TWENTY FIRST CENTURY science

GCSE Additional Applied Science

6 Materials and performance

Project directors Jenifer Burden
John Holman
Andrew Hunt
Robin Millar

Course editors Peter Campbell
Andrew Hunt

Module editor Michael Brimicombe

Authors Michael Brimicombe
David Brodie
Anna Grayson
David Sang

Contents

Materials in action
- Contact lenses — 3
- The building control surveyor — 4
- Dental braces — 6
- The lighting director — 8
- Sports performance — 10

People and organisations
- Setting standards — 12
- Putting toys to the test — 14
- Choosing materials — 16
- New structures from composites — 17
- Making it sound right — 18

The science
- Classes of material and their properties — 20
- Mechanical properties — 22
- Materials in collision — 26
- Thermal properties — 28
- Electric circuits — 30
- Electrical properties — 32
- A mouthful of materials — 33
- Room lighting — 34
- Lenses and images — 36
- Refractive index — 38
- Transparency — 39
- Stage lighting — 40
- Acoustic properties — 42
- Sound control — 44
- Bad vibrations — 45

Procedures and techniques
- Comparing stiffness — 46
- Comparing the strength of fibres — 47
- Finding the density of a solid — 48
- Comparing hardness — 49
- Measuring electrical conductance — 50
- Comparing thermal expansions — 51
- Comparing poor thermal conductors — 52
- Comparing good thermal conductors — 53
- Focal length of a converging lens — 54
- Measuring sound levels — 56
- Presenting data from standard procedures — 57

Your Work-related portfolio — 58
- Glossary — 60
- Index — 62

Introduction

The earliest humans used a variety of natural materials such as bone, wood, leather, and stone. Today's manufacturers use advanced materials in diverse products from long-lasting body implants to smaller and lighter computers and cars that are more energy efficient.

In this module you will meet some of the many people and organisations that develop, test, and use materials. You will extend the range of properties and materials that you learned about in the GCSE Science module Material choices. You will also practise some standard procedures and techniques used by scientists, engineers, and technicians.

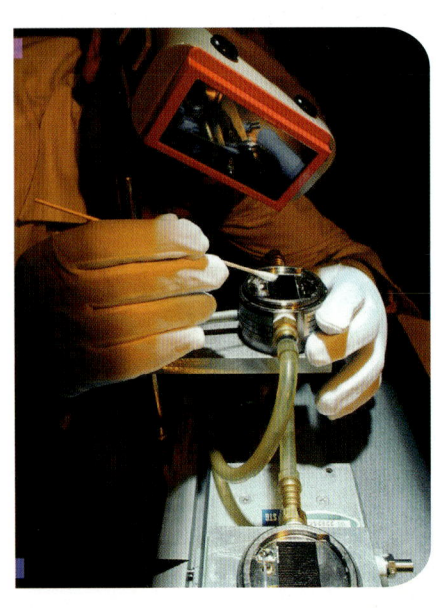

Materials in action

Contact lenses

Shelly Bensal did a three-year course as a dispensing optician. This means he can fit glasses. Then he went on to do a contact lens course. He gives contact-lens patients a choice – hard, gas-permeable lenses or soft lenses. Both of these are made from specially designed polymers.

'My patients need their contact lenses to give good vision and to settle well in the eye – in other words, to be comfortable. One of the most important things about the materials used for contact lenses is that they must be porous to oxygen. The front surface of your eye needs oxygen from the air.

'Contact lenses also need to react with the tears properly. Water solutions do not stick to most polymers but form little droplets and run off. The surfaces of contact lenses are made from a special polymer that lets the tears stick.

'The eye naturally produces proteins and lipids, and these stick to the lens. Using daily disposable soft lenses stops this problem. Hard gas-permeable lenses, which are worn for many days, need to be soaked regularly in a solution designed to remove any deposits.

'I love my job – I couldn't have taken a better path.'

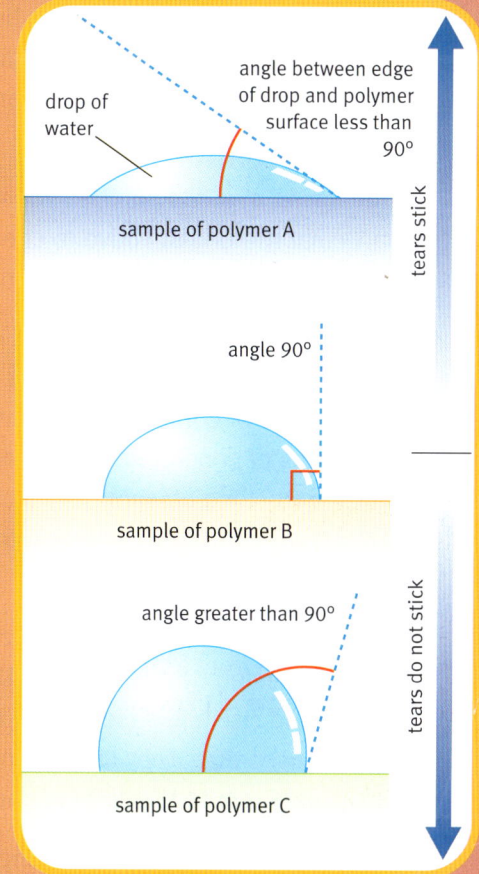

The wetting angle test on a polymer makes sure that tears will stick to the contact lens.

The building control surveyor

Hi, I'm Jaafer Merghani and I work for Camden Council in London.

Safety first

All local councils employ people like me. Buildings must be safe, and I make sure people stick to building regulations. I ask two main questions. First, is the building going to stand up? And second, could people get out in an emergency?

Fire prevention

I'm looking at some plans this morning. The architect has specified particular materials. I have to check that they are suitable. Preventing the spread of fire is important. For example, I might ask people to paint wood panelling with a fire-proof varnish.

Concrete – versatile and strong

The science of concrete is very interesting. You can change the properties of concrete by changing the proportions of sand, cement, water, and stone chippings. The chippings determine most of the strength. So the proportion and the size of the chippings have to be specified.

We'll often ask a building company to get a 150 mm cube of concrete tested by a laboratory. The block is crushed and the force required to crush it is measured. Builders often talk about '30 newton concrete'. They just mean that it takes at least 30 newtons per square mm to break the sample cube.

Concrete continues to gain strength after it sets. It will have three-quarters of its final strength after a month.

A typical day

My first call today is to a house where some of the downstairs walls are being knocked out to make it light and open-plan. Steel beams need to be put in to support the structure above. The gap is too wide and the load too heavy for a concrete beam.

My next visit is to a company that digs basements underneath houses. There are lots of things to think about here, not least keeping the rest of the house standing. I've got to know the builder, Adrian, quite well now.

Keeping out damp

Basements can be damp. To prevent that Adrian is lining the walls with a special membrane. It's made of thick polythene with dimples in it. The dimples go on the outside and any moisture will flow down between them. On the other side there's webbing, which sticks to the plaster inside the walls.

Keeping in heat

In this basement there's going to be a layer of expanded polystyrene and then plasterboard. This is to minimise heat loss. There are many new materials that do not conduct heat well and so help to keep a building warm.

That reminds me, I've had a query from someone about loft insulation. I need to give him a sketch of how to lay glass fibre in the roof space. Must dash – bye for now.

Dental braces

Clare Maidlow is a sales rep for 3M Unitek. This company makes everything you need to fit dental braces. Today she is visiting Watford General Hospital. She meets senior orthodontic nurse Lena Woods, who orders fittings and equipment there.

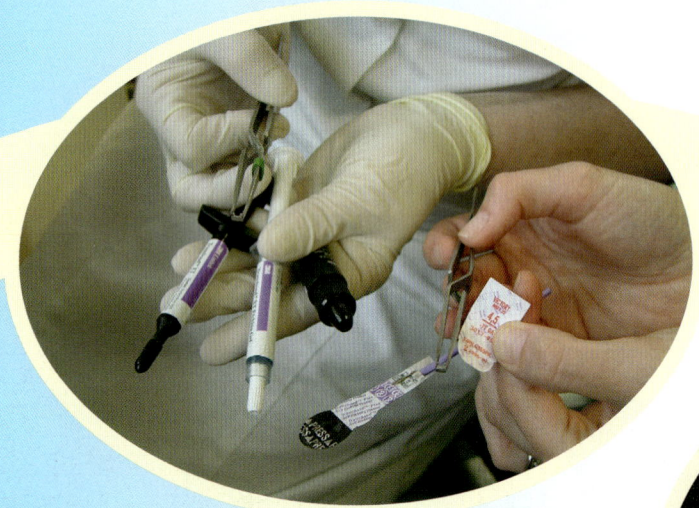

What are braces made of?

Clare   There's a little square bracket stuck to each tooth. The bracket determines the final position of the tooth. The brackets are linked together by an arch-shaped piece of wire. The wire provides a force to move the teeth. The brackets are linked to the wire with tiny elastic bands called ligatures.

How are they fitted?

Lena   The teeth must be really clean. All food debris and oil has to be removed so that the brackets will stick properly.

Clare   The teeth also need to be etched with a mild acid. This gives them a slightly rough surface so that the adhesive will stick. We sell brackets with the adhesive already applied. You just pop them on like a Post-it note! The adhesive is set using an ultraviolet light. It takes just a few seconds.

Lena   Most of the brackets are made of a metal mixture containing nickel. But some people are allergic to nickel.

Clare   We do have alternatives. We can supply gold-plated brackets. We also have white brackets made of a ceramic material. They hardly show on the teeth at all.

Lena Aren't ceramic brackets difficult to remove?

Clare They could be, yes. But ours have a groove and a slice of metal in them that makes them collapse when the orthodontist squeezes them with pliers.

Lena That's good, but they are more expensive.

Clare Once the adhesive has set, it's time to fix the archwire onto the brackets. Usually, wire made of an alloy of nickel and titanium is used for the first brace. We sell some really clever wires that have a 'shape-memory' effect. They are inserted cool. When they are in the mouth they warm up and change shape. This helps with the movement of the teeth.

Lena The archwire is fixed to the brackets with these elastic ligatures. Each ligature hooks over the four tiny wings at the corners of the brackets. This is where fashion and fun come in.

Clare Yes, we sell sets of ligatures in almost any colour.

Lena Pink and pale blue are the most popular for girls, deep blue for boys!

Clare Or you could have two colours to match your football team – yellow and red for Watford!

The lighting director

Mike Le Fevre is a lighting director for BBC Resources in West London.

I work in the studios in TV centre and on location – anywhere from the Cannes Film Festival to the Caribbean!

Sounds like a great job! What does a lighting director do?

Well, there are two sides to the job, really. First, I use my technical knowledge to get enough light from the actors and set to the TV or movie camera. I have to understand a lot about electrics and light to do that. Then the second part is artistic. I have to create the mood and use light to support what the performers and producers are trying to say.

What kind of lights do you use?

I could talk all night about different kinds of light. But there are two basic kinds of light source: soft and hard light.

Hard light?

Hard light is a single source of light. It's a bit like the rays from the Sun. I'll do you a sketch. See, with hard light the rays are all coming at the subject in parallel lines, and the shadows are clear and hard.

Mike's sketch

. . . and soft light?
Soft light is like the light coming from a cloudy sky. The rays are scattered and come at the subject in all directions. This makes a range of pale soft shadows.

These lights must get hot . . .
Yes, this unit has a tungsten filament which can reach around 3000 °C. The unit is made of aluminium with vents on top so hot air can escape.

It feels hot at the front!
Some units have special reflectors made of materials that absorb the infrared and only reflect the visible range. These ones are much cooler. The glass on the front of the units is vital because it absorbs ultraviolet. It stops actors getting sunburnt!

Do you use coloured lights?
Yes, all the time. For example, we might put a pale blue filter on these tungsten lights to make it look like daylight. Otherwise the camera picks it up as orange.

Do you use the filters for effects?
Yes, we use light to show emotion. There's a pale greeny-blue we call 'dismal blue'. And I might use a warm red or pink for a happy event.

So there's a whole range of colours?
The filters come in all sorts of colours, including the basic primary colours of red, green, and blue. I can mix any colour you want from these. But the position of the lights is just as important. It's not so much the light as where shade is that makes a scene.

Do you get to work with famous people?
Oh yes, all the time. And it's my job to make them feel comfortable with the look the lights are giving them. The most successful stars are often the nicest to work with!

Sports performance

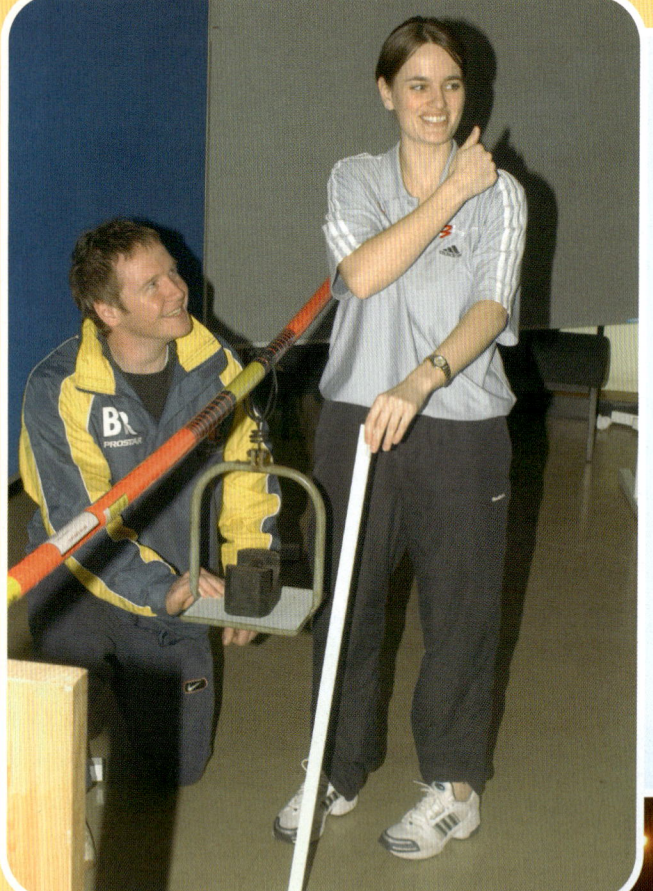

Students Amy Cleeton and Blake Raynor are at Birmingham University. After their course in Sports and Materials Science, they plan to help people in different sports to achieve outstanding performances.

They studied the properties of vaulting poles and the structure of the material they are made of. Blake explains, 'Pole-vault design turned out to be much more detailed than I'd expected. There's the internal structure of the composite material, for example. In a vaulting pole, the fibres are glass and the matrix is an epoxy (polymer) resin. Different composite materials have different amounts of fibres which can be woven together in different ways.'

International athletes use specially designed vaulting poles made from composite materials.

Want to be an international athlete capable of record-breaking performance? You'd better start training. Want to change what athletes can do, and how they do it? You should become a materials scientist. That's what Claire Davis is, and her work is international too.

In pole vaulting, the highest anybody could jump a hundred years ago was just over 3 metres. Today's athletes can get over 6 metres from the ground. That's almost as high as a house. The difference is mostly due to materials.

To jump as high as possible, a pole vaulter uses the energy of their run-up. They stick one end of the pole in the ground and bend it. This stores the energy from the run-up in the bent pole. The pole is elastic. It springs back and returns to its original shape. When the pole straightens, it transfers the energy back to the athlete again and pushes them up. A good pole must be not too flexible and not too rigid.

Around 1900, people used solid wooden poles. Then they switched to bamboo, which is still strong but more flexible. After that, records were smashed at nearly every Olympic Games. By 1956, pole vaulters could reach 4.5 metres, but the rate of improvement was levelling off.

Then, from 1960 onwards, records burst higher and higher again. That was the year when the top athletes started using composite poles. The material has low density, so the poles are light. And they are strong, for safety. If a pole breaks during a jump the athlete might get injured.

You can only win in modern sport if you have the best technology. The knowledge and skills of people like Claire, Amy, and Blake are behind winning performances in the sports stadium.

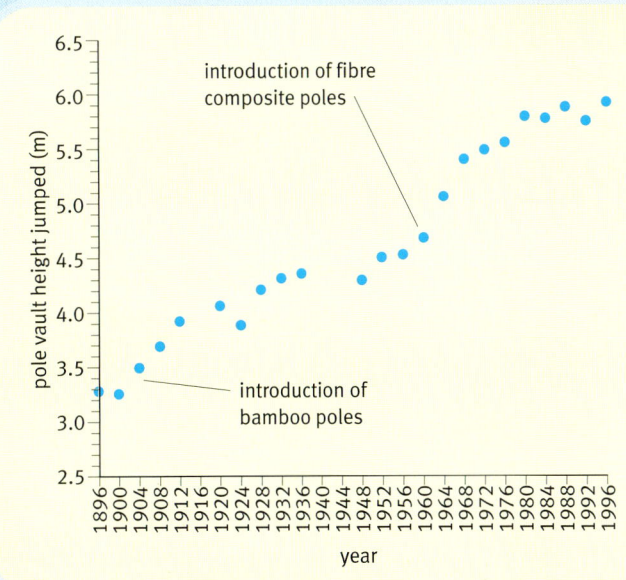

Athletes keep getting better. But when the materials technology improves, they get better faster.

Materials in action

People and organisations

People working in industrial, research, and standards organisations need to know about materials and their properties. This section introduces you to the challenges of their work, including how the use of materials is regulated.

Setting standards

New products have lots of codes and symbols on the packaging. Some of them tell you that samples of the product have been tested against a standard specification. That means it should be safe and reliable. But who sets the standard?

International standards

The International Organisation for Standardisation, ISO, is a network of standards organisations from 148 countries. The ISO brings people together from all over the world. They include

- experts from industries that make the products
- experts from the laboratories that test them
- people who represent consumers

They discuss a product and prepare a draft standard. Many people read this and comment on it. Eventually, the ISO creates a new International Standard. So, if you use a plastic bottle that meets standard BS ISO 16929, then you know that it is 'biodegradable'. It will rot away slowly when it is buried and won't clutter the environment forever.

All bank cards are the same size, so a British bank card will fit a French cash machine. It didn't just happen like that. The ISO had to create the standard.

European standards

In Europe, including the UK, you might find another mark on the product or packaging – CE marking. That means the product complies with the directives of the European Union. European standards are set by the European Committee for Standardisation (CEN). Their directives cover issues of reliability and safety.

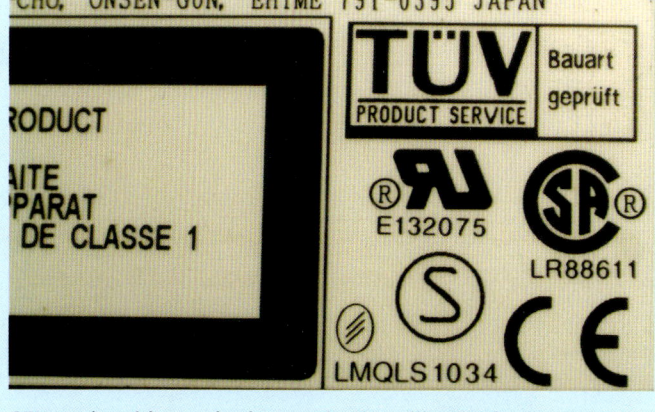

CEN works with standards organisations like BSI from every European country and creates European Standards that are all also British Standards.

British standards

Most countries have their own national organisation that sets standards for products and services. The British Standards Institution, BSI, was the first in the world. The BSI works with the ISO and many other organisations. It sets standards by consulting with experts. It says what should be tested, and how.

A manufacturer or an importer of a product can go to the British Standards Institution and pay a fee. Then the standard test will be carried out in a laboratory. If the product meets the test specifications, then the Kitemark can be used on the product.

The BSI shows that products match up to their standards using the 'Kitemark'. This is a kite-shaped symbol that includes the letters B and S, for 'British Standard'. Each Kitemark has a number, such as BS1877. You'll find this particular number on a pillow for a child's cot. It means air can flow through the pillow, so that the child can still breathe nose-down.

> **Questions**
> 1 What is a CE mark? What does it tell you?
> 2 Describe what the ISO does.
> 3 **a** Draw a Kitemark. What does it mean?
>   **b** If you were a manufacturer, why would you want a Kitemark on the goods that you make?
>   **c** Describe how you would get a Kitemark.
> 4 Suppose that standards organisations didn't exist. What difference would it make to people buying goods?

Putting toys to the test

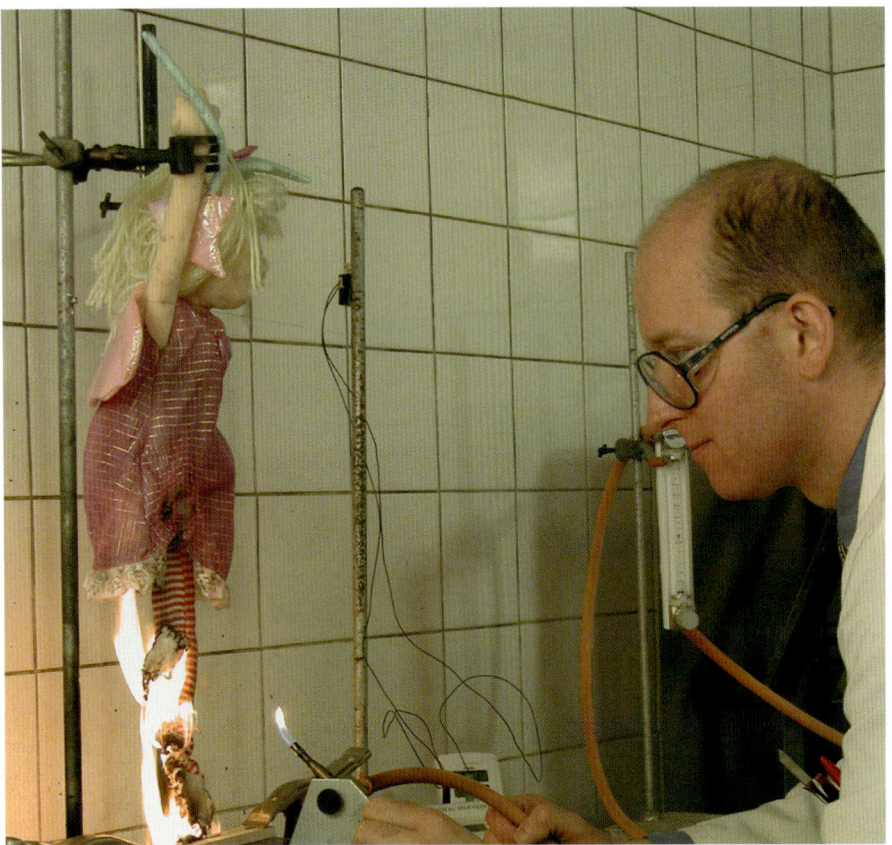

Testing flammability

If you walked into the laboratories of Scientific Services in Leicestershire on a day when they were testing soft toys, you might be a bit shocked. You would find them

- pulling the eyes off teddy bears
- dropping weights on plastic cars
- setting fire to dolls

But it's all in a good cause. They test products that you can buy, to make sure they are as safe and as reliable as they should be.

Anyone could take a product such as a teddy bear to the labs and pay them to carry out tests. But if you think that a product is poor in any way, then take it to your local Trading Standards Officers. They pay for firms like Scientific Services to carry out their tests.

Trading Standards Officers do not have to wait for someone to take a problem to them. They can gather goods from shops and market stalls, even Sunday car boot sales, and have them tested. If one sample of a kind of teddy bear fails the test, then all bears like this must be taken off sale. The shop and the manufacturer could face prosecution.

The people at Scientific Services must themselves work to standards set by a national or international organisation.

Those organisations might decide what hazards a teddy bear could present, for example. There should be no sharp points. The eyes must not come off too easily, because that would be a choking hazard. The fabric must not catch fire easily. And for simple quality of the product, the seams should not split. The list of what should be tested is called a **standard**.

For toys like teddy bears, the name of the standard is BS EN 71. At Scientific Services, they do all the tests exactly as the standard says they should, to make sure that the product is safe.

- An eye of a soft toy must stay attached when a 90 newton force is applied for 10 seconds.
- The seams must withstand a force of 70 newtons.
- The toy is put in the flame of a Bunsen burner, under safe conditions in the laboratory, to test flammability.

So, if you buy a soft toy for a new baby in the family, look out for the CE marking, which tells you that the toy should match BS EN 71. If it is on the label, then the baby will be safer.

Key word

standard

Testing that a toy can stand up to the right force

Questions

1. Why is it important that people at Scientific Services use standard procedures to test items?
2. Who sets the standard procedures used by Scientific Services?
3. Describe the standard procedure for testing BS EN 71.
4. Suppose that you bought an item which you thought was sub-standard. What should you do to get it tested?
5. There is a standard for chairs. Suggest a standard procedure for testing the suitability of a chair.

Choosing materials

Why do window designers work with glass, while hockey stick designers do not? Glass allows light to pass through it, but it is brittle. If you used it to hit a fast-moving ball, it would shatter.

What would be a good material for a hockey stick? Steel is stiff and strong. There are over 6000 varieties of steel, but they are all dense. A steel hockey stick would be too heavy. Polymers and wood are less dense. Again, there's a great range to choose from.

These professionals all choose the right material for the job:

- A product designer writes the product's specification (how it is expected to perform).
- A mechanical engineer knows that jet engine parts must keep their strength and toughness at high temperatures.
- An architect may specify energy-saving window glass to be used in a new building.
- A building control surveyor will make sure that thermal insulating materials comply with building regulations.

As well as performance, there are other factors to consider when choosing materials:

- Cost – is it affordable?
- Is it durable? How long will it last?
- Environmental impact – what damage might the material do, during manufacture, use, or as waste at the end of its lifetime?
- How easily can the material be shaped or joined to other materials?
- Aesthetic appeal – does it look good?

Questions

1 Imagine that you design tables for school dining halls. List all the factors which you need to consider. Explain the importance of each factor.

2 **a** Name an everyday object.

b Describe the materials from which it is made and explain why you think each of these materials was selected.

Hockey sticks are made of wood or composites. These materials are stiff, strong, and lightweight.

Some parts of this electrical machine are made of insulating materials.

New structures from composites

Sue Halliwell works for the Building Research Establishment (BRE) in Watford. BRE provides services for people who design and build all sorts of structures – such as houses, tower blocks, and bridges. Sue works on composites, and in particular on fibre-reinforced polymers, or FRPs. She explains her work:

'Here at BRE we test materials so that engineers and architects can use them for exciting designs of structures. We test properties like **compressive strength**, tensile strength, and density, and we make the data available in technical reports that people can buy.'

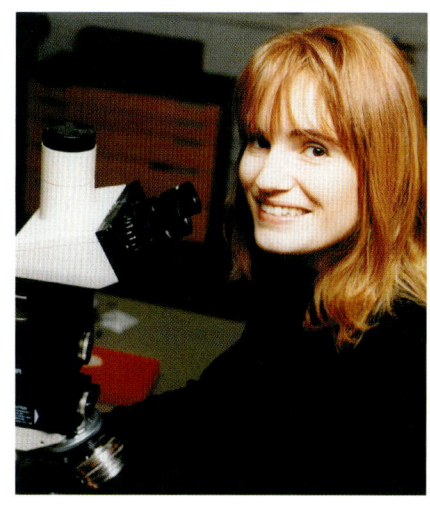

Sue Halliwell does research on materials. Her work tells designers of buildings and bridges which composite materials may be best for their designs.

Key words
compressive strength

Is this a tree killed by lightning? No, it's a phone mast. Composite materials can be moulded, and they have low density and high strength. They also allow radio waves to pass straight through them. So a composite mast can be made to look like a tree.

This footbridge was built in the Welsh mountains, a long way from a road. So the materials had to be delivered by helicopter. Lightness mattered. The low density of fibre-reinforced plastic made it ideal for the bridge deck.

Composites can be moulded to provide exciting design possibilities. Panels are light so they are easier to work with than traditional materials. This is the airport building in Harare, Zimbabwe.

A wind turbine has to take everything that the environment throws at it – howling gales and saltwater spray. Strength, lightness, and resistance to chemical corrosion are necessary properties.

Making it sound right

School classrooms are individually designed for their purpose. For example, the government and local councils publish guidelines for the design of music rooms in schools. These have to be followed by architects and builders of schools.

> **The design of music rooms**
> The acoustic environment is very important in music rooms. Sound quality inside the music room should be optimised. However, sound from a music room should not disturb neighbouring classrooms, and the music department should not suffer outside noise from a road or sports ground, for example. The two main objectives are good sound quality and good acoustic insulation.

Acoustically soft ceiling tiles help reduce the reverberation time.

Acoustic insulation

To keep music in music rooms, guidelines say schools should ideally

- put music rooms in their own block, separate from the rest of school
- build music rooms with materials that have the right acoustic properties
- think carefully about the shape and size of the rooms
- make sure that sound can't leak from one room to another

Acoustically hard materials should be used in the floors, walls, ceilings, and roofs of a music block. These materials reflect sound. Cavity walls and double glazing should be used for the outside of the block, to prevent noise getting in or the music getting out.

Panels and a moulded rail on the wall help prevent flutter echo.

> The quality of sound depends on a balance between how clear it is and fulness of the tone. This balance depends on reverberation time – the time taken in seconds for a loud sound to decay and no longer be heard.

Sound quality

Acoustically hard materials on the walls, ceiling, and floor keep the sound in the room. But this allows sound to bounce around the room for a long time before it dies away. The music sounds muddy, because new notes start before the previous ones have died away.

To solve this problem architects put acoustically soft materials inside music rooms. These materials absorb sounds. For example, each room may have a false ceiling made from foam plastic. The foam is fragile, but on the ceiling it will not be damaged.

Rooms the right shape

Square classrooms should be avoided. With parallel walls, sound bounces repeatedly between the walls, giving a ringing sound called flutter echo. Walls at right angles can also produce annoying reflections. If the room has to be square, then a curtain on one wall can solve the problem.

Noisy neighbours

Sounds may leak from one music room to the one next door. Here are some possible solutions:

- no keyholes in the doors
- acoustic breaks built in to any pipes and joists that go between rooms
- any gaps in walls for services such as water and electricity to be filled with foam

Designing a good suite of music rooms is a skilled job. It requires a lot of experience and a good understanding of the acoustic properties of building materials.

A steeply pitched ceiling prevents flutter echoes developing between the ceiling and the floor.

Questions

1. Explain, in your own words, what the two main design issues are for a school music room.
2. Why do the government and local councils publish design guidelines for school music rooms?
3. A school hall will be used for speeches and drama as well as music performances. Suggest how the acoustic design for a school hall may differ from the design of a music room.

The science

Classes of material and their properties

The many different materials used in buildings and manufactured products can be classed into three main types, shown in the table below. This is a general summary – materials vary greatly, and some materials in a class may behave differently.

Metals	Ceramics	Polymers
e.g. iron and steel	e.g. brick, tiles, and crockery	e.g. polythene, PVC, and Perspex
strong in tension and compression	strong in compression	often strong for their mass
often hard, stiff, and tough	weak in tension	often flexible
malleable, ductile	very hard	soft and easily scratched
	brittle	
good conductors of heat and electricity	thermal and electrical insulators	thermal and electrical insulators
shiny like a mirror	opaque	transparent or translucent

Properties of materials change with temperature. This table shows the properties at room temperature.

Composite materials

Composite materials are made of two or more materials combined. They have better mechanical properties than their component materials would alone. Bone, for example, is made of flexible collagen fibres held together with mineral crystals. Wood contains cellulose fibres held together by lignin.

Concrete is strong in compression but weak in tension (see page 22). Reinforced concrete is a composite material made by embedding steel rods in the concrete. These rods are very strong in tension.

The mast of this catamaran was reinforced with carbon fibres.

Modern composite materials are made with strong fibres held in a weaker matrix material. Glass-reinforced plastic is just one example. It is used to patch up car bodies. A glass fibre is strong but brittle. If one breaks, the plastic matrix shares out the extra stress amongst many other fibres. This stops cracks forming.

Material structure gives its properties

In the GCSE Science module Material choices, you learned how the very long molecules of polymers explain their mechanical behaviour. Structure and bonding provide the key to understanding the properties of *any* material. By controlling their internal structure, it is now possible to make 'designer materials'.

Imitating bone

Living materials such as bone and wood are 'smart' materials. They are able to grow and repair themselves in response to forces that they experience. Engineers learn a lot from looking at natural materials and structures.

Take bone, for example. Many bones are hollow. This enables them to be lightweight whilst remaining stiff and strong. This is especially important for birds. They wouldn't be able to fly if their bones were solid to the core.

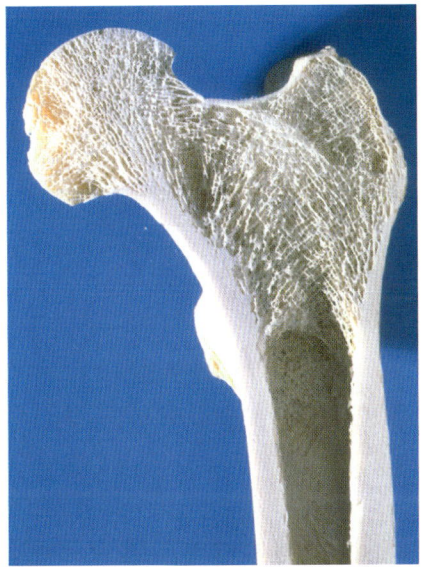

Within a bone, the hollow cavity may have struts that help to keep it rigid without adding much weight. Alternatively, the bone may be filled with a spongy honeycomb of supports. The spaces between are filled with bone marrow.

Racing-car bodies and aircraft wings make use of the same idea. They are made of an outer skin filled with a honeycomb material, giving a composite material that is both light and strong.

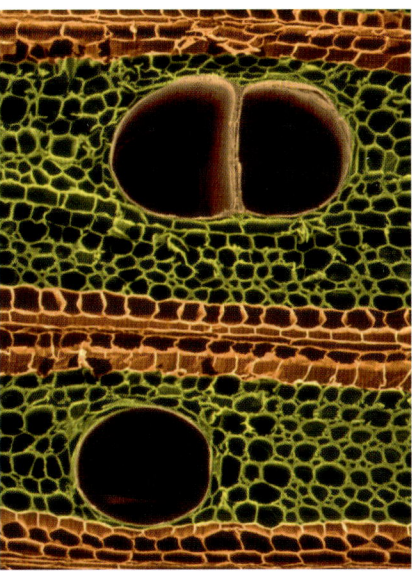

If you look at balsa wood using an electron microscope, you will see that it is made of cells that are long, thin tubes. Its density is about one-fifth that of water, which is not surprising, since it is mostly air. For its weight, balsa wood is a very stiff material. (Magnification ×70)

Designing weak points

Designers do not always want structures that are stiff and strong. Think of a bar of chocolate divided into segments. If you bend the bar, it will break at its thinnest point, and you will have a bite-sized portion.

> **Question**
> 1 Select one of the uses of composite materials from page 17. Suggest another material that could have been used. Discuss the advantages and disadvantages of this alternative to the composite material.

Key words
composite
metal
ceramic
polymer

Mechanical properties

Forces and materials

Mechanical properties describe the behaviour of materials when external forces act on them. **Forces** can **deform** (change the shape of) materials by stretching, squashing, or bending them.

- Stiff or flexible? – A material that holds its shape and resists being deformed is **stiff**. A **flexible** material is easily deformed.

- Elastic or plastic? – **Elastic** materials such as rubber spring back to their original shape after the force deforming them is removed. A **plastic** material does not recover its shape like this. It is deformed permanently, like Plasticine.

- Strong or weak? – **Strong** materials such as fishing line require a large force to break them. **Weak** materials such as liquorice lace break easily.

- Brittle or tough? – In an impact, **brittle** materials like glass snap cleanly and sharply. Tiny cracks open up and run through the material. **Tough** materials, like steel in a hammer head, resist the formation or spreading of cracks.

Key words
force
deform
stiff
flexible
elastic
plastic
strong
weak
brittle
tough
tension
compression
extension

Forces on a beam

Forces act on structures in different directions. Materials respond differently to these forces depending on their mechanical properties.

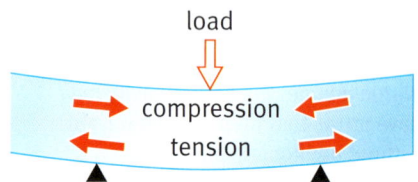

This beam is in **tension**. The forces are stretching the material. Steel cables are strong in tension – they hold up suspension bridges without breaking.

This beam is in **compression**. A pair of forces is squashing the material. A nail is strong in compression – it doesn't break when hammered.

This beam is in compression on the upper surface and tension on the lower surface.

The gymnast's beam is strong in both tension and compression.

Force–extension graphs

To find out how a material behaves in tension, you can hang different weights from it and measure the **extension** (how far it stretches). Plotting a graph of force against extension can tell you a lot about the material's behaviour.

Elastic behaviour

For small forces, the graph is always a straight line. The material is elastic – it returns to its original length when you remove the weights.

For an elastic material you can predict the extension produced by a particular force.

$$\text{force constant } k = \frac{\text{force}}{\text{extension}}$$

Example: A length of fishing line extends by 2.0 mm when it supports a fish of weight 0.5 N. What is the extension for a fish that weighs 1.2 N?

$$\text{force constant } k = \frac{0.5 \text{ N}}{2.0 \text{ mm}} = 0.25 \text{ N/mm}$$

for a fish that weighs 1.2 N, $k = 0.25 \text{ N/mm} = \frac{1.2 \text{ N}}{\text{extension}}$

and $\text{extension} = \frac{1.2 \text{ N}}{0.25 \text{ N/mm}} = 4.8 \text{ mm}$

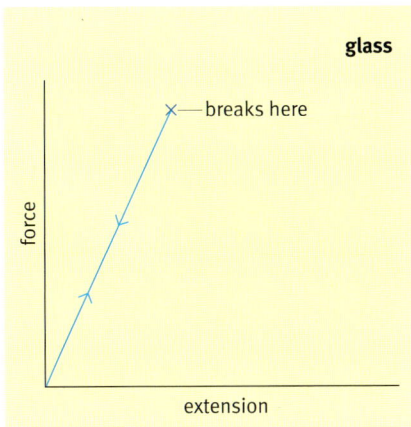

This force–extension graph is a straight line. The material is elastic. After being stretched it returns to its original size.

Plastic behaviour

Many materials become plastic if the force is big enough. They no longer return to their original size when you remove the weight. The force–extension graph becomes curved, as shown for copper.

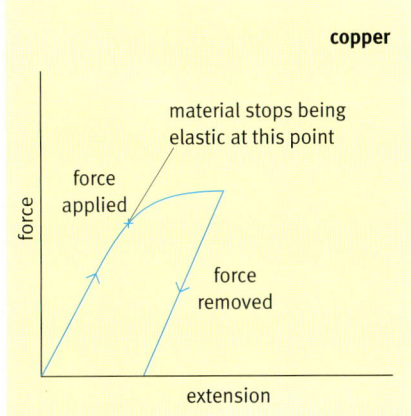

This material is elastic when the force is small. It becomes plastic with larger forces and does not return to its original size.

Stored energy

The area under a force–extension graph tells you how much energy is stored in the stretched material. In the graph opposite:

- Each small square is $10 \text{ N} \times 0.001 \text{ m} = 0.01 \text{ J}$.
- The total number of squares under the graph is about 10.
- So the amount of energy stored is $10 \times 0.01 \text{ J} = 0.1 \text{ J}$.

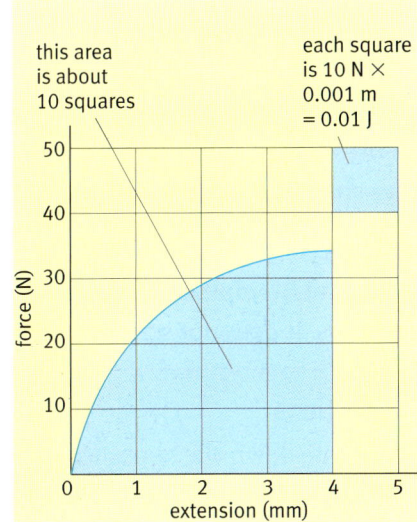

A stretched material stores energy. The graph allows you to calculate the energy stored.

Mechanical properties in action

Aircraft construction

The first aircraft were made of wood. People knew a lot about how to cut, shape, and join wood, and how to make use of its natural strength and stiffness. Wood is also a very lightweight material. Its **density** is low.

Today, aircraft are mostly made using a lightweight metal **alloy** called Duralumin. An alloy is a mixture of metals. Duralumin is mostly aluminium, with small amounts of copper, magnesium, manganese, and silicon to give the best possible combination of properties. These extra metals form a **solid solution**, with the different atoms spread evenly throughout the aluminium.

Steel

Steel is the alloy most commonly used today. Steels are mainly iron, with small amounts of carbon added. Specialised steels have tiny amounts of other elements added too, for example chromium, manganese, or nickel. Alloying can improve the **corrosion resistance** of a steel, and raise its **melting point**.

Metals are useful because they are **ductile**. They can be rolled into sheets, drawn into wire, or worked into other useful shapes without breaking.

Safety glass

Standard window glass is elastic. But it is also brittle. If it is bent too far, by wind or an impact, it shatters into thousands of sharp and dangerous fragments. For this reason, building regulations often specify that **toughened glass** must be used. Toughened glass absorbs lots of energy before shattering. And when it does shatter, the fragments are harmless.

Bullet-proof glass is tested with a 5 kg steel ball, moving at 30 mph. Steel is used because of its **hardness**. It does not scratch or dent easily.

Every kilogram saved when manufacturing an aircraft means that another kilogram of baggage can be carried. Low-density materials like aluminium and polymers make air travel possible.

Bullet-proof glass under test. It is made from two sheets of glass with a sheet of polymer sandwiched between them. It is tough stuff and is used in military vehicles, banks, and museums.

Questions

1. Write down the opposites for each of these terms: stiff, hard, strong, tough.
2. Use the key words lists here and on page 22 to describe the mechanical properties of these foods: biscuit, strawberry laces, boiled sweets.

Key words

density
alloy
solid solution
ductile
corrosion resistance
melting point
toughened glass
hardness

Making rigid structures

It's not only the material an object is made of that determines its mechanical properties. The way the material is shaped also plays an important part.

> **Key word**
> rigid

Folding paper

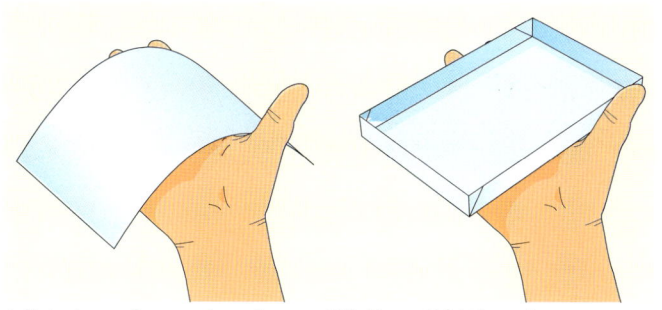

A flat piece of paper is not very stiff. If you fold the edge up as shown, it becomes much more **rigid** – it is stiffer and stronger.

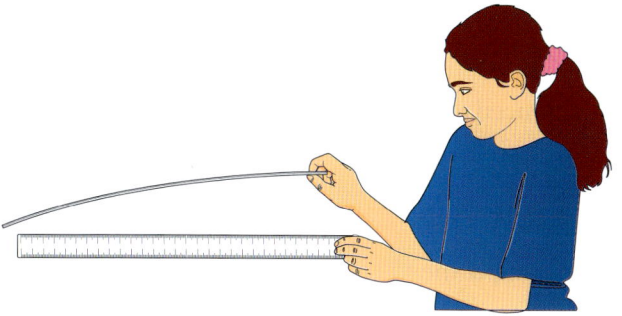

If you hold a metre rule flat, it bends. Hold it on edge and it's much stiffer. It's stronger in compression and tension. The folded edge of the paper is like this.

Paper and cardboard are folded to make them more rigid in corrugated card packaging, cardboard boxes, and egg boxes.

Tubes and triangles

Structures made of steel tubes joined as triangles are lighter than solid tubes, but still strong. Tubes of steel are

- much stiffer and stronger than flat sheets of steel
- much lighter and cheaper than solid rods of steel

The Forth railway bridge in Scotland was opened in 1890 and was a wonder of its age. It was the use of steel tubes joined as triangles that made such a long, high bridge possible.

Tubes and triangles allow bicycle frames to be very light but very strong.

> **Question**
> **3** Why is the frame of a bicycle made from hollow tubes rather than solid rods?

Materials in collision

Crashing cars

When a car has a head-on collision, it suddenly stops moving. The driver keeps moving forwards, until he meets something that provides a force to make him stop too. You can work out how big the force will be.

First you need to know the **momentum** of the driver before the collision:

original momentum = mass × velocity

Velocity simply means speed in a particular direction.

In a crash, the driver's momentum changes to zero.

change of momentum = 0 − original momentum

The change of momentum has a negative value.

In any collision,

force × collision time = change of momentum

So the force on the driver also depends on the **collision time**. The slower he stops, the better. The job of a seat belt is to increase the time it takes for him to stop, and reduce the force he feels.

Example: In a head-on collision at 30 m.p.h., a driver loses all of her momentum in just 0.20 s. In m/s, the collision speed is 14 m/s. The mass of the driver is 70 kg (so her weight is 700 N). How big is the force that stops her?

original momentum = mass × velocity = 70 kg × 14 m/s = 980 kg m/s
change of momentum = 0 − 980 kg m/s = −980 kg m/s
force × collision time = change of momentum
force × 0.20 s = −980 kg m/s

force = $\frac{-980 \text{ kg m/s}}{0.20 \text{ s}}$ = −4900 N

The minus sign shows that the force is in the opposite direction to the original momentum. This force is seven times the normal weight of the driver (700 N).

Force–time graphs

You can see the effect of a crumple zone by plotting force–time graphs for a test collision, as shown opposite.

Cars have safety features to protect the driver (and passengers) in a crash. These include
- a **seat belt** to slow his body down
- an **air bag** to slow down the forward motion of his head
- a **crumple zone** to make the car interior stop less quickly than its bumper does

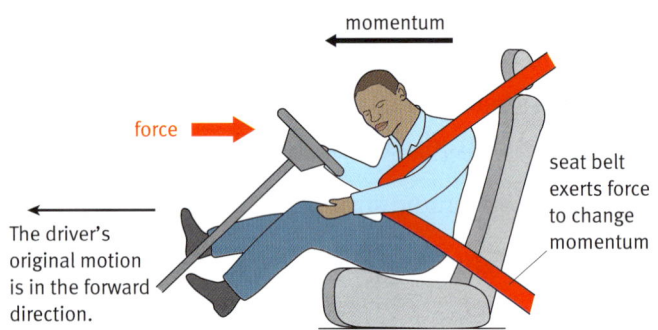

The force needed to stop the driver is opposite to the direction of his motion.

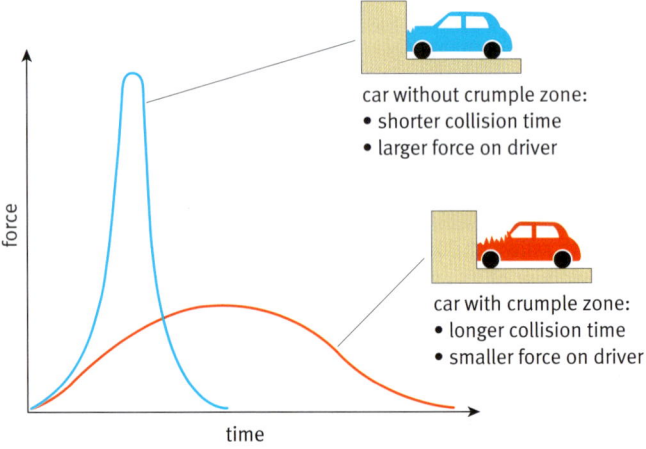

car without crumple zone:
• shorter collision time
• larger force on driver

car with crumple zone:
• longer collision time
• smaller force on driver

The force readings come from electronic devices in the dummy. The area under each graph is the change of momentum. It is the same for both cars.

Mind your skull

Your brain is soft, complex, and easily damaged. Your skull, which protects the brain, is much stronger, quite hard, and fairly rigid. But if you have a cycling accident, your head may suddenly experience too large a force. Head injuries can kill or cause disabling brain damage.

European standard EN1078, and others like it, guarantee that cycle helmets help protect cyclists in an accident. One lab test simulates what would happen if you fell off a stationary bike and hit your head against a kerbstone or some other hard object. There are also strap and buckle strength tests. Some standards include a test to see if the helmet will stay on when pushed to one side.

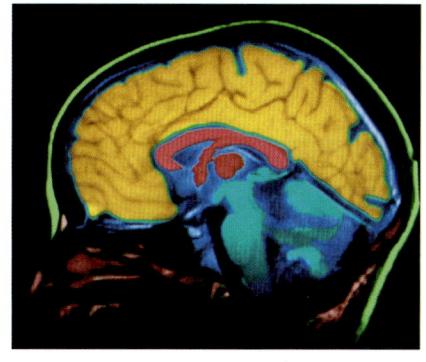

A scan of the brain inside the skull

How does a cycle helmet work?

In an accident, the helmet reduces forces acting on the head by spreading the impact over a larger area. It also increases the collision time. Doubling the collision time will reduce the force to half the size.

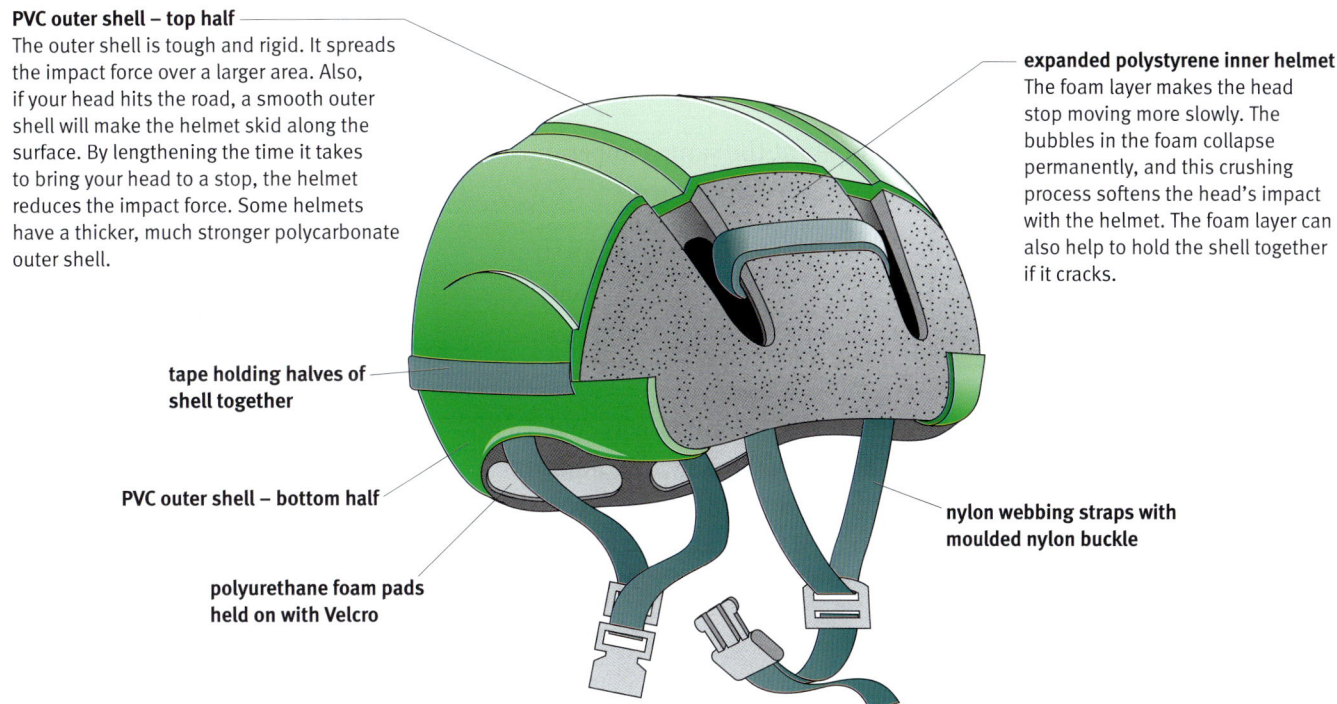

The mechanical properties of the shell and foam are complementary. But it's safer to ride carefully in the first place. You have to use your brain to protect your brain.

Questions

1 Describe three safety features of a car which protect the driver in a crash. Explain how they work.
2 A car travelling at 10 m/s crashes head-on into a wall, stopping in 0.3 s. Calculate the force needed to restrain a child of mass 20 kg in the car.

Key words

seat belt
air bag
crumple zone
momentum
velocity
collision time

Thermal properties

Thermal properties describe the behaviour of materials in response to temperature differences.

Ski clothing is lightweight and flexible, but it keeps you warm

Clothing for winter sports is designed to keep you warm in temperatures that might be 40 °C below your body temperature. You need to keep in the warmth of your body.

A ski jacket feels warm to the touch. It's really at the same temperature as the room, which is colder than you. So why does it feel warm?

Heat from your body escapes into the jacket fabric. However, it doesn't move through the fabric easily, because the fabric is a good thermal insulator. So when you touch the jacket, you are feeling the temperature of your own body.

If you touch a metal object, such as a car, on a cold day, it feels cold. Heat from your fingers travels into the metal and then flows away. The metal doesn't warm up. Instead, your finger cools down, so the metal feels cold. It's easy to be fooled by the sense of touch. You think you are finding out about the temperature of something, but you are really finding out how well it conducts heat.

Thermal conductance

The pattern for **thermal conductance** is the same as for electrical conductance:

- Metals are good conductors. Metal objects have high thermal conductance.
- Polymers (plastics) and ceramics (such as glass) are good thermal insulators and poor conductors. Objects made from polymers and ceramics have low thermal conductance.

This pan is made of metal so that heat conducts easily into the food inside. Its handle is made of plastic. Heat doesn't conduct easily up the handle, so it stays cool enough to be touched.

Expanding and contracting

This greenhouse has an aluminium framework that supports the glass panels. On a summer's day, the aluminium and glass both expand as they get warmer. Aluminium expands three times as much as glass. The glass might fall out as the frames get too big.

In winter, the aluminium and glass shrink when they get cold. Aluminium shrinks three times as much as glass. The frame may squeeze the glass until it cracks.

To avoid these problems, the glass is mounted with metal clips. It can move relative to the aluminium frame as temperatures change.

Most materials expand when they are heated and contract when they are cooled. You can predict the **thermal expansion** of an object if you know

- its original shape and size
- the temperature change
- what material it is made from

In other words, you can predict its new size.

A greenhouse close up. Glass panels are held in place with metal clips that allow the panels to move.

Stretching and bending

A useful type of switch is made using a **bimetal strip**. This is a strip made of two different metal alloys, chosen so that one expands more than the other. When the strip is heated, it bends. The alloy that expands more is on the outside of the curve.

Often product designers want to find materials that expand at the same rate. For example, it is important that the piston and cylinders of a car engine change size by the same amount as their temperatures rise and fall.

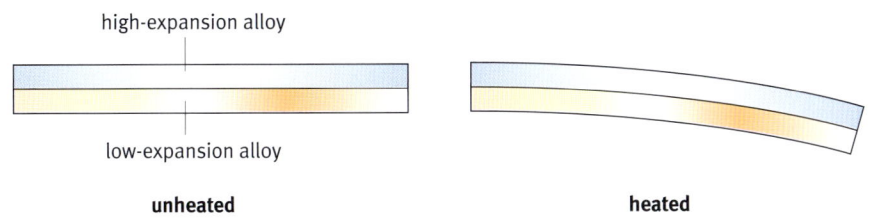

A bimetal strip is used in this iron to control the temperature.

> **Questions**
> 1. Explain why a metal spoon feels cold when you pick it up, but a wooden one feels warm.
> 2. Suggest why window frames are fixed into brick walls with a flexible sealant.

> **Key words**
> thermal conductance
> thermal expansion
> bimetal strip

Electric circuits

This technician at a car factory is learning about the electric circuits in a new model. It is easier on a test rig like this than on a real car. Look for the lights, switches, and connecting wires.

Electric circuits can be complicated. Electrical engineers and technicians learn how to look at a circuit and identify the important parts. They can then work out what the circuit does. They use circuit diagrams like maps, to find their way around circuits they are making or testing. It makes sense to use standard symbols for the different components.

You will already know about simple electric circuits. There must be a complete path that includes a power supply for any current to flow.

- In a **series circuit**, one component follows after another, and there is just one path for the current to follow.
- In a **parallel circuit**, there are one or more branches, and the current divides between them.

The table opposite shows some of the most important features to look for when you see a circuit or a circuit diagram.

power supply	cell	connecting wires	connecting wires joined	on/off switch	lamp	dimmer
A cell or power supply pushes the electric current around the circuit.		The connecting wires link up all the components in the right order. They are very good conductors of electricity.		When the switch is closed, it lets the current flow in the circuit.	The lamp lights when current flows in it.	To make a lamp brighter, you slide or turn the dimmer. Its resistance gets smaller, and so more current flows through the lamp.

Electrical measurements

You cannot see electricity. So engineers and technicians need to be able to make measurements to find out what's going on in a circuit. They use electrical meters to do this.

- An **ammeter** is used to measure the **current** in a circuit. The ammeter is connected in series. It gives readings in amps (A) or milliamps (mA).

- A **voltmeter** is used to measure **potential difference** (voltage). This is the energy a given current provides each second between two points in a circuit. A voltmeter is connected in parallel with a component in a circuit. It gives readings in volts (V) or millivolts (mV).

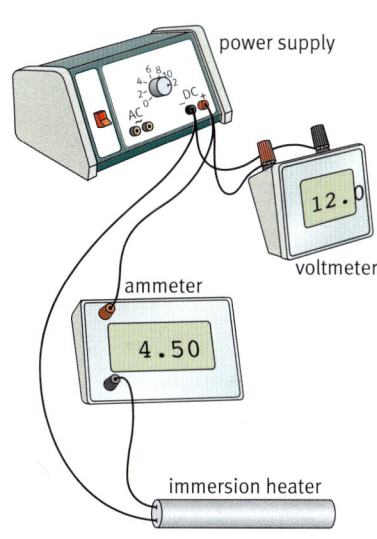

Questions

1 Draw a circuit diagram of a power supply, lamp, switch, and dimmer in series. What does the circuit do?

2 A standard test of a car headlamp bulb requires its current and voltage to be measured when it is connected to a 12 V battery. Draw a suitable circuit diagram for the procedure.

Key words

series circuit
parallel circuit
ammeter
current
voltmeter
potential difference

The science

Electrical properties

To make electrical equipment, you need to choose the best material for each part. The first thing to think about is how easily electricity flows through each material.

- Metals are good **electrical conductors**.
- Polymers and ceramics are good **electrical insulators** and poor conductors.

A look at a domestic electrical plug shows how these materials are put to work.

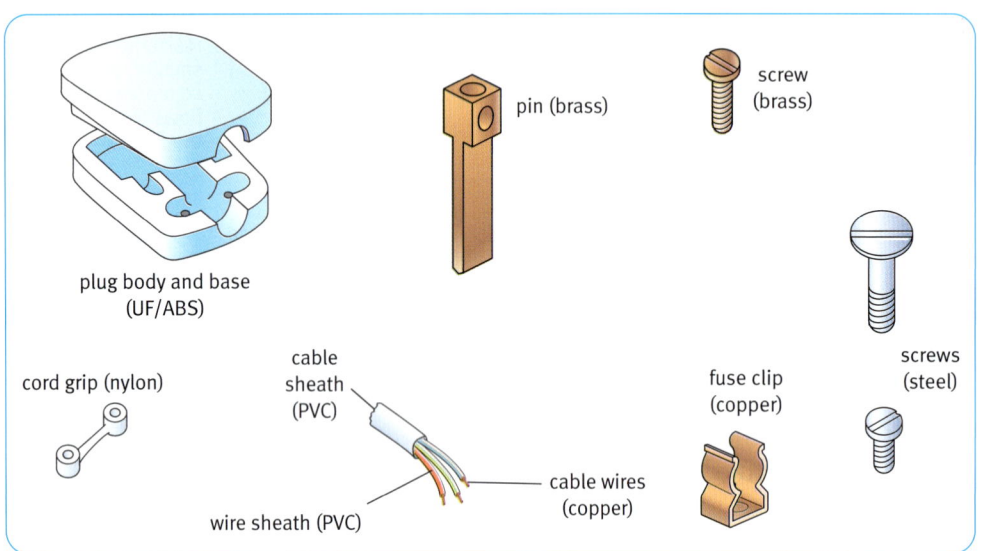

Electrical conductance

To see how well electricity flows through a particular component, you need to measure both current and potential difference. **Electrical conductance** is how much current passes for a given potential difference.

$$\text{conductance} = \frac{\text{current}}{\text{potential difference}}$$

In symbols:

$$G = \frac{I}{V}$$

The unit of conductance is the siemens. 1 siemens = 1 amp/volt

The range of electrical conductance for different materials is enormous. Conductance is compared in samples of material the same size and shape.

Example: Current of 3 A flows through a wire when there is a potential difference of 2 V across it. Here's how to calculate its conductance:

$$\text{conductance} = \frac{I}{V} = \frac{3A}{2V} = 1.5 \, S$$

Key words
electrical conductor
electrical insulator
electrical conductance

Question

1. The current in the heating element of an iron is 7.5 A when the voltage is 230 V. Calculate its conductance.

A mouthful of materials

If you damage your teeth, or you develop a dental cavity, your dentist can sort out the problem. Dentists choose materials with the right combination of properties.

Inside your mouth is a hostile environment. It's where saliva starts to break down your food. There are strong forces as your teeth bite, chew, and grind. The temperature can change suddenly, when you drink something hot or cold. If it is to do its job properly, a dental filling or a false tooth must be made of the right stuff.

By matching the properties of the filling to those of natural teeth, the dentist ensures that the filling lasts well and that the user does not notice that it is there.

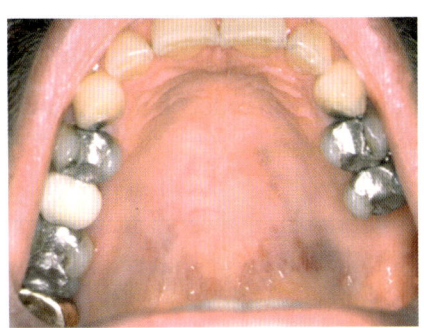

Amalgam dental fillings

Amalgam

Most metal fillings are made of amalgam, which is an alloy of metals including mercury. The metals are mixed to a paste which is put into a cavity. It soon hardens.

Advantages of amalgam	Disadvantages of amalgam
strong, so it resists the forces as your teeth crush together	like all metals, it conducts heat well, so drinking something hot can be painful, as heat reaches the nerve in your tooth
almost as hard as tooth enamel	thermal expansion is greater than that of tooth, so the filling can fall out or damage the tooth
	does not look like tooth enamel: it is silvery when new but later tarnishes

Composite fillings

Fillings, crowns, and false teeth made of composite look much better than amalgam. The composite is made of a plastic resin mixed with very fine particles of ground glass. Pigments are added to match the colour of your teeth. The dentist shines blue light on the filling to make it set hard.

Advantages of composites	Disadvantages of composites
as strong as natural teeth	thermal expansion is greater than that of a tooth, so the filling can fall out or damage the tooth
conducts heat poorly, like the tooth itself	
hardness similar to tooth enamel	

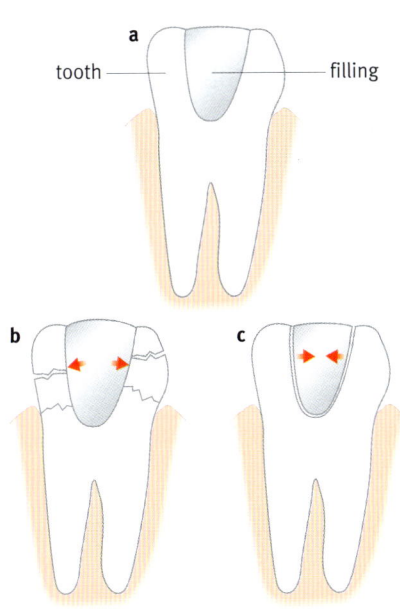

The material a filling is made of must have thermal properties similar to those of natural teeth (a). Otherwise, hot drink may cause the filling to expand too much and crack the tooth (b) or not expand enough and come loose (c).

Question

1 List the properties which are important for dental fillings. Justify each one.

Room lighting

A office or home needs lighting that suits a variety of activities. Lighting types include

- natural light – from windows or light pipes
- background lighting – for example, uplighters, or ceiling or wall lights
- accent and task lighting – for example, a table lamp or spotlights

A mirror will reflect light and can make a room seem larger.

A carefully placed mirror makes a room or a product seem larger.

Windows

A window is a sheet of **transparent** glass that will let light into a room. But did you know that ordinary window glass is not 100% transparent? It only transmits about 80% of the light that falls on it. Of the remaining 20%, most is reflected, and some is absorbed by the glass.

Frosted glass has an irregular surface. The light rays are jumbled as they pass through, and you don't see a clear image. It's described as **translucent**. Walls are **opaque** (no light passes through).

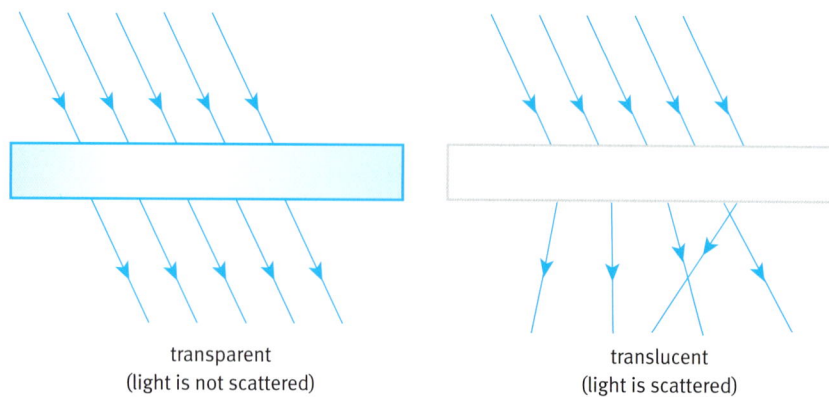

transparent
(light is not scattered)

translucent
(light is scattered)

opaque
(light is absorbed)

Designing windows

Sunlight contains both visible light and infrared (heat) radiation. You want these to get into your home, but you don't want the heat to get out again.

Metals are shiny materials that produce mirror reflections. The metal silver forms a **coating** on the back of most glass mirrors. Similarly, double-glazing glass has a thin coating on one surface. This lets heat and light through from the Sun, but reflects the heat back inside.

Self-cleaning glass is the latest technology. On the outer surface, it has a very thin coating of titanium dioxide. When dirt lands on the window, this coating helps to break it down, using the energy of ultraviolet light from the Sun. Any rain then washes the dirt off. Self-cleaning glass is expensive but ideal for buildings with inaccessible windows.

Key words
transparent
translucent
opaque
coating

34

Light pipes

Here is a new way of lighting a room: use a light pipe.

- On the roof, a glass dome collects the daylight that falls on it.
- A pipe with shiny metal walls reflects the light, guiding it towards a room below.
- A translucent diffuser in the ceiling spreads the light around the room.

This Dutch water tower was converted into smart offices with clear views, thanks to self-cleaning glass.

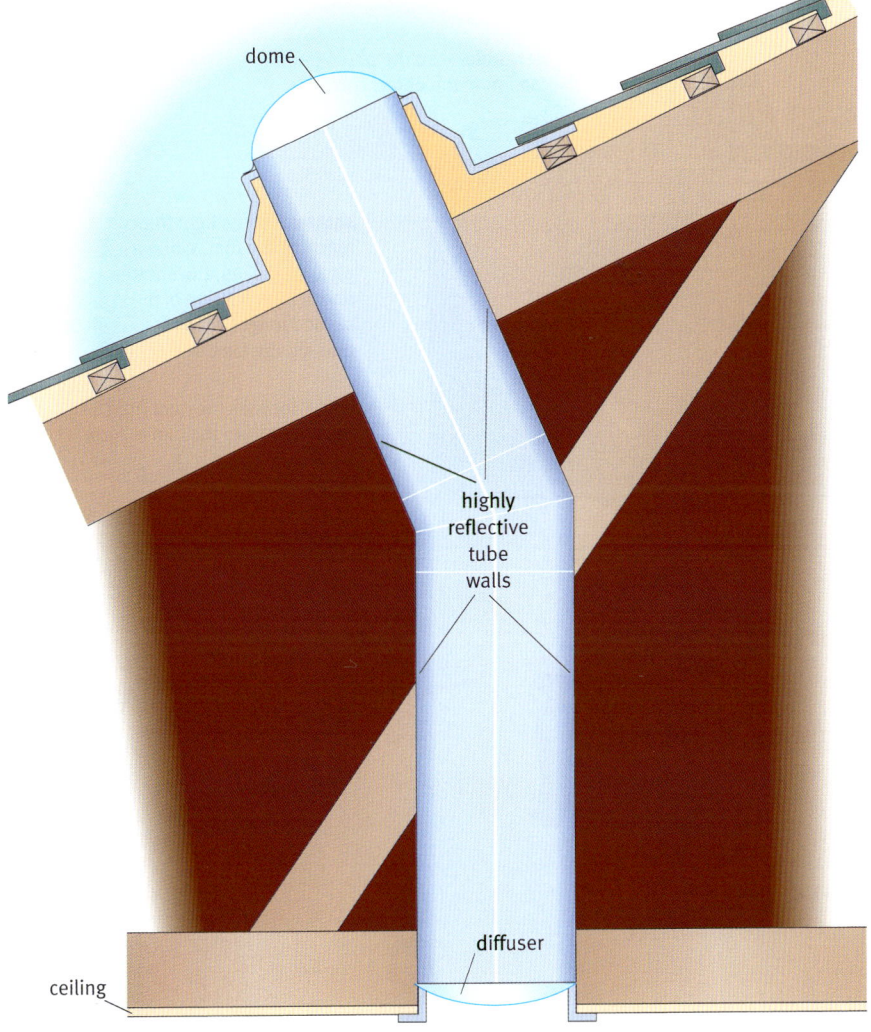

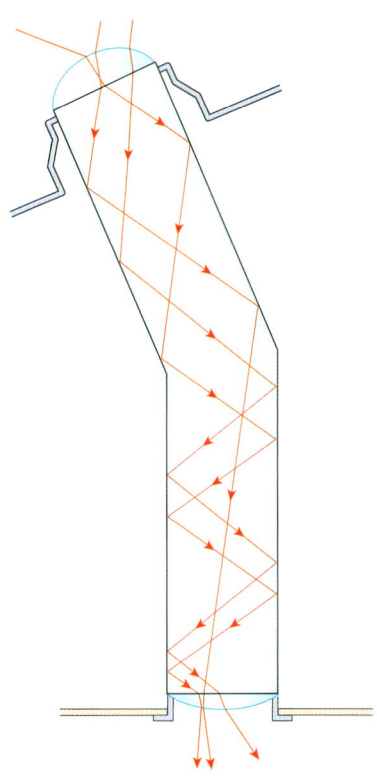

Questions

1. Explain the advantages of putting lots of mirrors in shops.
2. Give examples to explain the meaning of these terms: opaque, transparent, translucent, reflective.
3. Describe the properties needed for the ideal window glass. Justify each property.

Lenses and images

A simple camera

A camera lens refracts rays of light from an object to form an image on film (or an electronic screen, in a digital camera). This **real image** is of course much smaller than the object itself. And it is upside down. A camera lens is a **convex** or **converging lens**.

Some cameras also use a **concave** or **diverging lens**, in the viewfinder. This lens produces a small, upright image of the same object. This is called a **virtual image** because it cannot be formed on a screen.

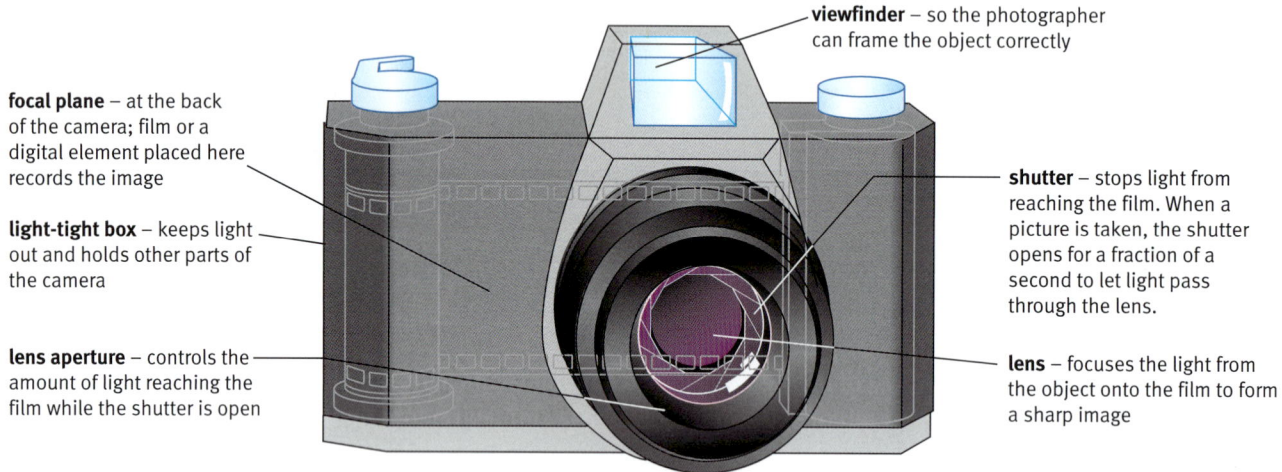

- **viewfinder** – so the photographer can frame the object correctly
- **focal plane** – at the back of the camera; film or a digital element placed here records the image
- **light-tight box** – keeps light out and holds other parts of the camera
- **lens aperture** – controls the amount of light reaching the film while the shutter is open
- **shutter** – stops light from reaching the film. When a picture is taken, the shutter opens for a fraction of a second to let light pass through the lens.
- **lens** – focuses the light from the object onto the film to form a sharp image

Focusing a camera

Some cameras can be adjusted to make a clear image whether the object is near or far away.

- For distant objects, you move the lens towards the film. The image is smaller.
- For near objects, you move the lens towards the object.

Other cameras have a fixed-focus lens, which cannot be moved. If an object is too near the camera, its image is blurred.

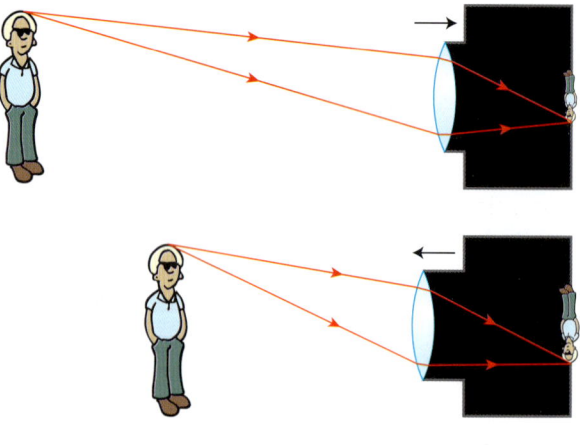

Focusing a camera by moving the lens

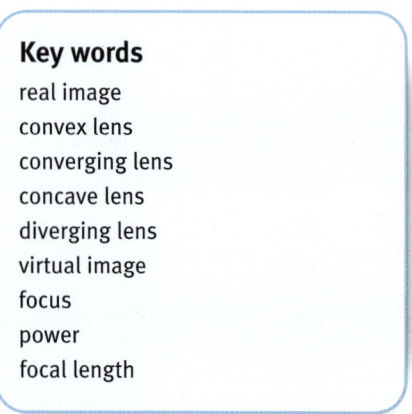

Key words
real image
convex lens
converging lens
concave lens
diverging lens
virtual image
focus
power
focal length

The power of a lens

wide-angle lens, short focal length	standard lens	telephoto lens, long focal length
f less than 35 mm	f between 40 and 60 mm	f greater than 70 mm

Camera lenses are labelled with their focal lengths. The focal length required depends on the size of the camera body.

The **focus** of a lens is the point to which it converges parallel rays of light, from a distant source. Using a magnifying glass to burn a hole in a piece of paper works like this.

Light from a source placed at the focus of a lens becomes a parallel beam. Searchlights work like this.

Opticians prescribe spectacle lenses by their **power**. The power P of a lens, measured in dioptres, is simply

$$P = \frac{1}{f}$$

where f is the **focal length** in metres. The power of a diverging lens is a negative number. The power of a converging lens is positive.

A fat lens has a shorter focal length and a larger power than a thin lens. The fat lens bends light more strongly.

If you stack several thin lenses together, their powers add up: $P = P_1 + P_2$. The optician can add the power of someone's contact lens to the power of the lens in their eye to get their combined power.

Lens coatings let in more light

Cameras collect light from an object and focus it onto a photographic film or electronic screen. The film or screen can only record the image if enough light gets through the lens. The lens is often coated with a thin film to make it transmit more light.

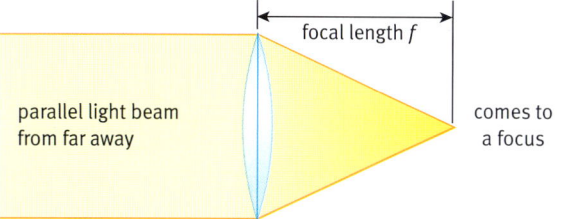

The focus of a lens is the point to which it converges rays of light parallel to the lens axis.

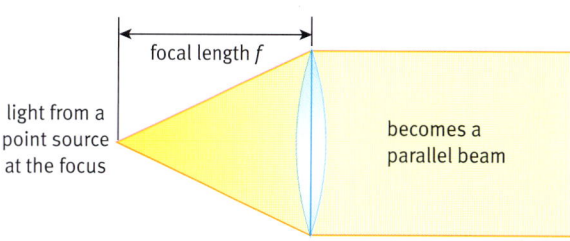

Light from a source placed at the focus of a lens becomes a parallel beam.

Questions

1. Describe how three different materials are used in a camera. Justify their use.
2. Name the type of camera lens that has each of these focal lengths:
   - **a** 135 mm
   - **b** 55 mm
   - **c** 28 mm
   - **d** 85 mm
   - **e** 35 mm
3. What is the difference between a real image and a virtual image?

Refractive index

This pencil in water looks broken, This is because light changes direction as it enters and leaves the water. The bending effect is called **refraction**. Glass and other transparent materials also refract light, though by different amounts.

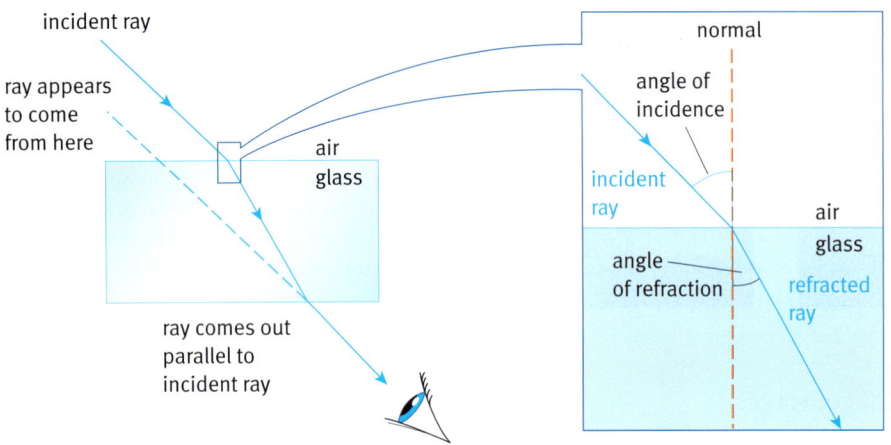

The broken pencil illusion.

The diagram above shows how a ray passes through a glass block. The line at right-angles to the side of the block is called the normal. The ray is refracted towards the normal when it enters the block, and away from the normal when it leaves it.

With a rectangular block like this one, the ray comes out parallel to its original direction. If the ray strikes the block at right-angles (square on), it goes straight through without being refracted.

The light ray is refracted when it enters glass because it slows down. A material that slows down light a lot has a high **refractive index**.

The greater the refractive index of a material, the more it can bend rays of light. So making a lens from glass with a high refractive index allows it to be thinner. This also makes it lighter, which is important for portable devices like binoculars or spectacles.

The refractive index of glass depends on what it is made of. Glass with a lot of lead has a higher refractive index than window glass, making it ideal for use in decorative cut glass.

Material	Refractive index
air	1.00
water	1.33
glass	1.4–1.6
acrylic (e.g. Perspex)	1.49
diamond	2.4

Some materials and their refractive indices

Questions

1. Explain why the best spectacle lenses are made from glass with a high refractive index.
2. Your eyes refract light that enters the eyeball to create a focused image. Use ideas of refraction to suggest why you need to wear goggles to see clearly under water.

Key words
refraction
refractive index

Transparency

Many homes have cable TV. The same cable often provides radio and telephone communications too. Light carries information down the cable.

Non-cable television signals are broadcast as radio waves through the air, and telephones work using electric currents in wires. Cable was made possible by the invention of very pure glass. The 'cable' is buried under the pavement, and each subscriber is linked up to it.

At the core of the cable is a very fine glass fibre, called an **optical fibre**. The cable company converts signals of the different TV channels into fast-changing pulses of laser light, on and off. That light is sent into the glass fibre.

The glass used to make this optical fibre must be of very high purity. Otherwise the laser light would be absorbed long before it reached subscribers. The glass has very high **transparency**.

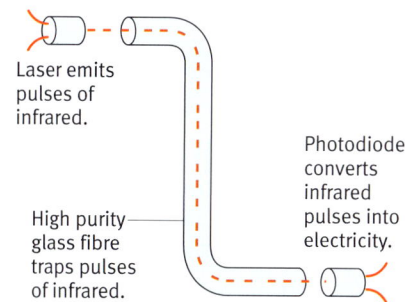

The laser sends pulses of infrared light into one end of the fibre. They cannot escape until they reach the other end, where they are detected by the photodiode. The high transparency of the glass allows the pulses to travel hundreds of kilometres through the fibre.

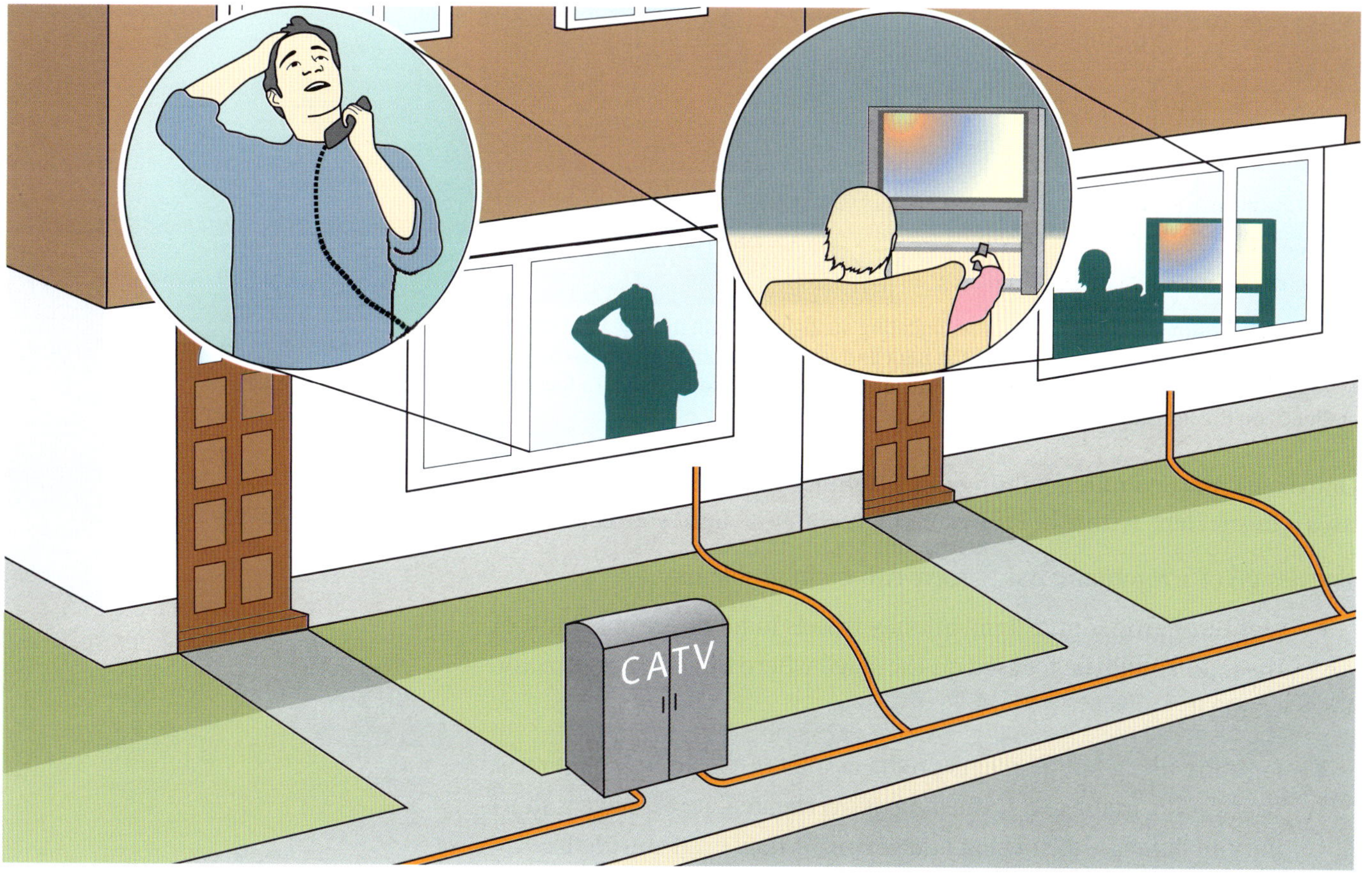

Question

1 Describe uses of optical fibres other than carrying TV signals.

Key words

optical fibre
transparency

Stage lighting

Stage lighting creates mood and interest, whether it's a huge pop concert or a drama production in a small studio. For film and TV productions, the lighting units must also provide enough light for the cameras to get good pictures of the action.

The colour of light and the angle at which it strikes surfaces make a big difference to how things appear. Next time you have a chance, notice how stage lighting is used.

Usually the performers are lit from above, imitating sunlight. A main or 'key' light will fill the whole set. Additional lights used at the back and front of the stage give a perception of depth and partially fill any shadows. 'Follow' spotlights may also be used, to focus the audience on the action.

One of the most common lighting units (luminaires) used in a film studio uses a fresnel lens with a hot tungsten filament as its light source.

The design of a luminaire has to solve two main problems:

- Light from the source lamp initially travels in all directions. Most of the light should be focused on the stage where the lighting director wants it.

- Lighting using hot filaments produces a lot of waste heat that must be carefully managed. If the lamp is overheated, it will have a short life and damage its housing and wiring. It may even start a fire.

> **Question**
> 1 You can read about TV studio lighting on pages 8–9. Explain why
>   a stage lighting units that use infrared reflectors produce 'cool' light
>   b it is important that stage lighting units absorb ultraviolet

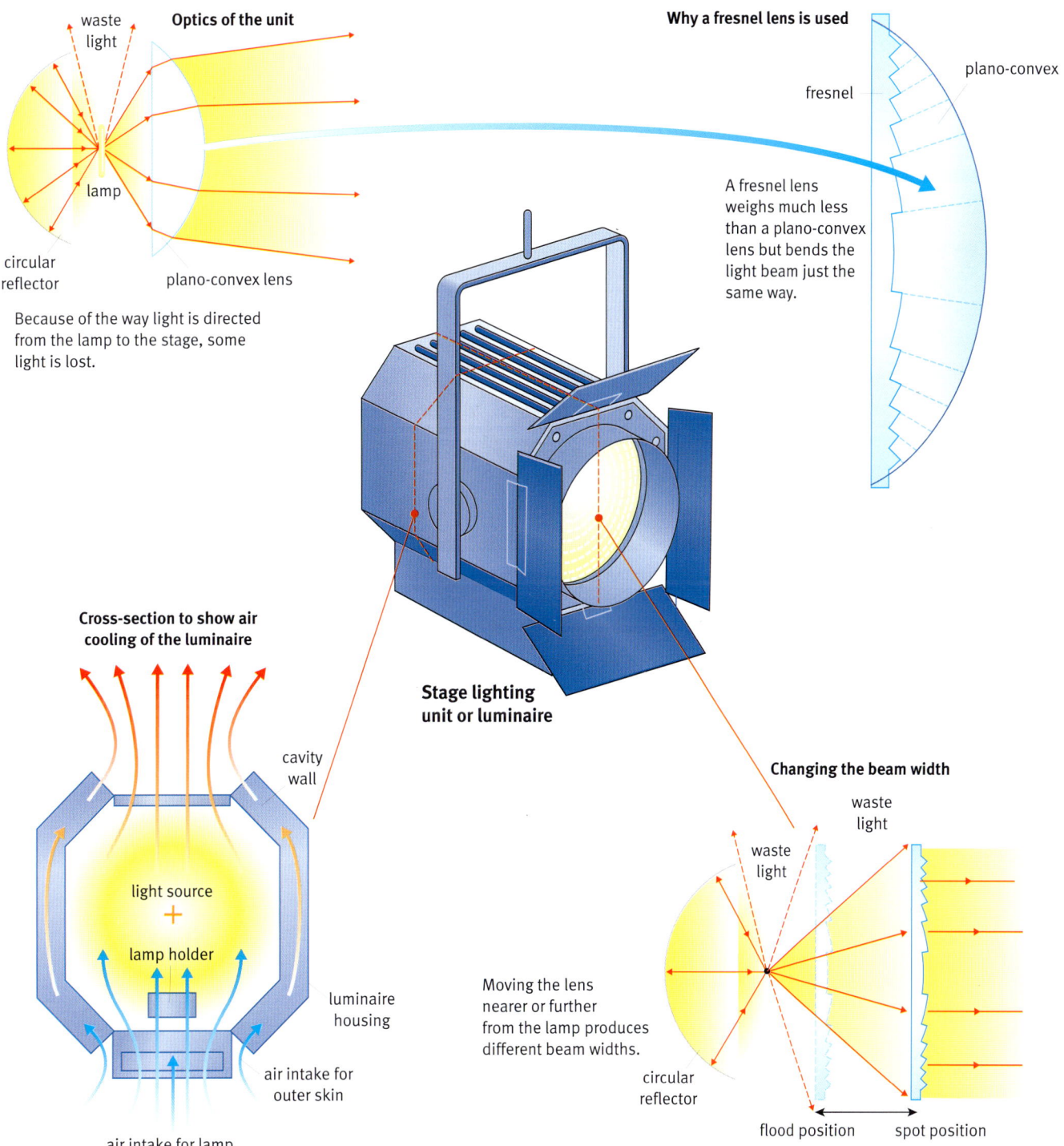

Acoustic properties

Noise is more than just a nuisance. It can also seriously damage your health. It is important to know how the materials in a building affect sound waves:

- Will they stop sound getting in from outside?
- Will they stop echoes interfering with people's conversations?
- How much sound will leak from one place to another?

Careful choice of building materials keeps unwanted sound away from people.

Sound is caused by vibrations. The energy of the vibrations is carried by pressure waves through solids, liquids, or gases in contact with the sound source. How far that energy travels depends on the **acoustic** properties of the material.

- Materials with lots of tiny cracks in them are acoustically soft. They **absorb** sound, because vibrations warm the material.
- Solid materials with no cracks are acoustically hard. They either **reflect** sound or **transmit** it.

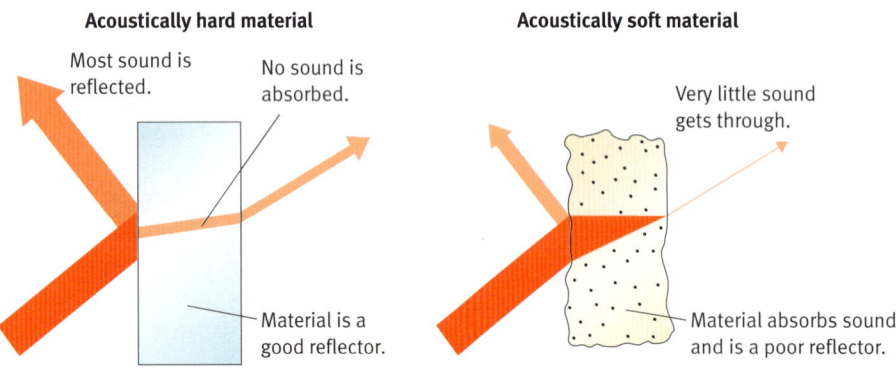

Unwanted sound can be dealt with in two ways. It can be reflected by hard surfaces, changing its direction. Or sound can be absorbed by soft surfaces, transferring its energy into heat.

Measuring sound

The pitch and intensity of a sound determine how much of a nuisance it is. The **pitch** depends on how rapidly its source is vibrating. It is usually given as the **frequency** of the sound. This is the number of vibrations of the source per second, measured in **hertz** (Hz).

The **intensity** of a sound is a measure of how much energy reaches your ear every second. The **loudness** of a sound is how you perceive it. Loudness depends not only on the intensity of a sound, but also its frequency and how long it lasts.

Humans can hear sounds only in the frequency range of 20 Hz to 20 000 Hz (20 kHz), and the ear is most sensitive to sounds at about 2 kHz. In other words, if the intensity of a sound stays the same while its frequency changes, it will sound loudest at about 2 kHz.

Intensity is measured in **decibels** (dB). A sound that increases by 10 decibels doubles its loudness.

Key words
acoustic
absorb
reflect
transmit
pitch
frequency
hertz
intensity
loudness
decibel

The science

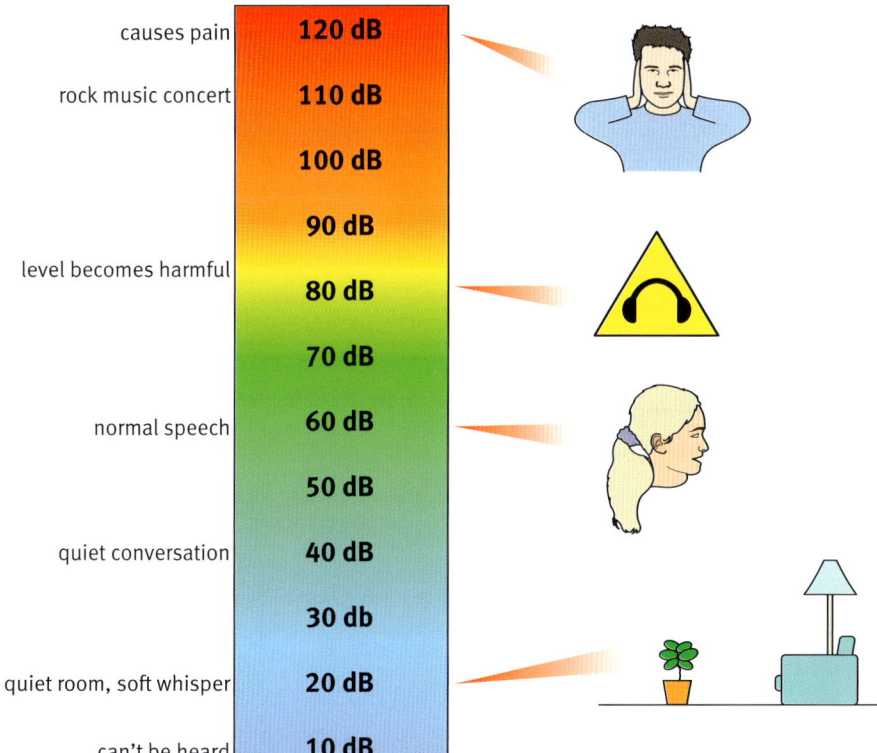

The decibel scale for sound intensity

Questions

1. A guitar string vibrates 256 times a second. What is the frequency of the sound it produces?

2. Describe the difference in structure between an acoustically soft material and an acoustically hard material.

3. Why are sound levels above 90 dB dangerous?

4. Describe and explain two ways of reducing sound levels.

Sound control

Ringing in the ears

Have you ever had a ringing sound in your ears after listening to music, perhaps at a rock concert? This was probably the result of exposure to sound levels above 85 dB. That level of sound often results in temporary loss of hearing too.

The damage may become permanent if you listen to sound at 90 dB or more for long periods of time, perhaps at work. You gradually lose the ability to hear high frequencies, eventually becoming deaf.

In some cases, the ringing sound doesn't go away but carries on and on. This distressing condition is called **tinnitus**.

There are three ways of reducing the level of unwanted sound in your ears:

- Move away from its source.
- Absorb the sound before it gets to you.
- Reflect it somewhere else.

Double glazing

Window glass is acoustically hard. Any sound reaching one side of it goes straight through, with almost no absorption. So single-pane windows transmit outside noise into buildings.

Using two panes of glass, as in double glazing, cuts out noise very effectively. Most of the sound that gets into the cavity is trapped there. It bounces off the inner surfaces of the glass until it has been absorbed by the air. Triple glazing is even more effective.

> **Questions**
> 1. Describe the consequences of prolonged exposure to high sound levels.
> 2. Explain why motorways near housing estates are often edged with high walls.
> 3. Explain how foam-filled cavity walls in buildings can provide sound insulation.

Key word
tinnitus

People who work in a dangerously noisy environment need to protect their ears.

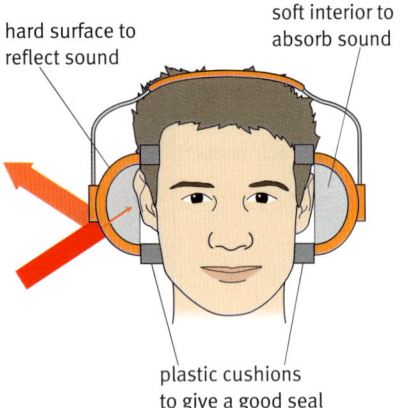

How ear defenders work

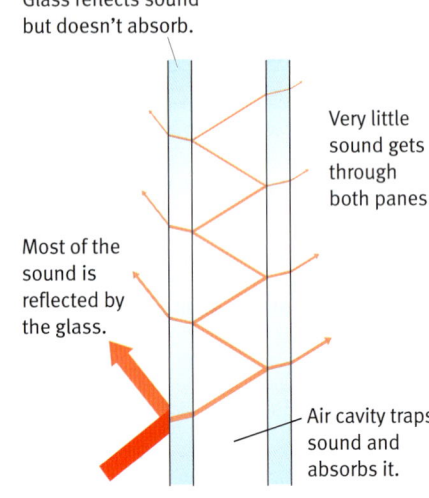

Double glazing not only keeps the heat in, it also helps keep noise out (or in).

Bad vibrations

Low frequency sound is difficult to manage. This is because acoustically soft materials absorb high frequency sound much better than low frequency sound. The graph opposite shows this. Frequencies below the audible range (under 20 Hz) are particularly troublesome. You feel them, instead of hearing them.

There are two ways of dealing with low frequency vibrations:

- Stop them happening in the first place.
- Isolate the source by mounting it on special absorbing material.

Sympathetic vibration

Annoying vibrations happen if the panels of a car vibrate in sympathy with the engine at a particular speed. Stiffening each panel by the right amount can suppress the vibration, making the car interior quieter.

Damping vibration

A car engine works by using rapidly repeating explosions, so you cannot stop it vibrating. However, it is possible to reduce the vibrations transmitted to the rest of the car by mounting the engine on pieces of rubber. This is because rubber is a good **damping** material. Rubber pads transfer the energy of vibrations into heat, as shown in this graph:

Rubber is also good at isolating vibrating machinery from the building that houses it. But it is not strong enough for very large structures, such as buildings and bridges. These use **fluid-filled dampers**. The vibrations force pistons to move inside cylinders of oil, transferring the energy of the vibration into heat.

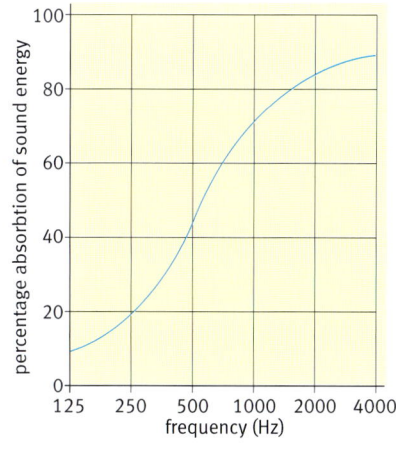

Foam tiles absorb high frequency sound much better than low frequency sound.

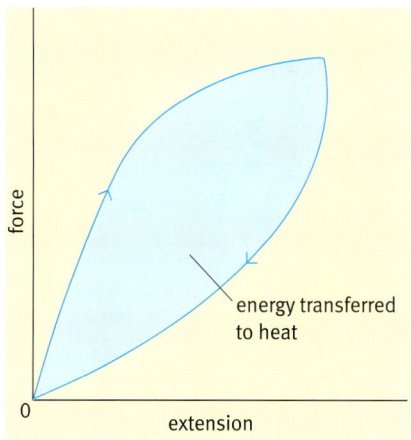

Rubber extends and contracts differently, heating up in the process.

> **Questions**
> 1 Describe two ways of damping vibrations of machinery.
> 2 Why are low frequency sounds troublesome?

> **Key words**
> damping
> fluid-filled dampers

This bridge was built without enough damping. Wind could set it vibrating. Shortly after this photograph was taken, the bridge fell to pieces.

Procedures and techniques

Comparing stiffness

CHECK SAFETY — Never work unsupervised

➥ Principle
The **stiffness** of a material describes how much it changes shape when an external force is applied.

➥ Equipment
- clamp, needle, clamp stand, ruler, slotted masses
- samples of different materials with the same dimensions (same length, width, and thickness)

➥ Procedure

➥ **Method 1 – material clamped at one end and loaded at the other**

1. Clamp one end of the material to the bench, so that 20 cm sticks out freely.

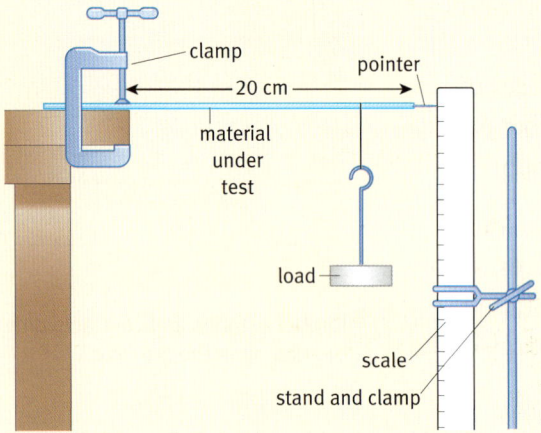

2. Tie the mass hanger securely to the end of the material using string. Keep feet clear!
3. Fix a needle to the end of the material using sticky tape.
4. Adjust the metre rule so that the needle is lined up with the zero on the scale.
5. Carefully add masses to the hanger until the needle has moved down exactly 10 mm. (Use larger and smaller masses to get as close as you can to 10 mm.) Record the mass.
6. Repeat with samples of other materials.

➥ **Method 2 – material supported at both ends and loaded at the mid-point**

1. Stand the material on two supports, as shown in the diagram, so that 20 cm is unsupported between them.

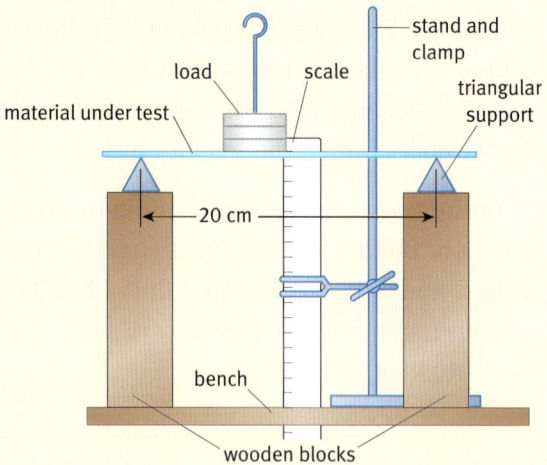

2. Tie the mass hanger securely to the middle of the material using string. Alternatively, place it on top of the material.
3. Adjust the metre rule so that the underside of the material is lined up with the zero on the scale.
4. Carefully add masses to the hanger until the material has moved down exactly 10 mm. (Use larger and smaller masses to get as close as you can to 10 mm.) Record the mass.
5. Repeat as above with samples of other materials.

➥ Interpreting the result
The greater the mass needed to bend the material, the greater its stiffness.

Comparing the strength of fibres

⇨ Principle

The **tensile strength** of a material describes the stretching force needed to break it.

⇨ Equipment

- G-clamp
- clamp stand
- metal slabs
- slotted masses and holder
- samples of different materials shaped as wires with the same length and thickness
- eye protection

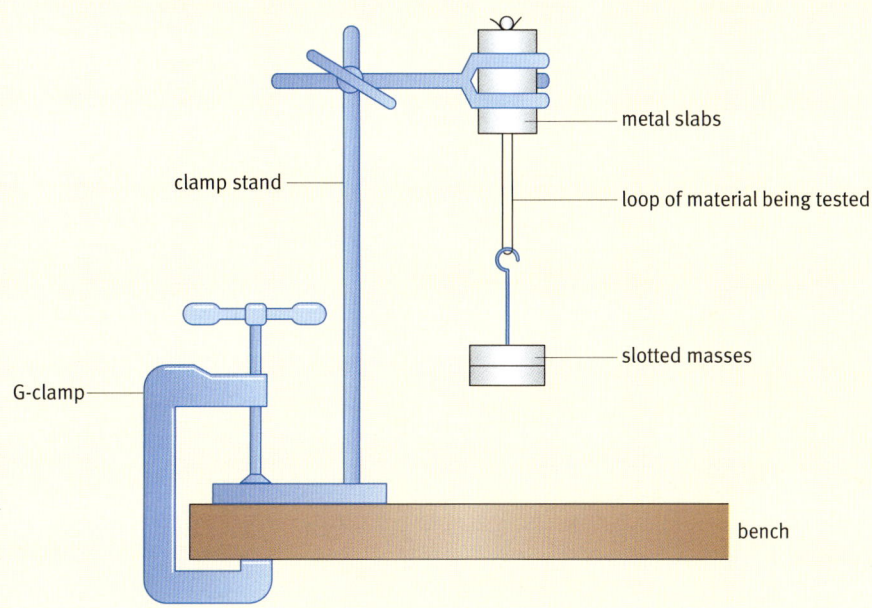

⇨ Procedure

1. Clamp the clamp stand to the bench. This stops it falling over.
2. Tie the ends of the material together to make a loop.
3. Tightly clamp the loop of material between the metal slabs.
4. Suspend the first slotted mass from the loop of material.
5. Add slotted masses until the material breaks.

⇨ Interpreting the result

The greater the mass needed to break the material, the greater its strength.

47

Finding the density of a solid

CHECK SAFETY
Never work unsupervised

Principle

A dense material has a lot of mass packed into a small volume. If you measure the mass and volume of an object, you can then calculate the **density** of the material from which it is made.

$$\text{density} = \frac{\text{mass}}{\text{volume}}$$

Units: g/cm^3 or kg/m^3

Example: A stone of mass 70 g has a volume of $20\,cm^3$.

$$\text{density of stone} = \frac{\text{mass}}{\text{volume}} = \frac{70\,g}{20\,cm^3} = 3.5\,g/cm^3$$

Equipment

- balance
- displacement can
- measuring cylinder
- object being measured

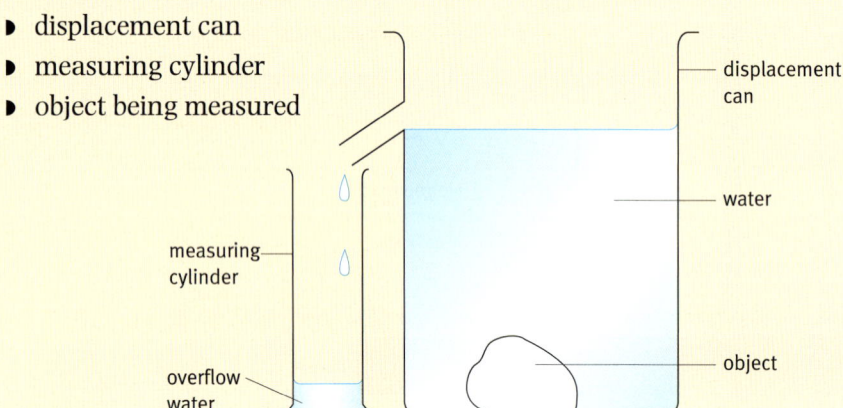

Procedure

An object may have an irregular shape. To find its volume, it is easiest to measure the volume of water it displaces.

1. Use a balance to weigh the object. Record its mass.
2. Fill the displacement can with water to the level of the spout.
3. Immerse the object and collect the water it displaces. Measure and record the volume of water displaced.
4. Calculate the density of the material.

Interpreting the result

The density of water is $1.0\,g/cm^3$ or $1000\,kg/m^3$. Materials with a density greater than this will sink in water. Those with a density less than this will float.

Comparing hardness

Principle

The **hardness** of a material is a measure of how difficult it is to dent or scratch.

Equipment

- plastic tubing
- metal punch
- wooden block guide
- 500 g slotted masses and holder
- string
- samples of materials being tested
- magnifier with scale

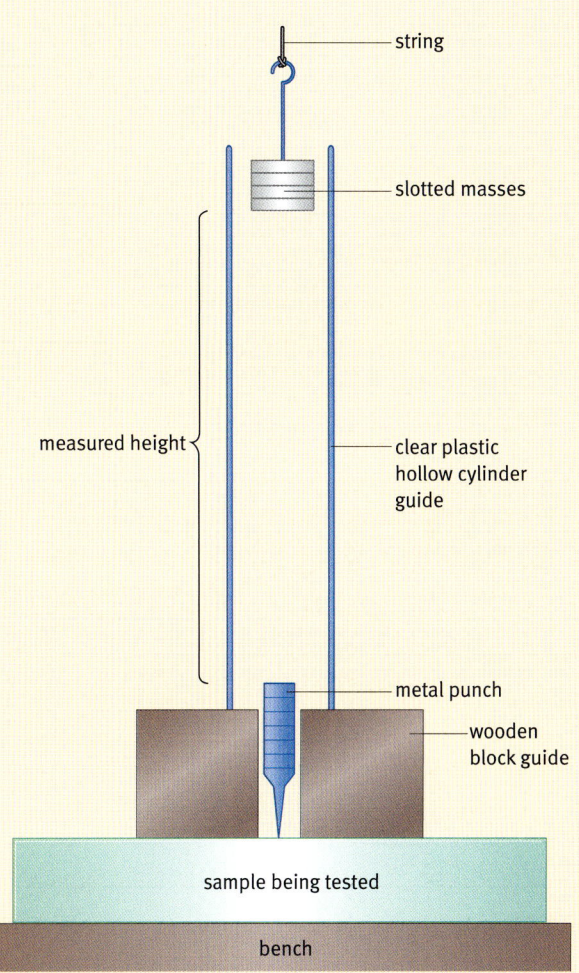

Procedure

1 Place the sample on a bench.

2 Clamp the length of clear plastic tubing above the sample. Clamp the ruler beside the tube. Place the metal punch on the sample, directly below the tube.

3 The slotted masses have a mass of 500 g. Tie a length of string to the hook.

4 Hold the string so that the bottom of the hanger is exactly 5 cm above the sample. Let go of the string.

5 Examine the sample to see if the metal punch has dented it. If not, try dropping the masses from a greater height. Record the height needed to produce a visible dent in the material.

6 Measure the diameter of the dent produced. You may need to use a magnifier with a scale in the field of view.

7 Repeat with samples of other materials.

Interpreting the result

The smaller the diameter of the dent (and the greater the height from which the masses are dropped), the harder the material.

Measuring electrical conductance

CHECK SAFETY
Never work unsupervised

➔ Principles

If a length of wire has a high **conductance**, it will let a large **current** flow even when the **potential difference** is small. If you *measure* the current flowing through the wire and the potential difference across it, you can *calculate* the wire's conductance.

$$\text{conductance} = \frac{\text{current}}{\text{potential difference}}$$

Unit: siemens (S)

➔ Equipment

- ammeter (or multimeter)
- voltmeter (or multimeter)
- resistance wire
- low-voltage power supply with variable output
- crocodile clips (× 2)
- leads

➔ Procedure

1. Set the power supply to 0 V. Connect the power supply, ammeter, and length of resistance wire to make a circuit. (Use crocodile clips to connect onto the wire.)
2. Connect the voltmeter to the ends of the wire.
3. Turn up the power supply until a current of about 1 A flows. Record the current and the potential difference. Switch off the power supply.
4. Calculate the conductance of the wire.

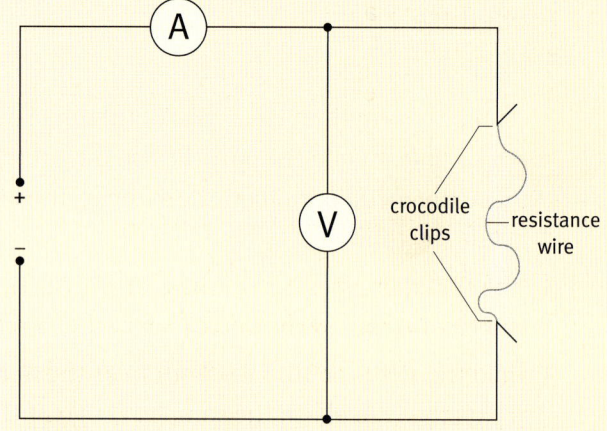

➔ Using a graph

1. Connect the circuit as before.
2. Turn up the power supply a little at a time. Record pairs of current and potential difference values in a table.
3. Draw a graph of current (*y* axis) against potential difference (*x* axis). Draw a smooth line close to the points of the graph, to show the pattern of the data.
4. Choose any convenient point on the graph. Read off values of current and potential difference, and calculate the conductance.

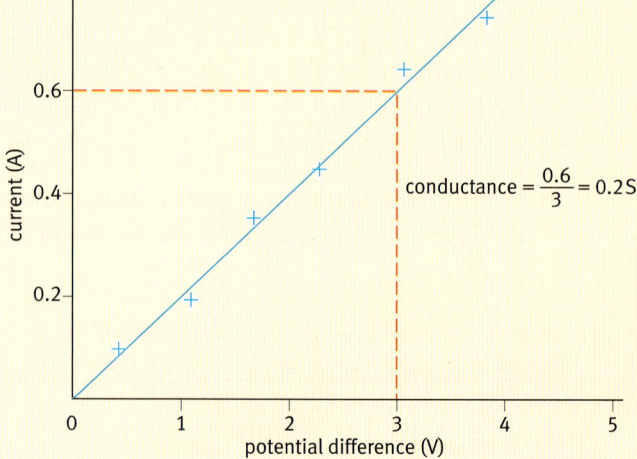

➔ Interpreting the result

For a given material,

- a shorter sample will have a higher conductance
- a sample with a larger cross-section will have a higher conductance

Comparing thermal expansions

CHECK SAFETY
Never work unsupervised

Principle

Any material will expand when its temperature rises, but some materials expand more than others.
You can compare their **thermal expansion** by using samples of the same cross-section and length.

Equipment

- Bunsen burner
- pin
- plastic straw
- metal block
- protractor
- two clamps
- two clamp stands
- stopclock
- rods of different metals of the same length and thickness
- G-clamp

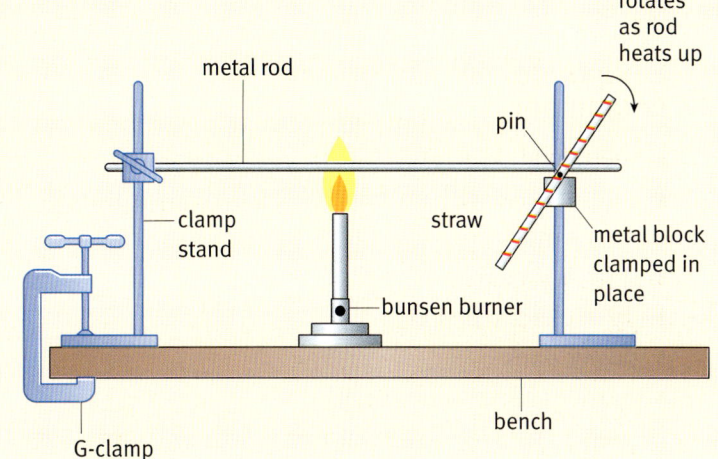

Procedure

1 Use one stand to hold the flat surface of the metal block above the bench.
2 Clamp the other stand to the bench, and firmly clamp one end of the metal rod to this stand.
3 Push the pin through the centre of the straw. The straw must rotate when the pin is twisted.
4 Place the pin between the metal rod and the flat metal block. The straw must be vertical.
5 Light the Bunsen burner and adjust it to a roaring flame.
6 Use the flame to heat the centre of the rod. Start the stopclock.
7 Stop the stopclock when the straw has rotated to horizontal.

CHECK SAFETY

Take care not to get burnt on the hot central portion of the rod.

Interpreting the result

As its temperature rises, the rod expands and pushes the pin round. With the same amount of heating, a material that expands more will push the pin round further. So the rod with the shortest time for a rotation of 90° has the largest thermal expansion.

This procedure is only approximate. It assumes that the temperature rise of the centre of the rod is the same for all of the metals tested. A more accurate procedure would take account of two other thermal properties: the thermal conductance of each material, and how much its temperature rises for a given amount of heating.

Comparing poor thermal conductors

CHECK SAFETY
Never work unsupervised

Principle

An object that conducts heat well has a high **thermal conductance**. Metal components have high thermal conductance. Polymers, glass, and wood components have low thermal conductance.

The thermal conductance of a poor conductor can be compared with a standard material whose thermal conductance is known.

Equipment

- metal plates (× 2)
- multimeters with probes to measure temperature (× 3)
- 100 g mass
- 250 cm³ beaker
- 12 V electric heater
- power pack
- hardboard reference plate
- specimens to test, of the same thickness as the reference plate
- eye protection

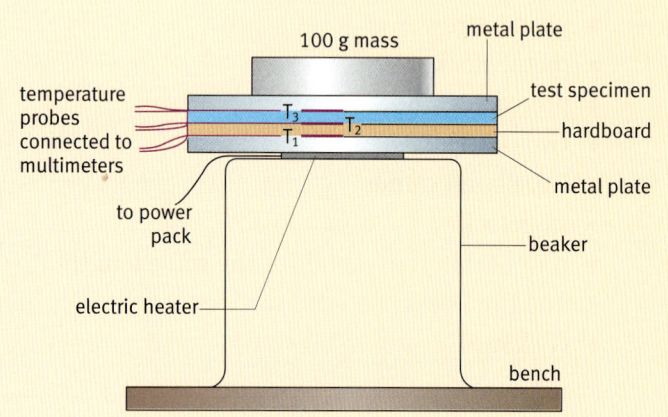

Procedure

1. Place on top of the upturned beaker, in order:
   - the heater
   - a metal plate
   - the hardboard
   - a test specimen
   - another metal plate
   - the 100 g mass
2. Insert the three temperature probes as shown in the diagram.
3. Connect the heater to the power pack.
4. Set the power pack to 12 V and switch on.
5. Wait until all three temperature readings are steady.
6. Once the readings have been steady for 5 minutes, record the values of T_1, T_2, and T_3.
7. Switch off the heater and let the whole apparatus cool down before dismantling it.

Interpreting the result

Hardboard has a thermal conductance, measured in tog values, of 0.3. The thermal conductance of the specimen is calculated using the formula

$$\text{tog value} = 0.3 \times \frac{T_2 - T_3}{T_1 - T_2}$$

Comparing good thermal conductors

CHECK SAFETY
Never work unsupervised

➔ Principle

An object that conducts heat well has a high **thermal conductance**. Metal components have high thermal conductance. Polymers, glass, and wood components have low thermal conductance.

You can get an approximate comparison of how easily heat travels along good conductors using the procedure below.

➔ Equipment

- metal rods of the same size (same length and diameter)
- Bunsen burner
- tripod
- Vaseline
- matchsticks
- stopclock

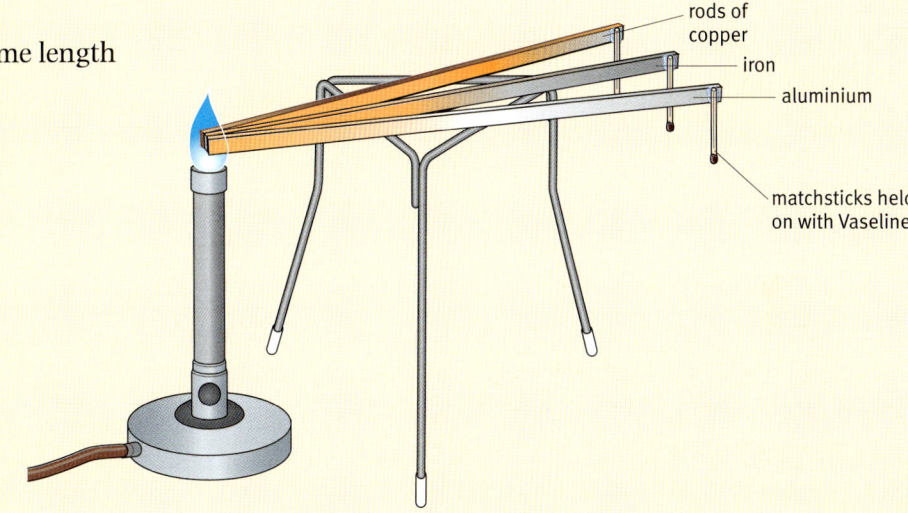

➔ Procedure

1. Arrange the rods on the tripod as shown. Use a small amount of petroleum jelly (Vaseline) to fix a matchstick to one end of each rod.
2. Adjust the Bunsen burner to give a gentle (not roaring) blue flame. Start heating the other end of the rods, and at the same time, start the stopclock.
3. Note the time at which the matchstick falls off the end of each rod.
4. **Take care not to burn yourself on the hot ends of the rods when you dismantle the apparatus.**

➔ Interpreting the result

The shorter the recorded time, the higher the thermal conductance of the metal.

Like the procedure used on page 51, this procedure is only approximate. A more accurate procedure would compare the rates at which energy must be supplied to maintain a fixed temperature difference along well-insulated rods.

Focal length of a converging lens

CHECK SAFETY
Never work unsupervised

⇨ Principle

The **focal length** of a lens is the distance from the lens to its focal point. A fat lens has a short focal length. A thin lens has a longer focal length.

There are several ways to measure the focal length of a convex lens.

⇨ Equipment

- convex lens
- metre rule
- light box with cross wires and screen
- plane mirror

⇨ Procedure

⇨ Method 1 – light from a distant object is focused at the focal point

1. Stand near the wall at the opposite side of the room to the window.
2. Hold up the lens and use it to focus an image of the window on the wall. Hold the lens so that the image is as clear as possible.
3. Measure the distance from the lens to the wall. This is the focal length f of the lens.

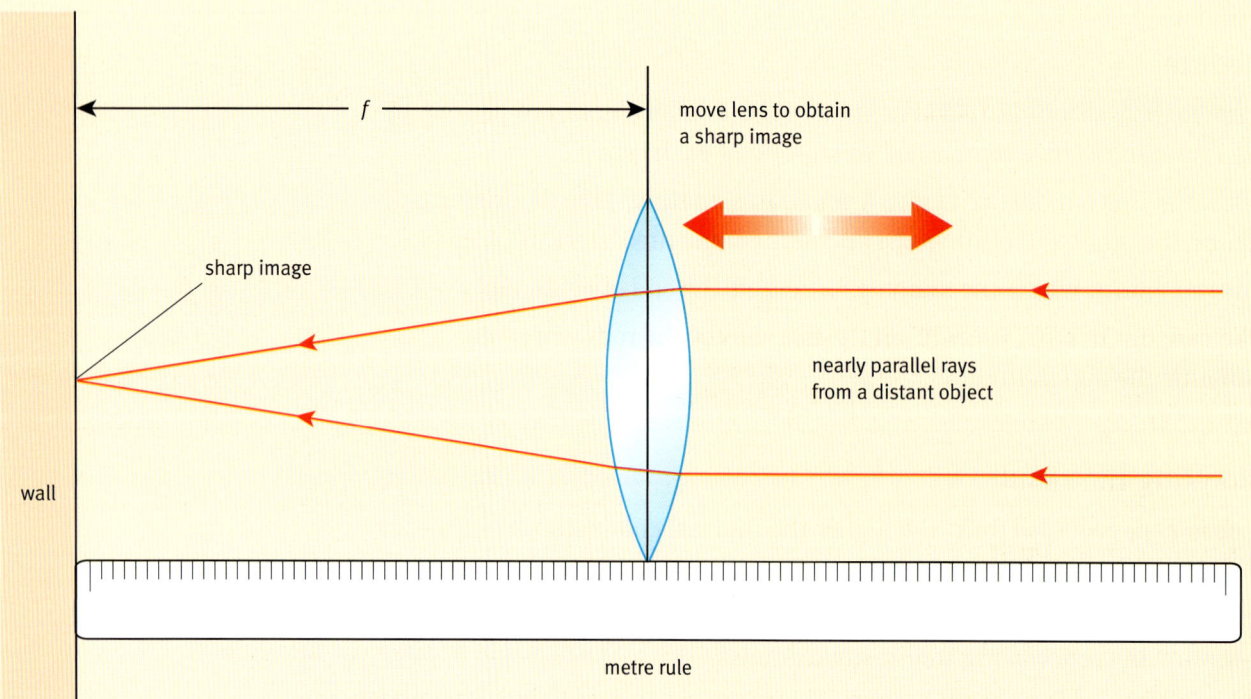

⇨ Method 2 – using a plane mirror

1. Stand the lens in front of the mirror. Place the light box behind the screen.
2. Check that the power supply is set at the correct voltage and switch it on. See how light is reflected back by the mirror onto the front of the screen. Look for an image of the cross wires on the screen.
3. Move the lens towards the screen or away from it, until the image of the cross wires is clearly focused.
4. Measure the distance from the screen to the lens. This is the focal length f of the lens.

CHECK SAFETY
Never work unsupervised

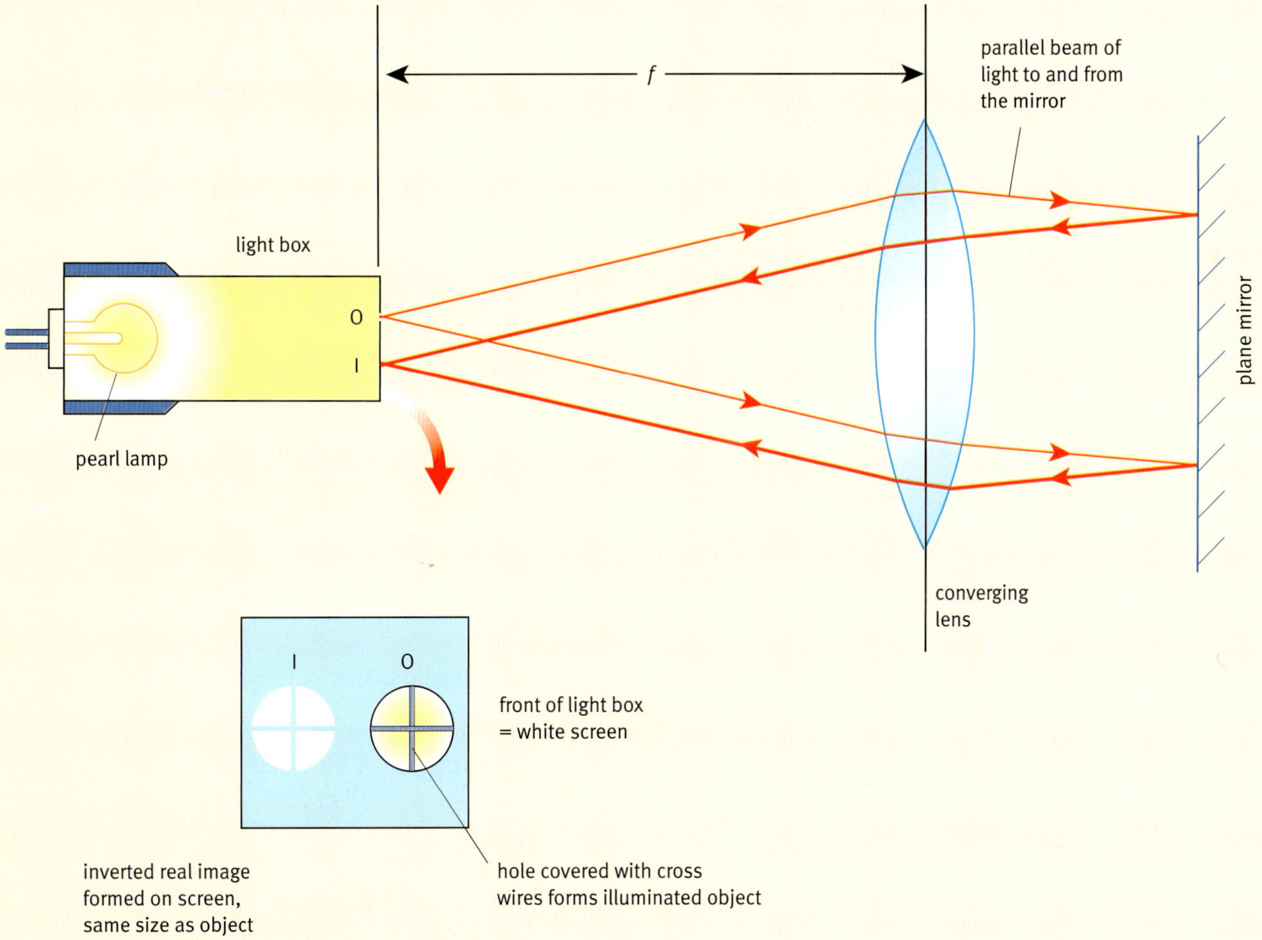

⇨ **Interpreting the result**

- A standard camera lens has a focal length of between 40 and 60 mm.
- Camera lenses with focal lengths less than 35 mm are considered to be wide-angle lenses.
- Lenses with focal lengths more than 70 mm are considered to be telephoto lenses.

Measuring sound levels

→ Principle

The sound level or **intensity** of a sound in **decibels** measures how much energy it delivers to surfaces which absorb it.

→ Equipment

- decibel meter
- sound sources

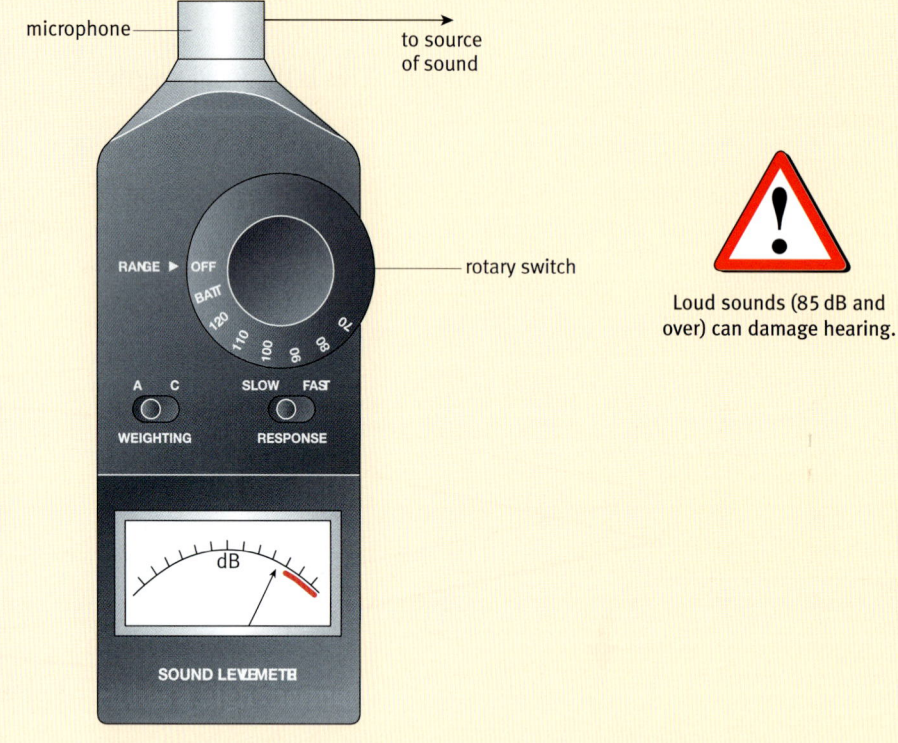

CHECK SAFETY
Never work unsupervised

Loud sounds (85 dB and over) can damage hearing.

→ Procedure

1. Use the rotary switch to set the meter to 'BATT' (battery test). If the needle is not in the red region, replace the battery.
2. Set the weighting switch to match the type of sound you are measuring. Use A for noise and C for music.
3. Set the response switch to SLOW if you are measuring noise or FAST to follow rapid changes.
4. Point the microphone at right angles to the source of sound, as shown in the diagram.
5. Rotate the switch to 120 dB and reduce it until the needle stays in the middle of the range.
6. Add the reading of the needle to the setting of the switch to obtain the sound level in dB. For example, if the needle reads 9 and the switch setting is 110, the sound level is 119 dB.

→ Interpreting the result

The intensity of a sound reduces as you get further away from its source, so to compare sound sources you must measure them from a fixed distance.

It is illegal to have sounds above 80 dB in the workplace without ear protection.

Presenting data from standard procedures

Principles

Suppose you have just performed a standard procedure for a client. How do you present your findings in the most efficient way? You need to do four things.

- State the procedure used.
- Present a table of your measurements, with units clearly shown.
- Draw a graph to present all of your results as a picture.
- State what you can conclude from the procedure.

By presenting data in this way, the client can instantly tell what you did, what you found out, and make a judgement about its reliability.

REPORT ON THE TENSILE STRENGTH OF ELASMAX FIBRES

Procedure

An Elasmax fibre of length 50 cm and thickness 0.76 mm was clamped at one end and the other end passed over a pulley wheel. The free end was tied to a scale pan. A piece of sticky tape was fastened 200 mm from the clamped end, over the 0 mm mark of a scale. The extension of the fibre was measured as weights were added to the scale pan. Weights were added until the fibre snapped. The procedure was repeated for three fibres of the same length and thickness.

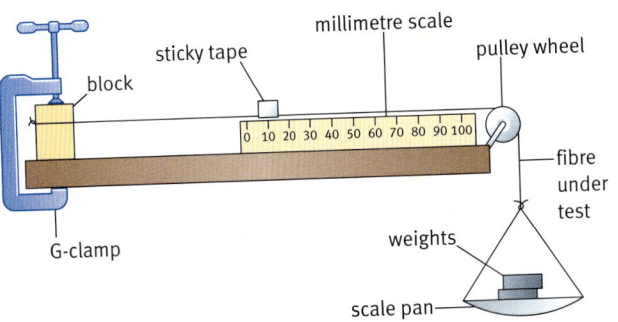

Table of measurements

Weight (N)	Extension 1 (mm)	Extension 2 (mm)	Extension 3 (mm)	Average extension (mm)
0	0	0	0	0
5	8	12	10	10
10	12	16	20	16
15	30	34	30	33
20	38	44	**75**	52
25	48	51	52	50
30	61	59	60	60
35	63	63	66	64
40	64	63	66	64
45	64	64	66	64
50	snaps	snaps	snaps	

Graph to show extension due to loading of fibres

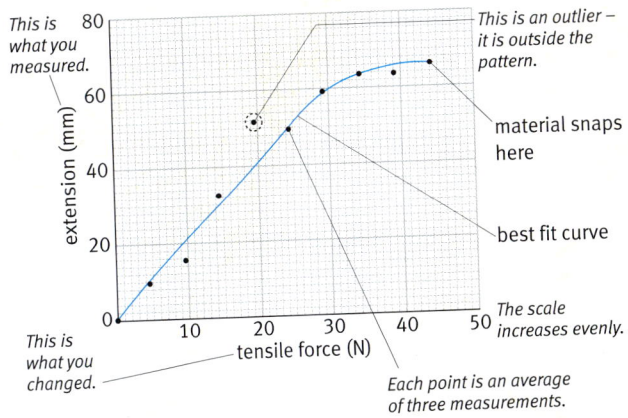

Summary

Although each fibre stretched slightly differently, they all snapped at a loading between 45 N and 50 N. One measurement, in bold in the table, is probably wrong due to a flaw in the sample.

Your Work-related portfolio

This Additional Applied Science course aims to help you

- carry out specific scientific procedures and understand that the results matter
- apply science knowledge and techniques, to solve problems
- learn about a variety of science-based workplaces
- select, organise, and communicate information clearly and logically

You will show your progress with these skills through your Work-related portfolio.

The Work-related portfolio counts for 50% of your total mark. Your school or college will give you details of the marking scheme for each part of it. This will help you check that your work meets the criteria for success.

Across the three modules that you study, the Work-related portfolio requires

- six **Standard procedures** (two from each module)
- one **Suitability test** (from any module)
- one **Work-related report** (from any module)

Work-related portfolio (50% of total marks)

Standard procedures (6 × 2% = 12% of total marks)

A Standard procedure is a series of practical steps, often including scientific techniques, that will achieve the same result no matter who carries it out. It involves following instructions, working safely, and making measurements or observations carefully. You will carry out six, each counting for 2%.

Suitability test (21% of total marks)

Suitability tests are another example of how science is used in the workplace. There are three types of test you might carry out:

- testing a material or comparing materials for a particular purpose
- comparing different procedures used for the same purpose
- testing the suitability of a device for a particular purpose

Work-related report (17% of total marks)

This gives you the opportunity to find out about some science-related activity carried out in a real workplace, such as a hospital or factory. You present your findings in a report. Choose your topic carefully, and don't let your report become too large. A good topic will

- interest you
- have information sources which you can identify, obtain, and understand

You will need to
- describe the workplace
- describe relevant applied science and skills

Standard procedures

Typically standard procedures will be carried out and assessed in a single lesson. They will involve a series of steps. Your teacher will tell you what, if anything, needs to be recorded.

Suitability test

You will need to

- describe desirable properties or characteristics
- follow or devise a suitable approach
- collect reliable data
- evaluate the suitability of the material, procedure, or device
- communicate through a structured report

Work-related report

Use any of the following sources of information:

- Internet
- school library
- local public library
- TV/video
- newspapers and magazines

You could also gather information from specific people or organisations:

- interview a practitioner (possibly a family member or a neighbour)
- write a letter to an organisation
- telephone to request a leaflet, or to find out who to write to

To obtain useful information from any of these, prepare detailed questions in advance. Explain who you are and what you are doing.

Use all information sources selectively, picking out parts relevant to your topic and writing in your own words. Keep a detailed record of any information sources you use.

Plan your report so that it has a clear and sensible structure. Use charts, graphs, or pictures that help convey the information.

Tip

The best advice is 'plan ahead'. Give your work the time it needs and work steadily and evenly over the time you are given. Your deadlines will come all too quickly, especially as you will have coursework to do in other subjects.

Glossary

absorb Sound, light, or vibrations are absorbed when they are converted to heat within a material, and are not reflected or transmitted.

acoustic To do with sound and hearing.

air bag Safety device in a vehicle which inflates in a crash and cushions the driver and passengers.

alloy A metallic material which is a solid solution of one metal in another, or of a non-metal in a metal.

ammeter A device for measuring the current in an electric circuit.

aperture The size of the opening that lets light into a camera when the shutter is open.

bimetal strip A strip of two metals joined together which have different thermal expansivities. When they are heated, one metal expands more than the other, causing the bimetal strip to bend.

brittle A brittle material snaps cleanly when it breaks as the body of the material cracks. A small deformation is enough to break it.

ceramic A material such as brick or pottery that is hard, strong in compression, and weak in tension.

coating A very thin layer of one material covering another material.

collision time The time it takes for the momentum to change to zero in a collision.

composite A material made up of two or more materials combined, which combines the properties of the constituent materials.

compression A material is in compression when it is being squashed by forces pushing together on it.

compressive strength The maximum load that a material of standard shape and size can withstand before crumbling.

concave lens A concave lens is thinner in the middle than at the edges. Also called a diverging lens, because of its effect on light rays.

conductance Another name for electrical conductance.

consistency (of a product) The proportion of a manufacturer's products that fully meet the specification (one aspect of product standards).

converging lens A lens that bends light rays so that they are closer together. Also called a convex lens, because of its shape.

convex lens A convex lens is thicker in the middle than at the edges. Also called a converging lens, because of its effect on light rays.

corrosion Damaging chemical reaction at the surface of a solid material, e.g. rusting of iron.

corrosion resistance The ability of a material not to take part in corrosion reactions, e.g. a steel alloy will resist rusting.

crumple zone Part of a car which easily deforms in a crash.

current The amount of electric charge that flows in an electric circuit each second. Measured in amps.

damping Using an absorbent material to stop vibrations.

decibel The unit used for the intensity of a sound.

deform To change the shape of something.

density The mass of a material per unit volume.

dioptre A unit describing the power of a lens. One dioptre (1 D) is the power of a lens with a focal length of 1 metre.

diverging lens A lens that bends light rays so that they spread further apart. Also called a concave lens, because of its shape.

ductile A ductile material can be worked, e.g. rolled into sheets or pulled into wires, without breaking.

durability Resistance to abrasive wear or weathering.

elastic An elastic material returns to its original shape after being deformed.

electrical conductance How much current passes through something for a given potential difference. Something made from a good conductor such as a metal generally has a high conductance.

electrical conductor A material through which electricity can pass easily.

electrical insulator A material through which electricity cannot pass easily.

extension The change in length of an object when loaded in tension (how far it stretches).

flexible A flexible material is easily deformed when a force is applied.

fluid-filled dampers Structures used on bridges and buildings to absorb vibrations.

focal length The distance from the centre of a lens to its focus.

focal plane The plane in which a camera focuses all parts of an image.

focus The point at which a lens forms an image from rays of light parallel to the lens axis.

force A push or a pull.

frequency The number of vibrations a second.

hardness The resistance of a material to scratches or dents.

hertz The unit used for frequency.

intensity The loudness of a sound, or how much energy reaches your ear a second. Intensity depends on amplitude.

loudness How you perceive a sound's intensity. As well as the energy of the sound, it also depends on the frequency and how long the sound lasts.

malleable A malleable material can be shaped or formed by hammering, without breaking.

melting point The temperature at which a solid melts and forms a liquid.

metal A material that is shiny and a good conductor of heat and electricity. Metals are elements on the left-hand side of the Periodic Table.

momentum The momentum of a moving object is its mass times its velocity.

opaque An opaque material absorbs light and does not allow it to pass through.

optical fibre A long and very thin fibre of very pure glass, used to transmit pulses of laser light over many kilometres.

parallel circuit An electric circuit that has two or more branches, and in which the current divides between them.

pitch How high or low a sound is. Pitch depends on frequency.

plastic A plastic material does not return to its original shape after being deformed.

polymer A material made up of very long molecules. The molecules are long chains of smaller molecules.

potential difference The difference in energy a given current provides each second between two points in a circuit, such as across a bulb or resistor. Also referred to as 'voltage' because it is measured in volts.

power (of a lens) Describes how strongly the lens bends light, measured in dioptres. The power of converging lenses is positive; the power of diverging lenses is negative.

real image An image that can be formed on a screen.

reflection Sound or light is reflected when it bounces off a material, and is not absorbed or transmitted.

refraction The bending of a ray of light as it goes from one medium to another and changes speed.

refractive index A measure of how much a material changes the direction of light rays at its surface. Refractive index is the speed of light in air divided by the speed of light in the material.

rigid A rigid structure is stiff and strong.

safety margin A specification standard beyond the minimum necessary to prevent an accident, e.g. ensuring the strength of a material or structure is greater than the expected maximum load.

seat belt Safety device in a vehicle which restrains the driver and passengers in a crash.

series circuit An electric circuit with one loop (no branches), so there is just one path for the current.

shutter A mechanism in a camera that operates when a picture is taken. It opens for a fraction of a second to let light onto the film or electronic screen.

solid solution A solid which has the atoms of one substance spread through another, such as an alloy.

standard A series of tests for assessing products or materials to make sure they are of an acceptable quality.

stiff A stiff material resists being deformed when bent.

strong A strong material withstands a large force before it breaks.

structure The way a material is shaped or put together, or the way its atoms or molecules are arranged.

tensile strength The stretching force needed to break a material.

tension A material is in tension when it is being stretched by forces pulling apart.

thermal conductance The rate of heat transfer through something for a given temperature difference. Something made from a good thermal conductor such as a metal generally transfers a lot of heat.

thermal expansion The way that solid objects increase in length (and volume) when heated. It can be defined quantitatively for a particular material as the ratio of the increase in length to its original length, for a temperature rise of 1 °C.

tinnitus A ringing sound in the ear.

tough A tough material is not easily broken and resists cracking.

toughened glass Glass that has been treated so that it absorbs a lot of energy before shattering.

translucent A translucent material scatters light as it passes through. Clear images cannot be seen through the material.

transmit Pass through.

transparency How much light can pass through a material.

transparent A transparent material transmits light – it allows light to pass through it. Clear images can be seen through the material.

velocity Speed in a certain direction.

viewfinder The device in a camera that enables the user to set up the desired image. Looking through the viewfinder, the user adjusts the camera direction and lens position.

virtual image An image that cannot be formed on a screen.

voltmeter A device for measuring the voltage (potential difference) in an electric circuit.

weak A weak material breaks with only a small force acting on it.

Index

absorb, sound 18–19, 42, 44, 60
absorb, light 34, 60
acoustic design 18–19
acoustic insulation 18–19, 42–3, 44
acoustic properties 18–19, 42–5, 60
air bags 26, 60
aircraft 24
alloy 7, 24, 33, 60
aluminium 24, 29
amalgam 33
ammeter 31, 60
angle of refraction 38
aperture, camera 36, 60
architect 4, 16, 18–19
archwire 7
athletics 10–11

beam 22
bimetal strip 29, 60
bone 20–1
braces, dental 6–7
British Standards Institution (BSI) 12–13
brittle 22, 24, 60
building control surveyor 4–5, 16
building regulations 4–5, 16
Building Research Establishment (BRE) 17

cable 39
cameras 36–7, 55
cars 26
CE marking 13
cell, electrical 31
CEN (European Committee for Standardisation) 13
ceramics 60
   in dental braces 6–7
   electrical conductance 20, 32
   general properties 20
   thermal conductance 20, 28
choosing materials 16
circuit diagrams 30–1
coating 60
   glass 34
   lens 37
collision time 26, 27, 60
collisions 26, 27
coloured lights 9
composites
   defined, with examples 20–1, 60
   living materials 21
   in dental fillings 33
   in hockey sticks 16
   in pole-vaulting 10–11
   research into 17
compression 22, 25, 60
compressive strength 4, 17, 60
concave lenses 36–7, 60
concrete 4, 20

conductance, electrical 32, 50
conductance, thermal 28, 52, 53
conductors, electrical 32
connecting wires 31
consistency, of a product 60
contact lenses 3
converging lens 36–7, 54–5, 60
convex lens 36–7, 54–5, 60
corrosion resistance 17, 60
   of a steel 24
crumple zone 26, 60
current, electric 31, 32, 50, 60
cycle helmets 27

damping 45, 60
deafness 44
decibel 43, 56, 60
deform 22, 60
density 16, 17, 24, 48, 60
dental braces 6–7
dimmer 31
dioptre 37, 60
diverging lens 36–7, 60
double glazing 18–19, 44
ductile 24, 60
durability 16, 60
Duralumin 24

elastic 11, 22, 23, 24, 60
electric circuits 30–1
electrical
   conductance 32, 50, 60
   conductor 32, 60
   insulation 16
   insulator 32, 60
   properties 31
energy
   and pole-vaulting 11
   stored in stretched material 23
   sound 43, 56
engineer
   electrical 30, 31
   mechanical 16
European Committee for Standardisation *see* CEN
European Standards 12–13
expansion, thermal 29, 51
extension 23, 57, 60

fibre-reinforced polymers (FRP) 17
fillings, dental 33
flexible 22, 60
fluid-filled dampers 45, 60
flutter echo 18–19
foam 18–19, 27, 44
focal length 36–7, 54–5, 60
focal plane 36–7, 60
focus 36–7, 60

force 22, 26, 27, 60
   force constant 23
   force–extension graphs 23
   force–time graphs 26
frequency 43, 45, 60
fresnel lens 41

glass 16
   bullet-proof 24
   double glazing 44
   mechanical properties 24
   optical fibres 39
   refraction in 38
   thermal expansion 29
   toughened 24
   windows 34, 44
glass-reinforced plastic (GRP) 20
greenhouse 29

hardness 24, 49, 60
hearing 43
hertz 43, 60

image 36–7, 54–5
incident ray 38
insulation
   acoustic 18–19, 42–3, 44
   electrical 16
   thermal 5, 28, 34
insulators, electrical 32
intensity, of sound 43, 56, 60
International Organisation for Standardisation (ISO 12–13)

Kitemark 13

lamp 31
lenses 36–7, 41, 54–5
   contact 3
ligatures 6–7
light and lighting 8–9, 34–5
   coloured 9, 40
   pipes 35
   refraction 38
   soft and hard 8–9
   stage 8–9, 40–1
loudness 43, 60
luminaire 40–1

malleable 20, 60
mass 26, 48
materials, summary of classes and properties 20
matrix, used in composites 20
mechanical engineer 16
mechanical properties 22–5
melting point, of a steel 24, 60
metals 7, 20, 24, 28, 32, 33, 61

mirrors 34, 55
momentum 26, 61
music rooms, design 18–19

noise 42–5
normal (in optics) 38

opaque 34, 61
optical fibres 39, 61
optician 3, 37

paper 25
parallel circuits 30, 61
pitch (of a sound) 43, 61
plastic behaviour 22–3, 61
polymers 28, 61
   and contact lenses 3
   polythene and polystyrene 5
properties 20, 32
potential difference 31, 32, 50, 61
power (of a lens) 36–7, 61
power supply 31
presentation, of data 57
procedures and techniques 46–57
product designer 16
PVC 27

real image 36, 61
reflection 61
   of light 34
   of sound 18–19, 42, 44
refraction 38, 61

refractive index 38, 61
rigid 25, 61
rubber 45

safety margin 61
seat belt 26, 61
series circuits 30, 61
shutter, camera 36, 61
siemens 32, 50
solid solution 24, 61
sound 42–5, 56
   insulation 18–19, 42–3, 44
   quality 18–19
specification 16
sports design 10–11
stage lighting 8–9, 40–1
standard procedures, presenting data
   from 57
standards 12–13, 14–15, 61
steel 16, 25
stiffness 22, 46, 61
strength 47
   see also tensile strength
strong 22, 61
structure 61
   made from composites 17
   triangular and tubular 25
switch 31

teeth 6–7, 33
tensile strength 17, 47, 57, 61
tension 22, 25, 61

thermal
   conductance 28, 52, 53, 61
   expansion 29, 51, 61
   insulation 5, 28, 34
   properties 28–9
thermostat 29
tinnitus 44, 61
tough 22, 61
toughened glass 24, 61
toys, testing 14–15
Trading Standards Officers 14
translucent 34, 35, 61
transmit 42, 61
transparency 39, 61
transparent 34, 38, 61
tungsten light filaments 9

vaulting pole design 10–11
velocity 26, 61
vibrations 42, 45
viewfinder 36, 61
virtual image 36–7, 61
voltage 31
voltmeter 31, 61
volume 48

water 38
weak 21, 22, 61
windows 34, 44
wood 20–1
Work–related portfolio 58–9

OXFORD
UNIVERSITY PRESS

Great Clarendon Street, Oxford OX2 6DP

Oxford University Press is a department of the University of Oxford.
It furthers the University's objective of excellence in research, scholarship,
and education by publishing worldwide in

Oxford   New York

Auckland   Cape Town   Dar es Salaam   Hong Kong   Karachi
Kuala Lumpur   Madrid   Melbourne   Mexico City   Nairobi
New Delhi   Shanghai   Taipei   Toronto

With offices in

Argentina   Austria   Brazil   Chile   Czech Republic   France   Greece
Guatemala   Hungary   Italy   Japan   Poland   Portugal   Singapore
South Korea   Switzerland   Thailand   Turkey   Ukraine   Vietnam

© University of York on behalf of UYSEG and the Nuffield Foundation 2006

The moral rights of the authors have been asserted

Database right Oxford University Press (maker)

First published 2006

All rights reserved. No part of this publication may be reproduced,
stored in a retrieval system, or transmitted, in any form or by any means,
without the prior permission in writing of Oxford University Press,
or as expressly permitted by law, or under terms agreed with the appropriate
reprographics rights organization. Enquiries concerning reproduction
outside the scope of the above should be sent to the Rights Department,
Oxford University Press, at the address above

You must not circulate this book in any other binding or cover
and you must impose this same condition on any acquirer

British Library Cataloguing in Publication Data

Data available

ISBN-13: 978-0-19-915031-1

10 9 8 7 6 5 4 3

Printed in Italy by Rotolito Lombarda

Acknowledgements

The publisher would like to thank the following for their kind permission to reproduce copyright material:

p2 Corbis UK Ltd.; **p3** Anna Grayson; **p4l** Anna Grayson, **p4r** Photodisc/Oxford University Press; **p5** Anna Grayson; **p6l&r** Anna Grayson; **p7l** Anna Grayson, **p7r** Photodisc/Oxford University Press; **p8** Anna Grayson; **p9** Steve Warren Avolites Ltd/Mike Le Fevre; **p10t** David Brodie, **p10b** Professor Harold Edgerton/Science Photo Library; **p12** images-of-france/Alamy; **p13t&b** Zooid Pictures; **p14** David Brodie; **p15** David Brodie; **p16l** Eliana Aponte/Reuters/Corbis UK Ltd., **p16r** Deep Light Productions/Science Photo Library; **p17bl** Faber Maunsell Ltd, **p17bc** Alifabs/White Young Green Consulting Ltd, **p17br** Martin Bond/Science Photo Library, **p17t** David Brodie, **p17c** Simon Emery/Rex Features; **p18t** Page One, **p18b** Photofusion Picture Library; **p19** Education Photos; **p20** Rex Features; **p21l** Lester V. Bergman/Corbis UK Ltd., **p21c** Pascal Goetgheluck/Science Photo Library, **p21r** Dennis Kunkel/Phototake Inc./Alamy; **p22** David Gregs/Alamy; **p24t** Pierre Perrin/Sygma/Corbis UK Ltd., **p24b** Volker Steger/Science Photo Library; **p25l** A.C. Waltham/Robert Harding Picture Library Ltd/Alamy, **p25r** Raleigh Cycles; **p26** David Woods/Corbis UK Ltd.; **p27** Corbis/Digital Stock/Oxford University Press; **p28t** Corel/Oxford University Press, **p28b** Andrew Lambert Photography/Science Photo Library; **p29t** Nigel Cattlin/Holt Studios International; **p29b** Bosch UK; **p30** Maximilian Stock Ltd/Science Photo Library; **p33** Medical-on-Line; **p34** Mark E. Gibson/Corbis UK Ltd.; **p35** Pilkington Plc; **p37tl&tc&tr** Canon (UK) Ltd, **p37bl&bc&br** Martyn F. Chillmaid; **p38** Andrew Lambert/Leslie Garland Picture Library/Alamy; **p42** B.S.P.I./Corbis UK Ltd.; **p44** Robert & Linda Mostyn/Eye Ubiquitous/Corbis UK Ltd.; **p45** Empics.

Illustrations by IFA Design, Plymouth, UK and Clive Goodyer

These resources have been developed to support teachers and students undertaking a new OCR suite of GCSE Science specifications, Twenty First Century Science.

Many people from schools, colleges, universities, industry, and the professions have contributed to the production of these resources. The feedback from over 75 Pilot Centres was invaluable. It led to significant changes to the course specifications, and to the supporting resources for teaching and learning.

The University of York Science Education Group (UYSEG) and Nuffield Curriculum Centre worked in partnership with an OCR team led by Mary Whitehouse, Elizabeth Herbert and Emily Clare to create the specifications, which have their origins in the Beyond 2000 report (Millar & Osborne, 1998) and subsequent Key Stage 4 development work undertaken by UYSEG and the Nuffield Curriculum Centre for QCA. Bryan Milner and Michael Reiss also contributed to this work, which is reported in: 21st Century Science GCSE Pilot Development: Final Report (UYSEG, March 2002).

Sponsors
The development of Twenty First Century Science was made possible by generous support from:
- The Nuffield Foundation
- The Salters' Institute
- The Wellcome Trust

THE MEMORIES OF A RUSSIAN YESTERYEAR

Volume 2

Compiled & Presented by

TONY ABBOTT

AUTHOR OF
'NICHOLAS II - TSAR TO SAINT'
'THE MEMORIES OF A RUSSIAN YESTERYEAR VOLUME 1'

65 ILLUSTRATIONS

New Angle Publishing

Copyright © Tony Abbott 2023
First Published October 2023
email@tonyabbott.co.uk

INDEPENDENT PUBLISHING NETWORK
ISBN 978-1-80517-078-5
Paperback Edition (cream paper)

CONTENTS

SECTION		PAGE
01	PUBLIC DOMAIN BOOKS	3
02	INTRODUCTION	4
	I Catherine Radziwill	6
	II Lily Dehn	9
	III Anna Viroubova	11
03	BOOK ONE	17

CONFESSIONS OF THE CZARINA
By PAUL VASSILI
(The pseudonym of Catherine Radziwill)

04	BOOK TWO	193

THE REAL TSARITSA
CLOSE FRIEND OF THE LATE
EMPRESS OF RUSSIA
By LILI DEHN

05	BOOK THREE	354

MEMORIES OF THE RUSSIAN COURT
By ANNA VIROUBOVA
LADY IN WAITING TO THE TSARINA

06	LIST OF ILLUSTRATIONS	626
07	INDEX OF NAMES	629
08	VOLUME INFORMATION	634
09	OTHER BOOKS BY THIS AUTHOR	635

PUBLIC DOMAIN BOOKS

Copyright applies for a further seventy years after the death of the originating author, this means that the rights to publish can be passed on. For some parts of the world the time length differs. It is also different for corporate rights such as in the United States of America, for Mickey Mouse for example, which is in place for 120 years; since the Sonny Bono Copyright Term Extension Act of 1998, which amended the duration of copyright. For non-corporate works published between 1923 and 1977, these enter the public domain 95 years after their creation (publication date), under the amendment act of 1998.

This means that the prolific memoirs coming out of Russia after the Russian Revolution 1917, mostly written in the 1920s and early 1930s, have been entering the public domain since 1 January 2019, when thousands of titles became available to download, reprint and repurpose, such as The Murder on the Links by Agatha Christie, Jacob's Room by Virginia Woolf, Kangaroo by D.H. Lawrence, and The Inimitable Jeeves by P.G. Wodehouse.

In 2023, those publications from 1927 that were protected by publishing enterprises, entered the public domain. This would include The Big Four by Agatha Christie, first published in the UK by William Collins & Sons on 27 January 1927 and in America by Dodd, Mead and Company later in the same year. This craze for books of old saw the rise of digital libraries such as the HathiTrust Digital Library which includes much of the content digitised by other repositories such as Google Books and Internet Archive.

It's hard to see why anyone would purchase a Public Domain ebook or hard copy version when almost any work out of copyright has been digitised and is freely available. For this to be viable, the new work must have added value, not just editorial by simply typesetting a work so that it looks neater, but reflected in the additional content; has updated news been provided about things that were not known about at the time? Have photographs been added to further explain the issues? In this volume, the three books are presented unabridged with very little added due to the bulging page count, although sharper images and the soothing cream and groundwood papers used, are reason enough to add this to the shelf.

* * *

INTRODUCTION

THE importance of the works within these covers is truly significant in the understanding of life in and around the court of Nicholas II. They share a commonality that they are centric to the Empress Alexandra Feodorovna. In the first book, the author's ties to the Russian court are through her aristocratic family. Her marriage to a Prussian officer developed further social circles in the European courts which she wrote about and during which time she had no less than seven children before returning to St Petersburg. We are talking about the aristocratic Catherine Radziwill whose maternal grandfather had served as Justice Minister to Emperor Nicholas I. Her connections to Alexander III and the inner circles were attained, in part, by her friendship with the Emperor's advisor Konstantin Petrovich Pobedonostsev and the affair she was having (some describing it as a civil marriage,) with the Emperor's court affiliate to the Okhrana police, Lieutenant-General Peter Alexander Aleksandrovich Cherevin. She is the professional author among the three books of memoirs herein, and therefore much higher expectations are held, but in truth her penmanship is matched by the compelling and sincere accounts of the Empress's closest friends as demonstrated in the other two books. These three prominent women make the revelations designed to sell their books which together hone down to a very intimate portrait of the Empress herself, told by those women who knew her best, at her best.

In comparison with Volume 1, there were hardly any typos at all found in the accounts of these authors. The only nuances are the rather long paragraph lengths. It appears that Victorian women did not appreciate that men require information to be spoon-fed in smaller chunks for the comprehension mechanism to activate. In this book format, 6,14" x 9.21", we can just about endure it but in their original format would have been rather taxing; Book III, for example, has a paragraph that begins on page 201 and ends on page 204. There's an underlying cloying erudition in the first two books, perhaps aiming for the erudite reader, which in fairness, doesn't take that long to get used to once acclimatised to the writing style. If we look behind the marketing mechanics these wonderful memories shine through as intended by their illustrious guardians.

Radziwill (Book I) was a contentious character and there can be no doubt that her motive for writing was solely for profit and fame.

INTRODUCTION

On the other hand, Lili Dehn (Book II) was a lucky survivor of the Bolshevik Revolution 1917 who came to live in England with her husband and son and wrote her book of memories. She had no official position at Court and was brought in to the Empress's close proximity by friendship; she writes from the unique perspective of an impartial confidant of the Empress and was making her case for publication as an eye-witness of the events in contrast to the less than well founded versions being circulated by the press. Madame Dehn was the first person to whom the Empress came in her grief with the news of the Emperor's abdication and she saw him return to his wife and face the supreme humiliation.

"There must surely be friends and relations in England who would welcome facts which proved that the Empress had been true to her English upbringing and to the traditional right living of the descendants of Queen Victoria. English people seem to have forgotten, when the Empress was vilified on the screen and in cold type, that she was the daughter of Princess Alice, a name which is associated with all that is noblest and best in a woman." — The World's News, 1 July 1922, Sydney

In Book III, Anna Viroubova promises the definitive version of the Empress's day to day life. It is certainly the lengthier and more detailed book, and like Dehn she defends the Empress without criticism of any sort, and disposes of the Rasputin legend. In a way, one gets the sense that it is an apologetic work, written as much to set the record straight, as to lay camouflage over her true role in much of the unfortunate events surrounding Rasputin. She was evidently in need of finances seeing that she sold six of her seven large photographic collections as well, nonetheless the extent of her work is a most significant first hand account. The Sun, on 13 January 1924, stated that "her book is politically insignificant, but it has a deep personal interest from the faithful picture which it gives of the Russian Imperial Family in the last years of Nicholas II's unhappy reign."

In *Nicholas II - Tsar to Saint*, I showed the parallelisms with the lives of Nicholas II and Alexandra to Louis XVI and Marie Antoinette; these authors strengthen that comparison. Succinctly, the essence of the Empress is given by Viroubova, "She was quick-tempered, and passionately devoted to her husband and to her children."

INTRODUCTION

I - CATHERINE RADZIWILL (b.1858-1941)

This first book of memoirs in Volume 2 is from a woman using a male pseudonym, certainly a talented literary agent of the time, who conjured interest over the author's true identity and perhaps some trepidation also around her obvious intimate knowledge of the inner workings of the European courts. Unlike many accounts from the 20s and 30s, Catherine Radziwill has not attracted the same scrutiny perhaps because of her mastery in the art of storytelling and her love for the written word – of which at times in her life she had come to depend on for her sole income, despite being a high noble.

She was born in St Petersburg, the only daughter of a tsarist General, Adam Adamowicz Rzewuski, from a Polish House of note. Her father was born and died in Kiev, and saw active service in the Crimean War. After the death of his first wife in 1853, he married Anna Dashkov and produced his only daughter, Catherine. On the maternal side therefore, Catherine Radziwill descended from the illustrious Russian Dashkov-Vorontsov House, among them Count Illarion Ivanovich, in the direct service of Alexander III and Nicholas II, until his death in 1916. Her mother died giving birth to her and her father returned to Kiev where they now lived together.

Catherine Maria Adamevna grew up in the Russian province of Ukraine and therefore considered herself to be Russian, despite her Polish/Lithuanian father. She was left alone to study and received a fair education. However, in 1873, when she was just fifteen years old, she married Wilhelm Radziwill, a Polish prince and Prussian officer, of whom little is known about, except that the couple circulated at the Prussian court in Berlin where they had moved to live with Wilhelm's family. They had seven children before their divorce in 1906, that did not all survive to adulthood. She continued to work as a secretary to the Empress consort Queen Victoria (the eldest daughter of Queen Victoria of Great Britain and mother to Wilhelm II,) after the divorce and also during her second marriage in 1909, until her first husband died in 1910.

The House of Radziwill had expanded their sphere of influence in to Poland and became a powerful House. The recent generations sought out English and American shores. Prince Stanisław Albrecht Radziwill, for example, was established in London, founded the Sikorski Museum for Polish immigrants there, and in 1959 married his third wife Caroline Lee Bouvier, the sister of former First Lady

INTRODUCTION

Jacqueline Kennedy. It was the surname that Catherine remained the proudest to hold up until her own death in 1941, notwithstanding her second marriage to Karl Emile Kolb-Danvin in 1909.

Both marriages exposed her directly to court life and the body of work she subsequently produced was a remarkable feat. Her second husband was of German birth but living in Stockholm when they married; where he exported engineering equipment between the English and Dutch. It was further exposure for her to yet another European court. As far as her extensive knowledge of the English court, she explains it in detail in her later book *Memories of Forty Years* (1915) in which the whole of Part I is devoted to describing, "the numerous journeys I have made to England."

These were dangerous times in Europe and loose talk amid the powerful families could see one removed from the four realms of existence. It was for her own safety that Catherine Radziwell used several pseudonyms. Letters from Count Paul Vasilli first appeared in a French magazine criticising the German court, which were also released in the book *La Société de Berlin: augmenté de lettres inédites* (1884). It drew speculation as to the identity of the true author, as there was no real Count Paul Vassili to account for. Under Vassili, as well as exposing the German court, she also produced *France From Behind The Veil* (1913), *Behind the Veil at the Russian Court* (1914) and *The Austrian Court From Within* (1916). Her sheer output during World War I is staggering, including a work of fiction and the book which is included in this volume: *Confessions of the Czarina* (1918). Her later works concentrated mainly on royal biographies with her autobiography, *It Really Happened,* following in 1932, ten years after the divorce from her second husband in 1922.

From 1915 when she released *The Royal Marriage Market of Europe*, her real name as the author appeared as; Princess Catherine Radziwill (Catherine Kolb-Danvin), acknowledging her current and former matrimonial surnames. But it wasn't until 1918, that she finally admitted that she was also the one and only Count Paul Vasilli - that revelation being made in the book *Confessions of the Czarina* - Book I of this volume – in which the credited author reverts to Count Paul Vassili to cement her claim to it, as it was the case that several works had sprung up using that author byline.

In this regard the reader will note on the title page that the book is by Count Paul Vassili but note also what is said in the 'Publishers'

INTRODUCTION

Note' at the start, and how straight away in the Introduction she admits to having written *La Société de Berlin* and *Behind the Veil of the Russian Court* – she is saying "I am Paul Vassili."

"Just before the war, when, indignant at the manner in which Nicholas II. was compromising the work of his great father, I wrote the book Behind the Veil of the Russian Court, I bethought myself of assuming once more the old pseudonym."

In Volume 1, I stated that Alexender Mossolov's memories were the perfect place to start as he began with a description of Nicholas II's parents . . . Likewise, Paul Vassili was publishing a full decade before Nicholas II became Tsar. Radziwill's view is comprehensive and in *Confessions of the Czarina* she brings the reader close and personal in to the workings of the Russian court where she was active for a time; estranged from her first husband before their divorce, in an affair with Peter Cherevin, a man who held Russia's highest military decoration, the Order of Saint George for bravery. He was described by the historian Alexander Polovtsov as being constantly drunk up until his death in 1896, encompassing the time Radziwill spent in St Petersburg. His passing meant she could not make ends meet and the mounting debts were more than her writing energies could cover. So, in 1899 she left everything and everyone behind and set out for South Africa to meet Cecil Rhodes, whom she eventually persuaded to pay off her debts. That expedition cost her sixteen months in a South African prison for various infractions, among them forging Rhodes' signature to obtain funds. But that's another leaf in the life of the woman who had seven children and exposed the workings of the European courts before she became 'bored' and headed out for an extramarital affair and an African adventure from which she was lucky to return, even if with scandal.

Here in Book I, she starts with, "The former Empress of All the Russias is to-day a prisoner, condemned to a horrible exile." It sets the timeframe perfectly as her book was also published whilst the ex-royal family were imprisoned in Siberia, (at Tobolsk before April and at Ekaterinburg from April 1918). How could she have suspected that in just a few months all of the prisoners would have been disposed of. She managed to escape the Bolsheviks and emigrated to America. She makes it clear in her book that she had not been impressed with the German court and indeed would have been glad to have been kicked out of Berlin had she been identified as the

INTRODUCTION

author of *La Société de Berlin*. Her experiences in the Prussian capital evidently had negative effects on her, not least as her son, Prince Vladislaw Radziwill was killed in action in East Prussia during World War I.

One might suspect when she mentions Marie Antoinette, "..., *she had also met on her path the devotion of a Fersen,...*" that she is using the Proto-Germanic word 'Fersen', the plural of Ferse, meaning 'heel', to imply a servant – but it is simply an innocent reference to Axel von Fersen (1755-1810), a Swedish aristocrat of her acquaintance. It is one of those books that you need to get one or two chapters in before the brain accepts the rather loquacious style, at times challenging and laborious. Yet it is deserving of its place in this volume of memories.

She was writing right up until a few hours before she died, having been admitted to hospital for a hip fracture and dying less than a month later, still in hospital. As early as 1915, in *Memories of Forty Years*, she had started to see herself as more than a writer, perhaps as a philosopher, considering the valuable knowledge she held of the workings of the European aristocratic society.

> "*People may call me a philosopher; they will never be able to think me a misanthrope, for indeed the faculty of enjoyment exists in me just as intensely as in the days of my youth. And, standing on the threshold of old age, I am glad to say that I have lived and loved, suffered and been merry; that my past has been sweet, though it has known bitter hours; but there is not a single page in it I would care to tear away.*"

II - LILI DEHN

Book II opens with a sonnet from Oswald Norman about the late Empress of Russia, in which he blames the mob and Revolution for her tragic fate and credits her enduring faith in a 'God of love' as the source of the fortitude that sustained her through the suffering. It is an apt opening in defence of the Empress's honour for a book from Lili Dehn, *née* Smolskaia, that aims to do the same. Norman published a book of 36 sonnets in 1897 of which the first one, written in 1884 was a dedication 'To My Father.' He credited his father for, '*The love of letters, and their priceless gain : Oh! would my verse did higher flights attain,*" and similarly Dehn tells us in the briefest ever foreword, that her Empress portrayed for her, the love of life, far removed from the caricatures that have been represented by the historical world record.

As Norman claims the peasant's perspective when noting the peculiarities of life, so too does Dehn make it clear that her notations of the Empress, unbiased by their long and intimate friendship, are not influenced by anything other than the innocent witness of her day to day existence as a kindly and simple-minded woman. In this respect we are fortunate to have an account that is totally devoid of the sensational which seems to have found currency in the lurid representations of the Empress by all other accounts, in which Dehn refutes any notion that the Empress was a hysterical neurotic woman – this, we are told, is the 'real' tsaritsa.

Indeed, it is in her attempt to paint the Empress whiter than white, that criticism of her book arises. It is to be expected that such a close friend would want to address imbalances against the Empress, and emphasise the undisputed facts that she was a loving mother, a deeply religious person and more British in her sympathies than German. But her insistence that Rasputin was an entertainer in the mysteries, a hypnotist, having no undue influence otherwise, reveals her level of ignorance, as it is well known that he condemned the hostilities of World War I directly to the Imperial couple – i.e. political, and he did have some effect on the attitudes of the Government and in the workings of the Church.

It is also well known that the Imperial couple abhorred the Duma. In this regard Dehn ascribes the Revolution to the Duma in that the catalyst for its inertia came from the bread famine of which she states was organised by the Duma and to a degree orchestrated by the Jews within. In the chapters devoted to the Revolution she provides no evidence to support these claims and the argument is therefore unpersuasive and simply a reflection of the Empress's views. Yet another writer of sonnets, the historian Hilaire Belloc, dismissed Jewish involvement in his book, The Jews, published in the same year, 1922. He highlighted that, "The Revolution was a Jewish movement, not not a movement of the Jewish race."

Madame Lili (Yulia) Dehn starts off by introducing her parents. Her father was a military engineer, and her mother in her second marriage. In 1907, she married captain Karl Akimovich von Dehn, an officer of the Emperor's personal guard, which was also during

INTRODUCTION

the time that she first met the Empress. Dehn proved her loyalty during the February Revolution 1917, remaining at Alexander Palace with the royal family, even though her own son was still in Petrograd, until the royal family were moved to Siberia,. There-fore, she is a first hand eye-witness to the harsh treatment accorded them by Mr Kerensky. A major interest of her book are the reproduction of letters received from the Tsaritsa when she was a prisoner in Siberia.

Lili Dehn died in Rome in 1963 and is buried in the Testaccio cemetery, sometimes referred to as the English cemetery; among its occupants are the English writers Keates and Shelley.

'The author succeeds in giving the reader more than a passing glimpse of one sort of autocracy, that came to an end with the Revolution. Bolshevism, perhaps, was a still worse autocracy of the other extreme, but it is impossible to read such a book as 'The Real Tsaritsa' without feeling that the Government of the Romanoffs did not crash before its time. Madame Dehn certainly tells a powerful and poignant story, and makes a loyal defence of her friends; but she shows that Russia was dead ripe for a change of a drastic kind.'
— The Queenslander, 22 July 1922; LITERATURE section

III - ANNA VIROUBOVA

For Book I we discussed the author in detail and for Book II, we acknowledged the author's sincerity but questioned her explanations for the Revolution and over Rasputin. The author of Book III should need no introduction as she is perhaps the best known associate of the inner royal family and was particularly close to the Empress from when they first met in 1905 up until their last days. In the Russian tradition there are variations of the spelling and usually Vyrubova is used, but the rule should be to adopt the spelling used in memoirs, which in this case is 'Viroubova' — Anna Alexandrovna Viroubova *née* Taneyeva.

She was just ten years old when Alexandra Feodorovna married Tsar Nicholas II. Her family were no strangers at court but she was mostly under the patronage of the Empress's sister Elizabeth Feodorovna and her husband Sergei Alexandrovich where she spent some of her childhood at their estate in Moscow. It was there that

INTRODUCTION

she also associated with playmate Felix Youssoupoff, the man that was to marry the Tsar's only niece in 1914 and murder Rasputin in 1916. (Chapter II) In 1903, at eighteen years old, she was moved closer to the court as a Maid-of-honour at the Winter Palace. It wasn't until 1905 that she moved to Tsarskoe Selo to replace the Empress's sickly Lady-in-waiting, Princess Orbeliani. As Rasputin had also arrived in St Petersburg in 1903 and appeared at the court in 1905, these were unfortunate coincidences that forever linked Viroubova to Rasputin. She moved in to a small house just opposite the palace gates and was seeing the Empress daily. Yet, the Empress would still visit her from time to time, sometimes with the Emperor.

Viroubova is conscious that she is seen as the person that may have betrayed the Empress under the influence of Rasputin as well as being thought of as having a German origin with an ulterior motive to spy on Russia. She is therefore keen to point out that neither theory were true and that her blood was every bit untainted Russian. This of course is demonstrated by her genealogy; born near St Petersburg to Alexander Taneyev, a high-ranking state official who served for twenty-two years as the head of His Imperial Majesty's Own Chancellery, and her maternal side being the Tolstoy family. It was mostly due to her father's connections that Viroubova drew nearer to the Imperial Court.

Another fallacy she is keen to address is the portrayal of her being a lonely child and the possessor of a rather low intellect, which Rasputin identified, being the reason he was able to easily manipulate her to gain access to the court. Firstly she reveals that she had a brother Serge and three sisters, all receiving an education, "rather more practical than was the average at that time." They were schooled in the arts and well grounded in academic studies, attending many court balls and official functions. She takes pride in explaining her parents' service to the Russian Court and her own introductions, first to the Dowager Empress Maria Feodorovna.

By way of addressing the question of her intellect, she mitigates any prospect of that being remotely true by admitting that when she and her brother were struck with Typhus, Serge had recovered well but she had been "at death's door," and was left with many complications, among them, "an affection of the brain whereby I lost both speech and hearing." Perhaps the quality of her memoirs do cast aside any shadows that would be cast by a wavering intellect and she has succeeded in maintaining her dignity and the reputation that was

INTRODUCTION

wrongfully tainted. She feels the need to attach at the end of the book, as Appendix A, the statement from Vladimir Michailovitch Roudneff, Kerensky's appointed investigator in to concerns around the royal family. In an important paragraph he absolves Viroubova of any wrong doing:

> "Such were the conditions which produced in the minds of persons ignorant of the nature of the intimacy between the Empress and Mme. Viroubova, belief in the exceptional influence of Mme. Viroubova on Court affairs. As has been said, Mme. Viroubova possessed no such influence, nor could she have possessed it. The Empress dominated the intelligence and the will of Mme. Viroubova, but the attachment between the two women was very strong. The religious instincts deeply rooted in their two natures explains the tragedy of their veneration of Rasputine. The relations between the Empress and Mme. Viroubova could be likened to those of a mother and daughter, nothing more."

The memories are notably biased and purposefully absent of any intricate conversation about Rasputin given that Viroubova was an ardent supporter of his. If indeed she had been manipulated to the extent that her actions played a part in the tragedy that befell the royal family, then one can understand why such details would be avoided but that is a scant criticism to consider considering the sheer volume of information she is providing. As a childhood friend of Felix Youssoupoff, had she inadvertently provided him information to assist his plot. Had she realised her role after the fact. Was it guilt and shame that was behind her staunch defence of the Empress.

It is a most loving legacy and written in a soft but commanding tone that leaves no doubt of her love for the royal family. Viroubova brings us in to the presence of the family members, "Every day at the same moment the door opened and the Emperor came in, sat down at the tea table, buttered a piece of bread, and began to sip his tea. He drank every day two glasses, never more, never less, and as he drank he glanced over his telegrams and newspapers." Indeed, up until World War I, Viroubova would have us believing that, "Monotonous though it may have been, the private life of the Emperor and his family was one of cloudless happiness."

At times it may seem she ascribes too much importance to her standing within the inner royal circle or that she is wilfully trying to

INTRODUCTION

right the foibles attributed to the royal family by other authors or attempting to smooth over the public record. For example, the well known tragedy at Khodynka Field during the coronation in 1896 is well documented. Many sources have denounced the royal couple for not reacting to the many deaths and for continuing with the celebrations. This is discussed in detail in *Nicholas II – Tsar to Saint*, in which I note that they could not have failed to be touched by the events, even if they did feel obliged to attend the evening ball. Viroubova attempts to right this by confirming that the royal couple were very severely emotionally affected by the events. The views on this are many, one account has it that the Emperor asked why he should cancel the evening ball; when he was advised to.

> "I think the Emperor liked to walk with me because he had need to talk to someone he trusted of purely personal cares which troubled his mind and which he could share with few. Some of these cares were of old origin, but had never been forgotten. I remember once he began to tell me, almost without any preface, of the dreadful disaster which attended his coronation, a panic, induced by bad management of the police, in the course of which scores of people were crushed to death. At the very hour of this fatal accident the coronation banquet took place, and the Emperor and Empress, despite their grief and horror, were obliged to take part in it exactly as though nothing had happened. The Emperor told me with what difficulty they had concealed their emotions, often having to hold their serviettes to their faces to hide their streaming tears."

Viroubova's memories are also invaluable in the corroboration of other accounts, such as with the often told story of Mlle Sophie Tutcheff, also given by Lili Dehn. And, with things like, "In his book M. Gilliard has recorded that he was never able to teach the Grand Duchesses to speak a fluent French. This is true because the languages used in the family were English and Russian, and the children never became interested in any other languages."

The main revelations the reader expects are about Rasputin. He is mentioned in Chapter V in connection to Mlle Tutcheff and thereafter in dribs and drabs, until Chapters XI and XII. It is more 'righting the record' it seems, and plays down Rasputin's behaviours. Her arguments are openly made and seem genuine enough so the

INTRODUCTION

reader alone is left to determine the validity of her account and that of Lily Dehn, against the many others that do not share their opinions on Rasputin.

Always looking for duplicity is a rule with the Romanov story. For example, there is the rather mysterious book *My Empress* by Madame Marfa Mouchanow, who claimed to have been the 'First Maid in waiting to her former Majesty the Czarina Alexandra of Russia', published in New York and London by John Lane Company. If the book is authentic it offers a new perspective on the royal family, The problem is that the book was published at the same time that the Bolshevik purge on the Romanov's began. It would have required time to write and get to print and it figures that within the aristocracy it may have been known by some that the royal family would perish.

So who was Madame Marfa Mouchanow? One theory is that it was an acquaintance of Princess Gedroitz during their time at the Tsarskoe Selo Infirmary. Despite Gedroitz's exemplary service record during the Russo-Japanese War and World War I, she is known to have turned Red during the Revolution and perhaps that exposé book was already in the making.

> "Alexandra Feodorovna did not make any real friends during the first years that followed upon her marriage. Indeed it was only after the Japanese war that she started the intimacies for which she was so much reproached by her subjects. The most notorious was that for Rasputin, but there were two others just as nefarious—that with Madame Wyroubieva and with the Princess Dondoukoff." — *My Empress*, chapter XVI

Another theory for the author of *My Empress* is given by the clues in Viroubova's book. A letter from Alexei dated 24 November 1917 states that Madeleine and Anna [VIROUBOVA] were still in Petrograd. Viroubova explains who Madeleine was and although there is no aroused suspicion, this trusted English person that Viroubova describes as 'invaluable' with access to the Empress's communique's and personal correspondence, could have been the candidate that was secretly writing the exposé book. Also described by Viroubova as 'tyrannical', Madeleine was 'often critical of her mistress's indolent habits'. This hidden author also appears to have disliked Viroubova:

> "Madame Wyroubieva was the daughter not of the Emperor's private secretary, as she represented herself to be, but of a State

INTRODUCTION

Secretary (which is quite a different thing, being a purely honorific position) called Tanieieff. She had been married to a navy officer with whom she could not agree, and they were divorced, not because he had grown mad, as she declared (divorce for insanity is not allowed in Russia), but because he had found reason to object to her conduct. The Empress, for reasons no one ever understood, took her part and invited her once or twice to the Palace of Czarskoi Selo. Madame Wyroubieva made the most of her opportunities and soon became quite indispensable to Alexandra Feodorovna. She it was who, with the Grand Duchess Elizabeth, introduced Rasputin into the Imperial household, and with him she established such control of the Czarina's actions that soon the latter became simply a tool in their hands."

— *My Empress*

"The chief maid of the Empress was Madeleine Zanotti, of English and Italian parentage, whose home before she came to Tsarskoe Selo was in England. Madeleine was a woman of middle age, very clever, and as usual with one in her position, inclined to be tyrannical. Madeleine had charge of all the gowns and jewels of the Empress, and as I think I have related, she was often critical of, in regard to correspondence, etc.

"Intimate letters, it is true, she answered promptly, but others she often left for weeks untouched. About once a month Madeleine, the principal maid of the Empress, would invade the boudoir and implore her mistress to clear up this heap of neglected correspondence."

— ANNA VIROUBOVA

Unravelling the mysteries is what is so rewarding about reading memoirs. In many cases one senses the answers are known, by someone, somewhere, just waiting for it all to be pieced together, and in this respect the amateur historian is much the same as the many astronomical novices that look up at the sky and now and then discover a galaxy, hitherto unseen by human eyes.

* * *

BOOK ONE

This is a book by the aristocratic writer Catherine Radziwill reprising her pseudonym byline count Paul Vassili. For many years the identity was not known and in this book she finally confirmed her claim to it. Her ambitions in life were matched by her talent in writing, she thought big and thought big of herself. Unfortunately her life choices were less than agreeable with her circumstances and she headed in to several scandals to address her finances.

With this work, Radziwill builds on the earlier interest from her book, *Behind the Veil at the Russian Court* (1914) for a much more intimate examination of the Russian Empress.

Original book published by Harper & Brothers, in 1918
Published in New York and London
310 printed pages

Confessions of the Czarina

by

COUNT PAUL VASSILI

Author of
"BEHIND THE VEIL AT THE RUSSIAN COURT"
"LA SOCIETE DE BERLIN"

ILLUSTRATED

Copyright, 1918, by Harper & Brothers
Printed in the United States of America
Published April, 1918

HARPER & BROTHERS PUBLISHERS
NEW YORK AND LONDON

CONTENTS

PART I

CHAP	PAGE
PUBLISHERS' NOTE	(ix) 21
INTRODUCTION	(xi) 22
I. BETROTHAL AND MARRIAGE	(1) 25
II. MARRIAGE AND LONELINESS	(18) 34
III. MY COUNTRY, MY BELOVED COUNTRY, WHY AM I PARTED FROM THEE?	(25) 38
IV. A SAD CORONATION	(34) 43
V. DAUGHTERS, DAUGHTERS, AND NO SON	(44) 49
VI. THE EMPRESS'S OPINIONS ABOUT RUSSIA	(53) 54
VII. WHAT THE IMPERIAL FAMILY THOUGHT ABOUT THE EMPRESS	(66) 61
VIII. SORROW AND UNEXPECTED CONSOLATION	(76) 67
IX. PHILIPPE AND HIS WORK	(88) 74
X. ANNA WYRUBEWA APPEARS ON THE SCENE AND HE SAW HER PASS	(99) 80
XI. AND HE SAW HER PASS	(112) 87
XII. LOVED AT LAST	(127) 95
XIII. HE DIED TO SAVE HER HONOR	(137) 100
XIV. A NATION IN REVOLT	(147) 106
XV. A PROPHET OF GOD	(157) 111
XVI. SHE SAW HIM ONCE MORE	(166) 116
XVII. MY SON! I MUST SAVE MY SON!	(177) 122
XVIII. ANOTHER WAR	(188) 129
XIX. MY FATHERLAND, MUST I FORSAKE THEE?	(199) 135
XX. IT IS YOUR HUSBAND WHO IS LOSING THE THRONE OF YOUR SON	(208) 140
XXI. PEACE, WE MUST HAVE PEACE	(219) 146

CHAP	PAGE
XXII. THE REMOVAL OF THE "PROPHET"	(229) 152
XXIII. ANNA COMES TO THE RESCUE	(240) 158
XXIV. YOU MUST BECOME THE EMPRESS	(251) 165
XXV. THE NATION WANTS YOUR HEAD	(261) 171
XXVI. A CROWN IS LOST	(271) 177
XXVII. A PRISONER AFTER HAVING BEEN A QUEEN	(281) 183
XXVIII. THE EXILE	(291) 189

THE CZARINA

From a photograph taken shortly before the Csar's downfall.

PUBLISHERS' NOTE

A FEW months before the great war broke out, there appeared a book, which, under the title *Behind the Veil of the Russian Court*, bearing the signature of Count Paul Vassili, a name that had become famous through the publication of the volume called *La Société de Berlin*. A lively interest was aroused by *Behind the Veil of the Russian Court*, dealing as it did with the intimate existence of four Russian Sovereigns and their respective Courts. The author of this book was declared to be already dead, out of a very natural feeling of precaution for his personal safety. Count Vassili was living in Petrograd at the time, and most certainly would have been banished to Siberia, and perhaps tried for *lèse-majesté*, if that fact had been discovered.

At the present moment the reasons for concealing it exist no longer, and Count Vassili is free to live once more and to publish another work of even greater interest—the life of the former Czarina Alexandra. In relating it, together with some most characteristic incidents which so far are but little known, Count Vassili remarks to the public what a small circle only have known; persons more or less interested in keeping the facts as secret as possible.

Count Vassili had known the Empress personally, in fact was regularly and most exactly informed by numerous friends as to all that went on at the Russian Court, and with all manner of intimate details concerning the existence led by the Czar and by his Consort in their Palace of Tsarskoye Selo. It is interesting to note that in *Behind the Veil of the Russian Court*, written at a time when but few people foresaw the fall of the dynasty of Romanoff, Count Vassili declared the event bound to take place in the then very near future.

INTRODUCTION

I AM not a coward, and it was not out of a feeling of uneasiness in regard to my personal safety, that I had not the courage to publish in my own name the book which, some thirty years ago, produced such a sensation when it appeared in the *Nouvelle Revue* of Madame Adam, under the title of "La Société de Berlin." But I was living in Germany at the time, and though I would have felt delighted had the publication of this volume driven me out of the Prussian capital, from which I was to shake the dust from my shoes with such joy, a few years later, I had there relatives who would most undoubtedly have fared very badly at Bismarck's hands, had my identity been disclosed. And once I am alluding to these distant times, it is just as well to say that the book in question had not at first been written for the benefit of the general public, but consisted of private letters addressed to Madame Adam, who, being happily still in the land of the living, can add many corroborative details. She suggested to me to publish some of these letters; I assented without suspecting the scandal which would follow, and which I do not regret in the very least, now that events have justified the mistrust with which the Prussian monster inspired me. The secret was well kept and one of the victims of it was poor Mr. Gérard, the secretary of Queen Augusta, who was accused of being the author of this book, an accusation that has clung to him ever since, and from which I am happy to relieve him.

The success of *La Société de Berlin* induced Madame Adam to publish other letters in the same style, devoted to other European capitals, with which, however, I had nothing to do, except those dealing with St. Petersburg life. The pseudonym of Count Paul Vassili remained a kind of public property divided between the *Nouvelle Revue* and my poor self. Just before the war, when, indignant at the manner in which Nicholas II. was compromising the work of his great father, I wrote the book *Behind the Veil of the Russian Court*, I bethought myself of assuming once more the old pseudonym. I was living at the time in St. Petersburg, as Petrograd was still called, and my brothers were in the Russian military service. I did not wish them to get into trouble. As it happened, my identity was suspected, and unpleasantness followed; but it is no stigma to have been ostracized by the Russian police under the old régime, so I did not mind or care.

I had not written the book out of any motives of revenge; on the contrary, I had many reasons to be personally grateful to Nicholas II. for various kindnesses I had met with at his hands; but it was impossible for any real Russian patriot to gaze unmoved at the German propaganda that was going on in the Empire, or to forgive its Sovereign Lady for disgracing herself together with the crown she wore, by the superstitious practices that had put her into the power of intriguing persons who ultimately brought about her own destruction, together with the ruin of the dynasty. It was impossible for any one who had known Russia during the reign of Alexander III., when the whole of Europe had its eyes turned upon her, and was clamoring for her alliance, not to feel deeply grieved in noticing the signs of the coming catastrophe which had been hovering in the air ever since the fatal Japanese war. The Monarch had become estranged from his people and his wife was the person responsible for it; or rather the people who had succeeded in getting hold of her mind. I do not wish here to throw stones at Alexandra Feodorowna, and in relating now what I know concerning her life, I will try not to forget that misfortune has got claims upon human sympathy, and that where a woman is concerned one is bound to be even more careful than in the case of a man.

The former Empress of All the Russias is to-day a prisoner, condemned to a horrible exile. She deserves indulgence; the more so that her follies, errors, and mistakes were partly due to a morbid state of mind, verging if not achieving actual insanity. Her existence, like that of the hero in the beautiful poem of Félix d'Arvers, had its secrets, and her soul its mysteries. The fact that she was a Sovereign did not shield her from feminine weaknesses, and, though she had always remained an innocent woman—a fact upon which one cannot sufficiently insist—in view of all the calumnies which have been heaped upon her, yet, like the unfortunate Marie Antoinette, to whom she has been more than once compared, she had also met on her path the devotion of a Fersen, as accomplished, as brave, and as handsome, as the Swedish officer whose name has gone down to posterity, thanks to his love for the poor Queen who perished on the scaffold of the Champs Elysées. While the latter was spared the sorrow of losing such a faithful friend, Alexandra Feodorowna was destined to be an unwilling witness of a cruel and unexpected tragedy, which ended brutally any dreams she might have nursed in the secret of her heart, and put her good name at the mercy of an infuriated man. Therein lies the drama of her life; a drama the remembrance of which probably haunts her to this day in the solitude of the lonely Siberian town, to which she has been banished by a triumphant Revolution.

This drama, which I am going to relate in the pages about to follow, was made the subject of a shameless exploitation that took advantage of the sorrow and despair to which it gave rise, that neither spared the woman nor respected the Sovereign, and that finally overthrew the Romanoff dynasty, and brought about the ruin of Russia. It seems to me that the revelation of it can harm neither its heroine, nor the country over which she reigned for twenty-two years; while, on the other hand, it may help the public to understand some of the causes of the great Revolution which was to be followed by such momentous consequences, not only for Russia, but also for the whole world.

Before relating it, I must, however, beg my readers to keep always in mind the fact that the Consort of Nicholas II. was not a normal woman; that madness was hereditary in the Hesse-Darmstadt family to which she belonged, twenty-two members of whom had, during the last hundred years or so, been confined in lunatic asylums; that consequently a different standard of criticism must be applied to Alexandra Feodorowna than to an ordinary person in full possession of all her intellectual faculties. The whole course of her history proves the truth of what I have just said, and claims indulgence for her conduct.

As for this history, I think that, such as it really was, few people have so far come to an exact knowledge of it, and that no one yet has related it as I am going to do. The information that has reached me has come almost day by day from sources which I have every reason to know are excellent. I have applied myself to eliminate many facts which appeared to me to be of too sensational a nature. I want also to point out to the reader that, though this book is called the *Confessions of the Czarina*, yet it does not contain one single word which I would like him to believe to have been uttered personally by the former Czarina. It is a story written ONLY by Count Paul Vassili, who accepts its responsibility in signing his name to it.

PAUL VASSILI.

February, 1918.

CONFESSIONS OF THE CZARINA

I

BETROTHAL AND MARRIAGE

TOWARD the close of February in the year 1894 the health of the Czar Alexander III. of Russia began to fail.

Those in the confidence of the inner circle of the Imperial Family, who constituted the small society which used to form the immediate surroundings of the Sovereign, whispered that the Emperor was taking a long time to rally from the attack of influenza which had prostrated him in the beginning of the winter, and that steps ought to be taken to ascertain whether or not he was suffering from something other than the weakness which generally follows upon this perfidious ailment. But they did not dare to mention openly their fears, because it was the tradition at the Russian Court that the Czar ought not, and could not, be ill; whenever any bulletins were public-shed concerning his health or that of any other member of the Imperial Family, it was immediately accepted by the general public as meaning that the end was approaching. In the case of Alexander III., his robust appearance, gigantic height and strength, seemed to exclude the possibility of sickness ever laying its grip upon him. In reality things were very different. The Czar had been suffering for years from a kidney complaint, which had been allowed to develop itself without anything being done to stop, or at least to arrest, its progress. He was by nature and temperament an indefatigable worker, accustommed to spending the best part of the day and a considerable portion of the night, seated at his writing-desk; he rarely allowed himself any vacations, except during his summer trips to Denmark, and he never complained when he felt unwell, or would admit that his strength was no longer what it had been. He had a most wonderful power of self-control and a very high idea of his duties as a Sovereign. On the day of his accession to the Throne, when, on his entering for the first time the Anitchkoff Palace, which was to remain his residence until his death, he was greeted by the members of his household with the traditional bread and salt, which is always offered in Russia upon occasions of the kind. When implored to show himself a father to his subjects, the giant's blue eyes had shone with even more kindness in their expression than was generally the case, and in a very distinct and quiet voice he had replied:

"Yes, I will try to be always a father to my people."

This promise, given in the solemn moment when the weight of his new duties and responsibilities was laid upon him, the late Czar had always kept faithfully, honestly, with a steadfast purpose and an indomitable will. He had put upon his program among other things the resolution never to complain at any personal ailment or misfortune that he might find himself obliged to bear. This resolution he kept up to the last moment, and he went on working at his daily task until at last the pen fell from his weary fingers and he had to own himself beaten. But during the last memorable year of his life he must have more than once felt that the end was drawing near, though he never spoke about it, with the exception of once, when finding himself alone with one of his intimate friends, General Tcherewine, he told him that he did not think he had long to live, adding, sadly:

"And what will happen to this country when I am no longer here?"

The General became so alarmed at this avowal of a state of things he had suspected, without daring to acknowledge, that he tried to open the eyes of Empress Marie as to the state of health of her husband. But the Czarina refused to see that anything was the matter, and angrily reproved the General for daring to suggest such a thing. The latter subsided, and sought one of the doctors who were generally in attendance on the Emperor, asking him to tell him honestly his opinion concerning the Czar. The doctor retrenched himself behind professional secrecy, and only replied vaguely. The truth of the matter was that he did not wish to own that he had been rebuffed by Alexander III. when he had asked the latter to allow him to make an examination of him, and that he had never dared to insist on its necessity.

At this time, when his father's life was trembling in the balance, the heir to the Russian Throne, the General Tcherewine, was twenty-six years old. If the traditions of the House of Romanoff had been adhered to in regard to him, he ought to have been married already, as it had been settled by custom that the eldest son of the Czar ought as early as possible to bring home a bride, so as to insure the succession to the crown. But the Empress Marie had never looked with favor at the possibility of seeing her family circle widened by the advent of a daughter-in-law, and whenever the question of the establishment of her eldest son was raised she always found objections to offer against any princess whose name was mentioned to her as that of a possible wife. The French party at the Imperial Court, which at that moment was in possession of great influence, tried hard to bring about the betrothal of the future Czar with the Princess Hélène of Orléans, and at one time it seemed as if it would be really possible to arrange such a marriage, in spite of the differrence of religion.

But another circumstance interfered; during one of his visits to Germany, where he often repaired as the guest of his aunt, the Grand Duchess Marie Alexandrowna of Coburg, the Grand Duke had fallen in love with the Princess Margaret of Prussia, the youngest daughter of the Empress Frederick, and the sister of William II., and had declared that he would not marry any one else. To this, however, Alexander III. decidedly objected, saying that he would never consent to a Prussian princess wearing again the crown of the Romanoffs. He expressed himself in such positive terms in regard to this matter that the Grand Duke did not dare to push it forward, and it was soon after this that he was sent on a journey round the world, while the Princess Margaret was hurried into a marriage with a Prince of Hesse by her brother, who, furious at her rejection by the Czar, decided to wed her offhand to the first eligible suitor who presented himself. The young girl wept profusely, but had to obey, and the Cesarewitsch for the first time in his life showed some independence, and declared to his friends that since he had not been allowed to marry the woman he loved, he would not marry at all.

Before this, however, there had been made by his aunt, the Grand Duchess Elisabeth, an attempt to betroth him with the latter's sister, the Princess Alix of Hesse, who had spent a winter season in St. Petersburg as her guest, and who was spoken of as likely to be considered an eligible bride for the future Emperor of All the Russias. She was not yet as beautiful as she was to become later on. The awkwardness of her manners had not impressed favorably St. Petersburg society. Smart women had ridiculed her and made fun of her dresses, all "made in Germany," and had objected to the ungraceful way in which she danced, and declared her to be dull and stupid. If one is to believe all that was said at the time, the Grand Duke Nicholas Alexandrowitch shared this opinion, and it was related that, one evening during a supper at the mess of the Hussar regiment of which he was captain, he had declared to his comrades that there was as much likelihood of his marrying the Princess Alix as there was of his uniting himself to the Krzesinska, the dancer who for some years already had been his mistress. But during the spring of the year 1894 things had changed. As the Czar's health became indifferent, his Ministers bethought themselves that it was almost a question of state to marry as soon as possible the Heir to the Throne.

Mr. de Giers, who was in possession of the portfolio of Foreign Affairs, and who (this by the way) had always been pro-German in his sympathies, gathered sufficient courage to mention the subject to Alexander III., saying that the nation wished to see the young Grand Duke married and father of a family. The Emperor understood, and a few days later, in despatching his son to Coburg to attend the nuptials of his cousin, the Princess Victoria Mélita, with the Grand Duke of Hesse, he told him that he would like him to ask for the hand of the Princess Alix, and to offer to the latter the diadem of the Romanoffs.

The Cesarewitsch did not object this time. For one thing, he did not think his father was really ill, and he was becoming very impatient at the state of subjection in which he was being kept by his parents. He imagined that, once he was married, he would be free to live his own life; what he had seen of the Princess Alix had not given him a very high opinion of her mental capacities, and therefore he believed that she would be contented with the grandeur that was being put in her way, and would shut her eyes to any little excursions he might make outside the beaten tracks of holy matrimony. The woman he had loved had been removed from his path, and perhaps in the secret of his soul he was not so very sorry, after all, to show her that he had consoled himself. It seems also that Miss Krzesinska, the Polish dancer by whom he had had two sons, had been won over to the marriage by means about which the less said the better, and had used her influence over her lover to persuade him that the Princess Alix was of so meek and mild a temperament that they would be able to continue their relations after his marriage with her, which perhaps would not be the case were he to wed some one gifted with more independence and more intellect. Nicholas has always been of the same opinion as that of the last person with whom he spoke. He therefore yielded, went dutifully to Coburg, and just as dutifully proposed to the young Princess whose arrival in

Russia was to herald so much misfortune to her new family, as well as to her new country.

The engagement was announced on the 20th of April, 1894, but was not made in Russia the subject of welcome it had been expected. Everybody felt that love had played no part in this union, which politics alone had inspired. The open repugnance which the bride displayed for everything that was Russian, and the hesitation she had shown before consenting to adopt the orthodox faith, had not predisposed in her favor St. Petersburg society. The Empress Marie, whose consent had been a matter of necessity, did not hide the want of sympathy with which this marriage inspired her; the Imperial Family did not care to see put over its head the insignificant Princess it had snubbed two years before; the nation, violently anti-German as it had become, wondered why it had not been possible to find for its future Sovereign a wife in some other country than the one which seemed to consider as its right the privilege of furnishing Russia with its Empresses.

By a curious anomaly, in Darmstadt, and in Berlin, the betrothal was exceedingly unpopular, and the press spoke of it as of an open scandal, on account of the change of religion imposed upon the Princess Alix. The only two people who rejoiced at her good luck were Queen Victoria, who always liked to see her daughters and granddaughters well married; and the Kaiser, who, since his earliest years, had been the particular friend of the future Czarina, and who had succeeded, at the time when she had shown herself reticent in regard to all her other relatives, in winning her confidence and her affection, perhaps out of gratitude, because he had been the only one who had troubled about her in general.

The first weeks which followed upon the engagement of the Cesarewitsch were spent by him in England, whither his fiancée had repaired, and while there he had been very much impressed with the grandeur of Great Britain, and with the kindness which Queen Victoria showed him. He would have liked nothing better than to be allowed to remain where he was for an indefinite time, forgetting all about Russia, which (this is unfortunately an uncontested fact) he never liked nor troubled about.

Events, however, were progressing, and very soon it became evident even to the most indifferent onlooker that the days of Alexander III. were numbered. The dying Sovereign was taken to Livadia in the Crimea, whither his son was hastily recalled. When the latter arrived there took place a small incident which, better, perhaps, than anything else, will give an idea of the young man's utter want of comprehension of the gravity of the events which went on around him. A few hours after he had reached Livadia his father's friend, General Tcherewine, called upon him, to make him a report concerning the health of the Czar. The Grand Duke listened to him in silence, then suddenly inquired:

"What have you been doing the whole time you have been here? Have you been at the theater, and are there any pretty actresses this year?"

The General, surprised, replied:

"But, Sir, I could not possibly go to the theater while the Emperor is so ill."

"Well, what has this got to do with going or not to the theater; one must spend one's evenings somewhere."

Tcherewine, who related to me himself this story a few weeks later, added:

"He will always remain the same; he will never understand anything that goes on around him."

It was during the last days of the useful life of Alexander III. that the plan of marrying immediately his son and future successor to the Princess Alix of Hesse, and of performing the ceremony at Livadia, was suggested, at the instigation, it is said, of the German Ambassador in St. Petersburg, General von Schweinitz, who had received instructions from Berlin to try and hasten the event as much as possible. But the Czar would not hear of it, declaring that the Heir to the Russian Throne could not be married privately. He consented, however, to a telegram being sent to the Princess Alix, inviting her to come at once to Livadia, to be presented to him. She obeyed the summons, but not without reluctance. She did not care for her future husband, and as she elegantly expressed it, to a lady whom she honored with her confidence, she "did not care to find herself in the Crimea at a time when no one would think of her, and when she would be compelled to be the fifth wheel to a coach." She was, how-ever, persuaded, and left for Warsaw, where her sister, the Grand Duchess Elisabeth, was to receive her, and to accompany her farther.

At Berlin she was met by William II., who traveled with her a part of the way, and during a long interview which lasted over five hours gave her his instructions as to what she ought to do in the future. As we shall see, she was to follow them but too well.

The Princess reached Livadia three days before the Czar breathed his last. He found sufficient strength to receive her, bless her, and wish her happiness in her new life. She replied (this must be conceded to her) with great tact to those solemn words of fare-well, and, suddenly surmounting her previous repugnance, she declared herself ready to abjure at once the Protestant faith, and to embrace that of her future husband and subjects. Some people say that she declared she wished to procure this last joy for Alexander III., but this is doubtful, considering the fact that her conversion took place only on the morrow of the death of the latter.

As soon as it had become an accomplished fact, she was given the title of a Russian Grand Duchess and of an Imperial Highness. Her name appeared in the liturgy, and she was treated with all the honors pertaining to a future Empress. But she found herself lonely and forsaken amid her newly acquired grandeur. The Dowager Empress was too entirely taken up by her grief to pay any attention to the haughty girl, who, already during those first few days of her new life, showed herself resentful when she thought that she was not awarded sufficient importance. The young Czar was so absorbed by the many duties and obligations which fell upon his shoulders that he had no time to remain with her as long as she would have wished, perhaps, and his family simply ignored her. Her days were spent in attending the many funeral services which, according to etiquette, took place twice, and sometimes thrice, daily beside the bier of the deceased Monarch. She found herself placed not only in an awkward, but also in an absurd, position, and if she did not realize other things, she understood this one but too well.

When the body of Alexander III. was brought back to St. Petersburg, the Princess Alix accompanied it, together with the other members of the Imperial Family, and one could see her, deeply veiled, during the funeral ceremonies which took place at the fortress, standing beside the Dowager Empress, silent and attentive to all that was going on around her, and making mental notes as to everything that was taking place. She began to assume a Sovereign's attitude, and she tried to take, as if accidentally, precedence over the Grand Duchesses. One of them, the Princess Marie Pawlowna, soon perceived the game, and one afternoon as the future bride was keeping close to her prospective mother-in-law, seeming to dance attendance upon the latter, the Grand Duchess pushed her aside most unceremoniously, saying as she did so:

"Not yet, not yet, Alix; this place belongs still to me."

Affronted, the young girl withdrew; but when she got home to the Palace belonging to her sister, where she had taken up her abode, she declared that she wished to return to Darmstadt because her position was too false in Russia.

Scene followed upon scene; and Nicholas II. was treated for the first time to the hysterics of which he was to see, later on, so many repetitions. At last the Prince of Wales, the future Edward VII., interfered, and it was partly at his instigation, and that of Queen Victoria, who wrote upon the subject to the Empress Marie, that it was at last decided that the marriage of the new Czar with the Princess Alix was to take place immediately after the funeral of the former's father.

I shall never forget that day. In the vast halls of the Winter Palace the whole of Russia was represented, eager to witness this unique ceremony, the marriage of a Reigning Emperor, an event which had never taken place before. The bride was on that day the object of great sympathy. One pitied her for finding herself so suddenly placed in a position for which she had not been at all prepared, and one felt disposed to grant her every indulgence in case she made a mistake of some kind or other, which was almost an unavoidable thing. Some people, whose English sympathies predisposed them in her favor, rejoiced openly to see the Throne occupied by a granddaughter of Queen Victoria, and hoped that the latter's influence and example would induce the new Empress to try and persuade her husband to renounce the principles of the tyrannous autocracy followed by his predecessors. The man in the street, however, remarked that nothing but bad omens surrounded this hurried marriage, and recalled the old Russian proverb, that "wedding-bells ought never to be heard in conjunction with funeral ones."

The most unconcerned person seemed to be the bride herself as, amid the hushed expectation of the crowd assembled on her passage, she entered the chapel of the Winter Palace on the arm of him who since a few days was Nicholas II., Emperor and Autocrat of All the Russias.

A murmur of admiration followed her as she passed. Seldom has anything more beautiful graced human eye than Alexandra Feodorowna in her wedding-dress, as she slowly walked along, with a diamond crown on her head and a long mantle of cloth of gold lined with ermine falling from her shoulders, and carried by Court officials in embroidered uniforms. She was a real vision of lovelyness, "divinely tall and divinely fair," and in the general feeling of admiration excited by her radiant beauty but few people noticed the thin, set lips, pressed together in firm determination, and the hard chin, which gave a disagreeable expression to what otherwise would have been a faultless face. Behind her, also in white attired, walked the Empress Marie, sobbing the whole time, and leaning on the arm of her aged father, the King of Denmark. Every heart went out to her in her widowhood and loneliness; while many wondered whether her successor, on the Throne she had graced so well, would ever become as popular as she had been during her short reign of thirteen years.

An hour later a State carriage with outriders drove the newly wedded couple from the Winter Palace to that of Anitchkoff where they were to take up their residence with the Dowager Empress until their own apartments were made ready for them. The bride was greeted with vociferous cheers by the crowds. It was the one solitary occasion in her life when she could have the illusion of being popular with her newly acquired subjects. Eighteen months later these were to show to her in an unmistakable manner that such was far from being the case, when she was making her entry into that old town of Moscow, where the Imperial Crown was to be put on her brow, to replace the orange flowers which had adorned her head on her wedding morning.

II

MARRIAGE AND LONELINESS

ONE must be fair. The first months of the wedded life of the young Empress Alexandra were not months of unmixed happiness. This, though partly her fault, was also due to circumstances and the people who surrounded her. Though the Consort of one of the mightiest monarchs in Europe, she yet found herself relegated to an absolutely secondary position; she discovered very quickly that no one considered her to be of any importance whatsoever beside her mother-in-law, the Dowager-Empress Marie. The latter had been one of the most popular Sovereigns who ever graced a throne, and from the very first days after her arrival in Russia she had applied herself to the task of pleasing the people. Like her sister, Queen Alexandra, she identified herself completely with the nation that now claimed her as its own, and she entered into all its interests and pursuits, without any exaggeration, but with that quiet, lovely dignity which never failed her, no matter in what position she found herself. Her influence over her husband had been immense, but no one had ever noticed it; on the contrary, she had persistently remained in the background and tried to pass for a pleasant, amiable, and just a little frivolous, woman who cared for balls, pretty clothes, fine jewels, and the pomp which surrounded her at every step she took. She held very properly the idea that it lowers a Sovereign to appear to be under the sway and influence of his wife, and so, though Alexander III. never took any decision of any importance without having first of all discussed it with her, in public she avoided not only talking politics, but even the appearance of being interested in them.

On the other hand, she had always been, not only conscious, but also very jealous, of her power. She did not in the least care to give it up after her widowhood. Her children, strange to say, had always stood in awe of her, much more than of the Czar, who was a most affectionate and loving father, while Marie Feodorowna had always treated them more from the point of view of Sovereign than mother.

This had been especially the case with the Grand Duke Nicholas, who, when he found himself Emperor, discovered that he could not avoid taking the Dowager Empress's opinion, especially in matters concerning his domestic life. He was told by her that the inexperience of his young wife made it imperative she should be guided by the advices of people older than herself.

This, however, did not suit at all Alexandra Feodorowna, and she found an unexpected support in the person of her own Mistress of the Robes, the Princess Galitzyne, who did not like Marie Feodorowna and was but too glad to put spokes in the latter's wheels. That was the cause of much trouble, and brought about strife in the Imperial Family, which might have been avoided by the exercise of a small amount of tact.

The young Empress, compelled to live in two badly furnished, poky little rooms on the ground floor of the Anitchkoff Palace, became impatient and fretful, and did not care to make a secret of the fact. She felt hurt, too, at several incidents which occurred about that time, the first one of which was connected with the introduction of her name in the liturgy. She wished it to figure immediately after that of the Emperor, while Marie Feodorowna pretended that hers ought not to be relegated to a secondary place, but be mentioned before that of her daughter-in-law.

The two ladies quarreled desperately on this subject, and at last the matter was referred to the Synod, which decided, in view of the existent precedents, that the name of the Consort of the Sovereign ought to be called before that of his mother. The Dowager was furious, while Alexandra Feodorowna was triumphant, and not wise enough to hide it from the world, expressing herself quite loudly in regard to the pleasure which she experienced in seeing defeated the attempt made by her mother-in-law to relegate her to an inferior place which she did not in the least wish to occupy.

Another cause of discontent arose in connection with the Crown Jewels. Marie Feodorowna had liked to wear them more often than any of her predecessors on the Throne, and, though her own private collection of pearls and diamonds was one of the most magnificent in Europe, yet she loved to put on the exceptional stones, tiaras, and necklaces which were the property of the State. Her husband, Czar Alexander III., also liked to see them adorn the person of his idolized wife, and in order to spare her the annoyance of going through the long ceremony associated with the demand of any *parure* it pleased her to require from the Treasury, he had had the jewels she cared for the most transferred to the Anitchkoff Palace, where they were kept in a special safe in the Empress's bedroom. After the latter's widowhood, the question arose as to whether she was to be allowed to retain the custody of all these precious stones, or whether, properly speaking, it was only the reigning Empress who had the right to wear them; had they not better be returned to the place which they had occupied before in the Imperial Treasury?

Some Court officials considered that this was the proper thing to do; the more so that, as it happened, the young Empress had not personal diamonds or pearls at all worthy of her new position. She had received some wonderful presents from her husband when they had become engaged, but the usual amount of jewels bestowed upon marriage on all the Grand Duchesses of Russia had not been offered to her, on account of the hurry with which this marriage had been achieved. It was therefore essential that she should be given the opportunity to adorn herself on all State occasions with the brilliants that the Crown held in reserve for the use of the Sovereign's Consorts. No one thought of subjecting the Empress to the ordeal of going to her mother-in-law, to beg from the latter the permission to use the things to which she was legally entitled, and one would have thought that the best way out of the difficulty would be to have the jewels returned to their original place of abode, and reinstated in the Treasury.

But one had not reckoned with the Dowager Empress! She absolutely refused to give up the ornaments she had been so fond of, and when driven out of her last intrenchments, and obliged to capitulate, she protested that it was not usual for an Empress to wear what belonged to the Crown, before that Crown had been officially laid upon her head, and said that she would relinquish the possession of the famous jewels only after the Coronation of her son and daughter-in-law. The Czar, weak as usual, yielded. Alexandra Feodorowna declared that she did not care for the "hateful things," and proceeded to buy out of her allowance the most gorgeous ornaments she could lay her hands upon, getting heavily into debt in consequence, a fact which did not help to make her popular with her subjects.

She had an unpleasant manner that told against her. Not affable by nature, timid to a certain extent, she imagined that her position as Empress of Russia required her to show herself haughty and disdainful with the people who were introduced to her. Her extremely indifferent knowledge of the French language, which was the only one in use in Court circles, also added to her unpopularity. Her mistakes in that respect were repeated every-where and ridiculed by the old ladies whom her want of politeness had con-tributed to offend, and before she had been married three months she found herself not only unpopular, but even disliked by almost every person she had met.

Then, again, Alexandra Feodorowna was possessed of a wonderful, but most unfortunate talent for drawing caricatures, of which she made no secret, but which, on the contrary, used against all those she disliked, and their name was legion! She found herself, of course, extremely lonely, without any friends of her own rank, and deprived of that liberty of going about she had enjoyed so much at Darmstadt. She had taken a violent dislike for all the Princesses belonging to her new family, and even the grace and liveliness of the Grand Duchess Xenia Alexandrowna, her sister-in-law, had failed to win her heart. She did not care for Russia; its climate did not agree with her, its language she could not learn; its religion she despised in those early days which followed upon her marriage, though she was later to become a fanatical adherent of the Greek Orthodox Church; its manners and customs she could not assimilate. All these circumstances put together made her sullen and angry, and added to her general discontent. She at last determined to try and assert herself, and, though secretly despising the weakness of character of her husband, whom she continually chaffed for his blind submission to his mother, she endeavored to supplant the latter in his heart and mind, and to substitute herself for Marie Feodorowna, not only in domestic, but also in political matters.

We shall presently see how this experiment was to be tried, and what were its ultimate consequences.

III

MY COUNTRY, MY BELOVED COUNTRY, WHY AM I PARTED FROM THEE?

THE spring of 1895 brought few changes in the existence of the young Empress.

For one thing, she contrived to influence the Czar to take up his residence in the small Palace of Tsarskoye Selo, which later on they were to inhabit permanently, but which at that time was still badly furnished and rather forlorn in appearance, owing to the fact that no one had ever lived there since the death of Alexander II. It had been a favorite resort of his, and of his morganatic wife, the Princess Youriewsky, and for that reason had been shunned by his successor, who had elected to establish himself in the huge castle of Gatschina. This place was left to the Dowager Empress for life, and thither she repaired in the beginning of the spring, not, however, without having made a feeble attempt to influence her son and daughter-in-law to accompany her. But for once Nicholas II. did not react, and ignored the invitation. His wife was expecting the birth of her first child, and this circumstance gave her more influence, and to her wishes more weight, than would perhaps have been the case under ordinary circumstances.

Though at Tsarskoye Selo Alexandra Feodorowna obtained more liberty than had been the case throughout the weary months of the preceding winter, yet she found that she had to keep in mind the necessity not to give any reason for the criticisms which she knew but too well were directed against her from every side. Needless to say, she might have avoided these criticisms by the display of some elementary notions of tact. In her way she was a very truthful woman; she even carried her love for veracity sometimes too far. She had no experience of the world, and her life at Darmstadt had not prepared her for the responsibilities of her position as Empress. She did not care for St. Petersburg society, which she considered frivolous, and she made no secret of this fact. Of course people resented it.

Her mother-in-law, the Empress Marie, though she had always kept herself very well informed as to all that was going on in the select circles of those privileged beings who were received at Court, yet had taken good care to appear to ignore the many love-affairs which were either known or suspected in regard to these people. She had so much tact that whenever anything she disapproved of occurred, among these Upper Ten Thousand of people, she let them see that such was the case, but never mentioned it in public.

The Empress Alexandra, on the contrary, spoke with acerbity of every small incident which came to her knowledge, and declared loudly that she would refuse to admit in her presence the persons guilty of indiscretions. During the second season which followed upon her marriage, when Court receptions interrupted by the mourning for the late Czar were once more resumed, the Empress struck off from the list of invitations submitted to her the names of some of the most prominent members of St. Petersburg society, giving her reasons for doing so. The result was that nothing but old frumps, or mothers with marriageable daughters, attended this particular ball, and that the Empress in her turn was boycotted by almost everybody of note in the capital, who did not care to have themselves or their relatives publicly branded as not worthy to be admitted within the gates of the Winter Palace. The effects of this ostracism became apparent on the New-Year's reception which followed upon this incident, which only four women attended, wives of Ministers, who, in virtue of their husbands' position, could not well do anything else. The Emperor, surprised at this absence of the feminine element, on an occasion when it was generally very conspicuous, inquired into the matter. When told the story which had given rise to it he forthwith consulted his mother, and the latter, profiting by the occasion, told her son that he had better have the names of the people about to be invited at Court balls submitted to her for inspection, and not to the young Empress. Of course this became known at once, with the result that the popularity of Marie Feodorowna increased, while that of her daughter-in-law, on the contrary, diminished with every day that passed.

Rebuffed on every side, Alexandra Feodorowna first sought comfort and advice from her sister, the Grand Duchess Elisabeth, who, by reason of her residing in Moscow, where her husband, the Grand Duke Sergius, occupied the position of Governor-General, did not often see her. The Grand Duchess, in response to an invitation which she received to come to Tsarskoye Selo, took the first train. When consulted by the Empress in regard to the difficulties with which she found her path beset, she could not find a solution for them, perhaps because she did not honestly seek it. Elisabeth, as well as her husband, was very ambitious, and they would not have been sorry to see Alexandra Feodorowna estranged from all her new family, in order to have her entirely under their influence and control, and to dominate through her the weak Nicholas II., whose character was already beginning to be known, with all its faults and defects, by his near relatives, as well as by his Ministers and advisers. Elisabeth, therefore, advised her sister to try and keep at arm's-length from her mother-in-law, uncles, aunts, and cousins, and especially to be suspicious of her two brothers-in-law, who were represented to her as being her natural enemies, notwithstanding the fact that one of them, the Grand Duke George, was consumptive and did not live in St. Petersburg, the climate of which he could not endure, while the second, the Grand Duke Michael, was a youth of sixteen, hardly out of school.

Alexandra Feodorowna, however, became suspicious of this advice, perhaps because she distrusted the Grand Duke Sergius just as much as her other relatives. Yet advice she felt she must have. It would have been natural for her to seek that of her brother, the Grand Duke of Hesse, and of her other sisters, the Princess Victoria of Battenberg and the Princess Henry of Prussia, but while the former had never been her favorite, the latter refused—at the instigation of her husband, most probably—to be mixed up in things which did not concern her, and intrenched herself behind her ignorance of Russian customs and Russian society. The Empress felt frantic, and it was then that she was seized with violent attacks of homesickness, which she did not attempt to conceal. More than once she was heard to say that she wished she were back in Germany, where at least she would find people capable of understanding her and of advising her well and soundly.

Germany has always, as is but too well known to-day, maintained an army of spies in Russia. Very quickly a report of what was going on in Tsarskoye Selo reached the ears of William II. He saw his opportunity and forthwith wrote to his cousin, reminding her of their former friendship and telling her that he was entirely at her disposal to help her, by his knowledge of Russian affairs, which he professed was very great, and by his experience of the world.

The Empress caught at the opportunity, and from that day there was established between them relations of the closest intimacy, linking the Empress and the Lord of Potsdam. She took the habit of sending him a kind of diary of what she was doing and of what went on at the Russian Court—a diary in which she did not spare her mother-in-law, or her husband, whom she reproached with not taking her part more openly.

Of course it was not easy to carry on such a correspondence. The young Empress was closely watched, a fact of which she was but too well aware. She tried the medium of the German Embassy, but apart from the fact that it would have seemed a suspicious thing to send there letters in a regular way, the Ambassador, Prince Radolin, refused to be the means of forwarding messages of which he did not know the import, and did not care to be involved in an intrigue that would inevitably have brought him to grief if discovered. Some other way, therefore, had to be devised, and for a time it seemed as if it would be next to impossible to find any. Once or twice the Princess Hohenlohe, wife of the Imperial Chancellor, who, through the fact that she was the owner of large estates in Lithuania, often visited St. Petersburg, brought with her messages from the Kaiser to the Empress Alexandra, and took back with her to Berlin the latter's replies. But this was not sufficient, and during the first visit paid by the Czar and his Consort to the German Court William and the young Czarina came to an understanding, after which their correspondence continued through the medium of friends of the Kaiser, who somehow appeared regularly in Russia whenever this was considered necessary.

People, and there were some, who happened to be in the secret of this intercourse pretended that one of the things which William II. urged upon his cousin was the necessity of getting rid of the influence of the Empress Marie, who, by reason of her avowed French sympathies, constituted a danger to German expansion and to German progress in the Muscovite Empire. The fact that for the present Alexandra Feodorowna was still considered a nonentity at the Russian Court was not of much importance because it was thought that if she were once to become the mother of a son she would immediately be raised to the position of an important personage in her husband's house and country. And it must not be forgotten that in the course of the summer of 1895 the Empress was known to be about to give birth to her first child, who of course had to be a boy and an Heir to the Russian Throne.

Alas, alas for these hopes!

It was a Grand Duchess, Olga Nicholaiewna, who saw the light of day on a November morning in the Imperial Palace of Tsarskoye Selo. The disappointment was intense and extended to all classes of the nation, except among the members of the Imperial Family, who made no secret of the fact that they were delighted the little Hessian Princess they all disliked so intensely had not fulfilled her husband's and her subjects' expectations. The news of their joy reached the ears of Alexandra Feodorowna through the channel of the Kaiser, and added to her bitterness against her Russian relatives, which made itself felt in the affected manner with which she continually made allusions in their presence to her regrets at having accepted the position of Empress of All the Russias. She openly spoke of her contempt for this "land of savages" as she called it, and more than once her attendants heard her give vent to the exclamation of "My country, my beloved country, why am I parted from Thee?"

THE CZARINA
When she was Princess Alix of Hesse Darmstadt, before her betrothal to the Czar, 1894.

IV

A SAD CORONATION

CONTRARY to the custom observed at the Imperial Court of Russia, the young Empress insisted herself on nursing her baby. This met with general disapproval, not only from Marie Feodorowna, who, never having thought of the possibility of such an infraction of the traditions of the House of Romanoff, felt considerably affronted at this piece of independence on the part of her daughter-in-law; also from all the dowagers of St. Petersburg, who considered the innovation as *infra dig.* and declared that such a breach of etiquette constituted a public scandal.

Some enterprising ladies, who, by virtue of their own unimpeachable positions, thought themselves entitled to express their opinions, ventured to say so to Alexandra Feodorowna herself. She was indignant at what she termed an insult, turned her back on those voluntary advisers, and flatly declared that she would refuse henceforward to admit into her presence people who had forgotten to such an extent the respect due to her and to her position as the wife of their Sovereign.

Matters assumed an acute form, and during the first ball which took place that season in the Winter Palace the incident was discussed most vehemently. One wondered what would happen later on, and how the Empress would behave in regard to those givers of unsought advice in the future. But Providence interfered in favor of Alexandra Feodorowna, because she suddenly was taken with an attack of the measles, not the German ones this time, but the real, authentic thing, and the Court festivities about to take place were immediately postponed in spite of the protestations of different Court officials, who urged that they could very well take place in the absence of the Empress, and that their abandonment would be a serious blow to trade, which already was very bad, and which had discounted the profits it generally made during a winter season when the gates of the Winter Palace were thrown open with the usual lavishness and luxury displayed there on such occasions. Trade and its requirements were about the last thing which troubled the mind of Alexandra Feodorowna. She was of the opinion prevalent in

Poland at the time of the Saxon dynasty that when Augustus was intoxicated the whole nation had to get drunk, and though she detested or pretended she detested Court balls and festivities, yet she was adverse to others enjoying them while she herself was debarred from doing so. Girls in their first season eager for showing off their pretty frocks, and lively young married women in quest of gaiety, were told to forego expectations of such pleasure, and the gates of the Palace remained closed for the first time in many years, to the general disappointment of St. Petersburg society and of its prominent members.

This disappointment, however, was soon forgotten in the expectation of the Coronation about to take place, the date of which had been fixed for the 15th of May. Great preparations were made for it. Those who remembered the pomp which had attended that of Alexander III., thirteen years before, wondered whether the ceremony about to be repeated would be as brilliant as the one which they had not yet forgotten. The whole of St. Petersburg society, with few exceptions, repaired to Moscow for the solemn occasion, and all the Foreign Courts sent representatives to attend the festival. One tried to guess how the young Empress would carry herself through the trying ordeal, and whether she would conde-scend for once to show herself amiable toward her subjects in the ancient capital of Muscovy, the population of which had always professed far more independence of opinions than that of St. Petersburg, where conversations were more restrained and guarded, in view of the constant presence of the Imperial Family within its walls. The one thing which everybody was looking forward to was the public entry of the young Sovereigns in the old town, an entry which was to be made with unusual pomp and solemnity.

I remember very well the day of the ceremony. I had a seat in a house situated on the great square opposite the residence of the Governor-General of the town, a position which was still occupied by the Grand Duke Sergius. Together with some friends, we watched the long line of troops, followed by representatives from all classes in the country; by Court officials on horseback, in gold-embroidered uniforms, behind whom rode, surrounded by a brilliant staff, the Czar himself, mounted on a gray charger; a small, slight figure, contrasting vividly with his father thirteen years before. Nicholas II. had already acquired the expression of utter impassebility which was never to change in the future. He surveyed with a grim look the vast crowds massed in the streets, who cheered him vociferously, but he did so with a look that expressed neither pleasure nor disappointment, but simply indifference mixed with tediousness.

Behind him came a long row of State carriages all gold and precious stones, the diamonds which glittered on them being valued at several millions of rubles. In the foremost, the carriage of Catherine the Great, with an immense Imperial Crown on its top, rode the Dowager Empress dressed in white and looking as young almost as she had done on the day of her own Coronation. Hurrahs without end greeted her appearance; the people cheered her with an enthusiasm such as had rarely been seen in Russia, while, pale and trembling, she bowed incessantly from right to left, with tears streaming down her cheeks. These hurrahs followed her all along her way from the distant Petrowsky Palace to the gates of the Kremlin, which she entered at last, amid the acclamations of the multitude assembled to see her pass.

Immediately behind her, divided only by a squadron of cavalry, drove her daughter-in-law, also dressed in a white gown, and sparkling with all the jewels belonging to the Crown, which she had assumed for the first time on that solemn day. A dead silence, contrasting painfully with the frenzied reception awarded to Marie Feodorowna, greeted her successor on the Throne of Russia. This contrast was so evident that everybody present was struck with it, and something like a presentiment of evil passed through the mind of most of the assistants of this strange scene. One remembered Marie Antoinette at Rheims during the Coronation of Louis XVI. when she also had been received with silence and contempt by the French nation, who a few years later was to send her to the scaffold.

Perhaps something of the kind crossed the mind of Alexandra Feodorowna herself, because it was evident that she was suffering from a violent desire to give vent to tears and rage. I saw her from the place where I stood, through the open large windows of the State carriage in which she sat quite alone, according to the requirements of etiquette, immovable like an Indian goddess, looking neither right nor left, but straight before her, her haughty head thrown back, two red spots on her cheeks, and a set expression on her thin lips closely joined together. She understood but too well the meaning of this strange reception she was awarded; too proud to complain, she seemed to ignore it. Once and once only did I see her start, and that was when, amid the profound silence which prevailed around her, a voice, that of a child, was heard exclaiming:

"Show me the German, mamma, show me the German!"

And with this cry in her ears and in those of other listeners, the big coach with Alexandra Feodorowna sitting in it, in all the splendor of her white dress and glorious jewels, vanished in the distance within the walls of that old fortress called the Kremlin, which, seen in the glamour of dusk already falling, looked more like a prison than a palace.

THE DOWAGER EMPRESS MARIE FEODOROWNA

Three days later I was to look once more on the slight and erect figure of the Consort of Nicholas II. as she emerged out of the bronze gates of the Cathedral of the Assumption walking under a canopy of cloth of gold and ermine, with ostrich plumes towering on its top, the Crown of the Russian Empresses standing high upon her small head and the long mantle of brocade embroidered with the black eagles of the Romanoffs trailing from her shoulders. She looked magnificent, but there was something in the expression of her haughty features which reminded one of the prophecy of the Italian sculptor in regard to Charles Stuart: "Something evil will befall that man; he has got misfortune written on his face."

Beside his wife, Nicholas II. looked the insignificant personage he was to remain until the end of his reign and very probably of his life. He could no more bear the weight of his Crown physically than he was able later on to carry the burden of his responsibilities. As he walked, he staggered and trembled; and one could distinctly notice the signs of the extreme fatigue under which he labored. Supported on either side by two attendants, who carried the folds of his Imperial mantle, he tried to keep erect the scepter which he held in his right hand, and the orb which reposed in his left.

And then occurred the memorable incident of that memorable day.

When the long procession reached the doors of the Cathedral of the Archangels where, according to custom, the newly crowned Czar was obliged to repair for a short service of thanksgiving, I saw Nicholas II. reel from right to left as would have done a drunken man, and suddenly the scepter which he grasped fell heavily from his hand to the stone floor, before the altar of the church.

It would be difficult to describe the emotion produced by this untoward incident, which was at once interpreted by the superstitious Russian people as a bad omen for the reign which had just begun. Strange though this may seem, yet it is absolutely true, that the faith of the Russian nation in Nicholas II. was shattered from that day when it had found him unable to carry the symbol of his supreme power and Imperial might and not strong enough to bear its weight.

This was not, however, the only unlucky incident which was connected with this sad Coronation, which in so many respects reminded one of several others that had marked the marriage festivities of Marie Antoinette, and the anointing of Louis XVI. at Rheims. I will not describe here the horrors which were enacted on the Khodinka field, when more than twenty thousand people were crushed to death during a popular festival given in honor of the Czar's assuming the Crown of his ancestors; I shall only mention the part played by Alexandra Feodorowna in the gruesome tragedy. As everybody knows, unfortunately for her reputation in history, she danced the night which followed upon it, at the French Embassy.

But what is not so well known is the fact that when she and the Emperor were asked by Count de Montebello, the French Ambassador, whether the ball which they had promised to attend had not better be postponed until the next day, which would have been an easy matter, Alexandra Feodorowna had exclaimed that she could not understand why such a fuss was made because "a few peasants had been victims of an accident likely to happen anywhere," while Nicholas II. had replied that he did not see any necessity to make any alteration in the program which had been officially sanctioned and adopted since a long time.

It was only on the third day following upon the catastrophe, when the clamors of public opinion reached even the deaf ears of the Czar and of his Consort, that they decided themselves at last to pay a visit to the various hospitals where the victims of the tragedy had been carried. They went there in great state and ceremony, the Empress dressed in lace and satin, holding in her hands a large bouquet of flowers which had been presented to her by the officials to whom had been deputed the charge of receiving her at the gates of the houses of suffering and death, whither her duties had called her, much against her will. It was related later on that a little girl ten years old or so, perceiving the roses held by the Sovereign, had exclaimed:

"Oh, the pretty roses!"

"Give them to her," said the Emperor.

"Certainly not. Flowers are most unwholesome in a sick-room," replied Alexandra Feodorowna, and she turned away without another word.

V

DAUGHTERS, DAUGHTERS, AND NO SON

IT was not generally known at the time of the Coronation that the Empress was about to become a mother for the second time. She had not mentioned the fact to her family and not to her mother-in-law, not wishing to be bothered with advice as to the manner in which she should take care of herself—advice which she was beforehand determined not to follow. But the strain of the Coronation festivities, with their attendant emotions and unavoidable fatigue, told upon her, and this was the principal reason which induced the Emperor to repair with her to Illinskoye, the country-seat of the Grand Duke Sergius, close to Moscow, immediately after the departure of the Foreign Envoys, who had been sent to Russia to represent their respective Governments.

The public wondered at this decision, the more so that it was openly said that the responsibility for the disaster of Khodinka rested with the Grand Duke, who had not known how to take the necessary precautions, which, if resorted to, would have prevented the catastrophe. No one suspected that the real reason for this determination of Nicholas II. to spend a few quiet weeks with his uncle and brother-in-law was due to the state of health of Alexandra Feodorowna.

The measure, however, was not to prove successful, because a very few days after the arrival of the Imperial pair at Illinskoye its hopes of an increase in the family were dashed to the ground, and an unlucky accident deprived the Empress of a son and the country of an heir, it having been proved that the child born too early was of the male sex. The fact was kept a close secret, as those in authority did not care for the nation to become aware of the disappointment which had overtaken its Monarch, and even Alexandra Feodorowna was not told of the full extent of the misfortune. She learned of it much later, after the birth of the only boy she ever had. To her anxious questions concerning the sex of the prematurely born infant she never got any satisfactory reply, and though she might have suspected the truth, yet it was not revealed to her at the time. She was only adjured to take care of herself and to avoid every kind of fatigue, a difficult thing to do, considering the fact that the Russian Sovereigns were about to start for a tour of visits at the different European Courts. These visits, with the exception of the stay in Paris where they were received with a burst of the most extraordinary enthusiasm ever witnessed in the French capital, did not turn out so successfully as had been hoped and expected. For one thing, Prince Lobanoff, who held the portfolio of Foreign Affairs and was by common consent considered as the ablest statesman in Russia and one of the cleverest in Europe, died suddenly on the Imperial train at a little station of the Southwestern Railway line, called Schepétowka, almost in the arms of the Emperor. Nicholas, seeing him stagger, rushed to his help. This sad event gave rise to many comments, and it was then that people began to whisper in Russia that the young Empress had got the evil eye and brought bad luck to all those who came into too close contact with her.

Nicholas and his Consort first proceeded to Breslau, where William II. with the Empress came to meet them and received them with the greatest cordiality. It was at that time that arrangements for his correspondence with the Czarina were made, much to the joy of the latter, who, as time went on, felt more and more in need of the help and advice of members of her own family. From Breslau, the Emperor and Empress proceeded to Vienna, but there a succession of unpleasant small incidents, insignificant in themselves, but destined in the course of time to bring about totally unexpected results, took place. Francis Joseph had decided to receive his Russian guests with all the pomp and splendor for which the Austrian Court had always been famous, and the Empress Elisabeth, after much pleading, had at last been persuaded to come to Vienna and to do the honors of the Hofburg to them. At the State banquet which was given there, she appeared, regal and magnificent, clothed in that deep mourning which she never gave up after the tragic death of her only son, the Archduke Rudolph, and she was far more observed and looked at than the young wife of Nicholas II., who resented the fact deeply. It is not generally known that at that time (later she outlived the feeling) Alexandra Feodorowna had a very high opinion of her own beauty and could not bear to play second fiddle in that respect to any one. She always hated pretty women whenever she saw them in a position to rival her, and the fact that Elisabeth of Bavaria, in spite of her fifty-seven years, eclipsed in many respects her own young and radiant beauty did not help to put the Czarina into a good temper. The interview, therefore, passed according to the rules of strict courtesy, but no cordiality permeated it. Wise politicians and diplomats began shaking their heads and murmuring that after this experiment it would become hard indeed to bring about pleasant relations between the Court of the Hofburg and that of Tsarskoye Selo.

From Vienna, the Russian Sovereigns went on to Copenhagen to pay to the aged King and Queen of Denmark their respects, but there also things did not go smoothly. The Russian Imperial Family had always been popular in Denmark, which the late Czar Alexander III. liked extremely, and where he used to spend happy weeks every summer. One had hoped that this tradition would continue, but after having seen Alexandra Feodorowna for three days Queen Louise had remarked that it would be just as well if she did not visit too often.

But what everybody in Russia looked forward to was the visits which Nicholas II. and his wife were about to pay to Balmoral and to Paris. In the first of these places they were made the objects of a warm and entirely homelike reception on the part of Queen Victoria. The latter had always been interested in the children of her favorite daughter, the Princess Alice, and had immensely rejoiced to see her youngest grandchild ascend the Throne of Russia. The Queen, however, was beginning to feel some misgivings as to the latter's fitness for the high position that she had been thrust into. She was perhaps the best informed person in Europe as to all that went on in Foreign Courts, and she had heard, not without serious apprehensions, of the growing unpopularity of Alexandra Feodorowna. She took the first opportunity which presented itself to talk seriously to her granddaughter and to try and persuade her that she ought to make some effort to win the respect and the affection of her subjects. To Victoria's surprise, the old lady never having been thwarted or contradicted, the Czarina replied that she did not know in the least what she was talking about, and that what Russians required was not amiable words but a sound administration of the whip. Under these circumstances the conversation very quickly came to an end, though the Queen, astounded as she was at Alexandra's impertinence, tried, nevertheless, to renew it with the Czar. The latter simply replied that his grandmother must have been misinformed, because everybody loved the Empress. After that Victoria gave up the subject, and she would probably never have mentioned it to any one had it not subsequently reached her ears that the Empress boasted among her friends about the way in which she had snubbed her grandmother. This was rather more than the equanimity of the Queen could stand, and in her turn she related her unsuccessful attempts to make the young Czarina listen to reason, not making any secret of the fact that the future of the latter filled her with the greatest apprehensions.

In Paris, the Empress found herself more at her ease. Flattery was poured down upon her in buckets. All the newspapers praised her looks, her jewels, her general demeanor, and it was only here and there that a dissenting voice was raised, as in the person of a dressmaker who remarked on the want of taste which had presided at the confection of the dresses with which Alexandra Feodorowna tried to astonish the Parisian natives. On the whole the visit was a success, and it inspired with new zeal all the promoters of the Franco-Russian alliance, among whom the Empress was most certainly not to be reckoned.

Very soon after this triumphal journey, a second child was born to Nicholas II. and his wife; another girl, to the intense disappointment of everybody. I am informed that the first words of Alexandra Feodorowna upon being told of the sex of the infant were:

"What will the nation say, what will it say?"

As a fact the nation said nothing; it had already begun to lose interest in the family affairs of its rulers.

As time went on this indifference to the joys and the woes of the Reigning House grew and grew, until at last it became a recognized fact in the whole of Russia that, as far as Nicholas II. was concerned, whatever happened to him or to his relatives was an object which presented no interest whatever to the millions of Russian men and women, who all of them were looking forward for a change in the destinies and the Government of their country. When he had ascended the Throne, any amount of expectations had been connected with him and with his name. These were very quickly dashed to the ground by his first public speech—the one which he made in reply to the congratulations of the zemstvos, or Russian local assemblies, on his accession and marriage, when he told the representatives of these institutions that they must not indulge "in senseless dreams" or hope that he would ever sacrifice the least little bit of his Imperial prerogatives or autocratic leanings. The Revolutionary committees, which had begun at that time and from the very day of the death of Alexander III. to renew their political activity, addressed to him a letter which, read to-day in the light of the events which have happened during the last twelvemonth, seems almost prophetic. They warned him that the struggle begun by him would only come to an end with his downfall, and the whole tone of this remarkable epistle, which I have reproduced in my volume, *Behind the Veil of the Russian Court*, reminds one at present that the prophesied blow has fallen, of the writing on the wall which appeared during the banquet of the Persian King, warning him of his approaching ruin.

Neither the Czar nor his Consort thought about these things. As time went on, the attention of the latter became more and more concentrated on the one fixed idea of having a son. She imagined that the secret of her unpopularity, which she had at last discovered, lay in the fact that she had not been able to give an Heir to the Russian Throne. Four times in succession daughters were born to her, each one received with increased disappointment, as the years went on, bringing into prominence the youngest brother of Nicholas II., the Grand Duke Michael, whom the Empress began hating with all her heart and soul. She imagined that wherever she went she was greeted with reproaches for having failed to fulfil the first duty of a Sovereign's Consort, that of assuring his succession in the direct line. The hysterical part of her temperament rose to the surface more and more with each day that passed. She locked herself up in her private apartments, refusing to see the members of her family and denying herself to all visitors, until at last it began to be whispered in Court circles that Alexandra Feodorowna's mind was getting unhinged and that she was suffering from religious mania, mixed up with the dread of persecution from her relatives. She used to sob for hours at a stretch, when no one could comfort her, and during those attacks of despair one cry continually escaped her lips, and was repeated until she could utter it no longer, out of sheer excitement and fatigue:

"Why, *why* will God not grant me a son?"

VI

THE EMPRESS'S OPINIONS ABOUT RUSSIA

ONE of the points about which there has been the most discussion in Russia is as to whether the Empress Alexandra had ever cared for the country which had become her own. Her friends have repeatedly asserted that she had become an ardent Russian patriot, and that her great, particular misfortune was that every action, word, or thought of hers had been misunderstood and this willingly.

As for her enemies, they declared, from the very first days which followed upon her unlucky marriage, that she had arrived in Russia imbued with the feelings of the deepest contempt for the country and its people, and that all her efforts had been applied toward making out of the Empire over which she reigned a vassal of her own native land.

It seems to me, who have had the opportunity to approach her personally, as well as that of hearing about her from persons who nourished no animosity against her, that neither the one nor the other of these two opinions was absolutely correct, though both were right, each in its way. When one attempts to judge the personality and the character of Alexandra Feodorowna, one must first of all take into account the fact that she belonged to that class of individuals who, while being fools, nevertheless think themselves clever. To this must be added a highly strung, hysterical temperament and the fact which was unknown in Russia at the time of her marriage, that madness was a family disease in the House of Hesse, to which she belonged by birth. The circumstances attending her rearing and education also had a good deal to do with the strangeness of her conduct after she had reached the years of discretion. She had been a mere baby, five or six years old, when she had lost her mother, the charming, clever, and accomplished Princess Alice of Great Britain, and she had been brought up partly at Windsor by Queen Victoria and partly at Darmstadt, where, however, she had not found any of the good examples its Court might have afforded her had her mother remained alive. She was the youngest member of her family, and as such treated with negligence and made to give way to her elder sisters, who were neither kind nor affectionate in regard to her—a fact which must have helped her a good deal to develop the haughty, disagreeable temper which was later on to play her so many bad tricks in life. On the other hand, the person who had charge of her education, as well as of that of the other Princesses, had conceived a great and most ill-advised affection for her; ill-advised in so far that she used to repeat to her that she was handsomer and cleverer than her sisters, and that she ought not to mind any slights which the latter might try to put upon her, because she was sure to make a better marriage than they.

When she was about twelve years old there occurred in the Grand-Ducal Palace of Darmstadt the tragedy or romance, call it as one likes, of the Grand Duke's morganatic union with a lovely Russian, Madame Kolémine, which came to such a sad end, owing to the interference of Queen Victoria and to the stupidity of the Grand Duke himself, who, in any case, ought first of all to have made careful inquiries as to the past life and conduct of his intended bride, and then—once he had plighted his troth to her—to have held the promises which he had made to her. He allowed her to be sent away from his Court and country in disgrace; the lady herself would have been but too willing to come to honorable terms with a man for whom she could no longer feel any esteem or affection, because in the whole long story of his intercourse with her Grand Duke Louis never showed himself otherwise than the true German he really was. Of course, the object of his transient affections was represented to his children as being merely an intriguing, base woman who had tried to make a great marriage and to supplant their mother. Whether the elder Darmstadt Princesses believed this calumny to have been the truth remains a matter of doubt. Judging impartially, this would seem to be hardly likely if one takes into consideration the fact that their ages hovered between eighteen and twenty-two, and that consequently one could reasonably assume that they knew what they were about when they showered one proof of affection after another on Madame Kolémine, and when they declared to her in many letters that there was nothing they wished for more than to see her become their father's wife.

This whole story, together with its heroine, is about one of the most perplexing affairs that ever occurred in any Royal House, and everything connected with it is to this very day shrouded in mystery. Madame Kolémine, who (this by the way) married again, after her divorce from the Grand Duke, a Russian diplomat, may or may not have been a bad woman. I hold no brief for or against her. Many people assert that in regard to certain scandals connected with the time of her early married life she was more sinned against than sinning, and that she became the victim of calumnies launched against her by unscrupulous enemies. But, true or not, the breath of suspicion had hovered around her good name to a sufficiently strong degree to have absolutely justified the objections of Queen Victoria to her becoming even the morganatic wife of the Grand Duke of Hesse.

It ought also to have influenced the latter into not admitting the fascinating Russian into the intimacy of his young daughters, which was precisely what he did. The girls could not be told every kind of gossip going about in the world, but they ought to have been shielded from the possibility of contracting friendships likely to lead them into unpleasantnesses in the future. On the other hand, considering the fact that this intimacy had once been established, one does not very well see how any of the Darmstadt Princesses could have been led to believe, after the three years or more that it had lasted, that Madame Kolémine was base and intriguing and cared only for a great marriage. Because this last accusation, leaving aside all others, was absolutely false, a fact no one was better able to know than themselves, who had repeatedly begged and implored her to accept their father's offer and to make him, together with themselves, happy people.

I have had some of these letters in my hands, and can therefore vouch for the truth of this last assertion, and to put an end to the questions of a suspicious public that may wonder how it came that such a correspondence was ever communicated to me, I will say at once that the reason for it was that I am a blood relation of Madame Kolémine, who after her divorce had thought I might be of some help to her in her troubles, and had herself asked me to read them. The impossibility in which I found myself to be of any use to the poor woman, whom I had never seen in my life before, and of interfering in a business which did not concern me in the very least, led her to take a most bitter attitude in regard to me and to become my enemy, so that in trying to take her part to-day I am doing so out of a feeling of justice and nothing else.

I have mentioned the story in general only because it explains to a certain degree the undisguised aversion of the Empress Alexandra for everything that was Russian or that had anything to do with Russia.

She had never shared her sisters' admiration for Madame Kolémine; on the contrary, she had always nourished a pronounced antipathy for the lady, and whatever the three other Darmstadt Princesses may have felt in regard to the woman whom their father was to marry and divorce on the same day, she, at least, had made no secret of her hatred for her. One of the first remarks which she made after she had become acquainted with St. Petersburg society was:

"I shall never like it; all the women remind me of Madame Kolémine."

This episode in the career of the Grand Duke of Hesse brought about, as might have been expected, a change in his relations with Queen Victoria, and he was no longer such a desired guest at Windsor or Balmoral as had been the case. His elder daughters married in quick succession, the second one wedding the Grand Duke Sergius of Russia. The little Alice was left alone at home, and though she was often requested by her grandmother to join her in England, she did not care so much for these invitations as formerly. The fact was that she was gradually acquiring a considerable influence over her father's mind, whose weakness of intellect rendered him an easy tool in his enterprising daughter's hands. She became the virtual mistress of his house, and developed during those years, where she remained absolutely without any feminine control over her, the imperious, disagreeable temper which was to play her such sorry tricks in the future. Small as was the Hessian Court, it yet was administered with that strict respect for etiquette always in vogue in Germany, and it pleased the Princess Alix to find herself the first lady in the land in her father's Dukedom. She preferred it to being the second in Rome.

It was during those years that she was taken on a visit to the Russian Court. This did not turn out a success, because no one in St. Petersburg was in the very least impressed by the beauty of the young girl. Russia, being celebrated for the loveliness of its women, would have required something more than she possessed to fall on its knees and worship her. Then, again, she was dressed with bad taste, her manners left much to be desired, and the rumor which began to circulate at the time of the possibility of her wedding the Heir to the Russian Throne did not appeal to public feeling. Alice thought herself slighted, and returned to her beloved Darmstadt more anti-Russian than she had ever been.

Two years went by, and the Grand Duke of Hesse died, carried off by a disease of the brain, difficult to account for if one takes into consideration the fact that he had never had any brains to lose. His son succeeded him, and, together with his sister, continued to inhabit the Darmstadt Palace, where nothing was changed except the master of the house, whom no one missed. For eighteen months Princess Alice reigned supreme, as she had done before; then one fine morning her brother announced to her that he was about to ask their cousin, the Princess Victoria Mélita of Coburg, to become his wife. A fit of hysterics followed upon this announcement. Alice could not resign herself to the necessity of playing second fiddle at her brother's Court, where she had been the center of attraction for such a long time. The fact that her future sister-in-law was just as young and more beautiful than she did not help her to get over her mortification. She was of a terribly jealous character and temperament, and she began from that very day to hate, with a ferocious hatred which went on increasing as time passed, the innocent girl for whom this Hessian marriage was to prove the source of so much sorrow. But about this I shall speak later on.

It was at this precise moment that talk about a Russian marriage for her began again. Many people wished for it. The Berlin Court was actively intriguing in favor of it, and during the whole of that winter of 1893-94 the newspapers were busy with it. The chancelleries of the different European capitals were very much preoccupied as to whether or not it would take place.

Perhaps few people will believe me when I say that had it not been for her brother's engagement nothing in the world would have ever decided the Princess Alice to give her consent to a union for which she did not feel the least sympathy. She was not at all dazzled by the prospect of becoming the Empress of Russia, because in her vanity and with her ideas of German grandeur she thought herself far superior to the Romanoffs, thanks to her long and unbroken line of ancestry. Her unimpeachable quarterings seemed to her to be so immeasurably above their doubtful ones that she considered it would be she who would do him an incommensurable honor by accepting as her wedded husband the Heir to the Throne of All the Russias. She would have infinitely preferred going on queening it in Darmstadt, or in any other small German town, than to have been chosen as the bride of the future Nicholas II., for whom she felt neither sympathy, affection, nor esteem.

But her brother's prospective marriage changed considerably her position. She would no longer occupy the position of the first lady of her beloved Hesse; she would find installed in the place which had been her own for so many happy years a woman younger than herself, with an independent character, a determined mind, a woman who would most probably grow very quickly to impose herself and her ways of thinking, not only on the whole Hessian Court, but also on the Grand Duke, whose sister was condemned beforehand to be neglected and treated as a negligible quantity.

This was gall and wormwood to the passionate, selfish girl, and this feeling of hers, which she allowed her cousin the Kaiser to guess, was very cleverly exploited by the latter in view of a marriage which none desired more ardently than himself. Next to his own sister, there was no one in the whole world whom he would have more ardently wished to become Empress of Russia than his cousin Alix. He invited himself to Darmstadt for a short visit, and while there took the first opportunity to discuss the subject with her. He told her what very few people knew at the time, and what the general public was entirely ignorant of—the serious nature of the illness with which Alexander III. was attacked, an illness which gave no hope whatever of recovery. By marrying the young Grand Duke Nicholas the Princess would find herself but for a short time in a so to say subordinate position. A few months would see her raised to one of the greatest thrones in Europe, from the height of which she would be enabled to look down with contempt and pity on the cousin who was about to take at the Darmstadt Court the place she had occupied so long she had grown to consider it as her very own.

Moreover, she would be able to win back Russia and its ruler to the cause of the German alliance, and thus accomplish one of her duties as a loyal German woman. He appealed to her worst instincts while seeming to call on her noblest ones to assert themselves, and once more he won the day.

In St. Petersburg, too, Hohenzollern influence and intrigues had worked actively, until at last the Czar, feeling perhaps that his days were numbered and perhaps also no longer strong enough to resist perfidious advice given to him by interested people, yielded the point, and when his eldest son started for Coburg to attend there the Princess Victoria Mélita's wedding, he authorized him to ask for the hand of Alix of Hesse.

As I have related, the marriage was at once announced, and we have seen already its first results. The reason why I have once more returned to its subject was to explain some of the causes which led the Empress Alexandra to conceive such a bad opinion about Russia, and to detest so cordially the Russian people. Her early dislike for Madame Kolémine had given her a natural antipathy for everything connected with the latter country; her visits there had strengthened this feeling; her vanity had been hurt by finding that St. Petersburg's society had paid absolutely no attention to her; and her slow, commonplace mind had been utterly unable to understand the refinement and high breeding of the Russian upper classes. Her natural coldness and ignorance had been repulsed instead of attracted by the simplicity but genuine kind-heartedness of the lower ones. She thought the nation one of savages and she made no secret whatever of that opinion, expressing her intention of correcting those "awful Russian manners," which had seemed to her young and inexperienced eyes so very dreadful when she had first become acquainted with them.

It is most likely that if she had married a small German Prince or Potentate she would have put herself out of the way to please his subjects. But she did not think the Russians worth her while. She considered that they ought to feel themselves highly honored by the fact that she had consented to come and reign over them, and in her own mind she did not attach any more importance to the judgments they might be inclined to bestow in regard to her person than she would have done to the criticisms of the first beggar in the street. She arrived in her new country despising it, together with its people, determined to ignore the wishes it might have or the necessities it might require. She arrived there prejudiced and bigoted, and so full of contempt for the land that hailed her as its Queen that she did not admit the possibility of treating it as one inhabited by human beings, but determined to apply to it some of the methods used by the Germans in their treatment of their Colonies.

For the opinion held by Alix of Hesse-Darmstadt in regard to Russia was simply that it ought to be nothing else but a Colony of the vast German Empire, and she felt more pride at the thought that she might reduce it to this condition than at the idea that she had been chosen out of so many other women to become the Empress of that Realm.

VII

WHAT THE IMPERIAL FAMILY THOUGHT ABOUT THE EMPRESS

IT would not have been human on the part of the Imperial Family to like the young wife of Nicholas II. in those early days which followed upon her marriage. The feminine portion of it especially could have been expected, before even the wedding of Alexandra Feodorowna had been solemnized, to look upon her with eyes full of criticism and with the desire to find fault with whatever she might say or do. Here she was, a young, lovely girl, in the full bloom of her beauty, put into the place of the first lady in the Realm, at a moment's notice, before even she had gone through that period of probation which falls as a general rule to the lot of every Consort of a Sovereign when she is but the wife of the Heir to the Throne. Had the haughty Imperial ladies, who for so many years had ruled according to their fancies St. Petersburg society, found themselves in presence of a Grand Duchess Czarevna whom they would have been able to advise, scold, or pet, according to their fancy, they might have taken, from the height of their own unassail-able positions, a more indulgent view of her unavoidable mistakes. They would have thought of her as of a young niece who owed them respect and submission, and whom it was their duty to train according to the exigencies of Russian etiquette. It must be remembered that Nicholas II. to the very day of his accession had been treated by his family like a mere boy without any importance. All of a sudden he found himself a Sovereign and, what was even worse, his wife, the little Hessian Princess, upon whom everybody had looked down with pity mixed with contempt, was the Empress of All the Russias. This was more than the Romanoffs could endure, especially when they remembered the cool, authoritative manner which the late Czar Alexander III. had always adopted in regard to them, and when they thought it might be possible his successor would imitate him in that respect at least, if not in others.

They need not have been in any apprehension as to this last point. Nicholas II., though he detested his uncles, yet stood in such awe of them that he would never have dared assert himself in their presence, far less contradict them. But the Empress had a different character, and she very quickly realized that all her relatives were furious at the fact of her being placed so far above them in rank and position. Fully conscious as she was of that rank, she determined that she would use its advantages to crush those in whom she saw but enemies, which in some cases was not quite exact, because there were then still some persons who, had she only appealed to them, would have responded to her call for sympathy and put themselves at her disposal, if only out of the motive that in rallying around her they were at the same time establishing their own influence.

Alexandra had no tact, and she never could hide her feelings in regard to the people who surrounded her. This explains the number of her enemies and the antagonism to which her mere presence anywhere gave rise. She knew very well that it would be very hard, if not impossible, for her to overcome certain prejudices existing against her. Instead, however, of trying to make for herself friends in other circles than purely aristocratic ones, she applied herself to wound those in whom she saw adversaries, and to discourage her friends by her utter disregard of the warnings that the latter sometimes thought it their duty to give to her. Her relations with the Empress Dowager had begun by being very cordial and affectionate, and it was she who had proposed to the Czar to take their abode in the Anitchkoff Palace with his mother, until their own apartments in the Winter Palace had been got ready for them. The arrangement had not been a successful one, and it is probable that Marie Feodorowna would have got on better later on with her daughter-in-law had the two ladies not lived under the same roof for about half a year. As it was they grew to know each other "not wisely, but too well," and the result was profound contempt on one side and sullen anger on the other. Servants' gossip did the rest; and the two incidents which I have already described, concerning the Crown Jewels and the liturgy, added the last drop of venom in a cup already full to overflowing. The Dowager began to criticize discreetly the young Empress, together with some of her intimate friends. These did not scruple to repeat what they had heard to their own near chums, and soon it became common property. The Grand Duchesses took their cue from Marie Feodorowna, and in an underhand way lamented over the failings of "dear Alexandra," her coldness, her want of politeness, and so forth, helping her in the meanwhile as much as they could to accentuate the shortcomings of an attitude which very soon came to displease everybody, even the people who had been the most enthusiastic about the young Empress.

As a proof of this fact I will relate a little incident which, at the time it occurred, proved the subject of much gossip in some select circles of St. Petersburg society. During one of the first receptions held at the Winter Palace, after the marriage of Nicholas II., there made her appearance an old lady who for the sake of convenience we shall call Madame A. She wished to be presented to the new Empress, an honor to which her own position, together with that of her late husband, gave her every right, besides the fact that she was one of the few ladies left in the capital who had adhered to the old Russian custom of keeping open house for her friends, and whose *salon* was a social authority in its way. The Empress, upon being shown the list of the people about to be presented to her, wanted to know who they were, and, seeing near her her aunt, the Grand Duchess Marie Pawlowna, the wife of the Grand Duke Wladimir, asked her whether Madame A. was or was not a person of importance. The Grand Duchess, who for reasons of her own disliked the latter, replied to her niece:

"Oh, she is an old frump. Give her your hand to kiss, and she will be satisfied."

Now this was the one thing which would not have satisfied Madame A. at all, who considered herself entitled to quite special consideration. Alexandra Feodorowna, believing her aunt, executed the latter's advice to the letter. She extended her much-bejeweled fingers to the astonished old lady, and then coolly turned her back upon her and passed on without having said one single word. The scandal was immense, so immense that the whole ballroom rang with it within a few minutes, and one of the Empress's ladies in waiting actually went up to her and tried to enlighten her as to the extent of the enormity which she had committed, advising her at the same time to seek out the irate Madame A. and to make her some kind of apology, under the pretext that she had not heard her name when it had been mentioned to her.

Alexandra Feodorowna in her turn, and with a certain amount of reason, became furious against the Grand Duchess Marie Pawlowna for having thus led her into a snare, and, boiling with rage, she crossed the room, went up to where Madame A. was discussing with volubility, together with some of her friends, the slight to which she had been subjected, and told her quite loudly:

"I am sorry, Madame, not to have treated you with the respect to which you are entitled, but it was the Grand Duchess Marie Pawlowna, my aunt, who had advised me to do it."

One may imagine the effect produced by this short sentence, which, instead of soothing the ruffled feelings of Madame A., added to her indignation. She turned round and replied quite distinctly, so that all the people standing near her heard her plainly:

"*Ce n'est pas à l'aide d'une trahison, Madame, que l'on excuse une impolitesse!*" ("It is not with the help of a treachery, Madame, that one can excuse a rudeness").

And making a deep courtesy to the discomfited Sovereign, Madame A. proudly retired and drove away from the Palace, leaving the Empress with the consciousness that in the space of five short minutes she had contrived to make for herself two mortal enemies.

The whole of the Imperial Family took up the cause of the Grand Duchess Marie Pawlowna. The latter's husband, the Grand Duke Wladimir, went to the Emperor and complained bitterly of the conduct of Alexandra Feodorowna. The other Grand Duchesses declared that, dating from that day, they would have nothing to do with her, except when the necessities of etiquette compelled them to appear at Court, but that personal relations with a person capable of such a grave piece of indiscretion were quite out of the question. The Grand Duchess Marie swore that she had never meant to advise her niece to show herself rude to such a respectable personage as Madame A.; that her words had been a mere joke, to which she had never imagined that any importance could be attached, and that it had been a cruel thing to denounce her in such a ruthless way to the worst gossip and most malicious tongue in St. Petersburg.

Even the Dowager Empress expressed herself as shocked beyond words at her daughter-in-law's behavior, but when she had tried to speak with the latter on the subject Alexandra Feodorowna had exclaim-med that she recognized the right of no one to criticize her actions, and forthwith produced for her mother-in-law's edification a caricature which she had drawn of the Emperor in swaddling-clothes, seated at a dinner-table in a high-backed chair, with his uncles and aunts standing around him, and threatening him with their fingers, adding that she was not going to follow the example of her spouse, and that if he chose to forget before his relatives that he was the Emperor of All the Russias, she would not do so for one single minute. After this the conversation came to an end, as was to be expected, but its consequences survived, with a vengeance into the bargain.

Of course incidents of the kind could not be productive of good relations. It did not take a long while before the general public, which, at that time, looked very much for its inspirations toward the Imperial Family, had come to the conclusion that the young Empress was a capricious, rude, and most disagreeable kind of person to whom it was preferable to give a wide berth. Once this legend had been transferred into the domain of history, every action, every word, every gesture of Alexandra Feodorowna was watched with attentive and critical eyes, always ready to make capital out of all her mistakes and to amplify all her errors into crimes. The fact of her having no son added to the resentful feelings of the nation against her, and that of her undisguised German sympathies did not contribute to make her popular. She in her turn, angry with her family, furious with St. Petersburg society, unable to seek friends among the Russian people, all of whom seemed in her inexperienced and prejudiced eyes to be more or less savage, set herself a task to show her contempt and dislike to those persons whom she had found so ready to throw stones against her on occasions when her conscience had told her that she did not deserve the insult. She retired more and more into the seclusion and privacy of her home at Tsarskoye Selo, and she announced to whoever wished to hear her that she did not see why she should spend her money in giving balls and entertaining a society that seemed to have made up its mind to insult her on every possible occasion. The words were repeated, and immediately taken up by the public in the light of another affront. One declared that for a penniless Hessian Princess to talk about "her money" was, to say the least, ridiculous, and one added that she ought to remember that it was part of the duties of a Russian Empress to entertain her subjects and to give them some pleasures in return for their fidelity.

Such was the position after Alexandra Feodorowna had been married three or four years. She might still at that time, had she attempted it in earnest, won back at least the respect if not the sympathies of the Russian nation. But to do so she would have had to bend down from the height of the Throne upon which she was seated, and to make some efforts to clear the misunderstandings which had arisen between her, her family, and her subjects. Unfortunately for her, the haughty Princess believed so firmly that she had been sinned against without having the least sin to her own credit that this "injustice," as she called it, in the world's judgments of her personality made her rebellious, and, not being clever enough either to forgive or to disdain it, she could find nothing else to do but to seek to revenge herself upon imaginary wrongs by making herself guilty of real ones.

GRAND DUCHESS MARIE PAWLOWNA
THE ELDER

Born Duchess Marie of Mecklenburg-Schwerin in 1854, she was the eldest daughter of a Prussian officer, the Grand Duke of Mecklenburg-Schwerin, by his first of three wives Princess Augusta Reuss of Köstritz. She married Alexander II's brother Vladimir and they produced six children; their second child and son, Kirill (aka Cyril), made himself pretender to the Russian throne after the death of Nicholas II was confirmed.

VIII

SORROW AND UNEXPECTED CONSOLATION

IT was not only her family and St. Petersburg society with whom the Empress could not agree. Her relations with her husband were also not of the best during the first years of her married life. Later on, when Alexandra Feodorowna had fallen into the hands of the clever gang of adventurers whose tool she was to remain until the final catastrophe which drove her from her Throne had taken place, she contrived to get hold of the feeble mind of Nicholas II., and to influence him absolutely, thanks to his love for his children, especially for his son.

During the first five years or so that followed upon his marriage the Czar, though he never quarreled with his wife, yet thought far less about her than he did about his mistress, the dancer Mathilde Krzesinska, a Pole of extreme intelligence, little beauty, but enormous attraction. Their friendship had begun when Nicholas was but a boy, or about that, rumor would have it, though I have reason for knowing that in this rumor was mistaken, as happens so often to the old lady, that the dancer had been chosen by the Empress Marie herself as a fit friend for her eldest son. The fact was that this liaison had started almost immediately after the Grand Duke's return from his journey round the world, which had had such a dramatic incident to enliven it in Japan, when a fanatic had attempted to take the life of the Heir to the Russian Throne, inflicting upon him a deep wound with his sword.

The Cesarewitsch had seen Mademoiselle Krzesinska on the stage of the Marinsky Theater, and had been very much impressed by her talent and grace. He had asked to be introduced to her, and had forthwith carried her off to supper at a fashionable restaurant called Cubat, where all the *jeunesse dorée* of St. Petersburg used to meet, eat, drink, and be merry. This supper, in which had taken part several of Nicholas's friends, officers in the same Hussar regiment where he was a captain, as well as one or two ladies of great beauty and doubtful reputation, had ended in a scandal, which for several weeks had been almost the only subject of discussion in the aristocratic *salons* of the capital. The company had been enjoying itself so much that glasses and plates had been broken; when, at two o'clock in the morning, the owner of the restaurant had ventured to suggest that it would be high time the entertainment came to an end, he had been sent to mind his own business. This the poor man would have been but too glad to do, but police regulations were very strict at that time, and he knew that if a patrol should see light in his windows from the outside that he would be fined heavily, no matter who had elected to remain in his establishment after the curfew had sounded.

This was precisely what happened.

A police officer walked up and knocked at the door of the private room where the Heir to the Russian Throne and his companions were disporting themselves, and ordered them to get out. The Grand Duke's aide-de-camp did not care to disclose the identity of his master, so he came out alone and tried to remonstrate with the man, asking him to give them another half-hour to finish their supper and pay for it. The officer refused and tried to force his way into the room, but was violently thrust aside. He had not the right to enforce his authority against a colonel in the army, which was the rank of the aide-de-camp, so he withdrew and telephoned to the Prefect of the town, General Wahl. The latter, who was an officious busybody, thought it a splendid occasion to assert his authority. He immediately proceeded himself to Cubat, where, in spite of the efforts made by the companions of the Grand Duke to keep him out, he rushed into the room, to find himself confronted by the Heir to the Throne. Nicholas became very angry and asked the General how he dared intrude upon his privacy. Wahl, furious in his turn, retorted that it was his duty to see that order was maintained in the capital, no matter who was troubling it, upon which, in one of the uncontrollable fits of rage to which he was sometimes subject, the Cesarewitsch seized hold of a dish full of caviar which stood on the table and threw its contents in the face of Wahl. A scene of indescribable disorder followed. At last Prince Bariatinsky, one of the generals in waiting on the Czar, who had accompanied the young Grand Duke during the latter's journey round the world, was sent for. He succeeded in putting an end to an incident which reflected credit upon none of those who had taken part in it.

The next day Alexander III. was apprised of what had taken place. History does not say what he told his son, but it was supposed that it had not been anything in the way of praise, because there was nothing that the Emperor hated more than a drunken brawl, and it must have been very painful for him to find that his Heir had become involved in one. But when General Wahl arrived, full of complaints and indignation at the treatment to which he had been subjected, the Monarch expressed to him his entire disapproval of his conduct, saying that he had had no right to intrude upon the privacy of the Grand Duke, and that he ought not to have forgotten the immense difference of rank which existed between him and the future Emperor of Russia. Wahl did not require to be told twice the same thing, and in the future he never attempted to interfere with the pleasures of any member of the Imperial Family.

People who were present at this ill-fated supper told afterward, when relating all the incidents which had made it a memorable one, that Nicholas wished to do something worse than pour the contents of a caviar-dish on General Wahl's head, but that Mademoiselle Krzesinska had thrown herself between them. True or not, it is certain that after this night the Grand Duke took to visiting the beautiful dancer in her home, and very soon their relations became an established fact. She bore him two sons, which gave her distinct advantages over all the other flirtations in which her Imperial lover indulged from time to time, flirtations which she was far too clever and careful to notice. What she aspired to afterward was to become a power in the land, a *Maîtresse de Roi*, such as had been seen at the French Court during the reigns of the last Bourbons. Her Polish propensity for intrigue coming to her help, she very soon contrived to make for herself an excellent position in the world as well as to earn a considerable fortune. She was a very reasonable, matter-of-fact woman; she knew very well that Nicholas had to marry, whether he liked it or not, and her only preoccupation, if we are to believe all that was related in St. Petersburg at the time, was whether he should marry a clever or a stupid woman. It is not difficult to guess the one she would have preferred had the choice been left to her discretion.

When the betrothal of the Cesarewitsch with the Princess Alix of Hesse was announced Mademoiselle Krzesinska, far from objecting to it, applied herself, on the contrary, to persuading him that he had done quite right and that he could not have chosen a better wife. She imagined that the placid German temperament of the bride-to-be would look with innocent eyes on the continuation of her intrigue with Nicholas, in which supposition she was vastly mistaken, because Alice, though she did not care for the husband she had been compelled to marry, did not mean to let him wander away from the conjugal home in search of a happiness she believed herself quite capable of alone procuring for him. She tried to separate the Grand Duke from the clever dancer who held him in her bondage, and of course she failed.

Nicholas kept up his former habits of going to see Mademoiselle Krzesinska whenever he had the time to do so; what was even worse, he continued to consult her on many matters which he never discussed with his wife. The latter became very unhappy, and it was then that even her affection for her children was not sufficient to prevent her from uttering aloud her despair at having been obliged to leave her dear Darmstadt for a country where everything and everybody conspired against her and her peace of mind, and where she could not even win the love of the husband who had been imposed upon her.

Among the few people whom she used to see more frequently than others was the Montenegrin Princess Stana, who had been married to Duke George of Leuchtenberg, with whom she had led a most unhappy, uncanny sort of existence. Stana, like all the Montenegrin daughters of King Nicholas, was a charming and attractive woman, clever into the bargain. In spite of her unhappy conjugal experiences she had grown very fond of Russia, and especially of her position as a member of the Russian Imperial Family. She was very willing to divorce the miserable husband to whom she had been united, who had insulted and outraged her without the least compunction from the very first day of their marriage; but she would have liked to find another one whose affection, and especially whose worldly situation, were such that her future would be assured on even more brilliant lines than the present. Her elder sister, Princess Militza, was the wife of the Grand Duke Peter Nicholaievitch, whose brother was that Grand Duke Nicholas who was later on to acquire such a reputation as Commander-in-chief of the Russian armies during the first months of the present war. Grand Duke Nicholas was not considered as a marriageable man, being bound by ties of close friendship since a good number of years with an attractive woman, Madame Bourénine. Nevertheless, Princess Stana made up her mind to marry him, an enterprise which seemed the more hopeless that it was against the canons of the Greek Orthodox Church for two sisters to marry two brothers. As we have seen, her sister was Grand Duke Nicholas's sister-in-law.

This, however, did not much trouble the determined Stana, but she knew very well that it would be quite impossible for her to succeed in her designs unless she managed to enlist on her side the sympathies of somebody strong enough to protect her and to lend her the support which she needed. It was useless to think of the Empress Dowager, because the latter had never looked kindly upon the Montenegrin Princesses, to whom she had been very good at the time that they were being brought up in the Smolny Convent in St. Petersburg, and who had rewarded her with the basest ingratitude later on. The Emperor was a mere puppet in the hands of his advisers, and these, Stana knew but too well, would be against any idea of her becoming the wife of Nicholas Nicholaievitch. Remained the young Empress, to whom no one to that day had ever dared to apply for anything, who had been considered by general consent as not being worthy of any attention or consideration. Stana imagined that any proofs of respect which she might give to her were bound to be more appreciated than they would have been under different circumstances. She forthwith proceeded to lay siege with great care and tact to the heart and the sympathies of Alexandra Feodorowna.

At first her advances were met with rebuff; then gradually, seeing how attentive and full of deference her cousin showed herself in respect to her person, the young Empress began to thaw; and soon a friendship, the more surprising that the two ladies did not seem to have anything whatever in common in their respective characters— even a close friendship—established itself between them, and the miserable wife of Nicholas II. poured out the sorrows which racked her heart to the willing ears of Stana Leuchtenberg, who, in her turn, related all her own misfortunes. At last Alexandra interested herself so much in the welfare of this other victim of an unhappy marriage that she exerted all her influence to persuade the Emperor to grant her the permission to sue for a divorce. At the same time she applied herself to invite the Grand Duke Nicholas as often as possible either at Tsarskoye Selo or at Livadia, and to make him meet there the beautiful Stana Leuchtenberg. The expected happened, and soon poor Madame Bourénine was forgotten, and the betrothal of the Empress's two protégés was announced, much to the indignation of the man in the street, who did not approve of it by any means.

The Grand Duke Nicholas was in his way just as ambitious a man as the fair Montenegrin he had married. To the Crimea they both repaired as soon as the divorce of the Princess had been pronounced. He knew very well the weakness which characterized his nephew, the Czar, and he would have dearly liked to become the latter's chief adviser and even his Prime Minister. He therefore favored his new wife's intimacy with the Empress, so that the couple were often seen at Tsarskoye Selo, much more so, in fact, than any other members of the Imperial Family.

Now the Grand Duke had one weakness. He believed in spiritualism, in turning tables, and all kinds of superstitious extravagances. The Empress's leanings had also since some time been directed toward the same subject, but she had felt afraid to speak about it, knowing very well that this would not be looked upon with lenient eyes by the Czar or by his mother. When she discovered, however, that Nicholas Nicholaievitch did not feel in the least ashamed if he were caught trying to communicate, through the medium of a table or of a pencil, with the inhabitants of the other world, she confided to him her great desire to do the same thing. The Grand Duke replied that nothing could be easier. They held several séances to which the Emperor also came, attracted by the descriptions which his cousin had made to him. Nicholas Nicholaievitch promised the Empress that he would bring to her a famous French medium called Philippe, who would most certainly make her witness most extraordinary performances in regard to the evocation of the spirits of people dead long before.

Alexandra Feodorowna was delighted. She had already derived great comfort from her intercourse with her cousins, and her feeling of affection for Stana had acquired considerable warmth since the beginning of their friendship. Moreover, she knew that the Grand Duke Nicholas was considered the strong man in the Romanoff family, and she realized that to have him on her side would be a distinct advantage for her, and that his support might help her to overcome many difficulties. Therefore she appreciated very much all the acts of attention which both Stana and her husband were fond of pouring upon her. When Nicholas told her that he would gratify her wish to see a real medium she was more than delighted. She did not foresee whither this fatal introduction was to lead her, nor realize the ill turn that her cousin was doing her by giving her an opportunity of indulging her tastes for the supernatural, to which she was to owe so many of the misfortunes which were to assail her in later years, and which were to play such an important part in the tragedy that ended with her downfall. She was looking for the consolation of the moment without thinking of the possibility of the catastrophe of the morrow. [imgage *Nicholas Nicholaievitch p101*]

THE CZAR

IX

PHILIPPE AND HIS WORK

THE Grand Duke Nicholas kept his word, and one afternoon he brought to Tsarskoye Selo the famous Philippe, about whom his wife had spoken so often and with such enthusiasm to the young Empress. Before relating what followed upon this hasty and ill-advised introduction of an adventurer in the family circle of the Czar, it may not be out of place to say a few words concerning this personage, as well as to give a short description of his person.

Philippe was a Frenchman who, if all that has been related about him is true, had come to grief in his native land, to which he had thought it wiser to bid good-by for a time at least. He had spent several years in Germany, studying at German universities (at least he said so) and had given a great deal of attention to occultism and everything connected with it. Why he came to St. Petersburg no one ever knew, and though he has been accused of having tried from the very first months of his arrival in Russia to get introduced to the Sovereigns, yet I do not personally believe in this part of the story, because at that time no one suspected Alexandra Feodorowna or Nicholas II. of being interested in the supernatural. What is more likely is that he only attempted to get acquainted with the aristocratic circles of the capital, some of which were known to be attracted by these manifestations which begin by turning tables and end in more or less genuine hysteria. Later on when it became known that the Emperor and Empress themselves had given a welcome to the spiritualistic doctrines which Philippe preached, it is probable that the idea was suggested to him, by people who realized what capital might be made out of this circumstance, that he might come to acquire political influence, if he would but make use of his science to enslave the weak persons who had come to believe in him.

Personal ambition and vanity did the rest, combined with a good deal of German money, cleverly and judiciously spent in the furtherance of deep schemes, the real purport of which he was never allowed to suspect. He was encouraged to consider himself as a personage of great importance, and one upon whose shoulders rested some of the responsibilities which, properly speaking, belonged to the Czar alone. He was clever, bright, and assimilated very quickly all that he heard or saw, and knew how to turn to the best advantage every possible circumstance with which his personal welfare or interest was connected. As soon as he found himself in the presence of Alexandra Feodorowna he understood how easy it would be for him to get hold of a mind which he judged at once, and this quite rightly, not to be well balanced. He therefore played upon it; he ministered to it; he took advantage of it and of its vagaries; and he soon acquired over the young Sovereign an influence such as no one before him had ever wielded, and such as no one in the future was to have, with the exception of the famous Raspoutine of evil memory.

At first Philippe proceeded with great caution—so great, indeed, as to elude even the suspicious eyes of the Grand Duke Nicholas, who, though he had been instrumental in bringing this impostor to Court, yet would not at all have liked to see him become influential there, and who watched him very carefully during the séances which were held every Saturday evening at Tsarskoye Selo. Even the suspicious eyes of the Grand Duke Nicholas could not detect anything the least dangerous in his manner of proceeding. Philippe acted the medium to perfection. He used to go into regular trances, during which he replied to the various questions put to him with more or less accuracy; he never could be detected, once he was awake again, as having the slightest knowledge or remembrance of what had taken place while he had been asleep. Several times he prophesied with such exactitude that it seemed marvelous.

On one of these occasions he announced to the small circle assembled to listen to him in the Empress's boudoir that a serious misfortune was threatening the State through the death of one of its most important functionaries. He was still plunged in the hypnotic sleep during which he made this startling announcement when Count Lamsdorff, who occupied the position of Under Secretary of State for Foreign Affairs, arrived at Tsarskoye Selo and asked to be received by the Emperor on urgent and important business. He had come to communicate to the Monarch the news of the sudden death that same evening of the Minister for Foreign Affairs, Count Mourawieff. Of course this set the seal to Philippe's reputation as a prophet. Afterward some meddlesome people assumed that he had become aware of the sad event before he had left St. Petersburg to proceed to the Imperial country Palace by one of those singular accidents which happen sometimes in life, and that he had very intelligently made use of this knowledge during the trance in which he had pretended to be plunged. True or not, the story circulated freely, and was repeated everywhere, but the people who ought to have been the most interested in it did not, of course, hear it. On the contrary, the influence of the impostor was considerably strengthened by the incident, even in regard to the Grand Duke Nicholas, who from that moment began himself to consult Philippe in various matters. Then the Grand Duke had to leave for the Crimea, where he usually spent part of the year, on account of the health of the Grand Duchess, which had never been of the strongest, and he left Philippe in possession of the field.

In spite of Nicholas Nicholaievitch's absence, the séances with the medium continued, and they became even longer and more frequent than had been the case before. The Empress developed more and more interest in their progress, and at last one day, when Philippe asked her whether she would not try to be sent to sleep by him in order to get rid of the cruel headaches from which she suffered, she did not object; on the contrary, expressed herself as quite willing to make the experiment. Philippe, however, insisted on one condition, which was that he should be left alone with her while it proceeded.

Here comes the surprising part of this singular business. Instead of protesting against this pretension of the adventurer, Alexandra Feodorowna accepted it as a matter of course, and, what is more surprising even, she induced the Czar to give his consent. Philippe sent her to sleep, with the result that her headaches really improved and that she began to get into the habit of talking with him, either willingly or unwillingly, about all the events of her daily life and of consulting him whenever she thought that she found herself confronted by any difficulty.

She confided to him—what he knew already—her passionate desire to become the mother of a son, as well as the many disillusions of her married life. Philippe encouraged her, and he was the first one who suggested to her the advisability of taking an interest in public affairs, instead of holding herself aloof from them, and to point out to her the necessity which existed for her, in order to consolidate her personal position, to try and acquire some influence over her husband's mind, and in this way to eliminate that of the Empress Dowager. When Alexandra Feodorowna protested, the adventurer declared to her that he had been sent from heaven to come to her help, that it had been suggested to him by the invisible spirits which always inspired him to go to Russia and to give her the benefit of his experience so as to deliver her from her numerous enemies.

When she declared that she understood nothing about politics, he replied that it was her duty to learn, and that if she did not find any one in Russia willing to teach her, there was her own family in Germany who would be but too glad to come to her rescue, together with their knowledge of the art of government and of handling men and facts. He added that this was the more indispensable that she was about to give birth at last to the son she had been longing for since so many years, and that this son would grow to be an honor to her, as well as the greatest Sovereign Russia had ever known.

Poor infatuated Alexandra believed the adventurer, believed in him so thoroughly that she imagined that she was really about to become a mother once more, and solemnly announced the fact to the Emperor and to the Imperial Family.

Great preparations were made for the auspicious event, and once more the hopes that Nicholas II. might have at last an Heir to his Throne and Crown were awakened. It is related that everything had been got ready, that even the guns which were to announce the birth of a new member of the Romanoff dynasty had been placed on the ramparts of the fortress of SS. Peter and Paul to be fired as soon as the event had taken place, when the suspicions of the Empress Dowager were awakened by the attitude of her daughter-in-law as well as by her physical appearance. She began to watch her, and to do so with the more care. The time for the latter's presumed confinement had passed without that confinement having occurred. Alexandra Feodorowna was observed to be in tears, and her nervous condition became almost alarming, but she refused to see a doctor, and declared that she felt sure she would be better as soon as the suspense under which she was laboring was over. She remained long hours closeted alone with Philippe, who seemed to be the only man capable of bringing some calm to her over-excited system.

Some member of the Court took upon himself the task of writing to the Grand Duke Nicholas Nicholaievitch in the Crimea, advising him that something had gone wrong with the Empress, and that Philippe was concerned in it. One must give the Grand Duke his due. He had never meant any harm to his cousin's wife when he had brought the impostor to Court. As we have seen, the latter had been most careful in his whole demeanor while the Grand Duke had remained at hand to control his conduct and his actions. This did not prevent him from rushing back to St. Petersburg as soon as he heard of the strange doings which were shaking the equanimity of the inhabitants of Tsarskoye Selo. He had no sooner seen the Emperor and the Empress than he guessed what had really occurred. He forthwith proceeded to tell the Monarch that the medical attendants of Alexandra Feodorowna must see and examine her, whether she liked it or not, because her state of health was a question which did not interest her alone, but was of the utmost importance to the whole country as well as to the dynasty. He hinted at certain gossip which was going about, to the effect that it was the intention of the Empress to palm off a supposed son on her husband and on his family. Altogether he spoke so strongly that Nicholas II. became seriously alarmed, and for once in his life asserted his authority and compelled his wife to submit to a medical examination.

The result stupefied him as well as other people, because it was ascertained that the hopes of motherhood of Alexandra Feodorowna had only existed in her imagination; that there was no prospect whatever of her giving to Russia that Heir for whose advent the whole country was so eager. Of course the scandal was great, though an attempt was made to soften it by the publication of an official bulletin stating that an unfortunate accident had destroyed the hopes of the Imperial Family. For those who had perforce to become aware of the true circumstances of this whole adventure, the Empress remained under the shadow of a ridicule which was to cling to her for a long time and was not forgotten even when the present war broke out.

The Grand Duke Nicholas had a stormy interview with Philippe. The impostor pretended that he was not to blame, that the Empress had misunderstood him altogether, and that he, together with the rest of the world, had honestly believed in her supposed hopes of maternity. But in the mean while it had been discovered that during the séances which he had held at Tsarskoye Selo he had mesmerized Alexandra Feodorowna, and abused the confidence she had reposed in him by trying to worm out of her State secrets she was believed to know. The Grand Duke kicked the man out of the Palace, and told him that if he ever dared to set his foot in it again he would have him sent to Siberia under escort. He proceeded to acquaint the Emperor with all that he had discovered and to request the latter to issue orders for the expulsion from Russia of the impostor who had thrown so much ridicule on him as well as on the whole dynasty, who had acquired, thanks to his underhand man-euvers, such a disastrous influence over the mind of his Imperial Consort.

Philippe disappeared and was never seen any more. No one knew what happened to him, or where he was sent, and no one troubled. He had been a nine days' wonder and he sank into oblivion, but the Empress's mad infatuation for him was not forgotten so easily. The more so because she did not attempt to hide her grief at his removal, and bitterly reproached the Grand Duke Nicholas for his interference in a matter which, as she declared, did not concern him. Angry words were exchanged and the old intimacy which had existed between the Grand Duchess Stana, her husband, and Alexandra Feodorowna not only came to an end, but was replaced by a hatred the more bitter that it had perforce to be concealed under the veil of politeness and amiability. The Empress's nature, as we know already, was essentially a vindictive one, and the insult, as she considered it, to which she had been subjected on the part of Nicholas Nicholaievitch was to be avenged by her many years later on the day when, thanks to her and to her new favorite, Raspoutine, he was deprived of his position as Commander-in-chief of the Russian armies in the field.

X

ANNA WYRUBEWA APPEARS ON THE SCENE
AND HE SAW HER PASS

AFTER the disastrous Philippe incident, the character of the Empress Alexandra changed considerably. She became a sullen, morose, melancholy woman, with a grudge against the world in general and the people with whom she lived in particular. Her sisters-in-law, the Grand Duchess Xenia Alexandrowna and the Grand Duchess Olga of Oldenburg, tried to come to her help and to enliven her by attempting to bring her out of the solitude in which she shut herself up, and if she would only have responded to these efforts it is possible that the whole course of her life might have run differently. But the Empress persisted in seeing enemies in everyone of her relatives, and, instead of trying to break through this wall of hostility with which she believed herself surrounded, she used all her powers of persuasion to induce her husband to take the same attitude of antagonism in regard to his family which she had adopted herself. Of course this was not forgiven her.

Nicholas II.'s sisters, who loved him dearly, were affronted when they discovered that their former intimate relations with their brother had come to an end, and that for some reason or other he looked upon them with suspicious eyes. Xenia simply shrugged her shoulders, and, being very wisely advised by her husband, the Grand Duke Alexander Michaylovitch, who, like all the members of that branch of the Romanoff family, was exceedingly intelligent, refrained from saying anything. But Olga, who was of a more enter-prising turn of mind, accosted the Czar one day and talked to him quite seriously about the conduct of the Empress, pointing out to him the harm which she was doing him by her rudeness toward the members of the Imperial Family, and expressing the conviction that times were sufficiently serious. This was during the Japanese war. The Emperor listened to her, as he listened to everybody who spoke to him, with courtesy and attention, but the only reply which she could obtain from him was to the effect that the Empress was in a bad state of health, that her nerves were quite unstrung, and that it would be wrong to take anything she said or did too seriously.

"But you are not nervous or ill," exclaimed the Grand Duchess. "How does it come, then, that you avoid us, your sisters, and even our mother just as much as does your wife. What have we done to you, except to love you, for you to treat us as if we were strangers?"

Nicholas II. pulled his mustache, but would not explain himself further, and Olga Alexandrowna had to own herself baffled.

The Empress heard of this conversation and it did not reconcile her to her sisters-in-law. She was in that morbid state of mind which gives an undue importance to the smallest incident which would not arrest for five minutes the attention of any normal person. The predisposition to insanity which existed in the Hesse-Darmstadt family had probably something to do with her condition, because she most certainly suffered from the mania of persecution; being a Sovereign, and a powerful one into the bargain, she imagined that the best use she could make of her unlimited power was to crush those in whom she persisted in seeing enemies bent on her destruction.

Rumors had reached her ears that some members of the Imperial Family (it had been the Grand Duke Nicholas Nicholaievitch, in fact) had said that her place ought to be in a convent rather than on the Throne, and she had immediately made out of the remark a desire on the part of her kinsman to shut her up in a monastery, as had been done in the Middle Ages with other Russian Czarinas, so as to give the Emperor the possibility to marry another woman who could bear him a son.

The supposition was a preposterous one, because such an idea had never crossed the Grand Duke's mind, but it could not be driven away out of the imagination of Alexandra Feodorowna. Hence her continual efforts to estrange her husband from his people, and to keep him entirely in her own hands, far away from any influence hostile to herself or to her daughters. There was, after all, some method in her madness. As things turned out, she was given several opportunities to exert her vengeful feelings in regard to the Imperial Family by the conduct of a few of its members.

I will here mention briefly two or three occasions when her intervention caused any amount of trouble and brought upon her head storms of abuse and indignation. The first one was the morganatic marriage of the Grand Duke Paul, the Emperor's uncle. This event was brought about principally through the want of tact and the stupidity of the people concerned in it, and it would have been far better for the Empress not to have interested herself in it at all, considering the fact that the personages concerned in this affair were certainly beneath her notice.

The Grand Duke had been upon terms of intimate friendship with a lady very well known in social circles of St. Petersburg, the wife of one of the officers of the regiment of which he was the commander. The thing had been going on for a number of years, and society had turned away its head and affected not to notice it; the more so that the husband of the lady in question seemed to ignore it, and to keep his eyes firmly closed as to her indiscretions. But one fine day the Grand Duke thought to make to Madame Pistolkors a present of some jewels which had belonged to his mother first, and to his wife afterward, and which had been locked up in a safe since the latter's death. This again might have passed unnoticed, had Madame Pistolkors not thought to put them on at a Court reception to which she was bidden. The Empress Dowager, who was present, recognized the unlucky ornaments, and, burning with wrath, forgot for once her strained relations with her daughter-in-law, and went up to her to draw her notice to the "scandal," as she termed it. Alexandra Feodorowna, as we know, had never been a tactful woman. She called a chamberlain and ordered him to invite Madame Pistolkors to leave the Palace immediately, and to escort her to her carriage. The next day Colonel Pistolkors, finding that matters had gone too far, introduced an action for divorce against his wife, and the latter, shunned by all her former friends, utterly disgraced before the world, had to flee abroad to hide her diminished head and her lost social prestige, in the solitude of a small Italian town. But then the unexpected, or rather the expected, occurred. The Grand Duke Paul took the only course left to him compatible with his honor as a gentleman. He followed the lady to Italy and married her there without asking anybody's leave, to the general scandal of St. Petersburg society, who declared that the incident with the diamond necklace that had been the primary cause of the catastrophe had been artfully engineered by its heroine in view of the result which was ultimately achieved.

The Emperor was furious; his mother equally so, but it is not likely that anything would have been done, or in general any notice taken of the action of the Grand Duke, had it not been for the intervention of the young Empress, who insisted on her uncle-by-marriage being deprived of his rank in the army and exiled abroad.

It was the first time that she had the opportunity to satisfy her instincts of hatred and of revenge in regard to a member of her husband's family, and she took a special delight, not only in doing so, but also in letting the world know that such was the case. Fate, for once kind to her, had delivered one of her enemies into her hands, and she was but too ready to seize this occasion for scoring her personal real or imaginary wrongs.

A few years later another incident of the same kind afforded her a second opportunity of exercising her powers of retaliation in regard to a Romanoff. The eldest son of the Grand Duke Wladimir, the young Grand Duke Cyril, the same who had nearly perished during the Japanese war in the catastrophe of the ship *Pétropawlosk*, married also without law or leave his first cousin, the divorced Grand Duchess of Hesse, the former sister-in-law of the Empress. The latter had always hated her, ever since the day that she had been obliged to play second fiddle to her at Darmstadt, and she had done her best to bring about an estrangement between her and her husband. This had not been difficult, because anything more brutal than the Grand Duke of Hesse had never existed. His young wife had had more to bear than the public knew, or that she cared herself to relate, but her own conduct had always been beyond reproach, and she had carried herself with remarkable tact and dignity. When at last she obtained her divorce, her only child, a little girl, was not even left entirely in her custody, but had to spend half of the year with the father. The latter did not well know what to do with the baby and most probably would never have availed himself of his rights had not his sister, the Empress Alexandra, interfered and persuaded him to confide to her own care the small Elisabeth, knowing very well that this would be about the most painful thing that could happen to the divorced Grand Duchess.

In accordance with this wish, the Grand Duke of Hesse brought his daughter to Spala in Poland, where the Russian Imperial Family were spending the autumn. The child sickened a few days later, and soon her condition became desperate. The doctors declared that the mother ought to be warned and asked to come, the more so that the little girl kept continually crying for her. But to this the Empress would never agree, until she knew it was positively too late. At last a telegram was sent to the Grand Duchess Victoria Mélita; it preceded but by a few hours the one advising her that her journey would be useless, as the end had come. One may imagine the feelings of the heartbroken mother and the natural resentment she must have felt at this piece of heartlessness on the part of her former sister-in-law. For a long time she would not be comforted, but at last she was induced to listen to her cousin, the Grand Duke Cyril, and she married him at Tegernsee in Bavaria, without the Czar's consent to this union having been so much as asked.

The rage of the Empress would be difficult to describe. Here was the sister-in-law whom she had hated for so many years the wife of a Russian Grand Duke, and of one, too, whose position put him very near to the succession to the Throne. One of those fits of hysterics to which Alexandra used to give way whenever she was crossed followed upon the news, and she insisted on the Czar declaring that he would never recognize the marriage and exiling the young couple. But here she met with an unexpected rebuff. Cyril's father, the Grand Duke Wladimir, was still alive at the time, and he was not a man to endure any slight offered either to him or to his children. He sought the Emperor and in a stormy interview reminded the latter that his new daughter-in-law was also the granddaughter of the Czar Alexander II., and asked him what he thought the latter would have said had he seen a Princess with Romanoff blood in her veins banished from the Russian Court. Nicholas was scared, and revoked the orders he had issued a few hours before, insisting only on the newly married pair not coming back to Russia for a few months, after which he left them free to do what they liked.

Alexandra Feodorowna was defeated, and this did not improve by any means her temper nor her feelings in regard to the Imperial Family. She then bethought herself to win over to her side that same Grand Duke Paul against whom she had been so incensed at the time he had married Madame Pistolkors. It must here be added that one of the reasons for her change of opinion in that respect lay in the fact that she had by that time struck up the extraordinary intimacy with Madame Wyrubewa which was to have such sinister consequences later on, and that this lady had always been one of the closest friends of the morganatic wife of Paul Alexandrowitch. The latter was therefore invited to return to Russia and given to understand that it depended on him to be reinstated in favor, if only he would take the Empress's part against their other relatives. Of course he promised he would do so, and we shall see presently what resulted of this intrigue in the years which followed.

Cyril and his wife returned to Tsarskoye Selo and to St. Petersburg in due course. They were received by both the Czar and Czarina coldly but civilly. Alexandra, however, persisted in her determination to keep her former sister-in-law at arm's-length, and the relations between the two ladies remained official, without the least attempt at any intimacy, until the Revolution sent the Empress into exile and threw into the arms of its leaders both Cyril Wladimirowitch and Victoria Mélita.

It was known already at the time that one of the persons who had the most contributed to excite Alexandra Feodorowna against her cousins had been Madame Wyrubewa. The latter was a new importation at Court, who, thanks to a very clever piece of strategy, had won the good graces of the Empress, whom she had met under rather peculiar circumstances. She was the daughter of a certain Mr. Tanieiew, who occupied important official functions at Court, and she had contrived to let the Czarina hear, through her father, that she was engaged in the occupation of writing a history of Hesse, which she meant to present to a public-school library or other institution of the same kind. Alexandra was immediately interested and asked to see the work. She sent for Madame Wyrubewa and soon the latter became her friend and confidante.

Madame Wyrubewa knew very well what she was about, even before circumstances turned out favorably in regard to her views and designs. She fully meant to become the Gray Eminence of the Empress, and, like the famous Père Joseph of Richelieu, to rule her, and through her the whole of Russia. We shall presently see how she proceeded to reach her aim, which in the mean while she knew very well she could never attain so long as there were near the Czar people whose close relationship with him allowed them to speak quite frankly with him on all subjects, even on that of the caprices and extraordinary behavior of his wife.

Anna Wyrubewa contrived to create a deadly feud between the Imperial pair and the whole clan of the Grand Duke Wladimir's family, who in a certain way was most powerful. The other members of the family were not dangerous in so far that the only thing they aspired to was to be left severely alone, and that they never cared to trouble with their presence the Emperor and Empress, for whom their dislike was only equaled by their contempt. There was only to be feared the Grand Duke Michael, the only brother of Nicholas II. and his Heir so long as the Empress had not given birth to a son. It was therefore against him that the new favorite turned her attention and against him that she excited the revengeful feelings of Alexandra Feodorowna.

What I wish to point out at present is that one of the secrets of the extraordinary influence which Anna Wyrubewa acquired over the mind of her Imperial mistress lay in the extreme ability which she displayed in appealing to all the bad sentiments of the latter, under the pretext of pitying her, and condoling with her on all the real or imaginary troubles of her life. She soon made herself indispensable to the Sovereign, who liked to visit her in her house, where she knew that no one would interfere with her and where she could meet the few people with whom she thoroughly sympathized, who in their turn were but too glad to have an opportunity of seeing almost in tête-à-tête the otherwise unapproachable Empress of Russia.

The small drawing-room full of flowers, where Alexandra Feodorowna was to spend so many happy and peaceful hours, and which was to witness in time such memorable events, filled itself with all manner of people, who, by common accord, never spoke of having been admitted within its precincts, or of having met one another there. It became also the meeting-place of a party, small at first, important later on; not, perhaps, on account of its number, but by the character of those who constituted it; a party that came to be known by the name of the "Empress's Party." It was to number among its adherents men like Mr. Sturmer, the latter's secretary, the too-famous Manassavitch-Maniuloff, Mr. Protopopoff, and, last but not least, the vagrant preacher who for a short time was to be the dominant figure in Russian politics, Grigory Raspoutine.

XI

AND HE SAW HER PASS

MADAME WYRUBEWA was a very clever woman, and an ambitious one into the bargain. Her ambition, however, was absolutely different from what might have been expected of a person brought up in the atmosphere of a Court and having been, if not actually mixed up, at least well posted, thanks to the position occupied by her father and family. She knew all the intrigues which always flourished and made the Court of St. Petersburg such a slippery ground for those who did not possess sufficient support to hold their own amid the rivalries and gossip which constituted the daily existence of the Imperial Family and of their friends. She did not care in the least for money, having got enough for her wants, nor for rank or position, which she knew too well could be lost or obtained according to circumstances, and which, besides, were never sufficient in Russia to make or mar an individual whose social worth depended only on the manner in which he was viewed by the Sovereign—the words of Paul I., when he said that the only persons deserving of any notice in his Empire were those "to whom he spoke, and only while he spoke with them." These words, about which one had laughed all through the three preceding reigns, had come to be absolutely true during that of Nicholas II., when favoritism assumed hitherto unknown proportions, as none knew better than Anna Wyrubewa, whose quick wit and ever-alert intelligence discovered very soon that she would become a far more important personage if she remained in the background content with being the Empress's friend, if she did not work toward obtaining for herself or for her husband a Court appointment or a lucrative official post. She aspired to something much more tangible, and at the same time much more amusing. She wanted to rule the Empress, and through her the whole of the vast Russian Empire. This young and delicate woman had the head of a statesman, and she might have risen to unheard-of might if she had not allowed those superstitious leanings which are inherent in the Russian character in so many cases to get the upper hand of her reason and lead her, together with her Imperial mistress, into the manifold mistakes which culminated in the catastrophe that destroyed the Throne of the Romanoffs.

At the same time Madame Wyrubewa sincerely loved the Empress. About this there is no doubt. She began by feeling sorry for the sad, miserable woman, so lonely amid her luxury and splendor, who stood friendless and defenseless among implacable enemies. She did not stop to consider whether this situation had arisen out of the personal fault of Alexandra Feodorowna, or out of other circumstances. She simply saw the fact, and hearing, as she did, all the different rumors concerning the Czarina which were going about in St. Petersburg society, she conceived the idea of coming to her help, and trying to be to her that friend in need she had never found since she came to Russia in quest of a Crown. This latter had certainly turned out to be, for her, one of thorns!

When her relations with the unfortunate Sovereign in whose life she was to play such an important part began, Anna Wyrubewa did not look beyond this simple fact, finding out how she could best be useful to her. The whole of St. Petersburg was discussing the question of a possible divorce which would send Alexandra Feodorowna into a convent, and bets had been made in select circles of Court society as to whether or not this would really take place. It was known that her relations with the Emperor were anything but tender, and that numerous quarrels had taken place between them.

Nicholas II., after an interval of several years, had resumed his former relations with Mademoiselle Krzesinska, and the dancer was contributing perhaps more than she herself suspected to sow dissentsion in the Imperial *ménage*. The Empress, as we know, was exceedingly proud, and as soon as she perceived, which did not take very long, that her husband was seeking amusement outside his home, she retired once more in haughty silence into the solitude of her own apartments and refused to fulfil the social duties required from her by her position, to the disgust of her friends and the joy of her numerous enemies. Matters had got to such a pass that sometimes days used to go by without the Czar and Czarina exchanging one single word beyond what was absolutely necessary during meals, and even these were not always taken together, Alexandra Feodorowna often putting forward her health as an excuse for having her dinner or lunch served in her own apartments. She was simply playing into her enemies' hands, and, whether consciously or unconsciously, herself tightening around her neck the rope which had been put within her reach.

It was this that made Anna Wyrubewa determined to come to the help of the unfortunate Sovereign whom she saw going with rapid steps toward ultimate destruction. She tried to reason with her, to speak to her of the necessity of not giving up the game, and of her imperative duty to remain upon good terms with her husband, so as to be able to bear him the son whose absence contributed so much to the bad relations that had taken the place of the affectionate ones which had undoubtedly existed at one time between her and Nicholas II. But the Empress would not listen, declaring that she was tired of always giving birth to girls, whose advent into the world only added to her unhappiness, and that, besides, she was sick of a husband whose deplorable weakness of character made him an easy prey for the first intriguing person who approached him. The only thing which she wished was to return to Darmstadt, together with her daughters; but as she knew very well that she would never be allowed to take them out of Russia, she preferred to be sent to a convent, where she could end her days in prayer, and where she could bring up her children without any interference from the outside world. The Emperor could divorce her and marry again; she did not care; all she wished for was a quiet life, far from those detestable Court intrigues that had wrecked all the hopes of happiness she had ever had.

Anna Wyrubewa listened, and very gently applied herself to reason with the sorely tried woman. She told her that it would be unworthy to throw up the game, but, on the contrary, that her duty toward her daughters required her to fight vigorously against destiny represented by the Empress Dowager, the Grand Dukes, the Court, and the nation, who judged her according to what it had been told of her. She repeated to her that if once she had a son her position would change immediately, and the affection of her husband would return to her, together with the popularity she had lost in the country. Alexandra only replied by floods of tears and complaints that she did not know how such a desirable event could happen. She loathed the Emperor and she knew that he did not care for her; that, in fact, no one cared for her; and that was the calamity which to her sensitive heart appeared the most terrible one among all those that had befallen her.

Madame Wyrubewa was at her wits' end, but she did not despair.

She felt, however, that she could not cope alone with the many difficulties which she found in her way, and so she looked round her to see whether she could not find any one in whom she could confide, and from whom she might, in her turn, seek advice.

I don't know whether I have related that the lady had always been a favorite in society. At that time she was going out a great deal, which was not the case later on, when her whole position changed and when she became the Empress's principal confidante, and had perforce to live in retirement. But twelve or fifteen years ago her house in Tsarskoye Selo was the meeting-place of a select circle, and especially of the officers of the regiments constituting the garrison of the Imperial Residence, who liked to drop in of an evening, and find a pleasant hostess, together with an excellent supper which was always waiting for them. Mr. Wyrubew, too, was a general favorite, and altogether the little house occupied by the young couple was very popular with the inhabitants of the Imperial Borough.

Among the special friends of Anna Wyrubewa was a dashing officer called Colonel Orloff. He had a commission in the regiment of Lancers of the Guard, the chief of whom was the Empress Alexandra. A wonderfully handsome man, he was also clever, brave, chivalrous, and altogether different from his comrades in so far that he had never cared for the boisterous pleasures which made up their daily existence. One day as he was going to call on Madame Wyrubewa he saw the Czarina leave her house in a state of evident agitation. Alexandra was alone and on foot, having walked from the Palace to her friend's house, and the Colonel, who, on recognizing the Sovereign, had respectfully stood aside, was much surprised to notice her red eyes and her general attitude of dejection. He waited until she had disappeared among the trees in the park and then rang the door-bell of Madame Wyrubewa.

He found her just as agitated as the Empress, and when he asked her what was the matter he was much surprised to see her begin to weep.

She related to him that she was terribly anxious about the fate of the unlucky Consort of Nicholas II., whose safety and person were threatened as much by her own stupidity as by the intrigues of her numerous enemies. Colonel Orloff listened in silence. He, too, was troubled by this unexpected revelation; the more so that for years he had nourished a secret adoration and worship for Alexandra Feodorowna, which he had hoped no one had, or would ever discover, and the news of her danger was terrible for him. His emotion was so evident that Anna noticed it at once, and an idea which was yet vague and misty began to take shape in her active brain, and induced her to seek the help of this unexpected ally whom circumstances and accident had brought to her. She started to discuss the situation seriously with the young officer, and together they determined to try and save the Empress, even against her own will, from the snares into which she was walking with an unconsciousness which was almost too pitiful to look upon otherwise than with a wild desire to snatch her away from the abyss whither she was sinking with what promised to become rapidity.

Colonel Orloff had a wonderful talent for music. On the very next day following upon the conversation which I have related, Madame Wyrubewa asked him to call on her in the afternoon, and to perform for her some melodies of Chopin which she knew were the favorite ones of the Empress. She also begged the latter to allow the Colonel to play for them, saying that it might interest her to hear him. Alexandra consented and, as in the case of David and Saul, she found a solace in listening to the wonderful music. Very soon she got into the habit of dropping in at her friend's whenever she had a spare moment, and then Orloff would be telephoned for, and he used to come and hold the two ladies under the spell of his rare talent. Of course no one was admitted to these meetings and no one knew anything about them. At that time people did not trouble about the Empress of All the Russias, and her actions did not offer the slightest interest to any one, to the Emperor least of all.

Colonel Orloff was something in character like the famous Count Fersen, the admirer and devoted friend of Marie Antoinette. He, too, had conceived a passion for his Sovereign, in whom he only saw the unfortunate, ill-treated, and misunderstood woman, and he conceived the thought to sacrifice everything for her service, to try and save her from the perils with which he saw her surrounded. And gradually, when his relations with her became more real and intimate, he, too, began to speak to her in the same sense as Anna Wyrubewa had done, of the necessity of trying to reconcile herself with her husband so as to be able to bring into the world that Heir after whom the whole of Russia had been longing for the last nine years or so.

One day Madame Wyrubewa, whether accidentally or intentionally, left the Colonel alone with the Czarina. He saw his opportunity, and began more seriously than he had ever done before to implore her to make an effort to save herself. The young man grew quite eloquent, until Alexandra, moved beyond words, started weeping in real earnest and asked him how he could suggest the possibility of a reconciliation between her and the Czar, in view of his own feelings for her, the nature of which she had guessed for some time. To her surprise, the Colonel fell on his knees before her and told her that it was because of these very feelings that he had felt himself justified in speaking to her as he had done. He was nothing beside her, and all he could do was to worship her from afar, and to try to come to her help, both for her own sake and for that of their country, that required from them both the supreme sacrifice he was asking of her. For once the cold and haughty Czarina was startled out of her usual indifference, and when they parted she had promised her devoted knight and admirer that, though she might not make an effort to win back the love of her husband, yet she would not repulse him, as she had done lately, if he made any attempt to return to her. She promised that on the love she owned to him that she felt for him, and on that of the one which they both had for this great Russia, which Orloff had never forgotten even amid the fervor of his passion. When Madame Wyrubewa came back to the room where she had left her two friends, she saw that something had happened, but she was far too clever to question them, and when the Empress said it was time for her to go home she simply offered to accompany her, hoping that something might be told to her during their walk back to the Palace. For once Alexandra was silent, and parted from Anna without betraying anything of what had passed during that half-hour when she had been left alone with the first man who had aroused some interest in her otherwise impassible heart.

Colonel Orloff was not so discreet, in the sense that he related to his friend all that had taken place between him and the Czarina—related it with such agitation and poignant regret that she saw at once that she was in the presence of a feeling capable of driving the man who was under its influence to any heights of personal sacrifice. She then communicated to him a plan out of which she hoped to find the solution of the troubles against which the Empress was struggling so bravely, but apparently so uselessly. It was a daring plan and it required much daring to accomplish it; but the future of the woman they both loved was at stake, and she thought they ought to risk it.

Nicholas II. was fond of Colonel Orloff, whom he had recently appointed one of his aides-de-camp. The Sovereign liked from time to time to go and dine or have supper at the mess of some regiment or other of the Guards, either at Tsarskoye Selo, Peterhof, or St. Petersburg. These entertainments used to last generally into the small hours of the morning, and ill-natured people said that the Czar when in this company of young men, which was more congenial to him than the one he was compelled to see generally, allowed himself to have more glasses of wine than were good for him, and to indulge in subjects of conversation he would have done better to avoid. Whether this was true or not, it is of course difficult to say, but the fact remains that Nicholas liked these "family festivities," as he used to call them, and that he always returned home in a good temper after having attended them. Colonel Orloff was aware of this weakness of the Sovereign, and one day he proposed to him to go and hear some regimental singers at the mess of his own Lancer regiment, stationed at Peterhof, the same regiment of which the Empress was Colonel-in-chief. Nicholas II. consented and a day was fixed. On the morning of that day Colonel Orloff sought Madame Wyrubewa; the two had a long conversation, the result of which was their reading together a certain page in French history relating how Louis XIII. had been compelled to seek the hospitality of his wife, Anne of Austria, on a stormy night when it had not been possible for him to return from Paris to St.-Germain, where he resided, an incident that had had world-wide consequences in the birth of the child who was to become in time Louis XIV. After that the Colonel returned to the Palace, where he was on duty that day, and his friend went to seek the Empress and to try to induce her to lend a helping hand to the plot which they had both engineered.

The supper took place, and it was nearly dawn when the Czar left the mess of the Lancers of the Guard, where he declared that he had spent a most pleasant evening. He drove in a motor-car back to Tsarskoye Selo in a very enjoyable frame of mind, which did not require the encouragement of his aide-de-camp, who sat next to him, to become a boisterous one. Lots of champagne had been drunk during the meal, and even after, and when some one in the gay assembly had ventured to say that the only pity of the whole thing was that no representatives of the fair sex had been invited to enliven the party with their presence, Nicholas II. had heartily echoed the regret expressed by the officer in question. Orloff, when alone with the Sovereign, had very cleverly turned the conversation into the same channel, and at last had wormed out of his Imperial Master the confession that he was very unhappy at the extreme coldness of character of his Consort, whose beauty he admired just as much as on the day he had married her. The Colonel, upon this, had ventured to express the conviction that this coldness was only assumed, and proceeded perhaps from jealousy more than from anything else. When at last Tsarskoye Selo was reached, instead of accompanying Nicholas to his own apartments, as it was part of his duties to do, he brought him to the door of the Empress's room, which he opened and closed upon him.

The next day a pale and haggard woman appeared in Anna Wyrubewa's house, coming to seek consolation in what she considered an overwhelming misfortune, and while she was sobbing out the agony of her soul with her head hidden in her friend's lap, a strong man who had borne many a misfortune without flinching, and who had stood calm and unmoved while his heart had been breaking, was sitting alone in his room, his head hidden in his hands, and hot tears dropping one by one between his fingers on the table over which he was leaning, in his overwhelming despair.

THE GRAND DUKE NICHOLAS

XII

LOVED AT LAST

AFTER a storm there comes, generally, so they say, at least, a great calm. And in a certain sense this happened in regard to the troubled mind of the Empress Alexandra. As time went on, she recognized the value of the good advice which she had received from Madame Wyrubewa as well as from Colonel Orloff. Her relations with the Czar, which had been more than strained for long months, became gradually better when she could at last tell him that she had once again, and this time without any mistake, the hope of giving him the Heir for which they had been longing. She saw his former confidence in her return, together with his affection; an affection to which she did not perhaps respond, but which she nevertheless appreciated, perhaps because she was told she ought to do so.

The fact was that her two friends were doing their best to get her to take a healthier view of her own position than had been the case until then. Intrigues at the Court were getting worse and worse, as the various events which finally brought about the Japanese war were slowly unfolding themselves, and it became every day more important for the security of the Empress that she should not disinterest herself from all that was going on around her, as had been her wont, since she had allowed disappointment and sorrow to overpower her.

It was an anxious and a critical time for the dynasty as well as for the country that was coming on, and Anna Wyrubewa with her clear mind was very well aware that such was the case. She used to hear all the gossip in the various circles of St. Petersburg society, and she knew very well that a war was wished for by the enemies of the existing order of things. They saw in it the possibility of overthrowing the dynasty, as the mistakes inevitable in dealing with such a corrupt administration as the Russian one would appear in a new, bold light before the horrified eyes of the public. She was also perfectly aware of the growing unpopularity of Nicholas II., and of the way in which he was daily losing what still remained of the former short-lived affection his subjects had felt for him. She would have liked the Empress to assert herself, and to claim as her right to be initiated in what was going on in the domain of public affairs, but it was still too early for Alexandra to avail herself of this advice. The Czarina did not feel sure of her ground as yet, and she only replied to her friend's adjurations that, if she were lucky enough to give birth to a son, she would follow her advice to the letter; in the mean while she felt afraid of being snubbed by the Emperor, who, though he treated her with far more consideration than he had done for a long time, still kept her in total ignorance of all questions relating to the affairs of the State. On the other hand, he did not hesitate to discuss them with his mother, the Dowager Empress, and even occasionally with his sisters and his brother-in-law, the Grand Duke Alexander Michaylovitch, who had always been his great friend and favorite.

The delicate condition of health of Alexandra Feodorowna furnished her with the pretext she required to isolate herself more than ever from her family, and she used to spend long hours with Madame Wyrubewa in the latter's small house, and whenever she went there she met, as if accidentally, Colonel Orloff, whose faithful, devoted eyes followed her with a love which she could not have helped noticing, even if she had not been aware of its existence. She was a woman gifted with a very pure mind, given to idealize the people she cared for and her own feelings in regard to them. She soon grew to think of the young officer as of a kind of guardian angel sent by Providence to help her in the various difficulties of her daily existence, and with a selfishness almost touching in its unconsciousness she took to confiding to him her various doubts and perplexities, and to initiate him into all the details of her married life, together with the constant disgust and struggles which attended it, not suspecting that by doing so she was breaking the heart of this one faithful friend who had sacrificed himself so entirely to her welfare.

In the mean while events had been rapidly unfolding themselves. The war with Japan had begun and was progressing, together with its long series of appalling disasters coming one on top of the other. Mukden had been fought, the *Pétropawlosk* had gone down in the waves of the Pacific, with brave Admiral Makharoff and its whole crew of officers and men, and the catastrophe of Tsu Shima had also taken place. These had been met by the utter indifference of Nicholas II., who had not even thought it worth while to interrupt the game of tennis he had been playing when the telegram with the news of this unprecedented misfortune had been brought to him. In the interior of the country trouble was also brewing. The Grand Duke Sergius, the uncle of the Czar and the husband of the Empress's eldest sister, Elisabeth, had fallen under the bomb of an assassin in Moscow, and the famous Minister of the Interior, Von Plehwe, whose very name was a horror to all the liberal elements in the land, had met with the same fate.

It was evident that grave events were at hand, and that unless something was attempted to meet them the very foundations of the Throne might come to be shaken by this rising tide of discontent which threatened to engulf the dynasty in its waves. It was high time something were done, and that some one should interfere to save Nicholas II. from impending calamity. Who could do so better than his wife and the mother of his children? Thus reasoned Anna Wyrubewa, and it was also what her friend, Colonel Orloff, thought; but that was not at all what was wished by the various other forces at work trying to dictate to the weak-minded Czar the conduct he ought to hold in the presence of these unexpected difficulties with which he found himself confronted, to his dismay and surprise. There had got about among the public an inkling as to the possibility of the Empress becoming all at once a factor to be reckoned with in the general situation. Immediately the efforts of all her enemies became concentrated on that one point—how best to eliminate this new element, which they understood but too well would necessarily counteract their own influence.

A careful watch was set on the person and the conduct of the young Sovereign. It did not bring any of the hoped-for results, because both Anna Wyrubewa and Colonel Orloff were prudent people, who contrived to arrange matters in such a way, that no one suspected they used to see Alexandra Feodorowna every day, and who had persuaded the latter to resort to all kinds of precautions whenever she visited her friend.

One day, however, an officer who was serving in that very same regiment of Laners to which Colonel Orloff belonged made a playful remark to the effect that he was believed to be a favorite with the lovely and cold Czarina, who had never hitherto allowed her glances to fall on any man whatsoever. The young Colonel became immediately alarmed, the more so that he could not discover the source whence this piece of gossip had arisen. He sought Madame Wyrubewa and told her that he had made up his mind to ask to be transferred to a regiment at the front, so as to put a quick end to any possible unpleasantness. She heartily agreed with him in the opinion that this was the best thing he could do, for the sake of everybody, and especially for that of the Empress.

The latter had to be told of Orloff's resolution. But when he broke to her his intention to request the favor of risking his life in distant Manchuria, she gave way to a fit of despair that absolutely frightened her two devoted friends, and implored him not to leave her, at least not until her child had been born, saying with sobs and tears that she would never be able to undergo the trial which awaited her if she did not know that he was there, as near to her as possible, and that she could see him after all was over, to wish her joy, if the expected babe were a son, and to comfort her if it turned out to be another girl, the one thing which she feared above all others.

At first the Colonel protested. He tried to explain to the despairing and over-excited woman that it was for her sake he wished to go away, at least for a while, and that it cost him more than he could say to come to such a resolution, but that he loved her far too much to let her run any risk. The Empress would not listen to anything, and at last she told him that if he went away she would consider it as a proof that he did not love her, and that all he had said to her had been nothing but empty phrases, such as no doubt he had repeated already to many more women than he even cared to remember. Orloff was stung to the quick, but he remained, nevertheless, firm until Alexandra Feodorowna exclaimed that unless he promised her to remain by her side she would make a scandal and depart for Darmstadt, whether the Emperor allowed her to do so or not. Man-like, he yielded, without suspecting whither this weakness was to lead him sooner than he could imagine.

While this drama was going on in the pretty little house whither Anna Wyrubewa received the Empress of All the Russias, unknown to the rest of the world, so she believed, at least, speculations were rife as to the eventual sex of the child expected by the Czar and Czarina. Everybody, with few exceptions, hoped that it would be another daughter, none more ardently than the Dowager Empress, who would have infinitely preferred the Throne passing to her youngest son than to any boy born to a daughter-in-law whom she made no secret of disliking, and whom she distrusted even more than she disliked. She realized very well that Alexandra Feodorowna, if she was the mother of an Heir to the Imperial Crown, would become a most important personage in the State, as well as in the eyes of her husband. This was not to be desired, in view of her strong German sympathies, which she had lately exhibited more than she had ever dared to do before.

The French alliance was very popular at the time I am talking about, and the Empress was considered as its principal and most bitter adversary. This was one more reason for not wishing her to acquire suddenly an importance that had never been awarded to her by the nation since she had become its Sovereign.

For months this kind of thing went on. Alexandra Feodorowna knew herself to be watched with anything but kind eyes, and this consciousness of the ill-will of which she was the object added to her anxiety and moral sufferings. As the weary months dragged on, she thought more and more of Orloff, and suddenly she realized that she loved him more than any one in the world, and she began to understand all that she must have cost him, in pain and vain regret.

But for her, at least, consolation was at hand. One July morning the Imperial Family were called together with the principal Court and State functionaries in all haste to Peterhof. The long-expected event was at hand, and a few hours would decide as to the future of the Romanoff dynasty. People with anxious faces thronged the vast halls of the Palace, waiting for news which seemed to be very long in coming.

At last, just as the clock struck noon, a doctor entered the room, and told the assistants that Nicholas II. was the father of a son.

There was one person present who listened to this announcement with an impassible face but with a breaking heart, and who could barely find sufficient strength to reach the little cottage where Anna Wyrubewa was sitting pale and anxious, in expectation of—she did not know well herself what. When she saw Colonel Orloff she extended toward him her two hands in a gesture of passionate greeting. But what was her surprise to see him fall on the sofa beside her and bury his head in the silk cushions, with such sobs as rarely shake the frame of a strong man. He had had the courage to sacrifice his personal happiness at the shrine of the woman whom he adored with such religious fervor, but it was more than he could bear to find how thoroughly this sacrifice had been accepted by Providence, and for just a few minutes he had hated this new-born child, whom he knew but too well was going to usurp the place he had hoped to keep forever in the heart and the affections of Alexandra Feodorowna.

XIII

HE DIED TO SAVE HER HONOR

THE christening of the little Grand Duke Alexis was solemnized with great pomp at Peterhof, and there is no doubt but that the position of his mother became, after his birth, quite different from what it had been before this much-wished-for baby had appeared. For one thing, the talk of a divorce between her and the Czar, which had been so frequently indulged in, came to an end, and it was felt, even by the most bitter enemies of the Empress, that it would be waste of time to think about the possibility of its ever taking place.

Nicholas II., in his joy at having at last an Heir, seemed to have returned to his former allegiance in regard to his wife, and he began to confide in her far more than he had done formerly, even consulting her on different occasions. She was the mother of the future Sovereign, and as such entitled to a consideration a childless Empress Dowager could never aspire to in the case of widowhood. It became, therefore, necessary to initiate her in matters concerning the government of the country, and the Czar did this the more willingly that at heart he distrusted his brother, and his numerous uncles and cousins, and feared that in case he died before the small Cesarewitsch had reached his majority the interests of the latter would not be looked after as well as would be necessary, unless his mother were there to protect them.

Alexandra Feodorowna, on the other hand, urged by her two friends, Madame Wyrubewa and Colonel Orloff, began to show far more interest in public affairs than she had ever done since her marriage, and she tried to establish between herself and her husband more intimate relations than she had cared to do formerly, when she used to spend her days lamenting over sorrows, imaginary most of the time, but sufficiently acute to render her intensely miserable. Her son became the principal preoccupation of her existence, and she would not intrust his care to any one, but transformed herself into his nurse, governess, and constant attendant, forgetting everything else, even the care of her daughters, in her nervous solicitude for him. Unfortunately the child was born excessively delicate, and had a curious and rare disease, a weakness of the blood-vessels, which were affected in such a way that he was attacked with hemorrhage at the slightest touch; the smallest of knocks or wounds would endanger his life. He might bleed to death from an ordinary bruise. An unfortunate accident which occurred when he was two years old, and which brought about a rupture that necessitated an operation from which he recovered only by a kind of miracle, only aggravated the chronic ailment with which he was afflicted.

One may imagine how terrible this state of things proved for the Empress, who very stupidly, as it seemed to some people, applied herself to hide from the public the state of physical health of her son, which had, among other results, that of people supposing him to be even more dangerously ill than was the case. The truth was that Alexandra feared that if it were known the boy was afflicted with an incurable disease, it might add to her own unpopularity. Her friends hoped that she might bear another son in time, but after the birth of Alexis she never had any more hopes of maternity, and so there remained nothing else to do but to try and rear this weak, frail, and puny infant, in whom were centered all the future hopes of the proud Romanoff dynasty.

Anna Wyrubewa did her best to comfort the sorrowing mother, and both she and Colonel Orloff agreed that the only thing to do in order to turn her thoughts into another channel than that of her child's state of health, over which she brooded until she had become absolutely morbid in her constant preoccupation of the painful subject, was to speak to her of the necessity of becoming the Czar's principal adviser and counselor. They tried to induce her to assert herself in the interest of Alexis, who they assured her would one day outgrow his native weakness and require her help in the numerous duties entailed upon him by his position as Heir to the Throne. In a certain sense they succeeded, and the Empress began to develop an independence of opinions and views in which she had never dared to indulge before. Ministers were surprised to hear the Czar say to them, when they pressed him for a reply to some decision or order they presented to him for confirmation, that he first wished to discuss the subject with his wife. Somehow there arose among the public, and especially among the Imperial Family, an impression that Alexandra had at last completely subjugated her husband, and that she was henceforward a factor to be reckoned with in every important

State affair which might arise in regard to foreign or home politics.

Of course people did not like it. One had been used for such a long time to consider the Czarina as a nonentity that it seemed a strange thing to have suddenly to take her into account; one began to wonder what could have brought about such an unexpected change in her whole conduct and demeanor. Maternal love was not sufficient to explain it, and the cause of it had to be looked for elsewhere, and one fine day her constant intercourse with Anna Wyrubewa was noticed. Once people were started on that path, there was but one step to take—to try and find out whether or not these suspicions were founded on anything tangible. Some inquisitive persons took to watching the actions of Anna Wyrubewa, and they were not long in discovering that her house served as a meeting-place for several people in whom Alexandra Feodorowna was interested, among others Colonel Orloff, whose hopeless passion for his Sovereign had been already suspected at different times.

Foremost among these voluntary observers, not to give them another name, figured members of the Imperial Family who had never taken kindly to the Consort of Nicholas II., and who hated the idea of her becoming a power in the State. They tried to find out something to her detriment, and who also attempted to enroll among their number the Dowager-Empress Marie, who, however, refused to listen to them, and whose affection for her eldest son induced her to make an effort to warn her daughter-in-law of the dangers which were threatening her. But the young Czarina would not hear anything, and haughtily refused the hand that was extended to her in sincere friendship. She snubbed Marie Feodorowna in such a manner that the latter, wounded to the quick in finding her good intentions misunderstood, swore that she would never again attempt to come to the help of a person who was so prejudiced against her.

In the mean while, ignorant of the conspiracy which was being engineered against her, Alexandra continued to spend her afternoons with Madame Wyrubewa, often taking her little boy with her. The two women watched the child sleeping in his cradle, and often Colonel Orloff shared their vigil with a bleeding heart, the baby reminding him of all that he had suffered for the sake of its mother, but with the consciousness of having done his duty to both. But one day rumors again reached his ears that his name had once more become associated with that of the Empress. This time he made up his mind to go away definitely, no matter how much she might ask him to stay. He realized, if neither she nor Anna Wyrubewa did so, that the position was becoming threatening, and that he ought to put an end to it in some way or other. Unfortunately, when he came to this conclusion it was already too late.

Madame Wyrubewa's husband was a naval officer, not gifted with a superabundance of brains, but honest in his way, and incapable of intrigues of any kind. He had troubled very little about his wife, and was perhaps the only man in St. Petersburg and in Tsarskoye Selo who was not aware of the high favor in which she stood with the Empress. His duties generally kept him far from his home most of the year, and when he was there he rarely troubled Anna with his presence. But he was known to be of a violent disposition, and as a fellow who would not suffer any stain to rest upon his honor. It was of this man that the enemies of Alexandra Feodorowna determined to make use in order to ruin her.

Anonymous letters were sent to him accusing his wife of carrying on a guilty intrigue with Colonel Orloff, intrigue which he was assured the Empress knew and favored. He was advised to return home unexpectedly any afternoon between four and five o'clock, when he would find proofs of the information vouchsafed to him by his unknown friend. The young man, instead of putting these denunciations in the fire, became so enraged that he determined to follow the advice of his anonymous correspondent. After having advised Anna that he was going away on a few days' cruise, he waited until the hour that had been indicated to him, and boldly walked back to his house.

He was met at the door by the Cossack in personal attendance on the Empress, who informed him that he could not get in. Wyrubew protested, and was quietly told that the Sovereign was visiting his wife, and that according to etiquette no one could be allowed to enter a place where she was unless by her special permission. The officer became furious, brushed the Cossack aside, and penetrated into the sitting-room, after having noticed that a military overcoat was hanging in his hall. He found the apartment empty, but in the adjoining one, which was Anna's boudoir, he could hear voices, one of which was distinctly masculine. He did not hesitate, but made his way inside, to find that his wife was not there, but that the Empress, pale and lovely, was standing by the mantelpiece, while Colonel Orloff, on his knees before her, was kissing passionately the hem of her skirt.

Alexandra Feodorowna gave one cry, which echoed through the whole building and brought Madame Wyrubewa to her help. Wyrubew himself remained silent and dazed by the unexpected sight. The only one not to lose presence of mind was Colonel Orloff, who, starting to his feet, went up to the intruder with the stern words:

"You are going to give me your word of honor to remain silent."

Wyrubew passed his hand over his eyes. He could hardly believe his own senses, and the terrible idea crossed his mind that his wife had been helping the Czarina in an amorous intrigue, and that very probably he would have to pay the penalty for this piece of complaisance, which he did not in the least care to do. He thought that insolence was the best way to get out of an impossible position with flying colors, and so he simply sneered in the face of Orloff, with the remark:

"Not I. If you have chosen to abuse my confidence, together with my wife, you cannot expect me to help you in your villainy."

Anna rushed to the Empress and took her in her arms, trying to lead her out of the room. Orloff made a movement forward as if he wanted to strangle Wyrubew; then he contained himself and said in a low voice:

"You know that you are not speaking the truth. Once more I implore you not to mention to any one what has taken place here, and I give you my word of honor to meet you whenever and wherever you like."

"You are not a man from whom one can expect satisfaction," replied Wyrubew, "and I will not claim it from you. There are other means at my disposal to punish you," and he turned away contemptuously.

The young Colonel's face became by turns deadly pale and fiery red. It was evident that he could hardly contain the tumultuous feelings which were racking him. Before him stood the Czarina looking at him with haggard eyes and trying to free herself from the encircling arms of her friend. Anna was weeping profusely and vainly struggling with an emotion she absolutely could not control. Orloff went up to the two women, and once more knelt before the Empress.

"Forgive me," he said. "I ought to have known better, but believe me, I shall atone."

He kissed once more the hem of her garment and went out of the room, without looking round, brushing past Wyrubew as if he had not seen him, and went back to his own house, calm and determined, but probably with the feelings of a man about to be taken to the scaffold.

Madame Wyrubewa seized her husband by the arm.

"Go now," she cried. "You have done enough evil for to-day, but remember that henceforth everything is at end between us."

He laughed sardonically, but obeyed her, and the couple never set eyes upon each other again after that terrible afternoon. The next day St. Petersburg was electrified by hearing that the popular Colonel Orloff had been found dead in his room, shot through the temple. He had atoned.

And two months later the Synod pronounced the decree of divorce between Anna Wyrubewa and her husband. The tragedy, like so many others of the same kind, had come to an end, by breaking two women's hearts.

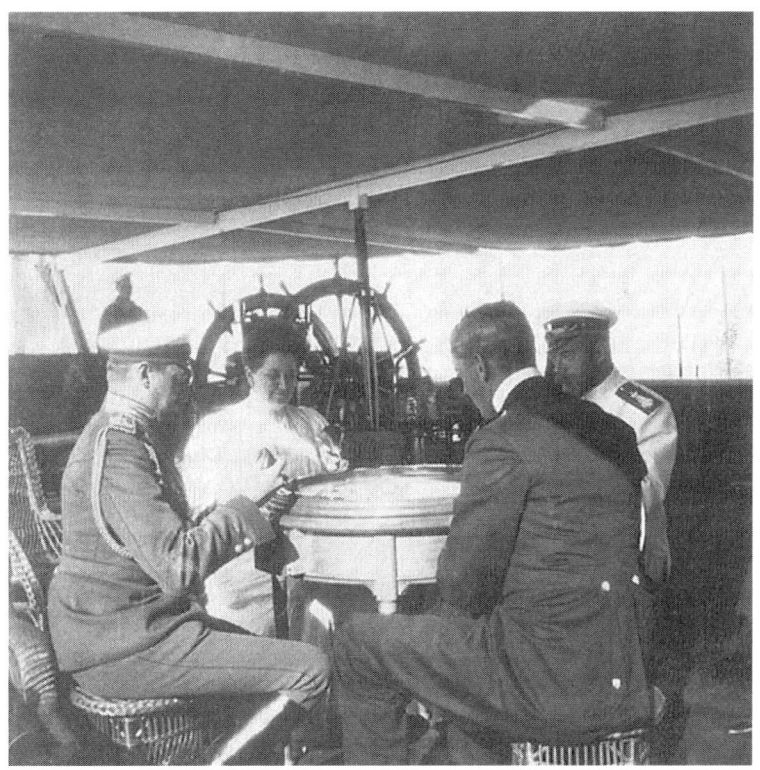

GENERAL ALEXANDER ORLOFF and ANNA VIROUBOVA sit to the left, relaxing with the Czar onboard Standart. (September 1906)

XIV

A NATION IN REVOLT

THE suicide of Colonel Orloff was perhaps one of the events which provoked the most sensation in St. Petersburg in recent years. Everybody had known him, and he had been a general favorite, not only in his regiment, but also among all the circles of society which he had frequented. The Czar, who had also liked him very much, was deeply affected by the catastrophe, and everybody kept wondering what could have induced a man who apparently had not a single thing in the world to trouble him to take his own life in such an unexpected manner.

The Empress alone said nothing. She was present at all the funeral services which were celebrated over the coffin of the young officer, but so was Nicholas II. Her attendance could not be considered as an extraordinary thing. No one, with the exception of Anna Wyrubewa, who had accompanied her, knew that on the night preceding the funeral of her friend Alexandra Feodorowna had proceeded alone and unattended, save for her, to the house where his mortal remains lay in state, and had spent an hour praying beside his dead body and weeping bitter tears. Outwardly, however, her calm had remained unshaken; and she had succeeded in quite a wonderful way in keeping her feelings under control. The only thing which she had insisted upon was to have Colonel Orloff buried in the cemetery of Tsarskoye Selo, where she had a simple monument, consis-ting of a large white marble cross, erected. She used to go every day to pray there, and to leave flowers on this tomb which represented for her so many hopes, and perhaps something else besides.

Of course these visits became known, but by a wonderful miracle they were not commented upon in the way they might have been. The reputation for eccentricity of Alexandra Feodorowna had by that time become so well established that people had left off wondering at anything she might attempt to do, and, besides, every one believed that the Colonel's death had been somehow connected with a love intrigue he had carried on with Anna Wyrubewa, whose divorce lent ground for such a theory. It was suspected or guessed that something had taken place in her house, but no one could exactly ascertain what this something had been, and Wyrubew himself had been for once thoroughly frightened, and had come to the conclusion that the best thing he could do for his own sake as well as for that of others was to hold his tongue, and to accept the divorce upon which his wife insisted. Later on, however, he unburdened his soul to some of his particular friends, but that happened at a time when people were thinking of other things than the tragical death of an officer whose existence was already forgotten by most of those who had known him.

As for the Empress, she had, as we have seen, borne herself wonderfully well in the first moments which had followed upon the tragedy, but afterward her nerves gave way entirely, and it was then that she had to be kept in strict seclusion, and under the care of trained nurses. It was said that her reason had given way under the load of her anxiety for her small son, and that the thought of his serious condition had weighed down so thoroughly on her mind that she had grown melancholy to an alarming extent. The story was believed perhaps because it suited so many people to think that it was true, and, besides, the political situation in Russia was becoming so alarming that it entirely absorbed public attention. The war with Japan had come to an ignominious end, and shown the many failings, as well as the thorough insufficiency, of the Government. The first symptoms of the Revolution were clearly appearing on the horizon, with its attendant horrors. Everybody felt that something had to be done in order to avert a catastrophe the extent of which it was impossible to foresee, but which was generally considered as being inevitable, unless the Czar made up his mind to grant the reforms for which his whole Empire was clamoring.

During those years, which were the prelude of other even more eventful ones that were later on to sweep away the Throne of the Romanoffs, Nicholas II. might still have regained the popularity which he had lost. If he had only bravely and courageously faced his people, and tried to get into direct contact with them, he could have secured for his dynasty a new lease of life. He was not liked, it is true, and he was not trusted, which was still worse; but nations are sometimes apt to be led by impulse, and it is certain that Russia would have felt grateful to him if he had only made an appeal to its loyalty for his person, and asked of her to help him in the task of repairing the wounds caused by the disastrous campaign that had come to an end with the signature of the Treaty of Portsmouth.

But the Czar ignored the wishes of his subjects and refused to acknowledge the justice of their claims to be taken into his confidence. He was narrow-minded, cruel by disposition, and though not at all an autocrat, yet every inch a tyrant. He was even something worse than that; he was a coward, and this is a defect which neither nations nor women forgive in those to whom they find their destinies intrusted.

The remembrance of that dreadful Sunday when a crowd of peaceful workmen, under the leadership of the afterward notorious priest, Gapone, marched toward the Winter Palace, to be met with the firing of machine-guns that laid them low by hundreds on the pavement—the remembrance of this bloody deed has never been effaced from the mind of the Russian nation. It traced between itself and its Czar a line of demarcation which could never be removed later on.

Many versions exist as to the conduct of Nicholas II. on that awful day. Some people have said that it was the Empress who had entreated him to fly to Tsarskoye Selo, where she thought that they would be in greater safety than in St. Petersburg; others have asserted that it was he who of his own accord had decided that it would be better for him to leave the capital and to abandon to his uncle, the Grand Duke Wladimir, the task of drowning in blood this attempt of his subjects to enter into direct communication with him. Probably both versions are right, in a sense, at least, because it is certain that Alexandra Feodorowna was always in fear something might happen to her son, and very likely she tried to induce her husband to consider how best to insure the safety of their only boy; on the other hand, the Emperor might, had he only come himself to take a sane view of the situation such as it presented itself at the time, have been able to reassure his wife and to explain to her that neither she nor their children were in any danger. Nicholas II., however, had only one thought in his small mind, and that was how to punish this "insolence," as he termed it, of his people. For him a mob was always a mob, except when it was ordered to cheer him, and lately he had had to acknowledge that, in regard to St. Petersburg, cheering had become rather a rare event.

I am not trying to relate here any of the numerous episodes which have made the unsuccessful Revolution of 1905 memorable. I am not writing the history of Nicholas II. Others have done so, and will do so, better than I could. What I only want to point out is the utter callousness shown by both the Czar and the Czarina in presence of the abominable repression which the police, together with some military commanders, inaugurated in regard to the people compromised even in a slight degree in the movement of emancipation which had shaken the existence of the dynasty. It was in vain that some wise people, like Count Witte, for instance, had tried to explain to Nicholas II. that unless he frankly granted some reforms without which it would be impossible to govern Russia in the future he might expect an explosion of wrath on the part of the nation which it would be almost impossible to subdue or to destroy. The Czar refused to listen, and when at last he yielded to the demands of his Ministry and signed the famous Manifest of the 17th of October, with its "simulacre" of constitution, it was with the firm intention not to keep any of the promises which it contained, and to try, on the contrary, to reduce to absolute powerlessness the National Assembly, or Duma, as it was called, the election of which he had allowed only because he could not help it, but not at all because he believed or hoped it might prove useful to him in the solution of the many problems which were waiting to be unraveled.

What followed upon the first convocation of the first Parliament Russia was to know is already a matter of history. It did not live for more than a few weeks, and very probably the Czar had never intended it to exist for any length of time. What he wished was to appear before the eyes of Europe as a Sovereign who had been willing to make any amount of sacrifices in order to insure the welfare of his subjects, who, instead of showing themselves grateful to him for his good intentions, had rewarded him with the basest ingratitude. He thought this a clever piece of policy, forgetting that any politician worthy of the name could see at once through his game, and that this game could have only one result—that of inspiring an utter contempt for his person as well as for his moral character.

Therein lies the great, the supreme, fault of Nicholas II. He never could bring himself to act frankly in regard to any serious matter in which his people were concerned. The Empress, in her strange way, was far more honest, because she did not hesitate to follow the instincts of her heart, and in her most mistaken actions she was at least sincere.

During the years that followed upon the insurrectionary movement of 1905 Alexandra Feodorowna was in such a state of health that it was almost impossible for her to take any part in what went on around her. Her reason had been seriously compromised by the shock caused by the tragical ending of the only romance she had known in her life, and she used to spend hours weeping in her room, absorbed in the contemplation of her own grief. It was in vain that Anna Wyrubewa, who had become more intimate with her than had even been the case before, had tried to induce her to fight the morbid ideas which were torturing her. The Empress would not listen to her friend, and insisted on secluding herself from the world and even from her own daughters, whose presence irritated her and made her give way to fits of impatience that were very nearly akin to madness. The girls were perfectly charming and had the luck to have an excellent governess, who tried to give them the love their own mother refused or was unable to award them; nevertheless their lives were blighted by the illness of the Empress, and it is not extraordinary that they came to care for their father more than for her, whom they were always more or less afraid to approach, whom they were constantly told they must not bother by questions of any kind or manifestations of affection.

It was only the little Cesarewitsch who was allowed to share his mother's solitude, whom she would never let out of her sight. He was the only preoccupation her diseased mind would admit, and when she saw that his state of health did not improve she became more and more desperate, until one day she confided to Anna Wyrubewa that she was sure God was punishing her for the affection which she acknowledged now that she had borne for Orloff, and that her boy would never get well. Her despair was so evident, and her mind was getting so unhinged, that at last the question of putting her in some retreat where she could be under a doctor's continual care was seriously considered by her medical attendants, who even informed the Czar of their fears in regard to the sanity of his Consort. Of course the fact that they had done so reached the knowledge of Madame Wyrubewa, and it was then that the latter began to consider whether it would not be possible to restore by some way or other the equanimity of Alexandra Feodorowna and to procure for her some kind of consolation for the seemingly incurable grief which was destroying her life and her reason. Unfortunately for all parties concerned, she was to make at that time the acquaintance of the notorious Raspoutine, whom she introduced, under the circumstances which I am going presently to relate, to the miserable, half-demented Empress, an introduction which was to prove so fatal not only to the unhappy Sovereign, but also through her to the whole of Russia.

XV

A PROPHET OF GOD

ANNA WYRUBEWA had always been inclined toward religious exaggeration, and this was perhaps one of the reasons why the Empress, who for years had buried herself in the exercise of all kinds of devotional practices, had taken to her so quickly. They were both of a mystical turn of mind, and never so happy as when enabled to spend long hours absorbed in prayer before some icon or other. And besides this, Anna was in the habit of frequenting certain circles of St. Petersburg society that were considered as the supporters of orthodoxy in its most rigid form, where all questions concerning the discipline of the Church were discussed and in some cases decided.

Such, for instance, was the house of the Countess Sophy Ignatieff, where the higher clergy used to meet at weekly assemblies, during which the laxity of the younger generation in regard to religious matters was discussed with many a sigh and many a shaking of wise heads, disposed to admit that this religious indifference, which was getting stronger and stronger every day, was bound to bring Russia to the brink of terrible misfortunes. Countess Ignatieff had traveled all over her native country in search of its sacred shrines and places, and was very well known personally in almost all the principal convents in the Empire. She had been suspected at one time of sympathies with dissenters, but this has never been proved; on the contrary, in her old age she gained the reputation of being fanatically orthodox, one who saw no salvation outside the fold of her own creed, who favored persecution of all others on account of her conviction that people ought to be brought back to the bosom of the Greek Church by any means, even through violence if other ones failed.

During one of the yearly pilgrimages in which so much of her time was spent she had had occasion to meet a kind of vagrant preacher whose wild eloquence had captivated her fancy and her imagination, and she had been partly instrumental in his coming to St. Petersburg, where she had arranged for him to hold religious meetings in her house, to which she had invited prominent church dignitaries, together with a few ladies of an enthusiastic turn of mind whom she believed would be inclined to listen to the wild ravings, for they were nothing else, of her new *protégé*.

At first people laughed at her, as well as at the uncouth appearance of the "Prophet of God," as she called him, who, while not blessed with the eloquence of a Savonarola, yet possessed sufficient persuasive gifts and talents to shake the equanimity of the hysterically inclined women who listened to him. This "Prophet" was none other than Grigory Raspoutine, who later on was to become such an important personage in Russia.

Madame Wyrubewa had heard about Raspoutine a long time before she ever came to hear him. But after she had had the opportunity of meeting him she thought that it would not be a bad thing to bring him to Tsarskoye Selo, where the poor Empress was eating away her heart in her grief at the loss of all that she had cared for in life, and to try to induce Alexandra to listen to him, and to pray together with him. He was supposed to perform wonders by the intensity and the fervor of his prayers, and it might just be possible that the very fact of his being a complete stranger to her, and moreover a man totally outside Court circles and Court intrigues, would influence the Czarina to give him her confidence and to permit him to cheer her up. At all events, she spoke about him several times, and pleaded hard with the Empress to allow him to be brought to her. This Alexandra Feodorowna absolutely refused, but she was induced at last to consent to see him at the house of Anna Wyrubewa, and thither came one winter evening the adventurer who was in time to become the Cagliostro of a reign which was not even worthy to have any one else but a common, uncouth peasant for its jester.

Now, as has been ultimately proved, Raspoutine was far from being the saintly man his admirers thought he was, but he was endowed with an unusual amount of cunning, and far more spirit of observation than he was credited with. When he was told that he would have the honor of meeting the Empress of Russia, and to pray in her presence for the health of her delicate little boy, he had at once perceived the advantages which might result for him out of this introduction, if only in regard to his personal prestige before his disciples and followers. He was above everything else a man who cared for his enjoyment as well as for the good things of life, and who, in the way of Paradise, only admitted the one described by Mohammed in his Koran. He had led a licentious, godless kind of existence, which he had contrived to persuade the weak women who had succumbed to his exhortations was in accord with the spirit of the doctrine which he preached, the principal points of which consisted in blind submission to his will and to his fancies. He had told them that they would be cleansed of their sins by a complete union with him, which he meant in the physical as well as in the moral sense of the word. It is probable that in his dealings with all the people who had grown to believe in him he had had recourse to his incontestable hypnotical powers and to practices of magnetic influence which he had learned amid some wild tribes of Siberia, where he had spent his childhood and early youth, who are to this day adepts in the art of witchcraft as well as in all kinds of magical rites and customs. At the same time the crafty adventurer knew very well that it would be unwise of him at the beginning of his intercourse with the Consort of his Sovereign, an intercourse which he was fully determined should be continued and not be limited to a single interview, to do aught else but assume the attitude of a man entirely absorbed in God and in the practices of religion. When he was introduced to Alexandra Feodorowna at the house of Anna Wyrubewa, he therefore remained standing before her, in an attitude of apparent humility, and he waited quietly until she should begin talking with him, which she immediately did, saying that she had heard so much about him that she had wished to see him and to ask him to pray for her little boy, whose state of health gave rise to so much anxiety and worry.

Raspoutine looked at her, then replied quietly that he would be happy to pray for the child, but that he thought she was just as much in need of prayer as her son because her state of moral health was far more alarming than Alexis's physical one.

The Empress was so amazed that she could not find a reply to what appeared to her in the first moment to be an unsurpassed piece of insolence. Anna Wyrubewa saw what was taking place in her mind, and, addressing her in English, a language which they always spoke together, implored her not to feel offended, as the man really did not know what he was saying, sometimes being urged by a strength superior to his own to give utterance to thoughts he would never have dared to express otherwise. She then urged the Czarina not to carry on the conversation further, but to ask Raspoutine to begin at once praying for her welfare, and also for that of Russia and of the Imperial Family.

Alexandra acquiesced, and the preacher proceeded to set himself before the icon which, as is usual in all Russian houses, was hanging in a corner of the room. He began long litanies which he recited in a peculiar deep tone of voice, that rose up louder and louder as gradually he worked himself up into a state of religious frenzy akin to the one displayed by the dancing and howling dervishes in Turkey. But whether or not his manner or the tone of his supplications or his personal influence was the cause of it, the Empress as she listened to him felt calmer and quieter than she had done for years. It seemed to her as if a great peace was stealing upon her after the despair and the sadness in which her days had been spent during the last months. When at last Raspoutine's orisons came to an end she was weeping silently, but all the nervous excitement under which she had been laboring at the beginning of the interview seemed to have disappeared and she looked more like a normal woman than she had done since the day when Orloff had said his last good-by to her in the boudoir of Anna Wyrubewa.

She silently extended her hand to the "Prophet," saying as she did so:

"You have done me a great deal of good, and I thank you with all my heart. I shall ask you again to pray for me."

It was thus that Alexandra Feodorowna met the man who was to have such a baneful influence over her whole life, whose fatal influence was to estrange her, still deeper than was already the case, from her subjects, and to give rise to the flood of calumnies in which she was ultimately to be drowned; and to perish, dragging along with her this mighty Russian Empire whose Crown she wore and whose people she had never understood nor even tried to understand.

Anna Wyrubewa was delighted to see that her beloved Czarina had really found some comfort in listening to Raspoutine's prayers. She believed in the "Prophet" who had found favor in the eyes of the Lord, and whose intercession in regard to the little Alexis would be crowned with success. The woman was superstitious to the backbone, and perhaps more mystically inclined even than most Russians are, which is saying a good deal. She thought, at all events, that, once the Empress got to be persuaded that she had to look to God alone for the recovery of her son from a disease that had been pronounced to be incurable by the best medical authorities, she would no longer fret as she had done, but begin to look at things from a religiously fatalistic point of view. She hoped also for another thing, and that was that the Czarina, once she had been taught to look above for comfort and consolation, would cease to lament over the "might have been" that has already caused so much heartburnings in this world, and that she would leave off reproaching herself, as she was constantly doing, for the death of the one man she had cared for, whom in all innocence she had sent to his destruction, and who had bravely preferred to disappear rather than allow a stain to rest upon her honor. She had guessed the agony of the self-reproach under which the soul of Alexandra Feodorowna had almost collapsed, and the remorse which had racked it until her intelligence had almost snapped, through the moral as well as through the physical pain which had clouded all her faculties. She hoped, therefore, seriously and earnestly, that the prayers of Raspoutine might ease this mental distress which had transformed the Empress of All the Russias into a half-demented woman. When she saw that his prayers had over the latter the beneficent influence she had expected, she determined to do her best to induce her to give her confidence to this man in whom her exalted imagination saw a savior as well as a friend.

This was the real beginning of the Raspoutine intrigue, and it would have been a lucky thing for all those who came afterward to be concerned in it if it had stopped at this stage, and not been transferred to a more dangerous one, the stage upon which European politics had to be played and, unfortunately for Russia, played by utterly unskilful hands. The comedy of Raspoutine did not last longer than a few months. Its drama dragged on for years, and is not yet over by a long way.

XVI

SHE SAW HIM ONCE MORE

AFTER she had made the acquaintance of Raspoutine the Empress changed considerably. For one thing, she became more cheerful and seemed once more to interest herself in what went on around her. She tried also to keep her mind away from the one morbid thought which had been haunting her, the thought that her son's bad health was a punishment which God had sent her on account of her conduct in regard to Colonel Orloff. She had most undoubtedly loved the young officer, and she realized with a painful but clear perspicacity that if she had allowed him to go away when he wished to do so the tragedy which had culminated with his suicide would never have taken place. Her mind, which was dimmed as to so many other points, was quite awake to the terrible one that the man to whom her whole heart had belonged had died to save her honor and to prevent her good name from being compromised. This was quite sufficient to fill her soul with acute remorse, but apart from this she missed the companionship of this faithful friend before whom she could allow herself to speak about her sorrows and her trials just as if she had been an ordinary woman and not an Empress.

There were times when her grandeur oppressed her, and then it was that she longed for a confidant and friend before whom she would not be ashamed to bare her heart and unburden it. She felt so lonely amid the pomp and splendor which surrounded her, so solitary in her great Palace which was so very different from the simple house in which her childhood and youth had been spent, and she was such a stranger in a land she had not learned to love and where she had found herself confronted with hostility from the very first day that she had set her foot in it. Of course her children, and especially her son, constituted a great interest and a great preoccupation in her life, but their existence was not sufficient to fill it entirely. In moments when she thought herself forsaken by the world she would have given ten years of her future existence to be able to see once more the man who had died for her because he had found it impossible to consecrate his whole life to her service.

Raspoutine was a keen observer of human nature. Lurking behind his hopeless ignorance there were immense cunning and a natural intuition of what was going on in other people's minds. Apart from this faculty, he always made it a point to try and find out as much as he could concerning the past of all persons with whom he happened to have dealings. He understood quite admirably the art of "drawing out" those with whom he conversed, and he could put together quite nicely the tangled threads which another man would never have gone to the trouble of trying to untwine. As soon as he had looked upon the Empress he had understood that she must have gone through some great grief which was not concerned with the state of health of her child alone, but which had deeper foundations. In the fashionable drawing-rooms where he was a welcome guest he had heard discussed more than once the personality as well as the conduct of Alexandra Feodorowna; he had come to the conclusion that the mystery which surrounded the death of Colonel Orloff was in some way connected with her, and not with Madame Wyrubewa alone. He applied himself, therefore, to dis-cover what had really taken place.

For some time he could learn nothing, as no one seemed to know anything more than the bare fact of the suicide of the young officer. It is true that when he had asked Anna for the true version the latter had angrily denied any connection implying guilt, but Raspoutine, peasant though he was, understood sufficiently the character of a woman of the world to know that such denials were not worth much. Altogether he was puzzled, but continued, however, to put in an appearance at Tsarskoye Selo whenever he was asked to do so, and he was shown several times the little Heir to the Throne. The Empress had brought the babe to Madame Wyrubewa's cottage several times for him to pray over. The "Prophet" had at once declared that the child would not die, and that there was every likelihood he would outgrow his weakness, a prophecy it had been relatively easy for him to make, considering the fact that before doing so he had taken good care to talk with a doctor of his acquaintance about the illness of Alexis, and had heard from him all that there was to hear on the subject, which was not much. The boy might live with care, and even get strong, once he had reached the years of adolescence; he might die from the effect of a hemorrhage, which the slightest accident might bring about. The whole thing was a matter of chance, and nothing else.

The Empress, however, became full of hope when Raspoutine told her not to worry unnecessarily, but to trust more to Providence than she had been doing. It happened just at that time that the little boy got stronger and better than he had been since his birth, and this fact inspired her with a hope such as she had never allowed herself to nurse since the day when she had realized to what a weak and frail piece of humanity she had given birth in the person of the only son and Heir of Nicholas II. She began to speak of the future, which she had hitherto not dared to do, and she seemed suddenly to think that this future might still hold some joys for her in reserve. As was but natural, she attributed this change in her feelings and mind to the influence of Raspoutine's prayers, and as was also natural she felt grateful to him for having brought it about.

The crafty peasant, however, was not so satisfied as the Empress. He had begun to make great plans concerning her and the influence he meant to acquire over her person. Somehow he could not bring them to realization. He might have gone on for a long time in this state of uncertainty if he had not made just at that moment the acquaintance of one of the cleverest secret police agents the Russian Government had in its pay, Manassavitch-Maniuloff. This personage, whom I have described at length in another book, knew more about what went on in the Imperial Palace of Tsarskoye Selo than any one else in the world. During the time when the famous Plehwe occupied the post of Minister of the Interior he had had the Palace watched just as much as the houses of the people whom he suspected of not favoring his views and policy. Among the agents whom he had intrusted with this task Manassavitch-Maniuloff had occupied a foremost place. He was one of the most unscrupulous men alive, and, as the future proved, had but one aim in his existence, that of enriching himself, thanks to all kinds of shady speculations and blackmail he practised on a large scale. He knew, if others did not, all that had taken place in the house of Anna Wyrubewa on the day when Colonel Orloff had left it for the last time, but he had never divulged this secret, and had been content with waiting patiently until the day when he might be able to turn it into account and to make capital out of it. Always on the alert, and just as keen an observer as Raspoutine himself of the weaknesses of human nature, with the additional advantage of being a very well educated and cultured man, he very quickly grasped the importance of what the "Prophet" confided to him when he started to relate to his friend the details of his first interview with the Empress of All the Russias.

Maniuloff was very well posted as to all the details of the Philippe incident, together with its ridiculous end. When he had heard how much Alexandra Feodorowna had been impressed by the fervor of Raspoutine's prayers, he suggested to the latter that he make use of the hypnotic faculties which he possessed in order to get the inexperienced and weak-minded Sovereign to become a tool in the hands of both. He gave him very detailed instructions as to how he was to proceed.

Armed with these instructions, Raspoutine started upon a campaign which brought Mr. Maniuloff to penal servitude, sent the Czarina to exile in Siberia, and himself to an untimely and bloody grave.

At the meetings at Anna Wyrubewa's house, during which the "Prophet" not only prayed himself for the prosperity of the House of Romanoff, he also persuaded the Empress to pray, too, in accordance with the particular rites which he declared were indispensable to a perfect communion of the human spirit with God, and which consisted in numerous genuflexions, and other things of the same kind; in long fasts and hours spent in meditation with the face on the floor, in what grew in time to be a hysterical state of ecstasy. These meetings went on undisturbed for a consider-able length of time, until one day Raspoutine informed Alexandra Feodorowna that he thought it wiser to discontinue them because certain things had been revealed to him by the Holy Ghost which had caused him to think that it would be better if he went away; otherwise he would be compelled to try and take her spiritually with him into regions whither perhaps she would not care to follow him. The Empress, of course, eagerly asked what he meant, upon which he replied that to perfect people such as he and she the Lord could grant the privilege of entering into relations with dead and gone people whom they had loved in this world; he did not know whether she would be able to go through this ordeal; therefore he thought it better to discontinue their meetings for the present.

The Czarina went home brooding upon what she had heard and with all her superstitious curiosity awakened. At first she tried not to think of what the "Prophet" had told her. Then she wondered whether she would be strong enough to face the ordeal of entering into communion with the other world, that world for which she had been longing, where had gone the one man she had loved beyond every other earthly thing. For some weeks she struggled against the temptation as it had been presented to her by Raspoutine; then at last she yielded to it, and asked Anna Wyrubewa to bring the "Prophet" once more to her house, as she wanted to speak with him again.

The adventurer demurred at first, finding one obstacle after another in order to decline the invitation which had been extended to him. At last he consented to an interview, but declared that he would insist that no one else be present at it, as the things which the "spirit" had commanded him to say to Alexandra Feodorowna were of such a secret nature that no one but herself could hear them. When he was introduced into the presence of the Sovereign he began by falling on his knees and praying with a fervor such as she had never seen him display before. At last he told the miserable, deluded woman that he had been commanded to say to her that there was one pure spirit now in another world who had been allowed to communicate with her through his medium; that he did not know who it was, but that if she wished to try the experiment she must, before attempting it, prepare herself for it, with long prayers and fastings, so as to be in a complete state of grace; otherwise the favor about to be conferred on her could not be awarded. By that time the Czarina had reached a nervous condition where anything Raspoutine told her would have been acceptable to her over-excited brain. She promised to conform herself to all the directions given to her, and three days later she met again the impostor in a place which he indicated to her, whither she went, accompanied by the faithful Anna. Madame Wyrubewa, however, was not admitted to the room where Raspoutine was waiting for the Empress. He stood before several holy images, with lamps burning before them.

The Empress had scarcely touched any food for three days; she had spent the time in long and almost continual orisons. She was just in a condition when any appeal to her superstition would be sure to meet with response. When she prostrated herself beside the "Prophet," she had reached a state of exhaustion and excitement which made her an easy prey to any imposture practised by the unscrupulous. For about an hour Raspoutine kept praying aloud, invoking the spirits of heaven in an impressive voice, every word of which went deep into the heart of Alexandra Feodorowna. Suddenly he seized her by the arm, exclaiming as he did so: "Look! look! and then believe!"

She raised her eyes, and saw distinctly on the white wall the image of Colonel Orloff, which, by a clever trick had been flashed on it by a magic lantern held for the purpose by Manassavitch-Maniuloff. The Empress gave one terrible cry and fell in a dead faint on the floor. Anna Wyrubewa, hearing her scream of agony, rushed into the room to find nothing but Raspoutine absorbed in deep prayer beside the inanimate form of his victim.

This was but the first scene of many of the same character. The Czarina recovered her scared senses with the full conviction that she had really seen the spirit of the man she had loved so dearly; she was very soon persuaded that he had been allowed to show himself to her and that he would henceforward watch over her and guide her with advice and encouragement in her future life. She quite believed that Raspoutine, whom she sincerely thought to be in total ignorance as to that episode in her life, was a real Prophet of God, and that, thanks to him, she would be able to communicate with the dead. Whether Anna Wyrubewa shared this conviction or not it is difficult to say, but it is not likely that either Raspoutine or Maniuloff confided in her. They knew too well the small reliance that, as a rule, can be placed upon feminine secrecy, and the game they were playing was far too serious for them to run the risk of compromising it by an indiscretion. It is therefore far more probable that they also played upon the superstitious feelings of the Empress's friend, and that they used both ladies for the furtherance of their own nefarious schemes with as much unscrupulousness as consummate art.

XVII

MY SON! I MUST SAVE MY SON!

AFTER the episode which I have just related, there was no longer any question of Raspoutine being allowed to leave the proximity of the Imperial Court. The Empress came to have such utter confidence in him that she even tried to induce the Czar to consult him; this he refused to do, but, seeing how much brighter his wife had become since her acquaintance with the "Prophet," he made no objection to her seeing him.

One must here remark that both Raspoutine and his chief adviser, Manassavitch-Maniuloff, played their cards wonderfully well by avoiding every appearance of mixing themselves up with politics. The "Prophet" talked with the Empress when he had the opportunity to do so, which, by the way, was not so frequent as might have been supposed. His conversations were always confined to religious subjects. He was very carefully coached by his accomplice every time he had to meet Alexandra Feodorowna, and he used to relate to her some sensational supernatural stories, which a man of his ignorance could not possibly have learned if he had not been inspired by the Almighty, as she fondly imagined. Her superstitious feelings had entirely taken the upper hand of her reason in all matters where Raspoutine was concerned, and she truly believed him to be a Prophet of God, whose every word was inspired by Heaven, whose intercession in her behalf had decided the Almighty to cure her son of a disease which all the doctors who had seen him had pronounced to be quite incurable.

In the mean while, although the relations of the Czarina with the crafty adventurer who had succeeded in captivating her confidence remained restricted to the purely religious ground, people were talking about them, trying to turn them into a vast agency where everything in the world could be bought and sold, providing the necessary money was forthcoming to do it. Manassavitch-Maniuloff, thanks to the numerous spies whose services he could command for a consideration, started to spread the rumor that Raspoutine had become all powerful in Court circles, and that if only one applied to him one could bring through the most difficult kind of business. It must be remembered that at the time I am referring to (the five years or so immediately preceding the war) Russia had been transformed into a vast stock-exchange, thanks to the mania for speculation which, since the Japanese war, had seized hold of the public. Industry always more or less neglected had suddenly taken a new and unexpected lease of life, and banks did a roaring business in selling and buying for the account of the innumerable speculators who rushed to invest their money. Nothing mattered in that respect save the quotation of yesterday and the one expected or hoped for to-morrow.

Government contracts for all kinds of things, especially contracts connected with the railway business and with factories of every sort, were eagerly sought for. In the fight which was taking place to obtain them every possible argument was employed. The art of Maniuloff and of his friends, because he was not alone in this detestable business, consisted in persuading others, even men in power who ought to have known better, that Raspoutine, through his connection with the powers who ruled at Tsarskoye Selo, could get for them such contracts that he expected in return a solid commission, which, of course, was never refused to him.

How long this kind of thing would have gone on it is difficult to say if Mr. Stolypine, who was at the time Prime Minister, had not had his attention drawn toward the activity of the "Prophet." Not knowing very well what to make of the conflicting reports which were brought to him, he expressed one day the desire to meet Raspoutine. After the interview he uttered his famous phrase:

"The only use the man could be put to was to light the furnace of the house he was living in."

The words were repeated, of course, to the person whom they concerned, and they proved the death sentence of Stolypine, because his "removal" by fair or by foul means was decided immediately after he had uttered them. Stolypine, however, in spite of his apparent disdain for the strange personality of Raspoutine, was far too clever not to realize that the constant presence of this man by the side of the Empress of Russia was likely to lead to gossip of a dangerous kind, if not to various complications. He tried at first to get rid of him by diplomatic means, and enrolled the sympathies of the Grand Duchess Elisabeth, the eldest sister of Alexandra Feodorowna, who, by reason of her having embraced a religious life, was in possession of great respect everywhere and could say what she liked to the Czar as well as to the Czarina. The Prime Minister explained to her that it was to the highest degree harmful for the reputation of the Imperial dynasty in general to see its heads give way to a superstition which only evoked ridicule on the part of reasonable people. Elisabeth Feodorowna promised that she would try what she could do, but after a while she had to acknowledge that at the first words she had spoken concerning the advisability of sending Raspoutine back to his native village of Pokrowskoye in Siberia the Empress had interrupted her so angrily that she had not been able to go on with the conversation.

Stolypine was not a man to stop at half-measures. He asked no one's law or leave, and in virtue of his powers as Prime Minister he had the "Prophet" exiled from the capital at twenty-four hours' notice.

Raspoutine wished to communicate with the Empress as soon as the order to leave St. Petersburg was signified to him, but he was prevented from doing so by his friend, Manassavitch-Maniuloff, who assured him that it would be far wiser not to murmur, and to accept the decree of banishment issued against him; because in that way he would acquire far more sympathy than would be the case if he rebelled; besides, in his absence it would be relatively easy to play upon the nervous temperament of the Empress to such an extent that after he had been recalled he would never stand again the risk of a second dismissal. This was accordingly done and Alexandra Feodorowna found herself alone, deprived of the possibility of going on with religious practices that had gradually assumed the character of those indulged in by that sect of the Khlystys to which Raspoutine belonged.

By a strange coincidence, which was nothing but a coincidence because, weak and foolish as was Anna Wyrubewa, she did not lend herself to the conspiracy which was so falsely attributed to her, which in reality did not exist, the conspiracy of drugging the little Cesarewitsch for the purpose of proving to his mother that he could not be well so long as Raspoutine was not there to pray for him—the child suddenly sickened in a more dangerous manner than ever before. The poor Empress again went out of her mind. She used to cry aloud that God was punishing her for not having known how to protect His "Prophet," and things of the same kind. At last the baby grew better, and the Court could remove to the Crimea, where it was hoped he would more rapidly recover than in the damp climate of St. Petersburg.

It was during this journey that Stolypine was murdered by secret police agents, a crime in which it was generally believed that Raspoutine, together with his accomplices, had been mixed up. The Empress, who had hated the Prime Minister ever since she had ascertained that it was he who had banished her favorite, did not disarm even in the presence of death, and it was related that she publicly prided herself upon having persuaded the Emperor not to attend the funeral of the man who had died for him, but to leave Kieff for Livadia on the eve of the day when it was to take place.

She had become very bitter just then, and she never missed any opportunity which presented itself to show her want of affection for the Imperial Family, as well as her contempt for the Russian people. The morganatic marriage of the only brother of Nicholas II., the Grand Duke Michael, which took place at about that time, procured her a new occasion to prove the unbounded influence which since the birth of her son she had acquired over the mind of the weak Emperor, and to exercise her revengeful feelings in an unexpected manner. This marriage, so much must be conceded, was of a nature to give rise to unpleasantness, and could not in any case have been viewed with favorable eyes either by the Czar or by the Imperial Family.

The lady had already been divorced twice, and the fact of her last husband having been an officer in the same regiment as the Grand Duke was also a reason why the match would have been disapproved of in any case. But, on the other hand, Michael Alexandrowitch, in uniting himself to the woman who had captivated his heart and his fancy, was acting as a man of honor, considering several facts which made it almost imperative for him not to forsake a person who had sacrificed much for his sake. It would certainly have been sufficient to oblige him to leave the army and to reside for some time abroad as a punishment, and no one imagined that worse could befall him.

The Empress had always intensely disliked her brother-in-law, who would have been Regent of the Empire in case the Czar had died before the Heir to the Throne had reached his majority, and she determined to make use of the opportunity which had arisen to vent her bad feelings on a man in whom she saw a rival to the claims of her own son. She induced Nicholas II. to deprive the Grand Duke of his fortune as well as of his civil rights, and to make out of him a ward in chancery. The scandal was immense, and it did not procure any friends for Alexandra Feodorowna.

In the mean while the Cesarewitsch sickened again, and the frantic mother implored Anna Wyrubewa to write to Raspoutine and to implore the latter to work a miracle of some kind in favor of her son. The "Prophet" replied that he would pray with all his heart for the child, but that he doubted very much whether this would avail, because the Empress had neglected her duties in regard to the Almighty and forgotten to continue the practices of mortification and of devotion she had been wrapped up in the whole time he had been near her to urge her to go on with them. Alexandra Feodorowna could not stand this last reproach, and she forthwith tarted to implore the Czar to recall the "Prophet." But Nicholas II. had been warned against him quite recently and refused to grant her request. This brought about a renewal of tears and hysterics on the part of the Czarina, and at last, one day that she was alone with Anna, she unburdened her soul to the latter, exclaiming that she knew her beloved boy was going to die and that it would be her fault, ending her confession with the agonized cry:

"My son! I must save my son!"

Madame Wyrubewa saw that the poor creature was in such an over-excited state that she might really be facing a collapse of her reason. She then proposed to the infatuated Alexandra to have recourse to a bold measure, which consisted in bringing back Raspoutine quite secretly to St. Petersburg, where he could stay at her house without any one getting to hear of it. If, then, his prayers brought about the amelioration required in the state of health of the little Alexis, the Empress would be able to tell the Czar what she had done, and perhaps to convince the latter of the efficacy of the holy man's intervention and intercession on behalf of their boy.

The Czarina caught eagerly at the idea, and after long negotiations, which very nearly failed because Raspoutine did not yield at once to the entreaties sent to him, he at last consented to return to St. Petersburg. He was secretly introduced into the room where the Heir to the Russian Throne was lying, in what every one thought were already the throes of death. He prayed for the child, he prayed for the Empress, and he urged the latter to submit to certain mysterious passes which he proceeded to perform over her head. A few days after this secret interview Alexis suddenly began to improve; not only this, but he became stronger and brighter than he had been for a long time.

Alexandra Feodorowna was radiant, and one day when Nicholas

II. was rejoicing at the happy change which had taken place in the condition of their son she informed him of what she had done and begged from him permission to bring Raspoutine to him and to allow him to remain in the vicinity of the Court in the future. Nicholas II. was convinced and granted the necessary authorization. After this the question of Raspoutine's return to Siberia was not raised again, and he never left, except for short vacations, the Sovereigns who had at last been persuaded to give to him their complete confidence.

He refused, however, to take up his abode in Tsarskoye Selo, and showed himself very discreet in his demeanor. He was admirably advised, and he prepared himself in silence for the part it was intended for him to play in the future. But at stated intervals, and upon stated days, he used to see the Empress, either in her own rooms or, most frequently, at the house of Anna Wyrubewa, when he evoked for her the spirit of Colonel Orloff and transmitted messages which he was supposed to have received.

Alexandra Feodorowna believed him, and this new understanding, which she firmly thought had, thanks to the prayers of the "Prophet," established itself between her and the man who had possessed her heart, proved to her the greatest consolation she had known. It induced her to come out of her retirement and to begin to take part in the management of public affairs, which she insisted upon the Czar communicating to her. The time was coming when it would become known in Russia that if the Sovereign was a weak man his Consort was trying to show herself a strong woman, and comparisons between Alexandra Feodorowna and Catherine the Great began to be heard in the yet small circle which affected to admire the new qualities it prided itself upon having discovered in the young Empress.

· · · · · · · ·

THE PALACE OF TSARSKOYE SELO

XVIII

ANOTHER WAR

THE years which followed upon Raspoutine's triumphant return to Tsarskoye Selo were most eventful ones for Russia as well as for the Imperial Family. Europe, too, went through political convulsions which were the preliminary of the disaster that was to sweep over it in 1914, but in which very few people in 1912 were able to discern danger. I am referring to the annexation by Austria of Bosnia and Herzegovina and to the two Balkan wars. When Servia was threatened by Bulgarian ambition there existed a powerful party in Russia which would have liked the Czar to interfere on her behalf, and to lend her his aid against King Ferdinand, on one side, and the Austrian spirit of conquest, on the other. Popular feeling was very much in favor of a Russian demonstration, and for some weeks St. Petersburg was the scene of a violent agitation which, in the opinion of many people, was destined to end in a war with the Austro-Hungarian monarchy. It was not a secret that the Servian Government would not have objected, had such a contingency presented itself, and during the whole of the summer and autumn of 1913 different Servian politicians came to Russia to discuss the situation. In Moscow, as well as in St. Petersburg, they applied themselves to the task of awakening in favor of their country the sympathies of all the Russian Slavophils. At one time it seemed as if they were going to succeed and as if the Czar would be compelled to yield to the general wishes of his subjects.

Here Raspoutine interfered, and, thanks to his influence over the Empress, he contrived to prevent the spread of a conflagration which threatened to extend itself far beyond the Balkan Peninsula. It must not be assumed, however, that in doing so he was actuated by any patriotic motives. He was a man for whom the word "patriotism" had absolutely no meaning. But his friends, as well as himself, were plunged head foremost in financial schemes which a war would in all probability have wrecked, and therefore he applied himself with all his energy to set hindrances in the path of the chauvinists who tried to induce the Emperor to assert the might of his Empire, to rush to the rescue of those Slav nationalities that had refused to conform themselves to the anti-Russian policy which Bulgaria had been pursuing ever since King Ferdinand had been put in control of her destinies.

This interference on the part of the "Prophet" in matters which did not concern him in the least became known very quickly, not only in Russia, but also abroad, and one of the most active members of the German Embassy in St. Petersburg, who was *persona grata* in the Wilhelmstrasse, wrote a whole report on the subject, raising at the same time the question as to whether it would not be worth while to try, with the help of substantial arguments, to win Raspoutine over to the idea of a *rapprochement* between Russia and Germany. The latter was steadily making preparations for the war which she was quite determined to provoke within a very few months. She had always worked toward the destruction of the Franco-Russian understanding, which stood in her way, which she feared might come to endanger her dreams of a world-wide Empire. Every effort had been made on the part of the Berlin Court to win over the Czar to the idea of renewing the intimate bonds which, during the whole time of his grandfather's reign, had united the Hohenzollerns and the Romanoffs. When Nicholas II. had repaired to Berlin for the marriage of the Kaiser's only daughter with the son of the Duke of Cumberland he had been made the object of one of the warmest welcomes he had ever received in his life, a welcome which had touched him so much that he had come back to Tsarskoye Selo full of enthusiasm for his Prussian relatives. If the truth need be told, he was also slightly disillusioned as to the advantages which his country might obtain through its alliance with the French Republic. This feeling of distrust which had thus been sown in his mind in regard to the good intentions of his Latin ally was of course at once reported to the Kaiser by the many friends which the latter had in St. Petersburg, and it made him doubly anxious to win over to his side the good-will as well as the sympathies of Nicholas II. At the same time William was very well aware that it was most difficult to rely on anything promised by a man with such a weak character, or rather with such a lack of character, as his Russian cousin. An ally who would continually whisper in the latter's ear all the advantages which a friendly treaty and understanding with Germany could bring to him, as well as to the whole Russian Empire, was indispensable; of course, when it was suggested to those who controlled the actions and the politics of the Wilhelmstrasse that he might be found in the person of the Empress Alexandra's favorite, the Kaiser came very quickly to the conclusion it would be worth while to obtain the good offices of this remarkable man.

This, however, would have proved difficult, even for the experienced spies which Prussia maintained in all circles of Russian society, as it was not easy to discover means of getting into contact with the formidable adventurer whose name had already become one of the most powerful to conjure with in the vast Russian Empire. At this juncture Mr. Manassavitch-Maniuloff interfered and volunteered his services to William II. The crafty fox had heard that the Czar's confidence in France was slightly shaken. Maniuloff at once bethought himself of the possibility of turning his knowledge to his personal advantage, and he managed, no one knows how, to impart to the German Ambassador in St. Petersburg, Count Pourtalès, his willingness to persuade Nicholas II., through Raspoutine, that he would do well to throw France overboard and to conclude a treaty with the Prussian Government, which eventually might prove of immense advantage to himself by assuring him of German protection in the not improbable case of a new Revolution taking place in his Empire.

This sort of thing went on for some time, and it is quite likely that if events had not precipitated themselves one upon the other with the most startling rapidity, the policy of Raspoutine and his friend might have borne fruit in some way or other, and the relations between the Cabinet of St. Petersburg and that of Paris, which had already sensibly cooled down, would have become even fresher than was already the case. In fact, the announced visit of President Poincaré had not appealed to the Czar, who, while unable to decline it, yet had expressed himself quite loudly as to the small amount of pleasure which he expected to get out of it. Of course Berlin heard about the remarks that had escaped the lips of the Russian Sovereign, and it was not slow to draw its own conclusions from them. In fact, if we are to believe all that was related at the time by persons well up as to what went on in European politics, it was confidently expected by the Kaiser that instead of drawing France and Russia closer together the journey of the French President, thanks to personal frictions he felt sure would arise, would, on the contrary, irritate Nicholas II. and make him look with more favorable eyes than he had done before on the possibility of a change in the conduct of

Russian Foreign Affairs.

Whether this would have taken place or not it is difficult to say, because at the last moment Germany lost her most devoted ally, and the influence of the man who had, more than any one else, worked in its interests was eliminated for the time being. A woman, who had just reasons for feeling revengeful against Raspoutine, stabbed him as he was coming out of church in his native village of Pokrowskoye in Siberia, whither he had gone on a short visit. He was ill for a long time, and during the weeks that he was laid up, to the intense consternation of the Empress, who was only with great difficulty prevented from going herself to nurse him, the Austrian ultimatum consequent on the assassination of the Heir to Francis Joseph's Throne was presented to Servia, and followed by the declaration of war launched by Germany almost simultaneously against Russia and France.

This proved for Alexandra Feodorowna the most terrible blow that had yet befallen her since the day when she had plighted her troth to the mighty Czar of All the Russias. During the eventful hours that preceded the initial act of the tragedy which was to change the face of the whole world she went about like a demented woman, crying and praying in turns, and imploring her husband to pause before he allowed the accomplishment of a calamity which she vaguely guessed would claim her for one of its first victims. But this time there was no Raspoutine at her side to play on the feelings of humanity of the weak-minded Nicholas, to persuade him that he ought rather to submit to the humiliation of Russian prestige than to allow another war to throw its shadow on his already too unfortunate reign. On the contrary, all the advisers of the Emperor, all his Ministers, public opinion, the press, and the army, eager to wipe out the remembrance of the Japanese disaster, poured into his ears their conviction that if he did not rush to the help of poor threatened Servia he would not only lose the last fragments of popularity which were left to him, but also put Russia before the whole world in a most shameful and dishonorable position.

As usual, the Czar yielded, with the results which we know and have seen. He could hardly have done anything else, if we take into consideration that Germany was absolutely determined to start the abominable war, from which she hoped to obtain the realization of her schemes of domination of the whole earth. But—and this must be told here—the Kaiser in letters far more authentic than the famous

Willy and Nicky correspondence, which personally I consider as subject to much doubt, in view of certain improbabilities which it contains, the Kaiser did propose at that time to his cousin to conclude with him a defensive and offensive alliance against France and England. In return for which he engaged himself to uphold any designs which Russia might nurse in regard to the Balkans and the Straits.

It may not be to the advantage of his intellectual faculties that Nicholas failed to see the vast political scheme which lay behind this offer; it is certainly to the honor of his moral character that he refused it, and this in spite of the supplications of his wife, who entreated him not to plunge their country into a war which, as she repeated, could only prove disastrous for its future, as well as for that of the dynasty. In spite of his natural defects, of his cruelty, harshness of heart, and utter disregard of the rights of others, the Czar was still a gentleman and he could not be induced to do anything capable of dishonoring him as a gentleman, though he may have lent himself to actions degrading for a Sovereign. During the terribly responsible days which preceded the declaration of war he behaved quite irreproachably. It was later on that he was influenced by Raspoutine and by the Empress to lend himself to political schemes unworthy of him, as well as of the nation over which he ruled.

On the 1st of August, 1914, twelve hours after Germany had thrown her gauntlet into his face, he showed himself for the last time to his people on the balcony of the Winter Palace. An immense crowd had gathered together in the big square which it faces, and for the last time, too, cheered him vociferously, forgetting in this solemn moment all the follies, mistakes, and errors which had saddened his reign and raised a barrier between him and this great Russia that his father had made so prosperous and so mighty. If in that supreme moment he had been able to find words capable of electrifying this crowd into believing in him again, who knows but that the reverses which were to crowd upon him could not have been avoided, or at least diminished! But Nicholas II. never knew how to speak to his subjects or how to touch their hearts. He remained impassible and indifferent in the most critical hours of his life and of theirs, and this incapacity to rise to the height of the situation of the moment was perhaps one of the things which contributed the most to his fall.

I remember him so well on that August afternoon, facing the multitude assembled to greet him as its Czar and leader, and I remember, too, the thought which swept through my mind, that it was a thousand pities it was not his father who stood there in his place. Alexander III. would have known how to address Russia in an hour of national danger. He was neither a brilliant nor an extremely intelligent man, but he was a man and a Sovereign, who realized the duties of a Monarch and of a man. He was, moreover, a Russian who thought and who felt as a Russian alone could think and feel, in questions where the honor and the future of the country were involved. Nicholas II. was simply an Emperor who wished to be an autocrat. It was too much and not enough at the same time, and many among those who looked upon him, as he appeared before his people on that historical balcony whence it was the custom to announce to the population of the capital the death of a Sovereign whenever it took place, many wondered whether they were not going to hear that another one had started on the long journey whence there is no return. His presence seemed to herald a funeral rather than the hope of a triumph, and this impression which he produced was so vivid that more than one acknowledged having experienced it when talking about this famous day which, though we knew it not, proved to be the last upon which a Russian Czar faced the Russian people before the latter overthrew the chief of the House of Romanoff from the Throne which he had disgraced.

XIX

MY FATHERLAND, MUST I FORSAKE THEE?

IT would not have been human on the part of the Empress Alexandra if she had not felt deeply aggrieved at the war which had so unexpectedly broken out between the country of her birth and that of her adoption. She had never really become a Russian at heart and her sympathies had remained exclusively German all through her married life. Apart from this, she had experienced from the intercourse which she had kept up with her own family the only pleasure which she had frankly enjoyed since the Crown of the Russian Czarinas had been put upon her head. She dearly loved her two sisters, the Princess Victoria of Battenberg and the Princess Irene of Prussia, far more, indeed, than she did her other one, the Grand Duchess Elisabeth, whom she considered more or less as a rival and whom in the secret of her heart she could not forgive for having won in Russia a popularity which had always been denied to her own self.

Then there was her brother, the Grand Duke of Hesse, with whom she had remained in correspondence, who paid her frequent visits in Tsarskoye Selo; there was also her cousin, the Kaiser, who had been the first person to point out to her the responsibilities which were inseparable from the exalted position she occupied as Empress of All the Russias, who had applied himself to persuade her that she had great political talents, and that she could undoubtedly, if she only wished it, become a most important factor in European politics. Strange to say, though she had been brought up partly in England, though her mother had been an English Princess, though she was the grandchild of Queen Victoria, she intensely disliked everything that was English, and had for English customs, English ambitions, and English politics the same hatred which characterized William II. Perhaps this common aversion was one of the reasons why they had always got on so well together, and why they had been able to be of so much use to each other. At all events, the fact that it existed in an equal degree in both of them had drawn them together, and at last, after she had contrived to eliminate the influence of her anti-German mother-in-law, Alexandra Feodorowna had been able to give herself up body and soul to the task of drawing together her husband and her own kindred. She had tried to persuade the former that the only means to insure the prosperity and the welfare of the Russian Empire in the future consisted in a closer union with Germany, with whom there existed absolutely no reason to quarrel, because there were no interests capable of clashing between the two people. She had represented to the weak-minded Nicholas that Russia had obtained from France all that she could hope to get, and that the latter had become weary of always being called upon to invest money in Russian bonds without any return being made for her generosity.

Nicholas II. had always detested republics, and though he had been made much of during his visits to Paris, which he had thoroughly enjoyed, he yet had never felt quite at home amid the Republican society he had been called upon to get acquainted with; in the secret of his heart he despised all French political men, whom he considered as much inferior to himself. But a natural inclination to dissimulation, which he carried so far that many people called it by quite another name, had made him carefully conceal the real state of his feelings in regard to his French ally. It is, however, quite certain that if the war had not broken out the Franco-Russian alliance would have died a natural death. As things occurred, it was for a short space of time to appear more complete than ever; this was not the merit of Nicholas, but the result of the honesty which the French Government brought to bear in all that happened in 1914. In Russian Court circles, which were all of them, more or less, given up to Germany, the news that the country was going to war was received with consternation, and there were many people who declared that it was a shame for Russia to be drawn into a struggle which was essentially a personal quarrel between France and Germany, with which she had nothing to do.

At first and before the anti-German feeling became fierce in St. Petersburg, the Empress, in spite of political complications, remained in private correspondence with her brother, and through him with the Kaiser, to whom she promised that she would spare no efforts to induce the Czar to conclude peace as soon as it became practicable. She had never been able to form an idea of the power which public opinion, especially in times of national danger, can exercise over a nation. She imagined that the authority wielded by the Crown would be sufficient to put an end to any manifestations of sympathy in regard to France on the part of the Russian people. She therefore felt confident that the struggle which had just begun would not last long, and that Russia could come out of it, if not with flying colors, at least without any serious losses.

No one during those early days of the war admitted for one moment the possibility that Warsaw and the line of fortresses which defended the Russian frontier on the side of the Niemen could fall into the hands of the enemy; all that the Empress expected was a defeat of the Russian armies which would not seriously compromise their prestige, but at the same time convince the country that an advantageous peace was, after all, the best way of getting out of a situation where all the time one adversary had either willingly or unwillingly misunderstood the good intentions of the other.

She was consequently working along this line when Raspoutine returned to Tsarskoye Selo. He did this as soon as the doctors had pronounced him fit to travel. She began once more to pray with him and to ask him to put her again into communication with that other world where she imagined that Colonel Orloff was waiting to advise her as to what she ought to do in regard to the war and to the necessity of putting an end to it as soon as possible. But while she believed that none outside the few people she had admitted into her confidence—one of whom was Anna Wyrubewa, and another Sturmer, who was later on to play such an important part in the tragedy of her fall—could guess what she was about, Sazonoff began to suspect that it was due to her influence that the Emperor was no longer so amenable to the advice which he ventured to offer. It was partly to put an obstacle in the way of any independent act of the Sovereign that might have been interpreted as not quite loyal in regard to Russia's Allies, that he had suggested the drawing up of the document known by the name of the Treaty of London, in which the Allied Powers engaged themselves not to conclude any individual or separate peace with Germany. He thought, and others did the same, that this would prove the best means to hold together the Entente without exposing it to mutual suspicion. He concluded this pact of his own authority, only acquainting the Czar with what he had done after it had become an accomplished fact.

Nicholas understood for once the significance of his Minister's bold action, but he could not disavow it; therefore he had to make the best of it. But he refrained from telling the Empress of this new complication which would surely interfere with her hopes of a prompt peace, and it was through a letter from her brother that she heard at last what had taken place in London. Her wrath was intense, the more so that her German relatives blamed her for a thing she had known nothing about and for which they tried to make her responsible. Alexandra Feodorowna had never under-stood what self-control meant, and she gave public vent to her indignation, accusing Sazonoff of having betrayed his Imperial Master's confidence, and vowing that he would be made to repent for this piece of audacity.

The Empress was still smarting under the sense of her personal defeat in a struggle against the people who were trying to control Russian politics and to lead them in a road she strongly objected taking, when the news of the defeat of the Russian army at Tannenberg came like a thunderbolt out of the blue, to stir up all the patriotic feelings of the Russian nation and to put an end to any idea of peace which may have existed in some timorous minds. The Empress had perforce to appear to share the general indignation against the ruthless conduct of Germany, and she had to acknowledge her momentary helplessness to speak what she considered to be the language of reason, and to try to persuade her subjects that it would be to their advantage to abandon their Allies to their fate, and to apply themselves to withdraw their own pawns out of the game.

In these days of suspense Raspoutine turned out to be the greatest comfort in the world to her. For one thing, he made it possible for her to begin again seeking in Berlin inspirations as to the course of conduct she ought to pursue. Thanks to him, Mr. Manassavitch-Maniuloff was persuaded to undertake a journey abroad, during which he was to see the leading political men in Europe and to ascertain their views on the subject of the conduct of the war in general, as well as of its chances of success. Ostensibly it was a newspaper on which he was assistant editor, the *Nowoie Wrémia*, that sent him on this perilous mission. In reality, he started as the agent of the Empress, and he saw several German officials in Stockholm, as well as in Copenhagen, where he spent a few days. He proceeded to London and to Paris, only to lend coloring to what otherwise would have been an impossible trip. When he returned to Russia he brought along with him a whole program drawn out by the Kaiser, which Alexandra Feodorowna proceeded at once to execute.

But here again she found obstacles in her path, the principal of which was the stubbornness of the Grand Duke Nicholas, who, in spite of the fact that he had to acknowledge that Russia had neither guns nor ammunition in sufficient quantity to be able to hold her own against the hordes of William II., yet refused to consider his country as beaten. The Grand Duke was popular in the army. The fact that it began to be known that he represented at Court the Russian party, in opposition to the hated Empress, who was supposed to head the German one, gave him considerable prestige. When the Czar had consulted him as to what ought to be done, he had replied:

"Do anything you like except conclude peace, because if you do I shall be the first one to lead the army against you, and to compel you to go on with the struggle."

Nicholas had repeated to the Czarina the threat of his cousin, and this had been sufficient to incense the latter, even more than she had been before, against a man whom she considered, perhaps not quite without reason, as her most formidable enemy.

Nevertheless, she tried to persuade him to change his mind, and made an appeal to his feelings of humanity, asking him whether it was right to go on with a war in which hundreds of thousands of Russian soldiers had already fallen, which would probably entail more sacrifices in the future than the country could afford. She spoke eloquently, but the Grand Duke remained unmoved, and at last Alexandra Feodorowna, worn out by the supreme effort which she had made, gave way to her uncontrollable grief, exclaiming in her deep distress:

"My country, my poor country, must I forsake thee?"

Nicholas Nicholaievitch turned round and said, with a withering contempt:

"To what country do you allude, Madam—to Russia or to Germany?"

The Empress jumped up, her eyes blazing with rage. She rang the bell, and told the lady in waiting who came in response to her call:

"Show the Grand Duke out. He must never be allowed to enter this room any more."

And Nicholas Nicholaievitch never did so again.

XX

IT IS YOUR HUSBAND WHO IS LOSING THE THRONE OF YOUR SON

THIS interview with the Grand Duke, Commander-in-chief of the armies in the field, could not fail to produce a deep impression on the troubled mind of the Empress. Her proud and unforgiving character had been goaded to the extreme by the irony with which her husband's cousin had received the overtures which she had made to him, and she could not bring herself to forgive him for the calm disdain with which he had asked her whether she considered Russia or Germany as her Fatherland.

Of course she flew to Anna Wyrubewa to seek consolation, but when the latter advised her to ask Raspoutine to pray for her in this crisis of her life, Alexandra Feodorowna for once did not accept this suggestion, saying that a man absorbed in religious practices like the "Prophet" could not be expected to take a sane view of a position which was getting so intricate that it would require a statesman of unusual ability to unravel it. But she expressed herself willing to talk to Mr. Sturmer about it, and to ask him what he thought of the Grand Duke's insolence, as she termed it, and what he would suggest as to the means of putting it down.

It is time here to say a word concerning Mr. Sturmer, who was so soon to play a prominent part in the drama of the Romanoffs' fate. He was a man of moderate intelligence, great ambition, and above everything else an opportunist—a perfect type of the class called in Russian Tchinownikis, who always and in everything it does approves the government of the day. He had for years paraded ultra-conservative opinions, and while he had performed the functions of Master of the Ceremonies at the Imperial Court, he had professed great sympathies for England and for everything British, playing the European, while at heart he was the personification of the Tartar hidden under the Russian flag. He was, moreover, an excellent talker and a well-read, well-educated man. His German origin had imbued him, as was to be expected, with considerable admiration for the Kultur, such as it was understood at the time I am referring to. The late

Czar Alexander III. had always abominated him and shown him that such was the case in an unmistakable manner. But Mr. Sturmer had the happy knack never to notice what it was inconvenient for him to be caught looking at; he stuck to his post until he contrived to get another appointment, that of President of the zemstwo of the province of Twer, where he possessed a large estate. This position, however, he had to abandon soon, because his colleagues happened all of them to be very ardent liberals who refused to accept his monarchical views.

Sturmer retired to private life, but at the time of the accession to the Throne of Nicholas II. he came to St. Petersburg, and managed to convey to the new Czar a detailed report as to the wave of liberalism that, to use his words, "infected" the province of Twer. If we are to believe a rumor which was persistently circulated in the capital, this had a good deal to do with the famous speech in which the Emperor told the deputies of the zemstwos (come to congratulate him on his marriage) that they need not in the future indulge in "senseless dreams," as it was his firm intention to uphold intact the principles of autocracy.

Sturmer was clever enough to conceal his extreme delight at the Sovereign's attitude, and he went on with his attempt to worm himself into the latter's favor. Very soon afterward he re-entered public life, was appointed Governor of that same province of Twer where he had met with such unsuccess, and proceeded steadily to work out for himself the reputation of being a first-rate statesman. He was shrewd enough to see what others had failed to perceive, and this was that, with the weak character of Nicholas II., it would require very little trouble on the part of the Empress to obtain complete mastery over his mind. He therefore applied himself to persuade the latter that it was her duty to make the attempt. He had always been a fanatical orthodox, perhaps because he had not been born one, and he was in great favor with several high Church dignitaries, including the new confessor of the Imperial Family, Father Schabelsky, whom the Czarina liked very much, and in whom she had great confidence. This made it relatively easy for him to carry to the ears of Alexandra Feodorowna his opinions on the current events of the day, and he did not fail to do so during the troubled times of the Revolution of 1905, and of the repression which followed upon it, in which he took an active part. He occupied then a post in the Ministry. However, he had to give up this upon his appointment as a member of the Council of State, which promotion had covered an attempt on the part of his colleagues to get rid of him. He took an important share in the deliberations of this Assembly, and very soon was recognized as one of the leaders of the ultra-conservative party there, and as a strong supporter of an alliance with Germany.

This attitude alone would have been sufficient to win for him the good-will of the Czarina, and when the war broke out she often talked with him over the sad consequences it was sure to bring; she discussed with this faithful friend the possibility of putting an end to it, in a sense favorable to Russian interests, not likely to harm Russian prestige abroad nor the dynasty at home.

Sturmer had been introduced to Raspoutine by the good offices of Manassavitch-Maniuloff, whose services he had had the opportunity to appreciate when they were both in the employ of the Government, and he soon played a prominent part in all the designs of these two sinister personages. It has even been related that it was due to his special suggestion that the comedy of the Empress being put into direct communication with the spirit of Colonel Orloff had been engineered; of this there exists so far no proof, and we must therefore accept the tale under the reserve that, according to the French proverb, it is only the rich to whom one lends money.

When Sturmer heard about the conversation which had terminated with such violence between Alexandra Feodorowna and the Grand Duke Nicholas he saw at once the capital that could be made out of the incident. He also disliked the Grand Duke; it was therefore easy for him to enter with alacrity and zeal into the plans of revenge that were being harbored by the Czarina, to whom he reported that Nicholas Nicholaievitch was trying to supplant the Czar, to get himself appointed Dictator of the Empire; that he had, moreover, the most sinister designs against the little Cesarewitsch, as well as against her, who, as he had openly declared, ought to be locked up in a convent. He pointed out further to the distracted Empress that the weakness of character of her husband might easily make him a prey to the ambitions of his cousin and cause him to lend himself to the latter's schemes. Besides this, it was against all the traditions of autocracy for a member of the Imperial Family to aspire to make for himself an independent position outside the Czar, and if the Grand Duke was allowed to work out the consolidation of his popularity among the army and the military party a Palace revolution could easily follow, which would overthrow Nicholas II. and dethrone him in favor of some other Romanoff, willing to become an easy tool in the hands of the Grand Duke Commander-in-chief.

After this it became the one object of Alexandra Feodorowna's ambition to deprive her cousin of his command, to have him exiled somewhere far from St. Petersburg, which by this time had been renamed Petrograd.

This, however, was a difficult piece of work to perform, precisely on account of the weakness of temperament of Nicholas II. and of the awe with which any violent decision to be taken in regard to any one whom he knew to be stronger than himself inspired him. Religious superstition was therefore brought to bear upon him; he was told by his wife, by a few people who were devoted to her, and last but not least by Raspoutine, that it was part of the duties of a Russian Czar to lead his nation in times of peril; that the enthusiasm which his presence at the head of the army would be sure to provoke would prove a great element in the achievement of a complete victory against a formidable foe, the strength of which had never been properly appreciated. At first Nicholas grew impatient and would not listen. At heart he had the vague consciousness of his own incapacity to command a big army in the field; he feared to take such a perilous responsibility upon his shoulders. He also knew that it was not the fault of the Grand Duke that he had been compelled to retreat before the invading German forces, but of the men who had failed to supply him with the necessary ammunition, artillery, and provisions. The Emperor did not care to make out of his cousin the scapegoat for all the sins of Israel. On the other hand, he dreaded the ascendency which Nicholas Nicholaievitch was undoubtedly acquiring in public opinion, and he did not care for any member of his family to become popular at his own expense. Still, he would not come to a decision. Even when the Grand Duke Commander-in-chief had objected to the presence of the Empress at headquarters, which she had wished to visit, he had refrained from insisting on the point. He had, on the contrary, applied himself to soothe his wife's ruffled feelings.

This hesitation on the part of the Sovereign did not please at all the small group of men who had entered into the schemes of the Empress. They knew very well that so long as Nicholas Nicholaievitch remained in power it would be impossible to bring to the front the question of a separate peace with Germany for which they were steadily working. It was therefore determined to force the Empress to extort from her husband the decision they wished for; consequently Raspoutine asked her to attend a prayer-meeting he wanted to hold, during which he said that it had been revealed to him that she would come to learn many things hitherto kept from her knowledge, but which it was time she should hear. What occurred at this meeting no one ever could ascertain exactly. It seems pretty certain that Raspoutine evoked the spirit of Colonel Orloff, and that the customary game of making a pencil write by itself was resorted to, with the result that Alexandra Feodorowna returned to the Palace fully convinced that, in resisting her demand for the removal of the Grand Duke Nicholas from his position as Commander-in-chief of the army, the Czar was endangering not only his own life, but also the Throne, and the chances of succession to it of his only son.

The Empress implored her husband to listen to her, telling him that if he really felt alarmed about taking any violent measures against the Grand Duke, he ought at least to dismiss the latter's head of the staff, General Januchevitch, to whose blunders all the disasters that had overpowered the Russian armies were due. She representted to her bewildered spouse that public opinion claimed some one should be punished for all the unsuccesses which had attended the war, and that it would be satisfied to a small degree if the General were removed from his command.

This was a compromise which Nicholas II. seized hold of with alacrity. It had been proposed to him because it was known very well that the Grand Duke would not consent to be parted from his faithful adviser with whom he had shared all the anxieties of the disastrous campaign that had been carried on amid such terrible difficulties, that he would rather resign his own command than give him up. The surmise proved quite correct. When Nicholas Nicholaievitch was informed of the change that had been made in the direction of the staff, without his having been consulted, he telegraphed to the Emperor, asking him to be also relieved as soon as possible from the duties of his responsible position. The Empress, Sturmer, and Raspoutine were jubilant. It was easy to persuade the Czar that his cousin, in thus resisting his orders, had rendered himself guilty of insubordination. It was decided not to accept his resignation, but simply to dismiss him and to appoint him at the same time Viceroy in the Caucasus, a position that had just been rendered vacant by the departure of Count Worontzoff-Daschkoff for reasons of health. This they thought would be a courteous way of getting rid once for all of a personality so strong and so encumbering at the same time as that of the Grand Duke, and of doing it in a manner to which no one could raise any objections.

The Emperor said yes to everything. He had been thoroughly frightened, and was no longer in a condition of mind capable of judging impartially of the events taking place around him. A solemn religious service was celebrated in the private chapel of the Imperial Palace of Tsarskoye Selo, to implore the protection of Heaven on the new Commander-in-chief of the Russian troops, after which Nicholas II. started for the headquarters of the army. He was received with great pomp and ceremony by the Grand Duke, and at once assumed the supreme command over demoralized regiments who were full of regret at the departure of their former leader.

Nicholas Nicholaievitch behaved with immense dignity. In this crisis of his life he only remembered that he was a Romanoff, and he showed an absolute submission to the decisions of the head of his dynasty. In words of incomparable nobility he issued an army order in which he thanked his soldiers for their good services, and expressed the hope that the presence of their Sovereign at their head would inspire them with a new energy in the struggle that lay before them. Then he left for his new post, accompanied to the railway station by the Czar himself, from whom he parted solemnly and respectfully, and whom he was never to see again, at least not as Emperor of All the Russias.

XXI

PEACE, WE MUST HAVE PEACE

THE removal of the Grand Duke Nicholas from the position of Commander-in-chief of the army did not meet with the general satisfaction that his enemies had hoped it would provoke. The sane elements of the nation understood quite well that, whatever mistakes he had been guilty of, they had proceeded more from the many difficulties which he had found in his way than from his own incapacity. No one liked the thought of his place having been taken by the Czar himself, who had long ago lost his personal prestige, whom no political party in the country trusted. The influence of the Empress was also dreaded, and the fact of her German leanings was openly discussed. The demand for a responsible Cabinet, from whom explanations could be demanded by the nation, was already to be heard everywhere. The Duma, when it had met, had been the scene of furious discussions during which the conduct of the Government had been severely censured. Russia was beginning to get tired of the tyrannous hand which was weighing it down and crushing every attempt at independence on the part of those who were in possession of her confidence.

The Ministry was neither respected nor considered, the Sovereign was despised, and his wife was hated. Dissatisfaction was spreading even in the spheres which out of old traditions and principles had kept it within bounds. The aristocracy had become weary of finding all its good intentions disdained or misconstrued; in all classes of society people were cursing the hidden "dark powers," as they were called, that disposed of the fate of the nation and that ruled the feeble and weak-minded Monarch who had been con-verted into a figurehead for whom no one cared except the un-scrupulous people who were abusing his credulity and who had contrived to get hold of his confidence.

The Czarina was openly accused of working hand in hand with her cousin, the Kaiser, and of assisting him in his dreams of a world-wide Empire into whose power the Russian one was to be delivered. And when the old, feeble, opinionated, but at any rate honest,

Gorémykine had been replaced as Prime Minister by the hated Sturmer, who by this time had risen to the position of leader of the ultra-conservative and reactionary party in the Council of State, the general indignation against the weakness of Nicholas II. could no longer be repressed, and the possibility of a Palace revolution came to be spoken of as the next thing likely to happen.

In the mean while Raspoutine and his friends were daily becoming more powerful. The "Prophet" had by that time completely mastered the details of the intrigue into which he had been drawn by the clever people of whom he had been the tool. These had been at first Count Witte, who in his hatred of the men who had driven him out of power had willingly lent himself to the conspiracy which transformed the Empress into one of the most active agents the Kaiser had ever had at his disposal in Russia. When this much-discussed statesman died at the very moment he might have been called again to play a part in the history of his country, his place had been taken by Sturmer, Manassavitch-Maniuloff, and other adventurers of the same kind, all eager to enrich themselves at the expense of their own Fatherland, all of them men who only looked for their personal financial advantage, who remained perfectly indifferent to the disasters which one after the other were crowding upon unfortunate Russia. Germany was clever enough to see through the game played by these sharks and she did not hesitate an instant in buying their services for all that they were worth.

Raspoutine had very accurately taken stock of the mental caliber of the half-demented Czarina, and while carefully avoiding discussing or even touching upon the subject of politics with her, he had contrived to persuade her to trust those so-called statesmen of whom he was but the instrument. As time went on she became more and more anxious to communicate with these spirits of the other world, in whose existence she had been led to believe as firmly as in that of the Divinity itself. Raspoutine, whenever he prayed in her presence, pretended to get into trances during which he told her things which he assured her he did not remember later on, but which he persuaded her he had been inspired by the celestial powers to tell. She was kept by him and by Anna Wyrubewa in a state of semi-hypnotism, which went so far that sometimes she was herself seized with attacks of convulsions bordering on epilepsy, during the long prayers in which she used to spend half of her days and most of her nights. The superstitious fears which had always haunted her were played upon by these clever adventurers whom she had admitted into the secret of her thoughts. She was finally convinced that her duty as a Russian Empress required of her to sacrifice herself for the welfare of her subjects, and to induce her husband to sign a peace that would put an end to the useless and terrible slaughter that had transformed the whole of Russia into one vast churchyard.

She still labored under the illusion that the dynasty was popular and that every decision of the Czar would be received with respect and gratitude by the nation. Though she knew that she was personally disliked, she did not imagine that this dislike extended itself to the Emperor, and she never supposed that, even in regard to her own person, the hatred of which she was the object existed anywhere else than among the aristocratic circles of Petrograd society. In one word, she believed in the power of autocracy, and she worked as hard as she could to consolidate it by getting Nicholas II. to appoint as his Ministers and advisers men who shared her opinions on this point, and who were ready to crush with the greatest vigor and the utmost severity every attempt to shake the prestige and the authority of the Crown.

Of course, the fact that the country was at war made her path most difficult; for this very reason she thought it was indispensable for the safety of the dynasty and of her son that peace should be concluded. She did not care in the least for the secret treaties or obligations Russia had assumed. To her, honor was but a question of opportunism. She set the existence of the Romanoffs before their self-respect. Her German blood made her lose sight of the real interests of her husband and of her children.

Here we must pause a moment and touch upon a point that has been as much discussed as it has remained mysterious to this day. Was Raspoutine a German agent directly employed by the Kaiser to persuade the half-demented Czarina that it was her duty to put an end to the war? Or was he simply the instrument of other people more in possession of the secret of Germany's schemes than himself? Personally I am inclined to believe this second version of his activity. Raspoutine was far too ignorant and uncouth to have been taken into the confidence of William II., but Mr. Sturmer, Mr. Manassavitch-Maniuloff, and Mr. Protopopoff undoubtedly were confidants of the Kaiser. They had been promised, most likely, large sums of money for their co-operation in this vile intrigue, which even after their fall was to be renewed and, as we have unfortunately seen, renewed with success.

I shall not repeat here the story of Mr. Protopopoff's famous journey to Sweden, where he got into direct touch with agents of the German Government. I shall not even return to the subject of the negotiations begun by him and continued by Mr. Sturmer. All this is now a matter of history, and what I am writing here only concerns the personal part played by the Empress in this dark plot, directed against all the Allies of Russia in the war as well as against Russia herself. I am only concerned with Alexandra Feodorowna and her share in the catastrophe which was to send her a captive and an exile to that distant Siberia whither so many innocent people had been banished by her husband.

I wish to explain how it could have become possible for her to be transformed into an active agent of German ambition on the Russian Throne. She was, as we have seen, only half-responsible for her actions. Her intelligence had never been properly balanced and self-control had never been taught her. She had, however, principles, and very strong ones, too, which had stood between her and temptation in the serious sentimental crisis of her life. But this resistance to what perhaps had been the one passion she had known, except her love for her son, had helped to overthrow her mental balance. She had given to God, represented by a Divinity of her own created by her imagination, all the affection she had not been allowed to expend on earth, and full of a spirit of self-sacrifice as stupid as it was devoid of any ground to stand upon. She had fancied that she could work out her personal salvation, together with that of her family and subjects, in restoring to the country whose Empress she happened to be the blessings of a peace that would stop the effusion of blood the thought of which robbed her of sleep at night and repose by day.

She was living in a state which most certainly was bordering on insanity, and she had entirely lost the faculty of discriminating between what was reality and what was a dream. Raspoutine held her in a kind of trance, which was further aggravated by the long fasts to which he obliged her to submit. She was told that she was the victim chosen by the Almighty to expiate all the sins of the Russian Empire, that it was only through constant prayer, combined with all kinds of other mortifications, that she could hope to see restored the peace of her mind and the health of her son. It is probable that she suffered from hallucinations during which she saw, as in a cloud, the rising shapes of soldiers killed in battle, clamoring to her to stop the useless massacres going on in the Polish plains where they had fallen. Is it a wonder that, unconscious of aught else than this condition of self-reproach to which she had been reduced, she tried to end her own sufferings, as well as the misery which had fallen upon her country, by disregarding all the advice she received from her real friends and making the most frantic efforts to induce her husband to accept the peace terms which the Kaiser had more than once caused to be secretly conveyed to him?

Nicholas II. was also weary of the struggle, but he realized better than his wife the impossibility which existed for him of acting independently of his Allies. He had Ministers who, in spite of their respect for his person and authority, would not have hesitated to point out to him the grave consequences which a defection of Russia would mean for the whole cause of the Allied nations, who, after all, had been entangled in this disastrous war because they had rushed to his help and to that of his people.

Sturmer, who had for a short time taken the conduct of Foreign Affairs in his hands, had been compelled to resign, owing to the opposition which he had encountered in the Duma, and especially owing to the masterful speech in which Professor Miliukoff had exposed all the vices and all the crimes of his administration. His retreat had not had for consequence a diminution of his favor or of his influence; he still remained the trusted adviser of both Czar and Czarina. Together with him were working Protopopoff, who pretended that he would be strong enough, with the help of the hundreds, nay thousands of police agents he had at his disposal, to crush every attempt at a revolution; Madame Wyrubewa; and, last but not least, the formidable Raspoutine, whose influence had proved wide enough to cause the postponement of the trial for blackmail of his confederate, Manassavitch-Maniuloff. A bank director from whom he had tried to extort 25,000 rubles had denounced the latter to the military authorities, and, in spite of the angry protest of Mr. Sturmer, whose confidential adviser he had become, he had been imprisoned and sent before a jury.

But even the efforts of these people combined could not move Nicholas II. to act in accordance with their wishes, because, as I have said, he still had Ministers unwilling to betray the country into the hands of its enemies. The head of the Cabinet was Mr. Trepoff, an honest man credited with liberal sympathies, who, at all events, would not lend himself to anything that could be interpreted into the light of a treason of Russia in regard to her Allies. Unfortunately, he could not hold out against the attacks that were directed against him by all the pro-German party, and after he had fallen the latter felt at last free to act as it liked, because Prince Galitzyne, who had accepted the difficult position of Prime Minister in a country already standing on the brink of ruin, was far too timid a man to dare express an opinion of his own, after the Sovereign had once spoken and signified his will to him.

RASPOUTINE

XXII

THE REMOVAL OF THE "PROPHET"

THERE is a well-known Latin proverb which says that the gods begin by depriving of their reason those whom they mean to destroy.

Never was its truth more forcibly illustrated than in the tragedy which brought about the fall of the autocratic system of government under which Russia had been suffering for centuries. Its last representative had incarnated in his person all the follies, the crimes, the mistakes, and the ruthless cruelty of his predecessors. Unlike them, he had not known how to temper them by personal authority or personal sympathy. He was an effeminate, degenerate descendant of strong ancestors; the whole atavism of a doomed race seemed to have become embodied in his weak individuality. If outside catastrophes had not occurred in his reign, it is still likely that he would have been compelled by a revolution of some kind or other to step back into an obscurity out of which he ought never to have emerged, because he was most certainly not able to bear the rays of the "fierce light which beats upon a throne." It is, however, possible that none of those supreme calamities which destroy the independence and self-respect of nations as well as of individuals would have been connected with his name and history. But destiny condemned him to remain forever, in the annals of the world, a living proof of the degeneracy which threatens all royal houses who do not possess sufficient energy to stand in perfect union with their people whenever a trial of some kind comes to threaten their mutual existence.

It would have been hard enough to be branded by the centuries to come as the last of the Romanoffs and as an unworthy Heir of Peter the Great. It was worse than hard for Russia, even more than for Nicholas II., to have to realize that, through stupidity, weakness of character, and an exaggerated opinion of his own power and might, he had been the direct cause of the ruin of his country and the means of plunging it into an abyss of distress and of anarchy from which it will take the work of several generations to redeem it.

His wife was the instrument of his destruction. About this last point there cannot exist any doubt whatever. She had a character stronger than his and she could speak to him in the name of the son to whom they were both so completely devoted. She could also appeal to his religious and superstitious feelings, which, though not as exaggerated as her own and not quite so foolishly carried to extremes, were yet also devoid of sound common sense. They were connected with the conviction that he had a mission to perform in regard to the future of his subjects, and to their welfare both in this world and in the next. Nicholas II. had in his character something of the traits of Caligula and other Roman emperors—a mixture of cruelty and theatrical sentimentality combined with cowardice in presence of danger and indecision before immediate peril. He never knew what it meant to play the game, and he perished because he refused to fight it out on the day that he discovered his adversary held all the trumps.

In the mean while the war was going on, claiming every day new victims. The insufficiency of the Government to face its various problems became more patent. Instead of applying himself to the task of coping with them, the Czar became absorbed, thanks to the remonstrances of his wife, in the one thought of how to consolidate his own authority, reduce to silence the protestations of the country and those of its representatives in the Duma, and conclude a peace with Germany which would allow him to make an appeal to his troops to help him to crush once more the Revolution which was hammering at his door, which he imagined he could subdue as easily as he had annihilated the one that had broken out after the Japanese campaign.

These were splendid plans indeed, and the Empress was already rejoicing at their success, in ignorance of the revolt which was shaking public opinion out of its previous apathy, a revolt which had extended itself to her own family. Bad as were most of the Grand Dukes, dissolute as their conduct had ever been, yet they had in their veins the blood of Catherine the Great and of all the dead and gone Romanoffs. They rose in rebellion against the gang of adventurers who were dishonoring the chief of their race and of their dynasty.

By that time the name of the Empress was being dragged in the dirt by every street boy, and open comments were made in public places in regard to her friendship, not to call it by another name, for Raspoutine—comments which were devoid of truth, because there was never any immorality in their relations, but which were generally believed, perhaps, because it would have been impossible for any one to guess that it was through superstitious practices that the "Prophet" had contrived to get absolute hold of her mind.

The Imperial Family felt the degradation to which this common peasant had reduced it, and though they had no reason in the world to like Nicholas II., yet they resented the humiliation which any slur upon the reputation of his wife conferred upon him as well. After all, Alexandra Feodorowna was the mother of the future Czar, and as such she ought to inspire respect in the Russian nation. If she did not realize this fact herself, others had to do it for her and rid her of a contact which was a slur. Besides, there was the hope that if once the adventurer was removed she could be brought to look upon the world from a more reasonable point of view. The principal thing was to deliver her from this evil adviser who was fast leading her, as well as the dynasty, to inevitable destruction and ruin.

The story of Raspoutine's assassination is too well known to be repeated here. At any time it would have broken the heart of the poor, misguided Czarina. But coming at the moment it took place, it did something more—it deprived her of what she considered to be her only moral support amid the troubles of her life, the possibility of communicating with the spirit of the man whom she had loved, who she felt sure was watching over her and over her child, from the heavens.

In the weeks preceding the murder of the "Prophet" he had subjected the Empress almost every evening to the agony of these prayer-meetings during which he communicated to her the so-called wishes of her dead friend, who, as he said, advised her, through his medium, as to what she ought to do to avert the dangers which were hovering over her head. The miserable woman used to listen to these revelations with anxious eagerness, and pray, pray, with a fervor she had never known before, for the strength to obey the commandments of a spirit who in death, as well as in life, had proved to be her best, and indeed her only, friend. Is it a wonder that the last remnants of sanity which were still left to her snapped under this terrible strain, and that at last she became the mere shadow of her former self, a fit subject for a lunatic asylum, where, indeed, she ought, for the good of everybody, to have been confined?

Her conduct after she had been told of the murder of the creature whom she revered as a Prophet of God was quite in accord with her character, such as it had developed itself through all the years during which she had allowed her mind to be invaded with superstitious notions, which would have been laughable if they had not been so pathetic. Her only thought was that of vengeance. She exercised it with a relentlessness which set against her the few people left in Petrograd who might have felt inclined to take her part and to pity her in this tragedy of her life. She left no peace for the Czar until he had exiled the persons whom she knew to have been the authors of the deed. When she was implored to take pity on the young Grand Duke Dmitry, and not have him sent to the Persian front, where there existed so many epidemics that it was hardly likely he would ever come back again, she had merely smiled and coldly said:

"Why should I pity him? He did not pity others."

And yet public feeling was so strong against her, and so entirely in favor of those who had had the courage to rid Russia of a man who had proved so fatal to it, that the schemes of revenge of Alexandra Feodorowna suffered a collapse. Mighty and powerful as Nicholas II. believed himself to be, yet he understood that the best thing he could do would be to let silence and oblivion fall over a crime that was eminently popular in the whole country. He had heard of the telegrams of congratulation, and of the flowers which had been sent to both his cousin Dmitry Pawlowitch and to the husband of his niece, young Prince Youssoupoff, as well as the joy to which the population of Petrograd had given way when it had become aware of the fate of the adventurer whose name had been so prominently and so sadly associated with that of the Empress of All the Russias. Perhaps at heart he was not so very sorry at an event which had certainly rid him of a great incumbrance.

Nicholas II. had always practised dissimulation to a considerable extent, and he had never allowed outsiders to guess what was going on in his mind. During the days which followed upon the disappearance of Raspoutine he certainly expressed great sympathy for the grief of his wife, but at the same time he did not, as she expected, cause the perpetrators of the murder of this low adventurer to be prosecuted publicly for their daring action. This apathy exasperated Alexandra Feodorowna.

During the last weeks of Raspoutine's life he had been working, conjointly with Sturmer and Protopopoff, toward convincing her to lend herself to a Palace revolution which would have overturned her husband and made little Alexis Czar under her own Regency. She had been told over and over again that she possessed all the great talents of Catherine II., that the Emperor was not a better man than Peter III. She had been made acquainted with his unpopularity, but at the same time persuaded that this unpopularity was a purely personal thing and that it did not extend itself to the person of the Heir to the Throne, nor even to her own. As Regent she could do any amount of good, and conclude peace with Germany the more easily that she was not bound by the terms of the agreement entered into by Mr. Sazonoff with the Entente, in the name of Nicholas II.

The foolish woman believed absolutely all the nonsense which was being constantly poured into her ears. Her ambition and lust for revenge over her enemies also played a part in this whole tragedy. She therefore began wondering whether, after all, she ought not to follow the advice which she had received from Heaven, as she fondly imagined, through the mouth of Raspoutine. She would have liked to be able to consult once again the spirit of Colonel Orloff so as to relieve her perplexity, because she had still sufficient scruples to hesitate before allowing those whom she considered to be her friends to use her name for the execution of a Palace revolution directed against her own husband, whom she may not have loved, but whom she still respected as the Czar of All the Russias.

It is at this juncture that a new incident occurred, the real details of which have never yet transpired. Raspoutine, just before he had been murdered, had introduced to the Empress a Tibetan doctor with whom he was on terms of intimate friendship, telling her that he was a man of great ability, devoted to occult sciences, had studied them in the convents of his country, and who was quite able to perform miracles. This man, whose name was Badmaieff, certainly saw Alexandra Feodorowna several times, and it was reported that he gave her certain drugs which he told her she ought to administer to the Emperor in secret, drugs which would make him quite subservient to her will. Whether she used them or not it is impossible to say. Young Prince Youssoupoff declared immediately after the Revolution that she had done so, and that in consequence of this experiment Nicholas II.'s will, which had always been a weak one, had completely disappeared, until he had been reduced to the condition of a puppet in the strong hands of his wife. But this assertion, coming as it did from a personage who could not have nursed kind feelings in regard to the Empress, must be accepted with caution.

It is a fact, however, that those in attendance on the Sovereign remarked more than once that he seemed at times to have lost the real consciousness of what was going on around him, that his eyes had acquired a vague, dazed look they had never worn before.

It is out of this introduction of Badmaieff into the intimacy of the Czarina that the rumor arose that Raspoutine, together with Anna Wyrubewa, had tried to administer slow poison to the small Grand Duke Alexis. Such a thing had never taken place, and indeed it could never have occurred if one considers the fact that the strongest trump in the game played by the pro-German agents who were leading Russia to its ruin was precisely the little Cesarewitsch, without whose existence it would have been impossible for them to think of making out of Alexandra Feodorowna a Regent of the Russian Empire. They had, therefore, the greatest interest in keeping the child in as good a state of health as possible, and he was far too delicate for them to try any experiment upon him. On the other hand, the necessity of getting rid by natural means of the Czar himself was so evident that it would not be surprising if the superstitious mind of his Consort had been influenced so as to persuade her to lend herself to what she had been told was nothing but a religious practice, but which in reality was an attempt to accomplish by this means what it would perhaps not have proved wise to try and bring about in another way.

PETER BADMAIEFF

XXIII

ANNA COMES TO THE RESCUE

IN the course of an interview which Anna Wyrubewa gave to a foreign newspaper correspondent a short while after she had been released from the fortress of SS. Peter and Paul, where she was confined for about three months following the outbreak of the Revolution, she said that the Empress Alexandra had never been so near to insanity as during the weeks which followed upon the murder of Raspoutine. What she failed to relate, however, was the manner in which she succeeded in preventing the half-balanced mind of the miserable woman from snapping altogether under the strain put upon it by circumstances.

When the outsider tries to form an opinion as to all the events which preceded the rebellion that destroyed the Throne of the Romanoffs, it is essential he should remember the state of mind of the Czarina at this particular time, as well as the condition of her nerves—a condition which was very nearly akin to the one into which a man falls when, after having been the victim of a pernicious drug habit, he finds himself unexpectedly and suddenly deprived of his favorite morphia or cocaine. The Empress had been living for months under the influence of these mysterious night sittings during which Raspoutine evoked for her, as she firmly believed, spirits of another world from whom she sought inspiration and in whom she found comfort. All at once this moral aid, which had helped her to live, was denied to her, and she did not know any longer what she was to do, surrounded as she felt herself to be by ever-increasing dangers which threatened not only her own person, but that of her beloved child, that son in whom she firmly believed Russia would find its salvation and who was destined to become one of the greatest and mightiest Sovereigns the country had ever seen reign over it since the days of Peter the Great. She felt absolutely at sea, like a ship deprived of its pilot and abandoned to inexperienced hands, ignorant of the first principles of navigation. Neither her husband, whom at heart she despised, nor her friend, Anna Wyrubewa, whom she had never entirely initiated into all the details of her secret intercourse with the dead, nor her faithful advisers, Sturmer and Protopopoff, could make up to her for the irreparable loss of the companionship which, thanks to Raspoutine, she believed she had succeeded in establishing between herself and the soul of the only man she had ever truly loved.

It is only after having grasped these essential facts in the life of the misguided Empress of Russia that it is possible to come to a reasonable appreciation of her person, character, and intrigues.

Once this has been done, it becomes relatively easy to understand the influence which Raspoutine had acquired over her mind, and not to share the general opinion that there existed something immoral in her relations with him. Immorality alone could not explain this entire submission on the part of a cultured, well-educated, elegant woman to the will of a dirty, uncouth, ignorant peasant. Besides that, Alexandra Feodorowna was far too proud to forget for one moment the social difference which separated her from the "Prophet." In her intercourse with him she remained the Empress, and on his side he was far too shrewd not to remember it also. He knew very well that one indiscreet word, one imprudent gesture, would have put an end at once to his influence, and the man as well as his accomplices were working for far too great and far too important an object to compromise its success by anything which might have savored of immoral intrigue.

The state of health of the little Cesarewitsch also was not the real reason why the latter's mother would not allow Raspoutine to leave her. She believed in the efficacy of his prayers for her son, but this belief alone would not have been sufficient to make her so entirely submissive to his will and to reduce her to the state of slavery into which she had been entranced. No, the secret of Raspoutine's influence lay in the simple fact that, thanks to the hypnotic faculties which he undoubtedly possessed, he had contrived to acquire an absolute dominion over her mind, and to persuade her that every time she prayed with him she was put into direct communication with her dead lover; that this lover had been allowed by the Almighty to come to her help in the troubles and perplexities of her life, to guide her in her conduct as a woman and a mother and in her duties as a Sovereign.

During the hours of agony which followed upon the news of the murder of that man whom she had considered as a holy creature and a real Prophet of God, Alexandra Feodorowna blurted out something of what lay on her mind to her devoted friend and companion, Anna Wyrubewa. The latter had removed from her own house to the Palace of Tsarskoye Selo, so as to be able to remain in constant attendance on the miserable Empress. Seeing her so forlorn and desolate, she bethought herself of rousing her faculties, and tried to persuade her that, though she had lost her advisers and counselors, she had yet a duty to perform, which consisted in going on with the work they had suggested to her to start. Peace was more than ever necessary to Russia, as well as to the dynasty, against which such fierce attacks were being launched. The sacred principles of autocracy that were being everywhere challenged ought to be

THE CESAREWITSCH

maintained, and how could this be done when the army which was the only force on which the Czar could rely was being kept at the frontier and uselessly butchered in battles it could not by any possibility win? There were other mothers besides herself in Russia who were crying over their dead sons and appealing to her to spare those who were still left to them. This war was a monstrous crime against humanity, as well as against the whole of the Russian nation. It must be stopped because otherwise worse calamities even than those that had already fallen on the country would occur. The performance of a duty was sometimes painful, but this ought not to prevent any right-minded person from trying to accomplish it. It was quite evident that the duty of the Empress required her to work toward the conclusion of peace with Germany, and this had been already suggested to her not only by the devoted friends she still had in the world, but by the spirit of the dead ones who had loved and honored her while they had been alive on earth.

Whether Anna Wyrubewa was sincere or not in thus pleading a cause which she knew her Imperial mistress had but too much at heart even without her interference, I shall not attempt to guess. Russia was most certainly going through a terrible crisis, and those who thought that the quick conclusion of a peace after which so many were secretly longing and sighing was indispensable were by no means a small minority in the country. It is quite likely that the Empress's confidante was sincere in her conduct, and it seems pretty certain that she had no pecuniary or material advantages in view when she lent herself to the dangerous scheme suggested to her by Sturmer and the latter's accomplices. *They* were not disinter-ested; they had decidedly ambitious views as to their own future, and they were most certainly in the employ of Germany, to which they had promised their co-operation. Protopopoff was a man who, in regard to the large fortune he was credited with possessing, was entirely self-made; he had never shown any hesitation as to the choice of the means by which he had acquired it. Sturmer thought himself endowed with the genius of a Bismarck or of a Richelieu, and dreamed of the glory of a peace that would leave Russia in appearance as strong as ever, but united by the closest of ties to the German Empire, of which he had been all through his political career a devoted admirer and servant. He had always preached the necessity of the renewal of the former alliance that in bygone times had united the Hohenzollerns and the Romanoffs. His vanity felt deeply flattered upon hearing from his friends in Berlin that the Kaiser, as well as the latter's Government, considered him the one great Minister Russia had ever possessed and were looking up to him to heal all the evils and all the miseries which the war had brought about. He did not care for the treaties that had been signed between Russia and her Allies, and probably shared the opinion of Mr. von Bethmann-Hollweg that all such documents were nothing but scraps of paper, not worthy of any notice on the part of intelligent people. He cared only for success, for titles, decorations, power, and a crowd of flatterers about him. Russia had ceased to be for him a matter for consideration. She would always fare well, in his opinion, if only he were allowed to direct her destinies. The war itself, with all the terrible breakage it had brought about, did not trouble him. It had begun with broken treaties and broken faith, broken honor and broken word; its result had been broken houses in broken lands, broken men, and broken hearts, but about these last Mr. Sturmer did not think at all.

And what about the third personage in this sinister tragedy? What about Manassavitch-Maniuloff, who had been all along the *Deus ex machina* of this dark intrigue, and the chief spy and accomplice of William II.? It was he who had engineered the conspiracy for peace which was being carried on by the Empress under his supervision. It was he who had been the real creator of the Raspoutine legend, and he was perhaps the one who at first suffered the most through the collapse of the adventurer. When the "Prophet" was murdered, Maniuloff was in prison under the accusation of blackmail. Once before he had escaped a trial that had been postponed on the personal order of Nicholas II. addressed to the president of the court. But after Raspoutine's disappearance the influence of Sturmer alone had not been able to help him. He was sent before a jury and sentenced to two years' penal servitude, which, however, he was never to undergo. The man had more than one arrow to his bow, and when the Revolution broke out he contrived to let Kerensky know that he could put at his disposal most incriminating documents in regard to the part played by the Empress in the peace negotiations which had taken place in the preceding February between Petrograd and Berlin. The bait probably took, because the spy who had for a long number of years cheated everybody was sent across the frontier to expiate his sins and most probably to go on for the benefit of the new masters of Russia with the nefarious game he had been playing in regard to all those who had had the misfortune to employ him.

After Sturmer had been compelled to resign his position of Prime Minister and leader of the Foreign Office he had, nevertheless, remained, as I have had already the occasion to tell, in close relations with the Court and with the Emperor and Empress. He had acquired a new ally in the person of the Metropolitan of Petrograd, Monseigneur Pitirim, a friend and favorite of Raspoutine, who now came to offer his consolations to the half-distracted Alexandra, and who also told her that it was henceforward her duty to go on doing all that the dead "Prophet" had suggested to her, no matter how much it might cost her. Between his preachings, the advice of Sturmer and Protopopoff, and the adjurations of Anna Wyrubewa, the Empress was at last persuaded to forget for a while the deep grief into which she had allowed herself to fall and to resume her political activity. But when she attempted to influence the Czar to approve of what she was about to do she found, to her surprise, that he did not show the same enthusiasm for her schemes as he had done before.

What had happened was this: The Imperial Family had once more tried to open the eyes of the Sovereign as to the folly of his wife's conduct. Nearly all the Grand Dukes and Grand Duchesses in Petrograd had sought his presence in succession, implored him to save the dynasty before it was too late, and to call together a responsible Ministry, chosen from among the men who had the confidence of the country and who represented it in the Duma. Their remonstrances had not convinced Nicholas II., but they had caused him to pause before consenting to the conclusion of a peace with Germany, which he began to fear he would not have the power or the strength to impose upon public opinion in Russia. He believed in his wife, and he felt convinced that she was the only disinter-rested friend left to him; at the same time he could not make up his mind to take a decision which—this much he knew—would be deeply resented by his Allies as well as by his own subjects. In his perplexity he preferred to wait for events to develop themselves in one sense or in the other, totally oblivious of the fact that there are periods in the life of nations when waiting is also a crime.

And while this struggle was going on in his mind, that of his wife was becoming more and more the prey of the evil advisers who had secured her sympathies and were abusing her confidence. They were becoming bolder and bolder as time went on, and at last they suggested to her to urge upon the Czar the necessity of returning to the front, where, they told her, he could come to a better understand-ding of the feelings of the army and be at last convinced that it was, like the rest of Russia, only longing for peace. Nicholas caught eagerly at the suggestion and departed, leaving the Empress mistress of the field and free to do what she liked, together with her friends.

METROPOLITAN PITIRIM (OKNOV PAVEL VASILIEVICH)
17th Metropolitan of Petrograd and Ladoga
b.1958 - d.1920

> Rasputin's influence on the Russian Orthodox Church through the Imperial Court meant that he controlled the Holy Synod. The Bishops opposed him but to no avail and this led to the exile of several bishops to Crimea or Siberia. It was in this way that the existing Metropolitan of Petrograd and Ladoga from 1912 to 1915, Vladimir Bogoyavlensky, who had been criticising Rasputin, found himself extricated to Kiev and replaced with Pitirim, a Rasputin supporter. Bogoyavlensky was killed by the advancing Bolsheviks in to Crimea in 1918.

XXIV

YOU MUST BECOME THE EMPRESS

WHEN the Czar left Tsarskoye Selo—for the last time, as it turned out, as a powerful, dreaded Sovereign—the Empress had not yet made up her mind as to what she ought to do. She was being urged by Sturmer and Protopopoff to come to a decision in regard to the future of the dynasty, which they declared to her was entirely in her hands; at the same time she lacked the moral courage to put herself boldly at the head of a movement to dethrone her husband. She had not the audacity of Catherine the Great, nor the latter's unscrupulousness, and, moreover, her mind was so weakened by the superstitious practices in which she had become absorbed that it is to be questioned whether or not she was given a true account of what was going on around her. She was entirely at the mercy of the first determined man who came along, audacious enough to compel her to sing according to his tune. But neither Sturmer nor Protopopoff were clever enough to be that. And they had no political party on whom they could rely to help them execute any plans they might form. They depended for their inspiration on the directions which they received from Berlin. By a lucky accident this inspiration failed them at the very moment they most needed it.

What had happened was this: The Allies had begun to get some inkling as to the intrigue which was going on under the Czar's own roof, an intrigue in which his wife held the foremost rôle. They contrived to put obstacles in the way of Mr. Protopopoff and of his friends, and to stop for a while the active correspondence which he was carrying on with the German Government *via* Stockholm. At the same time they arranged matters in such a way that the liberal leaders in the Duma became apprised of the negotiations pending between the Kaiser and his kinswoman at Tsarskoye Selo.

The story of the eventful days which preceded the Revolution have nothing to do with the present book, and I shall refer to them only in so far as they concern the Empress. She was mostly responsible for the rapidity with which rebellion spread and for the unexpected way in which it broke out. Had she remained quiet, it is likely that things might have dragged on for a few weeks, perhaps even for a few months, longer, because no one at this particular moment cared to see a change in the Government. But when it was ascertained that she had become a danger to the nation in general there was no longer any question of a delay, and events had to be forced on in some way or other.

What Sturmer proposed to the Czarina was to provoke a movement against the war in the garrisons of Petrograd and the towns in its neighborhood; this to be further accentuated by false news concerning the Czar, who would be represented as having died suddenly. The Government had at its disposal all the telegraph and telephone wires. It was, therefore, an easy matter to cut off the capital from all communication with the headquarters of the army. In the confusion inseparable from the consternation caused by the news of the Sovereign's demise it would have been but a matter of a few hours to get the little Grand Duke Alexis proclaimed Emperor under the Regency of his mother, who would thus have been left free to sign a peace which nothing and no treaty prevented *her* from concluding. Nicholas would be easily persuaded to accept accomplished facts and most likely would surrender with pleasure, or at least with absolute indifference, a Throne he had never cared for. So they thought that an act of formal abdication would not be difficult to obtain from him.

The country also would not feel sorry to be rid of a Monarch who had never been in possession of its affection or respect, and the army, glad to return to its homes, would most likely rally with alacrity around the Regent and the little Czar. The very fact that it was a woman and a delicate child upon whom the whole burden of an immense responsibility had fallen would predispose public opinion in their favor, and most likely this Palace revolution would end with complete success.

The Empress allowed herself to be won over to the conspiracy, and it was decided to put it into execution about the middle of the month of February. Protopopoff declared that he required that much time to gather together a sufficient number of police agents in Petrograd, without whom he did not dare to risk the adventure. Alexandra Feodorowna assented to everything that was proposed to her. She went about like one in a dream, unconscious of the abominable plot in which she had been induced to participate, thinking only of the time when she would be able at last to renew with her own family and with her own people the tender and intimate relations which the war had forcibly interrupted.

In the mean time the Emperor remained at the front, and if we are to believe all that was subsequently related about his conduct there, he changed considerably his opinion and point of view after having resumed direct contact with his troops. He convinced himself that they were not at all as anxious for peace as he had been led to expect, and that the feelings of the men in regard to Germany were revengeful more than anything else. His generals, and especially Aléxieieff, who was Head of the Staff, kept urging upon him the necessity of preparing a formidable offensive, this time on the Riga front. The General gave him hopes that it would turn out to be a successful one, provided (and this was the one everlasting and burning question) that the War Office sent sufficient ammunition to the front. The Emperor was persuaded that this could be done, but Aléxieieff was not so sanguine, and he started a private inquiry of his own as to what was going on in Petrograd in that respect. The result of it was that he was convinced that the Ministry had lately completely neglected this important item and had spent its time in arresting workmen whom it suspected of harboring democratic opinions, as well as in curtailing the hours of labor at the different factories where ammunition was manufactured. Protopopoff wanted the war to end, and he hoped that in limiting the output of shells and guns he would be able to place the country in such a position that a cessation of hostilities would become unavoidable.

A report to the Emperor, in which the situation such as it presented itself was exposed with great details, was brought to him by the Staff. As usual, it left him unmoved. He merely said that he would give orders to the War Office to take henceforward its orders from the Commander-in-chief of the Armies in the Field, meaning himself, but he refused to blame Protopopoff or to hear anything concerning the appointment of a liberal and responsible Cabinet from whom the Duma could require accounts. He did not mean to lessen his own prerogatives by the merest fraction, and he still thought that Russia might hold its own against her formidable foes without arms, provisions, shells, or big guns, and in general without means of defense capable of stopping the progress of the invaders in their triumphal march through his Empire.

The commanders of the different fronts held a consultation, and one of them, whose name I cannot mention at the present moment, first suggested the idea that it would not be a bad thing to try and bring about a military conspiracy which would overthrow the weak Monarch whom it was impossible to bring to take a sane view of the position in which the army found itself placed. Another general suggested that such an upheaval would only bring to the foreground the personality of the Empress, who would insist on being consulted in all matters in which the welfare of her son might be concerned. And no one wanted Alexandra Feodorowna to be raised to a position in which her voice might come to exercise an influence of any kind on the destinies of the country. It was by far preferable to let Nicholas II. remain where he was, and try to persuade him to allow the Staff, instead of the Cabinet, to have the last word to say in all questions connected with the national defense.

This secret, or rather not secret, conference, because its purport became known on the very same day it took place, thus accomplished nothing. In the mean while the object of its deliberations was communicated to the Ministry in Petrograd, and Protopopoff triumphantly informed the Empress of the fact that it had come to almost the same conclusions which he and his friends had arrived at long before. It was necessary to change the person of the Sovereign. He carefully refrained, however, from acquainting her with the knowledge of the opposition that the idea of a Regency had provoked.

It is a curious but certain fact that at this very time large sums of money were distributed to the troops quartered in Petrograd, Tsarskoye Selo, Peterhof, and Gatschina by unknown people in the name of the Empress. The latter declared, later on, when questioned on the subject by the Provisional Government, that she had known nothing about it; certainly it had not been her money which had been scattered about with such reckless generosity. I believe that in saying so she spoke the absolute truth. But then the question arises, by whose orders was this money thrown into the arena of the battle-field, where the fate of a nation and of a dynasty was about to be decided? Some people have declared that it was Protopopoff together with Sturmer who had hit upon the idea of making Alexandra Feodorowna popular among the army by appearances of a generosity with which no one had credited her before. But against this theory comes the probability that if either of the above mentioned gentlemen had been able to draw from the Treasury several millions of rubles to be applied to secret purposes, they would have begun by putting them into their own pockets and trusting to the future and to Providence for the success of any enterprise they embarked upon. Therefore the question arises again as to the origin of this money which was circulated with such a generous hand among the regiments considered as likely to lend themselves to a Palace revolution in favor of the delicate little boy who was the sole Heir to all the glory and the splendor of the Romanoffs.

I think that very few people, among those who knew how vital was Germany's interest at this particular moment to see a peace concluded, will doubt whence came these funds. They were certainly spent to favor the appointment of the Czarina as Regent of the Russian Empire. Who had procured them for the benefit of a vast conspiracy, the object of which was to deliver Russia, bound hand and foot, to the tender mercies of her formidable neighbor and enemy?

On the other hand, the liberal parties, now thoroughly awakened to the dangers of the situation, were also working earnestly toward the defeat of the plans conceived by Messrs. Sturmer, Protopopoff & Co. Several meetings of the leaders of the different factions in the Duma took place at the Tauride Palace, but none seemed to come to anything serious in the way of a revolution, which had been by that time recognized as absolutely inevitable.

The Cabinet saw this hesitation, and would undoubtedly have struck a serious blow at its adversaries if, just at the time, the children of the Empress had not sickened from the measles in a serious form. The mother forgot all her political intrigues in her anxiety; the plot about to be executed had perforce to be put off until a more favorable day. It must be here remarked that the Czar, when he heard about his son's and daughters' illness, telegraphed to his wife asking her whether she wished him to come back to Tsarskoye Selo. This did not suit in the least the people who were only waiting for a favorable opportunity to dethrone their Sovereign. Alexandra Feodorowna was easily persuaded to oppose herself to this desire of her husband and to wire back to him not to return. By a singular coincidence the presence of Nicholas II. at Tsarskoye Selo, which would without doubt have given quite another coloring to events which were going to happen within a few days, was desired neither by his friends nor by his foes nor even by his family. They all of them knew that something terrible was about to take place, but they also felt that, for the sake of everybody, it would be better he should be absent.

And in the silence of his study at Potsdam the Kaiser was secretly discounting this Russian Revolution which he saw quite clearly was approaching with quickening strides. He knew what he was about, and little did it matter to him if those whom he had used as pawns in the difficult game he had been playing would perish or not in the storm which his efforts had contributed to let loose.

ALEXANDER DMITRIEVICH PROTOPOPOFF
INTERIOR MINISTER (September 1916 – February 1917)

XXV

THE NATION WANTS YOUR HEAD

I FEEL personally sure—and others who were in Petrograd at the time of the fall of the Romanoffs told me the same thing—that in this whole history of the overthrow of one of the most formidable powers the world had ever known there are yet details which we do not know. In fact, no one knows them, but perhaps they will be explained to us later on. The catastrophe occurred with such startling rapidity that even those who were the most concerned in it were hardly able to realize its importance, even while recognizing its seriousness.

There is also another curious feature connected with the tragedy. All its principal actors, the men who were really instrumental in bringing about the change which transformed Russia from an autocratic—the most autocratic Government in the world, in fact—into a democratic Republic disappeared before even their task was done. It was the Duma in the person of its president, Mr. Rodzianko, it was the zemstwos who had taken up the cause of the liberal movement from the very beginning of the war, who really were responsible for the abdication of Nicholas II. And yet the Duma disappeared, melted into space with an unbelievable rapidity; Mr. Rodzianko has hardly been heard of since the activity of the zemstwos was suddenly interrupted.

How did all this happen? Who was responsible for the chaos into which Russia is plunged at the present moment? It is next to impossible to say to-day, though one may easily guess. All that the world knows is that chaos has supervened, and that, thanks to this chaos, Germans have once more re-entered Petrograd, by the back door, perhaps, but still re-entered it, and what does this detail matter to them! What they wanted was only to get there again; the rest would adjust itself as time went on, and the general confusion became even more complete than it was at the beginning.

Another feature in this extraordinary Revolution was the swiftness with which the country accepted it and accommodated itself to its consequences. In the space of a few hours the portraits of the Czar had disappeared from all public places, the Imperial arms, wherever these had graced a shop or concern of some kind, had followed suit. Ushers in the former Imperial theaters had discarded their liveries, sentinels at the Winter Palace had been removed, and the Red flag had taken the place of the Romanoff standard on top of the Imperial Residence. All this had been performed quietly, joyously, and in a perfectly orderly manner. It seemed almost as if people had been prepared for a long time for what was to come and had practised beforehand the various manifestations of their joy to which they gave vent as soon as it became known that the Guard regiments quartered in the capital had gone over to the Duma and sworn allegiance to Mr. Rodzianko, its president.

Of the war there was no longer any question. It seemed to be forgotten in the excitement of the hour, and somehow a general impression prevailed that, once the Czar had been overturned, peace was but a question of days. By one of those strange anomalies such as happen so often in life, the Czar had been accused of wishing to bring this peace about; yet when he was no longer there the world rejoiced at the thought that peace would surely be concluded before any unreasonable quantity of water had run through the Neva. It is also a singular feature of this singular time that while Petrograd was in the throes of revolution, while Ministers with Mr. Protopopoff at their head were being arrested and transferred to the fortress, the Czar at headquarters and the Empress at Tsarskoye Selo did not in the least suspect what was taking place in the capital. It was said later that the Grand Duke Paul had forced his way into the apartments of Alexandra Feodorowna and had acquainted her with the details of the upheaval which was to carry away her Throne.

I can hardly bring myself to believe this. For one thing, no one in the Imperial Family cared sufficiently for the Czarina to take the trouble to warn her of any peril in which she might find herself. And then she had not been upon good terms with the Grand Duke Paul in particular; it is to be questioned if she would, in view of the fact that it was his son who had helped to slay Raspoutine, have consented to receive him in general. I think it far more likely that it was only through the indiscretion of some of her attendants that the Empress heard of what was taking place. It is probable her first thought was that her friends had been working in her behalf, and that the insurrectionary movement which was shaking Petrograd was distinctly in her favor; that its aim was to make out of her the Regent of the Russian Empire.

It would be difficult, otherwise, to understand her apathy in the presence of this overwhelming catastrophe, or the resistance which she opposed to the advice which the few attendants who were still faithful to her and who had remained at Tsarskoye Selo, gave to her—to telegraph immediately to the Czar to return home. Up to the last minute she refused to do so, saying that she felt quite capable of resisting the mob in case it chose to invade the Imperial Residence. And at last it was not she, but the officer in command of the troops quartered in the town, who took it upon himself to inform General Woyeikoff, head of the Okhrana, or personal police service of the Czar, that it was high time for the Sovereign to return home, as he could no longer guarantee the safety of the Empress and of her children. All the regiments under his orders had gone over to the enemy.

Alexandra Feodorowna was waiting the whole time for Protopopoff and Sturmer; she was only wondering why they were so long in coming to her. When at last she was informed that they had been arrested by the mob and taken to the fortress, whither they had sent so many innocent people, she began to realize that things were not going so smoothly as she had fondly imagined, that something quite out of the common had taken place. Then she remembered certain words which Raspoutine had told her: so long as he was at her side no harm would befall her, but that, if he were once removed, misfortune upon misfortune would crowd on the House of Romanoff and sweep away the Crown to which she had become so attached.

In that acute moment when there flashed across her mind this prediction of a man in whom she had seen a Prophet of the Almighty, and the Empress realized the tragedy of her destiny, all the courage of which she had boasted earlier fell flat to the ground. She no longer thought of struggling against an implacable fate, and a complete indifference as to her possible future took the place of her previous energy and determination. The game was lost, absolutely lost, and she had better confess herself beaten before any more harm was done.

News of her husband's abdication reached her, and did not even rouse her sentiments of revolt at a piece of weakness which, under different circumstances, would have brought on one of those hysterical attacks to which she had been subject. She understood that she was alone, quite alone with the burden of her past sins and mistakes.

She accepted with stoical resignation the decrees of Destiny. Not one single feeling of pity for the miserable Monarch for whose fall she was so responsible, or for the children about to lose a glorious inheritance, moved her heart. She was thinking the whole time of the dead man who had loved her and of the murder-ed adventurer who had comforted her in the hours of her greatest moral agony.

Nothing seemed to make any impression on her blurred mind—not even the angry crowd when it appeared in the courtyard of the Palace where she was still staying, carrying before it great banners upon which were written the ominous words:

"Give us the head of Alexandra Feodorowna! We want the head of that German woman, Alexandra Feodorowna!"

When asked to leave the window and not to appear before this multitude clamoring for her blood, she merely shrugged her shoulders and remained where she was. She certainly was not courageous, but she did not lack bravery—the bravery born of fatalism or of indifference, which renders those who are endowed with it impassible before danger, because they fail to realize its importance or its imminence. This woman is a historical riddle which only history will be able to unravel, but not so soon as one imagines, because it is likely that she has not yet come to the end of her sinister and mischiefvous career.

While the life of his wife was threatened, while his Ministers were imprisoned, and while the nation was preparing to claim his abdication, Nicholas II., at Mohilew, where headquarters were stationed, remained just as indifferent to the convulsions which were shaking his country as the Empress watching by the bedside of her sick children. He also did not understand; he also failed to realize that what was taking place in Petrograd was the first act of a big game the stakes of which might easily come to be his own head and those of his family. The thought of Louis XVI. never once crossed his mind. At least it is allowed one to suppose so, because, when some officers of his suite remarked to him that the rebellion (the news of which had by that time reached him) bore many traits of resemblance to the premonitory riots that had heralded the introduction of the Terror in France, he simply replied:

"Oh, it is not at all the same thing. Russians are not Frenchmen—and the Romanoffs are not the Bourbons."

The Czar might at this early stage of the Revolution have returned to Tsarskoye Selo if he had only energetically insisted upon doing so. But he spent three days in complete indecision, and when at last he made up his mind to go home it was too late. By that time General Aléxieieff had been won over to the cause of the Duma, which was supposed to represent the only responsible authority in Russia; he put every possible obstacle in his way, going so far as to interfere with the arrangements made by General Woyeikoff for the departure of the Imperial train. It seems also that he sent telegrams asking for this train to be either stopped or at least delayed on its way.

No one at this stage wished Nicholas II. to go back to Petrograd, where it was feared his presence would prevent, if not stop, the establishment of the new Government; a useless fear, because, even if he had reached his former capital, he would never have found sufficient courage or energy to fight against an adverse fate or to do aught else but submit to the will of the multitude eager for his fall. The man who signed without one word of protest an abdication against which his whole soul ought to have protested, such a man was not to be feared, he could only be despised.

This was also the feeling which the whole nation began to entertain for him. People had pitied him in the beginning, but as the details of his conduct at Pskow became known, contempt took the place of any commiseration which the tragedy of his fate might have provoked. This opinion was so general that a friend of mine happening to discuss with one of the Deputies of the Workmen in those Soviets which were being organized just then the conduct of the former Czar, asked if he thought it likely the life of the deposed ruler was in danger. He received this characteristic reply:

"In danger? No. He is not worth a shot."

It is likely that the Empress, if she had been asked her opinion, would have agreed with this judgment. Though she had also thrown up her hands and renounced the game, she would not have given up her rights to the Crown that had been put upon her brow with such pomp and ceremony at Moscow twenty-one years before. She would have fought against the insolence of those who had come to demand it from her. Here I must say that, according to the words of one of the two Deputies sent by the Duma to interview Nicholas II. at Pskow, the prestige of the latter's personality as the anointed Czar of All the Russias was still so great that if he had mustered sufficient energy to throw out of the railway carriage the men audacious enough to claim his abdication, this gesture of Imperial rage would have brought back to him the allegiance of the troops. He was living through a terrible drama, and he was accepting it like a comedy. After having disgraced himself, he was dishonoring by his attitude the misfortunes which had fallen on his country, on his dynasty, and on his race.

MIKHAIL VLADIMIROVICH RODZIANKO
CHAIRMAN OF THE FIFTH DUMA

XXVI

A CROWN IS LOST

THE Monarchy of the Romanoffs had fallen like a house of cards which crumbles on the ground at the slightest touch. It had been considered one of the strongest, one of the most powerful, in Europe; yet its collapse had come with an amazing promptitude and there had not been found in the whole vast Empire over which it had ruled one single man or woman willing to arrest its downward course toward the abyss into which it finally disappeared. What the tyranny of Nicholas I., the selfishness of Alexander II., and the iron rule of Alexander III. had failed to produce, the weakness, indecision, and incapacity of Nicholas II. had made easy. What a German Princess, Catherine II., had maintained, another German Princess compromised and lost forever.

Without wishing to add to the faults and mistakes of Alexandra Feodorowna, it is nevertheless quite impossible to acquit her of blame in the catastrophe which finally wrecked poor, unfortunate Russia. Without her it is likely that the Crown would have kept some prestige, at least in the eyes of those whose family traditions were linked with the fate of the Monarchy in their country. She destroyed this prestige by the singularity of her conduct, the want of balance of her mind, and her proud, haughty, and totally false conception of the Russian character. She firmly believed that nothing she could do would be criticized and that even those who disliked her, whose number was legion, as she knew very well, would never dare to question her right to do whatever she pleased, or to choose her friends no matter in what circles or among what kind of people.

This woman, whom misfortune associated with the fate of the House of Romanoff at the very time when the latter ought to have had the aid of an intelligent, well-intentioned, and unselfish Princess to help it face the dangers which were threatening it, had never known how to put herself at the level of the persons by whom she found herself surrounded. She lacked not only tact, but also generosity, and she never could hold broad views about anything or about anybody. She was as scathing as she was hasty in her judgments.

From the very first day she was raised to the Throne of Russia she abused the privileges which her position conferred upon her, and either through stupidity—or willingly—because of her dislike for the nation whose Crown she wore, she applied herself to wound those whom she ought to have spared and to propitiate persons whom it would have been imperative for her to keep away as far as possible from her person and from that of her husband.

Of course she was in a certain sense a strong character, in so far, at least, as she never would yield to reason or accept any compromise. She had principles of her own, which, however, did not help her to win respect for herself or esteem for her conduct. Without ever allowing herself to be led by impulse, she failed to perceive that in most of her actions she was influenced by superstition of a most foolish kind. The fact that insanity existed in her family may excuse her to a certain point, but should not blind us to faults which might easily have been corrected had she only realized their existence.

She was a blameless wife; about this there cannot be any doubt. But she never loved her husband and she only cared for his position. She was a tender mother, at least to her son, whatever may have been her feelings in regard to her daughters, whom she most unjustly blamed for their sex, if we are to believe all that we have been told on the subject. But she lacked sympathy, which she never could give to others or win for herself. She was a cold, ambitious, stern creature, so convinced of her own perfection that she never could be brought to see good in anything with which she was not connected in some way or other. Her life certainly had tragedy entwined with its course. Perhaps the most cruel blow, until the final catastrophe that wrecked all her hopes, had been her unfortunate affection for the dashing officer, Colonel Orloff, who had died to save her honor and good name, whose post-mortem influence had been so cleverly made use of by unscrupulous adventurers in order to win her confidence. Alexandra had always at heart despised the weak man to whom she was married, but she had loved the high position which, thanks to her union with him, she had acquired; she would have liked to remain alone in control of it and to revenge her supposed wrongs at the hands of the Russian nation, by delivering it into the power of that German Fatherland of hers to which she had always remained attached. Her desire for peace was sincere (at least we must hope so), and everything we know about her and about her conduct during that momentous time when she kept working toward its conclusion points to the truth of this supposition. It had all along been a terrible trial for her to find the land of her birth at war with that of her adoption, and to this initial agony was added the superstitious terror which Raspoutine had inspired in her, thanks to the hypnotic practices in which he had induced her to participate—terror which ended by completely wrecking her already badly balanced mind.

But the supreme misfortune of the last Empress of Russia, a misfortune for which she was not responsible, was the fact of her having been married to a being who was too weak to lead her, too selfish to understand her, too cruelly inclined to sympathize with her; who at the same time did not acquire sufficient authority over her to inspire her with respect for his individuality as a man and for his position as a Sovereign. Had she been the wife of Alexander III., it is likely that she would have turned out entirely another woman from the one which she ultimately became; on the other hand, had Nicholas II. had for Consort a person different from the one to whom destiny had linked him, it is also probable that he would have contrived to avoid some of the mistakes into which he fell. He might have shown himself more plucky and more human in the different moments of crisis which made his reign such a sad and such an unfortunate one.

One of the most tragical things with which a student of history finds himself confronted when he analyzes any of the great catastrophes that come to change the fate of nations is the total insufficiency of the people who have to meet them or to handle them. There is no more pitiful spectacle than the vacillations of Louis XVI. during the first days of the great Revolution which sent him to the scaffold. Witness the want of character of the miserable Czar who is meditating at present in Tobolsk over the misfortunes that have landed him into exile; it is another of those sights one should have liked to see spared to posterity. During the twenty-two years he occupied the Russian Throne Nicholas II. constantly opposed himself to the wishes of his people, even the most reasonable ones, when he thought that they implied any diminution of his personal prerogatives or power. He sent hundreds of thousands of innocent people to the gallows or to horrible Siberian prisons under the slightest of pretexts. He had no hesitation at spilling the blood of his subjects either on the battle-field or on the scaffold. He allowed the detestable police system, which became, under his rule, stronger than it had ever been before, to interfere with private liberty and private opinions to an extent that had never been witnessed in his country even in the times of Nicholas I. or of Paul. While he reigned no one felt secure in his home or could go to bed with the consciousness that he would not be wakened in the middle of the night by an army of police agents come to search his drawers, or to arrest him under the most futile of pretexts or simply because he had refused to pay a bribe. And yet that man in whose name the most terrible injustices had been committed, who did not admit any resistance to his will, who believed in his unlimited power over one hundred and eighty millions of human beings—that man did not find sufficient courage to resist the only demand, among the many which were addressed to him during the course of his nefarious reign, that he ought never to have granted; and without one single thought of the future of his country or of his son he gave up without a murmur the Crown of which he was the bearer, when two determined men came to claim it from him, and he did so without even noticing that they were far more awed by the solemnity of the scene in which they found themselves unwilling actors than he was himself.

There never was a Throne relinquished with less dignity, there never was an act of abdication accomplished with less consciousness of the importance of its meaning. When one recapitulates all the details of the drama which was performed at Pskow, one can, when one is a Russian, feel but one passionate regret—that no one was found by the side of the last crowned Romanoff to drive a knife into his heart or put a bullet through his brain, and thus spare this haughty dynasty the shame of having been dragged into the gutter by its Head.

It is scarcely to be doubted that if Nicholas II. had only given more thought to the importance of the act he was invited to perform he might at least have saved his dynasty, if not himself. His brother, the Grand Duke Michael, who could easily renounce the Crown for himself, would hardly have been able to refuse the Regency on behalf of his little nephew. A man with the slightest political knowledge would have put the interests of his country before his own selfish feelings of paternal affection, and the Czar ought to have abdicated in favor of his son, and not have put forward this stupid pretext of lacking the courage to part from him. This very remark proves how little he understood the situation in which he found himself. It also shows us how utterly helpless he was when confronted by any difficulties of an overpowering and potential character. When one considers his whole conduct during those eventful hours when he lost not only his own, but his posterity's, Crown, it is impossible not to wonder as to whether or not there was any truth in the rumor that the Empress had been giving him drugs of some kind with the purpose of annihilating his will. It seems almost incredible that any man should have so quietly and so spontaneously lent a hand to his own degradation.

It is to be doubted, also, whether he regretted what he had done. Certainly he never imagined to what it would lead him. The idea that his people would have the courage to make him a prisoner does not seem to have crossed his mind, any more than did the fact that, once he had lost his position, he had become not only a useless, but an embarrassing factor in Russian politics. He went back to Mohilew, to the headquarters of that army of which he had been the Commander-in-chief as well as the Sovereign, quite naturally and in the same quiet manner he might have done in the days gone by. He did not even seem to yearn after his wife and children, and never once did he suggest the advisability of returning to Tsarskoye Selo. Of all the people assembled around him he appeared the most unconcerned. This indifference lasted even when he found himself faced with captivity and when the former Head of his Staff, General Aléxieieff, came to acquaint him with the decision of the Provisional Government to arrest him.

His wife, left alone in the Palace where she had spent so many happy days, did not perhaps share his indifference; she certainly displayed the same apathy. Alexandra Feodorowna, from the moment that she saw her schemes of personal grandeur frustrated, gave up the game; she gave it up with more dignity than her husband had ever shown—this much must be conceded to her. She never flinched before the insults that were poured down upon her; she never gave a sign that she was moved to anything else but disdain when General Korniloff read to her the orders of the Government in regard to her person, and acquainted her with the fact that she was a prisoner. She declared to the few people left with her that she considered herself only as a Sister of Charity in attendance on her sick children. The Empress had disappeared, outwardly at least, and perhaps it was just as well that she accepted the situation in this way, rather than attempt a useless resistance, which could only have added to her unpopularity.

But still the fact remained that the whole Russian Revolution had been conducted after the style of a comic opera of Offenbach. No one at first had recognized its serious character. No one had seemed to realize that it constituted the most portentous event of the last hundred years or so. Those who had carried it out had done it on the spur of the moment, without thinking of what would follow; and the Monarch who had bowed his head under its decrees also had not suspected that a morrow was there, waiting for the results of what was being done to-day. The historical stick that had been wielded by Peter the Great had been transformed into the ridiculous sword of the Grand Duchess of Gerolstein.

LA GRANDE DUCHESSE DE GÉROLSTEIN
An opéra bouffe in three acts and the greatest success of Offenbach's career. Originally banned in France after the Franco-Prussian War.

XXVII

A PRISONER AFTER HAVING BEEN A QUEEN

A NEW life began for Alexandra Feodorowna. Until that fatal day when she was taken into captivity her existence had been one of ease and luxury. She had been the Empress of All the Russias, being revered by some as almost a divinity, the absolute mistress of all her surroundings, with servants in attendance on her, eager to execute any commands it might please her to lay upon them. She had not a wish which was not instantly gratified; the misfortunes that had assailed her (I am not speaking now of those that fell upon Russia) had always left her indifferent; they had existed more in her imagination than in reality. Suddenly without warning and, what was even worse, at the very moment when she had expected to reach even loftier heights than the one upon which she was placed, she had been hurled down into an abyss of sorrow, of misery, and of pain such as she had never imagined she could ever know. She was no longer a Sovereign; her courtiers, servants, attendants, had all vanished with the exception of a very few, and those she had never cared for much, in the days of her prosperity. Her children were sick and she could not even obtain for them a doctor's help. Her friends had fled or were in prison; her Crown had been wrested from her; she was a prisoner, deprived of the means of communicating with her own people and relatives; the guards who surrounded her Palace were no longer placed there to protect her safety; they were intrusted with another mission, that of watching over every one of her movements and of preventing her from getting any news from the outside world. Instead of crowds gathered to cheer her, she saw assembled under her windows an angry multitude asking for her blood and calling out to her that she ought to be punished as a traitor. She had no friends, no money, no influence any longer. The dream had come to an end, and she found herself facing stern reality, a reality against which it was useless to struggle.

Her husband came back to her, a prisoner, likewise, but with perhaps less consciousness of the horror of their position than she had. They had to settle down to a new life entirely different from the previous one—a life of idleness, of inaction; an existence which made them realize with every step they took the awful change that had overtaken them. When they wished to go out they had to ask permission to do so from an officer who often refused it out of pure malice. They had to pass before sentinels who no longer presented arms to them, who only sneered in their faces as they saw them hurry through a room or a corridor, anxious to escape insult or outrage. No one was allowed to come near them. They were condemned to a solitude in which they were continually reminded of the days gone by forever.

    A few faithful attendants had been left them, it is true, but these last friends were just as badly off as themselves, and could do but very little to alleviate the miseries of a position which was an illustration of the famous verses of Dante, that there is nothing more dreadful during days of misery than to remember the past joyful ones. Even religion, which for such a long period of years had consoled the Empress in many sad and troubled hours, had ceased to be a comfort to her; divine service, during which her name and that of her husband were carefully omitted from the liturgy, was only one new source of torment for her. It seemed to her as if the Church as well as the Russian nation repulsed her and treated her as a pariah and an outcast. Another woman, with higher, loftier views, would have looked with more philosophy on these small sides in a great tragedy, might perhaps even have failed to notice them. But for Alexandra Feodorowna they constituted something far more tangible and real than the fact that the House of Romanoff had lost its Throne.

    She would most probably have wished to discuss with the Czar all the events which had brought about the catastrophe, but even this comfort was denied to her. The Provisional Government had issued orders that husband and wife should not be permitted to communicate with or see each other, except in presence of witnesses. Some people have said that this was an unnecessary cruelty, but it seems that there was some reason for this decision. A strong party at that time was clamoring for repressive measures in regard to the ex-Empress. Papers had been found in which her negotiations with the Kaiser had been revealed, and the question of bringing her to trial had been seriously discussed. But no one wished to see the former Czar mixed up with this business, as it was generally felt it would be a great political mistake to make a martyr out of him.

There was, however, ground to fear that if he were permitted to speak with his wife alone, she would contrive in some way or other to entangle him in her personal intrigues. This Mr. Miliukoff, then Minister for Foreign Affairs, wished to avoid, for reasons of a general political order, and Mr. Kerensky for other ones of a purely personal character. It seems that this leader of the Socialist party in the Duma had, before events had transformed him into a Minister, spoken also with certain agents of the Kaiser who had contrived to remain in the Russian capital. Nicholas II. had friends who, knowing this fact, warned the radical chief that if any harm was done to the former Sovereign his own participation in eventual peace negotiations with the enemy would be exposed. Can one imagine that when Nicholas was told of this fact he only blamed those who had thus attempted to save him, saying that he did not like blackmail of any kind, even when it was performed for his advantage? That man who had been one of the most important political factors of his time was not even shrewd enough to see that it was only politics which could save his life after they had dispossessed him of his Throne.

The Provisional Government, so long as decent men composed it, would have been willing to spare any unnecessary humiliations to the former Czar and his family. Unfortunately, the military men who had been put in charge of the Palace of Tsarskoye Selo and of its inhabitants did not share this opinion, and there is no doubt but that the deposed Monarch was subjected to insult, as well as to all kinds of small and petty annoyances calculated to make him feel bitterly the change in his position. I do not believe personally in the tales which were put into circulation as to his having been hustled about by the soldiers on guard at the castle the day he had returned there a State prisoner from Mohilew, a few short weeks after he had left it a powerful Sovereign. For one thing, his devoted aide-de-camp, Prince Dolgoroukoff, was with him, and he would most certainly have interfered had any violence been used in regard to his master. But the unfortunate Nicholas was made in other ways to drink the cup of humiliation to the dregs.

The troops were told not to salute him; the sentries were forbidden to present arms to him; he was addressed as Colonel Romanoff by his jailers; his letters were opened and his expenses controlled in a searching, insulting manner which must have been terribly bitter for him to bear. Every kind of newspaper containing insults addressed to him or to the Empress were sent to him or put in his way. When

THE ROYAL FAMILY

Seated, from left to right: The Grand-Duchess Marie; the Czarina, with the little Grand-Duke Alexis at her knee; the Czar; the Grand-Duchess Anastasia, the youngest of the four girls. Standing, from left to right: The Grand-Duchess Tatiana, and the Grand-Duchess Olga immediately behind her father.

he went out in the park he was often accosted by people who upbraided him for all the misfortunes that had fallen upon Russia, for which they made him responsible. I do not mention insignificant daily worries, such as the shutting off of the electric light, or of the water-pipes, so that the unfortunate Imperial Family was left without baths, and other small unpleasantnesses of the same kind. These would perhaps not have been noticed if the other ones had not been there to remind the once powerful Czar of All the Russias that he was at the mercy of the subjects whose rights he had not respected and whose cries for freedom he had quenched in blood.

But Nicholas, in the midst of all these miseries, preserved the same impassibility he had displayed when the news of the disasters of Mukden and Tsu Shima had been brought to him, or when he had heard that Warsaw and the long line of fortresses that had defended the Russian frontier on the Niemen and the Vistula had fallen into German hands. He accepted everything with stoicism; he expressed no surprise at the blows which were being hurled at his head. He simply remained indifferent, perhaps because he was too much of a fatalist to rebel, but most probably because he had not yet grasped the real significance of all that was happening to him.

The Empress was not so resigned, in spite of her apparent apathy. She had more reasons to fear for her personal safety than her husband, and she knew very well that in case of a rising of the anarchists in Petrograd she would be the first victim they would claim. This dread led her into another of the mistakes which she was continually perpetrating, the mistake of trying to call to her rescue her German cousin.

According to people whom I have reason to believe exceptionally well informed, she caused certain information to be carried to the Kaiser. In return for this she implored him to try and save her, together with her children. Of course this became known to the Provisional Government, but the latter wished to spare her, partly because it feared that if her new misdeeds were published nothing could save her from the wrath of the public, and it did not wish the Revolution to be dishonored by the murder of a defenseless woman, whatever that woman might have done. But the question of the transfer of Nicholas II. and of his family to a place where he could be guarded more closely than at Tsarskoye Selo was discussed seriously. It is likely that this would have been executed already during the first six weeks which followed upon his abdication if other things had not interfered, and if in rapid succession the men who had taken up the task he had been unable to fulfil had not in their turn disappeared one after the other, making room for Ministers more advanced in their opinions and more devoid of scruples as to the punishment which they believed ought to be inflicted on the former Emperor.

Alexandra Feodorowna had been subjected to a strict examination of her political activity by the military authorities in charge of the district of Petrograd, and particularly by General Korniloff, who had a personal grudge against her and who did not spare her in the scathing reproaches which he addressed to her. But nothing could shake the equanimity of the haughty Czarina. She sneered at the General, she scorned his threats, and proudly declared to him that she would not reply to any of his questions, as she did not recognize his right to address them to her. While her husband showed no sign of impatience under the affronts which were showered down upon him (on the contrary, he exhibited absolute submission to the will of those who had taken him captive), the Empress remembered the position which she had occupied a few days before, and simply smiled at her persecutors with a disdain that had certainly something exasperating about it if one considers the intellectual and moral standard of the people to whom this proof of her contempt was addressed.

Alexandra refused to show that she suffered from the change that had taken place in her position, while her husband hardly knew whether he was suffering from it or not. There lay the difference in their two characters and in their way of meeting the catastrophe which had changed their whole lives and destinies.

There came, however, a day when the composure of the Consort of Nicholas II. failed, when she at last gave way to despair. It was during the afternoon when her friend and the confidante of all her thoughts, Anna Wyrubewa, was taken away from her, and carried off to the fortress of SS. Peter and Paul in Petrograd. Until that day the Empress had not felt quite alone in her misery. There was at least near her one person with whom she could speak about all those dear dead ones whose memory she either cherished or worshiped. So long as that friend was there the miserable Empress could talk about Orloff, Raspoutine, and the prayer-meetings during which the latter evoked for her the spirit of the former. When Anna was taken away from her this last consolation came also to an end. Henceforward the solitude of Alexandra Feodorowna was to be complete; and nothing was left to her except her eyes to weep, and her memory to remind her of those whom she had loved and lost. The horrors which were to follow, the Siberian exile whither she was to be sent, were to leave her unmoved. She had inwardly died in that terrible hour when the last friend and the sharer of all the secrets of her life had been snatched away from her arms.

XXVIII

THE EXILE

THESE days in Tsarskoye Selo which seemed so hard to bear were Paradise compared with what awaited the previous masters of this Imperial place.

Soon there came one August morning when a man who a few months before had been known only as one of the leaders of that Socialist party which the Government of Nicholas II. had taken such trouble to suppress, and whom the tide of events had transformed into one of the Ministers of the new Russian Republic, Alexander Feodorovitch Kerensky, entered, uncalled for and unannounced, into the private apartments of the former Czar, and acquainted him with the new decision to which the Government of the day had come in regard to his person and to the fate of his family. How he did it, how he mustered the courage to inform the fallen Monarch that he was about to be exiled to that distant Siberia, where so many people had been sent during his reign and had gone cursing his name, no one knows. None of the actors in this scene ever revealed its details.

It is probable, however, that Kerensky had experienced far more emotion in signifying to the deposed Sovereign the horrible punishment to which he had been condemned than the latter had displayed when receiving the terrible news, the nature of which must have completely bewildered him. Of all the things he had expected, this was certainly the last. The possibility of a public trial, in imitation of the one of Louis XVI. in France, had been discussed between the Emperor and the Empress, and they had in a certain sense both schooled themselves for such a supreme ordeal. But exile in cold, bleak Siberia, in that land of mystery and of crime, of heroic deeds and ignoble deaths—this solution of the difficulties of a position which was daily growing more threatening had never presented itself to the mind of the last Russian Autocrat. Tobolsk, too! This dreariest spot in a dreary country; this accursed place from which all those who could do so fled away with alacrity! What could have been more awful than to have chosen it as the future residence of a Monarch who, if all was well considered, had by his own act rendered himself impossible as a ruler in the future? No political necessity required his being sent so far and there were many other places where he might have been just as safe, and not quite so unhappy, as in that small Siberian town. Can one wonder if despair took hold of the souls of the unhappy Czar and of his wife? Can one wonder if he exclaimed that he would have preferred to die rather than have to meet such an atrocious fate?

There were his children, too; his innocent children, who had done no harm and who were to share this miserable destiny to which he was condemned. Kerensky had told him that the young Grand Duchesses and their brother would be left free to follow their parents in that distant land whither they were to depart, or to remain in Russia if they preferred. But this was almost adding insult to an abominable injury, because the children could hardly do anything else than decide to accompany their father and mother. Verily nothing was to be spared to Nicholas II., not even the knowledge that his daughters and his idolized boy were about to be exposed to hardships which it was hardly likely they could survive.

Nevertheless, he took the news bravely, and neither he nor Alexandra Feodorowna murmured. The latter is credited with having remarked that the new Government had evidently wanted to please her by sending them to a place that could only be dear to her on account of its associations with Raspoutine, who had been born there. Whether this is exact or not I will not undertake to say. What is true, however, is that she left with dry eyes the place where she had spent so many happy years, and that not even during the religious service, which was celebrated in the private chapel of the Palace, when the blessings of Heaven were invoked in favor of people about to start on a long journey, did she shed a tear. All the other assistants, including her daughters, sobbed passionately and bitterly the whole time that it lasted.

Though the hour of the departure of Nicholas had been kept as secret as possible, the whole town of Tsarskoye Selo knew that he was about to leave it forever. The population for once was awed before the immensity of the disaster. One was used in Russia to people being sent to Siberia; one had seen a Menschikoff, a Biren, a Dolgorouky, and more recently a Prince Troubetskoy, and a Mourawieff-Apostol, start on this dreaded journey whence so few ever returned. But here was something different. Here was a

Romanoff about to go there where his ancestors and himself had deported so many, so many, entirely innocent people. Here was the dreaded Sovereign, whose name had been for such a long time mentioned only with reverence and with fear, exiled like one of those persons whom he had sent to Siberia with a mere stroke of his pen. Here was Nicholas II., formerly Emperor and Autocrat of All the Russias, here was the mighty Czar who had been crowned at Moscow, reduced to the condition of a common criminal by the subjects whose rights he had violated, whose consciences he had trampled upon, whose liberty he had taken away, and whose lives he had in so many instances tried to destroy. Truly the sight was appalling, and there were but few among those who in the early twilight of a summer morning looked at that mournful train which was carrying away toward the distant north so much dead grandeur and such awful misfortune, there were but few who did not realize that they were witnessing something more tragic and more solemn even than a funeral, that they were looking upon the end of a great chapter in Russian history, and that it was not only a Czar who lay buried under all these ruins, but also something of the past glories of the country.

They had been, after all, together with the future of the Romanoff dynasty, a great race that had produced great men and they deserved to have, as their last crowned descendant, some one better than Nicholas II. had proved to be. They had often been cruel, more frequently even unjust; they had never respected what the world is used to venerate and to esteem; they had shown themselves hard in regard to their subjects; but they had made out of dark, ignorant Russia an immense Empire on which even more immense hopes had been built. They had exercised on several occasions a restraining influence over the exaggerations of half-educated and half-instructed men with all the instincts of the savage and but few of the qualities of the civilized being. They had led their people through danger, through war, through many momentous days of rejoicing as well as of anguish. They had been a part of the Russian nation, and with their disgrace and fall something in that nation, too, had been dishonored and had perished.

With the personal failure of Nicholas II. this book has nothing to do. From the very first day that he had ascended the Throne of Russia it had been evident to any person gifted with the talent of observation that his reign was bound to end in disaster and in ruin;

that all the work performed by his late father, who, after all, when things are well considered, had been a great man and a wise Sovereign, would very soon be destroyed by his want of character and of principle and by his abominable selfishness. People, however, had hoped that during the years supreme power remained concentrated in his hands at least blood would not flow. No one's imagination had conceived the horrors of Mukden and of Tsu Shima, nor the massacres which took place at Tannenberg and in the Polish plains. No one had suspected that the rivers would be dyed red while he went on living unconscious and unconcerned about all the misery which would remain forever associated with his name. When destiny was at last fulfilled in regard to him, it was discovered that it had also been fulfilled in regard to Russia, and this was what the country could not bring itself to forgive him; this is what his ancestors also would not have pardoned him had they been able to arise out of their graves, and to

> Rend the gold brocade
> Whereof their shroud was made*

in horror at all the evil perpetrated by their unworthy descendant.

Yes, in the past they had been a strong race, these Romanoffs, about whom no one thinks any longer to-day in the vast realm that owned them once as its masters and lords. But it would be useless to deny that crimes without number were committed by them and that injustice flourished during the centuries when they could dispose absolutely of the fate of millions and millions of human creatures whom they killed and tortured at their will and according to their fancies. Perhaps it was a just punishment for the ruthless cruelty of some of them that the glories of their race came to an end and perished, together with all the traditions that had surrounded them for such a long time, because of the follies, sins, mistakes, blunderings, iniquities, and indecisions of a weak, characterless, and half-witted man and of a superstitious, intriguing, and half-demented woman.

* Longfellow, "The White Czar."

END OF BOOK I

BOOK TWO

Yulia Alexandrovna von Dehn, known as Madame Lily Dehn

There are some listings for this book published in 1919 but the widely available versions were published in London (April 1922) by Thornton Butterworth; the same publisher of *The Last Days of the Romanovs* by Robert Wilton in 1920 and The World Crisis 1911-1914 by Winston Churchill, in 1923. The other edition, *The Last Tsarista* was published in New York (September 1922) by Boston, Little, Brown, and Company.

— *The English edition was published by Thornton Butterworth Limited, London, 1922.*
— *Pages: 249 total pages*

THE
REAL TSARITSA

BY

MADAME LILI DEHN

CLOSE FRIEND OF THE
LATE EMPRESS OF RUSSIA

THORNTON BUTTERWORTH LTD.
15 BEDFORD STREET, LONDON, W.C.2

First Published – April, 1922

TSARKOE SELO

To

H.I.M. ALEXANDRA

THE LATE EMPRESS OF RUSSIA

Adieu, c'est pour un autre monde

The fate which destined thee for lofty power,
And crowned thee Sovereign o'er an Empire wide,
Placed too the cup of suffering by thy side
And sorrow gave thee for imperial dower :
How little did'st thou dream in Fortune's hour
Thy barque would founder on such tragic tide
Of blood as wrecks a mighty nation's pride,
While black the clouds of Revolution lower!
What force sustained thee in those days of stress
When death and ruin held their sombre court,
And frenzied mob set might all right above ?
What made thee still thy prayers to Heav'n address,
And solace to thy stricken spirit brought ?
'Twas faith unshaken in a God of love.

OSWALD NORMAN.

N.B. This list is from the London edition and varies from the U.S. edition published September 1922. It's included her for completion with the corresponding page numbers listed for this volume where applicable.

LIST OF ILLUSTRATIONS

H.I.M. Alexandra	(Frontispiece)	n/a
Anna (Ania) Virouboff	(to face page 48)	219
Her Imperial Majesty with the Tsarevitch	(56)	207
H.I.M. The Tsar with Officers of the Royal Yacht "Standart"	(96)	250
The Empress on board the "Standart"	(96)	250
H.I.M. on board the tender of the "Standart"	(96)	250
Grand Duchess Olga	(104)	n/a
Grand Duchess Tatiana	(104)	n/a
The Imperial Family	(152)	186
Royal Shooting Party	(160)	291
The Tsarevitch at G.H.Q.	(160)	291
The Tsarevitch and his Spaniel "Joy"	(160)	291
His Imperial Majesty and the Tsarevitch	(184)	308
H.I.M. Alexandra (end of 1915)	(184)	308
The Empress at Tobolsk	(208)	n/a
The Empress with Tatiana at Tsarkoe Selo	(208)	n/a
The Grand Duchesses Marie and Anastasie	(208)	n/a

FACSIMILIA

Part of letter of June 5/18, 1917	(to face page 240)	
Part of letter of March 2/15, 1918	(245)	350
Note from the Empress	(208)	n/a
Part of letter on day of departure for Siberia	(234)	n/a
Letter from the Empress (1916)	(235)	340
Part of letter of 30th July, 1917	(241)	n/a
Christmas Card drawn by the Empress	(242)	346

FOREWORD

In giving to the world my memories of the Empres
Russia, I do not wish to pose as one who is biased by a long and intimate friendship. I write of the Tsaritsa as I knew her: the real Tsaritsa. I was not acquainted with the heroine of the films, the hysterical devotee, or the pro-German who, it is asserted, betrayed both her country by adoption and the country which knew her as a granddaughter of Queen Victoria and the daughter of a much loved English Princess.

PART I – Old Russia

CHAPTER I

I WAS born on the beautiful estates in South Russia which belonged to my grandmother and my uncle. My father was Ismail Selim Bek Smolsky, whose ancestors hailed from Lithuanian Tartary, and my mother, before her marriage, was Mlle Catherine Horvat, whose grandfather had been invited by the Empress Elizabeth Petrovna to come from Hungary and assist in the colonization of South Russia. Colonel Horvat, who was half Serbian and half Hungarian by birth, was appointed general of the armies of the South by the Empress, and there is a story in our family that when he first arrived in Russia he was taken to the summit of a high mountain and told to look at the panorama of fields and forests lying beneath him.

Colonel Horvat dutifully admired the view, but an unexpected surprise awaited him. "Look well around you, M. le Colonel," said his guide, "the country, as far as you can see, is yours; it is the gift of the Empress!" Truly an Imperial gift, but all that remains of those great possessions are the estates where I was born. These properties were situated on the Dnieper, in the country known as "Little Russia," which in former times was the seat of the Ukranian Government. My forefathers became typical Russian noblemen; they were lavishly generous where their inclinations were concerned, and it is asserted that one of them once exchanged a large forest for a sporting dog which he especially coveted!

Revovka, my birthplace, was close to the other estates which came into our possession through Prince Goleniktcheff Koutousoff, the hero who saved Russia from falling into the hands of the French. It was a delightful old house, standing in a wellwooded park, with avenues of lime trees where the nightingales sang, and as I write, I can smell the unforgettable perfume of the limes, and recall the beauty and peace of the surroundings; it was, indeed, a real fairyland. All was prosperity and happiness at Revovka. The village nestled close to the Great House, and my ancestors were buried in the church. There were rows of little cottages which were whitewashed every week; the roofs were thatched with reeds, and the gardens were gay with flowers. A cherry tree stood in every garden (cherry trees are typical of South Russia), it was the country of cherry trees, spotless houses and simple joys.

The peasants were on the best of terms with my family, and they regarded my grandmother Horvat as a beneficent deity who replaced the reed roofs when they were destroyed by fire, and who supplied them with unlimited quantities of fuel. They were quite contented, and my grandmother still employed some of the peasants who had once been given to her as serfs. In the old days, it was customary to include a few serfs in a bride's *corbeille*, and the ten peasants who had been chosen to accompany my grandmother to Revovka adored her. "People say that we were unhappy as serfs," they would often remark, "but we were always well looked after—our landlord and our owner was also our father."

The peasant as master or mistress was invariably a tyrant, and I remember hearing about a beautiful girl who had become the mistress of a great nobleman, and who out-Heroded Herod in her arrogance. She employed her family to do her laundry work, and she always insisted upon her linen being rinsed in running water. If her petticoats were not sufficiently starched, the whole batch of her relatives was flogged. Personally, we did not resent the lack of starch, to this extent, but I suppose that this family flogging may be regarded as typical of the usual procedure of beggars on horseback!

My grandmother, Mme Horvat, *née* Baroness Pilar, was the sweetest of women, and I loved her with a child's passionate devotion. She used to tell me all kinds of stories, and our old nurse ably seconded her. Whenever we walked by the river, and I exclaimed at the beauty of the lilies, I was thrilled anew by hearing how, long ago, when the Tartar hordes descended on Beletskovka, the women and children used to wade into the water, and shelter under the broad green lily-leaves until the marauders had passed. The peasants at

Revovka were extremely superstitious, and they believed implicitly in witches and warlocks. It was common knowledge that certain women possessed tails and bewitched the cows, and woe betide the widow who mourned her husband too much! He would assuredly return in the likeness of a big snake, and make an unwelcome descent down the chimney. I was terribly scared by some of these narratives, and I much preferred the pretty customs prevalent at certain seasons, now vanished, alas! under the Bolshevik regime, since the teaching of Lenin would seemingly only include the ritual of blood in its category.

I chiefly remember the quaint methods of divination practised on New Year's Eve, when the girls of the village went out to listen at the closed doors, and those who heard a man's name mentioned were certain to marry within the year. They varied these proceedings by throwing their slippers over their heads, to see if they fell in the shape of anything that might be construed into an initial letter. Others preferred to try and catch the rays of the moon in a towel; all pretty gay conceits, dear to the heart of girlhood, and, on St. Catharine's Day, cherry tree branches were put in water, and, if the bare wood blossomed by Xmas, then marriage bells were about to ring.

Midsummer Day was sacred to the river, a survival doubtless of those pagan customs which are so difficult to destroy. Large fires were lighted along the river banks, and the village maidens, wearing wreaths, leapt into the water, across the fires, and left the wreaths in the river as an offering, perchance to the God of Streams. The next morning, they set out to look for their wreaths, and those who were lucky enough to find them discovered by the direction in which the wreath had been washed up the way by which marriage would come.

The storks brought luck, and they were invited to sojourn with us by means of wheels placed in the roofs on which they built their nests. The solemn birds were family friends, and, whenever a baby stork fell from its nest, everyone went to enormous trouble to put it back.

My grandmother had a passion for embroidery, and she employed from ten to fifteen girls constantly working for her. She believed that, as a typical industry, the art of embroidery in South Russia ought to be revived, and she spared no pains or expense over her hobby. She proved conclusively that the progress of the nations from East to West had left its traces even in embroidery patterns, as she often saw similar designs in antique carpets and Venetian work.

None of my grandmother's embroideries was ever sold: whenever a piece was finished, it was labelled with the date of its commencement and completion, and packed away in great presses, already nearly full of exquisite work. She presented a quantity of this embroidery to the Grand Duchess Elizabeth, the Tsaritsa's sister, when she was received into the Greek Church. My grandmother had the honour of acting as godmother to the Grand Duchess, and I believe her "christening" present was much appreciated. The embroideries were really wonderful: the designs were never drawn, the threads only were counted, and the pattern was evolved in this painstaking manner. Some of my grandmother's favourite designs were taken from Easter eggs, which were first covered with pinked-out wax, and colour inserted in them. Snow crystals formed another inspiration; my grandmother never tired of utilising anything decorative, and she was unusually successful. I like to think of those quiet days—the industrious girls, and the good feeling which existed between the employer and the employed. It is difficult to realise that the progress of Revolution has destroyed all this, that the great presses have been broken open and their contents dispersed to the four winds, and that to ask a peasant to pass her time profitably would be accounted a sin.

My grandmother, notwithstanding her patriarchal outlook, could be the "grande dame" when occasion warranted, and my old nurse used to relate how one of her neighbours, a certain Prince, came to ask her in marriage. This gentleman believed in the impressiveness of pomp and circumstance, so he arrived at Beletskovka in a carriage and six horses. He was most courteously received—and refused—by my grandmother, and, when he drove away, his horses, by some preconceived arrangement, cast their shoes in the avenue. These "cast off" shoes were solid silver, a mute testimony to his wealth, and, as he passed through the village, he and his postillions distributed undreamt-of largesse. The Prince was a haughty personage, who lived in a gorgeous mansion boasting fifty rooms. He gave two balls yearly, when an orchestra was specially sent for from Petrograd, a four days' journey from his estate. But in the Prince's opinion nobody, save my grandmother and our family, was good enough to associate (even as a dance partner) with him and his, so the balls were rather tame affairs, a few couples only taking the floor, but those who did were—like Cæsar's wife—entirely above suspicion.

Silver horse-shoes, expensive orchestras, and other unconsidered trifles cost money, and, as the male members of this super-aristocratic family were all in Hussar regiments, financial ruin eventually came as an uninvited and unwelcome guest: it closed the doors of the castle, the orchestra came no more, and the ladies of the house sought refuge in an institution for noble ladies of fallen fortunes!

My great-aunt, the Baroness Nina Pilar, was a romantic figure in my childhood's memories, as her name conjured up the fascination which surrounds those who breathe and have their being in the air of Courts. She was Lady-in-Waiting to the Empress Marie, wife of Alexander II, and she made her appearance at Court when she was sixteen, under the auspices of Countess Tizenhausen (another great-aunt), Grande Maîtresse de la Cour, who brought up Felix Soumarokoff, the grandfather of Prince Felix Yousopoff. There was a great deal of gossip about the paternity of old Soumarokoff, who had been confided, as a baby, to Countess Tizenhausen by an intimate friend, but nobody was ever any the wiser, and Soumarokoff's antecedents remained an unsolved mystery.

The Empress Marie loved Aunt Nina, and the Emperor was very kind to her until my innocent relative was the victim of chance, and a *costumière*. The Emperor had become infatuated with a certain Princess Dolgorouky, and one day, when my aunt was walking on the Quai, looking especially attractive in a new costume, she suddenly heard a voice addressing her in most endearing terms. She turned sharply round, and found to her dismay that the voice was the voice of the Emperor! Explanations followed, and my aunt discovered that Princess Dolgorouky possessed a duplicate of her new costume, and, as their heights and figures were similar, it was a case of mistaken identity.

The Empress was almost always ill, but her Court was distinguished by its elegance and refinement, and my aunt was one of the acknowledged leaders of fashion.

Like most pretty women, Aunt Nina had her love story, but she never married. Her Prince Charming was the Grand Duke Nicholas, to whom she was secretly engaged.

But, when the Grand Duke asked the Emperor's permission to marry his inamorata, the Emperor, who had never forgiven the contretemps on the Quai, refused his consent!

The unhappy lovers met in Switzerland when Aunt Nina was in attendance on the Empress, and there they bade each other farewell, and threw their engagement rings into the lake. The Grand Duke never forgot his broken romance, although he, like most lovers, eventually married someone else! But he was present at my aunt's funeral, and stood silently and sorrowfully looking at the coffin which held many of the dreams and much of the enchantment of his youth.

Aunt Nina practically sacrificed her life to save that of the Empress, although the latter died years later at Petrograd, when, it is asserted, a luminous Cross appeared over the Winter Palace, typical of her physical and mental sufferings.

It so happened that when the Empress and my aunt were driving in Switzerland, their carriage was run into by a cart, and, in order to prevent one of the shafts from striking the Empress, my aunt stood up to protect her, and was badly bruised in the chest. Some time afterwards cancer developed, but my aunt survived her Imperial mistress, and became Lady-in-Waiting to the Empress Dagmar, and Grande Maîtresse de la Cour to the Grand Duchess Elizabeth. The Grand Duchess was very much attached to her, and at her death she begged my grandmother to take her place. My grandmother, for family reasons, declined the honour, but she often used to visit the Grand Duchess and the Grand Duke Serge, and I remember hearing her describe the pathetic figure presented by the Grand Duchess after her husband's assassination, when she had relinquished the spleendours of life and had become a nun at Moscow.

My childhood was chiefly passed on my grandmother's estates. We led a somewhat patriarchal life at Revovka; a simple existence which will, I fear, never again return, and it is exceedingly difficult for me, as a Russian, to recognise the peasants of then and now. The average peasant was kindly by nature, entirely ignorant, and excessively difficult to educate. Whenever my grandmother tried to persuade her tenants to send their children to school, the answer was always the same: "Knowing how to read and write doesn't provide food. Our parents got on very well without education, our sons can do likewise." Their faith in the aristocratic class was boundless, they entirely depended on their landlords, but the Russian peasant has always, unfortunately for himself, been easily influenced by speeches and printed matter— hence the complete success of the Revolutionary Propaganda, and the belief in many of the false statements circulated in order to damage the Imperial family in the eyes of the people. I cannot defend our own attitude in not attempting to combat this danger; we were aware that it existed, but only one section, known as the Black Band, tried to destroy it by counter propaganda. Its efforts were unsuccessful, it received no support, for the very good reason that *nobody believed that the masses would rise*. The Russian aristocrat, secure in his class prejudices, and his optimistic faith in *himself*, was as loth as the French aristocrat of 1789 to realise that his position was, or could ever be, insecure!

The South Russian peasant, as I knew him, was a poetical, simple soul. After dinner we often used to watch the men turning their horses into our meadows for safety, and securing the animals' legs with chains, in order to prevent any inclination to roam. They invariably sang whilst making these nightly preparations, and they danced afterwards in the bright moonlight which flooded meadows and woodland with a white radiance. They had many quaint customs at Revovka, which may not be uninteresting to English readers who only know the Russia of to-day as a strange and poisonous growth, and not as the orchid which had its home in the eternal snows—a curious simile, perhaps, but in my mind a correct one. Our country, in many respects, was an exotic growth; super-refinement walked cheek by jowl with ignorance, and an almost oriental luxury brushed the skirts of poverty. It was a land of extreme contrasts, where emotions and passions either ran riot or else were suppressed to an undreamt-of extent. It was almost inconceivable at one time that the family coachman, who obstinately turned his horses' heads in the direction of home because he met a white dog in the road, could ever become the Bolshevik who would have murdered his employers instead of protecting them from the bad luck attendant on the unwelcome animal!

I must admit that my grandmother was as superstitious as her coachman. She believed implicitly in dreams, and an old woman from the village was always sent for to expound the more exciting ones. I remember that one of her dreams had a disastrous sequel, inasmuch as it involved the dismissal of a very devoted servant who, my grandmother dreamt, had attempted to kill her. She resolutely declined to see him again, and he was sent away to another estate. I supposed she was influenced in this by the knowledge that, on several occasions, she had "dreamed true."

Our peasants confided all their joys and sorrows in my grandmother, and, when any of them married, we were always invited to the wedding. This invitation was issued on set lines; the bride-to-be, dressed in full national costume, plentifully bedecked with flowers and ribbons, came with her bridesmaid to the servants' sitting-room, where she was received by my grandmother. The girl thereupon knelt, and bowed three times, informing my relation what an honour our presence would confer on her family, and, gratified by the assurance that we would promise to come, she withdrew, all smiles! After the ceremony, which always took place on a Sunday, the whole of the wedding party came back to our house and assembled on the terrace, where a village orchestra discoursed strange sweet sounds, and where unwearied dancing enlivened the music and singing. We always gave one kind of present—a cow! When I married, our employees surpassed themselves and gave me, not a cow, but two oxen!

We fasted on Christmas Eve until the first star appeared, when we partook of a heavy supper of which the fifteen courses always included fish. Hay was strewn under the tablecloth to remind us of the humility of the Manger, and it was customary for the children to carry the Christmas supper to their friends and relations. All the windows of the Chateau were darkened, but one was left open, and, when the first star appeared in the serene sky, this window was illuminated in honour of the Christ-Child. It was then that the children arrived "en masse," carrying revolving paper transparencies adorned with pictures of Christ; it was one illuminated stream of little children, and one of the prettiest sights imaginable.

New Year's Day was an occasion for general rejoicing, when the men of the village assembled on the terrace to congratulate us, throwing wheat in our pathway as a sign of prosperity. We then witnessed the procession of our servants, who filed past us, accompanied by their special charges. First, came the stablemen leading the horses, who, in addition to being superlatively well-groomed, were adorned with gilt crowns and many ribbons. Then came the herdsmen with their grave-eyed steers, whose horns were gilded in honour of the New Year; the sheep were accompanied by the shepherds, and the cortège was terminated by the poultry maid, who escorted a turkey smothered in ribbons.

On the first New Year's Day after the Revolution, the crowd came to the Chateau as usual, but there was no procession of animals, no smiling faces, and no wheat-strewn pathway. We were confronted by scowling peasants, who roughly informed us that henceforth nothing belonged to us, since they were masters. But to do our own people justice, the better minded amongst them absented themselves, and only the worst characters were in evidence—and these, in their turn, were under the evil influences rampant in towns. I have no hesitation in stating that the motive power in the destructtion of Russia emanated, and still emanates, from the Jews.

When the snows began to melt, the children and young people heralded the approach of Spring with song. Joining hands, they wandered singing in the twilight, a lovely, living chain of Youth in its Spring-time. They repeated these songs at Easter, that wonderful festival of Resurrection and the rebirth of Nature. On Holy Thursday the Gospels were read in the churches until midnight, and everyone carried a taper. My mother's estates were situated in the mountains, and it was a picturesque sight when the peasants wended their way churchwards at Easter. The church was half-way up a steep ascent, and the procession of taper-bearers could be traced by hundreds of lights, as two villages participated in the ceremony.

Revovka was an entrancing home for a child blessed, as I was, with an imaginative temperament. We had our particular White Lady, a tragic phantom who haunted the Park, and who used to swing in the branches of the lime trees. She had been the mistress of one of my great-uncles, and she was buried in the Park. No one seemed to know her fate, but it was said that she was beautiful and unhappy. Her grave was marked with a flat stone, without any inscription, as the poor little creature had sought refuge from love and life in self-destruction. But Nature was kinder to her than Man, and an enormous bush of wild roses threw out caressing arms towards the cold stones, and showered pink petal-tears on the unhonoured dead.

There was a similar forgotten grave on my father's property, formerly a huntingbox of the Kings of Poland. The occupant of this grave had been the mistress of a king, and, like the beauty of Revovka, she had killed herself; but she was a restless spirit, and she used to haunt the Park and the house in the summer, running swiftly across the greensward, wearing little scarlet slippers and darting up the staircase, her scarlet heels tap-tapping as she went her way, unsubstantial and fantastic as the morning mist.

I used to dream all kinds of dreams, but I never anticipated what Destiny held in store for me. I was, by nature, timid; I was to become courageous through force of another's shining example. I was to see and experience the real meaning of selfless love, and I was to know the comfort and beauty of religion. I do not say that I was irreligious—few Russians are really irreligious—our Belief is too deeply rooted—but I did not yet understand the meaning of the word Faith.

I always looked forward to our yearly pilgrimage to the Convent of Tchigrin, twenty-five miles away from Revovka. Custom ordained that we should proceed thither on foot, but the carriage invariably went with us! The convent contained a miraculous Virgin which, when the Turks pillaged Tchigrin, had been taken away by them. One day a disconsolate nun walking on the river's bank saw something floating on the surface of the water. The Virgin had returned to her convent, and from that time it became the scene of wonderful miracles, and many pilgrimages. I liked Tchigrin; it breathed an atmosphere of calm, standing alone in the midst of dense pine woods. But the wind, which respects neither convents nor humanity, was occasionally unkind to Tchigrin, as it swept away the sand which filled the crevices of the walls, almost like natural mortar, and the nuns daily brought bags of sand wherewith to repair the damage. This sand-carrying was an especial duty connected with Tchigrin, and occasionally it was a penance—but I think those simple creatures rarely deserved punishment.

I have perhaps devoted too much time to the festivals, ghosts and unexciting incidents of a country life. But I have done this in order to explain many subsequent happenings which would be otherwise incomprehensible to an English public. These events cannot, and must not, be judged entirely from an English standpoint. We are a race apart, our country is one of extreme mysticism and superstition. It is a land of miracles, where the holy pictures are believed to shed tears, and where every village possesses its seer and its saint. It would be possible to cover the length and breadth of England in a week's motoring tour, thus England is of necessity more circumscribed. One could not see Russia in such a manner. It is a country of vast distances, of densely populated cities, and lonely tracts which extend for thousands of miles. You cannot contrast the mode of life prevalent at Tooting with that of Tobolsk, or compare the customs of Moscow with those of Manchester. Our upbringing is entirely un-English. True, we are citizens of the world, we are indeed cosmopolitan, but—once a Russian, always a Russian. The Tsaritsa told me that, when she first came to Russia, she was greatly surprised to find that Russian servants did not understand the art of blackleading grates. She had always been accustomed to see shining grates in England when she stayed with her grandmother at Windsor—in Petrograd, shining grates were non-existent. We are miles apart from English ways in little things like these, and no Englishwoman worthy of the name has ever been known to be ignorant of the use of blacklead. But *we* ought not to be condemned for the non-recognition of its virtue. It is merely a question of outlook. In connection with these differences of outlook, I cannot do better than quote the words of a contributor to the "Daily Mail"; they will plead for my opinions, as the writer possesses the peculiar gift of racial and temperamental understanding:

"We have," he writes, "in England a cold fish-minded way of affecting to laugh at what we are prone to call local superstition. Let me tell you that this point of view will not work in Africa." (He is dealing, I fancy, with Morocco.) "What is obviously a childish hallucination in Hampstead or Newcastle is sober reality under this immense blue sky. You can disbelieve a lot of truths you do not understand as you strap-hang homewards, but you will learn to believe everything in Africa."

Might not this also apply to Russia?

HER IMPERIAL MAJESTY AND THE TSAREVITCH

CHAPTER II

MY childhood and early girlhood were passed quietly at Revovka and the Crimea. But I loved Revovka, and, whenever I went to stay with my uncle at Livadia, I took with me a little earth from the place which, to me, represented home. The great event at Revovka was the visit of my uncle Horvat, who came from Siberia to see my grandmother once a year. He was head of the Siberian railways, and occupied a political position which corresponded with that of a Viceroy of Ireland. He was a typical Horvat, tall, with deep, kind eyes, and he was also a very clever man. On the night of his arrival I never went to bed, and I remember that we saw the dawn together; he did not reach Revovka until 3 a.m. It was touching to witness his meeting with my grandmother. They were entirely "en rapport," and he was my greatest friend as well as my much loved uncle.

I never went to school. My first tutor was a priest, but, as I hardly knew Russian (we always spoke French at home) and he knew no French, I made little progress; afterwards Miss Ripe, an English governess, took me in hand, but I think she looked upon us as very much behind the times. The old house was protected at night by a watchman, and I regarded his intermittent coughing and his heavy tread somewhat as a lullaby. Whenever he went to the next town by boat, the watchman "called" my grandmother's maid in a very curious manner. He was an illiterate peasant, and time, as time, conveyed no meaning to him, so he would occasionally tap on the maid's window and tell her that such and such a star was in the sky. By this simple calculation she was enabled to judge how much longer it was permissible for her to remain in bed. Winter was a delightful season at Revovka, and I always wanted to be decorative, and drive out in the antique sledges which were painted with trails of flowers, and magnificently gilded. The modern sledges, covered with carpet, and piled up with bear skins, were not nearly so pretty. English people always associate sledges with wolves, and imagine that a winter's drive in Russia is fraught with desperate danger. The wolf terror is fast becoming a legend; wolves are now only found in districts far from the haunts of men, although an old custom at

Revovka ordained that lanterns were hung outside the stables at night to scare away the wolves! But I met a wolf unawares one evening when I was crossing the park. I had never seen one of our national animals face to face, so I thought that the big grey creature was a dog. I called it, and ran towards it, desirous of its better acquaintance, but it merely regarded me with furtive, unfriendly, green eyes, and then turned and trotted away in the opposite direction. When I reached the house, I described my encounter with the strange dog, but, greatly to my surprise, my story produced general excitement, and a search-party set forth to look for the footprints in the snow. These proved to be typical wolf marks, exactly like the print of a thumb, but our visitor had, by this time, completely disappeared.

When I was a young girl the disaffection in Russia was already well on the way to Revolution. In 1905, when I was staying with one of my uncles in Livadia who had charge of the Emperor's estates at Yalta, we were not left long in ignorance as to the methods which were employed by the Revolutionary Agents. It is now well known that most of the seeds of Revolution were sown at Yalta, but it was dreadful to see the boats smothered in red flags and to hear the Marseillaise sung defiantly from the water, since my uncle had prohibited all political meetings on land. One day, it was discovered that the golden eagles which marked the boundaries of the Emperor's estate had been broken and overthrown, but this act of vandalism was always attributed to the Jews and the more hot-headed of the students. There was general excitement in the Crimea at this time, and a few of the Revolutionary printing presses were secretly set up at the Grand Duke Constantine's Castle of Orianda, which for some reason had fallen into decay. It had always been my ambition to visit the ruins of Orianda, so one day I persuaded my cousins to accompany me thither. It was a forbidden expedition, but we considered the possible results of our disobedience would be amply compensated for by the mysteries of the underground passages, which we at once began to explore. As we neared the end of one of these the sound of distant voices broke the stillness, and, terrified out of our wits, we did not know whether to beat a retreat or to dare all and discover whence the sound proceeded. Curiosity eventually conquered cowardice, and we crept cautiously along until the darkness was lit up by a glow of a large fire. Thinking that we had now reached the entrance to the infernal regions we turned and fled precipitately, and, risking punishment, described the whereabouts of Hell to my uncle. And Hell, in a way, it proved to be, as it was discovered that secret printing presses existed underground, and that most of the evil propaganda had emanated from Orianda.

Although the Jews instigated much of the prevalent sedition, the biter was occasionally bit, and in 1905 there was serious trouble. Many people assert that the actual Revolution began by beating the Jews, as some of the soldiers returning from the war became very unruly, and set about the Jews most unmercifully.

My mother, who had married as her second husband an officer in a regiment stationed near us, received news of the trouble just at the moment when we were starting to drive into town. But she rather pooh-poohed the warning, until she saw for herself that the report was not exaggerated. We first encountered people fleeing through the fields, and, when at last we reached civilisation, we found the town in a state of confusion. Windows were broken, Jewish shops pillaged, and the leaders, regardless of the protesting Hebrews, seized their goods and distributed them broadcast to the mob. The black and white praying robes peculiar to the Jews were in special request, as pieces of these, worn next to the skin, were supposed to render the wearer immune from marsh fever.

Next day, when I was walking in the Park, I found myself close to the walled-in right of way which traversed it, and, to my surprise and horror, I heard the passers-by giving vent to most undreamt-of declarations. "It's the Jews *now*," said someone, uttering a curse, "but wait until the next time. We have our orders: soon it will be the turn of the landed proprietors!"

The speaker spoke the truth. Some days later fires and pillage broke out around my home, and, from the terrace at Revovka, we could see an ever widening circle of flame, and our peasants informed us that, most assuredly, Revovka would suffer next. But we escaped, although the house of Madame Tchebotaiff, a great landowner and Revolutionist, was one of the first to be destroyed. She was afterwards sent to Siberia, a rather ironical form of justice, I am inclined to think!

When all was calm, the Duma came into existence, in which representatives of every class met in Parliament for the first time. Troops were sent to punish the peasants, and many of them were flogged by the soldiers. Our peasants were not included amongst the offenders. The idea of whipping human beings was repellent to me, and, girl though I was, I felt that we, as a class, were responsible for the existence of many evils, and that it lay with us to try and remedy them. But whipping was applied to the guilty as the most effectual and the most easily understood antidote against rebellion: it is a barbarous punishment—in English eyes it must seem *utterly* so; but these whippings were as naught compared with the savagery and super-refinement of torture inflicted later by the whipped upon the whippers.

But my attention was soon to be diverted from rebellion and punishment. Shortly afterwards I went with my grandmother to Petrograd, where my marriage was arranged; in fact, I was already engaged when I was presented at Court. My fiancé was Captain Charles Dehn, of Swedish descent, whose ancestors had come into the northern provinces at the time of the Crusades, and the members of whose family were mostly generals or officers in the service of the State. Captain Dehn had taken part in quelling the Boxer Rebellion, and at the siege of Pekin he was the first officer to scale the walls of the Forbidden City in defence of the embassies. For this service he received the Order of St. George (the Russian Victoria Cross), and the Order of the Legion of Honour was awarded him by the ambassadors of the various nations represented in Pekin.

On his arrival at Petrograd he was presented to the Emperor, who appointed him an officer on the "Standart," and an officer of the Mixed Guard, whose members were chosen from various regiments, and many of whom were honoured by the personal friendship of the Emperor.

Captain Dehn was a great favourite with the little Tsarevitch and the Grand Duchesses, and he used to play with them in their nurseries, his nickname with the children being "Pekin Dehn." Both the Emperor and the Empress manifested the greatest interest in his engagement, and the Empress intimated to my grandmother that she wished to make my personal acquaintance.

My engagement was formally announced in 1907, but we waited in Petrograd for a month before we were received by the Empress. The Grand Duchess Anastasie was ill with diphtheria, and the Empress was nursing her at the Alexandria Palace, Peterhof, where, until all danger of infection had passed, she had isolated herself from the other members of the Imperial family.

How well I remember that first meeting with one whom I was to love so devotedly, and whose constant friendship has been one of my greatest joys. One summer morning in July, my grandmother and I arrived at the station at Peterhof, where my fiancé and a Court carriage were awaiting us. I was literally trembling with terror, and I was too excited to even notice Charles!

We duly reached the Alexandria Palace, but, as the Empress was still nervous about infection, it had been arranged that my presentation should take place in the Winter Garden attached to the Palace. We were received at the Palace by the Mistress of the Household, Princess Golitzin, who was exactly like an old picture, and whose adherence to regime made everyone dread being guilty of the smallest lapse of etiquette. But she was very kind and gracious to us, and I felt somehow that my simple white gown from Bressac's, and my rose-trimmed hat had met with her approval. As we walked through the Park to the Winter Garden I noticed a lady in one of the avenues, who stopped and looked at me intently. She was "petite," with an innocent baby face, and great appealing eyes, and so childish-looking in fact that she seemed only fit for boarding school. This lady was Anna Virouboff whose name was later to become associated with that of Rasputin, and whose friendship with the Empress has given rise to so many unwarrantable statements and scandalous stories.

I returned her scrutiny with interest, and we passed on with the Princess to the Winter Garden, a lovely tropical place, full of flowers and palms. It was exactly like a Garden of Dreams, at least I thought so until I saw the prosaically comfortable garden chairs, and noticed some toys and a child's dolls'-house. Then I decided that this beautiful garden must be real!

At last, advancing slowly through the masses of greenery, came a tall and slender figure. It was the Empress! I looked at her, admiration in my heart and in my eyes. I had never imagined her half so fair. And I shall never forget her beauty as I saw her on that July morning, although the Empress of many sorrows remains with me more as a pathetic and holy memory.

The Empress was dressed entirely in white, with a thin white veil draped round her hat. Her complexion was delicately fair, but when she was excited her cheeks were suffused with a faint rose flush. Her hair was reddish gold, her eyes—those infinitely tragic eyes—were dark blue, and her figure was as supple as a willow wand. I remember that her pearls were magnificent, and that diamond ear-rings flashed coloured fires whenever she moved her head. She wore a simple little ring bearing the emblem of the Swastika, her favourite symbol, and one which has given rise to so many conjectures, and been quoted triumphantly as proof positive of her leanings towards the occult by those who are ignorant of what it really meant to her.

Directly Princess Golitzin had left us alone, the Empress extended her hand for my grandmother and me to kiss; then, with a sweet smile, and a world of kindness in her eyes, "Sit down," she said, and, turning to Captain Dehn: "When is the marriage to take place?" she enquired.

My nervousness had vanished. I was no longer afraid; in fact it was the Empress who seemed shy, but she was, I found later, always shy with strangers, a trait peculiar to her and to her cousin, the Princess Royal, Duchess of Fife. However, this excessive shyness was not accounted as shyness in Petrograd, it was called German superciliousness! and as such it has even been described by some English writers.

The Empress talked to my grandmother for quite a long time, as she was anxious to hear the latest news of the Grand Duchess Elizabeth; she then chatted to my fiancé, and I noticed that she spoke Russian with a strong English accent. She afterwards addressed me as the blushing heroine of the morning, and she seemed quite pleased at the interest which I had displayed in the dolls'-house.

"Where are you going to spend your honeymoon?" she said, her blue eyes now mischievous. We told her. "Ah! . . . I do hope that I shall see you again very soon. I am quite alone, I cannot see my husband or my children, I shall be so glad when this tiresome quarantine is over, and we can be together again."

Our interview lasted well over half an hour. The Empress spoke French to my grandmother and me, she made no attempt to converse in German; then she rose to say good-bye, and we kissed hands. "I shall see you again very soon," she repeated. "Be sure you let me know when you return."

I went back to Petrograd almost beside myself with happiness. Mine was not the worldly pleasure of one who had been presented to an Empress. My happiness had its origin in another source. I felt instinctively that I had found a friend, someone I could love, and who, I dared hope, might love me! I was so tired out with my emotions that, on arriving home, I threw myself on my bed, regardless of my Bressac dress and my rose-wreathed hat, and I slept the sleep of exhaustion until four in the afternoon.

I was married two months later from my aunt's house in Livadia.

The Emperor received Captain Dehn before he left for the Crimea, blessed him, and gave him a beautiful ikon in a carved silver and gold frame. The Empress also presented him with an ikon, and, on our wedding day, we received a "wireless" from them, wishing us every happiness. This "wireless," so we heard afterwards, caused endless talk and many petty jealousies, as "wireless," then in its infancy, was only supposed to be used for important official communications.

We went to the Caucasus for our honeymoon and stayed three weeks in the mountains among the vines. It was the season of Autumn, and he had cast his flaming many-coloured mantle over everything. The wildness and luxuriance of that mountain region entranced me. I insisted upon being told all the legends connected with the locality, and I believed, with the peasants, that it was possible to hear the hoofs of the Centaurs, as they thundered down the passes in the silence of night. Gagree was an ideal place for a honeymoon, and I was actually sorry to return to my beloved Revovka, although we received a right royal welcome from my grandmother and her tenants.

Revovka was fifteen miles from the nearest railway station, but the whole of the way to our estate was illuminated with blazing tar barrels, and at every turn of the road we were offered bread and salt. Needless to say, the drive was a little protracted, and the *pièce de résistance* consisted in the two oxen which were presented to us at the journey's end.

My married life began under the most auspicious circumstances. Charles had promised me that he would always remain in the Emperor's Personal Guard, and I possessed a subconscious intuition that my future was to be closely connected with that of the Imperial family. This feeling did not arise from any worldly outlook, I never had any idea of the material benefits which might accrue to us through the Emperor's regard for my husband. My first meeting with the Empress had influenced me in an undreamt of manner. Although I felt it was ridiculous to associate any idea of sorrow with that radiant vision of the Winter Garden, I had, nevertheless, a strong feeling of fatality in connection with her. Time was destined to prove that my presentiment was right.

Our first home was in the Anitchkoff Palace, the residence of the

Dowager Empress Marie, where the Guards had their quarters, but afterwards we moved to Tsarkoe Selo. Our house was immediately opposite the Palace, and close to the barracks. The officers of the Personal Guard were most picturesque individuals, since each wore the uniform of the regiment from which he had been selected. There was no distinctive uniform; to be a member of the Guard was, in itself, an honour.

I used often to walk in the great Park of Tsarkoe Selo when my husband was on duty. The Palace dates from the time of Catherine the Great, and all the important receptions were held there. The Imperial family lived in the Alexander Palace, a white building in the style of the First Empire; the Palace had four entrances, the first was exclusively used by Their Majesties, two others were used for receptions, and the fourth was the entrance by which the Suite went to and fro. The Palace was entirely surrounded by the Park, in which was some beautiful ornamental water, a Chinese pavilion, and a bridge which connected the smaller park with that of the more important Palace.

As a young married woman, blessed with many kind relations and friends, it was not long before I took my place in Petrograd society. In 1907, one year after the Japanese war, life was not gay as many families were still in mourning, so those who looked for Court gaieties were disappointed—none being forthcoming. The Empress felt that the war was of too recent a date to warrant much entertaining; she was entirely sincere in this conviction, but her attitude did not meet with general approval. It was argued by the anti-Tsaritsa clique that an Empress of Russia belonged to Society, and not to herself. Her duty was merely to pose as a magnificent figure-head on the barque of pleasure—the war was over, and the world of Society wanted its ceaseless round of empty pleasures once again.

Petrograd Society was divided into many sets; each Grand Ducal Court had its own particular clique, and that of the Grand Duchess Marie, wife of the Grand Duke Vladimir, was perhaps specially joyous. The Grand Dukes, taken as a whole, led amusing lives; they were usually very handsome men—quite heroes of romance, many of them possessing a great admiration for the Imperial Ballet, in which they had various fair friends.

It was an expensive existence even in 1907, when Petrograd was supposed to be dull! People went every Sunday to the Ballet, and on Saturdays to the Théatre Français—this, a most fashionable rendezvous, where extremely decolleté toilettes were compensated for by an abundance of jewels! After the play, it was customary to adjourn to the Restaurant Cuba, or to that of L'ours, where a wonderful Roumanian orchestra enlivened supper; nobody thought of leaving the restaurants until three in the morning, and the officers usually remained until five! Occasionally, when I returned home in the early hours, I contrasted the dawn at Revovka with that of Petrograd; the same pearl, rose and silver tints painted the sky, but the dawn in South Russia witnessed no flight of human butterflies whose wings had been singed in the flame of pleasure. I was young enough to enjoy life, but at times our restless gaiety seemed to hold a hidden menace.

English was the medium of conversation in Society at Petrograd; it was invariably spoken at Court, and, although once fashionable to have German nurses, the fashion in 1907 was to have only English ones, and many Russians who could not speak English spoke French with an English accent! The great shopping centre was "Druce's" where one met one's friends, and bought English soaps, perfumery and dresses. The "Druce habit" primarily emanated from Court where everything English was in special favour—Jewish Society and that of the "haute finance" existed in Petrograd, but neither touched us.

The great enlivenments of the Season after the Japanese war were the Charity Bazaars. The Grand Duchess Marie always organised one in the Assemblée de la Noblesse, a huge building where an ultra-smart throng of Society leaders sold all kinds of pretty and expensive trifles. The Grand Duchess Marie (who was a German Princess) occupied the centre of the room, and sold at her own table. She was a tall, striking-looking woman, but not so handsome as the Grand Duchess Cyril at whose table I occasionally assisted. All the Grand Duchesses had tables, as was the case with the greater and lesser lights of Society. In fact the position of one's table was the index to one's position in Society. The bazaars were brilliant functions, the toilettes were wonderful, and it was quite the usual thing to change one's gown three times during the day. The air was heavy with perfume, flowers were lavishly displayed, and the tired vendors occasionally refreshed themselves with the best brands of champagne.

The Empress had her own table at the Assemblée de la Noblesse, and I sold at it once. She made quantities of things herself, instead of sending haphazard orders to Paris or London. The homely intimacy of her nature was very evident in this habit, nothing at her table was useless; she was true to type, the type of Queen Victoria's descendants, the Empress shared Queen Mary of England's love for needlework, and, like her, crocheted many pretty "woollies" for bazaars.

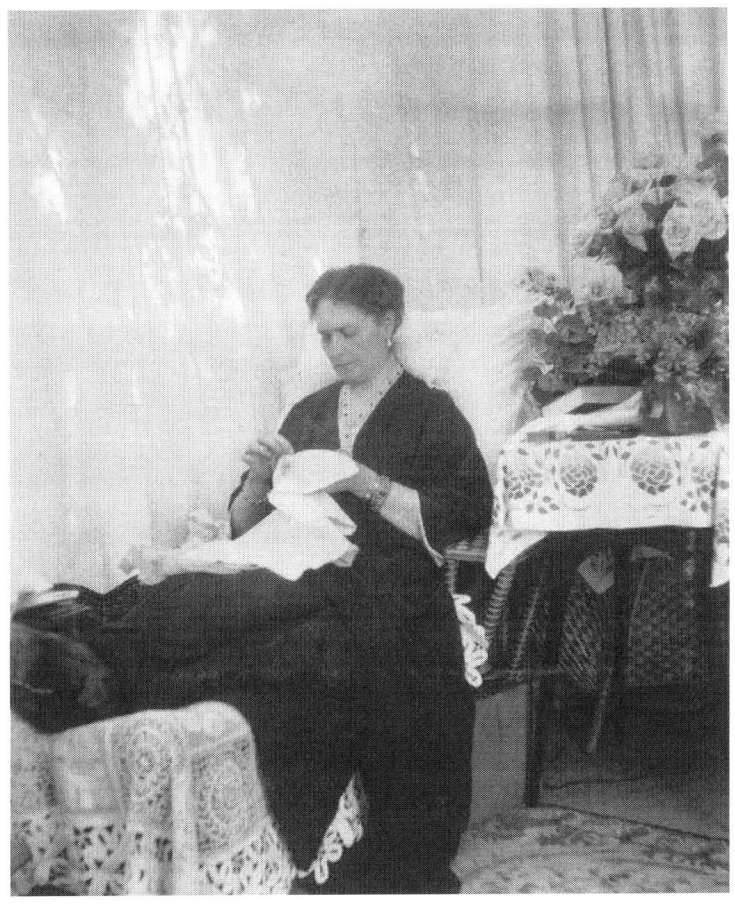

EMPRESS ALEXANDRA FEODOROVNA
Regularly worked on her needle point work

CHAPTER III

ALMOST immediately after my arrival at Tsarkoe Selo, I made the acquaintance of Anna Virouboff, the Lady of the Avenue, and my distant cousin, as her grandfather and my grandmother were related.

It is exceedingly difficult for me to discuss Anna Virouboff, as I am confronted with the tremendous prejudice which exists against her. In England she appears to be a Borgia-like heroine of the films, an hysterical sensualist, the mistress of Rasputin, and the evil genius of the Empress. Her political power is supposed to have been that of a Sarah Jennings and a Catherine Dashkoff, and her influence at Court paramount.

If I deny these charges, I shall lay myself open to the accusation of blind partisanship, and I shall be deemed an utterly untrustworthy chronicler; but, notwithstanding these possibilities, I can do no less than speak of Anna Virouboff as I knew her from 1907 until the day in March, 1917, when we were both removed from Tsarkoe Selo by order of Kerensky.

Anna's father, General Tanief, was Honorary Secretary of State, and all her family were connected with officers in the Imperial House. She married the same year as myself, but before her marriage she was deeply in love with General Orloff, who commanded the Lancers, and who was a great friend of the Empress. Rightly or wrongly, Her Majesty thought that General Orloff would be too old a husband for Anna, and, although the General loved her, and desired nothing better than to marry her, Anna yielded her will to that of the Empress, and accepted Lieutenant Virouboff, to whom she was married in the Palace Chapel at Tsarkoe Selo. The union turned out a complete failure, and I believe that the Empress's original interest in Anna was intensified by the fact that she was indirectly responsible for this unhappy marriage. The Empress accepted what she considered to be her responsibilities very seriously, as her salient characteristics were thoroughness and a fine sense of justice. It was not difficult for her to show more kindness to one whom she already loved, and whose unhappiness was now so poignant. Anna was one of those beings who always look as if someone has hurt them; one wanted to "mother" Anna, to amuse her, to hear her confidences, and to laugh at her exaggerated joys and sorrows.

This photograph of Anna Alexandrovna was used in *Nicholas II – Tsar to Saint*, on page 313. A different image is used here, of a slightly younger Anna.

ANNA ('ANIA') VIROUBOFF

In appearance, Anna is a person entirely different from the Anna Virouboff of the films and the novel, and she even dares to differ from more serious descriptions of her. She is of middle height, with brownish hair, large, appealing, long-lashed, grey-blue eyes, and a little turned-up nose. She has a baby face, all pink and white, and, alas for the Vampire the Anna of romance, she was then very fat. But her smile was charming, and her mouth pretty; she was weak as water, as clinging as the most obstinate ivy, and the Empress treated her much in the way that one treats a helpless child. Anna was excessively good-natured, always ready to help others, in whom she was never able to see evil. This virtue (for I suppose it is accounted a virtue) was the ultimate downfall of Anna. She was too credulous, and, therefore, too easily imposed on. She adored the Imperial Family with the devotion of an adherent of the Stuarts, but—and now I am about to make a statement which will be probably treated with derision—*she possessed no political influence whatever*; she could not influence the Empress one hair's breadth; the Empress petted her, teased her, and scolded her, but she never sought Anna's advice, save in questions of charity.

The Empress and her former Lady-in-Waiting were, however, one where religion was concerned; they shared the same religious sympathies in the midst of an unsympathetic and jealous entourage, and, as Anna did not get on well with the entourage, this fact gave the Empress an additional reason to protect her friend. Anna told me that some of the Ladies-in-Waiting disliked the Empress solely on account of her friendship with her, and, although she had told the Empress that, were she given an official position, all jealousies and comments would be silenced, the Empress had refused to entertain the suggestion.

Later on, when I became on intimate terms with the Empress, she gave me the reason for her refusal.

"I will never give Anna an official position," she said. "She is my friend, I wish to retain her as such. Surely an Empress is allowed the right of a woman to choose her friends. I assure you, Lili, I value my few real friends more than many of the persons in my entourage."

Four years after her marriage, Anna met with a train accident. She never again walked without crutches, her body was completely deformed, but even then slander did not spare her, and evil tongues in Petrograd asserted that, as well as being the friend of the Empress, Anna Virouboff was the mistress of the Emperor!! After her accident, the Empress gave Anna a carriage and pair, and often drove out with her. She lived in a pretty little house which had once belonged to Alexander I, and she usually lunched at home, after spending the morning at the Palace. "The children" liked her, everyone who really knew her liked her, and the best proof of her absolute harmlessness lies in the fact that after the Revolution she was never condemned to death. Surely, if she had been such an evil creature, the first action of those in authority would have been to destroy her? But Anna Virouboff lives, and perhaps one day she will defend herself.

One Monday, shortly after my marriage, I received a note from Anna, asking me to dine with her that evening. Captain Dehn had been in Petrograd for several days, and, as I was rather lonely, I was glad to accept. The dinner was very gay, several officers had been invited, and Emma Fredericks, the daughter of the Minister of the Court, was also a guest. At half-past nine, we heard the sound of wheels, and a carriage stopped outside the house. Anna instantly left the salon, and, a few minutes after, the door opened, and, to our great astonishment, the Emperor, the Empress and the Grand Duchesses entered. They were all laughing, as this surprise visit had been arranged by the Empress, who, seating herself, told us to do likewise, and motioned me to come to her.

"I told you that I should see you again very soon," she said, smiling, and thereupon she began to talk in the most friendly and simple manner.

Once again I had that curious, inexplicable foreboding of tragedy, but no tragedy lurked in that bright, gay room, and my gloomy thoughts were soon dispelled when I was presented to the Emperor.

This was the first occasion on which I had spoken to His Majesty, and I found him as charming and friendly as the Empress. His kind eyes, and his smile, struck me at once, he seemed to move in an aura of goodwill, and his peculiar fascinating charm of manner has been admitted even by his enemies, as M. Kerensky acknowledged that the Emperor possessed one of the noblest natures he ever met!

The Emperor, who bore a striking likeness to his cousin, King George of England, was a very amusing conversationalist, and blessed with a keen sense of humour. He instantly put me at my ease, and I made the acquaintance also of the Grand Duchesses, then quite girls, with whom I was later to become on terms of the closest friendship.

The Empress, having expressed a wish to play Halma, we had two or three games; she was greatly addicted to Halma, but she had one little lovable weakness in connection with it. She never liked to lose! The Emperor played dominoes in the next room, and afterwards Emma Fredericks sang, the Empress accompanying her. Her

Majesty was a very good pianist, and played with rare feeling, but her excessive shyness often precluded her from playing in the presence of others. At midnight the Imperial family took their departure, and the Empress whispered to me: "Au revoir, we shall meet to-morrow."

She did not forget. I was commanded to go to the Palace on the morrow. It was Tuesday, and I remember how pleased I was. "Everything nice happens on a Tuesday," I kept saying, for this was an old belief of mine.

After my meeting with the Empress at Anna's house, I often went to Tsarkoe Selo, and the Grand Duchesses and I used to ride on the wooden switchback, which was set up in one part of the Palace. It was tremendous fun, and we slid and played together for hours, but I quite forgot that I was a married woman and that I had hopes of becoming a mother in some months' time. However, the Empress had some idea of my condition, and one day, after she and Anna had been watching our performance on the switchback, Anna drew me aside.

"Lili," she said, "I've a message for you. The Empress wants you to be very careful just now." She held up a playful finger. "So no more switchback!"

During the months that followed, the Empress manifested the greatest kindness towards me. She insisted upon her own doctor attending me, and, when the Imperial family went yachting about a fortnight before the birth of my baby, my husband received orders to absent himself from the "Standart," and to remain with me instead.

This act of consideration was due to the Empress, and it caused, like the "wireless," much petty jealousy and a good deal of comment.

But the expected baby delayed his arrival, and, when the Imperial family returned to Tsarkoe Selo, the Emperor's first words to my husband were:

"Has the baby come?"

"No, Sire, not yet."

"Well, well, don't worry, Dehn, these things will happen, you know."

However, the baby arrived next morning, and shortly afterwards Anna Virouboff came to make enquiries on behalf of the Empress, bringing with her two lovely ikons, and a package done up in tissue paper and covered with masses of rambler roses. The package contained a thin, fleecy shawl, and my happiness was complete when Anna told me that the Empress wished to be my son's godmother.

This was a great honour, but it presented difficulties, inasmuch as the Dehns, in order to benefit from certain family monies, were obliged to be baptized as Lutherans. The Empress was told about this, and, although she made no objection at the time, I was to discover later how deeply she was imbued with the faith of her adopted country. At the first christening, the Empress attended in person, and held the baby, now known as Alexander Leonide. She gave me a beautiful sapphire and diamond brooch, and all kinds of presents, and for seven years the question of the child's religion was never mooted between us. But, at the end of that time, the Empress told me that her dearest wish was that "Titi" (as she called him) should be received into the Greek Church.

"It is more than a wish, Lili," she said earnestly, "it is a command. I insist upon my godson being Orthodox. He must be baptized before Christmas."

This quiet persistency seems to me to afford one of the most conclusive proofs of how Russian the Empress had become. It may be argued that most converts are usually fanatics, but this was not so in her case. With that "thoroughness" which I have mentioned as one of her chief characteristics, the Empress was now more Russian than most Russians, more Orthodox than the most Orthodox. She was intensely religious. Her love of God and her belief in His mercy came before her love of her husband and her children, and she found her greatest happiness in religion at a time when she was surrounded by the panoply of Imperial splendour. She was to derive consolation from her religion throughout the Via Dolorosa of the saddened years, and, if it is indeed true that she met death in the noisome cellar-room at Ekaterinburg, I am sure that the same ardent faith sustained her in that last moment of agony. She told me that she had hesitated to accept the Emperor's offer of marriage until she felt that her conscience would allow her to do so and she could say with truth: "Thy country shall be my country, thy people my people, and thy God my God."

Titi's second baptism took place during the war at the St. Theodor Cathedral. I had come to Tsarkoe Selo from Reval, and the ceremony took place at 8 in the morning. The Grand Duchesses Marie and Anastasie were present at the first service, but the Empress, previously indisposed, came with the Emperor and the suite to the second service, and afterwards took Holy Communion. Titi was obliged to remain during both services, but he was a good little boy, and he held his lighted candle carefully and firmly the whole time.

After the service we went back to the Palace, and the Empress displayed more emotion than she had done at the first christening. I could see how deeply the religious question had affected her all these years. She told me how relieved she was, how pleased, how she felt now that all was well with the child, and she gave her godson a wonderful ikon of St. Alexander and a Cross engraved with her initials.

But I must return to the earlier days—I have wandered from my narrative to give this example of how Russian the Empress was at heart; hers was no eye-service—to know her made it impossible to doubt her genuineness.

The Empress was always sweet with Titi. She adored children, and she often came to my house, when she nursed the baby and whistled to him. This amused her, and she declared that Titi knew her whistle and always opened his eyes whenever he heard it. I remember that on the morning after the "Lutheran" baptism the Empress paid me a surprise visit.

"I've come to see the baby," she said. "Let me go to the nursery and fetch him."

I followed her upstairs, and she took Titi out of his cot and carried him to the drawing-room, where she played with him for an hour, sitting on the carpet to do so.

I think I am right in saying that our affectionate friendship began from the birth of Titi. It was then that the Empress first called me "Lili," and as "Lili" I caused much mystification during the Revolution, when this signature was supposed to possess some cryptic meaning.

The Imperial Family spent part of that year in Finland, whither my husband accompanied them, and I and the baby went to stay with his parents. I was at Petrograd during the winter, and I saw a great deal of the Imperial Family, and learned to love them all. They led the simplest of lives; the Emperor often amused himself during the evening with a game of dominoes, and I worked with the Empress and her daughters. It was a real "vie de famille," the life which appealed to them as individuals, but not the life which appeals to the smart world, with which the Empress had so little in common. This was my first Christmas at Petrograd, and I deter-mined to have a little tree in Titi's honour. I came in from my shopping late in the afternoon of Christmas Eve, and at 6 o'clock a courier arrived with a large box full of all kinds of "surprises." This was a present from the Empress—she always sent a similar box at Easter, and it always arrived at 6 o'clock. Indeed, so punctual was this present, that my husband often used to hide the box and pretend that it had been forgotten—but I knew better!

We were invited to spend Christmas Day with the Imperial Family. There was a gigantic Christmas tree, the Grand Duchesses and the Tsarevitch thoroughly enjoyed themselves, and busied themselves in the distribution of friendship's offerings. The Empress had one curious fancy in connection with her Christmas trees: she always insisted upon blowing out the candles herself, and she was quite proud because she was able to extinguish the topmost candle by some extraordinary effort of breathing.

And now I feel I must speak of the real Tsaritsa, the Empress whose personality is known to so few—the Tsaritsa who was the most misjudged and unfortunate of human beings. I know in my heart that Time, the best historian, will make clear much that is dark. Even now, slowly, it is true, but none the less surely, people are beginning to wonder whether the Empress was in reality the pro-German and the hysterical *exaltée* she is supposed to have been. She did not deign to defend herself from the calumnies and lies which were scattered broadcast in Russia; to such a nature, these trials were sent by God—all that *she* had to do was to *endure*. But I saw her tears when she and the Emperor received the news of the loss of the "Hampshire" and the death of Kitchener. These were no Judas tears—hers was the grief of the woman and the Sovereign at the death of a brave soldier, and yet, whenever her name is mentioned in England, people say carelessly: "Oh, she saw to the torpedoing of the 'Hampshire,' and wasn't she the mistress of Rasputin?"

A pro-German, and the mistress of Rasputin!! Must this then, be the epitaph of the friend whom I knew, and the Empress to whom I owed the respect of a subject? I am not blind to the knowledge that any vehement defence may do her memory still more harm, but, nevertheless, I am impelled to write of her as she existed in her home, and in our hearts.

I have read and heard almost all that has been laid to her charge; I am no skilled writer, I know little or nothing of politics, but I can lay claim to some knowledge of my own sex. During the awful days of the Revolution, the Empress spoke to me as woman to woman. Her mind constantly dwelt on the days of her girlhood, her life with her grandmother, and the unhappiness of her childhood at Hesse

Darmstadt.

The Emperor was the love of her life. She told me herself that he was her first love, but, the greater her love, the greater her fear lest she would prove unworthy. She gave herself to Russia when she married, and she accepted Russia as a sacred trust; but she and the Emperor were always more husband and wife than Emperor and Empress— they lived the intimate life of happily married people, they liked simplicity, they shrank from publicity, and this love of retirement was the source of many of the evil reports which assailed the Imperial Family.

The Empress told me that when she cried at the marriage of her brother her tears were said to be tears of jealous rage at seeing herself dispossessed of authority.

"But, Lili, I was *not* jealous. I cried when I thought of my mother; this was the first festival since her death. I seemed to see her everywhere."

She described the dull Palace, its strict regime, her father's intermittent kindness, and how much she had looked forward to her visits to Windsor. I think that the intimacy with her grandmother unconsciously brought out the Early Victorian strain in the Empress's character. She undoubtedly possessed this strain, as in many ways she was a typical Victorian; she shared her grandmother's love of law and order, her faithful adherence to family duty, her dislike of modernity, and she also possessed the "homeliness" of the Coburgs, which annoyed Society so much. The Russian aristocracy could not understand why on all the earth their Empress knitted scarves and shawls as presents for her friends, or gave them dress-lengths. Their conception of an Imperial gift was totally different, and they were oblivious of the love which had been crocheted into the despised scarf or the useful shawl—but the Empress, with her Victorian ideas as to the value of friendship, would not, or could not, see that she was a failure in this sense. The Empress was in many ways as thrifty as her grandmother, but she did not share the miserly proclivities of her uncle, the late Duke of Saxe-Coburg. Her father was not a wealthy man, in fact life at Darmstadt was occasion-ally a question of ways and means. The Empress had been taught to be careful. She *was* careful.

"When I was engaged, Lili, I showed my grandmother some of the jewels which the Emperor had given me. What do you think she said?"

"I cannot imagine, Madame."

"Well . . . she looked at my diamonds and remarked: 'Now, Alix, don't get too proud!' The Queen was a tiny creature, and she wore such long trains . . . but she was very forceful." Then, reminiscently, "My sister Elizabeth and I always loved the little houses in England . . . dear little houses set in their pretty gardens. You'll see them one day, but I never shall."

Queen Victoria had instilled in the mind of her granddaughter the entire duties of a *Hausfrau*. In her persistent regard for these Martha-like cares, the Empress was entirely German and entirely English—certainly not Russian. I have mentioned her horror when she arrived at Petrograd and discovered that the servants were unaware of the use of blacklead. This was an actual worry to the Empress.

"I wanted my grates blackleaded every day," she said. "They were in a very bad condition, so I called one of my maids and told her to do the grate, only to discover that it was not within her province. Eventually a man-servant was sent for, but imagine, Lili, I had actually to show him how to blacklead a grate *myself*."

This practical side of the Empress was entirely distasteful to the entourage—they laughed at it equally as much as they criticised her friendships with people whom they did not consider in any way worthy of the friendship of an Empress of Russia. I and Anna came under the category of the unworthy, for, although we were well born, we were not of the "sang azur" of certain noble ladies who were desirous of admittance into the charmed circle. The Empress was accused of not being true to class, but on one point she was inflexible; she allowed no interference with her friendships. I sometimes wondered why she preferred "homely" friends to the more brilliant variety—I ventured to ask her this question, and she told me that she was, as I knew, painfully shy, and that strangers were almost repellent to her.

"I don't mind whether a person is rich or poor. Once my friend, always my friend." Yes, her loyalty was indeed worthy of the name of a friend, but she put friendship and its claims before material considerations. As a woman she was right, as an Empress perhaps she was wrong.

The aristocracy never tried to understand the real Tsaritsa. Their pride was up in arms against her—she found no favour in their eyes. I remember an incident which went to prove this, and which was widely discussed at the time.

Princess Bariatinsky, who then happened to be one of the Maids of Honour to the Empress, was a charming woman, but, like most of the aristocracy, she was excessively proud. One day, hearing that the Empress was about to go out, the Princess held herself in readiness to accompany her, but the Empress left the Palace by another entrance, accompanied by Mlle. Schneider, a Russian lady who gave the Empress lessons in Russian.

This unintentional slight was too much for the Princess. She, metaphorically and literally, put on her hat, and departed never to return, remarking as she did so: "*Quand une Bariatinsky met son chapeau, c'est pour sortir.*" The Empress detested any form of snobbism. One day, during the Japanese war, she was busy at one of her working parties at the Winter Palace; the windows of the salon opened on to the Neva Quai, and from where she sat the Empress could see the soldiers and officers passing to and fro. Suddenly she looked intently out of the window—an expression of distaste on her countenance—and she sighed impatiently. An officer ventured to ask her what was the matter. The Empress pointed to the Quai:

"*That* is the matter," she said, indicating an officer who had just been saluted by some soldiers, but who had not returned the salute. "Why cannot an officer recognise the men by whose side he may one day fall? I detest such snobbism," she added, coldly.

The scandals about the Empress, circulated by propaganda and rumour, will be believed, alas! for many years. She is credited with dabbling in occult practices, with a belief in Spiritualism, and of even attempting to call up the illustrious dead in order to influence the Emperor, who is supposed to have indulged in various dramatic séances at the Winter Palace. Perhaps these stories originated in the more or less retired life led by the Empress. This retirement was often enforced—she was a delicate woman, but, although many writers state that she suffered from the hereditary malady of her father's family, she never mentioned its existence to me. Her heart was weak, owing to rapid child-bearing, and at times she experienced great difficulty in breathing. I never saw the slightest trace of hysteria. The Empress was apt to get suddenly cross, but she usually kept her feelings well under control. Apart from her delicate health, there was another reason for these periods of retirement. The Tsarevitch and the Grand Duchesses were often ailing, the Empress was a devoted mother, and she insisted upon being with her children and sharing the duties of a nurse. The maternal element was strongly developed in her; the Empress was never so happy as when she was "mothering" somebody, and, whenever a person had gained her affection and her trust, she never failed to interest herself in the smallest details connected with him and his.

Her occultism has been grossly exaggerated. Her superstitions were of the most trivial description: she thought that a bright day was propitious for a journey, that the gift of an ikon to her was not propitious, but her fancy for the sign of the Swastika was not for the Swastika as a *charm*, only as a symbol. She told me that the ancients believed in the Swastika as the source of motion, the emblem of Divinity. The significance of it as a "luck bringer" never crossed her mind. "Faith, Love and Hope are *all* that matter," she would say. I will readily admit that she possessed a strong element of mysticism which coloured much of her life; this was akin to the "dreaming" propensities of her grandfather, the Prince Consort, and environment, and the Faith of her adoption fostered this mystic sense. English writers condemn this trait. I have before me a book in which the author quoted the opinion of one of the most bitter enemies of the Empress. "Alexandra Feodorovna," he says, "is an interesting type for future psychologists, historians and dramatic authors . . . a German Princess educated in England, on the Russian Throne, a convert to a peasant's religious sect, and an adept at occultism. She is made of the substance that those terrible, tyrannical Princesses of the XV-XVII centuries in the western countries of Europe were made of; those Princesses who united in their personality the despot Sovereign, bordering on the witch, and skirting the fanatical visionary, who were completely in the hands of their reactionary advisers, and their insinuating wily confessors."

I had read the book containing this extract before I began to write my memories of the real Tsaritsa. I read many passages with eyes half blinded with tears, sometimes I felt mine would be an impossible task. How could I, an unknown name in England, attempt to combat such statements? I am not assuming for one moment that the writer of the book was ill-disposed towards the Empress; he wrote for posterity, setting down his own opinion and that of others. But I am curious to know if he ever knew the Empress personally, and if he ever shared the intimate life of the Imperial Family. I did *both*—not only in the days before and during the war, but also in the days of despair, when murder and sudden death faced us at every turn. It was then no time for pretence—but the Empress never changed; she was the same unselfish soul, the same devoted mother and wife, the same loyal friend.

The material for another book which was largely circulated in England was supposed to have been "given" to the author by a lady well known, and in great favour at Court. This novel—for it was, in many respects, fiction pure and simple—was mentioned to me, and, upon reading it, I was amazed to find the names of persons who never existed, and who were, therefore, never at Court. There was no attempt to hide names under pseudonyms or initials—these imaginary beings lived, moved and had their being in the book as real individuals!

I was so much interested in the creative genius of the "Court Lady" that a friend of mine wrote to the part-author and asked him, on my behalf, to disclose her name. My request was refused: the part-author said that he was under an honourable vow of secrecy not to disclose the name of his collaborator!

But was this sporting? The book contained certain damning statements against the Empress, it bristled with inaccuracies; truly, anonymous Court histories cover a multitude of untruths! But surely those who profit thereby should have courage enough to come out in the open when certain questions arise. You either make a statement, or you do not. If you believe in its truth, you should not be ashamed to say why, and wherefore, and to acknowledge the source of its origin, but I am inclined to think that the words, "I gave my word not to say who told me," place little value on malicious gossip, either in books or in everyday life.

CHAPTER IV

THE Empress was an early riser. She had six dressers, of whom the chief, Madeleine Zanoty, was an Italian by birth, whose family had long been in the service of the Hesses. Louise Toutelberg, known as "Toutel," the second in authority, came from the Baltic, and there were four others. The dressers had three days' service, but none of them ever saw the Empress undressed or in her bath. She rose and went to her bath unassisted, and slipped on a Japanese kimono of silk or printed cotton over her undergarments when she was ready to have her hair arranged. The Empress was extraordinarily modest in her disarray, and in this the Victorian influence was again discernible, as her conception of the bedroom was à-la-mode de Windsor and Buckingham Palace in 1840. She did not countenance the filmy and theatrical, either in her lingerie or in her sleeping apartment; her underwear was of the finest linen, beautifully embroidered, but otherwise plain. Her red-gold hair was never touched with curling irons, and it was usually very simply dressed, except when great State functions called for a more elaborate coiffure.

The bedroom of the Emperor and the Empress was a large room with two tall windows opening on to the Park. It was on the ground floor, as, owing to the Empress's heart complaint, she found the exertion of ascending any stairs very exhausting. A lift in the corridor communicated with the nurseries, but during the Revolution the water supply was cut off, and the lift stopped working. Nevertheless the Empress insisted upon mounting the stairs to visit the invalid Grand Duchesses, and I always accompanied her, going behind her, and propping her up at each step. It brought tears to my eyes when I saw how ill she was, but she was determined not to miss a single chance of seeing her beloved children.

A large double bed made of lightish wood was near the windows, between which stood the Empress's dressing-table. At the right of the bed was a little door in the wall, leading to a tiny dark chapel lighted by hanging lamps, where the Empress was wont to pray. This chapel contained a table, and a praying-stand on which were a Bible and an ikon of Christ. This ikon was afterwards given to me by Her Majesty, in memory of the days which we spent together at Tsarkoe

Selo, and is one of my most treasured possessions to-day.

The furniture in the Imperial bedroom was in flowered tapestry, and the carpet was a plain coloured soft pile. The Emperor's dressing-room was separated from the bedroom by the corridor, and on the other side were the Empress's dressing-room and bathroom—but, alas! for her rumoured extravagances and her "odd" fancies! The bathroom was no luxurious place of silver and marble, but an old-fashioned bath set in a dark recess, and the Empress, with her Victorian love of neatness, insisted that the bath was hidden during the day under a loose cretonne cover. There was a fireplace in the dressing-room, and the dressers waited in the next room until the Empress required their services. The Empress's gowns were kept here, and another room full of large cupboards (half-way up the staircase leading to the nurseries) was given over to the use of those maids whose especial duty it was to iron and renovate Her Majesty's clothes.

The Empress favoured long, pointed footgear with very low heels: she usually wore suède, bronze or white shoes, never satin. "I can't bear satin shoes, they worry me," she would say. Her gowns, except those worn by her on State occasions, were very simple; she liked blouses and skirts, and she was greatly addicted to tea-gowns: her taste in dress was as refined as that of Queen Mary of England; like her she disapproved strongly of exaggerated fashions, and I shall not easily forget her condemnation when I once came to see her wearing a "hobble" skirt. "Do you really like this skirt, Lili?" asked the Empress.

"Well . . . Madame," I said helplessly, "c'est la mode."

"It is no use whatever as a skirt," she answered. "Now, Lili, prove to me that it is comfortable—run, Lili, run, and let me see how fast you can cover the ground in it."

Needless to say, I never wore a "hobble" skirt again.

The Empress has been accused of a mania for precious stones. I never saw any signs of it: true, she had quantities of magnificent jewels, but these possessions were consequent upon her position as Empress. She was fond of rings and bracelets, and she always wore a certain ring set with one immense pearl, and a jewelled cross. Some writers assert that this cross was set with emeralds, but I do not agree. I am sure that the stones were sapphires, and, as I saw it every day, I fancy I am correct. The Empress had soft, well-shaped hands, but they were neither small nor useless hands, and she never had her nails polished, as the Emperor detested highly polished and supermanicured nails.

At nine o'clock the Empress breakfasted with the Emperor; it was a simple meal à l'Anglaise, and after breakfast she went upstairs to see the children. Then Anna Virouboff arrived, and, if certain interviews were imperative, these were usually given during the morning, but, if the Empress found herself "free," she went to inspect her training college for domestic nurses, which was arranged entirely on English lines. She had great faith in the value of English-trained nurses for children, and she put all her usual "thoroughness" into the working and management of this institution.

Lunch was at one o'clock, and at twelve-thirty on Sundays; but when, as it often happened, the Empress was indisposed, she either lunched in her boudoir or alone with the Tsarevitch. After lunch the Empress walked, or drove herself in a little open carriage. Tea was at five, but sometimes receptions were held between lunch and tea. The Imperial Family all met at tea, which was quite "en famille"; and dinner, which was at 8 o'clock, was often a movable feast in the literal sense of the word. The Emperor disliked dining in one special room, so a table was carried to whichever room he happened to fancy that evening. Dinner over (and it was a very simple dinner) the Imperial Family spent the remainder of the evening together, and the Grand Duchesses, who had a *flair* for puzzles, usually indulged in puzzle-making: sometimes the Emperor read aloud whilst his daughters and their mother worked. It was the homely life of a united family—but a life with which the great world was not in sympathy; in fact a Russian writer did not hesitate to state openly that "it would have been better for Russia's felicity if the Empress had succumbed to the many frailties which were attributed to Catherine II." It is ironical to dwell on such an opinion when one remembers how the newspapers and the general public condemned her association with Rasputin. But had she been Catherine II, it is possible that this "frailty" might have been considered necessary for the "felicity" of Russia!

The Empress's boudoir, known as "Le Cabinet Mauve de l'Imperatrice," was a lovely room, in which the Empress's partiality for all shades of mauve was apparent. In spring-time and winter the air was fragrant with masses of lilac and lilies of the valley, which were sent daily from the Riviera. Lovely pictures adorned the walls—and one of the Annunciation, and another of St. Cecilia, faced a portrait of the Empress's mother, the late Princess Alice of England, Grand Duchess of HesseDarmstadt.

The furniture was mauve and white, Heppelwaite in style, and there were various "cosy corners." On a large table stood many family photographs, that of Queen Victoria occupying the place of honour.

The other private drawing-room was a large room, decorated and upholstered in shades of green, and the Empress had arranged in one corner a sort of tiny staircase and a balcony, which was always full of violets in the spring. In this room were pictures of herself and the Emperor, and some exquisite miniatures of the Grand Duchesses by Kaulbach, that of Marie being especially beautiful.

Books were everywhere; the Empress was a prolific reader, but she was chiefly addicted to serious literature, and she knew the Bible from cover to cover. The library was next the green drawing-room, and here all the newest books and magazines were placed on a round table, and constantly changed for others in the order of their publication.

The Empress was a great letter-writer, and she wrote her letters wherever she fancied. Her writing-table proper was in the room next her bedroom, but I have often seen her writing letters on a pad in her lap, and she invariably used a fountain-pen. Before the war she wrote daily to a great friend in Germany, and she always read this lady's letters to me. Her stationery, like her lingerie, was plain, but stamped with her cypher and the Imperial Crown.

Apropos of her fondness for lilac and lilies of the valley, I may mention that the Empress loved all flowers, her especial favourites being lilies, magnolias, wistaria, rhododendrons, freesias and violets. A love of flowers is usually akin to a love of perfumes, and the Empress was no exception to the rule. She generally used Atkinson's White Rose; it was, she said, "clean" as a perfume, and "infinitely sweet"—as an eau-de-toilette, she favoured Verveine.

When I first knew the Empress, she did not smoke, but during the Revolution she smoked cigarettes: I fancy they soothed her overwrought nerves.

The Empress always kept a diary, but I shall presently relate how it became my duty to burn her diaries, also those of Princess Sofia Orbeliani and Anna Virouboff; and last, but not least in sentimental interest, all the letters which the Emperor had sent her during their engagement and married life.

Dr. Botkin, the devoted friend and physician to the family, was introduced to me by Anna Virouboff, and I liked him exceedingly. He was a clever, liberal-minded man, and, although his political views were opposed to those of the Imperialists, he became so devoted to the Emperor that his once cherished views mattered little to him.

I think, from my description, which possesses the merit of accuracy, that it will be recognised what simplicity of life surrounded the rulers of one of the greatest Empires the world has ever known. Simplicity characterised all their doings, the simplicity which was to prove their undoing. The Imperial pair wished to lead the lives of private individuals; they imagined that it was possible. In Russia it has never been popular or possible for a Tsar to be human; he was an emblem, a representative of crystallised traditions; he united in himself the rôles of the Father of his people and the splendid, all-conquering, unapproachable Tsar. An Emperor or an Empress in mufti, so to speak, never yet appealed to popular imagination, and, just as the English cottager preserved and venerated the horrible "royal" oleographs of Queen Victoria, so did the Russian peasant venerate similar oleographs of the Emperor and his Consort. Neither cottager nor peasant would have understood or cared to possess "family" photographs of their rulers. Popular imagination has ever been appealed to by scarlet and ermine, golden crowns, and kingly sceptres. It doesn't understand or value anything else.

In the March following the birth of Titi, the Empress wrote and told me that she was anxious to see her godson, then nine months old. So I went with him to Tsarkoe Selo, where the Grand Duchesses made much of him, and used to take it in turns to bath him. We took up our quarters in Anna's house, where the Empress had personally superintended the arrangement of the baby's room, and she sent his cot, of which she crocheted the hangings and coverlet herself. She spent hours with the child, playing with him, "snap-shotting" him, and, after our first visit, I was constantly "commanded" to "come and bring the baby." I remember that, when I once missed the train, and arrived too late for lunch, the Empress, who was waiting for me, noticed my fatigue, and ordered tea. She took Titi on her lap, and saying, "Well . . . Lili, you do look hungry and tired," she fed me with pieces of sandwiches, pressing them on me much in the same way that a mother soothes a tired child. But she was ever "plus mère que mère, plus Russe que Russe," but her love of country was only for Russia and England. She had, and I say it with absolute conviction, no love for Germany as her "Motherland." She liked Darmstadt, because to her it represented home, but she manifested no interest in any other part of Germany.

My friendship with the Empress increased as the months passed. That autumn the Imperial Family went to Livadia, and I stayed with my uncle, going constantly to and from the Palace. The first day I saw the Empress in Livadia she gave me an entire layette for Titi which she had made herself. I had wondered why she had telegraphed for his measurements—now I knew! She would often call at my uncle's and take the baby with her for a drive. The little thing got to know her well, and one day, looking at her photograph, he said "Baby"; so after this the Empress of Russia was known to Titi by her own wish, *tout simplement,* in English, as "Aunt Baby." He always called her "Aunt Baby," and in many of her letters she alludes to herself by this pet name, but, needless to say, the favour shown to me and my child by the Imperial Family was the source of much comment at Court.

On one point my mind was made up. I determined never to allow any ideas of preferment or material advantage to spoil what was to me a condition of great happiness. My husband entirely agreed, and he declined to consider any mention of the posts which were from time to time spoken of in connection with him. As for myself, the Empress understood and appreciated my outlook. "You can always be my *friend* if matters remain as they are," she said. "I don't want to lose my Lili in an official personage."

We were very happy in those days. The Grand Duchesses were fast leaving childhood behind them and blossoming into charming girls; they did not greatly resemble one another, each was a type apart, but all were equally lovely in disposition. I cannot believe that any men so inhuman existed as those who, it is said, shot and stabbed those defenceless creatures in the house of death at Ekaterinburg. Apart from their beauty, their sweetness should have pleaded for them, but, if it is true that they have "passed," then surely no better epitaph could be theirs than the immortal words,

"Lovely and pleasant were they in their lives, and in their death they were not divided."

The Grand Duchess Olga was the eldest of these four fair sisters. She was a most amiable girl, and people loved her from the moment they set eyes on her. As a child she was plain, at fifteen she was beautiful. She was slightly above middle height, with a fresh complexion, deep blue eyes, quantities of light chestnut hair, and pretty hands and feet. She took life seriously, and she was a clever girl with a sweet disposition. I think she possessed unusual strength of character, and at one time she was mentioned as a possible bride for the Crown Prince of Roumania. But the Grand Duchess did not like him, and, as the Crown Prince liked the Grand Duchess Marie better than her sister, nothing came of the project. The sisters loved each other, and united in a passionate adoration for the Tsarevitch. In a recent book published in England, the Grand Duchesses have been described as Cinderellas, who were entirely subservient in family life owing to the attention paid the Tsarevitch. This is untrue. It is a fact that the Empress ardently desired a son, and that the birth of four daughters in succession was a disappointment to her, but she loved her daughters, they were her inseparable companions, and their plain and rather strict upbringing had nothing whatever of the Cinderella element.

The Grand Duchess Tatiana was as charming as her sister Olga, but in a different way. She has been described as proud, but I never knew anyone less so. With her, as with her mother, shyness and reserve were accounted as pride, but, once you knew her and had gained her affection, this reserve disappeared, and the real Tatiana became apparent. She was a poetical creature, always yearning for the ideal, and dreaming of great friendships which might be hers. The Emperor loved her devotedly, they had much in common, and the sisters used to laugh, and say that, if a favour were required, "Tatiana must ask Papa to grant it." She was very tall, and excessively thin, with a cameo-like profile, deep blue eyes, and dark chestnut hair . . . a lovely "Rose" maiden, fragile and pure as a flower.

All the Grand Duchesses were innocent children in their souls. Nothing impure was ever allowed to come into their lives—the Empress was very strict over the books which they read, which were mostly by English authors. They had no idea of the ugly side of life, although, poor girls, they were destined to see the worst side of it and to come in contact with the most debased passions of humanity! And yet it has been stated that the Empress, in her neurotic, religious exaltation, gave each of her daughters to Rasputin. Knowing her, knowing the Emperor, and knowing the daughters as I did, such an assertion savours of the monstrous; it has even been circulated that Mlle. Tutcheff objected to Rasputin being admitted to the Grand Duchesses' bedchamber to give them his nightly blessing after they had retired to bed, and that, as her protest was disregarded, she sent in her resignation. Mlle. Tutcheff was never governess to the Grand Duchesses, and she never witnessed Rasputin's nightly blessing, inasmuch as it never took place. The Emperor would never have permitted such a thing, even had the Empress wished it, and she certainly did not consider such a proceeding necessary for her daughters' salvation. Mlle. Tutcheff was the victim of her own spite and jealousy. She was not a very pleasant person, and, whenever the Imperial Family went to Livadia, she usually made herself very disagreeable, as she thoroughly disliked the Crimea. Continual grumbling wears away the patience of most people; the Empress was only human, and Mlle. Tutcheff was first given a holiday and then dismissed by the Grande Maîtresse de la Cour.

Mlle. Tutcheff did not hesitate to spread all kinds of vindictive rumours to account for her dismissal. She was too small-minded to state the real facts, and, as l'affaire Rasputin was generally spoken about, she decided to vent her spite on the Empress through this medium. I again assert that there is no truth in the legend of Rasputin's nightly blessing.

When I first knew the Grand Duchess Marie, she was quite a child, but during the Revolution she became very devoted to me, and I to her, and we spent most of our time together—she was a wonderful girl, possessed of tremendous reserve force, and I never realised her unselfish nature until those dreadful days. She too was exceeding fair, dowered with the classic beauty of the Romanoffs; her eyes were dark blue, shaded by long lashes, and she had masses of dark brown hair. Marie was plump, and the Empress often teased her about this; she was not so lively as her sisters, but she was much more decided in her outlook. The Grand Duchess Marie knew at once what she wanted, and why she wanted it.

Anastasie, the youngest Grand Duchess, might have been composed of quicksilver, instead of flesh and blood; she was most amusing, and she was a very clever mimic. She saw the humorous side of everything, and she was very fond of acting; indeed, Anastasie would have made an excellent comedy actress. She was always in mischief, a regular tom-boy, but she was not backward in her development, as M. Gilliard once stated. Anastasie was only sixteen at the time of the Revolution—no great age after all! She was pretty, but hers was more of a clever face, and her eyes were wells of intelligence.

All the sisters were utterly devoid of pride, and, when they nursed the wounded during the war, they were known as the Sisters Romanoff, and thus answered to the numbers 1, 2, 3 and 4.

The Grand Duchesses occupied two bedrooms; Olga and Tatiana shared one, Marie and Anastasie the other. These apartments were large and light, decorated and furnished in green and white. The sisters slept on camp beds—a custom dating back to the reign of Alexander I, who decreed that the daughters of the Emperor were not to sleep on more comfortable beds until they married. Ikons hung in the corners of the rooms, and there were pretty dressing-tables, and couches with embroidered cushions. The Grand Duchesses were fond of pictures and photographs—there were endless snapshots taken by themselves, those from their beloved Crimea being especially in evidence.

A large room, divided by a curtain, served as dressing-room and bathroom for the Grand Duchesses. One half of the room was full of cupboards, and in the other half stood the large bath of solid silver. The Grand Duchesses had departed from their mother's simple ideas, and, when they bathed at night, the water was perfumed and softened with almond bran. Like their mother, they were addicted to perfumes, and always used those of Coty. Tatiana favoured "Jasmin de Corse"; Olga, "Rose Thé"; Marie constantly changed her perfumes, but was more or less faithful to lilac, and Anastasie never deviated from violette.

The Grand Duchesses' attendants were a compromise between dressers, maids and nurses. They were all girls of good family, the most favoured being Mlle. Tegeleff, known as "Shoura"; the other two were "Elizabeth" and "Neouta." The Empress—once again Victorian—was very desirous for these girls to wear caps, but they declined respectfully but firmly to do so, and she did not press the matter. The Grand Duchesses liked their attendants, and often used to help them tidy the rooms and make the beds! Unlike their mother, but like most Russians, the four sisters showed a predilection for dress, but the Empress had her own ideas on the subject, and she chose and ordered all their clothes. As children, the girls were dressed alike, but later the two eldest wore similar gowns, and the next two were dressed, so to speak, "to match." The only frivolity which the Empress tolerated lay in her daughters' dressing-gowns, which carried out the colours of the regiments of which they were colonels, and the Grand Duchesses were very proud of their dressing-gowns and their regiments. They were always present at parades, when they wore the uniform of their regiments, and this excitement was one of their chief pleasures.

The sisters led most ordinary, uneventful lives; their exalted station never troubled them. With true courtesy they always made me pass out of a room before them, there was no ceremony, no fuss—they were the dearest, most affectionate girls, and I loved them all. The Grand Duchesses rose early, and were soon occupied with their lessons. After morning lessons they walked with the Emperor, and between lunch and tea they again went out with him. They spoke Russian, English or a little French, *never* German, and, although they danced well, they had not much chance to do so, unless the Imperial Family went to the Crimea, then Princess Marie Bariatinsky always arranged a series of dances for them.

The motive power in the lives of these charming children was family love. They had no thought apart from their home. Their affection was lavished on their father and mother, their brother and a few friends. Their parents were their paramount consideration. With the "children," as we called them, it was always a question of "Would Papa like it?" "Do you think this or that would please Mama?"—and they always alluded to their father and mother by the simple Russian words of Mama and Papa.

The Tsarevitch, that Child of many Prayers, one of the most pathetic figures in this tragedy of innocence, was born in 1904, and he was a healthy baby weighing eleven pounds at the time of his birth; many of the stories about his delicacy of constitution which have been given to the world are very exaggerated, especially the one which insists that the Nihilists mutilated the child when he was on the Imperial yacht. No such mutilation ever took place. The Tsarevitch certainly suffered from the hereditary trouble of thin blood-vessels, which first became apparent after a fall in Spala, but he was otherwise a normally healthy boy, and at the time of the Revolution he was really getting much stronger and much freer from the complaint. I know he was ailing at Tobolsk and Ekaterinburg, but that is hardly to be wondered at!

In appearance he resembled his sister Tatiana: he had the same fine features, and her beautiful blue eyes; he loved his sisters, and they adored him, and patiently submitted to his teasing. The Tsarevitch was a lively, amusing boy, with a wonderful ear for music, and he played well on the *balalika*: like Tatiana he was shy, but, once he knew and liked anyone, this shyness vanished.

The Empress insisted upon her son being brought up, like his sisters, in a perfectly natural way. There was no ceremonial in the daily life of the Tsarevitch: he was merely a son, and a brother to his family, although it was sometimes quaint to see him assume "grown up" airs. One day, when he was indulging in a romp with the Grand Duchesses, he was told that some officers of his regiment had arrived at the Palace and begged permission to be received by him.

The Tsarevitch instantly ceased his game, and, calling his sisters, he said very gravely: "Now, girls, run away. I am busy. Someone has just called to see me on business."

He adored his mother, and her passionate devotion to him is world-known, although, like many other things, this devotion has been used as a weapon against her. To the Empress, the Tsarevitch represented the direct result of prayer, the Divine condescension of God, the crowning joy of her marriage. Surely, if she manifested undue anxiety over him, she only did what all mothers have done, and will do until the end of time. There was certainly some subtle sympathy between mother and son: she was all that was lovely and beloved to him, and I especially remember one typical instance of this devotion:

My husband and I had been dining with the Imperial Family, and after dinner the Emperor suggested that we should accompany them to the Tsarevitch's bedroom, as the Empress always went thither to bid him good night and hear him say his prayers. It was a pretty sight to watch the child and his mother, and listen to his simple prayers, but, when the Empress rose to go, we suddenly found ourselves in complete darkness— the Tsarevitch had switched off the electric light over his bed!

"Why have you done this, Baby?" asked the Empress. "Oh," answered the child, "it's only light for me, Mama, when you are here. It's always quite dark when you have gone."

He loved his father, and the Emperor's great wish in the "happy days" was to undertake his son's education himself: this, for many reasons, was impossible, and Mr. Gibbs and M. Gilliard were his first tutors. Later, under very different conditions, the Emperor was enabled to carry out his wish. In the gloomy house at Tobolsk, he taught the Tsarevitch, and in the squalor and misery of Ekaterinburg the lessons still continued; but perhaps the greatest lesson learnt by the Tsarevitch and the other members of the unfortunate family was that of Faith: for faith sustained them, and strengthened them at a time when riches and friends had fled and they found themselves betrayed by the very country which had been all in all to them.

The Tsarevitch had various playmates—all sorts and conditions of boys shared his games: there were the two sons of his sailor-servant, two peasant boys with whom he was on friendly and affectionate terms, and my "Titi," who ran about with him, upsetting everything, and thoroughly enjoying himself. The Heir to the Throne was as courteous as his sisters. One day the Empress and I were sitting in the mauve boudoir, when we heard the excited voices of the Tsarevitch and Titi in the next room.

"I believe they're quarrelling," said the Empress, and she went to the door and listened to what the children were saying. Then she turned to me laughing. "Why they're not quarrelling, Lili. Alexis is insisting that Titi shall come into the mauve room first, and the good Titi won't hear of it!"

If the Tsarevitch had any peculiarities, I think the most striking was a decided penchant for hoarding. Many descendants of the Coburgs have been unusually thrifty, and perhaps the Tsarevitch inherited this trait. While thrifty he was really a most generous child, although he hoarded his things to such an extent that the Emperor often teased him unmercifully. During the sugar shortage he saved his allowance of sugar, which he gravely distributed among his friends. He was fond of animals, and his spaniel, "Joy," has happily found a home in England: his chief pet at Tsarkoe was an ugly sandy and white kitten, which he once brought from G.H.Q. This kitten he christened Zoubrovka, and bestowed a collar and a bell on it as a signal mark of affection. "Zoubrovka" was no respecter of palaces, and he used to wage war with the Grand Duchess Tatiana's bulldog "Artipo," and light-heartedly overthrow all the family photographs in the Tsaritsa's boudoir. But "Zoubrovka" was a privileged kitten, and I have often wondered what became of him when the Imperial Family were taken to Tobolsk.

All the children were fond of animals. The Grand Duchess Tatiana's pet was a bulldog called "Artipo," who slept in her bedroom, much to the annoyance of the Grand Duchess Olga, who disliked its propensity for snoring. The Grand Duchess Marie favoured a Siamese cat, and, the year before the Revolution, Anna Virouboff gave a little Pekinese dog to the Grand Duchess Anastasie.

This little creature had a tragic history. Curiously enough many people said that "Jimmi" seemed an unlucky dog; but he was a sweet little creature, whose tiny legs were so short that he could not walk up or down stairs. The Grand Duchess Anastasie always carried him, and "Jimmi" lavished a Pekinese devotion on her and her sisters. "Jimmi" went with the family to Tobolsk, and he is now identified in history with their fate. According to one account, his corpse was found, preserved in ice, at the top of the disused mine shaft; another writer has it that "Jimmi" defended his friends in the cellar at Ekaterinburg, barking defiance at the murderers, and guarding Tatiana's fainting body until they were both killed. His skeleton is said to have been discovered later in a clump of undergrowth, and subsequently identified by its size and by a bullet hole in the skull.

He was a dear little dog, and probably, could he have spoken, he would have desired no better fate than to perish with those in whose fortunes and affections he had equally participated.

The Emperor greatly resembled King George V in appearance, but his eyes were unforgettable; and those of his cousin, although fine, do not possess the expression peculiar to the eyes of the Emperor. It was a combination of melancholy, sweetness, resignation and tragedy: Nicholas II seemed as if he saw into the tragic future, but he also seemed to see the Heaven that lies beyond this earth. He was "God's good man." I can give no higher praise, render him no more fitting homage.

He was essentially charming: when you were with him you forgot the Emperor in the individual; he made formality impossible. He loved to tease people, and I came in for my full share of this propensity. One day when I was out walking at Livadia, several carriages passed me, but I did not especially notice their occupants. The next evening when I was dining at the Palace, the Emperor addressed me in grave tones: "*Lili—ce n'est pas bien, vous comprenez, mais ne pas reconnaitre vos amis.*"

"*Mais, Votre Majesté, qu'est que vous voulez dire?*"

"Well," said the Emperor, "you *cut* me yesterday."

"Votre Majesté, it's impossible!"

"Ah . . . it's quite possible, Lili. I drove past you, and bowed to you many times, but you wouldn't recognise me. Tell me in what I've offended you." And he continued to tease me until I felt ready to die with confusion. He loved his wife: no one has ever dared dispute the quality of the affection which existed between them; theirs was an ideal love-marriage, and when their love was tried in the furnace of affliction it was not found wanting.

Nicholas II had been reproached for his weakness of character, but this weakness was not weakness in the literal sense. The Empress, who was fully aware of what was said concerning the Emperor and herself, once told me how utterly people misunderstood her husband. "He is accused of weakness," she said bitterly. "He is the strongest—not the weakest. I assure you, Lili, that it cost the Emperor a tremendous effort to subdue the attacks of rage to which the Romanoffs are subject. He has learnt the hard lesson of self-control, only to be called weak; people forget that the greatest conqueror is he who conquers himself."

On another occasion she remarked that she knew that the Emperor and herself were blamed for not surrounding themselves with genuine people.

"It's an extraordinary thing, Lili," she said, "we've tried to find genuine advisers for the last twenty years, but we've never found them. I wonder whether any exist!"

The Empress always resented the cruel slanders which were circulated about the Emperor. "I wonder they don't accuse him of being too good: that, at least, would be true!" she cried.

As for herself, she troubled little.

"Why do people want to discuss me," she said. "Why *can't* they leave me alone!" Again: "Why will people insist that I am pro-German? I have spent twenty years in Germany, and twenty years in Russia. My interests, and my son's future lie in Russia:

how, therefore, can I be anything but Russian?"

The Empress has been censured for exerting undue influence over her husband, and this "pernicious" influence has made her the scapegoat for all the ills which have befallen Russia. But her "influence" was merely that of a good woman over a man. If she influenced the Emperor in any other way, it was done unconsciously. I will never believe otherwise, although, in making this assertion, I shall perhaps be confronted with all kinds of hostile criticism. It will be asked by what right I dare defend a woman who has been tried and found guilty. But I dare to do so. True, I am a person whose name is entirely unknown to the general public, but it cannot be disputed by those who knew life at Tsarkoe Selo and Petrograd that I was honoured by the Empress's friendship and confidence.

The Emperor shared his wife's "thoroughness"; he never believed anything until (were it possible) he had tried it for himself.

During the war, a new uniform was submitted for the Emperor's approval; he determined to test its qualities, and he walked for twenty miles wearing it, in order to see what weight was possible to carry with it. The sentinels failed to recognise the Emperor when he passed them wearing the sample "Tommy's" kit, a fact which greatly amused him; but, as a result of his practical experiment, the uniform (with certain alterations suggested by the Emperor) was "passed."

The Empress put her husband first in everything—it was always "The Emperor wishes it," "The Emperor says so"; she was very tender towards him, the maternal element was apparent in her love even for her husband: she took care of him, but perhaps this arose chiefly from a feeling that he suffered by reason of his love for her.

As husband and wife they were indeed one. They only asked happiness of life. The Emperor's tastes were of the simplest, the Empress was shy and retiring—both their dispositions were similar—and this similarity of tastes, ideal in the usual walks of life, was fatal to both of them as rulers. By this I do not for one moment wish to infer that they shirked their responsibilities: far from it, they were always ready to assume them, but they forgot that the times were out of joint, that it was their duty always to live in the fierce light that beats upon a throne. I do not think that by so doing they could have saved Russia. The case of Nicholas II and Alexandra of Russia is almost parallel with that of Louis XVI and Marie Antoinette. The Russian monarchs, like their French prototypes, were called upon to reign over a country ripe for Revolution, whose dragon's teeth had been sown by the vicious hands of their predecessors. France boasted as extravagant and exotic a society as that of Russia: the writing was already to be seen on the walls of Versailles and the Winter Palace, but the Sovereigns of Then and Now heeded it not. Louis XVI wanted to be left alone in his workroom, to make locks and to mend watches, and Marie Antoinette sighed for the simple pleasures of the Trianon and the pastoral joys of a farmer's wife.

Nicholas II did not care to be a locksmith, he merely wished to live the quiet life of a well-bred gentleman: chivalrous by nature, he (and here an English writer is correct) came nearer the British public-school idea than any other. The Empress did not require a Trianon, she wanted a home; but, although she loved Russia, Russia was always antagonistic to her. This she never realised, any more than she recognised the fact that the peasant class never wanted her to try and understand them.

The Emperor was a clever man, and he possessed that wonderful memory for faces peculiar to his uncle, King Edward VII. On one occasion when my husband was presented to the Emperor after receiving some special decoration, a colonel of a Siberian regiment also attended the Levée. The Emperor stretched out his hand to the colonel. "Surely I've seen you before?" he enquired. "Yes, Your Majesty." "Well, but *where*?" continued the Emperor, in puzzled tones; then brightening, "Ah, I know," he said, "I met you twelve years ago when I passed through Saratof."

The chief pleasures of the Emperor were those appertaining to an outdoor life. He was a good shot, fond of all kinds of sport, and his hands were exceptionally powerful. Boating was a favourite amusement; he liked to row in a small boat, or paddle a canoe, and the Emperor passed hours and hours on the water when the Imperial Family were staying at Shker, in Finland.

Both the Emperor and the Empress disliked the Kaiser. I say this with perfect sincerity, and in all truth. They rarely mentioned his name before the war, and I know that his love of theatrical displays appealed to neither of them. In 1903 the Emperor William arrived in his yacht at Reval to witness a military review. The "Standart" with the Emperor of Russia aboard was also at Reval. After the Kaiser had paid a formal call on the Emperor, signals passed between the two yachts.

"What's all this?" asked the Emperor. An officer enlightened him.

"Your Majesty," said he, "the signal from the 'Hohenzollern' says: 'The Emperor of the Atlantic salutes the Emperor of the Pacific.'" The Emperor looked cross.

"Oh, that's it—well reply 'Thank you'—that's quite enough."

The Kaiser did not shine as a visitor to the "Standart"; the first thing he did was to shake hands indiscriminately, a proceeding which caused much amusement and confusion, and everyone was heartily glad when the "Emperor of the Atlantic" took his departure.

The Grand Duchesses disliked any mention of the Kaiser, but some of the officers used to tease them about him. The usual question of any privileged arrival at Tsarkoe Selo was: "Well, how is Uncle Willie to-day?" And the invariable answer was: "No— no— he's not our Uncle Willie—we don't want to hear his name."

Russia has been described as a country of tears and misery during the war, but this is incorrect. The peasants were never so rich as at this time, and there was no discontent in the country districts;

the wives received big allowances, and they earned extra money for themselves without any difficulty. Every boy indulged in high patent-leather boots, every girl spent money on dress. There were certainly tears for the fallen, but there was no material misery in Russia.

The Emperor had made great plans to help those disabled in the service of their country. His idea was to give all wounded, disabled or decorated soldiers gifts of Crown Lands at the end of the war. He planned various land reforms, but the Revolutionaries incited the landlords against him by telling them that the Emperor was going to be generous at their expense, and not at his own!

It is impossible for an English public to realise the plots and counter-plots which existed in Russia. The Empress, on many occasions, barely escaped with her life; she was unpopular with all classes, but she was unable, mercifully, to estimate the quality of the hatred meted out to her. I do not think there is a single charge that has not been laid at her door; she is credited with hysteria, religious mania, pro-Germanism, the qualities of a Judas, the morals of a Messalina; she has been described as the intriguing, strong-minded consort of a weak man, a willing tool of an infamous sensualist, as well as being a half-witch, and a half-mystic. The real Tsaritsa, firm in her convictions, the devoted wife, mother and friend, is unknown. Her acts of charity have been misconstrued, her religion has been made her shame, the very nationality which she so willingly relinquished has become an unmerited reproach. She knew and read all the reports concerning her, but, although anonymous letters sought to vilify her, and journalism bespattered her with filth, nothing touched her serenity of soul.

I have seen her grow pale, and I have watched her eyes slowly fill with tears when something exceptionally vile came under her notice. But Alexandra Feodrovna was able to see the stars shining far above the mud of the streets.

CHAPTER V

I AM going to write of Gregory Rasputin as I knew him. My personal acquaintance with him lasted from 1910 to 1916, but I know that I might as well attempt to cleanse the Augean stables single-handed, as to be believed if I say one word in his defence. As a man, and as an infamous figure in history, he matters little to me, and, knowing the popular prejudice against him, I hesitated to mention his name in these pages. But I was urged to do so; it was represented to me that my silence might be equivalent to an acknowledgment, not only of his guilt, but also of that of the Empress. This last consideration decided me to forgo my resolution, and to write a faithful record of the man who was supposed to play such an important rôle during the last few years of the Russian Empire.

If I say that I never saw the evil side of Gregory Rasputin I shall be called a liar or a fool—perhaps, more chivalrously, the latter. It is, however, the truth when I say that we never saw the evil side of him. May I, therefore, plead for a hearing on the grounds that some men possess dual natures, and that they adapt these to the company in which they find themselves? I have heard of men who at home have led most moral lives, leading elsewhere existences before which an up-to-date French novel is as naught. Yet they never betrayed themselves to their nearest and dearest. Their friends were likewise deceived. Perhaps this dark side was never discovered, and they died and were buried as undefiled Christians. But even if something unforeseen had disclosed the man's secret orchard, his inner life, and his frailities (sic), their existence even then would most probably have been disbelieved by those who had known him intimately for years.

A person tells you that your dearest friend is a liar and a sensualist. Do you believe him? Rarely, I think, if you are worthy to call yourself a friend. You advise the traducer to make himself or herself scarce, and, if you allow your mind to become poisoned by slow dropping venom, you place yourself at once on a level with the slanderer.

The Empress refused to believe ill of Rasputin because she had never seen the evil side of him, and also because both she and the Emperor had extended the hand of friendship to him. There was no question of affection in her continual refusal to disown him, no phase of the passing passions which distinguished Catherine the Great, and which were so kindly tolerated by her subjects. The Empress inherited much of her illustrious grandmother's tenacity of purpose, and she refused to be dictated to. In this, she was the woman of character who resembled Queen Victoria. I do not wish to compare Rasputin with John Brown—they are as the poles apart—but what I wish to point out in connection with both of these persons, is that Queen Victoria and the Empress called John Brown and Gregory Rasputin their friends, and neither family disapproval nor public censure was a sufficient reason in their eyes to merit the sacrifice of a friend. There the similarity ends.

Gregory Rasputin arrived in Petrograd from Siberia on a pilgrimage, walking the entire way with irons on his body in order to make his progress more painful and difficult. If a pilgrim were to arrive in London from Edinburgh in similar circumstances he would be taken before a magistrate, and most probably sent to a lunatic asylum; these things do not happen in England, but they were of daily occurrence in Russia. We were so accustomed to the miraculous that I do not think the average Russian would have manifested any surprise if he had been accosted in the street by the Angel Gabriel! Rasputin had been introduced by certain people to Germogen, a priest and a friend of Elidor, who possessed great influence in the region of the Volga. Elidor's dominant idea was to found a particular sect of his own, but he failed to do so, and he was ultimately dismissed from authority. This, he attributed, rightly or wrongly, to Rasputin. Germogen was a firm believer in Rasputin's spiritual powers, and he was also much interested in his arduous pilgrimage. In fact, so greatly was he impressed that he decided to introduce the "staretz" to the Grand Duchess Peter, formerly Princess Meliza of Montenegro, and to her sister the Grand Duchess Anastasia, the wife of the Grand Duke Nicholas. Both these Princesses were addicted to mysticism; I may describe them as "soulful." Rasputin impressed them equally as much as he had impressed Germogen, and they talked everywhere about their wonderful "discovery."

At this time the two Grand Duchesses were on very friendly terms with the Empress, and it is not to be wondered that, little by little, her curiosity was aroused, and at last she and the Emperor expressed a wish to see Rasputin.

H.I.M. THE TSAR
Surrounded by the Officers of
The Royal Yacht 'Standart'

THE EMPRESS On board the Royal Yacht 'Standart'

H.I.M. THE TSAR WITH THE TSAREVITCH
On board the Tender going out to the Royal Yacht 'Standart'

The "staretz" was in due course presented to Their Majesties. Once again I repeat that such things could only happen in Russia, and it is therefore impossible to judge the Rasputin affair from an English standpoint. This uncouth peasant who came into the presence of Their Majesties barefooted, wearing the clumsy irons of penance, was in nowise impressed by his surroundings—he spoke freely to the Emperor, who was struck, like many others, by Rasputin's sincerity. The interview was not productive of any notable result, so far as Rasputin was concerned; it was merely an interesting incident, and when I first knew the Empress she never mentioned the name of Rasputin. In my opinion, and I speak in all sincerity, I believe that Rasputin was the unconscious tool of the Revolution. If John of Cronstadt had lived in 1910 to 1916, he would have been called another Rasputin. It was necessary for the Revolutionaries to find someone whose name they could couple with that of the Empress—a name whose connection with the Imperial Family would destroy their prestige with the higher classes, as well as nullifying the veneration of the peasant class. A member of the Duma once heckled one of the Revolutionary party on the question of Rasputin:

"Why," said he, "don't you kill Rasputin if you are so against him?" He received this surprising but wholly truthful reply:

"Kill Rasputin! Why, we should like him to live for ever! He represents our salvation!"

Rasputin's position was many-sided. One section of Society looked upon him as a "cult," and I have no doubt that there was a certain pathological interest in this. Another group formed a mystical conception of him as a "teacher," and a more material clique courted him, hoping thereby to gain influence with the Empress. The shame lies not so much with Rasputin as with those who "exploited" him.

At one time Rasputin was the guest of a well-known general, but, when this gentleman discovered that there was nothing to be gained by his hospitality, he quickly dropped his one-time acquaintance, and Rasputin took up his quarters in a small flat where he was supported by voluntary contributions. It was a humble abode, the "staretz" lived on the meanest food, and it was only during the last year of his life that he received presents of wine.

Anna Virouboff met Rasputin for the first time when she had just made up her mind to leave her husband. As I have said, her marriage with Lieutenant Virouboff had turned out disastrously, and their relations terminated in a most distressing manner. It so happened that once, when Anna was entertaining the Empress and General Orloff, Lieutenant Virouboff arrived unexpectedly from sea, and, as the police did not recognise him, he was refused admittance to his own house. There was a terrible scene between him and his wife after the Empress left, and Anna was beaten unmercifully. Anna then refused to live with him any longer, and returned to her parents. This affair created a great scandal, and, in order to console Anna, the "Montenegrin" Grand Duchesses took her to see Rasputin.

I cannot say whether or no this was a mistake. I am inclined to think that it was a well-meant error, as Anna Virouboff was a supersensitive, rather neurotic person, easily impressed by an effective *mise en scène*. And this *mise en scène* was amply provided for her. The heart-broken and insulted young wife was received at the Palace of the Grand Duchess Anastasia with immense ceremony, and what took place is best described as an emotional prayer meeting.

Suddenly a door opened and Gregory Rasputin made his appearance. He walked into the midst of the overwrought worshippers, untouched by their exaltation. He radiated peace, and he personified the Strong Man beloved as an ideal by the majority of women. To Anna, the shattered and the disillusioned, Rasputin typified the calm that comes after a great storm; he prayed with her, he consoled her, she felt that she could confide in him. She was utterly oblivious of the social gulf which separated them. Rasputin was something to lean on, and Anna always leant on somebody; this weak, lovable, credulous creature was unable to stand alone. And in this way their intimacy began. I am sure that Anna was never in love with the *man* (although she was always in love with someone), but his chief influence over her was that of the priest.

I believe that at this time the Empress saw Rasputin occasionally, but he was chiefly to be found in the company of the two Grand Duchesses who had "discovered" him, and who now reported that Rasputin was undoubtedly a "seer." This annoyed the Emperor, and, the next time he saw Rasputin, he asked him to tell him *how* he "saw" true. "Your Majesty, I know nothing of clairvoyancy," said Rasputin.

"Then why have the Grand Duchesses asserted that you possess clairvoyant gifts?" replied the Emperor, crossly; and, when the Empress put the same question to Rasputin, she received the same reply.

The real reason for this report will never be known. It was in all probability political, but, after Rasputin had disowned clairvoyancy, the two Grand Duchesses disowned their protegé and sided with Germogen against him. The commencement of endless intrigues dates from this period, as Elidor and Germogen were afraid that Rasputin would become more important than themselves.

I must now deal with Rasputin's alleged influence over the Empress. There is no doubt that her subconscious belief in his spiritual powers was confirmed by the long arm of coincidence. The Tsarevitch fell ill, the attack was severe and his parents were frantic. If any mother with an only son reads these pages, she will admit that the word "frantic" best describes the feelings of a mother at such a crisis. The Empress was literally beside herself; it was then that someone suggested that Rasputin should be sent for. When he arrived he bade the despairing parents hope. He prayed by the bedside of the Tsarevitch, and it seemed that directly he did so the child began to get better. There is not the slightest truth in the film and "novel" versions of the incident; coincidence, and coincidence alone, was responsible for the Tsarevitch's recovery at the moment of Rasputin's impassioned prayers.

I met Rasputin just before the Germogen scandals. My husband had gone to Copenhagen to escort the Empress Marie thither on the "Pole Star," and he was anxious for me to join him. To do this would have entailed leaving Titi with my mother, and I was reluctant to do so, although naturally desirous of acceding to my husband's wishes. Thus I was in somewhat of a dilemma. Anna noticed I was worried and unhappy.

"Look here, Lili, there's someone who can help you," she said.

"Who?" I asked.

"Gregory Rasputin," she answered.

I was not anxious to meet Rasputin—I did not possess the boundless belief in him which characterised Anna, but I agreed, to humour her, and she took me to Rasputin's eyrie (I say eyrie, since his flat was high up under the roof), and then left me.

I waited for some time alone in a little study until a man came in so noiselessly that I was almost unaware of his presence. It was Rasputin! Our eyes met, and I was instantly struck by his uncanny appearance. At a first glance, he appeared to be a typical peasant from the frozen North, but his eyes held mine, those shining steel-like eyes which seemed to read one's inmost thoughts. His face was pale and thin, his hair long, and his beard a lighter chestnut. Rasputin was not tall, but he gave one the impression of being so; he was dressed as a Russian peasant, and wore the high boots, loose shirt and long, black coat of the moujik. He came forward and took my hand.

"Ah . . . I see. Thou art worried." (He "tutoyed" everybody). "Well—nothing in life is worth worrying over—'tout passe'—you understand—that's the best outlook." He became serious.

"It is necessary to have Faith. God alone is thy help. Thou art torn between thy husband and thy child. Which of them is the weaker? Thou think'st that thy child is the more helpless. This is not so. A child can do nothing in his weakness—a man can do much."

Rasputin advised me to go to Copenhagen, but I did not go. I left Petrograd next day for the country—perhaps out of bravado! But the impression which Rasputin had produced on me was very vivid. I was at once attracted, repelled, disquieted and reassured; nevertheless, his eyes were productive of a feeling of terror and repugnance, and I made no answer when the Empress greeted me with the words: "So, Lili, you've seen our friend? He'll always help you."

My second meeting with Rasputin took place in the winter. Titi was seriously ill, it was thought that diphtheric conditions would set in, and the poor little boy lay tossing from side to side in delirium. Anna, who made constant enquiries, at last 'phoned. "Lili," she said, "my advice is—ask Gregory to come and pray." I hesitated—I knew my husband's distaste for anything touching the supernatural. But, when I saw how ill Titi was, I hesitated no longer. At any rate, no one could possibly condemn the prayers offered for a sick child. Rasputin promised to come at once, and he arrived in company with an old woman who was dressed as a nun. This quaint creature refused to enter the boy's bedroom, and sat on the stairs, praying.

"Don't wake Titi," I whispered, as we entered the nursery, for I was afraid that the sudden appearance of this strange peasant might frighten the child. Rasputin made no reply, but sat down by the bedside and looked long and intently at the sleeper. He then knelt and prayed. When he rose from his knees he bent over Titi.

"Don't wake him," I repeated.

"Silence—I *must.*"

Rasputin placed a finger on either side of Titi's nose. The child instantly awoke, looked at the stranger unafraid, and addressed him by the playful name which Russian children give to old people. Rasputin talked to him, and Titi told him that his head ached "ever so badly."

"Never mind," said Rasputin, his steel eyes full of strange lights. Then, addressing me: "To-morrow thy child will be well. Let me know if this is not so." And, bidding us farewell, he departed with his odd escort.

Directly Rasputin had gone the child fell asleep, and the next morning the threatened symptoms had disappeared, and his temperature was normal. In a few days, greatly to the doctor's amazement, he was quite well. After this, I could hardly dispute Rasputin's peculiar powers, and I always saw him whenever he came to the Palace— this, on an average, about once a month.

It is only fair to Rasputin to say that he derived no material benefits from these visits, in fact, he once complained to me that he was never even given his cab-fares! Rasputin's influence over the Empress was purely mystical. She had always believed in the power of prayer—Rasputin strengthened her in this belief, and I am sure that her perplexed soul was soothed by his ministrations. There was absolutely no sensual attraction. It gives me intense pain to touch on this subject, but I must not shrink from what I consider to be my duty. I have heard the most dreadful stories of the Empress—how, in the spirit of sacrifice she gave herself, and those dear children to Rasputin, in order to prove that the sacrifice of the body was acceptable to God. Such a monstrous thing never happened. But when I have defended her, and said that Rasputin was a common man, unpleasing to look on, dirty in his habits and uncouth in every respect, I have been told that these defects matter nothing in certain types of sensualism. I have put forward the indisputable fact that the Empress was an intensely fastidious woman, that she possessed no "animal" propensities, that her morals were the ultra-strict morals of her grandmother. The answer to this has been that many fastidious and super-moral women have been guilty of incomprehensible lapses, solely by reason of their fastidious and moral qualities. If such examples exist, why should not the Empress have done likewise?

I am confronted at every turn by these reports, and people say pityingly: "Well, of course, you *loved* the Empress." That is so ... but *I also knew* the Empress. The Emperor's attitude in the Rasputin scandal ought alone to destroy these accusations, as the Empress never saw Rasputin without the knowledge and consent of her husband. Even assuming Nicholas II to be a weak man, entirely under the domination of his wife, he would certainly have been man enough, husband enough, and father enough, never to have countenanced any immoral relations between Rasputin and his family. The Emperor was primarily a Christian and a gentleman, but he was likewise a Romanoff and an Emperor. In these capacities he would have meted out the only possible punishment for such an offence. When he was told the "outside" scandals concerning Rasputin, he would not credit them. And why not? *Simply because they were so bad*; had they been less so, the Emperor might have listened. It is a great mistake for anyone to attempt to destroy any friendship by describing the person whose ruin is contemplated as being entirely worthless. The desired result is obtained far more easily by damning him or her with faint praise!

When various people reproached the Empress for being on terms of friendship with a common peasant, and for believing that he was endowed with the attributes of holiness, she replied that Our Lord did not choose well-born members of Jewish society for His followers. All His disciples except St. Luke were men of humble origin. I am inclined to think that she placed Rasputin on a level with St. John . . . both were, in her opinion, mystics.

She was perfectly frank in her belief in Rasputin's powers of healing. The Empress was convinced that certain individuals possess this gift, and that Rasputin was one. When it was urged that the services of the most skilled physicians were at her disposal, she gave the invariable answer: "I believe in Rasputin." As for the stories that Rasputin and Anna Virouboff gave the Tsarevitch poisons and antidotes, I dismiss these with contempt—they belong solely to sensational fiction. Anna Virouboff would have been too frightened to give a kitten a dose of medicine, much less would she have tampered with the medicines given to the Tsarevitch.

The first grave scandal which assailed the Empress in connection with Rasputin was the discovery and publication of a letter written by her, in which she made use of the expression: "*Je veux reposer mon âme auprès de vous.*" The enemies of Rasputin were fully aware that he was guilty of the fatal habit of keeping interesting letters, so Rasputin (always desirous of popularity) was invited to meet certain influential people, and, on his way to the rendezvous, he was attacked and robbed, and all the correspondence which he carried on him was stolen.

In due time the contents of the Empress's letter were published, and this did her tremendous harm. Even the Duma took the worst view of the much quoted sentence, "*Je veux reposer mon âme auprès de vous.*" But that expression was not used at all in the physical meaning. The Empress merely wished to tell her friend that her soul was desirous of spiritual consolation.

Since I have lived in England, I have constantly met women who pin their faith in certain spiritual and physical advisers. Most Catholics have a special confessor to whom they invariably repair, just as most people have one particular doctor in whom they trust—most representatives of any denomination have their especial following. It is solely a question of one individual meeting the requirements of another.

The Emperor was very much troubled over the attacks which were made on the Empress. But both he and the Empress possessed a mistaken sense of their responsibilities in connection with Rasputin, and this mistaken sense of responsibility was to prove the ultimate destruction of both Rasputin and themselves. The Imperial couple resolutely refused to throw him over. In this decision the Emperor was as one with the Empress; perhaps they "humanly" declined to admit the right of anyone to dictate to them . . . but, be that as it may, Rasputin's position remained undisturbed.

It is well known that Rasputin condemned hostilities, but it is not equally well known that he tried to stop the declaration of war. Nevertheless, when mobilization began, he wired to Anna, saying: "The war must be stopped—war must *not* be declared; it will be the end of all things." No notice whatever was taken of this telegram, for the excellent reason that Rasputin's political influence was *nil*; he had, in fact, no influence in material matters, although many have thought otherwise.

General Beletsky once asked Rasputin to speak to the Emperor and suggest his name as Governor-General of Finland. Rasputin promised to do so, and mentioned the matter to the Emperor, in the presence of the Empress. The Emperor listened, but made no comment. General Beletsky was never appointed.

It seems impossible to obtain a logical hearing on behalf of either the Empress or Rasputin. All kinds of reports have been circulated in connection with the latter's excesses and debaucheries. There may have been some truth that Rasputin's private life was not all that it should have been, but I assert most solemnly that we never saw the slightest trace of impropriety in word, manner or behaviour when he was with us at Tsarkoe Selo.

Prince Orloff, the head of the Chancellerie Militaire, never made any pretence of liking or even tolerating the Empress. He experienced a sort of nervous repugnance to meeting her, and it was common knowledge that he took quantities of valerian in order to steady his nerves, whenever it was necessary for him to see her. The Empress was aware of this.

"I saw Prince Orloff to-day," she said to me, "he was reeking of valerian. Poor man, what an effort it must cost him to speak to me."

The Prince exercised no discretion whatever in his statements about the Empress and Rasputin; he seemed impelled to disparage her—his hatred amounted almost to a 'phobia—and at last the Emperor lost patience with him and sent him to the Caucasus. Princess Olga Orloff was received shortly afterwards by the Empress. The Empress was very fond of Olga, but it was a very unpleasant interview, as the Princess tried to explain that her husband had been grossly maligned. The Empress described the interview to me:

"I've had a dreadful time, Lili," she said, "Olga Orloff has just been. I'm very, very sorry for her, she's in a terrible state. When I rose, she began to speak most wildly, and to insist that her husband was devoted to me and to our interests. I knew that, if I were to sit down, I should burst into tears; so I kept standing. It was an awful moment."

Rasputin always had a presentiment of a violent death. He often remarked, with an air of profound conviction: "Whilst I'm alive all will be well, but, after my death, rivers of blood will flow. Nothing, however, will happen to 'Father' and 'Mother'"—this was his way of alluding to the Emperor and Empress. About this time an old woman, a disciple of Elidor's, came to see Rasputin one night, wearing a white dress plentifully trimmed with scarlet ribbons.

Rasputin reproved her for this display.

"How awful of you to wear these red ribbons," he said.

"Ah," replied the old woman. "I *know* why I wear red."

"And she knew full well," said Rasputin, gloomily, when describing the incident to me. "Red is the colour of blood—and blood will soon be as plentiful as her scarlet ribbons."

Everyone who loved the Imperial Family was horrified at the ever increasing scandals; the wildest reports, mostly lies, with a substratum of truth were current, and Rasputin was even said to have been sinning in Petrograd when he was actually in Siberia. It was impossible to persuade the Empress that popular feeling was against her. True, she heard what was said, and she occasionally read what was imputed to her, but she paid no attention to gossip or to mendacious paragraphs. She was obsessed by her religion, and she sent me and Anna Virouboff on a pilgrimage to Tobolsk in the summer of 1916. A new saint had been recently canonized at Tobolsk, and the Empress had made a vow to go thither herself, or to send a substitute. Anna asked me to consent, as she was afraid to travel alone, and, as the Empress begged me to go, I could do no less than prove my devotion to her wishes.

When I arrived at Petrograd I discovered that Rasputin was to travel with us. I could not help thinking that, in view of popular feeling, it was most ill-advised to advertise the expedition, but I dared not suggest this. We left Petrograd in the greatest publicity. ... A special saloon carriage was attached to the train . . . it was a progress of publicity, wires were sent in advance all along the line to announce our advent, and crowds thronged the stations to catch a glimpse of us.

At last, late in the evening, we arrived at Tumen, and from thence we took the steamer to Tobolsk. Little did I dream that, in a year's time, the Imperial Family were to make the same pilgrimage— of which the whole journey was to prove indeed a Via Dolorosa! They, too, were to see the black and swiftly flowing river, and the wild Tartar villages on its banks, and, like myself, they were to see the city on the mountain, with its churches and houses sharply silhouetted against the fast darkening sky.

We were received at Tobolsk by the Governor, the chief officials, and the Church dignitary, Varnava, and we were afterwards taken to our quarters in the Governor's house, where I slept in the little room which the Emperor, a year later, used as his study. The next day we visited the saint's grave, and attended a very impressive service in the Cathedral. Rasputin stayed with the priest, but, unfortunately, he quarrelled with Varnava, so matters became some-what strained, and I was not sorry when our two days' visit came to an end.

On the way back to Tumen, Rasputin made a point of us stopping at his village and seeing his wife. I was rather intrigued at this, as I had always wondered how and where he lived, and I felt quite interested when I saw the dark grey, carved wooden house which was the home of Rasputin. The village consisted of a group of small wooden houses built on two floors. Rasputin's house was, perhaps, a little larger than the others, and he said that he hoped one day

Their Majesties would visit him.

"But it's too far," I said—aghast at the proposal.

Rasputin was angry. "They *must*," he declared, and, a few minutes afterwards, he added the prophetic words: "Willing or unwilling, they will come to Tobolsk, and they will see my village before they die."

We remained one day at Rasputin's house. His wife was a charming, sensible woman, and the peasants were a fine type—honest, simple folk, who cultivated the fields belonging to Rasputin, and accepted no payment for so doing—working absolutely in the spirit of holiness.

Rasputin had three children—the two girls were being educated in Petrograd, but the boy was quite a peasant. Everyone was friendly, but most of the villagers were strongly against Rasputin's returning to Petrograd.

As we had decided to go on to Ekaterinburg, and from thence to the Convent of Verchoutouria, I thought it would be a good idea to persuade Rasputin to remain with his people. This he refused to do; I told Anna that there must be no more gossip, and that she must persuade Rasputin to leave us. She promised to do so, but at the last moment he went with us to Ekaterinburg.

I shall never forget my first impression of this fatal town. Directly we got out of the train, I felt a sense of calamity—we were all affected; Rasputin was ill at ease, Anna perceptibly nervous, and I was heartily glad when we reached the Convent of Verchoutouria, which is situated on the left bank of the river Toura. We stayed a night in the guest house attached to the Convent, and then Rasputin asked us to go into the woods with him and visit a hermit who was locally supposed to be a very holy man.

This pilgrimage must appear entirely foolish in the eyes of English readers. I try and put myself in their place, and imagine what the English public would think if the "Daily Mail" announced that Queen Mary had sent two of her friends on such an expedition. "This couldn't happen—Queen Mary is far too sensible," you will say.

No doubt Queen Mary *is* far too sensible . . . such a thing could never happen in England, and I am only relating it in order to prove that, once again, it is impossible to judge Russia from an English standpoint.

The hermit lived in the heart of the forest and his hermitage might easily have been taken for a poultry farm. He was surrounded by fowls of all sizes and descriptions. Perhaps he considered fowls akin to holiness; he gave quantities of eggs to the Convent, but we supped frugally off cold water and black bread. The hermit had no use for beds, so we slept miserably on the hard, unyielding floor of dried mud, and I must confess that I was glad when we returned to Verchoutouria and we were able to sleep and bath in comfort.

Rasputin decided to take leave of us at Verchoutouria, so we went on alone to Perm, where our saloon carriage was coupled to another train. Crowds came to stare at Anna, and some of their comments made me feel very uneasy. There was much dissatisfaction, and, when our saloon was uncoupled, it was done so forcibly that the carriage was almost derailed, and I was thrown from one end to the other. But we returned to Petrograd safely, there to be welcomed and thanked by the Empress.

"After all, Lili," said Anna, now prostrate with nerves and a heart attack, "we must believe that God *likes* us to endure."

I do not know whether this remark was reminiscent of the hermitage, or of the saloon carriage, but I was able honestly to thank God that I was once more within a civilized area.

Rasputin did not stay long in his village; he returned to Petrograd, and the brazen voice of scandal was again heard. One day, in 1916, when I was at Reval, the Empress telegraphed asking me to come and see her.

I obeyed, and found her alone, looking sad, and obviously much troubled in her mind. She did not, at first, touch on the subject nearest her heart; then, all at once, she told me how hard she thought it of people to speak against her so bitterly.

"I know *all*, Lili," she said. "Why does Gregory stop in Petrograd? The Emperor doesn't wish it. I don't. And yet we can't possibly discard him—he's done no wrong. Oh, why won't he see his folly?"

"I'll do all in my power, Madame, to make him do so," I replied. My heart overflowed with love for the Empress, she seemed so utterly broken, so tragically sad.

"I've already reproached Anna for not helping me in the matter," continued the Empress, and she gave me her permission to go at once to the house in Gorohovaya Street where Rasputin lived. I went with Anna.

We did not find Rasputin alone. It was tea time and he was surrounded by a little crowd of admirers. Next to him sat his *âme damnée*, Akilina Laptinsky, the secret agent, under whose skilful tutelage Rasputin unconsciously played the well-planned game of the Revolutionaries. Akilina posed as a Sister of Charity, and many people believed in her; she possessed great influence with Rasputin, and in his unguarded moments he made many deplorable confidences in Akilina, who used everything she heard in a way detrimental to the Imperial Family.

Akilina disliked me: she thought Anna was a weak fool, but I imagine that she regarded me as a foe more worthy of her steel. I acknowledged her presence, and I asked Rasputin if I could speak to him in private.

"But certainly," he answered, and we went into the next room, Akilina following us.

"And now?" enquired Rasputin, seating himself. I did not mince matters.

"Gregory," I said bluntly, "you must leave Petrograd at once. You can pray for Their Majesties equally well in Siberia. You *must* go—for their sakes, I implore you. Go—You know what is said—if you insist upon remaining, it will only mean danger for us all."

Rasputin considered me gravely—he did not speak. I could see Anna's "hurt child" look, I could feel Akilina's sinister scrutiny. Then Rasputin uttered these unexpected words:

"Perhaps thou art right. I'm sick and tired of it all. I'll go."

But a surprising interruption occurred. Akilina banged her clenched fist on the table, and confronted me with rage in her eyes.

"How *dare* you try and control the Father's spirit?" she screamed. "I say that he *must* stay. Who are you?—why, a nobody—you are too insignificant to judge what is best for anyone."

Silence, pregnant with meaning, fell in the little room. Anna was crying, Rasputin said nothing, but I still defied Akilina: the thought of the Empress gave me courage.

"Are you going to listen to the Sister?" I demanded coldly. Akilina recommenced her table-banging.

"If you leave Petrograd, Father, you'll have bad luck—you are *not* to go." "Well—well—" said Rasputin helplessly, "perhaps thou art right. I shall stay."

My efforts were unavailing. Rasputin could be as obstinate as a mule; and so, greatly distressed, I returned to the Palace. The Empress was very disappointed.

"I wonder why the Sister was so against my wishes," she said.

Later on we understood. I think that, despite her plotting and contriving, Akilina really had some affection for Rasputin, and she was occasionally ashamed of her Judaslike rôle. I remember that once, when Rasputin left Petrograd on a visit to his family, I went to see him off, and there, naturally, I encountered Akilina. As the train steamed out of the station she burst into tears—genuine tears; I saw there was no hypocrisy in her grief. Although I disliked Akilina, I felt sorry for her.

"You'd better let me drive you home," I said.

She accepted my offer, but in the car her tears recommenced.

"Whatever is the matter?" I enquired. "You'll see the Father again." Akilina raised her tear-drenched eyes.

"Ah—you know *nothing*—if you only knew—if you only knew what I know."

Surely this remark must have implied that she possessed some inner knowledge which terrified her, and which may have made her conscience-stricken.

Akilina nursed Anna at Tsarkoe Selo when she was ill with the measles, but on the second day of the Revolution she sent me a note, asking me to come over to the left wing of the Palace. She then informed me that Anna was delirious. . . .

"However, I can't do much for her. Will you tell Her Majesty that I must go into town for a day. I want to see Gregory's family."

I promised to deliver the message, but we never saw Akilina again. A fortnight later we were told that she was living in the family of one of the most prominent Revolutionaries.

Another "Sister," Voskoboinikova, equally associated with Rasputin, was head matron of Anna's hospital. She was, likewise, a great friend of M. Protopopoff, the Minister of the Interior, who used to spend hours in her company. Voskoboinikova possessed a certain fascination, but she was very inquisitive, and we equally disliked each other. Following the example of Akilina, she left Tsarkoe on the second day of the Revolution, but, the night before relinquishing her position at the hospital, she gave a dinner to the convalescent soldiers, when wine flowed freely and all sorts of seditious speeches were made. The soldiers were told to look to Petrograd for freedom, and that revolvers and bullets were fine things. Truly women had their uses during the Revolution!

But to return to Rasputin. The feeling against him daily assumed larger proportions. Elidor once sent a woman to kill him, and the Father was badly wounded in the stomach, but it is untrue to say that

Anna Virouboff nursed him during the illness which ensued. She never attempted to do so.

Prince Felix Yousopoff, whose name will always be connected with the tragedy of Rasputin, first met him at the house of Mme Golovina, a sister-in-law of the Grand Duke Paul. The demoiselle Golovina greatly admired Felix Yousopoff, in fact her "flamme" for him was well known. Some considerable time elapsed between the first meeting of Prince Felix and Rasputin: I spent the next two years chiefly in Reval, but I used to pay a fortnightly visit to the Empress, and, after my husband was sent to England, I went to Petrograd, where I saw the Empress daily. I was very surprised when she told me that Felix Yousopoff was a constant visitor at Rasputin's house; in fact I was so incredulous that I asked Rasputin whether this was true.

"Yes—it's quite true," he answered, "I have a great affection for Prince Yousopoff, I never call him anything else but 'Little One.'"

Mary Golovina, to whom also I expressed my astonishment, said that Prince Yousopoff declared that Rasputin's prayers benefited him: so there was nothing more to be said.

On December 16th, when I was at Tsarkoe Selo, I told the Empress that I wanted to see Rasputin on the morrow, but just before starting for his house—about five o'clock on the afternoon of December 17th—I was rung up from Tsarkoe Selo—the Empress wished to speak to me. Her voice seemed agitated.

"Lili," she said, "don't go to Father Gregory's to-day. Something strange has happened. He disappeared last night—nothing has been heard of him, but I'm sure it will be all right. Will you come to the Palace at once?"

Thoroughly startled by this disturbing news, I lost no time in taking the train to Tsarkoe Selo. An Imperial carriage was waiting for me, and I soon found myself at the Palace.

The Empress was in her mauve boudoir; once again I felt the premonition of coming disaster, but I endeavoured to disregard it. Never did the "cabinet mauve" look so home-like. The air was sweet with the fragrance of many flowers and the clean odour of burning wood; the Empress was lying down, the Grand Duchesses sat near her, and Anna Virouboff was sitting on a footstool close to the couch. The Empress was very pale—her blue eyes were full of trouble, the young girls were silent, and Anna had evidently been weeping. I heard all there was to tell me; Gregory had disappeared, but I believe the Empress never imagined for one moment that he was dead. She discountenanced any sinister conjectures; she soothed the ever weeping Anna, and then she told me what she wished me to do.

"You will sleep in Anna's house to-night," she said. "I want you to see people for me to-morrow—I am advised that it will be better for me not to do so."

I told the Empress that I was only too happy to be of service to her, and, after dinner, I went to Anna's house, which I was astonished to find in the occupation of the Secret Police!

The pretty little dining-room was full of police agents, who received me most courteously, explaining that their presence was accounted for by the fact that a plot to kill the Empress and Anna Virouboff had just been discovered. This was not reassuring, but I decided not to be nervous, and, bidding good night to the officers of justice, I went into Anna's bedroom.

The familiar room looked strangely unfamiliar—terror lurked in the shadows, and death seemed in the air. I am not by nature superstitious, but I must confess that I felt so when an ikon suddenly fell down with a crash, carrying a portrait of Rasputin with it in its fall. I hastily undressed and got into bed—I could not sleep; I lay awake for hours, and when, towards dawn, I dropped off in an uneasy slumber, I was suddenly aroused by what seemed a great noise outside. I heard in the distance the tread of countless feet, the sound of many voices; a mighty multitude was marching towards Tsarkoe Selo—and the dreadful thought flashed across my mind that perhaps there had been a rising at Petrograd. I jumped out of bed, threw on a wrapper, and rushed to the dining-room. There all was quiet; the police officers were sleeping on the floor. My entrance awakened them. "Why, madame, what's the matter?" they enquired.

"Cannot you hear for yourselves?" I said, impatiently, "the noise—the crowd— I'm sure something dreadful has happened at Petrograd."

"We have heard nothing. . . ."

"Oh, but I assure you it's correct."

The police opened the shutters, then the windows . . . outside all was still with the intense stillness of a winter's night. The officers made no comment, and closed the windows.

"Madame has perhaps been dreaming," said one, sympathetically. "She has had much to try her nerves."

But I knew differently. I had certainly experienced much to try my nerves, but what I heard was neither a nightmare nor a delusion. When I re-entered the sombre bedroom, with its fallen ikon and its fallen saint, I shuddered, for, although I knew it not, the veil had been lifted, and I had heard the fast approaching footsteps of Revolution and murder.

I was an early arrival at the Palace, but the Empress was already up and she greeted me most affectionately. She told me that M. Protopopoff had strongly urged her to receive no one: there was evidence of a plot to murder her, and, for the first time, she seemed to feel some misgivings concerning the fate of Rasputin. She manifested no anxiety about her own danger; she was utterly serene and fearless: I was so struck by this that I could not help saying:

"Oh, Madame, you don't seem afraid to die. I always dread death—I'm a horrible coward."

The Empress looked at me in astonishment.

"Surely, Lili, you are not *really* afraid to die?"

"Yes, Madame, I am."

"I cannot understand anyone being afraid to die," she said, quietly. "I have always looked upon Death as such a friend, such a *rest*. You mustn't be afraid to die, Lili."

I passed an anxious and exciting morning. I was besieged with visitors for Anna, and people who desired to see the Empress. I think my position gave rise to a great deal of jealousy in the Palace, as at this time the Empress made me the sole medium of her wishes and no official etiquette was observed.

Nothing was heard of Rasputin, but all kinds of disturbing rumours were current. A certain person paid twenty-two visits to Tsarkoe Selo in one day, hopeful to see the Empress, but, acting on the advice of Protopopoff, she absolutely declined to receive him.

Two days later, Rasputin's body was discovered under the ice in the Neva. It was taken to a hospital close by, where an autopsy was performed. Rasputin had been wounded in the face and side, and there was a bullet wound in his back. His expression was peaceful, and the stiff fingers of one hand were raised in a gesture of benediction; it was impossible to arrange the hand in a natural position! The autopsy proved without a doubt that Rasputin was alive when he was thrown into the Neva!

The news of the murder caused the greatest consternation at the Palace—Anna Virouboff was prostrated with grief, and the Imperial Family were deeply concerned. The reports that the Empress gave way to violent hysterics are incorrect. It would be untrue to say that she was not inexpressibly shocked and grieved, but she displayed no untoward emotion. The Emperor was troubled, but his feelings arose more from the significance of Rasputin's death than from the actual death of the man: he realised that this murder was the first definite blow against the hitherto absolute power of the Tsar!

Akilina Laptinsky came to the Palace immediately after the autopsy had been performed: she wished, so she said, to discuss the question of Rasputin's burial. She was received by the Empress; Anna and I were also present. The "Sister" first asked the Empress if she did not wish to see the corpse.

"Certainly not," replied the Empress—in a tone which admitted of no argument.

"But there is the question of the burial," said Akilina. "Gregory always wished to be buried at Tsarkoe Selo."

"Impossible . . . impossible . . ." cried the Empress. "The body had better be taken to Siberia and buried in the 'Father's' village."

Akilina wept. . . . She declared that Rasputin's spirit would never rest were he to be buried so far away from the Palace. The Empress hesitated. . . . I could see she was thinking that it would be equally as unfriendly to discard the dead as to discard the living. Anna, however, settled the question by proposing that Rasputin should be interred in the centre aisle of the new church adjoining her hospital for convalescents. The church and the hospital were being built on Anna's own property. . . . There could be no question of any scandal touching the Imperial Family. . . . This proceeding would only enable people to cast another stone at Anna's already shattered reputation.

"And . . . I care little for the opinion of the world," whimpered Anna, looking more than ever like a hurt baby.

So it was settled that Rasputin should be buried in Anna's church, and, as I attended the burial, I may say with absolute conviction that mine is a true account of the proceedings. I have been told, and I have read various wholly inaccurate reports—the most prevalent being that Rasputin was buried secretly at dead of night in the Park at Tsarkoe Selo. Nothing of the kind. Rasputin's burial took place at 8 o'clock on the morning of December 22nd. The Empress asked me, on the preceding evening, to meet the Imperial Family by the graveside, and I promised to do so.

It was a glorious morning, the sky was a deep blue, the sun was shining, and the hard snow sparkled like masses of diamonds; everything spoke of peace, and I could hardly believe that I was about to witness the closing scene of one of the greatest scandals and tragedies in history. My carriage stopped on the road some distance from the Observatory, and I was directed to walk across a frozen field towards the unfinished church. Planks had been placed on the snow to serve as a footpath, and when I arrived at the church I noticed that a police motor-van was drawn up near the open grave. After waiting several moments, I heard the sound of sleigh-bells, and Anna Virouboff came slowly across the field. Almost immediately afterwards, a closed automobile stopped, and the Imperial Family joined us. They were dressed in mourning, and the Empress carried some white flowers; she was very pale but quite composed, although I saw her tears fall when the oak coffin was taken out of the police van. The coffin was perfectly plain. It bore no inscription, and only a cross outside it testified to the faith of the departed.

The ceremony proceeded—the burial service was read by the chaplain to the hospital, and, after the Emperor and Empress had thrown earth on the coffin, the Empress distributed her flowers between the Grand Duchesses and ourselves, and we scattered them on the coffin.

When the last solemn words had been uttered, the Imperial Family left the church. Anna and I followed them. . . . Anna got into her sledge, I into my carriage. It was barely nine o'clock.

I looked back at the snowy fields, the bare walls of the unfinished church, and I thought of the murdered man who was sleeping there. I felt an immense pity for his fate, but, above all, I felt an immense pity and love for those who had believed in him and befriended him in defiance of the world, and on whose innocent shoulders the burden of his follies was destined to rest.

I have not attempted to introduce any picturesque imagery in my description of Rasputin's burial. I have stated the facts exactly as they occurred, and it now devolves upon me to contradict one of the most unjust accusations which have been made against the Empress in connection with the burial of Rasputin.

Several writers have asserted that, when Rasputin's remains were dug up after the Revolution, a holy image bearing the signatures of the Empress and the Grand Duchesses was discovered resting under the cheek of the dead man. The Empress has been credited with placing this image there herself, but this is not the case. The image (that of the Miraculous Virgin of Pskov) was one of several which the Empress brought back from Pskov when she and her daughters visited her hospital. The Empress purchased these images much in the same manner that visitors to Lourdes purchase souvenirs of Our Lady of Lourdes. The Imperial Family wrote their names and the date in pencil on the base of all these souvenirs, which were given to various friends. Rasputin received one, and, when his body was placed in the coffin, Akilina, with some sinister motive, insisted upon the image being placed under his cheek, and she was, doubt-less, responsible for the story that this was done by order of the Empress.

After Rasputin's death, his son and daughters came to Tsarkoe Selo and were received by the Empress. They related how, on the night of the murder, their father had received a message from Prince Yousopoff, asking him to come and see him. It appeared that Rasputin's daughters had some vague presentiment of ill, and begged their father to remain at home. He, however, insisted upon going to the "little one," and the finding of one of the goloshes which he wore on account of the deep snow was partly the means of discovering that foul play had taken place.

The family begged the Empress to avenge their father's death. She replied:

"I can promise you nothing. All rests with justice; we cannot possibly interfere in any way for or against that which has taken place."

These were her actual words, and they must surely discredit the story that Prince Yousopoff and the Grand Duke Dmitry were victims of the vindictive spirit of the Empress.

Rasputin, as I knew him, was, I repeat, not the villain of the novel and the films. In my eyes he was an uneducated man with a mission; he spoke an almost incomprehensible Siberian dialect, he could hardly read, he wrote like a child of four, and his manners were unspeakable. But he possessed both hypnotic and spiritual forces, he believed in himself and he made others do so. I am not ignorant of what has been said concerning his abnormal animalism, his satyr-like sensualities, the nameless orgies in which young women and young girls gave themselves as willing victims to his lust.

An English saying states that there is "no smoke without fire"—this may, perhaps, apply to Rasputin's sensual side, but never to the alleged extent. One woman in twenty may lose her sense of fitness and seek to mate with a man in an inferior station of life, but it is not an everyday occurrence. The reports about his dress and his extravagance are also very much exaggerated. Rasputin lived, and died, a poor man. He usually wore the dress of a peasant, and his wonderful jewelled cross only exists in the brains of novelists and journalists. Rasputin at first wore a simple copper cross, later he wore one of gold which he afterwards sent to the Emperor at the Stavka. This gift in Russia is usually unwelcome, as it signifies that you present with it the sorrows and sufferings synonymous with the Cross. The Emperor thought that Rasputin's cross was unlucky, so he gave it back to me, and asked me to give it to Anna. But Anna stubbornly refused to accept it, and I was at my wits' end to know what to do. I could not tell the Emperor that Anna would have none of Rasputin's cross—so I mislaid it, and I do not know what became of it. But I only saw the moral side of this apparently immoral man, and I was not alone in my conception of Rasputin's character. I know for a fact that many women of my world who had "affairs" and many demi-mondaines were not dragged further into the mire by Rasputin, for—incredible as it may appear—his influence in such cases was often for the best.

I remember that I once met Rasputin when I was walking on the Morskaya with a brother-officer of Captain Dehn's. He eyed me severely, and, when I returned home, I found a message telling me to come and see him. Partly out of curiosity I obeyed, and, when I saw Rasputin, he demanded an explanation.

"Of what?" I asked.

"Oh . . . thou know'st well enough. Art *thou* going to follow the example of these frivolous Society women? Why art thou not walking with thy husband?" He repeatedly said to women who sought his advice:

"If you mean to do wrong, first come and tell me."

So I can do no more than speak of Rasputin as I found him. If I had been a Rasputinière, or the victim of an abnormal passion, I should not be living happily with my husband, and Captain Dehn would never have countenanced any association with Rasputin if the latter had been guilty of immoralities at Tsarkoe Selo. His duty as a husband would have been greater than his devotion to the Imperial Family.

I cannot entirely defend the Empress's attitude. I love her, I reverence her memory, but I think she was, in many ways, perhaps, mistaken in her outlook. She argued, very rightly, that, even if she belonged to Russia, her soul belonged to God, and she had a perfect right to worship Him exactly in what manner most appealed to her.

I have mentioned her views as to position being no ban where the instruments of God were concerned. In a worldly sense this was impossible, especially in Russia, where humility appealed neither to the peasant nor to the higher classes. The religious "communism" of the Empress outraged their sense of fitness . . . the peasants could not understand one of their own class being on intimate terms with the Sovereigns . . . the higher classes were bitterly contemptuous.

Knowing the strong religious convictions of the Empress and the inborn characteristics of both classes, the Revolutionaries found in Rasputin a fitting agent of Imperial destruction.

The Greek Church is the most mediæval of religions . . . it is quite harmless, so to speak, when modern conditions are not introduced into its practice; but modernity, ever a fatal element in religion, is especially fatal to the Greek Church. The Empress would not understand this . . . her faith taught her to credit the existence of holy men, hermits, and seers—so, when Rasputin appeared in the character of one of these, she was not surprised, and she accepted the actuality of his heaven-sent mission, as the teachings of her Church bade her.

As I have stated, coincidence was largely responsible for the belief of the Empress in Rasputin's gift of healing. His prayers coincided with the recovery of the Tsarevitch—that child of many prayers. In her love for her son the Empress was *plus mère que mère.* I am likewise assured that there was no theatrical clap-trap in Rasputin's association with Anna Virouboff. Had Anna possessed the brains of Akilina, I might not be so positive—but Anna was no *intrigante*; in the face of possible denunciation as a Russian Sapphira, I repeat my estimate of Anna Virouboff, i.e., *childish, harmless, weak.*

If the Empress were guilty of any glaring weakness, it was, paradoxically, that of stubbornness. She did not allow any interference in what she considered her own province. Her grand-mother and the Prince Albert had tolerated none; her distant connection, Princess Clementine of Coburg, was ultra-obstinate; another of her connections, Ferdinand of Bulgaria, has also manifested the Coburg peculiarity. It is an interesting psychological study: in some of the family this trait is manifest in their undeviating pursuit of worldly ambition, in others it is apparent in their views of morality and domesticity. In the case of the Empress, morality, domesticity and religion were subjects in which she brooked no contradiction.

Had the Emperor been less religious, he might have (from a

worldly point of view) influenced his wife to have seen less of Rasputin. But he made no attempt to interfere with her on religious questions, remembering perhaps how wholly she had relinquished the faith of her fathers to embrace his own. The Empress has been accused of contributing to the downfall of Russia through her association with Rasputin. The finger of scorn and hatred has pointed at her, and an almost universal voice has cried, "Thou art the Woman." But history, if not always just, is at least generous, and it may be that Alexandra Feodorovna will one day be given the benefit of the doubt, and allowed to appeal against the sentence which has been passed on her. For many years prior to her advent as Empress of Russia, the movement for Freedom had been slowly but surely spreading over the entire country, and the creation of the Duma strengthened public opinion. But certain Revolutionaries—themselves as evil as their prototypes in the French Revolution—did not scorn to employ base agents in order to attain their base ends. These men used Rasputin—with what result is now apparent. But have the murders of Rasputin and the Empress cleansed Russia and enabled it to be rechristened Utopia?

The ashes of Rasputin are scattered to the four winds, the blood of the innocent cries aloud to Heaven for vengeance; but Russia—drunken with carnage, liberated from her ancient yoke, and delivered of her rulers—has as yet only produced Robespierres.

CHAPTER VI

I HAVE dealt with the subject of Rasputin before touching on that of the War, but his name is also connected with the War, as he is supposed to have been a German spy, and to have encouraged the alleged pro-German leanings of the Empress. Although I shall always adhere to my original belief that Rasputin was an unconscious agent of the Revolutionaries, I cannot deny that he was against the War, and always desirous of peace, but this attitude was due to his own wishes and convictions. I asked Rasputin in 1915 when he thought the war would be over. "Not yet. . . . Don't expect the war to be over yet," he answered; and in 1916, when I returned from Reval, I asked the Empress the same question. "Not yet, Lili, not yet," she said. Both these replies might serve to show how little was the political influence either of the Empress or of Rasputin. As an individual, doubtless the Empress desired peace: as a Russian, she could not possibly have desired the victory of Germany.

There was great excitement in 1914 throughout Russia; everyone hoped that England would come in, especially in naval circles, who were well aware of the weakness of the Russian fleet.

The excitement increased when Russia became the ally of France. The Imperial band played the hymns of the Allies daily; there was no question of pro-Germanism at Court—Russia, as befitting her great traditions, was fighting the good fight!

My husband was ordered to escort the Imperial Family to sea on the "Standart," and I knew that I must therefore spend my birthday without him. One evening, when we were sitting in the Park making plans for a belated celebration, my husband was accosted by one of the heads of his Department. "Dehn . . ." said he . . . "go at once to the Commander of the Port . . . you're wanted."

Upon his return my husband was very excited. "Lili," he cried, "I have received orders to join Admiral Essen's fleet. I must leave almost immediately." It was, indeed, "almost immediately," for at 3 a.m. my husband bade me good-bye.

The Empress sent me a note directly she knew that Charles had left. "I hope everything will be all right," she wrote. "Poor Lili, don't despair."

I tried *not* to despair, and, like most wives at this time, I kept a smiling face, although I was perilously near tears. Every day the Military Council was in consultation with the Emperor, and, on the evening before the declaration of war, I knew that mobilization had been decided upon.

The Emperor firmly believed that Russia was amply supplied with munitions. He had been assured on this point by the Grand Duke Nicholas and General Soukhomlinoff. Soukhomlinoff knew that the ammunition of the Russian army was insufficient, but he still continued to reassure the Emperor and the Allies. The Grand Duke Nicholas, who was far from blameless . . . instigated a Special Commission under the presidency of the Grand Duke Serge, with the declared object of providing the army with the requisite munitions. But three months passed, and nothing was done. Even when certain supplies of munitions arrived at the Front, these were useless, as they would not fit the guns and musketry which required them! The Emperor was most unjustly blamed for these calamities—but he was guiltless—the real offenders were the Grand Duke Nicholas, General Soukhomlinoff and their agents.

On the day following my husband's departure the Empress sent me a message asking me to go with her to the church usually attended by the Lancers (the Empress's Own). The service was very impressive; I stood behind the Empress, who was praying ardently, and, at the conclusion, she turned to me: "Don't look sad, Lili," she whispered. "This war *had* to be."

Whenever the regiments of which the Empress was colonel left for the front, she saw the officers and soldiers, and blessed them and spoke to them. A great deal has been said and written about the Empress's unpopularity with the soldiers. I have hardly heard a good word on her behalf, and yet I know how devotedly she was loved by many of the officers and men. It will be my privilege to show how, during the Revolution, she received many touching evidences of their affection, and I am determined not to allow the Sisyphus weight of calumny to deter me from telling what I know of the truth. After the declaration of hostilities the Empress at once instituted her own hospitals, and both she and her daughters went in for a medical course to qualify as Sisters of Charity. Princess Gedroits, herself a professor of surgery, instructed them, and the Imperial Family gave up most of their time to lectures and demonstrations.

Directly they had passed the necessary examinations, the Empress and "the four sisters Romanoff" started nursing, spending hours with the wounded and almost invariably being present at operations.

Society at once began to criticise this procedure. It argued that it was not the duty of an Empress of Russia to become a nurse. It failed to remember that at this time the illustrated papers were full of pictures of various crowned heads who were doing precisely the same thing for which they condemned the Empress! But she wore her rue with a difference. What was praiseworthy in others constituted a sin in her case. Without being accused of bitterness, I think I may be allowed to say that it makes me sad when I realise the persistent animosity displayed towards the Empress by all classes, from the prince to the peasant . . . "the evil that men do lives after them, the good is oft interred with their bones." In the case of the Empress, the good she undoubtedly did during her life was not only interred with her but it was never recognised during her life. Her innocent fault consisted (to quote the words of an English writer) in not being able to understand "that in the eyes of her subjects she must shine and be ornamental, but not useful in the trivial acceptance of the word." Perhaps the Empress erred in her conception of the mentality of the Russian peasant. As an impartial critic, I fear this was the case. When she wore the Red Cross, the sign of a universal Brotherhood of Pity, the average soldier only saw in the Red Cross an emblem of her lost dignity as Empress of Russia. He was shocked and embarrassed when she attended to his wounds and performed almost menial duties. His idea of an Empress was never as a woman, but only as an imposing and resplendent Sovereign.

The pro-German tendencies of the Empress were mentioned after our reverse at Brest, when the Emperor assumed command. Everyone was suspicious of her, and, when she spoke English at the hospitals to her daughters and her ladies-in-waiting, the soldiers declared she was speaking German, and this report once started was magnified exceedingly.

The actual dawn of Revolution occurred before the death of Rasputin, but during the war it was openly stated that the end of Tsardom was at hand. All our defeats were attributed to the pro-German influence of the Empress, who was spitefully alluded to as "The Colonel" in certain salons.

Protopopoff, the Minister of the Interior, was always reporting plots against the life of the Empress. One, it was said, had been disclosed in an intercepted letter from a Society woman to a friend in Moscow. The writer lamented that the murder of the Empress had not been a "fait accompli," and declared that, failing murder, the next best remedy was incarceration in a madhouse. Princess Vasiltchikoff sent a letter to the Empress, in the name of the women of Russia, telling her that all classes were against her, and daring her to mix further in Russian affairs.

It has been said that the Empress was equally furious at the contents of the letter, and the fact that it was written on paper torn off a letter-pad! But it was *not* the question of the breach of etiquette which writing to the Sovereign on a letter-pad implied, it was the horrible accusations, the virulent animosity of the missive which at first angered the Empress, and afterwards grieved her. She cried bitterly when she told me. "Of what am I accused?" she said. "Gregory is dead. Surely people might leave me alone!"

Princess Vasiltchikoff's letter gave rise to much excitement; her portrait was in all the newspapers, and public opinion was divided for and against her.

Another letter was sent to the Empress, this time anonymously, but it was equally reprehensible, and this letter and the preceding one caused the greatest indignation in the hospitals, as the officers who knew the Empress as she really was were very angry. Life in general was excessively difficult and painful, so much so that, when my husband arrived from Mourmansk, and asked Count Kapnist how things were going, the Count replied: "You'll soon see for yourself, and you'll be horrified. We have gone back to the days of Paul I. Ruin lies ahead of us."

The Empress saw a good many people at this time. Every Thursday there were musical evenings, where I met various friends—officers in the Artillery, the Emperor's A.D.C., Linavitch, Count Rabindar and his wife (who was a faulty likeness of the Empress), the officers of the "Standart," Prince Dolgouroki (who was after-wards murdered), Madame Voeikoff, the wife of the Commandant du Palais, Colonel Grotten, and many others.

A Roumanian orchestra, under the direction of the famous Goulesko, played on these Thursdays, and the Empress derived great pleasure in listening to the really exquisite music. A huge fire was always burning in the salon; the Empress sat near it, and a little seat immediately behind her was arranged for my exclusive use. If I happened to arrive after the Empress was seated, she always indicated the vacant place with a gesture and a sweet smile.

One evening, about a fortnight before the Revolution, when I was sitting in my usual place, listening to the Roumanian orchestra, I noticed that the Empress seemed unusually sad. So I ventured to bend forward and whisper, anxiously, "Oh, Madame, why are you so sad to-night?" The Empress turned and looked at me. . . . "Why am I sad, Lili?. . . I can't really say, but the music depresses me. . . . I think my heart is broken."

The same evening, Anna childishly observed: "We all seem out of sorts. What fun it would be to have some champagne!" The Empress was angry at the suggestion. "No . . ." she said, "the Emperor hates wine, he can't bear women to drink wine—but what matter his likes or his dislikes, when people will have it that he's a drunkard himself?" The Empress was in very indifferent health; mental worry had increased her heart trouble, but she endeavoured never to let her health interfere with her public duties. At an official reception following the departure of the Guards, the Empress told me that she hardly knew how to endure the strain. "Veronal is keeping me up. I'm literally saturated with it," she said.

When my husband came home on a few days' leave, the Emperor sent for him, and listened attentively to all that he had to say, questioning him very closely on certain subjects. We had never thought of or mentioned the subject of his preferment; he had now spent two strenuous years in the mine-fields, and the Emperor noticed how ill he looked. "Dehn must have a rest," remarked His Majesty. "I shall give him a post near my person."

But this kindly thought never matured. My husband was sent for by the Minister of the Marine, and left for England at twenty-four hours' notice, in company with General Meller-Zakomelsky, taking with them decorations destined by the Emperor for certain English officers. The news of the Revolution was not known by them or in England when they arrived, so an elaborate official reception was given them. Almost immediately afterwards the news was public property and it was impossible to use the Emperor's decorations. I often wonder what became of them.

Before leaving for England, my husband asked me to join him there. I could not promise. I loved him very dearly, but I felt that my duty lay with the Empress.

"No, Charles," I said, "I cannot promise anything at present, but, if things become better, I'll come."

When he had gone, I felt utterly unhappy, but I did not regret any sacrifice I was called upon to make for the Imperial Family. I loved them all far too much.

At this time the Emperor had every intention of remaining with his family, but, one morning, after having received General Gourko in audience, he suddenly announced:

"I'm going to G.H.Q. to-morrow." The Empress was surprised.

"Cannot you possibly stay with us?" she enquired.

"No," said the Emperor, "I must go."

Almost immediately after the Emperor's departure, the Tsarevitch fell ill with measles, and I used to spend every evening with the Empress, who was naturally much worried over her son's illness. In these days, our intimacy had increased so much that my time was mostly devoted to the Empress, and I saw few of my friends and relations. But my aunt, the Countess Kotzebue-Pilar, was a great Society leader, and I heard all that transpired in her salon. One evening before dinner my aunt (who was always furious at the rumours current about the Empress) 'phoned me to come to her house at once. I found her in an excessively agitated condition....

"It's awful what people are saying, Lili," she cried.... "And I must tell you—you *must* warn the Empress."

In somewhat calmer tones my aunt continued: "Yesterday I was at the Kotzebues'.... Many officers were present, and it was openly asserted that His Majesty will never return from G.H.Q. What are you going to do? You are constantly in the society of the Empress—you cannot allow her to remain in ignorance of these reports."

"She will not believe them," I said.

"Nevertheless," said my aunt, "it is your duty to warn her."

I returned to the Palace feeling very unhappy. I hardly knew what to do for the best. At last, after a struggle, I decided to tell the Empress. As I had anticipated, she made light of the story.

"It's all nonsense, Lili, I can't believe such a thing—it's nothing but malicious gossip. However, as you seem so apprehensive, send for Grotten (the Commandant du Palais) and tell him."

"Don't pay any attention to such a canard," cried Grotten angrily, when he heard my story. "It's a lie which stamps itself as the worst kind of lie."

"Well, General," I retorted, now thoroughly vexed with myself for having apparently made a mountain out of a molehill, "if God ordains my aunt's report to be a lie, so much the better."

"Don't be cross.... I'll most certainly get in touch with G.H.Q.," said Grotten reassuringly. THREE DAYS AFTER CAME THE REVOLUTION.

And now the funeral knell of Russia began to sound, at first muffled, but always insistently. Disorders broke out in Petrograd. The strikes began on February 21st (Old Style), and crowds clamoured for bread, of which the supplies had suddenly stopped.

No one could understand this, as Protopopoff's last words to the Emperor were: "There is plenty of flour, I'll pledge my word that we have enough flour to last us for a month, and after that fresh supplies will be coming in." The bread shortage was in reality due to the action of the Duma—it was an organised arrangement!!

Each day matters grew worse. Fighting took place in the streets, drunkards indulged in indescribable orgies, the police were murdered much in the same manner as they have been in Ireland. It was bitterly cold—snow lay in deep drifts, and Petrograd was in the iron grip of a black frost.

Protopopoff, the Minister of the Interior, was always ultra-optimistic—I never liked or trusted him; he did not seem the man to handle any great crisis. He was appreciated by the Duma until his deplorable interview in Stockholm, when he discussed the war in a very indiscreet manner; but, when the Emperor appointed Protopopoff Minister of the Interior, he was universally hated, and everyone blamed the Emperor for appointing a man so singularly devoid of merit. Protopopoff promised everything, without considering whether his promises were possible. It was the same with his statements: he disliked telling unpleasant truths, so he took refuge in pleasant evasions. He was the man who continually told the Imperial Family that nothing could possibly happen. "Trust in me," said Protopopoff, striking an attitude. And, whenever someone meekly remarked that the working classes were undoubtedly restive, Protopopoff struck another attitude which implied, "Did I fancy I heard you say '*restive*'?" and, aloud, in pained but hearty tones: "What? Are you actually troubling yourself about a little unrest? We'll soon crush them—Labour cannot stand up against *Me*."

It may be asked: Why did the Imperial Family, and especially the Empress, place so much reliance in M. Protopopoff's statements, as, since the Empress knew all that was written concerning her, she, at least, could have possessed no illusions? The answer is simple: The Empress knew that she was unpopular, but she never would believe that this unpopularity lay with the people—she attributed the scandals and calumnies to class-hatred, and to that craving for sensation without which a certain section of the Press would be unable to exist. When, made bold by my ever growing apprehendsions, I ventured to tell the Empress that in these days the "people" were not paragons of fidelity, she bade me remember the afternoon, not long distant, when we drove out to a little "Lett" village near Peterhof. I *did* remember. The automobile had stopped near the church, and, the moment the Empress alighted, she was surrounded by a crowd of peasants, who knelt before her, and, with tears in their eyes, prayed aloud for her happiness. After this the Empress was offered bread and salt, and it was with great difficulty that a passage was cleared to her waiting automobile. This incident occurred two years before the Revolution. "And yet you tell me, Lili, that these people wish me ill!"

"Madame, many things have happened during the last two years."

"*Nothing* has happened, Lili, to touch the real heart of Russia."

I do not profess to have any knowledge of politics, and I never wished to meddle in them, so it is impossible for me to attempt to discuss the so-called political influence of the Empress. We hardly ever spoke of politics, but I can truthfully state that I never once heard her utter one sentiment that might be described as even faintly pro-German. Her letters written after her arrest, which are reproduced for the first time, ought to plead for her more strongly than any words of mine. When the Empress wrote to me, neither she nor I had any idea that part of her correspondence would be read by the English public. The letters might never have reached *me*: they were smuggled out of the Palace and sent from Tobolsk in circumstances of much difficulty and danger. But they breathe sincerity of purpose in every line: they were written when the shadow of death was falling on the Imperial Family.... There is no trace of the hysterical, intriguing woman in any of them. The letter which contains the passage relating to the fleet will perhaps serve to vindicate the memory of the Empress more than anything else, at least so far as her alleged pro-Germanism is affected. Even now, Justice, blind, but nevertheless all-seeing, has decreed that Germany should acknowledge having laid the mines which destroyed the "Hampshire": Germany, brought to book, would not have scrupled to lay the guilt to the charge of the Empress, especially since she cannot defend herself. But Germany has not availed herself of the universal detestation which surrounds the name of Alexandra Feodorovna: so she has, at least, been spared *one* degradation.

Part II—The Revolution

CHAPTER I

ON Saturday, February 25th, 1917, the Empress told me that she wished me to come to Tsarkoe Selo on the following Monday, and I was (let me confess it) still in bed when the telephone rang at 10 a.m. I suppose my delay in answering must have amused the Empress, for her first words were: "I believe you have only just got out of bed, Lili. Listen, I want you to come to Tsarkoe by the 10.45 train. It's a lovely morning. We'll go for a run in the car, so I'll meet you at the station. You can see the girls and Anna, and return to Petrograd at 4 p.m. I'm certain you won't catch the train, but anyhow I'll be at the station to meet it."

I dressed at express speed, and, snatching up my gloves, a few rings, and a bracelet, I ran into the street in search of a fiacre. I had quite forgotten that there was a strike, and no conveyances were available! At this moment I saw M. Sablin's carriage: I hailed him, and begged for a lift to the station. On the way I questioned him.

"What news, Monsieur . . . ?"

"There's nothing fresh," he replied, "but everything is quite all right, although I must admit it is very strange about the bread shortage."

The train for Tsarkoe was just moving out of the station when I arrived on the platform, but I scrambled in, and found myself in the company of Madame Tanieff, Anna's mother, who was going to see her daughter, now ill, like the Grand Duchesses Olga and Tatiana, with the measles. Madame Tanieff, like M. Sablin, knew nothing fresh; she was chiefly concerned about Anna's illness; but the first words of the Empress, who, true to her promise, was awaiting me, were:

"Well, how is it in Petrograd? I hear things are very serious."

We said that there was apparently nothing alarming, and the Empress told Madame Tanieff to get into the car with us, and she would take her to the Palace.

It was a glorious morning: I remembered the splendour of the day long afterwards; the sky was an Italian blue, and snow lay everywhere. We were not able to drive in the Park on account of the drifts! On the way back, we met Captain Hvostchinsky, one of the Garde Equipage. The Empress intimated her wish to speak to him, and the car stopped.

Captain Hvostchinsky smiled at the notion of danger. "There is no danger, Your Majesty" he said; so, reassured, the Empress and I returned to the Palace. I went at once to see the Grand Duchesses. They were certainly very ill, suffering from bad pains in the ears; but they were pleased to see me, and I sat between the two camp beds, talking to them. After lunch I went up again, and presently the Empress joined us.

She beckoned me into the next room: I could see that she was agitated. "Lili," she said, breathlessly, "it is *very* bad. I have just seen Colonel Grotten, and General Resin, and they report that the Litovsky Regiment has mutinied, murdered the officers, and left barracks: the Volinsky Regiment has followed suit. I can't understand it. I'll never believe in the possibility of Revolution—why, only yesterday, everyone said it was impossible! The peasants love us . . . they adore Alexis! I'm sure that the trouble is confined to Petrograd alone. But I want you to go and see Anna . . . she may also have been told this, and you know how easily she is frightened!"

I found Anna ill, and light-headed, and, as I entered her bedroom, I thought what a contrast it presented to the cool, darkened room which I had just left. Olga and Tatiana were so patient, they lay so still, and were grateful for any attention. *This* sick room resembled a "lever du Roi" in the days of Louis XIV. Anna was surrounded by a crowd of "sisters" and three doctors were in attendance. Madame Tanieff was there, looking the picture of misery, and Anna's sister, who was almost hysterical, kept on exclaiming, "All is lost." They had expected General Tanieff to lunch, but he had not arrived . . . there was no news of him. What were they to do? General Tanieff entered in the midst of this confusion, breathless, and scarlet in the face. "Petrograd is in the hands of the mob," he exclaimed, "they are stopping all cars . . . they commandeered mine, and I've had to walk every step of the way."

At this intelligence, Allie Pistolkors (she had married the Grand Duke Paul's stepson) burst into tears and begged me to ask the Empress what she had better do. I promised to see the Empress at once, and, as the Grand Duchesses Anastasie and Marie had just come to fetch me, I returned to the private apartments with them.

The winter afternoon was fast drawing in, and I found the Empress alone in her boudoir. She could give me no message for Mme Pistolkors. "I don't *know* what to advise," she said, sadly. Then, turning to me, "What are *you* going to do, Lili? Titi is in Petrograd . . . had you not better return to him this evening?"

At the sight of the Empress, so tragically alone, so helpless in the midst of the signs and splendour of temporal power, I could hardly restrain my tears. Controlling myself with an effort, I tried to steady my voice:

"Permit me to remain with *you*, Madame," I entreated.

The Empress looked at me without speaking. Then she took me in her arms and held me close, and kissed me many times, saying as she did so:

"I *cannot* ask you to do this, Lili."

"But I must, Madame," I answered. . . . "Please, please let me stay. I can't go back to Petrograd and leave you here."

The Empress told me that she had tried to 'phone the Emperor, and that she had been unable to do so. "But I have wired him, asking him to return immediately. He'll be here on Wednesday morning."

After this conversation we went to see the Grand Duchesses, and the Empress lay down on a couch in their bedroom. I sat beside her, and we conversed in low tones so as not to awaken the sleeping girls. The Empress was still unable to believe in the reports, and she expressed a wish to see the Grand Duke Paul. "How I wish he would come," she said. She then asked me to go over to Anna's apartments, and say that she felt too unwell to come herself.

Anna's room still looked like a "lever du Roi"; Allie had taken her departure, so Mme Tanieff told me, and had gone to the Palace of the Grand Duke Paul. I lost no time in delivering the Empress's message, and quickly returned to her. The evening wore on. . . . News came that Petrograd was in a state of upheaval, and that crowds of mutineers were everywhere. The Empress begged me to 'phone Linavitch, the A.D.C. to the Emperor, and ask him to tell us what was happening. Linavitch was in command of a company of Horse Artillery at Pavlosk, two miles from Tsarkoe Selo, so it was not difficult to "get" him. "Tell Her Majesty," he said, "that I am here with my company, and that all will be well."

I spent the evening with the Empress in the mauve boudoir, and she told me how glad she was to have me near her. "I know the

Grand Duchesses want you to be somewhere close to their room, so I've decided that the red drawing-room will be the best place for you to sleep.* Come with me. Anastasie is waiting for us," she said.

The red drawing-room was a fine room; everything in it was upholstered in scarlet, and scarlet and white chintz covered the easy chairs. A bed had been arranged on one of the couches, and the two Grand Duchesses, with tender solicitude, had seen to the minor details themselves. Anastasie's nightgown lay outside the coverlet, Marie had put a lamp and an ikon on the table by the bed; and a snapshot of Titi, taken from their collection of photographs, had been hastily framed, and occupied a place next to the holy ikon. How dearly I loved them all . . . how glad I was that I was privileged to share their danger!

The Empress left me with Anastasie, as she wished to see Count Benckendorff, so Anastasie and I sat down comfortably on the red carpet, and amused ourselves with jigsaw puzzles until she returned.

The Empress came back from her interview with Count Benckendorff in a state of painful agitation, and, directly Anastasie had gone to bed, she told me that the reports were worse. "I don't want the girls to know anything until it is impossible to keep the truth from them . . ." she said, "but people are drinking to excess, and there is indiscriminate shooting in the streets. Oh, Lili, what a blessing that we have here the most devoted troops . . . there is the Garde Equipage . . . they are all our personal friends, and I place implicit faith in the tirailleurs of Tsarkoe."

I think that this thought comforted her: she seemed happier when she bade me good night.

I woke early on Tuesday morning. . . . Sleep had been almost impossible, but I had dropped into an uneasy slumber soon after dawn. I dressed at once, hoping to be ready for the Empress, but she was before me, and at half-past eight she entered the red drawing-room. We went at once to the Grand Duchesses, and drank our *café au lait* in their room. The Empress told me that she had wired repeatedly to the Tsar, but had received no reply. Later in the morning she received Count Benckendorff and Colonel Grotten, who informed her that matters were becoming more alarming and that the Garde Equipage had better remain inside the Palace, as there

---

* The apartments at Tsarkoe Selo reserved for guests and the suite were situated over the third and fourth entrances to the Palace. The red drawing-room was in the private apartments.—L. D.

THE REAL TSARITSA

was a report that the mob, supported by the Duma, was even now marching on Tsarkoe.

[*For the image that is inserted here, please see a very similar one already included in Book I on page 186.*]

THE IMPERIAL FAMILY
BACK ROW left to right: Grand Duchesses Marie, Olga, and Tatiana
CENTRE left to right: H.I.M. The Tsaritsa, Tsar Nicholas II. Grand Duchess Anastasia
FRONT The Tsarevitch

The Empress immediately consented; she was really delighted at the thought of having the Garde Equipage at the Palace, and the Grand Duchesses were frankly overjoyed. "It's just like being on the yacht again," they said. The Garde Equipage, which was now augmentted by the Mixed Guard, and by sentinels taken from the Cossack Convoi, took up its quarters outside the Palace and in the vast souterrains. One part of the Palace was arranged as an ambulance station. We were very busy, but the Grand Duchesses made light of danger and showed none of our agitation. The Empress was always awaiting a reply to her telegrams. None came.

Tuesday was a day of general unrest. It seemed as if the weather were in sympathy with man's savage mood. The blue sky of Monday had vanished, an icy blizzard swept around the Palace, and a north wind drove the deep snow into still deeper drifts. In the afternoon, on my way back from seeing Anna, I encountered Baroness Ysa Büxhoevgen on one of the corridors. She was almost running and she seemed very much disturbed. "I must see the Empress," she said. "I've just come from Tsarkoe Selo (the town): everything is awful—they say there is mutiny and dissatisfaction amongst the troops." Ysa's terror was general: panic seized the dwellers in the

Palace, but none of the servants left us. Mlle Schneider's maids, it is true, fled, but they came back again the next day.

The Empress was very anxious to see the Grand Duke Paul, but I believe that at first there was some misunderstanding, as the Grand Duke thought that etiquette demanded that the Empress should ask *him*, and he declared that he would not come unless she did. I had received a hint of this, so, when next I saw the Empress, I suggested that perhaps the Grand Duke was waiting for her invitation.... This had not occurred to the Empress; she told me to 'phone at once and ask the Grand Duke to come and see her after dinner.

I was placed, unwillingly, in a very awkward predicament. I had no official position at Court, but the Empress seemed to think that my duty was to act as her mouthpiece, and to assume an authority which I was far from desiring.

However, I 'phoned to the Palace of the Grand Duke, and, in the name of the
Empress, I asked him to come to Tsarkoe Selo. His son answerred the 'phone, and rather brusquely demanded to know who on all the earth was speaking.

"Lili Dehn," I said.

His "*Oh!*" was more eloquent than words!

During the afternoon the Empress called me into her boudoir. "Lili," she said, "they say that a hostile crowd of 300,000 persons is marching on the Palace. We shall not be, we *must* not be afraid. Everything is in the hands of God. To-morrow the Emperor is sure to come.... I *know* that, when he does, all will be well." She then asked me to 'phone to Petrograd, and get in touch with my aunt, Countess Pilar, and other friends. I 'phoned to several, but the news grew worse and worse. At last I 'phoned to my flat. The Emperor's A.D.C., Sablin, who lived in the same building, answered my ring. I begged him to take care of Titi, and, if it were possible, to join us at Tsarkoe, as the Imperial Family needed protection; but he replied that a ring of flames practically surrounded the building, which was well watched by hostile sailors. He managed, however, to bring Titi to the 'phone—and my heart ached when I heard my child's anxious voice:

"Mamma, when are you coming back?"

"Darling, I'll come very soon."

"Oh, *please* come; it's so dreadful here."

I felt torn between love and duty, but I had long since decided where my duty lay. I told the Empress what Sablin had reported; she listened in silence, and then, by some tremendous effort of will, she regained her usual composure. Her strength strengthened me. We had, indeed, every need for courage. The poor "children" were lying desperately ill. . . . They looked almost like corpses. . . . Anna was in high fever, the Palace was terror-stricken, and outside brooded the dread spectre of Revolution!

All at once the Empress was seized with an idea to talk to the soldiers. I begged to accompany her, in case of any unforeseen treachery, but she refused. "Why, Lili," she said, reproachfully, "they're all friends!" Marie and Anastasie went with her, and I watched them from a window. It was quite dark, and the great courtyard was illuminated with what appeared to be exceptionally powerful electric lights. The distant sound of guns was audible . . . the night was bitterly cold. From where I stood, I could see the Empress, wrapped in furs, walking from one man to another, utterly fearless of her safety. She was the calm, dignified Tsaritsa—the typical consort of the Tsar of all the Russias. Here was no hysterical religious maniac, no abandoned heroine of the novel! The Empress moved in this tragic *mise en scène*, protected by her own goodness; but, when the light fell on her fair, pale face, I trembled. I knew her weak heart, her delicacy of physique—suppose she were to faint?

When the Empress came back, she was apparently possessed by some inward exaltation. She was radiant; her trust in the "people" was complete, she was sustained by that, often, alas, broken reed of friendship. "They are our friends," she kept on repeating, "they are so devoted to us." She was, alas, presently to discover that the name of Judas is often synonymous with that of a friend.

One thing troubled her fleeting happiness. "I haven't seen a company in the basement. . . . It is such a pity, but I didn't feel well enough. Perhaps I can manage it tomorrow."

After her visit to the soldiers, the Empress received Count and Countess Benckendorff, who asked to be permitted to remain at the Palace. Their request was gladly granted, and rooms were arranged for them.

The Grand Duke Paul arrived later in the evening. He was a tall, imposing man, who was considered to be very fascinating, and, what was more to his credit, excessively kind at heart. He had a long conversation with the Empress, and we could hear their agitated voices in the next room. The Empress told me afterwards that almost her first words had been:

"What of the Guards?"

And the Grand Duke had replied in tones of fatality:

"I can do nothing. Nearly all of them are at the Front."

When we went to bid the Grand Duchesses good night, I was distressed to find that the firing was distinctly to be heard from their room. Olga and Tatiana did not appear to notice it, but, when their mother had gone, Olga asked me what the noise signified.

"Darling, I don't know—it's nothing. The hard frost makes everything sound much more," I said lightly.

"But are you *sure*, Lili?" persisted the Grand Duchess. "Even Mamma seems nervous, we're so worried about her heart; she's most certainly overtiring herself—*do* ask her to rest."

The Empress decided that Marie should sleep with her. "You, Lili, will sleep in the room with Anastasie, and have Marie's bed. Don't take off your corsets . . . one doesn't know what may happen. The Emperor arrives between 5 and 7 to-morrow morning, and we must be ready to meet him. Come to my room early, and then I'll tell you the train."

Neither the Grand Duchess nor I could sleep, and we lay awake in the darkness talking in low tones. Occasionally I was silent, but, when this was so, Anastasie never failed to ask: "Lili, are you asleep?"

During the night we got up and looked out of the windows. A huge gun had been placed in the courtyard. "How astonished Papa will be!" whispered Anastasie. We stood for a few minutes watching the weird scene. It was so bitterly cold that the sentinels were dancing round the gun in order to keep warm. Their figures were sharply defined against the arc-lights—it seemed like some new Carmagnole; in the distance we heard shouts of drunken voices and occasional shots—and so the night passed.

At 5 a.m. on Wednesday morning we went downstairs to the Empress's bedroom. She was awake, and as she opened the door she whispered: "Hush . . . Marie is asleep: the train is late. . . . Most probably the Emperor won't come until ten." The Empress was fully dressed, and she looked so sad that I could not help saying impulsively: "Oh, Madame, *why* is the train late?"

She smiled wanly, but did not reply. As we went back to our bedroom, Anastasie said in agitated tones: "Lili, the train is *never* late. Oh, if Papa would only come quickly. . . . I'm beginning to feel ill. What shall I do if I get ill? I can't be useful to Mamma. . . . Oh,

Lili, say I'm not going to be ill."

I tried to calm her, and I persuaded her to lie down on her bed and sleep; but the poor child was actually sickening for the measles. Anastasie was the sweetest-natured girl: she adored her mother, and delighted in running hither and thither on her errands.

The Empress always alluded to Anastasie as "my legs!"

When the Empress joined me in Olga's room a little before nine, she still hoped for the 10 o'clock train. "Perhaps the blizzard detains him," she said. She lay down on the couch, and I sat on the floor beside her; we spoke in undertones; but her chief anxiety was concerning my want of sleep. "Sit on a chair, Lili, and put your feet up on the couch," she said.

"No—no—Madame," I replied, "it is not to be thought of." But, at her request, I compromised matters by resting the tips of my shoes on the end of the couch.

Ten o'clock came, but we still heard nothing. It was the first of March, a month fatal to the Romanoffs—well might they "beware the Ides of March!" The Emperor Paul was suffocated on the first of March, and, thirty-six years previously, on this date, the Emperor's grandfather, Alexander II, was killed by a bomb. The March of 1917 is destined to be associated with the downfall of the dynasty.

We were living in a state of continual and unrelieved anxiety. Dr. Botkin and Dr. Direvenko were in constant attendance on the three Grand Duchesses, but the Tsarevitch was, fortunately, much better. Poor Anastasie could not reconcile herself to the idea of being ill: she cried and cried, and kept on repeating, "Please don't keep me in bed."

Service in the Palace was quite normal, but the water supply which worked the private lift used by the Empress had been cut off, and in consequence she was now obliged to walk upstairs. This sounds a trivial incident, but it entailed a great deal of suffering on the Empress, who was already overtired and overstrung. Her heart, always affected, now became much worse, owing to her having to go up and down stairs so often, but she insisted upon seeing her children, and she used to go up the staircase at times almost on the verge of fainting. I supported her—walking behind her and holding her underneath the arms.

We could not understand what had become of the Emperor: the Empress thought that the delay arose owing to the confusion on the railways, which were now in the hands of the Revolutionaries.

The dreary afternoon of March 1st was signalised by an unhappy occurrence. The Empress and I were standing at the window overlooking the courtyard, when we noticed that many of the soldiers had bound white handkerchiefs on their wrists. An enquiry as to the reason elicited the reply that the white handkerchiefs signified that upon the representation of a Member (who had come to Tsarkoe Selo) the troops had consented to act in unison with the Duma.

The Empress turned to me. "Well . . . so everything is in the hands of the Duma," she said, with a certain degree of bitterness. "Let us hope that it will bestir itself, and do something to remedy the disaffection."

She moved away from the window. I could see she was hurt and disappointed . . . but this was not destined to be the last of her many disillusions!

Count Appraxin, Secretary to the Empress, arrived later in the day: he had experienced the greatest difficulty in reaching Tsarkoe—and his news was not reassuring. We sat up late that evening—dinner had been a mere farce—our minds were too anxious and too preoccupied to think of food. The children were dangerously ill, the whereabouts of the Emperor were unknown, and the Revolution was at our gates. When at last I bade the Empress good night, she told me not to undress. "I'm not going to do so," she said, and her quiet tones were significant that she anticipated the worst!

DOCTOR DIREVENKO

SHOOTING PARTY IN FINLAND, AUTUMN, 1910
Centre—the Emperor: Right—Lieut.-Com. Dehn

THE TSAREVITCH AT G.H.Q.

THE TSAREVITCH AND HIS SPANIEL 'JOY'

CHAPTER II

EARLY on the morning of March 2nd the Empress came into the Grand Duchesses' bedroom. She was deathly pale—she seemed hardly alive. As I ran towards her I heard her agitated whisper: "Lili—the troops have deserted!"

I found no words with which to answer. I was stupefied. At last I managed to stammer:

"Why, Madame? In the name of God, why?"

"Their Commander-in-Chief, the Grand Duke Cyril, has sent for them." Then, unable to contain herself, the Empress said brokenly, "My sailors—my *own* sailors—I can't believe it."

But it was too true. The Garde Equipage had left the Palace at 1 a.m. and 5 a.m.— the "faithful friends," the "devoted subjects," were with us no longer. The officers of the Garde were received by the Empress in the mauve boudoir during the morning: I was present, and I heard from one of my husband's friends that the duty of taking the Garde to Petrograd had been carried out by a "temporary gentleman," Lieutenant Kouzmine. The officers were furious, especially their commandant, MiasocdoffIvanof, a big, burly sailor, whose kind eyes were full of tears. . . . One and all begged to be allowed to remain with the Empress, who, almost overcome by emotion, thanked them, saying: "Yes—yes—I beg you to remain: this has been a terrible blow, what *will* the Emperor say when he hears about it!" She then sent for General Resin, the Commander of the Mixed Guard, and instructed him to make room for the loyal officers in his regiment.

General Resin told me long afterwards that he was relieved when he knew that the cowardly Garde had actually left the Palace, as orders had been given for a detachment to go on one of the church towers which commanded a view of the courtyard, and if, by a certain time, the troops had not joined the Duma, to train two enormous field-guns on to the Palace!

There was still no news of the Emperor, although the Empress constantly telegraphed. It was reported that his train was returning to G.H.Q., and at the time many people thought that if it reached there the troops would have followed the Emperor. We 'phoned to the hospitals for news, and the Empress received a good many people. To all these she was her usual calm, dignified self. When I marvelled at her fortitude, she replied: "Lili, I must *not* give way. I keep on saying, '*I must not*'—it helps me."

In the late afternoon, Rita Hitrowo (one of the younger ladies-in-waiting, and a friend of the Grand Duchesses) arrived from Petrograd with the worst possible tidings, and, after the Empress had spoken to Rita, she received two officers of the Mixed Guard, who proposed to try and get a letter from her through to the Emperor: it was arranged that they should leave Tsarkoe the next evening. The Empress was always willing to hope. But the night passed, and still never a word came from the Emperor.

On March 3rd I took my *café au lait* with Marie, and we were joined by the Empress. It was a day of agony. The Grand Duchesses grew worse: their ears were badly inflamed, it seemed as if they might not recover. The Empress tried to snatch a little rest by occasionally lying on a couch: her feet had now become very painful, and her heart affection was, at times, alarming. Meals were silent and horrible affairs: I felt as though each morsel would choke me. But, as I had now grown desperate with anxiety, I conceived the notion of communicating with the Emperor by aeroplane. Might not his whereabouts be discovered in this way? The Empress welcomed the idea, and she sent for General Resin, and asked for an aeroplane to be despatched at once. He agreed, but even the weather was against us. . . . A blizzard set in; the dark sky was blotted out with scudding snow, and the wind howled dismally round the Palace.

The Grand Duke Paul arrived about 7 o'clock in the evening. The Empress was engaged in writing letters for the officers to convey to the Emperor, but she received the Grand Duke without a moment's delay.

The interview took place in the red drawing-room. Marie and I were in the adjoining study, and from time to time we heard the loud voice of the Grand Duke and the agitated replies of the Empress. Marie began to get apprehensive.

"Why is he shouting at Mamma?" she asked. "Don't you think I had better see what's the matter, Lili?"

"No, no," I said, "we had better remain here quietly."

"*You* can remain, but I'll go to my room," she answered. "I can't bear to think Mamma is worried."

Hardly had the Grand Duchess left the study when the door

opened and the Empress appeared. Her face was distorted with agony, her eyes were full of tears. She tottered rather than walked, and I rushed forward and supported her until she reached the writing-table between the windows. She leant heavily against it, and, taking my hands in hers, she said brokenly:

"*Abdiqué!*"

I could not believe my ears. I waited for her next words. They were hardly audible.

At last: "*Le pauvre . . . tout seul là bas . . . et passé . . . oh, mon Dieu, par quoi il a passé! Et je ne puis pas être près de lui pour le consoler.*"

"*Madame, très chère Madame, il faut avoir du courage.*"

She paid no attention to me, and kept on repeating, "*Mon Dieu, que c'est pénible. . . . Tout seul là bas!*" I put my arms around her and we walked slowly up and down the long room. At last, fearing for her reason, I cried: "*Mais Madame—au nom de Dieu— il vit!!*"

"Yes, Lili," she replied, as if new hope inspired her. "Yes, he lives."

"I entreat you, Madame, don't lose your courage, don't give way: think of your children and of the Emperor."

The Empress considered me with almost painful scrutiny. "And you, Lili, what of you?"

"Madame, I love you more than anything in this world."

"I know it—I see it, Lili."

"Well, Madame, *write* to him. Think how pleased he will be." I drew the Empress towards the writing-table, and she sank on a chair. . . . "Write, dear Madame, write," I repeated.

She obeyed almost like a child, murmuring, "Yes, Lili . . . how glad he'll be."

Feeling that I might venture to leave the Empress for a few minutes, I went in search of Dr. Botkin, who gave me a composing draught for her. . . . But the Empress did not wish to take it, and it was only when I said: "For *his* sake, Madame," that she complied.

The sound of bitter weeping now attracted my attention. In one corner of the room crouched the Grand Duchess Marie. She was as pale as her mother. She *knew* all! At this moment Volkoff, that faithful servant, entered, and in trembling tones announced that dinner was served. The Empress rose and endeavoured to regain her composure. . . .

I followed her into the next room. She looked round. "Where is Marie?" she said.

I went back to the red drawing-room. Marie was still crouching in the corner. She was so young, so helpless, so hurt, that I felt I must comfort her as one comforts a child. I knelt beside her, her head rested on my shoulder. I kissed her tear-stained face.

"Darling," I said, "don't cry. . . . You will make Mamma so unhappy. Think of *her*." At the words, "Think of *her*," the Grand Duchess remembered the unswerving devotion of the children towards their parents. Every thing was always subservient to Mamma and Papa.

"Ah . . . I'd forgotten, Lili. Yes, I must think of Mamma," she answered. Little by little her sobs ceased, her composure returned, and she went with me to her mother.

That night the Empress and I sat up very late: she had paid her usual visit to the Grand Duchesses, when she had tried outwardly to appear calm. But alone with me it was a different matter. The Empress told me that the Emperor had abdicated in favour of the Tsarevitch. "Now *he'll* be taken from me," she cried. "The people are to assume the Regency. What shall I do?" She started at every footfall; she trembled at the mere sound of a voice. . . . One idea obsessed her—someone might come at any moment to take away her son!

"But, Madame, nothing can be done until the Emperor returns."

"No, surely they will not dare; and he'll be with us very soon," she said. Then, with her usual unselfishness, the Empress insisted upon seeing Count Benckendorff. "I must console him and strengthen him. I can imagine his state of mind."

It was an affecting interview. . . . I do not know what actually transpired, but, when the Empress returned, she was crying. "*Le pauvre vieux*," she murmured, as if to herself.

I did not allow the Empress to see how apprehensive I was, how utterly despairing. I did not share her optimism. . . . The position was most precarious, and the desperate condition of the Grand Duch-esses augmented the general unhappiness. Our only hope lay in the Emperor's return—at any rate, his presence would afford us some moral protection! That night Marie and I slept in the red drawing-room. We lay awake for hours talking about the new developments. But one thought consoled us. The Emperor was still alive!

When the Empress paid her usual visit to the Grand Duchesses, she told us that her first idea was to see all those in the Palace, and console them as much as possible. Countess Gendrinkoff, her devoted lady-in-waiting, who was away visiting a sick relative, returned to Tsarkoe directly she heard of the Emperor's abdication, and her meeting with the Empress was most touching. At first neither of them spoke; and then the Countess, usually a most self-contained individual, broke into bitter weeping.

It was a tragic morning. Towards noon the Empress sent for me. "Lili," she said, "the Duma is losing no time. M. Rodziansko* has intimated that we must make our preparations for departure. He says we are to meet the Emperor somewhere *en route*. But we can't possibly go; how can we move the children? I've spoken to the doctors, and they say it would be fatal! I've told Rodziansko this, and he is returning later to acquaint me with the decision of the Duma."

Rodziansko and his colleagues returned at the time appointed. They were at once taken to the Empress.

"The decision of the Duma is unalterable," said Rodziansko curtly.

"But my children—my daughters . . ." pleaded the Empress.

"When a house is on fire, it is best to leave it," answered Rodziansko, with a sardonic smile.

There was apparently nothing to be done. We were at the mercy of Tiberius, and we commenced our preparations for departure. The Empress asked me if I would like to accompany them. I begged to be permitted to do so. "I *cannot* leave you, Madame," I said.

We endeavoured to 'phone to certain friends, but it was impossible. At last the operator, in frightened tones, whispered, "I can't give you the number; the telephone is not in our hands. I beg you, don't talk—I'll ring you up directly it is safe."

In the course of the afternoon a servant informed us that an officer of one of the Tartar regiments begged the Empress to receive him. The Empress asked me to interview him, as she felt too ill to do so, and accordingly I went over to the fourth wing of the Palace, where the officer was waiting. As I traversed the long corridors, I heard the sound of rough voices. I stopped, terrified, at the entrance of one of the salons—the Mixed Guards were just about to change the guard; but "changing the guard" was no longer the decorous proceeding of yester-year! When the fresh detachment entered the salon, they threw themselves literally into the arms of the other soldiers, shouting, "New-born citizens of freedom, we congratulate you."

I passed the "new-born citizens of freedom," and I found Lieu-

---

* M. Rodziansko, the President of the Duma, was an aristocrat who had turned Revolutionary: he was always antagonistic to the Imperial Family.

tenant Markoff, to whom I explained the reason of my "deputising." The poor boy had been wounded, he could scarcely stand; but his spirit was unconquerable. "Madame," he said, "I've fought my way through the mob in order to see the Empress, and assure her of my devotion.

The assassins wanted to tear off my epaulettes with HER cypher. I told them that the Empress had given them to me, and that it was her right alone to deprive me of them.

I've arrived here at last. . . . I entreat you to ask the Empress to allow me to remain somewhere near her. . . . I don't care if I wash up the dirty plates. I'll do anything—only let me stay!"

I promised Markoff to deliver his message, and on my way back I heard the soldiers laughing and singing. Sick at heart, and utterly disgusted at their behaviour, I reported it to the Empress. "*Les malheureux,*" she said, "*ce n'est pas leur faute, c'est la faute à ceux qui les trompent.*" She granted poor young Markoff's request, and told me to see General Resin, and arrange for Markoff to be included in his detachment.

I suppose the first idea of most people in the position of the Empress, faced with hurried flight, would have been to save their jewels. But jewels were a secondary consideration with the Empress; her chief treasures were those of sentiment, and, as I watched her collecting her favourite books and photographs, I thought that in this instance, as in all others, she was more of the woman than the Empress. And the idea of leaving the scene of many of her happiest associations must have been heart-rending to her. She had transformed the Palace into a home; here she had watched the beautiful growth of her four fair daughters and her adored son. And here she was destined to drink the uttermost dregs of the Cup of Sorrow.

Whilst she was gathering together her personal treasures, the Empress, recalled in imagination to Petrograd, by the sight of a photograph, asked me to telephone to Prince Ratief, the Commandant of the Winter Palace, and tell him that her thoughts were with them all. Fortunately I was enabled to do so; the Prince himself answered my call. "I thank Her Majesty from my heart. We are still alive, but crowds surround the Palace," he said.

After dinner, we went to see the Grand Duchesses, and then to the mauve boudoir—there was no news from the Emperor; all sorts of rumours were current, the most insistent being that he had returned to G.H.Q.

Sunday, the 5th of March, was for us another hopeless dawn. The Empress gave orders for a Te Deum to be sung, and the miraculous ikon from Znaminie* brought to the Palace and taken to the sick-rooms. The procession bearing the ikon passed through the Palace; the Empress walked in it, and, as I looked at the lovely representation of the Virgin and Child, the expression of the eyes seemed the same which I had often seen in those of the Empress—a combination of Faith, Hope and Tragedy!

It was a strange sight to witness the solemn little procession as it traversed the almost deserted splendours of the Palace. Incense wafted wreaths of perfume towards Heaven, the solemn chant rose and fell, the gold and blues of the Virgin's draperies glowed when the ikon passed one of the windows, the sacred symbol of the Cross raised its head above the tumult of Revolution. It seemed to me as if this were some last appeal to God, Who, we are told, is a God of Love and Pity.

The Empress was anxious that the ikon should be taken to Anna's room, so the procession wended its way thither. There, as usual, were the fuss and overcrowding which seemed inseparable from Anna's attack of measles; doctors, nurses and sisters took up all the available space, so, whilst the Empress was praying by the bedside, I stood by the door. One of the doctors from Anna's hospital was near, and, recognising me, he whispered: "I say, Madame Dehn, I think I shall say good-bye to the Palace. Things are getting too hot for *my* comfort." But, if he expected an answer, he received *none*. I simply stared at him.

The Empress was still kneeling by Anna's bed, and Anna, now thoroughly hysterical and *exaltée* by reason of much incense and many prayers, was crying and kissing the Empress's folded hands. It is quite impossible for English readers to imagine such a scene, but these religious processions in the case of illness were of common occurrence with us.

I went back to see Anna later in the evening, and, when I entered the bedroom, I was surprised to see the matron of Anna's hospital, who was praying—a taper in her hand. Directly she saw me, her prayers took unto themselves wings; we had always disliked each other, so our conversation was short and to the point.

"What, are *you* still here?" she exclaimed, meaningly.

---

* Znaminie is a little church adjacent to the Palace.

"Yes . . . I'm *here*," I replied, with equal emphasis.

Anna said nothing; she looked more childish than ever, and very ill at ease. The impression which I received was a bad one, and, when I related to the Empress what I had seen, she wrote to the doctor at the hospital, and asked him to send for the matron, as her presence was not required. Soon after this she resigned, and, like many others of her kind, she left Tsarkoe for an unknown destination.

On Monday, March 6th, all was in readiness for our departure. But one thing yet remained for us to do, and this was, in my eyes, of the utmost importance. During one of my restless nights, I suddenly remembered that the Empress had always kept a diary and that she possessed the diaries of her friend, Princess Orbelliany, which had been bequeathed to her by the Princess.

These contained most intimate accounts of various people, and events connected with the Court. I likewise remembered the Empress's sentimental habit of preserving correspondence with associations, and I dreaded the possibility of either letters or diaries falling into the hands of the Revolutionaries. I knew that the worst construction would be placed by the "Sons of Freedom" on anything unusual which these papers might contain. Even the Empress's habit of calling people by pet names might be construed as sensualism or treason!

I hardly dared suggest the wisdom of destroying this personal property, but my devotion triumphed over my nervousness. To my intense surprise, the Empress at once agreed to do as I proposed.

It may be argued that I was guilty of the worst Vandalism in persuading the Empress to destroy her diaries and correspondence. I may have been, in an historical and artistic sense—but I was right on the score of friendship. We had already experienced the misconstruction which had been put on *one* sentence in a letter: What might not be the fate of the contents of the Imperial diaries if they fell into the hands of censorious and "pure-minded" Revolutionaries?

Princess Orbelliany's diaries were burned first. They consisted of nine leatherbound volumes, and we experienced much difficulty in destroying them. This *auto-dafé* of sentiment took place in the red drawing-room, but we did not attempt to finish burning the diaries and correspondence in one day. It was at best a melancholy task, and we decided to spread it over a week—especially as the Grand Duchesses were very ill, and we had to be with them constantly. Olga was now suffering with inflammation in the head, and Anastasie made little or no progress.

After lunch, when the Empress and I were sitting in the mauve boudoir, we were startled by the sudden entrance of Volkoff. He was very agitated, his face was pale, he trembled in every limb. Without waiting to be addressed by the Empress, and utterly oblivious of etiquette, he cried: "The Emperor is on the 'phone!"

The Empress looked at Volkoff as if he had taken leave of his senses; then, as she realised the full import of his words, she jumped up with the alacrity of a girl of sixteen, and rushed out of the room.

I waited anxiously. I kept on praying that a little happiness might yet be hers . . . perhaps, for all we knew, the danger had passed.

When the Empress returned, her face was like an April day—all smiles and tears! "Lili," she exclaimed, "imagine what were his first words . . . he said: 'I thought that I might have come back to you, but they keep me here. However, I'll be with you all very soon.'" The Emperor added that the Dowager Empress was coming from Kieff to be with him, and that he had only received the Empress's wires *after* the abdication.

"The poor one!" said the Empress. "How much he has suffered! how pleased he'll be to see his mother!"

Thus the day which had begun so sadly ended happily . . . we went at once to tell the glad news to the Grand Duchesses and the Tsarevitch, who was much better, and greatly excited at the prospect of his father's return. M. Gilliard, a charming Swiss, who taught the children French, was with him, but Mr. Gibbs, his English tutor, was in Petrograd. I always remember Mr. Gibbs and his kindness to me. On one occasion upon going to Petrograd he put himself to great inconvenience to get news of Titi, and procure clothes for myself. Notwithstanding innumerable difficulties, he returned with reassuring tidings of Titi, and a clean nurse's uniform and lingerie for myself.*

---

* During this time the Empress and I wore nurses' uniforms. It has been erroneously stated that the Empress wore ordinary dress. This is not the case.

CHAPTER III

AFTER our usual visit to the children (March 7th) the Empress and I went into the red drawing-room, where a fierce fire was burning in the huge grate, and we recommenced our work of destruction.

A large oaken coffer had been placed on the table; this coffer contained all the letters written to the Empress by the Emperor during her engagement and married life. I dared not look at her as she sat gazing at the letters which meant so much. I think she re-read some of them, for at intervals I heard stifled sobs, and those sighs which have their origin in the heart's bitterness. Many of the letters had been written before she was a wife and a mother. They were the love-letters of a man who had loved her wholly and devotedly, who still loved her with the affection of that bygone Springtime. Little dreamt either the lover or the beloved that these letters were afterwards destined to be wet with tears.

The Empress rose from her chair, and, still weeping, laid her love-letters one by one on the heart of the fire. The writing glowed for an instant, as if desirous of burning itself into her very soul, then it faded, and the paper became a little heap of white ash. . . . Alas for Youth! Alas for Love!

When the Empress had destroyed her correspondence, she handed me her diaries to burn. Some of the earlier volumes were gay little books bound in white satin; others were bound in leather. She smiled bravely as I took them, and an immense disgust seized me when I thought that the country of my birth was responsible for her misery and the injustice meted out to her. "I can't bear Russia," I cried. "I hate it."

"Don't dare say such things, Lili," said the Empress. "You hurt me. . . . If you love me, don't ever say you hate Russia. The people are not to blame; they don't understand what they are doing."

A coloured post-card of South Russia fell out of one of the diaries. I picked it up. It was a pretty picture of young girls standing in a flower-starred meadow . . . and it brought Revovka back to me. "That's *home*," I murmured. But the Empress heard my words.

"What did you say? Repeat it, Lili. You said, 'That's *home*.' Now you must never say you hate Russia."

At this time, I am proud to say, the Empress relied on me as woman to woman. To her, I was always "Lili," or "My brave girl." I was her friend in trouble. The fact that I possessed no official position mattered nothing to her; every moment I was writing letters, taking messages, and seeing people on her behalf. I obeyed her absolutely, and her gentle influence gave me fresh strength to hope and to endure.

The burning of the diaries extended over Wednesday and Thursday . . . but on Thursday one of the Empress's dressers came to the red drawing-room and begged us to discontinue. "Your Majesty," said she, "the sweepers are searching for the half-charred pieces of paper, some of which have been carried up the chimney. I beg of you to cease. . . . These men are talking among themselves. ... They are utterly disloyal." But our task was completed—at any rate we had checkmated the curiosity of the Revolutionaries!

At 7 o'clock the Empress asked me to telephone again to the Winter Palace. As on the previous occasion, Prince Retief answered me.

"How are things with you?" I enquired.

"The mob is even now at the gates of the Palace," he replied with absolute unconcern. "I beg you, Madame, to present my assurances of fidelity and devotion to the Empress. . . . I may not be able to do so again. . . . Ah! . . . I thought as much. Madame, it distresses me to appear discourteous, but I fear I am about to be killed. . . . The doors of this room are being forced!" His voice ceased—there was a terrible crash. . . . I could bear no more, and the receiver slipped from my nerveless hands.

We remained in the mauve boudoir until quite late, but, just as we were about to go to bed, Volkoff entered in a state of painful agitation. He managed to tell us that M. Goutchkoff had arrived, and insisted upon seeing the Empress. It was then 11 o'clock.

"But, at this hour—it's impossible," said the Empress.

"Your Majesty, he *insists*," stammered Volkoff. The Empress turned to me—terror and pathos in her eyes. "He has come to arrest me, Lili," she exclaimed. "Telephone to the Grand Duke Paul, and ask him to come at once." Regaining her composure, the Empress rearranged the Red Cross head-dress which she had taken off, and stood waiting in silence for the Grand Duke. Neither Marie nor myself dared speak. At length, after what seemed an interminable agony of suspense, the Grand Duke entered, and the Empress told him in a few words about her ominous summons. The next moment, loud voices in the corridor, and the banging of a door, announced Goutchkoff's arrival in the adjoining room.

Goutchkoff, the Minister of War during the Revolution, was an openly avowed personal enemy of the Emperor, whom he had never forgiven for not having accepted him at his own valuation as the uncrowned king of Moscow. He had compelled the Emperor to abdicate through revenge; spiteful curiosity now urged him to gloat over the sufferings of a defenceless woman! He was a hideous creature, who wore big spectacles with yellow glasses, which partially disguised the fact that he was unable to look anyone straight in the face.

Marie and I clung desperately to the Empress; we were certain that all was now finished. She kissed us both tenderly, and passed out with the Grand Duke Paul, an infinitely tragic figure, recalling to my mind a vision of Marie Antoinette, whose troubles possessed so many similarities with those of the Empress. Volkoff, that loyal servant, true to the traditions of Imperial regime, informed us that Goutchkoff had brought two A.D.C.'s with him, and that one of these men had accosted him with the words: "Ha, ha! Here we are. You didn't expect us to-night, eh? But *we* are masters of the Palace *now*!"

Marie and I sat side by side on the sofa, the young girl shook with fear, but her terror was not for herself—Marie, like all the children, thought only of her beloved mother.

In this crisis of their fortunes, the Imperial Family manifested no sorrow at the loss of their rank and prestige. The only anxiety shown by them was the fear of parting one from the other. Theirs might have been the words inscribed upon the wall of a certain old prison in Italy: "Better death than life without you." And, if the report of their death be true, they most mercifully never knew the pain of separation.

At last footsteps sounded in the corridor—the door of the boudoir opened—and, to our unspeakable relief, we saw the Empress!

Marie ran towards her mother, half crying, and half laughing, and the Empress quickly reassured us.

"I am not to be arrested this time," she said. "But, oh! the humiliation of the interview! Goutchkoff was impossible—I could *not* give him my hand. He told me that he merely wanted to see how I was supporting my trials, and whether or no I was frightened." Her pale cheeks were rose-flushed, her eyes sparkled—at this moment the Empress was terrible in her anger. But she soon regained her calm dignity, and we bade her good night, thankful that she was spared to us.

Wednesday, March 8th, is a day momentous in the annals of new-born Russia, inasmuch as it witnessed the arrest of a woman and five sick children, and of those adherents who knew the meaning of the words Friendship and Duty.

In the morning Count Benckendorff came to inform us that the Emperor would arrive at Tsarkoe on the 9th, and that the Revolutionary authorities had decided to arrest everyone in the Palace by noon. The Count asked the Empress to give him a list of those of her suite who would be willing to remain, and the Empress at once addressed me: "Lili . . . do you understand what this order means? After it is enforced, nobody will be allowed to leave the Palace, all news from outside will be stopped. What do you wish to do? Think of Titi . . . Can you bear to be without tidings of him?" I did not hesitate. "My greatest wish is to remain with you, Madame," I replied. "I knew it!" exclaimed the Empress. "But . . . it will, I fear, be a terrible experience for you."

"Don't worry on my account, Madame," I answered. "We will share the danger together."

At noon, General Korniloff made his appearance at the Palace with the order for the arrest of the Imperial Family. The Empress received him wearing her Red Cross uniform, and she was genuinely pleased to see him, since she laboured under the mistaken idea that he was well disposed towards herself and the family. She was entirely mistaken, as Korniloff, thinking that the Empress disliked him, never lost an opportunity of spreading the most malicious reports concerning her.

Korniloff told the Empress that the Palace troops were to be replaced with those of the Revolution; there was no use for the Mixed Guard and the Cossack Convoi; the Palace was now thronged with Revolutionaries, who were walking about everywhere. When the officers of the Mixed Guard bade farewell to the Empress, many of them broke down and sobbed. She afterwards told me that it was also for her a most painful moment. The officers asked the Empress for a handkerchief, as a souvenir of her and the Grand Duchesses. . . . This handkerchief they proposed to tear in pieces, and divide between them; and later, to their great joy, we sent them some "initial" handkerchiefs.

It was a day of good-byes; many officers came in from Petrograd to bid farewell to the Imperial Family; the Tanieffs left, as the Empress had insisted upon them returning to the Palace of the Grand Duke

Michael, where they might reasonably hope to be in safety.

At last the Empress decided to tell the Grand Duchesses about the abdication . . . she could not bear this painful task to devolve upon her husband. She therefore made her way to their apartments, and was with them alone for a long time. Anantasie seemed to sense what had happened . . . and after her mother had left them she looked at me, and said, very quietly, "Mamma has told us everything, Lili; but, as Papa is coming, nothing else matters. However, you have known what was going on . . . how could you keep it from us? Why, you're usually so nervous . . . how is it you are so calm?"

I kissed her, and said that I owed all my fortitude to her mother. She had set such an example of courage that it was impossible for me not to follow it.

When the Empress broke the news to the Tsarevitch, the following conversation took place: "Shall I never go to G.H.Q. again with Papa?" asked the child.

"No, my darling—never again," replied his mother.

"Shan't I see my regiments and my soldiers?" he said anxiously.

"No . . . I fear not."

"Oh dear! And the yacht, and all my friends on board—shall we never go yachting any more?" He was almost on the verge of tears.

"No . . . we shall never see the 'Standart.' . . . It doesn't belong to us now."

The Empress and I took tea together, and she told me how glad she felt that the Garde Equipage had left their colours in the Palace. "I should be so sorry to think that the colours were in the possession of the Duma," she remarked. At that moment we heard the sound of voices, and a noise of singing and shouting. The Empress sprang off the couch on which she was lying, and rushed across to the window. "Oh, Madame, don't look, I implore you," I said, fearing the worst. But she did not hear me. Then I saw her grow pale, and she fell back half fainting on the couch. The sailors were leaving the Palace with the colours!

The Grand Duchess Marie was seized with measles late that evening. Like her sister, Anastasie, she dreaded being ill. "Oh, I did so want to be up when Papa comes," she kept on repeating, until high fever set in, and she lost consciousness . . . her last comprehensible words being, "Lili, can't you sleep with Mamma to-night?"

"Yes, darling," I told her. "I won't leave Mamma alone—I'll be somewhere near her, even if I have to sleep in the bath."

I went to the Empress. "Madame," I said, "will you permit me to remain near you to-night?"

"No, Lili, certainly not. If anything should happen, why should you be obliged to witness a tragedy?" she replied.

I returned to Olga and Tatiana, who, like Marie, were very anxious about their mother. "Lili, you *must* not leave Mamma alone. One of us has always slept with her*—she's not strong. Promise, promise us that you won't leave her alone;" and, when the Empress came to pay her last visit to the sick-room, the Grand Duchesses reiterated their request.

The Empress at first demurred . . . but, when she realised how much the Grand Duchesses dreaded her being left alone, she consented. "Well, Lili," she said reluctantly, "you see that the children must have their own way. But I will not allow anyone to think I am frightened. Undress upstairs, and, when my maids have left me, slip down the private staircase, bring your sheets and blankets, and you can make up a bed on the couch in my boudoir."

It was a bright moonlight night. Outside, the snow lay like a pall on the frost-bound Park. The cold was intense. The silence of the great Palace was occasionally broken by snatches of drunken songs and the coarse laughter of the soldiers. The intermittent firing of guns was audible. It was a night of beauty, defiled by the base passions of men.

I went quietly downstairs to the mauve boudoir. The Empress was waiting for me, and as she stood there I thought how girlish she looked. Her long hair fell in a heavy plait down her back, and she wore a loose silk dressing-gown over her night clothes. She was very pale, very ethereal, but unutterably pathetic. As I stumbled into the boudoir with my draperies of sheets and blankets she smiled—a little affectionate, mocking smile, which deepened as she watched me trying to arrange my bed on the couch. She came forward, still smiling. "Oh, Lili . . . you Russian ladies don't know how to be useful. When I was a girl, my grandmother, Queen Victoria, showed me how to make a bed. I'll teach *you*." And she deftly arranged the bedding, saying, as she did so: "Take care not to lie on this broken spring. I always had an idea *something* was amiss with this couch."

* From the time that the Emperor left for the Front, one of the Grand Duchesses always slept with the Empress.

The bed-making "à la mode de Windsor" was soon finished, and the Empress kissed me affectionately and bade me good night. "I'll leave my bedroom door open," she said; "then you won't feel lonely."

Sleep for me was impossible. I lay on the mauve couch—*her* couch—unable to realise that this strange happening was a part of ordinary life. Surely I must be dreaming; surely I should suddenly awake in my own bed at Petrograd, and find that the Revolution and its attendant horrors were only a nightmare! But the sound of coughing in the Empress's bedroom told me that, alas! it was no dream. . . . She was moving about, unable, like myself, to sleep. The light above the sacred ikon made a luminous pathway between the bedroom and the boudoir, and presently the Empress came back to me, carrying an eiderdown. "It's bitterly cold," she said. "I want you to be comfortable, Lili, so I've brought you another quilt." She tucked the quilt well round my shoulders, regardless of my protestations, and again bade me good night.

The mauve boudoir was flooded with moonlight, which fell directly on the portrait of the Empress's mother, and on the picture of the Annunciation. Both seemed alive. . . . The sad eyes of the dead woman watched the gradually unfolding tragedy of her daughter's life, whilst the radiant Virgin, overcome with divine condescension, welcomed the angel who hailed her as blessed among women.

Masses of lilac were arranged in front of the tall windows. It was customary for a fresh supply of lilac for the mauve boudoir to be sent daily to Tsarkoe Selo from the south of France; but, owing to the troublous times, no flowers had reached the Palace for a couple of days. Just before dawn, the dying lilac seemed to expire in a last breath of perfume . . . the boudoir was suddenly redolent of the perfume of Spring . . . tears filled my eyes. The poignant sweetness hurt me—winter was around us, and within our hearts. Should we ever know the joys of blue skies, and the glory of a world new-born? All was silent, save for the footsteps of the "Red" sentry as he passed and repassed up and down the corridor. At first the Revolutionaries had celebrated their sojourn in a Palace by singing seditious and obscene songs, but little by little these had ceased . . . the soldiers slept. My mind reverted constantly to the sick girls and to their brother, who, happily, unlike them, did not share their apprehendsions. What a contrast this night presented to the quiet, happy nights of long ago! I confess it was difficult to see the hand of God in this—to me—unnecessary suffering, and to accept all in the spirit of humility which the Empress manifested.

At seven o'clock the Empress told me I had better return to the red drawing-room, so I gathered my bedclothes together and slipped unperceived and unheard up the staircase.*

H.I.M. AND THE
TSAREVITCH, 1913

THE EMPRESS
(End of 1915)

* The remaining members of the suite occupied apartments in the fourth wing of the Palace. The Empress, who was afraid of infection for others, only saw them occasionally. I was quite alone with her and the children.

CHAPTER IV

ON the morning of Thursday, March 9th, the Empress came into the Grand Duchesses' bedroom; she was agitated and anxious, as she had been informed that the Emperor would arrive at the Palace between eleven and twelve. I went with her to see the Tsarevitch, and we sat by his bed talking to him. The little boy was very excited, and he kept on looking at his watch, and counting the seconds which must pass before his father's arrival.

Presently we heard the sound of an automobile, and Volkoff entered. The faithful servant had refused to accept the fact of the Emperor's abdication, and, in a manner worthy of Imperial traditions, he announced:

"His Majesty The Emperor!"

The Empress sprang from her chair, and ran out of the room. I, too, rose. The meeting between the reunited family must not, surely, be witnessed by any outsider! But the Tsarevitch seized my hand. "No, no, Lili, you're not to leave me," he insisted, so I sat down by him for five minutes, and eventually I managed to slip away and take refuge in Anna's room—where I remained until after lunch, when I was summoned to the Imperial presence.

Following my instructions, I went into the Grand Duchesses' room; the Empress was not there. Suddenly I heard the sound of footsteps. I knew to whom they belonged—but they were no longer the footsteps of a confident and happy man. They sounded as if the person who was advancing was very, very tired.

I trembled from head to foot—I dared not at first raise my eyes. When I did so, I encountered the tragic, weary eyes of the Emperor.

He advanced to where I was standing, and took my hands in his, saying, very simply:

"Thank you, Lili, for all you have done for us . . . and I? . . . what have I done for you? Absolutely nothing! Why, I have not even kept Dehn near you."

"Your Majesty," I answered, now unable to speak without crying . . . "it is for me to thank you for the privilege of being allowed to remain with you."

As we went into the red salon, and the light fell on the Emperor's face, I started. In the darkened bedroom I could not see clearly, but I now realised how greatly he had altered. The Emperor was deathly pale, his face was covered with innumerable wrinkles, his hair was quite grey at the temples, and blue shadows encircled his eyes. He looked like an old man; the Emperor smiled sadly when he saw my horrified expression, and he was about to speak, when the Empress joined us; he then tried to appear the light-hearted husband and father of the happy years; he sat with us and chatted on trivial matters, but I could see that he was inwardly ill at ease, and at last the effort was too much for him. "I think I'll go for a walk—walking always does me good," he said.

We passed through the corridors to Anna's apartments, where the Emperor left us, and went downstairs. The Empress and I entered the bedroom, and stood by one of the windows which looked out over the Park. Anna was very excited; she kept talking and crying, but we had eyes only for the Emperor, who by this time was outside the Palace. He walked briskly towards the Grande Allée, but suddenly a sentinel appeared from nowhere, so to speak, and intimated to the Emperor that he was not allowed to go in that direction. The Emperor made a nervous movement with his hand, but he obeyed, and retraced his steps; but the same thing occurred— another sentinel barred his passage, and an officer told the Emperor that, as he was now to all intents and purposes a prisoner, his exercise must be of the prison-yard description! . . . We watched the beloved figure turn the corner . . . his steps flagged, his head was bent, his whole aspect was significant of utter dejection; his spirit seemed completely broken. I do not think that until this moment we had realised the crushing grip of the Revolution, nor what it signified. But it was brought home to us most forcibly when we saw the passage of the Lord of All the Russias, the Emperor whose domains extended over millions of miles, now restricted to a few yards in his own Park.

The Empress said nothing, but I felt her hand grasp mine; it was, for her, an agonizing experience. After an interval, she spoke. . . . "We'll go back to the children, Lili; at any rate we can be together there."

The Grand Duchesses were delighted to know that their father had returned, and I think the knowledge of his safety acted on them like a tonic. Poor Marie, who had so longed to be the first to welcome the Emperor, was now delirious, with intervals of consciousness.

When I entered her room, she recognised me. "Well, Lili, where have you been?" she exclaimed. "I've been waiting and waiting for you. Papa is really *here*, isn't he?" The next moment she was back in the fantastic and terrible kingdom of fever. "Crowds of people ... dreadful people ... they're coming to kill Mamma!! Why are they doing these things?" Alas, poor child, others have since asked the same question.

That day the Emperor and the Empress dined and spent the evening together. The Empress told me afterwards that the Emperor lost his self-control when he was alone with her in the mauve boudoir; he wept bitterly. It was excessively difficult for her to console him, and to assure him that the husband and father was of more value in her eyes than the Emperor whose throne she had shared.

. . . . .

I cannot say that the Revolutionaries treated us with excessive discourtesy, but some of their methods were reprehensible. For instance, when certain complications ensued with Marie, it became necessary to have another medical opinion. This request was at first refused, but afterwards the authorities agreed, on condition that an officer and two soldiers were present at the medical examination! Colonel Kotzebue, the first Revolutionary commandant, had formerly been an officer in the Lancers, and, as he was a distant cousin of mine, I could hardly believe my eyes when I saw him in this official capacity, and I asked him to come and talk to me in Anna's room, as I considered he owed our family some explanation of his conduct.

"I can't imagine why I was nominated for the post," said Kotzebue. "All I can tell you, Lili, is that I was awakened in the middle of the night, and told to report myself at Tsarkoe Selo. Will you assure Their Majesties that there is nothing I will not try and do for them. This is really the happiest moment of my life, since it enables me to be of service to them."

When the Empress sent for me on the morning of March 10th, I found her lying on the couch in her boudoir. The Emperor was with her; she motioned me to come and sit beside her, and the Emperor talked to us.* (*footnote on next page*). He first described an incident which had impressed him most strongly that very morning.

"When I got up," he said, "I put on my dressing-gown and looked through the window which gives on the courtyard.† I noticed that the sentinel who was usually stationed there was now sitting on the steps—his rifle had slipped out of his hand—he was dozing! I called my valet, and showed him the unusual sight, and I couldn't help laughing—it was really absurd. At the sound of my laughter the soldier awoke, but he did not attempt to move—he scowled at us, and we withdrew. But what a conclusive proof of the general demoralisation! All must indeed be at an end for Russia, as without law, obedience and respect no empire can exist."

The Empress then questioned the Emperor about certain doings at G.H.Q. "Some occurrences were exceptionally painful," replied the Emperor. "My mother drove with me through the town, which was profusely decorated with red flags and a profusion of bunting. My poor mother couldn't bear to look at the flags . . . but the sight of them did not affect me; it seemed such a stupid and useless display! The behaviour of the crowd was in curious contrast to this exhibition of Revolutionary power, as they all knelt, as of yore, when our automobile passed."

"I could not bear to say good-bye to Voeikoff, Niloff and Fredericks. They didn't want to leave me. I had to insist at last. The Revolutionaries promised most faithfully not to harm them."‡

"One thing especially touched me," continued the Emperor. "When I got into the train, I noticed five or six schoolgirls who were standing on the platform trying to attract my attention. I went to the window, and, when they saw me, they began to cry, and made signs for me to write something for them. So I signed my name on a piece of paper, and sent it to the children. But they still lingered on the platform, and, as it was bitterly cold, I tried to make them understand that they had better go home. However, when my train left, two hours later, they were still there. They blessed me, poor children," said the Emperor, greatly moved by the recollection. "I hope their pure blessing will bring us happiness."

---

* (previous page)In all my descriptions of the conversations between the Emperor, the Empress and myself, I have endeavoured to describe what took place, almost word for word. I have not attempted to elaborate any of the statements, and my record may therefore be considered accurate.—L. D.

† The sleeping apartments of the Emperor and the Empress were situated on the ground floor of the Palace.—L. D.

‡ These faithful adherents were arrested at the next station and sent to Petrograd, where they were incarcerated in the Fortress of Peter and Paul.—L. D.

The Emperor told us that he had received countless telegrams after the news of his abdication was generally known. Many were abusive, but others breathed the concentrated spirit of loyalty. Count Keller sent a telegram informing the Emperor that he declined to recognise the existence of the Revolution.* The Count afterwards refused to sign the documents of allegiance, and he broke his sword and threw the pieces down.

"General Rousky was the first to broach the subject of my abdication," said the Emperor. "He boarded the train *en route*, and came into my saloon unannounced. 'Goutchkoff and Shoulgine are also coming to talk to you,' he informed me. These gentlemen made their appearance at the next station, and they were excessively impertinent. Rousky told them that he had already discussed matters with me. But I refused to be ignored. I struck the table with my fist. 'I'm going to speak, I *will* speak,' I cried.

"'You must abdicate in favour of the Tsarevitch, and the people will nominate a Regent,' said Goutchkoff and Shoulgine.

"'But,' I replied, 'are you sure—can you promise that my abdication will benefit Russia?' 'Your Majesty, it is the only thing to save Russia at the present crisis,' they replied.

"'But I must think it over. . . . I'll give you my answer in a couple of hours.'" "The delegates consented. I knew," continued the Emperor, looking with affection at his wife, "that their first idea was to separate Alexis from the Empress, so I spoke to Dr. Fedoroff, who was in the train, and I asked him whether he considered it advisable to allow the Tsarevitch to be taken from her.

"'It will shorten the Tsarevitch's life,' said Fedoroff bluntly.

"When Goutchkoff and Shoulgine returned, I intimated plainly that I would not part with my son. 'I am ready to abdicate,' I said, 'but not in favour of my son, only of my brother.'

"My decision appeared to trouble them: they asked me to think Better of it, but I was firm. Afterwards I signed the Act of Abdication. The train was then sent back to G.H.Q."

Such is the bare narrative of the abdication, related as nearly as possible in the Emperor's own words. Baron Stackelberg, a cousin of my husband's, who was travelling with the Emperor, afterwards told me that he and M. Voeikoff, the Commandant du Palais, met Rousky on the platform of the station where he joined the train. The

---

* Count Keller was killed at Kieff later.

two gentlemen were about to send some telegrams from the Emperor to Rodziansko, in which the Emperor replied to the former's request to give Russia a constitutional government. In the opinion of the Emperor, the moment had not arrived.

"Whose telegrams are these?" said Rousky.

"His Majesty's," answered Baron Stackelberg coldly.

Rousky snatched the telegrams from Baron Stackelberg, and put them in his pocket, remarking as he did so, "Useless!" So Rodziansko never received the Emperor's telegrams, and Baron Stackelberg, who is now in Finland, can confirm the truth of the story. M. Voeikoff and the Baron looked at each other, neither spoke, but each read in the other's eyes the unspoken thought—to kill Rousky then and there, and so avenge the insult to the Emperor. But Rousky had disappeared—the moment for righteous murder had passed!

. . . . . .

Life at first went on much as usual after the Emperor's return: he always insisted upon reading the daily papers, but the filth of the gutter press sickened and pained him. One evening I happened to come into the library where the Emperor was reading a newspaper: his expression showed that something had seriously displeased him. "Just look here, Lili," he said, showing me the portraits of the new Cabinet. "Look at these men. . . . Their faces are the real criminal type. And yet I was asked to approve of this Cabinet, and to agree to the Constitution," he added with a touch of bitterness. My time was now fully occupied. The Grand Duchess Marie was seriously ill, and I relieved the Empress in nursing her. . . . I had taken upon myself the task, formerly performed by the Empress, of sponging poor Marie's body, and, when the child was conscious, she liked me to brush and comb her lovely hair, which became sadly tangled as she tossed to and fro in her delirium. Marie was the first unmarried Grand Duchess to sleep on a "real" bed of her own, but, as she was so ill, we moved her from the narrow camp-bed to a more comfortable resting-place.

The Empress was a skilful nurse; she was especially expert in changing sheets and night-clothes in a few minutes without disturbing the patients. When I showed my surprise, she said quite simply: "I learnt to do useful things in England. . . . I've never forgotten what I owe to my English upbringing."

One day my cousin, Kotzebue, told me that an English gentleman, Mr. A. Stopford,* a friend of the Grand Duchess Marie Paul, was desirous of being of use to the Empress. He had, it appeared, a cult for the Imperial Family, and, as he was about to return to England, he asked Kotzebue whether the Empress would not like to send some letters by him to her relations. I told the Empress at once. It seemed such a wonderful chance.... Her first cousin, King George V, and his devoted consort, would surely welcome news from the Imperial Family!

The Empress was deeply touched by Mr. Stopford's offer. "I'll think about it, Lili," she said. But the next day she told me that she had decided not to communicate with King George and the Queen. "I *can't* write. What can I say? I'm too hurt and wounded by my country's behaviour.... But even with this I can't speak against Russia.... Besides, the Emperor is more worried than ever; he is so fearful that his abdication, and the unrest, may spoil the Great Offensive.... No ... we can't communicate with our cousins."

Both the Emperor and the Empress constantly referred to England. The first idea of the Duma had been to induce the Imperial Family to go to England, but certain powers there were antagonistic to the proposition, as it was considered likely to be unfavourably received by the Labour Party. But those who were fearful of sheltering a defenceless family, whose only crime consisted in being defenceless, need have had no apprehensions.

The Emperor and the Empress did not wish to leave Russia. "I'd rather go to the uttermost ends of Siberia," said the Emperor. Neither he nor the Empress could face the prospect of wandering about the Continent, and living at Swiss hotels as exRoyalties, snapshotted and paragraphed by representatives of the picture papers, and interviewed by amazing American journalists. Their retiring spirits shrank from cheap publicity; they considered that it was the duty of every Russian to stand by Russia, and face the common danger together.

Apart from their personal disinclination to go to England, the Soviets were opposed to the suggestion, and it was stated that, if any train left Tsarkoe with the Imperial fugitives, it would be stopped, and everyone murdered, as the Emperor knew too much to be allowed to leave Russia.

The Emperor brought me the newspaper which contained this statement. He was in a terrible rage.... He could scarcely contain

* If Mr. A. Stopford (1a St. James's Square) ever reads these lines, he may be glad to know that the Empress greatly appreciated his kindness.—L. D.

himself, and he almost threw the paper at me.

"Read this, Lili," he exclaimed, his face white with passion. "*Beasts!* How dare they say such things. . . . They judge others by themselves."

"Oh, Your Majesty," I answered, greatly troubled, "please don't read these horrible papers."

"I must, I must, Lili. I feel that I must know all," said the Emperor.

Occasionally he was in better spirits, and more like his old cheerful self. The Emperor was generally able to see the humour of any situation, and he would sometimes laugh at the idea of being, what he called, "an Ex." Everything was then "Ex." "Don't call me an Empress any more—I'm only an Ex," laughed the Empress; and one day, when some especially unpalatable ham was served at lunch, the Emperor remarked, "Well, this may have once been ham, but now it's nothing but an 'ex-ham.'" He was always amused by the likeness between him and his cousin, King George. One day he showed me a photograph of the latter, saying, "Have you seen my last photograph, Lili? Doesn't it flatter me?"

He had a great admiration for his cousin, and the Empress often spoke of Queen Alexandra, . . . her beauty, her sympathetic nature, and her boundless charity. "I would so much like to see my married sister in England," she invariably added, whenever she discussed her family. "Darmstadt is only a little spot in the garden of my memories," she would say, "but my mother died there, so I can't really be blamed for liking Darmstadt. . . . Isn't 'Home sweet Home' typically English?

"None of my daughters shall marry German Princes," she said on one occasion. It was suggested that Anastasie's future home might be in England, and the Empress welcomed the idea. . . . An English marriage would have been very near her heart. But "*l'homme propose, et Dieu dispose.*" If Russia had not betrayed herself, or if she had remained as solidly united as France, nothing would ever have been heard of the pro-Germanism attributed to the Empress. She was essentially English—English in her dress, her personal habits, her absolutely Victorian outlook; some of her ideas respecting a *ménage* were akin to those of the *Hausfrau*, but even these were English, as domesticity has always been a British attribute.

The Empress showed no special marks of favour to Germans who had settled in Russia. The reports of her having done so are untrue, or greatly exaggerated. There is no doubt that German agents were very active in Russia, and that the octopus of espionage put forth its tentacles in every direction. But in justice to a much defamed woman, surely it is unfair to credit her with being the instigator of this. Every European country was riddled with Germans, England more so than any other, and, although it was more intimately connected with Germany by marriage and consanguinity, no stones were ever hurled at the various personages, Royal and otherwise, who were really not as English as was the Empress. I remember, in connection with her impartial outlook, that, in 1910, a wealthy German named Faltsfein, was obsessed with the idea of becoming a Russian nobleman. A friend of his, an officer named Masloff, asked the Empress to make it possible for Herr Faltsfein to change his skin, but she was very disgusted, and told Masloff that nothing would induce her to put such a proposal before the Emperor!

One awful day a lorry full of soldiers, in charge of an excessively ill-favoured officer, arrived at the Palace. Kotzebue interviewed him.

"I've come to fetch the Emperor," said the officer, with an unprintable oath. "He's going to be imprisoned in 'Peter and Paul.'"

"You cannot remove the Emperor," answered Kotzebue. "I am commandant here. I refuse to give up the Emperor at your orders."

"Ah . . . ah . . . I knew it," shouted the officer. "The Emperor has fled! . . . we were told so in Petrograd. Let's search the Palace."

Kotzebue almost came to blows with the man. "I tell you the Emperor is *here* . . . I'll prove it." He then sent for Count Benckendorff and told him to ask the Emperor to pass through the corridor whilst the soldiers were looking. In a few moments the Emperor came slowly down the corridor . . . the officer rushed threateningly towards him, but Kotzebue restrained him, saying, "Well, you—, now you've seen the Emperor. Go back to the Soviet, tell them he's still here, and don't come again on a fool's errand."

The Emperor now walked in the Park every day, and each time he returned greatly depressed at some fresh mark of disrespect. "But," he said, "it's very foolish to think that this behaviour can affect my soul—how petty of them to seek to humiliate me by calling me 'Colonel' . . . after all, it's a very worthy appellation."

The Empress was a tragic figure, and, in her invariable Red Cross uniform, she symbolised Pity, in a world which knew not the meaning of the word. Every hour that I knew her, I loved her more.

One day, Kotzebue told me that Titi was ill; in fact, *very* ill, but I did not like to agitate the Empress until Kotzebue came to ask her to permit me to go with him and telephone from the basement of the Palace. She was greatly distressed to hear that her godson was ill, and equally concerned at not having been told before. "My poor girl, what you must have suffered!" she said.

Kotzebue and I descended into the basement: two soldiers guarded the telephone, and I was informed that I could only be allowed five minutes' conversation.

"How is the child?" was my first question. "Very ill, Madame," answered my maid.

"Please, please bring him to the 'phone.'" I waited impatiently, and then a little feeble voice whispered: "*Maman . . . c'est vraiment toi! quand viendras-tu?*" At that moment a soldier interposed.

"Your five minutes is up!"

I returned to the Empress, almost heart-broken, but I endeavoured to appear cheerful. The interminable day wore away, evening fell, and I assisted at what had now become a sort of nightly routine. Every evening the Emperor wheeled the Empress in her invalid-chair across the Palace in order to visit the suite. It was a melancholy pilgrimage. She first stopped to talk with the Benckendorffs, and afterwards passed from group to group of her faithful adherents, taking Anna's room on the way back— Anna, so to speak, representing the last word in dejection, as she was ever full of terrors and presentiments.

That night I was glad to seek refuge in the red drawing-room and find myself alone, and able to indulge in what is described as "a good cry." As I left the mauve boudoir, the Emperor and the Empress kissed me, and made the Sign of the Cross. I felt instinctively that they loved me, and were sorry for me.

A bright fire was burning in the red drawing-room, but I did not undress—I sat in front of the fire thinking of Titi. Yet even the knowledge that my son was seriously ill did not suffice to make me feel that my place was not here. I knew in my soul that the Empress came first, and would always be first where my duty was in question. I was well aware that I might never see my husband or my child again . . . but I knew that I should follow the Imperial Family wherever Destiny might beckon me. I confess I had my moments of weakness, when I longed for the security of home, and the peaceful existence which had hitherto been mine. To-night I felt more than usually despondent. The fire burnt low, and I sought to read the future in the red embers, just as I had done at Revovka in the long ago. Suddenly I heard the door of the salon open very softly, and a line of light pierced the darkness . . . someone was coming in!

I turned quickly to face the person who dared intrude upon the privacy of the apartments occupied by the Imperial Family. . . . Was it some fresh assumption of power on the part of the Revolutionaries?

But my visitor was no emissary of the Revolution—the slender figure standing in the doorway was that of the Empress. She looked more than usually fragile . . . she breathed with difficulty, her face was pale with fatigue, and, when I remembered the arduous ascent of the stairs, I was terrified lest a heart attack would ensue.

"Madame, Madame," I cried, "is anything amiss? Are you in danger?"

"Hush, Lili," said the Empress. "The Emperor and I are quite safe. But I couldn't rest without coming to see you. I know all about Titi, I quite realise what you feel." She took me in her arms just as a tender mother might have done, and she soothed me and caressed me. "My poor, dear child," she said. "Only God can help you. Trust in Him, as I do, Lili."

We mingled our tears, and she stayed with me for some considerable time. It was a strange scene, but I wish that those who revile the memory of the Empress could have seen her then, and experienced the pity, love and understanding which were so essentially her prerogatives. She strengthened and consoled me as no other could have done, and her last words of comfort before she left me were: "Perhaps they'll let us bring Titi from Petrograd to the Red Cross Hospital opposite the Palace, then you could always see him through one of the windows."

CHAPTER V

THE Tsarevitch was now almost well, and running about the Palace much as usual. I do not think he noticed many changes, the Revolution conveyed nothing to him except when he missed certain of his soldiers and his friends. He was still a happy, light-hearted child.

The Imperial Family had no presentiment of disaster for themselves, but they suffered untold agonies of mind over the fate of Russia. "Can you imagine what it means to the Emperor to know that he is cut off from active life?" said the Empress.

Soon after the episode of telephoning from the basement, Kotzebue went to Petrograd. I was anxious for his return, as he had promised to go and see Titi, and bring me the latest news from home. Days passed . . . I became apprehensive, and made enquiries, only to be told that we should not see him again at Tsarkoe! I saw in this an omen of coming trouble, so I went at once to the Emperor and acquainted him with what I had heard. The Emperor and the Empress were watching some of the ladies-in-waiting who were walking in the Park, followed by sentinels; the Empress noticed my agitation.

"Why, Lili, whatever is the matter?" she enquired.

"Madame . . . I hear that Kotzebue is to be replaced."

The Emperor looked at me. Then, shrugging his shoulders, he remarked: "Well— it can't be helped" and straightway changed the conversation . . . possibly to calm our fears, or more probably to show how unaffected he was by the mandates of the Revolutionaries.

The long, monotonous days passed—we endured them alternately with the calmness of despair and with gratitude for their dullness. Once we witnessed a sight of horror. Hearing a sound of military music, and the tramp, tramp of many people, we went to the windows, and saw a funeral procession wending its way across the snow-covered Park. But this was no ordinary funeral; the dead were some of the soldiers who had been killed at Tsarkoe Selo on the first day of the Revolution. It was a red burial—the coffins were covered in scarlet, the mourners were dressed in scarlet, and scarlet flags waved everywhere. Seen in the distance the procession looked like a river of blood flowing slowly through the Park. Everything was red and white, and the superstitious might have inferred from this a presage of the innocent blood so soon to be outpoured . . . since the snow was not whiter than the souls of the young and beautiful who are now safe in the keeping of a God of Justice, who most surely will repay!

None of us could forget the impression produced by this funeral; blood seemed everywhere, and terror lurked in the shadows. The soldiers were buried in the Park, within sight of the Palace—another refinement of torture for those whose imaginations were already overexcited. Our nerves were frayed, although I do not think that we were guilty of giving way to our emotions. But it was difficult to maintain our composure when insolent officers treated us in a shameful manner, or a soldier called the Empress by some filthy epithet. One soldier, however, was a Bayard. He possessed an English name, and his father taught in a school at Riga. This man was really extraordinary. He was not only polite, but he invariably tried to show us that he did not share the Revolutionary outlook. The two regiments which were at the Palace distinguished themselves by a series of petty thefts; not even the spoons were safe. I suppose they would have described these articles as "Souvenir spoons"!

. . . . .

We were no longer to complain of monotony. Even then, events unknown to us were moving quickly, and in my case definitely.

The Grand Duchess Marie was still very ill, and Anna, who knew this, decided to go and see her. The Empress was against the idea; Anna was ill, she said, and it was better for her health and her safety to keep as quiet as possible, and not to draw any undue attention to her presence in the Palace. So strongly did the Empress disapprove, that she was taken in her wheeled chair to see Anna, but she returned more nervous and apprehensive than before.

I spent the morning with the Empress, and I lunched with Anna, in the apparently forlorn hope of dissuading her from attempting to see Marie. After luncheon we discussed the burning question of Kotzebue's disappearance. Suddenly we were startled by hearing a noise in the corridor. . . . Anna instantly rang the bell. A servant answered it.

"Who is outside?" demanded Anna.

"I don't know," replied the man, who was evidently much disturbed; "the soldiers are here." At this moment a *skorohod** entered, and handed me a tiny folded note. I opened it. . . . Written in

pencil, in the Empress's handwriting, were these ominous words:
"*Kerensky passe par toutes nos chambres, pas avoir peur—Dieu est là. Vous embrasse toutes les deux.*"†

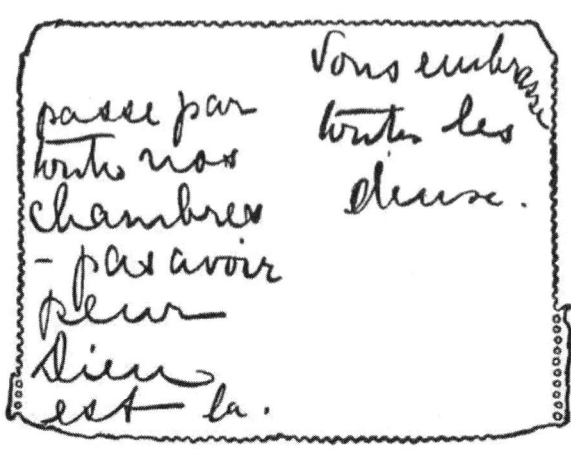

Heavy footsteps sounded in the corridor. I had barely time to slip the precious note inside my bodice when the door was flung open, and a man, followed by two others, came in. I stood up at once and looked at our visitor—it was Kerensky himself!

I saw a slight man with a pale face, thin lips, shifty eyes, seen under lowered lids, and a nondescript nose. Kerensky gave one the impression of being *mal soigné*. . . . He was not tall, but slight in figure, and his head drooped in a curious manner: he wore the blue jacket of an ordinary workman. Kerensky slowly considered us.

"Are you Madame Anna Virouboff?" he said, addressing Anna.

"Yes," replied Anna, faintly.

"Well, put on your clothes immediately and be ready to follow me." Anna made no answer.

"Why the devil are you in bed?" he demanded, staring at Anna's invalid *déshabillée*. "Because I'm ill," whimpered Anna, looking more childish than ever.

"Well" . . . said Kerensky, turning to an officer, "perhaps we had

---

* The skorohod were the confidential messengers of the Imperial Family. They wore a distinctive livery, and wonderful hats adorned with black and yellow ostrich feathers.

† The actual note is reproduced in these pages. Translation: "Kerensky is passing through all our rooms—Do not be afraid—God is present. I kiss you both."

better not move her. I'll have a chat with the doctors. In the meantime, isolate Madame Virouboff. Place sentinels before the door—she's to hold no communication with anyone. Nobody is to come into this bedroom or to leave it until I give the order."

He went out of the room, followed by the officers. Anna and I looked at each other, speechless with dismay. My first collected thought was for the Empress. I would not be separated from her.

"I *must* try and see Their Majesties," I said wildly.

"Yes, Lili, do. For God's sake see them," sobbed Anna.

I opened the bedroom door very softly: the sentinels had not yet arrived. I caught a glimpse of Kerensky entering the room occupied by the doctors; then, impelled by some desperate courage, I ran down the corridors, and arrived breathless in the Grand Duchesses' apartments. I found the Empress with Olga. I told her, in a few words, what had happened. Then distant footsteps warned us of Kerensky's approach.

"Run . . . Lili—hide in Marie's room—it's dark *there*," whispered the Empress.

I had barely time to crouch down behind a screen in Marie's room when Kerensky came in. He took no notice of the sick girl, but went in search of the Empress, who, with the Emperor, had now gone into the schoolroom. From where I was hiding I could hear Kerensky shouting. In a few moments the Empress entered; she was trembling visibly. . . . The Grand Duchesses Olga and Tatiana (now convalescent) rushed forward. "Mamma, Mamma, what is the matter?"

"Kerensky has insisted upon my leaving him alone with the Emperor," answered the Empress. . . . "They'll most probably arrest me."

The two girls clung to their mother, and slowly made their way back to Marie. I had now emerged from behind the screen, and I went into the schoolroom, where I determined to remain until I saw the Emperor.

After what seemed a very long time the Emperor came out—alone.

"Your Majesty," I cried, "tell me, I implore you, if there is anything dreadful in store for Her Majesty?"

The Emperor was painfully nervous. "No, no, Lili, and if Kerensky had uttered one word against Her Majesty, you would have heard me strike the table—thus—" and he struck the writing-table with his fist. "But I hear they've arrested Anna. Poor unfortunate woman, what will become of *her*?" At the sound of her husband's voice the Empress came out of Marie's bedroom. The Emperor told her that Kerensky had arrested Anna because he suspected that she was implicated in political plots. "If it's true, it's an awful thing," said Kerensky; "but I suppose everything will now be disclosed."

Their Majesties then related the particulars of their interview with Kerensky.

"His first words," said the Empress, "were, 'I am Kerensky. You probably know my name.'

"We made no answer.

"'But you must have heard of me?' he persisted.

"Still no reply.

"'Well,' said Kerensky, 'I'm sure I don't know why we are standing. Let's sit down—it's far more comfortable!'

"He seated himself," continued the Empress. "The Emperor and I slowly followed his example, and, finding that I still declined to speak, Kerensky insisted upon being left alone with the Emperor."

Shortly afterwards, to our great relief, we were informed that Kerensky had left the Palace and gone to the Town Hall. The new commandant, Colonel Korovichenko, was then presented to the Empress, who begged him to allow her to say good-bye to Anna. Korovichenko consented, and the Empress went, unaccompanied, to Anna's room. She sat very silent when she returned: she felt the parting keenly, as both the friends knew that, in all probability, it might be for ever!

The Emperor, the Grand Duchesses and myself now took up our position in "Orchie's room,"* from which the windows commanded a view of the entrance to Anna's apartments. I was sitting by the Empress near the window. . . . All at once she took my hand, and said in a voice choked with emotion:

"At least, God will allow you to remain, and. . . ."

Her sentence remained unfinished. . . . At this moment someone knocked at the door; it was Count Benckendorff, who had hurried along to tell the Empress that he still hoped better things for Anna. This was only a temporary respite. A little later we heard the sound of an automobile in the courtyard. I looked down, and saw two automobiles drawn up in front of the Imperial entrance to the Palace. Another knock! This time it was a servant who announced:

* Orchie was a pet name for Miss Orchard, the Empress's old governess, who had died at the Palace. Her room had been left undisturbed since her death.

"The new Commandant wishes to speak to Madame Dehn."

I went out; Korovitchenko, a fair-haired, common-looking man with a hard mouth, was standing at the end of the corridor.

"Madame Dehn?" he enquired brusquely.

"Yes . . . I am Madame Dehn."

"Well . . . get ready. Take as little as possible with you; you are going with Kerensky to Petrograd."

I nearly fainted, but I managed to run back to "Orchie's room." In a few hurried words I acquainted the Empress with Korovitchenko's orders. . . . I could not look at any of them. I tried to be calm, but at the sound of Tatiana's uncontrollable sobbing I broke down and wept in the arms of the Empress.

"*Eh bien . . .*" she said, releasing me gently from her embrace, "*il n'y rien à faire.*"

"Is Madame Dehn ready?" shouted someone outside.

The Empress called Zanoty (one of her dressers) and told her to put some things together in a suit-case. She did not speak to me—or I to her—our hearts were too full. It was like some terrible nightmare. At length I managed to go into Anastasie's room. . . . She was in bed. I kissed her many times, and told her that I would never forsake them. Poor Marie lay asleep in her darkened room. . . . I kissed her flushed cheek, blessed her, and went out quietly. There was no time to say good-bye to the Tsarevitch.

. . . . .

Zanoty had packed my suit-case, and the Empress now sent her to fetch a sacred medal, which she hung round my neck, blessing me as she did so. At the last moment Tatiana ran out of the room, and returned with a little leather case containing portraits of the Emperor and the Empress, which had stood on her especial table ever since she was a tiny child. "Lili . . ." she cried, "if Kerensky *is* going to take you away from us, you shall at least have Papa and Mamma to console you."

Another imperative summons told us that the moment of parting was at hand. I put on my hat, and we left "Orchie's room"; the Emperor and the Empress walked on either side of me, and the Grand Duchesses Olga and Tatiana followed us. I had never imagined in the "happy" days that it would ever be my lot to traverse this corridor with a breaking heart, or under such conditions. For ten years I had received nothing but affection from the Imperial

Family—I had watched the children grow up, I had been their playmate and their friend—now I had to leave them in hostile and menacing surroundings.

Russia had already deprived them of their Imperial state, their possessions and their liberty: surely she might not have deprived them of their friends!

We walked slowly towards the head of the great staircase . . . the moment for saying farewell had arrived . . . I tried to be brave . . . the silence was unbroken save by Tatiana's stifled sobbing. Olga and the Empress were quite calm, but Tatiana, who has been described by most contemporary historians as proud and reserved, made no secret of her grief.

Two soldiers were waiting on the staircase . . . the little group of the Imperial Family stopped, and surrounded me . . . then all pretence of self-control vanished. We clung together, but our unavailing tears made no impression on hearts harder than the marble staircase on which we stood.

"Come . . . Madame . . ." said one of the soldiers, seizing me by the arm.

I turned to the Empress. With a tremendous effort of will, she forced herself to smile reassuringly; then, in a voice whose every accent bespoke intense love and deep religious conviction, she said: "Lili, by suffering we are purified for Heaven. This goodbye matters little—we shall meet in another world."

The soldiers hurried me down the staircase, but I stopped halfway, and looked back. The Imperial Family was still where I had left them; with a rough gesture, my guards motioned me to descend. I could see my beloved Empress no longer.

I walked to the door of the second entrance where some officers and soldiers stood, laughing and talking. Two automobiles were waiting outside. It was bitterly cold, and a bleak wind howled round the Palace, and drove the snow in stinging dust against my face as I sat in the open automobile waiting for Anna. At last she appeared; she looked ghastly, and her eyes were swollen with crying. Two officers sat facing us, and a third took his place beside the chauffeur. In this manner we saw the last of Tsarkoe Selo . . . but I had left my heart behind.

We proceeded rapidly towards the private station, where the automobile stopped. I walked quickly inside. I held myself erect . . . I would *not* let our enemies think that I knew the meaning of the word FEAR. As I passed, some of the soldiers sneered . . . "See how haughty she is," they remarked; but I took no notice.

The Imperial train was waiting, and the thought flashed across my mind that the Revolutionaries were surely most inconsistent people, since Kerensky & Co. did not scruple to avail themselves of the luxuries appertaining to Imperial state. Anna and I made our way to the drawing-room compartment, where we seated ourselves— I say "ourselves," but, in reality, Anna was lying half fainting on a chair. I could just see the Palace through the window of the saloon, and I looked at nothing else until the train moved out of the station, and, even then, my straining eyes sought the familiar building which held so much that was dear to me.

Suddenly I became aware that someone was shouting, and thumping on the floor with a stick. I withdrew from the window to see what was the matter, and I encountered the angry gaze of Kerensky.

"Look here . . . you'd better listen when I'm talking to you," he raged.

I simply looked at him. Nobody had ever addressed me in such a manner! I am a tall woman; perhaps my height (I towered above him) and my unspoken contempt made him think better of continuing in this strain.

"I merely wanted to tell you that I am taking you to the prison of the Palais de Justice," said Kerensky. "From there you will be transferred (with deep meaning) *somewhere else*, and *that* will be the actual place of your imprisonment."

I still looked through him, and he beat a retreat into his own compartment. Ten minutes later we were at Petrograd!

The A.D.C.'s made Anna go first; I followed and as we walked down the train we passed through the saloon where Kerensky and another man were stretched out comfortably in the Emperor's easy chairs! When Kerensky saw me he sat up, and looked me up and down with a kind of half-fierce curiosity. I returned his appraising glance with one of disdain . . . the next moment Anna and I were told to get into a closed carriage (another relic of Imperialism), and we drove away in the company of the A.D.C.'s—mere boys—who were evidently keenly interested in us both.

I was horrified at the change which the Revolution had wrought in Petrograd. Its quiet, well-bred look had completely disappeared, it wore the aspect of a person just recovering from a drunken bout. Red flags were everywhere, and crowds of unrestful people were waiting in long queues outside the bakers' shops. This sight roused Anna from her lethargy of grief, and, childish as ever, she remarked, quite happily, "Well, Lili, it's no better *after* the Revolution than it was before." I silenced her further criticisms with a glance at the A.D.C.'s, and I felt quite relieved when our carriage sank first in one, and then in another of the dirty heaps of snow which cumbered the streets, and which had not been removed by the road sweepers. No policemen were visible; law and order had ceased to exist, but groups of odd-looking people hung about at the corners of the streets. These loungers were unmistakably Jews. . . . The Ghetto-like appearance of Petrograd was amply accounted for.

The carriage stopped outside the Palais de Justice, and we were conducted down seemingly endless corridors to a room on the fourth floor. This room was empty, save for two easy chairs, a small chair and a table on which stood a carafe of cold water. The aides-de-camp told us to ask the sentinels for anything we wanted, and they were about to leave us alone when I said to one of them: "Will you try and let my servants know that I'm here?"

"Impossible," he answered, "but in your next prison you'll be allowed to see your friends once a week." The young men then went away, and Anna at once began to cry. I tried to console her, but I was completely worn out—my powers of endurance had snapped, since there was no one to be brave for!

The room was bitterly cold, and we huddled together, wondering what next would happen. Suddenly shots rang out in the corridor ... were they harbingers of death? The firing was followed by coarse laughter, and a soldier ran into our room. "Ah . . . ha! . . . ha!! . . ." he mocked, "were you afraid . . . did you think you were going to be killed?"*

As I sat in the cheerless room, thinking over many things, I suddenly remembered that Anna had a great predilection for carrying letters and photographs about with her—my heart sank—supposing that she had done so now?

"Anna," I said, trying to speak lightly, "what papers have you

---

* General Knox was discussing certain matters with Kerensky at the moment when this shooting occurred, and he asked Kerensky what the shots signified. "Oh, it's only two friends of the Imperial Family who have just been brought here," answered Kerensky. I met General Knox after my escape to England, and when he related the incident I informed him that I was one of the "two friends."—L. D.

brought away with you?"

"Oh, lots, Lili," answered Anna. "I've some letters of the Empress, some letters from Gregory, and two photographs of him."

I suppose my expression must have betrayed me. Anna began to whimper. . . . "Oh, Lili, why do you look so grave? Surely they won't treat us badly? What *shall* we do?"

"You must give me every paper in your possession."

She demurred. "But *why*, Lili?"

"Because it's dangerous to retain anything connected either with Her Majesty or with Rasputin. The worst construction is likely to be placed on the most innocent expressions . . . you cannot surely wish to injure the Empress!"

Anna instantly handed over the letters, but the difficulty arose as to how best to destroy them. To burn them was impossible, as we had no stove; I therefore decided to tear the letters up in minute pieces, and throw them down the lavatory which we were permitted to use. In this way, I destroyed what might have been considered "compromising" documents!

After what seemed an interminable time, steps sounded in the corridor, the door was flung open, and Kerensky entered. He deliberately turned his back on Anna, but he surveyed me with the same appraising yet hostile scrutiny. We looked at each other without speaking. . . . At last, he shrugged his shoulders, and remarked to an officer:

"This place is damnably cold. Have the stove seen to immediately."

He left us without another word, and we heard him speaking at some length outside. The sentinels were then changed, and the soldier who was on duty in our room began to talk to me.

"Well, Mademoiselle," he said, "it's ten thousand pities to see you here . . . you *do* look sad. Whatever have you done?"

"Nothing."

"It's horrible . . . they've no right to arrest young ladies like you."

"Perhaps the new regulations are responsible for our arrest."

"The new regulations!" The man laughed loudly. "That's a good idea . . . I don't think they'll bring much luck. How can we get on without an Emperor? Don't imagine that *we* wanted this. Do you think we joined willingly? Why, they had to use force to get us . . . we were unarmed, it was no good attempting to resist them."

This kindly soul came from South Russia, and, when I told him who I was and where my estates were situated, he was ready to do anything for me.

"I'm on duty again to-morrow," he said, "so try and write a letter, and I'll see that it's delivered."

Night fell, and we were faint with hunger and fatigue. A little soup was brought us, but we could not swallow it. Every few minutes the door opened, and soldiers came in and made fun of us.

"We've two pretty girls now to look at," they mocked, but their laughter was better than their coarse jokes . . . some of these made me grow scarlet with, shame, and I trembled lest their coarseness might become something unspeakable. We wanted to wash . . . but washing was impossible—we had neither jug nor basin—the only water available was that in the carafe. I opened my suit-case, and as Zanoty had put some cotton-wool and lint with my things I quickly made a pad of some of the wool, and, pouring a little water into the glass, I damped the pad and mopped my face, drying it afterwards with some more cotton wool. At 1 a.m. we were surprised to see the two A.D.C.'s come in with some soldiers. One of the A.D.C.'s addressed Anna.

"Madame . . . we have orders to remove you."

Anna caught hold of my hand. "Oh, Lili, Lili," she moaned, "don't let them take me away. Can't you come with me?. . . I daren't go to another prison without you." "Cannot you let me accompany Madame Virouboff?" I said.

"The order is for *Madame Virouboff*," replied the A.D.C., and at this moment an officer entered.

"What's all the fuss about?" he demanded. The A.D.C. explained. "What . . . is Madame Virouboff really here?" cried the officer. "Well, I've always wanted to have a look at her . . . which one is it?" The A.D.C. indicated Anna, who was gazing from one to the other with frightened eyes.

"Get up," ordered the officer.

Anna meekly obeyed; as she did so, her crutch was visible.

"But . . . what's wrong?" asked the officer, now evidently greatly astonished. "I'm a cripple," faltered Anna.

"Good God," exclaimed the officer. He was silent, but he examined Anna much in the same way that a naturalist surveys a prehistoric beast. He could not reconcile the Anna of reality with the Anna of fiction. In common with many people, not only in Russia, but all the world over, he had imagined a totally different Anna

Virouboff. Perhaps he had visualised her as an adventuress of melodrama, a passionate *intrigante*, a subtle schemer, the masterful confidante of a weak Empress!

What did he actually see?

Rasputin's reputed *sorcière-en-chef* stood before him, a little trembling creature, with the prettiness and the plaintive voice of a child. The officer could not believe his eyes. "Do you mean to tell me that you are a cripple?" he stammered.

"I've always used a crutch since my railway accident," she said, helplessly, "I couldn't avoid being in an accident, could I?"

"Extraordinary, extraordinary," muttered the officer—he was still looking at her— "now, come along." But Anna threw herself on my neck, and refused to leave me. Her sobs were heart-breaking. To do them justice, the soldiers handled this butterfly broken on the wheel very gently. A group of journalists, male and female, all equally unkempt, were busy taking notes, and they glanced half-scornfully and half-pityingly at the shrinking figure of Anna Virouboff as she disappeared in the darkness.

CHAPTER VI

THE long days passed in their monotonous progress. I no longer seemed to belong to the outside world. I heard nothing, nobody came near me—I was as one dead. But, if my days were monotonous, my nights were full of horror. When darkness fell, and the authorities relaxed their incessant watchfulness, the soldiers became brutish . . . when I say that I dared not fall asleep, some idea may be gathered of my dread! I had never met the eyes of lust until now . . . but it was impossible not to understand the glances of many of the soldiers. And I was not under any false illusions about the morality of freedom, it might surely be called the Freedom of Immorality! I thought of my husband far away in England, of my child lying ill within a short distance of my prison, and of that dear family for whose sakes I would gladly suffer untold misery. Memory opened her book, and I saw within its pages people and scenes which stirred many bitter-sweet recollections in my heart. Once again I walked under the linden trees at Revovka, and listened to the nightingales. I saw the forgotten grave with the wild rose weeping her petal-tears over *la morte amoureuse*; once again I stood in the Winter Garden waiting to see the Empress, sometimes I played with Titi and the Grand Duchesses and heard the Empress's kind voice. The pale face and hypnotic eyes of Rasputin recalled my pilgrimage. . . . The church towers and houses of Tobolsk rose against the evening sky, the dark and sinister river flowed past me. . . .

Memory turned back more pages of her wonderful book, and I saw the Tsarkoe Selo of yesterday, the sick children, their fragile mother, and the Emperor, to whom Destiny had proved so cruel.

I endeavoured to preserve a calm mental outlook, it was useless. . . . I wondered whether escape might be possible, but my room was situated on the fourth floor, I dared not risk the descent from the window. One idea obsessed me. I *must* see Kerensky, and this idea grew more intense when I heard that I was shortly to be removed to another prison. "They are making enquiries about you," said the A.D.C.

"Well, I want you to do something, and inform the Minister Kerensky that I would like to see him."

The A.D.C. was evidently startled by my request.

"Hm . . . I'll do my best, but—" his gesture was significant of the hopelessness of such a request.

Upon his return, the A.D.C. said tersely:

"I've seen about your affair, but Kerensky sleeps; he has just dined." "Will you ask him to see me when he awakes?" "Yes. . . ." Again the significant gesture.

I waited impatiently. I felt that this interview with Kerensky would prove the critical point in my present desperate situation. I paced up and down the room, and my nervous agitation aroused the pity of one of the soldiers, who remarked kindly:

"Poor young lady! You *do* seem worried!"

Three hours passed. . . . They seemed like centuries, and then the A.D.C. entered.

"The Minister will receive you," he said.

I hastily arranged my sadly crumpled Red Cross uniform, and two soldiers with fixed bayonets stationed themselves on either side of me. The A.D.C. led the way down endless stairs and lengthy corridors. At last we halted before a half-open door, and, as I stood there, I smelt the delicate fragrance of roses. Surely no roses grew in this terrible prison soil? But the perfume was unmistakable, and I was not left long to wonder from whence it proceeded.

I was ushered into a large, well-furnished reception room, formerly occupied by some Minister under the Empire, and on a table stood an enormous basket of blood-red roses. On another table was a basket of scarlet carnations, the warm air was heavy with the mingled odours of roses and clove pinks. So the Ministers of the Revolution were able to indulge their taste for roses in March, whilst the Sons of Freedom clamoured in the snow for bread!

The door at the extreme end of the room was ajar; presently it opened, and Kerensky came in. He glanced at me, walked to the writing-table, where he seated himself, and indicated a place for me.

KERENSKY: "Well, what do you want. You asked to see me?"

MYSELF: "I want to ask you why I am under arrest. I have never meddled in politics, they are the last things that interest me. I can't regard myself as a political prisoner."

KERENSKY (taking a roll of paper off the desk, and perusing it): "Listen. . . . Firstly, you are accused of staying voluntarily with Their Majesties when you had no official position at Court. Can you deny this?"

MYSELF: "Certainly not, I have no wish to do so. I stayed with Their Majesties, as I could not possibly desert them at such a moment. I love the Imperial Family as individuals. Surely this cannot constitute a crime in your eyes."

KERENSKY: "Well . . . let it pass. . . . What is this close friendship between you and the Empress?"

MYSELF: "I am honoured with the friendship of the Empress. She knows my husband, she has been so good to us that we cannot be devoted enough to her."

KERENSKY (impatiently): "Enough of the Empress. What do you want?"

MYSELF: "What I ask is *not* freedom, but imprisonment in my own house. My child is ill. I want to be with him."

KERENSKY (laughing satirically): "You didn't consider your child when you left him alone in Petrograd in order to remain with your beloved Empress."

MYSELF (angrily): "I know best *why* I left him. You call yourself a patriot . . . I suppose you put the love of your country before family ties? I love the Imperial Family, they come before my family ties. You've taken me away from *them*—I haven't gone willingly. Why deprive me of my child?"

KERENSKY (with sinister emphasis): "Listen, Madame Dehn, *you know too much*. You have been constantly with the Empress since the beginning of the Revolution. You can, if you choose, throw quite another light on certain happenings which we have represented in a different aspect. You're DANGEROUS."

A long silence.

KERENSKY: "Can you explain why all orders from the Empress passed through you? You had no official position . . . it's a most suspicious occurrence."

MYSELF: "We were practically isolated in the private apartments through fear of contagion. Besides, what orders could the Empress give without their being known to *you*?"

KERENSKY: "The servants are witnesses that all orders came through you. Enquiries will reveal the truth . . . if you are honest ... well and good. If not . . . that's another matter."

I looked at him. Kerensky seemed absolutely implacable, but I decided to make one last appeal. He apparently loved flowers; this proved that, as his senses could be appealed to, why not his heart? "If *you* had a child of your own, you'd understand my feelings," I said.

Kerensky surveyed me with that now familiar appraising scrutiny. "I don't think much of you as a mother," he replied, smiling coldly, "but—how old is your child?"

"He is seven."

"Well, Madame, it so happens that I *have* a child, and he, too, is seven. I can decide nothing, but I am now going to a Council at which Prince Lvoff will be present. *He* must decide."

I looked him straight in the eyes. This time he met my gaze fully and squarely.

"I'm perfectly certain that you can do anything you like, without consulting anyone," I said. This tribute to his vanity appealed at once to Kerensky. With most men vanity is the most powerful factor. Wound a man's vanity and he will never forgive you; pander to it, and he is your friend for life. Kerensky was no exception: I had discovered the heel of this Russian Achilles.

"You are quite right. Of course I can do what I like. Go back to your room—I'll send you my answer later in the evening." He pressed an electric bell on his table. The A.D.C. entered.

"Has Madame Dehn a bed in her room?" asked Kerensky. "If not, see that one is placed there."

"Oh, I don't want a bed," I interrupted. "Please let me go to my child."

"I've already told you," said Kerensky, "that I'll let you know later. But . . . if I allow you to go home, you must give me your written promise not to act in any way against us."

The A.D.C. made a sign to the soldiers, Kerensky took no further notice of me, and I was hurried out of the warm flower-scented apartment into the icy corridor.

Black despair overcame me when I regained my room. Kerensky had been noncommittal; but I had hopes that my allusion to him as omnipotent might have some favourable effect; so I sat in the corner nearest the door, straining my ears to catch the sound of approaching footsteps.

Shortly after midnight my friend the A.D.C. made his appearance, and, with a theatrical gesture, indicative of boundless space, he advanced, saying:

"The Minister grants you permission to go home."

My feelings are better imagined than described. I sprang up, and made the Sign of the Cross, and my hand sought the beloved medal hidden in my dress. So I was really free! I could hardly believe it, surely I could not have heard aright!

The A.D.C. told me to put on my hat and cloak and follow him. . . . Before I did so he asked me to sign a paper agreeing not to leave Petrograd, and to hold myself in readiness to be interrogated. I did so; then, picking up my suit-case, I went downstairs. He left me in the hall. I had now apparently lost all interest for him, as he did not trouble to bid me farewell. . . . He merely pointed out the door, and disappeared. I looked round, hardly daring to move. I was not able to realize that I was free to go when, and where, I chose. I pushed open the heavy door, and found myself in the cold and darkness outside. Not a single fiacre was in sight; I felt too exhausted to move, but I made a supreme effort to walk. . . . Impossible! My feet slipped in all directions in the melted snow and slush of the road. Suddenly I noticed a man who was regarding me with evident curiosity. . . . My heart sank. What if this scrutiny meant that I was about to be rearrested?

The man made his way to where I was standing. "Are you Madame Dehn?" he enquired civilly.

"I am."

"I thought I recognised you, Madame. I've been at your house several times. I was formerly Madame Kazarinoff's footman. Poor, poor Madame, who would have believed this could happen to you. Let me help you. I know where I can find a fiacre."

He presently returned with a fiacre, and assisted me to get in with all the courtesy and deference of a well-trained servant. I thanked him many times. . . . He gave the direction to the driver, and we drove away.

It was one in the morning before I arrived home. I rang the bell, and after some delay the door was opened by my maid . . . who nearly fainted when she saw me. . . . I couldn't speak. My thoughts were concentrated on Titi. . . . I ran past her upstairs to his room. ... It was empty! What had happened—could he be dead? I hurried across the landing to my bedroom. . . . A light was burning. . . . Someone was in bed. . . . Thank God, I recognised the beloved dark head of my boy—he was safe. I fell on my knees beside him. With a little start, and a smile, which was like balm to my yearning heart, Titi awoke. . . .

"Mother, mother. . . ." He flung his arms round me. I covered his face with kisses.

"Where have you come from?" he enquired. "From prison."

The child began to cry. I realized the tactlessness of my reply. "If they ever take you away again I'll go too," he sobbed. "But where's 'Aunt Baby'? What has happened to her? And where is Papa? They say he's been killed."*

"Darling, darling, I can tell you nothing about Papa."

Hearing the sound of voices, my father now came into the room. He was greatly relieved to know that I was safe, as all sorts of stories were current respecting my fate and that of Anna Virouboff. But my one thought was for my child: he was much better, but the room struck cold, and I asked my father how it was that there was no fire. He shrugged his shoulders. "*Ma chère*," he replied, "the answer is quite simple—we have no wood! The servants manage to steal a little to burn during the day, but at night *c'est bien autre chose*."

I undressed as quickly as possible, and got into bed. I held Titi close. I kissed him passionately. I trembled with mingled joy and fear! . . . No one should separate us. I knew nothing as to our ultimate fate, but I had made up my mind, during these first hours of freedom, to escape as soon as possible to my estates in South Russia, and, if the Imperial Family were removed from Tsarkoe, to join them.

It was a strange home-coming. The whole house was disorganised. The servants were still devoted to my interests, but food and fuel were difficult to obtain. I spent the morning of the next day lying on a couch in my dressing-room. I was really ill; the long strain had told, and Nature was now exacting her toll in the shape of occasional heart attacks. The hours passed peacefully and slowly, but at ten o'clock in the evening the telephone rang, and my maid told me that the Commandant of the Equipage de la Garde wanted to speak to me.

I was surprised and vexed. After the way in which certain officers had treated the Imperial Family, it was not agreeable for me to continue their acquaintance. However, I went to the 'phone.

"Madame Dehn," said a well-known voice, "have you actually come back from the Palace?"

"Yes, I returned to Petrograd a few days ago."

"I heard that you had been placed under arrest. How is it then that you are at home?"

"Kerensky has given me permission to be with Titi. Cannot you,

---

* I heard later that it was reported that my husband had been killed and his body thrown overboard.

for my husband's sake, and as one of his brother-officers, come over and see me?"

"Impossible," answered the voice. "Look here, you can't stay where you are."

"Very well, since you order, I suppose I must obey. I'll try and find somewhere else, as soon as I am rested."

"You must go NOW."

"I haven't anywhere to go, and the child is ill."

"Take him to an hotel. I won't be responsible for your safety. Lots of things may happen during the night. . . . The sailors may come and murder you." The Commandant then rang off, and left me to face this new terror. But my mind was made up. I would not leave home at a moment's notice. If we had to die, we would die together. I was too exhausted, and the child was too ill, to contemplate a midnight flight.

I rang up my husband's nephew, who was in barracks, and he promised to keep me well advised; but fortunately the night passed peacefully. Nobody came near the house.

Weeks elapsed, and Kerensky seemed to have completely forgotten my existence. I led a quiet life, but my heart was torn with anxiety concerning my beloved friends. I received some letters from the Empress, and I wrote constantly to her, and to the Grand Duchesses. It was in connection with this correspondence that I was summoned to Tsarkoe Selo, by order of Commandant Kobilinsky.

I was instructed to leave Petrograd secretly, and to wear my Red Cross uniform. It was early in July, and the trees were bravely apparelled in their young verdure. It was very different to that bleak March afternoon when the snow lay thickly on the ground, and the wind had stung my face with its icy breath. Outwardly, at all events, everything was peaceful, but tears filled my eyes at the recollection of past Julys. . . . Surely God would not permit the innocent to suffer; surely Justice would awaken in the soul of misguided Russia, and all might yet be well.

As I approached the Palace I became sensible of an eerie change, both in it and in its immediate surroundings. I stopped to consider in what the change consisted. Then knowledge dawned upon me. Tsarkoe was a *dead* place. Its windows were almost hidden by the straggling branches of the unclipped trees, grass grew between the stones of its silent courtyard, and I instantly likened it to a famous Russian picture, "Le Chateau Oublié.". . . It was indeed a forgotten castle! I walked to and fro gazing up at the windows, but those within the Palace gave no sign of life. I wanted to call aloud that I was there, but I dared not imperil their safety or my own. I considered even now that I held my life in trust for the service of the Empress. . . . Who knew when she might require me?

PART OF LETTER FROM HER IMPERIAL MAJESTY WRITTEN ON THE DAY OF DEPARTURE FOR SIBERIA.
(*The note in centre is in the handwriting of the Tsarevitch.*)

Kobilinsky had taken up his quarters in the large building opposite the Palace, so I repaired thither. There were hardly any people visible, and I was directed to Kobilinsky's private room. He was a dark, shortish, nervous man, wearing military uniform, and, as the Empress had written that he was kind to them, I was naturally anxious to make a good impression. This interview is of some importance as I am enabled to contradict a part of Kobilinsky's deposition which appeared in a recent publication. In this deposition he queries the name of the writer of certain letters which came to Tsarkoe Selo, and attributes them to quite another person. The actual writer was myself, and the confusion respecting the signature

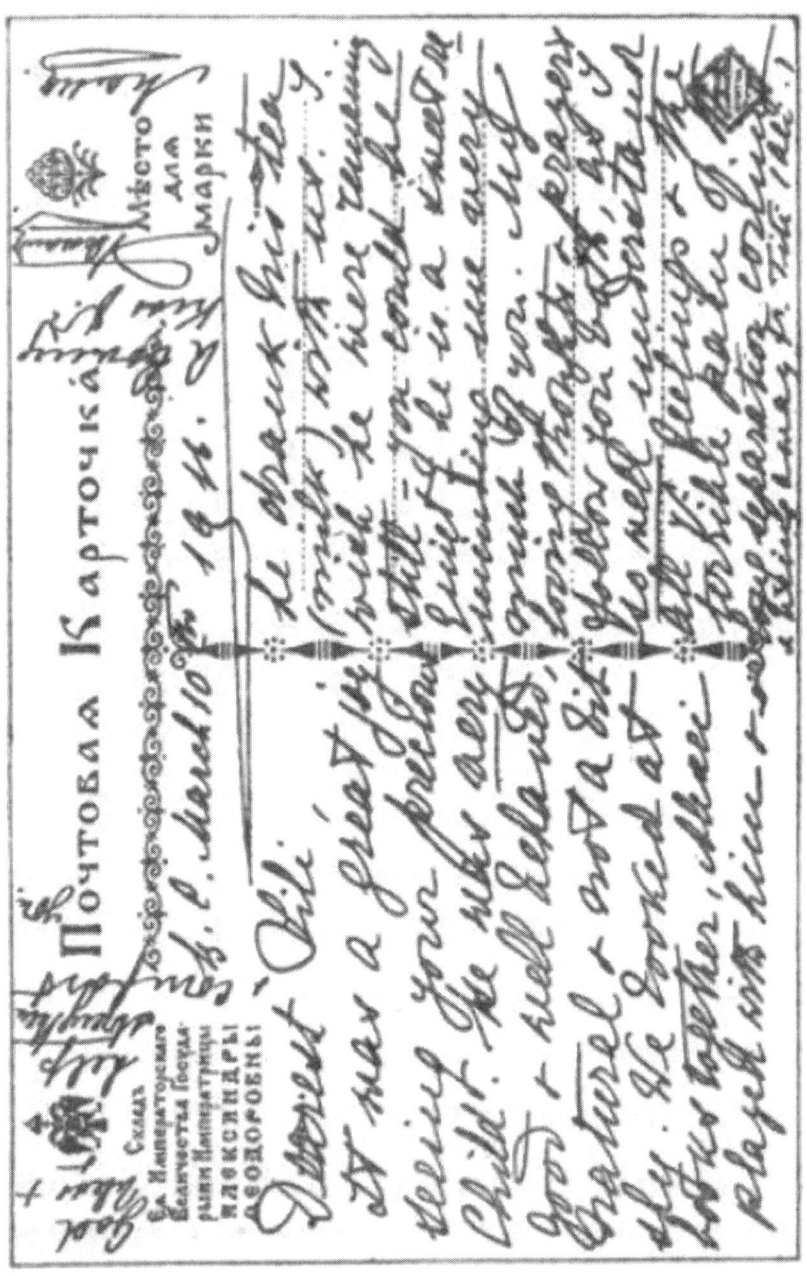

LETTER RECEIVED AT VLADIVOSTOK, IN 1916, WHEN I WAS ON MY WAY TO JAPAN WITH MY HUSBAND. HER IMPERIAL MAJESTY HERE GIVES ME A REPORT OF THE DOINGS OF MY LITTLE SON WHOM I HAD LEFT IN HER CHARGE.

arose from the fact that I had used a fanciful name composed of that of Titi and myself. There was not, and never has been, any "Mysterious Personage" as Kobilinsky's deposition leads one to suppose.

"Are you Madame Dehn?" asked Kobilinsky, eyeing me with some degree of curiosity.

"Yes, Commandant!"

"Are these from you?. . ." he continued, handing me a packet of letters.

"Most certainly. They are all in my handwriting," I said.

"Then why on earth don't you sign your full name when you write?" he queried testily.

"Because I've never been in the habit of doing so. 'Tili' is a fanciful name, a combination of 'Lili' and 'Titi.'"

"I don't believe you," he said bluntly. "It is the name of another lady."

"Why don't you make enquiries if you doubt my word?" I returned. "You'll easily find out that I'm telling the truth."

"Well, well," he grumbled. "I suppose I must believe you. But, see here, Madame, you've got to promise me something. You *must* agree to destroy all the letters which the Empress has sent you. If you don't, I won't allow you to write or to receive any more letters. I suppose," he added, "that such a devoted friend as yourself has not come to-day without bringing some letters for the Family?"

I acknowledged that such was the case. Kobilinsky smiled, and took the letters. He then signified that the interview was over.

Kobilinsky "passed" many letters to and from the Empress after this, but I was always haunted by the fear lest my precious corresponddence might be stolen, or else forcibly destroyed. Fortune favoured me, and an opportunity occurred to send my letters and certain private papers to England under the safe conduct of General Poole. These papers were ultimately deposited in a safe in London belonging to Prince George Shrinsky-Shihmatoff.

The Empress and the Grand Duchesses corresponded with me regularly after they left Tsarkoe, in fact up to a few weeks of their departure for Ekaterinburg. These letters were entrusted to confidential persons and smuggled by them out of the prison. Those who expect startling revelations of political importance will be sadly disappointed in these pathetic little leaves which have drifted from Friendship's tree across a passion-racked country, and, like the song, "have found their home" in the heart of a friend. But, for the student of psychology, the just man or woman, the curious seeker "behind the scenes" of Royalty, they will, I think, possess some interest. They will plead for a hearing far more effectively than any poor words of mine. Not one of them contains a sigh for the splendours of a throne. The woman who longed to be in the Crimea at a time of year when the acacias were like "perfumed clouds" made no allusion to the past glories of the Winter Palace, or the comfortable "English" life at Tsarkoe Selo. Perhaps the words of the writer who "being dead yet speaketh" may serve to efface some of the lies and scandals which have bespattered the name of an Empress who has been condemned so unmercifully.

The Empress and I have never met since that March afternoon when she bade me farewell. I cannot accept the almost overwhelming proofs of the tragedy of Ekaterinburg. From time to time reports of the safety of the Imperial Family have reached us, but the next moment we are faced with evidence that the whole of them have perished. God alone knows the truth, but I still permit myself to hope.

After my interview with Kobilinsky I returned to Petrograd, where I spent some uneventful weeks. Poor Anna was right when she said that things were no better after the Revolution than they were before! Existence was a difficult problem: a period of starvation set in, and we, like others, became familiar with the pangs of hunger. It was impossible to procure nourishing food for Titi; so, almost at my wits' end, I applied for permission to remove him to South Russia.

This permission was most unexpectedly granted. Two weeks later Kerensky's Government fell, and for the moment I was forgotten!

We lived very quietly at Beletskovka, and I was always planning the best way of escape to rejoin my beloved friends. *"L'homme propose, et Dieu dispose."* A wave of Bolshevism swept over South Russia, and our safety was menaced to such an extent that I was forced to escape with Titi to Odessa, and, as our adventures in no way touch on the subject of this book, I shall refrain from relating them. Suffice it to say that we managed to reach Odessa, and from thence, under the protection of the French, we went to Constantinople.

From Constantinople we made our way to Gibraltar, and from Gibraltar to England, where my husband was awaiting me after a three years' separation.

* * *

THE REAL TSARITSA

EXTRACT FROM THE LETTER OF 5 JUNE, 1917.
TSARKOE SELO.

Oh! how pleased I am that they have appointed a new Commander-in-Chief of the Baltic Fleet (Admiral Raswosoff). I hope to God it will be better now. He is a real sailor and I hope he will succeed in restoring order now. The heart of a soldier's daughter and wife is suffering terribly, in seeing what is going on. Cannot get accustomed and do not wish to. They were such hero soldiers, and how they were spoilt just at a time when it was necessary to start to get rid of the enemy (Germans). It will take many years to fight yet. You will understand how he (Tsar) must suffer. He reads, and tears stand in his eyes (newspapers), but I believe they will yet win (the War). We have so many friends in the fighting line. I can imagine how terribly they must suffer. Of course nobody can write. Yesterday we saw quite new people (new guard)—such a difference. It was at last quite a pleasure to see them. Am writing again what I ought not to, but this does not go by post, or you would not have received it. Of course, I have nothing of interest to write. To-day is a prayer at 12 o'clock. Anastasia is to-day 16 years old. How the time flies. . . .

I am remembering the past. It is necessary to look more calmly on everything. What is to be done? Once He sent us such trials, evidently He thinks we are sufficiently prepared for it. It is a sort of examination—it is necessary to prove that we did not go through it in vain. One can find in everything something good and useful—whatever sufferings we go through—let it be, He will give us force and patience and will not leave us. He is merciful. It is only necessary to bow to His wish without murmur and await—there on the other side He is preparing to all who love Him undescribable joy. You are young and so are our children—how many I have besides my own— you will see better times yet here. I believe strongly the bad will pass and there will be clear and cloudless sky. But the thunder-storm has not passed yet and therefore it is stifling—but I know it will be better afterwards. One must have only a little patience—and is it really so difficult? For every day that passes quietly I thank God. . . .

Three months have passed now (since Revolution)!! The people were promised that they would have more food and fuel, but all has become worse and more expensive. They have deceived everybody—I am so sorry for them. How many we have helped, but now it is all finished. . . .

It is terrible to think about it! How many people depended on us. But now? But one does not speak about such things, but I am writing about it because I feel so sadly about those who will have it more difficult now to live. But it is God's will! My dear own, I must finish now. Am kissing you and Titi most tenderly. Christ be with you.

"Most hearty greetings"—(from the Czar).

Yours loving,
AUNT BABY.

30th July, 1917.
TSARKOE SELO.

MY DEAREST,

Heartiest thanks for letter of the 21st. Cannot write—he has no time to read ("he"—Colonel Kobilinsky, Revolutionary Commandant of the Palace), the poor man is so busy all the time that he is often without lunch and dinner. Am pleased have made his acquaintance. E. S. has seen you ("E. S."—Doctor Botkin). I am so pleased that you know all about us.

Will remember your last year's trip. Do you remember? Have not been quite well lately—often had head and heartache. My heart was enlarged. Am sleeping very badly. But never mind—God gives me His strength. Have brought the ikon of Snameni (of God Mother). How thankful I am that this was possible, at this day dear to me (birthday of Tsarevitch). I prayed hard for you and remembered how we used to pray together before it. How Tina (Anna Virouboff) will now suffer—without anybody in the town and her sister in Finland and her friends going so far away (meaning herself)—how much people have to suffer—the path of life is so hard. Please write to A. W. (Colonel Siroboyarski—one of the wounded officers) and send him heartfelt greetings and blessings ✠—kiss you most tenderly and the darling Titi (my son). God preserve you and the Holy Mother.

Always yours,
AUNT BABY.

Kindest regards (meaning the Czar).

I remember—Faith, Hope, Love—that is all, all in life. You understand my feelings. Be brave. Thank you most heartily. All touched by your little ikons—will just put it on. Ask Rita (Miss Hitrovo) to write to the mother of your countryman (Colonel Siroboyarski).

Added by Tsarevitch:

Kiss you most tenderly. Thanks for congratulations.

ALEXEI.

Added by Grand Duchess Olga:

I also kiss you most tenderly and thank you Lili my heart, for post card, and little ikon. God preserve you.

OLGA.

Added by the Empress:

Thank you for your dear letters—we understand each other. It is hard to be separated. Greetings to R. Gor. ✠ I have learnt only now how you spent the first days (in prison). It is terrible, but God will reward. Am pleased that your husband has written.

PART OF LETTER OF 30TH JULY 1917.
(Day of removal from Tsarkoe Selo to Tobolsk. The upper portion is written by the Grand Duchess Olga, the postscript is in the handwriting of Her Imperial Majesty.)

29th November, 1917.
TOBOLSK.

MY DEAREST,

I am for such a very, very long time without news of you, and I feel sad. Have you received my post card of the 28th October?

CHRISTMAS CARD DRAWN SPECIALLY FOR ME
BY HER IMPERIAL MAJESTY WHILE AT TOBOLSK.

Everybody is well—my heart is not up to much, fit at times, but on the whole it is better. I live very quietly and seldom go out as it is too difficult to breathe in frozen air.

Lessons as usual. (News from Petrograd) "T" is as always. Zina has been to see her and O. V., who is very sad, she is always praying. Father Makari passed on on the 19th July.

Rumours have it that Gariainoff has married, but we do not know whether it is true. (Speaking of herself the Empress writes) Aunt Baby drew this herself. How is Titi?—Granny—I want to know such, such a lot. How is Count Keller? Have you seen him in Kharkoff? The present events are so awful for words, shameful and almost funny, but God is merciful, darling. Soon we shall be thinking of those days you passed with us. My God, what remembrances!

Matresha has married, they are now all in P., but the brother is at the front.

I read a lot, embroider and draw (I have to do it all with my spectacles, am so old). I think of you often and always pray fervently for you and love you tenderly.

I kiss you very, very much.

May Christ protect you.

Your countryman is at Vladivostok and Nicholas Jakovlevitch (one of the wounded) is, I think, also in Siberia. I am so lonely without you all. Where is your husband and his friends? We are still expecting Ysa and the others.

I kiss Titi tenderly. Write, I am waiting so. Verveine (toilet water) always reminds me of you.

2/15 March, 1918.
TOBOLSK.

MY OWN DEAR DARLING,

Best and tender thanks for your dear letter. At last we have received good news from you; it was an anxious time not to hear for so long, knowing that things are bad where you are living. I can imagine though what terrible mental agony you must be going through, and you are alone. My little godchild (Titi) is with you always—what he must see and hear! It is a hard school. My God, how sorry I am for you my little giant one; you have always been so brave. I think of those days of a year ago. I shall never forget that you were everything to me and believe that God will not leave you or forsake you. You left your son for "Mother" (meaning herself) and her family, and great will your reward be for this.

Thank God that your husband is not with you, for it would have been terrible, but not to know anything about him is more than awful. When I did not know for four days where mine was "then" (during the days of the Revolution), but what was that in comparison with you. But for us, in general, it is better and easier than for others—it hurts not to be with all our dear ones and not to be able to share their troubles. Yes, separation is a dreadful thing, but God gives strength to bear even this, and I feel the Father's presence near me and a wonderful sense of peaceful joy thrills and fills my soul (Tina feels the same), and one cannot understand the reason for it, as everything is so unutterably sad, but this comes from Above and is beside ourselves, and one knows that He will not forsake His own, will strengthen and protect.

Have news at last, two received new from K.; poor thing, she has a new sorrow, has buried her beloved father—her mother is with her. It is not easy for her to stay in town, though she has good friends and is not so cut off as you are, dearest. Be careful of certain of your friends—they are dangerous.

If you see dear Count Keller again, tell him that his ex-Chief (meaning herself) sends him her heartiest greeting (to her as well), and tell him that she prays constantly for him. I am anxious to know whether he has any news of his eldest son. Radionoff and his brother are in Kieff I hear that Gariainoff and his wife have been in Gagra and are now—so they say—at Rostoff. Am anxious about them, all last week have been *worrying* over it, and do not know why.

To-day we have 20 degrees of frost, but the sun is warm and we have already had real spring days. Godmother (meaning herself) does all the housekeeping now, looks through books and accounts—a lot to do, quite a real housewife. Everybody is well—only a few colds, and feet ached, not very badly, but enough to keep from walking. They have all grown, Marie is now much thinner, the fourth is stout and small. Tatiana helps everyone and everywhere, as usual; Olga is lazy, but they are all one in spirit. They kiss you tenderly—(stands for the Emperor) sends his hearty greetings. They are already sunburnt, they work hard, sew and cut wood, or we should have none. The court is full of timber, so we shall have enough to last.

We still are not allowed to go to church. A. V.'s mother (one of the Empress's wounded) is very sorry that you have not been to see her. She is living with some relatives of your mother's. Their estate has been taken away from them. The son has returned, he now looks, as they all do, pale and miserable.

They, poor things, can no longer keep M. S., and will probably be obliged soon to leave the house. She hardly ever gets a letter from her son; he too is complaining, so I copy what they write to me and send it on to them.

He is very upset not to hear from you, though he himself has written to you. He is going to Japan to learn English, he learnt more than 900 words in ten days and of course overtired himself and has been feeling ill. He was operated upon in December, in Vladivostok. Rita writes that Nicholas Jakovlevitch (one of the wounded) is at Simferopol with his friend, the brother of little M. Their splendid (good) friend (Alexandre Dumbadze) has been killed there, we loved him very much, he was one of our wounded.

I only write what I dare, for in the present days one never knows in whose hands the letter might fall. We hope to do our devotions next week if we are allowed to do so. I am already looking forward to those beautiful services—such a longing to pray in church. I dream of our church (at Tsarkoe Selo) and of my little cell-like corner near the altar. Nature is beautiful, everything is shining and brilliantly lighted up. The children are singing next door. There are no lessons to-day as it is Friday of Carnival week.

I relive in mind, day by day, through the year that has passed and think of those I saw for the last time. Have been well all along, but for the past week my heart has been bad and I do not feel well, but this is nothing. We cannot complain, we have got everything, we live well, thanks to the touching kindness of the people, who in secret send us bread, fish, pies, etc.

Do not worry about us, darling, dearly beloved one. For you all it is hard and especially for our Country!!! This hurts more than anything else—and the heart is racked with pain—what has been done in one year! God has allowed it to happen—therefore it must be necessary so that they might understand, that eyes might be opened to lies and deceits.

I cannot read the newspapers quietly, those senseless telegrams—and with the German at the door!!!

K. and everyone else looks at "brother" as a saviour—Great God, to what have they come to, to wait for the enemy to come and rid them from the infernal foe. And who is sent as the leader?

Aunt Baby's brother (meaning herself). Do you understand. They wished to act nicely, probably thinking that it would be less painful and humiliating to her—but for her (meaning herself) it is far worse—such an unbearable pain—but everything generally hurts now—all one's feelings have been trampled underfoot—but so it has to be, the soul must grow and rise above all else; that which is most dear and tender in us has been wounded—is it not true? So we too have to understand through it all that God is greater than everything and that He wants to draw us, through our sufferings, closer to Him. Love Him more and better than one and all. But my country—my God—how I love it, with all the power of my being, and her sufferings give me actual physical pain.

And who makes her (Russia) suffer, who causes blood to flow?... her own sons. My God, what a ghastly horror it all is. And who is the enemy? This cruel German, and the worst thing for Aunt Baby is that he (the enemy) is taking away everything as in the time of Tsar Alexsei Michailovich (meaning that frontiers of Russia would become again as during the reign of A. M.). But I am convinced that it will not remain so, help will come from Above, people can no longer do anything, but with God all things are possible, and He will show His strength, wisdom and all forgiveness and love—only believe, wait and pray.

This letter will, in all probability, reach you on the day of our parting (one year ago), it seems so near and yet again as if centuries had passed since then.

It is seven months that we have been here. We see Ysa* only through the windows, and Madeleine (the Empress's lady's-maid, Madeleine

Zanotti) too. They have been here for three or four months to-day, I am told. I must give that letter at once.

I kiss you and Titi tenderly, Christ be with you, my dearest ones. Greeting to Mother and Grandmother. The children kiss and love you, and he (the Emperor) sends his very best wishes.

YOUR OLD GODMOTHER.

PART OF LETTER DATED MARCH 2/15, 1918, WHICH REACHED ME THREE YEARS LATER IN ENGLAND

. . . . .

* (previous page) Baroness Büxhoevgen Lady-in-waiting to the Empress.

L'ENVOI

THE first idea of writing this book occurred to me some time after my arrival in England. I had always known that the Empress had been grossly misrepresented in Russia, but I had not attached much importance to the fact, as I had seen the Revolutionary propaganda, and I fully realized the methods of the Revolutionaries in relation to the Imperial Family.

I was, however, astonished and horrified to discover that the same ideas were current in the broad-minded and enlightened country which has afforded me and so many other fugitives such kindly sanctuary.

If possible, I think the Empress has been more universally condemned in England than in Russia. I have scarcely heard her name mentioned without its being coupled with the degrading attributes of treachery, sensualism, hysteria, and religious mania. To one who knew her intimately and who loved her devotedly, such a state of things is unspeakably painful. I accidentally saw a film which was the grossest libel on her character and her personality, the mind of the producer having been apparently bent upon presenting the Empress as a combination of the chief forms of lurid wickedness which appeal to patrons of the cinema. I have also read novels about her which, whilst enraging me as mendacious chronicles, have considerably enlightened me as to the capacity for invention of which the human imagination is capable. More serious works have condemned the Empress in a courteous manner, but they have been none the less scathing in their judgment. Some writers, after the story of Ekaterinburg was authentically given to the world, have been more tolerant and more pitying in their censure, but it has been always censure.

Therefore, in the face of such hatred and contempt for one at whose hands I have received nothing but kindness and love, I determined to write my impressions of the Empress as I knew her, both in the happy days and afterwards in those of war and unrest during the first dark weeks of the Revolution.

I reasoned, I trust with justice, that although the majority of people are always ready to believe the worst of anyone, there must be others who, in the spirit of fair play, would be willing to look on the reverse side of the picture. There must surely be friends and relations in England who would welcome facts which proved that the

Empress had been true to her English upbringing and to the traditional right living of the descendants of Queen Victoria. English people seem to have forgotten, when the Empress was vilified on the screen and in cold type, that she was the daughter of Princess Alice, a name which is associated with all that is noblest and best in woman, a name which alone, one might have thought, would have pleaded for that of her daughter. But nothing protected her, not even the facts that her first cousin was King of England and that one of her sisters was married and living in this country.

I knew the almost impossible task of rehabilitation which lay before me, but, as the task daily assumed greater proportions, love and pity for my beloved friend urged me to attempt it.

I knew that I might be accused of being a Rasputinière, since my photograph taken with him had appeared in one of the English illustrated papers; but my best reply to such a possible charge is that I am living in England with my husband and child, and that my husband has sanctioned my description of Rasputin as I and others knew him. If the Empress's association with Rasputin had been a guilty one, or if I had not been in a position to describe events exactly as they happened, this book would never have been written.

It is both unjust and untrue to ascribe the Revolution as directly consequent upon the Emperor's weakness, or the pro-Germanism and hysteric sensuality of the Empress. I have endeavoured to show that Rasputin was probably one of the unconscious tools of the Revolution against Imperialism: there is no doubt that German intrigues brought Lenin back from Switzerland to overthrow the milder rule of Kerensky, who was not ready to offer the country an efficient substitute for Tsardom, but the Empress was entirely innocent of pro-Germanism. Russia was ripe for Revolution; she had essayed Revolution years before the Empress or Rasputin saw the light. Her political history alone proves my statement, but War hurried the feet of Revolution toward her bloodstained goal. Other European kingdoms have tottered or fallen, but Russia is a land of extremes: hence the extreme methods of her ideas of equality, which are, in many respects, similar to those of the French Revolution.

I am well aware that certain "official" documents relative to the Empress were sent to England, and I know the shameful assertions which they contained. These documents emanated from the Duma, and were "arranged" by the Duma, in order to justify many things which would otherwise have been unjustifiable.

I have not attempted to give to the world any elaborate descripttions of Court festivities, and those happenings which are the common property of all European journalists. Mine is a very simple résumé of the daily life and personality of the Empress as I knew her. I have endeavoured to avoid anything in the nature of exaggeration, in the hope that the public, who have innocently lent a ready ear to those things which are untrue, and which have been exploited by people who never saw or spoke to the Empress, will give equal consideration to the testimony of one who both knew and loved The Real Tsaritsa.

PRINTED BY BURLEIGH LTD.,
AT THE BURLEIGH PRESS BRISTOL ENGLAND

END OF BOOK TWO

BOOK THREE

Anna Aleksandrovna Viroubova is a controversial figure with some painting her as an innocent friend of the Imperial Family and others as the nefarious mole in their camp for Rasputin. Her memories recorded here, are what she decided to leave for us, having had plenty of time, whilst living in Finland following the Revolution, to decide what to include. The reader must ascertain how sincere they are.

ANNA VIROUBOVA

Unlike other royal household staff, Viroubova was not permitted to accompany the royal family in to exile. She was imprisoned for some time in St Petersburg and interrogated by the Cheka police and the Provisional Government, of which she writes about in this book. She also wrote to the former Empress whenever possible, of which some correspondence is reproduced in this book.

She managed to escape Russia and lived out her life in relative safety in Finland where she joined a religious order. While there she wrote her memoirs, passed on her prized photo albums and even was interviewed.

Image: Viroubova stands with the walking cane she required for the rest of her life after a road traffic accident in St Petersburg.

First published in 1923 by the Macmillan Company – London, New York. Published in Australia 26 January 1924. Pages: 400

MEMORIES OF THE RUSSIAN COURT

THE EMPRESS OF RUSSIA IN HER HAPPY YEARS

TO MY EMPRESS,
WITH LOVE AND FIDELITY ETERNAL

MEMORIES OF
THE RUSSIAN COURT

BY

ANNA VIROUBOVA

THE MACMILLAN COMPANY
NEW YORK · BOSTON · CHICAGO · DALLAS
ATLANTA · SAN FRANCISCO

MACMILLAN & CO., Limited
LONDON · BOMBAY · CALCUTTA
MELBOURNE

THE MACMILLAN CO. OF CANADA, Ltd.
TORONTO

"When you are reproached—bless; when persecuted—be patient; when calumniated—comfort yourself; when slandered—rejoice; this is your road and mine." Words of St. Seraphine.

ALEXANDRA FEODOROVNA, from *Tobolsk*,
March 20, 1918

Yea, though I walk through the valley of the shadow of death I shall not fear. Thy rod and Thy staff shall comfort me.

ILLUSTRATIONS

THE EMPRESS OF RUSSIA IN HER HAPPY DAYS	*(Frontispiece)*	*355*
THE EMPRESS DRIVING HER PONY CHAISE	(8)	*364*
THE EMPRESS WITH GRAND DUCHESS TATIANA IN HER BEDROOM, TSARSKOE SELO	(8)	*364*
ALEXANDER SERGIEVITCH TANIEFF	(9)	*365*
THE WINTER PALACE, PETROGRAD	(20)	n/a
MILITARY REVIEW, TSARSKOE SELO	(21)	n/a
THE EMPEROR AND EMPRESS IN A QUIET HOUR ON BOARD THE IMPERIAL YACHT	(32)	*382*
THE EMPRESS DISTRIBUTING PRESENTS TO SAILORS AT THE END OF A CRUISE	(33)	*382*
LIVADIA, THE NEW PALACE OF THE TSARS IN THE CRIMEA	(38)	n/a
A CORNER OF THE COURT OF THE PALACE OF LIVADIA	(39)	n/a
THE IMPERIAL CHILDREN BATHING IN THE BLACK SEA AT LIVADIA	(39)	n/a
THE IMPERIAL YACHT ARRIVES AT LIVADIA, THE CRIMEA	(50)	*393*
THE TSAR, GRAND DUCHESSES OLGA AND TATIANA AND MME. VIROUBOVA AT HOMBURG	(51)	*394*
THE EMPRESS GIVING ALEXEI A LESSON ON THE TERRACE	(74)	n/a
ALEXEI PLAYING IN THE SNOW AT TSARSKOE SELO	(74)	n/a
THE EMPRESS IN BED WITH CONVALESCENT TSAREVITCH	(75)	n/a
GRAND DUCHESSES OLGA AND TATIANA ON BOARD THE "STANDERT"	(78)	n/a
THE TSAREVITCH WITH HIS COUSINS, CHILDREN OF GRAND DUKE ERNEST OF HESSE	(79)	n/a
NICHOLAS II AND THE TSAREVITCH ON BOARD THE IMPERIAL YACHT "STANDERT"	(80)	*412*
THE TSAREVITCH WITH HIS SAILOR, DEREVANKO	(81)	*413*
THE TSAR, TSARINA, AND ALEXEI IN THE GARDENS AT TSARSKOE SELO	(81)	*413*
THE EMPEROR AND EMPRESS IN OLD SLAVONIC DRESS, JUBILEE OF 1913	(98)	*427*

The Invalid Empress on the Balcony at Peterhof	(99)	428
The Guest Room in Rasputine's House in Siberia	(169)	472
The Three Children of Rasputine Before Their House in Siberia	(169)	471
The Empress and Young Grand Duke Dmitri, Afterwards One of Rasputine's Assassins	(184)	481
Minister of Court Count Fredericks, the Empress and Tatiana Taking Tea in Finnish Woods	(185)	482
Grand Duchesses Olga, Tatiana, Anastasie, and Marie, Prisoners at Tsarskoe Selo, 1917	(266)	534
Anna Viroubova Shortly After Her Release from the Fortress of Peter and Paul	(267)	535
Letters From Nicholas II to Anna Viroubova, from Tobolsk, 1917	(306)	n/a
Letters from Alexei, Tatiana, and Marie, Smuggled from Siberia in 1917	(307)	n/a
One of the Empress's Last Letters to Mme. Viroubova, Written in Old Slavonic, 1918	(320)	571
Smuggled Letter from the Empress on Birchbark After Paper Gave Out	(321)	572
The Ex-Emperor and Alexei Feeding Turkeys in the Barnyard, Tobolsk, 1918	(342)	n/a
The Last Photograph Taken of the Empress and Her Daughters, Olga and Tatiana, Shortly Before the Murder of the Imperial Family in Siberia	(343)	625

Publisher Note: With very few exceptions all these photographs were taken by members of the Imperial Family and by Mme. Viroubova, all of whom were experts with the camera.

Editing note: This list of illustrations from the London Edition 1923, of which some items are not included in certain versions; you can see n/a alongside these which were either very small or of very bad quality and not worth reproducing here.

CHAPTER I

It is with a prayerful heart and memories deep and reverent that I begin to write the story of my long and intimate friendship with Alexandra Feodorovna, wife of Nicholas II, Empress of Russia, and of the tragedy of the Revolution, which brought on her and hers such undeserved misery, and on our unhappy country such a black night of oblivion.

But first I feel that I should explain briefly who I am, for though my name has appeared rather prominently in most of the published accounts of the Revolution, few of the writers have taken the trouble to sift facts from fiction even in the comparatively unimportant matter of my genealogy. I have seen it stated that I was born in Germany, and that my marriage to a Russian officer was arranged to conceal my nationality. I have also read that I was a peasant woman brought from my native Siberia to further the ambitions of Rasputine. The truth is that I am unable to produce an ancestor who was not born Russian. My father, Alexander Sergievitch Tanieff, during most of his life, was a functionary of the Russian Court, Secretary of State, and Director of the Private *Chancellerie* of the Emperor, an office held before him by his father and his grandfather. My mother was a daughter of General Tolstoy, aide-de-camp of Alexander II. One of my immediate ancestors was Field Marshal Koutousoff, famous in the Napoleonic Wars. Another, on my mother's side, was Count Kontaisoff, an intimate friend of the eccentric Tsar Paul, son of the great Catherine.

Notwithstanding my family's hereditary connection with the Court our own family life was simple and quiet. My father, aside from his official duties, had no interests apart from his home and his music, for he was a composer and a pianist of more than national fame. My earliest memories are of home evenings, my brother Serge and my sister Alya (Alexandra) studying their lessons under the shaded lamp, my dear mother sitting near with her needlework, and my father at the piano working out one of his compositions, striking the keys softly and noting down his harmonies. I thank God for that happy childhood which gave me strength of soul to bear the sorrows and sufferings of after years.

Six months in every year we spent in the country near Moscow on an estate which had been in the family for nearly two hundred years. For neighbors we had the Princes Galatzine and the Grand

Duke and Grand Duchess Serge, the last named being the older sister of the Empress. I hardly remember when I did not know and love the Grand Duchess Elizabeth, as she was familiarly called. As small children she petted and spoiled us all, often inviting us to tea, the feast ending in a grand frolic in which we were allowed to search the rooms for toys which she had ingeniously hidden. It was at one of these children's teas that I first saw the Empress Alexandra. Quite unexpectedly the Tsarina was announced and the beautiful Grand Duchess Elizabeth, leaving her small guests, ran eagerly to greet her. The time was near the beginning of the reign of Nicholas II and Alexandra Feodorovna, and the Tsarina was at the very height of her youthful beauty. My childish impression of her was of a tall, slender, graceful woman, lovely beyond description, with a wealth of golden hair and eyes like stars, the very picture of what an Empress should be.

For my father the young Empress soon conceived a warm liking and confidence and she named him as vice president of the committee of *Assistance par le Travail*. During this time we lived in winter in the Michailovsky Palace in Petrograd, and in summer in a small villa in Peterhof on the Baltic Sea. From conversations between my mother and father I learned a great deal of the life of the Imperial Family. The Empress impressed my father both by her excessive shyness and by her unusual intelligence. She was above all a motherly woman and often combined baby-tending with serious business affairs. With the little Grand Duchess Olga in her arms she discussed all kinds of business with my father, and while with one hand rocking the cradle where lay the baby Tatiana she signed letters and papers of consequence. Sometimes while thus engaged there would come a clear, musical whistle, like a bird call. It was the Emperor's special summons to his wife, and at the first sound her cheek would turn to rose, and, regard less of everything, she would fly to answer it. That birdlike whistle of the Emperor I became very familiar with in later years, calling the children, signaling to me. It had a curious, appealing, resistless quality, peculiar to himself.

Perhaps it was a common love of music which first drew the Empress and our family into a bond of friendship. All of us children received a thorough musical education. From childhood we were taken regularly to concerts and the opera, and our home, especially on Wednesday evenings, was a rendezvous for all the musicians and composers of the capital. The great Tschaikovsky was a friend of my father, and I remember many others of note who were frequent guests at tea or dinner.

Apart from music we received an education rather more practical than was the average at that time. In the Russia of my childhood a girl of good family was supposed to acquire a few pretty accomplishments and nothing much besides. Accomplishments I and my sister were given, but besides music and painting, for which my sister had considerable talent, we were well grounded in academic studies, and we finished by taking examinations leading to teachers' diplomas. I may say also that even in our drawing-room accomplish-ments we were obliged to be thorough, and when my father ventured to show some of our work to the Empress she expressed warm approval. "Most Russian girls," she said, "seem to have nothing in their heads but officers."

The Empress, coming from a small German Court where everyone at least tried to occupy themselves usefully, found the idle and listless atmosphere of Russia little to her taste. In her first enthusiasm of power she thought to change things a little for the better. One of her early projects was a society of handwork com-posed of ladies of the Court and society circles, each one of whom should make with her own hands three garments a year to be given to the poor. The society, I am sorry to say, did not long flourish. The idea was too foreign to the soil. Nevertheless the Empress persisted in creating throughout Russia industrial centers, *maisons de travail*, where the unemployed, both men and women, and especially unfortunate women who, through errors of conduct, lost their positions, could find work.

Life at Court was by no means serious. In fact it was at that time very gay. At seventeen I was presented, first to the Empress Dowager who lived in a palace in Peterhof known as the Cottage. Extremely shy at first, I soon accustomed myself to the many brilliant Court functions to which my mother chaperoned my sister and myself. We danced that first winter, I remember, at no less than twenty-two balls besides attending many receptions, teas, and dinners. Perhaps it was partly the fatigue of all this social dissipation which made so serious the illness with which in the ensuing summer I was stricken. Typhus, that scourge of Russia, struck down at the same time my brother Serge and myself. My brother's illness ran a normal course and he made a rapid recovery, but for three months I lay at death's door. After the fever succeeded many complications, inflammation of the lungs and kidneys, and an affection of the brain whereby I lost both speech and hearing. In the midst of my suffering I had a vivid dream in which the saintly Father John of Kronstadt appeared to me and told me to have courage and that all would finally be well.

This Father John of Kronstadt, whom all true Russians reverence as a saint, I remembered as having thrice been at our house in my early childhood. The gentle majesty of his presence, the beauty of his benign countenance had so deeply impressed me that now, in my desperate illness, it seemed to me that he, more than the skilled physicians and the devoted sisters who attended me, had power of help and healing. In some way I managed to convey to my parents that I wanted Father John, and they immediately telegraphed begging him to come. It was some days before the message reached him, as he was away from home on a mission, but as soon as he received word of our need he hastened to Peterhof. As in a vision I sensed his coming long before he reached the house, and when he came I greeted him without astonishment with a feeble movement of my hand. Father John knelt down beside my bed, praying quietly, a corner of his long stole laid over my burning head. At length he rose, took a glass of holy water, and to the consternation of the nurses sprinkled it freely over me and bade me sleep. Almost instantly I fell into a deep sleep, and when I awoke next day I was so much better that all could see that I was on the road to recovery.

In September of that year I went with my mother first to Baden and afterwards to Naples. We lived in the same hotel with the Grand Duke and Grand Duchess Serge who were very much amused to see me in a wig, my long illness having rendered me temporarily almost bald. After a quiet but happy season in southern Italy I returned to Russia quite restored to health. The winter of 1903 I remember as a round of gaieties and dissipations. In January of that year I received from the Empress the diamond-studded *chiffre* of maid of honor, which meant that, following my marriage, I would have permanent entry to all Court functions. Not immediately but very soon afterwards I was called to duty to the person of the Empress, and there began then that close and intimate friendship which I know lasted with her always and which will remain with me as long as God permits me to live.

I would that I could paint a picture of the Empress Alexandra Feodorovna as I knew her before the first shadow of doom and disaster fell upon unhappy Russia. No photograph ever did her justice because it could reproduce neither her lovely color nor her graceful movements. Tall she was, and delicately, beautifully shaped, with exquisitely white neck and shoulders. Her abundant hair, red gold, was so long that she could easily sit upon it when it was unbound. Her complexion was clear and as rosy as a little child's. The Empress had large eyes, deep gray and very lustrous. It was only in later life that sorrow and anxiety gave her eyes the melancholy with which they are usually associated. In youth they wore an expression of constant merriment which explained her family nickname of "Sunny," a name by the way nearly always used by the Emperor. I began almost from the first day of our association to love and admire her, as I have loved her ever since and always shall.

The winter of 1903 was very brilliant, the season culminating in a famous ball in costumes of Tsar Alexis Michailovitch, who reigned in the seventeenth century. The ball was given first in the Hermitage, the great art gallery adjoining the Winter Palace, but so immense was its success that it had to be twice repeated, once in the *Salle de Concert* of the palace and again in the large ballroom of the Schermetieff Palace. My sister and I were two of twenty young girls selected to dance with twenty youthful cavaliers in an ancient Russian dance which required almost as much rehearsal as a ballet. The rehearsals were quite important society events, all the mothers attending, and the Empress often looking on as interested as any of us.

That summer I again fell ill in our villa in Peterhof, and I remember particularly that this was the first time the Empress ever visited our house. She drove in a low pony chaise, coming up to my sickroom all in white with a big white hat and in the best of spirits. Needless to say, her unexpected visit did me a world of good, as did her second visit at our home in the country when she left me a gift of holy water from Saroff, a place greatly venerated by Russians. That winter with its artless pleasures, and the pleasant summer which followed, marked the end of an era in Russia. Immediately afterwards came the catastrophe of the Japanese War, so needlessly entered into. This war was the beginning of a long line of disasters which ended in the supreme disaster of 1917.

I must confess that at the time the Japanese War made no very deep impression on young girls who, like myself, faced life lightly like happy children. We resigned ourselves to an almost complete cessation of balls and parties, and we put aside our pretty gowns for

THE EMPRESS DRIVING IN HER PONY CHAISE. PETERHOF, 1909.

THE EMPRESS WITH GRAND DUCHESS TATIANA IN HER
BEDROOM, TSARSKOE SELO. FAVORITE IKONS IN BACKGROUND

the sober dress of working sisters. The great salons of the Winter Palace were turned into workrooms and there every day society flocked to sew and knit for our soldiers and sailors fighting such incredible distances away, as well as for the wounded in hospitals at home and abroad. My mother, who was one of the heads of committees giving out work to be done at home, was constantly busy, and we obediently followed her example.

ALEXANDER SERGIEVITCH TANIEFF,
Director of the Tsar's Private Chancellerie, Father of Anna Viroubova.

Every day the Empress came to inspect the work, often sitting down at a table and sewing diligently with the others. This was shortly before the birth of the Tsarevitch and I have a clear picture in my mind of the Empress looking more than ever fine and delicate, her tall figure clad in a loose robe of dark velvet trimmed in fur. Behind her chair, bringing into splendid relief her bright gold hair, stood a huge negro servant, gorgeous in scarlet trousers, gold-embroidered jacket, and white turban. This negro, Jim, was one of four Abyssinians who stood guard before the doors of the private apartments. They were not soldiers and they had no functions except to open and close the doors, and to signify by a sudden, noiseless entrance into a state apartment that one of their Majesties was about to appear. The Abyssinians were in fact simply one of the left-overs from the days of Catherine the Great, in whose times dwarfs and negroes and other exotics figured as a part of Court ceremonials. They remained not because Nicholas II or the Empress wanted them, but because, as I shall later explain, it was practically impossible to change any detail of Russian Court life.

The following summer the heir was born amid the wildest rejoicings all over the Empire. I remember the Empress telling me with what extraordinary ease the child was brought into the world. Scarcely half an hour after the Empress had left her boudoir for her bedroom the baby was born and it was known that, after many prayers, there was an heir to the throne of the Romanoffs. The Emperor, in spite of the desperate sorrow brought upon him by a disastrous war, was quite mad with joy. His happiness and the mother's, however, was of short duration, for almost at once they learned that the poor child was afflicted with a dread disease, rather rare except in royal families where it is only too common. The victims of this malady are known in medicine as haemophiliacs, or bleeders. Frequently they die soon after birth, and those who survive are subject to frightful suffering, if not to sudden death, from slight injuries to blood vessels, internal as well as external.

The whole short life of the Tsarevitch, the loveliest and most amiable child imaginable, was a succession of agonizing illnesses due to this congenital affliction. The sufferings of the child were more than equaled by those of his parents, especially of his mother, who hardly knew a day of real happiness after she realized her boy's fate. Her health and spirits began to decline, and she developed a chronic heart trouble. Although the boy's affliction was in no conceivable way her fault, she dwelt morbidly on the fact that the disease is transmitted through the mother and that it was common in her family. One of her younger brothers suffered from it, also her uncle Leopold, Queen Victoria's youngest son, while all three sons of her sister, Princess Henry of Prussia, were similarly afflicted. One of these boys died young and the other two were lifelong invalids.

Everything possible, everything known to medical science, was done for the child Alexei. The Empress nursed him herself, as indeed, with the assistance of professional women, she had nursed all her children. Three trained Russian nurses were in attendance, with the Empress always superintending. She bathed the babe herself, and was with him so much that the Court, ever censorious of her, complained that she was more of a nurse than an Empress.

The Court, of course, did not immediately understand the serious condition of the infant heir. No parents, be their estate high or low, are ready all at once to reveal a misfortune such as that one. It is always human to hope that things are not as desperate as they seem, and that in time some remedy for the illness will be found. The Emperor and Empress guarded their secret from all except relatives and most intimate friends, closing their eyes and their ears to the growing unpopularity of the Empress. She was ill and she was suffering, but to the Court she appeared merely cold, haughty, and indifferent. From this false impression she never fully recovered even after the explanation of her suddenly acquired silence and melancholy became generally known.

CHAPTER II

IN one of the earliest days of 1905 my mother received a telegram from Princess Galatzine, first lady in waiting, saying that my immediate presence at Court was required. The Princess Orbeliani, also a lady in waiting, was seriously ill, and some one was needed to replace her in attendance on Her Majesty. I left at once for Tsarskoe Selo, then, as always, the favorite home of the Imperial Family, and on my arrival was conducted to the apartments in the palace known as the Lyceum. The rooms were small and dark with windows looking out on a little church. It was the first time I had ever been away from home, and in any surroundings I should have been homesick and forlorn, but in these unfriendly surroundings my spirits were with some excuse depressed.

The time of my coming to Court was unpropitious, the Imperial Family and all connections being in deep mourning for the Grand Duke Serge who, on the morning of February 4, had been barbarously assassinated. The Grand Duke Serge, uncle of Nicholas II, had been Governor of Moscow. He was undoubtedly a reactionary, and his rule was said to have been harsh. Certain it is that his administrative methods earned him the intense enmity of the Social Revolutionaries and he had long lived in danger of assassination. His wife, the Grand Duchess Elizabeth, was devoted to him in spite of his somewhat difficult temperament, and she never willingly allowed him to leave the palace of the Kremlin unaccompanied. Usually she went with him herself, but on this fatal February morning he, being in a dark mood, left the palace without her knowledge. Suddenly a great explosion shook all the windows, and the poor Grand Duchess, springing from her chair, cried out in an agonized voice: "It is Serge!"

Rushing out into the court she saw a horrible sight, the body of her husband scattered in a hundred bleeding fragments over the snow. The bomb had literally torn the unfortunate man to pieces, so that in the dismembered mass of flesh and blood there was nothing recognizable of what had been, only a few minutes before, a strong and dominating man.

The terrorist who threw the bomb was promptly arrested, tried, and sentenced to death. It was entirely characteristic of the Grand Duchess Elizabeth that in the midst of her grief and horror she still found room in her heart to pity the misguided wretch sitting in his cell waiting his miserable end. The Grand Duchess insisted on visiting the man in prison, assuring him of her forgiveness, and praying for him on the stone floor of his cell. Whether or not he joined in her prayers I do not know. The Social Revolutionaries prided themselves on being irreligious and very many of them were Jews.

The Court weighed down by this terrible tragedy was a sad enough place for a homesick girl like myself. Like all the other ladies in waiting I wore a black dress with a long veil, and when at length I was received by the Empress I found her, too, dressed in deep mourning. After this first formal reception I saw very little of the Empress, all her time being devoted to her sister, the Grand Duchess Elizabeth, and to Princess Henry of Prussia, who was visiting her. The Empress Dowager also came, so that the suite was thrown together in what for me was not altogether a pleasant association. My special duty, as I discovered, was attendance on the old Princess Orbeliani, whose illness, I am bound to admit, did not sweeten her disposition. But as she was dying of that terribly trying malady, creeping paralysis, I am ashamed, even now, to criticize her. For the other *dames d'honneur*, however, I have no hesitation to say that they were not on their best behavior. Being entirely a stranger at Court and unacquainted with insincerities which afterwards I came to know only too well, I suffered keenly from the cutting remarks of my colleagues. My French, which I own I spoke rather badly, came in for a great deal of ridicule. On the whole it was rather an unhappy period in my young life.

The one bright spot that I remember was a drive with the Empress to which I was summoned by telephone. It was a warm day in early spring and the snow around the tree roots along the road was thawing in the pale sunlight. We drove in an open carriage, a big Cossack, picturesquely uniformed, riding behind. It was my first public appearance with Royalty and I was a little confused as to how to behave in the presence of the low-bowing crowds that lined the way. The Empress, however, soon put me at my ease, chatting of simple things, talking of her children, especially of the infant heir, at that time about eight months old. Our drive was not very long because the Empress had to hurry back to superintend a dancing lesson of the young Grand Duchesses. I remember when I returned to the apartment of the invalid Princess Orbeliani, she commented rather maliciously on the fact that I was not invited to attend the dancing lesson. But by that time, alas! I knew that had I been invited her comment might have been more malicious still. Still I must not speak badly of the poor Princess, for in spite of her illness and approaching death she was very brave and kinder than most people in her circumstances would have been.

Lent came on and in the palace church there were held every Wednesday and Friday special services for the Imperial Family. I asked and was given permission to assist in these services and I found great solace in them. At that time also I became warmly attached to a maid of honor of the Grand Duchess Serge, Princess Senkovsky, a woman of rare character. She had recently lost her mother and was in a sad mood. Almost everyone, in fact, was sad at this time. The Grand Duchess Serge, although she bore her tragedy with dignity and courage, went about with a white face and eyes in which horror still lingered. On religious holidays she laid aside her black robes and appeared all in white like a madonna.

The Princess Irene of Prussia (Princess Henry) was still in mourning for her little son who had died of the same incurable disease which afflicted the Tsarevitch. She spoke to me with emotion of the child, to whom she had been deeply attached.

My duty came to an end in Holy Week, and I went to the private apartments to make my farewell of the Empress. She received me in the nursery, the baby Tsarevitch in her arms, and I cannot forget how beautiful the child appeared or how healthy and normal. He had a wealth of golden hair, large blue eyes, and an expression of intelligence rare in so young a child. The Empress was kindness itself. At parting she kissed me, and gave me as a souvenir of my first service a locket set in diamonds. Yet for all her gracious kindness how gladly I left that night for my beloved home.

The following summer, which as usual we spent at Peterhof, I saw much more of the Empress than in my month of attendance on her. With my mother and sister I again worked daily in the workrooms established for the wounded in the Japanese War, and there almost daily the Empress came to sew with the other women. Once every week she visited the hospitals at Tsarskoe Selo, and twice that summer, at her request, I accompanied her to her foundation hospital for training nurses. The Empress in the military hospitals was at her very best. Passing from bedside to bedside, speaking as tenderly as a mother to the sick and suffering men, sitting down to a game of checkers with convalescent officers, it was difficult to imagine how anyone could ever call her cold or shy. She was altogether charming and as she passed all eyes followed her with love and gratitude. To me she was everything that was good and kind, and into my heart there was born a great emotion of love and loyalty that made me determine that I would devote my whole life to the service of my Sovereigns. Soon after I was to know that they, too, desired that I should be intimately associated with their household. The first intimation came in the form of an invitation to spend two weeks on the Royal yacht which was about to leave for a cruise in Finnish waters. We left on the small yacht *Alexandria*, and at Kronstadt transferred to the larger yacht *Polar Star*. We were a fairly large company on board, among others Prince Obolensky, Naval Minister, Admiral Birileff, Count Tolstoy, Admiral Chagin of the Emperor's staff, and Mademoiselle Schneider and myself in attendance on Her Majesty. A little to my embarrassment I was placed at table next the Emperor with whom I was not at all acquainted. It is true that I had often seen him at Tsarskoe and at Peterhof riding, or walking with his kennel of English collies, eleven magnificent animals in which he took great pride. But this time, on the *Polar Star*, was the first time I had been brought into personal contact with him. With the Empress I felt more at home, and this he knew, for he began almost at once to speak to me of her and of her great help to him in the pain and anxiety of the Japanese War. "Without her," he said with feeling, "I could never have endured the strain."

The war was again recalled by a visit on board the yacht from

Count Witte, fresh from the Portsmouth Conference. As a reward for his work done there he received for the first time his title by which the world now knows him. During dinner he related with great gusto all his experiences in the United States, his triumph over the Japanese delegates, his popularity with the Americans, appearing very happy and satisfied with himself. The Emperor complimented him warmly, but Count Witte for all his talents was never a favorite with the Sovereigns.

Life on board the *Polar Star* was very informal, very lazy and agreeable. We sailed through the quiet waters of the Baltic, every day going ashore for walks, the Emperor and his staff sometimes shooting a little, but more often spending the time climbing rocks, hunting mushrooms and berries in the woods and meadows, and playing with the children to whom this country holiday was heavenly pleasure. Living long hours in the open air and indulging in so much vigorous exercise made me desperately sleepy so that I found myself drowsy at dinner and almost dead for sleep by the time the eleven o'clock tea hour came round. Everyone found my drowsiness a source of never-ending amusement, and once, after I had actually fallen asleep at tea and had nearly pitched out of my chair, the Emperor presented me with a silver matchbox with which he said I might prop my eyes open until bedtime.

There was, of course, a piano in the salon of the yacht, and the Empress and I found a new bond in our common love of music. We spent hours playing four-hand pieces, all our dearly loved classics, Bach, Beethoven, Tschaikovsky, and others. In our quiet hours with our music, and especially before going to bed, the Empress and I had many intimate conversations. As if to relieve a heart too much constrained to silence and solitude the Empress confided in me freely the difficulties of her life. From the first day of her coming to the Russian Court she felt herself disliked, and this was all the more a grief and mortification to her because her marriage with the Emperor was a true love match, and she ardently desired that their union should increase in the Russian people the loyalty and devotion they undoubtedly felt in those days for the House of Romanoff.

All the stories of the reluctance of Alexandra Feodorovna to marry Nicholas II are absurdly untrue. As a small child she had been taken to Petrograd to the marriage of her older sister Elizabeth and the Grand Duke Serge. With the Grand Duchess Xenia, sister of Nicholas, she formed a warm friendship, and with the young heir himself she was on the best of terms. One day he presented her with a pretty little brooch which from very shyness she accepted but afterwards repenting, she returned, squeezing the gift into his hand in the course of a children's party. The young Tsarevitch, much offended, or rather much hurt, passed the brooch on to his sister Xenia who, not knowing its history, cheerfully accepted it.

The attraction so early established increased with years and ripened into romantic love, yet Alexandra Feodorovna hesitated to accept Nicholas as her betrothed because of the change of religion which was necessary. Her home life at this time was not particularly happy. Her mother, Princess Alice of England, had died in her childhood, and now her father, the reigning Grand Duke of Hesse, died suddenly of a stroke of paralysis. Her brother Ernest, who inherited the title and who was of course her guardian, had made an unhappy marriage with Princess Victoria of Coburg, and the home life of the family was not particularly pleasant. Later this marriage was dissolved, and in 1908 Grand Duke Ernest was happily united to Princess Eleanor of Sohmslich. It was at his first marriage that Alexandra Feodorovna again met the Tsarevitch, and from this time on he became a suitor. After their formal betrothal the young pair spent some happy weeks with Queen Victoria in England, where the match met with the approval of all the English relatives.

Emperor Alexander III was at this time lying mortally ill in the Summer Palace Livadia, in the Crimea, and when his condition became hopeless Alexandra Feodorovna, as the future Tsarina, was summoned to join the Imperial Family at his bedside. The dying Tsar rose from his sickbed and, dressed in full uniform, gave her the greeting due her dignity as a royal bride. From the rest of the family, unfortunately, she had a less cordial reception. The Empress and her ladies in waiting, Princess Oblensky and Countess Voronzoff, were distant and formal, and the rest of the Court, as might be expected, followed their example. The whole atmosphere of the palace seemed to the young girl unwholesome and unsympathetic. Upstairs lay the dying Emperor, while below the suite lunched and dined and followed ordinary pursuits very much as though nothing untoward was happening. To Alexandra Feodorovna, accustomed to the intimacy of a small and much less formal Court, this behavior seemed unfeeling and unkind.

The end came suddenly one day when the Emperor, at the

moment almost free from pain or weakness, was sitting in his armchair. The Empress Marie, quite overcome, fainted in the arms of Alexandra, who in that hour of extreme sorrow, prayed sincerely that she and her future mother-in-law might be drawn together in bonds of affection. But this, alas! was never to be.

The days that followed were gray and desolate for the young bride. The funeral procession of Alexander III wound slowly and solemnly from the Crimea to Petrograd, a journey of many days. The young Emperor, absorbed in his new duties, had little time to devote to the lonely, homesick girl, and indeed they hardly met before the morning of their marriage, a few days after the state funeral of the dead Emperor. The marriage took place in the church of the Winter Palace, and those who witnessed it have said that the bride, in her rich satin robes, looked very pale and unhappy. As she herself told me, the wedding seemed only a continuation of the long funeral ceremonies she had so lately attended.

Thus came Alexandra Feodorovna to Russia, nor did the weeks that followed her arrival bring her any happiness. To her friend Countess Rantsau, lady in waiting to Princess Henry of Prussia, she wrote:

> I feel myself completely alone, and I am in despair that those who surround my husband are apparently false and insincere. Here nobody seems to do his duty for duty's sake, or for Russia, but only for his own selfish interests and for his own advancement. I weep and I worry all day long because I feel that my husband is so young and so inexperienced. He does not at all realize how they are all profiting at the expense of the State. What will come of it in the end? I am alone most of the time. My husband is all day occupied and he spends his evenings with his mother.

This was true, as Nicholas was very inexperienced and his mother's influence and, it must be said, her knowledge of affairs were very potent. All during the first year the Emperor and the two Empresses lived together in the Annitchkoff Palace on the Nevski Prospekt. Alexandra Feodorovna comforted herself with the thought that summer would bring her a real honeymoon in the Crimea. Meanwhile she and her young husband went for an occasional sledge ride together, about the only time granted them for confidences. Fortunately the first baby came soon and the second was soon expected. That autumn in the Crimea the Emperor was stricken with typhus and his wife insisted upon nursing him herself, hardly permitting his personal servant to assist her. Christmas was celebrated in his sickroom, his recovery having set in some weeks before. During these days of convalescence they went on solitary walks together, and the Emperor began to read with his wife, to confide in her with affection. When they went back to Petrograd it was with every cloud dispelled, and the Empress a radiantly happy wife. However, the somewhat cold and distant manner acquired in the first unhappy months of her stay in Russia remained with her. Russia seemed to her an unfriendly land, and she was never able to present to it her really sunny and amiable disposition.

Not all of these confidences did the Empress impart to me on that first cruise I was privileged to share with her on the *Polar Star*. Little by little, then and later, I learned the story of her unhappy youth. But what she told me that summer seemed to relieve her mind, and she was more cheerful at the ending of the cruise than at the beginning. The commander of the yacht was good enough to tell me that I had broken down the wall of ice that seemed to surround Her Majesty, and that now she could be more easily approached. At the close of the voyage the Emperor said: "You are to go with us every year after this."

But dearest of all in my memory were the words of the Empress at parting: "Dear Annia, God has sent me a friend in you." And so I remained ever afterwards, not a courtier, not long a lady in waiting, or even a maid of honor, or in any capacity an official member of the Court, but merely a devoted and an intimate friend of Alexandra Feodorovna, Empress of Russia.

CHAPTER III

SHORTLY after our return to Peterhof I went abroad with my family, stopping first at Karlsruhe, Baden, to visit my grandmother, and afterwards going on to Paris. The Empress had given me letters to her brother, the Grand Duke of Hesse, and to her eldest sister, Princess Victoria of Battenberg, both of whom I saw before leaving Germany. The seat of the Grand Duke of Hesse was

Wolfsgarten near Darmstadt, a beautiful place surrounded by extensive gardens laid out according to the Grand Duke's own plans. After my first luncheon at the palace, during which the Grand Duke asked me many questions about the Empress and her life at the Court of Russia, I walked in the gardens with Mme. Grancy, hofmistress of the Court of Hesse, a gracious and charming woman. She showed me the toys and other pathetic relics of the little Princess Elizabeth, only child of the Grand Duke's first marriage, who had died in Russia after an acute illness of a few hours. I also saw the white marble monument which the people of Hesse had raised to the memory of the child.

To the second luncheon I attended at the old Schloss came the Princess Victoria of Battenberg with her lovely daughter Louise. Etiquette at Hesse was of the severest order and I observed with some astonishment that the Princess Victoria curtsied deeply to her sister-in-law, Princess Eleanor, who though much younger than herself, was the wife of the reigning Grand Duke. The old Princess was a very clever woman and a brilliant conversationalist, although, to tell the truth, as she spoke very rapidly I lost a great deal of what she said. I remember her questioning me rather closely about the political situation in Russia, and although I was not very enlightening on the subject she was good enough to invite me and my sister to lunch with her at Jugenheim in the neighborhood of Darmstadt. Both the brother and the sister of the Empress entrusted me with letters to her, and I took them with me to Paris, not knowing that it would be a long time before I should be able to deliver them.

For in the midst of these pleasant days, all unknown to me, the tide of trouble and unrest was rising high in Russia. Beginning with a railroad strike in Finland, a succession of labor troubles and revolutionary demonstrations extending over a large territory brought about a serious crisis which for a time tied up most of the railroads and prevented our return to Russia. Of the cause of the trouble, and above all, of its ultimate consequences, I must say that I remained in complete ignorance. That the situation was grave of course I realized, and my heart went out to the Emperor on whom the responsibility of restoring order largely rested. But that this railroad strike, for that is all it seemed to amount to, was the beginning of a revolution never crossed my mind. I longed to get back to the Empress who I knew would be sharing the anxiety of the Emperor, but as a matter of fact I did not get back until after the manifesto of

October, 1905, had been signed and delivered to a startled world.

This October manifesto, relinquishing the principle of autocracy, creating for the first time a Duma of the Empire, was the result of many councils, some of them dramatic, not to say violent. Count Witte and Grand Duke Nicholas were determined that the Emperor should sign the manifesto, a thing which he was reluctant to do, not because he clung to his privileges as autocrat of all the Russias, though I know that this is the motive still attributed to him by almost all the world. The Tsar hesitated to create a house of popular representation because he knew how ill prepared the Russian people were for self-government. He knew the dense ignorance of the masses, the fanatical and ill-grounded socialism of the intelligentsia, the doctrinaire theories of the Constitutional Democrats. I can say with positive knowledge that Nicholas II fervently desired the progress of his country towards a high civilization, but in 1905 he felt very serious doubts of the wisdom of radical changes in the Russian system of government. At last, however, overborne by his ministers, he signed the manifesto. It is said that the Grand Duke Nicholas, in one of the last councils, lost all control of himself and drawing a revolver threatened to shoot himself on the spot unless the manifesto was signed. Whether this actually occurred or not I do not know, but from what was told me later by the Empress the scenes with the Grand Dukes and the ministers were painful in the extreme. When in one of the final councils the actual form of the national assembly was decided upon the Emperor, with a hand trembling with emotion, signed his name to the fateful document, all in the room rose and bowed to him in token of their continued fidelity.

The Empress told me that while these trying scenes were in progress she sat in her boudoir alone save for her near relative the Grand Duchess Anastasie, both of whom felt that in the stormy council chamber a child was being dangerously brought into the world. Yet all the prayers of the Empress, as well as those of the Emperor, were that the new policy of popular representation would bring peace to troubled Russia.

The Duma was elected, the Socialists alone of political parties repudiating it as too "bourgeois." I was present with all the Empress's household, in the Throne Room of the Winter Palace on the opening day of the Duma when the Tsar welcomed the deputies, and I remember with what a strong, steady voice, and with what clear enunciation, the opening speech was read. Of the proceedings of the first Duma I have no very definite recollections, because they were marked with endless and very wordy discussions rather than with any attempt at constructive action. Everyone knows that the Duma was dissolved by Imperial order after a short life of two months.

Of these momentous political events which rocked Russia and were featured prominently in every newspaper in the world only faint echoes reached the inner circle of the Russian Court. This may sound incredible to readers in republican countries where the press is entirely uncensored and where public opinion is educated in politics. In the Russia of 1906 the reading public was a comparatively small one and the press was poorly representative of the really intelligent people of the Empire. Few men and fewer women of my class attached any particular interest to the Duma, the best we hoped for it being that in time it would become an efficient working agency, like the parliaments of western European countries, adapted, of course, to Russian needs. The first Duma we thought of only as a rather foolish debating society.

The Empress and I were engaged, at that time, with singing lessons, our teacher being Mme. Tretskaia of the Conservatoire. The Empress was gifted with a lovely contralto voice, which, had she been born in other circumstances, might easily have given her a professional standing. My voice being a high soprano we sang many duets. Sometimes my sister joined us and as she also sang well we formed a trio singing many of the lovely arrangements for three voices by Schumann and others. Occasionally came also an English friend of the Empress, a talented violinist, and among us we arranged concerts which gave us the greatest pleasure, although we always had to hold them in another building of the palace called the Farm in order not to disturb the Emperor, who, for some strange reason, did not like to hear his wife sing.

When summer came and while the Duma was talking out its brief existence we again took up our sea life, this time on board the large royal yacht the *Standert*. We cruised for two months, the Emperor frequently going ashore for tennis and other amusements, but occupied two days of each week with papers and state documents brought to him by messenger from Petrograd. The Empress and I were almost constantly together walking on shore, or sitting on deck reading, or watching the joyful play of the children, each of whom had a sailor attendant to keep them from falling overboard or otherwise suffering mishap. The special attendant of the little Alexei was a big, good-natured sailor named Derevanko, a man seemingly devoted to the child. It was in fact Derevanko who taught Alexei to walk, and who during periods of great weakness following severe attacks of his malady carried the boy most tenderly in his arms. All of these sailors at the end of a cruise received watches and other valuable presents from the Emperor, yet most of them, even Derevanko, when the revolution came, turned on their Sovereigns with meanest treachery.

On my days of regular service, Wednesdays and Fridays, for I was then a regularly appointed lady in waiting, I dined with the Imperial Family, and at that time I formed a close friendship with General Alexander Orloff, an old companion in the Royal Hussars with the Emperor. After dinner the Emperor and General Orloff usually played billiards, while the Empress and I read or sewed under the warm lamplight. Those were happy evenings, full of bright talk and laughter, and I came to regard General Orloff as one of my best friends. Already the hateful hand of jealousy and gossip had been directed against me by people who could not understand, or who, from motives of palace politics, deliberately misunderstood the Empress's preference for my society. Practically every monarch has some close personal friend, absolutely disassociated with politics and social intrigue, but I have noticed that these friendships are always misunderstood and frequently bitterly resented. I used to take my small troubles to General Orloff, at least they seem small now after years of real trouble and affliction. But even after these bitter years of sorrow and affliction the kindly counsels of the good old general often come back to me, as they did then, like a friendly hand laid on my hot and resentful heart.

I was then, in 1906, a fully grown and mature young woman and, as I could not help knowing, I was the subject of many conversations in the family circle because of my indifference to marriage. I had, I suppose, the normal amount of attention from men, and the usual number of suitors, but none of the young officers and courtiers with whom I danced and chatted made any special appeal to my imagination. There was one young naval officer, Alexander Virouboff, who after December, 1906, came to our house almost every day, paying me the most marked attentions. One day at luncheon he spoke with pride of the very good service to which he had just been appointed, and very soon afterwards I found myself greeted on all sides as his affianced. In February there was a ball in which I was formally presented as a bride, and in the after whirl of dinners, presents, new gowns and jewels, I began to share the excitement, if not the happiness, of those around me. The Empress approved the match, my parents approved, and no one except my old friend General Orloff expressed even a faint doubt of the wisdom of the marriage. But on the day when he spoke to me frankly, advising me to think seriously before taking such a serious step, the Empress entered the room and said in a decided voice that I had given my word and that therefore I should not be given any discouragement.

I was married on the 30th of April, 1907, in the palace church at Tsarskoe Selo. The night before I slept ill and in the early morning I awoke in a mood of sadness and depression. The events of the day passed more like a dream than a reality. As in a dream I allowed myself to be dressed in my white satin wedding gown and floating veil, and still in a dream I knelt before their Majesties who blessed me, holding over my head a small ikon. Then began the marriage procession through the long corridors to the church. First walked Count Fredericks, master of ceremonies of the Court. Then came their Majesties, arm in arm, with my little boy cousin, Count Karloff, carrying a holy image. Then I, walking with my father. I must have shown by my excessive pallor the anxiety I felt, for on the stairs the Empress looked at me with concern and having caught my eye smiled brightly and glanced upward reassuringly at the bright sky.

During the ceremony I stood quite still like a manikin, gazing at my bridegroom as at some stranger. I had one moment of faint amusement when the officiating priest, who was very near-sighted, mistook the best man for the bridegroom addressing us affectionately as "my dear children." The Empress, as my matron of honor, stood at my left hand with the four young Grand Duchesses, and two others, the children of Grand Duke Paul. One of these was the Grand Duke Dmitri, who was destined to grow up to take part in the assassination of Rasputine. On the day of my marriage he was just a dear little boy, wide-eyed with the excitement of being one of a wedding party. After the ceremony there was tea with the Emperor and the Empress, and as usual when she and I parted there was an affectionate little note pressed into my hand. How like an angel she looked to me that day, and how hard it was for me to turn away from her and to go away with my husband. There was a family dinner that night in our home in Petrograd, and afterwards we went away for a month into the country.

It is a hard thing for a woman to tell of a marriage which from the first proved to be a complete mistake, and I shall say only of my husband that he was the victim of family abnormalities which in more than one instance manifested themselves in madness. My husband's nervous system had suffered severely in the rigors of the Japanese War, and there were many occasions when he was not at all responsible for what he did. Often for days together he kept his bed refusing to speak to anyone. One night things became so threatening that I could not forbear telephoning my fears to the Empress, and she, to my joy, responded by driving instantly to the house in her evening gown and jewels. For an hour she stayed with me comforting me with promises that the situation should, in one way or another, be relieved.

In August the Emperor and Empress invited us both to go for a cruise on the *Standert*, and sailing through the blue Finnish fjords it did seem for a time that I should find peace. But one day a terrible thing happened, possibly an accident, but if so a very strange one, as we had on board an uncommonly able Finnish pilot. We were seated on deck at tea, the band playing, a perfectly calm sea running, when we felt a terrific shock which shook the yacht from stem to stern and sent the tea service crashing to the deck. In great alarm we sprang to our feet only to feel the yacht listing sharply to larboard. In an instant the decks were alive with sailors obeying the harsh commands of the captain, and helping the suite to look to the safety of the women and children. The fleet of torpedo boats which always surrounded the yacht made speed to the rescue and within a few minutes the children and their nurses and attendants were taken off. Not knowing the exact degree of the disaster, the Empress and I hastened to the cabins where we hurriedly tied up in sheets all the valuables we could collect. We were the last to leave the poor *Standert*, which by that time was stationary on the rocks.

We spent the night on a small vessel, the *Asia*, the Empress taking Alexei with her in one cabin and the Emperor occupying a small cabin on deck. The little Grand Duchesses were crowded in a cabin by themselves, their nurses and attendants finding beds where they could. The ship was far from clean and I remember the Emperor, rather disheveled himself, bringing basins of water to the Empress and me in which to wash our faces and hands. We had some kind of a dinner about midnight and none of us passed an especially restful night. The next day came the yacht *Alexandria* on which we spent the next two weeks. A fortnight was required to get the ill-fated *Standert* off the rocks on which she had so mysteriously been driven. From the *Alexandria* and later to the *Polar Star*, to which we had been transferred, we watched the unhappy yacht being carefully removed from her captivity. We had not been very comfortable on the *Alexandria* because there was not nearly enough cabin room for our rather numerous company. The Empress occupied a cabin, the Tsarevitch and his sailor another one adjoining. The four little Grand Duchesses did as well as they could in one small cabin, while the Emperor slept on a couch in the main salon. As for me, I slept in a bathroom. Most of the suite found quarters on a Finnish ship which stood by.

After our return to Peterhof my husband became worse rather than better and his physician advised him to spend some time in a sanatorium for nervous patients in Switzerland. He left, but on coming back to Russia was noticeably in worse condition than before. In the hope that active service would be of benefit to his shattered nerves and disordered brain he was ordered to sea, but even this expedient proved of little benefit. After a year of intense suffering and humiliation my unhappy marriage, with the full approval of their Majesties and of my parents, was dissolved.

I kept my little house in Tsarskoe Selo, its modest furnishings beautified by many gifts from the Empress. Among these gifts were some charming pictures and six exquisitely embroidered antique chairs. A silver-laden tea table helped to make the salon cozy, and I have many happy memories of intimate teas to which the Empress sent fruit and the Emperor the cherry brandy which he especially affected.

The little house, however, was far from being the luxurious palace in which I have often been pictured as living. As a matter of fact, it was frightfully cold in winter because the house had no stone foundation but rested on the frozen earth. Sometimes when the Emperor and Empress came to tea we sat with our feet on the sofa to keep warm. Once the Emperor jokingly told me that after a visit to my house he kept himself from freezing only by going directly to a hot bath.

The summer of 1908 the Emperor and Empress paid an official visit to England, but on their return they sent for me and again I spent a happy holiday on the yacht. Not altogether happy, however, for

THE EMPEROR AND EMPRESS IN A QUIET HOUR ON BOARD
THE *STANDERT*. Photograph by Mme. Viroubova.

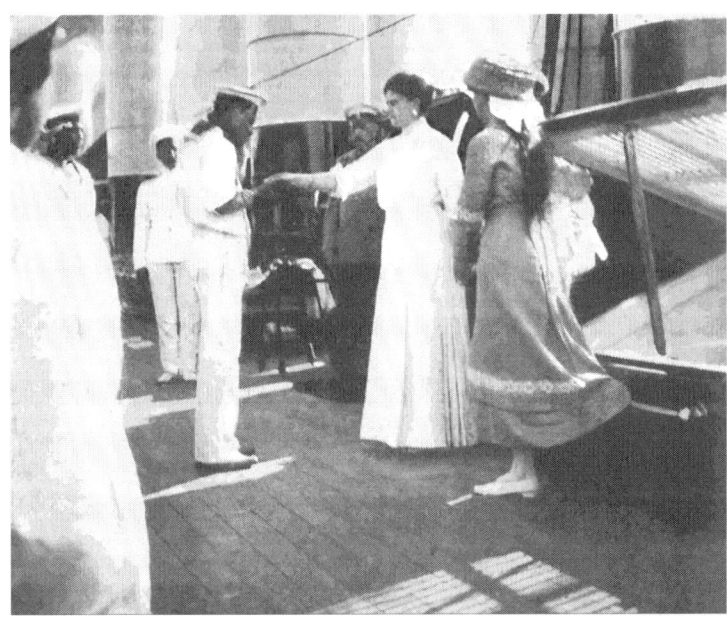

THE EMPRESS DISTRIBUTING PRESENTS AT THE END OF A CRUISE ON
THE IMPERIAL YACHT *STANDERT*.

towards the end of the cruise my poor friend General Orloff, then near his death from tuberculosis, came to say good-bye to his Sovereigns.

Correct in his uniform and all his orders the fine old soldier bade us all a brave farewell before leaving for Egypt, where he well knew that his end awaited him. Peace to his honored ashes. He lies buried at Tsarskoe Selo, where the Emperor and Empress often visited his grave. Poor Orloff, he too suffered from the malicious gossip of the Court where his honest admiration of the Empress was deliberately misinterpreted and assoiled. I can bear witness, and I do, that his greatest devotion was to the Emperor, his old comrade in arms, the friend of his youthful days.

CHAPTER IV

IN the autumn of 1909 I went for the first time to Livadia, the country estate of the Imperial Family in the Crimea. This part of Russia, dearer to all of the Tsars than any other, is a small peninsula, almost an island, surrounded on the west and south by the Black Sea and on the east by the Sea of Asov. A range of high hills protects it from the cold winds of the north and gives it a climate so mild and bland as to be almost sub-tropical. The Imperial estate, which occupies nearly half the peninsula, has always been left as far as possible in its natural condition of unbroken forests, wild mountains, and valleys. There was at the time of which I write but one short railroad in the whole of the Crimea, a short line running from Sevastopol, the principal port of the Black Sea, northward to Moscow. All other journeys had to be taken by carriage, motor cars, or on horseback.

The natural beauties of the Crimea would be difficult to exaggerate. The mountains, dark with pines, snow-covered during most of the year, make an imposing background for the profusion of flowering trees, shrubs and vines, making the valleys and plains one continuous garden. The vineyards of the Crimea are, or were previous to the Revolution, equal to any in Italy or southern France. What they became afterwards God knows. But certainly up to the summer of 1914, when I saw them last, the vine-clad hills and valleys of the Crimea were an earthly Paradise, as lovely and as peaceful as the mind can picture. From the grapes of the Crimea were distilled the best wines in Russia, among others an excellent champagne and a delicious sweet wine of the muscat variety.

Almost every kind of fruit flourished in the valleys, and in spring the wealth of blossoms, pink and white, of apples, cherries, peaches, almonds, made the whole countryside a perfumed garden, while in autumn the masses of golden fruit were a wonder to behold. Flowers bloomed as though they were the very soul of the fair earth. Never have I seen such roses. They spread over every building in great vines as strong as ivy, and they scattered their rich petals over lawns and pathways in fragrance at times almost overpowering. There was another flower, the glycinia, which grew on trailing vines in grapelike clusters, deep mauve in hue, the favorite color of the Empress. This flower, too, was intensely fragrant, as were the violets which in spring literally carpeted the plains. Imagine these valleys and plains, with their vineyards and orchards, their tall cypress trees and trailing roses, sloping down to a sea as blue as the sky and as gentle as a summer day, and you have a picture, imperfectly as I have painted it, of the country retreat of the Romanoffs. Here of all places in Russia they were loved and revered. The natives of the peninsula were Tartars, the men very tall and strong and the women almost invariably handsome. They were Mohammedans, and it was only within late years that the women had discarded their veils. Both men and women wore very picturesque dress, the men wearing round black fur caps and short embroidered coats over tight white trousers. It was the fashion for the women to dye their hair a bright red, over which they wore small caps and floating veils and adorning themselves with a wealth of silver bangles. These Tartars were an honest folk, absolutely loyal to the Tsar. They were wonderful horsemen, comparing favorably with the best of the Cossacks, and their horses, through long breeding and training, were natural pacers. To see a cavalcade of Tartars sweep by was to imagine a race of Centaurs come back to earth, so absolutely one was every horse and man.

The palace, as I saw it in 1909, was a large, old wooden structure surrounded by balconies, the rooms dark, damp, and unattractive. The only really sunny and cheerful room in the whole house was the dining room, where twice a day the suite met for luncheon and dinner. The Emperor usually presided at these meals, but the Empress being in bad health lunched privately with the Tsarevitch. The Empress had been for some time a victim of the most alarming heart attacks which she bravely concealed, not wishing the public to know her condition. Oftentimes when I remarked the blue whiteness of her hands, her quick, gasping breaths, she silenced me with a peremptory "Don't say anything. People need not know."

However, I was intensely relieved when at last she consented to have the daily attention of a special physician, this being the devoted Dr. Botkine, who accompanied the family in their Siberian exile, and shared their fate, whatever that fate may have been. Dr. Botkine, although a very able physician, was not a man of great social prominence, and when, at the Empress's request, I went to apprise him of his appointment as special medical adviser to their Majesties, he received the news with astonishment almost amounting to dismay. He began his administration by greatly curtailing the activities of the Empress, keeping her quietly in bed for long periods, and insisting on the use of a rolling chair in the gardens, and a pony chaise for longer jaunts abroad.

Life at Livadia in 1909 and in after years was simple and informal. We walked, rode, bathed in the sea, and generally led a healthful country life, such as the Tsar, eminently an outdoor man and a lover of nature, enjoyed to the utmost. We roamed the woods gathering wild berries and mushrooms which we ate at our al fresco teas, cooking the mushrooms over little campfires of twigs and dried leaves. The Emperor and his suite hunted a little, rode much, and played very good tennis. In this latter sport I was often the Emperor's partner and a very serious affair I had to make of each game. No conversation was allowed, and we played with all the gravity and intensity of professionals.

We had each year many visitors. In 1909 came sometimes to lunch the Emir of Bokhara, a big, handsome Oriental in a long black coat and a white turban glittering with diamonds and rubies. He seemed intensely interested in the comparative simplicity of Russian royal customs, and when he departed for his own land he distributed presents in true Arabian Nights' profusion, costly diamonds and rubies to their Majesties, and to the suite orders and decorations set with jewels. Nevertheless the souvenir of the Emir's visit to Livadia which I most prized was a photograph of himself for which he obligingly posed in the gardens. This photograph and hundreds of others which I took during the twelve years I spent with the Imperial Family I was obliged to leave behind me when I fled, a hunted refugee, across the Russian frontier. I have no hope of ever seeing any of them again.*

The 20th of October, the anniversary of the death of Alexander III, was always remembered by a solemn religious service held in the room where he died, the armchair in which he breathed his last being draped in heavy black. This death chamber was not in the main palace but in a smaller house adjoining, one which in 1909 was used as a lodging for the suite. The last part of our stay in the Crimea that year was not very gay. The Emperor left us for an official visit to the King of Italy, and on the day of his departure the Empress, greatly depressed, shut herself up in her own room refusing to see anyone, even the children. It was always to her an intolerable burden that she and the Emperor were obliged by etiquette to part from each other in public and to meet again after each absence in full view of the suite and often of the staring multitude.

This autumn was made sad also by one of the all too frequent illnesses of the unfortunate little Tsarevitch. The sufferings of the child on these occasions were so acute that everyone in the palace was rendered perfectly miserable. Nothing much could be done to assuage the poor boy's agony, and nothing except the constant love and devotion of the Empress gave him the slightest relief. We who could do nothing else for him took refuge in prayer and supplication in the little church near the palace. Mlle. Tutcheva, maid of honor to the young Grand Duchesses, read the psalms, while the Empress, the older girls, Olga and Tatiana, two of the Tsar's aides, and myself assisted in the singing. In the midst of our anxiety and distress during this illness of Alexei my father paid us a brief visit, bringing important reports to the Emperor, and this was at least a momentary bright hour in the sorrow of my existence. At Christmas time the Court returned to Tsarskoe Selo, both the Empress and the Tsarevitch by this time much improved in health.

The next time I went with their Majesties to the Crimea we found the estate transformed and greatly beautified by the substitution of a palace of white marble for the ancient and gloomy wooden buildings. The new palace was the work of the eminent architect, Krasnoff, who had also designed the palaces of the Grand Dukes Nicholas and George. In the two years Krasnoff had indeed worked marvels, not only in the palace, which was a gem of Italian Renaissance architecture, but in many smaller buildings, the whole constituting a town in itself, harmonious in material and design.

* Happily many of these photographs were later recovered and appear among illustrations of this volume.

I shall never forget the day we landed in Yalta, and the glorious drive through the bright spring sunshine to the palace. Before the carriage rode an old Tartar of the Crimea, one of the tribe I described earlier in this chapter. To ride before the Tsar's carriage was an ancient prerogative of these honest and loyal people, a prerogative which had to be resigned when carriages gave way to motor cars. No Tartar horse could have kept pace with, much less have preceded, a motor car of Nicholas II, for he always insisted on driving at a terrifying speed. But as late as 1911 he kept up the old custom of driving from Yalta to Livadia.

We drove, as I say, through the dazzling sunshine and under the fresh green trees of springtime until the white palace, set in gardens of blooming flowers and vines, burst on our delighted eyes. Russian fashion we proceeded first to the church, from whence in procession we followed the priests to the anointing and blessing of the new dwelling. The first day I spent with the Empress superintending the hanging of pictures and ikons, placing familiar and homely objects, photographs and souvenirs, so necessary to make a dwelling place out of an empty house, even though it be a royal palace. On the second floor were the private apartments of the family, including a small salon. The apartments of the Empress were furnished in light wood and pink chintzes and many vases and jars always kept full of the pink and mauve flowers she loved. From the windows of her boudoir one looked out on the wooded hills, and from the bedroom there was an enchanting view of the sparkling sea. To the right of the Empress's boudoir was the Emperor's study, furnished in green leather with a large writing table in the center of the room. On this floor also was the family dining room, the bedrooms of the Tsarevitch and of the Grand Duchesses and their attendants, a large day room for the use of the children, and a big white hall or ballroom, seldom used.

Below were the rooms of state, drawing rooms and dining rooms, all in white, the doors and windows opening on a marble courtyard draped with roses and vines which almost covered an antique Italian well in the center of the court. Here the Emperor loved to walk and smoke after luncheon, chatting with his guests or with members of the household. The whole palace, including the rooms of state, were lightly, beautifully furnished in white wood and flowered chintzes, giving the effect of a hospitable summer home rather than a palace.

That autumn was marked by a season of unusual gaiety in honor of the coming of age, at sixteen, of the Grand Duchess Olga, who received for the occasion a beautiful diamond ring and a necklace of diamonds and pearls. This gift of a necklace to the daughter of a Tsar when she became of age was traditional, but the expense of it to Alexandra Feodorovna, the mother of four daughters, was a matter of apprehension. Powerless to change the custom, even had she wished to do so, she tried to ease the burden on the treasury by a gradual accumulation of the jewels. By her request the necklaces, instead of being purchased outright when the young Grand Duchesses reached the age of sixteen, were collected stone by stone on their birthdays and name days. Thus at the coming-out ball of the Grand Duchess Olga she wore a necklace of thirty-two superb jewels which had been accumulating for her from her babyhood.

It was a very charming ball that marked the introduction to society of the oldest daughter of the Tsar. Flushed and fair in her first long gown, something pink and filmy and of course very smart, Olga was as excited over her début as any other young girl. Her hair, blonde and abundant, was worn for the first time coiled up young-lady fashion, and she bore herself as the central figure of the festivities with a modesty and a dignity which greatly pleased her parents. We danced in the great state dining room on the first floor, the glass doors to the courtyard thrown open, the music of the unseen orchestra floating in from the rose garden like a breath of its own wondrous fragrance. It was a perfect night, clear and warm, and the gowns and jewels of the women and the brilliant uniforms of the men made a striking spectacle under the blaze of the electric lights. The ball ended in a cotillion and a sumptuous supper served on small tables in the ballroom.

This was a beginning of a series of festivities which the Grand Duchess Olga and a little later on her sister Tatiana enjoyed to their utmost, for they were not in the least like the conventional idea of princesses, but simple, happy, normal young girls, loving dancing and parties and all the frivolities which make youth bright and memorable. Besides the dances given at Livadia that year, large functions attended by practically everyone in the neighborhood who had Court entrée, there were a number of very brilliant balls given in honor of Olga and Tatiana after the family returned to Tsarskoe Selo. Two of these were given by the Grand Dukes Peter and George and the girls enjoyed them so much that they begged for another before Christmas. This time it was Grand Duke Nicholas who provided a most regal entertainment, preceded by a dinner for the suite, to which I was invited. I went because the Empress wished it, but I went rather unwillingly knowing that the atmosphere was not a friendly one. Their Majesties were at that time particularly friendly with Grand Duke George and his wife who was Princess Marie of Greece, as formerly they had been with Grand Dukes Peter and Nicholas and their wives, the Montenegran princesses, Melitza and Stana, of whom more must be written later on.

In relating the events of the coming of age of Olga and Tatiana I must not forget to mention affairs of almost equal consequence which occurred in the Crimea in that season of 1911. The climate of the Crimea was ideal for tuberculosis patients, and from her earliest married life the Empress had taken the deepest interest in the many hospitals and sanatoria which nestled among the hills, some of them almost within the confines of the Imperial estate. Before the beginning of the reign of Nicholas II and Alexandra Feodorovna these hospitals existed in numbers but they were not of the best modern type. Not satisfied with these institutions the Empress out of her own private fortune built and equipped new and improved hospitals, and one of the first duties laid on me when I first visited the Crimea was to spend hours at a time visiting, inspecting and reporting on the condition of buildings, nursing and care of patients. I was particularly charged with discovering patients who were too poor to pay for the best food and nursing, and one of each summer's activities when the family visited the Crimea was a bazaar or other entertainment for the benefit of these needy ones. Four great bazaars organized and largely managed by the Empress I particularly remember. The first of these was held in 1911 and the others in 1912, 1913, and 1914. For all of these bazaars the Empress and her ladies worked very hard and from the opening day the Empress, however pre-carious the condition of her health, always presided at her own table, disposing of fine needlework, embroidery, and art objects with energy and enthusiasm. The crowds around her booth were enormous, the people pressing forward almost frenziedly to touch her hand, her sleeve, her dress, enchanted to receive their purchases from the hand of the Empress they adored, for she was adored by the real Russian people, whatever the intriguing Court and the jealous political rivals of her husband thought of her. Often the crowd at these bazaars would beg for a sight of Alexei, and smiling with pleasure the Empress would lift him to the table where the child would bow shyly but sweetly, stretching out his hands in friendly greeting to the worshipping crowds. Indeed the people loved all the Imperial Family then, whatever changes were made in the minds of the many by the horrible sufferings of the War, by propaganda, and by the mania of the Revolution. The great mass of the Russian people loved and were loyal to their Sovereigns. No one who knew them at all can ever forget that.

Perhaps they were more universally loved in the Crimea than elsewhere because of the simplicity of their lives and the close touch they were able to keep with the people of the country. We went to Livadia again in 1912, in 1913, and last of all in the spring and summer of 1914. We arrived in 1912 in the last week of Lent, I think the Saturday before Palm Sunday. Already the fruit trees were in full bloom and the air was warm with spring. Twice a day we attended service in the church, and on Thursday of Holy Week, a very solemn day in the orthodox Russian calendar, their Majesties took communion, previously turning from the altar to the congregation and bowing on all sides. After this they approached the holy images and kissed them. The Empress in her white gown and cap looked beautiful if somewhat thin and frail, and it was very sweet to see the little Alexei helping his mother from her knees after each deep reverence. On Easter eve there was a procession with candles all through the courts of the palace and on Easter Sunday for two hours the soldiers, according to old custom, gathered to exchange Easter kisses with the Emperor and to receive each an Easter egg. Children from the schools came to salute in like manner the Empress. For their Majesties it was a long and fatiguing ceremony, but they carried it through with all graciousness, while the Imperial household looked on.

Such was the intimate, the patriarchal relation between the Tsar and his people, and such was the real soul of Russia before the Revolution. I have often read, in books written by Western authors, that the Tsar and all the Imperial Family lived in hourly terror of assassination, that they knew themselves hated by their people and were righteously afraid of them. Nothing could possibly be farther from the truth. Certainly neither Nicholas II nor Alexandra Feodorovna feared their people. The constant police supervision under which they lived annoyed them unspeakably, and never were they happier than when practically unattended they moved freely among the Russian people they loved. In connection with the

Empress's care for the tuberculosis patients in the Crimea there was one day every summer known as White Flower Day, and on that day every member of society, unless she had a very good excuse, went out into the towns and sold white flowers for the benefit of the hospitals. It was a day especially delightful to the Empress and, as they grew old enough to participate in such duties, to all the young Grand Duchesses. The Empress and her daughters worked very hard on White Flower Day, spending practically the whole day driving and walking, mingling with the crowd and vending their flowers as enthusiastically as though their fortunes depended on selling them all. Of course they always did sell them all. The crowds surged around them eager and proud to buy a flower from their full baskets. But the buyers were no whit happier than the sellers, that I can say with assurance.

Of course life in the Crimea was not all simplicity and informality. There were a great many visitors, most of them of rank too exalted to be treated with informality. I remember in particular visits of Grand Duke Ernest of Hesse, brother of the Empress, and his wife, Princess Eleanor. I remember also visits of the widowed Grand Duchess Serge, who had become a nun and was now abbess of a wonderful convent in Moscow, the House of Mary and Martha. When she visited Livadia masses were said daily in the palace church. I ought not, while speaking of visitors, to omit mention of the old Prince Galitzin, a very odd person, but strongly attached to the Tsar, to whom he presented a part of his own estate, some distance to Livadia, and to which we made a special excursion on the royal yacht. Another memorable excursion was to the estates of Prince Oldenbourg on the coast of Caucasia. The sea that day was very rough and by the time we reached our destination the Empress was so prostrated that she could not go ashore. It was a pity because she missed what to all the others was a remarkable spectacle, a grand holiday of the Caucasians who, in their picturesque costumes, crowded down to the shore to greet their Sovereigns. The whole countryside was in festival, great bonfires burning in all the hills and on all the meadows wild music and the most fascinating of native dances.

Such was life in the Crimea in the old, vanished days. Simple, happy, kind, and loyal, all that was best in Russia.

CHAPTER V

THESE yearly visits to the Crimea were diversified with holiday voyages on the *Standert*, and visits to relatives and close friends in various countries. In 1910 their Majesties visited Riga and other Baltic ports where they were royally welcomed, afterwards voyaging to Finnish waters where they received as guests the King and Queen of Sweden. This was an official visit, hence attended with considerable ceremony, exchange visits of the Sovereigns from yacht to warship, state dinners and receptions. At one of these dinners I sat next the admiral of the Swedish fleet, who was much depressed because during the royal salute to the Emperor one of his sailors had accidentally been killed.

In the autumn of 1910 the Emperor and Empress went to Nauheim, hoping that the waters would have a beneficial effect on her failing health. They left on a cold and rainy day and both were in a melancholy state, partly because of separation from the beloved home, and partly because of the quite apparent weakness of the Empress. On her account the Emperor showed himself deeply disturbed. "I would do anything," he said to me, "even to going to prison, if she could only be well again." This anxiety was shared by the whole household, even by the servants who stood in line on the staircase saying their farewells, kissing the shoulder of the Emperor and the gloved hand of the Empress.

I heard almost daily from Frieberg, where the family were stopping, letters from the Emperor, the Empress, and the children, telling me of their daily life. At length came a letter from the Empress suggesting that I join my father at Hombourg, not far distant, that we might have opportunity for occasional meetings. As soon as I arrived I telephoned the château at Frieberg, and the next day a motor car was sent to fetch me. I found the Empress improved in health but looking thin and tired from the rather rigorous cure. The Emperor, in his civilian clothes, looked unfamiliar and strange, but he wore the conventional citizen's garb because he as well as the Empress wished to remain as far as possible private persons. When the health of the Empress permitted she, with Olga and Tatiana, enjoyed going unattended to Nauheim, walking unnoticed through the streets, and gazing admiringly into shop windows like ordinary tourists. Once the Emperor and the young Grand Duchesses motored over to Hombourg and for a short hour walked about quite happily

COMING IN AT YALTA, THE CRIMEA, 1911. THE EMPEROR, GRAND DUCHESSES OLGA AND ANASTASIE, MME. VIROUBOVA AND OFFICERS OF THE YACHT STANDERT. Photograph by the Empress.

THE EMPEROR, GRAND DUCHESSES OLGA AND MARIE STROLLING THROUGH HOMBURG IN 1910. THE EMPEROR OUT OF UNIFORM WAS ALMOST SAFE FROM RECOGNITION.

unobserved. Only too soon, however, the Emperor was recognized and our whole small party was obliged to flee precipitously before the gathering crowds and the ever enterprising news photographers. On some of our outings the Emperor was more fortunate. Once when we were wandering along a country road on the outskirts of Hombourg a wagon passing us dropped suddenly into the road a heavy box.

The carrier, try as he would, could not succeed in lifting the box back to its place until the Emperor went forward and, exerting all his strength, helped the man out of his difficulty. The carrier thanked his Majesty with every expression of respect and gratitude, recognizing him as a gentleman but never dreaming, of course, of his exalted station. To my expressions of amused enjoyment of the situation the Emperor said to me gravely: "I have come to believe that the higher a man's station in life the less it becomes him to assume any airs of superiority. I want my children to be brought up in this same belief."

Soon after this I returned to Russia to visit my sister, who had just borne her first baby, a little girl named for the Grand Duchess Tatiana, who acted as godmother for the child. My stay was not long, as letters from the Empress called me to Frankfort in order to be near her. On my arrival at Frankfort a surprise awaited me in the form of an invitation from the Grand Duke Ernest of Hesse to stay with his Imperial guests at his castle. At the castle gates I was welcomed by Mme. Grancy, the charming hofmistress of the Hessian Court, and by Miss Kerr, a bright and clever English girl, maid of honor to Princess Victoria. Miss Kerr took me at once to my apartments, near her own, and I quickly made myself at home. That night at dinner I sat between the Emperor and our host, the Grand Duke of Hesse. The company, which was most distinguished, included Prince Henry of Prussia, who that evening happened to be in rather a disagreeable mood, Princess Irene, Princess Victoria of Battenberg, and her beautiful daughter Princess Louise, Prince George of Greece, and the two semi-invalid sons of Prince and Princess Henry. The Empress was not present, being excused on account of her cure. Besides, it was understood that the Empress almost never appeared at state dinners.

The Grand Duke of Hesse I have always liked extremely both for his amiable disposition and for his many accomplishments. He was, and is still, an unusually gifted musician, a painter, and an artist craftsman seriously interested in the great pottery in Darmstadt, where his own designs are used. He has always been a man of liberal social ideals and his popularity among the people of Hesse not even the German Revolution has been powerful enough to overthrow. His wife, Princess Eleanor, when I knew her, was dignified and gracious and gifted with a genuine talent for dress. Prince Henry of Prussia, brother of the Kaiser and brother-in-law of the Empress, was a tall and handsome man, but inclined to be—let us say—temperamental. At times he was overbearing and very satirical, and at others friendly and charming. His wife was a small woman, simple in manner and of a kindly, unselfish nature. Princess Alice, daughter of Princess Victoria of Battenberg and wife of Prince Andrew of Greece, was a beautiful woman but unhappily quite deaf.

The Castle of Frieberg, which stands on a high hill overlooking a low valley and the little red-roofed town of Nauheim, is an ancient structure not particularly attractive either inside or out. There was nothing much for Grand Duke Ernest's guests to do in the way of amusement except to walk and drive. Of the Empress I saw rather less than we had planned, but sometimes late in the evenings the Emperor, the Empress, and myself met for Russian tea and for familiar talks before bedtime.

In October or November their Majesties returned to Tsarskoe Selo, the Empress greatly benefited by her cure. How happy we were to be once more at home, the Empress in her charming boudoir hung with mauve silk and fragrant with fresh roses and lilacs, I in my own little house which I dearly loved even though the floors were so cold. The opal-hued boudoir of the Empress, where we spent a great deal of our time, was a lovely, quiet place, so quiet that the footsteps of the children and the sound of their pianos in the rooms above were often quite audible. The Empress usually lay on a low couch over which hung her favorite picture, a large painting of the Holy Virgin asleep and surrounded by angels. Beside her couch stood a table, books on the lower shelf, and on the upper a confusion of family photographs, letters, telegrams, and papers. It was undeniably a weakness of the Empress that she was not in the least systematic about her correspondence. Intimate letters, it is true, she answered promptly, but others she often left for weeks un-touched. About once a month Madeleine, the principal maid of the Empress, would invade the boudoir and implore her mistress to clear up this heap of neglected correspondence. The Empress usually began by begging to be left alone, but in the end she always gave in to the importunities of the invaluable Madeleine. The Empress of course had a private secretary, Count Rostovseff, but it was one of her peculiarities that she preferred to handle her letters and telegrams before her secretary, and he seemed to accustom himself with ease to her dilatory ways.

It would be difficult to imagine two people more widely different on points of this kind than Nicholas II and Alexandra Feodorovna. Their private apartments were very close together, the Emperor's study, billiard and sitting room and his dressing room with a fine swimming bath, almost adjoining the apartments of the Empress. The big antechamber to the study, well furnished with chairs and tables and many books and magazines, looked out on a court, and here people who had business with the Emperor waited until they were summoned to his private room. The study was a perfect model of orderliness, the big writing table having every pen and pencil exactly in its place. The large calendar also with appointments written carefully in the Emperor's own hand was always precisely in its proper place. The Emperor often said that he wanted to be able to go into his study in the dark and put his hand at once on any object he knew to be there. The Emperor was equally particular about the appointments of his other rooms. The dressing table in the white-tiled bathroom, separated from the sitting room by a corridor and a small staircase, was as much a model of neatness as the study table, nor could the Emperor have tolerated valets who would not have kept his rooms in a condition of perpetual good order. Of course the ample *garderobes,* where the gowns, wraps, hats, and jewels of the Empress and the innumerable uniforms of the Emperor were kept, were always in order because they were in the care of experienced servants and were rarely if ever visited by others than their responsible guardians.

The Emperor's combined billiard and sitting room was not very much used because the Emperor spent most of his leisure hours in his wife's boudoir. But it was in the billiard room that the Emperor kept his many albums of photographs, records of his reign. These albums bound in green with the Imperial monogram, contained photographs taken over a period of twenty years. The Empress had her own albums full of equally priceless records, priceless from the historian's standpoint at any rate, and each of the children had their own. There was an expert photographer attached to the household whose only duty was to develop and print these photographs, which were, in almost every case, mounted by the royal photographer's own hand. This work used to be done, as a rule, on rainy days, either in the palace or on board the *Standert.* The Emperor, as usual, was neater about this work of pasting photographic prints than any other member of the household. He could not endure the sight of the least drop of glue on a table. As might be expected of so orderly a person the Emperor was slow about almost everything he did. When the Empress wrote a letter she did it very quickly, holding her portfolio on her knees on her *chaise longue.* When the Emperor wrote a letter it was a matter of hours before it was completed. I remember once at Livadia the Emperor retiring to his study at two o'clock to write an important letter to his mother. At five, the Empress afterwards told me, the letter remained unfinished.

The private life of the Imperial Family in these years before the War was quiet and uneventful. The Empress never left her room before noon, it being her custom, since her illness, to read and write propped up on pillows on her bed. Luncheon was at one o'clock, the Emperor, his aide-de-camp for the day, the children, and an occasional guest attending. After luncheon the Emperor went at once to his study to work or to receive visitors. Before tea time he usually went for a brisk walk in the open.

At half past two I came to the Empress, and if the weather was fine and she well enough, we went for a drive or a walk. Otherwise we read or worked until five, when the family tea was served. Tea was a meal in which there was never the slightest variation. Always appeared the same little white-draped table with its silver service, the glasses in their silver standards, and for the rest simply plates of hot bread and butter and a few English biscuits. Never anything new, never any surprises in the way of cakes or sweetmeats. The only difference in the Imperial tea table came in Lent, when butter and even bread made with butter disappeared, and a small dish or two of nuts was substituted. The Empress often used gently to complain, saying that other people had much more interesting teas, but she who was supposed to have almost unlimited power, was in reality quite unable to change a single deadly detail of the routine of the Russian Court, where things had been going on almost exactly the same for generations. The same arrangement of furniture in the state rooms, the same braziers of incense carried by footmen in the long corridors, the same house messengers in archaic costumes of red and gold with ostrich-feathered caps, and for all I know the same plates for hot bread and butter on the same tea table, were traditions going back to Catherine the Great, or Peter, or farther still perhaps.

Every day at the same moment the door opened and the Emperor came in, sat down at the tea table, buttered a piece of bread, and began to sip his tea. He drank every day two glasses, never more, never less, and as he drank he glanced over his telegrams and newspapers. The children were the only ones who found tea time at all exciting. They were dressed for it in fresh white frocks and colored sashes, and spent most of the hour playing on the floor with toys kept especially for them in a corner of the boudoir. As they grew older needlework and embroidery were substituted for the toys, the Empress disliking to see her daughters sitting with idle hands.

From six to eight the Emperor was busy with his ministers, and he usually came directly from his study to the eight o'clock family dinner. This was never a ceremonial meal, the guests, if any, being relatives or intimate friends. At nine the Empress, in the rich dinner gown and jewels she always wore, even on the most informal evenings, went to the bedroom of the Tsarevitch to hear him say his prayers and to tuck him into bed for the night. The Emperor worked until eleven, and until that hour the Empress, the two older Grand Duchesses, and I read, had a little music, or otherwise passed the time. Perhaps it is worth recording that bridge, or in fact any other card games, we never played. Nobody in the family cared at all for cards, and only a little, once in a while, for dominoes. At eleven the evening tea was served, and after that we separated, the Emperor to write his diary for the day, the Empress and the children to bed and I for home. All his life the Emperor kept a daily record of events, but like all the private papers of the Imperial Family, the diaries were seized by the Revolutionary leaders and probably (although I still hope to the contrary) destroyed. The diaries of Nicholas II, apart from any possible sentimental associations, should be possessed of great historical value.

Monotonous though it may have been, the private life of the Emperor and his family was one of cloudless happiness. Never, in all the twelve years of my association with them, did I hear an impatient word or surprise an angry look between the Emperor and the Empress. To him she was always "Sunny" or "Sweetheart," and he came into her quiet room, with its mauve hangings and its fragrant flowers, as into a haven of rest and peace. Politics and cares of state were left outside. Never were we allowed to speak of them. The Empress, on her part, kept her own troubles to herself. Never did she yield to the temptation to confide in him her perplexities, the foolish and spiteful intrigues of her ladies in waiting, nor even lesser troubles concerning the education and upbringing of the children. "He has the whole nation to think about," she often said to me. The only care she brought to the Emperor was the ever precarious health of Alexei, but this the whole family constantly felt, and it had to be spoken of very often. The Imperial Family was absolutely united in love and sympathy. I like to remember of the children, who adored their parents, that they never felt the slightest resentment of their mother's attachment for me. Sometimes I think the little Grand Duchess Marie, who especially worshipped her father, felt a little jealous when he invited me, as he often did, to accompany him on walks in the palace gardens. This may be imagination, and at all events the child's slight jealousy never interfered with our friendship.

I think the Emperor liked to walk with me because he had need to talk to someone he trusted of purely personal cares which troubled his mind and which he could share with few. Some of these cares were of old origin, but had never been forgotten. I remember once he began to tell me, almost without any preface, of the dreadful disaster which attended his coronation, a panic, induced by bad management of the police, in the course of which scores of people were crushed to death. At the very hour of this fatal accident the coronation banquet took place, and the Emperor and Empress, despite their grief and horror, were obliged to take part in it exactly as though nothing had happened. The Emperor told me with what difficulty they had concealed their emotions, often having to hold their serviettes to their faces to hide their streaming tears.

One of the happiest memories of my life at Tsarskoe Selo were the evenings when the Emperor, all cares past and present forgot, sat with us in the Empress's boudoir reading aloud from the work of Tolstoy, Tourgenieff, or his favorite Gogol. The Emperor read extremely well, with a pleasant voice and a remarkably clear enunciation. In the years of the Great War, so full of anguish and apprehension, the Emperor found relief in reading aloud amusing stories of Averchenko and Teffy, Russian humorists who perhaps have not yet been translated into foreign tongues.

Before the War the Emperor was pictured far and wide as a cruel tyrant deliberately opposed to the interests of his people, while the Empress appeared as a cold, proud woman, a *malade imaginaire*, wholly indifferent to the public good. Both of these pictures are cruelly misrepresentative. Nicholas II and his wife were human beings, with human faults and failings like the rest of us. Both had quick tempers, not invariably under perfect control. With the Empress temper was a matter of rapid explosion and equally sudden recovery. She was often for the moment furiously angry with her maids whom too often she discovered in insincerities and deceit. The Emperor's anger was slower to arouse and much slower to pass. Ordinarily he was the kindest and simplest of men, not in the least proud or over-conscious of his exalted position. His self-control was so great that to those who knew him little he often appeared absent-minded and indifferent. The fact is he was so reserved that he seemed to fear any kind of self-revealment. His mind was singularly acute, and he should have used it more accurately to gauge the characters of persons surrounding him. It was entirely within his mental powers to sense the atmosphere of gossip and calumny that surrounded the Court during the last years, and certainly it was within his power to put a stop to idle and malicious talk. But it was rarely possible to arouse him to its importance. "What high-minded person would believe such nonsense?" was his usual comment. Alas! he little realized how few were the really high-minded people who, in the last years of the Empire, surrounded his person or that of the Empress.

Sometimes the Emperor found himself obliged to take cognizance of the malicious gossip which made the Empress desperately unhappy and in the end poisoned the minds of thousands of really well-meaning and loyal Russians. Beginning as far back as 1909 the tide of treachery had begun to rise, and one of the earliest of those responsible for the final disaster, I regret to say, was a woman of the highest aristocracy, one long trusted and affectionately regarded by the Imperial Family. Mlle. Sophie Tutcheff, a protégée of the Grand Duchess Serge, and a lady who was a general over-governess to the children, was perhaps the first of all the intriguing courtiers of whom I have positive knowledge. Mlle. Tutcheff belonged to one of the oldest and most powerful families in Moscow, and she was strongly under the influence of certain bigoted priests, especially that of her cousin, Bishop Vladimir Putiata, who for ten years had lived in Rome as official representative of the Russian Church. It was he, I firmly believe, who inspired in Mlle. Tutcheff her antipathy to the Empress and her evil reports concerning the life of the Imperial Family. Mlle. Tutcheff, either of her own accord or encouraged by her relative, was continually opposed to what she called the English upbringing of the Imperial children. She wished to change the whole system, make it entirely Slav and free from any imported ideas.

Mlle. Tutcheff was, I believe, the first person to create what afterwards became the international Rasputine scandal. At the time of her residence in the palace at Tsarskoe Selo Rasputine's influence had scarcely been felt at all by the Emperor or Empress, although he was an intimate friend of other members of the Romanoff family. But Mlle. Tutcheff spread abroad a series of the most amazing falsehoods in which Rasputine figured as a constant visitor and virtually the spiritual guardian of the Imperial Family. I do not wish to repeat these stories, but merely to give an idea of their preposterous nature I will say that she represented Rasputine as having the freedom of the nurseries and even the bedchambers of the young Grand Duchesses. According to tales purported to have their origin with her, Rasputine was in the habit of bathing the children and afterward talking with them, sitting on their beds.

I do not think the Emperor believed all these rumors, but he did believe that Mlle. Tutcheff was guilty of malicious gossip of his family, and he therefore summoned her to his study and rebuked her severely, asking her how she dared to spread idle and untrue stories about his children. Of course she denied having done anything of the sort, but she admitted that she had spoken ill of Rasputine. "But you do not know the man," protested the Emperor, "and in any case, if you had criticisms to make of anyone known to this household you should have made them to us and not to the public." Mlle. Tutcheff admitted that she did not know Rasputine, and when the Emperor suggested that before she spoke evil of him it might be well for her to meet him she haughtily replied: "Never will I meet him."

For a short time after this Mlle. Tutcheff remained at Court, but being a rather stupid and very obstinate woman, she continued her campaign of intrigue. She managed to influence Princess Oblensky, long a favorite lady in waiting, until she entirely estranged her from the Empress. She even began to speak to the children against their own mother, until the Empress, who felt herself powerless against the woman, actually refused to visit the nurseries, and when she wanted her children near her sent for them to come to her private apartments. Too well she knew the Emperor's extreme reluctance to dismiss any person connected with the Court, and she waited in silent pain until the scandal grew to such proportions that the Emperor could no longer ignore it. Then Mlle. Tutcheff was summarily dismissed and sent back to her home in Moscow.

So powerful was the influence of the Tutcheff family that this incident was magnified beyond all proper proportions, and the former over-governess of the Imperial children was represented as a poor victim of Rasputine, a man whom she had never seen and who probably never knew of her existence. The last I ever heard of Mlle. Tutcheff, who, by the way, was a niece of the esteemed poet Tutcheff, she was living in Moscow, under the special protection of the Bolshevik Government. Her cousin, the former Bishop Vladimir Putiata, I understand has for several years been a great favorite of those Communists who have prosecuted such brave and fearless opponents of church despoilment as the unhappy Patriarch Tikhon and others.

Of the Emperor I think it ought to be said that his education, under his governor, General Bogdanovitch, was calculated to weaken the will of any boy and to encourage in Nicholas II his natural reserve and what might be called indolence of mind. But this I know of him that after his marriage he became much more resolute of temper and much more gentle of manner than other members of his family. It is certain that he loved Russia and the Russian people with his whole soul, and yet, under the political system for centuries in force, he had often to leave to people whom he knew only superficially many important details of government. Unquestionably it was a fault of the Emperor that he was over-confident, and only too ready to believe what was told him by people whom he personally liked. He was impulsive in most of his acts and sometimes made important nominations on the impression of a moment. It goes without saying that many of his officials took advantage of this overconfidence and sometimes acted in his name without his knowledge or authority.

Only too well for her own happiness and peace of mind did the Empress Alexandra Feodorovna understand her husband. She knew his kind heart, his love for his country and his people, but she knew also how easily influenced he could be by men in whom he reposed confidence. She knew that too often his acts were governed by the last person he happened to consult. But for all this I wish to say that the Emperor never appeared to his friends as a weak man. He had qualities of leadership with very limited opportunities to exercise those qualities. In his own domain he was "every inch an Emperor." The whole Court, from the Grand Dukes down to the last petty official and intriguing maid of honor, recognized this and stood in real awe of their Sovereign. I have a keen recollection of an episode at dinner in which a certain young Grand Duke ventured to utter an ill-founded grievance against a distinguished general who had dared to rebuke his Highness in public. The Emperor instantly recognized this as a mere display of temper and egoism, and his contempt and indignation knew no bounds. He literally turned white with anger, and the unfortunate young Grand Duke trembled before him like an offending servant. Afterwards the still indignant Emperor said to me: "He may thank God that the Empress and you were present. Otherwise I could not have held myself in hand." Towards the end of the Russian tragedy in 1917 the Emperor had learned to hold himself almost too well in hand, to subdue and to conceal the commanding personality of which he was naturally possessed. It would have been far better if he had used his personality and his great charm of manner to offset the tide of intrigue and revolution which in the midst of a world war overcame the Empire.

As long as I knew him, whether in the privacy of the palace at Tsarskoe Selo, in the informal life of the Crimea, on the Imperial yacht, in public or in private, I was always conscious of the strong personality of the Emperor. Everybody felt it. I can instance one occasion at a great reception of the Tauride Zemstvo when two men present were deliberately resolved to behave in a disrespectful manner to the Emperor. But the moment he entered the room these men found themselves completely overpowered. Their manner changed and they showed in every subsequent word and action their shame and regret. At one time a group of Social Revolutionaries were able to put on a cruiser which the Emperor was to visit a sailor charged with his Sovereign's assassination. But when the opportunity came the man literally could not do the deed. For his "weakness" this poor wretch was afterwards murdered by members of his party.

The character of the Empress was quite different from that of her husband. She was less lovable to the many, and yet of a stronger fiber. Where he was impulsive she was usually cautious and thoughtful. Where he was over-optimistic she was inclined to be a bit suspicious, especially of the weak and self-indulgent aristocracy. It was generally believed that the Empress was difficult to approach, but this was never true of sincere and disinterested souls. Suffering always made a strong appeal to the Empress, and whenever she knew of anyone sad or in trouble her heart was instantly touched. Few people, even in Russia, ever knew how much the Empress did for the poor, the sick, and the helpless. She was a born nurse, and from her earliest accession took an interest in hospitals and in nursing quite foreign to native Russian ideas. She not only visited the sick herself, in hospitals, in homes, but she enormously increased the efficiency of the hospital system in Russia. Out of her own private funds the Empress founded and supported two excellent schools for training nurses, especially in the care of children. These schools were founded on the best English models, and were under the general supervision of the famous Dr. Rauchfuss and of head nurse Miss Puchkine, a near relative of the great poet Puchkine. I could enlarge at length on the many constructive philanthropies of the Empress, paid for by herself, hospitals, homes, and orphanages, planned in almost every detail by herself, and constantly visited and inspected. After the Japanese War she built a *Hôtel des Invalides*, in which hundreds of disabled men were taught trades. She also built a number of cottages with gardens for wounded soldiers and their families, most of these war philanthropies being under the supervision of a trusted friend, Colonel the Count Shoulenbourg of the Empress's favorite Lancers.

The Empress possessed a heart and a mind utterly incapable of dishonesty or deceit, consequently she could never tolerate either in other people. This naturally got her heartily disliked by people of society to whom deceit was a matter of long practice. Another quality condemned in the Empress because entirely misunderstood, was her care as to expenses. Brought up in the comparative poverty of a small German Court, the Empress never lost the habit of a cautious use of money. Quite as in private families, where economy is an absolute necessity, the clothing of the young Grand Duchesses when outgrown by the elders were handed down to the younger girls. In the matter of selecting gifts for guests, for relatives, or at holidays for the suite, the Emperor simply selected from the rich assortment sent to the palace objects which best pleased him. The Empress, on the other hand, always examined the price cards and considered before choosing whether the jewel or the fur or the bijou, whatever it was, was worth what was asked for it. The difference between the Emperor and the Empress in regard to money was a difference in experience. The Emperor, all his life, had had everything he wanted without ever paying a single ruble for anything. He never had any money, never needed any money. I can recall but one solitary instance in which the Tsar of all the Russias ever even felt the need of touching a kopeck of his illimitable riches. It was in 1911 when their Majesties began to attend services at the Feodorovsky Cathedral at Tsarskoe Selo. In this church it was the custom to pass through the congregations alms basins into which everyone, of course, dropped a contribution, large or small. The Emperor alone was entirely penniless, and embarrassed by his unique situation he made a representation to the proper authorities, after which at exact monthly intervals he was furnished with four gold pieces for the alms basin of the Feodorovsky Cathedral. If he happened to attend an extra service he had to borrow his contribution from the Empress.

But if the Emperor carried no money in his pockets it was well enough known that he commanded vast sums, and it was characteristic of the sycophants who surrounded him that he was constantly importuned for "loans," for money to help out gambling or otherwise impecunious officers who, aware of the Emperor's great love for the army, played on it to their advantage. One day when the Emperor was taking his usual brisk walk through the grounds before tea a young officer who had managed to conceal himself in the shrubbery sprang out, threw himself on his knees, and threatened to kill himself on the spot unless the Emperor granted him a sum of money to clear the desperate wretch of some reckless deed. The Emperor was frightfully enraged—but he sent the man the money demanded.

The Empress had always handled money and knew quite well how to spend it wisely. From the depths of her honest soul she despised the use of money to buy loyalty and devotion. For a long time after my first formal service as maid of honor, with the usual salary, I received from her Majesty literally nothing at all. From my parents I had the income from my dowry, four hundred rubles a month, a sum entirely inadequate to pay the running expenses of my small establishment with its three absolutely indispensable servants, and at the same time to dress myself properly as a member of the Court circle. The Empress's brother, Grand Duke Ernest of Hesse, was one of the first of her intimates to point out to her the difficulties of my position, and to suggest to her that I be given a position at Court. The suggestion was not welcomed by Alexandra Feodorovna. "Is it not possible for the Empress of Russia to have one friend?" she cried bitterly, and she reminded her brother that her relation and mine were not without precedent in Russia. The Empress Dowager had a friend, Princess Oblensky; also the Empress Marie Alexandrovna, wife of Alexander III, had in Mme. Malzoff an intimate associate, and neither of these women had had any Court functions. Why should she not cherish a friendship free from all material considerations? However, after her brother and also Count Fredericks, Minister of the Court, had pointed out to her that it was scarcely proper that the Empress's best friend and confidante should wear made-over gowns and go home from the palace on foot at midnight because she had no money for cabs, the Empress began to relent a little. At first her change of attitude took the form of useful gifts bestowed at Christmas and Easter, dress patterns, furs, gloves, and the like. Finally one day she asked me to discuss with her the whole subject of my expenses. Making me sit down with pencil and paper, she commanded me to set forth a complete budget of my monthly expenditures, exactly what I paid for food, service, light, fire, and clothing. The domestic budget, apart from my small income, came to two hundred and seventy rubles a month, and at the orders of the Empress I was thereafter furnished monthly with the exact sum of two hundred and seventy rubles. It never occurred to her to name the amount in round numbers of three hundred rubles. Nor did it occur to me except as a matter of faint amusement. Of course I was often embarrassed for money even after I became possessed of this regular income, and even later when it was augmented by two thousand rubles a year for rent, and it often wrung my heart to have to say no to appeals for money. I knew that I appeared selfish and hard-hearted. The truth was that I was simply impecunious.

MLLE SOFIA TUTCHEFF

CHAPTER VI

THE year 1912, although destined to end in the almost fatal illness of the Tsarevitch, began happily for the Imperial Family. Peaceful and busy were the winter and spring, the Emperor engaged as usual with the affairs of the Empire, the Empress, as far as her health permitted, superintending the education of her children, and all of them busy with their books and their various tutors. Of the education and upbringing of the children of Nicholas II and the Empress Alexandra Feodorovna it should be said that while nothing was omitted to make them most loyal Russians, the educational methods employed were cosmopolitan. They had French, Swiss, and English tutors, but all their studies were under the superintendence of a Russian, the highly cultured M. Petroff, while for certain branches such as physics and natural science they were privately instructed in the gymnasium of Tsarskoe Selo. The first teacher of the Imperial children, she from whom they received their elementary education, was Miss Schneider, familiarly called "Trina," a native of one of the Baltic states of the Empire. Miss Schneider first came into service, years before the marriage of the Emperor and Empress, as instructor in the Russian language to Elizabeth, Grand Duchess Serge. Afterwards she taught Russian to the young Empress, and was retained at Court as reader to her Majesty. "Trina" was rather a difficult person in some ways, taking every advantage of her privileged position, but she was undeniably valuable and was heart and soul in her devotion to the family. She accompanied them to Siberia and there disappeared with them.

Perhaps the most valued of the instructors was M. Pierre Gilliard, whose book "Thirteen Years at the Russian Court" has been published in several languages and has been very well received. M. Gilliard, a Swiss gentlemen of many accomplishments, came first to Tsarskoe Selo as teacher of French to the young Grand Duchesses. Afterwards he became tutor to the Tsarevitch. M. Gilliard lived in the palace, and enjoyed to the fullest extent the confidence and affection of their Majesties. Mr. Gibbs, the English tutor, was also a great favorite. Both of these men followed the family into exile and remained faithful and devoted friends until forcibly expelled by the Bolsheviki.

In his book M. Gilliard has recorded that he was never able to teach the Grand Duchesses to speak a fluent French. This is true because the languages used in the family were English and Russian, and the children never became interested in any other languages. "Trina" was supposed to teach them German but she had less success with that language than M. Gilliard with French. The Emperor and Empress spoke English almost exclusively, and so did the Empress's brother, the Grand Duke of Hesse and his family. Among themselves the children usually spoke Russian. The Tsarevitch alone, thanks to his constant association with M. Gilliard, mastered the French language.

Every detail of the education of her children was supervised by the Empress, who often sat with them for hours together in the schoolroom. She herself taught them sewing and needlework, her best pupil being Tatiana, who had an extraordinary talent for all kinds of handwork. She not only made beautiful blouses and other garments, embroideries and crochets, but she was able on occasions to arrange her mother's long hair, and to dress her as well as a professional maid. Not that the Empress required as much dressing as the ordinary woman of rank and wealth. She had that kind of Victorian modesty that forbade any intrusion on the privacy of her dressing room. All that her maids were allowed to do was to dress her hair, fasten her boots, and put on her gown and jewels. The Empress had great taste in dress and always chose her jewels to finish rather than to ornament her costumes. "Only rubies to-day," she would command, or "pearls and sapphires with this gown."

The Empress and the children have been represented as surrounded by German servants, but this accusation is absolutely false. The chief woman of the household was Mme. Geringer, a Russian lady who came daily to the palace, ordered gowns, did all necessary shopping, paid bills, and attended to any business required by the Empress. The chief maid of the Empress was Madeleine Zanotti, of English and Italian parentage, whose home before she came to Tsarskoe Selo was in England. Madeleine was a woman of middle age, very clever, and as usual with one in her position, inclined to be tyrannical. Madeleine had charge of all the gowns and jewels of the Empress, and as I think I have related, she was often critical of her mistress's indolent habits in regard to correspondence, etc. A second maid was Tutelberg, "Toodles," a rather slow and quiet girl from the Baltic. She and Madeleine were mortal enemies, but they agreed on one thing at least, and that was that they would not wear caps and aprons. The Empress good-naturedly acquiesced and permitted simple black gowns and ribbon bows in the hair for her chief maids. There were three under maids, all Russians, and all perfectly devoted to the Imperial Family. These girls, who wore the regulation caps and white aprons, cared for the rooms of the Empress and the children. All the maids, when the Revolution came, remained faithful to the family, and one of them, as I shall tell later, performed the dangerous service of smuggling letters in and out of Siberia. One girl, Anna Demidoff, shared the fate of the family in 1918.

The Emperor had three valets, one of whom, Shalferoff, who had served Alexander III, turned spy during the Revolution. Another, old Raziesh, also a former servant of Alexander III, died in the service of Nicholas II, and was replaced by Chemoduroff, a fine and very loyal man. The third valet's name was Katoff. All three, as their names testify, were Russians, as were also the three men in the service of the Empress, Leo and Kondratief, both of whom died during the early days of the Revolution, and Volkoff, who followed the Royal exiles as far into Siberia as he was permitted by the Provisional Government.

The children's nurses were Russians, the head nurse being Marie Vechniakoff. Others I remember well were Alexandra, nicknamed "Shoura," a great favorite with the girls, Anna and Lisa, kind, faithful girls who spoke no word of any language except Russian. There were, of course, hundreds of house servants, and to my knowledge most, if not all of them, were Russians. The chef was a Frenchman, Cubat, a very great man in his profession. Sometimes, when an especially splendid dish had been prepared, Cubat was wont to introduce it, as it were, by standing magnificently in the doorway, clad in immaculate white linen, until the dish was served. Cubat became very wealthy in the Tsar's service, and now lives happily and luxuriously in his native France. He was, I believe, truly loyal to the Imperial Family, which is more than can be said for most of the servants. Their children were educated at the expense of the Emperor, and the majority, instead of choosing useful trades, elected to go to the universities, where they nearly all became Revolutionists.

In my father's opinion this was due to the fact that the Russian universities and higher schools offered little if any technical training. Recognizing this, the Empress created in Petrograd a technical school for boys and girls of the whole Empire. In this school the students were trained to become teachers in many useful handicrafts, and in addition to this normal academy the Empress established in many governments schools where boys and girls were perfected in the beautiful peasant arts of embroidery, dyeing, carving, and painting. I give these details because I think it only just to offset with facts the lying slanders of sensational writers who could not possibly have known anything of the intimate life of the Imperial Family of Russia but who have substituted propaganda for truth.

None of these sensational writers knew or tried to know how simple, not to say rigorous, was the régime followed by the Imperial children. All of them, even the delicate little Tsarevitch, slept in large, well-aired nurseries, on hard camp beds without pillows and with the least possible allowance of bedclothing. They had cold baths every morning and warm ones only at night. As a consequence of this simple life their manners were unassuming and natural without a single trace of *hauteur*. Although in 1912 the four girls were rapidly approaching womanhood—Olga was in her eighteenth year and Tatiana was nearly sixteen—their parents continued to regard them as children. The two older girls were spoken of as "the big ones," and were given many grown-up privileges, as for example, concerts and the theater to which the Emperor himself escorted them. The two younger Grand Duchesses and the Tsarevitch, "the little ones," were still in the nursery.

In the darkness of the mystery which surrounds the fate of these innocent children it is with poignant emotion that I recall them as they appeared, so full of life and joy, in those distant, yet incredibly near, days before the World War and the downfall of Imperial Russia. Of the four girls, Olga and Marie were essentially Russian, altogether Romanoff in their inheritance. Olga was perhaps the cleverest of them all, her mind being so quick to grasp ideas, so absorbent of knowledge that she learned almost without application or close study. Her chief characteristics, I should say, were a strong will and a singularly straightforward habit of thought and action. Admirable qualities in a woman, these same characteristics are often trying in childhood, and Olga as a little girl sometimes showed herself wilful and even disobedient. She had a hot temper which, however, she early learned to keep under control, and had she been allowed to live her natural life she would, I believe, have become a woman of influence and distinction. Extremely pretty, with brilliant blue eyes and a lovely complexion, Olga resembled her father in the fineness of her features, especially in her delicate, slightly tipped nose.

Marie and Anastasie were also blonde types and very attractive girls. Marie had splendid eyes and rose-red cheeks. She was inclined to be stout and she had rather thick lips which detracted a little from her beauty. Marie had a naturally sweet disposition and a very good mind. All three of these girls were more or less of the tomboy type. They had something of the innate brusqueness of their Romanoff ancestors, which displayed itself in a tendency to mischief. Anastasie, a sharp and clever child, was a very monkey for jokes, some of them at times almost too practical for the enjoyment of others. I remember once when the family was in their Polish estate in winter the children were amusing themselves at snowballing. The imp which sometimes seemed to possess Anastasie led her to throw a stone rolled in a snowball straight at her dearly loved sister Tatiana. The missile struck the poor girl fairly in the face with such force that she fell senseless to the ground. The grief and horror of Anastasie lasted for many days and permanently cured her of her worst propensities to practical jokes.

THE EMPEROR AND TSAREVITCH WALKING ON BOARD THE *STANDERT*. Photograph by the Empress.

MEMORIES OF THE RUSSIAN COURT

THE TSAR, TSARINA, AND ALEXEI IN THE GARDENS OF TSARSKOE SELO, 1911.

THE TSAREVITCH IN COSSACK UNIFORM WITH HIS SAILOR, DEREVANKO

Tatiana was almost a perfect reincarnation of her mother. Taller and slenderer than her sisters, she had the soft, refined features and the gentle, reserved manners of her English ancestry. Kindly and sympathetic of disposition, she displayed towards her younger sisters and her brother such a protecting spirit that they, in fun, nicknamed her "the governess." Of all the Grand Duchesses Tatiana was with the people the most popular, and I suspect in their hearts she was the most dearly loved of her parents. Certainly she was a different type from the others even in appearance, her hair being a rich brown and her eyes so darkly gray that in the evening they seemed quite black. Of all the girls Tatiana was most social in her tastes. She liked society and she longed pathetically for friends. But friends for these high born but unfortunate girls were very difficult to find. The Empress dreaded for her daughters the companionship of oversophisticated young women of the aristocracy, whose minds, even in the schoolroom, were fed with the foolish and often vicious gossip of a decadent society. The Empress even discouraged association with cousins and near relatives, many of whom were unwholesomely precocious in their outlook on life.

I would not give the impression that these young daughters of the Emperor and Empress were forced to lead dull and uneventful lives. They were allowed to have their little preferences for this or that handsome young officer with whom they danced, played tennis, walked, or rode. These innocent young romances were in fact a source of amusement to their Majesties, who enjoyed teasing the girls about any dashing officer who seemed to attract them. The Grand Duchess Olga, sister of the Emperor, sympathized with her nieces' love of pleasure and often arranged tea parties and tennis matches for them, the guests, of course, being of their own choice. We had some quite jolly tea parties in my little house also. In the matter of dress, so important to young and pretty girls, the Grand Duchesses were allowed to indulge their own tastes. Mme. Brisac, an accomplished French dressmaker, made gowns for the Imperial Family, and through her the latest Paris models reached the palace. The girls, however, inclined towards simple English fashions, especially for outdoor wear. In summer they dressed almost entirely in white. Jewels they were too young to wear except on very great occasions. Each girl received on her twelfth birthday a slender gold bracelet which was afterwards always worn, day and night, "for good luck." I have described in a previous chapter the Russian custom of presenting each Grand Duchess, on her coming of age, with a pearl and diamond necklace, but this was worn only at state functions or very formal balls.

Alexei, the only son of the Emperor and Empress, a more tragic child than the last Dauphin of France, indeed one of the most tragic figures in history, was, apart from his terrible affliction, the loveliest and most attractive of the whole family. Because of his delicate health Alexei began life as a rather spoiled child. His chief nurse, Marie Vechniakoff, a somewhat over-emotional woman, made the mistake of indulging the child in every whim. It is easy to understand why she did so, because nothing more heart-rending could be imagined than the little boy's moans and cries during his frequent illnesses. If he bumped his head or struck a hand or foot against a chair or table the usual result was a hideous blue swelling indicating a subcutaneous hemorrhage frightfully painful and often enduring for days or even weeks.

At five Alexei was placed in charge of the sailor Derevanko, who for a long term of years remained his constant body servant and companion. Derevanko, while devoted to the boy, did not spoil him as his women nurses had done, and the man was so patient and resourceful that he often did wonders in alleviating the child's pain. I can still in memory hear the plaintive, suffering voice of Alexei begging the big sailor to "lift my arm," "put my leg up," "warm my hands," and I can see the patient, calm-eyed man working for hours on end to give the maximum of comfort to the little pain-racked limbs.

As Alexei grew older his parents carefully explained to him the nature of his illness and impressed on him the necessity of avoiding falls and blows. But Alexei was a child of active mind, loving sports and outdoor play, and it was almost impossible for him to avoid the very things that brought him suffering. "Can't I have a bicycle?" he would beg his mother. "Alexei, you know you can't." "Mayn't I play tennis?" "Dear, you know you mustn't." Often these hard denials of the natural play impulse were followed by a gush of tears as the child cried out: "Why can other boys have everything and I nothing?"

Suffering and self-denial had their effect on the character of Alexei. Knowing what pain and sacrifice meant, he was extraordinarily sympathetic towards other sick people. His thoughtfulness of others was shown in his beautiful courtesy to women and girls and to his elders, and in his interest in the troubles of servants and dependents. It was a failing of the Emperor that even when he sympathized with the troubles of others he was rather slow to take action, unless indeed the matter was really serious. Alexei, on the contrary, was always for immediate action. I remember an instance when a boy in service at the palace was discharged for some reason which I have quite forgotten. The story somehow reached the ears of Alexei, who immediately took sides with the boy and gave his father no rest until the whole case was reviewed and the culprit was forgiven and restored to duty. Alexei usually defended all offenders, yet when the day came when his parents, in deep distress, told him that Father Gregory, that is, Rasputine, had been killed by members of his own family the boy's grief was swallowed up in rage and indignation. "Papa," he exclaimed, "is it possible that you will not punish them? The assassins of Stolypine were hanged."

I ask the reader to remember that the Imperial Family firmly believed that they owed much of Alexei's improving health to the prayers of Rasputine. Alexei himself believed it. Several years before Rasputine had assured the Empress that when the boy was twelve years old he would begin to improve and that by the time he was a man he would be entirely well. The undeniable fact is that after the age of twelve Alexei did begin very materially to improve. His illnesses became farther and farther apart and before 1917 his appearance had changed marvelously for the better. He resembled in no way the invalid sons of his mother's sister, Princess Henry of Prussia, who suffered from his own terrible malady. What the best physicians of Europe had been unable to do in their case some mysterious force had done in the case of the Tsarevitch. His parents to whom the young boy was as their very heart's blood believed that the healing hand of God had wrought the cure, and that it was in answer to the supplication of one whose spirit was able to rise in higher flight than theirs or any other's. They knew of course that the boy was not yet entirely well, but they believed that he was getting well. Alexei believed this also and it is certain that he looked forward to a healthy, normal manhood.

Alexei, like his father, dearly loved the army and all the pageants of military display. He had every kind of toy soldier, toy guns and fortresses, and with these he played for hours, with his sailor companion Derevanko, or "Dina" as the boy called him, and with the few boy companions he was allowed. Two of these boys were sons of "Dina," and a third was the son of one of the family physicians, by coincidence also named Derevanko. In the last years before the

Revolution a few carefully selected boys, cadets from the Military School, were called to the palace to play with Alexei. These boys were warned of the danger of any rough play, and all were extremely mindful of their responsibility. It was because no other type of boy could be trusted to play with Alexei that the Empress did not often invite to the palace the children of the Grand Dukes. They were Romanoffs, brusque and rude in their manners, thoughtless of the feelings of others, and the Empress literally did not dare to leave them alone with her son. But because of her caution she was bitterly assailed by her enemies who spoke sneeringly of her preference for "low born" children over the aristocratic children of the family.

The Emperor and Empress and all the children were passionately fond of pets, especially dogs. The Emperor's inseparable companion for many years was a splendid English collie named Iman, and when in the natural course of time this dog died the Emperor was inconsolable. After that he had a fine kennel of collies but he never made a special pet of any dog. The favorite dog of the Empress was a small, shaggy terrier from Scotland. This dog's name was Eira, and, to tell the truth, I did not like the little animal at all. His dis-agreeable habit of darting from under chairs and snapping at people's heels was a trial to my nerves. Nevertheless the Empress doted on him, carried him under her arm even to the dinner table, and amused herself greatly talking to and playing with the dour little creature. When he fell ill and had to be mercifully killed she wept in real grief and pity. Alexei's pets were two, a silky little spaniel named Joy and a beautiful big gray cat, the gift of General Voyeikoff. It was the only cat in the household and it was a privileged animal, even being allowed to sleep on Alexei's bed. There were two other dogs, Tatiana's French bull and a little King Charlie which I contributed to the menagerie. Both of these dogs went with the family to Siberia, and Jimmie, the King Charles spaniel, was found shot to death in that dreadful deserted house in Ekaterinaburg.

How far, how unbelievably far away now seem those peaceful days of 1912, when we were watching the Tsar's daughters growing towards womanhood, and even in our minds speculating on possible marriages for them. Their prospects as far as marriage was concerned, I must say, were rather vague. Foreign matches, because of religion and even more because of the girls' devotion to home and country, were almost out of the question, and suitable husbands in Russia seemed to be entirely lacking. There was a time in his boyhood when Dmitri, son of the Tsar's uncle, Grand Duke Paul, was a great favorite with the Imperial Family. But Dmitri as he grew older became so dissipated that he quite cut himself off from the prospect of an alliance with any of the Grand Duchesses. There had once been a faint possibility of an engagement between Olga and Crown Prince Carol of Rumania. As early as 1910 the beautiful Queen Marie and her son visited Russia for the purpose of introducing the young people, but nothing came of the visit. In 1914 the family made a return visit to Rumania on the *Standert*, the Rumanian Royal family, including the old Queen, "Carmen Sylva," meeting the yacht at Constanza, on the Black Sea, and making a splendid fête which lasted for three days. This time the matter was seriously broached to Olga who, in her usual quick, straightforward manner, declined the match. In 1916 Prince Carol again visited the Russian Court, and now his young man's fancy rested on Marie. He made a formal proposal for her hand, but the Emperor, declaring that Marie was nothing more than a schoolgirl, good-naturedly laughed the Prince's proposal aside.

Not all these proposals ended so merrily. One day coming as usual to Peterhof, I found the Empress in tears. A formal proposal had just been received from the old Grand Duchess, Marie Pavlovna, aunt of the Emperor, for a marriage between her son Boris Vladimirovitch and Grand Duchess Olga. This young man, Prince Boris, was much better known in questionable circles in Paris than in the Court of Russia and the mere suggestion of a marriage with one of her daughters was enough to reduce the Empress to mortified tears. Of course the proposal was rejected, greatly to the wrath of Grand Duchess Marie Pavlovna, a Russian *grande dame* of the old school in which the debauchery of young men was regarded as a perfectly natural phenomenon. She never forgave the slight, as she chose to consider it, and later became one of the most active of the circle of intriguers which, from the safety of a foreign embassy in Petrograd, plotted the ruin of the Imperial Family and of their country.

In the summer of 1912 the family and their immediate household, including myself, went on another long cruise in Finnish waters. During the cruise the yacht was visited by the Empress Dowager of whom previously I had seen but little. I write with some hesitation about the Empress Dowager, who is still living, and for whom I entertain all due respect. She was, as I remember her then, a small, slender woman, not beautiful certainly, not as attractive as her sister,

Queen Alexandra of England, but with a great deal of presence and, when she chose to exert it, considerable personal charm. The Emperor she apparently loved less than her other children, especially her son, Grand Duke Michail, and the Empress I fear she loved not at all. To the children she was affectionate but a trifle distant. I am sure that she resented the fact that the first four children were girls, and there is little doubt that she felt bitterly the affliction of the heir. Possibly she felt in her secret heart that it should have been her own strong son Michail who was the acknowledged successor of Nicholas II. I say this from my own conjecture and observations and not from positive knowledge. Yet after events, I think, confirmed my opinion.

The Dowager Empress after the death of Alexander III relinquished with rather bad grace her position of reigning Empress. In fact she never did relinquish it altogether, always taking precedence on public occasions of Alexandra Feodorovna. Just why the Tsar consented to this I never knew, but certain it is that always, when the Imperial Family made a state entrance the Tsar appeared first with his mother on his arm, the Empress following on the arm of one of the Grand Dukes. Society generally approved this procedure, the Empress Mother enjoying all the popularity which the Empress lacked. There were actually in Russia two Courts, a large one represented by society and the Grand Dukes, and a small one represented by the intimate circle of the Emperor and Empress. In the one everything done by the Empress Mother was right and by the shy and retiring Empress wrong. In the small Court it was exactly the other way around, except that even in the palace a certain amount of petty intrigue always existed.

The visit to Finnish waters by the Empress Mother in 1912 was marred by no coldness or disharmony. When we went ashore for tennis the Emperor admonished us all to play as well as we could, "because Mama is coming." We lunched aboard her yacht and she dined with us on the *Standert*. On the 22d of July, which was her name day, as well as that of the little Grand Duchess Marie, she spent most of the day on the Emperor's yacht, and after luncheon I took a photograph of her sitting with her arm around the Emperor's shoulders, her two little Japanese spaniels at their feet. She made us dance for her on deck, photographing us as we danced. After tea the children performed for her a little French playlet which seemed to delight her. Yet that evening at dinner I could not help noticing how her fine eyes, so kind and smiling towards most of the company, clouded slightly whenever they were turned to the Emperor or the Empress. Still I must record that later, passing the open door of Alexei's cabin, I saw the Empress Mother sitting on the edge of the child's bed talking gaily and peeling an apple quite like any loving grandmother.

I do not pretend to understand the Empress Dowager or her motives, but, as far as I can judge, her chief weakness was love of power. She carried her insistence on precedence so far that the *chiffres* of the maids of honor of both Empresses bore the initials M. A. instead of A. M., which was the proper order. She wanted to be first in everything and could not bear to abdicate either power or influence. She never, I believe, understood her son's preference for a quiet, family life, or the changed and softened manners he acquired under the influence of his wife.

CHAPTER VII

IN the autumn of 1912 the family went to Skernevizi, their Polish estate, in order to indulge the Emperor's love for big-game hunting. In the vast forests surrounding the estate all kinds of game were preserved and the sport of hunting there was said to be very exciting. During the war these woods and all the game were des-troyed by the Germans, but until after 1914 Skernevizi was a favorite retreat of the Emperor. I had returned to my house in Tsarskoe Selo but I was not allowed long to remain there. A telegram from the Empress conveyed the disquieting news that Alexei, in jumping into a boat, had injured himself and was now in a serious condition. The child had been removed from Skernevizi to Spala, a smaller Polish estate near Warsaw, and to Warsaw I accordingly traveled. Here I was met by one of the Imperial carriages and was driven to Spala. Driving for nearly an hour through deep woods and over a heavy, sandy road I reached my destination, a small wooden house, something like a country inn, in which the suite was lodged. Two rooms had been set apart for me and my maid, and here I found Olga and Tatiana waiting to help me get settled. Their mother, they said, was expecting me, and without any loss of time I went with them to the palace.

I found the Empress greatly agitated. The boy was temporarily improved but was still too delicate to be taken back to Tsarskoe Selo. Meanwhile the family lived in one of the dampest, gloomiest palaces I have ever seen. It was really a large wooden villa very badly planned as far as light and sunshine were concerned. The large dining room on the ground floor was so dark that the electric lights had to be kept on all day. Upstairs to the right of a long corridor were the rooms of the Emperor and Empress, her sitting room in bright English chintzes being one of the few cheerful spots in the house. Here we usually spent our evenings. The bedrooms and dressing rooms were too dark for comfort, but the Emperor's study, also on the right of the corridor, was fairly bright.

As long as the health of little Alexei continued fairly satisfactory the Emperor and his suite went stag hunting daily in the forests of the estate. Every evening after dinner the slain stags were brought to the front of the palace and laid out for inspection on the grass. The huntsmen with their flaring torches and winding horns standing over the day's bag made, I was told, a very picturesque spectacle. The Emperor and his suite and most of the household used to enjoy going out after dinner to enjoy this fine sight. I never went myself, having a foolish love of animals which prevents enjoyment of the royal sport of hunting. I even failed to appreciate, as the head of the estate, kind Count Velepolsky, thought I should, the many trophies of the chase with which the corridors and apartments of the palace were adorned.

What I did enjoy was the beautiful park which surrounded the palace, and the rapid little river Pilitsa that flowed through it. There was one leafy path through which I often walked in the mornings with the Emperor. This was called the Road of Mushrooms because it ended in a wonderful mushroom bench. The whole place was so remote and peaceful that I deeply sympathized with their Majesties' irritation that even there they could never stir abroad without being haunted by the police guard.

Although Alexei's illness was believed to have taken a favorable turn and he was even beginning to walk a little about the house and gardens, I found him pale and decidedly out of condition. He occasionally complained of pain, but the doctors were unable to discover any actual injury. One day the Empress took the child for a drive and before we had gone very far we saw that indeed he was very ill. He cried out with pain in his back and stomach, and the Empress, terribly frightened, gave the order to return to the palace. That return drive stands out in my mind as an experience of horror. Every movement of the carriage, every rough place in the road, caused the child the most exquisite torture, and by the time we reached home he was almost unconscious with pain. The next weeks were endless torment to the boy and to all of us who had to listen to his constant cries of pain. For fully eleven days these dreadful sounds filled the corridors outside his room, and those of us who were obliged to approach had often to stop our ears with our hands in order to go about our duties. During the entire time the Empress never undressed, never went to bed, rarely even lay down for an hour's rest. Hour after hour she sat beside the bed where the half-conscious child lay huddled on one side, his left leg drawn up so sharply that for nearly a year afterwards he could not straighten it out. His face was absolutely bloodless, drawn and seamed with suffering, while his almost expressionless eyes rolled back in his head. Once when the Emperor came into the room, seeing his boy in this agony and hearing his faint screams of pain, the poor father's courage complete-ly gave way and he rushed, weeping bitterly, to his study. Both parents believed the child dying, and Alexei himself, in one of his rare moments of consciousness, said to his mother: "When I am dead build me a little monument of stones in the wood."

The family's most trusted physicians, Dr. Rauchfuss and Professor Fedoroff and his assistant Dr. Derevanko, were in charge of the case and after the first consultations declared the Tsarevitch's condition hopeless. The hemorrhage of the stomach from which he was suffering seemed liable to turn into an abscess which could at any moment prove fatal. We had two terrible moments in which this complication threatened. One day at luncheon a note was brought from the Empress to the Emperor who, pale but collected, made a sign for the physicians to leave the table. Alexei, the Empress had written, was suffering so terribly that she feared the worst was about to happen. This crisis, however, was averted. On the second occasion, on an evening after dinner when we were sitting very quietly in the Empress's boudoir, Princess Henry of Prussia, who had come to be with her sister in her trouble, appeared in the doorway very white and agitated and begged the members of the suite to retire as the child's condition was desperate. At eleven o'clock the Emperor and Empress entered the room, despair written on their faces. Still the Empress declared that she could not believe that God had abandoned them and she asked me to telegraph Rasputine for his prayers.

His reply came quickly. "The little one will not die," it said. "Do not allow the doctors to bother him too much." As a matter of fact the turning point came a few days later, the pain subsided, and the boy lay wasted and utterly spent, but alive.

Curiously enough there was no church on this Polish estate, but during the illness of the Tsarevitch a chapel was installed in a large green tent in the garden. A new confessor, Father Alexander, celebrated mass and after the first celebration he walked in solemn procession from the altar to the sickroom bearing with him holy communion for the sick boy. The Emperor and Empress were very much impressed with Father Alexander and from that time on they retained him in their private chapel at Tsarskoe Selo. He was a good man but not a brave one, for when the Revolution came, and the Emperor and the Empress sent for him to come to them, he confessed himself afraid to go. Poor man! His caution, after all, did not save him. He was shot by the Bolsheviki a year or two afterwards, on what pretext I do not know.

The convalescence of Alexei was slow and wearisome. His nurse, Marie Vechniakoff, had grown so hysterical with fatigue that she had to be relieved, while the Empress was so exhausted that she could hardly move from room to room. The young Grand Duchesses were tireless in their devotion to the poor invalid, as was also M. Gilliard, who read to him and diverted him hours on end. Gradually the distracted household assumed a more normal aspect. The Emperor, in Cossack uniform, began once more to entertain the officers of his Varsovie Lancers, commanded by a splendid soldier, General Mannerheim, of whom the world has heard much. As Alexei's health continued to improve there was even a little shooting, and a great deal of tennis which the girls, after their long confine-ment to the house, greatly enjoyed. All of us began to be happy again, but one day the Emperor called me into his study and showed me a telegram from his brother, Grand Duke Michail, in which the latter announced his morganatic marriage to the Countess Brassoff, of whom the Emperor strongly disapproved. It was not the marriage itself that so strongly disturbed the Emperor, but that Michail had solemnly given his word of honor that it would never take place. "He broke his word—his word of honor," the Emperor repeated again and again.

Another blow which the Emperor received at this time was the suicide of Admiral Chagin, commandant of the *Standert* and one of the closest friends of the family. The Admiral shot himself on account of an unhappy love affair, and deeply as the Emperor mourned his death he was even more indignant at the manner of it. Russians, I know, are inclined to morbidity, and suicide with them is not an uncommon thing. But Nicholas II always regarded it as an act of dishonor. "Running away from the field of battle," was his characterization of such an act, and when he heard of Chagin's suicide he gave way to a terrible mood of anger and grief. Speaking of both Michail and Chagin he said bitterly: "How, in the midst of the boy's illness and all our trouble, how could they have done such things?" The poor Emperor, to whom every failure of those he loved and trusted came as an utterly unexpected blow, how near was his hour of complete and final disillusionment of nearly all earthly loyalties.

We had a few weeks of peaceful enjoyment before leaving Spala that autumn. The girls, bright and happy once more, rode every morning, the crisp air and the exercise coloring their cheeks and raising their spirits high. The Emperor tramped the woods, sometimes with me as his companion, and on one of these outings we both had a narrow escape from drowning. The Emperor took me for a row on the river which, as I have said, had a very rapid current. Intent on keeping the boat well into the current, the Emperor ran us into a small island, and for a few seconds escape from an ignominious upset seemed impossible. I was thoroughly frightened, the Emperor not a little embarrassed, and ardor for water sports was, for a time, rather lessened in both of us.

On October 21 (Russian Calendar) we celebrated the accession to the throne with high mass and holy communion, and a few days later the doctors decided that Alexei was well enough to be moved to Tsarskoe Selo. The Imperial train was made ready and their Majesties decided that I was to travel on it with the rest of the suite. This was, as a matter of fact, contrary to strict etiquette, and the announcement created among the ladies in waiting much consternation, not to say rancor. There is no question that being a regularly appointed lady in waiting to royalty and having nothing to do when a mere friend of the exalted one happens to be at hand is a bit irritating, so I cannot really blame the Empress's ladies for objecting to me as a traveling companion. The Imperial train, now used, one hears, by the inner circle of the Communists, was composed of a number of luxurious carriages, more like a home than a railway train. In the carriage of the Emperor and Empress the easy chairs and sofas were upholstered in bright chintz and there were books, family photographs, and all sort of familiar trinkets.

The emperor's study was in his favorite green leather, and adjoining their dressing rooms was a large and perfectly equipped bathroom. In this carriage also were rooms for the personal attendants of their Majesties. The Grand Duchesses and their maids had a similar carriage, and Alexei's carriage, which had compartments for the maids of honor and myself, was furnished with every imaginable comfort. The last carriage was the dining wagon with a small anteroom where the inevitable zakouski, the Russian table of *hors d'œuvres*, was served. At the long dining table the Emperor sat with his daughters on either hand, while facing him were Count Fredericks and the ladies in waiting. Throughout the journey of nearly two days the Empress was served in her own room or beside the bed where Alexei lay, very weak, but bright and cheerful once more.

This chapter may well close with one of the opening events of 1913, the Jubilee of the Romanoffs, celebrating the three hundredth anniversary of their reign. In February the Court moved from Tsarskoe Selo to the Winter Palace in Petrograd, a place they disliked because of the vast gloominess of the building and the fact that the only garden was a tiny space hardly large enough for the children to play or to exercise in. On reaching Petrograd the family drove directly across the Neva to Christ's Chapel, the little church of Peter the Great, where is, or was, preserved a miraculous picture of the Christ, very old and highly revered. The public had not been notified that the Imperial Family would first visit this chapel, but their presence quickly became known and they drove back to the Winter Palace through excited, but on the whole undemonstrative, masses of people, a typical Petrograd crowd.

The actual celebration of the Jubilee began with a solemn service in the Cathedral of Our Lady of Kazan, which everyone familiar with Petrograd remembers as one of the most beautiful of Russian churches. The vast building was packed to its utmost capacity, and that means a much larger crowd than in ordinary churches, since in Russia the congregation stands or kneels through the entire service. From my position I had a very good view of both the Emperor and the Tsarevitch, and I was puzzled to see them raise their heads and gaze long at the ceiling, but afterwards they told me that two doves had appeared and had floated for several minutes over their heads. In the religious exaltation of the hour this appeared to the Emperor a symbol that the blessing of God, after three centuries, continued to rest on the House of Romanoff. There followed a long series of functions at the palace, with deputations coming from all over the Empire, the women appearing at receptions and dinners in the beautiful national dress, which were also worn by the Empress and her daughters. The Empress, for all her weariness, was regal in her richly flowing robes and long-veiled, high *kokoshnik*, the Russian national headdress, set with magnificent jewels. She also wore the wide-ribboned order of St. Andrew, which was her sole privilege to wear, and at the most formal of the state dinners she wore the most splendid of all the crown jewels. The young Grand Duchesses were simply but beautifully gowned on all occasions, and they wore the order of Catherine the Great, red ribbons with blazing diamond stars. The crowds were enormous in all the great state rooms, the Imperial Family standing for hours while the multitudes filed past with sweeping curtsies and low bows. So long and fatiguing were these ceremonies that at the end the Empress was literally too fatigued to force a smile. Poor little Alexei also, after being carried through the rooms and obliged to acknowledge a thousand greetings, was taken back to his room in a condition of utter exhaustion.

There were state performances at the theater and the opera, Glinka's "Life for the Tsar" being sung to the usual tumult of applause and adulation, but for all that I felt that there was in the brilliant audience little real enthusiasm, little real loyalty. I saw a cloud over the whole celebration in Petrograd, and this impression, I am almost sure, was shared by the Empress. She told me that she could never feel happy in Petrograd. Everything in the Winter Palace reminded her of earlier years when she and her husband used to go happily to the theater together and returning would have supper in their dressing gowns before the fire talking over the events of the day and evening. "I was so happy then," she said plaintively, "so well and strong. Now I am a wreck."

Much as both she and the Emperor desired to shorten their stay in Petrograd, they were obliged to remain several weeks after the close of the official celebration because Tatiana, who unwisely had drunk the infected water of the capital, fell ill of typhoid and could not for some time be moved. With her lovely brown hair cut short, we finally went back to Tsarskoe Selo, where she made good progress back to health.

In the spring began the celebration of the Jubilee throughout the

THE EMPEROR AND EMPRESS IN OLD SLAVONIC DRESS
1913 JUBILEE.

Empire. The visit to the Volga, especially to Kostrama, the home of the first Romanoff monarch, Michail Feodorovnitch, was a magnificent success, the people actually wading waist deep in the river in order to get nearer the Imperial boat. It was the same through all the surrounding governments, crowds, cheers, acclamations, prayers, and great choruses singing the national hymn, very evidence of love and loyalty. I particularly remember when the cortège reached the town of Pereyaslovl, in the Vladimir Government, because it was from there that my father's family originated, and some of his relatives took part in the day's celebration. The Empress, to my regret, was not present, being confined to her bed on the Imperial train, ill and fatigued, yet under obligation to be ready for special ceremonies in Moscow. It would need a more eloquent pen than mine adequately to describe those days in Moscow, the Holy City of Russia. The weather was perfect, and under the clear sunshine the floating flags and banners, the flower-trimmed buildings, and the numberless decorations made up a spectacle of unforgettable beauty. Leaving his car at some distance from the Kremlin, the Emperor entered the great gate on foot, preceded by chanting priests with waving censers and holy images. Behind the Emperor and his suite came the Empress and Alexei in an open car through crowds that pressed hard against the police lines, while overhead all the bells of Moscow pealed welcome to the Sovereigns. Every day it was the same, demonstrations of love and fealty it seemed that no time or circumstance could ever alter.

THE EMPRESS ON HER BALCONY AT PETERHOF

CHAPTER VIII

NINETEEN-FOURTEEN, that year of fate for all the world, but more than all for my poor country, began its course in Russia, as elsewhere, in apparent peace and tranquillity. With us, as with other civilized people, the tragedy of Sarajevo came as a thrill of horror and surmise. I do not know exactly what we expected to follow that desperate act committed in a distant province of Austria, but certainly not the cataclysm of a World War and the ruin of three of the proudest empires of earth. Very shortly after the assassination of the Austrian heir and his wife the Emperor had gone to Kronstadt, headquarters of the Baltic fleet, to meet French and British squadrons then on cruise in Russian waters.* From Kronstadt he proceeded to Krasnoe, near Petrograd, the great summer central review center of the old Russian Army where the usual military maneuvers were in progress. Returning to Peterhof, the Emperor ordered a hasty departure to Finland because, he said, the political horizon was darkening and he needed a few days of rest and distraction. We sailed on July 6 (Russian Calendar) and had a quiet cruise, the last one we were ever destined to enjoy. Not that we intended it to be our last, for returning to Peterhof, from whence the Emperor hurried again to the reviews, we left nearly all our luggage on the yacht. The Empress, however, in one of her fits of melancholy, told me that she felt that we would never again be together on the *Standert*.

The political skies were indeed darkening. The Serbian murders and the unaccountably arrogant attitude of Austria grew in importance every succeeding day, and for many hours every day the Emperor was closeted in his study with Grand Duke Nicholas, Foreign Minister Sazonoff and other Ministers, all of whom urged on the Emperor the imperative duty of standing by Serbia. During the short intervals of the day when we saw the Emperor he seemed half dazed by the momentous decision he was called upon to make. A few days before mobilization I went to lunch at Krasnoe with a friend whose husband

---

* So little did any of the Allied rulers and statesmen anticipate the World War that in July, 1914, President Poincaré accompanied the French fleet on its cruise to the Baltic. Many festivities were arranged for him, and he was regally entertained by the Emperor. When receiving the ambassadors President Poincaré spoke gravely of the troubled political situation, but he said nothing to indicate that he expected war.

was on the Russian General Staff. In the middle of luncheon this officer, Count Nosstiz, burst into the room exclaiming: "Do you know what the Emperor has done? Can you guess what they have made him do? He has promoted the young men of the Military Academy to be officers, and he has sent the regiments back to their casernes to await orders. All the military attachés are telegraphing their Governments to ask what it means. What can it mean except war?"

From my friend's house I went almost at once back to Peterhof and informed the Empress what I had heard. Her amazement was unbounded, and over and over she repeated that she did not understand, that she could not imagine under what influence the Emperor had acted. He was still at the maneuvers, and although I remained late with the Empress I did not see him that night. The days that followed were full of suspense and anxiety. I spent most of my time playing tennis—very badly—with the girls, but from my occasional contacts with the Empress I knew that she was arguing and pleading against the war which apparently the Emperor felt to be inevitable. In one short talk I had with him on the subject he seemed to find a certain comfort in the thought that war always strengthened national feeling, and in his belief Russia would emerge from a truly righteous war stronger and better than ever. At this time a telegram arrived from Rasputine in Siberia, which plainly irritated the Emperor. Rasputine strongly opposed the war, and predicted that it would result in the destruction of the Empire. But the Emperor refused to believe it and resented what was really an almost unprecedented interference in affairs of state on the part of Rasputine.

I think I have spoken of the Emperor's aversion to the telephone. Up to this time none of his studies were ever fitted with telephones, but now he had wires and instruments installed and spent a great deal of time in conversations with Ministers and members of the military staff. Then came the day of mobilization, the same kind of a day of wild excitement, waving street crowds, weeping women and children, heartrending scenes of parting, that all the warring countries saw and ever will remember. After watching hours of these dreadful scenes in the streets of Peterhof I went to my evening duties with the Empress only to find that she had remained in absolute ignorance of what had been taking place. Mobilization! It was not true, she exclaimed. Certainly armies were moving, but only on the Austrian frontiers. She hurried from the room and I heard her enter the Emperor's study. For half an hour the sound of their excited voices reached my ears. Returning, the Empress dropped on her couch as one overcome by desperate tidings. "War!" She murmured breathlessly. "And I knew nothing of it. This is the end of everything." I could say nothing. I understood as little as she the incomprehensible silence of the Emperor at such an hour, and as always, whatever hurt her hurt me. We sat in silence until eleven when, as usual, the Emperor came in to tea, but he was distraught and gloomy and the tea hour also passed in almost complete silence.

The whole world has read the telegrams sent to Nicholas II by ex-Emperor William in those beginning days of the war. Their purport seemed to be sincere and intimate, begging his old friend and relative to stop mobilization, offering to meet the Emperor for a conference which yet might keep the peace. Historians of the future will have to decide whether those tenders were made in good faith or whether they were part of the sinister diplomacy of that wicked war. Nicholas II did not believe in their good faith, for he replied that he had no right to stop mobilization in Russia when German mobilization was already a matter of fact and that at any hour his frontiers might be crossed by German troops. After this interval the Emperor seemed to be in better spirits. War had come indeed, but even war was better than the threat and the uncertainty of the preceding weeks. The extreme depression of the Empress, however, continued unrelieved. Up to the last moment she hoped against hope, and when the German formal declaration of war was announced she gave way to a perfect passion of weeping, repeating to me through her tears: "This is the end of everything." The state visit of their Majesties to Petrograd soon after the declaration really seemed to justify the Emperor's belief that the war would arouse the national spirit, so long latent, in the Russian people. Never again do I expect to behold such a sight as the streets of Petrograd presented on that day. To say that the streets were crowded, thronged, massed, does not half express it. I do not believe that one single able-bodied person in the whole city remained at home during the hours spent in the capital by the Sovereigns. The streets were almost literally impassable, and the Imperial motor cars, moving at snail's pace from quay to palace through that frenzied sea of people, cheering, singing the national hymn, calling down blessings on the Emperor, was something that will live forever in the memories of all who witnessed it. The Imperial cortège was able, thanks to the police, to reach the Winter Palace at last, but many of the suite were halted by the crowds at the entrance to the great square in front of the palace and had to enter at a side door opening from the small garden to the west.

Inside the palace the crowd was relatively as great as that on the outside. Apparently every man and woman who had the right to appear at Court were massed in the corridors, the staircases, and the state apartments. Slowly their Majesties made their way to the great *Salle de Nicholas,* the largest hall in the palace, and there for several hours they stood receiving the most extraordinary tokens of homage from thousands of officials, ministers, and members of the *noblesse*, both men and women. Te Deums were sung, cheers and acclamations arose, and as the Emperor and Empress moved slowly through the crowds men and women threw themselves on their knees, kissing the hands of their Sovereigns with tears and fervent expressions of loyalty. Standing with others of the suite in the *Halle de Concert,* I watched this remarkable scene, and I listened to the historic speech of the Emperor which ended with the assurance that never would there be an end to Russian military effort until the last German was expelled from the beloved soil. From the *Salle de Nicholas* the Sovereigns passed to a balcony overlooking the great square. There with the Tsarevitch at their side they faced the wildly exulting people who with one accord dropped to their knees with mute gestures of love and obedience. Then as countless flags waved and dipped there arose from the lips and hearts of that vast assembly the moving strains of our great hymn: "God Save the Tsar."

Thus in a passion of renewed love and patriotism began in Russia the war of 1914. That same day the family returned to Peterhof, the Emperor almost immediately leaving for the casernes to bid farewell to regiments leaving for the front. As for the Empress, she became overnight a changed being. Every bodily ill and weakness forgotten, she began at once an extensive plan for a system of hospitals and sanitary trains for the dreadful roll of wounded which she knew must begin with the first battle. Her projected chain of hospitals and sanitary centers reached from Petrograd and Moscow to Charkoff and Odessa in the extreme south of Russia. The center of her personal activity was fixed in a large group of evacuation hospitals in and around Tsarskoe Selo, and there, after bidding farewell to my only brother, who immediately left for the southern front, I joined the Empress. Already her plans were so far matured that ten sanitary trains, bearing her name and the children's, were in active service, and something like eighty-five hospitals were open, or preparing to open, in Tsarskoe Selo, Peterhof, Pavlovsk, Louga, Sablino, and neighboring towns. The Empress, her two older daughters, and myself immediately enrolled under a competent woman surgeon, Dr. Gedroiz, as student nurses, spending two hours of every afternoon under theoretical instruction, and the entire hours of the morning in ward work in the hospitals. For the benefit of those who imagine that the work of a royal nurse is more or less in the nature of play I will describe the average routine of one of those mornings in which I was privileged to assist the Empress Alexandra Feodorovna and the Grand Duchesses Olga and Tatiana, the two last-named girls of nineteen and seventeen. Please remember that we were then only nurses in training. Arriving at the hospital shortly after nine in the morning we went directly to the receiving wards where the men were brought in after having first-aid treatment in the trenches and field hospitals. They had traveled far and were usually disgustingly dirty as well as blood-stained and suffering. Our hands scrubbed in antiseptic solutions we began the work of washing, cleaning, and bandaging maimed bodies, mangled faces, blinded eyes, all the indescribable mutilations of what is called civilized warfare. These we did under the orders and the direction of trained nurses who had the skill to do the things our lack of experience prevented us from doing. As we became accustomed to the work, and as both the Empress and Tatiana had extraordinary ability as nurses, we were given more important work. I speak of the Empress and Tatiana especially because Olga within two months was almost too exhausted and too unnerved to continue, and my abilities proved to be more in the executive and organizing than in the nursing end of hospital work. I have seen the Empress of Russia in the operating room of a hospital holding ether cones, handling sterilized instruments, assisting in the most difficult operations, taking from the hands of the busy surgeons amputated legs and arms, removing bloody and even vermin-infected dressings, enduring all the sights and smells and agonies of that most dreadful of all places, a military hospital in the midst of war. She did her work with the humility and the gentle tirelessness of one dedicated by God to a life of ministration. Tatiana was almost as skillful and quite as devoted as her mother, and complained only that on account of her youth she was spared some of the more trying cases. The Empress was spared nothing, nor did she wish to be. I think I never saw her happier than on the day, at the end of our two months' intensive training, she marched at the head of the procession of nurses to receive the red cross and the diploma of a certificated war nurse.

From that time on our days were literally devoted to toil. We rose at seven in the morning and very often it was an hour or two after midnight before we sought our beds. The Empress, after a morning in the operating room of one hospital, snatched a hasty luncheon and spent the rest of the day in a round of inspection of other hospitals. Every morning early I met her in the little Church of Our Lady of Znamenie, where we went for prayers, driving afterwards to the hospitals. On the days when the sanitary trains arrived with their ghastly loads of wounded we often worked from nine until three without stopping for food or rest. The Empress literally shirked nothing. Sometimes when an unfortunate soldier was told by the surgeons that he must suffer an amputation or undergo an operation which might be fatal, he turned in his bed calling out her name in anguished appeal. "Tsaritsa! Stand near me. Hold my hand that I may have courage." Were the man an officer or a simple peasant boy she always answered the appeal. With her arm under his head she would speak words of comfort and encouragement, praying with him while preparations for the operation were in progress, her own hands assisting in the merciful work of anesthesia. The men idolized her, watched for her coming, reached out bandaged hands to touch her as she passed, smiling happily as she bent over their pillows. Even the dying smiled as she knelt beside their beds murmuring last words of prayer and consolation.

In the last days of November, 1914, the Empress left Tsarskoe Selo for an informal inspection of hospitals within the radius of her especially chosen district. Dressed in the gray uniform of a nursing sister, accompanied by her older daughters, myself, and a small suite, she went to towns surrounding Tsarskoe Selo and southward as far as Pskoff, staff headquarters, where the younger Grand Duchess Marie Pavlovna was a hospital nurse. From there she proceeded to Vilna, Kovno, and Grodno, in which city she met the Emperor and with him went on to Dvinsk. The enthusiasm and affection with which the Empress was met in all these places and in stations along the route beggars description. A hundred incidents of the journey crowd my memory, each one worth the telling had I space to include them in this narrative. I remember, for example, the remarkable scene in the big fortress of Kovno, where acres of hospital beds were assembled and where the tall figure of the Empress, moving through those interminable aisles, was greeted like the visit of an angel. I never recall that journey without remembering the hospital at Grodno, where a gallant young officer lay dying of his wounds. Hearing that the Empress was on her way to the hospital, he rallied unexpectedly and declared to his nurses that he was determined to live until she came. Sheer will power kept life in the man's body until the Empress arrived, and when, at the door of the hospital, she was told of his dying wish to see her she hurried first to his bedside, kneeling beside it and receiving his last smile, his last gasping words of greeting and farewell.

After one very fatiguing day our train passed a sanitary train of the Union of Zemstvos moving south. The Empress, who should have been resting in bed at the time, ordered her train stopped that she might visit, to the surprise and delight of the doctors, this splendidly equipped rolling hospital. Another surprise visit was to the estate of Prince Tichkevitch, whose family supported on their own lands a very efficient hospital unit. It was impossible to avoid noticing how in the towns visited by the Empress, dressed as a simple sister of mercy, the love of the people was most manifest. In Grodno, Dvinsk, and other cities where she appeared with the Emperor there was plenty of enthusiasm, but on those occasions etiquette obliged her to lay aside her uniform and to dress as the wife of the Emperor. Much better the people loved her when she went among them in her nurse's dress, their devoted friend and sister. Etiquette forgotten, they crowded around her, talked to her freely, claimed her as their own.

Soon after returning from this visit of inspection the Empress accompanied by Grand Duchesses Olga and Tatiana, General Racine, Commander of the Palace Guards, a maid of honor and myself, set off on a journey to Moscow, where to my extreme sorrow and dismay I perceived for the first time unmistakable evidences of a spreading intrigue against the Imperial Family. At the station in Moscow the Empress was met by her sister, the Grand Duchess Serge and the latter's intimate friend and the executive of her convent, Mme. Gardieve. Welcome from the people there was none, as General Djounkovsky, Governor of Moscow, had announced, without any authority whatsoever, that the Empress was in the city incognito and did not wish to meet anyone. In con-sequence of this order we drove to the Kremlin through almost empty streets. Nevertheless the Empress began at once the inspection of hospitals, accompanied by General Racine and her maid of honor, Baroness Boukshoevden, daughter of the Russian Ambassador in Denmark. During our stay in Moscow I was not as constantly with the Empress as usual, our rooms in the Kremlin being far apart. However, General Odoevsky, the fine old Governor of the Kremlin, installed a telephone between our rooms, and on her free evenings the Empress often summoned me to sit with her in her dressing room, hung with light blue draperies and looking out over the river and the ancient roofs of Moscow. I lunched and dined with others of the suite in an old part of the immense palace known as the Granovita Palata, and here occurred one night a disagreeable scene in which General Racine, in the presence of the whole company, administered a stinging rebuke to General Djounkovsky, Governor of Moscow, for his responsibility for the cold welcome accorded her Majesty. The Governor turned very pale but made no answer to the accusation of General Racine. Already my mind was in a tumult of trouble, more and more conscious of the atmosphere of intrigue, plots, and conspiracies, the end of which I could not see. In the coldness of the Grand Duchess Serge, in my childhood such a friend to me and to my family, her chilly refusal to listen to her sister's denial of preposterous tales of the political influence exerted by Rasputine, by the general animosity towards myself, I began dimly to realize that there was a plot to strike at her Majesty through Rasputine and myself. There was absolutely nothing I could do, and I had to watch with tearless grief the breach between the sisters grow wider and deeper until their association was robbed of most of its old intimacy. I knew well enough, or I was convinced that I knew, that the dismissed maid of honor, Mlle. Tutcheff, was at the bottom of the whole affair, her family being among the most prominent in Moscow. But I could say nothing, do nothing.

With great relief we saw our train leave Moscow for a round of visits in surrounding territory, and here again the enthusiasm with which the people welcomed the Empress was unbounded. In the town of Toula, for example, and a little farther on in Orel, the people were so tumultuous in their greeting, they crowded so closely around their adored Empress, that our party could scarcely make our way to church and hospital. Once, following the Empress out of a church, carrying in my hands an ikon which had been presented to her, I was fairly overthrown by the crowding multitude and fell halfway down the high flight of steps before friendly hands could get me to my feet. I did not mind this, being only too rejoiced at evidences of love and devotion which the simple people of Russia felt for their Empress. In one town where there were no modern carriages she was dragged along in an old coach of state such as a medieval bishop might have used, the coach being quite covered with flowers and branches. In the town of Charkoff hundreds of students met the train bearing aloft portraits of her Majesty. In the small town of Belgorod, where the Empress wished to stop in order to visit a very sacred monastery, I shall never forget the joy with which the sleepy ischvostiks hurried through the darkness of the night to drive us the three or four versts from the railway to the monastery. Nor can I forget the arrival at the monastery, the sudden flare of lights as the monks hastened out to meet and greet their Sovereign Empress. These were the people, the plain people of Russia, and the difference between them and the plotting officials we had left behind in Moscow was a sad and a terrible contrast.

On December 6 (Russian Calendar), the birthday of the Emperor, we met his train at Voronezh, where our parties joined in visits to Tambov, Riasan, and other towns where the people gave their Majesties wonderful greetings. In Tambov the Emperor and Empress visited and had tea with a charming woman of advanced age, Mme. Alexandra Narishkin, friend of Alexander III and of many distinguished men of her time. Mme. Narishkin, horrible to relate, was afterwards murdered by the Bolsheviki, neither her liberal mind nor her long services to her country, and especially to her humble friends in Tambov, sparing her from the blood lust of the destroyers of Russia.

The journey of their Majesties terminated at Moscow, where the younger children of the family awaited them. I can still see the slim, erect figure of Alexei standing at salute on the station platform, and the rosy, eager faces of Marie and Anastasie welcoming their parents after their long separation. The united family drove to the Kremlin, this time not quite so inhospitably received. In the days following the Moscow hospitals and military organizations were visited in turn, and we included in these visits out of town activities of the Moscow Zemstvo (county council), canteens, etc. In one of these centers our host was Prince Lvoff, afterwards active in demanding the abdication of the Tsar, and I remember with what deference he received their Majesties, and the especial attention he paid to the Tsarevitch, whose autograph he begged for the visitors' book. Before we left

Moscow the Empress paid two visits, one to the old Countess Apraxin, sister of the former first lady in waiting, Princess Galatzine and, with the Emperor, to the Metropolitan Makari, a good man, but mercilessly persecuted during the Revolution.

There was one small but significant incident which happened after our return to Tsarskoe Selo, near the end of the year 1914. It failed of its intended effect, but had it not failed it might have had a far-reaching influence on world events at that time. Looking back on it now, I sometimes wonder exactly what lay back of the plot, and who was responsible for its inception. One evening late in the year I received a visit from two war nurses lately released from a German prison where they had been taken with a portion of a captured Russian regiment. In much perturbation of spirit these nurses told me of a third nurse who had been captured and imprisoned with them. This woman they had come to distrust as she had been accorded many special favors by the Germans. She had been given good food and even champagne, and when the nurses were released she alone was conveyed to the frontier in a motor car, the others going on foot. While in prison this woman had boasted that she expected to be received by the Emperor, to whom she proposed to present the flag of the captured regiment. The other nurses declared that in their opinion his Majesty should be warned of the woman's dubious character.

Hardly knowing what to think of such an extraordinary story, I thought it my duty to lay the matter before General Voyeikoff, Chief Commander of the Palace Guards, and when I learned from him that the Emperor had consented to receive the nurse I begged that the woman be investigated before being allowed to enter the palace. The Emperor showed some vexation, but he consented. When General Voyeikoff examined the woman she made a display of great frankness, handing him a revolver which she said it had been necessary for her to carry at the front. General Voyeikoff, thinking it strange that the weapon had not been taken away from her by the Germans, immediately ordered a search of her effects. In the handbag which she would certainly have carried with her to the palace were found two more loaded revolvers. The woman was, of course, arrested, and although I cannot explain why, her arrest caused great indignation among certain members of the aristocracy who previously had received her at their homes. The whole onus of her arrest was placed on me, although the Emperor declared his belief that she was a German spy sent to assassinate him. That she was a spy I have never doubted, but in my own mind I have never even tried to guess from whence she came.

CHAPTER IX

A VERY few days after the events chronicled in the last chapter I became the victim of a railroad accident which brought me to the threshold of death and for many months made it impossible for me to follow the events of the war, or the growing conspiracy against the Sovereigns. At a little past five o'clock of the afternoon of January 2, 1915, I took the train at Tsarskoe for a short visit to my parents in Petrograd. With me in my carriage was Mme. Shiff, a sister of a distinguished officer of Cuirassiers. We sat talking the usual commonplaces of travel when suddenly, without a moment's notice, there came a tremendous shock and a deafening crash, and I felt myself thrown violently forward, my head towards the roof of the carriage, and both legs held as in a vise in the coils of the steam-heating apparatus. The overturned carriage lurched and broke in two like an eggshell and I felt the bones of my left leg snap sharply. So intense was the pain that I momentarily lost consciousness. Too soon my senses returned to me and I found myself firmly wedged in the wreckage of wood and iron, a great bar of steel crushing my face, and my mouth so choked with blood that I could not utter a sound. All I could do in my agony was silently to pray that God would give me the relief of a quick death, for I could not believe that any human being could endure such pain and live.

After what seemed to me an interminable length of time I felt the pressure on my face removed and a kind voice asked: "Who lies here?" As I managed to breathe my name the rescuers exclaimed in astonishment and alarm, and immediately began to endeavor to extricate me from my agonizing position. By means of ropes passed under my arms and using great care and gentleness they ultimately got me free and laid me on the grass. In a moment's flash I recognized one as a Cossack of the Emperor's special guard, an excellent man named Lichatchieff, and the other as a soldier of the railway battalion. Then I fainted. Ripping loose one of the doors of the railway carriage, the men placed me on it and carried me to a nearby hut already crowded with wounded and dying. Regaining consciousness for a moment, I begged in whispers that Lichatchieff would telephone my parents in Petrograd and their Majesties at the palace. This the good fellow did without delay, and he also brought to my corner one of the surgeons summoned to the wreck. The man gave me a rapid examination and said briefly: "Do not disturb her. She is dying." He left to attend to more hopeful cases, but the faithful soldiers still knelt beside me, straightening my crushed and broken legs and wiping the blood from my lips. In about two hours another doctor, this time the surgeon Gedroiz, under whom the Empress, her daughters, and myself had taken our nurses' training, approached the corner where I lay. I looked with a kind of terror into the face of this woman, for I knew her to be no friend of mine. Simply giving my wounded head a superficial examination she said carelessly that I was a hopeless case, and left me without the slightest attempt to soothe my pain. Not until ten o'clock that night, four hours after the collision which had wrecked two trains, did any help reach me. At that hour arrived General Racine from the palace with orders from their Majesties to do everything possible in my behalf. At his imperative commands I was again placed on a stretcher and carried to a relief train made up of cattle cars. At the moment my poor father and mother arrived from Petrograd and the last things I remember were their sobs and a teaspoonful of brandy mercifully poured down my throat.

At the end of the journey to Tsarskoe Selo I dimly recognized the Empress and the four Grand Duchesses who had come to the station to meet the train. Their faces were full of sympathy and grief, and as they bent over me I found strength to whisper to them: "I am dying." I believed it because the doctors had said so, and because my pain was so great. Then came the ordeal of being lifted into the ambulance and the half-consciousness that the Empress was there too, holding my head on her knees and begging me to have courage. After that came an interval of darkness out of which I awoke in bed and almost free from pain. The Empress who, with my parents, remained near me, asked me if I would like to see the Emperor. Of course I replied that I would, and when he came I pressed the hand he gave me. Dr. Gedroiz, who was in charge of the ward, told everyone coldly to take leave of me as I could not possibly live until morning. "Is it so hopeless?" asked the Emperor. "She still has some strength in her hand."

Later on, I do not know exactly when, I opened my eyes quite clearly, and saw standing beside my bed the tall, gaunt form of Rasputine. He looked at me fixedly and said in a calm voice: "She will live, but will always be a cripple." A prediction which was literally fulfilled, for to this day I can walk only slowly and with the aid of a stout stick. I have been told that Rasputine recalled me from unconsciousness, but of his words I know only what I have recorded.

The next morning I was operated on and for the six weeks following I suppose I suffered as greatly as one can and live. My left leg which had sustained a double fracture, troubled me less than my back and my right leg which had been horribly wrenched and lacerated. My head wounds were also intensely painful and for a time I suffered from inflammation of the brain. My parents, the Empress, and the children came every day to see me, but despite their presence the neglect and unkindness of Dr. Gedroiz continued. The suggestion of the Empress that her trusted physician, Dr. Federoff, be brought into consultation was rudely repulsed by this woman, of whom I may finally say that she is now in high favor with the Bolsheviki whose ranks she joined in the autumn of 1917. Waited upon by none but the most inexperienced nurses, I do not know what might have become of me had not my mother brought to the hospital an old family nurse whom she absolutely insisted should take charge of me. Things went a little better after this, but happy was I when at the end of the sixth week, against the will of Dr. Gedroiz, I left that wretched hospital and was removed to my own home. There in the peace and security of my comfortable bedroom I enjoyed for the first time since my accident quiet and refreshing sleep.

It seems strange that the hostile and envious Court circle had deeply resented the daily visits of the Emperor and Empress to my bedside. To placate the gossipers the Emperor, before visiting me, used to make the rounds of all the wards. In spite of it all I had many visitors and many daily inquiries from the Empress Dowager and others. Very soon after my arrival home I was examined by skillful surgeons, among them Drs. Federoff and Gagentorn, who pronounced my crushed right leg to be in a very bad condition and placed it in a plaster cast, where it remained for two months. The Empress visited me daily, but the Emperor I seldom saw because, as I learned indirectly, the War was going very badly on the Russian front, and the Emperor was almost constantly with the armies. In the last week before Lent he came to my bedside with the Empress, in accordance with an old Russian custom, before confession, to beg my forgiveness for possible wrongs done me during the year past. Their pious humility and also the white and careworn face of the Emperor filled me with emotion which later events served only to increase, for very momentous and trying hours were even then crowding the destiny of Nicholas II, Tsar of all the Russias.

A soldier of the sanitary corps, a man named Jouk, had been assigned to duty at my house, and as soon as I was able to leave my bed he took me daily in a wheeled chair to church, and to the palace. This was the summer of 1915, a time of great tribulation for the Russian Army, as every student of the World War is aware. Grand Duke Nicholai Nicholaievitch was pursuing a policy which rightly disturbed the Emperor, who constantly complained that the commander in chief of his armies sent the men forward without proper ammunition, without artillery support, and with no adequate preparations for safe retreat. Disaster after disaster confirmed the Emperor's fears. Fortress after fortress fell to the Germans. Kovno fell. Novogeorgiesk fell, and finally Warsaw itself fell. It was a terrible day when the Emperor, white and trembling, brought this news to the Empress as we sat at tea on her balcony in the warm autumn air. The Emperor was fairly overcome with grief and humiliation as he finished his tale. "It cannot go on any longer like this," he exclaimed bitterly, and then he went on to declare that in spite of ministerial opposition he was determined to take personal command of the army himself. Only that day Krivosheim, Minister of Agriculture, had addressed him on the impossible condition of Russian internal affairs. Nicholai Nicholaievitch, not content with military supremacy, had assumed almost complete authority over all the business of the Empire. There were in fact two governments in Russia, orders being constantly issued from military headquarters without the knowledge, much less the consent, of the Emperor.

Very soon after the fall of Warsaw it became clear to the Emperor that if he were to retain any dignity whatever he would have to depose Nicholai Nicholaievitch, and I wish here to state, without any reservation whatever, that this decision was reached by the Emperor without advice from Rasputine, myself, or any other person. Even the Empress, although she approved her husband's resolution, had no part in forming it. M. Gilliard has written that the Emperor was forced to his action by bad advisers, especially the Empress and Rasputine, but in this he is absolutely mistaken. M. Gilliard writes that the Emperor was told that Grand Duke Nicholai Nicholaievitch was plotting to confine his Sovereign in a monastery. I do not believe for a moment that Rasputine ever made such a statement, but he did, in my presence, warn the Emperor to watch Nicholai Nicholaievitch and his wife who, he alleged, were at their old practices of table-tipping and spiritism, which he thought to be a highly dangerous way to conduct a war against the Germans. As for me, I repeat that never once did I say or do anything to influence the Emperor in state affairs. I wish I could here reproduce a letter written to my father by the Emperor in which all the reasons for taking the step he did were explained. The letter, alas! was taken from me by the Bolsheviki after my father's death, and I suppose was destroyed.

On the evening when the Emperor met his ministers to announce his great decision I dined at the palace, and I was deeply impressed with the firmness of the Emperor's decision not to be overborne by arguments or vain fears on the part of timid statesmen. As he arose to go to the council chamber the Emperor begged us to pray for him that his resolution should not falter. "You do not know how hard it has been for me to refrain from taking an active part in the command of my beloved army," he said at parting. Overcome and speechless, I pressed into his hand a tiny ikon which I had always worn around my neck, and during the long council which followed the Empress and I prayed fervently for the Emperor and for our distracted country.

As the time passed the Empress's anxiety grew so great that, throwing a cloak around her shoulders and beckoning me to follow, she went out on the balcony, one end of which gave on the council room. Through the lace of the window curtains we could see the Emperor sitting very upright, surrounded by his ministers, one of whom was on his feet speaking earnestly. Our eleven o'clock tea was served long before the Emperor, entirely exhausted, returned from the conference. Throwing himself in an armchair, he stretched himself out like a man spent after extreme exertion, and I could see that his brow and hands were wet with perspiration.

"They did not move me," he said in a low, tense voice. "I listened to all their long, dull speeches, and when all had finished I said: 'Gentlemen, in two days from now I leave for the Stavka.'" As he repeated the words his face lightened, his shoulders straightened, and he appeared like a man whose strength was suddenly renewed.

Yet one more struggle was before him. The Empress Dowager, whom the Emperor visited immediately after the ministerial conference, was by this time thoroughly imbued with the German-spy mania in which the Empress and Rasputine, not to mention myself, were involved. She believed the whole preposterous tissue of lies which had been built up and with all her might she struggled against the Emperor's decision to assume supreme command of the army. For over two hours a painful scene was enacted in the Empress Dowager's gardens, he trying to show her that utter disaster threatened the army and the Empire under existing conditions, and she repeating over and over again the wicked slanders of German plots which she insisted that he was furthering. In the end the Emperor left, terribly shaken, but with his resolution as strong as ever.

Before leaving for staff headquarters the Emperor and his family took communion together at the Feodorovsky Cathedral and at their last meal together he showed himself calm and collected as he had not been for some time; in fact, not since the beginning of the last disastrous campaign. From headquarters the Emperor wrote full accounts of the scenes which took place when he assumed personal command, and of the furious anger, not only of the deposed Nicholai Nicholaievitch but of all his staff, "Every one of whom," wrote the Emperor, "has the ambition himself to govern Russia."

I am not attempting to write a military history of those years, and I am quite aware of the fact that most published accounts of the Russian Army represent Nicholai Nicholaievitch as the devoted friend of the Allies and the Emperor as the pliant tool of German influences. It is undeniable, however, that almost as soon as Nicholai Nicholaievitch had been sent to the Caucasus and the Emperor took command of the Western Army a marked improvement in the general morale became apparent. Retreat at various points was stopped, the whole front strengthened, and a new spirit of loyalty to the Empire was manifest.

I wish to interpolate here, in connection with the Emperor's personal command of the army, a word on the immense service he rendered it at the beginning of the War in suppressing the manufacture and sale of vodka, the curse of the Russian peasantry. The Emperor did this entirely on his own initiative, without advice from his ministers or the Grand Dukes. The Emperor said at the time: "At least by this I will be remembered," and he was, because the condition of the peasants, the town workers, and of course the army became at once immeasurably better. In the midst of war-time privations the savings-banks accounts of the people increased enormously, and in the army there was none of the hideous debauchery which disgraced Russia in the Russo-Japanese War. As an eminent French correspondent long afterwards wrote: "It is to the dethroned Emperor Nicholas that we must accord the honor of having effected the greatest of all internal reforms in war-time Russia, the suppression of alcoholism."

In October the Emperor came to Tsarskoe Selo for a brief visit, and on his return he took with him to the Stavka the young Tsarevitch. This is the first time he had ever separated the boy from his mother, and the Empress was never happy except in the few minutes each day when she was reading the child's daily letter. At nine o'clock at night she went up to his bedroom exactly as though he were there and she was listening to his evening prayers. By day the Empress continued her tireless work in the hospitals from which, by reason of my accident, I had long been excluded. However, at this time, I received from the railroad as compensation for my injuries the considerable sum of eighty thousand rubles, and with the money I established a hospital for convalescent soldiers in which maimed and wounded men received training in various useful trades. This, it is needless to say, became a great source of happiness to me, since I knew as well as the soldiers what it meant to be crippled and helpless. From the first my hospital training school was a most gratifying success, and my personal interest in it never ceased until the Revolution, after which all my efforts at usefulness and service ended in imprisonment and persecution.

Not this action of mine, patriotic though it must have appeared, no amount of devotion of the Empress to the wounded, sufficed to check the rapidly growing propaganda which sought to convict the Imperial Family and all its friends of being German spies. The fact that in England the Empress's brother-in-law, Prince Louis of Battenberg, German-born but a loyal Briton, was forced to resign his command in the British Navy was used with effect against the Empress Alexandra Feodorovna. She knew and resented keenly this insane delusion, and she did everything in her power to overcome it. I remember a day when the Empress received a letter from her brother Ernest, Grand Duke of Hesse, in which he implored her to do something to improve the barbarous conditions of German prisoners in Russia. With streaming tears the Empress owned herself powerless to do anything at all in behalf of the unhappy captives. She had organized a committee for the relief of Russian prisoners in Germany, but this had been fiercely attacked, especially in the columns of *Novy Vremya*, an influential organ of the Constitutional Democratic Party. In this newspaper and in general society the Empress's committee was accused of being a mere camouflage gotten up to shield her real purpose of helping the Germans. Against such attacks the Empress had no defense. Her secretary, Count Rostovseff, indeed tried to refute the story concerning the Empress's prison-camp committee, but the editors of *Novy Vremya* insolently refused to publish his letter of explanation.

The German-spy mania was extended from the palace to almost every Russian who had the misfortune to possess a name that sounded at all German. Count Fredericks and Minister Sturmer were among those who suffered calumny, although neither spoke a single German sentence. But the greatest sufferers were those barons of the Baltic Provinces whose ancestors had bequeathed them names of quite certain German origin. Many of these men were arrested and sent to die, or to suffer worse than death in exile. The sons and relatives of many of these very Baltic proprietors were at the time fighting loyally in the Russian Army. That there were German spies at work in Russia all during the War I have no reason to doubt, but they were the men who after 1917 invited in and exalted Lenine and Trotzky, and not the Empress and her friends, nor yet the persecuted estate owners of the Baltic Provinces. Did the Emperor's family call upon the Germans to rescue them from Siberia? Did any of the Baltic Provinces at Versailles ask to be united to Germany?

The army and navy still remained loyal to the Sovereigns. On one of his home visits to Tsarskoe Selo the Emperor brought with him as a proof of this the Cross of the Order of St. George, the highest of all Russian military decorations, which none could bestow except the Emperor, or the chief command of one of the armies in the field. In this case it was the gallant Southern Army which had voted to bestow it on the Emperor, and his pride and joy in it were humbly great.

* * *

CHAPTER X

TO one who has always held the honor and faith of the Russian people very dear, who has never doubted that after the last hideous phase of revolution and anarchism has passed, the Russian nation will emerge stronger and better than ever before, the writing of these next chapters is a duty inexpressibly painful. I must tell the truth, otherwise it would have been better for me never to have written at all. Yet to picture in anything like its true colors the decadence of Petrograd society from 1914 onward is a task from which any loyal Russian must shrink. Without a knowledge of these conditions, however, students of the Russian Revolution will never be able to understand why the fabric of government slipped so easily from the feeble hands of the Provisional Government to the ruthless and bloody grasp of the Bolshevists.

During the entire winter of 1915, when the War was being waged on all fronts with such disaster to the Allies, when millions of men, Russians, Frenchmen, Belgians, Englishmen, were giving up their lives in the cause of freedom, the aristocracy of the Russian capital was indulging in a reckless orgy of dancing, sports, dining, yes, and wining also in spite of the Emperor's edict against alcohol, spending enormous sums for gowns and jewels, and in every way ignoring the terrible fact that the world was on fire and that civilization was battling for its very life. In the palace the most frugal régime had been adopted. Meals were simple almost to parsimony, no money was spent except for absolute necessities, and the Empress and her daughters spent practically every waking hour working and praying for the soldiers. But society, when it was not otherwise amusing itself, was indulging in a new and madly exciting game of intrigue against the throne. To spread slanders about the Empress, to inflame the simple minds of workmen against the state was the most popular diversion of the aristocracy. A typical instance of this mania was related to me by my sister, who one morning was surprised by an unexpected visit from her sister-in-law, daughter of a very great lady of the aristocracy. Bursting into the room, this woman exclaimed delightedly: "What do you think we are doing now? Spreading stories through all the factories that the Empress is keeping the Emperor constantly drunk. Everybody believes it." I mention this story as typical because the woman involved afterwards became very prominent in the Grand Ducal cabal that forced the abdication, and she was also one of two women present in the Yusupoff Palace on the night of Rasputine's assassination.

Every possible circumstance, no matter how inconsequential, was eagerly seized as capital by these plotters. A former lady in waiting, Marie Vassilchikoff, long retired from Court and living on her Austrian estates, came to Petrograd, I know not how, and asked for an audience with the Empress. Since Russia was at war with Austria this audience could not be granted, nor did the Empress even remotely desire it. Yet as the story was circulated Marie Vassilchikoff was represented as having been sent for by the Empress to negotiate a separate peace with Austria, and that this treachery was frustrated only by the vigorous intervention of the Grand Duchess Serge.

These stories were spread not only by Court and society people, but were made into a regular propaganda in the army, especially among the higher command. The propaganda was chiefly in the hands of members of the Union of Zemstvos, its most successful agent being the infamous Goutchkoff, who now, it is gratifying to know, has earned the contempt of every Russian political group, even including the Bolshevists. Thus in a whirl of heartless gaiety and an organized campaign against the Sovereigns and against the Empire passed the winter of 1915, the dark prelude of darker years to come.

In the spring of that year, my health being still very precarious, their Majesties sent me in charge of a sanitary train filled with invalid soldiers and officers to the soft climate of the Crimea. With me went a sister of mercy and the sanitary-corps man Jouk, of whom I have spoken. On the same train journeyed also three members of the secret police, ostensibly to protect, but really, as I well understood, to spy upon me. Their presence the Empress, who came in the pouring rain to see the train off from the station, was powerless to forbid, as she herself was constantly under the surveillance of the dread Okhrana. Our train traveled slowly, taking five days from Petrograd to the Black Sea. But this we did not mind as we were very comfortable, the weather became beautiful, and our frequent stops at Moscow and towns farther south were full of interest. Our destination was Evpatoria on the eastern shore of the Black Sea, and here all of us were cordially received, M. Duvan, the head man of the city, giving me for a residence his own flower-hung villa overlooking the sea. Here I spent two peaceful months, finding the mud baths wonderfully restoring, and meeting some unusually interesting people. I am sure that few people outside of Russia have ever heard of the Karaim, a racial group among the most ancient in the world and of whom, even then, a bare ten thousand existed. They were not Jews, although they worshipped in synagogues, because they acknowledged Christ as God, or at least a special prophet of God. They were, and are, if they still exist, a strange mixture of pious Jews and early Christians, left-overs from the days of the decaying Roman Empire when Judaism and Christianity were trying to unite in one faith. The head of the Karaim in Evpatoria was a fine black-bearded patriarch named Gaham, and with him I formed an almost immediate friendship. Dressed in the long black robe of his office, he used to sit with me for hours reading and reciting the legends of his people, many reaching back into the dim twilight of civilization. I liked the patriarch, not only for his simplicity and his kindness to me, but for his evident love and loyalty to the Imperial Family, a loyalty shared by all the people of the Karaim.

A telegram from the Empress told me that she was then leaving for the Stavka, from which she and the Emperor and the whole Imperial Family would proceed to the Crimea for an important military and naval review. Obeying her instructions I motored from Evpatoria to Sevastopol, through an enchanting landscape of hills and plains, the latter being literally carpeted with scarlet poppies. Arriving at Sevastopol, I had some difficulty in passing the guard, but the Empress's telegram, marked "Imperial," I had brought with me, and this proved the open sesame to the Emperor's special train. I lunched with the Empress and the Grand Duchesses, meeting the Emperor and Alexei when they came from the reviews at six o'clock. I spent that night in town, and the next day returned to Evpatoria, their Majesties promising to visit me within a few days. On May 16 they arrived and received a most enthusiastic welcome, not only from the townspeople but from the Tartars, who came in from the hills by thousands, from the people of the Karaim, and others as strange and as picturesque. The huge square before the cathedral was strewn with fragrant roses over which the Imperial Family walked to service. The next few hours were spent in a round of visits to churches, hospitals, and sanatoriums, and it was to a late luncheon at my villa that they finally arrived. After luncheon we walked and sat on the beach, but the gathering crowd became so large and so curious that the poor Emperor, who had looked forward to a sea bath and a swim, had to relinquish both. Alexei enjoyed the day, boy fashion, without regard to the crowds, playing on the beach and building a big sand fortress, which the schoolboys of the town next day surrounded by a high wall of stones to protect it from the ravages of the tide. We had tea in the garden, the Empress greatly enjoying the Oriental sweets sent her by the Tartars. In the evening I dined on the Imperial train and traveled with it a short distance on its way back to Petrograd.

In June I returned to Tsarskoe and resumed work in my beloved hospital training school. The weather was unusually hot but the Empress continued her constant duties in the hospitals and operating rooms. Often I accompanied her on her rounds, and it came to me as a painful shock that the surgeons and some of the wounded officers no longer regarded her, as before, with respect and veneration. Too often an officer would assume in her presence a careless and indifferent manner which even a professional nurse would have resented. The Empress never did. She must have noticed evidences of disrespect but no word of complaint ever passed her lips. When I ventured to suggest to her that it might be well to go less frequently to her hospital, she rewarded me with a look of reproach. Whatever other people did, whatever their attitude towards the War, Royalty knew its duty and would perform it faithfully to the end.

Both the Emperor and the Empress during all this rising tide of disaffection persisted in underestimating its importance. The Emperor especially treated the whole movement with the contempt which no doubt it merited but which as a national menace it was far too dangerous to ignore. I realized it keenly, but knowing how impossible it was to make their Majesties understand that everything that was said against me, against Rasputine, against the Ministers, was actually directed against themselves, I was obliged to keep my lips closed. My parents realized as well as I did what was going on. They had good reason, in fact, for my mother had received two most insulting letters, one from Princess Galatzine, sister of Mme. Rodzianko, whose husband was President of the Duma, and another from Mme. Timasheff, a woman of the highest aristocracy, letters which indicated a certain collusion between the writers. In them my mother was brutally informed that neither of the women desired any further acquaintanceship or association with her as she too undoubtedly belonged to the German-spy party. My parents at the time were living quietly in the little seaside town of Terioke, near Petrograd, and were studiously avoiding the vulgar orgies and intrigues of society.

In the midst of all these heart-breaking events I sought distraction in the enlargement and perfecting of my occupational hospital which was rapidly becoming overcrowded with invalids. I bought an additional piece of land and arranged for four portable houses to be brought from Finland. Two of these arrived duly, and I spent hours of absorbing interest watching them being put together on the newly acquired land. All these days I was constantly being bothered by people who, perhaps believing that the money I was investing in hospitals was another proof of my power over the Imperial treasury, tormented me with petitions of every kind and description, but all of them alike in the selfishness of their character. With cold hatred in their eyes, but with hypocritical words on their lips, these people besought my good offices with their Majesties on behalf of their sons, husbands, and relatives, all of whom were alleged to be worthy of promotions and of lucrative positions under the State. One woman of good social position invaded my hospital one day and treated me to a disgraceful scene because I had assured her that I was powerless to further her ambition to see her husband appointed head of a certain Government. Naturally it happened that some petitioners were poor and needy, and these, to the best of my ability, but without any political influence whatever, I did endeavor to help. I know now, after witnessing true sympathy and kindness to prisoners and persecuted, like myself in later days, that I never did half what I might have done in the time of my prosperity. If better days come to Russia in my lifetime God help me to devote all that remains of my years to the poor and especially to prisoners. Now that I have tasted poverty, now that I have known the hopelessness of captivity, I know better than I did what can be done for the lowly and unfortunate.

A number of very disquieting events occurred to us during the summer. On very hot days it was the custom of the Empress and the children to drive through the woods and shaded roads to Pavlovsk, a few versts from Tsarskoe Selo. One stifling afternoon we started out as usual in two carriages, the Empress and myself leading the way. The horses were magnificent animals, apparently in the very pink of condition, but suddenly one of the horses uttered a piercing scream and dropped dead in his harness. The other horse plunged sidewise in terror and for a few minutes it was all the coachman could do to avoid an overturn. The Empress, pale, but as always courageous, got out of the carriage and helped me, who was still on crutches, to alight. The carriage of the children drove up, and getting in, we returned without further incident to the palace. Whatever caused the sudden death of that horse, or what was the object of that carriage accident—if indeed it was an accident—we never knew, but it left behind in my mind, and I think also in the mind of the Empress, a strangely sinister impression. The Empress nevertheless went steadfastly on with her hospital work, arranging in the convalescent wards concerts and entertainments for the pleasure of the wounded. The best singers, the most accomplished musicians, were secured for these concerts, and the men seemed appreciative of them. Yet over the head of the Empress Alexandra Feodorovna drifted darker and darker the shadow of impending doom. The things I dared not say to her began to reach her from others. In August came from the Crimea the head man of the Karaim, of whom I have spoken. From the first he made an agreeable im-pression on the Empress and the children, especially upon Alexei, who never tired of listening to his stories. But Gaham had not made the journey from the Crimea to relate legends and tales. He had previously been connected with the Ministry of Foreign Affairs, serving in Persia and the East, and his acute mind was still occupied with the foreign affairs of the Empire on which he kept himself well informed.

Determined, if possible, to force the Empress to understand the gravity of the situation, he told her a number of extraordinary things which had come to his knowledge, among them an organized plot against the throne which was being carried on by near relatives of the Tsar in the seclusion of an allied foreign embassy in Petrograd. His story, involving, as it did, the ambassador of a friendly power, the trusted representative of an own cousin of the Emperor, seemed to the Empress too preposterous to be credited. Horrified, she ended the conversation, and a few days later she went, taking me with her, to visit the Emperor at the Stavka. What he had to comment on her report of an alleged ambassadorial plot against him I never knew, but I soon became aware that representatives of other foreign countries were undeniably hostile. At the Stavka were military commissions of practically every allied country, among them General Williams and his staff from Great Britain, General Janin from France, General Rikkel from Belgium, and high officers from Italy, Serbia, Rumania, Japan, and other countries, all accompanied by subordinate officers. One afternoon when the gardens were quite crowded by these men and men of our own army, and while the

Empress was making her customary circle, I chanced to overhear a conversation among officers of the foreign military missions, in which the most slanderous words against her Majesty were uttered. "She has come again, it appears," said one of these men, "to see her husband and give him the latest orders of Rasputine." "The suite hate to hear her arrival announced," said another officer. "They know it means changes."

Worse things were said, but without waiting to listen I managed to make my way to the Empress, and that night inviting, as I was well aware, her irritation and disbelief, I related something of what I had overheard. I went further and reminded her of what we both knew, the increasing demoralization of the Emperor's staff. The Grand Dukes and the commanding officers were, as a matter of course, invited each day to lunch with the Emperor, but with insolence and audacity hitherto unheard of, many of the Emperor's near kinsmen declined these invitations. They gave the most trivial and transparent excuses for their absence—headaches, fatigue, previous engagements, alleged duties. The Empress listened to what I said, silent and distraught. She knew, and I also knew, that nothing she could say to the Emperor would make the slightest impression. His eyes and ears were still closed to the gathering tempest.

General Alexieff, Chief of Staff, and undoubtedly a valuable officer, had, I soon learned, been drawn into the plot. The Emperor suspected him to be in correspondence with the traitor Goutchkoff, but when questioned General Alexieff denied this vehemently. He was soon, however, to prove his treachery to the Emperor. There was in attendance on his Majesty at the Stavka an old officer, General Ivanoff, a St. George Cross man, who formerly had held command of the Army of the South. This devoted and loyal old soldier General Alexieff knew he must get rid of, and this, had he been honest, he might have done by pleading age or decreased usefulness. Instead, he merely summoned General Ivanoff and informed him that to the regret of the whole staff he was removed. The Chief of Staff was not responsible for this, he declared, the order having come from the Empress and her accomplices, Rasputine and Mme. Virubova. What General Alexieff said to the Emperor on the subject I do not know, but when next the two met the Emperor turned his head aside. This sudden coldness on the part of the Emperor, whom old General Ivanoff loved dearly, made it impossible for him to seek an audience, and yet the general was valiantly determined not to leave the Stavka without presenting his case to the Sovereign. Calling on me that same day, he repeated to me, while tears rolled down his white beard, the lying words of General Alexieff against the Empress. Feeling it against reason and justice that the Emperor should remain in ignorance of this insult to his wife, I promised to speak to him about it, and this I did, but to little purpose. The Emperor's wrath against Alexieff was indeed kindled but he evidently felt that he could not, at that critical hour, dismiss an officer whose services were so urgently in demand. After-wards, however, his manner towards old General Ivanoff became conspicuously kind.

We remained for some time after this at the Stavka, days to me of such sad remembrance that I can scarcely endure the task of recording them. The Empress and her suite, the Grand Duchesses, and myself lived on board the Imperial train, motor cars coming each day at one o'clock to take us to staff headquarters to luncheon. Headquarters were in an ancient villa of the Governor of the Province, a rather old-fashioned and uncomfortable place. Even the huge dining room where the Emperor and Empress, the staff and the officers of the foreign missions met each day was a dull and gloomy room. When the weather became very warm this dismal apartment was abandoned, and luncheon was served in a large tent in a shady part of the grounds overlooking the town and farther away still the flowing tide of the mighty Dnieper. The only really bright circumstance of the time was the growing health and strength of the Tsarevitch. He was developing marvelously through the summer, both in bodily vigor and in gaiety of spirits. With his tutors, M. Gilliard and Petroff, he romped and played as though illness were a thing to him unknown. With, several of the allied officers, notably with the Belgian General Rikkel, he was also on the best of terms.

Every day after luncheon the maids came from the train with what gowns and other apparel we needed for the remaining functions of the day. There was little room in the house in which to change, but we managed to appropriate a few nooks and corners, and to make ourselves as presentable as possible in the circumstances. In the Emperor's scant hours of leisure he loved to walk with his family in the woods along the river brink, and sometimes when I saw the Empress sitting on the grass talking informally with the peasant women who crowded around her, I took comfort, believing then, as

I still believe, that the great mass of the Russian people were to the end faithful to their Sovereigns. As for the suite, most of them became increasingly indifferent, bound up in their foolish personal affairs, diverting themselves with whispered gossip and laughter, apparently quite indifferent to the calamitous progress of the War. People to whom religion is still in these cynical days a real refuge will understand me when I tell them what comfort I found in an ancient convent in the neighborhood, and in the poor little church which adjoined it. The one treasure of this church was an old and highly revered image of Our Lady of Mogiloff and almost every day of that distressful summer I managed to spend a few minutes on my knees before her dark and mystic image. One day, feeling in my heart the imminence of a danger I dared not name even to myself, I took off my diamond earrings and laid them at the foot of the shrine where I had sought and received peace of mind. I hope my poor offering was received with grace by the saint, who of course did not need it, but whose helpless ones always do. A little later the monks presented me with a small replica of the image, and strangely enough this was the one ikon I was permitted to take with me when I was sent to the Fortress of Peter and Paul.

Of that unhappy summer of 1916 I have only one or two more incidents to relate. One of these was a visit to the Stavka of the Princess Paley, wife of Grand Duke Paul. Coming from Kiev, where the Empress Dowager and the Grand Duke Nicholai Michailovitch were in residence, it appeared ominous to me that they too, all of them, seemed to be inoculated with the delusion of the German spy and the Rasputine influence. Neither the Princess nor the Grand Duke were in the least tactful in the expression of their opinions on the subject. Another visitor to the Stavka was Rodzianko, who came to demand the instant dismissal of Protopopoff, Minister of the Interior, once his friend and confidant, but now accused by the President of the Duma of being a lunatic. The Emperor received Rodzianko coldly, and did not even invite him to lunch. At tea that afternoon the Emperor said that the interview had angered him intensely as he knew quite well that Rodzianko's representations and motives were wholly insincere. Almost everything at the Stavka was growing worse and worse, the Grand Dukes being more insolent than ever and continually annoying General Voyeikoff by ordering trains and motors for themselves without any regard to the requirements of the Emperor. It was with feelings of unspeakable relief that in November, 1916, we left the Stavka for Tsarskoe Selo. In the Imperial train with us traveled young Grand Duke Dmitri Pavlovitch who even then was probably involved in a deadly plot against their Majesties. Yet this young man was able to keep up a pretense of friendship with the Empress, sitting beside her couch and entertaining her by the hour with amusing gossip and stories. Hearing the laughter the Emperor often opened his study door to listen and to join in the conversation. It was a merry journey home, yet within a few days after we arrived troubles again began to multiply. Entering the Empress's door one day, I found her in a passion of indignation and grief. As soon as she could speak she told me that the Emperor had sent her a letter from Nicholai Michailovitch, in which the Empress was specifically charged with the most mischievous political machinations. "Unless this is stopped," the letter concluded, "murders will certainly begin."*

Nicholai Michailovitch, it appears, had gone to the Stavka from the group in Kiev, with the express object of delivering this letter. Every member of the staff knew his errand and expected him to be ignominiously ejected from the Emperor's study. Nothing of the kind happened, and the Grand Duke stayed to luncheon in the most friendly manner. I do not know what he said to the Emperor, but I do know that the letter was laid on the Emperor's desk. Nothing was said or done to avenge this deadly insult to the wife of Nicholas II whom undoubtedly he loved dearer than his own life. The only explanation I can think of was the Emperor's complete absorption in the War, and in his unshaken conviction that the plotters' gossip was entirely harmless. He had the kind of mind which could concentrate on only one thing at a time, and at this period his whole heart and soul was with the fighting armies. I well remember scraps of conversation with him during those days which indicated that in the back of his mind were many plans for future internal reforms. He spoke of important social changes which must come after the War, social and constitutional reforms. "I will do everything necessary afterwards," he said in more than one of these conversations. "But I cannot act now. I cannot do more than one thing at a time."

The Empress, I think, for all her sensitiveness to the abominable

* Previous to the War and the impending revolution the Empress had had very little to do with politics, but it is true that when affairs became desperate she did what she rightly could to advise her husband.

accusations brought against her, tried to preserve the same waiting state of mind. Most disagreeable incidents she kept to herself, yet one day she showed me a letter written directly to her by a Princess Vassilchikoff, a letter so insulting that the Emperor was aroused to order the Princess and her husband, a member of the Duma, to their country estates. This letter was written on small scraps of paper evidently torn from a cheap writing tablet. "At least," said the Empress with faint sarcasm, "she might have used the stationery of a lady when addressing her Sovereign."

What had taken possession of Petrograd society? I often asked myself. Was it a mob delusion, contagious, like certain diseases? Was it a madness born of the War similar to other strange hysterias which arose during some of the wars of the Middle Ages? That the delusion was confined to Petrograd and a few other towns frequented by the aristocracy was perfectly apparent. In the last days of 1916 the Empress with Olga, Tatiana, and General Racine paid a brief visit to Novgorod to inspect military hospitals and to pray in the monastery and church of Sofisky Sobor, one of the oldest churches in Russia. Her visit was opposed, quite senselessly, by Petrograd society, which accused her of going for some bad purpose, God knows what. But at Novgorod the people poured out in throngs to greet her with peals of bells, music, and cheers. Before leaving the city the Empress paid a visit to a very old woman who had spent forty helpless years in bed, still wearing the heavy chains of penitence which as a pilgrim she had, almost a lifetime before, assumed. As her Majesty entered the old woman's cell a feeble voice uttered these words: "Here comes the martyred Empress, Alexandra Feodorovna." What could this aged and bedridden recluse have known or guessed of events which were to come?

CHAPTER XI

IN preceding chapters I have mentioned the name of Rasputine, that strange and ill-starred being about whom almost nothing is known to the multitude but against whom such horrible accusations have been made that he is universally classed with such monsters of iniquity as Cain, Nero, and Judas Iscariot. Even H. G. Wells, in whose "Outline of History" Joan of Arc and Abraham Lincoln are disposed of in a line, sacrifices valuable space to state as an established fact that in 1917 the Russian Court was "dominated by a religious impostor, Rasputin, whose cult was one of unspeakable foulness, a reeking scandal in the face of the world." I have no desire in this book to attempt an exoneration of Rasputine, for I am not so ambitious as to believe that I can change the collective mind of the world on any point. In the interests of historical truth, however, I believe it to be my simple duty to record the plain tale of how and why Rasputine came to be a factor in the lives of Nicholas II and of Alexandra Feodorovna, his wife, and exactly to what extent he did, or rather, did not, dominate the Russian Court. Those who expect from me secret and sensational disclosures will, I fear, be disappointed, for Rasputine's every movement for years was known to the Russian police, and the most sensational fact of his whole career, his assassination, has been described by practically every writer of the events of the Russian Revolution.

I will first explain the exact status of the man, for this does not appear to be generally understood. He has been called a priest, more often still a monk, but the truth is he was not in holy orders at all. He belonged to a curious species of roving religious peasant which in Russia were called *Stranniki*, the nearest English translation of the word being pilgrims. These wandering peasants, common sights in the old Russia, were accustomed to travel from one end of the Empire to the other, often walking with heavy chains on their bodies to make their progress more painful and difficult. They went from church to church, shrine to shrine, monastery to monastery, praying, fasting, mortifying the flesh, and their prayers were, by a very considerable population, eagerly sought and devoutly believed in. Once in a while a *Strannik* appeared who, by virtue of his extreme piety, gift of speech, or strong personality, acquired more than local reputation. Churchmen of high rank, estate owners, and even members of the nobility invited these men to their houses, listened with interest to their discourses, and asked for their prayers. Such a *Strannik* was Gregory Rasputine, who from the humblest beginnings in a remote Siberian village became known all over the Empire as a man of almost superhuman endowment.

Of the type of Russians to whom the *Stranniki* made a genuine appeal the Emperor and Empress undoubtedly belonged. The Emperor, like several of his near ancestors, was a born mystic, and the soul of Alexandra Feodorovna, either from natural inclination or from close association with him whom she so dearly loved, leaned also towards mysticism. By this I do not mean that the Emperor and the Empress were at all interested in spiritualism, table-tipping, or alleged materializations from the world beyond. Far from it. In the earliest days of my acquaintance with the Empress, as far back as 1905, she gave me a special warning against these things, telling me that if I wished for her friendship never to have anything to do with so-called spiritism. Both the Emperor and the Empress were profoundly interested in the religious life and expressions of the whole human race. They read with sympathy and understanding the religious literature not only of Christendom but of India, Persia, and the countries of the Far East. I remember in connection with the Empress's first warning against spiritism that she gave me a book, an obscure fourteenth-century missal called "*Les Amis des Dieu*" which, in spite of her warm recommendation, I found great difficulty in reading. This interest in religion and the life of the spirit was actually what constituted what Mr. Wells calls the "crazy pietism" of Nicholas II. It was simple Christianity lived and not merely subscribed to as a theory. They believed that prophecy, in the Biblical sense of the word, still existed in certain highly gifted and spiritually minded persons. They believed that it was possible outside the church and without the aid of regularly ordained bishops and priests to hold communion with God and with His Spirit. Before I came to Court there was a Frenchman, Dr. Philippe, in whom they reposed the greatest confidence, believing him to be one in whom the gift of prophecy existed. I never knew Dr. Philippe, hence I can speak of him only as a sort of a forerunner of Rasputine, because, as the Empress told me, his coming was foretold by Dr. Philippe. Very shortly before his death the French mystic told them that they would have another friend authorized to speak to them from God, and when Rasputine appeared he was accepted as that friend.

Rasputine, although very poor and humble and almost entirely illiterate, had acquired a great reputation as a preacher, and had especially attracted the attention of Bishop Theofan, a churchman of renown in Petrograd. Bishop Theofan introduced the *Strannik* to the wife of Grand Duke Nicholas, who immediately conceived a warm admiration for him, and began to speak to her friends of his marvelous piety and spiritual insight. At that time the Emperor was on very friendly terms with the Grand Duke Nicholas, or rather with his wife and her sister, two princesses of Montenegro who had married, not quite in conformity with the rules of the Orthodox Church, the brothers, Grand Dukes Nicholas and Peter. One of these sisters, Princess Melitza, Grand Duchess Peter, had something of a reputation as a mystic, and it was at her house that the Emperor and Empress met first Dr. Philippe and later Rasputine. In one of my first conversations with the Empress she told me this, and told me also how deeply the conversation of the Siberian peasant had interested both her husband and herself. In fact Rasputine, at that period, interested and impressed almost everyone with whom he came in contact. When the house of Stolypine was blown up by terrorist bombs and, among others, his beloved daughter was grievously wounded, it was Rasputine whom the famous statesman summoned to her bedside for prayer and supplication. I am aware that the public generally believes that it was I who introduced Rasputine into the Russian Court, but truth compels me to declare that he was well known to the Sovereigns and to most of the Court long before I ever saw him.

It was about a month before my marriage in 1907 that the Empress asked Grand Duchess Peter to make me acquainted with Rasputine. I had heard that the Grand Duchess was very clever and well read, and I was glad of the opportunity of meeting her in her palace on the English Quay in Petrograd. Interesting as I found her, I was nevertheless thrilled with excitement when a servant announ-ced the arrival of Rasputine. Before his entrance the Grand Duchess said to me: "Do not be astonished if I greet him peasant fashion," that is, with three kisses on the cheek. She did so greet him and then she presented us to each other. I saw an elderly peasant, thin, with a pale face, long hair, an uncared-for beard, and the most extraordinary eyes, large, light, brilliant, and apparently capable of seeing into the very mind and soul of the person with whom he held converse. He wore a long peasant coat, black and rather shabby from hard wear and much travel. We talked and the Grand Duchess, speaking in French, bade me ask him to pray for some special desire of mine. Timidly I begged him to pray that God would permit me to spend my whole life serving their Majesties. To this he replied: "Your whole life will be thus spent." We parted then, but shortly afterwards, just before my wedding day, when my heart was in a tumult of doubt and anxiety, I wrote to the Grand Duchess Peter and asked her to seek Rasputine's counsel in my behalf. His word to me was that I would marry as I had planned but that I should not find happiness in my marriage. It will be seen how little I regarded him as a prophet at this time since I paid no attention to his warning. A full year after my marriage I saw Rasputine for the second time. It was on a train going from Petrograd to Tsarskoe Selo, he being on his way there to visit friends who were in no way connected with the Court.

But, asks the bewildered reader, when and how did Rasputine acquire the dreadful, almost unprintable reputation which classes him with the arch-fiend himself? To answer the question satisfactorily I should have to reveal at great length the strangely abnormal and hysterical mentality of the Russian people of that epoch. I shall try to do this as I go farther, but here I shall give, as a sort of illustration of the lunacy of the hour, a little experience of my own. It was on the first occasion after my arrest by Kerensky in the spring of 1917, when I was brought before the High Commission of Justice of the Provisional Government. Weak and ill from my long imprisonment in the gloomy Fortress of Peter and Paul, I found myself facing an imposing group of something like forty judges, all learned in the law and clothed in such dignity of office that I gazed at them in a kind of awe. In my distracted mind I asked myself what questions these grave magistrates would ask me, and in what pro-found language would their questions be clothed. My heart almost stopped beating while I waited for the words of the chief judge. And this is what was said, in a deep and solemn voice: "Tell me, who was it at Court that Rasputine called a flower?" Sheer amazement held me speechless, but even had I been given time I could not have answered the question because there was no such person. The judges whispered together for a moment and then the same man, handing me a piece of cardboard, demanded impressively: "What is the meaning of this secret card which was found in your house by the soldiers?"

I took the piece of cardboard and almost instantly recognized it as a menu card of the yacht *Standert*, dated 1908. On the reverse side were written the names of war vessels present at that date at a naval review held near Kronstadt, Russian vessels all, among which the position of the Imperial yacht was marked by a crown. I handed the menu card back to the judge saying merely: "Look at it, and look at the date." He looked at it and in some confusion muttered: "It is true." One more question those giant intellects found to ask me. "Is it a fact that the Empress could not live without you?" To which I replied as any sensible person would have done: "Why should a happy wife and mother be unable to live without a mere friend?" The inquiry was then hastily closed and I was ordered back to prison, to be watched more closely than ever, *because I would not answer to judgment.*

This is a perfectly fair sample of the madness and confusion of the Russian mind, or rather the Petrograd mind, before and after the Revolution. That this madness, this unreasoning mania for the destruction of all institutions might have something to justify itself in the public mind, it was absolutely necessary to find and to persecute individuals who typified, in popular imagination, the things which were so bitterly hated. Rasputine, more than any one other individual in the Empire, did typify old and unpopular institutions, and I can readily see why some intelligent and fair-minded persons thus accepted him. Dillon, for example, in his book, "The Eclipse of Russia," says: "It is my belief that although his friends were influential Rasputine was a symbol."

Russia, like eighteenth-century France, passed through a period of acute insanity from which it is only now beginning to emerge in remorse and pain. This insanity was by no means confined to the ranks of the so-called Revolutionists. It pervaded the Duma, the highest ranks of society, Royalty itself, all as guilty of Russia's ruin as the most blood-thirsty terrorist. What had happened in these dark years between 1917 and 1923 is simply the punishment of God for the sins of a whole people. When His avenging hand has so plainly been laid upon all of the Russian people how dare any of us lay the calamity entirely at the doors of the Bolsheviki? We Russians look on the appalling condition of our once great country, we behold the famishing millions on the Volga and in the Ukraine, we count the fearful roll of the murdered, the imprisoned, the exiled, and we cry weakly that the Tsar was guilty, Rasputine was guilty, this man and that woman were guilty, but never do we admit that we were all guilty, guilty of blackest treason to our God, our Emperor, our country. Yet not until we cease to accuse others and repent our own sins will the white dawn of God's mercy rise over the starved and barren desert that was once mighty Russia.

Rasputine, it seems to be generally assumed, having been introduced to the Imperial Family, took up his residence in the palace of the Romanoffs and thereafter held in his hands the reins of government. Those who do not literally believe this are nevertheless persuaded that Rasputine lived very near their Majesties, saw them constantly, was consulted and obeyed by the Ministers, and with the aid and connivance of adoring women attached to the Court, ruled by fear and superstition the whole governing class of the Empire. If I denied that Rasputine ever lived at Court, ever had the smallest influence over governmental policies, ever ruled through adoring and superstitious women, I should not hope to be believed. I will then simply call attention to the fact that every move of Rasputine from the hour when he began to frequent the palaces of the Grand Dukes, especially from the day he met the Emperor and Empress in the drawing room of the Grand Duchess Melitza, to the midnight when he met his death in the Yusupoff Palace on the Moika Canal in Petrograd, is a matter of the most minute police record. The police know how many days of each year Rasputine spent in Petrograd and how much of his time was lived in Siberia. They know exactly how many times he called at the palace at Tsarskoe Selo, how long he stayed and who was present. They know when and under exactly the circumstances Rasputine came to my house, and who else came to the house at the same time. The police know more about Rasputine than all the journalists and the historians put together, and their records show that he spent most of his time in Siberia, and that when he visited Petrograd he lived in rather humble lodgings in an unfashionable street, 54 Gorochovaia. Rasputine never lived in the palace, seldom visited it, saw the Emperor less frequently than the Empress, and had among the women of the Court more enemies than friends.

The English-speaking reader may doubt the completeness and the accuracy of police records, knowing that in his own country only criminals and people of the underworld are really watched by the police. To know what police surveillance can mean it is necessary to have known Russia before 1917. I do not speak of the Bolshevik police. It is fairly well known what they are, but after all their methods, if not their motives, are founded on the Okhrana of the old days.

To give an idea of the ever-open and searching eye of the old Russian police I will describe what the situation was in the Imperial palace itself. In connection with the palace, or any of the Imperial residences, the persons of the Emperor and his family, the police force was organized in three sections. There were the palace police, a Cossack *convoi*, and a regiment of Guards known as the *Svodny Polk*. Besides the ranking officers of these organizations there was, over them all, a palace commandant, in the latest days of the Empire, General Voyeikoff. It was impossible for anyone to approach the palace, much less to be received by one of their Majesties, without the fact being known to scores of these police guards. Every soldier, every guard, in uniform or out, kept a notebook in which he was obliged to write down for inspection by his superiors the movements of all persons who entered the palace and even those who passed its walls. Moreover, they were obliged to communicate by telephone with their superior officers every event, however trivial, of which they were witness. This vigilance was extended even to the persons of the Emperor and his family. If the Empress ordered her carriage for two o'clock in the afternoon, the lackey receiving the order immediately informed the nearest police guard of the fact. The guard telephoned the news to the palace commandant's office and from there the information went by telephone to the offices of the separate police organizations: "Her Majesty's carriage has been ordered for two o'clock." This meant that from the time the Empress and her companion, or her children, drove from the palace doors to the hour when they re-turned the roads were lined with police, ready with their notebooks to record every single incident of the drive. Should the Empress stop her carriage to speak to an acquaintance, that unhappy individual would afterwards be approached by a guard standing in the road or behind trees or shrubbery, who would demand: "What is your name, and for what reason had you conversation with her Majesty?" With all her heart the Empress detested this system of police espionage, but it was one of the Russian ironclad traditions which neither she nor the Emperor could alter or abolish.

If the Imperial Family was thus subject to police surveillance the reader can easily imagine how closely the ordinary citizen and especially citizens of eminence were watched. I would not venture to declare on my own unsupported authority that Rasputine rarely visited the palace, at first two or three times a year, and but little oftener at the last, but I can state that these facts are on record in the police annals of Petrograd and Tsarskoe Selo. In the year of his death, 1916, Rasputine saw the Emperor exactly twice. There is one unfortunate fact in connection with these visits. I write it regretfully but it is true, and I can see how that circumstance served with some people to put a false emphasis on the visits of Rasputine to the Imperial household. In spite of the well-known fact that every visit of Rasputine was necessarily a public appearance, in full limelight, as it were, the Emperor and Empress attempted to throw over his visits a certain veil of secrecy. They had done the same thing with Dr. Philippe, and I suppose from the same motives. Every human being craves a little personal privacy. In the most loving family circle who does not at times want to be alone with his thoughts or his prayers behind closed doors? Thus it was with their Majesties. Rasputine represented to them hopes and aspirations far removed from earthly power and glory, and from earthly pain and suffering. They knew that he was a simple peasant and that many people of rank in official circles thought it strange, some even thought it undignified, for their Majesties of great Russia to listen to the counsels of so lowly and ignorant a man. For this reason, I know of no other, the Emperor and Empress vainly tried to make the visits of Rasputine as inconspicuous as possible. He was admitted into a side entrance instead of the main doorway; he went upstairs by a small staircase; he was received in the private apartments and never in the public drawing rooms. It was the same in Tsarskoe Selo and in the Crimea, in which latter place a day's visit served for a year's gossip throughout the entire estate. More than once I pointed out to the Empress the futility of the course pursued. "You know that before he reaches the palace, much less your boudoir, he has been written down at least forty times," I reminded her. The Empress always agreed. She knew that the police were everywhere, inside and outside the palace, in every corridor, at every door. She knew that there could be no secrets in the palace, and the Emperor knew it as well as she did, yet they persisted in trying to shield Rasputine from the publicity they knew to be inevitable for everyone.

It was generally in the evening that he was received, not because the eternal police vigilance was relaxed at that time, but because it was only in the evening that the Emperor found leisure for his personal friends. In the hour following dinner it sometimes happened that little Alexei came downstairs in his blue nightgown to talk with his father a few minutes before going to bed. When on these occasions Rasputine was present, the boy and his parents and any intimate friend who happened to be in the room would listen fascinated while the *Strannik* talked of Siberia and its peasants, of his wanderings through remote corners of Russia, and of his sojourn in the Holy Lands. His speech was simple, but strangely eloquent and uplifting. Their Majesties talked gladly to him of whatever happened to be on their minds, the ill health of their only son, principally, and he seemed to know how to comfort and to give them hope. They were always lighter of heart after his visits, and even had I conspired with him to gain their friendship the effort would have been quite useless and unnecessary. They liked him so well that when gossip or newspaper accusations of Rasputine's drunkenness and debauchery were brought to their attention they said only: "He is hated because we love him." And that ended the matter.

I will say for the Empress that although she had the fullest confidence in Rasputine's integrity she thought it worth while to make some inquiries into his private life in Siberia, where most of his time was spent. On two occasions she sent me, with others, to his distant village of Pokrovskoe to visit him. I wished then, and I do now, that she had selected someone wiser and more critical than myself. Of detective ability I possess not a trace. With me it is always, what I have seen I have seen. In company with Mme. Orloff, mother of General Orloff, and with two other women and our maids, I made the long journey to Siberia leaving the railroad at the little town of Toumean. Here Rasputine met us with a clumsy peasant cart drawn by two farm horses. In this springless vehicle we drove eighty versts across the steppes to the village where Rasputine dwelt with his old wife, his three children, and two aged spinsters who helped in the housework and in the care of the fields and the cattle. The household was almost Biblical in its bare simplicity, all the guests sleeping in an upper chamber on straw mattresses laid on the rough board floor. Except for the beds the rooms were practically without furniture, although on the walls were ikons before which faint tapers burned. We ate our plain meals in the common room downstairs, and in the evening there usually came four peasant men, devoted friends of Rasputine, who were called "the brothers." Sitting around the table they sang prayers and psalms with rustic faith and fervor. Almost every day we went down to the river to watch Rasputine and the brothers, fishermen all, draw in their nets, and often we ate our dinner by the river, cooking fish over little campfires on the shore, sharing in common our raisins, bread, nuts, and perhaps a little pastry. The season being Lent we had no meat, no milk, nor butter.

On my return to Tsarskoe Selo I described this pastoral existence to the Empress, and I had to add to my observations only that the clergy of the village seemed to dislike Rasputine, while the majority of the villagers merely took him for granted as one they had long been accustomed to. In a later year I was again sent to Siberia, this time with Mme. Julia (Lili) Dehn, wife of a naval officer on the yacht *Standert*, and several others, and a man servant as my special assistant as I was then very lame from the railroad accident which I have described. This time we went by boat from Toumean to Tobolsk on the River Toura, to view the relics of the Metropolitan John of Tobolsk, a sainted man of the time of Peter the Great. While in Tobolsk we were entertained in the house of the Governor of the Department, the same house where in the first days of their Siberian exile the Imperial Family were lodged. It was a large, very well furnished house on the river, but one could see that in winter it must have been extremely cold. On our way back we stopped for two days at Pokrovskoe, visiting Rasputine and finding him exactly as before, the old wife and the serving maids still occupied with household tasks and with field labor. I may add that in both of these visits I went to the famous monastery of Verchotourie, on the Ural River, where are kept some deeply venerated relics of St. Simeon. In the forests surrounding the monastery are many tiny wooden huts in which dwell solitary monks or anchorites, and among these was a celebrated old monk known as Father Makari. This aged and pious monk apparently held Rasputine in higher respect than did the village clergy, and they talked together like equals and friends, while we listened silently but with deep interest.

The wave of popular opposition against Rasputine began, I should say, in the last two and a half years of his life. Long after it began, long after his name was reviled and execrated in the press and in society, his lodgings in Petrograd, where he began to spend longer and longer intervals, were constantly crowded with beggars and petitioners. These were people of all stations who believed that whether he were good or evil his influence at Court was limitless. Every kind of petty official, every sort of poverty-stricken aspirant and grafting politician, and, of course, a whole crew of revolutionary agents, spies, and secret police haunted the place, pressing on Rasputine papers and petitions to be presented to the Emperor. To do Rasputine strict justice, he was forever telling the petitioners that it would be no good at all for him to present their papers, but he did not seem to have strength of mind to refuse point-blank to receive them. Often in pity for those who were sick and poor, or as he thought deserving, he would send them to one or another of his rich and influential acquaintances with a note saying: "Please, dear friend, receive him." It is very sad to reflect that his recommendation was the worst possible introduction a poor wretch could bear with him.

One of the hardest tasks which the Empress imposed upon me was the taking of messages, usually about the health of Alexei, to these crowded lodgings of Rasputine. As often as I appeared the people overwhelmed me with demands for money, positions, advancement, pardons, and what not. It was of no use to assure the people that I neither possessed nor desired to possess the kind of influence they believed to be mine. It was equally useless to assure them that their petitions, if I took them, would not be read by the Empress, but would merely be referred to her secretary, Count Rostovseff. Sometimes I encountered a case of great distress which if possible I tried privately to relieve. One day I met on the staircase a very poor young student who asked me if I could help him to a warm coat. I knew where I could get such a coat and I sent it to the student. Months afterwards when I was a prisoner in the fortress I received a note from this young man, telling me that he prayed daily for my safety and release. This almost unique instance of gratitude remains in my mind among memories much less agreeable of my visits to the lodgings of Rasputine.

CHAPTER XII

THERE is a photograph which, in the last days of the Empire, was published all over Russia, and was, I am informed, also published in western Europe and in America. It represents Rasputine sitting like an oracle in his lodgings, surrounded by ladies of the aristocracy. This photograph is supposed to illustrate the enormous hold which Rasputine possessed on the affections of the women of the Court. In plain language it is assumed to be a repre-sentation of Rasputine sitting in the midst of his harem. There has been no account published which, as far as I know, does not dwell on this phase of the Rasputine story, and there have been books published in which the most erotic letters, purporting to have been written him by the Empress herself and even by the innocent young Grand Duchesses, have been included, the publishers apparently never having inquired into their authenticity. Knowing that my evidence will be considered of little worth, I still have the temerity to state without any qualification whatsoever that these stories are without the slightest foundation. Rasputine had no harem at Court. In fact, I cannot remotely imagine a woman of education and refine-ment being attracted to him in a personal way. I never knew of one being so attracted, and although accusations of secret debauchery with women of the lower classes were made against him by agents of the Okhrana, the special inquiry instituted by the Commission of the Provisional Government failed to produce any evidence in support of the charges. The police were never able to bring forward a single woman of any class whom they could accuse with Rasputine.

The photograph, however, is authentic. I figure in it myself, therefore I am in a position to explain it. It shows a group of women and men who after attending early Mass sometimes gathered around Rasputine for religious discourse, for advice on all manner of things, and probably on the part of some for the gratification of idle curiosity. I do not know whether or not in western countries religion produces in the neurotic and shallow-minded a kind of emotional excitement which they mistake for faith, but in Russia there was a time when this was so. For the most part, however, it was really serious people, men and women, who went after Mass to listen to the discourses of Rasputine. He was, as I have said, an unlettered man, but he knew the Scriptures and his interpretations were so keen and so original that highly educated people, even learned churchmen, liked to listen to them. In matters of faith and doctrine he could never be confused or confounded. Moreover, his sympathy and his charity were so wide and tender that he attracted women of narrow lives whose small troubles might have been dismissed as trivial by ordinary confessors. For example, many lovelorn women (men too) used to go to those morning meetings to beg his prayers on their heart's behalf. He knew that unsatisfied love is a very real trouble, and he was always gentle and patient with such people, that is, if their souls were innocent. For irregular love affairs he had no patience whatever, and in this connection I remember an incident which illustrates this point, and also his remarkable powers of divination, or if you prefer, his keen intuition. A young married woman, harmless enough in her intentions, but rather frivolous nevertheless, came one morning to Rasputine's lodgings en route to a rendezvous with a handsome young officer who at the moment strongly attracted her. It was her idea to ask Rasputine's prayers in behalf of her special desire, but before she could say a word to him he gave her a keen glance and said: "I am going to relate to you a story. Once when I was traveling in Siberia I entered a small railroad station and beheld at a table a monk who recognized me and begged me to join him in a glass of tea. As I approached the table I saw him hastily conceal a bottle under the folds of his soutaine. He said: 'You are called a saint. Will you not help me to understand some of the troubled problems of my life?' I replied 'Ah! You call me a saint. But why do you at the time of asking me to help your troubled soul try to hide that bottle under your robe?'" The young woman turned deathly pale and without a word rose hastily and left the room.

This is only one of many similar incidents. Once at Kiev a Government functionary approached Rasputine and asked his prayers for one lying very ill. Rasputine's amazing eyes gazed into the eyes of the other and he said calmly: "I advise you to beseech not my prayers but those of Ste. Xenia." The functionary completely taken aback exclaimed: "How could you know that her name was Xenia?" I could relate many other such instances which can, of course, be attributed to intuition, thought transference, anything you like. But of true predictions of future events made by Rasputine what explanation can be given? What of his mysterious powers over the sick?

In behalf of the suffering little Tsarevitch the Emperor and Empress constantly asked the prayers of Rasputine, and the incident which I shall now relate will appeal to any mother or father of a suffering child and will render less childlike the faith of the afflicted parents of the heir to the throne. One day during the War the Emperor left Tsarskoe Selo for general headquarters, taking with him as usual the Tsarevitch. The child seemed to be in good condition, but a few hours after leaving the palace he was taken with a nosebleed. This is ordinarily a harmless enough manifestation, but in one suffering from Alexei's incurable malady it was a very serious thing. The doctors tried every known remedy, but the hemorrhage became steadily worse until death by exhaustion and loss of blood was threatened. I was with the Empress when the telegram came announcing the return of the Emperor and the boy to Tsarskoe Selo, and I can never forget the anguish of mind with which the poor mother awaited the arrival of her sick, perhaps her dying child. Nor can I ever forget the waxen, gravelike pallor of the little pointed face as the boy with infinite care was borne into the palace and laid on his little white bed. Above the blood-soaked bandages his large blue eyes gazed at us with pathos unspeakable, and it seemed to all around the bed that the last hour of the unhappy child was at hand. The physicians kept up their ministrations, exhausting every means known to science to stop the incessant bleeding. In despair the Empress sent for Rasputine. He came into the room, made the sign of the cross over the bed and, looking intently at the almost moribund child, said quietly to the kneeling parents: "Don't be alarmed. Nothing will happen." Then he walked out of the room and out of the palace.

THE THREE CHILDREN OF RASPUTINE BEFORE THEIR
HOUSE IN SIBERIA.

THE GUEST ROOM (THE ONLY LARGE ROOM) IN RASPUTINE'S HOUSE IN SIBERIA.

That was all. The child fell asleep, and the next day was so well that the Emperor left for his interrupted visit to the Stavka. Dr. Derevanko and Professor Fedoroff told me afterwards that they did not even attempt to explain the cure. It was simply a fact. For this and for other like services Rasputine never received any money from the Emperor or the Empress. Indeed he was never given any money by their Majesties except an occasional one-hundred-ruble note to pay cab fares and traveling expenses when he was sent for. In the last two years of his life the rent of his modest lodgings in Petrograd was paid. What money he had was received from petitioners who hoped through him to benefit in high quarters. Rasputine took this money, but he gave most of it to the poor, so that when he died his family was left practically penniless. That Rasputine, whatever his faults, was no mercenary is the simple truth. As far back as 1913 Kokovseff, Minister of Finance, who disliked and distrusted Rasputine, offered him 200,000 rubles if he would leave Petrograd and never return. Two hundred thousand rubles was a fortune beyond the dream of avarice to a Russian peasant, but Rasputine declined it, saying that he was not to be bought by anybody. "If their Majesties wish me to leave Petrograd," he said, "I will go at once, and for no money at all."

I know of many cases of illness where the prayers of Rasputine were asked, and had he been so minded he might have demanded and been given vast sums of money. But the fact is he often showed himself extremely reluctant to exert whatever strange power he possessed. In some instances where sick children were involved he would even object, saying: "If God takes him now it is perhaps to save him from future sins."

This indifference to money on the part of Rasputine was all the more conspicuous in a country where almost every hand was stretched out for reward, graft, or blackmail. The episode of one of Rasputine's bitterest enemies, the "mad" monk Illiador, is illuminating. Illiador was a person altogether disreputable, an unfrocked monk, and in my opinion a man mentally as well as morally irresponsible. He made friends with certain ministers, among them Chvostoff, one of several who, after the death of Stolypine, held for a time the portfolio of Minister of the Interior. Between Chvostoff and Illiador was concocted a plot to assassinate Rasputine. This was not successful because Illiador made the mistake of sending his wife to Petrograd with incriminating documents. But he was able to send a woman to Siberia, and she dealt Rasputine a knife wound from which he with difficulty recovered. This was in 1914.

After Rasputine the object of Illiador's greatest hatred was the Empress. His plot against Rasputine failing, he wrote against the Empress one of the most scurrilous and obscene books imaginable, but before attempting to publish it he sent her word that he would sell her the manuscript for sixty thousand rubles. Publishers in America, he wrote, would pay him a much higher price for the book, but he was willing to sacrifice something to save a woman's reputation. To this low blackmailer the indignant Empress returned no answer at all. Illiador lives in Russia now, a great favorite with the

Bolsheviki because of his bitter attacks on the clergy. But whether or not they permitted him to retain his profits on the book against the Empress I do not know.

But what of Rasputine's political influence, his treason with the Germans? The excuse for his murder was that he was leading the Emperor and Empress into the German net, persuading them to betray the Allies by making a separate peace. If I knew or suspected this to be true I would not hesitate to record it here. I would not dare to suppress such important historical evidence, if I had it, because all that I am writing in this book is for the future, not the present; for history, not for the ephemeral journalism of the day. Ministers, politicians, churchmen haunted the lodgings of Rasputine, and if any man ever had an opportunity to mingle in secret diplomacy he was that man. As a matter of plain justice to him, I do not believe such matters ever interested him. On two occasions of which I have knowledge he did give the Emperor political advice, and very shrewd advice, although it was received with irritation and resentment by his Majesty. One of these occasions was in 1912 when Grand Duke Nicholas, whose wife it will be remembered was a Montenegran, tried his every power of persuasion to bring Russia into the Balkan Wars. Rasputine implored the Emperor not to listen to this counsel. Only enemies of Russia, he declared, wanted to involve their country in that struggle, the inevitable outcome of which would be disaster to the Empire and to the house of Romanoff.

Rasputine always dreaded war, predicting that it would surely bring ruin to Russia and the monarchy. At the beginning of the World War he was lying wounded by Illiador's assassin in Siberia, but he sent a long telegram to the Emperor begging him to preserve peace. The Emperor, believing intervention in Serbia a point of honor, tore up the telegram and for a time appeared rather cold towards Rasputine. But as the War progressed they became friends again, for after it became inevitable Rasputine wanted the War fought through to a victorious end. The last time the Emperor saw him, about a month before his assassination, he gave a signal proof of this. The meeting took place in my house, and I heard every word of the conversation. The Emperor was depressed and pessimistic. Owing to heavy storms and lack of transportation facilities there had been difficulty in getting foodstuffs into Petrograd, and even some army battalions were lacking certain necessities. Nature itself, said the Emperor, seemed to be working against Russia's success in the War, to which Rasputine replied strongly advising the Emperor never, on any account, to be tempted to give up the struggle. The country that held out the longest against adverse circumstances, he said, would certainly win the War.

As Rasputine was leaving the house the Emperor asked him, as usual, for his blessing, but Rasputine replied: "This time it is for you to bless me, not I you." Finally at parting he humbly begged the Emperor to do everything he could in behalf of the wounded and of war orphans, reminding him that all Russia was giving its nearest and dearest for his sake. Did Rasputine on this day have a premonition of the fate that was so soon to overtake him? I cannot answer that question. It is impossible for me to know with any certainty whether or not this strange man was actually gifted with the spirit of prophecy or whether his frequent forecastings of truth were simply fruits of a mind more than normally keen and observant. All I can do, all I have attempted to do, is to picture Rasputine as I knew him. I never once saw him otherwise than I have described. I knew that he was reputed to drink and to indulge in other reprehensible practices. I heard, I suppose, every wild tale that was told of him. But no one ever presented to the Imperial Family or to myself any evidence, any facts in support of these accusations. It is a matter of record, and this the historians of the future will stress, that this man was called a criminal, but that he was never meted out the common justice which is supposed to be the right of the most abandoned criminal. He was accused of nameless crimes and he was executed for those crimes. But he was denied even the rough justice of a trial by his self-appointed judges. Did "Tsarist" Russia ever do such a thing to a man caught red-handed in the murder of an Emperor?

I have added as an appendix to this book a document which has been published in Russian and French, but which I believe appears here for the first time in English. It is the statement of Vladimir Michailovitch Roudneff, a judge of a superior court in Ekaterinoslav, one of a number of distinguished jurists appointed by Kerensky, when Minister of Justice in the Provisional Government, to a special High Commission of Inquiry and Investigation into the Acts of the Sovereigns and other prominent personages before the Revolution of 1917. Judge Roudneff, with great courage and honesty, made an effort to sift the evidence against Rasputine and to separate truth from mere rumor. That he was unable to treat the matter in a mood of perfect judicial calm, although he earnestly wished to do so, is proof enough of the madness of the Russian mind in that time of turmoil and bewilderment. Anyone at all familiar with rules of evidence will perceive how, with the best intentions, Judge Roudneff often offers opinion where facts alone are called for. A great many of his statements, if given in a court of justice, would in any civilized country be challenged and probably ruled out. However, the statement is valuable because it is the unique attempt of a justice-loving individual to escape from the mob mind of 1917 Russia and to present impartially the known facts about Rasputine. For his honesty in insisting that the facts be made public Judge Roudneff was ignominiously removed from the commission by its president, Judge Mouravieff. As far as I know and believe, none of the other members of the commission attempted to publish their findings.

I shall always feel that it was a great pity that Rasputine was not arrested, tried in the presence of his accusers and of all available witnesses, and if found guilty punished to the very limit of the law. As it was he was merely lynched and the question of his guilt or innocence will ever remain unsolved. Latest accounts certainly absolve the Empress of Russia from being his tool and his guilty partner, and death, whether by assassination or at the hand of public justice, has the same end, the righteous judgment of God, and from that perfect justice not the worst enemy of the man could bar the soul of Rasputine.

One thing more I deeply regret and that is that Judge Roudneff could not have tried Rasputine in person as he did try me. I appeared before him no less than fifteen times and I always found him studious at getting at the truth, separating facts from hysterical gossip, all in the interests of justice and of historical records. In his reports concerning me there are some errors, but not serious ones, some confusion of dates, but nothing important, and once or twice some trifling injustice for which I bear not the slightest malice. Judge Roudneff, for example, accuses me of loquacity, and in my testimony of jumping irrelevantly from one thought to another. I cannot help wondering if even a learned judge, after weeks of imprisonment, accompanied by inhuman insults and bodily injuries, and for the first time given an opportunity for explanation and self-defense, would have spoken in quite a calm and normal manner. However, I do not complain of anything Judge Roudneff says of me. I am grateful to the only Russian in a position of authority who has had the chivalry to give me the benefit of a reasonable doubt.

All others, including members of the Romanoff family who have known me from my earliest childhood, who in youth danced and chatted with me at Court balls, who knew my mother and my father, with his long and honorable record, have assailed me without a shred of mercy. They have represented me as a common upstart, an outsider in society who managed through unworthy schemes to worm her way into the confidence of the Empress. They have represented me as an abandoned woman, a criminal, a would-be poisoner of the Tsarevitch. They have been so loud in their denunciations of one defenseless woman that they have succeeded in concealing the fact of their own participation in events for which the Sovereigns were brought to ruin. They have thrown a blind before their responsibility for bringing Rasputine to the Court of Russia. Never do they allow it to be remembered that it was the Grand Dukes Nicholas and Peter and their Montenegran wives, Stana and Melitza, who introduced the Emperor and Empress to the poor peasant pilgrim who, had he never been taken up by these aristocrats, might have lived out an obscure, and perhaps valuable, existence in far Siberia. It was easier for these powerful ones, these sheltered women, these noble gentlemen, to avoid explanation of their part in the Russian tragedy and to take refuge behind the skirts of a woman who, after the overthrow of the Imperial Family, had not a friend on earth to defend or to protect her.

CHAPTER XIII

TWO days after the return of the Empress from her visit to Novgorod, in the earliest hours of December 17 (December 31, Western Calendar) was struck the first blow of the "bloodless" Russian Revolution, the assassination of Rasputine. On the afternoon of December 16 (December 30) I was sent by the Empress on an errand, entirely non-political, to Rasputine's lodgings. I went, as always, reluctantly, because I knew the evil construction which would be placed on my errand by any of the conspirators who happened to see me. Yet, as in duty bound, I went. I stayed the shortest possible time, but in that brief interval I heard Rasputine say that he expected to pay a late evening visit to the Yusupoff Palace to meet Grand Duchess Irene, wife of Prince Felix Yusupoff. Although I knew that

Felix had often visited Rasputine it struck me as odd that he should go to their house for the first time at such an unseemly hour. But to my question Rasputine replied that Felix did not wish his parents to know of his visit. As I was leaving the place Rasputine said a strange thing to me. "What more do you want?" he asked in a low voice. "Already you have received all." All that his prayers could give me? Did he mean that?

That evening in the Empress's boudoir I mentioned this proposed midnight visit, and the Empress said in some surprise: "But there must be some mistake. Irene is in the Crimea, and neither of the older Yusupoffs are in town." Once again she repeated thoughtfully: "There is surely a mistake," and then we began to talk of other things. The next morning soon after breakfast I was called on the telephone by one of the daughters of Rasputine, both of whom were being educated in Petrograd. In some anxiety the young girl told me that her father had gone out the night before in the Yusupoff motor car and had not returned. I was startled, of course, and even a little frightened, but I did not then guess the real significance of her news. When I reached the palace I gave the message to the Empress, who listened with a grave face but with little comment. A few minutes later there came a telephone call from Protopopoff in Petrograd. The police, he said, had reported to him that some time after the last midnight a patrolman standing near the entrance of the Yusupoff Palace had been startled by the report of a pistol. Ringing the doorbell, he was met by a Duma member named Puritchkevitch who appeared to be in an advanced stage of intoxication. In answer to the policeman's inquiry as to whether there was trouble in the house the drunken Puritchkevitch said in a jocular tone that it was nothing, nothing at all, only they had just killed Rasputine. The policeman, probably a none too intelligent specimen, took it as a casual joke of one of the high-born. They were always joking about Rasputine. The man moved on, but somewhat later he decided that he ought to report the matter to headquarters, which he did, but even then his superiors appear to have been too incredulous to act at once.

Protopopoff's message, however, so disquieted the Empress that she asked me to summon another of her trusted friends, Mme. Dehn, whose name I have mentioned before. Mme. Dehn came and we talked over the mystery together, but still without conviction that Puritchkevitch's reckless statement contained any real truth. Later in the day, however, came a telephone message from Grand

Duke Dmitri Pavlovitch, asking to be allowed to take tea with the Empress that afternoon at five. The message was conveyed to the Empress, who, pale and reflective, answered formally that she did not care just then to receive his Highness. Dmitri took the reply in bad grace, insisting that he must see the Empress as he had something special to tell her. Again the Empress refused, this time even more curtly. Almost immediately afterwards, almost as if the two men were in the same room, there came a telephone message from Felix Yusupoff asking if I would see him at tea, or later in the day if I so preferred. I answered that the Empress did not wish me to receive any visitors that day, whereupon Felix demanded an audience with the Empress that he might give her a true account of what had occurred. Her Majesty's reply was: "If Felix has anything to say let him write to me." Several times before the day ended telephone messages came from Felix to me, but none of these would the Empress allow me to answer.

Felix finally wrote a letter to the Empress. I cannot quote this letter verbatim, but I remember exactly its contents. By the honor of his house Prince Felix Yusupoff swore to his Sovereign Empress that the rumor of Rasputine's visit to his home was without any foundation whatever. He had indeed seen Rasputine in the interests of Irene's health, but he had never decoyed the man to his palace, as charged. There had been a party there, on the night in question, just a few friends, including Dmitri, to celebrate the opening of Felix's new apartments. All, he confessed, became drunk, and some foolish and reckless things were said and done. By chance, on leaving the house, one of the guests had shot a dog in the courtyard. That was absolutely all. This letter was not answered, but was turned over to the Minister of Justice.

Thoroughly aroused, the Empress now ordered Protopopoff to make an investigation of the whole affair. She called into council also Minister of War Belaieff, a good man, afterwards murdered by the Bolsheviki. The police, at their commands, went to the deserted Yusupoff palace, first searching for and finding the body of the dog which Felix said they had shot. But the bullet hole in the dog's head had let out little blood, and when the men entered the palace they found it a veritable shambles of blood and disorder. Evidences of a terrific struggle were found in the downstairs study of Prince Felix, on the stairs leading to an upper room, and in the room itself. Then, indeed, the whole power of the police was invoked, and somebody was found to testify that in the dead of night a motor car without any lights was seen leaving the Yusupoff Palace and disappearing in the direction of the Neva. Winter nights in Russia are very dark, as everyone knows, and the car was soon swallowed up in the shadows. The river was next searched, and by a hole in the ice, not far from Krestovsky Island, the police found a man's golosh. By Protopopoff's orders divers immediately searched the hole in the ice, and from it was soon dragged the frozen body of Rasputine. Arms and legs were tightly bound with cords, but the unfortunate man had managed to work loose his right hand which was frozen in a last attempt to make the sign of the cross. The body was taken to the Chesma Hospital, where an autopsy was performed. Although there were bullet holes in the back and innumerable cuts and wounds all over the body, the lungs were full of water, proving that they had thrown him alive into the icy river, and that death had occurred by drowning.

As soon as the news became public all Petrograd burst into a wild orgy of rejoicing. The "beast" was slain, the "evil genius" had disappeared never to return. There was no limit to the wild hysteria of the hour. In the midst of these demonstrations came a telephone message from Protopopoff asking the Empress's advice as to an immediate burial place for the murdered man. Ultimately the body would be sent to his Siberian village, but in the present circumstances the Minister of the Interior thought a postponement of this advisable. The Empress agreed, and she replied that a temporary interment might be arranged at Tsarskoe Selo. On December 29 (January 12) the coffin, accompanied by a kind-hearted sister of mercy, arrived at Tsarskoe. That same day the Emperor came home from the front, and in the presence of the Imperial Family and myself the briefest of services were held. On the dead man's breast had been laid an ikon from Novgorod, signed on the reverse by the Empress and her daughters as a last token of respect. The coffin was not even buried in consecrated ground, but in a corner of the palace park, and as it was being lowered a few prayers were said by Father Alexander, priest of the Imperial chapel. This is a true account of the burial of Rasputine, about which so many fantastic tales have been embroidered.

The horror and shock caused by this lynching, for it can be called by no other name, completely shattered the nerves of the family. The Emperor was affected less by the deed itself than by the fact that it was the work of members of his own family. "Before all

Russia," he exclaimed, "I am filled with shame that the hands of my kinsmen are stained with the blood of a simple peasant." Before this he had often shown disgust at the excesses of the Grand Dukes and their followers, but now he expressed himself as being entirely through with them all.

THE EMPRESS AND YOUNG GRAND DUKE DMITRI, AFTERWARDS ONE OF RASPUTINE'S ASSASSINS.

But Yusupoff and the others were by no means through with the Rasputine affair. Now that they had murdered and were applauded for the deed by all society, it seemed to them that they were in a position to claim full legal immunity. Grand Duke Alexander Michailovitch, the Emperor's brother-in-law, went to Dobrovolsky, Minister of Justice, and with a good deal of swagger told him that it was the will of the family—that is, of the Grand Dukes—that the whole matter should be quietly dropped. The next day, December 21 (January 5), Alexander Michailovitch drove with his oldest son to Tsarskoe Selo and, without the slightest assumption of deference or respect, entered the Emperor's study, demanding, in the name of the family, that no further investigation of the manner of Rasputine's death be made. In a voice that could easily be heard in the corridor outside the Grand Duke shouted that should the Emperor refuse this demand the throne itself would fall. The Emperor's answer to this insolence was an order of banishment to their estates of Nicholai Michailovitch, Felix, and Dmitri. At this the wrath of the Grand Dukes knew no bounds. A letter blazing with anger and impudence, signed by the whole family, was rushed to the Emperor, but his only comment was a single sentence written on the margin:

MINISTER OF COURT COUNT FREDERICKS, THE EMPRESS IN HER RECLINING CHAIR, AND GRAND DUCHESS TATIANA TAKING TEA IN THE WOODS IN FINLAND, 1911.

Nobody has a right to commit murder." Following this came a cringing letter from Dmitri who, like Felix, tried to lie himself out of all complicity in the crime. On his sacred honor, he declared he had had nothing to do with it. If the Emperor would only consent to see him he promised to establish his innocence. But the Emperor would not consent to see Dmitri. Pale and stern he moved through the rooms or sat so darkly plunged in thought that none of us ventured to disturb or even to speak to him. Into this troubled atmosphere a letter was brought to the Emperor by the Minister of the Interior, who had a right to seize suspicious mail matter. It was a letter written by the Princess Yusupoff to the Grand Duchess Xenia, sister of the Tsar and mother of Felix Yusupoff's wife. It was a most indiscreet letter to be sent at such a time, for it was a clear admission of the guilt of all the plotters. Although as a mother (she wrote) she felt deeply her son's position, she congratulated the Grand Duchess Xenia on her husband's conduct in the affair.

Sandro, she said, had saved the whole situation, evidently meaning that his demand for immunity for all concerned would have to be granted. *She was only sorry that the principals had not been able to bring their enterprise to its desired end.* However, there remained only the task of confining *Her*. Before the affair was finally concluded, she feared, they might send Nicholai Nicholaievitch and Stana to their estates. How stupid to have sent away Nicholai Michailovitch!

This was by no means the end of letters and telegrams seized by the police and brought to the palace. Many were written by relatives and close friends, people of the highest rank, and they all revealed a depth of callousness and treachery undreamed of before by the unhappy Sovereigns. When the Empress read these communications and realized that her nearest and dearest connections were in the ranks of her enemies, her head sank on her breast, her eyes grew dark with sorrow, and her whole countenance seemed to wither and grow old. A few days later the Grand Duchess Serge sent her sister several sacred ikons from the shrine of Saratoff. The Empress, without even looking at them, ordered them sent back to the convent of the Grand Duchess in Moscow.

I should add that from the day of the assassination of Rasputine my mail was full of anonymous letters threatening me with death. The Empress, perhaps more than any of us, instinctively aware of the endless ramifications of the Rasputine affair, commanded me in terms that admitted of no argument to leave my house and to take up residence in the palace. Sad as I was to leave the peace of my little home, I had no alternative than to obey, and with my maid I moved into two rooms in the Grand Ducal wing of the palace, occupied also by maids of honor and reached by the fourth large entrance to the palace. From that day, by command of their Majesties, every movement of mine was closely guarded. The soldier Jouk was assigned to my service and without him I never left the palace even to visit my hospital. When in the February following my only brother was married I was not allowed to attend the wedding.

Little by little, in spite of fears, the palace took on a certain air of tranquillity. In the evenings we sat in the mauve boudoir of the Empress; and as of old, the Emperor read aloud. At Christmas their Majesties saw that the customary trees and gifts were sent to the hospitals and that the usual presents were distributed to the servants. The children too had their Christmas celebration, but over us all hung a cloud of sorrow and of disillusionment. Never had the Emperor and Empress of Russia, rulers of nearly two hundred million souls, seemed so lonely or so helpless. Deserted and betrayed by their relatives, calumniated by men who, in the eyes of the outside world, seemed to represent the Russian people, they had no one left except a few faithful friends, and the Emperor's chosen ministers every one of whom was under the ban of popular obloquy. Most of them were accused of being the appointees of Rasputine, but this at least I am in a position to deny.

Sturmer, Minister of the Interior, and afterwards Prime Minister, was, according to Witte, recommended to the Tsar after the assassination of Pleve. The well-known fact that Sturmer was head of the nobility in the Government of Tver, that he was possessed of enormous estates, and that he had held several important positions at Court, ought to be sufficient proof that he needed no help from Rasputine or any other man. Sturmer was an old man, not brilliant perhaps, but certainly a man of high principles. He was arrested by the Provisional Government, and in the fortress suffered such frightful hardships that he died within a day after the Government, unable to fasten on him the slightest guilt, released him from prison. The Social Revolutionary Sokoloff, a just man, if wrong-headed, has declared publicly that had any Constitutional Assembly been held in Russia, the responsibility of Sturmer's death would have been laid upon Milukoff personally.

As for Protopopoff, he was appointed by the Emperor mainly on his record as a confidential agent of the Duma, and as a personal representative of Rodzianko, President of the Fourth Duma. After Protopopoff's return from an important foreign mission on behalf of the Duma he was presented to the Emperor at G. H. Q., and in a letter to the Empress a few days later, he expressed himself as delighted with the man. The appointment was made in one of those moments of impulse characteristic of Nicholas II, yet it must have been the result of some reflection, as it was the Emperor's expressed desire at this time to name a Minister of the Interior who could work in harmony with the Duma. Protopopoff, who, aside from his relations with Rodzianko, had for many years been a delegate from his own Zemstvo to the Union of Zemstvos, naturally appealed to the Emperor as an ideal popular candidate. No one could have been more astonished than he when, almost immediately after his appointment, Rodzianko and almost the entire majority party in the Duma joined in a clamor for Protopopoff's removal. The only charge I ever heard against him was that his mind had suddenly failed. Protopopoff, who was a man of high breeding, was nevertheless exceedingly nervous, and I always thought, somewhat weak-willed. He was not the infirm old man he has generally been represented, being about sixty-four years of age with white hair and mustache and young, bright black eyes. That he had plenty of physical and moral courage was proved by his conduct after the Revolution. Walking to the door of the council chamber of the Duma he announced himself thus: "I am Protopopoff. Arrest me if you like." He was arrested by orders of Rodzianko, but was released later, only to meet death by the bullets of the Bolsheviki. That Protopopoff was on friendly terms with Rasputine is true, but that Rasputine had anything to do with his appointment, or with his retention in office after the attack by the Duma, is simply absurd.

Maklakoff, Minister of the Interior before Protopopoff, was a former governor of Chernigoff. The Emperor met him in the course of a journey to the famous fête of Poltava, a jubilee of the wars of Peter the Great. The acquaintance was made in the leisure of a boat trip, and the Emperor, in another of his fits of impulsiveness, decided that he had found an ideal Minister of the Interior. Their friendship deepened with time, and the Emperor found great satisfaction in his new minister's reports, which he declared reflected his own point of view. Nothing against the administration of Maklakoff was ever even whispered until late in 1914, when Nicholai Nicholaievitch, as supreme commander of the Russian forces in the field, suddenly demanded his demission. Grand Duke Nicholas, it must be said, continually interfered with the affairs of the interior government, with which as military chief he had nothing whatever to do, but in the early days of the War the Emperor seemed to think it the part of wisdom to suffer this irregularity. Reluctantly he yielded to the request for Maklakoff's demission, saying to him with genuine regret: "They demand it, and at such a time I cannot stand against them."

In the place of Maklakoff was named Tcherbatkoff, a friend and protégé of Nicholai Nicholaievitch, a man whose former office had been head of the remount department of the State. Doubtless he knew a great deal about horses, but of the interior affairs of State he knew so little that even the influence of Grand Duke Nicholas was powerless to retain him in office longer than two months.

Tcherbatkoff was followed by Khvostoff who, previous to his appointment, was an entire stranger to Rasputine. Khvostoff had made a record as governor of Nizjni Novgorod, and afterwards as a vigorous anti-German orator in the Duma. He was also supposed to be a devoted friend of the Imperial Family. Soon after his appointment Khvostoff began sedulously to cultivate the friendship of Rasputine, and it is a matter of police record that this Minister of the Interior frequently played on Rasputine's unfortunate weakness for drink. Possibly he thought that by getting the poor man intoxicated he could worm from him the many Court secrets he was supposed to possess. Failing in this Khvostoff began, with the help of Chief of Police Belezky, a plot against Rasputine which nearly succeeded in the latter's assassination. This being discovered the demission of Khvostoff became imperative.

Soukhomlinoff, who when I knew him was an old man of seventy-five, was a former military governor of Kiev, and before his appointment as Minister of War, had been a great favorite of the Emperor. That he showed brilliant ability in the mobilization of the Russian Army in 1914 was admitted by the Allied Governments, and in fact no intrigue against him developed until some time after the beginning of the War. His principal enemies were Grand Duke Nicholas, General Polivanoff, and the notorious Goutchkoff. In my opinion their propaganda against him was instigated solely with the object of impairing the prestige of the Emperor. The crimes laid at the door of Soukhomlinoff were almost countless. He was accused of withholding ammunition from the armies, of harboring German spies in his house, and in general of being completely incapable of performing his duties of office. Of him the English historian Wilton says that time alone will prove whether the odium of the Russian war scandals rested on Soukhomlinoff or on Grand Duke Nicholas. At all events it was poor old Soukhomlinoff who was arrested, tried before a tribunal of the Provisional Government, and sentenced to life imprisonment. His young wife, who was arrested with him, occupied a cell next to mine in the Fortress of Peter and Paul, and without regard to the charges brought against her, I had reason constantly to admire the courage and self-possession with which she bore the hardships of prison life. So great was her dignity and self-command that she became universally respected by the soldiers, and I am confident that this alone saved us both from far worse indignities than those which we were called upon to bear. In prison Mme. Soukhomlinoff managed to keep herself constantly occupied. She wrote and read whenever writing materials and books were procurable, and her clever fingers fashioned out of scraps of the miserable prison bread really beautiful sprays of flowers. For coloring matter she used the paint from a moldering blue stripe on the walls of her cell, and scraps of red paper in which tea was wrapped. After months of imprisonment, bravely endured, Mme. Soukhomlinoff was brought to trial before a court of the Provisional Government. Her examination was of the most searching character, but at its close she left the courtroom fully acquitted, to the applause of the numerous spectators. Taking advantage of an amnesty pronounced some time later Mme. Soukhomlinoff got her aged husband released from prison and saw him safely to Finland. It is rather an anticlimax to the story that after so many trials borne together the marriage of the Soukhomlinoffs was dissolved, Mme. Soukhomlinoff marrying a young Georgian officer with whom she later perished under the Bolshevist terror.

One more person of whom I can speak with knowledge was, although not a minister, falsely alleged to be an appointee of Rasputine. This was the Metropolitan Pitirim, a man of impeccable honesty and very liberal views regarding Church administration. The Emperor met him in late 1914 on one of his visits to the Caucasus, Pitirim then being Exarch of Georgia. Not only the Emperor but his entire suite were enchanted by the charming manners, the piety, and learning of the Exarch, and when, a little later, the Empress met the

Emperor at Veronesh, he told her that he had Pitirim in mind for Metropolitan of Petrograd. Almost immediately after his appointment the propagandists began to connect his elevation with the Rasputine influence, but the truth is that the two men were never at any time on terms of more than formal acquaintanceship. As for their Majesties, they liked and respected Pitirim but he never was an intimate member of their household. Practically all their conversations which I overheard concerned the state of the Church in Georgia, which Pitirim insisted was lower than in other parts of the Empire. The Church of Georgia, Pitirim alleged, received too little support from the State, although it deserved as much if not more than others, because Georgian Christianity is the oldest in all Russia. According to tradition this Church was established by the Holy Virgin herself who, after a shipwreck off Mount Athos, visited Georgia, converted its chiefs and established the first Christian temple. Pitirim was essentially a churchman, yet he always advocated a certain separation of Church and State. That is, he desired the establishment of a parish system whereby the support of the Church should be the responsibility of the people rather than of the Imperial Government. Unworldly to the last degree, he nevertheless came in for his full share of slander and abuse. After my arrest by the Provisional Government my mother visited Kerensky in my behalf, and was astounded when he brutally told her that one of the charges against me was that all my diamonds were gifts from Pitirim, the inference being that we were on unduly intimate terms.

Another high personage to whom I wish to pay the tribute of just appreciation is Count Fredericks, chief minister of the Court. This honorable gentleman had spent almost his entire life in the service of the Imperial Family, having first been attached to the person of Alexander III. Nicholas II and his family he served with ability, discretion, and rare devotion. In virtue of his office he had to deal personally with the affairs of the Grand Dukes, their complicated financial transactions, their morganatic marriages, and other confidential affairs. Everyone, except those of the Grand Dukes who with reason had earned his contempt, loved this charming man whom their Majesties usually spoke of as "our old man." Count Fredericks, in his turn, always called them "*mes enfants.*" His house was to me for many years a second home, his daughters, the elder Mme. Voyeikoff, and the younger one, Emma, being among my dearest friends. Emma, who suffered a painful curvature of the spine, had the compensation of a rarely beautiful singing voice with which she often charmed the Emperor and Empress. Count Fredericks was arrested by the Provisional Government, but owing to his great age, was afterwards released.

The charge has often been brought against Nicholas II that he surrounded himself with inferior men. The fact of the case is that in the beginning of his reign he chose as his chief advisers men of ability and integrity who had been friends of his father, Alexander III. Later he chose men who in his opinion were the best ones available, and it must be admitted that there were few men of first-class ability among whom he could choose. The events of the War and the Revolution prove this, for neither of these two terrible emergencies produced in Russia a single man of conspicuous merit. Not one real leader appeared then nor in the years which have since elapsed. Truly has a distinguished American writer pointed out that never could Bolshevism and its insane philosophy have taken such strong roots in Russia, had not the soil been previously so well prepared. Every Russian who really loved his country must admit the truth of this statement. Too many exiled Russians, however, still cling to the delusion that some outside influence was the cause of their country's downfall. Let them acknowledge the truth that it was Russians themselves, especially Russians of the privileged classes, who principally are responsible for the catastrophe. For years before the Revolution the national spirit was in a state of decline. Few men or women cherished ideals of duty for duty's sake. Patriotism was practically extinct. Family life was weakened, and in the last days, the morale of the whole people was lower than in almost any other country of the civilized world.

May the blood of the thousands of innocents who have perished in War and Revolution wipe out the sins of the old hard-hearted and decadent Russia. May the millions still living, in exile and under Communist oppression, learn that only by repentance and by toleration of others' weaknesses can there be any possibility of a restoration of national life. Not by any outside help but by our own efforts, by loyal Russians coming together, not as political groups but as compatriots, can great Russia rise again out of her shame and desolation and become once more a nation among the nations of the earth.

* * *

CHAPTER XIV

FOR two months after the assassination of Rasputine the Emperor remained at Tsarskoe Selo, but he was by no means idle. In fact his whole heart and mind were occupied, not so much with the scandal that had reached its tragic climax in the Yusupoff Palace, but with the War which at that moment seemed to favor Russian arms. According to our advices the food shortage in Germany and in Turkey had become acute, and the Emperor believed that a vigorous spring offensive might bring the War to a speedy close. In his billiard room were spread out a large number of military maps which no one of the household, not even the Empress, was invited to inspect. The Emperor spent hours over these maps and his plan of a spring campaign, and when he left the billiard room he locked the door and put the key in his pocket. I had never seen him more completely the soldier, the commander in chief of a great army. All this time, from December, 1916, to February, 1917, the Russian front was comparatively quiet, furious snowstorms preventing the advance either of our own or the enemy's forces. Alas! The storms interfered also with railroad transport and Petrograd and Moscow were beginning to feel the pinch of hunger, a fact that gave their Majesties constant concern.

Meanwhile the Grand Duke Alexander Michailovitch persisted in his demand for an interview with the Empress, and as his letters to her failed of their object he began to write to the Grand Duchess Olga. The Empress, whose courage was great enough to enable her to ignore any possible danger to herself, decided to see the man and once for all let him have his say. In this decision the Emperor concurred, but he stipulated that he should be present in case the conversation should become unduly disagreeable. The Emperor's aide-de-camp for the day happened to be a spirited young officer, Lieutenant Linevitch, who after luncheon on the day set for the audience, lingered in the palace, apparently occupied in an amusing puzzle game with Tatiana. Afterwards Linevitch told me that so well did he know the extent of the Grand Ducal cabal, and especially the character of Alexander Michailovitch, that he had remained on purpose and that his sword had been ready at any moment to rescue the Empress from insult or from attempted assassination. As we expected the Grand Duke had nothing new to say to the Empress, but merely reiterated in more than usually violent terms the demand for Protopopoff's dismissal and for a constitutional form of government. The answer to these demands was as usual—everything necessary after the War, no fundamentally dangerous changes while the Germans remained on our soil. The Grand Duke, purple with anger, rushed out of the Empress's sitting room, but instead of leaving the palace, as he was expected to do, he entered the library, ordered pens and paper and began to write a letter to the Emperor's brother, Michail Alexandrovitch. No sooner had he begun his epistle than he perceived standing respectfully in the room the aide-de-camp Linevitch, whom, after a more or less civil greeting, he tried to dismiss. "You may go now," he said, coldly polite, but the astute Linevitch replied with ceremony: "No, your Highness, I am on service today and as long as your Highness is here it is not permitted for me to leave." In a fury Alexander Michailovitch got up and left the palace.

Men like Linevitch and many others, as faithful as ever to their Majesties, saw the threatening tempest more clearly than those within palace walls could possibly see it. The day after the visit of Alexander Michailovitch I received a call from one of the finest of the Romanoff connections, Duke Alexander of Luchtenberg. Pain-fully agitated, the Duke told me that he wanted me to help him to induce the Emperor to take a remarkable, indeed an unprecedented step. At the time of his accession to the throne every member of the family, it is well known, must make a solemn vow of fealty to the Tsar, and the Duke of Luchtenberg now begged me to persuade the Emperor, through the Empress, to exact from all the family a renewal of this vow. For the lives and safety of the Imperial Family the Duke believed this to be absolutely essential. "None of them are loyal, not one," he said earnestly. "And if the Emperor values the lives of his wife and children he must force the Grand Dukes and their families to declare themselves." Quite staggered, I replied that it was impossible for me to make such a proposition to their Majesties, but I added that the Duke himself, as a member of the family, might with entire propriety do so, and thus the matter was decided. Of the details of the conversation between the Emperor and his kinsman I know nothing, but I know that the conversation took place, because later the Emperor remarked in my hearing that "Sandro" Luchtenberg, in the kindness of his heart, had made a great matter out of a trifle, and he added, "Of course I could not ask of my own family the thing he suggested."

As one more indication of the gathering storm there came to me at my hospital from Saratoff an old man so feeble and so deaf that he had to bring with him a woman relative who through long familiarity was able to act as an interpreter in his conversations. This old man represented an organization known as the Union of the Russian People, a large group devoted to the Empire and to the persons of their Majesties. With intense emotion he told me that his organization had incontestable proofs of most treacherous propaganda which was being circulated by the Union of Zemstvos and Towns, under the personal direction of Goutchkoff and Rodzianko. He had brought with him documentary proofs of his assertions and he implored me to help him lay his proofs before the Emperor. I communicated his message to the Emperor, but as he was that day importantly engaged he suggested that the Empress might receive him instead. This she consented to do, but after an hour's conversation she sent the old man away, touched by his devotion but unconvinced of the gravity of the situation as he presented it.

To relieve somewhat the dullness and gloom that had settled on the palace we organized in those early winter days of 1917 a series of chamber-music recitals, the performers being Rumanian musicians who had been playing very beautifully in the convalescent wards of the Tsarskoe Selo hospitals. At the request of the Empress I arranged for performances in my own apartments in the palace, inviting, with their Majesties' approval, the Duke of Luchtenberg, Mme. Dehn, Count Fredericks, his daughters, my sister and her husband, and a few other intimate friends. The concerts were delightful, greatly cheering us all, including the somewhat lonely young Grand Duchesses and the much harassed Emperor. But something in the music, perhaps its wild and mournful tzigane numbers, moved the Empress to the depths of her sensitive soul. Her beautiful eyes became more than ever filled with melancholy and her heart seemed heavy with premonitions of disaster.

Partly because of her increased melancholy and partly moved by just anger against the propagandist press in which our innocent concerts were described as "palace orgies," the Emperor for the first time was awakened to consciousness that the safety of his family was indeed threatened. At least he became aware of the fact that despite the dangerous unrest of the times, Tsarskoe Selo and even Petrograd remained practically ungarrisoned. The capital was guarded by only a few regiments of reserves, while Tsarskoe Selo, the residence of the Imperial Family, had no regiments at all outside its peace-time quota of soldier and Cossack guards. At the command of the Emperor several additional regiments which had served for some time at the front were ordered to Tsarskoe for rest and recuperation, and, although naturally nothing of this was mentioned in the order, to augment if necessary the inadequate military force at hand. The first order was given for a strong detachment of naval guards, but after these men were actually entrained for Tsarskoe they were stopped by a counter order from General Gourko, who in the illness of General Alexieff was in command at G. H. Q. This counter order being at once communicated to the Emperor, he exercised his supreme authority and the regiment once more started for Tsarskoe Selo. But the audacity of General Gourko had not yet reached its limit. When the military train reached the station at Tsarskoe it was met by a telegram from General Gourko to the officer in command, ordering the regiment back to the front. The bewildered officer for a few moments was at a loss what to do, but fortunately news of his dilemma was telephoned to the palace, and the regiment, under the peremptory command of the Emperor, left the train and went into garrison at Tsarskoe. The Emperor next commanded that one of his favorite regiments of Varsovie Lancers be sent to Tsarskoe, but instead General Gourko left headquarters for the palace, where a long interview between the Emperor and the commander took place. By arguments of which I have no knowledge the Emperor was persuaded that the Lancers could not, for the time being, be spared from their front-line position, and he recalled his order.

However, it was clear that the Emperor was at last awake to the appalling menace of disaffection which was closing in like black cloud banks on every hand. The War was going badly, as every student of the times must remember. Brusiloff's brilliant offensive of the summer and autumn of 1916 had indeed made it plain that Russia was by no means out of the struggle, but although this famous drive had netted the Russians a gain of territory even larger than that which was yielded in the great Battle of the Somme, it had finally stopped leaving us with much lost territory still unredeemed. The Emperor knew this and it tormented his heart and soul. The intriguers knew it and resolved to use it as a weapon to get the Tsar away from his capital and from his family. It was on the 19th or 20th of February (Russian Calendar) that the Emperor's brother, Grand Duke Michail Alexandrovitch, visited the palace and told the Emperor that it was his immediate duty to return to the Stavka because of grave threats of mutiny in the army. Very reluctantly the Emperor consented to go. Mutiny in the army was a serious enough matter and demanded the presence of the commander in chief. But other things were at the same time occurring to cause keen anxiety. The Empress had acquainted me with the nature of these dis-quieting events, but because of the international character of the most serious I dislike even now to put them in writing. However, I am here repeating only what was then told me and I have no firsthand information to offer in verification of their truth. Their Majesties had been informed and finally from a source which they believed to be absolutely reliable, that the center of intrigue against the throne was not in any secret garret of disaffected workingmen but in the British Embassy, where the Ambassador, Sir George Buchanan, was personally aiding the Grand Dukes to overthrow Nicholas II and to replace him by his cousin Grand Duke Cyril Vladimirovitch. Sir George Buchanan's main purpose, it was said, was not so much to further the ambitions of the Grand Dukes as it was to weaken Russia as a factor in the future peace conference. Unable fully to believe that an ambassador of one of the Allied Powers would dare to meddle maliciously in the internal affairs of the Empire, the Tsar had nevertheless decided to communicate his information in a personal letter to his cousin King George of England. The Empress, deeply indignant, advised a demand on King George for the Ambassador's recall, but the Emperor replied that he dared not, at such a critical time, make public his distrust of an Ally's representative. Whether or not the Emperor ever wrote his letter to King George I never knew, but that his anxiety and depression of spirits persisted I can well testify. On the evening of February 29, the day before the Emperor's departure, I gave a small dinner to some intimate friends among the officers of the Naval Guard, Mme. Dehn helping me in my duties as hostess. A note from the Empress summoned us all to spend the end of the evening in her sitting room, and as soon as I saw the Emperor I knew that he was seriously upset. During the tea hour he spoke little, and when I tried to catch his eye he turned his head aside. The Empress murmured in my ear that all his instincts warned him against leaving Tsarskoe Selo at that time, and as this coincided exactly with my own judgment I ventured to tell him, on saying good night, that I should hope to the last moment that he would not go away until the worst of the uncertainties in Petrograd were removed. At this he smiled, almost cheerfully, and said that I must not allow myself to be frightened by wild rumors and idle gossip. Go he must, but within ten days he expected to be able to return.

The next morning I went to the door and watched his motor car drive out of the palace grounds, the Empress and the children going with it as far as the station. As usual on such occasions, there was a display of flags, of guards standing at salute, and bells from the churches pealing their farewell. Everything appeared the same, yet in that hour the flags, the soldiers, the pealing bells were speeding the Tsar of all the Russias to his doom.

I felt ill that morning, ill physically as well as mentally, yet as in duty bound I went to my hospital, where a soldier in whose case I took a special interest was to undergo an operation which he dreaded and at which he had implored me to be present. While the anesthetic was being administered I stood beside the poor man holding his hand, but at the same time I realized that I was becoming feverish and that my headache was almost unbearably increasing. Returning to the palace, I lay down in my bedroom, after writing a line to the Empress excusing myself from tea. An hour later Tatiana came in, sympathetic as usual, but troubled because both Olga and Alexei were in bed with high temperatures and the doctors suspected that they might be coming down with measles. A week or two before some small cadets from the military school had spent the afternoon playing with Alexei, and one of these boys had a cough and such a flushed face that the Empress had called the attention of M. Gilliard to the child, fearing illness. The next day we heard that he was ill with measles, but because our minds were so troubled with many other things none of us thought much of the danger of contagion. As for me, even after Tatiana had told me that Olga and Alexei were suspected cases, it did not at once occur to me that I was going to be ill. Still my temperature went on rising and my headache was unrelieved. I lay in bed all the next day until the dinner hour when Mme. Dehn came in and I made a futile effort to get up and dress. Mme. Dehn made me lie down again, and looking me over carefully she said: "You look very badly to me. I think you will have to have the doctor." The next instant, so it seemed to me, the doctor was in the room and I heard him say: "Measles. A bad case." Then I drifted off into sleep or unconsciousness.

That same day Tatiana fell ill, and now the Empress had four of us on her hands. Putting on her nurse's uniform, she spent all the succeeding days between her children's rooms and mine. Half conscious, I felt gratefully her capable hands arranging my pillows, smoothing my burning forehead, and holding to my lips medicines and cooling drinks. Already, as I heard vaguely, Marie and Anastasie had begun to cough, but this news disturbed me only as a passing dream. I was conscious of the presence of my mother and father and of my younger sister, and still as in a kind of nightmare I understood that they and the Empress spoke in hurried whispers of riots and disorders in Petrograd. But of the first days of Revolution, the strikes in Petrograd and Moscow, the revolt of the mobs and the hesitancy of the half-disciplined reserves to restore order, I know nothing except what was afterwards related to me. I do know, however, that through it all the Empress of Russia was completely calm and courageous, and that when my sister, hurrying to the palace after witnessing the wild scenes in Petrograd, had cried out to the Empress that the end had come, her fears were quieted by brave and reassuring words.

It was the devoted old Grand Duke Paul, as the Empress afterwards told me, who brought her the first official tidings, and made her understand that that most calamitous of all blunders, a political revolution in the midst of world war, had been accomplished. Even then she lost none of her marvelous courage. She did not call upon the Ministers or upon the Allied Ambassadors to protect her and her children. With dignity, unmoved she witnessed day by day the cowardly desertion of men who for years had lived at Court and who had enjoyed the faith and friendship of the Imperial Family. One by one they went, General Racine, Count Apraxine, officers and men of the bodyguard, servants the oldest and the most trusted, all with smooth excuses and apologies which translated meant only *sauve qui peut.*

One night came the noise of rioting and the sharp staccato of machine guns apparently approaching nearer and nearer the palace. It was about eleven o'clock and the Empress was sitting for a few minutes' rest on the edge of my bed. Getting up hastily and wrapping herself in a white shawl, she beckoned Marie, the last of the children on her feet, and went out of the palace into the icy air to face whatever threatened. The Naval Guard and the Konvoi Cossacks still remained on duty, although even then they were preparing to desert. It is altogether possible that they would have gone over to the rioters that night had it not been for the unexpected appearance of the Empress and her daughter. From one guard to another they passed, the stately woman and the courageous young girl, undaunted both in the face of deadly danger, speaking words of encouragement, and most of all of simple faith and confidence. This alone held the men at their posts during that dreadful night and prevented the rioters from attacking the palace. The next day the guards disappeared. The Naval Guards, led by Grand Duke Cyril Vladimirovitch,* marched with red flags to the Duma and presented themselves to Rodzianko as joyful revolutionists. The very men who in the previous midnight had hailed the Empress with the traditional greeting, "*Zdravie Jelaim Vashie Imperatorskoe Velichestvo!*" Health and long life to your Majesty! So loud had been their greeting that the Empress, not wishing me to know that she had left the palace, sent a servant to tell me that the Guards were waiting to meet the Emperor.

There was now in or about the palace practically no one to defend the Imperial Family in case the mob decided to attack. Still the Empress remained calm, saying only that she hoped no blood would have to be shed in their defense. A telegram from the Emperor revealed that the crisis had become known to him, for he implored the Empress to join him with the children at headquarters. At the same hour came an astounding message to the Empress from Rodzianko, now head of the Provisional Government, notifying her that she and her whole family must vacate the palace at once. Her answer to both messages was that she could not leave because all five of the children were dangerously ill. Rodzianko's reply to this appeal of an anguished mother was: "When the house is on fire it is time for everything to be thrown out." Desperately the Empress consulted doctors and nurses. Could the children possibly be moved? Could Anna? What was to be done in case the Provisional Government proved altogether pitiless?

Into this soul-racking dilemma of the mother came to the wife of the Emperor the terrible news of his abdication. I could not be with her in that hour of woe, nor did I even see her until the following morning. It was my parents who broke the news to me, almost too ill and too cloudy of mind to comprehend it. Mme. Dehn, who was with the Empress on the evening when Grand Duke Paul arrived with with the fatal tidings, has described the scene when the broken-hearted

* This is the same Cyril Vladimirovitch who has recently proclaimed himself "Head of the Romanoff Family and Guardian of the Throne."

Empress left the Grand Duke and returned to her own room.

"Her face was distorted with agony, her eyes were full of tears. She tottered rather than walked, and I rushed forward and supported her until she reached the writing table between the windows. She leaned heavily against it, and taking my hands in hers she said brokenly: '*Abdiqué!*'

"I could hardly believe my ears. I waited for her next words. They were scarcely audible. At last [still speaking in French, for Mme. Dehn spoke no English] 'Poor darling—alone there and suffering—My God! What he must have suffered!'"

In that hour of supreme agony there was not a word spoken of the loss of a throne. Alexandra Feodorovna's whole heart was with her husband, her sole fears that he might be in danger and that their boy might be taken from them. At once she began to send frantic telegrams to the Emperor begging him to come home as soon as possible. With the refinement of cruelty which marked the whole conduct of the Provisional Government in those days these telegrams were returned to the Empress marked in blue pencil: "Address of person mentioned unknown."

Not even this insolence nor all her fears broke the sublime courage of the Empress. When next morning she entered my sickroom and saw by my tear-drenched face that I knew what had happened her only visible emotion was a slight irritation that other lips than her own had brought me the news. "They should have known that I preferred to tell you myself," she said. It was only when gone her rounds of the palace and was alone in her own bedroom that she finally gave way to her grief. "Mama cried terribly," little Grand Duchess Marie told me. "I cried too, but not more than I could help, for poor Mama's sake." Never in my life, I am certain, shall I behold such proud fortitude as was shown all through those days of wreck and disaster by the Empress and her children. Not one single word of bitterness or resentment passed their lips. "You know, Annia," said the Empress gently, "all is finished for our Russia. But we must not blame the people or the soldiers for what has happened." Too well we knew on whose shoulders the burden of responsibility really rested.

By this time Olga and Alexei were decidedly better, but Tatiana and Anastasie were still very ill and Marie was in the first serious stage of the disease. The Empress in her hospital uniform moved tirelessly from one bed to another. Perceiving that from my floor of the palace practically every servant had fled, even my nurses and my once devoted Jouk having yielded to the general panic, she found people to move my bed upstairs to the old nursery of the Emperor. We were now almost alone in the palace. My father's resignation having been demanded and of course given, my parents were detained in Petrograd.

Days passed and still no word came from the Emperor. The Empress's endurance had almost reached its breaking point when there came to the palace a young woman, the wife of an obscure officer, who threw herself at the feet of the Empress and begged to be allowed the dangerous task of getting a letter through to the Emperor. Gratefully indeed did the Empress accept the offer, and within an hour the brave woman was on her way to Mogiloff. How she managed to reach headquarters, how she passed the cordon of soldiers and finally succeeded in delivering to the captive Emperor his wife's letter we never knew, but all honor to this heroic woman, she did it.

The palace was now full of Revolutionary soldiers, quite drunk with their new liberty. Their heavy boots tramped through all the rooms and corridors, and groups of dirty, unshaven men were constantly pushing their way into the nurseries bawling out hoarsely: "Show us Alexei!" For it was the heir who most of all aroused the interest and curiosity of the mob. Meanwhile, behind closed doors and anxiously awaiting the arrival of the Emperor, the Empress and her few faithful friends were at work forestalling the coming of Kerensky by burning and destroying letters and diaries, intimate personal records too precious to be allowed to fall into the ruthless hands of enemies.

CHAPTER XV

IN anxiety almost unbearable we waited until the morning of March 9 (Russian) the arrival of the Emperor. I was still confined to my bed and Dr. Botkine was making me his first visit of the day when my door flew open and Mme. Dehn, pale with excitement, rushed to my bedside exclaiming breathlessly: "He has come!" As soon as she could command words she described the arrival of the Emperor, not as of yore attended, but guarded like a prisoner by armed soldiers. The Empress was with Alexei when the motor cars drove into the palace grounds, and Mme. Dehn told how she sprang to her feet overjoyed and ran like a schoolgirl down the stairs and through the long corridors to meet her husband. For a time at least the happiness of reunion blotted out the suspense of the past and the gloomy uncertainty of the future. But afterwards, alone, behind their own closed doors, the emotion of the betrayed and deserted Emperor completely overcame his self-control and he sobbed like a child on the breast of his wife. It was four o'clock in the afternoon before she could come to me, and when she came I read in her white, drawn face the whole story of the ordeal through which she had passed. With prideful composure she related the events of the day. I tried to match her in courage but I am afraid I failed. I, who in all the twelve years of my life in the palace had but three times seen tears in the eyes of the Emperor, was entirely overwhelmed at her recital.

"He will not break down a second time," she said with a brave smile. "He is walking in the garden now. Come to the window and see." She helped me to the window and herself pulled aside the curtain. Never, never while I live shall I forget what we saw, we two, clinging together in shame and sorrow for our disgraced country. Below in the garden of the palace which had been his home for twenty years stood the man who until a few days before had been Tsar of all the Russias. With him was his faithful friend Prince Dolgorouky, and surrounding them were six soldiers, say rather six hooligans, armed with rifles. With their fists and with the butts of their guns they pushed the Emperor this way and that as though he were some wretched vagrant they were baiting in a country road. "You can't go there, Gospodin Polkovnik (Mr. Colonel)." "We don't permit you to walk in that direction, Gospodin Polkovnik." "Stand back when you are commanded, Gospodin Polkovnik." The Emperor, apparently unmoved, looked from one of these coarse brutes to another and with great dignity turned and walked back towards the palace. I had been a very sick woman, and I was now hardly fit to stand on my feet. The light went out suddenly and I fainted. But the Empress did not faint. She got me back to my bed, fetched cold water, and when I awoke it was to feel her cool hand bathing my head. From her calm and detached manner no one could have guessed that the scene we had just witnessed was part also of her own tragedy. Before leaving me she said as to a child: "If you will promise to be very good and not cry he shall come to see you this evening."

After dinner they came, the Emperor and Empress with our friend Lili Dehn. The two women sat down at a table with their needlework leaving the Emperor free to sit by my bed and talk to me privately. I have tried to show Nicholas II as a human person, with human emotions, and I have no desire now to represent him, in the hour of his humiliation, as other than a man feeling keenly and acutely the bitterness of his position. I had been unable until the day of his return to realize with any degree of clarity the full extent of his calamity. It was to me almost unbelievable that his enemies, who had so long plotted and schemed for his overthrow, had at last succeeded. It was beyond reason that the Emperor, the finest and best of the whole Romanoff family, should be allowed to fall under the feet of his decadent, treacherous kinsmen and subjects. But the Emperor, his eyes hard and glistening, told me that it was indeed true. And he added: "If all Russia came to me now on their knees I would never return."

With tears in his voice he spoke of the men, his most trusted relations and friends, who had turned against him and caused his downfall. He read me telegrams from Brusiloff, Alexieff, and other of his generals, others from members of the family, including a message from Nicholai Nicholaievitch, in which the writers "on their knees" begged his Imperial Majesty, for the salvation of Russia, to abdicate. In whose favor did they wish him to abdicate? The weak and ineffectual Duma? The great untaught masses of the people? No, to their own blind and self-seeking oligarchy, which, under a regent of its own choosing, would rule the boy Alexei and through him the people and the uncounted wealth of Russia. But this at least the Emperor could and did prevent. Both his heart and his mind forbade him to abdicate in favor of the Tsarevitch. "My boy I will not give to them," he said feelingly. "Let them get some one else, Michail, if he thinks he is strong enough."

I regret that I cannot remember every word the Emperor told me of the scenes in his train when the deputation from the Duma came to demand his abdication. I was trying too hard to obey the Empress's injunction to "be good and not cry." But I remember his telling me how arrogant and vain the deputies, especially Goutchkoff and Shoulgin, showed themselves. On their departure the Emperor's first words were addressed to the two tall Cossacks who stood guard at his door. "It is time now for you to tear my initials from your shoulder straps," he told them. The Cossacks saluted and one of them said: "Please your Imperial Majesty, please allow us to kill them." But the Emperor replied: "It is too late to do that now."

Of his mother, who hurried from Kiev, accompanied by Grand Duke Alexander Michailovitch, to see him, he said that he was vastly comforted to have her near him, but that the sight of the Grand Duke was unendurable. Driving away from the train with the Empress Dowager, the Emperor had been much moved to see the people along the whole distance of two versts fall on their knees to bid him farewell. There was a group of schoolgirls from the institute at Mogiloff who forced their way past the guards and surrounded their Sovereign, begging his handkerchief, his autograph on bits of paper, the buttons from his uniform, anything for a last souvenir. The Emperor's face grew sharply lined when he spoke of those brave girls and the kneeling people. "Why did you not appeal to them?" I asked. "Why did you not appeal to the soldiers?" But the Emperor answered gently: "The people knew themselves powerless, and as for appealing to the soldiers, how could I? Already I had heard threats of murdering my family." His wife and children, he said, were all on earth he had left to live for now. Their happiness and well-being were all his soul desired. As for the Empress, more than himself the real object of malice, only over his body should any hand be raised to injure her. Giving way once more for a brief moment to his grief the Emperor murmured half to himself: "But there is no justice, no justice on earth." Then as if in apology he said: "It has shaken me badly, as you see. For the first few days I was so little myself that I could not even write my diary."

As we talked it came over me for the first time in full force that all was indeed finished for Russia. The army was disrupted, the nation fallen. I could foresee, to some extent at least, the horrors we should have to meet, but in a kind of desperate hope I asked the Emperor if he did not think that the riots and strikes would now be put down. He shook his head. "Not for two years at least," he predicted. But what did he think was to become of him, of the Empress and the children? He did not know, but there was one prayer he should not be too proud to make to his enemies, and that was that they should not send him out of Russia. "Let me live here in my own country, as the humblest and most obscure proprietor, tilling the land and earning the poorest living," he exclaimed. "Send us to any distant corner of Russia, but only let us stay."

This was the only time I ever saw the Emperor in the least degree unmanned, or overcome with the bitterness of grief which I knew must have filled his spirit. After that first day in the palace gardens he gave his jailers no opportunity of insulting him. With Prince Dolgorouky he walked out daily but only along near pathways to the palace doors. The snow was heavy on the ground and the two men vigorously exercised themselves shoveling it from paths and roadways. Often the Emperor would look up from this strenuous work to wave a hand to those of us who were watching from the windows. In the solitude of my sick chamber I tormented myself with thoughts of what might be in store for the Emperor and the beloved family whose happiness and well-being were more to him than the most exalted throne. They were all prisoners of the Duma now, and what dark and hapless fate was the ruthless, irresponsible Duma preparing for them? Not a comforting question to haunt the mind of one ill in body and soul. From my first waking moment on I lived in anticipation of the daily visit of the Empress. She who had all at stake still kept her wonderful courage alive. She came in tall and stately, a smile on her gentle, melancholy face, bringing me the news of the nurseries, messages from the children, making me work, doing everything possible to cheer and to lighten my mind. In the evening the Emperor usually came, wheeling his wife in her invalid's chair, for by night her strength had all but gone. They stayed with me for an hour and then went on to say good night to the suite in the drawing room. Sadly diminished in numbers was that suite, but unchanged in fealty and affection for fallen majesty. Among those devoted friends who appeared almost like the survivors of a shipwreck were Count Benkendorff, brother of the former Russian ambassador to Great Britain, and his wife, who had boldly arrived at the palace when it was first surrounded by mutinous soldiers; two maids of honor, Baroness Buxhoevden and Countess Hendrikoff; the faithful Miss Schneider ("Trina"), Mme. Dehn, Count Fredericks, General Voyeikoff and the Hussar officer, General Groten. The two devoted aides-de-camp, Lieutenant Linevitch and Count Zamirsky, who had flown to the palace to be near the Empress after the abdication, had been forced to leave, or they too would have remained to the end. Of the household M. Gilliard and Mr. Gibbs, the French and English tutors of Alexei, had elected to remain. Madeleine, and several other personal attendants, including three nurses, also stayed. "In good times we served the family," said these honest souls, "never will we forsake them now."

Not once, after the very first of our conversations, and not at any time I believe to others in the palace did the Emperor or the Empress make the smallest complaint of their captivity. They seemed to suffer for Russia rather than for themselves, for they knew, and said so, that the army, suddenly in the midst of war released from all discipline, would soon cease to fight efficiently, or perhaps to obey orders at all. This of course the world knows is precisely what did happen. The Emperor, I must admit, sometimes betrayed a gruesome kind of humor over the fantastic blunders of the self-styled statesmen who were so rapidly making general shipwreck of their revolution. In every way they showed their weakness and bewilderment. Whether or not they feared to trust old officers of the Empire with the custody of the Imperial Family I cannot be sure, but the men they sent to Tsarskoe were a constant source of ironic mirth to the suite. Most of these men were young, raw, underbred, and inexperienced, the best of them being junior officers promoted since 1914. One day one of the guard officers, just to show how democratic Russia had become, swaggered up to the Emperor and offered to shake hands with him. Unfortunately, as he afterwards told me, the Emperor was so busy shoveling snow that he could not take advantage of the man's condescension.

The newly appointed commandant of the palace was a young man named Paul Kotzebou, before the War an officer of the lancers, but for some piece of misconduct cashiered from the service. I had long known Kotzebou and aside from his doubtful army record I was not sorry to see him in the palace, for I knew that if weak of character he was at least kind of heart. Kind indeed he proved himself, for he visited my sickroom in friendly fashion, risked arrest by consenting to smuggle letters to my parents in Petrograd, and was the first to warn me that the Provisional Government was contemplating my arrest. Many of the old friends and advisers of the Emperor were already in prison, but the proposal to arrest a woman whose sole crime had been devotion to the Empress and her children gave us all an uncomfortable, premonitory shock. The distress of the Empress was greater almost than her pride. The mercy she would have scorned to ask for herself she was ready to beg for me, and she did most earnestly implore Kotzebou to intercede in my behalf. "What possible good will it do them to arrest one helpless woman?"

she urged. "Parting with her would be like losing one of my own children." Kotzebou, whatever his feelings, could only reply: "If I could, Madame—but there is nothing I can do, nothing."

The Emperor alone refused to believe my arrest at all probable, but the others were badly frightened at the prospect. The sister of mercy who had worked in my hospital and was taking care of me, almost went on her knees to the Emperor and Empress. "Now is the time to show your real love for Anna Alexandrovna," she cried. "Take her into the rooms of your own children and never let anybody touch her." Cooler counsel came from Count Benkendorff, who advised the Emperor and Empress not to oppose my arrest if it were ordered. The only result of opposition, he pointed out, would be more arrests and perhaps increased hardship for the Empress. "I do not think they will detain her, unless it is in one of the rooms of the Tauride," he said, meaning that I might only be isolated for a time in the palace where the Duma held its sessions. Count Benkendorff was later to learn what kind of justice was being prepared by the criminal lunatics who were at Russia's throat.

One morning towards the 20th of March I had a hurried note from the Empress, the contents of which were enough to make me forget all my own troubles. Marie, who had been very ill and who now she feared was dying, was calling constantly for me. The servant who brought the note told me that Anastasie also was in a critical condition, lungs and ears being in a sad state of inflammation. Oxygen alone was keeping the children alive. Kotzebou was calling on me at the time, and as I sat up in bed wildly demanding to be dressed, he begged me not to leave my room. "They are only waiting until you are well enough to be arrested," he assured me. But though I feared arrest I feared still more letting the child I loved die with one single wish unfulfilled, and as soon as I could be sufficiently clothed it was Kotzebou himself who wheeled my chair through the long corridors to the nurseries. It was the first time in weeks that I had seen the children and our meeting was full of tears. We wept in each other's arms and then without wasting any time I went on into Marie's room. The child indeed seemed to be at the point of death, but when she saw me the suffering in her eyes turned to something like joy. Her weak hands fluttered on the bedclothes and with a feeble cry, "Annia, Annia," she began to weep. Long I sat beside her holding her hot and wasted body in my arms, and when I left her she was asleep. Shaken though I was with that experience, I had one more agony to bear. When my chair was being wheeled back along the corridor I passed the open door of Alexei's room, and this is what I saw. Lying sprawled in a chair was the sailor Derevanko, for many years the personal attendant of the Tsarevitch, and on whom the family had bestowed every kindness, every material benefit. Bitten by the mania of revolution, this man was now displaying his gratitude for all their favors. Insolently he bawled at the boy whom he had formerly loved and cherished, to bring him this or that, to perform any menial service his mean lackey's brain could think of. Dazed and apparently only half conscious of what he was being forced to do, the child moved about trying to obey. It was too much to bear. Hiding my face in my hands, I begged them to take me away from the sickening spectacle.

The next day, my last in the palace, I went again to the children, and for a few hours at least was a little bit happy. The Emperor and Empress had luncheon served in the nurseries, and we were all able to eat in some comfort because both Marie and Anastasie were showing signs of improvement. Still we were troubled because Kotzebou, as a reward for his too kindly treatment of the captives, had that morning been removed from the palace, and the doctors when they came brought with them newspapers, fair samples of the new "free" press of Russia, bristling with frightful stories, especially about me. For the first time I began to realize, with a sick heart, what an arrest might mean, what grotesque charges I might be called upon to face. For the first time, in these newspapers I read the amazing tale of how I had conspired with Dr. Badmieff to poison the Emperor and the Tsarevitch. Dr. Badmieff, that half mad old Siberian root and herb doctor, who never in his life had been admitted to the palace as a physician or even as a friend! It was too absurd to resent. Even the Empress who at first had shown anger, burst into mocking laughter. "Here, Annia," she cried, "keep this story for your collection."

The next day I was arrested. I awoke in a morning of storm and howling wind and in my soul a feeling of dread and foreboding. Immediately after my coffee I wrote a note to the Empress asking her not to wait until afternoon to see me. Her reply was kind and cheering, but she was busy in the nurseries and could not leave until after the arrival of the doctors. With luncheon came Lili Dehn, and scarcely had we finished the meal when we were aware of great noise and confusion in the corridor outside. An icy hand seemed to seize my heart. "They are coming," I whispered, and Mme. Dehn, springing from her chair cried: "Impossible. No—no—" and panic-struck fled the room. The door flew open to admit a frightened servant with a note from the Empress. "Kerensky is going through our rooms. Do not be frightened. God is with us." Hardly had the man retired when again the door opened and another frightened servant, a palace messenger in a feathered cap, announced in a drowned voice the arrival of Kerensky. In a moment the room seemed to fill up with men and walking arrogantly before them I beheld a small, clean-shaven, theatrical person whose essentially weak face was disguised in a Napoleonic frown. Standing over me in his characteristic attitude, right hand thrust into the bosom of his jacket, the man boomed out: "I am the Minister of Justice. You are to dress and go at once to Petrograd." I answered not a word but lay still on my pillows looking him straight in the face. This seemed to disconcert him somewhat for he turned to one of his officers and said nervously: "Ask the doctors if she is fit to go. Otherwise she must be arrested and isolated in the palace." Count Benkendorff, who stood in the back of the room near the door, volunteered to see the doctor, and when he returned it was with the message that Dr. Botkine gave them permission to take me. Afterwards I learned that the Empress reproached the doctor bitterly, saying over and over through her tears: "How can you? How can you? You who have children of your own." But Dr. Botkine was by this time a victim of craven fear, and he was incapable of refusing any request of the Provisional Government.

They gave me time to dress warmly, and I had a moment in which to reply briefly to a note from the Emperor and Empress, in which they enclosed small pictures of Christ and the Virgin, signed with their Majesties' initials, N. and A. When at last I was ready to go it suddenly surged over me that this might be the end of my long association with these dearly loved friends, my Sovereigns, whose intimate lives I had shared for twelve years. Ready to fall on my knees before him if necessary I made a final appeal to Colonel Korovitchinko, the new commandant of the palace, begging him to let me see them for one moment, just long enough to say good-bye. Colonel Korovitchinko, who afterwards died a cruel death at the hands of the Bolsheviki, at first refused, but moved by my tears he relented a little. The Emperor, he said, was outside and could not be summoned, but he would exert his authority far enough to send me under guard to say good-bye to the Empress. Under escort of two officers I was taken to the apartment of Mlle. Schneider, and very soon the pale Empress was wheeled into the room by her devoted attendant Volkov. We had time for only one long embrace and the hurried exchange of two rings. Then Tatiana, who came with her mother, embraced me, weeping, and as she too begged for a last memory gift I gave her the only thing I had to give, my wedding ring. Then the soldiers tore us apart but I saw that the man who gave the order did it with tears in his eyes. The last I remember was the white hand of the Empress pointing upward and her voice: "There we are always together." Volkov, weeping, cried out courageously: "Anna Alexandrovna, God will surely help."

They carried me downstairs to the motor, for I could neither walk nor stand, even with the help of my crutches. At the door stood several soldiers and Court servants, visibly distressed, but by this time I felt nothing, heard nothing. I was turned to stone. When I was lifted into the car I was startled to see there another woman, like myself swathed in wraps and veils. It was Lili Dehn, whose arrest had not before this day even been threatened. Dazed as I was, it was some comfort to hear her whisper that we were to travel to Petrograd together. I recovered myself a little, enough at least to recognize the frightened face of the servant who closed the door of the car. Killed a few months later, this good man had been for a long time a sailor on the Imperial yacht. "Take care of their Majesties," I managed to say to him. Then the motor car shot forward, and I left the palace at Tsarskoe Selo forever. Both Lili and I pressed our faces to the glass in a last effort to see those beloved we were leaving behind, and through the mist and rain we could just discern a group of white-clad figures crowded close to the nursery windows to see us go. In a moment of time the picture was blotted out and we saw only the wet landscape, the storm-bent trees, the rapidly creeping twilight. In another few moments we were at the station, the dear, familiar station of Tsarskoe, where so many, many times I had waited to greet or to say a short farewell to the Emperor and Empress. Ready for us was one of the small Imperial trains, now the special train of Kerensky. Our guards hurried us into a carriage, and the train immediately began to move. At the same time our carriage was invaded by Kerensky and a group of soldiers. Without even a pretense of decent politeness the new Minister of Justice began to shout at us: "Give your family names," and because we did not speak quickly enough the little man became insulted. "You will learn that when *I* ask a question you must answer promptly." We gave our names and Kerensky, turning triumphantly to the soldiers, ejaculated: "Well! Are you convinced now?" Apparently some of the men had expressed doubts as to whether they had bagged the right criminals. Sick and half fainting, I sank back into the cushions and closed my eyes on their departing figures. Lili bent over me with her salts bottle and soon I was able to sit up with some show of courage. It was the first time I had left the house since my illness and I was still very weak.

Arrived in Petrograd, Kerensky paraded us before his officers like barbarian captives of some Roman emperor, but this did not affect us seriously. Our eyes were busy gazing at the changed aspect of Petrograd, soldiers swanking around the streets proud of their slovenly appearance, the badge of their new freedom; mobs of people running aimlessly about, or pausing to listen to street-corner orators; and everywhere on walls and buildings masses of dirty red flags. An old-fashioned coach belong to the Imperial stables had been sent for us and still closely guarded we drove to the Ministry of Justice. There we climbed a long and very steep staircase—how I did it on my crutches I do not yet understand—and were shown into a room on the third floor, empty even of a wooden chair. Silently we stood and waited, and after a time men came in carrying two sofas. On one of these Lili sat down and on the other I lay prone. Again we waited, no one near us save the unkempt soldier who guarded the door. The evening lengthened and finally Kerensky honored us with another brief visit. He did not look at me at all but asked Lili if they had built us a fire. It was an unnecessary question, for he must have felt the icy chill of the room. A few minutes later, however, a servant did build a fire in the tiled stove, and another brought in a tray with eggs and tea. Left alone with the unkempt soldier, the man suddenly amazed us by breaking into a volley of speech in which he cursed most eloquently the new order of things. Nothing good would come of it, nothing, was his opinion. Somewhat reassured because we had a guard who was not at heart a Revolutionist, we lay down, but the night brought to neither of us any anodyne of sleep and rest.

* * *

CHAPTER XVI

MORNING dawned cold and gray, and so exhausted was I with sleeplessness and the discomfort of a hard bed without linen or blankets, that Lili was alarmed and when the tea arrived she begged the soldier who brought it to have a doctor sent me. But Kerensky replied that the doctor was engaged with War Minister Goutchkoff and could not be approached at present. Within a short time I was to be removed to a hospital, and as for Mme. Dehn, she might expect good news soon. As a matter of fact Mme. Dehn was released from custody the next day. Feeling confident that she would be let go, I gave her what jewels I had brought with me, asking her to turn them over to my mother. In return Lili gave me a few necessaries, including a pair of stockings for which later I was extremely grateful because the prison stockings were so coarse and heavy that they hurt my injured leg.

About three o'clock in the afternoon Colonel Peretz, who afterwards wrote a book on the Revolution, came into the room with a group of young boys, former cadets of the military academy, now commissioned officers of the new army. "Say good-bye to your friend and come along," I was ordered, and after a quick embrace I parted with Mme. Dehn, my last link with the past, and followed the men downstairs, where a large motor car was waiting. We all got in, the men's rifles considerably reducing the carrying capacity of the seats. As we drove off the colonel began a long and insulting monologue to which I tried not to listen. "Ah! You and your Grichka (Gregory)," I heard him saying, "what a monument you both deserve for helping us to bring about the Revolution." But all that I wanted to learn from him was my destination, and as if in answer to the unspoken question he said: "All night we were discussing the most appropriate lodgings for you, and we decided on the Troubetskoy Bastion in the fortress." At this point we passed a church and, after the invariable custom, I made the sign of the cross. Colonel Peretz flamed into anger at this. "Don't dare cross your-self," he cried with emphasis on the last word. "Rather pray for the souls of the martyrs of the Revolution." Then as I made no response he exclaimed: "Why don't you answer when I speak to you?" I replied coldly that I had nothing whatever to say to him, whereupon he began to revile the Emperor and Empress in coarsest terms, ending with the words:

"No doubt they are in hysterics over what has happened to them." Then I did speak. "If you knew with what dignity they are enduring what has happened you would not dare say what you have said." After which the monologue was for a moment or two halted.

Turning into the Liteiny, a street in which many barracks and ministries are located, the car stopped and Colonel Peretz dispatched one of the cadet officers on an errand into a Government building. On his return the colonel delayed matters long enough to make a bombastic speech on the great services to the Revolution performed by the cadets, and again we drove on. Realizing that we were not proceeding in the direction of the Fortress of Peter and Paul, I allowed my feminine curiosity to get the better of my pride and I asked whither we were bound. "To the Duma first," was the grim answer. "To the fortress afterwards." Arrived at the Tauride Palace we alighted at what is known as the Ministers' Pavilion and immediately went into the building. What a sight! Crowding the rooms and the corridors, men and women of all ages and conditions, prisoners of the Provisional Government! Looking about, I saw many people of my own class, among them Mme. Soukhomlinoff who for all her manner betrayed might have been a guest rather than a prisoner. We exchanged cheerful greetings and she introduced the two women beside her, Mme. Polouboiarenoff and Mme. Riman, wife of a well-known general. Mme. Polouboiarenoff, of whom I had heard as a brilliant writer on a conservative newspaper (murdered for this later by the Bolsheviki), was quite self-possessed, but Mme. Riman's face was wet with constantly flowing tears. A young girl student, a typical Revolutionist who seemed to be in some kind of authority, passed us in a hurry, pausing to say to Mme. Riman: "What are you crying about? You are going to be set free while these two"—Mme. Soukhomlinoff and myself—"are going to the fortress." Poor Mme. Riman was crying because her husband was already in prison, but the revolutionary student could not be expected to sympathize with that.

It really is easier to be calm over one's own than over another's fate, as I learned when I found myself, with Mme. Soukhomlinoff, once more in a motor car bound for that mysterious prison on the left bank of the Neva, directly opposite the Winter Palace, the Fortress of Peter and Paul. As we left the Tauride the girl student, who after all had some natural feelings, asked me for my father's telephone number that she might notify my parents where I had been sent. "No need to bother about that," broke in the chivalrous Colonel Peretz. "The newspapers will have a full report." "All the better," I rejoined, "for then many more will pray for me."

Rolling into the vast enclosure of the fortress, we stopped at the entrance of the Troubetskoy Bastion. A group of soldiers, dirty and wolfish of demeanor, rushed to meet us. "Now I am bringing you two very desperate political prisoners," shouted the colonel, as the men closed around us. But a stout Cossack, much more human than the rest, assumed authority saying that he was that day acting in place of the governor of the fortress. Preceded by this man, we traversed a long series of narrow, winding stone passages, so dark that I could see only a few feet ahead. Suddenly I was halted, hinges creaked, and I was roughly pushed into a pitch-dark cell the door of which was instantly bolted behind me.

No one who has not been a prisoner can possibly know the sickening sensation which possessed me, standing there in that dark hole, afraid to take a step forward, unable to touch with my groping hands either walls or furniture. My heart leaped and pounded in my breast and I clung desperately to my crutches lest I should fall into that unfathomed darkness. A few minutes of wild terror and then as my eyes grew accustomed to the dark I saw ahead of me a narrow iron cot towards which I moved with infinite caution. In my progress towards the bed my feet sank into pools of stagnant water which covered the floor, and soon I perceived that the walls of the cell were also dripping with moisture. The tiny window, high in the farthest wall, admitted little air, and the whole place was foul with dampness and the odor of years. It reeked with even worse smells as I quickly discovered, for close to the bed was an uncovered toilet connected with archaic plumbing. The bed was hard and lumpy and I do not think that the thin mattress had ever been cleaned or aired. However, that mattress was not to afflict me long. Within a few minutes my cell door was thrown open and several uniformed men entered. At their head was a black-bearded ruffian who told me that he was Koutzmine, representative of the Minister of Justice, and was authorized to arrange the régime of all prisoners. At his orders the soldiers tore from under me the ill-smelling mattress and the hard little pillow, leaving me only a rough bed of planks. Under his orders they tore off my rings and jerked loose a gold chain from which were suspended several precious relics. They hurt me and I cried out in protest, whereupon the soldiers spat at me, struck me with their fists and left, noisily clanging the iron door behind them. Wrapping my cloak around me, I crouched down on the bed shivering from head to foot and filled with such an agony of loathing and disgust and desolation that I thought I should die. Not a particle of food was brought me that day, and nothing broke the monotony of the dragging hours save now and again when the small grating in the door of my cell was pushed aside and a gaping soldier looked in. Then came night, hardly darker than the day, but more silent. Weak with hunger, spent with pain I clutched my aching head with my hands and asked God if He had forgotten me. At that moment of extreme misery I was startled and at the same time strangely comforted by a sudden low but distinct rapping on the other side of the wall. Instinctively I knew that it was Mme. Soukhomlinoff who was trying to speak to me in the only language prisoners have. I rapped back, almost happily, for I felt that with a friend so near I was not entirely deserted.

I must have slept after that, for the next thing I remember was a man entering the cell with a pot of hot water and a small piece of black bread which he placed on an iron shelf near the bed. "As soon as your money arrives you can have tea," he announced briefly. Tea would have been a priceless blessing in that cold place, but I was so thirsty that I drank every drop of the hot water and was thankful. I suppose I ate the black bread too, bad as it was, for I was very hungry.

How to describe the days that followed, slow-paced, monotonous, yet each one filled with its special meed of suffering? On one of the first days a grim woman came in and stripped me of my underclothes, substituting coarse and unclean garments marked with the number of my cell, which was 70. No prison dress seemed to be provided, so I was allowed to keep my own. But in the process of undressing the woman discovered a slender gold bracelet which I had worn day and night for many years and which was locked on my arm. She called Koutzmine and his guard of soldiers and they, indignant that they had overlooked a single article of value, began to force the bracelet over my hand. As the little circlet was not intended to go over my hand their efforts caused me such pain that I screamed in spite of myself. Touched, or perhaps merely annoyed at this, Koutzmine suggested to the soldiers that if I would promise not to give the bracelet to anyone I might be allowed to keep it. But his suggestion met with no sympathy and the bracelet was finally forced over my bruised hand.

The awful food and the still more awful solitude were daily afflictions, and I think they were really the worst of all. Twice a day a soldier brought in a nauseous dish, a kind of soup made of the bones and skin of fish, none too fresh. Sometimes, if the soldier happened to be in an especially vicious mood, he spat in the soup before giving it to me, and more than once I found small pieces of glass among the bones. Yet so ravenous was my hunger that I actually swallowed enough of the vile stuff to keep myself alive. Only by holding my nose with my fingers was I able to get a few spoonfuls down my throat. What was left I was careful to pour into the filthy toilet, for I had been told that unless I ate what was given me I would be left to starve. Hot water and black bread continued to be doled out in small quantities, but there was never any tea. No food was allowed to be given the prisoners even when it was brought to the fortress by relatives and friends. Neither was any kind of occupation given the wretched captives. We were not even allowed to clean our own cells, a soldier coming in once a week to wipe up the wet and slimy floors. When I begged the privilege of doing this myself the soldier replied: "A prisoner who works is not a prisoner at all." It is true that when he has absolutely nothing to do he is worse than a prisoner, he is a living corpse.

Actual death being too merciful for political prisoners, we were taken out, one by one, for ten minutes every day. The exercise ground was a small grassy court where a few shrubs and trees gave promise of green leaves later on. No words can describe the relief, the blessed joy that those few moments of light and air and the sight of the blue sky brought to my heart. It seemed to me that I lived only for those moments. Of course the walled court was well guarded by armed soldiers and never once did their fierce eyes ever leave me. Still it was a bit of God's beautiful world, a breath of His sweet air, and I breathed it deep into my soul, keeping it there for hope and comfort until the next day came. In the center of the court was a small and dingy bath house where, on Fridays and Saturdays, the prisoners were treated to a sort of a bath. On those days we were not permitted to walk, but I for one did not complain of this. Any respite from the gravelike existence of the cells was a blessing. It was still very cold and when I lay down for the night I never removed my clothes. I had two woolen handkerchiefs, or rather, head kerchiefs, and one of these I tied over my head and the other I wrapped around my shoulders for warmth. Usually I slept until about four o'clock when the bells of a church hard by broke into my slumbers. After that I tried to doze, but very soon came the tramp of boots on the stones of the corridors and the crash of wood which the soldiers brought in each day for their stoves. I always woke up shivering and my first move was towards a corner of my cell where the stones were dry and a little warm from the stove outside. Here I huddled and shook until the hot water and the black bread were thrust in. I had never fully recovered from my illness and the cold and damp brought on first a pleurisy and afterwards a racking cough. I was so weak that sometimes in crossing from the bed to what I called the warm corner I slipped and fell and lay on the wet floor unable to rise. The soldier who thus found me, if he were of the half decent sort, would pick me up and throw me on the plank bed. Otherwise he would merely kick me.

For the first two weeks I spent in the Troubetskoy Bastion the only attendants were men. The soldiers had the keys to the cells and the complete freedom of the corridors. The first lot were men of the 3rd Rifle Regiment of Petrograd, but within a few days some of them were shifted and their places were taken by a miscellaneous force from one of the most unruly of the mutinous reserves. Riots and fights between the two bands became an almost daily occurrence and the nerves of the prisoners were tortured by the yells and blows of the battle. My only comfort, aside from the ten minutes' respite of the exercise ground, was in the wall-tapping between my cell and Mme. Soukhomlinoff's. This had developed into a regular code and we managed to carry on, by alternately long and short taps, quite lucid conversations. Once to our fright the Governor of the bastion, Chkoni, caught us at this forbidden game and threatened us, if it happened again, with the dark cell, a place of unknown horrors, as we knew, for we had listened to the groans and cries of the former police chief Belezky while he suffered there. After the warning of Chkoni Mme. Soukhomlinoff and I communicated with each other only in the middle of the night when the snores of the soldiers in the corridors guaranteed a degree of safety. Without these cautiously tapped-out conversations I really do not know how I should have lived and kept sane.

The cough which had been afflicting me grew worse rather than better and the only relief that was offered me was a primitive kind of cupping which did the cough no good but covered my chest with black and blue bruises. Finally, at the request of the sanitary soldier who had done the cupping, the prison doctor was sent for. This man, whose name was Serebrianikoff, was one of the most dreadful persons I ever came in contact with. He had a red, malicious face, his clothes and person were revoltingly dirty, and to increase their effect he wore on his bulging waistcoat a huge red bow, emblem of his revolutionary ardor. When he came into my cell he literally tore the clothes from my back in a pretended examination, then turning to the soldiers in the doorway he shouted: "This woman is the worst of the whole lot; an absolute idiot from a life of vice." Slapping me on one cheek and then on the other, he began to ask me questions which I cannot repeat here of my alleged orgies with Rasputine, with Nicholas and "Alice" as he called the Empress. Even the soldiers looked disgusted and I shuddered away from him sick with repulsion. That night I was so far gone physically and mentally that I could not answer Mme. Soukhomlinoff when she tapped on the wall. All I could do was to cough and shiver and in an incoherent, half mad fashion pray: "My God, my God, hast Thou forsaken me?"

The next morning the soldier who brought my hot water and bread thought me dying and insisted in sending again for the unspeakable Serebrianikoff, although I begged him not to. "Send a woman, I implore you," I whispered. But there was no woman to send, and the prison doctor came instead. Declaring that I was merely shamming, this brute again struck me in the face and left saying: "I'll punish you for this. There'll be no exercise for you for two weeks after you think yourself well enough to go out." He kept his word, and for two weeks after I ceased to be acutely ill I remained all day in my cell weeping for the clean air and a sight of the blue sky. Little trickles of pale sunlight were beginning to steal through my barred windows, the cold was less intense and I knew that outside, in the world of freedom, the spring had come.

One little bit of good news came at this time. Women wardresses had been appointed to look after the special needs of the women prisoners. Two attendants from a women's prison were the first to arrive, but they were so shocked at the conditions they found in the fortress that they refused to stay. They were replaced by others, one a saucy young person whose sole energies went into flirtations with the soldiers, and an older woman with melancholy dark eyes and the best and kindest of hearts. I cannot tell her name because if she is still alive and in Russia she must be in the employ of the Bolsheviki.

I will call her simply the Woman. Her kindness to me I can never repay, but at least I shall never forget it, especially since I knew that every kind act she did was at her own personal risk. The Woman was on duty only until nine o'clock at night and was never allowed to enter my cell alone. Yet she often managed cleverly to follow slowly when she and the guard left the cell, and she frequently dropped on the floor behind her little pieces of sausage, chocolate, or bread nearly white. In the cell we dared not talk, but when she took me to the bath house we exchanged whispered conversations, and through her I got a little news of the exciting events of the time. The Provisional Government was tottering and the star of Kerensky was rising rapidly. The Imperial Family were still at Tsarskoe Selo, prisoners but alive, and that knowledge gave me a new impulse to live.

I must record one especially kind act my new friend did in my behalf. Easter Sunday came, and sitting on my hard bed I ventured to sing softly a verse or two of a well-remembered Easter hymn. On the Good Friday preceding we had been allowed to leave our cells one by one under guard and to confess to a good old priest, whose distress at our sorry plight so moved him that he heard our confessions with great tears in his eyes. Earnestly this old priest had begged Kerensky to allow him to visit prisoners in their cells and do what he could for their comfort, but Kerensky curtly refused.

I was thinking of him on this Easter morning. The soldiers had been running through the corridors calling to one another, perhaps in jest, perhaps as a matter of habit, the Russian greeting: "Kristos Voskrese," Christ is risen, to which the response is: "Voistino Voskrese," He is risen indeed. I could see that the soldiers had plates of the sugary cheese which everybody eats at Easter and which some of the prisoners received. Not I, because I was considered too wicked, too vile. Nevertheless, because of the trickle of sunshine that stole through the bars of my window, and because the old priest had really given me great comfort, I began to sing. Instantly the soldiers outside commanded me rudely to keep silent. It was too much. I laid my head down on the rags that formed my pillow and began to cry miserably. Then my hand strayed under the pillow, touching something. It was a little red Easter egg left there by the Woman, to make me feel that even in that place I was not entirely friendless. Never did a gift come as such a joyful surprise. I hugged it to my heart, kissed it and thanked God.

I was not forsaken. Indeed the worst was already passed for me, for the next day I was told that on every Friday after I was to receive a visit from my parents, whom I had feared I was never to see again on earth.

CHAPTER XVII

VISITORS in prison! Who but one who has spent days and nights of anguished loneliness behind bolted doors can possibly imagine the joy of such anticipation? I looked forward, almost as toward freedom itself, to the first Friday when I should see my beloved parents. I pictured myself running forward to embrace them, I could see my father's kind and loving smile, my mother's blue eyes full of happy tears. How we would sit, hand in hand, and talk over all that had happened since our parting! They would bring me news, messages, perhaps even letters from those other captives in Tsarskoe Selo. I should hear that the children were well again and the Empress's deepest anxieties were removed.

Alas! the harsh reality of my foolish dreams. When the day came I limped, between armed soldiers, through the long, gray corridors to the visitors' room, and there at the end of a long wooden table which divided us like an impassable gulf I saw my mother. There was no embrace allowed, not even a touch of hands. My mother tried to smile, tried to look at me with the love I craved, but in spite of herself her face paled and an expression of horror congealed her features. I stood there before her white with the pasty whiteness of prison, my uncombed, unkempt hair hanging about my shoulders, my dress dirty and wrinkled and an unhealed cut ploughing a bloody furrow across my forehead. To the question she dared not ask I touched the ugly wound and told her it was nothing, nothing. I could have told her that a soldier named Izotov, in a fit of animal temper, had knocked me against the edge of the cell door, and that the cut had received absolutely no attention since. Had we been alone I should have wept the whole story out on her breast, but we were not alone. Standing over us like inquisitors were the Procureur of Petrograd and the terrible Chkoni, governor of the Troubetskoy Bastion, and afterwards governor of the fortress itself. Ten minutes only were allowed us, and at the end of eight fleeting minutes Chkoni, watch in hand, roared out: "Two minutes left. Finish your talk." But we had no talk. Sobs choked our words, the few commonplace words that in such circumstances can be spoken. We could only bid each other be brave and trust in God's mercy. We could but gaze and gaze at each other through streaming tears. Then they separated us.

When the next Friday came I resolved to make myself a little more presentable. I had no mirror but I begged the Woman to loan me a small, cracked fragment. They had taken away all my toilet articles and every single hairpin, but the Woman gave me two hairpins of her own and, combing my hair with my fingers I arranged it more or less neatly. Every day I washed and cared for the cut on my forehead and when the visiting hour at last arrived I fancied that I looked rather more like myself. This time the precious ten minutes were spent with my father, and because he had been prepared in advance for the wretched object his daughter had become our brief interview was less emotional than that of the preceding Friday. Brave and erect my father held himself before those brutal jailers, and my heart glowed with love and pride to see him. We managed to exchange a few sentences and my father told me that he had obtained permission to send me money to buy tea and a few other comforts. He told me that he and my mother had waited three hours to see me and because it had been ruled that they could not both be admitted on the same day that my mother was standing close to the door of the next room just to catch the faint sound of my voice. These words roused Chkoni to a perfect fury. "So!" He fairly yelled. "But I'll spoil that game," and rushing out he slammed the door between the two rooms. My father flushed crimson but he spoke no word nor, of course, did I. A single protest might have meant punishment for me, and for us all no more visits.

I saw my father only three times, my mother a little oftener, as her health was the better of the two. The money my father sent me did not reach me except in very minute sums. By far the greater part of it was kept by the jailers, and gambled away. Not satisfied with that, the men warned my father that nothing except payment to the prison heads would save me from death, or worse still from assault by the soldiers. My father had long ago been deprived of his income, but he and my mother sold some valuables and gave it to the blackmailers who wanted it only for more gambling. Their sacrifice gave my parents a little peace of mind, but it did not save me from three of the most horrible nights I spent in the fortress. On each of these nights my cell was invaded by drunken soldiers who threatened me with unspeakable things. On the first occasion I simply groveled on the wet floor and prayed the man, in the name of his mother and mine, to let me alone, and, drunk as he was, my words actually penetrated his dark soul and shamed it. The next men were less drunk but were far more bestial. At the sight of them I threw myself against the wall and pounded frantically, screaming at the top of my lungs. Mme. Soukhomlinoff heard and understood. She screamed too, frightfully, and with all her might shook the heavy door of her cell. This brought the guard and once more I was saved. The third time I was so paralyzed with fright that I could not scream. I simply fell on my knees, holding up my little ikon, and begged like a trapped animal. The man hesitated a moment, spat on me contemptuously, and left. The next day, half dead with shame and fear, I managed to tell the Woman all that had passed. Indignantly she went to the Governor of the fortress, and after that even I, "the worst woman in prison," was spared the ultimate insult.

Although we could not know it, things were gradually changing for the better in the fortress. A little physical improvement was apparent. The cold had lessened and in our short walks in the prison yard we could see that lovely spring, with its fresh green leaves and springing flowers, had come to stay. I remember one day seeing in the grass a little yellow flower. It may have been a buttercup or a dandelion or something else we ordinarily call weeds, but to my eyes it was an exquisite thing. Audaciously I stooped and picked it, hiding it quickly in the bosom of my dress. The next visiting day I showed it to my father and dropped it on the table. On leaving the room he contrived to get hold of it and after his death in 1918 I found it, carefully preserved among his private papers. I never picked another flower in that prison yard, although once I tried. But this time a guard caught me, and struck the flower from my hand with the end of his rifle.

Things were improving under the surface, but aside from the welcome change in the weather conditions seemed for a time no better. In the cell adjoining that of Mme. Soukhomlinoff was my old friend General Voyeikoff, who was tortured almost as pitilessly as myself. My heart ached for him. In cell 69 was for some time the police detective Manouiloff, but when he was removed to another prison the writer Kolichko was placed in the cell. Kolichko, poor wretch, was so overcome by his arrest and imprisonment that during the first nights he sobbed so long and bitterly that I found it impossible to sleep. I was so unhappy that I began to pray for death, and once I even resolved to end my life. I had no weapon but a rusty needle which I had picked up and carefully concealed, but I had heard somewhere that there was a spot at the base of the brain which if punctured ever so little would cause death. Before seeking that spot I felt that I must say adieu to my brave little friend Mme. Soukhomlinoff, and so softly I rapped out a farewell message on the wall. Her quick mind instantly divined my intention and without losing any time she sent for the Woman and my rusty needle was taken away from me.

It began to be sultry in the Troubetskoy Bastion and the air in the cells became thick and foul. My small window, which looked out on a narrow court and a high wall, admitted little light and no breeze at all. I used to climb painfully up on the iron shelf which did duty for a table and pressing my face close to the bars I breathed in all the air possible. Instead of seeking the warm corner of my cell I now sat for hours together with my body against the wettest and coldest stones. My despondency increased every day, and I almost ceased to pray or to believe that the universe held any God to whom the prayers of captives could ascend. Yet all the time God was sending me help.

One day a soldier came to my cell and roughly bade me get up and go with two guards for examination. Not knowing exactly what that meant, I rose from my cot and followed the men to a room in the fortress where the High Commission of Inquiry appointed by Kerensky was then in session. Bewildered by the sudden transition from the bastion to a room full of comfortable furniture, and almost blinded by the brilliant light and sunshine, I had all I could do to answer their few inconsequential questions. I have described this first examination in another chapter, and I shall not repeat it here. It was so foolish that afterwards in my hot and ill-smelling cell I actually found myself laughing, and it had been a long time since I had laughed. Judge Roudneff, the only one of the commission who showed himself fair-minded or even capable of just judgment, was present at the inquiry, but I do not think he said a word. Afterwards he was charged with full responsibility of my case, and I appeared before him no less than fifteen times. At the close of the first of these personal interviews I thanked Judge Roudneff warmly. Astonished, he asked: "For what do you thank me?" And I answered: "For the happiness of four whole hours of sitting in a room with a window, and through it a glimpse of green trees." He did not reply except with a kind and sympathetic look, but I knew that his heart was touched, and that he received a new conception of what life meant to a prisoner.

Better things still were to come. Without our being aware of it the revolutionary mania had begun to subside a little and those men among our guards who had once been clean and decent were now getting back to their normal state of mind. Poor soldiers! Never let me forget that they were not to blame for the torments they inflicted on me and other prisoners. It was not they who invented the black calumnies that made me seem a creature undeserving of mercy or any clemency. It was not they who fashioned the cross on which I was crucified. The soldiers did only what they were incited to do by men and women far above them, people who conspired to crush me that they might crush the Empress. The soldiers I forgive, but I cannot yet forgive those others. The fate of the Imperial Family, the ruin of Russia, is on their souls. For what they did they have never shown any penitence, but those rough soldiers in the fortress repented and did what they could in atonement. One of the head guards was a man, handsome in a rustic sort of fashion, who at first had treated me with great insolence. One morning this man opened my door, hesitated for a moment, and then said in a low voice: "I am very sorry for you. Please take this," and vanished. "This" was an apple and a small piece of white bread. Another morning the soldier who brought my breakfast spoke in a grumbling aside but loudly enough for me to hear: "What idiocy to keep a poor sick woman in this place." One night the window in my cell door was pushed aside and in a trembling voice someone begged me to give him my hand. Tears fell on it while the unseen friend told me that he was a boy from Samara, and that it broke his heart to see women caged like beasts in such holes. He must have had a good mother, that boy. Perhaps they all had, for it became almost a habit for men passing through my corridor to slip me bits of bread, sausage, or sugar.

The most wonderful piece of good fortune came through the soldier in charge of the prison library. This man visited my cell one day, and after giving me a keen look which I could not understand he laid the library catalogue on my cot and went out. I had little interest in the dull books at our disposal, but when one sits hours in utter idleness he makes occupation out of almost nothing. I opened the catalogue and turned the leaves. To my astonishment out fell a folded paper. Cautiously I opened it and read these words: "Dear Anushka, I am sorry for you. If you have five rubles I can get a letter to your mother." For a long time after the incriminating paper had been destroyed I sat trembling in doubt and foreboding. I had barely five rubles, and if I gave them would they be gambled away? Was the letter a trap? Was it merely an effort to get me into trouble? I did not know, but on a bit of blank paper left in the catalogue I wrote with my stub of a pencil: "I have suffered so much already that I cannot believe that you wish to do me any more harm." Folding the five rubles and the paper into a tiny note, I tucked it into the catalogue and waited. After a while the librarian returned, and this time I read in his silent gaze that he was asking for my confidence. The next day he came back and again left the catalogue on my bed. This time I seized it eagerly and shook its leaves. A letter from my mother dropped out, a short letter, for she had been given only a few minutes to write, but I read and reread it until I knew every word by heart.

Then began a smuggled correspondence with my father and mother, they gladly giving money to the men who risked their own liberty by carrying the letters back and forth. The letters reached me in prison books, in the sheets of my bed, under the tin basin which held my food, and once even in a soldier's sock dropped carelessly on the floor. In this sock was concealed a note from Lili Dehn, free now and in correspondence with the family at Tsarskoe Selo. There was a slip of paper enclosed with a tiny white flower glued to it, and in the Empress's handwriting: "God keep you." Another precious souvenir of the Empress sent me by my mother was a little moonstone ring long ago given me at Tsarskoe. Tearing a rag from the lining of my coat, I made a bag for this jewel, and begging a safety pin from the Woman, I pinned it inside my dress. The poor librarian. This was the last favor he ever did me, for falling under the suspicion of the Governor, he was abruptly discharged. The letters, however, had done me so much good that I was in every way better and more cheerful. I felt in touch with the world again. I knew in a general way what was going on, and though not all the news was pleasant it gave me a sense of being alive and not altogether hopeless. I knew now what tireless efforts were being made in my behalf, and I felt that in the end something must come of them.

My parents had done everything humanly possible to move Kerensky but without any definite success. The first appointment with him was made through his secretary Chalpern, and although my parents were naturally exactly on time Kerensky kept them waiting for two hours. When at last they were received my parents were told that the Empress Alexandra Feodorovna, Rasputine, and Viroubova were responsible for the Revolution and would have to suffer for it. My parents had heard this before, but it was new to them to hear from Kerensky that he knew that I had had a great many diamonds from the Archbishop Pitirim and for that and other reasons nothing could be done for me. Later he softened a little and ended the interview by promising that my whole affair would be investigated. My parents then contrived an interview with the minister of Justice, Pereverzeff. They made two appointments in fact, for the first one Pereverzeff deliberately broke, going out for the day while my parents sat waiting in an ante-room. The next time my mother went to the Ministry she was received and was civilly treated. Pereverzeff also promised that a fair investigation would be made. By this time the Special Commission of Inquiry was sitting and my mother managed to see the president, Mouravieff. She took with her a letter from his brother to me before the abdication of the Emperor. In this letter I was warned of plots against me and was advised to leave the palace. I had replied to this letter, and my mother had a copy of my reply. I had written that I would never leave the Empress. My conscience was clean before God and man and I would remain to the end where God had placed me. I was astonished that a soldier should advise me to run away from a battlefield. Mouravieff who at first had been very harsh, changed after reading the letters. He even asked my mother to allow him to read them to the commission. They were significant, he said. As soon as my case had been referred to Judge Roudneff he called my parents to the Winter Palace, where he had his office, and talked with them, asking a great many questions, for nearly four hours. In this examination, for it was really that, my father and mother were allowed for the first time to defend me, to make explanations of obscure charges, to tell my life story to the man who was to judge me. No one else gave them such an opportunity, not even the Georgian deputy Cheidze, then very prominent in the Petrograd Soviet. Cheidze was kind and said that he would do anything in his power to help me to get justice, but I do not think he ever did anything. Members of the Provisional Government, Rodzianko and Lvoff, to whom, while they were still in power, my parents had written begging to be received, never even replied to the letters.

One day, sitting in my cell and remembering what had been written me in the smuggled letters, another wonderful thing happened. In the noon meal of fish soup which I must eat or starve I found a large piece of really decent meat. I ate it greedily, of course, and the next day I ate another piece which had mysteriously arrived. I took the first opportunity to ask the Woman where the food came from, and she told me that it was a cook, a poor man whose duty it was to carry food to our bastion. He too pitied me, she said, and she thought he might be willing to run almost any risk for me. So almost at once I was again in correspondence with my parents. This cook did more than carry letters, the brave man. He brought me food, chocolates, clean clothes, linen, stockings, and even a fresh frock. Growing bolder, he ventured regularly to take away my soiled linen and to replace it with clean things. All during those months in the fortress I had washed my linen and stockings in cold water, without soap, and in the night had hung them up in the warm corner on a hook improvised from a broken hairpin. Of course they were never clean, nor even, when I put them on, very dry, and now they were stiff with dirt. Can anyone imagine what it was to me to feel a clean, soft, smooth chemise against my skin?

I am sure the cook could never have done so much for me had not the guards closed their eyes to his activities. They were nearly all friendly now, and used to talk with me through the window in my door. In spring a number of pigeons flocked around the fortress and their constant sobbing voices got on my nerves. I spoke of this to one soldier who expressed surprise. "I was shut up here once," he said, "under the old Government, and I didn't find the birds bad at all. I used to feed them through the window." "You had a window in your cell," I exclaimed. "Then it couldn't have been as bad as this." And he assured me that it wasn't as bad under the Autocracy as under the beneficent Provisional Government and the Soviet. The prisoners had much better food and they could exercise two hours a day in the open.

Another prisoner of the Tsar's government, a non-commissioned officer named Diki, who had been very harsh to me in the beginning, now showed me kindness. Instead of robbing me, as of old, of every little privilege, he began to allow me an extra five minutes or so in the courtyard, he, too, saying that in the old days prisoners were better treated. Another of the guards in the courtyards, a man whom I had bitterly hated, and with cause, told the Woman that he wanted to speak to me. Afterwards while walking he approached me and I looked into his coarse face, deeply pitted with smallpox, and listened in fear at what he might have to say. Stammeringly he told me that he had just returned from a leave spent in his home in the Government of Saratoff. Visiting his sister's house, he was amazed to see, hanging under the ikon in the corner of the room, a photograph of me. "What!" he had exclaimed. "Do you have that shameless woman's picture in your house?" Whereupon his brother-in-law retorted: "Never dare to speak against her who was like a mother to me for two years in Tsarskoe. I was in her own hospital in the end, and it was like Heaven." The brother-in-law had charged the guard with all kinds of messages to me, telling him that they prayed for me daily in his family and hoped for my release. "Forgive me for being unjust to you," said the poor soldier, and offered me his hand. This was the first news I had of my hospital, and I learned with joy that the Provisional Government had not closed it.

Later I heard that the Government had not only carried on my work but had added five new buildings. None of my nurses or orderlies had left, though their openly expressed faith in me might easily have secured their dismissal. Some of the invalids had petitioned the Duma for my release, and another group, indignant because a revolutionary newspaper declined to publish their letter refuting the usual slanders about me, wanted to leave the hospital long enough to blow up the office building! They were good at heart, those misguided Russian soldiers, those poor ignorant children. I know them, and whatever they have been forced to do in these years of horror, I still believe them sound and good of soul. In the last days of my imprisonment in Peter and Paul the guards did not even lock my cell door. They used to linger and talk, and sometimes they brought paper and pencils that I might make sketches of them to take home. I was rather clever with a pencil in those days.

* * *

CHAPTER XVIII

THE prison had changed, and except for an occasional riot or a fight between two drinking soldiers, it was almost peaceful. For now there was a man attached to the fortress, a man so brave and kind, and above all so commanding that terrors fled before him—Dr. Ivan Manouchine. The gratitude and respect with which I write his name cannot be expressed in words. It was on the 23rd of April, the name day of the Empress, ever a day of memories to me, that this good man came into the house of pain where lay the prisoners of the Provisional Government. A few weeks before this the soldiers, gradually recovering from their first revolutionary blood lust, had begun to revolt against the needless brutality of the prison doctor, Serebrianikoff, and had finally sent in to the all powerful Kerensky a request for his demission. In those days Kerensky, whose ambition to be at the head of the government was maturing, made a special point of granting soldiers' petitions, and he really consented to replace Serebrianikoff with a physician of reputation. From the point of view of the Duma Dr. Manouchine was entirely a safe man to be appointed. He was a republican in politics, and he conformed to the popular superstition of "dark forces" surrounding the court. But what the Duma did not know about Dr. Manouchine was that he had a heart of gold and a mind that was ruled not by any political party but by principles of right and justice.

When the new prison doctor first came into my cell, accompanied by the retiring man looking frightened and ill at ease, I was lying on my cot in a mood of unusual rebellion. In a quiet, professional voice he asked me how I felt, and when he examined my poor chest and saw it black and blue and swollen from the clumsy cupping it had received, he frowned with displeasure. He gave some quick directions for my relief and in a gentle tone assured me that he intended to visit the bastion every day. It was the first time in many long weeks that I had been spoken to by the type of man we call a gentleman, and after the door closed behind him something in my frozen heart seemed to melt like icicles in the sun. Almost with the faith of childhood I fell on my knees and prayed, and after that I lay down and slept for several hours.

Every day soon after the booming of the noonday gun he came and every one among us stood up as close as possible to the cell doors, waiting to catch the first sound of his voice as he came down the corridor. At every door he stopped and asked the health of the prisoner. To him they were not prisoners but patients, and he treated them with all the skill and, above all, the courtesy he would have accorded the richest and most powerful of his patients. He examined our food and pronounced it entirely unsuited to our needs. He did not stop there, but in the end succeeded in greatly improving the ration and supplementing it for the sick with milk and eggs. How he did it in the Russia of those days I cannot imagine. I only know that Dr. Manouchine had a will of steel, and against that will and the staunch uprightness of his character malice and fanaticism broke like waves against a rock.

Little by little Dr. Manouchine instituted other reforms. The prisoners now received at least a part of the money furnished by their friends outside, and once a week the non-commissioned officer Diki went through the prison answering requests for such necessities as soap, tooth powder, and paper on which petitions to the Governor of the fortress might be written. Often when a prisoner lacked money to pay for these things the doctor supplied it out of his own pocket.

Meanwhile my examinations under the stern but just commissioner Roudneff were going on. Weary under the long and apparently pointless inquisition, I asked Dr. Manouchine one day how much longer he thought they intended to torment me. His reply was grave. "Not long, I think. But before it is over you may have to undergo a still more trying ordeal." A few day later he came to my cell alone; that is, he resolutely closed the door between us and his usual escort of soldiers, and told me in his kindest manner that the Special Commission of Inquiry had almost concluded that the charges against me were without foundation. One more proof, however, was necessary, a physician's sworn statement that the hideous accusations of vice made by enemies of the Emperor and Empress and their closest friends were false. Would I, for my own sake, for the sake of the Imperial Family, submit to a medical examination? Without at all knowing what was implied I gave an instant but rather frightened consent to any examination he thought necessary. It was a terrible ordeal for a woman to live through. Most of the questions asked me were of a nature which appalled me, and yet were beyond my understanding. I cannot here repeat even the least of them. I can only say that they opened up to me an abyss of wickedness and sin which I had not dreamed existed in the human soul. At the end of an hour— or many hours—of trial, I lay on my bed, hands clasped over my eyes, spent, exhausted, utterly incapable of speech. Up to the very end Dr. Manouchine's manner had been that of a physician, but now that it was over it was a friend beyond anything human and sympathetic who laid his hand on my quivering shoulder and said: "This clears you absolutely. They will take my word for it."

Towards the end of May, a hot and wearying season, the fortress was visited by the head of the Provisional Government's Commission of Inquiry, a pompous man, yet in his cautious way, rather kindly. Pausing before my cell, he told me that no crime had been fastened upon me and that I might hope soon to be transferred to a better place. Hope gave me new life momentarily, but as the days dragged on my hope gave way to bitter unbelief. My health always since my arrest indifferent, now began to decline and I could see that the doctor was seriously concerned for me. He came to the prison only four times a week now, and what ages seemed to elapse between his visits. All I had left of courage his voice and ministrations gave me.

One hot June day I was aroused from my sick lethargy by the tramping of heavy boots on the stones of the corridor. The heavy cell door swung open and I saw a crowd of strange men, several of whom unceremoniously invaded my cell and began an examination of my poor effects. Frightened, I watched them as they disdainfully picked up and threw aside the few rags a prisoner is allowed, but my fears were allayed when I saw in the background the tall figure of the doctor. "Do not be afraid, Anna Alexandrovna," he said. "This is only a committee of revision of prisoners." Later I heard him say to the committee: "This woman may have only a few days to live. If you are willing you may take on yourselves the responsibility of her death. As a physician I refuse to do so."

The next day he whispered to me that he was confident that I would be taken away, but that my release might be delayed a little because of renewed riots among the prison guards. He did not know where I was to be taken, and I feared it would be the Women's Prison, which the Woman had told me was almost as bad as the Troubetskoy Bastion. But soon I was relieved of that nightmare, for the doctor came again bringing me the good news that I should probably be taken to the House of Detention in a pleasant neighborhood on the other side of the river. In groups the friendly soldiers came to say good-bye and to assure me that even should the mutinous guards oppose my going they would see to it that I got safely way. Days went by, sleepless nights, and still no order of release arrived. I became almost hysterical with suspense. I gave way to dreadful fits of weeping until even the doctor grew stern and bade me control myself. I felt like a mouse under the teasing claws of a cat, and control was difficult even after I learned that the doctor had persuaded some members of the central committee of the Petrograd Soviet to visit the fortress and to reason with the mutinous guards.

Almost the last day of June, at six in the evening, I was standing barefooted and half dressed against the cool, wet wall of my cell thinking of my mother who, the day before, had visited me. Her face was brighter than usual and she had said to me: "The next time we meet it may be in better circumstances." At the moment my door opened and the hated Chkoni appeared. "Well," he said, with his usual sneer, "did you have hysterics after seeing your mother?" "Certainly not," I replied coldly. "No?" he commented, "I thought you might because to-morrow or the next day they may take you away." I fell against the wall too overcome to speak, too blind to see the hands of the guard pressing my limp hand in congratulation. Tomorrow or next day! The words repeated themselves in my brain countless times. But I was not even to wait until to-morrow, though Chkoni evidently wished me to think so. I heard the voice of the younger and less familiar wardress: "Dress yourself quickly. The doctor is bringing a deputation from the Soviet." I had nothing to put on except my ragged shoes and a torn gray woolen jacket, but these I rapidly seized while the wardress picked up and made a bundle of my small belongings.

On the opposite wall I heard brave Mme. Soukhomlinoff rapping out a farewell message to which I responded as well as I could. Then the deputation arrived, and the doctor. There was some confused talk. I cannot remember a word. I felt myself picked up and carried down the winding corridors. The great door of the bastion rolled open and we passed out into the cool, delicious evening air. There was a motor car into which I was lifted, another car into which the doctor climbed, there were soldiers, some friendly, some seemingly determined that the cars should not leave the courtyard. I remember very little until we drove out of the gates and over the Troizky Bridge. The wind, the brilliant twilight, the sight of water and the blue sky, blinded me so that I had to cover my face with both hands.

Within a short time the cars stopped at the Detention House in the Fourshkatskaia Ulitza, and I was carried into the office of the commissioner. He was an officer, rather short in stature, but dignified and efficient. Offering me his hand, he asked me if I would be seated while he made out the necessary papers. I had time to see that the House of Detention promised to be quite different from a prison. Indeed the soldiers of this house would not even permit the entrance of the fortress guards who had come with me. As if he divined that I was too weak to walk upstairs the commissioner gave orders that I was to be carried. It was into a large, light, clean room that they took me, and at my exclamation of joy at sight of windows the soldiers laughed heartily. But the doctor silenced them. "Go," he said, "see that her parents are telephoned, and send a woman to bathe and dress her." His own arms lifted me from the chair on which I half sat, half lay. On a bed softer and cooler than even existed in my memory he laid me, said good night, and gently left the room.

CHAPTER XIX

I SPENT a happy and peaceful month in the Detention House, the only disturbing event being the so-called July Revolution, the first serious attempt of Lenine's party to seize the government. The Soviet already transcended in power the old Provisional Government, most of whose original members had by this time disappeared from politics. Kerensky was premier, nominally, but only because a remnant of the Russian Army still resisted the separate peace propaganda and remained on duty at the front.

Persons in the Detention House were prisoners in the sense that they were under guard and were not allowed to leave the house. The guards were complacent, though, and visiting between the rooms was permitted. I soon found that I was the only woman in the place, and that some of the men there had suffered greater tortures than I. There were between eighty and ninety officers, almost the last remnant of the garrison of Kronstadt where in the first days of the Revolution the soldiers went quite mad and murdered, in ways too horrible to relate, a great many officers, and even young naval cadets against whom they could have had no possible grudge. The officers in the Detention House were in a sad state of body and mind. We talked together sometimes in the dining room, and learning that they longed for the consolation of Holy Communion, I remembered that my hospital in Tsarskoe Selo possessed a movable altar and holy vessels. With the consent of Nadjaroff, commandant of the Detention House, the altar and my own priest were brought from Tsarskoe and the sacred ceremony was twice celebrated, the last time on July 29, my birthday.

I ought to say of the commandant Nadjaroff that he was an excellent man, kind to the prisoners, and conscientious in his work. The poor man had one fatal weakness, gambling. So strong a hold had this vice on an otherwise good man that when his money ran short he was not above borrowing and even begging from the prisoners and their friends. It seems almost too bad to record this blot on the character of a man who was kind and courteous to me, but I am trying to give the psychology as well as to portray the events of the Russian Revolution, and I must emphasize the fact that it was the weakness and self-indulgence of the people themselves that made the Revolution and its frightful aftermath possible.

From my first day in the Detention House I began to recover my health and my self-control. My windows were not barred, and through the open casement I feasted my eyes on the beauty of grass and trees, on the familiar little church of Sts. Kosma and Damian which stood almost opposite, and, strangest of all to me, of people walking or driving through the streets below. It took a few days for me to get used to a normal state of life, and at first, when night grew near, I was seized with such nervousness that they had to let a maid sleep in the room with me. As the fresh air and sunshine began to bring color to my face and I felt strength returning to my limbs I forgot my fears, and became something like the woman I had been before I was caged like a beast in the Fortress of Peter and Paul. Visitors were admitted both morning and afternoon, and I had the happiness of talking privately with my father and mother and with friends who still remained faithful. They brought me clothes, toilet articles, books, flowers, writing materials, and, best of all, news of what had happened during the months of my imprisonment. I learned of the rapid disintegration of the army under the weak and ineffectual Provisional Government, the tottering state of Kerensky's régime, and the threatening domination of the Soviets. What was in store for Russia no one knew, happily for Russians. Of the fall of the

Soviets and the rise of Bolshevism no one yet had any premonition. The radical element was already in control, and there was a great deal of threatening talk of shooting the Emperor. However, the Imperial Family was still alive and in Tsarskoe Selo, which was as much as I had dared to hope.

Of the events of the July Revolution, the forerunner of the Bolshevist triumph of November, 1917, I know rather less than others who were at full liberty during that terrible week. It was about the 18th of the month, a brilliant summer day, when I was startled by long-continued shouting and bellowing of soldiers in a caserne not far from the house. In great excitement the men were running in and out of the yard calling on the *tovarishi* to arm themselves and join the uprising. As if by magic the streets filled with rough-looking people, singing wild songs, waving their arms, and forming processions behind huge scarlet banners on which I could read such inscriptions as "Down with the War!" "Down with the Provisional Government!" An endless line of these paraders passed and repassed, dirty, disorderly soldiers, equally dirty factory workers, yelling like crazed animals. Once in a while a gray motor truck would dash through the street, laden with shouting men and boys, rifles, and machine guns. In the distance we could hear shots and the ripping noise of the machine guns.

Of course we were all horribly frightened, especially the officers from Kronstadt, who knew that in case of invasion not one of them would be left alive. We were all advised to leave our rooms and take refuge in the corridors, as at any time the rioters might begin firing through the windows. But we were not out of danger even there because many of the guards openly sympathized with the rioters, and the head guard was so jubilant over the course of events that he went around boasting that he was quite prepared to surrender the house and all its inmates at the first demand of the Revolutionists. Some of the guards were better than this man, and one of them, a wearer of the St. George cross, said that in case of trouble he would try to get me to his sister's house, where I would be perfectly safe. For two nights nobody slept or even undressed. In the room next to mine was lodged old General Belaieff, former War Minister, whom imprisonment had left a sad wreck. He, like the other officers, fully expected death, and I found myself in the novel rôle of a cool and collected comforter. I, who had lately been afraid to sleep in a room by myself, now went from one old soldier to another urging them to keep up hope. The days passed, and the firing came no nearer, and within a week troops summoned from the front took possession of the city.

GRAND DUCHESSES OLGA, TATIANA, ANASTASIE, AND MARIE, PRISONERS AT TSARSKOE SELO, 1917.

ANNA VIROUBOVA.
A Photograph Taken Shortly after Her Release from the
Fortress of Peter and Paul, Petrograd.

My examination under the High Commissioner Roudneff not being entirely finished, he came once or twice to the Detention House bringing with him on one occasion Korinsky, Procurator of Petrograd, a courteous gentleman, who at parting expressed a hope that I would soon be free. A few days later, August 5, if I remember correctly, M. Korinsky himself telephoned that if my parents would call at his office they would be given my warrant of release. Alas, my parents happened to be in Terioke that day, but too impatient to wait until the morrow I telephoned my uncle Lachkereff, who immediately hastened to the Procurator's office for the coveted warrant. Trembling with excitement, I stood at my window with several of my good friends waiting the result of his errand. At six o'clock we heard a drosky driven at great speed over the cobbles, and as it came in sight we saw my uncle standing up and wildly waving the papers in his hand. "Free!" he called out. "Anna Alexandrovna, you are free!" The rest is confusion in my mind. There were laughter and sobs. People kissed and embraced me. I was in the drosky driving through Petrograd streets. I was in my uncle's house. The tea table was spread. It was like a dream.

After prison one gets used to freedom by slow degrees. It seems strange at first to be allowed to move about freely, to go to church, to walk, to drive, to go wherever one desires, through woods, along leafy country roads. Not that I was entirely free to go where I liked. I could not safely go to Tsarskoe Selo, even to my own house, which after my arrest had been taken over by the police,, and not only ransacked for evidence against me, but looted of every valuable. It was my faithful old servant Berchik who gave me the details of the search. He, honest soul, who had been forty-five years in the service of my family, was offered ten thousand rubles if he would testify against the Empress and myself. On his indignant refusal the police arrested him, while they tore up the carpets and even the floors of my rooms, demanding of Berchik the whereabouts of secret passages to the palace, the private telegraph and telephone wires to Berlin, my hidden writing desks, and all sorts of nonsense. Especially were they anxious to discover my wine cellar, and when they found that I possessed none they were angry indeed. They took possession of all the letters and papers they could find, and at the end of the search ordered my cook to prepare them an elaborate supper. Then they left taking with them the silverware.

If I could not visit Tsarskoe and those whom I loved and longed to see, I could at least, and I did, hear from the Empress. Just before the family were sent to their exile one of the maids smuggled out a letter which reached me safely and which I quote here, suppressing only the most intimate and affectionate passages.

"I cannot write much," the letter began, "my heart is too full. I love you, we love you, thank you, bless you, kiss the wound on your forehead. . . . I cannot find words. . . . I know what will be your anguish with this great distance between us. They do not tell us where we go (we shall learn only on the train), nor for how long, but we think it is where you were last" (Tobolsk). "Beloved, the misery of leaving! Everything packed up, empty rooms, such pain, the home of twenty-three years. Yet you have suffered far more. Farewell. Somehow let me know you received this. We prayed long before the Virgin of Znomenia, and I remembered the last time it was on your bed. My heart and soul are torn to go so far from home and from you. To be for months without news is terrible. But God is merciful. He won't forsake you, and will bring us together again in sunny times. I fully believe it."

With the letter the Empress sent me a box of my jewels which she had carefully guarded, and I heard a fairly full account of how the summer had been spent. For a time she and the Emperor had been kept apart, being allowed to speak to each other only at table and in the presence of guards. Revolutionary agents tried every possible means of incriminating the Empress, whom they hated even more than the Emperor, but finally failing in their efforts they allowed the family to be together once more. The day after they were sent to Siberia the maid visited me again with the story of their departure. Kerensky personally arranged every detail, and intruded his presence for hours together on the unhappy family. Under his orders everything was made ready for a midnight journey but actually they did not leave the palace until six o'clock in the morning. All night the prisoners sat in their traveling clothes and wraps in the round hall of the palace. At five a courageous servant brought them fresh tea, which gave them a little comfort, especially Alexei, who stood the night badly. They drove away from the palace with perfect serenity as if going on a holiday to Finland or the Crimea. Even the Revolutionary newspapers, with grudging admiration, had to admit this.

A day or two later Mr. Gibbs, Alexei's English tutor, came to see me, and he told me that although he was not permitted to accompany the Imperial Family with the other tutors, M. Gilliard and M. Petroff, he intended to follow them to Tobolsk. He took a photograph of me for the Empress, who was anxious to see for herself if the long imprisonment had impaired my health. As a matter of fact I was not very well just then, as I had something very like jaundice, so I am afraid my photograph was none too reassuring. At this time I was staying in the home of my sister's husband who was attached to the British Military Commission in Petrograd. It was a cool and comfortable apartment, and I should have been contented to stay on indefinitely. But one day my brother-in-law, in deep embarrassment, showed me a letter from his sister, who was expected on a visit. This lady expressed herself unwilling to live under the same roof with a person as notorious as myself, and I, equally unwilling to associate with her, moved back to my uncle's hospitable home. But even there I found no serenity. I had been acquitted of all the crimes charged against me by the Provisional Government,* but now the Government of Kerensky found new accusations to make of me. This time I was a counter-Revolutionist, and as papers served on me in the middle of the night of August 24 (Russian) ordered, I had to leave for an unknown destination within twenty-four hours. As I was without money and was really in need of a physician's care, my relatives began at once to petition every authority for a delay of at least twenty-four hours more. This was finally allowed, but two soldiers were immediately placed before my door and I was a prisoner in my uncle's apartment. Meanwhile my parents and friends continued to make every preparation for my comfort in exile, and two of my hospital staff, the director and a nurse, volunteered to go with me. The night before I left my poor parents stayed with me, none of us going to bed. Very early on a rainy morning two motor cars filled with police came for us. They were kind enough to let my parents accompany me almost to the Finnish side, and they explained that they had come so early because they feared street demonstrations.

At the station we found a miscellaneous company of alleged counter-Revolutionists including a few old acquaintances. Among these was former detective Manouiloff, a tall officer named Groten, the editor, Tanchevsky, and the curious little Siberian doctor Badmieff, with his equally curious wife and child and a young maid named Erika whom I came to know very well. Badmieff was the herb doctor who, it will be remembered, was supposed to purvey the deadly poisons which I was alleged to feed to the Tsarevitch. He was a small, round, shriveled man, excessively old—over a hundred, they said—and in appearance resembled a quaint carved Buddha out of an antiquarian shop. He had the smallest, blackest eyes imagi-nable, set in a face yellow and wrinkled, and his long, scraggly beard was as white as cotton. His wife, many years his junior, and his funny little child, Aida, were as Mongolian in appearance as himself. The maid, Erika, a girl of about eighteen, was not uncomely with her with her bright eyes and short, curly hair. All the "counter-Rev-olutionists" were herded together in one carriage, the one farthest from the engine, and in charge of us was a Jewish official of the Kerensky

* See Appendix B.

Government. At Terioke I parted with my father and mother, the train moving on quickly to the Finnish town of Belieovstrov. Here we were met by an enormous crowd of soldiers and working people, all hostile, demanding to see the dangerous counter-Revolutionists. Especially they demanded to see me, but I shrank back in my seat, fearing every moment that the shower of stones against the carriage would break the windows. But quickly the conductor's whistle was blown and the train moved beyond the reach of the mob.

Worse was to come. When we reached Rikimeaki we found waiting us a larger and a still more furious crowd. Our carriage was unfastened from the train and the mob rushed in yelling that we must all be given up and killed. "Give us the Grand Dukes!" they shouted. "Give us Gourko!" I sat with my face buried in the shoulder of my nursing sister fearing that my end had come. My fears were not imaginary, for several ruffians pitched on me shouting that they had found Gourko in women's clothes. Frantically the sister explained that I was not General Gourko but only a woman ill and lame. Refusing to believe her, they demanded that I be stripped, and I have no doubt that this would have happened had not a motor car opportunely dashed up carrying a sailor deputation from the Helsingfors Soviet. These men pushed their way into the carriage, and without ceremony booted the invaders out. One man, a tall, slender youth named Antonoff, made a speech at the top of his voice, commanding the mob to disperse and to leave things in the hands of the Soviet. So authoritatively did he speak that the crowd obeyed him and allowed our carriage to be attached to another train bound for Helsingfors. Antonoff remained with us, and in the friendliest fashion sat down beside me and bade me to be of good cheer. He did not know why we had been sent away from Petrograd, but the Soviet at Helsingfors, of which he was a member, had received a telegram, he thought directly from Kerensky, saying that we were being sent on, and when we arrived were to be placed under arrest. Doubtless there would be explanations, and after that we would surely be released. To my mind the thing seemed not quite so simple. Kerensky had sent us from Petrograd, but not to be imprisoned in Helsingfors. What he desired was that the mobs, notified of our arrival from his office, would kill us before we ever reached Helsingfors at all. No doubt he hoped at the same time to dispose of General Gourko and the Grand Dukes left in Petrograd.

But Gourko was too clever for Kerensky, and made good his escape to Archangel, where he took refuge with the British Occupational Force. As for the Grand Dukes, they were, for some reason, at this time left undisturbed by the Revolutionists.

It was night when we reached Helsingfors and we found the station practically deserted. The main body of the prisoners were taken away into the darkness, but Antonoff said that I and the nurse should spend the night in a hospital adjoining the station. We climbed several flights of steep stairs and passed through wards crowded with blue-gowned sick soldiers and sailors, not one of whom offered us the slightest rudeness. A skilled Finnish nurse undressed me and put me to bed, but unhappily not for long. Scarcely had I composed myself to sleep when the door opened, the lights flashed up, and Antonoff, red and very angry, entered the room. He had gone to the Soviet authorities, confident that he could persuade them to let me remain in the hospital, at least until word came from Petrograd of our exact status. But they refused his request and ordered him to take me at once to the ship on which the other prisoners were confined. There being no appeal I dressed and limped down the long stairs to the street where a dense mob had assembled, shouting, threatening, crowding dangerously around the motor car. It is a horrible thing to hear a mob shrieking for one's blood. One feels like a cornered hare in the face of yelping hounds. With the strength of desperation I clung to the arm of Antonoff, who for all I knew might yield suddenly and throw me to the crowd. Unworthy thought, for the man held me firmly, all the time demanding that the people give room and let us reach the car. When they saw me in the car their fury seemed to redouble. "Daughter of the Romanoffs," they yelled, "how dare she ride in a motor car? Let her get out and walk." Standing up in the car Antonoff repeated his commands that the mob disperse, and slowly at first and then more rapidly we got away. We reached the distant water front, and I was taken from the car to a ship. Picture my astonishment when I found myself standing on the deck of the *Polar Star*, the light and beautiful yacht on which I had so often sailed in Finnish waters with the Imperial Family. With all the Imperial property the *Polar Star* had been confiscated by the Provisional Government, and it was but another sign of the changing times that the yacht had later been taken away from the Provisional Government and was now the property of the Soviets, being the *Zentrobalt*, or headquarters of the

Baltic fleet.

From the deck I was hurried past the open door of the main dining salon, once a place of ceremony and good living, now a dingy, disordered apartment where crowds of illiterate workmen gathered to dispose of the rest of Russia's ruined fleet and the future of our unhappy country.* At least a hundred of these men were in the salon when I passed it first, and during the five days I spent on the yacht their voices seemed to go on in endless orations, ceaseless wrangling, twenty-four hours at a stretch. It was like nothing I can describe, like an ill-disciplined lunatic asylum. I was herded with the other "counter-Revolutionists" far below decks in what I conjectured had been the stokers' quarters. The stifling little cabins were filthy, like all the rest of the yacht, and they simply swarmed with vermin. It was so dark that night and day the electric lights burned, and I was thankful for that because somehow the bright light seemed to be a kind of protection against the swarm of grimacing, obscene sailors who infested the place, amusing themselves with discussions as to when and how we were likely to be killed. During the whole of the first night Antonoff stood guard over us and warned the sailors that no murder could be done without authority from the Soviet. Over and over again they suggested that he leave the place, but he always replied firmly that he was responsible for the prisoners and could not go. Finally towards morning the sailors left, and afterwards we learned that their blood lust towards us was not merely simulated. They had gone directly from the yacht to the *Petropavlovsk*, the flagship of the fleet, and had killed every one of the old officers left on board.

Antonoff left us early in the morning, left us expecting to return, but he never did return nor did we ever see or hear of him again. Such sudden disappearances were common enough even in those early days of the Russian Revolution, before murder became the fine art into which it has since developed. Five days we remained on the *Polar Star*, very miserable in our vermin-infested quarters below decks, but mercifully allowed part of each day in the open air. They might have allowed us longer time on deck had it not been for the hostile crowds that constantly thronged the quays. My time was spent in the shelter of the deckhouse near the main salon, a spot where in the old days the Empress and I loved to sit with our books

* Finland had not then separated from the old Russian Empire.

and work. Here five years before, when the Empress Dowager visited the yacht, I had taken a photograph of her with her arm around the shoulders of the Emperor, both smiling and happy in the sparkling light of the fjord. Every corner of the yacht had been exquisitely clean and white in those days. Dirty as the yacht's present crew appeared, I cannot say they starved their prisoners or were cruel to them. We had soup, meat, bread, and tea, luxurious fare compared to Peter and Paul. Our worst condition was suspense of mind as to our ultimate fate. At every change of guard we begged news from Petrograd, but always we received the same answer. The Kerensky Government gave no reason or justification for our arrest. Two of the sailors were especially friendly to me because, as they explained, they came from Rojdestino, our family estate near Moscow. "If we had known that you were going to be brought here," they said, "we might have done something. But now it is too late." That night I found in my cabin a tiny note, ill-spelled and badly written, warning me that all of us were about to be transferred to the Fortress of Sveaborg in the Bay of Helsingfors. "We are so sorry," the note concluded. Although it was unsigned, I knew the note must have been sent in kindness by one of the men from my old home. But at the prospect of another imprisonment my heart turned sick with dread.

Next evening came Ostrovsky, head of the Helsingfors Okhrana, accompanied by several members of the main committee of the Soviet. Ostrovsky was a very young man, scarcely eighteen I should judge, but he had fierce eyes and all the assurance of a born leader. Turning to my nurse, to Mme. Badmieff, Erika the maid, and her little Mongolian charge Aida, he said roughly that they were free but that all the rest would be taken at once to the fortress. In a sudden panic of alarm I threw myself into the arms of my nursing sister and begged her to accompany me. But she too was fear-stricken and drew back while all the men laughed heartlessly. "What's the difference?" asked Ostrovsky brutally. "You're all going to be shot anyhow." At which the dauntless Erika, putting Aida into her mother's arms, came over to me and tucking her hand under my arms said: "I'm not afraid. I'm going wherever the doctor goes and I'll stand by you both." I gave the trembling nurse a small box containing all the trinkets I had brought with me, gave her messages to my father and mother, and followed my fellow unfortunates to the deck, down a slippery gangplank to waiting motor boats on which we traveled the half hour's journey from the yacht to the fortress.

CHAPTER XX

SVEABORG before the War was one of three principal naval stations of the Russian Empire, the other two being Kronstadt and Reval. Sveaborg occupies a number of small islands in the Bay of Helsingfors. The bay itself, shaped like a rather narrow half moon, is so enclosed by these wooded islands that in winter the salt water freezes solidly. In summer the islands are green and lovely and a few of them, not under military control, are used by the Finns as pleasure resorts. Even in the darkness and in the unfortuitous circumstances of our arrival I could see that the main island might be a very attractive place. Up a steep hill we panted, past a white church surrounded with trees, and at last reached the place of our confinement, a long, dingy, one-storied stronghold. A young officer and several very dirty soldiers took our records, and Erika and I were pushed into a small cell with two wooden bunks covered with dust and alas, nothing else. The place smelled as only old prisons do smell, and the only air came in through a small window high in one of the walls. Wrapping ourselves in our coats, we lay down on the hard planks and tried to sleep. In the early dawn we got up, our backs aching and our throats choked with dust, but the irrepressible Erika laughed so heartily and sneezed so comically that I found it impossible to lament our surroundings. The place was a dreadful hole just the same, no proper toilet facilities at hand, and of course no opportunity of washing, to say nothing of bathing. We had to pay for our food at the rate of about ten rubles a day, at that time no small amount of money. The food was not very bad except that Stepan, the commissary, used to wipe our plates with a disgustingly dirty towel which he wore around his neck, the same towel being used in a laudable attempt to wipe the dust from our bunks.

Climbing on the bunks, we had a view through the window of a new building going up, the workmen being women as well as men. At the same time we got a glimpse of the detective Manouiloff who, ever pessimistic, held up three fingers as an expression of his belief that we had only that many days to live. We, however, ventured the guess that we would not remain at Sveaborg more than a month. It was a mere hazard but it turned out a fortunate one. We remained just about a month. It was a queer life we lived during that month, surrounded by tipsy and irresponsible men whose officers seemed to fear them too much to insist upon discipline. The officers, especially one fine young man, did everything they dared to make us comfortable. After the first ten days our plank beds were furnished with green leather cushions which might have made sleep a comfort if they had not persisted in slipping from under us about as soon as we dozed off. Somewhat later, a week perhaps before our liberation, these cushions were replaced by real mattresses stuffed with seaweed, wonderfully luxurious by comparison with the bare boards. The prisoners were exercised every day in the open under Sveaborg guards and the gaze of a crowd of Finnish Bolshevists. These people seemed at first immensely diverted by the pomposity of the Siberian doctor Badmieff who, in his long white robe, tall cap, and white gloves was certainly a curious spectacle. Soon they tired of him and turned their stolid, expressionless eyes on the other prisoners with what intentions we could only conjecture. Badmieff continued to be a center of interest in the prison. Erika, his faithful disciple, demanded the privilege of attending him, and this was granted. Every day he sat cross-legged like the Buddha he so much resembled, dictating endless medical treatises to Erika. In the evenings he used to put his lamp on the floor at the foot of his bunk, strew around it flowers and leaves brought from outside, burn some kind of ill-smelling herbs for incense, and generally create what I assumed to be the occult atmosphere of his beloved Thibet. Erika, scantily clad, always attended these séances and gradually they appeared to hypnotize the sailors, who thought highly of the doctor's professional powers. Indeed towards the end I often heard them swearing that whoever left the fortress, they would at least keep their highly esteemed *tovarish* Badmieff and his Siberian-Thibetan lore.

In sad contrast to the condition of Dr. Badmieff was that of the poor editor, Glinka Janchevsky, who being without money was treated with the utmost contempt. Housed in a wretched cell covered with obscene drawings, the miserable man spent most of his time lying on his wooden bed wrapped up, head and all, in his overcoat. He used to creep to our cell door with a glass of hot water in his hand begging for a pinch of tea and, if we had it, a little sugar. Every day he used to ask pathetically: "When do you think we shall be let go?" Like all journalists, he was famished for news, and whenever I got hold of a stray newspaper I used to read it to him from the first column to the last.

The vacillating conduct of the Bolshevist sailors toward the prisoners of Kerensky I can only ascribe to the increasingly bitter conflict going on between the weak Provisional Government and the Bolsheviki. The sailors hated us because we were "bourgeois," but they spared us because Kerensky desired our destruction. The officers good-naturedly brought me flowers from outside, an occasional newspaper, and even letters from people in Helsingfors who knew my history and pitied my fate. Sometimes I was even invited to tea with the officers, and twice I was taken out of prison, ostensibly for examination, but really to attend services at the little white church on the island. The guards were rough and kind by turns, sometimes uttering horrible threats against all the prisoners, sometimes bringing me a handful of the wild flowers they knew I loved to have near me. Discipline was lax, and we never knew from one day to another what might befall. For example, the padlock to my cell got lost and for several nights the door was left unlocked. One can imagine how I slept! On one of these unguarded nights the cell was invaded by a group of drunken and lustful men. Erika and I fought them, screaming at the top of our lungs, until a few sober and better-minded sailors came to the rescue. A day or two later, when a rumor spread that we were all to be hanged, I among the first, I for one felt less terror than relief. Anything, even hanging, seemed better than this lunatic prison where the guards drank, played cards, and wrangled all night, and where the men's attitude towards Erika and myself, the only women, was by turns dangerously savage and dangerously friendly.

Besides the Kerensky prisoners the fortress sheltered eight or nine prisoners charged with crimes ranging from theft to murder. Some of these whom we encountered in the exercise yard looked like very decent men, shining perhaps by contrast with the rowdy Revolutionists I had seen in the course of two imprisonments. For these unfortunates and for the guards we bought cigarettes, thus establishing more cordial relations. Nobody knew or could guess what was going to happen to us. One day appeared the president of the Helsingfors Soviet, a black-eyed Jew named Sheiman, who assured us that we were to be sent back to Petrograd, and that we might as well have our things ready by nine o'clock that night. Nothing happened that night, nor did we, for some reason, expect anything. The next day Sheiman came again with his bodyguard of soldiers and sailors, and told us that his Soviet refused for a time to release us. It appeared that telegrams had arrived from Kerensky and from Cheidze, the Georgian leader in the Petrograd Soviet, urgently demanding our return. The Helsingfors Soviet might have obliged Cheidze, but they would not honor any demand of Kerensky's, so there we were. The Provisional Government and the Petrograd Soviet sent over several deputies, Kaplan, a small, black-bearded man, who smilingly told us that there was no possible hope for us; Sokoloff, the famous, or rather infamous, author in the first instance of Order No. 1 which was principally responsible for the break-up of the army; and Joffe, the little Jew, who, a few years later, became influential enough to be included among the delegates to the Genoa Conference. After their visit, I don't know why, prison discipline became still further relaxed. We had visitors and the attention of physicians if we needed it. We were informed that henceforth we would not be regarded as prisoners at all, but only as persons temporarily detained. Two hours a day after this we were allowed in the open air, and I became very friendly with the Finnish women carpenters at work on the new building on our island. These good souls brought me bottles of delicious milk, and one day the building foreman, a Moscow Russian, invited me to his house to tea, and here I, a poor prisoner, was treated with such deference that I was actually embarrassed. Not one of the family would eat with me or even sit down in my presence.

At this time Erika and I were given a more commodious cell furnished with the seaweed mattresses of which I have spoken. But to our horror we found the walls covered with the most frightful scrawls and pictures. The sailor guards, however, brought water and sponges and with many apologies washed off the disgusting records as well as they could. I was thankful for this a few days later when all unexpectedly I received a visit from my dear mother. It had been some days after our parting at the frontier before she and my father learned that I was in prison. Immediately they had gone to Helsingfors to appeal to General Stachovitch, the Governor of Finland. But he advised them to avoid trouble for themselves, perhaps for me also, by going quietly back to Petrograd. My parents gave him money for me, which I never received, and despite the Governor's advice they stayed on in Helsingfors in faint hope of seeing me. Dr. Manouchine, my mother told me, had returned from a long visit in the Caucasus and was doing what he could to get me released. My mother also gave me news of the last struggle to maintain the army, the conflict between Korniloff and Kerensky, ending, as everyone knows, in the death of Korniloff. These two were about equally hated by the

Sveaborg sailors who would gladly have murdered them both. They had begun to speak with unbounded admiration of Lenine and Trotzky, especially of Lenine, who they declared was the coming saviour of Russia.

Bolshevism was in the air, and for a moment it assumed a really benevolent aspect. I remember a deputation of Kronstadt Bolshevists who came to Sveaborg to inspect us and to review our entire case. Some of these men were very civil to me, asking many questions about the Imperial Family and the life of the Court. At parting one said to me naïvely: "You are quite different from what I thought you'd be, and I shall tell the comrades so." The very next day another deputation came and, characteristic of the confused state of the public mind, these men were as brutal as the others had been kind. They stormed down the prison corridors roaring: "Where is Viroubova? Show us Viroubova!" I cowered in my cell, but when the guard came and admonished me, for my own safety, to show myself to the men I gathered courage to speak to them. Totally unprepared to see the terrible Viroubova merely a crippled woman in a shabby frock, the men suddenly quieted down and made civil response to my words. "We didn't know that you were ill," said one of the men as they prepared to move on.

Although we did not know it at the time, our fate really hung on the outcome of a Congress of Soviets which was then being held in Petrograd, and to which both Sheiman and Ostrovsky were delegates. Sheiman returned to Helsingfors and visiting my cell told me that both Trotzky and Lounacharsky were insistent on the release of Kerensky's prisoners. That evening, he said, would be held a secret session of the executives of the Helsingfors Soviet at which he would urge the recommendation of Trotzky and Lounacharsky. If the executives agreed the question would then be referred to the entire Soviet, made up principally of sailors of the old Baltic fleet. That evening I was invited to tea in the officers' quarters, and while sitting there the telephone rang. "It is for you," said the officer who answered the call. I picked up the receiver and heard Sheiman's voice saying briefly: "The executive has voted unanimously for the release of the prisoners."

There was little sleep for me that night, but tired as I was by morning, I greeted happily the unkempt cook and his messy breakfast plate. All day I waited with the dumb patience only prisoners know, and at early evening I was rewarded by the appearance of Sheiman and Ostrovsky. "Put on your coat and follow me," said Sheiman. "I have resolved to take you, on my own responsibility, to the hospital." To my nursing sister, who had spent the afternoon with me, he gave orders to go to Helsingfors and wait for further directions. At the prison gate Sheiman signed the necessary papers, and hurrying me past two gaping Bolshevist soldiers, he led the way down a bypath to the water. Boarding a small motor launch manned by a single sailor, we started off at high speed for Helsingfors. There was one bad moment when we approached a low bridge occupied by a strong guard, but at Sheiman's directions, uttered in a short whisper, I lay down flat in the launch and we passed unchallenged. The first stars were shining in the clear autumn sky as we reached the military quay of the town. We ran in under the lee of a huge warship and stepped ashore. There was a motor car waiting and the chauffeur, who evidently knew his business, started his engine without a word or even a turn of his head.

Sheiman spoke only one sentence. "Tovarish Nicholai, drive to—" naming a street and number. At once we were off, my head fairly swimming at the sight of electric lights, shaded streets, and people walking up and down. Turning into a quiet street we left the car, all three of us shaking hands with the discreet driver. Bidding Ostrovsky find my nurse and my small luggage, Sheiman conducted me to the door of the hospital where a nice clean Finnish nurse took me in charge and put me to bed in one of the freshest, airiest, most comfortable rooms I have ever occupied. "Take good care of this lady," were the last words of the President of the Helsingfors Soviet, "and let no one intrude on her." His words and the assured smile of the nurse were good soporifics and I fell almost instantly into a deep sleep.

Two days later, September 30 (Russian), Sheiman came to see me with the news that Trotzky had ordered all the Kerensky prisoners back to Petrograd, and that he, Sheiman, had personally seen to it that my nurse and my aunt, who was at that time in Helsingfors, were to accompany me. Sheiman himself, and also Ostrovsky, who was unfortunately very drunk, went with us in the train which left Helsingfors that same night about half past ten. It was an unpleasant journey, the prisoners being in a state of wild excitement, and many of the red-badged officers more or less tipsy. With my aunt and the nurse I sat in a corner of a dirty compartment praying for the day to come. At nine in the morning we reached Petrograd, and Sheiman, still solicitous of my welfare, escorted the three of us to the Smolny Institute, once an aristocratic school for girls, now the headquarters of the Petrograd Soviet. Here I had the happiness once more to embrace my mother, who, with relatives of other prisoners, waited our arrival. Many Soviet authorities were in the place, among others Kameneff, a small red-bearded man, and his wife, a sister of the renowned Trotzky. Both of the Kameneffs were extremely kind to us, seeing that my companions and I had tea and food, and expressing the hope that I should soon be out of trouble. Kameneff telephoned Kerensky's headquarters asking leave to send us home, but as it was a holiday nobody answered the call. "Well, go home anyhow," said Kameneff, leaving the telephone, but Sokolov stopped us long enough to make us understand that the prisoners all had to appear the next day before the High Commission in the Winter Palace. I never saw the Kameneffs again even to thank them for their kindness, but I read in the Kerensky news-papers that I was on terms of intimacy with them and was therefore a Bolshevist. It was even stated that I was a close friend of the afterwards notorious woman commissar Kolantai, whom I have never seen, and that Trotzky was a familiar visitor in my house.

Thus ended my second term of imprisonment. First I was arrested as a German spy and intrigant, next as a counter-Revolutionary. Now I was accused of being a Bolshevist and the name of Trotzky instead of Rasputine was linked with mine. Hardly knowing what next was in store for me, I reported at once to the High Commission. Here I was told that their inquiries concerning me were finished, and that I had better see the Minister of the Interior. At this ministry I was informed that I was in no immediate danger but that I would remain under police surveillance. I asked why, but got no satisfactory answer. Later I learned that the tottering Provisional Government wanted to send me and all the "counter-Revolutionists" to Archangel, but this move Dr. Manouchine, who was still very influential, was determined to prevent.

From my uncle's house, where I had first taken refuge, I moved to a discreet lodging in the heart of the city and from this place I never once in daylight ventured out. This was in late October, 1917, and the Bolshevist revolution had begun in deadly earnest. Day after day I sat listening to the sound of rifle shots and the putter of machine guns, the pounding of armored cars over the stone pavements, and the tramp, tramp, tramp of soldiers. Russia was getting ready for the long promised constitutional convention which turned out to be a

Communist *coup d'état.* Once in a while the husband of my landlady, a naval man, came to my lodgings, and it was he who gave me news of the arrest of the Provisional Government, the siege of the Winter Palace, and the ignominious collapse of Kerensky while women soldiers fought and died to hide his flight! The scenes in the streets, as they were described to me, were appalling, and soon it was decided that my retreat was too near the center of hostilities to be at all safe. About the end of October I was taken by night to a distant quarter of the town to the tiny apartment of an old woman, formerly a masseuse in my hospital. Here came our old servant Berchik, keen to protect me from danger, and here we stayed for a month, when my mother found me a still safer lodging on the sixth floor of a house in the Fourtchkatskaia, a cozy little apartment whose windows gave a pleasant view of roofs and church steeples. There for eight months I lived like a recluse, once in a great while venturing to go to church, well guarded by Berchik and the nurse. The Bolshevik Government seemed successfully established, and its policy of blood and terror and extermination was well under way. Yet in my hidden retreat it seemed to me that, for a time at least, I was forgotten, and my troubles were all over.

CHAPTER XXI

PARADOXICAL though it may appear, the last months of 1917 and the winter of 1918, spent in a hidden lodging in turbulent Petrograd, were more peaceful than any period I had known since the Revolution began. I knew that the city and the country were in the hands of fanatic Bolshevists and that under their ruthless theory of government no human life was at all secure. Food and fuel were scarce and dear, and there was no doubt that things were destined to grow worse long before they could, in any imaginable circumstances, grow better. The wreck of the army was complete, and while the war still waged in western Europe we, who had had so much to do with defiance of German militarism, were completely out of the final struggle. The peace of my soul was partly born of ignorance, I suppose, the ignorance of events shared by everyone not immediately in contact with the world catastrophe. I was free, I lived in a comfortable apartment, my dear father and mother came daily to see me, and two of my faithful old servants lived with me and were ready to protect me from all enemies.

Also, because the mind cannot fully realize the worst, I believed that the Russian chaos was a temporary manifestation. I thought I saw signs of a reaction in favor of the exiled Emperor. In this I was certainly encouraged by two of the oldest and most prominent Revolutionists known to the outside world, Bourtseff, a leader among the old Social Revolutionaries, and the novelist Gorky. It was in December, 1917, if I remember correctly, that I learned that Gorky was anxious to meet me, and as I preferred to keep my small corner of safety as free from visitors as possible, I made an appointment with the novelist in his own home, a modest apartment on the Petrograd side of the Neva, not far from the fortress. Gorky, whose gaunt features are familiar to all readers, is said to be a sufferer from tuberculosis, but as he has lived many years since the first rumors of this disease were circulated, there may be some reason to doubt his affliction. That he is a sick man none can doubt, for his high cheek bones seem almost to pierce his colorless skin and his darkly luminous eyes are deeply sunken in his head. For two hours of this first interview I sat in conversation with Gorky, strange creature, who at times seems to be heart and soul a Bolshevist and at other times openly expresses his loathing and disgust of their insane and destructive policies. To me Gorky was gentle and sympathetic, and what he said about the Emperor and Empress filled my heart with encouragement and hope. They were, he declared, the poor scapegoats of the Revolution, martyrs to the fanaticism of the time. He had examined with care the private apartments of the palace and he saw clearly that these unhappy ones were not even what are called aristocrats, but merely a bourgeois family devoted to each other and to their children, as well as to their ideals of righteous living. He expressed himself as bitterly disappointed in the Revolution and in the character of the Russian proletariat. Earnestly he advised me to live as quietly as possible, never reminding the Bolshevist authorities or any strangers of my existence. My duty, he told me, was to live and to devote myself to writing the true story of the lives of the Emperor and Empress. "You owe this to Russia," he said, "for what you can write may help to bring peace between the Emperor and the people."

Twice afterwards I saw and talked with Gorky, showing him a few pages of my reminiscences. He urged me to go on writing, suppressing nothing of the truth, and he even offered to help me with my work. But writing in Russia was at that time too dangerous a trade to be followed with any degree of confidence, and it was not until I was safely beyond the frontiers that I dared begin writing freely and at length. I wish to say, however, that it was principally due to Gorky's encouragement and to the encouragement of an American literary friend, Rheta Childe Dorr, that I ventured to attempt authorship, or rather that I undertook to present to the world, as they really were, my Sovereigns and my best beloved friends. My casual acquaintanceship with Gorky was naturally seized upon by certain foreign journalists as evidence that I had gone over to the Bolsheviki, and much abuse and scorn were hurled against me. How little those writers knew of Gorky and his half-hearted support of the Lenine policies! He held an important office under the Communists, it is true, and his wife, a former actress, was in the commissariat of theatricals and entertainments. But no man in Bolshevist Russia has ever been permitted more freedom of thought and speech than Gorky. He has done things which would have brought almost any other man to torture and death. I know, for example, that he sheltered under his roof at least one of the Romanoffs, and that the man was finally assisted by him across the Finnish frontier. Gorky interested himself also in the fate of several of the Grand Dukes, Nicholai Michailovitch, Paul and George, who were arrested and later shot to death in Peter and Paul. Gorky did everything in his power to save these men, in whom personally he had no interest whatever. He simply believed their murder to be unjustified, and it is said that he actually induced Lenine to sign an order for their release and deportation, but the order was signed too late, and the men were brutally executed.

At Christmas, 1917, I had a great happiness, nothing less than letters and a parcel of food from the exiles in Tobolsk. There were two parcels in fact, one containing flour, sugar, macaroni, and sausage, wonderful luxuries, and the other a pair of stockings knit by the Empress's own hands, a warm scarf, and some pretty Christmas cards illuminated in her well-remembered style. I made myself a tiny Christmas tree decorated with bits of tinsel and holly berries and hung with these precious tokens of affection and remembrance. Nor was this the only Christmas joy vouchsafed me after a year of sorrow and suffering. Under the escort of my good old servant Berchik I ventured to attend mass in the big church near the Nicholai station, a church built to commemorate the three hundredth anniversary of the Romanoff succession. After the service an old monk approached me and invited me to accompany him into the *réfectoire* of his monastery. I followed him, a little unwillingly, for one never knew what might happen. Entering I saw, to my astonishment, about two hundred factory women who almost filled the bare and lofty room. The old monk introduced me to the women, and to my bewilderment their leader came forward bowing, and holding in her outstretched hands a clean white towel on which reposed a silver ikon. It was an image of Our Lady of Unexpected Joy, and the kind woman told me that she and her fellow workers felt that after all that I had unjustly suffered in the fortress I ought to have from those who sympathized with me an expression of confidence and goodwill. She added that were I again in trouble I might feel myself free to take refuge in the lodgings of any one of them. Overcome with emotion, I could utter only a few stammering words of thanks. I kissed the good woman heartily, and all who could approached and embraced me. Knowing that I longed for more tangible expressions of gratitude, the good old monk pressed into my hands a number of sacred pictures and these I gave away, as long as they lasted, to my new friends. No words can tell how deeply I felt the kindness of these working women who, out of their scanty wages, bought a silver ikon to give to a woman of whom they knew nothing except that she had, as they believed, been persecuted for others' sake.

I needed the assurance that in the cruel world around me there were those who wished me well, for in the first months of the new year came one of the bitterest sorrows of my life, the death of my deeply loved and revered father. He died very suddenly, and without any pain, on January 25, 1918, leaving the world bereft of one of the kindest, most gifted, and sympathetic men of his generation in Russia. I have described my father as a musician and a composer, as well as a lifelong friend and functionary of the Imperial Family. His years of service as keeper of the privy purse might have made him a rich man, but so utterly honest was he that he accepted nothing except his moderate salary and he died leaving almost nothing, nothing but an unfading memory and the deep affection of my friends, including scores of poor students whose musical education and advancement he had furthered. At his funeral his own compositions were sung by volunteer choirs of his musician friends, and these followed his coffin in long procession the length of the Nevski Prospekt to the cemetery of the Alexandra Nevskaia Lavra, a monastic burial place where many of our greatest lie in everlasting repose. My mother came to live with me in my obscure lodgings, and together we faced our desolate future.

One thing alone lightened the darkness of those days. This was a correspondence daringly undertaken with my beloved friends in Siberia. Even now, and at this distance from Russia I cannot divulge the names of those brave and devoted ones who smuggled the letters and parcels to and from the house in Tobolsk, and got them to me and to the small group of faithful men and women in Petrograd. The two chiefly concerned, a man and a woman, of course lived in constant peril of discovery and death. Yet they gladly risked their lives that their Sovereigns might have the happiness of private communication with their friends. At this time their Majesties were permitted to write and receive a few letters, but every line was read by their jailers, and their list of correspondents was rigidly censored. Even in the letters smuggled out from Tobolsk the utmost precautions had to be observed, and the reader can see with what veiled and discreet phrases the sentences are couched.

I give these letters exactly as they were written, suppressing only certain messages of affection too intimate to make public. Most of the letters were written by the Empress, but one at least came from the Emperor, and a number are from the children. To me these letters are infinitely precious, not only as personal messages, but as proofs of the dauntless courage and deep religious faith of these martyrs of the Russian Revolution. Their patriotism and their love of country never faltered for a single moment, nor did they ever utter a complaint or a reproach against those who had so heartlessly betrayed them. It seems to me impossible that anyone, reading these letters, intended only for my own eyes, can continue to misjudge the lives and the characters of Nicholas II and the Empress Alexandra Feodorovna. What they reveal is their secret selves, unknown except to those who knew them best and knowing them loved them as they deserved to be loved.

The first communication to reach me was a brief message from the Empress, dated October 14, 1917, a short time after the news of my liberation from the fortress reached her in Siberia.

* * *

My darling: We are thinking constantly of you and of all the suffering you have had to endure. God help you in the future. How are your weak heart and your poor legs? We hope to go to Communion as usual if we are to be allowed. Lessons have begun again with Mr. Gibbs also. So glad, at last. We are all well. It is beautifully sunny. I sit behind this wall in the yard and work. Greetings to the doctors, the priest, and the nurses in your hospital. I kiss you and pray God to keep you.

A week later the Empress wrote me a long letter in which she ventures a few details of life in Tobolsk.

October 21, 1917.

My darling: I was inexpressibly glad to get news of you, and I kiss you fondly for all your loving thoughts of me. There are no real barriers between souls who really understand each other, but still it is natural for hearts to crave expressions of love. I wrote to you on the 14th, and now will try to send this to the same address, but I don't know how long you will remain. I wonder if you got my letter. I had hoped so much that you would see Zina and find comfort in her friendship. The expression in the eyes in the photograph which was brought me* has impressed me deeply, and I wept freely as I looked at it. Ah, God! Still He is merciful and will never forget His own. Great will be their reward in Heaven. The more we suffer here the fairer it will be on that other shore where so many dear ones await us. How are our Friend's† dear children, how well does the boy learn, and where do they live?

Dear little Owl, I kiss you tenderly. You are in all our hearts. We pray for you and often talk of you. In God's hands lie all things. From this great distance it is a difficult thing to help and comfort a loved one who is suffering. We hope tomorrow to go to Holy Communion, but neither today nor yesterday were we allowed to go to church. We have had services at home, last night prayers for the dead, tonight confession and evening prayer. You are ever with us, a kindred soul. How many things I long to say and to ask of you. It is strange to be in this house and to sleep in the dark bedroom.‡ I have heard nothing from Lili D. for some time. We are all well. I have been suffering from neuralgia in the head but now Dr. Kostritzky has come to treat me. We have spoken often of you.

They say that life in the Crimea is dreadful now. Still Olga A. is happy with her little Tichon whom she is nursing herself. They have no servants

---

* The snapshot taken of me by Mr. Gibbs soon after I was released from the fortress.
† Rasputine.
‡ This was the house and the room I occupied in my stay in Tobolsk on my second visit to Siberia.

so she and N. A. look after everything. Dobiasgin, we hear, has died of cancer. The needlework you sent me was the only token we have received from any of our friends. Where is poor Catherine? We suffer so for all, and we pray for all of you. That is all we can do. The weather is bad these last few days, and I never venture out because my heart is not behaving very well. I get a great deal of consolation reading the Bible. I often read it to the children, and I am sure that you also read it. Write soon again. We all kiss and bless you. May God sustain and keep you. My heart is full, but words are feeble things.

Yours, A.

The jacket warms and comforts me. I am surrounded by your dear presents, the blue dressing gown, red slippers, silver tray and spoon, the stick, etc. The ikon I wear. I do not remember the people you are living with now. Did you see the regimental priest from Peterhof? Ask the prayers of O. Hovari for us. God be with you. Love to your parents. Madeleine and Anna are still in Petrograd.

Card from Alexei, November 24, 1917.

I remember you often and am very sad. I remember your little house. We cut wood in the daytime for our baths. The days pass very quickly. Greetings to all.

On the same day the Empress wrote me a short letter in English.

Yesterday I received your letter dated November 6, and I thank you for it from my heart. It was such a joy to hear from you and to think how merciful is God to have given you this compensation. Your life in town must be more than unpleasant, confined in stuffy rooms, steep stairs to climb, no lovely walks possible, horrors all around you. Poor child! You know that in heart and soul I am near you, sharing all your pain and sorrow and praying for you fervently. Every day I read in the book you gave me seven years ago, "Day by Day," and like it very much. There are lovely passages in it.

The weather is very changeable, frost, sunshine, then darkness and thawings. Desperately dull for those who enjoy long walks and are deprived of them. Lessons continue as usual. Mother and daughters work and knit a great deal, making Christmas presents. How time flies! In two weeks more it will be eight months since I saw you last. And you, my little one, so far away in loneliness and sorrow. But you know where to seek consolation and strength, and you know that God will never forsake you. His love is over all.

On the whole we are all well, since I do not count chills and colds. Alexei's knee and arm swell from time to time, but happily without any pain. My heart has not been behaving very well. I read much, and live in the past, which is so full of rich memories. I have full trust in a brighter future. He will never forsake those who love and trust in His infinite mercy, and when we least expect it He will send help, and will save our unhappy country. Patience, faith and truth.

How did you like the two little colored cards? I have not heard from Lili Dehn for three months. It is hard to be cut off from all one's dear friends. I am so glad that your old servant and Nastia are with you, but where are the maids, Zina and Mainia? So Father Makari has left us. But he is really nearer than he was before.

Our thoughts will be very close together next month. You remember our last journey and what followed. After this anniversary it seems to me that God will show mercy. Kiss Praskovia and the children for me. The maid Liza and the girls have not come yet. All of us send tenderest love, blessings and kisses. God bless you, dearest friend. Keep a brave heart.

P. S. I should like to send you a little food, some macaroni for instance.

Up to this time, nearly the end of the year 1917, the Imperial Family in exile were treated with a certain degree of consideration. They had plenty of food and a limited freedom. In the next letter I received from the Empress, dated December 8, she speaks with gratitude of the fact that some of her favorite books were permitted to be retained by her, as a little later she overflows with gratitude to one of the Bolshevist Commissars who sent her a few familiar pictures and trinkets from the old home in Tsarskoe Selo. Little by little, however, privileges were taken from the family, and their status became that of criminal prisoners. I leave this to be shown in the letters which follow. On December 8, 1917, the Empress wrote me, in Russian, a letter which shows how poignantly she and the Emperor felt the desperate situation in Russia.

My darling: In thoughts and prayers we are always together. Still it is hard not to see each other. My heart is so full, there is so much I would like to know, so many thoughts I should like to share with you. But we hope the time will come when we shall see each other, and all the old friends who now are scattered in different parts of the world.

I am sorry you have had a misunderstanding with one of your best friends. That should never happen. This is no time to judge one's friends, every one of us being on such an unnatural strain.

We here live far from everybody and life is quiet, but we read of all the horrors that are going on. But I shall not speak of them. You live in their very center, and that is enough for you to bear. Petty troubles surround us. The maids have been in Tobolsk four days and yet they are not allowed to come to our house, although it was promised that they should. How pitiful this everlasting suspicion and fear. I suppose it will be the same with Isa.* Nobody is now allowed to approach us, but I hope they will soon see how stupid and brutal and unfair it is to keep them (the maids) waiting.

It is very cold—24 degrees of frost. We shiver in the rooms, and there is always a strong draught from the windows. Your pretty jacket is so useful. We all have chilblains on our fingers. (You remember how you suffered from them in your cold little house?) I am writing this while resting before dinner. Little Jimmy lies near me while his mistress plays the piano. On the 6th Alexei, Marie, and Gilik (M. Gilliard) acted a little play for us. The others are committing to memory scenes from French plays. Excellent distraction, and good for the memory. The evenings we spend together. He reads aloud to us, and I embroider. I am very busy all day preparing Christmas presents; painting ribbons for book markers, and cards as of old. I also have lessons with the children, as the priest is no longer permitted to come. But I like these lessons very much. So many things come back to my mind. I am reading with pleasure the works of Archbishop Wissky. I did not have them formerly. Lately also I have read Tichon Zadonsky. In spite of everything I was able to bring some of my favorite books with me. Do you read the Bible I gave you? Do you know that there is now a much more complete edition? I have given one to the children, and I have managed to get a large one for myself. There are some beautiful passages in the Proverbs of Solomon. The Psalms also give me peace. Dear, we understand each other. I thank you for everything, and in memory I live over again our happy past.

One of our former wounded men, Pr. Eristoff, is in hospital again. I don't know the reason. If possible give hearty greetings to him from us all. Give sincere thanks and greetings to Madame S. and her husband. God bless and comfort him.

Where are Serge (Mme. Viroubova's brother) and his wife? I received a touching letter from Zina. I know the past is all done with, but I thank God for all that we have received, and I live in the memory that cannot be taken from me. Still I worry often for my dearly loved, far distant, foolish little friend. I am glad that you have resumed your maiden name. Give greetings to Emma F., the English Red Cross nurse, and to your dear parents.

On the 6th we had service at home, not being allowed to go to church on account of some kind of a disturbance. I have not been out in the fresh

*Baroness Buxhoevden, lady in waiting.

air for four weeks. I can't go out in such bitter weather because of my heart. Nevertheless church draws me almost irresistibly.

I showed your photographs to Valia and Gilik. I did not want to show them to the ladies, your face is too dear and precious to me. Nastinka is too distant. She is very sweet, but she does not seem near to me. All my dear ones are far away. But I am surrounded by their photographs and gifts—jackets, dressing gowns, slippers, silver dish, spoons, and ikons. How I would like to send you something, but I fear it would get lost. I kiss you tenderly, love, and bless you. We all kiss you. He was touched by your letter of congratulation. We pray for you, and we think of you, not always without tears.

Yours.

The next day the Empress wrote again.

This is the feast day of the Virgin of Unexpected Joy. I always read the day's service, and I know that you, dear, do the same. It is the anniversary of our last journey together, to Saratoff. Do you remember how lovely it was? The old holy woman is dead now, but I keep her ikon always near me. Yesterday it was nine months since we were taken into captivity, and more than four months since we came here. Which of the English nurses was it who wrote to me? I am surprised to hear that Nini Voyeikoff and her family did not receive the ikons I sent them before leaving. Give kind regards to your faithful old servant and Nastia. This year I cannot give them anything for their Christmas tree. How sad. My dear, you are splendid. Christ be with you. Give my thanks to Fathers John and Dosifei for their remembrance. I am writing this morning in bed. Jimmy is sleeping nearly under my nose and interfering with my writing. Ortipo lies on my feet and keeps them warm.

Fancy that the kind Kommissar Makaroff sent me my pictures two months ago, St. Simeon Nesterofts, the little Annunciation from the bedroom, four small prints from my mauve room, five pastels of Kaulbach, four enlarged snapshots from Livadia; Tatiania and me, Alexei as sentry, Alexander III, Nicholas I, and also a small carpet from my bedroom.

My wicker lounge chair too is standing in my bedroom now. Among the other cushions is the one filled with rose leave given me by the Tartar women. It has been with me all the way. At the last moment of the night at Tsarskoe I took it with me, slept on it on the train and on the boat, and the lovely smell refreshed me. Have you had any news of Gaham (Chief of the Karaim)? Write to him and give him my regards. One of our former wounded, Sirobojarski, has visited him. There are 22 degrees of frost today, but bright sunshine. Do you remember the sister of mercy K. M. Bitner? She is giving the children lessons. What luck! The days fly. It is

Saturday again, and we shall have evening service at nine. A corner of the drawing room has been arranged with our ikons and lamps. It is homelike—but not church. I got so used to going almost daily for three years to the church of Znamenia before going on to the hospitals at Tsarskoe.

I advise you to write to M. Gilliard. (Now I have refilled my fountain pen.) Would you like some macaroni and coffee? I hope soon to send you some. It is so difficult for me here to take the vegetables out of the soup without eating any of it.* It is easy for me to fast and to do without fresh air but I sleep badly. Yet I hardly feel any of the ills of the flesh. My heart is better, as I live such a quiet life, almost without exercise. I have been very thin but it is less noticeable now, although my gowns are like sacks. I am quite gray too.

The spirits of the whole family are good. God is very near us, we feel His support, and are often amazed that we can endure events and separations which once might have killed us. Although we suffer horribly still there is peace in our souls. I suffer most for Russia, and I suffer for you too, but I know that ultimately all will be for the best. Only I don't understand anything any longer. Everyone seems to have gone mad. I think of you daily and love you dearly. You are splendid and I know how wonderfully you have grown. Do you remember the picture by Nesteroffs, Christ's Bride? Does the convent still attract you in spite of your new friend? God will direct everything. I want to believe that I shall see your buildings (my hospital) in the style of a convent. Where are the sisters of mercy Mary and Tatiana? What has become of Princess Chakoffskaia, and has she married her friend? Old Madame Orloff has written me that her grandson John was killed in the War, and that his fiancée killed herself from grief. Now they are buried beside his father.

My regards to my dear Lancers, to Jakoleff, Father John, and others. Pray for them all. I am sure that God will have mercy on our Russia. Has she not atoned for her awful sins?

My love, burn my letters. It is better. I have kept nothing of the dear past. We all kiss you tenderly and bless you. God is great and will not forsake those encircled by His love. Dear child, I shall be thinking of you especially

During that December I had the happiness of receiving letters from the Emperor, Alexei, and the Grand Duchesses Tatiana, Olga, during Christmas. I hope that we will meet again, but where and how is in His hands. We must leave it all to Him who knows all better than we. and Anastasie. The Emperor wrote acknowledging a note of mine written on his name day.

* The Empress Alexandra Feodorovna was always a strict vegetarian.

Tobolsk, 10 December, 1917.
Thank you so much for your kind wishes on my name day. Our thoughts and prayers are *always* with you, poor suffering creature. Her Majesty reads to us all your lines. Horrid to think all you had to go through. We are all right here. It is quite quiet. Pity you are not with us. Kisses and blessings without end from your loving friend, N.

Give my best love to your parents.

The children's letters were devoured because they gave so many details of the family life in Tobolsk. On December 9 Tatiana wrote:

My darling: I often think and pray for you, and we are always remembering and speaking of you. It is hard that we cannot see each other, but God will surely help us, and we will meet again in better times. We wear the frocks your kind friends sent us, and your little gifts are always with us, reminding us of you. We live quietly and peacefully. The days pass quickly. In the morning we have lessons, walk from eleven to twelve before the house in a place surrounded for us by a high board fence. We lunch together downstairs, sometimes Mamma and Alexei with us, but generally they lunch upstairs alone in Papa's study. In the after-noon we go out again for half an hour if it is not too cold. Tea upstairs, and then we read or write. Sometimes Papa reads aloud, and so goes by every day. On Saturdays we have evening service in the big hall at nine o'clock. Until that hour the priest has to serve in the church. On Sundays, when we are allowed, we go to a near-by church at eight o'clock in the morning. We go on foot through a garden, the soldiers who came here with us standing all around. They serve mass for us separately, and then have a mass for everybody. On holidays, alas, we have to have small service at home. We had to have home service on the 6th (St. Nicholas' day), and it was sad on such a big holiday not to be in church, but one can't have everything one wants, can one? I hope you at least can go to church. How are your heart and your poor legs? Do you see the doctor of your hospital? You remember how we used to tease you. Greetings to your old servants. Where are your brother and his wife? Have they got a baby? God bless you, my darling beloved. All our letters (permitted letters) go through the Kommissar. I am glad that the parents of Eristoff are kind to you. Him I remember well, but I never saw the parents. Isa has not come yet. Has she been to see you? I kiss you tenderly and love you.

Your T.

My darling dear Annia, How happy I was to hear from you. Thank you for the letter and the things. I wrote to you yesterday. It is so strange to be staying in the house where you stayed. Remember that we are sending this parcel secretly, so don't mention it. It is the only time probably that we can do it. Yesterday's letter I sent through the Kommissar. I am always thinking of you, my darling. We speak much of you among ourselves and also to Gilik, Valia, Prince Dolgorouky, and Mr. Gibbs. I wear your bracelet and never take it off, the one you gave me on January 12, my name day. You remember that cozy evening by the fireside? How nice it was. Did you ever see Groten and Linevitch?* Well, good-bye, my darling Annia. I kiss you tenderly and love you.

Your T.

From the Grand Duke Alexei, December 10, 1917.

My darling, I hope you got my postcard. Thank you very, very much for the little mushroom. Your perfumes remind us so much of you. Every day I pray God we shall live together again. God bless you.

Yours, A.

From the Grand Duchess Olga on the same date.

My darling, what joy it was to see your dear handwriting, and all the little things. Thanks awfully for all. Your perfumes reminded us so of you, your cabin on board, etc. It was very sad. I remember you often, kiss and love you. We four live in the corner blue room, arranged all quite cozily. Opposite to us in the little room is Papa's dressing room and Alexei's, then comes his room with Nagori. The brown room is Papa's and Mamma's bedroom. Then the sitting room, big hall, and beyond Papa's study. When there are big frosts it is very cold, and draughts blow from all the windows. We were today in church. Well, I wish you a peaceful and sunny Christmas. God bless you, darling. I kiss you over and over again.

Ever your own Olga.

From the Grand Duchess Anastasie.

My darling and dear: Thank you tenderly for your little gift. It was so nice to have it, reminding me especially of you. We remember and speak of you often, and in our prayers we are always together. The little dog you gave is always with us and is very nice. We have arranged our rooms comfortably and all four live together. We often sit in the windows looking at the people passing, and this gives us distraction. We have acted little plays for amusement. We walk in the garden behind high planks. God bless you.

AN.

* Groten and Linevitch were the two aides-de-camp who were so devoted to the family during the trying period before the Revolution. Afterwards they were denied entrance to the palace.

From the Empress.

My own precious child: It seems strange writing in English after nine weary months. We are doing a risky thing sending this parcel, but we profit through — who is still on the outside. Only promise to burn all we write as it could do you endless harm if they discovered that you were still in contact with us. Therefore don't judge those who are afraid to visit you, just leave time for people to quiet down. You cannot imagine the joy of getting your sweet letters. I have read and reread them over and over to myself and to the others. We all share the anguish, and the misery, and the joy to know that you are free at last. I won't speak of what you have gone through. Forget it, with the old name you have thrown away. Now live again.

One has so much to say that one ends by saying nothing. I am unaccustomed to writing anything of consequence, just short letters or cards, nothing of consequence. Your perfume quite overcame us. It went the round of our tea table, and we all saw you quite clearly before us. I have no "white rose" to send you, and could only scent the shawl with vervaine. Thanks for your own mauve bottle, the lovely blue silk jacket, and the excellent pastilles. The children and Father were so touched with the things you sent, which we remember so well, and packed up at Tsarskoe. We have none of such things with us, so alas, we have nothing to send you. I hope you got the food through — and Mme. ——. I have sent you at least five painted cards, always to be recognized by my signature. I have always to be imagining new things!

Yes, God is wonderful and has sent you (as always) in great sorrow, a new friend. I bless him for all that he has done for you, and I cannot refrain from sending him an image, as to all who are kind to you. Excuse this bad writing, but my pen is bad, and my fingers are stiff from cold. We had the blessing of going to church at eight o'clock this morning. They don't always allow us to go. The maids are not yet let in as they have no papers, so the odious commandant doesn't admit them. The soldiers think we already have too many people with us. Well, thanks to all this we can still write to you. Something good always comes out of everything.

Many things are very hard . . . our hearts are ready to burst at times. Happily there is nothing in this place that reminds us of you. This is better than it was at home where every corner was full of you. Ah, child, I am proud of you. Hard lessons, hard school, but you have passed your examinations so well. Thanks, child, for all you have said for us, for standing up for us, and for having borne all for our own and for Russia's sake. God alone can recompense you, for if He has let you see horrors He has permitted you to gaze a little into yonder world. Our souls are nearer now than before. I feel especially near you when I am reading the Bible. The children also are always finding texts suiting you. I am so contented with their souls. I hope God will bless my lessons with Baby. The ground is rich, but is the seed ripe enough? I do try my utmost, for all my life lies in this.

Dear, I carry you always with me. I never am separated from your ring, but at night I wear it on my bracelet as it is so loose on my finger. After we received our Friend's cross we got also this cross to bear. God knows it is painful being cut off from the lives of those dear to us, after being accustomed for years to share every thought. But my child has grown self-dependent with time. In your love we are always together. I wish we were so in fact, but God knows best. One learns to forget personal desires. God is merciful and will never forsake His children who trust Him.

I do hope this letter and parcel will reach you safely, only you had better write and tell — that you get everything safely. Nobody here must dream that we evade them, otherwise it would injure the kind commandant and they might remove him.

I keep myself occupied ceaselessly. Lessons begin at nine (in bed). Up at noon for religious lessons with Tatiana, Marie, Anastasie, and Alexei. I have a German lesson three times a week with Tatiana and once with Marie, besides reading with Tatiana. Also I sew, embroider, and paint, with spectacles on because my eyes have become too weak to do without them. I read "good books" a great deal, love the Bible, and from time to time read novels. I am so sad because they are allowed no walks except before the house and behind a high fence. But at least they have fresh air, and we are grateful for anything. He is simply marvelous. Such meekness while all the time suffering intensely for the country. A real marvel. The others are all good and brave and uncomplaining, and Alexei is an angel. He and I dine *à deux* and generally lunch so, but sometimes downstairs with the others.

They don't allow the priest to come to us for lessons, and even during services officers, commandant and Kommissar, stand near by to prevent any conversation between us. Strangely enough Germogene is Bishop here, but at present he is in Moscow. We have had no news from my old home or from England. All are well, we hear, in the Crimea, but the Empress Dowager has grown old and very sad and tearful. As for me my heart is better as I lead such a quiet life. I feel utter trust and faith that all will be well, that this is the worst, and that soon the sun will be shining brightly. But oh, the victims, and the innocent blood yet to be shed! We fear that Baby's other little friend from Mogiloff who was at M. has been killed, as his name was included among cadets killed at Moscow. Oh, God, save Russia! That is the cry of one's soul, morning, noon and night. Only not that shameless peace.*

I hope you got yesterday's letter through Mme. ——'s son-in-law. How

* Brest-Litovsk.

nice that you have him in charge of your affairs. Today my mind is full of Novgorod and the awful 17th.* Russia must suffer for that murder too. Dear, I am glad you see me in your dreams. I have seen you only twice, vaguely, but some day we shall be together again. When? I do not ask. He alone knows. How can one ask more? We simply give thanks for every day safely ended. I hope nobody will ever see these letters, as the smallest thing makes them react upon us with severity. That is to say we get no church services outside or in. The suite and the maids may leave the house only if guarded by soldiers, so of course they avoid going. Some of the soldiers are kind, others horrid.

Forgive this mess, but I am in a hurry and the table is crowded with painting materials. So glad you liked my old blue book. I have not a line of yours—all the past is a dream. One keeps only tears and grateful memories. One by one all earthly things slip away, houses and possessions ruined, friends vanished. One lives from day to day. But God is in all, and nature never changes. I can see all around me churches (long to go to them), and hills, the lovely world. Wolkoff wheels me in my chair to church across the street from the public garden. Some of the people bow and bless us, but others don't dare. All our letters and parcels are examined, but this one today is contraband. Father and Alexei are sad to think they have nothing to send you, and I can only clasp my weary child in my arms and hold her there as of old. I feel old, oh, so old, but I am still the mother of this country, and I suffer its pains as my own child's pains, and I love it in spite of all its sins and horrors. No one can tear a child from its mother's heart, and neither can you tear away one's country, although Russia's black ingratitude to the Emperor breaks my heart. Not that it is the *whole* country, though. God have mercy and save Russia.

Little friend, Christmas without me—up in the sixth story! My beloved child, long ago I took you to hold in my heart and never to be separated. In my heart is love and forgiveness for everything, though at times I am not as patient as I ought to be. I get angry when people are dishonest, or when they unnecessarily hurt and offend those I love. Father, on the other hand, bears everything. He wrote to you of his own accord. I did not ask him. Please thank everybody who wrote to us in English. But the less *they* know we correspond the better, otherwise they may stop all letters.

Ever your own, A.

The increasing poverty and hardships which surrounded the exiles, to say nothing of the lonely desolation of their lives, could not be kept out of the Empress's letters, although she tried to write cheerfully. I could read, in the growing discursiveness of her contra-

* Anniversary of Rasputine's assassination.

band letters, the disturbed and abnormal condition of her usual keen and concise mind. On December 15, 1917, she wrote:

> Dearest little one: Again I am writing to you, and you must thank — and reply carefully. My maids are not yet allowed to come to me, although they have been here eleven days. I don't know how it will come out. Isa (Baroness Buxhoevden, lady in waiting) is ill again. I hear that she will be allowed in when she arrives, as she has a *permis,* but I doubt it. I understand your wounded feelings when she did not go to see you, but does she know your address? She is timid, and her conscience in regard to you is not quite clear. She remembers perhaps my words to her last Autumn that there might come a time when she too would be taken from me and not allowed to return. She lives in the Gorochovaia with a niece. Zizi Narishkine (a former lady in waiting) lives in the Sergievskja, 54.
>
> I hope you will receive the things we sent for Christmas. Anna and Wolkoff helped me to send the parcels, the others I sent through —, so I make use of the opportunity to write to you. Be sure to write when you receive them. I make a note in my book whenever I write. I have drawn some postcards. Did you receive them? One of these days I shall send you some flour.
>
> It is bright sunshine and everything glitters with hoar frost. There are such moonlight nights, it must be ideal on the hills. But my poor unfortunates can only pace up and down the narrow yard. How I long to take Communion. We took it last on October 22, but now it is so awkward, one has to ask permission before doing the least thing. I am reading Solomon and the writings of St. Seraph, every time finding something new. How glad I am that none of your things got lost, the albums I left with mine in the trunk. It is dreary without them, but still better so, for it would hurt to look at them and remember. Some thoughts one is obliged to drive away, they are too poignant, too fresh in one's memory. All things for us are in the past, and what the future holds I cannot guess, but God knows, and I have given everything into His keeping. Pray for us and for those we love, and especially for Russia when you are at the shrine of the "All-Hearing Virgin." I love her beautiful face. I have asked Chemoduroff to take out a prayer (slip of paper with names of you all) on Sunday.
>
> Where is your poor old Grandmamma? I often think of her in her loneliness, and of your stories after you had been to see her. Who will wish you a happy Christmas on the telephone? Where is Serge and his wife? Where is Alexander Pavlovitch? Did you know that Linewitch had married, and Groten also, straight from the Fortress? Have you seen Mania Rebinder? This Summer they were still at Pavlovskoie, but since we left we have heard nothing of them. Where are Bishops Isidor and

Melchisedek? Is it true that Protopopoff has creeping paralysis? Poor old man, I understand that he has not been able to write anything yet, his experiences being too near. Strange are our lives, are they not? One could write volumes.

Zinaida Tolstoaia and her husband have been in Odessa for some time. They write frequently, dear people. Rita Hitrovo is staying with them, but she scarcely writes at all. They are expecting Lili Dehn soon, but I have heard nothing from her for four months. One of our wounded, Sedloff, is also in Odessa. Do you know anything of Malama?* Did Eristoff give you Tatiana's letter? Baida Apraxin and the whole family except the husband are in Yalta. He is in Moscow at the church conference. Professor Serge Petrovitch is also in Moscow. Petroff was, and Konrad is, in Tsarskoe. There too is Marie Rudiger Belaieff. Constadious, our old general, is dead. I try to give you news of all, though you probably know more than I do.

The children wear the brooches that Mme. Soukhomlinoff sent them. Mine I hung over a frame. Do you ever see old Mme. Orloff? Her grandson John was killed, and her Alexei is far away. It is sad for the poor old woman.

I am knitting stockings for the small one (Alexei). He asked for a pair as all his are in holes. Mine are warm and thick like the ones I gave the wounded, do you remember? I make everything now. Father's trousers are torn and darned, and the girls' under-linen in rags. Dreadful, is it not? I have grown quite gray. Anastasie, to her despair, is now very fat, as Marie was, round and fat to the waist, with short legs. I do hope she will grow. Olga and Tatiana are both thin, but their hair grows beautifully so that they can go without scarfs. Fancy that the papers say that Prince Volodia Troubetskoy has joined Kaledin with all his men. Splendid! I am sure that N. D.† will take part also now that he is serving in Odessa. I find myself writing in English, I don't know why. Be sure to burn all these letters as at any time your house may be searched again.

---

* A wounded officer and friend.
† A well-known marine officer.

CHAPTER XXII

THROUGH the winter and spring of 1918 I continued to receive letters and parcels, mostly contraband, from my friends in Siberia. I wish I dared to tell how and through whom these precious messages reached me, for it all belongs in the story of Revolutionary Russia. It illustrates the truth, often demonstrated, that tyranny and oppression can never kill the spirit of freedom in human beings. There are always a minority of people who hold their lives cheap by comparison with liberty, and in such people lives deathlessly the inspiration of fidelity to those they love, no matter how relentlessly the loved ones are persecuted. Poor as I was, poor as was the small group of friends who worked with me to communicate with the Imperial Family, we managed to get to them the necessities they lacked. Dangerous and difficult as travel was in those days, every traveler being almost certain to be searched several times along the way, there were three, two officers and a young girl, who at the risk of imprisonment and death by the most unspeakable tortures, calmly and fearlessly acted as emmissaries back and forth between Petrograd and remote Tobolsk. They had friends along the way, of course, but how they managed, through months of constant peril, to carry on their work is one of those mysteries which, to my mind, are not wholly earthly.

On January 9, 1918, I received the following Christmas letter from the Empress.

Thank you, darling, for all your letters which were a great joy to me and to us all. On Christmas Eve I received the letter and the perfume, then more scent by little ——. I regret not having seen her. Did you receive the parcels sent through the several friends, flour, coffee, tea, and lapscha (a kind of macaroni)? The letters and the snapshots sent through ——, did you get them? I am worried as I hear that all parcels containing food are opened. I begin today to number my letters, and you must keep account of them. Your cards, the small silver dish, and Lili's tiny silver bell I have not yet been able to receive.

We all congratulate you on your name day. May God bless, comfort, strengthen you, and give you joy. Believe, dear, that God will yet save our beloved country. He will not be unforgiving. Think of the Old Testament and the sufferings of the Children of Israel for their sins. And now it is we who have forgotten God, and that is why they* cannot bring any happiness. How I prayed on the 6th that God would send the spirit of good judgment and the fear of the Lord. Everyone apparently have lost their heads. The reign of terror is not yet over, and it is the sufferings of the innocent which nearly kills us. What do people live on now that everything is taken from them, their homes, their incomes, their money? We must have sinned terribly for our Father in Heaven to punish so frightfully. But I firmly and unfalteringly believe that in the end He will save us. The strange thing about the Russian character is that it can so suddenly change to evil, cruelty, and unreason, and can as suddenly change back again. This is in fact simply want of character. Russians are in reality big, ignorant children. However it is well known that during long wars all bad passions flame up. What is happening is awful, the murders, the persecutions, the imprisonments, but all of it must be suffered if we are to be cleansed, new born.

Forgive me, darling, that I write to you so sadly. I often wear your jackets, the blue and the mauve, as it is fearfully cold in the house. Outside the frosts are not often severe, and sometimes I go out and even sit on the balcony. The children are just recovering from scarletina, except Anastasie, who did not catch it. The elder ones began the new year by being in bed, Marie, of course, having a temperature of 39.5. Their hair is growing well. Lessons have begun again. Yesterday I gave three. Today I am free, and am therefore writing. On the 2nd of January I thought of you and sent a candle to be set before the Holy Seraphim. I have asked that prayers may be said in the cathedral where the relics lie, for all our dear ones. You remember the old pilgrim who came to Tsarskoe Selo. Fancy that he has been here. He wandered in with his big staff, and sent me a prosvera (holy bread).

I have begun your books. The style is quite different from the others. I have got myself some good books, too, but have not much time for reading. I embroider, knit, draw, and give lessons, but my eyes are getting weaker so that I can no longer work without glasses. You will see me quite an old woman! Did you know that the marine officer Nicolai Demenkoff has appendicitis? He is in Odessa. One of our wounded, Oroborjarsky, was operated on there a month ago. He is so sad and homesick, so far away. I correspond with his mother, a gentle, good, and really Christian soul. Lili Dehn went to see her.

I trust you received the painted cards that I put in the parcel of provisions. Not all were successful. If you receive my letters just write, thanks for No. I, etc. My three maids and Isa are still not allowed to come to us, and they are very much distressed, just sitting idle. But — is of better use on the outside. Little one, where are your brother Serge and his wife?

* Presumably the Soviet Government. (from previous page)

I know nothing of them. Your poor sister Alya, I hope she is not too sad; she has friends, but her husband, has he not become too sad away from her? How are the sweet children? Miss Ida is with her still, I hope. Did you know that sister Grekova is to be married soon to Baron Taube? How glad I am that you have seen A. P. Did he not seem strange out of uniform, and what did he say about his brother? Ah, all is past, and will never return. We must begin a new life and forget self. I must finish, my dear little soul. Christ be with you. Greetings to all. I kiss your mother. I congratulate you again. I want quickly to finish the small painting, and get it to you. I fear you are again passing through fearful days. Reports filter through of murders of officers in Sevastopol. Rodionoff and his brother are there.

Your own, A.

On the 16th of January the Empress wrote me a letter in Old Slavonic style to congratulate me on my name day. In this she addresses me as "Sister Seraphine." I should explain that my hospital in Tsarskoe Selo bore the name of that saint, because it was on her day that I suffered the terrible railway accident which left me lamed for life, but which gave me, in damages, the funds for founding the hospital.

Dearly beloved Sister Seraphine:
From a full heart I wish you well on your name day! God send you many blessings, good health, fortitude, meekness, strength to bear all punishments and sorrows sent by God, and gladness of soul. May the sun lighten the path you tread through life, warm all by your love, and let your light shine forth these sad, gloomy days. Do not despair, suffering sister, God will hear your prayers, all in good time. Also we pray for thee, sister chosen of the Lord. We have thee in fond remembrance. Your little corner is far away from us. All who love thee in this place send greetings. Do not misjudge the bad writing of thy sister. She is illiterate, an ailing lay sister. I am learning the writing of prayers, but weakness of sight prevents my striving. I read the works of Bishop Gr. Nissky, but he writes too much of the creation of the world. From our sister Zinaida I have received news, so much good will in every word, breathing peace of the soul.

The family known to thee are in good health, the children have suffered from the usual ills of the young, but are now restored to health. The youngest ill, but in good spirits however, and without suffering. The Lord has blessed the weather, beautiful and soft. Thy sister walks out and enjoys the sun, but when there is more frost she hides in her cell, takes a stocking, puts on her spectacles, and knits. Sister Sophia,* not long since arrived, has not been granted admittance, those in authority having refused it. She has found hospitality at the priest's with her old woman. The other sisters are all in different places. Dearly loved sister, art thou not weary reading this letter? All the others have gone to dinner. I remain on guard by the sick Anastasia. In the cells next to ours is sister Catherina† giving a lesson. We are embroidering for church, Sisters Tatiana and Maria with great zeal. Our father Nicholas gathers us around him in the evenings, and reads to us while we pass the time with needlework. With his meekness and good health he does not disdain to saw and chop wood for our needs, cleans the roads, too, with the children. Our mother Alexandra greets thee, sister, and sends her motherly blessings and hopes, sister, that thou livest in the Spirit of Christ. Life is hard but the spirit is strong. Dear sister Seraphine, may God keep thee. I beg for your prayers. Christ be with thee.

The Sinful sister FEODORA.

Prayers!

ONE OF THE EMPRESS'S LAST LETTERS, WRITTEN IN OLD SLAVONIC TO MME. VIROUBOVA IN 1918.

* Isa, Baroness Buxhoevden, lady in waiting.
† Miss Schneider.

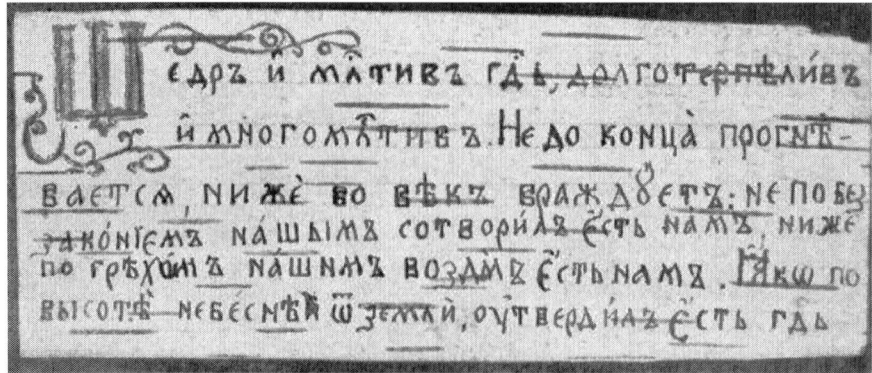

SMUGGLED LETTER FROM THE EMPRESS TO MME. VIROUBOVA WRITTEN IN OLD SLAVONIC ON BIRCH-BARK AFTER PAPER GAVE OUT.

22 of January.

So unexpectedly I received the letter of the 1st and the card of the 10th. I hasten to reply. Tenderly we thank through you Karochinsky. Really it is touching that even now we are not forgotten. God grant that his estates should be spared. God bless him. I am sending you some food but I do not know if it will ever reach you. Often we think of you. I wrote to you on the 16th through the hospital, on the 17th a card by Mr. Gibbs, and on the 9th two letters by ——. There! I have dropped my favorite pen and broken it. How provoking! It is fearfully cold, 29 degrees, 7 in the bathroom, and blowing in from everywhere. Such a wind, but they are all out. We hope to see the officer Tamarov if only from a distance. So glad you received everything. I hope you wear the gray shawl, and that it smells of vervaine, a well-remembered scent. Kind Zinoschka found it in Odessa, and sent it to me.

I am so surprised you have made the acquaintance of Gorky. He was awful formerly. Disgusting and immoral books and plays he wrote. Can it be the same man? How he fought against father and Russia when he lived in Italy. Be careful, my love. I am so glad you can go to church. To us it is forbidden, so service is at home, and a new priest serves. How glad I am that all is well with Serge. With Tina it will be difficult, but God will help her. It is true what they say about Marie Rebinder's husband? She wrote me, through Isa, that they are still in Petrograd, and that they threatened to kill him. It is difficult to understand people now. Sometimes they are with the Bolshevists outwardly, but in their hearts they are against them.

The cross we hung over the children's beds during their illness but during church service it lies on the table. Bishop Gerogene serves special prayers daily for father and mother—he is quite on their side, which is strange. I must hurry as one waits to take this letter. I am sending you a prayer I wrote on a piece of birchbark we cut. I can't draw much as my eyes are so bad, also my fingers are quite stiff from cold. Such a wind, and it blows so in the rooms. I am sending you a little image of the Holy Virgin. Thanks for the lovely prayer. I wear often the jackets you gave me. I send you all my soul-prayers and love. I believe firmly so I am quite calm. We are all your own and kiss you tenderly.

On the same day Grand Duchess Olga wrote a brief note.

Dearest, we were so glad to hear from you. How cold it is these days, and what a strong wind. We have just come back from a walk. On our window it is written—"Anna darling——" I wonder who wrote it. God bless you, dear. Be well.

Your OLGA.

Give my love to all who remember me.

Two other notes from Olga followed in February and just before Easter.

Darling, with all my loving heart I am with you these hard days for you. God help and comfort you, my darling. On Mamma's table stands the mauve bottle you sent her and which reminds us so much of you. There is much sun, but great frosts also and winds, and very cold in the rooms, especially in our corner room, where we live as before. All are well, and we walk much in the yard. There are many churches around here, so we are always hearing bells ringing. God bless you, darling. How sad your brother and sister are not with you.

Your own OLGA.

We all congratulate you tenderly with the coming Easter, and wish you to spend it as peacefully as anyone can now. I always think of you when they sing during mass the prayer we used to sing together on the yacht. I kiss you.

OLGA.

The other children also wrote me at this time. Grand Duchess Tatiana wrote two short but characteristic notes, the first one on my name day, January 12. In all these letters it will be seen how confidently the family looked forward to a future of freedom and happiness. This constant optimism in the midst of ever-increasing surveillance and cruelty is my excuse for including notes of slight general interest.

Tatiana wrote first:

"You remember the cozy evenings by the fireside? How nice it was. Did you again see Groten and Linevitch? (the faithful aides-de-camp). Well, good-bye, my darling Annia. God bless you. Good-bye—till when?

Your T.

Also—

My beloved darling. How happy we are to get news from you. I hope you got my letters. I think often of you and pray God to keep you from all harm and help you. I am glad you know the Eristoffs now. We get such good letters from Zina, she writes so well. There are many sadnesses in these days. God be with you. It is very cold. Papa wears his Cossack uniform and we remember how much you liked it. I kiss you tenderly, and love you, and congratulate you on your dear name day.

T.

From the Grand Duchess Marie Nicolaevna.

Good morning, my darling! What a long time since I have written to you, and how glad I was to get your little letter. It is very sad we don't see each other, but God will arrange for us to meet, and what joy it will be then. We live in the house where you have been. Do you remember the rooms? They are quite comfortable when a little arranged. We walk out twice every day. Some of the people here are kind. Every day I remember you, and love you very much. Mr. Gibbs gave us photographs he made of you—it was so nice to have them. Your perfumes remind us so much of you. I wish you every blessing from God, and kiss you tenderly. Don't be sad. Love to all yours.

Your loving MARIE.

My darling beloved, how are you? We are all well, walk much in the yard, and have a little hill down which we can slide. There is much frost these days so Mama sits at home. You will probably get this in February, so I congratulate you on your name day. God help you in future and bless you. We always remember and speak of you. May God guard all your ways. Don't be sad, dear. All will be well, and we shall be together again. I kiss you tenderly.

MARIE.

Alexei wrote that same month of January, 1918:

My darling Annia. We are so glad to have news from you, and to hear that you got all our things. Today there are 29 degrees of frost, a strong wind, and sunshine. We walked, and I went on skees in the yard. Yesterday I acted with Tatiana and Gilik a French piece. We are now preparing another piece. We have a few good soldiers with whom I play games in their rooms. Kolia Deravenko comes to me on holidays. Nagorini, the sailor, sleeps with me. As servants we have Wolkoff, Sednoff, Troup, and Chemoduroff. It is time to go to lunch. I kiss and embrace you. God bless you.

ALEXEI.

The remaining letters from the Empress, dating from the end of January to the last days of April, 1918, are uncomplaining, yet are full of suffering and the prescience of tragic events to come. I do not believe that the Empress ever lost faith in the ultimate happiness of her beloved family, but her keen mind fully comprehended the terrible march of events in the torn Empire, and she knew that trials and still greater trials had to be faced by the Emperor and herself. Her courage in the face of this certain conviction is beyond any praise of mine.

On the 23rd of January she wrote:

My precious child: There is a possibility of writing to you now as — leaves here on the 26th. I only hope no one robs him on the way. He takes you two pounds of macaroni, three pounds of rice, and a little ham. It is so well — does not live with us. I have knitted stockings, and have knitted you a pair. They are men's size but they will do under valenki and when it is cold in the rooms. Here we have 29 degrees of frost, and 6 in the big room. It is blowing terribly. I was keenly touched by the money you sent, but do not send any more as for the present we have all we need. There have been days when we did not know what to do. I wonder what you are living on. The little money you had I put in the box with your jewels. (My fingers are so stiff I can hardly hold my pen.) I am glad your rooms are so comfortable and so light, but it must be difficult for you to climb the long staircase. How are your poor back and legs?

I know nothing about Lili Dehn, and from my two sisters and my brother I have heard nothing for a year. Only one letter from my sister Elizabeth (Grand Duchess Serge) last summer. Olga Alexandrovna* writes long letters to the children all about her boy whom she adores and nurses herself. The grandmamma I think is getting very old, and is very sad.

Tudles has four in her room. They say that Marie P.† lives well in

---

* Sister of the Emperor.
† Grand Duchess Marie Pavlovna.

Kisslowdsk, both her sons are with her and she receives all the *beau monde* from Petrograd. Merika* lives there also and is expecting a baby. Marianna Ratkova has bought a house there, and receives on Thursdays. Mr. Gibbs asks often about you, also Tudles, and my big Niouta Demidoff. The little doggy lies on my knees and warms them. It is mortally cold, but in Petrograd there is probably worse darkness, hunger, and cold. God help you all to bear it patiently. The worse here the better in yonder world.

It hurts to think how much bloodshed will have to be before better days come. Darling, I send you all my love, and am so sad I can send you little else. I embroider for the church when my eyes allow me, otherwise I knit, but soon I shall have no more wool. We can't get any here—too dear, and very bad. I have had a letter from Shoura Petrovskaia, who is taking care of her brother's children. She sews boots and sells them. In October the children got a letter from their old nurse in England—the first one from there. What rot they publish about Tatiana in the newspapers! Do you see your new friend and saviour often? How is he? Love to your kind parents. I would love to write you certain things of interest, but just now there are many things one can't put in a letter. The little one has put on a sweater, and the girls wear valenki in their rooms. I know how sad you would feel.

The kind servant Sednoff has just brought me a cup of cocoa to warm me up. How do you pray with the rosary, and what prayers do you say on every tenth? I generally say Our Father and to the Holy Virgin, but should one say the same prayer to the end? I looked for it in the books but did not get any information. I long so to go to church but they allow us that only on great holidays (feasts). So we hope to go on the 2nd of February, and on the 3rd I shall order prayers at the relics for you. How is poor old Soukhomlinoff? Where is Sacha? I suppose one may completely trust the little officer you sent. I asked him to make the acquaintance of the priest who served us before, a most devoted and energetic man, a real fighting priest—more than spiritual perhaps—yet with a charming face, and a constantly sweet smile, very thin, long gray beard, and clever eyes. His feeling for us is known all over the country now by the good ones, therefore they took him away from us, but perhaps better so, as he can do more now. The Bishop is quite for father and mother, and so is the Patriarch in Moscow, and it seems most of the clergy. Only you must be careful what sort of people come to you. I am so anxious about your seeing Gorky. Be prudent, and don't have any serious conversations with him. People will try to get around you as before. I don't mean real friends, honest-meaning people, but others who for personal reasons will use you as their shield. Then you will have the brutes after you again.

I am racking my brains what to send you, as one can get nothing here at

* Princess Galatzine.

all. Our Christmas presents were all the work of our own hands, and now I must give my eyes a rest. How pleased I was that Princess Eristoff has spoken so kindly of us. Give her and also her son our love. Where does he serve now? The people here are very friendly—lots of Kirghise. When I sit in the window they bow to me, if the soldiers are not looking.

What dreadful news about the robbing of the sacristy in the Winter Palace. There were so many precious relics and many of our own ikons. They say it has been the same in the church of Gatchina. Did you know that the portraits of my parents and of father have been utterly destroyed? Also my Russian Court dresses and all the others as well? But the destruction of the churches is the worst of all. They say it was the soldiers from the hospital in the Winter Palace who did it. We hear that the soldiers in Smolny have seized all available food, and are quite indifferent to the prospect of the people starving. Why was money sent to us rather than having been given to the poor? True, there were for us some very difficult times when we could not pay any bills, and when for four months the servants had to go without any wages. The soldiers here were not paid, so they simply took our money to keep them quiet. All this is petty, but it makes great trouble for the commandant. The Hofmarshall Chancelerie is still in existence, but when they abolish it I really don't know what we shall do. Well, God will help, and we still have what we need.

I think often of Livadia and what may be happening there. They say that many former political prisoners are stationed there. Where is our dear yacht, the *Standert*? I am afraid to inquire about it. My God! How I suffered when I heard that you were imprisoned on the *Polar Star*. I cannot think of the yacht. It hurts too much.

It is said that our Kommissar is about to be removed, and we are so rejoiced. His assistant will leave with him. They are both terrible men, Siberian convicts formerly. The Kommissar was in prison for fifteen years. The soldiers have decided to send them away, but thank God they have left us our commandant. The soldiers manage absolutely everything here.

I am lying down, as it is six o'clock. There is a fire burning but it barely warms the room. Soon the little one will be coming in for a lesson. I am teaching the children the Divine Service. May God help me to teach it to them so that it will remain with them through their whole lives, and develop their souls. It is a big responsibility. It is such a blessing to live all together, and be so near to one another. Still you must know what I have to endure, having no news from my brother, nor any idea of what lies in the future. My poor brother also knows nothing of us. If I thought my own little old home and the family would have to suffer what we have—it is awful! Then it might begin also in England. However you remember that our Friend said that no harm would come to my old home.* I try to suppress all these thoughts that my soul may not be overwhelmed with despair. I trust all my dear ones to the Holy Virgin. May she shield them from all evil. I still have much to thank God for; you are well, and I can write to you; I am not separated from our own darlings. Thank God we are still in Russia (this is the chief thing), and we are near the relics of the Metropolitan John, and we have peace. Good-bye, my little daughter.

Old friends continued to be very dear to the exiled Empress, and she kept up her interest in all their affairs. Of my sister-in-law who had her first child while her husband was fighting on the Rumanian front the Empress wrote:

How much better it would have been if Tina could have gone to Odessa to have her baby, not far from Serge, and where kind Zinotchka could have looked after her and arranged everything. But now that the Rumanians have taken Kichineff Serge has probably left, and they are together again. Sharing hardships will cause their love to increase and strengthen. How is Alyas's (my sister) health? Was it Mariana's former husband, Derfelden, who was killed in the south? Her mother and family live in Boris's house.

I sometimes see Isa in the street (*i.e.* from the window). The sister of mercy Tatiana Andrievna is now in Petrograd taking care of her sister. Later she will return to Moscow. She seems rather nervous. Give our greetings to our confessor, father Afanasi, father Alexander, and my poor old Zio. I don't know anything about my second servant Kondratieff. What has become of our chauffeurs and the coachman Konkoff? Is old General Schwedoff still alive?

Holy Virgin, keep my daughter from all danger, bless and console her!

5th of February, 1918.

My own darling little one, How terribly sad I am for you about the death of your dear father, and that I could not be with you to help and console you in your great sorrow. You know that I am with you in my prayers. May Christ and the Holy Virgin comfort you, and wipe the tears from your eyes. May God receive his soul in peace. Tomorrow morning I will ask Anoushka to go and order service for him for forty days near the relics. Alas we can pray only at home. In him we both lost a true friend of many years. Father and the children suffer with you, tenderly kiss you, and know all that your sensitive heart feels.

As your telegram went by post I don't know what day God took him to

* (previous page) Rasputine foresaw this correctly and the Grand Duke of Hesse retains his old home in peace.

himself. Is it possible it was the same day you wrote to me? I am so glad you saw him daily, but how did it happen, your poor father? For himself one must thank God—so many hardships to live through—no home, and everything so bad. I remember how it was foretold to us (by Rasputine) that he would die when Serge married. And you two women are all alone now. I wonder if your brother-in-law was there to help you, or your kind uncle. I shall try to write to his address a long letter, and also to your mother. Tell her I kiss her tenderly, and how much we have always loved her and honored your father. He was a rare man. Don't cry. He is happy now, rests and prays for you at the Throne of God.

I am glad that you received my two letters. Now you will get two more. What your little messenger will tell you about your dear ones is for yourself alone. What horrors go on at Yalta and Massandra—My God! Where is the salvation for us all and for the poor officers? All the churches being ruined—nothing held sacred any more—it will finish in some terrible earthquake, or something like it as the chastisement of God. May He have mercy on our beloved country. How I pray for Russia.

They say that the Japanese are in Tomsk and keep good order there. I hope you got our little parcel. As we have no sugar I shall send you a little honey which you can eat during Lent. We live still by the old style, but probably shall have to change. Only I don't know how it will be then with Lent and all the services (festivals and fasts). The people may be very angry if two weeks are thrown out. That is why it was never done before.

The sun shines and even warms us in the day times. I feel that God will not forsake but will save us, though all is so dark and tears are flowing everywhere. My little one, don't suffer too much. All this had to be. Only My God, how sorry I am for the innocent ones killed everywhere. I can't write any more. Ask your mother to forgive the mistakes I shall make in writing to her in Russian, and that I cannot express myself as warmly as I would like to. Good-bye, my darling. I am sending you letters from father and the children.

2nd of March, 1918.

Darling child: Thanks for all from father, mother and the children. How you spoil us all by your dear letters and gifts. I was very anxious going so long without news from you, especially as rumors came that you were gone. Alas, I can't write you as I could wish for fear that this may fall into other hands. We have not yet received all that you have sent (contraband). It comes to us little by little. Dear child, do be careful of the people who come to see you. The way is so slippery, and it is so easy to fall. Sometimes a road is cleared through the snow on which one's true friends are to walk—and then the road becomes still more slippery!

We are all right, and I am now a real mistress of a household, going over accounts with M. Gilliard. New work and very practical. The weather is sunny—they are even sunburned, and even when the frost comes back it is warmer in the sun. I have sat twice on the balcony and sometimes sit in the yard. My heart has been much better, but for a week I have had great pains in it again. I worry so much. My God! How Russia suffers. You know that I love it even more than you do, miserable country, demolished from within, and by the Germans from without. Since the Revolution they have conquered a great deal of it without even a battle. If they created order now in Russia how dreadful would be the country's debasement—to have to be grateful to the enemy. They must never dare to attempt any conversations with father or mother.

We hope to go to Communion next week, if they allow us to go to church. We have not been since the 6th of January. I shall pray to the rosary you have written. Kiss your poor mother. I am glad you took some of your things from the hospital. Best love to poor G. Soukhomlinoff. What terrible times you are all living through. On the whole we are better off than you. Soon spring is coming to rejoice our hearts. The way of the cross first—then joy and gladness. It will soon be a year since we parted, but what is time? Life here is nothing—eternity is everything, and what we are doing is preparing our souls for the Kingdom of Heaven. Thus nothing, after all, is terrible, and if they do take everything from us they cannot take our souls. Have patience, and these days of suffering will end, we shall forget all the anguish and thank God. God help those who see only the bad, and don't try to understand that all this will pass. It cannot be otherwise. I cannot write all that fills my soul, but you, my little martyr, understand it better than I. You are farther on than I. We live here on earth but we are already half gone to the next world. We see with different eyes, and that makes it often difficult to associate with people who call themselves, and really are religious. My greatest sin is my irritability. The endless stupidity of my maid, for instance—she can't help being stupid, she is so often untruthful, or else she begins to sermonize like a preacher and then I burst—you know how hot-tempered I am. It is not difficult to bear great trials, but these little buzzing mosquitoes are so trying. I want to be a better woman, and I try. For long periods I am really patient, and then breaks out again my bad temper. We are to have a new confessor, the second in these seven months. I beg your forgiveness, too, darling. Day after tomorrow is the Sunday before Lent when one asks forgiveness for all one's faults. Forgive the past, and pray for me. Yesterday we had prayers for the dead, and we did not forget your father. A few days ago was the twenty-sixth anniversary of my father's death. I long to warm and to comfort others—but alas, I do not feel drawn to those around me here. I am cold towards them, and this, too, is wrong of me.

The cowardly yielding of the Bolshevist government to the triumphant Germans was a source of constant suffering to the Empress. In subsequent letters written me that spring she speaks almost indifferently of the cold and privations suffered in the house in Tobolsk, but she becomes passionate when she writes of the German invasion.

What a nightmare it is that it is Germans who are saving Russia (from Communism) and are restoring order. What could be more humiliating for us? With one hand the Germans give, and with the other they take away. Already they have seized an enormous territory. God help and save this unhappy country. Probably He wills us to endure all these insults, but that we must take them from the Germans almost kills me. During a war one can understand these things happening, but not during a revolution. Now Batoum has been taken—our country is disintegrating into bits. I cannot think calmly about it. Such hideous pain in heart and soul. Yet I am sure God will not leave it like this. He will send wisdom and save Russia I am sure.

It will always be to me an immense gratification that in the midst of her great pain and sorrow for Russia's piteous plight our small group of friends in Petrograd, and those brave souls who dared to risk their lives as message bearers, were able to get to the forlorn family in desolate Siberia at least the necessities of life of which a cruel and inefficient government deprived them. The Empress who all her life had but to command what she wanted for herself and her children was grateful, pathetically grateful, for the simple garments, the cheap little luxuries, even the materials for needlework we were able to convey to them. She thanks me almost effusively for the jackets and sweaters we sent her and the girls in their cold rooms. The wool was so soft and nice, but the linen, she feared, was almost too fine. This was early in March, but spring was already creeping across the steppes.

The weather is so fine that I have been sitting out on the balcony writing music for the Lenten prayers, as we have no printed notes. We had to sing this morning without any preparation, but it went—well, not too badly. God helped. After service we tried to sing some new prayers with the new deacon, and I hope it will go better tonight.*
On Wednesday, Friday and Saturday mornings we were allowed to go

to the eight o'clock morning service in church—imagine the joy and comfort! The other days we five women will sing during the home service. It reminds me of Livadia and Oreanda. This week we shall spend the evenings alone with the children, as we want to read together. I know of nothing new. My heart is troubled but my soul remains tranquil as I feel God always near. Yet what are they deciding on in Moscow? God help us.

"Peace and yet the Germans continue to advance farther and farther in," wrote the Empress on March 13 (Russian). "When will it all finish? When God allows. How I love my country, with all its faults. It grows dearer and dearer to me, and I thank God daily that He allowed us to remain here and did not send us farther away. Believe in the people, darling. The nation is strong, and young, and as soft as wax. Just now it is in bad hands, and darkness and anarchy reigns. But the King of Glory will come and will save, strengthen, and give wisdom to the people who are now deceived."

For some reason the Empress seemed to feel that the Lenten season of 1918 was destined to end in an Easter resurrection of the torn and distracted country. At least so her letters indicate. In a mood of fitful kindness and mercy the Bolshevist soldiers in authority in Tobolsk allowed their captives to go rather often to church and to Communion during this season, and the Empress was very happy in consequence. Her letters were full of prayers for the country, in which the whole family joined, and they appeared to look forward to Easter as the day when God would give some token that the sins of the Russian people, for which they were suffering, were forgiven. Yet never once did she speak of regaining power or the throne. All that was over and forgotten. Neither the Emperor nor the Empress ever indicated in any syllable that they expected to be returned to their former eminence. In fact they never spoke of what might actually happen to the Russian Empire, but they believed that God would hold it together and restore its people to wisdom and strength. For themselves they seemed to look forward to nothing better than an obscure existence with other Russian people. How uncomplainingly they accepted the hard terms of their lives, how grateful they were for the love of distant friends whom they might never see again, is shown in all the last letters I received from the

---

* (previous page) Western readers perhaps do not know how indispensable is vocal music in Russian church services where no organ is permitted. All priests are trained musicians, and there is much congregational singing.

Empress during March, 1918. After receiving one of our parcels of clothing she wrote me:

> We are endlessly touched by all your love and thoughtfulness. Thank everybody for us, please, but really it is too bad to spoil us so, for you are among so many difficulties and we have not many privations, I assure you. We have enough to eat, and in many respects are rich compared with you. The children put on yesterday your lovely blouses. The hats also are very useful, as we have none of this sort. The pink jacket is far too pretty for an old woman like me, but the hat is all right for my gray hair. What a lot of things! The books I have already begun to read, and for all the rest such tender thanks. He was so pleased by the military suit, vest, and trousers you sent him, and all the lovely things. From whom came the ancient image? I love it.
>
> Our last gifts to you, including the Easter eggs, will get off today. I can't get much here except a little flour. Just now we are completely shut off from the south, but we did get, a short time ago, letters from Odessa. What they have gone through there is quite terrible. Lili is alone in the country with her grandmother and our godchild, surrounded by the enemy. The big Princess Bariatinsky and Mme. Tolstoy were in prison in Yalta, the former merely because she took the part of the Tartars. Babia Apraxine with her mother and children live upstairs in their house, the lower floor being occupied by soldiers. Grand Duchess Xenia with her husband, children, and mother are living in Dülburg. Olga Alexandrovna (the Emperor's sister) lives in Haraks in a small house because if she had remained in Aitodor she would have had to pay for the house. What the Germans are doing! Keeping order in the towns but taking everything. All the wheat is in their hands, and it is said that they take seed-corn, coal, former Russian soldiers—everything. The Germans are now in Bierki and in Charkoff, Poltava Government. Batoum is in the hands of the Turks.
>
> Sunbeam (Alexei) has been ill in bed for the past week. I don't know whether coughing brought on the attack, or whether he picked up something heavy, but he had an awful internal hemorrhage and suffered fearfully. He is better now, but sleeps badly and the pains, though less severe, have not entirely ceased. He is frightfully thin and yellow, reminding me of Spala. Do you remember? But yesterday he began to eat a little, and Dr. Derevanko is satisfied with his progress. The child has to lie on his back without moving, and he gets so tired. I sit all day beside him, holding his aching legs, and I have grown almost as thin as he. It is certain now that we shall celebrate Easter at home because it will be better for him if we have a service together. I try to hope that this attack will pass more quickly than usual. It must, since all Winter he was so well.

I have not been outside the house for a week. I am no longer permitted to sit on the balcony, and I avoid going downstairs. I am sorry that your heart is bad again, but I can understand it. Be sure and let me know well in advance if you move again. Everyone, we hear, has been sent away from Tsarskoe. Poor Tsarskoe, who will take care of the rooms now? What do they mean when they speak of an "état de siège" there? . . .

Darling "Sister Seraphine":

I want to talk to you again, knowing how anxious you will be for Sunbeam. The blood recedes quickly—that is why today he again had very severe pains. Yesterday for the first time he smiled and talked with us, even played cards, and slept two hours during the day. He is frightfully thin, with enormous eyes, just as at Spala. He likes to be read to, eats little—no appetite at all in fact. I am with him the whole day, Tatiana or Mr. Gilliard relieving me at intervals. Mr. Gilliard reads to him tirelessly, or warms his legs with the Fohn apparatus. Today it is snowing again but the snow melts rapidly, and it is very muddy. I have not been out for a week and a half, as I am so tired that I don't dare to risk the stairs. So I sit with Alexei. A great number of new troops have come from everywhere. A new Kommissar has arrived from Moscow, a man named Jakovleff, and today we shall have to make his acquaintance. It gets very hot in this town in Summer, is frightfully dusty, and at times very humid. We are begging to be transferred for the hot months to some convent. I know that you too are longing for fresh air, and I trust that by God's mercy it may become possible for us all.

They are always hinting to us that we shall have to travel either very far away, or to the center (of Siberia), but we hope that this will not happen, as it would be dreadful at this season. How nice it would be if your brother could settle himself in Odessa. We are quite cut off from the south, never hear from anybody. The little officer will tell you—he saw me apart from the others.* I am so afraid that false rumors will reach your ears—people lie so frantically. Probably the little one's illness was reported as something different, as an excuse for our not being moved.† Well, all is God's will. The deeper you look the more you understand that this is so. All sorrows are sent us to free us from our sins or as a test of our faith, an example to others. It requires good food to make plants grow strong and beautiful, and the gardener walking through his garden wants to be pleased with his flowers. If they do not grow properly he takes his pruning knife and cuts,

* By this the Empress meant that the secret messenger would give me particulars she dared not write in her letter.
† To a convent as they desired.

waiting for the sunshine to coax them into growth again. I should like to be a painter, and make a picture of this beautiful garden and all that grows in it. I remember English gardens, and at Livadia you saw an illustrated book I had of them, so you will understand.

Just now eleven men have passed on horseback, good faces, mere boys—this I have not seen the like of for a long time. They are the guard of the new Kommissar. Sometimes we see men with the most awful faces. I would not include them in my garden picture. The only place for them would be outside where the merciful sunshine could reach them and make them clean from all the dirt and evil with which they are covered.

God bless you, darling child. Our prayers and blessings surround you. I was so pleased with the little mauve Easter egg, and all the rest. But I wish I could send you back the money I know you need for yourself. May the Holy Virgin guard you from all danger. Kiss your dear mother for me. Greetings to your old servant, the doctors, and Fathers John and Dosifei. I have seen the new Kommissar, and he really hasn't a bad face. Today is Sacha's (Count Voronzeff, aide-de-camp) birthday.

March 21.

Darling child, we thank you for all your gifts, the little eggs, the cards, and the chocolate for the little one. Thank your mother for the books. Father was delighted with the cigarettes, which he found so good, and also with the sweets. Snow has fallen again, although the sunshine is bright. The little one's leg is gradually getting better, he suffers less, and had a really good sleep last night. Today we are expecting to be searched—very agreeable! I don't know how it will be later about sending letters. I only hope it will be possible, and I pray for help. The atmosphere around us is fairly electrified. We feel that a storm is approaching, but we know that God is merciful, and will care for us. Things are growing very anguishing. Today we shall have a small service at home, for which we are thankful, but it is hard, nevertheless, not to be allowed to go to church. You under-stand how that is, my little martyr.

I shall not send this, as ordinarily, through ——, as she too is going to be searched. It was so nice of you to send her a dress. I add my thanks to hers. Today is the twenty-fourth anniversary of our engagement. How sad it is to remember that we had to burn all our letters, yours too, and others as dear.* (*see footnote on next page*) But what was to be done? One must not attach one's soul to earthly things, but words written by beloved hands penetrate the very heart, become a part of life itself.

I wish I had something sweet to send you, but I haven't anything. Why did you not keep that chocolate for yourself? You need it more than the children do. We are allowed one and a half pounds of sugar every month, but more is always given us by kind-hearted people here. I never touch sugar during Lent, but that does not seem to be a deprivation now. I was so sorry to hear that my poor lancer Ossorgine had been killed, and so many others besides. What a lot of misery and useless sacrifice! But they are all happier now in the other world. Though we know that the storm is coming nearer and nearer, our souls are at peace. Whatever happens will be through God's will. Thank God, at least, the little one is better.

May I send the money back to you? I am sure you will need it if you have to move again. God guard you. I bless and kiss you, and carry you always in my heart. Keep well and brave. Greetings to all from your ever loving. A.

This letter, written near the end of March, 1918, was the last I ever received written by her Majesty's own hand. A little later in the spring of that year she and the Emperor were hurriedly removed to Ekaterinaburg—the last place from which the world has received tidings of them. The children and most of the suite were left behind in Tobolsk, the poor little Alexei still ill and suffering, and cruelly deprived of the solace of his mother's love and devotion. In May I received a brief letter from Grand Duchess Olga who with difficulty managed to get me news of her parents and the family.

Darling, I take the first opportunity to write you the latest news we have had from ours in Ekaterinaburg. They wrote on the 23rd of April that the journey over the rough roads was terrible, but that in spite of great weariness they are well. They live in three rooms and eat the same food as the soldiers. The little one is better but is still in bed. As soon as he is well enough to be moved we shall join them. We have had letters from Zina but none from Lili. Have Alya and your brother written? The weather has become milder, the ice is out of the river Irtish, but nothing is green yet. Darling, you must know how dreadful it all is. We kiss and embrace you. God bless you.

OLGA.

(previous page)
* All purely personal letters were burned in the palace at Tsarskoe Selo as soon as the news of the Emperor's abdication reached us, the Empress being determined that her most sacred possessions should not be made public by the Provisional Government. She never recovered from the grief of destroying her youthful love letters, which were more to her than the most costly jewels she possessed, the richest of any sovereign in Europe. To me this is a singular revelation of the real character of the Empress.

After this short letter from Olga came a card from Ekaterinaburg written by one of the Empress's maids at her dictation. It contained a few loving words, and the news that they were recovering from the fatigue of their terrible journey. They were living in two rooms—probably, although this is not stated, under great privations. She hoped, but could not tell yet, that our correspondence could be continued. It never was. I had a card a little later from Mr. Gibbs saying that he and M. Gilliard had brought the children from Tobolsk to Ekaterinaburg and that the family was again united. The card was written from the train where he and M. Gilliard were living, not having been allowed to join the family in their stockaded house. Mr. Gibbs had an intuition that both of these devoted tutors were soon to be sent out of the country and such proved to be the case. This was my last news of my Empress and of my Sovereigns, best of all earthly friends.

In July short paragraphs appeared in the Bolshevist newspapers saying that by order of the Soviet at Ekaterinaburg the Emperor had been shot but that the Empress and the children had been removed to a place of safety. The announcement horrified me, yet left me without any exact conviction of its truth. Soviet newspapers published what they were ordered to publish without any regard whatever to facts. Thus when a little later it was announced that the whole family had been murdered—executed, as they phrased it—imagine "executing" five perfectly innocent children!—I could not make myself believe it. Yet little by little the public began to believe it, and it is certain that Nicholas II and his family have disappeared behind one of the world's greatest and most tragic mysteries. With them disappeared all of the suite and the servants who were permitted to accompany them to the house in Ekaterinaburg. My reason tells me that it is probable that they were all foully murdered, that they are dead and beyond the sorrows of this life forever. But reason is not always amenable. There are many of us in Russia and in exile who, knowing the vastness of the enormous empire, the remoteness of its communications with the outside world, know well the possibilities of imprisonment in monasteries, in mines, in deep forests from which no news can penetrate. We hope. That is all I can say. It is said, although I have no firsthand information on the subject, that the Empress Dowager has never believed that either of her sons was killed. The Soviet newspapers published accounts of the "execution" of Grand Duke Michail, and strong evidence has been presented that he was murdered in Siberia with others of the family, including the Grand Duchess Serge. These same newspapers, however, officially stated that Grand Duke Michail had been assisted to escape by English officers.

The most fantastic contradictions concerning all these alleged murders have from time to time cropped up. When I was in prison in the autumn of 1919 a fellow prisoner of the Chekha, the wife of an aide-de-camp of Grand Duke Michail, told me positively that she had received a letter from the Emperor's brother, safe and well in England.

Perhaps the strangest incident of the kind happened to me when I was hiding from the Chekha after my last imprisonment and my narrow escape from a Kronstadt firing squad. A woman unknown to me approached me and calling me by my name, which of course I did not acknowledge, showed me a photograph of a woman in nun's robes standing between two men, priests or monks. "This," she said mysteriously and in a whisper, "is one you know well. She sent it to you by my hands and asks you to write her a message that you are well, and also to give your address that she may write you a letter."

I looked long at the photograph—a poor print—and I could not deny to myself that there was something of a likeness in the face, and especially in the long, delicate hands. But the Empress had always been slender, and after her ill health became almost emaciated. This woman was stout. I might, had I had the slightest assurance of safety, have taken the risk of writing my name and address for this stranger. But no one in Russia takes such risks. The net of the Chekha is too far flung.

I have one word more to say about these letters of the Empress Alexandra Feodorovna. I have translated them as faithfully and as literally as possible, leaving out absolutely nothing except a few messages of affection and some religious expressions which seem to me too intimate to make public, and which might appear exaggerated to western readers. I have included letters which may be thought trivial in subject, but I have done it purposely because I yearned to present the Empress as she was, simple, self-sacrificing, a devoted wife, mother, and friend, an intense patriot, deeply and consistently religious. She had her human faults and failings, as she freely admits. Some of these traits can be described, as the French express it, as "the faults of her quality." Thus her great love for her husband, which never ceased to be romantic and youthful, caused her at times cruel heart pangs. Because this has nothing to do with her life or her story I should not allude to the one cloud that ever came between us—jealousy. I should leave that painful, fleeting episode alone, knowing that she would wish it forgotten, except that in certain letters which have been published she herself has spoken of it so bitterly that were I to omit mention of it entirely I might be accused of suppressing facts.

I have, I think, spoken frankly of the preference of the Emperor for my society at times, in long walks, in tennis, in conversation. In the early part of 1914 the Empress was ill, very low-spirited, and full of morbid reflections. She was much alone, as the Emperor was occupied many hours every day, and the children were busy with their lessons. In the Emperor's leisure moments he developed a more than ordinary desire for my companionship, perhaps only because I was an entirely healthy, normal woman, heart and soul devoted to the family, and one from whom it was never necessary to keep anything secret. We were much together in those days, and before either of us realized it the Empress became mortally jealous and suspicious of every movement of her husband and of myself. In letters written during this period, especially from the Crimea during the spring of 1914, the Empress said some very unkind and cruel things of me, or at least I should consider them cruel if they had not been rooted in illness, and in physical and mental misery. Of course the Court knew of the estrangement between us, and I regret to say that there were many who delighted in it and did what they could to make it permanent. My only real friends were Count Fredericks, Minister of the Court, and his two daughters, who stood by me loyally and kept me in courage.

That this illusion of jealousy was entirely dissipated, that the Empress finally realized that my love and devotion for her precluded any possibility of the things she feared, her letters to me from Siberia amply demonstrate. Our friendship became more deeply cemented than before, and nothing but death can ever sever the bond between us.

Other letters written by the Empress to her husband between 1914 and 1916 have within this past year found publication by a Russian firm in Berlin. Some of them have been reproduced in the London *Times*, and I have no doubt that they will also be published in America. These letters reveal the character of the Empress exactly as I knew her. It is balm to my bruised heart to read in the London *Times* that whatever has been said of her betrayal, or attempted betrayal of Russia during the war, must be abandoned as a legend without the least foundation. So must also be discarded accusations against her of any but spiritual relations with Rasputine. That she believed in him as a man sent of God is true, but that his influence on her, and through her on the Emperor's policies, had any political importance I must steadfastly deny. Both the Empress and Rasputine liked Protopopoff and trusted him. But that had nothing to do with his ministerial tenure. The Empress, and I think also Rasputine, disliked and distrusted Grand Duke Nicholas. But that had nothing to do with his demission. In these affairs the Emperor made his own decisions, as I have stated. The strongest proof of what I have written will be found in the letters of the Empress, those she wrote to the Emperor, to her relations in Germany and England, and those included in this volume. Nothing contradictory, nothing inconsistent has ever been discovered, despite the efforts of the Empress's bitter enemies, the Provisional Government and the Bolshevists. Before all the world, before the historians of the future, Alexandra Feodorovna, Empress of Russia, stands absolved.

CHAPTER XXIII

TOWARDS the close of the summer of 1918 life in Russia became almost indescribably chaotic and miserable. Most of the shops were closed, and only the few who could pay fantastic prices were able to buy food. There was a little bread, a very little butter, some meat, and a few farm products. Tea and coffee had completely disappeared, dried leaves taking their places, but even these substitutes were frightfully dear and very difficult to find. The trouble was that the Bolshevist authorities forbade the peasants to bring any food into Petrograd, and soldiers were kept on guard at the railway stations to confiscate any stocks that tried to run the blockade. Frequently the market stalls were raided, and what food was there was seized, and the merchants arrested. Food smuggling went on on a fairly large scale, and if one had money he could at least avoid starvation. Most people of our class lived by selling, one by one, jewels, furs, pictures, art objects, an enterprising class of Jewish dealers having sprung up as by magic to take advantage of the opportunity. There was also a new kind of merchant class, people of the intelligentsia, who knew the value of lace, furs, old china and embroideries, who dealt with us with more courtesy and rather less avarice than the Jews.

My mother and I fell into dire poverty. A home we had, and even a few valuable jewels, but we clung to everything we had as shipwrecked sailors to their life belts. We could not look far ahead, and we viewed complete bankruptcy with fear and dread. I recall one bitter day in that summer sitting down on a park bench weary and desolate as any pauper, for I had not in my pocket money enough to go home in a tram. I do not remember how I got home, but I remember that in that dark hour a former banker whom we had long known called at our lodgings and told us that he had a little money which he was about to smuggle to the Imperial Family in Siberia. He wanted us to accept twenty thousand rubles of this for our immediate needs, and gladly we did accept it. Very soon afterwards the banker suddenly and mysteriously disappeared, and his fate remains to this day a profound mystery. I do not even know if he succeeded in getting the money to Siberia. However, with the hope he inspired in me I began to think of possible resources which I might turn to account. My hospital in Tsarskoe Selo had been closed by the Bolsheviki, but its expensive equipment of furniture, instruments, horses and carriages still remained, and I employed a lawyer to go over the books and to estimate what money I could realize from a sale of the whole property. To my dismay I learned that the place with everything in it had been seized by my director and head nurse who, under the Bolshevist policy of confiscation, claimed all, ostensibly as state property but really as their own, for they had become ardent Bolshevists. I made a personal appeal to these old employees of mine to let me have at least one cow for my mother who, being very frail, needed milk. They simply laughed at me. My lawyer took steps to protect my rights, and the result of this rash action was that the former director denounced me to the Chekha as a counter-Revolutionist, and in the middle of an October night our home was invaded by armed men who arrested me and my nursing sister, and looted our rooms of everything that caught their fancy. Among other things they took was a letter from the Emperor to my father explaining the conditions which led him to assume supreme command of the army. This letter, treasured by me, seemed to them somehow very incriminating.

Driven ahead of the soldiers, I went downstairs and climbed into a motor truck which conveyed us to the headquarters of the Chekha in Gorohvaia Street. After my name had been taken by a slovenly official I followed the guard to one of two large rooms which formed the women's ward of the prison. There must have been close to two hundred women crowded in these rooms. They slept sometimes three to a narrow bed, they lay on the tables and even on the bare floor. The air of the place was, of course, utterly foul, for many of the women were of the class that never washes. Some were of gentle birth and breeding, accused of no particular offense, but held, according to Bolshevist custom, as hostages and possible witnesses for others who were under examination or who were wanted and could not be found. In the early morning all the prisoners got up from their narrow beds or the hard floor and made their way under soldier escort to a toilet where they washed their faces and hands. As I sat miserably on the edge of my bed a woman came up to me introducing herself as Mlle. Shoulgine, the oldest inhabitant of the place, and therefore a kind of a monitor. It was her business, she said, to see that each prisoner received food and to handle any letters or petitions the women might desire to send out. I told her that I desired to send a petition to the head of the Chekha, or to whatever committee was in charge of the prison, asking the nature of the charges against me, and begging for an early trial. This petition was duly dispatched, and very soon after a very large man, a Jew, came to see me and promised that my affair would be promptly investigated. The soldiers on guard spoke to me kindly and offered, if I had money, to carry letters back and forth from my home. I gave them money and was comforted to hear from my mother that Dr. Manouchine was once more working for my release. Although not a Bolshevist, the doctor's skill was greatly respected by the Communists, who had appointed him head physician of the old Detention House. There was a student doctor attached to our prison, and merely because he was a friend of Dr. Manouchine and knew that I was also, he was courteous and attentive to me. So potent is the influence of a truly great character.

The five days I spent in that filthy, crowded cell will never leave my memory. Every moment was a nightmare. Twice a day they served us with bowls of so-called soup, hot water with a little grease and a few wilted vegetables. This with small pieces of sour black bread was all the food vouchsafed us. Some of the prisoners got additional food from outside, and usually these fortunate ones divided what they had with the others. There was one beautiful woman of the half-world who daily received from some source ample food, and like most of the women of her class she was generous. I was told that she had been arrested because she had hidden and helped her lover, a White officer, to escape, and that she felt proud to be suffering for his sake. Perhaps it was from friends of his that she received the food, yet women of her kind, God knows, very seldom meet with gratitude even from those who owe it most.

Although I was accused of no crime and had no idea what accusations could be brought against me, I lived as all the others lived, in a state of constant anxiety and fear. All day and all night we heard the sound of motors and of motor horns, we saw prisoners brought in, and from our windows we could see great quantities of loot which the Bolshevist soldiers had collected, silver, pictures, rich wearing apparel, everything that appealed to them as valuable. In the courtyard we could see the men fighting like wolves over their spoils. It was like living in a pirates' den rather than a prison, and yet we were often enough reminded that we were prisoners. One day all the women in my room were roughly ordered into a larger room literally heaped with archives of the Imperial Government. With soldiers standing over us we set to work like charwomen to sort the papers and tie them up in neat bundles. Very often in the night when we were sleeping exhausted in our cell rooms the electric lights would suddenly be turned on, guards would call out names, and half a dozen frightened women would get up, gather their rags about them, and go out. Some returned, some disappeared. No one knew whose turn would come next or what her fate would be.

The name of my nursing sister was called before mine, and within a short time she returned smiling to say that she was to be sent home at once and that I should soon follow. Two hours later soldiers appeared at the grating and one called out my surname: "Tanieva, to Viborg Prison." I had spirit enough to demand the papers consigning me to this dread women's prison, but the soldiers merely pushed me back with the butts of their guns and bade me lose no time in obeying orders. I still had a little money with which I paid for a cab instead of walking the long distance to the prison, and I begged the soldiers to stop on the way and let me see my mother.

For this privilege I offered all the money remaining in my purse, which the soldiers took, also bargaining for the ring I wore on my hand. This I declined to give so they philosophically said: "Oh, well, why not?" And stopped the cab at the door of my mother's lodgings. Of course my poor mother was overjoyed to see me, even for a moment, and so was old Berchik, now almost at the end of his life. Both assured me that everything was being done in my behalf and that at the Viborg prison I would be in less danger of death than at the Chekha headquarters. I might even hope to be admitted to the prison hospital.

A little heartened in spite of myself I went on to Viborg, which lies in a far quarter of the town of what is known as the Viborg side of the Neva. A rather pretty Bolshevist girl was in charge of the receiving office, and when I pleaded ill health and asked to be sent to the hospital she promised to see what could be done. Viborg prison was one of many which during the first frenzied days of the Revolution were thrown open, the prisoners released, and the wardresses murdered. I do not know how other women were induced to take their places, but I do know that the women in whose charge I was placed were so kind and considerate that had any attempt been made against them the prisoners themselves would have fought in their defense. The wardress who locked me in my cell stopped to say a comforting word, and because she saw that I was shivering with cold as well as nervousness, she brought me bread and a little hot soup.

After some hours I had another visitor, Princess Kakouatoff, accused of being the ringleader of an anti-Bolshevist plot, who had been six months in Viborg and was regarded as a "trusty." Among other privileges she had the right to telephone friends of new prisoners, and at my request she telephoned messages to friends who could be of use to my mother if not to me. The princess brought me a little portion of fish which I ate hungrily, and I think she was also instrumental in finally getting me into the prison hospital. This was after I had fainted on the floor of my cell, and everyone in authority, including the prison doctor, knew that I was in no condition to endure the noisy confusion of the huge cell house. The hospital was a little cleaner than the rest of the prison, but it was a pretty dreadful place just the same. For nurses we had good-conduct prisoners, women of low type who stole food and everything else they could lay hands on. They stripped me of my clothes, substituting the prison chemise and blue dressing gown, and took away all my hairpins. I was given a bed in a room with six other women, one of them a particularly awful syphilis case, and two others, very dirty, who spent most of their time going over each other's heads for vermin. I stayed in this ghastly place a very short time, a woman doctor and a prisoner of my own class, Baroness Rosen, succeeding in getting me transferred to a better ward. Nevertheless the whole prison hospital was horrible. The trusties in charge of the wards were in the habit of eating the meat out of the prisoners' bowls, and fighting for food among prisoners throughout the institution was a daily occurrence. I can describe Viborg prison and most of its inmates in one word—beastly. Many of the women were syphilitic, most were verminous, some were half mad. One who slept near me had murdered her husband and burned his body. Nearly all sang the most obscene songs and held unrepeatable conversations. Mostly they were so depraved that the doctor in his rounds showed that he was afraid of them. Yet there were among them a few women who, like myself, had led sheltered and religious lives, and who were only now learning that such abandoned specimens of womanhood existed on the earth. There was no attempt at reforming the women. Once there had been a church attached to the prison, but this the Bolsheviki had closed, substituting a cinema to which on special occasions some of the prisoners were admitted. Not many political prisoners had this privilege because they were treated much more rigorously than common criminals. It was the common criminals also, the thieves, murderers, prostitutes, who were released in advance of "counter-Revolutionists," those accused, however vaguely, of political activities.

All the prisons of Petrograd by this time were so crowded with so-called political prisoners that even the women's prison was obliged to receive an overflow of sick men prisoners. This wholesale imprisonment of anti-Bolshevists naturally led to the shooting of thousands of citizens, shooting being simpler than feeding and housing, and in addition an economy of effort on the part of those charged with the mockery of trials. Later the Chekha dispensed with this mockery, but in those days prisoners were given the pretense of a hearing. I can testify to their futility, because I went through more than half a dozen trials and in no case was I accused of any crime, tried for any definite offense, or given anything like a fair hearing. On September 10, 1918, word was brought to the Viborg prison that on the next morning I was to be taken away not to return. This seemed to be a death sentence, and all that night I lay awake thinking of my poor mother and wondering what would become of her alone in the midst of the Bolshevist inferno. Silently and long I prayed for her and for the peaceful release of my own tried soul.

Very early in the morning I was summoned, my own clothes were given me, and I was led to the receiving office of the prison. Here two soldiers waited, and I was taken out between them and marched to the headquarters of the Chekha. In a small, dirty room I underwent an examination by two Jewish Communists, one of whom, Vladimirov—nearly all Jewish Communists assume Russian names—being prominent in the councils of the Communist central committee. For fully an hour these men did everything they could to terrorize me. They accused me of being a spy, of plotting against the Chekha, of being a dangerous counter-Revolutionist. They told me that I was to be shot at once and that they intended to shoot all the intellectuals and the "Bourju," leaving the proletariat in full possession of Russia. They continued this bluster until from sheer weariness they stopped, then one of the men leaned his elbows on the table and with a smile that was meant to be ingratiating said confidentially: "I tell you what. You relate the *true* story of Rasputine and perhaps we won't have you shot, at least not today." I assured the man that I knew no more about Rasputine than they did, perhaps not as much, since I had no access to police records and they had. Then they wanted to know all about the Czar and the life of the Court. As well as I could I satisfied their curiosity, which was that of ignorant children, and at the end of an exhausting interrogation they actually sent me, not to a wall and a firing squad, but back to the filthy cell in the Viborg prison. I dropped on my dirty bed, swallowed a little food brought me by a sympathetic fellow prisoner, and resigned myself for what next might happen to me. What happened was astonishing. A soldier came to the door and called out: "Tanieva, with your things to go home."

Within a short time I stood trembling and weak on the pavement in front of the prison. I could not have walked to my lodgings, in fact I felt incapable of walking at all, but a strange woman observing me and my piteous condition approached, put her arm around me, and helped me into a drosky. I had a little money, perhaps fifty rubles, and I gave it all to the ischvostik to drive me home. Here I found an amazing state of affairs, the general immorality and demoralization into which Bolshevism was driving the people having penetrated our own place. Everyone was turning thief, and my nursing sister, who had been with me since 1905, whom my mother had treated like a daughter, had become inoculated with the virus of evil. The woman had not only appropriated almost all the clothes I possessed, but had stolen all the trinkets and bits of jewelry she could lay hands on. She had even taken the carpets from the floors and stored them in her room. Not daring to attempt to regain any of this property I asked the nurse to please take what she wanted and leave the apartment. "Not at all," she replied. "This place suits me very well and as long as I choose I shall remain." She had embraced Bolshevism, not I am sure from principle, but as the safest policy, and in time she became rich in jewels, finery, and miscellaneous loot. It was months before we finally induced her to leave, and after her departure I have reason to believe that she did everything she could to keep me in trouble with the Bolshevists.

By this time the Communist régime was fully organized. The whole town was divided into districts, each one under command of a group of soldiers who had full license to search—and rob—houses, and to make arrests. Every night the search went on. At seven o'clock all electric lights were turned off, and when, two or three hours later, they suddenly flashed up again, every soul in the district was seized with fear, knowing that this was the signal for the invasion. Often women were included in the searching parties, terrible women dressed in silks and strung with jewelry, stolen of course from the hated "Bourju." Seven times our home was raided, once on the authority of an anonymous letter charging that we were in possession of firearms. Once more I was dragged off to an interminable examination, this time before the staff of the Red Army in a house in Gogol Street. The close connection between the Chekha and the Red Army was apparent because in the two hours during which I sat in the ante-chamber waiting examination a Lettish official of the Chekha passed freely in and out of the committee room, occasionally throwing me a reassuring word. My case would be settled favorably, he said, and it was, for the committee after bullying me for a length of time, dropped the subject of concealed firearms, assumed the snobbish and half cringing air with which I was becoming familiar to the point of nausea, and began asking questions about the Imperial household. They produced a large album of photographs and made me go through it and identify each picture. Finally the head inquisitor told me magnanimously that I could go home, cleared by the highest authority, but that soldiers would go with me and make sure that there were no revolvers or pistols in the house. The search was made anew, and then the men left, obviously disappointed that practically nothing worth stealing had come to light.

Two things of importance were happening in those days. The White Army was approaching Petrograd, and in all the streets soldiers were drilling in anticipation of a battle. Airplanes whirred overhead, and once in so often a shell screamed over the housetops. We prayed for the coming of the White Army, and at the same time dreaded the massacres we knew would precede its entry into the town. The second thing that marked this date was the Communist system of public feeding, free food being furnished by cards distributed according to the status of the individual. The Bolshevist authorities and the soldiers of course had the most food and the best. Next came the proletariat, so-called, and last of all the "Bourju" was provided for. These of the lowest strata in society got hardly anything at all and would have starved, most of them, had it not been for the food smuggling which constantly went on, the peasants from out of town boldly bringing in bulky parcels, and taking back in return for their food, not Bolshevist money, which they disdained, but everything they could accumulate in the way of furniture or dress materials. They even accepted window curtains and table linen, anything, in fact, that could be fashioned into clothing. These same peasants before the Revolution had been expert spinners and weavers, but now they scorned such plebeian occupations because it was easier to barter grains, milk, vegetables, and other produce for the last possessions of the townspeople.

We went on living, somehow, parting with clothing and furniture, burning boxes and even chairs for fuel, walking miles for stray bits of wood, praying for the success of the White forces, praying for protection against what must happen before that success could be achieved. My mother all these days was very ill with dysentery, which was rife in Petrograd, and I had that additional suffering, for I knew that it would take little to bring her frail life to an end.

* * *

CHAPTER XXIV

ON September 22 (October 6, New Style) I went in the evening to a lecture in a church. At that time every non-Bolshevist spent as many hours every day as possible in the churches, praying or listening to words of hope and comfort from the priests. The church was, in fact, the only home of peace and rest in the whole of the distracted country. That particular night in church I met some old friends who invited me to go home with them rather than walk the long and dreary, even the dangerous way back to my lodgings. I stayed with my friends that night, and the next morning early I went to mass in the little church where Father John of Kronstadt lies buried. I reached home about midday, and found the place in the possession of soldiers, two of whom had waited the entire night to arrest me, this time as a hostage, the White Army being reported within a few miles of Petrograd. My sick mother prepared me a little food, made a parcel of my scanty linen, and once more we bade each other the despairing farewell of two who knew that they might never meet again on earth. I was quickly conveyed to the headquarters of the Chekha where I was greeted with the exultant welcome: "Aha! Here we have the bird who has dared to stay out a whole night."

Thrust into the old filthy, ill-smelling cell room I found a spot near a dirty window from which I could get a far glimpse of the golden dome of St. Isaac's Cathedral. During my whole term in this place I kept my eyes and my whole mind on that golden dome, trying to forget the hell that whirled around me. The woman in charge of the room was a Finnish girl who had committed the crime of trying to run away to Finland. She was a stenographer and clerk, and the Chekha used her by night as an office assistant. Whether by nature or by association she had become as hard and as ruthless as her captors, and her imprisonment had many mitigations. It was her pleasant duty to make out the lists of those who, twice a week, were taken to Kronstadt to be shot, and her reports on the subject which she confided regularly to her chosen comrade, a Georgian dancer named Menabde, were enough to sicken even those of us who had become accustomed to wholesale slaughter of unoffending human beings. We heard little else except death and threats of death in this place. There was an official named Boze in the prison, and often we heard him screeching through the telephone to his wife that he would be late to dinner that night because he had a load of "game" to get off to Kronstadt. Under such conditions pity and sympathy become strangely dulled. On occasions when I was sent to the kitchens for hot water I used to get glimpses of the "game," huddled wretchedly in their seats or restlessly pacing their cells—waiting. Often when I returned with the water I found the seats and the cells empty, and although my heart sank and my senses swam, I never felt the screaming horror a normal person would have felt. This dulling of the emotions, I suppose, is nature's way of keeping the mind from giving way entirely. Of course nature took away all human dignity and self-respect, this, too, in mercy. Any prisoner who went to the kitchens was greeted with jeers and foul abuse from the cooks who threw us handfuls of potato parings and withered cabbage leaves, quite as one would throw bones to dogs. Like dogs we eagerly snatched at these leavings, because the prisoners' regular rations were nothing half as palatable, being mostly wormy dried fish and a disgusting substitute for bread.

One day I was called up for examination, and this time a real surprise awaited me. My judge was an Esthonian named Otto, not altogether a brutal man, as it turned out. As I approached his desk he regarded me grimly and without a word handed me a letter, unsigned, and reading about as follows: "To the Lady in Waiting, Anna Viroubova. You are the only one who can save us from this terrible Bolshevik administration, as you are at the head of a great organization fully equipped with guns and ammunition." Sternly the Esthonian judge commanded me to tell him the truth about the organization of which I was the head. Of course I told him that the whole thing was an invention, and he astonished me by saying that although the letter had been posted to my address he had very much doubted its verity. Then he asked, almost gently: "Are you very hungry?" Taken off my guard as much by the kindness as by the prospect of food, I fell against the desk murmuring only half aloud: "Hungry? Yes, oh, yes." Whereupon he opened a drawer of his desk and handed me a large piece of fresh, sweet bread. "Go now," he said, "and I will discuss your case with my colleague Vikman. In the evening we will see you again."

At eleven that night I was again summoned, this time before the two men. The Esthonian, still kind and courteous, gave me a glass of steaming tea, which did much to lend me courage. Both he and Vikman then put me through a searching examination especially about my relations, real and assumed, with the Imperial Family and with persons of the Court. At three in the morning they released me, more dead than alive with fatigue, Otto telling me heartily that he thought I would be set free within a few days. Vikman, however, declared that my case would have to be referred to Moscow and that I need not expect an early release. I went back to my evil cage expecting nothing. I knew, that the threat of the White Army advance filled with terror the whole Bolshevist population, and that in case of actual battle no life outside the slim Communist ranks would be worth the smallest scrap of their worthless paper money.

Very shortly after my return to the cell room I began to hear my name whispered from one wretched woman to another, and I accepted this without much emotion as a prelude to a boat journey to Kronstadt. Early on a certain morning a soldier approached the door and bawled out: "Tanieva, you to Moscow." I happened to be exceedingly ill that day, but mechanically I picked up my little handkerchief containing my few possessions, including a Bible, and followed the escort of two soldiers down the steep steps, as I believed, to my death. Perhaps they had orders to take me to Kronstadt, I cannot be sure of that, but I do know that the route we followed did not lead to the Moscow station. We had walked but a short distance when one of the soldiers said to the other: "What's the good of two of us bothering with one lame woman? I'll take care of her and you can go along. It will soon be over anyway." Nothing loath the other soldier, glad to get out of anything resembling work, took himself off while I, in charge of one armed man, mounted the crowded tram and rode on toward an unknown destination. At a certain point we had to change trams, and here occurred an incident so extraordinary that I almost hesitate to strain the credulity of a non-Russian reader by relating it. The second tram had been delayed for some reason, and a considerable crowd of passengers was waiting for it on the street corner. My soldier stood at my side waiting with the rest, but soon he became impatient. Ordering me not to move an inch in his absence, he ran down the street a short distance to see if the tram were in sight. As soon as he turned his back, people in the crowd began to speak to me. A girl in whom I recognized a former acquaintance asked me where I was going, and when I told her she took a bracelet I gave her and promised to carry it, with news of my fate, to my poor mother. An officer of the old army came up to me saying: "Are you not Anna Alexandrovna?" And when I said yes, he too asked me where I was being taken.

"Kronstadt, I think," I answered, but he said: "Who knows?" and pressed into my hands a roll of bills saying that they might be of use to me.

Other people surrounded me, mostly strangers, but two of them women whom I had often seen at mass in the small church of Father John. They said: "Why should you be shot? The soldier has not come back. Run while the chance is yours. Father John will surely help you." Encouraged by their sympathy, yet hardly knowing what I was doing, I limped off on my crutch much faster than I could have believed possible, the whole street-corner crowd spreading out to shield my flight. I limped and stumbled down Michel Street as far as the Nevski Prospekt weeping and praying all the time: "God save me! God save me!" until I reached the old shopping arcade known as the Gostiny Dvor. Here I caught sight of my soldier running in frantic pursuit of his escaped prisoner. It seemed all over with me then but I crouched in a corner of the deserted building and miraculously the soldier ran on without seeing me. As soon as I thought it at all safe I crept out of the old arcade and turned into the Zagorodny Prospekt, where I found a solitary cab. "Take me quickly," I cried to the ischvostik. "My mother is dying." The man replied indifferently that he had a fare waiting, but I thrust into his hands the entire roll of bills given me by the friendly officer, at the same time climbing into the drosky.

Said the ischvostik, "Where shall I drive you?" I gasped out the address of a friend in the suburbs of the city, and the man lashed his half-starved animal into a walk. After what seemed to me many hours we reached the place, I rang the doorbell and fell across the threshold in a dead faint.

My friend and her husband courageously took me in, fed, warmed me, and put me to bed. They even dared to send word to my mother that I was for the moment safe from pursuit, but they warned her not to come near the house as soldiers would certainly be watching her every movement. As a matter of fact my mother was visited by Red soldiers, arrested in her bed, and closely guarded for three weeks. Our maid also was arrested, as was everyone who came to the house. The old Berchick who had spent almost his entire lifetime in the service of our family was taken ill during this period and died. For five days his body lay uncoffined in the house, the Bolshevist authorities refusing him a burial permit. It was for my mother an interval of utter despair, since in addition to the death of

Berchick she lived in constant fear of my rearrest. In the opinion of the Bolshevist soldiers, however, I had escaped to the White Army, and photographs of me were posted conspicuously in all the railway stations.

The kind friends who had taken me in dared not for their lives keep me long, and wishing them nothing of harm I set out on a dark night without a kopeck in my pockets and with no certain idea where I could find a bed. I had in mind a religious hostel, a place where a few students, men and women, lived under the chaperonage of an old nun. There I went, begging them for Christ's sake to take me in, and there I was hidden for five perilous days. A girl student volunteered to go to see my mother, and go she did, but when hours passed, a day passed, and she did not return, a panic of fear seized all of us, and rather than expose these kind people to risk of imprisonment and death I voluntarily left the place. What else could I do?

How shall I describe the horrors of the next few months? Like a hunted animal I crept from one shelter to another, always leaving when it seemed at all possible that my protectors might be punished for their charity. Four nights I spent in the cell of an old nun whom I knew, but pitying her fears I put on the black head kerchief of a peasant woman and started in a cab, on borrowed money, for the house of a friend near the Alexandra Lavra on the outskirts of the town. All unknown to me a decree had that day been issued that no one could ride in a cab without written permission from the authorities. Consequently before we had traveled half the journey the cab was stopped by two women police, fierce creatures armed with rifles, who called out to the ischvostik: "Halt! We arrest you and your passenger." Hastily I crammed all the money I had into the ischvostik's hand and begged the women to let me go as I had just been discharged from hospital and knew nothing of the new rule. Oddly enough they let us drive on, but very soon the ischvostik, sick with terror, stopped his horse and told me that he would take me no further. I got out and staggered on through the muddy snow, for it was now late in the autumn of 1919. A former officer whom I had once known well met and recognizing me asked if he might not accompany me to my destination. "No, no," I cried. "It would be madness for you to be seen with me. I cannot explain, only go, go, as fast as you can." I staggered on, dripping with rain until I reached my friend's house. To my now customary greeting: "I am running away. Will you hide me?" she replied: "Come in. I have two others." Thus did brave Russians in those days risk their lives to save those of others. Under her protection I lived ten days, and in her house I met a woman, a servant in one of the Communist kitchens, who having access to food and supplies, afterwards more than once saved me from starvation.

From one such kindly haven to another I fled in the dead of night. Once I was received in the home of an English woman who out of her scanty stores gave me warm stockings, gloves, and a sweater. Another day or two I spent in the rooms of a dressmaker whose husband was an unwilling soldier in the Red Army. Once I ventured back to the student hostel, where they welcomed me and fed me well, one of their number having just returned from the country with a stock of smuggled food. Here I had news from my dear mother from the girl who had gone to her on my behalf, and had, after ten days' detention by the Chekha, got back to the hostel. Some members of the Chekha, she informed me, looked forward to shooting me instantly when I was caught, but others said that it was certain that I was with the White Army and would never be caught.

From the hostel I sought a paid lodging with the family of a former member of the orchestra of the Imperial Theater. These people, however, were very mercenary and would receive me only on advance payment of a large sum of money. Almost everything my mother and I had owned had been sold long before, but I retained a pendant of aquamarines and diamonds, a wedding present from the Empress, safely hidden in the house of a friend. This I had sold for fifty thousand rubles, giving half the money to the musician's wife in return for a few days' shelter in a wretchedly dirty, unheated room. Here I had to cut my hair short to get rid of vermin, and feeling unable to endure the hole I left it. Yet finding my next lodgings even worse, I returned, and here in the midst of discomfort and bitter cold, I had the joy of meeting my mother and also my aunt Lashkeroff, who brought me the welcome news that they thought they had at last found me a permanently safe retreat. It was miles from where I was staying, and I had to walk every step of the way, but when I arrived I found my hostess a lovely woman belonging to the Salvation Army. Gladly would I have stayed with her indefinitely but that was impossible as I had no passport and the police began to haunt the neighborhood. She did not abandon me for all that, but got me a new shelter in the home of a good priest and his wife. From here I was handed on from one to another of the priest's parishioners to whom he confided the story of my harried career. Once an Esthonian woman told me that her sister had found a Finnish woman who, for a good price, was willing to take fugitives over the frontier, and she strongly advised me to attempt the flight. Some instinct forbade, and it turned out a good instinct, for the Finnish woman, after taking the money, had abandoned the Esthonian's poor sister in the midst of a wood, from which she had to return, empty of purse and in deadly peril of arrest.

Cutting the story of my fugitive existence short, I finally found something like a permanent abode in the tiny and happily obscure woodland cottage of a working engineer, who kindly offered to take me in to his bachelor quarters a mile or two outside of Petrograd. Here I became once more the happy possessor of a passport, true not in my own name but perfectly legal otherwise. In Russia when a girl marries she gives up her passport to the priest, receiving a new one in the name of her husband. My kind old priest gave one of these maiden passports to the engineer, at the same time reporting to the Commissar of his neighborhood that such a passport had been lost. This was to prevent any possible trouble or inquiry. The Commissar obligingly gave the priest a duplicate, signed and sealed by Bolshevist authority. Now again I was a human being, for no one in Russia can be said to have any identity unless he is in possession of a passport. Mine described me as a teacher, and as such I was henceforth entitled to the Communist rations. For the time being I was less a teacher than an unskilled household servant, for naturally I wanted to do everything possible to repay the good engineer for affording me a safe shelter. I knew nothing whatever of cooking or housework, yet I attempted to do both. The engineer himself was absent all day, but when he returned at night he carried in wood enough to last twenty-four hours, and also water which had to be brought from a great distance. Food, of course, was very scarce. My mother and the friendly priest brought all they could, but even so I would often have suffered had it not been for my old acquaintance, the woman who worked in the Communist kitchen. And here I have to tell another incident which may seem impossible to some readers. One day I was sitting in the little house in the wood, feeling as secure as an escaped prisoner can feel, when I heard a sudden loud knocking at the door. There was no possible place where I could hide, but I sat absolutely still in my chair, hardly breathing for fear of disclosing the fact that the house was not empty. Again came the knocking at the door, this time louder and more peremptory than before. Realizing that it was useless to resist, I arose and with a prayer on my lips, I went to the door and opened it. No one was there. Nothing was in sight save the wintry trees and the frozen path that led to the highway. But yes! There almost at the end of the path stood the shivering figure of a little girl, the daughter of the woman in the Communist kitchen.

"Oh!" she cried, seeing me in the doorway. "I have been looking everywhere for your house and I could not find it."

"But you knocked," I said.

"No, I didn't," declared the child. "I haven't been near the house. I just this minute turned into the pathway to get out of the wind. I'm so glad I've found you. Mother has sent you something."

Who knocked at my door twice? The wind? It never did before or afterwards. If you believe in Providence, as I do, you may agree with me that God did not intend me at that time to starve in the depths of a desolate forest. If you prefer another explanation seek it.

In January, 1920, my kind friend the engineer told me reluctantly that he was about to marry and that the tiny room I occupied would have to be given up. I had not the remotest idea where I was to go. Above all things I desired to embrace a religious life, but in those perilous days no convent in Petrograd dared receive me. The convents were constantly being raided, and the younger nuns were frequently taken out and forced to work on the streets. No religious house could shelter a fugitive even though she possessed a false passport. Again I became a vagrant, spending a night here, a day there, sleeping in any refuge that opened to me. Towards the end of March I again found a home in the house of a priest and his wife who were as parents to me, and to whom I owe a lifetime of gratitude. Here I found not only safety but work, that blessed anodyne against all trouble. My passport, as I have said, described me as a teacher, and a teacher I now became, thanks to my new friends, who found me plenty of pupils among the working-class children of the neighborhood. I taught them the simple elements, and to children of the more intellectual classes languages and music. My pay was in food, but food in the Bolshevist paradise is worth much more than money, so I was completely satisfied.

By this time my appearance was so changed that I lost all fear of the police or the Chekha. One day when I was slowly walking the long distance across the river to my favorite church, the resting place of Father John, a motor car stopped in my path and I recognized as its occupant the Chekha inquisitor Boze, the man who had several times been my brutal jailer. "Grazhdanka (Citizeness)," he addressed me, "please tell me where to find —" he named a street and number whither he was bound, doubtless on some errand of terror. Giving him the direction, I moved on as fast as my crippled legs could carry me, but I need not have been afraid for he did not know me at all.

So went the year 1920, my mother and I and the good priest's family often discussing the possibilities of escape from the increasing starvation, death, and terror which everywhere surrounded us. People did escape, we knew, but how were we to do it—two women, one old and the other lame? It seemed altogether impossible. Besides, we had almost nothing with which to buy our way out of the country. My only shoes were homemade affairs of carpet, and I was so careful of them that often when walking I took them off and carried them in my hands to preserve them. Another thing, beset with dangers as we were in Russia we were no longer hungry, because I had an increasing number of pupils, and each one meant a tiny portion of food and firewood for my mother, my friends, and myself. But here is a strange and a universally human thing. Food and warmth do not bring content to prisoners, they create courage, and when one day in late October we received a letter from my sister, safe in a near-by country which I may not name, the flame of adventure blazed up in the soul of my brave little mother and in my own heart. My sister suggested the possibility of our getting out by one of the ways that persist in flourishing in spite of Bolshevism and the Chekha, and she offered us, if we succeeded in escaping, the shelter of her own home. I cannot reveal any detail of those secret ways of escape, because they still exist, and must not in any way be placed in jeopardy. Enough it is to say that Petrograd is separated from Finland by only a few versts of land, carefully guarded, and by a narrow arm of the Baltic Sea which cannot be quite as successfully guarded. In winter this water freezes, not as unsalted water freezes, smooth and thick and safe for passage, but in rough and treacherous hummocks of mixed ice and snow, with unexpected gaps of half-frozen water opening here and there between the ice masses. Still, the icy Baltic does at times admit of sledge passage, and there are men who make a business of taking over—for a price far beyond what most Russians can afford—refugees who have friends waiting for them in Finland or in countries to the west and south. Sometimes Red soldiers have to be bribed, and often they sell out the people whose money they accept. Sometimes also the men who contract to take refugees over the ice betray their passengers to the Bolshevik guards. Any way you look at it, escape from Bolshevik Russia is about as perilous as going unarmed into a tiger's cage. Yet people dare it, and we did.

It was about the first of December in our calendar, in the year 1920, when we received a second smuggled letter from my sister: "Be ready whenever we send for you." For that promised summons we waited in desperate suspense until two days after Christmas. Then to my mother's lodging came a fisherman and his little boy with the whispered news that we were to go with them on the day following. My mother found means of sending the news to our friend the priest, and he brought it to me. "Tomorrow at four o'clock you go abroad."

The next day at the appointed hour my mother and I, two shivering creatures facing death, but ready, met at a small railway station leading along the Baltic shores. The fisherman's son was also at the station, but obeying instructions, we did not notice him but simply followed wherever he led. Our train journey was short, and at five o'clock, pitch dark in the Russian winter, we alighted at a poor village, following the boy who carried on his back a bag of potatoes. Alas! In the darkness and confusion we lost him, and stood in the icy cold like lost souls, not knowing where to turn. Suddenly out of the shadows a peasant woman approached us. "Are you looking for a boy with a bag of potatoes?" she said in a low voice, and to our frightened assent she murmured: "Follow me." We followed, although, for all we knew, it was to a Chekha prison. Anybody in Russia may be Chekha, the friend who invites you to dinner, the man who buys your last jewel, the woman who offers to guide you over an unknown road. You can trust no one, consequently, when you must, you trust anyone. We followed the peasant woman into a dim hut, and there we found two fishermen who assured us that they were ready that night to take us across the frozen Baltic to a village on the Finnish side. Their horses and sledges, they told us, were safely hidden, but they would be ready to take us and three other fugitives, a lady, a child, and a maid, as soon as we could safely venture to leave the village. As luck would have it there was a festival and a dance going on that night, and we had to sit in that stifling hut in complete silence until two o'clock. Also we had to pay for our shelter and escape one hundred thousand rubles, which my mother had secured by selling her last treasure, a pearl necklace.

When the last peasant had gone to bed and silence wrapped the village, we stole out through the mud and the snow, and got into the rough sledge. Hardly had we struck the rough ice of the Baltic when the sledge overturned, waking the child who, silent before, now began to cry and to beg to go home. The little thing spoke only French and I can still hear him repeating over and over again in a high baby voice which he did not know imperiled the lives of all of us: "Maman, Maman, à la maison, à la maison." For six hours we drove thus, slowly and cautiously over the rotten ice, one of the men driving, and the other running ahead with a long pole testing the ice for a safe pathway. Often we stopped to listen for possible sentinels, and once in the neighborhood of Kronstadt we had such a fright that I wonder the men dared go farther. Plainly to our ears came the grinding of machinery, and we knew that where there was machinery there were men. We stopped long and listened, until our driver suddenly remembered that the noise was that of an ice breaker several miles out of our highway. By this time I was so stiff and drowsy with cold, so nearly frozen, in fact, that I hardly cared what happened to us. Seeing my wretched state, one of the men took off an extra pair of woolen socks he wore and slipped them on my feet. The unknown lady who accompanied us also spared me a warm wrap, and by rubbing and holding me close to their bodies they kept me alive. At eight o'clock of a pale winter morning they lifted me out of the sledge and with the others I stood trembling on the snowy shores of Finland.

"Now you are out of Sovdepia" (Soviet land), said the fishermen cheerfully, "but we are not safe yet, for the Finnish police may catch us and send us back." Hurriedly we climbed the hill to the cottage of one of the smugglers. Here we met his wife, who, gray with fear, came out to meet her husband after his night of peril on the ice. The woman gave us hot coffee, bread, and cheese, but she would not keep us long in her house. We knew that we must report as soon as possible at the quarantine station, and we knew, besides, that the sorely tried Finnish authorities would not be any too glad to see us coming. Do not blame the Finns for this. Every Russian refugee is a burden on their slender resources, and too often a pretended refugee is merely a Bolshevik agent sent to stir up trouble among disaffected workmen. However, on this occasion the Finns received our wretched group with infinite kindness, and made us comfortable during the required period we spent in the quarantine station. Then we went to our separate destinations, all of us to poverty, obscurity, homesickness, to that sunless clime which waits the exile wherever he may go. In the country where my mother and I finally arrived we found my sister, happier than ourselves, because she left Russia before the great horror began, thus saving part of her fortune. My sister gave us food, clothing, a lodging. Except for her bounty we had lost everything we ever owned, home, friends, possessions, country, for Russians now have no country, no flag, no place in the wide world. The best any of us can hope for is an obscure corner in some foreign land where we can earn enough to buy our daily bread, and a quiet place in which to pray every day of our lives: "God save Russia."

I am told, although I can hardly believe it, that in other lands, even in free America, there are beings so deluded that they wish to bring about revolution and Bolshevism. I do not wish for any of them the long nightmare of suffering that I, one of millions, have suffered under revolution and Bolshevism. I pray only that there may be revealed to them the fate of the betrayed who have died and are dying under the criminal administration of the Provisional Government and, later, of Lenine and his fanatical followers. If they can be made to know only in part what my poor, ravished country is today, they will forget their delusions and pray with the exiles: "God save Russia."

APPENDIX A

THE TRUTH CONCERNING THE RUSSIAN IMPERIAL FAMILY

Statement of Vladimir Michailovitch Roudneff, appointed by Minister of Justice Kerensky Special High Commissioner for Revision and Investigation of the actions of Ministers and other High Personages of the Imperial Government.

"I was acting as Procureur of the Court of Assizes of Ekaterinoslav when I received orders from Minister of Justice Kerensky to become a member of the High Commission of Inquiry charged with an examination of the acts and abuses of ministers and other high personages of the former Government. While working with this Commission in Petrograd I was especially assigned to examination of sources of secret influences at Court which were known as Dark Forces. My work with the Commission lasted until August, 1917, when I was forced to leave because the President, Mourvavieff, insisted upon my making reports of a plainly prejud-icial character.

"As an Attorney General (*juge d'instruction*) I had access to all documents, and the right to be present at the examination of all witnesses, with the view of establishing impartially the part played by persons accused by society and the public press of exerting influence on foreign and domestic politics. I was assigned to read all the papers and letters found in the Winter Palace, the palace at Tsarskoe Selo, and at Peterhof, especially the personal correspondence of the Emperor and Empress, certain of the Grand Dukes, and also the correspondence seized in the course of examination of the house of Archbishop Varnava, also of Countess S. S. Ignatieff, Dr. Badmaeff, Voyeikoff, and Anna Viroubova, and also to the relations existing between the Imperial family and the German Imperial family. Being aware of the importance of my inquiry in throwing light on historical events preceding and following the Revolution, I made copies of all documents and letters, *dossiers*, and statements of witnesses. In leaving Petrograd I took with me all these copies, concealing them in my home in Ekaterinoslav, but it is probable that these documents were destroyed when the Bolsheviki raided my house. If by happy chance I find that they still exist I shall certainly publish them in full, without any comments of my own.

"In the meantime I consider it my duty to write a short account of the principal persons who were accused of being Dark Forces. I must, however, warn the reader that as I write from memory some details may escape my mind. When I went to Petrograd to begin my work with the High Commission I admit that I was influenced by all the pamphlets and newspaper articles on the subject of the Rasputine influence, and other rumors and gossip, and I began my work under the domination of preconceived prejudices. But careful and impartial investigation soon forced me to the opinion that these rumors and newspaper accounts were based on slender foundations.

"The most interesting person charged with exercising a malign influence on political affairs was Gregory Rasputine, therefore this person was the central figure of my investigations. The account of the surveillance under which he lived, up to the very day of his death, is of great importance. This surveillance was exercised by the ordinary as well as the secret police, special agents noting all his goings and comings, some of these agents being disguised as policemen or as servants. Everything concerning the movements of Rasputine was carefully recorded every day. If he left his house, even for an hour or two, the moment of his departure and his return was noted, and also every person he met on the road.

"The secret agents kept strict account of all people he met and of all who visited him. In cases where the names of these persons were not known their full descriptions were taken. After having read all papers and examined many witnesses I reached the conclusion that Rasputine was a person more complex and less comprehensible than had been previously represented. In studying his personality I naturally paid attention to the chronological order of circumstances which finally opened to the man the doors of the Tsar's palace, and I discovered that the first preliminary was his acquaintance with the well known, pious, and learned churchmen Bishops Theofan and Hermogen. I noted also that it was afterwards due to the influence of Rasputine that these two great pillars of the Orthodox Church fell into disfavor. He was the cause of the relegation of Hermogen to the Monastery of Saratoff, and of the disgrace (demotion) of Theofan, after these two archbishops, discovering Rasputine's low instincts, openly turned against him. All the evidence pointed to the conclusion that in the inner life of Rasputine, a simple peasant of the Government of Tobolsk, there occurred suddenly a complete change transforming him and turning him toward Christ. Only in this way can I explain to myself his intimacy with these two remarkable bishops. This hypothesis is moreover confirmed by Rasputine's story of his journey to the Holy Land. This book is marked by extreme naïveté, simplicity, and sincerity. On the recommendation of the exalted churchmen mentioned, Rasputine was received by the Grand Duchesses Anastasie Nicholaevna and Melitza Nicholaevna, and it was through them that he made the acquaintance of Mme. Viroubova, *née* Tanieff, then maid of honor. He made a deep impression on this very religiously inclined woman, and gained at last an entry to the Imperial Palace. It was then that awoke in him his worst instincts, hitherto repressed, and it was then that he began adroitly to exploit the religious fervor possessed by very high personages. It must be admitted that he played his part with astonishing cleverness. Correspondence bearing on the subject and the testimony of various witnesses prove that Rasputine refused all subsidies, gratuities, and even honors which were freely offered him by their Majesties, indicating thus his integrity, his disinterestedness, and his profound devotion to the Throne, insisting that he was an intercessor for the Imperial family before God's throne. He alleged that everyone envied him his position, that he was surrounded by intriguers and slanderers, and that therefore evil reports concerning him were unworthy of belief. The only favor he accepted was the rental of his lodgings, paid by the personal Chancellor of his Majesty. He also accepted presents made by the hands of the Imperial family, such as shirts, waist-bands, etc.

"Rasputine had free entry to the apartments of the Emperor, saying prayers, addressing the Emperor and Empress with the familiar 'thou,' and greeting them in the Siberian peasant manner (with a kiss). It is known that he warned the Emperor, 'My death shall be thine also,' and that at Court he was regarded as a man gifted with the power of forecasting events. His predictions were couched in mysterious phrases like those of the Pythons of antiquity.

"Rasputine's income was derived from numerous persons who desired positions and money, and used Rasputine as their intermediary with the Emperor. Rasputine asked favors for his clients, promising, if these were granted, all kinds of blessings to the

Imperial family and to Russia.

"To this must be added that Rasputine possessed within himself a strange power by which he was able to exercise hypnotic suggestion. I have been able to establish the fact that he cured by hypnotism the disease of St. Vitus Dance which afflicted the son of one of his friends, Simanovitch. The young man was a student in the College of Commerce, and his malady completely disappeared after two séances in which Rasputine plunged the patient into hypnotic slumbers.

"Another case establishing the hypnotic power of Rasputine may be noted. During the winter of 1914-15 he was called to the house of the superintendent of railways in Tsarskoe Selo where lay, entirely unconscious, Anna Alexandrovna Vироubova, who had been seriously injured in a railroad accident. She was suffering from broken legs and a fracture of the skull. Their Majesties were in the room when Rasputine arrived, and he, simply raising his arms, said to the unconscious woman: 'Anushka, open your eyes,' which she instantly did, looking intelligently around her. This naturally made a deep impression on everyone present, including their Majesties, and it served to increase the prestige of Rasputine. Although Rasputine could barely read and write, he was far from being an inferior person. He had a keen and observant intellect, and a rare faculty of reading the character of any person with whom he came in contact. The rudeness and exaggerated simplicity of his bearing, which lent him the appearance of a common peasant, served to remind observers of his humble origin and his lack of culture.

"As so much was bruited in the public press about the immorality of Rasputine, the closest attention was given to this phase of his question. From the reports of the secret police it was proved that his love affairs consisted solely in night orgies with music-hall singers and an occasional petitioner. It is on record that when he was drunk he sometimes hinted of intimacies in higher circles, especially in those circles through which he had risen to power, but of his relations with women of high society nothing was established, either by police records or by information acquired by the commission. In the papers of the Bishop Varnava was found a telegram from Rasputine as follows: 'My dear, I cannot come, my silly women are shedding tears and won't let me go.' As for the accusation that in Siberia Rasputine was accustomed to bathe in company with women, and that he was affiliated with the 'Khlysty' sect, the Extraordinary Commission referred these charges to Gramoglassoff, professor in the Ecclesiastical Academy (of Moscow), who after examination of all the evidence, testified that among peasants of many parts of Siberia the common bath was a usual custom, and that he found no evidence in the writings or preachings of Rasputine of any affiliation with the 'Khlysty' doctrines.

"Rasputine was a man of large heart. He kept open house, and his lodgings were always crowded with a curiously mixed company living at his expense. To acquire the aureole of a benefactor, to follow the precepts of the Gospels according to which the generous hand is always filled, Rasputine took the money offered by his petitioners, but he gave generously to the poor and to people of the lower classes who begged his assistance. Thus he built up a reputation of being at once a generous and a disinterested man. Besides these alms Rasputine spent large sums in restaurants, cafés, music halls, and in the streets, so that when he died he left practically nothing. The investigation disclosed an immense amount of evidence concerning the petitions carried by Rasputine to Court, but all these, as has been said, referred merely to applications for positions, favors, railway concessions, and the like. Notwithstanding his great influence at Court not a single indication of Rasputine's political activity was disclosed.

"Many proofs of his influence were found in the papers of General Voyeikoff, Commandant of the Palace, as for example the following: 'My dear, Arrange this affair. Gregory.' These letters were annotated by Voyeikoff, with the names and addresses of the petitioners, the nature of their demands, the results of their applications, and the date of the replies. Many letters of the same kind were found among the papers of President of the Council of Ministers, Sturmer, and of other high personages. All the letters concerned themselves exclusively with favors and protection for the people in whom Rasputine interested himself. He had special names for various persons with whom he was in frequent contact. Sturmer was called 'The Old Man,' Archbishop Varnava 'Butterfly,' the Emperor 'Papa,' and the Empress 'Mama.' The nickname of Varnava, 'Butterfly,' was found in a letter to Mme. Viroubova.

"The inquiry into the influence of Rasputine on the Imperial family was intensive, but it was definitely established that that influence had its source in the profound religious sentiments of their Majesties, joined to their conviction that Rasputine was a saint, and was the sole intermediary between God and the Emperor, as well as of all Russia. The Imperial family believed that they saw proofs of his sanctity in his psychic power over certain persons of the Court, such as bringing back to life and consciousness the desperately injured Mme. Viroubova, whose case has been described; also in his undoubtedly benign influence on the health of the heir, and on a whole series of fulfilled forecasting of events.

"It is evident that sly and unscrupulous people did everything in their power to profit by Rasputine's influence on the Imperial family, thus waking up in the man his worst instincts. This is particularly true of the former Minister of the Interior, A. N. Khvostoff and of Belezky, Director of the Police Department. To consolidate their position at Court they came to an understanding with Rasputine whereby they agreed to pay him, out of the private funds of the Police Department, the sum of three thousand rubles monthly, besides other sums, that he might require, provided he helped them to place candidates agreeable to them. Rasputine accepted these conditions, and for three months filled his engagements, but finding that the arrangement was not advantageous to himself, returned to his independent manner of work. Khvostoff, fearing that Rasputine would betray him, began openly to oppose him. He knew that he stood well with the Imperial family, and he counted also on the coöperation of the Duma, of which he was a member, and in which Rasputine was cordially hated. This put Belezky in a difficult position, because he doubted Khvostoff's power at Court, and he had no doubt at all concerning Rasputine's power. Belezky decided therefore to betray his chief, and range himself on the side of Rasputine. His object was, to use the words of Rasputine himself, to throw down the Khvostoff ministry. The struggle between these two officials culminated in the famous plot against the life of Rasputine, which created such a sensation in the press during the year 1916. The plot was laid by Belezky in the following manner. An engineer named Heine, owner of several private gambling houses in Petrograd, was hired to go to Christiania to meet the unfrocked monk Illiador Troufanoff, a former friend of Rasputine. The result of this journey was a series of telegrams addressed to Heine and signed by Illiador covertly alluding to a conspiracy against the life of Rasputine. In one of these telegrams it was stated that the forty men engaged in the conspiracy were dissatisfied to wait longer, and it was necessary to send them immediately thirty thousand rubles. These telegrams, coming in war time from a neutral country, were delivered to the police, only after having been read being passed on to the person addressed. Finally, after receiving all the telegrams, Heine presented himself to Rasputine in the guise of a repentant sinner, giving him full details of the plot, in which he owned himself concerned, but which he vowed Khvostoff to be the leading spirit. The result was that Rasputine took the story to the Imperial family, and the dismissal of Khvostoff quickly followed. It is an interesting fact that Heine's telegrams from Christiania mentioned a number of names of persons living in Tsaritzine, former friends of Illiador, who were supposed to be in Christiania busy with the details of the plot. The evidence given at the inquiry proved beyond doubt that the persons concerned had never left their homes.

"Personally the official Khvostoff was highly esteemed by both the Emperor and the Empress, they believing him to be sincerely religious, and devoted to the interests of the Imperial family and to Russia, but the evidence shows that he was really devoted only to his personal interests. He once invited the head of the Gendarmerie, General Komissaroff, to go with him in civilian dress, and to introduce Rasputine to the Metropolitan Pitirim. They were received by a novice who went to the Metropolitan's study to announce them. When the Metropolitan appeared Rasputine introduced General Komissaroff, and disagreeable as it was to see a gendarme officer in his house, his Eminence invited the men to follow him into his study. There they discovered Khvostoff sitting on a sofa. Seeing Rasputine Khvostoff laughed rather nervously, but continued his conversation with the Metropolitan, then, rising to take his departure, asked General Komissaroff to drive home with him. Komissaroff found himself in an awkward position, and when Khvostoff suddenly asked him if he understood the affair he answered in the negative. 'Well,' said Khvostoff, 'it is now clear in what relation Pitirim stands with Rasputine. When you were announced he was just telling me that he had nothing in common with Rasputine, and that the person who was waiting to see him was an eminent Georgian. "Permit me," he said, "to leave you for a few minutes." Now we see who the "eminent Georgian" really was.' This was testified to by Komissaroff himself.

"Of all the ministers Khvostoff was the closest to Rasputine. Rumors of the intimate relations between Sturmer and Rasputine were found to be without foundation. There was between them, it is true, a friendship. Sturmer understood Rasputine's great influence, and did what he could to advance the interests of his clients. He sent fruit, wine, and delicacies to Rasputine, but there is no evidence that he allowed him to influence political affairs. The relations between Rasputine and Protopopoff, who, for some reason, Rasputine called 'Kalinine' were no more intimate, although Protopopoff liked Rasputine, and it is certain that Rasputine defended Protopopoff when the position of the latter was menaced. This was done usually in the absence of the Sovereigns, Rasputine addressing himself to the Empress, at the same time uttering predictions.

"Protopopoff distinguished himself by an extraordinary lack of will power, representing at different times quite opposing organizations. He was even at one time elected vice-president of the Duma. Protopopoff has publicly been accused of initiating and carrying out an attempt to put down the popular uprising of the first days of the Revolution. He is accused of having placed machine guns on the roofs of houses to shoot down the armed insurgents. However, the *juge d'instruction* Jousvik-Kompaneitz, after having interrogated many witnesses, and examining all the machine guns found in the streets of Petrograd in the first days of the Revolution, has testified that all the machine guns belonged to different regiments, and none, not even those found on the roofs of houses, to the police. Generally speaking, there were no machine guns on roofs, except those placed there at the beginning of the war as a defense against airplane attacks. It must be said that during the critical days of February, 1917, Protopopoff showed a complete incapacity, and from the legal point of view, his absolutely criminal weakness. Among his papers were found intimate and even affectionate letters from Rasputine, but not one letter contained anything more than recommendations in favor of his protégés. Nor in the papers of any other high personages were found letters of different tenor signed by Rasputine. Both press and public seem to have been persuaded that Rasputine was very intimate with two political adventurers, Dr. Badmaeff and Prince Andronnikoff, and that through him these men were able to exercise wide political influence. Evidence has established, however, that these rumors were without any foundation. The two adventurers were, in fact, nothing more than the hangers-on of Rasputine, glad to gather up the crumbs from his table, and falsely representing to their clients that they had influence over Rasputine, and through him influence at Court."

(Here follows at some length the result of the High Commission's inquiry into the activities of Dr. Badmaeff and Prince Andronnikoff, but as they have nothing whatever to do with this history they are omitted. A. V.)

"Badmaeff was the physician of Minister Protopopoff, but the Imperial family had no confidence in his methods—any more than had Rasputine—and in an examination of the servants of the Imperial household, it was demonstrated clearly that the Thibetan doctor had never been called in his professional capacity to the apartments of the Emperor's children.

"General Voyeikoff, Commandant of the Palace, I examined many times in the Fortress of Petropavlosk where he was imprisoned. He did not play a very powerful rôle at Court, but according to letters from his wife, daughter of Court Minister Fredericks, covering the years 1914-15, and found in his house, he was esteemed by the Imperial family as a man devoted to the throne, an impression which I, after several interviews with him, did not share. From letters of Voyeikoff to his wife it is plain that he was hostile to Rasputine. In certain of the letters he calls Rasputine the evil genius of the Imperial family and of Russia, and he believed that his intimacy at Court discredited the throne and gave strength to humors and opinions and slanderous stories by which the anti-Government party profited. Nevertheless he took full advantage of the influence of Rasputine. He had not the courage to reject his petitions, which is proved by the annotations in his handwriting on the letters of Rasputine."

(High Commissioner Roudneff adds that, in his opinion, Voyeikoff thought badly of Rasputine, and that his wife hated the man, but that neither of them communicated their views to the Imperial family. A. V.)

"Having heard a great deal of the exceptional influence at Court of Mme. Viroubova, and of her relations with Rasputine, and having read and believed what was said about her in society and the press, I must admit that when I went to examine her in the Fortress of Petropavlosk I was frankly prejudiced against her. This hostility remained with me up to the moment of her entrance into the office of the Fortress under the escort of two soldiers. As she entered the room I was struck with the expression of her eyes, an expression of more than earthly gentleness and meekness. This first impression was confirmed in all my subsequent interviews with her. From the first conversation which I had with her I became convinced that, given her individuality and her character, she could never have had any influence on politics either foreign or domestic. I believe this in the first place because of the essentially feminine point of view shown by her on all political matters of which we talked, and in the second place because of her loquacity and her complete incapacity to keep secret even facts which might reflect on herself. I became convinced that to ask Mme. Viroubova to keep anything a secret was equivalent to proclaiming it from the housetops, because anything that she thought important she felt impelled to communicate, not only to friends but to possible foes. Noting these two characteristics of Mme. Viroubova, I asked myself two questions—why she stood in close relations with Rasputine, and what was the secret of her intimacy with the Imperial family.

"I found the answer to the first question in conversations with the parents of Mme. Viroubova, M. Tanieff, chief of the private Chancellory of his Majesty, and his wife, *née* Countess Tolstoy. From them I learned of an episode in the life of their daughter which, in my opinion, explained why Rasputine obtained later such an influence over the will of the young woman. At the age of thirteen Mme. Viroubova fell gravely ill of typhus, the illness being complicated with peritonitis, and her condition, according to the physicians, was desperate. Her parents called to her bedside the famous priest, Father John of Kronstadt. Following his prayers the illness took a favorable turn, and the young girl was soon pronounced out of danger. This made a deep impression on her mind, and thereafter strongly inclined her to a religious life.

"Mme. Viroubova first met Rasputine in the house of the Grand Duchess Melitza Nicholaevna (wife of Grand Duke Peter), and that meeting was not a happy event. The Grand Duchess had prepared Mme. Viroubova for the meeting by conversations on the subject of religion, and had given her certain French books on occult subjects. Later the Grand Duchess invited Mme. Viroubova to her house, promising to introduce her to a great intercessor before God in favor of Russia, a man who possessed gifts of prophecy, and the faculty of curing the sick. This interview by Mme. Viroubova, then Mlle. Tanieff, made a great impression on the young woman who was then on the eve of marriage with Lieutenant Viroubova. Rasputine spoke only on religious subjects, and when the young girl asked him if he approved her marriage he answered allegorically saying that the pathway of life was strewn not only with roses but with thorns, and that man progressed towards perfection only through sufferings and trials.

"The marriage of Mme. Viroubova was from the first unhappy. According to the testimony of Mme. Tanieff, the man was completely impotent, addicted to perverted practices and saddistic habits, causing her daughter the most frightful moral sufferings and physical disgust. Nevertheless, believing in the Biblical injunction 'Whom God hath joined let no man put asunder,' Mme. Viroubova for a time kept her sufferings a secret even from her parents, and only after she had been nearly killed by her husband did she reveal to them the tragedy of her marriage. The result was, of course, a divorce. The testimony of Mme. Tanieff concerning the moral character of her son-in-law was confirmed by a medical examination of Mme. Viroubova, ordered by the Commission of Inquiry, and by which was established the virginity of the young woman. This examination was held in May, 1917. In consequence of her shocking marital experience the religious inclinations of Mme. Viroubova were increased and were developed into something approaching religious mania. She became the purest and most sincere admirer of Rasputine, who, up to the last day of his life, she considered a holy man, and one completely disinterested from every worldly point of view.

"In regard to the question of the intimacy of Mme. Viroubova with the Imperial family, I concluded that it had its roots in the wholly different mentalities of the Empress and Mme. Viroubova, that attraction of opposites which so often seems necessary to complete a balance. The two women were entirely different, and yet they had many things in common. Both, for example, were devotedly fond of music, and as the Empress possessed an agreeable contralto voice and Mme. Viroubova a good soprano, they occupied many leisure hours singing duets.

"Such were the conditions which produced in the minds of persons ignorant of the nature of the intimacy between the Empress and Mme. Viroubova, belief in the exceptional influence of Mme. Viroubova on Court affairs. As has been said, Mme. Viroubova possessed no such influence, nor could she have possessed it. The Empress dominated the intelligence and the will of Mme. Viroubova, but the attachment between the two women was very strong. The religious instincts deeply rooted in their two natures explains the tragedy of their veneration of Rasputine. The relations between the

Empress and Mme. Viroubova could be likened to those of a mother and daughter, nothing more.

"My opinions regarding the moral qualities of Mme. Viroubova, resulting from interviews with her in the Fortress of Petropavlosk and in the Winter Palace were entirely confirmed by the forgiving and Christian spirit displayed by her towards those who had caused her, in the course of her imprisonment, the most horrible suffering. Of the insults and tortures to which she was subjected in the Fortress I did not learn, in the first instance, from Mme. Viroubova herself, but from her mother. Only on direct examination did Mme. Viroubova confirm her mother's testimony, and even then she spoke calmly and with astonishing meekness, saying that her persecutors should not be blamed too severely because they did not realize what they were doing. These tortures of the prison guards, such as spitting in her face, dealing her blows on the head and body, accusing her of being the mistress of the Emperor and of Rasputine, tearing off her clothes and threatening to murder a sick woman who could walk only with the aid of crutches, caused the Commission of Inquiry to transfer the prisoner to a house formerly occupied by the Director of the Gendarmerie (House of Detention). The testimony of Mme. Viroubova presented a complete contrast to that of Prince Andronnikoff. Her statements were all candid and sincere, and their truth was subsequently established beyond doubt by documentary evidence. The only fault I found with Mme. Viroubova was her tendency to wordiness, and her amazing habit of skipping from one subject to another, without regard to the fact that she might be hurting her own cause. Mme. Viroubova appears to have interceded at Court for various persons, but her petitions were received with a certain distrust because of her known goodness and her simplicity of mind.

"The character of the Empress Alexandra was shown clearly in her correspondence with the Emperor and with Mme. Viroubova. This correspondence, in French and English, is filled with sentiments of affection for her husband and children. The Empress occupied herself personally with the education of her children, and she often indicates in her letters that it is desirable not to spoil them or to give them habits of luxury. The correspondence reveals also the deep piety of the Empress. In her letters to her husband she often describes her emotions during religious services, and speaks of the peace and tranquillity of her soul after prayer. Hardly ever, in the course of this long correspondence, are any allusions made to politics. The letters concern intimate and family affairs only. In passages in which Rasputine is mentioned she speaks of him as 'that holy man,' and shows that she considers him one sent of God, a prophet, and a man who prays sincerely for the Imperial family. Through the whole correspondence, which covers a period of ten years, I found not one single letter written in German. According to the testimony of Court adherents I have proof that before the War German was never spoken at Court. Because of public rumors of the sympathy of the Empress for Germany and of the existence in the Palace at Tsarskoe Selo of private wires to Berlin, I made a careful examination of the apartments of the Imperial family, and I found no indications at all of communications between the Imperial household of Russia and the Imperial household of Germany. I also examined the rumors concerning the beneficence of the Empress towards the German wounded and prisoners of War, and I found that the Empress showed compassion for the sufferings of Germans and Russians alike, without distinction, desiring to fulfill the injunction of Christ who said that whoever visited the sick and suffering also visited Himself.

"For these reasons, and above all on account of the frail health of the Empress, who suffered from a disease of the heart, the Imperial family led a very retired life, which favored the development, especially in the Empress, of extreme piety. Inspired by her devotion the Empress introduced into certain churches attached to the Court a régime of monastic services, and followed with delight, in spite of her ill health, up to the very end, masses which lasted for hours on end. This same excessive religious zeal was the foundation for her admiration for Gregory Rasputine, who, possessing an extraordinary power of suggestion, exercised an undeniably salutary effect on the invalid Tsarevitch. Because of her extreme piety the Empress was in no proper state of mind to understand the real source of the amazing influence of Rasputine on the health of the Heir, and she believed the explanation to be due, not at all to hypnotism, but to the celestial gifts which Rasputine owed to the sanctity of his life.

"A year and a half before the Revolution of 1917, the former monk, Illiador Troufanoff, sent his wife from Christiania to Petrograd with the proposal that the Imperial family purchase the manuscript of his book, which later appeared under the title of 'The Holy Devil,' in which the relations of the Imperial family with Rasputine were scandalously represented. The Police Department interested itself in the matter, and at its own imminent risk entered into negotiations with the wife of Illiador concerning the purchase of the manuscript for which Illiador demanded, I am assured, sixty thousand rubles. The affair was finally submitted to the Empress Alexandra who repudiated with indignation the vile proposition of Illiador, saying that 'white could never be made black, and that an innocent person could never be assoiled.'

"In terminating this inquiry I believe it necessary to repeat that Bishops Theofan and Hermogen contributed importantly to the introduction of Rasputine at Court. It was because of their recommendations that the Empress, in the beginning, received Rasputine cordially and confidently. Her sentiments towards him were fortified only by the reasons indicated in the course of this document."

APPENDIX B

Copy of certificate of acquittal of Anna Viroubova issued by the High Commission of Inquiry, August, 1917.

Ministry of Justice

The High Commission of Inquiry into the acts and abuses of Ministers and other High Personages of the Former Government.
25th of August, 1917.
No. 3285
Petrograd
Winter Palace
Tel. 1-38-20 and 186.

(Seal)

Testimonial

This testimonial delivered to Anna Alexandrovna Viroubova at the end of the investigation of the High Commission of Inquiry, certifies that she was found not guilty and that she will not again be called to judgment. This statement is given under the signature and seal of the President of the High Commission.

(*Signed*) N. MOURVAVIEFF.

THE LAST PHOTOGRAPH TAKEN OF THE EMPRESS AND HER
DAUGHTERS, OLGA AND TATIANA, 1918

END OF BOOK THREE

LIST OF ILLUSTRATIONS

All the images used for this volume are rights free, i.e. in the public domain, and were sourced from either Wikimedia or Yandex LLC.

Empress Alexandra Feodorovna with Alexander Leonide, 1909. Lili Dehn's son born 9 August 1908, affectionately called 'Titi'.

PAGE

Cover C. Radziwill; L. Dehn; A. Viroubova
17 Book One cover and author Catherine Radziwill
20 Book I frontispiece portrait of Alexandra Feodorovna
42 Shoulder portrait of Princess Alix of Hesse and by Rhine
46 Shoulder portrait of Dowager Maria Feodorovna
66 Shoulder portrait of Grand Duchess Marie Pawlowna
73 Shoulder portrait of the Czar Nicholas II
94 Shoulder portrait of Grand Duke Nicholas Nikolaevich
105 General Orloff and Anna Viroubova with Czar on Standart

LIST OF NAMES

128 The Palace of Tsarskoye Selo (i.e. Catherine Palace)
151 Portrait of Raspoutine
157 Portrait of Peter Badmaieff
160 Full portrait in uniform of Cesarewitsch Alexei
170 Shoulder portrait of Alexander Protopopoff
176 Portrait of Mikhail Rodzianlo
182 Poster for La Grande Duchesse De Gérolstein
186 Royal Family portrait
193 Book Two shoulder portrait of author Lili Dehn
207 Portrait of Alexandra Feodorovna and son Alexei
217 Empress Alexandra Feodorovna doing needle work
219 Inset: Anna Viroubova
219 Standing portrait of Anna Viroubova
250 H.I.M. Nicholas II on Standart portrait with officers
250 Empress Alexandra Feodorovna aboard Standart
250 The Emperor with Tsarevitch on Tender
285 Thumbnail portrait of royal family
290 Shoulder portrait of Doctor Direvanko
291 Emperor out in the field with Lieutenant Dehn
291 The Tsarevitch at attention at G.H.Q.
291 Tsarevitch and his dog 'Joy'
308 The Emperor with Tsarevitch Alexei on the balcony
308 Empress Alexandra Feodorovna sitting with flowers
322 Drawing of Empress's note
339 Drawing of part of Empress's letter
340 Drawing of Empress's letter to Lily Dehn
345 Drawing of part of letter 30th July 1917
346 Drawing of Christmas card made by the Empress
350 Drawing of letter from Empress to Lily Dehn
354 Book Three full portrait of Anna Viroubova in habit
355 Sitting portrait of Empress Alexandra Feodorovna
364 Empress Alexandra on pony chaise
364 Empress and Tatiana in bed
365 Side portrait of Alexander Tannieff
382 Emperor and Empress relaxing on deck of Standart
382 Empress on deck of Standart handing gifts to sailors
393 Royal Family on jetty alongside Standart, in Yalta
394 Nicholas II, with Mdm Viroubova walking in Hamburg

ILLUSTRATIONS

407 Shoulder portrait of Mlle Sofia Tutcheff
412 Emperor and Tsarevitch in naval uniforms on Standart
413 Tsarevich Alexei with his minder Dr. Derevanko
413 The Emperor, Empress and Tsarevich in the gardens
427 Emperor and Empress in Slavic dress at ball of 1903
428 A sickly Empress on the Peterhof balcony
471 Rasputin's three children
472 Rasputin's guestroom
481 Empress with Grand Duke Dmitri
482 Empress Alexandra taking tea in the woods of Finland
534 The four Grand Duchesses (Princesses)
535 Shoulder portrait of Anna Viroubova
571 Drawing of last letter from Empress to Anna Viroubova
572 Drawing Smuggle letter from Empress to Anna Viroubova
625 Last known photograph of Empress and daughters, 1918
626 The Empress holding young Tatiana, 1909
628 Princess Victoria Melita

PRINCESS VICTORIA MELITA
Of Saxe-Coburg and Gotha

INDEX OF NAMES

A

Alexander II, 84, 177, 201, 289, 359
Alexander III, 4, 6, 23, 25-31, 34-35, 44, 51-53, 59, 61, 68, 134, 141, 177, 179, 372-373, 386, 406, 410, 419, 437, 488-489, 559
Alexandra, Feodorovna, *passim*
Alexis Michaelovitch, (Tsar), 363
Anastasie, Grand Duchess (aka Anastasia), 211, 223, 238-239, 242-243, 252, 282, 284-285, 287-289, 300, 305, 316, 343, 373, 411-412, 437, 496, 498, 505-506, 560, 562, 564, 567, 569, 571, 613
Antoinette, Marie, 5, 9, 23, 45, 47, 91, 245
Antonoff, 539-541
Appraxin (aka Apraxine), Count, 290
Apraxin, Countess Baida, 438, 567

B

Badmaieff (aka Badmieff), 156-157, 506, 538, 542, 544
Bariatinsky, Princess, 68, 228, 240, 583
Belaieff, (Minister of War), 479, 533, 567
Beletsky, General, 257
Benckendorff, Count, 284, 287, 295, 304, 317-318, 324
Berchik, 536, 550, 553, 594
Birileff, Admiral, 370
Botkin (aka Botkine), Dr. 234, 289, 344, 385, 499
Bouvier, Caroline Lee, 6
Brassoff, Countess, 423
Brisac, Mme., 414
Buchanan, Sir George (Ambassador), 494
Buxhoevden, Baroness (aka Buxhoeveden), 503, 558, 566, 571

C

Catherine the Great, 45, 127, 153, 165, 215, 249, 366, 398, 426
Chagin, Admiral, 370, 423-424
Cheidze, Deputy, 524, 545-546
Chemoduroff, 410, 566, 575

Cherevin, Lieutenant-General Peter Alexander Aleksandrovich, 4
Chkoni, 515, 518-519, 530
Clementine, Princess (of Coburg), 271

D

Dehn Charles, 211, 213-214, 221, 270, 273, 277, 291
Dehn, Madame – 5, 9-11, 14-15, 222, 286, 298, 325, 334-337, 341, 467, 478, 492, 494-495, 497-501, 503, 506-508, 510, 523, 557, 567, 569, 575
Demenkoff, Nikolai, 569
Demidoff, Anna (aka Niouta), 410, 576
Derevanko (aka Direvenko), Kolia, 00
Direvenko (aka Derevanko), Dr, 289-290, 378, 415-416, 422, 472, 506, 583
Djounkovsky, General (Governor of Moscow), 435-436
Dobiasgin, 556
Dolgorouky, Prince, 190, 500, 503, 562
Dolgorouky, Princess, 201,
Dorr, Rheta Childe, 552
Duvan, M., 448

E

Edward VII, 32, 246
Elidor, 249, 253, 258, 264
Elizabeth Feodorovna, 360, 362, 369, 391, 401, 435-436, 448, 575, 588
Etiquette, Louise of Battenberg (see Princess Louise)

F

Fedoroff, Professor, 313, 422, 472
Feodorovna, Tsaritsa Alexandra (aka Alix of Hesse and by Rhine) – See Alexandra Feodorovna
Feodorovna, GD Elizabeth (aka GD Serge) – See Elizabeth Feodorovna
Feodorovna, Maria (aka Marie Feodorowna) – See Marie Dowager-Empress
Feodorovnitch, Michail (First Tsar), 427

INDEX

Ferdinand, King (of Bulgaria), 129-130, 271
Fredericks, Count (Minister of the Court), 221-222

G

Gagentorn, Dr., 441
Galitzyne (aka Golitzin), Princess, 35, 151, 212-213
Gapone, (priest), 108
Gardieve, Mme., 435
Gedroiz, Dr., 433, 440-441
Gendrinkoff, Countess, 296
General:
 Aléxieieff (aka Alexieff), 167, 175, 181, 453-454, 493, 501
 Belaieff, 479, 533, 567
 Bogdanovitch, 402
 Gourko, 278, 493, 539-540
 Groten, 503, 538, 562, 566, 574
 Ivanoff, 453-454, 486
 Januchevitch, 144
 Komissaroff, 617
 Korniloff, 181, 187, 304, 546
 Mannerheim, 423
 Odoevsky, 436
 Poole, 541
 Polivanoff, 486
 Racine, 435-436, 440, 457, 496
 Rikkel, 452, 454
 Tcherewine, 26, 30
 Williams, 452
 Woyeikoff, 173, 175
George V, (of Great Britain) 221, 243, 315-316, 494
Germogen (see Hermogen)
Gibbs, Mr. (English tutor), 241, 300, 408, 503, 537, 555, 562, 572, 574, 576, 587
Gilliard, M. Pierre (French tutor), 14, 238, 241, 300, 408-409, 423, 442-443, 454, 495, 503, 537, 558, 560, 580, 584, 587
Golovina, Mary, 264
Gorémykine, Count (Prime Minister), 147
Gorky, (novelist), 551-552, 572, 576
Goutchkoff, M. (Minister of War), 302-303, 313, 448, 453, 486, 492, 501, 510
Gramoglassoff, professor, 614
Grancy, Mme (hofmistress of the Court of Hesse), 375, 395

Grand Duchesses:
 Anastasia Nicholaiewna (see Anastasie)
 Elizabeth (Elizabeth Feodorovna) of Gerolstein, 182
 Marie Alexandrowna of Coburg, 27
 Olga of Oldenburg, 80
 Olga Nicholaiewna, 42
 Marie (aka Maria) Nicholaiewna, *passim*
 Marie Paul, 315
 Melitza Nicholaevna, 389, 460, 463, 477, 613, 620
 Tatiana Nicholaiewna, 237, 239-240, 282, 288, 323, 325, 348, 360, 364, 388-389, 411, 414, 420, 433, 482, 495, 498, 561, 564, 567, 571, 573-576, 578, 584
 Xenia Alexandrowna, 37, 80, 371-372, 470, 483
Grand Duke:
 Alexander Michaylovitch, 80, 96
 Alexis, Tsarevich, 100-101, 103, 114, 117, 126, 155, 157, 166, 242, 282, 313
 Cyril, Vladimirovitch, 83-85, 216, 292, 494, 497
 Dmitri Pavlovitch, 379, 456, 479, 481-483
 Ernest, of Hesse, 391, 395, 406
 George Alexandrovich, 386, 389, 552
 George of Leuchtenberg, 40, 70
 Michael, 40, 53, 85, 125, 180, 305
 Nicholas (Nicholai) Nicholaievitch, 71-73, 76-77, 79, 87, 139, 142-145, 442-444, 455-456, 483, 486, 501
 Nikolai Michailovich, 482, 552-553
 Louis, (aka Prince Louis of Battenberg), 55, 445
 Paul, Alexandrowitch, 81-81, 84, 172, 180, 264, 282-283, 286-287, 293, 302-303, 379, 418, 455, 496-497, 552
 Sergius, 39-40, 44, 49, 57, 97
Grotten (aka Groten), Colonel, 276, 278, 282, 284, 503, 538, 562, 566, 574

H

Hélène of Orléans, Princess, 27
Hendrikoff, Countess, 503
Hermogen, Bishop (aka Germogen, Germogene, Gerogene), 249, 253, 564, 572, 612, 624
Hitrowo, Rita, 293

INDEX

Hohenlohe, Princess, (wife of Imperial Chancellor), 41
Horvat,Colonel, 197-198, 208
Hvostchinsky, Captain, 282

I

Ignatieff, Countess Sophy S., 111, 611
Illiador, 473-474, 616-617, 623-624
Irene, Princess, (of Prussia -formerly of Hesse, aka Princess Henry of Prussia), 135, 366, 368-369, 373, 395, 416, 422

J

Janchevsky, Glinka, 544
John, Father (of Kronstadt), 362, 560, 599, 602, 607, 620
Joseph, Francis, 50, 132

K

Kapnist, Count, 276
Karloff, Count, 379
Katoff, 410
Keller, Count, 313, 346, 348
Kennedy, Jaqueline, 7
Kerensky, Alexander Feodorovitch, 322-325, 327-329, 332-335, 337-338, 342, 352, 461, 475, 488, 499, 507-510, 517, 521, 524, 527, 531-532, 537, 540, 542, 544-550, 611
Kerr, Miss, 395
Khvostoff, A. N. (Minister of the Interior), 486, 616-617
Kitchener, Lord, 225
Kobilinsky, Commandant, 339-339, 341-342, 344
Kolantai, Mme. (Commisar), 549
Kolémine, Madame, 55-57, 60
Kolichko, (writer), 521
Korinsky, M., 535
Kontaisoff, Count, 359
Korovichenko (aka Korovitchinko), Colonel, 324
Kostritzky, Dr., 555
Kotzebue (aka Kotzebou), Colonel Paul, 278, 311, 315, 317-318, 320, 321, 504-506
Kotzebue-Pilar, Countess, 278

Koutousoff, Prince (Field Marshal), Goleniktcheff, 198, 359
Kouzmine, Lieutenant, 292
Krivosheim, (Minister of Agriculture), 442
Krzesinska, Mlle (Polish dancer), 28, 67, 69-70, 88

L

Laptinsky, Akilina, 262, 267
Lamsdorff, Count, 75
Leopold, (QV's youngest son), 366
Lenin (aka Lenine), 199, 352, 446, 531, 547, 552, 610
Lichatchieff, Mr, 439-440
Linavitch, 276, 283
Linevitch, Lieutenant, 490-491, 503, 562, 574
Lounacharsky, 547
Louis XVI, 5, 45, 47, 174, 179, 189, 245
Lvoff, Prince, 335, 437, 525

M

Makari, (priest), 346, 438, 467, 557
Makharoff, Admiral, 96
Malzoff, Mme., 406
Manassavitch-Maniuloff, GD Wladimir, 86, 118, 121-122, 124, 131, 138, 142, 147-148, 150, 162
Manouchine, Dr., 527-529, 546, 549, 592
Manouiloff, Detective, 520, 538, 543
Margaret of Prussia, Princess, 27
Marie, Dowager-Empress, (aka Maria Feodorovna, Feodorowna) 12, 34-35, 37-38, 45-46, 62-65, 102
Markoff, Lieutenant, 297
Mary, Queen (of England), 217, 232, 260
Melita, Princess Victoria, (of Saxe-Coburg and Gotha) 628
Menabde, (dancer), 599
Miasocdoff-Ivanof, 292
Miliukoff, Professor, 150, 185
Mourawieff (aka Mouravieff), Judge (Minister for Foreign Affairs), 75, 190

N

Nagorini, 575
Narishkin, Mme. Alexandra, 437

Narishkine, Zizi, 437, 566
Nicholai Michailovitch, 455
Nicholas II, 442, 456, 458-459, 485, 488-489, 494, 501, 554, 587
Nissky, Bishop, 570
Nosstiz, Count, 430

O

Olga, Grand Duchess (Alexandrovna), 81
Olga, Grand Duchess (Nicholaiewna), 42, 236-237, 239, 242, 281-282, 285, 288-289, 306, 323, 325-326, 345, 348, 360, 386, 388-389, 392, 394, 411, 414, 418, 420, 433, 435, 457, 490, 495, 498, 555, 560, 562, 567, 573, 575, 586-587, 625
Olga, Grand Duchess (of Oldenburg), 80
Orbelliany, Princess, 299
Orchard, Miss, 324, 384
Orianda, 209, 210
Orloff, General (aka Colonel) Alexander, 90-93, 95-106, 114, 116-118, 121, 127, 137, 144, 156, 178, 188, 218, 252, 258, 378-379, 383
Orloff, Princess Olga, 258

P

Paul I, 87, 276, 289, 359
Pavlovitch, Alexander, 566
Pavlovitch (Pawlowitch), GD Dmitry, 155, 456, 479
Pawlowna, Princess Marie (wife of GD Wladimir – aunt to Emperor), 31, 63-64, 66
Peretz, Colonel, 510-512
Pereverzeff, Judge, 524
Peter the Great, 152, 158, 182, 425, 467, 485
Petroff, M., 408, 454, 537, 567
Petrovskaia, Shoura, 576
Philippe, M., 72, 74-80, 119, 459-460, 465
Pistolkors, Madame Allie, 82, 84, 282-283
Pitirim, Archbishop, 163, 487-488, 524, 617
Pleve, von (aka Plehwe) (Minister of the Interior), 97, 118, 484
Pobedonostsev, Konstantin Petrovich, 4
Poincaré, President (of France), 131, 429
Polouboiarenoff, Mme., 511

Potsdam, Lord of, 41, 109
Pourtalès, Count, (German Ambassador to Russia), 131
Prince:
  Andrew, Prince (of Greece), 395
  Andronnikoff, 618-619, 622
  Boris, 418, 578
  Carol (of Rumania), 418
  Dolgoroukoff, 185
  George, Prince (of Greece), 395
  George Shrinsky-Shihmatoff, 341
  Lobanoff, 50
  Radolin, 41
  Stanislaw Albrecht Radziwill, 6
  Troubetskoy, Volodia, 190, 510, 512, 515, 518, 529, 567
Princess:
  Alice of Great Britain, 5, 51, 57-58, 69, 234, 372, 395
  Bariatinsky, Marie, 68, 228, 240, 583
  Eleanor of Sohmslich, 372, 375, 391, 395
  Kakouatoff, 594
  Louise, (of Battenberg - daughter of Victoria Battenberg), 375, 395
  Marie (of Greece), 389
  Oblensky, 372, 402, 406
  Orbeliani, 12, 234, 367, 368-369
  Scnkovsky, 369
  Youriewsky, 38
Protopopoff, Alexander Dmitrievich, (Minister of the Interior), 86, 148-150, 155, 159, 162-163, 165-170, 172-173, 263, 266, 275, 279, 455, 478-480, 485, 491, 567, 590, 618-619
Puchkine, Miss (Head Nurse), 404
Puritchkevitch, M., 478
Putiata, Bishop Vladimir, 401-402

Q

Queen Louise of Denmark, (formerly Louise of Hesse-Kassel), 51
Queen Alexandra of Great Britain, 34, 316, 419

R

Rabindar, Maria (Marie) 276
Rasputin (aka Raspoutine), 5, 10-16, 212, 218,

INDEX

225, 233, 237-238, 248-249, 251-273, 275, 329, 331-332, 352, 354, 359, 379, 401-402, 416, 422, 430, 441-444, 448, 450, 453, 455, 457-488, 490, 516, 524, 549, 579, 590, 596, 612-624
Raswosoff, Admiral, 343
Ratief, Prince, 297, 410, 578
Rauchfuss, Dr., 404, 422
Resin, General, 282, 292-293, 297
Retief, Prince, 302
Rheta, (American writer), 552
Riman, Mme., 511
Ripe, Miss, 208
Rodzianko (aka Rodziansko), M., 171-172, 176, 296, 314, 455, 485, 492, 497, 525
Rodzianko, Mme., 450
Rostovseff, Count, (Empress's secretary), 396, 446, 468
Roudneff, Judge, 13, 475-476, 521-522, 524, 528, 535, 611, 619
Rousky, General, 313-314

S

Sablin, M., 281, 286-287, 433
Sandro, (husband to GD Xenia), 483, 491
Sazonoff, Mr (Foreign Minister), 138, 156, 429
Schabelsky, (priest), 141
Schneider, Mlle, 228, 286, 370, 408, 503, 508, 571
Schweinitz, General von, 30
Sednoff, 573
Shiff, Mme., 539
Shoulenbourg, Colonel, 405
Shoulgine, Mme., 313, 592
Shrinsky-Shihmatoff, Prince, 341
Sokoloff, M., 546
Soukhomlinoff, General, 274, 486-487, 576, 580
Soukhomlinoff, Mme., 487, 511, 515-516, 576 520-521, 530
Stackelberg, Baron, 313-314
Stolypine, Mr, 123-125, 416, 460, 473
Stopford, A., 315
Sturmer, Mr, 86, 137, 140-142, 144, 147-148-150, 155, 159, 162-163, 165-166, 168-169, 173, 446, 484, 615, 617

T

Tanieff, Madame (aka Tanieva - later Viroubova), (see Virouboff, Madame Anna)
Tanieff (aka Tanieiew), General Alexander Sergievitch, 85, 282, 359, 365, 620
Tanchevsky, 538
Tcherbatkoff, M., 486
Theofan, Bishop, 459, 612, 624
Tikhon, Patriarch, 402
Timasheff, Mme., 450
Tolstoaia, Zinaida, 567, 570
Tolstoy, Count (General), 359, 370,
Tolstoy, Countess, 583, 620
Toutelberg, Louise, 231
Trepoff, Mr (Head of Cabinet), 150
Tretskaia, Mme, 377
Trotzky, Leon, 446, 547-549
Troup, 575
Tutcheff, Mlle Sophie, 14, 237-238, 401-402, 407, 436

V

Varnava, Bishop, 259, 611, 614-615
Vasiltchikoff, Princess, 276
Vassilchikoff, Mme. (Lady in waiting), 448, 457
Vechniakoff, Marie (Head Nurse), 410, 415, 423
Velepolsky, Count, 421
Victoria, Princess (of Battenberg), 40, 135, 374-375, 395
Victoria, Queen, 5-6, 29, 32, 51, 54-57, 135, 197, 217, 327, 234-235, 249, 306, 352, 366, 372
Virouboff, Alexander, 218, 251-252, 378
Virouboff, Madame Anna, (aka Wyrubewa, Viroubova), *passim*
Volkoff (aka Wolkoff), 294, 300, 302-303, 309-310
Voronzoff (Voronzeff), Countess, 372

W

Wilhelm II, Kaiser, (aka William II), 6, 27, 29-30, 40-42, 50, 59, 130-131, 133, 135-138, 146-148, 150, 162, 165, 169, 184-185
Wissky, Archbishop, 558

633

INDEX

Witte, Count Alexander, 109, 147, 371, 376, 484
Worontzoff-Daschkoff, Count, 145

Y

Yousopoff (yusupoff), Prince Felix, 201, 264, 269
Yousopoff (yusupoff), Princess Irene, 477-479

Z

Zadonsky, Tichin, 558

BOOK INFORMATION

65 Illustrations

Word Count

Introduction: 5,734
Book I: 63,854
Book II: 62,095
Book III: 115,098
Total word count: 251,831

Comparisons: Ulysses   (265, 222)
               Pride and Prejudice (122,685)

Volume II editions
Black & White images

Ebook   ISBN: 978-1-80517-208-6
Paperback ISBN: 978-1-80517-078-5 (cream paper, matt cover)
Hardback   ISBN: 978-1-80517-207-9 (cream paper, gloss cover)
Hardback   ISBN: 978-1-80517-079-2 (groundwood paper with dust jacket)

Series editions

Nicholas II – Tsar to Saint
The Memories of a Russian Yesteryear Volume 1
The Memories of a Russian Yesteryear Volume 2
The Memories of a Russian Yesteryear Volume 3

OTHER BOOKS BY THIS AUTHOR

NICHOLAS II - TSAR TO SAINT
THE RULER THAT LOST A DYNASTY

ebook
ISBN: 978-1-80352-908-0

Paperback
B&W on Groundwood paper
ISBN: 978-1-80352-799-4

Paperback
B&W on Cream paper
ISBN: 979-8-85465-243-8

Hardback
B&W on Groundwood paper
with dust jacket
ISBN: 978-1-80352-911-0

Hardback
B&W on Cream paper
ISBN: 979-8-85221-513-0

Pages: 546 / Illustrations: 203

From the back Cover:

Why was Russia's Imperial Family murdered? What part did they play in the revolution that gave rise to the Bolsheviks and transformed Russian Orthodoxy and State in to the Soviet Union? Tony Abbott takes us back to the pivotal moment in 1894 when Nicholas Alexandrovich Romanov became Tsar and there existed a real opportunity for Russia to modernise and become an industrial power and world leader. This is an incredible story of two rulers very much in love, who unwittingly created the circumstances that collapsed their dynasty.

Through first hand commentaries and with over 200 images, the events around the twenty-three year reign of Nicholas II are presented with clear and acute observations. What emerges is a fascinating insight of the man who made history by helping to bring about the revolution of an empire.

Printed in Great Britain
by Amazon

Cook on Costs
2014

Simon Middleton is a District Judge and Regional Costs Judge. He has sat on both the Midland and Western regions. Prior to judicial appointment he was a solicitor in private practice for 17 years and had higher court rights of advocacy (civil).

Simon has been a member of the Judicial College tutor team for 6 years and is now one of the Course Directors responsible for the delivery of civil law education to the judiciary. His primary responsibilities in these roles have been related to the teaching of costs and costs/case management. He was one of the small cadre of judges charged with the development and delivery of judicial training on the 'Jackson' reforms in the run up to implementation.

Simon has been interested in costs throughout his career and has lectured and written on the topic of costs extensively over the years. He would say that he never had a chance as he was an articled clerk at Messrs Ward Bowie when the senior partner of that firm was none other than Michael Cook. He is delighted to have the opportunity to maintain the 'Ward Bowie' link with this text.

Jason Rowley is a Costs Judge, or Taxing Master, at the Senior Courts Costs Office, having previously been a Deputy Master since 2006. He has been involved in costs matters since 1993 when he attended a conference chaired by Michael Cook and which left him with a desire to know far more about the world of costs.

Jason was the Forum of Insurance Lawyers' inaugural Chair of its Special Interest Group on costs and intervened on FOIL's behalf in cases such as *Callery v Gray* before running test cases on behalf of clients involving Claims Direct and The Accident Group. This led to stints on the Law Society's Civil Litigation Committee, including drafting the 2005 model CFA, as well as participation in the 'Big Tents' convened to broker industry agreements on success fees.

Jason started as a solicitor in an insurance practice dealing mainly with personal injury claims. He was a partner for 11 years, the last 5 of which were as Managing Partner. Thereafter he was the CEO of a barristers' chambers in the Inner Temple and a senior underwriting manager at a legal expenses insurance company.

Cook on Costs 2014

A guide to legal remuneration in civil contentious and non-contentious business

Simon Middleton
Jason Rowley

Members of the LexisNexis Group worldwide

United Kingdom	LexisNexis Butterworths, a Division of Reed Elsevier (UK) Ltd, Lexis House, 30 Farringdon Street London, EC4A 4HH, and London House, 20–22 East London Street, Edinburgh EH7 4BQ
Australia	LexisNexis Butterworths, Chatswood, New South Wales
Austria	LexisNexis Verlag ARD Orac GmbH & Co KG, Vienna
Benelux	LexisNexis Benelux, Amsterdam
Canada	LexisNexis Butterworths, Markham, Ontario
China	LexisNexis China, Beijing and Shanghai
France	LexisNexis SA, Paris
Germany	LexisNexis GmbH, Dusseldorf
Hong Kong	LexisNexis Hong Kong, Hong Kong
India	LexisNexis India, New Delhi
Italy	Giuffrè Editore, Milan
Japan	LexisNexis Japan, Tokyo
Malaysia	Malayan Law Journal Sdn Bhd, Kuala Lumpur
New Zealand	LexisNexis NZ Ltd, Wellington
Poland	Wydawnictwo Prawnicze LexisNexis Sp, Warsaw
Singapore	LexisNexis Singapore, Singapore
South Africa	LexisNexis Butterworths, Durban
USA	LexisNexis, Dayton, Ohio

© Reed Elsevier (UK) Ltd 2013

Published by LexisNexis Butterworths

All rights reserved. No part of this publication may be reproduced in any material form (including photocopying or storing it in any medium by electronic means and whether or not transiently or incidentally to some other use of this publication) without the written permission of the copyright owner except in accordance with the provisions of the Copyright, Designs and Patents Act 1988 or under the terms of a licence issued by the Copyright Licensing Agency Ltd, Saffron House, 6–10 Kirby Street, London EC1N 8TS. Applications for the copyright owner's written permission to reproduce any part of this publication should be addressed to the publisher.
Warning: The doing of an unauthorised act in relation to a copyright work may result in both a civil claim for damages and criminal prosecution.

Crown copyright material is reproduced with the permission of the Controller of HMSO and the Queen's Printer for Scotland. Parliamentary copyright material is reproduced with the permission of the Controller of Her Majesty's Stationery Office on behalf of Parliament. Any European material in this work which has been reproduced from EUR-lex, the official European Communities legislation website, is European Communities copyright.
A CIP Catalogue record for this book is available from the British Library.

Printed and bound by CPI Group (UK) Ltd, Croydon, CR0 4YY

Visit LexisNexis UK at www.lexisnexis.co.uk

Preface

I remember a letter from a client being circulated amongst the staff during my articles praising the work of a particular fee earner (not me I hasten to add). On it one of the partners had written 'I taught her all she knows'. Below that the senior partner had written 'And I taught you all you know and somebody else taught me all I know'. Those words have remained with me and the more experienced (older) I become the more I realise that knowledge is something to be shared and not hoarded in the hope that one knows something more than the next person. A text such as this offers the opportunity to do so.

For the record the senior partner in question was Michael Cook. In his preface to the first edition of this text he was at pains to attribute his 'understanding of, and enthusiasm for, costs' to his late partner Lionel Cranfield. Whilst Michael's understanding and enthusiasm remain as acute as ever, he has chosen to move on to pastures new. It would be wrong if I did not mark the occasion by thanking him both for the assistance that the previous editions has provided to me and, more importantly, for instilling in me 30 years ago that same enthusiasm for costs. It is a privilege to have the opportunity to follow in his footsteps. However, acknowledging that, despite his best efforts, my understanding does not match Michael's, it is an enormous relief and a great pleasure to be joined by Jason Rowley, who for many years has sat as an assessor on costs hearings and is now, of course, a master of the Senior Courts Costs Office. Together we take on responsibility for editing the text.

Our timing may leave something to be desired. We take over the text in a year that has seen the most radical overhaul of costs since the introduction of the CPR in 1999. As a result some of the views we express are speculative. We await with interest decisions of the Higher Courts that enable more authoritative comment on key concepts such as costs management and the new proportionality test. In the meantime, though, as lawyers seek to persuade the court to define and refine the April amendments to the CPR, Lionel Cranfield's description of costs as 'the interesting part' resonates as clearly as it did all those years ago.

Some of you may note that the actual text is a slimmed down version from that of recent vintages with the relevant parts of the CPR as appendices. Our response is that it is deliberately so and that this, inevitably, is our acknowledgement of the primacy of proportionality.

The law is stated as at 1 October 2013.

Simon Middleton

Foreword to the First Edition

It is my privilege as President of the Law Society to commend Michael Cook's very practical guide to all aspects of solicitors' costs (other than criminal costs).

He modestly suggests that the book does not contain any facts and figures that cannot be found elsewhere. Recognising the ultimate and total dependence of solicitors on the costs which they receive, he presents all those facts and figures with clarity borne of experience. This book adds to the major contribution which Michael Cook has already made to the profession on this vital topic.

It is also a clear response to the need for a definitive and manageable guide to an area which requires such practical and positive skills.

I hope that it will find a place in every solicitor's office.

October 1991

P T Ely
President
The Law Society

Preface to the First Edition

When the publishers asked me to write a preface I cavilled on the two grounds that no-one reads prefaces and they serve no useful purpose. I was told that the purpose is to explain the aims and intention of the work; in other words, why I have written the book and why you should read it. Fair enough. My late partner, Lionel Cranfield, at the end of some heavy litigation used to say, 'Now we come to the interesting part – the costs!'. In the quarter of a century since Lionel was killed in the hunting field, the profession has become increasingly aware that money is the life-blood of a solicitor's practice, but I continue to be perplexed that, in spite of this realisation, most solicitors and their clients lose out financially because of the profession's basic lack of understanding and interest in all aspects of costs. In a College of Law survey of the areas of skills on which solicitors wished the Law Society's new Legal Practice course to be based, costs did not even appear in the list.

And that is why I have written this book. It does not contain any facts and figures you cannot find elsewhere – apart from it being illustrated from my collection of unreported cases – but I have tried to write about costs in an interesting, coherent and digestible way. My aim in writing the book is therefore to impart to you the understanding of, and enthusiasm for, costs that Lionel Cranfield aroused in me all those years ago.

I much appreciate the kindness of Philip Ely in writing a foreword to the book in his year as President of the Law Society. I am seriously indebted to my old friend Master Michael Devonshire of the Supreme Court Taxing Office for reading the manuscript of the book and making many useful suggestions on the condition that I did not mention his name.

The law is stated as at 1 October 1991.

Michael Cook

Limpsfield
October 1991

Acknowledgments

The joint editors would like to thank Helen Wood for Chapter 39 and James Laughland and Joanna Hughes for their contribution to Chapters 34 to 38.

Contents

Preface	v
Foreword to the first edition	vii
Preface to the first edition	ix
Acknowledgments	xi
Table of statutes	xxv
Table of statutory instruments	xxix
Table of Civil Procedure Rules	xxxi
Table of non-statutory provisions	xxxvii
Table of cases	xxxix

PART I SOLICITOR AND CLIENT

Chapter 1 The retainer	3
Introduction	3
The contract	4
Quotations and estimates	7
Client care	14
Complaints	21
Ending the agreement	21
Contingency agreement issues	23
Writing off/waiving costs	23
Chapter 2 Billing the client	25
Introduction	25
Terminology	25
Who can you bill?	25
Interim bills	27
Final bills	30
Value Added Tax	33
Chapter 3 Recovery of costs from clients	39
Introduction	39
Before issuing	39
Court procedure	41
Applications by the client	44
Third parties	47
Recovery without proceedings	49
Statutory demand	49
Bankruptcy petition	50

Contents

Solicitor's lien	51
Bankruptcy petition	50

PART II   FUNDING

Chapter 4   Creating Conditional Fee Agreements — 59
- The Golden Rule — 59
- Requirements — 60
- Drafting considerations — 63
- Differential rates — 65
- Assessing the risk — 67
- Risk and the cost of funding — 70
- Cost of setting up a funding arrangement — 71

Chapter 5   Running cases with conditional fee agreements — 73
- During the case — 73
- No notice of funding — 74
- Notice of funding — 74
- Estimates of costs — 74
- Counsel — 75
- Interim applications — 78
- Changing arrangements — 79
- Security for costs — 80
- Ending the retainer before the case concludes — 82
- After the case ends — 84
- Calculation of the cap — 85
- QOCS — 86
- Notification to insurers — 87
- Interest on costs — 87
- Where unsuccessful — 89

Chapter 6   'Old' conditional fee agreements and ATE insurance — 91
- Introduction — 91
- Compliance with the primary legislation — 92
- Compliance with the secondary legislation — 95
- CFA Regulations 2000 — 95
- CFA (Miscellaneous Amendments) Regulations 2003 — 100
- Collective conditional fee agreements — 102
- Membership organisations — 103
- Court approach to legislative compliance — 104
- Professional conduct compliance — 110
- ATE insurance — 121
- Notification of additional liabilities — 124

Transitional provisions | 129

Chapter 7 Damages-based agreements | 131
Introduction | 131
The Ontario Model | 131
Statutory framework | 132
Maximum level of the recoverable percentage | 137
Reduction in damages or costs | 139
Recoverability | 139
Termination | 140
Operation of the indemnity principle | 141
Conflict of interest | 142
Challenges by the client | 142

Chapter 8 Contentious and non-contentious business agreements | 145
Introduction | 145
Defining contentious and non-contentious business | 145
Business agreements | 146
Restricting challenges to fees | 150
Procedure | 152
Non-contentious work prior to 2009 | 153
Solicitor mortgagee's costs | 153

Chapter 9 Insurance arrangements | 155
Introduction | 155
Before the event insurance | 156
BTE usage | 156
BTE coverage | 157
After the event insurance | 160
Risks covered | 160
Premiums still recoverable | 163
Can the client challenge the premium level? | 166
Contents of an ATE policy | 166
Paying adverse costs | 168
ATE and prospective costs control | 169
Solicitor self insurance | 170

Chapter 10 State and commercial funding | 173
Introduction | 173
State funding | 173
Commercial funding | 176
The funding agreement | 179
Profile of funded cases | 180
The Association of Litigation Funders | 180

Contents

Fee sharing in personal injury cases	181

PART III BETWEEN THE PARTIES

Chapter 11 An overview of the key costs amendments to the Civil Procedure Rules	185
Introduction	185
The overriding objective	185
Relief from sanction	186
Part 44	186
Part 45	188
Part 46	189
Part 47	189
Part 48	189
Other transitional provisions	190
The Costs Practice Direction	190
Chapter 12 The indemnity principle	191
No profit	191
Possible pitfalls and some exceptions	192
The future of the indemnity principle	200
Chapter 13 Prospective costs control – introduction	201
Chapter 14 Prospective costs control – Proportionality	203
Introduction	203
The case for reform	203
The reform	204
Proportionality and justice	208
Chapter 15 Prospective costs control – Costs and cases management	211
Introduction	211
What is costs management?	211
Costs management pilot schemes	212
The scope of the costs management regime under CPR 3.12–CPR 3.18 and CPR PD 3E	213
The procedural code	215
Form H	222
Completing the Form H	223
The interaction between case and costs management	227
Setting the budget	231
The case law so far	234
Conclusion	237

Contents

Chapter 16	Prospective control of costs – The relevance of costs budgets in cases not costs managed	239
Chapter 17	Prospective costs control – Costs capping	243

A brief history – and possibly an even briefer future 243
Jurisdiction 244
Some of the case law 244
The procedural code 246
Conclusion 247

Chapter 18	Prospective control of costs – protective costs orders	249

Introduction 249
Case law 249
The future 252
Aarhus Convention cases 253
Beddoe orders 254

Chapter 19	Prospective costs orders – Security for costs	255

Introduction 255
Under the Civil Procedure Rules, rules 25.12–25.14 255
The Conditions under CPR 25.13(2) 258
Security for costs of an appeal 261
Is it just to make an order for security of costs having regard to all the circumstances of the case? 263
Security for costs under CPR Part 3 266
The amount of security 268
Variation of an order for security 268
Security for costs under the Arbitration Act 1996 269

Chapter 20	Costs inducements to settle – Part 36 offers and other admissible offers	271

Introduction 271
The content of a valid Part 36 offer 271
Who may make a Part 36 offer? 273
When may a Part 36 offer be made? 273
Time periods for acceptance and withdrawal 274
Clarification of a Part 36 offer 278
Part 36 and interest 278
Disclosure of Part 36 offers 278
Part 36 offers and counter offers 280
The consequences of acceptance of a Part 36 offer within time 281
Acceptance after the expiry of the relevant time period 283
The consequences of acceptance of a Part 36 offer made by only one of two or more defendants 285
The consequences of not accepting a Part 36 offer 285

xvii

Contents

Defendants' Part 36 offers and qualified one way costs shifting	287
Part 36 and success in part	287
Part 36 and future pecuniary loss claims	288
Part 36 and provisional damages claims	289
Part 36 and recoupment of state benefit	289
Part 36 and fixed success fees	291
Part 36 and the 'tactical offer'	291
Part 36 and the Pre-Action Protocol for Low Value Personal Injury Claims in Road Traffic Accidents and the Pre-Action Protocol for Low Value Personal Injury (Employers' Liability and Public Liability) Claims	292
Part 36 and the small claims track	294
Part 36 and VAT Tribunal	294
Pre-6 April 2007 Part 36 payments and offers	295
Other admissible offers	295
Conclusion	299

Chapter 21 Costs inducements to settle – ADR 301
Introduction 301
Failure to engage in ADR leading to a costs sanction 301
Failure to engage in ADR – no costs sanction 302
The *Halsey* test 303
The voluntary nature of ADR 304
Conclusion 306

Chapter 22 Costs awards between the parties 309
Introduction 309
No order as to costs 309
The menu of costs orders available to the court 310
Solicitor's duty to notify the client 311
The time for compliance with a costs order 312
Deemed costs orders 312
Costs orders by consent 314
Contested awards between the parties 316
The menu of costs orders available to the court to give effect to its conclusions under CPR 44.2–CPR 44.5 324
Costs issues for the trial judge in the award of costs or for the costs judge on assessment 327
Costs of pre-action disclosure applications 330
Small claim track costs 332
Costs following allocation and re-allocation 333
Costs only proceedings 334
Agreement in issued proceedings in respect of everything except costs 336

Claim and counterclaim and set off generally	337
No winner or loser	340
Bullock and *Sanderson* orders	342
State funded parties	343

Chapter 23   Wasted costs orders 345
The law	345
The procedure	345
Satellite litigation	347
The appropriate three stage test	348
Two stage discretion	351
Problems with privilege	351
Threatening an application or putting the other side on notice?	352
Wasted costs and public funding	352
Wasted costs and advocacy	353
Wasted costs where the solicitor places reliance on counsel	353
Wasted costs against a barrister for his advice	356
Some successful applications	356
Some unsuccessful applications	357
Tribunals	363
Conclusion	364

Chapter 24   The bases of costs 365
Introduction	365
The indemnity basis – when is it appropriate?	367
Conclusion	377
The future	377

Chapter 25   Payment on account of costs 379

PART IV   QUANTIFICATION OF COSTS

Chapter 26   Fixed costs 385
Introduction	385
Structure of this chapter	386
Part 45 Section I (CPR 45.1–CPR 45.8)	386
Part 45 Section II (CPR 45.9–CPR 45.15)	388
Part 45 Section III (CPR 45.16–CPR 45.29)	392
Part 45 Section IIIA (CPR 45.29A–CPR 45.29L)	398
Part 45 Section IV (CPR 45.30–CPR 45.32)	401
Part 45 Section V (CPR 45.33–CPR 45.36)	405
Part 45 Section VI (CPR 45.37–CPR 45.40)	405
Part 45 Section VII (CPR 45.41–CPR 45.44)	407

Contents

The small claims track 408

Chapter 27   Summary assessment 411
Introduction 411
Who conducts a summary assessment? 411
The benefits of a summary assessment 412
When is a summary assessment appropriate? 412
When is a summary assessment inappropriate? 414
The extent of the costs to be assessed 419
The statement of costs 419
The procedure 420
The summary assessment hearing 421
Hourly rates 422
Counsel's fees 423
Proportionality 423
Appeals from summary assessment 424
Conclusion 424

Chapter 28   Detailed assessment – procedure 425
Introduction 425
Method of assessment 425
Preliminary issues 426
Entitlement to start detailed assessment proceedings 426
Venue 428
Reaching agreement 429
The detailed assessment procedure 430
Commencing detailed assessment 430
Statements of case 434
Obtaining a default costs certificate 436
Setting aside a default costs certificate 437
Requesting a hearing date 440
Interim costs certificate 441
Provisional assessment 442
The detailed assessment hearing 442
Costs of the detailed assessment proceedings 450
Final costs certificate 451
Appeal from a detailed assessment hearing 452
Authorised court officers 453
Other detailed assessment proceedings 454

Chapter 29   Detailed assessment – statements of case 457
Introduction 457
The bill of costs: form and content 457

The title page	459
The narrative	459
Heads of costs	461
The detail	468
Points of dispute	477
Reply	479
Part 18 requests	480
Chapter 30 Detailed assessment – provisional assessment	**483**
Introduction	483
The pilot scheme	483
The provisional assessment procedure	484
Court of Protection	491
Legal aid	491
Trusts and other funds	491
Chapter 31 Time and value	**493**
Introduction	493
Time as a management tool	494
Time as a charging tool	495
Calculating the hourly charging rate	496
The Seven Pillars of Wisdom	501
Amount or value	502
Skill etc	503
Place/circumstances	503
What level of B factor to claim?	504
The A and B test	507
The SCCO Guide to Summary Assessment	508
Locality of fee earners	512
Assessing the time spent	515
Counsel's fees	521
Value billing in non-contentious business	522
Alternatives to time as a management tool	525
Alternatives to time as a charging tool	526
Chapter 32 Interest on costs	**529**
Introduction	529
The incipitur rule (as opposed to the allocatur rule)	529
Orders deemed to have been made	530
Enhanced interest	530
Backdating and post-dating of interest – the court's discretion	531
The rate of interest	533

Contents

Chapter 33 Appeals against assessments	535
Introduction	535
The importance of reasons	535
A simple formulation of reasoning when assessing items and overall proportionality	536
Permission to appeal	537
Grounds for appeal	537
The time for appealing	538
Route of appeals	539
Review not re-hearing	539
The role of assessors	540
Appeals from the decisions made on detailed assessment by authorised court officers	541
Conclusion	542

PART V SPECIAL CASES

Chapter 34 Children and protected parties	545
Introduction	545
Litigation friend	545
Expenses incurred by a litigation friend	545
Detailed assessment	546
Chapter 35 Litigants in person	547
Introduction	547
Who is a litigant in person?	547
Professional litigants in person	548
Financial loss	548
Quantification	552
Chapter 36 Costs payable under a contract	555
Introduction	555
Mortgages	555
Landlord and tenant	556
Contractual costs	556
Chapter 37 Group (multi-party) litigation orders	559
Introduction	559
Discontinuance	560
Individual costs	560
Generic costs	561
Costs capping	562
Funding	563

Contents

Chapter 38 Trustees and personal representatives	565
Entitlement to costs out of the trust or estate	565
Costs against trustees	565
Pre-emptive orders	566
Prospective costs orders	566
Detailed assessment of costs from the trust or estate	567

Chapter 39 Family proceedings	569
Family Procedure Rules 2010	569
Applications other than in financial remedy proceedings	574
Children Act 1989 applications	575
Appeal	578
Calderbank offers	578
Financial remedy proceedings where FPR 28.3 applies	579
The financial effect on the parties of any costs order	580
The overriding objective	586
Being creative with costs orders	589
Costs information	590
Funding family costs	591
Applications in matrimonial and civil partnership proceedings	591
Applications under Schedule 1 to the Children Act 1989	593
Solicitor and client	594
Curbing costs	594
Summary assessment	597
Enforcement	598

Chapter 40 Costs against non-parties	601
Introduction	601
The general approach	601
The application procedure	602
Requirements for notice of an application	603
The approach of the court to managing an application	604
Is it a prerequisite that the non-party has provided funding for the claim?	605
Is it a necessary pre-requisite of an order that the non-party has caused the other party to incur costs over and above those that would have been incurred anyway?	606
Specific categories of non-parties traditionally in the firing line	609
Funders	616
Solicitors	618
Family cases	622

Chapter 41 Arbitration	623
Introduction	623

Contents

Costs of the arbitration defined	623
Agreement as to payment of costs	623
The award of costs	624
Recoverable costs	625
Failure to give reasons	629
Applications to the court	629
The costs of enforcement	629
Security for costs	630
Conclusion	630
Appendix	631
Index	829

Table of Statutes

References in the right-hand column are to paragraph numbers. Paragraph references printed in **bold** type indicate where the statute is set out in part or in full.

A

Access to Justice Act 1999
.... 6.17
s 11 .... **22.35**, 27.11, 39.15
29 .... 6.22, 6.43, 6.46, 9.20, 9.22
30 .... 6.19, 6.20, 6.22, 6.39
31 .... 6.18
Sch 2 .... 10.3
Administration of Justice Act 1990
Sch 8 .... 39.49
Arbitration Act 1996
.... 19.23, 41.16
s 24, 25 .... 41.10
28 .... 41.2
(2) 41.10, 41.13
33 .... 41.11
(1)(a) 41.6
37(2) 41.2, 41.10
38(3) **19.24**
52(3) 41.12
53(3) 41.12
56(2) 41.13
57(3)(a) 41.11
59 .... 41.1, 41.2
60 .... 41.1, 41.3
61 .... 41.1, 41.4, 41.6
62 .... 41.1, 41.3
(2)(a) 41.11
(i) 41.11
63 .... 41.1, 41.9
(3) 41.7
(4) 41.7, 41.13
(5) 41.7
64 .... 41.1
(1) 41.10
(2) 41.10, 41.13
(4) 41.10
65 .... 17.2, 41.1
(2) 41.11
68 .... 41.12, 41.13
(2)(a) 41.6
69 .... 41.13
70(4) 41.13
Attorneys and Solicitors Act 1870
s 4 .... 12.3

C

Child Abduction and Custody Act 1985
.... 39.15
Children Act 1989
.... 39.14
s 37 .... 40.20
Sch 1 .... 39.11, 39.13, 39.40
Civil Partnership Act 2004
.... 39.39
s 48(2) 39.9
Sch 5
Pt 9 (paras 39–45)............. 39.11
para 69..................... 39.11
Sch 6 .... 39.11
Companies Act 1985
.... 19.3
s 726(1) 19.13
Consumer Credit Act 1974
.... 1.7, 10.15
County Courts Act 1984
s 49(1) **19.10**
74 .... 32.3
(1) **32.1**
Courts and Legal Services Act 1990
.... 4.7, 6.2, 12.9
s 3 .... 8.14
17 .... 9.9
28(5) 35.6
58 ... 4.2, **4.14**, 6.11, 6.18, 6.25, 7.3,
8.7, 9.30, 9.31
(1) 6.3
(3) 40.18
(b) 4.5
(c) 4.9
(a) 6.6
(c) 6.5
(4B) **4.8**, 5.25
58A(1) 4.5
58AA 7.3, 7.4
58C 9.16
59(1) **8.7**
(2) **8.7**
60(3) **8.7**
98 .... **8.7**
119(1) 6.3

xxv

Table of Statutes

D

Damages Act 1996 .... 20.27
Domestic Proceedings And Magistrates Courts Act 1978
   Pt 1 (ss 1–35) 39.11

E

Environment Protection Act 1990 .... 6.6
   s 82 .... 4.5

F

Fatal Accidents Act 1976 .... 11.5
Financial Services and Markets Act 2000 .... 9.32
   s 26(1) 9.31

H

Human Rights Act 1998
   Sch 1
      art 2 .... 29.18
      . 6 .... 39.39

I

Insolvency Act 1986 .... 9.17
   s 382(1) 39.47
Inheritance Tax Act 1984 .... 15.21

J

Judgments Act 1838
   s 17 .... 32.3
      (1) **32.1**, 32.5

L

Landlord and Tenant Act 1925
   s 146(1) 36.3
Land Registration Act 2002 .... 28.59
Law Reform (Miscellaneous Provisions) Act 1934 .... 11.5
Legal Aid Act 1988
   s 17 .... 27.11
    31 .... 3.34
Legal Aid, Sentencing and Punishment of Offenders Act 2012
   .... 4.0, 4.7, 5.12, 6.15, 9.1, 9.13, 9.15, 9.19, 10.3, 39.40
   s 46(1) 6.56

Legal Aid, Sentencing and Punishment of Offenders Act 2012 – *cont.*
   s 48 .... 9.17
    61 .... 12.4
Legal Services Act 1990 .... 1.19
Legal Services Act 2007
   s 194 .... 12.4, 28.28

M

Matrimonial Causes Act 1973
   s 10(2) 39.9
    22ZA, 22ZB **39.39**
    25 .... 39.42
    27 .... 39.11
    35 .... 39.11
Mental Capacity Act 2005 .... 28.56

P

Pneumoconiosis etc. (Workers' Compensation) Act 1979 .... 9.17
Proceeds of Crime Act 2002
   Pt 5 (ss 240–316) 28.60

S

Senior Courts Act 1981
   s 18(1)(f) 23.31
    35A **32.1**
    51 .... 12.3, 12.13, 19.20, 22.2, 23.2, 23.9, 23.25, 23.27, 23.32, 29.18, 40.1, 40.16
      (1) 22.24, 40.18
      (2) 12.11
      (3) 17.2, 36.4, 40.17, 40.18
      (6) **23.1**, 23.26, 40.18
      (7) **23.1**, 23.4, 40.18
      (13) 23.14, 23.26
Solicitors Act 1974
   ... 1.1, 1.2, 2.6, 2.7, 8.1, 8.8, 8.14, 30.4
   s 25(1) 41.8
   Pt II (ss 31–55) 7.19
   Pt III (ss 56–75) 8.13
   s 56 .... 8.10
    57 .... 8.3
    58 .... 8.16
    59–61 .... 8.3
    62 .... 8.3, 8.4
    63 .... 8.3, 8.9
    64(2) 2.9, 3.17
    65(1) 3.32
      (2) 1.4, 1.28, 2.5, 2.10
    67 .... 2.11
    68 .... 2.5, 3.16, 3.30
    69 .... 3.4, 3.28

Solicitors Act 1974 – *cont.*
  s 69(1) 3.6, 3.27
      (2)(b) 3.5
      (3) 3.11
    70 .... 3.11, 3.19, 3.20
      (1), (2) 3.7
      (6) 3.15, 19.24
      (9), (10) 3.15
    71 .... 3.11, 3.25
      (1) 3.23, 3.24
    73 .... 3.31, 19.24
    74(3) 1.17, 3.10
    75 .... 19.24
    87 .... 8.2
State Immunity Act 1978
  .... 23.5
Supply of Goods and Services Act 1982
  s 15 .... 1.10

T

Third Parties (Rights against Insurers) Act 1930 .... 9.24
Third Parties (Rights against Insurers) Act 2010 .... 9.24
Trusts of Land and Appointment of Trustees Act 1996 .... 15.21

V

Value Added Tax Act 1994
  s 24 .... 29.24

W

Water Resources Act 1991
  s 165 .... 22.7

Table of Statutory Instruments

References in the right-hand column are to paragraph numbers. Paragraph references printed in **bold** type indicate where the statutory instrument is set out in part or in full.

A

Access to Justice (Membership Organisation) Regulations 2005, SI 2005/2306 .... 6.19, 6.21

C

Cancellation of Contracts made in a Consumer's Home or Place of Work etc. Regulations 2008, SI 2008/1816 .... 1.27, 4.13, 5.16
    reg 7(5) 1.7
Civil Jurisdiction and Judgments Act 1982 .... 26.7
Civil Legal Aid (General) Regulations 1989, SI 1989/339 .... 22.19, 30.24
    Pt XI (regs 87–99)
      reg 64 .... 3.34
          112 .... 29.23
          121 .... 29.23
Civil Legal Aid (Merits Criteria) (Amendment) Regulations 2013, SI 2013/772
    reg 42 .... 10.6
        43 .... 10.5
Civil Proceedings Fee Amendments (No 2) Order 2013, SI 2013/1410 .... 28.35
Collective Conditional Fee Agreements Regulations 2000, SI 2000/2988 .... 6.17
Conditional Fee Agreements Order 2000, SI 2000/823 .... 4.6
    art 4 .... 6.5
Conditional Fee Agreements Order 2013, SI 2013/689 .... 4.9
    art 5(1), (2) **4.8**
Conditional Fee Agreements Regulations 2000, SI 2000/692 .... 6.7, 6.21, 6.24, 7.5, 7.7, 12.11
    reg 1 .... 6.11
        2 .... 6.13, 7.6
          (1)(a)(b) **6.9**
        3 .... 6.13
          (1)(b) 4.22, 6.10, 6.23
        3A(5) 6.13
        4 .... 6.11, 6.12, 6.13, 6.27

Conditional Fee Agreements Regulations 2000, SI 2000/692 – *cont.*
    reg 4(2)(c), (d) 6.25
          (e) 6.26
        5(1) 6.18
Conditional Fee Agreements (Miscellaneous Amendments) Regulations 2003, SI 2003/1240 .... 4.11, 6.9
Consumer Protection (Cancellation of Contracts Concluded away from Business Premises) Regulations 1987, SI 1987/2117 .... 1.7
County Court (Interest on Judgment Debts) Order 1991, SI 1991/1184 .... 5.32, 32.1
    art 4 .... 32.6
Court of Protection Rules 2005, SI 2005/1744 .... 28.56

D

Damages Based Agreement Regulations 2010, SI 2010/1206 .... 7.4, 7.12
Damages-Based Agreements Regulations 2013, SI 2013/609
    reg 1 .... 7.5, 7.7
        3 .... 7.3
        4 .... **7.15**
          (1) 7.11
          (4) 7.13
        5 .... 7.5
        6 .... 7.6
        8 .... 7.16

F

Family Procedure Rules 2010 SI 2010/2955 .... 39.44
    Pt 1 .... 39.30
    r 1.1(2) 39.31
    Pt 2
    r 2.69 .... 39.24
      2.69A–269E 39.22
      2.71 .... 39.24
    Pt 9
    r 9.17(3), (4) 39.25
      9.27 .... 39.37

Table of Statutory Instruments

Family Procedure Rules 2010 SI 2010/2955 – *cont.*
Pt 23
r 23.8(7) 39.20
Pt 28 .... 39.1, 39.7
r 28.1 .... **39.2**, 39.8, 40.20
28.2 . **39.3**, 39.8, 39.10, 39.13, 39.31
28.3 . **39.4**, 39.8, 39.10, 39.11, 39.18, 39.20, 39.21, 39.24, 39.31, 39.36
   (6) **39.19**, 39.30
   (7) 39.30
     (a)–(e) **39.19**, 39.21
     (f) . **39.19**, 39.21, 39.34, 39.37
28.4 .... **39.5**
PD 28A 39.1, **39.6**
   (b) 39.11
4.4 .... **39.20**
4.5 .... 39.37
Family Proceedings Rules 1991, SI 1991/1247
2.62 .... 39.42
7.4(1) 39.49
   (3A) 39.49
   (7B) **39.49**
   (7C) 39.49
Financial Services and Markets Act 2000 (Regulated Activities) Order 2001, SI 2001/544 .... 9.31

H

High Court Enforcement Officers Regulations 2004, SI 2004/400
Sch 3 .... 28.61

I

Insolvency Rules 1986, SI 1986/1925
r 12.3 .... **39.47**
Insolvency (Amendment) Rules 2005, SI 2005/527 .... 39.48

P

Proceeds of Crime Act 2002 (Legal Expenses in Civil Recovery Proceedings) Regulations 2005, SI 2005/3382 .... 28.60

Proceeds of Crime Act 2002 (Legal Expenses in Civil Recovery Proceedings) (Amendment) Regulations 2008, SI 2008/523 .... 28.60

R

Recovery of Costs Insurance Premiums in Clinical Negligence Proceedings (No 2) Regulations 2013, SI 2013/739
.... 6.56, 9.16

S

Solicitors' (Non-Contentious Business) Remuneration Order 1994, SI 1994/2616 .... 8.13, 8.15, 31.35
Solicitors' (Non-Contentious Business) Remuneration Order 2009, SI 2009/1931 .... 8.13, 31.35
art 3 ... **8.10**, 31.8, 31.10, 31.13, 31.34
Supreme Court Rules 2009 SI 2009/1603
r 46(1) 28.58
47 .... 28.58
49(1) 28.58

T

Tribunal Procedure (First-tier Tribunal) (Health, Education and Social Care Chamber) Rules 2008, SI 2008/2699
.... 23.33
Tribunal Procedure (First-tier Tribunal) (Property Chamber) Rules 2013, SI 2013/1169 .... 28.59
Tribunal Procedure (Upper Tribunal) Rules 2008, SI 2008/2698
.... 23.33

Table of Civil Procedure Rules

References in the right-hand column are to paragraph numbers. Paragraph references printed in **bold** type indicate where the CPR is set out in part or in full.

Civil Procedure Rules 1998, SI 1998/3132 .. 6.1, 6.18, 10.13, 11.14, 12.1, 14.2, 15.15, 15.22, 17.5, 23.33, 24.2, 24.22, 28.14, 28.59, 28.61, 29.1, 29.25, 30.7, 31.7, 31.21, 33.2, 39.1, 39.13
Pt 1
  r 1.1 .... **14.3**
    (2) 15.34
    (c) 14.1
  1.4(2)(e) 21.5
Pt 2
  r 2.3(1) 19.2, 35.6
Pt 3 .... 15.8
  r 3.1 .... 19.12
    (1) 25.1
    (2)(m) 17.2
    (3) 19.21
    (3)(a) 25.1
    (5) 19.1, 19.21
    (6A) 19.1
    (7) 20.18
    .... 32.3
  3.7(6)(b) 22.8
  3.9 .... 6.52, 11.3, 15.34, 28.25
    (1) **15.28**
  3.12 .. 12.8, 14.3, 15.8, 25.1, 41.11
    (1) 16.1
    (c) 15.4
    (2) **15.1**
  3.13 ... 12.8, 14.3, 15.1, 15.4, 16.1, 41.11
  3.14 12.8, 14.3, 15.1, 15.4, 15.6, 15.34, 41.11
  3.15 .. 12.8, 14.3, 15.1, 15.4, 41.11
    (3) 15.8
  3.16 .. 12.8, 14.3, 15.1, 15.4, 15.14, 41.11
  3.17 ... 12.8, 14.3, 15.1, 15.4, 15.8, 41.11
  3.18 .. 12.8, 14.3, 15.1, 15.4, 15.11, 15.12, 15.31, 24.5, 27.16, 41.11
  3.19 .... 11.4, 12.8
    (5) 17.6
  3.20, 3.21 .... 11.4, 17.6
  PD 3E 15.4, 16.1
  para 1 .... 15.18
    2.1 .... **15.10**

Civil Procedure Rules 1998, SI 1998/3132 – cont.
  para 2.3 12.8, **14.3**, 15.31, **15.32**, 15.33, 27.16
    2.4 .... 15.8
    2.6 .... 15.8, 15.14
    2.7 .... 15.7
    2.8 .... 15.14
    2.9 .... 14.3, 15.9
  PD 3F
    para 1.1, 1.2 .... 17.6
Pt 4
  PD 4
  Table 1
    Form N161 .... 28.53, 33.10
      N235 .... 19.10
      N251 5.2, 6.47, 6.48, 6.49, 6.52, 9.27, 11.4, 27.9
      N252 .... 28.15
      N254 .... 28.27
      N255 .... 28.27
      N258 28.34, 28.35, 28.56, 30.3, 30.4, 30.5, 30.6, 30.14, 30.16
      N258B 30.14, 30.16
      N258C 3.13
      N260 27.9, 27.16, 27.21, 27.23, 39.37
      N460 .... 28.54
Pt 6
  r 6.8 .... 23.3
  PD 6A
    Section 4 .... 20.6
Pt 7 .... 22.27, 22.29, 26.39
  r 7.4(2) 15.34
Pt 8 . 3.11, 3.31, 8.14, 15.5, 17.1, 18.5, 22.27, 22.29, 26.8, 26.16
  PD 8B 26.16, 26.17
  para 10.1 .... 20.32
Pt 13
  r 13.3 .... 19.21
Pt 16
  r 16.3 .... 26.43
Pt 18 .... 29.31
Pt 19 .... 19.10
  r 19.2 .... 19.19
  PD 19B
    para 1.6 .... 37.1
Pt 20
  r 20.3 .... 19.2
Pt 21 .... 6.12, 19.10, 26.13

Table of Civil Procedure Rules

Civil Procedure Rules 1998, SI 1998/3132 – cont.
- r 21.1(2) 34.1
- 21.2(1) 34.2
- (3) 34.2
- 21.5, 21.6 .... 34.2
- 21.9(6) 34.2
- 21.12(3) 34.3
- PD 21
  - para 11.1 .... 34.3
- Pt 22
- PD 22 .... 15.21
  - para 2.2A 12.8
- Pt 23 .. 3.11, 28.19, 28.38, 28.55, 38.3
- r 23.10 .... 22.3
- Pt 24
- r 24.6 .... 19.21
- Pt 25 .... 3.12, 38.3
- r 25.1(1)(m) 3.32
- 25.12 ...... 19.1, 19.2, 19.3, 19.21
- (2) 19.22
- 25.13 .... 19.21
- (1)(a) 19.4, 19.13
- (2) 19.4, 19.13
- (a) 19.12
- (ii) 19.5
- (c) 19.1, 19.6, 19.7
- (g) 19.12
- 25.14 .... 19.1, 19.3, 19.10
- 25.15 ..... 19.1, 19.3, 19.12, 19.21
- Pt 26
- r 26.3 .... 16.1
- (1) 15.5
- 26.7(3) 15.23, 22.25
- Pt 27 .... 22.23, 23.2
- r 27.1 .... 26.46
- 27.2 .... 20.33
- 27.14 .... 20.33, 26.46
- (2) 22.25
- (h) 20.32
- (g) 26.47
- (5), (6) 22.25, 26.49
- PD 27
  - para 7 .... 26.46
- Pt 29
- r 29.1(2) 15.24
- Pt 30 .... 26.32
- Pt 31
- r 31.5(7) 15.26
- 31.6 .... 22.24
- Pt 32
- r 32.2(3) 15.25
- Pt 34 .... 40.15
- Pt 35
- r 35.4(2), (3) 15.27
- 35.6 .... 15.27
- r 35.15(3) 33.9
- PD 35
  - para 10 .... 33.9
  - 11.1–11.4 .... 15.27

Civil Procedure Rules 1998, SI 1998/3132 – cont.
- Pt 36 . 6.9, 6.29, 6.32, 11.5, 11.9, 13.1, 15.4, 17.2, 20.16, 20.33, 20.37, 22.17, 22.18, 22.20, 24.8, 28.48, 29.6, 30.10
- r 36.1 .... 20.2, 20.32
- (2) 20.36
- 36.2 ...... 20.2, 20.3, 20.13, 20.32
- 36.3 .... 20.32
- 36.3(2) 20.2
- (b) 20.4
- (4) 20.4
- (5) 20.5
- (7) 20.5
- 36.4 .... 20.32
- (2) 20.20
- 36.5 .... 20.2, 20.27, 20.32
- 36.6 .... 20.2, 20.32
- (1) 20.28
- 36.7 .... 20.6, 20.32
- 36.8 .... 20.13, 20.32
- 36.9 .... 20.32
- (3)(d) 20.10, 20.12
- (5) 20.11
- 36.10 20.2, 20.4, 20.26, 20.32, 20.36, 20.38, 22.6, 24.7
- (1) 20.18, 22.29
- (2)(b) 20.18
- (3) 20.17
- (4) 20.21
- (a) 20.7
- (5) 20.21
- 36.11 .... 20.2, 20.32, 20.36
- (1), (2) 20.18
- (3)(a) 20.18
- (4) 20.19
- (6)–(8) 20.20
- 36.12 .... 20.32
- (2), (3) 20.22
- 36.13 .... 20.32
- (2) 20.15
- 36.14 20.2, 20.3, 20.9, 20.11, 20.13, 20.14, 20.32, 20.34, 20.36, 20.38, 20.41, 24.20, 32.4
- (1A) 20.1
- (2) .. 20.7, 20.23, 20.24, 20.25
- (3) 20.7, 20.24
- (d) 20.1
- (4) 20.21
- 36.14A 20.39, 22.22
- (3)(c) 20.32
- 36.15 .... 20.2, 20.29, 20.32
- 36.16–36.20 .... 20.32
- 36.21(2)–(4) 20.32
- 36.22 .... 20.32
- PD 36A
  - para 2.2 .... 20.15
  - 3.2 .... 20.15
- PD 36B 20.35, 20.38
- Pt 38 .... 29.6, 29.18

Table of Civil Procedure Rules

Civil Procedure Rules 1998, SI 1998/3132 – cont.
  r 38.6 .... 24.14
    (1) 22.7
Pt 39
  r 39.3 .... 36.3
Pt 40
  r 40.8 .... 5.32
    (1) **32.2**
Pt 41
  r 41.5(2), (3) 20.27
    41.8 .... 20.27
Pt 43 .... 11.1
  43.2(1)(k) 11.10
    (3), (4) 6.16
Pt 44 .... 6.13, 6.16, 11.1, 11.8, 22.34,
      24.20, 26.46, 41.16
  r 44.1 .... 6.44, 28.2, 41.4
    (1) .. 27.18, 28.52, **28.55**, 33.10
    (2) 41.7
  44.2 .. 6.44, 11.4, 20.2, 20.9, 20.23,
    20.24, 20.38, 22.23, 22.33, 23.2,
      36.4
    (1) 22.1
    (2) .. 22.16, 22.21, 22.32, 24.3,
      39.11
      (a) 22.18
    (3) 22.21, 39.11
    (4) .. 20.36, 22.17, 22.21, 24.7,
      24.8, 24.12, 24.25
      (c) 20.39
    (5)14.3, 22.17, 22.18, 22.21, 22.
      32, 24.7, 24.25
      (a) 24.10
      (d) 22.18
    (6)(a) 22.21, 22.32
      (b)–(f) 22.21
      (g) 21.3, **32.5**
    (7) 22.21, 22.22
    (8) 3.12, 5.**23**, 5.24, 25.1, 28.33,
      29.22
  44.3 7.17, 11.4, 20.24, 20.30, 20.38,
    22.17, 22.23, **39.11**
    (1) 24.1
    (2)11.11, **14.3**, 15.4, 15.30, 16.1,
    24.4, 27.18, 27.21, 39.8, 39.23,
      41.7
      (a) 22.17, 37.4
      (b) 37.4
    (3) 15.12, 27.18
    (4) 24.1
      (c) 39.18
    (5) 11.2, 11.11, **14.3**, 15.31, 15.
    33, 16.1, 24.4, 27.21, 33.3, 40.5,
      41.7
    (6) **39.36**, 39.37
      (a) 21.3
      (g) **32.5**
      .... **39.36**
    (8) 25.1, **39.36**
    (9) 22.32, **39.36**

Civil Procedure Rules 1998, SI 1998/3132 – cont.
  r 44.3A 27.9
    44.3B(1)(a) 6.10
      (c) 6.51
      (d), (e) 6.53
  44.4 8.10, 27.18, 31.8, 31.10, 31.26
    (1) 20.18, 22.23
      (a) 37.4
    (3) 14.3, 24.3, 31.13
  44.5 11.4, 17.7, 22.23, 26.24, 27.10,
      36.1, 36.2, 36.4
    (1) 6.45
    (3) 6.44
  44.6 .... 27.1
  44.7 .... 22.5
  44.8 .... 22.4
  44.9(1) 22.9
    (4) 22.10, 32.3
  44.10(2) 39.11
    (c) 22.2
    (3) 22.2, 39.11
  44.11 .... 22.23, 28.20
  44.12 .... 22.32
    (a) 26.37
  44.12A 22.27, 26.8
  44.12A(4A) 22.29
  44.13 .... 11.5, 20.24
  44.14 .... 11.5, 20.24, 28.20
  44.15 .... 11.5, 20.24, 26.29
  44.16 6.10, 11.5, 20.24, 20.25, 26.29
  44.17 .... 11.5, 11.12
    (5) 22.5
  44.18 .... 11.6, 17.6, 18.2
    (1), (2) 7.17
    (5) 9.29
PD 44 .... 27.5
para 3 .... 1.16
  3.1 .... 16.1
  4.2 .... 22.2, 22.3
  .7 .... 36.1, 36.4
  7.2, 7.3 .... 27.10
  9.1 .... 27.4
  9.2 .... 27.4, 27.6
  9.3 .... 27.10
  9.4 .... 27.14
  9.5(1) 27.16
    (4) 27.17
  9.6 .... 27.17
  9.7 .... 27.2
  9.8 .... 27.11
  9.9(2) 27.12
  9.10 .... 27.13
  10.2, 10.3 .... 22.4
  13.2 .... 39.45
Pt 45 ..... 4.16, 6.13, 6.16, 6.29, 6.39,
    11.1, 11.8, 41.4, 41.7, 41.16
Section I (rr 45.1–45.6)
  .... 2.19
  r 45.1 .... 26.3
    45.2 .... 26.3, 26.4

xxxiii

Table of Civil Procedure Rules

Civil Procedure Rules 1998, SI 1998/3132 – cont.

Rule	Reference
r 45.3	26.3
45.4	26.3, 26.5
45.5, 45.6	26.3
Section II (rr 45.7–45.14)	22.29, 26.15, 26.17, 26.19, 26.21, 26.25, 26.26, 26.28, 34.4
r 45.7	12.11, 26.3, 26.6
45.8	12.11, 26.3, 26.7
45.9	12.11
45.10	12.11, 26.11
45.11	12.11, 26.9, 26.11
45.12	12.11
(2)(c)	26.11
45.13	12.11, 26.31
45.14	12.11, 26.31
(2)	26.14
45.15	12.11, 26.31
Section III (rr 45.16–45.29)	6.38, 26.15, 26.25, 26.26, 26.28, 34.4
r 45.16	6.42, 20.30
45.18	6.33, 6.40
45.19	26.19
45.20	20.32
45.21	26.22
45.22	26.22, 26.24
45.23	26.22
45.23B	26.18
45.24	26.21
45.25	26.24
45.26	20.32
45.28	26.8
45.29	12.11, 20.32, 26.24, 26.28, 26.31
Section IIIA (rr 45.29A–45.29L)	26.25, 34.4
45.29A–45.29E	11.7
45.29F	11.7, 26.29
(2)–(7)	27.15
45.29G, 45.29H	11.7
45.29I	11.7, 26.28
45.29J–45.29L	11.7
Section IV (rr 45.30–45.32)	6.38, 26.26, 26.32
45.30	26.33
(3)	26.35
45.31	26.34
(2)	26.37
45.32	26.38
Section V (rr 45.33–45.36)	6.38, 26.26
45.36	26.39
Section VI (rr 45.37–45.40)	12.11, 26.26, 26.40
45.37, 45.38	11.7
45.39	11.7
(5)	35.7
45.40	11.7
45.41	11.7, 18.4

Civil Procedure Rules 1998, SI 1998/3132 – cont.

Rule	Reference
r 45.41(2)	26.45
45.42	11.7, 18.4
45.43	11.7, 18.4
(1)	26.45
45.44	18.4
(3)(a), (b)	26.45
PD 45	26.28
para 1.3	26.4
2.6	26.10
2.10	26.11
.5	18.4
5.1, 5.2	26.45
Pt 46	6.13, 6.16, 11.1, 11.4, 11.7, 11.8, 39.11, 41.4, 41.7, 41.16
r 46.1	22.24, 40.15
46.2	40.1, 40.4, 40.5, 40.15
(1)(b)	40.3
46.3	38.1
46.4	33.10, 34.4
(2)	27.12
46.5	26.42
(2)	35.5
(3)(a)	35.8
(4)	35.8
(a), (b)	35.4
(3)	37.1
46.8(2)–(4)	23.2
46.9	8.6
(2)	1.17, 3.10
(3)	31.27
46.10	2.9
(2)	2.12
46.11(2)	22.26, 26.49
46.13	22.26
(1), (2)	26.49
46.14	20.4, 22.29, 26.8
(5)	22.27
PD 46	
para 1	38.1
2.1	27.12
3.1	35.6
3.4	35.4
.5	23.3
5.2, 5.3	23.2
5.4(b)	23.24
5.5(b), (c)	**23.4**
5.6	23.2, 23.27
5.7	23.8
5.8, 5.9	23.24
6.6(b)	**2.12**
9.4	**22.29**
9.6	22.28
9.9	22.28
9.10	22.27
9.12	22.27
Pt 47	6.13, 6.16, 11.1, 39.11, 41.4, 41.7, 41.16
Section I (rr 41A–40.4)	
r 47.1	5.23, 28.6

Table of Civil Procedure Rules

Civil Procedure Rules 1998, SI 1998/3132 – cont.
r 47.3	33.10
47.4	28.7
47.7	28.19, 28.21, 28.34, 28.60
47.8	28.21
(1)	28.20
(3)	28.25
(b)	28.19
	28.43, 29.28
47.9(3)	3.14, 28.24
(4)	28.27
(5)	28.24
47.10	28.12
47.11	3.12
(4)	28.28
47.12(1)	28.24, 28.29
(2)	28.30
(3)	28.28
47.13	28.26, 29.30
47.14	6.24, 28.60
(1), (2)	28.34
(6)	29.28
(7)	28.44
47.15	11.9, 22.21, 27.3, 28.2, 28.40, 30.3, 33.11
(1)	11.13
(3)	30.5
(5)	30.9
(6)	30.4
(7)	30.11
(10)	15.30, 30.13
47.16	25.1
(4)	28.28
Section VI (rr 47.17–47.17A)	
r 47.17	28.50
(4)	30.5
(6)	28.28
47.19	28.48, 30.16, 38.5
47.20	11.9, 12.2
(3)	28.47
(4)	28.48
47.21–47.24	28.53, 33.10
PD 47	
para 1.1	28.5
1.2–1.4	28.6
3.1	33.10
.4	28.7
5.2	28.17
5.5(2)	28.15
5.6	28.23
5.8	29.19
5.11	29.7
5.13, 5.14	29.19
5.15	29.19
5.16	29.9
5.18	29.15
5.19	29.19
5.20	29.20
5.22(2)–(4)	29.19
(6)	29.17

Civil Procedure Rules 1998, SI 1998/3132 – cont.
para 8.2	**29.28**
8.3	28.25, 30.5
9.4	28.12
10.7	26.6, 28.27
11.1	28.29
11.2(1), (2)	28.31
(3)	28.32
11.3	28.33
12	29.30
13.1	28.42
13.2	28.36, 28.42, 28.56, 30.5
(a)–(h)	28.35
(i)	12.9, 28.35
(j), (k)	28.35
(l)(v)	28.35
13.3	**28.45**, 29.6
13.5	28.37
13.10	29.24, 29.30
13.12	30.14
14	28.40, 30.3
14.1	11.9
14.3	30.5
14.4(1)	30.6
(2)	30.11
14.6	30.10
15	28.38
16	28.50
18	30.16, 38.5
19	28.48
20	28.53
20.1–20.6	33.10
39.10	29.24
Pt 48	6.13, 6.16, 11.1
r 48.1	5.3, 6.54
(2)	22.24
48.2	5.3, 6.55, 11.10
(2)	9.17
48.3	11.8
48.6	37.1
(4)	35.8
PD 48	6.55
Pt 51	
PD 51D	15.6
PD 51G	15.8
para 1.3	14.3
.2	15.12
3.2	15.12
4.2	14.3
.8	15.12
Pt 52	28.54
r 52.3(1)	33.4
(6)	33.4
(7)	19.12
52.4	33.6
52.9(1)(c)	19.12
52.11(3)	33.5
PD 52A	
para 3.6	33.7
3.8(2)	33.7

Table of Civil Procedure Rules

Civil Procedure Rules 1998, SI 1998/3132 – cont.
  Pt 61 .... 24.20
  Pt 62 .... 41.4, 41.16
    r 62.18 .... 41.14
  Pt 63 .... 26.32
    r 63.26(2) 26.38
  Pt 64
    r 64.2(a) 18.5, 38.4
  PD 64A
    para 6 .... 38.4
  PD 64B 18.5

Civil Procedure Rules 1998, SI 1998/3132 – cont.
  Sch 1 RSC
    Ord 62
      r 18(1) 35.8
    Ord 106 .... 3.31
  Sch 2 CCR
    Ord 11
      r 10 .... 39.22
  Pre-Action Protocols
    Protocol for Low Value Personal Injury
      .. 26.15, 26.16, 26.17, 26.25, 26.30

Table of Non-Statutory Provisions

References in the right-hand column are to paragraph numbers. Paragraph references printed in **bold** type indicate where the Provisions are set out in part or in full.

Admiralty and Commercial Courts Guide
    para F14.2–F14.4 .... **27.7**
Bar Code
    para 209 .... 23.12
        506 .... 23.12
Guideline Hourly Rates 2010
    .... 29.7
Law Society's Practice Rules
    r 15 .... 6.18
Senior Courts Costs Office Guide to the Summary Assessment of Costs
    .... 31.22
Solicitors' Accounts Rules 1991
    .... 2.5
Solicitors' Code of Conduct 2007
    .... 1.1
    r 2 .... 1.20, 6.18
    2.03 .... 1.18, **1.19**
    Guidance to Rule 2
    .... **1.29**
    r 9 .... 10.16
Solicitors' Code of Conduct 2011
    ..... 1.1, 1.5, 1.8, 1.10, 1.18, 1.20, 1.21, 3.8, 6.28
    Outcomes
    r 1.1 .... **1.22**
    1.6 .... **1.22**
    1.9–1.11 .... 1.26
    1.12 .... **1.22**
    1.13, 1.14 .... **1.22**
    1.27 .... **1.22**

Solicitors' Code of Conduct 2011 – *cont.*
    Indicative Behaviours
    r 1.1 .... **1.23**
    1.3 .... 1.25
    1.4, 1.5 .... **1.23**
    1.13–1.21 .... **1.23**
    1.22–1.24 .... 1.26
Solicitors' Costs Information and Client Care Code 1999
    .... 1.10
Solicitors' Practice Rules 1990
    r 15 .... 1.18, 1.19, 1.20, 6.18
Standard Contractual Terms for the Supply of Legal Services by Barristers to Authorised Persons 2012
    .... 5.7
Supreme Court Costs Office Guide 2003 .... 31.24
Supreme Court Costs Office Guide 2005 .... 31.24, 31.32
Supreme Court Costs Office Guide 2010 .... 31.25
Supreme Court Costs Office Guide 2013 .... 28.10, 31.24
Technology and Construction Court Guide
    s 6.4.4 .... 27.4
Terms of work on which Barristers Offer their Services to Solicitors and the Withdrawal of Credit Scheme 1988 .... 5.7

Table of Cases

A

A (wasted costs order), Re [2013] EWCA Civ 43, [2013] 2 FLR 1, [2013] Fam Law 533 .... 23.30
A v A (Maintenance pending suit: provision for legal costs) [2001] 1 WLR 605, [2001] 1 FCR 226, [2001] Fam Law 96, [2000] All ER (D) 1627 .... 39.6
A v A (Maintenance pending suit: provision for legal fees) [2001] 1 FLR 377 .... 39.39
A v B (No 2) [2007] EWHC 54 (Comm), [2007] 1 All ER (Comm) 633, [2007] Bus LR D59, [2007] 1 Lloyd's Rep 358, [2007] ArbLR 1, [2007] All ER (D) 157 (Jan) 24.15
A v BCD (a firm) (5 June 1997, unreported), QBD 3.20
A v Chief Constable of South Yorkshire Police [2008] EWHC 1658 (QB), [2008] All ER (D) 226 (Jul) 31.30
A L Barnes Ltd v Time Talk (UK) Ltd [2003] EWCA Civ 402, [2003] BLR 331, (2003) Times, 9 April, 147 Sol Jo LB 385, [2003] All ER (D) 391 (Mar) 22.17
AB v CD [2011] EWHC 602 (Ch), [2011] IP & T 504, [2011] All ER (D) 25 (Apr) 20.15
AEI Rediffusion Music Ltd v Phonographic Performance Ltd (No 2). See Phonographic Performance Ltd v AEI Rediffusion Music Ltd
Aaron v Shelton [2004] EWHC 1162 (QB), [2004] 3 All ER 561, [2004] NLJR 853, [2004] 3 Costs LR 488, [2004] All ER (D) 347 (May) 22.23
Abedi v Penningtons (a firm) [2000] NLJR 465, CA 2.7
Abraham v Thompson [1997] 4 All ER 362, [1997] 37 LS Gaz R 41, 141 Sol Jo LB 217, CA 19.3
Adams v Mackinnes (Case No 13 of 2001), unreported, SCCO 2.7
Adrian Allen Ltd v Fuglers (Case No 13 of 2003), unreported, SCCO 1.25
Adris v Royal Bank of Scotland (Cartel Client Review Ltd, additional parties) [2010] EWHC 941 (QB), [2010] NLJR 767, [2010] 4 Costs LR 598, [2010] All ER (D) 156 (May) 40.7
Agassi v Robinson (Inspector of Taxes) (Bar Council intervening) [2005] EWCA Civ 1507, [2006] 1 All ER 900, [2006] 1 WLR 2126, [2006] STC 580, [2005] NLJR 1885, (2005) Times, 22 December, 150 Sol Jo LB 28, [2006] 2 Costs LR 283, [2005] All ER (D) 40 (Dec) 35.6
Agrimex Ltd v Tradigrain SA [2003] EWHC 1656 (Comm), [2003] 2 Lloyd's Rep 537, [2003] NLJR 1121, (2003) Times, 12 August, [2003] ArbLR 1, [2003] All ER (D) 151 (Jul) 41.10
Aiden Shipping Co Ltd v Interbulk Ltd, The Vimeira (No 2) [1986] AC 965, [1986] 2 All ER 409, [1986] 2 WLR 1051, [1986] 2 Lloyd's Rep 117, 130 Sol Jo 429, [1986] LS Gaz R 1895, [1986] NLJ Rep 514, HL 40.2, 40.13
Al-Koronky v Time Life Entertainments Group Ltd [2005] EWHC 1688 (QB), [2005] All ER (D) 457 (Jul); affd sub nom Al-Koronky v Time-Life Entertainment Group Ltd [2006] EWCA Civ 1123, [2007] 1 Costs LR 57, [2006] All ER (D) 447 (Jul) 5.17, 19.15, 19.22
Alex Sherred (a child suing by his mother and litigation friend Denise Sherred) v David Carpenter (5 March 2009, unreported) 26.13
Ali (Saif) v Sydney Mitchell & Co (a firm) [1980] AC 198, [1978] 3 All ER 1033, [1978] 3 WLR 849, 122 Sol Jo 761, HL 23.5
Alison Jones v Alcom UK Ltd (24 February 2011, unreported) 22.29
Amber Construction Services Ltd v London Interspace HG Ltd [2007] EWHC 3042 (TCC), [2008] Bus LR D46, [2008] 1 EGLR 1, [2008] BLR 74, [2008] 09 EG 202, [2008] 5 Costs LR 715 .... 26.3
Amin v Amin [2007] EWHC 827 (Ch) 22.32
Amoco (UK) Exploration Co v British American Offshore Ltd (No 2) [2002] BLR 135, [2001] All ER (D) 327 (Nov) 24.9, 24.18, 32.5
Andrews v Dawkes (2006), unreported, QBD Birmingham 6.27
Antonelli v Allen [2001] EWCA Civ 1563 .... 19.12

Table of Cases

Antonelli v Wade Gery Farr (a firm) [1994] Ch 205, [1994] 3 WLR 462, CA 23.12
Aoun v Bahri [2002] EWHC 29 (Comm), [2002] 3 All ER 182, [2002] All ER (D) 104 (Feb) 19.8, 19.9
Apex Frozen Foods Ltd v Ali [2007] EWHC 469 (Ch), [2007] 6 Costs LR 818, [2007] BPIR 1437, [2007] All ER (D) 158 (Mar) 40.13
Aquila Design (GRP Products) Ltd v Cornhill Insurance plc [1988] BCLC 134, 3 BCC 364, CLY 2952, CA 19.18
Arkin v Borchard Lines Ltd [2001] CP Rep 108 .... 7.15
Arkin v Borchard Lines Ltd (No 5) [2003] EWHC 2844 (Comm), [2004] 1 Lloyd's Rep 88, [2003] NLJR 1903, [2004] 2 Costs LR 231, [2003] All ER (D) 406 (Nov); revsd sub nom Arkin v Borchard Lines Ltd [2005] EWCA Civ 655, [2005] 3 All ER 613, [2005] 1 WLR 3055, [2005] 2 Lloyd's Rep 187, [2005] NLJR 902, (2005) Times, 3 June, [2005] 4 Costs LR 643, [2005] All ER (D) 410 (May) ... 10.10, 10.13, 40.7, 40.18
Arkin v Borchard Lines Ltd. See Arkin v Borchard Lines Ltd (No 5)
Arrowfield Services v BP Collings (Case No 15 of 2003), unreported, SCCO 3.20
Arthur J S Hall & Co (a firm) v Simons [2002] 1 AC 615, [2000] 3 All ER 673, [2000] 3 WLR 543, [2000] 2 FCR 673, [2000] 2 FLR 545, [2000] Fam Law 806, [2000] 32 LS Gaz R 38, [2000] NLJR 1147, [2000] BLR 407, [2000] EGCS 99, 144 Sol Jo LB 238, [2001] 3 LRC 117, [2000] All ER (D) 1027, HL 23.12
Ashia Centur Ltd v Barker Gillette LLP [2011] EWHC 148 (QB), [2011] 4 Costs LR 576, [2011] All ER (D) 265 (Feb) 1.31
Ashley Cole v News Group (18 October 2006, unreported), SCCO 6.24
Ashworth v Berkeley-Walbrood Ltd (1989) Independent, 9 October, CA 19.2
Astrazeneca UK Ltd v International Business Machines Corpn [2011] EWHC 3373 (TCC), [2012] NLJR 95, [2012] All ER (D) 22 (Jan) 36.4
Atack v Lee [2004] EWCA Civ 1712, [2005] 1 WLR 2643, [2006] RTR 127, [2005] NLJR 24, (2004) Times, 28 December, 149 Sol Jo LB 60, [2005] 2 Costs LR 308, [2004] All ER (D) 262 (Dec) 6.29, 6.38
Atrium Training Services Ltd (in liq), Re, Smailes v McNally [2013] EWHC 1562 (Ch), [2013] All ER (D) 105 (Jun) 15.34
Automotive Latch Systems Ltd v Honeywell International Inc [2006] EWHC 2340 (Comm), [2006] All ER (D) 121 (Sep) 19.7, 19.18
Automotive Latch Systems Ltd v Honeywell International, Inc [2010] EWHC 1031 (Comm) 40.17

B

B (costs), Re [1999] 3 FCR 586, [1999] 2 FLR 221, CA 39.16
B v Pendlebury [2002] EWHC 1797 .... 23.27
BCCI v Ali (13 April 2000, unreported), ChD 37.1
BOS GmbH & Co KG v Cobra UK Automotive Products Division Ltd (in admin) [2012] EWPCC 44, [2012] 6 Costs LR 1083, [2012] All ER (D) 146 (Sep) 26.37
Bailey v IBC Vehicles Ltd [1998] 3 All ER 570, 142 Sol Jo LB 126, CA 12.9, 29.21
Baker v Quantum Clothing Group Ltd, Meridian and Pretty Polly [2008] EWCA Civ 823 .... 18.3
Balmoral Group Ltd v Borealis (UK) Ltd [2006] EWHC 2531 (Comm), [2006] All ER (D) 183 (Oct) 24.24
Bani K/S A/S and K/S A/S Havbulk I v Korea Shipbuilding and Engineering Corpn [1987] 2 Lloyd's Rep 445, CA 19.24
Bank of Credit and Commerce International SA v Ali (No 3) [1999] NLJ 1734 ...... 22.17
Bank of Tokyo-Mitsubishi UFJ Ltd v Baskan Gida Sanayi Ve Pazarlama As [2009] EWHC 1696 (Ch), [2009] NLJR 1066, [2010] 5 Costs LR 657, [2009] All ER (D) 159 (Jul) 22.18, 22.32
Barker (Kim) Ltd v Aegon Insurance Co (UK) Ltd (1989) Times, 9 October, CA 19.7
Barndeal Ltd v Richmond-upon-Thames London Borough Council [2005] EWHC 1377 (QB), [2006] 1 Costs LR 47, [2005] All ER (D) 369 (Jun) 40.4
Barr v Biffa Waste Services Ltd [2009] EWHC 2444 (TCC), [2009] NLJR 1513, [2010] 3 Costs LR 317, [2009] All ER (D) 176 (Oct) 9.29, 17.6, 37.5

Table of Cases

Barr v Biffa Waste Services Ltd [2011] EWHC 1003 (TCC), [2011] 4 All ER 1065, 137 ConLR 125, [2011] NLJR 661, [2011] All ER (D) 25 (May); revsd [2012] EWCA Civ 312, [2013] QB 455, [2012] 3 All ER 380, [2012] 3 WLR 795, [2012] 2 P & CR 99, [2012] PTSR 1527, [2012] HLR 415, 141 ConLR 1, [2012] 14 LS Gaz R 21, [2012] BLR 275, [2012] 13 EG 90 (CS), 156 Sol Jo (no 12) 31, [2012] All ER (D) 141 (Mar) 17.6
Barrister, a (wasted costs order No 1 of 1991), Re [1993] QB 293, [1992] 3 All ER 429, [1992] 3 WLR 662, 95 Cr App Rep 288, [1992] 24 LS Gaz R 31, [1992] NLJR 636, 136 Sol Jo LB 147, (1995) Times, 21 April, CA 23.22
Barrister, a, (wasted costs order), Re [2001] EWCA Crim 1728, [2002] 1 Cr App Rep 207, (2001) Times, 31 July, [2001] All ER (D) 296 (Jul) 23.2
Beasley (by his litigation friend) v Alexander [2012] EWHC 2715 (QB), [2013] 1 WLR 762, (2012) Times, 02 November, [2012] 6 Costs LR 1137, [2012] All ER (D) 75 (Oct) 20.15
Beddoe, Re, Downes v Cottam [1893] 1 Ch 547, 62 LJ Ch 233, 2 R 223, 41 WR 177, 37 Sol Jo 99, 68 LT 595, [1891–4] All ER Rep Ext 1697, CA 18.5, 38.3
Bell Electric Ltd v Aweco Appliance Systems GmbH & Co KG [2002] EWCA Civ 1501, [2003] 1 All ER 344, (2002) Times, 20 November, [2002] All ER (D) 460 (Oct) 19.12
Bensusan v Freedman [2001] All ER (D) 212 (Oct), SC 6.30
Berkeley Administration Inc v McClelland [1990] 2 QB 407, [1990] 1 All ER 958, [1990] 2 WLR 1021, [1990] 2 CMLR 116, [1990] FSR 381, [1990] BCC 272, [1990] NLJR 289, CA 19.5
Bermuda International Securities Ltd v KPMG (a firm) [2001] EWCA Civ 269, 145 Sol Jo LB 70, [2001] All ER (D) 337 (Feb) 22.24
Berry v Spousals (BM 309007) (2007), unreported, Birmingham CC 6.27
Bevan v Power Panels Electrical Systems Ltd [2007] EWHC 90073 (Costs) 6.11, 6.27
Bexbes LLP v Beer [2009] EWCA Civ 628, [2009] All ER (D) 273 (Jun) 6.9
Blackham v Entrepose UK [2004] EWCA Civ 1109, [2004] 35 LS Gaz R 33, (2004) Times, 28 September, 148 Sol Jo LB 945, [2005] 1 Costs LR 68, [2004] All ER (D) 478 (Jul) 20.14
Blackmore v Cummings [2009] EWCA Civ 1276, [2010] 1 WLR 983, [2010] All ER (D) 154 (Mar) 25.1
Blair v Danesh [2009] EWCA Civ 516 .... 6.9, 7.6
Blue Sky One Ltd v Mahan Air [2011] EWCA Civ 544, [2011] All ER (D) 107 (May) 19.18
Blue Sphere Global Ltd v Revenue and Customs Comrs [2010] EWCA Civ 1448, [2011] STC 547, [2011] 2 Costs LO 182, [2010] All ER (D) 207 (Dec) 20.34
Booker Belmont Wholesale Ltd v Ashford Developments Ltd (18 July 2000, unreported), CA 22.31
Booth v Britannia Hotels Ltd [2002] EWCA Civ 579, [2002] All ER (D) 422 (Mar) 28.46
Botham v Khan [2004] EWHC 2602 (QB), [2005] 2 Costs LR 259, [2004] All ER (D) 222 (Nov) 28.20
Bournemouth & Boscombe Athletic Football Club Ltd v Lloyds TSB Bank plc [2004] EWCA Civ 935, [2004] All ER (D) 323 (Jun) 40.12
Bowring (C T) & Co (Insurance) Ltd v Corsi & Partners Ltd [1994] 2 Lloyd's Rep 567, [1995] 1 BCLC 148, [1994] BCC 713, [1994] 32 LS Gaz R 44, 138 Sol Jo LB 140, CA 19.2
Brampton Manor (Leisure) Ltd v McLean [2007] EWHC 3340 (Ch), [2009] BCC 30, [2007] All ER (D) 258 (Dec) 40.4
Brawley v Marczynski (No 2) [2002] EWCA Civ 1453, [2002] 4 All ER 1067, [2003] 1 WLR 813, (2002) Times, 7 November, [2002] All ER (D) 288 (Oct) 24.21
Bray Walker Solicitors (a firm) v Silvera [2010] EWCA Civ 332, [2010] 15 LS Gaz R 18, 154 Sol Jo (no 13) 29, [2010] All ER (D) 261 (Mar) 6.10, 6.35
Brennan v Associated Asphalt (0507905) (2006), unreported, SCCO 6.10
Brierley v Prescott (Ref: 0504718) (2006), unreported, SCCO 6.9
British Cash and Parcel Conveyors Ltd v Lamson Store Service Co Ltd [1908] 1 KB 1006, 77 LJKB 649, [1908–10] All ER Rep 146, 98 LT 875, CA 10.9
British Waterways Board v Norman (1993) 26 HLR 232, [1993] NPC 143, [1994] COD 262, [1993] EGCS 177, (1993) Times, 11 December 4.11, 12.3, 12.5
Brown v Bennett (Wasted Costs) (No 1) (2002), unreported 23.6, 23.9, 23.14

xli

Table of Cases

Brown v Russell Young & Co [2007] EWCA Civ 43, [2007] 2 All ER 453, [2008] 1 WLR 525, [2007] NLJR 222, (2007) Times, 13 February, [2007] 4 Costs LR 552, [2007] All ER (D) 287 (Jan) 37.4
Brown-Quinn v Equity Syndicate Managment Ltd [2011] EWHC 2661 (Comm), [2012] 1 All ER 778, [2012] Lloyd's Rep IR 248, [2011] 44 LS Gaz R 20, [2012] 1 Costs LR 1, [2011] All ER (D) 243 (Oct); revsd in part [2012] EWCA Civ 1633, [2013] 3 All ER 505, [2013] 1 WLR 1740, [2013] 2 CMLR 514, [2013] Lloyd's Rep IR 371, [2013] 1 Costs LR 1, [2012] All ER (D) 104 (Dec) 9.8
Budgen v Andrew Gardner Partnership [2002] EWCA Civ 1125, (2002) Times, 9 September, [2002] All ER (D) 528 (Jul) 22.21
Bullock v London General Omnibus Co [1907] 1 KB 264, 76 LJKB 127, [1904–7] All ER Rep 44, 51 Sol Jo 66, 95 LT 905, 23 TLR 62, CA 22.34
Burchell v Bullard [2005] EWCA Civ 358, [2005] NLJR 593, [2005] BLR 330, 149 Sol Jo LB 477, [2005] 3 Costs LR 507, [2005] All ER (D) 62 (Apr) 21.5, 22.32
Burstein v Times Newspapers Ltd [2002] EWCA Civ 1739, [2003] 03 LS Gaz R 31, (2002) Times, 6 December, 146 Sol Jo LB 277, [2002] All ER (D) 442 (Nov) ... 6.23, 12.3
Byrne v South Sefton (Merseyside) Health Authority [2001] EWCA Civ 1904, [2002] 1 WLR 775, [2002] 01 LS Gaz R 19, (2001) Times, 28 November, 145 Sol Jo LB 268, [2001] All ER (D) 351 (Nov) 23.26

C

C v C (maintenance pending suit: legal costs) [2006] 2 FLR 1207, [2006] Fam Law 739 .... 39.6
C v D [2011] EWCA Civ 646, [2012] 1 All ER 302, [2012] 1 WLR 1962, 136 ConLR 109, [2011] 2 EGLR 95, [2011] NLJR 780, [2011] 5 Costs LR 773, [2011] All ER (D) 287 (May) 20.5
C v W [2008] EWCA Civ 1459, [2009] 4 All ER 1129, [2009] RTR 199, [2009] NLJR 73, [2009] 1 Costs LR 123, [2008] All ER (D) 239 (Dec) 6.29, 6.30, 6.33
CMCS Common Market Commercial Services AVV v Taylor [2011] EWHC 324 (Ch), [2011] NLJR 365, (2011) Times, 15 April, [2011] 3 Costs LO 259, [2011] All ER (D) 269 (Feb) 23.28, 23.29
Calderbank v Calderbank [1976] Fam 93, [1975] 3 All ER 333, [1975] 3 WLR 586, [1976] Fam Law 93, 119 Sol Jo 490, CA 20.37, 39.18, 39.21, 39.22, 39.24, 39.25
Callery v Gray [2001] EWCA Civ 1117, [2001] 3 All ER 833, [2001] 1 WLR 2112, [2001] NLJR 1129, (2001) Times, 18 July, [2001] 2 Costs LR 163, [2001] All ER (D) 213 (Jul); affd sub nom Callery v Gray [2001] EWCA Civ 1246, [2001] 4 All ER 1, [2001] 1 WLR 2142, [2001] 35 LS Gaz R 33, (2001) Times, 24 October, 145 Sol Jo LB 204, [2001] 2 Costs LR 205, [2001] All ER (D) 470 (Jul); affd [2002] UKHL 28, [2002] 1 WLR 2000, [2002] 3 All ER 417, [2003] RTR 71, (2002) Times, 2 July, [2002] NLJR 1031, [2002] 2 Costs LR 205, [2002] All ER (D) 233 (Jun) . 4.16, 4.18, 6.25, 6.29, 6.30, 6.43, 6.44, 9.20
Camertown Timber Merchants Ltd v Sidhu [2011] EWCA Civ 1041, [2011] All ER (D) 47 (Sep) 22.33
Carr v Leeds City Council (1999) 32 HLR 753, [2000] COD 10, (1999) Times, 12 November, [1999] All ER (D) 1117, DC 12.3
Carver v BAA plc [2008] EWCA Civ 412, [2008] 3 All ER 911, [2009] 1 WLR 113, [2008] PIQR P288, (2008) Times, 4 June, [2008] 5 Costs LR 779, [2008] All ER (D) 295 (Apr) 20.1
Chamberlain v Boodle and King (a firm) [1982] 3 All ER 188, [1982] 1 WLR 1443, 125 Sol Jo 257, CA 2.8, 8.6, 8.8, 31.1
Chapman (TGA) Ltd v Christopher [1998] 2 All ER 873, [1998] 1 WLR 12, [1998] Lloyd's Rep IR 1, CA 40.9
Cheesman, Re [1891] 2 Ch 289, 60 LJ Ch 714, 39 WR 497, 64 LT 602, CA 3.20
Chemistree Homecare Ltd v Roche Products Ltd [2011] EWHC 1579 (Ch), [2011] All ER (D) 115 (Jun) 19.18
Choudhury v Kingston Hospital Trust [2006] EWHC 90057 (Costs) 6.27
Church Comrs v Ibrahim [1997] 1 EGLR 13, [1997] 03 EG 136, CA 2.3

Table of Cases

Cie Noga D'Importation et D'Exportation SA v Abacha (as personal representatives of Sani Abacha (dec'd)) (No 4) [2004] EWHC 2601 (Comm), [2004] All ER (D) 292 (Nov) 19.19
Citation plc v Ellis Whittam Ltd [2012] EWHC 764 (QB), [2012] NLJR 504, [2012] 5 Costs LR 826, [2012] All ER (D) 227 (Mar) 22.24
Claims Direct Test Cases, Re [2003] EWCA Civ 136, [2003] 4 All ER 508, [2003] 2 All ER (Comm) 788, [2003] Lloyd's Rep IR 677, [2003] 13 LS Gaz R 26, (2003) Times, 18 February, 147 Sol Jo LB 236, [2003] 2 Costs LR 254, [2003] All ER (D) 160 (Feb) 6.29, 6.43, 6.56
Clarke v Maltby [2010] EWHC 1856 (QB), [2010] All ER (D) 253 (Jul) 24.13
Cobbett v Wood [1908] 2 KB 420, 77 LJKB 878, 52 Sol Jo 517, 99 LT 482, 24 TLR 615, CA 2.3
Coles v Barnsley Metropolitan BC (1999) 32 HLR 753, [2000] COD 10, (1999) Times, 12 November, [1999] All ER (D) 1117, DC 12.3
Colour Quest Ltd v Total Downstream UK plc [2009] EWHC 823 (Comm), [2010] 2 Costs LR 140, [2009] All ER (D) 152 (Apr) 32.5
Company, a (No 006798 of 1995), Re [1996] 2 All ER 417, [1996] 1 WLR 491, [1996] 2 BCLC 48 .... 23.16
Condliffe v Hislop [1996] 1 All ER 431, [1996] 1 WLR 753, [1996] EMLR 25, CA 19.3
Connaughton v Imperial College Healthcare NHS Trust [2010] EWHC 90173 (Costs) 6.4
Continental Assurance Co of London plc (in liq), Re (13 June 2001, unreported) 24.22
Contractreal Ltd v Davies [2001] EWCA Civ 928, [2001] All ER (D) 231 (May) 29.18
Cooper v P & O Stena Line Ltd [1999] 1 Lloyd's Rep 734 .... 24.16
Cox v MGN Ltd [2006] EWHC 1235 (QB), [2006] All ER (D) 396 (May) 6.33, 31.14, 31.29
Crabtree (B J) (Insulation) Ltd v GPT Communications Systems Ltd (1990) 59 BLR 43, CA 19.2
Crane v Canons Leisure Centre [2007] EWCA Civ 1352, [2008] 2 All ER 931, [2008] 1 WLR 2549, [2008] NLJR 103, [2008] 1 Costs LR 132, [2007] All ER (D) 281 (Dec) 29.17, 35.6
Crema v Cenkos Securities plc [2011] EWCA Civ 10, [2011] 4 Costs LR 552 .... 5.23
Crook v Birmingham City Council [2007] EWHC 1415 (Admin), [2007] NLJR 939, [2007] 5 Costs LR 732, [2007] All ER (D) 191 (Jun) 6.23
Currey v Currey (No 2) [2006] EWCA Civ 1338, [2007] 1 FLR 946, [2006] NLJR 1651, (2006) Times, 3 November, 150 Sol Jo LB 1393, [2007] 2 Costs LR 227, [2006] All ER (D) 218 (Oct) 39.6, 39.39
Cuthbert v Gair (t/a Bowes Manor Equestrian Centre) (2008) 152 Sol Jo (no 38) 29 .... 35.6

D

DB Casqueiro (in a matter of wasted costs) v Barclays Bank plc (UKEAT/0085/12MAA), unreported 23.33
Daniels v Metropolitan Police Comr [2005] EWCA Civ 1312 .... 21.5
Darby v The Law Society [2003] EWHC 2270 (Admin), [2003] All ER (D) 210 (Oct) 1.19
Dardana Ltd v Yukos Oil Co [2002] EWCA Civ 543, [2002] 1 All ER (Comm) 819, [2002] 2 Lloyd's Rep 326, [2002] ArbLR 48, [2002] All ER (D) 126 (Apr) 19.24
Data Delecta Aktiebolag v MSL Dynamics: C-43/95 [1996] ECR I-4661, [1996] All ER (EC) 961, [1996] 3 CMLR 741, ECJ 19.5
David Smith v Countrywide Farmers plc (11 February 2010, unreported), QBD . 6.24, 6.27
David Truex (a firm) v Kitchin [2007] EWCA Civ 618, [2007] 2 FLR 1203, [2007] Fam Law 903, [2007] NLJR 1011, (2007) Times, 29 August, [2007] 4 Costs LR 587, [2007] All ER (D) 53 (Jul) 10.1
David Warner and SMP Trustees Ltd v Merriman White (A Firm) [2008] EWHC 1129 (Ch) 40.19
Davidsons v Jones-Fenleigh (1980) 124 Sol Jo 204, (1980) Times, 11 March, CA 2.7

Table of Cases

Davy-Chiesman v Davy-Chiesman [1984] Fam 48, [1984] 1 All ER 321, [1984] 2 WLR 291, 127 Sol Jo 805, [1984] LS Gaz R 44, CA 23.13, 23.15
Dawson v First Choice (5BM15907) (12 March 2007, unreported) 15.8
Day v Day [2006] EWCA Civ 415, [2006] All ER (D) 184 (Mar) 22.17
De Beer v Kanaar & Co (a firm) [2001] EWCA Civ 1318, [2002] 3 All ER 1020, [2003] 1 WLR 38, [2001] All ER (D) 40 (Aug) 19.5
Debtor, a (No 88 of 1991), Re [1993] Ch 286, [1992] 4 All ER 301, [1992] 3 WLR 1026, [1992] LS Gaz R 34, [1992] NLJR 1039, 136 Sol Jo LB 206 .... 3.27, 3.28
Dempsey v Johnstone [2003] EWCA Civ 1134, [2003] All ER (D) 515 (Jul) ... 23.9, 23.25
Denton v Denton [2004] EWHC 1308 (Fam), [2004] 2 FLR 594, [2005] Fam Law 353 .... 39.41
Dickinson (t/a Dickinson Equipment Finance) v Rushmer (t/a FJ Associates) [2002] NLJR 58, ChD 12.9
Dickinson (t/a Dickinson Equipment Finance) v Rushmer (t/a FJ Associates) [2002] Costs LR 128, [2001] All ER (D) 369 (Dec) 28.45
Dickinson v Tesco plc [2013] EWCA Civ 226, CA 22.19
Digicel (St Lucia) Ltd (a company registered under the laws of St Lucia) v Cable & Wireless plc [2010] EWHC 888 (Ch), [2010] 5 Costs LR 709, [2010] All ER (D) 166 (Apr) 24.17
Dix v Townend [2008] EWHC 90117 (Costs) 9.31
Dole v ECT (17 September 2007, unreported), SCCO 6.27
Dolphin Quays Developments Ltd v Mills [2007] EWHC 1180 (Ch), [2007] 4 All ER 503, [2008] 1 BCLC 1, [2008] 2 Costs LR 220, [2007] BPIR 1482, [2007] All ER (D) 270 (May); affd [2008] EWCA Civ 385, [2008] 4 All ER 58, [2008] 1 WLR 1829, [2008] Bus LR 1520, [2008] 2 BCLC 774, [2008] All ER (D) 257 (Apr), sub nom Mills v Birchall [2008] BCC 471, [2008] 4 Costs LR 599, [2008] BPIR 607 .... 40.13
Drake v Fripp [2011] EWCA Civ 1282, [2012] 2 Costs LR 264 .... 6.37
Drew v Whitbread plc [2010] EWCA Civ 53, [2010] 1 WLR 1725, [2010] NLJR 270, (2010) Times, 30 March, [2010] 2 Costs LR 213, [2010] All ER (D) 104 (Feb) 22.23, 26.41
Duffy v Port Ramsgate (JOH 0306901) (2004), unreported, SCCO 6.18
Dumrul v Standard Chartered Bank [2010] EWHC 2625 (Comm), [2010] NLJR 1532, [2010] All ER (D) 216 (Oct) 19.2
Dunn v Mici [2008] EWHC 90115 (Costs) 6.12
Dunnett v Railtrack plc (in railway administration) [2002] EWCA Civ 303, [2002] 2 All ER 850 .... 21.2
Dymocks Franchise Systems (NSW) Pty Ltd v Todd [2004] UKPC 39, [2005] 4 All ER 195, [2004] 1 WLR 2807, [2004] NLJR 1325, 148 Sol Jo LB 971, [2005] 1 Costs LR 52, [2005] 3 LRC 719, [2004] All ER (D) 420 (Jul) 40.7
Dyson Ltd v Hoover Ltd [2003] EWHC 624 (Ch), [2003] 2 All ER 1042, [2004] 1 WLR 1264, [2003] IP & T 591, [2003] 15 LS Gaz R 25, (2003) Times, 18 March, [2003] All ER (D) 252 (Jul) 25.1
Dyson Technology Ltd v Strutt [2007] EWHC 1756 (Ch), [2007] 4 Costs LR 597, [2007] All ER (D) 381 (Jul) 22.21, 22.32

E

E C-L v D M (costs in Hague Convention proceedings) [2005] EWHC 588 (Fam), (2005) Times, 10 May, [2005] All ER (D) 187 (Apr), sub nom E C-L v D M (child abduction: costs) [2005] 2 FLR 772, [2005] Fam Law 606, [2005] 4 Costs LR 576 .... 39.15
E Ivor Hughes Education Foundation v Leach [2005] EWHC 1317 (Ch), [2005] All ER (D) 127 (Jun) 20.37
East West Corpn v Dampskibsselskabet AF, 1912, Aktieselskab (a body corporate) [2003] EWCA Civ 174, [2003] 1 Lloyd's Rep 265n, [2003] 12 LS Gaz R 31, [2003] 14 LS Gaz R 27, 147 Sol Jo LB 234, [2003] All ER (D) 249 (Feb), sub nom East West Corp v Dampskibsselskabet AF, 1912, Aktieselskab (a body corporate); Utaniko Ltd v P (2003) Times, 21 February 20.4
Eastwood, Re, Lloyds Bank Ltd v Eastwood [1975] Ch 112, [1973] 3 All ER 1079, [1973] 3 WLR 795, 117 Sol Jo 487; revsd [1975] Ch 112, [1974] 3 All ER 603, [1974] 3 WLR 454, 118 Sol Jo 533, CA 8.7, 31.8

Table of Cases

Easyair Ltd (t/a Openair) v Opal Telecom Ltd [2009] EWHC 779 (Ch), [2009] 6 Costs LR 882, [2009] All ER (D) 104 (Apr) 24.22
Edwards v Smiths Dock Ltd [2004] EWHC 1116 (QB), [2004] 3 Costs LR 440, [2004] All ER (D) 181 (May) 6.30
Electricity Supply Nominees Ltd v Farrell [1997] 2 All ER 498, [1997] 1 WLR 1149, CA 32.5
Ellerton v Harris [2004] EWCA Civ 1712, [2005] 1 WLR 2643, [2006] RTR 127, [2005] NLJR 24, (2004) Times, 28 December, 149 Sol Jo LB 60, [2005] 2 Costs LR 308, [2004] All ER (D) 262 (Dec) 6.29, 6.38
Elliott v Pensions Ombudsman [1997] 45 LS Gaz R 27, 142 Sol Jo LB 19, [1998] OPLR 21 .... 40.14
Elvanite Full Circle Ltd v AMEC Earth & Environmental (UK) Ltd [2013] EWHC 1643 (TCC), [2013] All ER (D) 187 (Jun) 15.12, 15.34, 24.5, 25.1
Emezie v Secretary of State for the Home Department [2013] EWCA Civ 733, [2013] All ER (D) 243 (Jun) 22.31
English v Emery Reimbold & Strick Ltd [2002] EWCA Civ 605, [2002] 3 All ER 385, [2002] 1 WLR 2409, [2003] IRLR 710, (2002) Times, 10 May, [2002] All ER (D) 302 (Apr) 33.2
Envis v Thakkar (1995) Times, 2 May, [1997] BPIR 189, CA 19.10
Epsom College v Pierse Contracting Southern Ltd (formerly Biseley Construction Ltd) (in liq) [2011] EWCA Civ 1449, 142 ConLR 59, [2012] 3 Costs LR 451, [2011] All ER (D) 153 (Dec) 20.5
Equitas Ltd v Horace Holman & Co Ltd [2008] EWHC 2287 (Comm), [2009] 1 BCLC 662, [2008] All ER (D) 35 (Oct) 40.3
Eschig v UNIQA Sachversicherung AG: C-199/08 [2009] ECR I-08295, [2010] 1 All ER (Comm) 576, [2010] 1 CMLR 131, [2010] Bus LR 1404, [2009] All ER (D) 78 (Sep), ECJ 9.8
Eurocross Sales Ltd v Cornhill Insurance plc [1995] 4 All ER 950, [1995] 1 WLR 1517, [1995] 2 BCLC 384, [1995] BCC 991, [1996] LRLR 1, CA 19.19
Europa Holdings Ltd v Circle Industries (UK) plc [1993] BCLC 320, CA 19.18
Europeans Ltd v Revenue and Customs Comrs [2011] EWHC 948 (Ch), [2011] STC 1449, [2011] 4 Costs LO 447, [2011] All ER (D) 147 (Apr) 40.4
Evans v Evans [1990] 2 All ER 147, [1990] 1 WLR 575n, [1990] FCR 498, [1990] 1 FLR 319, [1990] Fam Law 215, 134 Sol Jo 785, [1990] 12 LS Gaz R 40, [1990] NLJR 291 .... 39.42
Everglade Maritime Inc v Schiffahrtsgesellschaft Detlef von Appen mbH, The Maria [1993] QB 780, [1993] 3 All ER 748, [1993] 3 WLR 176, [1993] 2 Lloyd's Rep 168, CA 41.5
Eweida v British Airways plc [2009] EWCA Civ 1025, 153 Sol Jo (no 40) 37, [2009] All ER (D) 161 (Oct) 18.2
Excelsior Commercial & Industrial Holdings Ltd v Salisbury Hamer Aspden & Johnston (Costs) [2002] EWCA Civ 879 .... 24.8
Ezair v Ezair [2012] EWCA Civ 893, [2013] 1 FLR 281 .... 39.27, 39.35

F

F & C Alternative Investments (Holdings) Ltd v Barthelemy [2012] EWCA Civ 843, [2012] 4 All ER 1096, [2013] 1 WLR 548, [2013] Bus LR 186, [2012] NLJR 872, [2013] 1 Costs LR 35, [2012] All ER (D) 145 (Jun) 20.38, 22.17
Falmouth House Freehold Co Ltd v Morgan Walker LLP [2010] EWHC 3092 (Ch), [2010] NLJR 1686, [2011] 2 Costs LR 292, [2010] All ER (D) 256 (Nov) 3.20
Fattal v Walbrook Trustees (Jersey) Ltd [2009] EWHC 1674 (Ch), [2009] 4 Costs LR 591, [2009] All ER (D) 190 (Jul) 32.5
Fenton v Holmes [2007] EWHC 2476 (Ch), [2008] 2 Costs LR 238, [2007] All ER (D) 12 (Jun) 6.12
Filmlab Systems International Ltd v Pennington [1994] 4 All ER 673, [1995] 1 WLR 673, [1993] NLJR 1405 .... 23.2
Findlay v Cantor Index Ltd [2008] EWHC 90116 (Costs) 6.24
Finley v Glaxo Laboratories Ltd [1997] Costs Law Rep 106 .... 31.9, 31.19

xlv

Table of Cases

Firle Investments Ltd v Datapoint International Ltd [2001] EWCA Civ 1106, [2001] NPC 106, [2001] All ER (D) 258 (Jun) 21.3
Fisher Meredith v JH and PH (financial remedy: appeal: wasted costs) [2012] EWHC 408 (Fam), [2012] 2 FCR 241, [2012] 2 FLR 536, [2012] All ER (D) 157 (Mar) 39.35
Flatman v Germany [2013] EWCA Civ 278, [2013] 4 All ER 349, [2013] 1 WLR 2676, [2013] NLJR 16, 157 Sol Jo (no 15) 31, [2013] All ER (D) 41 (Apr) 2.12, 9.30, 12.3, 40.18
Flender Werft AG v Aegean Maritime Ltd [1990] 2 Lloyd's Rep 27 .... 19.2
Flynn v Scougall [2004] EWCA Civ 873, [2004] 3 All ER 609, [2004] 1 WLR 3069, 148 Sol Jo LB 880, [2005] 1 Costs LR 38, [2004] All ER (D) 226 (Jul) 20.8
Fons HF v Corporal Ltd [2013] EWHC 1278 (Ch), [2013] 4 Costs LO 646, [2013] All ER (D) 292 (May) 15.22, 15.34
Forcelux Ltd v Binnie [2009] EWCA Civ 1077, [2010] HLR 340, [2009] 5 Costs LR 825, [2009] All ER (D) 234 (Oct) 36.3
Forde v Birmingham City Council [2009] EWHC 12 (QB), [2010] 1 All ER 802, [2009] 1 WLR 2732, [2009] 2 Costs LR 206, [2009] All ER (D) 64 (Jan) 5.14, 6.34
Fortune v Roe [2011] EWHC 2953 (QB), [2012] RTR 480, [2012] 2 Costs LR 288, [2011] All ER (D) 91 (Nov) 6.9, 6.29
Fosberry v Revenue and Customs Comrs [2006] EWHC 90061 (Costs) 6.3
Fox v Foundation Piling Ltd [2011] EWCA Civ 790, [2011] 6 Costs LR 961, [2011] All ER (D) 61 (Jul) 20.39, 20.41, 22.17, 22.18, 22.20, 22.31
Franses (Ian) v Al Assad [2007] EWHC 2442 (Ch), [2007] All ER (D) 415 (Oct) 24.11
Frascati (in chambers), Re (2 December 1981, unreported), QBD 31.12, 31.26
Fred Perry (Holdings) Ltd v Brands Plaza Trading Ltd [2012] EWCA Civ 224, [2012] FSR 807, [2012] 6 Costs LR 1007, [2012] All ER (D) 77 (Jun) 15.28
French v Carter Lemon Camerons LLP [2012] EWCA Civ 1180, [2012] NLJR 1155, 156 Sol Jo (no 34) 27, [2012] 6 Costs LO 879, [2012] All ER (D) 14 (Sep) 3.33
French v Groupama Insurance Co Ltd [2011] EWCA Civ 1119, [2011] 4 Costs LO 547, [2011] All ER (D) 91 (Oct) 20.38
French (A Martin) v Kingswood Hill Ltd [1961] 1 QB 96, [1960] 2 All ER 251, [1960] 2 WLR 947, 104 Sol Jo 447, CA 20.37
Freudiana Holdings, Re (1995) Times, 4 December, CA 23.8

G

G (costs: child case), Re [1999] 3 FCR 463, [1999] 2 FLR 250, [1999] Fam Law 381, CA 39.16
G v G [1985] 2 All ER 225, [1985] 1 WLR 647, [1985] FLR 894, [1985] Fam Law 321, 129 Sol Jo 315, HL 33.5
G v G (maintenance pending suit: legal costs) [2002] EWHC 306 (Fam), [2002] 3 FCR 339, [2003] 2 FLR 71, [2003] Fam Law 393, [2002] All ER (D) 306 (Feb) 39.6
G v G [2009] EWHC 2080 .... 39.40
GS v L (No 2) (financial remedies: costs) [2011] EWHC 2116 (Fam), [2013] 1 FLR 407, [2012] Fam Law 802 .... 39.27, 39.30, 39.31
GW v RW [2003] EWHC 611 (Fam), [2003] 2 FCR 289, [2003] 2 FLR 108, [2003] Fam Law 386, [2003] All ER (D) 40 (May) 39.22
Gamlen Chemical Co (UK) Ltd v Rochem Ltd [1980] 1 All ER 1049, [1980] 1 WLR 614, 123 Sol Jo 838, CA 3.33
Gannet Shipping Ltd v Eastrade Commodities Inc [2002] 1 All ER (Comm) 297, [2002] 1 Lloyd's Rep 713, [2001] ArbLR 27, [2001] All ER (D) 74 (Dec) 41.11
Garbutt v Edwards [2005] EWCA Civ 1206, [2006] 1 All ER 553, [2006] 1 WLR 2907, [2005] 43 LS Gaz R 30, (2005) Times, 3 November, [2006] 1 Costs LR 143, [2005] All ER (D) 316 (Oct) 1.15, 1.19, 6.28
Gardiner v FX Music Ltd [2000] All ER (D) 144, ChD 40.11
Garrett v Halton Borough Council; Myatt v National Coal Board [2006] EWCA Civ 1017, [2007] 1 All ER 147, [2007] 1 WLR 554, (2006) Times, 25 July, [2006] All ER (D) 239 (Jul) 6.5, 6.12
Gaynor v Central West London Buses Ltd [2006] EWCA Civ 1120, [2007] 1 All ER 84, [2007] 1 WLR 1045, [2006] NLJR 1324, (2006) Times, 25 August, [2007] 1 Costs LR 33, [2006] All ER (D) 453 (Jul) 6.3

Table of Cases

Gazley v Wade [2004] EWHC 2675 (QB), [2005] 1 Costs LR 129, [2004] All ER (D) 291 (Nov) 31.30
Gbangbola v Smith & Sherriff Ltd [1998] 3 All ER 730 .... 41.6
Gemma Ltd v Gimson [2005] EWHC 69 (TCC), 102 ConLR 87, [2005] BLR 163, [2005] All ER (D) 79 (Feb) 40.7
General Mediterranean Holdings SA v Patel [1999] 3 All ER 673, [2000] 1 WLR 272, [1999] NLJR 1145 .... 23.17
General of Berne Insurance Co v Jardine Reinsurance Management Ltd [1998] 2 All ER 301, [1998] 1 WLR 1231, 142 Sol Jo LB 86, [1998] Lloyd's Rep IR 211, [1998] All ER (D) 47, CA 12.8
Ghadami v Lyon Cole Insurance Group Ltd [2010] EWCA Civ 767, [2010] 6 Costs LR 903, [2010] All ER (D) 126 (Jul) 12.12
Giambrone v JMC Holidays Ltd (formerly Sunworld Holidays Ltd) [2002] EWHC 2932 (QB), [2003] 1 All ER 982, [2003] NLJR 58, [2003] 2 Costs LR 189, [2002] All ER (D) 359 (Dec) 28.46
Gibbon v Manchester City Council [2010] EWCA Civ 726, [2011] 2 All ER 258, [2010] 1 WLR 2081, [2010] 3 EGLR 85, [2010] 36 EG 120, [2010] 27 EG 84 (CS), [2010] 5 Costs LR 828, [2010] All ER (D) 218 (Jun) 20.2, 20.5, 20.16
Gibson's Settlement Trusts, Re, Mellors v Gibson [1981] Ch 179, [1981] 1 All ER 233, [1981] 2 WLR 1, 125 Sol Jo 48 .... 19.20, 22.14, 29.18
Giles v Thompson [1994] 1 AC 142, [1993] 3 All ER 321, [1993] 2 WLR 908, [1993] RTR 289, [1993] 27 LS Gaz R 34, 137 Sol Jo LB 151, HL 10.9
Gimex International Groupe Import Export v Chill Bag Co Ltd [2012] EWPCC 34, [2012] NLJR 1259, [2012] 6 Costs LR 1069, [2012] All ER (D) 117 (Sep) 26.34
Gliddon v Lloyd Maunder Ltd [2003] NLJR 318, [2003] All ER (D) 225 (Feb), SC 6.18
Global Marine Drillships Ltd v La Bella [2010] EWHC 2498 (Ch), [2011] 2 Costs LR 183 .... 31.24
Gloucestershire County Council v Evans [2008] EWCA Civ 21, [2008] 1 WLR 1883, [2008] NLJR 219, [2008] 2 Costs LR 308, [2008] All ER (D) 284 (Jan) 6.5
Gojkovic v Gojkovic (No 2) [1992] Fam 40, [1992] 1 All ER 267, [1991] 3 WLR 621, [1991] FCR 913, [1991] 2 FLR 233, [1991] Fam Law 378, CA 39.8, 39.12
Goldman v Hesper [1988] 3 All ER 97, [1988] 1 WLR 1238, [1989] 1 FLR 195, [1989] Fam Law 152, [1988] NLJR 272, CA 28.45
Gomba Holdings (UK) Ltd v Minories Finance Ltd (No 2) [1993] Ch 171, [1992] 4 All ER 588, [1992] 3 WLR 723, [1993] BCLC 7, [1992] BCC 877, 12 LDAB 217, CA 2.3, 36.2
Goodwood Recoveries Ltd v Breen [2005] EWCA Civ 414, [2006] 2 All ER 533, [2006] 1 WLR 2723, 149 Sol Jo LB 509, [2007] 2 Costs LR 147, [2005] All ER (D) 226 (Apr) 40.12
Gordano Building Contractors Ltd v Burgess [1988] 1 WLR 890, 132 Sol Jo 1091, [1988] NLJR 127, CA 19.23
Gower Chemicals Group Litigation v Gower Chemicals Ltd [2008] EWHC 735 (QB), [2008] 4 Costs LR 582, [2008] All ER (D) 233 (Apr) 28.45
Grand v Gill [2011] EWCA Civ 902, [2011] NLJR 1100, [2011] 6 Costs LR 977, [2011] All ER (D) 249 (Jul) 35.5
Gray v Buss Merton (a firm) [1999] PNLR 882 .... 1.3
Gray v Going Places Leisure Travel Ltd [2005] EWCA Civ 189, [2005] 3 Costs LR 405, [2005] All ER (D) 94 (Feb) 23.34
Gray v Toner (11 November 2010, unreported), Liverpool County Court 5.32
Great Future International Ltd v Sealand Housing Corp [2003] EWCA Civ 682, [2003] All ER (D) 365 (May) 19.21
Greville v Sprake [2001] EWCA Civ 234 .... 35.8
Griffiths v Metropolitan Police Comr [2003] EWCA Civ 313, [2003] All ER (D) 32 (Mar) 22.2
Griffiths v Solutia UK Ltd (formerly Monsanto Chemicals Ltd) [2001] EWCA Civ 736, [2001] 2 Costs LR 247, [2001] All ER (D) 196 (Apr) 17.2, 33.5
Gulf Azov Shipping Co Ltd v Idisi [2004] EWCA Civ 292, [2004] All ER (D) 284 (Mar) 40.16
Gundry v Sainsbury [1910] 1 KB 645, 79 LJKB 713, 54 Sol Jo 327, 102 LT 440, 26 TLR 321, [1908–10] All ER Rep Ext 1170, CA 12.3

xlvii

Table of Cases

H

HB (mother) v PB (father), OB (a child by his guardian) & Croydon London Borough Council (respondent on issues of costs only) [2103] EWHC 1956 (Fam) 40.20
HH v BLW (appeal: costs: proportionality) [2012] EWHC 2199 (Fam), [2013] 1 FLR 420 ...... 39.16
HLB Kidsons (a firm) v Lloyds Underwriters subsribing to Lloyd's Policy No. 621/PKIDOO101 [2007] EWHC 2699 (Comm), [2008] 3 Costs LR 427, [2007] All ER (D) 341 (Nov) 24.24
Haji-Ioannou v Frangos [2006] EWCA Civ 1663, [2007] 3 All ER 938, [2006] NLJR 1918, [2007] 2 Costs LR 253, [2006] All ER (D) 72 (Dec) 28.20
Hallam-Peel & Co v Southwark London Borough Council [2008] EWCA Civ 1120, [2009] 2 Costs LR 269, [2008] All ER (D) 200 (Oct) 23.27
Halloran v Delaney [2002] EWCA Civ 1258, [2003] 1 All ER 775, [2003] 1 WLR 28, [2003] RTR 147, [2002] NLJR 1386, [2003] PIQR P 71, [2002] All ER (D) 30 (Sep) 6.29
Halsey v Milton Keynes General NHS Trust; Steel v Joy [2004] EWCA Civ 576, [2004] 4 All ER 920, [2004] 1 WLR 3002, 81 BMLR 108, [2004] 22 LS Gaz R 31, [2004] NLJR 769, (2004) Times, 27 May, 148 Sol Jo LB 629, [2004] 3 Costs LR 393, [2004] All ER (D) 125 (May) 21.4, 21.5
Hamilton v Al Fayed [2002] EWCA Civ 665, [2003] QB 1175, [2002] 3 All ER 641, [2003] 2 WLR 128, [2002] 25 LS Gaz R 34, (2002) Times, 17 June, 146 Sol Jo LB 143, [2002] EMLR 931, [2002] All ER (D) 266 (May) 40.7, 40.16
Hammersmatch Properties (Welwyn) Ltd v Saint-Gobain Ceramics and Plastics Ltd [2013] EWHC 2227 (TCC), [2013] All ER (D) 303 (Jul) 20.39, 22.22
Hanley v Smith and MIB [2009] EWHC 90144 (Costs) 6.9, 6.32
Harold v Smith (1860) 5 H & N 381, 29 LJ Ex 141, 6 Jur NS 254, 8 WR 447, 2 LT 556 ...... 12.2, 12.3
Harrington v Wakeling [2007] EWHC 1184 (Ch), 151 Sol Jo LB 744, [2007] 5 Costs LR 710, [2007] All ER (D) 317 (May) 6.34
Harris v Moat Housing Group South Ltd [2007] EWHC 3092 (QB), [2008] 1 WLR 1578, [2008] NLJR 67, [2008] 2 Costs LR 294, [2007] All ER (D) 323 (Dec) 29.21
Harris v Wallis [2006] EWHC 630 (Ch), (2006) Times, 12 May, [2006] All ER (D) 158 (Mar) 19.11
Harrison v Tew [1990] 2 AC 523, [1990] 1 All ER 321, [1990] 2 WLR 210, 134 Sol Jo 374, [1990] NLJR 132, HL 3.21
Hart v Aga Khan Foundation (UK) [1984] 2 All ER 439, [1984] 1 WLR 994, 128 Sol Jo 531, [1984] LS Gaz R 2537, CA 35.8
Hawksford Trustees Jersey Ltd v Stella Global UK Ltd [2012] EWCA Civ 987, [2012] 1 WLR 3581, [2013] Lloyd's Rep IR 337, [2012] 5 Costs LR 886, [2012] All ER (D) 337 (Jul) 6.46, 9.22
Hawley v Luminar Leisure Ltd [2006] EWCA Civ 18, [2006] Lloyd's Rep IR 307, [2006] IRLR 817, 150 Sol Jo LB 163, [2006] All ER (D) 158 (Jan) 20.11
Hay v Szterbin [2010] EWHC 1967 (Ch), [2010] 6 Costs LR 926, [2010] All ER (D) 336 (Jul) 22.32
Haydon v Strudwick [2010] EWHC 90164 (Costs) 6.52
Hazlett v Sefton Metropolitan Borough Council [2000] 4 All ER 887, [1999] NLJR 1869, [2001] 1 Costs LR 89, DC 12.9
Henderson v All Around the World Recordings Ltd [2013] EWPCC 19, [2013] All ER (D) 301 (Mar) 26.34
Hendry v Chartsearch Ltd (1998) Times, 16 September, [1998] CLC 1382, CA 19.3
Henry v BBC [2005] EWHC 2503 (QB), [2006] 1 All ER 154, [2005] NLJR 1780, [2006] 3 Costs LR 412, [2005] All ER (D) 149 (Nov) 17.5
Henry v News Group Newspapers Ltd [2013] EWCA Civ 19, [2013] 2 All ER 840, [2013] IP & T 660, [2013] NLJR 140, [2013] 2 Costs LR 334, [2013] All ER (D) 192 (Jan) 15.13, 15.34, 15.35
Heron v TNT (UK) Ltd [2013] EWCA Civ 469, [2013] 3 All ER 479, [2013] NLJR 21, [2013] All ER (D) 28 (May) 12.3, 40.18
Hextalls (a firm) v Al-Sami [2009] EWHC 3678 (QB) 25.1
Hickman v Blake Lapthorn [2006] EWHC 12 (QB), [2006] 3 Costs LR 452, [2006] All ER (D) 67 (Jan) 21.5

Higgins v Ministry of Defence [2010] EWHC 654 (QB), 154 Sol Jo (no 14) 28, [2010] 6 Costs LR 867, [2010] All ER (D) 281 (Mar) 31.30
Higgs v Camden and Islington Health Authority [2003] EWHC 15 (QB), 72 BMLR 95, [2003] 2 Costs LR 211, [2003] All ER (D) 76 (Jan) 31.21
Hill v Archbold [1968] 1 QB 686, [1967] 3 All ER 110, [1967] 3 WLR 1218, 111 Sol Jo 543, CA 40.18
Hobbs v Marlowe [1978] AC 16, [1977] 2 All ER 241, [1977] 2 WLR 777, 120 Sol Jo 838, CA; affd [1978] AC 16, [1977] 2 All ER 241, [1977] 2 WLR 777, [1977] RTR 253, 121 Sol Jo 272, HL 26.46
Hodgson v Imperial Tobacco Ltd [1998] 2 All ER 673, [1998] 1 WLR 1056, [1998] 15 LS Gaz R 31, [1998] NLJR 241, 142 Sol Jo LB 93, [1998] All ER (D) 48, CA ... 40.18
Hogan v Hogan (No 2) [1924] 2 IR 14 .... 19.16
Hoist UK Ltd v Reid Lifting Ltd [2010] EWHC 1922 (Ch), [2011] Bus LR D58, (2010) Times, 5 October, [2011] 1 Costs LR 36 .... 22.7
Hollins v Russell [2003] EWCA Civ 718, [2003] 28 LS Gaz R 30, [2003] 4 All ER 590, [2003] 1 WLR 2487, [2003] NLJR 920, (2003) Times, 10 June, 147 Sol Jo LB 662, [2003] All ER (D) 311 (May) 4.2, 6.23, 6.24, 6.25, 6.26, 8.7, 12.9, 28.45, 29.21
Holmes v Alfred McAlpine Homes (Yorkshire) Ltd [2006] EWHC 110 (QB), [2006] 3 Costs LR 466, [2006] All ER (D) 68 (Feb) 5.14
Horth v Thompson [2010] EWHC 1674 (QB), [2011] 3 Costs LO 249, [2010] All ER (D) 47 (Jul) 6.9
Huck v Robson [2002] EWCA Civ 398, [2002] 3 All ER 263, [2003] 1 WLR 1340, [2002] All ER (D) 316 (Mar) 20.31
Hughes v George Major Skip Hire [2009] EWHC 90147 (Costs) 6.10, 6.23
Hughes v Newham London Borough Council (0502314) (2005), unreported, SCCO 6.25
Hullock v East Riding of Yorkshire County Council [2009] EWCA Civ 1039, [2009] All ER (D) 04 (Dec) 22.17
Hunt v RM Douglas (Roofing) Ltd [1990] 1 AC 398, [1988] 3 All ER 823, [1988] 3 WLR 975, 132 Sol Jo 1592, [1989] 1 LS Gaz R 40, HL 32.2, 32.5
Hurst v Leeming [2002] EWHC 1051 (Ch), [2003] 1 Lloyd's Rep 379, [2003] 2 Costs LR 153, [2002] All ER (D) 135 (May) 21.3
Huscroft v P&O Ferries Ltd [2010] EWCA Civ 1483, [2011] 2 All ER 762, [2011] 1 WLR 939, [2011] NLJR 28, (2011) Times, 11 February, [2010] All ER (D) 263 (Dec) 19.21
Hutchings v British Transport Police Authority [2006] EWHC 90064 (Costs) 29.31
Hutchison Telephone (UK) Ltd v Ultimate Response Ltd [1993] BCLC 307, CA 19.2

I

Igloo Regeneration (GP) Ltd v Powell Williams Partnership [2013] EWHC 1859 (TCC), 157 Sol Jo (no 28) 31, [2013] All ER (D) 59 (Jul) 24.20
Ilangaratne v British Medical Association [2007] EWHC 920 (Ch), [2008] 3 Costs LR 367, [2007] All ER (D) 133 (May) 12.5
Individual Homes Ltd v Macbreams Investments Ltd (2002) Times, 14 November, [2002] All ER (D) 345 (Oct) 40.15
Innovare Displays plc v Corporate Broking Services Ltd [1991] BCC 174, CA 19.22
Irvine v Metropolitan Police Comr [2005] EWCA Civ 129, [2005] 3 Costs LR 380, [2005] All ER (D) 46 (Feb) 22.34
Isaac v Isaac [2005] EWHC 435 (Ch), [2005] All ER (D) 379 (Mar) 22.7

J

Jackson v Ministry of Defence [2006] EWCA Civ 46 .... 22.18
Jemma Trust Co Ltd v Liptrott [2003] EWCA Civ 1476, [2004] 1 All ER 510, [2004] 1 WLR 646, [2003] 45 LS Gaz R 29, [2003] NLJR 1633, (2003) Times, 30 October, 147 Sol Jo LB 1275, [2003] All ER (D) 405 (Oct) ... 8.7, 31.26, 31.35, 31.36, 31.37, 33.2

Table of Cases

Jenkins v Young Brothers Transport Ltd [2006] EWHC 151 (QB), [2006] 2 All ER 798, [2006] 1 WLR 3189, [2006] NLJR 421, [2006] 3 Costs LR 495, [2006] All ER (D) 270 (Feb) 5.15
John v Price Waterhouse (t/a PricewaterhouseCoopers) [2002] 1 WLR 953, [2001] 34 LS Gaz R 39, (2001) Times, 22 August, 145 Sol Jo LB 188, [2001] All ER (D) 145 (Jul) 24.24
Johnsey Estates (1990) Ltd v Secretary of State for the Environment, Transport and the Regions [2001] EWCA Civ 535, [2001] NPC 79, [2001] 2 EGLR 128, [2001] All ER (D) 135 (Apr) 33.5
Johnson v Reed Corrugated Cases Ltd [1992] 1 All ER 169 .... 31.19, 31.26
Jones v Caradon Catnic Ltd [2005] EWCA Civ 1821, [2006] 3 Costs LR 427 .... 6.5, 7.10
Jones v Environcom Ltd [2009] EWHC 16 (Comm), [2010] Lloyd's Rep IR 190, [2010] 1 BCLC 150, [2009] All ER (D) 115 (Jan) 19.2
Jones v Secretary of State for Energy and Climate Change [2012] EWHC 3647 (QB), [2013] 2 Costs LR 230, [2013] All ER (D) 81 (Feb) 37.4
Jones v Secretary of State for Energy and Climate Change [2013] EWHC 1023 (QB), [2013] 3 All ER 1014, [2013] All ER (D) 65 (May) 32.5
Jones v Secretary of State for Wales [1997] 2 All ER 507, [1997] 1 WLR 1008 ...... 31.29
Jones v Wrexham Borough Council [2007] EWCA Civ 1356, [2008] 1 WLR 1590, (2008) Times, 21 January, 152 Sol Jo (no 2) 33, [2008] 1 Costs LR 147, [2007] All ER (D) 300 (Dec) 4.6, 7.9

K

KMT v Kent City Council [2012] EWHC 2088 (QB), [2012] 6 Costs LR 1039, [2012] All ER (D) 245 (Jul) 31.7, 31.21
KR v Bryn Alyn Community (Holdings) Ltd (in liq) [2003] EWCA Civ 383, [2003] PIQR P562, 147 Sol Jo LB 387, [2003] All ER (D) 347 (Mar) 20.4
Kashmiri v Ejaz [2007] EWHC 90074 (Costs) 6.27
Kasir v Darlington & Simpson Rolling Mills Ltd [2001] 2 Costs LR 228 .... 33.6
Khans Solicitor (a firm) v Chifuntwe [2013] EWCA Civ 481, [2013] 4 All ER 367, [2013] All ER (D) 103 (May) 1.27
Kiam v MGN Ltd (No 2) [2002] EWCA Civ 66, [2002] 2 All ER 242, [2002] All ER (D) 65 (Feb) 24.8
Kilby v Gawith [2008] EWCA Civ 812, [2009] 1 WLR 853, [2009] RTR 8, (2008) Times, 13 June, 152 Sol Jo (no 21) 28, [2008] All ER (D) 248 (May) 6.41
King v Telegraph Group Ltd [2004] EWCA Civ 613, [2005] 1 WLR 2282, [2004] 25 LS Gaz R 27, [2004] NLJR 823, (2004) Times, 21 May, 148 Sol Jo LB 664, [2004] 3 Costs LR 449, [2004] EMLR 429, [2004] All ER (D) 242 (May) 31.30
King v Telegraph Group Ltd [2005] EWHC 90015 (Costs) 5.14, 6.34
Kingsley, Re (1978) 122 Sol Jo 457 .... 2.9, 31.26
KOO Golden East Mongolia v Bank of Nova Scotia [2008] EWHC 1120 (Admin), [2008] All ER (D) 254 (May) 23.5, 23.29
Kopel v Safeway Stores plc [2003] IRLR 753, [2003] All ER (D) 05 (Sep), EAT 20.40
Kralj v Birkbeck Montague (18 February 1988, unreported), CA 3.20
Kris Motor Spares Ltd v Fox Williams [2010] EWHC 1008 (QB), [2010] 4 Costs LR 620, [2010] All ER (D) 87 (May) 6.44
Kris Motor Spares Ltd v Fox Williams LLP [2009] EWHC 2813 (QB), [2009] 6 Costs LR 931, [2009] All ER (D) 156 (Nov) 5.16, 5.18
Kunaka v Barclays Bank plc [2010] EWCA Civ 1035, [2011] 2 Costs LR 179 .... 20.21
Kutsi v North Middlesex University Hospital NHS Trust [2008] EWHC 90119 (Costs) 6.52
Kuwait Airways Corpn v Iraqi Airways Co (No 2) [1995] 1 All ER 790, [1994] 1 WLR 985, [1994] 8 LS Gaz R 38, 138 Sol Jo LB 39, CA 32.5

Table of Cases

L

L/M International Construction Inc v Circle Partnership Ltd [1995] CLY 4010, CA 19.2
Lahey v Pirelli Tyres Ltd [2007] EWCA Civ 91, [2007] 1 WLR 998, [2007] NLJR 294, [2007] PIQR P292, (2007) Times, 19 February, [2007] 3 Costs LR 462, [2007] All ER (D) 165 (Feb) 20.17
Lamb v Khan [2004] EWHC 2602 (QB), [2005] 2 Costs LR 259, [2004] All ER (D) 222 (Nov) 28.20
Lamont v Burton [2007] EWCA Civ 429, [2007] 3 All ER 173, [2007] 1 WLR 2814, [2008] RTR 58, [2007] NLJR 706, [2007] PIQR Q166, 151 Sol Jo LB 670, [2007] 4 Costs LR 574, [2007] All ER (D) 131 (May) 6.41, 20.30
Land and Property Trust Co plc (No 4), Re [1994] 1 BCLC 232, sub nom Land and Property Trust Co plc (No 2), Re [1993] BCC 462, [1993] 11 LS Gaz R 43, CA 40.11
Langsam v Beachcroft LLP [2011] EWHC 1451 (Ch), [2011] 3 Costs LO 380, [2011] All ER (D) 82 (Jun); affd [2012] EWCA Civ 1230, [2013] 1 Costs LO 112, [2012] All ER (D) 68 (Oct) 6.11, 6.27
Law Society v Persaud (1990) Times, 10 May 35.8
Lazarus (Leopold) Ltd v Secretary of State for Trade and Industry (1976) 120 Sol Jo 268, [1976] Costs Law Rep 62 .... 31.8
Leadbeater v Leadbeater [1985] FLR 789, [1985] Fam Law 280 .... 39.27
Lee v Birmingham City Council [2008] EWCA Civ 891, [2008] NLJR 1180, [2009] 2 Costs LR 191, [2008] All ER (D) 423 (Jul) 22.26
Leigh v Michelin Tyre plc [2003] EWCA Civ 1766, [2004] 2 All ER 175, [2004] 1 WLR 846, (2003) Times, 16 December, [2003] All ER (D) 144 (Dec) 1.10, 14.1, 16.1
Less v Benedict [2005] EWHC 1643 (Ch), [2005] 4 Costs LR 688, [2005] All ER (D) 355 (Jul) 28.21
Leverton (H) Ltd v Crawford Offshore (Exploration) Services Ltd (in liq) (1996) Times, 22 November 40.12
Levy v Legal Services Commission [2001] 1 All ER 895, [2000] All ER (D) 1775, [2001] 1 FCR 178, [2001] Fam Law 92, [2000] BPIR 1065, [2000] All ER (D) 1775, CA 39.47
Lifeline Gloves Ltd v Richardson [2005] EWHC 1524 (Ch), [2006] 1 Costs LR 58, [2005] All ER (D) 36 (Jul) 24.12
Lingfield Properties (Darlington) Ltd v Padgett Lavender Associates [2008] EWHC 2795 (QB), [2008] All ER (D) 162 (Nov) 40.6
Little Olympian Each Ways Ltd, Re [1994] 4 All ER 561, [1995] 1 WLR 560, [1995] 1 BCLC 48, [1994] BCC 959 .... 19.5
Lobster Group Ltd v Heidelberg Graphic Equipment Ltd [2008] EWHC 413 (TCC), [2008] 2 All ER 1173, [2008] Bus LR D58, 117 ConLR 64, [2008] 1 BCLC 722, [2008] BLR 314, [2008] 5 Costs LR 724, [2008] All ER (D) 88 (Mar) 19.20
Locabail (UK) Ltd v Bayfield Properties Ltd (No 3) [2000] 2 Costs LR 169, ChD 40.17
Locke v Camberwell Health Authority [2002] Lloyd's Rep PN 23, [1991] 2 Med LR 249, [1990] NLJR 205, CA 23.13
London and Regional (St George's Court) Ltd v Ministry of Defence [2008] EWHC 526 (TCC), 121 ConLR 26, 152 Sol Jo (no 14) 28, [2008] All ER (D) 249 (Mar); affd [2008] EWCA Civ 1212, 121 ConLR 26, [2009] BLR 20, [2008] 45 EG 100 (CS), [2008] All ER (D) 52 (Nov) 10.12, 10.16
London Scottish Benefit Society v Chorley (1884) 13 QBD 872, 53 LJQB 551, 32 WR 781, [1881–5] All ER Rep 1111, 51 LT 100, CA 35.2, 35.3
Longbotham & Sons, Re [1904] 2 Ch 152, 73 LJ Ch 681, 52 WR 660, [1904–7] All ER Rep 488, 48 Sol Jo 546, 90 LT 801, CA 3.23
Longman v Feather and Black (18 March 2008, unreported) 28.49
Loveday v Renton (No 2) [1992] 3 All ER 184 .... 31.19, 31.20
Lownds v Home Office [2002] EWCA Civ 365, [2002] 4 All ER 775, [2002] 1 WLR 2450, (2002) Times, 5 April, [2002] 2 Costs LR 279, [2002] All ER (D) 329 (Mar) 14.1, 14.2, 14.3, 24.3, 28.42, 28.46, 29.15, 31.14
Lumb v Hampsey [2011] EWHC 2808 (QB), [2012] All ER (D) 18 (Feb) 20.21
Lumos Skincare Ltd v Sweet Squared Ltd [2012] EWPCC 28, [2012] 4 Costs LR 735, [2012] All ER (D) 160 (Jun) 26.35

Table of Cases

M

M (Local Authority's Costs), Re [1995] 1 FCR 649, [1995] 1 FLR 533, [1995] Fam Law 291 .... 39.14
M, Re [2009] EWCA Civ 311 .... 39.17
M (A child), Re (5 July 2012, unreported), CA 23.30
M v Croydon Borough of London [2012] EWCA Civ 595, [2012] 3 All ER 1237, [2012] 1 WLR 2607, [2012] LGR 822, [2012] 3 FCR 179, [2012] 4 Costs LR 689, [2012] All ER (D) 73 (May) 22.31
M v M [2009] EWHC 1941 (Fam), [2010] 1 FLR 256, [2009] Fam Law 1029 ...... 39.30
MIOM 1 Ltd v Sea Echo E.N.E. (No 2) [2011] EWHC 2715 (Admlty), [2012] 1 Lloyd's Rep 140, [2011] NLJR 1596, [2011] All ER (D) 51 (Nov) 24.20
M/S Alghanim Industries Inc v Skandia International Insurance Corpn [2001] 2 All ER (Comm) 41.9
M-T v T [2006] EWHC 2494 (Fam), [2007] 2 FLR 925, [2007] Fam Law 1066 ..... 39.40
McCarthy v Essex Rivers Healthcare NHS Trust (Case No HQ06X03686) [2010] 1 Costs LR 59 .... 6.30
MacDonald v Taree Holdings Ltd [2001] 1 Costs LR 147 .... 27.17
McFarlane v McFarlane [2004] EWCA Civ 872, [2005] Fam 171, [2004] 3 All ER 921, [2004] 3 WLR 1480, [2004] 2 FCR 657, [2004] 2 FLR 893, (2004) Times, 9 July, [2004] All ER (D) 105 (Jul); sub nom Miller v Miller [2005] EWCA Civ 984, [2005] 2 FCR 713, [2005] Fam Law 766, [2005] All ER (D) 467 (Jul); sub nom Miller v Miller [2006] UKHL 24, [2006] 2 AC 618, [2006] 3 All ER 1, [2006] 2 WLR 1283, [2006] 2 FCR 213, [2006] 1 FLR 1186, [2006] Fam Law 629, [2006] NLJR 916, (2006) Times, 25 May, 150 Sol Jo LB 704, [2006] All ER (D) 343 (May) 39.6
McGlinn v Waltham Contractors Ltd [2005] EWHC 1419 (TCC), [2005] 3 All ER 1126, 102 ConLR 111, [2006] 1 Costs LR 27, [2005] All ER (D) 145 (Jul) ... 19.20, 22.23
McGlinn v Waltham Contractors Ltd [2007] EWHC 698 (TCC), 112 ConLR 148, [2007] All ER (D) 475 (Mar) 22.34
Mackay (as court-appointed receivers) v Ashwood Enterprises Ltd [2013] EWCA Civ 959, [2013] All ER (D) 51 (Aug) 22.2
Magical Marking Ltd v Ware & Kay LLP [2013] EWHC 636 (Ch), [2013] All ER (D) 218 (Apr) 22.17
Mahmood v Penrose [2002] EWCA Civ 457, [2002] All ER (D) 227 (Mar) 27.3
Mainwaring v Goldtech Investments Ltd (No 2) [1999] 1 All ER 456, CA 28.51
Malkinson v Trim [2002] EWCA Civ 1273, [2003] 2 All ER 356, [2003] 1 WLR 463, [2002] 40 LS Gaz R 33, [2002] NLJR 1484, (2002) Times, 11 October, [2002] All ER (D) 66 (Sep) 35.2
Maltby v D J Freeman & Co [1978] 2 All ER 913, [1978] 1 WLR 431, 122 Sol Jo 212 .... 31.11, 31.36
Manches LLP v Freer [2006] EWHC 991 (QB), [2006] All ER (D) 428 (Nov) 1.3
Manning v King's College Hospital NHS Trust [2011] EWHC 2954 (QB), [2011] NLJR 1634, [2012] 1 Costs LR 105, [2011] All ER (D) 97 (Nov) 6.9, 6.52
Maqsood v Mahmood [2012] EWCA Civ 251, [2012] All ER (D) 100 (Mar) 15.22
Maria, The. See Everglade Maritime Inc v Schiffahrtsgesellschaft Detlef von Appen mbH, The Maria
Marren v Dawson Bentley & Co Ltd [1961] 2 QB 135, [1961] 2 All ER 270, [1961] 2 WLR 679, 105 Sol Jo 383 .... 3.9
Mars UK Ltd v Teknowledge Ltd [1999] 2 Costs LR 44 .... 25.1
Mastercigars Direct Ltd v Withers LLP [2007] EWHC 2733 (Ch), [2008] 3 All ER 417, [2009] 1 WLR 881, [2008] IP & T 946, [2007] NLJR 1731, [2008] 1 Costs LR 72, [2007] All ER (D) 385 (Nov) 1.10
Mastercigars Direct Ltd v Withers LLP [2009] EWCA Civ 1526, [2010] 3 Costs LR 374 .... 1.10
Mayer v Harte [1960] 2 All ER 840, [1960] 1 WLR 770, 104 Sol Jo 603, CA 22.34
Medcalf v Mardell (Weatherill) [2002] UKHL 27, [2003] 1 AC 120, [2002] 3 All ER 721, [2002] 3 WLR 172, [2002] 31 LS Gaz R 34, [2002] NLJR 1032, (2002) Times, 28 June, [2002] All ER (D) 228 (Jun) 23.3, 23.9, 23.14
Medway Oil and Storage Co Ltd v Continental Contractors Ltd [1929] AC 88, 98 LJKB 148, [1928] All ER Rep 330, 140 LT 98, 45 TLR 20, HL 22.32

Table of Cases

Medway Primary Care Trust v Marcus [2011] EWCA Civ 750, 123 BMLR 112, [2011] 5 Costs LR 808, [2011] All ER (D) 219 (Jun) 22.17, 22.19
Mehjoo v Harben Barker [2013] EWHC 1669 (QB), [2013] All ER (D) 162 (Jun) ... 20.11, 20.13
Melchior v Vettivel [2001] All ER (D) 351 (May) 23.2
Mengi v Hermitage [2012] EWHC 2045 (QB), [2012] 5 Costs LO 641, [2012] All ER (D) 293 (Jul) 5.17
Merchantbridge & Co Ltd v Safron General Partner 1 Ltd [2011] EWHC 1524 (Comm), [2012] 2 BCLC 291, [2011] All ER (D) 39 (Jul) 10.10
Messih v McMillan Williams [2010] EWCA Civ 844, [2010] 6 Costs LR 914, [2010] All ER (D) 254 (Jul) 22.7
Metal Distributors Ltd, Re [2004] EWHC 2535 (Ch), [2004] All ER (D) 486 (Jul) 3.20
Metalloy Supplies Ltd (in liq) v MA (UK) Ltd [1997] 1 All ER 418, [1997] 1 WLR 1613, [1997] 1 BCLC 165, CA 40.13
Michael Phillips Architects Ltd v Riklin [2010] EWHC 834 (TCC), [2010] NLJR 943, [2010] BLR 569, [2010] All ER (D) 164 (Jun) 5.17
Microsoft Corpn v Datel Design [2011] EWHC 1986 (Ch) 22.24
Miller v Miller. See McFarlane v McFarlane
Miller v Miller. See Parlour v Parlour
Mills v Birchall. See Dolphin Quays Developments Ltd v Mills
Milne v David Price Solicitors (04/P8/340) (2005), unreported, SCCO 6.9
Minkin v Cawdery Kaye Fireman and Taylor [2011] EWHC 177 (QB), [2011] NLJR 256, 155 Sol Jo (no 7) 35, [2011] 3 Costs LR 465, [2011] All ER (D) 82 (Feb); revsd sub nom Minkin v Cawdery Kaye Fireman & Taylor [2012] EWCA Civ 546, [2012] 3 All ER 1117, [2013] 2 FCR 125, [2012] NLJR 681, [2012] 19 EG 94 (CS), 156 Sol Jo (no 18) 31, [2012] 4 Costs LR 650, [2012] All ER (D) 35 (May) 1.4, 1.10
Minotaur Data Systems Ltd, Re, Official Receiver v Brunt [1998] 4 All ER 500, [1999] 1 WLR 449, [1998] 2 BCLC 306, (1998) Times, 25 June, [1998] BPIR 756, [1998] All ER (D) 236; revsd [1999] 3 All ER 122, [1999] 1 WLR 1129, [1999] 2 BCLC 766, [1999] NPC 27, [1999] NLJR 415, 143 Sol Jo LB 97, [1999] BPIR 560, [1999] All ER (D) 215, CA 35.2
Mitchell v James [2002] EWCA Civ 997, [2003] 2 All ER 1064, [2004] 1 WLR 158, [2002] 36 LS Gaz R 38, (2002) Times, 20 July, [2002] All ER (D) 200 (Jul) .. 20.38, 28.48
Mitchell v News Group Newspapers Ltd [2013] EWHC 2179 (QB) 15.6, 15.34
Mitchell v News Group Newspapers Ltd [2013] EWHC 2355 (QB) 15.6, 15.34
Moon v Garrett [2006] EWCA Civ 1121, [2007] ICR 95, [2007] PIQR P30, [2006] BLR 402, (2006) Times, 1 September, [2007] 1 Costs LR 41, [2006] All ER (D) 429 (Jul) 22.34
Moore's (Wallisdown) Ltd v Pensions Ombudsman [2002] 1 All ER 737, [2002] 1 WLR 1649, [2002] ICR 773, (2002) Times, 1 March, [2001] All ER (D) 372 (Dec) 40.14
Morgan v Spirit Group Ltd [2011] EWCA Civ 68, [2011] 3 Costs LR 449, [2011] All ER (D) 24 (Feb) 27.1, 27.18
Morgan v UPS Ltd [2008] EWCA Civ 1476, [2008] All ER (D) 100 (Nov) 22.17
Morris v Dennis [2008] EWHC 90112 (Costs) 6.5
Morris v London Borough of Southwark [2010] EWHC 901 (QB), [2010] 4 Costs LR 526, [2010] All ER (D) 103 (Apr); affd sub nom Morris v Southwark London Borough Council (Law Society intervening); Sibthorpe v same [2011] EWCA Civ 25, [2011] 2 All ER 240, [2011] 1 WLR 2111, [2011] HLR 295, [2011] 06 LS Gaz R 18, [2011] NLJR 173, (2011) Times, 14 February, [2011] 3 Costs LR 427, [2011] All ER (D) 183 (Jan) 9.31
Morris v Wiltshire & Woodspring District Council (16 January 1998, unreported), QBD 35.5
Morrison v Buckinghamshire County Council [2011] EWHC 3444 (QB), [2012] All ER (D) 10 (Feb) 15.31
Mortgage Agency Number Four Ltd v Alomo Solicitors (a firm) [2011] EWHC B22 (Mercantile) 24.22
Morton v Portal Ltd [2010] EWHC 1804 (QB), [2010] All ER (D) 167 (Jul) 22.18
Moses-Taiga v Taiga [2005] EWCA Civ 1013, [2008] 1 FCR 696, [2006] 1 FLR 1074, [2006] Fam Law 266, [2005] All ER (D) 57 (Jul) 39.6

liii

Table of Cases

Motto v Trafigura Ltd [2011] EWCA Civ 1150, [2012] 2 All ER 181, [2012] 1 WLR 657, [2011] 42 LS Gaz R 21, [2011] NLJR 1483, (2011) Times, 22 November, [2011] 6 Costs LR 1028, [2011] All ER (D) 138 (Oct) 4.26, 6.44
Motto v Trafigura Ltd [2011] EWHC 90201 (Costs) 37.4
Mubarak v Mubarak [2001] 1 FLR 673, [2000] All ER (D) 1797; revsd [2001] 1 FLR 698, [2001] Fam Law 178, [2000] All ER (D) 2302, CA 39.49
Multiplex Constructions (UK) Ltd v Cleveland Bridge UK Ltd [2008] EWHC 2280 (TCC), 122 ConLR 88, [2009] 1 Costs LR 55, [2008] All ER (D) 04 (Oct) 22.22
Murphy v Young & Co's Brewery plc and Sun Alliance and London Insurance plc [1997] 1 All ER 518, [1997] 1 WLR 1591, [1997] 1 Lloyd's Rep 236, CA 40.9
Murray v Neil Dowlman Architecture Ltd [2013] EWHC 872 (TCC), 148 ConLR 256, [2013] NLJR 18, [2013] 3 Costs LR 460, [2013] All ER (D) 92 (Apr) ... 9.27, 15.8, 15.15, 15.34
Murray Lewis v Tennants Distribution Ltd [2010] EWHC 90161 (Costs) 9.31
Murray (Edmund) Ltd v BSP International Foundations Ltd (1992) 33 ConLR 1, CA 19.18
Myatt v National Coal Board [2007] EWCA Civ 307, [2007] 4 All ER 1094, [2007] 1 WLR 1559, (2007) Times, 27 March, [2007] 4 Costs LR 564, [2007] All ER (D) 301 (Mar) 40.18
Myers v Bonnington [2007] EWHC 90077 (Costs) 6.27

N

Nasser v United Bank of Kuwait [2001] EWCA Civ 556, [2002] 1 All ER 401, [2002] 1 WLR 1868, [2001] All ER (D) 146 (Apr) 19.5
National Westminster Bank plc v Rabobank Nederland [2007] EWHC 1742 (Comm), [2008] 1 All ER (Comm) 243, [2008] 3 Costs LR 396, [2007] All ER (D) 331 (Jul) 24.10
NatWest Bank v Feeney [2006] EWHC 90066 (Costs) 29.16
Neale v Hutchinson [2012] EWCA Civ 345, [2012] 5 Costs LO 588, [2012] 2 P & CR D1, [2012] All ER (D) 151 (Mar) 22.18
Nederlandse Reassurantie Groep Holding NV v Bacon & Woodrow [1998] 2 Costs LR 32, QBD 12.8
Neil v Stephenson (8 December 2000, unreported), QBD 27.6
Newall v Lewis [2008] EWHC 910 (Ch), [2008] 4 Costs LR 626, [2008] All ER (D) 426 (Apr) 22.14
Nichia Corpn v Argos Ltd [2007] EWCA Civ 741, [2007] Bus LR 1753, [2007] FSR 895, [2007] IP & T 943, [2007] All ER (D) 299 (Jul) 14.4
Nizami v Butt [2005] EWHC 159 (QB), [2006] 2 All ER 140, [2006] 1 WLR 3307, [2006] RTR 315, [2006] 09 LS Gaz R 30, [2006] NLJR 272, [2006] 3 Costs LR 483, [2006] All ER (D) 116 (Feb) 6.41, 12.11
Noorani v Calver [2009] EWHC 592 (QB), 153 Sol Jo (no 13) 27, [2009] All ER (D) 274 (Mar) 24.8
Norjarl K/S A/S v Hyundai Heavy Industries Co Ltd [1992] 1 QB 863, [1991] 3 All ER 211, [1991] 3 WLR 1025, [1991] 1 Lloyd's Rep 524, [1991] NLJR 343, CA 41.10
North West Holdings plc, Re, Secretary of State for Trade and Industry v Backhouse [2001] EWCA Civ 67, [2001] 1 BCLC 468, [2002] BCC 441, (2001) Independent, 9 February, 145 Sol Jo LB 53, [2001] All ER (D) 184 (Jan) 40.12
Northampton Coal, Iron and Waggon Co v Midland Waggon Co (1878) 7 Ch D 500, 26 WR 485, 38 LT 82, CA 19.7
Northstar Systems Ltd v Fielding [2006] EWCA Civ 1660 .... 22.18, 22.23
Nossen's Patent, Re [1969] 1 All ER 775, [1968] FSR 617, sub nom Nossen's Letter Patent, Re [1969] 1 WLR 638, 113 Sol Jo 445 .... 35.3, 35.6
Nykredit Mortgage Bank plc v Edward Erdman Group Ltd (No 2) [1998] 1 All ER 305, [1997] 1 WLR 1627, 75 P & CR D28, 14 LDAB 67, [1998] 01 LS Gaz R 24, [1998] 05 EG 150, 142 Sol Jo LB 29, HL 32.5

Table of Cases

O

O v Ministry of Defence [2006] EWHC 990 (QB), [2006] All ER (D) 203 (May) 37.3
O'Beirne v Hudson [2010] EWCA Civ 52, [2010] 1 WLR 1717, (2010) Times, 9 April, [2010] 2 Costs LR 204, [2010] All ER (D) 91 (Feb) 22.23
Ochwat v Watson Burton (a firm) [1999] All ER (D) 1407 .... 37.1
O'Driscoll v Liverpool City Council (2007), unreported, Liverpool County Court 6.27
Olatawura v Abiloye [2002] EWCA Civ 998, [2002] 4 All ER 903, [2003] 1 WLR 275, [2002] 36 LS Gaz R 39, [2002] NLJR 1204, (2002) Times, 24 July, [2002] All ER (D) 253 (Jul) 19.21
Oliver v Doughty [2011] EWCA Civ 1584, [2012] 2 All ER 825, [2012] 1 WLR 1048, [2012] RTR 316, [2012] NLJR 68, [2012] 2 Costs LR 314, [2011] All ER (D) 148 (Dec) 22.29
Oliver (executor of the estate of Oliver) v Whipps Cross University Hospital NHS [2009] EWHC 1104 (QB), 108 BMLR 181, 153 Sol Jo (no 21) 29, [2009] All ER (D) 199 (May) 6.31
Orchard v South Eastern Electricity Board [1987] QB 565, [1987] 1 All ER 95, [1987] 2 WLR 102, 130 Sol Jo 956, [1986] LS Gaz R 412, [1986] NLJ Rep 1112, CA 23.27
Oriakhel v Vickers [2008] EWCA Civ 748, [2008] All ER (D) 69 (Jul) 40.4
Oyston v Royal Bank of Scotland [2006] EWHC 90053 (Costs) 4.6, 4.16

P

P v P. See Priestley v Priestley
PR Records Ltd v Vinyl 2000 Ltd [2007] EWHC 1721 (Ch), [2008] 1 Costs LR 19, [2007] All ER (D) 265 (Jul) 40.3
Painting v University of Oxford [2005] EWCA Civ 161, (2005) Times, 15 February, [2005] 3 Costs LR 394, [2005] All ER (D) 45 (Feb) 22.18
Palmer v Palmer [2008] EWCA Civ 46, [2008] Lloyd's Rep IR 535, [2008] 4 Costs LR 513, [2008] All ER (D) 71 (Feb) 40.9
Palmier plc (in liq), Re [2009] EWHC 983 (Ch), [2009] All ER (D) 93 (May) 6.23
Pamplin v Express Newspapers Ltd [1985] 2 All ER 185, [1985] 1 WLR 689, 129 Sol Jo 188 .... 28.45
Pankhurst v White [2010] EWCA Civ 1445, [2011] 3 Costs LR 392, [2010] All ER (D) 184 (Dec) 5.31, 6.32
Parker (Kenneth Ronald) v Seixo (Joel Carlos) [2010] EWHC 90162 (Costs) 6.45
Parker-Tweedale v Dunbar Bank plc (No 2) [1991] Ch 26, [1990] 2 All ER 588, [1990] 3 WLR 780, 134 Sol Jo 1040, [1990] 29 LS Gaz R 33, CA 36.2
Parkinson (Sir Lindsay) & Co Ltd v Triplan Ltd [1973] QB 609, [1973] 2 All ER 273, [1973] 2 WLR 632, 117 Sol Jo 146, 226 Estates Gazette 1393, CA 19.14
Parlour v Parlour [2004] EWCA Civ 872, [2005] Fam 171, [2004] 3 All ER 921, [2004] 3 WLR 1480, [2004] 2 FCR 657, [2004] 2 FLR 893, (2004) Times, 9 July, [2004] All ER (D) 105 (Jul); sub nom Miller v Miller [2005] EWCA Civ 984, [2005] 2 FCR 713, [2005] Fam Law 766, [2005] All ER (D) 467 (Jul); sub nom Miller v Miller [2006] UKHL 24, [2006] 2 AC 618, [2006] 3 All ER 1, [2006] 2 WLR 1283, [2006] 2 FCR 213, [2006] 1 FLR 1186, [2006] Fam Law 629, [2006] NLJR 916, (2006) Times, 25 May, 150 Sol Jo LB 704, [2006] All ER (D) 343 (May) 39.6
Peacock v MGN Ltd [2009] EWHC 769 (QB), [2009] 4 Costs LR 584, [2009] All ER (D) 88 (Apr) 17.6
Peacock (Matthew) v MGN Ltd [2010] EWHC 90174 (Costs) 6.29
Pearless de Rougemont & Co v Pilbrow. See Pilbrow v Pearless De Rougemont & Co
Pendennis Shipyard Ltd v Magrathea (Pendennis) Ltd (in liq) [1998] 1 Lloyd's Rep 315, [1997] 35 LS Gaz R 35 .... 40.9
Perry v Lord Chancellor (1994) Times, 26 May 31.26
Persaud v Persaud [2003] EWCA Civ 394, 147 Sol Jo LB 301, [2003] All ER (D) 80 (Mar) 23.25, 23.27, 23.31
Persimmon Homes Ltd v Great Lakes Reinsurance (UK) plc [2010] EWHC 1705 (Comm), [2011] Lloyd's Rep IR 101, [2010] All ER (D) 114 (Jul) 9.24

Table of Cases

Petromec Inc v Petroleo Brasileiro SA [2007] EWHC 1589 (Comm), 115 ConLR 11, [2007] All ER (D) 102 (Jul); affd sub nom Petromec Inc v Petroleo Brasileiro SA Petrobras [2007] EWCA Civ 1371, [2008] 1 Lloyd's Rep 305, 115 ConLR 11, [2007] All ER (D) 378 (Dec) 40.17
Petromec Inc v Petroleo Brasileiro SA Petrobras [2006] EWCA Civ 1038, 150 Sol Jo LB 984, [2007] 2 Costs LR 212, [2006] All ER (D) 260 (Jul) 40.6
Petromin SA v Secnav Marine Ltd [1995] 1 Lloyd's Rep 603 .... 19.2
Petursson v Hutchison 3G UK Ltd [2004] EWHC 2609 (TCC) 17.5
Phillips v Princo [2003] All ER (D) 99 .... 40.7
Phillips v Symes [2004] EWHC 2330 (Ch), [2005] 4 All ER 519, [2005] 1 WLR 2043, [2005] 2 All ER (Comm) 538, 83 BMLR 115, (2004) Times, 5 November, [2005] 2 Costs LR 224, [2004] All ER (D) 270 (Oct) 40.15
Phoenix Finance Ltd v Federation International De L'Automobile [2002] EWHC 1028 (Ch), (2002) Times, 27 June, [2002] ArbLR 32, [2002] All ER (D) 347 (May) 24.19
Phonographic Performance Ltd v AEI Rediffusion Music Ltd [1999] 2 All ER 299, [1999] RPC 599, sub nom AEI Rediffusion Music Ltd v Phonographic Performance Ltd (No 2) [1999] 1 WLR 1507, 143 Sol Jo LB 97, [1999] EMLR 335, CA 22.33
Pilbrow v Pearless De Rougemont & Co (a firm) [1999] 3 All ER 355, [1999] 2 FLR 139, [1999] NLJR 441, sub nom Pearless de Rougemont & Co v Pilbrow 143 Sol Jo LB 114, CA 1.25
Piper Double Glazing Ltd v DC Contracts (1992) Ltd [1994] 1 All ER 177, [1994] 1 WLR 777 .... 41.8
Platinum Controls Ltd v Aleris Recycling (Swansea) Ltd [2012] EWHC 1675 (Ch), [2012] NLJR 912, [2012] All ER (D) 224 (Jun) 19.18
Plymouth & South West Co-operative Society Ltd v Architecture, Structure & Management Ltd [2006] EWHC 3252 (TCC), 111 ConLR 189, [2006] All ER (D) 248 (Dec) 40.9
Polak v Marchioness of Winchester [1956] 2 All ER 660, [1956] 1 WLR 818, 100 Sol Jo 470, CA 3.17
Porzelack KG v Porzelack (UK) Ltd [1987] 1 All ER 1074, [1987] 1 WLR 420, 131 Sol Jo 410, [1987] LS Gaz R 735, [1987] NLJ Rep 219 .... 19.15
Powell v Herefordshire Health Authority [2002] EWCA Civ 1786, [2003] 3 All ER 253, (2002) Times, 27 December, [2003] 2 Costs LR 185, [2002] All ER (D) 415 (Nov) 32.5
Practice Direction [1982] 2 All ER 800 .... 39.42
Practice Direction (family proceedings: court bundles) [2000] 2 All ER 287, Fam 39.42
Practice Note: Case Management [1995] 1 All ER 586, Fam 39.43
Preece v Caerphilly (15 August 2007, unreported), Cardiff CC 6.12
Priestley v Priestley [1989] FCR 657, sub nom P v P [1989] 2 FLR 241, [1989] Fam Law 313 .... 39.42
Procter & Gamble Co v Svenska Cellulosa Aktiebolaget SCA [2012] EWHC 2839 (Ch), [2013] 1 WLR 1464, [2013] 1 Costs LR 97, [2012] All ER (D) 282 (Oct) 20.38
Property and Reversionary Investment Corpn Ltd v Secretary of State for the Environment [1975] 2 All ER 436, [1975] 1 WLR 1504, 119 Sol Jo 274 .. 8.11, 31.13, 31. 15, 31.16, 31.17
Providence Capitol Trustees Ltd v Ayres [1996] 4 All ER 760 .... 40.14

Q

Q v Q (Family Division: costs: summary assessment) [2002] 2 FLR 668 (Fam) . 27.4, 39.45

R

R (on the application of Goodson) v Bedfordshire and Luton Coroner (Luton and Dunstable Hospital NHS Trust, interested party) [2004] EWHC 2931 (Admin), [2005] 2 All ER 791, [2006] 1 WLR 432, 84 BMLR 72, [2004] All ER (D) 298 (Dec) 18.2
R v Cardiff City Council, ex p Brown (11 August 1999, unreported), QBD 27.6

R v Common Professional Examination Board, ex p Mealing-McCleod (2000) Times, 2 May, CA	35.8
R (on the application of Scrinivasans Solicitors) v Croydon County Court [2013] EWCA Civ 249, [2013] All ER (D) 223 (Feb)	22.18
R (on the application of Edwards) v Environment Agency (Cemex UK Cement Ltd, intervening) [2010] UKSC 57, [2011] 1 All ER 785, [2011] 1 WLR 79, [2011] NLJR 101, [2011] NLJR 65, (2011) Times, 06 January, 154 Sol Jo (no 48) 35, [2011] 2 Costs LR 151, [2010] All ER (D) 183 (Dec)	28.58
R (on the application of Wulfsohn) v Legal Services Commission (formerly Legal Aid Board) [2002] EWCA Civ 250, [2002] All ER (D) 120 (Feb)	35.5
R v Lord Chancellor, ex p Child Poverty Action Group [1998] 2 All ER 755, [1999] 1 WLR 347, [1998] NLJR 205	18.2
R v Miller (Raymond) [1983] 3 All ER 186, [1983] 1 WLR 1056, 78 Cr App Rep 71, [1983] Crim LR 615, 127 Sol Jo 580, DC	12.5
R v Oxfordshire County Council, ex p Wallace [1987] NLJ Rep 542	23.13
R (on the application of the Campaign for Nuclear Disarmament) v Prime Minister [2002] EWHC 2777 (Admin), (2002) Times, 27 December, [2003] 3 LRC 335, [2002] All ER (D) 245 (Dec)	18.2
R v R (financial remedies: needs and practicalites) [2011] EWHC 3093 (Fam), [2013] 1 FLR 120	39.27
R v Sandhu (29 November 1984, unreported)	31.12
R (A) (Disputed Children) v Secretary of State for the Home Department [2007] EWHC 2494 (Admin), [2007] All ER (D) 20 (Nov)	18.3
R (on the application of Corner House Research) v Secretary of State for Trade and Industry [2005] EWCA Civ 192, [2005] 4 All ER 1, [2005] 1 WLR 2600, (2005) Times, 7 March, 149 Sol Jo LB 297, [2005] 3 Costs LR 455, [2005] All ER (D) 07 (Mar)	18.2
R (on the application of Factortame) v Secretary of State for Transport, Environment and the Regions (No 2) [2002] EWCA Civ 932, [2003] QB 381, [2002] 4 All ER 97, [2002] 3 WLR 1104, [2002] 35 LS Gaz R 34, [2002] NLJR 1313, [2003] BLR 1, (2002) Times, 9 July, [2002] All ER (D) 41 (Jul)	9.31, 10.11
R (on the application of Mendes) v Southwark London Borough Council (2009) Times, 7 April, [2009] All ER (D) 231 (Mar), CA	22.33
R (on the application of Hide) v Staffordshire County Council [2007] EWHC 2441 (Admin), [2007] NLJR 1543, [2007] All ER (D) 402 (Oct)	23.4
R v Supreme Court Taxing Office, ex p John Singh & Co [1997] 1 Costs LR 49 (CA)	28.46
R (on the application of Buglife, The Invertebrate Conservation Trust) v Thurrock Thames Gateway Development Corpn [2008] EWCA Civ 1209, [2008] 45 EG 101 (CS), (2008) Times, 18 November, 152 Sol Jo (no 43) 29, [2009] 1 Costs LR 80, [2008] All ER (D) 30 (Nov)	18.2
R (on the application of Boxall) v Waltham Forest London Borough Council (2001) 4 CCL Rep 258	22.31
R (on the application of Ministry of Defence) v Wiltshire and Swindon Coroner [2005] EWHC 889 (Admin), [2005] 4 All ER 40, [2006] 1 WLR 134, (2005) Times, 5 May, [2005] All ER (D) 242 (Apr)	18.2
R (on the application of Compton) v Wiltshire Primary Care Trust [2008] EWCA Civ 749, [2009] 1 All ER 978, [2009] 1 WLR 1436, [2009] PTSR 753, 152 Sol Jo (no 27) 29, [2008] All ER (D) 12 (Jul)	18.2
R v Wiseman Lee (Solicitors) (wasted costs order) (No 5 of 2000) [2001] EWCA Crim 707	23.2
RH v RH (costs) (adjustment for gross disparity) [2008] EWHC 347 (Fam), [2008] 2 FLR 2142, [2008] Fam Law 841, [2008] All ER (D) 67 (Apr)	39.30
RSA Pursuit Test Cases [2005] EWHC 90003 (Costs)	6.44
Radnor's (Earl) Will Trusts, Re (1890) 45 Ch D 402, 59 LJ Ch 782, 6 TLR 480, CA	38.2
Ralph Hume Garry (a firm) v Gwillim [2002] EWCA Civ 1500, [2003] 1 All ER 1038, [2003] 1 WLR 510, [2002] 45 LS Gaz R 35, [2002] NLJR 1653, (2002) Times, 4 November, 146 Sol Jo LB 237, [2002] All ER (D) 316 (Oct)	2.9, 8.2
Ratcliffe Duce and Gammer v Binns (UKEAT/100/08), unreported	23.33

Table of Cases

Redwing Construction Ltd v Wishart [2011] EWHC 19 (TCC), [2011] Lloyd's Rep IR 331, [2011] 1 EGLR 13, [2011] NLJR 137, [2011] BLR 186, [2011] 2 Costs LO 212, [2011] All ER (D) 101 (Jan) 6.30
Reed Executive plc v Reed Business Information Ltd [2004] EWCA Civ 887, [2004] 4 All ER 942, [2004] 1 WLR 3026, [2005] FSR 16, [2004] IP & T 1087, 148 Sol Jo LB 881, [2004] 4 Costs LR 662, [2004] All ER (D) 233 (Jul) 20.37, 28.49
Regent Leisuretime Ltd v Skerrett [2006] EWCA Civ 1032, [2006] All ER (D) 34 (Jul) 23.24
Reid Minty (a firm) v Taylor [2001] EWCA Civ 1723, [2002] 2 All ER 150, [2002] 1 WLR 2800, [2002] 1 CPLR 1, [2002] EMLR 347, [2001] All ER (D) 427 (Oct) 24.8
Remnant, Re (1849) 11 Beav 603, 18 LJ Ch 374, 14 LTOS 265 .... 2.12
Republic of Kazakhstan v Istil Group Inc [2005] EWCA Civ 1468, [2006] 1 WLR 596, [2006] 2 All ER (Comm) 26, (2005) Times, 17 November, [2005] ArbLR 35, [2005] All ER (D) 120 (Nov) 19.23
Revenue and Customs Comrs v Blue Sphere Global Ltd [2011] EWHC 90217 (Costs) 6.5
Reynolds v Stone Rowe Brewer (a firm) [2008] EWHC 497 (QB), 152 Sol Jo (no 14) 29, [2008] 4 Costs LR 545, [2008] All ER (D) 250 (Mar) 1.11
Richard Buxton (a firm) v Mills-Owens [2010] EWCA Civ 122, [2010] 4 All ER 405, [2010] 1 WLR 1997, [2010] 2 EGLR 73, [2010] 10 LS Gaz R 15, [2010] 17 EG 96, [2010] 09 EG 166 (CS), (2010) Times, 4 June, [2010] 3 Costs LR 421, [2010] All ER (D) 242 (Feb) 1.28
Richardson Roofing Co Ltd v Colman Partnership Ltd [2009] EWCA Civ 839, [2009] 4 Costs LR 521 .... 22.14
Ridehalgh v Horsefield [1994] Ch 205, [1994] 3 All ER 848, [1994] 3 WLR 462, [1994] 2 FLR 194, [1994] Fam Law 560, [1994] BCC 390, CA 23.2, 23.3, 23.4, 23.13, 23.25
Ridler v Walter [2001] TASSC 98, Supreme Court of Tasmania 41.12
Riniker v University College London [2001] 1 WLR 13, [2001] 1 Costs LR 20, CA 33.4
Roach v Home Office [2009] EWHC 312 (QB), [2010] QB 256, [2009] 3 All ER 510, [2010] 2 WLR 746, [2009] NLJR 474, [2009] 2 Costs LR 287, [2009] All ER (D) 164 (Mar) 29.18
Robertson Research International Ltd v ABG Exploration BV (1999) Times, 3 November, [1999] All ER (D) 1125 .... 40.5
Roche v Newbury Homes [2009] EW Misc 3 (EWCC) 6.9
Rogers v Merthyr Tydfil County Borough Council [2006] EWCA Civ 1134, [2007] 1 All ER 354, [2007] 1 WLR 808, 150 Sol Jo LB 1053, [2007] 1 Costs LR 77, [2006] All ER (D) 471 (Jul) 6.44, 6.45, 9.16, 33.8
Rogers v Mouchel (Case No OMY01382) (2010), unreported 6.4
Romer and Haslam, Re [1893] 2 QB 286, 62 LJQB 610, 4 R 486, 42 WR 51, 69 LT 547, CA 2.7
Ross v Bowbelle (Owners) [1997] 1 WLR 1159, [1997] 2 Lloyd's Rep 196, CA 29.18, 32.5
Royal Bank of Canada Trust Corpn Ltd v Secretary of State for Defence [2003] EWHC 1479 (Ch), [2004] 1 P & CR 448, [2003] 2 P & CR D50, [2003] All ER (D) 171 (May) 21.2
Ryan v Tretol Group Ltd [2002] EWHC 1956 (QB), [2002] All ER (D) 156 (Jul) 31.30

S

SES Contracting Ltd v UK Coal plc [2007] EWCA Civ 791, (2007) Times, October 16, [2007] 5 Costs LR 758, [2007] All ER (D) 410 (Jul) 22.24
SG (a minor by his mother and litigation friend) v Hewitt [2012] EWCA Civ 1053, [2013] 1 All ER 1118, [2012] 5 Costs LR 937, [2012] All ER (D) 16 (Aug) 20.21
Samco Europe v MSC Prestige [2011] EWHC 1656 (Admlty), [2011] NLJR 988, [2012] BLR 267, [2011] All ER (D) 55 (Jul) 20.9
Samonini v London General Transport Services Ltd [2005] EWHC 90001 (Costs) 6.25
Sampla v Rushmoor Borough Council [2008] EWHC 2616 (TCC), [2008] All ER (D) 335 (Oct) 20.16

Sanderson v Blyth Theatre Co [1903] 2 KB 533, 72 LJKB 761, 52 WR 33, 89 LT 159, 19
TLR 660, CA 22.34
Sandhu v Sidhu [2009] EWHC 983 (Ch), [2009] All ER (D) 93 (May) 6.23
Sarwar v Alam [2001] EWCA Civ 1401, [2001] 4 All ER 541, [2002] 1 WLR 125,
[2002] Lloyd's Rep IR 126, [2001] NLJR 1492, [2001] All ER (D) 44 (Sep) 6.25
Satellite (2003) Ltd, Re (17 November 2003, unreported), Ch D 24.15
Saxton v Bayliss (20 March 2013, unreported), Ch Div 40.15
Sayers v Merck SmithKline Beecham plc [2001] EWCA Civ 2017, [2003] 3 All ER 631,
[2002] 1 WLR 2274, 146 Sol Jo LB 31, [2001] All ER (D) 365 (Dec) 37.2
Scott v Duncan [2012] EWHC 1792 (QB), [2012] 4 Costs LR 787 .... 6.52
Scott v Harrogate Borough Council (20 January 2003, unreported), Harrogate County
Ct 34.3
Scott v Transport for London (December 2009, unreported), (Hastings County Court)
.... 6.9
Sears Tooth (a firm) v Payne Hicks Beach [1998] 1 FCR 231, [1997] 2 FLR 116, [1997]
Fam Law 392, [1997] 05 LS Gaz R 32, 141 Sol Jo LB 37 .... 39.41
Seaspeed Dora, The. See Slazengers Ltd v Seaspeed Ferries International Ltd, The
Seaspeed Dora
Seventh Earl of Malmesbury v Strutt & Parker [2007] EWHC 2199 (QB), [2007] 42 EG
294 (CS), [2007] All ER (D) 103 (Oct) 21.2
Shah v Karanjia [1993] 4 All ER 792, [1993] NLJR 1260 .... 40.16
Shahrokh Mireskandari v Law Society [2009] EWHC 2224 (Ch), 153 Sol Jo (no 34)
29 .... 22.9
Sharratt v London Central Bus Co and other cases (No 2), The Accident Group Test
Cases [2003] NLJR 790, 147 Sol Jo LB 657, [2003] All ER (D) 232 (May), SC; affd
[2004] EWCA Civ 575, [2004] 3 All ER 325, [2004] 23 LS Gaz R 32, [2004] NLJR
891, 148 Sol Jo LB 662, [2004] 3 Costs LR 422, [2004] All ER (D) 285 (May) ... 6.8, 6.9,
6.11
Shepherds Investments Ltd v Andrew Walters [2007] EWCA Civ 292, [2007] 6 Costs LR
837, [2007] All ER (D) 40 (Apr) 20.15
Sheppard v Essex Strategic Health Authority [2005] EWHC 1518 (QB), [2006] 1 Costs
LR 8 .... 17.4
Shirley v Caswell [2000] Lloyd's Rep PN 955, (2000) Independent, 24 July, [2001]
1 Costs LR 1, [2000] All ER (D) 807, CA 22.23
Sibley & Co v Reachbyte Ltd and Kris Motor Spares Ltd [2008] EWHC 2665 (Ch),
[2009] 2 Costs LR 311, [2008] All ER (D) 15 (Nov) 1.3, 33.5
Simaan General Contracting Co v Pilkington Glass Ltd [1987] 1 All ER 345, [1987] 1
WLR 516, 131 Sol Jo 297, [1987] LS Gaz R 819, [1986] NLJ Rep 824, 3 Const LJ
300, CA 19.17
Simcoe v Jacuzzi UK Group plc [2012] EWCA Civ 137, [2012] 2 All ER 60, [2012] 1
WLR 2393, 156 Sol Jo (no 11) 31, [2012] 2 Costs LR 401, [2012] All ER (D) 107
(Feb) 5.32, 32.2
Simms v Law Society [2005] EWCA Civ 849, [2005] NLJR 1124, [2006] 2 Costs LR
245, [2005] All ER (D) 131 (Jul) 24.24
Simpsons Motor Sales (London) Ltd v Hendon Corpn [1964] 3 All ER 833, [1965] 1
WLR 112, 109 Sol Jo 32 .... 31.32
Sims v Hawkins [2007] EWCA Civ 1175, [2008] 5 Costs LR 691, [2007] All ER (D) 247
(Nov) 40.5
Sisu Capital Fund Ltd v Tucker [2005] EWHC 2170 (Ch), [2006] BCC 463, [2006] BPIR
154, [2005] All ER (D) 200 (Oct) 35.3
Sitapuria v Khan (10 December 2007, unreported) 6.42
Skylight Maritime SA v Ascot Underwriting [2005] EWHC 15 (Comm), [2005] NLJR
139, [2005] All ER (D) 114 (Jan) 40.19
Slatter v Ronaldsons (a firm) [2002] 2 Costs LR 267, [2001] All ER (D) 251 (Dec)
.... 1.31
Slazengers Ltd v Seaspeed Ferries International Ltd, The Seaspeed Dora [1987] 3 All ER
967, [1988] 1 WLR 221, [1988] 1 Lloyd's Rep 36, 132 Sol Jo 23, [1987] LS Gaz R
3577, [1987] NLJ Rep 1085, CA 19.5
Slick Seating Systems v Adams [2013] EWHC 1642 (QB), [2013] 4 Costs LR 576,
[2013] All ER (D) 66 (Jun) 25.1, 27.4, 27.16
Smart v East Cheshire NHS Trust [2003] EWHC 2806 (QB), 80 BMLR 175 .... 17.4, 17.6

Table of Cases

Societe Internationale de Telecommunications Aeronautiques SC v Wyatt Co (UK) Ltd, (Maxwell Batley (a firm), Pt 20 defendant) [2002] EWHC 2401 (Ch), [2002] All ER (D) 189 (Nov) 20.24, 21.3
Solicitor, A, Re [1993] 2 FLR 959, [1993] Fam Law 627, [1993] 24 LS Gaz R 40, 137 Sol Jo LB 107, CA 23.21
Solomon v Cromwell Group plc [2011] EWCA Civ 1584, [2012] 2 All ER 825, [2012] 1 WLR 1048, [2012] RTR 316, [2012] NLJR 68, [2012] 2 Costs LR 314, [2011] All ER (D) 148 (Dec) 22.29
South Coast Shipping Co Ltd v Havant Borough Council [2002] 3 All ER 779, [2002] NLJR 59, [2001] All ER (D) 382 (Dec) 12.9, 28.45
Southern Counties Fresh Foods Ltd, Re, Cobden Investments Ltd v Romford Wholesale Meats Ltd [2011] EWHC 1370 (Ch), [2011] NLJR 882, [2011] 3 Costs LO 343, [2011] All ER (D) 66 (Jun) 5.32
Southwark London Borough Council v IBM UK Ltd [2011] EWHC 653 (TCC), [2011] NLJR 474, [2011] All ER (D) 261 (Mar) 24.20
Spencer v Gordon Wood t/a Gordons Tyres (a firm) [2004] EWCA Civ 352, 148 Sol Jo LB 356, [2004] 3 Costs LR 372, [2004] All ER (D) 275 (Mar), sub nom Spencer v Wood (2004) Times, 30 March 6.23
Square Mile Partnership Ltd v Fitzmaurice McCall Ltd [2006] EWHC 236 (Ch), [2006] All ER (D) 84 (Jan) 22.32, 22.33
Straker v Tudor Rose (a firm) [2007] EWCA Civ 368, 151 Sol Jo LB 571, [2008] 2 Costs LR 205, [2007] All ER (D) 224 (Apr) 22.16
Stringer v Copley [2012] Lexis Citation 68 .... 26.12
Stubbs v Board of Governors of the Royal National Orthopaedic Hospital [1997] Costs Law Rep 117 .... 31.9
Styler v Ingham & Hilton (31 July 2008, unreported) 6.42
Sullivan v Co-operative Insurance Society Ltd [1999] 2 Costs LR 158, CA 31.30
Supperstone v Hurst [2008] EWHC 735 (Ch), [2008] 4 Costs LR 572, [2008] BPIR 1134, [2008] All ER (D) 211 (Apr) 6.52
Sutherland v Turnbull [2010] EWHC 2699 (QB), [2010] All ER (D) 02 (Nov) 20.18
Swedac Ltd v Magnet & Southerns plc [1989] FSR 243; on appeal [1990] FSR 89, CA 23.13
Swift v Robertson [2012] EWCA Civ 1794, [2013] Bus LR 479, 177 JP 169, 157 Sol Jo (no 3) 31, [2013] All ER (D) 62 (Jan) 4.13
Sycamore Bidco Ltd v Breslin [2013] EWHC 583 (Ch), [2013] All ER (D) 173 (Mar) 22.21
Symphony Group plc v Hodgson [1994] QB 179, [1993] 4 All ER 143, [1993] 3 WLR 830, [1993] 23 LS Gaz R 39, [1993] NLJR 725, 137 Sol Jo LB 134, CA .. 40.2, 40.8, 40.9
Szekeres v Alan Smeath & Co [2005] EWHC 1733 (Ch), [2005] 32 LS Gaz R 31, [2005] 4 Costs LR 707, [2005] All ER (D) 08 (Aug) 3.19

T

T (a child) (order for costs), Re [2005] EWCA Civ 311, [2005] 1 FCR 625, [2005] 2 FLR 681, [2005] Fam Law 534, [2005] All ER (D) 335 (Mar) 39.16
T (children) (Costs: Care Proceedings: Serious Allegation Not Proved), Re [2012] UKSC 36, [2012] 4 All ER 1137, [2012] 1 WLR 2281, [2012] PTSR 1379, [2012] 3 FCR 137, [2013] 1 FLR 133, [2012] Fam Law 1325, [2012] NLJR 1028, (2012) Times, 14 August, [2012] 5 Costs LR 914, [2012] All ER (D) 254 (Jul) 39.14
TL v ML (ancillary relief: claim against assets of extended family) [2005] EWHC 2860 (Fam), [2006] 1 FCR 465, [2006] 1 FLR 1263, [2006] Fam Law 183 .... 39.39
Tait v Cataldo [2010] EWHC 90166 (Costs) 6.52
Tanfern Ltd v Cameron-MacDonald [2000] 2 All ER 801, [2000] 1 WLR 1311, [2000] All ER (D) 654, CA 33.5
Tankard v John Fredricks Plastics Ltd [2008] EWCA Civ 1375, [2009] 4 All ER 526, [2009] 1 WLR 1731, (2009) Times, 16 January, [2009] 1 Costs LR 101, [2008] All ER (D) 126 (Dec) 6.26
Taylor v Pace Developments Ltd [1991] BCC 406, CA 40.11
Tearle & Co v Sherring (29 October 1993, unreported), QBD 2.11

Table of Cases

Teasdale v HSBC Bank plc [2010] EWHC 612 (QB), [2010] 4 All ER 630, [2010] NLJR 878, [2010] 4 Costs LR 543, [2010] All ER (D) 34 (Jun) 22.7
Ted Baker plc v Axa Insurance UK plc [2012] EWHC 1779 (Comm), [2012] 6 Costs LR 1023 .... 20.15
Thai Trading Co (a firm) v Taylor [1998] QB 781, [1998] 3 All ER 65, [1998] 2 WLR 893, [1998] 2 FLR 430, [1998] Fam Law 586, [1998] 15 LS Gaz R 30, 142 Sol Jo LB 125, CA 12.3
Thames Chambers Solicitors v Miah [2013] EWHC 1245 (QB), [2013] 4 Costs LR 582, [2013] All ER (D) 249 (May) 23.18
Thevarajah v Riordan [2013] EWHC 3179 (Ch), [2013] All ER (D) 239 (Oct) 15.34
Thewlis v Groupama Insurance Co Ltd [2012] EWHC 3 (TCC), 142 ConLR 85, [2012] BLR 259, [2012] 5 Costs LO 560, [2012] All ER (D) 09 (Jan) 20.2
Thomson v Berkhamsted Collegiate School [2009] EWHC 2374 (QB), [2009] NLJR 1440, [2009] 6 Costs LR 859, [2009] All ER (D) 39 (Oct) 40.17
Thornley v Lang [2003] EWCA Civ 1484, [2004] 1 All ER 886, [2004] 1 WLR 378, [2003] 46 LS Gaz R 24, [2003] NLJR 1706, (2003) Times, 31 October, 147 Sol Jo LB 1277, [2003] All ER (D) 466 (Oct) 6.18, 6.22
Thornley (by his litigation friend Lavinia Thornley) v Ministry of Defence [2010] EWHC 2584 (QB), [2011] 3 Costs LR 335, [2011] All ER (D) 177 (Jan) 6.32
Three Rivers District Council v Bank of England [2006] EWHC 816 (Comm), [2006] All ER (D) 175 (Apr) 22.23
Tim Martin Interiors Ltd v Akin Gump LLP [2011] EWCA Civ 1574, [2012] 2 All ER 1058, [2012] 1 WLR 2946, [2012] 1 EGLR 153, [2012] NLJR 66, [2012] 2 Costs LR 325, [2012] All ER (D) 02 (Jan) 3.25
Tinseltime Ltd v Roberts [2012] EWHC 2628 (TCC), [2012] NLJR 1290, 156 Sol Jo (no 38) 31, [2012] 6 Costs LR 1094, [2012] All ER (D) 19 (Oct) 9.30
Tolstoy-Miloslavsky v Lord Aldington [1996] 2 All ER 556, [1996] 1 WLR 736, [1996] 01 LS Gaz R 22, 140 Sol Jo LB 26, [1996] PNLR 335, CA 23.13, 23.15, 40.18
Total Spares & Supplies Ltd v Antares SRL [2006] EWHC 1537 (Ch), [2006] BPIR 1330, [2006] All ER (D) 314 (Jun) 40.7
Tramountana Armadora SA v Atlantic Shipping Co SA [1978] 2 All ER 870, [1978] 1 Lloyd's Rep 391 .... 41.5
Tranter v Hansons [2009] EWHC 90145 (Costs) 6.27
Treasury Solicitor v Regester [1978] 2 All ER 920, [1978] 1 WLR 446, 122 Sol Jo 163 .... 31.1, 31.15
Trident International Freight Services Ltd v Manchester Ship Canal Co [1990] BCLC 263, CA 19.18
Trill v Sacher (No 2) [1992] 40 LS Gaz R 32, CA 23.20
Troy Foods v Manton [2013] EWCA Civ 615, [2013] 4 Costs LR 546 .. 14.3, 15.32, 15.34
Truex v Toll [2009] EWHC 396 (Ch), [2009] 4 All ER 419, [2009] 1 WLR 2121, [2009] 2 FLR 250, [2009] Fam Law 474, [2009] NLJR 429, [2009] 5 Costs LR 758, [2009] BPIR 692, [2009] All ER (D) 98 (Mar) 3.28
Truscott v Truscott [1998] 1 All ER 82, [1998] 1 WLR 132, [1998] 1 FCR 270, [1998] 1 FLR 265, [1998] Fam Law 74, CA 31.30
Trustees of Stokes Pension Fund v Western Power Distribution (South West) plc [2005] EWCA Civ 854, [2005] 3 All ER 775, [2005] 1 WLR 3595, [2006] 2 Costs LR 226, [2005] All ER (D) 107 (Jul) 20.38
Turner & Co v O Palomo SA [1999] 4 All ER 353, [2000] 1 WLR 37, CA . 2.5, 3.22, 3.27
Turner Page Music v Torres Design Associates Ltd (1998) Times, 3 August, CA 23.23

U

U v Liverpool City Council [2005] EWCA Civ 475, [2005] 1 WLR 2657, (2005) Times, 16 May, [2005] 4 Costs LR 600, [2005] All ER (D) 381 (Apr) 6.29, 6.38, 33.8
Ultraframe (UK) Ltd v Fielding [2006] EWCA Civ 1660, [2007] 2 All ER 983, (2007) Times, 8 January, [2007] 2 Costs LR 264, [2006] All ER (D) 81 (Dec) 22.23
Unisoft Group (No 2), Re [1993] BCLC 532 .... 19.22
University of Nottingham v Eyett (No 2) [1999] 2 All ER 445, [1999] ICR 730, [1998] All ER (D) 643 .... 40.14

Table of Cases

V

Valentine v Allen [2003] EWCA Civ 915, [2003] All ER (D) 79 (Jul) 21.3
Various Claimants v Corby Borough Council [2008] EWHC 619 (TCC) 37.5
Various Claimants v Gower Chemicals (CW05054225) (2007), unreported, SCCO
.... 6.18
Vaughan v Jones [2006] EWHC 2123 (Ch), [2006] BPIR 1538, [2006] All ER (D) 62
(Aug) 40.16
Venture Finance plc v Mead [2005] EWCA Civ 325, [2006] 3 Costs LR 389,
[2005] All ER (D) 376 (Mar) 36.4
Venulum Property Investments Ltd v Space Architecture Ltd [2013] EWHC 1242 (TCC),
[2013] 4 Costs LR 596, [2013] All ER (D) 276 (May) 15.34
Verrecchia (t/a Freightmaster Commercials) v Metropolitan Police Comr [2002] EWCA
Civ 605 .... 22.21, 22.33
Villa Agencies SPF Ltd v Kestrel Travel Consultancy Ltd [2012] EWCA Civ 000, CA
.... 22.32
Vimeira (No 2), The. See Aiden Shipping Co Ltd v Interbulk Ltd, The Vimeira (No 2)
Virani Ltd v Manuel Revert y CIA SA [2003] EWCA Civ 1651, [2004] 2 Lloyd's Rep 14,
[2003] All ER (D) 324 (Jul) 21.2

W

W (a child), Re (26 August 2000, unreported), CA 39.16
Wagstaff v Colls [2003] EWCA Civ 469, (2003) Times, 17 April, 147 Sol Jo LB 419,
[2003] All ER (D) 25 (Apr) 23.2
Walker Windsail Systems Ltd, Re, Walker v Walker [2005] EWCA Civ 247, [2006]
1 All ER 272, [2006] 1 WLR 2194, (2005) Times, 3 March, [2005] 3 Costs LR 363,
[2005] BPIR 454, [2005] All ER (D) 277 (Jan) 22.7
Wallersteiner v Moir (No 2) [1975] QB 373, [1975] 1 All ER 849, [1975] 2 WLR 389,
119 Sol Jo 97, CA 40.18
Walsh v Shanahan [2013] EWCA Civ 675, [2013] 5 Costs LO 738, [2013] All ER (D)
180 (Jun) 20.26
Walton v Egan [1982] QB 1232, [1982] 3 All ER 849, [1982] 3 WLR 352, 126 Sol Jo
345 .... 8.11
Wates Construction Ltd v HGP Greentree Allchurch Evans Ltd [2005] EWHC 2174
(TCC), 105 ConLR 47, [2006] BLR 45, [2005] All ER (D) 170 (Nov) 24.14
Watts (Thomas) & Co (a firm) v Smith [1998] 2 Costs LR 59, CA 3.22
Webb v Environment Agency (5 April 2011, unreported), QBD 22.7
Webster v Ridgeway Foundation School [2010] EWHC 318 (QB), [2010] 6 Costs LR
859, [2010] All ER (D) 97 (Mar) 24.24
Weir v Secretary of State for Transport [2005] EWHC 2192 (Ch), [2005] All ER (D) 160
(Oct) 18.2
West African Gas Pipeline Co Ltd v Willbros Global Holdings Inc [2012] EWHC 396
(TCC), 141 ConLR 151, [2012] NLJR 682, [2012] All ER (D) 60 (May) 15.26
Westwood v Knight [2011] EWPCC 11, [2011] FSR 847, [2011] 4 Costs LR 654 ... 26.33,
26.35, 26.36
Wethered Estate Ltd v Davis [2005] EWHC 1903 (Ch), [2006] BLR 86, [2005] All ER
(D) 336 (Jul) 21.5
Wetzel v KBC Fidea [2007] EWHC 90079 (Costs) 6.41
Widlake v BAA Ltd [2009] EWCA Civ 1256, 153 Sol Jo (no 45) 29, [2010] 3 Costs LR
353, [2009] All ER (D) 246 (Nov) 22.17, 22.18
Wild v Simpson [1919] 2 KB 544, 88 LJKB 1085, [1918–19] All ER Rep 682, 63 Sol Jo
625, 121 LT 326, 35 TLR 576, CA 1.4
Wilkinson v Kenny [1993] 3 All ER 9, [1993] 1 WLR 963, [1993] NLJR 582, CA
.... 23.31
William Dronsfield (a protected party who proceeds by his litigation friend Alison
Millington) v Benjamin Street (25 June 2013, unreported) 6.40
Willis v Nicolson [2007] EWCA Civ 199, [2007] PIQR P318, 151 Sol Jo LB 394,
[2007] All ER (D) 205 (Mar) 17.5
Wilson v GP Haden t/a Clyne Farm Centre [2013] EWHC 1211 (QB) 22.21

Table of Cases

Wilson v Ministry of Defence (23 April 13, unreported), QC 20.12
Wilson v William Sturges & Co [2006] EWHC 792 (QB), [2006] 16 EG 146 (CS), [2006] 4 Costs LR 614, [2006] All ER (D) 110 (Apr) 2.8
Wilsons Solicitors v Johnson (2011) UKEAT/515/10, [2011] ICR D21, EAT 23.33
Wong v Vizards (a firm) [1997] 2 Costs LR 46, QBD 1.4, 1.13
Wood v Worthing and Southlands Hospitals NHS Trust (9 July 2004, unreported) 31.29
Woollard v Fowler [2005] EWHC 90051 (Costs) 26.12
Wraith v Sheffield Forgemasters Ltd [1998] 1 All ER 82, [1998] 1 WLR 132, [1998] 1 FCR 270, [1998] 1 FLR 265, [1998] Fam Law 74, CA 31.30
Wright v Bennett [1948] 1 KB 601, [1948] 1 All ER 410, [1948] LJR 1019, 92 Sol Jo 167, 64 TLR 149, CA 29.18
Wright v Michael Wright Supplies Ltd [2013] EWCA Civ 234, [2013] 4 Costs LO 630, [2013] All ER (D) 02 (Apr) 21.6
Wyche v Careforce Group plc (25 July 2013, unreported), QBD 15.26, 15.34
Wyld v Silver (No 2) [1962] 2 All ER 809, [1962] 1 WLR 863, 106 Sol Jo 409, CA 19.22

X

XYZ v Schering Health Care (Costs Appeal No 9 of 2004), unreported, SCCO 31.33

Y

Yao Essaie Motto v Trafigura [2008] EWCA Civ 1150 .... 6.29
Yao Essaie Motto v Trafigura Ltd (15 February 2011, unreported), SCCO 4.26
Yonge v Toynbee [1910] 1 KB 215, 79 LJKB 208, [1908–10] All ER Rep 204, 102 LT 57, 26 TLR 211, CA 40.19

Z

Zakirov v Newmans Solicitors [2012] EWHC 90222 (Costs) 35.2
Zissis v Lukomski [2006] EWCA Civ 341, [2006] 1 WLR 2778, [2006] 2 EGLR 61, [2006] 15 EG 135 (CS), (2006) Times, 24 April, [2006] All ER (D) 63 (Apr) 24.24
Zuliani v Veira [1994] 1 WLR 1149, PC 3.4

Decisions of the European Court of Justice are listed below numerically. These decisions are also included in the preceding alphabetical list.

C-43/95: Data Delecta Aktiebolag v MSL Dynamics [1996] ECR I-4661, [1996] All ER (EC) 961, [1996] 3 CMLR 741, ECJ 19.5
C-199/08: Eschig v UNIQA Sachversicherung AG [2009] ECR I-08295, [2010] 1 All ER (Comm) 576, [2010] 1 CMLR 131, [2010] Bus LR 1404, [2009] All ER (D) 78 (Sep), ECJ 9.8

Part 1

Solicitor and client

CHAPTER 1

THE RETAINER

INTRODUCTION

[1.1]

The relationship between the client and his solicitor is at the heart of the law of costs. The giving of instructions by a client to a solicitor constitutes the solicitor's retainer by the client. It is a contract. It creates the solicitor's right to be paid. The rights and liabilities of the parties are governed by the ordinary law of contract, but the relationship is also subject to the special provisions which govern contracts between a solicitor and his client. These come in two forms – the statutory kind, such as the provisions of the Solicitors Act 1974: and the regulatory kind, in particular the Codes of Conduct introduced by the Law Society and then the Solicitors Regulation Authority. The statutory provisions affecting the retainer are dealt with throughout the book. They are particularly prominent in this Part and Part II in relation to funding agreements. The regulatory provisions are dealt with in this chapter, after we have looked at the general contractual provisions and before we reach the termination issues.

'Solicitor' and 'client'

[1.2]

It will not have escaped your attention that the lawyer representing a client can no longer simply be described as a solicitor. Chartered legal executives and costs lawyers have acquired rights of audience since this book was first written. Barristers are rapidly acquiring rights to conduct litigation and many who could call themselves solicitors use other titles, whether they work in traditional partnerships or in some form of Alternative Business Structure.

This book has always been written directly to the legal representative and that provides an immediacy that is often helpful, particularly when dealing with the solicitor and client aspects of civil remuneration. The word 'you' refers to the legal representative and 'we' refers to the writers of this work. Some of that immediacy would be lost if we were to use some all embracing nomenclature such as 'legal representative' or 'fee earner'. Rather than do this, we have used the word solicitor as shorthand for the party's lawyer with due apology to any other legal representative who reads this book.

Similarly, but rather more discretely, we have used the word client rather than 'party chargeable' when considering Solicitors' Act 1974 assessments. Hopefully we have made it clear where a third party interest applies rather than a client in the strict sense, but that is a fairly rare occurrence and the word client hugely aids readability.

THE CONTRACT

[1.3]

It is an implied term of the contract that the client will pay the solicitor's charges and disbursements. Other than for certain contingency based arrangements, the retainer need not be in writing but, if the true construction of an arrangement is that the solicitor's costs are guaranteed by a third party, then the requirement that a guarantee must be in writing applies to the arrangement between the solicitor and the third party. For example, in *Manches LLP v Freer* [2006] EWHC 991 (QB), solicitors claimed payment of their outstanding fees from the defendant for work done for a company of which he was a director. The solicitors claimed that the defendant was personally liable because of a provision in their terms of business that stated the directors would be directly liable for fees and disbursements if the company failed to pay them. The defendant accepted that he had signed the engagement letter but said it was on behalf of the company and not so as to make him personally liable. The court agreed with the defendant and held that for the defendant to be personally liable as a guarantor he had to have signed the letter both as a director and in his personal capacity, or preferably he should have signed two letters. A potential trap which many a solicitor may have been fortunate to avoid.

Where there is a dispute between a solicitor and his client about the terms of an oral retainer, the word of the client is to be preferred to the word of the solicitor, or, at least, more weight is to be given to it. The reason is plain. It is because the client is ignorant and the solicitor is, or should be, learned in the law. If the solicitor does not take the precaution of getting a written retainer, he has only himself to blame for being at variance with his client over it and must take the consequences. The onus is on the solicitor to establish the terms of the retainer and in the absence of persuasive evidence the court should prefer the client's version. 'It is up to the solicitor to take the appropriate steps to clarify precisely the extent of his retainer' (*Gray v Buss Merton (a firm)* [1999] PNLR 882 & 892) 'because the client, through ignorance of the correct terminology, may not be able to express his instructions clearly' (*Sibley & Co v Reachbyte Ltd and Kris Motor Spares Ltd* [2008] EWHC 2665 (Ch)).

An entire contract

[1.4]

A retainer is normally a contract under which the solicitor is to do certain work for the client and cannot seek any remuneration until that work has been completed or the retainer has been terminated in some other way. This is described by the law of contract as an 'entire contract.' The solicitor is not entitled to any payment on account of his costs other than disbursements.

The only circumstances in which the retainer is not an entire contract is if the parties have agreed to interim payments; there is a request from the client for a final bill; or it is contentious business and s 65(2) of the Solicitors Act 1974 applies.

If the solicitor wrongfully terminates the retainer he is not entitled to any payment at all for the work he has done, either on a quantum meruit or any other basis (*Wild v Simpson* [1919] 2 KB 544, 88 LJKB 1085, CA).

The previous paragraphs are worth re-reading as they set out a reality that is often not grasped by solicitors. The subject of interim bills is dealt with in detail in the next chapter but time and again solicitors consider withdrawing their labour as a result of tardy or non-payment of interim bills by the client.

As will be seen from the two following cases, the Court of Appeal at least, has started to soften the strict contractual approach. But bills of costs are regularly thrown out in full for the solicitor repudiating the contract and the message to take away is twofold. First, if this situation happens to you, alarm bells should ring and you should look very carefully at the steps you take to encourage your client to pay up. Secondly, time spent by solicitors on their terms of business is seldom wasted.

In *Wong v Vizards (a firm)* [1997] 2 Costs LR 46, QBD, where solicitors declined to represent their client at a hearing unless he made a substantial payment on account of a disputed bill the judge held that because the amount claimed by the solicitors was unreasonable they had wrongfully terminated the retainer on the grounds of non-payment and were therefore not entitled to any payment at all for the work they had done in preparation for the hearing.

In *Minkin v Cawdery Kaye Fireman and Taylor (a firm)* [2011] EWHC 177 (QB), [2011] NLJR 256, 155 Sol Jo (no 7) 35 the solicitors had to go up to the Court of Appeal to recover their costs. The costs judge had held that the solicitors were in repudiatory breach when they delivered an interim bill substantially in excess of their estimate and refused to continue to act until it was paid and therefore they were not entitled to any payment at all. The solicitors' appeal to a High Court judge was dismissed.

The solicitors obtained permission to go to the Court of Appeal which held that the complaint that the bill exceeded the estimate could not stand in the face of the fact that the letter enclosing the terms of business made plain that estimates were not intended to be fixed or binding and that other factors might have meant that the estimate would be varied from time to time. The terms of business further informed the client that estimates were given 'as a guide' only. There was no guarantee that the final charge would not exceed the estimate because there were many factors outside the solicitors' control which might affect the level of costs. Refusal to pay could not be justified on that ground. Moreover, the client was fully informed of his right to challenge in court any bill which he felt was excessive. The reality was that the client was short of money and could not readily pay for his solicitors' services in coping with a new and unexpected turn of events. The comparatively simple case had become more complicated and as a result more expensive. The unexpected complication did not justify a refusal to pay a bill which became payable on presentation. The client was obliged by the payment terms set out in the terms of business to pay bills on presentation. He did not do so and that put him in breach.

The client had no reasonable justification for not meeting the bill presented to him. It was the client who repudiated the contract, not the firm. Not being prepared to act until money was paid showed a willingness to act when there was money on account.

The retainer came to an end when the client communicated to the solicitors that he had lost confidence in them. The client's termination of the retainer absolved the solicitor from any further performance of the contract but it did not absolve the client from the payment of the costs properly incurred to that date. The costs judge was therefore wrong not to order the payment of the costs as they were assessed and he was wrong to further order the solicitors to make repayment to the client.

In delivering judgment Lord Justice Ward said 'The client was short of money and could not pay for his solicitors in coping with a new and unexpected turn of events that complicated the case. Every solicitor will encounter, in one way or another, the kind of problem which gives rise to this appeal'. Lord Justice Elias added: 'Any other view would compel a solicitor to carry on working for a client even though there may be little realistic prospect of payment'.

Increases in the hourly rate

[1.5]

For charging rates to be increased the solicitor must have reserved the contractual right to do so in the retainer. It has nothing to do with the Code of Conduct even though successive versions have required the client to be kept informed of any increases as a matter of professional conduct. In any event, a favourite paying party challenge on assessment is the question of whether the client was aware of any increases in the hourly rate during the course of the case. So, you must make sure that the client care letter provides for a periodic review and increase where appropriate and that you have a system which remembers to notify your client of the increase promptly when that periodic review has taken place.

Client's instructions

[1.6]

Where instructions are received not from a client but from a third party purporting to represent that client, it is prudent for the solicitor to obtain written instructions from the client that he or she wishes him or her to act. Similarly where instructions are given by only one client on behalf of others in a joint matter, it is unwise to proceed without checking that all clients agree with the instructions given.

It is a wise solicitor who knows his own client. Is the client the trade union or its member? The driver of the car or his insurers? The employer or his insurers? The insured or the legal expenses insurers? The limited company or its director personally? A husband and wife or just one of them? The solicitor who does not ascertain clearly at the outset who his client is may repent at leisure if it transpires he has taken instructions from the wrong person or finds his bill rejected by the person to whom he renders it.

Cancellable Agreements

[1.7]

Since 1 October 2008 the seven day cooling-off period for contracts made at a consumer's home or place of work has been extended to legal services by the Cancellation of Contracts made in a Consumer's Home or Place of Work etc,

Regulations 2008, SI 2008/1816. The predecessor regulations (the equally snappily entitled 'Consumer Protection (Cancellation of Contracts Concluded away from Business Premises) Regulations 1987) applied in similar circumstances except in one important respect. Where a meeting had been solicited by the client, the 1987 Regulations did not apply. The prospect of an unsolicited visitor by a solicitor (presumably therefore an 'unsolicitor') to a potential client is sufficiently rare for it to be excluded for the purposes of this book.

The 2008 Regulations apply even where the retainer is entered into at the solicitor's office following a visit to the client's home or place of work. Where the regulations apply, reg 7(5) requires the client to be notified in writing of the right to cancel 'set out in a separate box with the heading Notice of the Right to Cancel'. This notice is said to be 'incorporated in the same document' but that has been widened somewhat by case law only requiring the notice to be in a document 'inextricably linked' to the contract, rather than the contract itself.

As with the Consumer Credit Act 1974 itself, a failure to get the documentation and procedure correct can lead to the entire agreement becoming unenforceable. This is a regular source of enquiry between the parties by defendants facing claimants using CFAs where the explanation and signature process has been at least partially dealt with at the client's house (or place of work).

Where the client decides to cancel the agreement, then unlike termination of the retainer by the client generally (see [1.27] below), cancellation has the effect of the retainer being treated as if it had never existed. Consequently, no fees would be payable and this may well be a deterrent to a solicitor carrying out any work or making any disbursement during the cooling off period. However in a case of urgency the client care letter may include a provision that the client accepts liability for any costs or disbursements incurred during the cooling off period.

QUOTATIONS AND ESTIMATES

[1.8]

As we shall see, the 2011 Code of Conduct looks for, amongst other things, a clear explanation by the solicitor of the fees to be charged. There is no specific reference to quotations and estimates, which are, of course, governed by the law of contract. We will start by looking at how the solicitor should describe his 'fee indications' so as not to fall foul of the law before going on to see how they should be phrased so that the Regulatory Gods are equally pleased.

Quotations

[1.9]

Do you really know the difference between a quotation and an estimate in respect of solicitors' costs? A quotation is a fixed price for doing the work which cannot be exceeded in any circumstances except with the freely given consent of the client. Common examples are the buying and selling of houses and the drawing of wills.

Estimates

[1.10]

Estimates have a chameleon quality. They have a tendency to turn into quotations when you are not looking. An unqualified estimate, for example, is for all practical purposes a fixed price quotation. On 20 February 2003, the Law Society Gazette recorded a case where a firm of solicitors had informed the client at the outset of his matrimonial proceedings that he could expect a bill in the region of £2,500. The firm regularly rendered interim bills, the penultimate one leaving the total cost just a little short of the original estimate. Nine months later, after the fee earner who had dealt with the client had left the firm, the solicitors sent the client an additional bill that took his costs £750 above the estimate he had been given originally. The Regulator ordered the firm to refund to the client the costs in excess of the estimate.

Successive Codes of Conduct have required solicitors to give the best possible information to their clients at the outset of the likely costs to be incurred. That inevitably means estimating the likely future cost (and for this purpose it matters not whether it is described as an estimate or a budget or anything else.)

When doing so, you should make it clear that the estimate is not intended to be fixed. This makes the fee indication into a qualified estimate. Solicitors should inform the client that if there are unforeseen developments and/or complications the estimate may need to be revised and updated and the client will be advised of this as the matter progresses. In protracted matters such as litigation, it may be advisable for solicitors to give staged estimates, with an overall estimate at a later stage once all the issues have been identified.

In *Mastercigars Direct Ltd v Withers LLP* [2007] EWHC 2733 (Ch), [2008] 3 All ER 417, the relevant wording of Withers' qualified estimate is worth quoting: 'When you instruct us we will do our best to tell you the likely level of our fees. Unless we tell you otherwise, this will be an estimate only, not a fixed quotation. If you ask for a fixed quotation we will try to provide one. However it may not be possible to predict the amount of time we will need to deal with the matter. You may set an upper limit on our costs. We will not do any work that will take our fees over this limit without your permission . . . ' The SCCO Master had held that in the circumstances Withers had failed to update their estimate and were bound by it but Withers' appeal was allowed. Mr Justice Morgan held that their estimate was not a fixed quotation, nor was it an upper limit on costs, nor did it define the work to be done. The retainer was subject to the Supply of Goods and Services Act 1982, s 15 and it was therefore an implied term that the solicitors would be paid reasonable remuneration for their services. Although the solicitors had given a contractual promise to update the costs estimate, that was not a condition precedent to them recovering any sum in addition to the sums set out in the estimate. He held that where a costs estimate is given but the costs subsequently claimed exceeded the estimate, it does not follow that the solicitor would be restricted to recovering the sum in the estimate. The question was 'What, in all the circumstances, is it reasonable for the client to be expected to pay?' And the estimate was one, but only one, of those circumstances. As explained in *Leigh v Michelin Tyre plc* [2003] EWCA Civ 1766, [2004] 2 All ER 175, [2004] 1 WLR 846 the court could have regard to the estimate and it was a factor to be taken into consideration as a yardstick in determining what was reasonable. The greater the difference between an estimate and the final bill, the greater the explanation called for. However, if there is a satisfactory explanation of the difference an estimate may cease to be a useful yardstick. Reliance on the estimate by the client is another factor. It was not necessary to imply into the contract of retainer a term that the solicitors had to comply with the Solicitors Costs Information and Client Care Code in respect of updating costs information.

The case went back to the SCCO Master for consideration of whether the client relied on the estimate (this is dealt with in more detail below.) The Master's decision on this point was also appealed. Morgan J overturned this decision and the claimant sought to appeal his judgment. When refusing permission to appeal ([2009] EWCA Civ 1526, [2010] 3 Costs LR 374) the Court of Appeal endorsed paragraph 54 of the second judgment of Morgan J which included the sentence 'it is not the proper function of the court to punish the solicitor for providing a wrong estimate or for failing to keep it up to date as events unfolded.'

[The case of *Minkin v Cawdery Kaye Fireman and Taylor* referred to above is also an example of an effective qualification of an estimate.]

Quotations that used to be estimates

[1.11]

Even a qualified estimate is not carte blanche to charge whatever sum is justified by the time recorded. A qualified estimate is an indication of price which is qualified by a statement that the solicitor may have to charge more if the matter involves more work than he expects. There are circumstances in which what started out in life as a qualified estimate can finish up as binding on the solicitor as a quotation.

In particular, the final amount payable should not vary substantially from the estimate unless there has been a change in circumstances of which the client has been informed. The qualified estimate should have stated the circumstances which might give rise to an increase and the client should have been informed when any of those circumstances arose. The fact that the solicitor has seriously underestimated the work or disbursements involved will not necessarily be a changed circumstance – indeed it could be a strong indication of negligence on the part of the solicitor in preparing the original estimate. A factor in considering whether any upward revision of an estimate is reasonable is whether the client instructed the solicitor after shopping around and taking the lowest estimate he had been given.

In *Reynolds v Stone Rowe Brewer (a firm)* [2008] EWHC 497 (QB), 152 Sol Jo (no 14) 29, [2008] 4 Costs LR 545 the claimant instructed the defendant solicitors to represent her in a dispute with a building contractor and they informed her that the estimated cost of taking the matter forward and through to trial would be in the region of £10,000–£18,000 plus VAT. Throughout the course of the litigation, the defendants rendered a number of invoices and they then wrote to the claimant saying that their estimate of the likely cost of the case had to be revised to around £30,000 plus VAT. The court upheld the finding of the costs judge that the revised estimate had been an attempt to correct an 'earlier under-estimate' and was not attributable to any change in the facts. The solicitors were not saved by the provision in their client care letter that 'This is only of course an estimate which could be increased depending on how strenuously the matter is defended'. There had been no significantly unusual developments before the revised estimate such as to explain the difference between the £18,000 estimate and the £30,000 revised estimate.

Increasing the estimate

[1.12]

The best time to rectify an errant estimate is obviously before things get out of hand. Whilst the client is relying on you in respect of the case, he will be much more receptive to revising the likely cost than once the case has come to an end. The best approach is always to address the issue before the costs are incurred which will cause the earlier estimate to be exceeded. Indeed, previous Law Society guidance stated that 'clients should be informed immediately if it appears that the estimate will be, or is likely to be, exceeded. In most cases this should happen before undertaking the work that exceeds the estimate'.

This guidance was in place long before the concept of budgets in the manner that we have now came to light. But it remains good advice. Clients, as much as opponents and the court, want to be appraised of increases in the cost of the litigation before it occurs and not afterwards. It means that some alternatives to the current plan might be considered – fewer or different experts; alternative cases abandoned and so forth. Almost as importantly, it gives the client the impression that you are on top of the financial aspects of the case and not simply the legal aspects. Solicitors may think that it is acceptable to dismiss overspends with generic comments about the behaviour of the opponent, but clients are often less than impressed that they have incurred costs unknowingly that will cause problems in settlement or assessment in due course.

Exceeding the estimate

[1.13]

In *Wong* v *Vizards* (a firm) [1997] 2 Costs LR 46, the solicitors in a letter to their client allowed for profit costs of £9,955 stating 'the fee proposal hopefully sets out the fullest extent of your liability to this firm for costs likely to be incurred in the future'. Although this did not amount to a binding agreement that in no circumstances would the solicitors' fees exceed their fee proposal, it was a clear and considered indication of the maximum, upon which the client was likely to rely and did rely. The solicitors in fact delivered bills exceeding £45,000. In considering whether a reasonable amount for the work done should exceed what the fee-payer had been led to believe was a worst case assessment, regard should be had to any explanation of the divergence. It had not been suggested there was any unexpected development between the date of the solicitors' letter and the date of trial, and no satisfactory explanation had been given why the solicitors should be entitled to profit costs exceeding the amount put forward as their worst case assessment. The judge limited the solicitors' costs to £9,955.

But, that was not the end of the story. The trial had taken two days less than estimated by the solicitors at a charge of £660 a day. The client argued that the estimate should therefore be reduced by the resultant saving of £1,320. This was about 15% of the total estimate and the judge thought it not unreasonable to give the solicitors the benefit of this margin and therefore allowed them to recover the full amount of their estimate. And it is from this that the myth has arisen that *Wong* v *Vizards* is authority for the proposition that solicitors may exceed their estimates by a margin of 15%.

The judge in *Mastercigars*, whilst allowing the solicitors to exceed their estimate, confirmed there is no 15% margin. The judge in *Reynolds*, whilst not allowing any increase, said it was not possible to say that the sum he would have allowed represented any particular margin over the estimates. However, in at least two SCCO cases costs judges have found it helpful to apply it, and this was one of the matters on which the assessors disagreed with the judge in *Mastercigars*. On appeal from the costs judge, who had applied a 20% margin, the High Court judge addressed margins in these terms: 'The adoption of an approach which involves adding a margin, usually expressed as a percentage, to the estimate as the conventional approach and the majority of cases would pay scant, if any, regard to that legal process. While there are advantages to the margin approach, it should not be systematically endorsed. The adoption of a margin approach greatly simplifies the steps which a costs judge needs to take when carrying out a detailed assessment of the bill which has been preceded by a lower estimate. If the margin approach became the permissible conventional approach, then the costs of the detailed assessment could be reduced and the outcome would be more predictable. But even where a court had followed the proper assessment process, it can never be right for costs to be expressed by reference to a margin.'

The client's reliance on estimates

[1.14]

The longer an estimate is left unrevised, the more risk there is that the client will come to rely heavily on the estimate. After all, why shouldn't the client believe that your estimate holds good if you do not revisit it?

In *Mastercigars*, the judge said the following in respect of how the court should look at the client's apparent reliance on an estimate. He said:

> 'In my judgment, the legal process involved in a case where a client contends that its reliance on an estimate should be taken into account in determining the figure which it is reasonable for the client to pay is as follows. The court should determine whether the client did rely on the estimate. The court should determine how the client relied on the estimate. The court should try to determine the above without conducting an elaborate and detailed investigation. The court should decide whether the costs claimed should be reduced by reason of its findings as to reliance and, if so, in what way and by how much. Whether there should be a reduction, and if so to what extent, is a matter of judgment. Specific deductions can be made from the costs otherwise recoverable to reflect the impact which an erroneous and uncorrected estimate had on the conduct of the client. Such an approach requires the court to form an assessment of the impact of the estimate on the conduct of the client. The court should consider the deductions which are needed in order to do justice between the parties. It is not the proper function of the court to punish the solicitor for providing a wrong estimate or for failing to keep it up to date as events unfolded.

In terms of the sequence of the decisions to be made by the court, it has been suggested that the court should determine whether, and if so how, it will reflect the estimate in the detailed assessment before carrying out the detailed assessment. The suggestion as to the sequence of decision making may not always be appropriate. The suggestion is put forward as practical guidance rather than as a legal imperative. The ultimate question is as to the sum which it is reasonable for the client to pay, having regard to the estimate and any other relevant matter.'

If you don't want to leave it to the judgment of a judge, make sure that the estimate (and the client) is kept up to date.

Not giving an estimate

[1.15]

Garbutt v Edwards [2005] EWCA Civ 1206, [2006] 1 All ER 553, [2006] 1 WLR 2907 held that, between the parties, failure to give an estimate was merely a matter for the costs judge, who should consider whether this has in any way increased the costs over what they would have been if an estimate had been given.

Between solicitor and client, the position appears to be that if a solicitor gives a costs estimate he is bound by it, but there is still no effective sanction for failing to do so. If so, we reach the bizarre position that solicitors who provide estimates which prove to be inaccurate are in a worse position than those who provide no estimates at all. The regulatory sanctions for such a failure ought to be significant.

The effect of budgets on estimates

[1.16]

The great majority of cases which appear to be multi-track in value are going to require a costs budget to be prepared once proceedings are commenced. The recoverable fee for creating a budget is limited to 1% of the budget or £1,000, whichever is the greater. If the solicitor has to start from scratch, such sum is not going to be adequate to cover the costs of completing the precedent form in a sizeable case.

It was said, when the requirements for budgeting were being unveiled, that the fee was based on an assumption that most of the work had already been carried out in providing the client with a costs estimate at the outset. As such, those figures (and their assumptions) could simply be transposed on to the Precedent H spreadsheet.

The costs estimates provided to clients in client care letters to date have been broad figures – or brackets of broad figures – with generally no explanation of how they have been reached. Explanatory comments have tended to be no more than 'in our experience' fees up to, for example the issue of proceedings or disclosure, 'for this type of work' are in the order of £X,000, or between £X,000 and £Y,000. Often there is a comment that 'hopefully' the other side will see sense and the case will be settled earlier, just in case the client takes fright at the figures that have been set out.

It will be interesting to see whether firms bite the bullet and start preparing Precedent H type budgets as the costs estimates for their clients. As budgets become more common, each firm's figures may well become more standardised, and therefore quicker to prepare for clients and the court alike.

Assuming this is the case, there will be further case law on the circumstances in which it is appropriate for a firm to exceed the costs budget/estimate that it originally provided. As with budgets and estimates generally, the setting out of the underlying assumptions is absolutely key to later variation.

The wording of subsection 3 of the Practice Direction to Part 44 leaves open the question of the relevance of estimates in the form of budgets generally. The received wisdom seemed to be that once a budget had been agreed between the parties or approved by the court, it could be revised during the course of proceedings but the court would have regard to the last such budget when the final costs came to be considered. The need to be within 20% of the budgeted figure did not seem to be relevant where a Case Management Order had been made (and as part of which the budget would be agreed or approved). However the Practice Direction suggests that there may still be some relevance to the 20% threshold even where a CMO has been made. That suggests a double test for the errant budgeter. Not only will a 20% deviation call for an explanation but any deviation will need to have good reasons to go with it. Budgets are going to be subject to the Goldilocks test – not too much, not too little but just right – or else they will cause trouble for the client and/or his solicitor.

The effect of recoverable fixed costs on estimates

[1.17]

Section 74(3) of the Solicitors Act 1974 states that a solicitor may only recover from his client on assessment the costs which are recoverable on a between the parties assessment. This statement is made with the proviso that it only relates to the county court and not High Court. Furthermore it can be displaced by rules of court. CPR 46.9(2) disapplies s 74(3) if there is a written agreement between the solicitor and his client.

Why is this important? The implementation of fixed costs on the small claims track and, sooner or later, the fast track means that solicitors will be limited to these fixed figures unless they have set out in writing that s 74(3) is not to apply to their agreement. Indeed, it is prudent for the solicitor at the outset of litigation to identify the probability of a shortfall to the client as the result of fixed costs (or the proportionality test), and to obtain the client's signature to an agreement that the solicitor and client costs may exceed those recoverable between the parties.

1.18 *Chapter 1 The Retainer*

CLIENT CARE

Old Regulatory Position

[1.18]

The Solicitors Practice Rules came into effect as far back as 1936. The last version of the SPR was produced in 1990 and included Practice Rule 15 (costs information and client care). It was revised periodically until it was replaced by the Solicitors' Code of Conduct 2007. During the 17 years up to 2007 solicitors gradually became used to the concept of providing their clients with prospective costs estimates as well as information on how their case was to be run. The phrase 'Rule 15 letter' is still regularly heard as a synonym for a client care letter: not least because since 2007 there have been several further revisions of the Code so that the relevant rule number has not stayed long enough with practitioners for the number 15 to have been displaced.

The 2007 Code of Conduct was not the Law Society's Code but the Solicitors' Code because it was prescribed not by the Law Society but by the Solicitors' Regulation Authority. The new Code contained 25 rules, with Rule 2 – promoted from 15 – dealing with client relations, including information about the cost in Rule 2.03. In 2011 a further Code of Conduct was introduced with its focus being on the outcome for the client of the service provided rather than whether prescriptive regulations had been met (a so-called 'tick-box' approach.) The latest regulations are considered in detail later in this Chapter. But their lack of prescription means that there is some benefit of reminding ourselves of what was seen to be good practice prior to their introduction on 6 October 2011.

[1.19]

'2.03 Information about the cost

(1) You must give your client the best information possible about the likely overall cost of a matter both at the outset and, when appropriate, as the matter progresses. In particular you must:
(a) advise the client of the basis and terms of your charges;
(b) advise the client if charging rates are to be increased;
(c) advise the client of likely payments which you or your client may need to make to others;
(d) discuss with the client how the client will pay, in particular:
 (i) whether the client may be eligible and should apply for public funding; and
 (ii) whether the client's own costs are covered by insurance or may be paid by someone else such as an employer or trade union;
(e) advise the client that there are circumstances where you may be entitled to exercise a lien for unpaid costs;
(f) advise the client of their potential liability for any other party's costs; and
(g) discuss with the client whether their liability for another party's costs may be covered by existing insurance or whether specially purchased insurance may be obtained.

(2) Where you are acting for the client under a conditional fee agreement, (including a collective conditional fee agreement) in addition to complying with 2.03(1) above and 2.03(5) and (6) below, you must explain the following, both at

the outset and, when appropriate, as the matter progresses:
(a) the circumstances in which your client may be liable for your costs and whether you will seek payment of these from the client, if entitled to do so;
(b) if you intend to seek payment of any or all of your costs from your client, you must advise your client of their right to an assessment of those costs; and
(c) where applicable, the fact that you are obliged under a fee sharing agreement to pay to a charity any fees which you receive by way of costs from the client's opponent or other third party.

(3) Where you are acting for a publicly funded client, in addition to complying with 2.03(1) above and 2.03(5) and (6) below, you must explain the following at the outset:
(a) the circumstances in which they may be liable for your costs;
(b) the effect of the statutory charge;
(c) the client's duty to pay any fixed or periodic contribution assessed and the consequence of failing to do so; and
(d) that even if your client is successful, the other party may not be ordered to pay costs or may not be in a position to pay them.

(4) Where you agree to share your fees with a charity in accordance with 8.01(k) you must disclose to the client at the outset the name of the charity.

(5) Any information about the cost must be clear and confirmed in writing.

(6) You must discuss with your client whether the potential outcomes of any legal case will justify the expense or risk involved including, if relevant, the risk of having to pay an opponent's costs.

(7) If you can demonstrate that it was inappropriate in the circumstances to meet some or all of the requirements in 2.03(1) and (5), you will not breach 2.03.'

The accompanying Guidance, which it was emphasised was not part of the Code and was not mandatory explained that a breach of 2.03 would not invariably render the retainer unenforceable. This was a nod to the arguments in the 'CFA costs wars' where non-compliance with the Courts and Legal Services Act 1990 (as amended) did have this effect. For a short while, defendants argued that a breach of the Code, which has been found to have the force of secondary legislation by *Swain v The Law Society* [1983] 1 AC 598, was similarly catastrophic in effect. The wording of the Guidance was a regulatory sandbag with which to bolster the wall of resistance against such attacks.

Similarly, in *Garbutt*, the Court of Appeal held that the requirement in Solicitor's Practice Rule 15 that a solicitor 'shall' give to the client the information prescribed in the Costs Code was not mandatory, and failure to comply with it did not render the retainer unlawful.

The Court of Appeal was at pains to emphasise that the rules were not designed to relieve paying parties of their obligations but to protect clients.

The purpose of Rule 2.03 according to the Guidance was said to be to ensure that the client was given relevant costs information which was clearly expressed. The information had to be worded in a way that was appropriate for the client. It needed to be given in writing and regularly updated. The Guidance accepted that it was often impossible to tell at the outset what the overall cost would be. The requirement was simply that the client was provided with as much information as possible at the start and was kept updated thereafter. If (as was usually the case) a precise figure could not be given at the outset, the Guidance proposed that this should be explained to the client and agreement reached on a ceiling figure or review dates.

1.20 *Chapter 1 The Retainer*

In the 1990 Code the requirement had been to provide the client with the 'best information possible about the likely overall cost' at the outset and to update the cost information 'at least every six months'. The time period was watered down to 'when appropriate' in 2007 but many kept to the 6 month period as a guideline.

The Guidance went on to recognise that clients could not all be seen to have the same requirements. Some would require more information than others; some aspects would be more important to some clients than others. Consequently, the Guidance said, this rule would be enforced in a manner which was 'proportionate' to the seriousness of the breach.

But, beware of the warning by the Court of Appeal in *Darby v Law Society* [2003] EWHC 2270 (Admin):

> 'There is . . . a heavy onus on a solicitor to establish that his client is so sophisticated [about costs] that the rule may be disregarded.'

Other exceptions are where compliance may be insensitive or impractical, such as the taking of instructions for a deathbed will and emergencies. While an emergency might be a sufficient reason for not giving costs information at the outset of the retainer and before work is started, it does not excuse complete failure to give any information at any stage. Where solicitors took over the conduct of proceedings from another firm only five days before the final hearing, they inadvertently overlooked the matter of costs information. This may have been understandable, but unfortunately the retainer continued for a further 18 months, during which time the client was given no information as to costs nor were any interim bills rendered. The outcome was that the solicitors were ordered to pay compensation to the client amounting to almost one-fifth of their fees (Law Society Gazette, 25 April 2002).

Current Regulatory Position

[1.20]

The SRA Code of Conduct 2011 forms part of the SRA Handbook which came into effect on 6 October 2011. In addition to the Code are the 10 mandatory SRA Principles which are at the heart of the new regime. These apply to all solicitors and to all firms that are regulated by the SRA and everybody who works in them – including owners in an Alternative Business Structure who are not lawyers.

The principles state that 'you must:
(1) uphold the rule of law and the proper administration of justice,
(2) act with integrity,
(3) not allow your independence to be compromised,
(4) act in the best interests of each client,
(5) provide a proper standard of service to your clients,
(6) behave in a way that maintains the trust the public places in you and in the provision of legal services,
(7) comply with your legal and regulatory obligations and deal with your regulators and ombudsmen in an open, timely and co-operative manner,

(8) run your business or carry out your role in the business effectively and in accordance with proper governance and sound financial and risk management principles,
(9) run your business or carry out your role in the business in a way that encourages equality of opportunity and respect for diversity, and
(10) protect client money and assets.'

The Handbook gives the following guidance on the Principles: 'They define the fundamental ethical and professional standards that we expect of all firms and individuals (including owners who may not be lawyers) when providing legal services. You should always have regard to the Principles and use them as your starting point when faced with an ethical dilemma. Where two or more Principles come into conflict, the Principle which takes precedence is the one which best serves the public interest in the particular circumstances, especially the public interest in the proper administration of justice.'

Client care moved up the batting order from Rule 15 in 1990 to Rule 2 in 2007. In the latest Code it has reached Chapter 1 which is a testament to the continuing complaints received by clients of the care they receive – generally in respect of service rather than advice – as much as anything else. Chapter 1 concerns providing a proper standard of service, which takes into account the individual needs and circumstances of each client. This includes providing clients with the information they need to make informed decisions about the services they need, how these will be delivered and how much they will cost. It confirms solicitors are generally free to decide whether or not to accept instructions in any matter, provided they do not discriminate unlawfully.

Outcomes Focused Regulation

[1.21]

The introductory text of Chapter 1 is short (as with the other chapters). Most of the text relates to 'Outcomes' which must be achieved and by 'Indicative behaviours' which 'may' tend to show that the outcomes have been achieved and as such you have complied with the Principles set out above.

This approach is a veritable sea change to the regulation of legal services. Previously, a conscientious solicitor could be assured of complying with his professional obligations if he read through the Rules and noted any with which he did not seem to be complying and took steps to address them. An exhaustive list could be produced and each one ticked off as they were completed.

The variation in solicitors' practices from multinational firms with thousands of partners to sole practitioners, not to mention Alternative Business Structures, meant, at least as far as the SRA were concerned, that one set of prescriptive Rules, even with Guidance, could not fit all circumstances. Consequently the 2011 Code looks from the other end of the telescope to see how things turned out, particularly for the client. If your client has a good outcome (not necessarily the same as having been successful) then you have adhered to the Principles. How do you know whether your client has had a good outcome? A good question. The indicative behaviours are meant to show how good practice might achieve the right outcome but they are no more than examples and may not be relevant in your situation. They are, in practice, likely to be something that is only looked at when trying to demonstrate that the outcomes were achieved following a complaint by the client.

Outcomes Focused Regulation and costs

[1.22]

This is a book about costs not regulatory compliance. You need to be aware that non-compliance may affect your remuneration but it is beyond the scope of this book in our view to set out all of the Outcomes and Indicative behaviours with commentary. There are other books where that can be obtained and the Handbook itself can be obtained from the SRA website (www.sra.org.uk).

Outcomes

The Outcomes particularly relevant to the issue of costs are as follows:-

'The Outcomes

The outcomes in this chapter show how the Principles apply in the context of client care. They are mandatory.

You must achieve these outcomes:

O(1.6) you only enter into fee agreements with your clients that are legal, and which you consider are suitable for the client's needs and take account of the client's best interests;

Entering into an agreement which is legal (O(1.6)) sounds as if it is so obvious that it does not need saying. Nevertheless, there is an indicative behaviour (IB(1.27)) to show what would not demonstrate achieving a good outcome namely 'entering into unlawful fee arrangements such as an unlawful contingency fee' so it must have been concluded by the SRA that this was not obvious. However, there is certainly case law to suggest that a solicitor foolish enough to enter into an agreement that is unenforceable because it is unlawful unintentionally provides the client with free legal services which may well be a very good outcome for the client.

O(1.12) clients are in a position to make informed decisions about the services they need, how their matter will be handled and the options available to them;

O(1.12), and its compatriot indicative behaviour IB(1.1) (below) presage a change in working practice which is quite new. Traditionally, solicitors have provided the same level of service to all their clients, at least in theory. This Outcome, however, provides the solicitor with the opportunity to agree different service levels with different clients which may, or may not, simply reflect the ability of the client to pay for the solicitors' full service. It is but a short step to see the so-called 'unbundling' of the solicitor's assistance to his client so that certain stages or elements are dealt with by the client himself, or by using the assistance of a third party, and the solicitor is only involved in those parts that really require his expertise. The effect of such arrangements on time-hallowed concepts such as being 'on the court record' will have to be worked out in due course.

O(1.13) clients receive the best possible information, both at the time of engagement and when appropriate as their matter progresses, about the likely overall cost of their matter;
O(1.14) clients are informed of their right to challenge or complain about your bill and the circumstances in which they may be liable to pay interest on an unpaid bill;

Client care **1.23**

Providing the client with the best possible information at the outset (O(1.13)) and information on how to challenge bills (O(1.14)) follow on from previous Codes and highlight the client's statutory entitlement (see [3.16]ff) to such information.

Indicative Behaviours

[1.23]

The most relevant Indicative behaviours are set out below. As you will see there is an entire section headed 'fee arrangements with your client' in addition to the more general indicative behaviours. They are all worth studying. You will find some of the first twelve appear later in this Chapter rather than immediately below.

> Dealing with the client's matter
>
> IB(1.1)   agreeing an appropriate level of service with your client, for example the type and frequency of communications;
> IB(1.4)   explaining any arrangements, such as fee sharing or referral arrangements, which are relevant to the client's instructions;
> IB(1.5)   explaining any limitations or conditions on what you can do for the client, for example, because of the way the client's matter is funded;
>
> 'Fee arrangements with your client
>
> IB(1.13)   discussing whether the potential outcomes of the client's matter are likely to justify the expense or risk involved, including any risk of having to pay someone else's legal fees;
> IB(1.14)   clearly explaining your fees and if and when they are likely to change;
> IB(1.15)   warning about any other payments for which the client may be responsible;
> IB(1.16)   discussing how the client will pay, including whether public funding may be available, whether the client has insurance that might cover the fees, and whether the fees may be paid by someone else such as a trade union;
> IB(1.17)   where you are acting for a client under a fee arrangement governed by statute, such as a conditional fee agreement, giving the client all relevant information relating to that arrangement;
> IB(1.18)   where you are acting for a publicly funded client, explaining how their publicly funded status affects the costs;
> IB(1.19)   providing the information in a clear and accessible form which is appropriate to the needs and circumstances of the client;
> IB(1.20)   where you receive a financial benefit as a result of acting for a client, either:
> •   paying it to the client;
> •   offsetting it against your fees; or
> •   keeping it only where you can justify keeping it, you have told the client the amount of the benefit (or an approximation if you do not know the exact amount) and the client has agreed that you can keep it;
> IB(1.21)   ensuring that disbursements included in your bill reflect the actual amount spent or to be spent on behalf of the client;'

In-house and Overseas practice

[1.24]

There are variations on the outcomes if you work in-house or overseas. For the former, some of the outcomes are not applicable where the work is carried out for the employer. All of them apply if work is done for someone else 'unless it is clear that the outcome is not relevant to your particular circumstances.' For the latter, regard is to be had in respect of prevailing local customs including the need to avoid contingency fee arrangements where they are unlawful locally.

Status of personnel

[1.25]

One of the other general indicative behaviours is to ensure that the client is told in writing the name and status of the people dealing with the case and whoever is responsible for its supervision (IB(1.3)). A similar provision in the 1990 Code enabled the Court of Appeal to uphold the client's refusal to pay his solicitor's bill on the grounds that he had asked for an appointment with a solicitor but it transpired that the person to whom he was referred by the receptionist was neither a solicitor nor a qualified legal executive. In *Pilbrow v Pearless De Rougemont & Co (a firm)* [1999] 3 All ER 355, [1999] 2 FLR 139, CA, the work had been done to the standard of a competent solicitor. Nevertheless, the contract was one to provide legal services by a solicitor and therefore the firm had not performed that contract at all. The firm should have trained its receptionist when faced with a request to see a solicitor to do one of the following:

(i) refer the client to a solicitor;
(ii) refer the client to someone who was not a solicitor but inform the client that that person was not a solicitor; or
(iii) refer the client to someone who the receptionist knew was not a solicitor, refrain from telling the client that fact and alert the referee to the fact that the client had asked for a solicitor.

If the last course were adopted then it would be the duty of the referee straight away to make clear to the client that he was not a solicitor.

The indicative behaviour talks of setting out the name and status in writing rather than on meeting. Compliance with this behaviour is usually achieved in the client care letter and arguably therefore the situation in *Pilbrow* would be rectified by that subsequent letter even if the client was still under a misapprehension when he left the building after his consultation with his 'solicitor.'

In *Adrian Allen Ltd v Fuglers* (SCCO Case No 13 of 2003) a struck-off solicitor (employed by the defendant firm with the permission of the Law Society) informed the claimant, who was a client of the firm, that he was a solicitor and that he could assist in the litigation in which it was engaged. The claimant instructed him, but he proved to be incompetent, and subsequently disappeared. The solicitors were not entitled to any payment for the work done because the terms of the retainer were for the work to be done by a solicitor. Presumably the work would have been disallowed on assessment for being incompetently done in any event.

COMPLAINTS

[1.26]

It has long been a grumble of many solicitors that the client has to be told about his rights to complain about the solicitor in (more or less) the very first letter (now O(1.9)). It may make solicitors feel better to know that when, much more recently, similar provisions became applicable to barristers so that they had to notify the client directly about the right to complain (so-called 'signposting' provisions required by the Legal Ombudsman), they were just as vociferous about the poor impression that this gave. Nevertheless the number of complaints dealt with by the Legal Ombudsman (in respect of service) and the SRA (in respect of a breach of the SRA Principles) suggest that there is a very real need for clients to understand their rights in this area.

Clients are encouraged to contact their solicitor directly initially but if that proves unsuccessful, the client can then contact the Legal Ombudsman (O(1.10)). Clients' complaints need to be dealt with promptly, fairly, openly and effectively (O(1.11)). Such an approach does not sit naturally with most professionals and needs to be worked at constantly. For those in need, further study of the complaints handling indicative behaviours can be found at IB(1.22) to IB(1.24).

ENDING THE AGREEMENT

By the client

[1.27]

The client may terminate a retainer at any time for any reason.

The retainer may also be terminated by the effluxion of time if it was for a fixed period, or by the death, bankruptcy or insanity of the solicitor or the client, or if its continuance becomes unlawful.

The client may have the right to cancel under the Cancellation of Contracts made in a Consumer's Home or Place of Work etc Regulations 2008 (see **[1.7]** above).

But a client who seeks to end the retainer in order to deprive his solicitor of fees by claiming them directly from the opponent will be prevented by the solicitor's lien if the money is the solicitor's client account. If it goes directly to the client, the Court of Appeal's decision in *Khans Solicitors (a firm) v Chifuntwe* [2013] EWCA Civ 481, [2013] 4 All ER 367, makes clear that the paying party makes payment to your client at his own risk if he is on notice of your interest in the money (or is colluding with the client to deprive you of your fees). If the paying party is unsure as to whom he should pay the costs, he can apply to the court for guidance or to enable him to pay the money into court to discharge his debt. The Court of Appeal suggested that solicitors' terms of business should be altered as necessary to take this situation into account. The Court suggested that a clause requiring the paying party to pay monies into court in such circumstances might be appropriate.

1.28 Chapter 1 The Retainer

By the solicitor

(i)   Under the Solicitors Act 1974, s 65(2)

[1.28]

In respect of contentious costs only, a solicitor can request a client to pay a reasonable sum of money on account of costs incurred and to be incurred. If the client fails to make that payment within a reasonable time, this shall be a good cause upon which the solicitor may, on giving reasonable notice to the client, terminate the retainer.

(ii)   For good cause

[1.29]

There is an implied term that the solicitor may terminate the retainer upon reasonable notice for good cause, such as a failure to provide funds for disbursements or give adequate instructions; the client requiring the solicitor to behave unlawfully or unethically; obstructing the solicitor, or preventing him from dealing with the matter, and where there is a serious breakdown in confidence between them.

The main issue in *Richard Buxton (a firm) v Mills-Owens* [2010] EWCA Civ 122, [2010] 4 All ER 405, [2010] 1 WLR 1997 was whether a solicitor had 'good reason' for terminating the retainer if the client insisted on his putting forward a case and instructing counsel to argue a case that was 'doomed to disaster' or which the solicitor believed was 'bound to fail'.

There is no comprehensive definition of what amounts to a 'good reason' to terminate because it is a fact-sensitive question. It was wrong to restrict the circumstances in which a solicitor could lawfully terminate his retainer to those in which he was instructed to do something improper. Solicitors should not lightly be able lawfully to terminate their retainers, but the desirability of protecting a client from an arbitrary and unreasonable termination was not a sufficient justification for a narrow interpretation of the phrase 'good reason'.

It would be improper in a statutory planning appeal to advance an argument based on the merits of the decision by the planning inspector, which was hopeless and not genuinely arguable. As the client had insisted that such arguments be advanced, the solicitors had good reason for terminating the retainer. The retainer was an entire contract that had not been completed, but as it had been terminated for good reason the solicitors were entitled to their proper costs and disbursements for work done prior to the termination.

(iii)   Where the client loses capacity

[1.30]

What should you do if your client loses capacity to enter into a contract or (more likely) continue to give instructions?

Guidance to the Solicitors Code of Conduct 2007 gave this helpful advice: 'If your client loses mental capacity after you have started to act, the law will automatically end the contractual relationship. However, it is important that

Writing off/waiving costs **1.32**

the client, who is in a very vulnerable situation, is not left without legal representation. Consequently, you should notify an appropriate person (eg the Court of Protection), or you may look for someone legally entitled to provide you with instructions, such as an attorney under an enduring power of attorney, or take the appropriate steps for such a person to be appointed, such as a receiver or a litigation friend. This is a particularly complex legal issue and you should satisfy yourself as to the law before deciding on your course of action.'

CONTINGENCY AGREEMENT ISSUES

[1.31]

A private paying client who decides to terminate the retainer with his solicitor can do so for any reason and simply needs to pay whatever sum is outstanding in respect of the solicitors' charges. To the extent they are in dispute, they can be assessed by the court (see Chapter 3) and any lien the solicitor has will be resolved, or an undertaking given by subsequent solicitors instructed by the claimant. This was the position for hundreds of years.

However, when conditional fee agreements were allowed in 1995 and now damages-based agreements can be used in contentious matters, the position becomes more complicated. At the time the retainer comes to an end, the determination of whether the case has been successful or not will be a matter of conjecture if the case is still proceeding. Is a success fee or slice of the damages going to be paid? Or is there in fact no liability to the solicitor at all?

The Law Society has published a model CFA whenever significant changes have been made to the regime. Each set of Law Society Conditions sets out what happens if the agreement ends before the claims for damages ends. These cover most, if not all, eventualities in respect of the outcome of the case. But these have not dealt with issues such as the firm of solicitors becoming insolvent or merging with another firm or being intervened by the SRA etc. This seems to excite paying parties to argue that there is a flaw in the receiving party's retainer for some or all of the time covered by the case on regular occasions. To avoid this, if you decide to draft your own CFA, you should take considerable care to make sure that every eventuality that you can conceive is dealt with in the termination provision.

There is no Law Society DBA and so you are on your own in drafting suitable provisions, although the CFA terms are the obvious starting point.

WRITING OFF/WAIVING COSTS

[1.32]

In *Slatter v Ronaldsons (a firm)* [2002] 2 Costs LR 267 the defendant solicitors had acted for the claimant in matrimonial proceedings for which they had outstanding costs. When the claimant became unemployed they concluded that it would be uneconomic to pursue him for the balance of their costs and accordingly 'wrote them off', obtaining the appropriate VAT refund. Subsequently the client instructed other solicitors who wrote to the defendants requesting delivery up of all relevant papers which the defendants refused to do, contending that they had a lien over the papers until their bills were discharged in full. The judge, on appeal, upheld the lien (see [**3.29**] below for more on liens). The contractual liability to pay the balance of the bills survived their 'writing off' and there was no evidence that the claimant acted to his detriment entitling him to promissory estoppel, on the basis that the balance of the bills had been written off.

Similarly, where solicitors had agreed not to charge for work done after judgment was handed down they were not precluded from claiming costs for work done after that date because there was no consideration for their waiver (*Ashia Centur Ltd v Barker Gillette LLP* [2011] EWHC 148 (QB), [2011] 4 Costs LR 576).

CHAPTER 2

BILLING THE CLIENT

INTRODUCTION

[2.1]

This Chapter is split into three sections. The first covers the surprisingly complex area of interim bills. The second deals with the final bill rendered to the client and is the introduction to the next Chapter regarding suing the client where necessary for payment. The final section concerns VAT which is something which catches people out all too often and, like most things, is much easier to get right first time than to have to remedy errors that have been made.

Before getting on to these sections we should say a few words about two subjects which impact on this Chapter and indeed the next – terminology which is particularly important here and the fundamental question of to whom can you render your bill?

TERMINOLOGY

[2.2]

When communicating with clients, solicitors are apt to call their bill by a plethora of different names – an invoice, an account, a fee note, 'a note of my charges' etc. In each case, the solicitor is referring to a document on which is set out the sums payable to his firm for the firm's efforts together with those of other people engaged or payments made to third parties. As will soon be seen, there is more than one kind of interim bill that can be rendered. There is more than one kind of final bill as well. An interim bill may in fact only be a request for a payment on account, rather than a formal bill, anyway. We use the word 'bill' in this Chapter to cover most if not all of these documents. You need to be clear about the concepts behind the distinctions (and their effect) rather than slavishly following the correct terminology. In our experience, even the most skilled advocates trip over the correct terminology from time to time.

WHO CAN YOU BILL?

[2.3]

Save for those with a contract with the Legal Aid Agency, there is only one source from which a solicitor can be remunerated and that is by the client. It is fundamental to an understanding of the law of costs to appreciate this but it is regularly said by solicitors that they are seeking an order from a court for 'their costs' and not the client's. This is completely wrong, even where the client has been told that the solicitor will take whatever can be recovered from the other side and render no additional bill to the client.

It may well be that the client has rights against other persons or funds, which will ultimately pay the solicitor's fees. The trustee looks to a trust fund for reimbursement; a successful party in litigation can expect an order for costs against the loser; the landlord may provide in the lease that his legal costs be paid by the tenant, but none of this affects the solicitor and client relationship between the trustee, the successful litigant or the landlord and their respective solicitors. The solicitor on behalf of his client seeks to recover money due to the client from the trust fund, the unsuccessful litigant or the tenant, but he does not have any personal relationship with them. The solicitor looks to his client for payment of the costs whether or not the client is reimbursed by a third party. The solicitor must therefore deliver a bill to his client for the whole of his costs and give credit for any sum that has been received. This is the strict legal position and has been since at least 1908. In *Cobbett v Wood* [1908] 2 KB 420, 77 LJKB 878, CA, the Court of Appeal held that a bill of costs which excluded the between the parties' items and simply referred to the excess chargeable as between solicitor and own client was not a proper bill. Whether a court would take such a strict line now must be open to question. But the fact that the costs are the client's rather than the solicitor's is beyond doubt.

Misunderstandings about this arise frequently when a third party who is paying the client's costs wishes to have an invoice on which he can recover the VAT. For example, leases regularly contain clauses where the landlord has obtained an undertaking from the tenant to pay his costs. Such a clause also regularly provides that the landlord should be 'compensated fully' or 'indemnified' for costs and expenses incurred as a result of the tenant's breach of the lease. Similar clauses appear in most mortgage deeds. The solicitor is providing legal services to his client, not the third party tenant (for example), and that tenant cannot obtain a VAT invoice even though he has ultimately funded payment of the client's bill.

The Court of Appeal decided in *Gomba Holdings (UK) Ltd v Minories Finance Ltd (No 2)* [1993] Ch 171, [1992] 4 All ER 588, CA, that mortgagees were entitled to recover costs on the indemnity basis following a review of the meaning of 'indemnity costs' clauses. The Court of Appeal in *Church Comrs v Ibrahim* [1997] 1 EGLR 13, [1997] 03 EG 136, CA, confirmed that the principles in *Gomba Holdings* are not confined to mortgage cases. It held that in general the landlord is not to be deprived of a contractual right to indemnity costs. Although the court always retains a discretion on costs, that discretion should be used to reflect the contractual agreement between the landlord and the tenant unless the landlord's conduct is improper or unreasonable. This decision clarified the landlord's position on costs and made it easier either to negotiate or obtain from the court adequate compensation for the costs of litigation against defaulting tenants. However, the wording of the indemnity clause must be clear and a claim to costs should be fully pleaded in the particulars of claim.

INTERIM BILLS

[2.4]

Solicitors have always been free to agree the terms of their retainer with their clients in respect of both non-contentious and contentious business. It is only in recent years that solicitors have realised what has long been appreciated in every other walk of life: that without stage payments by the client the entire burden of financing the work falls upon he who is doing it. 'Cash is King' as every managing partner, chief executive and their bank manager can tell you without hesitation. Without cash flow, any business is in difficulties.

Litigation in particular can be protracted, complicated and lingering, as can some non-contentious work. An agreement with a client that the firm will render interim bills at monthly, three-monthly or six-monthly intervals will transform a firm's cash flow. Any basic system of time recording and costing will enable simple interim bills to be produced based on time spent and hourly rates. If done correctly (see below) any anomalies or inequities can be rectified in the final bill. The clients will be grateful. Solicitors live in fear of offending their clients with requests for payments on account of costs and disbursements, but it is nothing compared with the fear of the clients of the ever-growing size of an unknown bill which they know they will inevitably receive. (This is true now even of CFA clients who used to have the luxury of being told that they would receive all their compensation and not have to pay any costs to their solicitor or their opponent, whatever the circumstances.) Clients welcome knowing the amount of costs they have incurred to date, even if it may mean them crying 'Halt!' before any more costs are incurred; they will (generally) also welcome the opportunity of making stage payments. Furthermore, clients appreciate that a solicitor who is efficient in the conduct of his own affairs is likely to be no less efficient in looking after theirs.

There are two kinds of interim bill, and the difference between them is crucial. When deciding what sort of interim bills you want to send out, you need to consider in particular whether you think you might need:
- to sue the client on such bills (and not simply on a final bill); and/or
- to seek a different amount from your client at the end of the case for the period the interim bill covers

Keep these questions in mind as we now look at interim bills on account and interim statute bills.

(a) Interim bills on account

[2.5]

A bill on account is really nothing more than a request for payment on account in fancy dress. Not being a statute bill it cannot be sued on by the solicitor, the client cannot apply for a detailed assessment of it and, therefore, the time limits for applying for a detailed assessment do not run.

In *Turner & Co v O Palomo SA* [1999] 4 All ER 353, [2000] 1 WLR 37, CA, five bills rendered during the course of litigation had been headed 'on account of charges and disbursements incurred or to be incurred'. It was held these could not be construed as final or statute bills in respect of the work covered by them, and accordingly the time limits for applying for a detailed assessment under the Solicitors Act had not started to run. If the client does not pay a bill on account the solicitor should give him 'reasonable notice' according to s 65(2), that unless payment is made within a stipulated (reasonable) time, the solicitor will withdraw from the retainer. But you should note that if the client regards the amount requested on account as excessive he can invite the solicitor to render a statute bill which he may then have assessed. If the solicitor then fails to render a statute bill the client can obtain an order from the court that he should do so, pursuant to the Solicitors Act 1974, s 68.

One advantage to the solicitor of rendering a bill on account is that it need not be the final quantification of all the work included in it, but is merely the minimum amount of his charges to date. It also avoids the risk of limiting any between-the-parties costs recoverable in respect of this period to the amount of the bill on account under the indemnity principle. Furthermore, unless the sum charged is based on the unsuccessful rate set out in a CFA, there is a risk of appearing to charge by results without the benefit of a statutorily compliant CFA. For this reason it is important to make it clear to the client that the bill on account is simply a request for payment on account of the final (statute) bill which will be delivered later. Wording on the following lines will achieve this:

> 'There is statutory provision for various discretionary factors to be taken into account when calculating solicitors' fees, some of which cannot be assessed until all the work is completed; these will be taken into account in our final bill when we shall be able to make an overall evaluation of the matter.'

A bill on account constitutes a written intimation to the client of the amount of costs incurred to date and therefore entitles a solicitor to transfer from his client account into his office account money received from the client in accordance with the Solicitors' Accounts Rules 1991. Such a transfer is restricted to the amount of costs already incurred and must not cover anticipated future costs. A bill on account must therefore be restricted to costs incurred. Any request for money on account of future costs must, in respect of contentious business, comply with the requirements of the Solicitors Act 1974, s 65(2) and in respect of non-contentious costs must be pursuant to an agreement with the client.

Therefore if you are looking to improve cash flow and leave flexibility as to the final bill, a bill on account is the interim bill for you. It does mean that you cannot pursue your client on such bills but a separate final bill can be delivered and pursued where necessary. You need to be careful that the payment on account does not turn into a statute bill by mistake and which inadvertence may cause you many problems.

(b) Interim statute bills

[2.6]

These are called statute bills because they comply with all the requirements of the Solicitors Act 1974 and result in all the consequences which flow from such compliance – the solicitor can enforce payment by suing the client, the client can obtain an order for a Solicitors Act assessment and the various time limits relating to the client's rights to an assessment run from the date of their delivery. Although they are interim bills they are also final bills in respect of the work covered by them. There can be no subsequent adjustment in the light of the outcome of the business. They are complete self-contained bills of costs to date.

Interim statute bills during the currency of the retainer can arise in only two ways: by natural break or agreement. These days it is rare for a client care letter not to cater for the making of interim payments.

By Agreement

[2.7]

'Before a solicitor is entitled to require a bill to be treated as a complete self-contained bill of costs to date, he must make it plain to the client expressly or by implication that that is the purpose of his sending in that bill for that amount at that time. Then of course one looks to see what the client's reaction is. If the client's reaction is to pay the bill in its entirety without demur, it is not difficult to infer an agreement that the bill is to be treated as a self-contained bill of costs to date.'

That was how Roskill LJ put it in *Davidsons v Jones-Fenleigh* (1980) 124 Sol Jo 204, (1980) Times, 11 March, CA. In that case the court found that each of four bills delivered was complete and final in its own right and that the time for challenging three of them had expired. In *Abedi v Penningtons (a firm)* [2000] NLJR 465, CA, the solicitors had not agreed with their client that they could deliver interim bills and there were no natural breaks in the litigation, but the solicitors did in fact deliver bills on a monthly basis, each purporting to be a final bill for the period in question. The client at first paid regularly, but then stopped, leaving five bills unpaid and four bills partially paid. The client then alleged that she had been overcharged and sought a detailed assessment under the Solicitors Act. She was so long out of time in making her application that if each of the bills was treated as a final bill she was not entitled to have any of them assessed under the Act. The Court of Appeal upheld the award of summary judgment to the solicitors on the grounds that the possibility of interim bills being statute bills could arise by virtue of an inferred as well as an express agreement. The client, far from disputing the bills, had paid them regularly and had promised to pay the outstanding bills. The case was distinguished from *Re Romer and Haslam* [1893] 2 QB 286, 62 LJQB 610, CA, because in that case the solicitors had never asked for payment of any of their bills, but merely sought and obtained payment on account.

In *Adams v Mackinnes* SCCO Case No 13 of 2001 the retainer commenced in 1990 with a client care letter, no copy of which survived. In August 1994 the solicitors sent a further client care letter, which was available, stating they would apply an uplift for care and conduct but probably not until the end of the action, when they could decide what mark-up was merited. However, they did not change the format of their bills until May 1996, when they bore a prominent message that the bill was an interim bill. Prior to then the bills were drawn in a way which made them look like final bills. They set out exactly what work had been done, the rate being charged and the period covered by the bill, and each carried a notice informing the client of his rights to detailed assessment under the Solicitors Act. In these circumstances the solicitors were not entitled to any uplift for care and conduct prior to their client care letter of August 1994. It was not mentioned in this short report, but on the face of it, it would be arguable that the format of the bills until May 1996 resulted in them being final bills on which no mark-up could subsequently be claimed.

Natural break

[2.8]

If there is no agreement in the client care letter or other document regarding the delivery of interim bills, the solicitor has to rely on the concept of a natural break in protracted litigation. There is authority for the rendering of an interim statute bill at such points but unfortunately, there is little authority to help to identify what is a natural break. In *Chamberlain v Boodle and King (a firm)* [1982] 3 All ER 188, [1982] 1 WLR 1443, CA, Lord Denning said: 'It is a question of fact whether there are natural breaks in the work done by a solicitor so that each portion of it can and should be treated as a separate and distinct part in itself, capable of and rightly being charged separately and taxed separately'. In that case the Court of Appeal held that there had been no natural breaks justifying treating a series of accounts rendered during litigation as final accounts and that they should accordingly be treated as one bill all of which could be assessed.

In *Wilson v William Sturges & Co* [2006] EWHC 792 (QB), [2006] 16 EG 146 (CS), [2006] 4 Costs LR 614 a bill delivered at the end of the first stage of proceedings was held to be a statute bill. Even though the court held it to be 20% in excess of the proper amount, the solicitors' insistence on it being paid before proceeding further did not terminate the retainer and disentitle the solicitors to their reasonable costs. The Law Society's advice is not to rely on the 'natural break' principle as a ground for delivering an interim statute bill except in the clearest circumstances.

So if you take the view that your clients may need suing and/or are likely to challenge your bill, you should consider rendering formal interim statute bills rather than a bill on account. The all-important time limits for challenging the bill (see [**3.19**]) start running from when the statute bill is delivered. The limitation period may well have run out on at least some bills if the litigation carries on for a long period of time. Bills that were paid quite happily by the client originally have a nasty habit of appearing far less reasonable to them at a later stage, particularly if the litigation has not gone so well in the meantime.

FINAL BILLS

Form and content

[2.9]

The formalities governing the delivery of a bill are surprisingly few. The Solicitors Act 1974 is silent as to the form of a bill for non-contentious work

Final bills **2.9**

and in respect of contentious business s 64 merely prescribes that the bill of costs may, at the option of the solicitor, be either a bill containing detailed items or a gross sum bill.

A bill of costs must contain sufficient particulars to enable the client to judge the fairness of the charges. Furthermore, it needs to be sufficiently particularised for (another) solicitor to advise on it and a costs officer to judge the propriety of the various items of which it is composed. Unsurprisingly, in *Re Kingsley* (1978) 122 Sol Jo 457 a bill merely stating 'for professional services' was held to be void as inadequate.

This problem has been superseded in the main by two practical answers arising from s 64 Solicitors Act 1974. First, in respect of contentious business, the client can require the solicitor to replace the uninformative gross sum bill with a detailed one (for more on this see Chapter 3 at [**3.17**]). The client has to be within various time limits to do so. Sometimes clients ask for some clarification which may (or may not) amount to a request for a fully detailed bill in substitution for the gross sum bill. It is prudent when receiving such a request to write to the client and ascertaining precisely what the client wants. At the same time, you can point out that s 64(2) (which gives the client the right to request the detailed bill), entitles you to submit a replacement detailed bill that exceeds the amount of the gross sum bill, if the work done justifies it.

Second, the solicitor is required to 'furnish the costs officer with such details of any of the costs covered by the bill as the costs officer may require.' This is translated by the court rules which require, at rule 46.10(2) the solicitor to provide a breakdown of the costs claimed so that the client and the court can see what is being claimed in more detail.

In non-contentious business, a client used to be able to request a Remuneration Certificate to check that the fee charged was fair and reasonable but that route was removed when the relevant Order was repealed.

The law in this area was reviewed by the Court of Appeal in *Ralph Hume Garry (a firm) v Gwillim* [2002] EWCA Civ 1500, [2003] 1 All ER 1038, [2003] 1 WLR 510. A balance must be struck between protection of the client's right to seek a Solicitors Act assessment and the solicitor's right to payment not being defeated by 'opportunist resort to technicality'. To establish that a bill was not in 'bona fide compliance' with the Act, a client must establish (i) it contained no sufficient narrative to identify what he was being charged for; and (ii) he did not have sufficient knowledge from documents in his possession, or from what he had been told, to take advice about challenging the bill. The more the client knows, the less the need for the bill to spell it out.

In these days of computerised time recording and standard documents generated via case management systems, it is tempting simply to create a bill which merely sets out a mathematical calculation of the time spent because it can be done virtually automatically. But, in the absence of at least a sufficient summary of the work done, such a bill might well be treated by the court as not being a properly delivered bill. Nevertheless, a computer print out can help to provide clients with additional information to a suitable summary as Ward LJ's cri de coeur at the end of Ralph Hume Garry makes clear:

> 'I add this postscript for the profession's consideration so that an unseemly dispute of this kind does not happen again. Surely in 2002 every second of time spent, certainly on contentious business, is recorded on the Account Department's com-

puter with a description of the fee-earner, the rate of charging and some description of the work done. A copy of the print-out, adjusted as may be necessary to remove items recorded for administrative purposes but not chargeable to the client, could so easily be rendered and all the problems that have arisen here would be avoided. In these days where there seems to be a need for transparency in all things, is a print-out not the least a client is entitled to expect?'

A full narrative could include details embarrassing to the client who wishes to use the bill as a VAT invoice and there are various methods of preserving this confidentiality, the most usual being to set out the narrative either on a separate sheet or on a tear-off portion at the end of the bill.

Taking account of interim bills

[2.10]

Where interim bills on account have been rendered pursuant to s 65(2) of the Solicitors Act 1974 these should be ignored for the purposes of the final bill and treated as requests for payment of money on account. The final bill should therefore be for the total amount of the solicitor's charges and disbursements with any additional mark-up (or mark-down) on the interim bills and give credit for all payments received as a result of bills on account.

If pursuant to an agreement with your client (or where there have been natural breaks) you have been rendering interim statute bills, each bill is a self-contained account for the work done during the period which it covers. Therefore the final bill will only be for the fees incurred since the period covered by the last interim statute bill.

Unpaid disbursements

[2.11]

The bill may include unpaid disbursements, as well as those that have been paid, but only if they are described in the bill as not yet paid. The defect is fatal to these items if they are challenged on a subsequent assessment (Solicitors Act 1974, s 67). The only remedy would be to ask the costs judge for an adjournment, apply to the court for leave to withdraw the entire bill, re-deliver it and then start again. Even if you were not ordered to pay all the costs thrown away by the client, the economics of starting again will often be questionable.

In *Tearle & Co v Sherring* (29 October 1993, unreported, QBD), Wright J held that where a solicitor has acted in good faith but by inadvertence has omitted to describe disbursements as unpaid, the court not only had power to give him leave to withdraw his bill and deliver another one, to save costs it could in an appropriate case give leave to amend his bill by adding the words 'unpaid'. He tentatively expressed the view that a costs officer might have the same powers.

Section 67 says that, if the bill is assessed, the unpaid disbursement 'shall not be allowed' unless it is paid before the assessment is completed. If the assessment is lengthy, one practical answer would of course be simply for the solicitor to pay the disbursement before the assessment is completed.

Cash account

[2.12]

The first question most solicitors ask in reference to a cash account is 'what is it?' It is defined at sub-paragraph 6.6 in the Practice Direction to Part 46 as an account 'showing money received by the solicitor to the credit of the client and sums paid out of that money on behalf of the client but not payments out which were made in satisfaction of the bill or of any items which are claimed in the bill.'

The breakdown required by CPR 46.10(2) (see [2.9]) includes a cash account so that the court can see the whole picture between the solicitor and client. In effect the cash account deals with money expended by the solicitor on behalf of the client for anything which is not in the bill (or which ought to be there).

Items which the solicitor is not bound by law or custom to make, such as purchase money, interest, sums paid into court, or damages, costs paid to an opponent, estate duty and Land Registry fees, are properly charged in the cash account.

Items which should have been included in the bill as disbursements are, for example court fees, counsel's fees, expenses of witnesses, agents and stationers'.

It is important to know what disbursements should appear in the bill itself and what should be charged in the cash account because, perhaps obviously, only those items in the bill can be recovered from the client. Many solicitors lose money on assessments of costs by regarding as cash account entries which should have been disbursements.

Matters are not helped by the fact that definitions of disbursements are only to be found in Victorian cases such as *Re Remnant* (1849) 11 Beav 603. They are said to be 'such payments as the solicitor in the due discharge of his duty is bound to make whether his client furnishes him with money for the purpose, with money on account, or not'. Rather more recent cases such as *Flatman v Germany* [2013] EWCA Civ 278, [2013] 4 All ER 349, [2013] 1 WLR 2676 have skirted round whether there is any worthwhile distinction to be drawn between disbursements and expenses when many of the examples given above are financed by the solicitor rather than paid from money received from the client.

VALUE ADDED TAX

[2.13]

A solicitor's services to his client are subject to VAT at the standard rate unless the solicitor is not registered for VAT or the services are zero-rated.

VAT Chargeability Chart

Service Rendered by	Capacity of Client	Residence of Client	VAT Chargeability

2.14 Chapter 2 Billing the client

Solicitors	(See Note 1)	(See Note 2)	
1. General legal services other than those below	Private or Business	UK	Full
		EEC (Non UK)	Full
		World (Non EEC)	Zero
		UK	Full
		EEC (Non UK)	Zero
		World (Non EEC)	Zero
2. Services relating to land outside UK	Private Or Business	UK	Zero
		World (Non UK)	Zero
3. Services relating to land inside UK i.e.: conveyancing, litigation and advice associated with land	Private Or Business	UK	Full
		World (Non UK)	Full

Note 1. Capacity of client

It should be clearly established in every case whether a client gives instructions in a private or in a business capacity.

Note 2. Residence of client

Private individuals: country where they are usually resident.

Business client with premises: the country where it has any business establishment including a branch or agency. If this would give it more than one country or residence then it is to be treated as resident in the country where its establishment is located which is most directly concerned with the services rendered.

Business client without premises: for corporate clients, the country or incorporation or legal constitution; for individual clients the country of the individual's usual residence.

What rate to charge?

[2.14]

The standard rate of VAT has changed regularly during the last few years. For ease of reference the rates were/are
  17.5% prior to 1 December 2008
  15% from 1 December 2008 to 31 December 2009
  17.5% from 1 January 2010 to 3 January 2011

20% from 4 January 2011

A VAT invoice is normally raised when the service is completed. However, in circumstances where work is yet to be completed, the rate of VAT to be charged will depend on the date at which the supply of service, the tax point, occurs.

Where the client can claim the VAT as input tax, the difference in rates will matter only modestly to the client. The rate changes do not make much difference to the solicitor either. But if the client is not VAT registered and is liable to pay all the costs at the end, for example, as in a CFA, the difference can be significant. When parties seek to recover costs against their opponents in such circumstances, the bill should be split to reflect the work done in different VAT periods. To do otherwise means the paying party is being asked to pay more than would have been the case if the receiving party had been billed as the case progressed. The Practice Direction to Part 44 assumes that the client will have elected to receive bills at the lower rate and puts the receiving party to the burden of justifying his course of action where this has not been done.

The Tax Point

[2.15]

The basic rule is that the tax point for services is the date on which their performance is completed. VAT becomes payable on that date irrespective of when the bill is delivered.

For solicitors there is a minor exception. If a tax invoice is issued within three months of the basic tax point, the date of the invoice becomes the actual tax point. VAT is therefore payable by a solicitor when the work is completed or when his bill is delivered provided that this is within three months of the completion of the work (HM Revenue and Customs Notice 700.)

Disbursements

[2.16]

In respect of which litigation disbursements should a client be charged VAT? The simple answer is 'none'. Unfortunately this gives rise to the much more complicated question 'what is a disbursement?' The detailed answer is to be found in the VAT Guide (HM Revenue and Customs Notice 700) on the HMRC website, which sets out eight criteria. The test is whether the goods or services are supplied to the solicitor to enable him to render his service to the client, in which event they are not a disbursement and VAT is chargeable. For example, postage, telephone calls, travelling and hotel expenses are not disbursements and VAT must be charged on them. A disbursement is a payment which relates to goods or services supplied to the client, even if it is, in the first place, paid for by the solicitor. Examples are oath and court fees. A method of avoiding VAT on a payment which would not otherwise attract VAT, such as an air fare, is for the client to pay for it direct. The solicitor need not then include it in his account.

VAT on medical reports?

[2.17]

In 2008 the Manchester Tax Tribunal accepted the proposition that although the use of medical reports and records is part of the legal services provided and upon which VAT was already charged as part of the tax payable on the lawyer's fees, the 'obtaining' of the reports was a separate service carried out, as an agent, on behalf of the injured party, and any expense incurred was merely a disbursement and so not subject to VAT.

Counsel's fees

[2.18]

The fees of counsel are part of the service rendered by the solicitor and should therefore be included in the solicitor's bill and attract VAT even if the barrister is not registered for VAT. However, there is a long-standing concession under which HM Revenue & Customs permit solicitors to re-address counsel's fee notes to the lay client, who then makes payment direct (or is requested by the solicitor to return the fee note to him with a separate cheque payable to counsel) so that counsel's fees do not pass through the solicitor's books at all. Where this is done the fees of the unregistered barrister do not attract VAT, and the solicitor avoids incurring VAT liability on counsel's fees because he need not include them in his VAT quarterly return.

VAT on fixed costs recovered

[2.19]

Fixed costs recovered under CPR Rule 45 Part I contain no element of VAT. The solicitor may either charge the client an amount equal to the fixed costs or charge on the normal solicitor and client basis, in both cases giving credit for the sum recovered. By either method he must charge the client VAT in addition to the fixed costs. VAT is recoverable from the opponent in the normal way under the remainder of CPR Rule 45, for example on fast track trial fixed costs provided the client is not registered for VAT.

Insurance claims

(a) Policy holder can recover VAT

[2.20]

If the insurance relates to the client's business and it is VAT registered the client will be able to take a full input tax credit for the VAT on the solicitor's bill. In these circumstances the insurers will only be responsible to their policy holder for the amount of the solicitor's bill exclusive of VAT.

Where the policy holder does not pay the VAT element, the insurer will generally meet this part of the solicitor's fee as well because the only alternative, the solicitor writing off the VAT output tax, disadvantages the solicitor and so is not really an alternative.

(b) Policy holder cannot recover VAT

[2.21]

Where the client is partly exempt, or the insurance does not relate to his business, or he is not VAT registered, the solicitor's bill must still be addressed to the client and include VAT. In the circumstances there is no objection to the client sending the entire bill to the insurers for payment, as previously. If, for internal accounting purposes, the insurers wish solicitors to issue invoices directly to them, this may be done but the invoice so delivered will not be a VAT invoice.

CHAPTER 3

RECOVERY OF COSTS FROM CLIENTS

INTRODUCTION

[3.1]

Litigation is meant to be the last resort and never a truer statement was made in the context of solicitors and their clients. Things have normally gone badly wrong if you are contemplating the issuing of proceedings or taking other forceful steps to recover your fees from your client. The irrecoverable cost of pursuing the client – in terms of lost fee earning and management time – should make you stop and consider whether a deal, any deal, with this client might be better in the long run. Learning lessons on what to do (or not do) next time will be more palatable if a large amount of the costs of pursuing the client are not thrown away in addition. In particular, is your client going to be able to pay the sum assessed by the court or is his non-payment of bills to date a clue that he cannot pay you even if he wanted to do so? Many a complaint about negligent work has been made to cover up a lack of cash.

In this Chapter we look at what needs to be done before commencing proceedings as well as during them; what rights the client has to request further and better information as well as an assessment itself; and what steps you might take outside assessment proceedings, including enforcement options to recoup your fees.

BEFORE ISSUING

Credit Checks

[3.2]

Unless you know your client's finances very well, now is the time to make some enquiries. Many accounts departments have direct access to online checks as part of the Know Your Client procedures.

Final bill

[3.3]

It is essential that a final bill in the form and content described in Chapter 2 has been delivered. Where there has been a series of interim bills on account no proceedings may be commenced until the delivery of a final statute bill. This bill should be for the entirety of the work done and disbursements incurred throughout, either at the agreed charging rate or the direct cost with an appropriate mark-up in light of all the factors including the outcome of the matter, giving credit for payments received.

Signature

[3.4]

Although not essential to the validity of the bill, a solicitor may not sue for his costs unless he has complied with the requirements of the Solicitors Act 1974, s 69 as to signature.

Until 7 March 2008, s 69 of the Solicitors Act 1974 provided that solicitor's bill of costs to the client must be signed by the solicitor or one of his partners, either in his own name or in the name of the firm or be accompanied by a letter which is so signed and refers to the bill. For bills delivered after 7 March 2008, The Legal Services Act 2007 amended s 69 to require a bill to be signed by the solicitor or on his behalf by an employee of the solicitor authorised by him to sign.

The practical difference between a bill which is valid but unenforceable by court proceedings because it is not signed and an invalid bill, is that the former entitles a solicitor to appropriate money paid on account in settlement of the bill or to exercise a lien until the bill is paid, while the latter is of no effect for any purpose. However, if an action on a defective bill gets before the court, the judge has the discretion in appropriate circumstances to give leave for the bill to be withdrawn and replaced without commencing fresh proceedings (*Zuliani v Veira* [1994] 1 WLR 1149, PC).

Delivery of bill

[3.5]

The Solicitors Act 1974, s 69(2)(b) also provides that the bill must be delivered to the party to be charged either personally or by being sent to him by post to, or left for him at, his place of business, dwelling house or last known place of abode.

Elapse of one month

[3.6]

One month must have elapsed since the delivery of the bill unless the party chargeable is about to: quit England and Wales; to become bankrupt or to compound with his creditors or do any other act which tends to prevent or delay the solicitor obtaining payment. In these circumstances the solicitor can seek leave to issue proceedings within a month (Solicitors Act 1974, s 69(1)).

No application or order for a detailed assessment

[3.7]

The client must not have either made an application to the court for a detailed assessment of the costs within one month of delivery of the bill or obtained an order for the bill to be assessed, in either of which events no action may be commenced on the bill or proceeded with until the assessment is completed: Solicitors Act 1974, s 70(1), (2).

Client information

[3.8]

Under the SRA Code 2011 (see **[1.26]**) it is mandatory that clients are informed of their right to challenge or complain about a solicitor's bill. The requisite information nowadays is invariably printed on the paper used for creating the firm's bills so that it is there without having to remember it. For those who use multi-functional devices in conjunction with case management templates, the same information is stored in the templates and printed out on to plain paper.

Limitation Act 1980

[3.9]

As a matter of basic contract law, proceedings must be commenced within six years from the day after the cause of action arose (*Marren v Dawson Bentley & Co Ltd* [1961] 2 QB 135, [1961] 2 All ER 270). This is when the solicitor becomes entitled to his fees under the retainer whether or not they have been quantified and irrespective of the requirement that one month must have expired since the bill was delivered before proceedings may be commenced unless the debt has been acknowledged since the cause of action arose.

The recoverable costs trap

[3.10]

Unless the solicitor and client have entered into a written agreement pursuant to CPR Rule 46.9(2) expressly permitting payment to a solicitor of an amount of costs greater than that which the client could have recovered from another party in the proceedings, the Solicitors Act 1974, s 74(3), limits the solicitor's entitlement to that amount (in county court proceedings).

In other words, there is no point pursuing a client for bills of costs which exceed the sums recovered from the other side if you do not have an express written agreement with your client allowing you to do so.

COURT PROCEDURE

Where to start

[3.11]

A solicitor can sue for his fees in the High Court using a Part 7 Claim Form. But if the client disputes only the amount of the bill, usually on an application for summary judgment, the order will direct that the bill be referred to a costs judge and that the solicitor be entitled to sign judgment for the costs as assessed, together with the costs of the action. The client can also obtain such an order by making an application in an existing action under CPR Part 23.

3.11 *Chapter 3 Recovery of costs from clients*

More usually a solicitor brings proceedings to have his costs assessed under the Solicitors Act 1974 by using CPR Part 8 (as modified by CPR 67.3) if it is an originating application (as is usually the case) or CPR Part 23 in existing proceedings.

The application is normally made in the Senior Courts Costs Office. But, where a solicitor's bill of costs relates wholly or partly to contentious business done in a county court and the amount of the bill does not exceed £5,000, then the powers of the court relating to assessment of the solicitor's bill under the Solicitors Act 1974, ss 70 and 71 may be exercised and performed by the county court (s 69(3)).

The Claim Form is accompanied by the bill (or bills) to be assessed. There are model precedents in the Schedule of Costs Precedents to the CPR to assist. Precedent J is the model Claim Form. If the costs relate to a CFA, a copy of that agreement also needs to accompany the Claim Form.

Default judgments and interim payments

[3.12]

There is no mechanism for a default judgment or costs certificate in solicitor and client assessment because CPR 47.11 is disapplied. But there is no similar disapplication of CPR 44.2(8) and so the court ought to consider making an interim payment order unless there is good reason not to do so. If such an order is not made, the solicitor can seek an interim payment on account of costs using the procedure set out in Part 25.

Directions

[3.13]

Upon receipt of the court file the judge will give directions for a breakdown of the bill to be served by the solicitor. The breakdown includes both 'details of the work done' and a cash account (see **[2.12]** above). Thereafter the directions will deal with service of Points of Dispute and any Replies. Standard directions also include a stay on any proceedings for the bills in question and a resolution of the solicitors lien on papers once the bills have been assessed. The judge may also give directions regarding inspection of the solicitors' files by the client or his representative.

Once the directions have been complied with the parties may then request a hearing date if no settlement has been reached. There is a specific Notice to be used (N258C). The court will then fix a date for the hearing. In practice at the SCCO a date will already have been provisionally fixed as part of the directions being given.

Basis of assessment

[3.14]

This is the indemnity basis used for some between the parties' assessments. It is sometimes described as being a 'modified indemnity basis' because CPR 46.9(3) says that:

' . . . costs are to be assessed on the indemnity basis but are to be presumed-

– to have been reasonably incurred if they were incurred with the express or implied approval of the client:
– to be reasonable in amount if their amount was expressly or impliedly approved by the client;
– to have been unreasonably incurred if-
– they are of an unusual nature or amount: and
– the solicitor did not tell the client that as a result the costs might not be recovered from the other party.'

You might question what is left of the indemnity basis after these significant presumptions have been imposed. The answer in a nutshell is that there is no proportionality test to be imposed, as there would be if the costs were assessed on the standard basis. The other distinction between standard and indemnity basis costs is the reversal of the benefit of the doubt. But 'where there's no doubt, there's no difference' and the presumptions in CPR 46.9(3) go a long way to dispelling any doubt the court might have on most items in the bill.

Costs of the assessment

[3.15]

The award of costs of the proceedings is not left to the discretion of the judge in the way of between the parties' assessments. Section 70(9) expects the order made to be based on the outcome. If one-fifth of the amount of the bill is disallowed, the solicitor pays the costs, if not, the client pays. There is some discretion left to the costs officer by s 70(10) but he has to certify that there are special circumstances relating to the bill, or the assessment of it, to vary the one-fifth rule. Where some of the costs claimed are outside the scope of the solicitor's retainer, such costs are to be ignored when calculating the one fifth. If a significant amount of such costs were originally claimed from the client this may amount to a special circumstance entitling the court to award costs to the client even though the 'in scope' costs were reduced by less than one-fifth (see *Bentine and Bentine v The Official Solicitor and Wilsons Solicitors LLP* [2013] EWHC 3098 (Ch)).

A client may seek an order for detailed assessment (see [3.19]). Where he does so, and with this provision in mind, the client may think it prudent to limit the order to the profit costs only if there are substantial disbursements which he does not wish to challenge or, conversely, limited to counsel's or expert's fees if these are all that he wishes to challenge (s 70(6)).

Where the bill being assessed is a gross sum bill, it is important to appreciate that any breakdown ordered by the court is for the purposes of the assessment only and it is still the original gross sum bill delivered to the client that at the end of the assessment will be either upheld or reduced. Accordingly, although the detailed breakdown may quite properly justify a figure considerably higher than the amount of the gross sum bill, it is of no relevance to the question of whether or not the gross sum bill has been reduced by one-fifth.

One slightly peculiar aspect to s 70(9) is that if the client does not attend the detailed assessment (and the proceedings had been started by the solicitor rather than the client) the one-fifth rule does not apply. It may be that no costs are payable but it is more likely that the court will consider the imposition of a costs order based on special circumstances where the client has not settled the costs prior to a hearing and then not turned up for it.

APPLICATIONS BY THE CLIENT

Application for delivery of a bill

[3.16]

Section 68 of the Solicitors Act 1974 empowers the court on an application by the client to order the solicitor to deliver a bill. The jurisdiction in the High Court extends to all cases, including those in which no business has been done by the solicitor in the High Court, while the county court has similar jurisdiction where the bill of costs relates wholly or partly to contentious business done by the solicitor in that county court. There are two common circumstances which may give rise to such an application. The first is where the solicitor attempts to retain money paid on account without having delivered an adequate bill. The second is where the solicitor makes what the client regards as an excessive demand for an interim payment on account.

Application for further information

[3.17]

Section 64(2) provides that the client may, before he is served with proceedings and within three months of the date on which the bill was delivered to him (whichever is the earlier), require the solicitor to deliver a bill containing detailed items in lieu of the original bill. This has the dramatic effect that the gross sum bill is of no effect. Accordingly, the solicitor is no longer bound by the amount of the gross sum bill and he is free to deliver a detailed bill for a higher amount if his new calculations justify this (*Polak v Marchioness of Winchester* [1956] 2 All ER 660, [1956] 1 WLR 818, CA).

The court's invariable practice of requiring a breakdown to be provided – now enshrined in the Practice Direction to Part 46 – means that applications under s 64(2) are perhaps less important than when *Polak* was decided. Nevertheless, it is an important right for a client to use for the understandable wish to resolve matters outside any formal court proceedings.

Application to set aside a Contentious Business Agreement

[3.18]

Contentious Business Agreements are considered in Chapter 8. Their aim is to make all of the terms of the agreement clear and definite. The attraction for the solicitor is that the agreement reached as to fees within the CBA is then enforceable without the rigmarole of a detailed assessment. The CBA can only be set aside if the client can show that it is unfair or unreasonable as a whole. It is intended to be a difficult test to achieve. But, the result is that in practice they are set aside for a lack of certainty in their terms more often than because they are inherently unfair. In order to create room for argument an application to set aside a CBA needs to be made when the solicitor applies for leave to enforce the agreement. It can also be made as a free standing application by the client.

Application for detailed assessment

[3.19]

The Solicitors Act 1974 tries to reward those clients who actively deal with their solicitors' costs and puts to the test those who sit on their hands and wait to be sued. The solicitor has to wait for one month from the delivery of his bill before commencing proceedings (see [3.6] above). If the client applies to the court within this time for the costs to be assessed, the court will automatically grant such application. Furthermore there will be no requirement for the client to pay any money into court and no other proceedings can be commenced by the solicitor until the bill has been assessed.

If the client makes the application between one and twelve months, the court may order the assessment on such terms as it thinks fit and order a stay on any proceedings that the solicitor has commenced in the meantime.

If the client waits until after 12 months have expired, the court will only make an order if 'special circumstances' can be shown (see below).

These time limits apply if there has been no payment of the bill and suggest that there is a dispute between the solicitor and client. Where the client has paid the bill and then seeks to have the court assess it, the requirements are more strict. If the request is made within 12 months of payment of the bill, the client will need to show 'special circumstances' for the court to order an assessment. If the request is made after 12 months then the court has no power to make such an order. These provisions of the Solicitors Act s 70 may be presented in the following table.

Bill not paid	Bill paid	The Court . . .
Application made with 1 month of receiving the bill.	-	. . . will automatically make an order for assessment.
Application made more than 1 month and less than 12 months of receiving the bill.	-	. . . may make an order for assessment on such terms as it sees fit.
Application made after 12 months of receiving the bill.	Application made within 12 months of paying the bill.	. . . will only make an order for assessment if 'special circumstances' are demonstrated.
-	Application made more than 12 months after paying the bill.	. . . has no power to make an order for assessment.

The client may well be a litigant in person and the court will be more concerned with substance than form where the client has made the application. In *Szekeres v Alan Smeath & Co* [2005] EWHC 1733 (Ch), [2005] 32 LS Gaz R 31, [2005] 4 Costs LR 707, defects in the client's claim form seeking the detailed assessment of eight bills of costs did not prevent the proceedings having been validly commenced within one month. Accordingly the costs judge had been in error by refusing to make an order for detailed assessment of the bills because of the formal deficiencies.

What are special circumstances?

[3.20]

Where 12 months have expired from the delivery of the bill or after a judgment has been obtained or where the bill has been paid within 12 months of the application, no order shall be made except in 'special circumstances'.

The cases are inevitably fact specific and as such previous case law simply provides examples of what was considered to be a special circumstance in that particular case. A recent example is the case of *Falmouth House Freehold Co Ltd v Morgan Walker LLP* [2010] EWHC 3092 (Ch), [2010] NLJR 1686, [2011] 2 Costs LR 292, Mr Justice Lewinson reviewed a number of older cases when considering how the Costs Judge had dealt with the issue in 2010. He stated that whether special circumstances exist is essentially a value judgment. It depended on comparing the particular case with the run-of-the-mill case, in order to decide whether a detailed assessment was justified despite the restrictions contained in s 70(3). The Court of Appeal had said previously (in *Re Cheesman* (1891) 2 Ch 289 CA) that it would not interfere with the decision of the first-instance judge on whether special circumstances existed except in a strong case. That was especially so, according to Mr Justice Lewinson, where the value judgment had been made by a specialist costs judge.

In *Falmouth House* the costs judge had taken into account the fact that an invoice called for an explanation. It was also for a large sum which was also something that could be taken into account.

In *Kralj v Birkbeck Montague* (18 February 1988, unreported, CA) the Court of Appeal found special circumstances where the solicitors had dissuaded their client from having their costs assessed and had charged her substantially more than was recovered between the parties, including such items as time spent with law reporters. In *A v BCD (a firm)* (5 June 1997, unreported), QBD, the court observed that the word used was 'special' and not 'exceptional'. The court also discounted the practice of a client paying a bill in full while unilaterally 'reserving the right' to have the bill assessed. Although the solicitor could, by agreement in return for payment, waive the right to resist an order for assessment out of time, the most the client could do was to pay under protest and deploy the protest as a special circumstance if possible. In *Arrowfield Services v BP Collings* SCCO Case No 15 of 2003 the Court held that it was difficult to conceive of a more powerful special circumstance for assessing a bill out of time than where the solicitor has agreed to such an assessment, because by his agreement that the assessment should take place, he had apparently agreed that the protection given to him by the time limits should not apply.

Interestingly, in *Re Metal Distributors (UK) Ltd* [2004] EWHC 2535 (Ch), it was held that the refusal of solicitors to provide a breakdown of their costs, whether or not it this was good client management, did not amount to special circumstances justifying an application being made after the expiration of the 12-month period.

No power after 12 months following payment

[3.21]

Even where there are special circumstances, the court has no power to order an assessment on an application by the party chargeable with the bill after the expiration of 12 months from the payment of the bill. It is important to remember that this provision also affects third parties liable to meet the solicitor's bill to his client, such as mortgagors and lessors. This cannot be circumvented by resort to the inherent jurisdiction of the court: *Harrison v Tew* [1990] 2 AC 523, [1990] 1 All ER 321, HL.

A third party should be aware of the risk of losing their right to an assessment as the result of this provision. If a mortgagee has paid his solicitor's costs in respect of possession proceedings which resulted in a suspended order, and added those costs to the mortgage debt, the mortgagor may not discover this until long after the 12-month period has expired. It would therefore be prudent for the mortgagor after any such litigation to ascertain the amount of the mortgagee's solicitor's costs so that an application for an assessment may be made within the requisite time.

Challenging the level of costs only

[3.22]

However, all is not lost if you are acting for the client. Whilst the right to a detailed assessment has been lost, the common law right to make a claimant (here, the solicitor) proved the quantum of an unspecified claim remains. Accordingly, if a solicitor sues on its invoices, the client can defend quantum on the basis that, like other forms of damages, the claimant has to prove its loss. The obvious judges to ascertain the loss are costs judges and so the case can be put before them for a judicial assessment. This may be more along the lines of a quantum meruit trial than a detailed assessment but it prevents a solicitor claimant from obtaining a default judgment for whatever sum the solicitor saw fit to bill. This line of authority comes from the Court of Appeal decision in *Watts (Thomas) & Co (a firm) v Smith* [1998] 2 Costs LR 59, CA). The Court of Appeal heard a further case the following year (*Turner & Co v O Palomo SA* [1999] 4 All ER 353, [2000] 1 WLR 37, CA) which followed *Watts*. The bill was ordered to be sent for assessment (but not for detailed assessment) by a costs judge.

THIRD PARTIES

[3.23]

Section 71(1) of the Solicitors Act 1974 provides that a person, other than the client, who is liable to pay a bill either to the solicitor or to the client may apply to the High Court for an order for the assessment of the bill as if he were the client, and the court may make the same order (if any) as it might have made if the application had been made by the client.

Although the assessment must be conducted as between the solicitor and his client, a third party does not, by obtaining an order to assess, increase his liability to the solicitor's client. For example, if a mortgagor has the costs of the mortgagee's solicitor assessed, items which the mortgagor would not be liable to pay as between himself and the mortgagee will be disallowed even if the solicitor is entitled to charge them against his client, the mortgagee (*Re Longbotham & Sons* [1904] 2 Ch 152, 73 LJ Ch 681, CA).

Residuary beneficiaries

[3.24]

Such beneficiaries are able to complain and to expect the solicitor to respond to the matter under the solicitors' complaints handling procedure. This is in line with a residuary beneficiary's ability to seek third party assessment of costs under s 71(1) of the Solicitors Act 1974, whether or not the Solicitor is an executor of the estate.

It is obviously good practice for solicitors to provide relevant client care information to residuary beneficiaries for reference at the outset, together with costs estimates and any later revisions. If a residuary beneficiary complains to the firm at any point in the administration, then this should be dealt with in the same manner as one would handle a complaint from a client.

Mortgagee's costs (and similar)

[3.25]

What is the position where one party is contractually obliged to meet the costs of the other party? For example, a bank takes steps to enforce a mortgage against its borrower, using solicitors in the process. It claims to be entitled to recover all of its costs from the borrower, including those for which it is liable to the solicitor. It can do so directly or sometimes, more easily, by deduction from the proceeds of the sale of the mortgaged property. It may have no particular incentive to query the amount of the solicitor's bill of costs. It agrees those costs, and pays them. The borrower wishes to challenge the amount recoverable from it by way of the solicitor's costs.

At first sight, this looks to be the sort of situation for which s 71 was drafted. The borrower is the party liable to pay the costs rather than the party chargeable, namely the bank. If the third party has not yet paid anything in respect of the bill (or only sums on account which are less than the amount properly allowable) then the s 71 assessment may be useful to the third party, because he should not be liable to pay more than the amount so certified.

But the situation is not so straightforward where the bank (in this case) has already paid the solicitors' charges from the sale proceeds. Any reductions on assessment will not actually reduce the solicitors' fees because they have been paid. Indeed they have been paid by the borrower from the sale proceeds which would otherwise go to the borrower. To the extent that there is any repayment to be made (because a particular item is not recoverable as against the borrower) the shortfall would need to be paid by the bank as the client.

In these circumstances, the case of *Tim Martin Interiors Ltd v Akin Gump LLP* [2011] EWCA Civ 1574 suggests that the third party ought to bring proceedings against the client to establish how much was due from him to the client. In a mortgage case such as in *Tim Martin Interiors*, the proceedings would be conventional proceedings for an account of what was due under the mortgage. Such proceedings would enable the court to determine the correct issue as between the correct parties, and if appropriate to order repayment by the mortgagee to the mortgagor. In such proceedings it would be possible for the court to do what cannot be done under a s 71 assessment, namely to disallow part of an amount claimed on the basis that something was due, but not as much as is claimed – for example by substituting a lower hourly rate.

Instead of seeking an assessment under s 71, therefore, in almost all cases a mortgagor or other party seeking to challenge the costs claimed and received by a mortgagee should bring a claim for an account of the sums due under the mortgage.

Lord Justice Lloyd concluded his judgment in *Tim Martin Interiors* by saying:

> 'In the light of this judgment it may be anticipated that third party assessments will become rare, whereas claims for an account, and like proceedings in other types of case, where the real issue is as to the reasonableness of legal costs, best resolved by those experienced in the assessment of costs, may become much more frequent. With that in mind, it seems to me that it might be sensible for a dispute which is only, or mainly, about legal costs to be able to be commenced as an application for an account directly in the SCCO, rather than having to go via the Chancery Division.'

A claim for an account may be the right approach for several situations which can throw up this sort of problem, for example in the case of a trust or the administration of an estate. In other cases that may not be the right approach, and it may be necessary to claim a declaration as to the amount properly due, especially if the amount claimed has had to be paid by the third party, no doubt under protest.

RECOVERY WITHOUT PROCEEDINGS

[3.26]

There are two strands to this part of the Chapter. The first is to look at the use of Statutory Demands and Insolvency Petitions as an aggressive manoeuvre outside the general civil court approach. The second is to consider the use of the solicitors' lien as a method by which a client may be forced to compromise with his solicitor over fees in order to deal with his case.

STATUTORY DEMAND

[3.27]

Service of a statutory demand for payment of a solicitor's costs does not constitute the bringing of an action for the purposes of the Solicitors Act 1974, s 69(1) which prohibits solicitors from bringing any action to recover costs within one month of delivery of their bill without special leave of the court. The consequence, to which a statutory demand leads if not complied with, is a presumption that the debtor is unable to pay the debt in question. This, in turn, enables the creditor to present a bankruptcy or winding up petition. In general the court should exercise its discretion to set aside the statutory demand if, but only if, it would not be just for those consequences to apply in the circumstances. Arguments based on s 69 do not amount to sufficient grounds for setting aside the demand. Accordingly, a solicitor may serve a statutory demand for payment of his costs before the expiration of a month from the date of delivery of his bill of costs (*Re A Debtor (No 88 of 1991)* [1993] Ch 286, [1992] 4 All ER 301).

The Law Society has recommended that solicitors should be wary of following this course of action because of the power to set aside a statutory demand on the grounds of injustice (*Re A Debtor (No 88 of 1991)*. Another reason for caution is that the decision in *Turner & Co v O Palomo SA* [1999] 4 All ER 353, [2000] 1 WLR 37, CA could be interpreted as precluding statutory demands based on a solicitor's bill because the amount of the bill is always open to challenge in proceedings brought to recover it even after the time limit for a detailed assessment has expired and therefore it is not a liquidated demand.

BANKRUPTCY PETITION

[3.28]

In *Truex v Toll* [2009] EWHC 396 (Ch) the defendant client presented a list of perceived complaints about the solicitor's services in respect of her matrimonial proceedings. The solicitor served a statutory demand for his outstanding fees and followed it up with a bankruptcy petition. It was held on appeal that as the costs did not form the subject of a judgment, assessment or agreement, they were not a liquidated sum for the purpose of founding a bankruptcy petition; the bill as a whole was capable of challenge as to quantum and was thus for an unliquidated sum. It was not possible to say that any part of the work done by the solicitor had been quantified, or was quantifiable by the bankruptcy court as a mere matter of arithmetic. The sum claimed only became a liquidated sum once the fees had been assessed by a costs judge, determined in an action, or agreed. Whether a sum is liquidated and whether there is a defence to the claim are separate issues, and the first must be determined before the second is addressed.

If the statutory demand is not complied with within 21 days of service the solicitor may present a bankruptcy petition, but because this is an action within the meaning of s 69, a petition may not be issued within a month without leave (*Re A Debtor (No 88 of 1991)*).

The considerations regarding a winding up petition in respect of a corporate client are, for the purposes of this commentary at least, exactly the same as for an individual client.

SOLICITOR'S LIEN

General Lien

[3.29]

A solicitor has, at common law, a general lien to retain any money, papers or other property belonging to his client which properly come into his possession until payment of his costs, whether or not the property was acquired in connection with the matter for which the costs were incurred. The solicitor may retain, until payment of his costs, property other than money to any value even if it greatly exceeds the amount due, but he cannot hold money in excess of the amount due. A solicitor is not entitled to sell property held under a lien or to transfer it into his ownership without an order from the court.

Particular Lien

[3.30]

A solicitor also has at common law a particular lien on property recovered or preserved by him in litigation which extends to all costs incurred, both billed and unbilled. Unlike the general lien, a particular lien covers property not in the solicitor's possession and gives him an equitable right to have the property transferred into his possession. The costs are incurred when work is done under the retainer entitling the solicitor to exercise a general or a particular lien until the costs are paid. If some of the costs are unbilled the client's remedy is to request a bill for any outstanding costs. If the solicitor does not comply with the request the client may apply to the court under section 68 of the Solicitors Act 1974 for an order that he does so (see [**3.16**]).

Solicitors Act Charging Order

[3.31]

The lien on property recovered is in effect extended by the Solicitors Act 1974, s 73, which provides that any court in which a solicitor has been employed to prosecute or defend any suit, matter or proceeding may at any time 'declare the solicitor entitled to a charge on any property recovered or preserved . . . ; and make such orders for the assessment of those costs and for raising money to pay or for paying them out of the property recovered or preserved as the court thinks fit'.

A Solicitors Act Charging Order differs from a general lien in two particular ways. First, it applies to real as well as personal property and, second, it does not apply where the claim for costs is statute barred (a solicitor's lien on the other hand cannot become statute barred.) The application is made under CPR Part 8.

Loss of lien

[3.32]

A solicitor will lose his lien if he takes other forms of security from his client. The solicitor may do this under s 65(1) but if he does so, this will preclude him from claiming a lien over subsequent money, property or documents in his possession unless that lien has been expressly reserved.

CPR 25.1(1)(m) provides that where the defendant to a claim for the recovery of personal property does not dispute the title of the party making the claim but claims to be entitled to retain the property by virtue of a lien, the court may make an order permitting the claimant to pay money into court pending the outcome of the proceedings and directing that if he does so, the property shall be given up to him.

If at any stage the solicitor takes alternative security from a client with the intention of satisfying his claim for fees by this alternative means his lien will have been waived. This is because the taking of security for costs generally is inconsistent with the lien and the solicitor who does not preserve his lien when taking the security will be taken to have abandoned it.

Who ended the retainer?

[3.33]

If the retainer is terminated by the client other than for misconduct by the solicitor, the solicitor's lien is virtually absolute. He cannot be required to hand over or produce for inspection any papers in his possession and he is entitled to keep them until his costs have been paid. In *French v Carter Lemon Camerons LLP* [2012] EWCA Civ 1180, [2012] NLJR 1155, 156 Sol Jo (no 34) 27 the claimant instructed the defendant solicitors to act for her in litigation against an insurance company but complained about their conduct of the litigation. Ultimately she stated that she was left with no choice but to represent herself in the insurance litigation and requested a copy of her file. The defendant exercised its purported right to a lien in respect of its unpaid fees. The claimant issued proceedings seeking recovery of those files. The application was dismissed in the lower courts and the claimant appealed to the Court of Appeal who upheld the decision. The retainer spelled out the rights of each party to terminate the retainer and the solicitors' entitlement to be paid their fees in the event of termination.

But the court went on to say that if the solicitor terminates the retainer, he may be ordered to hand over the papers to the new solicitor on the new solicitor's undertaking to hold them without prejudice to his lien, to return them intact after the action is over and to allow the former solicitor access to them in the meantime and if necessary to prosecute the proceedings in an active manner. The case of *Gamlen Chemical Co (UK) Ltd v Rochem Ltd* [1980] 1 All ER 1049, [1980] 1 WLR 614, CA is the authority for the extent of the undertakings which a solicitor can expect from a client's new solicitors.

Substituting the lien for an undertaking is not automatic and the court needs to exercise its discretion. In approaching the matter, the overriding principle is that the order made should be that which would best serve, or at least not frustrate, the interests of justice. The principle that a litigant should not be deprived of material relevant to the conduct of his case and so driven from the judgment seat is to be weighed against the principle that litigation should be conducted with due regard to the interests of the court's own officers, who should not be left without payment for what was justly due to them. Where the solicitors have behaved impeccably and of whose conduct there has been no criticism, whilst their clients, without any excuse, have not paid the costs and there is a default judgment against them, the balance of hardship would be far greater on the solicitors if the lien were not enforced, because they would then probably recover nothing, whereas it was open to the clients to preserve their position by paying the solicitor's costs.

Where the retainer is discharged by the solicitor, he has only a qualified lien over the papers. Unless there are exceptional circumstances, the fact that the solicitor had reasonable cause to end the retainer would not justify modifying the overriding principle that a solicitor discharging himself should not be allowed to exert his lien so as to interfere with the course of justice. He would be entitled to the appropriate undertakings from the new solicitors.

State funding

[3.34]

A solicitor's lien arises in respect of costs due for work done on the instructions of the client, for which the client has undertaken personal liability. Pre-certificate costs and disbursements will fall within this category. However, once a certificate has been issued the situation is altered, since the assisted person's solicitor has a statutory right to be paid out of the fund, and may not take any payment other than from the fund (Civil Legal Aid (General) Regulations 1989, reg 64).

When a certificate is amended to enable a new solicitor to have the conduct of a legally aided person's case, it does not appear logical that a common law lien can arise in respect of costs and disbursements payable under the statutory certificate. Indeed, the Law Society has taken the view that a solicitor's costs are secured by an order for assessment or certificate and as such it would be inappropriate even to call for a professional undertaking from the successor solicitor to pay the costs except in respect of any outstanding pre-certificate costs.

A solicitor should not part with the papers on a legally aided matter until the certificate is transferred to the successor solicitor, although they should be made available for inspection in the meantime or copies provided.

In order to make sure the solicitor is not prejudiced in claiming costs due from the fund it is quite proper to ask for an undertaking requiring the successor solicitor to
(i) return the papers promptly on completion to enable a bill of costs to be drawn up; or
(ii) have the first solicitor's costs included in the successor solicitor's bill, collect those costs and pay them to the first solicitor.
Accordingly, where, under a certificate, a change of solicitor is authorised, subject to (a) there being no lien in respect of pre-certificate costs and disbursements, and (b) an undertaking being given by the new solicitor as to the eventual assessment of costs, there is no reason why the papers should not be expeditiously transferred to the new solicitor.

Documents to be handed over

[3.35]

What papers should be handed over by a solicitor on the termination of his retainer and lien? A full answer is complicated and outside the scope of this book. Suffice to say, the client is not strictly speaking entitled to the whole file but as records become more heavily computerised, the ability to store documents without hardship means that it is often easier to give the client/successor solicitor the entire paper files rather than spend unproductive time weeding things out. Furthermore, an incomplete file is likely to result in further correspondence with the client/new solicitor which is time wasted by all concerned.

Client's remedy

[3.36]

Where there is a valid and enforceable lien, the only practical remedy for the client who wishes to obtain without delay documents on which his previous solicitor is claiming a lien is to pay his costs. Sometimes the solicitor is willing to agree to the money being paid into a joint account to await the outcome of a Solicitors Act assessment. If not, the client should obtain either the original solicitor's consent to a Solicitors Act assessment or an order from the court before making the payment, as otherwise the court might not order an assessment. This course not only has the advantage of securing the immediate release of the papers, but it also enables the new solicitors to inspect and consider them at their leisure in order to advise whether the costs should be challenged – and, if so, on what grounds – before proceeding with the assessment.

Part II

Funding

INTRODUCTION

[4.0]

The old businessman's adage drummed into managing partners is that 'turnover is vanity, profit is sanity, but cash flow is king.' The amount of revenue your firm makes is of very little value unless it outstrips your expenditure. Moreover, if you cannot afford to keep the business running whilst the revenue is generated you simply cannot trade.

This part of the book concerns the funding of cases, but not in the manner that any businessman would recognise. The solicitor actually funds the case in the sense of providing the cash flow. The 'funding' agreement with the client is really only a mechanism to establish how much the client will eventually pay for the solicitor's services. The concept of an 'entire contract' (see **[1.4]** above) demonstrates the idea that the solicitor will only be paid once the job has been completed. Primary legislation and the development of case law eventually allowed for interim payments to be made as the case progressed. But essentially, the solicitor funded his client's case and was paid at the conclusion. That payment then funded future clients' cases.

The agreements that we consider in this Part are consistent in that they try to make clients more comfortable with embarking on the expensive business that is involved in using a lawyer to create, enforce or defend rights. These agreements try to spell out the client's potential liability – either to a specific sum or to a percentage or similar calculation – if things go badly. The client's liability in the event of a successful outcome is sometimes, but not always, less explicitly described because the assumption is that the opponent will be liable for the majority if not all of the costs. If the Woolf and Jackson Reforms tell us anything about clients' wishes, it is that they would prefer as much certainty as possible as to costs, win or lose. The fixing of fees in the small and fast tracks and prospective budget setting in the multi-track all point towards a drive for certainty for the client. The risk of under reward for the work done is transferred squarely to the lawyers. Consequently, it is not just managing partners who need to keep a close eye on whether the work is profitable and capable of being turned into cash sufficiently quickly to keep the practice viable – you do too.

The enactment of LASPO 2012 and its subordinate legislation has changed the landscape more in the area of funding than almost anywhere else. This part of the book is largely written on the basis of agreements entered into after 1 April 2013. Readers concerned with agreements before that date – in particular CFAs – will need to look at Chapter 6 as well.

We have taken the view that state funding via the Legal Aid Agency and its predecessors is no longer an area that this book should cover directly. It has become so specialised that those who practice with it, will know the regulations inside out and will not gain benefit from a further chapter here. Most readers will have no dealings with legal aid any longer and so will not be

4.0 Introduction

interested in such a chapter. We have covered eligibility etc sufficiently to make sure that professional obligations on funding advice are satisfied. We also still deal with situations that arise where the opponent is legally aided in Chapter 10.

We start with Conditional Fee Agreements because they are the most prevalent of the funding agreements currently. We then deal with Damages-Based Agreements which may become more popular over time, followed by Contentious and Non-Contentious Business Agreements which, in an age of increasing certainty may have a comeback in popularity. Then we look at insurance, both before and after the event. Traditionally such insurance has been used with a particular funding agreement, most obviously ATE with CFAs. But the need to be more client specific with funding arrangements may well mean that the boundaries are blurred and that insurance is used in different ways with different arrangements: so we deal with the insurance options on their own. At the end of the insurance chapter (Chapter 9), we look at the position where the solicitor decides to take on the funding responsibility rather than insuring it. Finally in this part we look at litigation funding arrangements by commercial third parties.

CHAPTER 4

CREATING CONDITIONAL FEE AGREEMENTS

[4.1]

On 1 April 2013 the landscape changed considerably for those who wanted to use CFAs to fund their cases. In order to keep this book clear and concise, the law and practice in relation to CFAs is split into 3 chapters. This Chapter deals with the setting up of the CFA and concentrates on drafting requirements and risk assessment. The next (Chapter 5) deals with issues during the life of the CFA and in particular when the outcome of the case is known. These two chapters deal with CFAs taken out on or after April 1 2013 and so are based on the current regime. Much of what is said is relevant to all CFAs, whenever they have been taken out. As time goes by, the current practice will take over from the earlier regime(s). But clearly there are many cases going through the system which use a pre-1 April 2013 CFA and issues which relate to those agreements, particularly in respect of recoverability of success fees is dealt with in the third chapter (Chapter 6.)

THE GOLDEN RULE

[4.2]

Traditionally the idea that a solicitor would be interested in the outcome of his client's case (other than professionally) was seen as an undesirable state of affairs. The lawyer should be disinterested in order to give impartial and proper advice. He would therefore need to be paid the same whether the case was won or lost. An agreement which depended upon the outcome of events would run counter to this requirement. Consequently, agreements which did so were considered to be 'maintaining' the action and if they included a sharing of the spoils of the litigation they were said to be 'champertous'. Maintenance and champerty were crimes up to the middle of the last century. Even when abolished as crimes, they remained as torts and still rendered solicitors' agreements unenforceable when CFAs came to be considered at the end of the 1980s.

(None of this applied to non-contentious business such as conveyancing where agreements which were dependent on say, completion taking place, were seen as quite proper. The prohibition on contingency agreements related solely to contentious business. (See chapter 8 regarding the definitions of contentious and non-contentious business in more detail.)

The Courts and Legal Services Act 1990 ('CLSA') created CFAs. The wording of s 58 is not the simplest to understand because it recognises that agreements contingent on the outcome of events are generally unenforceable. So, it carved out an area of contingency fee agreements which would be enforceable, but did not otherwise alter the general position. As Ian Burnett QC (now Burnett J) poetically put it in *Hollins v Russell* [2003] EWCA Civ 718, CFAs are 'islands of legality in a sea of illegality.' They have now been joined in this situation by Damages-Based Agreements as of April 2013.

So the Golden Rule to remember in respect of CFAs is that they need to comply with the Courts and Legal Services Act 1990 and any subordinate legislation. If they do not, they become unenforceable contingency fee agreements. They are then unenforceable against the client and, by operation of the indemnity principle, fees generated under such an agreement cannot be recovered from the opponent.

This effect is the root cause of the so-called 'Costs Wars' in the early part of this century. Non-compliance was comparatively easy to demonstrate based on the original regulations. Much of those have been swept away but new Regulations and a new CFA Order came into being on 1 April 2013 and there may be more attempts by paying parties to render successful parties' agreements unenforceable.

REQUIREMENTS

[4.3]

The requirements to create a valid CFA are now quite limited. A CFA needs to:
- be in writing;
- not be in relation to family or criminal proceedings.

Any success fee:
- must be no more than 100% of the base fees; and
- may need to be limited to a percentage of the client's damages.

Let us look at each one of these in turn

In writing

[4.4]

All of the terms need to be in writing. There have been various decisions at first instance to the effect that other documents, such as the client care letter, can be looked at in addition to the CFA itself to ascertain the terms. But agreements which are partly in writing and partly oral do not comply with the requirements and such agreements are unenforceable.

Not in relation to family or criminal proceedings

[4.5]

The precise wording of the requirement is that the CFA 'must not relate to proceedings which cannot be the subject of an enforceable conditional fee agreement' (s 58(3)(b)). The provisions of s 58A(1) of the CLSA 1990 describes those proceedings as 'criminal proceedings, apart from proceedings under s 82 of the Environmental Protection Act 1990' or 'family proceedings.' There then follow 11 sub-sections defining the relevant statutes which are considered to be family proceedings. The impact in a nutshell is that, for public policy reasons, it is inappropriate for a solicitor advising his client in criminal or family matters to be paid depending upon the outcome.

Any success fee must be no more than 100% of the base fees

[4.6]

A CFA does not need to have any success fee but if there is one, it must not be more than a doubling of the base fees.

So in *Oyston v Royal Bank of Scotland* [2006] EWHC 90053 (Costs) an agreement for the solicitors to receive a £50,000 bonus if the damages recovered were over £1m rendered the CFA invalid where the solicitor was already entitled to a 100% success fee.

This case came hard on the heels of *Jones v Wrexham Borough Council* [2007] EWCA Civ 1356 where the Court of Appeal held that a success fee set at 120% rendered the agreement unenforceable even though on assessment the claim was limited to 100%. The Court of Appeal was clear that claiming a success fee over and above the maximum allowed by the CFA Order 2000 (now replaced by the CFA Order 2013) was contrary to the administration of justice even if not so obviously contrary to the interest of the client (who would never pay it).

Any success fee may be limited to a percentage of the client's damages

[4.7]

When the subordinate legislation bringing the CLSA 1990 into effect was originally introduced in 1995, the success fee in a CFA was not recoverable from the opponent and there was no limit on the extent to which the client's damages could be depleted to pay for that success fee by the client. The Law Society gave professional conduct guidance that the limit should be 25% of the client's damages but there was nothing in the legislation to require this. When recoverability of success fees was introduced in April 2000 the need to seek payment of any fees by the client dissipated. The 'market' in personal injury and debt recovery work (the main areas for using CFAs) was very much the need to offer the client a 100% recovery of damages with the solicitor taking whatever fees could be recovered. Some firms did still seek payments from their clients but that was relatively rare and the sums sought limited.

With the ending of recoverability from the opponent by LASPO 2012, the issue of payment of the success fee has reappeared. This time, the secondary legislation has prescribed limits to the amount that can be claimed from the client in personal injury and clinical negligence cases, albeit not in any other case.

Personal injury and clinical negligence cases

[4.8]

A new provision (s 58(4B)) has been added to CLSA 1990 by LASPO 2012 with which the CFA needs to comply:

'(4B) The additional conditions are that—
(a) the agreement must provide that the success fee is subject to a maximum limit,
(b) the maximum limit must be expressed as a percentage of the descriptions of damages awarded in the proceedings that are specified in the agreement,

(c) that percentage must not exceed the percentage specified by order made by the Lord Chancellor in relation to the proceedings or calculated in a manner so specified, and
(d) those descriptions of damages may only include descriptions of damages specified by order made by the Lord Chancellor in relation to the proceedings.'

As with all Parliamentary drafting, it is done in a style which needs to be read more than once. The 'maximum limit' is a reference to how much of the client's damages can be taken in fees by the solicitor. The limit has to be expressed as a percentage of the damages. Since many people see personal injury damages as sacrosanct, some of the heads of damage claimed cannot be reduced by the agreement, particularly future losses.

The order of the Lord Chancellor referred to in (4B)(c) above is the CFA Order 2013 and the description of damages in that order from which a percentage can be deducted is set out at Article 5(2) of the Order as follows
(a)   general damages for pain, suffering, and loss of amenity; and
(b)   damages for pecuniary loss, other than future pecuniary loss

So, future losses are untouchable which makes the largest personal injury and clinical negligence claims less attractive from this stand point than they would be otherwise. Where the Compensation Recovery Unit is entitled to recover sums for benefits paid or treatment rendered, those sums are also removed from the pot before a percentage can be taken.

What is the maximum percentage that can be deducted? Article 5(1) says:
(a)   in proceedings at first instance, 25%; and
(b)   in all other proceedings, 100%.

The 25% figure is not a surprise given the previous limit expected by the Law Society. This figure is also reflected in the Damages-Based Agreements Regulations considered later in this part of the book. The increase to 100% in relation to appeals is perhaps more surprising. The widely used model Law Society CFA expects an appeal by the opponent to be covered under that agreement. As a result the step change in the potential liability to the client has to be explained and which almost inevitably will reduce the simplicity of the explanation to the client and with it the client's understanding of his liability.

The Law Society model does not cover appeals by the client so that the solicitor can take stock before deciding whether to continue under a CFA and without any risk of breaching the entire contract. If it is the opponent who is appealing the theory is that the client must have had a good result at first instance and is in a strong position to defend the opponent's appeal. It may be better to remove the automatic right for the client to be covered on an appeal by any party and enter into a new agreement in the future. That way, the 25% cap can be explained at the outset. If the case reaches an appeal, a more case-specific cap of up to 100% can be put on the extent of the damages that might be risked at that point.

DRAFTING CONSIDERATIONS

General approach

[4.9]

The sting in the tail to the amendment to s 58 of the CLSA 1990 is not that the client will now start to pay some of his solicitor's fees with his damages. It is the risk that a badly drafted CFA will be found to be unenforceable. S 58(3)(c) requires the CFA to comply with such requirements as are prescribed by the Lord Chancellor. This must therefore include the CFA Order 2013. If the wording does not comply with that set out above, the defendant will be seeking to avoid any liability for costs under the 'Golden Rule' described earlier in this Chapter.

It is not surprising therefore that the statutory requirements, such as those of the CFA Order 2013 are written into the Law Society model agreement more or less verbatim. It may not help the client to understand the terms of the agreement – when compared with the use of plainer English – but it has the undoubted value of making the agreement more robust to challenges. If you decide to draft your own CFA, it is something to which you must give some thought. One option that superficially is attractive is to draft the agreement in the language of the legislation and provide an explanatory note or leaflet to go with it. Be careful. That way leads to inconsistencies ripe for exploitation by paying parties seeking to allege uncertainty and non-compliance. Even copious phrases explaining that the wording of the CFA is to take precedence over such a leaflet if there is any inconsistency can be grist to the challenger's mill. 'What is the purpose of a leaflet which concedes it is (or may be) inconsistent with the document it is explaining? There must be more to it, probably an oral explanation which creates terms (and thereby a partly oral CFA which is inevitably unenforceable.)'

Given that a CFA is fundamentally an agreement between a solicitor and his client, it has always seemed odd to go to counsel to get that agreement right. But if you are looking to use a CFA in standard terms on many cases (or perhaps a few cases but of potentially significant value), there may be some comfort in using another professional's professional indemnity insurance as a potential backstop.

Relationship with DBAs

[4.10]

As we will see in Chapter 7 on Damages-Based Agreements, the aim is to make the agreement simpler. The current regime may not do that, but it may well be in time that CFAs can use some DBA terminology to create a shorter, simpler agreement. Furthermore, where the fees are fixed if the case is won because the recoverable fixed fees are all that will be taken by the solicitor and no fees will be charged if unsuccessful, the agreement ought to be capable of being expressed in many fewer words. The Law Society model seeks to give options and cover various situations. Your CFA may not need to do so, depending upon the case on which you are instructed.

4.11 *Chapter 4 Creating Conditional Fee Agreements*

Speculative agreements

[4.11]

Arguably the simplest agreement is one where the client is told that you will accept whatever is received from the other side, but nothing otherwise. This used to be called 'speccing' because the case was brought entirely speculatively by the solicitor, at least insofar as his own costs were concerned. Such agreements offended the indemnity principle and impecunious clients who retained solicitors to pursue claims were often seen as being parties to a speccing agreement as in the case of *British Waterways Board v Norman* (1993) 26 HLR 232, [1993] NPC 143.

Once CFAs became lawful, the need for speccing agreements reduced. Those who drafted CFAs which altered the definition of a 'win' to the recovery of costs rather than simply the award of costs found themselves met with indemnity principle arguments as to the client's liability. These CFAs were called CFA Lite(s) because they reduced the client's liability for costs to whatever sum was recovered from the opponent. If the case failed, the client was not liable for his solicitor's costs and essentially the client had achieved a costs free environment. The CFA (Miscellaneous Amendments) Regulations 2003) established that such agreements do not offend the indemnity principle.

It is not just impecunious clients who wish to benefit from a CFA Lite. Businesses which pursue debt collection claims and Insolvency Practitioners seeking to make sure funds are not dissipated in legal fees are amongst those who are keen to agree no recovery, no fee arrangements.

Termination provisions

[4.12]

These are fundamental to all contracts. Where payment depends upon the outcome of the case, as in CFAs, particular care needs to be taken on how to deal with the ending of the retainer before that outcome is known. We deal with these situations in more detail in the next Chapter but you should be aware of the issues when drafting the agreement. The Law Society model has always covered problems of the client's making such as a lack of co-operation or wishing to continue against the lawyer's advice. It did not originally deal with any issues affecting the law firm's existence. The change of many firms to LLP status called into question the effectiveness of the original CFA thereafter and various ways were tried by firms to get around perceived problems. The latest Law Society model includes the following provision.

> 'Cessation of Business
>
> If we stop carrying on business then you must pay us or any successor to our business (or that part of our business which takes over conduct of your claim) our basic charges and our expenses and disbursements including barristers' fees and success fees if you go on to win your claim for damages.'

On the face of it, this provision is wide enough to cover most situations that can be contemplated. How it fits together with the specific provisions about success fees may be tested. Interventions, for example, have a nasty habit of happening more or less overnight. There may be a very limited (or no) opportunity for the law firm to elect whether to risk waiting for payment of all fees including success fees at the end of the case or to require payment of the base fees upon termination.

Cancellable agreements

[4.13]

You will also need to consider at the outset the 'Cancellation of Contracts made in a Consumer's Home or Place of Work etc, Regulations 2008' if you intend (a) to use CFAs with 'consumers' ie natural persons and (b) see them outside of your offices at any stage during the signing up process. Whilst the regulations may be aimed at doorstep salesmen, they undoubtedly cover solicitors as 'traders' and Cancellation Notices ought to be used if there is any doubt at all. The cooling off period of 7 days enables the client to terminate the agreement during that period without incurring any costs, unless there has been a specific agreement that work done during this period will still be payable in the event of cancellation.

Where work needs to be done in a hurry, this requirement can cause problems as was demonstrated in the case of *Swift v Robertson* [2012] EWCA Civ 1794, [2013] Bus LR 479, 177 JP 169 where a removal man fell foul of the regulations. He went to the claimant's house at the claimant's request to size up the job which was to take place a week later. He prepared a quotation after the visit which was accepted by the claimant. However, the claimant then shopped around and, having obtained a cheaper quotation, cancelled the agreement the day before the move and refused to pay the cancellation fee in the agreement that he signed. The Court of Appeal reluctantly found in favour of the claimant based on the Regulations. In the general course of business the 7 day cooling off period is not going to be a huge factor to solicitors in terms of time spent. But there may be cases where, for example, limitation is in issue, where not following the regulations may be an expensive mistake in respect of recovering costs from the client.

DIFFERENTIAL RATES

[4.14]

The definition of a CFA in s 58 refers to an agreement 'which provides for his fees and expenses, or any part of them, to be payable only in specified circumstances.' There are three points to make about this part of the definition.

Fees and expenses

[4.15]

The agreement can deal with expenses and not just fees in different ways depending upon the outcome. This is an important consideration in respect of drafting agreements. It is dealt with later in this part of the book at Chapter 9 regarding solicitors taking more of the funding risk.

Specifying circumstances

[4.16]

The specification of circumstances can be as simple or as complicated as the solicitor and his client wish. The great majority of CFAs have been used in personal injury cases where an hourly rate and success fee would be payable if the case won and no payment at all if the case lost. Initially, the success fee would be a single figure whenever the case settled. But with judicial encouragement from the Court of Appeal in landmark cases such as *Callery v Gray* [2001] EWCA Civ 1117, [2001] 3 All ER 833, [2001] 1 WLR 2112 and the industry agreements under the auspices of the Civil Justice Council, a number of success fees were potentially payable depending upon the exact circumstances of settlement. ('Staged' success fees were enshrined in Part 45 for personal injury cases arising out of road traffic accidents and claims against employers (see **[6.38]** in Chapter 6).

In commercial cases, different base fees and not just the success fee have been specified to be applicable in certain circumstances. In other cases a fixed sum for a successful outcome has been agreed in addition to a percentage uplift. This has not always ended happily – see for example the case of *Oyston v Royal Bank of Scotland* [2006] EWHC 90053 (Costs) – but the theory is sound, the client and his solicitor are able to specify the circumstances in infinite variety should they wish to do so.

Fees where unsuccessful

[4.17]

The market in personal injury and to some extent clinical negligence claims prior to 1 April 2013, was that of 'no win, no fee'. Whilst inaccurate in some respects, this description did clearly represent the position that there would be no fee if the claimant lost his case. That outcome was not attractive in other areas where the prospect of an unsuccessful outcome was much more likely.

So, defendants' solicitors in personal injury cases (who usually do not achieve 'success' in obtaining an order for costs from the claimant), have agreed collective CFAs with insurance companies which have higher (usually court guideline) rates where successful and lower rates where unsuccessful. The lower rates reflect the general agreements reached between the law firm and insurer. Such agreement for unsuccessful cases might be a fixed figure rather than a lower hourly rate. In commercial cases, the unsuccessful fee would generally be a lower hourly rate with figures in the region of 50% of the successful rate not uncommon.

These agreements are described in myriad ways. They are differential fee agreements, or hybrid agreements or 'No win, low(er) fee' agreements. Those which have no fee where unsuccessful are sometimes called 'pure CFAs' but perhaps fortunately those where there is a lower rate have not been called 'impure CFAs' or any other antonym to pure.

ASSESSING THE RISK

The limits of risk assessment

[4.18]

'We do not consider that it can ever be said that a case is without risk' said Lord Woolf in *Callery v Gray* [2001] EWCA Civ 1117, [2001] 3 All ER 833, [2001] 1 WLR 2112. All litigators know of cases that seemed to be stone cold certainties that failed to succeed when they went to court. Equally, most can point to cases which they ran as much in hope as in expectation but which triumphed at trial. Where the case was paid for privately, the outcome did not matter to the lawyer financially. As such, if the client wished to risk triumph or disaster on an obviously risky case, it was very much their choice. Those twin imposters would have been considered by the client's lawyers at the outset when advice on prospects was rendered. Those prospects would have been reconsidered as the case progressed, and reviewed prior to preparing for trial. No matter how many times the risks were assessed, however, the essential prospects of the case would be unlikely to alter significantly. A risky case remains a risky case however often those risks are assessed. Parties and their lawyers may strive to find evidence and arguments to improve the case's prospects but a case that starts as a risky one rarely becomes a very strong case by the time it gets to trial.

The benefits of risk assessment

[4.19]

Consequently, there is a school of thought that says that there are only three kinds of case – good ones; bad ones; all the ones in the middle. The first category are the ones you would always wish to run, whether on a CFA or otherwise. The second category contains those that you should jettison unless the client wishes to pay you irrespective of the result. The third category has those cases where you need to consider the risk and decide whether you wish to take it on or not. If you do decide to do so, you must remember that the risks have not disappeared simply because you have assessed those risks and decided to accept them. It is surprising how many litigators treat a case with an accepted risk as being as good as a strong case and pursue it as if all the time and effort invested is almost certainly going to be recouped at the end of the case.

There is another school of thought that says the risks can be weighed in a much more sophisticated and scientific manner. There have been many books written on the assessment of risk generally. There have even been some solely on assessing risk in relation to conditional fee agreements. Risk assessment is ultimately, like faith, very much a personal thing. You either think the risks can be gone through with the fine tooth comb or just the broad brush, or perhaps some implement in between.

Wherever you stand on this issue, there are sound management reasons for assessing the risk to see if the rewards are commensurate to the risks that your firm is running. Therefore even if you think assessing the prospects of success is just a gut reaction, you should consider a more formal risk assessment to

4.20 *Chapter 4 Creating Conditional Fee Agreements*

consider whether taking on the case stacks up generally. For example, will it involve a huge amount of resource or tie up unpaid fees for an inordinate amount of time? Furthermore, your firm will only benefit properly from the risk assessments if you record information centrally to be able to build up a picture of your success and profitability in any given area.

The other reason for creating a risk assessment regardless of your belief in its relevance has ended with the change in regime. Whilst success fees were recoverable, a contemporaneous risk assessment was a key document to support the risks as seen by the litigator when the CFA was created. It was a nervous litigator's charter because every risk that could be imagined was put down since there was no reason not to do so. Even if the risks suggested that a success fee well over 100% could be justified, they were still worth setting down because they might just persuade the assessing judge that the success fee actually claimed was a very reasonable one given the risks. This is no longer required but the risk assessments are still valuable in cases going through the courts and which are considered in Chapter 6.

Forms of risk assessment

[4.20]

The Risk Assessment forms perused by the courts have often been described as unhelpful or designed to establish a high success fee in every case. There is no Law Society model risk assessment to go with the model CFA. The model CFA simply takes the 'assessment of the risks in your case' as being one of the factors to justify the success fee claimed in Schedule 1 to the agreement.

Most risk assessments are variations on a table which sets out all the risks the solicitor can think of down one side and levels of risk along the other axis. So, the solicitor may consider 'expert evidence' to be a risk and have the options of low, medium or high to categorise that risk. Variations include many more options to each risk so that expert evidence might be broken down into risks regarding the client's expert as distinct from the opponent's; or from the risk of single joint expert. Instead of low, medium and high, there might be several categories to distinguish the level of risk. There may be a method of using figures rather than words so that risks might be 1 to 9 to give more refinement to the perceived level. Whatever amount of detail is used, the resulting figures (or impression if ticks are used against the level of risk) produce a risk percentage which shows the prospects of success or failure.

This percentage is then used, conventionally, in a purely arithmetical way to be converted to a percentage success fee by using a ready reckoner first made widely known by the Law Society book on Conditional Fees written by Michael Napier and Fiona Bawdon when CFAs were first introduced. The ready reckoner's simplicity has found favour with the judiciary on many occasions.

Assessing the risk **4.21**

The Ready Reckoner

[4.21]

The simplest way of explaining its use is by an example. Supposing that your risk assessment suggested that you considered that there was a 75% prospect of success (or 25% prospect of failure.) The ready reckoner below would give you a 33% success fee to be claimed on your base costs.

The logic for the ready reckoner is this. A 75% prospect of success is the same as saying that if several cloned cases of this one were run to trial, 3 out of every 4 would succeed. Therefore there is a risk that this particular case is the other 1 of the 4. In order to guard against that risk, a success fee is claimed. Mathematically the calculation is to divide the prospects of failure (25%) by the prospect of success (75%) and multiply by 100. This calculation has been done for each prospect of success or 'chance of winning' of 50% or more as follows

Chance of winning	Success Fee	Chance of winning	Success Fee
50	100%	75	33%
51	96%	76	32%
52	92%	77	30%
53	89%	78	28%
54	85%	79	27%
55	82%	80	25%
56	79%	81	23%
57	75%	82	22%
58	72%	83	20%
59	69%	84	19%
60	67%	85	18%
61	64%	86	16%
62	61%	87	15%
63	59%	88	14%
64	56%	89	12%
65	54%	90	11%
66	52%	91	10%
67	49%	92	9%
68	47%	93	8%
69	45%	94	6%
70	43%	95	5%
71	41%	96	4%
72	39%	97	3%
73	37%	98	2%
74	35%	99	1%

4.21 *Chapter 4 Creating Conditional Fee Agreements*

It will not take you long to realise that there is a certain amount of artificiality in this approach. It assumes in particular that the costs of each case, whether won or lost, are the same (which is why we have described them as 'cloned'). Nevertheless it is relatively easy to explain to clients and, since it is mathematical, is seen to be objective rather than a subjective figure plucked from the air. What most clients do not appreciate is that the success fee depends upon the original risk assessment and so the gloomier the risk assessment, the higher the success fee is bound to be. The level of gloom is entirely in the hands of the assessor of the risk.

RISK AND THE COST OF FUNDING

[4.22]

The payment of fees (and, in some cases, disbursements) will be postponed until the outcome of the case. The fact that the cost of this delay is effectively carried by the solicitor is discussed at the very beginning of this part of the book.

Regulation 3(1)(b) of the CFA Regulations 2000 raised the concept of the cost of funding being reflected in the success fee. The original Law Society model drafted in 1995 made no reference to such a concept. The success fee was simply a reflection of the risk as calculated by the ready reckoner.

Having seen reg 3(1(b), the revised Law Society model contained a schedule for calculating the success fee and the first two factors to be considered were:

'(a) the fact that if you win we will not be paid our basic charges until the end of the claim;
(b) our arrangements with you about paying disbursements'

The schedule went on to say that 'the matters set out at paragraphs (a) and (b) above together make up [   ]% of the increase on basic charges.'

With minor amendments and a demotion in the batting order to (d) and (e) these two factors continue in the latest model agreement.

An unnecessary complication

[4.23]

The main difference between then and now is that in 2000 the funding factors were considered to justify a separate success fee from the risk factors. This made the schedule more complicated and resulted in a number of solicitors seeking a composite success fee above 100%. In particular, a 100% risk success fee became invariable for cases going to trial (with a lower fee for earlier settlement). In addition a further 5% or 10% was claimed for the cost of funding resulting in a total success fee claimed of 105% or 110% which was starkly in breach of the CFA Order 2000 maximum of 100% and rendered the agreement unenforceable.

Why make this system more complicated? It is a good question. There was never any explanation why a percentage applied to the base costs was a good method of seeking compensation for funding the case. It would have been much more obvious to seek a sum based on interest charged on the base fees and disbursements that were outstanding. The need for separation from the risk factor success fee was precisely because the risk factor element was recoverable from the losing party but the cost of the funding element was not. This was provided for by the CPR in response to reg 3(1)(b). But those solicitors who kept the success fee purely for the risk made completion of the CFA, and explanations to the client, much simpler. Now, whilst the schedule reflects the fact of a delay in payment until the end of the case, it is included within the single success fee claimed in the CFA.

What to do now

[4.24]

If you intend to charge your client for the delay in payment, you need to consider how this is to be charged. If the Law Society model wording is used, it is included in the success fee. If you decide to make a separate calculation paragraphs (d) and (e) need to be removed from the standard schedule. In either event, the charge will need to be included in the 25% cap if you are dealing with a personal injury case.

Funding disbursements

[4.25]

As mentioned above, and dealt with in detail below, the client's expenses can be part of the CFA, and not just the solicitor's fees. Accordingly, they will be paid by the opponent if the client succeeds, and paid by the solicitor if unsuccessful. If you are considering taking this approach, you should consider what if any charge you propose to make for funding those disbursements in the interim. Presumably, such a charge would usually only be made in the event of success.

COST OF SETTING UP A FUNDING ARRANGEMENT

[4.26]

The cost of setting up the client retainer is a matter between the solicitor and his client. Where there is a funding arrangement included in that retainer the time spent in setting everything up may take rather longer. For example, in commercial litigation there may be discussions with (third party) litigation funders (see Chapter 10) and after the event insurers (Chapter 9) as well as a conditional fee agreement to sort out. In group litigation there may be these factors and more.

It is perhaps surprising therefore that it was not until 2011 that we had a decision concerning the recovery of the cost of explaining CFAs to clients and of putting in place ATE insurance. In *Motto v Trafigura Ltd* SCCO (2011) 15 February 2011 the Senior Costs Judge answered the following question 'yes': 'Is the work undertaken by solicitors, counsel, costs draftsmen and insurers in establishing and setting up (1) the conditional fee arrangements and

4.26 *Chapter 4 Creating Conditional Fee Agreements*

(2) the insurance policy recoverable in principle?' On appeal, the Court of Appeal overturned that decision ([2011] EWCA Civ 1150, [2012] 2 All ER 181, [2012] 1 WLR 657). Claimants cannot recover the cost of preparing and advising on CFAs nor can they recover costs incurred in discussing litigation with or taking instructions from ATE insurers.

The court's reasoning was that the expertise and effort devoted by solicitors to identifying a potential claimant, and negotiating the terms on which they were to be engaged by the claimant, in connection with litigation, could not be properly described as an item incurred by the client for the purposes of the litigation. Until the CFA was signed, the potential claimant was not a client let alone a claimant. Liaising with ATE insurers after the insurance was in place was collateral to the action and liaising with insurers was designed to ensure the claimant was protected against costs.

CHAPTER 5

RUNNING CASES WITH CONDITIONAL FEE AGREEMENTS

DURING THE CASE

Payments on account

[5.1]

Where you have agreed a CFA with your client that is not a 'No win, no fee' agreement, you are bound to receive some fees regardless of the outcome. In many commercial cases, the unsuccessful hourly rate is in the region of 50% of the successful rate. As such, half of your fees ought to be paid by your client as the case progresses. Your client will be reimbursed in the event of a win from the recovered costs.

If you had been instructed on a traditional private paying arrangement, you would expect to render regular bills. You should do so using a CFA wherever possible. The client is never going to be delirious about paying your fees, but unless he really cannot do so, it will help your cash flow and make your client understand the costs involved as they are incurred. It is usually the case that clients are more positive about paying your fees whilst the case is proceeding since they believe they will win and therefore recover them anyway.

In order to be able to do this, you need to have express wording agreed with your client otherwise you have to rely on natural breaks etc discussed at [2.9]. There is no need for this. You should have it in your client care letter anyway. But if you do not, make sure it is in your CFA.

The interim bill could be either a payment on account or an interim statute bill. The differences are discussed in more detail at [2.4]. For the purposes of this chapter, it is simply worth considering the risk of an interim statute bill providing an argument to your opponent that the amount billed to your client is the full extent of your client's liability for the period covered by each individual statute bill. If you are billing say 50% of the successful fees as the amount you will receive win or lose, you are risking the other 50% if your opponent establishes the effect of your statute bills is to limit your client's liability to the statute bills. You need to be very careful about the wording of your interim bills and unless there is a benefit to the certainty created by an interim statute bill there is plenty to be said for the flexibility of a payment on account.

NO NOTICE OF FUNDING

[5.2]

One of the main matters which solicitors needed to keep in mind prior to 1 April 2013 was the notification of the CFA to the opponent. This might be in the letter before action or subsequently by a Notice of Funding in form N251. There is more detail on this requirement in the next chapter. Such notification was required wherever a CFA had a success fee and so the opponent needed to be told of the 'additional liability' that was being incurred over and above base costs. Where a CFA did not have a success fee, but merely different hourly rates depending upon the outcome, there was no need to serve a notice. So differential CFAs, as were often used by defendants in personal injury cases, did not have to be alerted to the opponent. Now that success fees are no longer recoverable (save for certain discrete areas discussed below) there is no longer any need to notify the opponent of the existence of the CFA.

Traditionally, and save for Legally Aided cases, each party's financing of the case was a matter for the client and not discussed between the parties. The use of Notices of Funding in some forms of litigation has become so widespread that parties appear to believe that they are entitled to know their opponent's arrangements. It will be interesting to see whether this prurience dies down or whether a more open approach will become the norm.

NOTICE OF FUNDING

[5.3]

Cases where the CFA was entered into prior to 1 April 2013 still need to follow the old rules in accordance with CPR 48.1 and 48.2. This includes the need for notification.

Furthermore those cases where recoverability of additional liabilities has not been withdrawn for the time being – mesothelioma claims and those involving Insolvency Practitioners or against the media for publication and privacy actions – will also need to be notified by virtue of the same provisions.

ESTIMATES OF COSTS

[5.4]

CFA clients are just as entitled to be kept up to date in respect of the costs their solicitors have incurred as any other client. That would seem obvious, particularly where the client is simply paying a different hourly rate depending upon the outcome. But where the client was on a 'pure' CFA and so nothing would be paid if the case was unsuccessful, there have been many who took the view that the client simply would not be interested in the costs involved and keeping mum would save unnecessary conversations and correspondence with the client (which may well not be recoverable from the opponent.) This was a particularly prevalent view in personal injury work when the success fee was recoverable and the client was guaranteed to receive his full damages.

The removal of recoverability has meant that many clients will now end up parting with some of their damages to pay for their lawyer's fees and it becomes obvious once more that they ought to be kept informed. The eventual introduction of fixed fees in the fast track along with budgeting in the multi-track will make estimating costs easier. Such figures are of course only those that are recoverable from the opponent. To the extent that the solicitor's fees will not be fully met from such sums, the client will still be liable to his solicitor in the event of success.

COUNSEL

[5.5]

For hundreds of years, solicitors instructed counsel in contentious matters based on the 'honorarium' approach. In other words, the solicitor would send instructions to the barrister who would advise in writing or in conference or attend court on the client's behalf. The fee for each piece of work would be agreed, usually in advance of the work being done, and the solicitor would pay for the work in due course. If the client did not put the solicitor in funds to pay counsel, the solicitor was still required as a matter of professional conduct to pay counsel's fees. Counsel could not sue on the agreement and if there was a disagreement between the solicitor and barrister regarding the fee a Joint Tribunal (see below) would be convened consisting of solicitors and barristers to adjudicate. Whatever fee was deemed payable by the Tribunal then had to be paid by the solicitor, again as a matter of professional conduct.

Three things have happened in the last few years which have altered this arrangement.

Solicitor's professional obligation

[5.6]

The first is that from 2007 the solicitor stopped having a professional obligation to meet counsel's fees if his client did not put him in funds to pay counsel. This change took some time to be widely noticed and there are many solicitors who still believe they are professionally obligated. But, a solicitor's continued use of the chambers in which counsel is based is unlikely in such circumstances and that has been of sufficient concern to many solicitors to encourage them to pay counsel even if they have to do so themselves.

The Joint Tribunal has continued to be in existence even though the solicitors' professional obligations have evaporated. The Tribunal only comes into being if the Bar Council convenes it and the speed with which notifications of non-payment have been processed (and which are the precursor to a Tribunal being convened) has left this process to be unwieldy and time consuming.

5.7 Chapter 5 Running Cases with Conditional Fee Agreements

New Contractual Terms

[5.7]

The Bar's answer to this unsatisfactory situation has been to bring in new contractual terms on which counsel carries out work. The Bar Council and the Law Society spent several years trying to agree terms but were unable to do so. Consequently the Bar has gone it alone. As from January 2013, the default position is no longer an honorarium arrangement (set out in 'The Terms of work on which Barristers Offer their Services to Solicitors and the Withdrawal of Credit Scheme 1988') but a contract based on the contract terms (the equally wordy 'Standard Contractual Terms for the Supply of Legal Services by Barristers to Authorised Persons 2012'). There were earlier contractual terms which barristers could use but they were very rarely used. The 2012 version will be much more widely implemented because of the default position change. The parties are free to vary these terms and many chambers have varied the standard provisions to a greater or lesser extent. But the variation must be agreed in writing and that should prevent most disputes as to the agreed variation.

Now if a solicitor does not pay, he and/or his client, can be sued in the courts in the ordinary way. Interestingly, the Joint Tribunal can be used as an alternative and which presumably is seen as a slightly more gentlemanly form of fighting over fees.

A solicitor who did not pay up following a Joint Tribunal would find himself unable to instruct any barrister on the usual credit terms but would have to send money with the brief. A list of such solicitors was kept by the Bar Council and regularly circulated amongst chambers to make sure that no services on credit were being provided. The contractual terms are seen by the Bar Council as sufficiently robust for this additional sanction to be otiose. So a solicitor who finds himself on the List of Defaulting solicitors (previously the Withdrawal of Credit list) is able to go to other counsel on the same case and seek credit with that counsel/chambers. This has been an unnecessary shot straight through the metatarsal by the Bar Council in the view of many chambers whose methods of enforcement are limited.

These first two changes, the lack of a professional obligation and then the new contractual terms, have not really altered the way in which solicitors and counsel interact during the life of a case. Counsel is still paid win or lose and so is prepared to draft pleadings, advise on merits etc and 'do his best' in court regardless of the merits. It is only if, for example, counsel is asked to plead fraud without evidence or otherwise argue a point so hopeless that he will incur the wrath of the court or the censure of the Bar Standards Board, that he will decline instructions.

Use of CFAs

[5.8]

The third change is the use of conditional fee agreements (and from 1 April 2013, potentially Damages-Based Agreements). It is here that the change in the relationship between solicitors and counsel has altered. It is surprising that it has not altered more in fact but, ironically, some changes that should have taken place have not done so because seemingly neither the solicitor nor the barrister have read the CFA closely enough and acted upon it.

Instructions to advise using a CFA

[5.9]

The first time many counsel see a CFA backed case is when they are asked to advise on the merits. If this is requested in writing, counsel can take the view that the case does not have sufficient merits for him to take it on under a CFA. But if the advice is in conference with the papers arriving relatively shortly beforehand, counsel and his instructing solicitor can be embarrassed if counsel has to decline the instructions at the last minute because he does not have confidence in the case. In practice, counsel have often taken the view that the better course is to conduct the conference and see if any merits emerge. If they do not, the preparation and conference time will have to be written off.

Brief to appear

[5.10]

Similarly, if counsel attends at a trial having only received the papers shortly beforehand, he either has to return the papers and cause professional embarrassment all round or prepare and appear in the expectation that he will not be paid for so doing. The choice is often governed by the value placed on the relationship with the instructing solicitor.

Since counsel is being paid on the outcome of the case, he cannot be compelled to take the brief and the 'cab rank' rule specifically does not apply to CFA cases for this reason.

Counsel can also decline the brief on a CFA if he thinks that he is too senior to deal with it on the basis that he is unlikely to be able to recover the sort of fee he usually commands on that hearing.

Being kept informed

[5.11]

The biggest change ought to be in relation to informing counsel of the progress of the case. The agreements drafted by the Personal Injury Bar Association (with APIL), the Chancery Bar Association and the Commercial Bar Association all contain numerous requirements during the life of the case which are honoured in the breach by many solicitors. The requirements are not simply updates regarding progress in the court timetable. They expect papers to be sent to counsel for advice at specific stages and before certain decisions are taken. Nevertheless, it is a common complaint that cases on which counsel have outstanding fees, for example, for an early advice on the merits, are abandoned by the solicitor without counsel knowing, let along having any input on the decision.

Post-April 2013 Agreements

[5.12]

The restrictions on the recoverability of success fees and the deductions from the client's damages apply equally to counsel as they do to solicitors. Therefore, all of the solicitor and counsel agreements need to be rewritten. This has meant that some agreements were rewritten in January to take account of the new contractual terms and then three months later redrafted again to deal with LASPO 2012. The venerable APIL / PIBA agreement for personal injury and clinical negligence cases has been updated. It is now version 9. It has lost the tick box success fee calculator and has acquired a section with options regarding any fees not recovered on settlement or assessment. Both alterations reflect the end of recoverability for success fees and the issue of who will now pay counsel's fees. The other regularly seen agreement, the Chancery Bar Association's CFA has also been updated, although it requires rather more effort on the part of the drafter in using its precedent.

The central question is on what basis will counsel be instructed? Anecdotally, it appears that many counsel are attempting to return to a disbursement basis to avoid taking some of the client's damages for a success fee.

INTERIM APPLICATIONS

[5.13]

The drafting of most CFAs seeks to cater for the situation where an order for costs is made on the way to a successfully concluded case. The definition of a win is generally based on an award of damages and it is sometimes argued by defendants that until such time as an award for damages has been made, the claimant has not been successful as defined in the CFA. That means that the claimant does not have to pay his solicitor any fees and so the defendant should not have to do so either.

The Law Society model CFA seeks to get round the problem in this way:

> 'If on the way to winning or losing you are awarded any costs, by agreement or court order, then we are entitled to payment of those costs, together with a success fee on those charges if you win overall.'

What does 'if you win overall' mean in this sentence? It seems clear that it is intended to refers solely to the success fee. The entitlement to payment of the base costs is meant to be crystallised when the client is awarded costs. However it can quite plausibly be argued that the base costs, together with the success fee, only become an entitlement if and when the case is won overall.

Most courts in fact seem to consider this to be a satisfactory form of words. They take the view that the ultimate outcome does not matter, other than for the success fee. With the end of recoverability of the success fee, it may well be that the general argument outlined above, is not pursued by paying parties since courts are not going to be very keen to deprive a party of his base costs following an application unless there is a clear problem with the wording of the agreement.

CHANGING ARRANGEMENTS

Dating of CFAs

[5.14]

In order to begin work under a CFA it is normally prepared and dated as soon as possible. The work done is usually described as being from 'now' until the agreement ends. But in some litigation, it is not always possible to sign up the client to a CFA at the outset because the prospects are too uncertain. In other situations, a new CFA is required because the original party needs to be substituted, through death, insolvency, incapacity etc.

The question therefore was always going to crop up of 'can I make my CFA cover work done before the CFA was signed?' The first judicial pronouncement on this was from the Senior Costs Judge, Master Hurst in *King v Telegraph Group Ltd* [2005] EWHC 90015 (Costs) where he approved the concept as a matter of contract in relation to the base fees but not in respect of the success fees.

Master Hurst described the process as 'back dating' but this was disapproved by Stanley Burnton J. in *Holmes v Alfred McAlpine Homes (Yorkshire) Ltd* [2006] EWHC 110 (QB), [2006] 3 Costs LR 466. The correct approach was for the agreement to be dated on the date it was signed but for the wording to be revised to make clear that it had retrospective effect. The judge put it like this:

> 'Mr Wilkinson submitted that the agreement was on its face, retrospective. That is incorrect. It was not retrospective: it was back-dated, which is a very different thing. A properly drafted agreement would have borne the date on which it was executed, but would have expressly provided for its application to work done from the prior date agreed by the parties. The written agreement in this case was misleading.'

The claimant was lucky in *Holmes* because he succeeded on other grounds to show the CFA was enforceable. He, or perhaps more accurately his solicitors got the essential point wrong regarding back dating rather than making agreements properly retrospective. Parties sometimes also refer to the case of *Forde v Birmingham City Council* [2009] EWHC 12 (QB), [2010] 1 All ER 802, [2009] 1 WLR 2732 where the position described in *Holmes* was confirmed.

How should you make your CFA retrospective? The standard amendment to the Law Society wording is made under the heading 'Basic Charges' where the original wording is 'These are for work done from now until this agreement ends.' Amendments are usually either to remove the word 'now' in order to say 'These are for work done from xx/yy/zzzz . . . ' (being the date on which the client was first seen) or 'These are for work done from when you first instructed us . . . .' Wording along these lines was used in *King* and has been widely used thereafter.

Assignment of CFAs

[5.15]

In *Jenkins v Young Brothers Transport Ltd* [2006] EWHC 151 (QB), [2006] 2 All ER 798, [2006] 1 WLR 3189 the claimant instituted proceedings funded

5.15 *Chapter 5 Running Cases with Conditional Fee Agreements*

by a CFA. When the solicitor acting for him changed firms, the CFA was assigned from the first firm to the second. The solicitor changed firms again and once again the CFA was assigned. The case was settled on terms of damages of £445,000 with costs to be agreed. The defendant challenged the lawfulness of the CFA on the basis that it could not be assigned as a matter of general contract law.

The court held that it would be a novel approach to the administration of justice if the court were to seek to interfere with a professional relationship whose propriety and worth had never been challenged. In these circumstances the benefit and burden of the CFA could be assigned as an exception to the general rule.

Cancellation of CFAs

[5.16]

If the client terminates a CFA through his cancellation rights under the Cancellation of Contracts made in a Consumer's Home or Place of Work etc, Regulations 2008, SI 2008/1816, that is the end of the CFA and the client has no other liability to the solicitor.

However, if the CFA is cancelled otherwise, for example, under the provisions requiring the client not to mislead the solicitor, residual fees may be payable under an underlying retainer. In *Kris Motor Spares Ltd v Fox Williams LLP* [2009] EWHC 2813 (QB), [2009] 6 Costs LR 931 the client misled the solicitors as to the independence of an expert witness whose evidence was crucial. The solicitors terminated the CFA on terms that they would continue to provide services to conclude the case on ordinary fee terms, the opponents having offered a 'drop hands' outcome. The CFA provided for full fees to be paid in the event of such termination. Had the solicitors not terminated at that stage the case would have concluded as a loss under the CFA and the solicitors would have only recovered 70% of their fees. It was held that the CFA had been validly terminated, the retainer remained and therefore the client was not left unrepresented but had a liability for costs. It should be kept in mind that termination of a retainer at common law requires a good reason to terminate the retainer and also reasonable notice to cease acting.

SECURITY FOR COSTS

[5.17]

We deal with security for costs applications in detail in Chapter 19. But where the client is using a CFA, there are some considerations which do not apply to other forms of funding. In particular, it is sometimes possible to purchase ATE insurance, essentially in the form of a bond, in order to persuade the court that the opponent's concerns about the client being good for the money are exaggerated.

On many occasions an opponent will accept the existence of an ATE policy as being sufficient not to proceed with a security for costs application. But to do so, in our view, suggests that they have not really considered the relationship between the client and the ATE insurer sufficiently closely. All ATE policies have their policy terms requiring the client to comply with various matters regarding the conduct of the litigation. If the client fails to do so, the insurer will be at liberty to cancel the policy. This may mean that adverse costs incurred after the date of cancellation will not be met. In a more serious case, the policy might be avoided ab initio and as such the opponent will not be able to take any benefit of the ATE policy indemnity unless the client is made insolvent and the opponent successfully sues the ATE insurer in the client's shoes (which would involve demonstrating that the policy should not have been cancelled or avoided as the case may be.)

The opponent ought to have been notified of the cancellation of the policy when Notices of Funding were obligatory, but that did not always happen. Now that such Notices are no longer required there is very little prospect of such notification taking place.

The potential inadequacy of ATE policies in security for costs applications was considered in *Al-Koronky v Time-Life Entertainment Group Ltd* [2005] EWHC 1688 (QB). The court held that the existence of a satisfactory ATE policy could help in resisting an application. But since the policy did not pay out if the claimant had been untruthful, and as this was a libel claim where the only real issue was the truthfulness of the claimant, the policy was in effect worthless.

In *Michael Phillips Architects Ltd v Riklin* [2010] EWHC 834 (TCC), [2010] NLJR 943, [2010] BLR 569 Mr Justice Akenhead set out the following common sense guide to ATE in security for costs cases:

(a) There is no reason in principle why an ATE insurance policy could not provide some or some element of security.
(b) It will be a rare case where the ATE insurance policy can provide as good security as a payment into court or a bank bond or guarantee.
(c) A claimant must demonstrate that the policy actually does provide some security: there must not be terms pursuant to which or circumstances in which the insurers can readily but legitimately and contractually avoid liability to pay out for the defendant's costs.
(d) There is no reason in principle why the amount fixed by a security for costs order could not be somewhat reduced to take into account any realistic probability that the ATE insurance would cover the costs of the defendant.

The question that arose in *Mengi v Hermitage* [2012] EWHC 2045 (QB), [2012] 5 Costs LO 641, was whether an order for security ought to take into account a defendant's potential success fee. Tugendhat J included an allowance for a 100% success fee without seeing the CFA, reasoning that to take into account that costs would involve an uplift for the CFA did not involve any illegitimate speculation. Obviously this is a case that is only relevant where the case is sufficiently old for the defendant to have entered into a CFA or CCFA prior to 1 April 2013 (and therefore to have a recoverable success fee.)

ENDING THE RETAINER BEFORE THE CASE CONCLUDES

[5.18]

As is always said, a client can end a retainer at any time, a solicitor only for good reason and with reasonable notice. The position changes with a funding arrangement in respect of terminating that arrangement. A client can still effectively end it at any time because he is free to instruct the solicitor of his choice. If he decides to move his case to another firm, that is the end of the funding arrangement as well as the retainer. The client may have to pay costs to retrieve his papers but is not limited otherwise. The difference comes with the solicitor's entitlement to end the CFA. As we have seen in *Kris Motor Spares Ltd v Fox Williams LLP* above, the CFA funding arrangement can be ended without the underlying retainer of the solicitor bringing proceedings on the instruction of the client concluding.

Why would you want to end the CFA? The usual reasons for ending a CFA before the end of a case are:
(a) a different form of funding is more appropriate
(b) the prospects of success are not sufficient to continue
(c) the client is not co-operating

Let us take these in turn

Other funding

[5.19]

Alternative funding enquiries are meant to have been carried out at the beginning of the case so a decision that another option is more appropriate during the case ought to be rare. It is likely to be the case that BTE insurance has come to light either through more diligent searching or a belated response from a potential insurer. It could possibly be – though this is now extremely rare – that the client is now eligible for legal aid funding but was not previously. Even if this is so, it is not necessarily the case that to continue with a CFA would be an unreasonable choice and the decision would be very much case specific. The same is true if a different form of funding, such as union backing, became available part way through a case.

It was not always the case that the use of CFAs and ATE insurance was always to be the last resort purely because they would be more expensive for the opponent. Nevertheless, that was generally the approach taken on assessment by the courts where the party could have used Legal Aid or BTE insurance (or had started off doing so but changed to a CFA and ATE insurance during the case). Now that these additional liabilities are no longer recoverable from the opponent the concept of a CFA being a last resort has presumably ended. It may still have been an unreasonable choice on its facts, but that would be a matter for the client to take up, not the opponent.

Lack of Prospects

[5.20]

Much CFA litigation is conducted with ATE insurance firmly in tow. If the prospects drop below the insurer's minimum threshold, the diminution in prospects has to be reported and the likelihood is that the ATE insurance will be cancelled from that point. The consequence for the solicitor is that any further disbursements will not be recoverable from the ATE insurer if they are not recovered from the opponent. This is a sizeable hurdle to surmount even if the solicitor is prepared to spend more time and effort in trying to improve the prospects, difficult though that almost always is.

Where the prospects have deteriorated slowly, particularly in ways about which the opponent is not yet aware, the solicitor will be expected to try to bring a resolution to the claim so that the policy does not have to be formally cancelled and potentially notification given to the opponent. Acceptance of a nuisance offer, even out of time, or a drop hands settlement has been preferable to a discontinuance from the ATE insurer's point of view and they are likely to allow the solicitor a little time in which to try to bring this about. The arrival of Qualified One Way Costs shifting will affect some of the assumptions made to date since the ATE insurer will not have a liability in respect of a discontinuance unless there is some suggestion of fraud, but the ATE insurer will not want to be backing cases with slim prospects of success in any event.

Since payment is based on the outcome of the case, all CFAs contain a clause allowing the solicitor to bail out if the prospects have reduced. To do otherwise would be to lock in the legal team to a case that could not win but would involve the lawyers in (potentially considerable) irrecoverable costs. The existence of this bail out clause has been used by paying parties to argue that the risk assessment of a case is skewed in the solicitor's favour because the case will not always reach the end of the litigation, ie a trial. Accordingly, the prospects are improved in the solicitor's favour. See [6.31] for a discussion of the case of *McCarthy v Essex Rivers Healthcare NHS Trust (Case No HQ06X03686)* (13 November 2009, unreported) QBD on this point.

Non-cooperation

[5.21]

Some forms of non-cooperation are not always the client's fault eg, death or insolvency but they still impact on the solicitor's ability to run the case.

The provision of inadequate instructions is problematic whether looked at through the prism of the CFA or the retainer as a whole. If the situation is bad enough to end the CFA through it, the relationship with the client is likely to be at end in any event.

Which rate?

[5.22]

If the CFA comes to an end at a point where it can be said that the client may still win, the solicitor should be entitled to seek his costs at the successful rate. The usual provisions allow for such base costs to be paid out upon termination. Alternatively the base costs and any success fee will be paid if and when the client has been successful.

If the CFA comes to an end because the case is unlikely to win, then the unsuccessful rates will apply. In a personal injury case, this rate may well be nothing. But in a commercial matter, there is likely to be a lower rate payable

5.22 Chapter 5 Running Cases with Conditional Fee Agreements

and which should be charged, if it has not already been charged as the case progressed. The drafting of the CFA is much more complicated in relation to the termination provisions if there is a lower fee when unsuccessful than where it is a no win, no fee agreement.

AFTER THE CASE ENDS

Payment on account following a hearing

[5.23]

Where your client has succeeded at a final hearing, you have succeeded within the terms of your CFA. If you have not yet satisfied the definition of success at that point, then let us hope that it is because your client has only won on liability, to use that term in a relatively loose sense for some forms of litigation. As such, you still need to agree the damages and/or have a hearing to determine how much. You may have obtained an order for costs of the liability trial already. If it is a forthwith order then you could have the costs assessed, but not otherwise (CPR 47.1). So if your opponent can easily establish that you have not really 'won' yet according to your CFA then better to wait. In that case you can come back to this chapter later.

If you have agreed a form of CFA Lite so that recovery is a requirement and not just an award of costs, you should still be in a position to seek an interim payment because logically that is only hastening the payment that your client has agreed with you anyway. Furthermore, the recovery is usually based on a recovery of damages rather than costs which is the purpose of the interim payment here.

If there is some other reason why the definition has not been triggered you may be in some difficulties in recovering anything under that agreement.

CPR 44.2(8) states that 'where the court orders a party to pay costs subject to detailed assessment, it will order that party to pay a reasonable amount on account of costs, unless there is good reason not to do so.'

This rule came into force on 1 April 2013. Before then, the court might make an order for an interim payment. Now, as with much of the Jackson inspired legislation the court is exhorted to make a positive order. The theory is that if a sizeable interim payment is made by the loser to the winner, the remaining sum will be more easily agreed because the overall playing field for the parties has been reduced. Make sure that you or your advocate have worked out how much you will be seeking on an interim payment application at the end of the hearing. You will need an estimate or your latest approved budget for this purpose. Whilst it will take time to be sure how judges will react, our assumption is that trial judges will be willing to make an award of a larger percentage of the overall costs where there is a budget than if there is only an estimate before the court.

What should a court do when faced with a sizeable success fee claimed as part of the interim payment? The practical answer is that the judge will undoubtedly take a view about how risky he thought the case was and will factor that into his decision. Whether he will pronounce this publicly is a different matter. Some judges are quite willing to assess the success fee at the time of the trial and record it rather than leaving it for the judge on assessment. In *Crema v Cenkos Securities plc* [2011] EWCA Civ 10, [2011] 4 Costs LR 552 the Court of Appeal considered this issue in the light of a 100% success fee amounting to £250,000. Neither side could point to any previous case law on whether a large uplift should be taken into account when ordering an interim payment. On the basis that the court should order what will almost certainly be recovered the figure arrived at was £300,000. As such Crema would almost certainly get some uplift.

Payment on account following an agreement

[5.24]

Where agreement has been reached by acceptance of a Part 36 offer, there is no scope to seek an interim payment of costs until a detailed assessment hearing is requested (via an interim costs certificate). Consequently, these circumstances should prove a powerful incentive for you to get your breakdown or bill drafted and submitted to the other side as soon as possible.

Where there has been a mediation or other form of negotiation which requires a consent order to be prepared for the court's approval, there is scope for you to include an order for an interim payment. Given the wording of CPR 44.2(8) it will be difficult for your opponent to justify any outright opposition to an interim payment.

Don't forget, cash (flow) is king. If you do not get an order as discussed here, you will have to wait until you have requested a detailed assessment hearing before being able to apply for an interim costs certificate. The laudable aim of provisional assessments is that they will be completed within six weeks of the request for detailed assessment. It is unlikely that courts will entertain interim costs certificate applications if the delay in getting a provisionally assessed bill is well under two months. If the courts do not manage to hold the six week target, they are likely to be sufficiently overwhelmed with work that they are not necessarily going to be dealing with interim costs applications speedily in any event. So if your costs are under £75,000 in total you definitely need to make sure that you seek an interim payment when the case concludes and do not leave it until later.

CALCULATION OF THE CAP

Personal injury cases

[5.25]

The requirements of s 58(4B) of the Courts and Legal Services Act 1990 added by LASPO 2012 were discussed in Chapter 4. In essence, a personal injury client's damages cannot be reduced by more than 25% of his PSLA award and past losses. This cap only applies to personal injury cases and does not apply in respect of an appeal where the limit is 100% of these damages.

By defining the percentage as being based on various heads of damage, the effect of any finding of contributory negligence will be to reduce the damages and therefore the success fee available to the solicitor. Deductions to the damages also have to be made for any recoupable CRU benefits with the same shrinking effect to the potential pot.

The evil of delay

[5.26]

Clients have often complained about the lack of speed with which their claims have been brought. These complaints may well have more weight to them in the future. The longer the case takes to get to trial, the more that future losses will become past losses and thereby become susceptible to deduction. Meticulous records of who caused what delay could well prove to be very valuable if a claim of dilatoriness is made by the client at the end of the case.

The need for communication

[5.27]

A superficial glance at this new structure is enough to see that communication with the client prior to settlement is paramount. The more the client understands the likely damages he may receive before the defendant makes a Part 36 offer or round table proposal (or the case gets to trial), the more that his expectations can be managed. Defendants are not going to make life easy for the claimants and offers are unlikely to be broken down so that easy calculations can be made of the costs consequences to the client of acceptance. Court judgments will hopefully be clear on how a damages figure is calculated but if it is not, it will be very difficult to get further clarification from the court. A client who knows in round terms what his past pecuniary losses and PSLA awards are can make an informed choice. A client who comes upon the options for the first time when the offer is made will almost certainly feel pressured into settlement to some degree however balanced and objective the advice may be.

QOCS

[5.28]

The issue of QOCS is dealt with in Chapter 11 but it is easy to see how it may impact in the situations being discussed here. If the claimant does not beat the defendant's Part 36 offer, a proportion of his damages will be kept by the defendant to meet his own costs. If the damages are extinguished by such costs then at least there is no need to discuss this point further with your client because 25% of nothing needs no discussion. But if the expected damages are halved by QOCS, for example, the profitability of the case may be called into question. (Cases involving fundamental dishonesty or struck out for lack of progress are not going to be problematic in the same way because there will not be any damages in the first place.)

Which costs have to fit into the cap? The legislation is clear that it is the success fee which is subject to the maximum limit. Therefore any disbursements, including any ATE premium fall outside that limit and can be sought separately. If counsel is being paid by an ordinary retainer, his fees will be dealt with separately as well. But if he is using a CFA as well as the solicitor, there is a further success fee which needs to fit within the maximum limit.

Will there be room? It must be doubtful. Even a success fee of 12.5% of the costs may well be a significant proportion of the damages and since the standard figures for success fees in personal injury cases have been removed from the rules for new cases, there is no reason to expect continued adherence to those figures.

If there is not enough room, will counsel be prepared to continue to act on a no win, no fee basis but without a success fee? Or will the solicitor or client be prepared to pay a smaller fee – but still some fee – in the event of a loss? A number of counsel would wish to go back to being a disbursement paid win or lose and therefore without any success fee. Some have suggested increasing their hourly rate so that they build in the success fee into that rate. Why that increased rate is going to seem reasonable on either a between the parties or solicitor and client assessment is not immediately obvious. Perhaps the agreement of the client at the outset is thought to change an unreasonable rate into a reasonable one.

NOTIFICATION TO INSURERS

[5.29]

Where the case is backed by any form of legal expenses insurance, you should make sure that the insurers are notified of the outcome. In many cases this is now possible online. It is far easier to remember to do this straightaway than to be chased for information months or years later when the files have been archived.

If you make sure you do this regularly, it also affords an opportunity to make any claims for unrecovered disbursements that can be paid under the policy but will not be paid very willingly if claimed a long time after the case has ended.

INTEREST ON COSTS

[5.30]

There are two aspects to interest on costs that need to be considered. The first is the court admonishing the defendant for failing to accept the offer your claimant client made which your client subsequently beat at the hearing. The second is the question of interest on costs which have been 'fructifying in the wrong pocket' to quote one of the old authorities.

Punitive interest

[5.31]

The decisions in this area are very much fact sensitive. In *Crema v Cenkos Securities* the Court of Appeal considered that the winning claimant had taken a number of time consuming, bad points and so only allowed the claimant 75% of his costs. Nevertheless, since the claimant beat his own Part 36 offer, the costs were not only to be assessed on the indemnity basis but also would attract interest at the rate of 5% above base rate.

By contrast, the fee arrangements in *Pankhurst v White* [2010] EWCA Civ 1445, [2011] 3 Costs LR 392 came in for severe criticism from Jackson LJ who was moved to describe them as 'grotesque'. The chronology of the facts is sufficiently complicated for both parties to have been able to claim that their opponent would have been better off taking their offer than continuing. The judge at first instance declined to give the claimant indemnity costs or interest on the costs. The Court of Appeal took the view that the level of the success fee claimed where liability was admitted, the fact that the client still paid the solicitor in the event of not beating a Part 36 offer, and that there was ATE insurance against Part 36 adverse costs meant that the decision not to award interest on the costs was fully justified.

Out of pocket interest

[5.32]

A number of first instance decisions took the view that a CFA funded client should not get interest on his costs until they had been assessed because the client had not paid them. This assumed that the claimant had not been billed once the case had been successful and had settled the solicitor's fees pending the detailed assessment. But this assumption certainly followed the general practice of not asking clients to pay out for anything if at all possible. The best known of these cases is called *Gray v Toner* (11 November 2010, Liverpool County Court).

The Court of Appeal found an opportunity to consider this issue in *Simcoe v Jacuzzi UK Group plc* [2012] EWCA Civ 137, [2012] 2 All ER 60, [2012] 1 WLR 2393. The Court held that the starting point under CPR 40.8 is the same as the rule under the County Courts (Interest on Judgment Debts) Order 1991, namely that interest runs from the date ordered not the date assessed or agreed. This is often referred to as the 'incipitur' rule rather than the 'allocatur' rule. The Court then considered whether the existence of a CFA was relevant in deciding whether to make an order on cases to which CPR 40.8 applied. It was held that the existence of a CFA was not relevant and so the usual order would be that interest runs from the date costs are ordered.

In *Re Southern Counties Fresh Foods Ltd, Cobden Investments Ltd v Romford Wholesale Meats Ltd* [2011] EWHC 1370 (Ch), [2011] NLJR 882, [2011] 3 Costs LO 343 the client had paid substantial interim costs before entering into a CFA. The judge held that he had a discretion to allow interest on the costs incurred prior to the date of the costs order in order to compensate the claimant for being out of pocket.

WHERE UNSUCCESSFUL

[5.33]

If you have been operating under a pure no win, no fee agreement, your fees are unfortunately written off since there is no-one to pay them. The same will be the case for expenses if you have agreed with your client that both fees and expenses will be conditional upon the outcome. Alternatively, you or your client may have insured the case using either BTE or ATE insurance and in which case it will be important to get your claim form in before any claim for adverse costs comes in from your opponent. If it is a case to which costs protection under QOCS applies, that risk is obviously lower.

If, on the other hand, your client has agreed to a CFA with differential fee rates, you will need to render a final bill. You will need to check whether you have sought and obtained money on account through interim bills. If they are not statute bills, you need to look at whether there was any additional work carried out during the case that has not been charged for and include it in the final invoice. Similar considerations apply if you have agreed a fixed figure in the event of an unsuccessful outcome. If you have had payments on account they will need to be offset against the fixed sum. Hopefully, this will not mean a reimbursement to the client but if it does it will no doubt inform your estimation of the fee (or the appropriate interim bills) next time.

CHAPTER 6

'OLD' CONDITIONAL FEE AGREEMENTS AND ATE INSURANCE

INTRODUCTION

[6.1]

In this chapter we look at the issues that arise specifically in respect of CFAs and ATE insurance taken out prior to 1 April 2013 and which, as a result, generally involve 'additional liabilities' to be paid by the losing opponent. The comments in this chapter also apply to those discrete areas where claims can still be brought with a recoverable insurance policy – mesothelioma, privacy and publication, some insolvency and to a certain extent, clinical negligence claims. For simplicity, this chapter assumes that 'pre-commencement' means any date prior to 1 April 2013.

Prior to April 2000, additional liabilities were not recoverable and we have taken the view that it is unnecessary to cover the regime in place at the time. Indeed the number of CFAs which pre-date 30 November 2005 is also probably small but much of the case law comes in the period April 2000 to November 2005 and continues to apply to cases thereafter. In some areas, the relaxation of the regime in November 2005 changed the position significantly. We have taken the approach of discussing each issue in turn. If there is a substantial change in November 2005, it is highlighted. If there is no mention of a change, you can assume that the 'old' case law continued to apply up to 31 March 2013.

All of the following issues raised are woven together by the desire of the paying party – invariably a defendant liability insurer in a personal injury case – to persuade a court that, for one reason or another, the additional liability should not be recoverable. Some of the arguments raised have the effect of making all fees irrecoverable because of the indemnity principle (see the Golden Rule at [4.2]) and this has meant that the consequences of losing a generic point have driven a number of firms out of business.

The issues in this chapter revolve around:-
(a) the various legislative provisions that have applied during the period from 1 April 2000 to 31 March 2013
(b) compliance with those requirements, as considered by the courts
(c) professional conduct compliance
(d) the level of the success fees claimed as dealt with by the courts and by the CPR
(e) Challenges to the level of ATE premiums
(f) the giving of notice of the existence of the CFA or ATE policy.

The issues are dealt with in this order below.

COMPLIANCE WITH THE PRIMARY LEGISLATION

[6.2]

The relevant primary legislation is the Courts and Legal Services Act ('CLSA') 1990 as amended by the Access to Justice Act 1999.

Advocacy or Litigation Services

[6.3]

Section 58(1) applies to the provision of litigation services which are defined in s 119(1) as 'any services which it would be reasonable to expect a person who is exercising, or is contemplating exercising, a right to conduct litigation in relation to any proceedings, or any contemplated proceedings, to provide'. In *Gaynor v Central West London Buses Ltd* [2006] EWCA Civ 1120, [2007] 1 All ER 84, [2007] 1 WLR 1045 the retainer letter included the provision: 'If your claim is disputed by your opponent and you decide not to pursue your claim then we will not make a charge for the work we have done to date'. The claim was pursued and the client awarded her costs. Did this provision make the agreement a CFA and thus unenforceable because it did not comply with the CFA regulations? 'No' said the Court of Appeal. The work done before a decision was made not to pursue the claim was pre-litigation work which did not constitute the provision of litigation services under section 119(1). The solicitors were not exercising their right to conduct litigation and could not be said to be contemplating exercising that right until the potential defendant disputed the claim. Advising on the merits and writing a letter before action did not amount to litigation services. Therefore the agreement was not a CFA and the costs were recoverable.

It is possible to make either too much or too little of this decision. Clearly the use of the disputed provision does not render a retainer unenforceable. Although pre-proceedings work is by definition non-contentious business this does not prevent the parties entering into a CFA in respect of it at the outset. Most CFAs are entered into at the non-contentious business stage and many of them relate to claims which are settled without the issuing of proceedings (and which would retrospectively convert the work into contentious business (see Chapter 8)). The curiosity of this decision is that although the work in question was converted into contentious business by the subsequent commencement of proceedings it was not litigation services for the purposes of s 119(1). The Court of Appeal appears to have identified the new category of 'pre-litigation contentious business'. In the absence of a valid CFA the successful claimant cannot of course recover any success fee between the parties, nor may a success fee be recovered for non-contentious business which has not been converted into contentious business by the commencement of proceedings.

Section 58 also applies to the exercise of a right of audience and in *Fosberry v Revenue & Customs Comrs* [2006] EWHC 90061 (Costs) the parties conceded that the VAT Tribunal was a Court for the purposes of the section. It was held that the tribunal had granted a non-legal professional advisor a right of audience and the right to conduct litigation. Accordingly his fee arrangement was governed by s 58, and was, unhappily, an invalid CFA, a decision upheld on appeal to the High Court.

Pre-Action applications

[6.4]

The standard Law Society wording in its model CFA is as follows:

> "If on the way to winning or losing you are awarded any costs by agreement or Court order, then we are entitled to payment of those costs together with a success fee on those charges if you win overall."

Does the client have a liability to pay the costs of a pre-action application for disclosure based on this provision? 'Yes' said Master Haworth in *Connaughton v Imperial College Healthcare NHS Trust* [2010] EWHC 90173 (Costs) and there was therefore no breach of the indemnity principle in ordering the opponent to pay the costs of a pre-action application. A lay person's reasonable expectation would be that non-compliance with a pre-action protocol which was part of the pre-litigation process and necessitated an application to the court would be covered under the terms of this CFA.

A twist in this case was that the Claimant was not now proceeding against the Defendant, but against another party – the Defendant's cleaner. Did that mean the PAD application could not be 'within the claim' by virtue of the fact that proceedings had now been brought against another party? No. The fact that proceedings had not been issued against the Defendant, did not mean that the Claimant was not claiming against the Defendant in accordance with the definition of 'claim' within the CFA.

Similar arguments concerning pre-action work failed again in *Rogers v Mouchel* (unreported – 2010 Case No. OMY01382 HHJ Birtles). It was held that the words 'your claim' contained in the Law Society Model CFA were to be given a broad and purposive intention so as to include pre-action conduct even if no proceedings were ever issued.

Maximum success fee of 100%

[6.5]

The CFA Order 2000 (article 4) states that the maximum success fee that can be claimed in a CFA is 100%. The order is made in accordance with s 58(4)(c) of the CLSA 1990 and so a CFA which seeks more than 100% breaches the provisions of both the primary and secondary legislation.

In *Jones v Caradon Catnic Ltd* [2005] EWCA Civ 1821, [2006] 3 Costs LR 427 the Court of Appeal held that a statement in the risk assessment to a CFA of a 120% success fee was such a stark departure from the 100% maximum that the CFA could not be saved by a 100% cap elsewhere in the agreement even though neither party would be the loser. If this breach were held to be immaterial, all breaches would be held to be immaterial and the administration of justice would suffer. The robust language of the judgment was quoted with approval in *Garrett* (below).

Following swiftly on the heels of *Jones* was the case of *Oyston v Royal Bank of Scotland* [2006] EWHC 90053 (Costs) where the CFA provided for a 100% success fee and a £50,000 bonus if more than £1 million was recovered. A deed of variation that removed the reference to the bonus payment was then made to try to avoid invalidity. The senior costs judge held against the claimant. No retrospective rectification could be made once the substantive proceedings had concluded. The breach would inescapably have a materially adverse effect on the proper administration of justice and there should be no severance of the offending term from the remainder of the agreement.

In *Gloucestershire County Council v Evans* [2008] EWCA Civ 21, [2008] 1 WLR 1883, [2008] NLJR 219 the local authority had a Collective CFA providing for an hourly rate of £95 payable irrespective of outcome and a rate of £145 for a win. A success fee of 100% was to apply to the higher rate in the event of a win. It might be assumed that the percentage increase is an increase on base costs – model CFAs take that approach. But the statute talks not of the base costs you are charging in the CFA but a fee that you would have charged had there been no CFA at all – of course no one ever states what that fee would have been. The CFA in question referred to 'basic charges' (£145) and 'discounted charges' (£95) and applied the success fee only to the former.

It was argued that the success fee sought by the local authority really amounted to 290% based on the fact that the costs at risk were only £50 per hour. (Clearly even referring to the £95 rate would have left a success fee in excess of 100% anyway but that was not the argument run.) The Court held that the success fee applied to the basic charges and did not offend the statute. The agreement provided for basic charges of £145 per hour. That was the amount of the fees that would be payable if the agreement was not a CFA and so no more than 100% had been claimed in the CFA.

A similar result was reached in the SCCO case of *Morris v Dennis* [2008] EWHC 90112 (Costs) where the CFA defined the base charges to which a 100% success fee was applied plus an administration charge of £150. It was held that the charge (although only payable on success) was not part of the uplift which accordingly did not exceed the 100% maximum.

Commissioners for HMRC v Blue Sphere Global Ltd [2011] EWHC 90217 (Costs) provides a useful example of an assessment of the reasonableness of a success fee where only part of the solicitor's fees are at risk. The CFA provided for a discounted fee arrangement of two hourly fee levels, one payable in any event and a higher one if an appeal by HMRC failed. There was in addition a success fee of 80% based on the higher hourly rate. Applying *Gloucestershire County Council v Evans* the solicitors were only at risk for the difference between the discounted fee rate and the higher fee rate. That difference was £200 per hour – the higher rate being £500 per hour. The success fee had to reflect that limited risk and to apply a success fee to the higher rate fee would be wholly unreasonable. On the facts it was held that the risk was 50/50 which would of course justify a 100% success fee. To reflect the real risk of only £200 per hour the correct calculation was 0.5 x 200 divided by 500. That produced a success fee of 20% applied to the higher hourly rate.

Environmental Protection Act 1990, s 82

[6.6]

For the purposes of s 58(4)(a) of the CLSA 1990 provision for a success fee may be made in all enforceable conditional fee agreements except in proceedings under this provision of the Environmental Protection Act 1990. This is a halfway house provision since it is the only form of arguably 'criminal' proceedings where a CFA can be used at all. Someone must have thought that allowing a success fee would be one step too far.

COMPLIANCE WITH THE SECONDARY LEGISLATION

[6.7]

This mainly relates to the CFA Regulations 2000 and so we shall start with these. But even when things were made simpler, this did not prevent new challenges springing up.

CFA REGULATIONS 2000

Regulation 1

[6.8]

This regulation includes in the definition of 'client' a person who is liable to pay for litigation services under a CFA, for example an insurance company or trade union. The same regulation specifies that a 'legal representative' is the person providing the advocacy or litigation services – a definition of considerable importance when applying the requirements to provide information to clients, as was demonstrated in the *Sharratt v London Central Bus Co and other cases (No 2), The Accident Group Test Cases* [2003] NLJR 790, 147 Sol Jo LB 657, where TAG's use of unqualified representatives to carry out the obligations of the legal representative was challenged. The Court of Appeal held that the obligations can be delegated to an unqualified person ('outsourced'), provided there is adequate supervision by a solicitor. The issue of the adequacy of supervision has not yet been considered by a court.

Regulation 2

[6.9]

This regulation requires a CFA to specify 'the proceedings, or parts of them, to which it relates' and particularly 'whether it relates to any counterclaim or proceedings for enforcement'. This paragraph caused difficulty in *Roche v Newbury Homes* [2009] EW Misc 3 (EWCC) with respect to pre-action work. In *Blair v Danesh* [2009] EWCA Civ 516 the CFA referred to a claim for refund of monies given under undue influence but four years later the action ranged widely. Costs were awarded for the successful undue influence claim but it was argued that the CFA was invalid because the proceedings as they turned out were not specified. The Court of Appeal refused leave on that point seemingly because the costs had only been awarded for undue influence and the consumer protection purpose of the regulations had been satisfied.

The Law Society Model CFA includes an appeal brought against the client. The Court of Appeal in *Bexbes LLP v Beer* [2009] EWCA Civ 628, where BexBes had given notice before trial of a CFA which provided for a success fee, was required to consider whether notice of the CFA had been given in relation to the appeal. The conclusion, having consulted with the Senior Costs Judge, was that there was no separate requirement for giving notice of a CFA in an appeal and opponents and advisers should know that a standard CFA would continue if the opponent brings an appeal.

In *Brierley v Prescott* 2006 (SCCO Ref: 0504718) the CFA referred to a claim against an insurer. In the proceedings the driver's name was later substituted for that of the insurer. On assessment it was argued that there was no liability under the CFA in respect of the claim against the driver and that therefore the driver could not be liable to indemnify those costs. It was held that the CFA covered a claim arising out of the accident and costs were therefore payable by the client and recoverable. A similar decision was reached in *Scott v Transport for London* (December 2009 unreported) *(Hastings County Court)* where the CFA named a local authority and not the defendant. No harm was done because the CFA Regulations did not require the defendant to be named in the CFA.

Paragraph 2(1)(b) requires 'the circumstances in which the legal representative's fees and expenses, or part of them are payable.' The phrase 'no-win, no-fee' is an attractive sound-bite, but it is too simplistic. In the House of Lords, the Lord Chancellor warned that 'it would be wise' for lawyers and clients to discuss at the outset the likely proceeds from the enterprise and that 'it might be possible to meet the point by putting in some requirement about the situation that should obtain with regard to the net amount that the client could expect to recover as a measure of success in terms of the agreement'.

In *Milne v David Price Solicitors* 2005 (SCCO) 04/P8/340) the client agreed a settlement with his opponent that did not amount to a win which was defined in the CFA as settlement at a specified sum. His acceptance of the settlement was held to be a termination of the CFA with the result that he had to pay solely his base costs.

There are potential snags also once an opponent is bound to pay costs. The definition of 'win' given in the Law Society model CFA is as follows: 'Your claim for damages is finally decided in your favour whether by a court decision or an agreement to pay your damages.' In *Fortune v Roe* [2011] EWHC 2953 (QB), [2012] RTR 480, [2012] 2 Costs LR 288 the CFA in model form had been entered into after judgment had been entered with damages to be assessed. It was argued that the definition of win had therefore already been satisfied. It was held that the defining words required an agreement to have been reached as to the amount of damages and not merely that some damages will be paid.

The same wording was significant also in *Manning v King's College Hospital NHS Trust* [2011] EWHC 2954 (QB), [2011] NLJR 1634, [2012] 1 Costs LR 105. An original claim for clinical negligence brought by the widower of the patient of the defendant trust was taken over by the widower's executors upon his death. The widower had been represented by solicitors and leading counsel under a CFA which provided for the lawyers to take the risk on Part 36. New CFAs were entered into with the executors but the wording did not put the lawyers at risk on Part 36. The executor CFAs were made after trial but before judgment. Spencer J held that although the executor CFAs had in strict terms not put the Part 36 risk onto the lawyers, unlike the original CFAs, it was clearly the intention of the executors and the lawyers to have done so. Accordingly the CFAs did place the Part 36 risk onto the lawyers, and until judgment was given that risk remained. Given trial had already taken place and the trial judge had yet to make up his mind at the time of the widower's death, the risk was 50/50. You may think it quite a step for the court to have interpreted the new CFAs to have the same effect as the originals, even though the necessary words were not there at all.

In *Hanley v Smith and MIB* [2009] EWHC 90144 (Costs) the CFA was made two years after a standard retainer and by then the first defendant had admitted liability albeit not its extent. It was argued that the definition of win in the Law Society model had already been satisfied. That argument failed with the court holding that the claim against the MIB was an essential ingredient of the claim, a reasonable person would have understood that the central purpose of the litigation would not be satisfied merely by obtaining a worthless (but necessary) judgment against the driver. A win though, (however defined) is surely not enough. What if none, or only part of, the damages awarded and between the parties' costs are recovered because, for example, the defendant is, or becomes, insolvent? Unless the formula is 'No recovery, no pay' the empty-handed client will have nothing to pay with. The lawyers will be entitled to their costs and success fee, so the client will, in fact, be substantially worse off than if he had lost the action.

Also, what effect is a successful counterclaim, perhaps arising out of an allegation of contributory negligence, to have on the specified circumstances? The result in *Horth v Thompson* [2010] EWHC 1674 (QB), [2011] 3 Costs LO 249, is not for the faint-hearted. The defendant pursued a counter-claim in respect of a collision. The defendant was found to be 65% responsible. His own damage was far greater than that of the claimant but he was still over £300 down after apportioning the blame. The judge ordered each party to pay the other's costs of the entire claim and counterclaim combined. The defendant was represented under a CFA so his costs were seriously greater than those of the claimant. The defendant who was far more to blame and had received more damage than he had himself caused the claimant ended up receiving costs from the claimant.

The Law Society's Model CFA defines a win in terms only of a final decision on the claim in the client's favour. A win might be defined in terms of actual recovery and the Senior Costs Judge expressed the view in *Sharratt v London Central Bus Co and other cases (No 2), The Accident Group Test Cases* [2003] NLJR 790, 147 Sol Jo LB 657 that a standard CFA could in any event define win in terms of recovery. The issue was put beyond doubt in 2003 when simplified CFAs were allowed by the Conditional Fee Agreements (Miscellaneous Amendment) Regulations 2003(see below).

Regulation 3

[6.10]

This regulation contains the extra requirements for CFAs which provide for a success fee and these can be tricky. The agreement must give brief reasons for setting the success fee at the agreed percentage. What if the reasons are that the client is extremely demanding or difficult, or that he and his witnesses are of doubtful credibility? That will require tactful wording, but you cannot afford to be too tactful because the reasons given will have to be relied upon if the percentage increase is challenged.

One use for the success fee that was not immediately apparent was as a compensatory mechanism for being kept out of any money (usually) until the end of the case. This 'postponement' element is not recoverable according to CPR 44.3B(1)(a), unlike the risk element. Consequently a differentiation between the two was required (reg 3(1)(b)) so that the client would know how much he was liable for if he was successful in his case. (This issue is also discussed in terms of drafting CFAs at [4.22].)

Following *Bray Walker Solicitors (a firm) v Silvera* [2010] EWCA Civ 332, [2010] 15 LS Gaz R 18, 154 Sol Jo (no 13) 29 it appears that 'brief reasons' can be very brief indeed. In that case merely stating that the reasons were the unspecified sufficed. The client must consent in the agreement to the reasons being disclosed if required by the court, which, of course, it will do if there is a dispute between the parties. It was held in *Brennan v Associated Asphalt* 2006 (SCCO 0507905) that failure to specify that the postponement element was nil was a breach of reg 3. On the basis however that no postponement fee could therefore be charged there was no materially adverse effect on the protection of the client. Nor was there a materially adverse effect on the administration of justice. A similar result was arrived at in *Hughes v George Major Skip Hire* [2009] EWHC 90147 (Costs) where the client had been certain from the solicitor's explanation of the CFA that she would not be liable for any costs not recovered from the defendant. The CFA had a confusing reference to the postponement element such that regulation 3 had been breached but no material adverse effect was shown because of the client's certainty from the explanation given.

Also under 'brief reasons' the agreement must provide that any between the parties' reduction of the success fee shall be binding on the solicitor unless the court orders otherwise. The court will only 'order otherwise' if the solicitor can show that there were exceptional circumstances. The conflict between a solicitor and his client could not be more starkly illustrated. The client is unlikely to have been separately advised before entering into the agreement; will he be separately advised when the solicitor seeks to recover the shortfall? For the procedure see the old CPR 44.16. To the best of our knowledge, such an application has never been made. The potential for a conflict is stark but the prospect of claiming a percentage where successful because of some problem with the client is too unattractive for most to contemplate. (If the judge has assessed a lower than claimed success fee on the case generally, he is unlikely to think that the client should pay any of the unreasonable element in the absence of any blame on the client's part.)

Regulation 4

[6.11]

This regulation sets out the information to be given to clients before they enter into a CFA. The regulation contains the unusual requirement that four elements of information, whilst they may be given in writing, must also be given *orally*: what is a win; the right to a detailed assessment; has the client pre-purchased legal expenses insurance and what other methods of financing the costs are available. The solicitor's views and advice on any particular method of funding and insurance together with an explanation of the effect of the agreement and a disclosure if there is an interest in recommending the policy must be given to the client both *orally* and *in writing*. There must clearly be some careful drafting and record keeping (perhaps even literally an audio recording of the oral advice, or perhaps a tape recording to be played to each client) to ensure that compliance with these provisions could be proved when necessary.

An accurate oral explanation failed to cure an inaccurate written one in *Bevan v Power Panels Electrical Systems Ltd* [2007] EWHC 90073 (Costs). All of this must be given by the 'legal representative' which by Regulation 1 means the 'person' providing the advocacy or litigation services. In *The Accident Group Test Cases* the explanations were given by non-legally qualified persons not directly employed by the solicitor. The Court of Appeal held that the regulations did permit delegation, the important safeguard being supervision by the solicitor.

Regulation 4 has been a fertile ground for paying parties to unearth 'technical challenges' demonstrating non-compliance. The consequences of failing to comply with any of the regulations are dire and are discussed at [4.2].

If a CFA has to be replaced by a subsequent CFA the regulations will need to be complied with in respect of the replacement. The failure to explain to the client the effect of a replacement CFA was fatal to its validity in *Langsam v Beachcroft LLP* [2011] EWHC 1451 (Ch), [2011] 3 Costs LO 380, resulting in the solicitors being unable to enforce the CFA against their own client. A quantum meruit claim also failed on the basis that it would significantly undermine the operation of CLSA 1990, s 58 if a solicitor who is unable to claim his fees for the legal services provided because of material non-compliance with the statutory requirements could nonetheless recover payment for those services from the client on the basis of a quantum meruit claim.

Regulation 5

[6.12]

This regulation requires the agreement to be signed by both parties unless it is a CFA between one legal representative and an additional legal representative. This usually means counsel although it includes any other person providing advocacy or litigation services such as a solicitor advocate or agent. Following the draconian decision in *Garrett v Halton Borough Council* [2006] EWCA Civ 1017, [2007] 1 All ER 147, [2007] 1 WLR 554 failure of one or both parties to sign will be regarded as a material breach. In *Fenton v Holmes* [2007] EWHC 2476 (Ch), [2008] 2 Costs LR 238, the client had not signed and in *Preece v Caerphilly* (2007) Cardiff CC 15/8/2007 the solicitors had not signed. In each case the CFA was invalid. In *Dunn v Mici* [2008] EWHC 90115 (Costs) the CFA stated that the agreement covered Mr Dunn's claim for damages and personal injury 'by his mother and litigation friend'. The issue for decision was whether Mr Dunn alone was the client; if so, did he receive advice which complied with reg 4; if not and both he and his mother were clients, what was the effect on the enforceability of the CFA if one had received compliant advice but the other had not? Master Campbell held that no formal appointment of a litigation friend under CPR 21 was ever made and therefore

Mrs Dunn was not, and could not, have been the client. As to the reg 4 enquiry regarding legal expense insurance, the solicitors discharged the tasks which the regulations required them to undertake in relation to Mr Dunn, there being no requirement that these should be repeated in relation to Mrs Dunn.

CFA (MISCELLANEOUS AMENDMENTS) REGULATIONS 2003

Simpler CFAs

[6.13]

From 2 June 2003 these regulations amended the CFA Regulations 2000 by inserting reg 3A into them to provide that a CFA is still enforceable even though the client is liable to pay his legal representative's fees and expenses only if, and to the extent that, he recovers damages or costs in the proceedings. These are sometimes referred to as a 'CFA Lite'. Amendments made to the CPR provide that costs payable under such a CFA are recoverable under the old Parts 44 to 48. Thus the indemnity principle is in effect abrogated in relation to this type of CFA.

Simplification was achieved because regs 2, 3 and 4 of CFA Regulations 2000 did not apply to simplified CFAs, although they must still state the circumstances in which fees and expenses are payable – they must define a 'win'. Brief reasons for the percentage success fee must still be given and the client should in the agreement waive his privilege in those reasons should the court order disclosure. Regulation 3A(5) permits the solicitor to include a provision imposing liability on the client for fees and expenses in specified circumstances of non-cooperation or misbehaviour by the client or the death or insolvency of the client.

Simplified Collective CFAs

[6.14]

The CFA (Miscellaneous Amendments) Regulations 2003 also introduced a simplified version of a CCFA which restricts own costs liabilities to any sums recovered. Where a CCFA is being used by a membership organisation it is probable that the member is intended to retain all his damages even under a standard CCFA. The difference is that this regulation means the agreement can actually say so, but it is subject to the same wording as the simplified individual agreement 'sums are recovered . . . whether by way of costs or otherwise'.

Costs or otherwise

[6.15]

A client's liability for his own solicitor's 'fees and expenses' under a simplified CFA is limited to 'sums recovered . . . whether by way of costs or otherwise' – not just to 'costs' recovered. The limitation does not include the client's liability for any ATE insurance premium. Problems arise with the wording 'or otherwise'. These envisage circumstances in which a solicitor deducts his solicitor and client costs from the damages. Their inclusion was influenced by the belief of Lord Phillips MR that they were needed to produce a logical scheme. He said that as a matter of principle there was no reason why one should allow this kind of agreement and then limit the client's liability to costs recovered. It is still leaving the client better off than having a liability to pay the solicitor simply because the claim succeeded, even though nothing is recovered where, for example, damages are recovered but not costs. Is that what is intended and, if so, does the client understand that? If the damages are intended to be protected, the CFA should use the word 'costs' only, provided the solicitor is willing to write off the disbursements if they are not recovered.

Ten years after these regulations came into force (and eight years since they were revoked) LASPO 2012 has brought the concept of clients paying costs from their damages to the fore.

Part 36 offers

[6.16]

Where the client has won, the simplified CFA rules allow a solicitor to say to the client that he will keep his damages. But that does not deal with the situation where the client has lost or has failed to beat a Part 36 offer. In these situations the simplified CFA produces a very different result. Where the client loses and recovers nothing he may have no liability at all, not for fees, not for disbursements and not for counsel. Strictly speaking, if he has no liability at all he probably has no insurable interest on which to base an ATE insurance policy – therefore the whole risk lies with the solicitor. In practice, many ATE insurers would countenance this sort of arrangement. Where there is a failure to beat a Part 36 offer, some damages will still be recovered and the client can therefore be liable for his own solicitor's costs and disbursements up to the amount of the sums recovered. Presumably that amount is what will be left after the client has paid his opponent's costs since the Part 36 offer was made. If it is intended to protect the client's damages, the client's liability for his opponent's costs after the Part 36 offer must be insured against under a policy which covers such costs without setting off the pre-Part 36 offer from own costs recovered. Otherwise any shortfall must either come out of the client's damages or be borne by his solicitor. The regulations may be simpler, but explaining them to a client is certainly not. The arrival of Qualified One-way Costs Shifting ('QOCS') does not affect these agreements because QOCS does not apply to pre-commencement CFAs.

CPR 43.2(3) and (4) provide that costs whose recovery is limited by a simplified CFA are nevertheless recoverable from an opponent for the purposes of the old CPR Parts 44 to 48.

COLLECTIVE CONDITIONAL FEE AGREEMENTS

CCFA Regulations 2000

[6.17]

Collective conditional fee agreements (CCFAs) were a response to concerns expressed during the passage of the Access to Justice Act 1999 that the individual CFA regime was not administratively suitable for the bulk purchase of legal services. In particular commercial organisations and membership organisations such as the AA and RAC and the trade unions were not able to use the existing CFA provisions because of the practical and physical difficulties of administering the rules on an individual basis. As with individual CFA regulations, the CCFA Regulations 2000 were revoked from 1 November 2005.

The CCFA Regulations 2000 made provision for all CCFAs irrespective of the context in which they were to be used. The regulations applied to bulk purchasers of legal services, such as the legal department of a multi-national company, and to bulk providers of legal services such as the legal representatives retained by a trade union to act for its members. There is a crucial difference between these two categories in that the bulk purchaser is also the client, whereas the bulk provider will involve numerous clients who are not funding the litigation. The regulations, however, apply consumer protection provisions to all CCFAs, including those where the client is also the funder. The regulations are not set out so as to have provisions applying according to whether the CCFA is for a bulk purchaser or not.

Indemnity principle challenges

[6.18]

There are no provisions in the CPR for CCFAs and no changes have been made following the commencement of s 31 to disapply the indemnity principle. There was therefore a question as to the validity of CCFAs where the party to the proceedings was not also the funder. Two explanations of CCFAs emerged from costs assessments. In *Gliddon v Lloyd Maunder Ltd* [2003] NLJR 318 Master O'Hare took the view that the CCFA Regulations 2000 did not abrogate the indemnity principle. He found, however, that there was a link between the individual client and the solicitor such that there was a retainer between them. In *Thornley v Lang* [2003] EWCA Civ 1484, [2004] 1 All ER 886, [2004] 1 WLR 378, the Court of Appeal reviewed the line of authorities dealing with trade union funding and held that the member here was liable on the basis of the CCFA. It reached that result by alternative routes. Either the union agreed with the authority of its member or the member ratified the union's agreement. On either footing there was a contract under which the member was liable and that contract was a CCFA. The Court expressly rejected the argument that the member had entered into a CFA on an individual basis which would be subject to the CFA Regulations.

Several challenges to a trade union CCFA were made in *Duffy v Port Ramsgate* (2004) SCCO (SCCO JOH 0306901). It was argued that because the member's liability was, under the terms of the CCFA, limited to the liability of the losing opponent the whole agreement was invalid – an indemnity principle argument. It was also argued that a member represented under such a CCFA still required the same explanation as to costs under Rule 15 of the Law Society Rules as a private client. Both arguments were rejected. The CCFA also covered work done on the case before 30 November 2000 when CCFAs were introduced. No success fee was claimed in respect of the earlier work and the court rejected the argument that this earlier work had been conducted under an invalid agreement. The challenge in *Various Claimants v Gower Chemicals* (2007) SCCO CW05054225 failed where Field J sitting in the county court held that Regulation 5(1) is complied with where the CCFA contains a requirement to prepare and retain a risk assessment – actual performance is not a further requirement.

As with the individual version, a simplified CCFA can still impose an own costs liability where the client fails to co-operate, attend a medical or expert examination or court hearing, fails to give instructions or withdraws instructions and upon the death or insolvency of the client. The burden of explanation in the case of CCFAs was always rather less than for individual CFAs and no changes were made to the level of explanation required.

The revocation of the Regulations from 1 November 2005 means that a CCFA made after 1 November 2005 need only comply with Courts and Legal Services Act 1990, s 58 and it can of course be worded to achieve the result intended by the simplified regulations. On the basis of *Duffy* rule 15 of the Professional Conduct Rules (or indeed rule 2 of the SRA Code of Conduct for later cases) plays no significant role, at least where the CCFA relates to a Trade Union and the realistic position is that the member receives free legal services.

MEMBERSHIP ORGANISATIONS

Access to Justice (Membership Organisations) Regulations 2000

[6.19]

These regulations came into force on 1 April 2000 and defined the bodies which are prescribed for the purpose of AJA 1999, s 30. They are as able to recover a sum as part of legal costs from unsuccessful opponents to reflect the provision of legal help for members and their families. The definition is also contained in the Access to Justice (Membership Organisations) Regulations 2005.

Limit on amount claimed

[6.20]

By s 30 of the Access to Justice Act 1999 and the Access to Justice (Membership Organisations) Regulations 2000, provision is made for the recoupment of a sum no greater than the equivalent of the cost to a member of taking out a personal insurance policy covering adverse costs only. This is imported into the CPR by Rule 44.3B(1)(b). It states that any provision for an

6.20 *Chapter 6 'Old' Conditional Fee Agreements and ATE insurance*

additional liability by a membership organisation which exceeds the likely cost of a commercial ATE insurance premium shall not be recoverable between the parties.

Access to Justice (Membership Organisations) Regulations 2005

[6.21]

These regulations replaced the 2000 Regulations from 1 November 2005. This regulatory scheme still leaves the organisation responsible for the administration of the litigation and for ensuring that an agreement with the member exists which can give rise to the recoupment of the costs. It also leaves own costs in lost cases with the organisation.

CCFAs, ATE and Membership Organisations

[6.22]

A CCFA can be used to transfer the own costs risk to its lawyers who can seek to cover that risk by success fees in successful cases. A membership organisation can also make use of an ATE policy as an alternative to going along the s 30 route. Alternatively, on the basis of *Thornley v Lang* (above) the individual has a costs liability for his own costs as well as his opponent's costs and could therefore insure by use of an individual ATE policy and seek then to recover the premium under s 29. Such a route deals with own disbursements where the case loses, s 30 does not. It is possible that a membership organisation may fund the ATE premium or negotiate deferred premiums. If the organisation were to take out its own insurance the premium would be the 'provision' made by the organisation but that would lead to seemingly insurmountable problems in apportioning any part of such provision to an individual case. Nevertheless, increasingly s 30 has been abandoned in practice in favour of a CFA with ATE insurance.

COURT APPROACH TO LEGISLATIVE COMPLIANCE

'Technical challenges'

[6.23]

In view of the Golden Rule (see **[4.2]**) paying parties wasted no time in challenging CFAs as soon as success fees became recoverable in April 2000. The two Court of Appeal decisions in *Callery v Gray* in 2001 together with *Sarwar v Alam* (see below) later the same year gave the higher courts a taste of the warfare being conducted in courts of first instance. Seemingly endless allegations of non-compliance with the legislation came in 2002. In *Burstein v Times Newspapers Ltd* [2002] EWCA Civ 1739, [2003] 03 LS Gaz R 31, (2002) Times, 6 December Latham LJ ended the judgment of the court with these words:

' . . . The deputy costs judge is to be commended for ensuring that the detailed assessment did not become an excuse for further expensive litigation at the behest of a disappointed but persistent litigant. Satellite litigation about costs has become a growth industry, and one that is a blot on the civil justice system. Costs Judges should be astute to prevent such proceedings from being protracted by allegations that are without substance. In future district judges and costs judges must be equally astute to prevent satellite litigation about costs from being protracted by allegations about breaches of the CFA Regulations where the breaches do not matter. They should remember that the law does not care about the very little things, and that they should only declare a CFA unenforceable if the breach does matter and if the client could have relied on it successfully against his solicitor.'

So by the time the case of *Hollins v Russell* [2003] EWCA Civ 718 reached the Court of Appeal, the judiciary were fully aware that a war was being conducted, particularly in the personal injury field. The court in *Hollins* (and several conjoined cases) considered several alleged breaches of the secondary legislation. The defendants' arguments were described by the court as being 'as unattractive as they were unmeritorious' and held that a departure from the regulations did not affect the validity or enforceability of a CFA unless the departure was both material and adversely affected the client or the administration of justice. None of the challenges considered by the Court of Appeal were successful. The court summarised its 228-paragraph judgment as: 'The court should be watchful when it considers allegations that there have been breaches of the regulations. The parliamentary purpose is to enhance access to justice, not to impede it, and to create better ways of delivering litigation services, not worse ones. These purposes will be thwarted if those who render good service to their clients under CFAs are at risk of going unremunerated at the culmination of the bitter trench warfare which has been such an unhappy feature of the recent litigation scene.'

Rightly or wrongly the profession interpreted *Hollins* as heralding the end of technical challenges unless the client had suffered some material detriment. The euphoria did not last long.

In *Spencer v Gordon Wood t/a Gordons Tyres (a firm)* [2004] EWCA Civ 352, 148 Sol Jo LB 356, [2004] 3 Costs LR 372, reg 3(1)(b) was breached because the CFA failed to specify how much of the 75% success fee related to postponement. There was a separate risk assessment on file which included an item 'Deferment of costs until conclusion of case 50%', but, as the circuit judge commented, '50% of what? Of the success fee? Of the profit costs? How is the postponement charge reflected in the overall success fee of 75%? The risk assessment is silent as to that, as are the agreement and the schedule.' The Court of Appeal agreed. The breach materially affected the client because he did not know how much of the 75% would be recoverable from the defendant and the punishment (no fees at all) was not objectionable – the words 'shall be unenforceable' meant what they said.

A CFA using the Law Society Model caused similar problems but a different outcome in *Hughes v George Major Skip Hire* [2009] EWHC 90147 (Costs). The CFA referred to a 50% success fee. Schedule 1 to the CFA stated that the postponement element was 50%, that the risk elements amounted to 50% and that the total success fee was 50%. That confusion and breach of reg 3 was rescued by the oral explanation that left the client certain that they would not be paying any costs out of damages so there was no material breach. (This oral clarification should be distinguished from oral terms of a CFA which will inevitably render it unenforceable since it needs to be in writing.)

Burnton J in *Re Palmier plc (in liquidation); Sandhu v Sidhu* [2009] EWHC 983 (Ch), upheld a CFA that stated there was no charge for postponement although the schedule containing the risk assessment made no reference to postponement. That absence of postponement in the schedule made it clear that there was no charge for that element.

In *Crook v Birmingham City Council* [2007] EWHC 1415 (Admin), [2007] NLJR 939, [2007] 5 Costs LR 732 the solicitor agreed with a number of clients of modest means in housing litigation that if they recovered damages of £3,000 or less but no costs they would be charged at the solicitors' usual hourly rate with a cap of £1,000. The defendants contended that the reduced level of fees represented what the market would bear and therefore the increment amounted to a success fee. It was held that there was no success fee but simply a discount from the 'normal fee'. If the defendant's analysis were right it could be argued in every case and would obliterate the distinction in conditional fee agreements between the 'base fee' and the 'success fee'.

Disclosure

[6.24]

The Court in *Hollins* said that it should become normal practice for a CFA to be disclosed for the purpose of costs proceedings in which a success fee is claimed, subject to the provision in paragraph 40.14 of the Costs Practice Direction that the judge may ask the receiving party to elect whether to disclose the CFA to the paying party in order to rely on it or whether to decline disclosure and instead rely on other evidence (see [**28.45**]). If the CFA contains confidential information relating to other proceedings, it may be suitably redacted before disclosure takes place. A party given that option who then consistently refuses to provide the CFA or sufficient other evidence of compliance will suffer the consequences of failing to recover costs, the exact outcome in *David Smith v Countrywide Farmers plc* (unreported 11 February 2010 QBD Cardiff District Registry).

Attendance notes and other correspondence should not ordinarily be disclosed, but the judge conducting the assessment may require the disclosure of material of this kind if a genuine issue is raised. A genuine issue is one in which there is a real chance that the CFA is unenforceable as a result of failure to satisfy the applicable conditions (*Pratt v Bull, Worth v McKenna* [2003] EWCA Civ 718.

Since the revocation of the 2000 Regulations from 1 November 2005, a CFA needs only to be in writing (seemingly it need not be signed) but that did not stop an application for disclosure in *Ashley Cole v News Group* (2006) October 18 SCCO. That application failed simply because no points of dispute had been served hence CPR 47.14 and CPD 40.14 did not apply. The applicability of *Hollins* to post 1 November 2005 CFAs was considered in *Findlay v Cantor Index Ltd* [2008] EWHC 90116 (Costs) where disclosure was sought of the CFA, the reasons for the success fee and counsel's opinion. Master Campbell held that Costs Practice Direction 32.5 could have no application to post 1 November 2005 CFAs (where there was no requirement to state reasons for the success fee) and thus the paying party has no right to information concerning the setting of the success fee, and presumably no proof other than the solicitor's assertion as to what it was. Nonetheless, arguing for a success fee without showing how it was set is likely to be a fraught endeavour. As to the CFA, *Hollins* still required it to be disclosed at the costs stage although counsel's opinion was privileged despite being referred to in a disclosed risk assessment. This gap was closed by amendments to the Costs Practice Direction in 2009 so that a statement of reasons for the success fee together with either the CFA or at least certain parts of the CFA (regarding the definition of a 'win'; the payments provisions following a Part 36 Offer etc.) has to be disclosed when setting the case down for a detailed assessment hearing.

Alternative forms of funding

[6.25]

The Court of Appeal heard *Sarwar v Alam* [2001] EWCA Civ 1401, [2001] 4 All ER 541, [2002] 1 WLR 125 on September 11, 2001. It was mentioned by the court at the end of *Callery v Gray (No 2)* as being a case which was in the process of being fast tracked to the Court of Appeal. Whilst it was reconsidered in detail by the case of *Myatt* five years later, it is still of importance in considering whether the use of an ATE policy could be seen as reasonable rather than an existing BTE policy.

Lord Justice Brooke provided the following helpful summary:

> 'This is another appeal concerned with the new arrangements for financing the costs of personal injury litigation which came into effect last year. Legal Aid is now no longer available for most litigation of this type. In *Callery v Gray*, the Court of Appeal was concerned two months ago with issues relating to the appropriate size of a success fee in a conditional fee agreement made in connection with a small claim for personal injuries suffered in a road traffic accident which was settled quite quickly without any need to bring court proceedings. In that case a passenger in a car had made a claim against the driver of the car involved in an accident. The court was also concerned with the appropriateness of taking out 'after the event' ('ATE') insurance in connection with such a claim, and the reasonableness of the ATE premium claimed in that case.

> The present appeal is concerned with a similar claim brought by a passenger against the driver of the car in which he was travelling. The court below had disallowed the recovery of an ATE premium on the grounds that the claimant ought to have enquired into the availability of 'before the event' ('BTE') legal expenses insurance which formed part of the cover provided by the driver's insurance policy, and then made use of that cover. This policy covered the costs and expenses of both sides in a claim brought by a passenger in the car against the insured driver himself up to a limit of £50,000.

> The Court of Appeal allowed the appeal on the grounds that the policy did not provide the claimant with appropriate cover in the circumstances of this case. Representation arranged by the insurer of the opposing party, to which the claimant had never been a party, and of which he had no knowledge of the time it was entered into and where the opposing insurer through its chosen representative reserved to itself the full conduct and control of the claim, was not a reasonable alternative to representation by a lawyer of the claimant's own choice, backed by an ATE policy.

> The court suggested that the position might be different if BTE insurers financed some transparently independent organisation to handle such claims, and made it clear in the policy that this is what they were doing.
>
> The court said, however, that if a claimant making a relatively small (ie less than about £5,000) claim in a road traffic accident had access to pre-existing BTE cover which appeared to be satisfactory for a claim of that size, that in the ordinary course of things he/she should be referred to the relevant BTE insurer.'

The court gave guidance as to the nature of the enquiries a solicitor should make in this class of case into the availability of BTE cover and the insurance policies and other documents the solicitor should ask the client to produce. A solicitor should normally invite a client, by means of a standard form letter, to bring to the first interview any relevant motor and household insurance policy, as well as any stand-alone LEI policy belonging to the client and/or spouse or partner. However, regard has always to be had to the amount at stake, and a solicitor was not obliged to embark upon a treasure hunt. The court emphasised that this guidance should not be treated as an inflexible code, and that the overriding principle was that the claimant, assisted by his/her solicitor, should act in a manner that was reasonable. Now that this sort of claim would go into the Portal, it must be arguable that the extent of the 'treasure hunt' is now limited by the fixed recoverable costs involved.

In *Samonini v London General Transport Services Ltd* [2005] EWHC 90001 (Costs) the client said he had no existing legal expenses cover and the solicitor took his word for it without further enquiry. The client was correct — he had no BTE and however diligently the solicitor had searched he would have found nothing because there was nothing to find. Nevertheless, the Senior Costs Judge held that the solicitor's failure to comply with reg 4(2)(d) (which required him to 'consider' the availability of legal expenses insurance) was a material breach invalidating the CFA.

Hughes v Newham London Borough Council 2005 SCCO 0502314 was a housing disrepair case where legal aid was available and the client would have been eligible. The failure of the solicitors to consider legal aid was a material breach of reg 4(2)(d) fatal to the CFA.

In *Garrett v Halton Borough Council; Myatt v National Coal Board* [2006] EWCA Civ 1017, [2007] 1 All ER 147, [2007] 1 WLR 554, the Court of Appeal gave what it described as 'guidance' on CFAs concerning a solicitor's duty to consider existing legal expenses insurance under CFA reg 4(2)(c). However, as the regulation had been revoked on 1 November 2005, the decisions were not so much guidance but more of a checklist for those wishing to embark on satellite litigation in respect of CFAs that had been created prior to that date. The Court held that enforceability of a CFA was to be determined as at the date of its commencement and not in the light of its consequences. It was not necessary for there to be any actual material detriment to the client or to the administration of justice to constitute a breach. The language of s 58 CLSA 1990 was clear and uncompromising. The statutory scheme provided that if any of the conditions were not satisfied, the CFA would not be enforceable and the solicitor would not be paid. That was clear and stark. Such a policy was tough but not irrational: it was designed to protect clients and to encourage solicitors to comply with the statutory requirements. *Hollins* had done no more than deal a fatal blow to challenges which were based on literal but trivial grounds and immaterial departures from the statutory requirements.

In *Myatt* the solicitor asked the wrong question in her attempt to comply with reg 4(2)(c). She should not simply have asked unsophisticated clients whether they had credit cards, household or motor insurance policies or trade union membership which would entitle them to legal expenses insurance in respect of the contemplated claim. The solicitor is required to take reasonable steps to ascertain whether the client's risk for costs is already insured and those steps will depend on a variety of circumstances such as the nature of the client, the circumstances in which the solicitor is instructed, the nature of the claim, the cost of the ATE premium may be a relevant factor and whether a referring body has already investigated the availability of BTE. The requirement in *Sarwar v Alam* that the client should be invited to bring all relevant policy and other documents to the first interview should be treated with considerable caution in high-volume low-value litigation – in which the solicitor might never have a first interview. In any event, *Sarwar* related not to reg 4 but to the reasonableness of entering into a CFA or ATE insurance.

Disclosing an interest

[6.26]

In *Garrett* the solicitors were members of a panel which referred work to them provided they recommended a particular policy of insurance. This was clearly an interest disclosable under reg 4(2)(e) and it was not sufficient merely to inform the client of membership of the panel without explaining its implications. Further consideration was given to reg 4(2)(e) in the context of the Accident Line Protect (ALP) insurance scheme. In *Tankard v John Fredricks Plastics Ltd* [2008] EWCA Civ 1375, [2009] 4 All ER 526, [2009] 1 WLR 1731, the court said a solicitor has an interest if a reasonable person with knowledge of the relevant facts would think that the existence of the interest might affect the advice given by the solicitor to his client. The ALP scheme did not involve such an interest. The court went on to consider, albeit obiter, what level of disclosure would be required where a solicitor did have an interest. The client should have been told what the nature of the interest is – it was not enough just to say that there was an interest. The regulation says the solicitor must when recommending insurance inform the client 'whether he has an interest in doing so'. Once the solicitor has an interest, using the test set out above, he must explain the nature of that interest to the client. The court in *Garrett* did not dissent from its decision in *Hollins* that it is not necessary to state that a solicitor has no interest in the insurance he is recommending – it is only necessary to state when he has.

Challenges after *Garrett* and *Myatt*

[6.27]

In the light of the Court of Appeal's guidance, it seemed inevitable that the spate of satellite costs litigation would continue unabated for so long as the indemnity principle survived. It did survive and there was no abatement for some time. A robust approach was taken in *David Smith v Countrywide*

6.27 *Chapter 6 'Old' Conditional Fee Agreements and ATE insurance*

Farmers plc (11 February 2010, unreported), QBD Judge Seys-Llewellyn QC sitting as a deputy judge of the High Court declaring that it was in principle and in practice wrong for a judge to trawl through a CFA to check compliance. Here is a sample of the cases on breach of the regulations. In *Bevan v Power Panels Electrical Systems Ltd* [2007] EWHC 90073 (Costs) both *Garrett* and *Myatt* points were taken. *Myatt* also tripped up solicitors in *Andrews v Dawkes* (2006) (QBD Birmingham) where the solicitors did not check that their client was unlikely to have access to before the event insurance (BTE). In *Berry v Spousals* (2007) Birmingham CC BM 309007 solicitors did not at any time check claims management company arrangement to see if it would cover costs rather than use a CFA. *Myatt* featured in *Choudhury v Kingston Hospital Trust* [2006] EWHC 90057 (Costs) but there the client consultant anaesthetist wrote to the solicitors confirming she had no BTE – a sophisticated client answering the *Myatt* ultimate question. The same result occurred in *Kashmiri v Ejaz* [2007] EWHC 90074 (Costs) with a commercial client able to provide all the *Myatt* answers himself. BTE has caused particular difficulties in cases involving buses with no consistency in the lower courts as to whether a solicitor before 2005 could have been expected to check the bus company's insurance to see if it covered injured passengers. Paying parties should consult *Tranter v Hansons* [2009] EWHC 90145 (Costs), receiving parties *Dole v ECT* (17 September 2007, unreported). Most recent is the successful own client challenge in *Langsam v Beachcroft LLP* [2011] EWHC 1451 (Ch), [2011] 3 Costs LO 380, where a replacement CFA did not contain the cap on total fees that the original CFA had contained. The failure to explain the change was a breach of Regulation 4(3) and the CFA was unenforceable.

Membership of Accident Line was the basis of a *Garrett* challenge in *Myers v Bonnington* [2007] EWHC 90077 (Costs). The firm had received 24 referrals over six years – a de minimis factor and not enough to amount to an interest. Finally, in *O'Driscoll v Liverpool City Council* (2007) (Liverpool County Court) there had been no 'positive averment' that solicitors would lose TAG panel membership if their client had not taken a policy with TAG, and therefore no basis for an inference to be drawn as that in *Myatt* that the solicitors had an interest in recommending an insurance policy.

PROFESSIONAL CONDUCT COMPLIANCE

[6.28]

The transferring of regulation to the SRA from the secondary legislation managed to put an end to costs satellite litigation between the parties. The paying party briefly sought to argue that a breach of the code of conduct, since it has statutory force, meant that the client could still contend that the retainer was unenforceable (or at least that the solicitor's costs entitlement is reduced). Consequently, the paying party ought to be able to rely on the indemnity principle to avoid liability for payment. These arguments were addressed in *Garbutt v Edwards* [2005] EWCA Civ 1206, [2006] 1 All ER 553, [2006] 1 WLR 2907, where the Court of Appeal confirmed that the practice rules and care code did have statutory effect, but held that the particular breach in this case did not make the agreement unenforceable. In terms of providing costs information such as an estimate, the word 'shall' in Rule 15 was not mandatory. The sanctions for non-compliance were disciplinary and a client could not be penalised for not initiating disciplinary proceedings against his own solicitor. Between the parties the costs judge should consider whether and, if so, to what extent the costs claimed would have been significantly lower had an estimate been given. The judgment contains clear comments that the court did not want to see paying parties seeking to take advantage of any such 'technical' points based on the SRA code of conduct.

Level of success fee – court decisions

[6.29]

In *Callery v Gray* [2001] EWCA Civ 1117, [2001] 3 All ER 833, [2001] 1 WLR 2112, the court dealt with the defendants' argument that a CFA should not be entered into until a risk assessment could be carried out by the claimant in possession of (at least most of) the facts by turning the argument on its head. Instead of there being little or no success fee recoverable until the risk assessment could be carried out, the court took the view that a higher success fee should be set with the facility for that fee to be reduced if the case settled early, or at least the risk reduced (which generally would mean the admission of liability.) The court gave no guidance as to how the two figures should be calculated but it envisaged both figures being set at the outset. If the case settled in the Protocol period the lower fee would apply to the whole case; if the case did not settle, the higher figure would apply to the whole case. The court gave an example of a 100% fee rebated to 5% if the claim settled within the Protocol period.

The defendants took *Callery* to the House of Lords ([2002] 3 All ER 417). The House largely washed its hands of the problem by asserting that the appropriate forum to monitor the new funding regime and resolve its teething problems was the Court of Appeal. Their Lordships did comment, however, that a 20% success fee for simple cases (as the Court of Appeal had allowed 'as a maximum' in *Callery*) looked too high and that two-stage success fees might indeed be the way ahead.

In *Halloran v Delaney* [2002] EWCA Civ 1258, [2003] 1 All ER 775, [2003] 1 WLR 28, Lord Justice Brooke stunned the costs industry when he said:

> 'After taking advice from our assessor, and after considering the arguments in the present case, we consider that judges concerned with questions relating to the recoverability of a success fee in claims as simple as this which are settled without the need to commence proceedings should now ordinarily decide to allow an uplift of 5% on the claimant's lawyers' costs (including the costs of any costs only proceedings which are awarded to them) pursuant to their powers contained in CPD 11.8(2) unless persuaded that a higher uplift is appropriate in the particular circumstances of the case. This policy should be adopted in relation to all CFAs, however they are structured, which are entered into on and after 1st August 2001, when both *Callery* judgments had been published and the main uncertainties about costs recovery had been removed.'

6.29 *Chapter 6 'Old' Conditional Fee Agreements and ATE insurance*

So, no sooner had the ink dried on the House of Lords decision not to interfere in *Callery*, than the Court of Appeal had not only replaced the 20% success fee benchmark with 5%, but also backdated its effect to 1 August 2001. All was not, however, as it seemed. Lord Justice Brooke in the *Claims Direct Test Cases* [2003] EWCA Civ 136 provided the following 'clarification' of his judgment in *Halloran*:

> 'Subsequent events have shown that I should have expressed myself with greater clarity. The type of case to which I was referring was a case similar to *Callery v Gray* and *Halloran v Delaney* in which, to adopt the 'ready reckoner' in *Cook on Costs* (2003), p 545, the prospects of success are virtually 100%. The two-step fee advocated by the court in *Callery v Gray (No 1)* is apt to allow a solicitor in such a case to cater for the wholly unexpected risk lurking below the limpid waters of the simplest of claims. It did not require any research evidence or submissions from other parties in the industry to persuade the court that in this type of extremely simple claim a success fee of over 5% was no longer tenable in all the circumstances. The guidance given in that judgment was not intended to have any wider application.'

Yet more clarification and further encouragement for two-stage success fees was provided in *Atack v Lee* [2004] EWCA Civ 1712, [2005] 1 WLR 2643, [2006] RTR 127. In *Atack* the Court upheld the reduction of a success fee from 100% to 50% in a case which went to trial and then settled for £30,000 after a finding of liability. In *Ellerton* the court reduced the success fee from 30% to 20% on the basis that it was a straight forward case and therefore 20% was the maximum to be awarded. The only significant risk was the possibility of the claimant accepting her solicitor's advice and then not beating a payment-in, which was just one of the rare risks which justified a success fee set as high as 20% in the simplest of claims.

Lord Justice Brooke, who to this point had been in all of the landmark decisions, provided this summary:

> 'Because there seems to be some lingering uncertainty about the combined effect of *Callery v Gray* and *Halloran v Delaney* we feel that we ought to restate for the benefit of district judges and costs judges the principles in cases governed by the old regime The reasonableness of the success fee has to be assessed as at the time the CFA was agreed. It is permissible for any CFA to include a two-stage success fee, and this is to be encouraged. In other words the success fee may be a higher percentage (up to 100% in an appropriate case) in the event that a claim does not settle within the protocol period, and a lower success fee (down to 5% in the very simplest of cases) in claims which do settle within that period. Further statistical evidence is now available to which it will be legitimate for parties to refer in relation to success fees agreed in an old regime case after the date of this judgment.'

The reference to the 'old regime' was to cases which did not fall into the fixed success fee rules that by now had started to come into Part 45 (see below.)

The case of *U v Liverpool City Council* [2005] EWCA Civ 475, [2005] 1 WLR 2657, (2005) Times, 16 May considered the reasonableness of a single stage success fee of 100% set in October 2001 in a personal injury claim involving a four year old child who had stepped into a hole in a grass verge. The Court allowed 50% as a single stage success fee but set out the reasoning behind the use of two stage fees:

Professional conduct compliance **6.29**

'[21] When deciding upon a success fee [the claimant's solicitor] had two choices. He could have taken the view that this claim would probably settle without fuss at a reasonably early stage, but he wished to protect himself against the risk that the claim might go the full distance and might eventually fail. In those circumstances he could select the two-stage success fee discussed by this court in *Callery v Gray* [2001] EWCA 1117 at [106]–[112]. In this situation he would be willing to restrict himself to a low success fee if the case settled within the protocol period – or within such other period, perhaps until the service of the defence, as he might choose – and to have the benefit of a high success fee for the cases that did not settle early. As things turned out, he would have benefited on the facts of this case if he had adopted this course: a high two-stage success fee would have been more readily defensible in a case which did not settle until proceedings were quite far advanced.

[22] Alternatively, he could have selected, as he did in fact, a single-stage success fee, being a fee which he would seek to recover at the same level however quickly or slowly the claim was resolved. In those circumstances it would not be possible to justify so high a success fee.'

The judgment endorsed Lord Woolf's encouragement to lawyers in *Callery v Gray* to take seriously the possibility of agreeing an initial success fee of, say 100%, on the basis that if the claim settled within the protocol period (or some other period identified by the parties to the CFA) a lower success fee would be recoverable under the CFA. At the assessment of costs attention would then be paid to the reasonableness of the success fee which was recoverable as things turned out, and this type of arrangement would lead to a greater chance of establishing the reasonableness of a higher success fee where the claim did not settle within the agreed period.

Costs judges should therefore be more willing to approve what appear to be high success fees in cases which have gone a long distance towards trial if the maker of the CFA has agreed that a much lower success fee should be payable if the claim settles at an early stage. The *Claims Direct Test Cases* [2003] EWCA Civ 136 at [101], [2003] 4 All ER 508, [2003] 2 All ER (Comm) 788 was an earlier exposition of this principle.

The message from the Court of Appeal was that where the CFA was entered into before the *Callery* judgment on 1 August 2001 the guidance given in *Callery* applies so that even a simple road traffic case would attract a single step success fee of 20%. Where the CFA was made after 1 August 2001 and a two stage fee was provided, then, again in simple cases, the lower figure should be 5%. There had been repeated encouragement to use a two-stage fee and the carrot of the award of a substantially higher second stage fee has been dangled.

But, even the Court of Appeal has not always fully embraced these messages. In *C v W* [2008] EWCA Civ 1459, [2009] 4 All ER 1129, [2009] RTR 199 the court decided that a 20% success fee in a road traffic passenger claim where liability had been admitted was about right. (No two stage fee in this case). The court did say that if a solicitor agrees to forgo all fees post Part 36 if the offer is not beaten then the success fee could be reviewed during the life of the CFA – quite how that is to be achieved was not explained. The statute requires the percentage to be stated. A two stage fee states the percentage for each stage – how a reviewable fee would fit the statutory requirement must be in some doubt.

6.29 *Chapter 6 'Old' Conditional Fee Agreements and ATE insurance*

The most recent re-iteration of its preference for staged success fees by the Court of Appeal was in *Motto v Trafigura* [2008] EWCA Civ 1150. A single stage success fee of 100% was reduced to 58% with the court again stressing that a more sympathetic view might be taken of relatively high success fees where the level is 'reviewable as the case progresses'. Again the Court of Appeal seems to have in mind a success fee that changes in response to changes in prospects once the case is up and running. It is hard to see how such a provision in a CFA could satisfy the statutory requirement to state the success fee at the time the CFA is made. The concept of a staged success fee is that different percentages are fixed at the outset to reflect the progress of the case. (This is the approach taken by the fixed figures set out in the old Part 45).

We do now at least have recognition that the mere fact that a success fee is staged does not remove the need for it also to be reasonable. In *Fortune v Roe* [2011] EWHC 2953 (QB), [2012] RTR 480, [2012] 2 Costs LR 288 the CFA was entered into after an admission of liability and judgment entered on liability in a catastrophic injury case. The staged success fee was 100% payable if the case settled within three months of the trial date and 25% for any earlier settlement. The case settled less than a month before trial. The risk assessment identified Part 36 as the major risk. The claimant pointed to the fact that if the case had been within the Part 45 scheme a 100% success fee would be recoverable as the fixed success fee at a trial. On appeal it was held that the mere fact that a success fee is staged does not mean it is reasonable. The question still remains as to what the level of risk was and what success fee was justified. There was an admission on liability and no risk of any substance until any Part 36 offer was made. Such an offer was likely to be made only close to the trial in this particular case. The Claimant's solicitors' costs up to that time, which would have been substantial, were secured. A success fee of 100% in such circumstances was unreasonable. A reasonable success fee, whether single or second stage, in the circumstances which pertained when the CFA was entered into, was 20%.

For an example of a staged success fee in defamation see *Peacock (Matthew) v MGN Ltd* [2010] EWHC 90174 (Costs). The success fee was in three stages: 100% of the basic charges, where the claim proceeds to 28 days after service of the defence; 50% if the case settles after proceedings are issued but before 28 days after the defence is served; or 25% if the case settles before proceedings are issued. Given MGN continued with a reasoned defence to stage three Master Campbell allowed the 100% success fee.

Routine or straight forward cases

[6.30]

In *Bensusan v Freedman* [2001] All ER (D) 212 (Oct), SCCO the Senior Costs Judge equated a simple clinical negligence case, which involved a dental tool being dropped in the claimant's mouth who then swallowed it, with a personal injury action arising out of a rear-end shunt. While the costs judge accepted that the risk of failure in a clinical negligence action can be greater than in a personal injury case, this particular case was 'simple and straightforward'. In *Callery v Gray* the Court of Appeal indicated that in a routine road traffic accident action the success fee should not exceed 20% and the Senior Costs Judge took the same view in this case, reducing a success fee of 50% to 20%. Using *Callery* as 'no more than a starting point', he said the effect of the Costs Practice Direction was to prevent excessive claims for success fees in cases which settle without the need for proceedings when it was clear, or ought to have been clear, from the outset that the risk of having to commence proceedings was minimal.

In *Edwards v Smiths Dock Ltd* [2004] EWHC 1116 (QB), [2004] 3 Costs LR 440, the court, in approving an 87% success fee, distinguished a complicated assessment of quantum from a simple claim as in *Callery*. These early cases can be contrasted with the decision in *C v W* where a 20% success fee was allowed in a road traffic claim where liability had been admitted.

Nonetheless, similar sentiments to those expressed in *Bensusan* appear in the field of construction adjudications in *Redwing Construction Ltd v Wishart* [2011] EWHC 19 (TCC), [2011] Lloyd's Rep IR 331, [2011] 1 EGLR 13 where Akenhead J held that CFAs and ATE could be used in enforcement proceedings provided there was no exemption in the CPR from the usual rules relating to funding arrangements. It needed to be borne in mind however that the large majority of reported cases on adjudication enforcements are successful and are usually pursued (as here) via application for summary judgment because there is no realistic defence. It was important that claimants did not use CFAs and ATE insurance primarily as a commercial threat to defendants. It was legitimate for the Court to ask itself whether, in any particular case, a CFA or ATE Insurance was a reasonable and proportionate arrangement to make.

CFAs taken out at an early stage

[6.31]

The essence of *Callery v Gray* was that a client could be offered a CFA at the outset of a case. The question then arises as to how the success fee is to be calculated when at such an early stage little is known about the risks of the case. According to the judgment, *McCarthy v Essex Rivers Healthcare NHS Trust (Case No HQ06X03686)* [2010] 1 Costs LR 59 is the fourth case in which the same clinical negligence firm's CFA has been reviewed by a court. This time the point taken was that the CFA (as do most) provided that it could be terminated ' . . . if we believe that you are unlikely to win'. That, said Mackay J, was relevant to the level of a single stage success fee (claimed at 100%). Also relevant was the fact that this was not a two stage success fee. The success fee had been reduced to 80% and this appeal against that decision failed. It was accepted that the case was, at the time taken on, a 50:50 risk. Mackay J took the view that the termination clause meant that at a fairly early stage cases below a 50% chance can be removed leaving claims falling into the range of 50% to 80% prospects. In *Oliver (executor of the estate of Oliver) v Whipps Cross University Hospital NHS* [2009] EWHC 1104 (QB), 108 BMLR 181, 153 Sol Jo (no 21) 29, John Frederick Oliver had died of septicaemia in a hospital run by the defendants. The same solicitors agreed to act under the same conditional fee agreement as in *McCarthy* with a success fee of 100%, which represented a notional 50% prospect of success in the action.

The agreement noted that the solicitors had not yet had the opportunity to test the credibility of the evidence or assess the relevant facts of expert witnesses. Accordingly the prospects of success were uncertain and it was impossible to assess the percentage chance of success with any mathematical precision. The proceedings were settled. The costs judge reduced the success fee to 67% which represented a 60% chance of success on the grounds that the solicitors must have thought there was more than a 50% chance of success, otherwise the claim would not have been accepted. On appeal, the judge held that the claim was of a kind that faced difficulties and had uncertain prospects, and based on what the solicitors knew when the conditional fee agreement was made, it was one that could easily have been assessed as having chances of success lower than 50%. Accordingly, the costs judge was wrong to take the view that he did. The judge also rejected the view that there had to be a two stage fee if a 100% figure was ever to be approved. The result was he approved a single stage 100% fee.

Liability admitted cases

[6.32]

C v W considered the success fee that could be set where liability had been admitted prior to the CFA being made. The figure arrived at was 20%. In *Hanley v Smith and MIB* [2009] EWHC 90144 (Costs) the CFA was made two years after a standard retainer and by then the first defendant had admitted liability. The issues of liability, quantum and the liability of the MIB were all still live. The solicitor had a 100% success fee of which 90% was for risk. It was held that apart from a double counted element in the risk assessment the risks were properly assessed and 82% was allowed. Leading counsel whose CFA was made after the MIB had admitted liability but before quantum was agreed had sought a success fee of 82% but it was reduced to 54%. The court said that there was a Part 36 risk and the earlier an offer was likely to be made the greater the risk was, but not great enough to justify 82%.

Faring even less well was the success fee in *Thornley (by his litigation friend Lavinia Thornley) v Ministry of Defence* [2010] EWHC 2584 (QB), [2011] 3 Costs LR 335, where counsel entered into a CFA with a 100% success fee at a time when liability and causation had been admitted. Counsel's CFA provided for fees to be paid even if counsel advised rejection of a part 36 offer which was subsequently not beaten. In all of those circumstances there was no realistic possibility that counsel would not get her fees. No success fee was allowed. See also Jackson LJ's trenchant observations as to the fee arrangements in *Pankhurst v White* [2010] EWCA Civ 1445, [2011] 3 Costs LR 392.

High value

[6.33]

In *Cox v MGN* [2006] EWHC 1235 (QB), a media privacy action, the solicitor claimed a 95% success fee and the paying party offered 5%. Unfortunately for the solicitor, leading counsel's pre-CFA opinion was 'bullish' and the success fee was reduced on assessment to 40%, a figure held on appeal to be within the range of reasonable assessments. The CFA was entered into in 2002 at a time when the law on privacy was not as clear as it is now. Today the success fee allowed would probably be lower.

In *C v W* Moore-Bick LJ queried the assumption that a higher value case made the risks greater rather than simply requiring more work. He said (at paragraph 15):

> 'It is probably true in general that high value claims tend to be more complex and to involve a greater amount of work than claims of lower value, but that does not of itself increase the risk of losing.'

However, he did accept that there might be a larger number of potential pitfalls in a bigger case and in which case this could be reflected in the risk assessment and from there into the success fee. It was not correct simply to add, as in that case, a further 20% to the success fee solely to reflect the size of the claim. As is often argued on assessment, a larger value case will encourage the potential paying party to look under more stones for possible defences than would be the case in a smaller value case.

The approach set out in *C v W* contrasts with the provisions in the old Part 45 (rule 45.18) where the escape clause is triggered by value rather than anything else.

Retrospective success fees

[6.34]

Forde v Birmingham City Council [2009] EWHC 12 (QB), [2010] 1 All ER 802, [2009] 1 WLR 2732 was an appeal from the decision of Master Campbell that a CFA can be retrospective but the success fee, following the decision of the Senior Costs Judge in *King v Telegraph Group Ltd* [2005] EWHC 90015 (Costs), could not. On appeal Christopher Clarke J took a different view. Retrospective success fees are permitted. There were at least three situations where such fees might arise: an ordinary retainer later turned into a CFA with a success fee; a CFA on no win, no fee terms and, after the work has converted after some time to include a success fee; CFA with a success fee at x% initially and later changed, retrospectively to y%. The only limit then appears to be that, following *Harrington v Wakeling* [2007] EWHC 1184 (Ch), 151 Sol Jo LB 744, [2007] 5 Costs LR 710 the CFA must be made before the close of the case – otherwise there is no element of futurity.

Multiple claims and global success fees

[6.35]

In *Bray Walker Solicitors (a firm) v Silvera* [2010] EWCA Civ 332, [2010] 15 LS Gaz R 18, 154 Sol Jo (no 13) 29, one CFA covered two claims. The first claim was for the costs of litigation negligently pursued by solicitors. The prospects there were 70% to 80%. The second claim was for loss of chance of success in that negligently pursued litigation. There the prospects were rather lower. The CFA provided for a single success fee of 70% triggered if any recovery at all was made. Wilson LJ took the view (obiter) that the global success fee had to reflect the split on risk and that 70% was too high. The level of the success fee did not fall to be determined but these observations leave the very difficult question of how to set a success fee where different heads of claim carry different risks.

Disbursement liability

[6.36]

Section 11.8(1)(b) of the Costs Practice Direction expressly stated that the legal representative's liability for any disbursements is a factor to be taken into account in assessing the reasonableness of the success fee. The success fee is applied to the fees charged and not to the disbursements.

Nervous litigator's charter

[6.37]

Immediately after *Trafigura*, the then Master of the Rolls, Lord Neuberger, was concerned with the setting of unreasonably high success fees in *Drake v Fripp* [2011] EWCA Civ 1282, [2012] 2 Costs LR 264. This was an unsuccessful appeal in a boundary dispute which had lasted less than a day. Junior and leading counsel for the respondent had 100% success fees. The 100% success fee was reduced to 50%. Lord Neuberger expressed himself in these terms:

> 'I believe that there may be a regrettable, if understandable, tendency to charge the maximum success fee of 100% in every case. The client with whom the fee is negotiated by the lawyer has no interest in the level of success fee (at least in a case such as this, where he has to pay no more than he is entitled to recover from the paying party), and the lawyer has an obvious and strong interest in the success fee being as high as possible. In many cases, it is easy for a lawyer, acting in complete good faith, to persuade himself that the prospects of his client's case succeeding are no better than 50% when it is in his interest to do so, and when he has no negotiations with the party who will or may have to pay the success fee. The court has a particular duty, therefore, to be vigilant in considering the reasonableness of the level of success fee agreed, but, as I have said, this does not mean that the court can invoke the wisdom of hindsight or should adopt an unduly harsh approach.'

It has always been thus. The more risks the solicitor can reasonably recite, the lower the prospects of success and therefore the higher the success fee that can be justified.

Assessing levels of success fee – Part 45

Statistics

[6.38]

Lying behind the fixed success fee regime in the old Part 45 Sections III, IV and V is a set of statistics produced for the Civil Justice Council and which used to

Professional conduct compliance 6.39

be available on the costs debate page of its website. The Court of Appeal made it clear in *Atack v Lee* that the fixed success fees themselves cannot be used for cases that do not fall under that regime but the underlying statistics can be used as guidance.

The Court of Appeal in *U v Liverpool City Council* referred to other statistical material obtained by the Association of Personal Injury Lawyers. A full table was published in *Litigation Funding* of August 2003, but here is a summary:

Success Rates – Personal Injury Cases CRU Statistics

	RTA	EMPLOYER	PUBLIC	CLIN NEG
2001-2	89%	74%	61%	46%
2002-3	87%	77%	60%	46%

Fixed success fees

[6.39]

It should be kept in mind that costs belong to the client. Part 45 fixes the success fee which can be ordered against a paying party. The CFA will govern the level of fees payable to the solicitor and thought must be given to drafting the CFA to reflect the fixed fee steps. These multiple variations are not easy to explain to the client but if they are not included then the client will keep the difference between the costs allowed and those specified in the CFA. The Law Society Model CFA sets out the success fee in steps for fixed success fee cases.

For ease of reference the following table summarises the recoverable success fees in personal injury claims arising from road traffic accidents, EL/PL accidents or EL disease claims. It assumes a degree of knowledge of the working of the system and is provided as a reference point. Otherwise, you will need to consult the rules fully.

Type of Case	Standard Percentage (solicitor or barrister)	Settlement near to trial (barrister only)*	Percentage at trial (solicitor or barrister)
RTA	12.5	50/75	100
EL Accident	25**	50/75	100
Type A Disease	27.5**	50/75	100
Type B Disease	100	100	100
Type C Disease	62.5***	62.5/75	100

* If a fast track case settles within 14 of the trial, the standard percentage is increased to the first figure. If the case is in the multi-track, the trigger point is 21 days of trial and the increase is to the second figure.

** A further 2.5% may be claimed if a s 30 of the AJA 1999 undertaking is involved

*** A further 7.5% may be claimed if a s 30 of the AJA 1999 undertaking is involved

Escape

[6.40]

The figures in the table can be varied where the damages arising from the accident are over £500,000 or from the disease are over £250,000. The escape mechanism is set out at old Rule 45.18 and was considered by HHJ Wood QC sitting as a High Court judge in *William Dronsfield (a protected party who proceeds by his litigation friend Alison Millington) v Benjamin Street* (Manchester District Registry 25 June 2013).

Challenges

[6.41]

The practical effect of the fixed success fees has, perhaps inevitably, been tested by the parties. The courts have clearly been determined to make the arrangements work, even to the extent of some surprising outcomes. Certainty has arguably trumped everything else, particularly in the case of *Lamont v Burton* [2007] EWCA Civ 429, [2007] 3 All ER 173, [2007] 1 WLR 2814, where the claimant (or perhaps that should be his solicitors) still achieved a 100% success fee because the case went to trial, even though the claimant failed to beat the defendant's Part 36 offer. The base costs were divided between pre- and post- offer in the usual way: but the success fee was unaffected. This led to a number of commentators with handy examples demonstrating that in some instances it was better for the solicitor to tell the client to take his chance at court regardless of the offer because the solicitor's increased success fee of 100% over, say 12.5%, made up for the potential loss in base costs.

In *Kilby v Gawith* [2008] EWCA Civ 812, [2009] 1 WLR 853, [2009] RTR 8 the court decided that the defendant having admitted liability before the claimant had entered into a CFA had no effect on the recoverable success fee as the court had no discretion to vary it. Challenges brought on questions of potential alternative funding *(Wetzel v KBC Fidea* [2007] EWHC 90079 (Costs) and the validity of the CFA itself (*Nizami v Butt* [2006] EWHC 159 (QB), [2006] 2 All ER 140, [2006] 1 WLR 3307 Simon J) were given similarly short shrift by the courts.

Concluding at trial

[6.42]

The defendants' challenges to the meaning of 'concludes at trial' in CPR 45.16 have been more successful. A number of judges have concluded that the phrase can only be interpreted as meaning that the trial must have started and not simply settled on the day of the hearing (see eg *Sitapuria v Khan* December 10, 2007, Liverpool CC, HHJ Stewart Q.C; *Styler v Ingham & Hilton* July 31, 2008, Harrogate CC, HHJ Shaun Spencer Q.C.). This may seem hard on claimants who arrive at court ready to proceed only to settle at the door of the court. But that does chime with the deliberations of the stakeholders who originally brokered the fixed success fees. It is only when the claimant (and his solicitors) take the risk of a judge deciding the outstanding issues that the trial success fee can be claimed. If the judge opens and then the parties settle that will be enough but that is an occurrence that is rather rarer than has sometimes been suggested.

ATE INSURANCE

Challenging the contents of the premium

[6.43]

The level of premium may be challenged on the basis that not all of the sum paid was in fact premium. This type of challenge featured early before the courts. In *Callery v Gray (No 2)* [2001] EWCA Civ 1246 the court considered the make up of an ATE policy based on submissions received from intervenors and which were largely encompassed in the report of Master O'Hare who was also sitting as an assessor. The Court of Appeal reconsidered the peripheral benefits that might come within a premium the following year in *Re Claims Direct Test Cases* [2002] All ER (D) 76 (Sep) and in challenges to the scheme known as The Accident Group (TAG) see *The Accident Group (TAG) test cases* 2003 SCCO Case No: PTH 0204771 15 May. Since then there have been few if any challenges to the level of the premium based upon its contents and whether they all comprise insurance which is recoverable within the terms of s29 AJA 1999.

Challenging the level of the premium

[6.44]

The provisions of CPR 44.4(1) and (2) and 44.5(3) (the seven pillars of wisdom) and paras 11.1, 11.5, 11.7 and 11.10 of the Costs Practice Direction are especially relevant to the assessment of premiums.

In the *RSA Pursuit Test Cases* [2005] EWHC 90003 (Costs) the Senior Costs Judge set out the test:

' . . . the court should look both at the costs risks and at the size of the claim when considering the premium' [261]

On the basis that no other ATE policy had been obtainable, exceptionally high premiums were allowed once the insurer's flawed method of calculation had been corrected. Master Hurst concluded that the method was flawed because it was based on estimates of costs with no provision for reflecting true costs. A formula based on the opponent's actual costs was accepted. That same formula was used in arriving at the £9m premium in *Motto v Trafigura* [2011] EWCA Civ 1150, [2012] 2 All ER 181, [2012] 1 WLR 657 with a prospects of success figure of 65% being used by the insurer. The premium rate thus arrived at (62% including administration and profit) was applied to the totality of the defendant's costs, irrespective of the risk level at the time those costs were incurred.

6.44 *Chapter 6 'Old' Conditional Fee Agreements and ATE insurance*

General guidance has also been provided by the Court of Appeal as to how a costs assessor can hope to approach the question of the level of premium. Brooke LJ had this to say in *Rogers v Merthyr Tydfil County Borough Council* [2006] EWCA Civ 1134, [2007] 1 All ER 354, [2007] 1 WLR 808:

> "District Judges and Costs Judges do not as Lord Hoffmann observed in *Callery v Gray (Nos 1 and 2)* have the expertise to judge the reasonableness of a premium except in very broad brush terms and the viability of the ATE market will be in peril if they regard themselves (without the assistance of expert advice) as better qualified than the underwriter to rate the financial risk the insurer faces. Although the Claimant very often does not have to pay the premium himself this does not mean that there are no competitive or other pressures at all in the market. The evidence before this court shows it is not in an insurer's interest to fix a premium at a level which will attract frequent challenges.'

That passage is now often cited in decisions on recovery of premium and was the focus of the judgment of Simon J in *Kris Motor Spares Ltd v Fox Williams* [2010] EWHC 1008 (QB), [2010] 4 Costs LR 620 who made it clear that, without reversing the burden of proof, the paying party must produce some evidence that the premium is unreasonable, remembering that doubt must be resolved in its favour. He provided the following guidance:

> '. . . challenges must be resolved on the basis of evidence and analysis, rather than by assertion and counter-assertion. The issue should be identified promptly and, where necessary, there should be directions for the proper determination of specific issues. This may involve the Costs Judge looking at the proposal; and in the receiving party providing a note for a one-off ATE premium and not just for a staged premium.' [46]

The requirement for 'a note' is a reference to the decision in *Rogers v Merthyr Tydfil County Borough Council*. Simon J also dismissed the argument that taking ATE insurance late in the proceedings was unreasonable. The solicitor defendants had insured their increasing down side only days before a hearing.

Staged premiums

[6.45]

In *Rogers v Merthyr Tydfil County Borough Council* the court considered the stepped premium model offered by the DAS 80e policy. The court allowed the premium in full, even though it was twice the amount of the damages in a simple tripping claim. There was no objection to staged premiums in principle. The evidence showed that it was necessary for solicitors to sign up to provide products from particular insurers. DAS imposed such an obligation. The solicitor explained why he chose DAS as a provider, and his reasons were legitimate. Although the premium appeared disproportionate to the damages, the *Lownds v Home Office* test of necessity was to be applied. Given the need for the solicitor to subscribe to DAS in all of his cases, it was therefore necessary to incur the premium in the present claim. Therefore it was not in fact disproportionate. 'Necessity . . . may be demonstrated by the application of strategic considerations which travel beyond the dictates of the particular case. Thus it may include the unavoidable characteristics of the market in insurance of this kind. It does so because this very market is integral to the means of providing access to justice in civil disputes in what may be called the post-legal aid world.' It was wrong to consider proportionality only with reference to size of the damages. The court had to take 'all the circumstances' into account (CPR 44.5(1)), and this included the risk to which the insurer was exposed. Here the costs exposure of about £6,500 justified the premium claimed in any event, even though the damages claimed did not exceed half this amount.

The court gave the following guidance:

(a) A party who has an ATE insurance policy incorporating two or more staged premiums should inform his opponent that the policy is staged, and should set out accurately the trigger moments at which the second or later stages will be reached. If this is done, the opponent has been given fair notice of the staging, and unless there are features of the case that are out of the ordinary, his liability to pay at the second or third stage a higher premium than he would have had to pay if the claim had been settled at the first stage should not prove to be a contentious issue.

(b) If an issue arises about the size of a second or third stage premium, it will ordinarily be sufficient for a claimant's solicitor to write a brief note for the purposes of the costs assessment explaining how he came to choose the particular ATE product for his client, and the basis on which the premium is rated – whether block rated or individually rated.

In practice, a bald statement that, for example, 'this premium is individually rated' does little to quell the paying party's desire for further information. As the case law stands, the receiving party is well within his rights to refuse to provide any further information until such time as the paying party produces some credible, alternative insurance which the receiving party would have taken out instead.

The challenge to a staged premium in *Parker (Kenneth Ronald) v Seixo (Joel Carlos)* [2010] EWHC 90162 (Costs) came in the context of a £120,000 personal injury claim where the staged premium began at £551 with an additional £9,550 if the case reached service of a defence. The challenge failed again on the basis that without expert evidence the court was in no position to second guess the underwriter.

ATE and appeals

[6.46]

The Court of Appeal was divided on the answer to the question of whether an appeal was to be regarded as the same proceedings as the trial when considering the costs order to be made following the appeal. In *Hawksford Trustees Jersey Ltd v Stella Global UK Ltd* [2012] EWCA Civ 987, [2012] 1 WLR 3581, [2013] Lloyd's Rep IR 337 the majority held that the appeal was separate from the trial:

(1) The word 'proceedings' in s 29 of the Access to Justice Act 1999 should be given its traditional meaning which distinguishes between proceedings at trial and on appeal.

(2) The risk that the incidence of costs at trial might be changed by the costs order of the appeal court may be a new risk of the appeal, but the costs liability and costs order in question remain those of the trial: the risk insured against is a risk of incurring a liability in the trial proceedings not in the appeal proceedings.
(3) The costs liability in respect of which the premium has been taken out remains a costs liability in the trial proceedings, not in the appeal proceedings.

It followed that an ATE policy taken out to defend the appeal, but which gave cover for trial costs (which were at risk had the appeal succeeded), gave rise to a recoverable premium only in respect of the cover for the appeal and not for the trial costs element of the premium. The court made it clear that where a claimant takes out ATE insurance before trial and loses at trial but wins on appeal that premium would be allowed as costs of the trial. In the current case the policy had only been taken out after the trial and therefore at trial there was no premium.

NOTIFICATION OF ADDITIONAL LIABILITIES

The need to notify

[6.47]

Traditionally, a party's funding of his case was a private matter between the client and his solicitor. If the client had to take out a loan or incur some other loss in order to fund the case, he would not be able to claim that additional interest or other sum from his opponent even if he was successful. As such the opponent had no entitlement to know about any such arrangement.

This position was breached with the introduction of Legal Aid. The granting of a certificate, and the likelihood that the opponent would not be able to recover any costs from the legally aided party, was of sufficient moment that the opponent was entitled to be made aware of it so that he could consider whether to continue to pursue or defend the claim as the case may be.

When additional liabilities became recoverable in April 2000, the opponent's potential exposure to costs more or less doubled. The prospect of a 100% success fee and an ATE premium were sufficient to persuade the rule makers that the opponent should be notified in a similar way to a legally aided party. Indeed some argued that the extent of the additional liabilities needed to be notified since otherwise there was a potential breach of ECHR rights. But the argument that such information might prejudice a party's strategy or give away the strength of belief in the case held sway. Consequently only the existence, and not the amount, was to be notified. This means that a party who changes solicitor needs to notify his opponent of the CFA with each solicitor if that is to be the case. This will confirm to the opponent whether he remains at risk of an additional liability, albeit not whether the percentage uplift has gone up or down as a result of the change. If a party's solicitor decides to replace his CFA with a new one for any of a number of reasons, he will need to make sure that the opponent is notified of the existence of the new CFA and, strictly speaking, the ending of the old CFA (there is scope on the N251 to provide both pieces of information at the same time). What the opponent is meant to make of such notification is not clear but it will provide him with some ammunition for the detailed assessment proceedings in all probability. Neither of these situations may lead to an increase in the opponent's liability. There are several situations where that liability will be increased but no notice needs to be given (see below).

Notice of Funding

[6.48]

In order to ensure that parties provided the requisite information in CPD Section 19.4, a notice of funding (N251) was produced. Section 19.4, and therefore the form, changed to allow for more information to be provided from 1 October 2009 in the light of experience but it is rare for any serious point to be taken on the information on the form (unless it is an error as to dates): issues invariably arise from a failure to serve the notice.

Where the information changes so that the notice of funding is no longer accurate, notice of the change must be filed and served on the opponent. The practice direction confirms that the notice of change is in fact a further N251 rather than any other form of notice. The fact that more information had to be provided in notices of funding served after 1 October 2009 does not render notices served before that date inaccurate since the provision is not retrospective.

Service of the Notice

[6.49]

(a) Pre-proceedings – until October 2009, the pre-action protocols only said that notice of funding, in whatever format, 'should' be given. This was interpreted in several decisions to mean that no sanction would apply if notice was not given until the commencement of proceedings. From 1 October 2009, the provision was changed so that notice 'must' be given and that change has brought pre-proceedings notification into line with the post-commencement provisions.

(b) At the beginning of proceedings – The claimant needs to serve his N251 when commencing proceedings. This may be by providing sufficient copies with the proceedings when they are issued and served by the court or by adding a copy to the other documents when serving if this is to be done by the solicitors. The defendant needs to serve his N251 with the unimaginatively entitled 'first document.' What this document is depends upon the circumstances of the case. It may be an acknowledgment of service. It may be a defence. If a default judgment has been obtained it may be the application to set judgment aside. The rationale is clear. It needs to be served at the earliest opportunity by each party.

(c) At any other time – If the funding arrangement is not entered into until part way through proceedings, for example because a new solicitor has been instructed or a top up ATE policy has been purchased, the N251 clearly cannot be served with the proceedings or with the first

document. The rationale of notifying the opponent as soon as possible still applies and the party has seven days within which to file and serve the requisite notice.
(d) Notice of change of funding arrangement – As with notification 'at any other time' a party has seven days within which to file and serve the N251 showing the altered information.

No need to notify

[6.50]

In the following situations there is no need to notify the opponent of the existence of, or any change to, a party's arrangements:
(1) If the CFA has no success fee – but simply different hourly rates payable depending upon the outcome – there is no need to give the other party any notice of the existence of the CFA. The rationale is simple. The opponent does not face any additional liability if he loses than if the party with the CFA was instructing his solicitor on a private paying basis.
(2) If the CFA is between a solicitor and 'an additional legal representative' – this will usually be counsel but could be a solicitor agent or solicitor advocate. CPD 19.3(2)(a) is clear that no notice needs to be given though why that should be so is hard to fathom since counsel's fees may be significant and therefore so too may be his success fee. The wording of the practice direction does suggest that notice of the solicitors own CFA with his client has to have been given already for counsel's CFA not to be notified. It would be a rare case where counsel used a CFA but the solicitor did not do so. In any event, there would still be little logic in counsel's CFA being notified in one circumstance but not another.
(3) If an ATE policy has been 'topped up' – if the limit of indemnity has been increased from the original policy limit the level of premium is likely to have increased as well thereby increasing the opponent's potential liability. Nevertheless, CPD 19.3(2)(b) indicates that where notice of 'some insurance cover' has been notified already, there is no need to notify the opponent of any change in that cover unless it is that the policy has been cancelled. If further insurance cover is obtained by taking out a policy with another insurer, notification of that new policy does need to be notified.

Failure to notify

[6.51]

The contents of CPR 44.3B are stark. With the usual proviso that the court may order otherwise, a party who fails to notify his opponent in accordance with court rule, practice direction or order cannot recover any additional liability for the period of default: CPR 44.3B (1)(c).

Where there has been no notification, the opponent is by definition unaware of the existence of the additional liability. Accordingly, he will not usually be in any position to take a point about this failure until he serves his points of dispute (or raises it in correspondence where a schedule of costs has been put forward by the receiving party for negotiation.) Equally, a party who has overlooked serving notice usually remains oblivious to the failure until the paying party raises it. Why this does not become apparent when the bill of costs is being drafted is not clear. Perhaps receiving parties take the view that they will see whether the paying party will take the point. If so, that is optimistic to say the least since the bill will be considerably higher than would otherwise be the case and is one of the first points that any drafter of points of dispute is likely to take.

Applications for relief from sanctions

[6.52]

The result is that applications for relief from the sanction imposed by this section have generally been made in the detailed assessment proceedings. Indeed, there is a specific provision in the costs practice direction (10.1) in relation to such applications. Consequently, most of the reported decisions are those of costs judges at the SCCO and appeals from those decisions. The change in emphasis in applications for relief from sanctions under CPR 3.9 is covered fully in Part 3 of this book. That change in emphasis means that the reported decisions to date may be of only limited guidance for the future. The cases seem to divide in general terms between those where no notification was given and those where the paying party knew something but not everything he was entitled to know. Where there has been no notice, prejudice to the paying party almost goes without saying and so applications face an uphill task in front of most of the judiciary. If the paying party was aware that, for example, the receiving party was using a CFA, but not exactly when that started, it will be more difficult for the paying party to demonstrate any prejudice, not least with CFAs potentially being retrospective. With these cautionary comments, the following paragraphs set out the relevant case law to date on applications for relief in respect of funding arrangements.

In *Tait v Cataldo* [2010] EWHC 90166 (Costs) a Notice of Funding served in February 2008 referred only to a CFA made in November 2006 with no reference to two earlier CFAs or to an ATE policy. In respect of the ATE the failure was an error in transcribing a written N251 into the typed version sent to the defendants. As to the two earlier CFAs those had come to an end by the time the notice had to be given and the solicitor took the view that those earlier CFAs, being spent, did not need to be referred to. That understanding was not however relied upon before Master O'Hare who took the view that counsel was right not to argue that the earlier CFAs need not be referred to. The explanations given for the mistakes did not count in favour of granting relief but Master O'Hare did grant relief in all respects based on the view that the CFA mistakes were of little significance, the ATE mistake had caused no prejudice to the defendants and the mistakes had in terms of substance been remedied informally before the settlement process commenced.

In *Supperstone v Hurst* [2008] EWHC 735 (Ch), [2008] 4 Costs LR 572, [2008] BPIR 1134 Floyd J said that relief from sanctions should not be granted lightly (albeit that he decided to do so here) and a party who fails to comply with the CPR runs a significant risk that he would be refused relief. If a party does not have a very good explanation, or the other side is prejudiced by his failure, relief from sanctions would usually be refused. That decision was followed in *Kutsi v North Middlesex University Hospital NHS Trust* [2008]

EWHC 90119 (Costs). Contrary to CPR 44.15 and CPR 44.3B, the defendant had not been notified of the existence of the policy until after the claim was settled, as a result of which the claimant needed the court to grant relief from sanctions before she could attempt to recover the premium of £80,325.00 from the defendant on detailed assessment. The costs judge was critical of the firm's failure to be aware of rudimentary CPR principles and there was no good explanation for the complete failure to give any notice of the premium at all. It followed that irrespective of any prejudice to the paying party relief from sanction would be refused. *Haydon v Strudwick* [2010] EWHC 90164 (Costs) provides a further example of the application of CPR 3.9. Here an initial CFA with success fee was entered into whilst the client was a minor. Upon gaining majority the client's position became one of patient and a second CFA was entered into. ATE insurance was also taken out. No notice of funding for any of these additional liabilities was ever sent. Both CFAs were at various stages referred to in correspondence. Early correspondence did refer also to the first CFA having additional liabilities. At no point was the ATE policy referred to. Master Haworth in the SCCO granted relief from sanction in respect of the two CFAs but not in respect of the ATE policy. His reasoning was that having applied all of the factors of CPR 3.9 for the granting of relief none of those factors indicated that relief ought to be withheld. In particular the opponent was aware that there was a CFA with additional liabilities (but not the ATE) and was not prejudiced by a failure to provide the technical detail that form N251 sets out.

Applications for relief from sanction should be made quickly. Relief was not sought in *Manning v King's College Hospital NHS Trust* [2011] EWHC 2954 (QB), [2011] NLJR 1634, [2012] 1 Costs LR 105 until detailed assessment, over 12 months after the opponents had referred to the breach of the rules. On appeal Spencer J reversed the costs judge's decision to refuse relief. It had been based on inconsistent views expressed by the costs judge as to the prejudice suffered by the defendants as a result of the breach of the rules. Spencer J held that there was no prejudice on the substantive proceedings in the failure to refer to a success fee. He then held that and any prejudice at the assessment of costs stage, because of the delay in seeking relief from sanction, could be dealt with by the costs order relating to that assessment and not by refusing relief from sanction. That meant the success fee and ATE premium were recovered. The second judgment in this case found Spencer J critical of the very long delay in making an application for relief. The penalty was a 25% reduction in the claimant's costs of the assessment. The same judge confirmed on appeal in *Scott v Duncan* [2012] EWHC 1792 (QB), [2012] 4 Costs LR 787 that it was appropriate to grant relief from sanctions where a paying party had always been aware that the claimant was represented under a CFA in circumstances where there had been a failure to give notice of a replacement CFA.

Related failures

[6.53]

The CPD Section 32.5 sets out the information that needs to be included with the notice of commencement in respect of additional liabilities. If you are serving a notice of commencement, it is worth checking this lengthy provision in the practice direction because a failure to comply with its terms brings CPR rule 44.3B into play. Rules 44.3B(1)(d) and 44.3B(1)(e) relate specifically to failures to provide the required information in detailed assessment proceedings for CFAs and ATE policies respectively. Some paying parties will invite receiving parties to make a formal application for relief from sanctions to rectify this default. If this is not done, or not done promptly, applications made at detailed assessment hearings have been known to result in the additional liability being disallowed notwithstanding the fact that it was notified within the substantive proceedings themselves. So, if you are invited to make an application, it is generally better to be safe and a little poorer in terms of a modest adverse costs order than sorry with no success fee or an awkward conversation with your ATE provider.

In relation to CFAs, disclosure of a copy of the CFA will ensure that there are no difficulties under this section. If you do not wish to serve even a redacted version of the CFA, you will need to serve a statement setting out certain provisions in any event (CPD 32.5(1)(d). This requirement is in addition to the risk assessment or statement of reasons regarding the level of the success fee. It is worth noting that these provisions make no exception in respect of personal injury cases where the level of success fee is prescribed by Part 45. The fact that the paying party can ascertain the recoverable success fee without sight of anything from the receiving party will only be relevant in terms of prejudice on an application for relief from sanctions.

For the ATE policy, the key document is the 'insurance certificate' (CPD 32.5(2)). Many ATE providers do not produce a document with the word 'certificate' upon it. The closest document is the schedule to the policy and which will generally be accepted by the paying party albeit that it does not always cover all of the information required by CPD 32.5(2). The only sure way to cover this aspect is to disclose the full wording of the policy but that is something that may need the prior agreement of the ATE insurer.

TRANSITIONAL PROVISIONS

[6.54]

Shortly prior to April 2013, it seemed possible to attend a seminar every day on how the Jackson inspired reforms would play out in relation to many aspects of civil litigation but particularly in respect of costs management and the ending of recoverability of additional liabilities. At the time of writing this book, most of the queries regarding the transitional provisions are yet to be tested. We have not added to the speculation in this book. But you may wish to consult the Jackson Review supplement to the 2013 edition which, as the name suggests, deals specifically with the change in regime and raises many issues for which there are as yet no answers.

The transitional provisions are contained in Part 48 and are relatively brief. The old costs rules and practice directions are preserved, as if in aspic, by rule 48.1 for 'pre-commencement' CFAs, CCFAs, ATE policies and Membership Organisation undertakings. So if you have a pre-commencement additional liability, you can continue to rely on the provisions set out in this chapter regardless of when the costs come to be assessed.

'Pre-commencement'

[6.55]

The phrase 'pre-commencement' applies to all CFAs and ATE policies entered into before 1 April 2013. It also applies to CFAs and ATE policies in the three discrete areas of mesothelioma claims; privacy and publication claims; and insolvency related claims. The precise wording of these three areas can be found at rule 48.2 and PD 48. The primary and secondary legislation used to carve out these three areas suggests a lack of permanence in any of them. Elsewhere, some form of Portalesque fixed fee regime is proposed for mesothelioma claims. Some extension of the Qualified One way Costs Shifting may be an answer in privacy and publication claims.

Clinical negligence claims

[6.56]

Section 46(1) of LASPO 2012 carved out one seemingly permanent exception to the ending of recoverability of ATE premiums. As fleshed out by the Recovery of Costs Insurance Premiums in Clinical Negligence Proceedings (No 2) Regulations 2013 (SI 2013/79), it enables ATE policies which insure the costs of certain experts reports to be recoverable. The limitations on the policies are considerable. They can only relate to experts dealing with breach of duty or causation rather than quantum. They only relate to the reports and not to the costs of any attendance upon either counsel or the judge at trial (or even possibly the opposing witness for the purpose of a joint statement). Section 46 allowed for the regulations to deal with the cost of such premiums but the Government decided to leave that to the courts to determine. Undoubtedly the NHSLA and other defendant organisations will put the reasonableness of the premiums to the test. It may be that 'destruction' of the premiums in the manner carried out by the Senior Costs Judge in the *Claims Direct Test Cases* [2003] EWCA Civ 136, [2003] 4 All ER 508, [2003] 2 All ER (Comm) 788 will be required. It is certainly the case that there will be court decisions to consider in this area in the future.

CHAPTER 7

DAMAGES-BASED AGREEMENTS

INTRODUCTION

[7.1]

As we have discussed at the outset of this part of the book, there is more than one kind of agreement where the entitlement to fees depends upon the outcome of the case. In other words, the extent of the fees is 'contingent' upon the outcome. All of these agreements can therefore be described as contingency fee agreements. However, the law of costs tends to treat the phrase contingency fee agreement as meaning a particular type of agreement, namely one where the lawyer agrees to act for the client on the basis of receiving a share of the damages as his payment. Whilst these are not uncommon in non-contentious business, eg conveyancing, such agreements are unenforceable at common law as being champertous where they relate to contentious business. It is only an agreement that complies with a statutory exception which can legitimately depend upon the outcome. So, conditional fee agreements are lawful contingency fee agreements. So too, are retainers which comply with the Legal Aid Agency requirements (because the hourly rates recoverable if the client wins are far higher than those recoverable from the Agency if the client loses.) But the purest form of contingency fee agreement is a damages-based agreement and these have only just become available in contentious business.

There have been many who have championed the use of contingency fees in litigation for some considerable time. The simplicity of the concept of the client agreeing to pay over a share of his damages for his representation has seemed very desirable when compared with some of the tortuous descriptions required to be given by those using conditional fee agreements. But there is only room for one use of the acronym CFA and since conditional fees arrived on the Statute Books first, contingency fee agreements have had to be branded damages-based agreements so that they are DBAs rather than a further form of CFA.

THE ONTARIO MODEL

[7.2]

Since the Statute of Gloucester in the thirteenth century, the basic rule has been that costs follow the event. In other words, the winning party will have at least a contribution towards his costs paid by the losing party. This basic tenet causes a difficulty with the concept of a DBA where the client pays his lawyer out of his share of the damages – what does the client/solicitor do with the order for costs payable by the other side? To ignore it would be to give a windfall to opponents at the expense of the client. But to capitalise upon it would be to destroy the inherent simplicity of the model by importing rules regarding the taking into account of the opponent's contribution to the client's costs.

As we shall see, these rules do not just cause headaches as to whether the agreement is in the client's best interest (though that is a significant concern). They also require the agreement to deal with matters such as hourly rates which would not be required under a pure contingency fee agreement and which inevitably turn the agreement into something resembling a CFA.

Now that recoverability of success fees has ended, a CFA might be described as an agreement based on hourly rates which are recoverable from the opponent in the event of success with the client meeting any shortfall from his damages, up to a capped limit. The form of DBA now available can be described as an agreement based on a share of the damages up to a capped limit but which is reduced by such sums as are recoverable from the opponent in the event of success calculated using hourly rates. Confused? Some figures may help to explain the last two sentences but at this point you should simply be aware that to a large extent a DBA and a CFA have similar mechanisms even if in certain cases, the outcome can be very different. We will look at some figures later in the chapter but first we shall look at the statutory requirements of a DBA and which is very much based on a Canadian arrangement described by Sir Rupert Jackson as the 'Ontario model.'

STATUTORY FRAMEWORK

[7.3]

The wording of s 58AA of the Courts and Legal Services Act 1990 (brought in by LASPO 2012) is very similar to s 58 in respect of CFAs. An agreement created under s 58AA is not unenforceable simply because it is a DBA. This double negative definition nods at the common law position that a DBA is unenforceable at common law. Consequently, as with CFAs, the Act creates an island of legality in the sea of illegality that is the common law. In order to comply with s 58AA a DBA must be:
- entered into on or after 1 April 2013;
- in writing;
- in accordance with the Damages-Based Agreements Regulations 2013.

Regulation 3 of the DBA Regulations requires the DBA to:
- specify the claim or proceedings to which it relates;
- the circumstances in which the solicitors fees are payable;
- the reasons for setting the level of the percentage.

Let us look at these in turn.

Entered into on or after 1 April 2013

[7.4]

Proceedings brought in the employment tribunal have always been considered to be non-contentious business which just goes to show that you cannot always rely on the ordinary meaning of words. The effect of this is that contingency fees have always been available in employment cases and indeed are widely used by solicitors. Counsel, for professional conduct reasons, were not allowed to use a contingency fee agreement. They were able to use a CFA and, since there is no costs recovery generally in the employment tribunal, the fees were almost always paid out of the client's damages which made the prohibition a distinction without a difference. Such contingency fee agreements between solicitors and clients were regulated by the Damages-Based Agreements Regulations 2010. These regulations continue to apply to agreements entered into before April 2013 but are otherwise revoked.

In non-employment matters, however, a DBA pre April 2013 cannot comply with s 58AA and is unenforceable. Whilst it is probable that the same ability to make retrospective agreements will apply to DBAs as it does to CFAs, the contractual abilities of the parties will not overcome the fact that the statute is not retrospective in effect.

In writing

[7.5]

A degree of formality is required to make sure that everyone knows the bargain they have struck. It may well be that the client understands the cap on DBAs more easily than on CFAs but, for the reasons set out below on tactical and professional considerations, it is imperative that the terms of the bargain, and the explanation of it, are clearly set out in writing. The fact that it is a statutory requirement, as with CFAs, ought to be irrelevant. However, agreements reached in a rush, because circumstances have changed or limitation is imminent, can easily be reduced only partly in writing. Such agreements will be vulnerable to challenge if the paying party gets wind of the oral elements, particularly if they tend to disadvantage the claimant.

In Regulation 1 'costs' are defined as 'the total of the representative's time reasonably spent, in respect of the claim or proceedings, multiplied by the reasonable hourly rate of remuneration of the representative.' In order to be able to quantify the costs recoverable from the opponent, it is very important that the hourly rates are set out. Provisions for revising the hourly rates on a periodical basis ought to be included. A simple alternative to bespoke rates might be to refer to the hourly rates set from time to time for your local court by virtue of the Guideline Rates for Summary Assessment.

Where to put the hourly rate information? If the DBA is going to include client care type information anyway, then the obvious answer is 'in the DBA itself.' But many firms send a separate client care letter which sets out hourly rates amongst many other pieces of information. It is easy for the wording of the client care letter and the DBA to get out of synchronisation, including the hourly rates, if they are not regularly reviewed.

There is another issue regarding a separate client care letter which sets up a retainer with the client in addition to the DBA. The law of costs is ambivalent about the idea of there being concurrent retainers. But it would seem to be the only method which potentially will allow a DBA to be used and still be able to charge some fees if the case does not succeed. This is set out in detail below but mention is made here so that you can consider what you wish to include in your client care letter when also using a DBA.

In employment matters, reg 5 sets out the information that needs to be given in writing to the client. There is nothing which says that such information must be given prior to the commencement of the DBA. But the wording is very similar to the CFA Regulations 2000 and which required the information to be given at the outset. From a client care point of view, the information is the sort that ought to be provided at the beginning of the matter anyway. The information to be provided is set out as follows:

- The circumstances in which the client may seek a review of the representative's costs and expenses and the procedure for doing so;
- The dispute resolution service provided by ACAS regarding actual and potential claims (presumably in relation to the underlying proceedings rather than a putative problem between the solicitor and client);
- Other methods of funding that may be available such as Legal aid, legal expenses insurance, trade union funding or pro bono advice. If any are available, information on how they apply to the particular case;
- The point at which expenses become payable and a reasonable estimate of the amount likely to be spent on expenses (including VAT).

If the client asks for further explanation, advice or information, the *representative* needs to provide it. Civil litigators could do worse than use these categories as part of their template for providing clients with costs information.

Specifying the proceedings

[7.6]

The need to specify the proceedings, or parts of them, to which the agreement relates has unfortunate echoes of reg 2 of the CFA Regulations 2000. Cases such as *Blair v Danesh* [2009] EWCA Civ 516 were brought by the paying party in an attempt to show non-compliance with the regulations and therefore a lack of enforceability of the agreement against either the claimant or the defendant.

Problems have tended to crop up where there is more than one defendant. If there is only a single defendant, then a general description of a claim against that defendant, even if it changes its name or entity through merger etc, will usually be adequate. Where, however a defendant brings a Part 20 claim against a third party and the claimant joins that party as second defendant, it is easy to overlook amending the agreement with the client to include the second defendant in the specification of the proceedings in the DBA. There are an infinite variety of such possibilities. One option is to make the definition of the proceedings wide enough to encompass 'any additional defendants brought into the claim' or similar but you should be careful not to make it so vague that the reverse argument of the description being so wide as to lack any contract certainty becomes available.

In employment matters, reg 6 requires any amendment to a DBA to cover additional causes of action must be in writing and signed by the client and the representative. There is no mention of amendment to a non-employment DBA in the Regulations.

The circumstances in which fees and expenses are payable

[7.7]

This is another concept brought in from the CFA Regulations and to which old cases may be relevant. Generally the idea of payment in the event of success but not otherwise has been relatively easy to define. However, the meaning of payment in the DBA Regulations is rather more complicated than in the CFA Regulations.

A DBA cannot require the client to pay any more than the agreed 'payment'. This is the percentage of damages that has been agreed can be deducted and which is subject to statutory limits (see below). The payment is the shortfall after any costs and expenses have been recovered from the opponent. In civil cases, it includes both solicitors' costs and counsel's fees but excludes expenses, and which would include After the Event insurance if that was used in conjunction with a DBA. In employment matters, counsel's fees are treated as an expense and so are not part of the payment itself.

The definition of payment is in reg 1 of the DBA Regulations and starts off with the phrase 'that part of the sum recovered in respect of the claim or damages awarded the client agrees to pay . . . ' From this it is clear that the client can only be someone who is recovering damages, ie the claimant and not a defendant (unless there is a counterclaim).

As this is currently drafted, it appears to preclude any form of differential or 'no win, low fee' fee agreement because the absence of success will mean there is nothing recovered from which the lawyers could be paid. It is said by those who want to use such agreements that this provision is going to prevent a good deal of litigation being carried on under DBAs. If it is altered so that a lower hourly rate or fixed fee is payable in the event of failure, it will mean that the distinctions between a DBA and a CFA continue to be more and more blurred. At the time of writing this chapter there is a good deal of dispute about whether a so called hybrid DBA is allowed under the existing framework. The only method for doing so is by having a separate retainer – a 'failure retainer' – which enables the solicitor to receive a payment based on hourly rates (or a fixed fee) on the basis of a retainer that has nothing to do with the DBA itself. However, this immediately causes a concern regarding the ability of a party to have two retainers at the same time – a situation which occurs from time to time and which is not entirely clear as to whether it is acceptable or not (save for legally aided cases where it is certainly not permissible). Given the various statutory instruments being put before Parliament to amend bugs in the system, it would seem a reasonable bet that further legislative clarification will be provided on this point.

Reasons for setting the level of the percentage

[7.8]

The attraction to clients of DBAs is the certainty with which they are provided. Whatever the outcome they know that they are not going to be liable to pay out more than X% of their damages for their legal fees. (This figure might well be increased to pay expenses and it does not deal with adverse costs but they can be dealt with separately.) On the other side of this coin is the realisation on the part of the solicitor that there is a finite sum that can be recovered in payment of your fees. The ability to recover costs from the opponent based on an hourly rate at first sight seems to guarantee that a reasonable sum will be recovered for a reasonable amount of work. But, as we shall see below, the workings of the Ontario Model often militate against this. It will be the naïve solicitor who does not consider the sum likely to be achieved via a percentage of the client's damages before starting any work.

Ideally, you should have some idea at the outset of:
(a) the overall value of the claim (or the relevant heads of claim if it is a personal injury case (see below);
(b) the overall cost of reaching a trial and proving the case there;
(c) the prospects of the case settling successfully earlier than at trial;
(d) the likely recoverable costs on assessment or by agreement;
(e) the firm's appetite for taking on such work and any guidelines in place which affect the cost of doing so.

These factors can all be summed up by the phrase 'risk reward.' If the case is not very valuable in terms of damages, but is likely to take a good deal of time to prepare and is unlikely to settle short of trial, it does not present a very attractive picture. Even if the case is successful, the amount of money recovered in fees may be nowhere near the sort of return expected based on a calculation of time spent multiplied by your hourly rate.

Conversely a strong case on liability with an easily calculated quantum, such as is set out on invoices, may well suggest that the case will settle quickly and without a great deal of work. In such cases, the prospects of taking a percentage of the costs for limited effort sounds promising.

Then of course there are all the cases in the middle. Some, such as those involving personal injury or negligence, are unlikely to crystallise early in terms of quantum even if liability or breach of duty is clear cut. Others, such as commercial disputes will often suffer from a lack of any objective third party evidence (such as is often available in personal injury cases) to be able to be confident on liability for some time even if the quantum is clear from the beginning.

How much it will cost to get the case 'home' at trial?

[7.9]

Factor (b) is perhaps more important than any other. It is certainly the one that is most in danger of being overlooked. It is true that very few cases get to trial, but if you start from the assumption that it will usually settle early, you will almost certainly take a bath on a number of occasions. You will also be prone to feeling that you are going to be under rewarded during the life of the case. If you have a long track record in the area in which you are considering using DBAs you may be able to take a more bullish approach. But even so, there are uncertainties as to opponent's tactics once they are aware of the fact that you are using a DBA. This will become clear to them relatively soon if you are involved in repeat litigation with regular opponents, notwithstanding the fact that there is no need to serve any form of notice of funding. The need to keep statistics; to take heed of judicial dicta such as that in *Stevens v Watts* (cited with approval by Lord Woolf in both *Jefferson v NFC* and *Home Office v Lownds*); as well as budget carefully have never been greater, nor as likely to provide significant advantages to your calculation of profitability.

Likely recoverable costs on assessment

[7.9A]

Factor (d) is relevant if you charge an hourly rate for your work which does not generally find favour with the court on a between the parties' assessment. This may be perfectly proper. For example, in *Jones v Wrexham Borough Council* [2007] EWCA Civ 1356, [2008] 1 WLR 1590, (2008) Times, 21 January the rates claimed by a Reading firm were based on City of London rates because many of the lawyers had come from City firms to do similar work but in a different environment. The firm was successful in that case in charging higher rates. You might consider your expertise to justify something similar but have not yet persuaded the local judiciary that this is so. There is nothing to prevent you charging such rates to your client in ordinary retainer cases. If you do this using a DBA, you will know that there is very likely to be a shortfall on assessment as a result and such sum is limited to what can be claimed from the client as the payment in addition to the costs recovered from the opponent. Similar considerations would apply if you use specialist counsel in your area of work and their fees are also considered to be vulnerable to challenge on assessment.

If you are instructed in relation to an employment matter, the reasons for setting the level of the percentage have to have regard (where appropriate) to whether the claim or proceedings is one of several similar claims or proceedings. Perhaps peculiarly if you are instructed on a group action, that does not seem to be a relevant consideration when setting the percentage level.

MAXIMUM LEVEL OF THE RECOVERABLE PERCENTAGE

[7.10]

You need to make sure that your agreement does not inadvertently exceed the maximum level of recoverable fees from your client. Similar provisions in CFA legislation rendered a number of agreements unenforceable even though the parties to the agreement were perfectly content with the terms of the bargain struck. Once the opponent was able to demonstrate that the maximum level had been exceeded, it was but a short step for the Court of Appeal to consider that the administration of justice was affected, even if the client could not really be said to be prejudiced. In *Jones v Caradon Catnic Ltd* [2005] EWCA Civ 1821, [2006] 3 Costs LR 427 Lord Justice Laws was particularly vehement in his condemnation of agreements which flouted the statutory requirements. They went 'flat against the grain' against the will of Parliament, amongst other comments.

The headline maximum levels are as follows. Each needs some explanation of the finer detail:
- Personal injury 25%

- Employment 35%
- All other civil cases 50%
- Appeals 100%

Personal injury

[7.11]

In personal injury cases the 25% deduction can only come from the damages for pain, suffering and loss of amenity and past pecuniary losses (net of the recoupment of any CRU benefits). This is the same arrangement as for the deductible element in CFAs.

The 25% includes VAT on the solicitors' charges as well as counsel's fees (including VAT if any). It does not include expenses (reg 4(1) of the DBA Regulations 2013).

These provisions are bound to cause problems. By limiting the heads of damage from which sums can be deducted, a solicitor and his client will always have divergent interests in allocating offers to those heads of claim where a global offer is made by the defendant. There is little prospect of a defendant offering to breakdown a Part 36 offer or similar. Discussions between the solicitor, counsel and the client will need to be carried whilst the lawyers walk a professional tightrope in providing appropriate advice whilst keeping their commercial interests at bay.

Employment

[7.12]

The figure of 35% in employment cases continues from the DBA Regulations 2010. Unlike personal injury and civil cases, counsel's fees are treated as being expenses and so are not included within the 35%. The client is therefore liable to hand over a rather higher percentage of his damages than the headline figure suggests. Many solicitors, particularly since their own remuneration is capped by the percentage, will use counsel for advocacy rather than do it themselves.

Other civil cases including appeals

[7.13]

The limit of 50% was a surprise to many people. The rationale for finally allowing contingency fees to be used in contentious work was largely based on allowing parties the freedom to contract on whatever terms they wished. As such, the parties' options were increased from the existing methods of funding. By restricting parties to a 50/50 split between the solicitor and client, the freedom of contract argument has rather been sidelined. Presumably the idea of the lawyers taking the majority of the spoils of litigation was a newspaper headline the Ministry of Justice did not wish to see. But that does not sit very happily with reg 4(4) which states that the limitation to 50% only relates to cases at first instance. If it is accepted that there may be sufficient risk for a party justifiably to give most of his damages away if successful on appeal, there does not appear to be much logic in preventing this from happening earlier. Not all clients are consumers needing protection. Many are sophisticated and large businesses, often with their own in-house legal department well able to agree terms with external solicitors.

REDUCTION IN DAMAGES OR COSTS

[7.14]

The hinge on which DBAs work is the sum recovered by way of damages and costs. Anything which reduces either element of this sum will have the effect of reducing the pot from which the solicitor can claim his fees.

Accordingly, consideration of the risk reward potential needs to include the possibility that the damages or costs may be reduced by court findings. There are a number of possibilities here. Findings of contributory negligence will obviously reduce the claim itself. Counterclaims or other forms of set off may have exactly the same effect (see below). In personal injury cases, the operation of the Compensation Recovery Unit recoupment provisions will also need to be 'netted off.' The one consoling factor is that the spectre of a reduction in the damages in any of these ways should be relatively obvious from an early stage. The extent of the reduction will not be clear in most situations but the possibility will be.

By comparison, the likelihood of an issues based costs order or a percentage order will not become clear (other than a theoretical possibility in any case) until later. It is unusual for issues to be run without any real expectation that they will be successful. Properly alternative cases do not necessarily result in only a percentage of costs being allowed. The conduct of the parties often seems to be the catalyst for orders which are less than a full order for costs. Such conduct invariably occurs during the currency of the case and so can only be considered by a regular and objective monitoring of the conduct of the litigation.

RECOVERABILITY

[7.15]

Further consideration also needs to be given to whether the sums will be physically recovered from the opponent. If there is no insurance, or the opponent is not of any obvious substance, this may well be a concern. Does the word 'recovered' denote a settlement short of a court hearing or does it in fact require money to be received by the client before he is liable to discharge his solicitor's fees?

Regulation 4, when defining 'payment', refers to amounts 'that have been paid or are payable by another party to the proceedings by agreement or order.' So at first blush that suggests that it is only the award of costs that is required. But in fact it is the opposite because the clause is written in a negative fashion. A DBA must not require a payment by the client to his lawyer other than one that has taken into account the sums that have been paid or are payable by the opponent. If the opponent has not paid the between the parties element, the lawyer will only be able to claim from his client the additional sum. To put this into some very simple figures – the client damages are £12,000 and he has agreed to pay 25% to his solicitor net of any recovery from the opponent. The between the parties costs are assessed at £2,000. The client is liable to pay his solicitor £3,000 as the 25% share. He can offset the £2,000 and pay his solicitor the £1,000 balance. If the opponent is not good for the £2,000 it appears that the risk is on the solicitor.

Legal Aid practitioners of any vintage appreciate that the statutory charge bites on any property 'recovered (or preserved)'. In the absence of anything physically recovered, whether in cash or property, there would be nothing on which the statutory charge could bite. The wording of the relevant regulations certainly appears to suggest that recovered means actually received in that context.

However, in the commercial context, the decision in *Arkin v Borchard Lines Ltd* [2001] CP Rep 108 considered the meaning of the word 'recover' in relation to a CFA. Mr Justice Colman decided that there was a distinction between 'recover' and 'paid'. He decided that counsel had entered into a CFA because the client was unable to pay under an ordinary retainer. The idea that counsel would make payment of their fees also contingent upon the defendant's means was one he found to be 'completely inconceivable.'

It is dangerous to assume that parties and their lawyers will always do what would seem to be logical. The idea of solicitors buying in work for which they would use a 'pure' CFA with no success fee would also have seemed unlikely at the time Mr Justice Colman was hearing the *Arkin* case. However, that is what solicitors running the Accident Group cases agreed to, in great and small numbers, before the scheme fell apart in 2003. There is likely to be further argument on this point even if the solicitors are able to take enforcement proceedings in the client's name as of right.

TERMINATION

[7.16]

Regulation 8 of the DBA Regulations deals with the question of early termination of the agreement in an employment matter. The general law of contract is preserved but two particular issues of concern in respect of a DBA are addressed. First, a solicitor may only charge fees on a time basis for his fee and expenses if the agreement ends early. In other words, he cannot seek a sum equivalent to a percentage of the likely damages. Secondly, the client cannot validly terminate the agreement after a settlement has been agreed or within a week of the tribunal hearing. No doubt some unscrupulous clients will seek to use their lawyers for as long as possible and then try to sack them just before the damages become available. These provisions help to prevent that from occurring.

It is perhaps disappointing therefore that there are no parallel provisions in relation to non-employment matters. On the face of it, the client could seek to remove his obligations to his solicitor in exactly the same way in a civil case as for an employment matter. There will also be cases where either the solicitor or client decides that a case ought not to be pursued further, whether on a lack of prospects or on a risk reward basis. Why the use of a quantum meruit approach is not equally good for civil cases in these circumstances is hard to fathom. There was a storm of protest when the original (overly detailed) DBA Regulations were unveiled. It may be that which made the Government's draftsman decide to put the bare minimum into the Regulations instead, thereby allowing the parties to contract on any further aspects. If that is so, civil litigation lawyers would be well advised to consider following the employment matter provisions when drafting their DBA.

OPERATION OF THE INDEMNITY PRINCIPLE

[7.17]

When CFAs were first introduced in 1995, they had no impact on the opponent in the sense that they did not make it more expensive for a losing opponent than was already the case (based on a traditional retainer.) When recoverability was introduced, the success fee made cases potentially almost twice as expensive and consequently paying parties' interest was engaged in attacking the validity of such arrangements. Knowledge of the impact of the indemnity principle became widespread for the reasons described as the Golden Rule and discussed at [**4.2**]. Removal of recoverability may mean that the paying parties' interest will wane. It is likely to lead courts to consider any inadequacy of drafting to be a matter between the solicitor and his own client.

The introduction of DBAs has no detrimental impact on the opponent. CPR 44.18(1) specifically says that the fact of the DBA 'will not affect the making of any order for costs which otherwise would be made in favour of the [successful] party'. Rule 44.18(2) confirms that the assessment of costs under that order will be carried out in accordance with rule 44.3 (as would any other order for costs). Furthermore, whatever is the extent of the DBA client's liability to his solicitor (in other words, the percentage share of damages) is the most that the opponent can be required to pay when calculated on a time spent basis. In some situations the opponent's liabilities will be decreased by the operation of the indemnity principle. The comments in this paragraph can be demonstrated with a simple example.

A commercial client successfully sues for damages in the sum of £50,000. His agreement with his solicitor is for the lawyer to take 30% of the damages (ie £15,000) as payment. The costs based on the time spent are £10,000. The opponent has to pay this £10,000 and the client pays the remaining £5,000 in order to make up the 30% share.

But if the time costs had amounted to £20,000 instead, the opponent's liability would still only be £15,000 since that is the client's maximum liability. The opponent would therefore only hand over £15,000. Notably, there would be no shortfall for the client to pay and so the client pays nothing from the damages to his solicitor and the solicitor will have to write off the extra £5,000 of fees that he has incurred in the case. This in a nutshell, is why the risk of a case becoming protracted is entirely the solicitor's. Once his fees have reached the value of the percentage share, the solicitor is effectively providing his client with free representation and the opponent is also at no further costs risk. Hence the need to consider the risk reward very carefully at the outset and keep it in mind thereafter (see [7.8]). This exacerbates the potential for solicitor and client conflicts as to whether a case should be settled or not.

CONFLICT OF INTEREST

[7.18]

It is obvious that if there is a finite amount of money as the potential reward, the longer the case takes to bring to a conclusion, the less remunerative it is for the solicitor. The reverse is true for the client. Giving the solicitor 25% of the claim may be a good deal if the litigation is likely to be hard fought. But if not much more than a stiff letter before action is required, it does not look to be such a bargain.

The client care provisions discussed in Chapter 1, and the natural wish to look after a client's best interests can easily lead to muddled thinking and a tying of your professional soul in knots. No client will expect you to offer a choice of agreements at the beginning of a case and then to select whichever is least advantageous to you at the end of it as some manifestation of professional ethics. As long as the pros and cons of agreements are spelled out to clients at the beginning of the case, they will be content to follow the chosen agreement regardless of the outcome. Moreover, the Legal Ombudsman is unlikely to criticise any agreement which ostensibly dealt reasonably with the client's circumstances. You may well decide not to offer all of the options that could be used to run a case. That is a matter for you. As long as the client is aware that there are other options, but that you do not offer them, he cannot validly criticise you later if, in the eventual circumstances, one of those other options might have suited him better.

By way of illustration, there are many practitioners who are currently unwilling to offer DBAs to clients because of the 'hybrid' issue. There is no reason for them to offer an agreement that either does not suit their practice (because there are no fees at all if unsuccessful) or risks being found unenforceable (if a 'low fee' based on an hourly rate is agreed). There is no suggestion that this unwillingness is unprofessional. It is a perfectly rational choice and if you do not want to run cases on any form of contingent agreement you are entitled to do so.

CHALLENGES BY THE CLIENT

[7.19]

Embarking on a practice using DBAs is not for the faint-hearted, though some traps can be avoided by looking at the evolution of CFAs and the challenges to them. One final concern to bear in mind is one which did not occur with CFAs with recoverable success fees. It applied before the age of recoverability but

Challenges by the client 7.19

was not a point that was taken at the time. It may occur now and applies equally to DBAs. It goes back to the beginning of this chapter and the concept that all contingency fees are unenforceable unless they comply with the statutory provisions that created them. A shrewd client will appreciate that if his lawyer is using a DBA (or CFA) and that agreement is unenforceable, there will be no deduction from his damages because the solicitors' fees will be irrecoverable. Therefore, such a client would seek to find some non-compliance with the legislation once the case has concluded and take the solicitor to an assessment under Part III of the Solicitors Act 1974 to determine the point. Obviously, a client who regularly used a solicitor would not take such steps but it is wise to be aware of this possibility where DBAs are offered to clients on a one off basis.

CHAPTER 8

CONTENTIOUS AND NON-CONTENTIOUS BUSINESS AGREEMENTS

INTRODUCTION

[8.1]

The difference between contentious and non-contentious business has been said to be fundamental to the remuneration of solicitors for decades if not centuries. But there are more similarities than differences in the two forms of business. Furthermore, events over the last few years have reduced the relevance of the distinction to the point where there is little practical difference between the two in most circumstances. Consequently, this Chapter, having discussed the meaning of contentious and non-contentious work, deals with the two together in considering the benefit and detriment to solicitors and their clients of formal business agreements to which provisions of the Solicitors Act 1974 apply.

DEFINING CONTENTIOUS AND NON-CONTENTIOUS BUSINESS

[8.2]

There is a drafting convention of defining something by saying that it is everything other than something else. That is the approach used to describe non-contentious business (it is everything other than contentious business) in s 87 of the Solicitors Act 1974. It is off-putting to the extent that contentious business is not a phrase that is necessarily obvious in its meaning. But this approach does make sure that nothing falls between the cracks. It is either one thing or the other. Matters are not helped by the definition of contentious business (which also appears at s 87) concluding with the phrase 'not being business which falls within the definition of non-contentious or common form probate business'. It is easy to have sympathy with Lord Justice Ward in *Ralph Hume Garry (a firm) v Gwillim* ([2002] EWCA Civ 1500, [2003] 1 All ER 1038, [2003] 1 WLR 510) where he described s 87 as 'a fairly useless circular definition of contentious and non-contentious costs'.

The distinguishing part of the definition of contentious business is that it 'means business done, whether as solicitor or advocate, in or for the purposes of proceedings begun before a court or before an arbitrator'. Consequently, all proceedings before a court, once litigated are contentious business. Work done in anticipation of court proceedings is non-contentious unless and until court proceedings are actually commenced. When that happens, the work converts to becoming contentious business retrospectively. The only exception to this may be where the work has already been billed with a statute bill and as such has crystallised the work done. The point will be an academic one in most circumstances.

Non-contentious business usually covers work that is transactional rather than litigious in nature. Private client work such as conveyancing and wills and probate (other than contentious probate naturally) is non-contentious. So too is commercial work in mergers and acquisitions, construction projects and contracts generally. Often the words transactional and non-contentious are used as synonyms in these areas. Less obviously, the bringing of claims in tribunals has been considered to be non-contentious work, even if tribunal proceedings are commenced. As such, employment matters are non-contentious business. However, that position is likely to change in our view with the reorganisation of the Tribunals structure. There are now two 'tiers' for the majority of tribunals. There are seven 'chambers' in the first tier and four chambers in the upper tier. The upper tier includes the Upper Tribunal (Lands Chamber) (formerly the Lands Tribunal). The Employment Appeals Tribunal is not within this structure. These two tribunals were previously considered to be courts of record and so proceedings before them would be considered to be contentious business. It appears that all of the tribunals are starting to become costs (and court fee) bearing, and it will be odd if they are not considered to be contentious business in the relatively near future.

BUSINESS AGREEMENTS

[8.3]

Section 57 of the Solicitors Act 1974 deals with non-contentious business agreements. Sections 59 to 63 deal with contentious business agreements. Whilst using different terms in places, there is a great deal of overlap in these sections. So, to avoid the reader searching for differences that do not exist simply because of some of the language, we have dealt with them together under the following headings. Obviously where there are differences, these are pointed out.

Representative Party

[8.4]

The agreement will of course usually be between the solicitor and the client. But there are some circumstances where the person contracting with the solicitor is in fact representing the ultimate client. In such circumstances, apart from any drafting issues, the representative party needs to be aware of the provisions of s 62. Such representative capacity would occur if the 'client' signs the agreement as guardian; trustee under a deed or will; Court of Protection deputy or other person authorised under that Act.

The crucial point is that the agreement needs to be put before a costs officer before any payment is made. If the representative party pays the solicitor's fee without doing this, he runs the risk of being held to account for any shortfall that the ultimate client would have achieved if the agreement had gone before the court.

Timing

[8.5]

A solicitor may make an agreement with his client as to his remuneration before, during or after doing the work. It is therefore explicit that such agreements can have retrospective effect.

In writing

[8.6]

The business agreement must be in writing. A non-contentious business agreement must also be signed by the client or his agent. According to Lord Denning in *Chamberlain v Boodle and King (a firm)* ([1982] 3 All ER 188, [1982] 1 WLR 1443, CA) a contentious business agreement can be created by an exchange of letters which would mean it was in writing but not signed by the client as if it were a single document. Given that the benefit of these agreements is very largely in the solicitor's favour and therefore to the detriment of the client, it is imperative that you obtain a signature from your client to a document setting out the agreed terms. It may be that your business agreement is essentially a client care letter. If so, the client should be supplied with an extra copy of the letter with a request to countersign it and return it.

In passing, there is another advantage from the solicitor's point of view in getting the letter (or equivalent) signed. The letter will constitute express authority from the client in respect of the matters contained in it and these agreed terms are therefore, pursuant to CPR 46.9, presumed to be reasonable (see the end of this Chapter). Whatever the view of the costs officer may be as to the agreement, on a Solicitors Act assessment of costs the client is bound by it.

Methods of Charging

[8.7]

A business agreement may provide for remuneration by:
- reference to an hourly rate
  Until s 98 of the Courts and Legal Services Act 1990 was enacted, a business agreement could not be made based on hourly rates. The use of hourly rates brings a risk of uncertainty to the agreement and, as can be seen below, the trade-off for limiting the client's rights to assessment is based on the certainty as to fees which a business agreement gives them.
  Where an hourly rate is agreed, the client is bound by that agreement. Only the number of hours spent can be challenged subsequently (see [**8.12**]).

8.7 Chapter 8 Contentious and Non-Contentious Business Agreements

A contentious business agreement using an hourly rate is expressly permitted to be 'at a higher or lower rate than that at which he would otherwise have been entitled to be remunerated' (s 59(1)). There is no reason in principle why the same should not apply to a non-contentious business agreement.

- a gross sum

These days we tend to talk of a fixed fee rather than a gross sum. It is the perfect type of fee arrangement to encompass within a business agreement because it is the most certain of all the possibilities. Even if there are in fact a number of fixed fees depending upon the circumstances, this remains the case. So a building block approach or a menu of fees as a case progresses would fit the bill. So too, would an agreement that set out different figures depending upon the stage at which the case concluded. One which set out different figures depending upon success or failure however would fall foul of s 59(2) and would probably also be difficult to convert into a CFA (see 'a percentage' below).

- a commission

In Chapter 31 we look at the concept of value billing as an alternative to charging based solely on the time spent. The case of *Jemma Trust Co Ltd v Liptrott* [2003] EWCA Civ 1476, [2004] 1 All ER 510, [2004] 1 WLR 646 is the most recent exposition of the appropriateness of this idea. In essence the solicitor charges on an hourly rate and then additionally claims a percentage of the estate or property involved. The idea is to reflect the value and therefore skill of the solicitor. The hourly rate is often suppressed to acknowledge the separate value element. This is an approach used in non-contentious work, generally property or probate based. It is not strictly contingent upon the outcome because there is no success or failure as such. In any event, it has always been acceptable to carry out non-contentious work on a results basis.

- a percentage

Section 59(2) expressly forbids a contentious business agreement to found a contingency fee agreement, particularly a champertous one. So the agreement must not give the solicitor any interest in the proceedings, nor depend upon success for payment.

There are now two ways around the statutory prohibition. The first is that a conditional fee agreement can be entered into in accordance with s 58 of the CLSA 1990 stipulating for payment only in the event of success and which will make it legal. The agreement may also be capable of fulfilling the terms of a contentious business agreement (the Court of Appeal found this to be so in *Hollins v Russell* [2003] EWCA Civ 718). But if the agreement does not comply with the requirements of s 58 of the CLSA it will be unenforceable against the client by virtue of s 59(2) and therefore irrecoverable against the opponent by s 60(3) (a statutory version of the indemnity principle.) Secondly, as discussed in Chapter 7, DBAs became lawful in April 2013 for use in court work ie contentious business.

- a salary

There is nothing wrong in principle with your client employing you to carry out work on his behalf. There are a whole host of regulatory issues to consider and economically it might not make sense in many situations, but it is an option for you to look at. If, at the end of the case, your client has a costs order in his favour, your fees would be assessed at an hourly rate in the usual way without the need for any calculations about the actual cost of your employment. In *Re Eastwood, Lloyds Bank Ltd v Eastwood* [1975] Ch 112, [1973] 3 All ER 1079, CA the Court of Appeal made it clear that the costs of an employed lawyer should be assessed in the same way as for an independent solicitor in private practice.

- or otherwise

    An example of an alternative to the named options is payment by shares in new companies which proved popular during the 'dot.com' era. This approach is sometimes the only way a start-up company can realistically pay for legal services.

Achieving Certainty

[8.8]

This is the fundamental point of a business agreement. If the terms as to payment cannot be set out with certainty there is very little point attempting to create a business agreement, whether contentious or non-contentious.

In order to be a binding contract, it must show all the terms of the bargain. The leading case on this point remains *Chamberlain v Boodle and King (a firm)* [1982] 3 All ER 188, [1982] 1 WLR 1443, CA) which is an indication that such agreements have fallen out of favour. When solicitors were used to receiving a fixed sum for their work, such agreements were much more attractive than when hourly rates took hold of the legal profession. As clients demand more certainty and the rules of court require prospective budgets where fees are not fixed, it may be that the idea of certainty is one which will encourage the use of business agreements once more.

The following extract from Lord Denning's judgment in *Chamberlain* shows how much certainty is required. In effect any purely hourly rate based agreement is going to have difficulty coming within the requirements.

> 'Further, the agreement must be sufficiently specific – so as to tell the client what he is letting himself in for by way of costs. It seems to me that the letters in this case did not give the client the least idea of what he is letting himself in for . . . .Take for instance the rate . . . It is £60 to £80 an hour. What rate is to be charged? And for what partner? Of what standard? [etc.] . . .
>
> I only make those observations because it seems to me that this is not an agreement as to remuneration at all. It is simply an indication of the rate of charging on which the solicitors propose to make up the bill. It is by no means an agreement in writing as to the remuneration.'

More recent cases have tended to fall down on the basis of brackets of hourly rates for a particular grade of fee earner (as it did in *Chamberlain*.) Even an open ended right to review the hourly rates annually has created a degree of uncertainty for the courts. If hourly rates are to be used they need to be specified figures, preferably linked to the main fee earners. Any review needs to reflect a specified percentage increase or a link to an objective figure such as the Retail Prices Index.

The courts' attitude is perhaps not surprising. The Solicitors Act gives solicitors an opportunity to make a contract with their client which, if done correctly, prevents the court from having virtually any say in the reasonableness of the remuneration sought thereafter. If, on the other hand, the client's liability is not certain, the court is going to want to protect the lay client from the professional adviser. If the agreement is certain, it may be a bad bargain, but that is a matter for the client and would cut across the client's freedom to contract if the court could interfere.

Early Termination

[8.9]

Where some work has been carried out under a contentious business agreement, but before completing it, the client decides to change solicitor (or indeed if the solicitor dies or becomes incapacitated), s 63 comes into play. Unlike the usual position (see next heading), the court may order an assessment of the fees even if it considers the terms of the agreement to be fair and reasonable.

Where the client has decided to change solicitor, the court will look to see if there has been any default, negligence, improper delay or other conduct such as to justify allowing less than the full amount of the remuneration agreed. If there are none of these aggravating factors, then the fee ought to be allowed in full. This provision does not apply where the solicitor is not able to keep to his side of the bargain.

RESTRICTING CHALLENGES TO FEES

Invalid Business Agreements

[8.10]

Where the business agreement has been found wanting in its construction, the solicitors' costs will be assessed via a solicitor and client assessment as described in Chapter 3.

If the work was contentious the costs will be assessed in accordance with the rules of court and in particular the factors set out at CPR 44.4. If the work is non-contentious then obviously there has been no litigation, save for the Part 8 proceedings bringing the question of costs before the court. The various Remuneration Orders produced under the powers in s 56 of the Solicitors Act 1974 set out the factors that the court should take into account when assessing non-contentious work. Article 3 of the Solicitors' Remuneration (Non-Contentious Business) Order 2009 sets them out as follows:

> 'A solicitor's costs must be fair and reasonable having regard to all the circumstances of the case and in particular to-
> (a)   the complexity of the matter or the difficulty or novelty of the questions raised;

(b) the skill, labour, specialised knowledge and responsibility involved;
(c) the time spent on the business;
(d) the number and importance of the documents prepared or considered, without regard to length;
(e) the place where and the circumstances in which the business or any part of the business is transacted;
(f) the amount or value of any money or property involved;
(g) whether any land involved is registered land within the meaning of the Land Registration Act 2002;
(h) the importance of the matter to the client; and
(i) the approval (express or implied) of the entitled person or the express approval of the testator to-
(i) the solicitor undertaking all or any part of the work giving rise to the costs; or
(ii) the amount of the costs.'

Apart from the specific property and probate factors, this list essentially sets out the same 'seven pillars of wisdom' as are seen in the court rules. It is noteworthy that the time spent on the matter is a factor even if the bill is rendered as a result of a fixed fee, salary, commission etc and not solely when by reference to an hourly rate.

Valid Business Agreements

[8.11]

Where the agreement is a valid one, however, then the fees ought to be payable without difficulty by the client. If the client changes his mind about the reasonableness of the bargain he originally struck he will have to persuade a court that the fee he is being asked to pay is not a 'fair and reasonable' one. Only if he can do this, will the agreement be overturned so that a detailed assessment can take place. The test is the same whether or not the business agreement relates to contentious or non-contentious work.

How does the court come to consider whether the fees are fair and reasonable? It is not a precise science. According to Donaldson J in *Property and Reversionary Investment Corpn Ltd v Secretary of State for the Environment* [1975] 2 All ER 436, [1975] 1 WLR 1504, 119 Sol Jo 274:

> 'It is an exercise in assessment, an exercise in balanced judgment — not an arithmetical calculation. It follows that different people may reach different conclusions as to what sum is fair and reasonable, although all should fall within a bracket which, in the vast majority of cases, will be narrow.'

Later in the judgment he described it as a 'value judgment based on discretion and experience.' As such, 'it might not be the right figure, and indeed such a figure probably does not exist, but we hope that it will be a right figure.'

You would be forgiven for thinking that the bulwark against the opening of the agreement is therefore less than certain. That would be ironic given the need for certainty in the original agreement. In order to give any effect to the provisions of the Solicitors Act, the bar has to be set fairly high. In practice, it is often clear that either the solicitor or the client is on the moral high ground and it is the other party who is trying to set aside (or uphold) the agreement purely on tactical grounds.

Although the court requires only prima facie evidence of unfairness or unreasonableness in order to intervene, in the opinion of Mustill J in *Walton v Egan* [1982] QB 1232, [1982] 3 All ER 849, [1982] 3 WLR 352 'from a practical point of view the agreement of the client is the strongest evidence that the fee is reasonable'.

Hourly rates

[8.12]

If the client does not allege that the business agreement is unfair or unreasonable, he can still challenge the amount of hours spent on his case if the agreement is based on hourly rates rather than, for example, a fixed fee. The amount of the hourly rate will be binding on the client, but the costs officer may inquire into (a) the number of hours worked by the solicitor, and (b) whether the number of hours worked by him or her was excessive.

PROCEDURE

Non-contentious business

[8.13]

Where a solicitor seeks payment of fees due under a non-contentious business agreement, he will begin proceedings under Part III of the Solicitors Act 1974 as described in Chapter 3. If the proceedings are contested, the solicitor can rely upon the terms of the agreement to defeat arguments raised about the fees (other than the number of hours spent if it's an hourly rates based agreement) that are challenged.

If the terms are challenged, the costs officer will enquire into the facts surrounding the agreement and certify whether the fees are fair and reasonable or not. If they are, then they are payable under the terms of the agreement. If they are not, then the court will give consequential directions for a detailed assessment.

One reason for a solicitor to consider the use of non-contentious business agreement was that the provisions of the Solicitors' Remuneration (Non-Contentious Business) Order 1994 did not apply to it and therefore the client was precluded from applying for a remuneration certificate. That reason no longer applies unless the work done was before August 2009 when the new Order came into force. This procedure is described in more detail at [8.15].

Contentious business

[8.14]

The effect of a contentious business agreement is to preclude a Solicitors Act assessment of the costs as between the solicitor and the client except in respect of agreements by reference to hourly rates. The agreement itself does not give a cause of action and before a solicitor can rely on it, he must apply to the court for leave to enforce the agreement. Equally, the client may apply to the court to set it aside. Both applications are made under CPR Part 8. The outcome will depend on whether or not the court is of the opinion that the agreement is fair and reasonable. Applications in respect of bills less than £25,000 must be made in the county court (CLSA 1990, s 3). An application may be made at any time before the bill is paid or before the work has been done. An application can still be made after the bill has been paid for the court to re-open the agreement and assess the costs. The court will need to find that there are special circumstances in order to do so. Such applications are to be made within 12 months of payment although the court may allow further time if it appears to be reasonable to do so.

NON-CONTENTIOUS WORK PRIOR TO 2009

[8.15]

On 11 August 2009, the Solicitors' Remuneration (Non-Contentious Business) Order 2009 replaced the 1994 version. However, the earlier Order still applies to any work carried out prior to 11 August 2009. The 1994 Order provides for a procedure whereby the client can challenge the costs charged by his solicitors without having the costs assessed by the court. Instead the client can request a Remuneration Certificate to be provided by the Law Society as to what sum is a fair and reasonable one to pay. The client has to have paid at least half of value of the bill(s) he is seeking to challenge before he can use the procedure. Having done so, a solicitor in another practice will, on behalf of the Law Society, look at the file and determine whether the work done was fair and reasonable. If it was not reasonable, a different figure will be substituted.

If the client does not like the decision, he could still require the costs to be assessed through the court. But, as mentioned earlier, this procedure is not available, even for pre-2009 work, if there is a valid non-contentious business agreement in existence.

SOLICITOR MORTGAGEE'S COSTS

[8.16]

Section 58 of the Solicitors Act 1974 contains specific provisions to deal with the circumstance of where a solicitor is a mortgagee, or is one of a number of mortgagees. In such circumstances the solicitor is entitled to charge as if he were the solicitor to the mortgagee and employed by him to investigate title, negotiate the loan etc.

CHAPTER 9

INSURANCE ARRANGEMENTS

INTRODUCTION

[9.1]

We began this part of the book by pointing out that the word 'funding' used in the context of lawyers was not one which necessarily accorded with accepted wisdom in other spheres. Much of the so-called funding is really no more than a fee arrangement with the client, particularly where it limits the client's liability in the event of failure, and typically avoids the need for a client to pay fees in the meantime (the very antithesis of funding as most people would see it.) The use of legal expenses insurance ('LEI') to cover the risk of litigation is a prime example of a concept misdescribed as a funding option but which is undoubtedly something which you need to consider with your client as part of providing the best information as to costs generally.

Much of the effort of rule makers, judges, parties and their lawyers in respect of insurance has concentrated on the vexed question of 'recoverability.' That subject, together with the related provisions concerning solicitors' success fees, can be found in detail at Chapter 6. This chapter only deals with recoverability to the extent that LASPO 2012 and its subordinate legislation has allowed certain niche areas to remain, at least for the time being.

It is conventional to deal with Before the Event ('BTE') insurance and After the Event ('ATE') insurance separately as if the twain would never meet. Whilst there are separate headings in this chapter, you will soon realise that custom and practice has meant that they are more intertwined than is often realised. The virtual end to recoverability of ATE premiums is in our view likely to enhance this situation rather than reduce it.

One of the questions arising from LASPO 2012 is whether or not a solicitor should in fact advise his client not to trouble to use any insurance at all. This might be because the risks have diminished following the implementation of Qualified One way Costs Shifting (QOCS). It might be because the lawyer is prepared to take on some of the adverse risk himself. Possibilities here include that the solicitor wants to charge for 'funding' and make some money on doing so; the client cannot afford to pay for insurance but the solicitor is keen to run the case; or the client is sufficiently important to the solicitor that he can require the latter to take the risk anyway. Whatever is the situation, the role of the solicitor as a quasi-insurer is considered at the end of this chapter.

Chapter 9 Insurance Arrangements

BEFORE THE EVENT INSURANCE

[9.2]

Chronologically, if not alphabetically, BTE insurance comes before ATE insurance. As the name implies it pre-dates the cause of action at the centre of the litigation. Like motor or property insurance it is bought in the hope that it will never be required and statistically, it will not be required in the great majority of cases. Consequently the BTE insurer does not pay out very often and the cost is limited because 'the many pay for the few' as the insurance maxim has it.

Funding via BTE insurance

[9.3]

There will be some who read the introduction and who frowned at the idea that BTE insurance does not fund litigation. In theory fees and disbursements have been payable as the case progressed. But in practice fees in particular and disbursements to a lesser extent have not been paid by BTE insurers as part of the panel arrangements. Indeed some agreements have had the expectation, if not necessarily the explicit wording, that disbursements would not be paid, win or lose. As with much of insurance arrangements, they differ from one insurer to the next.

There are a number of motor and household insurance policies which have included BTE cover for free. That practice has reduced for two reasons. The first is that its inclusion was, for some insurers, simply a tactic at the height of the 'costs wars', to cause claimants some difficulty in recovering the cost of ATE premiums (and success fees). That imperative appears to have waned. More recently, the comparison websites have obliged insurers to place a nominal cost against add-ons such as BTE insurance so that the policies can be compared against other providers. Anecdotally, this has caused the take up of BTE insurance to drop considerably where potential policyholders look to minimise the cost of their insurance.

This development would suggest that the encouragement given by Sir Rupert Jackson in his Final Report to the BTE industry as a way to underwrite litigation at a proportionate cost will fall on stony ground.

BTE USAGE

[9.4]

Prior to the advent of recoverability, BTE insurance tended to be used by claimants via firms of solicitors with whom the insurer had panel arrangements. The ferocity with which potential claimants are now courted by various organisations to sign up with them did not exist in the 1990s and clients would generally be content to use the insurer's recommendation if they did not have their own solicitor already. In the latter event, the existence of the BTE insurance was largely ignored.

Once recoverability arrived in April 2000, it became a considerably cheaper option from the defendant liability insurers' point of view and had to be considered by all solicitors as an alternative form of funding to using a CFA (see Chapter 6). As a result, BTE insurers found themselves being asked to indemnify non-panel firms against costs, many of whom were unknown to them. Whilst approaches varied, most BTE insurers initially took a hard line and refused to cover these firms and required the case to be transferred to panel solicitors if the client wanted the benefit of the BTE cover. Most BTE insurers gradually softened this line but some still relied on the Ombudsman's decision that the Insurance Companies (Legal Expenses Insurance) Regulations 1990 did not give the client freedom of choice of solicitor until proceedings had been commenced.

Since April 2013, the world has turned again. Many solicitors have spent a decade trying to justify not using a client's BTE cover in order to recover success fees and ATE premiums. Now a client's pre-paid insurance is a valuable commodity when compared with the cost of ATE insurance. It is difficult to justify not using BTE cover unless it is clearly insufficient for the case or is otherwise unsuitable. If the BTE insurer continues to refuse to indemnify a particular solicitor as to costs prior to issue, that may no longer be a reason not to use it. This is particularly so in lower value personal injury work (where most of the BTE insurance exists) since the introduction of the Portals which significantly limit the costs risks to the claimant even if proceedings are commenced (see Chapter 26).

BTE COVERAGE

[9.5]

BTE insurance traditionally covers all sides' costs. Therefore it is unlike (most) ATE insurance which does not cover the own solicitor's costs, but merely disbursements and the other sides' costs. This can result in the level of indemnity in a BTE policy being exhausted relatively quickly because all sides are burning the indemnity at the same time.

The fact that the claimant's solicitor's fees were underwritten scuppered any arguments that panel firms had in using a collective CFA on cases sent via the BTE insurer. In particular they could not claim a success fee to offset the cost of the losers when they were paid even if the client lost. Consequently, at their panel firms' urging, some BTE insurers reduced the level of coverage so that they no longer covered the claimant's solicitor's costs but just the disbursements. As such the policy mirrored an ATE policy. The BTE insurers were content with this as it reduced their potential exposure and the solicitors were free to pursue claims using a CFA but without the need for any ATE cover. Some BTE insurers extended this arrangement to non-panel firms who sought to use the BTE cover on behalf of their clients. Some were prepared to reduce their level of cover on an ad hoc basis when requested to do so by non-panel firms in order to be able to use the cover. This approach tended to mean that the BTE insurer's objection to using a non-panel firm – that they might run a poor case at great expense – largely disappeared.

9.5 *Chapter 9 Insurance Arrangements*

Now that recoverability of ATE premiums has ended, BTE insurers are receiving more requests to use the insurance to fund cases. Some will agree to do so based on the terms of the existing policy. Others are prepared to do so only if it is altered to mirror ATE insurance. Some will do neither. In any event, if your client has BTE insurance, you will need to ask the BTE insurer what, if anything, it is prepared to do in your particular case.

Top up cover

[9.6]

In the infancy of ATE insurance, a request for a policy which provided additional cover to a client who had BTE insurance was met with grave suspicion. The whole ethos of insurance is to insure a large basket of cases so that the odd loser is paid for by the premiums from the much larger number of winners. A one-off proposal for insurance was therefore a clear example of 'adverse selection' ie that the solicitor was only seeking to insure his riskiest cases without insuring the basket of less risky cases. A one-off proposal where the case was sufficiently advanced to have exhausted the BTE cover was an aggravated form of adverse selection. Not surprisingly such proposals were invariably rejected.

As time passed, the concept of topping up ATE policies became understood as one which often resulted in sizeable premium income rather than huge losses. The need for additional cover did not always mean that the case was a difficult one. Often, the defendant would fight hard and expensively, even though his battle was very much uphill. This knowledge enabled many ATE insurers to start providing additional cover on cases initially being run with the benefit of BTE insurance. Providing an excess layer of cover was if anything less risky than cover from the ground up. Any modest adverse costs order would probably be met by the underlying BTE cover. The same applied where one ATE insurer was giving additional cover to an ATE policy provided by a different ATE insurer.

In the current environment, it may take ATE insurers a little while to become comfortable with topping up BTE insurance, whether or not it has been converted into something resembling ATE insurance. But if there is sufficient premium income achievable for such products, top up cover is likely to continue.

The client's indemnity

[9.7]

Sometimes the BTE insurer will agree to you running the case with the benefit of the BTE insurance, but upon reading the insurer's protocol agreement, you decide that it is not workable. What do you do?

If the stumbling block relates to reporting provisions, level of indemnity etc it will probably require a discussion with your client about the pros and cons of using the insurance. The cover is ultimately no more than an indemnity against expense and if your client would rather instruct you than use the insurance cover, that is a perfectly proper decision to make as long as the potential repercussions have been clearly explained and understood beforehand. As with all such matters, obtaining the client's agreement in writing is a must.

Hourly rates

[9.8]

The most awkward stumbling block – and it is a regular one – is the question of your fees and in particular your hourly rate. BTE insurer panel agreements generally limit the hourly rates payable based on the volume of work being provided to the firm. There is often also a non-panel agreement containing rates which the insurer considers to be reasonable to pay on individual cases. The differential can be catered for by using a CFA where your client is a claimant. Your usual hourly rate is the successful rate. The insurer's lower rate is the unsuccessful rate. But if your client is not the claimant (and so is less likely to obtain an order for costs), or is in a regime, such as the employment tribunal, where costs are generally not recoverable, a CFA is not a solution.

In *Brown-Quinn v Equity Syndicate Management Ltd* [2011] EWHC 2661 (Comm), [2012] 1 All ER 778, [2012] Lloyd's Rep IR 248 the clients (there were three conjoined cases) had been involved in employment tribunal matters. The defendant had a non-panel costs rate that was below the rate the solicitors would ordinarily charge. It argued originally that it was not liable to meet any costs if it did not agree to the insured's choice of solicitor. By the time of the hearing before Mr Justice Burton the defendant had revised its position (rightly in the judge's view). The defendant no longer contended that it could prevent the choice of solicitor: merely that the rate it was liable to pay should be based on the non-panel rate where there had been no agreement as to the applicable rate to charge. The judge however decided that any Solicitors Act assessment of costs due should use the insurer's non-panel costs rate, not as a starting point, but as a comparator. It would thus be necessary and right for the court assessing the costs to take into account the availability of any other suitable firms on lesser rates negotiated with the insurers.

This case has been held up by some commentators as a great victory for non-panel firms. The same it has to be said was true of the case of *Eschig v UNIQA Sachversicherung AG: C-199/08* [2009] ECR I-08295, [2010] 1 All ER (Comm) 576, ECJ. However that case opened no floodgates and this one also appears to be quite dependent upon its facts. One, for example, was the agreement to the non-panel solicitors being the 'appointed representative' to the claimant under the terms of the policy. Having agreed that, but not having agreed fee rates, the insurer put itself in a difficult position. If this case gains any traction it will be a simple matter for insurers to make sure they do not agree one part without the other. Furthermore, there is a considerable difference between civil cases where the solicitor is likely to recover his costs and an employment case where he is unlikely to do so. Finally, if solicitors do manage to charge higher rates to the BTE insurer (and which cannot be passed on to the opponent) the premiums will inevitably increase, or the sales will reduce, or quite possibly both.

9.9 *Chapter 9 Insurance Arrangements*

AFTER THE EVENT INSURANCE

The price isn't right

[9.9]

ATE insurance has had a curious existence. At its inception as an adjunct to the Law Society Model CFA it provided the answer to the question of paying for the defendant's costs and subsequently the client's disbursements. But the price of £85 for £100,000 of cover soon resulted in the underwriters leaving the market licking their wounds. With the advent of recoverability, the ATE insurers became 'entrepreneurs doing the Government's business' when providing the statutory objective of new and better legal services according to CLSA 1990, s 17. Almost immediately the premiums charged were said to be too high and were only ever to be seen on cases that couldn't lose. The underwriting of this oft-called 'fragile market' was a mystery to all but the underwriters and the courts were told by the higher courts that they should not try to second guess that underwriting. However, when Master Hurst did 'deconstruct' the policies sold by Claims Direct and the Accident Group he found policies in which non-recoverable aspects were embedded so the premiums had to be reduced. Moreover, the cases told a tale of underwriters' fingers being burned by claims management companies and seeking to take an ever larger share of the premium to offset their mounting losses. At the time the premium levels could quite properly be described as being both too low and too high.

The size of the policy premiums, particularly in cases where the damages were not high, remained an issue ever after. When Sir Rupert Jackson proposed removing recoverability, he was pushing at an open door with the judiciary and many parts of the civil litigation world. He was vehemently opposed by the ATE insurers and those who used their products, particularly claimant personal injury and clinical negligence practitioners, but to no avail.

The end of recoverability is a watershed moment for ATE insurers. If they are right that they provide security for claimants (and it is invariably claimants) to access the courts, they will flourish because they will no longer have to justify their premiums to third parties whose only interest is to reduce the figure in front of them. Instead the policyholders will pay for their insurance because they value the transfer of risk that it represents. If, on the other hand, the security provided is unnecessary, this form of insurance will wither on the vine. Early indications suggest that reports of ATE insurance's demise are exaggerated.

RISKS COVERED

[9.10]

Initially, ATE insurance solely covered adverse costs risks. In other words the opponents' costs if the client lost his case, whether at trial or by earlier discontinuance. This left a gap in the funding arrangements on unsuccessful cases. The opponents' costs were covered by the ATE policy. The solicitor's own fees were not payable by virtue of the CFA. But the client's disbursements still had to be met by the client and in some cases, particularly clinical negligence, that could amount to a considerable sum.

ATE insurance quickly evolved to cover 'own disbursements'. Often this did not include counsel's fees because counsel was expected to share in the risk with a CFA as the solicitor had already done. But some ATE insurers did cover counsel's fees, albeit that then made it difficult for counsel to recover success fees if they positively did wish to act under a CFA.

In the early days of ATE insurance, there was a dichotomy of view between the insurers as to the appropriate shape of policies. A number of insurers offered a policy which, like BTE insurance, covered both sides' costs and disbursements. The selling point of such policies was that the solicitor did not need to use a CFA because he would be paid win or lose. That was understandably attractive to a profession which did not generally wish to move to a position of gambling its fees on the outcome of the case. The difficulty with it however, was that solicitors who used it did not necessarily alter their previous practice and so continued to take cases whose prospects were comparatively modest because their fees did not depend upon the result. Consequently the claims experience of insurers offering both side's cover tended to be poor and the premium levels were high.

Claims Direct was the main exponent of running cases on this model and its influence at the time was such that a provision in the Costs Practice Direction at the commencement of the CPR was directly aimed at making sure that the cost of such policies did not greatly exceed the use of a CFA and the more standard ATE insurance (see Section 11.10(1) CPD).

Personal injury cases

[9.11]

Trades Unions, and other Membership Organisations, did not cover own disbursements, but rather left this to their panel firms. Consequently, the statutory provisions making the cost of their undertakings recoverable (in the same way as if they had provided a formal insurance policy) did not make the cost of covering own disbursements recoverable (see Chapter 6 for more detail).

When Sir Rupert Jackson recommended the ending of recoverability of ATE premiums, he proposed Qualified One Way Costs Shifting as a method of removing the adverse costs risk from the client who might otherwise have to pay for ATE insurance himself. QOCS does not deal with the problem of the client's own disbursements. Sir Rupert's view was that this was a matter for the claimant and his solicitor. According to his Final Report, it would be 'perverse' for a vindicated defendant to have to pay for them (via other cases where claimants were successful) and defendants were making a sufficient contribution to the system by having to bear their own costs as a result of QOCS.

So the risks covered in personal injury cases by ATE policies are the own disbursement risk and the adverse costs risk in the event that QOCS protection is removed. That would occur without limit in relation to a claim to be found 'fundamentally dishonest' but is of little significance to ATE policies because they would not respond if the insured was found to have been fraudulent in this fashion. Where the claim had been struck out, then depending upon the consequences, the policy may provide cover. The opponents' claim in such circumstances is again unlimited. Where the claimant has failed to beat a Part 36 offer, the adverse costs risk is limited to the extent of the damages that were awarded. This gives the appearance of ATE insurance protecting the client's damages rather than covering the opponents' costs in some circumstances.

It is unquestionably the case that the risk to ATE insurers has significantly decreased where QOCS applies and premium levels have reduced as a result. The introduction of fixed costs for cases that 'escape' the Portals is likely to reduce the levels further. In personal injury cases, more than any other area of litigation, the question for each client is whether there is sufficient benefit in the protection afforded to justify the irrecoverable cost of the ATE premium.

Clinical negligence cases

[9.12]

Much of what is said in respect of personal injury cases applies equally to clinical negligence cases. As discussed below, part of the premium remains recoverable, but a significant part does not. The advent of QOCS reduces the risk to the insurer because a loss or discontinuance is no longer likely to trigger any payment for adverse risks under the policy.

But the parties in clinical negligence cases are heavily dependent upon their expert evidence. The risk of the expert changing his view, or simply not being accepted by the court, is ever present. The fees for an expert attending conferences with counsel and at any hearing are not covered by the recoverable insurance policy and it is a rare client who will be prepared to risk being liable for such fees without insurance. As such, ATE insurance is likely to continue to be taken up by most clinical negligence claimants, whether individually or by some firm-wide scheme.

All other cases

[9.13]

Arguably the financial imperative for taking out ATE insurance in civil and commercial cases has not altered a great deal by the implementation of LASPO 2012. Most cases were compromised anyway and they would often be settled via a global deal which left a pot for damages, costs and disbursements including the premium. For all practical purposes, the premium was therefore already payable by the client in many cases.

Outside the emotive area of damages for personal injury, civil cases are really about money and the costs risk as to obtaining/denying that claim for money. The transfer of the costs risk to an insurer for the payment of a premium is a classic insurance arrangement and many clients see the benefit of transferring that risk. The risks remain those of own disbursements and adverse costs. There are no QOCS issues (although QOCS may well spread to other civil cases) to consider. The place for ATE insurance in such disputes is likely to remain. It may or may not be in conjunction with CFAs since there is no actual need for a CFA to be in place to operate an ATE policy. It is simply that most ATE insurers like the solicitors to be taking some risk even if the CFA is of the 'hybrid' variety where a lower fee is payable if the case loses.

PREMIUMS STILL RECOVERABLE

[9.14]

The general position on recoverability after 1 April 2013 is that, in principle:
- Policies taken out before 1 April 2013 are recoverable from a losing opponent: but policies taken out thereafter are not;
- Policies in respect of certain experts' reports in clinical negligence claims are recoverable;
- Policies in respect of insolvency proceedings, publication and privacy proceedings and mesothelioma claims are recoverable for the time being

Let us look at these in turn.

Policies taken out prior to 1 April 2013

[9.15]

These are described as being 'pre-commencement' policies because they were taken out before the effect of LASPO 2012 on 1 April 2013. They continue to be recoverable even though the order for costs obtained by the client post-dates April 2013. The issue of recoverability generally is dealt with in detail in Chapter 6. There are at least a couple of areas where judicial input is very likely to be required on how the transitional provisions relate to ATE premiums.

One issue that may well prove to be more illusory than real is whether a staged premium policy is affected by the change of regime. Some commentators have suggested that the increased levels in the later stages contravene the legislation by seeking additional sums post April 2013. This is an argument we find difficult to follow. The agreement was taken out 'pre-commencement' and specified the premium that would be payable depending upon the point at which the case successfully settled. Once the case does settle, there seems to us to be no reason in principle why the level agreed by the client at the outset cannot be claimed from the opponent. If the premiums decreased rather than increased as the case progressed there would presumably be no argument from a paying party about a later stage being claimed.

The more contentious issue is likely to be whether a pre-commencement policy which is 'topped up' to provide additional cover becomes wholly or partly a post-commencement policy as a result of the varied terms. Has the agreement been novated in some way rather than simply varied by consent? Naturally, the opponent is unlikely to object to a change in the policy terms which simply increased the level of cover because that potentially would have been of advantage to the opponent if things had turned out differently. It will only be where the additional cover has attracted an additional premium that arguments will be raised.

Policies in Clinical Negligence proceedings

[9.16]

Section 58C was added to the CLSA 1990 by LASPO 2012. It deals solely with ATE insurance recovery and starts by ending recoverability generally. It then goes on to carve out a specific exception in relation to clinical negligence

9.16 *Chapter 9 Insurance Arrangements*

proceedings. This is contrary to the methods used to continue recoverability in other areas discussed below. It would appear that the clinical negligence arrangement is intended to be permanent.

Section 58C allows for the costs of a 'costs insurance policy' e an ATE policy, to be included in a successful party's order for costs. The risk covered by the policy is in respect of 'incurring a liability to pay for one or more expert reports in respect of clinical negligence.' The fact that the cover only relates to the report of the expert(s) is significant. Much of the costs of expert evidence in clinical negligence cases relates to further work done by the expert in conferences with counsel and attending at court. It is not clear that a joint report following a meeting of the experts – a pre-requisite in most proceedings – is necessarily covered by this description.

In order to flesh out the detail of this section, it allows for the use of subordinate regulations. These took two goes to be passed by Parliament and so the relevant ones are called 'The Recovery of Costs Insurance Premiums in Clinical Negligence Proceedings (No 2) Regulations 2013.' The prescribed description of eligible policies is that:

(a) the underlying damages claim must exceed £1,000;
(b) the expert report or reports must relate to 'liability or causation'.

Presumably the minimum figure in (a) will go up if the small claims limit for such cases is increased beyond £1,000. The limitation in (b) shows the rationale of the Government behind this exception to the ending of recoverability. Invariably, the questions of breach of duty (rather than 'liability') and causation are investigated first in potential clinical negligence claims. If they can be established then issues regarding quantum can be pursued. The idea of being able to insure the early reports is intended to assist parties to investigate claims without having to expose themselves to a costs risk of unsupportive medical evidence. The theory therefore is that everyone can take out insurance, investigate their claim and then either pay for the medical evidence from the insurance if it cannot be pursued or claim the cost of the insurance from the defendant in due course if it is pursued.

The regulations might have prescribed a maximum amount for the costs of this premium either as a monetary figure or as some form of percentage of the wider policy premium, but the Government decided after consultation not to do this. Instead, the assessment of the premium level has been left to the courts. This may just be a postponement of the problem. ATE insurers have not generally underwritten clinical negligence in the way described above. They have tended to want some evidence from a claimant before insuring the case. This has usually meant the claimant has had to risk the cost of an unsupportive report. The new scheme expects insurance to be in place before the report is commissioned. If ATE insurers are to insure cases in this manner, the burning cost (ie the money paid out on policy claims) may be so high that the premium levels claimed from the opponents will always be the subject of challenge. Paying parties will be arguing that such premiums are disproportionate and that cases such as *Rogers v Merthyr Tydfil County Borough Council* [2006] EWCA Civ 1134, [2007] 1 All ER 354, [2007] 1 WLR 808 are no longer good law given the Jackson Review and the introduction of LASPO 2012. Faced with this prospect, the ATE insurers may only make such insurance be available to firms with a good track record. For everyone else, the likelihood is that policies will still only be incepted once reports have been obtained. The cover will be retrospective to cover these reports in the event that the case does not succeed.

On the other hand the cover provided by the recoverable element of the insurance is going to be insufficient for most claimants since they cannot guarantee that their case will win without incurring additional expert fees in the manner outlined above. It is just enough cover to get the claimant into serious difficulties. Almost inevitably, claimants are going to have to pay for additional insurance themselves. Since the Government's approach might be considered to be either too much or too little (but not 'just right') you might wonder why clinical negligence schemes should be treated differently from any other case.

Policies in respect of insolvency proceedings, publication and privacy proceedings and mesothelioma claims

[9.17]

The fifth commencement order under LASPO 2012 brought the end-of-recoverability provisions into force. It 'saved', ie prevented, this occurring in relation to certain types of proceedings in accordance with written answers previously provided by Government Ministers. Presumably it would only take a further commencement order to lift the brake on these proceedings and end recoverability thereafter. As such, there is not a great deal of permanence about this mechanism. It reflects the intention of dealing with these different types of claim in new ways in the near future. The definition of each set of proceedings is as follows:

- Insolvency – proceedings brought by a liquidator of a company or an administrator under the Insolvency Act 1986. They also include proceedings brought by a company being wound up or which has entered administration under that Act.
- Publication and privacy – proceedings for defamation, malicious falsehood, breach of confidence involving publication to the general public, misuse of private information or harassment where the defendant is a news publisher.
- Mesothelioma – a claim for damages in respect of diffuse mesothelioma (within the meaning of the Pneumoconiosis etc. (Workers' Compensation) Act 1979 (this is the definition in CPR 48.2(2) to which the regulations refer. It might have made more sense to refer directly to the 1979 Act.)

The rules of court to back up the Commencement Order can be found in CPR 48.2.

The fate of insolvency-related proceedings is not clear. The justification for its inclusion in the first place was opaque. Cynics suggested that HM Revenue & Customs wished to benefit from recoverable premiums in proceedings it brought. But in the absence of any clear reason as to why such proceedings were saved, it must be open to doubt as to how long that position can remain.

The exception for publication and privacy proceedings was a late change and followed the views expressed by Lord Justice Leveson in his report at the end of his inquiry into press ethics. The holding of the existing position is meant to last until some other mechanism is found. Lord Justice Leveson favoured the use of QOCS but the asymmetric relationship (individual claimant versus large, insured, corporation) is not always present and indeed can be quite the reverse (conspicuously wealthy celebrity versus small publisher).

Mesothelioma claims have benefited from a bespoke procedure which has meant that claims are dealt with very quickly in the hope of concluding before the claimant dies. It is not at all obvious why this category of personal injury claim should benefit from the saving on recoverability but a specific provision (LASPO 2012, s 48) was included well before the end of the Parliamentary process. Other options, such as an online process, are being considered as an alternative to the current proceedings. If a different option is available that reduces or eliminates the claimant's adverse costs risk, presumably ATE premiums will cease to be recoverable.

CAN THE CLIENT CHALLENGE THE PREMIUM LEVEL?

[9.18]

Can the client challenge the premium level? In a word, no. The premium, as with all of the other policy terms, is a matter of contract between the insurer and insured and the law does not prevent a party from entering into a bad bargain. So, in the absence of misrepresentation or any other breach of contract point, the premium agreed is the premium payable. The days of courts trying to grapple with burning costs, profit and administration elements and so forth are at an end for 'post-commencement' policies.

CONTENTS OF AN ATE POLICY

[9.19]

Notwithstanding the following comments on some specific matters, the wording of ATE policies has not changed greatly with the arrival of LASPO 2012. Furthermore, each insurer has its own wording and a detailed discussion of the different options is outside the scope of this book.

Collateral benefits

[9.20]

The Court of Appeal in *Callery v Gray (No 2)* [2001] EWCA Civ 1246 originally clarified the limits of insurance premiums made recoverable by s 29 of the Access to Justice Act 1999. At the time there was some concern by defendants that items outside those contemplated by the Parliamentary draftsman would be included and their cost would be reflected in the ATE premium sought. They were described as 'collateral benefits'. Perhaps the defendant's shot across the ATE insurers' bows prevented this from occurring. In any event the contents of ATE policies were largely uniform amongst the major players.

The ending of recoverability means that the policy contents are no longer of any interest to the defendant. If a claimant wishes to have a bells and whistles policy then he can pay for the additional items. It is only the recoverable policies – particularly the new clinical negligence policy – where the defendant will be keen to make sure additional benefits have not been loaded into the wording and additional premium charged as a result. Otherwise a party is free to contract with his insurer on whatever terms he wants.

Counsel's fees

[9.21]

One aspect which may change is the coverage of counsel's fees. Some insurers have always covered these fees but many have not. The capping of success fees against the client's damages has caused some barristers to propose returning to a 'disbursement' basis so that the success fee can be foregone. This would allow the solicitor more room to agree his own success fee with the client. If counsel is insured under an ATE policy, the client does not need to worry about payment of counsel's fees if the case does not go well.

Retrospective cover

[9.22]

Coverage under some policies is retrospective. This allows for cases to be worked up at no original risk to the insurer in terms of disbursements. If the case appears to have good prospects, it can then be insured and any disbursements already incurred can be covered in case the proceedings do not turn out as well as was hoped. However, this is a world away from the situation in *Hawksford Trustees Jersey Ltd v Stella Global UK Ltd* [2012] EWCA Civ 987, [2012] 1 WLR 3581, [2013] Lloyd's Rep IR 337 where a retrospective policy completely altered the balance of the costs claimed. In *Hawksford* a party who had not taken out insurance in the first instance proceedings sought to buy cover during the appeal that covered not just the appeal but also the costs of the first instance proceedings. The result was that a premium of £394, 638 was claimed by the respondent even though the appellant's estimate for the appeal was only £68,502. The Court accepted that if the appellant had won, the respondent was likely to be liable to pay both the cost of the appeal and of first instance. Nevertheless, by a majority decision the Court concluded that they were separate proceedings and that the proper construction of s 29 was that it related to the proceedings in which the policy was taken out and that other proceedings were separate and therefore not covered.

PAYING ADVERSE COSTS

[9.23]

Where the insured party loses, the opponent has no direct right against the insured party's insurers. If the insurer refuses to indemnify the insured and the insured cannot meet his liability to the opponent, the opponent has two choices if he wishes to take matters further.

Proceedings against the insurer

[9.24]

If the opponent believes the insurer has been actively involved in conducting the litigation he can seek a non-party costs order against the insurer under s 51 of the Senior Courts Act 1981. Otherwise he needs to step into the shoes of the insured to exercise the insured's rights against the insurer in accordance with The Third Parties (Rights against Insurers) Act 1930. In order to take that step, the insured has to be made insolvent and so this is not a step to be taken lightly. Furthermore, the opponent has no better right to make a claim than the insured. Therefore, if the insurer has avoided the policy in accordance with its terms, the opponent will not be able to seek payment any more than the insured could. In order to see how this works in practice, a good example is the decision of Mr Justice David Steel in *Persimmon Homes Ltd v Great Lakes Reinsurance (UK) Plc* [2010] EWHC 1705 (Comm), [2011] Lloyd's Rep IR 101.

The Third Parties (Rights against Insurers) Act 2010 was enacted to bring into effect changes in practice since the 1930s but as yet this has not been brought into force. When it is, it will end any lingering doubt as to whether voluntarily purchased insurance, such as LEI, comes within the provisions of the legislation. It will also allow a claim to be brought before the amount of the claim has been crystallised. Currently the amount of the costs has to be assessed (or agreed) before proceedings under the 1930 Act can be brought. Under the new Act, the quantification exercise can be brought as part of the claim against the insurer so that expense does not have to be incurred in a situation where no recovery is going to be made in any event.

Avoidance of a policy

[9.25]

The fact that the ATE insurance is based on a contract has been regularly overlooked by successful opponents, particularly in personal injury cases. It may be that the compulsory nature of motor and employer's liability insurance leads to the belief that all insurance will pay claims even if the insured has acted in contravention of the policy terms. But this is clearly not the case. If a claimant is found to be fundamentally dishonest in a personal injury claim, and therefore he loses his QOCS protection, it should be no surprise to the opponent (who must have been alleging and proving the deception) if the ATE insurer avoids the policy for material misrepresentations as to the underlying facts of the case on which the proposal for insurance was founded.

Order of payment?

[9.26]

Not all policies specify the order in which payments under the indemnity need to be made by the insurer. If it is not clear to you that there is any ranking, you should make sure you make a claim for unrecovered disbursements promptly if there is any prospect of adverse costs having to be paid and thereby reducing the indemnity available.

ATE AND PROSPECTIVE COSTS CONTROL

Costs Budgeting

[9.27]

If there ever was a need to include in a party's budget any additional liabilities – success fee or ATE premium – that effectively ceased when recoverability ended because the budget only deals with recoverable costs. There are of course many cases where proceedings have commenced after 1 April 2013 (and so need to be budgeted) where there is a pre-commencement CFA and/or ATE policy. Furthermore, all insolvency, privacy and mesothelioma cases continue to be 'pre-commencement' for the reasons explained earlier in this chapter. As such there are a number of situations where the existence of the ATE premium needs to be alerted to the court and the opponent in order to make sure that it is recoverable in principle at the end of the case. The method of doing so continues to be the Notice of Funding in form N251. Precedent H expressly states that any recoverable success fee or ATE premium is not to be included in the budget.

Unfortunately, the pilot schemes did not deal with this situation so clearly. In the case of *Murray v Neil Dowlman Architecture Ltd* [2013] EWHC 872 (TCC), 148 ConLR 256, [2013] NLJR 18, [2013] 3 Costs LR 460, the TCC pilot scheme in CPR PD 51G the solicitor did not specify that the additional liabilities were excluded from the budgeted figures and had to seek relief from sanctions to make sure that the budget was not read in this fashion (as the defendant contended.) It may be unfair to the drafters of the pilot scheme precedent to criticise it too severely since the solicitors in the *Murray* case do not appear to have used it in any event.

Security for Costs

[9.28]

We deal with security for costs applications in detail in Chapter 19. ATE insurance is essentially an unsatisfactory substitute for a bond or guarantee because of the insurer's contractual rights to deny coverage where the insured has not kept to the policy conditions.

9.29 *Chapter 9 Insurance Arrangements*

Costs Capping

[9.29]

In *Barr v Biffa Waste Services Ltd* [2009] EWHC 2444 (TCC), [2009] NLJR 1513, [2010] 3 Costs LR 317 Coulson J thought it entirely random to link the amount at which a claimant's costs could be capped to the amount that a defendant could recover against the claimants under an ATE policy, particularly where the latter figure was outside the control of the defendant and, at least directly, outside the control of the court. What mattered were the criteria in CPR 44.18(5) namely:

(1) Is there a risk that costs will be disproportionately incurred?
(2) If so, can that risk be adequately controlled by case management and/or detailed assessment of costs?
(3) In all the circumstances, is it in the interests of justice to make a costs capping order?

No costs capping order was made. The case is also a good example of the point made at the end of the section headed 'paying adverse costs.' The court always seems to consider that the limit of indemnity is always available to meet the opponent's costs and does not take any note of the client's own disbursements. This can often be a significant sum.

SOLICITOR SELF INSURANCE

Client's Disbursements

[9.30]

A solicitor can carry the liability for disbursements in the event that the case fails without taking himself outside the normal practice of a solicitor and turning himself into a funder. For the reasons discussed at **[10.10]** regarding commercial litigation funders, this is fortunate. If it were not so, solicitors could find themselves not only left out of pocket in respect of disbursements in an unsuccessful case, but also having to pay the opponent's costs to the value of twice the disbursements that he agreed to meet based on the decision in *Arkin v Borchard Lines*.

This position was recently confirmed by the Court of Appeal in *Flatman v Germany* [2013] EWCA Civ 278, [2013] 4 All ER 349, [2013] 1 WLR 2676. This decision overturned the judgment of Eady J which said that a solicitor who agreed to fund disbursements as the case progressed and not to seek them from the client if they could not be obtained from the opponent at the end had put himself into the position of a commercial funder. The subsequent High Court decision in *Tinseltime Ltd v Roberts* [2012] EWHC 2628 (TCC), [2012] NLJR 1290, 156 Sol Jo (no 38) 31 doubted Eady J and held that the funding of disbursements by a solicitor was insufficient reason to make a third party costs order. The Court of Appeal in *Flatman* agreed with the judge in *Tinseltime*. The Court of Appeal had the benefit of an intervention from the Law Society who pointed out that CLSA 1990, s 58, the starting point for CFAs, referred to agreements regarding 'fees and expenses' being paid dependent upon the outcome of the case. Notwithstanding a spirited argument from the defendant that there was a material difference between expenses and disbursements, the Court of Appeal (and the judge in *Tinseltime* who had seen the Law Society's skeleton argument) considered the solicitor was simply providing legal services in accordance with the statute.

There are postscripts to both *Flatman* and *Tinseltime* that are worth noting. In *Flatman* the appeal overall failed. It was an appeal against an order for disclosure of information in the detailed assessment proceedings. The Court of Appeal concluded that, although Eady J had fallen into error, there were other circumstances revealed on the appeal that justified the orders that Eady J had made. In *Tinseltime*, the solicitor thought that the disbursements would be modest and that, together with the fact that the company was on its uppers, persuaded him to act on the company's behalf. He also thought that the prospects were good but that was proved wrong when the claims were struck out. The solicitor was left with £22,000 to pay in disbursements: not a modest sum at all. This case is well worth reading if you are contemplating taking on any liability for your client to see just how easy it is to be led into difficulty.

Opponent's Costs

[9.31]

Offering to indemnify the client against the opponent's costs is a much more fraught undertaking than offering to meet disbursements. For a start you have much less say in the extent of the liability being incurred. Furthermore, the bargain of indemnifying the client in order to get the claim going, tends to have the appearance of champerty about it. There is also the question of whether you are acting as a quasi insurer without the necessary, regulatory approval.

The SCCO decision in *Dix v Townend* [2008] EWHC 90117 (Costs) raised the profile of champerty when the costs judge found that a solicitor undertaking to indemnify the client against adverse costs gave rise to a champertous agreement. The risk had been uninsured and the retainer was successfully challenged. Even if not champertous it was in any event contrary to public policy as explained by Lord Phillips MR in *R (Factortame Ltd) v Secretary of State for Transport, Local Government and the Regions (No 2)* [2002] EWCA Civ 932 (see [**10.10**]). Dix was not followed however in the decisions of MacDuff J in *Morris v London Borough of Southwark* [2010] EWHC 901, [2010] 4 Costs LR 526 and of Master O' Hare in *Murray Lewis v Tennants Distribution Ltd* [2010] EWHC 90161 (Costs). In the former the risk to the solicitor in having to pay opponent's costs under the arrangement made with their client was small and far outweighed by the advantages of the arrangements as a whole. In the latter it was simply not accepted that an agreement to shoulder the risk of adverse costs orders ought to be regarded as threatening the integrity of any solicitor.

The decision of MacDuff J in *Morris* was the subject of a conjoined appeal in *Sibthorpe v Southwark London Borough Council* [2011] EWCA Civ 25 and in which the Law Society intervened. The claimants' solicitors did not have a legal aid contract for housing work and the claimants could not afford the premium for after-the-event insurance, even if a policy could be found. The solicitors therefore offered to provide cost protection for their two housing disrepair claims to enable the tenants to proceed. In return the clients agreed to use CFAs which provided for success fees of 10%, limited in respect of the solicitors' costs to costs recovered. The defendants contended that the indemnity against the liability to pay the opponent's cost was tainted by champerty or maintenance because it gave the solicitors a financial interest, and therefore the whole CFA became unenforceable. The Court of Appeal held that although CLSA 1990, s 58 permitted some previously champertous agreements to be entered into, the arrangements in the present case were not covered by the Act. However, in no case cited to a court had it been held to be champertous for a person to agree to run the risk of a loss should the action fail without enjoying any gain should the action succeed. To hold the indemnity in the present case champertous would involve extending the law of champerty at a time when its scope was being curtailed rather than extended. The indemnity was accordingly not champertous.

As a side note, the defendant also sought to argue that the arrangement was a provision of insurance within the terms of the Financial Services and Markets Act 2000 (Regulated Activities) Order 2001. Such insurance could only be provided by an entity authorised by the FSMA 2000 and since the solicitors were not so authorised the CFA was void in accordance with s 26(1) of that Act. This ground of appeal was not given leave to appeal but it was dealt with in any event at the end of the Court of Appeal judgment. The CFA was for the provision of legal services. Whilst it provided an indemnity, it could not be categorised as a contract of insurance because its principal object was to provide legal services, not insurance.

Practical considerations

[9.32]

It is tempting with the fixing of recoverable court fees in various areas and the introduction of QOCS to consider this option rather than take out insurance. If nothing else it avoids a conversation about some (more) of your client's damages being taken as a result of the litigation process. It may be that, in the course of time, LEI insurers will be prepared to provide some form of excess layer or catastrophe insurance to allow solicitors to underwrite themselves up to a certain limit, but that day has not arrived. If there is no insurance in place and an indemnity given to the client, your liability is only limited by what can be agreed or is assessed by the court.

Given that *Dix* is the only decision in which the solicitor has been held to fall foul of the Financial Services and Markets Act 2000 in providing quasi-insurance it appears that this is a theoretical risk rather than a practical one. So you may only be struck off for being bankrupted, not for failing to comply with the Prudential Regulation Authority's code.

CHAPTER 10

STATE AND COMMERCIAL FUNDING

INTRODUCTION

[10.1]

In this Chapter we look at funding via the Legal Aid Agency and by Third Party Funders (aka Litigation Funders). We have said at the beginning of this part of the book that the use of Legal Aid is now so specialised that there is little point in dealing with it in detail. Those who might find such a chapter useful already know the detail: most people have no need to use a detailed exposition. Similarly, those who regularly use commercial funding will be au fait with the mechanics involved. Anecdotally, the commercial funders' evidence to Sir Rupert Jackson was that fewer than 100 cases had been backed by such funding at the time. On that basis, unless the take up has increased exponentially, most readers of this book do not currently need to know about the detail of such funding either.

But do not let that stop you reading this brief chapter. There are two very good reasons why you should do so. The first is that litigation funding is likely to increase over time and therefore it will become more relevant to more people. The second is that you need to be able to consider whether your potential client is eligible for Legal Aid or might benefit from commercial funding, when giving the best information possible to him at the outset of the case (and thereafter). This is so, even if you do not work with either form of funding. The correct advice to your client might be to go elsewhere so that the client can benefit from such funding. That sort of advice is at the core of being a professional advisor and is easy to forget in the ever more commercialised world in which the law is practised. In the case of *David Truex (a firm) v Kitchin* [2007] EWCA Civ 618, [2007] 2 FLR 1203, [2007] Fam Law 903 the Court of Appeal dismissed a solicitor's claim for costs against a client who was eligible for public funding on the grounds that solicitors are bound at the outset to consider whether a client might be eligible for public funding, rather than continue to take instructions and run up costs while they gathered information before considering public funding eligibility.

STATE FUNDING

[10.2]

In April 2013 the Legal Services Commission was replaced by the Legal Aid Agency, a new Executive Agency of the Ministry of Justice, and its powers and functions have been transferred to the Lord Chancellor. Whether that makes much practical difference remains to be seen but at least it means that everyone can go back to using the phrase 'legally aided' rather than attempting various less attractive phrases based on the Legal Services Commission.

Not for everyone

[10.3]

In February 1999 the Legal Services Commission (as it then was) started to contract with legal service providers to the exclusion of any lawyers who did not have such a contract. It was the beginning of the end of any solicitor wishing to act for a client with the benefit of state funding being able to do so. The restriction started in clinical negligence cases, but in 2000 it extended to the giving of any initial help in any civil matter or provide any level of service (ie including full blown representation) in family and immigration cases. In the same year most personal injury cases (except clinical negligence) were excluded from public funding. On 1 April 2001, the civil contracting scheme was extended to cover all levels of service for all types of case, which can now only be supplied under a contract. Further amendments have occurred since then, most recently with the enactment of LASPO 2012 to limit further the use of public funds in clinical negligence matters.

In short, unless you have a contract with the Legal Aid Agency (as it now is), you cannot act for your client with the benefit of public funds. Even if you do have such a contract, there are many types of claim which are no longer eligible for assistance. For example, Schedule 2 to LASPO 2012 provides that excluded from funding are civil legal services provided in relation to:

(a) personal injury or death;
(b) conveyancing;
(c) damage to property;
(d) the making of wills;
(e) matters of trust law;
(f) defamation or malicious falsehood;
(g) matters of company or partnership law;
(h) other matters arising out of the carrying on of a business;
(i) claims in tort in respect of negligence, assault, battery or false imprisonment, trespass to goods or to land;
(j) claims in tort in respect of breach of statutory duty.

Criteria

[10.4]

So, depending upon the nature of your client (for example, a company) or the type of case, you may well be able to advise your client at this point that they are not eligible for public funding. But if they are still potentially eligible, there are three aspects of the case and the client to consider.

- The merits of the case
- The cost benefit analysis
- The client's financial eligibility

In practice, you may want to start with the third aspect first since it will determine (negatively) the prospects more quickly than the other two. But let us consider them in order.

Merits

[10.5]

This is covered in reg 43 of the Civil Legal Aid (Merits Criteria) Regulations 2013. In a nutshell the merits have to be high and not the borderline prospects that might still justify a CFA or modest prospects which would be pursued under an ordinary pay as you go retainer. Whilst there are several classifications – very good (80% or more); good (60%–80%); moderate (50%–60%); borderline (less than 50%) – the cost benefit analysis will mean in most cases that only those with very good prospects are going to be funded sufficiently to get to trial if needs be. The percentages in the brackets above are not in reg 43 but come from the LSC Funding Code which had the same criteria and give a useful indication of the prospects of success required to fit within these categorisations.

Costs benefit analysis

[10.6]

One man's costs benefit might be described, somewhat contentiously perhaps, as another man's justice at proportionate cost. Regulation 42 of the Civil Legal Aid (Merits Criteria) Regulations 2013 sets outs the test for establishing whether the merits of the case justify its likely costs. The better the prospects of success (as set out above), the more costs in proportion to those prospects will be sanctioned. In itself it is not a bad approach and it might benefit you in considering your firm's approach generally to the investment in the case relative to its prospects. Of course if your payment is not dependent upon the outcome you can continue to consider prospects with a certain amount of equanimity.

Where the prospects of success are:
- Very Good – the anticipated damages must exceed the costs.
- Good – the estimated damages must exceed the costs by a ratio of two to one.
- Moderate – the estimated damages must exceed the costs by a ratio of four to one.
- Borderline – will only qualify for funding if it is of significant wider public interest or is of 'overwhelming' importance to the individual.

Financial eligibility

[10.7]

The acronym MINELAS (Middle Income Not Eligible for Legal Aid Support) was coined many years ago to highlight the many people who were not sufficiently wealthy to litigate without financial concern but who could not obtain Legal Aid. That position has steadily increased as the qualifying means have decreased. The criteria defy any brief description which would be of benefit to the practitioner. But it is safe to say that if you have a client who has a good case which can be taken to trial within the costs benefit analysis, they will have to be extremely lacking in income and capital to be able to take up Legal Aid funding in most circumstances. You may need to develop a relationship with a firm who has a contract if you regularly come across such clients so that they can assess the client's financial eligibility. If you need to look at something for yourself you can go to the website www.gov.uk/legal-aid/eligibility.

COMMERCIAL FUNDING

[10.8]

In order to deal with the current arrangements for litigation funding it is necessary to look briefly at the history of attempts by third parties to become involved in the dispute of others. There are many cases on this subject but we will look at just a few to give you the flavour of the issues involved. The common law restrictions on maintenance and champerty still remain, and the courts therefore still have to decide on the facts of each litigation funding agreement whether the contract is unenforceable on the grounds of public policy.

Maintenance and Champerty

[10.9]

These two concepts are at the heart of why funding by third parties has proved problematic over the years.

Maintenance is said to be 'the procurement, by direct or indirect financial assistance, of another person to institute, or carry on or defend civil proceedings without lawful justification' (The Law Commission 1966). More flamboyantly, the definition in the cases comes from the case of *British Cash and Parcel Conveyors Ltd v Lamson Store Service Co Ltd* [1908] 1 KB 1006, 77 LJKB 649, CA: 'Maintenance is the wanton and officious intermeddling with the disputes of others in which the maintainer has no interest whatever, and where the assistance he renders to the one or other party is without justification or excuse'.

Champerty, as described by the House of Lords judges in the case of *Giles v Thompson* [1994] 1 AC 142, [1993] 3 All ER 321, HL is (a) 'an aggravated form of maintenance. The distinguishing feature of champerty is the support of litigation by a stranger in return for a share of the proceeds' or (b) 'Champerty is maintenance with the addition of a division of the spoils of the litigation'.

On the face of it therefore, an agreement for a third party to fund the cost of a case in return for (say) a third of the damages is both champertous and maintenance.

The extent of a funder's liability

[10.10]

In *Arkin v Borchard Lines Ltd (No 5)* [2003] EWHC 2844 (Comm), [2004] 1 Lloyd's Rep 88, [2003] NLJR 1903 the Part 20 defendant (MPC), a professional funding company, entered into a funding agreement with the claimant.

MPC agreed to fund the employment of expert witnesses, the preparation of their evidence and the organisation of the enormous quantities of documents which it became necessary to investigate before the trial.

When the claimant lost, the defendants applied for a costs order against MPC. The defendants laid stress on the very substantial proportion of any recoverable damages or settlement payments (25% of the first £5 million and 23% of any excess) which MPC was to receive under its funding agreement. The amount of the claim, including exemplary damages, eventually reached $160 million. That would have meant a payment of some $40 million to the funders. The defendants also drew attention to the absence of any undertaking by MPC to pay the defendants' recoverable costs or to take out after-the-event (ATE) insurance cover in respect of such costs. They submitted that, in principle, professional funders, as distinct from pure funders (such as friends or family who back a case without expecting any reward), and who are maintaining litigation for their profit, should be liable for the costs of the defendants if their claim fails, which in this case it did.

MPC argued that funding agreements with professional funders which have the purpose of enabling impecunious claimants to pursue claims of real substance which, but for such funding, they could not have done, should not be visited with costs orders against the funders if the claim fails.

The High Court favoured MPC's public policy arguments and refused to make an order for costs against the Part 20 defendant. On appeal the Court of Appeal held that a professional funder, who finances part of a claimant's costs of litigation, should be potentially liable for the costs of the opposing party *to the extent of the funding provided*. In its judgment the court said:

> 'The effect of this will, of course, be that, if the funding is provided on a contingency basis of recovery, the funder will require, as the price of the funding, a greater share of the recovery should the claim succeed. In the individual case, the net recovery of a successful claimant will be diminished. While this is unfortunate, it seems to us that it is a cost that the impecunious claimant can reasonably be expected to bear. Overall justice will be better served than leaving defendants in a position where they have no right to recover any costs from a professional funder whose intervention has permitted the continuation of a claim which has ultimately proved to be without merit.'

The decision is very much policy driven – but unfortunately the Court clearly did not think that either policy objective completely outweighed the other. As such, the result is something of a half-way house. A professional funder who finances litigation to the cost of £500,000 will, if the client loses, expect to pay up to a further £500,000 in costs to the opponent. If the access to justice arguments had won, the funder would have paid nothing to the opponent. If the causation arguments had won (i.e. the case was only run because of the funding) the funder would have had to pay all of the opponent's costs.

In the chapter of his report on litigation funding by a third party, Sir Rupert Jackson suggested that this rule should be revisited by judges asked to consider awards of costs against funders – his steer suggesting that they should consider whether *Arkin* should be varied to make the funder liable for the whole amount. An example of that steer being exercised may be seen in the case of *Merchantbridge & Co Ltd v Safron General Partner 1 Ltd* [2011] EWHC 1524 (Comm). In that case the defendant was funded by third party investment

10.10 *Chapter 10 State and Commercial Funding*

banks who had an interest in the proceedings while not being a party to it. The Claimant succeeded and the judge determined that the investment banks should be liable to the full extent of the claimants' costs, in a departure from *Arkin*.

Funding of expert evidence

[10.11]

In *R (on the application of Factortame) v Secretary of State for Transport, Environments and the Regions (No 2)* [2002] EWCA Civ 932 impecunious Spanish trawler owners had obtained judgment against the British Government for damages for breaches of their fishing rights. Unfortunately, they could not afford to proceed with the assessment of damages. Accountants agreed to provide litigation support in the form of handling documents and programming services, as well as undertaking to pay the fees of expert witnesses, in exchange for 8% of any amount recovered. The Court of Appeal held that the agreement was not champertous, saying:

> 'Where the law expressly restricts the circumstances in which agreements in support of litigation are lawful, this provides a powerful indication of the limits of public policy in analogous situations. Where this is not the case, then we believe one must today look at the facts of the particular case and consider whether those facts suggest that the agreement in question might tempt the allegedly champertous maintainer for his personal gain to inflame the damages, to suppress evidence, to suborn witnesses or otherwise to undermine the ends of justice.'

The Court bore in mind the fact that the share of the damages taken by Grant Thornton was only 8%. They were a firm of accountants and so were members of a respectable and regulated profession. Whilst they played an important role in the preparation of the computer model on which the damages claims were based, this was subject to checking by the other side and was transparent. The claimants were in any event represented by highly experienced solicitors and counsel, and the solicitors had very properly insisted on remaining in control of the conduct of the litigation. In these circumstances, there was no realistic prospect of there being any undermining of justice by the 8% agreement. It was also clear to the Court that there was no other realistic way of the accountants being paid for work they had already done, other than to support the litigation in the manner they had chosen to do.

Current position

[10.12]

In *London and Regional (St George's Court) Ltd v Ministry of Defence* [2008] EWHC 526 (TCC), 121 ConLR 26, 152 Sol Jo (no 14) 28 Coulson J summarised the present state of the authorities as:
- the mere fact that litigation services have been provided in return for a promise in the share of the proceeds is not by itself sufficient to justify that promise being held to be unenforceable;

- in considering whether an agreement is unlawful on grounds of maintenance or champerty, the question is whether the agreement has *a tendency to corrupt public justice*, and such a question requires the closest attention to the nature and surrounding circumstance of a particular agreement;
- the modern authorities demonstrate a flexible approach where courts have generally declined to hold that an agreement under which a party provided assistance with litigation in return for a share of the proceeds was unenforceable;
- the rules against champerty, so far as they have survived, are primarily concerned with the protection of the integrity of the litigation process by the limitation of control of the conduct of the action by a third party.

THE FUNDING AGREEMENT

[10.13]

A litigation funder covers all the costs (or such costs as the funded party seeks to have covered by the funder) of the litigation in return for a share of the proceeds of the action, including its own legal costs, expert's and court fees and any adverse costs orders.

There is no need for a CFA (though funders will no doubt look more favourably on cases where the solicitor is prepared to act on at least a partial conditional fee agreement as a demonstration of their belief in the merits of the case in question). It is also said that there is no need for ATE insurance although this is usually sought to limit the funder's risk of an *Arkin* payment having to be made. The price is typically 20%–50% of the amount recovered, depending on how long this takes to achieve. This form of funding is not champertous provided the funder does not (and is not entitled to) control the conduct of the action by the client.

Agreements must be structured so that the client retains full control over the way in which they conduct their action. This includes the funder not being permitted to set minimum settlement levels when signing up a case, or interfering in the day to day conduct of the matter. Withdrawal of funding if the claim's merits plummet does not amount to interference.

Once the case is signed up, the client is then left to run his litigation in the usual way. The funding agreement sets out the responsibilities and liabilities of the parties. If at any stage the claim's merits suffer a 'material adverse decline' (which will include the ability of any successful judgment to be enforced against the defendant, the position on liability or claim value, or the conduct of the claimant, amongst others), the funder has the right to terminate the funding, while retaining the liability for all own side and adverse costs up to that date.

Litigation funders will say that the fact that they are only interested in funding good claims sends a powerful message to the opponent that a dispassionate and commercially focused third party also thinks the claim is good. Where the involvement of funders in accelerating settlement discussions has occurred, this has inevitably been a welcome development for third party funded claimants. ATE insurers used to say very much the same thing about the signal given to the opponent by their independent backing of the claim. To the extent that this was so, it lessened in importance over time as the use of ATE funding became more common. It will be interesting to see whether there is any impact in the announcement of the existence of third party funding (which is not something required by the CPR) and if so, whether its influence dissipates if it becomes more common place.

PROFILE OF FUNDED CASES

[10.14]

Litigation funding by a third party is not currently available to everyone. The minimum sizes of claim litigation funders will fund ranges from £350,000 to £25,000,000. There need to be sufficient damages available to make the time and effort invested in considering and setting up the claim worthwhile. In addition to value, the minimum eligibility criteria for considering funding a claim are:
- a defendant who can pay the amount claimed;
- good legal merits ie both in relation to liability and with a demonstrable (not aspirational) minimum claim value;
- where the costs of pursuing the matter are proportionate to the size of the claim; and
- the lawyer who it is proposed will run the claim is demonstrably experienced in the area to which the claim relates.

If you cannot afford to litigate otherwise, foregoing a percentage of the damages on success under a litigation funding agreement is of course a lot better than nothing. The alternatives of CFAs and ATE tend to have more value for the lower value claims. But now that success fees and premiums have ceased to be recoverable, litigation funding by a third party may be a more attractive option in some cases. It is certainly the case that in all substantial litigation, whatever the financial position of the client, the solicitor now has a duty to advise on this new method of finance.

THE ASSOCIATION OF LITIGATION FUNDERS

[10.15]

In November 2011 the Association of Litigation Funders (ALF) was formed (www.associationoflitigationfunders.com) as a forum for funders and non-funders alike to discuss matters relating to funding and provide a contact point for those using funding. At the same time a Code was launched and the ALF is responsible for future developments of the Code.

The Code sets out the standards of best practice and behaviour for litigation funders in the UK. It provides transparency to claimants and their solicitors and requires litigation funders to provide satisfactory answers to certain key questions before entering into relationships with claimants.

Under the Code, litigation funders are required to give assurances to claimants that, among other things, the litigation funder will not try to take control of the litigation, the litigation funder has the money to pay for the costs of the funded litigation and the litigation funder will not terminate funding absent a material adverse development.

The Code was originally encouraged and ultimately approved by Sir Rupert Jackson and commended by Lord Neuberger. A copy of the Code can be found on the ALF website.

At present the bulk of funding occurs in commercial litigation. If there are new entrants who operate solely consumer based litigation, there may well need to be a separate code and association to deal with the issues arising there. In the meantime of course, the consumer has the protection of the Consumer Credit Act 1974.

FEE SHARING IN PERSONAL INJURY CASES

[10.16]

Up to, and including, the SRA Code 2007, it was contrary to the solicitor's code of conduct for him to share any fees with a third party funder (or indeed anyone else other than another lawyer) in personal injury cases. The final incarnation of the relevant rule was Rule 9 but that has not been replicated in the 2011 version discussed in Chapter 1. The change in emphasis from prescriptive regulation to a concentration on the outcome means that a direct successor to Rule 9 would have been problematic. Furthermore, the ending of recoverability in CFAs and the introduction of DBAs all point towards the idea of the claimant parting with a share of his damages in order to secure representation. Although a commercial funder is not strictly part of the legal team, the same considerations apply.

As set out in the *London and Regional* decision, the modern view is that an agreement of this sort is not, without more, inimical to the client's best interests or those of the administration of justice. Consequently, an unhappy client who subsequently complains about the solicitor's conduct in using a third party funder, is unlikely to succeed unless the outcome of his litigation was prejudiced in some way as a result.

Part III

Between the parties

CHAPTER 11

AN OVERVIEW OF THE KEY COSTS AMENDMENTS TO THE CIVIL PROCEDURE RULES

INTRODUCTION

[11.1]

The costs landscape has altered dramatically in 2013. 1 April 2013 saw the introduction of the much trailed 'Jackson' reforms, with the ink still drying on the new rules, and the swingeing cuts to public funding. Before the summer was over the Low Value Personal Injury Claim Protocol had been the subject of radical revision and extension, fixed costs had been introduced to certain fast track personal injury claims and as this edition is being written the first decisions on costs, and linked case, management are starting to filter through from the higher courts. Most of these changes impact upon the recoverability of costs between the parties. It has been a busy year for practitioners and the judiciary alike. The supplement to the 2013 edition set out the major changes in some detail. The changes to funding arrangements have been considered in Part II. Those that concern the position between the parties will be discussed in context in this part. Some things, though, remain reassuringly familiar.

In simplistic terms CPR Part 43 is no more, CPR Parts 44, 45 and 47 have been revised, CPR Part 48 has become CPR Part 46 and Part 48 becomes transitional provisions. Well that is all clear then! However, in case it is not, what follows in this chapter will be a brief look at the revised costs provisions in more detail. This is not intended to be an exhaustive look at where all the costs provisions can now be found, but instead merely to highlight some key additions, deletions and movements. Before that though, it is necessary to consider two key amendments to the CPR that underpin the ethos of the new post Jackson era.

THE OVERRIDING OBJECTIVE

[11.2]

As we shall see later (see Chapter 14 – Control of Costs – Proportionality), proportionality is the key to the reforms. It is no surprise, then, to find that the 'Holy Grail' that is the overriding objective has been amended – so that the rules are designed to enable the court to deal with cases justly *and at proportionate cost*. Justice has been qualified and this qualification resonates at every procedural stage. Given the definition of proportionality in CPR 44.3(5) it is clear that rather than neatly sidestepping costs issues in most claims, leaving them to be resolved with, if necessary, in the words of Brooke LJ, an axe on assessment, the court is now charged with placing costs at centre stage throughout the claim. Every procedural decision must be made against a consideration of the cost implications.

There have been two changes to the overriding objective and the second of these reflects the other key amendment – the provision for relief from sanction.

RELIEF FROM SANCTION

[11.3]

What has relief from sanction to do with costs? The answer is everything. The amendments to CPR 3.9 to place the emphasis on the need for the efficient and proportionate conduct of litigation, whilst enforcing compliance with orders, rules and practice directions (both of which provisions are reinforced by the altered overriding objective), make it clear that dilatory conduct, whether deliberate or by mere oversight, which inevitably increases the costs, will no longer be tolerated. Failure to comply with orders or rules drives a coach and horses through carefully constructed case management, which will have been informed in the multi track by agreed and approved or set budgets and in the fast track by reference to the overriding objective – in other words by consideration of the reasonable and proportionate costs for the litigation. It will no longer be enough to suggest that the consequences of breach on the 'innocent party' can be compensated for by an order for costs.

PART 44

[11.4]

Whilst sections of Part 44 remain familiar, even those have been subject to a numbering change to ensure that those who like to reel off references have work to do. Surely the most irritating one is that the award of costs provision at CPR 44.3 is now found at CPR 44.2.

A curious deletion from Part 44 is what was CPR 44.5, relating to the provision of information about funding arrangements – the need to serve an N251. In respect of funding arrangements entered in to after 31 March 2013 there will be no requirement to notify the other party of the existence of a conditional fee agreement ('CFA'). Presumably the logic for this is that any success fee under the agreement is no longer recoverable between the parties. However, the existence of such an agreement is still relevant to both the other party and the court in respect of interim hearings and entitlement to payment. If the CFA does not define 'success' as success on an interim application, then the client has no liability to pay the solicitor at the time of the interim order and so payment of any costs assessed should not be ordered at that stage – a simple application of the indemnity principle. However, how is either the court or the paying party to know this if no notice of funding has to be given?

Costs only proceedings have found a new home in Part 46 as have the provisions relating to the specific costs limitations under small claims and fast track cases and costs in cases reallocated. The other significant departure from Part 44 is cost capping. This has been sensibly relocated to CPR 3.19–CPR 3.21 where it follows on immediately from the new provisions in that part in respect of costs management, creating a seamless section on costs control.

The major additions to Part 44 are the introduction of the rules relating to 'qualified one way costs shifting' ('QOCS') and to Damages Based Agreements.

Qualified one way costs shifting

[11.5]

During Sir Rupert Jackson's preliminary investigations it was suggested to him by an unnamed insurance company that the cost recovered by the insurance industry in those personal injury/fatal accident cases that it successfully defended was less than the amount it paid out in after the event insurance premiums on those claims it lost. A suggestion was made that if the insurance industry waived its right to costs in those claims it defended successfully, then there would be no risk to claimants of adverse costs orders and therefore no need to insure against that risk (justifying the simultaneous abolition of the recoverability of insurance premiums between the parties).

This suggestion has found its way in to the CPR, but in qualified terms. There are four qualifications, two requiring permission of the court and two not:

(i) An adverse costs order may be enforced against the claimant *without court permission* BUT only up to the level of any damages and interest that the claimant has recovered (ie as an offset). This provision is intended to ensure that there is some sanction in respect of any interim applications upon which the defendant is the successful party and some risk to a claimant in respect of CPR Part 36.

(ii) An adverse costs order may be enforced in full against the claimant *without court permission* if the claim has been struck out on the basis that it discloses no reasonable grounds, if it is an abuse of court or if the conduct of the claimant or a person acting on the claimant's behalf, and of whose conduct the claimant has knowledge, is likely to obstruct the just disposal of proceedings. Without this exception there would be no risk to deter a claim that is so flawed or conduct that is so culpable that the claim merits being struck out.

(iii) An adverse costs order may be enforced in full *with court permission* where the claim is found on the balance of probabilities to be fundamentally dishonest. This should prove to be an interesting battleground, raising many a question. What is fundamentally dishonest? The rule refers to the claim, suggesting the entirety of the claim. What of those cases where there is an injury for which compensation is recovered, but there has been substantial exaggeration, the claimant has not beaten a Part 36 offer, but the defendants costs dwarf the amount they can offset against the damages under i) above? Given that most claims where there is exaggeration seem to settle soon after service of surveillance evidence, and the practice direction supporting the rules suggests that issues relating to allegations that a claim is fundamentally dishonest will normally be determined at trial and only in exceptional circumstances will the court order that the issues are determined when the claim has settled, does this exception have any real teeth? No doubt costs judges breathed a sigh of relief that determination of 'fundamentally dishonest' will not fall at their door in the course of a detailed assessment. Inevitably, at this early stage there are far more questions than answers.

(iv) An adverse costs order may be enforced in full *with court permission*:
 (a) where the claim is made for the financial benefit of someone other than the claimant or a dependant under the Fatal Accidents Act 1976. This does NOT include claims for gratuitous care, for medical expenses or for an employer; and
 (b) where the claim is made for the benefit of the claimant other than a claim to which this section applies

CPR 44.13–CPR 44.17 is not retrospective. The transitional arrangements provide that 'QOCS' does not apply to claims under funding arrangements entered in to before 1 April 2013. A funding arrangement refers to:

- agreements for advocacy or litigation services which provide for payment of a success fee or under which work has already been done; or
- where an insurance policy was taken out prior to 1 April 13; or
- where an agreement has been entered in to with a member ship organisation to meet costs prior to 1 April 13.

There is a sting in the tail with a trap for the unwary as this provision does *not* apply to applications for pre-action disclosure. This is not at all surprising as, of course, a claim for pre-action disclosure is a free standing claim and is not a claim for personal injury, a claim under the Fatal Accidents Act 1976 or a claim under the Law Reform (Miscellaneous Provisions) Act 1934.

Damages Based Agreements

[11.6]

CPR 44.18 provides the procedural support for the statutory introduction of this method of funding (see Chapter 7 – Funding for more detail on Damages Based Agreements).

PART 45

[11.7]

Apart from minor alterations to the particular sections there have been few changes to Part 45. Many of the numbering changes have been caused as a result of the abolition of recoverable additional liabilities necessitating the deletion from the rules of the various fixed success fee provisions.

Sensibly the fixed trial cost provisions of what was Part 46 have been moved to Part 45 (at CPR 45.37–CPR 45.40) to bring all fixed cost rules within the same part.

The only major additions are those of limits on recoverability of costs in Aarhus Convention cases and fixed costs in certain fast track personal injury claims. These provisions are dealt with at CPR 45.41–CPR 45.43 and CPR 45.29A–CPR 45.29L.

PART 46

[11.8]

With fixed trial fees in fast track moving to Part 45 then Part 46 has altered completely. It is, in fact, Part 48 as was, save that the former Part 48.3 – 'Amount of costs where costs are payable pursuant to a contract' – has moved to Part 44 and by way of a trade the provisions relating to pro bono costs, costs only proceedings and reallocation have moved in the opposite direction.

PART 47

[11.9]

Part 47 remains concerned with assessments. The only changes of any significance are those that introduce i) the procedure for provisional assessment of between the parties bills of £75,000 or less (CPR 47.15) and ii) the sanctions of Part 36 to a receiving party's costs offer (CPR 47.20). In respect of the former it had been widely anticipated that the £75,000 figure would be exclusive of VAT. However, CPR PD 47, para 14.1 simply gives a total figure and so, at first blush, fewer bills than expected will fall within the provisional scheme. However, it would seem that the Senior Court Costs Office is interpreting this as a VAT exclusive figure and this seems correct as the definition of 'costs' in CPR 44.1 does not include VAT.

PART 48

[11.10]

Do not throw out all the old rules yet! As stated in the introduction to this chapter Part 48 contains the transitional provisions. These provide that the pre-1 April 2013 rules apply to funding arrangements in place before that date. CPR 48.2 defines such a funding arrangement. It should come with an ice pack and a darkened room. It must come with the old rules as CPR 43.2(1)(k) is essential if the provision is to be understood. However, in broad terms, if there was a pre-1 April 2013 funding agreement, as defined in CPR 43.2(1)(k) or, in respect of a collective conditional fee agreement, services were provided to the party under that agreement, then the old rules apply.

Beware, though, for Part 48 only deals with transitional provisions in respect of funding arrangements in place before 1 April 2013. Other transitional provisions crop up in different locations.

11.11 *Chapter 11 Overview of Key Costs Amendments to the CPR*

OTHER TRANSITIONAL PROVISIONS

Proportionality

[11.11]

Given the significant changes caused by CPR 44.3(2) and (5) it will have come as a relief to practitioners to find that the new proportionality test and 'proportionality trumping necessity' (see Chapter 14 Control of Costs – Proportionality) do not apply to cases commenced before 1 April 2013 or, if the case had not been issued by that date, but work had been done on it, then to the cost of the work undertaken prior to that date. However, as will be seen later that does mean that for a period the court will have to apply two different proportionality tests to separate parts of work done on the same case at assessment.

QOCS

[11.12]

Part 44.17 makes it clear that QOCS only applies where the claim is one where no funding arrangement was in place before 1 April 2013.

Provisional assessments

[11.13]

This procedure only applies to those cases where the detailed assessment commenced after 31 March 2013. In other words all cases where the Notice of Commencement was served after 31 March 2013 and the bill is for £75,000 or less will be subject to the provisional assessment scheme (CPR 47.15(1)).

THE COSTS PRACTICE DIRECTION

[11.14]

It would be wrong not to mention the costs practice direction. Instead of one practice direction the amendments have introduced specific practice directions that relate to the individual cost parts of the CPR. Gone are the days of trying to match up the individual parts of the CPR with one PD, they are now self-contained and specific as with the rest of the CPR.

CHAPTER 12

THE INDEMNITY PRINCIPLE

[12.1]

From the new, the April 2013 CPR amendments, where else can any consideration of costs between the parties go than to the old and the indemnity principle? In England and Wales it is usual for the loser of litigation (whether that is at a discrete interim hearing or a trial or an appeal) to be ordered to pay the winner's costs. Costs are awarded to indemnify the winning party for the costs and expenses incurred. Although this is called an indemnity it is rarely, if ever, a full indemnity for a variety of reasons: the costs that the client has agreed to pay to the solicitor, the barrister or for disbursements are unreasonably incurred or are reasonably incurred but unreasonable in amount or are reasonably incurred and reasonable in amount but are disproportionate. All of these reflect situations over which the loser has no control and it would be wrong for the winner to be indemnified for them.

NO PROFIT

[12.2]

Although it is extremely unusual for the costs ordered between the parties to be as much as the receiving party's solicitor and client costs, what is certain as a general rule is that costs between the parties can never exceed the solicitor and client costs. Every general rule is qualified and the indemnity principle is no exception. The obvious exceptions arise as a result of the application of one of the fixed cost schemes (although given the concerns over the economic viability of undertaking work under such schemes it is hard to conceive of a situation where the fixed cost would now exceed the solicitor and client cost) and where the receiving party has made an effective offer under CPR 47.20 and is entitled to an extra percentage on the costs assessed. Apart from the exceptions the receiving party is entitled only to be indemnified for the actual liability to his solicitor and cannot make a profit out of the costs recovered from the other party. The first record of this principle being expounded was in *Harold v Smith* (1860) 5 H & N 381:

> 'Costs as between party and party are given by the law as an indemnity to the person entitled to them: they are not imposed as a punishment on the party who pays them, nor given as a bonus to the party who receives them. Therefore, if the extent of the indemnification can be found out, the extent to which costs ought to be allowed is also ascertained.'

It is clear from this wording that CPR 47.20 clearly offends the principle as the extra percentage is one or both of a punishment or a bonus. However, it is an expressly permitted exception.

12.3 *Chapter 12 The Indemnity Principle*

POSSIBLE PITFALLS AND SOME EXCEPTIONS

(a) Agreements to charge no fee and impecunious clients

[12.3]

If a solicitor agrees to work for nothing then the client has no liability for costs and, therefore, has no need for, nor entitlement to, an indemnity from the other party even if successful in the litigation. The seminal case on this is *Gundry v Sainsbury* [1910] 1 KB 645, 79 LJKB 713, CA in which the Court of Appeal held that to award costs to a client whose solicitor had agreed not to charge costs would have been giving a bonus to the party receiving them. This is a simple application of the law as laid down in *Harold v Smith*.

However, it is, perhaps, more understandable that a solicitor may agree to act on the basis that there will be no charge unless the claim succeeds. If a potential client has no money, but a good claim the solicitor will wish to obtain the instruction. The client cannot afford to pay unless the claim is successful and so a creative retainer is required. Sadly what may seem creative may also fall foul of the law. In *Gundry v Sainsbury* itself, the successful client had said in cross examination 'I could not pay costs and I had arranged with my solicitor not to pay the costs of the action'. Whilst the solicitor disagreed, he was precluded by the Attorneys and Solicitors Act 1870, s 4 from giving evidence and stating his version of the retainer. As a result there was no recovery of costs between the parties.

More recently in *British Waterways Board v Norman* (1993) 26 HLR 232, [1993] NPC 143, the court held that where solicitors agreed to act for a client whose financial circumstances were such that there was no prospect of the client paying the solicitors' costs unless the claim succeeded, in the absence of a specific agreement to the contrary, there must have been an understanding between the solicitors and the client that they would not look to her for any costs if she lost.

This case is often referred to as having been overruled by *Thai Trading Co (a firm) v Taylor* [1998] QB 781, [1998] 3 All ER 65, CA. However, *Thai Trading* was concerned about the lawfulness of a 'no win, no fee' agreement and it was only this aspect of the judgment in *British Waterways Board v Norman* that was overruled by *Thai Trading*. The basic tenet remains that, if a solicitor expressly or impliedly agrees that he will not in any circumstances charge his client, no costs are recoverable from the other party.

Two subsequent cases in which the question of impecuniosity was invoked were *Carr v Leeds City Council* and *Coles v Barnsley Metropolitan BC* (1999) 32 HLR 753, [2000] COD 10, DC, in which the local authorities sought to avoid paying costs ordered against them on the grounds that private prosecutions in respect of defective properties were funded by the solicitors on the basis that their clients would not be liable to pay any costs to them, win or lose, and that any purported agreement to the contrary was a sham. In both cases the court found there was no sufficient basis for the allegation.

In *Burstein v Times Newspapers Ltd (No 2)* [2002] EWCA Civ 1739 the claimant was awarded his costs of a successful libel action against the defendants. The defendants contended that when the claimant's solicitors became aware that their client was no longer able to pay their costs, the agreement in their retainer that he would pay their costs became a sham and unenforceable. Accordingly, the defendants contended that the claimant had no liability towards his own solicitors for costs. The judge and the Court of Appeal rejected that argument, holding that the material produced by the defendants did not undermine the evidence of the claimant's solicitors that the agreements were proper. The defendants' proposition that a retainer was, or became, champertous and therefore unlawful and unenforceable if a solicitor became aware at any time that their client could not have afforded to pay his costs, was a proposition which could not on the authorities be supported.

The problems with the impecunious client have reappeared, albeit in a different context, recently. In *Heron v TNT (UK) Ltd* [2013] EWCA Civ 469, [2013] 3 All ER 479, [2013] NLJR 21 it was suggested, amongst other submissions, that the failure of the paying party's solicitors to seek after the event insurance demonstrated that the firm had become a party to the action and, as such, a non-party costs order under s 51 of the Senior Courts Act 1981 and CPR 46.2 should be made. In *Flatman v Germany* [2013] EWCA Civ 278, [2013] 4 All ER 349, [2013] 1 WLR 2676 the successful defendants argued that where a claimant pursued a claim under a CFA, with no after the event insurance, was impecunious and where disbursements might have been paid by the solicitors on behalf of the client, then this, too, might form the basis for a non-party costs order. In both cases the receiving party was unsuccessful. Neither the failure to obtain after the event insurance, nor the funding of disbursements rendered a solicitor a 'real party' to the claim.

(b) Pro bono

[12.4]

A topical example of working for no fee, whatever the outcome of the litigation, is that of the professional pro bono groups. In its consultation paper, '*Costs Recovery in Pro Bono Assisted Cases*', the Department for Constitutional Affairs (now the Ministry of Justice) said:

> 'Pro bono work is an important adjunct to the main strands of the provision of legal services. However, there is an injustice caused by the fact that costs cannot be recovered from the losing party if the successful party is represented pro bono, even though a pro bono assisted party would be liable for his opponent's costs if his opponent won. In such cases, the opponent is aware that if he loses, he will not be asked to pay the other party's costs because that party is represented pro bono. It is intended that the sums recovered will go, not to those providing the representation but, to a prescribed charitable body that will administer and distribute monies received to voluntary organisations that provide free legal support to the community.'

The result was s 194 of the Legal Services Act 2007 and the abrogation of the indemnity principle in pro bono assisted cases. Section 61 of the Legal Aid, Sentencing and Punishment of Offenders Act 2012 extends the previous provision by permitting the Supreme Court to make pro bono awards. The procedure is set out at CPR 46.7 and the prescribed charity is the 'Access to Justice Foundation'. The consensus appears to be that the possibility of pro bono orders being made in the Supreme Court may serve to publicise the possibility of such orders and lead to an increase in their number.

(c) Payment by a third party

[12.5]

Litigation is increasingly being funded under arrangements whereby the solicitor's costs are to be paid by a third party, perhaps a before the event insurer, a trade union, an insurance company or a pure litigation funder. In order to recover costs from another party in such a situation it must be shown that the client had a primary, or dual, liability for the solicitor's costs, as the client was able to do in *R v Miller (Raymond)* [1983] 3 All ER 186, [1983] 1 WLR 1056, DC. It is therefore important expressly to incorporate in the client care letter the right to look to the client for payment of the solicitor and client costs if the case is won, and, indeed, if it is lost, unless there is a valid conditional fee or damages based agreement. Otherwise, on the basis of *British Waterways Board v Norman*, it may be argued that there is an implication to the contrary and that the client has no liability to the solicitor for which there is an entitlement to be indemnified between the parties.

In *Ilangaratne v British Medical Association* [2007] EWHC 920 (Ch), [2008] 3 Costs LR 367 the insurers of the BMA, who were successful in the action, had instructed the solicitors and the claimant alleged there was therefore no retainer between the defendant and the solicitors – another breach of the indemnity principle. However, there was a standing arrangement between the insurers and the solicitors, as to costs evidenced by a letter between them. Although the letter itself was not a written contract or retainer it was evidence of a standing arrangement that any instructions to the solicitors on behalf of the insurer's customers would give rise to a retainer between the customer and the solicitor on terms already advised and agreed, including the charging rate previously agreed with the insurers. There was therefore a retainer.

(d) Fixed fee

[12.6]

Where a solicitor agrees to work for a fixed fee, then that is the maximum amount that can be recovered from another party. As will become apparent when considering 'item by item' at (f) below and 'Costs Control', this may present some interesting challenges when a case is costs managed.

(e) Interim bill

[12.7]

If, during the course of litigation, the solicitor delivers a bill to the client for work done to date – a stage payment – that bill is a statute bill and not merely a request for payment on account and the solicitor has fixed costs at that amount for the work done during the period covered by the bill and cannot recover a higher amount from another party for that period.

(f) Item-by-item

[12.8]

A dramatic development in the application of the indemnity principle was the decision of the Court of Appeal in *General of Berne Insurance Co v Jardine Reinsurance Management Ltd* [1998] 2 All ER 301, [1998] 1 WLR 1231, CA. Two Supreme Court Taxing Office masters (now costs judges) had reached conflicting decisions on whether the application of the indemnity principle should be on an item-by-item basis, or whether it only provides a global cap, so that the receiving party may recover on assessment uplifted hourly expense rates which are judged to be reasonable, even if they exceed the rates that the solicitors were entitled to receive from their client under a contentious business agreement, provided that the total amount allowed between the parties did not exceed the total amount the solicitors were entitled to recover from their client. There were arguments both ways and, in the words of the Court of Appeal, 'each of the parties' cases has its problems'. The court came down in favour of the item-by-item approach, even though it accepted that examples could arise where complicated and painstaking reductions could be required, commenting 'taxation [assessment] of costs can be a laborious procedure in any event, and can be expensive in taxing [assessing] fees'. No one would quarrel with that. *Nederlandse Reassurantie Groep Holding NV v Bacon & Woodrow* ([1998] 2 Costs LR 32, QBD), a review of a taxation [*detailed assessment*] by Tucker J, confirmed that the *General of Berne v Jardine* interpretation of the indemnity principle applied to all retainers, whether or not a formal contentious business agreement was in place.

The full consequences of these decisions are still emerging, for example: where interim costs have been awarded against the receiving party or where parts of the receiving party's costs have been excluded; if the receiving party's solicitor was acting for a fixed fee, there cannot be an item-by-item comparison; if the retainer provides that the solicitor will not charge the client more than the amount of costs recovered between the parties what is the amount for which the client is entitled to be indemnified? Nil? Or perhaps the solicitor should agree not to charge the client *less* than the amount recovered between the parties! How is a retainer for a fixed expense rate with a flexible mark-up to be compared between the parties? How does the indemnity principle apply where there is a financial limit on the funding certificate? How does the amount of the success fee under a conditional fee agreement affect the application of the indemnity principle? How does the principle apply where a party is entitled to fixed costs? In any event, the Law Society recommends the inclusion of a 'slip clause' when acting on an hourly charge-out basis to permit an additional percentage in the final bill in certain circumstances to cover the item-by-item approach if necessary.

This problem becomes more pronounced when the claim in which the solicitor is representing the client on a global fixed fee comes within the costs management provisions of CPR 3.12–CPR 3.18. Form H is broken down into phases with the budget to be set by reference to a sum for each phase (CPR PD 3E, para 2.3). How is a global fixed fee to be split between the phases? Will solicitors undertaking fixed fee work have to agree how the overall sum is to be split between the phases with the client at the time of creating the retainer? If not how will the solicitors complete the Form H?

One suggestion is that the Form H is completed as though there were no fixed fee in place. In other words the solicitors prepare the Form H using a more conventional time cost approach. There are myriad difficulties with this approach. The first is if the total budget calculated this way exceeds the fixed fee. This is because CPR PD 22, para 2.2A provides that the statement of truth verifying a costs budget requires the solicitors to confirm that 'the costs incurred do not exceed the costs which my client is liable to pay in respect of such work. Even if the costs incurred at the time of the Form H certification are within the fixed fee, can the solicitor honestly give this certificate when there was never a breakdown of that fixed fee? Arguably estimated fees to be incurred in the budget present less of a problem if the total of the budget is less than the fixed fee as the certification is less stringent 'The future costs stated in this budget are a proper estimate of the reasonable and proportionate costs which my client will incur in this litigation'. However, even if the problems at the time of completing the Form H are overcome the assessment presents challenging issues. On standard basis assessments in such cases the court has regard to the receiving party's budget *for each phase*. At this stage the solicitors will have to accept that the absolute cap is the fixed fee and this will open the door to arguments over how this has been attributed per phase. In addition merely being within the overall fixed fee which is higher than the budget does not satisfy the requirement to be within the budget per phase. With an increase in the number of clients demanding fixed fee work this is not merely an academic point.

(g) Disclosure

[12.9]
A matter which has bedevilled assessments of costs for some years is how is the paying party to discover the terms of the retainer to ascertain the level of the indemnity to which the receiving party is entitled? Almost invariably the first challenge in the Points of Dispute relates to whether or not there is a valid retainer. There can be no doubt that the onus should be on the receiving party to satisfy the court that the terms of the retainer provide an entitlement to the indemnity for costs sought. This view received support in *Bailey v IBC Vehicles Ltd* [1998] 3 All ER 570, 142 Sol Jo LB 126, CA, in which the court held that in future, any client care letter setting out the terms of the client's financial obligations to the solicitors should be attached to the bill of costs together with any contentious business agreement or other relevant documents.

However, Henry LJ suggested that the signing of the between-the-parties bill of costs by the solicitor as an officer of the court was effectively a certificate that the receiving party's solicitors were not seeking to recover in relation to any item more than they had agreed to charge their client and was, usually, sufficient confirmation that there had been no breach of the indemnity principle. There was a presumption of trust and any breach of that trust should be treated as a most serious disciplinary offence.

This suggestion was rejected by the Court of Appeal in *Hollins v Russell* [2003] EWCA Civ 718, in respect of challenges to conditional fee agreements, which it distinguished from conventional challenges, as in *Bailey*. In *Bailey* the paying party was not saying that there was no liability at all to pay any costs to the receiving party, but was challenging the hourly rate and mark-up being applied to it.

This *Hollins* type challenge to retainer and the principle of paying anything at all increased the stakes significantly from a mere challenge to the figures produced. The conditional fee agreement (CFA) regulations introduced a new level of complexity. The solicitor's certificate as to accuracy might not be sufficient evidence of there being no breach of the indemnity principle where the quality and quantity of the information served on the paying party about the success fee was less than would be made available in respect of the other aspects of the bill in the case of an assessment where there was no additional liability claimed. The question of whether a CFA complies with the Courts and Legal Services Act 1990 is principally a matter of law, while challenges to conventional bills are generally questions of fact.

This approach has been exemplified in both the Supreme Court Costs Office's *Guide to the Summary Assessment of Costs* and in the case of *Hazlett v Sefton Metropolitan Borough Council* [2000] 4 All ER 887, [1999] NLJR 1869, DC in which it was held that for the purpose of making an order for costs between the parties, there is a presumption that the client would be personally liable for the solicitor's costs and it would not normally be necessary for the client to have to adduce evidence to that effect. However, where there was a genuine issue raised by the paying party as to whether the receiving party had properly incurred costs in the proceedings, the position would be different. If it were alleged that the receiving party was not liable to pay the solicitor's costs, whether because the client had entered into an unlawful and an unenforceable CFA with the solicitor or for any other reasons, the client would be at risk if continuing to rely upon the presumption that there was a liability for the solicitor's costs. If the client did not then adduce evidence to prove that costs had properly incurred in the proceedings or the paying party could show by evidence or argument that the client had not, it would be most unlikely that the costs would be successfully recovered. The need for a claimant to give evidence to prove the entitlement to costs rather than relying upon the presumption would not, however, arise if the paying party simply put the claimant to proof of the entitlement to costs. Then the claimant would be justified in relying on the presumption.

In practice post *Hollins* there tends to have been voluntary disclosure of CFAs in detailed assessment proceedings. If that does not happen then the court ought not to take it upon itself to examine the CFA to determine the dispute without disclosure to the paying party unless the parties are happy for this to be the approach. In *Dickinson (t/a Dickinson Equipment Finance) v Rushmer (t/a FJ Associates)* [2002] NLJR 58, ChD there was an appeal from a detailed assessment by a costs judge to Rimer J (as he was) sitting with assessors. On the detailed assessment the paying party (the defendant) invoked the indemnity principle, asserting that the receiving party (the claimant) could not have assumed a personal responsibility to pay costs of the amount claimed in his solicitor's bills. The claimant's solicitors produced to the costs judge documents to prove the terms of the retainer and demonstrate there was no breach of the indemnity principle. The costs judge refused to allow the defendant to see the documents on the grounds they were privileged. It was held that the costs judge was wrong. It is one of the most basic principles of natural justice that each side is entitled to know what the other side's case is and to see the documentary material on which he is relying. The receiving party, without producing any documents, could ask the costs judge to direct whether he regarded the paying party as having raised a genuine issue which needed to be met by evidence, or if he accepted the signature certification on the bill of costs that the indemnity principle had not been offended. If the receiving party pre-empts any such decision by the costs judge by producing the documents, the paying party is entitled to see them, even though they are privileged.

In *South Coast Shipping Co Ltd v Havant Borough Council* [2002] 3 All ER 779, [2002] NLJR 59 the costs judge had been shown documents which the paying party had not been allowed to see on the grounds that they were privileged. The appeal raised human rights issues, in particular the conflict between the right to privilege and the right to a fair trial. If the costs judge, having seen the documents in question, required the receiving party to elect between giving secondary evidence of the retainer and waiving the privilege, there was no incompatibility with the principles articulated by the European Convention for Human Rights. That was not intended to suggest that the costs judge should put the receiving party to its election in respect of every document relied on, regardless of its degree of relevance. In the great majority of cases the paying party should be content to agree that the costs judge alone should see privileged documents. Only where it was necessary and proportionate should the receiving party be put to his election. Otherwise the redaction and production of privileged documents, or the adducing of further evidence, would lead to additional delay and increase costs. Again in practice frequently the parties are content for the retainer issue to be resolved by the court examining the relevant documentation (which must be filed where there is a dispute pursuant to CPR PD 47, para 13.2 (i)) and indicating whether or not it is satisfied there is a valid retainer.

The issue of whether or not there has been a breach of the indemnity principle may become less common following the abolition of recoverable additional liabilities. Where it is raised it will be interesting to see how the court deals with it when undertaking a provisional assessment on paper without the parties present.

(h) Unlawful retainer

[12.10]

Worst of all, if a solicitor enters into an invalid retainer with the client it is unenforceable against the client and the client can therefore recover nothing from the other party. This has been the case where the court has found CFAs to be unenforceable. With CFA funding still available and still subject to some statutory regulation and with the jury still out on Damages Based Agreements this remains a genuine risk to solicitors undertaking work on these types of retainer.

(i) Fixed costs

[12.11]

In *Butt v Nizami* [2006] EWHC 159 (QB), sometimes reported as *Nizami v Butt*, the claimants both suffered whiplash injuries when the car they were travelling in was hit by the defendant's car. Their claims were settled, but costs could not be agreed. They commenced costs-only proceedings under CPR Part 8 claiming fixed recoverable costs under CPR 45.9, disbursements under CPR 45.10 and a fixed success fee under CPR 45.11. The defendants alleged that the claimants' solicitors had failed to make appropriate enquiries about the availability of before-the-event insurance and sought a direction for the solicitors to certify compliance with the Conditional Fee Agreement Regulations 2000. The costs judge held that the claimants' entitlement to costs depended not on the existence of a valid and enforceable conditional fee agreement, but on their entitlement under the fixed recoverable costs rule. He was upheld on appeal.

The intention underlying CPR 45.7–CPR 45.14 (now CPR 45.9–CPR 45.15) is to provide an agreed scheme of recovery that is certain and easily calculated by providing fixed levels of remuneration, which might over – reward in some cases and under – reward in others, but which were regarded as fair when taken as a whole. There was a change in the law effected by the amendment to s 51(2) of the Senior Courts Act 1981 which significantly modified the indemnity principle and permitted changes in the rules to give effect to the modification.

It was clear that it was intended that the indemnity principle should not apply to the figures that were recoverable, and accordingly there was little reason why the indemnity principle should have any application to CPR 45.9 (now CPR 45.11) and CPR 45.11(no longer relevant), and good reasons why it should not. The CPR had successfully disapplied the indemnity principle in relation to the predictable costs scheme.

CPR Part 45 now encompasses all the prescribed fixed costs regimes. The decision in *Nizami* illustrates that whatever type of fee retainer solicitors are acting under it is implicit in all fixed costs regimes that the indemnity principle is disapplied. In Section 1, Table I gives the fixed costs on commencement, Table II costs on entry of judgment, Table III the fixed costs on commencement of a claim for the recovery of land or a demotion claim, Table IV miscellaneous fixed costs, which are concerned mainly with service of documents and Table V fixed enforcement costs. The prescribed amounts are payable without any enquiry into the terms of the retainer between the receiving party and their solicitor. Section II is concerned with predictable costs in road traffic accident claims. Section IIIA applies to certain injury claims that no longer remain in the low value injury protocols. A further form of fixed costs is the fixed trial costs on the fast track, which as seen in the preceding chapter are now at Section VI, which prescribes that the court shall not award less than the amount shown in the table. Again this clearly precludes an investigation into the terms of the successful party's retainer and, for example, it is accepted that the amount of counsel's brief fee is of no relevance, even if it is substantially less than the prescribed amount of the fixed costs. The attraction of the certainty and consistency of fixed, or predictable costs, would be undermined if they were subject to the indemnity principle. Fixed costs are an incentive to efficiency and a sanction against inefficiency.

(j) BTE insurance

[12.12]

In *Ghadami v Lyon Cole Insurance Group Ltd* [2010] EWCA Civ 767, [2010] 6 Costs LR 903, the claimant pursued a claim against the defendant who had a before-the-event indemnity insurance policy covering the costs of the proceedings subject to an excess of £1,000. The defendant paid its solicitors the excess and the insurers paid the balance of their costs. The claimant lost and was ordered to pay the defendant's cost. The claimant contended that as the defendant's liability for costs was limited to the excess of £1,000, the indemnity principle prevented the defendant recovering more than the excess from the claimant. On appeal it was held that there was an implicit agreement that the solicitors would act for the defendant in relation to the claim without any express terms as to costs. There was therefore no agreement whereby the liability of the defendant for its solicitor's fees and disbursements was expressly limited to £1,000 and they were entitled to recover the full between the parties costs.

THE FUTURE OF THE INDEMNITY PRINCIPLE

[12.13]

For years there has been a clamour for abolition of the indemnity principle. Sir Rupert Jackson described in his final report how he had received forceful submissions from both pro and anti-abolitionists. He concluded that the principle should be abrogated. It was thought that the introduction of damages based agreements, which are no more than contingency agreements by another name, might sound the final death knell. However, the indemnity principle has survived the latest reforms. This seems curious when:
- there are the numerous permitted exceptions to which reference has already been made;
- costs management removes the risk of utterly unchecked costs; and
- the statutory framework already exists – s 31 of the Access to Justice Act 1999, which has not been implemented, amends s 51 of the Senior Courts Act 1981 to provide that the amount recovered by way of costs may not be limited to 'what would have been payable by him (the client) to them (the solicitors) if he had not been awarded costs.'

For the time being, though, the indemnity principle survives, albeit with an ever increasing number of permitted exceptions.

CHAPTER 13

PROSPECTIVE COSTS CONTROL – INTRODUCTION

[13.1]

It has long been acknowledged that the major failure of the Woolf reforms, introduced as the CPR in 1999, has been the failure to control costs. In part this may be an unfair criticism as the introduction of recoverable 'additional liabilities' in 2000, which led to the beginning of over a decade of 'costs wars', and the failure of the government to implement a fixed costs scheme in the fast track, have certainly exacerbated the situation. Certainly the focus on costs has not been in the direction that Lord Woolf intended when stating in his Final Report:

> 'I recognise that my reforms involve learning new skills. These will have to be learned not only by judges but by members of the profession generally. The profession as well as the judiciary must pay more attention to and be better informed about costs than they are at present. My objective is to require greater attention to be focused on costs throughout the process of resolving disputes by everyone involved: judges, litigants and lawyers.'

Instead the emphasis, by dint of restrictive rules on costs capping, a lack of policing of the then existing powers under the CPR to require costs estimates (and limited use of those estimates that were filed) and a reliance on ex post facto cost assessment, has served only to shift the gaze very much to the end of the court process. As a result, as courts at all levels have repeatedly said, by the time many claims reach trial the case is no longer about the original dispute, but instead is all about the costs because neither party can afford to lose. The other inevitable consequence of this approach is that prospective litigants are deterred from pursuing legitimate claims because of their inability to fund the costs.

It was in recognition of this failing that Sir Rupert Jackson focussed much of his attention on prospective costs management by the court, not only to curb expenditure, but also to ensure that litigants are able to make informed decisions knowing the cost consequences of pursuing or defending claims and to enable the court to allocate increasingly restricted resources proportionately between cases.

As a result formal costs management has been introduced for most multi track cases, after two pilot schemes trail blazed the path. Arguably costs management renders costs capping redundant, but for the moment that, too, remains an option. In addition amendments to CPR Part 36 are designed to offer parties greater incentives to settle claims, and these changes are extended to the detailed assessment process to try to discourage retrospective arguments on costs.

At the heart of the costs reforms is the concept of proportionality – familiar in name, but with centre stage comes a different meaning and application.

CHAPTER 14

PROSPECTIVE COSTS CONTROL – PROPORTIONALITY

INTRODUCTION

[14.1]

The concept of proportionality permeates throughout the CPR, from the amended overriding objective, through case and costs management to costs assessment at the conclusion of a claim. In theory this is nothing new. Prior to April 2013, CPR 1.1(2)(c) included in its definition of 'dealing with a case justly', the requirement to deal 'with the case in ways which are proportionate'. Case managers, with an eye to the overriding objective, ought to have been giving directions that were proportionate. In addition, the case of *Lownds v Home Office* [2002] EWCA Civ 365, [2002] 4 All ER 775, [2002] 1 WLR 2450 introduced the test of proportionality to the assessments of costs – if the global costs claimed by the receiving party were disproportionate then individual items would only be recoverable if *necessarily* incurred and reasonable in amount. The need to keep control over costs, linked to proportionality, was reinforced in *Leigh v Michelin Tyre plc* [2003] EWCA Civ 1766, [2004] 2 All ER 175, [2004] 1 WLR 846 and subsequently codified in Section 6 of the Costs Practice Direction. So why has proportionality taken on such importance within the reforms and why has it attracted so much comment – much of which is highly critical? Is the criticism justified?

THE CASE FOR REFORM

[14.2]

Whilst Lord Woolf went on to qualify his definition of proportionality in *Lownds*, his earlier comments in that case set out the purists' position:

> 'If, because of lack of planning or due to other causes, the global costs are disproportionately high, then the requirement that the costs should be proportionate means that no more should be payable than would have been payable if the litigation had been conducted in a proportionate manner.'

However, it became accepted wisdom that applying the two – fold test of *Lownds* did not necessarily achieve this desired outcome. Even after preliminary findings that the global costs of cases were not proportionate in amount, there was a perception that assessments could, and frequently did, still lead to assessed costs, on an item by item basis, that remained disproportionate (See for example the comments of May LJ set out at Chapter 3, para 4.7 of the '*Review of Civil Litigation Costs: Final Report*')

14.2 *Chapter 14 Prospective Costs Control – Proportionality*

This was a theme picked up Sir Rupert Jackson in his Final Report. His remit was, as he described, 'to promote access to justice at proportionate cost'. He devoted an entire chapter to consideration of proportionality and returned to what he saw as the causes of disproportionate costs at various points in his analysis and conclusions. His clear verdict was that:

> 'Access to justice is only practicable if the costs of litigation are proportionate.'

It was with this desire to preserve access to justice in mind that Sir Rupert Jackson set out the proposals that have become enshrined in the Civil Procedure Rules from 1 April 2013.

THE REFORM

[14.3]

As we have already seen, in order to leave no one in any doubt as to the prominence of proportionality within the reforms, the overriding objective', has been amended to provide at CPR 1.1(1) that:

> 'These Rules are a new procedural code with the overriding objective of enabling the court to deal with cases justly and at proportionate cost.'

In simple terms this means that when making procedural decisions on any case the court must ensure that the outcome enables the claim to be determined at proportionate cost. This has led to concern being expressed across the legal community that pursuit of justice has been qualified – that claims will no longer be investigated fully. The court will be compelled to dispense a lesser, cost constrained, version of justice. This debate will be considered in more detail below.

The court must determine proportionality by reference to a new definition. This is set out at CPR 44.3(5) as follows:

> 'Costs incurred are proportionate if they bear a reasonable relationship to—
> (a) the sums in issue in the proceedings;
> (b) the value of any non-monetary relief in issue in the proceedings;
> (c) the complexity of the litigation;
> (d) any additional work generated by the conduct of the paying party; and
> (e) any wider factors involved in the proceedings, such as reputation or public importance.'

What does this mean? Like many statutory and procedural checklists this does not lead to a clear and precise specific outcome. One person's view of proportionality will differ from that of another. The answer in any given case is that this definition will lead to whatever the judge considering the checklist, in the context of the case before him, decides is proportionate within the parameters of a reasonable exercise of judicial discretion. Unsurprisingly, this, too, has come in for sustained criticism. Different judges considering the same case might come to different views. The easy answer is that this is nothing new! Up and down the country every day this happens – not just on a procedural level, but also in terms of the final decisions made. Exercises of discretion and particular factual findings inevitably lead to different conclusions.

The specific concern with this checklist, though, is that, when applied prospectively to case management decisions and to budget setting within the costs management regime, different decisions on proportionality may result in a more thorough investigation of the claim being permitted in one court than in another. Again this concern will be addressed below.

As stated, proportionality is central to the new costs management regime. CPR PD 3E, para 2.3 sets out the way in which the court will determine the budget. It provides:

' . . . The court's approval will relate only to the total figures for each phase of the proceedings . . . When reviewing budgets, the court will not undertake a detailed assessment in advance, but rather will consider whether the budgeted costs fall within the range of reasonable and proportionate costs.'

In other words the court will consider the CPR 44.3(5) factors and will then determine the reasonable and proportionate costs for each phase of the proceedings. As the budget figure for each phase will be set as a total sum under CPR PD 3E, para 2.3, it is the consideration of proportionality that ultimately will be determinative of that amount.

At first blush, it seems curious that one of the five factors seemingly cannot inform that decision. At the time of setting the budget, the identity of the paying party will not be known. Does this mean that the court must assume, for the purposes of setting a particular party's budget, that it will be the receiving party? There is nothing in CPR 3.12–CPR 3.18 or CPR PD 3E that supports this broad interpretation. Accordingly, even if the conduct of one party has already increased the costs of the other, it seems that this cannot be taken in to account at this stage as that party may not ultimately be the paying party.

However, on reflection, there is clear logic to this. The court is not able to budget costs that have already been incurred and is not, at this stage, undertaking an assessment of those costs. Any arguments on conduct increasing incurred costs can be taken when considering the reasonableness of costs and the overall proportionality at any subsequent assessment. By robust case management the court will control the procedural conduct of parties going forward. Any subsequent conduct issues can either be dealt with by free standing applications (the costs of which fall outside the budget – CPR PD 3E, para 2.9) or by subsequent arguments on assessment that the conduct complained of is 'good reason' to depart from the budget set. As such there is no reason to consider past conduct when setting the budget. Indeed to do so may itself lead to disproportionately time consuming argument.

As an aside there has been some concern that the conduct of the receiving party is omitted from the proportionality checklist. This can be readily explained – it would be superfluous. In respect of those costs not budgeted, then any conduct issues will be dealt with, as appropriate, on the award of costs (CPR 44.2(5)), whether in respect of free standing applications or the overall claim, and at any subsequent assessment of the reasonable costs under such orders (CPR 44.4(3)). Both these provisions contain clear obligations on the court to consider conduct. In respect of budgeted costs, as has already been stated, these are set by reference to reasonableness as well as proportionality. Accordingly there is already provision for the consideration of conduct.

14.3 Chapter 14 Prospective Costs Control – Proportionality

The final proportionality procedural change is undoubtedly the most pervasive and seismic. It results in the demise of the *Lownds* test and its replacement by something altogether more consequential. *Lownds* has received sustained criticism from the Court of Appeal and it was no surprise to find that Sir Rupert Jackson recommended its reversal in the final report:

> 'In other words, I propose that in an assessment of costs on the standard basis, proportionality should prevail over reasonableness and the proportionality test should be applied on a global basis. The court should first make an assessment of reasonable costs, having regard to the individual items in the bill, the time reasonably spent on those items and the other factors listed in CPR rule 44.5(3). The court should then stand back and consider whether the total figure is proportionate. If the total figure is not proportionate, the court should make an appropriate reduction.'

This recommendation has found its way in to the CPR at CPR 44.3(2):

> 'Where the amount of costs is to be assessed on the standard basis, the court will—
>
> (a) only allow costs which are proportionate to the matters in issue. Costs which are disproportionate in amount may be disallowed or reduced even if they were reasonably or necessarily incurred; and
> (b) resolve any doubt which it may have as to whether costs were reasonably and proportionately incurred or were reasonable and proportionate in amount in favour of the paying party.'

The potential consequence of this for the receiving party ought to be enough to persuade parties to sign up willingly to prospective costs management! The court will undertake an assessment of the non-budgeted costs on a reasonableness test, stand back at the end and, if the overall costs are not proportionate (budgeted plus non-budgeted), reduce the overall amount to the figure that the court determines to be proportionate by application of the CPR 44.3(5) factors. The irreducible minimum will be the budgeted costs as, of course, unless there is a 'good reason' to depart from the budget, CPR 3.18 protects the budgeted costs.

We say that the provisions of CPR 3.18 protect the budgeted costs because they have already been subjected to a proportionality (and indeed reasonableness) analysis under CPR PD 3E, para 2.3 when the budget was set. However, it may be that this certainty has to be tempered to take account of the decision of Moore-Bick LJ in *Troy Foods v Manton* [2013] EWCA Civ 615, [2013] 4 Costs LR 546. Moore-Bick LJ was concerned with an application for permission to appeal a costs management decision under CPR PD 51G – the Costs Management in Mercantile Courts and Technology and Construction Courts – Pilot scheme. The defendant sought to challenge the decision to approve the claimant's costs for witness statements and the hourly rate for counsel because it submitted that the judge had not had sufficient concern for the effect that setting the budget has on subsequent detailed assessments on the standard basis. The defendant suggested that concern was necessary as it was likely that costs judges would treat the approval of the budget as establishing that costs then incurred would be reasonable if they fell within the approved budget. Moore-Bick LJ granted permission to appeal and in so doing commented:

'It follows that I do not accept that costs judges should or will treat the court's approval of a budget as demonstrating, without further consideration, that the costs incurred by the receiving party are reasonable or proportionate simply because they fall within the scope of the approved budget.'

Whilst an expression of regret that the claim subsequently settled without the appeal being heard is not a particularly proportionate sentiment, it is a shame that this decision has created some uncertainty at a time when the new rules are only just bedding in and the Court of Appeal has not had an opportunity to clarify the position. However, we suggest that this decision is, in fact, of limited impact for the following reasons:

- The appeal arose in the context of a pilot, rather than under the generally applicable costs management provisions. Interestingly the pilot scheme under CPR PD 51G is opaque when prescribing how a budget is to be set. CPR PD 51G, para 1.3 contains the only reference to 'reasonable and proportionate' costs, but does so when dealing with the relevance of costs that have already been incurred and which cannot be budgeted. The only other assistance in the pilot scheme is to be found at CPR PD 51G, para 4.2 which describes the objective of costs management as being 'to control the costs of litigation in accordance with the overriding objective'. It is certainly arguable that under this pilot the costs have not been subjected to a prospective determination of reasonableness. This is in marked contrast to costs management under CPR PD 3E, para 2.3 which clearly states that the costs for each of the prescribed phases of the proceedings will be set by reference to the range of reasonable and proportionate costs. If a phase of a budget is set after court determination that it is both reasonable and proportionate, then the costs officer/judge on any subsequent assessment is constrained unless there is 'good reason'. The onus is on the paying party to establish 'good reason' rather than the court having to consider reasonableness and proportionality again as a matter of course.
- There is a suggestion that the judge approved the contentious part of the budget despite expressing doubts about whether the figures were reasonable.
- The decision relates only to permission to appeal. The respondent did not attend and to that extent the court only heard one side of the argument.
- The effect of this decision would be to encourage paying parties to go to detailed assessment to have a second bite of the cherry – quite the reverse of Sir Rupert Jackson's stated goal to reduce both the frequency of and the time spent on assessments.

So, whilst *Troy* was settled and the Court of Appeal was denied the opportunity to clarify the position, we remain firmly of the view that revisiting reasonableness and proportionality on assessment will be extremely rare and can only be done within an argument that there is 'good reason' to depart from the budget – eg where decisions on reasonableness and proportionality have been made by the costs managing judge on the basis of a high value claim that transpires to have been exaggerated. However, fuelled by the words of Moore-Bick LJ we anticipate that it will not be long before the Higher Courts have another opportunity to consider this.

14.3 *Chapter 14 Prospective Costs Control – Proportionality*

In any event the wording of CPR 44.3(2) has consequences that extend far further than merely to an assessment of costs. It is simply here that it becomes glaringly apparent that proportionality trumps necessity. It is the practical consequences of this provision at a far earlier stage on case and costs management that support the argument of those who believe justice has been compromised.

PROPORTIONALITY AND JUSTICE

[14.4]

The concerns expressed by the legal community, which became more vocal as 1 April 2013 approached, have been referred to above – cut price justice not being justice at all, unpredictability as one court may take a different view of proportionality from another (even one sitting in the court/room next door) and a victory of process over outcome.

Even the staunchest supporters of the reforms accept that in some cases proportionate case management will result in less just outcomes than had there been a lesser form of costs control. The analysis set out in Chapter 15 – Prospective Control of Costs and Case Management below of how directions may have to be tailored out of the cloth available, renders this an inevitability.

However, those supporters are keen to stress that in the majority of cases there will be no discernible difference in outcome. In the large part the reforms are aimed to take waste out of the justice system. To target statements to essential material by limiting the length of the statement, to limit disclosure to fewer, key documents, to avoid waste of valuable resources by ensuring that parties comply with court orders and rules and to reduce the length of the trial, will avoid prolix and unfocused hearings where the judge is referred to only a few pages of the plethora of paper paginated in many lever arch files and ensure that cases are pursued efficiently or run the risk of being struck out.

Those advocates of reform advance the case that without change justice had become the right of all, but the privilege of only the few – those with sufficient financial resources or those under a funding scheme that gave them total immunity from risk (and therefore no interest in controlling cost). As such, an imperfect system that re-opens the court doors to a greater number, results in better justice than where many cannot afford the entry fee to open the court door, which is no system at all.

Whatever the force of the competing views, the reforms are in place. Proportionality, in its new guise, is centre stage and it is time for practitioners and judges to move from theory to practice. The final word goes to Jacob LJ in*Nichia Corpn v Argos Ltd* [2007] EWCA Civ 741, [2007] Bus LR 1753, [2007] FSR 895 who, perhaps, best expresses the inevitable compromise that will be the changes:

> ' "Perfect justice" in one sense involves a tribunal examining every conceivable aspect of a dispute . . . No stone, however small, should remain unturned . . . But a system which sought . . . "perfect justice" in every case would actually defeat justice. The cost and time involved would make it impossible to decide all but the most vastly funded cases. The cost of nearly every case would be greater than what it is about. Life is too short to investigate everything in that way. So a

compromise is made: one makes do with a lesser procedure even though it may result in the justice being rougher. Putting it another way, better justice is achieved by risking a little bit of injustice.'

CHAPTER 15

PROSPECTIVE COSTS CONTROL – COSTS AND CASES MANAGEMENT

INTRODUCTION

[15.1]

Sir Rupert Jackson's Final Report identified that, in conjunction with the requirement for the court to case manage proceedings, there was also need for costs management to become part of that process. His view was that case and costs management are inextricably linked – that no claim should be case managed without consideration of the costs implications of any given step in those proceedings. This view has become central to the costs management scheme introduced in CPR Part 3. Indeed CPR 3.12(2) could not make this any clearer:

> 'The purpose of costs management is that the court should manage both the steps to be taken and the costs to be incurred by the parties to any proceedings so as to further the overriding objective.'

However, the link between the two exists even if we take a broader view of costs management than that within the formal scheme at CPR 3.12–CPR 3.18. This is because the amendments to the overriding objective make it plain that proportionality of costs lies at the heart of every case – whatever track it may be on and in whatever division it may be issued. This is reinforced in multi track cases by CPR 3.17 which leaves no room for doubt that even in non-budgeted cases every case management decision comes with a price tag.

WHAT IS COSTS MANAGEMENT?

[15.2]

A costs management order ('CMO') enables the court to control the parties' expenditure throughout the proceedings. Strictly this is qualified to the extent that all that the court does by a CMO is set the amount that, in the absence of a 'good reason' to depart from the prescribed budget, will be recoverable between the parties on an assessment on the standard basis. However, as we shall see when looking at some specific examples of routine case management decisions, the inextricable tie between case and costs management means that the effect of a CMO may trespass into, and influence, solicitor and client costs. This is because the requirement for the parties to provide information as to their estimated costs, as well as costs already incurred, will enable the court to consider these estimates alongside the directions to be given for the case management of the proceedings. Where a court considers the estimated costs to be disproportionate, it will tailor the directions to bring the costs down to a reasonable and proportionate level; this is particularly likely to affect those phases of the litigation where costs have traditionally been disproportionate such as disclosure, experts' reports and trial.

A CMO will record the extent to which the budgets are agreed between the parties. Where no agreement has been reached it will record the court's approval of the budgets after the court has made appropriate revisions. The court may also set a timetable, give directions for future reviews of budgets or make provision for parties to notify it of any agreed changes to the budget. In addition, the court may consider holding costs management conferences to consider revising the budgets.

COSTS MANAGEMENT PILOT SCHEMES

[15.3]

The feasibility of implementing the costs management provisions was tested in two pilot schemes:
(a) 'Costs Management in the Birmingham Mercantile and TCC courts' (Costs management scheme). This was subsequently extended to all Mercantile and TCC courts.
(b) 'Defamation Proceedings Costs Management Scheme' (the Defamation costs scheme). This ran in the Royal Courts of Justice and the Manchester District Registry.

There was a marked difference in the use of the two schemes. The Costs management scheme was voluntary and used by commercial clients who, in a difficult economic climate were already making stringent costs demands on their lawyers. Those lawyers, in the main, already have relatively sophisticated time recording/case management systems which assisted in monitoring the costs expenditure in line with the budgets prescribed by the court. Indeed for many of them the Form H appears primitive when compared with the case/cost management matrices already being demanded by clients. In contrast the defamation pilot proved to be more controversial, no doubt in part for the simple reason that it was compulsory, but also because of the confrontational nature of, and the historic concerns over costs expenditure in, such litigation.

As a result feedback messages have been mixed. There have been both critics and supporters of the pilots. These are considered in the 'Costs Management Final Report'. The summary of the Final Report provides an indication of the benefits that costs management may provide to practitioners, and, more importantly, their clients. These are:
- it makes the parties focus on the issues early on, and more thoroughly analyse what is necessary to prosecute the action;
- it helps to focus on the costs of the future conduct of the case;
- it informs the parties about each other's budgets for the litigation and provides an insight into the opponent's tactics;
- it introduces a degree of certainty to the planned amount of work and costs for the client, and provides a strong incentive to keep within the budget;

- it may avoid lengthy detailed assessments of costs at the end of the litigation; and it informs the parties about the costs consequences of not settling at an early stage and thus can encourage settlement.

Despite the differing viewpoints, the simple fact is that from 1 April 2013 practitioners involved in proceedings subject to the new costs management regime need to understand how the regime works and embrace the change. Indeed whilst the regime is in its infancy astute practitioners actually have a chance to influence the way it develops. The range of opinions from panellists and the audience at the 2013 Association of Costs Lawyers conference suggests creative minds are already turning to fertile areas for dispute.

THE SCOPE OF THE COSTS MANAGEMENT REGIME UNDER CPR 3.12–CPR 3.18 AND CPR PD 3E

[15.4]

The formal costs management scheme applies to all multi-track cases commenced on or after 1 April 2013. A number of exceptions to this, set out in CPR 3.12, provide that it will not apply:

- In the Admiralty and Commercial Courts. This is regardless of the value of the claim.
- In any proceedings subject to fixed costs, eg claims for the recovery of land.
- In any proceedings subject to scale costs, eg patent cases in the county courts.
- If the court itself orders that costs budgeting will not apply. There is no provision within the rules as to why the court should make such an order and therefore it will be interesting to see the extent to which such orders are made. There is a school of thought that costs management may be unnecessary for defendants in those claims subject to the QOCS provisions. Why put the defendant to the expenditure of budget production when, in the vast majority of cases, the maximum recovery would be limited to the level of damages? There are many answers. Two obvious ones are that if there is a valid offer under CPR Part 36 in a high value case the defendant might still be able to recover all the costs by way of set off. Another is that the court needs the costs information to fulfil its function to case manage proportionately.

At the eleventh hour, two further exceptions were added and as a consequence the costs management regime will not apply to any cases in either the Chancery Division of the High Court or the Technology and Construction Court (TCC) and Mercantile Courts if the sums in dispute, as at the date of the first CMC, exceed £2 million.

The late addition of these two exceptions was due, in part, to a concern that parties and their legal advisers might seek to issue in the Commercial Court in cases that could just as comfortably be in that court as the Chancery Division, Mercantile Courts and TCC, in an attempt to avoid being subject to the new costs management regime – at least until any anticipated satellite litigation as to the interpretation of the new CPR provisions had been forthcoming. This perceived risk of forum shopping led to the exceptions.

15.4 *Chapter 15 Prospective Costs Control – Costs and Cases Management*

The exceptions in the Chancery Division and the Queen's Bench Division met with mixed reviews in any event, including criticisms that they have resulted in a clear watering down of the attempt by Sir Rupert Jackson to bring about a change in the culture of costs management. Another legitimate criticism was made by those practitioners who had invested time and money preparing unnecessarily for implementation.

In addition, by definition, the exceptions apply to high value cases which are more likely to incur substantial and, often, out of proportion costs and are surely the very types of claim at which the provisions are aimed. At the time of writing a sub-committee of the Civil Procedure Rules Committee has produced a consultation paper suggesting the £2 million exemption may be 'unnecessary and inappropriate. The consultation will consider whether commercial litigants (including international ones) would be deterred from using our courts if they were subject to costs management provisions. The conclusion will make interesting reading. We, too, wonder what the commercial client users of the courts that have been exempted from the regime make of all this. They have legal budgets to manage and reserves to place on claims. It is difficult to believe that they would not want the increased certainty that costs management brings, particularly if they are aware of the provisions of CPR 44.3(2) and the proportionality cross check on retrospective assessment of costs. However, it seems from the comments of those that represent them that they resist the straight jacket that will be compulsory clothing under a budget.

We understand that the TCC will be conducting its own review of the £2 million exception by examining cases which have gone through the first CCMC since 1 April 2013. What remains to be seen is whether this is to enable a decision to be made to remove the exception so that all cases in the TCC are subject to costs budgeting or whether it may result in the exemption being applied across all cases in the TCC. Ramsay J, who took over the reins from Sir Rupert Jackson in guiding the reforms through their latter stages up to implementation, has publicly stated that he does not consider the exceptions for the various courts to be sustainable and considers that they will be removed. The provision in CPR 3.12(1)(c), which leaves it to the discretion of the President of the Queen's Bench Division to determine which cases should be exempt, means that any change could simply be done through a court guidance notice rather than requiring any amendment to the CPR provision itself. For anyone with matters proceedings in the TCC, or looking to commence proceedings there, it will be important to be aware of any change to the remit of the current exemption.

The £2 million exception is stated to exclude interests and costs, although again the court has the discretion to order otherwise. On the surface this seems a simple enough basis for an exception but is bound to raise its own issues, for example:

- The exceptions are based on the sums in dispute rather than the level of costs and consequently parties with claims around this value may succumb to a temptation, at least in the early days, artificially to inflate the value of the claim to beyond the £2 million mark so as to avoid the costs budgeting regime altogether.
- The reference is to the 'sums in dispute' and so will include not only the claims by the claimant but also any counterclaims by the defendant.

- Is the value of the 'sums in dispute' to be that at the start of the CMC or at the end of the CMC after any applications have been dealt with. A successful application may have a big impact on the amount of the sums in dispute, eg an application to strike out part of the claimant's claim. If for example a £2.5 million claim was reduced to a £1.8 million claim following a successful application would the courts then seek to adjourn the CMC to enable the parties to put together costs budgets and seek to agree them? If so, parties may take a strategic decision not to make an application for, eg strike out until after the CMC given that there is no provision for the court to introduce costs budgeting at a later date in the proceedings once the value of the sums in dispute fall below the £2 million level.

Even in those matters proceeding in the Chancery Division, the Technology and Construction Court or the Mercantile Courts which are not subject to costs budgeting, the court is still required under CPR 3.17 to consider the costs when taking any case management decision. It will be interesting to see how judges seek to implement this requirement without the aid of costs budgets, given that the courts' failure to do so in the past under the then overriding objective which required 'dealing with the case in ways which are proportionate', in part, contributed to the need for reform and implementation of the costs budgeting regime.

THE PROCEDURAL CODE

Filing and service of the budget

[15.5]

Apart from litigants in person who are excluded from the requirement to prepare costs budgets, all parties in cases falling within the regime must file and exchange budgets. In general terms in Part 7 claims these should be filed and served with the directions questionnaire. In those claims where there is no notice under CPR 26.3(1), being a notice of provisional allocation, then the budgets must be filed and exchanged 7 days before the first case management conference. Already teething problems are emerging. What of those CPR Part 8 claims automatically allocated to the multi track (and so within the regime) in which there is no case management conference? What of claims where a judgment is entered for damages to be decided in default of a defence being filed, the claim is allocated to multi track without the filing of directions questionnaires and paper directions are appropriate? These oddities have not gone unnoticed and details are being collated. Expect the rules committee to address them in further updates.

The consequence of not filing and serving the budget

[15.6]

However, the timing of the trigger for filing and exchange of costs budgets is not a matter of idle academic speculation. Failure to comply with the rules has a draconian consequence. CPR 3.14 provides that a party failing to comply will be treated as though having filed a budget limited to court fees only! Whilst the court has the discretion to order otherwise, in this brave new world where it is more difficult to obtain relief from sanction, parties may well find that the court is unsympathetic. This view is borne out by the decision of Master McCloud in the case of *Mitchell v News Group Newspapers Ltd* [2013] EWHC 2179(QB) when the court limited the claimant's budget to court fees only as the claimant had failed to file and serve his costs budget in accordance with the Defamation Pilot at CPR PD 51D. The master subsequently also refused relief from this sanction – *Mitchell v News Group Newspapers Ltd* [2013] EWHC 2355 (QB). At the time of going to press the appeals against these decisions are only weeks away. It will be interesting to see whether the decisions are restricted to the Defamation Pilot or of broader application.

Approval of the budget

[15.7]

Once the court has made a costs management order, whether by recording the agreed costs or, if not agreed, by recording the court's approval after making appropriate revisions, each party must re – file and re – serve the budget in the form approved annexed to the order approving it (CPR PD 3E, para 2.7). For reasons considered later concerning clarity about what work has been included within an agreed or approved budget, our view is that if the budget revisions have been made on the basis that the claim will proceed on different assumptions than are in the budget, then the obligation to file the amended budget in the form approved extends to revising the assumptions section – see **Completing the Form H** below.

Agreement and revision of the budget

[15.8]

Remember that budgets cannot be approved retrospectively – both CPR 3.12 and CPR 3.15 refer to budgeting costs 'to be incurred' and CPR PD 3E, para 2.4 is clear that 'the court may not approve costs incurred before the date of any budget'. The astute will notice that the rules and the PD are not quite the same – with the latter suggesting some jurisdiction in respect of costs incurred between the date of the budget and the costs management conference. The rule clearly takes precedence.

The inability to address expenditure that has already occurred by the time of the first case management conference is a failing in the scheme, as in some cases the costs incurred during the pre-action phase and through to the CCMC can be considerable. An extreme example was seen in *Dawson v First Choice* (unreported) Birmingham District Registry 5BM15907 12.3.07; a costs capping case decided in 2007. In that case significant costs had been incurred prior to the first CMC. With the requirements for parties to comply with pre-action protocols, expensive front loading applications such as those for freezing injunctions and, in technical cases, early assistance from experts, costs can escalate rapidly and unchecked before the first CMC. It is no surprise that the regional Costs Judge who sat as an assessor in Dawson responded to the interim report of Sir Rupert Jackson with the suggestion that budgets should be agreed/set once there was a response to the letter of claim and before the real expenditure had started.

With the introduction of costs budgeting there is a risk that parties will front load work to a greater extent, increasing the incurred costs and limiting the scope for prospective court intervention. However, this approach may backfire. The costs already incurred still have a significant role to play in budget setting the future costs. The court can record its comments on those incurred costs and these costs will be considered when determining whether the budgeted costs going forward are reasonable and proportionate, eg if the reasonable and proportionate costs of the disclosure phase are, say, £20,000 and £20,000 has already been spent will the court allow no further expenditure on that phase going forward? In addition front loading work exposes a greater level of costs to the overall proportionality cross check on a detailed assessment. Our suspicion is that parties may find out rather quickly that it is better to have the certainty of knowing in advance what the recoverable costs expenditure will be, rather than find out once the money has been spent that a disproportionate amount has been spent with a significant proportion unrecoverable.

What is clear, though, is that if a party wishes to revise a budget then he must follow the procedure set out in CPR PD 3E, para 2.6. An attempt should be made to agree the revision with the other party and in default of agreement he must make an application to the court. If the revision is agreed there is no requirement in the rules for the amended budget to be filed. This entire provision seems curious. CPR 3.15(3) places a duty on the court where it has made a costs management order to control the parties' budgets in respect of recoverable costs thereafter. How can it do so when its approval is not required to the variation and, worse still, it may never see the revised budget?

For those thinking that agreement of budgets generally is the way to avoid the risk that the court may not take the same view as the parties of what constitutes proportionate case management, then pause and reflect. The parties are, of course, free to agree their directions and budgets and free to agree revisions to them. However, these agreements are only effective if the court makes the case management orders that justify the agreed budgets, eg – the parties agree that they will each have four experts and agree the budget for that and the court permits that number of experts. Of course the court is not bound by the extent of the case management agreement between the parties – indeed quite the contrary. As already stated, the court will take account of costs in making any case management decision (see CPR 3.17) and the mere agreement to the costs does not mean that the court is compelled to order the agreed directions if that renders the expenditure unreasonable and/or disproportionate. Any compulsion in such a situation is to tailor the directions to ensure reasonableness and proportionality of costs.

This is as true on agreed revision of the budget as it as when first setting the budget and perhaps the concerns that an approved budget may be varied by agreement without the court knowing are more theoretical than practical. This is because CPR PD 3E, para 2.6 only permits revision where there has been a significant development in the litigation. It is hard to imagine such a development that does not require further court directions. Accordingly the court may still control the budget in that situation indirectly as it may decide that the further directions are not proportionate and refuse to order them.

Beware of authorities that may have merged from the pilots suggesting that there can be retrospective approval of budget variations (such as in *Murray v Neil Dowlman Architecture Ltd* [2013] EWHC 872 (TCC), 148 ConLR 256, [2013] NLJR 18, [2013] 3 Costs LR 460, which is considered in more detail below). They are of no relevance to costs management under Part 3. The pilot scheme under CPR PD 51G specifically permitted the court to approve or disapprove of departures that had occurred from the previous budget. There is no such provision in Part 3.

Costs outside the scope of the budget

[15.9]

CPR PD 3E, para 2.9 states that if there are interim applications during the claim that were not provided for in the budget, then any costs that arise from them are treated as additional to the budget. In other words the court should assess what order for costs to make on any such applications and then, as appropriate, undertake a summary assessment of those costs in the usual way. A classic example would be in respect of an application to enforce compliance with the directions timetable. Obviously some applications may have been sufficiently likely at the time of approving the budget that they fall within already allowed contingent costs.

The costs of costs management

[15.10]

The costs of the costs process has become the source of much judicial criticism. Inevitably the costs management process adds an additional expense to litigation. Rather than allow for protracted argument about how much, the rules prescribe the sums that will be recoverable. CPR PD 3E, para 2.1 provides that:

'2.2 Save in exceptional circumstances –
(1) the recoverable costs of initially completing Precedent H shall not exceed the higher of £1,000 or 1% of the approved budget;
(2) all other recoverable costs of the budgeting and costs management process shall not exceed 2% of the approved budget.'

It will be interesting to see the circumstances in which parties seek to test the exception. It may be that the costs of repeated revision of the budget caused by the other party's conduct of the litigation would satisfy the test. However, the court could actually recompense for that by use of interim costs orders rather than store up an argument for later in the proceedings (presumably at assessment). To do so in this way would be both a fairer and a more proportionate approach. Fairer because the party causing the revisions may not ultimately be the paying party and more proportionate because even if the offending party is the paying party, deferring the argument simply increases the chance of a detailed assessment.

The relevance of the budget set to ultimate costs recovery under a standard basis assessment

[15.11]

Notwithstanding the views of Lord Justice Moore-Bick in *Troy* (considered above), the importance of the budget on recoverable costs on standard basis assessments is profound. Unless there is *'good reason'* to do so the court will not depart from the last approved or agreed budget – CPR 3.18. The rule seems clear.

The relevance of the budget set to ultimate costs recovery under an indemnity basis assessment

[15.12]

An inevitable consequence would appear to be that the budget has no part to play in an assessment ordered on the indemnity basis. CPR 3.18 is expressly limited to standard basis assessments for an entirely logical reason. Budgets are set by reference to reasonable and proportionate costs. CPR 44.3(3) makes it plain that costs on an indemnity basis are set by reference to reasonableness only. Proportionality is not mentioned. Accordingly to extend the relevance of a budget to an indemnity basis assessment appears to import a proportionality test that is not relevant to the assessment.

However, in *Elvantine Full Circle Ltd v AMEC Earth and Environmental (UK) Ltd* [2013] EWHC 1643 (TCC) Coulson J. suggests that the budget does still have a relevance to an indemnity basis assessment stating that:

> 'Prima facie, whether under PD 51G paragraph 8, or CPR 3.18, the costs management order (with its approval of the costs budget) is expressed to be relevant only to an assessment of costs on a standard basis. However, as a matter of logical analysis, it seems to me that the costs management order should also be the starting point of an assessment of costs on an indemnity basis, even if the 'good reasons' to depart from it are likely to be more numerous and extensive if the indemnity basis is applied.
>
> The first reason for this is that, as set out in paragraphs 2 and 3.2 of PD 51G (paragraph 10 above), the costs budgets represent the parties' estimate of all the costs that they think that they will incur. It is not an estimate based on any particular form of costs assessment; it is just an estimate of likely costs. If it is an accurate estimate of all the costs that will be incurred, then it seems to me that it should be the relevant starting point for an assessment of costs on an indemnity basis as well as for an assessment on the standard basis.
>
> Secondly, this would provide the benefits of both consistency and certainty. There is a concern that, if an order for indemnity costs allows a receiving party to ignore the costs management order, then that will encourage successful parties to argue for indemnity costs every time. That would be unfortunate, and would leave an unacceptable doubt hanging over even approved costs budgets, all the way through to judgment and beyond. A paying party will have fought the trial assuming that, even if it loses, its opponent will be unlikely to recover more than the amount recorded in the costs management order, unless there is good reason for any departure. That is the certainty that the new regime provides. Even if the paying

15.12 *Chapter 15 Prospective Costs Control – Costs and Cases Management*

> party has to pay costs on an indemnity basis, that does not seem to me automatically to justify an abandonment of that certainty, and the encouragement of a costs free-for-all.
>
> Of course, in any given case, it might be said that an award of indemnity costs – which does not require any assessment of proportionality – might be a 'good reason' to depart from the costs budget approved by the court pursuant to paragraph 8 of PD 51G. I can well see that, in particular factual circumstances, an award of indemnity costs might be a good reason to permit such a departure. But that would be fact-specific, and it would not detract from the principle of at least starting the costs assessment by reference to the approved budget.'

Apart from the fact that this decision was obiter, as the judge had already decided that this was not a case where he was prepared to make an order for indemnity costs, we would urge caution in placing too much reliance on this decision for the following reasons:

- It is important to stress that this decision was under the TCC pilot scheme. That is particularly pertinent because, as we shall see below, the wording of the statement of truth on Form H <u>does not</u> lead to an assumption that the budget includes 'all the costs' that a party thinks it will incur. Form H should include only those costs 'that are an estimate of the reasonable and proportionate costs' that will be incurred. There may be a quite legitimate difference between the Form H estimate and all the costs a party thinks it will incur. The wording of the statement limits the costs by specific reference to proportionality, which has no relevance on the indemnity basis
- An indemnity costs order is intended to be a sanction against a paying party. In such circumstances it seems entirely appropriate to us that the certainty of the costs liability of the offending party should be lost. If that is not the case, what is the purpose of an indemnity order and what incentive is there for parties to conduct claims reasonably if they know that their potential liability for costs is capped by the budget in any event? We accept that the substantial difference for a receiving party between costs on the standard or on the indemnity basis may lead to an increase in applications for the latter. However, the time spent scotching those speculative applications is surely a proportionate price well worth paying to leave in place a sanction with teeth to deter proceedings and conduct that are an abuse.
- It seems unlikely to us that the limited possibility of obtaining an order for indemnity costs and escaping the constraints of the budget will encourage parties to ignore the costs management order. The incidence of indemnity costs orders is so infrequent that most parties will be advised that the budget will limit the cost recovery. To take a risk knowing this would be a throw of the dice of an inveterate gambler only.
- If the onus is on the party who has obtained the indemnity basis costs award to argue 'good reason' to depart from the budget that has two obvious implications. The first, and most obvious, is that this is not what CPR 3.18 states and it places a burden on the receiving party that appears in neither CPR 3.18 nor CPR 44.3(3). The second is that this is likely to have the effect of increasing the number of claims proceeding to detailed assessment, because instead of recognising that he has a

higher cost liability, the paying party may decide to make the receiving party take the risk on establishing 'good reason'. We venture to suggest that the extra time and expense spent on an increased number of detailed assessments will dwarf that spent on an increased number of applications for indemnity costs.

Good reason

[15.13]

What the courts consider to constitute a good reason remains to be seen, although it is likely to be very limited. Whilst many practitioners and academic commentators saw the Court of Appeal decision in *Henry v News Group Newspapers Ltd* [2013] EWCA Civ 19, [2013] 2 All ER 840, [2013] IP & T 660 as marking the end of costs management before it had properly begun, we disagree with that interpretation. It was a case decided under the defamation pilot and, of much greater significance, before the amendments to both the overriding objective and the relief from sanction provision were introduced. The qualification to the overriding objective by insertion of 'at proportionate cost' and the more restrictive approach to relief from sanctions, suggests to us that the 'good reason' hurdle bar will be set high. Lord Justice Moore-Bick hinted at this in Henry when he said, in relation to the rules in place from 1 April 2013, at paragraph 19 that they will:

> ' . . . impose greater responsibility on the court for the management of the costs of proceedings and greater responsibility on the parties for keeping budgets under review as the proceedings progress. Read as a whole they lay greater emphasis on the importance of the approved or agreed budget as providing a prima facie limit on the amount of recoverable costs. In those circumstances, although the court will still have the power to depart from the approved or agreed budget if it is satisfied that there is good reason to do so, and may for that purpose take into consideration all the circumstances of the case, I should expect it to place particular emphasis on the function of the budget as imposing a limit on recoverable costs.'

One of aims of costs budgeting is to avoid costly and time consuming detailed assessments – a process more likely to be avoided if costs budgeting is used correctly and the court discourages argument on 'good reason'.

Having predicted that instances of 'good reason' will be few and far between, some seem reasonably obvious, eg budgeted contingency costs when the contingency does not become reality and costs for the trial phase of the budget when the claim settles without a trial.

Litigants in person

[15.14]

Whilst litigants in person are excluded from the obligation to prepare costs budgets they must be provided with a copy of the budget of any represented party (CPR PD 3E, para 2.8). However, what of the situation where a party is unrepresented at the time of the first case management conference, but subsequently instructs solicitors? On the face of it the rules make no provision for that party's costs to be budgeted. It may be that in such a situation the other side may make an application for the court to fix a costs management hearing pursuant to CPR 3.16. Alternatively it may be that some judges instigate internal practices that in any claims where notices of acting are filed where previously a party was unrepresented, the file is referred to the judge to consider convening a costs management conference. There is also an argument that this is a significant development in the litigation and as such the newly represented party should follow the procedure in CPR PD 3E, para 2.6 and submit a budget to the other party for agreement and in default of agreement apply to the court. Indeed, the budgeted party may also take the view that representation represents a significant development and wish to vary his budget adopting the same procedure, which may put the matter before the court anyway. Our view is that the purpose of costs management is such that, whether there is a specific provision or not, the prudent solicitor in such a situation ought at the very least to try to agree a budget with the other party and in default apply to the court for a costs management hearing.

FORM H

[15.15]

The costs budget must be in the form of Precedent H although there is no requirement for a specific font or font size; the requirement is simply for an easily legible typeface in landscape format. From a practical perspective it is sensible to use Precedent H, despite the current issues with it. The Mercantile and TCC pilot scheme was less prescriptive, requiring the budget in a form substantially following Form HB. However, the dangers of adapting the form were revealed in *Murray v Neil Dowlman Architecture Ltd* [2013] EWHC 872 (TCC), 148 ConLR 256, [2013] NLJR 18, [2013] 3 Costs LR 460, a case within that pilot. The claimant used its own pro forma costs budget which did not contain all of the information required; it excluded the tick box it should have ticked to state that the success fees under its CFA and the ATE insurance premium were to be excluded from the budget. Consequently these sums were not included in the budget set, but were not specifically excluded and the claimant, if successful, would have been limited during costs recovery to the amount set out in the approved budget which would be taken to include, rather than exclude, the success fee and insurance premium. The claimant sought relief from sanctions to rectify this mistake.

Mr Justice Coulson did not consider that to be an appropriate application and determined the issue on the basis of an application to amend the costs budget. In this case he allowed the amendment. In doing so the critical issue for the court was that the current Precedent H, in place since 1 April 2013, contains a default setting so that success fees under a CFA and ATE insurance premiums are expressly stated to be excluded from the budget, consequently a party is not required to specifically state this. As there is now no requirement to tick a box, Coulson J was 'uncomfortable' with the thought that the claimant would be penalised by a failure to tick a box which he would now no longer be required to do. In addition, the claimant had raised the problem with its costs budget very soon after the CCMC and the defendant had not suffered any prejudice; it had not been misled or confused as it had been aware of the CFA and ATE insurance as a result of the separate funding notification requirements. Remember that *Murray* was decided under the TCC pilot rules that permit revision of budgets. The post-31 March 2013 CPR provisions *do not* permit retrospective budgeting and the claimant would now have been left arguing 'good reason'.

COMPLETING THE FORM H

[15.16]

The costs budget is intended to provide the other party and the court with a detailed breakdown of both the costs a party has incurred to date and the costs he estimates he will incur from the time of the budget onwards. The Form is divided in to a summary page, costs incurred and estimated for 10 specific phases of litigation and a contingency section. There is a guide to completion of the Form H. In an age of proportionality it is best described as pithy with obvious areas of uncertainty (eg one of the phases is headed '*Settlement/ADR*' and yet the guidance note suggests that mediation does not go in this phase, but instead as a contingency). Until experience provides the answers it seems sensible to be clear on the face of the Form H precisely what work has been included under which phase. This reduces the scope for argument later as to precisely what an agreed or approved budget includes. Fortunately Form H provides such an opportunity

Page 1 – The assumptions and the summary

[15.17]

Apart from making sure that the correct form is used, it is vital that the form contains accurate details of the costs, both incurred and estimated for the future and sets out clearly the assumptions upon which the budget has been based. We cannot stress enough the importance that should be attached to ensuring that the assumptions underlying the sums in each phase of the budget are set out in detail. This is important for a number of reasons:
- The assumptions page is, in effect, a case plan. It should illustrate to the court that thought has been given to how to progress the claim. If it does not, then the impression that it will convey is that the case has not been adequately planned. This is bound to have an adverse effect on the view that the court takes of the costs
- The assumptions justify the costs. If, for example, the expenditure on witness statements is high, but for a valid reason, eg there are a number of witnesses and they are hard to trace, then say so and use the assumptions as an opportunity to inform the court of the reasons for the expenditure. The figures in isolation mean nothing. It is only when they are viewed with the assumptions of how the claim will progress that they have any purpose.
- As the court will be linking proportionate case management to the budget it will need to understand the basis upon which the costs have been calculated to assist in the determination of what is reasonable and proportionate.

- As any subsequent revision of the budget can only be where there has been a significant development in the litigation, the court will look back to the original assumptions upon which a budget was agreed or approved to check that the claim has indeed taken a different course from that originally costed.

Remember that the Form H will be considered in conjunction with the directions questionnaire, the proposed directions and the disclosure report, in all but personal injury multi track claims. If these documents are contradictory (eg the estimate of costs for the expert evidence in the directions questionnaire differs from that in the budget) that will immediately raise concerns about the amount of control being exercised over costs. Linked to this is the position where a party provides insufficient information in the directions questionnaire. We have all seen allocation questionnaires that answer key questions, such as what witness are to be called and on what issues, with unhelpful answers. If in doubt 'TBA' (to be advised) seemed to be the stock answer. Not only is the court unlikely to accept directions questionnaires filled in with incomplete information, but it will also highlight serious issues with the proposed budget. How can a party who can neither name his witnesses, nor identify the issues that those witnesses will address, have possibly drafted detailed assumptions about the evidence to be called and included anything other than a speculative figure for the witness statement phase of the Form H?

If the budgeted costs do not exceed £25,000 the only requirement is to complete page 1 of the costs budget. It seems logical that as contingencies are included in the Form H, then if the amount of any budgeted contingencies takes the overall budget over £25,000 the entire Form H must be completed.

The phases

[15.18]

There are ten phases: Pre-action costs, issue/pleadings, CMC, disclosure, witness statements, expert reports, PTR, trial preparation, trial and settlement/ADR. These are followed by 'contingency A' and 'contingency B'.

Such has been the uncertainty surrounding completion that some thought that as CPR PD 3E, para 1 insists that the budget must be in the format of Form H, that meant that there must be two contingencies. Plainly that is nonsensical, but other concerns over completion are more valid. Sometimes work done is hard to categorise. An example often given is work on valuing a personal injury claim. Legitimately this may fall within both 'pleadings' and 'trial preparation' if it involves preparing or revising a schedule of loss. It might also happily sit in the 'settlement' phase if it is part of preparation for a round table meeting even though it is still a pleading. Where should this work feature in the budget? Until the position becomes clearer the answer seems straightforward to us. Use the assumptions column to explain what has been included within each phase. Any later analysis of the agreed or approved budget is then clearly measurable against specific tasks undertaken within a particular phase. This also gives the court the opportunity to move that expenditure to another phase if it approves it, but believes the work to have been categorised wrongly.

Further problems arise in ensuring that the time spent by fee earners is recorded against the correct phase. This occurs where work done on a particular occasion is in part on one phase and in part on another – eg consideration of documents as part of the process of preparing witness statements. Is the time to be recorded against disclosure or witness statements? Is there a temptation to record against one of the phases to ensure the work stays within the budget? In reality we suspect that it will be clear to the fee earner involved what the purpose was of the task he was undertaking at any given time. In the example given, we suspect that the fee earner should record that time against the witness statement phase.

One area where the guidance is superb is on the question of where to insert costs pre-action. At first blush this may seem an odd conundrum when there is a phase titled 'Pre-action costs'. However, there is also a clear logic to thinking that these should, where they relate to another phase, be included as incurred costs in that phase. Without the guidance there was a clear risk that parties would think that the pre-action costs on a particular phase should go in at both relevant points. Form H is a self-calculating spreadsheet and to insert the figures twice would result in duplication. The guidance helpfully directs parties to include in the 'pre-action' phase only those costs incurred pre-action that are not already incurred in another phase of the budget. In other words the pre-action phase is always likely to be relatively small as most work will be incurred work on disclosure, witness statements, experts, preparing statements of case for issue and settlement. If there is what seems a large pre-action phase total in the context of the claim, expect the court to ask for a breakdown of how this has been calculated as it will wish to ensure that the work has been correctly allocated. This does serve a legitimate purpose, for, as we have said, the court wants to use the already incurred figure for a phase to inform its decision about what further expenditure on that phase is reasonable and proportionate.

Counsel

[15.19]

This is a pithy point. Do no forget to allow for counsel's fees. It is as well to obtain an early estimate of these so that it does not become hard to retain counsel later, because the budgeted fees are too low, counsel is constrained to do the work for a low fee out of a sense of commercial loyalty or necessity or the client has a large portion of counsel's fees that are irrecoverable win or lose the claim.

Specifically excluded costs

[15.20]

Form H also stipulates that certain costs are specifically excluded from the budget. These are set out on page 1 of Precedent H and are VAT (if applicable), success fees and ATE insurance premiums (if applicable), costs of detailed assessment, costs of any appeals and the costs of enforcing any judgment.

An issue arose with court fees. Precedent H originally stated on page 1 that court fees were to be excluded. This was inconsistent with the requirement to provide them on pages 2 onwards when dealing with the incurred and estimated costs for each phase of the litigation. This discrepancy was picked up quickly by the judiciary and Precedent H was amended to remove court fees from the list of exclusions on page 1.

15.21 *Chapter 15 Prospective Costs Control – Costs and Cases Management*

Statement of truth

[15.21]

The costs budget must contain a statement of truth. This must be signed by a senior legal representative of the party. The wording for this verifies the budget and is set out in CPR PD 22 which provides:

> 'Statement of truth
>
> The costs stated to have been incurred do not exceed the costs which my client is liable to pay in respect of such work. The future costs stated in this budget are a proper estimate of the reasonable and proportionate costs which my client will incur in this litigation.'

We have already considered the potential difficulty that this may present when a client pursues a claim on a fixed fee that does not lend itself to division between specific phases of the budget in Chapter 12 – The Indemnity Principle.

There is no provision in the costs budget for any statement by the client that the costs budget has been discussed with him and it has been approved; notwithstanding that it is the client that will be liable to pay the costs. From a practical perspective discussions must be undertaken with the client prior to completion of the budget. The client may well have a very different idea as to what budget he will require to pursue, or defend a case and this must be addressed at the outset. An explanation should be provided as to the costs that will need to be incurred; incurred costs should already have been discussed. It is also essential to explain that the costs budget will provide the basis for determining the recoverable costs at the end of the proceedings. At the end of the day the client may decide that he wants to spend in excess of what the court determines to be a reasonable budget and there is nothing to prevent him from doing so (other than court case management limiting disclosure, witness evidence, number of experts etc. which is considered below under 'The interaction between case and costs management'). The client should also be aware of the assumptions on which the budget is based so that he understands the practical effect of the financial constraint.

One consequence of the silence in the rules on the need to serve the Form H on the client is that judges may decide to hold more CCMCs with the parties present, rather than by telephone. This way the court may be sure that the client knows the costs that are being bandied around and the court's concerns about proportionality. The thinking is that this may lead to earlier compromise. The sorts of cases suitable might be boundary and right of way disputes, Inheritance Act claims and Trusts of Land and Appointment of Trustees Act claims – where often the case management conference represents the last chance for settlement before the costs escalate to such an extent that the claim becomes about the costs rather than the original dispute. However, it is essential that the client understands that the costs the court is budgeting are those between the parties. Any solicitor client amount on top will be irrecoverable.

As a final thought, it will be interesting to see how comfortable solicitors are when signing a statement of truth that the costs stated in the budget are a reasonable estimate of proportionate costs. It will be even more interesting, when budgets have been agreed, to see how willing to stick to such a certification solicitors will be once the court has declined to make the case management directions upon which the agreed budget was based. This control that the court is still able to exert by robust case management, and the ways in which this may arise requires more detailed consideration.

THE INTERACTION BETWEEN CASE AND COSTS MANAGEMENT

[15.22]

Let there be no doubt that the judiciary sees robust and effective case management as the vital ingredient essential to the success of the 'Jackson' reforms. This is plain from:

- The comments of Sir Rupert Jackson in the executive summary to his final report:

    'One of the points that was impressed upon me during the Costs Review was that judges should take a more robust approach to case management, to ensure that (realistic) timetables are observed and that costs are kept proportionate. Case management can and should be an effective tool for costs control.'

    and in his eagerness to highlight the experience of Singapore when implementing its own procedural reforms in the 1990s. The impact of the case management reforms there was 'electric'. There, as here, there was deep discontent about the reforms within the legal profession. However, as the profession came to terms with the new provisions 'it was generally recognised that the long term effect of these reforms was highly beneficial.'

- The clear indications that have been given to the judiciary that it can expect the support of the Court of Appeal where robust, but proportionate, case management decisions have been reached. This has been reinforced by the announcement by the Master of the Rolls, Lord Dyson, of a small cadre of Lord Justices designated to hear costs and case management appeals. These are the Master of the Rolls himself, the Deputy Head of Civil Justice, Lord Justice Stephen Richards and Lord Justices Jackson, Davis and Lewison. There will be at least one of these designated judges sitting on each appeal.

- The approach taken by the Court of Appeal in the run up to implementation of the reforms and subsequent decisions of lower courts. In *Maqsood v Mahmood* [2012] EWCA Civ 251 the Court of Appeal upheld the decision at first instance to strike out the Claimant's claim on the basis of a failure to comply with the CPR. Since implementation of the new case management provisions we have seen this approach being continued in cases such as *Fons HF v Corporal Ltd* [2013] EWHC 1278 (Ch), [2013] 4 Costs LO 646 and *Biffa Waste Services Ltd v Ali Dinler* (QBD) 10.10.13. In the former the court extended time for exchange of witness statements, but clearly stated that non-compliance would result in a party being debarred from relying upon 'any' evidence at trial. In the latter Swift J overturned the decision of a district judge to grant relief from sanctions – the sanction having been that the claim was struck out – where the claimant filed his statements 2 hours outside the time prescribed in an unless order.

Whilst only time will tell if this robust approach will continue, all the early signs are that a new culture has arrived – one in which timetables set by the court should no longer be viewed as merely targets, but as dates written in tablets of stone and in which proportionate claim disposal trumps 'perfect justice'.

How does this approach to case management sit with costs management? We envisage that the procedural issues raised below illustrate clearly the impact of costs management on case management decisions.

(i) Allocation

[15.23]

The most likely noticeable effect will be the demise of the 'fast track' value claim that makes its way in to the multi track due to the time estimate based on the number of witnesses. In future parties may expect the court to limit the number of witnesses (see below) and be prescriptive about the trial timetable to ensure that the claim is disposed of within 5 hours. However, the deletion of CPR 26.7(3) in respect of claims issued after 31 March 2013, which had prevented the court allocating a claim to a track with a financial limit lower than the value of the claim without the consent of the parties, means that the court may decide that a claim over £10,000 should still be allocated to the small claims track as a proportionate case management decision. (see Chapter 22 Costs Awards Between the Parties – Small Claims Track Costs, below)

(ii) Use of standard direction templates

[15.24]

Although this may seem like small beer in the overall drive to proportionality, there is no doubt that the introduction of standard directions and the mandatory filing requirement with directions questionnaires in multi track, ought to assist by forcing parties to address case management at an earlier stage and avoid the expense for both parties and the court in preparing bespoke orders. Early anecdotal evidence suggests that the professions have yet to embrace the standard directions – see CPR 29.1(2).

(iii) Witnesses and witness statements

[15.25]

Whilst the court has always had powers under the CPR to control evidence, the amendments in CPR 32.2(3) take this further. All the indications are that the judiciary are primed to use these powers. Directions questionnaires will be scrutinised to see upon what witness evidence the parties propose to rely. In multi track claims this will be cross referenced to the proposed budget for that phase. If the budget is not reasonable and proportionate then parties may have to select their best evidence. In other cases the court will be looking to limit the witness evidence to that which will enable the trial to be completed within a proportionate time estimate. Parties should expect to see orders limiting the number of witnesses, the issues that are to be addressed by those witnesses and the length of the statements.

(iv) Disclosure

[15.26]

Sir Rupert Jackson made no secret of his view that disclosure is one of the main drivers of cost in litigation. The proliferation of e-disclosure has only served to heighten the difficulties and increase this cost, as can be seen from a cursory glance at *West African Gas Pipeline Co Ltd v Willbros Global Holdings Inc* [2012] EWHC 396 (TCC), 141 ConLR 151, [2012] NLJR 682 at paragraph 65 and *Wyche v Careforce Group Plc* QBD (Comm) 25 July 2013. As a result CPR 31.5(7) now offers a menu of disclosure options, with, in multi track non personal injury claims, standard disclosure being the last, and least attractive, of these. The requirement to file and serve disclosure reports 14 days before the first case management conference and for discussion between the parties about the method of disclosure 7 days before the first case management conference, means that the court will be far better informed to make a proportionate disclosure order from the menu. In multi track cases expect the court to compare this with the phase budget proposed in Form H.

One option not listed, which may well be attractive to the court, is that parties should disclose those documents upon which reliance is to be placed by them in support of the case <u>and</u> any documents of which they are aware that are adverse to that case. This, perhaps, better balances the conflict between justice and proportionality. Given the high cost of disclosure and the desire of lawyers to follow the paper trail to the end in the (all too often forlorn) hope of discovering a contaminatory document, it seems likely that this phase will be at the forefront of the 'needs v proportionality' conflict.

(v) Experts

[15.27]

Another battleground between necessary work and proportionality is that of expert evidence. Parties may expect to encounter more use of single joint experts, tighter prescription by the court of the specific issues upon which the expert is to report (shorter, focused reports will be the order of the day) and limits on experts' recoverable fees. Amendments to CPR 35.4(2) and (3) require up front estimates of an expert's fee (which will be necessary if the Form H is to be accurately completed in any event) and identification of the issues to be addressed by that expert. Practitioners who do not follow pre – action protocols for expert nomination are likely to find difficulty in obtaining permission to rely upon the expert evidence then obtained (and for which the client has already incurred a cost liability).

The jury is still out on whether concurrent expert evidence produces savings in costs. The relevant provisions are found in CPR PD 35, para 11. 1–para 11.4. The limited experience from the pilot at the Manchester Construction and Technology and Mercantile courts is that, in the right case, hearing the evidence, or part of it, of opposing experts concurrently can save court time and therefore be a proportionate approach.

By way of aside it is worth noting that the court has had some ability to restrict the recoverability of experts' fees under the CPR prior to the introduction of costs management. CPR 35.4 has long provided for the court to limit the amount of an expert's fees and expenses that are recoverable between the parties and CPR 35.6 provides that where an expert fails to answer a question posed by a party other than the one which instructed him, the court may order that the expert's fees are irrecoverable.

(vi) Relief from sanctions

[15.28]

We have already considered the new CPR 3.9 in Chapter 11 – An Overview of the Key Costs Amendments to the Civil Procedural Rules. Failure to comply with orders, the CPR and practice directions is a drain on the resources of the court and parties alike; the court because non-compliance may result in an increase in the number of interim hearings or paper case management or, at worst, adjournment of a trial leaving a valuable slot of court time that cannot be filled: parties because efforts to compel compliance inevitably cost money. Rapidly the best laid plans for a claim to be resolved proportionately can be left in tatters. Not so in future. As we have said the court will be quick to make 'unless orders' to compel compliance and keep claims on the rails and slow to grant relief from sanction for those who fail to comply. For those still in any doubt the words of Jackson LJ in *Fred Perry (Holdings) Ltd v Brands Plaza Trading Ltd* [2012] EWCA Civ 224, [2012] FSR 807, [2012] 6 Costs LR 1007 will come as a chilling reinforcement:

> 'I should however draw attention to the forthcoming amendments to Rule 3.9. There is a concern that relief against sanctions is being granted too readily at the present time. Such a culture of delay and non-compliance is injurious to the civil justice system and to litigants generally. The Rule Committee has recently approved a proposal that the present rule 3.9(1) be deleted and the following be substituted:
>
>> 'On an application for relief from any sanction imposed for a failure to comply with any rule, practice direction or court order, the court will consider the circumstances of the case, so as to enable it to deal justly with the application including the need –
>> (a) for litigation to be conducted efficiently and at proportionate cost; and
>> (b) to enforce compliance with rules, practice directions and court orders.'
>
> It is currently anticipated that this revised rule will come into force on 1 April 2013. After that date litigants who substantially disregard court orders or the requirements of the Civil Procedure Rules will receive significantly less indulgence than hitherto. As I say, that rule amendment lies in the future. In the present case, on the rules as they stand, relief from sanction must be refused.'

(vii) Trial

[15.29]

The court has always had the power to set the trial timetable and control the way in which evidence is presented. However, the court now has, as part of the overriding objective, the obligation to keep the trial proportionate. With civil court time at a premium the judiciary is increasingly likely to invoke this power.

The increase in the small claim track limit to £10,000 will see more substantial claims being conducted by litigants in person. Many will have had no previous experience of the court process. Litigants in person often arrive at court having no idea what to expect. If the extended jurisdiction is not to lead to the court allocating a disproportionate amount of a finite resource, then judges are likely to impose strict timetables for evidence in chief and cross examination, to ensure these hearings are proportionately constrained.

This will be as true of the fast track – particularly, if, as predicted above, claims that might previously have been allocated to the multi track solely on time estimate are to be kept within the fast track.

(viii) Assessments of costs

[15.30]

The effect on the outcome of assessments of costs with proportionality overreaching all else at CPR 44.3(2) has already been highlighted. However, on a more rudimentary note, parties may find the court eager to undertake even more assessments on a summary basis. Why? Well, the decision of whether or not to undertake a summary or detailed assessment is a case management one. All case management decisions are taken with consideration of cost and proportionality. Detailed assessment costs more than summary assessment and results in more court time being allotted to a case. Accordingly, summary assessment must be the preferred option whenever possible.

Where detailed assessment is ordered the new provisional assessment procedure under CPR 47.15, which limits the recoverable costs within the assessment and is designed to reduce the use of court time on the assessment, has been introduced to ensure that the process itself is proportionate. The high threshold required for a successful challenge to the provisional assessment set out at CPR 47.15(10) will deter most unsatisfied parties.

SETTING THE BUDGET

Introduction

[15.31]

We know that proportionality is the key to the budget, that every direction has a price tag and that the court will use robust case management to assist in the budget setting process. This all seems quite clear. What is less certain is how this translates in to actual figures. The pilot schemes and the various implementation lectures give some pointers – but not always in the same direction.

One of the first, if not the first, decisions emerging from the defamation pilot is informative in that it illustrates the learning curve involved for the judiciary. In that case, *Morrison v Buckinghamshire County Council* (unreported) the case managing judge set the directions and convened another appointment when a budget would be set by the SCCO for the claim based on those directions. The difficulties with this approach are obvious:

- The directions have been set before the cost implications of them are considered and as such the directions determine the budget
- In consequence it will be a matter of chance whether the budgeting of these directions leads to a proportionate expenditure rather than this being a considered and tailored exercise
- If a robust approach were subsequently taken to the costing exercise to set a proportionate budget, that would still leave the parties to comply with the directions with inadequate funds to do so

In fact this case is extremely helpful as a learning tool. It reinforces what we have stressed throughout this chapter – namely that case and costs management go hand in hand. But how does that translate in to practice?

Many of us thought that what Sir Rupert Jackson sought was overall proportionality: in other words that he was not concerned with micro management of particular aspects of expenditure provided that a party's overall costs were within a global budget set. Of course, from the outset, the pilots required completion of a budget in broadly the same format as the Form H, giving a steer that something more was required and that the court would be interested in the costs associated with the various procedural steps to be undertaken. So it has proven. CPR PD 3E, para 2.3 is plain that the court is required to budget by the phases of the proceedings. This is reinforced in the same part by the fact that the parties can plainly agree parts of the budget, even if they cannot agree the entire budget. CPR 3.18 completes this approach by providing that on an assessment on the standard basis the court will have regard to the last approved or agreed budget 'for each phase of the proceedings' and not depart from that budget unless there is good reason.

So the budget will be set by phases. However, our view is that this cannot be without a clear idea of overall proportionality. It would surely defeat the object of the exercise if the court budgeted a claim by reference to the phases, added the total up, concluded that the overall cost was disproportionate when judged against the criteria at CPR 44.3(5) and could do nothing about it. This would involve the court approving a budget that permitted the litigation to be undertaken disproportionately.

The answer is surely that, as Sir Rupert Jackson envisaged, the court will begin by taking an overall view of proportionality (in general terms), then budget the individual phases to bring the claim to trial within that overall budget with the ability to tweak the overall view of proportionality slightly in the light of the information that emerges when setting the spend on the phases. This approach involves the court hearing brief initial submissions on overall proportionality, if the budgets are not agreed, and forming this initial view. If the budgets are agreed this presents a more challenging approach. It will be for

Setting the budget **15.32**

the court to raise its concerns on proportionality and confront parties who are united in opposition to court intervention! It will be interesting to see the extent to which the court does intervene. In *CRM Trading Ltd v Chubb Electronic Security Ltd* QBD (Merc) 4.9.13 Judge Mackie QC suggested that it would be rare for budgets to be contested and that the court had to be cautious about altering budgets. Early reports from the county court indicate that both these suggestions are proving unfounded.

The relevance of hourly rates

[15.32]

So far this discussion has been in general terms, but once the court starts to set a budget this will involve specific sums. There appear to be two clear camps forming on how it will do so. On the one hand are those who see it as inevitable that the court will set the budget by reference to hourly rates. After all, Form H, which must be completed, makes provision for hourly rates and time to be inserted and, by dint of the self-calculating nature of the form, this produces the resultant figures per phase. On the other hand are those who believe that hourly rate has no, or at best a limited, role in budgeting. We are firmly in this latter camp. We take this view on the bases that:

- Hourly rate is only of relevance if it is multiplied by an amount of time. Consideration of time chimes of 'need' and, as we have seen, proportionality trumps need. If the court does become sucked in to rate setting, then it must consider time (for rate alone is purposeless). What does the court do if the figure that emerges from this multiplication is disproportionate? Revisit rate? Revisit time? The only certainty is that the budget setting exercise itself will then take a disproportionate amount of court time.
- CPR PD 3E, para 2.3 specifically states that 'The court's approval will relate only to the total figures for each phase of the proceedings . . . '.
- CPR PD 3E, para 2.3 also steers the court away from undertaking a prospective detailed assessment. What could be more akin to a detailed assessment than lengthy arguments on appropriate hourly rate?
- CPR PD 3E, para 2.3 defines the task of the court when setting the phase budget to consider ' . . . whether the budgeted costs fall within the range of reasonable and proportionate costs'. This is not language that resonates with descending to the detail that is required to determine hourly rates
- Setting a budget by hourly rates creates later uncertainty. Who is to say that the work will then be undertaken by a fee earner commanding that hourly rate, as the mere setting of the budget cannot compel that? As the budgeted sum is what will be recovered on a standard basis assessment unless there is a good reason to the contrary, this raises the spectre of either a possible breach of the indemnity principle or an increase in detailed assessments with paying parties seeking to ascertain who has actually done the work. This immediately undermines one of the aims of budgeting, which is to reduce the number of assessments.
- Setting hourly rate opens the door to forum shopping in those cases that are not geographically constrained to be issued in a specific court. If word emerges that in Court A budgets are set by reference to hourly

15.32 *Chapter 15 Prospective Costs Control – Costs and Cases Management*

rate and the rate currently allowed for a grade A fee earner is £100 per hour more than Court B, which also sets budgets by reference to hourly rate, allows, before too long Court A will be inundated with claims whilst Court B sits empty.

We accept that this is an area ripe for argument. We also note that in some of the cases that have progressed through the pilots hourly rate has been a factor – indeed in *Troy* (above) an integral part of the grounds of appeal related to the hourly rate of counsel. We acknowledge that Form H appears to demand detailed information about hourly rate and time spent and to be spent. Interestingly the Form will work without such information simply with totals being inserted for each phase. However, aside from the reasons set out above, avoiding protracted argument about hourly rate provides the simplicity to the process that we think Sir Rupert Jackson intended and enables parties to have the freedom to determine how the budget is spent, ie to make decisions about how they want to model the expenditure – whether by less time at a higher fee earner level or more time at a lower level with the inevitable difference in service provision that flows from that decision. We shall return to this topic in future editions when practice becomes clearer and shall, if necessary, eat humble pie.

Setting the figure

[15.33]

Having turned our backs on the traditional costs calculation of hourly rate multiplied by an amount of time we still have to establish how the precise budget figures will be calculated.

Our view is that the court will adopt a far broader approach. It is required to set the budget for each phase by total figure. The approach we advance is that the court will identify the salient features of proportionality under the CPR 44.3(5) checklist. It will then apply those features to each phase and determine an appropriate total sum. In practice this will involve the court giving a brief judgment identifying which of the CPR 44.3(5) factors are relevant, and in what way, to a particular case and then adopting the wording of CPR PD 3E, para 2.3 and using a formula of words for each phase along the lines of 'Having identified the relevant considerations on proportionality in this claim, the appropriate case management order is [y] and the reasonable and proportionate budget for this phase is [£x]'. Critics of this approach may see this as too opaque a process. However, it seems to be precisely what the rules require, linking costs and case management by reference to proportionality. Anything more complicated can only be trespassing in to the territory of a detailed assessment.

THE CASE LAW SO FAR

[15.34]

We have already referred to the cases of *Henry*, *Mitchell*, *Elvanite*, *Fons*, *Wyche*, *Murray*, *Troy* and *Biffa Waste* above. The case law so far suggests that there will be more to follow attempting to clarify:

- what constitutes good reason for the receiving party to depart from its budget if the budget is exceeded;
- what constitutes good reason for the paying party to depart from the receiving party's budget if it is not exceeded;
- the relevance of the budget if costs are awarded on the indemnity basis;
- the relevance of hourly rate;
- what happens if the budget is late and the party in breach seeks to escape the sanction of CPR 3.14.

So far the messages have been mixed and seem to rail against the simplicity of the CPR itself – surely not what is necessary if arguments over costs are to be kept proportionate and if there are to be fewer detailed assessments (which are certainly two of the outcomes that Sir Rupert Jackson envisaged).

However, the court already seems to have embraced the amendments to the overriding objective and the procedural provisions to use robust case management as a method of curbing cost expenditure. The jurisprudence emerging all points one way – compliance with the CPR, Practice Directions and court orders is not optional. However, there is a degree of ambiguity as to how severe will be the consequences of non-compliance.

In *Fons HF v Corporal Ltd* (above) the court was concerned with the second defendant's failure to comply with two previous orders for the exchange of witness statement evidence. The problem was compounded because it appeared that the second defendant's case had altered from that pleaded, which would have implications in respect of the trial that had already been listed. Unusually the alteration would reduce the time estimate with the consequence as the judge stated ' . . . that while this trial continues to be listed with a time estimate of five says it will, in truth, be completed in a fraction of the time with the result that a valuable national resource, namely court sitting days, will be wasted as a result . . . '. The court concluded that it had come 'very close to refusing an extension' but was 'persuaded to extend the time . . . .because this hearing is taking place only a very short while after the amendment of the CPR'. The judge extended the time until 4 pm the day after the hearing. Both the breached orders pre-dated the CPR amendments and the court left little doubt what the order would have been had the case been decided a few months later.

In *Venulum Property Investments Ltd v Space Architecture Ltd* [2013] EWHC 1242 (TCC), [2013] 4 Costs LR 596 the court was even less forgiving. Here the claimant sought an extension of time within which to serve the Particulars of Claim, having thought that it had 14 days from service of the claim form to serve, but having overlooked the provisions of CPR 7.4(2), the claim form had been served on the last day of its validity. For two of the defendants the point was not academic as any new claim would be statute barred against them. The application required the court to consider CPR 1.1 and CPR 3.9 in their new guise. Interestingly both parties made submissions addressing the old CPR 3.9 factors as well as addressing the new rule – agreeing that whilst these would still be covered by 'all the circumstances' and therefore of relevance, it was the emphasis of the rule that had changed. Ultimately the judge concluded:

> 'In my judgment, when the circumstances are considered as a whole, particularly in the light of the stricter approach that must now be taken by the courts towards those

who fail to comply with rules following the new changes to the CPR, this is a case where the court should refuse permission to extend time.'

This acceptance of the fact that the CPR now represents a stricter regime was confirmed by Henderson J in *Re Atrium Training Services Ltd (in liq), Smailes v McNally* [2013] EWHC 1562 (Ch). The court's consideration involved a long history of difficulties with disclosure and a request for a further extension of time for the liquidators to complete the process. The court was clear that it was not dealing with an application for relief from sanction, the application for extension having preceded the expiry of the time for disclosure: instead the court had to determine the application by reference to the amended overriding objective. The court granted the extension, but accompanied by cautionary words:

> 'The matters set out in rule 1.1(2) now include, of course, the enforcement of compliance with orders. To that extent, it is no doubt the case that the court will scrutinize an application for an extension more rigorously than it might have done before 1st April, and that it must firmly discourage any easy assumption that an extension will be granted if it would not involve any obvious prejudice to the other side.'

It would be wrong not to make the point that the rules still contain provision for extensions of time and for relief from sanctions and that the parties should adopt realistic positions in the context of each case. As Henderson J observed in the context of an extension of time and not relief from sanction:

> 'Everything will always depend on the circumstances of the particular case, and the stage in the proceedings when the order is made . . . It would, I think, be unfortunate if the new and salutary emphasis on compliance with orders were to lead to a situation where, in cases of the general type I have described, a reasonable request for an extension were to be rejected in the hope that the court might be persuaded to refuse any extension at all.'

This was a theme picked up in *Wyche v Careforce Group Plc* (above) in which Walker J granted relief from sanctions in respect of failure to comply with e-disclosure. In granting relief the court stated that it was not a 'martinet' and should accept that there would be human error on occasions. In this case the first of two breaches had led to more disclosure than was necessary and the second had led to a temporary failure to disclose 24 documents in a complex e-disclosure exercise. Preparation for trial had only been delayed for one week. The granting of relief was held to be consistent with the principles of the new regime.

In contrast relief was not granted by Hildyard J where there had been breach of an unless order in respect of discovery in *Thevarajah v Riordan* [2013] EWHC 3179 (Ch). However, it was granted on a second application in that case by Andrew Sutcliffe QC (Ch D) 10.10.13 who concluded that there had been a material change in circumstances and that consideration of the 'old' CPR 3.9 factors remained relevant and this was not trumped by the fact relief would be disproportionate.

Perhaps the position so far then may be that it seems that the court is adopting a more robust regime certainly, but one where decisions remain contextual and not subordinate to an inflexible mantra of compliance with no prospect of relief. Instead, as Edwards-Stuart J summarised counsel for the defendant's position in *Venulum Property Investments Ltd v Space Architecture Ltd* a change of emphasis so that the court is required to take a much stronger and less tolerant approach to failures to comply with matters such as time limits.

CONCLUSION

[15.35]

What the review of the case law to date, and with the new rules in their infancy, demonstrates is that it is impossible to be prescriptive as to how the court will approach costs management and proportionate case management. As it will be the court approach that dictates how parties must engage, there will be an inevitable period of uncertainty. For the reasons that we have set out above, jurisprudence emerging from the pilot schemes should be treated cautiously – Moore-Bick LJ suggested as much in *Henry*. Inevitably this creates uncertainty. It will also lead to a rash of appeals as practitioners and parties test the parameters of judicial discretion in this area, attempt to clarify what is meant by proportionality and seek some form of definition of 'good reason'. Sir Rupert Jackson always recognised that there would be a 'bedding-in' period similar to that which occurred in Singapore and that there would be some casualties along the way. The consistency of decision that will result from the Court of Appeal by dint of the designation of a limited number of Lord Justices to sit on costs and case management decisions, means that a clearer picture should emerge relatively swiftly – nipping any further costs wars in the bud.

The debate concerning the extent to which certain areas of litigation remain exempt from the costs management regime is also worth monitoring. As we have indicated the Civil Procedure Rule Committee has a consultation in process about the exemptions. Part of its remit is also to consider whether there are any other areas of litigation where mandatory costs management is causing difficulties. All this at a time when there was also initial uncertainty over the extent to which the Queen's Bench Division had granted its own exemption to clinical negligence claims, which in turn led practitioners to believe that any such exception also extended to the District Registries, when many practitioners in the field of lease renewal seek exclusion and when Insolvency practitioners are seeking to extend their temporary exclusion to a permanent one. If we are not careful the rule will need to be re-written for simplicity's sake to exclude all claims from case management except the few that remain within the scheme.

CHAPTER 16

PROSPECTIVE CONTROL OF COSTS – THE RELEVANCE OF COSTS BUDGETS IN CASES NOT COSTS MANAGED

[16.1]

In the 16th Costs implementation lecture Mr Justice Ramsey noted that:

> 'As costs management is a necessary adjunct to proper case management and to the furtherance of the overriding objective there will, in most cases, be a presumption in favour of making a costs management order.'

It seems clear, then, that multi track cases where a costs management order is not made will be few and far between. However, even in these cases the Form H that has been filed and served pursuant to CPR 3.13 remains relevant for reasons known to those familiar with the old Costs Practice Direction Section 6.

Section 6 of the old Costs Practice Direction required parties to provide the court with estimates of the base costs already incurred and those to be incurred with allocation questionnaires and pre-trial checklists. Despite the requirement for such estimates they were generally regarded as an unsuccessful tool in managing the costs of litigation. The major reasons for this were:
- that parties often failed to provide an estimate at all;
- the estimate provided was one in an incomplete form;
- save where the costs met the stringent test for cost capping there was not a lot that the court could do with the estimate at the time it was given; and
- the court routinely failed to police the requirement to provide proper estimates.

This was curious as the Court of Appeal in *Leigh v Michelin Tyre plc* [2003] EWCA Civ 1766, [2004] 2 All ER 175, [2004] 1 WLR 846 had deprecated the circuit judge's description of costs estimates as being 'damp squibs'. In that case the court had stressed the importance of costs estimates and had given the following guidance:

(i) First, the estimates made by solicitors of the overall likely costs of the litigation should usually provide a useful yard-stick by which the reasonableness of the costs finally claimed may be measured. If there was a substantial difference between the estimated costs and the costs claimed, that difference called for an explanation. In the absence of a satisfactory explanation, the court may conclude that the difference itself was evidence from which it could conclude that the costs claimed were unreasonable.

(ii) Second, the court may take the estimated costs into account if the paying party showed that it relied on the estimate in a certain way. An obvious example would be where the paying party concluded that it had only a relatively slim chance of winning but as the estimated costs

16.1 *Chapter 16 Prospective Control of Costs – Relevance of Costs Budgets*

of the receiving party were low it was worth risking paying those to take the chance. Having relied on the estimate, fought the claim and lost, the receiving party then claimed costs in excess of the estimate. Here the paying party could point to reliance on the estimate and the fact that had the costs been estimated at the level at which they were subsequently claimed, he would not have run the case to trial and would have settled it.

(iii) Third, the court may take the estimate into account in cases where it decided that it would probably have given different case management directions if a realistic estimate had been given. It might, for example, have trimmed the number of experts who could be called, and taken other steps to slim down the complexity of the litigation in the interests of controlling costs in a reasonable and proportionate manner – an early steer towards the proportionate case management that is now required.

Section 6 of the old Costs Practice Direction effectively codified the views expressed in *Leigh*. It provided that where there was a 20% or more difference between an estimate and the costs claimed that required an explanation. The paying party then had the opportunity to argue reliance upon the estimate. At assessment the court could have regard to the estimate when assessing reasonableness and proportionality of the costs. If the difference was 20% or more and either the receiving party had not provided a satisfactory explanation for this or the paying party showed reliance, then the court could regard the difference as evidence that the costs claimed were unreasonable or disproportionate.

These provisions have largely been imported in to CPR PD 44, sub-section 3. However, there are important changes.

Sub-section 3.1 provides that the section only relates to budgets filed under CPR PD 3E and only where the court has not made a costs management order. CPR 3.12 expressly limits CPR PD 3E to multi track claims (with the exceptions already considered). In other words sub-section 3 only covers claims in the multi track. This may seem obvious as there is no longer an obligation to file an estimate in the fast track. However, there will be cases where the notice of provisional allocation under CPR 26.3 refers to the multi track. As such the parties must file and exchange Forms H at the time of filing the directions questionnaire. The case managing judge may then decide that the provisional allocation was incorrect and formally allocate to fast track. Do the provisions of sub-section 3 then apply? The logical answer is that the claim is not in the multi track and is therefore excluded. However, it is hard to imagine that the court on assessment will ignore the budget altogether if there are significant departures from it.

It is odd that fast track claims are excluded entirely from the estimating process. We accept that to have full blown costs management in the fast track may itself be a disproportionate process, but the absence of any requirement to provide even an estimate seems unfortunate. Experience shows that it is often fast track claims that are the epitome of disproportionality. We suspect the exclusion was because the intention was that all fast track claims should be dealt with under fixed fee schemes. Fast track personal injury claims arising out of road traffic, employer liability and public liability claims have, from 31 July 2013 become subject to such a regime and maybe the hope is that the

Prospective Control of Costs – The Relevance of Costs Budgets in Cases Not Costs Managed **16.1**

remainder of the fast track will follow in due course. The effect in the meantime is the curious situation where the court has only retrospective costs assessment tools in its armoury in these cases, unless the court, of its own volition, requires estimates or decides to take the bold step of using the proviso at CPR 3.12(1) to costs manage a fast track claim. Whilst we can see the attraction of that, case selection is paramount to ensure that the costs management exercise is not, itself, disproportionate. Otherwise, for those conducting fast track claims, the sting is very much in the tail. Remember the provisions of CPR 44.3(2) and (5). The last thing that the court will do on an assessment is the proportionality cross check. It is at this point, when the court having assessed the reasonable amount of costs cuts those still further by reference to the proportionality factors, that the certainty of prospective budgeting may seem more attractive.

The major amendment in the new provision is to add a clearer sanction in the situation where the paying party can show that he placed reliance on the budget. Instead of this simply enabling the court to treat this as evidence that the costs claimed are unreasonable or disproportionate (which remains the position where the receiving party has not provided a satisfactory explanation for the difference), the court may now restrict the recovery to the amount it is reasonable for the paying party to pay in the light of the reliance *even* if that is a lesser amount than the reasonably and proportionately incurred costs of the receiving party. At last a genuine sanction and a much clearer steer for the assessing judge and a greater degree of certainty for the parties as to the likely outcome of a finding of reliance.

If there is a difference of less than 20% then there is no change. Costs Practice Direction 6.6(1) has simply become sub-section 3.4. At the assessment the court will assess the reasonableness and proportionality of the costs claimed and in doing so may consider the budgets filed by any of the parties.

CHAPTER 17

PROSPECTIVE COSTS CONTROL – COSTS CAPPING

A BRIEF HISTORY – AND POSSIBLY AN EVEN BRIEFER FUTURE

[17.1]

A primary objective of the CPR even before the April 2013 amendments was to restrict the costs through case management, with the court controlling the work to be done by the parties to ensure it was proportionate to the issues. For various reasons, including the front-loading of work by the CPR regime and the introduction of recoverable additional liabilities, this approach had limited, if any, success.

Sir Rupert Jackson recognised that the other side of the coin is to limit the work by restricting costs in advance. On the small claims track this has always been achieved by not awarding between-the-parties profit costs save where a party has behaved unreasonably. Both parties know at the outset what, in normal circumstances, the recoverable costs will be, whatever the outcome of the litigation. On the fast track, control has gradually been achieved by the court by fixed trial costs, the predictable costs scheme for road traffic pre-action costs, the CPR Part 8 Low Value Personal Injury Claims in Road Traffic Accidents ('RTA') pre-action protocol, the extension of this to employers' liability ('EL') and public liability ('PL') claims and a continuing move towards fixed pre-trial costs, including the introduction of fixed recoverable costs in claims which exit the RTA, EL and PL protocols. On the multi-track an important development in court control of costs was costs capping, which a former Master of the Rolls called 'rough justice; which has to be contemplated to rein in costs'. As costs capping was, in part, an acknowledgement of the failure to restrict the costs at the start of the proceedings through case management and at the end of the proceedings by failing to award between the parties only those costs which were reasonable and proportionate it is hard to see what role it has now that costs management, robust case management and a proportionality cross check at the end of an assessment are the order of the day. The CPR still provides for costs capping, but we cannot help but think that its days are almost numbered. However, as it remains an option, and whilst there are still exceptions to the costs management scheme, it merits further comment.

17.2 *Chapter 17 Prospective Costs Control – Costs Capping*

JURISDICTION

[17.2]

Initially jurisdiction was founded on Section 6 of the Costs Practice Direction as was and on s 51(3) of the Senior Courts Act 1981 which provides: 'The court shall have full power to determine by whom and to what extent costs are to be paid' without specifying that this shall be done at any particular stage of the proceedings. Further support was drawn from CPR 3.1(2)(m) which empowers the court to 'take any other step or make any other order for the purpose of managing the case and furthering the overriding objective'. However this is subject to the provision 'except where these rules provide otherwise' and CPR 44.3 (as was) provides that the court must have regard, amongst other things, to CPR Part 36 offers and the conduct of the parties, neither of which could be done until after the conclusion of the proceedings. Accordingly the requirements of CPR 3.1(2)(m) were apparently infringed. This problem was avoided by making cost capping orders that permitted both prospective and retrospective variation on these lines.

One of the first cases to embrace and justify the costs capping regime was *Griffiths v Solutia UK Ltd (formerly Monsanto Chemicals Ltd)* [2001] EWCA Civ 736, [2001] 2 Costs LR 247. The court said that although the CPR did not confer upon the court an express power to place an advance limit on the costs which would be recoverable for all or part of the litigation (equivalent to the power in the Arbitration Act 1996, s 65), case management powers surely allowed a judge to exercise the power of limiting costs, either indirectly or even directly, so that they are proportionate to the amount involved. Judges conducting cases should make full use of their powers under Section 6 of the Costs Practice Direction to obtain estimates of costs and to exercise their powers in respect of costs and case management to keep costs within the bounds of the proportionate in accordance with the overriding objective.

However, establishing the jurisdiction proved to be only the start of the debate. Other issues arose. Fortunately some of them, such as whether the cap could extend to additional liabilities, are no longer of any concern and need not be considered. The two most important issues, which remain relevant to both costs capping and costs management, were:
- In what circumstances should the court cap costs?
- At what stage in the proceedings should the court exercise it jurisdiction?

SOME OF THE CASE LAW

[17.3]

A wealth of case law ensued. This established many of the principles that have formed the basis for the costs management regime that was introduced in April 2013. As a result a handful of those cases justify further consideration.

In what circumstances should the court cap costs – the exception or the common place?

[17.4]

In *Smart v East Cheshire NHS Trust* [2003] EWHC 2806 (QB), 80 BMLR 175 Gage J acknowledged the legality of costs capping by the procedural judge but was less than enthusiastic about it. He concluded that:

'In my judgement, the court should only consider making a costs cap order in such cases where the applicant shows by evidence that there is a real and substantial risk that without such an order costs will be disproportionately or unreasonably incurred; and that this risk may not be managed by conventional case management and a detailed assessment of costs after a trial and it is just to make an order. It seems to me that it is unnecessary to ascribe to such a test the general heading of exceptional circumstances. I would expect that in the run of ordinary actions it will be rare for this test to be satisfied but it is impossible to predict all the circumstances in which it may be said to arise'.

In other words Gage J firmly set out the stall of those who believed that costs capping orders should be a rarity.

Hallett J (as she then was) favoured a wider application indicating in *Sheppard v Essex Strategic Health Authority* [2005] EWHC 1518 (QB), [2006] 1 Costs LR 8 that the courts were moving towards a system of pre-emptive strikes in order to avoid the costs of litigation spiralling out of control and becoming unreasonable or disproportionate. She determined that it was far better for the court to attempt to control and budget for costs where appropriate, than to allow costs to be incurred and then submitted to detailed assessment after the event.

At what stage in the proceedings should the court exercise it jurisdiction?

[17.5]

In *Henry v BBC* [2005] EWHC 2503 (QB), [2006] 1 All ER 154, [2005] NLJR 1780, the judge refused a costs capping order on the ground that the application was made too late. This reinforced the view expressed by HHJ Kirkham when refusing an application for a costs capping order in *Petursson v Hutchison 3G UK Ltd* [2004] EWHC 2609 (TCC) The appropriate time to consider a costs cap was at an early stage of an action when the parties and the court could together plan the steps needed to bring the matter to trial, the costs implications of those steps and whether a cap was appropriate.

It appeared that definitive answers were to be provided in *Willis v Nicolson* [2007] EWCA Civ 199, [2007] PIQR P318, 151 Sol Jo LB 394. It was confidently predicted that the Court of Appeal would give general guidance on costs capping. Lord Justice Buxton said:

'The very high costs of civil litigation in England and Wales is a matter of concern not merely to the parties in a particular case, but for the litigation system as a whole. One element in the high cost of litigation was undoubtedly the expectation as to annual income of the professionals who conducted it. The costs system could not do anything about that, because it assessed the proper charge for work on the basis of the market rates charged by professionals. It had been hoped when the CPR came into force that that practice might change. However, no change has occurred. The

17.5 *Chapter 17 Prospective Costs Control – Costs Capping*

reasonable amount per hour of a professional's time continues to be determined by the market. Therefore, the focus of costs limitation has to be on the way in which the professionals intended to conduct the case, because the amount recoverable on assessment was fixed, as to rates, by the standard amounts allowed. To limit the way in which professionals intended to conduct a case was a delicate matter. The court would have to be careful to select the right moment in the litigation process for the consideration of a costs cap . . . With all these factors in mind we drafted a comprehensive set of principles to be applied in personal injury cases, which are the most obvious candidates for costs capping; which could also be considered for application to other types of case.'

It seemed that the long awaited guidance was imminent. It was not to be for as Lord Justice Buxton continued:

'However, further discussion with members of the court, including the Master of the Rolls and the Deputy Head of Civil Justice, has demonstrated that, despite the terms in which permission to appeal was granted in this case and the observations in this court, there remain serious doubts as to whether further guidance on costs capping, if it is to be given at all, should emanate from a constitution of the court as opposed to being formulated by the Civil Procedure Rules Committee, after extensive consultation. We are bound to recognise the imperative of that view. We therefore do not pursue the question further. It will be for the Rules Committee to decide whether, and if so with what degree of urgency, to take up the issues that we have identified earlier in this judgment.'

THE PROCEDURAL CODE

[17.6]

However, the wait for clarity was not much prolonged, because, with commendable alacrity, the Civil Procedure Rules Committee drafted amendments to CPR Part 44 and the Costs Practice Direction on which it invited comments. Less encouraging was the accompanying statement that the proposals 'are drawn from current case law' and 'do not propose new policy' when the 'current case law' was in conflict and there appeared to be no 'present policy'.

The rules which were introduced into the Civil Procedure Rules by the Civil Procedure (Amendment No 3) Rules 2008 on 6 April 2009 were, perhaps understandably as a result of the conflicting authorities, extremely conservative in the view of those who saw costs capping as a way to curb excessive and disproportionate expenditure.

Those rules, which appeared at CPR 44.18–CPR 44.20 and at Costs Practice Direction Section 23A, have been moved, almost unchanged, to follow the costs management provisions and can now be found at CPR 3.19–CPR 3.21 and at CPR PD 3F.

In summary the rules provide that:
- The court will make a costs capping order only in exceptional circumstances (CPR PD 3F, para 1.1) and may do so if it is in the interests of justice to do so, there is a substantial risk that without such an order costs will be disproportionately incurred and this risk cannot be managed by case management decisions and detailed assessment of costs (CPR 3.19(5)).

- The application should be made as soon as possible – preferably before or at the first case management conference CPR PD 3F, para 1.2)
- Costs capping cannot be retrospective (CPR 3.19)
- The court does not have to cap the costs of all parties (CPR 3.19)

In other words the result was an absolute triumph for those from the *Smart v East Cheshire NHS Trust* (above) school of thought. Costs capping was virtually dead in the water immediately. Costs capping orders can now be made only in exceptional cases where (i) it is in the interests of justice to do so, (ii) there is a substantial risk that without the imposition of a cap, disproportionate costs will be incurred and (iii) where conventional case management and a detailed assessment are not sufficient to control costs adequately.

That these are virtually insuperable hurdles was demonstrated by Coulson J in *Barr v Biffa Waste Services Ltd* [2009] EWHC 2444 (TCC), [2009] NLJR 1513, [2010] 3 Costs LR 317 and Eady J in *Peacock v MGN Ltd* [2009] EWHC 769 (QB), [2009] 4 Costs LR 584. In the words of Coulson J:

> 'It would be a very unusual case in which a High Court judge did not feel able to utilise one or both of [the tools of case management and detailed assessment] to control disproportionate costs. That is, after all, what they are there for'.

Although it should be noted that he also felt moved to say at the end of the case in *Barr v Biffa Waste Services Ltd* [2011] EWHC 1003 (TCC), [2011] 4 All ER 1065, 137 ConLR 125 'The Group Litigation has, in the end, not been of any benefit to anyone at all except the lawyers.'

CONCLUSION

[17.7]

Costs capping orders will be rare. Whilst in theory they remain a tool of costs control in those cases where a costs management order is not made, the exceptionality test, coupled with renewed court focus on proportionate case management, conspire to read the last rites over an innovative and interesting harbinger of the rationale behind the CPR amendments of April 2013.

Those with an eye for detail will note that the rule may have been moved as part of the amendments, but it has escaped the continuity cross check. Anyone involved in that rare case where the court has decided to cap the costs and is determining the amount of that cap, will be surprised that it is to be set by taking account of the factors at CPR 44.5 (the seven pillars under the old rules). CPR 44.5 now deals with cases determining the amount of costs payable under a contract. Perhaps this is deliberate and it is intended to be the final nail in the coffin.

CHAPTER 18

PROSPECTIVE CONTROL OF COSTS – PROTECTIVE COSTS ORDERS

INTRODUCTION

[18.1]

It is important to define our terminology at the outset of this chapter. Many of the procedures that we have considered in previous chapters under the umbrella heading of *'Prospective costs control'* lead, to a greater or lesser extent, to the provision of some form of costs protection. At the end of this chapter we shall consider *Beddoe orders* – although strictly these are better described as a pre-emptive costs order. That leaves something else as *protective costs orders*. That something is an advance order that if a party, usually a claimant, but as we shall see below it can be a defendant, is unsuccessful either he will not be ordered to pay the costs of the successful party/ies or that his liability under any such costs orders is limited to a specific amount.

CASE LAW

[18.2]

One of the first cases where guidance was given in respect of protective costs orders was in *R v Lord Chancellor ex p Child Poverty Action Group* [1998] 2 All ER 755, [1999] 1 WLR 347. Dyson J (as he then was) suggested that such orders should only be made in exceptional circumstances and gave the following guidance:
(i) The court must be satisfied that the issues raised are truly ones of general public importance;
(ii) The court must be satisfied, following short argument, that it has a sufficient appreciation of the merits of the claim that it can be concluded that it is in the public interest to make the order;
(iii) The court must have regard to the financial resources of the applicant and respondent, and the amount of costs likely to be in issue;
(iv) The court will be more likely to make an order where the respondent clearly has a superior capacity to bear the costs of the proceedings than the applicant, and where it is satisfied that, unless the order is made, the applicant will probably discontinue the proceedings, and will be acting reasonably in so doing.

An early illustration of the application of a protective costs order came in the case of *R (on the application of the Campaign for Nuclear Disarmament) v Prime Minister* [2002] EWHC 2777 (Admin), (2002) Times, 27 December, [2003] 3 LRC 335. In the case the Campaign for Nuclear Disarmament (CND) sought permission to apply for a declaration that the United Nations Security Council Resolution 1441, on the basis of which weapons inspectors had gone to Iraq, did not authorise the use of force in the event of there being a breach, and that a further Security Council resolution would be required to authorise such force. In the course of proceedings CND applied for and obtained an order that in the event of costs being awarded against it in the High Court, such costs be limited to £25,000, on the grounds that CND's modest resources were such that if they were not given the comfort sought by the order they would be unable to proceed with the challenge.

The authorities and philosophy of protective costs orders were considered in R *(on the application of Corner House Research) v Secretary of State for Trade and Industry* [2005] EWCA Civ 192, [2005] 4 All ER 1, [2005] 1 WLR 2600. The Court of Appeal, having considered the statutory framework for awarding costs and the historical perspective on protective costs orders offered clear guidelines as follows:

- Such orders will only be made in exceptional cases.
- The governing principles are:
  - A protective costs order may be made at any stage of the proceedings, on such conditions as the court thinks fit, provided that the court is satisfied that:
    - the issues raised are of general public importance;
    - the public interest requires that those issues should be resolved;
    - the applicant has no private interest in the outcome of the case;
    - having regard to the financial resources of the applicant and the respondent and to the amount of costs that are likely to be involved it is fair and just to make the order;
    - if the order were not made the applicant would probably discontinue the proceedings and would be acting reasonably in so doing.
- If those acting for the applicant were doing so *pro bono* that would be likely to enhance the merits of the application for a protective costs order.
- It is for the court, in its discretion, to decide whether it is fair and just to make the order in the light of the above considerations.
- A protective costs order can take a number of different forms and the choice of form is an important aspect of the judge's discretion. Where an applicant is seeking an order for costs if it is successful, the court should prescribe by way of a capping order a total amount of the recoverable costs, which allows for modest legal representation.
- A claimant should apply for a protective costs order in his claim form, which should include a schedule of the claimant's future costs, of and incidental to, the full judicial review application. A defendant should set out any reasons for resisting an order in its acknowledgment of service. The judge should consider making an order on paper with any actual hearing being brief and, recognising the costs of the process itself, proportionate.

R (on the application of Compton) v Wiltshire Primary Care Trust [2008] EWCA Civ 749, [2009] 1 All ER 978, [2009] 1 WLR 1436 confirmed that the principles for making a protective costs order set out in *Corner House* are to be followed in the court of first instance and in the Court of Appeal. The criterion that the issues raised had to be of 'general' public importance did not mean that they had to be of interest to all the public nationally. In the present case it had been open to the judge to find that there was a public interest in the resolution of issues as to the closure of parts of a hospital that affected a wide community and that the issues were therefore of general public importance. 'Exceptionality' was not an additional criterion but a prediction of the effect of applying the principles of *Corner House*.

Procedurally, if the recipient of the protective costs order in the court below wished to appeal, an application for an order should be lodged with an application for permission. The respondent should have an opportunity of providing written reasons why an order was inappropriate. The decision would be taken on paper by the single Lord Justice. If an order was refused the applicant could apply orally. If it was granted then a respondent would need compelling reasons to set it aside. Appeals from the refusal to grant an order, or against an order, should be dealt with by a single Lord Justice on paper with the normal order being no order for costs.

In *R (on the application of Buglife, The Invertebrate Conservation Trust) v Thurrock Thames Gateway Development Corpn* [2008] EWCA Civ 1209, [2008] 45 EG 101 (CS), (2008) Times, 18 November, the applicant Conservation Trust applied for a protective costs order capping its liability in costs in a dispute with the respondent local planning authority, which in turn applied for an order capping their own liability in costs to the applicant. The Court of Appeal stressed that the courts should do their utmost to dissuade the parties from engaging in expensive satellite litigation on the question of whether protective costs orders and cost capping orders should be made. In *Buglife* the guidelines had not been followed. The local authority's written reasons were not put before the court on paper before the application for permission to appeal and for a protective costs order were considered by the judge. It was of great importance that issues relating to permission to appeal and to a protective costs order and a consequent cost capping order should all be considered at the same time and on paper. In the present case, because the claimant had been granted permission to appeal, it should have some protection but it would be unfair for it to have total protection especially given the fact that there was a significant risk that it would lose. The just order was to limit the claimant's costs in the Court of Appeal to a further £10,000 making its potential total liability £20,000. It was right to cap the authority's liability to the claimant in an appropriate sum, which was also £10,000.

The Master of the Rolls said:

'In the rare case in which it is necessary to have an oral hearing, it should last a short time as contemplated in Corner House and it should take place in good time before the hearing of the substantive application for judicial review, so that the parties may know the position as to their potential liabilities for costs in advance of incurring the costs.'

18.2 *Chapter 18 Prospective Control of Costs – Protective Costs Orders*

A number of cases have demonstrated the difficulty in satisfying the dual requirements of there being a public interest but there being no private interest. In *R (on the application of Goodson) v Bedfordshire and Luton Coroner (Luton and Dunstable Hospital NHS Trust, interested party)* [2004] EWHC 2931 (Admin), [2005] 2 All ER 791, [2006] 1 WLR 432 although there was an issue of general importance, the applicant had an undoubted private interest in the proceedings and the public interest in having the issues decided was not so great as to require a decision from the Appeal Court at the inevitable expense to the Luton and Dunstable Hospital NHS Trust, the second respondent, as to its own costs. In *Weir v Secretary of State for Transport* [2005] EWHC 2192 (Ch), the claimant, a member of the Railtrack private shareholders action group who had commenced proceedings against the Secretary of State alleging misfeasance in public office and breach of the European Convention on Human Rights, failed to obtain a protective costs order limiting his costs liability to £1.35 million because he personally was seeking compensation.

R (on the application of Ministry of Defence) v Wiltshire and Swindon Coroner [2005] EWHC 889 (Admin), [2005] 4 All ER 40, [2006] 1 WLR 134, demonstrated that a protective costs order can, in theory be made in favour of a defendant. The court granted an application by the Ministry of Defence that the Coroner for Wiltshire and Swindon produce digital recordings of his summing up at an inquest, but refused the application of the defendant coroner for a protective costs order because, provided he acted reasonably, he would be indemnified by the council as his active involvement in the case was reasonable. Therefore a protective costs order, whilst available to a defendant, was inappropriate in that case.

In *Eweida v British Airways Plc* [2009] EWCA Civ 1025, 153 Sol Jo (no 40) 37, the claimant failed to obtain either a protective costs order or a costs cap. She worked for the employer on its check-in desks and complained that she had not been allowed to wear a cross denoting her Christian faith in such a way as to be visible outside her uniform. The tribunal rejected all aspects of the claim and the EAT dismissed her appeal. When she obtained permission to appeal to the Court of Appeal she applied for a protective costs order but her application was refused not only because this was not public law litigation but also the employee's private interest was too significant to make an order appropriate even if it were public law litigation. Her alternative of a costs capping order was also refused on an application of the appropriate test under the then CPR 44.18–CPR 44.20 on the basis that the rates proposed to be charged were not unreasonable and there was no reason to suppose that any risk of disproportionate costs could not be adequately controlled by the costs judge on the detailed assessment.

THE FUTURE

[18.3]

The Civil Justice Council recommended building on the protective costs order as explained in *Corner House Research* (above) to permit access to justice in public law cases by further consideration being given to the wider import of the judgment (*'Improved Access to Justice – Funding Options and Proportionate Costs August 2005'- Recommendation 12*). However, in *R (A) (Disputed Children) v Secretary of State for the Home Department* [2007] EWHC 2494 (Admin) while the Court accepted that the issues raised were of general importance and that the public interest required them to be resolved, it did not accept that the public interest required them to be resolved on terms that exposed the Secretary of State to a substantial costs order if she failed, but which prevented her from recovering her costs if she were successful. Although the Secretary of State's pocket was deep, her resources were not unlimited and money spent on litigation was money which could otherwise be spent on her ordinary operations.

On the other hand, in *Baker v (1) Quantum Clothing Group Ltd. (2) Meridian (3) Pretty Polly* [2008] EWCA Civ 823, when the claimant appealed against the dismissal of her claim against her employer for damages for industrial injury suffered by her in the course of her employment, it was apparent that Meridian and Pretty Polly could be affected by the judgment on appeal and they were joined as parties to the proceedings. The claimant had after – the – event insurance which covered the first defendant's costs but not those of the added defendants. Accordingly, the claimant applied for an order that the 2nd and 3rd defendant companies should bear the burden of their own costs of her appeal in any event. The claimant stated that without the protection of an order the risks of an adverse costs order would mean that she was unable to pursue her appeal. The costs that would be incurred on the appeal were, by comparison with the costs liabilities accrued in the case to date, relatively small and it was important that those prior costs should not have been incurred pointlessly by bringing the matter to a premature end. In those circumstances and in the interests of justice, the fact that the claimant's appeal would be stifled if her application was refused outweighed the injustice that would be caused to Meridian and Pretty Polly in compelling them to bear the burden of their own costs of the appeal. Accordingly the court made a protective costs order.

AARHUS CONVENTION CASES

[18.4]

Although arguably something that falls within either the later comments on various fixed costs regimes or costs capping (above), the addition to the CPR in the April 2013 amendments to include recoverable costs limits in Aarhus convention claims is founded on the need to prevent the risk of prohibitive costs acting as a bar on proceedings. As such, whilst not a pure protective costs order as we know it, the purpose is that of providing prospective costs protection. The rules are set out at CPR 45.41–CPR 45.44 and the limit is to be found in CPR PD 45, para 5. If the claimant is the paying party the limit is £5,000 where the claimant is an individual or £10,000 in all other cases. If the paying party is the defendant the limit is £35,000.

BEDDOE ORDERS

[18.5]

Until the advent of the type of prospective cost orders considered above, the only form of pre-emptive costs orders were to be found in the Chancery Division. There they remain an option and if the wording in the practice direction to CPR Part 64 is to be adopted we must call these prospective costs orders as well.

If trustees of personal representative are concerned that in commencing or defending a claim it might subsequently be suggested that they have acted unreasonably, they may apply to the court prospectively for an order that, win or lose, the costs that they incur will come out of the fund or the estate. As this applies as much to defending a claim as pursuing one, it is not unusual to see a claim stayed whilst trustees or personal representatives make this application in the Chancery Division. For more than a hundred years these applications and the ensuing orders have been known as '*Beddoe applications*' and '*Beddoe orders*' after the case of *Re Beddoe, Downes v Cottam* [1893] 1 Ch 547, 62 LJ Ch 233, CA .

The procedure is set out at CPR PD 64B, which comes very loosely from the general provision at CPR 64.2(a) that enables the court to determine any question arising in the administration of an estate or the execution of a trust. In summary:

- The application should be made under CPR Part 8.
- The application must be supported by evidence and, to ensure that the trustees or personal representatives are properly protected, that evidence must give full disclosure of all relevant matters.
- The evidence in support of the application must cover the advice of a lawyer as to the prospects of success, the value to the fund of the dispute, details of the likely costs the trustees or personal representative and other parties.
- The evidence must provide any known information about the means of the other parties.
- The evidence must identify any other relevant factors that may influence the court's decision.
- The evidence must indicate the extent and outcome of discussion with beneficiaries.
- In respect of any litigation, information about whether any relevant pre-action protocols have been complied with and whether ADR has been proposed or will be (and if not, why not).
- If a beneficiary opposes the application he should be a defendant to the Part 8 claim

Ordinarily the court will dispose of the application on paper. If either the claimant or the defendant believes a court hearing is required then they must expressly state this and give reasons. The court is not obliged to hold a hearing, but where it makes an order without one then it will give the parties an opportunity to apply to vary or discharge the order at an oral hearing. Any order sanctioning proceedings or the defence of proceedings can be limited to a particular stage and be subject to review on the material then available. See Chapter 38 –Trustees for further consideration of this.

CHAPTER 19

PROSPECTIVE COSTS ORDERS – SECURITY FOR COSTS

INTRODUCTION

[19.1]

One party to litigation may be ordered to provide security for the costs of the other party in the very limited circumstances prescribed by CPR 25.12–25.15 and the Arbitration Act 1996, s 38. In addition there is a limited discretion under CPR 3.1(5) to order a payment in to court which CPR 3.1(6A) refers to as security for the other party's costs. There was also provision to apply for security under the Companies Act 1985, s 726(1). That provision has been repealed (October 2009) but much of the case law that emerged from it and the principles established by it have a resonance with the provisions in CPR Part 25, particularly under CPR 25.13(2)(c).

UNDER THE CIVIL PROCEDURE RULES, RULES 25.12–25.14

Who may make an application?

[19.2]

The procedural code sets out at CPR 25.12 that only a defendant may apply for security. However, although there can be no order for security for costs against a defendant who is exercising his right to defend himself, even though resident out of the jurisdiction, there may be such an order in respect of a counterclaim. This is because under CPR 20.3 an additional claim (the definition of which includes a counterclaim) is treated as a claim and CPR 2.3(1) defines a defendant as 'a person against whom a claim is made'.

As a general rule, where a counterclaim can properly be relied upon as a set-off and where it arises out of the same subject matter as the claim, the counter-claiming defendant ought not to be required to give security for the costs of that counterclaim unless there are exceptional circumstances (*Ashworth v Berkeley-Walbrood Ltd* (1989) Independent, 9 October, CA). Security against a defendant was also refused where the counterclaim raised the same issues as the claim in *BJ Crabtree (Insulation) Ltd v GPT Communications Systems Ltd* (1990) 59 BLR 43, CA.

A counterclaim is more than a mere defence to a claim: it is a claim which has a 'vitality of its own' is how it was put in *Jones v Environcom Ltd* [2009] EWHC 16 (Comm), [2010] Lloyd's Rep IR 190, [2010] 1 BCLC 150. The inference that the defendant has a claim which has a vitality of its own can be drawn where (i) the defendant would have issued proceedings himself and it was a matter of chance which party issued proceedings first and (ii) the defendant would continue with the proceedings pursuing his own claim even if the claimant discontinued its claim.

The marked discrepancy in size between the amount claimed in the action and the very much greater amount claimed by the counterclaim is relevant to the consideration of whether a counterclaim is a mere defence or a cross-claim with 'vitality of its own', which might well stand alone and be pursued even if the original claim were abandoned (*Hutchison Telephone (UK) Ltd v Ultimate Response Ltd* [1993] BCLC 307, CA). Another example of the practical application of this was in *L/M International Construction Inc v Circle Partnership Ltd* [1995] CLY 4010, CA, where the claimant's claim for costs and fees of about £1 million was met with a counterclaim for breach of contract totalling £15 million. The court, perhaps unsurprisingly, took the view that the amount of the counterclaim put the defendant in the character of a claimant and considered that it ought to be ordered to give security irrespective of the defence to the original action. Interestingly, in *Petromin SA v Secnav Marine Ltd* [1995] 1 Lloyd's Rep 603, QBD where both parties in a case involving a counterclaim were making substantial claims based on the same set of facts, the order for security for costs of the counterclaim in favour of the claimant was for the full amount of those costs and not merely for the amount by which the claimant's costs were increased in defending the counterclaim. The claimant was entitled to be secured in respect of costs no less fully than if it were merely a defendant to the claim advanced in the counterclaim notwithstanding that it was also claimant in the action.

The fact that some form of protection may be held as a result of other proceedings does not necessarily preclude an application for security. In *Flender Werft AG v Aegean Maritime Ltd* [1990] 2 Lloyd's Rep 27, QBD it was held that obtaining a freezing injunction against the claimant in an arbitration as security for a counterclaim did not prevent the defendant obtaining an order for security for costs of the claim. The freezing of the funds relating to the counterclaim was not relevant to the issue of security for costs in the main action, in respect of which the defendant was entitled to a separate security in the event that the claimant's claims against them failed. The defendant had offered to provide security for the costs of its counterclaim. In the circumstances it was fairer to make an order to the effect that each side should secure the other, rather than to make no order as to security at all. Neither party should be at risk in costs if successful.

However, there is no jurisdiction to order a defendant seeking an enquiry as to damages arising out of the claimant's interim injunction to provide security for the claimant's costs arising from the defendant's application (*Bowring (C T) & Co (Insurance) Ltd v Corsi & Partners Ltd* [1994] 2 Lloyd's Rep 567, [1995] 1 BCLC 148, CA).

There are two exceptions to the general rule that the court will not exercise its discretion under CPR 25.12 to order security for costs if the same issues arise on the claim and the counterclaim and the costs incurred in defending the claim would also be incurred in prosecuting the counterclaim. These are:

(1) Where the claim raises substantial factual enquiries which are not the subject of the counterclaim, an order for security might be appropriate notwithstanding the fact that the claim provided a defence to the counterclaim. However, in such a situation the security may be limited to the costs of the additional issues raised by the claim;

(2) Where the claim and counterclaim both raise additional issues it might also be relevant to consider whether the quantum of the claim in respect of which security is sought is substantially greater than the counterclaim (*Dumrul v Standard Chartered Bank* [2010] EWHC 2625 (Comm), [2010] NLJR 1532).

Security for costs other than from the claimant

[19.3]

In *Condliffe v Hislop* [1996] 1 All ER 431, [1996] 1 WLR 753, CA, Kennedy LJ observed that if the circumstances suggested that if the litigating party were to lose, an order for costs would be difficult to enforce against a maintainer of that party, even though there was no power to order security for costs against the maintained party (who did not fall within RSC Ord 23 (now CPR 25.12–CPR 25.15)) nevertheless a stay of the proceedings could be ordered as a protection for the defendant as protection against this risk

In *Abraham v Thompson* [1997] 4 All ER 362, [1997] 37 LS Gaz R 41, CA, the Court of Appeal disagreed. The right of a claimant to bring a properly pleaded and constituted claim in good faith took precedence over the interest of a defendant who might be unable to recover costs against an impecunious claimant. It was preferable that a successful defendant should suffer the injustice of irrecoverable costs than that a claimant with a genuine claim should be prevented from pursuing it. The defendant's application for an order requiring the claimant to disclose details of any funding he was receiving from third parties to enable the defendant to make an application for an order for security for costs on the grounds that it might be difficult to enforce an order for costs against the funder was rejected.

In *Hendry v Chartsearch Ltd* (1998) Times, 16 September, [1998] CLC 1382, CA, the claimant company was in financial difficulties and could not afford to give security for costs. Accordingly, it transferred and assigned its causes of action to its chairman and majority shareholder who commenced proceedings. The contract giving rise to the cause of action precluded assignment without the prior consent of the defendant, which was not to be unreasonably refused. The court concluded that The Companies Act 1985 gave the defendant a statutory right which protected it from being sued by an impecunious limited company without security for its costs. It could not be regarded as unreasonable for the defendant to insist upon that statutory right and refuse consent to assignment. If the individual assignee were willing to provide security on the same terms as if he were the company then the defendant could not reasonably refuse its consent on that ground. However, that was not the position in this case.

However, the introduction of CPR 25.14 in 2000 has codified the position with clarity. Under this rule an order may be sought against someone who has assigned a claim to the claimant with a view to avoiding an adverse order for costs or who has contributed or agreed to contribute to the claimant's costs in exchange for a share of any money or property the claimant may recover. Given the increase in 'litigation funding' agreements this may prove to be a more frequently relied upon provision. The very fact that a claimant has sought external funding in exchange for a financial interest in the claim is a clear indication that the claimant does not have the means to meet an adverse costs order.

General conditions to be satisfied for applications under Part 25.12

[19.4]

These are set out at CPR 25.13. In effect there is a twofold test. The court has to be satisfied, having regard to all the circumstances of the case, that it is just to make an order AND at least one of the conditions set out at CPR 25.13(2) must be met or some other statutory provision permits such an order. In practice the tests tend to be taken the other way around, for if the defendant cannot satisfy the court under CPR 25.13(2), then the court does not need to exercise its discretion under CPR 25.13(1)(a). These are six conditions in CPR 25.13(2).

THE CONDITIONS UNDER CPR 25.13(2)

(1) Residence outside the jurisdiction but not resident in a Brussels Convention state or a state bound by the Lugano Convention

[19.5]

CPR 25.13(2)(a) provides that an individual claimant or company ordinarily resident out of the jurisdiction in a country which is not a party to the Brussels or Lugano Conventions, may be ordered to give security. The test of whether or not a corporation is ordinarily resident outside the jurisdiction requires the court to locate its central management and control (*Re Little Olympian Each Ways Ltd* [1994] 4 All ER 561, [1995] 1 WLR 560). This provision covers claimants resident in the Isle of Man and the Channel Islands.

Although residence abroad is a condition precedent to the application, this alone is not sufficient to justify an order for security (*Berkeley Administration Inc v McClelland* [1990] 2 QB 407, [1990] 1 All ER 958, CA). Such an approach would be discriminatory and contrary to art 14 of the European Convention for the Protection of Human Rights. The court has now moved to a more flexible approach and the discretion should be exercised on objectively justified grounds relating to obstacles to or the burden of enforcement in the context of the particular country concerned. In *Nasser v United Bank of Kuwait* [2001] EWCA Civ 556, [2002] 1 All ER 401, [2002] 1 WLR 1868, the Court of Appeal limited the security to the additional costs that would arise as a result of enforcement abroad. It is important to note that the additional costs arise in the country where the claimant's assets are held, which may not be the same as the country in which he resides.

In *Data Delecta Aktiebolag v MSL Dynamics Ltd: C-43/95* [1996] ECR I-4661, [1996] All ER (EC) 961, ECJ, it was held that a provision that security could be ordered against an individual claimant resident in the EC, even if there was cogent evidence of substantial difficulty in enforcing the judgment, was discriminatory. There is therefore now no basis upon which an order can be made against an individual claimant who resides in the EC on the grounds of his residence abroad. However, the fact that the claimant, a Dutch national residing in Florida, had assets in Holland and Switzerland did not protect him from an order for security for costs under CPR 25.13(2)(a)(ii) as being 'a person against whom a claim [could] be enforced under the Brussels Conventions or the Lugano Convention'. The rule is aimed at the juridical characteristics of the claimant, regardless of the assets that he owned or where those assets might be situated. A claimant who was not ordinarily resident in the UK or a Convention state could not escape liability to give security for costs merely by placing an asset in a Convention state (*De Beer v Kanaar & Co (a firm)* [2001] EWCA Civ 1318, [2002] 3 All ER 1020, [2003] 1 WLR 38).

There is also no settled rule of practice that no order will be made against a foreign claimant if there are co-claimants resident in England (*Slazengers Ltd v Seaspeed Ferries International Ltd, The Seaspeed Dora* [1987] 3 All ER 967, [1988] 1 WLR 221, CA).

(2) The claimant is a company or other body (whether incorporated inside or outside Great Britain) and there is reason to believe that it will be unable to pay the defendant's costs if so ordered

[19.6]

CPR 25.13(2)(c) covers the impecunious company, whether within the jurisdiction or not. It is important to note that there has to be reason to believe that the claimant 'will', as opposed to 'may' not be able to pay the defendant's costs.

Inability to pay

[19.7]

The fact that a company is in liquidation is, on the face of it, evidence that it is unable to pay the defendant's costs unless evidence to the contrary is given (*Northampton Coal, Iron and Waggon Co v Midland Waggon Co* (1878) 7 Ch D 500, CA). Otherwise, an application for security must be supported by a statement which credibly and reasonably shows the inability of the company to pay the costs if the defendant was successful. The mere issuing of a debenture charging all the company's assets is not a sufficient reason to order security. Where a company's accountant deposed to there being sufficient cash-flow to meet an order for costs, despite a shortage of assets, the court accepted this in the absence of expert evidence to the contrary (*Kim Barker Ltd v Aegon Insurance Co (UK) Ltd* (1989) Times, 9 October, CA).

In *Automotive Latch Systems Ltd v Honeywell International Inc* [2006] EWHC 2340 (Comm), the court rejected the claimant's submission that in respect of orders for security for costs, CPR 25.13(2)(c) looked to the ability to pay costs at the time an order to pay was made and that as a matter of jurisdiction, the defendant could not show that in two years' time or so when any litigation would be likely to be the subject of a judgment and a costs order that the claimant's finances would not be such as to enable it to pay the defendant's costs. The court also rejected the claimant's submission that the defendant had caused its financial difficulties because to do so would be to pre-judge one of the major issues in dispute in the proceedings.

(3) The claimant has changed address since the litigation was commenced with a view to evading the consequences of litigation

[19.8]

Although it seems obvious in a rule that relates to the provision of security for costs, one of the 'consequences of litigation' is the possibility of being ordered to give security (*Aoun v Bahri* [2002] EWHC 29 (Comm), [2002] 3 All ER 182). Accordingly if a change of address can be linked to the evasion of any consequence of litigation then this ground is satisfied – Moore-Bick J (as he then was) stated that once a change of address was established then 'the only question is whether' a party 'has done so with a view to evading the consequences of the litigation.'

(4) The claimant failed to give his address in the claim form, or gave an incorrect address on that form

[19.9]

One might think that this situation is unlikely to arise. However, we are aware of one case at least in the county court in the last year where a plainly accidental omission of an address opened the door for an expensive argument about security. Indeed in *Auon v Bahri* (above) this argument also arose. Of course, the rationale behind the provision is obvious. If no address or an incorrect address is given then that may make enforcement of any costs order harder and more expensive.

(5) The claimant is acting as a nominal claimant, other than as a representative claimant under Part 19, and there is reason to believe he will be unable to pay the defendant's costs

[19.10]

CPR 2.3 defines a claimant as 'a person who brings a claim'. It does not offer any definition of a nominal claimant. Further consideration suggests that this condition will be of limited, if any, application. CPR Part 19 representative claimants are specifically excluded. A defendant bringing a counterclaim is not a nominal claimant, but is clearly a claimant in the counterclaim for the reasons already considered. A trustee in bankruptcy bringing a claim in that capacity is also clearly a claimant in his own right. We have already considered the provisions of CPR 25.14 in respect of assigned debts and litigation funders. This leaves only limited scope for a claimant to be nominal only, eg, a trustee in bankruptcy continuing a claim brought by the bankrupt. However, even then there is already a separate provision for security under s 49 of the County Courts Act 1984, which provides:

'49(1) The bankruptcy of the plaintiff (claimant) in any action in a county court which the trustee might maintain for the benefit of the creditors shall not cause the action to abate if, within such reasonable time as the court orders, the trustee elects to continue the action and to give security for the costs of the action.'

This leaves the rather unattractive possibility that a litigation friend under CPR Part 21 may be treated as a 'nominal claimant'. The N235 'Certificate of suitability' already provides for the litigation friend of a claimant to undertake to pay costs, but this is subject to a right to recover these from the claimant. Strictly in such claims the claimant is the child or protected party anyway.

In *Envis v Thakkar* (1995) Times, 2 May, [1997] BPIR 189, CA the claimant was in financial difficulties when he began his action, but if he succeeded he would be able to discharge his liabilities. His creditors would benefit but so would he. Before a person could be branded a 'nominal claimant' there had to be some element of deliberate duplicity or window-dressing operating to the detriment of the defendant. In this case the claimant could not be regarded simply as a nominal claimant suing for the benefit of some other person.

(6) The claimant has taken steps in relation to his assets that would make it difficult to enforce an order for costs against him.

[19.11]

This provision does not require the defendant to show the claimant's intent to avoid the consequences of an adverse costs order. The rule is not concerned with the claimant's motivation for the disposal of an asset (*Aoun v Bahri*, above). There does not have to be a subjective intention; all the defendant need do is show that the steps taken make enforcement more difficult (*Harris v Wallis* [2006] EWHC 630 (Ch), (2006) Times, 12 May).

However, if intent to defeat any subsequent adverse costs order can also be shown then there is no doubt that this would influence the court's consideration of whether it is just to make an order.

SECURITY FOR COSTS OF AN APPEAL

[19.12]

CPR 25.15 provides that an order for security for costs may be made against an appellant and a cross – appealing respondent on the same grounds as appear in CPR 25.13.

In *Antonelli v Allen* [2001] EWCA Civ 1563 the claimant paid £100,000 to her solicitor, who had subsequently been struck off the roll of solicitors, in dubious circumstances. She retrieved £70,000 of the money but lost her claim for the balance of £30,000 from the solicitor's partner. On the claimant obtaining permission to appeal, the defendant sought an order for security for costs under CPR 25.13(2)(a) (residence outside the jurisdiction) and CPR 25.13(2)(g) (against a claimant who 'has taken steps in relation to his assets that would make it difficult to enforce an order for costs against him'). At the time of proceedings the claimant was living in Israel, but had, by the appellate stage, moved to New York, which made the case for security stronger, because there is no reciprocal enforcement of judgments between this country and the United States. The claimant had failed to respond to any questions about the fate of the £70,000 she had recovered, and there was an inference that she had dealt with that sum in such a way as to put it beyond the reach of any creditors. The respondent to the appeal was therefore entitled to security. The court was not prepared to assume in the absence of evidence that the costs of enforcing a judgment against the claimant in New York would not be substantial and would not, for the most part, be irrevocable. In these circumstances the court fixed the amount of security at £10,000 with a provision that if payment was not made by a fixed date the appeal would be struck out automatically and that the appeal be stayed in the meantime.

In *Bell Electric Ltd v Aweco Appliance Systems GmbH & Co KG* [2002] EWCA Civ 1501, [2003] 1 All ER 344, (2002) Times, 20 November, the defendant appealed against a judgment ordering it to pay to Bell within 14 days £100,000 by way of interim damages and £35,000 on account of costs. Aweco was in deliberate breach of the order to pay the judgment sum and its application for a stay had been refused. The failure or delay in making the payment was due not to any financial difficulty but was cynically based upon the practical difficulties for the respondent in seeking enforcement in a foreign jurisdiction. Accordingly there was a compelling reason for the Court of Appeal to order that the appeal be stayed unless within 14 days Aweco paid into court £135,000 to abide its outcome.

A problem peculiar to appeals is that until permission to appeal has been granted there is no provision under CPR 25.15 to enable the court to order security. This is notwithstanding that sometimes the respondent to the appeal may incur not insubstantial costs during that period. A possible solution is under the court's powers under CPR 3.1 (see below).

Rule 36 of the Supreme Court Rules 2009 provides the Supreme Court with the power to order security for costs. The Supreme Court may, on the application of the respondent, order an appellant to give security for the costs of the appeal, and will set out the amount of the security and the manner in which it must be given. Security for costs will not generally be required for appellants who have been granted state funding, for Ministers or Government departments, or where the appeal is under the Child Abduction and Custody Act 1985. No security for costs is required in cross-appeals. Failure to provide security as required will result in the appeal being struck out by the Registrar although the appellant may apply to reinstate the appeal.

As with CPR 25.13 the court must still consider the justice of making an order for security for costs having regard to all the circumstances of the case.

Both CPR 52.3(7) and CPR 52.9(1)(c) may operate to provide an order security for costs by the court imposing some form of payment in to court in respect of costs as a condition of granting permission to appeal.

IS IT JUST TO MAKE AN ORDER FOR SECURITY OF COSTS HAVING REGARD TO ALL THE CIRCUMSTANCES OF THE CASE?

[19.13]

Having considered the CPR 25.13(2) conditions, let us turn our attention to CPR 25.13(1)(a) and the second part of the twofold test. Even if the defendant has satisfied the court that one of the conditions under CPR 25.13(2) is met, the court retains a discretion and must be satisfied that it is just to order security. It is here that many of the authorities that emerged under s 726(1) of the Companies Act 1985 remain relevant and informative.

Discretion

[19.14]

In *Parkinson (Sir Lindsay) & Co Ltd v Triplan Ltd* [1973] QB 609, [1973] 2 All ER 273, CA, Lord Denning identified the following circumstances which the court might take into account in exercising its discretion:

(i) Whether the claimant's claim is bona fide and not a sham.
(ii) Whether the claimant has a reasonably good prospect of success.
(iii) Whether there is an admission by the defendant on the pleadings or elsewhere that the money is due.
(iv) Whether there is a substantial payment into court or an 'open offer' of a substantial amount.
(v) Whether the application for security is being used oppressively, eg so as to stifle a genuine claim.
(vi) Whether the claimant's want of means is being brought about by any conduct by the defendants, such as delay in payment or in doing their part of the work.
(vii) Whether the application for security is made at a late stage of the proceedings.

Prospects of success

[19.15]

If it can clearly be demonstrated that the claimant has a very high probability of success, that is a matter that can properly be weighed in the balance. Similarly, if it can be shown that there is a very high probability that the defendant will succeed that also is a matter that can be weighed. The court deplores attempts to go into the merits of the case, unless it can be clearly demonstrated one way or the other that there is a high degree of probability of success or failure (*Porzelack KG v Porzelack (UK) Ltd* [1987] 1 All ER 1074, [1987] 1 WLR 420).

This approach was confirmed in *Al-Koronsky v Time Life Entertainment Group Ltd* [2005] EWHC 1688 (QB), which also held that a defendant should not be denied security merely because the claimant had succeeded in previous litigation.

Admission

[19.16]

If a defendant admits so much of the claim as would be equal to the amount for which security would have been ordered, the court may refuse him security, for he can secure himself by paying the admitted amount into court (*Hogan v Hogan (No 2)* [1924] 2 IR 14).

Negotiations

[19.17]

A defendant should not be adversely affected in seeking security merely because he has attempted to reach a settlement. Evidence of negotiations conducted 'without prejudice', should not be admitted without the consent of the parties (*Simaan General Contracting Co v Pilkington Glass Ltd* [1987] 1 All ER 345, [1987] 1 WLR 516, CA).

No order if that would be oppressive

[19.18]

Where an order for security for costs against the claimant company might result in oppression in that the claimant company would be forced to abandon a claim which had a reasonable prospect of success, the court is entitled to refuse to make that order notwithstanding that the claimant company, if unsuccessful, will be unable to pay the defendant's costs (*Aquila Design (GRP Products) Ltd v Cornhill Insurance plc* [1988] BCLC 134, 3 BCC 364, CA). It is not necessary for the company to produce evidence of its inability to pursue the proceedings if an order for security is made for the application to be dismissed; it is sufficient if the company shows that there is a probability that it will be unable to pursue the proceedings. Unless it is clearly demonstrable one way or the other, it is not appropriate to go into the merits of the claim in such an application (*Trident International Freight Services Ltd v Manchester Ship Canal Co* [1990] BCLC 263, CA).

Europa Holdings Ltd v Circle Industries (UK) plc [1993] BCLC 320, CA also supports this approach. In that case the claimant was a small company operating on a limited turnover with a net deficit in a time of great depression. It was a solvent and prudently managed company with a genuine claim for payments for work done. In such circumstances it was found that it would be oppressive to force the claimant to abandon its claim by ordering it to give security for costs. The same conclusion was reached in *Murray (Edmund) Ltd v BSP International Foundations Ltd* (1992) 33 ConLR 1, CA in which the Court of Appeal set aside an order for security for costs on the ground that the claimant was simply unable to meet it and would be forced to abandon its claim, which was acknowledged to be a credible one, and this would represent a denial of justice. This was a balancing exercise and the detriment to the claimant would be worse if the order was made than to the defendant if it were not.

On the other hand in *Automotive Latch Systems Ltd v Honeywell International Inc* (above) the court observed that it was a common, if not inevitable, feature of any order for security that the paying party would expect to find a better use for the money if it did not have to pay it as security. However, there is a significant difference between finding a better use for money and not being able to find the money to fund the security.

In *Blue Sky One Ltd v Mahan Air* [2011] EWCA Civ 544, the appellant contended that any substantial order for security would stifle its appeal for which permission had been granted, which would be wrong as a matter of domestic law and would also infringe its rights under Article 6 of the Human Rights Act 1998. The court concluded that if such a submission were relied upon, the onus was upon the party alleging that its appeal would be stifled if a condition of security was imposed to put before the Court full and frank evidence as to its means. The Court rejected the submission that the requirement for such evidence was incompatible with rights under Article 6, concluding that it is for the party seeking to establish its impecuniosity to prove its financial position.

In *Chemistree Homecare Ltd v Roche Products Ltd; Blackbay Ventures Ltd v Roche Products Ltd; Zanrex Ltd v Roche Products Ltd* [2011] EWHC 1579 (Ch), the claimant sought an interim injunction. The defendant estimated that its costs of defending the claim would be substantial and sought an order for 75% of the costs it had already incurred and 75% of costs it would incur up to the end of the injunctive period. The court held: (1) Having regard to in-depth financial reports concerning the health of the claimant's businesses and their respective profitability, it was clear that there was a real risk that it would not be able to meet an award of costs if one was ordered against it. (2) The defendant only sought costs to the end of the injunctive period and there was no evidence that the claimant could not seek to raise the necessary sums from a third party. The court ordered security of £450,000, being 60% of the estimated costs.

In *Platinum Controls Ltd v Aleris Recycling (Swansea) Ltd* [2012] EWHC 1675 (Ch), [2012] NLJR 912, the court re-iterated that its discretion whether to order security for costs or not in the case of a company is a balancing one, weighing the injustice to the defendant if no security was ordered and the claimant failed at trial against the need to avoid oppressive suppression of litigation. All the circumstances of the case must be taken into account, including the claimant's prospects of success, but the court should not go into the merits in detail unless it could clearly be demonstrated that there would be a high degree of probability of success or failure. Further, it would be necessary for the claimant to investigate all possibilities of obtaining financial support, including its directors, shareholders, other backers or interested persons, to put up security for costs before the court could satisfy itself that the claim would be stifled.

(See also 'Amount' below)

Co-claimant

[19.19]

Where the defendants obtained an order for security for costs against a claimant company which was not complied with, this did not entitle them to an order for security against an individual claimant as a condition of him being joined in the proceedings. The law offered the defendants no protection against costs which could not be paid by impoverished personal claimants (*Eurocross*

19.19 *Chapter 19 Prospective Costs Orders – Security for Costs*

Sales Ltd v Cornhill Insurance plc [1995] 4 All ER 950, [1995] 1 WLR 1517, CA). Similarly, where a claimant company had failed to comply with an order to provide security but had executed a deed of assignment of its equitable interests in the claims to a shareholder justifying his joinder as a party under CPR 19.2, an order for security for costs could not be made against the shareholder under the guise of a condition imposed on joinder, except to the extent of security for any additional costs caused by or wasted as a result of his joinder: *Cie Noga D'Importation et D'Exportation SA v Abacha (as personal representatives of Sani Abacha (deceased)) (No 4)* [2004] EWHC 2601 (Comm).

Pre-action costs

[19.20]

In *Lobster Group Ltd v Heidelberg Graphic Equipment Ltd* [2008] EWHC 413 (TCC), [2008] 2 All ER 1173, [2008] Bus LR D58 lengthy pre-action mediation had failed and the first defendant sought security for the costs of the mediation from the claimant, which was now in administration. The court held that although as a matter of principle pre-action costs can be the subject of an application for security (*Re Gibson's Settlement Trusts, Mellors v Gibson* [1981] Ch 179, [1981] 1 All ER 233 and *McGlinn v Waltham Contractors Ltd* [2005] EWHC 1419 (TCC), [2005] 3 All ER 1126, 102 ConLR 111) the court should be slow to exercise its discretion in favour of an applicant as there was a risk that if the pre-action period was lengthy the costs could be extensive and any subsequent attempt to obtain security might become penal in nature. Under the terms of the mediation the parties had agreed to bear their own costs. As a consequence the costs of the mediation were unlikely to be recoverable in any event and, even if they were, they should not form part of the security ordered. Costs of separate pre-action mediation were not 'costs of and incidental to the proceedings'. Both the course of the mediation and the reasons for its unsuccessful outcome were privileged matters and, as a matter of general principle, the costs incurred in respect of such procedure were not recoverable under the Senior Courts Act 1981, s 51. The pre-action period was very prolonged covering a period of over two years and to order the claimant to provide security for costs incurred during that period would be draconian. Instead the claimant was ordered to provide security for costs from the commencement of proceedings up until the exchange of witness statements.

SECURITY FOR COSTS UNDER CPR PART 3

[19.21]

This provision specifically permits the court to order a party to pay sums into court if there has been a failure to comply with a rule, practice direction or any relevant pre-action protocol. Unlike Part 25 it is not restricted to a claimant being ordered to pay security: any party can be ordered to make a payment into court. The provision is specifically referred to in CPR 13.3 as the court can impose conditions on setting aside a default judgment and CPR 24.6 when conditions can be imposed on the disposal of an application for summary judgment. Interestingly the wording almost mirrors that of the amended overriding objective. It will be interesting to see whether, in this age of more robust case management enforcement, the court is more inclined to use the powers under CPR Part 3. We suspect that the answer will be no, for the reasons set out in the two cases considered below.

In *Olatawura v Abiloye* [2002] EWCA Civ 998, [2002] 4 All ER 903, [2003] 1 WLR 275, the court stated that before ordering security for costs in any case the court should be alert and sensitive to the risk that by making such an order it might be denying the party concerned the right to access to the court.

In *Huscroft v P&O Ferries Ltd* [2010] EWCA Civ 1483, [2011] 2 All ER 762, [2011] 1 WLR 939, the claimant lived in Portugal and the defendant applied for an order under CPR 3.1(3) and CPR 3.1(5) that he pay £20,000 into court as security for costs with conditions, on the grounds that the claim did not have a reasonable prospect of success, the claimant had failed to comply with court orders and did not have the financial resources to meet a judgment for costs against him. At a case management hearing, the district judge found that the claimant had not failed to comply with any rule, practice direction or pre-action protocol so he could not make an order under CPR 3.1(5), but instead he made an order under CPR 3.1(3)) that the claimant was to pay £5,000 into court as security for the costs and that in default of payment the claim was to be struck out. The district judge was overruled on appeal. The court stressed that litigants should not be encouraged to regard CPR 3.1(3) as providing a convenient means of circumventing the requirements of Part 25 and thereby of providing a less demanding route to obtaining security for costs. When a court was asked to consider making an order under CPR 3.1(3) or CPR 3.1(5) which was, or amounted to, an order for security for costs, or when it considered doing so of its own motion, it had to bear in mind the principles underlying CPR 25.12 and CPR 25.13. An order should not be made under CPR 3.1(3) to protect a defendant from being unable to enforce a judgment for costs against a personal claimant, who was resident within the jurisdiction or in one of the other member states of the European Union and had been impecunious and whose conduct of the proceedings might be open to criticism, unless one or more additional factors were present which made it appropriate to impose a burden of that kind on one party and a corresponding benefit on the other party. The order had enabled the defendant to obtain, on the back of case management directions, an order for security for costs which it could not have obtained under Part 25 and which was unrelated to the orders being made. It was inappropriate to make such an order.

The difference between rule CPR 3.1 and CPR 25.15 was illustrated in *Great Future Ltd v Sealand Housing Corpn* [2003] EWCA Civ 682. The defendant, who had not complied with interim orders to pay £1 million on account of costs, sought permission to appeal against the judgment. The claimant sought orders that the interim orders be complied with and for security for its costs of the appeal before the application was heard. Although it could not rely on CPR 25.15 because there was no appeal until permission was granted, the Court of Appeal could exercise case management powers under CPR 3.1. There was no evidence that making the order sought would stifle the appeal and accordingly the orders were made. Accordingly it is here that the lacuna in the CPR 25.15 provisions for security on appeals is filled.

19.22 Chapter 19 Prospective Costs Orders – Security for Costs

THE AMOUNT OF SECURITY

[19.22]

The application must be supported by evidence (CPR 25.12(2)). Obviously this should include details of the costs already incurred and an estimate of the future costs. As these applications will invariably arise in claims in the multi track this process ought to be simplified and the expense reduced by making use of existing Forms H and any costs budgets approved by the court.

Similarly the court's consideration of the costs will be far simpler as a result of any costs management orders. Even then the court may discount the amount by the possibility of early settlement or the reduction of costs that are not within the approved budget on assessment. Of course, the court has power to order security in a sum less than the total potential order for costs. It may, for example, only be ordering security for the additional costs of enforcement. The entire process is a discretionary one. A balancing exercise is required (*Re Unisoft Group (No 2)* [1993] BCLC 532, ChD (Companies Ct). As we have seen, the court will not make an order which would prevent a claimant from proceeding with a legitimate claim. An example of this was *Innovare Displays plc v Corporate Broking Services Ltd* [1991] BCC 174, CA, where the Court of Appeal held that 'sufficient security' for costs does not mean complete security, but security of a sufficiency in all the circumstances of the case as to be just. Accordingly, although the defendant sought to obtain an order for security of costs in the sum of £147,655, which it estimated would be its costs, the court held that a sum of £10,000 would be appropriate! A state-funded claimant cannot be ordered to give greater security than the amount he would be likely to be ordered to pay if the defence were to succeed (*Wyld v Silver (No 2)* [1962] 2 All ER 809, [1962] 1 WLR 863, CA).

The fact that a claimant has either before (BTE) or after-the-event (ATE) insurance, is not in itself a sufficient reason for not making an order for security for costs. Nearly all BTE and ATE policies contain wide terms enabling an insurer to repudiate if a claim fails because of the insured's own conduct. Accordingly there is no certainty that in the situation where the defendant would have an order for costs against the claimant (usually where the claim has been unsuccessful), the insurance will still be available as a source of meeting the defendant's costs. (*Al-Koronky v Time-Life Entertainment Group Ltd* [2006] EWCA Civ 1123. (see Chapter 5 – Running Cases with Conditional Fee Agreements for further consideration of the link between funding and security).

VARIATION OF AN ORDER FOR SECURITY

[19.23]

In *Gordano Building Contractors Ltd v Burgess* [1988] 1 WLR 890, 132 Sol Jo 1091, CA, the claimant was ordered to give £20,000 security for the defendant's costs with 'liberty to both parties to apply'. Subsequently, the claimants obtained a letter from their bank stating that it would be willing to lend the claimants £20,000 to pay the costs if required. The claimants applied to have the order set aside but the judge held he had no jurisdiction to hear the application.

It was held on appeal that there were two questions: Can a claimant return to court in respect of the order if he can show a material change of circumstances? Can a claimant return if he produces fresh evidence as to the state of affairs extant at the date of the original order? The answer to the second question is simple: 'No'. If a claimant wishes to have an opportunity to produce further evidence he should, no doubt at some penalty as to costs, apply for an adjournment of the original security hearing. The answer to the first question must be that it is open to a claimant to apply to get an order for security set aside or varied in the light of changed circumstances. In *Gordano* the judge had not considered whether there was a material change of circumstances and, if so, what as a matter of discretion he would do about it. The appeal was allowed to the extent of remitting the matter to the judge for reconsideration.

In an application for relief under the Arbitration Act 1996 where the parties had agreed £30,000 security for costs and had also agreed that it would not be increased even if there were a material change of circumstances, the court still retained a residual discretion to vary the agreement if there were wholly exceptional circumstances. In *Republic of Kazakhstan v Istil Group Inc* [2005] EWCA Civ 1468, [2006] 1 WLR 596, [2006] 2 All ER (Comm) 26, there were such circumstances and accordingly the judge had been wrong to refuse to order further security for costs

SECURITY FOR COSTS UNDER THE ARBITRATION ACT 1996

[19.24]

The Arbitration Act 1996, s 38(3) provides:

> 'The Tribunal may order a claimant to provide security for the costs of the arbitration. This power shall not be exercised on the ground that the claimant is:
> (a) An individual ordinarily resident outside the United Kingdom, or
> (b) A corporation or association incorporated or formed under the law of a country outside the United Kingdom, or whose central management and control is exercised outside the United Kingdom.'

This gives to the arbitrator a wide general discretion, which he must nevertheless exercise judicially – especially as the 'costs of the arbitration' include his own fees and expenses! No doubt the various factors relevant to applications for security for costs in other litigation will be taken into account by arbitrators. Also still of relevance is the decision in *K/S A/S Bani v Korea Shipbuilding and Engineering Corpn* [1987] 2 Lloyd's Rep 445, CA, in which it was held that where an international arbitration is of a type regularly conducted in London for many years and is not a one-off arbitration, security for costs should be ordered; all the more so if the litigation costs are very high.

Another practical difference between security for costs in arbitration and in other litigation is the tradition in litigation that applications for security for costs are not heard by the trial judge, because privileged and prejudicial matters, such as offers for settlement, may have to be considered on the application. Parliament must have taken the view that arbitrators are made of sterner stuff and can more readily put such matters out of their minds, because arbitrators themselves deal with applications for security.

Where a party is challenging the enforcement of a foreign arbitration award, it is right for the court to treat that party as a defendant, with the consequence that the court has jurisdiction to grant security for costs against the holder of the award (*Dardana Ltd v Yukos Oil Co* [2002] EWCA Civ 543, [2002] 1 All ER (Comm) 819, [2002] 2 Lloyd's Rep 326).

Section 70(6) of the Act empowers the court to order the applicant or appellant in an appeal against an award to provide security for the costs of the application or appeal, although such an order cannot be made on the grounds of residence outside the United Kingdom.

Section 75 empowers an arbitrator to make the same declarations and orders as the court charging property recovered with the payment of the solicitor's costs under s 73 of the Solicitors Act 1974.

CHAPTER 20

COSTS INDUCEMENTS TO SETTLE – PART 36 OFFERS AND OTHER ADMISSIBLE OFFERS

INTRODUCTION

[20.1]

On 5 April 2007 the Civil Procedure (Amendment No 3) Rules 2006 introduced a new Part 36 to replace in its entirety the previous rule. In particular it dispensed with the requirement that for a defendant's offer of settlement to have costs consequences it had to be accompanied by a payment into court. Since then there have been two key additions.

The first was the introduction in October 2011 of CPR 36.14(1A). This overturned the unhelpful decision of *Carver v BAA plc* [2008] EWCA Civ 412, [2008] 3 All ER 911, [2009] 1 WLR 113. CPR 36.14(1A) confirms what we had all thought always to be the case – namely that in any claim involving money, i) a judgment more advantageous than a defendant's Part 36 offer is a judgment better in money terms, however small the difference and ii) a judgment at least as advantageous as a claimant's Part 36 offer is one that equals or exceeds the offer in amount.

The second, also to CPR 36.14, was the introduction in respect of claimants' offers made after 31 March 2013 of an additional sanction against a defendant who fails to accept a claimant's Part 36 offer, which the claimant subsequently equals or betters at trial. CPR 36.14(3)(d) is intended to provide that in such a situation the claimant will receive an extra 10% of the damages on awards of up to £500,000 and an extra 10% on the first £500,000 and 5% on the next £500,000 on awards of over £500,000. We say intended for reasons to which we shall return.

Two clear themes emerge from recent authorities and rule changes. The first is that 14 years after its introduction Part 36 is still causing practitioners and parties significant problems. The second is that Part 36 is at the very centre of attempts to encourage settlement in a greater number of cases and to avoid the expense of trial. Both mean Part 36 merits close investigation.

THE CONTENT OF A VALID PART 36 OFFER

[20.2]

It is important to note that CPR 36.1(2) provides that a party may make an offer to settle in whatever way he chooses, but if the offer is not made in accordance with it will not have the costs consequences specified in CPR 36.10, CPR 36.11 and CPR 36.14. It emphasises that in deciding what order to make about costs CPR 44.2 requires the court to have regard to any offer whether it has been made in accordance with Part 36 or not. Given the clarity of this rule it is odd how many cases are reported where the offering party has failed to make a valid offer.

To satisfy the requirements of CPR 36.3(2) an offer must be in writing, must specifically state that it is intended to be a Part 36 offer, specify a period of not less than 21 days within which, if the offer is accepted, the defendant pays the claimant's costs ('the relevant period'), confirm whether the offer relates to all the claim or to only part of it and whether it takes into account any counterclaim. If a trial is listed to start less than 21 days after the offer then the requirement to specify the period of not less than 21 days does not apply and instead the period must be up to the end of the trial or such other period as the court determines. In certain circumstances the offer must contain more information – eg in a personal injury claim involving a claim for future pecuniary loss (CPR 36.5), where the claim involves a claim for provisional damages (CPR 36.6), and, more generally, in relation to the deduction of recoverable benefits (CPR 36.15).

Part 36.4 provides that an offer must be for a single lump sum which is payable within 14 days; if it is not it is not effective for the purposes of Part 36 unless the offer is accepted.

On the face of it the provisions seem straightforward. However, repeatedly the court has been presented with arguments over whether or not an offer falls within the Part 36 scheme. The starting and, one would have thought, the finishing point must be the wording of Part 36 because it is clear from the judgment of Moore-Bick LJ in *Gibbon v Manchester City Council* [2010] EWCA Civ 726, [2011] 2 All ER 258, [2010] 1 WLR 2081 that this part of the CPR contains a free standing procedural code outside the rules of law governing the formation of contracts:

> 'In my view, Part 36 was drafted with these considerations in mind and is to be read and understood according to its terms without importing other rules derived from the general law, save where that was clearly intended.'

The Court of Appeal concluded that Part 36 is a self-contained code, prescribing both the manner in which an offer might be made and the consequences flowing from accepting or failing to accept it. Although basic concepts of offer and acceptance clearly underpinned Part 36, it was not to be understood as incorporating all the rules governing the formation of contracts. Indeed, it was not desirable that it should so do. Certainty was to be commended in a procedural code which had to be understood by ordinary citizens, and it was with that in mind that Part 36 had been drafted. It was to be read and understood according to its terms without importing other rules derived from the general law, save where that was clearly intended.

We might be forgiven then, for wondering quite how so many arguments have arisen over the wording adopted by parties. We suspect that the reason is that the wording in CPR 36.2 still comes as a surprise to those regularly making what they believe to be valid Part 36 offers. In *Thewlis v Groupama Insurance Co Ltd* [2012] EWHC 3 (TCC), 142 ConLR 85, [2012] BLR 259 HHJ Behrens sitting as a Judge of the High Court determined that an offer that did not state that it was intended to have the consequences of section 1 of Part 36 could not be a valid Part 36 offer. He concluded that the regime was a free standing one and that compliance with its terms was mandatory.

WHO MAY MAKE A PART 36 OFFER?

[20.3]

Nowhere in Part 36 does it state specifically that an offer may be made by any party. The wording of CPR 36.2 creates further uncertainty by referring to the offer providing the period during which the defendant will be liable to pay the claimant's costs if the offer is accepted. However, CPR 36.14, which sets out the consequences after judgment if an offer has not been accepted, leaves no room for argument. It is clear that the rule contemplates offers from both claimants and defendants.

WHEN MAY A PART 36 OFFER BE MADE?

[20.4]

A Part 36 offer may be made before proceedings. Indeed in terms of a defendant seeking to maximise his costs protection and a claimant seeking to make the most of the possible sanctions if he receives an outcome at least as advantageous as his offer, the earlier that the offer is made the better. However, although a Part 36 offer may be made before proceedings, if it is accepted in the same period the claimant cannot claim costs under CPR 36.10 because there are no proceedings, unless the offer is worded 'the defendant will be liable for the claimant's costs, including the costs pre-issue of proceedings in accordance with CPR Part 36.10' (*Udogaranya v Nwagw* [2010] EWHC 90186 (Costs)). This creates no practical difficulties because if the costs cannot be claimed under CPR 36.10, then they certainly may under the costs only proceedings provisions of CPR 46.14.

CPR 36.3(2)(b) also extends the scope of Part 36 to appeal proceedings. However, there is a trap for the unwary. In *East West Corpn v Dampskibsselskabet AF, 1912, Aktieselskab (a body corporate)* [2003] EWCA Civ 174, [2003] 1 Lloyd's Rep 265n, [2003] 12 LS Gaz R 31 and *KR v Bryn Alyn Community (Holdings) Ltd (in liq)* [2003] EWCA Civ 383, [2003] PIQR P562, 147 Sol Jo LB 387, the claimants had all made offers to settle before the trial under the previous Part 36 which were not accepted by the defendants. In each case the defendants were unsuccessful at the trial and the trial judge ordered that they should pay the claimants' costs on the indemnity basis. After the defendants' unsuccessful appeals, the claimants sought a similar order in respect of their costs of the appeal. However, they had made no further Part 36 offer in respect of the appeal, perhaps thinking that the offer in the original proceedings provided sufficient protection. They were unsuccessful. The court concluded that they ought to have made a specific offer in the appeal proceedings. CPR 36.3(4) reflects these decisions. An offer only has Part 36 consequences in the proceedings in which it is made and <u>not</u> in relation to the costs of any appeal from the final decision.

TIME PERIODS FOR ACCEPTANCE AND WITHDRAWAL

General

[20.5]

Once again the provisions of Part 36 seem clear. If a party wishes to withdraw an offer during the period in which he has said that he will be responsible for the other party's costs, then he may only do so with the permission of the court (CPR 36.3(5)). After the expiry of that period an offer may be withdrawn or the terms of it changed to be less advantageous at any time and without the permission of the court. However, this can only be done by <u>written notice</u> of the withdrawal or change of terms to the other party. So a Part 36 offer is open for acceptance unless, and until, it is formally withdrawn or varied in writing. The following sections suggest that this is not all quite so straightforward as it appears.

In *Gibbon v Manchester City Council* [2010] EWCA Civ 726, [2011] 2 All ER 258, [2010] 1 WLR 2081 the Court of Appeal concluded that where a number of Part 36 offers have been made, but none have been withdrawn, then all remain open for acceptance – in other words there may be a number of Part 36 offers available for acceptance at any given moment. The Court of Appeal made it clear that rejection of a Part 36 offer does not prevent its later acceptance, if the offer is not formally withdrawn in the meantime. In addition the fact that the claimant rejected an offer by the defendant, which was in the same sum as an earlier Part 36 offer made by the claimant herself (which she had not withdrawn), could not amount to an implied withdrawal of the her earlier Part 36 offer. A Part 36 offer can **ONLY** be withdrawn in accordance with CPR 36.3(7) above. The consequence in *Gibbon* was that the defendant was able to accept the claimant's own offer 3 months or so after it had been made at a time when the claimant had made it clear that she now valued her claim at a higher level. This was undoubtedly an illustration of the law of unintended consequences in action.

As a result even if an offer has been made before the commencement of proceedings it survives the commencement of proceedings. There is no need to repeat it or do anything. It will last until it is withdrawn or accepted.

In proceedings arising out of a commercial dispute over the sale of development land the claimant made an offer headed 'Offer to settle under CPR Part 36', which provided that 'the offer will be open for 21 days from the date of this letter ('the relevant period')' and expressly stated that it was intended to have Part 36 consequences. The offer was not accepted within 21 days or the slight extension agreed by the parties, but much later and shortly before trial the defendant purported to accept it. The judge held that, despite its wording, the offer had not been a Part 36 offer because it is not possible to make a time-limited offer under Part 36 and this offer was time limited. He therefore rejected the suggestion that the case had been compromised by acceptance of the offer. Each side appealed. The parties were anonymised in case there was eventually a trial.

The Court of Appeal held that although the judge had been right to hold that a time-limited offer cannot be a Part 36 offer, the words 'the offer will be open for 21 days', had to be construed in the context of Part 36, and in that context, the words meant no more than that the offer would not be withdrawn for at least 21 days (as required by Part 36). As that in turn suggested that the offer could be withdrawn at a later date the offer fell squarely within Part 36. In the circumstances the offer, not having been withdrawn in writing, had remained open and was validly accepted (*C v D* [2011] EWCA Civ 646, [2012] 1 All ER 302, [2012] 1 WLR 1962). The explanation given by Rix LJ at paragraph 54 sets out his reasoning as follows:

> 'In the context of Part 36, it seems to me to be entirely feasible and reasonable to read the words 'open for 21 days' as meaning that it will not be withdrawn within those 21 days. Part 36 permits withdrawal within the 21 day relevant period, but only with the permission of the court. It seems to me that 'open for 21 days' is an obvious way of saying that there will be no attempt to withdraw within those 21 days. It is also a warning that after the expiry of those 21 days, a withdrawal of the offer is on the cards. Such a construction would save the Part 36 offer as a Part 36 offer and would also give to both parties the clarity and certainty which both Part 36 itself, and the offer letter with its reference to 'open for 21 days', aspire to. It would leave the offeror entirely free to withdraw the offer immediately upon expiry of the stated period, or to let it roll on for as long as it wished. At the same time it would assure the offeree that it had 21 days to consider what it wanted to do, but was at risk if it had not accepted within that period. There might be an issue, had the offeror wished to withdraw within the relevant period, as to whether the court would permit it to do so where it had stated that it was open for 21 days: but that issue does not affect the current question.'

Rix LJ confirmed, and indeed slightly extended, this construction in *Epsom College v Pierse Contracting Southern Ltd (formerly Biseley Construction Ltd) (in liq)* [2011] EWCA Civ 1449, 142 ConLR 59, [2012] 3 Costs LR 451. In this case he concluded that an offer stated to 'remain open for acceptance 21 days from the date of receipt of this letter by you . . . and as such diarise expiry as 4pm on Friday 3 April 2009 . . . ' was not a time limited offer, even though, on the face of it, the offer prescribed a finite period.

The message from these two cases is twofold. First, the court appears to be adopting a purposive interpretation of the wording used in offers and, second, that an offer must be expressly withdrawn if the offeror no longer wishes it to be accepted, even if the offeree has already formally rejected the offer.

When is a Part 36 offer made?

[20.6]

CPR 36.7 provides, unsurprisingly, that an offer is made when it is served on the offeree and a change in the terms is made when that is served on the offeree. The rule is silent about when a withdrawal of an offer is made, but that, too, must be only when it is served. The fastest way to make a written offer or acceptance under the rules other than by personal service is by fax or other electronic method. A fax or email message sent before 4.30 pm is deemed to be served on that very same day. This is, of course, on the basis that the recipient is willing to accept service by that method – for details on this see CPR PD 6A, section 4.

Offers made less than 21 days before trial.

[20.7]

As we have seen the usual requirement that a Part 36 offer must provide a period for acceptance of at least 21 days during which the offeror accepts liability for the offeree's costs does not apply to offers made less than 21 days before trial. This has two consequences as follows:
- If such an offer is accepted, then under CPR 36.10(4)(a) the parties must either agree the liability for costs or the court will make an order.
- If the offer is not accepted, but at trial the offer transpires to be a good one, then the costs consequences provided in CPR 36.14(2) and (3) (see below) do not apply unless the court has abridged the time.

Permission to withdraw a Part 36 offer

[20.8]

In *Flynn v Scougall* [2004] EWCA Civ 873, [2004] 3 All ER 609, [2004] 1 WLR 3069 without waiting for a medical report on the claimant the defendant paid £24,500 into court. Six days later the report arrived saying that the claimant's injuries were not as serious as had been thought. The defendant applied for permission to withdraw £14,500, leaving only £10,000. The application was refused. There had been no material unforeseen change in circumstances. The defendant had taken a risk in making an early payment-in and it would not be justice to deprive the claimant of his right to accept the payment within 21 days.

The effect of withdrawal of a Part 36 offer

[20.9]

Apart from ensuring that an offer is no longer available for acceptance, a withdrawal of an offer means that the costs and other sanctions which the court may impose after judgment under CPR 36.14 do not apply. However, CPR 44 2 provides that in deciding what order to make about costs the court must have regard to an admissible offer which is not an offer to which costs consequences under Part 36 apply.

In *Samco Europe v MSC Prestige* [2011] EWHC 1656 (Admlty), [2011] NLJR 988, [2012] BLR 267, although the claimant had withdrawn its offer and it was no longer available for acceptance it retained its costs potency. There was no suggestion that the defendant could not reasonably have been expected to accept it during the 21 day period. Had it been accepted within the period of 21 days no further costs would have been incurred by either party. The reason why the claimant had incurred costs thereafter was because its offer had not been accepted.

Offers accepted at trial

[20.10]

A party may only accept a Part 36 offer once the trial has commenced with the permission of the court (CPR 36.9(3)(d).

Acceptance after trial but before the court has given judgment

[20.11]

CPR 36.9(5) enshrines the decision in *Hawley v Luminar Leisure Ltd* [2006] EWCA Civ 18, [2006] Lloyd's Rep IR 307, [2006] IRLR 817, in which after the judge had reserved judgment the defendant purported to accept a Part 36 offer the claimant had made earlier in the proceedings and had not withdrawn. The court held that an offer was probably only open for acceptance until the hearing commenced and certainly could not be accepted after the court had reserved judgment.

This was also the conclusion of Silber J in *Mehjoo vHarben Barker* [2013] EWHC 1669 (QB). In this case the trial ended in March 2013, but judgment was not handed down until June 2013. The claimant had made a Part 36 offer, which he subsequently bettered at trial. However, he had withdrawn that offer in May 2013. The judge considered that there had been no need for the claimant to withdraw his Part 36 offer after trial because it was no longer capable of acceptance anyway without his consent, which he would not have given. Accordingly, as the claimant could not withdraw his offer his attempt to do so could not deprive him of the benefit that followed under CPR 36.14.

When does a trial commence?

[20.12]

In *Wilson v Ministry of Defence* (unreported HHJ Hughes QC 23.4.13) the court had to grapple with this very question. The claimant's claim arose out of indirect asbestos exposure (from washing asbestos contaminated clothing). The defendant alleged that the claim was statute barred and a preliminary trial was ordered on the limitation issue. Prior to this trial the claimant made a Part 36 offer. The preliminary trial concluded in the claimant's favour. The defendant sought to accept the offer. The claimant argued that once the trial on limitation had commenced, then under CPR 36.9(3)(d) the defendant required the permission of the court to accept. The defendant submitted that no permission was required in the period between the end of the limitation trial and the beginning of the quantum trial (breach having been admitted). The court concluded that the trial began when the first of two or more trials of contested issues commenced. CPR 36.9(3)(d) did not seek to differentiate between a trial undertaken in stages and one where all issues were determined. Accordingly the defendant needed permission to accept the offer.

CLARIFICATION OF A PART 36 OFFER

[20.13]

CPR 36.8 contains what must be one of the least used provisions in the CPR. It provides that an offeree may seek clarification of the Part 36 offer provided that this is done within 7 days of receipt of the offer. If the offeror fails to provide the clarification, then the offeree can apply to the court for an order that the offeror does so and the court will set the date when the offer is then deemed to have been made. We can see no reason why the court would set that date as any other than the date by which the clarification is to be provided.

Our comment on the under use of this provision finds support in the conclusions of Silber J in *Hossein Mehjoo v Harben Barker* [2013] EWHC 1669 (QB). Here the defendants sought to rely upon a perceived lack of information about the claimant's costs when trying to avoid the consequences of CPR 36.14. Apart from stating that there was no requirement in CPR 36.2 for the claimant to have provided costs information as part of the offer, the judge stated that instead of an offeree complaining about a failure by the offeror to provide information at the time of the offer, that party should have invoked the provisions of CPR 36.8 to seek clarification of the offer.

PART 36 AND INTEREST

[20.14]

Both a claimant's and defendant's Part 36 offers will be deemed to be inclusive of interest until the expiry of the period during which the defendant has indicated he will be liable for costs or the claimant has indicated he will accept standard basis costs without any CPR 36.14 sanctions or, if the offer is made less than 21 days before the trial, until a date 21 days after the offer.

This does mean that the court may be involved in some calculations of the net offer when determining whether or not an offer exceeds an award at trial. The older the offer the more challenging the arithmetic! In *Blackham v Entrepose UK* [2004] EWCA Civ 1109, [2004] 35 LS Gaz R 33, (2004) Times, 28 September, the trial judge was wrong to have regarded his award of £40,854 including interest to the date of judgment as being less advantageous to the claimant than the claimant's Part 36 offer of £40,000 made two years earlier. The damages element of the offer was £39,644, the balance being interest, which exceeded the damages element of the judgment.

DISCLOSURE OF PART 36 OFFERS

[20.15]

Part 36.13 provides that a Part 36 offer will be treated as 'without prejudice' and must not be communicated to the trial judge or the to the judge allocated to conduct the trial until the case has been decided. This is particularly pertinent given the post March 2013 aim of judicial docketing: providing for the same judge to see a claim through from case management to trial. Parties will have to be astute to ensure that any Part 36 offers made by any party are not communicated at interim hearings. So, for example, any application for security for costs should be listed before a different judge for, as we have seen in the previous chapter, the existence of a settlement offer is a factor relevant to the exercise of discretion when the court considers whether or not to order security for costs.

Similarly if there is an application for permission to withdraw a Part 36 offer within the relevant time period for acceptance CPR PD 36A, para 2.2 is a reminder to the parties not to list this before the judge who will be the trial judge unless all parties agree that the trial judge may hear that application. This is also the position when permission of the court is required to accept a Part 36 offer (CPR PD 36A, para 3.2).

The position where there is split trial between liability and quantum and the claimant succeeds on liability where there is a Part 36 offer presents more challenging problems. Under the pre 2007 rule, if the claimant applied for costs at that stage the judge could be told there was a Part 36 payment which invariably resulted in the costs being reserved. The new rule includes no such provision. It could well be wrong that a claimant who subsequently fails to beat a Part 36 offer should have been awarded his costs of the liability hearing. Either the old rule needs restoring or the claimant's costs on a split trial should always be reserved until all the issues have been resolved.

Support for this approach came in *Shepherds Investments Ltd v Walters* [2007] EWCA Civ 292, [2007] 6 Costs LR 837, where the Court of Appeal dismissed the claimant's appeal against an order reserving the issue of costs until after the outcome of an account of profits in its claim against the defendants, holding that a judge is not required by the CPR to make an immediate decision on costs and has a discretion to postpone it until quantum had been finally determined.

A literal reading of CPR 36.13(2) suggests that even the existence of a Part 36 offer must not be communicated to the trial judge until the second stage of the case has been decided as the wording refers simply to 'the fact that a Part 36 offer has been made'. Accordingly, in nearly all split trial cases where an offer has been made all questions of costs will have to be reserved to the conclusion of the second stage. However, it is clearly desirable for the costs of the liability hearing to be dealt with at its conclusion. A possible solution might be, in appropriate cases, to construe the words 'until the case has been decided' in CPR 36.13(2) as referring to the conclusion of the first part of a split trial. Even then the court could only be told of the existence of the offer, so the question of costs would still have to be reserved (*AB v CD* [2011] EWHC 602 (Ch), [2011] IP & T 504 – a case where the parties' names were removed to prevent potential prejudice at any later hearing). However, the subsequent case of *Beasley (by his litigation friend) v Alexander* [2012] EWHC 2715 (QB), [2013] 1 WLR 762, (2012) Times, 02 November doubts this approach, construing 'until the case has been decided' as being a reference to final determination.

The conundrum was wrestled with in two other cases that appear to accept the approach in *Beasley*, but without any enthusiasm.

20.15 *Chapter 20 Costs Inducements to Settle – Part 36 Offers*

In *Ted Baker plc v Axa Insurance UK plc* [2012] EWHC 1779 (Comm), [2012] 6 Costs LR 1023 there had been a trial of a preliminary issue and the question of the costs of that trial arose for decision. Unless the parties were to agree to disclosure to the court of any Part 36 offer, CPR 36.13 would preclude the court knowing whether there had been any such offer.

Eder J thought that to be an unsatisfactory position:

'It seems to me . . . that there is a real problem here. In my view, there is an urgent need for CPR 36.13 to be reviewed and possibly reformulated in order to deal in particular with the question of split trials and the kind of difficulties which have arisen in the present case.'

Earlier in the judgment he commented on the wording of CPR 36.13:

' . . . although CPR Part 36 (2) prohibits the fact of any CPR Part 36 offer being communicated to the court in the circumstances there specified, it does not on its language appear to prohibit the fact that a CPR Part 36 offer has not been made being communicated to the court. Again, I did not understand the parties to suggest otherwise although this may seem somewhat odd if only because if that is right and the court is not told that a CPR Part 36 has not been made then the inference would seem to be that a CPR Part 36 offer must have been made; and this would appear to undermine the prohibition in CPR Part 36.13.'

The court was in fact informed without objection that it could and should proceed on the basis that no Part 36 offer had been made specifically in relation to the preliminary issues but could not know if any wider Part 36 offer had been made.

The second case was *Jean Scene Ltd v Tesco Stores Ltd* [2012] EWHC 1275 (Ch) in which Judge Pelling QC held:

'It is undesirable for a judge to make even a partial costs order if it is in relation to part of the costs of the action generally, as opposed to a freestanding application, where there is a Part 36 offer that is intended to be dispositive of the whole claim, particularly where the Part 36 offer was made at a very early stage in the proceedings, the parties are unable or unwilling to disclose the terms of the offer to the judge to whom the costs application is made and where the quantum issues on which the effectiveness of the Part 36 offer depends remain to be decided. In such circumstances I regard it as almost inevitable that the court will only in the most exceptional of circumstances be prepared to entertain a costs application in relation to even part of the general costs of a claim.'

PART 36 OFFERS AND COUNTER OFFERS

[20.16]

As we have seen in *Gibbon v Manchester City Council* (above) because Part 36 is its own self-contained scheme, unlike in the law of contract, a counter-offer does not terminate the original offer which will still exist if the counter-offer is refused and may be accepted at any time before it is withdrawn. This should always be borne in mind, but never more so that where a defendant insurer has made a Part 36 offer in a substantial claim and then obtains surveillance evidence of exaggeration. The wise practitioner should formally withdraw the

Part 36 offer before serving the evidence to avoid the claimant having an opportunity to accept the original offer. Failure to consider this can have disastrous effects. *Sampla v Rushmoor Borough Council* [2008] EWHC 2616 (TCC), provides a dramatic illustration of this. Mr Crowley settled the claim and sought a 20% contribution from his co-defendant (RBC). This was robustly refused but in the course of a two day hearing RBC offered a contribution of 33.5% which Mr Crowley rejected. RBC then accepted his original offer of 20% which had not been withdrawn.

THE CONSEQUENCES OF ACCEPTANCE OF A PART 36 OFFER WITHIN TIME

[20.17]

CPR 36.10 sets out the effect of accepting an offer within the relevant period – the claimant will be entitled to the costs of the proceedings up to the date on which the notice of acceptance was served on the offeror.

These costs will be on the standard basis (CPR 36.10(3)). In *Lahey v Pirelli Tyres Ltd* [2007] EWCA Civ 91, [2007] 1 WLR 998, [2007] NLJR 294, the claimant accepted a Part 36 payment by the defendant and thereby became entitled to his costs of the proceedings up to the date of serving notice of acceptance. However, at the outset of the detailed assessment, the defendant asked the district judge to order, before embarking on the detailed assessment, that the claimant should be only awarded 25% of the assessed costs. The defendant contended that in determining whether costs had been 'unreasonably incurred or are unreasonable in amount' (within the meaning of CPR 44.4(1) as was), the court was not constrained only to look at items of cost individually. It might conclude that a whole stage of the proceedings was unreasonable. It could look at the conduct of the parties in the round and not only by reference to specific items of costs. The district judge held he had no jurisdiction to order any such reduction, he was upheld by the judge on appeal and again by the Court of Appeal. The effect of the CPR is that, upon acceptance of a Part 36 payment, 'a costs order is deemed to have been made on the standard basis'. This means the claimant is entitled to 100% of the assessed costs, those being the amount that the cost judge decided was payable at the conclusion of the detailed assessment. The district judge had no power to vary that order. The power to vary or revoke an order given by CPR 3.1(7) is only exercisable in relation to an order that the court has previously made, and not to an order that is deemed to be made by operation of the rules.

The same automatic provisions also apply where the defendant's offer relates only to part of the claim, but at the time of accepting it the claimant abandons the balance of the claim (CPR 36.10(2)). However, in this situation the court has a residual discretion to order otherwise and make different costs provisions.

Settlement of all or part of the claim

[20.18]

In *Sutherland v Turnbull* [2010] EWHC 2699 (QB) the claimant accepted the defendant's Part 36 offer to pay 30% of the damages to be assessed on an agreed list of specific injuries arising out of a road traffic accident. The defendant sought an order, pursuant to CPR 36.10(2), for the court to exercise its discretion in relation to costs in respect of liability and causation. This gave rise to the issue of whether the offer related to part only of the claim, and, therefore, came within CPR 36.10(2). An objective reading of the offer led to the conclusion that it was an offer to settle all outstanding liability issues. The fact that the only money on offer was referable to some injuries did not carry with it the consequence that the offer as a whole was confined to the claim for damages in respect of those injuries leaving the remaining injuries as a discrete and fully contested part of the proceedings. The court resolved the matter by reference to the terms and effect of CPR 36.11(1), (2) and (3). If the offer had related to part only of the claim, then CPR 36.11(3)(a) would apply and the claim would be stayed only in relation to the claim in respect of some injuries. It would follow that notwithstanding the claimant's acceptance of that offer there would have been no stay in respect of the claimant's claim for damages for other injuries. That was plainly not what had been intended nor contemplated within the Part 36 offer made by the defendant and not what would have been understood by any reasonable solicitor reading it. CPR 36.10(2)(b) contemplated a separate distinct abandonment carried out at the same time as the act of accepting the Part 36 offer. There was in the present case no such act by the claimant or his solicitors. In any event no such act could be implied by the acceptance of the Part 36 offer. By accepting the offer the claimant was not accepting an offer to settle the claims in respect of some injuries but the whole of her liability claim, leaving only damages to be assessed. The claimant was entitled to the costs of the proceedings up to the date on which notice of acceptance of the defendant's offer was served pursuant to CPR 36.10(1).

As a result any party making a Part 36 offer which is not intended to be a global offer to conclude a claim must clearly state that on the face of the offer. Even if it is clear from the amount offered that nothing has been included in monetary terms in respect of a discrete part of the claim (for example future loss of earnings or care), unless there is a clear indication that the offer is in part only and specific issues have been excluded so that they remain live before the court, then the offer will not fall within CPR 36.10(2) but instead will fall within CPR 36.10(1). The claimant will not be treated as implicitly abandoning parts of the claim. Instead he will be treated as accepting a global offer in respect of the entirety of the claim. The consequences will be that:

- The entire claim will be stayed
- The claimant cannot pursue further sums
- The claimant is entitled to his costs on the standard basis and the defendant cannot seek a different costs award. Instead the defendant is left to raise the discrepancy between sums sought and sums accepted within arguments on reasonableness and proportionality of the costs at any subsequent assessment of costs.

A stay of proceedings

[20.19]

CPR 36.11 confirms that an acceptance of a Part 36 offer stays the claim. An exception is in claims where the approval of the court is concerned, such as those involving children and protected parties. In those cases the stay only operates once the court has approved the terms of settlement (CPR 36.11(4)). It is surprising how many parties still try to conclude such claims by lodging a Tomlin order. The Practice Direction to Part 36 requires the party accepting the offer to file a copy of the notice of acceptance with the court at the same time as serving it so that the court knows that the claim has been settled.

If the acceptance is of part of the claim only, then only that part is stayed.

The stay does not operate to prevent any action to enforce the Part 36 offer or to resolve the costs of the proceedings – including issues of interest on costs.

Payment

[20.20]

Unless the parties have agreed otherwise in writing, then acceptance of an offer that is to pay by one single sum of money, means this must be paid within 14 days of acceptance or the date of the order for provisional damages or the order including periodical payments (CPR 36.11(6)). Under CPR 36.4(2)) an offer that includes terms for payment of all or part of the sum later than 14 days is not an offer within the Part 36 regime anyway unless the offer is accepted.

If payment is not made within this period the claimant may enter judgment for the unpaid sum. Curiously the rule at CPR 36.11(7) refers to the offeree entering judgment. However, if the Part 36 offer was a claimant's one that would be a nonsense as it would be the offeror who was the one with the incentive to enter judgment.

If the offer was one that did not provide for payment of all the sum within 14 days then CPR 36.11(8)) provides that if there is a dispute about what was agreed and one party alleges that the other has not complied with the agreement, then an application can be made to the court for enforcement without needing to begin a fresh action based on breach of contract.

ACCEPTANCE AFTER THE EXPIRY OF THE RELEVANT TIME PERIOD

[20.21]

If a party wishes to accept an offer after the time period within which a liability for costs has been accepted, then he must either agree the costs consequences of the late acceptance on acceptance with the other party or the court will make an order as to the costs (CPR 36.10(4)). The usual provision in this situation under CPR 36.10(5), subject to any other order by the court, is that the claimant recovers his costs up to the last day he could have accepted the offer with his costs liability accepted by the defendant and with a liability for the defendant's costs after that date. Interestingly this second part of this rule refers to offeror and offeree, recognising that whether the offer is a defendant's or claimant's one, the claimant will recover costs up to the date that the offer could have been accepted, but after that there is a difference depending upon who made the offer.

When might the court order otherwise?

In *Lumb v Hampsey* [2011] EWHC 2808 (QB) Lang J confirmed that there was no guidance in the rule itself or the practice direction to assist the court in how it might exercise this discretion. The only previous authority had seen a departure from the usual adverse cost consequences on the party accepting late because of 'exceptional circumstances'. Lang J concluded

> 'In my view the test which I should adopt in deciding the issue under Rule 36.10(5) is similar to that set out in Rule 36.14. The test which I apply is whether the usual costs order set out in Rule 36.10(5) should be departed from because it would be unjust for the claimant to pay the defendant's costs after the expiry of the relevant period, in the particular circumstances of the case. Such a departure would be the exception rather than the rule.' (Para 6)

Lang J then made specific reference to the factors at CPR 36.14(4). Having done so, and having reviewed the grounds put forward, she refused to disapply the normal cost consequences of late acceptance.

Contrast *Lumb* to *SG (a minor by his mother and litigation friend) v Hewitt* [2012] EWCA Civ 1053, [2013] 1 All ER 1118, [2012] 5 Costs LR 937. SG was injured at the age of six, and sustained 'facial scarring and a severe head injury'. The court heard that the defendant made a Part 36 offer of £500,000 in 2009, when the boy was 12, which was ultimately accepted by the claimant two years later. The defendant agreed to pay the claimant's costs until 21 days after the Part 36 offer was made, and the costs involved in approving the settlement. After the offer, the claimant was advised by counsel that it was impossible to put a 'definitive value' on the claim at that stage. He advised that further investigations should be carried out, but the offer should not be rejected.

Following counsel's advice, the claimant's solicitors wrote to the defendant's solicitors inviting attention to the evidence that the claimant was still too young for final conclusions to be drawn, explaining that they were not able to advise on the reasonableness of the offer and setting out the further investigations they intended to carry out.

They neither rejected the offer nor asked for the defendant's agreement to extend the time during which it would remain open for acceptance. Equally, the defendant did not withdraw it.

Black LJ said that the court had to make the normal order for costs, 'unless it considered it unjust to do so' and, in deciding whether it was unjust, had to take into account all the circumstances of the case. Awarding the claimant his costs throughout and departing from the usual order she continued:

> 'Some words of caution: as I have already said, costs decisions are particularly fact sensitive. The view I have formed of this case, albeit a clear one, is an amalgam of all of its features. It is unlikely that they would be replicated precisely in another case. The various factors interact with each other and just as Stanley Burnton J identified in Matthews that a particular change in the circumstances there may have

led to a different result, so differences between the facts of this case and the facts of other cases may mean that the result in the other case should differ from the result in this one.'

Kunaka v Barclays Bank plc [2010] EWCA Civ 1035, [2011] 2 Costs LR 179 is another example of the court ordering otherwise and reiterates that no hard and fast rule can be prescribed. The discretion of the court in each case will depend upon the facts. Mr Kunuka was in person. When during negotiations the defendant bank reminded Mr Kunuka that their Part 36 offer of £35,000 made previously remained open, he accepted it. The bank only then informed him that under CPR 36.10(5) they were entitled to their costs from 21 days after the offer. The Court of Appeal observed that this was not litigation between trained lawyers and the bank had not stated the consequence of the late acceptance when reminding the claimant of the offer. The court had discretion to assess what the fairness of the situation demanded and awarded the claimant his costs to 21 days after the offer with no order for costs thereafter.

THE CONSEQUENCES OF ACCEPTANCE OF A PART 36 OFFER MADE BY ONLY ONE OF TWO OR MORE DEFENDANTS

[20.22]

The situation differs depending upon whether or not the claimant has sued the defendants jointly or in the alternative or has sued them severally. If it is the former then under CPR 36.12(2) the claimant may only accept the Part 36 offer of one if:
- he discontinues his claim against the remaining defendants who have not made the offer; and
- those remaining defendants give their written consent to the acceptance.

If it is the latter then under CPR 36.12(3) the claimant may:
- accept the offer; and
- continue with the claim against the remaining defendants who did not make the offer.

THE CONSEQUENCES OF NOT ACCEPTING A PART 36 OFFER

A defendant's offer not accepted by the claimant

[20.23]

Clearly if the claimant recovers more than the defendant's best Part 36 offer, then Part 36 does not apply and the court determines costs under CPR 44.2 (see below).

If the claimant does not recover as much as the defendant's Part 36 offer then CPR 36.14(2) applies, the defendant is entitled to his costs from the date upon which the relevant period for acceptance expired and interest on his costs. This is subject to a financial limit on recovery of those costs if the claimant has 'QOCS' protection – see below. The rule is silent about the costs prior to that date as those will fall to be determined within the analysis under CPR 44.2 (see Chapter 22 Costs Awards between the Parties).

A claimant's offer not accepted by the defendant

[20.24]

Obviously if the defendant does not accept a claimant's offer, but at trial the claimant is awarded less than the amount of his offer (ie does not receive a judgment at least as advantageous as the offer) and there is no relevant defendant's offer, then Part 36 does not apply and the court determines costs under CPR 44.2 (see Chapter 22 Costs Awards between the Parties).

However, under CPR 36.14(3) a claimant who obtains a judgment at least as advantageous as his offer is entitled to interest at a rate not exceeding 10% above base rate, his costs on the indemnity basis and interest on the costs also at a rate not exceeding 10% above base rate. In addition the CPR amendments of April 2013 introduced a further incentive for claimants to make realistic Part 36 offers. In respect of claimants' Part 36 offers made after 31 March 2013 where the judgment against the defendant is at least as advantageous to the claimant as the claimant's Part 36 offer then, in addition to the other benefits listed, the claimant will recover an extra 10% of the amount awarded by the court on awards of up to £500,000 and on awards of up to £1 million, 10% of the first £500,000 and an extra 5% of the amount awarded above £500,000. All these sanctions are 'unless the court considers it unjust' to impose them.

We promised to return to this provision in the introduction to this chapter and we do so now. In our view there seems to have been a problem with the final drafting of this provision, if the intention was that on awards of over £1 million, the claimant should receive an extra £75,000 (which seemed to be what was intended in early drafts). We believe that the wording now used, if applied with ordinary meaning, results in a claimant recovering an award of over £1 million who has made a relevant Part 36 offer, recovering no extra percentage because, strictly, the court has not awarded a sum up to £500,000 or a sum between £500,000 and £1 million. This problem seems, in part, to be because the final wording has been taken from 'The offers to settle in civil proceedings Order 2013' (SI 2013/93). However, that order contained a specific provision for awards over £1 million that the CPR 36.14 does not – namely that for awards over £1 million the claimant receives 7.5% of the first £1 million and 0.001% of the amount awarded above that figure. As the potential sum in dispute could be £75,000, and therefore worth the candle, we expect either an amendment to the rule to say clearly what it meant to say or a High Court decision on this in due course.

CPR 36.14(2), considered above, makes no similar provision for sanctions in favour of a defendant who has made an offer which is not beaten. A defendant wishing to obtain costs on the indemnity basis must establish that there are circumstances which take the case out of the norm (Excelsior Commercial and *Industrial Holdings Ltd v Salisbury Hamer Aspden & Johnson (a firm)* [2002] EWCA Civ 879). In *Société Internationale de Telecommunications Aeronautiques v Wyatt Co (UK) Ltd (Maxwell Batley (a firm), Part 20 defendant* [2002] EWHC 2401 (Ch), Maxwell Batley had made a token

Part 36 payment into court of £1,000 and, because it wholly succeeded in the action, sought its costs on the indemnity. After all it was entitled to its costs on the standard basis under the analysis of the then CPR 44.3 factors as it was the successful party. Surely there should have been some additional reward in recognition that it had also made a Part 36 offer that had not been beaten? The court concluded that for Maxwell Batley to have had a good claim for indemnity costs it would have been necessary for it to show more than just that Watson Wyatt refused its Part 36 offer. It would have had to have satisfied the usual requirements for an indemnity order. As it could not it was not appropriate to award indemnity costs.

This discrepancy of sanction between claimants who fail to beat defendants' offers and defendants who fail to beat claimants' offers is further accentuated in those claims to which the qualified one way costs shifting ('QOCS') provisions of CPR 44.13–CPR 44.16 applies.

DEFENDANTS' PART 36 OFFERS AND QUALIFIED ONE WAY COSTS SHIFTING

[20.25]

In a case in which a claimant has the benefit of 'QOCS' (see Chapter 11 An Overview of the Key Costs Amendments to the Civil Procedure Rules), but fails to obtain a result more advantageous than a defendant's Part 36 offer, then the claimant will have to pay the defendant's costs from the date on which the relevant period for acceptance expired (under CPR 36.14(2)). However, the amount that the claimant has to pay under this costs order cannot, when added to any other interim costs orders against him which he has paid or has to pay, be more than the aggregate of orders for damages and interest in the claimant's favour unless one of the exceptions in CPR 44.16 applies. In other words the pendulum of sanctions under Part 36 has swung even further in favour of the claimant.

PART 36 AND SUCCESS IN PART

[20.26]

In *Walsh v Shanahan* [2013] EWCA Civ 675, [2013] 5 Costs LO 738, the Court of Appeal held that where the claimant had succeeded on a claim for breach of fiduciary duty, had been unsuccessful on a claim for an account and had failed to recover more than the defendants' Part 36 offer it was entirely appropriate for the judge to have commenced his consideration of costs by asking himself what costs order he would have been made if there had been no Part 36 offer.

Adopting this approach the judge had concluded that he would have found the defendants to recover 90% of their costs. However, having then taken into account the Part 36 offer, the claimant was ordered to pay: (i) 90% of the defendants' costs until the expiry of the relevant period for acceptance of the Part 36 offer and (ii) all of their costs thereafter.

20.26 *Chapter 20 Costs Inducements to Settle – Part 36 Offers*

The Court of Appeal considered this approach was both orthodox and rational. The costs after the expiry of the relevant period for acceptance were covered by the provisions of Part 36, but it was entirely logical for the judge to ignore the Part 36 offer when approach overall success. The defendants were the substantially successful parties, although their success was reduced by their unsuccessful resistance of the damages claim. This approach to the award of costs was entirely unexceptionable. The fact that the claimant would have been entitled to his costs incurred in the relevant period had he accepted the Part 36 offer was irrelevant because he had not accepted it and had taken the risk of continuing the claim. The judge had correctly assessed that the claim was about an account of profits, the claimant had lost this issue and had made a costs order accordingly.

This case illustrates the enormous uncertainty and risk that rejecting a Part 36 offer can create. From a position where the claimant would have recovered damages and, under CPR 36.10 his costs up to acceptance, the claimant went to a situation where he recovered none of his costs and paid virtually all those of the defendants.

PART 36 AND FUTURE PECUNIARY LOSS CLAIMS

[20.27]

Under the Damages Act 1996 the court has the power to award losses for future pecuniary loss by way of periodical payments ('PPs'). Strictly CPR 41.5 requires each party to state its case on PPs in the respective statements of case. In reality this is often overlooked and in anticipation CPR 41.5(2) and (3) provide that the court can order parties to make such statements and provide further information about them. One reason for this is that the court has to give an indication as soon as practicable as to whether a conventional lump sum award or PPs is likely to be the most appropriate form for the damages.

CPR 36.5 sets out detailed provisions for offers where future pecuniary losses arise. In reality the vast majority of claims are not suitable for PPs and settlement is on a lump sum basis with no application of these provisions. However, in those claims where PPs remain a live issue the following specific rules apply:

- An offer may be that all future pecuniary losses are dealt with by lump sum, all are dealt with by PPs or that part of the damages are dealt with by lump sum and part by PPs and that the whole or part of any other damages by lump sum.
- An offer must state the amount of the lump sum offered for all or part of any damages
- An offer may state what part of the lump sum, if any, relates to future pecuniary loss and what part relates to other damages to be accepted in the form of a lump sum
- An offer must state what part of the offer relates to damages for future pecuniary loss to be paid in the form of PPs and must specify the amount and duration of the PPs, the amount of any payments for substantial capital purchases and when those are to be made and whether each amount is to vary by reference to the retail prices index

('RPI'), some other index or not at all (remember when making such an offer that the court order under CPR 41.8 must specify variation annually by reference to the RPI unless the court orders otherwise).
- An offer must state that any part of the damages in the form of PPs will be funded to ensure continuity of payment is reasonably secure.
- If an offer is a mixture of lump sum and PPs, the acceptance can only be to the offer as a whole. In other words the acceptance cannot be to part being funded in a specific way and leave alive an argument on the fashion of payment of the rest of the claim.
- Whoever makes the offer, if it includes payment of some part of the damages in the form of PPs and is accepted, it is from the claimant to apply to the court within 7 days of acceptance for a formal court order under CPR 41.8

PART 36 AND PROVISIONAL DAMAGES CLAIMS

[20.28]

Any party to a claim for provisional damages can make a Part 36 offer (CPR 36.6(1)). The offer must state:
- Whether or not the proposed settlement includes an award of provisional damages
- Where it does the offer must specify
  (i) that the sum offered is on the assumption that the claimant will not develop the disease or suffer the type of deterioration specified in the offer
  (ii) that the offer is subject to the condition that any application for further damages must be made within a particular time
  (iii) What that time period is to be.

Regardless of who makes the offer, if it is accepted, the onus is on the claimant to apply to the court under CPR 41.2 for an award of provisional damages.

PART 36 AND RECOUPMENT OF STATE BENEFIT

[20.29]

On this occasion the rule says it all:

'Part 36.15: Deduction of benefits and lump sum payments

(1) In this rule and rule 36.9—
(a) 'the 1997 Act' means the Social Security (Recovery of Benefits) Act 1997;
(b) 'the 2008 Regulations' means the Social Security (Recovery of Benefits)(Lump Sum Payments) Regulations 2008;
(c) 'recoverable amount' means—
(i) 'recoverable benefits' as defined in section 1(4)(c) of the 1997 Act; and
(ii) 'recoverable lump sum payments' as defined in regulation 4 of the 2008 Regulations;
(d) 'deductible amount' means—
(i) any benefits by the amount of which damages are to be reduced in accordance with section 8 of, and Schedule 2 to the 1997 Act ('deductible benefits'); and
(ii) any lump sum payment by the amount of which damages are to be reduced in accordance with regulation 12 of the 2008 Regulations ('deductible lump sum payments'); and
(e) 'certificate'—
(i) in relation to recoverable benefits is construed in accordance with the provisions of the 1997 Act; and
(ii) in relation to recoverable lump sum payments has the meaning given in section 29 of the 1997 Act as applied by regulation 2 of, and modified by Schedule 1 to the 2008 Regulations.

(2) This rule applies where a payment to a claimant following acceptance of a Part 36 offer would be a compensation payment as defined in section 1(4)(b) or 1A(5)(b) of the 1997 Act.

(3) A defendant who makes a Part 36 offer should state either—
(a) that the offer is made without regard to any liability for recoverable amounts; or
(b) that it is intended to include any deductible amounts.

(4) Where paragraph (3)(b) applies, paragraphs (5) to (9) of this rule will apply to the Part 36 offer.

(5) Before making the Part 36 offer, the offeror must apply for a certificate.

(6) Subject to paragraph (7), the Part 36 offer must state—
(a) the amount of gross compensation;
(b) the name and amount of any deductible amount by which the gross amount is reduced; and
(c) the net amount of compensation.

(7) If at the time the offeror makes the Part 36 offer, the offeror has applied for, but has not received a certificate, the offeror must clarify the offer by stating the matters referred to in paragraphs (6)(b) and (6)(c) not more than 7 days after receipt of the certificate.

(8) For the purposes of rule 36.14(1)(a), a claimant fails to recover more than any sum offered (including a lump sum offered under rule 36.5) if the claimant fails upon judgment being entered to recover a sum, once deductible amounts identified in the judgment have been deducted, greater than the net amount stated under paragraph (6)(c).

(Section 15(2) of the 1997 Act provides that the court must specify the compensation payment attributable to each head of damage. Schedule 1 to the 2008 Regulations modifies section 15 of the 1997 Act in relation to lump sum payments and provides that the court must specify the compensation payment attributable to each or any dependant who has received a lump sum payment.)

(9) Where—
(a) further deductible benefits have accrued since the Part 36 offer was made;

(b) and
the court gives permission to accept the Part 36 offer,

the court may direct that the amount of the offer payable to the offeree shall be reduced by a sum equivalent to the deductible amounts paid to the claimant since the date of the offer.

(Rule 36.9(3)(b) states that permission is required to accept an offer where the relevant period has expired and further deductible amounts have been paid to the claimant).'

PART 36 AND FIXED SUCCESS FEES

[20.30]

In *Lamont v Burton* [2007] EWCA Civ 429, the claimant in a road traffic accident claim was represented under a conditional fee agreement, but although liability was admitted he failed to beat a Part 36 payment. Accordingly, the claimant was awarded his costs up until the latest date on which the Part 36 payment could have been accepted. When those costs were assessed he claimed a 100% success fee under the fixed uplift provisions of what was CPR 45.16 on the ground that the claim had been concluded at trial. The defendant argued that although CPR 45.16 did not itself give the court jurisdiction to allow a different percentage increase, it contained a lacuna in that it did not deal with situations in which a claimant failed at trial to better a Part 36 offer, which could be remedied by the court exercising its discretion under the then CPR 44.3 to award a success fee no greater that it would have been under the then CPR 45.16 had the offer been accepted in time, namely 12.5%. The court held the 100% success fee to be mandatory in all cases. There was no discretion to award the claimant a success fee of less than 100%, Part 44 was excluded by the very words used. It would be a matter for the Rules Committee and the Civil Justice Council whether to amend Part 45 to make special provision to deal with the Part 36 issue. Quite whether the court had in mind that the amendment should remedy this apparent injustice by abolishing the recoverability of success fees at all is another matter. However, this has certainly removed this problem once those older claims with recoverable success fees have worked their way through the court process.

PART 36 AND THE 'TACTICAL OFFER'

[20.31]

Huck v Robson [2002] EWCA Civ 398, [2002] 3 All ER 263, [2003] 1 WLR 1340 concerned a road traffic accident in which it was likely the finding on liability would be either 50/50 or 100%, the claimant made a pre-action offer to accept a 95%–5% split on liability. The defendant rejected the offer and the trial judge refused to take it into account in his award of costs on the grounds that the apportionment was 'illusory'. The Court of Appeal disagreed. Although a judge would be entitled to exercise his discretion and refuse indemnity costs where an offer was purely tactical, for example, to settle for 99.9% of the full value of the claim, that could not be said of the claimant's offer. The reduction of 5% provided the defendant with a real opportunity of settlement and did not represent the court's probable decision on liability. After applying the prescribed factors the court, by a majority, awarded the claimant her costs on the indemnity basis. We find the concept of a 'purely tactical' offer difficult. All Part 36 offers are tactical, and they are none the worse for it.

PART 36 AND THE PRE-ACTION PROTOCOL FOR LOW VALUE PERSONAL INJURY CLAIMS IN ROAD TRAFFIC ACCIDENTS AND THE PRE-ACTION PROTOCOL FOR LOW VALUE PERSONAL INJURY (EMPLOYERS' LIABILITY AND PUBLIC LIABILITY) CLAIMS

[20.32]

Under both the Pre-action protocol for low value personal injury claims in road traffic accidents ('RTA protocol') and the Pre-action protocol for low value personal injury (employers' liability and public liability) claims ('EL/PL protocol') procedure, if the parties are unable to agree settlement at Stage 2, then they move to Stage 3 and proceedings are issued under CPR Part 8. Those proceedings are dealt with by the court either as a paper exercise at a non-attended hearing or, if either party so requests, at an attended hearing. In these proceedings Part 36.16 specifically excludes CPR 36.1–CPR 36.15. Instead offers in the proceedings are governed by CPR 36.16–CPR 36.22.

When commencing proceedings the claimant must include the 'court proceedings pack'. This includes the parties' final offers and should be filed in a sealed envelope. These offers are defined by CPR 36.17 as protocol offers and are deemed made on the first business day after the 'court proceedings pack form' is sent to the defendant (CPR 36.18). A protocol offer is treated as <u>exclusive</u> of interest and is relevant only to the fixed costs in the stage 3 procedure and not to any costs arising on any appeal (CPR 36.19).

Once the court has determined the claim, whether by a written judgment at a non-attended hearing or by judgment given in the usual way in court at an attended hearing, then, without any fanfare but in the case of one of the editors, with some excitement, the envelope containing the relevant protocol offers is opened. It is obviously vital that the offer must not be communicated to the court until the award has been made.

The contents of the envelope determine the outcome of the stage 3 costs. The possible outcomes are as follows:
- If the award is more than the protocol offer of the defendant, but less than that of the claimant, CPR 36.21(3) applies and the defendant simply pays the claimant the fixed costs in CPR 45.20.
- If the award is the same as or less than the defendant's RTA protocol offer, CPR 36.21(2) applies and the claimant pays the defendant's fixed costs in CPR 45.26 and interest on those costs.
- If the award is for the same as or more than the claimant's RTA protocol offer, CPR 36.21(4) applies and the defendant pays the claimant's fixed costs in CPR 45.20, interest on the entire damages

awarded at a rate not exceeding 10% above base rate for some or all of the period commencing under CPR 36.18 (above) and interest on the fixed costs at a rate not exceeding 10% above base rate. No start date is specified for interest on the fixed costs. It may be that this is because there is a risk for the claimant, unlike the defendant where the start date for interest is specified, that some of the Stage 1 or Stage 2 costs may remain outstanding and the period of interest on those should run from before the Stage 3 proceedings and because the claimant will have incurred disbursements and may wish to argue when interest should run from in respect of those payments. The upshot, whatever the rationale for it, is that the period of interest is a matter for court discretion in this situation.

Once Stage 3 proceedings have started a party may only withdraw a protocol offer with the permission of the court (CPR PD 8B, para 10.1). If the court gives permission then the claim ceases to be Stage 3 proceedings and the court must give directions for its determination. However, the court will only give permission if there is a good reason for the claim not to continue under the Stage 3 procedure.

Despite the significant percentage reduction in fees under the RTA protocol, stage 3 has remained unscathed and the fees are as they were upon introduction of the scheme. Whilst both the protocols prescribe different stage 1 and 2 fees based on whether the claim is under or over £10,000 (the limit for claims in the RTA protocol having increased to £25,000 and the EL/PL protocol having been introduced with effect from 31 July 2013) and the EL/PL protocol stage 1 and 2 fees being higher in any event, the stage 3 costs for all types of claim caught within the protocols is the same for stage three as follows:

TYPE A	Legal representative's costs	£250
TYPE B	Advocate's costs	£250
TYPE C	Costs for the advice on damages where the claimant is a child	£150

The court's involvement with the protocols is not solely limited to stage 3 proceedings. Under CPR 27.14(2)(h) the court has the power, in cases allocated to the small claims track to award the stage 1 and stage 2 costs to the claimant where the claim began in the protocol, the claimant reasonably believed that the claim was valued at more than the small claim track limit and the defendant admitted liability under the protocol process, but has not paid the stage 1 and 2 fixed fees.

In addition where a claim ceases to be dealt with within the relevant protocol then the provisions under CPR 36.10A apply, which take account that from 31 July 2013 fixed costs now apply. The amount to which a claimant is entitled in such claims where the Part 36 offer is accepted within the relevant time period depends upon the stage at which the proceedings have reached at the date of notice of acceptance (CPR 36.10A(3)). Tables 6B and 6C Section

IIIA of Part 45 'Claims which no longer continue under the RTA or EL/PL pre-action protocols – Fixed recoverable costs' set out the stages and the sums then payable.

If, though the claimant does not accept in the relevant time period and subsequently receives a judgment that is not more advantageous than the Part 36 offer the claimant will recover the fixed costs relevant to the stage that had been reached at the time of expiry of the relevant period as above, but will be responsible for the defendant's costs from the date on which the relevant period expired.

The position is even worse for a claimant who fails to better a protocol offer where the claim does not continue under the protocol. In that situation the claimant is limited to the Stage 1 and Stage 2 fixed costs as per Table 6A accompanying CPR 45.29, but will be liable for the defendant's costs from the date on which the protocol offer is deemed to be made until judgment. Even with the protections of qualified one way costs shifting, the claimant may be left with little or nothing of his damages after off-setting. For the avoidance of doubt CPR 36.14A(3)(c) makes it plain that the amount of the judgment is less than the protocol offer where the judgment is less <u>after</u> deductible amounts are deducted (ie after the deduction of deductible benefits and lump sum payments as per CPR 36.15).

PART 36 AND THE SMALL CLAIMS TRACK

[20.33]

Although Part 36 does not apply to the small claims track (CPR 27.2) there is no reason why a party to a small claim should not make an offer which could be taken into account when awarding costs and disbursements. The increase of the small claims track limit to £10,000 for claims issued after 31 March 2013 means that in some cases a substantial amount of work may need to be done. Whether this will influence the court to exercise its jurisdiction to make adverse costs order under CPR 27.14 more and to take into account attempts to settle when doing so is another matter. Bear in mind, of course, that the court will be applying the overriding objective to these claims and proportionality will still be the order of the day.

PART 36 AND VAT TRIBUNAL

[20.34]

In *Blue Sphere Global Ltd v Revenue and Customs Comrs* [2010] EWCA Civ 1448, [2011] STC 547, [2011] 2 Costs LO 182, Blue Sphere Global had made a Part 36 offer in proceedings in which they sought to recover from HMRC sums paid by way of VAT. HMRC had refused repayment on the grounds that the transactions were connected with fraud. That refusal of repayment was upheld by a VAT Tribunal but reversed by a High Court Judge. The Revenue then appealed to the Court of Appeal. BSG made a Part 36 offer in relation to the appeal, and in the event the appeal failed. BSG therefore sought orders under CPR 36.14 for indemnity costs, enhanced interest on costs and enhanced interest on damages.

The Court of Appeal rejected the HMRC's contentions that:
(i) Part 36 had no application to cases that started in tribunals, such as the VAT tribunal, where the CPR costs regime did not apply;
(ii) that if the rule did apply then it would be unjust to make an order because this was a test case;
(iii) it was in the public interest to clarify the law and therefore the HMRC was acting in the public interest and should not face punitive orders under CPR 36.14; and

that no order should be made as BSG's offer was only just beaten.

Instead the Court of Appeal ordered that full effect should be given to the offer and ordered indemnity costs and enhanced interest on both damages and costs.

PRE-6 APRIL 2007 PART 36 PAYMENTS AND OFFERS

[20.35]

Whilst we would like to think that 6 years on all cases where a Part 36 offer or payment into court was made have now worked there way through the court process, in case we are wrong (and because this is still contained in the CPR), we would simply refer to the contents of CPR PD 36B which explains the position in some detail. We wonder when it will finally be deemed safe to remove this provision from the rules?

OTHER ADMISSIBLE OFFERS

Introduction

[20.36]

CPR Part 36 is not the full story of offers to settle. It is important to remember that CPR 36.1(2) provides that a party may make an offer to settle in whatever way he chooses, but, and this is the important bit, if the offer is not made in accordance with Part 36 it will not have the costs consequences specified in CPR 36.10, CPR 36.11 and CPR 36.14. This emphasises that in deciding what order to make about costs under CPR 44.2(4) (see below) the court will have regard to any admissible offer which does not have the costs consequences of Part 36.

'Calderbank' letters

[20.37]

Since 1975 a frequently used alternative to a payment into court or Part 36 offer has been an offer contained in a Calderbank letter. It is called 'Calderbank' because it is a form of offer first approved by the Court of Appeal in *Calderbank v Calderbank* [1976] Fam 93, [1975] 3 All ER 333, [1975] 3 WLR 586, CA, and this is still a useful description. It is an offer made 'Without prejudice save as to costs'. The magic words are 'save as to costs' as was demonstrated in *Reed Executive plc v Reed Business Information Ltd* [2004] EWCA Civ 887, [2004] 4 All ER 942, [2004] 1 WLR 3026. The defendant had made offers in privileged meetings and in 'without prejudice' correspondence on which it sought to rely on the question of costs. Disaster! The court could not look at the correspondence. In its judgment the Court of Appeal said: 'Parties may negotiate in the faith and expectation that the negotiations cannot be used against them even on the question of costs unless the negotiations are expressly stated to be 'without prejudice save as to costs'. Magic words indeed.

Even though offers no longer have to be accompanied by a payment into court there are still many circumstances in which a defendant may consider it advantageous to make a Calderbank offer rather than one which complies with Part 36. Some of these are:
- if the defendant cannot raise the money in a single sum or within the prescribed 14 days to enable him to make a Part 36 offer;
- to make a walk-away offer;
- to make a package of proposals unsuitable for Part 36 offer;
- to offer to do remedial work;
- to make an apology;
- to offer to do further business; and
- in a multi-party action.

A perfect illustration of when a Part 36 offer is, in principle, inappropriate is where there is a monetary claim, but in fact something far more important underlies the proceedings. A paradigm example of this is *E Ivor Hughes Educational Foundation v Leach* [2005] EWHC 1317 (Ch). The defendant was accused, among other things, of fiddling his expenses to the tune of £87,000, which he denied but paid £5,000 into court in order to dispose of this head of claim as an economic settlement. What he wished to do was to reach a commercial settlement, but accompany it with a denial. Part 36 does not permit qualification, other than implicitly in the amount of money offered. The claimant accepted the offer, but the judge directed that the employment tribunal in the defendant's claim for wrongful dismissal could construe the payment as an admission of liability in relation to making false expenses claims. Whilst this may have been a step too far as Lord Devlin in *French (A Martin) v Kingswood Hill Ltd* [1961] 1 QB 96, [1960] 2 All ER 251, CA had said that a payment in should not be equated to an admission, an offer outside Part 36 in which there was also an express denial might better have served the defendant's non-monetary objective.

The different consequences – to be a Part 36 offer or not to be a Part 36 offer

[20.38]

The key distinction is that an offer under CPR 44.2 cannot open the door for the imposition of the provisions of Part 36. As we have already seen Part 36 is its own self-contained procedural code. An offer is either within or without it. Just before the April 2007 changes from payments into court to offers, the

distinction between the two had blurred thanks to cases like *Trustees of Stokes Pension Fund v Western Power Distribution (South West) plc* [2005] EWCA Civ 854, [2005] 3 All ER 775, [2005] 1 WLR 3595. *Stokes* determined that an offer 'should usually' be treated as having the consequences of the then Part 36 even in the absence of a payment into court if four conditions were satisfied:
- the offer had to be expressed in clear terms so that there could be no doubt what was being offered;
- the offer had to be open for acceptance for at least 21 days and otherwise accord with the substance of a Calderbank offer;
- the offer had to be genuine (not a sham or non-serious in some way); and
- the defendant should clearly have been good for the money when the offer was made.

Just when it seemed safe to transfer *Stokes* to history text books along came a timely reminder that perhaps CPR PD 36B is safe for a few years yet. The case of *French v Groupama Insurance Co Ltd* [2011] EWCA Civ 1119, [2011] 4 Costs LO 547 saw the Court of Appeal grappling with Calderbank letters and offers that failed to comply with CPR Part 36 as it was at the time that the offer was made.

In this case the claimant brought a claim based on breach of contract against her insurers in respect of subsidence at her home. At trial she was awarded £126,963.53. There had been pre-action offers on 22 December 2006 and 15 February 2007 of £115,000. The trial judge accepted that of the award about £20,000 was in respect of additional damages that the claimant suffered after the date of the offers and that, as such, the amount offered (which was inclusive of costs and interest) was more advantageous to her than what she subsequently recovered. The trial judge applied *Stokes* even though the defendant accepted that neither offer was within the CPR Part 36 provisions of the time, accorded the second offer letter the same status as a payment in (not the first letter because it did not satisfy all the requirements of *Stokes*) and ordered that the claimant should meet all the defendant's costs.

On appeal the Court of Appeal determined that the offer was not akin to a Part 36 offer because:
(a) an offer inclusive of costs could not be a Part 36 offer (*Mitchell v James* [2002] EWCA Civ 997, [2003] 2 All ER 1064, [2004] 1 WLR 158); and.
(b) the parties' agreement that when the offer was made it was privileged (and not qualified with the words 'save as to costs') meant it was not a Calderbank offer and,

therefore, condition ii) of *Stokes* was not met.

Having concluded that *Stokes* and the transitional provisions of Part 36 did not apply and that the trial judge had erred, the Court of Appeal approached the issue of costs afresh under CPR 44.2, taking account of the offer as an admissible offer (but not one akin to a Part 36 offer) and concluded that there should be no order as to costs from 21 days after the February 2007 offer and that the defendant should pay the claimant's costs up to that date.

The Court of Appeal returned to this topic in *F & C Alternative Investments (Holdings) Ltd v Barthelemy* [2012] EWCA Civ 843, [2012] 4 All ER 1096, [2013] 1 WLR 548, where offers had been made 'without prejudice save as to costs' explicitly not as Part 36 offers but stating that the court would be invited to apply the same costs consequences as for Part 36 offers.

The trial judge referred to the 'infelicity in the wording' of Part 36 not making its use sensible in the present case where Part 7 proceedings meant the offeror would have to agree to pay the costs of those proceedings. The trial judge then made a costs order in the same terms as if Part 36 had applied. He was reversed by the Court of Appeal which held:

- This was not a Part 36 offer and consequently the judge had no jurisdiction to make a costs order under CPR 36.14;
- The judge's jurisdiction as to costs fell to be exercised under CPR 44.3 (now CPR 44.2);
- Under Part 44 the costs regime of Part 36, whether indirectly or by analogy, cannot properly be invoked;
- There is no reason or justification, for indirectly extending Part 36 beyond its expressed ambit. To do so would tend to undermine the requirements of Part 36 and the repeated insistence of the courts that intended Part 36 offers should be very carefully drafted so as to comply with the requirements of Part 36.

This case was considered in *Procter & Gamble Co v Svenska Cellulosa Aktiebolaget SCA* [2012] EWHC 2839 (Ch), [2013] 1 WLR 1464, [2013] 1 Costs LR 97, where the claimant sought to minimise its liability to make payment to Svenska. Procter & Gamble made a claimant's offer expressed to have the consequences of Part 36. The offer included an offer to pay the defendant's costs if it was accepted within 21 days. The effect of this offer was that the claimant was reversing the costs consequences under CPR 36.10 by giving up its entitlement to its costs and offering to pay the defendant's costs.

Having considered the decision in *F & C Alternative Investments (Holdings) Ltd v Barthelemy*, Hildyard J concluded as follows:

> 'In my view, the issue in the F & C case was really whether an offer accepted not to be within Part 36 could be given, by analogy, the same consequences as would have followed if it had been compliant and intended to be so. Here, the issue is whether CPR 36.2(2), and thus the gateway to CPR 36.10 and 36.14, is to be so strictly construed that it requires (by rule 36.2(2)(c)) the offer made to provide for the defendant to be liable for the claimant's costs even if the claimant expresses his offer to be a Part 36 offer, but as part of that offer, agrees to forsake that entitlement and instead pay the defendant his costs. Put another way, I do not accept that it is impossible for a claimant to comply with Part 36 unless he requires to be paid his costs and such payment to be made within a period of not less than 21 days.'

Part 44 'near miss' offers

[20.39]

As we stated in the introduction, CPR 36.14A made clear that in money claims *'more advantageous'* means better in money terms. In *Hammersmatch Properties (Welwyn) Ltd v Saint Gobain Ceramics & Plastics Ltd* [2013] EWHC 2227 (TCC) the claimant recovered £1,058,768, which exceeded the Part 36 offer by a mere £3637.90, Ramsey J concluded that where a claimant had obtained a judgment more advantageous than the best Part 36 offer of the opposing party it would be wrong to revisit this 'near miss' when considering costs more generally under CPR 44.2(4)(c). He concluded that to determine otherwise would simply introduce by Part 44 the very uncertainty that had been removed by CPR 36.14A. He doubted whether a *'near miss'* offer added anything to broader conduct arguments that might flow based on an unreasonable refusal to negotiate.

The final words on the subject should go to Jackson LJ, who, in *Fox v Foundation Piling Ltd* [2011] EWCA Civ 790, [2011] 6 Costs LR 961, said:

> ' . . . parties are quite entitled to make Calderbank offers outside the framework of Part 36. Where a party makes such an offer and then achieves a more advantageous result, the court's discretion is wider. Nevertheless it may well be appropriate to order the party which has optimistically rejected the Calderbank offer to pay all costs since the date when that offer expired.'

Employment tribunals

[20.40]

An offer is a factor that the Employment Tribunal may take into account in deciding whether to make a costs order in accordance with Rule 14 of the Employment Tribunals Rules of Procedure, but failure to achieve an award in excess of the offer is not conclusive. The Employment Tribunal must also be satisfied that the conduct of the applicant in rejecting the offer was unreasonable (*Kopel v Safeway Stores plc* [2003] IRLR 753, EAT).

CONCLUSION

[20.41]

There can be little doubt that the additional benefit for the claimant under CPR 36.14 is intended:
- to make claimants think of a figure, deduct something from it and make an offer in the hope of recovering an additional percentage of the damages
- to make defendants think of a figure and add to it either in accepting claimants' offers or in making their own offers to buy off the risk of paying an additional amount of damages

This is to encourage settlement and avoid the situation that Jackson LJ described in *Fox v Foundation Piling Ltd* in these terms:

> 'A not uncommon scenario is that both parties turn out to have been over-optimistic in their Part 36 offers. The claimant recovers more than the defendant has previously offered to pay, but less than the claimant has previously offered to accept . . .'

Whether or not it will work remains to be seen. One of his recommendations was that interested parties in the personal injury litigation field should try to devise a neutral computer programme that valued general damages in smaller personal injury claims to avoid a costly process where the difference between

20.41 *Chapter 20 Costs Inducements to Settle – Part 36 Offers*

the parties was only a small fraction of the costs incurred in having determination through the court. It was one of the few recommendations not to come to fruition (indeed not to get off the ground at all).

What does seem certain, though, is that 6 years after the last radical changes to Part 36, its provisions are still catching practitioners by surprise and generating an unhealthy amount of jurisprudence.

CHAPTER 21

COSTS INDUCEMENTS TO SETTLE – ADR

INTRODUCTION

[21.1]

In his *'Review of Civil Litigation Costs; Final Report'* Sir Rupert Jackson dwelt on the merits of Alternative Dispute Resolution ('ADR') at some length. He considered that:

> 'ADR is relevant to the present Costs Review in two ways. First, ADR (and in particular mediation) is a tool which can be used to reduce costs. At the present time disputing parties do not always make sufficient use of that tool. Secondly, an appropriately structured costs regime will encourage the use of ADR. It is a sad fact at the moment that many cases settle at a late stage, when substantial costs have been run up. Indeed some cases which ought to settle (because sufficient common ground exists between the parties) become incapable of settlement as a result of the high costs incurred. One important aim of the present Costs Review is to encourage parties to resolve such disputes at the earliest opportunity, whether by negotiation or by any available form of ADR.'

Whilst, as we shall see from Sir Rupert's conclusions below, his view was that the court has a greater role to play than it has done to date, there can be no doubt that the court has made attempts to steer parties, whether by stick, carrot or both, towards ADR. This is clear from the case law that has emerged.

FAILURE TO ENGAGE IN ADR LEADING TO A COSTS SANCTION

[21.2]

In *Dunnett v Railtrack plc (in railway administration)* [2002] EWCA Civ 303, [2002] 2 All ER 850, on granting the claimant permission to appeal against the dismissal of his action for damages for negligence by Railtrack, the judge told both parties they should attempt alternative dispute resolution, but Railtrack refused to do so, on the grounds that it was not prepared to make any further payment to the claimant and was confident that it would succeed on the appeal. Railtrack did indeed succeed in having the appeal dismissed, but the Court of Appeal demonstrated its displeasure at Railtrack's outright refusal to consider ADR by depriving it of its costs, noting that parties and their lawyers should ensure that they are aware that it is one of their duties fully to consider ADR, especially when the court has specifically suggested it, and not merely flatly to turn it down. The court warned that to adopt that approach placed the party doing so at risk of adverse consequences in costs regardless of the outcome of the litigation.

21.2 *Chapter 21 Costs Inducements to Settle – ADR*

A failure to mediate led to similar consequences in *Royal Bank of Canada Trust Corpn Ltd v Secretary of State for Defence* [2003] EWHC 1479 (Ch), [2004] 1 P & CR 448, [2003] 2 P & CR D50, where in a lease dispute the Ministry of Defence refused to mediate because the dispute turned on a point of law, was between commercial parties and (unlike previous ADR cases involving costs penalties) the matter was not one where emotions played a significant part in the case. The MoD won but was deprived of its costs because it had ignored the Government's ADR pledge (given by the Lord Chancellor's Department in March 2001) that ADR would be considered and used in all suitable cases wherever the other party was prepared to adopt it. The judge said that the MoD's reasons did not make the matter unsuitable for mediation.

The court maintained the same line in *Virani Ltd v Manuel Revert y CIA SA* [2003] EWCA Civ 1651, [2004] 2 Lloyd's Rep 14, when it ordered an unsuccessful appellant to pay costs on the indemnity basis because he had refused the offer of the Court of Appeal's own mediation service on being granted permission to appeal, failed to negotiate or to enter into any form of mediation or ADR.

A party who agrees to mediation, but then causes the mediation to fail by reason of his unreasonable position in the mediation is in reality in the same position as a party who unreasonably refuses to mediate. This is something which the court can and will take account of in the costs order. In *Seventh Earl of Malmesbury v Strutt & Parker* [2007] EWHC 2199, [2007] 42 EG 294 (CS) the court determined that had the claimant made an offer which better reflected its true position, the mediation might have succeeded. The judge said 'It would be wrong to say more', presumably because he could only say what he did because both parties waived privilege for the mediation. Taking account of unreasonable conduct in privileged mediation is not easy. The judge reduced the claimant's costs to 80% to reflect the unreasonable attitude in the mediation.

FAILURE TO ENGAGE IN ADR – NO COSTS SANCTION

[21.3]

Hurst v Leeming [2002] EWHC 1051 (Ch), [2003] 1 Lloyd's Rep 379, [2003] 2 Costs LR 153, was the first of a series of cases where the court refused to penalise a failure to mediate. Mr Leeming was held to have been justified in taking the view that mediation was not appropriate because it had no realistic prospect of success as, viewed objectively, the mediation had no real prospect of success. It was plain that Mr Hurst had been so seriously disturbed by the tragic course of events resulting from the dissolution of his partnership that he was incapable of a balanced evaluation of the facts. He was out to obtain a substantial sum in the mediation process and was not likely to accept any mediation which did not achieve that result, although his claim plainly entitled him to nothing (as he had conceded by the time of this decision).

In *Firle Investments Ltd v Datapoint International Ltd* [2001] EWCA Civ 1106, [2001] NPC 106, the claimant had undoubtedly been at fault in being unwilling to negotiate and by persisting with hopeless issues, thereby lengthening the trial, but ultimately it had beaten the payments in and had succeeded by a substantial amount in the period up to a second payment in. The claimant recovered all of its costs up to the date of the second payment and 70% of its costs thereafter. The judge retained a discretion under CPR 44.3 (now CPR 44.2) to disapply the normal rule relating to costs and could take into account such factors as payments in, offers to settle, the conduct of the parties and whether a party had lost an issue. However, the judge should refrain from speculating on whether hypothetical offers to settle would have been accepted.

Société Internationale de Telecommunications Aeronautiques SC v Wyatt Co (UK) Ltd (Maxwell Batley (a firm), Part 20 defendant) [2002] EWHC 2401 (Ch), was another case in which the court refused to impose a sanction for refusing to mediate. The judge refused on a number of grounds to deprive Maxwell Batley of any of its costs because of its refusal to mediate. First, the only reason why Watson Wyatt wanted Maxwell Batley to take part in the mediation was so that pressure could be brought on it to make a large contribution to whatever Société Internationale was willing to accept from Watson Wyatt. Second, the purpose of the invitation to mediate was not with a view to resolving the liability of Maxwell Batley without litigation: there was no suggestion that, with influence by the mediator, Watson Wyatt might not pursue their claim against Maxwell Batley. Third, Watson Wyatt had tried to browbeat and bully Maxwell Batley into the mediation in a manner which the judge found disagreeable and off-putting, even suggesting that Maxwell Batley's solicitor's reputation would suffer as a result of the way in which it was conducting the claim. Finally, Watson Wyatt told Maxwell Batley that the mediator had told them that they could get $10 million from Maxwell Batley and that he was 'motoring' against Maxwell Batley. The judge found that to be an astonishing way of trying to persuade Maxwell Batley to join the mediation. The invitation, or rather demand, of Watson Wyatt to Maxwell Batley to participate in the mediation had been self-serving and it would be a grave injustice to Maxwell Batley to deprive it of any part of its costs because it had declined to mediate.

In *Valentine v Allen* [2003] EWCA Civ 915, the claimant lost his proceedings seeking to restrain alleged trespass by his neighbours over his land and he lost his appeal. He endeavoured to resist an order that he should pay the costs of the appeal because the defendant had refused his offers to mediate before the trial. The court held that the defendant had made reasonable and generous offers to resolve the matter but these were refused by the claimant who was seeking a settlement by way of payment of a large sum. In these circumstances, the defendant had not acted unreasonably by refusing the offer to mediate, and in any event the claimant had been refused permission to appeal the judge's order as to costs, in which the mediation issue had already been considered.

THE *HALSEY* TEST

[21.4]

In *Halsey v Milton Keynes General NHS Trust; Steel v Joy* [2004] EWCA Civ 576, [2004] 4 All ER 920, [2004] 1 WLR 3002, the Court of Appeal considered previous decisions to set out clear guidelines. In doing so it held that there was no presumption that a party to a dispute should agree to mediation or other alternative dispute resolution processes. The general rule is that costs of litigation should follow the event. It concluded that refusal to agree to ADR does not justify departure from the general rule, unless it is shown that the successful party acted unreasonably in refusing to do so. To oblige truly unwilling parties to refer their disputes to mediation would be to impose an unacceptable obstruction on their right to access the court, and indeed a court order to mediate could itself be a violation of article 8 of the European Convention on Human Rights. The court set out the following factors that the court ought to consider when determining if a refusal to engage in ADR was reasonable or unreasonable:

- The nature of the dispute – although the court suggested a few categories of case that might be unsuitable for ADR it concluded that 'most cases are not by their very nature unsuitable for ADR'.
- The merits of the case – clearly the strength of a case is relevant to whether a refusal to engage in ADR is reasonable. The court acknowledged that if this were not accepted a party with a weak case could almost compel the other party to engage in ADR and make some concession by threat of an adverse costs order. However, as we know, a party's belief in the impregnability of his position is often as much self-righteous hope as it is reality. As the court summarised the position:

    'The fact that a party unreasonably believes that his case is watertight is no justification for refusing mediation. But the fact that a party reasonably believes that he has a watertight case may well be a sufficient justification for a refusal to mediate.'

- Whether other settlement methods have already been attempted – whilst the fact that previous offers have been made may illustrate that one party may be making efforts to settle and the other is blithely pressing on regardless and justify the refusal of the offering party to enter into ADR, the court was keen to stress that ADR can prove successful even in cases where previous offers have not led to counter-offers.
- Whether the costs of ADR would be disproportionately high – clearly this factor weighs heavier in justifying a refusal of ADR when the sums in dispute are smaller.
- Delay – the stage at which ADR is offered is relevant, as if it is suggested late in the day acceptance of it may have the effect of delaying the trial of the action.
- Whether the mediation has a reasonable prospect of success – objectively viewed does the ADR have any real prospect of success or would it simply have added an extra tier of costs to no avail?

THE VOLUNTARY NATURE OF ADR

[21.5]

What remains clear, and was reinforced in Sir Rupert's final report is that compulsory ADR continues to be a contradiction in terms. The pendulum of case law has swung from the assumption that it is unreasonable to refuse to mediate, to the assumption that such refusal is reasonable unless the other party can demonstrate otherwise. ADR is to be achieved by persuasion and encouragement from the court and not force. The court's obligation to consider ADR as part of its case management duty has been an ever present in the CPR – CPR 1.4(2)(e) is clear that active case management includes 'encouraging the parties to use an ADR procedure if the court considers that appropriate and facilitating the use of such a procedure'.

However, the emphasis has changed. The message is clear that parties should routinely consider whether their disputes are suitable for ADR and appreciate that if the unsuccessful party can demonstrate that the refusal to mediate was unreasonable this could rebut the presumption that costs follow the event.

In *Burchell v Bullard* [2005] EWCA Civ 358, [2005] NLJR 593, [2005] BLR 330, the Court of Appeal endorsed *Halsey* saying that the case made it plain there was a high rate of success achieved by mediation and also established its importance as a route to a just result running parallel with that of the court system. The court had given its stamp of approval for mediation and it was now the legal profession which had to become fully aware of and acknowledge its value. The profession could no longer with impunity shrug aside reasonable requests to mediate. Claimants and defendants alike in the future could expect little sympathy if they blithely battled on regardless of the alternatives. In particular, a party could not ignore a proper request to mediate simply because it had been made before the claim had been issued.

Despite the warnings in *Halsey* and *Burchall* these cases were followed by another three cases in which a refusal to mediate was upheld: *Daniels v Metropolitan Police Comr* [2005] EWCA Civ 1312 (it was reasonable for a public body to contest what it reasonably considered to be an unfounded claim in order to deter similarly unfounded claims); *Wethered Estate Ltd v Davis* [2005] EWHC 1903 (Ch) (it was not unreasonable to refuse mediation until the true nature of the dispute had been defined) and *Hickman v Blake Lapthorn* [2006] EWHC 12 (QB), [2006] 3 Costs LR 452 (a barrister who was not prepared to compromise a negligence claim against him had legitimately and reasonably refused to mediate). In *Hickman* the court gave the following guidance, reinforcing and refining the *Halsey* guidelines:

'– A party cannot be ordered to submit to mediation as that would be contrary to article 6 of the European Convention on Human Rights.
– The burden is on the unsuccessful party to show why the general rule of costs following the event should not apply, and it must be shown that the successful party acted unreasonably in refusing to agree to ADR.
– A party's reasonable belief that he has a strong case is relevant to the reasonableness of his refusal otherwise the fear of costs sanctions may be used to extract unmerited settlements.
– Where a case is evenly balanced a party's belief that he would win should be given little or no weight in considering whether a refusal was reasonable. The belief must be unreasonable.
– The cost of mediation is a relevant factor.
– Whether the mediation had a reasonable prospect of success is relevant to the reasonableness of a refusal to agree to mediation.
– In considering whether the refusal to agree to mediation was unreasonable, it is for the unsuccessful party to show that there was a reasonable prospect that the mediation would have been successful.

21.5 *Chapter 21 Costs Inducements to Settle – ADR*

- Where a party refuses to take part in mediation despite encouragement from the court to do so, that is a factor to be taken into account.
- Public bodies are not in a special position.'

CONCLUSION

[21.6]

That ADR has a significant role in the control of costs is clear from the recommendations made by Sir Rupert at the end of his consideration of ADR. These are that:
- There should be a serious campaign:
  - (a) to ensure that all litigation lawyers and judges are properly informed about the benefits which ADR can bring; and
  - (b) to alert the public and small businesses to the benefits of ADR.
- An authoritative handbook should be prepared, explaining clearly and concisely what ADR is and giving details of all reputable providers of mediation. This should be the standard handbook for use at all Judicial College seminars and CPD training sessions concerning mediation

Unsurprisingly, within weeks of the April 2013 amendments to the CPR the handbook recommended by Sir Rupert was published, endorsed by, amongst others, the Judicial College (the JSB as was). Copies have been sent to all full time judges. This coupled with a discrete phase in the Form H budget for ADR/settlement could not make the position clearer – the judiciary will be expecting parties to consider ADR at an early stage and a party not considering this as part of the assumptions on page 1 of the Form H and not providing a budget for this phase will have some explaining to do. This could lead to the curious situation where the court actually encourages an increase to a budget at the costs management stage to enable a party who has made no costs provision for ADR to have a pot of money to spend on it. At a time of great professional concern that the court's sole objective is to cut costs and slash budgets, this may prove to be the crunchiest carrot on offer.

The ADR Handbook has already received endorsement by the Court of Appeal in *PGF II SA v OMFS CO 1Ltd* [2013] EWCA Civ 1288. In this case the Court of Appeal upheld the trial judge's decision not to award the defendant its costs for a period after a valid Part 36 offer in part because it had failed to respond to an offer to enter into ADR. Briggs LJ referred to the ADR Handbook and stated that silence in the face of a invitation to enter into ADR should, as a general rather than invariable rule, itself be treated as unreasonable conduct.

Ward LJ in *Wright v Michael Wright Supplies Ltd* [2013] EWCA Civ 234, [2013] 4 Costs LO 630 even raised the spectre of revisiting *Halsey* and in particular the rule expressed by the Court of Appeal in that case that 'to oblige truly unwilling parties to refer their disputes to mediation would be to impose an unacceptable obstruction on their right to access to the court'. He expressed the concern that there are always cases that illustrate that parties determined to have their day in court cannot be compelled to try to resolve their dispute through ADR despite the most persuasive judicial cajoling. With customary eloquence he identified the problem in this way:

> 'You may be able to drag the horse (a mule offers a better metaphor) to water, but you cannot force the wretched animal to drink if it stubbornly resists. I suppose you can make it run around the litigation course so vigorously that in a muck sweat it will find the mediation trough more friendly and desirable. But none of that provides the real answer. Perhaps, therefore, it is time to review the rule in *Halsey* . . . .'

CHAPTER 22

COSTS AWARDS BETWEEN THE PARTIES

INTRODUCTION

[22.1]

CPR 44.2(1) confers on the court a discretion as to whether costs are payable by one party to another, the amount of those costs and when they are to be paid. This power extends both to interim and final orders. In certain situations there is a deemed costs order, subject usually to a residual power for the court to order otherwise, and no court adjudication is required. Where there is no deemed order then the court must determine whether there is to be a costs order and, if there is, in whose favour it will be made. Finally there are some unusual situations where discrete rules apply. However, let us start with some of the easy bits.

NO ORDER AS TO COSTS

[22.2]

CPR PD 44, para 4.2 sets out, in tabular form, those costs orders that the court will commonly make. It links 'no order as to costs' and 'each party to pay own costs' together, and points out that such orders mean precisely what they say- the costs covered by the order are not to be recovered from the other party and each party has to pay his own costs.

CPR 44.10 takes this a step further and provides that where a court order makes no mention of costs at all, then the general rule is that no party is entitled to costs (including a pro bono award of costs). This was demonstrated in *Griffiths v Metropolitan Police Comr* [2003] EWCA Civ 313, when the court held not only that where an interim order was silent no party was entitled to their costs, but also that the trial judge had no jurisdiction to vary that order.

There are exceptions and these are at CPR 44.10(2) which states that where the court gives permission to appeal, permission to apply for judicial review or makes an order on a not on notice application, but is silent on costs, then an order for 'the applicant's costs in the case' is deemed included. A party may apply to vary the deemed order under CPR 44.10(3).

It would be wrong to leave CPR 44.10(2) without making reference to *Mackay and Busby v Ashwood Enterprises Ltd* [2013] EWCA Civ 959 in which the Court of Appeal stated that the deemed *'costs in the case'* order provided for under CPR 44.10(2)(c) supported the contention that the court could in fact make an order for costs on a not on notice application against the party who had not had notice (founding the general jurisdiction under the broad provisions of s 51 of the Senior Courts Act 1981). This is, of course, subject to the right of a party not served to apply to vary or set aside the order under CPR 23.10.

THE MENU OF COSTS ORDERS AVAILABLE TO THE COURT

[22.3]

Reference has been made above to the table at CPR PD 44, para 4.2. This table is a useful point of reference in respect of costs orders following interim and appeal hearings and merits inclusion here:

Term	Effect
Costs Costs in any event	The party in whose favour the order is made is entitled to that party's costs in respect of the part of the proceedings to which the order relates, whatever other costs orders are made in the proceedings.
Costs in the case Costs in the application	The party in whose favour the court makes an order for costs at the end of the proceedings is entitled to that party's costs of the part of the proceedings to which the order relates.
Costs reserved	The decision about costs is deferred to a later occasion, but if no later order is made the costs will be costs in the case.
Claimant's/ Defendant's costs in case/ application	If the party in whose favour the costs order is made is awarded costs at the end the proceedings, that party is entitled to that party's costs of the part of the proceedings to which the order relates. If any other party is awarded costs at the end of the proceedings, the party in whose favour the final costs order is made is not liable to pay the costs of any other party in respect of the part of the proceedings to which the order relates.
Costs thrown away	Where, for example, a judgment or order is set aside, the party in whose favour the costs order is made is entitled to the costs which have been incurred as a consequence. This includes the costs of – preparing for and attending any hearing at which the judgment or order which has been set aside was made; preparing for and attending any hearing to set aside the judgment or order in question; preparing for and attending any hearing at which the court orders the proceedings or the part in question to be adjourned; any steps taken to enforce a judgment or order which has subsequently been set aside.

Costs of and caused by	Where, for example, the court makes this order on an application to amend a statement of case, the party in whose favour the costs order is made is entitled to the costs of preparing for and attending the application and the costs of any consequential amendment to his own statement of case.
Costs here and below	The party in whose favour the costs order is made is entitled not only to that party's costs in respect of the proceedings in which the court makes the order but also to that party's costs of the proceedings in any lower court. In the case of an appeal from a Divisional Court the party is not entitled to any costs incurred in any court below the Divisional Court.
No order as to costs Each party to pay own costs	Each party is to bear that party's own costs of the part of the proceedings to which the order relates whatever costs order the court makes at the end of the proceedings.

It is worth pointing out that an order for 'costs reserved' becomes an order for 'costs in the case', if there is no later determination of where the responsibility for those costs lies.

Of course, with the exception of 'no order for costs' this table is of no assistance when the court makes an order that a party's costs are to be paid by another, whether on an interim application or after a final hearing.

SOLICITOR'S DUTY TO NOTIFY THE CLIENT

[22.4]

CPR 44.8 provides that where a costs order is made against a represented party and that party is not present at the hearing at which the order is made the solicitor must inform his client of the order within 7 days of receipt by the solicitor of notice of the order. CPR PD 44 defines a party to include anyone who has instructed a solicitor to represent a party, eg a trade union, an insurer or the legal aid authority. In fact most professional instructors will make it a contractual term of instruction that notification takes place in these circumstances anyway. CPR PD 44, para 10.3 enables the court to require evidence of the steps taken by the solicitor to notify the client of the adverse costs order. This is a watered down version of original proposals for enforcement of this rule that provided for a solicitor who failed to comply to pay the costs. We have never seen an order that required a solicitor to provide proof of efforts to comply.

CPR PD 44, para 10.2 requires the solicitor to provide the client with an explanation of how the order came to be made when serving the notification.

THE TIME FOR COMPLIANCE WITH A COSTS ORDER

[22.5]

CPR 44.7 provides that unless the court specifies another date, a party must comply with an order for the payment of costs within 14 days of the date of the order specifying the amount of costs or, if the amount is quantified at a detailed assessment, in accordance with the final costs certificate (CPR 47.17(5)).

DEEMED COSTS ORDERS

Part 36

[22.6]

Part 36.10 provides for a deemed costs order in favour of the claimant on the acceptance of a Part 36 offer. This has been considered in more detail in Chapter 20 Costs Inducements to Settle – Part 36 Offers and other Admissible Offers, above. Just a reminder that this does not apply on acceptance of a pre-issue Part 36 offer and the costs only procedure should be used.

Discontinuance

[22.7]

CPR 38.6 provides that a claimant who discontinues is liable for the costs of the defendant against whom he discontinues for the costs up until the date of service of the notice of discontinuance, unless the court orders otherwise. If the discontinuance is of part of the proceedings only then the liability is only in respect of the costs referable to that part and those cannot be assessed until the conclusion of the remained of the proceedings.

Where claimants deleted a claim by amendments to the particulars of claim this was in effect a discontinuance of that claim with the same costs consequences (*Isaac v Isaac* [2005] EWHC 435 (Ch)). The burden is on a claimant who seeks to avoid the costs consequences of discontinuance to persuade the court that some other order is appropriate, perhaps because of some unavoidable and unforeseeable change of circumstance (*Re Walker Windsail Systems Ltd, Walker v Walker* [2005] EWCA Civ 247, [2006] 1 All ER 272, [2006] 1 WLR 2194). However, changing circumstances are part and parcel of litigation and mere change alone will not suffice. In *Teasdale v HSBC Bank plc* [2010] EWHC 612 (QB), [2010] 4 All ER 630, [2010] NLJR 878 the court concluded that it would be difficult to see how any change in circumstances could amount to good reason unless connected with some conduct on the part of the defendant which merited a departure from the general rule.

In *Messih v McMillan Williams* [2010] EWCA Civ 844, [2010] 6 Costs LR 914, the claimant brought proceedings against two firms of solicitors alleging different causes of action arising out of his tenancy of commercial premises. The proceedings against the first solicitors were settled on the terms that the

Deemed costs orders **22.7**

solicitors paid damages and the claimant's costs. The claimant wished to discontinue against the second firm of solicitors on the basis that each party would pay their own costs, but the solicitors would not agree. Accordingly the claimant served notice of discontinuance and applied for an order under CPR 38.6(1) that he should not be required to pay the second firm's costs. The judge made the order requested against which the defendant appealed.

The Court of Appeal held that the judge was wrong concluding that if CPR 38.6 had been intended to create a general discretion as to costs on discontinuance it would have said so. The rule made it clear that the defendant started from the position of being entitled to his costs and it was for the claimant to justify the making of some other order. Accordingly the claimant was ordered to pay the second solicitors' cost up to the date of discontinuance.

An example of the court exercising its discretion, and affirming the decisions on the defendant's conduct is *Webb v Environment Agency* (2011) QBD 5 April. In *Webb* the claimants applied for a costs order against the defendant agency even though they had discontinued their claim for damages against the agency for negligence, breach of duty and/or nuisance for property damage as a result of flooding caused by the installation of a grate in a watercourse maintained by the agency. The agency in response had contended that there had been a failure to act which did not give rise to any liability. On the issue of proceedings the agency maintained the same defence but later amended its defence to contend that the installation of the grate was a positive act under its powers under the Water Resources Act 1991, s 165, which meant that any dispute had to be heard by the Lands Tribunal. The parties agreed for the matter to be heard by the Lands Tribunal and the claimants agreed to discontinue the court proceedings, but sought their costs. The judge adjourned the costs hearing until after the tribunal decision. The tribunal proceedings were settled and the agency agreed to pay 80% of the claimants' costs. In the main proceedings the court after argument held this was one of those rare cases where the usual costs consequences should not apply. Until the amended defence, the agency had represented a factual situation that it abandoned and then relied on a statutory defence as being conclusive. Justice would be met by an order that the agency pay 80% of the claimants' costs.

It is clear from *Hoist UK Ltd v Reid Lifting Ltd* [2010] EWHC 1922 (Ch), [2011] Bus LR D58, (2010) Times, 5 October that one of the factors the court will take into account on an application to depart from the general rule is the fact that such an application was made some time after discontinuance. The court also concluded that the fact that proceedings were discontinued before the application for a different costs order did not mean that there was no longer a claim in which to make the application (as the defendant ingeniously tried to argue). It mattered not: the application that there be no order for costs was nevertheless refused.

Strike out for non-payment of fees

[22.8]

If the claimant party fails to pay or obtain fee exemption from a fee payable for either the allocation of the claim or on the filing of the pre-trial checklist, then pursuant to CPR 3.7(6)(b) there is a deemed order that claimant pays the defendant's costs unless the court orders otherwise

The basis of assessment under deemed orders.

[22.9]

CPR 44.9(1) makes it clear that assessments under the deemed orders set out above are conducted on the standard basis. However, where the court has a residual discretion that does extend to making an order for indemnity costs in place of standard basis costs on the application of the party against whom the claim has been discontinued, if it is appropriate to do so. It did just this in *Shahrokh Mireskandari v Law Society* [2009] EWHC 2224 (Ch), 153 Sol Jo (no 34) 29, where the discontinued claim had always been utterly speculative.

Interest on the costs assessed under a deemed order

[22.10]

CPR 44.9(4) is clear that any interest on costs runs from the date when the deemed costs order is made.

COSTS ORDERS BY CONSENT

[22.11]

The parties may agree where the liability for costs rests between themselves, whether in respect of interim of final orders. This can be done by way of consent order, consent judgment or Tomlin order. However, the court retains the discretion whether or not to approve these orders and there a number of potential pitfalls to avoid.

Tomlin orders

[22.12]

A Tomlin order stays a claim on terms. The terms are usually contained in a schedule to the order. All too often parties still include the agreement in respect of costs in the schedule. This is acceptable if it is an agreement to pay a specified sum for costs. However, if the agreement is that the costs will be assessed if not agreed then the order must be in the body of the order. If it is not then there is no order giving a right to assessment proceedings and a failure to agree costs can present a major problem in recovering them.

Agreement in interim orders for costs to be assessed if not agreed

[22.13]

As we have considered earlier, an order for a detailed assessment is a case management decision. It must be made by the court in accordance with the overriding objective. This requires the court to consider proportionality. In most cases ordering a detailed assessment of the costs relating to an interim hearing is not proportionate. Whilst this may have been included to avoid an argument at this stage on the amount of the costs, the likelihood is that the court will be increasingly reluctant to approve such an order insisting that either the amount of costs is agreed or that there is a summary assessment of them there and then.

Uncertain and unclear terms

[22.14]

It may sound obvious, but it is important that the parties consider carefully the costs order that is drafted before consenting to it. This is not only to ensure that it contains enforceable costs provisions, but also to be satisfied that the order states precisely what has been agreed after consideration of all potential consequences.

In *Richardson Roofing Co Ltd v Colman Partnership Ltd* [2009] EWCA Civ 839, [2009] 4 Costs LR 521 under a consent order the claimant was to pay a fourth party's costs incurred and thrown away by the adjournment of the trial. Three years later, the fourth party served a draft bill of cost for its entire costs until the hearing ordering a preliminary issue. The fourth party issued proceedings seeking an order that the costs judge dealing with the assessment should be directed that the costs included the fourth party's preparation for trial because there was no prospect that the claim would be revived. The judge made an order directing the costs judge to carry out the assessment using the guidance set out in six specified paragraphs of his judgment.

The Court of Appeal thought it highly debatable whether the judge had had any jurisdiction to hear the application. The order in dispute was a consent order with no application to vary it. As the jurisdiction point had not been fully argued before the Court of Appeal, the point could not be decided and the court therefore assumed jurisdiction, despite its serious reservations. Even if the judge had jurisdiction, he should not have exercised it because the matter would ordinarily go before a costs judge for a detailed assessment. It was undesirable for judges to make this sort of order referring to guidance set out in their judgment, which was extremely diffuse. Judgments provide the reasons for the subsequent orders and any order made at the end of a judgment should stand on its own. The judge had failed to answer the issue in relation to the construction of the consent order and his consequent order was of no assistance to a costs judge. Accordingly, his order could not stand.

In *Newall v Lewis* [2008] EWHC 910 (Ch), [2008] 4 Costs LR 626, the SCCO master decided that all that the parties had meant by 'incidental costs' in a consent order were the investigative costs pre-issue, which would be subject to detailed assessment and referral to a judge, and that all costs claimed post-issue, including incidental ones, were potentially recoverable and did not need to be referred to a judge. On appeal it was held, contrary to the master's conclusion, the order imposed the requirement on the costs judge of separating into distinct categories the costs of the Part 8 proceedings and the costs incidental to them. The correct approach was that if disputes on which costs had been expended pre-issue were relevant to the eventual proceedings and the other party's attitude had made it reasonable to expect them to be included in the litigation, those costs should be recoverable, *Re Gibson's Settlement Trusts, Mellors v Gibson* [1981] Ch 179, [1981] 1 All ER 233 applied. The matter was one of fact and legal analysis rather than discretion

Reserved costs

[22.15]

Again do not expect the court to be too enamoured of the suggestion that costs are reserved to be dealt with later. If there is a good reason to do so then an explanation of this may be sensible.

CONTESTED AWARDS BETWEEN THE PARTIES

The starting point

[22.16]

CPR 44.2(2) contains the rebuttable presumption that the unsuccessful party pays the costs of the successful parties (often referred to as 'costs follow the event'). However, like all 'general rules' in the CPR, CPR 44.2(2) is then followed by reasons why the court might depart from the starting point. However, it is only after the successful party has been identified that the court moves on to consider reasons to depart. To undertake the exercise as a combined one, whilst it may still lead to the same end result, does not accord with the rule. In *Straker v Tudor Rose (a firm)* [2007] EWCA Civ 368, 151 Sol Jo LB 571, [2008] 2 Costs LR 205 Waller LJ gave this helpful guidance on the approach to 'between the parties' costs:
- First is it appropriate to make an order for costs?
- Second, if it is, the general rule is that the unsuccessful party will pay the costs of the successful party.
- Third, identify the successful party
- Fourth, consider whether there are any reasons for departing from the general rule in whole or in part. If so the court should make clear findings of the factors justifying the departure.

Who is the successful party?

[22.17]

In *A L Barnes Ltd v Time Talk (UK) Ltd* [2003] EWCA Civ 402, [2003] BLR 331, (2003) Times, 9 April Longmore LJ set out at para 28 a formulation that the trial judge ought to adopt to determine the identity of the successful party:

'In deciding who is the successful party the most important thing is to identify the party who is to pay money to the other. That is the surest indicator of success and failure.'

He indicated that if the trial judge had asked himself this question:

' . . . he would in my judgment have had to answer that it was the claimants who recovered more than the defendants had ever offered and thus it must be the claimants who were the successful party.'

This approach was endorsed by Ward LJ in *Day v Day* [2006] EWCA Civ 415. This was a case concerning the beneficial interest of parties in the proceeds of the sale of a property. The trial judge found that the shares were 2/5th and 3/5th respectively, but made only a time limited costs order declaring that after that time there was a 'no score draw'. In substituting an award for the claimant to recover her costs from the defendant Ward LJ stated at para 17:

'I would go further and say that in a case like this, the question of who is the unsuccessful party can easily be determined by deciding who has to write the cheque at the end of the case.'

The judgment of Lightman J (as he was) in *Bank of Credit and Commerce International SA v Ali (No 3)* [1999] NLJ 1734 Vol 149 had been cited to the Court of Appeal in Day. Lightman J had formulated a definition of success as being 'For the purposes of the CPR success is not a technical term but a result in real life, and the question as to who has succeeded is a matter for the exercise of common sense.' Perhaps this simply puts into words what all practitioners know – that a look at the reactions of the parties in court when judgment is given is the surest indicator of success and failure. However, Ward LJ felt that this did not go far enough and adopted the simple approach set out above.

The logic of this simple formulation of success is patent. After all, it only answers the first question posed of who was the successful party and engages the general rule. The court then has the task of considering the factors at CPR 44.2(4) and CPR 44.2(5) to determine whether there is a reason to depart from that general rule. Many a cost order has been lost on that journey through the rules. There is adequate provision within them to ensure that a just outcome is received. Adopting a clear approach to identification of the successful party has the additional benefit of avoiding protracted and costly debate upon this point. Sadly that has not always proved to be the case

In *AL Barnes* Longmore LJ did qualify the position by stating that it should apply to what 'might generally be called commercial litigation' and this appears to be broadly what has transpired in the commercial arena. *F&C Alternative Investments (Holdings) Ltd v Barthelemy* [2011] EWHC 2807 (Ch), held that often it would be appropriate for the loser to pay the winner's costs, even where there had been issues on which the overall winner had lost. The court re-iterated that in commercial litigation, the starting point in working out who the winner is for the purposes of making costs orders would usually be to look at what money had been ordered to be paid as the payee would be the successful party.

Indeed this approach seemed to be mirrored in non-commercial litigation, although much is often made of the decision in *Hullock v East Riding of Yorkshire County Council* [2009] EWCA Civ 1039, which, on first reading, seems to suggest that in a personal injury quantum only claim, things may be different. However, whilst the Court of Appeal judgment raises consideration of the identification of the *'real issue'* and how that is determined as 'central . . . in determining who should pay the costs' in fact this consideration is limited to a period after an interim payment was made for a sum which ultimately proved to be the settlement sum. It is our view that this case helps little in the overall consideration of success.

However, the 'commercial' qualification led to a busy few days for the Court of Appeal in the summer of 2011 and it seems that the debate may still be raging.

In *Medway Primary Care Trust v Marcus* [2011] EWCA Civ 750, 123 BMLR 112, [2011] 5 Costs LR 808, the claimant commenced proceedings alleging clinical negligence resulting in his left lower leg having to be amputated. The defendants admitted breach of duty but denied causation. Shortly before the trial as to liability, the appropriate quantum of the claim was agreed as being £525,000. At trial the issue of causation in respect of the amputation was decided against the claimant, but he was awarded damages of £2,000 for pain and suffering over a limited period of time relating to admitted breaches of duty in respect of the provision of pain killing medication. The trial judge decided that the claimant was the successful party, but then reduced the award of costs to 50% of his costs.

It was held on appeal by a majority that the £2,000 recovered did not constitute vindication for the claimant. The £2,000 was scant consolation for the claimant whose only real claim was for the amputation. The action was in reality all about the cause of the amputation and the costs were spent in advancing and defending that. The defendants were therefore the successful parties and so the starting point should be a costs order in their favour. There should then be a reduction for the fact that the claimant did succeed to a very small extent; the fact that the trust did not concede liability until a very late stage and because the trust's case as to breach of duty was not withdrawn until just before trial. The fact that there was no offer made under CPR Part 36 was not a ground for a reduction but it was relevant that the defendants had not written a 'Calderbank letter' offering, say, £3,000 plus costs proportionate to the recovery. However, the real claim failed and no rational person would have issued the proceedings and pursued these proceedings to recover only £2,000. The claimant was therefore ordered to pay 75% of the defendants' costs.

However, the dissenting judgment was given by Jackson LJ. It was on the basis that the defendants should have made a Part 36 offer, the claimant had a good claim for the £2,000 and the only way he could have recovered that sum was by pursing a claim through the court as the defendants had refused to pay anything. He was obviously attracted to the clarity of the 'who writes the cheque' evaluation of success and failure and keen to stress that it is the first stage in the CPR 44.2 (then CPR 44.3) test only, stating at paragraph 30

'In my view, in a personal injury case where (a) the claimant has pursued his claim in a reasonable manner, (b) the claimant recovers damages (other than nominal damages) and (c) there is no or no sufficient Part 36 offer, the starting point should be that the claimant recovers his costs. That flows from rule 44.3(2)(a). The next question to consider is whether any adjustment should be made to reflect the issues on which the claimant has lost.'

Jackson LJ concluded that he regarded the claimant as the successful party, but that the trial judge's award to the claimant of 50% of his costs was 'on the generous side'.

Just eight days later he had the opportunity to revisit the topic in *Fox v Foundation Piling Ltd* [2011] EWCA Civ 790, [2011] 6 Costs LR 961. The claimant had suffered personal injuries at work, having fallen carrying heavy equipment. He made a claim put at over £280,000. Only quantum was in issue. Surveillance evidence revealed him to have been less seriously affected than he alleged. On disclosure of the evidence he accepted a revised net offer of about £31,000 and the claim settled, but the parties were unable to agree on an appropriate costs order. The court was asked to determine the costs award. At first instance the court concluded that the claimant should pay the costs from the date of the first offer of about £23,000 made about 13 months before settlement. The court concluded that after that date the defendant was the successful party. It stated that even if it was wrong about the effect of the offer (and the claimant had subsequently accepted an offer that gave him a higher net receipt), the claimant's own conduct justified this order. Perhaps unsurprisingly, in the light of the pre-*Medway* authorities, by the time of the appeal the parties had agreed that the claimant was in fact the successful party for the purpose of CPR 44.2 (as is) and the argument was only about whether, and to what extent, the court should depart from that position.

Jackson LJ, giving the lead judgment of a unanimous Court of Appeal, which substituted an order that the defendant pay the claimant's costs on the standard basis, offered some timely reminders (both to practitioners and to the appellate judiciary).

- Where both parties are over optimistic with their Part 36 positions, the claimant should normally be regarded as the 'successful party', because s/he has been forced to bring proceedings in order to recover the sum awarded. (para 46)
- A defendant in possession of surveillance evidence should make a prompt and realistic Part 36 offer – see *Morgan v UPS* [2008] EWCA Civ 1476. Here the defendant had this evidence but delayed in making a realistic offer. Its remedy was to have made an early, modest Part 36 and its failure to do so prevented it seeking costs protection (paras 58–60).
- The fact that the successful party has won and lost some issues may be a good reason for modifying the usual order under CPR 44.2 AND this is commonly achieved by awarding the successful party a specified proportion of her/his costs (paras 47-49)
- The growing tendency of Courts at all levels (including the Court of Appeal) to depart from the starting point in CPR 44.2 too far and too often was an unwelcome trend which had itself increased costs by arguments at first instance and a 'swarm of appeals'. (para 62)

A definitive statement from an impeccable source appeared to have concluded the debate. There the matter seemed to lie. However, it would be wrong to leave the topic without reference to the decision of Briggs J (as he then was) in *Magical Marking Ltd & Phillis v Ware & Kay Ltd & 10 ors* [2013] EWHC 636 (Ch) which distinguished *Fox* on the basis that 'success' had been conceded by the time that the appeal was heard and was not a live issue before the Court of Appeal and that it did not, therefore, undermine the *Medway* line of authority on 'substantial success'. Strictly it is undeniable that by the time of the appeal, the defendant had conceded that the claimant was the successful party in *Fox*, but Jackson LJ's view could not have been expressed in clearer terms:

> 'In my view, there is no justification for departing from the usual starting point as set out in rule 44.3(2)(a) [now 44.2(2)(a)], namely that the unsuccessful party should pay the successful part's costs. The judge exercised his discretion on the wrong basis, namely the assumption that the defendant was the successful party. It therefore falls to this court to re-exercise that discretion.'

For our part we prefer the *A L Barnes*, *Day* and *Fox* approach. At a time when the court has no desire or resource for further rounds of the 'costs wars' it has the virtue of clarity and simplicity. Concerns over conduct, partial success etc. can be addressed in any event when looking at whether there are reasons to depart from this starting point as we shall see below with cases such as *Widlake v BAA Ltd* [2009] EWCA Civ 1256, 153 Sol Jo (no 45) 29, [2010] 3 Costs LR 353 illustrating how to achieve the desired outcome within the provisions of CPR 44.2. In that case Ward LJ reverted to the template of Longmore LJ in *Straker* (above) stating:

> 'Waller LJ was surely right to endorse Longmore LJ's views that the most important thing is to identify the party who is to pay money to the other <u>even in a case of personal injury</u>.' (Our emphasis)

Reasons to depart from the general rule

(i) Conduct of the parties

[22.18]

CPR 44.2(5) makes it plain that the conduct concerned may be:
- both before and during proceedings;
- the unreasonable pursuit of, raising or contesting of a particular issue or allegation;
- the manner in which the case, an allegation or an issue has been pursued;
- the fact that a successful claimant has exaggerated the claim in whole or in part.

The starting point on conduct is the case of *Painting v University of Oxford* [2005] EWCA Civ 161, (2005) Times, 15 February, [2005] 3 Costs LR 394. Judgment had been entered for the claimant in her personal injury claim and all that remained was the determination of quantum. The defendant paid £184,000 into court, but then secured video surveillance taken of the claimant. This showed she was able to walk normally without aid and was able to bend and straighten herself looking at display items in shops, all completely contrary to the level of disability she alleged. The University obtained permission to withdraw all the money in court save for £10,000 and contested quantum on the basis that the claimant had been exaggerating her claim. On the assessment of damages the judge found the claimant had indeed exaggerated her claim and awarded her only £22,000 but also her costs on the basis that she had beaten the payment into court. The Court of Appeal held that the judge should have taken into account <u>all</u> the provisions of CPR 44.2. The judge had only taken into account the inadequacy of the Part 36 payment. The court stated:

> 'Mrs Painting had been deliberately misleading in the course of the claim, and the fact that the exaggeration is intended and fraudulent is, to my mind, a very important element which needs to be addressed in any assessment of costs'

The Court of Appeal was also singularly unimpressed that the claimant had not attempted to negotiate settlement. The upshot was that the claimant was ordered to pay the costs of the action from the date of the reduced payment into court, which was only about 3 weeks after the original sum had been offered. Accordingly the claimant would have very little, if any money, left over from her damages.

We pause to note that Kay LJ made the point, subsequently re-iterated by Jackson LJ in *Fox v Foundation Piling Ltd* (which we have considered in the Chapter 20 Costs Inducements to Settle – Part 36 Offers and other Admissible Offers, above), about a defendant's salvation being by way of a realistic Part 36 offer, stating:

> 'What the University chose to do was to make a Part 36 payment which amounted to a rock-bottom figure even on the basis that it established exaggeration to the maximum extent. If it had chosen to do so, it could have pitched the payment higher without for a moment weakening its position on the central issue in the case.'

The following year the Court of Appeal had a further opportunity to consider conduct and exaggeration in *Jackson v Ministry of Defence* [2006] EWCA Civ 46, [2006]. The claimant had issued proceedings for personal injury against the MoD for injury suffered during a training exercise. He advanced substantial claims for damages for future loss of earnings and for specially adapted accommodation based on his account of his residual disability. The medical evidence did not support the claim of residual disability and those claims were eventually abandoned, reducing his claim from over £1 million to £240,000. The MoD made a Part 36 payment into court in the sum of £150,000. Damages of £155,000 were awarded and the Court of Appeal upheld the judge's reduction of the claimant's costs by 25% to reflect the fact that the award had only just beaten the payment into court and the fact the claimant had exaggerated his evidence. The reduction that the judge made in costs and the likely reduction at the detailed assessment were likely to act as a disincentive to claimants who sought to make exaggerated claims. The order made was well within the judge's wide discretion.

What these cases go to show is that the ultimate outcome is fact specific, but have a common theme that the starting point is that the successful party recovers his costs and the court only then looks at departure from that position.

This point was never better illustrated than in *Widlake v BAA* (see above). The claimant brought an employers' liability claim. Her claim was put at around £34,000. The trial judge awarded £5,522.38 plus interest. More tellingly he found that the claimant had deliberately concealed her history of back pain 'in the hope of increasing the amount of compensation which she would recover'. He also noted that the claimant had not made any counter offers to the defendant's Part 36 offer and made no attempt to negotiate.

22.18 *Chapter 22 Costs Awards Between the Parties*

Notwithstanding this the claimant still recovered more than the defendant's Part 36. The trial judge declared the defendant to be the real winner and ordered the claimant to pay the defendant's costs. The Court of Appeal disagreed and in so doing restated the correct approach to award costs is first to identify the successful party. Here that was the claimant. However, the court then considered the exaggeration and conduct and concluded:

> 'I start with the claimant getting her costs because she beat the payment in and was the successful party. That is the starting point. Those costs should not include costs related to Miss Porter's reporting and the costs judge must be directed to exclude those matters. Pursuing her claim in the exaggerated way she did had the result that this became heavily contested litigation whereas it might have settled. The defendant has been put to unnecessary expense. But an order for costs against the claimant is less justified where, as here, the defendant failed to alleviate its predicament by making a proper Part 36 offer and so lost the opportunity provided by the rules of recovering those costs from the claimant. The claimant's dishonesty must be penalised. The claimant's failure to negotiate a claim which was clearly capable of being settled must also be recognised. When I balance those factors, and attempt to do justice to both parties and to be fair to them, I conclude that the right order in this case is that there be no order for costs.' (para 44)

The Court of Appeal has been clear that there is no general rule that a finding of dishonest conduct will replace the general starting point under CPR 44(2)(a). An evaluation of the effect it has on the issues in the trial will be required. In *Neale v Hutchinson* [2012] EWCA Civ 345, [2012] 5 Costs LO 588, [2012] 2 P & CR D1 the appellants, who had been successful at the trial of a boundary dispute, were found to have changed a measurement on a document and had sought to suggest that this had in fact been done by the respondents. The trial judge had been explicit in his condemnation of the appellants' dishonesty. The trial judge made a number of separate costs orders and in his reasoning made specific reference to the appellants' dishonesty. The Court of Appeal accepted and re-iterated the principle set out by Waller LJ in *Northstar Systems Ltd v Fielding* [2006] EWCA Civ 1660 that:

> 'There is no general rule that a losing party who can establish dishonesty must receive all his costs of establishing this dishonesty, however, disproportionate they may be.'

with Pitchford LJ concluding that:

> 'In my judgment, the judge erred in his unreserved acceptance of the sweeping proposition that the defendants (the appellants) should not expect to be able to fabricate documents and lie under oath in support of their case and still recover their costs if they succeed at trial.'

Instead the Court of Appeal stressed the importance of starting with the general rule at CPR 44.2(2) and then considering conduct as a reason to depart from that point.

It is important to stress in this consideration of conduct and exaggeration that the fact that a claimant accepts a sum very much less than he had originally claimed does not itself show there to have been exaggeration within CPR 44.2(5)(d). That paragraph could not have been intended to be satisfied merely because a genuine claim was overestimated. Exaggeration for the purposes of that rule must indicate conduct meriting criticism. This was different from a party merely stating his best case (*Morton v Portal Ltd* [2010] EWHC 1804 (QB)).

Remember, too, that pre-action conduct is also relevant. Gone are those pre-CPR days where parties could conduct themselves without concern prior to issue provided that they then were seen to be paragons of virtue once the claim had been issued.

> 'In my judgment it would be wrong to conclude, if there ever was a strict rule that pre-action conduct was relevant to costs only if causative of the bringing of an unsuccessful claim, or of increased expense in the subsequent litigation, that such a rule survives the introduction of the CPR. The language of Part 44 requires the court to have regard to the conduct of all the parties, both before as well as during the proceedings, and is otherwise wholly unqualified.' (*Bank of Tokyo-Mitsubishi UFJ Ltd v Baskan Gida Sanayi Ve Pazarlama As* [2009] EWHC 1696 (Ch), [2009] NLJR 1066, [2010] 5 Costs LR 657)

Finally, the conduct of the proceedings themselves can also lead to a departure from the general rule. In *R (on the application of Scrinivasans Solicitors) v Croydon County Court* [2013] EWCA Civ 249 the claimant firm of solicitors was successful on a judicial review. However, the solicitors had abandoned and re-shaped points at the 11th hour and had pursued wrong criteria. The first instance judge indicated that the claimant had been successful but there was a considerable amount of failure to pursue the right submissions on the right point at the right time. Rather than make an issue based order the judge made no order for costs to reflect the poor conduct. The Court of Appeal concluded, rather damningly, that given how the litigation had been conducted the order in respect of costs was both justified and sensible!

(ii)   Whether a party has succeeded on part of its case, even if that party has not been wholly successful

[22.19]

Mixing up partial success with the definition of the successful party can be easy. A party can win on a number of issues, but remain unsuccessful in the litigation. The pre-CPR days when the court did not look at the component parts of a claim, but simply the overall outcome, when determining where the costs liability lay, are long gone. However, even now there seems to be an acceptance that a successful party is likely to suffer reverses on certain issues along the way and that these do not necessarily sound in costs. As Jackson LJ said in *Fox*:

> 'In a personal injury action the fact that the claimant has won on some issues and lost on other issues along the way is not normally a reason for depriving the claimant of part of his costs.'

There is also a clear overlap between partial success and conduct in those cases that we have considered under the preceding ground for departure from the general rule.

However, there does not have to be. In *Dickinson, Simmonds, Verley and Moonsam v Tesco PLC, Stewart Alexander Group Ltd, O'Neil and Axa Corporate Solutions Assurances SA* (2013) [2013] EWCA Civ 226, the Court of Appeal was concerned with a dispute over whether to permit retrials in credit hire claims where fresh evidence had come to light. A car hire company had been joined for the purposes of costs determination. Axa contested every issue before the court. Whilst it won on two issues relevant to the retrials these issues did not occupy much court time. The successful party was the car hire company. However, to reflect the partial success of *Axa* the court concluded that the car hire company should recover only 70% of its costs of the appeals.

In our view it is under this heading that the Court of Appeal could have recognised that the claimant in *Marcus v Medway* was the successful party (as he had recovered £2,000), but taken account of the fact that he had lost the substantial argument before the court on causation of the amputation. Whether, by doing so, it would have been sufficient to displace the general rule to the extent that justified an order for costs against the claimant is another matter.

(iii) *Any admissible offer to settle made by a party which is drawn to the court's attention, and which is not an offer to which costs consequences under Part 36 apply*

[22.20]

See the chapter on Part 36 and other admissible offers above. In particular note the comments of Jackson LJ in *Fox v Foundation Piling Ltd (above)* making specific reference to the relevance of such offers:

' . . . parties are quite entitled to make Calderbank offers outside the framework of Part 36. Where a party makes such an offer and then achieves a more advantageous result, the court's discretion is wider. Nevertheless it may well be appropriate to order that party which has optimistically rejected the Calderbank offer to pay all costs since the date when that offer expired.'

THE MENU OF COSTS ORDERS AVAILABLE TO THE COURT TO GIVE EFFECT TO ITS CONCLUSIONS UNDER CPR 44.2–CPR 44.5

[22.21]

CPR 44.2(6) sets out a menu of the various orders the court may make giving a successful party less than the whole of his costs if that is what consideration of CPR 44.2(2)–CPR 44.2(5) demands. In order of practicality CPR 44.2(6)(a) to (f) are in descending order of desirability for the purposes of ease of any assessment of costs. Indeed CPR 44.2(7) states that (f) is to be avoided if it is possible to order either (a) or (c). Why is this? Well whilst it may seem clear to the trial judge that there have been obvious discrete issues in the case, the chances are that those will not be so transparent when costs have to be apportioned to the specific issues. This may not simply be an attempt to load the costs towards recovery. It is just as likely to be because the work done was of generic value to the claim as well as touching on the discrete issue in respect of which costs are recoverable. This was a problem highlighted in the case of *Dyson Technology Ltd v Strutt* [2007] EWHC 1756 (Ch), [2007] 4 Costs LR 597, which confirmed that where the costs of the action are awarded to one party with the exception of costs relating to a particular matter or issue, the party in whose favour the costs of that issue were awarded is not entitled to recover anything except the extra costs generated by that issue and not to costs that were incurred and were equally attributable to other elements of the claim. Patten J (as he then was) expressed his concern with issue based orders as follows:

> 'The CPR make no special provision for dealing with costs of this type and some of the difficulties in the assessment of these costs arise directly from a common failure by judges to appreciate the complexities which can be created by orders which seek to split the responsibility for costs between the parties other than by an order for the payment of a simple percentage or proportion of the total costs bill.'

The difficulties envisaged by Patten J are obvious to those familiar with detailed assessments. The costs judge will have to master the issue in detail to determine what costs were properly incurred in dealing with it. Invariably the costs judge will probably be without the assistance of anyone who was present at the trial to clarify precisely how the trial judge expressed his formulation of the issue and without a transcript of the judgment – particularly as more detailed assessments will be dealt with under the provisional assessment procedure (CPR 47.15). All this adds to the cost of detailed assessment and to the amount of time absorbed in dealing with costs on this basis. The costs incurred on assessment may then not be proportionate to the benefit gained. In all the circumstances, contrary to what may seem at trial to be thought to be the case, a 'percentage' order under CPR 44.2(6)(a) made by the judge will often produce a fairer result than a 'distinct parts' order under CPR 44.2(6)(f). Moreover such an order is consistent with the amended overriding objective of the CPR to deal with cases proportionately.

In *Verrecchia (t/a Freightmaster Commercials) v Metropolitan Police Comr* [2002] EWCA Civ 605, [2002], the court expressly emphasised that the CPR places an obligation on the court to make an issue based order which allows or disallows costs by reference to certain issues only if other forms of order cannot be made which sufficiently reflect the justice of the case (CPR 44.2(7)). This was the approach taken in *Budgen v Andrew Gardner Partnership* [2002] EWCA Civ 1125, (2002) Times, 9 September, where the trial judge ordered the defendant to pay only 75% of the claimant's costs of the action because the claimant had lost on one issue, which had taken up a substantial amount of the trial. The defendant's appeal against the judge's refusal to make a 'distinct parts' order in its favour was dismissed.

The court will probably start with (a) and try *not* to work its way down the list. A recent illustration of this approach was that adopted by Mann J in *Sycamore Bidco Ltd v Breslin and Dawson* [2103] EWHC 583 (Ch). This was a claim concerning alleged misrepresentation and breach of warranty in a share purchase transaction. The claimant failed on misrepresentation, but was successful on breach of warranty. Mann J concluded that the claimant had lost on significant matters, those matters had led to a vast amount of evidence and it was appropriate to reflect this by a costs deduction. He was satisfied that this could be treated as a separate issue and reflected in an issue based costs order. However, in pursuit of a proportionate costs order the correct order was a percentage one and he awarded the claimant 60% of its costs.

This methodology was also endorsed in a personal injury context by Swift J in *Wilson v GP Haden t/a Clyne Farm Centre* [2013] EWHC 1211 (QB). Where the claimant had been successful overall, but had been unsuccessful on a causation issue, it was appropriate to order a percentage reduction in costs rather than to make an issue-based order. Swift J identified the pitfalls of an issue based costs order considered above, stating:

> 'It seems to me therefore, that in the circumstances of this case it would be right for me to depart to some extent from the usual rule that costs follow the event. Having said that, I must decide how that is to be done. I am not attracted by the defendant's submission that it should be done by way of requiring the costs judge to separate out all those costs relating to the issue of impact attenuation. It seems to me that such a course is likely to involve an element of complexity and undoubtedly would add to the costs.'

So in most cases a 'distinct parts' order is not to be made. Wherever practicable, the judge should endeavour to form a view as to the percentage of costs to which the winning party should be entitled, or alternatively, whether justice would be sufficiently done by awarding costs from or until a particular date only, as suggested by CPR 44.2(6)(c).

Conclusion

[22.22]

If CPR 44.2 offers a clear template to the way in which the court will approach the award of costs between the parties, except in the specific situations considered below, then the comments of Jackson J (as he then was) in *Multiplex Constructions (UK) Ltd v Cleveland Bridge UK Ltd* [2008] EHWC 2280 (TCC), 122 ConLR 88, [2009] 1 Costs LR 55 act as a useful step by step guide. There were eight steps. We have taken out one of the steps as it dates back to the time when *Carver* was good law before the change to CPR 36.14A and the decision of Ramsey J in *Hammersmatch Properties Welwyn Ltd v Saint Gobain Ceramics & Plastics Ltd* (see Chapter 20 Costs Inducements to Settle – Part 36 offers and Other Admissible Offers, above). Here are the remaining seven steps:

- The party which ends up receiving payment should generally be characterised as the overall successful party in respect of the entire action.
- The starting point is the general rule that the successful party is entitled to an order for costs.
- The judge must then consider what departures are required from that starting point, having regard to all the circumstances of the case.
- Where the circumstances of the case require an issue-based costs order, that is what the judge should make. However, the judge should hesitate before doing so, because of the practical difficulties which this causes and because of the steer given by CPR 44.2(7).
- In many cases the judge can and should reflect the relative success of the parties on different issues by making a proportionate costs order.

- In considering the circumstances of the case the judge will have regard not only to any part 36 offers made but also to each party's approach to negotiations (insofar as admissible) and general conduct of the litigation.
- In assessing a proportionate costs order the judge should consider what costs are referable to each issue and what costs are common to several issues. It will often be reasonable for the overall winner to recover not only the costs specific to the issues which he has won but also the common costs.

COSTS ISSUES FOR THE TRIAL JUDGE IN THE AWARD OF COSTS OR FOR THE COSTS JUDGE ON ASSESSMENT

[22.23]

Aaron v Shelton [2004] EWHC 1162 (QB), [2004] 3 All ER 561, [2004] NLJR 853 was a case much loved by those charged with assessing costs. This was because it could be relied upon to head off at the pass some challenging arguments relating to the receiving party's conduct (the sort we have already considered under 'Award of costs' above). The court held that where a losing party considers that he should not be liable to pay the whole of the costs of the claim by reason of the receiving party's conduct, he should make an application to the trial judge, raising this argument when the court is considering what orders as to costs should be made under CPR 44.2.

In fact there was also an authority suggesting quite the reverse, that of *Shirley v Caswell* [2000] Lloyd's Rep PN 955, (2000) Independent, 24 July, CA. In this case the claimant recovered about 1/12th of the value of the claim and had abandoned certain items of claim. The trial judge made a split costs order – 60% of the claimant's costs against the defendant and 40% of the defendant's costs against the claimant. The Court of Appeal concluded that:

> 'In the light of his findings the judge had been entitled to conclude in the instant case that a special order for costs in favour of the defendant was justified. However where, as here, the successful party had pursued issues which were later abandoned but had incurred costs in the process, such costs were not to be deducted by the trial judge from the costs recoverable by the successful party. The costs of abandoned issues were prima facie to be disallowed by a costs judge as costs unnecessarily incurred upon detailed assessment. To adopt the approach taken by the judge below would result in the successful party being penalised twice for pursuing unnecessary issues.'

In other words make one costs order and let the assessing judge determine whether costs were reasonably incurred or not. Apart from seeking to deter the making of multiple costs order, the rationale seemed to be that the costs judge would be in a far better position to make a determination, given the additional information that would be available to him on a detailed assessment. The main reason that assessing judges disliked this decision was because the assumptions behind it were flawed. These assumptions were that the assessing judge would know what the trial judge had said and the reasons why he felt an inquiry was better held at assessment. However, invariably those who attend to conduct assessments have not attended the trial of the claim, the order of the trial judge is usually silent as to the specific concerns of the trial judge or his particular findings on facts/issues that go to conduct, there is no transcript of the trial to enlighten the costs judge and those appearing before him have different notes of what the trial judge may have said! The overriding concern of the costs judge under *Shirley v Caswell* was that he would have to undertake a mini re-trial on conduct at the assessment.

To be fair the major concern of the Court of Appeal, as set out in the final sentence above, was to avoid a party being doubly penalised.

The Court of Appeal clarified the position in *Ultraframe (UK) Ltd v Fielding* [2006] EWCA Civ 1660, [2007] 2 All ER 983, (2007) Times, 8 January. It was held that the principle stated in *Aaron v Shelton* was too unqualified in application. Where the paying party had not sought an order from the judge reflecting the misconduct (in this case dishonest) of the receiving party in the award of costs, that failure should not deprive the paying party from referring to it on the assessment of costs, or prevent the assessing judge from considering whether the costs incurred by the dishonest party were reasonable. Indeed consideration of a party's conduct should normally take place both when the trial judge was considering what order for costs he should make, and then when the costs judge was assessing costs under the award. The court would want to ensure that dishonesty/misconduct was penalised, but that the dishonest party was not placed in double jeopardy. Ultimately, the key is one of the proper construction of the order for costs made by the trial judge.

In *Ultraframe* the court concluded that:

- The consideration of conduct should take place both under CPR 44.2 and CPR 44.4 (which is, after all, no more than these rules provide)
- The court must be astute to avoid double jeopardy.
- As a result the trial judge making the order should consider the effect of this order on the subsequent assessment of amount
- The trial judge ought to help the assessing judge understand the residual task left to him when considering conduct under CPR 44.4
- Judges may wish to consider whether to make an order under CPR 44.11 before they make the costs order!

As is apparent it is not, therefore, in those cases where conduct is not taken before the trial judge that problems really emerge. *Ultraframe* is clear that the issue may still be raised before the trial judge. The more challenging position is where conduct is raised before the trial judge and he either makes an order that discounts to reflect the conduct or decides that the conduct does not merit any departure from the general rule that the successful party recovers its costs (CPR 44.2). In both the assessing judge must ensure there is no double penalty – which is particularly hard where no deduction has been made and the costs judge may not even know that conduct was argued before the trial judge as the order will simply be for one party to pay the costs of another party.

Accordingly, it is important for the judge who is asked to take conduct into account at the end of the trial when considering the order as to costs, to consider what is likely to occur on the assessment. Where conduct is being reflected in an order made by the trial judge, it must be wise for him to make clear, ideally on the face of the order, whether he is making the order on the basis that on the assessment the paying party would still be entitled to raise the conduct to argue that costs incurred in supporting the particular conduct were unreasonably incurred. Where the trial judge concludes that there is no conduct that justifies a departure from the general rule, it would be sensible for that to be recited in the order. However, the parties must also be vigilant to ensure that the intended effect of the trial judge's order is clear, rather than leaving the burden exclusively to him. Otherwise the parties are storing up an argument for a later date and may well find, as anecdotally we know has happened, that the subsequent detailed assessment is adjourned for clarification to be sought from the trial judge or for a transcript of the costs judgment of the trial judge to be obtained.

In *Three Rivers District Council v Governor and Company of the Bank of England* [2006] EWHC 816 (Comm) the trial judge not only awarded indemnity costs to the defendants but offered to give the costs judge such assistance as he reasonably could, including answering his written questions and sitting with him on the assessment if necessary.

It is fair to say that following *Northstar* there still seemed to be some confusion about how the issue of conduct might be addressed by both the trial judge and the costs judge. This was clarified in *Drew v Whitbread plc* [2010] EWCA Civ 53, [2010] 1 WLR 1725, [2010] NLJR 270. This claim was allocated to the multi-track but the claimant only recovered damages well within the limits of the fast track. The judge awarded him his costs to be assessed on the standard basis. At the assessment, the district judge ruled that the claimant could never have recovered the damages he was claiming, and that the claim should be treated as if it were a fast track one in order to ensure that costs were proportionate. The claimant appealed, contending that the district judge did not have the power to impose fast track costs, as in effect that rescinded the award of standard basis costs by the trial judge and was a retrospective re-allocation. The Court of Appeal held that the district judge had not been entitled simply to rule that she was going to assess the costs of trial as if the case were on the fast track. That would be to rescind the trial judge's order. The permissible approach was to assess the costs on the standard basis taking into account that the case ought to have been allocated to the fast track. A costs judge is entitled, as part of the process of assessment, to hold that a case should, if reasonably presented, have been allocated to the fast track, and to assess costs accordingly. The fact this point was not raised at trial did not preclude it. The approach in *Aaron v Shelton* was too narrow, and was disapproved. There might be some points which could not be raised at an assessment, because it would in effect require the costs judge to re-try the case. But that did not mean there was a general rule that a failure to raise a matter at trial for the purposes of CPR 44.2 precluded the raising of the matter at assessment for CPR 44.4 purposes.

Waller LJ set out the link between the two procedural provisions as follows:

'In my view 44.3 and 44.5 [as were] are intended to work in harmony and it is intended that the parties' conduct (for example) may have to be considered under both. If what is sought is a special order as to costs which a costs judge should follow that obviously should be sought from the trial judge. If it is clear that a costs judge would be assisted in the assessment of costs by some indication from the trial judge about the way in which a trial has been conducted, a request for that indication should be sought'

Earlier he had also set out how, in practice, the trial judge and costs judge might both consider conduct without that necessarily representing a risk of double penalty and scotched, once and for all, the *Aaron* argument that failure to raise conduct at trial precluded it being raised at assessment:

> 'On the face of the two provisions, in fulfilling their different functions, the trial judge under 44.3 and the costs judge under 44.5 are enjoined to take into account many similar factors. That may mean that if a factor has been raised before the trial judge and the trial judge has ruled on that factor, that will bind the costs judge but (and it is important to emphasise this) more often than not the costs judge has material which the trial judge did not have, and thus will not be bound. But the notion that if a party has not raised a matter under 44.3 he should be precluded from raising it under 44.5 does not sit easily with the express provisions.'

In other words provided that the costs judge considers the factors at CPR 44.4 as part of the exercise of determining what costs have been reasonably incurred and are reasonable in amount (and now by final cross check against proportionality), that will not undermine the trial judge's costs order. An example arose in *O'Beirne v Hudson* [2010] EWCA Civ 52, where the consent order provided: 'The defendant do pay the claimant's reasonable costs and disbursements on the standard basis, to be subject to detailed assessment if not agreed.' The defendant argued that had the matter proceeded to allocation it would have been allocated to the small claims track and therefore it was liable for only fixed costs under CPR Part 27. The Court of Appeal held that although a costs judge cannot vary a costs order he could exercise his discretion in considering whether costs were reasonably incurred, and whether it was reasonable for the paying party to pay more than would have been recoverable in a case that would have been allocated to the small claims track. A costs judge cannot simply apply small claims track costs, but this would still be a highly material circumstance in considering what by way of assessment was reasonably incurred and reasonable in amount. Under CPR 44.4(1) the costs judge is required to take into account all of the circumstances of the case and that includes the fact that the case would almost certainly been allocated to the small claims track.

COSTS OF PRE-ACTION DISCLOSURE APPLICATIONS

[22.24]

Section 51(1) of the Senior Courts Act 1981 provides that the costs 'of and incidental to all proceedings' shall be in the discretion of the court. *McGlinn v Waltham Contractors Ltd* [2005] EWHC 1419 (TCC), [2005] 3 All ER 1126, 102 ConLR 111, held that costs incurred in complying with any pre-action protocol are capable of being costs 'incidental to' any proceedings which are subsequently commenced. However, only in exceptional circumstances could costs incurred by a defendant at the stage of a pre-action protocol, in dealing with and responding to issues which were subsequently dropped from the action when the proceedings were commenced be costs 'incidental to' those proceedings. It would be contrary to the whole purpose of the pre-action protocols if claiming parties were routinely penalised if they decided not to pursue claims in court which they had originally included in their protocol claim letters. The whole purpose of a pre-action protocol procedure is to narrow issues and to allow a prospective defendant, wherever possible, to demonstrate to a prospective claimant that a particular claim is doomed to failure. It would be wrong in principle to penalise a claimant for abandoning claims which the defendant has demonstrated are not going to succeed because to do so would be to penalise the claimant for doing the very thing which the protocol was designed to achieve. Even so, the defendant in *McGlinn* was £20,000 out of pocket as a result of complying with the protocol and responding to claims which did not subsequently form part of the proceedings.

Citation plc v Ellis Whittam Ltd [2012] EWHC 764 (QB), [2012] NLJR 504, [2012] 5 Costs LR 826, found it to be settled law that:

(1) if no claim form is issued, then there is no litigation and so there are no costs of litigation for a defendant, whatever costs might have been incurred in complying with a pre-action protocol; however

(2) if a claim form is issued, the costs incurred in complying with a pre-action protocol might be recoverable as costs incidental to any subsequent proceedings

By contrast pre-action disclosure is one area of a pre-proceedings skirmish between parties that comes with its own specific costs provisions. Notwithstanding this anyone considering the rule and observing the standard orders that are routinely made up and down the country on a daily basis may ponder the relevance of the CPR provisions.

CPR 46.1 provides that the general rule is that the court will award the party against whom disclosure is sought his costs of:

(a) The application; and

(b) Compliance with the order for disclosure.

We hazard a guess that the vast majority of pre-action disclosure orders made (virtually always by consent) contain a provision that the party giving disclosure pays the costs of the party applying for that disclosure and is utterly silent on the costs of compliance. This may be because the general rule is followed by an exception. This is that the court may make a different order having regard to all the circumstances, including the extent to which it was reasonable to oppose the application and whether the parties have complied with any pre-action protocol governing the claim. As the majority of these applications are made in personal injury claims where protocol disclosure has not been provided within the prescribed time limits, this may go some way to explaining the myriad departures from the general rule. However, even then case law suggests that these routine departures from the general rule may be a step too far.

In *SES Contracting Ltd v UK Coal plc* [2007] EWCA Civ 791, the claimant made a successful application for pre-action disclosure against the first named defendant under CPR 31.16. The application had been opposed and the judge ordered the first named defendant to pay the costs of the application. The Court of Appeal held the judge had failed to have sufficient regard to the general rule that the respondent to such an application was normally entitled to his costs. There was ample material to justify a departure from the general rule in this case, but not to the extent of ordering the first defendant to pay the whole of the claimant's costs. In the circumstances, the right order was no order as to costs. In the words of Moore-Bick LJ:

> 'If one is starting from the position set out in rule 48.1(2) [now 46.1] one would expect an order of this kind to be made only in a case where it was clearly unreasonable for the respondent to oppose the application or where the manner of his opposition was so unreasonable as to make it appropriate to require him to bear the whole of the parties' costs.'

This seems in keeping with the earlier decision of *Bermuda International Securities Ltd v KMPG (a Firm)* [2001] EWCA Civ 269, 145 Sol Jo LB 70, although the Court of Appeal in *SES Contracting* stated that it did not find this decision of much assistance. In this case KPMG had resisted disclosure 'root and branch', but the departure from the general rule was only to the extent that there should be no order as to costs.

Bermuda International Securities also contained the order that if subsequently substantive proceedings were issued between these parties then the costs of the pre-action disclosure would be costs in the case in those proceedings. Arnold J took this approach a step further in *Microsoft Corporation v Datel Design* [2011] EWHC 1986 (Ch). In that case the pre-action disclosure application had to be adjourned. By the time it returned to court it had been rendered otiose because, amongst other reasons, substantive proceedings had been issued. Arnold J reserved the costs of the pre-action disclosure proceedings to the trial judge in the substantive proceedings.

SMALL CLAIM TRACK COSTS

[22.25]

In the small claims track the court cannot order one party to pay any other anything towards that party's costs, fees and expenses except that which is prescribed in CPR 27.14(2). We shall consider this under the 'fixed costs' part of the next section for the detail. However, one important point to note is that the previous provisions at CPR 27.14(5) and CPR 27.14(6) are no more. They were deleted in the April 2013 amendments to the CPR and do NOT apply to any claim issued after that date. This coupled with the amendment to delete CPR 26.7(3), which had prevented the court from allocating a claim to a track with a lower limit than the value of the claim without the consent of the parties, means the court may now allocate a claim to the small claims track despite the value of the claim and if it does so the parties are subject to the small claim costs regime. Expect to see a significant number of credit hire claims in excess of £10,000 where the only issue is quantum and the only evidence is in the form of generic and ubiquitous basic rate reports allocated to the small claims track (with those same rate reports limited in length by specific order) as to do otherwise may not be a proportionate case management decision.

COSTS FOLLOWING ALLOCATION AND RE-ALLOCATION

[22.26]

Under CPR 46.11 the limitations on costs recovery in the small claims track apply to work done both before and after allocation to that track (save where the allocation is by way of re-allocation, in which case the provisions of CPR 46.13 apply – see below).

In *Lee v Birmingham City Council* [2008] EWCA Civ 891 the secure tenant's solicitors had sent to the defendant a letter of claim in respect of disrepair, invoking the Pre-action Protocol for Housing Disrepair Cases. In response most of the repairs were carried out and the council offered a global sum for damages and costs. It was agreed any action would have been allocated to the small claims track as it was for less than £5,000 and there was no claim for specific performance – it no longer being necessary. However, the Court of Appeal concluded that since the introduction of the protocol it was no longer the position that a claim only began on issue. The protocol included a warning that there was likely to be a costs penalty if a claim was not first pursued in accordance with its terms. This clearly evidenced that its object was to achieve settlement of disrepair claims without recourse to litigation. If the effect of making the claim was to compel the defendant to undertake the required work, then providing that the landlord was liable for the disrepair, the tenant should recover the reasonable costs of achieving that result. Under CPR 46.13 pre-allocation costs are unaffected by subsequent allocation anyway. Accordingly the court's powers are unrestricted in respect of pre-allocation costs and it was perfectly proper to make an order in respect of them under CPR 46.11(2) if to do so was necessary to ensure that the protocol did not operate to prevent recovery of costs reasonably incurred in achieving the repair. We are somewhat surprised that there have not been more reported cases following this, as plainly the ruling affects all cases where there is a pre-action protocol and where partial settlement pre-issue reduces the value of the residual claim below the small claims threshold.

CPR 46.13 provides that where a party has obtained a costs order before allocation to any track, subsequent allocation will not affect the earlier award. This is true even if the subsequent allocation is to the small claims track. Accordingly, if, for example, either party makes an application for summary judgment before the case is allocated and defers allocation pending the outcome of that application, then either party may recover the costs of that application, depending upon the outcome.

If the court allocates a claim and subsequently re-allocates to a different track then the general rule is that any special costs rules that apply to the initial track will apply up to re-allocation and any special rules that apply to the subsequent track apply from the date of re-allocation. As noted before this is a general rule and so, inevitably, comes with an exception – that the court may order otherwise

COSTS ONLY PROCEEDINGS

Procedure

[22.27]

From 3 July, 2000 the Civil Procedure (Amendment No 3) Rules 2000 introduced a useful procedure to be followed where parties to a dispute have reached an agreement on all issues, including which party is to pay the costs, but have failed to agree the amount of the costs where no proceedings have been started. Previously the only possible courses of action were to institute proceedings based upon the agreement seeking an order that the costs be assessed and paid pursuant to the agreement or the issue of substantive proceedings on the basis that in the absence of agreement about costs there was no concluded agreement and the entire dispute remained unresolved.

Under what was CPR 44.12A and is now CPR 46.14 either party may start proceedings by issuing a claim form in accordance with Part 8. The claim form must contain details of the agreement and attach copies of documents that evidence it. Until April 2013 the court had either to make an order for costs or dismiss the claim. This is no longer the case and the court is not now so constrained. Under CPR 46.14(5) the court may make an order for costs. What it does if it does not make an order for costs appears now to be a matter of discretion – it is no longer obliged to dismiss the claim. CPR PD 46, para 9.10 refers the defendant opposing the claim to CPR Part 8 and the requirement to file a witness statement. The court will then give directions, including, if appropriate the possibility of ordering the claim to continue but as a Part 7 claim

CPR PD 46, para 9.12 confirms that this rule does not prevent a party from issuing a claim form under Part 7 or Part 8 to sue on an agreement made in settlement of a dispute where that agreement makes provision for costs, nor from claiming in that case an order for costs or a specified sum in respect of costs. However, the Practice Direction has been altered with effect from 1 April 2013 so that if the sole issue in dispute is the amount of costs, then the CPR 46.14 procedure must be used. Other Part 7 or Part 8 proceedings may only be used whether there are other issues as well.

Summary or detailed assessment

[22.28]

In previous editions it was suggested that in no circumstances should a district judge or costs judge attempt to dispose of the application for an order for costs and then immediately embark upon a summary assessment of the costs in dispute. That is no longer the position. CPR PD 46, para 9.9, whilst retaining a general rule that the costs order under this procedure should be subject to detailed assessment, also states that if the order for assessment is made at a hearing and the court is then in a position to undertake a summary assessment it may do so. As these orders are invariably made as a paper work exercise without a hearing the proviso appears of limited effect. However, given the desire for proportionality we wonder whether parties will now be asking the court to list these applications for hearing to enable the court to undertake a summary assessment. As CPR PD 46, para 9.6 gives the district judge or costs judge jurisdiction 'to hear and decide any issue which may arise in a claim issued under this rule . . . ' there seems no reason, in principle, why the court cannot accede to such a request.

On the assumption that the general rule will apply more often than not (as after all it is the general rule) it is accepted that for bills of modest size or where there is a challenge to just one item then a detailed assessment may be an unnecessarily cumbersome and disproportionate method of resolution even under the provisional assessment scheme. However, it appears anecdotally, that this problem has often been resolved by the parties and the court creatively agreeing some short form of bill and Points of Dispute and Replies to the Points of Dispute dealing with the only issue in dispute.

CPR 46.14 and fixed recoverable costs scheme

[22.29]

Solomon v Cromwell Group; Oliver v Doughty [2011] EWCA Civ 1584, were low value RTA claims which were settled pre-action by the acceptance of Part 36 offers. The claimants obtained orders for assessment under CPR 44.12A (now CPR 46.14) but then put in detailed bills for assessment on the ordinary standard basis. They contended that they were entitled to do so under CPR 36.10(1) and that they had the benefit of deemed costs orders. The defendants denied this, contending that the claimants' entitlement was confined to fixed recoverable costs by virtue of CPR 44.12A(4A). In *Solomon* the claimant succeeded before the district judge but lost on appeal to the circuit judge who held that CPR 36.10 simply had no application, as it only applied where there were 'proceedings'. In *Oliver* the defendant succeeded before the district judge, who held that fixed recoverable costs were a form of costs on the standard basis. The claimants appealed in each case and the *Oliver* appeal was leap-frogged to the Court of Appeal which held:

(1) The judge had been wrong in *Solomon* to decide that there were no 'proceedings' for the purpose of CPR 36.10. The terms of Part 36 as a whole made it clear that steps taken in contemplation of proceedings were to be regarded as 'proceedings' for the purpose of CPR 36.10(1).
(2) However, there could be no deemed costs order under CPR 44.12 in the absence of actual court proceedings. An order for costs could not exist in a vacuum. The contrary view of the costs judge in *Alison Jones v Alcom UK Ltd* (24 February 2011) was not correct.
(3) An assessment in accordance with Part 45 Section II could not properly be regarded as an assessment on the standard basis. Therefore there was a degree of conflict between CPR 36.10(1) and CPR 44.12A(4A).
(4) However, the Rule Committee would not have intended that a claimant in a low value RTA case who accepted a Part 36 offer pre-proceedings should recover costs on the standard basis whereas such a claimant who accepted an offer other than under Part 36 should be limited to fixed recoverable costs. Accordingly the rule that the general gave way to the particular applied, and CPR 44.12A (4A) and Part 45, Section II prevailed.

(5) Although a claimant had the right to start proceedings under either Part 7 or Part 8, it was doubtful that he could recover more than the fixed costs for which Section II of Part 45 provides. It was, though, not necessary to decide that as both claimants had issued under CPR 44.12A.
(6) It is possible to contract out of Part 45 Section II, at least in part, but on a construction of the correspondence in these cases that had not been done.

Both appeals were dismissed accordingly. The situation has not subsequently been clarified by amendment

Basis of costs

[22.30]

CPR PD 46, para 9.4 states that 'unless the court orders otherwise or Section II of Part 45 applies the costs will be treated as being claimed on the standard basis.' This replaces the previous position which required the claimant to state in the claim form whether costs were claimed on the indemnity or standard basis with silence indicating the standard basis.

AGREEMENT IN ISSUED PROCEEDINGS IN RESPECT OF EVERYTHING EXCEPT COSTS

[22.31]

If the parties have agreed who will pay the costs but have not agreed the amount then, of course, they have reached agreement and there is no difficulty about the court ordering a detailed assessment of the costs. But what if the parties are agreed on everything but the incidence of costs: can they ask the court to resolve the issue? The simple answer is yes, but the court position when asked to do so and the ultimate outcome is far from straightforward.

In *Booker Belmont Wholesale Ltd v Ashford Developments Ltd* (18 July 2000, unreported), CA, the parties reached an agreement whereby the fourth party should 'pay such proportion of the [claimant's] costs of the action as the court shall determine and the reasonable costs of the [claimant's] expert . . . '. The judge was held to have been wrong to refuse to make an order determining the proportion of costs to be paid by the fourth party on the grounds that both as a matter of construction and jurisdiction he had no power to do so. As a matter of construction, this was a valid order for costs in a proportion to be fixed by the court. As a matter of jurisdiction, the court always had jurisdiction to decide the costs of an action. The correct approach for the judge would have been to determine the extent to which having regard to the degree of lack of success of the claimant, there should be some reduction in the amount of costs awarded.

The appropriate test is set out in *M v Croydon Borough of London* [2012] EWCA Civ 595. The Court of appeal held that where a claim has been settled, there is a sharp difference between (i) a case where a claimant has been wholly successful whether following a contested hearing or pursuant to a settlement, and (ii) a case where he has only succeeded in part following a contested hearing, or pursuant to a settlement, and (iii) a case where there has been some compromise which does not actually reflect the claimant's claims. While in every case, the allocation of costs will depend on the specific facts, there are some points which can be made about these different types of case.

- In case (i), it is hard to see why the claimant should not recover all his costs, unless there is some good reason to the contrary. Whether pursuant to judgment following a contested hearing, or by virtue of a settlement, the claimant can, at least absent special circumstances, say that he has been vindicated, and, as the successful party, that he should recover his costs.
- In case (ii), when deciding how to allocate liability for costs after a trial, the court will normally determine questions such as how reasonable the claimant was in pursuing the unsuccessful claim, how important it was compared with the successful claim, and how much the costs were increased as a result of the claimant pursuing the unsuccessful claim.
- In case (iii), the court is often unable to gauge whether there is a successful party in any respect, and, if so, who it is. In such cases, therefore, there is an even more powerful argument that the default position should be no order for costs. However, in some such cases, it may well be sensible to look at the underlying claims and inquire whether it was tolerably clear who would have won if the matter had not settled. If it is, then that may well strongly support the contention that the party who would have won did better out of the settlement, and therefore did win.

This approach was confirmed in *Emezie v Secretary of State for the Home Department* [2013] EWCA Civ 733. The Court of Appeal was at pains to stress that the previous test distilled from *R (on the application of Boxall) v Waltham Forest London Borough Council* (2001) 4 CCL Rep 258 had been superseded by *M v Croydon LBC*. The correct starting point is to ask whether the claimant achieved by the settlement what was sought in the proceedings. The Court of Appeal concluded that in this case the answer to that question was yes and substituted an order for costs in the claimant's favour in place of no order for costs.

If there remained any doubt as to whether the court can step in on the question of costs only, it is worth remembering that the original decision in *Fox v Foundation Piling Ltd* was one where the parties had settled all but the costs and invited the court to determine where the liability lay and the court accepted that invitation.

CLAIM AND COUNTERCLAIM AND SET OFF GENERALLY

[22.32]

For those worried that CPR 44.3(9) as was had not followed the rest of that part into CPR 44.2 do not panic. This rule is now accorded a subsection of its own and can be found at CPR 44.12 under the heading of 'Set-off'. In simplistic terms it allows the court to order that the costs assessed against a

22.32 *Chapter 22 Costs Awards Between the Parties*

party who is also entitled to recover costs are to be set off against that entitlement. The alternative is that the court delays the issue of a final costs certificate for the costs to which that party is entitled until he has paid the costs assessed against him.

The rule covers, amongst other situations, the position where the claimant succeeds on the whole or on part in his claim and the defendant succeeds on the whole or on part on his counterclaim. However, as its heading suggests the rule extends to a far more general consideration of setting off respective costs orders.

In the context of cross claims it is infinitely preferable and proportionate for there to be only one order for costs adjusted appropriately by allowing a proportion or some other partial order as to costs in favour of one party rather than making cross-orders for costs in favour of the successful claimant and the successful defendant on his counterclaim.

If that is not possible, it appears that the lengthy, tedious and often unfair investigation of the costs attributable to the claim and those attributable to the counterclaim prescribed in the case of *Medway Oil and Storage Co Ltd v Continental Contractors Ltd* [1929] AC 88, 98 LJKB 148, HL will still apply. Under this approach, if both claim and counterclaim are successful, the counterclaim will have attributed to it only the increase in the costs which it had brought about, so that the result will be that the balance of the costs will almost certainly be in favour of the claimant and be unfair to the defendant. Of course the reverse is true if the claim and counterclaim are both unsuccessful. There the defendant will recover all the costs save those that are attributable to the counterclaim only, for which he will be responsible. *Dyson Technology Ltd v Strutt* [2007] EWHC 1756 (Ch) demonstrated the importance of time records distinguishing between work done on the claim and work done on the counterclaim in cases where *Medway Oils* might apply.

In *Medway Oils*, the court also concluded that unless there was a specific order to this effect, costs could not be apportioned between claim and counterclaim. The case of *Hay v Szterbin* [2010] EWHC 1967 (Ch), [2010] 6 Costs LR 926 concerned the difference between apportionment and division of costs. Under the terms of a Tomlin order settling the claim between the claimant and the third defendant, Green Wright Charlton (a firm) agreed to pay the costs of the claimant's action against it only. The costs judge, when asked to construe this agreement, determined that this meant the firm only had to pay those costs that related exclusively to the claim against it and nothing in respect of any costs common to the claim against all the defendants. The claimant had suggested that those costs could be apportioned between the defendants. On appeal the *Medway* distinction between division and apportionment of costs was confirmed. The third defendant would be liable by a process of division for those common costs that could be attributed exclusively to the claim against it. Items that did not relate solely to the claim against the third defendant were not susceptible to division and were irrecoverable under the order.

The potential for injustice created by *Medway Oils* is clear. Take the example of a road traffic accident in which both parties sustain injury and loss and both blame each other. If there is a claim and counterclaim and ultimately liability is apportioned between them, the defendant will not recover any liability costs (as those arose within the claim anyway) and will have to pay the claimant's costs of those because the claimant issued proceedings first. This creates a race to issue proceedings.

Fortunately this approach is not always followed.

The Court of Appeal in *Burchell v Bullard* [2005] EWCA Civ 358, [2005] NLJR 593, [2005] BLR 330 found that the trial judge had been correct to dismiss this approach because of the difficulty in the preparation of the bill of costs and the enormous complication of the process of detailed assessment in this building dispute. However, he had not found a better solution by resorting to an order that costs followed the event, ordering the defendants to pay the costs of the claim and the claimant to pay the defendant's costs of the counterclaim. In doing so he had fallen into the error of fettering his discretion and not considering what alternatives were available under CPR 44.2(6). The most obvious and frequently most desirable option should be that in CPR 44.2(6)(a) (considered above), namely that of ordering a proportion of the party's costs to be paid. An order for 'Costs following the event' was the general rule and in this kind of litigation, the event was determined by establishing who had written the cheque at the end of the case. In this case the defendants had done so and they therefore were the unsuccessful party. The starting point was that the claimant was entitled to the costs of the proceedings, claim and counterclaim taken together. The specific aspects of conduct identified in CPR 44.2(5) had then to be taken into account when considering a departure from the general rule. This approach resulted in the Court of Appeal ordering the defendants to pay 60% of the claimant's costs of the claim, counterclaim and the Part 20 proceedings and 60% of the claimant's liability to pay the Part 20 defendant's costs.

By departing from the *Medway Oils* approach and instead treating the exercise as one of set off the court is better able to recognise the justice of the situation. A helpful illustration of this was the case of *Square Mile Partnership Ltd v Fitzmaurice McCall Ltd* [2006] EWHC 236 (Ch). The court found that the claimant had to issue proceedings to compel the defendant to admit its claim and had to come to court in order to recover the money it obtained. However, most of the evidence and pre-trial work went to the main issue of the counterclaim upon which the defendant had been successful. Accordingly, both parties could be said to have succeeded on their respective claims in accordance with CPR 44.2. To award the claimant a significant proportion of its costs would not be just, as it would fail to reflect the defendant's successful counterclaim. To give the defendant its costs would be harsh, as the claimant had succeeded in recovering significant sums from the defendant despite losing on the counterclaim. Taking all the factors into account, the appropriate order was no order as to costs.

In *Villa Agencies SPF Ltd v Kestrel Travel Consultancy Ltd* [2012] EWCA Civ 000 the Court of Appeal stripped the vexed issue of costs on claim and counterclaim right back to the general rule at CPR 44.2(2). In this case the claimant sued for outstanding rent and the defendant counterclaimed for the same amount for repair work for which it had paid. The outcome was that there was a substantial deduction applied to the rent to take account of the repair work. The trial judge had decided that the case was 'all about the counterclaim' and made one costs order – namely that the claimant should pay

2/3rds of the defendant's costs. The Court of Appeal regarded the starting point as an error in principle. The claimant had ended up recovering a significant amount of its claim and was plainly the successful party. In reality the case was about whether the defendant owed anything to the claimant and the court had concluded that it did. The starting point was that the claimant was the successful party but there should be departures from the general rule to recognise that the counterclaim occupied most of the trial and the claimant had only succeeded in part. The Court of Appeal awarded the claimant 25% of its costs.

However, sometimes attempting to undertake the set off at the award of costs stage without knowing the respective costs of the parties would, itself, create an injustice. This was the conclusion reached by the court in *Amin v Amin and 17 others (costs)* [2007] EWHC 827 (Ch D) where it decided that where the costs on each side are very large it is not safe to carry out an exercise of setting-off by reference to percentages of one side's costs against the other. Instead, it is appropriate to ascertain a percentage of each side's costs which is payable to the other and leave the parties to turn those percentages into money amounts, whether by agreement or assessment, and to effect set-off at that level. The litigation concerned two actions concerning a partnership dispute and an unfair prejudice petition against a company in which the court decided that overall 35% of each side's total costs were to be attributed to the company action, and 65% to the partnership action. Although it was not appropriate to set off these figures against each other because of the disparity, the parties were of course free to agree to such a set-off should they wish to do so.

Set-off, although not under CPR 44.12, can take place in a broader sense, in the award of the trial judge under CPR 44.2 in connection with success on issues. We have included it here because this offers a contrast to the position in *Amin*. In *The Bank Of Tokyo-Mitsubishi UFJ, Ltd Baskan Gida Sanayi Ve Pazarlama AS* (see above) the parties were aware of their overall costs and the portion of these that was attributable to one very specific issue upon which the successful party had lost. Concessions were also made as to the likely percentage of those costs that would be recovered on an assessment – either under the standard or indemnity basis. Accordingly the trial judge was able to calculate the overall expenditure of the parties on this issue, work out what percentage that represented of the likely cost recovery of the successful party and make an order for costs in its favour but reduced by that percentage with a clear direction to the assessing costs judge to make no further deduction in respect of that issue. Whilst this is a set-off, what the trial judge was in effect doing was a combination of the award of and quantification of the costs of that issue. He would not have been able to undertake the set off without the detailed costs information and concessions before him.

NO WINNER OR LOSER

[22.33]

We have seen above in *Square Mile Partnership Ltd* that the court can conclude that both parties have been successful on their respective claim and counterclaim and make no order for costs by way of a reasoned set-off.

However, that is different from the court determining that neither party had been successful. This was precisely the situation in *Phonographic Performance Ltd v AEI Rediffusion Music Ltd* [1999] 2 All ER 299, CA, an appeal from an order for costs by the Copyright Tribunal, in which the Court of Appeal held that the Tribunal had erred in principle by seeking to find a winner and a loser and that in reality neither side had won. Accordingly, there should be no order for costs.

This was also the outcome in *Verrechia (t/a Freightmaster Commercials) v Metropolitan Police Comr* (above) where the claimant sought damages of £141,500 plus aggravated and exemplary damages under the Torts (Interference with Goods) Act 1977 in respect of 65 items which the police had failed to return to him. The claimant had offered to settle for £98,000 which was followed by a payment into court by the police of the sum of £5,500 in full settlement of the whole of the claimant's claim. The judge awarded damages in the sum of £37,300 plus interest with the result that neither party succeeded to the extent of their prior offers. The judge clearly thought the action had resulted in a 'draw'. Although when the case started there were some 65 items, by the time of closing speeches, there were only 40 items in issue. The appellant won in respect of half in number of these items. There were no criticisms relevant to costs in the conduct of either party. The appellant had won half of his case and lost on the rest. The police, as well as the appellant, had come to court to win that part of the case in which they succeeded. To this extent the judge was entitled to take the view that each side was the winner. Alternatively, it was open to the judge, in the light of the wide powers conferred by the CPR, to conclude that, in any event, the appellant should only have part of his costs as he had been successful in part only of his case, and that the police should have the cost of the part of the case on which they had been successful. On either basis the judge could properly conclude that the proportion of costs which each party should receive was 50% and that the net result was nil when these two percentages were set against each other.

In *Camertown Timber Merchants Ltd v Sidhu* [2011] EWCA Civ 1041, when deciding costs, the judge clearly took account of conduct. The defendant had tried to bolster his claim and the claimant had exaggerated his. Each party had won and lost on various points and both were unsatisfactory witnesses. The judge's decision to make no order as to costs was one within his discretion. The order for costs following such a trial was a paradigm example of the exercise of discretion. It could not be said that the judge misdirected himself or that he exceeded the ambit of his discretion.

However, there is a risk of injustice in making no order for costs. The court must still undertake an assessment of whether there is a successful party and if there is, then follow the provisions of CPR 44.2. In *R (on the application of Mendes) v Southwark London Borough Council* [2009] EWCA Civ 594 the Court of Appeal warned judges against being tempted too readily to adopt the fall-back position of 'no order for costs' where judicial review claims are withdrawn on settlement and confirmed that costs can and should in appropriate cases be awarded in favour of publicly funded claimants where claims are settled outside of court. The judge had been wrong to make no order as to costs where clearly the claimant had strong prospects of success and the decision to bring proceedings had been a reasonable one in the face of the defendant's refusal to withdraw its incorrect decision before proceedings were commenced.

BULLOCK AND SANDERSON ORDERS

[22.34]

These orders arise where an action founded on either contract or tort against two separate defendants is successful against one and unsuccessful against the other. The court has a discretion (see *Mayer v Harte* [1960] 2 All ER 840, CA for the principles upon which such discretion is exercised) to order the unsuccessful defendant to pay the successful defendant's costs. This can be done in one of two ways:

(i) An order that the unsuccessful defendant pay directly to the successful defendant the latter's costs (known as a Sanderson order because it was first made in *Sanderson v Blyth Theatre Co* [1903] 2 KB 533, CA).
(ii) An order that the claimant pay the successful defendant's costs, permitting the claimant to add them to the costs ordered to be paid to him by the unsuccessful defendant (a Bullock order – *Bullock v London General Omnibus Co* [1907] 1 KB 264, CA).

Where the claimant has no means, then, subject to the court's discretion, the Sanderson order (ie (i) above) is the fairer way of dealing with the justice of the case. Neither order is appropriate in favour of a claimant who has sued both defendants because of a doubt as to the law as opposed to the facts, or where the causes of action against each are quite distinct, nor where the respective claims are not alternative or are based upon quite distinct sets of facts.

Irvine v Metropolitan Police Comr [2005] EWCA Civ 129, [2005] 3 Costs LR 380 confirmed that the jurisdiction to make a Sanderson order survived the introduction of the CPR. The exercise of discretion in deciding whether to make such a costs order has to be guided by the overriding objective and CPR Part 44 and it has to be recognised that it is capable of working injustice against a successful defendant. The court had a wide discretion over costs. In determining whether to order an unsuccessful defendant to pay the costs of a successful defendant the relevant factors included whether the claim against the successful defendant had been made 'in the alternative', whether the causes of action had been connected with those on which the claimant had been successful and whether it had been reasonable for the claimant to join and pursue a claim against the successful defendant. Although a significant factor was likely to be whether one defendant blamed another, whether it had been reasonable to join that defendant and pursue the claim would depend on the facts of the case and whether the claimant could in fact sustain such a claim.

In this case although the claimant had succeeded against the first defendant he had failed to establish a sustainable claim against the second defendant and in relation to the third defendant it was only sued in the alternative 15 months after the proceedings had started: the claimant had produced no cogent evidence in support of his claim against it. Accordingly, the judge had been entitled to conclude that although the first defendant should pay the costs of the claimant, the claimant should pay the costs of the two defendants against which his claims in negligence had failed.

Moon v Garrett [2006] EWCA Civ 1121, (2006) Times, 1 September confirmed the decision in *Irvine* (above) that there are no hard and fast rules as to when it is appropriate to make a *Bullock* or a *Sanderson* order. If a claimant has acted reasonably in suing two defendants, it will be harsh if he ends up paying the costs of suing the defendant against whom he has failed. Whether one defendant has blamed another will always be a factor. The fact that claims are not truly alternative does not mean that the court does not have the power to order one defendant to pay the costs of another.

Inevitably only in truly exceptional circumstances will a claimant who has lost against two separate defendants be able to recover the costs of pursuing one defendant from the other. Where claims made by the claimant against both the second and fourth defendants had failed a *Bullock* order was not appropriate and the fourth defendants' costs were ordered to be borne by the claimant. It had been unreasonable for the claimant to pursue the fourth defendants but, even if it had been reasonable, it would be unjust to make the second defendant pay the costs of the unsuccessful pursuit of allegations against the fourth defendant where the second defendant had not blamed the fourth defendant or encouraged pursuit of the fourth defendant by the claimant (*McGlinn v Waltham Contractors Ltd* [2007] EWHC 698 (TCC), 112 ConLR 148).

A final point, solely for aficionados, is that in *Sanderson* although the court approved of the jurisdiction to order an unsuccessful defendant to pay directly the costs of a successful defendant, the court actually made a *Bullock* type order.

STATE FUNDED PARTIES

[22.35]

This has been considered within the section on funding. However, by way of reminder:
- If an order for costs is made in favour of a public funded party there must be a detailed assessment of those costs
- If an order for costs is made against a public funded party the amount of costs payable is also subject to a determination under s 11 of the Access to Justice Act 1999. This section provides that such costs:

> ' . . . shall not exceed the amount (if any) which it is reasonable for him to pay having regard to all the circumstances including –
> (a) the financial resources of all the parties to the proceedings; and
> (b) their conduct in connection with the dispute to which the proceedings relate.'

Given the extremely limited availability of public funding after the last series of cuts to civil legal aid, we suspect that s 11 determinations will soon become obsolete.

CHAPTER 23

WASTED COSTS ORDERS

THE LAW

[23.1]

From 1 October 1991, s 4 of the Courts and Legal Service Act 1990 inserted a new s 51 into the Senior Courts Act 1981 relating to both the High Court and the County courts introducing the concept of 'wasted costs', as follows:

'(6) In any proceedings mentioned in sub-section (1) [in the Court of Appeal, High Court and County court], the court may disallow or, (as the case may be) order the legal or other representative concerned to meet the whole of any wasted costs or such part of them as may be determined in accordance with Rules of Court.
(7) In sub-section (6) 'wasted costs' means any costs incurred by a party: (a) as a result of any improper, unreasonable or negligent act or omission on the part of any legal or other representative or any employee of such a representative; or (b) which in the light of any such act or omission occurring after they were incurred, the court considers it unreasonable to expect that party to pay.'

THE PROCEDURE

[23.2]

The relevant procedure is set out at CPR 46.8. Whenever the court is considering making such an order under s 51 it has to afford the legal representative against whom the order is being considered the opportunity to make written submissions or, if this is what the legal representative prefers, attend a court hearing where this will be considered, before any order is made. This may sound obvious, but the possibility of a wasted costs order may arise at a hearing where the relevant representative is not in attendance (for example counsel is in court and the solicitor is in the firing line) and/or where the representative is not in a position that day to explain his position.

The possibility of the legal representative responding by written submissions is an addition to the previous rule and was introduced in April 2013. Another amendment is that previously the rules suggested that the onus was on the legal representative to 'show cause' why the wasted costs order should not be made. That was not the case. The onus was on the party seeking a wasted costs order (a good reason why the court should be shy of commencing such an enquiry). Under CPR 46.8 there is no such suggestion. The only obligation is on the court to afford the legal representative the opportunity before making any order. Gone, too, are the provisions empowering the court to order an inquiry by and report from a costs judge or district judge and the option of referring the question of wasted costs to a costs judge or district judge, instead of making a wasted costs order (as opposed to referring to such a judge for assessment of amount which is permissible – see below). In other words the court considering the possibility of a wasted costs order should grasp the nettle and, having given the legal representative the opportunity to make submissions (whether written or oral), should decide whether or not to make an order. Clearly the intention is to make the procedure more efficient and proportionate.

Although *Re Wiseman Lee (Solicitors) (wasted costs order) (No 5 of 2000)* [2001] EWCA Crim 707, was a criminal case the principle applies equally to civil matters. The court made a wasted costs order against the defendant solicitors in their absence with permission to make representations against the order by a given date. When no representations were received, the order was drawn up. The solicitors' appeal was allowed on the grounds that a legal representative must be allowed to make representations before a wasted costs order is made against him under what is now CPR 46.8(2).

An application for a wasted costs order may be made by any party or pursuant to CPR PD 46, para 5.3 of the court's own initiative. We have referred above to one good reason that a court should be wary of commencing this process. Another is that if the court ultimately concludes that a wasted costs order is not justified then it will have put in chain a process that has caused the parties to incur costs for no purpose. Who then pays the costs incurred? If the court allows one party take the initiative, then the award of the costs of the process fall to be considered between the parties under CPR 44.2 in the usual way.

CPR PD 46, para 5.2 indicates that an application for an order may be made at any stage up to and including detailed assessment proceedings, although in general, applications are best left until after the trial. Indeed in *FilmLab Systems International Ltd v Pennington* [1994] 4 All ER 673 the court indicated that it '*would rarely be wise or right to seek to obtain such an order until after the trial*'. In *Melchior v Vettivel* [2001] All ER (D) 351 the court made a wasted costs order after the conclusion of the trial and the sealing of the costs order. Indeed even where the proceedings have been settled and stayed under a Tomlin order, there is no need to apply to lift the stay to make the application: see *Wagstaff v Colls* [2003] EWCA Civ 469. This only goes to emphasise that a wasted costs application is genuinely free standing.

It is important to note as a result that a party does not lose the right to apply for wasted costs by waiting until the trial or even the detailed assessment proceedings. The court is not being asked to overturn a previous costs order on such an application. The statutory power to make a wasted costs order is a discrete jurisdiction conveyed on the court.

Despite this clear indication that these applications are best left until trial, many are in respect of specific hearings (eg where one party's representative fails to attend and the hearing has to be adjourned with wasted costs) and relatively straightforward. These can be, and frequently are dealt with at an interim stage. In practice what often happens is the court puts the representatives on notice that this will be considered at the next hearing, by which time no order is required because the representatives have accepted 'fault' and already discharged the other party's wasted costs. That this is an acceptable approach was endorsed by the Court of Appeal in *Ridehalgh v Horsefield* [1994] Ch 205 (see below).

Applications are either to be made under CPR Part 23 or by oral application at any hearing. If the application is made by Part 23 application then the notice and any evidence in support must identify what the legal representative is alleged to have done of failed to do and the costs that the legal representative may be ordered to pay or which are sought against him.

CPR PD 46, para 5.6 gives the court discretion as to how to case manage the procedure to be adopted. The two imperatives are that:
- the process is fair;
- the process is as simple (and proportionate) as possible.

The court should determine the procedure to be followed to meet the requirements of the individual case. In *Re a Barrister (wasted costs order)* [2001] EWCA Crim 1728, [2002] 1 Cr App Rep 207, (2001) Times, 31 July the court gave the following guidance:
- Elaborate pleadings should in general be avoided.
- No formal process of disclosure is appropriate.
- The court could not imagine any circumstances in which the applicant should be permitted to interrogate the respondent lawyer.
- On the other hand, the respondent must be entitled to present a full defence and must be informed of the conduct complained of, the amount claimed and the alleged causal link between the two.
- Hearings should be measured in hours, not days or weeks

If the court does decide to disallow some costs or order the representative to meet some costs the court must set the amount or direct a costs judge or district judge to set the amount (CPR 46.8(3)). The court may also direct that notice of any proceedings for wasted costs and/or any order made in those proceedings is given to the legal representative's client (CPR 46.8(4)).

SATELLITE LITIGATION

[23.3]

As stated above the court must look to resolve these applications in the simplest way possible. This includes ensuring that claims do not become side tracked by tangential litigation. When introduced the wasted costs provisions started as a trickle and then became a steady stream, many of which were aimed at circumventing the limited cost recovery against state funded parties by seeking orders against their representatives. Understandably this was a development that the Court of Appeal was eager to deter. Accordingly, it addressed all aspects of wasted costs applications in *Ridehalgh v Horsefield*, where, in appeals backed by the Bar Council, The Law Society and the Solicitors Indemnity Fund, the Court of Appeal set aside wasted costs orders against two solicitors and a barrister and in the lead case declined to make an order in a case referred to them by a different division of the Court of Appeal. In delivering the judgment of the court, the Master of the Rolls said that while judges must not reject the weapon which Parliament intended to be used for the protection of those injured by the unjustifiable conduct of the other side's lawyers, they must be astute to control what threatened to become a new and costly form of satellite litigation.

Of course this definition of purpose is too restrictive. There is nothing to stop a party seeking a wasted costs order against his own representative as the House of Lords concluded in *Medcalf v Mardell (Weatherill)* [2002] UKHL 27. However, this may cause difficulties with privilege (see below). In this case Lord Bingham re-iterated the concerns of the Court of Appeal in approving the approach of the Privy Council in a New Zealand case, that wasted costs orders should be confined to questions which are apt for summary disposal by the court, such as failures to appear; conduct which leads to an otherwise avoidable step in the proceedings; the prolongation of a hearing by gross repetition or extreme slowness in the presentation of evidence or argument. Such matters can be dealt with summarily on agreed facts or after a brief enquiry. Any hearing to investigate the conduct of a complex action is itself likely to be expensive and time-consuming. Compensating litigating parties who have been put to unnecessary expense was only one of the public interests to be considered. The robust case management that we expect from April 2013 ought to limit the circumstances in which wasted costs may arise once parties and their representatives realise that when the court prescribes a step to be taken by a set date that is not negotiable.

In summary anyone considering applying for a wasted costs order should think twice.

The Practice Direction supplementing CPR 46.8 largely embodies the major findings of the Court of Appeal in *Ridehalgh*. As wasted costs orders are made under the statute and, therefore CPR 6.8 and CPR PD 46, para 5 govern merely the practice and procedure, not the principles, this is an area in which decisions made prior to the April amendments and the introduction of the CPR are still of relevance and assistance.

THE APPROPRIATE THREE STAGE TEST

[23.4]

Section 51(7) of the Senior Courts Act 1981 defines wasted costs as 'any costs incurred by a party - (a) as a result of any improper, unreasonable or negligent act or omission on the part of any legal or other representative or any employee of such a representative . . . .' This is repeated in CPR PD 46, para 5.5 which sets out the limited situation in which the court may make such an order. However, in addition CPR PD 5.5 also provides that the court may only make a wasted costs order:

- if the legal representatives conduct has caused a party to incur unnecessary costs, or has meant that costs incurred by a party prior to the unreasonable or negligent act or omission have been wasted;
- if it is just in all the circumstances to order the legal representative to compensate that party for the whole or part of these costs.

This three stage test was identified by the Court of Appeal in *Re Barrister (wasted costs order) No 1 of 1991*

A striking illustration of the three stage test in action and the residual breadth of the court's discretion is the case of *R (on the application of Hide) v Staffordshire County Council* [2007] EWHC 2441 (Admin), [2007] NLJR 1543. The claimant's solicitor advocate had engaged in behaviour which could properly be regarded as improper, unreasonable and/or negligent. The proceedings were completely unnecessary. They were doomed to failure and a reasonably competent solicitor should have known as much. Inevitably the other party had incurred wholly unnecessary costs. Nevertheless, the judge reached the conclusion that he should not make an order against the solicitor. The reason for that conclusion related to the difficult financial circumstances in which the solicitor advocate found herself. The third stage of the test was engaged. In this case, the evidence was that an order for wasted costs would carry a significant risk of causing the solicitor to become bankrupt. That would be a disproportionate consequence of her unreasonable and negligent conduct in the litigation. On that discrete basis, a wasted costs order was not made.

Having looked at a practical example of the three stage test from *Ridehalgh* let us look at the component parts in more detail.

Stage 1 Improper, unreasonable or negligent conduct

[23.5]

The definitions of improper, unreasonable and negligent are as follows:

> 'Improper' covers, but is not confined to, conduct which would ordinarily be held to justify disbarment, striking off, suspension from practice or other serious professional penalty. It also covers conduct which according to the consensus of professional, including judicial, opinion could be fairly stigmatised as being improper whether it violated the letter of a professional code or not. (Lord Bingham)

> 'Unreasonable' includes conduct which is vexatious, designed to harass the other side rather than advance the resolution of the case and it makes no difference that the conduct is the product of excessive zeal and not improper motive. Legal representatives cannot lend assistance to proceedings which are an abuse of process and they are not entitled to use litigious procedures for purposes for which they are not intended, as by issuing or pursuing proceedings for purposes unconnected with success in the litigation or pursuing a case known to be dishonest nor are they entitled to evade rules intended to safeguard the interests of justice, such as by knowingly failing to make full disclosure on an not on notice application or knowingly conniving with incomplete disclosure of documents. However, conduct is not unreasonable simply because it leads to an unsuccessful result or because other more cautious legal representatives would have acted differently. The acid test is whether the conduct permitted of a reasonable explanation. It is not unreasonable to be optimistic. (Lord Bingham)

> 'Negligent' does not mean conduct which is actionable as a breach of the legal representative's duty to his own client. There is of course no duty of care to the other party. Negligence should be understood in an untechnical way to denote failure to act with the competence reasonably expected of ordinary members of the profession. However, the court firmly discountenanced any suggestion that an applicant for a wasted costs order needed to prove under the negligence head anything less than he would have had to prove in an action for negligence (Lord Bingham).

The court adopted the test in *Ali (Saif) v Sydney Mitchell & Co (a firm)* [1980] AC 198, [1978] 3 All ER 1033, HL: 'advice, acts or omissions in the course of their professional work which no member of the profession who is reasonably well-informed and competent would have given or done or omitted to do'; an error 'such as no reasonably well-informed and competent member of that profession could have made'. This approach was confirmed in *KOO*

Golden East Mongolia v Bank of Nova Scotia [2008] EWHC 1120 (Admin), where the defendant bank claimed a wasted costs order against the solicitors who had acted for the unsuccessful claimant in a claim to trace and recover missing gold. It contended that it was entitled to a wasted costs order because the solicitors' conduct was persistently negligent and unreasonable in making and continuing to have the Central Bank of Mongolia as a party to the action because the solicitors should have appreciated that the bank was immune from suit because of the provisions of the State Immunity Act 1978. In dismissing the claim the court gave the following reasons:

- a bill of costs should have been served before making the application;
- the claimant should have been given a proper opportunity to pay the costs;
- the absence of an application to strike out the claim was hardly consistent with the submission that it was misconceived;
- the suggestion that 'no reasonable solicitor could have been optimistic' was well below the threshold for sufficient negligence or unreasonableness to justify a wasted costs order;
- even a binding authority fatal, or almost fatal, to the client's case might not justify a wasted costs order.

Stage 2 Causation

[23.6]

The causal link must be established by the applicant and failure to do so will result in no award being made. This rarely presents a problem. The misconduct required to satisfy the first stage is such that it almost invariably has inevitably caused unnecessary costs. As most applications for wasted costs are made by the other party there is no difficulty with problems of privilege (to which we shall return below) that would prevent that party evidencing the loss alleged to have been caused. However, the court has been clear that the burden is on the party seeking a wasted costs order. The alleged conduct must, in fact, have led to identifiable wasted costs. In *Brown v Bennett (Wasted Costs)* (No 1) [2002] the defendant contended that the court must be satisfied that there was a real prospect that the applicant would not have incurred all the costs that he did incur if the lawyers had not acted and advised as they did, and if the court were not so satisfied any uncertainty would have to be taken into account when assessing the level of costs or order against the lawyers. The court held that although there was a powerful argument in logic for awarding only a proportion of the total costs on the 'loss of chance' basis, this was not the appropriate approach to take. The court should ask itself whether, on the balance of probabilities, the applicant would have incurred the costs that he claimed from the lawyers if they had not acted or advised as they did.

Stage 3 It is just in all the circumstances to make an order

[23.7]

This is one of the two entirely discretionary elements of the process (the other being at the outset as to whether to embark on the wasted costs proceedings): see below 'two stage discretion'.

TWO STAGE DISCRETION

[23.8]

Jurisdiction to make an order for wasted costs depends upon the exercise of the court's discretion at two stages. These stages are set out at CPR PD 46, para 5.7.

First, when the initial application is made. The court should not embark on the process automatically. The court must be satisfied that it has before it the evidence or other material which, if unanswered, would be likely to lead to a wasted costs order being made. It is at this stage where the conduct complained of arises on an interim application that it may be justifiable not to defer application or, indeed, determination, as the judge at that hearing is likely to be in the best position to decide this. The Court of Appeal has stated that this discretion must be exercised judicially, but judges might, not infrequently, decide that further proceedings were not justified (*Re Freudiana Holdings Ltd* (1995) Times, 4 December, CA).

Second, even if the court is satisfied that legal representatives had acted improperly, unreasonably or negligently so as to waste costs, the court is not bound to make an order, but has to give sustainable reasons for the exercise of its discretion in that way. The Court of Appeal has emphasised that judges should approach their task with caution and where possible consider the applicability of other sanctions of a disciplinary nature.

PROBLEMS WITH PRIVILEGE

[23.9]

As we have already noted in *Medcalf v Mardell (Weatherill)* [2002] UKHL 27, the House of Lords rejected the argument that s 51 conferred no right on a party to seek a wasted costs order against any legal representative other than his own. It did, however, hold that in these circumstances if the party for whom the legal representative was acting refused to waive legal privilege, thereby preventing the practitioner from telling the whole story, the court should not make a wasted costs order unless (a) satisfied there is nothing the practitioner could say, if unconstrained, to resist the order and (b) it is in all the circumstances fair to make the order. However, it was held in *Brown v Bennett*, on a wasted costs application, that it was permissible for the respondent barristers to be asked whether they saw or knew of non-privileged documents, provided that the purpose of the question was not to discover what was in the barristers' briefs or instructions, even though answering the question might reveal the contents of the instructions or brief. That distinction might appear to be a very narrow and technical one. However, it was self-evident that it would be quite impermissible for the barristers to answer the question if the purpose of putting the question was to find out if a particular document was in the instructions, even if the document was open. But, it was by no means self-evident that if the purpose of asking the question was solely to discover whether the document was seen by or known to counsel that the question should not be answered. The question was permissible even if it happened to reveal that open documents, as opposed to privileged documents, were included in counsels' briefs or instructions, provided that no prejudice was thereby caused to the client.

In *Dempsey v Johnstone* [2003] EWCA Civ 1134, the claimant was a publicly funded bankrupt whose claim was dismissed at the trial on the grounds that he had no arguable prospect of success. The defendant sought and obtained a wasted costs order against the claimant's solicitors on the grounds that no reasonably competent legal representative would have continued with the action. The Court of Appeal held that the judge had applied the right test (see below under 'Some unsuccessful cases – negligence') but had come to the wrong conclusion because he took the view that the facts pleaded could not support the claim made as a matter of law. The Court of Appeal disagreed. The statement of claim settled by leading counsel was capable of supporting the claim: it was a matter for evidence. In determining that question, the judge could only come to a conclusion adverse to the solicitors if he had the opportunity of seeing counsel's advice which was privileged. In the absence of waiver of privilege, it could not be inferred that the evaluation of the claim was negligent in the relevant sense within s 51 of the Senior Courts Act 1981.

THREATENING AN APPLICATION OR PUTTING THE OTHER SIDE ON NOTICE?

[23.10]

The threat of a wasted costs order should not be used as a means of intimidation. However, if one side considers that the conduct of the other has been improper, unreasonable or negligent and likely to cause a waste of costs it is not objectionable to alert the other side to that view. There appears to be a fine line between 'threatening' and 'alerting'.

WASTED COSTS AND PUBLIC FUNDING

[23.11]

It is incumbent on courts to bear prominently in mind the peculiar vulnerability of legal representatives acting for assisted persons. It would subvert the benevolent purposes of state funding legislation if such representatives are subject to any unusual personal risk and their advice and conduct is not to be tempered by the knowledge that their client is not their paymaster and so not, in all probability, liable for the costs of the other side. This is clearly a reference to what the Master of the Rolls called a new and costly form of satellite litigation arising out of attempts by successful litigants to obtain costs against counsel and solicitors for unsuccessful state-funded litigants on the grounds that the Legal Aid Board (replaced by the Legal Services Commission and now, in turn the Legal Aid Authority) should have been advised either at the outset or after disclosure that the action had no realistic prospect of success.

WASTED COSTS AND ADVOCACY

[23.12]

Although the legislation intended to encroach upon the traditional immunity of the advocate by subjecting him to the wasted costs jurisdiction, full allowance must be made for the fact that an advocate in court often has to make decisions quickly and under pressure. Mistakes will inevitably be made and things done which the outcome may show with hindsight to have been unwise. Advocacy is more an art than a science: it cannot be conducted according to formulae. It is only when, with all allowances made, an advocate's conduct of court proceedings is quite plainly unjustifiable that it is appropriate to make a wasted costs order against him.

In *Antonelli v Wade Gery Farr (a firm)* [1994] Ch 205, CA the Court of Appeal set aside an order that counsel should pay the costs of one day of a trial that had been made on the grounds that her acceptance of an unseen brief at very short notice was unreasonable and amounted to improper conduct as it was improbable that she had time to grasp properly the issues involved in the matter. She had been unclear about the issues involved, her submissions had been rambling, had contained many embarrassing pauses and she had failed to prepare written submissions when requested by the judge. The judge had failed to take into account para 209 of the Bar Code, known as the 'cab-rank' rule, which in the opinion of the Court of Appeal precluded the barrister from refusing the brief. She did not then know how inadequate her instructions would be, but even if she had known, she would not have been entitled to refuse. Even when the inadequacy of her instructions became only too plain, para 506 of the Bar Code precluded her from returning the brief or withdrawing from the case in such a way or in such circumstances that the client might be unable to find other legal assistance in time to prevent prejudice being suffered by him. There was no reason to think anyone else would have been better placed to conduct the case than this barrister.

The question of an advocate's immunity from suit generally was re-visited by the House of Lords in *Arthur JS Hall & Co (a firm) v Simons* [2002] 1 AC 615, HL when it was decided that advocates no longer enjoy immunity from suit in respect of their conduct of civil and criminal proceedings.

WASTED COSTS WHERE THE SOLICITOR PLACES RELIANCE ON COUNSEL

[23.13]

To what extent is a solicitor able to avoid a wasted costs order by pointing to his instruction of reliance on counsel? Case law seems clear. A solicitor plainly cannot delegate or abrogate his professional responsibility simply by seeking the assistance of counsel. He must supply counsel with appropriate papers and apply his mind to the advice received. However, the more specialist the nature of the advice the more reasonable it is likely to be for him to accept it. In more specific terms the court has supplied a number of answers to the question, some more encouraging to solicitors than others.

23.13 *Chapter 23 Wasted Costs Orders*

The low watermark was *Davy-Chiesman v Davy-Chiesman* [1984] Fam 48, CA, where counsel had advised in writing that the husband should not seek a lump sum payment from his wife payable directly to him as otherwise it would be taken immediately by his trustee in bankruptcy. At a subsequent conference, counsel advised that the husband should seek a lump sum payable to him direct and abandon all his other claims. The court concluded that it ought to have been glaringly apparent to any reasonable solicitor that the only form of relief for which counsel was going to ask fell foul of the fundamental requirement that because of the bankruptcy any capital sum should not go direct to the husband. The duty the solicitor owed to inform the legal aid fund of any change of circumstances was not just to pass on any views expressed by counsel but to consider the effect of any change of circumstances. The solicitor was at that stage guilty of 'a serious dereliction of duty' or 'serious misconduct' and it was not sufficient to absolve him that he acted in accordance with the advice of counsel. The court acknowledged that in many circumstances a solicitor is protected from personal liability if he has acted on the advice of experienced counsel properly instructed. However, the protection to the solicitor is not automatically total. A solicitor is highly-trained and expected to be experienced in his particular fields of law and accordingly he did not and could not abdicate all responsibility whatever by instructing counsel. In this particular case, the court concluded that the solicitor had allowed his own skill and ability to be entirely subordinated to the dominant and forceful personality of counsel. Obviously, the Legal Aid Committee was not to be bombarded with notifications of every minute fluctuation in the estimate of the percentage prospects of success but only when it appeared, or should appear to a reasonable solicitor, that the assisted person no longer had any reasonable chance of success. As a result The Law Society's application that the solicitor should pay both the husband's and wife's costs personally was allowed.

The prime culprit was counsel and yet it was the solicitors who were ordered to pay the costs of both parties. Counsel was not even named in the law report!

By way of contrast is the case of *R v Oxfordshire County Council, ex p Wallace* [1987] NLJ Rep 542, QBD. The applicant commenced judicial review proceedings against the local education authority and the local health authority. On the first day of the hearing he was given leave to discontinue the proceedings against the health authority who at the end of the hearing applied for an order for their costs to be paid personally by the applicant's solicitors.

Counsel for the health authority argued that counsel's advice that the health authority should be joined as a party was based quite clearly on his stating the statutory position inaccurately and his advice that the decision of the health authority had been perverse was patently wrong, or ought to have been recognised as such by anyone with experience in this field, such as the applicant's solicitors. He relied on the decision in *Davy-Chiesman v Davy-Chiesman* that because the solicitors should have realised that there was not an arguable case for judicial review against the health authority they should be personally liable for costs in spite of the advice of counsel. The solicitors argued that they had obtained counsel's specific advice and it was clear and unequivocal on the merits.

It was held that the case could be distinguished on its facts on *Davy-Chiesman*. In that case, counsel propounded a claim which he had earlier advised was bound to fail, as the solicitors in the case knew, and it was not the one for which, on his advice, the client had been granted legal aid. In *R v Oxfordshire County Council*, counsel had been specifically asked to advise against whom the proceedings should begin and the form they should take. A mere mistake or error of judgment is insufficient to attract liability of a solicitor for the costs of the other parties, for which purpose it was necessary to establish gross or serious negligence. Accordingly an order for costs on behalf of the health authority against the solicitors was refused.

The solicitors also emerged unscathed from *Swedac Ltd v Magnet and Southerns plc* [1989] FSR 243, QBD. When striking out a claim the trial judge criticised counsel and his solicitors. However, he refused to order the claimant's solicitors to pay the costs personally. It was held on appeal ([1990] FSR 89, CA) that although the solicitors could not automatically shelter behind counsel, in this particular case they had been justified in relying upon his advice.

In *Locke v Camberwell Health Authority* [1991] 2 Med LR 249, CA, it was held at first instance that the failure of a solicitor to ensure that pertinent information was before counsel advising on the merits in a negligence action constituted a gross dereliction of the solicitor's duty as an officer of the court and made him liable for the defendant's costs, thrown away. In advising on the merits, counsel did not have photocopies of the hospital notes, which it was his duty to request and the solicitor's duty to supply, and as a result his advice was short and inadequate. However, when it was established on appeal that the defendant had not in fact disclosed the relevant hospital notes, counsel and the solicitors were exonerated and the order set aside. The court concluded that as a general rule, a solicitor should be entitled to rely upon the advice of counsel properly instructed, but must not do so blindly and must still exercise independent judgment.

However, the pendulum swung back again in *Count Tolstoy-Miloslavsky v Lord Aldington* [1996] 2 All ER 556, (CA) (after the decision in *Ridehalgh*), where the Court of Appeal, in upholding an order that Count Tolstoy should pay 60% of Lord Aldington's costs of an action brought by Count Tolstoy to set aside on the grounds of fraud the monumental libel damages awarded against him, said that:

> ' . . . although the solicitors relied on fully instructed and very experienced, respected leading counsel who put his name to a statement of claim . . . counsel having extensive knowledge of the background of the case . . . this does not absolve the solicitors exercising their independent judgment nor allow them to close their eyes to the blindingly obvious . . . '.

This notwithstanding that the document the solicitors were expected to have prevented had been signed by both leading and junior counsel. It would seem that independent and courageous judgment is required.

WASTED COSTS AGAINST A BARRISTER FOR HIS ADVICE

[23.14]

In *Brown v Bennett* and *Medcalf v Mardell* (both above), it was submitted that a wasted costs order could only be made against a barrister by virtue of his conduct when actually exercising a right of audience, in other words, when actually conducting a case in court. The basis for this argument was that a wasted costs order could only be made as a result of inappropriate conduct on the part of any 'legal or other representative' which s 51(13) limited to 'any person exercising a right of audience or right to conduct litigation'. As a barrister, unlike a solicitor, did not 'conduct litigation' he could only be liable when he was exercising a right of audience. Rejecting this argument the court held that although it was true that the concept of conducting litigation would in many circumstances be understood to involve the traditional litigation activities of a solicitor, 'a right to conduct litigation' was not defined in the Act and it was quite permissible to give that expression a meaning which was less technical and more vernacular. There was no reason why it should not extend to activities such as drafting or settling of documents and advising on prospects or procedure (in other words undertaking tasks that someone conducting litigation would ordinarily do)

SOME SUCCESSFUL APPLICATIONS

[23.15]

Successful examples we have already considered are the decisions in *Davy-Chiesman* and *Tolstoy-Miloslavsky v Aldington*. Here are a few others to give a flavour to how the court exercises its discretion.

Re a Company (006798 of 1995) [1996] 2 All ER 417, ChD

[23.16]

A solicitor swore an affidavit in support of a winding-up petition, asserting on oath a belief that a company was insolvent on the ground that a debt was owing and that the company was unable to pay its debts as they fell due. There were in fact no grounds upon which a competent solicitor could have reached that view on the material available to him and he was ordered to pay the costs of the company personally.

General Mediteranean Holdings SA v Patel [1999] 3 All ER 673, QBD (Comm Ct)

[23.17]

The defendants denied an allegation of fraud until shortly before the trial when they admitted it. The action was compromised on terms that there be no order for costs, but the claimant obtained a wasted costs order against the defendants' solicitor on the grounds that the solicitor knew the defendants had admitted the existence of the fraud in other proceedings.

Thames Chambers Solicitors v Azad Miah [2013] EWHC 1245 (QB)

[23.18]

The solicitors against whom a wasted costs order was made had represented a client in debt recovery proceedings. They had known on receipt of instructions that their client was bankrupt and had been put on notice that they needed to obtain the trustee in bankruptcy's consent to continuation of the proceedings and failed to do so. The claim was struck out. The court was satisfied that the three stage test was satisfied.

SOME UNSUCCESSFUL APPLICATIONS

[23.19]

And to balance the consideration, a few examples of cases where the court declined to make a wasted costs order. There are more examples of these as they are more informative of the approach that the court takes.

(i) Problems with public funding

Trill v Sacher (No 2) [1992] 40 LS Gaz R 32, CA

[23.20]

Where a statement of claim was struck out for want of prosecution the defendants sought wasted costs orders against the solicitors for the legally – aided claimant. The solicitors' explanation for the delay related to the difficulties experienced in complicated cases such as the present, where the claimant had only a limited legal aid certificate. The solicitors' hands were tied in dealing with the Legal Aid authorities. The slow action by the authorities and by counsel in preparing written opinions had contributed to the delay. An order for wasted costs was set aside.

Re a Solicitor (wasted costs order) [1993] 2 FLR 959, CA

[23.21]

In this case a solicitor who was having difficulty resolving the position regarding his client's Legal Aid certificate had not sought an adjournment until the day before the fixed hearing date, causing expense to the other side and to its witnesses. His failure to warn the court or his opponent in sufficient time was an error of judgment. However, the material now available showed that the Law Society accepted that the solicitor had found himself in an extremely difficult position. It would not be right to go further than to say the solicitor had been guilty of an error of judgment: he had not acted in dereliction of his duty. A wasted costs order against him personally was not warranted.

(ii) Unrealistic time estimates

Re a Barrister (wasted costs order No 4 of 1993) (1995) Times, 21 April, CA

[23.22]

It is important for a judge considering making a wasted costs order, which is a draconian order, to remember that he is removed from the daily demands of practice and to make allowance for difficulties with time estimates. The appellant had accepted a brief for a two-day trial listed at Derby Crown Court immediately prior to another trial in which he was to appear at Nottingham Crown Court. The first trial was late starting and progressed more slowly than anticipated. In granting an adjournment of the second trial due to the unavailability of the barrister, the judge ordered him to pay the consequential wasted costs. In quashing the order, the court said that although the barrister had been over-optimistic in failing to anticipate delays in the first trial, his conduct could not be described as unreasonable.

(iii) Cost benefit analysis

Turner Page Music v Torres Design Associates Ltd (1998) Times, 3 August, CA

[23.23]

The ability of the court to make a wasted costs order can have advantages, but it will be of no advantage if it is going to result in complex proceedings which involve detailed investigation of fact. If the situation involves detailed investigation of fact and indeed allegations of dishonesty then it may well be that the wasted costs procedure is largely inappropriate to cover the situation, save in what would be an exceptional case. If the situation involved breach of a solicitor's professional duty to a client that too might make it unsuited to a summary procedure. The claimant's application for a wasted costs order against the defendant's solicitors was dismissed.

(iv) Oral application or Part 23 application supported by evidence?

Regent Leisuretime Ltd v Skerrett [2006] EWCA Civ 1032

[23.24]

Although an oral application in the course of a hearing is possible pursuant to CPR PD 46, para 5.4(b), that is only likely to be sensible if the scope of the application to the costs said to have been wasted is narrow and clear; for example, if an adjournment is necessary because of a solicitor's or counsel's conduct, as regards the costs thrown away by the adjournment. In this case the scope and nature of the costs claimed was wholly unclear at the time of the hearing before the judge and the Court of Appeal held that as a result he should neither have allowed the application to be made orally nor even considered it at that stage. Instead he should have told the defendant litigants in person that if they wanted to apply for costs they should issue a Part 23 application notice supported by evidence as required by CPR PD 46, para 5.9. There could have then have been either a first stage hearing or, if CPR PD 46, para 5.8 was satisfied, a hearing at which the first stage and, if relevant, the second stage were both considered.

The judge's approach had been wrong and was outside the scope of the admittedly flexible discretion given to him as to how to proceed. He had not had enough material before him to form even a preliminary view, and, indeed, had not formed such a view, that the solicitor had acted improperly, unnecessarily or negligently, and he had no material on which he could form a view as to whether any significant unnecessary costs had been caused to be incurred by reason of the solicitor's conduct. Nor could he save any time or money by proceeding to a first-stage assessment on that day even if he had had before him the necessary material. The judge was plainly wrong to order the defendants' claim for wasted be investigated because he was unable to be satisfied of either CPR PD 46, para 5.7(a)(i) or (ii). On the further information that Court of Appeal had available it was clear that any costs lost by the defendants as litigants in person did not justify the time and expense of an investigation.

(v) Negligence

Dempsey v Johnstone (above under 'Problems with privilege')

[23.25]

In this case the Court of Appeal reviewed authorities after *Ridehalgh* and concluded that the meaning of 'negligence' had not been modified as had been suggested in *Persaud v Persaud* [2003] EWCA Civ 394 where Peter Gibson LJ had suggested that something more than *'negligence'* was required: something akin to 'abuse of process'. Instead, the Court of Appeal concluded that where it is alleged that a legal representative has pursued a hopeless case, the question is whether no reasonably competent legal representative would have continued with the action. Latham LJ stated 'that negligence could be the appropriate word to describe a situation in which it is abundantly plain that the legal representative has failed to appreciate that there is a binding authority fatal to his client's case'. In this case in the absence of seeing counsel's advice, which was privileged, it could not be inferred that the evaluation of the claim was negligent in the relevant sense within s 51 of the Senior Courts Act 1981.

(vi) Causation of loss

Byrne v South Sefton (Merseyside) Health Authority [2001] EWCA Civ 1904

[23.26]

The solicitors had acted for the claimant in connection with his allegations of clinical negligence by the defendant, but ceased to act for the claimant before he instructed other solicitors, who commenced proceedings. When the action was dismissed as out of time the defendant applied for a wasted costs order against the previous solicitors on the grounds that its costs had been incurred as a result of the solicitors' failure to bring proceedings within the limitation period. Section 51(6) of the Senior Courts Act 1981 empowers costs to be awarded against 'a legal or other representative', while s 51(13) defines a legal or other representative as a person who has issued proceedings, exercised rights of audience or performed ancillary functions in relation to the conduct of litigation. The defendant's case against the solicitors rested on the very opposite, namely the fact that they had failed to do any of these things. Furthermore there was no causative link between the conduct of these solicitors and the costs incurred by the defendant. The defendant's costs had not been incurred by the claimant's previous solicitors' failure to act, but by the claimant's current solicitors' decision to bring an action outside the limitation period.

(vii) Wasted costs and a legal representative's duty to the court

B v Pendlebury [2002] EWHC 1797

[23.27]

On an application for wasted costs in a case in which there has been no adjudication of the primary facts, the court should be reluctant to enquire into a state of affairs in which it has no solid foundation from which the process of analysis essential to the wasted costs procedure can proceed. The process and procedure envisaged in the Practice Direction is not readily to be equated with what was involved on a trial of issues. The phrases 'simple and summary as the circumstances will permit' (CPR PD 46, para 5.6) and 'after giving the legal representative an opportunity to give reasons' (now 'make representations') did not lend themselves as appropriate to a disputed trial involving consideration and resolution *of complex* and disputed evidence. The court found it to be axiomatic that a solicitor is bound by the instructions of his client. He is not obliged to act as a filter between the instructions provided by the client and the opposing party.

Quite simply, a solicitor owes no duty to the opposing party although he does, of course, owe such a duty to the court: *Orchard v South Eastern Electricity Board* [1987] QB 565, CA.

This duty to the court was confirmed in *Hallam-Peel & Co v Southwark London Borough Council* [2008] EWCA Civ 1120. A firm of solicitors had not acted unreasonably in raising a new point in possession proceedings, which led to further adjournments and its conduct did not involve a breach of duty to the court. Accordingly, a wasted cost order should not have been made against them and their appeal was allowed.

In *Persaud v Persaud* (above) the court explained that its jurisdiction to make a wasted costs order against a solicitor is founded on breach of the duty owed by the solicitor to the court to perform his duty as an officer of the court.

There is no doubt that the jurisdiction under s 51 of the Senior Courts Act 1981 to make a wasted costs order has now been extended to barristers, but before a wasted costs order can be made against a member of the Bar there must have been a breach of duty to the court by the barrister. It is not enough that the court considers the advocate has being arguing a hopeless case. The litigant is entitled to be heard; to penalise the advocate for presenting his client's case to the court would be contrary to constitutional principles. The position is different if the court concludes that there has been improper time-wasting by the advocate or that the advocate has knowingly lent himself to an abuse of process. However it is relevant to bear in mind that, if a party is raising issues or is taking steps which have no reasonable prospects of success or are scandalous or an abuse of process, both the aggrieved party and the court have powers to remedy the situation by invoking summary remedies – striking out, summary judgment, other peremptory orders etc. The making of a wasted costs order should not be a primary remedy; by definition it only arises once the damage has been done. It is a last resort. There must be something more than negligence for the wasted costs jurisdiction to arise; there must be something akin to an abuse of process if the conduct of the legal representative is to make him liable to a wasted costs order. In the present case the conduct of counsel did not involve any breach of duty to the court, nor was there an abuse of process.

(viii) Disclosure

CMCS Common Market Commercial Services AVV v Taylor [2011] EWHC 324 (Ch)

[23.28]

There is no difference in principle between the ambit of a solicitor's duty in the conduct and supervision of disclosure and the conduct and supervision of redaction of disclosable documents before they are offered for inspection. The wasted costs jurisdiction is compensatory rather than punitive in nature and an applicant is required to establish that they would not have incurred the additional costs if the alleged breach of duty had not occurred. The reality in this case was that the redactions made had cried out for challenge, which was eventually successfully mounted. Had the solicitor made the more complete disclosure incumbent on him, it would have marginally increased the already strong prospects of a successful challenge but would not have led to a challenge being made earlier than it was. It was the client's decision not to comply with an 'unless order' that caused, or at least contributed to, his decision to abandon his defence and he had been debarred by then from defending the claim. It had not been shown that the breach of duty had caused any wasted costs.

(ix) Earlier remedy

[23.29]

In *CMCS* and *Koo Golden East Mongolia (A Body Corporate) v (1) Bank Of Nova Scotia (2) Scotia Capital (Europe) Ltd (3) Central Bank Of Mongolia (t/a Mongolbank)* (both above) the court stressed that instead of waiting and applying retrospectively for a wasted costs order, the aggrieved party should make a prospective application – whether it be for an 'unless order' or to strike out – highlighting in *CMCS* that the particular abuse was one which was always within the ability of the other party to bring to an end by an appropriate application for a strike out or debarring order. Even if a solicitor's conduct did improperly enable his client to continue his claim unreasonably, it would still be a difficult and finely balanced question whether a wasted costs order was appropriate. Where there was an available remedy against the abusive party, the wasted costs jurisdiction was to be treated as a last resort.

(x) **Wasted costs and family proceedings**

[23.30]

There has been a spate of cases in recent years involving wasted costs order applications in care proceedings. In *Re M (A child)* 5 July 2012 (CA) the Local Authority failed at the second stage of the three stage test – being unable to show that the conduct complained of had caused it to incur additional costs. However, in the same case the solicitors were ordered to pay the costs of the guardian in respect of the costs of a hearing. This was by way of concession by the solicitors, but one which the court accepted was well made. In *Re A (wasted costs order)* [2013] EWCA Civ 43 the Court of Appeal found the solicitors conduct to be woefully short of that which was required in pursuing an application for permission to appeal, but again concluded that the conduct was not the cause of any wasted costs

(xi) **Appeals**

[23.31]

Appeals against wasted costs orders or the failure to make them are possible. An appeal against a wasted costs order does not relate only, or indeed primarily, to costs: it relates to the conduct of the solicitor, and accordingly does not fall within the ambit of s 18(1)(f) of the Senior Courts Act 1981 which prohibits appeals against orders for costs only, unless the judge has mis-exercised his discretion (*Wilkinson v Kenny* [1993] 3 All ER 9, CA – which was in fact an appeal not against the wasted costs order but the costs of the application for wasted costs and as such the appeal did fall foul of s 18(1)(f)).

However, that does not mean that such appeals have been encouraged. There have been numerous cautionary statements warning against appeals from judges who have refused to make a wasted costs order. The Court of Appeal stressed in *Persaud* (above) that it will only be in a very rare case that it will interfere with the decision by the judge as to whether or not to make a wasted costs order. The rationale for this is that the judge who has conducted the trial will be fully aware of the conduct of legal representatives in the case before him.

Wall v Lefever [1998] iFCR 605, CA

[23.32]

The wasted costs provisions of s 51 involve the tension between two important public interests. First, lawyers should not be deterred from pursing their clients' interests by fear of incurring a personal liability to their clients' opponents and second litigants should not be financially prejudiced by the unjustifiable conduct of litigation by their or their opponents' lawyers. Before launching an appeal against a refusal to make a wasted costs order by a judge at first instance who had heard the evidence, parties should exercise great care. If the judge concluded that the conduct complained of had not fallen within that proscribed by s 51 an appeal was only justified if some point of principle indicated that the judge's approach had been wholly wrong. Unsurprisingly, given the court used this case to reinforce this message, the Court of Appeal did not find that the judge had been 'wholly wrong'.

TRIBUNALS

[23.33]

From 3 November 2008 the Tribunal Procedure (Upper Tribunal) Rules 2008 and the Tribunal Procedure (First-Tier Tribunal) (Health, Education and Social Care Chamber) Rules 2008 provide that a tribunal shall not make an order in respect of costs other than for wasted costs or if the tribunal considers that a party or its representative has acted unreasonably in bringing, defending or conducting the proceedings. The Upper Tribunal may also make an order in respect of cost in proceedings on appeal from another tribunal, to the extent and in the circumstances that the other tribunal had the power to make an order in respect of costs. Either tribunal may make an order for costs on an application or on its own initiative. The amount of cost to be paid may be ascertained by:
(a) summary assessment;
(b) agreement of a specified sum by the paying person and the person entitled to receive the cost;
(c) assessment as the whole or a specified part of the costs incurred by the receiving person, if not agreed.

Following an order for assessment, the paying person or the receiving person may apply to the High Court in the Upper Tribunal and to the county court in the First-Tier Tribunal for a detailed assessment of costs in accordance with the Civil Procedure Rules 1998 on the standard basis or, if specified in the order, on the indemnity basis.

Wilsons Solicitors v Johnson (2011/ UKEAT/515/10), [2011] ICR D21, EAT. It is necessary for employment judges where wasted costs orders are sought to remind themselves of the Employment Tribunals (Constitution and Rules of Procedure) Regulations 2004, r 48 and the relevant authorities and do so explicitly in their reasoning. As was made clear by the decision of Elias P in *Ratcliffe Duce and Gammer v Binns* (UKEAT/100/08), in applying rule 48 tribunals should apply the principles developed in relation to the equivalent High Court jurisdiction.

The distinction drawn in *Ridehalgh* (see above) between a hopeless case where proper advice absolved a lawyer from responsibility and an abuse of process did not apply to a suggestion of defective preparation or presentation. The only question in such a case was whether, applying ordinary professional standards, a representative had acted negligently. The purpose of a case management hearing is to achieve a final definition of the issues. Because of the solicitors' failures that did not happen and the other party's costs of attendance were wasted. Wasted costs were purely a matter of case management which could not give rise to an issue of law. It was a self – contained issue and there was nothing irrational in the tribunal dealing with it at any convenient point in the proceedings. The solicitors' appeal was dismissed. This confirms the practice that we have referred to above of dealing with wasted costs at an interim stage if there is a discrete issue involved.

The case of *DB Casqueiro (in a matter of wasted costs) v Barclays Bank PLC* UKEAT/0085/12MAA clarified that the employment tribunal has no power to refer the assessment of the amount to be paid under a wasted costs order that it has made to the county court.

CONCLUSION

[23.34]

The best summary of the wasted costs jurisdiction is that set out in *Gray v Going Places Leisure Travel Ltd* [2005] EWCA Civ 189 where the Court of Appeal considered the link between the award of costs and the wasted costs process as follows:

- The making of an order as to who should bear the costs and on what basis, in respect of proceedings which go to trial, are in principle part of the overriding order made by the court at the conclusion of the trial.
- In the absence of at least a good reason to the contrary, the costs of proceedings should be dealt with by the tribunal which determines the issue which disposes of the case immediately after the judgment in disposing of the case
- In principle there is no difference between a costs order against a party and a costs order against a non-party, they are all part of the judicial function involved in disposing of a case
- It is not mandatory that the application for wasted costs be made at the end of the trial. In many cases a party considering an application for a wasted costs order will ask the judge for time to consider whether to make such an application and, even if such an application is made, the normal course is for the court to give directions in relation to the disposal of the application rather than to deal with it straightaway.
- The application for a wasted costs order can be made after the order in relation to the proceedings has been drawn up, although, in the absence of good reason for the delay the court hearing the late application will not necessarily grant it.

CHAPTER 24

THE BASES OF COSTS

INTRODUCTION

[24.1]

When the court is to assess the amount of costs it must do so on either the 'standard' or 'indemnity' basis (CPR 44.3(1)). CPR 44.3(4) provides that where an order for costs to be assessed is silent or, and sadly this does still happen, the court makes an order for assessment other than on the standard or indemnity basis, then the costs will be assessed on the standard basis. As a result of this most orders are actually silent on the basis, which means standard basis assessment.

The relevance of the different bases

[24.2]

Both bases allow costs reasonably incurred and of a reasonable amount. Prior to the CPR the only difference between the two bases was in respect of the burden of proof of reasonableness. On the standard basis any doubt as to reasonableness is to be resolved in favour of the paying party, whilst on the indemnity basis any doubt is to be resolved in favour of the receiving party. As former Chief Taxing Master Matthews succinctly put it: 'If there is no doubt, there is no difference'.

The introduction within the CPR of the concept of proportionality introduced a further distinction between the bases in 1999. This remains, but, since the April 2013 CPR amendments, may now be of seismic importance. As there are still claims within the court process that do not fall within the proportionality regime from April 2013, we shall look at the difference caused by proportionality in respect of both the pre- and post-31 March 2013 rule provisions.

Pre-April 2013

[24.3]

In respect of those claims which were either commenced before 1 April 2013 or commenced after that date in respect of costs incurred at that date, then the previous CPR provision at what was CPR 44.4(2) applies. This provides that on the standard basis the costs must not only be reasonably incurred and reasonable in amount, but also proportionate to the matters in issue and proportionate in amount.

In practice this will often involve the court having to make an initial assessment of overall proportionality as per *Lownds v Home Office* [2002] EWCA Civ 365, [2002] 4 All ER 775, [2002] 1 WLR 2450. If the overall costs

24.3 Chapter 24 The Bases of Costs

are found to be disproportionate by reference to the overriding objective and the factors now found at CPR 44.4(3) with a new eighth factor (last approved or agreed budget), then the court will assess the costs against a test of necessarily incurred and reasonable and proportionate in amount.

The provision in respect of an indemnity basis assessment for these cases is what it remains now – unqualified by reference to proportionality.

So for these cases the proportionality distinction relating to proportionality is that the court may on the standard basis find the overall costs disproportionate at the outset of an assessment and undertake the assessment only allowing necessarily incurred costs. If it finds the overall costs proportionate then the court permits all reasonably incurred costs which are and reasonable and proportionate in amount. This, together with where the benefit of the doubt rests, inevitably makes a difference. Ask any costs judges and we suspect that they will tell you the difference is anything between 10%-15% on average.

Post-31 March 2013

[24.4]

In respect of those costs that are governed by the post-March 2013 provisions the difference between standard and indemnity basis may be significantly more marked. In these cases CPR 44.3(2) inserts this pithy provision:

> 'Where the amount of costs is to assessed on the standard basis, the court will-
>
> (a) only allow costs which are proportionate to the matters in issue. Costs which are disproportionate in amount may be disallowed or reduced even if they were reasonably or necessarily incurred . . . '

Pithy it may be, but of quite extraordinary potency. This means that on a standard basis assessment a costs judge can undertake the assessment allowing what is reasonably incurred and reasonable in amount, add up the total allowed and then determine by reference to the new proportionality test at CPR 44.3(5) that this still leaves a figure that is disproportionate and reduce still further until satisfied that the figure is proportionate. (See Chapter 14 Prospective Costs Control – Proportionality, above)

In contrast there is no test of proportionality on the indemnity basis. As the new approach to proportionality is commonly accepted as likely to see costs reduced on the standard basis to lower levels than under the previous provisions, then the benefit of an indemnity basis assessment increases.

Costs bases and budgets

[24.5]

In those case in which a costs management order has been made by the court then on a standard basis assessment CPR 3.18 requires the court to have regard to the last approved or agreed budget and not to depart from it unless there is good reason to do so. There is no such constraint on an indemnity basis assessment. The case management order does not impact on the assessment of the costs. Whilst we are aware that the court in *Elvantine Full Circle Limited v AMEC Earth & Environment (UK) Ltd* (see above) suggested that the budget would still be the starting point, for the reasons we have already set out above in Chapter 15 Prospective Costs Control – Costs and Case Management we disagree. The budget has been set by an analysis of what costs are reasonable and proportionate per phase of the budget going forward. An indemnity basis assessment cannot be constrained by figures that have been subjected to a proportionality scrutiny when this has no place on such an assessment and where the budget submitted was one of, amongst other matters, an estimate of future proportionate costs.

Conclusion

[24.6]

For the reasons we have set out above the difference between the amount assessed on a standard and indemnity basis assessment for those costs subject to the post-31 March 2013 regime, may now be sizeable. Proportionality lies at the heart of the recent reforms. An indemnity costs order bypasses the assessment of costs by reference to this. Suddenly there may be more than a few percentage points difference in outcome. We certainly expect there to be an increase in the number of applications for such orders.

THE INDEMNITY BASIS – WHEN IS IT APPROPRIATE?

Introduction

[24.7]

With the exception of CPR Part 36, which specifically provides for an order for costs on the indemnity basis in a defined situation (CPR 36.10 – see below) and costs under a contract where that contract specifically provides for costs on an indemnity basis, indemnity costs can only be awarded by the court exercising its discretion under CPR 44.2. It does so by reference to the overriding objective and the factors that it is obliged to consider when deciding what order to make about costs (CPR 44.2(4) and CPR 44.2(5)).

How does the court exercise its discretion?

[24.8]

The discretion to make such an order is wide, indeed so wide that the Court of Appeal has shied away from setting a prescriptive list of circumstances where such an order would be appropriate. In *Excelsior Commercial & Industrial Holdings Ltd v Salisbury Hamer Aspden & Johnston (Costs)* [2002] EWCA Civ 879 Lord Woolf explained why guidance was of limited assistance:

> 'In my judgment it is dangerous for the court to try and add to the requirements of the CPR which are not spelt out in the relevant parts of the CPR. This court can do no more than draw attention to the width of the discretion of the trial judge and re-emphasise the point that has already been made that, before an indemnity order can be made, there must be some conduct or some circumstance which takes the case out of the norm.'

This approach was adopted by Coulson J in *Noorani v Calver* [2009] EWHC 592 (QB)). However, he went a stage further extracting from the authorities to summarise the position as follows:

> 'Indemnity costs are no longer limited to cases where the court wishes to express disapproval of the way in which litigation has been conducted. An order for indemnity costs can be made even when the conduct could not properly be regarded as lacking in moral probity or deserving of moral condemnation. However, such conduct must be unreasonable "to a high degree". "Unreasonable" in this context does not mean merely wrong or misguided in hindsight'

Coulson J then confirmed why specific guidance in this area is so difficult, saying:

> 'In any dispute about the appropriate basis for the assessment of costs, the court must consider each case on its own facts . . . '

What does seem clear is that where the court has made an award of indemnity costs the cases in which it has done so can be divided between those where there has or has not been culpability and abuse of process. This seems to be a distillation of the decisions of the Court of Appeal in the immediate aftermath of the introduction of the CPR

One of the first cases to emerge on indemnity costs was *Reid Minty (a firm) v Taylor* [2001] EWCA Civ 1723, The Court of Appeal determined that a party can be ordered to pay costs on the indemnity basis under CPR 44.2 even though there has been no moral lack of probity or conduct deserving of moral condemnation on its part. The provision specifically included a discretion to decide whether some or all of the costs awarded should be on the standard or indemnity basis. If costs were awarded on the indemnity basis, in many cases there would be some implicit expression of disapproval of the way in which the litigation had been conducted, but that would not necessarily be so in every case. Litigation could be conducted in a way which was unreasonable and which justified an award of costs on the indemnity basis, but which could not properly be regarded as lacking moral probity or deserving moral condemnation. It would not be right, however, that every defendant in every case could put themselves in the way of claiming indemnity costs simply by inviting the claimant at an early stage to give up and pay the defendant's costs (as the defendant had done here). It might be different if the defendant offered to move some way towards the claimant's position and the result was more favourable to the defendant than that.

This was followed by *Kiam v MGN Ltd (No 2)* [2002] EWCA Civ 66. Here the claimant had been awarded £105,000 damages in a jury libel case against which the defendant appealed. Before the appeal was heard the claimant's solicitors wrote 'without prejudice save as to costs' offering to accept £75,000 and to return to the defendant £30,000 plus appropriate interest, an offer which the defendant simply ignored. When the appeal was dismissed the claimant sought costs on the indemnity basis under the court's general discretion under CPR Part 44. The court considered *Reid Minty* and concluded that where litigation was conducted in a way that was unreasonable, even though the conduct could not properly be regarded as lacking moral probity or deserving moral condemnation, CPR 44.2(4) requires the court when deciding what order to make about costs, to have regard to all the circumstances, including any admissible offer to settle made by a party.

The court held it would be a rare case where a refusal of a settlement offer would attract, under CPR Part 44, not merely an adverse order for costs, but an order on the indemnity rather than on the standard basis. Although conduct falling short of misconduct deserving of moral condemnation could be so unreasonable as to justify an order for indemnity costs, such conduct would need to be unreasonable to a high degree, not merely wrong or misguided in hindsight. An indemnity costs order made under Part 44, unlike one made under Part 36, was intended to carry at least some stigma. The court was keen to reiterate that it was not generally appropriate to condemn in indemnity costs those who declined reasonable settlement offers. In the instant case, it was quite impossible to regard the defendant's refusal of the claimant's offer as unreasonable, let alone unreasonable to so pronounced a degree as to merit an award of indemnity costs.

The Court of Appeal revisited the question of conduct in *Excelsior Commercial & Industrial Holdings Ltd v Salisbury Hamer Aspden & Johnson* (see above) Lord Woolf was at pains to stress that 'an indemnity costs order may be justified not only because of the conduct of the parties, but also because of other particular circumstances of the litigation.'

(a) **Culpability and abuse of process**

[24.9]

Traditionally costs on the indemnity basis have only been awarded where there has been some culpability or abuse of process such as:
- deceit or underhandedness by a party;
- abuse of the court's procedure;
- failure to come to court with open hands;
- the making of tenuous and hopeless claims;
- reliance on utterly unjustified defences;
- the introduction and reliance upon voluminous and unnecessary evidence; or
- extraneous motives for the litigation (an example of which is the use of litigation for an ulterior commercial purpose – see *Amoco (UK) Exploration v British American Offshore Limited* [2002] BLR 135 below)

What seems clear is the exercise of the discretion by the court is best considered by reference to specific examples of where the court has made indemnity costs orders. It is one of those instances where it is hard to pinpoint specific conduct, but one knows it when one sees it!

Unreasonableness

[24.10]

National Westminster Bank plc v Rabobank Nederland [2007] EWHC 1742 (Comm) found that the minimum nature of the conduct required to justify an

24.10 *Chapter 24 The Bases of Costs*

order for costs on the indemnity basis was, except in very rare cases, that there had been a significant level of unreasonableness or otherwise inappropriate conduct in its widest sense. This could be pre-litigation conduct or in relation to the commencement or conduct of the litigation itself (mirroring the conduct provisions of CPR 44.2(5)(a)). The conduct must be looked at in the context of the entire litigation and a view taken as to whether the level of unreasonableness or inappropriateness is, in all the circumstances, high enough to engage such an order. In this case the entire underlying foundation of the defendant's core allegation of fraudulent design by senior officials of the claimant bank showed itself from the very commencement of the counterclaim to be deeply flawed and the allegation involved an assumption so improbable as to be far-fetched. The defendants had vigorously pursued claims of dishonesty throughout a long trial despite the fact that such allegations were highly speculative, if not doomed from the start. The defendants had crossed the frontier by conducting litigation in a manner so unreasonable and/or so unsatisfactory as to justify an order for costs against it on the indemnity basis in respect of its counterclaim.

Unreasonable behaviour and underhandedness

[24.11]

In *Franses (Ian) v Al Assad* [2007] EWHC 2442 (Ch), the claimant obtained a freezing order improperly without notice, there were severe procedural flaws (including insufficient documentary evidence and a defective affidavit), and the duty of full and frank disclosure was breached in two respects. The injunction was discharged. The court determined that the award of costs on the indemnity basis will not be justified unless the conduct of the paying party can be said in some respects to have been unreasonable. In this case the decision to make the application without notice was unreasonable, and at least some of the procedural deficiencies were serious matters of which the court should mark its disapproval. These features taken together were sufficient to remove the case from the ordinary run, or to take it 'out of the norm'.

Conduct and hopeless defences

[24.12]

In *Lifeline Gloves Ltd v Richardson* [2005] EWHC 1524 (Ch), the defendants successfully opposed the claimant's application for summary judgment with the costs being reserved. The defendants then withdrew their defence. The claimant was awarded its costs on the indemnity basis on the grounds that the defendants' conduct and the lack of merit in their defence amounted to unreasonable behaviour under CPR 44.2(4) justifying an award of indemnity costs.

Failure to come to court with open hands and misconduct

[24.13]

In *Clarke v Maltby* [2010] EWHC 1856 (QB), the claimant was awarded general damages and damages for her loss of earnings and sought an order that the defendant pay her costs on the indemnity basis in the light of the defendant's conduct at the quantum hearing. The manner in which the case was conducted went far beyond testing the claimant's evidence. Critically, there was simply no support for an allegation of deliberate exaggeration furthermore the counter-schedule had implied serious professional impropriety on the part of the claimant's solicitors, an implication that was unreservedly withdrawn at trial. The claimant was ordered to pay the defendant's costs on the indemnity basis.

Unreasonable conduct and continued pursuit of a hopeless claim

[24.14]

In *Wates Construction Ltd v HGP Greentree Allchurch Evans Ltd* [2005] EWHC 2174 (TCC), the claimant informed the defendant on the day of the trial that it was discontinuing its claim and accepted in accordance with CPR 38.6 that it had to pay the defendant's costs. The defendant contended that the claimant's conduct of the case had been so unreasonable that the court should award the defendant all its costs on the indemnity basis, alternatively on the indemnity basis from the time of exchange of witness statements, when it was apparent that the claimant's claim was hopeless.

The court agreed. The witness statements and in particular an agreement reached between the experts, made it clear beyond doubt that the claimant had no case against the defendant and should have discontinued its claim at that stage. From then on its conduct was so unreasonable that it justified an order for costs on the indemnity basis.

Abuse of court process and misconduct

[24.15]

A v B (No 2) [2007] EWHC 54 (Comm), concluded that provided that it can be established by a successful application for a stay, or an anti-suit injunction, as a remedy for breach of an arbitration or jurisdiction clause, that the breach had caused the innocent party reasonably to incur legal costs, then those costs should normally be recoverable on the indemnity basis. If costs were confined to the standard basis, there would necessarily be part of the successful applicant's costs of the application which had been properly incurred but could not recover by such an order because of the restricted process of assessment.

The unidentified portion of costs would then be a loss which could only be recoverable as damages for breach of the jurisdiction or arbitration agreement, if such a damages claim were permissible. Authority suggested it would not be

24.15 *Chapter 24 The Bases of Costs*

permitted, which would lead to a fundamentally unjust outcome. The court found that there is no policy argument which precludes costs on the indemnity basis in this situation.

In *Re Satellite (2003) Ltd* (17 November 2003, unreported), Ch D, the petitioning creditor served a number of statutory demands on the company seeking sums allegedly due from the company. The company not only denied that the sums were due, but had a cross-claim. Nevertheless, the creditor presented a winding-up petition, which was dismissed. The winding-up jurisdiction should not be used as a means of debt collection. There was no evidence of the company's insolvency or any proper basis for the presentation of a petition. Accordingly, the petitioning creditor was ordered to pay the company's costs on the indemnity basis.

Unjustified defence

[24.16]

Cooper v P&O Stena Line Ltd [1999] 1 Lloyd's Rep 734, QBD (Admiralty Ct).

In an action for personal injuries, an allegation of malingering is a serious allegation of fraud and has to be pleaded. If the defendant had undertaken a proper investigation it was unlikely it would have defended on liability at all and in terms of quantum there never had been sufficient material on which to base the allegation of fraud. The unusual circumstances of the case justified the award of costs on the indemnity basis.

The introduction and reliance upon voluminous and unnecessary evidence

[24.17]

In *Digicel (St Lucia) Ltd (a company registered under the laws of St Lucia) v Cable & Wireless plc* [2010] EWHC 888, Ch D, the claimant had alleged bad faith and conspiracy, which were allegations of serious wrongdoing. The claims were very wide, which meant that the defendant had to respond to allegations that virtually everything it had done was unlawful. The claimant's witness statements had not been confined to evidence of facts within the witnesses own knowledge. The claimant had significantly overstated the quantum of its claim. On the basis of those findings, it was just that the successful defendant should recover its costs, provided those were reasonably incurred and reasonable in amount, without being subject to the possibility that some part of those costs should be disallowed on the grounds of proportionality, as would be the case under the standard basis. There was no injustice in denying the claimant the benefit of an assessment on a proportionate basis when it had showed no interest in proportionality by casting its claim disproportionately wide and requiring the defendant to meet such a claim. The claimant had also forfeited its right to the benefit of the doubt on reasonableness.

Extraneous motive for litigation

[24.18]

In *Amoco (UK) Exploration Co v British American Offshore Ltd* (see above), the claimant sought to avoid a contract which had become unprofitable by putting pressure on the defendant to renegotiate and, when this failed, terminating the contract. The grounds of termination were different to those subsequently put forward at the trial. The claimant lost heavily and was ordered to pay costs on the indemnity basis because it had conducted itself throughout on the basis that its commercial interests took precedence over the rights and wrongs of the matter, and had then sought to justify its stance by reference to a constantly changing case.

Indemnity costs and causation of increased costs

[24.19]

Phoenix Finance Ltd v Federation International de l'Automobile [2002] EWHC 1028 (Ch), held that there is no need for a party when seeking an indemnity costs order to show that the conduct complained of has increased the costs. The question is the reasonableness or otherwise of the conduct and is not dependent upon whether the conduct, whether reasonable or unreasonable, has increased the costs payable. Letters before action are required in all proper cases and if letters before action are not written and if in the event the claimant loses then he can hardly complain if he ends up paying indemnity costs, particularly in the circumstances of this case where the evidence suggested strongly that there was no justification for interim relief in the first place.

Refusal of offers under CPR Part 44

[24.20]

This does seem to be an area of some uncertainty. In *Kiam* above the court was clear that a failure to beat an admissible offer under CPR Part 44 was unlikely in itself to be a reason to make an award of indemnity costs. However, in *Southwark London Borough Council v IBM UK Ltd* [2011] EWHC 653 (TCC) an order for indemnity costs was made when the claimant failed to accept an offer for settlement that was a good one. The court concluded that given the difficulties that the claimant faced it ought to have appreciated there was no good reason why the offer should not be accepted. Accordingly it was fair for the defendant to have costs on the indemnity basis after the offer.

However, as we have seen when considering Part 44 offers in Chapter 20 Costs Inducements to Settle – Part 36 Offers and other Admissible Offers, above, the court has been at pains to stress that these offers should not be given the status of Part 36 offers and the sanctions imposed by Part 36.14 should not be applied to Part 44 offers.

This seems to have informed the decision of the court in *MIOM 1 Ltd v Sea Echo E.N.E. (No 2)* [2011] EWHC 2715 (Admlty), [2012] 1 Lloyd's Rep 140, [2011] NLJR 1596. Here the court observed that CPR Part 61 (which deals with costs in admiralty claims) does not provide for costs on the indemnity basis where a CPR Part 61 offer is successful, whereas CPR Part 36 does provide for such costs when a Part 36 offer was successful. That was a clear indication that the drafters of CPR Part 61 did not intend that indemnity costs should be awarded merely because a Part 61 offer had been successful. In those circumstances, it was not appropriate in a collision action in the Admiralty Division governed by CPR Part 61 to order costs on the indemnity basis merely because an offer had been successful.

More recently Akenhead J in *Igloo Regeneration (GP) Ltd v Powell Williams Partnership (Costs)* [2013] EWHC 1859 TCC, 157 Sol Jo (no 28) 31 made an award of costs against the claimants on the indemnity basis where they declined to accept an offer that was exactly in the terms that they had offered to accept a few days previously, but which had expired. This was against a backdrop where there had been no change of circumstances between the expiry of the claimants' offer and the making of the defendant's offer. We can see that a specific factual context arose here and the order was not made simply on the back of a failure to accept a Part 44 offer

Indemnity costs and public funding

[24.21]

In *Brawley v Marczynski (No 2)* [2002] EWCA Civ 1453, costs were awarded on the indemnity basis to a publicly funded party in order to penalise the losing party's unreasonable conduct in the case. It is no impediment to an award of indemnity costs that the only beneficiaries of penalising the defendant's solicitors would be the claimant's lawyers.

(b) **No culpability or abuse of process**

[24.22]

In *Re Continental Assurance Co of London plc (in liquidation)* (13 June 2001), unreported Park J was concerned with the situation where conduct had definitely increased the costs. He expressly disavowed his disapproval of some of the conduct of the liquidators as being the reason for his award of costs against them being on the indemnity basis. He did take into account what the consequences of the choice would be and where it would be appropriate for the benefit of any doubt to lie. He accepted that there is a prima facie steer in the CPR that in the normal case costs should be awarded on the standard basis so that the benefit of any doubt should go to the paying party. However, this was not a normal, but a wholly exceptional case where it would be wrong and unacceptable for matters to be resolved against the respondents. The liquidators had pursued all issues at full length and right to the end, with the result that the amount of costs incurred must have been enormous. It was appropriate the costs of the successful respondents should be assessed on the indemnity basis so that they, and not the liquidators, received the benefit of any doubts. In other words the order was made looking forwards to the assessment outcome and not back to the conduct.

In future, with parties eager to escape the constraints of a costs budget, it will be interesting to see if a new line of awards on the indemnity basis emerges on the grounds that, although the costs were disproportionate to the matters in issue, there nevertheless were good reasons for incurring them and therefore the award should be on the indemnity basis, to avoid the proportionality assessment undertaken when setting the budget at the detailed assessment. Already the seeds of such arguments have germinated with the decision of HHJ Simon Brown QC in *Mortgage Agency Number Four Limited v Alomo Solicitors (a firm)* [2011] EWHC B22 (Mercantile). This was a case that had been subject to costs management in the pilot scheme in the Mercantile Court at Birmingham Civil Justice Centre. The claimants were entitled to indemnity costs in a respect of some of the claim as a result of a Part 36 offer. They sought the balance of their costs on the same basis. Whilst making it clear that there had been deplorable and unreasonable conduct by the defendant that had led to an unnecessary trial, the judge gave this as a further reason for an award of indemnity costs:

> 'Furthermore, I am satisfied that it is only fair on the Claimants that that should be the case, putting the burden of proof on a detailed assessment on the Defendants to show, if they dare to do so, that the Claimants costs – apparently disproportionately high and in excess of approved budget as they are – are "unreasonable", rather than vice versa i.e. having to prove that their own costs are "reasonable".'

On the opposite side of this coin is that proportionality represents a substantial check on costs and, as the court stressed in *Easyair Ltd (t/a Openair) v Opal Telecom Ltd* [2009] EWHC 779 (Ch), [2009] 6 Costs LR 882 that is not something that should be removed from the consideration of amount of costs lightly.

Indemnity costs under Part 36

[24.23]

Costs on the indemnity basis may also be awarded as the result of a Part 36 offer (see Chapter 20 Costs Inducements to Settle – Part 36 Offers and other Admissible Offers, above).

Indemnity costs refused

[24.24]

Again a review of instances where the court has declined to make an order is informative.

In *John v Price Waterhouse (t/a PricewaterhouseCoopers)* [2002] 1 WLR 953, Ch D, [2001] 34 LS Gaz R 39 the defendant had successfully resisted the claimant's claim arising out of its conduct as auditor of the claimant companies. The defendant sought an order for costs on the indemnity basis on the grounds that as auditor it was entitled to be indemnified out of the assets of the companies against all its costs. The court held that in its role of a litigant the defendant was entitled only to an order for costs on the standard basis.

There was nothing to prevent the auditor from seeking to recover the difference between standard costs and indemnity costs in separate proceedings to enforce the relevant contractual terms.

In *Simms v Law Society* [2005] EWCA Civ 849, the judge fell into error. After making a between the parties order for costs on the indemnity basis against the claimant, the judge rejected the claimant's attack on the proportionality of the Law Society's response to his application to the High Court, which he claimed had generated 'enormous costs', saying the matter could be dealt with on the detailed assessment. On the indemnity basis lack of proportionality is of no relevance and the judge's response was therefore inconsistent with the order he had made. On appeal the order was varied to the standard basis.

In *Zissis v Lukomski* [2006] EWCA Civ 341, the district judge had awarded costs on the indemnity basis on the ground that if parties litigate only as to costs, then it seemed to him that they must bear a greater risk that if they are unsuccessful they will be ordered to pay costs on the indemnity basis. In allowing an appeal against the order, the Court of Appeal held that there is no reason why parties litigating over costs alone should be at any greater risk of an award of indemnity costs than those litigating over other matters.

In *Balmoral Group Ltd v Borealis (UK) Ltd* [2006] EWHC 2531 (Comm), the successful defendant contended that the claimant's behaviour had been so unreasonable both before and during the proceedings, including the presentation of a grossly exaggerated claim, unreasonable failure to make efforts to settle and by the character of the technical evidence adduced, that costs should be awarded on the indemnity basis.

The judge held that justice does not demand that a resounding defeat should always carry with it an award of indemnity costs. The claimant's pre-action activity had not over-stepped the mark and although the claim for damages (loss of profits) was put very high and the explanation given by the claimant rested more on wishful thinking than evidential support, the court was not persuaded that continuing with the claim was so unreasonable that costs should be awarded on the indemnity basis. However, the deficient expert evidence adduced by the claimant had led to unnecessary costs incurred by the defendant and accordingly the costs incurred by the defendant in that respect were awarded on the indemnity basis.

In *(HLB Kidsons (a firm) v Lloyds Underwriters subscribing to Lloyd's Policy No. 621/PKIDOO101* [2007] EWHC 2699 (Comm) the court concluded that a claimant's rejection of a defendant's Part 36 offer (which does not carry the same entitlement as a claimant's Part 36 offer does as to indemnity costs) does not, as of right, take the case out of the norm even though, in light of the outcome of the case, it may prove wrong to reject it. As Gloster J explained:

> 'In my judgment it is not appropriate, in the circumstances of this case, to make an order for indemnity costs. . . . there has been nothing in the conduct of the litigation of the claim by Kidsons that takes their rejection of the Part 36 offer out of the norm. In the light of my judgment, they were wrong to have done so, but no doubt they were advised by their legal advisors not to accept the offer. In a complex piece of commercial litigation of this nature, I cannot say they were unreasonable to have done so.'

Indeed, the simple position is that had the rule makers wished to impose indemnity costs as a matter of course on a claimant who fails to obtain a judgment more advantageous than the defendant's Part 36 offer it could have done so.

Although the claims in *Webster v Ridgeway Foundation School* [2010] EWHC 318 (QB), failed, they were not hopeless, except one in respect of which indemnity costs was awarded. Although a collateral purpose in pursuing the proceedings was to bring the defendant to book, there was nothing improper in pursuing litigation for this purpose. A desire to obtain compensation can often co-exist with a wish to demonstrate that the defendant has been at fault. It is only where an over-zealous claimant engages in unreasonable conduct that indemnity costs are justified.

CONCLUSION

[24.25]

Whilst it may seem a step too far to tread where the Court of Appeal has refused to and offer some guidelines, there are certainly some clear themes that emerge from the case law considered. These are:

- the discretion of the court is broad and the exercise of it is entirely fact sensitive;
- the starting point is that costs on the standard basis is the usual order and therefore to justify an award of indemnity costs requires something out of 'the norm';
- the court must have regard to the factors set out at CPR 44.2(4) and CPR 44.2(5) when considering what costs order to make;
- this means that the court may take into account conduct before the issue of proceedings in reaching its decision (including compliance with pre-action protocols);
- it is not always necessary for there to have been deliberate misconduct, but there must be a substantial level of unreasonableness;
- the mere fact that parties may discontinue all or part of their claims, withdraw defences, be unsuccessful on a summary disposal or at trial does not represent something out of 'the norm' – something more is required;
- similarly a claimant's failure to accept a Part 36 offer or a party's failure to accept a Part 44 offer does not inevitably result in an order for indemnity costs – it depends on the context;
- the intent behind an award of indemnity costs is to ensure that the receiving party does not have to surmount the test of proportionality and that the burden in respect of reasonableness is imposed on the paying party.

THE FUTURE

[24.26]

We have already stated that because the rewards of an indemnity costs order are now greater (avoidance of the strait jacket of a court approved budget etc) we believe that there are likely to be more applications for such an order. We have also suggested that the factors that the court must take into account have not changed and, at first blush, there is no reason to suppose that an increased number of applications will result in an increased number of awards. However, this really depends upon two things:

- How the court chooses to interpret 'unreasonableness' in the light of the amendment of the CPR. In an era where we are told that court orders really are meant to be complied with, will repeated failure to do so be treated as sufficiently unreasonable conduct to merit the visitation of an indemnity costs award? Will being struck out for breach of court orders and being refused relief from sanction suffice or will that become 'the norm'?
- Whether the court will see the consequential constraint on recovery of costs between the parties that follows a costs management order as a reason to remove proportionality (and the budget) from the assessment by making an indemnity costs order.

CHAPTER 25

PAYMENT ON ACCOUNT OF COSTS

[25.1]

The April 2013 amendments to the CPR made a substantial alteration to the position in respect of payments on account of costs pending the quantification of the overall costs liability. Prior to April 2013 the rule was contained in CPR 44.3(8). It stated that the court may make an order for payment of costs on account once it had made a costs order that was to be subject to later assessment of amount. The rule is now found at CPR 44.2(8) and has undergone a shift in emphasis. Now, in such a situation, the court will order the paying party to pay a reasonable sum on account unless there is good reason not to do so. It would appear that *Blackmore v Cummings* [2009] EWCA Civ 1276, [2010] 1 WLR 983 which held there is no legal presumption that a party is entitled to realise the benefit of a costs order in his favour by having an interim payment on account can be consigned to the history books. What was not even to be elevated to a presumption is now mandatory in the absence of good reason.

This provision, when linked with the fact that in multi track claims the likelihood is that there will be a costs management order and so an approved or agreed budget, makes the entire process much simpler. Not only is a payment on account mandatory unless there is a good reason not to order one, but the assessment of the amount has become a far easier exercise because reference can be made to the budgeted figure.

It is interesting to note that in a judgment that covered many of the issues that arise on award of costs and costs management (including whether there should be an order for indemnity basis costs, the relevance of a budget where there is an award of indemnity costs and good reason to depart from a budget) and which ran to 69 paragraphs, Coulson J in *Elvantine Full Circle Ltd v AMEC Earth and Environmental (UK) Ltd* (see above) dealt with payment on account in just twelve lines. The main conclusion brooking no debate:

> 'The defendant is entitled to an interim payment on account of costs. Since, for the reasons that I have given, the costs management order is likely to be the benchmark for the costs to be recovered by the defendant, any interim award of costs should operate on the basis that the defendant's recoverable costs in this case are unlikely to be much less than the costs management order but unlikely to very much greater either.'

Changing the emphasis also has the effect of making a detailed assessment less attractive. One advantage of going to detailed assessment was to defer the date for payment of the costs. If, as seems likely, the court will order something close to the budgeted amount as a payment on account any benefit of delay is lost.

There will still be cases under the pre-April 2013 regime which have not been subject to costs management, any multi track cases where no costs management order has been made (those cases currently exempted under CPR

25.1 *Chapter 25 Payment on Account of Costs*

3.12 and multi track claims where the court decides not to make a costs management order) and fast track claims where no summary assessment is undertaken, where the court may find it harder to determine the amount to be paid on account.

The benchmark case is still *Mars UK Ltd v Teknowledge Ltd* [1999] 2 Costs LR 44 Ch D. Albeit that this was a case governed by the discretionary power to order a payment on account the court identified factors that might still be relevant as 'good reason' not to make an order eg an unsuccessful party's wish to appeal and the relative financial position of each party

Mars is cited to the court regularly on the quantification of the payment on account. Curiously if the submissions are to be believed it is at different times, and depending on whether the paying party or the receiving party is advancing the submission, authority for a payment of a sum as low as 14.4% or as high as 66.66% of the costs claimed. In fact Jacob J ordered the defendants to make an interim payment of £80,000. The total costs were in the order of £550,000. However, the costs award only provided for the receiving party to recover 60% of it costs. The judge assessed that only about 40% of the costs were in fact recoverable. The % awarded really depends on how one does the arithmetic. If one takes £80,000 as a percentage of £550,000 it does equal 14.4%. If one recognises that the costs order was for 60% only then the starting point is £330,000, which equals 24.2%. The way that Jacob J calculated the sum was to start by reducing the £550,000 by 60% (as that took account of his assessment that only about 40% of the costs would be recoverable). This left a figure of £220,000 (although oddly this was said to be about £200,000). The costs award was for 60% of costs and so the starting point was £120,000 of which £80,000 – 66.66% – was awarded. In fact this arithmetical exercise is not particularly relevant because what it shows is the judge undertook his own approximate assessment of what he thought the likely recovery would be and awarded a significant proportion of that sum. Each case will be fact specific. Indeed in *Mars* Jacob J took into account the claimant's pre-action heavy handedness and misconduct during the proceedings which informed his conclusion that he thought it unlikely that on a detailed assessment the claimant would recover more than 40% of the claimed costs.

Days Healthcare UK Ltd v Pihsiang Machinery Manufacturing Co Ltd [2006] EWHC 1444 (QB), raised the interesting issue of the effect on the detailed assessment proceedings where a defendant has failed to comply with an order to make an interim costs payment (in this case of £2 million plus interest). The receiving party claimant applied for an order that unless the payment was made, a final costs certificate should be issued in the full amount it had claimed. The court concluded that quite apart from any specific rule, the court has an inherent jurisdiction to control its own processes sufficiently enough to enable it to make the order sought. CPR 3.1(1) expressly preserves the inherent powers of the court, while CPR 3.1(3)(a) provides that where the court makes an order in the course of its general powers of management, it can do so subject to conditions, including a condition to pay a sum of money into court. However, as points of dispute had been properly served, the court decided that the appropriate order should provide for there still to be an assessment, but that the paying party defendants should not be permitted to participate further unless they made the interim payment.

Conditional orders will never be made if the effect would be to stifle the litigant's access to the court and will usually only be made against a litigant without assets in England and against whom enforcement is likely to be difficult or impossible. This paragraph was considered in *Hextalls (a firm) v Al-Sami* [2009] EWHC 3678 (QB) in which the costs judge described it as 'an unwarranted gloss', but on appeal the High Court judge expressly endorsed it. It will be interesting to see how this will survive the change that is CPR 44.2(8).

The amended CPR maintains the distinction between a payment on account (CPR 44.2(8)) and an interim costs certificate (CPR 47.16). The former, as we have seen, obliges a court that has made a costs award to order a payment on account. The latter may only be requested after the receiving party has requested a detailed assessment. What of the situation where between trial and the award of costs and filing N258 to request a detailed assessment hearing the receiving party wishes to pursue further costs over and above those ordered by the trial judge on an interim basis? In *Dyson Ltd v Hoover Ltd* [2003] EWHC 624 (Ch), the claimant obtained judgment for damages to be assessed for infringement of its registered patent by the defendant. Two weeks before the hearing of an enquiry into damages the claimant accepted a payment into court of £4 million. The payment in was more advantageous than the claimant's own Part 36 offer and the judge awarded the claimant the cost of the enquiry to be assessed on the standard basis. Before the costs had been assessed the claimant applied to another judge for an order under the then CPR 44.3(8) that an amount be paid on account of its costs before they were assessed. The application was opposed on the grounds that because the court had not had the benefit of hearing the full trial or of hearing the enquiry, there was no normal rule applicable and the court was ill-placed to make any order for an interim payment. The court agreed, saying it had to exercise particular caution when invited to make an interim payment in these circumstances, because the judge was blind to the issues between the parties and the sums of money in dispute were considerable. The court stressed that it was important to bear in mind that the costs judge was empowered to issue an interim costs certificate at any time after the receiving party had filed a request for a detailed assessment hearing. This provision operated on the basis that the costs judge's power to order interim costs had been preceded by steps placing him in the position to make an accurate assessment of the likely costs and be able to decide whether an interim payment was appropriate and, if so, for how much it should be.

The April 2013 amendments to the CPR all point to a reduction, if not cessation, of arguments about payments on account. Summary assessment being more proportionate than detailed assessment, will mean fewer cases being deferred for later quantification of costs. The obligation on the court to make an order for payment on account unless there is good reason, shifts the burden to the paying party to satisfy the court why no such order should be made. The easier assessment of the amount by reference to a budget that has been set by phase expenditure that the court has already determined to be reasonable and proportionate means that a significant proportion of costs is likely to be ordered at this stage. Indeed if docketing works as envisaged and there is judicial continuity it will be the same judge deciding whether to undertake a summary assessment or ordering a detailed assessment and the payment on account as the one who set the budget in the first place. This point

25.1 *Chapter 25 Payment on Account of Costs*

is amply illustrated by the decision of HHJ Simon Brown QC in *Slick Seating Systems v Adams* [2013] EWHC 1642 (QB), [2013] 4 Costs LR 576. He had had exclusive involvement in the management of the claim, had entered an interim judgment and was concerned with assessing losses under it. Having done so the judge turned to consider the costs issues that arose. Whilst he ultimately concluded that an award of indemnity costs was appropriate (and duly assessed those summarily), his comments on the link between budget and assessment bear repetition:

> 'By running this case with a costs budget, I approved a budget of a grand total of £359,710.35 pence for doing this case through to trial. In my judgment that budget was proportionate to what was at stake: the £4.4million sum that I have just awarded. The claimants have laudably kept within that budget and have exercised due control over their activities and expenditure in an exemplary fashion. The statement of costs on 13.5.13 (which is today) is favourably compared with the costs estimate of 22.5.12. The form is signed by the partner of the solicitors and a member of the client company as well, Mr Beasley; the grand total is £351,267.35 pence. In my judgment that is a sum which is, looking at each of the phases, within the budget that was set and the claimants are to be commended with controlling the budget throughout this particular period.
>
> That will be the sum that I would award to be paid within 14 days without the need for detailed assessment, detailed assessment becoming otiose . . . '

Part IV

Quantification of costs

CHAPTER 26

FIXED COSTS

INTRODUCTION

[26.1]

Why is it that litigation solicitors always assume that fixed costs are the road to penury? Those dealing with non-contentious work have always worked on fixed or percentage figures, scale rates or some other easily definable and quotable figure from the outset. Before hourly billing became the dominant method, litigation in the county court was also based on scale figures. Whilst they ultimately failed to keep pace with inflation the 'main preparation item' was a fixed sum as far back as the 1950s (when it was £40) to take a case to trial.

Perhaps it is the assumption that the case is bound to reach trial and will be bitterly fought along the way thereby exhausting the fixed fee. But that assumption can be disapproved by a quick look at any litigator's past cases. Most cases settle prior to litigation. Most that litigate settle well before trial. Those that get to a final hearing are a small, single figure percentage of the overall case load of most solicitors and of those many will only be fighting over quantum and not liability. For this reason the fixed fees are not based on the worst case scenario. To do so would be to overcompensate in most cases. But, the fact that the fee expects a settlement along the way should not give you the mindset that cases which do not settle early are simply to be seen as loss makers. They are the roundabouts to the early settlers' swings.

It is difficult to deal with fixed fees if you only have the odd civil litigation case. But if you have a basket, they should be capable of making a profit overall. If you cannot, then you have to look at the way your cases are run. The Portal (see [26.15] below) is challenging traditional providers of legal services with fees which are low by historic standards. They are the pinch point in the transition from hourly rates. But the rates coming into the fast track, as well as those applicable to the ever expanding small claims track, require you to consider your approach in such cases. It is generally a question of doing all, but only, the amount of work required to get the case 'home.' That may well entail revising your view of what actually needs to be done, compared with what has always been done and paid for by the opponent.

To the extent that the fixed fees do not cover your costs, including profit, you must remember that these are only the recoverable costs. There is nothing to stop you from claiming an additional sum from your client. This may be a further fixed figure. It may be a share of your damages via a DBA which is offset by the recoverable fee. It may be a CFA which also caters for a payment from the client. The 'direction of travel' in cases outside the multi-track is clearly towards one of fixing the recoverable element of fees. Better to work out a way for you to embrace such fees rather than adopt a Canute-like posture.

STRUCTURE OF THIS CHAPTER

[26.2]

The rearrangement of the sections of the CPR from April 2013 has brought all of the fixed costs provisions into Part 45 with the exception of those in the small claims track. There are now eight sections to Part 45 (including IIIA) and we deal with them in order before turning to the small claims track. As with the rest of this book, we have not set out the full rules in the Chapter but you will find them, including the tables at the end of this book.

PART 45 SECTION I (CPR 45.1–CPR 45.8)

[26.3]

This section contains the traditional fixed costs regarding the commencement of the claim, entry of judgment and enforcement of judgments. It is the only section where VAT is not recoverable in addition to the fees set out. The amounts set out are those which will apply 'unless the court orders otherwise' (CPR 45.1 and CPR 45.3). It will be rare for a court to order otherwise but an example of a case where this happened is *Amber Construction Services Ltd v London Interspace HG Ltd* [2007] EWHC 3042 (TCC). Mr Justice Akenhead took the view that the defendant had brought the proceedings on its own head by failing to pay an adjudication award in the context of a construction dispute. The claimant had followed the TCC Guide in such circumstances which required an application to be made in conjunction with the issue of proceedings. The application sought to abridge the time for acknowledging service and applying for summary judgment including a witness statement with contractual documents exhibited. It may well be this factor that particularly influenced the judge to allow considerably more than the £100 usually recovered (he awarded over £6,000). It seems debatable that some of the other reasons given, for example the fact that the defendant had argued in correspondence that it had a good defence, and indeed a counterclaim, should weigh heavily on whether fixed costs should be exceeded. If it were so, greater sums would be sought and paid with much greater regularity than does in fact take place.

Commencement costs

[26.4]

The fixed costs on commencement are the figures to be set out on the Claim Form. Relevant claims are those for specified damages over £25 or relate to claims for the recovery of goods or the possession of land (including under the accelerated procedure for assured shorthold tenancies). If a claim does not fall within these categories the words 'to be assessed' are to be entered on the claim form instead (CPR PD 45, para 1.3).

For damages claims the fixed costs are set out in Table 1 (CPR 45.2) and range between £50 and £100. Costs for possession or demotion claims are set out in Table 3 (other than under the accelerated procedure). They are £69.50 regardless of value.

In either case, there are additional costs for personal service of a defendant (£10/£7.50) and £15 for personal service on any further defendants. It is not obvious why the figures are not the same for dealing with this documentation and its service. The sums involved hardly justify this differentiation.

Judgment costs

[26.5]

Table 2 (CPR 45.4) deals with costs on the entry of judgment in damages claims. There are various possibilities as to the route to judgment – in default of acknowledgment of service or defence; on admission; by summary judgment; or on delivery of goods – and so there is a range of costs recoverable. They also depend on whether the judgment figure exceeds £5,000. The range is between £22 and £85 other than for summary judgment where the figures are £175/£210.

There is no table in respect of possession claims because a standard amount of £57.25 applies to all possession or demotion claims save for the accelerated procedure cases (where a figure of £79.50 applies for both commencement and judgment upon an order for possession being made.)

For all other cases the judgment costs are added to the commencement costs.

Miscellaneous costs

[26.6]

Table 4 (CPR 45.7) sets out what are described as miscellaneous fixed costs but which are in fact costs of service. They cover the situations regarding personal service of a document at any point (£15); service by an alternative method or place approved by the court (£53.25); and service out of the jurisdiction (£68.25 if elsewhere in Great Britain or £77 otherwise).

No reference is made to the fixed costs recoverable in detailed assessment proceedings for a default costs certificate although this might have found a home here. Instead that is referred to in CPR PD 47, para 10.7 where £80 is allowed together with any court fee. Such sum is subject to the case not having been in the small claims track and to the court's power to order otherwise. These figure are specified in the Notice of Commencement.

Enforcement costs

[26.7]

Table 5 (CPR 45.8) sets out the fixed costs recoverable upon various methods of enforcement being used. Without exception they are very unlikely to cover the costs involved, not least where an attendance at court is required to achieve a final order. They are presumably intended to strike a balance between compensating the judgment creditor and not penalising too heavily a possibly already impecunious judgment debtor.

Some of the nomenclature of the enforcement methods has changed – never for the shorter – but the methods have not changed. A third party debt order (previously a garnishee order) and a charging order have similar procedures but have different fixed costs (£98.50 and £110 respectively). The charging order allows for reasonable disbursements as well which reflects the need to register the charging order with HM Land Registry for it to have any effect.

Where the court does all of the work on receipt of the judgment creditor's application, the costs are particularly miserly. Attachment of earnings and warrants of execution are both under £10. There are no costs at all for a writ of execution (formerly a writ of fieri facias) in the High Court where the High Court Enforcement Officer (formerly the Sheriff) levies his own fees on the debtor.

For oral examinations under CPR 71.2, a half hourly rate of £15 is allowed. The questions will usually be posed by a court officer or, if complex by a district judge, but it is not unknown for the judgment creditor to ask questions of a debtor to see where any assets may lie.

Where an award is made elsewhere than in the High or county courts, fixed costs ranging between £30.75 and £75.50 are recoverable for an application to enforce such an award. There are no fewer than four bands here but as the top band is for £2,000 or more, that is presumably the most commonly used band in any event. If a judgment in one part of the United Kingdom is certified and registered in another part under the Civil Jurisdiction and Judgments Act 1982, the costs of that registration may also be claimed.

PART 45 SECTION II (CPR 45.9–CPR 45.15)

[26.8]

This section deals with the costs recoverable in personal injury cases arising out of road traffic accidents. It was the first of the sections in Part 45 which arose out of agreements hammered out by the personal injury industry under the auspices of the Civil Justice Council. It has to some extent been superseded by the implementation of the Portal (see Section III below). Originally any claim which settled pre-proceedings for under £10,000 would be subject to the provisions of this section. Now, many if not most of such claims are settled into the Portal and to which different provisions apply. Therefore in order for this section to apply a claim needs:

- To arise from an RTA occurring on after 6 October 2003
- To comprise a total value of agreed damages of £10,000 or less
- To have fallen out of the Portal
- To have settled pre-proceedings
- Not to involve a litigant in person
- Not to involve an untraced driver
- Not to be within the small claims limit

If a case comes within these criteria there is a self-contained code for calculating the fixed recoverable costs and disbursements as well as a mechanism for the claimant to apply for a higher sum in an appropriate case. Since the case must have settled pre-proceedings, 'costs only' proceedings under CPR 46.14 are required. (This provision used to be CPR 44.12A and some cases refer to CPR 44.12A proceedings as a result). Where a claimant was a child or protected party, an approval hearing would not take a case out of the criteria if the approval hearing was commenced under Part 8 in the usual way.

If the case started in the Portal and fell out after liability was admitted, there may be fees paid by the defendant to the claimant even though a settlement was not achieved. Where this has happened the interim payment needs to be deducted from the fixed costs calculated in accordance with this section (CPR 45.28).

Fixed costs

[26.9]

The solicitors' costs are calculated by reference to the damages received. Care has to be taken to calculate the correct level of damages. Interim payments need to be included. Any contributory negligence needs to be deducted and so too does any CRU deduction.

The formula set out in CPR 45.11 is:
- £800 as a base fee; plus
- 20% of the damages up to £5,000; plus
- 15% of any damages between £5,001 and £10,000
- 12.5% London weighting (see below)
- VAT on these sums

This formula, excluding London weighting, makes for a maximum recoverable sum of £2,550 (£800 + £1,000 + £750) for damages of exactly £10,000.

London weighting

[26.10]

Where the claimant lives or works in London (as defined in CPR PD 45, para 2.6) and the legal representative also works in London, the additional 12.5% can be claimed. This was a hotly contested provision originally. Defendants thought that claimant lawyers would route work through London offices to benefit from the increase. Claimant lawyers thought that clients would find 'advice deserts' if they happened to live in London since solicitors would not deal with these cases without some reflection of the increased costs of practising in London.

Fixed disbursements

[26.11]

Unless a specific application is made to escape the fixed provisions, CPR 45.10 only allows the disbursements set out in CPR 45.11 to be recoverable. This provision was meant to simplify the issue of disbursements and therefore not prevent settlement being held up. However, a combination of factors has meant that often it is the disbursements which have proved the sticking point. The defendants' fear that solicitors work would be repackaged as disbursements started the process. Some defendant negotiators have taken a very robust line with the aid of cases on medical report fees. Both sides have

26.11 *Chapter 26 Fixed Costs*

sometimes taken little notice of what is meant to be recoverable when putting claims forward or for challenging them. PD45 para 2.10 specifically provides for disbursement only proceedings under CPR 46.14 where the parties cannot agree the level of recoverable disbursements.

According to CPR 45.12 the court may allow any of the following disbursements but will not allow any other type. However, the final 'type' is described as 'any other disbursement that has arisen due to a particular feature of the dispute' (CPR 45.12(2)(c)) which leaves a large door open for the parties to disagree about a particular disbursement. The specific disbursements are the costs of obtaining:
- medical records;
- a medical report;
- a police report;
- an engineer's report;
- a DVLA search;
- counsel and court fees for an approval hearing if the claimant is a child or protected party.

Medical fees

[26.12]

In *Stringer v Copley* [2012] Lexis Citation 68 HHJ (Michael) Cook considered the use of medical agencies in routine personal injury cases. He said that it had become common practice to instruct a medical agency to arrange a medical examination of the claimant; undertake the collation and obtaining of relevant medical reports; arrange the appointment with the medical expert and the claimant; deal with any cancellations or rearrangements and deliver the resultant medical report to the solicitors. Because of the specialisation, experience, expertise and contacts of the medical agency they were able to do this administrative work at least as efficiently, expeditiously and economically as most firms of solicitors using their own fee earners. There could be no objection in principle to the fees of the medical agency being recoverable between the parties, provided it could be demonstrated that their charges did not exceed the reasonable and proportionate costs of the work if it had been done by the solicitors. Where the invoice, or 'fee note', from the medical agency showed the medical consultant's fees and their own charges separately, it was possible for the costs officer to assess them both. But where a medical agency charged a composite fee, without differentiating between the amount of the medical consultant's fee and their own charges, the medical fees and the agency's charges were not sufficiently particularized. As such, the costs officer could not be satisfied that they did not exceed the reasonable and proportionate cost of a solicitor's fee earner doing the work.

From a law of costs point of view this could not be faulted. Unfortunately, the practical implications of this (or perhaps simply the reluctance of medical reporting organisations to do as HHJ Cook stipulated) meant that the settlement of many cases was held up by a wait for further information. The same problem arose in relation to medical records which were often obtained by the medical agency for the benefit of the clinician. The decision of Master Hurst in *Woollard v Fowler* ([2005] EWHC 90051 (Costs)) did not resolve matters. The temperature was reduced by an industry agreement reached between many of the liability insurers and the Association of Medical Reporting Organisations ('the AMRO agreement') which set standard figures for reports and which, with subsequent updates, have generally been adopted and allowed by the courts.

Counsel's fees

[26.13]

The formal wording in respect of the approval hearing fees is 'where they are necessarily incurred by reason of one or more of the claimants being a child or protected party as defined in Part 21 – (i) fees payable for instructing counsel; or (ii) court fees payable on an application to the court.' You might think that the very fact that the claimant is a child would 'necessarily' incur the cost of counsel attending the approval hearing so that the judge could approve the settlement. After all, the court fee was necessarily incurred by the making of the application. However, a number of judges at first instance have taken the view that this is not so. A solicitor receives a standard amount of costs for running the case. If he attends the approval hearing, no further costs will be incurred because there is no provision for any such costs to be paid. Therefore, the instruction of counsel is an additional cost and unless the case is a difficult one, it is an unreasonable one to incur. The same argument is not raised about a written advice for the benefit of the court and a fee is usually agreed for this (fixed figure?). It is merely the attendance. The case of *Alex Sherred (a child suing by his mother and litigation friend Denise Sherred) v David Carpenter* (Taunton CC) (5 March 2009) (HHJ O'Malley) is one example of this line of judicial thinking. It does not seem to take much note of the geographical dislocation that often occurs these days in personal injury cases. If the solicitor attends (and somebody obviously has to do so), the cost of him doing so could wipe out a large part of the overall recoverable fee. In practice counsel tends to be sent along and is instructed to seek to persuade the court that the additional fees should be allowed.

Seeking to exceed the fixed recoverable costs

[26.14]

CPR 45.13 contains the optimistic opening that the court 'will entertain a claim for an amount of costs . . . greater than the fixed recoverable costs' but then largely dashes hopes of escape by continuing that it will only do so 'if it considers there are exceptional circumstances' which would make it appropriate to do so. This has proved to be a high hurdle on the relatively rare occasions when an application has been made. The opponent's arguments tend to revolve around 'swings and roundabouts' and the extensive data that was available to the stakeholders and subsequently the Government when this section of Part 45 was brought in. Consequently, it will be a rare case whose facts are so extreme as to take it out of the data set originally used to calculate the figures. Arguments that the defendant has given the claimant 'the run around' and thereby increased the claimant's costs hugely have occasionally been successful but generally the courts have not been hugely entertained by the applications made.

26.14 *Chapter 26 Fixed Costs*

In order to make sure parties think twice before a Part 8 claim is made, CPR 45.13 to CPR 45.15 spell out the costs consequences of such an application. There are two separate hurdles to overcome:
(1) The claimant must persuade the court that exceptional circumstances exist so that it is appropriate to exercise its discretion at all.
If the claimant fails to get past this hurdle, the court must make an order for fixed recoverable costs and disbursements only. The court may make an order for the claimant to pay the costs of the defendant in defending the proceedings.
(2) The claimant must persuade the court to allow increased costs at a figure at least 20% above the fixed figure (excluding VAT).

The assessment may be by summary or detailed assessment. If the claimant fails to achieve the 20% increase threshold, the court must order the defendant to pay the lesser of the fixed recoverable costs or the assessed costs. This provision (CPR 45.14(2)) makes it clear that assessed costs will not necessarily be higher than the fixed costs. As with hurdle 1, if the claimant does not overcome it, the court may make an order for the claimant to pay the defendant's costs of the Part 8 claim.

A little peculiarly, there is no specific provision for the claimant who surmounts both hurdles to be awarded costs. We do not think that is because the drafters assumed no one would ever manage this. It simply implies that costs will follow the event in the usual way and the claimant will receive his costs of the Part 8 claim to be assessed at the hearing.

PART 45 SECTION III (CPR 45.16–CPR 45.29)

[26.15]

The title of this section is 'The Pre-Action Protocols for low value personal injury claims in road traffic accidents and low value personal injury (employers' liability and public liability) claims.' The title gives away two things. The first is that the employers' liability ('EL') and public liability ('PL') claims have been grafted on to the original RTA Protocol making everything more awkwardly drafted than would probably have been the case if the protocol was drafted from scratch. But that could not happen because thousands of personal injury cases were already progressing through 'the Portal' as it is known. (The increase in the value of the scheme and its inclusion of EL and PL accident claims relates to all claims where the Claims Notification Form is sent on or after 31 July 2013. For EL disease claims the trigger point is whether a letter of claim has been sent by 31 July 2013).

The second is that the Protocols, together with the Part 8 procedure documentation run to an inordinate length given that the idea was meant to be a simple and cheap system to deal with low value cases where liability was not in dispute.

The length of the wording is a reflection of the drafters' intention to cover every eventuality. If you imagine for a moment, a cartoon where a wooden box contains two creatures who are experts at escaping from such boxes. Along comes the hero of the cartoon who is charged with keeping these creatures in the box. There then exists a scene where the hero is constantly nailing more and more pieces of wood on to the original box in order to plug gaps being created by the two creatures. Soon the box looks nothing like the original creation and is probably odds on to survive anything short of a nuclear explosion. That, in a way, is the situation here. Both claimant lawyers and defendant insurers have become experts at seeking to gain advantage on the pre-action playing field on which the vast majority of personal injury cases settle. Once the predictable costs regime came into play (see Part 45 Section II above), claimants' lawyers would escape its provisions and commence proceedings where they thought the defendant insurers were dragging their heels. The defendant insurers (or their lawyers) would then dispute the level of costs claimed as being caused by premature issuing of proceedings and so the allowed sums should be limited to those under the predictable costs regime. With this background, it is no surprise that the Protocol would need careful drafting to avoid manipulation by the parties. Thankfully it is outside the scope of this book to dissect the enlarged protocol in terms of procedure. It is sufficient to look at the costs' provisions. But its importance should not be underestimated. The use of online settlement procedures may well be the future sooner than most litigators think.

The three stages

[26.16]

For both protocols there are three stages. The first two occur prior to any court proceedings. The third stage requires a hearing by the court (but nothing else) and therefore is commenced by a Part 8 claim, hence the relevant Practice Direction is 8B.

Stage 1 is all about liability for the accident. The claimant puts forward his case via an online system known as the Portal. If the defendant denies liability (or does not respond) the case falls out of the Portal (see below). If the defendant admits liability, Stage 1 ends and the Stage 1 fee becomes payable. It originally had to be paid within 10 days of the admission but this had the effect of a number of claims stopping at this point which was considered undesirable. Consequently, the Stage 1 fee is now only payable at the end of Stage 2.

Stage 2 is concerned with quantum and negotiation. The claimant obtains medical evidence to support the claim and sends this through to the defendant together with a Part 36 offer. If the defendant does not accept the claimant's valuation, he can put forward a counter offer and there is a short window for negotiation. If settlement is reached the claimant receives his damages. If not, the claimant has to initiate the Stage 3 procedure. In either event, his solicitor now receives the Stage 1 and Stage 2 costs. Even though the case has not concluded, the claimant is at no risk as to costs to this point and therefore the recoverable costs are to be paid over. The cost of the only significant disbursement, the medical report, is met directly by the defendant.

Stage 3 is about obtaining a court determination. There is nothing to prevent the parties negotiating formally after the window in Stage 2 has closed. (A good many cases settle in this way and are catered for within Section IIIA.) But in theory both parties have made their best offers by now. The claimant then completes a court proceedings pack and begins Part 8 proceedings. The court will either give a determination on paper as to the appropriate quantum

26.16 *Chapter 26 Fixed Costs*

or at a hearing if either party wants to take that latter step. Sealed offers are provided to the court as part of the court pack and so the judge is in a position to determine damages and costs on paper or at a hearing. In the first year of the RTA Protocol (as it then was) existing, hardly any cases reached Stage 3.

A word should be said here about claims needing approvals. More is said below because the options are varied. But for the moment, it is worth noting that an approval hearing could come at Stage 2 in order to approve a settlement between the parties or at Stage 3 where it is more akin to an assessment of damages hearing.

The fixed costs

[26.17]

In the Portal, the only recoverable costs allowed are the fixed costs and disbursements set out in this section. There is no opportunity to obtain assessed costs and therefore issues of seeking indemnity costs for the opponent's conduct do not arise. The same is true in relation to Part 36 offers in the Portal and which we deal with in Chapter 20. The recoverable costs are contained in Tables 6 and 6A in the Practice Direction. There are rather more options than might be expected for a three stage procedure. The tables distinguish between RTA on the one hand and EL /PL on the other. They also break at £10,000, which was the original limit for the RTA protocol. These distinctions are predicated on the basis that EL/PL claims are perceived to be more difficult than RTA claims and that added value brings added complexity. There are further complications because the fees for stage 3 are broken down into the uninformative Type A, B and C costs. Type A fixed costs refer to the 'legal representative's costs' for running the third stage. Type B refers to the 'advocate's costs' where a hearing in person takes place and Type C refers to the written advice required for an approval hearing. The tables are set out at the end of the book but they are condensed as follows:

	Stage 1	Stage 2	Stage 3		
			A	B	C
RTA up to £10,000	£200	£300	£250	£250	£150
RTA above £10,000	£200	£600	£250	£250	£150
EL/PL up to £10,000	£300	£600	£250	£250	£150
EL/PL above £10,000	£300	£1300	£250	£250	£150

Where the parties reach agreement after the Part 8 proceedings have been sent to the court, but before they have been issued, the claimant will be entitled to a Type A payment so long as the agreed figure is more than the defendant's formal protocol offer.

'London weighting' applies here as it does in Section II. If the claimant lives or works in London and the legal representative also practises in London, the recoverable costs are increased by 12.5%. This provision does not apply to Type B or C costs in Stage 3. The definition of London is set out in Practice Direction 45.

'Counsel's' advice

[26.18]

Before the Portal was expanded vertically (by increasing the damages to £25,000) and horizontally (to include EL and PL claims), the only use of counsel which could be recovered was for the provision of a written advice and / or representation at an approval hearing or as an advocate at the Stage 3 hearing. The increased value of claims persuaded the rule makers that seeking advice on quantum was appropriate for the larger cases. Consequently CPR 45.23B allows for a Type C costs claim to be made where the damages are over £10,000. This provision looks fraught with difficulty. What happens if the solicitor thinks it is worth more than £10,000 but counsel disagrees? Which is the 'value' of the claim? It would seem fairly clear that the Advice would generally be taken as representing the correct figure, but what if a Stage 3 hearing decides damages are indeed more than £10,000? CPR 45.23B requires the advice to be 'reasonably required' and that will undoubtedly be viewed differently in different quarters. Finally, the advice can be obtained by 'a specialist solicitor or counsel.' Will defendants seek to argue that the solicitor was not a specialist and so the fee is not payable? (Why is it not specialist counsel anyway?) Will local arrangements between solicitors come to pass to earn further income by providing opinions on a reciprocal basis? This provision looks to be something of a sop to the junior Bar but it may simply lead to a lot of work being done and billed but never recovered.

The disbursements

[26.19]

The disbursements which may be recovered are set out at CPR 45.19. They are in very similar terms to those in Section II (**[26.11]**). As with that Section, the rule says that the court may allow certain prescribed disbursements but will not allow any other type. It then finishes with the catch-all phrase that it may allow 'any other disbursement that has arisen due to a particular feature of the dispute' which rather destroys the strength of the earlier wording.

The disbursements which are recoverable are the costs of obtaining medical records and a medical report. As mentioned above, in most cases the defendant has paid for the medical report and associated records directly. This is likely to change, at least for the time being, as EL and PL practitioners can be expected to take a less collaborative approach than some RTA lawyers whose clients have very largely needed standard medical evidence to support soft tissue claims.

Engineer's report fees and those for searching the DVLA and MID databases can be claimed. So too can court fees where an approval is needed, a Part 8 claim has to be issued for Stage 3 or where protective proceedings have been issued to avoid falling foul of limitation. This would be rare in RTA proceedings but may be more common in EL and PL cases. It is an area of potential disagreement, particularly where the defendant could agree to extend the limitation period rather than have proceedings issued to safeguard the claimant's position.

Offers

[26.20]

These are compulsory in the Portal in order to encourage settlement and have to be made before the Stage 3 procedure can be invoked. If they are not made, then the case falls out of the Portal. They are dealt with along with other Part 36 offers in Chapter 20.

Avoidance behaviour

[26.21]

If the defendant wishes the case to exit the Portal he has numerous ways of doing so, mostly simply by inaction on his part, and there are no specific sanctions if he decides to take this step. The expectation is that defendants, or more particularly their insurers, will always want to keep cases in the Portal and therefore will not act in this way. That was a debatable assumption when the options were either the RTA Portal or the predictable costs regime in Section II because for many cases that started in the Portal, it was cheaper in terms of costs for the defendant to pay under the Section II formula. However, the revised Portal rates introduced in April 2013 have now probably placed the defendant in the position that the rule makers previously thought was the case all the time.

There is and always has been a temptation for claimant lawyers to exit their client's case from the Portal in order to make the case more profitable. Whilst that may sound pejorative, there is no obvious improvement in the evidence before a court to assess damages between a Stage 3 hearing and a disposal or other assessment of damages hearing so it does not appear to be a question of the client's damages as to why cases exited with alacrity. Consequently, CPR 45.24 spells out the costs consequences if a court considers that the claimant (or his lawyer) has not engaged properly with the Portal in one of the following ways:

- providing insufficient information in the Claim Notification Form;
- by unreasonably valuing the claim at more than £25,000 so that there is no need to comply with the Protocol;
- discontinuing the process set out in the Protocol unreasonably and starting Part 7 proceedings;
- commencing proceedings on the basis that an interim payment made is insufficient;
- acting unreasonably in any other way that causes the Protocol to be discontinued.

In any of these situations the court may demonstrate its displeasure by ordering the defendant to pay no more to a successful claimant than the appropriate Portal fixed costs and disbursements.

Approval hearings

[26.22]

Where the claimant is a child and the parties have agreed liability and damages the claimant applies for an approval hearing using a Part 8 claim in the traditional way. If the court approves the settlement the claimant will be awarded a full house of Stage 1, Stage 2 and Type A, B and C costs from Stage 3 for the hearing (even though it is not a 'Stage 3 hearing' in the sense of a court assessing the damages) and disbursements.

However, if the court does not approve the settlement, it will only award Stage 1 and Stage 2 costs (and disbursements) at the time and the parties will be sent away to prepare for another hearing. This might be once a larger sum has been agreed or it might be an assessment of damages hearing because the parties cannot agree any larger figure. At the second hearing, the claimant will be awarded the Type A and C costs for the preparation work and the advice. The court will also award the Type B (advocacy) costs for one of the hearings. The court has the discretion to award an additional amount for the extra Type A and Type B fees incurred by having two hearings. The court can order either the claimant or defendant to pay such extra costs or it could decide to make no order in this respect.

Where the parties are apart at the end of Stage 2 but, having applied for a Stage 3 hearing the parties reach a settlement, the provisions of CPR 45.22 apply rather than CPR 45.21 (described in the previous paragraphs). The rules are identical in terms of what will be paid when and with what discretion. The only difference under this rule is that the Stage 3 hearing is used as the second hearing if the approval hearing is not originally successful. This rule is also built on the basis that the settlement is more than the defendant's protocol offer. If it were not, presumably the defendant would want to alter the incidence of costs.

In the unlikely event that the court considers that the claim is not suitable for approval to be determined by the court under the Stage 3 procedure, CPR 45.23 decrees that the court will order the defendant to pay the 'full house' costs of Stage 1, Stage 2 and Stage 3 A, B and C but there is no mention of disbursements being paid at this point.

Adjournment

[26.23]

Where the court adjourns a hearing – whether Stage 3 or approval – it may, in its discretion, order a party to pay an additional amount of Type B costs and any court fee for the adjournment.

Costs of the costs only proceedings

[26.24]

Should it be the case that a Part 8 claim is started for an approval hearing or for a Stage 3 hearing and then the case settles, it is to be hoped that the parties can agree the costs of these proceedings as well. But if the amount of costs is the sticking point (and not the principle of who should pay) either party may apply to the court for it to determine the costs by virtue of CPR 45.29. Where this happens the court will assess the costs in accordance with CPR 45.22 and CPR 45.25 and not CPR 44.5 which has no application here.

CPR 45.22 deals with costs based on the parties reaching agreement where the claimant is a child (see above). CPR 45.25 deals with two situations depending on the level of settlement agreed by the parties. If the settlement is more than the defendant's protocol offer, the claimant will be awarded any Stage 1 and 2 payments not already received. He will also receive Type A costs and his disbursement. But where the claimant has not beaten the offer the claimant will not get the Type A costs (but will get the other items) and the court may, in its discretion, order either party to pay the costs of the application.

PART 45 SECTION IIIA (CPR 45.29A–CPR 45.29L)

[26.25]

It used to be the case that a claimant could escape any fixed costs strictures if he could leave the Portal (which many cases did) and get to issue proceedings so as to leave the predictable costs regime (in Section II) behind as well. Thereafter, with the exception of the trial costs (which were generally there simply to pay counsel's fees anyway) the solicitor was on to payment by hourly rate. That is no more. This section deals with the fixed recoverable costs for 'Claims which no longer continue under the RTA or EL/PL Pre-Action Protocols.' These changes apply to all cases where the CNF was sent on or after on 31 July 2013 and which is when this section came into the CPR. Although this section does not apply to disease claims, it does leave the great majority of personal injury claims in an entirely fixed costs world.

This section borrows drafting from sections II and III and references to comments on those sections are set out below rather than repeating the comments here.

General points

[26.26]

(1) The 'London weighting' provision referred to in section II [26.10] and III [26.17] appears here as well. So if the party and his solicitor work in London (as defined) or the claimant lives in London and the solicitor works there, a sum equivalent to 12.5% is added to the figures in the various tables. The only element which does not get a London weighting is the recoverable disbursements.

(2) VAT is payable in addition to the figures set out for costs.

(3) The only recoverable costs are the fixed costs and disbursements set out in this section. There are no outside costs or orders that can be brought in to these claims.

(4) The fixed costs are generally a combination of a base fee plus a figure equivalent to a percentage of the damages. Whilst some parts of Tables 6C, 6D and 6E are internally consistent, the individual figures defy any summary and need to be viewed in full at the end of the book.

(5) Nevertheless the parameters of the tables are consistent amongst the three types of personal injury claim. In each case:
   (a) There are five stages:
      (i) pre-issue;
      (ii) from issue but prior to allocation;
      (iii) from allocation but prior to the date of listing;
      (iv) from the date of listing but prior to the date of trial
      (v) trial
   (b) Where the case settles pre-issue the costs vary based on three value bands:
      (i) £1,000 up to £5,000;
      (ii) Over £5,000 up to £10,000
      (iii) Over £10,000 up to £25,000.
   (c) Once proceedings have been issued, the value of the case does not matter in terms of calculating the level of costs; it simply depends on the stage of settlement, if settled before trial.
   (d) There is no increase in either the base fee or the percentage figure between stages (iv) and (v) set out above. Therefore there is no incentive to run a case to a hearing as any form of costs building exercise. The only increase to the recoverable costs is the addition of the trial advocacy fee.
   (e) The trial advocacy fees are the same whether it is an RTA, EL or PL claim. They are contained in four bands which does not fit with the rest of these costs but does tie in with the fixed costs in section VI. With the exception of the top band, the figures appear to have been increased by roughly 3% since they were last up-rated in October 2007. The fees are:

Value of the claim	Trial advocacy fee
No more than £3,000	£500
£3,001 – £10,000	£710
£10,001 – £15,000	£1,070
£15,001 – £25,000	£1,705

Interim Applications

[26.27]

Where an interim application is made, the recoverable costs are one half of the applicable Type A and Type B costs in Table 6 and 6A: in other words, £125 each. London Weighting, disbursements and VAT are all potentially payable in addition.

Disbursements

[26.28]

These are set out in CPR 45.29I. The wording follows sections II and III where some disbursements may be allowed; any others must not be; but case specific disbursements can in theory at least be made an exception. The categories that may be allowed are:
- Medical records and reports (see **[26.12]** in section II for more detail)
- Non-medical expert reports
- Advice from a specialist solicitor or counsel (see **[26.18]** in section III)
- Court fees
- Expert's trial attendance fees where the court has given the expert permission to attend
- Travelling and subsistence expenses for a party or their witness attending the hearing
- Loss of earnings or leave for a party or their witness attending the hearing (see Practice Direction 45 for the limits on this)
- In an RTA case, an engineer's report (on the condition of the car and any repairs rather than as contemplated in the second bullet point above) and DVLA or MID searches

Defendant's costs

[26.29]

It appeared for a long time as if the defendant's costs were not going to be subject to fixed costs which would have been a surprising outcome. Whilst defendants do not win many personal injury claims and QOCS make enforcement of any costs order impossible in most cases, there are defendants with counterclaims as well defendants who successfully uphold Part 36 offers. There are also defendants who are successful in interim applications. CPR 45.29F deals with defendant's costs. As with claimants, defendants are entitled to London weighting, VAT and disbursements where appropriate.

The general rule is that, unlike claimants, the defendant will not necessarily be awarded the fixed recoverable costs set out in Tables 6C, 6D and 6E. The court is required to have regard to those Tables and is not to make an award that exceeds them. In order to be able to refer to the Tables the court has to decide what the value of the claim is. This will usually be the amount specified on the claim form but excluding any contributory negligence; amounts not in dispute; interest or costs; or vehicle damage (if an RTA claim). If the claim form does not state the value, the court will look at the statement of value required by CPR 16.3. If the claimant has stated he cannot reasonably say how much is likely to be recovered the value will be taken to be the maximum i.e. £25,000.

Part 36, and in particular CPR 36.10A and CPR 36.14A, apply rather than this section where the claimant accepts the defendant's offer or fails to beat the defendant's offer at trial. If the claimant is struck out or found to be fundamentally dishonest under CPR 44.15 and CPR 44.16, the court will assess the defendant's costs without reference to CPR 45.29F and therefore the fixed recoverable costs in this section.

Counterclaims

[26.30]

Where a defendant successfully brings a counterclaim which includes a claim for personal injury to which the RTA Protocol applies, the order for costs will be assessed in the same way as if the defendant was the claimant bringing a claim.

However, where there is no claim for personal injuries within the counterclaim, the defendant will only be entitled to costs amounting to one half of the applicable Type A and Type B costs in Table 6 together with London Weighting, VAT and disbursements where appropriate.

Escaping the fixed fees

[26.31]

Unlike CPR 45.13 in section II, the court does not ask to be entertained in this section. Nevertheless the two hurdles that need to be jumped – exceptionality and a 20% increase on the fixed figures – are the same as in CPR 45.13 to CPR 45.15 (see **[26.14]**).

An application may be made by the claimant or the defendant. If the court is not persuaded of the exceptional nature of the case, the claimant will be limited to the fixed recoverable costs set out in this section. The defendant may also get those figures but will not necessarily get that much. CPR 45.29J says the court will allow an amount that 'has regard to' the fixed recoverable costs. It makes it clear that the assessed figure will not exceed the fixed figure.

If the court agrees that there are exceptional circumstances but, having assessed the costs, does not award a sum that is 20% or more above the fixed figures, the court will allow the lower of the assessed figure or the fixed figure. As with the rules in section II, it is clear that the court may allow a lower figure than the fixed costs in an appropriate case.

In either of these failed eventualities, the court may decide not to award the costs of the proceedings to the receiving party. Indeed the court may make the receiving party pay the costs of the paying party in respect of the proceedings and/or assessment.

PART 45 SECTION IV (CPR 45.30–CPR 45.32)

[26.32]

The Patents county court underwent a package of reforms in October 2010 in relation to procedure, transfers and costs dealt with by Parts 63, 30 and 45 of the CPR respectively. From 1 October 2013 it also underwent a change of name so that it is now the Intellectual Property Enterprise Court ('IPEC'). The reforms (other than the change of name) might seem to be esoteric to many but in our view, they hold the seeds of future rules and practice, particularly in the fast track and bear some scrutiny as a result.

HHJ Birss QC (now Birss J) was responsible for implementing the new rules and so has been responsible for tackling a number of issues arising from the fixed, or scale, fee structure imposed. Some of those decisions can easily be transferred into costs budgeting issues already and if the Jackson Review proposals for fast track costs come into play these decisions can certainly be 'read across' there as well.

Scale management

[26.33]

The guiding principle of HHJ Birss QC in a series of judgments is to allow the court to 'facilitate access to justice for smaller litigants in intellectual property cases.' The key to doing this is to ensure that there is certainty that the cost capping figures will not be exceeded except in the most unusual case. Every judgment makes this comment. In *Westwood v Knight* [2011] EWPCC 11, [2011] FSR 847, [2011] 4 Costs LR 654 he puts it this way

> 'The correct approach must be to apply the limits if they can possibly be applied, recognizing however that in the end the court always has a discretion as to costs and that includes the amount of costs. It is a discretion which in my judgment will very rarely (if ever) be exercised to exceed the limits set by Section [IV]. For one thing specific exceptions are provided for [r. 45.30]. Furthermore to exercise discretion on a wider basis in all but the most rare and exceptional case would undermine the very object of the scale in the first place. For the scale to give a measure of certainty to litigants it has to be possible to be sure that the limits will apply well before any costs are incurred and most likely before any action has even commenced.'

The exceptions to which the judge refers relate to a finding that one party's conduct amounts to an abuse of process or where there is a certificate of contested validity relating to a patent or registered design. Other than these exceptions the scale costs apply to all cases in the IPEC although where a case has transferred from the High Court, the provisions do not relate to the pre-transfer costs incurred.

The scales

[26.34]

There are two tables set out in the Practice Direction. Table A relates to activities undertaken when dealing with the question of liability. Table B relates to quantum activities and which will take the form either of an inquiry as to damages or an account of profits. Against each activity is a scale figure which is in fact a cap since it is the maximum amount of costs allowed for that stage unless the court exercises its discretion.

Overarching these individual stages and caps are global caps set out in CPR 45.31. In relation to liability, the cap is £50,000. For quantum it is £25,000. These sums are net of VAT where recoverable. They are inclusive of any additional liabilities claimed (*Henderson v All Around the World Recordings Ltd* [2013] EWPCC 19). They are also the maximum claimed even if there is more than one opponent. In *Gimex International Groupe Import Export v The Chill Bag Company Ltd* [2012] EWPCC 34 the successful claimant sought to claim £45,000 from each of two defendants. HHJ Birss QC was clear that if Gimex had lost the case it would not have been required to pay more than £50,000 even though there were two defendants. On that basis it would be wrong to allow it to 'share out its single costs bill as between two sets of defendants and thereby recover more than £50,000 in costs in these proceedings.'

Assessment

[26.35]

The default method of assessment of costs in the IPEC is by summary assessment. This is so even if the trial lasted more than a day (unlike the fast track). Consequently, the rules as to interim payments and detailed assessment proceedings are specifically disapplied by CPR 45.30(3).

Where there are costs from High Court proceedings pre-transfer, these are assessed summarily in the normal way (see Chapter 27). Otherwise, the assessment of costs is something of a hybrid between the usual summary assessment and the application of costs capping. In order to carry this out the receiving party needs to break down its costs claimed into the stages in Tables A and B. Where this was not done, the court did its best to carry out a summary assessment but disallowed costs relating to disclosure entirely for a period. Recorder Campbell, in *Lumos Skincare Ltd v Sweet Squared Ltd* [2012] EWPCC 28, [2012] 4 Costs LR 735 said, having found that he had no clear information on the defendant's disclosure, 'I am not prepared either to make assumptions in the defendant's favour about matters which the defendant could have proved, or to award some arbitrary sum, since to do so would be unfair to the claimant. I therefore make no award for the defendant's costs relating to disclosure over this period.'

In *Westwood* the receiving party did provide a breakdown of costs by stages and furthermore a summary schedule setting out totals for each stage. HHJ Birss QC thought the schedule was a useful document and commended it to other litigants in the future. It is easy to see a similarity with the revised precedent H used for costs budgeting.

Once the judge has looked at each stage's breakdown and established a summarily assessed figure, he then compares that figure with the scale maximum and will allow the lower of the two. HHJ Birss QC's description of this procedure in *Westwood* is careful to say that if this procedure is followed correctly the lower of the two figures is 'very likely' to be the right figure, rather than definitely will be that figure. However, the robustness of his decisions to date leave little room for doubt that they will be the final figure, subject to the overall cap. If the assessed figure is above the scale maximum the parties and the court can be confident that there is more than sufficient work to justify the maximum recoverable sum. If the figure is lower, the paying party has had his liability specifically assessed on that case's own circumstances.

The final stage is to compare the total of the figures allowed for each stage, whether the scale maximum or a lower assessed figure, with the global caps of £50,000 or £25,000 as the case may be. The approach is very similar to the comparison of assessed and scale figures for each stage. The lower of the totalled figures and the global maximum will be allowed. This limit does not apply to any High Court costs summarily assessed and which are payable in addition (see *Westwood* as an example of this.)

Costs outside the scales?

[26.36]

Again, there is a similarity to the costs management approach on display here. There is no scale figure for pre-action costs which would militate against parties using the pre-action protocol. However, where the costs can properly be claimed against a post proceedings stage, for example the drafting of witness statements or considering disclosure, such costs should be claimed under that stage rather than as pre-action costs. As such, the only irrecoverable costs ought to be those dealing specifically with the protocol itself.

As with costs management, there is no 'general costs of the action,' or similar stage, into which costs can be placed. Unlike costs management, there is no scope for amending the stages to fit the particular case. As such the approach according to HHJ Birss QC is *Westwood* is not to take 'too narrow a view of the scope of the stages in Table A. I say that because even if a broad view of the scope of those stages is taken, this would not undermine the system.'

The order for costs

[26.37]

The award of costs between the parties is carried out in the same way as in any other court, ie in accordance with Part 44 with costs following the event as the starting point.

If the court decides to reflect the parties' success and failure with a percentage order or issues based order, how does that fit in with the scale figures? HHJ Birss QC considered this in *BOS GmbH & Co KG v Cobra UK Automotive Products Division Ltd* [2012] EWPCC 44. He came to the view that 25% of the claimant's costs should be deducted as a result of a failure on certain issues. He thought that a further 10% should be deducted to reflect the issues on which the defendant should have his costs. On the basis that the parties' costs were of similar amounts, the better way to reflect this was to reduce the percentage of the claimant's costs rather than allow the claimant their costs of the particular issue. (No doubt he had CPR 44 in his mind although this is not specifically stated.) So, he ended up with deducting the 25% and the 10% from the starting point of a 100% order and awarded the claimant 65% of its costs.

The key issue then was the order in which the summary assessment should be carried out. Should the assessed figure for each stage be reduced to 65% and then compared with the scale maximum as the claimant contended, or should the assessed figure be compared with the scale maximum and then 65% of whichever was the lower figure allowed (as the defendant said)? HHJ Birss QC referred to CPR 45.31(2) which states that the global caps apply 'after the court has applied the provision on set off in accordance with rule [r 44.12(a)].' In his view, the correct approach was therefore to reduce the assessed figure to reflect the percentage order and then compared that figure with the scale maximum for the relevant stage. He accepted that if the originally assessed figure is considerably higher than the maximum there may be little effect in the percentage order because the party would only receive the scale maximum

Part 45 Section VI (CPR 45.37–CPR 45.40) **26.40**

whether that was lower than the 100% or 65% figure. But this was more in accord with CPR 45.31(2) in its operation. He also considered that the other approach would have a disproportionate effect since it would never allow more than 65% of the scale maximum which was too low in this particular case.

Unreasonable conduct

[26.38]

To the extent that a party has behaved unreasonably, the court may make an award of costs against him in accordance with CPR 63.26(2). Such costs will be assessed summarily at the end of the hearing and such sums will, in accordance with CPR 45.32, be payable in addition to the total costs payable for the proceedings generally.

PART 45 SECTION V (CPR 45.33–CPR 45.36)

[26.39]

It just goes to show how many Part 7 claims that the HM Revenue and Customs issue that it has its own section in Part 45. This section closely follows Part 1 in its format. There is a table (Table 7) of commencement costs ranging from £33 to £180 based on bands which run from £25 to £500 near the bottom to £300,000 and above at the top. There is also a table (Table 8) of judgment costs but that barely justifies being a table since the fixed costs are either £15 or £20 depending upon whether the judgment sum is under or over the £5,000 threshold.

Other provisions in this section, such as 45.36, mimic provisions in section I (CPR 45.3 in this example), to confirm that if payment of the damages and fixed costs is made within 14 days of service of the proceedings, there are no further costs payable.

PART 45 SECTION VI (CPR 45.37–CPR 45.40)

[26.40]

This section was a free transfer from Part 46 as part of the reorganisation of the costs section in the CPR. The trial costs in the fast track were fixed when the Woolf Reforms were brought into play on 26 April 1999. They have continued to be the only fixed fees on the Fast Track notwithstanding the view of Lord Woolf that this was an important remedy for the evil of costs that he identified in his report. They have only been changed on two occasions since then. The first, in October 2007, was an uplifting of the original figures. Since this took a good eight years to come through, the delay remains an example to point to of what happens when the Government decides to fix costs. The second was to tie in with the increase in the scope of the fast track from

26.40 *Chapter 26 Fixed Costs*

£15,000 to £25,000 in April 2009. The original three bands were supplemented with a fourth to cater for proceedings between £15,000 and £25,000 where the proceedings were issued on 6 April 2009 or later.

Scope of 'fast track trial costs'

[26.41]

The costs awarded under this section are for the preparation and appearance at a fast track trial. They do not include any other disbursements and nor do they include VAT. As such, if the party cannot recover VAT, it can be charged in addition to the relevant figure for costs.

The word 'trial' includes an assessment of damages hearing where liability has already been admitted. It does not include summary judgment hearings or approval hearings.

If the case has not been allocated to the fast track then the fees do not apply. This does not prevent a judge assessing costs on a case allocated to the multi-track from taking note of what would have been allowed on the fast track in an appropriate case *(Drew v Whitbread* [2010] EWCA Civ 53, [2010] 1 WLR 1725, [2010] NLJR 270).

Amount of costs

[26.42]

The figures are set out in Table 9 in PD 45. They are as follows

Value of the claim	Costs
No more than £3,000	£485
£3,001 – £10,000	£690
£10,001 – £15,000	£1,035
£15,001 – £25,000	£1,690

The only circumstances in which the court can award any other sum instead of one of these figures are if:
(a) the court decides:
    (i) not to award any costs at all;
    (ii) to apportion the costs to reflect the respective degrees of success of the parties;
    (iii) to set off the costs of the parties' claims and make an award of the difference to one party;
(b) one or more of the parties have behaved improperly during the trial:
    (i) if it is a winning party, the trial costs may be reduced as the court considers appropriate (and this may be if the behaviour is 'only' unreasonable rather than improper);
    (ii) if it is a losing party, the trial costs may be supplemented by such additional amount as the court considers appropriate;
(c) a party has a legal representative present in addition to the advocate and the court considers it was 'necessary' for the legal representative to attend. If so the allowable amount is £345;

(d) a successful party is a litigant in person in which case he can claim costs in accordance with CPR 46.5. Where the litigant in person can show financial loss, the two-thirds figure is based on the appropriate trial costs for the value of the case;
(e) it is necessary to have a separate trial of an issue in which case the court must award at least £485 but will otherwise not award more than two thirds of the standard costs (of any of the higher bands).

Calculating the value of the claim

[26.43]

For the purposes of calculating into which band a claim should fall, the value of the claim for a successful claimant is the total amount of the judgment excluding any interest, costs or reduction for contributory negligence. For the defendant it is the amount specified in the claim form (excluding interest and costs). If there is no specification in the claim form of the level of damages, the statement of value on the claim form required by r 16.3 will be used unless the claimant states that he cannot reasonably say how much the claim is worth and in which case the value will be treated as being more than £15,000. This means that the defendant will receive the highest level of costs if he is successful.

If both parties have claims and the defendant's counterclaim is the more valuable, the calculation of value for either party will be based on the defendant's counterclaim.

If the claim is for a remedy other than money, the case will be treated as having a value between £3,000 and £10,000 unless the court otherwise orders. If there is a non-monetary claim and a money claim, this provision will apply unless the money claim is for more than £10,000 and in which case the relevant higher bracket applies.

If the same advocate acts for more than one claimant, and each one has a separate claim against the defendant, the value of the claims is based on the aggregate amount of the damages; whether it is the claimants or the defendant who is successful.

Multiple parties

[26.44]

Where the same advocate acts for more than one party, the court will only make one award in his favour and the parties instructing him are only jointly entitled to such costs. Similarly, where there is a single defendant against multiple opponents, the court may only make one order in favour of the defendant.

PART 45 SECTION VII (CPR 45.41–CPR 45.44)

[26.45]

This section deals with Aarhus Convention claims. What is the Aarhus Convention you might say? CPR 45.41(2) defines a claim arising from the Convention as being:

'. . . a claim for judicial review of a decision, act or omission all or part of which is subject to the provisions of the UNECE Convention on Access to Information, Public Participation in Decision-Making and Access to Justice in Environmental Matters done at Aarhus, Denmark on 25 June 1998, including a claim which proceeds on the basis that the decision act or omission, or part of it, is so subject.'

Part of the Convention requires proceedings to be capable of being brought in the Convention countries without ruinous risk. As such there are special provisions in this section to limit the claimant's costs. The general rule (CPR 45.43(1)) is that a claimant cannot be ordered to pay a greater amount in costs to a defendant than is the amount prescribed in the Practice Direction to Part 45. The amount payable depends upon the nature of the claimant. If the claimant is an individual and is not acting on behalf of a business or other legal entity, the sum is £5,000. In any other case it is £10,000 (CPR PD 45, para 5.1).

Where the claimant has not stated that the Convention applies, or has specifically stated that it does not apply, the costs protection provisions of this section do not apply either.

The defendant's potential liability in costs in a Convention claim is limited to £35,000 (CPR PD 45, para 5.2). However, where the defendant states that the Convention does not apply in the acknowledgment of service and defence, he risks increasing his potential liability. The court will decide whether the Convention applies at the earliest opportunity so that the parties know where they stand. If the court accepts the defendant's position that the proceedings are not a Convention claim, it will usually make no order as to costs (CPR 45.44(3)(a)). But if the court decides against the defendant and holds that it is a Convention claim, it will normally order the defendant to pay the claimant's costs on the indemnity basis. This may well cause the recoverable costs to be more than the prescribed fixed costs of £35,000 but that is no bar to their recovery (CPR 45.44(3)(b)).

THE SMALL CLAIMS TRACK

[26.46]

The intention of the small claims track is made clear in CPR 27.1 where it talks of providing a special procedure for dealing with claims and limiting the costs that can be recovered by that procedure. This is done by stripping out many of the procedural requirements of civil litigation as far as possible and enabling and not just requiring that a party 'should act as his own lawyer' as Lord Diplock put it in *Hobbs v Marlowe* [1978] AC 16 when discussing a predecessor system for small claims.

Unlike all of the other fixed costs provisions, the costs in relation to small claims track cases are set out in CPR 27.14 and Practice Direction 27, sub-para 7. There are also one or two relevant references to Part 44 in relation to allocation as discussed below.

The small claims track **26.49**

Recoverable fixed costs

[26.47]

By removing the procedural requirements so that a party can represent himself, the court rules justify there being no recoverability of the costs of legal representation. This includes any fee or reward charged by a party's lay representative.

There are four exceptions to this general point:
(1) The fixed costs on commencement of a claim under Part 45 Section I can be claimed, whether or not the case actually comes within the relevant definition in that section for the type of case involved.
(2) If advice is taken in respect of an injunction or specific performance then legal fees can be claimed. Such costs are limited to £260.
(3) If a party has behaved unreasonably, the court can order such further costs as it thinks fit and will summarily assess these (CPR 27.14(2)(g)). Rejecting an offer will not of itself amount to unreasonable behaviour but it may be taken into account by the court.
(4) Where a personal injury case started in the Portal and the defendant admitted liability for the accident but it then transpired that the level of damages were below £1,000 and so the case fell out of the Portal. In such circumstances, and so long as the claimant reasonably believed the value was more than £1,000, the Stage 1 and, if appropriate Stage 2, costs are payable by the defendant.

Recoverable disbursements

[26.48]

There are various disbursements that can be claimed:
(1) any court fees paid;
(2) travelling, accommodation and subsistence expenses incurred by a party or witness in attending the hearing limited to a maximum of £90;
(3) loss of earnings or leave incurred by a party or witness in attending the hearing limited to a maximum of £90;
(4) Experts' fees limited to £750 per expert (this was £200 prior to 1 April 2013);
(5) In an appeal, the cost of any reasonably incurred approved transcript.

Allocation and re-allocation to track

[26.49]

As with all tracks, the allocation procedure will consider the small claims track to be the 'normal' track where certain criteria are met, particularly relating to the value of the claim. In such cases the costs restrictions set out above automatically apply to all costs incurred before and after allocation.

There are two qualifications to this point.
(1) Any costs orders made prior to allocation will continue to be effective and entitle the party to assessed costs (CPR 46.13(1) overriding CPR 46.11(2))

26.49 *Chapter 26 Fixed Costs*

(2) Where a case is allocated to one track and then re-allocated to another, any 'special rules' about costs in either track will apply up to/after the re-allocation as the case may be. This is always subject to the court ordering otherwise (CPR 46.13(2)).

Prior to 1 April 2013, the court could allocate a case to the small claims track even if its value was above the limits which would have made it the normal track. In those circumstances, the costs restrictions would also apply unless the parties agreed that the fast track costs provisions should apply instead. Where that happened, the fixed trial costs (see [26.42] above) would not apply but they would not be exceeded when the court came to assess the costs overall. This provision (CPR 27.14(5) and (6)) was removed on 1 April 2013 and so if the court now allocates a larger value case to the small claims track, the costs restrictions will apply regardless of the parties' wishes.

CHAPTER 27

SUMMARY ASSESSMENT

INTRODUCTION

[27.1]

Since 26 April 1999 whenever the court makes an order that one party is to pay costs to another party, then unless the costs are fixed costs, the court must either make a summary assessment of those costs or order a detailed assessment. This remains the position after the April 2013 amendments to the CPR and the rule is found at CPR 44.6.

In his preliminary report Sir Rupert Jackson put forward 3 options in respect of the summary assessment procedure as follows:

– Option 1: make no change to the present rules governing summary assessment.
– Option 2: abolish the summary assessment procedure and instead encourage judges to order interim payments on account of costs, alternatively provisional assessments.
– Option 3: restructure the summary assessment procedure

His conclusion was that:

'I am quite satisfied that option 3 is the proper way forward. Summary assessment is a valuable tool which has made a substantial contribution to civil procedure, not least by deterring frivolous applications and reducing the need for detailed assessment proceedings. The summary assessment procedure should be retained and improvements should be made in order to meet the criticisms which have been expressed . . . '

So the summary assessment procedure remains. But how summary is summary? In *Morgan v Spirit Group Ltd* [2011] EWCA Civ 68, when overturning the decision of the recorder, the Court of Appeal made the obvious point that the court must undertake either a summary or detailed assessment of the costs. The recorder had not done so. Instead, in a paragraph, the recorder had simply determined a proportionate figure and added something to that to allow for the existence of what he had, erroneously, described as 'a contingent fee agreement'. Summary though the assessment is intended to be, the court concluded that what had taken place here was not a summary assessment. So, summary though the procedure may be, it must still recognisably be an assessment.

WHO CONDUCTS A SUMMARY ASSESSMENT?

[27.2]

A summary assessment of the costs of a hearing or a trial must be carried out by the judge who conducted that hearing. CPR PD 44, para 9.7 precludes the

27.2 *Chapter 27 Summary Assessment*

court that awards costs from ordering a summary assessment before a costs officer. If a summary assessment is not undertaken at the hearing at which costs are awarded then the court must give directions for its later hearing before the same judge unless a detailed assessment is ordered.

THE BENEFITS OF A SUMMARY ASSESSMENT

[27.3]

There are benefits to both the parties and to the court in conducting a summary, as opposed to a detailed, assessment. These are as follows:
- A summary assessment is more proportionate than a detailed assessment – both in terms of the costs of the parties and the court time involved. This is true even with the fixed cost of the provisional costs regime introduced in CPR 47.15.
- The judge who has awarded the costs has a better understanding of the issues/case than the assessing judge will ever have. Indeed, with the greater continuity that comes with 'docketing', the judge may well have been involved throughout the court proceedings. In *Mahmood v Penrose* [2002] EWCA Civ 457, the Court explained the reasoning behind summary assessment was that the person who has actually heard the case and knows about it is in a position to make a summary assessment.
- In cases that have been costs managed the assessment process should be much simplified as the court is likely to be dealing with the pre-budget costs only. Accordingly the uncertainty of the outcome of a summary assessment is much reduced.
- There is no delay. The receiving party will receive the costs far earlier than waiting for a detailed assessment. The paying party's liability is crystallised earlier, avoiding a prolonged period when interest on costs at 8% falls due.
- The court is spared dealing with the usual standard objections that fill the first few pages of Points of Dispute and which invariably occupy court time to little or no avail. The assessment is focused on the key issues.

WHEN IS A SUMMARY ASSESSMENT APPROPRIATE?

[27.4]

The simple answer, following on from the benefits listed above, is that summary assessments should be carried out in preference to detailed assessment as often as possible. CPR PD 44, para 9.2 sets out a general rule that the court should conduct a summary assessment unless there is good reason not to do so:
- at the conclusion of a trial within the fast track: and
- at the conclusion of any other hearing that has lasted no more than a day.

The general rule applies to all hearings that have lasted – the word is 'lasted', not 'listed', so the test is hindsight not foresight – not more than one day

There was some debate about whether or not the apparent distinction between 'a trial' and 'a hearing' precludes a summary assessment after a multi track trial. The debate was short lived. The court may and, whenever possible, should, conduct summary assessments after multi track trials. We expect this to occur more often now that most multi track claims will have been costs managed (see Chapter 25 Payments on Account of Costs, above and the decision of HHJ Simon Brown QC in *Slick Seating Systems v Adams* [2013] EWHC 1642 (QB), [2013] 4 Costs LR 576) and now that the judiciary is astute to the key concept of proportionality when making case management decisions (of which a decision whether to have a summary or detailed assessment is certainly one). Indeed 'other hearings' includes not only multi-track trials, but also appeals in the Court of Appeal which do not last more than a day. This expectation is notwithstanding one of the conclusions of the Review of Civil Litigation Costs: Final Report that recognised that in larger cases and with judges less familiar with costs an increase in referral to detailed assessment would be appropriate in these terms:

> 'If any judge at the end of a hearing within Costs PD paragraph 13.2 (now PD 44, para 9) considers that he or she lacks the time or the expertise to assess costs summarily (either at that hearing or on paper afterwards), then the judge should order a substantial payment on account of costs and direct detailed assessment.'

Although the general rule applies to cases that last no more than one day it was held in *Q v Q (Family Division: costs: summary assessment)* [2002] 2 FLR 668 (Fam) that there is no presumption against summary assessment in relation to costs where hearings last more than one day. Indeed CPR PD 44, para 9.1 requires summary assessment to be considered in every case where a costs award is made and fixed costs do not apply. Section 6.4.4 of the Technology and Construction Court Guide provides that if the hearing of an application has to be adjourned because of delays by one or other of the parties in serving evidence, the court is likely to order that party to pay the costs straight away, and to make a summary assessment of those costs. Frequently the insurmountable difficulty for the court in undertaking a summary assessment in a case that lasts in excess of one day is that the parties have not prepared costs statements, because they do not anticipate that there will be a summary assessment and they are not obliged by the rules to file and serve them in such cases. However, there is nothing in the rules to prevent a party preparing, filing and serving such a statement voluntarily. Nor is there any reason why the court as a case management decision, either at the pre – trial review or on the first day of the trial, should not order the parties to file and serve a costs statement. Again we would expect efficient parties and courts to adopt these practices

WHEN IS A SUMMARY ASSESSMENT INAPPROPRIATE?

[27.5]

CPR PD 44 sets out a number of situations where the court may decide not to undertake a summary assessment or must not do so.

When the court decides not to undertake a summary assessment

(a) Exception to the general rule

[27.6]

The general rule at CPR PD 44, para 9.2 provides the court with the option of not undertaking a summary assessment where there is 'good reason' not to do so. It gives as an example where the paying party shows substantial grounds for disputing the sum claimed for costs that cannot be dealt with summarily. The most obvious example of this might be a challenge to the overall retainer. However, do not expect that the court will simply rubber stamp 'good reason' and order a detailed assessment. For the reasons already specified the court should be reluctant to defer the assessment. Accordingly it is likely to delve behind a bold assertion of substantive issue to ascertain whether there is genuinely an issue that requires more detailed investigation or whether the paying party is speculating in the hope that on detailed assessment he may stumble upon something.

Another possible reason for not having a summary assessment could be legal arguments which could not properly be dealt with on a summary assessment. In *R v Cardiff City Council, ex p Brown* (11 August 1999, unreported, QBD) it was alleged that the hourly rate claimed was in breach of the indemnity principle because the work had been done by the receiving party's own legal department and the hourly cost of doing the work could not have been as much as the rate claimed between the parties. The issue required a consideration of the legal department's costings and the various relevant authorities, which Harrison J concluded was not a suitable exercise for a summary assessment.

Another reason, and one which the court may have to grapple with more often with the withdrawal of public funding and the increase in the incidence of unrepresented parties, is whether a litigant in person receiving party has suffered a financial loss. In *Neil v Stephenson* (2000) CLW, 8 December, QBD although the claimant, acting in person, had his action struck out, he had been awarded his costs of an interim hearing. When the claim was struck out, the judge acceded to the claimant's request for a detailed assessment to give him an opportunity to demonstrate that he should be entitled to recover the value of a contract which his business had allegedly lost as a result of his attendance at the interim hearing. The fairest course was to order an expedited detailed assessment of the costs of both parties to give the litigant in person the opportunity to consider his arguments and put the case before an experienced costs judge.

(b) Admiralty and Commercial Courts

[27.7]

Paragraph F14.2 of the Admiralty and Commercial Courts Guide provides 'active consideration will generally be given by the court to adopting the summary assessment procedure in all cases where the schedule of costs of the successful party is no more than £100,000, but the parties should always be prepared for the court to assess costs summarily even where the costs exceed this amount'. There is further guidance in the next two paragraphs:

> 'In carrying out a summary assessment of costs, the court may have regard amongst other matters to:
> (i) advice from a Commercial Costs Judge or from the Chief Costs Judge on costs of specialist solicitors and counsel;
> (ii) any survey published by the London Solicitors Litigation Association showing the average hourly expense rate for solicitors in London;
> (iii) any information provided to the court at its request by one or more of the specialist associations (referred to at section A4.2) on average charges by specialist solicitors and counsel.

The figure is an interesting one. Under the Regional Costs Judge scheme, which was introduced in 2005/6, one of the criteria for reference to a regional costs judge by a district judge is that the bill exceeds £50,000 (soon to be increased to £100,000). So, notwithstanding that assessment of costs is part and parcel of their daily diet of cases, district judges have been referring cases to a specialist where the costs exceed £50,000, whilst there is an expectation that commercial judges will undertake summary assessments of all cases where the costs are double that limit!

(c) Costs in the case

[27.8]

Originally, the costs practice direction provided that generally there would be no summary assessment of costs if an order has been made for costs in the case. However, this restriction was removed. There was a time when it was thought that the court would summarily assess the costs of all parties where an order for 'costs in the case' was made, to obviate the need for a detailed assessment of these later when it became known in whose favour the ultimate costs order had fallen. We have never known this done. Neither the parties nor the court seem to have the will to undertake what will largely prove to be a fruitless exercise.

(d) Where there is a deferred payment retainer in place.

[27.9]

Until those claims governed by the pre-April CPR have concluded there will be cases where parties are entitled to recover success fees. Indeed, even after that date there will be cases funded under conditional fee agreements ('CFAs') and damages based agreements ('DBAs') where a client's liability to his solicitor is dependent upon success (however defined). Special provisions apply to such cases where interim costs orders are made. In both instances (both pre- and post-31 March 2013) there should still be a summary assessment. However, where the receiving party's liability to his solicitor has not arisen (for example where success under the CFA is defined as obtaining an award of damages, rather than success at an interim hearing along the way) then plainly under the indemnity principle the paying party cannot be ordered to make any payment unless and until a liability does arise. What does the court do?

In respect of pre-April CFAs then under the transitional arrangements the old Costs PD sections 9, 13.12(1) and 14 and old CPR 44.3A still apply. In essence these provide that the court should make a summary assessment of the base costs (splitting the disbursements allowed between counsel's fees and others to facilitate the later application of any success fee to which counsel may be entitled to these sums), but not order payment of that sum and make provision for the later payment of that sum and assessment and payment of the relevant success fees that might apply once the trigger for the client's liability to his solicitor has arisen. The old Costs PD, para 14.4 suggested that the summarily assessed base costs be paid into court or that the order assessing them prevented enforcement pending further order or postponed the payment in some other way. We always took the view that this was a rather laborious provision and what the court should do was provide one order that put the onus on the parties to resolve when payment was due and deferred the time for assessment of the additional liabilities (otherwise commencement within 3 months of the order might still be before any liability of the receiving party to his solicitor had crystallised). Our preferred form of order in this situation is:

'1(a) The (claimant/defendant) shall pay the (defendant's/claimant's) costs of and incidental to the application dated the (insert). The base costs are summarily assessed as follows:

profit costs	v
disbursements (non counsel)	w
disbursements (counsel)	x
VAT	y
Total	z

These costs shall be paid by the (defendant/claimant) to the (claimant/defendant) within 14 days of the (defendant/claimant) receiving written notice from the (claimant/defendant) or his solicitors confirming that the (claimant/defendant) has a liability under the terms of his fee arrangement with his solicitors to pay these costs. Liberty to apply

1(b) Unless the additional liabilities due under this order are agreed there shall be a detailed assessment of these. The time for commencement of any such assessment shall be no earlier than the date upon which notice is given under paragraph 1(a) above and no later than 3 months after that date.'

In respect of orders after 31 March 2013 the court is not concerned with the success fee and need make no provision in respect of it, as that falls to be dealt with exclusively between the solicitor and the client. However, the question of when liability for the base costs (under CFAs) or any costs under DBAs arises is still relevant. In those circumstances we suggest that paragraph 1(a) of the order above is used with removal of the word 'base'. The order has the benefit of there being an uncomplicated and proportionate procedure to obtain the assessed costs once the client has a liability to his solicitor, without the application having to return to the court.

Many solicitors have sought to avoid the restriction on immediate payment under a CFA, by excluding pre-action and interim applications from the substantive CFA that covers the claim and which defines success in overall terms. On each occasion of a pre-action or interim application a separate CFA is written with success defined as a variant on obtaining the order sought, including an order for costs from the other party/parties. Others have sought to draft the substantive CFA so that immediate payment of costs is due on each and every successful pre-action or interim application. Our experience is that this is a much more difficult exercise as there is a danger of triggering the client's liability at an earlier stage than that which has been sold to the client under the 'no win, no fee' sales pitch.

Interestingly, for CFAs entered into after 31 March 2013 and for DBAs there is no longer any need to serve and file a notice of funding (FORM N251). The rationale is obvious – additional liabilities are no longer recoverable between the parties and so notice that they may exist is only of relevance to the client. However, this does mean that the only check on whether an immediate entitlement to costs has arisen is by the receiving party's solicitor's accurate completion of the certificate on the FORM N260 (see below). We wonder how many orders for costs to be paid within 14 days will be made on interim applications with the court accepting the accuracy of the signed certificate, when in fact the funding mechanism agreed does not, at that stage, create any liability for the client to pay the solicitor.

Where the court must not undertake a summary assessment

(a)   Mortgagee's costs

[27.10]

CPR PD 44, para 9.3 disapplies the general rule to the mortgagee's costs of mortgage possession proceedings or other proceedings relating to a mortgage unless the mortgagee waives this exclusion and asks for an award of costs against another party. The rationale for this provision is that usually a mortgagee does not seek an award of costs because it is entitled to recover the costs under the terms of the mortgage deed. CPR PD 44, paras 7.2 and 7.3 sets out more detailed provisions in respect of costs under a mortgage, ultimately directing that where there is a dispute as to the amount the court may direct an assessment under CPR 44.5 (costs under a contract).

(b)   A public funded party

[27.11]

CPR PD 44, para 9.8 provides that where one of the parties is in receipt of state funding there cannot be a summary assessment of any costs ordered to be paid to that party by the other party. This is because the Legal Aid Act 1988 and Legal Service Commission regulations require those costs to be the subject of a detailed assessment. There is nothing to prevent a summary assessment against a state funded party, although there will need to be a determination of the assisted person's liability to pay those costs for the purposes of the Legal Aid Act 1988, s 17 or the Access to Justice Act 1999, s 11 and the exercise may prove to be an academic one only (see Chapter 10 State Funded Parties).

(c)   Children or protected parties

[27.12]

CPR PD 44, para 9.9 prevents a summary assessment of the costs due to a child or a protected party, unless their solicitor agrees not to make any further charge to the client beyond that recovered from the other party. However, there is a conflict between CPR PD 44, para 9.9(2) which indicates that the court may make a summary assessment of any costs due from a child or protected party and CPR 46.4(2) which provides that 'the general rule is that (a) the court must order a detailed assessment of the costs payable by, or out of money belonging to, any party who is a child or protected party . . . '. CPR PD 46, para 2.1 qualifies the strict requirement for detailed assessment of the costs payable by a child or protected party, but not to the extent that the conflict is necessarily resolved. Our experience has been that there is an assumption that the obligation to order a detailed assessment under Part 46 has been understood to arise only where the child or protected party is the recipient of costs and his solicitor has not waived the entitlement to any further costs. However, CPR 46.4(2) clearly also deals with the situation where the child or paying party is the paying party (see Chapter 34 Part 21/Protected Parties).

(d)   Agreed costs

[27.13]

CPR PD 44, para 9.10 provides that the court will not endorse disproportionate and/or unreasonable costs. Accordingly the Practice Direction states that when the amount of the costs to be paid has been agreed, the order should make it clear that it was made by consent.

(e) Agreement 'save as to costs'

[27.14]

Where the parties have agreed the outcome of an application and wish to avoid a court hearing they should seek to agree a figure for costs or agree that there is to be no order for costs. It is not for the court to assess costs summarily in their absence (see CPR PD 44, para 9.4).

THE EXTENT OF THE COSTS TO BE ASSESSED

[27.15]

The summary assessment at the end of a multi track or fast track trial will deal with the costs of the whole of the claim, remembering that the advocacy costs of the fast track trial are fixed. Of course from 31 July 2013 all fast track road traffic accident, employers' and public liability claims where the claim notification form was sent after that date are subject to fixed costs provisions. No assessment will be necessary in those cases if the award of costs is in the claimant's favour. If it is in the defendant's favour then an assessment will be required but given that the court cannot exceed what the claimant would have recovered under the fixed costs provisions, this is likely to be far simpler than prior to this (see CPR 45.29F(2)–(7)).

A summary assessment at the end of an interim application will deal with the costs of the application or matter to which it related.

THE STATEMENT OF COSTS

[27.16]

CPR PD 44, para 9.5(1) makes it the duty of the parties and their legal representatives to assist the judge in undertaking a summary assessment. As such the parties must provide a written statement of costs that should follow as closely as possible FORM N260. This must particularise the information specified in the Practice Direction and be signed by the party or his legal representative and include the certificate contained on that form (save in certain public funding situations and where the party is represented by someone in his employment).

The FORM N260 was altered at the time of the April 2013 amendments to the CPR, in line with one of the recommendations made by Sir Rupert Jackson in the final report. However, experience to date shows that many solicitors are using the old form, or, where they adapted the old form for their own purposes, have failed to adapt that to meet the new requirements. The new form was introduced as a result of the almost universal criticism of the lack of information on the old form (endorsed by Sir Rupert Jackson) and the potential that this created for 'broad brush' injustice. The new form now has an additional sheet that provides for a breakdown of the time claimed for documents so that the judge has a far clearer idea of what work the receiving

27.16 *Chapter 27 Summary Assessment*

party has done. However, do not expect the assessment to become a quasi-detailed assessment with the court going through each individual item listed in the document breakdown. The assessment remains a summary one and the breakdown is only informative of how time has been spent. Indeed the very reason that one of us dislikes the new form is that the breakdown smacks of, and encourages, something more than a summary process!

One area where the new FORM N260 is deficient in respect of assessments of the costs of the entire claim is in its link to costs management orders. As we have seen earlier under CPR PD 3E, para 2.3 the court sets the budget by total sum per phase and under Part 3.18 on assessment (and this includes a summary assessment) has regard to the amount budgeted for each phase of the proceedings and only departs from the last approved or agreed budget when satisfied that there is 'good reason' to do so. Accordingly one might be forgiven for thinking that the new form would list expenditure under the phase headings to make for easy comparison, so that both the paying party and the court can readily see whether the receiving party has kept within the budget sum set for each phase. In fact the new FORM N260 largely maintains the old form of breakdown between time spent on client, on opponents, on others etc. and still requires details of the number of letters written, phone calls made etc. by each grade of fee earner. Whilst this is fine for interim applications and final assessment in non-budgeted claims, this is not true of final assessments in costs managed cases. As such direct comparison with the budgeted sums for the phases of the claim is impossible. All that the form reveals is whether the overall claim for costs is within or without the overall budget sum. Will this mean that the court will be more reluctant to undertake summary assessments on the basis that more information to enable budget comparison will be available on detailed assessment? We think not because:

- There is no new form of bill and so there is no guarantee that any more information as to the costs of each phase will be available on a detailed assessment
- The 'broad brush' of summary assessment makes determination of the issues far easier and, therefore, more proportionate for the court.
- The summary assessment procedure has the attraction of being a much simpler one than it was for those judges less familiar with costs where the case has been costs managed – see *Slick Seating Systems v Adams* above

THE PROCEDURE

[27.17]

The statement must be filed at court and copies served on any party against whom an order for payment is intended to be sought not less than 24 hours before the time fixed for the hearing or 2 days before a fast track trial. The beauty of this procedure is that when the solicitor is filling in his statement of costs he does not know whether his client is going to be the receiving or the paying party, which should assist him to take a balanced view of the exercise. CPR PD 44, para 9.6 says that failure to comply will be taken into account in the decision about what order to make in respect of costs (so it is relevant to the award as well as the amount of assessed costs) and the defaulter will be at the risk of an order for costs if his failure necessitates an adjournment or a detailed assessment. Even worse, at first blush that party might not recover any costs at all because the Practice Direction makes the preparation, lodging and serving of a statement of costs a condition precedent to applying for costs (CPR PD 44, para 9.5 refers to the fact that a party intending to seek costs must prepare a written statement and CPR PD 44, para 9.5(4) states that this must be filed and served the prescribed time before the hearing).

However, in *MacDonald v Taree Holdings Ltd* [2001] 06 LS Gaz R 45, ChD it was held that, despite the use of the word 'must', the provision is not mandatory and a deputy district judge had been wrong to refuse the successful party's application for summary assessment of his costs on the grounds that he had not served a statement of costs upon the respondent 24 hours in advance. The court held that it has a wide discretion when deciding whether or not to award costs under Part 44 and in applications for award of costs and summary assessments of those costs, the failure to serve a schedule of costs was often being used for grounds for depriving a party of his costs or for curtailing a party's costs. Where, however, the only factor against awarding costs was merely a failure to serve a statement without aggravating factors, a party should not be deprived of all his costs. The court should take the matter into account but its reaction should be proportionate. The question the court should ask itself was what, if any, prejudice there had been to the paying party and how should that prejudice be dealt with. The court should consider: first, whether it would be appropriate to have a brief adjournment for the paying party to consider the statement and then proceed to a summary assessment of the costs. In that event, the judge should err in favour of awarding a lighter figure; secondly, whether the matter should be stood over for a detailed assessment; thirdly, whether the matter should be stood over for a summary assessment at a later date or for summary assessment to be dealt with in writing.

Where there is an adjournment for a summary assessment at a later date or an order for a detailed assessment, the receiving party will, of course, be at the risk of paying the consequential additional costs.

THE SUMMARY ASSESSMENT HEARING

[27.18]

Whilst *Morgan v Spirit Group Ltd* (above) is clear that the court must conduct a summary assessment process, what that is will vary from judge to judge. Some will take longer than others. Some will prefer to deal with the challenges one by one and make a ruling before moving on to the next area of dispute. Others will hear all the challenges and then produce one judgment dealing with them all. Neither approach is wrong. This permissible variation is not surprising when one considers the actual definition of summary assessment, which is less than enlightening. It appears at CPR 44.1(1) and states that a summary assessment 'means the procedure whereby costs are assessed by the judge who has heard the case or application.'

27.18 *Chapter 27 Summary Assessment*

In essence the court must consider the costs in the same way that it would within a detailed assessment, but does so within a much condensed time span – for example the White Book carries a specimen timetable for a fast track trial that allows 30 minutes for costs (both the award and assessment of these). Advocates will raise the same points routinely seen within Points of Dispute, such as challenges to:
- The indemnity principle
- The hourly rate/grade of fee earner
- The amount of time spent
- The sums spent on disbursements
- The amount of counsel's fees (where not fixed)
- The overall proportionality of the costs

The court must determine these challenges by reference to CPR 44.3(2) on a standard basis assessment and CPR 44.3(3) on an indemnity basis (see Chapter 24 Bases of Costs, above). The court will also have regard to the factors at CPR 44.4 (the seven pillars of wisdom as was – now eight with the introduction of reference to the last agreed or approved budget). These factors are examined fully in Chapters 28-30 Detailed Assessment (below).

HOURLY RATES

[27.19]

Whilst hourly rates will be considered in far more detail in Chapter 00, the Senior Courts Costs Office ('SCCO') guideline hourly rates included within its *Guide to the Summary Assessment of Costs* is, if not peculiar to summary assessments, certainly more particularly directed at these than at detailed assessments. In summary assessments it is rare to see arguments about the guideline rates, other than as to the appropriate grade of fee earner suitable to conduct the work carried out and whether the location of the solicitor (rather than the client and the court) determines the appropriate rate.

The guidelines set out the hourly rate for different levels of fee earner in different regions of the country. The title of the document gives a number of clues – namely that the rates are guideline only and are for use in summary assessments. However, the broader application of the rates is a topic for discussion elsewhere as already noted.

After their first introduction the guidelines were updated regularly. Most procedural text books will show the relevant rates back to 2003 and most courts have their own tables of the rates to assist judges when dealing with older cases. Until 2005 rates were set for each court. After that the country was divided in to 4 bands (one being London, although that is in turn divided into 'City of London', 'Centre' and 'Outer London') and different rates were set for each band. From 2009 onwards, the minimal differential between Band 2 and 3 led to them being merged with the same rates and to the then remaining outside London bands being re-titled National 1 and National 2. (What also seemed to happen that year and has previously gone unremarked upon, was that East Sussex became lost in the relocation and, having been in band 1 previously, has escaped the clutches of the guideline rates ever since.)

However, the rates have not been updated since January 2010, when they were broadly increased in line with inflation. This has been deliberate. The combination of implementation of the recommendations from the final report of Sir Rupert Jackson and the 'will they be, won't they be' approach of the government to the introduction of fixed fees in the fast track, meant that the future of guideline hourly rates became uncertain. However, a working party is now looking at the rates again.

The current rates are divided between four categories of fee earner:
- Solicitors with over eight years' post qualification experience including at least eight years' litigation experience
- Solicitors and legal executives with over four years' post qualification experience including at least four years' litigation experience
- Other solicitors and legal executives and fee earners of equivalent experience
- Trainee solicitors, paralegals and fee earners of equivalent experience

Note that the word 'partner' does not appear – the rates are set by experience and not status. Members of the Institute of Legal Executives who are not qualified as legal executives either come within the lowest category or must argue a case based on equivalent experience. It is surprising how often arguments emerge on assessment as to whether or not a particular fee earner meets what appear to be clearly defined criteria – and not just in terms of equivalent experience, but as to whether they have the period of post qualification experience necessary to come within a particular category. The level of offended indignation can be high.

COUNSEL'S FEES

[27.20]

Counsel's fees are prescribed in fast track trials under CPR 45.38. Beyond this there are no set rates for counsel. The multiplicity of scenario under which the court is assessing counsel's fees militates against this. However, the SCCO Guide to the Summary Assessment of Costs does set out some figures based on the SCCO's statistics for what it describes as 'run of the mill proceedings'. The commentary stresses that these are not recommended rates, but reference points for judges looking for starting points upon which to assess counsel's fees.

PROPORTIONALITY

[27.21]

Whilst the FORM N260 requires the provision of information by reference to hourly rate and time spent by solicitors, remember that once the court has gone through the challenges and determined what is reasonably incurred and reasonable in amount it must, on standard basis assessments, then step back (whether or not the parties request it to do so) and determine whether the sum

27.21 *Chapter 27 Summary Assessment*

so assessed is proportionate. If it is not then under CPR 44.3(2) it will then reduce the costs further by reference to the definition in CPR 44.3(5) to such sum as it deems to be proportionate. Increasingly, therefore, an in depth consideration of hourly rate may not determine the ultimate outcome of the summary assessment.

APPEALS FROM SUMMARY ASSESSMENT

[27.22]

See Chapter 33 Appeals against Assessments (below)

CONCLUSION

[27.23]

It is clear from the final report of Sir Rupert Jackson that summary assessments are here to stay. Indeed with the introduction of costs budgets and the increased emphasis on proportionality we believe that there will be an increased incidence of these assessments. Parties should be prepared for this. Failure to comply with the procedural requirements for filing and serving the statement of costs and departures from the FORM N260 format are likely to result in less lenient treatment than before April 2013. Compliance is likely to result in far earlier recoupment of costs from the paying party.

There is still more work to be done. Guideline hourly rates are being reconsidered. Work is under way to link the format for bills for detailed assessments to the budget phases, which should mean that something offering a better comparison between costs claimed and costs budgeted ought to be available even on a summary assessment. As Sir Rupert stated in the final report:

> 'A new software system should be developed, which will be capable of generating bills of costs at different levels of generality. An intermediate level of generality should be used for the purpose of assisting the court to carry out summary assessment of costs.'

Finally, the introduction of costs management means, inevitably, that the judiciary across the board will have to develop a level of understanding of costs never previously required. The experience acquired will be brought to bear at a summary assessment and the number of judges feeling uncomfortable with the process, and so the occasions of deferring costs to a detailed assessment, should reduce.

CHAPTER 28

DETAILED ASSESSMENT – PROCEDURE

INTRODUCTION

[28.1]

Detailed assessment proceedings in many ways look like the baby brother of the proceedings in which an order for costs was originally created. The parties exchange Statements of Case (see Chapter 29) and then have a hearing (sometimes now on paper) if they cannot resolve their differences. In larger cases, there may also be the need for witness statements and even expert evidence. One area in which detailed assessment proceedings are unique is that of disclosure. It is usually only the documents held by the party whose bill is being assessed – the receiving party – that are of interest. Some documents, such as the inter-party correspondence and exchanged witness statements, are obviously in the hands of both parties. Some documents, such as receipts to vouch for the larger disbursements claimed, need to be disclosed to the opponent –the paying party. Most others, however, are protected by privilege and so do not have to be disclosed to the opponent. Nevertheless the court will have seen them and the receiving party will seek money in relation to the work done that is revealed by them. This looks completely contrary to all notions of natural justice. The judge takes on the role of protecting the paying party by raising matters of concern, eg on the retainer terms or as part of assessing whether the time claimed for an activity in the bill is reasonable as evidenced by, for example, an attendance note. It is a position which gives rise to certain issues and these are discussed at **[28.41]** below.

The structure of this Chapter is to deal with some specific considerations first, then to deal with the procedure in detail and finally to look at assessments in some discrete areas.

METHOD OF ASSESSMENT

[28.2]

CPR Rule 44.6 provides that

'The amount of any costs ordered to be paid by one party is ascertained either by
(a) A summary assessment by the judge at the end of a hearing [see Chapter 27]; or
(b) A detailed assessment by a costs officer. The provisions for detailed assessment are contained in CPR Part 47.'

A 'costs officer' encompasses costs judges, district judges and authorised court officers (CPR 44.1). For more information on authorised court officers see [28.55]).

28.2 Chapter 28 Detailed Assessment – Procedure

A provisional assessment may take place if the bill is valued at £75,000 or less in accordance with CPR 47.15. It is essentially a detailed assessment carried out by the judge without the parties attending. As the name implies, it is not necessarily the final word and the parties can seek what amounts to a detailed assessment of some or all of the bill if they wish to challenge a provisional assessment (see Chapter 00).

So, in fact there are really only two forms of assessment, summary or in detail. A provisional assessment is a subsidiary form of detailed assessment for smaller bills.

PRELIMINARY ISSUES

[28.3]

This part of the chapter looks at the following questions. Can I start the detailed assessment proceedings yet? If so, where should that be? How do I stop the procedure if I reach agreement with my opponent?

ENTITLEMENT TO START DETAILED ASSESSMENT PROCEEDINGS

[28.4]

There are two points to consider, and the first one will not take very long in the great majority of cases.

Stayed by appeal?

[28.5]

If your client has an order for costs at the end of proceedings, he is entitled to have those costs assessed regardless of whether the opponent seeks to appeal that order as part of an appeal against the judgment (or even if it is the only thing he is seeking to appeal). CPR 47.2 is admirably brief and clear on this point. If the opponent wants a stay of the detailed assessment proceedings, he needs to apply to the court for an order. Do not be seduced by arguments that the assessment will be a 'sterile exercise' and therefore pointless. Most appeals fail so the chances are the assessment will stand and your client will be able to recover his money more quickly than if he had waited.

Of course what can be done is not necessarily what should be done. If you consider the judgment may well be overturned on appeal you may wish to ponder the advisability of having your client's bill drawn in the meantime. If your fears are realised, the only person paying the costs of having the bill drawn will be your client.

If you contemplate waiting until the opponent's appeal has been heard, make sure you protect yourself against subsequent arguments about the delay in commencing detailed assessment proceedings. It is remarkable how often a paying party is prepared to argue that CPR 47.2 is clear and his opponent should have got on with the detailed assessment proceedings because the appeal proceedings were 'obviously hopeless' from the outset. At the very least, you should offer to delay the proceedings only upon condition that no arguments regarding delay can be brought for the period that the appeal proceedings are on foot.

The same position applies in the rare situation in personal injury cases where provisional damages have been ordered. The proceedings are at an end when that order is made and the costs can be assessed (CPR PD 47, para 1.1).

'Forthwith' orders

[28.6]

The general rule is that costs of proceedings will be assessed at the end of the proceedings, rather than part way through (CPR 47.1). This obviously does not apply in relation to the costs of interim applications which are dealt with summarily at the end of the application.

It does mean that for this general rule to be usurped, the court has to make an order that the costs are to be assessed 'forthwith' or 'immediately' or a similar term. The usual order that costs are 'to be assessed on the standard basis' will not do. Detailed assessment proceedings brought under such an order whilst the substantive case is still going can be challenged immediately by the paying party and the notice of commencement may be set aside (CPR PD 47, para 1.3).

What happens if your client has an order for the costs regarding liability (or similar) at the end of a split trial? Do you have to wait until the loss has been quantified? The answer is no. Rule 47.1 talks of the costs of 'proceedings, or any part of the proceedings' being concluded. Whilst the provision can be read either way, it is generally assumed to mean that a costs order in respect of a discrete part of the proceedings can be assessed whilst the remainder continues. There is, in any event, nothing to prevent the parties agreeing to treat the proceedings as concluded even though they are continuing (CPR PD 47, para 1.2). Appeal proceedings are treated as separate proceedings. As such, the costs of the appeal can be dealt with even though the case continues at first instance following the appeal.

You may wonder why this rule exists. Its purpose is to ensure that all costs are assessed by the same person at the same time and this continues notwithstanding the introduction of summary assessments at the end of interim hearings. The assessing judge may have bills from both parties for different periods of the case. There may be several different parties who have bills to be assessed and some of whom were released from the proceedings before the end. Considering whether bills are globally disproportionate, whether at the beginning of the assessment or at the end, is much more difficult where the costs involved are only for part of the case.

PD 47 para 1.4 contains the useful provision that where there has been an order for a detailed assessment of interim costs and there is no prospect of the claim continuing, the court may authorise the commencement of detailed assessment proceedings.

VENUE

[28.7]

The rule regarding where to commence detailed assessment proceedings (CPR 47.4) is, at first sight, a triumph of opaque drafting. You may not be surprised to learn that proceedings should be commenced at 'the appropriate office.' This may be the SCCO or a county court which is different from the one in which the case was proceeding. This is so, even if a formal transfer of the case is not actually made.

In fact the provision for the correct venue is quite detailed and so the definition of 'the appropriate office' is better left to the practice direction (CPR PD 47, para 4) than trying to draft an unwieldy rule in Part 47. The arrangements are as follows:

- High Court and Court of Appeal cases – the SCCO;
- District Registries or county court cases:
  - The Registry/court in which the case proceeded when the order for costs was made; or to which it has subsequently been transferred; unless it is a
  - 'London' county court – who refer cases to the SCCO (see below)
- Tribunal, body or other person – a county court (unless in 'London').

'London' cases

[28.8]

Cases which have been run in any of the following county courts do not deal with detailed assessment hearings. As such, when a request for a detailed assessment hearing is made (see below) the request needs to be sent to the SCCO. If it is sent to the original court, it will be returned with a brief message to that effect. The original court will deal with any interlocutory stages before the request, for example, setting aside a default costs certificate. It will also deal with enforcement proceedings arising from the sum assessed by the SCCO. In some ways, the SCCO acts as a trial centre for detailed assessment proceedings. The courts affected are set out in the practice direction as follows:-

Barnet	Ilford	Central London
Bow	Kingston	Clerkenwell & Shoreditch
Brentford	Lambeth	Mayors and City of London
Bromley	Romford	West London
Croydon	Uxbridge	Willesden
Edmonton	Wandsworth	Woolwich

'Difficult' cases

[28.9]

In addition to taking on London cases, the SCCO regularly has files referred to it by a court that would otherwise be 'the appropriate office.' The criteria according to the practice direction (para 4.3(2)) are:

- The size of the bill
- The difficulty of the issues involved
- The likely length of the hearing
- The cost to the parties
- Any other relevant matter

A case is only meant to be referred if it is appropriate to do so based on these criteria. In practice, most county courts will take little persuading to transfer a case which might require several days of court time to be dealt with, particularly if both parties are agreed that this should happen.

Regional Costs Judges

[28.10]

In order to limit the need to refer cases physically to the SCCO, an alternative was introduced of 'regional costs judges' who comprise district judges with a particular interest or expertise in costs and who could deal with the case more locally. Such judges can have cases transferred to them if they (the cases) contain complex arguments on points of law or an issue affecting a group of similar cases are identified in the points of dispute or reply, or indeed are referred to in argument at a detailed assessment hearing itself. Alternatively, cases may by referred if they are of sufficient value or will take a long time to deal with. The current criteria thresholds are £100,000 and at least two days respectively. Cases concerning costs involved in Court of Appeal proceedings can only be dealt with by the SCCO even if the case originated in the county court. The most recent list of regional costs judges can be found in the SCCO Guide 2013.

REACHING AGREEMENT

Prior to requesting a hearing

[28.11]

Most cases issued in the courts settle before they reach a final hearing. So too do most detailed assessment proceedings. Where agreement is reached during the exchange of the statements of case (see Chapter 29), there is no court involvement – documents are only served, not filed – and so there is no need to do anything formally to bring the detailed assessment proceedings to an end. It is, in effect, a settlement pre-proceedings.

Formalising the agreement

[28.12]

The court is engaged once the receiving party seeks a detailed assessment hearing. The case will be given a hearing date, albeit that in larger cases, it may only be a directions hearing. Rule 47.10 enables either party to seek a costs certificate to confirm the agreed sums for the bill and the costs of the detailed assessment proceedings themselves. Such a certificate may be an interim or final certificate. The benefits of the certificate are (a) to protect either party from any subsequent suggestion that the agreement reached was on different terms and (b) to enable the receiving party to enforce the agreement if payment is not forthcoming. (If the paying party had paid more on account than was agreed in settlement, the protection would be for the paying party but that would be a very rare occurrence indeed.)

A receiving party may discontinue the detailed assessment proceedings in accordance with Part 38. Once a request for a hearing date has been made the receiving party needs the court's permission to do so. This provision in the practice direction (CPR PD 47, para 9.4) aims to protect the paying party from being left without a mechanism to seek any costs he has incurred in the detailed assessment proceedings. However, given that the receiving party has already been awarded costs in the substantive proceedings, it will be a rare case where he decides to discontinue the detailed assessment proceedings and thereby not receive any costs at all.

Informal agreement

[28.13]

In practice, most parties reach agreement on the quantum payable to the receiving party and are content to deal with payment without the benefit (and cost) of obtaining a formal certificate. An exchange of letters will suffice for the parties. A letter from the receiving party to the court will almost always be sufficient for a hearing to be vacated, regardless of whether the paying party also writes to the court.

THE DETAILED ASSESSMENT PROCEDURE

[28.14]

This section, as the title suggests, deals with the procedure of detailed assessment proceedings. The issues regarding the drafting of the bill of costs and subsequent documents – what we have called the 'statements of case' – are dealt with in Chapter 00. For the purposes of this Chapter, the documents are assumed to be compliant with the requirements of the CPR and case law.

COMMENCING DETAILED ASSESSMENT

[28.15]

Detailed assessment proceedings are commenced by the service of a notice of commencement in form N252 together with a bill of costs. These documents obviously need to be served on the paying party. They also need to be served on any other 'relevant person' as specified in CPR PD 47, para 5.5. Any such relevant person becomes a party to the detailed assessment proceedings in addition to the receiving party(ies) and paying party(ies). Who might be a relevant person? The practice direction describes three broad categories:

'Any person who has already taken part in the proceedings and who is directly liable under a court order for costs. (This would normally describe the paying party.)
(a) Any person who tells the receiving party that he has a financial interest in the outcome of the assessment and wishes to be a party.
(b) Any other person whom the court orders to be treated as such.'

If a receiving party is in doubt about the relevance of a person, he can ask the court for directions (CPR PD 47, para 5.5(2)), or simply serve on that person anyway and leave them to decide whether to take any part. The court is unlikely to make that person a party to the proceedings unless he applies to do so in accordance with Part 19.

Notice of Commencement

[28.16]

The notice of commencement contains the bare bones of information. The order for assessment (or equivalent) and its date needs to be set out. So too does the total claimed in the bill of costs and, separately, a further sum which will be added to the total of the bill if a default costs certificate is obtained in the absence of the paying party's points of dispute. This sum comprises fixed costs together with the court fee and currently amounts to £140 (see [28.27]).

The time by which points of dispute need to be served also has to be inserted into the form. The general rule is 21 clear days and is looked at in more detail at [28.24]. If the notice is to be served outside England and Wales you will need to consult the rules in Section IV of Part 6 to make sure that sufficient time is allowed. The notice is treated as if it were a claim form and the points of dispute as if they were a defence.

Additional documents

[28.17]

In addition to the notice of commencement and bill of costs, the receiving party needs to serve any or all of the following documents, where appropriate:
- copies of the fee notes of counsel and any expert whose fees are claimed in the bill;
- written evidence as to any other disbursements which are claimed and which exceed £500;
- where recoverable success fees or ATE premiums are concerned, the further documents and information required by Section 32 of the (old) Costs Practice Direction;
- a statement of parties on whom the notice of commencement etc is being served. (Although CPR PD 47, para 5.2 suggests this needs to be done in all cases, this is often ignored where the paying party or parties are clear and there is no other relevant person.)

No filing

[28.18]

It is well worth noting that the first part of the detailed assessment procedure is entirely a matter between the parties; there is no involvement of the court at this stage and if, as often happens, agreement is reached between the parties, the court will not have been troubled by the detailed assessment proceedings at all.

Delay in commencing proceedings

[28.19]

The period for commencing detailed assessment proceedings is three months after the event giving rise to the right of assessment (CPR 47.7). The most common event is a court order for costs but it could also be, for example, an acceptance of a Part 36 offer or a notice of discontinuance.

If the receiving party does not commence detailed assessment proceedings within the time prescribed by the rules (or by any direction of the court), the paying party's remedy is to make an application under Part 23 for an 'unless' order. CPR 47.8 empowers the court to order that unless the receiving party commences proceedings within a specified period all or part of the costs will be disallowed. Such applications are in fact rarely made, perhaps because the paying party does not wish to bring forward the day when he will have to hand over money to the receiving party. But where the application is made, the receiving party invariably ends up serving a notice of commencement and bill of costs to begin the detailed assessment proceedings before the application is heard so the application has a galvanising effect. Furthermore, the receiving party almost always ends up paying the costs of the application since there are few cases where the receiving party could not feasibly have a bill drawn and served a notice within three months of the right to do so having arisen. Moreover, the application is the gateway to seeking any sanction greater than interest being disallowed.

Where the paying party has not made an application under CPR 47.8, there is nothing to prevent the receiving party from commencing the assessment proceedings at any time, the only sanction being the court's power to disallow interest for the delayed period (CPR 47.8(3)(b)). Whilst the court's power is entirely discretionary as to the period of interest to be disallowed, conventionally it is for the period from three months after the order for assessment until the notice of commencement is served, which can be several years later.

'Misconduct' in the form of delay

[28.20]

Where a paying party makes an application under CPR 47.8, it is sometimes accompanied by an application under CPR 44.11 which empowers the court to disallow all or part of the costs where there has been misconduct. If the application is only made as part of the points of dispute to be heard at the final detailed assessment hearing, rather than at the beginning of the proceedings, the case law does not hold out much hope for applications under CPR 44.11 based on costs delays.

In *Botham v Khan* and *Lamb v Khan* [2004] EWHC 2602 (QB) the court held that disallowance of costs pursuant to CPR 44.11 (then CPR 44.14) on the ground of misconduct would be a disproportionate sanction. Although there was culpable delay by the defendant, the claimants had not availed themselves of the right to apply under CPR 47.8(1) for an order requiring the defendant to commence detailed assessment within a specified period. The delay had not prevented a fair assessment of the defendant's costs although the process would be more difficult than if it had been carried out on a timely basis.

The paying party was equally unsuccessful under CPR 44.11 in *Haji-Ioannou v Frangos* [2006] EWCA Civ 1663, even though the delay was for more than five years. The court decided that CPR 47.8 and CPR 44.14 (now CPR 44.11) were not inconsistent and there was no case for imposing upon the defendants the further sanction of disallowing any part of their costs under CPR 44.14 in addition to loss of interest under CPR 47.8. Their delay was not deliberate or wilful and the claimant had not himself been a model of expedition. Where the relevant rule not only gave the paying party the option of preventing further delay by himself taking the initiative but also spelled out the normal sanction for such delay, the court should be hesitant to impose further penalties by way of reducing otherwise allowable costs.

Delay no breach of Article 6

[28.21]

The fact that the remedy for delay is in the hands of the paying party was clearly demonstrated in *Less v Benedict* [2005] EWHC 1643 (Ch). The claimants had been ordered to pay the defendant's costs and the defendant served notices of commencement within the time limit specified in CPR 47.7. But, as a result of confusion and oversight it was not until three and half years later that the defendant re-served the notices on the claimants at their last known addresses. The claimants submitted their right to a hearing within a reasonable time under Article 6 of the European Convention on Human Rights would be breached if the costs assessment was allowed to continue; the excessive and unreasonable delay without explanation was an abuse of process; the claimants could not have a fair hearing as they no longer had access to the relevant files. The court held there had been no violation of the claimant's rights under Article 6 because CPR 47.8 provided a mechanism for the claimants to bring the matter to the attention of the court to obtain a hearing within a reasonable time. If a party failed to take advantage of that mechanism it could not be said that he had thereby been deprived of his rights under Article 6. The court would not, by proceeding to a hearing, be sanctioning a continuance of a breach of any rights, but would be taking remedial action to correct the consequences of such a breach.

STATEMENTS OF CASE

[28.22]

You will find all of the drafting requirements of the bill of costs, points of dispute and replies in Chapter 29.

Serving bills of costs

[28.23]

As we have seen, the bill of costs is served with the notice of commencement. Sometimes, it is informally served upon the paying party as part of a negotiation of costs. Whilst that is sometimes seen as an aggressive form of negotiation, particularly if no schedule of costs has been previously provided, it certainly concentrates the mind of the paying party. Where the bill is for a substantial sum, it will take some time to draft and there is plenty to be said for going straight for a bill rather than producing a schedule. The cost of preparing the bill is conventionally the last item of the bill itself and so is recoverable in negotiation. The cost of producing a schedule is sometimes disputed by the paying party and there appears to be no direct authority on the point.

If the paying party asks for an electronic copy of the bill at any time prior to the detailed assessment hearing, the receiving party must provide it free of charge within 7 days of the request, always assuming that it can be copied electronically. This is a new provision in the 2013 recasting of the rules and practice directions (CPR PD 47, para 5.6). Previously a copy could be requested on a disk but time has moved on. Most parties emailed a copy of the document rather than sent it on a disk anyway. The new wording introduces us to the concept of a 'native format' which is explained in parentheses as 'for example, in Excel or an equivalent.' It is not hard to see the Rules Committee grappling with a suitable phrase which is explanatory but which is not prescriptive. Most bills are created using proprietary costs drafting software rather than Excel. That is likely to increase if and when the style of bills alter (see [29.25]. Perhaps, like the BBC, the Rules Committee is not meant to advertise any particular brand of such software.

Serving points of dispute

[28.24]

The points of dispute need to be served within the time period set out on the notice of commencement. This is a minimum of 21 days after the date of service of the notice. (Whilst the rules provide for the parties to agree to shorten this period, in practice a paying party who wants to get on with proceedings will simply serve his points of dispute as soon as possible). In order to deal with the vagaries of the post, some receiving parties add a day or two to the minimum period. Others will allow for a longer period, especially on larger bills, rather than have to deal with a request for further time.

The points of dispute need to be served on every other party in the detailed assessment proceedings. This is the purpose of the statement of parties referred to above. It is surprisingly common that paying parties overlook serving points of dispute on the other paying parties.

According to CPR 47.9(3) a party who serves his points of dispute outside the 21 day period may not be heard further in the detailed assessment proceedings unless the court gives him permission to do so. This is very much a backstop provision aimed at enabling a court to deal with one of several paying parties who serves his points of dispute late and takes little or no part until much later in proceedings. It would be a very rare case indeed that a court would decline to give permission for a party liable to pay the costs assessed by the court to be heard on that assessment. Nevertheless, Akenhead J considered this provision to amount to a sanction in *Baker v Hallam Estate Ltd* [2013] EWHC 1046 (QB) and so an application for relief from sanctions is required.

The real risk in not serving the points of dispute promptly is the issuing of a default costs certificate (see [28.27] below). If the points of dispute are served out of time but still arrive before the court has issued a default costs certificate, the court must not do so (CPR 47.9(5)) and if it has done so in ignorance of that service, the receiving party should inform the court that he is not entitled to the certificate and the court will set it aside without an application (CPR 47.12(1)).

Offer with Points of Dispute

[28.25]

Prior to April 2013, a paying party needed to make an offer (then a Part 47 offer) within 14 days of being served with the bill of costs for that offer to have maximum costs protection when the court came to consider the costs of the detailed assessment proceedings.

When the rules and practice directions were recast, this provision was removed. Instead, CPR PD 47, para 8.3 brought in a new requirement which appears destined to require judicial consideration to interpret its precise requirements.

The new provision requires the paying party to state in an open letter accompanying the points of dispute *'what sum, if any, that party offers to pay in settlement of the total costs claimed.'* Such offer appears to be separate from any Part 36 offer since the practice direction specifically states that a paying party may also make a Part 36 offer.

What is the purpose of making a paying party make an open offer? Presumably, it will generally be for a lesser sum than any Part 36 offer since otherwise there would be no purpose in Part 36 offers. But if it is lower, the paying party will generally wish to rely on any separate Part 36 offer anyway. Since the provision qualifies the requirement with 'if any' it suggests that there may not have to be any offer made anyway. It is easy to see that the Rule Committee was likely to have a few qualms at obliging a paying party to make an obligatory offer. As we have seen elsewhere in this book, it is not impossible for the entire bill to be disallowed, particularly where a CFA or DBA is used. If the paying party is putting forward indemnity principle arguments, it may well not want to make any offer at all.

With the recasting of CPR 3.9 regarding relief from sanctions and the general increase in robust case management, a failure by a paying party to make an offer with the points of dispute, affords the receiving party with an apparent opportunity to compel an offer to be made, or a sanction to be applied. However, it would seem to be a disproportionate sanction to allow a default costs certificate on the basis that no points have been served if no offer has accompanied them (as some receiving parties have sought to persuade the courts).

The pre-April 2013 arrangements were never entirely satisfactory, particularly where offers were made rather later in the detailed assessment proceedings. Whilst time may tell otherwise, it does not appear that this provision is going to be an improvement. In practice, the paying parties' Part 36 offers will be used to consider the costs of the detailed assessment proceedings. These open offers are more likely to be referred to solely in relation to issues of conduct.

Optional Replies

[28.26]

CPR 47.13 entitles the receiving party to serve replies within 21 days of service of the points of dispute. The heading in the rules is 'Optional Reply' and there is no compulsion on the receiving party to serve any such document. The benefits and limitations of serving a reply are dealt with in more detail in Chapter 29.

The assessing judge will generally expect to see a reply, particularly where a provisional assessment is being conducted. If, for example, a retainer issue is raised in the points of dispute, the reply is the perfect place in which to dispel any doubts raised by the paying party. The judge will not thank a receiving party for failing to provide such an answer if it would have been simple to do so and instead he has had to spend time trawling through the receiving party's papers to find the answer.

OBTAINING A DEFAULT COSTS CERTIFICATE

[28.27]

The ability to obtain a default costs certificate is a continuation of the philosophy of not involving the court until it is clear that there is a dispute for the court to resolve.

Where points of dispute are not served within 21 days of service of the notice of commencement, the receiving party is entitled to apply to the court for a default costs certificate under CPR 47.9(4). He does so by using forms N254 and N255 (form N255 (HC) for the High Court) which are a request for a Default Costs Certificate and the Certificate itself respectively. The request provides sufficient information regarding the service of the notice of commencement and the amount in the bill to enable the court to process the request in a stand-alone fashion. Where a request is made to the SCCO, there will be no court file at that point so the necessary information and paperwork has to be provided. Even where there is an existing court file, the court will be unaware of the amount of the costs in the bill and when the paying party's 21 day period expired, hence the need for a formal request document.

In the same manner as a request for default judgment on a specified damages claim, the only costs recoverable for seeking judgment are fixed by the court rules. The fixed costs are set out in CPR PD 47, para 10.7 at £80; a sum which has not altered since the CPR came in to force in 1999. The court fee is currently £60.

The default costs certificate requires payment within 14 days as with other court orders. The paying party's time from the moment he is served with the notice of commencement to receiving a default costs certificate may therefore be less than six weeks.

A paying party faced with a default costs certificate therefore either has to seek to set that certificate aside, if he wants to challenge the sums claimed, or to deal with payment. If he cannot pay, he can make an application to the court to stay enforcement proceedings. Such applications will be heard by either the court which issued the certificate or one which has general jurisdiction to enforce the certificate (if different).

Enforcement proceedings based on the certificate will similarly be brought by the receiving party in such courts, save that no enforcement proceedings can be brought in the SCCO.

Pro Bono Representation

[28.28]

Where the receiving party is represented by a legal team acting under s 194 of the Legal Services Act 2007, a copy of the default costs certificate (or interim or final costs certificate) needs to be sent to the nominated charity.

Pro bono representation under the Act is distinct from a solicitor or counsel agreeing to act without fee for a party. The pro bono lawyer needs to nominate one of the prescribed charities who will receive the benefit of the costs that would otherwise be payable by the client to that lawyer. In effect, the costs are received from the opponent in a successful case, rather than the client, and the indemnity principle issue that arises is suspended by the provisions of the Act.

Once the pro bono agreement is set up, the charity is clearly entitled to know when an order in its favour is made. Similarly, if a certificate is set aside for any reason, it is entitled to know about that as well. The relevant rules dealing with this notification are:
- default costs certificates – CPR 47.11(4) and CPR 47.12(3);
- interim costs certificates – CPR 47.16(4);
- final costs certificates – CPR 47.17(6).

SETTING ASIDE A DEFAULT COSTS CERTIFICATE

Mandatory setting aside

[28.29]

The court must set aside a default costs certificate if the receiving party was not entitled to it according to CPR 47.12(1). That is not a surprising situation you might think. But it helps to set it out directly so that receiving parties do not seek to oppose a set aside application on grounds similar to those set out below. It would be all too simple for a receiving party to argue that the court should not set aside an invalidly obtained certificate unless the paying party had demonstrated some 'good' reason to do so. Arguments that the same result was inevitable and that the courts' resources should not be used to set aside something that was ultimately going to be achieved anyway can be envisioned. You might think it a rare case that all of the costs in a bill are allowed on an assessment, but it is not unheard of. Moreover, if the paying party did not provide points of dispute with the application, it could well lead to the receiving party to argue that this demonstrated the paying party had no real arguments anyway.

The rules prior to April 2013 placed an obligation on the receiving party to set aside an invalidly obtained default certificate. This would usually be where the paying party had not been given the minimum period of time to serve points of dispute, or there had been a problem with service of the notice of commencement, or a combination of both. That provision has gone and so it would appear possible for the paying party to apply to the court for a mandatory setting aside as well as for the normal discretionary setting aside. Only a receiving party's request can be dealt with by a court officer (CPR PD 47, para 11.1) but a paying party's request could be dealt with by a judge without a hearing if the circumstances were sufficiently clear cut. In practice, such applications are likely to be listed in the expectation that the receiving party will agree to the application if the circumstances are as the paying party alleges.

Discretionary setting aside

[28.30]

In order to set aside a validly obtained certificate, a paying party must demonstrate to the court that there is 'some good reason' why the certificate should be set aside so the proceedings can continue (CPR 47.12(2)).

Reason for the default

[28.31]

There is a need to make the application swiftly because one of the factors the court will take into account is whether the application has been made 'promptly' (CPR PD 47, para 11.2(2)). But, it is equally important that the application is made with evidence to support it (CPR PD 47, para 11.2(1)). Parties often concentrate on the reasons why they had failed to serve the points of dispute in time, even though such 'reasons' are no more than over-work, lack of organisation etc which, it ought to be clear objectively, are not reasons but simply a description of the occasional lapse in a busy professional's life. Whilst each party must decide for themselves as to how much detail they need to put in to describe the oversights, he should realise that the judge will have heard such comments many times before and have his own view about how acceptable such conduct is. It might be thought better simply to take the blame on the chin and move on to the 'good reason' part of the application as quickly as possible. Obviously if there is an issue about any service of the document at

Setting aside a Default Costs Certificate 28.33

all (or simply delay so the paying party has not been given the requisite time), full detail is required because, if proved, the certificate would have to be set aside regardless of any other merit as discussed under the previous heading.

Good reason to continue

[28.32]

In addition to the application notice form N244, the application should exhibit a copy of the bill and default costs certificate (which the court will not otherwise have). Furthermore, there should be a copy of the points of dispute which the paying party proposes to serve if the certificate is set aside. CPR PD 47, para 11.2(3), which sets out these requirements, says that certificates will only be set aside 'as a general rule' if these documents are filed and served with the application. Preparing points of dispute costs money and if the application is not granted, that money will have been wasted. But do not let that deter you from having them drafted for they are fundamental to the application. They show that the paying party is serious about wanting to take part in the detailed assessment proceedings notwithstanding the default certificate. They also provide the court with a clear explanation of why the case should continue. If they do not do this, it can only be because there is little that is worth challenging in the bill. At that point, overriding objective considerations come into play regarding allocating further court resources to the case.

Points of dispute also take time to prepare. In order to make your application promptly, we would suggest that, where necessary, the application is issued whilst the points are being drafted, so that they can be filed and served prior to the application being heard. There may be some comment from your opponent about being unable to consider your application until those points are received, but unless your opponent is prepared to confirm that no point will be take about how promptly the application was made, the safer course of action is to issue pending the points of dispute being prepared.

Setting aside on terms

[28.33]

As with setting aside default judgments in substantive proceedings, the court can attach conditions to the terms on which the certificate is set aside. This power comes from the court's case management powers in Part 3 and as such its scope is wide. In particular, a court which has some doubts about the merits of setting aside the certificate may decide to require the defaulting party to make an interim payment of costs to the receiving party. CPR PD 47, para 11.3 refers to the expectation in CPR 44.2(8) for a court to make an interim payment of costs when it makes a relevant order. The fact that the judge hearing the application did not make the original order for costs does not detract from this power.

REQUESTING A HEARING DATE

Time for requesting a hearing date

[28.34]

CPR 47.14(1) requires the receiving party to request a detailed assessment hearing date within three months of the last date on which a notice of commencement could be served based on CPR 47.7 or any court order. Therefore, in theory at least, the interlocutory phases of the detailed assessment proceedings should take no longer than six months from the order of costs (etc) being made before the court is asked to fix a hearing.

In practice CPR 47.14(1) is treated in much the same way as CPR 47.7 regarding the commencement of proceedings. The rule is expressed in mandatory fashion ('must') but there is no need to seek relief from any sanction if the request is made outside the stipulated three months. The court will still provide a hearing date whenever the request in form N258 is made. Unless the paying party makes an application under CPR 47.14(2) to compel the receiving party to file a request(on pain of disallowing some or all of the bill), the only sanction which the court can impose for a breach of CPR 47.14(1) is one of interest.

For some reason, paying parties tend to take this point rather less often than in respect of delay in actually commencing detailed assessment proceedings. Perhaps it reflects the fact that any delay is usually related to the parties negotiating with a view to avoiding a court hearing.

Documents to accompany the request

[28.35]

CPR PD 47, para 13.2 sets out items (a) to (l(v)) which depending upon the circumstances, may be required to accompany the form N258 request. In order to assist the receiving party, form N258 provides a checklist of the documents required.

One of the documents is the 'document giving the right to detailed assessment'. This is usually the order from the court at the end of a contested hearing showing the order for costs. It may be the notice of acceptance of a Part 36 offer, or the notice of discontinuance. Less often it may be the arbitration award or an order or award from a statutory tribunal or other body.

Notwithstanding the introduction of the new precedent G in the costs precedents which combines the points of dispute and replies, the practice direction continues to talk of annotated points of dispute. The purpose of annotation is to show the judge which items have been agreed and their value and which items remain in dispute. Such annotation is rarely done, unless it is part of the reply that the receiving party is serving anyway. The use of Precedent G should mean that annotations become more consistent as to concessions made and their value.

The final document to go with the request is the fee which is payable by reference to the Civil Proceedings Fee Amendments (No 2) Order 2013, SI 2013/1410, which came into force on 1 July 2013. The fee begins at £325 for a bill that does not exceed £15,000 and increases to a maximum of £5,455 for bills of more than £500,000. The most that can be charged for a bill going to provisional assessment is £980.

What the court does

[28.36]

Having safely banked your money, the court office will consider whether the documents have been received in accordance with CPR PD 47, para 13.2. If they have not, a request for the missing documentation will be made or the papers will be returned depending upon how lacking the papers are initially.

The court will then look to see if the case should be dealt with by provisional assessment. The key document is the notice of commencement. If it is dated on after 1 April 2013 and the bill is said to be for no more than £75,000, it will usually be referred for provisional assessment. If there is a clear need for evidence to be given it will not be listed for a provisional assessment. (See Chapter 30 for more details.)

If the case is outside the terms of the provisional assessment criteria, it will be listed for a detailed assessment hearing, unless the court decides to give some directions or fix a preliminary appointment. Practices vary between courts. At the SCCO the costs judge or costs officer will review the file before giving directions (which may simply be to list the case for hearing.) In some other courts the parties are required to discuss the case to make sure that the issues have been narrowed as far as possible.

The court will give the parties (and anyone else who is relevant) at least 14 days' notice of the hearing unless they have agreed to short notice.

Disbursement only assessment

[28.37]

If the dispute only concerns disbursements, the court will ordinarily deal with the case on the papers, ie without the need for either party to attend. (There remains of course the party's entitlement to ask the court to order 'otherwise.') In such circumstances the judge will give a written reasons to the parties as to his decision (CPR PD 47, para 13.5).

INTERIM COSTS CERTIFICATE

[28.38]

Once a hearing date has been requested, a party may seek an interim costs certificate. In order to do so, an application notice is required in accordance with Part 23 (CPR PD 47, para 15).

The court has wide powers to issue a certificate for such sum as it considers appropriate. It may require the money to paid into court rather than to the receiving party. Where there is pro bono representation (see **[28.28]**), a copy of the certificate needs to be sent to the nominated charity. Similarly, if an existing certificate is varied or cancelled, the charity needs to be notified of the change.

28.39 *Chapter 28 Detailed Assessment – Procedure*

Provisional assessments

[28.39]

Some courts will issue interim costs certificates without a hearing but practice varies. Where the case will be heard by provisional assessment, the interim certificate application is unlikely to be heard before the assessment takes place. If there is no challenge to the provisional assessment, there will be a final certificate shortly after the hearing and the need for an interim certificate will be rendered otiose. However, if a post-provisional hearing is required, particularly if the paying party is the challenger, an interim certificate is likely to be issued at the point when the notice of challenge is received if an application has previously been filed with the court.

PROVISIONAL ASSESSMENT

[28.40]

Once a request for a detailed assessment hearing has been filed with the court, cases which fit within the requirements of CPR 47.15 and CPR PD 47, para 14 will be assessed provisionally. This procedure is dealt with fully at Chapter 30. If the provisional assessment is challenged, a hearing will be convened to deal with the item or items challenged. There are specific costs provisions in relation to such a hearing, but otherwise it will take place akin to a detailed assessment hearing.

THE DETAILED ASSESSMENT HEARING

[28.41]

One of the reasons that practitioners read textbooks and commentaries is, of course, to familiarise themselves with the procedure in a particular court and not simply to understand the applicable law. It is perhaps strange therefore, that such books do not normally describe how a hearing actually takes place. Sometimes, the practitioner has the benefit of seeing the process for themselves as part of their training or attending someone else's case, but that is not always so. There are certain peculiarities in the detailed assessment hearing which are likely to catch out even seasoned practitioners from other spheres. In that spirit, we offer the following description of a detailed assessment hearing, accepting that each judge will deal with things in his own way.

Prior to the hearing

[28.42]

The judge, at least in an ideal world, will have read the receiving party's papers that have been lodged with the court. In practice, this is more likely to have occurred at the SCCO, or where a regional costs judge (see **[28.10]**) is hearing the case. Practices vary between judges, but any time will have been used to consider specific points raised by the paying party, eg in respect of the retainer, or to get a flavour of how the receiving party's solicitor has conducted the file. For example, has there been a lot of time spent on activities that do not justify that time? Or has there been an economy of effort shown that more than justifies the time spent? Does the correspondence suggest endless letters to the client or others simply copying them into others' correspondence to keep them 'updated' or has the paying party's correspondence increased the time and effort required to deal with things than should have been the case?

Where there are points of principle, skeleton arguments are sometimes prepared and filed and served shortly before the hearing.

The paying party will be expected to have reviewed the points of dispute in the light of any replies to see if explanations or concessions have narrowed the dispute between the parties. Whilst this should have been done when the replies were served, the paying party may not have made any firm decisions on whether to accept any concessionary figures offered by the receiving party for example. The judge at the hearing is going to expect the paying party to set out his position clearly in respect of such matters.

The receiving party needs to remember to prepare for the hearing before lodging papers with the court. While some solicitors copy their files to send to the court (and therefore retain a copy to prepare from), most people lodge the original papers and so cannot prepare at the last minute. CPR PD 47, para 13.11 requires the papers to be lodged between 7 and 14 days before the hearing. (It is worth checking with the court due to hear the case as they vary in the enthusiasm with which they wish to receive papers beforehand. Cases at the SCCO and before regional costs judges will always require the lodgement of papers.) CPR PD 47, para 13.12 expects the papers to be lodged in the following order:

(i) instructions and briefs to counsel arranged in chronological order together with all advices, opinions and drafts received and response to such instructions;
(ii) reports and opinions of medical and other experts;
(iii) any other relevant papers;
(iv) a full set of any relevant statements of case;
(v) correspondence, file notes and attendance notes.

Whether it is the transit to the court that is the problem is unknown but it is rare to see the receiving party's papers properly in this order.

The theory is that the paying party is limited to raising the issues set out in the points of dispute. As a result the receiving party can prepare for the arguments it has to meet. That theory is subject to the three following points:

(1) If the explanation for one issue at the hearing leads to another point being legitimately taken by the paying party, the court will invariably allow that further point to be dealt with at the hearing.
(2) As discussed in Chapter 29 on statements of case, any of the major documents can be amended without permission prior to the hearing. It is a matter for the court to decide whether to allow those variations at the hearing itself. Sometimes the amended documents are served late in the day entitling the opponent to an adjournment but that is not always the case.

28.42 *Chapter 28 Detailed Assessment – Procedure*

(3) If the court considers the costs to be 'globally disproportionate' under the test arising from *Lownds v Home Office* [2002] EWCA Civ 365, [2002] 4 All ER 775, [2002] 1 WLR 2450, many judges will then consider each and every item in the bill regardless of whether the item was challenged in the points of dispute. If 'proportionality' is challenged in the points of dispute, the prospect of every item being considered is at least a theoretical possibility. (see *The* Lownds *test* below).

During the hearing

[28.43]

Whilst detailed assessment hearings are no longer held 'in chambers' but are in public, the only people able to attend the hearing and take part (rather than observe) are the receiving party, the paying party and any other party who has served points of dispute in accordance with rule 47.9. Of course, in an appropriate case, the court can decide to allow others to take part but that is only likely to be the case where it assists the resolution of the case and not simply because they may be affected by its result in an indirect fashion.

As mentioned at the very beginning of this chapter, certain notions of natural justice are apparently offended by the conduct of a detailed assessment hearing. There is no concept of an agreed bundle for the advocates to work from. The papers lodged by the receiving party are protected by privilege so they cannot be seen by the paying party unless privilege is waived. Nevertheless, they have been (or at least could be) seen by the judge and may conclusively determine points in issue between the parties. There are times when the privileged documents are of such importance to an issue which itself is important in the context of the assessment that the judge considers the document needs to be disclosed if it is to be relied upon by the receiving party. The judge cannot compel disclosure of a privileged document. He can, however, put that party to his election as to whether to disclose it or whether to rely on other evidence to support his position. This is discussed in more detail below under the heading *'Elections'*.

Usually any points of principle or points generally applicable to the bill are dealt with first at the assessment. It is sometimes the case that counsel, or other advocate, appear solely to deal with such matters before stepping aside for another person to deal with the detail of the bill. For larger bills, the court will sometimes list a hearing solely to deal with these preliminary issues. This may help the parties with questions of representation. More importantly, it minimises the extent of the preparation that needs to be undertaken by the court and in particular the parties, thereby minimising the cost of the hearing all round.

Where there are preliminary points, the court will decide as to which party it wishes to hear first depending upon the issue raised, although generally it will be for the paying party to go first. Such points are dealt with in a very similar fashion to an application before any other court. At the end of the parties' submissions the court will give its ruling. If there is a lot of material put before the court, including case law, the decision may be reserved.

When the detail of the bill comes to be considered, it is invariably for the paying party to make his submission first. Often there is little more to be said than is encapsulated by the particular point of dispute. The nature of a detailed assessment is to some extent explanatory. The paying party challenges an aspect as being irrecoverable for one of a number of reasons; the receiving party explains the context in which the item has been incurred and the judge rules on whether the item is allowed, disallowed or allowed in part. This process is repeated for each of the items in dispute. If any items are no longer pursued by the paying party, or have been conceded by the receiving party, the relevant item is marked as being allowed or disallowed as the case may be on the bill.

The final item on the bill is conventionally the cost of preparing the bill itself and any time claimed for the solicitor who signs the bill checking its accuracy and completing the certificates on the bill. These costs are claimed as being within the scope of the original order for costs. However, the costs involving in any costs negotiations, serving the notice of commencement and dealing with the statements of case etc are the costs of the detailed assessment proceedings and the receiving party needs to obtain a further order of the court to receive these (see next heading). Consequently, once the challenged items have all been determined, the parties need to be in a position to address the court on the costs of the detailed assessment proceedings. Those experienced in attending detailed assessment proceedings tend to be able to keep a running score so that they know how much the assessed costs are as soon as the final item has been ruled upon. Unless this is the case, the parties need to 'make up' the bill either precisely or at least sufficiently to be able to ascertain whether any offers are relevant to the question of the costs of the detailed assessment proceedings. Sometimes the parties make their calculations in the courtroom. Often they step outside so that they can clarify any figures without the pressure of the judge watching.

The parties then make any submissions about the costs of the detailed assessment proceedings. These may be on the principle of who should meet the costs (see next main heading for more detail). They may be only on the amount. Invariably the costs are summarily assessed rather than in detail (which would require a further hearing).

After the hearing

[28.44]

Traditionally, the original bill would be returned to the receiving party at the end of the hearing for it to be completed. It would then accompany a request for a final costs certificate (previously called the allocatur). This reflected the old High Court practice of there being no court file. Now that all courts have their own files, the practice of returning the original to the receiving party has decreased and the parties usually complete their own copies of the bill with the original remaining with the court. A request for a final costs certificate is dealt with below; so too are provisions regarding appeals. (It is worth noting that the rule on the time limit for appeals has been altered so that time does not start to run until the last day of the detailed assessment where that assessment takes place on more than one day (CPR 47.14(7)).

The receiving party needs to make arrangements to remove his files from the court if he is not able to carry them away at the end of the hearing. (Since the SCCO moved to the Thomas More building within the Royal Courts of Justice, access by couriers and outdoor clerks to retrieve papers is complicated by the secure car parking arrangements.)

Elections

[28.45]

In addition to making submissions as to the context in which the item claimed in the bill came about, the receiving party may wish to refer the judge to a document such as a file note or attendance note to support, for example, the time claimed. Alternatively the judge may require the receiving party to make the document available to him. In either event, the receiving party does not waive his privilege to the document simply by producing it to the court. In many cases, the judge will allow or disallow the item having considered the proffered document. If the item is disallowed the paying party is unlikely to be too concerned about what the document said in most cases. If the amount is allowed, however, it is incumbent upon the judge to confirm that he is satisfied that the item is necessary or reasonable as the case may be and preferably give some reason, however brief, as to why that is so. Otherwise, the paying party has little idea why his challenge has failed.

In some situations a judge is not prepared to deal with the item on this basis. An obvious example is the use of a CFA. At the height of the 'costs wars' extremely technical challenges were raised as to the construction of CFAs and their compliance with both primary and secondary legislation. The paying party clearly needed to be afforded an opportunity to look at the CFA in order to raise any points he wished about its compliance. Equally, receiving parties understandably wished to avoid disclosing their CFAs because any minor non-compliance would have fatal consequences to the recoverability of their client's costs.

If the judge is not prepared to deal with an item as set out above, he invokes a procedure sometimes called the rule in *Goldman v Hesper* [1988] 1 WLR 1238 or the 'Pamplin procedure' (*Pamplin v Express Newspapers Limited* [1985] 1 WLR 689) and which was enshrined in the court rules when the CPR came into being. Its current incarnation can be found at CPR PD 47, para 13.13 which says:

'The court may direct the receiving party to produce any document which in the opinion of the court is necessary to enable it to reach its decision. These documents will in the first instance be produced to the court, but the court may ask the receiving party to elect whether to disclose the particular document to the paying party in order to rely on the contents of the document, or whether to decline disclosure and instead rely on other evidence.'

The workings of this rule are helpfully discussed in *Hollins v Russell* [2003] EWCA Civ 718. The case concerned, amongst other things, the disclosure of the CFAs in *Hollins* and the various conjoined cases. The Court took the view that the CFA was of such fundamental importance that a receiving party would have to be put to its election if the case reached a hearing. The delicacy of balancing the receiving party's right to legal professional privilege and the paying party's right to natural justice can be seen in the wording of this judgment. In our view, the judgment clearly says that a receiving party has to make an election if the case gets to a hearing but not before. Some paying parties argue that it should be disclosed earlier: some receiving parties say that it is not authority for an election at all.

Many of the cases in this area relate to the disclosure of the solicitors retainer. In the cases of *Dickinson v Rushmer* [2002] Costs LR 128 Mr Justice Rimer required the receiving party to disclose retainer documents if he intended to rely upon them. He considered it to be 'almost self-evident' that the most basic aspects of fairness required the paying party to be able to see documents in order to make submissions upon them. In *South Coast Shipping v Havant BC* [2002] 3 All ER 779 Mr Justice Pumfrey concluded that the rule, as set out in the Practice Direction, was compatible with the requirements of the European Convention on Human Rights.

All of the cases we have referred to are at pains to point out that putting a party to his election is only going to be required in rare cases. Most of the time, the parties should be content with the judge looking at the documents himself and making his assessment accordingly. This was described as 'an informal, sensible, pragmatic and time and cost saving procedure' by Mr Justice Davis in *Gower Chemicals* (see below). In *Goldman*, Taylor LJ said that the 'contents of documents will almost always be irrelevant to considerations of taxation which are more concerned with time taken, the length of documents, the frequency of correspondence and other aspects reflecting on costs.' So, only 'where it is necessary and proportionate should the receiving party be put to his election. The redaction and production of privileged documents, or the adducing of further evidence, will lead to additional delay and increase costs' (Pumfrey J in *South Coast Shipping*.)

One area where disclosure is likely to be required is in respect of unused expert evidence. If the receiving party contends that, notwithstanding that it had not been disclosed, it was reasonable for him to have incurred the cost of such evidence, he may well be put to his election. If the case had not advanced sufficiently for that evidence to be disclosed in accordance with a court timetable, for example, there may be no issue. But if the case concluded at or near trial and it is clear the evidence was not going to be relied upon, then an issue is raised that means the judge may well put the party to his election. In the *Gower Chemicals* litigation (*Various claimants in the Gower Chemicals Group Litigation v Gower Chemicals* [2008] EWHC 735 (QB)) the defendant paying party took issue with a claim for the cost of a number of expert reports which had not been disclosed. The costs judge refused to look at the report before putting the claimants to their election. The claimants appealed this decision although did not explain to Mr Justice Davis specifically why they did not wish to disclose the evidence to support its cost. The appeal was dismissed with the judge following the rationale of the other cases set out above.

He did cause one issue by supporting the costs judge's decision not to look at the evidence at all before putting the receiving party to their election. This is not how the Practice Direction describes the procedure but understandably Mr Justice Davis thought it easier for the costs judge not to consider something and then have to put it out of his mind. Practically speaking, it does make for a potential problem with the judge's pre hearing preparation if he is not clear on which documents he can see and which he cannot. If it were simply a matter of undisclosed reports, it would be relatively easy to avoid any pre-reading assuming that the point was raised in the points of dispute. However, there is nothing to suggest that there is any such limitation and it is inevitable that on some occasions a judge will already have seen the document before considering whether to put the receiving party to his election.

The *Lownds* Test

[28.46]

In *Lownds v Home Office* [2002] EWCA Civ 365, [2002] 4 All ER 775, [2002] 1 WLR 2450 the Court of Appeal considered how to apply the principle of proportionality to costs reasonably incurred and reasonable in amount. A standard basis assessment has the additional requirement of proportionality, but it is absent from an indemnity basis assessment.

In the lead judgment, Lord Woolf set out a two-stage test. First the judge was to look at the costs claimed in the bill on a 'global' basis. Did he consider the overall costs to be disproportionate, or at least have the appearance of being so? If the judge answered this in the negative, he would assess the costs to see whether they had been reasonably incurred and were reasonable in amount. But if the judge answered his question positively, he was to allow only those costs which were 'necessarily' incurred (and reasonable in amount.)

It is clear that the test of necessity was intended to be a higher hurdle than simply one of reasonableness. But the *Lownds* judgment itself cautioned against too high a hurdle being erected by the court. At paragraph 37, the Court said

> 'Although we emphasise the need, when costs are disproportionate, to determine what was necessary, we also emphasise that a sensible standard of necessity has to be adopted. This is a standard which takes fully into account the need to make allowances for the different judgments which those responsible for litigation can sensibly come to as to what is required. The danger of setting too high a standard with the benefit of hindsight has to be avoided. While the threshold required to meet necessity is higher than that of reasonableness, it is still a standard that a competent practitioner should be able to achieve without undue difficulty.'

The result has been that it can be difficult to spot the difference between assessments conducted on the basis of necessity and those simply on the basis of reasonableness. As Sir Rupert Jackson's Final Report pointed out, judges would get to the end of an assessment on the necessity basis and still consider that they had allowed a disproportionate amount of costs overall.

Another problem with the effect of the *Lownds* test approach was that it appeared to condemn the assessment to become one where each and every item had to be considered as to its necessity. The paying party was unlikely to have challenged every item and indeed may have been sparing in its challenges. The time estimate would have been based on the number of challenges made in the points of dispute. By its own decision at the beginning of the assessment, the court would often cause a hearing to go part-heard because it could not possibly be completed in time. This, at least anecdotally, led to a judicial reluctance to consider bills to be globally disproportionate when otherwise they may well have been considered to be so.

The alternative approach by many judges was to consider the bill to be globally proportionate but certain items to be disproportionate, for example the documents item. This enabled the court to concentrate solely on the items that appeared to be problematic. This approach was supported by the decision of Mr Justice Morland in *Giambrone v JMC Holidays Ltd (formerly Sunworld Holidays Ltd)* [2002] EWHC 2932 (QB), [2003] 1 All ER 982, [2003] NLJR 58.

In any event it is not, in our view, clear that *Lownds* does require every single item to be scrutinised by the court if the costs are found to be globally disproportionate. The judgment refers to considering each item but in the same way as judgments often refer to an 'item by item' approach without actually expecting every single item to be considered. The assessment is essentially based on the items which the paying party wishes to challenge. If the paying party does not challenge a particular item it is not for the court (unless the court's conscience is offended in any way) to raise additional issues. Obviously, the court may be sympathetic to allowing additional challenges to be raised by the paying party once the finding of global disproportionality has been made, but that is a very different matter from automatically considering each and every item in the bill.

There are some items which do not easily lend themselves to the concept of necessity. For example, what is a necessary hourly rate, rather than a reasonable one? If it is considered to be the rate necessary to get an appropriate solicitor (however that is considered) to do the work it is essentially the same as being a reasonable rate for the solicitor who did do the work to receive anyway.

At virtually the same time as *Lownds* was heard by the Court of Appeal, it also heard the case of *Booth v Britannia Hotels* [2002] EWCA Civ 579. That case suggested a different approach, namely conducting the detailed assessment based on reasonableness and then standing back at the end of the assessment to consider the question of proportionality. If the assessed figure was considered to be disproportionate, the court would have to take further steps to render it proportionate. This might have been to revisit certain items to reduce the overall costs. It might have been simply to reduce the assessed figure by a further sum or percentage. The latter approach is used on appeals from costs assessed on criminal matters following the case of *R v Supreme Court Taxing Office, ex p John Singh & Co* [1997] 1 Costs LR 49 (CA).

The Court in *Booth* did not specify the approach to take. The conclusion of the Final Report is that judges should stand back in a manner akin to that expressed in *Booth*. In the 15th implementation lecture regarding the Jackson reforms, the assumption is that the *Lownds* approach should be reversed for costs incurred after 1 April 2013. But the *Lownds* approach was never included in the rules and it has not been expressly displaced in the new rules and so the case must presumably be still relevant, at least for the time being. It has only disapproved of by judicial commentary outside of court decisions and so cannot of themselves overturn a Court of Appeal decision. The wording of CPR 44 does suggest support for the approach in the judicial commentary but it does not prescribe any particular procedure. It must be a matter of time before the Court of Appeal revisits the issue of proportionality and puts this point beyond doubt.

COSTS OF THE DETAILED ASSESSMENT PROCEEDINGS

Award of costs

[28.47]

The detailed assessment procedure is a quantification exercise of the costs awarded to a (receiving) party in previous proceedings. If the paying party puts the receiving party to the task of justifying the costs claimed before a court, the presumption in CPR 47.20 is that the receiving party will receive the costs of having to do so. The only exceptions are where the provisions of any Act, or a rule or practice direction provide otherwise or the court decides to make a different order. The court will only make a different order if:
- the conduct of the parties warrants it;
- the bill has been reduced by an amount which justifies it; or
- a party has unreasonably claimed the costs of a particular item (or occasionally unreasonably disputed it) (CPR 47.20(3)).

Where the court decides to reflect such circumstances in its order, it may reduce the costs of assessment by a percentage or specific sum, it may decide to make no order as to costs or award them to the paying party.

Offers

[28.48]

Until April 2013, a paying was able to make an offer under CPR 47.19 which, if not beaten, would afford protection against the receiving party seeking the costs of the detailed assessment proceedings. It was an echo of Part 36 but did not have the same rigidity and it left open the question of whether a Part 36 offer on costs could be made in any event.

The recasting of the costs rules has clarified this situation by making clear (CPR 47.20(4)) that Part 36 offers by both parties can be made. The terminology of Part 36 is adapted so that, for example, the claimant in Part 36 is the receiving party in Part 47.

It is to be hoped that the rate at which the Part 36 regime appears to be reconsidered by the senior courts does not affect its operation in Part 47. The scope for making complicated offers on costs which will afford any protection is probably limited and the occasions when offers will be withdrawn is also likely to be rare in practice.

One of the problems with CPR 47.19 offers was whether such offers included the costs of the detailed assessment proceedings themselves, or whether they simply related to the costs in the bill. Where offers did include such costs they would wrap everything up if accepted, but were difficult to disentangle if they were relied upon at an assessment. The opposite occurred if the detailed assessment costs were left outside the offer. Parties may still take either option under the new rules. It would seem to be unlikely that an all-inclusive offer can properly be made under Part 36 given the decision of the Court of Appeal in *Mitchell v James* [2002] EWCA Civ 997.

CPR PD 47, para 19 confirms that any offer, whether under Part 36 or not, ought to specify exactly what it contains, in particular whether it includes the costs of the preparation of the bill, VAT or interest. If the offer is silent on these aspects, it will be taken to include all of them.

Without prejudice (save as to costs)

[28.49]

Negotiations on a 'without prejudice save as to costs' basis are, as the description suggests, admissible on the question of the costs of the relevant proceedings. However, negotiations on a completely 'without prejudice' basis are not. The parties may negotiate in the faith and expectation that purely without prejudice negotiations could not be used against them even on the question of costs. Although there were some exceptions to that rule, costs was not one of them (*Reed Executive Plc v Reed Business Information Ltd* [2004] EWCA Civ 887, [2004] 4 All ER 942, [2004] 1 WLR 3026).

If the receiving party seeks to recover the costs of negotiating the costs of the substantive proceedings, it should make clear that those negotiations will be dependent upon the paying party bearing the costs of the negotiations. Alternatively a claimant could accept the final offer for costs conditionally upon the defendant paying his costs of the negotiations. What must be avoided is what was described as the Russian doll analogy in the case of *Longman v Feather and Black* (18 March 2008, unreported) in Southampton County Court where an appeal was allowed against an order that the defendant pay the claimant's costs of the costs negotiations to be assessed if not agreed. Under the order, if the defendant had opted to negotiate the costs of the costs negotiations, and if these were agreed, the claimant would have then been entitled to demand the costs of those negotiations, and so on ad infinitum.

FINAL COSTS CERTIFICATE

[28.50]

Following the detailed assessment hearing, the receiving party calculates the costs allowed on assessment by completing or 'making up' the bill. In many cases, once the figure has been agreed between the parties it is paid by one to the other and no further step is required.

However, if the receiving party wishes to be in a position to enforce the costs as assessed, he needs a final costs certificate which is the quantified order for costs. CPR 47.17 governs the procedure. It expects the completion of the bill and request for a final certificate to be done within 14 days but there is no effective penalty if it is done later. The practice direction enables the paying party to make an application to compel the certificate to be produced but such an application must be the rarest of beasts. It takes a good deal of conjecture to think of any situation in which the paying party wishes to be able to demonstrate his indebtedness with a formal certificate; unless he is contemplating insolvency (and in which case there will be problems with enforcement anyway). Apart from this, there is no need to worry if the lack of payment by your client's opponent comes as a surprise, you can still obtain a certificate. The only danger is that the court file may have been archived but that is not fatal.

Unlike a default costs certificate there is no request form: the completed bill acts as the request. The key part of the bill is the summary which needs to show the figures that are claimed so that they can be transposed onto the final certificate. The practice direction (CPR PD 47, para 16) has a number of requirements to be satisfied before a certificate will be issued. For example, receipted fee notes from counsel and receipted fee notes for any other disbursement over £500 (and so not covered by Certificate (5) on the bill). Given that these disbursements will have been allowed by the assessing judge – and quite possibly disputed in detail – these provisions are rather cautious. They are also perhaps a little anachronistic. In many cases counsel will be paid from the fees recovered from the opponent. If that recovery cannot take place in the absence of a final certificate, there is little purpose in requiring a receipted fee note. These restrictions may also help to explain why most receiving parties do not obtain a final costs certificate after a detailed assessment.

Setting aside a final costs certificate

[28.51]

Since this certificate is only produced after a detailed assessment hearing, the circumstances in which a setting aside (other than for mathematical error) will be few and far between. However, in *Mainwaring v Goldtech Investments Ltd (No 2)* [1999] 1 All ER 456, CA a receiving party delayed for about six years in serving a bill of costs on one of two paying parties. The first paying party's assessment had been completed and a certificate issued. When the second party successfully applied to set aside the certificate and had the costs against them disallowed in their entirety because of the delay the first paying party succeeded in having the costs award against her also set aside in its entirety, because there could not be two certificates in different amounts arising from a single costs order imposing joint and several liability. There was only one bill and there could therefore be only one penalty.

APPEAL FROM A DETAILED ASSESSMENT HEARING

[28.52]

Detailed assessments are carried out by a 'costs officer' which, according to CPR 44.1(1) comprises costs judges, district judges and authorised court officers.

Appeals from authorised court officers

[28.53]

There is a self-contained code for such appeals set out in CPR 47.21 to CPR 47.24. A would-be appellant does not require any permission to appeal an authorised court officer's decision. The appeal is to a costs judge (or district judge of the High Court) and the appeal hearing will be a rehearing of all issues that the parties wish to raise. It is not a review of whether the costs officer's decision was 'plainly wrong' etc.

The appellant needs to file an appeal notice in form N161 within 21 days of the decision which is being appealed. The court will then serve the notice on the other parties and fix a date for the appeal hearing.

CPR PD 47, para 20 sets out some surprisingly complicated provisions regarding the production of a transcript or at least note of the court officer's decision. Given that the appeal hearing is a review, it is questionable as to the benefit of this requirement.

(The nature of an authorised court officer is discussed below.)

Appeal from costs judges or district judges

[28.54]

An appeal from a decision of a costs judge or district judge requires permission to appeal and is dealt with under Part 52 as with any other appeal.

The judge needs to complete a form N460 regarding any application for permission to appeal so that it is on the file in the event that an appellate court needs to see it. Since the nature of detailed assessments is comparatively informal and a formal order is not generally required at the end of the hearing, parties can find themselves in some difficulty in demonstrating the decision that they are appealing. It may be necessary to request the judge to prepare an order, sometimes declaratory in nature, so that it can set out what it is the party wishes to appeal. Since that is not necessarily top of a busy judge's list of things to do, it may be helpful to suggest that you will draft an order for him to approve.

AUTHORISED COURT OFFICERS

[28.55]

An authorised court officer is defined in CPR 44.1(1) as 'any officer of:
(i)   a county court;
(ii)   a district registry;
(iii)   the Principal Registry of the Family Division; or
(iv)   the Senior Courts Costs Office whom the Lord Chancellor has authorised to assess costs.'

Since costs in family proceedings are now dealt with in the SCCO, there are no authorised court officers in the PRFD, and their presence is really only seen at the SCCO. They are called Costs Officers rather than authorised court officers, presumably on the basis that they inhabit the Costs Office. This phrase is used notwithstanding it has a separate, specific meaning in the CPR.

Authorised court officers do not have any formal judicial training but are civil servants employed by the Ministry of Justice. As a matter of policy they do not deal with solicitor and client bills, nor cases where the Ministry of Justice is one of the parties.

The parties may agree that the detailed assessment should not be dealt with by an authorised court officer; if one party only objects he may make an application under Part 23 to a costs judge or district judge. In practice an application by letter will usually suffice.

In addition to dealing with provisional and detailed assessment of between the parties' costs, they also deal with many Court of Protection bills and which are generally provisionally assessed.

OTHER DETAILED ASSESSMENT PROCEEDINGS

Court of Protection

[28.56]

Deputies are entitled by virtue of the Mental Capacity Act 2005 and the Court of Protection Rules 2007 to payment of their charges, whether as a professional deputy or not. There is a clear regard for keeping the cost of assessment to a minimum since payment of the fees is likely to come out of the protected person's property. Consequently, there are provisions for the payment of fixed costs where the work is limited and the option of a short form bill where the fees are below £3,000. Furthermore, most Court of Protection bills are assessed provisionally to avoid the need for attendance and further detail can be found in Chapter 30 on such assessments.

The costs of the application for the appointment of the deputy include all work up to the date of the appointment. All costs thereafter are treated as general management charges and are assessed if an order for such charges has been made. That order needs to be included with the N258 requesting an assessment (along with the other papers usually lodged in accordance with CPR PD 47, para 13.2) for the first year's management charges. Thereafter it is not required because the SCCO keeps a record of such orders and the costs allowed.

Legal Aid

[28.57]

A legally aided party who has an order for costs to be assessed against a non-legally aided party will use the same procedure as for any other between the parties' bill. The party's solicitors may decide to waive its entitlement to any costs from the Legal Aid Agency (perhaps because they would form a statutory charge on their client's property) and in which case the assessment will be dealt with in exactly the same way as any other assessment. If costs unrecovered from the opponent are going to be sought from the Legal Aid Agency, there are two differences from the usual procedure. The bill will be drawn in a six column format (see [29.3]) so that unrecovered items can potentially be transferred to the legal aid columns. At the end of the detailed assessment, the receiving party's advocate will stay behind to seek assessment of any purely legal aid items, eg reporting to the Legal Aid Agency and which does not concern the opponent.

Supreme Court Proceedings

[28.58]

Part 7 of the Supreme Court Rules 2009 deals with costs in the Supreme Court accompanied by Practice Direction No 13. Costs may be awarded on either the standard or the indemnity basis with the same definitions as in the CPR. There is provision for the paying party to file and serve points of dispute and for the receiving party to respond.

Rule 47 provides that submissions as to costs are to be made *before* judgment unless before the close of oral argument a party applies to defer making submissions until after judgment. The court could direct oral submissions immediately after judgment, or the simultaneous or sequential filing of written submissions or written submissions followed by oral submissions.

Under rule 49 either party may request a provisional assessment of costs by the Registrar. There are no similar provisions to those set out in Chapter 30 regarding the costs consequences of an 'appeal' of a provisional assessment. If either or both parties are dissatisfied with that assessment the registrar will try to resolve it in correspondence. If that fails he will appoint an oral hearing before two costs officers, one of whom will be a costs judge and the other an appointee of the President – in practice, this is usually the Registrar.

The function of Supreme Court costs officers is to carry out the detailed assessment (rule 49(1) and SCPD No 13, para 16.1). That is the limit of their jurisdiction. Decisions as to whether the receiving party is to receive less than 100% of the assessed costs are reserved to the court in the exercise of the discretion given to it by rule 46(1). Costs officers must confine their attention to the basis of assessment prescribed by rule 50, subject to any directions that might be given to them by the court: *R (on the application of Edwards) v Environment Agency (Cemex UK Cement Ltd, intervening)* [2010] UKSC 57.

There is no appeal from a decision of the costs officers on the quantum of any sums allowed. An appeal on a point of principle lies to a single Supreme Court Justice. He may affirm the costs officers' decision or he may refer the issue to a panel of Justices to decide it and who may or may not require an oral hearing to do so.

Tribunal (Property Chamber)

[28.59]

The Land Registration Act 2002 empowers the Adjudicator to award costs in respect of cases referred to him by the Land Registry in relation to disputes over land. The procedure for dealing with costs was originally set out in the Adjudicator Regulations 2003 (and then 2007). The position has been simplified as part of the reorganisation of the Tribunals structure so that the Adjudicator and his deputies have become part of the property chamber. The Tribunal Procedure (First-tier Tribunal) (Property Chamber) Rules 2013 came into force on 1 July 2013 and have replaced the earlier Adjudicator Regulations.

The new regulations import the provisions of the CPR (reg 13(8)) regarding a detailed assessment and so replace the CPR lite arrangements set out previously and to which most people appearing before it had little or no knowledge. The amount of costs may be determined by a summary assessment, in which case the Adjudicator will deal with the costs. They may be agreed by the parties or they may be dealt with by a detailed assessment. If it is the last of these options, the hearing may well be conducted by a Deputy Master from the SCCO sitting as a Deputy Adjudicator.

Civil Recovery Orders

[28.60]

The Director of the Assets Recovery Agency may apply to the High Court for a property freezing order or an interim receiving order in an appropriate case under Part 5 of the Proceeds of Crime Act 2002. The property is then vested in a trustee for civil recovery. Any person who has incurred legal expenses in the proceedings and the court has ordered their costs to be paid from the property is meant to try to agree them initially with the Director. If agreement is reached, the Director will authorise the trustee to pay those costs from the property in trust. If they cannot be agreed, the costs will have to be assessed in accordance with the Proceeds of Crime Act 2002 (Legal Expenses in Civil Recovery Proceedings) Regulations 2005 (and 2008).

The regulations specify that CPR Part 47 applies save that the three month time limits for commencing detailed assessment (CPR 47.7) and for requesting a hearing (CPR 47.14) are both reduced to two months.

Costs are assessed on the standard basis but the hourly rates allowable are subject to restrictions set out in regulation 17. There are two columns of figures which are based on seniority in respect of both solicitors and counsel. In order to get into the second, higher column the case need to involve 'substantial novel or complex issues of law or fact'. All of the rates set out in the table can be enhanced if the legal representatives are based in London as defined by postcodes. An increase of 20% is allowed for 'Central London' (postcodes EC1-4, SW1, W1 and WC1-2). An increase of 10% is allowed for 'Outer London' (BR, CR, DA, E, N, NW, SE, SW, UB and W.) The rates are not particularly high and so it may not matter that there appears to be no discretion for the court to award a lower hourly rate if to do otherwise would offend the indemnity principle.

High Court Enforcement Officers

[28.61]

The High Court Enforcement Officers Regulations 2004 set out fees which may be charged by HCEOs (formerly Sheriffs) in respect of the executions against goods which they carry out on behalf of judgment creditors. Not all of the items listed in Schedule 3 are fixed and so it is open for the party liable to pay the HCEO's fees to challenge them in accordance with the Regulations.

An application needs to be made to the SCCO which will be listed for a directions hearing. The costs judge will firstly consider whether to make an order for detailed assessment of the costs in dispute. If he does, then he will also give directions regarding the need for documents to expand upon the parties' positions, if they have not already been set out sufficiently, together with directions for the hearing. The provisions of the CPR will apply to such proceedings.

CHAPTER 29

DETAILED ASSESSMENT – STATEMENTS OF CASE

INTRODUCTION

[29.1]

You will not find the description 'Statements of Case' used in the CPR to describe the documents in a detailed assessment but we have used the description which normally refers to Particulars of Claim, Defence and so forth for two reasons. The first is aesthetic. It is ungainly and unnecessarily lengthy to refer regularly to 'bill of costs, points of dispute and reply' and so where we talk about these documents together we have used the phrase statements of case and have headed this chapter appropriately.

Second, and more importantly, it is no bad thing to echo the ethos as well as the terminology of the underlying case's pleadings. In detailed assessment proceedings the bill of costs is intended to set out detailed information upon which the receiving party relies and the solicitor has to sign several certificates at the end of the bill to confirm the accuracy of the document. Doing so is 'no empty formality.' Similarly the points of dispute are intended, concisely, to set out the objections to the bill in a way that the receiving party (and the court) can understand. Replies are meant to answer questions raised by the points of dispute to narrow the issues. The fact that all three documents came in for criticism during the Jackson Review suggests that parties have not necessarily taken the care that parties take when drafting the statements of case in the original proceedings.

THE BILL OF COSTS: FORM AND CONTENT

[29.2]

If you have never seen a bill of costs, you may find it helpful to look at the model bills annexed to Practice Direction 47 (precedents A to E) before going further. Even if you have, they are a useful resource and highlight, for example, the difference between bills drafted for costs solely between the parties, and those where a claim is to be made against the Legal Aid fund, or both an opponent and the Legal Aid fund.

3 and 6 column bills

[29.3]

Sometimes reference is made to '3 column bills' or '6 column bills.' Model bills A, B and E demonstrate the 3 column version with columns for profit costs, disbursements and VAT. Model bill C demonstrates the 6 column version. Where some costs are claimed against an opponent and some claimed against the Legal Aid Fund, two sets of 3 columns are set out, hence the 6 column format. Between the parties' items are set out on one set of the 3 columns whilst items such as the costs of reporting to the Legal Aid Agency, are set out on the other set. Where, on assessment, the court decides that an item is not recoverable from the opponent, it may transfer it over to the Legal Aid columns if it thinks that the item should be paid by the Legal Aid Fund. The Fund is essentially in the place of the client and as the client is liable for the solicitors reasonably incurred costs, so too will the Legal Aid Fund, even if they are not recoverable from the opponent. (Some items may be considered unreasonably incurred and not be allowed against either the opponent or the Fund.)

4 column bills

[29.4]

The format of a bill takes a while to master. It has evolved over time and the various Reviews and reforms have not altered the basic format hugely. It is still, as was said in the interim Jackson report, modelled on a 'Victorian account book.' Model Bill D is drawn in a different way in respect of the columns and was an attempt to update the standard 3 column approach. It is a 4 column bill with two sets of two columns. The first set is headed, 'amount claimed' and the second set, 'amount allowed.' When the bill is drafted, figures for profit costs and disbursements share one column with VAT having a separate column. The costs claimed would all be set out in the first set of the two columns. If the bill reached an assessment, the amount allowed for each item would then be recorded in the other two columns.

However, this did not accord with most courts' approach of striking through disallowed items and writing in the revised figure (if any). Moreover it made it very difficult to separate the disbursements from the profit costs. As such it is rarely used but there could be no formal objection to it should you decide that it is the one for you.

A new format?

[29.5]

Moves are afoot to bring the format of a bill into the computer age so that it is easier to get the time recorded as the case progresses to populate the bill created at the end of the proceedings without a costs draftsman having to start from scratch. Computer print outs to date have been treated with some scepticism by the courts in terms of demonstrating the time that has been spent. That is likely to change as computerised time recording is used more directly. But, although the format of the bill may change in the way aspects are grouped together, it will not alter the building blocks on which a bill is drafted and ultimately the assessment takes place. The format of the bill can be broken down into the following sections – the Title page; the narrative; the heads of costs and their detail; the summary; and the certificates.

THE TITLE PAGE

[29.6]

As with all court pleadings the bill starts by setting out the full title of the proceedings. There are three further items that need to be included on this page. The first applies in all circumstances; the second and third only apply where relevant.

(1) The order for assessment – The front page needs to set out, usually in 'the tramlines', the order or provision which entitles the party to have the bill assessed. This will usually be a court order but could be the acceptance of a Part 36 offer or on the discontinuance of a claim where the provisions in Part 36 and Part 38 respectively provide for a party's entitlement to costs. The various options for this are set out in CPR PD 47, para 13.3. The order for assessment was formerly known as the reference to tax.

(2) VAT registration number – this is the solicitor's registration number rather than the client's. It is only required if VAT is to be claimed from the paying party. Where the receiving party can recover the VAT he has paid to his legal representatives as input tax, that VAT cannot be recovered from the opponent. See **[2.14]** for more detail on this point.

(3) Legal Aid information – where the party has been legally aided, certain information as to the certificates provided by the LSC/LAA need to be set out.

THE NARRATIVE

[29.7]

The practice direction (CPR PD 47, para 5.11) describes this section as being 'background information' but it is universally known as the narrative because most of what is put down is a retelling of the subject matter of the proceedings and then the course of the proceedings. As the practice direction puts it, the information that should be set out covers three areas.

(1) A brief description of the proceedings up to the date of the Notice of Commencement – note the word 'brief.' The best narratives are ones which describe the subject matter and then the proceedings in a concise fashion. But many drafters are tempted to throw in endless detail which tests the resolve of the judge as to whether to get to the end of the narrative or not. In literary terms, it should be a pacy page turner, not some turgid and dense prose. Just as importantly, it should show some sense of objectivity in describing events. All too often the description of what happened and why tends to descend into partisan mud throwing. Why is this? The belief must be that setting out, for example, all the terms of the shareholders' agreement or all of the injuries sustained in a severe accident of themselves demonstrate how important and complex the litigation was. Similarly, describing the endless delay and obfuscation of the opponent (in the receiving party's eyes) in every interlocutory skirmish clarifies why the time spent was spent and guards against allegations of disproportionate cost. But such things will

become clear to the judge in his pre-reading (if he has chance) or during the detailed assessment (if he hasn't). Filling pages of the bill with inordinate detail sets the receiving party off on the wrong foot because it does the opposite of what it intended.

Although it is said that the narrative should continue up to the date of the notice of commencement (see [28.15] but essentially up to the date of the bill itself), in practice the narrative rarely goes this far. It usually ends either with the event that created the order for assessment, eg accepting the Part 36 offer or carries on whilst the conclusion is put into effect, eg perfecting the order of the court based on a judgment given at trial.

(2) A statement of the status of the fee earners involved and the hourly rates claimed – the introduction of the Guideline Hourly Rates (see [31.22]) has led to some misconceptions in this requirement. The description of the fee earners may be along traditional lines – partner, solicitor, trainee solicitor – in accordance with the client care letter and any agreement such as a contentious business agreement. There is no requirement for such descriptions to tie in with the four bands that the Guideline uses: just as there is no requirement for the rates claimed to tie in with the Guideline rates. The hourly rates are allowed at whatever the court considers to be the reasonable rate for the fee earner concerned.

Some firms have specialised in niche areas and have taken the view that they should describe some of their employees by reference to similarly niche, specialist titles. There is nothing wrong with doing this and as the traditional retainer becomes more flexible in work being done by outsourced companies as well as solicitor's firms, such titles are likely to proliferate.

What is important is that the bill gives some information about the experience and expertise of such fee earners so that the paying party can form his own view of the reasonableness of the rate claimed. Where the fee earner has a professional qualification such as a solicitor, chartered legal executive or barrister, the number of years of Post Qualification Experience is usually sufficient.

The wording of the practice direction points out that hourly rates do not need to be given if the solicitor is not claiming via hourly rates. This will only apply to a solicitor and client assessment. On any between the parties' assessment, even where a Damages-Based Agreement is used, the work done will be claimed using hourly rates.

(3) A brief explanation of any agreement or arrangement between the receiving party and his legal representatives which affects the costs claimed in the bill. The effect of any agreement on the costs claimed is a manifestation of the indemnity principle, meaning that the receiving party cannot claim any more costs from his opponent than he is liable to pay his own legal representatives. If we carry on the DBA example mentioned above, the limit of costs payable by the receiving party to his lawyers will be a percentage of the damages recovered (see Chapter 7 for more detail.) If the costs set out in a bill calculated by the amount of time spent exceed that percentage sum the bill must confirm that the amount sought from the paying party is the percentage sum and not the arithmetical total of the figures in the bill.

HEADS OF COSTS

[29.8]

Up to the end of the narrative, a novice viewer of a bill would be feeling relatively comfortable with the format used. It is at the point that the detail of the bill arrives, and to some extent its order, that many people start to become flummoxed. The bill is meant to be split into the following headings and if there is more than one part to the bill (see below) then these headings, or at least the relevant ones, will be repeated in each part. We discuss the issues raised by each one in turn.

(1) Attendances at court and upon counsel

[29.9]

This heading essentially produces a chronology of the action. Pre-proceedings there may be advice received from counsel in writing, at the telephone or in person. The date of such advice will fix the chronology which is useful to all parties and the judge in getting a flavour of the case. Once proceedings have started, the dates of the pleadings, case management conferences and interim applications will continue the chronology until the trial, or earlier settlement.

The time spent at hearings and attending conferences with counsel will be included here and so too will be counsel's fees for advocacy at court or advice in conference or in writing. One of the peculiarities about the arrangement of the headings is that the letters and telephone calls between the solicitor and counsel to arrange these attendances at court and conferences are invariably only captured under heading (6). It makes sense in that it is neater to place the number of letters, etc in one place rather than a smaller number against each activity under this heading. But if the costs of an attendance at court or counsel are disallowed, it would be simpler to strike out the communications relevant to that activity at the same time.

The Practice Direction (at para 5.16) requires the bill to aid the chronology by including orders which do not assist the receiving party. This may be an application where there was no order for costs or the costs might have been payable by the receiving party to the paying party in any event. It might simply be a relevant event from which no specific charge is made. These entries assist the court to get an overall picture but regularly upset paying parties, particularly litigants in person, who think that costs are being claimed in relation to such hearings etc. even if phrases such as 'no fees claimed' are prominently displayed.

(2) Attendances on and communications with the client

[29.10]

This heading generally comes second only to work done on documents in terms of being the largest item. The nature of the attendances and communication is changing over time. The traditional model of the client visiting the solicitor in his offices to provide instructions periodically and with the occasional letter in between to keep the client abreast of developments is disappearing. The traditional arrangement required the solicitor to be located near to the client but that is no longer the case in many situations. In such circumstances the number of personal attendances now tends to be much lower but in their place will be more letters, telephone calls or increasingly emails, or a combination of all three. Electronic communications are discussed at [29.19] below. The relative informality of emails and the ease with which they can be sent does mean that they can mount up when dealing with the client.

Should the solicitor attend upon the client or the client upon the solicitor? Where personal attendances do take place, it has been the position for a long time that the client is to visit the solicitor. This was on the basis that the client could be expected to put himself out in the cause of his action sufficiently to call on the solicitor. To do otherwise, would be seen as a 'luxury' which the paying party, certainly on a standard basis assessment, ought not to meet. With ever increasing client care and competition for some clients, solicitors are regularly visiting their clients rather than vice versa. Such travelling time is likely to continue to be seen as a luxury, particularly when electronic communication is often a perfectly good substitute. The cost of travelling to see the client may be recoverable where the client could not realistically attend upon the solicitor. Seriously injured claimants or those suffering from terminal diseases are often visited at home or in hospital to take instructions and such travelling is often allowed on assessment.

(3) Attendances on and communications with witnesses including any expert witnesses

[29.11]

In contrast with the client, travelling time to visit a witness is often recoverable, particularly if the witness has no connection with the party. The cost of drafting and finalising witness statements falls into the documents item (heading (7)) but the letters or emails forwarding drafts etc will be seen here.

At one time correspondence with expert witnesses would be little different from lay witnesses. But the rise of medical reporting organisations in the medical field has changed the approach quite considerably. A pro-forma is often all that is required to obtain a report. The agency will organise the examination and the medical records and deliver the report in time scales that used to appear to be wildly optimistic. The agency's cost has proved a contentious issue at certain times (see [26.12]) but if the cost of the report is in line with the AMRO agreement, it is paid more often than not without issue.

The more esoteric medical experts are often still instructed directly in a similar way to non-medical experts. As with lay witnesses, the time spent reviewing draft reports and suggesting amendments and additions falls into the work done heading (and so too does the letter of instruction). It is the more routine communication regarding matters such as the willingness to accept instructions in the first place, dates of availability for trial and payment of fees that are set out here.

(4) Attendances to inspect any property or place for the purposes of the proceedings

[29.12]

This heading speaks for itself and is not one that is present in most bills.

(5) Attendances on and communications with other persons including offices of public records

[29.13]

The obvious 'other person' is the paying party and any other parties to the proceedings. Given that the paying party was obviously present during telephone calls with the paying party and received the receiving party's letters it is surprising how often the amount of time on attendances or the number of routine letters or emails are challenged. If there is one item that ought to pass without argument it ought to be this one. The only exception is where there are regularly several letters written each day where there was no particular urgency and could seemingly have been rolled up into one letter rather than the several claimed for.

Unlike the other items in the bill, the paying party has as much information as the receiving party here. Accordingly, any challenge to the reasonableness of the items claimed ought to be set out with precision rather than a general 'some of the correspondence appears to be non-progressive.' Anything less may be given short shrift from the court.

Where the attendances or communications have resulted in a fee being paid, for example for a search, the fee will be set out under this heading as a disbursement as well.

(6) Communications with counsel

[29.14]

As mentioned in (1) above the rest of the engagement with counsel is set out separately. This means that a telephone advice might be set out under heading (1) but a routine telephone call with counsel falls under this heading. If it was not advice that was being proffered it is often argued that the time was not reasonably spent. It is often difficult to distinguish between attendances or correspondence upon counsel and the same activities with his clerk. Both are recoverable in principle but can lead to a higher level of communications than might otherwise be the case.

The traditional use of counsel at specific points of the case with discrete instructions appears to be on the wane. With counsel becoming more of a team player in many areas of litigation and also having a keener interest in progress where his fees depend upon the outcome, he often wishes to have more regular contact. It is now often the case that counsel is sent documents to keep him updated as to developments regardless of whether he is expected to do anything with them at the time. This can lead to significant numbers of communications being claimed under this heading. It also renders the receiving party's solicitor vulnerable to the argument that he is being over reliant on counsel and as such should not be entitled to the hourly rate being claimed.

29.15 *Chapter 29 Detailed Assessment – Statements of Case*

(7) Work done on documents

[29.15]

This is the largest and in some ways most unsatisfactory item in the bill. The Practice Direction (para 5.18) requires any of the headings from (2) to (10) to use a schedule where any item has 20 or more entries for attendances or non-routine communications. In practice it is the documents item which invariably requires a schedule. Attendance upon the client (heading 2) is the only other heading where schedules are regularly seen.

The work done on documents, whether set out in the bill or in a schedule, is listed chronologically. At a busy period in a case, the solicitor may well be dealing with several different aspects at the same time. This can mean that a discrete area of work, for example dealing with disclosure, may have entries spread over several different pages of the documents schedule which can lead to some difficulty in taking an overall view of the reasonableness of the time spent. The revision of the bill format mentioned at the beginning of this chapter and discussed further at the end is aimed at altering the documents item in particular. The idea is to group items together so an overall view on say disclosure can be taken. There is no easy answer, however, because the costs often have to be split into time periods, and a gathering of all the work by area (or phase to use the cost budgeting terminology) may fall foul of this. If there is a change in the legal representation from one firm to another and the work done straddles the two firms' involvement some assessment is going to have to be employed to determine how much of an overall figure is payable to each firm. Even a simple change in the VAT rate would cause complications.

In addition to the problems caused by the chronological approach to the documents item is the effect of an item by item approach to each item claimed in the schedule. Where a bill was considered to be disproportionate under the *Lownds* test (see [28.4]) the theory was that every item in the bill would then have to be considered. That would often mean, ironically enough, a disproportionate amount of court time being used to consider each and every item. Even on an ordinary standard basis assessment, some paying parties seek to challenge virtually every entry in the documents item. If the paying party wants to have each item considered in detail, or, less often, the receiving party wants that approach, the court has little option but to oblige the parties once they have been reminded of the overriding objective and in particular the need to allocate an appropriate amount of the court's resources to a particular case.

Many parties wish the court to exercise a broad brush approach to the documents item. Others will want part of the documents item to be considered in detail and then will extrapolate the judge's findings to the whole documents item, or those in later parts, without going through them in the same detail. Whichever approach is adopted, the paying party is relying on the judge having had the chance to look through the receiving party's papers to some extent to get a feeling for the amount of time claimed for the work done.

(8) Work done in connection with negotiations with a view to settlement if not already covered in the previous headings

[29.16]

Cases regularly settle at a Joint Settlement Meeting ('JSM') or Round Table Meeting ('RTM') and such time is clearly recoverable in principle. If a third party were added to assist the parties to reach a settlement, then as a general principle, such costs would also amount to work done in negotiations with a view to settlement. The third party would be a mediator and you need to be careful about the terms of a mediation if you expect to be recovering the costs of the mediation from your opponent in the proceedings.

All formal mediations have paperwork to deal with the costs of the mediation as well as many other factors such as confidentiality. In the optimistic air of a potential settlement, it is usually the case that each party agrees to share the costs of the mediator, the venue and the catering (etc) equally. Some agreements also stipulate that the party's own costs of attending the mediation will be borne by the party. It is not unusual for one or more parties to be reticent about attending a mediation at all and so agreements as to the costs of it sometimes have to be adjusted as a result.

What happens if a party agrees to bear his own costs of the mediation, does not reach a settlement and then subsequently wins the litigation? In *NatWest Bank v Feeney* [2006] EWHC 90066 (Costs) Master Campbell considered an agreement in these terms and decided that the Tomlin Order which concluded the case did not overturn that agreement. The case was appealed to Eady J who, in an unreported decision, dismissed the appeal. The mediation agreement had a specific provision enabling the parties to amend the standard wording and the issue was raised in the guidance notes but the parties had made no amendment. Consequently the successful defendants could not claim their modest attendance costs nor the more significant fees of their counsel.

At the hearing Master Campbell also disallowed the defendants' share of the mediators' fee. When providing a written judgment on this point he took a different view but had already given his judgment and so the fee continued to be disallowed. Despite the explanation of why he had come to different conclusion in the written judgment it appears that Eady J preferred the original approach and dismissed the appeal on this aspect as well. As such, it seems that the mediator's fee is only recoverable if either the agreement allows for it to be subsequently recoverable or is silent on the issue. It is only if the agreement says that each party will pay a proportion of the fee will it be irrecoverable. Even then this could be overturned by the final order expressly incorporating such costs.

In *Lobster Group Ltd v Heidelberg Graphic Equipment Ltd* [2008] EWHC 413 (TCC) Mr Justice Coulson confirmed the approach in NatWest Bank. However he drew a distinction between mediations which take place after proceedings have commenced (as in NatWest Bank) and those which occur pre-proceedings. In respect of the latter he said:

> 'First, unlike the costs incurred in a pre-action protocol, I do not believe that the costs of a separate pre-action mediation can ordinarily be described as 'costs of and incidental to the proceedings.' On the contrary, it seems to me to be clear that they are not. They are the costs incurred in pursuing a valid method of alternative dispute

resolution . . . As a matter of general principle, therefore, I do not believe that the costs incurred in respect of such a procedure are recoverable under s 51 [Senior Courts Act 1981].'

The question of costs of and incidental to proceedings is considered in detail under heading (10) below.

(9) Attendances on and communications with London and other agents and work done by them

[29.17]

The idea of a 'country' solicitor instructing a London solicitor to act as his agent has something of John Major's warm beer and cricket on the village green quotation about it. It is easy to forget that prior to the CPR coming into existence in 1999 the High Court had no files and so the original application notice had to be produced before the Master for the hearing date and the eventual order to be endorsed upon it (by way of example). In that sort of environment, last minute faxes and emails for the court file could not take place and a local physical presence was required. Some solicitors still use London agents as a cost effective alternative to them travelling to a hearing in the High Court and such costs are added to the bill as if the work was done by the solicitor on the record (CPR PD 47, para 5.22(6)).

If some commentators are correct and the legal process unravels so that work we would currently call solicitors' work is done by a variety of entities, it may be under this heading that such work would be placed. Surprisingly, given how long they have been involved in the detailed assessment process, it is the use of costs draftsmen that has relatively recently gained the Court of Appeal's interest in the use of outsourced resources to do work that a solicitor would traditionally carry out.

In *Crane v Canons Leisure* [2007] EWCA Civ 1352, the Court of Appeal wrestled with this problem. The receiving party's solicitors had been instructed via a Collective Conditional Fee Agreement with a success fee. For the detailed assessment, the solicitors instructed a firm of costs consultants. They claimed the work, under the agency principle as their own work and so sought a success fee on the base costs. The paying party argued that, based on the definition in the CCFA, the work was properly described as a disbursement because the work was carried out by external costs draftsmen rather than employees of the solicitors and as such no success fee should be payable. By a majority decision, the Court of Appeal took the view that the work carried out by the costs consultants was the type of work that the solicitors were retained to do. The solicitors may have chosen to delegate the work, but they never relinquished control of and responsibility for it. The classification of the work carried out could not sensibly depend on whether the solicitors did the work themselves or whether they delegated it to another solicitor or whether they delegated it to costs draftsmen who were not solicitors. Accordingly, the costs consultant's work was properly described as 'solicitors' work' done on behalf of the solicitors and so the consultant's fees were properly described as base costs within the terms of the CCFA thereby attracting a success fee.

(10) Other work which was of or incidental to the proceedings
[29.18]

Section 51 of the Senior Courts Act 1981 is the foundation stone of the court's power to award costs. It is usually mentioned in the context of the court awarding costs against non-parties (see Chapter 40) but its terms are also prayed in aid where the costs of peripheral work to the case are sought to be claimed by the winner of proceedings. This is because the terms of s 51 talk of costs 'of and incidental' to the proceedings.

In *Contractreal Ltd v Davies* [2001] EWCA Civ 928, Arden LJ reviewed various authorities and considered that the phrase 'of and incidental to' is a 'time-hallowed phrase in the context of costs' and that it had received a limited meaning so that it denoted 'some subordinate costs to the costs of the action.'

In *Re Gibson's Settlement Trusts, Mellors v Gibson* [1981] Ch 179, [1981] 1 All ER 233, Sir Robert Megarry considered that there were three features of work which the court should consider which rendered it 'of and incidental to' an action. These strands were that:

(1) The work involved was 'of use and service' to the action.
Later cases confirmed that this was not an absolute test in the sense that the receiving party had to show that it was actually used at the hearing. Otherwise any case that settled short of a hearing could not possibly meet that test. The work needs to be 'likely' to be of use and service.
(2) It was relevant to the action.
(3) It could fairly be attributed to the conduct of the defendant.

The case was concerned with, amongst other things, whether pre-proceedings work could be considered to be of and incidental to the costs of the proceedings. Based on the three strands of reasoning above, Sir Robert Megarry confirmed that it could.

Costs incurred in one set of proceedings cannot, as a general rule, be recovered in another. For example, see *Wright v Bennett* [1948] 1 KB 601, [1948] 1 All ER 410, CA where papers prepared for counsel were used at first instance and then used again before the appeal court. The court in the second proceedings is entitled to make its own order as to costs and the recoverable costs for one party will flow from the terms of that order and not as a result of the order in the first set of proceedings.

But what happens where the costs are incurred in the context of proceedings where no costs can be awarded? In a line of cases starting with *Ross v Bowbelle (Owners)* [1997] 1 WLR 1159, [1997] 2 Lloyd's Rep 196 courts have decided that work done in respect of inquests can be recovered in a civil claim for damages as long as they come within the three strand test set out in *Re Gibson's Settlement Trusts*. The most recent confirmation of this point is set out in *Roach v Home Office* [2009] EWHC 312 (QB). The logic of these decisions is that they are evidence gathering opportunities and as the work could be claimed if the evidence was obtained by separate interviews of witnesses, why should that evidence not be obtained by attending and cross examining witnesses at the inquest? The inquest costs are not automatically recoverable and are case specific. You should consider what issues are still live if you are going to attend at an inquest and expect the costs to be recoverable in a civil claim. If the deceased died immediately there can be no evidence as to quantum (unlike in the *Bowbelle* case). So, if the intended defendant admits liability for civil damages prior to the inquest taking place, it is difficult to see what 'use and service' attending will be, even if the family understandably wishes to be represented. To what extent is attendance at preliminary hearings an evidence gathering opportunity? Is the attendance at the inquest to make submissions to the coroner as to the verdict (in the hope of getting a more favourable verdict with which to pursue a civil claim) recoverable? There are first instance decisions going both ways on these questions. The case law is clear however, that the existence of a potential civil claim will not necessarily enable the funding of attending an inquest to be achieved by a roundabout method. Given the dicta in *Contractreal* above, inquest costs will regularly be challenged by paying parties as they are often significantly greater than the costs of the rest of the claim which does not sit comfortably with the concept of being subordinate to the costs of the action itself.

The advent of inquests conducted under Article 2 of the European Convention on Human Rights give rise to their own problems. The one year time limit for bring a Human Rights Act claim can mean that proceedings are issued protectively prior to the inquest taking place. If the case continues based on the evidence at the inquest all is well. But what happens if the inquest does not demonstrate any liability on the part of the public body or any other potential defendant? The civil claim will have to be discontinued and the defendant could bring a claim for costs in accordance with Part 38. If the costs of the inquest are in principle of and incidental to the civil claim for the claimant, then presumably they are equally so for the defendant. As these incidental costs often dwarf the costs of the remainder of the claim, the claimant might be in for a nasty shock upon discontinuance. The QOCS shield may save the day in such circumstances but it is 'Qualified' and so it cannot be taken as read that this would necessarily occur.

THE DETAIL

The building blocks

[29.19]

(1) *Consecutive numbering* – each item in the bill needs to be consecutively numbered (para 5.15). Practices vary to some extent on this but the guiding principle ought to be that it is difficult to break the time claimed into too many different items. The precedent bills set out the sort of method to be adopted. They can be found in the Schedule of Costs Precedents annexed to Practice Direction 47. They are not mandatory in their format but it is desirable that they should be followed where possible.

The use of consecutive numbers enables the paying party to identify the items which he seeks to challenge in the points of dispute (see below).

(2) *Dividing the bill into parts* – the bill is essentially a chronological document running from the first instruction by the client to the conclusion of the case. There may be circumstances during the life of the case which mean that it would be appropriate to divide the bill into the work done prior to the

circumstance occurring and the work done thereafter. The possibilities are almost endless (see 5.8 for further examples) but perhaps the four most common reasons are:-

(i) The client has had the benefit of Legal Aid for some of the case but not all of it; or has been a litigant in person for some of the case but not all of it.

(ii) The case has transferred from one firm of solicitors to another. As a result there is a new retainer, new fee earners (unless the transfer has taken place because the person dealing with the case has moved), new hourly rates etc to take into account.

(iii) The VAT rate has changed. If there is no division of the bill into two (or more) parts, the allocation of VAT to work done during different periods is extremely difficult.

(iv) The solicitor has rendered interim statute bills during the life of the case (see Chapter 2 in respect of interim bills). The rendering of these bills creates a limit to the costs that can be recovered between the parties. In order for the paying party and the court to understand whether the indemnity principle is in danger of being breached the between the parties' costs claimed need to be set out in respect of the same periods as are covered by the interim statute bill(s). For reasons explained in Chapter 2, requests for payments on account do not have the same effect.

The requirements of costs budgeting are likely to mean that bills are drafted in parts to mirror the phases in the budgets so that it can be seen whether the budgets for the individual phases have been exceeded or not.

(3) *Six minute units* – not everyone records their time in six minute units (indeed not everyone records their time at all). Often attendance notes will show time claimed in 5 minute units, particularly multiples of 5 such as 15 minutes and 30 minutes. Some people record their time to the minute and will regularly have 11 minute telephone calls or a meeting time at 2 hours and 14 minutes. There is nothing inherently wrong with charging in any of these methods as long as the client knows the basis on which you calculate your time.

The court expects to allow the time for routine letters and emails out and telephone calls (in or out) to be allowed on a unit basis of 6 minutes each. The relevant hourly rate for the fee earner involved will enable the value to be charged in the bill. There is always at least one person in a firm (and often many more) who cavil at the idea of this rounding up of time. But such arguments are only valid in relation to the time claimed as a charge to the client. The court sets out the method in which costs are to be recoverable from the opponent and the party's bill needs to comply with the Practice Direction (para 5.22). In any event, the requirement regarding 6 minute units only applies to routine communications and client care documentation invariably explains this charging method to clients nowadays anyway. But if you feel particularly strongly about this, you can always point to the fact that the court will allow 6 minute units 'in general' and so you could have your bill drawn in a different manner. You should not expect a court to entertain your arguments in respect of the opponent's bill with any enthusiasm however.

29.19 *Chapter 29 Detailed Assessment – Statements of Case*

The six minute unit charge for letters and emails out 'will include perusing and considering the routine letters or emails in.' This reflects the fact that routine correspondence can often be deal with very quickly, for example a letter sending a cheque in respect of a fee note received. To be able to claim 6 minutes for considering the letter (fee note) received and a further 6 minutes for sending a two line letter in response would encourage the time claimed to mount up very quickly on very routine matters. Hence, the inclusion of the consideration of the letter in. Sometimes a separate 6 minute item is claimed for considering a routine communication on the basis that there was no need for any response and as such the time considering the incoming document is otherwise lost. Such arguments do not generally succeed on a between the parties' assessment but if the client agreed to pay the costs of letters in then there is no reason in principle why the cost should not be sought. Traditionally letters in were claimed at half the rate of letters out, presumably on the basis that they would often not require any specific action.

We all know that an email may not take very long to create and send. Should such brief emails count as a 6 minute unit? The original draft of the Costs Practice Direction in 1999 set the rate at 3 minutes. That suggested that no email took very long to write and experience has shown that longer and longer emails are now sent and fewer formal letters are sent, particularly to clients. The amount of fee earner input, even in a short email, is often comparable with standard letters produced by case management systems or by the fee earner asking their secretary or assistant to send 'the usual letter' which is then charged at the fee earner's rate.

Emails and indeed text messages and other 'electronic communications' can be claimed if they can be demonstrated to have occurred and their contents are available for the court to scrutinise. The Practice Direction (para 5.22(2)) is a little wary of such new-fangled things. These electronic communications – other than emails – are allowed at the court's discretion, which is how emails used to be allowed before they became equal to letters.

(4) *Attendances* – these are sometimes called personal attendances. They may be face to face but also may be by telephone or video conference. If they are by telephone they need to be for longer than six minutes since otherwise they would fall into the 'routine communications' item instead.

(5) *Communications* – these are defined at para 5.13. They include 'letters out', ie ones that you send elsewhere, emails out and telephone calls. The Practice Direction says that non-routine communications (ie ones that are of substance (see routine communications by contrast below) 'must be set out in chronological order.' In practice, the time taken to write longer letters or emails tends to be included in the documents item rather than under the heading of, for example, the client. Longer telephone calls however, do appear under the heading of the person to whom the call was made (eg the client) and as such are really treated as an attendance (see previous heading).

(6) *Routine communications* – these are defined at para 5.14. They comprise the same activities as longer communications but, 'because of their simplicity should not be regarded as letters or emails of substance or telephone calls which properly amount to an attendance.' They are simply counted up and set out as two figures, one for telephone calls and one for letters or emails out.

(7) *Travelling (and waiting) costs* – where these costs are claimed, for example in attending court, they should be charged at the rate agreed with the client, always assuming that such rate is no higher than the rate agreed in respect of the other work done. Traditionally, this rate would be half or two-thirds of the hourly rate generally claimed because of the absence of any 'B' factor for care and conduct (see Chapter 31) and this is still the case in some agreements with clients. In default of any specific agreement however, the time will be charged at the hourly rate claimed for other activities in the bill.

(8) *Local travelling expenses* – the expenses incurred by legal representatives in attending the local court (or indeed anywhere else locally) is not allowed. The rule of thumb set out in the Practice Direction (para 5.22(3)) as to what is local is 10 miles from the court dealing with the case at the time. That rule is becoming increasingly anachronistic where proceedings are issued centrally and then transferred to a court which may be many miles away from the solicitor.

(9) *The cost of postage, etc* – whether documents are sent by post, courier, DX; messages are sent by fax or telex; or outgoing telephone calls are made, the costs associated with the use of these methods is not generally recoverable. Where the cost is unusually heavy or the circumstances unusual, the court may exceptionally allow sums in respect of these items. The use of couriers because of urgency, eg the imminent expiry of a limitation period will only be allowed if the urgency does not appear to have been created by delay on the part of either the client or his legal representatives. The use of couriers because of a lack of trust in the postal system, etc will not be recoverable between the parties.

This provision of the Practice Direction (para 5.22(4)) is sometimes used by paying parties to challenge the cost of conference calls arranged by commercial organisations so that the calls can be recorded. If such a call is required for a telephone hearing, the cost is usually recoverable. If it is for the purpose of a conference between the client and his legal representatives, then it is less likely to be so.

(10) *The cost of making copies* – as with postage, etc, the cost of copying is not usually recoverable because it is seen to be included in the overheads of running a firm of solicitors. The test is whether the copying is unusually numerous for the nature of the case or the copying is carried out in unusual circumstances. Claims tend to be made following a trial where numerous copies of the trial bundles have had to be produced externally and there is an invoice to be met. It is difficult to suggest that a trial can be considered to be an unusual circumstance in litigation. Consequently the cost of copying trial bundles is unlikely to be recoverable in most circumstances.

If a claim is to be made, the number of copies, their purpose and the costs claimed for them must be set out in the bill. There is no going rate for the cost of copying and the rate chosen will have to be justified on the assessment if challenged. Equally the paying party will have to justify any contrary figure put forward. The cost should not include the time spent in the sum put forward. To the extent that the creation of trial bundles is a fee earner task, the time claimed should be included in the documents item and not here.

(11) *The cost of preparing the bill* – this is usually the last item in the bill. It is incurred before the Notice of Commencement is served and, in accordance with the Practice Direction (para 5.19), is an item taken to fall under the

29.19 *Chapter 29 Detailed Assessment – Statements of Case*

order for costs giving rise to the existence of the bill. Costs incurred thereafter however are costs of the detailed assessment proceedings and a further order is needed to recover these (unless they are agreed between the parties). Therefore entries included for considering the points of dispute, preparing replies, making up (completing) the bill after the detailed assessment hearing are not recoverable in the bill itself, nor is a provisional sum for such matters.

At one time, this item would be a disbursement and often calculated by reference to a percentage of the costs 'as drawn'. That practice has largely died out and the cost is usually calculated as with other work by reference to an hourly rate and the time spent. The hourly rate is often claimed by reference to the Guideline rates for either a Grade C or Grade D fee earner.

The summary

[29.20]

As you would expect, this is the shortest section of the bill. It is one which can often be the most useful section. It gives the reader an overview of the extent of the profit costs, disbursements and VAT which have been claimed (para 5.20). Where there is more than one part to the bill, a summary for each part needs to be provided. If there are totals at the foot of each page, these also need to be included in the summary. Most summaries fit on a single page as a table, but in larger bills, or ones that have a lot of parts (such as bills for group actions) the summary can run for several pages. The summary can also be the simplest way of locating certain items, particularly large disbursements, which might be placed in several different areas of the bill.

The main purpose of the summary however is to enable the receiving party, and then the court, to be clear as to the amount allowed on assessment and which figure will then be transferred to a final certificate if the receiving party requests it (see [28.44]). Since it is not completed until decisions at the assessment have been made, it cannot be confidently completed at any earlier point. Traditionally it would be filled in using pencil so that it could be rubbed out and completed in ink. With computerisation, including self-totalling software, it is more common for the summary to be completed at the outset and a reprinted bill produced after the assessment. Why complete the summary in any manner prior to the assessment? It gives all the parties, and the court, an overview of the size of the bill from the outset. It therefore gives the figure for the notice of commencement. It is often helpful when the parties are negotiating and it helps the court consider the time estimate for listing when a hearing is requested.

Certificates

[29.21]

The practice direction perhaps plays down the importance of the certificates in simply saying that the bill must contain such certificates set out in Precedent F of the Schedule of Costs Precedents as are appropriate. As mentioned at the outset of this chapter, the signing of the certificates is 'no empty formality' according to the Court of Appeal in *Bailey v IBC Vehicles Ltd* [1998] 3 All ER 570, 142 Sol Jo LB 126, CA. To reinforce the strength of that comment, the Court of Appeal in *Hollins v Russell* [2003] EWCA Civ 718 felt the need to spell out no fewer than five reasons why the solicitor's signature should not be taken as sacrosanct in relation to CFAs. The effect of *Bailey* is that, unless a paying party can point to a case specific reason as to why there should be some doubt over the veracity of the certificates signed by the receiving party's solicitor, the court need not enquire into the matters certified. By way of example, the court does not need to see receipts or other documents to vouch for the payment of disbursements under £500. The trade-off for this position, therefore, is that if the certificates have been wrongly signed, it is a serious matter indeed and can lead to professional conduct sanctions which traditionally would have meant the court referring the solicitor to the Law Society.

It was for this reason that the Court of Appeal in *Hollins* was concerned about the effect CFAs were having on the certificate signing process. Numerous cases had come before the courts where the indemnity principle certificate had been signed (as must happen on all bills) to say that the claimant was not seeking more from the defendant than he was liable to pay to his solicitor. However, those courts had then concluded that the CFA had not complied with the legislative requirements (see Chapter 6) and as a result the CFA had become unenforceable against the client. Since the client was no longer liable for his solicitor's costs, neither was the defendant and the certificate signed by the solicitor was false as a result. What should the court do in circumstances where the solicitor thought that his CFA was compliant and so signed the bill certificate, but was found to be wrong as a result of the court's decision? Reporting the solicitor to the Law Society seemed a draconian step but the need for the parties and the courts to be able to rely on the certificates was not to be underestimated.

The Court concluded that, in relation to CFAs, the position in *Bailey* could not stand. This did not mean that a different certificate (or even no certificate) should be signed. Instead, the paying party would be able to see the CFA and decide whether it wanted to take any points which might render the CFA unenforceable. Interestingly, it was not for the court to be the watchdog on this issue. The court could rely on the paying party to raise a point if there was one. The Court did not go so far as to say that the receiving party had to disclose his CFA to the opponent since that was a privileged document. But it did require the court to put the receiving party 'to his election' about either disclosing it or relying on something else to prove his retainer (see [28.45] for more detail on this procedure.)

The introduction of DBAs may herald a similar accommodation for these agreements if their terms are regularly challenged but in any event the need for certificates to be accurate is absolute. If, as sometimes occurs, the bill certificates are not signed prior to a detailed assessment hearing taking place, the court will at the very least, not allow a final costs certificate to be produced until the certificates have been signed.

All certificates must be signed by the receiving party or by his solicitor (apart from certificate (6) see below). Where the bill claims costs in respect of work done by more than one firm of solicitors, certificate (1), appropriately completed, should be signed on behalf of each firm. The remaining certificates only need to be completed by the firm on record. There can be practical difficulties in getting solicitors who have ceased trading or fallen out with the client to co-operate in the production of the bill with the later firm of solicitors. Where the conducting solicitor has moved from one firm to another he may be able to sign all the certificates on behalf of both firms by amending the certificates appropriately.

Bear in mind that the client could always sign the certificate which would be a lot easier logistically in many cases. Where the earlier solicitor's fees are based on an agreement contingent on the outcome (CFA or DBA) it would be wise to think twice before taking this approach. If there is anything wrong with the enforceability of the agreement, the client would not want to have affirmed anything in the bill which would prevent him otherwise being the beneficiary of an unenforceable agreement. But, in this situation, any solicitor who does not co-operate in the recovery of fees from the opponent is risking leaving himself without any payment whatsoever.

The risk is not all on the earlier solicitor's side however. Parties who, having instructed two firms of solicitors, commenced detailed assessment proceedings with the bill of one firm alone, were given a bitter pill to swallow by the court in *Harris v Moat Housing Group South Ltd* [2007] EWHC 3092 (QB). The appellants ('either oblivious or heedless of the provisions of the practice direction,' said Christopher Clarke J) failed to include the costs of their previous solicitors in their bill of costs, and the costs of their second solicitors were compromised by agreement with the paying party. If the appellants had, either in their notice of commencement or in the bills or otherwise, made clear that the amount claimed was only part of their claim to costs and that they would be claiming later in respect of the work of their first solicitors; and the agreement was that the respondents would pay a sum in respect of the costs claimed, recognising that the costs in respect of the first solicitors were still to be dealt with, the appellants would not have been prevented from making a claim in respect of those costs. There would have been a failure to comply with the Practice Direction, but subject to any sanction that the court thought fit to impose there would be no reason in principle why the court should not assess the remaining costs in dispute. However, the position here was different, because what had been settled was the amount of the receiving party's costs pursuant to particular orders. If they had left out of their bill part of what they should have claimed and there had been a settlement of the bill, they could not recover more than the amount agreed. The omission was their misfortune.

The Certificates cover the following issues:

A. *Mandatory Certificates*

[29.22]

(1) *Accuracy*: This certificate starts with the overarching confirmation that 'this bill is both accurate and complete'. It then confirms that the indemnity principle has not been breached. In other words that 'the costs claimed herein do not exceed the costs which the receiving party is required to pay me/my firm'.

Additionally, this certificate confirms that any employed legal representative who carried out work for which costs are claimed is employed by the receiving party. This certificate supported the rule that an employed solicitor could not carry out litigation for anyone other than his employer. That restriction, to the extent that it still applies, is likely to be removed with the arrival of Alternative Business Structures etc.

Finally this certificate confirms that any work done for a legally aided client was carried out pursuant to a certificate from the Legal Aid Agency (or earlier equivalent.)

(2) *Interest and interim payments*: This certificate deals with activities that have happened during the case and since the case has ended but before it has come back to court to be assessed.

Both certificates are written in the alternative. Either there has been a ruling affecting the receiving party (or his solicitor)'s entitlement to interest (and in which case details need to be given of the date and text of the ruling together with the identity of the judge); or there has been no such ruling.

Similarly, there has either been one or more interim payments (and in which case the date, amount and the identity of the person making the payment are needed for each payment); or there have been no such payments.

The presumption in the rules at CPR 44.2(8) that an interim payment will be awarded will mean that this certificate will be answered more often in the positive than has previously been the case.

B. Legal Aid cases

[29.23]

(3) *Position of a Legally Aided client*: This certificate is made pursuant to regulation 119 of the Civil Legal Aid (General) Regulations 1989. It deals with the issue of whether the client has an interest in the detailed assessment proceedings. For example, if the client may have to meet any shortfall in costs received from the opponent through the operation of the statutory charge.

Where the client does have an interest, the certificate confirms that a copy of the bill has been sent to the client with an explanation of the nature of his interest and the steps that can be taken to safeguard his interest in the assessment. Furthermore, the client has not requested that the costs officer be informed of his interest and has not requested a copy of the notice of hearing to be sent to him. If the client has no interest, then the certificate will record this and of course there is no need to send the bill to the client with the explanation.

The purpose of this certificate therefore is to deal in a simple way with clients who do not need or wish to attend the hearing being left in peace. Otherwise, they would need to be sent a Notice of Hearing on the basis that they might have an interest and would turn up at the hearing, more often than not, for little or no purpose. Where a client has an interest and does wish to receive notice of the hearing, this certificate cannot be completed and the client's details should appear on the statement of parties lodged with the Request for a Detailed Assessment Hearing.

(4) *Notification re counsel's fees and consent to FCC*: This certificate is made pursuant to regulations 112 and 121 of the Civil Legal Aid (General) Regulations 1989. It gives consent to the signing of the final costs certificate within 21 days of detailed assessment.

As part of this, the certificate confirms that either counsel has been notified in writing of any fees reduced or disallowed on the detailed assessment (and if so, when) or that there were no such reductions rendering notice to counsel to be unnecessary.

Disbursements and VAT

[29.24]

(5) *Lower value disbursements paid*: This certificate saves a considerable amount of effort on the part of the receiving party's solicitor and the court. If it cannot be completed, the receiving party will have to produce vouchers for every single disbursement in the bill, no matter how small. The only reason for not signing this certificate is where the smaller disbursements have not been paid and that might raise alarm bells as to how the case has been financed and / or the state of the receiving party's file if the solicitor cannot properly certify that everything has been paid. Why counsel does not have to be paid is something of a peculiarity. It may stem from the honorarium approach to his fees discussed in Chapter 2. If so, the introduction of the new contractual terms may render this less likely in the future in any event.

(6) *VAT position*: The VAT certificate is a forest of square brackets and opportunities to cross out the various options envisaged by the drafters of the certificate. In short, the solicitors or auditors certify, supposedly with the benefit of the client's last completed VAT return that the receiving party is entitled to claim all / a percentage / none of the VAT charged on costs and disbursements as input tax in accordance with Section 24 of the Value Added Tax Act 1994.

If the client can recover all the VAT he has paid (or will pay) as input tax then he will not be able to recover it from the opponent. If he can recover a percentage, then he can only claim the remaining percentage. If he cannot recover any VAT then of course he can seek his VAT from the opponent.

This certificate is different from the others in two respects. The first is that the client's auditors can sign it. Every other certificate can only be signed by the party or the solicitor. Second, according to the pre-amble to Schedule F of the Costs Precedents

> 'Certificate (6) may be included in the bill, or, if the dispute as to VAT recoverability arises after service of the bill, may be filed and served as a supplementary document amending the bill under paragraph 39.10 of Practice Direction 47.'

The one flaw with this certificate is that there is no paragraph 39.10 of CPR PD 47. Nor, at least in recent times, was such a paragraph in the old Costs Practice Direction. No doubt sooner or later it will be amended so that it refers to the new practice directions (presumably CPR PD 47, para 13.10).

The future for bills of costs?

[29.25]

In September 2010 a working group was formed by the Association of Costs Lawyers with a core remit to design a model bill of costs which satisfied Sir

Rupert Jackson's recommendations and a wider remit to redesign the format of bills of costs so that they could be used throughout the whole costs process, from instruction of a legal representative to the final costs certificate. In October 2011 the working group produced an interim report, which recommended a change of format to provide for bills to be prepared by reference to phases, tasks and activities, based on the litigation code set in the Uniform Task-Based Management System (UTBMS) already well-established in the USA. The group saw no point in trying to reinvent the wheel when there is a tested system which could be tailored to the jurisdiction of England and Wales and the requirements of the Civil Procedure Rules. A bill of costs for detailed assessment would become just one of the reports at different levels of generality that could be generated from the same data entries. One of the benefits of using the UTBMS codes would be the facility, in the long run, to import data directly from law firms' practice management software.

How long is the long run? The ACL group's report envisaged a pilot study and evaluation over a medium term of 12-18 months, followed by three phases: two years when either the old format or the new format could be used; two years when the new format would become compulsory for larger bills; and a third phase when it would be mandatory for all bills.

We do not think you should hold your breath in respect of these developments. The use of UTBMS for the panel firms of liability insurers was popular briefly at the beginning of the century. It faded out when the cost to the insurers of the licences was not matched by the cost reductions that greater transparency of fee earners' work was expected to achieve. This is not the driver for the ACL group and there is no reason in principle why time should not be recorded in terms of phases and tasks or activities. However, the prescriptive nature of the data entry – particularly the narrative – would in our view be entirely prohibitive. The alternative would be for standard narrative entries but they would then destroy the purpose of these changes which is to provide meaningful information with which to assess the time spent. Added to these difficulties is the general reluctance of fee earners to record time in many cases and to do so with bespoke narrative input in almost all cases.

In addition to the practical issues, there are also issues of confidentiality and relevance. The data entered into time-recording software is privileged and confidential; it could not simply be made available to the court and litigation opponents without some process of vetting to preserve confidentiality. Work which is not recoverable under the particular terms of an order for costs would have to be filtered out. More prosaically, so too would any client items. In each case, human intervention and judgment will be required.

POINTS OF DISPUTE

[29.26]

In comparison with the bill of costs, points of dispute have to provide very little information and as such they have few procedural requirements. More time is spent in the CPR in dealing with the consequences of a party not serving his points of dispute than in what should be in them in the first place.

29.27 *Chapter 29 Detailed Assessment – Statements of Case*

Structure

[29.27]

There is one model points of dispute and that can be found at Precedent G of the Schedule of Costs forms annexed to Part 47. It has been revised as of April 2013 but it is very largely, and almost literally, a matter of form over substance. Points of dispute would tend to be on landscape oriented paper but now they are on portrait. Instead of the points made by the paying party being side by side with the receiving party's replies, they are placed one above the other. This change of orientation has allowed for a space to be included after each reply to a point and before the next point with the anticipatory heading of 'costs officer's decision.' But as there are no requirements as to how large the box is for that decision, the boxes tend to be sized to fit comfortably on the paper. As such, there is often insufficient room to write anything resembling a reasoned decision and just occasionally there is far too much. The drafter of the points should at least consider whether it is a general point of principle which might require considerable space or an individual item where enough space to write a figure is all that is required.

In the rules regarding Provisional Assessments (see Chapter 30), there appears to be an expectation that the Precedent G document is the one that contains the costs officer's decisions. In practice that does not often seem to occur but it is one more reason to consider allowing more space for the judge if you are usually cheese-paring in this respect.

Format

[29.28]

Other than stating that the points of dispute may challenge any item of the bill, CPR 47.9 says nothing on how the points should be written. Such guidance as there is, is contained in CPR PD 47, para 8 and the model document in the Precedents.

According to para 8.2, 'Points of dispute must be short and to the point.' They must follow Precedent G so far as practicable. In particular they must:
(a) Identify any general points or matters of principle which require decision before the individual items in the bill are addressed; and
(b) Identify specific points, stating concisely the nature and grounds of dispute.
Sub-section 8.2 concludes that 'once a point has been made it should not be repeated but the numbers where the point arises should be inserted in the left hand box as shown in Precedent G.'

So to recap: the points should be short and to the point, stated concisely and not repeated. Anyone would think that they had previously been verbose. In his final report Sir Rupert Jackson observed:

> 'Points of dispute are said to be overlong, therefore expensive to read and expensive to reply to. Points of reply are similarly prolix. Both of these pleadings are in large measure formulaic and are built up from standard paragraphs held by solicitors on their databases. In addition, there are lengthy passages in the points of dispute and the points of reply dealing with time spent on documents.'

There should be no need to plead to every individual item in a bill of costs, but this practice stems to some extent from the wording of CPR 47.14(6): 'Only items specified in the points of dispute may be raised at the hearing, unless the court gives permission.' The alternative would be to allow the paying party to bring points in at the hearing without warning but such trial by ambush tactics went out when the Woolf Reforms brought in the cards on the table approach.

Early indications are in fact that many paying parties are using the reforms as an opportunity to keep points of dispute brief and not deal with the objections in detail. Whilst this means the end of long repetitive comments about the documents item in particular, it does take the parties back towards a trial by ambush.

The introduction of provisional assessments militates against brevity and selectivity. The points of dispute are probably going to be the only opportunity the paying party is going to have to influence a judge carrying out a provisional assessment. On that basis, the kitchen sink is almost bound to be pleaded along with everything else.

Calculating offers

[29.29]

The exercise of setting out the challenges to the bill of costs should, amongst other things, enable the paying party to come to a view as to by how much he can realistically expect to reduce the receiving party's bill on assessment. That figure should inform the offers made to settle the case short of a hearing. The more realistic the figure is, the more helpful it will be. It is unusual to succeed (or fail) on every argument put before a judge and there are bound to be grey areas where some experience is going to be beneficial. Offers made should logically exceed the best case scenario set out in the points of dispute. The need for open offers with the points of dispute and the effect of such offers, as well as Part 36 offers is considered elsewhere (see [28.25]).

REPLY

[29.30]

The heading above CPR 47.13 is 'Optional Reply' and, as that suggests, the receiving party does not have to put in a reply to the points of dispute. It may be very clear without a further document as to what the argument is between the parties.

If the receiving party decides to serve a reply, he should do so within 21 days of receipt of the points of dispute according to CPR 47.13. But there is no sanction if they are not served within this period and they are often served later, especially in the run up to the hearing itself. There is no obvious need to seek the court's permission to rely upon them if they are served late. The court's permission is never expressly required in CPR PD 47, para 13.10 although it can refuse to allow variations that have been made to the reply (or other statements of case.) If the reply is the original version, but simply served late, it is almost always bound to be allowed in before the court. This is especially so since the purpose of the reply is either to explain any confusing parts of the bill raised in the points of dispute or it is to narrow the issues and therefore is concessionary in nature which is always a good thing.

Sub-section 12 of PD 47 contains three stipulations, the first of which is that the reply must follow Precedent G whenever practicable (see points of dispute above for the overall look of Precedent G). The second and third requirements are:
- it must be limited to responses on the points of principle or to concessions on the individual items; and
- it must not contain general denials, specific denials or standard form responses.

The quotation regarding the prolixity of points of dispute above applies as much to replies as to points of dispute. The purpose of a document which simply 'maintains' the claimed figure or 'rejects' all offers made without any counter proposal was always a waste of time. If there was no reply, the receiving party would be taken to be defending the challenge in the points of dispute anyway. That is why they are described as being optional.

The point made above regarding drafting points of dispute on bills which will be provisionally assessed also applies to replies. The receiving party gets two goes at providing information to the costs officer with the bill and the replies. There is something to be said for keeping the bill narrative relatively brief and saving all the ammunition for the reply but the key thing is to make sure the information is in front of the provisional assessor in some shape or form. Having said this, the practice of serving replies only when a receiving party knows the bill will be provisionally assessed (and thereby adding extra detail that would have been given orally at a detailed assessment hearing) is more likely to result in them not being allowed in at all. They should be served in accordance with the rule so that they are included within the documents enclosed with the request for a detailed assessment hearing.

PART 18 REQUESTS

[29.31]

The frequency and extent of Part 18 requests in detailed assessment proceedings increased during the last decade as paying parties sought to challenge additional liabilities, ie recoverable CFA success fees and ATE premiums. Such challenges, if successful, might bring a considerable windfall (see the 'Golden Rule' at the beginning of Chapter 4). Consequently, bills which would otherwise have settled were taken to a court hearing to see if a knockout blow could be delivered. In order to try to reduce the cost of having to pay for the assessment hearing where no such blow was landed, paying parties would try to obtain further information than would usually be received from the bill in order to weigh up the prospects of running such an argument. In *Hutchings v British Transport Police Authority* [2006] EWHC 90064 (Costs), the senior costs judge, sitting as a recorder, gave guidance on requests for further information under para 35.7 of the Costs Practice Direction. The starting point was the overriding objective, and in particular the requirement of proportionality. The court should not willingly do anything which is likely to promote further satellite litigation. The defendant's Part 18 request originally ran to 13 questions all of which the deputy district judge had refused on the grounds it was a fishing expedition. On appeal the senior costs judge described the application as a brash and ill-considered attempt to uncover information which would enable the defendant to challenge the claimant's bill on a technical point but he allowed three of the requests as being reasonable and proportionate. The other questions could be raised as part of the points of dispute and argued on the assessment.

The genie is out of the bottle in relation to Part 18 requests and so they remain an option even though the era of recoverability has come to an end. Whilst there is no equivalent of para 35.7 in the new CPR PD 47, there is no suggestion that a Part 18 request, as a matter of principle, could only be made in respect of additional liabilities. The nature of points of dispute is that they are often a request for information where the wording of the bill is not understood or seems to contradict itself or does not appear to be consistent with other things about which the paying party is aware. The main reason for converting such requests into something more formal under Part 18 was the receiving party's refusal to answer the questions posed. If, or when, another situation arises where requests for information are routinely rebuffed, the use of Part 18 may rise again.

CHAPTER 30

DETAILED ASSESSMENT – PROVISIONAL ASSESSMENT

INTRODUCTION

[30.1]

Litigators of a certain age may recall having bills of costs provisionally assessed by the court in days gone by. In some areas, such as Court of Protection work, this method has continued. As from April 2013, the use of provisional assessment for a large swathe of bills in civil cases, is likely to change the nature of detailed assessment of costs in a fundamental way.

In this Chapter we look at the new scheme together with some comments on the provisional assessments in Legal Aid only and Court of Protection bills. Whilst this Chapter is comparatively short, early practical and procedural problems suggest that it is destined to grow if not like topsy, then certainly as if the costs equivalent of John Innes No 1 had been applied to it, over time.

It is worth making the point (also made in Chapter 28) that provisional assessment is essentially just a stage in the detailed assessment procedure. It is not a third form of assessment to go with summary and detailed assessment. If a case is not suitable for a provisional assessment, it will go straight to a detailed assessment. If a party does not like the outcome of the provisional assessment, he can require there to be a detailed assessment, albeit the rules as to who pays the costs of that hearing are altered from the normal rules. The method by which the bill is assessed is essentially the same as if the parties are there. The purpose of the provisional assessment scheme is to make the detailed assessment process quicker and cheaper; not to create a new form of assessment.

THE PILOT SCHEME

[30.2]

The new arrangements arise from the Jackson sponsored pilot scheme which ran from 1 October 2010 until 31 March 2013 in the Leeds, York and Scarborough County Courts. Under the pilot scheme the provisional assessment of parties' costs took place without an oral hearing where the base costs were under £25,000. Practice Direction 51E (now revoked) set out the details.

Leeds County Court dealt with 119 cases entering the pilot during the first year, most of which were personal injury claims. Of the 100 cases which proceeded to provisional assessment, 17 led to requests for subsequent oral hearings (9 by paying parties and 8 by receiving parties). Only two cases went as far as the oral hearing. In a report based on the Leeds data, Sir Rupert Jackson recommended that the provisional assessment procedure should be incorporated into the Civil Procedure Rules.

THE PROVISIONAL ASSESSMENT PROCEDURE

Requesting a provisional assessment

[30.3]

All of the procedural requirements for provisional assessments can be found at CPR 47.15 and CPR PD 47, para 14.

The journey towards a provisional assessment starts from the same place as any other detailed assessment, namely a notice of commencement and bill of costs. The parties serve points of dispute and any replies and if the case has not settled, the receiving party will request a hearing date by filing an N258 and supporting documents. The N258 was changed in April 2013 to cater for provisional assessments. The receiving party has to request a provisional assessment if the value in the Notice of Commencement is £75,000 or less. If it is more, he requests a detailed assessment hearing and gives a time estimate in the usual way.

Suitability for a provisional assessment

[30.4]

Upon receipt of the N258, the court will consider the suitability of the case for the provisional assessment procedure. Practices will vary from court to court. In the SCCO a costs officer or Master will look at the file.

What makes a case suitable? That is the wrong question. All cases will be considered suitable unless they have a case specific reason for not being so. Having said this, the procedure does not apply to Solicitors Act 1974 assessments between solicitors and clients.

When considering why a provisional assessment might not be suitable, two things need to be borne in mind. The first is that matters such as the importance of the case to the client or the complexity of the facts or issues in the proceedings giving rise to the detailed assessment are not going to be good reasons for an oral hearing. The assessor is going to be able to get to grips with the issues etc just as well by reading the papers at a provisional assessment as he would be if he had read some papers before the detailed assessment hearing.

The second issue is whether evidence is going to be required. Most detailed assessments are successfully completed without any formal evidence being given. Advocates often tread a thin line between submissions and evidence in detailed assessment hearings but that is inevitable where explanations are being requested on a variety of matters. But where there is an evidential dispute over the facts, for example regarding the terms or formation of the retainer, there will be little option but to have a hearing so that such evidence as may be required can be given and tested.

Such issues may not be apparent on an initial sift of the case for non-suitability. If the case gets to the point of being provisionally assessed, there is no reason why the judge cannot halt that assessment and list the case for a hearing instead (CPR 47.15(6).)

Supporting papers

[30.5]

The court will not undertake a provisional assessment until it is in receipt of 'the relevant supporting documents' specified in CPR PD 47, para 14. These documents should in fact have been lodged with the N258 in the first place. But this provision (CPR 47.15(3)) enables the court to decline to carry out any assessment until the minimum amount of paperwork has been filed. The practice direction stipulates the following documents in addition to the N258.

- The notice of commencement, bill of costs, annotated points of dispute from each paying party, any replies, the order for assessment and the other relevant documents that should be filed with the N258 for any detailed assessment (CPR PD 47, para 13.2)
- An additional copy of the bill, including a statement of the costs claimed in respect of the detailed assessment drawn on the assumption that there will not be an oral hearing following the provisional assessment
- All Part 36 and similar offers which must be placed in a sealed envelope and marked 'Part 36 or similar offers' but making sure that there is no indication as to which party made the offer(s). This should include the paying party's open letter served with the points of dispute in accordance with CPR PD 47, para 8.3.
- A completed Precedent G (combining the points of dispute and any reply)

In the pilot scheme, these were all the papers that were filed with the court and there was no provision for any further documents to be lodged to try to influence the assessor. It is perhaps surprising in those circumstances that there were no reports of the points of dispute or replies becoming longer as a result. (The cynic would suggest it was perhaps difficult for that to happen.) Furthermore, the fact that the limit was £25,000 might well have meant that there was less need to provide any further documentation.

The tripling of the limit to £75,000 caught many by surprise and it may well be that a certain amount of tweaking of the procedure will be required to accommodate the larger bills that will be assessed. In the SCCO, which used to have a separate provision in the old costs rules regarding the lodging of papers, when other courts did not, there is an expectation that some papers will be lodged in addition to those specified by the rule. But any requirement to lodge papers would be a local practice direction and therefore to be avoided if at all possible. Consequently, the parties receive with the notice of hearing (see below) a sheet with directions which requires the paying party to confirm whether he intends to lodge any further papers. He need do nothing further if he says that he does not intend to lodge any papers. If, however, he does wish to do so, there is a window of between 2 and 10 days before the hearing for this to be done. The expectation is that there will be a box or two of core papers, but not a van load. Be careful, if you say that you will lodge further papers but then overlook doing so in the time period mentioned, the provisional assessment is likely to be adjourned until the papers have arrived. The court is unlikely to chase for such papers.

CPR 45.17(4) states clearly that the provisional assessment will be based on the information contained in the bill and supporting papers and the contentions set out in Precedent G. Will the additional box load of papers render the assessment contrary to this rule or will they be defined as 'supporting papers' even though they are not specified in the rule or practice direction? The list of documents set out in CPR PD 47, para 14.3 and referred to above are ones that it is said 'must' be filed. Presumably this does not limit what the court can see and prevent it from looking at other documents that may be filed?

Some parties only appreciate the prospect of a provisional assessment rather than a detailed assessment hearing when they get notice of the date of the provisional assessment. That notice spurs them on to file and serve replies that would not otherwise be prepared (if they are the receiving party) or simply to write to the court to expand upon their client's position. Such correspondence is always prefaced with a suggestion that it is written to assist the court but it is debatable that this will ever be the case. If the judge finds himself unable to reach a decision on a particular issue he is perfectly entitled to end the provisional assessment process and list the case for a detailed assessment hearing. If he makes a decision which is wrong in some way, the party can always seek a post-provisional hearing. Attempting to deal with arguments in numerous letters and other documents as well as those raised in the points of dispute and replies is generally going to be unhelpful rather than helpful. It also leads to a position where each party feels compelled to respond to whatever is written by their opponent.

The six week limit

[30.6]

According to CPR PD 47, para 14.4(1) the court will use its best endeavours to carry out the assessment within six weeks of receipt of the N258 and supporting papers. This is a very unusual use of a 'best endeavours' clause. Its purpose appears to be to focus judicial and court manager minds on the need to get on with these assessments rather than to put them at the end of the box work. There is not the slightest hint of what would happen if the six week time period cannot be met.

Practices vary as to the best way to get through this work. In the pilot, the cases were listed for a 'hearing' even though the parties were not able to attend. The Masters in the SCCO have followed this approach so that cases should be dealt with in a finite time. Listing the case also gives the parties an understanding of when the assessment is going to take place and therefore when they should receive the result. There is definitely something strange in a notice of hearing which states in underlined script that no party is to attend (CPR PD 47, para 14.4). The costs officers at the SCCO have taken the opposite approach, ie not listing the cases. As long as they can be dealt with expeditiously this may well result in a speedier turn around for parties who lodge their papers (or confirm they are not going to do so) promptly.

Carrying out the provisional assessment

[30.7]

The rule and practice direction are silent on exactly how the bill should be assessed. They merely deal with how the results of the assessment are recorded. The reason for the silence is two-fold. First, the method of assessing the bill is essentially the same for a provisional assessment as it is for a detailed assessment (see [28.43]). The points of dispute and any replies stand in the place of any oral submissions but as long as the point is raised, the court is seized of the issue and will consider the relevant documents and information before coming to a decision on that point.

The second reason can be highlighted by asking what does the CPR have to say about the conduct of a detailed assessment hearing or indeed any application or trial? In fact there is nothing in the rules regarding how a hearing is run. That is left to the judge and will be entirely case specific. However, since the provisional assessment is carried out without the parties being present, there is an understandable wish to know how it is done.

One of the main differences is the need for the assessor to make an intelligible note of his decision on each point so that the parties can understand that decision when they receive the decision. During a detailed assessment hearing, the judge may, for example decide to reduce the hourly rate claimed and will record this by simply striking out the hourly rate set out in the narrative, or at the beginning of a subsequent part of the bill, and substitute the allowed rate in pen. Where the parties are not present to hear the assessor's reasoning, they are left with the decision (eg the reduced hourly rate) but no explanation. That is not a satisfactory exercise of judicial involvement because part of justice being done is to have reasons for the decisions reached. The following options are open to the judge to explain the decisions reached:

(a) he can set out reasons on the bill;
(b) he can set out reasons on the Precedent G;
(c) he can set out reasons on a separate schedule to go in or with a covering letter;
(d) he can decide to let the amendments speak for themselves on the minor aspects and give reasons as per (a) to (c) above for the points of principle.

The attraction of (c) over (a) or (b) is that there is no issue of space. As described in Chapter 29, the size of the box in which to give reasons in Precedent G seems to depend on how well the box fits on to the page rather than necessarily leaving an adequate amount of room for the decision to be set out. Similar considerations apply on the bill itself in certain areas. Nevertheless, annotating the bill is likely to be the most common option, particularly as it does not require the judge or member of the court staff to type out the reasons.

Option (d) does not really deal with the justice point raised above but will probably be seen on smaller items. At most there will be a very terse comment about it such as 'not recoverable' which arguably is no better than there being no comment at all.

Notifying the parties of the assessment

[30.8]

The change in the scope of provisional assessments means that the rule and practice direction have essentially been drafted in a vacuum, notwithstanding the running of the pilot scheme. This is evident in the provisions regarding annotation.

CPR 47.15(7) explains that a copy of the bill, as provisionally assessed, will be sent to each party. CPR PD 47, para 14.2(2) envisages that the court will return a copy of the Precedent G with the court's decisions noted upon it. This suggests that the only markings on the bill will be the reduced figures where that is appropriate together with whatever sum is allowed for the costs of the provisional assessment (see below). The reasons will be set out separately on the Precedent G. For the reasons given above, this may not be a viable option in some cases. In any event it does seem odd that the rule refers to one document being returned and the practice direction refers to another. It seems probable that many assessors will put all the information on the bill and it will only be that document that is sent to the parties together with the notice regarding what to do next. Those judges who wish to limit the number of challenges to the detailed assessment may consider it worthwhile to have a schedule of reasons to use to explain why some items are not recoverable and which could be rather longer than manuscript reasons given on the bill or precedent G.

Judges usually try to use a pen that does not contain black ink when marking a bill so that the alterations can clearly be seen. If the bill is photocopied, that ploy is rendered useless unless the court copies the bill in colour, which is unlikely. It may well be that the annotated bills are in fact scanned and emailed to parties to preserve the colour as well as save photocopying.

Costs of the provisional assessment

[30.9]

CPR 47.15(5) originally stated simply that the court would not award more than £1,500 to any party in respect of the provisional assessment. This was subsequently clarified to confirm that the £1,500 figure does not include either VAT, where appropriate, nor the court fee. There was a good deal of consternation initially that the capped figure would be largely eaten up by the court fee. That has now passed but it is worth considering how much the court is likely to award by way of provisional assessment costs since CPR 47.15(5) confirms that it is to be done on the basis that there will be no subsequent, oral hearing. The receiving party will have to:
- consider the points of dispute;
- draft any replies;
- give consideration to the open offer and any Part 36 offers made.
- draft the N258 and collate the papers to be sent to the court.
- The cost of preparation of the bill is already catered for on the bill itself.

However, the cost of preparation of the bill is already catered for on the bill itself and, a detailed assessment there is no time to be claimed for:

- preparing the papers for lodging with the court (or at least not as much);
- preparing the advocate for the hearing;
- attending the hearing and travelling to and from it.

How much will be allowed for the costs of negotiations between the parties? The schedules which are lodged with the court which do exceed the recoverable limit generally have considerable time claimed for (unsuccessful) negotiations. Absent such time, these activities set out above will, in the general run of things, not be sufficient to reach the capped limit.

Considering Part 36 offers

[30.10]

Contrary to the implications raised by the filing of sealed envelopes, the court is not going to look at any Part 36 or similar offers at this stage. A sum will be allowed on the bill on the assumption that the receiving party is entitled to the costs of the provisional assessment. This is the purpose of the additional bill of costs referred to in the 'supporting papers' section above.

If a paying party is content with the provisional assessment, other than the costs of the assessment itself which have been awarded to the receiving party, there is a specific procedure set out at CPR PD 47, para 14.6. In short, the court will invite the parties to make written submissions and the issue will be finally determined (ie not provisionally determined) without a hearing. The precise procedure may vary depending upon the circumstances of the case.

When would this circumstance occur? It may simply be that the paying party considers that the award has been too generous and so the amount should be made more reasonable. Such arguments will be entirely case specific. It may also be because the paying party has made an offer which has not been beaten and the paying party wishes to have an order for the costs of the provisional assessment. It would seem probable that in these circumstances the receiving party would be asking for an oral hearing in any event and so the costs would be considered after that hearing.

Concluding the provisional assessment

[30.11]

Upon receipt of the notice of provisional assessment and annotated bill and/or Precedent G, the parties must agree the total sum due to the receiving party based on the court's decisions. If the parties cannot agree the arithmetic they must refer the disputed sums to the court and a decision will be made upon the written submissions of the parties (CPR PD 47, para 14.4(2)).

Unless one of the parties challenges the provisional assessment (see next heading), the provisional assessment will be binding on the parties save in exceptional circumstances (CPR 47.15(7)). There is no requirement to obtain a final costs certificate to complete the process but the procedure is available if a certificate is required for enforcement proceedings, for example.

Challenging the provisional assessment

[30.12]

The parties will receive a notice with the provisionally assessed bill telling them what they may do next. Their options are:
(a) to accept the provisional assessment (see previous section);
(b) to challenge all of the items assessed;
(c) to challenge some of the items assessed;
(d) to challenge the costs of the assessment (see section headed 'considering Part 36 offers').

Where a party decides to challenge some or all of the items assessed in the bill, he must file and serve a written request for an oral hearing. This has to be done within 21 days of receipt of the notice otherwise the provisional assessment will be binding in all but exceptional circumstances.

The request needs to identify the item or items which the challenger wishes to be reviewed at the hearing. It must also provide a time estimate for the hearing.

Upon receipt of the request, the court will fix a date and time for the hearing and will give the parties at least 14 days' notice of the time and place of that hearing.

The post-provisional hearing

[30.13]

At the hearing the court will consider afresh the issues raised. Whilst there is nothing to prevent one judge from hearing a case which another had provisionally assessed, this is unlikely to occur in practice. Human nature will dictate that a judge essentially hearing an appeal from himself will generally only move from his original decision if there are factors of which he was unaware originally.

The would be challenger needs to consider very carefully the costs provisions in respect of a post-provisional hearing set out at CPR 47.15(10). The rule is written so that the challenger will pay the costs of and incidental to the post-provisional hearing unless he 'achieves an adjustment' in his favour of 20% or more of the sum provisionally assessed. This means that if the bill is assessed at 80% of the bill as originally drawn, the receiving party would need to get the total up to at least 96% or the paying party down to at most 64% to avoid paying the costs of the oral hearing. So, unless the assessment is particularly harsh or particularly generous, it will be very difficult to challenge the bill without having to pay the costs of doing so. For a bill at the top end of the provisional assessment limit it may be worthwhile to do so in any event but that would be rare.

One of the options above was to challenge some of the items but not all of them. The costs hurdle is the same however many items are challenged so if only one item is challenged it will have to be very significant for it not to be inevitable that the challenger will have to pay the costs even if successful.

There is a saving provision of the court ordering otherwise and the practice direction (CPR PD 47, para 14.5) specifically refers to the conduct of the parties and the existence of any offers made. It may be that a Part 36 offer made prior to the detailed assessment proceedings by the paying party could alter the incidence of costs completely. However, a Part 36 offer made after the provisional assessment simply to reduce the hurdle that would otherwise have to be overcome at the post-provisional hearing is unlikely to be received favourably by the court.

COURT OF PROTECTION

[30.14]

Requests for detailed assessment of the costs of proceedings in the Court of Protection should be directed to the Senior Courts Costs Office using form N258 (or N258B if, as is likely, the costs are being paid out of a fund.) The SCCO will normally deal with Court of Protection assessments on a provisional basis by post. If the solicitor is not satisfied with the assessment, the costs officer must be informed within 14 days of receipt of the provisional assessment. The costs officer will usually carry out an informal review of any items with which the deputy is dissatisfied. If any informal review does not remedy the dissatisfaction, the SCCO will then fix a date for an oral hearing. In practice the costs officer will deal with any enquiries by telephone or letter.

Authorised court officers (see **[28.55]**) generally deal with any bills below £100,000 and a Master will deal with any larger bills. The documents to be lodged in support of the bill are those set out in CPR PD 47, para 13.12.

LEGAL AID

[30.15]

In days gone by, the courts would regularly assess 'legal aid only' bills where the solicitor would be paid from the Legal Aid fund (in its various incarnations.) That practice has dwindled as assessment of such costs has gradually been subsumed by the Legal Aid authority itself.

Where a legally aided party settles the between the parties' element of his costs without a detailed assessment hearing, it is still possible for the court to assess the legal aid only elements as if the detailed assessment had been reached.

TRUSTS AND OTHER FUNDS

[30.16]

CPR 47.19 is described, in a rather wordy fashion, as setting out the 'detailed assessment procedure where costs are payable out of a fund other than the community legal service fund' (which will no doubt be re-named soon.) It is accompanied by CPR PD 47, para 18.

30.16 *Chapter 30 Detailed Assessment – provisional assessment*

The procedure is that the receiving party needs to request an assessment within three months of the event giving rise to the entitlement to assessed costs. The request is in form N258B. The court may require the bill to be served on other parties if it considers they have a financial interest in the outcome. This would usually be a beneficiary of the trust but may be the trustee in some cases.

The court will provisionally assess the bill and return it to the receiving party who has 14 days to request an oral hearing should he wish to do so. If a hearing is requested, notice will be given to any person with a financial interest as well as the receiving party. If no request is made (or the oral hearing has taken place) the receiving party will be expected to complete the bill so that the final certificate can be produced, always assuming the receiving party is legally represented.

CHAPTER 31

TIME AND VALUE

INTRODUCTION

[31.1]

The use of time as the yardstick by which the size of a solicitor's bill is measured is embedded within both contentious and non-contentious work. The time taken on a matter is part of assessing the reasonableness of the fee charged even where it is not based on an hourly rate. Even Damages-Based Agreements, whose rationale is an agreement of a fee based on the damages, have to be able to be viewed from an hourly rate basis in order to recover the client's costs.

Not only is time one of the prescribed factors for both contentious and non-contentious business, and in routine matters it is the most important, it also runs like a thread through the other factors. The importance and complexity of the matter, the difficulty or novelty of questions raised can all affect the amount of time spent. An hourly cost rate applied to recorded time, in the words of Donaldson J:

> '. . . if calculated accurately, informs a solicitor of the minimum figure which he must charge, if he is not to make an actual loss on the transaction. Second, it gives him an idea of the relationship between the overheads attributable to the transaction and the profit accruing to him. This latter point is plainly relevant in the broad sense that the nature of some transactions will justify much larger profits than others of a more routine type. But we must stress that it is only one of a number of cross-checks on the fairness and reasonableness of the final figure. The final figure will result from an exercise in judgment, not arithmetic, whatever arithmetical cross-checks may be employed.'
>
> (*Treasury Solicitor v Regester* [1978] 2 All ER 920, QBD.)

This chapter looks at time from all angles since it should be used as a management tool as well as for charging. As you would expect, there have been many cases which have considered the issues raised. At the beginning of Chapter 26 on fixed costs we queried the legal fraternity's desire to use hours rather than fixed fees as the basis of charging. There are other alternatives, particularly value which we also look at here. But we cannot get away from the point that paying for a reasonable amount of time, and no more, for a particular piece of work is in our psyche. As Lord Denning said in the 1980s when deciding the case of *Chamberlain v Boodle and King (a firm)* [1982] 3 All ER 188, [1982] 1 WLR 1443, CA:

> 'These rates per hour are over a pound a minute. It would seem there must be a very good system of timing – almost by stopwatch – if that is to be the rate of payment.'

31.2 *Chapter 31 Time and Value*

TIME AS A MANAGEMENT TOOL

[31.2]

Time is a solicitor's raw material, his stock in trade. In the same way that businesses keep control over materials and stock, a solicitor should keep control over his own and his employees' time. The first relevance of time is therefore to the good running and management of a solicitor's practice. Could the time be put to better use?

Time recording and time costing were introduced to solicitors' firms in the days when it was slowly dawning on them that they were no longer certain to make a profit simply by working all hours that God sent and charging as much as the matter or client would stand. It was for that reason that the profession introduced time costing – to see what work was being run at a profit and at a loss. It had nothing to do with charging – indeed conveyancing was still on a fixed scale but conveyancers were exhorted to cost their time to see what profit, if any, they were making. The object, therefore, was to check whether work which was being charged on a scale or some other basis was in fact being done at a profit. As a result many Central London firms stopped doing domestic conveyancing. In the present climate of competitive charging in both non-contentious and contentious work, with costs for the latter being increasingly restricted by being fixed, capped or subject to a predictable matrix (see Chapter 26), it is more important than ever for a solicitor to know what each area of work is costing and whether it can be charged for at a profit.

The Expense of Time

[31.3]

For the system to work you have to know how much each job costs and to do this you have to calculate the cost to the firm of each fee earner in reasonable units of time. There have been a variety of different methods of costing work but the method suggested by the Law Society in its booklet *The Expense of Time* has seemed to most firms using it simple and satisfactory. It has also been extolled in previous editions of this book as including a simplified back-of-an-envelope approach upon which it is difficult to improve. Unfortunately, it is now out of print and very difficult to get hold of, even online. As such you will have to make do with the following summary and should also look ahead to [31.23] for an alternative. *The Expense of Time* basically divides the firm into fee earners and non-fee earners and then divides the projected overheads of the firm, including the salaries of the non-fee earners, between the fee earners, after building in various adjustments such as notional salaries for partners, deduction of interest on the client account and applying the retail price index and the estimated rate of inflation.

Recording non-chargeable time

[31.4]

If fee earners record their non-chargeable time under various categories as well as their chargeable time, the firm's management will have an overall picture of each fee earner's day and can identify how he spent his time. This 'big brother' aspect of time recording does not commend itself to everyone, and it is often not very effective since non-chargeable time is often nebulous and so can lead to 'time-dumping' on difficult to measure activities such as marketing.

TIME AS A CHARGING TOOL

[31.5]

We have seen that the recording and costing of time was introduced for the purpose of knowing whether work was being done at a profit or a loss and that it had the additional advantage of affording management information for those who desired it. It was not intended as a basis of charging. What then happened was that costs judges applying the rules for the assessment of costs; those responsible for devising those rules; and indeed the profession itself, all realised that the cost of doing the work was a useful method of quantifying both contentious and non-contentious work, particularly in routine matters.

Nevertheless, this was using time costing for a purpose for which it was never intended, so it was not really surprising when complaints began to ring out that the formula was too unreliable and too unsophisticated to bear the burden that had been imposed on it. As an aid to business efficiency, the costing of time was a matter of only domestic concern to each firm costing its work. If the formula they used, or their arithmetic, was wrong, it was only they who were misled into thinking they were making a bigger or smaller profit than they really were.

The painful plodder's charter

[31.6]

Another objection to time charging is the same as the objection to the historical quantitative methods of assessment. We now laugh at the thought of payment calculated in relation to the height of the file or the weight of papers, but payment by the metre, the kilo or the hour are all equally flawed: they all relate to quantity not quality. Payment based on time means that the longer a solicitor takes to do the work, the more he is paid. The greater his skill and expertise, the greater his expedition, the less time he will record and the less will be his reward. Where a solicitor brings to bear years of experience, or has a flash of inspiration, it happens so quickly that the time recording clock has barely moved. High quality work which should be encouraged is in fact severely penalised in any system of charges based solely on time. Charges based on the time spent reward the painful plodder for his slowness and inefficiency while penalising the speed and efficiency of others. In contending that this basis of charging is against the client's interests an American commentator referred to the "misaligned interests of the hourly rate".

31.7 *Chapter 31 Time and Value*

CALCULATING THE HOURLY CHARGING RATE

A fee earner's hourly rate

[31.7]

For those who have only practised since the Civil Procedure Rules came into being in 1999, the 'hourly rate' which they charge tends to be a single figure which is either based on the guideline hourly rates (see [**31.22**]) or is a figure provided by the firm's management as the rate to be charged. In some firms, there may be different rates for fee earners to charge depending upon the type of work undertaken but that is relatively rare. Even rarer, is the concept of charging a bespoke hourly rate for an individual piece of work. Where this happens, it tends now to be a reduction from a fee earner's published or 'rack' rate depending upon the negotiating power of the client, rather than a rate designed to reflect the complexity, urgency or other material factor of the work in question.

Many fee earners simply understand that they have a single hourly rate that they charge on all matters and which is increased annually or upon achieving some further qualification or status and notified to clients. But this is a blunt approach to achieving an appropriate hourly rate and the courts have long been uncomfortable with the simplicity of the CPR's approach. Consequently, cases are periodically reported which show that the pre-CPR approach is alive and well. The most recent case is that of *KMT V Kent County Council* [2012] EWHC 2088 (QB). We shall now look at the pre-CPR approach which can safely be said is also a post CPR approach.

A and B factors

[31.8]

This approach to assessment was explained by Brightman J in *Re Eastwood, Lloyds Bank Ltd v Eastwood* [1975] Ch 112, [1973] 3 All ER 1079, CA, in the following extract from his judgment:

> 'During the hearing of the argument I was given certain advice by the assessors as to the manner in which a taxation proceeds at the present day . . . The advice given to me is this:
>
>> "At the present day, on the taxation of a bill of costs of a firm of solicitors in private practice which has been engaged in litigation on behalf of a client . . . the taxation invariably proceeds on the following basis. The firm informs the taxing master of the period of time that has been spent by any partner or employee of the firm on any 'relevant' aspect of the case: the word 'relevant' is intended to exclude time spent on a part of the case for which there is a fixed charge prescribed by statute or rule. The firm submits (a) what is the proper cost per hour of the time so spent, having regard to a reasonable estimate of the overhead expenses of the solicitors' firm including (if the time spent is that of an employee) the reasonable salary of the employee or (if the time spent is that of a partner) a notional salary. The firm will also submit (b) what is a proper additional sum to be allowed over and above (a) by way of further profit costs."'

This philosophy was further expounded in *Lazarus (Leopold) Ltd v Secretary of State for Trade and Industry* (1976) 120 Sol Jo 268, [1976] Costs Law Rep 62 in which Kerr J promoted '(a)' and '(b)' to 'A' and 'B' and ever since they have been known as the A factor and the B factor. This was also the first time that the expression 'direct cost' was used:

> 'In his Answer the master helpfully summarised the practice concerning the computation of Item 26 in cases such as this. Having considered the weight of the proceedings and the responsibility and skill involved, the practice is then to arrive at a total figure consisting of two elements which are referred to as A and B in the judgment of the Court of Appeal in *Eastwood*'s case.
>
> The computation of the A figure involves an assessment of the reasonable direct cost, that is to say the grade of person (senior solicitor, assistant solicitor, legal executive, etc) whom it was reasonable to employ at each stage; an approximation of the cost of employment of each individual by considering the number of hours for each of them to be reasonably engaged; and assessing a rate per hour sufficient to cover the salary and the appropriate share of the general overheads of each such person. The assessment of the appropriate rate per hour would be based on the taxing master's knowledge and experience of the average solicitor or executive employed by the average firm in the area concerned. The total hours of each person multiplied by an approximate cost per hour, together with an allowance for letters, telephone calls and telex messages, then produces what the Court of Appeal referred to as the A figure. The B figure is then conventionally assessed by adding a percentage to the A figure which is appropriate in all the circumstances to cover matters which cannot be calculated on an hourly basis, that is to say supervision and other indirect expenses, together with what the master referred to in his Answer as "imponderables", which reflect the degree of skill, responsibility and the other factors set out in [what is now CPR Rule 44.4]. As mentioned above, the increase for the B figure claimed in the present case was just under 50 per cent, which would be perfectly normal and certainly not excessive in cases of this type. This is the figure which, in the case of an independent firm of solicitors, is expected to make a contribution to the profits of the firm. The appropriateness of the total of A and B, arrived at in this way, is then considered against the background of the proceedings as a whole and rounded off to a convenient sum which appears right in all the circumstances.'

The A factor was therefore identified as the hourly expense rate, and the B factor as the profit, A and B together giving the hourly charging rate. The A factor is arrived at by ascertaining the amount and cost of the time spent, whilst the B factor is arrived at by the application of the factors prescribed in the Solicitors' (Non-Contentious Business) Remuneration Order 2009, art 3 or in CPR 44.4.

Ascertaining the Direct Cost ('A' Factor)

[31.9]

Basing your rates on what other firms are charging or by adopting the going rates being allowed on between the parties' assessments of costs will not tell you whether you are making a profit or loss on a particular kind of work. There is no satisfactory substitute for each firm of solicitors undertaking an *Expense of Time* type calculation at least annually. On an assessment of costs between a solicitor and his client in both contentious and non-contentious costs, the costs officer starts with the retainer. In days gone by, the letter of retainer (if any) might be silent as to the hourly expense rates charged and in which case the costs officer would consider the actual cost to the firm of doing the work and whether that cost was reasonable in the context of the nature of the work. Now that virtually all retainer letters and agreements make express provision for hourly rates, the costs officer starts from the position of knowing the composite hourly charging rate and, where appropriate, working backwards to establish the direct cost. As between the solicitor and his client, the hourly rate set out in the retainer letter or agreement, particularly where that document is signed, is going to be very persuasive, if not conclusive, evidence of the reasonable hourly rate.

However, it does not follow that this amount will be allowed between the parties, regardless of how meticulously you may have calculated your hourly expense rate. If you choose to be a prestigious firm with luxurious offices and expensive cars – as you are perfectly entitled to do – and your client chooses you and your firm – as he is perfectly entitled to do – it is not reasonable that the loser in litigation should have to pay your above-average expense rates because of your sybaritic lifestyle. Between the parties' costs assessed on the standard basis must be both reasonable and proportionate.

How are judges to decide what the hourly cost rate should be? By sitting day in and day out, hearing solicitors disputing or accepting the rates of others and forming a view as to market rates. The judges do not lay down the rates. They adjudicate on the opposing contentions of the parties. Guidance on hourly cost rates was given in two cases:

Hirst J in *Stubbs v Board of Governors of the Royal National Orthopaedic Hospital* [1997] Costs Law Rep 117:

> 'Stress is placed on the average cost of the average solicitor or legal executive – in the particular area concerned . . . In arriving at a figure, the essence of the exercise was an appropriate apportionment of the estimated overhead expenses of the solicitors' firm having regard to the position, status, and likely rate of remuneration of the persons who were employed on the case under consideration. This is of its very nature suitable for assessment by the taxing master in the light of his very wide experience; it is also a matter which is properly approached by reference to averages since it is unlikely that there will be a very wide divergence between comparable firms of solicitors operating in similar fields of work in similar geographical areas.'

In *Finley v Glaxo Laboratories Ltd* [1997] Costs Law Rep 106, Hobhouse J said:

> ' . . . The next point – and it is the one which formed the main part of the argument before me – is the rate per hour which should be allowed. It is clear that the rate which should be allowed is the actual cost, assessed on an objective basis. In other words, it is not answered merely by reference to what has been the cost to the solicitor in question of doing the relevant work on an hourly basis. It has to be assessed on an objective basis having regard to what is reasonable. Therefore, one must consider the position of other solicitors in question. One has to consider whether it is the appropriate level of fee earner that is claimed for. In the present matter it has been accepted, and I accept, that the appropriate level of fee earner for a case of this character was a senior litigation partner; and the solicitor concerned matched that description.'

The introduction of composite hourly rates has meant that solicitors very rarely are in a position to provide the court with the sort of information that these two cases describe. Once in a while, an expense of time type calculation is contained within a witness statement from the receiving party's solicitor but it is far from common. That is surprising given that it forces paying parties' advocates to use the guideline rates as indicative of the direct cost and which (for reasons discussed below) can mean a ceiling on the recoverable hourly rates which is below the rates actually agreed with the client.

Ascertaining the Profit Element ('B' Factor)

[31.10]

So far we have been concerned only with the cost of doing the work (the A factor), to which must, of course, be added the solicitor's reasonable profit (the B factor) to ascertain the hourly charging rate. This is an exercise in evaluating the prescribed factors other than time in CPR 44.4 or art 3. It is not an approach that is unique to solicitors. Businesses of all shapes and sizes regularly price work on the basis of cost plus profit.

The profit percentage in routine matters is generally accepted as 50%, in other words, one-third of the composite charging rate. In pre-CPR assessments travelling and waiting would not justify any B factor and attendance upon counsel in conference or trial would generally be allowed at £35%. Solicitor advocacy would often justify more than 50%. All of these percentages were based on the theory that the solicitor was taking more or less responsibility for the conduct of the case during the particular activity than the standard 50%. In a non-routine case the receiving party can seek an increased charging rate because of the B factor, and indeed the paying party could seek a reduction of the profit element. This is discussed below.

The use of a composite rate in the CPR has risked removing all of the controls, checks and balances built into between the parties' costs over the years. A combined rate in no way reduces the amount of costs, it merely makes the calculation less transparent. Whether or not the A and B factors are identified or concealed in a charging rate, they are still there in terms of costs plus profit. It should also be remembered that the B factor is not solely concerned with profit, it also covers such matters as unrecordable time, supervision and other indirect expenses which would otherwise be unremunerated.

Unrecordable time

[31.11]

Time not capable of being recorded and not included in the A factor, was described by Walton J in *Maltby v D J Freeman & Co (a firm)* [1978] 2 All ER 913, ChD:

> 'No professional man, or senior employee of a professional man, stops thinking about the day's problems the minute he lifts his coat and umbrella from the stand and sets out on the journey home. Ideas, often very valuable ideas, occur in the train or car home, or in the bath, or even whilst watching television. Yet nothing is ever put down on a time sheet, or can be put down on a time sheet adequately to reflect this out of hours devotion of time.'

Supervision

[31.12]

We will see in *Re Frascati* (in chambers) (2 December 1981, unreported, QBD) ([31.26]) that time which should have been recorded but wasn't, should be taken into account in the A factor. What sort of time are we dealing with here? Supervision is one example.

R v Sandhu (29 November 1984, reported in the Lord Chancellor's Department's *Taxing Compendium*) identified three categories of supervision:

(i) The ordinary day-to-day supervision which is part of the overheads of the firm reflected in the basic mark-up – or indeed in the partners' non-chargeable time.
(ii) Supervision where a senior fee earner takes a direct hand in the case, in which event he should record his time and charge for it in the direct costs in the A factor.
(iii) Where a junior fee earner competently conducts the case with the assistance of regular but unquantified supervision from a senior fee earner whose time is not charged in the direct cost. This supervision is covered by the B factor.

Starting point for assessing the B factor

[31.13]

In *Property and Reversionary Investment Corpn Ltd v Secretary of State for the Environment* [1975] 2 All ER 436, [1975] 1 WLR 1504, 119 Sol Jo 274, Donaldson J. said that when looking at the B factor:

> ' . . . it is wrong always to start by assessing the direct and indirect expense to the solicitor, represented by the time spent on the business. This must always be taken into account, but it is not necessarily, or even usually, a basic factor to which all others are related. Thus, although the labour involved will usually be directly related to, and reflected by, the time spent, the skill and specialised knowledge involved may vary greatly for different parts of that time. Again not all time spent on a transaction necessarily lends itself to being recorded, although the fullest possible records should be kept.
>
> This error is compounded if, as an invariable rule, the figure representing the expense of recorded time spent on the transaction is multiplied by another figure to reflect the other factors. The present case provides an illustration of this error. The responsibility and value of the property involved were linked factors, but neither was affected by whether the recorded time spent was 30 hours or 60 hours. Yet the application of a multiplier would double the responsibility/value factor, if the recorded time spent had happened to be 60 rather than 30 hours.
>
> In my judgment the proper approach is to start by taking a broad look at "all the circumstances of the case" and in particular the general nature of the business. This should be followed by a systematic consideration of the factors specified in the paragraphs of [art 3] of the order.'

All the circumstances of the case are intended to be covered by the factors represented by the Seven Pillars of Wisdom.

THE SEVEN PILLARS OF WISDOM

[31.14]

In *Cox v MGN* [2006] EWHC 1235 (QB), The court explained that it is 'required to take into account all the circumstances including, in particular, the factors listed at CPR Part [44.4(3)], which are sometimes referred to as the "Seven Pillars of Wisdom". It is necessary to have regard to the solicitor's particular skill, effort, specialised knowledge and responsibility. Obviously, also, the case in hand must be assessed for importance, complexity, difficulty or novelty. All the while the court will apply the test of proportionality.'

The phrase 'seven pillars of wisdom' comes originally from the Book of Proverbs and there is something almost biblical about the way it has stood firm against attempts to bring those pillars down over time. The seven pillars contained in Order 62 of the Rules of the Supreme Court were revised by the CPR to include the issue of conduct which lay centrally in the Woolf Reforms that led to the CPR. Consequently conduct became the first pillar and the issues of complexity and novelty had to share a bed in a later pillar. The Jackson reforms have added a further pillar by making reference to the receiving party's last approved or agreed budget. Whether the phrase 'the eight pillars of wisdom' will take hold is a matter for conjecture. These pillars were, until the recasting of the rules in April 2013, also used for considering the issue of proportionality by the *Lownds v Home Office* test ([28.46]).

But the new pillars – conduct of the parties and the level of a budget set during proceedings – do not assist with considering the factors in the case which establish the appropriate B factor to include in the hourly charging rate. So for our purposes, the seven pillars that need to be considered are the longer standing ones, namely:

(1) the amount or value of any money or property involved
(2) the importance of the matter to all the parties
(3) the complexity of the matter
(4) the difficulty or novelty of the questions raised
(5) the skill, labour, specialised knowledge and responsibility involved
(6) the place where and the circumstance in which the work was done
(7) the time spent on the case

Together with the value of the case, the time spent is one of the only two factors that are quantifiable. The remaining factors are ones which the judge weighs in the balance, but cannot ascribe any particular figure to them. Let us look at some of them in more detail. The relevant case law comes largely from decisions on non-contentious business where the other factors have tended to weigh more heavily in bills which often have been fixed fee (gross sum) bills. When challenged by the client, the courts have been faced with deciding whether the fees charged are "fair and reasonable" without always having the crutch of hourly rates to fall back on.

But these cases are of value when trying to justify (or challenge) an hourly rate on detailed assessment for reasons that will become clear and so they should not be dismissed as only relevant in non-contentious work, even if the value bands discussed immediately below do not appear to be relevant. They

31.14 *Chapter 31 Time and Value*

may, if nothing else, give food for thought for 'hybrid' DBA arrangements where the mix of hourly rates and value relates to the successful outcome and does not run into the difficulty that the description 'hybrid DBA' usually connotes (see Chapter 7).

AMOUNT OR VALUE

[31.15]

In *Treasury Solicitor v Regester* Donaldson J said:

> 'Turning now to value, we remind ourselves that scale fees have been abolished. Nevertheless, it is reasonable and fair to the client that the remuneration should not be disproportionate to the value of the property involved. It was therefore useful to employ a yardstick to assess that relationship. Various yardsticks, with a regressive basis, can be suggested, and an example will be found in the Oyez Practice Notes [No 20, 6th edn, at p 31]. The fact is that there is no right yardstick, although some may be wrong. For our part, we would consider that 1/2% on the first £250,000 in a major transaction, and thereafter regressing, provides a reasonable method of assessment.'

The yardstick to which Donaldson J was giving cautious approval followed the bands suggested by Donaldson J himself in the *Property and Reversionary* case. The Council of the Law Society suggested revised bands in 1980 and took the opportunity to make further adjustments in July 1987 in their *An Approach to Non-Contentious Costs*, as follows (revised in 1994):

Band	£ Percentage
Up to £400,000	½% (0.5%)
On next £600,000 (maximum total value £1,000,000)	⅜% (0.375%)
On next £1,500,000 (maximum total value £2,500,000)	¼% (0.25%)
On next £2,500,000 (maximum total value £5,000,000)	¼% (0.125%)
On next £5,000,000 (maximum total value £10,000,000)	1/10% (0.1%)

What the Law Society booklet does not tell you, and what you may be desperate to know, is that Donaldson J suggested a percentage charge of 0.05% for values of over £10,000,000. Do not however assume that this provides the profession with a yardstick, because Donaldson J continued:

> 'We must also make it clear that we disagree with the suggestion in the Oyez Notes that the purpose of a scale is to arrive at remuneration for the responsibility/risk element, which is then to be added to the remuneration for the other elements. This is not the case. Remuneration has to be assessed for all the circumstances taken together, and the purpose of a regressing yardstick is, as we have said, to check that the provisional figure bears a reasonable relationship to the value of the property. This is not only reasonable but also fair to both parties and in particular to the client.'

SKILL ETC

[31.16]

Where a solicitor brings to bear great skill and experience far in excess of that of the average solicitor, or where one of the other factors such as complexity, documents, place, title or importance to the client may result in the amount of time spent being an irrelevance.

Although the *Property and Reversionary* and *Treasury Solicitor v Regester* cases dealt with non-contentious commercial matters, there is no reason why there should not be an identical approach in non-routine contentious matters, where the value, the adrenalin or another factor overwhelms the time involved. A gross sum bill to the client can be calculated on these principles. Should the client request a detailed bill or details be ordered on a Solicitors Act assessment the solicitor will be expected to produce a breakdown in terms of hourly rates. Even if the rates appear to be ludicrously high, the overall total may be justified on the approach of *Property and Reversionary* and *Treasury Solicitor v Regester*.

PLACE/CIRCUMSTANCES

[31.17]

In *Treasury Solicitor v Regester*, Donaldson J identified a ninth factor, 'the adrenalin factor'. It is the increased responsibility when time becomes important in a different sense to the rest of this chapter – there is little time to play with and the solicitor has to get things right first time. It seems to us that this conveniently falls within the factor of 'the circumstances in which work or any part of it was done'. In any event, the judge described it in this way:

> 'We then looked at the eight factors and asked ourselves, what was the factor or factors, if any, which distinguished this transaction from the general run of such transactions? The answer was clearly "the adrenalin factor". By this, we mean that the solicitor had not only to work fast but had absolutely no margin for error. The transaction had to be completed by 31 July come what might, or their client had lost not only this deal but all possibility of avoiding the effects of the development land tax. In a different case we might have found that there was plenty of time and that the transaction was very similar to one with which the solicitors had previously been concerned for the same client. This would have caused us to look in the reverse direction . . . recorded time does not provide an arithmetical basis for a charge in cases such as this. Its relevance is to check whether the provisional figures for remuneration bear a reasonable relationship to the overheads attributable to the transaction. In this case, the figures which we had in mind did not seem to bear an unusual relationship to the overheads in a transaction of this type, and we therefore obtained no positive assistance from considering this factor.'

31.18 Chapter 31 Time and Value

WHAT LEVEL OF B FACTOR TO CLAIM?

Starting point

[31.18]

In practice the starting point is to expect to claim 50% for the B factor as being appropriate for the run of the mill case. Most people will readily appreciate that if there is something that is out of the ordinary, it may increase the percentage towards, and sometimes over, 100%. But the beauty of the A and B approach is that the percentage allowed for the B factor could go down as well as up to bring the overall costs to an appropriate figure. So, where someone has made a meal or a mess of a case, the percentage increase for care and conduct for that work could be nothing.

In other cases, it could be much higher than average. For example, where a junior fee earner has done work which could justifiably have been done by a senior partner he may be rewarded with a 200% mark-up for care and conduct on his hourly expense rate. Similarly, if someone has been brilliant, or had extraordinary responsibility, or has been very expeditious, this is the place to take it into account – otherwise the fee earner will be penalised by his own expedition.

In these examples the paying party is not penalised by the A and B approach. The fee earner who made a meal of the case would find the extra work, if allowed, would do no more than make up for the lack of any B factor on the work that should have been required. The junior fee earner with a £125 an hour expense rate who is given a 200% mark-up receives no more than would a solicitor with a £250 an hour expense rate and a 50% mark-up. Similarly, a solicitor who did all the work himself quickly and without using counsel, even with a 200% mark-up, would be unlikely to cost the paying party any more than if the average solicitor had taken an average length of time and incurred average counsel's fees. A solicitor who is expeditious and efficient is entitled to benefit from, and not be penalised by, the lack of time he has spent on the matter. The result is reward for merit – and that is the difference between time-costing and time-charging.

Just one caveat to that. The higher the level of fee earner, the more he is expected to know. Therefore the higher the expense rate the lower the mark-up – you cannot expect to be rewarded twice.

Case law examples

[31.19]

There was consideration of the B factor mark-up in *Finley v Glaxo Laboratories Ltd* where Hobhouse J said:

> 'The solicitor claimed 125%. The Registrar allowed 85% so as to give a composite figure of £65 per hour in conjunction with the £35 per hour rate. The solicitor was, at the material times, effectively a sole practitioner. He had another solicitor with him for part of the time but that is not material to the present case. He had a special experience and knowledge of vaccine and similar matters. He brought to the consideration of the plaintiff's case a familiarity with the subject matter and a

measure of expertise appropriate to a lawyer dealing with that class of case. It must also be borne in mind that he had acted for the plaintiff previously in the tribunal matter, so that he was aware of the background through that source. He is a practitioner in Newcastle . . . That takes me to the second half of the calculation, which is the question of uplift and the "B" factor. One of the matters on which the district registrar was criticised in relation to the "A" factor, in my judgment correctly, was that he got too much involved, in considering the cost assessment, with questions of profit and with the question of the uplift that he was later going to apply. The simplest way to illustrate this point is by referring to the concluding passage in his reasons, where he said:

> "I cannot see, on the evidence before me, any compelling reasons why I should on this evidence increase the currently allowed hourly rate. The allowed rate is £35 per hour. This plus an 85% mark-up gives a remuneration of £65 per hour, which I consider to be both fair and reasonable having regard to the weight of the case and Mr Deas' seniority."

'The district registrar is quite right that at the end of any assessment of this kind he should stand back for a moment and consider the implications and the overall picture presented by his decision on the detail. That is what he is doing there. But it also shows that there is a relationship between the percentage that he chose to allow and the hourly rate. There are many other passages in his reasons where he allows the overall profit assessment to colour his views about the hourly rate. I consider – and counsel has not sought to argue to the contrary – that, having substantially altered the hourly rate, it is appropriate and proper for me to reconsider the uplift that has been allowed.

The uplift factor was a subject that was referred to Kerr J in the *Leopold Lazarus* case at page 4 of the transcript. In view of the hour I will not read out what he says. He refers to Appendix 2 to Order 62 in the earlier version, which is now a slightly different version in the current Appendix. That requires the court, and indeed the taxing officer, in assessing the costs, to take into account, among other things, the complexity of the item or of the cause or matter in which it arises; the difficulty or novelty of the questions involved; the skills, specialised knowledge and responsibility required; the time and labour expended by the solicitor; the number and importance of documents; the importance of the cause or matter to the client; the amount of money involved, and soon.

This present case did involve matters of responsibility for the solicitor, because he was taking upon himself the responsibility of assessing the evidence that was available on a preliminary basis and advising the client. He advised against the continuation of proceedings and, far from his deserving to be marked down for that, he deserves credit for taking upon himself that responsibility and also being prepared to say, of his own judgment, that the proceedings should not continue as to do so would not be in the interests of the client and would, indeed, be a waste of time and money. It was an important case for the client. It potentially involved a very large sum of money. But the medical issues were ones which, although complicated and technical, were not outside the expertise of the solicitor; and it is because of this expertise that the matter was being dealt with by a senior partner.

One must also bear in mind that this case was at a very early stage: it was at a stage of, essentially, perusal and advice. The solicitor was not involved in the preparation of complicated documents, nor was he involved in the conduct of any negotiations. Still less was he actually involved in the conduct of any litigation. All those factors might have put this case in a different light.

I have to consider what is the appropriate uplift for a senior partner in the conduct of a case of this character at the stage which it had reached. The starting point for this exercise is 50%; that is the advice I received, and it is also the practice in the North East. If one is concerned with a High Court or potential High Court action, that is the appropriate starting point for the uplift. Likewise I am satisfied that 125% was far too high for a case of this kind. One of the reasons, I suspect, why 125% was even being considered by the solicitor is the too low figure that he might have feared he was going to be allowed as an hourly rate. As I have already made clear, I would not lend support to the adoption of an unduly low hourly rate and then seeking to put it right by applying a higher uplift percentage. The right approach is that which I have emphasised, namely to adopt a realistic approach to the hourly rate to reflect the actual cost of the fee earner involved, and then to apply an appropriate but not excessive uplift.

The advice that I have had from my assessors in this matter is that no more than 75% is the maximum justifiable uplift in a case of this character, at the stage that it was at, and involving the work which it did involve at that stage. I am advised, and I have also formed the view, that 85% is too high and cannot be supported.'

Evans J also dealt with mark-up in *Johnson v Reed Corrugated* as follows:

' . . . the range for normal, i.e. non-exceptional cases, starts at 50% . . . an appropriate figure for "run-of-the-mill" cases. The figure increases above 50% so as to reflect a number of possible factors — including the complexity of the case, any particular need for special attention to be paid to it, and any additional responsibilities which the solicitor may have undertaken towards the client, and others, depending upon the circumstances—but only a small percentage of accident cases results in an allowance of over 70%. To justify a figure of 100% or even one closely approaching 100% there must be some combination of factors which mean that the case approaches the exceptional.'

He too allowed 75%.

The major test case of *Loveday v Renton (No 2)* [1992] 3 All ER 184, QBD relating to the whooping cough vaccine was adjudged to merit a 125% B factor. The action ran between 1982 and 1988 and culminated in a 300-page judgment which took two days to read. The trial itself lasted for 65 working days. The documentation ran to some 100 lever arch files and The Wellcome Foundation had prepared a 'library' of some 50 files containing literature on the subject. The case was truly exceptional in terms of weight, complexity and responsibility. In delivering his judgment Hobhouse J emphasised that the court deplored any artificially depressed rates being compensated for by artificially inflated mark-up figures. He continued:

'To justify an uplift in excess of 100% it is necessary, as has recently been re-stated by Evans J in Johnson v Reed Corrugated Cases Ltd to demonstrate that the case is exceptional. There has been a tendency among some firms of solicitors to put forward grossly inflated percentages by way of uplift and a failure to appreciate that to justify an uplift even as high as 100% requires the demonstration that the case is exceptional. In the present case, having regard to the features to which I have referred, I am satisfied (not without some hesitation) that the taxing master was wrong to alter his own original assessment on bill A of an uplift of 125%. I consider that 125% is at the top end of the bracket of uplift which would be proper for this case overall and I would not have interfered with a figure which lay somewhere between 125% and 100%. But I do accept the solicitors' submission before me that 100% was too low for bill A and I therefore shall reinstate the original allowance of 125%.'

Self-imposed ceiling?

[31.20]

Based on the case law, solicitors cannot expect to obtain mark-ups in excess of 75% except in the heaviest and most complex cases and probably only those which involve a contested hearing. This leads to a problem where the receiving party cannot demonstrate to the court its A factor direct cost by any method other than dividing the guideline hourly rate into a notional A and B (invariably by deducting a third to represent a 50% uplift). Let's take the National Regional 2 Band A fee earner rate of £201 as our example. The notional A factor will be two thirds of this figure – £134 – and the B factor will be £67. If the B factor is increased to 75% the figure will be £100.50 which when added to the A factor achieves an hourly rate of £234.50. This is the rate that may be the maximum achievable for almost any case run by a solicitor in National Region 2. Whichever band is appropriate, receiving parties regularly find themselves in difficulty in justifying hourly rates on detailed assessment. This is because the A and B calculation requires them to justify 100% or more for the B factor on what cannot be described as exceptional cases of the *Loveday v Renton* type.

The issue is particularly acute where CFAs have meant that the client has had little interest in the hourly rates sought because he has been given an indication by his solicitor that the hourly rates are payable by the opponent and if the court reduces them on assessment the solicitor will live with that reduction. In such circumstances there is little incentive to put anything but an aspirational hourly rate in the client care and/or CFA documentation. If the rates you are seeking in your agreements with clients exceed the guideline rates in the table at **[31.24]**, you should consider making sure you have an Expense of Time type calculation discussed at the beginning of this chapter to justify your charging rates rather than having to rely on the guideline rates. If, as some suggest, the guideline rates may come down in certain areas, action on this issue ought to be all the more pressing.

THE A AND B TEST

[31.21]

The first case following the introduction of the CPR to confirm that the A and B factors were alive and well was *Higgs v Camden and Islington Health Authority* [2003] EWHC 15 (QB), 72 BMLR 95, [2003] 2 Costs LR 211. The nine-year-old claimant was awarded £3.5 million damages for inadequate care during the first 35 minutes of his life, resulting in severe, acquired, hypoxic brain injury; dyskinetic quadriplegic cerebral palsy causing major and permanent motor disabilities, but no intellectual impairment. He would be dependent for his, probably lengthy, lifetime for all daily activities, such as eating, toileting, washing and dressing. It was held that the guideline hourly rates had only limited significance in circumstances such as these, given in particular that brain damage at birth is a particularly sensitive subject matter for litigation and that the specific demands placed upon solicitors by clients and litigation friends will vary widely from case to case). The guidelines expressly state that costs and fees exceeding the guidelines 'may well be justified' in an appropriate case at the discretion of the court. Further, the guideline figures were not supposed to replace the experience and knowledge of those familiar with the local area and the field generally. Accordingly, the guideline figures were not of great value in this instance. The paying party on appeal criticised the costs judge for carrying out an old 'A plus B' calculation in order to test the reasonableness of the rate charged by the partner. However, the bill of costs did in fact claim an inclusive rate of £300 per hour and the court found that the costs judge did no more than use the A plus B method as one of the measures and indicators to ensure that he was able to gauge the propriety or otherwise of a figure of £300 per hour. He used the A plus B type analysis to inform his consideration of the reasonableness of the inclusive rate sought; he had in mind the solicitor's expense rate and he took the seven pillars of wisdom into account in reaching the final figure. The court upheld the award of £300 an hour which represented £150 an hour with a 100% mark-up. (It could, of course, have been £200 an hour with a 50% mark-up.)

In *KMT v Kent County Council* [2012] EWHC 2088 (QB), the claimant, together with her sisters were the subject of sexual, physical and emotional abuse whilst in the care of the defendant and brought claims via the Official Solicitor who instructed Irwin Mitchell. The costs judge decided that it was reasonable to instruct a Central London (now London 2) firm and allowed £335, £255, £180 and £120 for fee earners corresponding to guideline bands A to D respectively. The costs judge applied a 'cross-check' by taking the relevant guideline figure (£263 for the Grade A, by way of example) and reducing it by a third to deduct the notional 50% B factor said to be included in the guideline rates. That left him with £175 and to which he added a 75% B factor given the complexities of the case. This gives £307 in round terms and was sufficient to confirm to the costs judge that his assessment of a reasonable figure was supported by the cross check. Eady J on appeal described the approach adopted as "one that falls very much within the remit of a costs judge."

THE SCCO GUIDE TO SUMMARY ASSESSMENT

[31.22]

The Senior Courts Costs Office Guide to the Summary Assessment of Costs includes guideline hourly rates for different levels of fee earner in different parts of the country. It is not the fault of the Guide that many ignore its title and most of its content, and focus entirely on the tables of hourly rates. As the title and the content state, the Guide is specifically limited to summary assessments of costs and is intended to provide a simple collation of hourly rates applicable for routine costs to be assessed summarily at the end of a hearing which has lasted no more than a day. It was created as a resource for judges who were not used to assessing costs, whether summarily or at all, particularly given the increasing geographical dislocation between solicitors' firms and the courts in which their clients appeared. 'Going' rates for local solicitors were no longer necessarily of any use. The Guide was not intended to be used in detailed assessments where it was assumed that the judge would have more experience of appropriate hourly rates and the parties would have the opportunity of providing more evidence to support their positions if the rates were challenged.

However, the Guide is regularly treated as if it prescribes hourly rates and is appropriate for detailed assessments of costs. The figures were originally produced by collating figures largely provided by district judges and local law societies. The description of them as being the 'SCCO' Guide perhaps gave them an unintended stamp of approval. The more recent revisions of the rates have had more to do with arithmetical increases for inflation than anything else. So to regard these rates as a substitute for solicitors calculating their own rates and justifying them on assessment is really to put the cart before the horse.

In 2005 the Guide ceased to give hourly rates approved for each court. It massaged the rates into three groups for the entire country plus three further groups for London. As a result the rates are no longer approved by any member of the judiciary, do not refer to any particular court and to that extent have become a bureaucratic and not a judicial exercise.

Civil Justice Council Costs Committee

[31.23]

In its report *Improved Access to Justice – Funding Options and Proportionate Costs* in 2005, the Civil Justice Council recommended the creation of a Costs Council 'to oversee the introduction, implementation and monitoring of the reforms we recommend and in particular to establish and review annually the recoverable fixed fees in the fast track and guideline hourly rates between the parties in the multi-track'.

In 2008 the Ministry of Justice transferred the task of collating hourly rates to an Advisory Committee on Civil Costs under the chairmanship of Professor Stephen Nickell. In announcing the Advisory Committee the Ministry of Justice minister said it was 'in response to a recommendation from the Civil Justice Council', but it was not quite the response the Council expected. Professor Nickell wrote to the Master of the Rolls on 9 December 2008 in the following terms:

> 'The Advisory Committee on Civil Costs recommends the attached Table of Guideline Hourly Rates to apply from 1st January, 2009. As you know these guideline rates are broad approximations to be used only as a starting point for judges carrying out summary assessment. These rates are interim in nature in the sense that there remain some unresolved issues which are made clear in the enclosed document entitled "The Derivation of New Guideline Hourly Rates", from which you will understand that at least one member was pressing for an immediate reduction in rates. The unresolved issues include the extent of work done by solicitors outside the region in which they are located and the extent to which referral fees can account for the gap between the hourly rates charged by claimants', as opposed to defendants', solicitors. We hope to have looked at these specific issues by 2010.
>
> Our new interim Guideline Hourly Rates are based on data collected in a survey of solicitors and other interested parties as well as both written and oral evidence provided by representatives of the main interest groups and others. The information

collected refers to the calendar year 2007 and, as last year, we have used the rise in the ONS Average Earnings Index (AEI) for Private Sector Service industries, excluding bonuses, seasonally adjusted, from 2006 Q3 to 2008 Q3 to uprate the 2007 numbers.

I should emphasise that the Committee sees this as unfinished business and that when the outstanding issues have been resolved, we shall revisit the question.'

The committee was concerned that the figures were skewed because an increasing number of firms have offices both in London and the provinces and by the payment of referral fees.

The Master of the Rolls accepted the recommendation of the Advisory Committee that the guideline hourly rates for Summary Assessment should be increased in line with inflation by 1.7% with effect from 1 April 2010. The rates for London 3, Bands A and B are presented as ranges which are said to go some way towards reflecting the wide range of work types transacted in these areas. The Advisory Committee recommended a similar increase for 2011 but that did not come to pass.

As part of the Jackson Review, a Costs Council was proposed to replace the Advisory Committee. However, that did not come to pass either and, in an ironic closing of the circle, the job of reviewing the guideline rates has been handed to the Civil Justice Council by the Government. A new subcommittee has been created with Foskett J in the chair and the Senior Costs Judge as his vice-chairman. Its terms of reference are to make evidence based recommendations and a month long survey took place in November 2013 to seek to gather such evidence. We will have to see how long it will be before that evidence translates into any revised rates to replace the 2010 figures. In the absence of the Law Society's *Expense of Time* the parameters of the CJC's survey provides a very useful indication of the sort of information that a court will find relevant when considering a calculation to support an hourly rate.

Bands of Fee Earner

[31.24]

In the original Guide in 2003, fee earners were divided into four levels of seniority and lettered A to D. The different levels were described as grades. For some reason this was changed in 2010 to bands but to date the phraseology of a 'Grade A fee earner' has remained. (The word 'band' had denoted, until 2010, the geographical grouping of courts which first took place in the 2005 Guide. Those bands have become regions instead.)

The grades were originally agreed between representatives of the SCCO, the Association of District Judges and the Law Society. The categories are as follows:

(A) Solicitors with over eight years post qualification experience including at least eight years litigation experience.
(B) Solicitors and legal executives with over four years post qualification experience including at least four years' litigation experience.
(C) Other solicitors and legal executives and fee earners of equivalent experience.
(D) Trainee solicitors, para legals and other fee earners.

The SCCO Guide to Summary Assessment 31.24

It is worth noting that the word 'partner' does not appear. The bands are based upon experience, not status. 'Legal executive' means a Chartered Fellow of the Institute of Legal Executives. Although fee earners of an equivalent experience may be entitled to similar rates it must be borne in mind that Fellows of the Institute of Legal Executives generally spend two years in a solicitor's office before passing their Part 1 general examinations, a further two years before passing the Part 2 specialist examinations and then complete a further two years in practice before being able to become Fellows. They have therefore acquired considerable practical and academic experience. It is semantically unfortunate that lesser beings, although they are members of the Institute of Legal Executives, may not call themselves legal executives and must be described by some other name such as, fee earner, clerk or paralegal and treated as being in the bottom grade unless the court is satisfied they have the equivalent experience.

The Guide states that:

' . . . many High Court cases justify fee earners at a senior level. However the same may not be true of attendance at pre-trial hearings with counsel. The task of sitting behind counsel should be delegated to a more junior fee earner in all but the most important pre-trial hearings. As with hourly rates the costs estimate supplied by the paying party may be of assistance. What grade of fee earner did they use?

In some proceedings solicitors appear as advocates more frequently than they used to. It must be borne in mind that, especially in substantial hearings, it may be more economical if the advocacy is conducted by counsel rather than a solicitor. In all cases the court should consider whether the decision to instruct counsel has led to an increase in costs and whether that increase is justifiable."

The Guide also points out that an hourly rate in excess of the guideline figures may be appropriate for Band A fee earners in substantial and complex litigation where other factors, including the value of the litigation, the level of complexity, the urgency or importance of the matter as well as any international element would justify a significantly higher rate to reflect higher average costs.

For example, in *Global Marine Drillships Ltd v La Bella* [2010] EWHC 2498 Ch D, Mr Justice Smith on a summary assessment held that although the guideline figure for a Band (he called it 'Grade') A fee earner was £317, given the nature of the application and the urgency, a figure higher than the guideline figure was justified and he allowed £400. It was an application for committal and the defendant faced being sent to prison.

The following table sets out the guideline hourly rates set out in the SCCO Guides from 2003 to date. In order to understand the geographical bands, you will need to read the commentary after this table. 'Region' means 'National Region' for the avoidance of any doubt.

Year	2003				2005			
Bands	A	B	C	D	A	B	C	D
London 1	342	247	189	116	359	259	198	122
London 2	263	200	163	105	276	210	171	110

31.24 *Chapter 31 Time and Value*

London 3	189-221	142-189	137	100	198-232	149-198	144	105
Region 1	175	155	130	95	184	163	137	100
Region 2	165	145	120	90	173	152	126	95
Region 3	150	135	115	85	158	142	121	90
Year	2007				2008			
Bands	A	B	C	D	A	B	C	D
London 1	380	274	210	129	396	285	219	134
London 2	292	222	181	116	304	231	189	121
London 3	210-246	158-210	152	111	219-256	165-219	158	116
Region 1	195	173	145	106	203	180	151	110
Region 2	183	161	133	101	191	168	139	105
Region 3	167	150	128	95	174	156	133	99
Year	2009				2010			
Bands	A	B	C	D	A	B	C	D
London 1	402	291	222	136	409	296	226	138
London 2	312	238	193	124	317	242	196	126
London 3	225-263	169-225	168	124	229-267	172-229	165	121
Region 1	213	189	158	116	217	192	161	118
Region 2/3*	198	174	144	109	201	177	146	111

* National regions 2 and 3 were amalgamated from 2009.

LOCALITY OF FEE EARNERS

[31.25]

Until 31 December 2008 the guideline figures showed Bands One, Two and Three outside the London Area, and a further three localities within the London area (City, Central and Outer London). The difference in the rates

Locality of Fee Earners 31.25

between Bands Two and Three were perceived to be so small that Band Three was promoted to the same level as Band Two and the bands re-titled National Regional 1 and National Regional 2. The London bands are simply described as London 1, 2 and 3.

As the Guide says:

'. . . the guideline figures have been grouped according to locality by way of general guidance only. Although many firms may be comparable with others in the same locality, some of them will not be. For example, a firm located in the City of London which specialises in fast track personal injury claims may not be comparable with other firms in that locality and vice versa.

In any particular case the hourly rate it is reasonable to allow should be determined by reference to the rates charged by comparable firms. For this purpose the costs estimate supplied by the paying party may be of assistance. The rate to allow should not be determined by reference to locality or postcode alone.

The guideline rates for solicitors provided here are broad approximations only. In any particular area the Designated Civil Judge may supply more exact guidelines for rates in that area. Also the costs estimate provided by the paying party may give further guidance if the solicitors for both parties are based in the same locality.'

In some ways, one of the main uses for the regional bands has been to be a tool for paying parties to use to argue about the use of solicitors from a more expensive region by the receiving party (see 'Distant solicitor' [31.30]).

The following table sets out the regions as set out in the SCCO Guide for 2010 onwards.

National		London		
Regional 1	Regional 2	1	2	3
Aldershot, Farnham, Bournemouth (including Poole)	Birmingham (Outer), Bradford (Dewsbury, Halifax, Huddersfield, Keighley & Skipton), Bath, Cheltenham and Gloucester, Taunton, Yeovil	EC1 EC2 EC3 EC4	W1 WC1 WC2 SW1	All other London postcodes: W, NW, N, E, SE, SW and
Birmingham Inner	Bury			Bromley, Croydon, Dartford, Gravesend and Uxbridge
Cambridge City, Harlow	Chelmsford North, Cambridge County, Peterborough, Bury St Edmunds, Norfolk, Lowestoft			
Canterbury, Maidstone, Medway and Tunbridge Wells	Chester and North Wales			

31.25 *Chapter 31 Time and Value*

National		London		
Regional 1	Regional 2	1	2	3
Cardiff (Inner)	Coventry, Rugby, Nuneaton, Stratford and Warwick, Cumbria, Devon, Cornwall			
Chelmsford South, Essex and East Suffolk	Exeter, Plymouth, Grimsby, Skegness			
Fareham, Winchester	Hull (City), Hull (Outer), Kidderminster			
Hampshire, Dorset, Wiltshire, Isle of Wight	Leeds (Outer), Wakefield & Pontefract			
Kingston, Guildford, Reigate, Epsom	Leigh			
Leeds Inner (within 2 kilometres radius of the City Art Gallery)	Lincoln			
Lewes	Luton, Bedford, St Albans, Hitchin, Hertford			
Liverpool, Birkenhead	Manchester (Outer), Oldham, Bolton, Tameside			
Manchester Central	Newcastle (other than City Centre), Northampton & Leicester, Nottingham & Derbyshire			
Newcastle City Centre (within a 2-mile radius of St Nicholas Cathedral)	Scarborough & Ripon			
Norwich City	Sheffield, Doncaster and South Yorkshire			
Nottingham City	Stafford, Stoke, Tamworth			
Oxford, Thames Valley	St Helens, Shrewsbury, Telford, Ludlow, Oswestry, South & West Wales, Southport			
Southampton, Portsmouth	Stockport, Altrincham, Salford, Swansea, Newport, Cardiff (Outer)			

National		London		
Regional 1	Regional 2	1	2	3
Swindon, Basingstoke	Teesside			
Watford	Wigan			
	Wolverhampton, Walsall, Dudley and Stourbridge, Worcester, Hereford, Evesham and Redditch			
	York, Harrogate			

ASSESSING THE TIME SPENT

Proving the time was spent

[31.26]

The costs officer must be satisfied that the amount of time claimed was actually spent. Properly kept and detailed time records are helpful in support of a bill provided they explain the nature of the work as well as recording the time involved. In *Jemma Trust Co Ltd v Liptrott* [2003] EWCA Civ 1476, [2004] 1 All ER 510, [2004] 1 WLR 646, the court held that there is no obligation on a solicitor to keep attendance notes in respect of either contentious or non-contentious work. The true position is that in both kinds of work the burden is on the solicitor not only to show that the time claimed has been spent, but that it had been reasonable to spend that time. The keeping of an attendance note is one way, but not the only way, in which this can be demonstrated. The failure to keep such notes exposes the solicitor to the risk of being unable to prove the reasonableness of the time spent. In the *Jemma Trust* case the costs judge after examining all the files was satisfied that the time recorded had been reasonably spent.

While the absence of records could result in the disallowance or diminution of the charges claimed, the records themselves will not be accepted as conclusive evidence that the time recorded either has been spent, or if spent, is 'reasonably' or 'proportionately' chargeable. Always keep in mind the words of Payne J in *Re Kingsley* (1978) 122 Sol Jo 457:

> 'I ought to add that this case illustrates the dangers which are present if reliance is placed on a modern system of recording, without at the same time retaining the old and well tried practice of keeping attendance notes showing briefly the time taken and the purport of the work done day by day. It may be that this case will invite attention to the importance of appreciating the limits to which the computer system can be used in cases where taxation of costs must follow litigation and to the necessity of preserving as well the use of the traditional systems.'

31.26 *Chapter 31 Time and Value*

It is unlikely that the stage will ever be reached of producing a computer print-out as incontrovertible evidence of the number of hours spent. For the time being paperless offices, or perhaps more accurately, documents all being held electronically by solicitors is a problem on detailed assessment. Generally the papers have to be printed out and if there are any gaps (as often seems to be the case) will generally mean the cost of work done is disallowed. The provision of CDs and USB drives or "memory sticks" containing papers for the court's use is still in its infancy and is subject to the inevitable teething problems.

Letters received and copies of letters sent, instructions to counsel, statements of case, witness statements and all other documents are tangible records and evidence of the work involved. All other work, such as attendances on the client, counsel, court and witnesses, telephone conversations, perusals, considering the facts and law, does not automatically create records of itself and a solicitor must be able to prove that he has done work of this nature. The best way is to create and keep records of it. Inferences can be drawn and oral evidence given but it is in the solicitor's interests to have available on assessment the best possible evidence of work done, and this can most easily be provided by making contemporaneous records. However, here are some words of comfort from Parker J in *Re Frascati (in chambers)* (2 December 1981, unreported, QBD):

> 'The right to charge cannot depend upon the question whether discussions are recorded or unrecorded. It must depend, initially, upon whether they in fact took place and occupied the time claimed. If they are recorded in attendance notes this will no doubt ordinarily be accepted as sufficient evidence of those facts. If they are not so recorded it may well be that the claimant is unable to satisfy the taxing (costs) officer or master (costs judge) as to the facts. But neither the presence nor the absence of an attendance note is conclusive. It may well be for example that it is wholly impractical in some instances to keep such notes. In an exceptionally complex case, such as this which is occupying two fee earners there may be short but important discussions in respect of which it would be wholly unreasonable to expect attendance notes to be kept. In such cases an estimate of the time involved is inevitable. The question which then arises for decision is whether the estimate given is reasonable. This is a matter wholly for the taxing authorities. In general, however, all such discussions involving any substantial period of time should be recorded and an estimated addition should only be allowed for short discussions which it would be impracticable to record.'

A number of points arise:
(1) We are dealing here with time which could have been recorded but for some reason was not, and not with unrecordable time, such as supervision, which is covered by general care and attention, ie the 'B' factor (see [**31.12**]).
(2) How the time was recorded does not, for these purposes, matter. It can be on the most sophisticated computer system or on the back of old envelopes. If the latter contains greater detail of the work which was done it is preferable to the former. The problem with most computer systems is that they do not require, and the fee earner regularly fails to provide, a sufficient narrative. A separate file note is essential. With a mark-up for care and control on all contentious work it is not only the

(3) Experience shows that many solicitors have now trained themselves adequately to record the time spent on attendances but they are not so well disciplined in respect of time spent on the preparation and perusal of documents such as pleadings, instructions to counsel and, particularly in long actions, refreshing their memory from the statements of case, affidavits, attendance notes, correspondence and opinions. These are the danger areas.

(4) Time spent considering the law and procedure is usually non-chargeable – and the higher the expense rate, the more law and procedure the fee earner is expected to know. In a review of criminal costs it was held that leading and junior counsel can be assumed to be fully up-to-date with the law in the field in which they hold themselves out as practising in and they will not be paid for researching the law unless the case is unusual or infrequent (*Perry v Lord Chancellor* (1994) Times, 26 May, QBD). The same principle of course applies to civil work and to solicitors.

(5) Where there has been an omission to record time, the costs draftsman – and the costs officer – must rely on their own experience to assess from the documents in the file what time a competent solicitor would have spent on their preparation and perusal. The documents prove the work has been done but someone has to put a time and value on that work.

(6) In *Johnson v Reed Corrugated Cases Ltd* [1992] 1 All ER 169, QBD Evans J was less enthusiastic about giving credit for unrecorded time, saying:

> 'In my judgment, the submission that there were unrecorded occasions when chargeable time was spent on these cases must be rejected. This leaves the registrar's decision that in practice not all time will be fully recorded, even for those items of work in respect of which a claim is made. The claims invariably are for global figures, mostly to the nearest five minutes. No doubt there were some occasions when the periods spent were slightly more, others when it was slightly less. There is no evidence that any substantial items were not recorded at all. In my judgment, therefore, this item must be disallowed.'

Was the time reasonably spent?

[31.27]

This is an objective test between the parties. Was it reasonable to do all the work that was done and was it done within a reasonable time? If on the standard basis, was it proportionate to the matters in issue? If it was excessive or irrelevant it will be unreasonable or disallowed. In contentious costs on the standard basis any doubt as to reasonableness will be resolved in favour of the paying party. Between a solicitor and his client the same objective tests apply, together with, in respect of contentious work, the express and implied subjective assumptions prescribed by CPR 46.9(3). Since solicitor and client costs are assessed on the indemnity basis, any doubt will be resolved in favour of the solicitors.

The level of fee earner

[31.28]

Has a senior partner done work which could have been done by a legal executive? If so, only the expense rate of a legal executive can be allowed. Has a legal executive done the work of a senior partner? Again only the expense rate of a legal executive can be allowed. However, in both examples, the hourly rate can be appropriately adjusted by the application of the other prescribed factors – perhaps to award the partner for his expertise and expedition and the legal executive for punching above his weight.

Is the hourly rate too high?

[31.29]

As already mentioned, rates above the going rate can be claimed on the basis of the prescribed factors other than time, but there are other circumstances in which the going rate for a particular court may be exceeded on a detailed assessment.

In *Jones v Secretary of State for Wales* [1997] 2 All ER 507, QBD, a provincial firm of solicitors costs were allowed at a higher figure than the local rate because it demonstrated that it was more specialised than the norm for the area. Pitmans of Reading were in effect a London firm who had moved to Reading and were doing work that only a London firm could do. The judge was of the view that it would be odd and undesirable if the higher London rate could be recovered by a London firm, but not the somewhat lower rate claimed by Pitmans for doing a case which otherwise would probably have to be handled by a London firm.

This approach was followed by Master Wright in the SCCO in *Wood v Worthing and Southlands Hospitals NHS Trust* (9 July 2004, unreported). In a complex clinical negligence case he allowed central London rates for the specialist partner in the outer London office of a national firm on the grounds that it would be inappropriate to apply the rates recoverable if the work had been done by a small Hampstead firm. A note of caution about this approach was sounded in *Cox v MGN*:

> 'If you wish to take yourself out of the norm you have to provide the court with evidence to enable you to do so. You may have a niche practice, and you may be able to persuade celebrities that you are the solicitor to go to at whatever rate you choose to charge them, but without evidence that your overheads are out of the ordinary there is no basis for holding that a Jones [*Jones v Secretary of State for Wales* [1997] 2 All ER 507[increase should apply.'

In Jones the salary costs were similar to a London firm rather than a Reading firm thereby enabling Pitmans to explain the increased overheads in the manner referred to in Cox.

Distant solicitor

[31.30]

In *Truscott v Truscott, Wraith v Sheffield Forgemasters Ltd* [1998] 1 All ER 82, CA and *Sullivan v Co-operative Insurance Society Ltd* ([1999] 2 Costs LR

158, CA, the Court of Appeal addressed the question of whether the liability of an unsuccessful party ordered to pay costs should be restricted to what a reasonably competent solicitor practising in the area of the court (or in the area where the successful party lived) might have been expected to charge, or whether the successful party should be entitled to recover the sums claimed by the solicitor who was in fact instructed to act on his behalf. The court held that a costs judge had to consider whether, having regard to all the relevant considerations, the successful party had acted reasonably in instructing the particular solicitors. In *Truscott* the claimant had been ill-served by local solicitors in the Brighton County Court and had instructed London solicitors who rectified the matter and obtained a wasted costs order against the original solicitors. The court held that it was reasonable for a London firm to have been instructed in these circumstances and that as their charges were reasonable by London standards they were allowed. The same principles applied in *Wraith* but in that case the only reason why the work went to London solicitors was that the claimant's trade union had adopted the practice of sending all their work to these solicitors. The action was proceeding in the Sheffield District Registry and the court was satisfied there were firms of solicitors in Sheffield or Leeds well qualified to do the work. The trade union knew, or ought to have known, what sort of legal fees it would have to expend to obtain competent services and their connection with one firm of solicitors in London was of limited relevance. *Sullivan* applied the same test over the Pennines, holding there were in Manchester, and in many other centres outside London, legal practitioners who conducted cases of substantially greater weight and complexity than the present case, every day of their working lives. Although the fact that a trade union or other organisation habitually used a particular firm of solicitors was a relevant factor, it was of limited relevance in an individual case. There were no weighty factors in the context of costs justifying the trade union in employing London solicitors to conduct litigation in Manchester. Subjectively the choice may have been entirely reasonable, but between the parties the various circumstances had to be balanced objectively.

Wraith was followed in *A v Chief Constable of South Yorkshire Police* [2008] EWHC 1658 (QB) where in proceedings in Sheffield the court disallowed the costs of specialist London solicitors on the grounds that a reasonable person in the position of the claimant would have enquired about solicitors in Sheffield with experience of bringing claims against the police. Such enquiries would have brought to his attention a law firm in Sheffield that undertook police misconduct cases. He would have appreciated that there was a substantial difference in rates and would have instructed the Sheffield firm.

In *Ryan v Tretol Group Ltd* [2002] EWHC 1956 (QB), the claimant in an asbestosis claim instructed local solicitors, but after the legal executive dealing with the matter left the firm, the claimant was dissatisfied with the service he was receiving. The claim was legally and factually complex and in the circumstances it was reasonable for the claimant to change his legal representatives to a London firm with considerable asbestosis experience in the absence of other firms with similar experience in the claimant's locality.

Subjective factors also applied in *Higgins v Ministry of Defence* [2010] EWHC 654 (QB). At the age of 82 the claimant, who lived in Broadstairs in Kent, was diagnosed with asbestosis by a local consultant who told Mr Higgins that his condition was advanced and there was no treatment. By this

31.30 *Chapter 31 Time and Value*

time Mr Higgins was able to walk very little, and his daughter had moved in to care for him. The consultant mentioned a firm of solicitors in Central London, Field Fisher Waterhouse (FFW), who had experience in these matters and whom the claimant consulted.

The claimant had been exposed to asbestos during the course of his work for the Ministry of Defence in Davenport and the solicitors were able to reach a settlement, but could not agree the costs. The solicitors claimed an hourly rate of £345 while the defendants argued that according to the guidelines for summary assessment the recoverable hourly rate would have been up to £200 for Kent and up to £250 for outer London firms. The judge could find no point of principle saying: 'It is not in dispute that a reasonable litigant would normally be expected to investigate the hourly rates of solicitors whom we might instruct, and that he will normally be expected to consider a number of other factors, including the time and costs associated with geographical location, before choosing whom to instruct and to take advice on these and other matters before he does so'. It would not be objectively reasonable to expect an 82 year old man who had just been informed that he was incurably ill, to undertake a trawl of local solicitors, in circumstances where an experienced consultant had given him the name of FFW as solicitors who specialised in this field. None of the alternative firms whose names were put forward were markedly more accessible from Broadstairs than is London. The costs judge had considered all the relevant factors including the claimant's age and the urgency of the case before deciding it reasonable for Mr Higgins to have instructed FFW and it would not be appropriate for the court to interfere with his decision on appeal.

Gazeley v Wade [2004] EWHC 2675 (QBD) went the other way. The appellant appealed against a decision that he should recover costs from the respondents on the basis of rates applicable to solicitors in the Norwich area although his solicitors were London libel specialists. The costs judge was not wrong to find that it was unreasonable for the appellant to instruct London libel specialists rather than a local firm in Norwich in respect of his defamation claim against a national newspaper even though he erred in deciding that the case 'was obviously a Norfolk case'. The fact that the claimant was from Norfolk did not make it a Norfolk case. At the relevant time the likelihood was that the hearing would have been in London. Moreover, it was a grave libel published nationwide by defendants based in London. However, it was important to recognise that in order to have the necessary or proportionate expertise available, it was not always necessary to instruct London specialist solicitors. An important factor was that any competent litigation solicitor in the country could call upon specialist members of the Bar at very short notice. The costs judge's conclusion of the unreasonableness of instructing London solicitors was based on the particular facts of the case, in the light of his very wide experience of litigation generally, including the role of specialist practitioners.

That the nature of the work is relevant to the level of fees was demonstrated by the senior costs judge in *King v Telegraph Group Ltd* [2004] EWCA Civ 613. City rates for City solicitors are recoverable only where the City solicitor is undertaking City work, which is normally heavy commercial or corporate work. Defamation is not in that category, and, particularly given the reduction in damages awards for libel, is never likely to be. A City firm which undertakes work, which could be competently handled by a number of Central London solicitors, is acting unreasonably and disproportionately if it seeks to charge City rates. Newspapers and insurers control defendants' lawyer's fees and are not prepared to pay the level of fees which the lawyers may wish to charge.

COUNSEL'S FEES

Guideline application fees

[31.31]

At the end of the SCCO Guideline rates are benchmark rates for counsel's attendance at short hearings. Unlike the solicitors' rates, the benchmark rates for counsel are rarely relied upon in detailed assessments. The Guide states:

> 'The following table sets out figures based on SCCO statistics dealing with run of the mill proceedings in the Queen's Bench and Chancery Division and in the Administrative Court. The table gives figures for cases lasting up to an hour and up to half a day, in respect of counsel up to five years call, up to ten years call and over ten years call. It is emphasised that these figures are not recommended rates but it is hoped that they may provide a helpful starting point for judges when assessing counsel's fees. The appropriate fee in any particular case may be more or less than the figures appearing in the table, depending upon the circumstances.
>
> The table does not include any figures in respect of leading counsel's fees since such cases would self evidently be exceptional. Similarly, no figures are included for the Commercial Court or the Technology & Construction Court.'

Table of Counsel's Fees

QUEENS BENCH	1 hour hearing	½ day hearing
Junior up to 5 years call	£259	£450
Junior 5–10 years call	£386	£767
Junior 10+ years call	£582	£1,164
CHANCERY DIVISION	1 hour hearing	½ day hearing
Junior up to 5 years call	£291	£556
Junior 5–10 years call	£497	£931
Junior 10+ years call	£757	£1,397
ADMINISTRATIVE COURT	1 hour hearing	½ day hearing
Junior up to 5 years call	£381	£582
Junior 5–10 years call	£698	£1,164
Junior 10+ years call	£989	£1,746'

Non guideline fee applications

[31.32]

The SCCO Guide continues:

> 'if the paying parties were represented by counsel, the fee paid to their counsel is an important factor but not a conclusive one on the question of fees payable to the receiving party's counsel.
>
> In deciding upon the appropriate fee for counsel the question is not simply one of counsel's experience and seniority but also of the level of counsel which the particular case merits.
>
> Counsel's fees should not be allowed in cases in which it was not reasonable to have instructed counsel, but it must be borne in mind that, especially in substantial hearings, it may be more economical if the advocacy is conducted by counsel rather than a solicitor. In all cases the court should consider whether or not the decision to instruct counsel has led to an increase in costs and whether that increase is justifiable.'

The test for counsel's fees is usually said to come from *Simpsons Motor Sales (London) Litd v Hendon Corporation* [1964] 3 All ER 833, [1965] 1 WLR 112. The fee is based on a reasonably competent counsel being instructed rather than someone pre-eminent in the field.

Brief fees

[31.33]

The traditional approach is that the time spent is a relevant factor in assessing a brief fee but it is not appropriate to determine a brief fee simply by having regard to an hourly rate. In *XYZ v Schering Health Care* SCCO Costs Appeal No 9 of 2004 leading counsel had been required to prepare large amounts of material, and to undertake complex cross-examinations which required highly specialised skills. It could not be said that his brief fee was unreasonable.

There is a trend towards brief fees becoming more regularly the product of hours multiplied by hourly rate. Large corporations such as insurance companies as well as Government departments appear to be wedded to the idea that hourly rates provide certainty and transparency. How they could be any more certain or transparent than a previously agreed fixed fee is not immediately obvious.

VALUE BILLING IN NON-CONTENTIOUS BUSINESS

[31.34]

The assessment of fair and reasonable remuneration is an art not a science. It is arrived at by a commonsense feel of the matter – not by the application of a rigid formula by a computer. The charging options available in non-contentious matters are hourly rate only; hourly rate plus value; fixed fee; contingency fee agreement; or non-contentious business agreement. The last three categories are dealt with elsewhere in Part 2 of this book. Hourly rates have been dealt with earlier in this chapter so the following comments concentrate on the idea of charging the client an hourly rate based fee with a value charge in addition.

Hourly rate plus value

[31.35]

In certain areas of work, such as probate, it may be appropriate to include a value element in the method of charging. However, the question of whether a value element is charged is a matter of judgement for the solicitor and for agreement between the solicitor and the client. It is certainly not mandatory.

Where a value element is used, the overall consideration must always be that the charges are fair and reasonable, having regard to all the circumstances of the matter, in accordance with Article 3 of the Solicitors' (Non-Contentious Business) Remuneration Order 2009, as amended.

The leading case in this area is *Jemma Trust Co Ltd v Liptrott* [2003] EWCA Civ 1476, [2004] 1 All ER 510, [2004] 1 WLR 646. The Court of Appeal held that a value charge can either be made in addition to an hourly rate, or it can be included in the hourly rate, but the value element must not be reflected in both charges. Where appropriate, charges may consist of two elements:

(1) *Hourly rate.* This should be an inclusive figure incorporating the fee earner's expense rate and any appropriate care and conduct uplift. [So if the second element is not charged, the solicitor is remunerated in the normal way.]

(2) *Value element.* Account may be taken of the value of the assets in the estate. In calculating the value element of the charge, the following approach may be helpful:

First, consider the value nature and number of assets. It is usual to divide the estate (i.e. total value of the assets left after death) into two parts:

(a) *The deceased's residence.* The value of the deceased's home, or as much of it as he or she owned, if it was shared with another person. For example, where the property is jointly owned by two people, the value is reduced by half.

(b) Value of rest of the estate.

Second, apply an appropriate percentage. An appropriate percentage should be considered in the light of the circumstances of the matter but the following may be helpful:

(a) Solicitor not acting as an executor – Value of gross estate less residence 1%; Value of residence 0.5%;

(b) Solicitor acting as sole executor or joint executor with another person – Value of gross estate less residence 1.5%; Value of residence 0.75%.

The final figure should always be reviewed to ensure that the charges are fair and reasonable having regard to all the circumstances.

31.36 *Chapter 31 Time and Value*

High value estates

[31.36]

When dealing with high value estates, consideration should be given to reducing the value element percentage charged in order to ensure that the overall level of charge is fair and reasonable.

In *Jemma Trust*, the costs judge had held that a solicitor's charges for administering a large estate should no longer be based on both hourly rates and an additional element in respect of value. The estate in question was worth nearly £10 million for the administration of which the solicitor had elected to charge on the basis of an hourly rate for work carried out and in addition he charged 1.5% of the gross value of the estate (save for the house, which was charged at 0.5%) as a value element. This resulted in a value element of £227,000 in addition to the hourly rate charges of £386,000, totalling some £613,000 plus VAT. The master concluded that now hourly rates are calculated invariably on the basis of sophisticated time recording material, it is anachronistic and wrong to include an additional element in respect of value.

The Court of Appeal held that he was wrong on two counts. First, it is still appropriate for solicitors administering an estate and charging the time spent on administration to charge a separate fee based on the value of the estate, provided of course it is fair and reasonable remuneration in the light of all the circumstances. Second, by disallowing a charge based on the value element and only allowing a miserly hourly rate, the costs judge had in effect failed to have any regard to the value element. However, the Law Society's advice in its 1995 publication 'An Approach to Non-Contentious costs', and its subsequent 1999 booklet, that in high value estates 'consideration should be given to reducing the value element percentage' was insufficiently firm. The regressive scale adopted in *Maltby v D J Freeman & Co* [1978] 2 All ER 913, [1978] 1 WLR 431 should be used, resulting, when updated for inflation, in the following figures for work done in 2003:

(1) Up to £1 million   1.5%
(2) Over £1 million up to £4 million   ½%
(3) Over £4 million up to £8 million   ⅙%
(4) Over £8 million up to £12 million   1/12%

The Court of Appeal gave this guidance:

'(1) Much the best practice is for a solicitor to obtain prior agreement as to the basis of his charges not only from the executors but also, where appropriate, from any residuary beneficiary who is an entitled third party under the 1994 Order [*now the 2009 Order*]. This is encouraged in the 1994 booklet and letter 8 of Appendix 2 to the 1999 booklet provides a good working draft of such agreement. We support that encouragement;
(2) in any complicated administration, it will be prudent for solicitors to provide in their terms of retainer for interim bills to be rendered for payment on account; this is, of course, subject to the solicitor's obligation to review the matter as a whole at the end of the business so as to ensure that he has claimed no more than is fair and reasonable, taking into account the factors set out in the 1994 Order;
(3) there should be no hard and fast rule that charges cannot be made separately by reference to the value of the estate; value can, by contrast, be taken into account

as part of the hourly rate; value can also be taken into account partly in one way and partly in the other. What is important is that:
(a) it should be transparent on the face of the bill how value is being taken into account; and
(b) in no case should it be taken into account more than once;
(4) in many cases, if a charge is separately made by reference to the value of the estate, it should usually be on a regressive scale. The bands and percentages will be for the costs judge in each case; the suggestions to the costs judge set out in paragraph 30 may be thought by him to be appropriate for this case but different bands and percentages will be appropriate for other cases and figures set out in paragraph 30 cannot be any more than a guideline;
(5) it may be helpful at the end of the business for the solicitor or, if there is an assessment, for the costs judge, when a separate element of the bill is based on the value of the estate, to calculate the number of hours that would notionally be taken to achieve the amount of the separate charge. That may help to determine whether overall the remuneration claimed or assessed is fair and reasonable within the terms of the 1994 Order;
(6) it may also be helpful to consider the Law Society's Guidance in cases where there is no relevant and ascertainable value factor which is given in the 1994 booklet at paragraph 13.4. If the time spent on the matter is costed out at the solicitor's expense rate (which should be readily ascertainable from the Solicitor's Expense of Time calculations) the difference between that sum (the cost to the solicitor of the time spent on the matter) and the final figure claimed will represent the mark-up. The mark-up (which should take into account the factors specified in the 1994 Order including value) when added to the cost of the time spent must then be judged by reference to the requirement that this total figure must represent "such sum as may be fair and reasonable to both solicitor and entitled person".'

Routine non-contentious work

[31.37]

In all other routine non-contentious matters, where there is an ascertainable value element, consider whether the non-value factors are subsumed by value. It used to be only in exceptional circumstances that there would be a mark-up expressed as a percentage of time in addition to the value factor. However, the fall in the value of money since the value yardsticks were set means that there are now probably fewer cases where a value factor in itself is sufficient. If there is a value factor the usual mark-up on time might be less, perhaps between 25% and 33% in matters other than routine domestic conveyancing, or none at all following *Jemma Trust Co Ltd v Liptrott*, above.

ALTERNATIVES TO TIME AS A MANAGEMENT TOOL

Budgeting

[31.38]

You can prepare a budget of expenses for the next year. You can divide that by 12. You now know that if in each month you do not bill at least that amount you are running at a loss. You can take it a stage further by adding to your budgeted expenses for the forthcoming year the profit you hope to make during that period and dividing that total by 12. This gives you a total target figure for bills delivered. Having calculated the figure for bills to deliver during the year, it can be apportioned among the fee earners so that the total targets of the fee earners equal your overall target. If each fee earner accepts his target as realistic, you know you should reach your overall target for the year. As a result of monthly checks you will have an early warning if any fee earner is falling behind his target. Each month you will know whether you are on, ahead of or behind your target.

Fee earner targets

[31.39]

There is no point in one solicitor (the principal) employing another solicitor or legal executive (the assistant), paying him a salary, supplying him free of charge with accommodation, secretarial services and all the other tools of his profession, unless that assistant either makes a profit for his principal or relieves the principal of routine work, thereby enabling the principal to undertake more profitable work. A rough formula which has stood the test of time is that a fee earner's share of the overheads is not less than one-and-a-half times his salary. An assistant with a salary of £40,000 a year will have a share of the overheads of £60,000 a year, meaning that he must bill £100,000, or two-and-a-half times his salary, to break even. If to this is added a modest profit of half his salary, this produces a target of £120,000, or three times his salary. A survey of surveyors identified what they called a factor of 2.7–3.0; they had reached the same conclusion – that unless an assistant earns at least three times his salary, his employment must be justified by some means other than the profit he is making for the firm.

These two aspects of management control depend solely upon the calculation of expenses and projected profit. They do not of themselves require any recording or costing of time.

ALTERNATIVES TO TIME AS A CHARGING TOOL

[31.40]

There are various ways in which solicitors' charges could be calculated, other than on the basis of time. Examples are according to value, scale, commission, salary, fixed fees, a contingency basis, and a quantum meruit.

Not only does time not enter into any of these bases, the less time a solicitor takes to do the work where time is not the basis of calculation, the bigger the profit he makes because he has more time to spend on other work.

In an article in *The Lawyer*, Anne Gallagher considered various methods of charging in the United States. She identified four basic billing methods: hourly, fixed-fee and contingency, all of which determine in advance the billing method that will be used, and 'value billing', where the lawyer retrospectively determines what the fee will be, based upon either subjective or agreed-upon criteria. Variations of billing methods included:

(i)	Blended hourly rate	All services are charged at a common hourly rate whether for a senior partner, junior partner or associate.
(ii)	Fixed fee plus hourly rate	Certain defined services are charged on a fixed-fee basis, while the remainder of the time expended is billed on an hourly basis.
(iii)	Hourly rate plus premium	The established hourly rate is standard or lower than standard with a bonus paid if certain results are achieved.
(iv)	Percentage fees	The actual amount of the fee depends on the value of the total transaction.
(v)	Unit charges	For a defined segment of service, a unit charge is established, usually in combination with hourly billing.
(vi)	Relative value fees	A fee is established based on the relative value of the activity. For example, time spent answering correspondence may be assigned a lower value than time spent in research or negotiation.
(vii)	Availability only retainer	No services are included in this arrangement, which is established to ensure that legal counsel will be available when needed and not available to represent an adverse party. Fees are then paid as rendered.

CHAPTER 32

INTEREST ON COSTS

INTRODUCTION

[32.1]

The entitlement to interest arises under statute as follows:
- Section 35A of the Senior Courts Act, sub-s (1), provides:

    ' . . . subject to rules of court, in proceedings (whenever instituted) before the High Court for the recovery of a debt or damages there may be included in any sum for which judgment is given simple interest, at such rate as the court thinks fit or as rules of court may provide, on all or any part of the debt or damages in respect of which judgment is given, or payment is made before judgment, for all or any part of the period between the date when the cause of action arose . . . '

- Section 74(1) of the County Courts Act 1984 is similar:

    ' . . . the Lord Chancellor may by order made with the concurrence of the Treasury provide any sums to which this subsection applies shall carry interest at such rate in between such time as may be prescribed by the order.'

- Section 17(1) of the Judgments Act 1838 continues the provision:

    ' . . . every judgment debt shall carry interest at the rate of . . . from such time as shall be prescribed by rules of court until the same shall be satisfied, and such interest may be levied under a writ of execution on such judgment.'

The Judgments Act 1838, s 17 provides for the recovery of interest on costs in the High Court, whilst the County Court (Interest on Judgment Debts) Order 1991, made under the County Courts Act 1984, s 74, provides for interest to be paid on 'relevant judgments'. A relevant judgment is one for not less than £5,000 or in respect of a 'qualifying debt' as defined by the Late Payment of Commercial Debts (Interest) Act 1998, into which we need not and shall not delve.

THE INCIPITUR RULE (AS OPPOSED TO THE ALLOCATUR RULE)

[32.2]

The dispute as to the date from which interest on costs runs from went all the way up to the House of Lords where, in *Hunt v R M Douglas (Roofing) Ltd* [1990] 1 AC 398, it was held that interest on costs is a judgment debt for the purpose of s 17 above and that it runs from the date on which judgment is pronounced and not from the date of the costs officer's certificate. The former is known as the *incipitur rule* and the latter as the *allocatur rule*. Neither rule is entirely satisfactory but it was held that the balance of justice favoured the incipitur rule because the unsuccessful party had unnecessarily caused the costs to be incurred and that as neither rule covered costs incurred before the judgment, the application of the allocatur rule, generally speaking, does greater injustice than the operation of the incipitur rule, even though this means the successful party may recovers interest on disbursements before they are paid and costs after judgment before they are incurred. This has now been enshrined in CPR 40.8(1) as follows:

'(1) Where interest is payable on a judgment pursuant to section 17 of the Judgments Act 1838 or section 74 of the County Courts Act 1984, the interest shall begin to run from the date that judgment is given unless—
(a) a rule in another Part or a Practice Direction makes different provision; or
(b) the court orders otherwise.
(2) The court may order that interest shall begin to run from a date before the date that judgment is given.'

In *Simcoe v Jacuzzi UK Group plc* [2012] EWCA Civ 137, [2012] 2 All ER 60, [2012] 1 WLR 2393, the claimant appealed against a judgment that interest of 8% should be added to damages only from the allocatur date. It was held on appeal that the fact the claimant's solicitors were acting under a conditional fee arrangement did not justify departing from the general rule that interest on the costs runs from the incipitur date. The court concluded that the solicitors will have done the work which was reflected in the costs awarded to the claimant and will have incurred overheads on which those costs were based.

The effect of the claimant not paying anything to his solicitors until after the costs have been recovered from the defendant is that those solicitors finance their clients' litigation, and they should not be expected to continue to do so until the costs awarded are agreed or assessed.

ORDERS DEEMED TO HAVE BEEN MADE

[32.3]

CPR 44.9(4) provides that interest payable pursuant to of the Judgments Act 1838, s 17 or the County Courts Act 1984, s 74 on costs deemed to have been ordered (namely where a claim has been struck out for non-payment of fees under CPR 3.7, where a CPR 36 offer is accepted or where a claimant discontinues under CPR 38) shall begin to run from the date on which the event which gives rise to the entitlement to costs occurred.

ENHANCED INTEREST

[32.4]

CPR 36.14 provides that where at trial a defendant is held liable for more, or the judgment against a defendant is more advantageous to a claimant, than the proposals contained in a claimant's Part 36 offer, the court may order that the claimant is entitled to his costs on the indemnity basis and interest on those costs at a rate not exceeding 10% above base rate.

BACKDATING AND POST-DATING OF INTEREST – THE COURT'S DISCRETION

[32.5]

Where the costs judge has made various orders for the costs of the receiving party during the course of the assessment, the interest on those costs also runs from the date of the order for costs to be assessed under the incipitur rule not from the dates of the costs officer's certificates (*Ross v Bowbelle (Owners)* [1997] 1 WLR 1159, [1997] 1 WLR 1159, CA). Similarly, in *Electricity Supply Nominees Ltd v Farrell* [1997] 2 All ER 498, CA, where a consent order provided that the costs 'shall be taxed [assessed] upon the standard basis if not agreed and such costs when taxed [assessed] or agreed shall be paid by the defendants' it did not operate as a genuine postponement until after assessment of the obligation to pay the costs so as to avoid interest running from the date of the order pursuant to the rule in *Hunt v RM Douglas (Roofing) Ltd* (above). The court concluded that the words in italics were probably verbiage derived from a precedent and were no more than a recitation of what in fact was to occur. Although it is open to the parties to a consent order to make a side agreement as to interest or to take account of the effect of the court's order on interest, altering the wording of costs orders is unlikely to achieve the desired result. The court indicated that it was doubtful if even in a consent order the court could directly alter the incidence of interest on the judgment because that is governed by statute and rules of court.

In *Nykredit Mortgage Bank plc v Edward Erdman Group Ltd (No 2)* [1998] 1 All ER 305, HL the House of Lords said that the Court of Appeal had been 'lured into error' when in *Kuwait Airways Corpn v Iraqi Airways Co (No 2)* [1995] 1 All ER 790, CA it had held that the Rules of the Supreme Court empowered it to backdate its order for costs to enable interest to run from the original judgment. Statute apart, courts have no power to award interest on costs, however desirable it might be for the court to have power to order the payment of interest on costs from a date earlier than the date on which the court gave judgment. However, although the court had no general inherent power to order the payment of interest, the result of the appeal had been that orders in the courts below should not have been made and that some of the money previously paid by the defendants to the plaintiffs as damages and costs pursuant to orders of the trial judge and the Court of Appeal had fallen to be repaid to them. This could have been an idle exercise unless the court was able to make consequential orders that achieved, as near as reasonably practicable, the necessary restitution which included interest on the money to be repaid. The power to do so was derived from the inherent jurisdiction of the House of Lords, which was also possessed by the Court of Appeal.

32.5 *Chapter 32 Interest on Costs*

To enable lower courts to do the same, the Civil Procedure (Modification of Enactments) Order 1998 amended the Judgments Act 1838, s 17(1) to provide that interest will run from such time as is prescribed by rules of court. The resultant rule is CPR 44.2(6)(g) which empowers any court to award 'interest on costs from or until a certain date, including a date before judgment'.

Powell v Herefordshire Health Authority [2002] EWCA Civ 1786, was a clinical negligence claim on behalf of a child who suffered severe brain damage soon after birth. Liability was admitted and judgment entered by consent in April 1994. The damages were not assessed until June 2001, when there was an award of £2,175,000. The claim for interest on the costs almost exceeded the amount of the bill itself, being dated back to April 1994. The delay had arisen from uncertainly about the medical prognosis rather than any sloth on the part of the claimant or his lawyers. The defendant contended that interest should run from June 2001, the date of the actual assessment of the quantum of damages. On the assumption that these were the only two dates available to him, the costs judge ordered interest from the date of the judgment rather than the date of the quantification of damages. In overturning the costs judge the Court of Appeal noted that neither party had referred the costs judge to the power conferred on the court by CPR 44.2(6)(g) which gave him discretion to select any date which fitted the justice of the case. By the time of the case the parties had settled the interest point and so we shall never know what would have been the outcome of a decision under CPR 44.2(6)(g).

In *Amoco (UK) Exploration Co v British American Offshore Ltd (No 2)* [2002] BLR 135, QBD (Comm Ct) the court exercised this power to order interest to run from a date prior to judgment to recognise the reality of when the costs expenditure actually took place with the judge stating:

> 'For my part, I think it may well be appropriate for at least in substantial proceedings involving commercial interests of significant importance both in balance sheet and reputational terms, that the court should award interest on costs under CPR Part 44.3(6)(g) [now 44.2(6)(g)] where substantial sums have inevitably been expended perhaps a year or more before the award of costs is made and interest begins to run on it under the general rule.'

By contrast In *Colour Quest Ltd v Total Downstream UK plc* [2009] EWHC 823 (Comm), the unusual nature of the case and the substantial costs involved justified an order that interest should only run from six months after the judgment. At a hearing which Steel J noted was said to costing £250 per minute, he concluded:

> '... justice requires a postponement of the liability for the interest until a later date. This was indeed a case very much out of the norm where the costs are very large indeed. Indeed the claimants themselves wish to double the time allowed for the presentation of a detailed account for assessment. The disparity between the claimants' costs and those assessed as due may, it is contended, run to millions. It follows that payments on account are exposed to an enormous margin of error. In my judgement the starting date should be extended to 6 months from today.'

As Clarke J held in *Fattal v Walbrook Trustees (Jersey) Ltd* [2009] EWHC 1674 (Ch):

> 'The ability of the High Court to depart from the incipitur rule was conferred in order that the court could take account of the fact that money would often be

expended before any judgment, Conversely, where money has not been expended, for example where the bulk of the costs have been paid at a date long after the relevant judgment, justice requires that the date of commencement of the interest is postponed beyond the date of that judgment'

More recently in *Jones v Secretary of State for Energy and Climate Change* [2013] EWHC 1023, QB, [2013] 3 All ER 1014 the court held that the claimants were entitled to interest on disbursements for a pre- judgment period at the rate agreed in a disbursement funding agreement under which their solicitors agreed to provide the claimants with sums to discharge the disbursements as they arose.

What these authorities show is that the court has a broad discretion. Whilst the incipitur rule may be appropriate in the majority of cases there is nothing that requires the court to find any exceptional circumstances to depart from this when exercising its discretion under CPR 44.2(6)(g). As Clarke J stated in *Fattal* (above) 'The most important criterion is that any order should reflect what justice requires' and it is the pursuit of that goal which should inform the exercise of discretion.

THE RATE OF INTEREST

[32.6]

Since April 1993 the rate of interest on judgment debts has been 8%, as prescribed by the Judgment Debts (Rate of Interest) Order 1993. This is despite the reduction of interest rates on funds invested in court which since July 2009 has seen the special account rate reduced to 0.5%. The basis court funds rate is an even less appealing 0.3%. Judgment rate interest therefore remains a good investment – provided the judgment and costs are eventually paid. Indeed where there is a paying party good for the money and there are no cash flow problems for the receiving party we can understand why there may have been a reluctance to seek payments on account of costs before the detailed assessment as this is a rate of return on funds to be envied.

In respect of county court judgments where the judgment creditor applies to enforce payment of the judgment by execution or some other means, interest will not accrue if the enforcement process is wholly or partly successful. This is set out in the County Court (Interest on Judgment Debts) Order 1991, SI 1991/1184, at art 4. Presumably, this provision was made because, if the enforcement process is wholly successful the judgment will have been satisfied, whilst if it is only partly successful it will be too complicated to calculate the interest on the various amounts outstanding from time to time.

CHAPTER 33

APPEALS AGAINST ASSESSMENTS

INTRODUCTION

[33.1]

One of the least illuminating ways to spend a few hours is to read the transcript of an assessment hearing (particularly a detailed assessment). However, this is the task that falls to an appellate court on an appeal from an assessment. In the course of an assessment, where the court is dealing with a myriad of disputes (often challenges to six minute units), judgments come thick and fast. The key is that the court gives adequate reasons for its decision on each challenge (even if it is simply to resolve whether the six minute unit is allowed or is assessed off the bill). However, it is the failure to give such reasons that often leads to an appeal.

THE IMPORTANCE OF REASONS

[33.2]

In *English v Emery Reimbold & Strick Ltd* [2002] EWCA Civ 605 the Court of Appeal provided some clear guidance on reasons and appeal. For these purposes references to substantive decisions may just as easily be to decisions made within an assessment. The court held (amongst other matters):

(1) Where an application for permission to appeal on the ground of lack of reasons is made to the judge at first instance, he should consider whether his judgment is defective for lack of reasons, adjourning for that purpose if necessary. If he concludes that it was, then he should remedy the defect by providing additional reasons, and then refuse permission to appeal. If he concludes that his reasons were adequate, then he should refuse permission to appeal.

(2) Where an application for permission to appeal on the ground of lack of reasons is made to an appellate court, and it appears to that court that the application is well – founded, it should consider adjourning the application and remitting to the trial judge with an invitation to provide reasons, or additional reasons. Where the appellate court is in doubt as to the adequacy of reasons it may be appropriate to adjourn the application to an oral hearing on notice to the respondent.

(3) Where permission to appeal is granted on the ground of inadequate reasons, the appellate court should review the judgment in the light of the evidence and submissions at trial in order to determine whether it was apparent why the judge had reached the decision he did. If the

(4) If the reasons for a judge's order as to costs are plain, as where costs followed the event, there is no need for the judge to give reasons for a costs order. However, the CPR sometimes require a more complex approach to costs, and judgments dealing with costs will more often need to identify the provisions of the rules which have been in play and why those have led to the order made. Where no express reason for a costs order is given the appeal court will approach the material facts on the assumption that the judge has a good reason for the order made, and where there is a perfectly rational explanation for the order the appeal court is likely to draw the inference that this is what motivated the judge in making the order. Accordingly it is only in cases where an order for costs is made with neither reasons nor any obvious explanation for the order that it is likely to be appropriate to give permission to appeal on the ground of lack of reasons.

The Court of Appeal had to consider this in the context of assessments of costs in *Jemma Trust Co Ltd v Liptrott & Forrester (No 2)* [2003] EWCA Civ 1476. It concluded that the first question to answer was: 'Was the costs judge's judgment unsustainable for want of sufficient reasons?' The court held that it will often be impossible, and sometimes undesirable, for a costs judge to spell out the exact process of reasoning which led to the final figure – that figure will frequently be the result of triangulation, based very much on expert 'feel' between a variety of relatively unfixed possible positions. In that case the court concluded that despite the criticism which could be made of the judgment for its lack of reasoning at the crucial point, it was not persuaded that it was right to interfere with the judgment on that ground.

A SIMPLE FORMULATION OF REASONING WHEN ASSESSING ITEMS AND OVERALL PROPORTIONALITY

[33.3]

In practice when dealing with actual sums in the bill, as opposed to substantive preliminary points, the judge at first instance ought to be able to make himself virtually immune from appeal by the adoption of a simple formulation of words, eg 'the item is/is not reasonably incurred' and if reasonably incurred 'is/is not reasonable in amount' and if not reasonable in amount 'the reasonable amount is £x' and by way of proportionality cross check at the end 'Having reached the sum that is reasonably incurred and reasonable in amount it is necessary to view that sum against the factors at CPR 44.3(5) to ensure that this sum is proportionate. The relevant considerations under CPR 44.3(5) are (insert as relevant). In the light of these factors in my judgment the sum is/is not proportionate' and if not proportionate 'and I assess the proportionate sum by reference to the factors identified as £y'. Similarly parties wishing to challenge/defend a particular item or the amount of time spent on it ought to be focussing on the same wording, which is no more than an expression of the tests that the court must adopt under the rules.

PERMISSION TO APPEAL

[33.4]

In all costs appeals (save those from an authorised costs officer – see below), the appellant will require permission to appeal (CPR 52.3(1). As with other appeals, the appellant may seek permission from the judge conducting the assessment and (if the first instance judge refuses permission)/or (if no application is made to the first instance judge) to the appeal court. CPR 52.3(6) provides that permission to appeal will only be granted where the court considers that the appeal would have a real prospect of success or there is some other compelling reason why the appeal should be heard.

The Court of Appeal has no jurisdiction to entertain an appeal against the refusal of the judge to grant permission to appeal against the decision of the costs judge. It also has no jurisdiction to review the judge's decision (see *Riniker v University College London* [2001] 1 WLR 13, CA).

An appeal does not act as a stay and a separate application for a stay of enforcement must be made. It is likely that a court granting a stay will make it dependent upon some form of payment (if only to ensure that the appellate process is not being used simply as a device to defer payment).

GROUNDS FOR APPEAL

[33.5]

The only grounds of appeal are those prescribed in CPR 52.11(3), namely that the decision was wrong or unjust because of a serious procedural or other irregularity in the proceedings.

'Wrong' may be an error of i) law, ii) fact or iii) the exercise of discretion. While the appellate court will readily entertain an appeal on a question of law, it will need considerable persuasion to interfere with a finding of fact or the exercise of discretion.

The meaning of 'wrong' was considered in *Griffiths v Solutia UK Ltd (formerly Monsanto Chemicals Ltd)* [2001] EWCA Civ 736, [2001] 2 Costs LR 247, the relevant part of the judgment being:

> 'The question before us is whether the deputy judge was entitled to conclude that the decision of the costs judge was wrong. The test which the deputy judge had to apply, in determining whether or not the costs judge's decision was wrong, was explained by this court in Tanfern v Cameron-McDonald [2000] 1 WLR 1311, para 32, where Brooke LJ says as follows:
>
>> "The first ground for interference speaks for itself. The epithet 'wrong' is to be applied to the substance of the decision made by the lower court. If the appeal is against the exercise of a discretion by the lower court, the decision of the House of Lords in G v G (minors: custody appeal [1985] 1 WLR 647 warrants attention. In that case Lord Fraser of Tullybelton said, at page 652
>>
>>> 'Certainly it would not be useful to inquire whether different shades of meaning are intended to be conveyed by words such as "blatant error" used by the President in the present case, and words such as "clearly wrong", "plainly wrong" or "simply wrong" used by other judges in other cases. All these various

expressions were used in order to emphasise the point that the appellate court should only interfere when they consider that the judge of first instance has not merely preferred an imperfect solution, which is different from an alternative imperfect solution which the Court of Appeal might or would have adopted, but has exceeded the generous ambit within which a reasonable disagreement is possible'

The task that faces us is to apply that same test. Essentially the test requires the appellate court to consider whether or not, in a case involving the exercise of discretion, the judge has approached the matter applying the correct principles, has taken into account all relevant considerations and has not taken into account irrelevant considerations, and has reached a decision which is one which can properly be described as a decision which is within the ambit of reasonable decisions open to the judge on the facts of the case."'

In *Johnsey Estates (1990) Ltd v Secretary of State for the Environment, Transport and the Regions* [2001] EWCA Civ 535, the Court of Appeal held that an appellate court should not interfere with the judge's exercise of discretion merely because it takes the view that it would have exercised that discretion differently. This requires an appellate court to exercise a degree of self – restraint. It must recognise the advantage which the trial judge enjoys as a result of his 'feel' for the case which he has tried. Indeed, it is not for an appellate court even to consider whether it would have exercised the discretion differently unless it has first reached the conclusion that the judge's exercise of his discretion is flawed. That is to say, that he has erred in principle, taken into account matters which should have been left out of account, left out of account matters which should have been taken into account or reached a conclusion which is so plainly wrong that it can be described as perverse.

A practical illustration of this was in *Sibley & Co v Reachbyte Ltd & Kris Motor Spares Ltd* [2008] EWHC 2665 (Ch). This was an appeal against the reduction of counsel's fees on a detailed assessment. The appeal court determined that it could not be said that the deputy master could not reasonably have come to the conclusion that he had based on the material before him. It stressed that the appeal court should not interfere with a decision without there being any valid criticism of the first instance judge. To allow an appeal merely because a different view was entertained on appeal on the same material that was before the first instance judge would fail to give true effect to the primacy of the factual findings of that judge.

THE TIME FOR APPEALING

[33.6]

As the appeals are governed by CPR Part 52, the time period for an appeal is prescribed within that rule at CPR 52.4. This provides that an appellant must file notice of the appeal within 21 days of the date of the decision of the lower court against which an appeal is pursued or such other period, being either longer or shorter, that the lower court may have directed.

What does this mean in the context of a detailed assessment where the court is making many decisions that together lead to an overall conclusion? In *Kasir v Darlington & Simpson Rolling Mills Ltd* [2001] 2 Costs LR 228, QBD, at the end of a four day detailed assessment it was left to the respective costs draftsmen to agree the calculations arising out of the costs judge's findings and to submit the resultant assessed bill to him. Unfortunately they were unable to agree the result, which necessitated another hearing, which in turn led to further correspondence so that the final figure was not agreed until 5 months after the original assessment. The claimant submitted notice of appeal within 14 days of the final agreement. The court held that this was out of time, as the time period runs from the decisions on the items under appeal. As was pointed out to the judge, this means that in lengthy assessments the final date for applications for appeal will vary depending upon which day of the assessment the decision being appealed was made. This might result in the parties having to file several appellant's notices. It also results in parties having to make decisions on interim items before the overall outcome of the assessment is known. The practical solution offered by the judge in this case was that the parties could agree and/or the costs judge could order at the start of an assessment that the time for any appeals is extended to a date either 21 days or some other agreed and ordered period after the last day of the assessment.

ROUTE OF APPEALS

[33.7]

It is hard to imagine that any decision in a claim is more final than a summary or detailed assessment of the costs at the conclusion of a claim. However, for the purpose of the destination of any appeal, CPR PD 52A, para 3.8(2) states that a summary or detailed assessment of costs are examples of decisions that are not final. Indeed the definition of a 'final decision' in Part PD 52A, para 3.6 specifically excludes detailed assessment as otherwise it would be difficult to see how that could not be regarded as 'a decision of a court that would finally determine the entire proceedings, whichever way the court decided the issues before it.' There is a rationale to this. If the assessment takes place in a multi-track claim, then by defining the assessment as an interim decision this avoids making the Court of Appeal the destination court for any appeal – (see the Destination tables in CPR PD 52A Section 3),

Appeals from summary or detailed assessments conducted by a district judge are to a circuit judge. Appeals from summary or detailed assessments of a costs judge and against summary assessments by a circuit judge are to a High Court judge. Appeals from summary assessments by a High Court judge are to the Court of Appeal.

REVIEW NOT RE-HEARING

[33.8]

Appeals against assessment decisions are not full re-hearings, unless the appellate court so orders. Instead they are limited to a review of the decision under appeal. Two recent examples of where the appellate court has decided to adopt a re-hearing were *U v Liverpool City Council* [2005] EWCA Civ 475,

[2005] 1 WLR 2657, (2005) Times, 16 May (whether an appeal should be permitted on an issue not raised at the detailed assessment) and *Rogers v Merthyr Tydfil County Borough Council* [2006] EWCA Civ 1134, [2007] 1 All ER 354, [2007] 1 WLR 808 (the quantum of ATE insurance premiums affecting the insurance industry generally).

THE ROLE OF ASSESSORS

[33.9]

The costs rules used to provide for the appointment of assessors to sit with the judge on a review of a decision at a taxation of costs. These were usually a costs officer and a practising lawyer (generally a solicitor unless the appeal was in connection with counsel's fees when a barrister would be used). There is no longer such a specific provision. However, courts may, and regularly still do, avail themselves of the assistance of assessors under their powers deriving from the Senior Courts Act 1981, s 70 or the County Court Act 1984, s 63 as provided for in CPR 35.15. Regularly at county court level the appellate court will involve the local Regional Costs Judge. In the High Court the use of masters from the Senior Court Costs Office is customary. In the High Court, the judge may well wish to sit with more than one assessor, the other one still, traditionally, being drawn from the professions.

CPR 35.15(3) prescribes that it is a matter for the court to determine the extent to which the assessor is involved. This may simply be to prepare a report (such as on the availability and cost of after the event insurance) or to attend the entire hearing and advise the court. However, it is important to stress that:
- if the assessor prepares a report for the court that must be sent to the parties before the hearing and the parties may use it at the hearing;
- the assessor is not treated like an expert and cannot be cross examined;
- the final decision remains that of the court (and the court may, and has, disagreed with its assessor/s);
- the court must give at least 21 days' notice to the parties that it proposes to sit with an assessor (naming that assessor). Any party may object to the appointment provided that notice of the objection is made to the court in writing within 7 days of receiving the notice and the court will take the objection into account when determining whether to sit with the assessor (CPR PD 35, para 10).

Save where the assessor is a salaried member of the judiciary (eg a costs judge or regional costs judge) the assessor's remuneration shall be determined by the court and shall form part of the costs of the proceedings. The court may order any party to deposit in the court office a specified sum in respect of the assessor's fees and may defer the appointment of the assessor until this has been done.

APPEALS FROM THE DECISIONS MADE ON DETAILED ASSESSMENT BY AUTHORISED COURT OFFICERS

[33.10]

An authorised costs officer is 'any officer of the county court, district registry, principal registry of the family division or costs office whom the Lord Chancellor has authorised to assess costs (CPR 44.1(1)). Part PD 47, para 3.1 sets out the financial limits on their jurisdiction (base costs excluding Vat do not exceed £35,000, save in respect of principal officers where the limit is £110,000.)

CPR 47.3 sets out the powers of authorised costs officers stating that they have the same powers as the court on detailed assessment proceedings save that they cannot make 'wasted costs orders', orders in relation to misconduct, orders imposing sanctions for delay in commencing detailed assessment and cannot undertake assessments of the amount that a client should pay his solicitor save in respect of assessments where the money is payable by a child or protected party under CPR 46.4. The parties do have the right to object to a detailed assessment by an authorised costs officer and the court may then order the assessment to be conducted by a costs judge or district judge.

As already stated the appellate route prescribed by CPR Part 52 does not apply to appeal from authorised court officers. Instead the relevant rules relating to appeals from costs officers are set out at CPR 47.21-CPR 47.24. These provide that:

- any party to detailed assessment proceedings may apply to appeal without needing permission;
- the appeal lies to a costs judge or a district judge of the High Court;
- the appeal notice must be filed within 21 days after the date of the decision being appealed against;
- the court will serve the notice and give notice of the hearing date;
- the court will re – hear the detailed assessment proceedings giving rise to the appeal and in so doing may give such directions and make such order as it sees fit. In other words the jurisdiction on appeal extends beyond the customary review of the decision being appealed;
- interestingly the previous exclusion of a Legal Services Commission funded client or an assisted person from the definition of a party to the detailed assessment for the purpose of an appeal against the decision of an authorised court officer has been removed from the rules. CPR 47.21 no longer contains any qualification on the definition of 'party';

The relevant procedural requirements are set out at CPR PD 47, paras 20. 1–20.6 and include:

- prescription of the relevant appeal notice (FORM N 161);
- provisions for filing a suitable record of the judgment appealed against, whether by transcript, the bill annotated with reasons or agreed advocates notes of the reasons as approved by the authorised court officer. Specific obligations are imposed on the respondent to the appeal where the appellant was unrepresented before the authorised court officer and there is no official record;

- provision for amending the notice of appeal where there is no record of the judgment or reasons for it available by the time the notice has to be filed. In this situation the notice must be completed as best as is possible. Any later application to amend will require the permission of the judge hearing the appeal,

CONCLUSION

[33.11]

Appeals from assessments tend to be on substantive issues as opposed to challenges to individual amounts of time. The provisional detailed assessment process (see Chapter 30 Detailed Assessment – Provisional Assessment, above) introduced by CPR 47.15, with its own review process, may reduce the number of appeals with the parties being satisfied by the opportunity to challenge the court's initial decision. However, we wait with bated breath for appeals aimed at challenging the concepts underpinning the April 2013 costs reforms. In particular appeals relating to 'good reason' to depart from budgets and the effect of proportionality cross checks at the end of an assessment are likely battlegrounds. It will be interesting to see how far parties seek to obtain guidelines and parameters on these concepts – with the judiciary astute to the fact that both are key to the drive to control costs.

Part V

Special cases

CHAPTER 34

CHILDREN AND PROTECTED PARTIES

INTRODUCTION

[34.1]

Phrases to describe a party who lacks the capacity to litigate have come and gone with some rapidity. A 'party under a disability' and a 'patient' have given way to a 'protected party', which CPR 21.1(2) defines as a party, or an intended party, who lacks capacity to conduct the proceedings. Similarly, a 'minor' or an 'infant' has been replaced by 'child' even though that word is not age restricted in common parlance. According to the CPR however it means a person under 18.

LITIGATION FRIEND

[34.2]

A protected party must have a litigation friend to conduct proceedings on his behalf (CPR 21.2(1)) and a child must have one unless the court permits the child to conduct his own proceedings (CPR 21.2(3)).

Where the child or protected party is a claimant, the litigation friend must give an undertaking to pay any costs which the child or protected party may be ordered to pay in relation to the proceedings, subject to any right he may have to be repaid from the assets of the child or protected party whether the litigation friend has become one without a court order (CPR 21.5) or has been appointed under a court order (CPR 21.6). The litigation friend's liability for costs continues until the child or protected party serves notice that the litigation friend's appointment has ceased (giving his address for service and stating whether or not he intends to carry on the proceedings) or the litigation friend serves notice on the parties that his appointment to act has ceased (CPR 21.9(6)).

EXPENSES INCURRED BY A LITIGATION FRIEND

[34.3]

Under CPR 21.12 and CPR PD 21, para 11.1 a litigation friend who incurs expenses on behalf of a child or protected party in any proceedings is entitled to recover the amount paid or payable out of any money recovered or paid into court to the extent that it has been reasonably incurred and is reasonable in amount. Items which have been disallowed on a between the parties' assessment, cannot be claimed from the child or protected party (CPR 21.12(3)).

Expenses may include all or part of an ATE policy premium and interest on a loan taken out to pay an insurance premium or other recoverable disbursement. In *Scott v Harrogate Borough Council* (20 January 2003, unreported, Harrogate County Ct) the district judge ordered a deduction from the child's damages of a conditional fee agreement administrative success fee of 10% and interest on a bank loan obtained to cover expenses, both of which were not recoverable from the defendant. (There is therefore a distinction between items which are disallowed on assessment and those which are simply not recoverable between the parties.)

Where the claim settles for £5,000 or less, the litigation friend's expenses can only exceed 25% of the settlement with the court's permission. It cannot be more than 50% in any event.

Whilst the foregoing paragraphs accurately summarise the contents of this Rule, they cannot hide the fact that this is a peculiarly worded rule. It seems obviously to be meant to cover the Litigation Friend's personal expenses rather than solicitor's costs but the reference to ATE premiums certainly clouds the issue.

DETAILED ASSESSMENT

[34.4]

CPR 46.4 is designed to protect the interests of a child or protected party and to prevent exploitation by unscrupulous legal representatives or litigation friends. Generally, detailed assessment is required of any costs payable by a child or protected party to his solicitor and when such an assessment takes place the court must also assess any costs payable to that party in the proceedings, unless a default costs certificate has been issued.

The court is given the discretion to dispense with a detailed assessment where it is not needed to protect the interests of the child or protected party:

(a) where another party has agreed to pay a specified sum in respect of the costs and the solicitor has waived the right to claim any further costs from the child or protected party;

(b) where the court has decided the amount of costs payable by way of summary assessment; and

(c) where an insurer or other person is liable to discharge the costs which the child or protected party would otherwise be liable to pay and the court is satisfied that the person liable is financially able to discharge the costs.

If the claim is (or was) proceeding through the Portal or the Predictable Costs Regime in CPR Part 45 Sections II, III or IIIA, the recoverable costs are fixed and so there is no prospect of a detailed assessment of such costs. A detailed assessment would still be required if the solicitor sought to claim any costs from his client.

CHAPTER 35

LITIGANTS IN PERSON

INTRODUCTION

[35.1]

The Litigants in Person (Costs and Expenses) Act 1975 provides that where any costs of a litigant in person ('LIP') are ordered to be paid by any other party to proceedings there may be allowed (subject to any rule of the court) on assessment or determination of the costs, sums in respect of any work done and any expenses and losses incurred by the LIP in connection with the proceedings. In simple terms the court has the power to award an LIP his costs quantified by either summary or detailed assessment. The provisions of the Act apply to all civil proceedings in the Supreme Court, the Senior Courts, the Lands Tribunal and county courts in which any order is made that the costs of a LIP are to be paid by any other party to those proceedings or in any other way.

Prior to the coming into force of the Act in April 1976, LIPs (other than those who were practising solicitors) whose costs were ordered to be paid were entitled to recover from the paying party no more than such out-of-pocket expenses as had been properly incurred. The Act enabled LIPs to recover the same category of costs as would have been allowed if the work had been done by a solicitor on the litigant's behalf, limited in amount to such sum as was required to compensate the litigant for the time he had reasonably spent on preparing and conducting his case, subject to any rules of court.

The term 'self represented' party briefly gathered attraction from those who considered the term litigant in person to be unwieldy. But, as the rules of court stem from the 1975 Act, the terminology employed by that enactment has prevailed. The Master of the Rolls has issued a statement confirming this to be the case.

WHO IS A LITIGANT IN PERSON?

[35.2]

CPR 46.5(6) provides that a litigant in person includes a company or other corporation which is acting without a legal representative, a barrister, solicitor, solicitor's employee or other authorised litigator who is acting for himself. It was established in *London Scottish Benefit Society v Chorley* (1884) 13 QBD 872, CA that where an action was brought against a solicitor who defended it in person and obtained judgment, he was entitled to the same costs as if he had employed a solicitor, except in respect of items which the fact of his acting directly rendered unnecessary. The exception within CPR 46.5(6) for individuals who are represented by a firm in which they are a partner means that the principle applies where a defendant solicitor does not expend his own time and skill in defending the claim, because the defence was undertaken by one of his partners and by others within his firm (*Malkinson v Trim* [2002] EWCA Civ 1273). Accordingly a solicitor who instead of acting in person is represented by his or her firm or acts in his or her firm's name and is not regarded as a LIP. This interpretation was endorsed by Master Leonard in *Zakirov v Newmans Solicitors* [2012] EWHC 90222 (Costs) when the issue was directly in point.

It is therefore important for a solicitor litigant, especially if he is a sole practitioner, to make it clear that he is acting through his firm and not in person, for example by using his firm's, and not his private, notepaper.

In *Re Minotaur Data Systems Ltd, Official Receiver v Brunt* [1998] 4 All ER 500, ChD the judge had with some regret come to the conclusion that the Official Solicitor acting as an applicant in proceedings under the Company Director Disqualification Act 1986, s 6 without the aid of a solicitor was not a LIP and therefore not entitled to an allowance in respect of costs. The judge therefore would have been pleased that his decision was reversed by the Court of Appeal on 2 March 1999 ([1999] 3 All ER 122, CA), which held that although the Official Solicitor could not be said to be acting for the Crown as a LIP, he was himself a LIP for the purposes of recovering his costs.

PROFESSIONAL LITIGANTS IN PERSON

[35.3]

In *Re Nossen's Patent* ([1969] 1 WLR 638) it was held that when it was appropriate a corporate litigant should recover, on a between-the-parties basis, a sum in respect of expert services performed by its own staff, the amount must be restricted to a reasonable sum for the actual and direct costs of the work undertaken, and not a proportion of the corporation's overheads, no part of such expenditure being occasioned by the litigation. *London Scottish Benefit Society v Chorley* ((1884) 13 QBD 872) established the rule of practice that a LIP who was a solicitor could recover costs as if he had employed a solicitor, applied only in the context of LIPs who were solicitors and that generally speaking, a LIP could not recover for his time. Although the decision was before the CPR, it would be an inadmissible extension of that case to treat the principle established by it to other professionals such as accountants. An accountant LIP could at most recover for work he had done himself, what the cost of obtaining the expert advice of an independent professional would have been, but he is not entitled to recover the costs of general assistance to the expert in the conduct of the litigation (*Sisu Capital Fund Ltd v Tucker* [2005] EWHC 2170 (Ch)).

FINANCIAL LOSS

[35.4]

The recoverable costs of a LIP depend upon whether he can show that he has suffered any financial loss in dealing with the proceedings. If he can proves a loss, he may claim a sum which does not exceed the amount of his actual financial loss (CPR 46.5(4)(a)). If he cannot show any loss, he may still make a claim for the amount of time he reasonably spent at the prescribed rate (CPR 46.5(4)(b)). This rate is set out in CPR PD 46, para 3.4. It increased from £9.25 to £18.00 per hour on 1 October 2011.

The Two Thirds Rule

[35.5]

Rules of court have always limited the costs recoverable by a LIP to two-thirds of the amount which would have been allowed to a legal representative. The reasoning behind this figure is that a solicitor's charges have usually included a 50% profit mark-up on his expense rate, but as a LIP may not make a profit out of the costs of litigation, the 50% is deducted, leaving two-thirds.

Since there is no profit mark up on disbursements, the two thirds rule does not apply to them.

The provision that a LIP may only recover two-thirds of what a notional solicitor would have charged limits the amount the LIP may claim for any particular item. The correct approach therefore for the assessment of a LIP claiming at the specified rate is to ascertain what the total of that item is. Then compare that with two-thirds of the notional solicitor's rate and allow the lower of the two items. This means that the bill of costs drawn by the LIP must be gone through in some detail, item by item (*Morris v Wiltshire & Woodspring District Council* (16 January 1998, unreported, QBD).

In *R (on the application of Wulfsohn) v Legal Services Commission* [2002] EWCA Civ 250, a successful LIP produced a rough costs schedule purporting to show that he had been engaged for over 1,200 hours on research and he also gave oral evidence, which the court had no reason to disbelieve, that he had spent well in excess of 1,200 hours. In these circumstances the right course was to start with the cap imposed by CPR 46.5(2) which provides that the costs of a litigant in person shall not exceed two-thirds of the amount which would have been allowed if the litigant in person had been represented by a legal representative. The litigant in person produced a letter to him from a firm of solicitors saying: 'On the limited information that we have been provided by yourself and the Citizens Advice Bureau in the Royal Courts of Justice and having seen at a very preliminary stage the documentation with regards the matter we would estimate that the legal costs would be in the region of £15,000 to £20,000 plus VAT'. Doing the best it could on the information in front of it the court 'being extremely rough-and-ready' about it took the figure of £15,000 which resulted in a cap of £10,000. In addition to that the court allowed photocopying, postage and travelling totalling £460, resulting in a total award of costs of £10,460, instead of the £120 awarded by the trial judge.

The same considerations, particularly in respect of proportionality, will apply to the assessment of a litigant in person's costs as to costs incurred by a legal representative. In *Grand v Gill* [2011] EWCA Civ 902 the claimant had limited success on appeal in reducing the level of an award of damages made against her. She then made an application for her costs in preparing for the appeal as a litigant in person. The Court of Appeal agreed in principle that costs should be awarded but considered the sum claimed to be 'unjustified, unjustifiable and disproportionate'. The Court of Appeal determined that the claimant ought only to be compensated for her time spent that was fairly referable to the issues upon which she had ultimately succeeded. There would be no compensation for work done or expense incurred which was misdirected or wasted.

Legal Costs

[35.6]

Although a litigant may have acted in person without a solicitor on the record, he is nevertheless entitled to recover payments reasonably made for legal services in two respects:
(1) Legal services relating to the conduct of the proceedings
(2) The costs of obtaining expert assistance in connection with assessing the claim for costs, including representation on a detailed assessment.
There is no guidance given in respect of (1). It will usually be a solicitor but with the relaxation of the Bar's prohibition on effectively conducting proceedings, it may be a barrister via the direct access scheme.

By contrast, there is a list of no fewer than 7 categories of expert set out in CPR PD 46, para 3.1 including 3 different kinds of 'costs draftsmen'.

The price of a LIP not using a solicitor was dramatically illustrated in *Agassi v Robinson (Inspector of Taxes)* [2005] EWCA Civ 1507. Former Wimbledon champion, Andre Agassi in his battle with the taxman retained a tax expert who was a member of the Chartered Institute of Taxation licensed to instruct counsel directly. No solicitors were involved. Mr Agassi was awarded his costs as a LIP. Were the tax expert's fees recoverable as costs under the general costs provisions of CPR 46.5 (as it is now)? No. Although Mr Agassi could recover counsel's fee as a disbursement he was not entitled to recover as a LIP costs in respect of work done by the tax expert which would normally have been done by a solicitor. That meant he was not entitled to recover the costs of the tax expert providing general assistance to counsel. However, it could be appropriate to allow Mr Agassi at least part of the expert's fees as a disbursement. It might be possible to argue that the cost of discussing the issues with counsel, and assisting with the preparation of the skeleton argument etc was allowable as a disbursement, because the provision of that kind of assistance in a specialist esoteric area was not the kind of work that would normally be done by the solicitor instructed to conduct the appeals. Another way of making the same point is that it might be possible to characterise the specialist services as those of an expert.

The court observed that the solution to the problems raised by the case was for an organisation such as the Chartered Institute of Taxation to become an 'authorised body' within the meaning of s 28(5) of the Courts and Legal Services Act 1990, and for those members who wished to conduct litigation to become authorised litigators and thereby 'legal representatives' within the meaning of CPR 2.3(1).

The unhappy postscript for Mr Agassi was that the Revenue appealed on the merits and succeeded in the House of Lords, which awarded them their costs both in the House and in the Court of Appeal; so his inability to recover the tax expert's costs was in any event academic. Game, set and match to the Inspector of Taxes.

Agassi was followed in the SCCO in *Cuthbert v Gair (t/a Bowes Manor Equestrian Centre)* (2008) 152 Sol Jo (no 38) 29. The claimant issued proceedings for damages for personal injuries suffered while attending an equestrian event, but served notice of discontinuance. On the detailed assessment of the defendants' costs, the defendants sought to recover payments made to a loss adjuster by the defendants' insurers, both before and after solicitors were instructed. The work of the loss adjuster under the first invoice included corresponding with the claimant's solicitor, investigating the accident, obtaining witness statements and dealing with documentation. That work was work that would normally be carried out by a solicitor and the defendants were not entitled to recover costs in respect of it.

In respect of a second invoice, for work done by the loss adjuster after solicitors had been instructed, it was necessary to assess the relationship between the defendants' solicitors and the loss adjuster. If the defendant's solicitors had sought assistance from the loss adjuster on an agency basis, then they would have been entitled to recover their costs, not as a disbursement, but as a profit cost, following *Crane v Canons Leisure Centre* [2007] EWCA Civ 1352. However no true agency agreement existed between the solicitors and the loss adjuster: there was no letter of instruction and no terms of engagement. On that basis, it was not possible for the defendants to recover the loss adjuster's fees after the solicitors had been instructed.

Furthermore, the work undertaken by the loss adjuster did not fall within the category of 'expert assistance' that otherwise might have rendered the costs recoverable (*Re Nossen's Letter Patent* [1969] 1 WLR 638). The present was a simple case of an insurer contracting out part of its work in order to investigate claims made against the insured. It was routine work, which many insurers would have undertaken in-house. The mere fact that the defendants' insurer chose to contract out that work did not render the costs recoverable.

Fast Track Trial Costs

[35.7]

Where a LIP appears at a Fast Track trial, CPR 45.39(5) makes express provision for how costs are to be awarded. It maintains the distinction between LIPs who can demonstrate a financial loss and those that cannot:
- LIPs who can prove financial loss will receive two thirds of the amount that would have been award as advocates' fees;
- LIPs who cannot prove financial loss will receive an amount in respect of the time spent reasonably doing the work at the rate of £18.00 per hour.

The amount of the advocate's fees is set in 4 bands as discussed in Chapter 26. Those figures are only varied if there has been unreasonable or improper behaviour by one or more of the parties during the hearing. Absent such behaviour, the court will be obliged to award the LIP two thirds of the advocate's fee regardless of the extent of the loss actually sustained by the LIP.

QUANTIFICATION

[35.8]

In *Law Society v Persaud* (1990) Times, 10 May, QBD, the defendant, who lived in South Africa, successfully contested proceedings brought by the Law Society and obtained an order for costs against them. On the assessment of his costs two items were disallowed: the cost of travelling from South Africa to England to defend the action in person of £1,391.25 and the cost of travelling between Birmingham and London of £74.

The claimant did not suggest that the defendant had been extravagant or acted in bad faith but that the disbursements did not come within the terms of RSC Ord 62, r 18(1) (now CPR 46.5(3)) because no solicitor would have been allowed to charge as a disbursement the cost of travel from South Africa to England because the solicitor would already be in England and similarly the cost of travelling from Birmingham to London would not be incurred because the solicitor would already be in London.

The defendant justified his disbursements on the grounds that they were less than would have been the cost of instructing solicitors and counsel. The costs officer had assessed the costs of solicitors at £2,457 and allowed the defendant £1,643, being two-thirds of this, resulting in a saving to the claimant of £821 of the amount if the defendant had instructed solicitors. To this, the defendant argued, must be added the estimated cost of instructing counsel. The Law Society and the costs officer took the view that this was wrong, following the decision in *Hart v Aga Khan Foundation (UK)* [1984] 2 All ER 439, CA, where the Court of Appeal had held that the notional cost of counsel should not be allowed where counsel had not in fact been instructed. To apply that decision in *Persaud* was too literal an approach; it did not preclude the costs officer from allowing reasonable disbursements which had in fact been incurred. The costs sought to be recovered by the defendant did not exceed what would have been the actual cost and disbursements allowable to a London solicitor, including counsel's fees. There was still an overall saving to the other side as a result of the relevant party having chosen to represent himself, and therefore, both sets of disallowed disbursements were restored.

The *New Law Journal* of 1 July 1994, p 890 reported that a LIP who, after a three-and-a-half-day hearing, was granted leave to proceed under the Mental Health Act against seven defendants with the costs of the application awarded against those defendants, was allowed on taxation her costs in excess of £59,000 for obtaining legal assistance in the preparation and presentation of a case from solicitors who were not on the record.

In *R v Common Professional Examination Board, ex p Mealing-McCleod* (2000) Times, 2 May, CA it was held:
(a) CPR 48.6(4) (now CPR 46.5(4)) suggests that more time should be allowed to a litigant in person than to a solicitor doing the same task.
(b) When a litigant in person is restricted to the litigant in person charging rate of £9.25 per hour (now £18.00), it is likely that the costs of the other side, if legally represented are going to be very much higher. Costs incurred by a party must be proportionate to the amount of the claim and the litigation generally. This does not mean that the costs of a

legally represented party must be reduced below a reasonable level of remuneration because the other side is a litigant in person restricted to payment of £18.00 per hour.

However, in *Greville v Sprake* [2001] EWCA Civ 234, the Court of Appeal, without referring to *Mealing-McCleod*, held that a LIP is limited to the time which would reasonably have been spent by a solicitor on the preparation of his or her case based on CPR 46.5(3)(a).

CHAPTER 36

COSTS PAYABLE UNDER A CONTRACT

INTRODUCTION

[36.1]

CPR 44.5 and CPR PD 44, para 7 deal with the amount of costs where costs are payable pursuant to a contract. The most obvious example of this situation relates to mortgage agreements but it is possible for general contractual arrangements to come within these provisions, notwithstanding the old maxim that the law does not protect a party from a bad bargain (including provisions as to payment of costs).

MORTGAGES

[36.2]

Many contracts, particularly mortgages, contain provisions to the effect that one or other party will be liable for the costs incurred pursuant to the contract. Under the terms of a mortgage deed, the mortgagee is usually entitled to add to his security his usual and proper costs of proceedings between himself and the mortgagor or any surety. He does not require an order from the court to do so. However, the court has an equitable jurisdiction to disallow all or part of a mortgagee's costs as being unreasonably incurred or of an unreasonable amount, in fixing the terms of redemption. In addition to his costs of proceedings between himself and the mortgagor, a mortgagee may recover the reasonable and proper costs of proceedings between himself and a third party where what is impugned is the title to the estate. But where a third party impugns the title to a mortgage or the enforcement or exercise of some right or power accruing to the mortgagee under it, the mortgagee's costs of the proceedings, even though reasonable and proper, are not recoverable from the mortgagor (*Parker-Tweedale v Dunbar Bank plc (No 2)* [1991] Ch 26, CA).

There is a presumption that costs payable under a mortgage or any other contract are reasonably and properly incurred and such costs are to be paid on the indemnity basis (*Gomba Holdings (UK) Ltd v Minories Finance Ltd (No 2)* [1993] Ch 171, [1992] 4 All ER 588, CA).

CPR 44.5 is a summary of the law as explained above, supplemented by the provisions of sub-section 7 of the Part 44 Costs Practice Direction. To reflect the advantageous position of a mortgagee with such a contractual entitlement, paragraph 9.3 of the Part 44 Costs Practice Direction varies the general rule otherwise usually applying for the timing of summary assessment. Paragraph 9.3 states:

'The general rule in paragraph 9.2 does not apply to a mortgagee's costs incurred in mortgage possession proceedings or other proceedings relating to a mortgage unless the mortgagee asks the court to make an order for the mortgagee's costs to be paid by another party.'

In other words, the costs will not be summarily assessed by the court unless the mortgagee wishes this to take place.

LANDLORD AND TENANT

[36.3]

In *Forcelux Ltd v Binnie* [2009] EWCA Civ 1077 the tenant had fallen into arrears with payment of ground rent and charges and the landlord obtained a default judgment against him. The landlord obtained an order for possession which a district judge set aside in the exercise of his discretion under CPR 39.3 and granted relief from forfeiture on terms as to payment of outstanding monies. The Court of Appeal confirmed the setting aside of the possession order. The landlord claimed to be entitled to its costs of the entirety of the proceedings on the basis that the tenant had covenanted in the lease to pay 'all costs charges and expenses (including legal costs . . .) which may be incurred by the lessor in or in contemplation . . . of any steps or proceedings under Section 146 of the Landlord and Tenant Act 1925'.

The Court of Appeal held that possession proceedings brought to enforce a right of re-entry following a notice under s 146(1) of the Act were proceedings 'under' that section. The possession action was within the scope of the words 'any statutory proceedings' within the scope of the covenant. It followed that the application to set aside the possession order was as much an application as the application for relief and forfeiture coupled with it, and was also within the scope of those words. However, the tenant had been the substantial winner of the appeal. Even assuming that the contractual provisions in the lease covered the costs of the appeal, which was a matter of construction, the contractual right was not an absolute one and did not oust the jurisdiction of the court to make another order if there were good reason for doing so. In the circumstances, the tenant was entitled to a costs order which departed from the contract in the exercise of the court's discretion.

CONTRACTUAL COSTS

[36.4]

In *Venture Finance plc v Mead* [2005] EWCA Civ 325, the claimant had obtained judgment by consent for less than one-third of the amount claimed but there was nothing that enabled the judge to decide why it had been willing to settle for that sum. It was impossible to say that one party had obviously won and the other had obviously lost. The judge had erred in principle. He had wrongly thought that the only order for costs that gave effect to the parties' contractual rights was that each defendant should be liable for only 50% of the whole costs of the proceedings. The contractual obligation on each defendant was to pay all costs and expenses arising out of the recovery of monies from that defendant under the guarantee. The right course for the judge was to consider, in relation to each defendant, the extent to which the whole costs of the proceedings could be said to arise out of the claim to recover under the guarantee obligations of that defendant. That approach might have led to a conclusion that a proper proportion of the whole costs of proceedings to be awarded against each defendant was 100%, or some lower proportion. It was difficult to see how the appropriate proportion could be as low as 50%. If the claimant could not recover 100% of its costs from the second defendant, it would be left with a shortfall as the first defendant was bankrupt.

The judge had also erred in thinking that he was required to apply what is now CPR 44.5 (Amount of costs where costs are payable under a contract) when making the costs order. That rule applied only at the stage when the court was assessing costs and not when it was deciding by whom costs should be paid. The judge should have been exercising his discretion under what is now CPR rule 44.2 and the Senior Courts Act 1981, s 51(3).

The case of *Astrazeneca UK Ltd v International Business Machines Corpn* [2011] EWHC 3373 (TCC), confirmed that while, in principle, there might be two alternative bases for obtaining costs, namely under the terms of an express contractual indemnity or by the exercise of the court's discretion pursuant to s 51 of the Senior Courts Act 1981 and the CPR, the fact that the court made an order pursuant to s 51 did not detract from any contractual right to claim indemnity costs. It is clear that in exercising its discretion under CPR 44.2, the court should ordinarily exercise that discretion so as to reflect the contractual right.

Equally, if the court were giving effect to a contractual right to costs, then the provisions of CPR 44.5 and CPR PD 44, para 7 would provide, first, that the costs recoverable were those which had been reasonably incurred and reasonable in amount, and, second, that the costs payable should be disallowed if the court was satisfied by the paying party that costs had been unreasonably incurred or were unreasonable in amount. The fact that there was a contractual obligation to pay costs meant that the court ought to exercise its discretion so as to reflect those contractual rights and also be consistent with the requirements of CPR 44.5.

CHAPTER 37

GROUP (MULTI-PARTY) LITIGATION ORDERS

INTRODUCTION

[37.1]

CPR 46.6 was originally introduced on 3 July 2000 (as CPR 48.6) in an attempt to codify the guidance in the authorities on costs issues arising in group litigation. Some of the problems occur because a few claims are selected out of a large number as test cases and issues crop up regarding how the costs of those cases are to be apportioned among all of the claimants in a variety of circumstances. For example, when a between the parties' costs award in their favour is obtained which is less than the solicitor and client costs, or where there is an adverse costs order.

A frequent complication is where some, but not all, of the claimants are state-funded. In *BCCI v Ali* (13 April 2000, unreported, ChD) of some 300 claimants, five were chosen as test cases. They were not awarded any between the parties' costs, (although not ordered to pay any either) but the non-test case claimants argued that the solicitor and client costs should be paid solely by the test case claimants, on the grounds that they had lost because of their own dishonesty. Surprisingly, no costs sharing order or agreement had been made. The court found that when the test cases were chosen it was the common expectation of all the claimants that the test case costs would be shared equally, and it so ordered.

In *Ochwat v Watson Burton (a firm)* [1999] All ER (D) 1407 the test case claimants won on the general issues of duty of care and breach of that duty, but failed on causation. The judge awarded the defendants 75% of their costs of which he ordered the test case claimants to pay 75% and the remaining claimants 25%. On appeal it was held that this was unfair to the test case claimants and that the costs awarded to the defendants should be paid by all the claimants.

There is no supplementary practice direction to this rule, but CPR PD19B, para 16 provides that the costs judge shall apportion the amounts of common and individual costs at or before the commencement of detailed assessment, if the court has not already done so.

CPR 46.6(3) provides that, unless the court orders otherwise, orders against group litigants impose several liability for an equal proportion of common costs and that a group litigant will be responsible for his own individual solicitor and client costs and an equal share of the common costs. The court may make provision for the costs contribution of a party who joins the group late, or leaves it early.

DISCONTINUANCE

[37.2]

In *Sayers v Merck SmithKline Beecham plc* [2001] EWCA Civ 2017, appeals arose in relation to the details of costs sharing in three separate multi-party actions. However likely it might be that, if common issues were directed at trial, the costs of those issues would be ordered to follow the determination of those issues rather than await the individual fate of each claimant's action, it would be wrong to say that that should always be the presumption. Parties who settle their cases did not usually need any presumptive order as to the incidence of costs since costs would be part of the discussion leading to a settlement in any event. Discontinuers, however, gave rise to a more difficult problem. The order usually made in these circumstances was too blunt an instrument and was unnecessarily favourable to defendants, at a stage of the proceedings when it was as yet unknown whether the claimants as a whole were to be successful in the common issues which were to be tried. The orders would be amended to read:

> 'If in any quarter a claimant discontinued his claim against any one or more of the defendants or it is dismissed by an order of the court whereby the claimant is ordered to pay such defendant's costs, then he will be liable for his individual costs incurred by such defendants up to the last day of that quarter; liability for common costs and disbursement to be determined following the trial of common issues, with permission to apply if such trial does take place.'

INDIVIDUAL COSTS

[37.3]

In *O v Ministry of Defence* [2006] EWHC 990 (QB), the claimant in group litigation was awarded personal damages for clinical negligence but the MoD sought to set off the claimant's proportionate share of the costs of trial of the generic issues, against either his costs of his individual action or against his damages. Although the claimant might have benefited in the pursuit of his individual action had the generic issues been resolved differently the fact remained that in his own individual action the claimant proved that he sustained injury and consequential loss and damage, as a consequence of negligent treatment. He would have succeeded in that action whether or not the generic issues had been litigated. The formation of the group litigation and his involvement in it, as to which he had no real choice, inevitably resulted in a very substantial delay in the resolution of his claim. The justice of the case required that neither the claimant's own order for costs nor his damages should be subject to a set-off of his share of the generic costs.

GENERIC COSTS

[37.4]

Where a client is a member of a group involved in litigation which is awarded costs against the other party he is entitled to recover the costs for which he would have been liable to his solicitor. He is liable to his solicitor for all costs properly incurred whether they were incurred solely on his behalf or whether they were incurred for the benefit of the group and he has only to pay an appropriate proportion. There is nothing fundamentally different or special about generic costs; they are simply costs that have been shared for the sensible purpose of keeping the costs of each claim down.

A defendant's protection against an inflated claim is to be found in CPR 44.4(1)(a) where it provides that when assessing costs on the standard basis, the court has to consider whether, in the circumstances the costs are proportionately and reasonably incurred and are proportionate and reasonable in amount. That is an objective test. Furthermore, any doubts about the reasonableness or proportionality of any item on the bill has to be resolved in favour of the paying party (CPR 44.3(2)(b)). Since 1 April 2013, it is now also the case that costs which are disproportionate in amount may be disallowed or reduced even if they were reasonably or necessarily incurred: CPR 44.3(2)(a).

The protection provided by detailed assessment under CPR 44.4(1) would not in any way be enhanced by the existence of a letter sent at the time of the retainer whereby the solicitor told the client that some of the costs would be expended for the benefit of other claimants besides himself, so that he would only be asked to pay a share of those costs. In the real world, that would be a statement of the obvious. In practice, it is highly likely that the claimant would be aware that there were others making similar claims, through the same solicitors. Why would the solicitor do anything other than share costs, which were capable of being shared and thereby save money?

It would be good practice for a solicitor to mention in a client care letter that some of the work to be done would be for the benefit of a group of clients and that individuals would be liable only for their share. It would be sensible for a firm to keep records of the number of clients for whom it was acting at any time. Such records would help to demonstrate, if need be, that the proportion claimed for any individual client was justified. Where solicitors intend to claim a share of costs incurred by other solicitors, it would be wise for them to ensure that the terms of the agreement between the solicitors is clearly defined, to demonstrate more easily that the bill under scrutiny is reasonable and proportionate. In short, such records are desirable because they would be an aid to proving the reasonableness of a bill; they are not required as a pre-requisite to the recovery of a share of generic costs.

Where the litigation has been funded under CFAs entered into by each client, it would only be necessary for the individual client to demonstrate that there is an agreement between him and his solicitor collateral to the CFA specifically relating to generic costs if generic costs were in some way different from costs incurred solely for the individual client. There is therefore no requirement for any additional or collateral agreement relating to generic costs in a CFA for a successful claimant to recover such costs in an action where no group litigation order has been made (*Brown v Russell Young & Co* [2007] EWCA Civ 43).

No discussion of the assessment of costs in GLOs could fail to make some mention of *Motto & Others v (1) Trafigura Ltd (2) Trafigura Beheer BV* [2011] EWHC 90201 (Costs) where the Senior Costs Judge dealt with preliminary issues raised on assessment of the £104 million bill of costs incurred. These issues largely concerned the usual arguments on detailed assessment such as proportionality and reasonableness rather than GLO specific arguments. But, as well as being the largest bill to be seen in the SCCO it was also noteworthy for the bravura argument of the paying party regarding hourly rates. 30,000 French speaking Ivorian clients using CFAs was apparently still a 'run of the mill' case.

In *Jones v Secretary of State for Energy and Climate Change* [2012] EWHC 3647 (QB), [2013] 2 Costs LR 230, (the Phurnacite Litigation) Mrs Justice Swift had to consider the appropriate costs order to make where the lead claimants had lost at trial on some of the issues; namely causation of skin and bladder cancer. The parties were in full agreement on one point, namely that whatever she did, the Judge should not make an issue-based order for costs as such would present 'extremely complex problems for the judge carrying out the detailed assessment of costs and would in all probability add significantly to the costs of the assessment process. The Judge noted how the general rule was that the successful party will usually get their costs and that there was no reason to treat GLO cases any differently. The Defendants' arguments as to what percentage reduction should be made to reflect the issues upon which the Claimants had lost concentrated too much, in the Judge's eyes, on what time had been spent on such issues at trial. She felt, no doubt correctly, that she must consider work done on the litigation as a whole. She accepted that the interests of justice demanded that there must be some reduction in the Claimants' entitlement to costs to reflect the issues on which they had lost, this being more than simply a case of the inevitable loss of one or two issues on the way to ultimate success at trial, and adopting an admittedly 'broad brush approach' awarded the Claimants 80% of their common costs.

COSTS CAPPING

[37.5]

In *Various Claimants v Corby Borough Council* [2008] EWHC 619 (TCC), the claimants were children born to mothers who had lived in or close to Corby during the 1980s and 1990s. They alleged that as a result of negligent work, toxins had escaped and had affected pregnant women so as to cause birth defects in their offspring.

The parties accepted that an overall costs cap should be fixed. In fixing such a cap, the court had to have regard to the constituent elements making up each party's submitted costs estimates. It was impossible to predict with absolute precision or accuracy how much individual heads of costs would ultimately cost and an overall costs cap allowed an element of flexibility. Each party had agreed a CFA with their solicitors and the claimants had taken out After the Event insurance. It was accepted that the costs capping should not be on the basis of what their CFA would allow the successful party. Nor was the claimants' insurance taken into account. It was necessary to take an informed but broad-brush approach to the amount of the costs cap to be fixed upon each of the parties.

The parties agreed that a contingency was a fair and sensible allowance to make given the nature of the litigation and the likely encountering of expenditure or increase in levels of expenditure which was probably inevitable even if it could not be specifically foreseen which the judge fixed at 5%. The parties had liberty to apply to the Court for adjustment of the costs caps if, unforeseeably and beyond the reasonable control of the party in question, circumstances so changed or new circumstances arose such that there was a genuine need to adjust the figures.

In *Barr v Biffa Waste Services Ltd* [2009] EWHC 2444 (TCC), [2009] NLJR 1513, [2010] 3 Costs LR 317, the Judge was asked to impose a costs-capping order on the Claimants' costs equivalent to the £1 million limit of their ATE insurance. The Judge declined to do so, saying that it was entirely random to try and link the amount of any appropriate costs capping order to the amount of available ATE cover when such was beyond the scope of any control by either the Defendant or the Court. The Judge further refused to make a costs capping order, holding that he was not satisfied that it would be impossible to control the Claimants' costs through a combination of case management and detailed assessment. However, the Judge did recognise the potential unfairness of the Defendant's wider commercial position when faced with a group of Claimants whose costs liability, if the claim failed, was likely to exceed the value of the ATE cover and made an order noting that the Claimants' recently provided estimate of future costs 'was to be taken as a reasonable estimate of such costs and therefore their likely maximum recovery at the end of trial', albeit with leave to apply to modify that estimate if later events showed such would be appropriate.

FUNDING

[37.6]

Time will tell as to whether the introduction from 1 April 2013 of Damages-Based Agreements will lead to much of a change in the attractiveness, or otherwise, to legal representatives and clients of seeking to consolidate multiple claims into one piece of group litigation.

CHAPTER 38

TRUSTEES AND PERSONAL REPRESENTATIVES

ENTITLEMENT TO COSTS OUT OF THE TRUST OR ESTATE

[38.1]

Trustees and personal representatives have a duty to protect the trust fund or the estate. If, in so doing, they are able to recover any costs from another party or source, then they will claim them in the ordinary way. But if this is not possible, they are entitled to their costs of any proceedings undertaken in their capacity of trustee or personal representative out of the trust fund or estate.

Such costs will be assessed on the indemnity basis to the extent that they have been properly incurred. Whether that has happened depends upon all the circumstances, but in particular whether they have acted in the interests of the trust or estate rather than for some other benefit (including their own.) The fact that the trustee has defended a claim which includes a claim against the trustee personally is not necessarily determinative that the trustee has not acted in the interest of the trust. The reasonableness of bringing or defending proceedings (and the conduct of them) will be considered and so too will the question of whether the trustee obtained directions from the court before bringing or defending those proceedings (CPR 46.3 and CPR PD 46, para 1).

COSTS AGAINST TRUSTEES

[38.2]

A trustee or personal representative who has acted outside his duty or has acted in his own interests or otherwise unreasonably not only will be unable to recover his costs out of the fund or estate, but he may be ordered to pay the costs of another party personally. Similarly, although trustees and personal representatives will generally be justified in making an application to obtain the opinion of the court on a matter of construction or difficulty, if they appeal against the court's decision unsuccessfully, they may expect to be ordered to pay the costs of the appeal personally (*Re Earl of Radnor's Will Trusts* (1890) 45 Ch D 402, CA). Only in exceptional circumstances, say, for example, where large interests are at stake or where the interests of unborn persons are affected, will the costs be ordered to be paid out of the estate.

PRE-EMPTIVE ORDERS

[38.3]

Trustees or personal representatives may apply to the court before commencing or defending proceedings for an order that, win or lose, they will be entitled to their costs out of the property in dispute, before the facts have been fully investigated and before the law has been fully argued. This may be appropriate if there is a risk of a suggestion that the trustee has acted unreasonably or in his own interest. This is known as a 'Beddoe application' (*Re Beddoe, Downes v Cottam* [1893] 1 Ch 547, 62 LJ Ch 233, CA). Such a pre-emptive costs order affords the trustees the comfort of knowing that they will be indemnified against the costs of the claim. In addition to asking for an indemnity in respect of the costs, a *Beddoe* application usually asks the court for directions as to whether the claim should be continued or defended. Applications are made under CPR Part 25 if made before proceedings are commenced or by application notice in accordance with CPR Part 23 if made after proceedings have commenced.

PROSPECTIVE COSTS ORDERS

[38.4]

CPR 64.2(a) and CPR PD 64A, para 6 apply to claims for the court to determine any question arising in the administration of a deceased person or the execution of a trust. This includes a direction that the beneficiaries' costs should be paid in advance out of the trust fund, 'a prospective costs order'. The procedure for such costs applications is found within the Practice Direction to Part 64 Estates, Trusts and Charities.

Where a trustee has the power to enter into a contract no prospective order is required as the trustee is entitled to recover any monies from the trust that he has paid out. But if the trustee does not have this power, or simply decides not to exercise it, he may apply to the court for a prospective costs order.

To the extent that the order relates to the trustee's own costs, the court can simply authorise those costs to be paid from the fund. If there are any other party's costs the court can direct that the agreed (or assessed) costs be paid either on the standard or indemnity basis. Interim payments may be made on account of such costs. A model form of order is annexed to the practice direction.

Applications will be dealt with by the court on paper where possible. If the trustee considers that an oral hearing is required, he needs to cover this in the evidence supporting his application. If the court is minded to refuse the application on paper, it will inform the trustee so that he can ask for a hearing if he wishes to do so.

DETAILED ASSESSMENT OF COSTS FROM THE TRUST OR ESTATE

[38.5]

The procedure for this is set out at CPR 47.19 and CPR PD 47, para 18. Since most assessments are carried out provisionally, it is discussed in Chapter 30 at [30.16].

CHAPTER 39

FAMILY PROCEEDINGS

FAMILY PROCEDURE RULES 2010

[39.1]

Since the introduction of the CPR there have been successive attempts to unify the treatment of costs in civil and family proceedings, most recently by the Family Procedure Rules 2010. These rules came into force on 5 April 2011. The costs of all family proceedings are now governed by FPR 2010 Part 28 and the Practice Direction 28A which supplements it. Here they are:

[39.2]

'FPR Rule 28.1: Costs

The court may at any time make such order as to costs as it thinks just.'

[39.3]

'FPR Rule 28.2: Application of other rules

(1) Subject to rule 28.3 and to paragraph (2), Parts 44 (except rules 44.2(2) and (3), and 44.10(2) and (3)) 46 and 47 and rule 45.8. of the CPR apply to costs in proceedings, with the following modifications –
(a) in rule 44.1(1)(f)(ii), 'district judge' includes a district judge of the principal registry;
(b) in rule 46.8(1) after 'section 51(6) of the Senior Courts Act 1981' insert 'or section 145A of the Magistrates' Courts Act 1980';
(c) in accordance with any provisions in Practice Direction 28A; and
(d) any other necessary modifications.
(2) Part 47 and rules 45.8 and 46.7 of the CPR do not apply to proceedings in a magistrates' court.'

[39.4]

'FPR Rule 28.3: Costs in financial remedy proceedings

(1) This rule applies in relation to financial remedy proceedings.
(2) Rule 44.2(1), (4) and (5) of the CPR do not apply to financial remedy proceedings.
(3) Rule 44.2(6) to (8) and 44.12 of the CPR apply to an order made under this rule as they apply to an order made under rule 44.3 of the CPR.
(4) In this rule –
(a) 'costs' has the same meaning as in rule 44.1(1)(c) of the CPR; and
(b) 'financial remedy proceedings' means proceedings for –
(i) a financial order except an order for maintenance pending suit, an order for maintenance pending outcome of proceedings, an interim

periodical payments order or any other form of interim order for the purposes of rule 9.7(1)(a), (b), (c) and (e);
(ii) an order under Part 3 of the 1984 Act;
(iii) an order under Schedule 7 to the 2004 Act;
(iv) an order under section 10(2) of the 1973 Act;
(v) an order under section 48(2) of the 2004 Act.

(5) Subject to paragraph (6), the general rule in financial remedy proceedings is that the court will not make an order requiring one party to pay the costs of another party.
(6) The court may make an order requiring one party to pay the costs of another party at any stage of the proceedings where it considers it appropriate to do so because of the conduct of a party in relation to the proceedings (whether before or during them).
(7) In deciding what order (if any) to make under paragraph (6), the court must have regard to –
(a) any failure by a party to comply with these rules, any order of the court or any practice direction which the court considers relevant;
(b) any open offer to settle made by a party;
(c) whether it was reasonable for a party to raise, pursue or contest a particular allegation or issue;
(d) the manner in which a party has pursued or responded to the application or a particular allegation or issue;
(e) any other aspect of a party's conduct in relation to proceedings which the court considers relevant; and
(f) the financial effect on the parties of any costs order.
(8) No offer to settle which is not an open offer to settle is admissible at any stage of the proceedings, except as provided by rule 9.17.'

[39.5]

'FPR Rule 28.4: Wasted costs orders in the magistrates' court: appeals

A legal or other representative against whom a wasted costs order is made in the magistrates' court may appeal to the Crown Court.'

[39.6]

'Practice Direction 28A Costs

Application and modification of the CPR

1.1 Rule 28.2 provides that subject to rule 28.3 of the FPR and to paragraph (2) of rule 28.2, Parts 43, 44 (except rules 44.3(2) and (3), 44.9 to 44.12C, 44.13(1A) and (1B) and 44.18 to 20),47 and 48 and rule 45.6 of the CPR apply to costs in family proceedings with the modifications listed in rule 28.2(1)(a) to (d). Rule 28.2(1)(c) refers to modifications in accordance with this Practice Direction.
1.2 In addition to the modifications to the CPR listed in rule 28.2(1),in rule 48. 1(1)(b) after paragraph (ii) insert '(iii) section 68A of the Magistrates' Courts Act 1980.'.
1.3 Rule 28.2(2) provides that Part 47 and rules 44.3C and 45.6 of the CPR do not apply to proceedings in a magistrates' court.

Application and modification of the Practice Direction supplementing CPR Parts 43 to 48

2.1 For the purpose of proceedings to which these Rules apply, the Practice Direction about costs which supplements Parts 43 to 48 of the CPR ('the costs practice direction') will apply, but with the exclusions and modifications explained below to reflect the exclusions and modifications to those Parts of the CPR as they are applied by Part 28 of these Rules.

2.2 Rule 28.2(1) applies, with modifications and certain exceptions, Parts 43 to 48 of the CPR to costs in family proceedings. Paragraph 1.2 of this Practice Direction modifies rule 48.1(1)(b) when it applies to family proceedings. Rule 28.2(2), by way of exception, disapplies Part 47, rules 44.3C and 45.6 of the CPR in the case of family proceedings in a magistrates' court. Rule 28.3, again by way of exception, additionally disapplies CPR rule 44.3(1), (4) and (5) in the case of financial remedy proceedings, regardless of court.

2.3 The costs practice direction does not, therefore, apply in its entirety but with the exclusion of certain sections reflecting the non-application of certain rules of the CPR which those sections supplement.

2.4 The costs practice direction applies as follows –
— to family proceedings generally, other than in magistrates' courts, with the exception of sections 6, 15, 16, 17 and 23A;
— to family proceedings generally, in magistrates' courts only, with the exception of sections 6, 15, 16, 17, 23A and sections 28–49A;
— to financial remedy proceedings, other than in magistrates' courts, with the exception of section 6, paragraphs 8.1 to 8.4 of section 8 and sections 15, 16, 17 and 23A;
— to financial remedy proceedings in magistrates' courts only, with the exception of section 6, paragraphs 8.1 to 8.4 of section 8, sections 15, 16, 17, 23A and sections 28–49A.

2.5 All subsequent editions of the costs practice direction as and when they are published and come into effect shall in the same way extend to all family proceedings.

2.6 The costs practice direction includes provisions applicable to proceedings following changes in the manner in which legal services are funded pursuant to the Access to Justice Act 1999. It should be noted that although the cost of the premium in respect of legal costs insurance (section 29) or the cost of funding by a prescribed membership organisation (section 30) may be recoverable, family proceedings (within section 58A(2) of the Courts and Legal Services Act 1990) cannot be the subject of an enforceable conditional fee agreement.

2.7 Paragraph 1.4 of section 1 of the costs practice direction shall be modified as follows –
in the definition of 'counsel' for 'High court or in the county courts' substitute 'High Court, county courts or in a magistrates' court'.

General Interpretation of references in CPR

3.1 References in the costs practice direction to 'claimant' and 'defendant' are to be read as references to equivalent terms used in proceedings to which these Rules apply and other terms and expressions used in the costs practice direction shall be similarly treated.

3.2 References in CPR Parts 43 to 48 to other rules or Parts of the CPR shall be read, where there is an equivalent rule or Part in these Rules, to that equivalent rule or Part.

Costs in financial remedy proceedings

4.1 Rule 28.3 relates to the court's power to make costs orders in financial remedy proceedings. For the purposes of rule 28.3, 'financial remedy proceedings' are defined in accordance with rule 28.3(4)(b). That definition, which is more limited

39.6 *Chapter 39 Family proceedings*

than the principal definition in rule 2.3(1), includes –
(a) an application for a financial order, except –
  (i) an order for maintenance pending suit or an order for maintenance pending outcome of proceedings;
  (ii) an interim periodical payments order or any other form of interim order for the purposes of rule 9.7(1)(a),(b),(c) and (e);
(b) an application for an order under Part 3 of the Matrimonial and Family Proceedings Act 1984 or Schedule 7 to the Civil Partnership Act 2004; and
(c) an application under section 10(2) of the Matrimonial Causes Act 1973 or section 48(2) of the Civil Partnership Act 2004.
4.2 Accordingly, it should be noted that –
(a) while most interim financial applications are excluded from rule 28.3, the rule does apply to an application for an interim variation order within rule 9.7(1)(d),
(b) rule 28.3 does not apply to an application for any of the following financial remedies –
  (i) an order under Schedule 1 to the Children Act 1989;
  (ii) an order under section 27 of the Matrimonial Causes Act 1973 or Part 9 of Schedule 5 to the Civil Partnership Act 2004;
  (iii) an order under section 35 of the Matrimonial Causes Act 1973 or paragraph 69 of Schedule 5 to the Civil Partnership Act 2004; or
  (iv) an order under Part 1 of the Domestic Proceedings and Magistrates' Courts Act 1978 or Schedule 6 to the Civil Partnership Act 2004.
4.3 Under rule 28.3 the court only has the power to make a costs order in financial remedy proceedings when this is justified by the litigation conduct of one of the parties. When determining whether and how to exercise this power the court will be required to take into account the list of factors set out in that rule. The court will not be able to take into account any offers to settle expressed to be 'without prejudice' or 'without prejudice save as to costs' in deciding what, if any, costs orders to make.
4.4 In considering the conduct of the parties for the purposes of rule 28.3(6) and (7) (including any open offers to settle), the court will have regard to the obligation of the parties to help the court to further the overriding objective (see rules 1.1 and 1.3) and will take into account the nature, importance and complexity of the issues in the case. This may be of particular significance in applications for variation orders and interim variation orders or other cases where there is a risk of the costs becoming disproportionate to the amounts in dispute.
4.5 Parties who intend to seek a costs order against another party in proceedings to which rule 28.3 applies should ordinarily make this plain in open correspondence or in skeleton arguments before the date of the hearing. In any case where summary assessment of costs awarded under rule 28.3 would be appropriate parties are under an obligation to file a statement of costs in CPR Form N260.
4.6 An interim financial order which includes an element to allow a party to deal with legal fees (see *A v A (maintenance pending suit: provision for legal fees)* [2001] 1 WLR 605; *G v G (maintenance pending suit; costs)* [2002] EWHC 306 (Fam); *McFarlane v McFarlane, Parlour v Parlour* [2004] EWCA Civ 872; *Moses-Taiga v Taiga* [2005] EWCA Civ 1013; *C v C (Maintenance Pending Suit: Legal Costs)* [2006] Fam Law 739; *Currey v Currey (No 2)* [2006] EWCA Civ 1338) is an order made pursuant to section 22 of the Matrimonial Causes Act 1973 or an order under paragraph 38 of Schedule 5 of the 2004 Act, and is not a 'costs order' within the meaning of rule 28.3.
4.7 By virtue of rule 28.2(1), where rule 28.3 does not apply, the exercise of the court's discretion as to costs is governed by the relevant provisions of the CPR and in particular rule 44.3 (excluding r 44.3(2) and (3)).'

[39.7]

Part 28 is not entirely self-contained. Most of the costs provisions of the CPR are adopted and so must be referred to separately, with appropriate modifications as provided in the rules.

[39.8]

From the outset, the presumption in civil cases that costs follow the event that is enshrined in CPR Rule 44.3(2) was disapplied in family proceedings by the Family Proceedings (Miscellaneous Amendment) Rules 1999.

Nevertheless, in *Gojkovic v Gojkovic (No 2)* [1992] Fam 40, CA it was held that even in the Family Division the starting point has to be that costs on the face of it follow the event, although the presumption might be more easily displaced and in circumstances which do not apply in other Divisions of the High Court. Indeed, in family matters it is often difficult to identify an 'event' which costs might follow. An important example is that it was unusual to order costs in Children Act cases.

FPR 2010 Rule 28.1 sets out the guiding principle that in family proceedings the court may at any time make such order as to costs as it thinks just. However, this is qualified by rule 28.3 as far as financial remedy proceedings are concerned. Rule 28.3 retains the presumption that there will be no order for costs between the parties. The presumption can be displaced where appropriate because of the conduct of a party in relation to the proceedings.

There are therefore two costs regimes that apply in family proceedings.

Where the application is not for a financial remedy or financial order FPR 2010, rule 28.2 will always apply, in other words, CPR as modified for use in family proceedings.

Where the application is for a financial remedy or order, whether it is FPR 2010, rules 28.2 or 28.3 that will apply will depend upon whether the application is in financial remedy proceedings. This is not necessarily the same thing as proceedings for a financial remedy! The potential for confusion can be avoided by using as a starting point the applications to which FPR 2010, rule 28.3 *will* apply.

[39.9]

Here they are:
- avoidance of disposition
- periodical payments
- lump sum
- property adjustment
- variation
- pension sharing
- section 10(2) Matrimonial Causes Act 1973/section 48(2) Civil Partnership Act 2004

[39.10]

However, it is important to note that for the purposes of rule 28.3, the following applications are excluded from that costs regime, and fall to be dealt with under the costs provisions that apply in all other cases, whether for a financial order or not, namely FPR 2010, rule 28.2.

- maintenance pending suit;
- legal services order
- maintenance pending outcome of the proceedings;
- interim periodical payments;
- any other form of application for an interim order apart from an interim variation order;
- appeals;
- preliminary issue applications;
- setting aside an order made in financial remedy proceedings;
- enforcement.

[39.11]

In addition to that list of exceptions, FPR PD 28A, para 4.2(b) also excludes the following applications for a financial remedy from those to which rule 28.3 applies:
- Children Act 1989, Sch 1;
- Matrimonial Causes Act 1973, s 27;
- Civil Partnership Act 2004, Sch 5, Part 9;
- Matrimonial Causes Act 1973, s 35;
- Civil Partnership Act 2004, Sch 5, para 69;
- Domestic Proceedings and Magistrates' Courts Act 1978, Part 1;
- Civil Partnership Act 2004, Sch 6.

APPLICATIONS OTHER THAN IN FINANCIAL REMEDY PROCEEDINGS

[39.12]

The costs of all applications other than those in financial remedy proceedings, to which Rule 28.3 apply, are governed by those provisions of the CPR that are specified in Rule 28.2.

The whole of Part 44 (except CPR 44.2(2)(3) and CPR 44.10(2),(3)), CPR 46 and CPR 47 and CPR 45.8 apply. The most significant of the exceptions to Part 44 is CPR 44.2(2)). The effect of this is to disapply in family proceedings the default position for civil costs that costs follow the event. The operative part of the Rule for family proceedings, excluding financial remedy proceedings to which FPR 28.3 applies, is therefore:

'Rule 44.3

(1) The court has a discretion as to–
(a) whether costs are payable by one party to another;
(b) the amount of those costs; and
(c) when they are to be paid.'

The Rule goes on to set out the matters that the court must consider when exercising its discretion

'(4) In deciding what order (if any) to make about costs, the court must have regard to all the circumstances, including –
(a) the conduct of the parties;
(b) whether a party has succeeded on part of his case, even if he has not been

(c) any payment into court or admissible offer to settle made by a party which is drawn to the court's attention.

(5) The conduct of the parties includes

(a) conduct before, as well as during, the proceedings and in particular the extent to which the parties followed the Practice Direction (Pre-Action Conduct) or any relevant pre-action protocol;

(b) whether it was reasonable for a party to raise, pursue or contest a particular allegation or issue;

(c) the manner in which a party has pursued or defended his case or a particular allegation or issue;

(d) whether [an applicant] who has succeeded in his claim, in whole or in part, exaggerated his claim.'

Accordingly in family proceedings that are not financial remedy proceedings, to which FPR 28.3 applies, the court is neither bound by the presumption that there will be no order for costs as it is where the rule applies, nor is it bound by the civil presumption that costs follow the event. The court therefore has a wide discretion. However, family judges do like to have a starting point as we saw in *Gojkovic (No 2)* [39.00]. In that case the Court of Appeal was concerned with a similar provision in the pre-CPR Rules of the Supreme Court on an appeal in a family case that disapplied the general rule that costs follow the event. In that case Butler-Sloss LJ said there needed to be a starting point and in her judgment the starting point should be that costs follow the event. Having said that, the presumption may be more easily displaced in a family case.

CHILDREN ACT 1989 APPLICATIONS

(a) Schedule 1 applications

[39.13]

Schedule 1 of the Children Act 1989 provides three forms of financial relief for a child: (1) a maintenance order, settlement and transfer of property order to be made in favour of a child against either or both of its parents; (2) a periodical payments order and lump sum order to be made in favour of a child who has reached 18; and (3) the court may vary a maintenance agreement containing financial arrangements for the child either during the lifetime of the parent or after the death of one of them. The application may be made by a parent or guardian or the child itself if over 18.

These are family proceedings, but not financial remedy proceedings. Any issue as to costs therefore falls to be dealt with under the relevant provisions of the CPR, as provided for by FPR 2010, rule 28.2 Even without litigation misconduct the parties are at risk of a between-the-parties costs order but they are able to protect their costs position by making a 'Calderbank' offer.

39.14 *Chapter 39 Family proceedings*

(b) Other Children Act 1989 applications

[39.14]

Although the court has the power to make such order for costs as it thinks fit in family proceedings, it is unusual in a case involving children for an order for costs against a party to be made.

The recent decision in *Re T (Costs: Care Proceedings: Serious Allegation Not Proved)* [2013] 1 FLR 133 restated the principle set out by Cazalet J in *Re M (Local Authority's Costs)* [1995] 1 FLR 533 that the general practice of not awarding costs in the absence or reprehensible behaviour or an unreasonable stance accorded with the interests of justice.

In private law Children Act 1989 applications too, orders for costs against a party will be rare. However, where the court is satisfied that a party's conduct of the proceedings has been unreasonable, that party will be at risk of an adverse costs order.

[39.15]

In an application under the Child Abduction and Custody Act 1985 the mother had been unreasonable in the conduct of the case through her persistent pursuit of uncorroborated false allegations of abduction and serious dishonesty involving the forgery of documents. At the conclusion of the hearing the father, a man of limited means who was bringing up the parties' child on his own without any financial support from the mother and who had been wholly successful in the outcome of the proceedings, had asked for his costs against the mother. The mother, who was publicly funded, contended there should be no order for costs in view of Article 26 of the Convention on the Civil Aspects of International Child Abduction 1980 and that as a matter of public policy, so as not to deter genuine abduction applicants or the acceptance of instructions by solicitors from the Child Abduction Unit of the Official Solicitor's Office. The court held It would be inconsistent to say that the rules generally and the costs rules in particular applied to the 1985 Act but that they should not have effect because the power to make a costs order against a claimant was not expressly set out on the face of the Hague Convention. There was nothing inimical to the operation of the Hague Convention in the application of ordinary funding costs principles against an unsuccessful claimant. Accordingly there was jurisdiction to make a costs order against a claimant in proceedings under the 1985 Act. Deliberate and persistent falsification of a case in an attempt to deprive a child of his habitual residence or to render ineffective the custody, access and Article 8 rights of the child and the other parent were relevant when exercising the court's discretion as to costs. Accordingly, since in each case where a costs application was made there should be a costs enquiry on the merits, having regard to the statutory test in s 11 of the Access to Justice Act 1999, it was appropriate to make an order for costs against the mother and adjourn for detailed assessment on the standard basis by a costs judge (*EC-L v DM (child abduction: costs)* [2005] EWHC 588 (Fam), [2005] 2 FLR 772, [2005] Fam Law 606).

[39.16]

Re T (a child) (order for costs) [2005] EWCA Civ 311, [2005] 1 FCR 625 concerned an application by a father for contact with his child.

Although after a fact-finding hearing the judge found that the mother with whom the child resided need have no concerns about the child's contact with its father the mother remained obdurate and made further unfounded allegations against the father necessitating a total of four hearings. There is a limit to which allowance could be made for a parent who deliberately and unreasonably obstructs contact and the Court of Appeal upheld the judge's order that the mother should pay the costs of all four hearings (*Re T (a child) (order for costs)* [2005] EWCA Civ 311, [2005] 1 FCR 625).

In *Re G (costs: child case)* [1999] 3 FCR 463, CA, [1999] 2 FLR 250 the Court of Appeal held that 'it was unusual to order costs in a family case, although it would be appropriate to order costs where a parent, even a litigant in person, went beyond the limit of what was reasonable to pursue the application before the court'. Even though the judge had found the father's case to be hopeless, the court held that hopelessness and unreasonableness were not necessarily the same thing. Furthermore, a greater degree of generosity might be appropriate in ruling that pursuing an application had become unreasonable where the litigant was acting in person rather than where he was receiving legal advice. In *Re W (a child)* (26 August 2000, unreported), CA on the father's application to vary an order for residence with the mother, the judge confirmed the existing arrangements, extended staying contact and ordered the father to pay the costs, commenting that the proceedings had been long and protracted and that the mother had been the successful party. In allowing the father's appeal against the order for costs, the court said the judge had not criticised the father's motives or provided any other readily discernible reason for departing from the normal rule in private family law cases that each party should pay its own costs. It is essential that between-the-parties costs orders should only ever be made in exceptional cases where a party had been guilty of manipulating the litigation for their own ends rather than advancing a case which they genuinely believed to be in the best interests of the child.

In *Re B (costs)* [1999] 3 FCR 586, [1999] 2 FLR 221 the Court of Appeal demonstrated that although it is rare to make costs orders in Children Act cases, the rule is not invariable. The mother had suffered from a serious manic depressive illness as a result of which the child resided with the father, who opposed the mother's application for staying contact because of his fear the mother might have a relapse which could put the child at risk. The mother produced evidence from a psychiatric consultant that the child would not be at any risk, but the father was unwilling to accept this even when a psychiatrist instructed by him agreed with the mother's psychiatrist. The Court of Appeal held the judge was right to take the view that from the date that the two psychiatrists reached agreement the father should have withdrawn his opposition to staying contact and therefore ordered him to pay the mother's costs from that date, and that the judge was also correct in assessing the costs of the father's continued and unjustified opposition at 80% of the costs of the hearing.

In *HH v BLW (Appeal: Costs: Proportionality)* [2013] 1 FLR 420 the court at first instance made an order for costs in the sum of £2,468 against the applicant father who withdrew his application for contact to his 15 year old

39.16 *Chapter 39 Family proceedings*

daughter. The decision not to proceed was made after the father had heard from the Cafcass officer that the child was firmly expressing an unwillingness to have contact.

The father appealed against the decision to award costs to the mother. In considering the application for permission to appeal, Holman J disagreed with the view of the district judge that the father's application had been misconceived and the outcome a foregone conclusion. His view was that the father had been entitled to await the recommendation of the Cafcass officer before deciding what course to take and it was entirely reasonable of him to do so. Had he decided to pursue the application in the face of the Cafcass report he would have been foolish and a costs order would have been justified. Nonetheless, Holman J was unable to say that the high hurdle in order to succeed on an application for permission to appeal had been cleared. Coupled with that, the application of the overriding objective, in particular proportionality, determined that the application should fail.

APPEAL

[39.17]

Costs are much more likely to be awarded on an appeal than at first instance. At first instance nobody knows what the judge is going to find. On an appeal both parties have the chance to take stock and make an offer. In *Re M* [2009] EWCA Civ 311 the Court of Appeal refused permission to appeal from the High Court judge's order that the husband should pay the wife's costs of her successful appeal from the district judge. The husband through his counsel had opposed the appeal 'root and branch' and announced that he intended to apply for costs if the appeal were successfully opposed. Such a litigant, if he lost as here, could not complain when the judge took the view that he should contribute to or pay, the appellant's costs.

CALDERBANK OFFERS

[39.18]

A 'Calderbank' offer is so-called because it is a form of offer first approved by the Court of Appeal in *Calderbank v Calderbank* [1976] Fam 93, [1975] 3 All ER 333, [1975] 3 WLR 586 CA, and this is still a useful description. It is an offer made 'Without prejudice save as to costs'. It may not be seen by the court except in the context of costs between the parties once the substantive issues have been determined.

Wherever there is a risk that costs may be ordered between the parties it is sensible to make offers to settle, if necessary using *Calderbank* terms, 'without prejudice save as to costs'.

Calderbank offers are not admissible before the court in financial remedy proceedings to which Rule 28.3 applies, except at the Financial Dispute Resolution appointment. However, in all other family proceedings they are.

CPR 44.3(4)(c) specifically requires the court to consider any admissible offers made as one of the circumstances of the case to be taken into account in the exercise of its discretion.

FINANCIAL REMEDY PROCEEDINGS WHERE FPR 28.3 APPLIES

[39.19]

In financial remedy proceedings, to which FPR 28.3 applies, the court's costs discretion is limited. The general rule for such applications is that the court will not make an order requiring one party to pay the costs of another party. This however is subject to FPR 28.3(6):

> 'The court may make an order requiring one party to pay the costs of another party at any stage of the proceedings where it considers it appropriate to do so because of the conduct of a party in relation to the proceedings (whether before or during them).'

Therefore, the court may depart from the general rule if there is litigation conduct which justifies the making of an order for costs.

The matters relevant to deciding whether there has been any relevant conduct are set out at FPR 28.3(7) and the court is required to have regard to them:

(a) any failure by a party to comply with these rules, any order of the court or any practice directions which the court considers relevant;
(b) any open offer to settle made by a party;
(c) whether it was reasonable for a party to raise, pursue or contest a particular allegation or issue;
(d) the manner in which a party has pursued or responded to the application or a particular allegation or issue;
(e) any other aspect of a party's conduct in relation to proceedings which the court considers relevant; and although not an aspect of conduct the court must also weigh in the balance;
(f) the financial effect on the parties of any costs order.

[39.20]

This does no more than preserve the position as it had been for this type of application since the Family Proceedings (Amendment) Rules 2006 came into effect from 3rd April 2006. However, Practice Direction 28A adds something to it:

> '4.4 In considering the conduct of the parties for the purposes of rule 28.3(6) and (7) (including any open offers to settle), the court will have regard to the obligation of the parties to help the court to further the overriding objective (see [FPR 2010]) rules 1.1 and 1.3) and will take into account the nature, importance and complexity of the issues in the case. This may be of particular significance in applications for variation orders and interim variation orders or other cases where there is a risk of the costs becoming disproportionate to the amounts in dispute.'

39.20 *Chapter 39 Family proceedings*

So although the list in FPR 23.8(7) is an exhaustive list, and the court cannot avail itself of its broad brush and sweep up all the circumstances of the case, there are additional factors set out in the Practice Direction to which the court must have regard.

The final sentence of paragraph 4.4 of the Practice Direction gives a clue as to what it may primarily be aimed at. This provision highlights the position that is frequently encountered in applications to vary a periodical payments order, where costs will frequently outweigh any financial benefit that may be obtained even from a successful application. This clear indication that the court will be looking particularly at cases of that sort should encourage both parties to such an application to take a pragmatic view before disproportionate costs are incurred. The corollary is the exclusion of maintenance pending suit applications from FPR 28.3 thus disapplying the 'no order' presumption.

THE FINANCIAL EFFECT ON THE PARTIES OF ANY COSTS ORDER

[39.21]

As we have seen, when deciding whether or not to make an order for costs in financial remedy proceedings to which FPR 28.3 applies the court will look at litigation conduct as set out in FPR 28.3(7)(a) to (e). The court will then go on to consider the financial effect on the parties of any costs order under FPR 28.3(f). This is unrelated to conduct and involves an assessment of where any costs order would leave the parties financially. It is a necessary exercise in applications for a financial remedy since it may be the case that if an adverse costs order were to be made it may have the effect of undermining the basis upon which the court's substantive order was made.

This of course was the case under the rule that FPR 28.3 substantially replaces, rule 2.71 of the FPR 1991. That rule came into force on 3rd April 2006 and its effect was to make *Calderbank* offers inadmissible in what used to be known as ancillary relief proceedings.

Before *Calderbank* offers were rendered inadmissible, costs were dealt with at the conclusion of the case as they would be in a civil case. So once the court had determined the substantive application the question of costs was addressed. In order to protect their position on costs, parties would make offers to settle that were expressed to be 'without prejudice save as to costs'. These were produced after judgment was given and the judge was then invited to make an order for costs against the party who had been less successful in accurately predicting the outcome.

There had been disquiet for some time about how the system of *Calderbank* offers operated in ancillary relief applications.

(a) False teeth

[39.22]

As far back as 1992, with a view to 'giving teeth' to *Calderbank* offers, CCR Order 11, rule 10, which related to *Calderbank* offers in civil proceedings, was applied to ancillary relief by a new FPR 2.69. On 5 June 2000 in the light of eight years' experience the procedure was codified and expanded in new FPR 2.69A to E. Whether because of defects in the new rules or the way in which they were applied, or perhaps because they highlighted underlying flaws in the system itself, they have not worked. In the words of Thorpe LJ: 'The innovation was not well received and did not stand up to hard wear and tear'. Nicholas (now Mr Justice) Mostyn QC did not believe 'these rules say what we intended them to say'. In *GW v RW* [2003] EWHC 611 (Fam), [2003] 2 FCR 289, Mr Mostyn, sitting as a deputy High Court judge, held that the rules were incomprehensible and that where both parties had made *Calderbank* offers, became unworkable. He identified two problems in particular. The first was the destabilising effect that costs can have on financial settlements that have been carefully constructed by the court. Having considered the facts and circumstances of a case the court arrives at a settlement that, in its judgment, does justice between the parties. At the conclusion of some cases it is revealed to the court that one party has failed to 'beat' a *Calderbank* offer, the consequences of which could undermine completely the substantive order for ancillary relief that the court had just made. This stems from *Calderbank* offers not being brought to the attention of the court until all matters have been determined and the court is considering the issue of costs. The second problem identified by Mr Mostyn was that the system of closed offers had introduced a degree of procedural gamesmanship which lead to uncertainty and had introduced an undesirable element of gambling into ancillary relief proceedings. *Calderbank* offers had been likened to a form of spread betting. The result could be disproportionate and produce unfairness not only in big money cases but also where modest sums were involved.

(b) The Senior Costs Judge writes

[39.23]

The view that the approach of the courts in family financial matters needed reconsideration was reinforced by a letter to the President from the Senior Costs Judge shortly after the SCCO took over the assessment of costs in family proceedings. He wrote: "It is apparent that whereas in non-family civil proceedings the resolution of the substantive dispute frequently takes the heat out of any animosity between the parties, and enables settlement of the costs to be achieved in a significant number of cases, in family proceedings that animosity, which is in any event likely to be at a very high level, continues unabated during the assessment proceedings. The successful spouse on one side vows to bleed the other dry of every penny if at all possible, whilst the paying spouse goes out of his or her way to deny the other the possibility of any recovery. The number of settlements in assessments arising out of family proceedings is very low . . . It may be worth giving serious thought to doing away with fee shifting in family proceedings. The Family Proceedings (Miscellaneous Amendment) Rules 1991 disapply CPR 44.3(2) (costs follow the event). It is therefore a relatively short step to providing that in family proceedings no order for costs will be made unless a particular party has behaved in such an unreasonable manner that the court feels that a sanction should be imposed."

(c) The matrimonial pot

[39.24]

In October 2004, the Department for Constitutional Affairs, in collaboration with the President's Ancillary Relief Advisory Group published a Consultation Paper *Costs in Ancillary Relief Proceedings and Appeals in Family Proceedings*. The Consultation Paper included a draft statutory instrument making amendments to FPR 2.69 on offers to settle and inserting a new FPR 2.71 on costs in ancillary relief proceedings. The principal objective was to establish a clear general rule in ancillary relief proceedings that the court should make no order for costs, unless there had been "unreasonable conduct" in relation to the proceedings, by one of the parties. It was envisaged that as a result, the role of *Calderbank* offers in ancillary relief proceedings would be ended. The new philosophy was to move the costs regime in ancillary relief proceedings further away from the classic 'costs follow the event' approach in civil litigation and to enable the court to include consideration of costs as part of the overall settlement of the parties' financial affairs. Justice would be better served by dealing with costs as part of the substantive application rather than treating costs as a separate issue. Absent misconduct, costs would be paid out of the matrimonial 'pot' and the court would then divide the remainder between the parties. The new regime was introduced by the Family Proceedings (Amendment) Rules 2006 on 3 April 2006 and is substantially replicated in FPR 28.3.

Calderbank offers in financial remedy proceedings

[39.25]

FPR 28.2(8) specifically excludes the reliance upon *Calderbank* offers in financial remedy proceedings, save for the purposes of Financial Dispute Resolution hearings under the provisions in FPR 9.17:

'(3) Not less than 7 days before the FDR appointment, the applicant must file with the court details of all offers and proposals, and responses to them.
(4) Paragraph (3) includes any offers, proposals or responses made wholly or partly without prejudice, but paragraph (3) does not make any material admissible as evidence if, but for that paragraph, it would not be admissible.'

Calderbank offers therefore have a very limited role in financial remedy proceedings. They carry no costs sanction and therefore if privilege is sought the mere heading 'Without Prejudice' will suffice. The district judge is able to comment on the privileged offers, and to observe perhaps, that a wholly inappropriate open offer to settle may amount to litigation misconduct that could result in a departure from the general rule of no costs order. At this stage, and indeed during any subsequent proceedings, there may be two separate strands of negotiation with more generous proposals contained in privileged correspondence which will be admissible before the district judge at the FDR hearing but not before any other judge. Parties may therefore continue to make privileged offers in negotiations more favourable than those known to the court but if settlement is not reached, the offers will have no costs conse-

(i) Between-the-parties costs orders

1. No misconduct

[39.26]

If there is no litigation misconduct the general rule that there will be no order for costs will apply. So, does the court look at the total solicitor and client costs of each party, add them both together, deduct their total from the family assets and then proceed to share between the parties what the lawyers have left for them?

There are two schools of thought.

[39.27]

(a) Costs out of the pot. In the red corner is the consultation paper on *Costs in Ancillary Relief Proceedings* which it is important to remember was produced not only by the Department for Constitutional Affairs but also by the illustrious President's Ancillary Relief Advisory Group and its Costs sub-committee. Paragraph 27 is unambiguous: 'The purpose of applying a 'no order for costs' principle in ancillary relief proceedings is to stress to the parties, and to their legal advisers, that running up costs in litigation will serve only to reduce the resources that the parties will have left to support them in their new lives apart. The proposed amendments to the costs rules are designed to establish the principle that, in the absence of litigation misconduct, the normal approach of the court to costs in ancillary relief proceedings should be to treat them as part of the parties' reasonable financial needs and liabilities. *Costs will have to paid from the matrimonial 'pot' and the court will then divide the remainder between the parties.* (My italics)

On this approach there are problems:
- there is no control over the costs a party could claim out of the matrimonial pot. Solicitor and client costs are on the indemnity basis to which the test of proportionality does not apply and if the client accepts their solicitor's costs, those costs cannot be challenged on the grounds of being unreasonable. The client is contractually bound to pay them. So how could they be reduced even if either the court or the other party objects to the amount claimed coming out of the pot?
- where is the incentive for a party to control their legal costs if half are in effect to be paid by the other spouse?
- what if one party is economical and the other is not? Should the economical party have to subsidise the other? David Burrows in *The New Ancillary Relief Costs Regime* gives the example of family assets of £1 million, with the husband having incurred costs of £100,000 and the wife £50,000. If the costs are deducted before the net assets of £850,000 are divided 50/50, the wife will pay £25,000 of the husband's costs. And if the wife is to receive 60% of the assets she will on this basis pay £40,000 of the husband's costs.

- is there any incentive for a party to act in person to reduce costs if the other party has legal representation paid out of the pot?
- each of these factors gives rise to what have been described as 'back-door adverse costs orders' and 'standing justice on her head'. Only if you adopt what has been called the *Leadbeater* technique – Mr Justice Balcombe in *Leadbeater v Leadbeater* [1985] FLR 789, [1985] Fam Law 280 – and deal with the gross assets – notionally writing back any costs already paid – do you arrive at the position of each party paying their own costs out of their share of the assets.

However, recent decisions have discouraged the *Leadbeater* approach. In *R v R (financial remedies: needs and practicalities)* [2011] EWHC 3093 (Fam), [2013] 1 FLR 120, the wife's costs were approximately double those of her husband. Although the judge, Coleridge J, was sympathetic to the wife he was 'driven to say that he would discourage the pursuit of this add-back principle or approach, which inevitably leads to a quasi-taxation or assessment of costs during the hearing, but without the court having all the material which would be available to, for instance, a costs judge. It also rather flies in the face of the non-order starting point and leads to debates about costs by the back door, which the new rules were designed to try and reduce or prevent.'

This was followed in *GS v L (No 2) (Financial Remedies: Costs)* [2011] EWHC 2116 (Fam), [2012] Fam Law 802, in which it was held that the judge should first decide whether or not it was appropriate to make an issue-based costs order and, if it were appropriate, make the order for costs expressing the order by way of a percentage, or for the costs to be from a particular date.

In *Ezair v Ezair* [2012] EWCA Civ 893 the Court of Appeal said that the circuit judge should not have adjusted the lump sum award to effectively reflect a costs element He should have made an assessment of the lump sum by reference to the section 25 criteria and then made a separate costs order to reflect conduct within the litigation.

[39.28]

(b) Individual debts. In the blue corner we have Judge Martin Cardinal and District Judge Simon Middleton in the second edition of *Matrimonial Costs* who quote Lord Justice Wilson, as he then was, that the proper treatment of liabilities for costs under the new regime will generally be that they are debts to which the judge should have regard in making his substantive award. In other words, the court must first apply the s 25 criteria to all the available assets and it is then for each party to discharge their own costs together with any other debts out of their share of the family assets, subject, of course, to the court taking into account all of a party's liabilities in determining what is a fair division of the family assets according to traditional practice.

But even this approach has its difficulties. If a party's liability for costs is to be taken out of the equation altogether and treated as a personal debt to be paid out of their share of the assets, this can drive a coach and horses through a needs-based order. Cardinal and Middleton give the example of a mother who needs £150,000 to re-house herself and her two children but has a costs liability of £20,000. Does she get the £170,000 she needs? If she does it is another back-door order for costs in the guise of a needs-based asset division.

If liability for costs is to be taken into account as a debt in sharing the assets, what if one party has incurred twice the costs of the other? Should the court look behind the solicitor and client screen and hold a mini Solicitors Act assessment and for the purposes of s 25 ignore any costs above the amount it regards as reasonable – perhaps the amount of the costs of the other party? If so we are into retrospective costs capping.

[39.29]

(c) Half-way house? There is, I suggest, a possible halfway house between the two positions of either the costs of both parties being deducted from the family assets before applying the s 25 factors on the one hand and on the other hand applying the s 25 factors to the whole of the assets, leaving each party to pay their legal fees out of their own share. At the outset of the proceedings the court could give an indication of (or the parties might even agree) the amount of costs each party is likely to receive out of the pot before the remaining assets are shared between them. In that way, the parties and their legal advisers would know from the start how much of their costs they will recover out of the pot and that any costs above that amount will have to come out of their share of the family assets.

This approach could even lead to a matrix of predictable or benchmark costs to be charged against the matrimonial assets for each stage of the ancillary relief proceedings. The amount could be geared to the value of the matrimonial assets and the complexity of the issues, leaving each party free to incur additional solicitor and client costs out of their own share of the assets if they so choose.

Until the Court of Appeal declares the red corner or the blue corner or some half-way house to be the winner, the 'no order for costs' principle should be treated with great caution.

2. *Misconduct in the proceedings*

[39.30]

The concept of litigation conduct, or more particularly, misconduct, is common to all family proceedings, whether financial remedy proceedings to which FPR 28.3 applies, or all other applications to which the relevant provisions of CPR apply.

Notwithstanding the general rule that there will be no order for costs in financial remedy proceedings, FPR 28.3(6) provides that the court may make such an order if it considers the conduct of a party in relation to the proceedings (litigation, not matrimonial, misconduct), both before (for example failure to comply with the pre-action protocol) and during them, makes it appropriate to do so. FPR 28.3(7) then lists the matters to which the court must have regard in deciding whether to make an order (para [39.2]).

In all the other family proceedings to which CPR apply litigation conduct is one of the circumstances of the case to be taken into account in the exercise of the court's discretion as to costs.

The relevant rules spell out the sort of conduct that might justify a costs order under either regime and there are some reported cases that assist.

RH v RH (Costs) (Adjustment for Gross Disparity) [2008] EWHC 347 (Fam), [2008] All ER (D) 67 (Apr) was a rare report of a decision to make a between-the-parties costs order based on litigation misconduct.

In protracted ancillary relief proceedings brought by the husband, the wife had argued that her £1.55m inheritance should be disregarded from the calculation, while the husband sought the equalisation of their capital situation. The husband further contended that in light of the wife's conduct in attempting to conceal the ownership of several properties he was entitled to more than half.

The husband formulated an offer whereby settlement would be reached through a property exchange in his favour, or a £750,000 payment being made to him. The wife neither accepted nor rejected the husband's offer. The court rejected the respective arguments of the wife in relation to the disregarding of the inheritance, and of the husband in respect of entitlement to more than a half share. In awarding the husband £670,000, the court had had regard to the considerable disparity in the costs that each side had run up. At the date of the award the wife's costs were £265,000 and the husband's £486,000. Accordingly, the court adjusted the assets subject to division by, in effect, writing back £225,000 of the husband's costs. The husband appealed and contended that the judge erred in discounting £225,000 and that the award payable should have been £782,500 to achieve parity. On appeal the court ruled that the husband's offer represented a genuine offer that the wife should have accepted. However, the husband had been unreasonable to expend so much more than the wife in litigating the case. As such, the husband should not recoup more than that which was reasonable for the wife to pay and the wife should pay the husband's costs in the fixed sum of £150,000.

In *M v M* [2009] EWHC 1941 (Fam), [2010] 1 FLR 256, [2009] Fam Law 1029 a complete change of tack by the wife only weeks before the trial necessitated the whole focus of the husband's trial preparation to shift, only to shift again a matter of days before the trial when the wife abandoned her attempt to have the shares in the husband's company transferred to her and instead claimed periodical payments. In the witness box the wife admitted that the inflated budget annexed to her Form E was devised 'in revenge' and her expert's report was disclosed only on the first day of the hearing. She was ordered to pay £175,000 to the husband which represented about 20% of his costs.

In *GS v L (No 2)(Costs)* [2013] 1 FLR 407 Eleanor King J suggested that the starting point for an assessment of conduct must be the overriding objective contained in FPR 2010, Part 1.

THE OVERRIDING OBJECTIVE

[39.31]

Part 1 of FPR 2010 sets out the overriding objective that the court is expected to further, with the help of the parties. The overriding objective to deal with a case justly must be applied by the court whenever it exercises any power given to it by the rules. This includes the power to make orders for costs, whether that power is exercised under FPR 28.2 or FPR 28.3.

Accordingly, a party's conduct in the litigation must be viewed in the context of what is expected from the parties by way of assistance in the furthering of the overriding objective to deal with cases justly.

This is what the rules say:

'Rule 1.1(2) Dealing with a case justly includes, so far as is practicable –

(a) ensuring that it is dealt with expeditiously and fairly;
(b) dealing with the case in ways which are proportionate to the nature, importance and complexity of the issues
(c) ensuring that the parties are on an equal footing;
(d) saving expense; and
(e) allotting to it an appropriate share of the court's resources, while taking into account the need to allot resources to other cases.'

In *GS v L* Eleanor King J made an issues-based costs order against the husband who doggedly pursued his erroneous contention that the proceedings raised issues of Spanish law and should therefore be heard in Spain. As a result, a case that could have been dealt with in two days instead became protracted and unduly complicated, all this incurring disproportionate costs. The judge ordered the husband to pay a fixed sum towards the wife's costs, roughly equating to one third.

(ii) The matrimonial pot

[39.32]

A fundamental difference between civil and matrimonial litigation, is that in civil litigation no order for costs means each party must pay its own costs out of its own resources, but in matrimonial litigation there are no outside resources and the costs of both parties must come out of the family assets. Abolishing between-the-parties costs orders does not alter the basic fact that all the costs must come out of the matrimonial pot, nor does it answer the basic question of how the costs are to be allocated fairly between the parties. An immediate consequence of the abolition of costs orders will be that many aspects of costs which have hitherto been dealt with on a detailed assessment by a costs judge must now be considered by the judge conducting the final hearing before he or she can make an order. Two aspects of costs in particular must be resolved before a partisan order can be made: quantum and penalties for litigation misconduct.

(iii) Quantification

[39.33]

The judge at any hearing of family proceedings has to apply the relevant criteria in exercising the discretion given to make between the parties costs orders.

Litigation conduct is relevant to all family proceedings. If it is established, one approach is for the court to quantify the costs of both parties arising out of the misconduct, and then order that the total of those costs be paid out of the share of the offending party, otherwise the sanction has no teeth.

The judge must also decide whether a costs order should be on the standard or the indemnity basis, there are different kinds and levels of what may be called litigation misconduct in family matters, not all of them justifying the award of indemnity costs.

And because the costs have to be quantified before the family assets can be apportioned, this will again involve the family judge in a task which hitherto has been undertaken by a costs judge on a detailed assessment.

(iv) **Compensation or sanction?**

[39.34]

A second area of difficulty arises out of the dual role of costs as both compensation and as a sanction for litigation misconduct.

A fee shifting order in a big money case where the family assets exceed the aggregate needs of the family creates no problem, but what about the great majority of cases where there is simply not enough money to go round to start with? Take Cardinal and Middleton's example of a wife who needs £150,000 to re-house herself and the children but has £20,000 solicitor and client costs. What if the court finds she has been guilty of litigation misconduct causing the husband to incur £5,000 of costs? If you do not increase her award by £20,000 to pay her solicitor and client costs and you also order her to pay the husband £5,000 between-the-parties costs, she is £25,000 short of her needs. Where does that leave her and the children? Out in the street?

The dilemma arises whenever a between-the-parties costs order for litigation misconduct would result in the needs of one of the parties not being met. A fee-shifting order could in most cases upset the delicate balance between meeting the needs of both parties and meeting the needs of only one of them.

Which is to take priority – the *needs* of a party or *punishing* them for litigation misconduct? FPR 28.3(7)(f) requires the court to have regard to the financial effect on the parties of making a costs order. If a party's and the children's needs are in any event going to override the making of a between-the-parties costs order then what is the point of investigating misconduct in the proceedings in the first place?

(v) **Wasted costs**

[39.35]

There is one way for an order for costs based on litigation misconduct not to be met out of the matrimonial pot – that is an order that the solicitors pay them personally. CPR 44.14 which enables the court to order the legal representative of one party to pay the costs of the other for misconduct has not been disapplied from ancillary relief proceedings nor have the wasted costs provisions of s 51(7) of Senior Courts Act 1981. I have a fantasy of both parties getting together and agreeing each to claim wasted costs against the other's solicitors in order to preserve the matrimonial pot!

In *Ezair v Ezair* (para **[39.00]**) the Court of Appeal held that it was inappropriate to inflate a lump sum award in financial remedy proceedings in order to compensate for wasted costs. This may have been one of those unusual ancillary relief cases which justified an order for costs against a litigant in person.

In *Fisher Meredith v JH and PH (financial remedy: appeal: wasted costs)* [2012] EWHC 408 (Fam), [2012] 2 FCR 241, [2012] 2 FLR 536 the wife's solicitors applied to adjourn to allow third parties to be joined to the proceedings where the husband alleged that family members were in fact the beneficial owners of shares in a company held in his name.

The district judge allowed the adjournment but ordered the wife to pay wasted costs and her solicitors to show cause why they should not be personally liable for these.

At the wasted costs hearing the district judge ordered the wife's solicitors to pay the costs, stating that insufficient thought had been given as to the enforceability of any orders against third parties.

On appeal the judge distinguished between a case where a wife states that property or shares in the name of the third party belong to the husband or where (as here) it was alleged that shares in which the husband clearly had legal title, belonged to a third party.

In the former case the obligation to join the third party would be on the claimant. In the latter case, the duty to clarify the position for the court, by joining a third party, was not clear cut at all. Indeed, there might be good reasons why the husband would join the third parties in order to reinforce his position. For those reasons the wasted costs order was set aside.

Writing in the New Law Journal, Kim Beatson and Lehna Hewitt observed that all the recent cases indicate that an order for wasted costs is an order of last resort. It is likely to involve another full and extensive hearing where independent legal advice will be necessary for the legal advisers defending the application and it is likely to be inappropriate for the case to be heard until conclusion of the trial in order that the conduct can be fully assessed and usually by the same trial judge.

BEING CREATIVE WITH COSTS ORDERS

[39.36]

Whenever the court makes an order for costs, whether under FPR 28.3 in financial remedy proceedings, or under the CPR in any other family proceedings CPR 44.3(6) to (9) apply. These are as follows:

'CPR 44.3

(6) The orders which the court may make under this rule include an order that a party must pay –
- (a) a proportion of another party's costs;
- (b) a stated amount in respect of another party's costs;
- (c) costs from or until a certain date only;
- (d) costs incurred before proceedings have begun;
- (e) costs relating to particular steps taken in the proceedings;
- (f) costs relating only to a distinct part of the proceedings; and
- (g) interest on costs from or until a certain date, including a date before judgment.

(7) Where the court would otherwise consider making an order under paragraph (6)(f), it must instead, if practicable, make an order under paragraph (6)(a) or (c).

(8) Where the court has ordered a party to pay costs, it may order an amount to be paid on account before the costs are assessed.

(9) Where a party entitled to costs is also liable to pay costs the court may assess the costs which that party is liable to pay and either –

(a) set off the amount assessed against the amount the party is entitled to be paid and direct him to pay any balance; or

(b) delay the issue of a certificate for the costs to which the party is entitled until he has paid the amount which he is liable to pay.'

These provisions enable the court to be creative when making a costs order in order to better reflect the outcome of the case. So for example, one party might be ordered to pay the other's costs only in relation to a particular issue, or the costs from, or up to, a certain date.

COSTS INFORMATION

[39.37]

In the light of the requirement upon the court imposed by FPR 28.3((7)(f) to consider the effect of any costs order on the parties, and the powers that the court has under CPR 44.3(6) to make orders for any part of the costs to be paid, it is important that it is provided with the necessary material with which to work.

Practitioners are accustomed to producing Forms H and H1 with details of their costs to the date of any hearing. This requirement, introduced by the President's Practice Direction of 20 February 2006, is retained and can be found in FPR 2010, Part 9, that part of the FPR 2010 that deals with all applications for a financial remedy:

'**FPR 9.27: Estimates of costs**

(1) Subject to paragraph (2), at every hearing or appointment each party must produce to the court an estimate of the costs incurred by that party up to the date of that hearing or appointment.'

This is to be in Form H.

There are special provisions in relation to the final hearing of an application for a financial remedy:

'(2) Not less than 14 days before the date fixed for the final hearing of an application for a financial remedy, each party ('the filing party') must (unless the court directs otherwise) file with the court and serve on each other party a statement giving full particulars of all costs in respect of the proceedings which the filing party has incurred or expects to incur, to enable the court to take account of the parties' liabilities for costs when deciding what order (if any) to make for a financial remedy.'

The form prescribed for the purpose is Form H1 and it is a detailed and accurate statement of costs rather than an estimate.

In financial remedy proceedings that are governed by FPR 28.3, PD 28A, para 4.5 provides that where it is intended to ask the court to exercise its discretion to make an order for costs because of the litigation conduct of the other party, this should ordinarily be made plain either in open correspondence or a skeleton argument before the hearing.

In any case in which it is likely that the court will wish to make a summary assessment of costs in the event of a successful application, a statement of costs in CPR Form N260 must be filed.

FUNDING FAMILY COSTS

[39.38]

In these days when fairness is equated to equality in family finance, fairness must surely extend to equality of arms in financing legal advice and representation. It is a simple question of cash-flow. If all the liquid assets are under the control of one of the parties, funds should be made equally available to both parties for paying the lawyers during the proceedings, whatever the final distribution of the family assets may be.

APPLICATIONS IN MATRIMONIAL AND CIVIL PARTNERSHIP PROCEEDINGS

[39.39]

The enactment of The Legal Aid, Sentencing and Punishment of Offenders Act 2012 with effect from 1 April 2013 is designed to address the requirements of the overriding objective for the parties to be on an equal footing, and the Article 6 ECHR right to a fair trial. An amendment to the Matrimonial Causes Act 1973 in the form of new sections 22ZA and 22ZB introduces a distinct form of order known as a legal services order. This replaces the allowance for costs that for well over a decade judges have been making as part of an order for maintenance pending suit.

Here are the new provisions:

MATRIMONIAL CAUSES ACT 1973

22ZA Orders for payment in respect of legal services

(1) In proceedings for divorce, nullity or marriage or judicial separation, the court may make an order or orders requiring one party to the marriage to pay to the other ("the applicant") an amount for the purpose of enabling that applicant to obtain legal services for the purposes of the proceedings.

(2) The court may also make such an order or orders in proceedings under this Part for financial relief in connection with proceedings for divorce, nullity of marriage or judicial separation.

(3) The court must not make an order under this section unless it is satisfied that, without the amount, the applicant would not reasonably be able to obtain appropriate legal services for the purposes of the proceedings or any part of the proceedings.

(4) For the purposes of subsection (3), the court must be satisfied in particular, that-

(a) the applicant is not reasonably able to secure a loan to pay for the services, and

(b) the applicant is unlikely to be able to obtain the services by granting a charge over assets recovered in the proceedings.

(5) An order under this section may be made for the purpose of enabling the applicant to obtain legal services of a specified description, including legal services provided in a specified period or for a specified part of the proceedings.
(6) An order under this section may-
(a) provide for payment of all or part of the amount by instalments of specified amounts, and
(b) required the instalments to be secured to the satisfaction of the court.
(7) An order under this section may direct that payment of all or part of the amount is to be deferred.
(8) The court may at any time in the proceedings vary an order made under this section if it considers that there has been a material change of circumstances since the order was made.
(9) For the purposes of the assessment of costs in the proceedings, the applicant's costs are to be treated as reduced by any amount paid to the applicant pursuant to an order under this section for the purposes of those proceedings.
(10) In this section "legal services", in relation to proceedings, means the following types of services-
(a) providing advice as to how the law applies in the particular circumstances,
(b) providing advice and assistance in relation to the proceedings,
(c) providing other advice and assistance in relation to the settlement or other resolution of the dispute that is the subject of the proceedings, and
(d) providing advice and assistance in relation to the enforcement of decisions in the proceedings or as part of the settlement or resolution of the dispute, and they include, in particular, advice and assistance in the form of representation and any form of dispute resolution, including mediation.
In subsections (5) and (6) "specified" means specified in the order concerned.'

The new provisions apply only to proceedings for decrees under Matrimonial Causes Act 1973 and dissolution under Civil Partnership Act 2004, together with financial relief under both statutes. To a large extent the new provisions do no more than reflect the approach taken by judges since Holman J started the ball rolling in *A V A (maintenance pending suit: provision for legal fees)* [2001] 1 FLR 377. In particular, s 22ZA(4)(a) and (b) require the applicant to satisfy the court that he or she is not reasonably able either to secure a loan or grant a charge over any assets recovered, in order to fund the proceedings. In *Currey v Currey (No 2)* [2006] EWCA Civ 1338 Wilson LJ, as he then was, regarded these criteria as central to the exercise of the court's discretion whether to make a costs allowance.

As to the evidence that the applicant would be required to adduce in order to discharge the burden of proving a negative, it is likely that this will be no different from that suggested by the judge in *TL v ML (ancillary relief: claim against assets of extended family)* [2005] EWHC 2860 (Fam), [2006] 1 FCR 465, [2006] 1 FLR 1263. This was that the applicant should provide correspondence with at least two banks declining the request for a loan, and a letter from her solicitors stating that they were not prepared to enter into a *Sears Tooth* agreement.

If the court is not satisfied as to either of these matters it must not make an order. If it is satisfied, before making an order the court must exercise its discretion in accordance with the criteria set out at s 22ZB:

'**22ZB Matters to which the court is to have regard in deciding how to exercise power under section 22ZA**

(1) When considering whether to make or vary an order under section 22ZA, the

court must have regard to-
(a) the income, earning capacity, property and other financial resources which each of the applicant and the paying party has or is likely to have in the foreseeable future,
(b) the financial needs, obligations and responsibilities which each of the applicant and the paying party has or is likely to have in the foreseeable future,
(c) the subject matter of the proceedings, including the matters in issue in them,
(d) whether the paying party is legally represented in the proceedings,
(e) any step taken by the applicant to avoid all or part of the proceedings, whether by proposing or considering mediation or otherwise,
(f) the applicant's conduct in relation to the proceedings,
(g) any amount owed by the applicant to the paying party in respect of costs in the proceedings or other proceedings to which the applicant and the paying party are or were party, and
(h) the effect of the order or variation on the paying party.
(2) In subsection (1)(a) "earning capacity", in relation to the applicant or the paying party, includes any increase in earning capacity which, in the opinion of the court, it would be reasonable to expect the applicant or the paying party to take steps to acquire.
(3) For the purposes of subsection (1)(h), the court must have regard, in particular, to whether the making or variation of the order is likely to-
(a) cause undue hardship to the paying party, or
(b) prevent the paying party from obtaining legal services for the purposes of the proceedings.
. . .
(7) In this section "legal services" has the same meaning as in section 22ZA.'

If the court exercises its discretion in favour of the applicant, the legal services order may take the form of payment by instalments (which may be secured), lump sum, or both, and may be limited either as to time, as to scope, or both. The payments may be deferred in whole or in part. The court has the power to vary the order in the light of any material change of circumstances on either side.

The good news for the paying party is that any sums paid to the applicant under a legal services order will stand to his or her credit against any costs that are assessed to be paid to the applicant in the proceedings.

APPLICATIONS UNDER SCHEDULE 1 TO THE CHILDREN ACT 1989

[39.40]

Although The Legal Aid, Sentencing and Punishment of Offenders Act 2012 fails to make statutory provision for legal services orders in cases of this sort, the precedent has been set for costs allowances to be made in orders for interim provision under Schedule 1.

In *M-T v T* [2006] EWHC 2494 (Fam), [2007] 2 FLR 925, [2007] Fam Law 1066 Charles J made an allowance for legal costs in an application for interim periodical payments. This was on the basis that since the application was brought by the mother effectively on behalf of, and for the benefit of the child, she was acting in a quasi-fiduciary or representative capacity and should not have to bear the costs personally.

Although a principle has therefore been established that the court has power under Schedule 1 to make an allowance for costs by way of an interim periodical payments or lump sum order, in exercising its discretion the court will always be mindful that the pursuit of the substantive claim must be for the benefit of the child and not to satisfy the applicant's desire to litigate.

In *G v G* [2009] EWHC 2080 Moylan J made an award of interim periodical payments of £40,000 as a contribution to the costs of the applicant mother. The court held that there was jurisdiction to make such an order under Schedule 1 if it was considered appropriate to do so for the benefit of the children. In that case Moylan J took the view that it would be of benefit to the children if the mother were properly represented. It was relevant to his decision that there was a significant financial imbalance between the parents, the applicant was conducting the litigation in an entirely responsible way, and the respondent could well afford to pay.

SOLICITOR AND CLIENT

[39.41]

The Conditional Fee Agreement Rules and Regulations (see Section VII) precludes solicitors entering into conditional fee agreements in respect of matrimonial matters. A SCCO costs judge was in error when he found that a client care letter in matrimonial proceedings sent to the wife by her solicitor containing the clause 'we have agreed that a claim for costs will not be made until money is received at the end of the case' was a conditional fee agreement which did not comply with the regulations, and was therefore unenforceable. The Court, on appeal, was satisfied that the letter taken as a whole showed that the wife was to be liable for her own costs and the phrase meant that the solicitor agreed not to claim his costs until the conclusion of the case, a position wholly in line with the manner in which family proceedings were often undertaken (*Denton v Denton* [2004] EWHC 1308 (Fam), [2004] 2 FLR 594, [2005] Fam Law 353).

Similarly, an agreement by a wife to pay her solicitor's fees in divorce proceedings out of such sums as the court might award her in the proceedings was not champertous and invalid (*Sears Tooth (a firm) v Payne Hicks Beach (a firm)* [1997] 2 FLR 116).

CURBING COSTS

(a) Guidelines

[39.42]

There is nothing new under the sun.

Concern over increasingly heavy legal costs was expressed by Booth J as long ago as 1990 in *Evans v Evans* [1990] 2 All ER 147, [1990] 1 WLR 575n when she said that the court was at times unable to make appropriate provision orders for wives and children because of the liability for costs. She issued the following general guidelines to family law practitioners in the preparation of substantial ancillary relief cases:

'(1) Affidavit evidence should be confined to relevant facts and should not be prolix or diffuse. Each party should normally file one substantive affidavit dealing with matters to which the court should have regard under s 25 of the Matrimonial Causes Act 1973 [as substituted by s 3 of the Matrimonial and Family Proceedings Act 1984] and matters which are material to the application. If any further affidavit is necessary it should be confined to matters as answering a serious allegation made by the other party dealing with any serious allegation made by the other party dealing with any serious issue raised or setting out any material change of circumstances.

(2) Inquiries made under [rule 2.62 of the Family Proceedings Rules 1991] should, as far as possible, be contained in one comprehensive questionnaire and should not be made piecemeal at different times.

(3) Wherever possible valuations of properties should be obtained from a valuer jointly instructed by both parties. Where each party instructs a valuer then reports should be exchanged and the valuers should meet in an attempt to resolve any differences between them or otherwise to narrow the issues.

(4) While it may be necessary to obtain a broad assessment of the value of a shareholding in a private company it is inappropriate to undertake an expensive and meaningless exercise to achieve a precise valuation of a private company which will not be sold: see *P v P (financial provision)* [1989] 2 FLR 241.

(5) All professional witnesses should be careful to avoid a partisan approach and should maintain proper professional standards.

(6) Care should be taken in deciding what evidence, other than professional evidence, should be adduced and emotive issues which are not material should be avoided. Where affidavit evidence is filed the deponents must be available for cross-examination on notice from the other side.

(7) Solicitors on both sides should together prepare bundles of documents for use at the hearing and should reach agreement as to what should be included and what excluded; duplication of documents should always be avoided. [This is now covered by Practice Direction (family proceedings: court bundles) [2000] 2 All ER 287, Fam.]

(8) A chronology of material facts should be agreed and made available to the court.

(9) In a substantial case it might be desirable to have a pre-trial review to explore the possibility of settlement and to define the issues and to ensure readiness for hearing if a settlement cannot be reached.

(10) Solicitors and counsel should keep their clients informed of the costs at all stages of the proceedings and, where appropriate, should ensure that they understand the implications of the legal aid charge: the court will require an estimate of the approximate amount of the costs on each side before it can make a lump sum award (see Practice Direction [1982] 2 All ER 800).

(11) The desirability of reaching a settlement should be borne in mind throughout the proceedings. While it is necessary for the legal advisers to have sufficient knowledge of the financial situation of both parties before advising their client on a proposed settlement, the necessity to make further inquiries must always be balanced by a consideration of what they are realistically likely to achieve and the increased costs which are likely to be incurred by making them.'

(b) Practice Note: case management

[39.43]

On 31 January 1995 the President took matters further by issuing the *Practice Note: Case Management* [1995] 1 All ER 586, Fam. Here it is:

[39.44]

'(1) The importance of reducing the cost and delay of civil litigation makes it necessary for the court to assert greater control over the preparation for the conduct of hearings than has hitherto been customary. Failure by practitioners to conduct cases economically will be visited by appropriate orders for costs, including wasted costs orders.

(2) The court will accordingly exercise its discretion to limit—
(a) discovery;
(b) the length of opening and closing oral submissions;
(c) the time allowed for the examination and cross-examination of witnesses;
(d) the issues on which it wishes to be addressed;
(e) reading aloud from documents and authorities.

(3) Unless otherwise ordered, every witness statement or affidavit shall stand as the evidence in chief of the witness concerned. The substance of the evidence which a party intends to adduce at the hearing must be sufficiently detailed, but without prolixity; it must be confined to material matters of fact, not (except in the case of the evidence of professional witnesses) of opinion; and if hearsay evidence is to be adduced, the source of the information must be declared or good reason given for not doing so.

(4) It is a duty owed to the court both by the parties and by their legal representatives to give full and frank disclosure in ancillary relief applications and also in all matters in respect of children. The parties and their advisers must also use their best endeavours—
(a) to confine the issues and the evidence called to what is reasonably considered to be essential for the proper presentation of their case;
(b) to reduce or eliminate issues for expert evidence;
(c) in advance of the hearing to agree which are the issues or the main issues.

(5) Unless the nature of the hearing makes it necessary, and in the absence of specific directions, bundles should be agreed and prepared for use by the court, the parties and the witnesses, and shall be in A4 format where possible, suitably secured. The bundles for use by the court shall be lodged with the court (the Clerk of the Rules in matters in the RCJ in London) at least two clear days before the hearing. Each bundle should be paginated, indexed, wholly legible and arranged chronologically. Where documents are copied unnecessarily or bundled incompetently, the cost will be disallowed.

(6) In cases estimated to last for five days or more and in which no pre-trial review has been ordered, application should be made for a pre-trial review. It should when practicable be listed at least three weeks before the hearing and be conducted by the judge or district judge before whom the case is to be heard, and should be attended by the advocates who are to represent the parties at the hearing. Whenever possible, all statements or evidence and all reports should be filed before the date of the review and in good time for them to have been considered by all parties.

(7) Whenever practicable, and in any matter estimated to last five days or more, each party should, not less than two clear days before the hearing, lodge with the court, or the Clerk of the Rules in matters in the RCJ in London, and deliver to other parties, a chronology and a skeleton argument concisely summarising that party's submissions in relation to each of the issues, and citing the main authorities relied upon. It is important that skeleton arguments should be brief.

(8) In advance of the hearing upon request, and otherwise in the course of their opening, parties should be prepared to furnish the court, if there is no core bundle, with a list of documents essential for a proper understanding of the case.
(9) The opening speech should be succinct. At its conclusion other parties may be invited briefly to amplify their skeleton arguments. In a heavy case the court may in conjunction with final speeches require written submissions, including the findings of fact for which each party contends.
(10) This Practice Direction, which follows the directions handed down by the Lord Chief Justice and the Vice-Chancellor to apply in the Queen's Bench and Chancery Divisions, shall apply to all family proceedings in the High Court and in all Care Centres, Family Hearing Centres and divorce county courts.
(11) Issued with the concurrence of the Lord Chancellor.'

Since then the court's case management powers have been further sharpened and the Overriding Objective rules. There are pre-action protocols. Parties are expected to attend ADR and the court is required at every stage to consider whether a case should be adjourned for ADR to take place.

All this and more is set out in the FPR 2010.

SUMMARY ASSESSMENT

(a) No presumption of detailed assessment

[39.45]

The general rule in para 13.2 of the Practice Direction in Part 44 of the CPR did not imply a general rule requiring detailed assessment of costs in hearings lasting longer than a day. On the contrary, the exercise of the power to make a summary assessment must be considered in every case. In family cases in particular, the aggravation of detailed assessment ran counter to the satisfactory resolution of disputes (*Q v Q (Family Division costs: summary assessment)* [2002] 2 FLR 668, Fam).

(b) Discrete issues

[39.46]

Because so many aspects of family matters, such as the divorce, children and ancillary relief, are dealt with at the same time, often in the same attendance, telephone call or letter, it is particularly important to be able to distinguish between them for the purposes of the summary assessment of the costs of an interim application. Otherwise, there is the risk that the statement of costs may include either more or less than the work done for that particular hearing. In substantial matters you should therefore consider having separate folders or dividers within the file, or even separate files, relating to each aspect, or to a specific hearing such as an application for maintenance pending suit. Where appropriate, separate letters should be written on each aspect of the matter, with the time spent on attendances and telephone calls being similarly apportioned. By these means you will be able to satisfy the judge that you are not over-charging – and yourself that you are charging enough.

ENFORCEMENT

(a) **Statutory demand**

[39.47]

Rule 12.3 of the Insolvency Rules 1986 provides:

'(1) Subject as follows, in bankruptcy . . . all claims by creditors are provable as debts against . . . the bankrupt . . . The following are not provable – (a) in bankruptcy, any fine imposed for an offence and any obligation arising under an order made in family proceedings or under a maintenance assessment made under the Child Support Act 1991 . . . '

On this basis, Mr Levy, who was a bankrupt, applied to set aside a statutory demand served by the Legal Aid Board, now the Legal Services Commission, claiming £62,732 costs incurred by his former wife in her family proceedings. Although there was jurisdiction to make a bankruptcy order on a petition based upon a non-provable debt, the jurisdiction was discretionary and would not be exercised except in special circumstances. It was puzzling that s 382(1) contemplated that a non-provable debt might be regarded as a bankruptcy debt, but it was difficult to conceive of any circumstances where that would occur (*Levy v Legal Services Commission* [2001] 1 All ER 895, CA, [2001] 1 FCR 178).

(b) **Bankruptcy**

[39.48]

The Insolvency (Amendment) Rules 2005 which came into force on 1 April 2005 provided that lump sum orders and costs orders made in family proceedings are now provable in bankruptcy in respect of bankruptcy orders made on or after that date. Arrears of periodical payments and under child support assessments are still not provable.

(c) **Judgment summons**

[39.49]

Schedule 8 to the Administration of Justice Act 1990, which prescribes the matrimonial orders enforceable by judgment summons, does not specifically mention orders for costs, but rule 7.4 of the Family Proceedings Rules 1991 contemplates enforcement of costs orders by judgment summons by providing that on the hearing of a judgment summons the judge may make a new order for payment of the amount due where the original order is for a lump sum provision or costs.

However, in *Mubarak v Mubarak* [2001] 1 FLR 673, [2000] All ER (D) 1797, CA, the Court of Appeal held that the existing judgment summons procedure under s 5 of the Debtors Act 1869 was not human rights compliant. Judgment summons proceedings for contempt are classified as criminal proceedings and the requirement that the debtor shall attend before the court to show cause why he should not be sent to prison amounts to an unacceptable reversal of the burden of proof and removes the protection against self-incrimination. Accordingly rule 7.4(1) was amended so that instead of the judgment debtor being ordered 'to appear and be examined on oath as to his means' he is now simply ordered 'to attend court'. A new rule 7.4(3A) requires the judgment creditor to file with the request, copies of all written evidence on which he (in practice, invariably 'she') intends to rely. A new rule 7.4(7B) provides:

> 'No person may be committed on an application for a judgment summons unless—(a) where the proceedings are in the High Court, the court has summoned the debtor to attend, he has failed to do so, and he has also failed to attend the adjourned hearing; (b) where the proceedings are in the county court, an order is made under section 110(2) of the County Courts Act 1984; or (c) the judgment creditor proves that the debtor— (i) has or has had since the date of the order the means to pay the sum in respect of which he is made default; and (ii) has refused or neglected, or refuses or neglects, to pay that sum.'

And for the avoidance of doubt Rule 7.4(7C) now provides that the debtor may not be compelled to give evidence.

Acknowledgment

The section dealing with Applications under Schedule 1 to the Children Act 1989 under the main heading Funding Family Costs at para [39.40] is reproduced from *Cohabitation Law Practice and Precedents* 5th edition Wood *et al* by kind permission of Jordan Publishing Limited.

CHAPTER 40

COSTS AGAINST NON-PARTIES

INTRODUCTION

[40.1]

The position in respect of costs orders against non-parties is set out at CPR 46.2. It is a rule without an accompanying practice direction. It sets out a reminder that the relevant jurisdiction under which the court can make such an order is s 51 of the Senior Courts Act 1981 and prescribes the procedure that must be adopted. In essence whenever the court is considering making such an order it must join the person against whom such an order is contemplated as a party and provide an opportunity to that person to attend court when the position will be considered. The rule does not provide any guidance on how the court might interpret the statutory provision which is itself broad – the court has the power to determine by whom and to whom costs are to be paid and to what degree. Further guidance has come from the case law that has emerged on the subject.

THE GENERAL APPROACH

[40.2]

The seminal case on non-party costs is *Symphony Group plc v Hodgson* [1993] 4 All ER 143, CA. The claimant obtained injunctive relief against the defendant who had left its employment to join Halvanto, a competitor, in breach of restrictive covenants in his contract of employment. The claimant neither added Halvanto as a defendant nor initiated proceedings against it nor told it that it might seek to make it liable for costs. Halvanto's managing director gave evidence for the defendant. The defendant was protected against a cost liability to the claimant by a legal aid certificate. As a result the aggrieved claimant sought and successfully obtained an order that Halvanto should pay its costs of the action. In allowing Halvanto's appeal the court said that since *Aiden Shipping Co Ltd v Interbulk, The Vimeira (No 2)* [1986] AC 965, [1986] 2 All ER 409, HL the courts had entertained claims for costs against non-parties where a person who was not a party:
- had some management of or financed an action;
- had caused the action;
- was a party to a closely related action which had been heard at the same time but not consolidated; or
- was involved in group litigation.

The court identified the following material considerations to be taken into account, which remain the guiding principles when considering costs orders against non-parties:

(a) An order for the payment of costs by a non-party will always be exceptional. The judge should treat any application for such an order with considerable caution.
(b) It will be even more exceptional for such an order to be made where the applicant has a cause of action against the non-party and could have joined him to the substantive claim. Joinder as a party gives the person concerned all the protection conferred by the rules, eg the framing of issues, involvement in the disclosure process and the knowledge of what the issues are before giving evidence.
(c) Even if the applicant can provide a good reason for not joining the non-party against whom he has a valid cause of action, he should warn the non-party that he might apply for costs against him and should do so at the earliest opportunity.
(d) An application should normally be determined by the trial judge.
(e) The fact that the trial judge might have expressed views on the conduct of the non-party neither constitutes bias nor the appearance of bias (and is not, therefore, a basis for recusal of that judge).
(f) The procedure for the determination of costs is a summary procedure, not necessarily subject to all the rules that would apply in an action. Thus, subject to any statutory exceptions, judicial findings are inadmissible as evidence of the facts on which they are based in proceedings between one of the parties to the original proceedings and a stranger. However, in the summary procedure for a solicitor to pay the costs of an action to which he was not a party, the judge's findings of fact might be admissible. This departure from basic principles could only be justified if the connection of the non-party with the original proceedings was so close that he would not suffer any injustice by allowing the exception to the general rule.
(g) The normal rule is that witnesses in either civil or criminal proceedings enjoy immunity from any form of civil action in respect of evidence given during the proceedings. In so far as the evidence of a witness in proceedings might lead to an application for costs against him or his company, it introduces an exception to a valuable general principle.
(h) The fact that an employee or even a director or managing director of a company gives evidence in an action does not normally mean that the company is taking part in that action.
(i) The judge should be alert to the possibility that an application for costs against a non-party is motivated by resentment or an inability to obtain an effective order for costs against a state-funded litigant.

THE APPLICATION PROCEDURE

[40.3]

An application for a non-party costs order should be made to the trial judge, even if he has expressed a view about the conduct of the non-party. It is not wrong for a judge to mention the possibility of an application for a non-party costs order at the outset of the trial (*Equitas Ltd v Horace Holman & Co Ltd* [2008] EWHC 2287 (Comm)).

An application for an order to join a party for the purpose of seeking an order for costs against him would normally be expected to explain the nature of the claim against the intended party and the purpose to be served by joining that party. If it is clear that a joinder of an intended party is an abuse of process, then the court will dismiss the application. In *PR Records Ltd v Vinyl 2000 Ltd* [2007] EWHC 1721 (Ch), it was held that a Master had been wrong to refuse to join the second defendant as a party to the proceedings and not to permit the matter to proceed to the second stage hearing envisaged under CPR 46.2(1)(b). The application had not involved an abuse of process of the court. At the stage of considering joinder, it is not appropriate to attempt a preliminary assessment of the merits in order to see whether an application for a non-party cost order has a real prospect of success. The right order in this case should have been to permit the joinder and to allow the matter to proceed to the second stage hearing.

REQUIREMENTS FOR NOTICE OF AN APPLICATION

[40.4]

A relevant consideration in the exercise of the court's discretion whether to make an order for costs against a non-party is whether the non-party has received notice, before or during the litigation, that he may be made subject to an order for costs. Although CPR 46.2 contains no requirement for notice to be given, the court has repeatedly taken account of whether notice has been given and when. It is a denial of the fundamental right of non-parties to be heard on serious allegations if the court was to decide them at a hearing of the sort deliberately intended to be of a summary nature without the court embarking on an enquiry of the type that justice would require without adequate notice (*Barndeal Ltd v Richmond-upon-Thames London Borough Council* [2005] EWHC 1377 (QB)). This point was re-iterated in *Brampton Manor (Leisure) v McClean* [2007] EWHC 3340 (Ch) where the court said that the non-party must have notice and a minimum period of time to prepare and reflect upon his position before the court determination.

In *Oriakhel v Vickers* [2008] EWCA Civ 748, the second defendant ('G') alleged that the claimant ('O') was involved in a fraudulent conspiracy with the first defendant ('V') under which O brought a road traffic accident claim for damages against V. V's insurers contended that the claim was bogus and were joined as second defendant by G to contest the claim and to counterclaim. The judge dismissed the claim as fraudulent. At the trial a Mr Khan ('K') gave evidence which the judge disbelieved concluding that he too was involved in the conspiracy. G then joined K as a party for the purpose of seeking a costs order against him, which the judge refused to make under the mistaken belief that it had to be shown that the non-party was a funder or controller of the litigation before a non-party costs order could be made. Nevertheless on appeal the Court of Appeal refused to make a non-party costs order against K on the grounds it would be exceptional for an order to be made against a non-party where the applicant had a separate cause of action against the non-party and could have joined him as a party to the original proceedings. Prior to the trial G had contended that K was a dishonest conspirator and

40.4 *Chapter 40 Costs against Non-Parties*

could have joined him at that stage. If he had been made a defendant to the counterclaim, he would have had a full opportunity of taking legal advice, adducing such evidence and documents as might support his defence and considering his own position in the litigation. K had not been given notice of the claim at any time when he could have taken legal advice and, if necessary, deployed further material by way of his defence and, if so advised, applied to be joined as a party. Where a non-party has, in effect, controlled the primary litigation it would be bound by the result. But K did not have such a close connection with the primary claim that he was bound by the result. Witness immunity was yet another reason why the court should not exercise its discretion to award costs against K. In any event G remained free to sue K for his part in the dishonest conspiracy and if it succeeded, the costs of successfully defending the primary claim would be recoverable damages flowing from the conspiracy.

In *Europeans Ltd v Revenue and Customs Comrs* [2011] EWHC 948 (Ch), [2011] STC 1449, [2011] 4 Costs LO 447, the Revenue and Customs Commissioners applied for a non-party costs order against the respondent, who had been the Managing Director of, and the sole shareholder in, Europeans Ltd. In September 2008 Europeans' appeal to the VAT and Duties Tribunal was dismissed. At a hearing in May 2009, the judge dismissed an appeal and ordered Europeans to pay the commissioners' costs. In November 2010, the commissioners made the non-party application, seeking the recovery of those costs from the respondent, Europeans having entered voluntary liquidation in May 2009. The respondent argued that notice of the commissioners' intention to seek an order against him personally had been given too late. The first time he was told of an intention to pursue him personally was in July 2009, over a month after the appeal had been dismissed and Europeans had entered into voluntary liquidation. The delay was compounded by the fact that this application was then not made for a further 16 months. Such delay seriously weighed in the respondent's favour. However, the failure to give an early warning was not a stand alone requirement which would operate conclusively against the applicant; it was no more than a material consideration, albeit a highly material one. It was hard to see what the respondent would have done differently given that the appeal was withdrawn at the last minute and on the face of it had little or no merit in any event. Further, the underlying finding of the tribunal was of fraud. In all the circumstances, the absence of an early warning was insufficient to deny the commissioners a non-party costs order to which they were plainly otherwise entitled.

THE APPROACH OF THE COURT TO MANAGING AN APPLICATION

[40.5]

Whether such jurisdiction should be exercised is, of course, another matter entirely and the extent to which a respondent has, in fact, funded any proceedings may be very relevant to the exercise of discretion. There is a danger that the exercise of the jurisdiction to order a non-party to proceedings to pay the cost of those proceedings becomes over-complicated by reference to authority. In addition the parties must be astute to the proportionality issues that such an application raises.

In *Robertson Research International Ltd v ABG Exploration BV* (1999) Times, 3 November, QBD (Pat Ct), after directions for trial had been given in a robustly defended action, the defendant company gave notice that it was ceasing to trade, its solicitors came off the record and it consented to judgment for the entire sum claimed. By then it was an empty shell. The claimant joined as parties a director and the financial controller of the defendant company alleging that they had known there was no defence and seeking an order for costs against them. The defendants argued that in view of the strictures in wasted costs matters that such applications will not be heard unless they can be dealt with summarily, the present applications should be dismissed. The master accepted these submissions, but on appeal, Laddie J held that unlike the wasted costs jurisdiction, applications against non-parties were solely compensatory and it was irrelevant whether there was any impropriety on the part of those responsible. Although such orders would always be exceptional, it was not necessary that the applications had to be capable of being dealt with summarily. The matter could be dealt with appropriately by, for example, limiting it to principle issues, limiting or dispensing with cross examination and limiting the length of the hearing.

However, what does seem clear is that the procedure should not be adopted where the sums involved are relatively small (see the postscript on disproportionate costs in *Sims v Hawkins* [2007] EWCA Civ 1175). This consideration has become even more important after the CPR amendments of April 2013. The decision to join a party and embark upon a non-party costs application is a case management one to which the amended overriding objective applies. If the preliminary determination is that the exercise is disproportionate by reference to the definition at CPR 44.3(5) then the court is unlikely to permit joinder and move to the second stage of CPR 46.2.

IS IT A PREREQUISITE THAT THE NON-PARTY HAS PROVIDED FUNDING FOR THE CLAIM?

[40.6]

Although the non-party has provided funding in most of the reported cases, it is not, essential, in the sense of being a jurisdictional pre-requisite to the exercise of the court's discretion. If the evidence is that a respondent to the application (whether director or shareholder or controller of a relevant company) has effectively controlled the proceedings and has sought to derive potential benefit from them, that will be enough to establish the jurisdiction.

Whilst the absence of provision of funding is not a bar to an application, the mere fact that the non-party has provided funding also does not, itself, justify an order. In *Petromec Inc v Petroleo Brasileiro SA Petrobas* [2006] EWCA Civ 1038 the Court of Appeal stated that in a situation where a non-party director can be described as the 'real party', seeking his own benefit, controlling and/or funding the litigation and even where he has acted in good faith or without any impropriety, justice may well demand that he be liable in costs on a fact sensitive and objective assessment of the circumstances,

In *Lingfield Properties (Darlington) Ltd v Padgett Lavender Associates* [2008] EWHC 2795 (QB), the court refused to join a company secretary personally as a party to the proceedings for the purposes of a non-party costs order. Although he was no longer a director he had been the principal person involved in managing the company's claims and this particular claim had depended on his evidence. He was the only person through whom the company had acted. He had also arranged the funding of the claim, although he had not funded it personally. Nevertheless, it was clear he was not the real party to the claim. The board probably applied its mind independently to the matters to be decided in the litigation. The company risked its own assets and made arrangements for funding the litigation that did not materially depend upon the financial contribution of the company secretary. The company remained the real party to the litigation.

IS IT A NECESSARY PRE-REQUISITE OF AN ORDER THAT THE NON-PARTY HAS CAUSED THE OTHER PARTY TO INCUR COSTS OVER AND ABOVE THOSE THAT WOULD HAVE BEEN INCURRED ANYWAY?

Causation of costs

[40.7]

In *Dymocks Franchise Systems (NSW) Pty Ltd v Todd* [2004] UKPC 39, the Privy Council held that costs would be awarded against a non-party who had not merely funded the proceedings but had substantially controlled them, or was to benefit from them or who promoted and funded proceedings by an insolvent company solely or substantially for his own financial benefit. The Privy Council also took the opportunity to review the relevant case law and amongst other matters defined 'exceptional' in the context of non-party costs orders as conveying nothing more than something outside the 'norm' where a party funds a claim or defence for his own benefit at his own expense. It also confirmed that once a court has determined that a case is 'exceptional' what it must do is determine whether, in all the circumstances, it is just to make a non-party costs order.

Arkin v Borchard Lines Ltd [2005] EWCA Civ 655, picking up on *Dymocks Frachise Systems*, is seen as the first really clear illustration that causation of costs is not strictly required. In this case MPC, a professional funding company, entered into a funding agreement with the claimant, whereby it funded the employment of expert witnesses, the preparation of their evidence and the organisation of the enormous quantities of documents which became necessary to investigate before the trial. The defendant sought a non-party costs order arguing that, in principle, professional funders, as distinct from pure funders, who are maintaining litigation for their profit, should be liable for the costs of the defendants if the claim fails, which in this case it did.

Pre-Requisite that Non-Party has Caused the other Party to Incur Costs 40.7

MPC resisted the claim, submitting that conditional fee agreements with professional funders which have the purpose of enabling impecunious claimants to pursue claims of real substance which, but for such funding, they could not have done, should not be visited with costs orders against the funders if the claim fails.

At first instance the court held that in these circumstances the public policy objectives of the deterrence of weak claims and of the protection of the due administration of justice from interference by those who fund litigation must yield to the objective of making access to the courts available to impecunious claimants with claims of sufficient substance. An order for costs against MPC would, no doubt, operate as a strong deterrent to professional funders to provide support for impecunious claimants with large and complex claims.

On appeal the Court of Appeal, relying on the summary of the jurisdiction set out in *Dymocks Franchise Systems* (above), held that a professional funder, who finances part of a claimant's costs of litigation, ought potentially to be liable for the costs of the opposing party *to the extent of the funding provided*. The likelihood that the decision would result in professional funders taking a larger cut of any award to reflect this heightened risk was deemed preferable in the pursuit of 'overall justice' to leaving a successful defendant in such a claim without the ability to recover costs. Interestingly, as a harbinger of the April 2013 reforms, the court felt that this would act as a natural check on costs, as the non-party funder would want to keep costs proportionate to limit its potential exposure and this potential costs exposure would also act as a deterrent against speculative claims.

Total Spares & Supplies Ltd v Antares SRL [2006] EWHC 1537 (Ch), confirmed that whilst causation of costs is no longer a pre-requisite to a non-party costs order it is still a consideration. In this case the principal defendant had transferred most of its assets to a new company a week before the trial and then allowed itself to be struck off the Italian company register leaving the claimant unable to recover its costs. The claimant sought an order that Francisco Gargani, who controlled both companies, and the new company, Antares, should pay the costs. Following *Arkin v Borchard Line Ltd* (see above), the court concluded that it could no longer be said that causation is a necessary pre-condition to an order for costs against a non-party. Causation would often be a vital factor but there could be cases where, in accordance with principle, it was just to make an order for costs against a non-party who could not be said to have caused the costs in question. In the circumstances, it was just to make an order. The facts of the case were exceptional and justified the making of an order that Antares should pay 55% of the claimant's costs incurred after the transfer. Mr Gargani was directly responsible for the transfer and it was just that he should be responsible for the costs originally ordered to be paid by the defendant.

In the earlier case of *Phillips v Princo* [2003] All ER (D) 99, the judge dealt with causation in a far broader way, holding that there is a clear distinction to be drawn between disinterested, in the sense of commercially disinterested, funders on the one hand and those whose interests stood to be advanced by a successful outcome of the litigation on the other. Phillips commenced patent infringement proceedings against Aventi in respect of unlicensed discs made by Princo. Phillips made a deliberate decision not to join Princo in the litigation. Princo undertook payment of the costs of Aventi's defence which enabled

40.7 *Chapter 40 Costs against Non-Parties*

Aventi to defend the action. Aventi went into insolvent liquidation and Phillips was permitted to join Princo as a party for the purpose of seeking an order for costs against it. It was clear that Princo was interested in the outcome of the proceedings and was not a pure funder. The action would not have been defended but for its funding. Even though Princo did not involve itself in the action and did not control it, Phillips ran up costs in an action which, but for Princo's funding of Aventi for its own commercial benefit, would not have needed to be incurred. Accordingly Princo was ordered to pay the entire costs of the action. In other words the mere fact of funding a claim that could not otherwise be funded was treated as causing all the costs.

Similarly in *Gemma Ltd v Gimson* [2005] EWHC 69 (TCC), it was clear that the defendant company could not have afforded its role in the litigation without funding provided by the non-party and that funding directly led to the claimant incurring substantial costs. The non-party would have been the sole or principal beneficiaries of any recovery from the claimant and had failed to inform the company's professional advisors of material facts that, if disclosed, would have inevitably led to advice that the company's case was hopeless. The proceedings were only pursued out of the non-party's vendetta against the claimant. Fairness and justice required that a non-party costs order should be made and accordingly the funding company was joined in the proceedings and was made jointly and severally liable for all the costs the defendant was ordered to pay to the claimant.

Whilst this review of causation is of academic interest, in practical terms, as the last two cases suggest, it is difficult to imagine how the funding provided by a non-party will not inevitably cause the non-funded party to incur further costs. This was the point made by the court in *Adris v Royal Bank of Scotland (Cartel Client Review Ltd, additional parties)* [2010] EWHC 941 (QB). Here the allegations before the court were the more usual ones of control and a consideration of who was the 'real party'. However, the court also concluded that it was hard to see how a party could 'control' litigation without it following that such conduct had a bearing on the incurrence of costs by the other side. The same was true of funding – funding was necessary and it followed that in its absence the litigation may not have started or continued – with the inevitable consequence that some or all of the costs would not have been incurred. On the facts, a non-party order was justified in respect of CCLS, who was running the claims, but not as against the sole shareholder in that company. He had not funded or promoted proceedings by a claimant which was an insolvent company, nor were the proceedings solely or substantially for his own financial benefit, nor was he trading under his own name with a blurred distinction between the personal and corporate businesses.

The conclusions that we draw from this review of the case law are:
- whilst it was once stated to be a pre-condition of a non-party costs order that additional costs had been incurred (see *Hamilton v Al Fayed* [2002] EWCA Civ 665 per Simon Brown LJ at para 54), that is no longer the case;
- the fact that the non-party has caused additional costs to be incurred is an additional factor supporting an order against the non-party;

- In most cases (certainly those where the non-party has provided funding to an otherwise impecunious party) causation is, in any event, easy to establish – but for the funding provided the claim or defence could not have been maintained and the other party, inevitably, would have avoided incurring the costs of a contested claim.

SPECIFIC CATEGORIES OF NON-PARTIES TRADITIONALLY IN THE FIRING LINE

[40.8]

As set out above the court identified in *Symphony Group plc v Hodgson* the type of conduct on behalf of a non-party that might lead to an order for costs. A review of the case law readily reveals categories of non-party with a specific risk of exposure to a non-party costs order. We shall look at them in turn, then consider separately the position of funders and solicitors where very specific considerations arise and finally look briefly at non-party costs orders in the family jurisdiction where normally a 'no costs' regime applies.

Insurers

[40.9]

Inevitably the position of insurers has come under scrutiny given their interest in the outcome of claims where they are providing an indemnity to an insured. Certain themes have emerged.

Insurers who take over the defence of an action and conduct it for their own benefit may properly be said to be the real defendants. In such a situation in *Pendennis Shipyard Ltd v Magrathea (Pendennis) Ltd (in liq)* [1998] 1 Lloyd's Rep 315 (QB) they were ordered to pay the costs although they were not a party to the action.

However, in *Murphy v Young & Co's Brewery plc and Sun Alliance and London Insurance plc* [1997] 1 All ER 518, [1997] 1 WLR 1591, CA legal expenses insurers were not liable to pay more than the limit of their cover merely on the grounds that they had funded the litigation under a commercial agreement.

In *TGA Chapman Ltd v Christopher* and *Sun Alliance* [1998] 2 All ER 873, CA, the Court of Appeal held that the defendant's liability insurers were liable for the full amount of a judgment of £1,100,000 plus costs even though their liability under the insurance policy was limited to £1,000,000 inclusive of opponents costs. The insurers were in a different position from those in *Murphy* because, as in *Pendennis* above, the litigation was funded, controlled and directed by the insurance company motivated entirely by its own interest. The same reasoning was adopted in *Plymouth and South West Co-operative Society Ltd v Architecture, Structure and Management Ltd* [2006] EWHC 3252 (TCC), where the insured had virtually no assets and had ceased trading, but the insurers had fought the claim to protect their own liability to the defendant of £2 million under the policy.

In *Palmer v Palmer* [2008] EWCA Civ 46, [2008] Lloyd's Rep IR 535, [2008] 4 Costs LR 513, the appellant Royal & Sun Alliance ('RSA') insured the second defendant ('PZ') and had financed its unsuccessful liability defence to the claim. The other defendants had admitted liability and one of the co-defendants, the MIB, had made a Part 36 Offer to accept a contribution of £300,000 from PZ and to meet the whole of the remainder of the claim. This offer had been rejected.

PZ were ordered to pay the claimant's and defendants' costs of the liability proceedings. The damages were likely to exceed £2 million. RSA's liability under the policy was limited to £500,000, which would be exhausted by the claims for damages and costs. PZ was in financial difficulties and would not be able to make a substantial contribution toward the sums claimed and, as a result, PZ's co-defendants would have to pay the balance. The MIB sought an order under s 51 of the Senior Courts Act 1981 that RSA should pay the costs incurred by the other parties which would free the whole of the £500,000 payable under the policy to be applied towards the damages.

The court concluded that the Part 36 offer ought to have been accepted to save PZ from further risk and that RSA had rejected that offer without consulting PZ, demonstrating that RSA had been motivated either exclusively, or at least predominantly, by its own interest in the manner in which it had conducted the defence, and that that was a circumstance that pushed the case into an exceptional situation of the type identified by Balcombe J in *Symphony Group plc* in which it was appropriate to make the order sought. RSA could be regarded as the true defendant in all but name. PZ had no commercial interest in pursuing the claim and RSA should have been aware of this. This was evidenced by the fact that when the opportunity to settle had arisen, RSA had turned it down without reference to PZ. The inference was that the defence of the claim was conducted on the basis that the only real interest being protected was that of RSA. On that basis it was proper to make an order for costs against RSA.

Company Directors

[40.10]

As with insurers there has been a plethora of cases that concern the liability of the officers of limited companies for costs of claims involving those companies where the specific company subject to a potential liability is impecunious. A review of cases where orders have not and have been made gives a flavour of the approach which the court has adopted in such a situation. This review comes with the inevitable reminder that the discretion is a broad one and that the exercise of it will be fact specific.

(a) Orders not made

[40.11]

In *Taylor v Pace Developments Ltd* [1991] BCC 406, CA, the Court of Appeal held that the controlling director of a one-man company should not be liable personally for the costs of defending an action even though he knew the company would not be able to meet the claimant's costs should the company's defence prove unsuccessful. The court found that to impose a liability in such a situation would be far too great an inroad into the principle of limited liability. However, a sole director might still be made liable if the company's defence transpired not to be bona fide (eg where a company had been advised that there was no defence).

Gardiner v FX Music Ltd [2000] All ER (D) 144, ChD confirmed that an order for costs against a non-party is always an exceptional order. In the case of a sole or guiding director of an insolvent company, such an order is not normally made unless it can be shown that the director caused the company to bring or defend proceedings improperly.

Re Land and Property Trust Co plc (No 4) [1994] 1 BCLC 232, CA provided a good illustration of the dangers inherent in treating an application for costs against a non-party in the same manner as one against a party to the proceedings. On refusing a petition by a company for an administration order, the judge had ordered the costs incurred by a creditor to be paid by the directors personally. Although he had jurisdiction to make such an order for costs, he had refused the directors' application for an adjournment to enable them to have a proper opportunity of putting in evidence and his failure to do so caused him to err in principle. On the fresh evidence that was available before the Court of Appeal the order for costs ought not to have been made.

(b) Orders made

[40.12]

In *H Leverton Ltd v Crawford Offshore (Exploration) Services Ltd (in liquidation)* (1996) Times, 22 November, QBD, an application succeeded where a Mr Crawford was found to be the defendant company for all practical purposes. He had acted in bad faith by concocting false claims supported by forged documents, had destroyed and suppressed documents, had given false evidence, and had substantially financed the proceedings. These were exceptional circumstances making it appropriate to order costs against him.

In *North West Holdings plc, Re, Secretary of State for Trade and Industry v Backhouse* [2001] EWCA Civ 67, [2001] 1 BCLC 468, [2002] BCC 441, two companies were ordered to be wound up, the judge having found that no proper books of account had been kept, that the businesses of the companies were intertwined with the defendant personally, who treated the companies' moneys as his own, and that a savings scheme marketed through the companies lacked intrinsic merit. After judgment, the Secretary of State warned the defendant for the first time that he intended to apply for a costs order against him personally. The judge ordered the defendant to pay the costs personally because he had defended the proceedings solely to protect his own reputation and position without seriously considering the interests of the companies or their creditors. The crucial question was whether the relevant director held a bona fide belief that the companies had arguable defences and that it was in the companies' interests to advance those defences. If he did so believe, to make a non-party pay the costs would constitute an unlawful inroad into the principle of limited liability. However, on the facts, the defendant had not considered the companies' interests but only his own; there was ample evidence to justify the costs order despite the absence of an early warning concerning costs.

In *Bournemouth and Boscombe Athletic Football Club Ltd v Lloyds TSB Bank plc* [2004] EWCA Civ 935, the claimant football club was in severe financial trouble and was unable to meet its obligations to its bankers, the defendant, who appointed administrative receivers. The club entered into a company voluntary arrangement and its business was sold to a new company. One of the directors initiated proceedings against the bank in the name of the company which were struck out. He then initiated further proceedings making the same allegations. That action too was struck out and an appeal against the striking out also failed. The director was ordered to be added as a defendant to enable an application to be made for a non-party costs order against him. The court concluded that the director had sought to play fast and loose with the civil justice system by commencing and presenting hopeless claims and by presenting a hopeless appeal on behalf of an insolvent company. The circumstances were sufficiently exceptional to justify the court making him face the financial consequences of his conduct by ordering him personally to pay the costs of the company's unsuccessful appeal.

In *Goodwood Recoveries Ltd v Breen: Breen v Slater* [2005] EWCA Civ 414, a Michael Slater controlled the claimant debt recovery company and was also a consultant solicitor in a firm which funded proceedings brought by the company under a conditional fee agreement. The costs of the litigation brought by the claimant company would not have been incurred without Mr Slater's involvement. He formulated the claim and brought it, albeit in the name of the company. He was the real party for whose benefit the litigation was brought. He did not fund it, but only because it was funded under a CFA by the firm of solicitors for whom he acted as a consultant and in whose name he undertook the conduct of the litigation. A lack of *bona fides* is not necessarily a condition of claiming against a non-party, if that non-party was really the party for whose benefit the case was conducted. In any event the whole of the costs of the litigation were caused by Mr Slater's dishonesty or impropriety, irrespective of whether he had any *bona fide* belief in the claim. Therefore it was not necessary to decide whether there had to be a causal link between all the costs and the alleged impropriety: it was appropriate to order Mr Slater to pay the whole of the costs and not merely the additional costs which could be attributed directly to improper conduct on his part.

Liquidators and receivers

[40.13]

Although the court has jurisdiction to order a liquidator, as a non-party to proceedings brought by an insolvent company, to pay costs personally, it will only exercise that jurisdiction in exceptional circumstances where there has been impropriety on the part of the liquidator, particularly in view of the fact that the prospective remedy of obtaining an order for security for costs is available to the defendant (see Chapter 19 – Prospective Costs Control – Security for Costs, above). The caution necessary in all cases when an attempt is made to render a non-party liable for costs is all the greater in the case of a liquidator having regard to public policy considerations (*Metalloy Sup-*

Specific Categories of Non-Parties Traditionally in the Firing Line 40.13

plies Ltd (in liquidation) v MA (UK) Ltd [1997] 1 All ER 418, CA). Once again an examination of relevant case law offers pointers to the key components necessary to establish a personal liability on the part of the receivers or liquidators.

The court has been clear that the decision in *Aiden Shipping Co Ltd v Interbulk Ltd, The Vimeira (No 2)* (1986) AC 965, H, [1986] 2 All ER 409, HL does not justify the judicial creation of a substantive rule that receivers should be personally responsible for the costs of a successful party: indeed quite the contrary. However, each case will be determined on its own facts.

In *Dolphin Quays Developments Ltd (In Administrative and Fixed Charge Receivership) v Mills* [2007] EWHC 1180 (Ch), the court set out the importance of drawing a distinction between the inevitable functions of receivers and liquidators and the type of conduct necessary to trigger a non-party costs order. Here the applicant applied for an order that the receivers should pay the costs of unsuccessful litigation that had been brought by the claimant company against him. He was a substantial creditor of the claimant company's parent company, and contended that the receivers were the real parties to the unsuccessful litigation and had conducted the litigation for their own benefit and the benefit of the Royal Bank of Scotland which had appointed them. The application was refused with a reminder that non-party costs orders are only to be made in exceptional circumstances and this was an entirely normal case of receivers seeking to enforce a contractual right forming part of the security. The fact that the claim failed might be unusual did not mean it was exceptional. This claim could hardly be classified as 'exceptional'. If an order were made in this case then it would have to be made in all such cases. There was no impropriety or unreasonableness in pursuing the litigation on which a non-party costs order could be founded. Neither the receivers nor the bank could be viewed as the real parties to the litigation. Again the court made the point that the applicant could have applied for security for costs from the company when the litigation was first instigated.

However, that such orders can be made is confirmed by the decision in *Apex Frozen Foods Ltd (in liquidation) v Abdul Ali* [2007] EWHC 469 (Ch). A freezing order was granted subject to the liquidator giving a personal undertaking to the effect that he would be liable for any loss to the claimant if the court found that the order had occasioned such loss and that the claimant should be compensated for it.

The freezing order was subsequently discharged on the basis that there had been improper disclosure of the material facts at the time that the freezing order was granted. The purpose of the undertaking was to ensure that a mechanism was available to make good any detriment suffered by the claimant through the grant of the freezing order if it was subsequently established that there should not have been an injunction. The claimant had incurred costs in relation to the injunction proceedings as a result of the freezing order being obtained in the absence of proper disclosure of material facts. The fact that the liquidator was innocent of any personal conscious failure was insufficient to absolve him from liability when such absolution would produce precisely the injustice which the undertaking was designed to guard against. Accordingly, the claimant was entitled to recover its costs as recoverable damages on the standard basis. The undertaking here was the key element, making out the required 'exceptionality'.

40.13 *Chapter 40 Costs against Non-Parties*

Accordingly there is still scope for a non-party costs order against a receiver or against a creditor in appropriate cases, especially where it can be shown that the non-party was the 'real party'. Costs orders against receivers will more readily be made where a company is in liquidation and the receiver's agency has terminated, or where the successful party has not been able to obtain security for costs.

It is clear that the court regards a failure to obtain security for costs in appropriate cases as an important factor in the exercise of its discretion. However, even then this must be viewed in context and is merely a factor to which the court will have regard. The lack of an application for security is utterly irrelevant if the applicant cannot make out a case for a non-party order in the first place. This was the position in *Mills v Birchall* [2008] EWCA Civ 385. This was an entirely normal case of receivers seeking to enforce a contractual right forming part of the security. The absence of any element of impropriety or unreasonableness confirmed that the claim was not exceptional and also underlined the fact that it was not an alternative justification for making the order sought. The receivers had directed the proceedings on behalf of the company without any direction or interference by the bank and the funding of the proceedings by the company was derived from the realisations in the receivership. The receivers had not funded the claim, nor had they had any interest in the monies from which the claim was funded or in the outcome of the claim. Neither the receivers nor the bank were to be regarded as the real party. The position of receivers as agents was found to be analogous to the position of directors and liquidators. In an ordinary case in which a director or liquidator causes a company to bring proceedings that are unsuccessful, a personal costs order will not be made. Some additional element will be required. In this case the judge did not place too much weight on the absence of an application for security of costs as other factors determined the outcome. There were no factors that, in accordance with established principles, justified interference with the judge's exercise of discretion.

Tribunals

[40.14]

The court has power to order costs against a tribunal whose decision was overturned on appeal, but only if the tribunal effectively makes itself a party by appearing on the appeal or taking steps to defend its determination (*Providence Capitol Trustees Ltd v Ayres* [1996] 4 All ER 760, ChD). Where, in a successful appeal against a decision of the Pensions Ombudsman, the ombudsman had appeared at the appeal and made representations in support of his determination, it was held that the costs recoverable from him were not limited to the amount by which they had been increased by his appearance but could, in principle, extend to the whole of the successful party's costs in making the appeal (an early in road into the 'causation' argument). This approach was followed in *Moore's (Wallisdown) Ltd v Pensions Ombudsman* [2002] 1 All ER 737, ChD, where it was held that the decisions in *Elliott v Pensions Ombudsman* [1998] OPLR 21, ChD and *University of Nottingham v Eyett (No 2)* [1999] 2 All ER 445 ChD, that the ombudsman could be ordered to pay the costs of a successful appeal only to the extent that those costs were increased by his participation, were wrong and would not be followed.

Witnesses

[40.15]

In *Phillips v Symes* [2004] EWHC 2330 (Ch), the court had to consider the extent to which witnesses could incur a liability for costs. The particular case concerned an expert evaluation of the capacity of one party. A cost order was made against the expert witness, who by his evidence had caused significant expense to be incurred in a flagrant reckless disregard of his duties to the court. The court was clear that it would be wrong to exclude a power for it to award costs against witnesses in appropriate circumstances. However, the court should be astute to prevent any unfair questioning of witnesses designed solely to set up the possibility of a later application for non-party costs. This case revisited the issue of notice in the context of possible orders against witnesses, stating that a party contemplating applying for a non-party costs order against a lay witness would inevitably have to give early notice to that person, in contrast to the position of legal representatives who would need no reminding of the possibility of wasted costs orders.

More recently in *Saxton v Bayliss* Ch Div 20.3.13 the court set aside a non-party costs order made at first instance against a witness (the wife of the first appellant) in boundary dispute proceedings. The court reminded itself that such an order was always exceptional, that the jurisdiction should be exercised with caution and that notice of an impending application should be given to the non-party. The court also took account of the fact that the additional costs incurred as a result of this particular evidence were minimal in the context of the overall costs.

On the other side of the coin is the position where a witness incurs costs in compliance with court procedure and seeks to recover those costs. It is worth remembering that CPR 46.2 is headed '*Costs orders in favour of or against non-parties*'. In *Individual Homes Ltd v Macbream Investments Ltd* (2002) Times, 14 November the claimant had issued and served a witness summons under CPR 34 on an employee of the Halifax Bank of Scotland plc requiring his attendance at the hearing of the claim and the production by him of the bank's documentation. The court held that it was an anomaly that if the application had been made under s 34 of the Senior Courts Act 1981 (non-party disclosure of documents), the rules would have provided that the person against whom the order had been made could recover the costs of compliance with the order (pursuant to CPR 46.1), but that there was no equivalent provision in Part 34 to enable a witness to recover his costs of complying with a witness summons. The conclusion reached was that the courts of first instance have jurisdiction to resolve this anomaly under CPR 46.2 to award costs where it is just and reasonable. However, the court also held it was inappropriate to order that the bank be joined to the proceedings for the purpose of pursuing these costs. The claimant was ordered to pay the costs incurred by the bank in complying with the summons.

40.16 *Chapter 40 Costs against Non-Parties*

FUNDERS

[40.16]

Whilst it is glaringly obvious, as a starting point to an application based on the non-party having provided funding, the applicant must be able to establish that the respondent to the application actually funded the litigation. In *Shah v Karanjia* [1993] 4 All ER 792, ChD, an application against a non-party failed both because he had been given no adequate warning of the application and because of lack of evidence that he had actually funded the proceedings.

The case of *Hamilton v Al Fayed (No 2)* [2002] EWCA Civ 665, sought to define the type of funding that would be needed as part of the requirement for a non-party costs order. The court distinguished between 'pure' funders, who made donations in support of a litigant as an act of charity, and 'professional' funders, such as insurers, who were almost always contractually bound to fund the litigation. It concluded that:

- An order for costs would rarely be made against a charitable, philanthropic, altruistic or merely sympathetic donor who, on the information before him, had reasonable grounds for believing that the litigant had reasonable grounds for asserting his right or a defence to the claim, and who wished to ensure that a genuine dispute was not lost or inadequately contested by default due to financial constraint.
- If pure funders were regularly exposed to liability under s 51, such funds would dry up and access to justice would thereby on occasions be lost. They should not, therefore, ordinarily be held liable.
- So long as the law continued to allow impoverished parties to litigate without having to provide security for their opponents' costs, those sympathetic to their plight should not be discouraged from assisting them to secure representation. In the light of the recent further reductions in the availability never has this sentiment had greater resonance.

In *Gulf Azov Shipping Co Ltd v Chief Humphrey Irikefe Idisi* [2004] EWCA Civ 292, the court allowed an appeal by a Nigerian lawyer against a costs order made against him for personally intervening on behalf of one of the defendants against whom damages had been awarded for the wrongful detention of the claimant's vessel. That defendant's assets were subject to a worldwide freezing order in an attempt to enforce the award. The lawyer could not be criticised for assisting the defendant financially in instructing solicitors of high standing and for assisting the defendant in attempting to discharge his costs liabilities. There was no suggestion that the lawyer had been personally interested in the outcome of the litigation.

It is clear that the benefit derived by the funder does not have to be a direct financial one and funding motivated by personal animosity may also justify a non-party costs order (*Vaughan v Jones and Fowler* [2006] EWHC 2123 (Ch))

Orders made

[40.17]

In *Locabail (UK) Ltd v Bayfield Properties Ltd (No 3)* [2000] 2 Costs LR 169, ChD, the first husband of one of the defendants had given evidence on her behalf at the trial when he had been found to be an unsatisfactory witness, prone to exaggeration. He admitted to funding his former wife's action and recommending a solicitor to her. The claimants sought costs against him and he was joined in the action solely in relation to the claim for costs. It was not appropriate to punish the former husband by awarding costs solely on the basis of his funding the litigation of another and behaving unsatisfactorily as a witness. However he was ordered to pay the claimants' costs on the following bases:

(a) he had funded proceedings knowing that his former wife would be unable to satisfy a costs order if unsuccessful;
(b) his intense identification with his former wife's position in his own evidence;
(c) his indifference to the legal and factual issues in the case; and
(d) the court's rejection of the factual basis of his former wife's case.

Petromec Inc v Petroleo Brasileiro SA Petrobras [2007] EWHC 1589 (Comm), held that actual funding by a non-party is not a jurisdictional pre-requisite to the exercise of the court's discretion under s 51 of the Senior Courts Act 1981 (although it did find that there was funding in this case). If the evidence is that a person, whether a director or shareholder or controller of a relevant company, had effectively controlled the proceedings and sought to derive potential benefit from them, that was enough to establish the jurisdiction. Whether the jurisdiction should be exercised is another matter, and the extent to which a person has, in fact, funded any proceedings might be very relevant to the exercise of discretion.

In *Thomson v Berkhamsted Collegiate School* [2009] EWHC 2374 (QB), the claimant was a 25-year-old unemployed university graduate who sued the private school he had attended between 1994 and 2002 for injury, loss and damages of nearly £1 million for failing to prevent him being bullied. He discontinued the proceedings two weeks into the trial. He was unable to meet any costs order and the defendant wished to seek a non-party costs order against his parents. Pursuant to that application the school sought orders requiring the parents to file and serve disclosure statements setting out correspondence between them and their son's solicitors, experts and counsel, and orders against the son with respect to disclosure and his claim of legal professional privilege.

The parents were not merely funders but were directly concerned with the facts of the claim and played an active role in the litigation. It was doubtful that it would have been funded if the parents had not made funds available themselves. Accordingly, an application for non-party costs had a reasonable prospect of success. The only doubt was over whether the parents gained a benefit from the litigation and sought to control its course. The defendant could only demonstrate the element of control if it knew what communications the parents had had with the solicitors, counsel and experts in the case. Accordingly the school was entitled to disclosure.

In *Automotive Latch Systems Ltd v Honeywell International, Inc* [2010] EWHC 1031 (Comm), the claimant was ordered to pay the defendant's costs, some of them on the indemnity basis, and the defendant applied under the Senior Courts Act 1981, s 51(3) for a non-party to pay those costs. The non-party was in control of both the claimant and the litigation; the claimant only had continued in existence in order to pursue the litigation; the non-party had lent money to the claimant and in doing so was funding the litigation. The non-party had also sought and procured funding for the litigation from external investors, in addition to his own funding contribution. He was the principal witness for the claimant and without whose evidence the claim would not have been sustainable. The non-party had a 74% holding in the claimant and stood to benefit personally to a very considerable extent if the claim succeeded. Applying the relevant principles, the present case was a clear one for the imposition of a non-party costs order.

SOLICITORS

As funders

[40.18]

In a previous edition Michael Cook warned that *Arkin v Borchard* (above) opened the door for the argument that in certain situations a solicitor who invests his time, and perhaps funds disbursements, becomes a 'commercial funder'. His particular concern was where lawyers agree substantial success fees under conditional fee agreements when acting for impecunious clients without insurance cover for the defendant's costs if the claim fails. As he put it:

> 'Quite simply, the financing of litigation by a solicitor is maintenance and because he seeks to profit out of it, it is champertous maintenance. Success fees are the solicitor's share of the proceeds of litigation. Champerty is unlawful and therefore a champertous retainer is unenforceable: that is why it was necessary in s 58(3) of the Courts and Legal Services Act 1990 to provide that a CFA should not be unenforceable for that reason alone. Section 58(3) did not purport to abolish a solicitor's liability to a third party as the maintainer of litigation.'

Lawful, or justifiable, maintenance is no longer against public policy. In the words of Lord Scarman in *Wallersteiner v Moir (No 2)* [1975] 1 All ER 849. 'The maintenance of other people's litigation is no longer regarded as a mischief: trade unions, trade protection societies, insurance companies and the State do it regularly and frequently'. However, as was said by Lord Denning in *Hill v Archbold* [1968] 1 QB 686, and confirmed in case after case thereafter: 'It is perfectly justifiable and is accepted by everyone as lawful, provided always that the one who supports the litigation, if it fails, pays the costs of the other side'

Hodgson v Imperial Tobacco Ltd [1998] 2 All ER 673 is often quoted as authority to the contrary. It is not. The court in *Hodgson* simply held that the existence of a CFA did not alter the circumstances in which a legal adviser could be personally liable for the costs of a party. The position was not any different than other retainer arrangements. The court was concerned with the solicitor's liability as a solicitor and not as a maintainer of litigation. The words 'maintenance' and 'champerty' are nowhere mentioned in the judgment, nor did the Court of Appeal consider any of the authorities. It did not say that a CFA put a solicitor in a better position than any other maintainer of litigation, and why should it? Why should a successful defendant be in a worse position against a maintainer of litigation because he is a member of the legal profession who, like other professional maintainers, is making a profit out of it?

The consequences of the abolition of recovery of additional liabilities, but preservation of conditional fee agreements with success fees (but being a matter of solicitor and client costs) and the introduction of Damages Based Agreements ('DBAs'), which are, in effect, no more than a form of contingency fee agreement, demand further examination of the position of the legal representative.

In *Myatt v National Coal Board* [2007] EWCA Civ 307 the Court of Appeal had dismissed the claimants' appeals against the finding of the costs judge that the conditional fee agreements they had entered into with their solicitors were unenforceable and therefore although the four claimants had each succeeded in their claim for damages for personal injuries the indemnity principle precluded them from recovering costs from the defendants.

The claimants had no insurance against any liability for costs because it was a condition precedent to the liability of the insurers under the claimants' after the event policies that enforceable conditional fee agreements were in place. Accordingly, the claimants had contested the appeal without any valid funding mechanism in place to meet the appeal costs of the successful defendant. In those circumstances the defendant sought an order for the costs of the appeals against the solicitors under s 51 of the Senior Courts Act 1981.

There were 60 other cases where clients had entered into CFAs with the solicitors in similar circumstances, who had a total a sum in the region of £200,000 at stake.

These four claimants did have a financial interest in the appeals because they were liable for their own disbursements which, including the now irrecoverable insurance premium, averaged £2,500 to be paid out of their modest damages of £3,000–£4,000.

Of course the Court of Appeal identified the starting point as the statutory provision itself – s 51 of the Senior Courts Act. The court referred extensively to the judgment in *Tolstoy-Miloslavsky v Aldington* [1996] 1 WLR 736 quoting Rose LJ at page 743:

> 'Sections 51(1) and (3) of the Supreme Court Act 1981 do not confer jurisdiction to make an order for costs against legal representatives when acting as legal representatives . . . '

but highlighting that he had in the same case qualified this statement respectively as follows:

> 'There are only three categories of conduct which can give rise to an order for costs against a solicitor:
> 1. It is within the wasted costs jurisdiction of section 51(6) and (7);
> 2. It is otherwise a breach of duty to the court . . . e.g. if he acts even unwittingly without authority or in breach of an undertaking;
> 3. If he acts outside the role of solicitor e.g. in a private capacity or as a true third party funder for someone else.' (per Rose LJ)

40.18 *Chapter 40 Costs against Non-Parties*

In *Myatt* the court found that the third category described by Rose LJ included a solicitor who is 'a real party . . . in very important and critical respects' and who 'not merely funds the proceedings but substantially also contributes, or at any rate, is to benefit from them'. The mere fact that the same solicitor is formally on the court record advancing or defending proceedings for his client does not, itself, act as a bar to an application by the successful opposing party under s 51(1) and (3) of the Senior Courts Act 1981, that the solicitor should pay some or all of the costs.

Were it not for the relatively minor financial interest of the claimants in the disbursements (minor in the context of the solicitors' overall exposure to costs that were potentially irrecoverable), only the claimants' solicitors would have had an interest in the appeal, which the Court of Appeal concluded took them outside the role of solicitors. Indeed Dyson LJ (as he then was) thought it unlikely that the individual claimants would have pursued an appeal solely for the minimal disbursements. In these circumstances the court held that it would be very surprising if the fact that the claimants had a modest financial interest meant that the solicitor's financial interest counted for nothing when deciding what order for costs it was just to make.

The court reiterated that the non-party need not be the only real party to the litigation, provided that he is 'a real party . . . in very important and critical respects'.

The Court of Appeal held that there was no good reason why those observations should not apply with equal force to solicitors as to non-solicitors and was in no doubt that there is jurisdiction to make an order under s 51(3) against a solicitor where litigation is pursued by the client for the benefit or to a substantial degree for the benefit of the solicitor.

The Court of Appeal held that the fair and just order to make in this case was to order the solicitors to pay 50% of the defendant's costs of the appeal. In arriving at this percentage the court took into account the fact that the claimants had a residual real financial interest in the success of the appeals; their disbursements represented approximately one third of the total costs incurred by them before their claims were settled. It also took into account the fact that the solicitors were not given a warning until the appeals had been dismissed that an application for costs might be made against them.

In *Myatt* the court accepted that the authorities did not set out in definitive terms where the line in the sand is drawn between the case where a solicitor acts purely as the legal representative in the ordinary way on behalf of a client and is therefore immune from the jurisdiction of the court under ss 51(1) and (3), and, on the other hand, a case where the solicitor's acts are such that he becomes 'a real party' to the claim. However, Lloyd LJ also chose to refer to a passage from the judgment of Roth LJ in *Tolstoy-Miloslavsky v Aldington* which rather neatly sets out the line that the court will be looking for on the facts of each individual case where arguments as to the extent of the solicitor's role arise:

> 'The legal representative who acts as a legal representative does not make himself a quasi party and no jurisdiction to make an order for costs against him under s.51(1) and (3) arises. However, a legal representative who goes beyond conducting proceedings as a legal representative and behaves as a quasi party will not be immune from a costs order under s 51(1) and (3) merely because he is a barrister or a solicitor'.

The court was clear, though, that there is a distinction to be made between privately paying clients who retain a costs liability to the solicitor – even if it is notional – and those represented under conditional fee agreements where the clients do not have and have not had any personal liability to pay the solicitors' costs.

The court concluded in *Myatt* that it was correct to regard the solicitors in relation to the conduct of the appeal as having acted in part for the sake of their own benefit in a respect which was of no interest or concern to their clients, and as having acted as a matter of business to seek to establish their right to be paid, not by their own clients in practice, the profit costs on these four cases and all the others of which these were representative. In those circumstances, which could be common in relation to cases where the enforceability of a CFA is at stake, but would be most unusual in any other situation, it was proper to regard the solicitors as having acted in respect of the appeal in a dual capacity; acting for their clients, certainly and with a real interest of those clients to protect, but primarily acting for their own sake. In other words they had crossed the line drawn by Roth LJ.

The introduction of damages based agreements where the client may, in some circumstances, have no liability to the solicitor other than in respect of non-advocate disbursements (where he is unsuccessful) again creates the potential for solicitors to become quasi parties. If a claim is unsuccessful the solicitors will not be paid. Whilst it may indeed be in the client's interest to appeal, there can be no doubt that a successful appeal would also be in the interests of the solicitors, who would then be paid. The advice and the decision making process will need to be clear to protect solicitors in such a situation.

However, no doubt, of enormous relief to solicitors at a time when they are having to be more creative in terms of the funding arrangements into which they are prepared to enter are the decisions in the cases of *Heron v TNT (UK) Ltd* [2013] EWCA Civ 469, [2013] 3 All ER 479, [2013] NLJR 21 and *Flatman v Germany* [2013] EWCA Civ 278, [2013] 4 All ER 349, [2013] 1 WLR 2676 (see Chapter 12 The Indemnity Principle, above). In the former case the Court of Appeal upheld the judge's decision to refuse an order under s 51 on the basis that a solicitor retained under a conditional fee agreement without any backing after the event insurance, had not become 'a real party' to the claim. In the latter case the Court of Appeal again concluded that a solicitor in the same situation who also funded the disbursements was, similarly, not the 'real party' or even 'a real party' to the litigation.

Lack of authority

[40.19]

In *Skylight Maritime SA v Ascot Underwriting* [2005] EWHC 15 (Comm), proceedings were brought against solicitors on the grounds of breach of warranty of authority in commencing proceedings on behalf of a one-yacht Panamanian company against insurance brokers without authority from the client to begin proceedings. In such circumstances the general rule is that the court has jurisdiction to make a summary order against the solicitor for the costs incurred by the opposite party caused by the solicitor's unauthorised conduct (*Yonge v Toynbee* [1910] 1 KB 215). However, in this case there were substantial issues of fact that could not be resolved in the course of the usual summary procedure and therefore the application for a summary determination was refused.

In *(1) David Warner (2) SMP Trustees Ltd v Merriman White (A Firm)* [2008] EWHC 1129 (Ch) a firm of solicitors issued a petition purporting to act on behalf of the two petitioners, when it had, in fact, only obtained instructions from one of them and the other had no knowledge of the proceedings until it was ordered to give security for costs. In striking the other party from the petition and ordering the solicitors to pay the costs of the other parties the court drew attention to the Guide to the Professional Conduct of Solicitors, which was explicit about a solicitor's duty to ensure that there were clear instructions from a client where instructions are given by a third party.

FAMILY CASES

[40.20]

Whilst FPR 28.1 affords the court a wide discretion to make such order for costs as it thinks just, the prevailing position in the family courts is that the starting point is 'no order for costs' between the parties. However, it is important to note that the family court also has the jurisdiction to make non-party costs awards. In *HB (mother) v (1) PB (father) (2) OB (a child by his guardian) & Croydon London Borough Council (respondent on issues of costs only)* [2013] EWHC 1956 (Fam) Cobb J concluded that there was limited authority on these awards in the family court. However, he found that the clear failure of the local authority to consider guidance that existed on fabricated illness when compiling its report under s 37 of the Children Act 1989 had led to fruitless hearings at which the father had incurred wasted costs. He referred to the 'exceptionality test' in non-family cases, concluded that a local authority charged with preparing a report was sufficiently closely connected with the court proceedings, that its failures in the case were exceptional and had caused costs to be incurred that would not otherwise have arisen and ordered the local authority to pay the father's costs of the abortive hearings.

CHAPTER 41

ARBITRATION

INTRODUCTION

[41.1]

The law relating to arbitration is contained in the Arbitration Act 1996, which was greeted with universal acclaim as a model statute in both form and content when it came into force on 1 January 1997. Sections 59–65 set out the key costs provisions, but these are supplemented by specific provisions elsewhere in the statute.

COSTS OF THE ARBITRATION DEFINED

[41.2]

Section 59 provides that the costs of the arbitration include the arbitrator's fees and expenses, the fees and expenses of any arbitral institution involved and the legal and other costs of the parties to the arbitration. Section 63 makes it clear that costs include those of and incidental to any proceedings to determine the amount of the recoverable costs

Section 28 of the Act makes the parties jointly and severally liable for such of the arbitrator's reasonable fees and expenses as are appropriate in the circumstances. It also makes provision for the parties to agree with the arbitrator his fees and expenses, including capping and restrictions, or to apply to the court for them to be considered and adjusted.

Section 37(2) provides that the arbitrator's expenses include the fees and expenses of any expert, legal adviser or assessor appointed by the arbitrator.

AGREEMENT AS TO PAYMENT OF COSTS

[41.3]

Section 60 provides that any agreement between the parties that a party is to meet the whole or part of the costs of the arbitration whatever the outcome of it may be is only valid if that agreement was made after the dispute being arbitrated arises. In other words s 60 of the Act preserves the unusual prohibition in respect of arbitrations that the parties may not in the arbitration agreement provide that each party shall pay his own costs in any event, and extends the prohibition to agreements that one party shall pay the other party's costs whatever the outcome of the arbitration. The purpose is to try to create a level playing field between large contractors and small contractors, and not deter a party from commencing arbitration proceedings because he will be liable for his own costs in any event. This consideration does not apply to any post-dispute agreement the parties may wish to enter into.

Section 62 provides that any agreement between the parties as to liability for the costs of the arbitration extends only to those costs that are recoverable, unless the parties agree otherwise.

THE AWARD OF COSTS

[41.4]

The general rules that apply to the award of costs in arbitrations are to be found in s 61. This empowers the arbitrator to award the whole or part of the costs of the arbitration to either party, subject to (i) a valid agreement between the parties as to the liability for costs and (ii) the general principle that costs follow the event except where it appears to the tribunal that in the circumstances this is not appropriate in relation to the whole or part of the costs. Matters such as exaggeration, conduct, failure on particular issues, reasonableness, proportionality and sealed offers were relevant to the award of arbitration costs long before these concepts were introduced into civil litigation by the CPR. CPR 44.1 applies the CPR rules CPR Parts 44–47 to the costs of arbitration proceedings and to this extent the practice has been codified in the CPR both in respect of the principles on which costs are both awarded and quantified, but also in respect of offers to settle. CPR Part 62, and its supplementary practice direction, apply to arbitration proceedings but do not specifically mention costs.

Offers

[41.5]

If a sealed offer is made in arbitration, being the arbitral equivalent of a CPR Part 36 offer, a respondent is normally entitled to payment of costs from the date of the offer if the award in respect of the claim and interest is less than the offer. The arbitrator is not entitled to take into account whether an award of costs would be made in favour of the claimant, because that would require the claimant to assess not only the likelihood of achieving an award on his claim and interest exceeding the offer, but also, if there was a risk of an order that the claimant pay the respondent's costs, the chance of obtaining an award greater than the offer and the respondent's costs. Such a result would hinder settlement and introduce complications inconsistent with the principle that the costs should follow the event (*Everglade Maritime Inc v Schiffahrtsgesellschaft Detlef von Appen mbH* [1993] QB 780, CA).

As it is customary for an award to deal at one and the same time both with the parties' claims and with the question of costs, the existence of a sealed offer has to be brought to the attention of the arbitrator either before he has reached a decision or he may have to revise that decision on costs in the light of the terms of the sealed offer when he sees it. However, obviously, if adopting the former approach the offer should remain sealed and the content unknown until the substantive decision has been reached as it would be wholly improper for the arbitrator to look at it before he has reached a final decision on the matters in dispute other than as to costs.

There are arbitrators and umpires who feel that the former procedure is not satisfactory. They take the view that respondents will feel that their defence is weakened if the arbitrator knows that they have made a sealed offer, even if the figure is concealed. If this is so, respondents may be deterred from making a sealed offer. The solutions are either to adopt a procedure where the award of costs is not dealt with at the same time as the award or for an arbitrator to require the respondents to give him at the end of the hearing a sealed envelope which is to contain either a statement that no sealed offer has been made or the sealed offer itself. If this procedure is adopted, the existence or otherwise of a sealed offer is concealed from the tribunal until the moment at which it has to consider that part of the award which relates to costs, the delivery of a sealed envelope of itself being devoid of all significance (*Tramountana Armadora SA v Atlantic Shipping Co SA* [1978] 2 All ER 870, [1978] 1 Lloyd's Rep 391).

Fairness

[41.6]

An arbitrator's power to award costs under s 61 of the Act or under the applicable procedural rules is subject to the general duty under s 33(1)(a) of the Act to act fairly and impartially as between the parties. Accordingly, if the arbitrator has been troubled by matters not raised by the parties and on which he wishes to rely in making his order as to costs, he must bring them to the attention of the parties before so doing so that the parties have an opportunity to address these issues. It was held in *Ghangbola v Smith & Sherriff Ltd* [1998] 3 All ER 730 that failure to do so constituted a serious irregularity within the meaning of s 68(2)(a) of the Act.

RECOVERABLE COSTS

The basis of costs and proportionality

[41.7]

Prior to the introduction of the CPR arbitrators were empowered and encouraged to assess costs themselves where possible and otherwise to refer them to the court, subject always to the right of the parties to agree what costs were recoverable. Costs were in effect on the standard basis as it was defined before 26 April 1999 unless the arbitrator ordered otherwise. However, the application of the CPR now incorporates the specific additional test of proportionality.

Section 63(3) of the Act permits the arbitrator to award costs on such basis 'as he thinks fit'. Where costs awarded between the parties are to be determined by the court the effect of sections 63(4) and (5) is that they are assessed on the standard basis as it was defined before the introduction of the CPR unless the arbitrator orders otherwise, namely costs of a reasonable amount in respect of those costs reasonably incurred. However, CPR 44.1(2) requires the court to apply CPR Parts 44–47 to arbitration proceedings and these include the principle of proportionality at CPR 44.3(2) and (5). We therefore appear to have the unsatisfactory, and no doubt unintentional, position that the arbitrator may, if he thinks fit, ignore the test of proportionality but the court must apply it (and as we have seen the new CPR 44.3(2) heightens the implications of a proportionality cross check on costs).

Non-solicitors

[41.8]

Where a party is represented in an arbitration by a person who is not qualified as a barrister or solicitor, but who provides similar services, and an award is made providing for payment of that party's costs by the other party or for such costs to be assessed in the High Court if not agreed, the court has power to allow the costs of the unqualified person in relation to the conduct of the arbitration. The prohibition in s 25(1) of the Solicitors Act 1974 against the recovery of costs in respect of anything done by any unqualified person 'acting as a solicitor' does not apply to an unqualified person representing a party in an arbitration since an unqualified person does not 'act as a solicitor' within the meaning of s 25(1) merely by doing acts of a kind commonly done by solicitors. Acts prohibited by s 25(1) are limited to acts which are lawful only for a qualified solicitor to do and which only a solicitor may perform, or acts purportedly done in that capacity. They do not include acts commonly done by a solicitor but which do not involve a representation that the person so acting is acting as a solicitor. A person acting as an advocate for a party in arbitration proceedings who is not qualified as a barrister or solicitor and does not hold himself out as such is not acting as a barrister or solicitor and accordingly the party employing him is not precluded from entitlement to payment of his costs (*Piper Double Glazing Ltd v DC Contracts (1992) Ltd* [1994] 1 All ER 177 – in which Potter J found that no part of the arbitral process involved acts which could be said to be 'acting as a solicitor').

Recoverability

[41.9]

Section 63 provides three methods of resolving the question of costs recoverability, namely by:
- agreement;
- determination by the arbitrator. In which case the arbitrator must specify the basis on which it has acted, the items of recoverable costs and the amount allowed for each item;
- application to the court in the absence of either of the above. Either party may apply to the court.

Does an award of costs 'to be agreed or assessed in default of agreement' permit the arbitrator to determine the costs, or does it amount either to a reference to the court or to a failure by the arbitrator to determine the costs, enabling the receiving party to apply to the court? In *M/S Alghanim Industries Inc v Skandia International Insurance Corpn* [2001] 2 All ER (Comm) 30 it was held that the phrase is not to be construed as a determination of the proceedings with a reference of assessment to the court in the event that the parties could not agree. The arbitrator had not expressly declined to assess and settle the costs, requiring the matter to go to the court. An award of costs to be agreed or assessed in default of an agreement is not a refusal to assess but is neutral in its language: it leaves open the possibility of an application by either party to the arbitrator. This was consistent with the spirit of the Act that as far as possible matters should be resolved by arbitration rather than application to the court.

The arbitrator's fees and expenses

[41.10]

Section 64 permits the recovery within the recoverable costs of the arbitration such reasonable fees and expenses of the arbitrator as are appropriate in the circumstances – again with the proviso that the parties can agree otherwise. Section 64(1) is also subject to any order of the court made under ss 24 and 25 of the Act on the removal or resignation of an arbitrator.

As already noted s 28 of the Act makes the parties jointly and severally liable for such appropriate reasonable fees and expenses. It also makes provision for the parties to agree with the arbitrator his fees and expenses, including capping and restrictions, or to apply to the court for them to be considered and adjusted.

In addition to the provisions of s 64, s 37(2) provides that 'expenses' may include the fees and expenses of experts, legal advisers and assessors appointed by the arbitrator. Of course, for these to be recoverable, they too must be reasonable.

Any dispute about the reasonableness of an arbitrator's fees and expenses may be resolved by the court under s 64(2). However, this is subject to the right of the parties and the arbitrator set out in s 64(4) to agree the amount of his fees and expenses at the outset, creating a contract with which the court cannot later interfere. If the agreement relates only to hourly or daily rates, the court can investigate the reasonableness of the time spent, but not the agreed amounts.

If an application is made to the court to determine the fees and expenses, s 28(2) makes provision for the court to order that the amount of an arbitrator's fees and expenses shall be considered and adjusted by such means and upon such terms as it may direct. This course was successfully adopted by the paying party in *Agrimex Ltd v Tradigrain SA* [2003] EWHC 1656 (Comm), in respect of the fees of a legal draftsman employed by the arbitrators to assist them in drafting the award. The draftsman was a solicitor of only three years' post-qualification experience, but charged £9,300 out of total costs of under £20,000. After deprecating the use of legal draftsmen except in very special circumstances, the court held that the charges were disproportionate and that no competent lawyer could have spent the amount of time upon which the claim was based. The solicitor's fee was allowed at £5,000 – which would have been even lower had not this been the amount proposed by the paying party!

It is not improper for a party appointing an arbitrator to agree his fees before appointment since the appointing party and the proposed arbitrator are respectively free to appoint someone else or not to accept the appointment if the terms are not mutually acceptable. However, once an appointment had been made it is contrary to the arbitrator's quasi-judicial status for him to bargain unilaterally with only one party for his fees and any agreement made between the appointing party and his arbitrator after he had accepted appointment for the payment of his fees without the consent of the other party, probably constitutes misconduct, and is in any event liable to render the arbitrator vulnerable to the imputation of bias. Although it is not improper for an arbitrator to stipulate at the time of his appointment for a commitment fee to be made payable in any event even if the arbitration does not take place, once appointed an arbitrator is not entitled unilaterally to change the terms of his contract by demanding a commitment fee unless there is a significant and substantial change in the commitment required of him which justifies the payment of a further fee (*Norjarl K/S A/S v Hyundai Heavy Industries Co Ltd* [1992] 1 QB 863).

Section 56 empowers the arbitrator to refuse to deliver his award unless his fees and expenses are paid in full. This is no doubt designed to prevent the situation where the arbitrator fears that all parties are likely to be so dissatisfied with the decision that obtaining payment from either will prove difficult.

Limits on the recoverable costs

[41.11]

Adopting a philosophy with which the Civil Procedure Rules are only now adopting, s 65 of the Arbitration Act 1996 empowers the arbitrator, unless the parties have agreed otherwise, to limit the costs recoverable in the arbitration either as a whole or in respect of a specified part. This enables the arbitrator either on his own initiative or, more likely, on the application of one of the parties to specify prospectively the maximum liability for costs of the arbitration. Section 65(2) provides that any such direction may be made at any stage but this must be done sufficiently far in advance of the costs to which the limit relates being incurred for the limit to be of effect. A clear harbinger of the prospective costs management orders introduced in CPR 3.12–3.18 seventeen years later.

However, as proportionality does not feature in costs dealt with by the arbitrator (and as a consequence in the relevant procedures adopted in the arbitration), this will not necessarily act as a brake to prevent either party spending a disproportionate amount on costs (as it might under the overriding objective and costs management within the CPR), but it will put a cap on the amount they can recover from the other party.

Any cap imposed may be varied at any time under s 65(2), but again subject to the qualifications that the variation cannot be retrospective and must be made sufficiently in advance of the costs under the exiting cap being incurred for the lower limit sought to be taken into account. In other words if the costs have already been spent or the attempt to vary is late in the process when the parties are committed to a course from which it is not possible to deviate, there should be no variation.

Under s 57(3)(a) of the 1996 Act, an arbitrator has the default power to 'correct an award so as to remove any clerical mistake or error arising from an accidental slip or omission'. In *Gannet Shipping Ltd v Eastrade Commodities Inc* [2002] 1 All ER (Comm) 297, the court held that the arbitrator also had jurisdiction to vary the costs order at the same time. The costs error was an error 'arising from' the 'accidental slip' in the amount awarded, and accordingly it could be corrected. Furthermore, the court found that the costs award could also have been amended under the jurisdiction to intervene conferred by both s 62(2)(a) (failure by the tribunal to comply with the natural justice principle in s 33 of the 1996 Act) and s 62(2)(i) (irregularity in the conduct of the proceedings or in the award which is admitted by the tribunal). The only limiting factor on the intervention is the need to show substantial injustice.

FAILURE TO GIVE REASONS

[41.12]

Under s 52(3) of the act, an award 'shall contain reasons for the award'. In the light of this statutory requirement it is odd that there is no sanction in s 53(3) for a failure to include reasons. A failure to include reasons for a costs award does not render the award void but simply allows a party to seek reasons by means of an application to the court under s 68 of the 1996 Act on the basis that the arbitrator has failed to comply with the requirements as to the form of the award (s 68(2)(h); *Ridler v Walter* [2001] TASSC 98, Supreme Court of Tasmania).

APPLICATIONS TO THE COURT

[41.13]

There are a number of provisions in the Act under which costs issues arising from an arbitration may be put before the court. We have considered some of them but by way of summary these are:
- s 28(2) to have the amount of the arbitrator's fees and expenses considered and adjusted;
- s 56(2) for a review of the arbitrator's fees and expenses;
- s 63(4) to determine the recoverable costs of the arbitration;
- s 64(2) to determine the reasonable fees and expenses of the arbitrator;
- s 68 on grounds of serious irregularity;
- s 69 on the grounds of a serious error of law in the award of costs;
- s 70(4) for an order that the arbitrator provides proper reasons for his award of costs.

THE COSTS OF ENFORCEMENT

[41.14]

Enforcement of an award involves an application under CPR 62.18 for permission to enforce the award as a judgment with the option of entering

41.14 *Chapter 41 Arbitration*

judgment in the terms of the award. The application should be for costs to be included in the order giving permission and if judgment is to be obtained 'for the costs of any judgment to be entered'

SECURITY FOR COSTS

[41.15]

(See Chapter 19 – Prospective Costs Control Security for Costs, above).

CONCLUSION

[41.16]

Whilst any decision about costs relating to an arbitration remains with the arbitrator, the Arbitration Act 1996 provides a statutory code for determination of any issues arising. Once matters have been put before the court, then the provisions of either CPR Part 62 or, if the matter relates to costs, CPR Parts 44–47 apply.

Appendix

Civil Procedure Rules and Practice Directions

Part 3	The court's case and costs management powers
Part 36	Offers to settle and payments into court
Part 44	General rules about costs
Part 45	Fixed costs
Part 46	Costs Payable by or to Particular Persons
Part 47	Procedure for detailed assessment of costs and default provisions
Part 48	Part 2 of the Legal Aid, Sentencing and Punishment of Offenders Act 2012 relating to civil litigation funding and costs: transitional provision in relation to pre-commencement funding arrangements

Solicitors Act 1974

Part III	Remuneration of solicitors

CPR 3.1 *Appendix*

PART 3   THE COURT'S CASE AND COSTS MANAGEMENT POWERS

I
CASE MANAGEMENT

Rule 3.1	The court's general powers of management	CPR 3.1
Rule 3.2	Court officer's power to refer to a judge	CPR 3.2
Rule 3.3	Court's power to make order of its own initiative	CPR 3.3
Rule 3.4	Power to strike out a statement of case	CPR 3.4
Rule 3.5	Judgment without trial after striking out	CPR 3.5
Rule 3.5A	Automatic transfer	CPR 3.5A
Rule 3.6	Setting aside judgment entered after striking out	CPR 3.6
Rule 3.7	Sanctions for non-payment of certain fees	CPR 3.7
Rule 3.7A		CPR 3.7A
Rule 3.7B	Sanctions for dishonouring cheque	CPR 3.7B
Rule 3.8	Sanctions have effect unless defaulting party obtains relief	CPR 3.8
Rule 3.9	Relief from sanctions	CPR 3.9
Rule 3.10	General power of the court to rectify matters where there has been an error of procedure	CPR 3.10
Rule 3.11	Power of the court to make civil restraint orders	CPR 3.11

II
COSTS MANAGEMENT

Rule 3.12	Application of this Section and the purpose of costs management	CPR 3.12
Rule 3.13	Filing and exchanging budgets	CPR 3.13
Rule 3.14	Failure to file a budget	CPR 3.14
Rule 3.15	Costs management orders	CPR 3.15
Rule 3.16	Costs management conferences	CPR 3.16
Rule 3.17	Court to have regard to budgets and to take account of costs	CPR 3.17
Rule 3.18	Assessing costs on the standard basis where a costs management order has been made	CPR 3.18

III
COSTS CAPPING

Rule 3.19	Costs capping orders - General	CPR 3.19
Rule 3.20	Application for a costs capping order	CPR 3.20
Rule 3.21	Application to vary a costs capping order	CPR 3.21
	Practice Direction 3E—Costs Management	CPR PD 3E

I CASE MANAGEMENT

[CPR 3.1]

3.1 The court's general powers of management

(1)   The list of powers in this rule is in addition to any powers given to the court by any other rule or practice direction or by any other enactment or any powers it may otherwise have.

(2)   Except where these Rules provide otherwise, the court may—

    (a)   extend or shorten the time for compliance with any rule, practice direction or court order (even if an application for extension is made after the time for compliance has expired);

    (b)   adjourn or bring forward a hearing;

    (c)   require a party or a party's legal representative to attend the court;

    (d)   hold a hearing and receive evidence by telephone or by using any other method of direct oral communication;

    (e)   direct that part of any proceedings (such as a **counterclaim**) be dealt with as separate proceedings;

    (f)   **stay** [GL] the whole or part of any proceedings or judgment either generally or until a specified date or event;

(g) consolidate proceedings;
(h) try two or more claims on the same occasion;
(i) direct a separate trial of any issue;
(j) decide the order in which issues are to be tried;
(k) exclude an issue from consideration;
(l) dismiss or give judgment on a claim after a decision on a preliminary issue;
(ll) order any party to file and exchange a costs budget;
(m) take any other step or make any other order for the purpose of managing the case and furthering the overriding objective.

(3) When the court makes an order, it may—
(a) make it subject to conditions, including a condition to pay a sum of money into court; and
(b) specify the consequence of failure to comply with the order or a condition.

(4) Where the court gives directions it will take into account whether or not a party has complied with the Practice Direction (Pre-Action Conduct) and any relevant **pre-action protocol** (GL).

(5) The court may order a party to pay a sum of money into court if that party has, without good reason, failed to comply with a rule, practice direction or a relevant pre-action protocol.

(6) When exercising its power under paragraph (5) the court must have regard to—
(a) the amount in dispute; and
(b) the costs which the parties have incurred or which they may incur.

(6A) Where a party pays money into court following an order under paragraph (3) or (5), the money shall be security for any sum payable by that party to any other party in the proceedings.

(7) A power of the court under these Rules to make an order includes a power to vary or revoke the order.

(8) The court may contact the parties from time to time in order to monitor compliance with directions. The parties must respond promptly to any such enquiries from the court.

[CPR 3.2]

3.2 Court officer's power to refer to a judge

Where a step is to be taken by a court officer—
(a) the court officer may consult a judge before taking that step;
(b) the step may be taken by a judge instead of the court officer.

[CPR 3.3]

3.3 Court's power to make order of its own initiative

(1) Except where a rule or some other enactment provides otherwise, the court may exercise its powers on an application or of its own initiative.
(Part 23 sets out the procedure for making an application)
(2) Where the court proposes to make an order of its own initiative—
(a) it may give any person likely to be affected by the order an opportunity to make representations; and
(b) where it does so it must specify the time by and the manner in which the representations must be made.

CPR 3.4 Appendix

(3) Where the court proposes—
  (a) to make an order of its own initiative; and
  (b) to hold a hearing to decide whether to make the order,
it must give each party likely to be affected by the order at least 3 days' notice of the hearing.

(4) The court may make an order of its own initiative, without hearing the parties or giving them an opportunity to make representations.

(5) Where the court has made an order under paragraph (4)—
  (a) a party affected by the order may apply to have it **set aside** (GL), varied or **stayed** (GL); and
  (b) the order must contain a statement of the right to make such an application.

(6) An application under paragraph (5)(a) must be made—
  (a) within such period as may be specified by the court; or
  (b) if the court does not specify a period, not more than 7 days after the date on which the order was served on the party making the application.

(7) If the court of its own initiative strikes out a statement of case or dismisses an application (including an application for permission to appeal or for permission to apply for judicial review), and it considers that the claim or application is totally without merit—
  (a) the court's order must record that fact; and
  (b) the court must at the same time consider whether it is appropriate to make a civil restraint order.

[CPR 3.4]

3.4 Power to strike out a statement of case

(1) In this rule and rule 3.5, reference to a statement of case includes reference to part of a statement of case.

(2) The court may **strike out** (GL) a statement of case if it appears to the court—
  (a) that the statement of case discloses no reasonable grounds for bringing or defending the claim;
  (b) that the statement of case is an abuse of the court's process or is otherwise likely to obstruct the just disposal of the proceedings; or
  (c) that there has been a failure to comply with a rule, practice direction or court order.

(3) When the court strikes out a statement of case it may make any consequential order it considers appropriate.

(4) Where—
  (a) the court has struck out a claimant's statement of case;
  (b) the claimant has been ordered to pay costs to the defendant; and
  (c) before the claimant pays those costs, he starts another claim against the same defendant, arising out of facts which are the same or substantially the same as those relating to the claim in which the statement of case was struck out,
the court may, on the application of the defendant, **stay** (GL) that other claim until the costs of the first claim have been paid.

(5) Paragraph (2) does not limit any other power of the court to **strike out** (GL) a statement of case.

(6) If the court strikes out a claimant's statement of case and it considers that the claim is totally without merit—

(a) the court's order must record that fact; and
(b) the court must at the same time consider whether it is appropriate to make a civil restraint order.

[CPR 3.5]

3.5 Judgment without trial after striking out

(1) This rule applies where—
    (a) the court makes an order which includes a term that the statement of case of a party shall be struck out if the party does not comply with the order; and
    (b) the party against whom the order was made does not comply with it.
(2) A party may obtain judgment with costs by filing a request for judgment if—
    (a) the order referred to in paragraph (1)(a) relates to the whole of a statement of case; and
    (b) where the party wishing to obtain judgment is the claimant, the claim is for—
        (i) a specified amount of money;
        (ii) an amount of money to be decided by the court;
        (iii) delivery of goods where the claim form gives the defendant the alternative of paying their value; or
        (iv) any combination of these remedies.
(3) Where judgment is obtained under this rule in a case to which paragraph 2(b)(iii) applies, it will be judgment requiring the defendant to deliver the goods, or (if he does not do so) pay the value of the goods as decided by the court (less any payments made).
(4) The request must state that the right to enter judgment has arisen because the court's order has not been complied with.
(5) A party must make an application in accordance with Part 23 if he wishes to obtain judgment under this rule in a case to which paragraph (2) does not apply.

[CPR 3.5A]

3.5A Automatic transfer

If—
    (a) a claimant files a request for judgment which includes an amount of money to be decided by the court in accordance with rule 3.5; and
    (b) the claim is a designated money claim,
the court will transfer the claim to the preferred court upon receipt of the request for judgment.

[CPR 3.6]

3.6 Setting aside judgment entered after striking out

(1) A party against whom the court has entered judgment under rule 3.5 may apply to the court to set the judgment aside.
(2) An application under paragraph (1) must be made not more than 14 days after the judgment has been served on the party making the application.
(3) If the right to enter judgment had not arisen at the time when judgment was entered, the court must **set aside** (GL) the judgment.
(4) If the application to **set aside** (GL) is made for any other reason, rule 3.9 (relief from sanctions) shall apply.

CPR 3.7 Appendix

[CPR 3.7]

3.7 Sanctions for non-payment of certain fees
(1) This rule applies where—
   (a) a directions questionnaire or a pre-trial check list (listing questionnaire) is filed without payment of the fee specified by the relevant Fees Order;
   (b) the court dispenses with the need for a directions questionnaire or a pre-trial check list or both;
   (c) these Rules do not require a directions questionnaire or a pre-trial check list to be filed in relation to the claim in question; *or*
   (d) the court has made an order giving permission to proceed with a claim for judicial review; or
   (e) the fee payable for a hearing specified by the relevant Fees Order is not paid.

(Rule 26.3 provides for the court to dispense with the need for a directions questionnaire and rules 28.5 and 29.6 provide for the court to dispense with the need for a pre-trial check list)

(Rule 54.12 provides for the **service** of the order giving permission to proceed with a claim for judicial review)

(2) The court will serve a notice on the claimant requiring payment of the fee specified in the relevant Fees Order if, at the time the fee is due, the claimant has not paid it or made an application for full or part remission.
(3) The notice will specify the date by which the claimant must pay the fee.
(4) If the claimant does not—
   (a) pay the fee; or
   (b) make an application for full or part remission of the fee,
by the date specified in the notice—
      (i) the claim will automatically be struck out without further order of the court; and
      (ii) the claimant will be liable for the costs which the defendant has incurred unless the court orders otherwise.

(Rule 44.9 provides for the basis of assessment where a right to costs arises under this rule and contains provisions about when a costs order is deemed to have been made and applying for an order under section 194(3) of the Legal Services Act 2007)

(5) Where an application for—
   (a) full or part remission of a fee is refused, the court will serve notice on the claimant requiring payment of the full fee by the date specified in the notice; or
   (b) part remission of a fee is granted, the court will serve notice on the claimant requiring payment of the balance of the fee by the date specified in the notice.
(6) If the claimant does not pay the fee by the date specified in the notice—
   (a) the claim will automatically be struck out without further order of the court; and
   (b) the claimant will be liable for the costs which the defendant has incurred unless the court orders otherwise.
(7) If—
   (a) a claimant applies to have the claim reinstated; and
   (b) the court grants relief,

the relief will be conditional on the claimant either paying the fee or filing evidence of full or part payment or remission of the fee within the period specified in paragraph (8).

(8) The period referred to in paragraph (7) is—
- (a) if the order granting relief is made at a hearing at which a claimant is present or represented, 2 days from the date of the order;
- (b) in any other case, 7 days from the date of service of the order on the claimant.

[CPR 3.7A]

3.7A

(1) This rule applies where—
- (a) a defendant files a counterclaim without—
  - (i) payment of the fee specified by the relevant Fees Order; or
  - (ii) making an application for full or part remission of the fee; or
- (b) the proceedings continue on the counterclaim alone and—
  - (i) a directions questionnaire or a pre-trial check list (listing questionnaire) is filed without payment of the fee specified by the relevant Fees Order;
  - (ii) the court dispenses with the need for a directions questionnaire or a pre-trial check list or both;
  - (iii) these Rules do not require an allocation questionnaire or a pre-trial checklist to be filed in relation to the claim in question; or
  - (iv) the fee payable for a hearing specified by the relevant Fees Order is not paid.

(2) The court will serve a notice on the defendant requiring payment of the fee specified in the relevant Fees Order if, at the time the fee is due, the defendant has not paid it or made an application for full or part remission.

(3) The notice will specify the date by which the defendant must pay the fee.

(4) If the defendant does not—
- (a) pay the fee; or
- (b) make an application for full or part remission of the fee,

by the date specified in the notice, the counterclaim will automatically be struck out without further order of the court.

(5) Where an application for—
- (a) full or part remission of a fee is refused, the court will serve notice on the defendant requiring payment of the full fee by the date specified in the notice; or
- (b) part remission of a fee is granted, the court will serve notice on the defendant requiring payment of the balance of the fee by the date specified in the notice.

(6) If the defendant does not pay the fee by the date specified in the notice, the counterclaim will automatically be struck out without further order of the court.

(7) If—
- (a) the defendant applies to have the counterclaim reinstated; and
- (b) the court grants relief,

the relief will be conditional on the defendant either paying the fee or filing evidence of full or part remission of the fee within the period specified in paragraph (8).

(8) The period referred to in paragraph (7) is—

CPR 3.7B *Appendix*

    (a)   if the order granting relief is made at a hearing at which the defendant is present or represented, 2 days from the date of the order;

    (b)   in any other case, 7 days from the date of service of the order on the defendant.

[CPR 3.7B]

3.7B Sanctions for dishonouring cheque

(1) This rule applies where any fee is paid by cheque and that cheque is subsequently dishonoured.

(2) The court will serve a notice on the paying party requiring payment of the fee which will specify the date by which the fee must be paid.

(3) If the fee is not paid by the date specified in the notice –

    (a)   where the fee is payable by the claimant, the claim will automatically be struck out without further order of the court;

    (b)   where the fee is payable by the defendant, the defence will automatically be struck out without further order of the court,

and the paying party shall be liable for the costs which any other party has incurred unless the court orders otherwise.

(Rule 44.9 provides for the basis of assessment where a right to costs arises under this rule)

(4) If–

    (a)   the paying party applies to have the claim or defence reinstated; and

    (b)   the court grants relief,

the relief shall be conditional on that party paying the fee within the period specified in paragraph (5).

(5) The period referred to in paragraph (4) is–

    (a)   if the order granting relief is made at a hearing at which the paying party is present or represented, 2 days from the date of the order;

    (b)   in any other case, 7 days from the date of service of the order on the paying party.

(6) For the purposes of this rule, 'claimant' includes a Part 20 claimant and 'claim form' includes a Part 20 claim.

[CPR 3.8]

3.8 Sanctions have effect unless defaulting party obtains relief

(1) Where a party has failed to comply with a rule, practice direction or court order, any sanction for failure to comply imposed by the rule, practice direction or court order has effect unless the party in default applies for and obtains relief from the sanction.

(Rule 3.9 sets out the circumstances which the court will consider on an application to grant relief from a sanction)

(2) Where the sanction is the payment of costs, the party in default may only obtain relief by appealing against the order for costs.

(3) Where a rule, practice direction or court order—

    (a)   requires a party to do something within a specified time, and

    (b)   specifies the consequence of failure to comply,

the time for doing the act in question may not be extended by agreement between the parties.

[CPR 3.9]

3.9 Relief from sanctions

(1) On an application for relief from any sanction imposed for a failure to comply with any rule, practice direction or court order, the court will consider all the circumstances of the case, so as to enable it to deal justly with the application, including the need—
    (a)   for litigation to be conducted efficiently and at proportionate cost; and
    (b)   to enforce compliance with rules, practice directions and orders.

(2) An application for relief must be supported by evidence.

[CPR 3.10]

3.10 General power of the court to rectify matters where there has been an error of procedure

Where there has been an error of procedure such as a failure to comply with a rule or practice direction—
    (a)   the error does not invalidate any step taken in the proceedings unless the court so orders; and
    (b)   the court may make an order to remedy the error.

[CPR 3.11]

3.11 Power of the court to make civil restraint orders

A practice direction may set out—
    (a)   the circumstances in which the court has the power to make a civil restraint order against a party to proceedings;
    (b)   the procedure where a party applies for a civil restraint order against another party; and
    (c)   the consequences of the court making a civil restraint order.

II COSTS MANAGEMENT

[CPR 3.12]

3.12 Application of this Section and the purpose of costs management

[(1) This Section and Practice Direction 3E apply to all multi-track cases commenced on or after 1st April 2013 except—
    (a)   cases in the Admiralty and Commercial Courts;
    (b)   such cases in the Chancery Division as the Chancellor of the High Court may direct; and
    (c)   such cases in the Technology and Construction Court and the Mercantile Court as the President of the Queen's Bench Division may direct,

unless the proceedings are the subject of fixed costs or scale costs or the court otherwise orders. This Section and Practice Direction 3E shall apply to any other proceedings (including applications) where the court so orders.

(2) The purpose of costs management is that the court should manage both the steps to be taken and the costs to be incurred by the parties to any proceedings so as to further the overriding objective.

CPR 3.13   *Appendix*

[CPR 3.13]

3.13 Filing and exchanging budgets

Unless the court otherwise orders, all parties except litigants in person must file and exchange budgets as required by the rules or as the court otherwise directs. Each party must do so by the date specified in the notice served under rule 26.3(1) or, if no such date is specified, seven days before the first case management conference.

[CPR 3.14]

3.14 Failure to file a budget

Unless the court otherwise orders, any party which fails to file a budget despite being required to do so will be treated as having filed a budget comprising only the applicable court fees.

[CPR 3.15]

3.15 Costs management orders

(1)   In addition to exercising its other powers, the court may manage the costs to be incurred by any party in any proceedings.
(2)   The court may at any time make a "costs management order". By such order the court will—
   (a)   record the extent to which the budgets are agreed between the parties;
   (b)   in respect of budgets or parts of budgets which are not agreed, record the court's approval after making appropriate revisions.
(3)   If a costs management order has been made, the court will thereafter control the parties' budgets in respect of recoverable costs.

[CPR 3.16]

3.16 Costs management conferences

(1)   Any hearing which is convened solely for the purpose of costs management (for example, to approve a revised budget) is referred to as a "costs management conference".
(2)   Where practicable, costs management conferences should be conducted by telephone or in writing.

[CPR 3.17]

3.17 Court to have regard to budgets and to take account of costs

(1)   When making any case management decision, the court will have regard to any available budgets of the parties and will take into account the costs involved in each procedural step.
(2)   Paragraph (1) applies whether or not the court has made a costs management order.

[CPR 3.18]

3.18 Assessing costs on the standard basis where a costs management order has been made

In any case where a costs management order has been made, when assessing costs on the standard basis, the court will—

(a) have regard to the receiving party's last approved or agreed budget for each phase of the proceedings; and
(b) not depart from such approved or agreed budget unless satisfied that there is good reason to do so.

(Attention is drawn to rule 44.3(2)(a) and rule 44.3(5), which concern proportionality of costs.)

III COSTS CAPPING

[CPR 3.19]

3.19 **Costs capping orders – General**

(1) A costs capping order is an order limiting the amount of future costs (including disbursements) which a party may recover pursuant to an order for costs subsequently made.

(2) In this rule, "future costs" means costs incurred in respect of work done after the date of the costs capping order but excluding the amount of any additional liability.

(3) This rule does not apply to protective costs orders.

(4) A costs capping order may be in respect of –
 (a) the whole litigation; or
 (b) any issues which are ordered to be tried separately.

(5) The court may at any stage of proceedings make a costs capping order against all or any of the parties, if—
 (a) it is in the interests of justice to do so;
 (b) there is a substantial risk that without such an order costs will be disproportionately incurred; and
 (c) it is not satisfied that the risk in subparagraph (b) can be adequately controlled by–
  (i) case management directions or orders made under this Part; and
  (ii) detailed assessment of costs.

(6) In considering whether to exercise its discretion under this rule, the court will consider all the circumstances of the case, including—
 (a) whether there is a substantial imbalance between the financial position of the parties;
 (b) whether the costs of determining the amount of the cap are likely to be proportionate to the overall costs of the litigation;
 (c) the stage which the proceedings have reached; and
 (d) the costs which have been incurred to date and the future costs.

(7) A costs capping order, once made, will limit the costs recoverable by the party subject to the order unless a party successfully applies to vary the order. No such variation will be made unless—
 (a) there has been a material and substantial change of circumstances since the date when the order was made; or
 (b) there is some other compelling reason why a variation should be made.

[CPR 3.20]

3.20 **Application for a costs capping order**

(1) An application for a costs capping order must be made on notice in accordance with Part 23.

CPR 3.21   *Appendix*

(2) The application notice must –
- (a) set out –
  - (i) whether the costs capping order is in respect of the whole of the litigation or a particular issue which is ordered to be tried separately; and
  - (ii) why a costs capping order should be made; and
- (b) be accompanied by a budget setting out –
  - (i) the costs (and disbursements) incurred by the applicant to date; and
  - (ii) the costs (and disbursements) which the applicant is likely to incur in the future conduct of the proceedings.

(3) The court may give directions for the determination of the application and such directions may –
- (a) direct any party to the proceedings –
  - (i) to file a schedule of costs in the form set out in paragraph 3 of Practice Direction 3F – Costs capping;
  - (ii) to file written submissions on all or any part of the issues arising;
- (b) fix the date and time estimate of the hearing of the application;
- (c) indicate whether the judge hearing the application will sit with an assessor at the hearing of the application; and
- (d) include any further directions as the court sees fit.

[CPR 3.21]

3.21 Application to vary a costs capping order

An application to vary a costs capping order must be made by application notice pursuant to Part 23.

PRACTICE DIRECTION 3E—COSTS MANAGEMENT

THIS PRACTICE DIRECTION SUPPLEMENTS SECTION II OF CPR PART 3

Budget format

[CPR PD 3E.1]

1. Unless the court otherwise orders, a budget must be in the form of Precedent H annexed to this Practice Direction. It must be in landscape format with an easily legible typeface. In substantial cases, the court may direct that budgets be limited initially to part only of the proceedings and subsequently extended to cover the whole proceedings. A budget must be dated and verified by a statement of truth signed by a senior legal representative of the party. In cases where a party's budgeted costs do not exceed £25,000, there is no obligation on that party to complete more than the first page of Precedent H.
(The wording for a statement of truth verifying a budget is set out in Practice Direction 22.)

Costs management orders

[CPR PD 3E.2]

2.1 If the court makes a costs management order under rule 3.15, the following paragraphs apply.
2.2 Save in exceptional circumstances:

(1) the recoverable costs of initially completing Precedent H shall not exceed the higher of £1,000 or 1% of the approved budget;
(2) all other recoverable costs of the budgeting and costs management process shall not exceed 2% of the approved budget.

2.3 If the budgets or parts of the budgets are agreed between all parties, the court will record the extent of such agreement. In so far as the budgets are not agreed, the court will review them and, after making any appropriate revisions, record its approval of those budgets. The court's approval will relate only to the total figures for each phase of the proceedings, although in the course of its review the court may have regard to the constituent elements of each total figure. When reviewing budgets, the court will not undertake a detailed assessment in advance, but rather will consider whether the budgeted costs fall within the range of reasonable and proportionate costs.

2.4 As part of the costs management process the court may not approve costs incurred before the date of any budget. The court may, however, record its comments on those costs and should take those costs into account when considering the reasonableness and proportionality of all subsequent costs.

2.5 The court may set a timetable or give other directions for future reviews of budgets.

2.6 Each party shall revise its budget in respect of future costs upwards or downwards, if significant developments in the litigation warrant such revisions. Such amended budgets shall be submitted to the other parties for agreement. In default of agreement, the amended budgets shall be submitted to the court, together with a note of (a) the changes made and the reasons for those changes and (b) the objections of any other party. The court may approve, vary or disapprove the revisions, having regard to any significant developments which have occurred since the date when the previous budget was approved or agreed.

2.7 After its budget has been approved, each party shall re-file and re-serve the budget in the form approved with re-cast figures, annexed to the order approving it.

2.8 A litigant, in person, even though not required to prepare a budget, shall nevertheless be provided with a copy of the budget of any other party.

2.9 If interim applications are made which, reasonably, were not included in a budget, then the costs of such interim applications shall be treated as additional to the approved budgets.

CPR 36.A1 *Appendix*

PART 36   OFFERS TO SETTLE AND PAYMENTS INTO COURT

Rule 36.A1   Scope of this Part .... CPR 36.A1

I
PART 36 OFFER TO SETTLE

Rule 36.1	Scope of this Section	CPR 36.1
Rule 36.2	Form and content of a Part 36 offer	CPR 36.2
Rule 36.3	Part 36 offers—general provisions	CPR 36.3
Rule 36.4	Part 36 offers—defendants' offers	CPR 36.4
Rule 36.5	Personal injury claims for future pecuniary loss	CPR 36.5
Rule 36.6	Offer to settle a claim for provisional damages	CPR 36.6
Rule 36.7	Time when a Part 36 offer is made and accepted ..	CPR 36.7
Rule 36.8	Clarification of a Part 36 offer	CPR 36.8
Rule 36.9	Acceptance of a Part 36 offer	CPR 36.9
Rule 36.10	Costs consequences of acceptance of a Part 36 offer	CPR 36.10
Rule 36.10A	Costs consequences of acceptance of a Part 36 offer where Section IIIA of Part 45 applies	CPR 36.10A
Rule 36.11	The effect of acceptance of a Part 36 offer	CPR 36.11
Rule 36.12	Acceptance of a Part 36 offer or a Part 36 payment made by one or more, but not all, defendants	CPR 36.12
Rule 36.13	Restriction on disclosure of a Part 36 offer or a Part 36 payment ..	CPR 36.13
Rule 36.14	Costs consequences following judgment	CPR 36.14
Rule 36.14A	Costs consequences following judgment where Section IIIA of Part 45 applies	CPR 36.14A
Rule 36.15	Deduction of benefits and lump sum payments	CPR 36.15

II
RTA PROTOCOL OFFERS TO SETTLE

Rule 36.16	Scope of this Section	CPR 36.16
Rule 36.17	Form and content of an RTA Protocol offer	CPR 36.17
Rule 36.18	Time when an RTA Protocol offer is made	CPR 36.18
Rule 36.19	General provisions	CPR 36.19
Rule 36.20	Restrictions on disclosure of an RTA Protocol offer	CPR 36.20
Rule 36.21	Costs consequences following judgment	CPR 36.21
Rule 36.22	Deduction of benefits	CPR 36.22
	Practice Direction 36A—Offers to Settle	CPR PD 36A
	Practice Direction 36B—	CPR PD 36B

[CPR 36.A1]

36.A1 Scope of this Part

(1)   This Part contains rules about—

  (a)   offers to settle; and

  (b)   the consequences where an offer to settle is made in accordance with this Part.

(2)   Section I of this Part contains rules about offers to settle other than where Section II applies.

(3)   Section II of this Part contains rules about offers to settle where the parties have followed the Pre-Action Protocol for Low Value Personal Injury Claims in Road Traffic Accidents ("the RTA Protocol") or the Pre-action Protocol for Low Value Personal Injury (Employers' Liability and Public Liability) Claims ("the EL/PL Protocol") and have started proceedings under Part 8 in accordance with Practice Direction 8B.

SECTION I PART 36 OFFERS TO SETTLE

[CPR 36.1]

36.1 Scope of this Section
(1) This Section does not apply to an offer to settle to which Section II of this Part applies.
(2) Nothing in this Section prevents a party making an offer to settle in whatever way he chooses, but if the offer is not made in accordance with rule 36.2, it will not have the consequences specified in rules 36.10, 36.11 and 36.14.
(Rule 44.2 requires the court to consider an offer to settle that does not have the costs consequences set out in this Section in deciding what order to make about costs)

[CPR 36.2]

36.2 Form and content of a Part 36 offer
(1) An offer to settle which is made in accordance with this rule is called a Part 36 offer.
(2) A Part 36 offer must—
    (a) be in writing;
    (b) state on its face that it is intended to have the consequences of Section I of Part 36;
    (c) specify a period of not less than 21 days within which the defendant will be liable for the claimant's costs in accordance with rule 36.10 if the offer is accepted;
    (d) state whether it relates to the whole of the claim or to part of it or to an issue that arises in it and if so to which part or issue; and
    (e) state whether it takes into account any counterclaim.
(Rule 36.7 makes provision for when a Part 36 offer is made)
(3) Rule 36.2(2)(c) does not apply if the offer is made less than 21 days before the start of the trial.
(4) In appropriate cases, a Part 36 offer must contain such further information as is required by rule 36.5 (Personal injury claims for future pecuniary loss), rule 36.6 (Offer to settle a claim for provisional damages), and rule 36.15 (Deduction of benefits).
(5) An offeror may make a Part 36 offer solely in relation to liability.

[CPR 36.3]

36.3 Part 36 offers—general provisions
(1) In this Part—
    (a) the party who makes an offer is the "offeror";
    (b) the party to whom an offer is made is the "offeree"; and
    (c) "the relevant period" means—
        (i) in the case of an offer made not less than 21 days before trial, the period stated under rule 36.2(2)(c) or such longer period as the parties agree;
        (ii) otherwise, the period up to end of the trial or such other period as the court has determined.
(2) A Part 36 offer—
    (a) may be made at any time, including before the commencement of proceedings; and
    (b) may be made in appeal proceedings.

CPR 36.4 *Appendix*

(3) A Part 36 offer which offers to pay or offers to accept a sum of money will be treated as inclusive of all interest until—
    (a) the date on which the period stated under rule 36.2(2)(c) expires; or
    (b) if rule 36.2(3) applies, a date 21 days after the date the offer was made.

(4) A Part 36 offer shall have the consequences set out in this Section only in relation to the costs of the proceedings in respect of which it is made, and not in relation to the costs of any appeal from the final decision in those proceedings.

(5) Before expiry of the relevant period, a Part 36 offer may be withdrawn or its terms changed to be less advantageous to the offeree, only if the court gives permission.

(6) After expiry of the relevant period and provided that the offeree has not previously served notice of acceptance, the offeror may withdraw the offer or change its terms to be less advantageous to the offeree without the permission of the court.

(7) The offeror does so by serving written notice of the withdrawal or change of terms on the offeree.

(Rule 36.14(6) deals with the costs consequences following judgment of an offer that is withdrawn)

[CPR 36.4]

36.4 Part 36 offers—defendants' offers

(1) Subject to rule 36.5(3) and rule 36.6(1), a Part 36 offer by a defendant to pay a sum of money in settlement of a claim must be an offer to pay a single sum of money.

(2) But, an offer that includes an offer to pay all or part of the sum, if accepted, at a date later than 14 days following the date of acceptance will not be treated as a Part 36 offer unless the offeree accepts the offer.

[CPR 36.5]

36.5 Personal injury claims for future pecuniary loss

(1) This rule applies to a claim for damages for personal injury which is or includes a claim for future pecuniary loss.

(2) An offer to settle such a claim will not have the consequences set out in rules 36.10, 36.11 and 36.14 unless it is made by way of a Part 36 offer under this rule.

(3) A Part 36 offer to which this rule applies may contain an offer to pay, or an offer to accept—
    (a) the whole or part of the damages for future pecuniary loss in the form of—
        (i) a lump sum; or
        (ii) periodical payments; or
        (iii) both a lump sum and periodical payments;
    (b) the whole or part of any other damages in the form of a lump sum.

(4) A Part 36 offer to which this rule applies—
    (a) must state the amount of any offer to pay the whole or part of any damages in the form of a lump sum;
    (b) may state—
        (i) what part of the lump sum, if any, relates to damages for future pecuniary loss; and

(ii) what part relates to other damages to be accepted in the form of a lump sum;
(c) must state what part of the offer relates to damages for future pecuniary loss to be paid or accepted in the form of periodical payments and must specify—
  (i) the amount and duration of the periodical payments;
  (ii) the amount of any payments for substantial capital purchases and when they are to be made; and
  (iii) that each amount is to vary by reference to the retail prices index (or to some other named index, or that it is not to vary by reference to any index); and
(d) must state either that any damages which take the form of periodical payments will be funded in a way which ensures that the continuity of payment is reasonably secure in accordance with section 2(4) of the Damages Act 1996 or how such damages are to be paid and how the continuity of their payment is to be secured.

(5) Rule 36.4 applies to the extent that a Part 36 offer by a defendant under this rule includes an offer to pay all or part of any damages in the form of a lump sum.

(6) Where the offeror makes a Part 36 offer to which this rule applies and which offers to pay or to accept damages in the form of both a lump sum and periodical payments, the offeree may only give notice of acceptance of the offer as a whole.

(7) If the offeree accepts a Part 36 offer which includes payment of any part of the damages in the form of periodical payments, the claimant must, within 7 days of the date of acceptance, apply to the court for an order for an award of damages in the form of periodical payments under rule 41.8.

(Practice Direction 41B contains information about periodical payments under the Damages Act 1996)

[CPR 36.6]

36.6 Offer to settle a claim for provisional damages

(1) An offeror may make a Part 36 offer in respect of a claim which includes a claim for provisional damages.

(2) Where he does so, the Part 36 offer must specify whether or not the offeror is proposing that the settlement shall include an award of provisional damages.

(3) Where the offeror is offering to agree to the making of an award of provisional damages the Part 36 offer must also state—
(a) that the sum offered is in satisfaction of the claim for damages on the assumption that the injured person will not develop the disease or suffer the type of deterioration specified in the offer;
(b) that the offer is subject to the condition that the claimant must make any claim for further damages within a limited period; and
(c) what that period is.

(4) Rule 36.4 applies to the extent that a Part 36 offer by a defendant includes an offer to agree to the making of an award of provisional damages.

(5) If the offeree accepts the Part 36 offer, the claimant must, within 7 days of the date of acceptance, apply to the court for an order for an award of provisional damages under rule 41.2.

[CPR 36.7]

36.7 Time when a Part 36 offer is made

(1) A Part 36 offer is made when it is served on the offeree.

CPR 36.8 *Appendix*

(2) A change in the terms of a Part 36 offer will be effective when notice of the change is served on the offeree.
(Rule 36.3 makes provision about when permission is required to change the terms of an offer to make it less advantageous to the offeree)

[CPR 36.8]

36.8 Clarification of a Part 36 offer
(1) The offeree may, within 7 days of a Part 36 offer being made, request the offeror to clarify the offer.
(2) If the offeror does not give the clarification requested under paragraph (1) within 7 days of receiving the request, the offeree may, unless the trial has started, apply for an order that he does so.
(Part 23 contains provisions about making an application to the court)
(3) If the court makes an order under paragraph (2), it must specify the date when the Part 36 offer is to be treated as having been made.

[CPR 36.9]

36.9 Acceptance of a Part 36 offer
(1) A Part 36 offer is accepted by serving written notice of the acceptance on the offeror.
(2) Subject to rule 36.9(3), a Part 36 offer may be accepted at any time (whether or not the offeree has subsequently made a different offer) unless the offeror serves notice of withdrawal on the offeree.
(Rule 21.10 deals with compromise etc. by or on behalf of a child or protected party)
(3) The court's permission is required to accept a Part 36 offer where—
    (a) rule 36.12(4) applies;
    (b) rule 36.15(3)(b) applies, the relevant period has expired and further deductible amounts have been paid to the claimant since the date of the offer;
    (c) an apportionment is required under rule 41.3A; or
    (d) the trial has started.
(Rule 36.12 deals with offers by some but not all of multiple defendants)
(Rule 36.15 defines "deductible amounts")
(Rule 41.3A requires an apportionment in proceedings under the Fatal Accidents Act 1976 and Law Reform (Miscellaneous Provisions) Act 1934)
(4) Where the court gives permission under paragraph (3), unless all the parties have agreed costs, the court will make an order dealing with costs, and may order that the costs consequences set out in rule 36.10 will apply.
(5) Unless the parties agree, a Part 36 offer may not be accepted after the end of the trial but before judgment is handed down.

[CPR 36.10]

36.10 Costs consequences of acceptance of a Part 36 offer
(1) Subject to rule 36.10A and to paragraphs (2) and (4)(a) of this rule, where a Part 36 offer is accepted within the relevant period the claimant will be entitled to the costs of the proceedings up to the date on which notice of acceptance was served on the offeror.
(2) Where—
    (a) a defendant's Part 36 offer relates to part only of the claim; and

(b) at the time of serving notice of acceptance within the relevant period the claimant abandons the balance of the claim,

the claimant will be entitled to the costs of the proceedings up to the date of serving notice of acceptance unless the court orders otherwise.

(3) Costs under paragraphs (1) and (2) of this rule will be assessed on the standard basis if the amount of costs is not agreed.

(Rule 44.3(2) explains the standard basis for assessment of costs)

(Rule 44.9 contains provisions about when a costs order is deemed to have been made and applying for an order under section 194(3) of the Legal Services Act 2007.)

(4) Where—
   (a) a Part 36 offer that was made less than 21 days before the start of trial is accepted; or
   (b) a Part 36 offer is accepted after expiry of the relevant period,

if the parties do not agree the liability for costs, the court will make an order as to costs.

(5) Where paragraph (4)(b) applies, unless the court orders otherwise—
   (a) the claimant will be entitled to the costs of the proceedings up to the date on which the relevant period expired; and
   (b) the offeree will be liable for the offeror's costs for the period from the date of expiry of the relevant period to the date of acceptance.

(6) The claimant's costs include any costs incurred in dealing with the defendant's counterclaim if the Part 36 offer states that it takes into account the counterclaim.

[CPR 36.10A]

36.10A Costs consequences of acceptance of a Part 36 offer where Section IIIA of Part 45 applies

(1) This rule applies where a claim no longer continues under the RTA or EL/PL Protocol pursuant to rule 45.29A(1).

(2) Where a Part 36 offer is accepted within the relevant period, the claimant will be entitled to the fixed costs in Table 6B, Table 6C or Table 6D in Section IIIA of Part 45 for the stage applicable at the date on which notice of acceptance was served on the offeror.

(3) Where—
   (a) a defendant's Part 36 offer relates to part only of the claim; and
   (b) at the time of serving notice of acceptance within the relevant period the claimant abandons the balance of the claim,

the claimant will be entitled to the fixed costs in paragraph (2).

(4) Subject to paragraph (5), where a defendant's Part 36 offer is accepted after the relevant period—
   (a) the claimant will be entitled to the fixed costs in Table 6B, Table 6C or Table 6D in Section IIIA of Part 45 for the stage applicable at the date on which the relevant period expired; and
   (b) the claimant will be liable for the defendant's costs for the period from the date of expiry of the relevant period to the date of acceptance.

(5) Where the claimant accepts the defendant's Protocol offer after the date on which the claim leaves the Protocol—
   (a) the claimant will be entitled to the applicable Stage 1 and Stage 2 fixed costs in Table 6 or Table 6A in Section III of Part 45; and

CPR 36.11   *Appendix*

  (b) the claimant will be liable for the defendant's costs from the date on which the Protocol offer is deemed to be made to the date of acceptance.
(6) For the purposes of this rule a defendant's Protocol offer is either—
  (a) defined in accordance with rules 36.17 and 36.18; or
  (b) if the claim leaves the Protocol before the Court Proceedings Pack Form is sent to the defendant—
    (i) the last offer made by the defendant before the claim leaves the Protocol; and
    (ii) deemed to be made on the first business day after the claim leaves the Protocol.
(7) A reference to the 'Court Proceedings Pack Form' is a reference to the form used in the Protocol.
(8) Fixed costs shall be calculated by reference to the amount of the offer which is accepted.
(9) Where the parties do not agree the liability for costs, the court will make an order as to costs.
(10) Where the court makes an order for costs in favour of the defendant—
  (a) the court will have regard to; and
  (b) the amount of costs ordered shall not exceed,
the fixed costs in Table 6B, Table 6C or Table 6D in Section IIIA of Part 45 applicable at the date of acceptance, less the fixed costs to which the claimant is entitled under paragraph (4) or (5).
(11) The parties are entitled to disbursements allowed in accordance with rule 45.29I incurred in any period for which costs are payable to them.

[CPR 36.11]

36.11 The effect of acceptance of a Part 36 offer
(1) If a Part 36 offer is accepted, the claim will be stayed (GL).
(2) In the case of acceptance of a Part 36 offer which relates to the whole claim the stay (GL) will be upon the terms of the offer.
(3) If a Part 36 offer which relates to part only of the claim is accepted—
  (a) the claim will be stayed (GL) as to that part upon the terms of the offer; and
  (b) subject to rule 36.10(2), unless the parties have agreed costs, the liability for costs shall be decided by the court.
(4) If the approval of the court is required before a settlement can be binding, any stay (GL) which would otherwise arise on the acceptance of a Part 36 offer will take effect only when that approval has been given.
(5) Any stay (GL) arising under this rule will not affect the power of the court—
  (a) to enforce the terms of a Part 36 offer;
  (b) to deal with any question of costs (including interest on costs) relating to the proceedings.
(6) Unless the parties agree otherwise in writing, where a Part 36 offer by a defendant that is or that includes an offer to pay a single sum of money is accepted, that sum must be paid to the offeree within 14 days of the date of—
  (a) acceptance; or
  (b) the order when the court makes an order under rule 41.2 (order for an award of provisional damages) or rule 41.8 (order for an award of periodical payments), unless the court orders otherwise.

(7) If the accepted sum is not paid within 14 days or such other period as has been agreed the offeree may enter judgment for the unpaid sum.
(8) Where—
- (a) a Part 36 offer (or part of a Part 36 offer) which is not an offer to which paragraph (6) applies is accepted; and
- (b) a party alleges that the other party has not honoured the terms of the offer,

that party may apply to enforce the terms of the offer without the need for a new claim.

[CPR 36.12]

36.12 Acceptance of a Part 36 offer made by one or more, but not all, defendants

(1) This rule applies where the claimant wishes to accept a Part 36 offer made by one or more, but not all, of a number of defendants.
(2) If the defendants are sued jointly or in the alternative, the claimant may accept the offer if—
- (a) he discontinues his claim against those defendants who have not made the offer; and
- (b) those defendants give written consent to the acceptance of the offer.

(3) If the claimant alleges that the defendants have a several liability (GL) to him, the claimant may—
- (a) accept the offer; and
- (b) continue with his claims against the other defendants if he is entitled to do so.

(4) In all other cases the claimant must apply to the court for an order permitting him to accept the Part 36 offer.

[CPR 36.13]

36.13 Restriction on disclosure of a Part 36 offer

(1) A Part 36 offer will be treated as "without prejudice$^{(GL)}$ except as to costs".
(2) The fact that a Part 36 offer has been made must not be communicated to the trial judge or to the judge (if any) allocated in advance to conduct the trial until the case has been decided.
(3) Paragraph (2) does not apply—
- (a) where the defence of tender before claim (GL) has been raised;
- (b) where the proceedings have been stayed (GL) under rule 36.11 following acceptance of a Part 36 offer; or
- (c) where the offeror and the offeree agree in writing that it should not apply.

[CPR 36.14]

36.14 Costs consequences following judgment

(1) Subject to rule 36.14A, this rule applies where upon judgment being entered—
- (a) a claimant fails to obtain a judgment more advantageous than a defendant's Part 36 offer; or
- (b) judgment against the defendant is at least as advantageous to the claimant as the proposals contained in a claimant's Part 36 offer.

(1A) For the purposes of paragraph (1), in relation to any money claim or money element of a claim, "more advantageous" means better in money terms by any

CPR 36.14 *Appendix*

amount, however small, and "at least as advantageous" shall be construed accordingly.

(2) Subject to paragraph (6), where rule 36.14(1)(a) applies, the court will, unless it considers it unjust to do so, order that the defendant is entitled to—
- (a) costs from the date on which the relevant period expired; and
- (b) interest on those costs.

(3) Subject to paragraph (6), where rule 36.14(1)(b) applies, the court will, unless it considers it unjust to do so, order that the claimant is entitled to—
- (a) interest on the whole or part of any sum of money (excluding interest) awarded at a rate not exceeding 10% above base rate (GL) for some or all of the period starting with the date on which the relevant period expired;
- (b) costs on the indemnity basis from the date on which the relevant period expired;
- (c) interest on those costs at a rate not exceeding 10% above base rate (GL); and
- (d) an additional amount, which shall not exceed £75,000, calculated by applying the prescribed percentage set out below to an amount which is—
    - (i) where the claim is or includes a money claim, the sum awarded to the claimant by the court; or
    - (ii) where the claim is only a non-monetary claim, the sum awarded to the claimant by the court in respect of costs—

Amount awarded by the court	Prescribed percentage
up to £500,000	10% of the amount awarded;
above £500,000 up to £1,000,000	10% of the first £500,000 and 5% of any amount above that figure

(4) In considering whether it would be unjust to make the orders referred to in paragraphs (2) and (3) above, the court will take into account all the circumstances of the case including—
- (a) the terms of any Part 36 offer;
- (b) the stage in the proceedings when any Part 36 offer was made, including in particular how long before the trial started the offer was made;
- (c) the information available to the parties at the time when the Part 36 offer was made; and
- (d) the conduct of the parties with regard to the giving or refusing to give information for the purposes of enabling the offer to be made or evaluated.

(5) Where the court awards interest under this rule and also awards interest on the same sum and for the same period under any other power, the total rate of interest may not exceed 10% above base rate$^{(GL)}$.

(6) Paragraphs (2) and (3) of this rule do not apply to a Part 36 offer—
- (a) that has been withdrawn;
- (b) that has been changed so that its terms are less advantageous to the offeree, and the offeree has beaten the less advantageous offer;

(c) made less than 21 days before trial, unless the court has abridged the relevant period.

(Rule 44.2 requires the court to consider an offer to settle that does not have the costs consequences set out in this Section in deciding what order to make about costs)

[CPR 36.14A]

36.14A Costs consequences following judgment where Section IIIA of Part 45 applies

(1) Where a claim no longer continues under the RTA or EL/PL Protocol pursuant to rule 45.29A(1), rule 36.14 applies with the following modifications.

(2) Subject to paragraph (3), where an order for costs is made pursuant to rule 36.14(2)—

 (a) the claimant will be entitled to the fixed costs in Table 6B, 6C or 6D in Section IIIA of Part 45 for the stage applicable at the date on which the relevant period expired; and

 (b) the claimant will be liable for the defendant's costs from the date on which the relevant period expired to the date of judgment.

(3) Where the claimant fails to obtain a judgment more advantageous than the defendant's Protocol offer—

 (a) the claimant will be entitled to the applicable Stage 1 and Stage 2 fixed costs in Table 6 or Table 6A in Section III of Part 45; and

 (b) the claimant will be liable for the defendant's costs from the date on which the Protocol offer is deemed to be made to the date of judgment; and

 (c) in this rule, the amount of the judgment is less than the Protocol offer where the judgment is less than the offer once deductible amounts identified in the judgment are deducted.

("Deductible amount" is defined in rule 36.15(1)(d).)

(4) For the purposes of this rule a defendant's Protocol offer is either—

 (a) defined in accordance with rules 36.17 and 36.18; or

 (b) if the claim leaves the Protocol before the Court Proceedings Pack Form is sent to the defendant—

  (i) the last offer made by the defendant before the claim leaves the Protocol; and

  (ii) deemed to be made on the first business day after the claim leaves the Protocol.

(5) A reference to the 'Court Proceedings Pack Form' is a reference to the form used in the Protocol.

(6) Fixed costs shall be calculated by reference to the amount which is awarded.

(7) Where the court makes an order for costs in favour of the defendant—

 (a) the court will have regard to; and

 (b) the amount of costs ordered shall not exceed,

the fixed costs in Table 6B, 6C or 6D in Section IIIA of Part 45 applicable at the date of judgment, less the fixed costs to which the claimant is entitled under paragraph (2) or (3).

(8) The parties are entitled to disbursements allowed in accordance with rule 45.29I incurred in any period for which costs are payable to them.

CPR 36.15   *Appendix*

[CPR 36.15]

36.15 Deduction of benefits and lump sum payments

(1) In this rule and rule 36.9—
 (a) "the 1997 Act" means the Social Security (Recovery of Benefits) Act 1997;
 (b) "the 2008 Regulations" means the Social Security (Recovery of Benefits)(Lump Sum Payments) Regulations 2008;
 (c) "recoverable amount" means—
  (i) "recoverable benefits" as defined in section 1(4)(c) of the 1997 Act; and
  (ii) "recoverable lump sum payments" as defined in regulation 4 of the 2008 Regulations;
 (d) "deductible amount" means—
  (i) any benefits by the amount of which damages are to be reduced in accordance with section 8 of, and Schedule 2 to the 1997 Act ("deductible benefits"); and
  (ii) any lump sum payment by the amount of which damages are to be reduced in accordance with regulation 12 of the 2008 Regulations ("deductible lump sum payments"); and
 (e) "certificate"—
  (i) in relation to recoverable benefits is construed in accordance with the provisions of the 1997 Act; and
  (ii) in relation to recoverable lump sum payments has the meaning given in section 29 of the 1997 Act as applied by regulation 2 of, and modified by Schedule 1 to the 2008 Regulations.

(2) This rule applies where a payment to a claimant following acceptance of a Part 36 offer would be a compensation payment as defined in section 1(4)(b) or 1A(5)(b) of the 1997 Act.

(3) A defendant who makes a Part 36 offer should state either—
 (a) that the offer is made without regard to any liability for recoverable amounts; or
 (b) that it is intended to include any deductible amounts.

(4) Where paragraph (3)(b) applies, paragraphs (5) to (9) of this rule will apply to the Part 36 offer.

(5) Before making the Part 36 offer, the offeror must apply for a certificate.

(6) Subject to paragraph (7), the Part 36 offer must state—
 (a) the amount of gross compensation;
 (b) the name and amount of any deductible amount by which the gross amount is reduced; and
 (c) the net amount after deduction of compensation.

(7) If at the time the offeror makes the Part 36 offer, the offeror has applied for, but has not received a certificate, the offeror must clarify the offer by stating the matters referred to in paragraphs (6)(b) and (6)(c) not more than 7 days after receipt of the certificate.

(8) For the purposes of rule 36.14(1)(a), a claimant fails to recover more than any sum offered (including a lump sum offered under rule 36.5) if the claimant fails upon judgment being entered to recover a sum, once deductible amounts identified in the judgment have been deducted, greater than the net amount stated under paragraph (6)(c).

(Section 15(2) of the 1997 Act provides that the court must specify the compensation payment attributable to each head of damage. Schedule 1 to the 2008 Regulations modifies section 15 of the 1997 Act in relation to lump sum payments and provides that the court must specify the compensation payment attributable to each or any dependant who has received a lump sum payment.)
(9) Where—
    (a)   further deductible amounts have accrued since the Part 36 offer was made; and
    (b)   the court gives permission to accept the Part 36 offer,
the court may direct that the amount of the offer payable to the offeree shall be reduced by a sum equivalent to the deductible amounts paid to the claimant since the date of the offer.
(Rule 36.9(3)(b) states that permission is required to accept an offer where the relevant period has expired and further deductible amounts have been paid to the claimant)

SECTION II RTA PROTOCOL AND EL/PL PROTOCOL OFFERS TO SETTLE

[CPR 36.16]

36.16 Scope of this Section
(1) Where this Section applies Section I does not apply.
(2) This Section applies to an offer to settle where the parties have followed the RTA Protocol or the EL/PL Protocol and started proceedings under Part 8 in accordance with Practice Direction 8B ("the Stage 3 Procedure").
(3) A reference to the "Court Proceedings Pack Form" is a reference to the form used in the RTA Protocol.
(4) Nothing in this Section prevents a party making an offer to settle in whatever way that party chooses, but if the offer is not made in accordance with this Section, it will not have any costs consequences.

[CPR 36.17]

36.17 Form and content of a Protocol offer
(1) An offer to settle which is made in accordance with this rule is called a Protocol offer.
(2) A Protocol offer must—
    (a)   be set out in the Court Proceedings Pack (Part B) Form; and
    (b)   contain the final total amount of the offer from both parties.

[CPR 36.18]

36.18 Time when a Protocol offer is made
The Protocol offer is deemed to be made on the first business day after the Court Proceedings Pack (Part A and Part B) Form is sent to the defendant.

[CPR 36.19]

36.19 General provisions
A Protocol offer—
    (a)   is treated as exclusive of all interest; and
    (b)   has the consequences set out in this Section only in relation to the fixed costs of the Stage 3 Procedure as provided for in rule 45.29,

CPR 36.20   *Appendix*

and not in relation to the costs of any appeal from the final decision of those proceedings.

[CPR 36.20]

36.20 Restrictions on disclosure of a Protocol offer

(1)   The amount of the Protocol offer must not be communicated to the court until the claim is determined.

(2)   Any other offer to settle must not be communicated to the court at all.

(3)   Once the claim is determined, the court will examine the Protocol offer.

[CPR 36.21]

36.21 Costs consequences following judgment

(1)   This rule applies where, on the determination by the court, the claimant obtains judgment against the defendant for an amount of damages that is—
- (a)   less than or equal to the amount of the defendant's Protocol offer;
- (b)   more than the defendant's Protocol offer but less than the claimant's Protocol offer; or
- (c)   equal to or more than the claimant's Protocol offer.

(2)   Where paragraph (1)(a) applies, the court will order the claimant to pay—
- (a)   the fixed costs in rule 45.26; and
- (b)   interest on those fixed costs from the first business day after the deemed date of the Protocol offer under rule 36.18.

(3)   Where paragraph (1)(b) applies, the court will order the defendant to pay the fixed costs in rule 45.20.

(4)   Where paragraph (1)(c) applies, the court will order the defendant to pay—
- (a)   interest on the whole of the damages awarded at a rate not exceeding 10% above base rate for some or all of the period starting with the date specified in rule 36.18;
- (b)   the fixed costs in rule 45.20;
- (c)   interest on those costs at a rate not exceeding 10% above base rate; and
- (d)   an additional amount calculated in accordance with rule 36.14(3)(d).

[CPR 36.22]

36.22 Deduction of benefits

For the purposes of rule 36.21(1)(a) the amount of the judgment is less than the Protocol offer where the judgment is less than that offer once deductible amounts identified in the judgment are deducted.

('Deductible amount' is defined in rule 36.15(1)(d).)

PRACTICE DIRECTION 36A—OFFERS TO SETTLE

THIS PRACTICE DIRECTION SUPPLEMENTS CPR PART 36

Formalities of Part 36 offers and other notices under this Part

[CPR PD 36A.1]

1.1 A Part 36 offer may be made using Form N242A.
1.2 Where a Part 36 offer, notice of acceptance or notice of withdrawal or change of terms is to be served on a party who is legally represented, the document to be served must be served on the legal representative.

Application for permission to withdraw a Part 36 Offer

[CPR PD 36A.2]

2.1 Rule 36.3(4) provides that before expiry of the relevant period a Part 36 offer may only be withdrawn or its terms changed to be less advantageous to the offeree with the permission of the court.
2.2 The permission of the court must be sought-
(1) by making an application under Part 23, which must be dealt with by a judge other than the judge (if any) allocated in advance to conduct the trial, unless the parties agree that such judge may hear the application;
(2) at a trial or other hearing, provided that it is not to the trial judge or to the judge (if any) allocated in advance to conduct the trial, unless the parties agree that such judge may hear the application.

Acceptance of a Part 36 offer

[CPR PD 36A.3]

3.1 Where a Part 36 offer is accepted in accordance with rule 36.9(1) the notice of acceptance must be served on the offeror and filed with the court where the case is proceeding.
3.2 Where the court's permission is required to accept a Part 36 offer, the permission of the court must be sought-
(1) by making an application under Part 23, which must be dealt with by a judge other than the judge (if any) allocated in advance to conduct the trial, unless the parties agree that such judge may hear the application;
(2) at a trial or other hearing, provided that it is not to the trial judge or to the judge (if any) allocated in advance to conduct the trial, unless the parties agree that such judge may hear the application.
3.3 Where rule 36.9(3)(b) applies, the application for permission to accept the offer must-
(1) state-
    (a) the net amount offered in the Part 36 offer;
    (b) the deductible amounts that had accrued at the date the offer was made;
    (c) the deductible amounts that have subsequently accrued; and
(2) be accompanied by a copy of the current certificate.

PRACTICE DIRECTION 36B

THIS PRACTICE DIRECTION SUPPLEMENTS CPR PART 36

CPR PD 36B.1 *Appendix*

From 6th April 2007, new rules came into force concerning offers to settle, and Part 36, as it was in force immediately before 6th April 2007, was substituted by a new Part 36.

Rule 7 of the Civil Procedure (Amendment No.3) Rules 2006 that brought those new rules into force and replaced the previous rules contained some provisions that dealt with how the rules are to apply to offers and payments into court made before 6th April 2007.

This Practice Direction explains how those provisions will operate.

Offers and Payments made before 6th April 2007

[CPR PD 36B.1]

1.1 Paragraph (2) of rule 7 provides that where a Part 36 offer or Part 36 payment was made before 6th April 2007, if it would have had the consequences set out in the rules of court contained in Part 36 as it was in force immediately before 6th April 2007, it will have the consequences set out in rules 36.10, 36.11 and 36.14 after that date.

1.2 This provision makes clear that a Part 36 offer or Part 36 payment that was valid before 6th April 2007, will continue to be a valid Part 36 offer under the rules in force from 6th April 2007, and will have the consequences set out in those rules, specifically in relation to costs and the effect of acceptance.

Permission of the court

[CPR PD 36B.2]

2.1 Paragraph (3) of rule 7 provides that where a Part 36 offer or Part 36 payment was made before 6th April 2007, the permission of the court is required to accept that offer or payment, if permission would have been required under the rules of court contained in Part 36 as it was in force immediately before 6th April 2007.

2.2 This provision preserves the requirement to obtain the permission of the court to accept an offer as it existed under the rules in force immediately before 6th April 2007. Therefore, if permission would have been required before 6th April 2007, it will be required after that date. But, if permission would not have been required because the parties have been able to agree liability for costs, or if a further offer has been made triggering a new period for acceptance, permission will not be required after 6th April 2007.

Payments into court made before 6th April 2007

[CPR PD 36B.3]

3.1 Paragraph (4) of rule 7 provides that rule 37.3 will apply to a Part 36 payment made before 6th April 2007 as if that payment into court had been made under a court order.

3.2 Rule 37.3 applies to all payments under Part 37, including payments into court under order, and permission is required to take the money out of court.

3.3 By applying rule 37.3 to payments into court made before 6th April 2007, this provision preserves in particular the requirement that permission be obtained to withdraw such payment.

3.4 But, rule 37.3 also provides that money may be taken out of court without the court's permission where a Part 36 offer (including an offer underlying a Part 36 payment) is accepted without needing the permission of the court and the defendant agrees that the sum may be paid out in satisfaction of the offer. Paragraph 3.4 of the Practice Direction to Part 37 makes provision about how to take money out of court.

3.5 This exception to the permission requirement preserves the right under rule 37.2, as it was in force immediately before 6th April 2007, to treat a payment into court made under order or by way of a defence of tender before claim as a Part 36 payment.

3.6 This provision has the effect that a Part 36 payment made before 6th April 2007 may be taken out of court simply by filing a request for payment if the offer underlying the Part 36 payment is accepted without needing permission. In those circumstances, it may be assumed that the defendant agrees to the money being used in satisfaction of the sum offered, and the requirement in paragraph 3.4 of the Practice Direction to Part 37 to file a Form 202 will not apply.

Offers remaining open for acceptance

[CPR PD 36B.4]

4.1 Paragraph (5) of rule 7 provides that the rules of court contained in Part 36 as it was in force immediately before 6th April 2007 shall continue to apply to a Part 36 offer or Part 36 payment made less than 21 days before 6th April 2007.

4.2 This provision preserves those rules in their entirety in relation to offers and payments made less than 21 days before 6th April 2007 for the period that they are expressed to remain open for acceptance.

4.3 Paragraph (6) of rule 7 provides that paragraph (5) ceases to apply at the expiry of 21 days from the date that the offer or payment was made, unless the trial has started within that period.

4.4 This provision has the effect that once the 21 day period has expired, the new regime (including the modifications at paragraphs (2), (3) and (4) of rule 7) will apply to the offer or payment.

4.5 If the trial has started within the 21 day period, the rules that were in force before 6th April 2007 will continue to apply to the offer or payment.

Offers made before commencement of proceedings

[CPR PD 36B.5]

5.1 Paragraph (7) of rule 7 deals with the position where, before 6th April 2007, a person made an offer to settle before commencement of proceedings which complied with the provisions of rule 36.10 as it was in force immediately before 6th April 2007.

5.2 The court will take that offer into account when making any order as to costs. This preserves the discretion of the court to take into account an offer made before commencement of proceedings as it existed before 6th April 2007.

5.3 The permission of the court will be required to accept such an offer after proceedings have been commenced. This preserves the position under rule 36.10(4) as it was in force immediately before 6th April 2007.

5.4 If proceedings are commenced after 6th April 2007, the requirement to pay money into court in respect of a defendant's money offer under rule 36.10(3)(a) (as it was in force before 6th April 2007) will not an apply to a defendant's money offer made before the proceedings were commenced.

CPR PD 36B.5 *Appendix*

Notice of offer to settle
(Section 1 - Part 36)

Name of court *(If proceedings have started)*	
Claim No. *(or other ref)*	
Claimant *(including ref)*	
Defendant *(including ref)*	

To the Offeree ('s Solicitor) *(Insert name and address)*

Take notice the (defendant)(claimant) offers to settle the claim. This offer is intended to have the consequences of Section 1 Part 36. If the offer is accepted within _____ days (must be at least 21 days) of service of this notice the defendant will be liable for the claimant's costs in accordance with Rule 36.10 of the Civil Procedure Rules.

The offer is to settle:

(tick as appropriate)
- [ ] the whole of the claim
- [ ] part of the claim *(give details below)*
- [ ] a certain issue or issues in the claim *(give details below)*

The offer is:

(Insert details - expand box as necessary)

Note: Rule 36.5 specifies details that must be included in an offer including periodical payments of damages for future pecuniary loss.

Rule 36.11 requires that an offer by a defendant to pay a sum of money (other than periodical payments) must be paid within 14 days of acceptance.

- [ ] It (does)(does not) take into account all(part) of the following counterclaim:

(give details of the counterclaim)

N242A Notice of offer to settle (Section 1 Part 36) (08.11) © Crown Copyright 2011
http://www.hmcourts-service.gov.uk/cms/index.htm © Crown Copyright Published by LexisNexis 2011 under the Open Government Licence

Civil Procedure Rules 1998 **CPR PD 36B.5**

Include only if claim for provisional damages

☐ The offer is made in satisfaction of the claim on the assumption that the claimant will not [develop (state the disease)   ] **OR** [suffer (state type of deterioration)   ]. But if that does occur, the claimant will be entitled to claim further damages at any time before (insert date)   .

OR

☐ This offer does not include an offer in respect of the claim for provisional damages.

To be completed by defendants only

☐ This offer is made without regard to any liability for recoverable benefits under the Social Security (Recovery of Benefits Act) 1997.

OR

☐ This offer is intended to include any relevant deductible benefits for which I am liable under the Social Security (Recovery of Benefits Act) 1997.

The amount of [£   ] is offered by way of gross compensation.

[I have not yet received a certificate of recoverable benefits]
OR
[The following amounts in respect of the following benefits are to be deducted (insert details).

Type of benefit	Amount

The net amount offered is therefore [£   ]]

Signed		Position held (If signing on behalf of a firm or company)	
Offeror('s solicitor)			

Date

661

CPR 44.1 *Appendix*

PART 44 GENERAL RULES ABOUT COSTS

SECTION I
GENERAL

Rule 44.1	Interpretation and application	CPR 44.1
Rule 44.2	Court's discretion as to costs	CPR 44.2
Rule 44.3	Basis of assessment	CPR 44.3
Rule 44.4	Factors to be taken into account in deciding the amount of costs	CPR 44.4
Rule 44.5	Amount of costs where costs are payable under a contract	CPR 44.5
Rule 44.6	Procedure for assessing costs	CPR 44.6
Rule 44.7	Time for complying with an order for costs	CPR 44.7
Rule 44.8	Legal representative's duty to notify the party	CPR 44.8
Rule 44.9	Cases where costs orders deemed to have been made	CPR 44.9
Rule 44.10	Where the court makes no order for costs	CPR 44.10
Rule 44.11	Court's powers in relation to misconduct	CPR 44.11
Rule 44.12	Set-off	CPR 44.12

SECTION II
QUALIFIED ONE-WAY COSTS SHIFTING

Rule 44.13	Qualified one-way costs shifting: scope and interpretation	CPR 44.13
Rule 44.14	Effect of qualified one-way costs shifting	CPR 44.14
Rule 44.15	Exceptions to qualified one-way costs shifting where permission not required	CPR 44.15
Rule 44.16	Exceptions to qualified one-way costs shifting where permission required	CPR 44.16
Rule 44.17	Transitional provision	CPR 44.17

SECTION III
DAMAGES-BASED AGREEMENTS

Rule 44.18	Award of costs where there is a damages-based agreement	CPR 44.18
	Practice Direction 44—General Rules About Costs	CPR PD 44

SECTION I GENERAL

[CPR 44.1]

44.1 Interpretation and application

(1) In Parts 44 to 47, unless the context otherwise requires—
"authorised court officer" means any officer of—

        (i)   a county court;
        (ii)   a district registry;
        (iii) the Principal Registry of the Family Division; or
        (iv) the Costs Office,

whom the Lord Chancellor has authorised to assess costs;
"conditional fee agreement" means an agreement enforceable under section 58 of the Courts and Legal Services Act 1990;
"costs" includes fees, charges, disbursements, expenses, remuneration, reimbursement allowed to a litigant in person under rule 46.5 and any fee or reward charged by a lay representative for acting on behalf of a party in proceedings allocated to the small claims track;
"costs judge" means a taxing master of the Senior Courts;
"Costs Office" means the Senior Courts Costs Office;
"costs officer" means—

        (i)   a costs judge;
        (ii)   a district judge; or
        (iii) an authorised court officer;

"detailed assessment" means the procedure by which the amount of costs is decided by a costs officer in accordance with Part 47;

"the Director (legal aid)" means the person designated as the Director of Legal Aid Casework pursuant to section 4 of the Legal Aid, Sentencing and Punishment of Offenders Act 2012, or a person entitled to exercise the functions of the Director;

"fixed costs" means costs the amounts of which are fixed by these rules whether or not the court has a discretion to allow some other or no amount, and include—

    (i)   the amounts which are to be allowed in respect of legal representatives' charges in the circumstances set out in Section I of Part 45;

    (ii)   fixed recoverable costs calculated in accordance with rule 45.11;

    (iii)   the additional costs allowed by rule 45.18;

    (iv)   fixed costs determined under rule 45.21;

    (v)   costs fixed by rules 45.37 and 45.38;

"free of charge" has the same meaning as in section 194(10) of the 2007 Act;

"fund" includes any estate or property held for the benefit of any person or class of person and any fund to which a trustee or personal representative is entitled in that capacity;

"HMRC" means HM Revenue and Customs;

"legal aid" means civil legal services made available under arrangements made for the purposes of Part 1of the Legal Aid, Sentencing and Punishment of Offenders Act 2012;

"paying party" means a party liable to pay costs;

"the prescribed charity" has the same meaning as in section 194(8) of the 2007 Act;

"pro bono representation" means legal representation provided free of charge;

"receiving party" means a party entitled to be paid costs;

"summary assessment" means the procedure whereby costs are assessed by the judge who has heard the case or application;

"VAT" means Value Added Tax;

"the 2007 Act" means the Legal Services Act 2007.

("Legal representative" has the meaning given in rule 2.3).

  (2)   The costs to which Parts 44 to 47 apply include—

    (a)   the following costs where those costs may be assessed by the court—

        (i)   costs of proceedings before an arbitrator or umpire;

        (ii)   costs of proceedings before a tribunal or other statutory body; and

        (iii)   costs payable by a client to their legal representative; and

    (b)   costs which are payable by one party to another party under the terms of a contract, where the court makes an order for an assessment of those costs.

  (3)   Where advocacy or litigation services are provided to a client under a conditional fee agreement, costs are recoverable under Parts 44 to 47 notwithstanding that the client is liable to pay the legal representative's fees and expenses only to the extent that sums are recovered in respect of the proceedings, whether by way of costs or otherwise.

[CPR 44.2]

44.2 Court's discretion as to costs

  (1)   The court has discretion as to—

    (a)   whether costs are payable by one party to another;

CPR 44.3 *Appendix*

      (b)   the amount of those costs; and
      (c)   when they are to be paid.
(2)   If the court decides to make an order about costs—
      (a)   the general rule is that the unsuccessful party will be ordered to pay the costs of the successful party; but
      (b)   the court may make a different order.
(3)   The general rule does not apply to the following proceedings—
      (a)   proceedings in the Court of Appeal on an application or appeal made in connection with proceedings in the Family Division; or
      (b)   proceedings in the Court of Appeal from a judgment, direction, decision or order given or made in probate proceedings or family proceedings.
(4)   In deciding what order (if any) to make about costs, the court will have regard to all the circumstances, including—
      (a)   the conduct of all the parties;
      (b)   whether a party has succeeded on part of its case, even if that party has not been wholly successful; and
      (c)   any admissible offer to settle made by a party which is drawn to the court's attention, and which is not an offer to which costs consequences under Part 36 apply.
(5)   The conduct of the parties includes—
      (a)   conduct before, as well as during, the proceedings and in particular the extent to which the parties followed the Practice Direction – Pre-Action Conduct or any relevant pre-action protocol;
      (b)   whether it was reasonable for a party to raise, pursue or contest a particular allegation or issue;
      (c)   the manner in which a party has pursued or defended its case or a particular allegation or issue; and
      (d)   whether a claimant who has succeeded in the claim, in whole or in part, exaggerated its claim.
(6)   The orders which the court may make under this rule include an order that a party must pay—
      (a)   a proportion of another party's costs;
      (b)   a stated amount in respect of another party's costs;
      (c)   costs from or until a certain date only;
      (d)   costs incurred before proceedings have begun;
      (e)   costs relating to particular steps taken in the proceedings;
      (f)   costs relating only to a distinct part of the proceedings; and
      (g)   interest on costs from or until a certain date, including a date before judgment.
(7)   Before the court considers making an order under paragraph (6)(f), it will consider whether it is practicable to make an order under paragraph (6)(a) or (c) instead.
(8)   Where the court orders a party to pay costs subject to detailed assessment, it will order that party to pay a reasonable sum on account of costs, unless there is good reason not to do so.

[CPR 44.3]

44.3 Basis of assessment

(1)   Where the court is to assess the amount of costs (whether by summary or detailed assessment) it will assess those costs—

(a) on the standard basis; or
(b) on the indemnity basis,

but the court will not in either case allow costs which have been unreasonably incurred or are unreasonable in amount.

(Rule 44.5 sets out how the court decides the amount of costs payable under a contract.)

(2) Where the amount of costs is to be assessed on the standard basis, the court will—
 (a) only allow costs which are proportionate to the matters in issue. Costs which are disproportionate in amount may be disallowed or reduced even if they were reasonably or necessarily incurred; and
 (b) resolve any doubt which it may have as to whether costs were reasonably and proportionately incurred or were reasonable and proportionate in amount in favour of the paying party.

(Factors which the court may take into account are set out in rule 44.4.)

(3) Where the amount of costs is to be assessed on the indemnity basis, the court will resolve any doubt which it may have as to whether costs were reasonably incurred or were reasonable in amount in favour of the receiving party.

(4) Where—
 (a) the court makes an order about costs without indicating the basis on which the costs are to be assessed; or
 (b) the court makes an order for costs to be assessed on a basis other than the standard basis or the indemnity basis,
the costs will be assessed on the standard basis.

(5) Costs incurred are proportionate if they bear a reasonable relationship to—
 (a) the sums in issue in the proceedings;
 (b) the value of any non-monetary relief in issue in the proceedings;
 (c) the complexity of the litigation;
 (d) any additional work generated by the conduct of the paying party; and
 (e) any wider factors involved in the proceedings, such as reputation or public importance.

(6) Where the amount of a solicitor's remuneration in respect of non-contentious business is regulated by any general orders made under the Solicitors Act 1974, the amount of the costs to be allowed in respect of any such business which falls to be assessed by the court will be decided in accordance with those general orders rather than this rule and rule 44.4.

(7) Paragraphs (2)(a) and (5) do not apply in relation to—
 (a) to cases commenced before 1st April 2013; or
 (a) costs incurred in respect of work done before 1st April 2013,
and in relation to such cases or costs, rule 44.4(2)(a) as it was in force immediately before 1st April 2013 will apply instead.

[CPR 44.4]

44.4 Factors to be taken into account in deciding the amount of costs

(1) The court will have regard to all the circumstances in deciding whether costs were—
 (a) if it is assessing costs on the standard basis—
  (i) proportionately and reasonably incurred; or
  (ii) proportionate and reasonable in amount, or
 (b) if it is assessing costs on the indemnity basis—

CPR 44.5   *Appendix*

        (i)   unreasonably incurred; or
        (ii)   unreasonable in amount.
(2) In particular, the court will give effect to any orders which have already been made.
(3) The court will also have regard to—
    (a)   the conduct of all the parties, including in particular—
        (i)   conduct before, as well as during, the proceedings; and
        (ii)   the efforts made, if any, before and during the proceedings in order to try to resolve the dispute;
    (b)   the amount or value of any money or property involved;
    (c)   the importance of the matter to all the parties;
    (d)   the particular complexity of the matter or the difficulty or novelty of the questions raised;
    (e)   the skill, effort, specialised knowledge and responsibility involved;
    (f)   the time spent on the case;
    (g)   the place where and the circumstances in which work or any part of it was done; and
    (h)   the receiving party's last approved or agreed budget.
(Rule 35.4(4) gives the court power to limit the amount that a party may recover with regard to the fees and expenses of an expert.)

[CPR 44.5]

44.5 Amount of costs where costs are payable under a contract
(1) Subject to paragraphs (2) to (4), where the court assesses (whether by summary or detailed assessment) costs which are payable by the paying party to the receiving party under the terms of a contract, the costs payable under those terms are, unless the contract expressly provides otherwise, to be presumed to be costs which—
    (a)   have been reasonably incurred; and
    (b)   are reasonable in amount,
and the court will assess them accordingly.
(2) The presumptions in paragraph (1) are rebuttable. Practice Direction 44 – General rules about costs sets out circumstances where the court may order otherwise.
(3) Paragraph (1) does not apply where the contract is between a solicitor and client.

[CPR 44.6]

44.6 Procedure for assessing costs
(1) Where the court orders a party to pay costs to another party (other than fixed costs) it may either—
    (a)   make a summary assessment of the costs; or
    (b)   order detailed assessment of the costs by a costs officer,
unless any rule, practice direction or other enactment provides otherwise.
(Practice Direction 44 – General rules about costs sets out the factors which will affect the court's decision under paragraph (1).)
(2) A party may recover the fixed costs specified in Part 45 in accordance with that Part.

[CPR 44.7]

44.7 Time for complying with an order for costs

(1) A party must comply with an order for the payment of costs within 14 days of—
    (a) the date of the judgment or order if it states the amount of those costs;
    (b) if the amount of those costs (or part of them) is decided later in accordance with Part 47, the date of the certificate which states the amount; or
    (c) in either case, such other date as the court may specify.
(Part 47 sets out the procedure for detailed assessment of costs.)

[CPR 44.8]

44.8 Legal representative's duty to notify the party

Where—
    (a) the court makes a costs order against a legally represented party; and
    (b) the party is not present when the order is made,
the party's legal representative must notify that party in writing of the costs order no later than 7 days after the legal representative receives notice of the order.
(Paragraph 10.1 of Practice Direction 44 defines "party" for the purposes of this rule.)

[CPR 44.9]

44.9 Cases where costs orders deemed to have been made

(1) Subject to paragraph (2), where a right to costs arises under—
    (a) rule 3.7 (defendant's right to costs where claim is struck out for non-payment of fees);
    (a1) rule 3.7B (sanctions for dishonouring cheque);
    (b) rule 36.10(1) or (2) (claimant's entitlement to costs where a Part 36 offer is accepted); or
    (c) rule 38.6 (defendant's right to costs where claimant discontinues),
a costs order will be deemed to have been made on the standard basis.

(2) Paragraph 1(b) does not apply where a Part 36 offer is accepted before the commencement of proceedings.

(3) Where such an order is deemed to be made in favour of a party with *pro bono* representation, that party may apply for an order under section 194(3) of the 2007 Act.

(4) Interest payable under section 17 of the Judgments Act 1838 or section 74 of the County Courts Act 1984 on the costs deemed to have been ordered under paragraph (1) will begin to run from the date on which the event which gave rise to the entitlement to costs occurred.

[CPR 44.10]

44.10 Where the court makes no order for costs

(1) Where the court makes an order which does not mention costs—
    (a) subject to paragraphs (2) and (3), the general rule is that no party is entitled—

CPR 44.11 *Appendix*

        (i)   to costs; or
        (ii)   to seek an order under section 194(3) of the 2007 Act,

in relation to that order; but

    (b)   this does not affect any entitlement of a party to recover costs out of a fund held by that party as trustee or personal representative, or under any lease, mortgage or other security.

(2) Where the court makes—
    (a)   an order granting permission to appeal;
    (b)   an order granting permission to apply for judicial review; or
    (c)   any other order or direction sought by a party on an application without notice,

and its order does not mention costs, it will be deemed to include an order for applicant's costs in the case.

(3) Any party affected by a deemed order for costs under paragraph (2) may apply at any time to vary the order.

(4) The court hearing an appeal may, unless it dismisses the appeal, make orders about the costs of the proceedings giving rise to the appeal as well as the costs of the appeal.

(5) Subject to any order made by the transferring court, where proceedings are transferred from one court to another, the court to which they are transferred may deal with all the costs, including the costs before the transfer.

[CPR 44.11]

44.11 Court's powers in relation to misconduct

(1) The court may make an order under this rule where—
    (a)   a party or that party's legal representative, in connection with a summary or detailed assessment, fails to comply with a rule, practice direction or court order; or
    (b)   it appears to the court that the conduct of a party or that party's legal representative, before or during the proceedings or in the assessment proceedings, was unreasonable or improper.

(2) Where paragraph (1) applies, the court may—
    (a)   disallow all or part of the costs which are being assessed; or
    (b)   order the party at fault or that party's legal representative to pay costs which that party or legal representative has caused any other party to incur.

(3) Where—
    (a)   the court makes an order under paragraph (2) against a legally represented party; and
    (b)   the party is not present when the order is made,

the party's legal representative must notify that party in writing of the order no later than 7 days after the legal representative receives notice of the order.

[CPR 44.12]

44.12 Set Off

(1) Where a party entitled to costs is also liable to pay costs, the court may assess the costs which that party is liable to pay and either—
    (a)   set off the amount assessed against the amount the party is entitled to be paid and direct that party to pay any balance; or

(b) delay the issue of a certificate for the costs to which the party is entitled until the party has paid the amount which that party is liable to pay.

SECTION II QUALIFIED ONE-WAY COSTS SHIFTING

[CPR 44.13]

44.13 Qualified one-way costs shifting: scope and interpretation
(1) This Section applies to proceedings which include a claim for damages—
 (a) for personal injuries;
 (b) under the Fatal Accidents Act 1976; or
 (c) which arises out of death or personal injury and survives for the benefit of an estate by virtue of section 1(1) of the Law Reform (Miscellaneous Provisions) Act 1934,
but does not apply to applications pursuant to section 33 of the Senior Courts Act 1981 or section 52 of the County Courts Act 1984 (applications for pre-action disclosure), or where rule 44.17 applies.
(2) In this Section, "claimant" means a person bringing a claim to which this Section applies or an estate on behalf of which such a claim is brought, and includes a person making a counterclaim or an additional claim.

[CPR 44.14]

44.14 Effect of qualified one-way costs shifting
(1) Subject to rules 44.15 and 44.16, orders for costs made against a claimant may be enforced without the permission of the court but only to the extent that the aggregate amount in money terms of such orders does not exceed the aggregate amount in money terms of any orders for damages and interest made in favour of the claimant.
(2) Orders for costs made against a claimant may only be enforced after the proceedings have been concluded and the costs have been assessed or agreed.
(3) An order for costs which is enforced only to the extent permitted by paragraph (1) shall not be treated as an unsatisfied or outstanding judgment for the purposes of any court record.

[CPR 44.15]

44.15 Exceptions to qualified one-way costs shifting where permission not required
Orders for costs made against the claimant may be enforced to the full extent of such orders without the permission of the court where the proceedings have been struck out on the grounds that—
 (a) the claimant has disclosed no reasonable grounds for bringing the proceedings;
 (b) the proceedings are an abuse of the court's process; or
 (c) the conduct of—
  (i) the claimant; or
  (ii) a person acting on the claimant's behalf and with the claimant's knowledge of such conduct,
is likely to obstruct the just disposal of the proceedings.

CPR 44.16 *Appendix*

[CPR 44.16]

44.16 Exceptions to qualified one-way costs shifting where permission required

(1) Orders for costs made against the claimant may be enforced to the full extent of such orders with the permission of the court where the claim is found on the balance of probabilities to be fundamentally dishonest.

(2) Orders for costs made against the claimant may be enforced up to the full extent of such orders with the permission of the court, and to the extent that it considers just, where—

 (a) the proceedings include a claim which is made for the financial benefit of a person other than the claimant or a dependant within the meaning of section 1(3) of the Fatal Accidents Act 1976 (other than a claim in respect of the gratuitous provision of care, earnings paid by an employer or medical expenses); or

 (b) a claim is made for the benefit of the claimant other than a claim to which this Section applies.

(3) Where paragraph (2)(a) applies, the court may, subject to rule 46.2, make an order for costs against a person, other than the claimant, for whose financial benefit the whole or part of the claim was made.

[CPR 44.17]

44.17 Transitional provision

This Section does not apply to proceedings where the claimant has entered into a pre-commencement funding arrangement (as defined in rule 48.2).

SECTION III DAMAGES-BASED AGREEMENTS

[CPR 44.18]

44.18 Award of costs where there is a damages-based agreement

(1) The fact that a party has entered into a damages-based agreement will not affect the making of any order for costs which otherwise would be made in favour of that party.

(2) Where costs are to be assessed in favour of a party who has entered into a damages-based agreement—

 (a) the party's recoverable costs will be assessed in accordance with rule 44.3; and

 (b) the party may not recover by way of costs more than the total amount payable by that party under the damages-based agreement for legal services provided under that agreement.

PRACTICE DIRECTION 44—GENERAL RULES ABOUT COSTS

THIS PRACTICE DIRECTION SUPPLEMENTS PART 44 OF THE CIVIL PROCEDURE RULES

Section I — General

Subsection 1 of this Practice Direction

Documents and forms

[CPR PD 44.1]

1.1 In respect of any document which is required by Practice Directions 44 to 47 to be signed by a party or that party's legal representative, the provisions of Practice Direction 22 relating to who may sign apply as if the document in question was a statement of truth. Statements of truth are not required in assessment proceedings unless a rule or Practice Direction so requires or the court so orders.
(Practice Direction 22 makes provision for cases in which a party is a child, a protected party or a company or other corporation and cases in which a document is signed on behalf of a partnership.)
1.2 Form N260 is a model form of Statement of Costs to be used for summary assessments.
(Further details about Statements of Costs are given in paragraph 9.5 below.)
Precedents A, B and C in the Schedule of Costs Precedents annexed to this Practice Direction are model forms of bills of costs to be used for detailed assessments. A party wishing to rely upon a bill which departs from the model forms should include in the background information of the bill an explanation for that departure.
(Further details about bills of costs are given in Practice Direction 47.)

Subsection 2 of this Practice Direction — Special provisions relating to VAT

Scope of this subsection

[CPR PD 44.2]

2.1 This subsection deals with claims for VAT) which are made in respect of costs being dealt with by way of summary assessment or detailed assessment.
VAT Registration Number
2.2 The number allocated by HMRC to every person registered under the Value Added Tax Act 1994 (except a Government Department) must appear in a prominent place at the head of every statement, bill of costs, fee sheet, account or voucher on which VAT is being included as part of a claim for costs.

Entitlement to VAT on Costs
2.3 VAT should not be included in a claim for costs if the receiving party is able to recover the VAT as input tax. Where the receiving party is able to obtain credit from HMRC for a proportion of the VAT as input tax, only that proportion which is not eligible for credit should be included in the claim for costs.
2.4 The receiving party has responsibility for ensuring that VAT is claimed only when the receiving party is unable to recover the VAT or a proportion thereof as input tax.
2.5 Where there is a dispute as to whether VAT is properly claimed the receiving party must provide a certificate signed by the legal representatives or the auditors of the receiving party substantially in the form illustrated in Precedent F in the Schedule of Costs Precedents annexed to Practice Direction 47. Where the receiving party is a litigant in person who is claiming VAT, evidence to support the claim (such as a letter from HMRC) must be produced at the hearing at which the costs are assessed.
2.6 Where there is a dispute as to whether any service in respect of which a charge is proposed to be made in the bill is zero rated or exempt from VAT, reference should be made to HMRC and its view obtained and made known at the hearing at which the

CPR PD 44.2 *Appendix*

costs are assessed. Such enquiry should be made by the receiving party. In the case of a bill from a solicitor to the solicitor's legal representative's own client, such enquiry should be made by the client.

Form of bill of costs where VAT rate changes

2.7 Where there is a change in the rate of VAT, suppliers of goods and services are entitled by sections 88 (1) and 88(2) of the Value Added Tax Act 1994 in most circumstances to elect whether the new or the old rate of VAT should apply to a supply where the basic and actual tax points span a period during which the rate changed.

2.8 It will be assumed, unless a contrary indication is given in writing, that an election to take advantage of the provisions mentioned in paragraph 2.7 and to charge VAT at the lower rate has been made. In any case in which an election to charge at the lower rate is not made, such a decision must be justified to the court assessing the costs.

Apportionment

2.9 Subject to 2.7 and 2.8, all bills of costs, fees and disbursements on which VAT is included must be divided into separate parts so as to show work done before, on and after the date or dates from which any change in the rate of VAT takes effect. Where, however, a lump sum charge is made for work which spans a period during which there has been a change in VAT rates, and paragraphs 2.7 and 2.8 above do not apply, reference should be made to paragraphs 30.7 or 30.8 of the VAT Guide (Notice 700) (or any revised edition of that notice) published by HMRC. If necessary, the lump sum should be apportioned. The totals of profit costs and disbursements in each part must be carried separately to the summary.

Change in VAT rate between the conclusion of a detailed settlement and the issue of a final certificate

2.10 Should there be a change in the rate between the conclusion of a detailed assessment and the issue of the final costs certificate, any interested party may apply for the detailed assessment to be varied so as to take account of any increase or reduction in the amount of tax payable. Once the final costs certificate has been issued, no variation under this paragraph will be permitted.

Disbursements not classified as such for VAT purposes

2.11
(1) Legal representatives often make payments to third parties for the supply of goods or services where no VAT was chargeable on the supply by the third party: for example, the cost of meals taken and travel costs. The question whether legal representatives should include VAT in respect of these payments when invoicing their clients or in claims for costs between litigants should be decided in accordance with this Practice Direction and with the criteria set out in the VAT Guide (Notice 700).
(2) Payments to third parties which are normally treated as part of the legal representative's overheads (for example, postage costs and telephone costs) will not be treated as disbursements. The third party supply should be included as part of the costs of the legal representatives' legal services and VAT must be added to the total bill charged to the client.
(3) Disputes may arise in respect of payments made to a third party which the legal representative shows as disbursements in the invoice delivered to the receiving party. Some payments, although correctly described as disbursements for some purposes, are not classified as disbursements for VAT purposes. Items not classified as disbursements for VAT purposes must be shown as part of the services provided by the legal representative and, therefore, VAT must be added in respect of them whether or not VAT was chargeable on the supply by the third party.

(4) Guidance as to the circumstances in which disbursements may or may not be classified as disbursements for VAT purposes is given in the VAT Guide (Notice 700, paragraph 25.1). One of the key issues is whether the third party supply—
   (a) was made to the legal representative (and therefore subsumed in the onward supply of legal services); or
   (b) was made direct to the receiving party (the third party having no right to demand payment from the legal representative, who makes the payment only as agent for the receiving party).
(5) Examples of payments under subparagraph (4)(a) are: travelling expenses, such as an airline ticket, and subsistence expenses, such as the cost of meals, where the person travelling and receiving the meals is the legal representative. The supplies by the airline and the restaurant are supplies to the legal representative, not to the client.
(6) Payments under subparagraph (4)(b) are classified as disbursements for VAT purposes and, therefore, the legal representative need not add VAT in respect of them. Simple examples are payments by a legal representative of court fees and payment of fees to an expert witness.

Litigants in person

2.12 Where a litigant acts in person, that litigant is not treated for the purposes of VAT as having supplied services and therefore no VAT is chargeable in respect of work done by that litigant (even where, for example, that litigant is a solicitor or other legal representative). Consequently in such circumstances a bill of costs should not claim any VAT.

Government Departments

2.13 On an assessment between parties, where costs are being paid to a Government Department in respect of services rendered by its legal staff, VAT should not be added.

Payment pursuant to an order under section 194(3) of the 2007 Act

2.14 Where an order is made under section 194(3) of the 2007 Act, any bill presented for agreement or assessment pursuant to that order must not include a claim for VAT.

Subsection 3 of this Practice Direction – Costs budgets

Costs budgets

[CPR PD 44.3]

3.1 In any case where the parties have filed budgets in accordance with Practice Direction 3E but the court has not made a costs management order under rule 3.15, the provisions of this subsection shall apply.

3.2 If there is a difference of 20% or more between the costs claimed by a receiving party on detailed assessment and the costs shown in a budget filed by that party, the receiving party must provide a statement of the reasons for the difference with the bill of costs.

3.3 If a paying party—
(a) claims to have reasonably relied on a budget filed by a receiving party; or
(b) wishes to rely upon the costs shown in the budget in order to dispute the reasonableness or proportionality of the costs claimed,

the paying party must serve a statement setting out the case in this regard in that party's points of dispute.

3.4 On an assessment of the costs of a party, the court will have regard to the last approved or agreed budget, and may have regard to any other budget previously filed

CPR PD 44.4   *Appendix*

by that party, or by any other party in the same proceedings. Such other budgets may be taken into account when assessing the reasonableness and proportionality of any costs claimed.
3.5 Subject to paragraph 3.4, paragraphs 3.6 and 3.7 apply where there is a difference of 20% or more between the costs claimed by a receiving party and the costs shown in a budget filed by that party.
3.6 Where it appears to the court that the paying party reasonably relied on the budget, the court may restrict the recoverable costs to such sum as is reasonable for the paying party to pay in the light of that reliance, notwithstanding that such sum is less than the amount of costs reasonably and proportionately incurred by the receiving party.
3.7 Where it appears to the court that the receiving party has not provided a satisfactory explanation for that difference, the court may regard the difference between the costs claimed and the costs shown in the budget as evidence that the costs claimed are unreasonable or disproportionate.

Subsection 4 of this Practice Direction — Court's discretion as to costs: Rule 44.2

Court's discretion as to costs: rule 44.2

[CPR PD 44.4]

4.1 The court may make an order about costs at any stage in a case.
4.2 There are certain costs orders which the court will commonly make in proceedings before trial. The following table sets out the general effect of these orders. The table is not an exhaustive list of the orders which the court may make.

Term	Effect
Costs Costs in any event	The party in whose favour the order is made is entitled to that party's costs in respect of the part of the proceedings to which the order relates, whatever other costs orders are made in the proceedings.
Costs in the case Costs in the application	The party in whose favour the court makes an order for costs at the end of the proceedings is entitled to that party's costs of the part of the proceedings to which the order relates.
Costs reserved	The decision about costs is deferred to a later occasion, but if no later order is made the costs will be costs in the case.
Claimant's/Defendant's costs in case/application	If the party in whose favour the costs order is made is awarded costs at the end the proceedings, that party is entitled to that party's costs of the part of the proceedings to which the order relates. If any other party is awarded costs at the end of the proceedings, the party in whose favour the final costs order is made is not liable to pay the costs of any other party in respect of the part of the proceedings to which the order relates.
	Where, for example, a judgment or order is set aside, the party in whose favour the costs order is made is entitled to the costs which have been incurred as a consequence. This includes the costs of –
	preparing for and attending any hearing at which the judgment or order which has been set aside was made;
Costs thrown away	preparing for and attending any hearing to set aside the judgment or order in question;

Term	Effect
	preparing for and attending any hearing at which the court orders the proceedings or the part in question to be adjourned;
	any steps taken to enforce a judgment or order which has subsequently been set aside.
Costs of and caused by	Where, for example, the court makes this order on an application to amend a statement of case, the party in whose favour the costs order is made is entitled to the costs of preparing for and attending the application and the costs of any consequential amendment to his own statement of case.
Costs here and below	The party in whose favour the costs order is made is entitled not only to that party's costs in respect of the proceedings in which the court makes the order but also to that party's costs of the proceedings in any lower court. In the case of an appeal from a Divisional Court the party is not entitled to any costs incurred in any court below the Divisional Court.
No order as to costs Each party to pay own costs	Each party is to bear that party's own costs of the part of the proceedings to which the order relates whatever costs order the court makes at the end of the proceedings.

Subsection 5 of this Practice Direction — Fees of Counsel

Fees of Counsel

[CPR PD 44.5]

5.1
(1) When making an order for costs the court may state an opinion as to whether or not the hearing was fit for the attendance of one or more counsel, and, if it does so, the court conducting a detailed assessment of those costs will have regard to the opinion stated.
(2) The court will generally express an opinion only where—
 (a) the paying party asks it to do so;
 (b) more than one counsel appeared for a party; or
 (c) the court wishes to record its opinion that the case was not fit for the attendance of counsel.
5.2
(1) Where the court refers any matter to the conveyancing counsel of the court the fees payable to counsel in respect of the work done or to be done will be assessed by the court in accordance with rule 44.2.
(2) An appeal from a decision of the court in respect of the fees of such counsel will be dealt with under the general rules as to appeals set out in Part 52. If the appeal is against the decision of an authorised court officer, it will be dealt with in accordance with rules 47.22 to 47.24.

CPR PD 44.6 *Appendix*

Subsection 6 of this Practice Direction – Basis of assessment: Rule 44.3

Costs on the indemnity basis

[CPR PD 44.6]

    6.1 If costs are awarded on the indemnity basis, the court assessing costs will disallow any costs—
    (a)   which it finds to have been unreasonably incurred; or
    (b)   which it considers to be unreasonable in amount.

Costs on the standard basis

    6.2 If costs are awarded on the standard basis, the court assessing costs will disallow any costs—
    (a)   which it finds to have been unreasonably incurred;
    (b)   which it considers to be unreasonable in amount;
    (c)   which it considers to have been disproportionately incurred or to be disproportionate in amount; or
    (d)   about which it has doubts as to whether they were reasonably or proportionately incurred, or whether they are reasonable and proportionate in amount.

Subsection 7 of this Practice Direction — Amount of costs where costs are payable pursuant to a contract: Rule 44.5

Application of rule 44.5

[CPR PD 44.7]

    7.1 Rule 44.5 only applies if the court is assessing costs payable under a contract. It does not—
    (a)   require the court to make an assessment of such costs; or
    (b)   require a mortgagee to apply for an order for those costs where there is a contractual right to recover out of the mortgage funds.

Costs relating to a mortgage

    7.2
    (1)   The following principles apply to costs relating to a mortgage.
    (2)   An order for the payment of costs of proceedings by one party to another is always a discretionary order: section 51 of the Senior Courts Act 1981 ("the section 51 discretion").
    (3)   Where there is a contractual right to the costs, the discretion should ordinarily be exercised so as to reflect that contractual right.
    (4)   The power of the court to disallow a mortgagee's costs sought to be added to the mortgage security is a power that does not derive from section 51, but from the power of the courts of equity to fix the terms on which redemption will be allowed.
    (5)   A decision by a court to refuse costs in whole or in part to a mortgagee may be—
        (a)   a decision in the exercise of the section 51 discretion;
        (b)   a decision in the exercise of the power to fix the terms on which redemption will be allowed;
        (c)   a decision as to the extent of a mortgagee's contractual right to add the mortgagee's costs to the security; or
        (d)   a combination of two or more of these things.
    (6)   A mortgagee is not to be deprived of a contractual or equitable right to add costs to the security merely by reason of an order for payment of costs made

without reference to the mortgagee's contractual or equitable rights, and without any adjudication as to whether or not the mortgagee should be deprived of those costs.

7.3
(1) Where the contract entitles a mortgagee to—
    (a) add the costs of litigation relating to the mortgage to the sum secured by it; or
    (b) require a mortgagor to pay those costs,
    the mortgagor may make an application for the court to direct that an account of the mortgagee's costs be taken.
(Rule 25.1(1)(n) provides that the court may direct that a party file an account.)
(2) The mortgagor may then dispute an amount in the mortgagee's account on the basis that it has been unreasonably incurred or is unreasonable in amount.
(3) Where a mortgagor disputes an amount, the court may make an order that the disputed costs are assessed under rule 44.5.

Subsection 8 of this Practice Direction — Procedure for assessing costs: Rule 44.6

Procedure for assessing costs: rule 44.6

[CPR PD 44.8]

8.1 Subject to paragraph 8.3, where the court does not order fixed costs (or no fixed costs are provided for) the amount of costs payable will be assessed by the court. Rule 44.6 allows the court making an order about costs either—
(a) to make a summary assessment of the amount of the costs; or
(b) to order the amount to be decided in accordance with Part 47 (a detailed assessment).

8.2 An order for costs will be treated as an order for the amount of costs to be decided by a detailed assessment unless the order otherwise provides.

8.3 Where a party is entitled to costs some of which are fixed costs and some of which are not, the court will assess those costs which are not fixed. For example, the court will assess the disbursements payable in accordance with rules 45.12 or 45.19. The decision whether such assessment should be summary or detailed will be made in accordance with paragraphs 9.1 to 9.10 of this Practice Direction.

Subsection 9 of this Practice Direction — Summary assessment: General provisions

When the court should consider whether to make a summary assessment

[CPR PD 44.9]

9.1 Whenever a court makes an order about costs which does not provide only for fixed costs to be paid the court should consider whether to make a summary assessment of costs.

Timing of summary assessment

9.2 The general rule is that the court should make a summary assessment of the costs—
(a) at the conclusion of the trial of a case which has been dealt with on the fast track, in which case the order will deal with the costs of the whole claim; and
(b) at the conclusion of any other hearing, which has lasted not more than one day, in which case the order will deal with the costs of the application or

CPR PD 44.9 *Appendix*

matter to which the hearing related. If this hearing disposes of the claim, the order may deal with the costs of the whole claim,
unless there is good reason not to do so, for example where the paying party shows substantial grounds for disputing the sum claimed for costs that cannot be dealt with summarily.

Summary assessment of mortgagee's costs

9.3 The general rule in paragraph 9.2 does not apply to a mortgagee's costs incurred in mortgage possession proceedings or other proceedings relating to a mortgage unless the mortgagee asks the court to make an order for the mortgagee's costs to be paid by another party.
(Paragraphs 7.2 and 7.3 deal in more detail with costs relating to mortgages.)

Consent orders

9.4 Where an application has been made and the parties to the application agree an order by consent without any party attending, the parties should seek to agree a figure for costs to be inserted in the consent order or agree that there should be no order for costs.

Duty of parties and legal representatives

9.5
(1) It is the duty of the parties and their legal representatives to assist the judge in making a summary assessment of costs in any case to which paragraph 9.2 above applies, in accordance with the following subparagraphs.
(2) Each party who intends to claim costs must prepare a written statement of those costs showing separately in the form of a schedule—
   (a) the number of hours to be claimed;
   (b) the hourly rate to be claimed;
   (c) the grade of fee earner;
   (d) the amount and nature of any disbursement to be claimed, other than counsel's fee for appearing at the hearing;
   (e) the amount of legal representative's costs to be claimed for attending or appearing at the hearing;
   (f) counsel's fees; and
   (g) any VAT to be claimed on these amounts.
(3) The statement of costs should follow as closely as possible Form N260 and must be signed by the party or the party's legal representative. Where a party is—
   (a) an assisted person;
   (b) a LSC funded client;
   (c) a person for whom civil legal services (within the meaning of Part 1 of the Legal Aid, Sentencing and Punishment of Offenders Act 2012) are provided under arrangements made for the purposes of that Part of that Act; or
   (d) represented by a person in the party's employment,
   the statement of costs need not include the certificate appended at the end of Form N260.
(4) The statement of costs must be filed at court and copies of it must be served on any party against whom an order for payment of those costs is intended to be sought as soon as possible and in any event—
   (a) for a fast track trial, not less than 2 days before the trial; and
   (b) for all other hearings, not less than 24 hours before the time fixed for the hearing.

9.6 The failure by a party, without reasonable excuse, to comply with paragraph 9.5 will be taken into account by the court in deciding what order to make about the costs of the claim, hearing or application, and about the costs of any further hearing or detailed assessment hearing that may be necessary as a result of that failure.

No summary assessment by a costs officer

9.7 The court awarding costs cannot make an order for a summary assessment of costs by a costs officer. If a summary assessment of costs is appropriate but the court awarding costs is unable to do so on the day, the court may give directions as to a further hearing before the same judge.

Assisted persons etc

9.8 The court will not make a summary assessment of the costs of a receiving party who is an assisted person or LSC funded client or who is a person for whom civil legal services (within the meaning of Part 1 of the Legal Aid, Sentencing and Punishment of Offenders Act 2012) are provided under arrangements made for the purposes of that Part of that Act.

Children or protected parties

9.9
(1) The court will not make a summary assessment of the costs of a receiving party who is a child or protected party within the meaning of Part 21 unless the legal representative acting for the child or protected party has waived the right to further costs (see Practice Direction 46 paragraph 2.1).
(2) The court may make a summary assessment of costs payable by a child or protected party.

Disproportionate or unreasonable costs

9.10 The court will not give its approval to disproportionate or unreasonable costs. When the amount of the costs to be paid has been agreed between the parties the order for costs must state that the order is by consent.

Subsection 10 of this Practice Direction — Legal representative's duty to notify party: Rule 44.8

Legal representative's duty to notify party: rule 44.8

[CPR PD 44.10]

10.1 For the purposes of rule 44.8 and paragraph 10.2, "party" includes any person (for example, an insurer, a trade union or the LSC or Lord Chancellor) who has instructed the legal representative to act for the party or who is liable to pay the legal representative's fees.
10.2 A legal representative who notifies a party of an order under rule 44.8 must also explain why the order came to be made.
10.3 Although rule 44.8 does not specify any sanction for breach of the rule the court may, either in the order for costs itself or in a subsequent order, require the legal representative to produce to the court evidence showing that the legal representative took reasonable steps to comply with the rule.

Subsection 11 of this Practice Direction — Court's powers in relation to misconduct: Rule 44.11

Court's powers in relation to misconduct: rule 44.11

[CPR PD 44.11]

11.1 Before making an order under rule 44.11, the court must give the party or legal

CPR PD 44.12 Appendix

representative in question a reasonable opportunity to make written submissions or, if the legal representative so desires, to attend a hearing.

11.2 Conduct which is unreasonable or improper includes steps which are calculated to prevent or inhibit the court from furthering the overriding objective.

11.3 Although rule 44.11(3) does not specify any sanction for breach of the obligation imposed by the rule the court may, either in the order under rule 44.11(2) or in a subsequent order, require the legal representative to produce to the court evidence that the legal representative took reasonable steps to comply with the obligation.

Section II – Qualified one-way costs shifting

Subsection 12 of this Practice Direction – Qualified one-way costs shifting

Qualified one-way costs shifting

[CPR PD 44.12]

12.1 This subsection applies to proceedings to which Section II of Part 44 applies.

12.2 Examples of claims made for the financial benefit of a person other than the claimant or a dependant within the meaning of section 1(3) of the Fatal Accidents Act 1976 within the meaning of rule 44.16(2) are subrogated claims and claims for credit hire.

12.3 "Gratuitous provision of care" within the meaning of rule 44.16(2)(a) includes the provision of personal services rendered gratuitously by persons such as relatives and friends for things such as personal care, domestic assistance, childminding, home maintenance and decorating, gardening and chauffeuring.

12.4 In a case to which rule 44.16(1) applies (fundamentally dishonest claims)—
(a) the court will normally direct that issues arising out of an allegation that the claim is fundamentally dishonest be determined at the trial;
(b) where the proceedings have been settled, the court will not, save in exceptional circumstances, order that issues arising out of an allegation that the claim was fundamentally dishonest be determined in those proceedings;
(c) where the claimant has served a notice of discontinuance, the court may direct that issues arising out of an allegation that the claim was fundamentally dishonest be determined notwithstanding that the notice has not been set aside pursuant to rule 38.4;
(d) the court may, as it thinks fair and just, determine the costs attributable to the claim having been found to be fundamentally dishonest.

12.5 The court has power to make an order for costs against a person other than the claimant under section 51(3) of the Senior Courts Act 1981 and rule 46.2. In a case to which rule 44.16(2)(a) applies (claims for the benefit of others)—
(a) the court will usually order any person other than the claimant for whose financial benefit such a claim was made to pay all the costs of the proceedings or the costs attributable to the issues to which rule 44.16(2)(a) applies, or may exceptionally make such an order permitting the enforcement of such an order for costs against the claimant.
(b) the court may, as it thinks fair and just, determine the costs attributable to claims for the financial benefit of persons other than the claimant.

12.6 In proceedings to which rule 44.16 applies, the court will normally order the claimant or, as the case may be, the person for whose benefit a claim was made to pay costs notwithstanding that the aggregate amount in money terms of such orders exceeds the aggregate amount in money terms of any orders for damages, interest and costs made in favour of the claimant.

12.7 Assessments of costs may be on a standard or indemnity basis and may be subject to a summary or detailed assessment.

Civil Procedure Rules 1998 **CPR PD 44.12**

PART 45 FIXED COSTS

I
FIXED COSTS

Rule 45.1	Scope of this Section	CPR 45.1
Rule 45.2	Amount of fixed commencement costs in a claim for the recovery of money or goods	CPR 45.2
Rule 45.3	When defendant only liable for fixed commencement costs	CPR 45.3
Rule 45.4	Costs on entry of judgment in a claim for the recovery of money or goods	CPR 45.4
Rule 45.5	Amount of fixed commencement costs in a claim for the recovery of land or a demotion claim	CPR 45.5
Rule 45.6	Costs on entry of judgment in a claim for the recovery of land or a demotion claim	CPR 45.6
Rule 45.7	Miscellaneous fixed costs	CPR 45.7
Rule 45.8	Fixed enforcement costs	CPR 45.8

II
ROAD TRAFFIC ACCIDENTS—FIXED RECOVERABLE COSTS

Rule 45.9	Scope and interpretation	CPR 45.9
Rule 45.10	Application of fixed recoverable costs	CPR 45.10
Rule 45.11	Amount of fixed recoverable costs	CPR 45.11
Rule 45.12	Disbursements	CPR 45.12
Rule 45.13	Claims for an amount of costs exceeding fixed recoverable costs	CPR 45.13
Rule 45.14	Failure to achieve costs greater than fixed recoverable costs	CPR 45.14
Rule 45.15	Costs of the costs-only proceedings or the detailed assessment	CPR 45.15

III
PRE-ACTION PROTOCOL FOR LOW VALUE PERSONAL INJURY CLAIMS IN ROAD TRAFFIC ACCIDENTS

Rule 45.16	Scope and interpretation	CPR 45.16
Rule 45.17	Application of fixed costs, disbursements and success fee	CPR 45.17
Rule 45.18	Amount of fixed costs	CPR 45.189
Rule 45.19	Disbursements	CPR 45.19
Rule 45.20	Where the claimant obtains judgment for an amount more than the defendant's RTA Protocol offer	CPR 45.20
Rule 45.21	Settlement at Stage 2 where the claimant is a child	CPR 45.21
Rule 45.22	Settlement at Stage 3 where the claimant is a child	CPR 45.22
Rule 45.23	Where the court orders the claim is not suitable to be determined under the Stage 3 Procedure and the claimant is a child	CPR 45.23
Rule 45.24	Failure to comply or electing not to continue with the RTA Protocol – costs consequences	CPR 45.24
Rule 45.25	Where the parties have settled after proceedings have started	CPR 45.25
Rule 45.26	Where the claimant obtains judgment for an amount equal to or less than the defendant's RTA Protocol offer	CPR 45.26
Rule 45.27	Adjournment	CPR 45.27
Rule 45.28	Account of payment of Stage 1 fixed costs	CPR 45.28
Rule 45.29	Costs-only application after a claim is started under Part 8 in accordance with Practice Direction 8B	CPR 45.29

IIIA
CLAIMS WHICH NO LONGER CONTINUE UNDER THE RTA AND EL/PL PRE-ACTION PROTOCOLS – FIXED RECOVERABLE COSTS

Rule 45.29A	Scope and interpretation	CPR 45.29A
Rule 45.29B	Application of fixed costs and disbursements – RTA Protocol	CPR 45.29B
Rule 45.29C	Amount of fixed costs – RTA Protocol	CPR 45.29C
Rule 45.29D	Application of fixed costs and disbursements – EL/PL Protocol	CPR 45.29D
Rule 45.29E	Amount of fixed costs – EL/PL Protocol	CPR 45.29E
Rule 45.29F	Defendants' costs	CPR 45.29F
Rule 45.29G	Counterclaims under the RTA Protocol	CPR 45.29G
Rule 45.29H	Interim applications	CPR 45.29H

681

CPR 45.1 *Appendix*

Rule 45.29I   Disbursements .... CPR 45.29I
Rule 45.29J   Claims for an amount of costs exceeding fixed recoverable costs .. CPR 45.29Jl
Rule 45.29K   Failure to achieve costs greater than fixed recoverable costs .... CPR 45.29K
Rule 45.29L   Costs of the costs-only proceedings or the detailed assessment .. CPR 45.29L

IV
SCALE COSTS FOR CLAIMS IN THE INTELLECTUAL PROPERTY ENTERPRISE COURT

Rule 45.30   Scope and interpretation .... CPR 45.30
Rule 45.31   Amount of scale costs .... CPR 45.31
Rule 45.32   Summary assessment of the costs of an application where a party has behaved unreasonably .... CPR 45.32

V
FIXED COSTS: HM REVENUE AND CUSTOMS

Rule 45.33   Scope, interpretation and application .... CPR 45.33
Rule 45.34   Amount of fixed commencement costs in a county court claim for the recovery of money .... CPR 45.34
Rule 45.35   Costs on entry of judgment in a county court claim for recovery of money .... CPR 45.35
Rule 45.36   When the defendant is only liable for the fixed commencement costs . CPR 45.36

VI
FAST TRACK TRIAL COSTS

Rule 45.37   Scope of this Section .... CPR 45.37
Rule 45.38   Amount of fast track trial costs .... CPR 45.38
Rule 45.39   Power to award more or less than the amount of fast track trial costs . CPR 45.39
Rule 45.40   Fast track trial costs where there is more than one claimant or defendant .... CPR 45.40

VII
COSTS LIMITS IN AARHUS CONVENTION CLAIMS

Rule 45.41   Scope and interpretation .... CPR 45.41
Rule 45.42   Opting out .... CPR 45.42
Rule 45.43   Limit on costs recoverable from a party in an Aarhus Convention claim .... CPR 45.43
Rule 45.44   Challenging whether the claim is an Aarhus Convention claim ..... CPR 45.44
               Practice Direction 45—Fixed Costs .... CPR PD 45

SECTION I FIXED COSTS

[CPR 45.1]

45.1 Scope of this Section

(1) This Section sets out the amounts which, unless the court orders otherwise, are to be allowed in respect of legal representatives' charges.

(2) This Section applies where—

    (a)   the only claim is a claim for a specified sum of money where the value of the claim exceeds £25 and—

        (i)   judgment in default is obtained under rule 12.4(1);

        (ii)   judgment on admission is obtained under rule 14.4(3);

        (iii)   judgment on admission on part of the claim is obtained under rule 14.5(6);

        (iv)   summary judgment is given under Part 24;

        (v)   the court has made an order to strike out a defence under rule 3.4(2)(a) as disclosing no reasonable grounds for defending the claim; or

        (vi)   rule 45.4 applies;

(b) the only claim is a claim where the court gave a fixed date for the hearing when it issued the claim and judgment is given for the delivery of goods, and the value of the claim exceeds £25;

(c) the claim is for the recovery of land, including a possession claim under Part 55, whether or not the claim includes a claim for a sum of money and the defendant gives up possession, pays the amount claimed, if any, and the fixed commencement costs stated in the claim form;

(d) the claim is for the recovery of land, including a possession claim under Part 55, where one of the grounds for possession is arrears of rent, for which the court gave a fixed date for the hearing when it issued the claim and judgment is given for the possession of land (whether or not the order for possession is suspended on terms) and the defendant—
  (i) has neither delivered a defence, or counterclaim, nor otherwise denied liability; or
  (ii) has delivered a defence which is limited to specifying his proposals for the payment of arrears of rent;

(e) the claim is a possession claim under Section II of Part 55 (accelerated possession claims of land let on an assured shorthold tenancy) and a possession order is made where the defendant has neither delivered a defence, or counterclaim, nor otherwise denied liability;

(f) the claim is a demotion claim under Section III of Part 65 or a demotion claim is made in the same claim form in which a claim for possession is made under Part 55 and that demotion claim is successful; or

(g) a judgment creditor has taken steps under Parts 70 to 73 to enforce a judgment or order.

(Practice Direction 7B sets out the types of case where a court will give a fixed date for a hearing when it issues a claim.)

(3) No sum in respect of legal representatives' charges will be allowed where the only claim is for a sum of money or goods not exceeding £25.

(4) Any appropriate court fee will be allowed in addition to the costs set out in this Section.

(5) The claim form may include a claim for fixed commencement costs.

[CPR 45.2]

45.2 Amount of fixed commencement costs in a claim for the recovery of money or goods

(1) The amount of fixed commencement costs in a claim to which rule 45.1(2)(a) or (b) applies—
  (a) will be calculated by reference to Table 1; and
  (b) the amount claimed, or the value of the goods claimed if specified, in the claim form is to be used for determining the band in Table 1 that applies to the claim.

(2) The amounts shown in Table 4 are to be allowed in addition, if applicable.

CPR 45.3 *Appendix*

Table 1
Fixed costs on commencement of a claim for the recovery of money or goods

Relevant Band	Where the claim form is served by the court or by any method other than personal service by the claimant	Where – the claim form is served personally by the claimant; – there is only one defendant	Where there is more than one defendant, for each additional defendant personally served at separate addresses by the claimant
Where— — The value of the claim exceeds £25 but does not exceed £500	£50	£60	£15
Where— — The value of the claim exceeds £500 but does not exceed £1,000	£70	£80	£15
Where— — The value of the claim exceeds £1,000 but does not exceed £5,000; or — the only claim is for delivery of goods and no value is specified or stated on the claim form	£80	£90	£15
Where— — the value of the claim exceeds £5,000	£100	£110	£15

[CPR 45.3]

45.3 When defendant only liable for fixed commencement costs
   Where—
   (a)   the only claim is for a specified sum of money; and
   (b)   the defendant pays the money claimed within 14 days after being served with the particulars of claim, together with the fixed commencement costs stated in the claim form,
the defendant is not liable for any further costs unless the court orders otherwise.

[CPR 45.4]

45.4 Costs on entry of judgment in a claim for the recovery of money or goods
   Where—

Civil Procedure Rules 1998 **CPR 45.4**

(a) the claimant has claimed fixed commencement costs under rule 45.2; and
(b) judgment is entered in a claim to which rule 45.1(2)(a) or (b) applies in the circumstances specified in Table 2, the amount to be included in the judgment for the claimant's legal representative's charges is the total of—
 (i) the fixed commencement costs; and
 (ii) the relevant amount shown in Table 2.

Table 2
Fixed costs on entry of judgment in a claim for the recovery of money or goods

	Where the amount of the judgment exceeds £25 but does not exceed £5,000	Where the amount of the judgment exceeds £5,000
Where judgment in default of an acknowledgment of service is entered under rule 12.4 (1) (entry of judgment by request on claim for money only)	£22	£30
Where judgment in default of a defence is entered under rule 12.4 (1) (entry of judgment by request on claim for money only)	£25	£35
Where judgment is entered under rule 14.4 (judgment on admission), or rule 14.5 (judgment on admission of part of claim) and claimant accepts the defendant's proposal as to the manner of payment.	£40	£55
Where judgment is entered under rule 14.4 (judgment on admission), or rule 14.5 (judgment on admission on part of claim) and court decides the date or times of payment	£55	£70
Where summary judgment is given under Part 24 or the court strikes out a defence under rule 3.4 (2)(a), in either case, on application by a party	£175	£210
Where judgment is given on a claim for delivery of goods under a regulated agreement within the meaning of the Consumer Credit Act 1974 and no other entry in this table applies	£60	£85

685

CPR 45.5 Appendix

[CPR 45.5]

45.5 Amount of fixed commencement costs in a claim for the recovery of land or a demotion claim

(1) The amount of fixed commencement costs in a claim to which rule 45.1(2)(c), (d) or (f) applies will be calculated by reference to Table 3.

(2) The amounts shown in Table 4 are to be allowed in addition, if applicable.

Table 3
Fixed costs on commencement of a claim for the recovery of land or a demotion claim

Where the claim form is served by the court or by any method other than personal service by the claimant	Where the claim form is served personally by the claimant; there is only one defendant	Where there is more than one defendant, for each additional defendant personally served at separate addresses by the claimant
£69.50	£77	£15

[CPR 45.6]

45.6 Costs on entry of judgment in a claim for the recovery of land or a demotion claim

(1) Where—
   (a) the claimant has claimed fixed commencement costs under rule 45.5; and
   (b) judgment is entered in a claim to which rule 45.1(2)(d) or (f) applies, the amount to be included in the judgment for the claimant's legal representative's charges is the total of—
      (i) the fixed commencement costs; and
      (ii) the sum of £57.25.

(2) Where an order for possession is made in a claim to which rule 45.1(2)(e) applies, the amount allowed for the claimant's legal representative's charges for preparing and filing—
   (a) the claim form;
   (b) the documents that accompany the claim form; and
   (c) the request for possession,
is £79.50.

[CPR 45.7]

45.7 Miscellaneous fixed costs

Table 4 shows the amount to be allowed in respect of legal representative's charges in the circumstances mentioned.

Table 4
Miscellaneous fixed costs

For service by a party of any document required to be served personally including preparing and copying a certificate of service for each individual served	£15
Where service by an alternative method or at an alternative place is permitted by an order under rule 6.15 for each individual served	£53.25
Where a document is served out of the jurisdiction—	
(a) in Scotland, Northern Ireland, the Isle of Man or the Channel Islands	£68.25
(b) in any other place	£77

[CPR 45.8]

45.8 Fixed enforcement costs

Table 5 shows the amount to be allowed in respect of legal representatives' costs in the circumstances mentioned. The amounts shown in Table 4 are to be allowed in addition, if applicable.

Table 5
Fixed enforcement costs

For an application under rule 70.5(4) that an award may be enforced as if payable under a court order, where the amount outstanding under the award:	
exceeds £25 but does not exceed £250	£30.75
exceeds £250 but does not exceed £600	£41.00
exceeds £600 but does not exceed £2,000	£69.50
exceeds £2,000	£75.50
On attendance to question a judgment debtor (or officer of a company or other corporation) who has been ordered to attend court under rule 71.2 where the questioning takes place before a court officer, including attendance by a responsible representative of the solicitor:	
	for each half-hour or part, £15.00
	(When the questioning takes place before a judge, he may summarily assess any costs allowed.)
On the making of a final third party debt order under rule 72.8(6)(a) or an order for the payment to the judgment creditor of money in court under rule 72.10(1)(b):	
if the amount recovered is less than £150	one-half of the amount recovered
Otherwise	£98.50

CPR 45.9 *Appendix*

	On the making of a final charging order under rule 73.8(2)(a):	£110.00
		The court may also allow reasonable disbursements in respect of search fees and the registration of the order.
	Where a certificate is issued and registered under Schedule 6 to the Civil Jurisdiction and Judgments Act 1982, the costs of registration	£39
	Where permission is given under RSC Order 45, rule 3 to enforce a judgment or order giving possession of land and costs are allowed on the judgment or order, the amount to be added to the judgment or order for costs—	
(a)	basic costs	£42.50
(b)	where notice of the proceedings is to be to more than one person, for each additional person	£2.75
	Where a writ of execution as defined in the RSC Order 46, rule 1, is issued against any party	£51.75
	Where a request is filed for the issue of a warrant of execution under CCR Order 26, rule 1, for a sum exceeding £25	£2.25
	Where an application for an attachment of earnings order is made and costs are allowed under CCR Order 27, rule 9 or CCR Order 28, rule 10, for each attendance on the hearing of the application	£8.50

SECTION II ROAD TRAFFIC ACCIDENTS – FIXED RECOVERABLE COSTS

[CPR 45.9]

45.9 Scope and interpretation

(1) Subject to paragraph (3), this Section sets out the costs which are to be allowed in—
    (a) proceedings to which rule 46.14(1) applies (costs-only proceedings); or
    (b) proceedings for approval of a settlement or compromise under rule 21.10(2),
in cases to which this Section applies.

(2) This Section applies where—
    (a) the dispute arises from a road traffic accident occurring on or after 6 October 2003;
    (b) the agreed damages include damages in respect of personal injury, damage to property, or both;
    (c) the total value of the agreed damages does not exceed £10,000; and
    (d) if a claim had been issued for the amount of the agreed damages, the small claims track would not have been the normal track for that claim.

(3) This Section does not apply where—
    (a) the claimant is a litigant in person; or
    (b) Section III or Section IIIA of this Part applies.
(4) In this Section—

"road traffic accident" means an accident resulting in bodily injury to any person or damage to property caused by, or arising out of, the use of a motor vehicle on a road or other public place in England and Wales;

"motor vehicle" means a mechanically propelled vehicle intended for use on roads; and

"road" means any highway and any other road to which the public has access and includes bridges over which a road passes.

[CPR 45.10]

45.10 Application of fixed recoverable costs

Subject to rule 45.13, the only costs which are to be allowed are—
    (a) fixed recoverable costs calculated in accordance with rule 45.11; and
    (b) disbursements allowed in accordance with rule 45.12.

(Rule 45.13 provides for where a party issues a claim for more than the fixed recoverable costs.)

[CPR 45.11]

45.11 Amount of fixed recoverable costs

(1) Subject to paragraphs (2) and (3), the amount of fixed recoverable costs is the total of—
    (a) £800;
    (b) 20% of the damages agreed up to £5,000; and
    (c) 15% of the damages agreed between £5,000 and £10,000.
(2) Where the claimant—
    (a) lives or works in an area set out in Practice Direction 45; and
    (b) instructs a legal representative who practises in that area,
the fixed recoverable costs will include, in addition to the costs specified in paragraph (1), an amount equal to 12.5% of the costs allowable under that paragraph.
(3) Where appropriate, VAT may be recovered in addition to the amount of fixed recoverable costs and any reference in this Section to fixed recoverable costs is a reference to those costs net of any such VAT.

[CPR 45.12]

45.12 Disbursements

(1) The court—
    (a) may allow a claim for a disbursement of a type mentioned in paragraph (2); but
    (b) will not allow a claim for any other type of disbursement.
(2) The disbursements referred to in paragraph (1) are—
    (a) the cost of obtaining—
        (i) medical records;
        (ii) a medical report;
        (iii) a police report;
        (iv) an engineer's report; or

CPR 45.13   *Appendix*

        (v)   a search of the records of the Driver Vehicle Licensing Authority;
    (b)   where they are necessarily incurred by reason of one or more of the claimants being a child or protected party as defined in Part 21—
        (i)   fees payable for instructing counsel; or
        (ii)   court fees payable on an application to the court; or
    (c)   any other disbursement that has arisen due to a particular feature of the dispute.

[CPR 45.13]

45.13 Claims for an amount of costs exceeding fixed recoverable costs
(1) The court will entertain a claim for an amount of costs (excluding any success fee or disbursements) greater than the fixed recoverable costs but only if it considers that there are exceptional circumstances making it appropriate to do so.
(2) If the court considers such a claim appropriate, it may—
    (a)   summarily assess the costs; or
    (b)   make an order for the costs to be subject to detailed assessment.
(3) If the court does not consider the claim appropriate, it will make an order for fixed recoverable costs (and any permitted disbursements) only.

[CPR 45.14]

45.14 Failure to achieve costs greater than fixed recoverable costs
(1) This rule applies where—
    (a)   costs are assessed in accordance with rule 45.13(2); and
    (b)   the court assesses the costs (excluding any VAT) as being an amount which is less than 20% greater than the amount of the fixed recoverable costs.
(2) The court must order the defendant to pay to the claimant the lesser of—
    (a)   the fixed recoverable costs; and
    (b)   the assessed costs.

[CPR 45.15]

45.15 Costs of the costs-only proceedings or the detailed assessment
Where—
    (a)   the court makes an order for fixed recoverable costs in accordance with rule 45.13(3); or
    (b)   rule 45.14 applies, the court may—
        (i)   decide not to make an award of the payment of the claimant's costs in bringing the proceedings under rule 46.14; and
        (ii)   make orders in relation to costs that may include an order that the claimant pay the defendant's costs of defending those proceedings.

SECTION III THE PRE-ACTION PROTOCOLS FOR LOW VALUE PERSONAL INJURY CLAIMS IN ROAD TRAFFIC ACCIDENTS AND LOW VALUE PERSONAL INJURY (EMPLOYERS' LIABILITY AND PUBLIC LIABILITY) CLAIMS

[CPR 45.16]

45.16 Scope and interpretation

(1) This Section applies to claims that have been or should have been started under Part 8 in accordance with Practice Direction 8B ("the Stage 3 Procedure").

(2) Where a party has not complied with the relevant Protocol rule 45.24 will apply.

The "relevant Protocol" means—

    (a) the Pre-Action Protocol for Personal Injury Claims in Road Traffic Accidents ("the RTA Protocol"); or

    (b) the Pre-action Protocol for Low Value Personal Injury Claims (Employers' Liability and Public Liability) Claims ("the EL/PL Protocol").

(3) A reference to "Claim Notification Form" or Court Proceedings Pack is a reference to the form used in the relevant Protocol.

[CPR 45.17]

45.17 Application of fixed costs, and disbursements

The only costs allowed are—

    (a) fixed costs in rule 45.18; and

    (b) disbursements in accordance with rule 45.19; and

    (c) where applicable, fixed costs in accordance with rule 45.23A or 45.23B.

[CPR 45.18]

45.18 Amount of fixed costs

(1) Subject to paragraph (4), the amount of fixed costs is set out in Tables 6 and 6A.

(2) In Tables 6 and 6A —

"Type A fixed costs" means the legal representative's costs;

"Type B fixed costs" means the advocate's costs; and

"Type C fixed costs" means the costs for the advice on the amount of damages where the claimant is a child.

(3) "Advocate" has the same meaning as in rule 45.37(2)(a).

(4) Subject to rule 45.24(2) the court will not award more or less than the amounts shown in Tables 6 and 6A.

(5) Where the claimant—

    (a) lives or works in an area set out in Practice Direction 45; and

    (b) instructs a legal representative who practises in that area,

the fixed costs will include, in addition to the costs set out in Tables 6 and 6A, an amount equal to 12.5% of the Stage 1 and 2 and Stage 3 Type A fixed costs.

(6) Where appropriate, VAT may be recovered in addition to the amount of fixed costs and any reference in this Section to fixed costs is a reference to those costs net of any such VAT.

CPR 45.19 *Appendix*

Table 6 – Fixed costs in relation to the RTA Protocol

Where the value of the claim for damages is not more than £10,000		Where the value of the claim for damages is more than £10,000, but not more than £25,000	
Stage 1 fixed costs	£200	Stage 1 fixed costs	£200
Stage 2 fixed costs	£300	Stage 2 fixed costs	£600
Stage 3—		Stage 3—	
Type A fixed costs	£250	Type A fixed costs	£250
Type B fixed costs	£250	Type B fixed costs	£250
Type C fixed costs	£150	Type C fixed costs	£150

Table 6A – Fixed costs in relation to the EL/PL Protocol

Where the value of the claim for damages is not more than £10,000		Where the value of the claim for damages is more than £10,000, but not more than £25,000	
Stage 1 fixed costs	£300	Stage 1 fixed costs	£300
Stage 2 fixed costs	£600	Stage 2 fixed costs	£1300
Stage 3—		Stage 3—	
Type A fixed costs	£250	Type A fixed costs	£250
Type B fixed costs	£250	Type B fixed costs	£250
Type C fixed costs	£150	Type C fixed costs	£150

[CPR 45.19]

45.19 Disbursements

(1) The court—
    (a) may allow a claim for a disbursement of a type mentioned in paragraphs (2) or (3); but
    (b) will not allow a claim for any other type of disbursement.

(2) In a claim to which either the RTA Protocol or EL/PL Protocol applies, the disbursements] referred to in paragraph (1) are—
    (a) the cost of obtaining—
        (i) medical records;
        (ii) a medical report or reports or non-medical expert reports as provided for in the relevant Protocol;
        (iii) ...
        (iv) ...
    (b) court fees as a result of Part 21 being applicable;
    (c) court fees payable where proceedings are started as a result of a limitation period that is about to expire;
    (d) court fees in respect of the Stage 3 Procedure; and
    (e) any other disbursement that has arisen due to a particular feature of the dispute.

(3) In a claim to which the RTA Protocol applies, the disbursements referred to in paragraph (1) are also the cost of—
    (a)   an engineer's report; and
    (b)   a search of the records of the—
        (i)   Driver Vehicle Licensing Authority; and
        (ii)   Motor Insurance Database.

[CPR 45.20]

45.20 Where the claimant obtains judgment for an amount more than the defendant's relevant Protocol offer

Where rule 36.21(1)(b) or (c) applies, the court will order the defendant to pay—
    (a)   where not already paid by the defendant, the Stage 1 and 2 fixed costs;
    (b)   where the claim is determined—
        (i)   on the papers, Stage 3 Type A fixed costs;
        (ii)   at a Stage 3 hearing, Stage 3 Type A and B fixed costs; or
        (iii)   at a Stage 3 hearing and the claimant is a child, Type A, B and C fixed costs; and
    (c)   disbursements allowed in accordance with rule 45.19.

[CPR 45.21]

45.21 Settlement at Stage 2 where the claimant is a child

(1) This rule applies where—
    (a)   the claimant is a child;
    (b)   there is a settlement at Stage 2 of the relevant Protocol; and
    (c)   an application is made to the court to approve the settlement.
(2) Where the court approves the settlement at a settlement hearing it will order the defendant to pay—
    (a)   the Stage 1 and 2 fixed costs;
    (b)   the Stage 3 Type A, B and C fixed costs; and
    (c)   disbursements allowed in accordance with rule 45.19.
(3) Where the court does not approve the settlement at a settlement hearing it will order the defendant to pay the Stage 1 and 2 fixed costs.
(4) Paragraphs (5) and (6) apply where the court does not approve the settlement at the first settlement hearing but does approve the settlement at a second settlement hearing.
(5) At the second settlement hearing the court will order the defendant to pay—
    (a)   the Stage 3 Type A and C fixed costs for the first settlement hearing;
    (b)   disbursements allowed in accordance with rule 45.19; and
    (c)   the Stage 3 Type B fixed costs for one of the hearings.
(6) The court in its discretion may also order—
    (a)   the defendant to pay an additional amount of either or both the Stage 3—
        (i)   Type A fixed costs;
        (ii)   Type B fixed costs; or
    (b)   he claimant to pay an amount equivalent to either or both the Stage 3—
        (i)   Type A fixed costs;
        (ii)   Type B fixed costs.

CPR 45.22 *Appendix*

[CPR 45.22]

45.22 Settlement at Stage 3 where the claimant is a child
(1) This rule applies where—
    (a) the claimant is a child;
    (b) there is a settlement after proceedings are started under the Stage 3 Procedure;
    (c) the settlement is more than the defendant's relevant Protocol offer; and
    (d) an application is made to the court to approve the settlement.
(2) Where the court approves the settlement at the settlement hearing it will order the defendant to pay—
    (a) the Stage 1 and 2 fixed costs;
    (b) the Stage 3 Type A, B and C fixed costs; and
    (c) disbursements allowed in accordance with rule 45.19.
(3) Where the court does not approve the settlement at the settlement hearing it will order the defendant to pay the Stage 1 and 2 fixed costs.
(4) Paragraphs (5) and (6) apply where the court does not approve the settlement at the first settlement hearing but does approve the settlement at the Stage 3 hearing.
(5) At the Stage 3 hearing the court will order the defendant to pay—
    (a) the Stage 3 Type A and C fixed costs for the settlement hearing;
    (b) disbursements allowed in accordance with rule 45.19; and
    (c) the Stage 3 Type B fixed costs for one of the hearings.
(6) The court in its discretion may also order—
    (a) he defendant to pay an additional amount of either or both the Stage 3—
        (i) Type A fixed costs;
        (ii) Type B fixed costs; or
    (b) the claimant to pay an amount equivalent to either or both of the Stage 3—
        (i) Type A fixed costs;
        (ii) Type B fixed costs.
(7) Where the settlement is not approved at the Stage 3 hearing the court will order the defendant to pay the Stage 3 Type A fixed costs.

[CPR 45.23]

45.23 Where the court orders that the claim is not suitable to be determined under the Stage 3 Procedure and the claimant is a child
Where—
    (a) the claimant is a child; and
    (b) at a settlement hearing or the Stage 3 hearing the court orders that the claim is not suitable to be determined under the Stage 3 Procedure,
the court will order the defendant to pay—
        (i) the Stage 1 and 2 fixed costs; and
        (ii) the Stage 3 Type A, B and C fixed costs.

[CPR 45.23A]

45.23A Settlement before proceedings are issued under Stage 3
Where—
- (a) there is a settlement after the Court Proceedings Pack has been sent to the defendant but before proceedings are issued under Stage 3; and
- (b) the settlement is more than the defendant's relevant Protocol offer,

the fixed costs will include an additional amount equivalent to the Stage 3 Type A fixed costs.

[CPR 45.23B]

45.23B Additional advice on the value of the claim
Where—
- (a) the value of the claim for damages is more than £10,000;
- (b) an additional advice has been obtained from a specialist solicitor or from counsel;
- (c) that advice is reasonably required to value the claim,

the court will order the defendant to pay—
the fixed costs may include an additional amount equivalent to the Stage 3 Type C fixed costs.

[CPR 45.24]

45.24 Failure to comply or electing not to continue with the relevant Protocol – costs consequences

(1) This rule applies where the claimant—
- (a) does not comply with the process set out in the relevant Protocol; or
- (b) elects not to continue with that process,

and starts proceedings under Part 7.

(2) Where a judgment is given in favour of the claimant but—
- (a) the court determines that the defendant did not proceed with the process set out in the relevant Protocol because the claimant provided insufficient information on the Claim Notification Form;
- (b) the court considers that the claimant acted unreasonably—
  - (i) by discontinuing the process set out in the relevant Protocol and starting proceedings under Part 7;
  - (ii) by valuing the claim at more than £25,000, so that the claimant did not need to comply with the relevant Protocol; or
  - (iii) except for paragraph (2)(a), in any other way that caused the process in the relevant Protocol to be discontinued; or
- (c) the claimant did not comply with the relevant Protocol at all despite the claim falling within the scope of the relevant Protocol,

the court may order the defendant to pay no more than the fixed costs in rule 45.18 together with the disbursements allowed in accordance with rule 45.19.

(3) Where the claimant starts proceedings under paragraph 7.28 of the RTA Protocol or paragraph 7.26 of the EL/PL Protocol and the court orders the defendant to make an interim payment of no more than the interim payment made under paragraph 7.14(2) or (3) of the RTA Protocol or paragraph 7.17(2) or (3) of the EL/PL Protocol the court will, on the final determination of the proceedings, order the defendant to pay no more than—
- (a) the Stage 1 and 2 fixed costs; and

CPR 45.25 *Appendix*

  (b) the disbursements allowed in accordance with rule 45.19.

[CPR 45.25]

45.25 Where the parties have settled after proceedings have started
(1) This rule applies where an application is made under rule 45.29 (costs-only application after a claim is started under Part 8 in accordance with Practice Direction 8B).
(2) Where the settlement is more than the defendant's relevant Protocol offer the court will order the defendant to pay—
  (a) the Stage 1 and 2 fixed costs where not already paid by the defendant;
  (b) the Stage 3 Type A fixed costs; and
  (c) disbursements allowed in accordance with rule 45.19.
(3) Where the settlement is less than or equal to the defendant's relevant Protocol offer the court will order the defendant to pay—
  (a) the Stage 1 and 2 fixed costs where not already paid by the defendant; and
  (b) disbursements allowed in accordance with rule 45.19.
(4) The court may, in its discretion, order either party to pay the costs of the application.

[CPR 45.26]

45.26 Where the claimant obtains judgment for an amount equal to or less than the defendant's relevant Protocol offer
Where rule 36.21(1)(a) applies, the court will order the claimant to pay—
  (a) where the claim is determined—
    (i) on the papers, Stage 3 Type A fixed costs; or
    (ii) at a hearing, Stage 3 Type A and B fixed costs;
  (b) any Stage 3 disbursements allowed in accordance with rule 45.19.

[CPR 45.27]

45.27 Adjournment
Where the court adjourns a settlement hearing or a Stage 3 hearing it may, in its discretion, order a party to pay—
  (a) an additional amount of the Stage 3 Type B fixed costs; and
  (b) any court fee for that adjournment.

[CPR 45.28]

45.28 Account of payment of Stage 1 and Stage 2 fixed costs
Where a claim no longer continues under the relevant Protocol the court will, when making any order as to costs including an order for fixed recoverable costs under Section II or Section IIIA of this Part, take into account the Stage 1 and Stage 2 fixed costs that have been paid by the defendant.

[CPR 45.29]

45.29 Costs-only application after a claim is started under Part 8 in accordance with Practice Direction 8B
(1) This rule sets out the procedure where—

(a) the parties to a dispute have reached an agreement on all issues (including which party is to pay the costs) which is made or confirmed in writing; but
(b) they have failed to agree the amount of those costs; and
(c) proceedings have been started under Part 8 in accordance with Practice Direction 8B.

(2) Either party may make an application for the court to determine the costs.

(3) Where an application is made under this rule the court will assess the costs in accordance with rule 45.22 or rule 45.25.

(4) Rule 44.5 (amount of costs where costs are payable pursuant to a contract) does not apply to an application under this rule.

SECTION IIIACLAIMS WHICH NO LONGER CONTINUE UNDER THE RTA OR EL/PL PRE-ACTION PROTOCOLS – FIXED RECOVERABLE COSTS

[CPR 45.29A]

45.29A Scope and interpretation

(1) Subject to paragraph (3), this section applies where a claim is started under—
(a) the Pre-Action Protocol for Low Value Personal Injury Claims in Road Traffic Accidents ("the RTA Protocol"); or
(b) the Pre-Action Protocol for Low Value Personal Injury (Employers' Liability and Public Liability) Claims ("the EL/PL Protocol"),
but no longer continues under the relevant Protocol or the Stage 3 Procedure in Practice Direction 8B.

(2) This section does not apply to a disease claim which is started under the EL/PL Protocol.

(3) Nothing in this section shall prevent the court making an order under rule 45.24.

[CPR 45.29B]

45.29B Application of fixed costs and disbursements – RTA Protocol

Subject to rules 45.29F, 45.29G, 45.29H and 45.29J, if, in a claim started under the RTA Protocol, the Claim Notification Form is submitted on or after 31st July 2013, the only costs allowed are—
(a) the fixed costs in rule 45.29C;
(b) disbursements in accordance with rule 45.29I.

[CPR 45.29C]

45.29C Amount of fixed costs – RTA Protocol

(1) Subject to paragraph (2), the amount of fixed costs is set out in Table 6B.

(2) Where the claimant—
(a) lives or works in an area set out in Practice Direction 45; and
(b) instructs a legal representative who practises in that area,
the fixed costs will include, in addition to the costs set out in Table 6B, an amount equal to 12.5% of the costs allowable under paragraph (1) and set out in Table 6B.

(3) Where appropriate, VAT may be recovered in addition to the amount of fixed recoverable costs and any reference in this Section to fixed costs is a reference to those costs net of VAT.

(4) In Table 6B—

CPR 45.29C *Appendix*

    (a)   in Part B, "on or after" means the period beginning on the date on which the court respectively—
            (i)   issues the claim;
            (ii)   allocates the claim under Part 26; or
            (iii)   lists the claim for trial; and
    (b)   unless stated otherwise, a reference to "damages" means agreed damages; and
    (c)   a reference to "trial" is a reference to the final contested hearing.

Table 6B – Fixed costs where a claim no longer continues under the RTA Protocol

A. If Parties reach a settlement prior to the claimant issuing proceedings under Part 7			
Agreed damages	At least £1,000, but not more than £5,000	More than £5,000, but not more than £10,000	More than £10,000, but not more than £25,000
Fixed costs	The greater of— (a) £550; or (b) the total of— (i) £100; and (ii) 20% of the damages	The total of— (a) £1,100; and (b) 15% of damages over £5,000	The total of— (a) £1,930; and (b) 10% of damages over £10,000

B. If proceedings are issued under Part 7, but the case settles before trial			
Stage at which case is settled	On or after the date of issue, but prior to the date of allocation under Part 26	On or after the date of allocation under Part 26, but prior to the date of listing	On or after the date of listing but prior the date of trial
Fixed costs	The total of— (a) £1,160; and (b) 20% of the damages	The total of— (a) £1,880; and (b) 20% of the damages	The total of— (a) £2,655; and (b) 20% of the damages

C. If the claim is disposed of at trial	
Fixed costs	The total of— (a) £2,655; and (b) 20% of the damages agreed or awarded; and (c) the relevant trial advocacy fee

Civil Procedure Rules 1998 CPR 45.29E

D. Trial advocacy fees				
Damages agreed or awarded	Not more than £3,000	More than £3,000, but not more than £10,000	More than £10,000, but not more than £15,000	More than £15,000
Trial advocacy fee	£500	£710	£1,070	£1,705

[CPR 45.29D]

45.29D Application of fixed costs and disbursements – EL/PL Protocol

Subject to rules 45.29F, 45.29H and 45.29J, in a claim started under the EL/PL Protocol the only costs allowed are—
    (a)   fixed costs in rule 45.29E; and
    (b)   disbursements in accordance with rule 45.29I.

[CPR 45.29E]

45.29E Amount of fixed costs – EL/PL Protocol

(1)   Subject to paragraph (2), the amount of fixed costs is set out—
    (a)   in respect of employers' liability claims, in Table 6C; and
    (b)   in respect of public liability claims, in Table 6D.
(2)   Where the claimant—
    (a)   lives or works in an area set out in Practice Direction 45; and
    (b)   instructs a legal representative who practises in that area,
the fixed costs will include, in addition to the costs set out in Tables 6C and 6D, an amount equal to 12.5% of the costs allowable under paragraph (1) and set out in Table 6C and 6D.
(3)   Where appropriate, VAT may be recovered in addition to the amount of fixed recoverable costs and any reference in this Section to fixed costs is a reference to those costs net of VAT.
(4)   In Tables 6C and 6D—
    (a)   in Part B, "on or after" means the period beginning on the date on which the court respectively—
        (i)   issues the claim;
        (ii)   allocates the claim under Part 26; or
        (iii)   lists the claim for trial; and
    (b)   unless stated otherwise, a reference to "damages" means agreed damages; and
    (c)   a reference to "trial" is a reference to the final contested hearing.

Table 6C – Fixed costs where a claim no longer continues under the EL/PL Protocol – employers' liability claims

A. If Parties reach a settlement prior to the claimant issuing proceedings under Part 7

CPR 45.29E *Appendix*

Agreed damages	At least £1,000, but not more than £5,000	More than £5,000, but not more than £10,000	More than £10,000, but not more than £25,000
Fixed costs	The total of— (a) £950; and (b) 17.5% of the damages	The total of— (a) £1,855; and (b) 12.5% of damages over £5,000	The total of— (a) £2,500; and (b) 10% of damages over £10,000

B. If proceedings are issued under Part 7, but the case settles before trial			
Stage at which case is settled	On or after the date of issue, but prior to the date of allocation under Part 26	On or after the date of allocation under Part 26, but prior to the date of listing	On or after the date of listing but prior the date of trial
Fixed costs	The total of— (a) £2,630; and (b) 20% of the damages	The total of— (a) £3,350; and (b) 25% of the damages	The total of— (a) £4,280; and (b) 30% of the damages

C. If the claim is disposed of at trial	
Fixed costs	The total of— (a) £4,280; and (b) 30% of the damages agreed or awarded; and (c) the relevant trial advocacy fee

D. Trial advocacy fees				
Damages agreed or awarded	Not more than £3,000	More than £3,000, but not more than £10,000	More than £10,000, but not more than £15,000	More than £15,000
Trial advocacy fee	£500	£710	£1,070	£1,705

Table 6D – Fixed costs where a claim no longer continues under the EL/PL Protocol – public liability claims

Civil Procedure Rules 1998 **CPR 45.29E**

A. If Parties reach a settlement prior to the claimant issuing proceedings under Part 7			
Agreed damages	At least £1,000, but not more than £5,000	More than £5,000, but not more than £10,000	More than £10,000, but not more than £25,000
Fixed costs	The total of— (a) £950; and (b) 17.5% of the damages	The total of— (a) £1,855; and (b) 10% of damages over £5,000	The total of— (a) £2,370; and (b) 10% of damages over £10,000

B. If proceedings are issued under Part 7, but the case settles before trial			
Stage at which case is settled	On or after the date of issue, but prior to the date of allocation under Part 26	On or after the date of allocation under Part 26, but prior to the date of listing	On or after the date of listing but prior the date of trial
Fixed costs	The total of— (a) £2,450; and (b) 17.5% of the damages	The total of— (a) £3,065; and (b) 22.5% of the damages	The total of— (a) £3,790; and (b) 27.5% of the damages

C. If the claim is disposed of at trial	
Fixed costs	The total of— (a) £3,790; and (b) 27.5% of the damages agreed or awarded; and (c) the relevant trial advocacy fee

D. Trial advocacy fees				
Damages agreed or awarded	Not more than £3,000	More than £3,000, but not more than £10,000	More than £10,000, but not more than £15,000	More than £15,000
Trial advocacy fee	£500	£710	£1,070	£1,705

CPR 45.29F   *Appendix*

[CPR 45.29F]

45.29F Defendants' costs

(1) In this rule—
   (a) paragraphs (8) and (9) apply to assessments of defendants' costs under Part 36;
   (b) paragraph (10) applies to assessments to which the exclusions from qualified one way costs shifting in rules 44.15 and 44.16 apply; and
   (c) paragraphs (2) to (7) apply to all other cases under this Section in which a defendant's costs are assessed.

(2) If, in any case to which this Section applies, the court makes an order for costs in favour of the defendant—
   (a) the court will have regard to; and
   (b) the amount of costs order to be paid shall not exceed,
the amount which would have been payable by the defendant if an order for costs had been made in favour of the claimant at the same stage of the proceedings.

(3) For the purpose of assessing the costs payable to the defendant by reference to the fixed costs in Table 6, Table 6A, Table 6B, Table 6C and Table 6D, "value of the claim for damages" and "damages" shall be treated as references to the value of the claim.

(4) For the purposes of paragraph (3), "the value of the claim" is—
   (a) the amount specified in the claim form, excluding—
      (i) any amount not in dispute;
      (ii) in a claim started under the RTA Protocol, any claim for vehicle related damages;
      (iii) interest;
      (iv) costs; and
      (v) any contributory negligence;
   (b) if no amount is specified in the claim form, the maximum amount which the claimant reasonably expected to recover according to the statement of value included in the claim form under rule 16.3; or
   (c) £25,000, if the claim form states that the claimant cannot reasonably say how much is likely to be recovered.

(5) Where the defendant—
   (a) lives, works or carries on business in an area set out in Practice Direction 45; and
   (b) instructs a legal representative who practises in that area,
the costs will include, in addition to the costs allowable under paragraph (2), an amount equal to 12.5% of those costs.

(6) Where an order for costs is made pursuant to this rule, the defendant is entitled to disbursements in accordance with rule 45.29I.

(7) Where appropriate, VAT may be recovered in addition to the amount of any costs allowable under this rule.

(8) Where, in a case to which this Section applies, a Part 36 offer is accepted, rule 36.10A will apply instead of this rule.

(9) Where, in a case to which this Section applies, upon judgment being entered, the claimant fails to obtain a judgment more advantageous than the claimant's Part 36 offer, rule 36.14A will apply instead of this rule.

(10) Where, in a case to which this Section applies, any of the exceptions to qualified one way costs shifting in rules 44.15 and 44.16 is established, the court will assess the defendant's costs without reference to this rule.

[CPR 45.29G]

45.29G Counterclaims under the RTA Protocol
(1) If in any case to which this Section applies—
    (a) the defendant brings a counterclaim which includes a claim for personal injuries to which the RTA Protocol applies;
    (b) the counterclaim succeeds; and
    (c) the court makes an order for the costs of the counterclaim,
rules 45.29B, 45.29C, 45.29I, 45.29J, 45.29K and 45.29L shall apply.
(2) Where a successful counterclaim does not include a claim for personal injuries—
    (a) the order for costs of the counterclaim shall be for a sum equivalent to one half of the applicable Type A and Type B costs in Table 6;
    (b) where the defendant—
        (i) lives, works, or carries on business in an area set out in Practice Direction 45; and
        (ii) instructs a legal representative who practises in that area,
the costs will include, in addition to the costs allowable under paragraph (a), an amount equal to 12.5% of those costs;
    (c) if an order for costs is made pursuant to this rule, the defendant is entitled to disbursements in accordance with rule 45.29I; and
    (d) where appropriate, VAT may be recovered in addition to the amount of any costs allowable under this rule.

[CPR 45.29H]

45.29H Interim applications
(1) Where the court makes an order for costs of an interim application to be paid by one party in a case to which this Section applies, the order shall be for a sum equivalent to one half of the applicable Type A and Type B costs in Table 6 or 6A.
(2) Where the party in whose favour the order for costs is made—
    (a) lives, works or carries on business in an area set out in Practice Direction 45; and
    (b) instructs a legal representative who practises in that area,
the costs will include, in addition to the costs allowable under paragraph (1), an amount equal to 12.5% of those costs.
(3) If an order for costs is made pursuant to this rule, the party in whose favour the order is made is entitled to disbursements in accordance with rule 45.29I.
(4) Where appropriate, VAT may be recovered in addition to the amount of any costs allowable under this rule.

[CPR 45.29I]

45.29I Disbursements
(1) The court—
    (a) may allow a claim for a disbursement of a type mentioned in paragraphs (2) or (3); but
    (b) will not allow a claim for any other type of disbursement.
(2) In a claim started under either the RTA Protocol or the EL/PL Protocol, the disbursements referred to in paragraph (1) are—
    (a) the cost of obtaining medical records and expert medical reports as provided for in the relevant Protocol;

CPR 45.29J *Appendix*

  (b) the cost of any non-medical expert reports as provided for in the relevant Protocol;
  (c) the cost of any advice from a specialist solicitor or counsel as provided for in the relevant Protocol;
  (d) court fees;
  (e) any expert's fee for attending the trial where the court has given permission for the expert to attend;
  (f) expenses which a party or witness has reasonably incurred in travelling to and from a hearing or in staying away from home for the purposes of attending a hearing;
  (g) a sum not exceeding the amount specified in Practice Direction 45 for any loss of earnings or loss of leave by a party or witness due to attending a hearing or to staying away from home for the purpose of attending a hearing; and
  (h) any other disbursement reasonably incurred due to a particular feature of the dispute.
(3) In a claim started under the RTA Protocol only, the disbursements referred to in paragraph (1) are also the cost of—
  (a) an engineer's report; and
  (b) a search of the records of the—
    (i) Driver Vehicle Licensing Authority; and
    (ii) Motor Insurance Database.

[CPR 45.29J]

45.29J Claims for an amount of costs exceeding fixed recoverable costs
(1) If it considers that there are exceptional circumstances making it appropriate to do so, the court will consider a claim for an amount of costs (excluding disbursements) which is greater than the fixed recoverable costs referred to in rules 45.29B to 45.29H.
(2) If the court considers such a claim to be appropriate, it may—
  (a) summarily assess the costs; or
  (b) make an order for the costs to be subject to detailed assessment.
(3) If the court does not consider the claim to be appropriate, it will make an order—
  (a) if the claim is made by the claimant, for the fixed recoverable costs; or
  (b) if the claim is made by the defendant, for a sum which has regard to, but which does not exceed the fixed recoverable costs,
and any permitted disbursements only.

[CPR 45.29K]

45.29K Failure to achieve costs greater than fixed recoverable costs
(1) This rule applies where—
  (a) costs are assessed in accordance with rule 45.29J(2); and
  (b) the court assesses the costs (excluding any VAT) as being an amount which is in a sum less than 20% greater than the amount of the fixed recoverable costs.
(2) The court will make an order for the party who made the claim to be paid the lesser of—
  (a) the fixed recoverable costs; and
  (b) the assessed costs.

[CPR 45.29L]

45.29L Costs of the costs-only proceedings or the detailed assessment

(1) Where—
    (a)   the court makes an order for costs in accordance with rule 45.29J(3); or
    (b)   rule 45.29K applies,

the court may—
    (i)   decide not to award the party making the claim the costs of the costs only proceedings or detailed assessment; and
    (ii)   make orders in relation to costs that may include an order that the party making the claim pay the costs of the party defending those proceedings or that assessment.

SECTION IV SCALE COSTS FOR CLAIMS IN THE INTELLECTUAL PROPERTY ENTERPRISE COURT

[CPR 45.30]

45.30 Scope and interpretation

(1) Subject to paragraph (2), this Section applies to proceedings in the Intellectual Property Enterprise Court.

(2) This Section does not apply where—
    (a)   the court considers that a party has behaved in a manner which amounts to an abuse of the court's process; or
    (b)   the claim concerns the infringement or revocation of a patent or registered design the validity of which has been certified by a court or by the Comptroller-General of Patents, Designs and Trade Marks in earlier proceedings.

(3) The court will make a summary assessment of the costs of the party in whose favour any order for costs is made. Rules 44.2(8), 44.7(b) and Part 47 do not apply to this Section.

(4) "Scale costs" means the costs set out in Table A and Table B of the Practice Direction supplementing this Part.

[CPR 45.31]

45.31 Amount of scale costs

(1) Subject to rule 45.32, the court will not order a party to pay total costs of more than—
    (a)   £50,000 on the final determination of a claim in relation to liability; and
    (b)   £25,000 on an inquiry as to damages or account of profits.

(2) The amounts in paragraph (1) apply after the court has applied the provision on set off in accordance with rule 44.12(a).

(3) The maximum amount of scale costs that the court will award for each stage of the claim is set out in Practice Direction 45.

(4) The amount of the scale costs awarded by the court in accordance with paragraph (3) will depend on the nature and complexity of the claim.

(4A) Subject to assessment where appropriate, the following may be recovered in addition to the amount of the scale costs set out in Practice Direction 45 – Fixed Costs—
    (a)   court fees;

CPR 45.32 *Appendix*

  (b) costs relating to the enforcement of any court order; and
  (c) wasted costs.
(5) Where appropriate, VAT may be recovered in addition to the amount of the scale costs and any reference in this Section to scale costs is a reference to those costs net of any such VAT.

[CPR 45.32]

45.32 Summary assessment of the costs of an application where a party has behaved unreasonably
Costs awarded to a party under rule 63.26(2) are in addition to the total costs that may be awarded to that party under rule 45.31.

SECTION V FIXED COSTS: HM REVENUE AND CUSTOMS

[CPR 45.33]

45.33 Scope, interpretation and application
(1) This Section sets out the amounts which, unless the court orders otherwise, are to be allowed in respect of HM Revenue and Customs charges in the cases to which this Section applies.
(2) For the purpose of this Section—
  (a) "HMRC Officer" means a person appointed by the Commissioners under section 2 of the Commissioners for Revenue and Customs Act 2005[1] and authorised to conduct county court proceedings for recovery of debt under section 25(1A)[2] of that Act;
  (b) "Commissioners" means commissioners for HMRC appointed under section 1 of the Commissioners for Revenue and Customs Act 2005;
  (c) "debt" means any sum payable to the Commissioners under or by virtue of an enactment or under a contract settlement; and
  (d) "HMRC charges" means the fixed costs set out in Tables 7 and 8 in this Section.
(3) HMRC charges must, for the purpose of this Section, be claimed as "legal representative's costs" on relevant court forms.
(4) This Section applies where the only claim is a claim conducted by an HMRC Officer in the county court for recovery of a debt and the Commissioners obtain judgment on the claim.
(5) Any appropriate court fee will be allowed in addition to the costs set out in this Section.
(6) The claim form may include a claim for fixed commencement costs.

[1] 2005 c.11.
[2] 2005 c.11. Section 25(1A) was inserted by the Finance Act 2008 (c.9) section 137(1)(a).

[CPR 45.34]

45.34 Amount of fixed commencement costs in a county court claim for the recovery of money
The amount of fixed commencement costs in a claim to which rule 45.33 applies—
  (a) will be calculated by reference to Table 7; and
  (b) the amount claimed in the claim form is to be used for determining which claim band in Table 7 applies.

Table 7 – Fixed Costs on Commencement of a County Court Claim Conducted by an HMRC Officer

Where the value of the claim does not exceed £25	Nil
Where the value of the claim exceeds £25 but does not exceed £500	£33
Where the value of the claim exceeds £500 but does not exceed £1,000	£47
Where the value of the claim exceeds £1,000 but does not exceed £5,000	£53
Where the value of the claim exceeds £5,000 but does not exceed £15,000	£67
Where the value of the claim exceeds £15,000 but does not exceed £50,000	£90
Where the value of the claim exceeds £50,000 but does not exceed £100,000	£113
Where the value of the claim exceeds £100,000 but does not exceed £150,000	£127
Where the value of the claim exceeds £150,000 but does not exceed £200,000	£140
Where the value of the claim exceeds £200,000 but does not exceed £250,000	£153
Where the value of the claim exceeds £250,000 but does not exceed £300,000	£167
Where the value of the claim exceeds £300,000	£180

[CPR 45.35]

45.35 Costs on entry of judgment in a county court claim for recovery of money

Where—
- (a) an HMRC Officer has claimed fixed commencement costs under Rule 45.34; and
- (b) judgment is entered in a claim to which rule 45.33 applies,

the amount to be included in the judgment for HMRC charges is the total of—
- (i) the fixed commencement costs; and
- (ii) the amount in Table 8 relevant to the value of the claim.

Table 8 – Fixed Costs on Entry of Judgment of a County Court Claim Conducted by an HMRC Officer

Where the value of the claim does not exceed £5,000	£15
Where the value of the claim exceeds £5,000	£20

[CPR 45.36]

45.36 When the defendant is only liable for fixed commencement costs

Where—
- (a) the only claim is for a specified sum of money; and
- (b) the defendant pays the money claimed within 14 days after service of the particulars of claim, together with the fixed commencement costs stated in the claim form,

the defendant is not liable for any further costs unless the court orders otherwise.

CPR 45.37 *Appendix*

SECTION VI FAST TRACK TRIAL COSTS

[CPR 45.37]

45.37 Scope of this Section
(1) This Section deals with the amount of costs which the court may award as the costs of an advocate for preparing for and appearing at the trial of a claim in the fast track (referred to in this rule as "fast track trial costs").
(2) For the purposes of this Section—
"advocate" means a person exercising a right of audience as a representative of, or on behalf of, a party;
"fast track trial costs" means the costs of a party's advocate for preparing for and appearing at the trial, but does not include—
    (i)   any other disbursements; or
    (ii) any value added tax payable on the fees of a party's advocate; and
"trial" includes a hearing where the court decides an amount of money or the value of goods following a judgment under Part 12 (default judgment) or Part 14 (admissions) but does not include –
    (i)   the hearing of an application for summary judgment under Part 24; or
    (ii) the court's approval of a settlement or other compromise under rule 21.10.

[CPR 45.38]

45.38 Amount of fast track trial costs
(1) Table 9 shows the amount of fast track trial costs which the court may award (whether by summary or detailed assessment).

Table 9

Value of the claim	Amount of fast track trial costs which the court may award
No more than £3,000	£485
More than £3,000 but not more than £10,000 [but not more than £15,000]	£690
More than £10,000	£1,035
For proceedings issued on or after 6th April 2009, more than £15,000	£1,650

(2) The court may not award more or less than the amount shown in the table except where—
    (a) it decides not to award any fast track trial costs; or
    (b) rule 45.39 applies,
but the court may apportion the amount awarded between the parties to reflect their respective degrees of success on the issues at trial.
(3) Where the only claim is for the payment of money—
    (a) for the purpose of quantifying fast track trial costs awarded to a claimant, the value of the claim is the total amount of the judgment excluding—
        (i)   interest and costs; and
        (ii) any reduction made for contributory negligence; and

(b) for the purpose of quantifying fast track trial costs awarded to a defendant, the value of the claim is—
  (i) the amount specified in the claim form (excluding interest and costs);
  (ii) if no amount is specified, the maximum amount which the claimant reasonably expected to recover according to the statement of value included in the claim form under rule 16.3; or
  (iii) more than £15,000, if the claim form states that the claimant cannot reasonably say how much is likely to be recovered.

(4) Where the claim is only for a remedy other than the payment of money, the value of the claim is deemed to be more than £3,000 but not more than £10,000, unless the court orders otherwise.

(5) Where the claim includes both a claim for the payment of money and for a remedy other than the payment of money, the value of the claim is deemed to be the higher of—
  (a) the value of the money claim decided in accordance with paragraph (3); or
  (b) the deemed value of the other remedy decided in accordance with paragraph (4),
unless the court orders otherwise.

(6) Where—
  (a) a defendant has made a counterclaim against the claimant;
  (b) the counterclaim has a higher value than the claim; and
  (c) the claimant succeeds at trial both on the claim and the counterclaim,
for the purpose of quantifying fast track trial costs awarded to the claimant, the value of the claim is the value of the defendant's counterclaim calculated in accordance with this rule.

[CPR 45.39]

45.39 Power to award more or less than the amount of fast track trial costs

(1) This rule sets out when a court may award—
  (a) an additional amount to the amount of fast track trial costs shown in Table 9 in rule 45.38(1); or
  (b) less than those amounts.

(2) If—
  (a) in addition to the advocate, a party's legal representative attends the trial;
  (b) the court considers that it was necessary for a legal representative to attend to assist the advocate; and
  (c) the court awards fast track trial costs to that party,
the court may award an additional £345 in respect of the legal representative's attendance at the trial.

(3) If the court considers that it is necessary to direct a separate trial of an issue then the court may award an additional amount in respect of the separate trial but that amount is limited in accordance with paragraph (4) of this rule.

(4) The additional amount the court may award under paragraph (3) will not exceed two-thirds of the amount payable for that claim, subject to a minimum award of £485.

(5) Where the party to whom fast track trial costs are to be awarded is a litigant in person, the court will award—

CPR 45.40 *Appendix*

      (a)   if the litigant in person can prove financial loss, two-thirds of the amount that would otherwise be awarded; or

      (b)   if the litigant in person fails to prove financial loss, an amount in respect of the time spent reasonably doing the work at the rate specified in Practice Direction 46.

(6) Where a defendant has made a counterclaim against the claimant, and—

      (a)   the claimant has succeeded on his claim; and

      (b)   the defendant has succeeded on his counterclaim,

the court will quantify the amount of the award of fast track trial costs to which—

           (i)   but for the counterclaim, the claimant would be entitled for succeeding on his claim; and

           (ii)   but for the claim, the defendant would be entitled for succeeding on his counterclaim,

and make one award of the difference, if any, to the party entitled to the higher award of costs.

(7) Where the court considers that the party to whom fast track trial costs are to be awarded has behaved unreasonably or improperly during the trial, it may award that party an amount less than would otherwise be payable for that claim, as it considers appropriate.

(8) Where the court considers that the party who is to pay the fast track trial costs has behaved improperly during the trial the court may award such additional amount to the other party as it considers appropriate.

[CPR 45.40]

45.40 Fast track trial costs where there is more than one claimant or defendant

(1) Where the same advocate is acting for more than one party—

      (a)   the court may make only one award in respect of fast track trial costs payable to that advocate; and

      (b)   the parties for whom the advocate is acting are jointly entitled to any fast track trial costs awarded by the court.

(2) Where—

      (a)   the same advocate is acting for more than one claimant; and

      (b)   each claimant has a separate claim against the defendant,

the value of the claim, for the purpose of quantifying the award in respect of fast track trial costs is to be ascertained in accordance with paragraph (3).

(3) The value of the claim in the circumstances mentioned in paragraph (2) or (5) is—

      (a)   where the only claim of each claimant is for the payment of money—

           (i)   if the award of fast track trial costs is in favour of the claimants, the total amount of the judgment made in favour of all the claimants jointly represented; or

           (ii)   if the award is in favour of the defendant, the total amount claimed by the claimants,

and in either case, quantified in accordance with rule 45.38(3);

      (b)   where the only claim of each claimant is for a remedy other than the payment of money, deemed to be more than £3,000 but not more than £10,000; and

      (c)   where claims of the claimants include both a claim for the payment of money and for a remedy other than the payment of money, deemed to be—

(i) more than £3,000 but not more than £10,000; or
(ii) if greater, the value of the money claims calculated in accordance with subparagraph (a) above.

(4) Where—
(a) there is more than one defendant; and
(b) any or all of the defendants are separately represented,

the court may award fast track trial costs to each party who is separately represented.

(5) Where—
(a) there is more than one claimant; and
(b) a single defendant,

the court may make only one award to the defendant of fast track trial costs, for which the claimants are jointly and severally liable.

(6) For the purpose of quantifying the fast track trial costs awarded to the single defendant under paragraph (5), the value of the claim is to be calculated in accordance with paragraph (3) of this rule.

SECTION VII COSTS LIMITS IN AARHUS CONVENTION CLAIMS

[CPR 45.41]

45.41 Scope and interpretation

(1) This Section provides for the costs which are to be recoverable between the parties in Aarhus Convention claims.

(2) In this Section, "Aarhus Convention claim" means a claim for judicial review of a decision, act or omission all or part of which is subject to the provisions of the UNECE Convention on Access to Information, Public Participation in Decision-Making and Access to Justice in Environmental Matters done at Aarhus, Denmark on 25 June 1998, including a claim which proceeds on the basis that the decision, act or omission, or part of it, is so subject.

(Rule 52.9A makes provision in relation to costs of an appeal.)

[CPR 45.42]

45.42 Opting out

Rules 45.43 to 45.44 do not apply where the claimant—
(a) has not stated in the claim form that the claim is an Aarhus Convention claim; or
(b) has stated in the claim form that—
(i) the claim is not an Aarhus Convention claim, or
(ii) although the claim is an Aarhus Convention claim, the claimant does not wish those rules to apply.

[CPR 45.43]

45.43 Limit on costs recoverable from a party in an Aarhus Convention claim

(1) Subject to rule 45.44, a party to an Aarhus Convention claim may not be ordered to pay costs exceeding the amount prescribed in Practice Direction 45.

(2) Practice Direction 45 may prescribe a different amount for the purpose of paragraph (1) according to the nature of the claimant.

CPR 45.44 *Appendix*

[CPR 45.44]

45.44 Challenging whether the claim is an Aarhus Convention claim
(1) If the claimant has stated in the claim form that the claim is an Aarhus Convention claim, rule 45.43 will apply unless—
    (a) the defendant has in the acknowledgment of service filed in accordance with rule 54.8—
        (i) denied that the claim is an Aarhus Convention claim; and
        (ii) set out the defendant's grounds for such denial; and
    (b) the court has determined that the claim is not an Aarhus Convention claim.
(2) Where the defendant argues that the claim is not an Aarhus Convention claim, the court will determine that issue at the earliest opportunity.
(3) In any proceedings to determine whether the claim is an Aarhus Convention claim—
    (a) if the court holds that the claim is not an Aarhus Convention claim, it will normally make no order for costs in relation to those proceedings;
    (b) if the court holds that the claim is an Aarhus Convention claim, it will normally order the defendant to pay the claimant's costs of those proceedings on the indemnity basis, and that order may be enforced notwithstanding that this would increase the costs payable by the defendant beyond the amount prescribed in Practice Direction 45.

PRACTICE DIRECTION 45—FIXED COSTS

THIS PRACTICE DIRECTION SUPPLEMENTS PART 45

Section I of Part 45 – Fixed costs

Fixed costs in small claims

[CPR PD 45.1]

1.1 Under Rule 27.14 the costs which can be awarded to a claimant in a small claim include the fixed costs payable under Part 45 attributable to issuing the claim.
1.2 Those fixed costs are the sum of—
(a) the fixed commencement costs calculated in accordance with Table 1 of Rule 45.2;
(b) the appropriate court fee or fees paid by the claimant.

Claims to which Part 45 does not apply

1.3 In a claim to which Part 45 does not apply, no amount shall be entered on the claim form for the charges of the claimant's legal representative, but the words "to be assessed" shall be inserted.

Section II of Part 45 – Road traffic accidents: Fixed recoverable costs in costs-only proceedings

Scope

[CPR PD 45.2]

2.1 Section II of Part 45 ('the Section') provides for certain fixed costs to be

Civil Procedure Rules 1998 **CPR PD 45.2**

recoverable between parties in respect of costs incurred in disputes which are settled prior to proceedings being issued. The Section applies to road traffic accident disputes as defined in rule 45.9(4)(a), where the accident which gave rise to the dispute occurred on or after 6th October 2003.
2.2 The Section does not apply to disputes where the total agreed value of the damages is within the small claims limit or exceeds £10,000. Rule 26.8(2) sets out how the financial value of a claim is assessed for the purposes of allocation to track.
2.3 Fixed recoverable costs are to be calculated by reference to the amount of agreed damages which are payable to the receiving party. In calculating the amount of these damages—
(a) account must be taken of both general and special damages and interest;
(b) any interim payments made must be included;
(c) where the parties have agreed an element of contributory negligence, the amount of damages attributed to that negligence must be deducted;
(d) any amount required by statute to be paid by the compensating party directly to a third party (such as sums paid by way of compensation recovery payments and National Health Service expenses) must not be included.
2.4 The Section applies to cases which fall within the scope of the Uninsured Drivers Agreement dated 13 August 1999. The section does not apply to cases which fall within the scope of the Untraced Drivers Agreement dated 14 February 2003.

Fixed recoverable costs formula

2.5 The amount of fixed costs recoverable is the sum of –
(a) £800;
(b) 20% of the agreed damages up to £5,000; and
(c) 15% of the agreed damages between £5,000 and £10,000.
For example, agreed damages of £7,523 would result in recoverable costs of £2,178.45 i.e.
£800 + (20% of £5,000) + (15% of £2,523).

Additional costs for work in specified areas

2.6 The area referred to in rules 45.11(2) and 45.18(5) consists of (within London) the county court districts of Barnet, Bow, Brentford, Central London, Clerkenwell and Shoreditch, Edmonton, Ilford, Lambeth, Mayors and City of London, Romford, Wandsworth, West London, Willesden and Woolwich and (outside London) the county court districts of Bromley, Croydon, Dartford, Gravesend and Uxbridge.

Multiple claimants

2.7 Where two or more potential claimants instruct the same legal representative, the provisions of the section apply in respect of each claimant.

Information to be included in the claim form

2.8 Costs only proceedings are commenced using the procedure set out in rule 46.14. A claim form should be issued in accordance with Part 8. Where the claimant is claiming an amount of costs which exceed the amount of the fixed recoverable costs the claim form must give details of the exceptional circumstances to justify the additional costs.
2.9 The claimant must also include on the claim form details of any disbursements. The disbursements that may be claimed are set out in rule 45.12(1). If the disbursement falls within 45.12(2)(c) (disbursements that have arisen due to a particular feature of the dispute) the claimant must give details of the particular feature of the dispute that made the disbursement necessary.

Disbursements

2.10 If the parties agree the amount of the fixed recoverable costs and the only dispute is as to the payment of, or amount of, a disbursement, then proceedings should be issued under rule 46.14.

CPR PD 45.2A *Appendix*

Section IIIA — Claims Which No Longer Continue Under The Rta Or EI/PI Protocols – Disbursements

Claims for loss of earnings: Rule 45.29I(2)(g)

[CPR PD 45.2A]

2A.1 Where, under rule 45.29I(2)(g) (loss of earnings), the court allows a claim for any loss of earnings or leave by a party or witness due to attending a hearing or staying away from home for the purpose of attending a hearing, the specified sums, per day, for each person are—
(a) £90, where the value of the claim for damages is not more than £10,000; and
(b) £135, where the value of the claim for damages is more than £10,000.

Section IV of Part 45 – Scale costs for proceedings in the Intellectual Property Enterprise Court

Tables A and B

[CPR PD 45.3]

3.1 Tables A and B set out the maximum amount of scale costs which the court will award for each stage of a claim in the Intellectual Property Enterprise Court.
3.2 Table A sets out the scale costs for each stage of a claim up to determination of liability.
3.3 Table B sets out the scale costs for each stage of an inquiry as to damages or account of profits.

Table A

Stage of a claim	Maximum amount of costs
Particulars of claim	£7,000
Defence and counterclaim	£7,000
Reply and defence to counterclaim	£7,000
Reply to defence to counterclaim	£3,500
Attendance at a case management conference	£3,000
Making or responding to an application	£3,000
Providing or inspecting disclosure or product/process description	£6,000
Performing or inspecting experiments	£3,000
Preparing witness statements	£6,000
Preparing experts' report	£8,000
Preparing for and attending trial and judgment	£16,000
Preparing for determination on the papers	£5,500

Table B

Stage of a claim	Maximum amount of costs
Points of claim	£3,000

Stage of a claim	Maximum amount of costs
Points of defence	£3,000
Attendance at a case management conference	£3,000
Making or responding to an application	£3,000
Providing or inspecting disclosure	£3,000
Preparing witness statements	£6,000
Preparing experts' report	£6,000
Preparing for and attending trial and judgment	£8,000
Preparing for determination on the papers	£3,000

Section VI of Part 45 – Fast track trial costs

Scope

[CPR PD 45.4]

4.1 Section VI of Part 45 applies to the costs of an advocate for preparing for and appearing at the trial of a claim in the fast track.
4.2 It applies only where, at the date of the trial, the claim is allocated to the fast track. It does not apply in any other case, irrespective of the final value of the claim.
4.3 In particular it does not apply to a disposal hearing at which the amount to be paid under a judgment or order is decided by the court (see paragraph 12.4 of Practice Direction 26)).

Section VII of Part 45 — Costs limits in Aarhus Convention claims

Limit on costs recoverable from a party in an Aarhus Convention claim: Rule 45.43

[CPR PD 45.5]

5.1 Where a claimant is ordered to pay costs, the amount specified for the purpose of rule 45.43(1) is—
(a) £5,000 where the claimant is claiming only as an individual and not as, or on behalf of, a business or other legal person;
(b) in all other cases, £10,000.
5.2 Where a defendant is ordered to pay costs, the amount specified for the purpose of rule 45.43(1) is £35,000.

CPR 46.1 *Appendix*

PART 46   COSTS — SPECIAL CASES

SECTION I
COSTS PAYABLE BY OR TO PARTICULAR PERSONS

Rule 46.1	Pre-commencement disclosure and orders for disclosure against a person who is not a party	CPR 46.1
Rule 46.2	Costs orders in favour of or against non-parties	CPR 46.2
Rule 46.3	Limitations on court's power to award costs in favour of trustee or personal representative	CPR 46.3
Rule 46 4	Costs where money is payable by or to a child or protected party	CPR 46.4
Rule 46 5	Litigants in person	CPR 46.5
Rule 46 6	Costs where the court has made a group litigation order	CPR 46.6
Rule 46 7	Orders in respect of pro bono representation	CPR 46.7

SECTION II
COSTS RELATING TO LEGAL REPRESENTATIVES

Rule 46 8	Personal liability of legal representative for costs – wasted costs orders	CPR 46.8
Rule 46 9	Basis of detailed assessment of solicitor and client costs	CPR 46.9
Rule 46 10	Assessment procedure	CPR 46.10

SECTION III
COSTS ON ALLOCATION AND RE-ALLOCATION

Rule 46 11	Costs on the small claims track and fast track	CPR 46.11
Rule 46 12	Limitation on amount court may allow where a claim allocated to the fast track settles before trial	CPR 46.12
Rule 46 13	Costs following allocation, re-allocation and non-allocation	CPR 46.13

SECTION IV
COSTS-ONLY PROCEEDINGS

Rule 46 14	Costs-only proceedings	CPR 46.14
	Practice Direction 46—Costs — Special Cases	CPR PD 46

SECTION I COSTS PAYABLE BY OR TO PARTICULAR PERSONS

[CPR 46.1]

46.1 Pre-commencement disclosure and orders for disclosure against a person who is not a party

(1)   This paragraph applies where a person applies—
  (a)   for an order under—
    (i)   section 33 of the Senior Courts Act 1981; or
    (ii)   section 52 of the County Courts Act 1984,
  (which give the court powers exercisable before commencement of proceedings); or
  (b)   for an order under—
    (i)   section 34 of the Senior Courts Act 1981; or
    (ii)   section 53 of the County Courts Act 1984,
  (which give the court power to make an order against a non-party for disclosure of documents, inspection of property etc.).

(2)   The general rule is that the court will award the person against whom the order is sought that person's costs—
  (a)   of the application; and
  (b)   of complying with any order made on the application.

(3)   The court may however make a different order, having regard to all the circumstances, including—

(a) the extent to which it was reasonable for the person against whom the order was sought to oppose the application; and
(b) whether the parties to the application have complied with any relevant pre-action protocol.

[CPR 46.2]

46.2 Costs orders in favour of or against non-parties

(1) Where the court is considering whether to exercise its power under section 51 of the Senior Courts Act 1981 (costs are in the discretion of the court) to make a costs order in favour of or against a person who is not a party to proceedings, that person must—
    (a) be added as a party to the proceedings for the purposes of costs only; and
    (b) be given a reasonable opportunity to attend a hearing at which the court will consider the matter further.
(2) This rule does not apply—
    (a) where the court is considering whether to—
        (i) make an order against the Lord Chancellor in proceedings in which the Lord Chancellor has provided legal aid to a party to the proceedings;
        (ii) make a wasted costs order (as defined in rule 46.8); and
    (b) in proceedings to which rule 46.1 applies (pre-commencement disclosure and orders for disclosure against a person who is not a party).

[CPR 46.3]

46.3 Limitations on court's power to award costs in favour of trustee or personal representative

(1) This rule applies where—
    (a) a person is or has been a party to any proceedings in the capacity of trustee or personal representative; and
    (b) rule 44.5 does not apply.
(2) The general rule is that that person is entitled to be paid the costs of those proceedings, insofar as they are not recovered from or paid by any other person, out of the relevant trust fund or estate.
(3) Where that person is entitled to be paid any of those costs out of the fund or estate, those costs will be assessed on the indemnity basis.

[CPR 46.4]

46.4 Costs where money is payable by or to a child or protected party

(1) This rule applies to any proceedings where a party is a child or protected party and—
    (a) money is ordered or agreed to be paid to, or for the benefit of, that party; or
    (b) money is ordered to be paid by that party or on that party's behalf.
("Child" and "protected party" have the same meaning as in rule 21.1(2).)
(2) The general rule is that—
    (a) the court must order a detailed assessment of the costs payable by, or out of money belonging to, any party who is a child or protected party; and

CPR 46.5 *Appendix*

      (b)   on an assessment under paragraph (a), the court must also assess any costs payable to that party in the proceedings, unless—
          (i)   the court has issued a default costs certificate in relation to those costs under rule 47.11; or
          (ii)   the costs are payable in proceedings to which Section II or Section III of Part 45 applies.

(3)   The court need not order detailed assessment of costs in the circumstances set out in Practice Direction 46.

(4)   Where—
      (a)   a claimant is a child or protected party; and
      (b)   a detailed assessment has taken place under paragraph (2)(a),

the only amount payable by the child or protected party is the amount which the court certifies as payable.

(This rule applies to a counterclaim by or on behalf of a child or protected party by virtue of rule 20.3.)

[CPR 46.5]

46.5 Litigants in person

(1)   This rule applies where the court orders (whether by summary assessment or detailed assessment) that the costs of a litigant in person are to be paid by any other person.

(2)   The costs allowed under this rule will not exceed, except in the case of a disbursement, two-thirds of the amount which would have been allowed if the litigant in person had been represented by a legal representative.

(3)   The litigant in person shall be allowed—
      (a)   costs for the same categories of—
          (i)   work; and
          (ii)   disbursements,

which would have been allowed if the work had been done or the disbursements had been made by a legal representative on the litigant in person's behalf;
      (b)   the payments reasonably made by the litigant in person for legal services relating to the conduct of the proceedings; and
      (c)   the costs of obtaining expert assistance in assessing the costs claim.

(4)   The amount of costs to be allowed to the litigant in person for any item of work claimed will be—
      (a)   where the litigant can prove financial loss, the amount that the litigant can prove to have been lost for time reasonably spent on doing the work; or
      (b)   where the litigant cannot prove financial loss, an amount for the time reasonably spent on doing the work at the rate set out in Practice Direction 46.

(5)   A litigant who is allowed costs for attending at court to conduct the case is not entitled to a witness allowance in respect of such attendance in addition to those costs.

(6)   For the purposes of this rule, a litigant in person includes—
      (a)   a company or other corporation which is acting without a legal representative; and
      (b)   any of the following who acts in person (except where any such person is represented by a firm in which that person is a partner)—
          (i)   a barrister;

(ii) a solicitor;
(iii) a solicitor's employee;
(iv) a manager of a body recognised under section 9 of the Administration of Justice Act 1985; or
(v) a person who, for the purposes of the 2007 Act, is an authorised person in relation to an activity which constitutes the conduct of litigation (within the meaning of that Act).

[CPR 46.6]

46.6 Costs where the court has made a group litigation order

(1) This rule applies where the court has made a Group Litigation Order ("GLO").
(2) In this rule—
"individual costs" means costs incurred in relation to an individual claim on the group register;
"common costs" means—
    (i) costs incurred in relation to the GLO issues;
    (ii) individual costs incurred in a claim while it is proceeding as a test claim, and
    (iii) costs incurred by the lead legal representative in administering the group litigation; and
'group litigant' means a claimant or defendant, as the case may be, whose claim is entered on the group register.
(3) Unless the court orders otherwise, any order for common costs against group litigants imposes on each group litigant several liability for an equal proportion of those common costs.
(4) The general rule is that a group litigant who is the paying party will, in addition to any liability to pay the receiving party, be liable for—
    (a) the individual costs of that group litigant's claim; and
    (b) an equal proportion, together with all the other group litigants, of the common costs.
(5) Where the court makes an order about costs in relation to any application or hearing which involved—
    (a) one or more GLO issues; and
    (b) issues relevant only to individual claims,
the court will direct the proportion of the costs that is to relate to common costs and the proportion that is to relate to individual costs.
(6) Where common costs have been incurred before a claim is entered on the group register, the court may order the group litigant to be liable for a proportion of those costs.
(7) Where a claim is removed from the group register, the court may make an order for costs in that claim which includes a proportion of the common costs incurred up to the date on which the claim is removed from the group register.
(Part 19 sets out rules about group litigation.)

[CPR 46.7]

46.7 Orders in respect of pro bono representation

(1) Where the court makes an order under section 194(3) of the 2007 Act—
    (a) the court may order the payment to the prescribed charity of a sum no greater than the costs specified in Part 45 to which the party with pro

CPR 46.8 *Appendix*

bono representation would have been entitled in accordance with that Part and in respect of that representation had it not been provided free of charge; or

(b) where Part 45 does not apply, the court may determine the amount of the payment (other than a sum equivalent to fixed costs) to be made by the paying party to the prescribed charity by—

(i) making a summary assessment; or

(ii) making an order for detailed assessment,

of a sum equivalent to all or part of the costs the paying party would have been ordered to pay to the party with pro bono representation in respect of that representation had it not been provided free of charge.

(2) Where the court makes an order under section 194(3) of the 2007 Act, the order must direct that the payment by the paying party be made to the prescribed charity.

(3) The receiving party must send a copy of the order to the prescribed charity within 7 days of receipt of the order.

(4) Where the court considers making or makes an order under section 194(3) of the 2007 Act, Parts 44 to 47 apply, where appropriate, with the following modifications—

(a) references to "costs orders", "orders about costs" or "orders for the payment of costs" are to be read, unless otherwise stated, as if they refer to an order under section 194(3);

(b) references to "costs" are to be read as if they referred to a sum equivalent to the costs that would have been claimed by, incurred by or awarded to the party with pro bono representation in respect of that representation had it not been provided free of charge; and

(c) references to "receiving party" are to be read, as meaning a party who has pro bono representation and who would have been entitled to be paid costs in respect of that representation had it not been provided free of charge.

SECTION II COSTS RELATING TO LEGAL REPRESENTATIVES

[CPR 46.8]

46.8 Personal liability of legal representative for costs – wasted costs orders

(1) This rule applies where the court is considering whether to make an order under section 51(6) of the Senior Courts Act 1981 (court's power to disallow or (as the case may be) order a legal representative to meet, "wasted costs").

(2) The court will give the legal representative a reasonable opportunity to make written submissions or, if the legal representative prefers, to attend a hearing before it makes such an order.

(3) When the court makes a wasted costs order, it will—

(a) specify the amount to be disallowed or paid; or

(b) direct a costs judge or a district judge to decide the amount of costs to be disallowed or paid.

(4) The court may direct that notice must be given to the legal representative's client, in such manner as the court may direct—

(a) of any proceedings under this rule; or

(b) of any order made under it against his legal representative.

[CPR 46.9]

46.9 Basis of detailed assessment of solicitor and client costs

(1)   This rule applies to every assessment of a solicitor's bill to a client except a bill which is to be paid out of the Community Legal Service Fund under the Legal Aid Act 1988 or the Access to Justice Act 1999.

(2)   Section 74(3) of the Solicitors Act 1974 applies unless the solicitor and client have entered into a written agreement which expressly permits payment to the solicitor of an amount of costs greater than that which the client could have recovered from another party to the proceedings.

(3)   Subject to paragraph (2), costs are to be assessed on the indemnity basis but are to be presumed—
    (a)   to have been reasonably incurred if they were incurred with the express or implied approval of the client;
    (b)   to be reasonable in amount if their amount was expressly or impliedly approved by the client;
    (c)   to have been unreasonably incurred if—
        (i)   they are of an unusual nature or amount; and
        (ii)   the solicitor did not tell the client that as a result the costs might not be recovered from the other party.

(4)   Where the court is considering a percentage increase on the application of the client, the court will have regard to all the relevant factors as they reasonably appeared to the solicitor or counsel when the conditional fee agreement was entered into or varied.

[CPR 46.10]

46.10 Assessment procedure

(1)   This rule sets out the procedure to be followed where the court has made an order under Part III of the Solicitors Act 1974 for the assessment of costs payable to a solicitor by the solicitor's client.

(2)   The solicitor must serve a breakdown of costs within 28 days of the order for costs to be assessed.

(3)   The client must serve points of dispute within 14 days after service on the client of the breakdown of costs.

(4)   The solicitor must serve any reply within 14 days of service on the solicitor of the points of dispute.

(5)   Either party may file a request for a hearing date—
    (a)   after points of dispute have been served; but
    (b)   no later than 3 months after the date of the order for the costs to be assessed.

(6)   This procedure applies subject to any contrary order made by the court.

SECTION III COSTS ON ALLOCATION AND RE-ALLOCATION

[CPR 46.11]

46.11 Costs on the small claims track and fast track

(1)   Part 27 (small claims) and Part 45 Section VI (fast track trial costs) contain special rules about—
    (a)   liability for costs;
    (b)   the amount of costs which the court may award; and
    (c)   the procedure for assessing costs.

CPR 46.12 *Appendix*

(2) Once a claim is allocated to a particular track, those special rules shall apply to the period before, as well as after, allocation except where the court or a practice direction provides otherwise.

[CPR 46.12]

46.12 Limitation on amount court may allow where a claim allocated to the fast track settles before trial

(1) Where the court—
- (a) assesses costs in relation to a claim which—
  - (i) has been allocated to the fast track; and
  - (ii) settles before the start of the trial; and
- (b) is considering the amount of costs to be allowed in respect of a party's advocate for preparing for the trial,

it may not allow, in respect of those advocate's costs, an amount that exceeds the amount of fast track trial costs which would have been payable in relation to the claim had the trial taken place.

(2) When deciding the amount to be allowed in respect of the advocate's costs, the court will have regard to—
- (a) when the claim was settled; and
- (b) when the court was notified that the claim had settled.

(3) In this rule, "advocate" and "fast track trial costs" have the meanings given to them by Part 45 Section VI.

[CPR 46.13]

46.13 Costs following allocation, re-allocation and non-allocation

(1) Any costs orders made before a claim is allocated will not be affected by allocation.

(2) Where—
- (a) claim is allocated to a track; and
- (b) the court subsequently re-allocates that claim to a different track,

then unless the court orders otherwise, any special rules about costs applying—
- (i) to the first track, will apply to the claim up to the date of re-allocation; and
- (ii) to the second track, will apply from the date of re-allocation.

(3) Where the court is assessing costs on the standard basis of a claim which concluded without being allocated to a track, it may restrict those costs to costs that would have been allowed on the track to which the claim would have been allocated if allocation had taken place.

SECTION IV COSTS-ONLY PROCEEDINGS

[CPR 46.14]

46.14 Costs-only proceedings

(1) This rule applies where—
- (a) the parties to a dispute have reached an agreement on all issues (including which party is to pay the costs) which is made or confirmed in writing; but
- (b) they have failed to agree the amount of those costs; and

(c) no proceedings have been started.
(2) Where this rule applies, the procedure set out in this rule must be followed.
(3) Proceedings under this rule are commenced by issuing a claim form in accordance with Part 8.
(4) The claim form must contain or be accompanied by the agreement or confirmation.
(5) In proceedings to which this rule applies the court may make an order for the payment of costs the amount of which is to be determined by assessment and/or, where appropriate, for the payment of fixed costs.
(6) Where this rule applies but the procedure set out in this rule has not been followed by a party—
- (a) that party will not be allowed costs greater than those that would have been allowed to that party had the procedure been followed; and
- (b) the court may award the other party the costs of the proceedings up to the point where an order for the payment of costs is made.

(7) Rule 44.5 (amount of costs where costs are payable pursuant to a contract) does not apply to claims started under the procedure in this rule.

PRACTICE DIRECTION 46—COSTS SPECIAL CASES

THIS PRACTICE DIRECTION SUPPLEMENTS PART 46

Awards of costs in favour of a trustee or personal representative: Rule 46.3

[CPR PD 46.1]

1.1 A trustee or personal representative is entitled to an indemnity out of the relevant trust fund or estate for costs properly incurred. Whether costs were properly incurred depends on all the circumstances of the case including whether the trustee or personal representative ("the trustee")—
- (a) obtained directions from the court before bringing or defending the proceedings;
- (b) acted in the interests of the fund or estate or in substance for a benefit other than that of the estate, including the trustee's own; and
- (c) acted in some way unreasonably in bringing or defending, or in the conduct of, the proceedings.

1.2 The trustee is not to be taken to have acted for a benefit other than that of the fund by reason only that the trustee has defended a claim in which relief is sought against the trustee personally.

Costs where money is payable by or to a child or protected party: Rule 46.4

[CPR PD 46.2]

2.1 The circumstances in which the court need not order the assessment of costs under rule 46.4(3) are as follows—
- (a) where there is no need to do so to protect the interests of the child or protected party or their estate;
- (b) where another party has agreed to pay a specified sum in respect of the costs of the child or protected party and the legal representative acting for the child or protected party has waived the right to claim further costs;
- (c) where the court has decided the costs payable to the child or protected party by way of summary assessment and the legal representative acting for the child or protected party has waived the right to claim further costs; and

CPR PD 46.3   *Appendix*

(d)   where an insurer or other person is liable to discharge the costs which the child or protected party would otherwise be liable to pay to the legal representative and the court is satisfied that the insurer or other person is financially able to discharge those costs.

Litigants in person: Rule 46.5

[CPR PD 46.3]

3.1 In order to qualify as an expert for the purpose of rule 46.5(3)(c) (expert assistance in connection with assessing the claim for costs), the person in question must be a—
(a)   barrister;
(b)   solicitor;
(c)   Fellow of the Institute of Legal Executives;
(d)   Fellow of the Association of Costs Lawyers;
(e)   law costs draftsman who is a member of the Academy of Experts;
(f)   law costs draftsman who is a member of the Expert Witness Institute.
3.2 Where a self represented litigant wishes to prove that the litigant has suffered financial loss, the litigant should produce to the court any written evidence relied on to support that claim, and serve a copy of that evidence on any party against whom the litigant seeks costs at least 24 hours before the hearing at which the question may be decided.
3.3 A self represented litigant who commences detailed assessment proceedings under rule 47.5 should serve copies of that written evidence with the notice of commencement.
3.4 The amount, which may be allowed to a self represented litigant under rule 45.39(5)(b) and rule 46.5(4)(b), is £18 per hour.

Orders in respect of pro bono representation: Rule 46.7

[CPR PD 46.4]

4.1 Where an order is sought under section 194(3) of the Legal Services Act 2007 the party who has pro bono representation must prepare, file and serve a written statement of the sum equivalent to the costs that party would have claimed for that legal representation had it not been provided free of charge.

Personal liability of legal representative for costs – wasted costs orders: Rule 46.8

[CPR PD 46.5]

5.1 A wasted costs order is an order—
(a)   that the legal representative pay a sum (either specified or to be assessed) in respect of costs to a party; or
(b)   for costs relating to a specified sum or items of work to be disallowed.
5.2 Rule 46.8 deals with wasted costs orders against legal representatives. Such orders can be made at any stage in the proceedings up to and including the detailed assessment proceedings. In general, applications for wasted costs are best left until after the end of the trial.
5.3 The court may make a wasted costs order against a legal representative on its own initiative.
5.4 A party may apply for a wasted costs order—
(a)   by filing an application notice in accordance with Part 23; or
(b)   by making an application orally in the course of any hearing.
5.5 It is appropriate for the court to make a wasted costs order against a legal representative, only if—
(a)   the legal representative has acted improperly, unreasonably or negligently;

(b) the legal representative's conduct has caused a party to incur unnecessary costs, or has meant that costs incurred by a party prior to the improper, unreasonable or negligent act or omission have been wasted;
(c) it is just in all the circumstances to order the legal representative to compensate that party for the whole or part of those costs.

5.6 The court will give directions about the procedure to be followed in each case in order to ensure that the issues are dealt with in a way which is fair and as simple and summary as the circumstances permit.

5.7 As a general rule the court will consider whether to make a wasted costs order in two stages—
(a) at the first stage the court must be satisfied—
 (i) that it has before it evidence or other material which, if unanswered, would be likely to lead to a wasted costs order being made; and
 (ii) the wasted costs proceedings are justified notwithstanding the likely costs involved;
(b) at the second stage, the court will consider, after giving the legal representative an opportunity to make representations in writing or at a hearing, whether it is appropriate to make a wasted costs order in accordance with paragraph 5.5 above.

5.8 The court may proceed to the second stage described in paragraph 5.7 without first adjourning the hearing if it is satisfied that the legal representative has already had a reasonable opportunity to make representations.

5.9 On an application for a wasted costs order under Part 23 the application notice and any evidence in support must identify—
(a) what the legal representative is alleged to have done or failed to do; and
(b) the costs that the legal representative may be ordered to pay or which are sought against the legal representative.

Assessment of solicitor and client costs: Rules 46.9 and 46.10

[CPR PD 46.6]

6.1 A client and solicitor may agree whatever terms they consider appropriate about the payment of the solicitor's charges. If however, the costs are of an unusual nature, either in amount or the type of costs incurred, those costs will be presumed to have been unreasonably incurred unless the solicitor satisfies the court that the client was informed that they were unusual and that they might not be allowed on an assessment of costs between the parties. That information must have been given to the client before the costs were incurred.

6.2 Costs as between a solicitor and client are assessed on the indemnity basis. The presumptions in rule 46.9(3) are rebuttable.

6.3 If a party fails to comply with the requirements of rule 46.10 concerning the service of a breakdown of costs or points of dispute, any other party may apply to the court in which the detailed assessment hearing should take place for an order requiring compliance. If the court makes such an order, it may—
(a) make it subject to conditions including a condition to pay a sum of money into court; and
(b) specify the consequence of failure to comply with the order or a condition.

6.4 The procedure for obtaining an order under Part III of the Solicitors Act 1974 is by a Part 8 claim, as modified by rule 67.3 and Practice Direction 67. Precedent J of the Schedule of Costs Precedents is a model form of claim form. The application must be accompanied by the bill or bills in respect of which assessment is sought, and, if the claim concerns a conditional fee agreement, a copy of that agreement. If the original bill is not available a copy will suffice.

6.5 Model forms of order, which the court may make, are set out in Precedents K, L and M of the Schedule of Costs Precedents.

6.6 The breakdown of costs referred to in rule 46.10 is a document which contains the following information—
(a) details of the work done under each of the bills sent for assessment; and
(b) in applications under Section 70 of the Solicitors Act 1974, a cash account showing money received by the solicitor to the credit of the client and sums

paid out of that money on behalf of the client but not payments out which were made in satisfaction of the bill or of any items which are claimed in the bill.

6.7 Precedent P of the Schedule of Costs Precedents is a model form of breakdown of costs. A party who is required to serve a breakdown of costs must also serve–
(a) copies of the fee notes of counsel and of any expert in respect of fees claimed in the breakdown, and
(b) written evidence as to any other disbursement which is claimed in the breakdown and which exceeds £250.

6.8 The provisions relating to default costs certificates (rule 47.11) do not apply to cases to which rule 46.10 applies.

6.9 The time for requesting a detailed assessment hearing is within 3 months after the date of the order for the costs to be assessed.

6.10 The form of request for a hearing date must be in Form N258C. The request must be accompanied by copies of—
(a) the order sending the bill or bills for assessment;
(b) the bill or bills sent for assessment;
(c) the solicitor's breakdown of costs and any invoices or accounts served with that breakdown;
(d) a copy of the points of dispute;
(e) a copy of any replies served;
(f) a statement signed by the party filing the request or that party's legal representative giving the names and addresses for service of all parties to the proceedings.

6.11 The request must include the estimated length of hearing.

6.12 On receipt of the request the court will fix a date for the hearing, or will give directions.

6.13 The court will give at least 14 days notice of the time and place of the detailed assessment hearing.

6.14 Unless the court gives permission, only the solicitor whose bill it is and parties who have served points of dispute may be heard and only items specified in the points of dispute may be raised.

6.15 If a party wishes to vary that party's breakdown of costs, points of dispute or reply, an amended or supplementary document must be filed with the court and copies of it must be served on all other relevant parties. Permission is not required to vary a breakdown of costs, points of dispute or a reply but the court may disallow the variation or permit it only upon conditions, including conditions as to the payment of any costs caused or wasted by the variation.

6.16 Unless the court directs otherwise the solicitor must file with the court the papers in support of the bill not less than 7 days before the date for the detailed assessment hearing and not more than 14 days before that date.

6.17 Once the detailed assessment hearing has ended it is the responsibility of the legal representative appearing for the solicitor or, as the case may be, the solicitor in person to remove the papers filed in support of the bill.

6.18 If, in the course of a detailed assessment hearing of a solicitor's bill to that solicitor's client, it appears to the court that in any event the solicitor will be liable in connection with that bill to pay money to the client, it may issue an interim certificate specifying an amount which in its opinion is payable by the solicitor to the client.

6.19 After the detailed assessment hearing is concluded the court will –
(a) complete the court copy of the bill so as to show the amount allowed;
(b) determine the result of the cash account;
(c) award the costs of the detailed assessment hearing in accordance with Section 70(8) of the Solicitors Act 1974; and
(d) issue a final costs certificate.

Costs on the small claims and fast tracks: Rule 46.11

[CPR PD 46.7]

7.1

(1) Before a claim is allocated to either the small claims track or the fast track the court is not restricted by any of the special rules that apply to that track but see paragraph 8.2 below.
(2) Where a claim has been so allocated, the special rules which relate to that track will apply to work done before as well as after allocation save to the extent (if any) that an order for costs in respect of that work was made before allocation.
(3) Where a claim, issued for a sum in excess of the normal financial scope of the small claims track, is allocated to that track only because an admission of part of the claim by the defendant reduces the amount in dispute to a sum within the normal scope of that track; on entering judgment for the admitted part before allocation of the balance of the claim the court may allow costs in respect of the proceedings down to that date.

Costs following allocation, re-allocation and non-allocation: Rule 46.13

[CPR PD 46.8]

8.1 Before reallocating a claim from the small claims track to another track, the court must decide whether any party is to pay costs to the date of the order to re-allocate in accordance with the rules about costs contained in Part 27 If it decides to make such an order the court will make a summary assessment of those costs in accordance with that Part.

8.2 Where a settlement is reached or a Part 36 offer accepted in a case which has not been allocated but would, if allocated, have been suitable for allocation to the small claims track, rule 46.13 enables the court to allow only small claims track costs in accordance with rule 27.14. This power is not exercisable if the costs are to be paid on the indemnity basis.

Costs-only proceedings: Rule 46.14

[CPR PD 46.9]

9.1 A claim form under rule 46.14 should not be issued in the High Court unless the dispute to which the agreement relates was of such a value or type that proceedings would have been commenced in the High Court.

9.2 A claim form which is to be issued in the High Court at the Royal Courts of Justice will be issued in the Costs Office.

9.3 Attention is drawn to rule 8.2 (in particular to paragraph (b)(ii)) and to rule 46.14(3). The claim form must—
(a) identify the claim or dispute to which the agreement relates;
(b) state the date and terms of the agreement on which the claimant relies;
(c) set out or attach a draft of the order which the claimant seeks;
(d) state the amount of the costs claimed.

9.4 Unless the court orders otherwise or Section II of Part 45 applies the costs will be treated as being claimed on the standard basis.

9.5 The evidence required under rule 8.5 includes copies of the documents on which the claimant relies to prove the defendant's agreement to pay costs.

9.6 A costs judge or a district judge has jurisdiction to hear and decide any issue which may arise in a claim issued under this rule irrespective of the amount of the costs claimed or of the value of the claim to which the agreement to pay costs relates. The court may make an order by consent under paragraph 9.8, or an order dismissing a claim under paragraph 9.10 below.

9.7 When the time for filing the defendant's acknowledgement of service has expired, the claimant may request in writing that the court make an order in the terms of the claim, unless the defendant has filed an acknowledgement of service stating the intention to contest the claim or to seek a different order.

9.8 Rule 40.6 applies where an order is to be made by consent. An order may be made by consent in terms which differ from those set out in the claim form.

9.9 Where costs are ordered to be assessed, the general rule is that this should be by detailed assessment. However when an order is made under this rule following a hearing and the court is in a position to summarily assess costs it should generally do so.

9.10 If the defendant opposes the claim the defendant must file a witness statement in accordance with rule 8.5(3). The court will then give directions including, if appropriate, a direction that the claim shall continue as if it were a Part 7 claim. A claim is not treated as opposed merely because the defendant disputes the amount of the claim for costs.

9.11 A claim issued under this rule may be dealt with without being allocated to a track. Rule 8.9 does not apply to claims issued under this rule.

9.12 Where there are other issues nothing in rule 46.14 prevents a person from issuing a claim form under Part 7 or Part 8 to sue on an agreement made in settlement of a dispute where that agreement makes provision for costs, nor from claiming in that case an order for costs or a specified sum in respect of costs but the "costs only" procedure in rule 46.14 must be used where the sole issue is the amount of costs.

PART 47 PROCEDURE FOR DETAILED ASSESSMENT OF COSTS AND DEFAULT PROVISIONS

I
GENERAL RULES ABOUT DETAILED ASSESSMENT

Rule 47.1	Time when detailed assessment may be carried out	CPR 47.1
Rule 47.2	No stay of detailed assessment where there is an appeal	CPR 47.2
Rule 47.3	Powers of an authorised court officer	CPR 47.3
Rule 47.4	Venue for detailed assessment proceedings	CPR 47.4

II
COSTS PAYABLE BY ONE PARTY TO ANOTHER—COMMENCEMENT OF DETAILED ASSESSMENT PROCEEDINGS

Rule 47.5	Application of this section	CPR 47.5
Rule 47.6	Commencement of detailed assessment proceedings	CPR 47.6
Rule 47.7	Period for commencing detailed assessment proceedings	CPR 47.7
Rule 47.8	Sanction for delay in commencing detailed assessment proceedings .	CPR 47.8
Rule 47.9	Points of dispute and consequence of not serving	CPR 47.9
Rule 47.10	Procedure where costs are agreed	CPR 47.10

III
COSTS PAYABLE BY ONE PARTY TO ANOTHER—DEFAULT PROVISIONS

Rule 47.11	Default costs certificate	CPR 47.11
Rule 47.12	Setting aside default costs certificate	CPR 47.12

IV
COSTS PAYABLE BY ONE PARTY TO ANOTHER—PROCEDURE WHERE POINTS OF DISPUTE ARE SERVED

Rule 47.13	Optional reply	CPR 47.13
Rule 47.14	Detailed assessment hearing	CPR 47.14
Rule 47.15	Provisional Assessment	CPR 47.15

V
INTERIM COSTS CERTIFICATE AND FINAL COSTS CERTIFICATE

Rule 47.16	Power to issue an interim certificate	CPR 47.16
Rule 47.17	Final costs certificate	CPR 47.17

VI
DETAILED ASSESSMENT PROCEDURE FOR COSTS OF A LSC FUNDED CLIENT OR AN ASSISTED PERSON WHERE COSTS ARE PAYABLE OUT OF THE COMMUNITY LEGAL SERVICE FUND

Rule 47.18	Detailed assessment procedure for costs of a LSC funded client or an assisted person where costs are payable out of the Community Legal Service Fund	CPR 47.18
Rule 47.19	Detailed assessment procedure where costs are payable out of a fund other than the Community Legal Service Fund	CPR 47.19

VII
COSTS OF DETAILED ASSESSMENT PROCEEDINGS

Rule 47.20	Liability for costs of detailed assessment proceedings	CPR 47.20

VIII
APPEALS FROM AUTHORISED COURT OFFICERS IN DETAILED ASSESSMENT PROCEEDINGS

Rule 47.21	Right to appeal	CPR 47.21
Rule 47.22	Court to hear appeal	CPR 47.22
Rule 47.23	Appeal procedure	CPR 47.23
Rule 47.24	Powers of the court on appeal	CPR 47.24
	Practice Direction 47—Procedure for Detailed Assessment of Costs and Default Provisions	CPR PD 47

CPR 47.1   *Appendix*

SECTION I GENERAL RULES ABOUT DETAILED ASSESSMENT

[CPR 47.1]

47.1 Time when detailed assessment may be carried out
The general rule is that the costs of any proceedings or any part of the proceedings are not to be assessed by the detailed procedure until the conclusion of the proceedings, but the court may order them to be assessed immediately.
(Practice Direction 47 gives further guidance about when proceedings are concluded for the purpose of this rule.)

[CPR 47.2]

47.2 No stay of detailed assessment where there is an appeal
Detailed assessment is not stayed pending an appeal unless the court so orders.

[CPR 47.3]

47.3 Powers of an authorised court officer
(1)   An authorised court officer has all the powers of the court when making a detailed assessment, except—
    (a)   power to make a wasted costs order as defined in rule 46.8;
    (b)   power to make an order under—
        (i)   rule 44.11 (powers in relation to misconduct);
        (ii)   rule 47.8 (sanction for delay in commencing detailed assessment proceedings);
        (iii)   paragraph (2) (objection to detailed assessment by authorised court officer); and
    (c)   power to make a detailed assessment of costs payable to a solicitor by that solicitor's client, unless the costs are being assessed under rule 46.4 (costs where money is payable to a child or protected party).
(2)   Where a party objects to the detailed assessment of costs being made by an authorised court officer, the court may order it to be made by a costs judge or a district judge.
(Practice Direction 47 sets out the relevant procedure.)

[CPR 47.4]

47.4 Venue for detailed assessment proceedings
(1)   All applications and requests in detailed assessment proceedings must be made to or filed at the appropriate office.
(Practice Direction 47 sets out the meaning of "appropriate office" in any particular case)
(2)   The court may direct that the appropriate office is to be the Costs Office.
(3)   A county court may direct that another county court is to be the appropriate office.
(4)   A direction under paragraph (3) may be made without proceedings being transferred to that court.
(Rule 30.2 makes provision for any county court to transfer the proceedings to another county court for detailed assessment of costs.)

SECTION II COSTS PAYABLE BY ONE PARTY TO ANOTHER – COMMENCEMENT OF DETAILED ASSESSMENT PROCEEDINGS

[CPR 47.5]

47.5 Application of this Section
This Section of Part 47 applies where a cost officer is to make a detailed assessment of—
    (a)   costs which are payable by one party to another; or
    (b)   the sum which is payable by one party to the prescribed charity pursuant to an order under section 194(3) of the 2007 Act.

[CPR 47.6]

47.6 Commencement of detailed assessment proceedings
(1) Detailed assessment proceedings are commenced by the receiving party serving on the paying party—
    (a)   notice of commencement in the relevant practice form; and
    (b)   a copy of the bill of costs.
(Rule 47.7 sets out the period for commencing detailed assessment proceedings.)
(2) The receiving party must also serve a copy of the notice of commencement and the bill on any other relevant persons specified in Practice Direction 47.
(3) A person on whom a copy of the notice of commencement is served under paragraph (2) is a party to the detailed assessment proceedings (in addition to the paying party and the receiving party).
(Practice Direction 47 deals with—
    other documents which the party must file when requesting detailed assessment;
    the court's powers where it considers that a hearing may be necessary;
    the form of a bill; and
    the length of notice which will be given if a hearing date is fixed.)
(Paragraphs 7B.2 to 7B.7 of the Practice Direction - Civil Recovery Proceedings contain provisions about detailed assessment of costs in relation to civil recovery orders.)

[CPR 47.7]

47.7 Period for commencing detailed assessment proceedings
The following table shows the period for commencing detailed assessment proceedings.

Source of right to detailed assessment	Time by which detailed assessment proceedings must be commenced
Judgment, direction, order, award or other determination	3 months after the date of the judgment etc
	Where detailed assessment is stayed pending an appeal, 3 months after the date of the order lifting the stay.
Discontinuance under Part 38	3 months after the date of service of notice of discontinuance under rule 38.3; or
	3 months after the date of the dismissal of application to set the notice of discontinuance aside under rule 38.4

CPR 47.8 *Appendix*

    Acceptance of an offer to settle   3 months after the date when the right to costs arose.

[CPR 47.8]

47.8 Sanction for delay in commencing detailed assessment proceedings
(1) Where the receiving party fails to commence detailed assessment proceedings within the period specified—
    (a)   in rule 47.7; or
    (b)   by any direction of the court,
the paying party may apply for an order requiring the receiving party to commence detailed assessment proceedings within such time as the court may specify.
(2) On an application under paragraph (1), the court may direct that, unless the receiving party commences detailed assessment proceedings within the time specified by the court, all or part of the costs to which the receiving party would otherwise be entitled will be disallowed.
(3) If—
    (a)   the paying party has not made an application in accordance with paragraph (1); and
    (b)   the receiving party commences the proceedings later than the period specified in rule 47.7,
the court may disallow all or part of the interest otherwise payable to the receiving party under—
        (i)   section 17 of the Judgments Act 1838; or
        (ii)   section 74 of the County Courts Act 1984,
but will not impose any other sanction except in accordance with rule 44.11 (powers in relation to misconduct).
(4) Where the costs to be assessed in a detailed assessment are payable out of the Community Legal Service Fund, this rule applies as if the receiving party were the solicitor to whom the costs are payable and the paying party were the Legal Services Commission.

[CPR 47.9]

47.9 Points of dispute and consequence of not serving
(1) The paying party and any other party to the detailed assessment proceedings may dispute any item in the bill of costs by serving points of dispute on—
    (a)   the receiving party; and
    (b)   every other party to the detailed assessment proceedings.
(2) The period for serving points of dispute is 21 days after the date of service of the notice of commencement.
(3) If a party serves points of dispute after the period set out in paragraph (2), that party may not be heard further in the detailed assessment proceedings unless the court gives permission.
(Practice Direction 47 sets out requirements about the form of points of dispute.)
(4) The receiving party may file a request for a default costs certificate if—
    (a)   the period set out in paragraph (2) for serving points of dispute has expired; and
    (b)   the receiving party has not been served with any points of dispute.

(5) If any party (including the paying party) serves points of dispute before the issue of a default costs certificate the court may not issue the default costs certificate.
(Section IV of this Part sets out the procedure to be followed after points of dispute have been served.)

[CPR 47.10]

47.10 Procedure where costs are agreed

(1) If the paying party and the receiving party agree the amount of costs, either party may apply for a costs certificate (either interim or final) in the amount agreed. (Rule 47.16 and rule 47.17 contain further provisions about interim and final costs certificates respectively)

(2) An application for a certificate under paragraph (1) must be made to the court which would be the venue for detailed assessment proceedings under rule 47.4.

SECTION III COSTS PAYABLE BY ONE PARTY TO ANOTHER – DEFAULT PROVISIONS

[CPR 47.11]

47.11 Default costs certificate

(1) Where the receiving party is permitted by rule 47.9 to obtain a default costs certificate, that party does so by filing a request in the relevant practice form.
(Practice Direction 47 deals with the procedure by which the receiving party may obtain a default costs certificate.)

(2) A default costs certificate will include an order to pay the costs to which it relates.

(3) Where a receiving party obtains a default costs certificate, the costs payable to that party for the commencement of detailed assessment proceedings will be the sum set out in Practice Direction 47.

(4) A receiving party who obtains a default costs certificate in detailed assessment proceedings pursuant to an order under section 194(3) of the 2007 Act must send a copy of the default costs certificate to the prescribed charity.

[CPR 47.12]

47.12 Setting aside a default costs certificate

(1) The court will set aside a default costs certificate if the receiving party was not entitled to it.

(2) In any other case, the court may set aside or vary a default costs certificate if it appears to the court that there is some good reason why the detailed assessment proceedings should continue.
(Practice Direction 47 contains further details about the procedure for setting aside a default costs certificate and the matters which the court must take into account)

(3) Where the court sets aside or varies a default costs certificate in detailed assessment proceedings pursuant to an order under section 194(3) of the Legal Services Act 2007, the receiving party must send a copy of the order setting aside or varying the default costs certificate to the prescribed charity.

CPR 47.13   *Appendix*

SECTION IV COSTS PAYABLE BY ONE PARTY TO ANOTHER – PROCEDURE WHERE POINTS OF DISPUTE ARE SERVED

[CPR 47.13]

47.13 Optional Reply
(1)   Where any party to the detailed assessment proceedings serves points of dispute, the receiving party may serve a reply on the other parties to the assessment proceedings.
(2)   The receiving party may do so within 21 days after being served with the points of dispute to which the reply relates.
(Practice Direction 47 sets out the meaning of "reply".)

[CPR 47.14]

47.14 Detailed assessment hearing
(1)   Where points of dispute are served in accordance with this Part, the receiving party must file a request for a detailed assessment hearing within 3 months of the expiry of the period for commencing detailed assessment proceedings as specified—
   (a)   in rule 47.7; or
   (b)   by any direction of the court.
(2)   Where the receiving party fails to file a request in accordance with paragraph (1), the paying party may apply for an order requiring the receiving party to file the request within such time as the court may specify.
(3)   On an application under paragraph (2), the court may direct that, unless the receiving party requests a detailed assessment hearing within the time specified by the court, all or part of the costs to which the receiving party would otherwise be entitled will be disallowed.
(4)   If—
   (a)   the paying party has not made an application in accordance with paragraph (2); and
   (b)   the receiving party files a request for a detailed assessment hearing later than the period specified in paragraph (1),
the court may disallow all or part of the interest otherwise payable to the receiving party under—
      (i)   section 17 of the Judgments Act 1838; or
      (ii)   section 74 of the County Courts Act 1984,

but will not impose any other sanction except in accordance with rule 44.11 (powers in relation to misconduct).
(5)   No party other than—
   (a)   the receiving party;
   (b)   the paying party; and
   (c)   any party who has served points of dispute under rule 47.9,
may be heard at the detailed assessment hearing unless the court gives permission.
(6)   Only items specified in the points of dispute may be raised at the hearing, unless the court gives permission.
(7)   If an assessment is carried out at more than one hearing, then for the purposes of rule 52.4 time for appealing shall not start to run until the conclusion of the final hearing, unless the court orders otherwise.

(Practice Direction 47 specifies other documents which must be filed with the request for hearing and the length of notice which the court will give when it fixes a hearing date.)

[CPR 47.15]

47.15 Provisional Assessment

(1) This rule applies to any detailed assessment proceedings commenced in the High Court or a county court on or after 1 April 2013 in which the costs claimed are the amount set out in paragraph 14.1 of the practice direction supplementing this Part, or less.

(2) In proceedings to which this rule applies, the parties must comply with the procedure set out in Part 47 as modified by paragraph 14 Practice Direction 47.

(3) The court will undertake a provisional assessment of the receiving party's costs on receipt of Form N258 and the relevant supporting documents specified in Practice Direction 47.

(4) The provisional assessment will be based on the information contained in the bill and supporting papers and the contentions set out in Precedent G (the points of dispute and any reply).

(5) In proceedings which do not go beyond provisional assessment, the maximum amount the court will award to any party as costs of the assessment (other than the costs of drafting the bill of costs) is £1,500 together with any VAT thereon and any court fees paid by that party.

(6) The court may at any time decide that the matter is unsuitable for a provisional assessment and may give directions for the matter to be listed for hearing. The matter will then proceed under rule 47.14 without modification.

(7) When a provisional assessment has been carried out, the court will send a copy of the bill, as provisionally assessed, to each party with a notice stating that any party who wishes to challenge any aspect of the provisional assessment must, within 21 days of the receipt of the notice, file and serve on all other parties a written request for an oral hearing. If no such request is filed and served within that period, the provisional assessment shall be binding upon the parties, save in exceptional circumstances.

(8) The written request referred to in paragraph (7) must—
   (a) identify the item or items in the court's provisional assessment which are sought to be reviewed at the hearing; and
   (b) provide a time estimate for the hearing.

(9) The court then will fix a date for the hearing and give at least 14 days' notice of the time and place of the hearing to all parties.

(10) Any party which has requested an oral hearing, will pay the costs of and incidental to that hearing unless—
   (a) it achieves an adjustment in its own favour by 20% or more of the sum provisionally assessed; or
   (b) the court otherwise orders.

SECTION V INTERIM COSTS CERTIFICATE AND FINAL COSTS CERTIFICATE

[CPR 47.16]

47.16 Power to issue an interim certificate

(1) The court may at any time after the receiving party has filed a request for a detailed assessment hearing –

CPR 47.17 *Appendix*

      (a)   issue an interim costs certificate for such sum as it considers appropriate; or
      (b)   amend or cancel an interim certificate.

(2) An interim certificate will include an order to pay the costs to which it relates, unless the court orders otherwise.

(3) The court may order the costs certified in an interim certificate to be paid into court.

(4) Where the court –
      (a)   issues an interim costs certificate; or
      (b)   amends or cancels an interim certificate,
in detailed assessment proceedings pursuant to an order under section 194(3) of the 2007 Act, the receiving party must send a copy of the interim costs certificate or the order amending or cancelling the interim costs certificate to the prescribed charity.

[CPR 47.17]

47.17 Final costs certificate

(1) In this rule a "completed bill" means a bill calculated to show the amount due following the detailed assessment of the costs.

(2) The period for filing the completed bill is 14 days after the end of the detailed assessment hearing.

(3) When a completed bill is filed the court will issue a final costs certificate and serve it on the parties to the detailed assessment proceedings.

(4) Paragraph (3) is subject to any order made by the court that a certificate is not to be issued until other costs have been paid.

(5) A final costs certificate will include an order to pay the costs to which it relates, unless the court orders otherwise.

(Practice Direction 47 deals with the form of a final costs certificate.)

(6) Where the court issues a final costs certificate in detailed assessment proceedings pursuant to an order under section 194(3) of the 2007 Act, the receiving party must send a copy of the final costs certificate to the prescribed charity.

SECTION VI DETAILED ASSESSMENT PROCEDURE FOR COSTS OF A LSC FUNDED CLIENT OR AN ASSISTED PERSON WHERE COSTS ARE PAYABLE OUT OF THE COMMUNITY LEGAL SERVICE FUND

[CPR 47.18]

47.18 Detailed assessment procedure where costs are payable out of the Community Legal Services Fund

(1) Where the court is to assess costs of a LSC funded client or an assisted person which are payable out of the Community Legal Services Fund, that person's solicitor may commence detailed assessment proceedings by filing a request in the relevant practice form.

(2) A request under paragraph (1) must be filed within 3 months after the date when the right to detailed assessment arose.

(3) The solicitor must also serve a copy of the request for detailed assessment on the LSC funded client or the assisted person, if notice of that person's interest has been given to the court in accordance with community legal service or legal aid regulations.

(4) Where the solicitor has certified that the LSC funded client or that person wishes to attend an assessment hearing, the court will, on receipt of the request for assessment, fix a date for the assessment hearing.

(5) Where paragraph (3) does not apply, the court will, on receipt of the request for assessment provisionally assess the costs without the attendance of the solicitor, unless it considers that a hearing is necessary.

(6) After the court has provisionally assessed the bill, it will return the bill to the solicitor.

(7) The court will fix a date for an assessment hearing if the solicitor informs the court, within 14 days after receiving the provisionally assessed bill, that the solicitor wants the court to hold such a hearing.

[CPR 47.19]

47.19 Detailed assessment procedure where costs are payable out of a fund other than the community legal service fund

(1) Where the court is to assess costs which are payable out of a fund other than the Community Legal Service Fund, the receiving party may commence detailed assessment proceedings by filing a request in the relevant practice form.

(2) A request under paragraph (1) must be filed within 3 months after the date when the right to detailed assessment arose.

(3) The court may direct that the party seeking assessment serve a copy of the request on any person who has a financial interest in the outcome of the assessment.

(4) The court will, on receipt of the request for assessment, provisionally assess the costs without the attendance of the receiving party, unless the court considers that a hearing is necessary.

(5) After the court has provisionally assessed the bill, it will return the bill to the receiving party.

(6) The court will fix a date for an assessment hearing if the receiving party informs the court, within 14 days after receiving the provisionally assessed bill, that the receiving party wants the court to hold such a hearing.

SECTION VII COSTS OF DETAILED ASSESSMENT PROCEEDINGS

[CPR 47.20]

47.20 Liability for costs of detailed assessment proceedings

(1) The receiving party is entitled to the costs of the detailed assessment proceedings except where—
  (a) the provisions of any Act, any of these Rules or any relevant practice direction provide otherwise; or
  (b) the court makes some other order in relation to all or part of the costs of the detailed assessment proceedings.

(2) Paragraph (1) does not apply where the receiving party has pro bono representation in the detailed assessment proceedings but that party may apply for an order in respect of that representation under section 194(3) of the 2007 Act.

(3) In deciding whether to make some other order, the court must have regard to all the circumstances, including—
  (a) the conduct of all the parties;
  (b) the amount, if any, by which the bill of costs has been reduced; and
  (c) whether it was reasonable for a party to claim the costs of a particular item or to dispute that item.

CPR 47.21 *Appendix*

(4) The provisions of Part 36 apply to the costs of detailed assessment proceedings with the following modifications—
- (a) "claimant" refers to "receiving party" and "defendant" refers to "paying party";
- (b) "trial" refers to "detailed assessment hearing";
- (c) in rule 36.9(5), at the end insert "or, where the Part 36 offer is made in respect of the detailed assessment proceedings, after the commencement of the detailed assessment hearing.";
- (d) for rule 36.11(7) substitute "If the accepted sum is not paid within 14 days or such other period as has been agreed the offeree may apply for a final costs certificate for the unpaid sum.";
- (e) a reference to "judgment being entered" is to the completion of the detailed assessment, and references to a "judgment" being advantageous or otherwise are to the outcome of the detailed assessment.

(5) The court will usually summarily assess the costs of detailed assessment proceedings at the conclusion of those proceedings.

(6) Unless the court otherwise orders, interest on the costs of detailed assessment proceedings will run from the date of default, interim or final costs certificate, as the case may be.

(7) For the purposes of rule 36.14, detailed assessment proceedings are to be regarded as an independent claim.

SECTION VIII APPEALS FROM AUTHORISED COURT OFFICERS IN DETAILED ASSESSMENT PROCEEDINGS

[CPR 47.21]

47.21 Right to appeal
Any party to detailed assessment proceedings may appeal against a decision of an authorised court officer in those proceedings.

[CPR 47.22]

47.22 Court to hear appeal
An appeal against a decision of an authorised court officer lies to a costs judge or a district judge of the High Court.

[CPR 47.23]

47.23 Appeal procedure
(1) The appellant must file an appeal notice within 21 days after the date of the decision against which it is sought to appeal.
(2) On receipt of the appeal notice, the court will—
- (a) serve a copy of the notice on the parties to the detailed assessment proceedings; and
- (b) give notice of the appeal hearing to those parties.

[CPR 47.24]

47.24 Powers of the court on appeal
On an appeal from an authorised court officer the court will—
- (a) re-hear the proceedings which gave rise to the decision appealed against; and

(b) make any order and give any directions as it considers appropriate.

PRACTICE DIRECTION 47—PROCEDURE FOR DETAILED ASSESSMENT OF COSTS AND DEFAULT PROVISIONS

THIS PRACTICE DIRECTION SUPPLEMENTS PART 47

Time when assessment may be carried out: Rule 47.1

[CPR PD 47.1]

1.1 For the purposes of rule 47.1, proceedings are concluded when the court has finally determined the matters in issue in the claim, whether or not there is an appeal, or made an award of provisional damages under Part 41.
1.2 The court may order or the parties may agree in writing that, although the proceedings are continuing, they will nevertheless be treated as concluded.
1.3 A party who is served with a notice of commencement (see paragraph 5.2 below) may apply to a costs judge or a district judge to determine whether the party who served it is entitled to commence detailed assessment proceedings. On hearing such an application the orders which the court may make include: an order allowing the detailed assessment proceedings to continue, or an order setting aside the notice of commencement.
1.4 A costs judge or a district judge may make an order allowing detailed assessment proceedings to be commenced where there is no realistic prospect of the claim continuing.

No stay of detailed assessment where there is an appeal: Rule 47.2

[CPR PD 47.2]

2. An application to stay the detailed assessment of costs pending an appeal may be made to the court whose order is being appealed or to the court which will hear the appeal.

Powers of an authorised court officer: Rule 47.3

[CPR PD 47.3]

3.1 The court officers authorised by the Lord Chancellor to assess costs in the Costs Office and the Principal Registry of the Family Division are authorised to deal with claims where the base costs excluding VAT do not exceed £35,000 in the case of senior executive officers, or their equivalent, and £110,000 in the case of principal officers.
3.2 Where the receiving party, paying party and any other party to the detailed assessment proceedings who has served points of dispute are agreed that the assessment should not be made by an authorised court officer, the receiving party should so inform the court when requesting a hearing date. The court will then list the hearing before a costs judge or a district judge.
3.3 In any other case a party who objects to the assessment being made by an authorised court officer must make an application to the costs judge or district judge under Part 23 setting out the reasons for the objection.

Venue for detailed assessment proceedings: Rule 47.4

[CPR PD 47.4]

4.1 For the purposes of rule 47.4(1) the 'appropriate office' means—

(a) the district registry or county court in which the case was being dealt with when the judgment or order was made or the event occurred which gave rise to the right to assessment, or to which it has subsequently been transferred;
(b) where a tribunal, person or other body makes an order for the detailed assessment of costs, a county court (subject to paragraph 4.2); or
(c) in all other cases, including Court of Appeal cases, the Costs Office.

4.2
(1) This paragraph applies where the appropriate office is any of the following county courts: Barnet, Bow, Brentford, Bromley, Central London, Clerkenwell and Shoreditch, Croydon, Edmonton, Ilford, Kingston, Lambeth, Mayors and City of London, Romford, Uxbridge, Wandsworth, West London, Willesden and Woolwich.
(2) Where this paragraph applies—
   (a) the receiving party must file any request for a detailed assessment hearing in the Costs Office and, for all purposes relating to that detailed assessment (other than the issue of default costs certificates and applications to set aside default costs certificates), the Costs Office will be treated as the appropriate office in that case;
   (b) default costs certificates should be issued and applications to set aside default costs certificates should be issued and heard in the relevant county court; and
   (c) unless an order is made under rule 47.4(2) directing that the Costs Office as part of the High Court shall be the appropriate office, an appeal from any decision made by a costs judge shall lie to the Designated Civil Judge for the London Group of County Courts or such judge as the Designated Civil Judge shall nominate. The appeal notice and any other relevant papers should be lodged at the Central London Civil Justice Centre.

4.3
(1) A direction under rule 47.4(2) or (3) specifying a particular court, registry or office as the appropriate office may be given on application or on the court's own initiative.
(2) Unless the Costs Office is the appropriate office for the purposes of rule 47.4(1) an order directing that an assessment is to take place at the Costs Office will be made only if it is appropriate to do so having regard to the size of the bill of costs, the difficulty of the issues involved, the likely length of the hearing, the cost to the parties and any other relevant matter.

Commencement of detailed assessment proceedings: Rule 47.6

[CPR PD 47.5]

5.1 Precedents A, B, C and D in the Schedule of Costs Precedents annexed to this Practice Direction are model forms of bills of costs for detailed assessment.

5.2 The receiving party must serve on the paying party and all the other relevant persons the following documents—
(a) a notice of commencement in Form N252;
(b) a copy of the bill of costs;
(c) copies of the fee notes of counsel and of any expert in respect of fees claimed in the bill;
(d) written evidence as to any other disbursement which is claimed and which exceeds £500;
(e) a statement giving the name and address for service of any person upon whom the receiving party intends to serve the notice of commencement.

5.3 The notice of commencement must be completed to show as separate items—
(a) the total amount of the costs claimed in the bill;
(b) the extra sum which will be payable by way of fixed costs and court fees if a default costs certificate is obtained.

5.4 Where the notice of commencement is to be served outside England and Wales the date to be inserted in the notice of commencement for the paying party to send points of dispute is a date (not less than 21 days from the date of service of the notice) which must be calculated by reference to Section IV of Part 6 as if the notice were a claim form and as if the date to be inserted was the date for the filing of a defence.

5.5
(1) For the purposes of rule 47.6(2) a "relevant person" means—
   (a) any person who has taken part in the proceedings which gave rise to the assessment and who is directly liable under an order for costs made against that person;
   (b) any person who has given to the receiving party notice in writing that that person has a financial interest in the outcome of the assessment and wishes to be a party accordingly;
   (c) any other person whom the court orders to be treated as such.
(2) Where a party is unsure whether a person is or is not a relevant person, that party may apply to the appropriate office for directions.
(3) The court will generally not make an order that the person in respect of whom the application is made will be treated as a relevant person, unless within a specified time that person applies to the court to be joined as a party to the assessment proceedings in accordance with Part 19 (Parties and Group Litigation).

5.6 Where—
(a) the bill of costs is capable of being copied electronically; and
(b) before the detailed assessment hearing,
a paying party requests an electronic copy of the bill, the receiving party must supply the paying party with a copy in its native format (for example, in Excel or an equivalent) free of charge not more than 7 days after receipt of the request.

Form and contents of bills of costs — general

5.7 A bill of costs may consist of such of the following sections as may be appropriate—
(1) title page;
(2) background information;
(3) items of costs claimed under the headings specified in paragraph 5.12;
(4) summary showing the total costs claimed on each page of the bill;
(5) schedules of time spent on non-routine attendances; and
(6) the certificates referred to in paragraph 5.21.

If the only dispute between the parties concerns disbursements, the bill of costs shall be limited to items (1) and (2) above, a list of the disbursements in issue and brief written submissions in respect of those disbursements.

5.8 Where it is necessary or convenient to do so, a bill of costs may be divided into two or more parts, each part containing sections (2), (3) and (4) above. Circumstances in which it will be necessary or convenient to divide a bill into parts include the following—
(1) Where the receiving party acted in person during the course of the proceedings (whether or not that party also had a legal representative at that time) the bill must be divided into different parts so as to distinguish between;
   (a) the costs claimed for work done by the legal representative; and
   (b) the costs claimed for work done by the receiving party in person.
(2) Where the receiving party had pro bono representation for part of the proceedings and an order under section 194(3) of the Legal Services Act 2007 has been made, the bill must be divided into different parts so as to distinguish between—
   (a) the sum equivalent to the costs claimed for work done by the legal representative acting free of charge; and
   (b) the costs claimed for work not done by the legal representative acting free of charge.
(3) Where the receiving party was represented by different legal representatives during the course of the proceedings, the bill must be divided into different parts so as to distinguish between the costs payable in respect of each legal representative.

Civil Procedure Rules 1998 CPR PD 47.5

(4) Where the receiving party obtained legal aid or LSC funding or is a person for whom civil legal services (within the meaning of Part 1 of the Legal Aid, Sentencing and Punishment of Offenders Act 2012) were provided under arrangements made for the purposes of that Part of that Act in respect of all or part of the proceedings, the bill must be divided into separate parts so as to distinguish between—
  (a) costs claimed before legal aid or LSC funding was granted or before civil legal services were provided;
  (b) costs claimed after legal aid or LSC funding was granted or after civil legal services were provided; and
  (c) any costs claimed after legal aid or LSC funding ceased or after civil legal services ceased to be provided.
(5) Where the bill covers costs payable under an order or orders under which there are different paying parties the bill must be divided into parts so as to deal separately with the costs payable by each paying party.
(6) Where the bill covers costs payable under an order or orders, in respect of which the receiving party wishes to claim interest from different dates, the bill must be divided to enable such interest to be calculated.

5.9 Where a party claims costs against another party and also claims costs against the LSC or Lord Chancellor only for work done in the same period, the costs claimed against the LSC or Lord Chancellor only can be claimed either in a separate part of the bill or in additional columns in the same part of the bill. Precedents B and C in the Schedule of Costs Precedents annexed to this Practice Direction show how bills should be drafted when costs are claimed against the LSC only.

Form and content of bills of costs: Title page

5.10 The title page of the bill of costs must set out—
(1) the full title of the proceedings;
(2) the name of the party whose bill it is and a description of the document showing the right to assessment (as to which see paragraph 13.3 of this Practice Direction);
(3) if VAT is included as part of the claim for costs, the VAT number of the legal representative or other person in respect of whom VAT is claimed;
(4) details of all legal aid certificates, LSC certificates, certificates recording the determinations of the Director of Legal Aid Casework and relevant amendment certificates in respect of which claims for costs are included in the bill.

Form and content of bills of costs: Background information

5.11 The background information included in the bill of costs should set out—
(1) a brief description of the proceedings up to the date of the notice of commencement;
(2) a statement of the status of the legal representatives' employee in respect of whom costs are claimed and (if those costs are calculated on the basis of hourly rates) the hourly rates claimed for each such person.
(3) a brief explanation of any agreement or arrangement between the receiving party and his legal representatives, which affects the costs claimed in the bill.

Form and content of bills of costs: Heads of costs

5.12 The bill of costs may consist of items under such of the following heads as may be appropriate—
(1) attendances at court and upon counsel up to the date of the notice of commencement;
(2) attendances on and communications with the receiving party;
(3) attendances on and communications with witnesses including any expert witness;
(4) attendances to inspect any property or place for the purposes of the proceedings;

CPR PD 47.5 *Appendix*

(5) attendances on and communications with other persons, including offices of public records;
(6) communications with the court and with counsel;
(7) work done on documents:
(8) work done in connection with negotiations with a view to settlement if not already covered in the heads listed above;
(9) attendances on and communications with London and other agents and work done by them;
(10) other work done which was of or incidental to the proceedings and which is not already covered in the heads listed above.

5.13 In respect of each of the heads of costs—
(1) 'communications' means letters out e-mails out and telephone calls;
(2) communications, which are not routine communications, must be set out in chronological order;
(3) routine communications must be set out as a single item at the end of each head;

5.14 Routine communications are letters out, e-mails out and telephone calls which because of their simplicity should not be regarded as letters or e-mails of substance or telephone calls which properly amount to an attendance.

5.15 Each item claimed in the bill of costs must be consecutively numbered.

5.16 In each part of the bill of costs which claims items under head (1) in paragraph 5.12 (attendances at court and upon counsel) a note should be made of—
(1) all relevant events, including events which do not constitute chargeable items;
(2) any orders for costs which the court made (whether or not a claim is made in respect of those costs in this bill of costs).

5.17 The numbered items of costs may be set out on paper divided into columns. Precedents A, B and C in the Schedule of Costs Precedents annexed to this Practice Direction illustrate various model forms of bills of costs.

5.18 In respect of heads (2) to (10) in paragraph 5.12 above, if the number of attendances and communications other than routine communications is twenty or more, the claim for the costs of those items in that section of the bill of costs should be for the total only and should refer to a schedule in which the full record of dates and details is set out. If the bill of costs contains more than one schedule each schedule should be numbered consecutively.

5.19 The bill of costs must not contain any claims in respect of costs or court fees which relate solely to the detailed assessment proceedings other than costs claimed for preparing and checking the bill.

5.20 The summary must show the total profit costs and disbursements claimed separately from the total VAT claimed. Where the bill of costs is divided into parts the summary must also give totals for each part. If each page of the bill gives a page total the summary must also set out the page totals for each page.

5.21 The bill of costs must contain such of the certificates, the texts of which are set out in Precedent F of the Schedule of Costs Precedents annexed to this Practice Direction, as are appropriate.

5.22 The following provisions relate to work done by legal representatives—
(1) Routine letters out, routine e-mails out and routine telephone calls will in general be allowed on a unit basis of 6 minutes each, the charge being calculated by reference to the appropriate hourly rate. The unit charge for letters out and e-mails out will include perusing and considering the routine letters in or e-mails in.
(2) The court may, in its discretion, allow an actual time charge for preparation of electronic communications sent by legal representatives, which properly amount to attendances provided that the time taken has been recorded.
(3) Local travelling expenses incurred by legal representatives will not be allowed. The definition of 'local' is a matter for the discretion of the court. As a matter of guidance, 'local' will, in general, be taken to mean within a radius of 10 miles from the court dealing with the case at the relevant time. Where

travelling and waiting time is claimed, this should be allowed at the rate agreed with the client unless this is more than the hourly rate on the assessment.

(4) The cost of postage, couriers, out-going telephone calls, fax and telex messages will in general not be allowed but the court may exceptionally in its discretion allow such expenses in unusual circumstances or where the cost is unusually heavy.

(5) The cost of making copies of documents will not in general be allowed but the court may exceptionally in its discretion make an allowance for copying in unusual circumstances or where the documents copied are unusually numerous in relation to the nature of the case. Where this discretion is invoked the number of copies made, their purpose and the costs claimed for them must be set out in the bill.

(6) Agency charges as between principal legal representatives and their agents will be dealt with on the principle that such charges, where appropriate, form part of the principal legal representative's charges. Where these charges relate to head (1) in paragraph 5.12 (attendances at court and on counsel) they must be included in their chronological order in that head. In other cases they must be included in head (9) (attendances on London and other agents).

Period for commencing detailed assessment proceedings: Rule 47.7

[CPR PD 47.6]

6.1 The time for commencing the detailed assessment proceedings may be extended or shortened either by agreement (rule 2.11) or by the court (rule 3.1(2)(a)). Any application is to the appropriate office.

6.2 The detailed assessment proceedings are commenced by service of the documents referred to. Permission to commence assessment proceedings out of time is not required.

Sanction for delay in commencing detailed assessment proceedings: Rule 47.8

[CPR PD 47.7]

7 An application for an order under rule 47.8 must be made in writing and be issued in the appropriate office. The application notice must be served at least 7 days before the hearing.

Points of dispute and consequences of not serving: Rule 47.9

[CPR PD 47.8]

8.1 Time for service of points of dispute may be extended or shortened either by agreement (rule 2.11) or by the court (rule 3.1(2)(a)). Any application is to the appropriate office.

8.2 Points of dispute must be short and to the point. They must follow Precedent G in the Schedule of Costs Precedents annexed to this Practice Direction, so far as practicable. They must:

(a) identify any general points or matters of principle which require decision before the individual items in the bill are addressed; and

(b) identify specific points, stating concisely the nature and grounds of dispute.

Once a point has been made it should not be repeated but the item numbers where the point arises should be inserted in the left hand box as shown in Precedent G.

8.3 The paying party must state in an open letter accompanying the points of dispute what sum, if any, that party offers to pay in settlement of the total costs claimed. The paying party may also make an offer under Part 36.

CPR PD 47.9 *Appendix*

Procedure where costs are agreed and on discontinuance: Rule 47.10

[CPR PD 47.9]

9.1 Where the parties have agreed terms as to the issue of a costs certificate (either interim or final) they should apply under rule 40.6 (Consent judgments and orders) for an order that a certificate be issued in the terms set out in the application. Such an application may be dealt with by a court officer, who may issue the certificate.

9.2 Where in the course of proceedings the receiving party claims that the paying party has agreed to pay costs but that the paying party will neither pay those costs nor join in a consent application under paragraph 9.1, the receiving party may apply under Part 23 for a certificate either interim or final to be issued.

9.3 Nothing in rule 47.10 prevents parties who seek a judgment or order by consent from including in the draft a term that a party shall pay to another party a specified sum in respect of costs.

9.4
(1) The receiving party may discontinue the detailed assessment proceedings in accordance with Part 38 (Discontinuance).
(2) Where the receiving party discontinues the detailed assessment proceedings before a detailed assessment hearing has been requested, the paying party may apply to the appropriate office for an order about the costs of the detailed assessment proceedings.
(3) Where a detailed assessment hearing has been requested the receiving party may not discontinue unless the court gives permission.
(4) A bill of costs may be withdrawn by consent whether or not a detailed assessment hearing has been requested.

Default costs certificate: Rule 47.11

[CPR PD 47.10]

10.1
(1) A request for the issue of a default costs certificate must be made in Form N254 and must be signed by the receiving party or his legal representative.
(2) The request must be accompanied by a copy of the document giving the right to detailed assessment and must be filed at the appropriate office. (Paragraph 13.3 below identifies the appropriate documents).

10.2 A default costs certificate will be in Form N255.

10.3 Attention is drawn to Rules 40.3 (Drawing up and Filing of Judgments and Orders) and 40.4 (Service of Judgments and Orders) which apply to the preparation and service of a default costs certificate. The receiving party will be treated as having permission to draw up a default costs certificate by virtue of this Practice Direction.

10.4 The issue of a default costs certificate does not prohibit, govern or affect any detailed assessment of the same costs which are payable out of the Community Legal Service Fund or by the Lord Chancellor under Part 1 of the Legal Aid, Sentencing and Punishment of Offenders Act 2012.

10.5 An application for an order staying enforcement of a default costs certificate may be made either–
(a) to a costs judge or district judge of the court office which issued the certificate; or
(b) to the court (if different) which has general jurisdiction to enforce the certificate.

10.6 Proceedings for enforcement of default costs certificates may not be issued in the Costs Office.

Default costs certificate: Fixed costs on the issue of a default costs certificate

10.7 Unless paragraph 1.2 of Practice Direction 45 (Fixed Costs in Small Claims) applies or unless the court orders otherwise, the fixed costs to be included in a default costs certificate are £80 plus a sum equal to any appropriate court fee payable on the issue of the certificate.

10.8 The fixed costs included in a certificate must not exceed the maximum sum specified for costs and court fee in the notice of commencement.

Setting aside default costs certificate: Rule 47.12

[CPR PD 47.11]

11.1 A court officer may set aside a default costs certificate at the request of the receiving party under rule 47.12. A costs judge or a district judge will make any other order or give any directions under this rule.

11.2
(1) An application for an order under rule 47.12(2) to set aside or vary a default costs certificate must be supported by evidence.
(2) In deciding whether to set aside or vary a certificate under rule 47.12(2) the matters to which the court must have regard include whether the party seeking the order made the application promptly.
(3) As a general rule a default costs certificate will be set aside under rule 47.12 only if the applicant shows a good reason for the court to do so and if the applicant files with the application a copy of the bill, a copy of the default costs certificate and a draft of the points of dispute the applicant proposes to serve if the application is granted.

11.3 Attention is drawn to rule 3.1(3) (which enables the court when making an order to make it subject to conditions) and to rule 44.2(8) (which enables the court to order a party whom it has ordered to pay costs to pay an amount on account before the costs are assessed). A costs judge or a district judge may exercise the power of the court to make an order under rule 44.2(8) although he did not make the order about costs which led to the issue of the default costs certificate.

Optional reply: Rule 47.13

[CPR PD 47.12]

12.1 A reply served by the receiving party under Rule 47.13 must be limited to points of principle and concessions only. It must not contain general denials, specific denials or standard form responses.

12.2 Whenever practicable, the reply must be set out in the form of Precedent G.

Detailed assessment hearing: Rule 47.14

[CPR PD 47.13]

13.1 The time for requesting a detailed assessment hearing is within 3 months of the expiry of the period for commencing detailed assessment proceedings.

13.2 The request for a detailed assessment hearing must be in Form N258. The request must be accompanied by—
(a) a copy of the notice of commencement of detailed assessment proceedings;
(b) a copy of the bill of costs,
(c) the document giving the right to detailed assessment (see paragraph 13.3 below);
(d) a copy of the points of dispute, annotated as necessary in order to show which items have been agreed and their value and to show which items remain in dispute and their value;
(e) as many copies of the points of dispute so annotated as there are persons who have served points of dispute;

CPR PD 47.13 *Appendix*

- (f) a copy of any replies served;
- (g) copies of all orders made by the court relating to the costs which are to be assessed;
- (h) copies of the fee notes and other written evidence as served on the paying party in accordance with paragraph 5.2 above;
- (i) where there is a dispute as to the receiving party's liability to pay costs to the legal representatives who acted for the receiving party, any agreement, letter or other written information provided by the legal representative to the client explaining how the legal representative's charges are to be calculated;
- (j) a statement signed by the receiving party or the legal representative giving the name, address for service, reference and telephone number and fax number, if any, of—
  - (i) the receiving party;
  - (ii) the paying party;
  - (iii) any other person who has served points of dispute or who has given notice to the receiving party under paragraph 5.5(1)(b) above;

  and giving an estimate of the length of time the detailed assessment hearing will take;
- (k) where the application for a detailed assessment hearing is made by a party other than the receiving party, such of the documents set out in this paragraph as are in the possession of that party;
- (l) where the court is to assess the costs of an assisted person or LSC funded client or person to whom civil legal services (within the meaning of Part 1 of the Legal Aid, Sentencing and Punishment of Offenders Act 2012) are provided under arrangement made for the purposes of that Part of that Act—
  - (i) the legal aid certificate, LSC certificate, the certificate recording the determination of the Director of Legal Aid Casework and relevant amendment certificates, any authorities and any certificates of discharge or revocation or withdrawal;
  - (ii) a certificate, in Precedent F(3) of the Schedule of Costs Precedents;
  - (iii) if that person has a financial interest in the detailed assessment hearing and wishes to attend, the postal address of that person to which the court will send notice of any hearing;
  - (iv) if the rates payable out of the LSC fund or by the Lord Chancellor under Part 1 of the Legal Aid, Sentencing and Punishment of Offenders Act 2012 are prescribed rates, a schedule to the bill of costs setting out all the items in the bill which are claimed against other parties calculated at the legal aid prescribed rates with or without any claim for enhancement: (further information as to this schedule is set out in paragraph 17 of this Practice Direction);
  - (v) a copy of any default costs certificate in respect of costs claimed in the bill of costs.

13.3 "The document giving the right to detailed assessment" means such one or more of the following documents as are appropriate to the detailed assessment proceedings—
- (a) a copy of the judgment or order of the court or tribunal giving the right to detailed assessment;
- (b) a copy of the notice served under rule 3.7 (sanctions for non-payment of certain fees) where a claim is struck out under that rule;
- (c) a copy of the notice of acceptance where an offer to settle is accepted under Part 36 (Offers to settle);
- (d) a copy of the notice of discontinuance in a case which is discontinued under Part 38 (Discontinuance);
- (e) a copy of the award made on an arbitration under any Act or pursuant to an agreement, where no court has made an order for the enforcement of the award;

(f) a copy of the order, award or determination of a statutorily constituted tribunal or body.

13.4 On receipt of the request for a detailed assessment hearing the court will fix a date for the hearing, or, if the costs officer so decides, will give directions or fix a date for a preliminary appointment.

13.5 Unless the court otherwise orders, if the only dispute between the parties concerns disbursements, the hearing shall take place in the absence of the parties on the basis of the documents and the court will issue its decision in writing.

13.6 The court will give at least 14 days' notice of the time and place of the detailed assessment hearing to every person named in the statement referred to in paragraph 13.2(j) above.

13.7 If either party wishes to make an application in the detailed assessment proceedings the provisions of Part 23 apply.

13.8
(1) This paragraph deals with the procedure to be adopted where a date has been given by the court for a detailed assessment hearing and—
  (a) the detailed assessment proceedings are settled; or
  (b) a party to the detailed assessment proceedings wishes to apply to vary the date which the court has fixed; or
  (c) the parties to the detailed assessment proceedings agree about changes they wish to make to any direction given for the management of the detailed assessment proceedings.
(2) If detailed assessment proceedings are settled, the receiving party must give notice of that fact to the court immediately, preferably by fax.
(3) A party who wishes to apply to vary a direction must do so in accordance with Part 23.
(4) If the parties agree about changes they wish to make to any direction given for the management of the detailed assessment proceedings—
  (a) they must apply to the court for an order by consent; and
  (b) they must file a draft of the directions sought and an agreed statement of the reasons why the variation is sought; and
  (c) the court may make an order in the agreed terms or in other terms without a hearing, but it may direct that a hearing is to be listed.

13.10
(1) If a party wishes to vary that party's bill of costs, points of dispute or a reply, an amended or supplementary document must be filed with the court and copies of it must be served on all other relevant parties.
(2) Permission is not required to vary a bill of costs, points of dispute or a reply but the court may disallow the variation or permit it only upon conditions, including conditions as to the payment of any costs caused or wasted by the variation.

13.11 Unless the court directs otherwise the receiving party must file with the court the papers in support of the bill not less than 7 days before the date for the detailed assessment hearing and not more than 14 days before that date.

13.12 The papers to be filed in support of the bill and the order in which they are to be arranged are as follows—
(i) instructions and briefs to counsel arranged in chronological order together with all advices, opinions and drafts received and response to such instructions;
(ii) reports and opinions of medical and other experts;
(iii) any other relevant papers;
(iv) a full set of any relevant statements of case
(v) correspondence, file notes and attendance notes;

13.13 The court may direct the receiving party to produce any document which in the opinion of the court is necessary to enable it to reach its decision. These documents will in the first instance be produced to the court, but the court may ask the receiving party to elect whether to disclose the particular document to the paying party in order to rely on the contents of the document, or whether to decline disclosure and instead rely on other evidence.

13.14 Once the detailed assessment hearing has ended it is the responsibility of the receiving party to remove the papers filed in support of the bill.

CPR PD 47.14 *Appendix*

Provisional assessment: Rule 47.15

[CPR PD 47.14]

14.1 The amount of costs referred to in rule 47.15(1) is £75,000.

14.2 The following provisions of Part 47 and this Practice Direction will apply to cases falling within rule 47.15—
(1) rules 47.1, 47.2, 47.4 to 47.13, 47.14 (except paragraphs (6) and (7)), 47.16, 47.17, 47.20 and 47.21; and
(2) paragraphs 1, 2, 4 to 12, 13 (with the exception of paragraphs 13.4 to 13.7, 13.9, 13.11 and 13.14), 15, and 16, of this Practice Direction.

14.3 In cases falling within rule 47.15, when the receiving party files a request for a detailed assessment hearing, that party must file—
(a) the request in Form N258;
(b) the documents set out at paragraphs 8.3 and 13.2 of this Practice Direction;
(c) an additional copy of the bill, including a statement of the costs claimed in respect of the detailed assessment drawn on the assumption that there will not be an oral hearing following the provisional assessment;
(d) the offers made (those marked "without prejudice save as to costs" or made under Part 36 must be contained in a sealed envelope, marked "Part 36 or similar offers", but not indicating which party or parties have made them);
(e) completed Precedent G (points of dispute and any reply).

14.4
(1) On receipt of the request for detailed assessment and the supporting papers, the court will use its best endeavours to undertake a provisional assessment within 6 weeks. No party will be permitted to attend the provisional assessment.
(2) Once the provisional assessment has been carried out the court will return Precedent G (the points of dispute and any reply) with the court's decisions noted upon it. Within 14 days of receipt of Precedent G the parties must agree the total sum due to the receiving party on the basis of the court's decisions. If the parties are unable to agree the arithmetic, they must refer the dispute back to the court for a decision on the basis of written submissions.

14.5 When considering whether to depart from the order indicated by rule 47.15(10) the court will take into account the conduct of the parties and any offers made.

14.6 If a party wishes to be heard only as to the order made in respect of the costs of the initial provisional assessment, the court will invite each side to make written submissions and the matter will be finally determined without a hearing. The court will decide what if any order for costs to make in respect of this procedure.

Power to issue an interim certificate: Rule 47.16

[CPR PD 47.15]

15. A party wishing to apply for an interim certificate may do so by making an application in accordance with Part 23.

Final costs certificate: Rule 47.17

[CPR PD 47.16]

16.1 At the detailed assessment hearing the court will indicate any disallowance or reduction in the sums claimed in the bill of costs by making an appropriate note on the bill.

16.2 The receiving party must, in order to complete the bill after the detailed assessment hearing make clear the correct figures agreed or allowed in respect of each item and must re-calculate the summary of the bill appropriately.

16.3 The completed bill of costs must be filed with the court no later than 14 days after the detailed assessment hearing.

16.4 At the same time as filing the completed bill of costs, the party whose bill it is must also produce receipted fee notes and receipted accounts in respect of all disbursements except those covered by a certificate in Precedent F(5) in the Schedule of Costs Precedents annexed to this Practice Direction.

16.5 No final costs certificate will be issued until all relevant court fees payable on the assessment of costs have been paid.

16.6 If the receiving party fails to file a completed bill in accordance with rule 47.17 the paying party may make an application under Part 23 seeking an appropriate order under rule 3.1.

16.7 A final costs certificate will show—
(a) the amount of any costs which have been agreed between the parties or which have been allowed on detailed assessment;
(b) where applicable the amount agreed or allowed in respect of VAT on such costs.

This provision is subject to any contrary statutory provision relating to costs payable out of the Community Legal Service Fund or by the Lord Chancellor under Part 1 of the Legal Aid, Sentencing and Punishment of Offenders Act 2012.

16.8 A final costs certificate will include disbursements in respect of the fees of counsel only if receipted fee notes or accounts in respect of those disbursements have been produced to the court and only to the extent indicated by those receipts.

16.9 Where the certificate relates to costs payable between parties a separate certificate will be issued for each party entitled to costs.

16.10 Form N257 is a model form of interim costs certificate and Form N256 is a model form of final costs certificate.

16.11 An application for an order staying enforcement of an interim costs certificate or final costs certificate may be made either—
(a) to a costs judge or district judge of the court office which issued the certificate; or
(b) to the court (if different) which has general jurisdiction to enforce the certificate.

16.12 An interim or final costs certificate may be enforced as if it were a judgment for the payment of an amount of money. However, proceedings for the enforcement of interim costs certificates or final costs certificates may not be issued in the Costs Office.

Detailed assessment procedure where costs are payable out of the community legal service fund or by the lord chancellor under part 1 of the legal aid, sentencing and punishment of offenders act 2012: Rule 47.18

[CPR PD 47.17]

17.1 The time for requesting a detailed assessment under rule 47.18 is within 3 months after the date when the right to detailed assessment arose.

17.2
(1) The request for a detailed assessment of costs must be in Form N258A. The request must be accompanied by—
(a) a copy of the bill of costs;
(b) the document giving the right to detailed assessment (see paragraph 13.3 above);
(c) copies of all orders made by the court relating to the costs which are to be assessed;

CPR PD 47.17 Appendix

       (d)   copies of any fee notes of counsel and any expert in respect of fees claimed in the bill;
       (e)   written evidence as to any other disbursement which is claimed and which exceeds £500;
       (f)   the legal aid certificates, LSC certificates, certificates recording the determinations of the Director of Legal Aid Casework, any relevant amendment certificates, any authorities and any certificates of discharge, revocation or withdrawal; and
       (g)   a statement signed by the legal representative giving the representative's name, address for service, reference, telephone number, fax number, e-mail address where available and, if the assisted person has a financial interest in the detailed assessment and wishes to attend, giving the postal address of that person, to which the court will send notice of any hearing.

(2) The relevant papers in support of the bill as described in paragraph 13.12 must only be lodged if requested by the costs officer.

17.3 Where the court has provisionally assessed a bill of costs it will send to the legal representative a notice, in Form N253 annexed to this practice direction, of the amount of costs which the court proposes to allow together with the bill itself. The legal representative should, if the provisional assessment is to be accepted, then complete the bill.

17.4 If the solicitor whose bill it is, or any other party wishes to make an application in the detailed assessment proceedings, the provisions of Part 23 applies.

17.5 It is the responsibility of the legal representative to complete the bill by entering in the bill the correct figures allowed in respect of each item, recalculating the summary of the bill appropriately and completing the Community Legal Service assessment certificate (Form EX80A).

Costs payable by the legal services commission or lord chancellor at prescribed rates

17.6 Where the costs of an assisted person or LSC funded client or person to whom civil legal services (within the meaning of Part 1 of the Legal Aid, Sentencing and Punishment of Offenders Act 2012) are provided under arrangements made for the purposes of that Part of that Act are payable by another person but costs can be claimed against the LSC or Lord Chancellor at prescribed rates (with or without enhancement), the solicitor of the assisted person or LSC funded client or person to whom civil legal services are provided must file a legal aid/ LSC schedule in accordance with paragraph 13.2(l) above. The schedule should follow as closely as possible Precedent E of the Schedule of Costs Precedents annexed to this Practice Direction.

17.7 The schedule must set out by reference to the item numbers in the bill of costs, all the costs claimed as payable by another person, but the arithmetic in the schedule should claim those items at prescribed rates only (with or without any claim for enhancement).

17.8 Where there has been a change in the prescribed rates during the period covered by the bill of costs, the schedule (as opposed to the bill) should be divided into separate parts, so as to deal separately with each change of rate. The schedule must also be divided so as to correspond with any divisions in the bill of costs.

17.9 If the bill of costs contains additional columns setting out costs claimed against the LSC or Lord Chancellor only, the schedule may be set out in a separate document or, alternatively, may be included in the additional columns of the bill.

17.10 The detailed assessment of the legal aid/ LSC schedule will take place immediately after the detailed assessment of the bill of costs but on occasions, the court may decide to conduct the detailed assessment of the legal aid/ LSC schedule separately from any detailed assessment of the bill of costs. This will occur, for example, where a default costs certificate is obtained as between the parties but that certificate is not set aside at the time of the detailed assessment of the legal aid costs.

17.11 Where costs have been assessed at prescribed rates it is the responsibility of the legal representative to enter the correct figures allowed in respect of each item and to recalculate the summary of the legal aid/ LSC schedule.

Detailed assessment procedure where costs are payable out of a fund other than the community legal service fund or by the lord chancellor under part 1 of the legal aid, sentencing and punishment of offenders act 2012: Rule 47.19

[CPR PD 47.18]

18.1 Rule 47.19 enables the court to direct under rule 47.19(3) that the receiving party must serve a copy of the request for assessment and copies of the documents which accompany it, on any person who has a financial interest in the outcome of the assessment.

18.2 A person has a financial interest in the outcome of the assessment if the assessment will or may affect the amount of money or property to which that person is or may become entitled out of the fund. Where an interest in the fund is itself held by a trustee for the benefit of some other person, that trustee will be treated as the person having such a financial interest unless it is not appropriate to do so. 'Trustee' includes a personal representative, receiver or any other person acting in a fiduciary capacity.

18.3 The request for a detailed assessment of costs out of the fund should be in Form N258B, be accompanied by the documents set out at paragraph 17.2(1) (a) to (e) and the following—
(a) a statement signed by the receiving party giving his name, address for service, reference, telephone number,
(b) a statement of the postal address of any person who has a financial interest in the outcome of the assessment; and
(c) if a person having a financial interest is a child or protected party, a statement to that effect.

18.4 The court will decide, having regard to the amount of the bill, the size of the fund and the number of persons who have a financial interest, which of those persons should be served and may give directions about service and about the hearing. The court may dispense with service on all or some of those persons.

18.5 Where the court makes an order dispensing with service on all such persons it may proceed at once to make a provisional assessment, or, if it decides that a hearing is necessary, give appropriate directions. Before deciding whether a hearing is necessary, the court may require the receiving party to provide further information relating to the bill.

18.6
(1) The court will send the provisionally assessed bill to the receiving party with a notice in Form N253. If the receiving party is legally represented the legal representative should, if the provisional assessment is to be accepted, then complete the bill.
(2) The court will fix a date for a detailed assessment hearing, if the receiving party informs the court within 14 days after receiving the notice in Form N253, that the receiving party wants the court to hold such a hearing.

18.7 The court will give at least 14 days notice of the time and place of the hearing to the receiving party and to any person who has a financial interest and who has been served with a copy of the request for assessment.

18.8 If any party or any person who has a financial interest wishes to make an application in the detailed assessment proceedings, the provisions of Part 23 (General Rules about Applications for Court Orders) apply.

18.9 If the receiving party is legally represented the legal representative must complete the bill by inserting the correct figures in respect of each item and must recalculate the summary of the bill.

Costs of detailed assessment proceedings – rule 47.20: Offers to settle under part 36 or otherwise

[CPR PD 47.19]

19. Where an offer to settle is made, whether under Part 36 or otherwise, it should specify whether or not it is intended to be inclusive of the cost of preparation of the bill, interest and VAT. Unless the offer states otherwise it will be treated as being inclusive of these.

Appeals from authorised court officers in detailed assessment proceedings: Rules 47.22 to 47.25

[CPR PD 47.20]

20.1 This Section relates only to appeals from authorised court officers in detailed assessment proceedings. All other appeals arising out of detailed assessment proceedings (and arising out of summary assessments) are dealt with in accordance with Part 52 and Practice Directions 52A to 52E. The destination of appeals is dealt with in accordance with the Access to Justice Act 1999 (Destination of Appeals) Order 2000.

20.2 In respect of appeals from authorised court officers, there is no requirement to obtain permission, or to seek written reasons.

20.3 The appellant must file a notice which should be in Form N161 (an appellant's notice).

20.4 The appeal will be heard by a costs judge or a district judge of the High Court, and is a re-hearing.

20.5 The appellant's notice should, if possible, be accompanied by a suitable record of the judgment appealed against. Where reasons given for the decision have been officially recorded by the court an approved transcript of that record should accompany the notice. Where there is no official record the following documents will be acceptable—

(a) the officer's comments written on the bill;
(b) advocates' notes of the reasons agreed by the respondent if possible and approved by the authorised court officer.

When the appellant was unrepresented before the authorised court officer, it is the duty of any advocate for the respondent to make a note of the reasons promptly available, free of charge to the appellant where there is no official record or if the court so directs. Where the appellant was represented before the authorised court officer, it is the duty of the appellant's own former advocate to make a note available. The appellant should submit the note of the reasons to the costs judge or district judge hearing the appeal.

20.6 Where the appellant is not able to obtain a suitable record of the authorised court officer's decision within the time in which the appellant's notice must be filed, the appellant's notice must still be completed to the best of the appellant's ability. It may however be amended subsequently with the permission of the costs judge or district judge hearing the appeal.

Schedule of costs precedents

A: Model form of bill of costs (receiving party's solicitor and counsel on conditional fee agreement terms)

B: Model form of bill of costs (detailed assessment of additional liability only)

C: Model form of bill of costs (payable by Defendant and the LSC)

D: Model form of bill of costs (alternative form, single column for amounts claimed, separate parts for costs payable by the LSC only)

Civil Procedure Rules 1998 **CPR PD 47.20**

E: Legal Aid/ LSC Schedule of Costs

F: Certificates for inclusion in bill of costs

G. Points of Dispute and Reply

H: Costs Budget

J: Solicitors Act 1974: Part 8 claim form under Part III of the Act

K: Solicitors Act 1974: order for delivery of bill

L: Solicitors Act 1974: order for detailed assessment (client)

M: Solicitors Act 1974: order for detailed assessment (solicitors)

P: Solicitors Act 1974: breakdown of costs

Schedule of costs precedents
Precedent A

IN THE HIGH COURT OF JUSTICE	2011-B-9999
QUEEN'S BENCH DIVISION	
BRIGHTON DISTRICT REGISTRY	
BETWEEN	
AB	Claimant
- and -	
CD	Defendant

Claimant's bill of costs to be assessed pursuant to the order dated 2nd April 2013

V.A.T. No 33 4404 90

In these proceedings the claimant sought compensation for personal injuries and other losses suffered in a road accident which occurred on 1st January 2011 near the junction between Bolingbroke Lane and Regency Road, Brighton, East Sussex. The claimant had been travelling as a front seat passenger in a car driven by the defendant. The claimant suffered severe injuries when, because of the defendant's negligence, the car left the road and collided with a brick wall.

The defendant was later convicted of various offences arising out of the accident including careless driving and driving under the influence of drink or drugs.

In the civil action the defendant alleged that immediately before the car journey began the claimant had known that the defendant was under the influence of alcohol and therefore consented to the risk of injury or was contributorily negligent as to it. It was also alleged that, immediately before the accident occurred, the claimant wrongfully took control of the steering wheel so causing the accident to occur.

The claimant first instructed solicitors, E F & Co, in this matter in July 2011. The claim form was issued in October 2011 and in February 2012 the proceedings were listed for a two day trial commencing 25th July 2012. At the trial the defendant was found

CPR PD 47.20 *Appendix*

liable but the compensation was reduced by 25% to take account of contributory negligence by the claimant. The claimant was awarded a total of £78,256.83 plus £1,207.16 interest plus costs.

The claimant instructed E F & Co under a retainer which specifies the following hourly rates.

Partner — £217 per hour plus VAT

Assistant Solicitor — £192 per hour plus VAT

Other fee earners — £118 per hour plus VAT

Success fees exclusive of disbursement funding costs: 40%

Success fee in respect of disbursement funding costs: 7.5% (not claimed in this bill)

Except where the contrary is stated the proceedings were conducted on behalf of the claimant by an assistant solicitor, admitted November 2008.

E F & Co instructed Counsel (Miss GH, called 1992) under a conditional fee agreement dated 5th June 2001 which specifies a success fee of 75% and base fees, payable in various circumstances, of which the following are relevant

Fees for interim hearing whose estimated duration is up to 2 hours: £600

Brief for trial whose estimated duration is 2 days: £2,000

Fee for second and subsequent days: £650 per day

Civil Procedure Rules 1998 **CPR PD 47.20**

Item No	Description of work done	V.A.T.	Disbursements	Profit Costs
	8th July 2011 – EF & Co instructed			
1	7th October 2011 – Claim issued Issue fee	—	£685.00	
2	21st October 2011 – Particulars of claim served			
	25th November 2011 – Time for service of defence extended by agreement to 14th January 2012			
3	Fee on allocation	—	£220.00	
	20th January 2012 – case allocated to multi-track			
	9th February 2012 – Case management conference at which costs were awarded to the claimant and the base costs were summarily assessed at £400 (paid on 24th February 2012)			
	23rd February 2012 – Claimant's list of documents			
	12th April 2012 – Payment into court of £25,126.33			
	13th April 2012 – Filing pre-trial check list			
4	Paid listing fee	—	£110.00	
5	Paid hearing fee		£1,090.00	
	25th July 2012 – Attending first day of trial: adjourned part heard Engaged in Court 5.00 hours	£960.00		

CPR PD 47.20 *Appendix*

Item No	Description of work done	V.A.T.	Disbursements	Profit Costs
	Engaged in conference 0.75 hours	£144.00		£1,392.00
	Travel and waiting 1.5 hours	£288.00		
6	Total solicitor's fee for attending			
7	Counsel's base fee for trial (Miss GH)	£400.00	£2,000.00	
8	Fee of expert witness (Dr. IJ)	—	£850.00	
9	Expenses of witnesses of fact	—	£84.00	
	26th July 2012 – Attending second day of trial when judgment was given for the claimant in the sum of £78,256.53 plus £1207.16 interest plus costs			
	To Summary	£ 400	£5,039.00	£1,392.00
	Engaged in Court 3.00 hours	£576.00		
	Engaged in conference 1.5 hours	£288.00		
	Travel and waiting 1.5 hours	£288.00		
10	Total solicitor's fee for attending			£1,152.00
11	Counsel's fee for second day (Miss GH)	£190.00	£950.00	
	Claimant			
12	8th July 2011 – First instructions: 0.75 hours by Partner			£162.75
	Other timed attendances in person and by telephone – See Schedule 1			
13	Total base fee for Schedule 1 – 7.5 hours			£1,440.00

Civil Procedure Rules 1998 **CPR PD 47.20**

Item No	Description of work done	V.A.T.	Disbursements	Profit Costs
14	Routine letters out and telephone calls – 29 (17 + 12) total fee			£556.80
	Witnesses of Fact			
	Timed attendances in person, by letter out and by telephone – See Schedule 2			
15	Total base fee for Schedule 2 – 5.2 hours			£998.40
16	Routine letters out, e mails and telephone calls – 8 (4 + 2 + 2)total base fee			£153.60
17	Paid travelling on 10th October 2011	£4.59	£22.96	
	Medical expert (Dr. IJ)			
18	11th September 2011 – long letter out 0.33 hours: fee			£63.36
19	31st January 2012 – long letter out 0.25 hours: fee			£48.00
20	23rd May 2012 – telephone call 0.2 hours fee			£38.40
21	Routine letters out and telephone calls – 10 (6 + 4) total fee			£192.00
22	Dr. IJ's fee for report	—	£500.00	
	Defendant and his solicitor			
23	8th July 2011 – timed letter sent 0.5 hours: fee			£96.00
24	19th February 2012 – telephone call 0.25 hours: fee			£48.00
25	Routine letters out and telephone calls – 24 (18 + 6) total fee			£460.80

CPR PD 47.20   *Appendix*

Item No	Description of work done	V.A.T.	Disbursements	Profit Costs
26	**Communications with the court** Routine letters out and telephone calls – 9 (8 + 1) total fee			£172.80
	To Summary	£194.59	£1,472.96	£5,582.91
27	**Communications with Counsel** Routine letters out, e mails and telephone calls – 19 (4 + 7 + 8) tota fee			£364.80
	Work done on documents Timed attendances – See Schedule 3			
28	Total fees for Schedule 3 – 0.75 hours at £217, 44.5 hours at £192, 12 hours at £118			£10,122.75
	Work done on negotiations 23rd March 2012 – meeting at offices of Solicitors for the Defendant			
	Engaged – 1.5 hours	£288.00		
	Travel and waiting – 1.25 hours	£240.00		
29	Total fee for meeting			£528.00
	Other work done Preparing and checking bill			
	Engaged: Solicitor – 1 hour	£192.00		
	Engaged: Costs Draftsman – 4 hours	£480.00		
30	Total fee on other work done			£672.00
31	VAT on solicitor's fees (20% of £18,662.46)	£3,731.49		

Item No	Description of work done	V.A.T.	Disbursements	Profit Costs
32	VAT on Counsel's base fees (20% of £3,950)"	£790.00		
	To Summary	£4,521.49		£11,687.55
	SUMMARY			
	Page 2	£400.00	£5,039.00	£1,392.00
	Page 3	£194.59	£1,472.96	£5,582.91
	Page 4	£4,521.49		£11,687.55
	Totals:	£5,116.08	£6,511.96	£18,662.46
	Grand total:			£30,290.50

CPR PD 47.20 *Appendix*

**Schedule of costs precedents
Precedent B**

IN THE HIGH COURT OF JUSTICE QUEEN'S BENCH DIVISION BRIGHTON DISTRICT REGISTRY BETWEEN	2000 - B - 9999
AB	Claimant
- and -	
CD	Defendant

Claimant's bill of costs to be assessed pursuant to the order dated 26th July 2001

V.A.T. NO 33 4404 90

In these proceedings the claimant sought compensation for personal injuries and other losses suffered in a road accident which occurred on Friday 1st January 1999 near the junction between Bolingbroke Lane and Regency Road, Brighton, East Sussex. The claimant had been travelling as a front seat passenger in a car driven by the defendant. The claimant suffered severe injuries when, because of the defendant's negligence, the car left the road and collided with a brick wall.

The defendant was later convicted of various offences arising out of the accident including careless driving and driving under the influence of drink or drugs.

In the civil action the defendant alleged that immediately before the car journey began the claimant had known that the defendant was under the influence of alcohol and therefore consented to the risk of injury or was contributorily negligent as to it. It was also alleged that, immediately before the accident occurred, the claimant wrongfully took control of the steering wheel so causing the accident to occur.

The claimant first instructed solicitors, E F & Co, in this matter in July 2000. The claim form was issued in October 1999 and in February 2000 the proceedings were listed for a two day trial commencing 25th July 2001. At the trial the defendant was found liable but the compensation was reduced by 25% to take account of contributory negligence by the claimant. The claimant was awarded a total of £78,256.83 plus £1,207.16 interest plus costs, and the base costs were summarily assessed

The claimant instructed E F & Co under a conditional fee agreement dated 8th July 2000 which specifies the following base fees and success fees.

Partner – £180 per hour plus VAT

Assistant Solicitor – £140 per hour plus VAT

Other fee earners – £85 per hour plus VAT

Success fees exclusive of disbursement funding costs: 40%

Success fee in respect of disbursement funding costs: 7.5% (not claimed in this bill)

Except where the contrary is stated the proceedings were conducted on behalf of the claimant by an assistant solicitor, admitted November 1999.

Civil Procedure Rules 1998 **CPR PD 47.20**

E F & Co instructed Counsel (Miss GH, called 1992) under a conditional fee agreement dated 5th June 2001 which specifies a success fee of 75% and base fees, payable in various circumstances, of which the following are relevant.

Fees for interim hearing whose estimated duration is up to 2 hours: £600

Brief for trial whose estimated duration is 2 days: £2,000

Fee for second and subsequent days: £650 per day

CPR PD 47.20 *Appendix*

Item No	Description of work done	V.A.T.	Disbursements	Profit Costs
1	8th July 2000 – EF & Co instructed 22nd July 2000 – AEI with Eastbird Legal Protection Ltd Premium for policy	—	£120.00	—
2	9th February 2001 – Case management conference at which costs were awarded to the Claimant and the base costs were summarily assessed at £400 Success fee on costs of case management conference (40% of £400) plus VAT 28th June 2001 – Pre trial review: costs in the case (base costs included base costs at trial) 25th July 2001 – First day of trial 26th July 2001 – Second day of trial at which judgment was given for the claimant as follows: Compensation: £78,256.83 Interest thereon: £1,207.16 Base costs to trial Solicitor's fees: £12,500.00 plus £2187.50 VAT thereon Counsel's fees: £3,200.00 plus £560.00 VAT thereon Other disbursements: £2,300.00 plus £4.02 VAT thereon	£28.00		£160.00

Item No	Description of work done	V.A.T.	Disbursements	Profit Costs
3	Success fee on solicitor's base costs awarded at trial (40% of £12,500) plus VAT	£875.00		£5,000.00
4	Success fee on Counsel's base costs awarded at trial (75% of £3,200) plus VAT	£420.00	£2,400.00	
	Other work done Preparing and checking bill Engaged: Solicitor – 0.25 hours £ 35.00 Engaged: Costs draftsman – 1.75 hours £ 148.75	£32.16		£183.75
5	Total base fee for other work done plus VAT	£12.87		£73.50
6	Success fee for other work done (40% of £183.75) plus VAT	£1,368.03		
	Totals: Profit Costs Disbursements VAT Grand total:		£2,520.00	£5,417.25 £5,417.25 £2,520.00 £1,368.03 £9,305.28

CPR PD 47.20 *Appendix*

Schedule of costs precedents
Precedent C

> IN THE HIGH COURT OF JUSTICE 1999 - B - 9999
> QUEEN'S BENCH DIVISION
> BRIGHTON DISTRICT REGISTRY
> BETWEEN
>
>   AB   Claimant
>   - and -
>   CD   Defendant

Claimant's bill of costs to be assessed pursuant to the order dated 26th July 2000 and in accordance with Regulation 107A of the Civil Legal Aid (General) Regulations 1989

Legal Aid Certificate No 01. 01. 99. 32552X issued on 9th September 1999.

V.A.T. No 33 4404 90

In these proceedings the claimant sought compensation for personal injuries and other losses suffered in a road accident which occurred on Friday 1st January 1999 near the junction between Bolingbroke Lane and Regency Road, Brighton, East Sussex. The claimant had been travelling as a front seat passenger in a car driven by the defendant. The claimant suffered severe injuries when, because of the defendant's negligence, the car left the road and collided with a brick wall.

The defendant was later convicted of various offences arising out of the accident including careless driving and driving under the influence of drink or drugs.

In the civil action the defendant alleged that immediately before the car journey began the claimant had known that the defendant was under the influence of alcohol and therefore consented to the risk of injury or was contributorily negligent as to it. It was also alleged that, immediately before the accident occurred, the claimant wrongfully took control of the steering wheel so causing the accident to occur.

The claimant first instructed solicitors, E F & Co, in this matter in July 1999. The claim form was issued in October 1999 and in February 2000 the proceedings were listed for a two day trial commencing 25th July 2000. At the trial the defendant was found liable but the compensation was reduced by 25% to take account of contributory negligence by the claimant. The claimant was awarded a total of £78,256.83 plus £1,207.16 interest plus costs.

The proceedings were conducted on behalf of the claimant by an assistant solicitor, admitted November 1998. The bill is divided into two parts.

Part 1
Costs payable by the defendant to the date of grant of legal aid

This covers the period from 8th July 1999 to 8th September 1999. In this part the solicitor's time is charged at £140 per hour (including travel and waiting time) and letters out and telephone calls at £14.00 each.

Part 2
Costs payable by the defendant and L.S.C. from the date of grant of legal aid

This part covers the period from 9th September 1999 to the present time, the client

Civil Procedure Rules 1998 CPR PD 47.20

having the benefit of a legal aid certificate covering these proceedings. In this part, solicitor's time in respect of costs payable by the defendant has been charged as in Part 1 plus costs draftsman's and trainee's time charged at £85 per hour. Solicitor's time in respect of costs payable by the LSC only are charged at the prescribed hourly rates plus enhancement of 50%.

Preparation: £74

Attending counsel in conference or at court: £36.40

Travelling and waiting: £32.70

Routine letters out: £7.40

Routine telephone calls: £4.10

CPR PD 47.20 *Appendix*

Item No	Description of work done	Payable by L.S.C. only		Payable by Defendant		
		V.A.T.	Disbursements	Profit Costs	V.A.T.	Disbursements	Profit Costs
	Part 1: COSTS TO DATE OF GRANT OF LEGAL AID.						
	Claimant						
1	8th July 1999 – First Instructions – 0.75 hours						£105.00
2	Routine Letters out – 3						£42.00
	Witnesses of Fact						
3	Routine Letters out – 2						£28.00
	The Defendant						
4	8th July 1999 – Timed letter sent – 0.5 hours						£70.00
5	VAT on total profit costs (17.5% of £245)				£42.88		
	To Summary				£42.88	£ -	£245.00
	Part 2: COSTS FROM DATE OF GRANT OF LEGAL AID						
	7th October 1999 – Claim issued						
6	Issue fee				—	£400.00	
	21st October 1999 – Particulars of claim served						

Civil Procedure Rules 1998 **CPR PD 47.20**

	Description of work done	Payable by L.S.C. only	Payable by Defendant
7	25th November 1999 – Time for service of defence extended by agreement to 14th January 2000		
	17th January 2000 – Filing allocation questionnaire		
	Fee on allocation	—	£80.00
	20th January 2000 – Case allocated to multi-track		
	9th February 2000 – Case management conference		
	Engaged 0.75 hours	£105.00	
	Travel and waiting 2.00 hours	£280.00	
8	Total solicitor's fee for attending		£385.00
	23rd February 2000 – Claimant's list of documents		
	12th April 2000 – Payment into court of £25,126.33		
	13th April 2000 – Filing pre-trial check list		
9	Fee on listing		—
	28th June 2000 – Pre-trial review		
	Engaged 1.5 hours	£210.00	£400.00

	Description of work done	Payable by L.S.C. only		Payable by Defendant	
	Travel and waiting 2.00 hours	£280.00			
10	Total solicitor's fee for attending				£490.00
11	Counsel's brief fee for attending pre-trial review (Miss GH)		£105.00	£600.00	£875.00
	To Summary	£ —	£ —	£105.00 £1,480.00	
	25th July 2000 – Attending first day of trial: adjourned part heard				
	Engaged in court 5.00 hours	£700.00			
	Engaged in conference 0.75 hours	£105.00			
	Travel and waiting 1.5 hours	£210.00			
12	Total solicitor's fee for attending				£1,015.00
13	Counsel's brief fee for trial (Miss GH)			£350.00 £2,000.00	
14	Fee of expert witness (Dr IJ)			— £850.00	

	Description of work done	Payable by L.S.C. only		Payable by Defendant	
15	Expenses of witnesses of fact			—	£84.00
	26th July 2000 – Attending second day of trial when judgment was given for the claimant in the sum of £78,256.83 plus £1,207.16 interest plus costs				
	Engaged in court 3.00 hours	£420.00			
	Engaged in conference 1.5 hours	£210.00			
	Travel and waiting 1.5 hours	£210.00			
16	Total solicitor's fee for attending				£840.00
17	Counsel's fee for second day (Miss GH)			£113.75	£650.00
	Claimant ~ **(1)** Payable by Defendant				
	Timed attendances in person and by telephone – see Schedule 1				
18	Total fees for Schedule 1 – 7.50 hours				£1,050.00
19	Routine letters out and telephone calls – 26 (14 + 12)				£364.00

CPR PD 47.20 *Appendix*

	Description of work done		Payable by L.S.C. only	Payable by Defendant	
	Claimant ~ (2) Payable by LSC only				
	11th September 1999 – telephone call				
	Engaged 0.25 hours	£18.50			
	Enhancement 50%	£9.25			
20	Total solicitor's fee		£27.75		
	10th April 2000 – telephone call				
	Engaged 0.1 hours	£4.10			
	Enhancement 50%	£2.05			
21	Total solicitor's fee		£6.15		
	Witnesses of fact				
	Timed attendances in person, by letter out and by telephone – see Schedule 2				
22	Total fees for Schedule 2 – 5.2 hours				£728.00
23	Routine letters out (including e mails) and telephone calls – 6 (4 + 2)				£84.00
24	Paid travelling on 9th October 1999			£4.02	£22.96

Civil Procedure Rules 1998 **CPR PD 47.20**

	Description of work done	Payable by L.S.C. only		Payable by Defendant		
		£ —	£ —	£33.90	£467.77	£3,606.96
	To Summary					£4,081.00
	Medical expert (Dr IJ)					
25	11th September 1999 – long letter out 0.33 hours					£46.20
26	30th January 2000 – long letter out 0.25 hours					£35.00
27	23rd May 2000 – telephone call 0.2 hrs				£28.00	£140.00
28	Routine letters out and telephone calls – 10 (6 + 4)					
29	Dr IJ's fee for report				—	
	Solicitors for the defendant					
30	19th February 2000 – telephone call 0.25 hours				£350.00	£35.00
31	Routine letters out and telephone calls – 24 (18 + 6)					£336.00
	Communications with the court					
32	Routine letters out and telephone calls – 9 (8 + 1)					£126.00
	Communications with Counsel					
33	Routine letters out (including e mails) and telephone calls – 19 (11 + 8)					£266.00

CPR PD 47.20 *Appendix*

	Description of work done	Payable by L.S.C. only		Payable by Defendant
	Legal Aid Board and LSC ~ Payable by LSC only 2nd August 2000 – Report on case			
	Engaged 0.5 hours	£37.00		
	Enhancement 50%	£18.50		
34	Total solicitor's fee		£55.50	
	Routine letters out and telephone calls			
	Letters out – 2	£14.80		
	Telephones call – 4	£16.40		
35	Total solicitor's fee		£31.20	
	Work done on documents Timed attendances – see Schedule 3			
36	Total fees for Schedule 3 – 45.25 hours at £140 + 12 hours at £85			£7,355.00
	Work done on negotiations 23rd March 2000 – meeting at offices of solicitors for the Defendant			
	Engaged – 1.5 hours	£210.00		
	Travel and waiting 1.25 hours	£175.00		

774

	Description of work done	Payable by L.S.C. only		Payable by Defendant	
37	Total solicitor's fee for meeting				£385.00
	Other work done ~ (1) Payable by Defendant Preparing and checking bill				
	Engaged: Solicitor – 1 hour	£140.00			
	Engaged: Costs Draftsman – 4 hours	£340.00			
38	Total on other work done (1)				£480.00
	To Summary	£ —	£ —	£86.70	£ —
				£350.00	£9,232.20
	Other work done ~ (2) Payable by LSC only Preparing and checking bill				
	Engaged: Solicitor – no claim				
	Engaged: Costs Draftsman – 1 hour	£74.00			
39	Total on other work done (2)	£74.00			

CPR PD 47.20 *Appendix*

	Description of work done	Payable by L.S.C. only	Payable by Defendant	
40	VAT on total profit costs payable by Defendant (17.5% of £14,176.20)			£ —
41	VAT on total profit costs payable by LSC only (17.5% of £205.60)	£35.98		
	To summary	£35.98 £ —	£74.00 £2,480.84	£ —
	SUMMARY			£245.00
	Part 1 – Pre Legal Aid			
	Page 3	£ —	£ — £42.88	£ —
	Part 2 – Costs since grant of legal aid			
	Page 3	£ —	£ — £105.00	£1,480.00 £875.00
	Page 4	£ —	£33.90 £467.77	£3,606.96 £4,081.00
	Page 5	£ —	£86.70 £ —	£350.00 £9,232.20
	Page 6	£35.98 £ —	£74.00 £2,480.84	£ —
	Totals	£35.98 £ —	£194.60 £3,096.49	£5,436.96 £14,433.20
	Grand totals			
	Costs payable by Defendant			£22,966.65
	Costs payable by LSC only			£230.58
	Grand total:			£23,197.23

Civil Procedure Rules 1998 **CPR PD 47.20**

**Schedule of costs precedents
Precedent D**

IN THE HIGH COURT OF JUSTICE	1999 – B – 9999
QUEEN'S BENCH DIVISION	
BRIGHTON DISTRICT REGISTRY	
BETWEEN	
AB	Claimant
- and -	
CD	Defendant

Claimant's bill of costs to be assessed pursuant to the order dated 26th July 2000 and in accordance with Regulation 107A of the Civil Legal Aid (General) Regulations 1989

Legal Aid Certificate No 01. 01. 99. 32552X issued on 9th September 1999.

V.A.T. No 33 4404 90

In these proceedings the claimant sought compensation for personal injuries and other losses suffered in a road accident which occurred on Friday 1st January 1999 near the junction between Bolingbroke Lane and Regency Road, Brighton, East Sussex. The claimant had been travelling as a front seat passenger in a car driven by the defendant. The claimant suffered severe injuries when, because of the defendant's negligence, the car left the road and collided with a brick wall.

The defendant was later convicted of various offences arising out of the accident including careless driving and driving under the influence of drink or drugs.

In the civil action the defendant alleged that immediately before the car journey began the claimant had known that the defendant was under the influence of alcohol and therefore consented to the risk of injury or was contributorily negligent as to it. It was also alleged that, immediately before the accident occurred, the claimant wrongfully took control of the steering wheel so causing the accident to occur.

The claimant first instructed solicitors, E F & Co, in this matter in July 1999. The claim form was issued in October 1999 and in February 2000 the proceedings were listed for a two day trial commencing 25th July 2000. At the trial the defendant was found liable but the compensation was reduced by 25% to take account of contributory negligence by the claimant. The claimant was awarded a total of £78,256.83 plus £1,207.16 interest plus costs.

The proceedings were conducted on behalf of the claimant by an assistant solicitor, admitted November 1998. The bill is divided into three parts.

Part 1
Costs payable by the defendant to the date of grant of legal aid

This covers the period from 8th July 1999 to 8th September 1999. In this part the solicitor's time is charged at £140 per hour (including travel and waiting time) and letters out and telephone calls at £14.00 each.

Part 2
Costs payable by the defendant from the date of grant of legal aid

This part covers the period from 9th September 1999 to the present time, the client

CPR PD 47.20 *Appendix*

having the benefit of a legal aid certificate covering these proceedings. In this part, solicitor's time in respect of costs payable by the defendant has been charged as in Part 1 plus costs draftsman's and trainee's time charged at £85 per hour.

Part 3
Costs payable by the LSC only

This part covers the same period as Part 2. In this part solicitor's time in respect of costs payable by the LSC only are charged at the prescribed hourly rates plus enhancement of 50%.

Preparation: £74

Attending counsel in conference or at court: £36.40

Travelling and waiting: £32.70

Routine letters out: £7.40

Routine telephone calls: £4.10

Civil Procedure Rules 1998 **CPR PD 47.20**

Item No	Item	Amount claimed	VAT	Amount allowed	VAT
	Part 1: COSTS PAYABLE BY THE DEFENDANT				
	Claimant				
1	8th July 1999 – First Instructions – 0.75 hours	£105.00	£18.38		
2	Routine Letters out – 3	£42.00	£7.35		
	Witnesses of Fact				
3	Routine Letters out – 2	£28.00	£4.90		
	The Defendant				
4	8th July 1999 – Timed letter sent – 0.5 hours	£70.00	£12.25		
	To Summary	£245.00	£42.88		
	Part 2: COSTS PAYABLE BY THE DEFENDANT				
	7th October 1999 – Claim issued				
5	Issue fee	£400.00	—		
	21st October 1999 – Particulars of claim served				
	25th November 1999 – Time for service of defence extended by agreement to 14th January 2000				
	17th January 2000 – Filing allocation questionnaire				
6	Fee on allocation	£80.00	—		

779

CPR PD 47.20 *Appendix*

Item No	Item	Amount claimed	VAT	Amount allowed	VAT
	20th January 2000 – Case allocated to multi-track				
	9th February 2000 – Case management conference				
	Engaged 0.75 hours	£105.00			
	Travel and waiting 2.00 hours	£280.00			
7	Total solicitor's fee for attending	£385.00	£67.38		
	23rd February 2000 – Claimant's list of documents				
	12th April 2000 – Payment into court of £25,126.33				
	13th April 2000 – Filing pre-trial check list				
8	Fee on listing	£400.00	—		
	28th June 2000 – Pre-trial review				
	Engaged 1.5 hours	£210.00			
	Travel and waiting 2.00 hours	£280.00			
9	Total solicitor's fee for attending	£490.00	£85.75		
10	Counsel's brief fee for attending pre-trial review (Miss GH)	£600.00	£105.00		

Item No	Item	Amount claimed	VAT	Amount allowed	VAT
	25th July 2000 – Attending first day of trial: adjourned part heard				
	Engaged in court 5.00 hours	£700.00			
	Engaged in conference 0.75 hours	£105.00			
	Travel and waiting 1.5 hours	£210.00			
11	Total solicitor's fee for attending	£1,015.00	£177.63		
12	Counsel's brief fee for trial (Miss GH)	£2,000.00	£350.00		
13	Fee of expert witness (Dr IJ)	£850.00			
14	Expenses of witnesses of fact	£84.00			
	26th July 2000 – Attending second day of trial when judgment "was given for the claimant in the sum of £78,256.83 plus" "£1,207.16 interest plus costs"				
	Engaged in court 3.00 hours	£420.00			
	Engaged in conference 1.5 hours	£210.00			
	Travel and waiting 1.5 hours	£210.00			

Item No	Item	Amount claimed	VAT	Amount allowed	VAT
15	Total solicitor's fee for attending	£840.00	£147.00		
16	Counsel's fee for second day (Miss GH)	£650.00	£113.75		
	Claimant				
	Timed attendances in person and by telephone – see Schedule 1				
17	Total fees for Schedule 1 – 7.50 hours	£1,050.00	£183.75		
18	Routine letters out and telephone calls – 26 (14 + 12)	£364.00	£63.70		
	Witnesses of fact				
	"Timed attendances in person, by letter out and by"				
	telephone – see Schedule 2				
19	Total fees for Schedule 2 – 5.2 hours	£728.00	£127.40		
20	Routine letters out (including e mails) and telephone calls – 6 (4 + 2)	£84.00	£14.70		
21	Paid travelling on 9th October 1999	£22.96	£4.02		
	Medical expert (Dr IJ)				
22	11th September 1999 – long letter out 0.33 hours	£46.20	£8.09		

Civil Procedure Rules 1998 **CPR PD 47.20**

Item No	Item	Amount claimed	VAT	Amount allowed	VAT
23	30th January 2000 – long letter out 0.25 hours	£35.00	£6.13		
24	23rd May 2000 – telephone call 0.2 hours	£28.00	£4.90		
25	Routine letters out and telephone calls – 10 (6 + 4)	£140.00	£24.50		
26	Dr IJ's fee for report	£350.00	—		
	Solicitors for the defendant				
27	19th February 2000 – telephone call 0.25 hours	£35.00	£6.13		
28	Routine letters out and telephone calls – 24 (18 + 6)	£336.00	£58.80		
	Communications with the court				
29	Routine letters out and telephone calls – 9 (8 + 1)	£126.00	£22.05		
	Communications with Counsel				
30	Routine letters out (including e mails) and telephone calls – 19 (11 + 8)	£266.00	£46.55		
	Work done on documents				
	Timed attendances – see Schedule 3				

783

CPR PD 47.20 *Appendix*

Item No	Item		Amount claimed	VAT	Amount allowed	VAT
31	Total fees for Schedule 3 – 45.25 hours at £140 + 12 hours at £85		£7,355.00	£1,287.13		
	Work done on negotiations					
	23rd March 2000 – meeting at offices of solicitors for the Defendant					
	Engaged – 1.5 hours	£210.00				
	Travel and waiting – 1.25 hours	£175.00				
32	Total solicitor's fee for meeting		£385.00	£67.38		
33	**Other work done**					
	Preparing and checking bill					
	Engaged: Solicitor – 1 hour	£140.00				
	Engaged: Costs Draftsman 4 hours	£340.00				
	Total on other work done		£480.00	£84.00		
	To summary		£19,625.16	£3,055.70		
	Part 3: COSTS PAYABLE BY LSC ONLY					
	Claimant					
	11th September 1999 – telephone call					
	Engaged 0.25 hours	£18.50				
	Enhancement 50%	£9.25				
34	Total solicitor's fee		£27.75	£4.86		

Item No	Item	Amount claimed		VAT	Amount allowed	VAT
	10th April 2000 – telephone call					
	Engaged 0.1 hours	£4.10				
	Enhancement 50%	£2.05				
35	Total solicitor's fee		£6.15	£1.08		
	Legal Aid Board and LSC					
	2nd August 2000 – Report on case					
	Engaged 0.5 hours	£37.00				
	Enhancement 50%	£18.50				
36	Total solicitor's fee		£55.50	£9.71		
	Routine letters out and telephone calls					
	Letters out – 2	£14.80				
	Telephone calls – 4	£16.40				
37	Total solicitor's fee		£31.20	£5.46		
	Other work done					
	Preparing and checking bill					
	Engaged: Solicitor – no claim					
	Engaged: Costs Draftsman – 1 hours	£74.00				
38	Total on other work done		£74.00	£12.95		
	To summary		194.60	34.06		
	SUMMARY					
	Costs payable by the Defendant					

Civil Procedure Rules 1998 **CPR PD 47.20**

CPR PD 47.20 *Appendix*

Item No	Item		Amount claimed	VAT	Amount allowed	VAT
		Part 1	£245.00	£42.88		
		Part 2	£19,625.16	£3,055.70		
	Total costs payable by the Defendant		£19,870.16	£3,098.58		
	Costs payable by LSC only					
	Part 3		£194.60	£34.06		
	Grand Totals					
	Costs payable by the Defendant		£19,870.16	£3,098.58		
	Costs payable by LSC only		£194.60	£34.06		
	Grand total		£20,064.76	£3,132.63		

Schedule of costs precedents
Precedent E

Legal Aid/LSC Schedule of Costs
IN THE HIGH COURT OF JUSTICE   1999 - B - 9999
QUEEN'S BENCH DIVISION
BRIGHTON DISTRICT REGISTRY
BETWEEN
           AB   Claimant
           - and -
           CD   Defendant

CPR PD 47.20 *Appendix*

Claimant's bill of costs: Legal Aid/LSC Schedule

Item No	Description of work done		V.A.T.	Disbursements	Profit Costs
6	Issue fee		—	£400.00	
7	Allocation fee		—	£80.00	
8	Solicitor's fee for hearing				
	Engaged 0.75 hours	£55.50			
	Enhancement thereon at 50%	£27.75			
	Travel and waiting 2.00 hours	£65.40			
	Total solicitor's fee for attending				£148.65
9	Fee on listing		—	£400.00	
10	Solicitor's fee for hearing				
	Engaged 1.5 hours	£111.00			
	Enhancement thereon at 50%	£55.50			
	Travel and waiting 2.00 hours	£65.40			
	Total solicitor's fee for attending				£231.90
11	Counsel's fee		£105.00	£600.00	
12	Solicitor's fee for trial				
	Engaged in court 5.00 hours	£182.00			
	Engaged in conference 0.75 hours	£27.30			
	Enhancement thereon at 50%	£104.65			
	Travel and waiting 1.50 hours	£49.05			
	Total solicitor's fee for attending				£363.00
13	Counsel's brief fee for trial		£350.00	£2,000.00	
14	Expert's fee for trial		—	£850.00	

Civil Procedure Rules 1998 **CPR PD 47.20**

Item No	Description of work done		V.A.T.	Disbursements	Profit Costs
15	Witnesses' expenses			£84.00	
	To summary		£455.00	£4,414.00	£743.55
16	Solicitor's fee for trial (second day)				
	Engaged in court 3.00 hours	£109.20			
	Engaged in conference 1.50 hours	£54.60			
	Enhancement thereon at 50%	£81.90			
	Travel and waiting 1.50 hours	£49.05			
	Total solicitor's fee for attending				£294.75
17	Counsel's fee for second day of trial		£113.75	£650.00	
18	Timed attendances on Claimant (1) 7.5 hours	£555.00			
	Enhancement thereon at 50%	£277.50			£832.50
19	Routine communications with Claimant (1)				
	Letters out – 14	£103.60			
	Telephone calls – 12	£49.20			£152.80
22	Timed attendances on and communications with witnesses of fact 5.2 hours	£384.80			
	Enhancement thereon at 50%	£192.40			
23	Routine communications with witnesses of fact				£577.20

CPR PD 47.20 *Appendix*

Item No	Description of work done		V.A.T.	Disbursements	Profit Costs
	Letters out – 4	£29.60			£37.80
	Telephone calls – 2	£8.20			
24	Paid travelling		£4.02	£22.96	
25	Timed attendance on medical expert				
	0.33 hours	£24.42			£36.63
	Enhancement thereon at 50%	£12.21			
26	Timed communications with medical expert				
	0.25 hours	£18.50			£27.75
	Enhancement thereon at 50%	£9.25			
27	Timed communications with medical expert				
	0.2 hours	£14.80			£22.20
	Enhancement thereon at 50%	£7.40			
28	Routine communications with medical expert				
	Letters out – 6	£44.40			£60.80
	Telephone calls – 4	£16.40			
29	Expert's fee for report		—	£350.00	
30	Timed communications with solicitors for Defendant				
	0.25 hours	£18.50			£27.75
	Enhancement thereon at 50%	£9.25			
31	Routine communications with solicitors for Defendant				
	Letters out – 18	£133.20			

Civil Procedure Rules 1998 **CPR PD 47.20**

Item No	Description of work done		V.A.T.	Disbursements	Profit Costs
32	Telephone calls – 6	£24.60			£157.80
	Routine communications with the court				
	Letters out – 8	£59.20			
	Telephone calls – 1	£4.10			£63.30
	To summary		£117.77	£1,022.96	£2,291.28
33	Routine communications with Counsel				
	Letters out – 11	£81.40			
	Telephone calls – 8	£32.80			£114.20
36	Work done on documents				
	57.25 hours	£4,236.50			
	Enhancement thereon at 50%	£2,118.25			£6,354.75
37	Work done on negotiations				
	Engaged – 1.5 hours	£111.00			
	Enhancement thereon at 50%	£55.50			
	Travel and waiting – 1.25 hours	£40.88			£207.38
38	Other work done (1)				
	Preparing and checking bill				£370.00
40	VAT on total profit costs set out above (17.5% of £10,216.86)		£1,787.95		
	To summary		£1,787.95	£ —	£7,046.33
	SUMMARY				
	Page 1		£455.00	£4,414.00	£743.55
	Page 2		£117.77	£1,022.96	£2,291.28
	Page 3		£1,787.95	£ —	£7,046.33

CPR PD 47.20 *Appendix*

Item No	Description of work done	V.A.T.	Disbursements	Profit Costs
	Totals:	£2,360.72	£5,436.96	£10,081.16
	Grand total:			£17,878.84

Civil Procedure Rules 1998 **CPR PD 47.20**

Schedule of costs precedents
Precedent F: Certificates for inclusion in bill of costs

- Appropriate certificates under headings (1) and (2) are required in all cases. The appropriate certificate under (3) is required in all cases in which the receiving party is an assisted person or a LSC funded client. Certificates (4), (5) and (6) are optional. Certificate (6) may be included in the bill, or, if the dispute as to VAT recoverability arises after service of the bill, may be filed and served as a supplementary document amending the bill under paragraph 39.10 of Practice Direction 47.
- All certificates must be signed by the receiving party or by his solicitor. Where the bill claims costs in respect of work done by more than one firm of solicitors, certificate (1), appropriately completed, should be signed on behalf of each firm.

(1) Certificate as to accuracy

I certify that this bill is both accurate and complete [and]

☐ *(where the receiving party was funded by legal aid)*

[in respect of Part(s) .... of the bill] all the work claimed was done pursuant to a certificate/contract issued by the Legal Services Board/ LSC granted to the assisted person.

☐ *(where costs are claimed for work done by an employed solicitor)*

[in respect of Part(s) .... of the bill] the case was conducted by a legal representative who is an employee of the receiving party.

☐ *(other cases where costs are claimed for work done by a solicitor)*

[in respect of Part(s) .... of the bill] the costs claimed herein do not exceed the costs which the receiving party is required to pay me/my firm.

(2) Certificate as to interest and payments

I certify that:

☐ No rulings have been made in this case which affects my/the receiving party's entitlement (if any) to interest on costs.

or

793

CPR PD 47.20   *Appendix*

☐   The only rulings made in this case as to interest are as follows:

[*give brief details as to the date of each ruling, the name of the Judge who made it and the text of the ruling*]
and

☐   No payments have been made by any paying party on account of costs included in this bill of costs.

or

☐   The following payments have been made on account of costs included in this bill of costs:

[*give brief details of the amounts, the dates of payment and the name of the person by or on whose behalf they were paid*]

(3) Certificate as to interest of assisted person/ LSC funded client pursuant to Regulation 119 of the Civil Legal Aid (General) Regulations 1989

I certify that the Legally Aided person/LSC funded client has no financial interest in the detailed assessment.

or

I certify that a copy of this bill has been sent to the Legally Aided person/LSC funded client pursuant to Regulation 119 of the Civil Legal Aid General Regulations 1989 with an explanation of his/her interest in the detailed assessment and the steps which can be taken to safeguard that interest in the assessment. He/she has/has not requested that the costs officer be informed of his/her interest and has/has not requested that notice of the detailed assessment hearing be sent to him/her.

(4) Consent to the signing of the certificate within 21 days of detailed assessment pursuant to Regulation 112 and 121 of the Civil Legal Aid (General) Regulations 1989

I certify that notice of the fees reduced or disallowed on detailed assessment has been given in writing to counsel on [date].

or

I certify that: there having been no reduction or disallowance of counsel's fees it is not necessary to give notice to counsel.

I/we consent to the final costs certificate being issued immediately.

(5) Certificate in respect of disbursements not exceeding £500

I hereby certify that all disbursements listed in this bill which individually do not exceed £500 (other than those relating to counsel's fees) have been duly discharged.

(6) Certificate as to recovery of VAT

With reference to the pending assessment of the [claimant's/defendant's] costs and

Civil Procedure Rules 1998 **CPR PD 47.20**

disbursements herein which are payable by the [claimant/defendant] we the undersigned [legal representative of] [auditors of] the [claimant/defendant] hereby certify that the [claimant/defendant] on the basis of its last completed VAT return [would/would not be entitled to recover would/be entitled to recover only percent of the] Value Added Tax on such costs and disbursements, as input tax pursuant to section 24 of the Value Added Tax Act 1994.

Schedule of costs precedents
Precedent G

IN THE HIGH COURT OF JUSTICE   2000 – B – 9999
QUEEN'S BENCH DIVISION
OXBRIDGE DISTRICT REGISTRY
BETWEEN
     WX   Claimant
     - and -
     YZ   Defendant

Points of dispute served by the defendant

Point 1 General point	Rates claimed for the assistant solicitor and other fee earners are excessive. Reduce to £158 and £116 respectively plus VAT.
	Receiving Party's Reply:
	Costs Officer's decision:
Point 2 Point of principle	The claimant was at the time a child/protected person/insolvent and did not have the capacity to authorise the solicitors to bring these proceedings.
	Receiving Party's Reply:
	Costs Officer's decision:
Point 3 (6), (12), (17), (23), (29), (32)	(i) The number of conferences with counsel is excessive and should be reduced to 3 in total (9 hours). (ii) There is no need for two fee earners to attend each conference. Limit to one assistant solicitor in each case.
	Receiving Party's Reply:
	Costs Officer's decision:
Point 4 (42)	The claim for timed attendances on claimant (schedule 1) is excessive. Reduce to 4 hours.
	Receiving Party's Reply:
	Costs Officer's decision:
Point 5 (47)	The total claim for work done on documents by the assistant solicitor is excessive. A reasonable allowance in respect of documents concerning court and counsel is 8 hours, for documents concerning witnesses and the expert witness 6.5 hours, for work done on arithmetic 2.25 hours and for other documents 5.5 hours. Reduce to 22.25 hours.

CPR PD 47.20 *Appendix*

	Receiving Party's Reply:
	Costs Officer's decision:
Point 6 (50)	The time claimed for preparing and checking the bill is excessive. Reduce solicitor's time to 0.5 hours and reduce the costs draftsman's time to three hours.
	Receiving Party's Reply:
	Costs Officer's decision:

Served on . . . . [date] by. . . . . [name] [legal representative of] the Defendant.

Civil Procedure Rules 1998 **CPR PD 47.20**

Schedule of costs precedents
Precedent H

PRECEDENT H

In the: [to be completed]
Parties: [to be completed]
Claim number: [to be completed]

Costs budget of [Claimant / Defendant] dated [ ]

Work done / to be done	Assumptions	Incurred		Estimated		Total (£)
		Disbursements (£)	Time costs (£)	Disbursements (£)	Time costs (£)	
Pre-action costs						
Issue / pleadings						
CMC						
Disclosure						
Witness statements						
Expert reports						
PTR						
Trial preparation						
Trial						
ADR / Settlement discussions						
Contingent cost A: [explanation]						
Contingent cost B: [explanation]						
Contingent cost [explanation]						
GRAND TOTAL (including both incurred costs and estimated costs)						

This estimate excludes VAT (if applicable), success fees and ATE insurance premiums (if applicable), costs of detailed assessment, costs of any appeals, costs of enforcing any judgment and [complete as appropriate]

[Statement of truth]

Signed _____ Date

Position

www.justice.gov.uk   © Crown Copyright. Published by LexisNexis 2013 under the Open Government Licence

Page 1 of 5

797

CPR PD 47.20 *Appendix*

In the: [to be completed]
Parties: [to be completed]
Claim number: [to be completed]

	RATE (per hour)	PRE-ACTION COSTS			ISSUE / STATEMENTS OF CASE				CMC			
		Incurred costs	Estimated costs		TOTAL	Incurred costs	Estimated costs		TOTAL	Incurred costs	Estimated costs		TOTAL
		£	Hours	£		£	Hours	£		£	Hours	£	
Fee earners' timecosts (fee earner description)													
1													
2													
3													
4													
5 Total Profit Costs (1 to 4)													
Expert's costs													
6 Fees													
7 Disbursements													
Counsel's fees [indicate seniority]													
8 Leading counsel													
9 Junior counsel													
10 Court fees													
11 Other Disbursements													
12 Explanation of disbursements [details to be completed]													
13 Total Disbursements (6 to 11)													
14 Total (5 + 13)													

In the: [to be completed]
Parties: [to be completed]
Claim number: [to be completed]

	RATE (per hour)	DISCLOSURE				WITNESS STATEMENTS				EXPERT REPORTS			
		Incurred costs	Estimated costs		TOTAL	Incurred costs	Estimated costs		TOTAL	Incurred costs	Estimated costs		TOTAL
		£	Hours	£		£	Hours	£		£	Hours	£	
Fee earners' timecosts (fee earner description)													
1													
2													
3													
4													
5 Total Profit Costs (1 to 4)													
Expert's costs													
6 Fees													
7 Disbursements													
Counsel's fees [indicate seniority]													
8 Leading counsel													
9 Junior counsel													
10 Court fees													
11 Other Disbursements													
12 Explanation of disbursements [details to be completed]													
13 Total Disbursements (6 to 11)													
14 Total (5 + 13)													

CPR PD 47.20 *Appendix*

In the: [to be completed]
Parties: [to be completed]
Claim number: [to be completed]

	RATE (per hour)	PTR			TRIAL PREPARATION			TRIAL		
		Incurred costs	Estimated costs		TOTAL	Incurred costs	Estimated costs		TOTAL	Incurred costs	Estimated costs		TOTAL
		£	Hours	£		£	Hours	£		£	Hours	£	
Fee earners' timecosts (fee earner description)													
1													
2													
3													
4													
5 Total Profit Costs (1 to 4)													
Expert's costs													
6 Fees													
7 Disbursements													
Counsel's fees [indicate seniority]													
8 Leading counsel													
9 Junior counsel													
10 Court fees													
11 Other Disbursements													
12 Explanation of disbursements [details to be completed]													
13 Total Disbursements (6 to 11)													
14 Total (5 + 13)													

Civil Procedure Rules 1998 **CPR PD 47.20**

In the: [to be completed]
Parties: [to be completed]
Claim number: [to be completed]

	RATE (per hour)	SETTLEMENT / ADR			CONTINGENT COST A: [EXPLAIN]				CONTINGENT COST B: [EXPLAIN]			
		Incurred costs	Estimated costs		TOTAL	Incurred costs	Estimated costs		TOTAL	Incurred costs	Estimated costs		TOTAL
Fees		£	Hours	£		£	Hours	£		£	Hours	£	
Fee earners' timecosts (fee earner description)													
1													
2													
3													
4													
5 **Total Profit Costs (1 to 4)**													
Expert's costs													
6 Fees													
7 Disbursements													
Counsel's fees [indicate seniority]													
8 Leading counsel													
9 Junior counsel													
10 Court fees													
11 Other Disbursements													
12 Explanation of disbursements [details to be completed]													
13 Total Disbursements (6 to 11)													
14 Total (5 + 13)													

801

CPR PD 47.20 *Appendix*

1. This is the form on which you should set out your budget of anticipated costs in accordance with CPR Part 3 and Practice Directions 3E and 3F.
2. This table identifies where within the budget form the various items of work, **in so far as they are required by the circumstances of your case**, should be included. Allowance must be made in each phase for advising the client, taking instructions and corresponding with the other party/parties and the court in respect of matters falling within that phase.

Phase	Includes	Does NOT include
Pre-action	• Pre-Action Protocol correspondence	• Any work already incurred in relation to any other phase of the budget
	• Investigating the merits of the claim and advising client	
	• Considering ADR, advising on settlement and Part 36 offers	
	• All other steps taken and advice given preaction	
Statements of case	• Preparation of Claim Form	• Amendments to statements of case (see below)
	• Issue and service of proceeding	
	• Preparation of Particulars of Claim, Defence, Reply, including taking instructions, instructing counsel and any necessary investigation	
	• Considering opposing statements of case and advising client	
	• Part 18 requests (request and answer)	
	• Any conferences with counsel primarily relating to statements of case	
CMC	• Completion of AQs	• Subsequent CMCs
	• Arranging a CMC	
	• Preparation of costs budget for first CMC and reviewing opponent's budget	
	• Correspondence with opponents to agree directions and budgets, where possible	
	• Preparation for, and attendance at, the CMC	
	• Finalising the order	
Disclosure	• Obtaining documents from client and advising on disclosure obligations	• Applications for specific disclosure

Phase	Includes	Does NOT include
	• Reviewing documents for disclosure, preparing disclosure report or questionnaire response and list	• Applications and requests for third party disclosure
	• Inspection	
	• Reviewing opponent's list and documents, undertaking any appropriate investigations	
	• Correspondence between parties about the scope of disclosure and queries arising	
	• Consulting counsel, so far as appropriate, in relation to disclosure	
Witness Statements	• Identifying witnesses	• Arranging for witnesses to attend trial (include in trial preparation)
	• Obtaining statements	
	• Preparing witness summaries	
	• Consulting counsel, so far as appropriate, about witness statements	
	• Reviewing opponent's statements and undertaking any appropriate investigations	
	• Applications for witness summaries	
Expert Reports	• ☐Identifying and engaging suitable expert(s)	• Obtaining permission to adduce expert evidence (include in CMC or as separate application)
	• Reviewing draft and approving report(s)	
	• Dealing with follow-up questions of experts	• ☐Arranging for experts to attend trial (include in trial preparation)
	• Considering opposing experts' reports	
	• Meetings of experts (preparing agenda etc)	
PTR	• Bundle	• Assembling and/or copying the bundle (this is not fee earners' work)
	• ☐Preparation of updated costs budgets and reviewing opponent's budget	
	• Preparing and agreeing chronology, case summary and dramatis personae (if ordered and not already prepared earlier in case)	
	• Completing and filing pre-trial checklists	

803

CPR PD 47.20   *Appendix*

Phase	Includes	Does NOT include
Trial Preparation	• Correspondence with opponents to agree directions and costs budgets, if possible • Attendance at the PTR • ☐Trial bundles • Witness summonses, and arranging for witnesses to attend trial • Any final factual investigations • Supplemental disclosure and statements (if required) • Agreeing brief fee • Any pre trial conferences and advice from Counsel • Pre-trial liaison with witnesses	• Assembling and/or copying the trial bundle (this is not fee earners' work) • Counsel's brief fee and any refreshers
Trial	• ☐Solicitors' attendance at trial • All conferences and other activity outside court hours during the trial • Attendance on witnesses during the trial • Counsel's brief fee and any refreshers • Dealing with draft judgment and related applications	• Preparation for trial • Agreeing brief fee
Settlement	• ☐Settlement negotiations, including Part 36 and other offers and advising the client • Drafting settlement agreement or Tomlin order • Advice to the client on settlement (excluding advice included in the pre-action phase)	• Mediation (should be included as a contingency)

3. The 'contingent cost' sections of this form should be used for **anticipated costs** which do not fall within the main categories set out in this form. Examples might be the trial of preliminary issues, a mediation, applications to amend, applications for disclosure against third parties or (in libel cases) applications re meaning. **Costs which are not anticipated** but which become necessary later are dealt with in paragraph 4.7 of the Practice Direction.

4. Any party may apply to the court if it considers that another party is behaving oppressively in seeking to cause the applicant to spend money disproportionately on costs and the court will grant such relief as may be appropriate.

Civil Procedure Rules 1998 **CPR PD 47.20**

Schedule of costs precedents
Precedent J

IN THE HIGH COURT OF JUSTICE   Claim No
SUPREME COURT COSTS OFFICE
IN THE MATTER OF [name of solicitor or solicitors' firm]
Claimant
Defendant(s)

Claim form (CPR Part 8)

Details of claim (see also overleaf)

The following orders are applied for:

() An order in standard form for the delivery of a bill of costs in [all cases and matters] [the following causes and matters. . . . .] in which
the Defendant has acted for the Claimant(s).
() An order in standard form for the detailed assessment of the bill(s) dated . . . . [and] [bearing the invoice numbers . . . .
delivered by the [claimant/Defendant] to the [Defendant/Claimant/person named . . . .]
() An order dealing with the costs of this application

Defendant's name and   £
address
        Court fee
        Solicitor's costs
        Issue date

Claim No	

Details of claim (continued)

Statement of Truth

*(I believe) (The Claimant believes) that the facts stated in these particulars of claim are true. *I am duly authorised by the Claimant to sign this statement.

Full name . . . .   Name of Claimant's solicitor's firm

Signed . . . . position or office held . . . . *(Claimant) (Litigation friend) (Claimant's solicitor) (if signing on behalf of firm or company)

*delete as appropriate

Claimant's or claimant's solicitor's address to which documents should be sent if different from overleaf. If you are prepared to accept service by DX, fax or e-mail, please add details.

CPR PD 47.20 *Appendix*

Schedule of costs precedents
Precedent K: Order for delivery of a Bill

Solicitor's Act: order for delivery of bill

DATED the [DATE]

IN THE HIGH COURT OF JUSTICE [Claim No]

[DIVISION]

[JUDGE TYPE] [JUDGE NAME]

[CLAIMANT]

BETWEEN:

Claimant

- and -

[DEFENDANT]

Defendant

UPON THE APPLICATION OF THE [PARTY]

[the parties and their representatives who attended]

AND UPON HEARING

AND UPON READING the documents on the Court File

IT IS ORDERED THAT

(1) The [PARTY] must within [NUMBER OF DAYS] deliver to the [PARTY] or to his solicitor, a bill of costs in all causes and matters in which he has been concerned for the [PARTY]

(2) The [PARTY] must give credit in that bill for all money received by him from or on account of the [PARTY]

Civil Procedure Rules 1998 **CPR PD 47.20**

Schedule of costs precedents
Precedent L: Order for Detailed Assessment (Client)

Order on Client's Application for Detailed Assessment of Solicitor's Bill

DATED the

IN THE HIGH COURT OF JUSTICE

BETWEEN:

Claimant

- and -

Defendant

UPON THE APPLICATION OF THE [PARTY]

[the parties and their representatives who attended]

AND UPON HEARING

AND UPON READING the documents on the Court File

IT IS ORDERED THAT

(1) A detailed assessment must be made of the bill dated [   ] delivered to the claimant by the defendant.

(2) On making the detailed assessment, the court must also assess the costs of these proceedings and certify what is due to or from either party in respect of the bill and the costs of these proceedings.

(3) Until these proceedings are concluded the defendant must not commence or continue any proceedings against the claimant in respect of the bill mentioned above.

(4) Upon payment by the claimant of any sum certified as due to the defendant in these proceedings the defendant must deliver to the claimant all the documentation in the defendant's possession or control which belong to the claimant.

CPR PD 47.20 *Appendix*

Schedule of costs precedents
Precedent M: Order for Detailed Assessment (Solicitor)

SCHEDULE OF COSTS PRECEDENTS
PRECEDENT M
Order on Solicitor's Application for Assessment Under the Solicitor's Act 1974 Part III

Upon hearing ... upon reading ...

IT IS ORDERED THAT

(1) A detailed assessment must be made of the bill dated [   ] delivered to the defendant by the claimant.

(2) If the defendant attends the detailed assessment the court making that assessment must also assess the costs of these proceedings and certify what is due to or from either party in respect of the bill and the costs of these proceedings.

(3) Until these proceedings are concluded the claimant must not commence or continue any proceedings against the defendant in respect of the bill mentioned above.

(4) Upon payment by the defendant of any sum certified as due to the claimant in these proceedings the claimant must deliver to the defendant all the documentation in the claimant's possession or control which belong to the defendant.

© Crown Copyright. Reproduced by permission of the Controller of Her Majesty's Stationery Office. Published by LexisNexis Butterworths.

Civil Procedure Rules 1998 **CPR PD 47.20**

Schedule of costs precedents
Precedent P: Solicitors Act: Breakdown of Costs

IN THE HIGH COURT OF JUSTICE   Claim Number
QUEEN'S BENCH DIVISION
SUPREME COURT COSTS OFFICE
BETWEEN
      EF   Claimant
      - and -
      GH & Co   Defendants

Breakdown of defendant's bill of costs dated 27th February 2012
to be assessed pursuant to the order dated 27th April 2012

The claimant instructed the defendants in connection with a summons for careless or inconsiderate driving which had been served upon him. By letter dated 21st October 2011 the defendants wrote to the claimant setting out their terms of business including the hourly rates of the fee earners who would act on his instructions. On 23rd October 2011 the claimant dated and signed a copy of that letter and returned it to the defendants so indicating his acceptance of the terms set out.

The proceedings were of the highest importance to the claimant who feared losing his licence and who wished to defend any civil proceedings that might be taken against him as a result of the prosecution. The defendants entered into correspondence with the CPS and eventually obtained their witness statements and invited them to consent to an adjournment because of the absence overseas of an important witness for the claimant (Mr LM). Eventually the defendants successfully applied to the court for an adjournment, and also applied for and obtained a witness summons. At the trial the claimant was found guilty and was fined £300 and 4 points were endorsed on his driving licence.

Proceedings were conducted by an assistant solicitor admitted in 2000 whose time is charged at an agreed rate of £192.00 per hour with routine letters out and telephone calls at an agreed rate of £19.20 each. At the trial a trainee attended with counsel. The trainee's time is charged at an agreed rate of £118. per hour.

Cash account for client: EF

Received		**Paid**	
From client on account generally:		Refund to client:	
24th October 2011	1,500.00	27th February 2012	385.00
From client on account generally:			
19th February 2012	1,000.00	Balancing Item	2,115.00
	2,500.00		2,500.00
Balance due to client EF:	£2,115.00		

CPR PD 47.20 *Appendix*

Item no	Item	V.A.T. £	Disbursements £	Profit Costs £
	Attendances on Court and Counsel			
	5th November 2011 — application made for an adjournment for the convenience of a witness			
	13th November 2011 — contested hearing of the application			
	Engaged 10 minutes £ 32.00			
	Travel and waiting 30 minutes £ 96.00			
1	Total solicitor's fee for hearing	£25.60		£128.00
	17th November convenience of a witness — conference with counsel			
	Engaged 45 minutes £ 144.00			
	Travel and waiting 1 hour £ 192.00			
2	Total solicitor's fee for conference	£67.20		£336.00
3	Counsel's fee for conference paid (Miss JK)	£30.00	£150.00	
	15th February 2012 — Brief to counsel — 4 pages A4			
	23rd February 2012 — attending the trial			
	Trainee engaged in court 1 hour £ 118.00			
	Engaged in conference 30 minutes £ 59.00			
	Travel and waiting (apportioned) 45 minutes £ 88.50			
4	Total trainee solicitor's fee for trial	£53.10		£265.50

Civil Procedure Rules 1998 **CPR PD 47.20**

Item no	Item		V.A.T. £	Disbursements £	Profit Costs £
5	Counsel's brief fee (paid: Miss JK)		£60.00	£300.00	
	Claimant				
6	21st October 2011 — first instructions 1 hour		£38.40		£192.00
7	16th November 2011 — finalising proof of evidence 45 minutes		£28.80		£144.00
8	Routine letters out and telephone calls — 12 (9 + 3)		£46.08		£230.40
	Witness (Mr LM)				
	Personal attendance by trainee solicitor				
	Engaged 45 minutes	£ 88.50			
	Travel and waiting 1 hour	£118.00			
9	Total trainee solicitor's fee for attendance		£41.30		£206.50
10	Routine letters out — 6		£35.04		£175.20
	Other persons				
11	Routine letters out and telephone calls to the CPS — 8 (6 + 2)		£30.72		£153.60
	Communications with the court				
12	Routine letters out — 6		£23.04		£115.20
	Communications with Counsel				
13	Routine letters out and telephone calls — 8 (4 + 4)		£30.72		£153.60
	Work done on documents				

CPR PD 47.20 *Appendix*

Item no	Item	V.A.T. £	Disbursements £	Profit Costs £
14	Instructions to counsel — 30 minutes	£19.20		£96.00
15	Attendance note of conference — 15 minutes	£9.60		£48.00
16	Brief to counsel — 30 minutes	£19.20		£96.00
17	Attendance note of trial (trainee) — 45 minutes	£17.70		£88.50
		£575.70	£450.00	£2428.50

Summary

Total costs claimed in breakdown £2,878.50 + £575.70 VAT.

(Total costs billed £2,500.00 + £500.00 VAT)

PART 48 PART 2 OF THE LEGAL AID, SENTENCING AND PUNISHMENT OF OFFENDERS ACT 2012 RELATING TO CIVIL LITIGATION FUNDING AND COSTS: TRANSITIONAL PROVISION IN RELATION TO PRE-COMMENCEMENT FUNDING ARRANGEMENTS

Rule 48.1 .... CPR 48.1
Rule 48.2 .... CPR 48.2
    Practice Direction 48—Part 2 of the Legal Aid, Sentencing and Punishment of Offenders Act 2012 Relating to Civil Litigation Funding and Costs: Transitional Provision and Exceptions .... CPR PD 48

[CPR 48.1]

48.1.

(1) The provisions of CPR Parts 43 to 48 relating to funding arrangements, and the attendant provisions of the Costs Practice Direction, will apply in relation to a pre-commencement funding arrangement as they were in force immediately before 1 April 2013, with such modifications (if any) as may be made by a practice direction on or after that date.

(2) A reference in rule 48.2 to a rule is to that rule as it was in force immediately before 1 April 2013.

[CPR 48.2]

48.2.

(1) A pre-commencement funding arrangement is—
    (a) in relation to proceedings other than insolvency-related proceedings, publication and privacy proceedings or a mesothelioma claim—
        (i) a funding arrangement as defined by rule 43.2(1)(k)(i) where—
            (aa) the agreement was entered into before 1 April 2013 specifically for the purposes of the provision to the person by whom the success fee is payable of advocacy or litigation services in relation to the matter that is the subject of the proceedings in which the costs order is to be made; or
            (bb) the agreement was entered into before 1 April 2013 and advocacy or litigation services were provided to that person under the agreement in connection with that matter before 1 April 2013;
        (ii) a funding arrangement as defined by rule 43.2(1)(k)(ii) where the party seeking to recover the insurance premium took out the insurance policy in relation to the proceedings before 1 April 2013;
        (iii) a funding arrangement as defined by rule 43.2(1)(k)(iii) where the agreement with the membership organisation to meet the costs was made before 1 April 2013 specifically in respect of the costs of other parties to proceedings relating to the matter which is the subject of the proceedings in which the costs order is to be made;
    (b) in relation to insolvency-related proceedings, publication and privacy proceedings or a mesothelioma claim—
        (i) a funding arrangement as defined by rule 43.2(1)(k)(i) where—
            (aa) the agreement was entered into before the relevant

CPR 48.2 *Appendix*

       date specifically for the purposes of the provision to the person by whom the success fee is payable of advocacy or litigation services in relation to the matter that is the subject of the proceedings in which the costs order is to be made; or

    (bb) the agreement was entered into before the relevant date and advocacy or litigation services were provided to that person under the agreement in connection with that matter before the relevant date;

  (ii) a funding arrangement as defined by rule 43.2(1)(k)(ii) where the party seeking to recover the insurance premium took out the insurance policy in relation to the proceedings before the relevant date.

(2) In paragraph (1)—

 (a) "insolvency-related proceedings" means any proceedings—

  (i) in England and Wales brought by a person acting in the capacity of—

   (aa) a liquidator of a company which is being wound up in England and Wales or Scotland under Parts IV or V of the Insolvency Act 1986; or

   (bb) a trustee of a bankrupt's estate under Part IX of the Insolvency Act 1986;

  (ii) brought by a person acting in the capacity of an administrator appointed pursuant to the provisions of Part II of the Insolvency Act 1986;

  (iii) in England and Wales brought by a company which is being wound up in England and Wales or Scotland under Parts IV or V of the Insolvency Act 1986; or

  (iv) brought by a company which has entered administration under Part II of the Insolvency Act 1986;

 (b) "news publisher" means a person who publishes a newspaper, magazine or website containing news or information about or comment on current affairs;

 (c) "publication and privacy proceedings" means proceedings for—

  (i) defamation;

  (ii) malicious falsehood;

  (iii) breach of confidence involving publication to the general public;

  (iv) misuse of private information; or

  (v) harassment, where the defendant is a news publisher.

   (d) "a mesothelioma claim" is a claim for damages in respect of diffuse mesothelioma (within the meaning of the Pneumoconiosis etc. (Workers' Compensation) Act 1979; and

   (e) "the relevant date" is the date on which sections 44 and 46 of the Legal Aid, Sentencing and Punishment of Offenders Act 2012 came into force in relation to proceedings of the sort in question.

Civil Procedure Rules 1998 CPR PD 48.2

PRACTICE DIRECTION 48—PART 2 OF THE LEGAL AID, SENTENCING AND PUNISHMENT OF OFFENDERS ACT 2012 RELATING TO CIVIL LITIGATION FUNDING AND COSTS: TRANSITIONAL PROVISION AND EXCEPTIONS

THIS PRACTICE DIRECTION SUPPLEMENTS PART 48

Transitional Provisions: General

[CPR PD 48.1]

1.1 Sections 44 and 46 of the Legal Aid, Sentencing and Punishment of Offenders Act 2012 ("the 2012 Act") make changes to the effect that a costs order may not include, respectively, provision requiring the payment by one party of all or part of a success fee payable by another party under a conditional fee agreement or of an amount in respect of all or part of the premium of a costs insurance policy taken out by another party. These changes come into force on 1 April 2013.

1.2 Sections 44(6) and 46(3) of the 2012 Act make saving provisions to the effect, respectively, that these changes do not apply so as to prevent a costs order including such provision where the conditional fee agreement in relation to the proceedings was entered into (or, in relation to a collective conditional fee agreement, services were provided to the party under the agreement), or the costs insurance policy in relation to the proceedings taken out, before the date on which the changes come into force.

1.3 The provisions in the CPR relating to funding arrangements have accordingly been revoked (either in whole or in part as they relate to funding arrangements) with effect from 1 April 2013; but they will remain relevant, and will continue to have effect notwithstanding the revocations, after that date for those cases covered by the saving provisions.

1.4 The provisions in the CPR in force prior to 1 April 2012 relating to funding arrangements include—
(a) CPR 43.2(1)(a), (k), (l), (m), (n), (o), 43.2(3) and 43.2(4);
(b) CPR 44.3A, 44.3B, 44.12B, 44.15 and 44.16;
(c) CPR 45.8, 45.10, 45.12, 45.13, Sections III to V (45.15 to 45.19, 45.20 to 22 and 45.23 to 26), 45.28 and 45.31 to 45.40;
(d) CPR 46.3;
(e) CPR 48.8.

Mesothelioma claims

[CPR PD 48.2]

2.1 By virtue of section 48 of the 2012 Act, the changes relating to recoverable success fees and insurance premiums which are made by sections 44 and 46 of the Act may not be commenced, and accordingly will not apply, in relation to mesothelioma claims (defined by section 48(2) of the Act as having the same meaning as in the Pneumoconiosis etc. (Workers' Compensation) Act 1979) until such time as a review has been carried out and the conclusions of that review published. It will accordingly remain possible for a costs order in favour of a party to such proceedings to include provision requiring the payment of success fees and premiums under after the event costs insurance policies, and so the provisions of the CPR relating to funding arrangements as in force immediately prior to 1 April 2013 will continue to apply in relation to such proceedings, whether commenced before or after 1 April 2013. This will include the provision for fixed recoverable success fees in respect of employers' liability disease claims in Section V of Part 45 (CPR 45.23 to 45.26), which will otherwise cease to apply other than to claims in which a CFA was entered into or a costs insurance policy taken out before 1 April 2013).

2.2 On the later date when sections 44 and 46 are brought into force in relation to mesothelioma claims, the saving provisions of sections 44(6) and 44(3) will have

815

effect in relation to funding arrangements in such claims as they do more generally, save that the operative date for the saving provisions will not be 1 April 2013 but the later date.

Insolvency-related proceedings and publication and privacy proceedings

[CPR PD 48.3]

3.1 Sections 44 and 46 of the 2012 Act are not being commenced immediately in relation to certain proceedings related to insolvency. Until such time as those sections are commenced in relation to those proceedings, therefore, they are in a similar position as regards funding arrangements to mesothelioma claims.
3.2 Similarly, sections 44 and 46 of the 2012 Act are not being commenced immediately in respect of publication and privacy proceedings, which will accordingly be in a similar position as regards funding arrangements to mesothelioma claims and insolvency-related proceedings until such time as those sections are commenced in relation to them.

New provision in relation to clinical negligence claims

[CPR PD 48.4]

4.1 Section 46 of the 2012 Act enables the Lord Chancellor by regulations to provide that a costs order may include provision requiring the payment of an amount in respect of all or part of the premium of a costs insurance policy, where—
(a) the order is made in favour of a party to clinical negligence proceedings of a prescribed description;
(b) the party has taken out a costs insurance policy insuring against the risk of incurring a liability to pay for one or more expert reports in respect of clinical negligence in connection with the proceedings (or against that risk and other risks);
(c) the policy is of a prescribed description;
(d) the policy states how much of the premium relates to the liability to pay for such an expert report or reports, and the amount to be paid is in respect of that part of the premium.
4.2 The regulations made under the power are the Recovery of Costs Insurance Premiums in Clinical Negligence Proceedings Regulations 2013 (S.I. 2013/92). The regulations relate only to clinical negligence cases where a costs insurance policy is taken out on or after 1 April 2013, so the provisions in force in the CPR prior to 1 April 2013 relating to funding arrangements will not apply.

SOLICITORS ACT 1974
(c 47)

s 56	Orders as to the remuneration for non-contentious business	SOL 56
s 57	Non-contentious business agreement	SOL 57
s 58	Remuneration of a solicitor who is a mortgagee	SOL 58
s 59	Contentious business agreements	SOL 59
s 60	Effect of contentious business agreement	SOL 60
s 61	Enforcement of contentious business agreement	SOL 61
s 62	Contentious business agreements by certain representatives	SOL 62
s 63	Effect on contentious business agreement of death, incapability or change of solicitor	SOL 63
s 64	Form of bill of costs for contentious business	SOL 64
s 65	Security for costs and termination of retainer	SOL 65
s 66	Taxations with regard to contentious business	SOL 66
s 67	Inclusion of disbursements in bill of costs	SOL 67
s 68	Power of court to order solicitor to deliver bill etc	SOL 68
s 69	Action to recover solicitor's costs	SOL 69
s 70	Taxation on application of party chargeable or solicitor	SOL 70
s 71	Taxation on application of third parties etc	SOL 71
s 72	Supplementary provisions as to taxations	SOL 72
s 73	Charging orders	SOL 73
s 74	Special provisions as to contentious business done in county courts	SOL 74
s 75	Wording	SOL 75

[SOL 56]

56. Orders as to remuneration for non-contentious business

(1) For the purposes of this section there shall be a committee consisting of the following persons—
- (a) the Lord Chancellor;
- (b) the Lord Chief Justice;
- (c) the Master of the Rolls;
- (d) the President of the Society;
- (da) a member of the Legal Services Board nominated by that Board;
- (e) a solicitor, being the president of a local law society, nominated by the Lord Chancellor to serve on the committee during his tenure of office as president; and
- (f) for the purpose only of prescribing and regulating the remuneration of solicitors in respect of business done under the Land Registration Act 2002, the Chief Land Registrar appointed under that Act.

(2) The committee, or any three members of the committee (the Lord Chancellor being one), may make general orders prescribing the general principles to be applied when determining the remuneration of solicitors in respect of non-contentious business.

(3) The Lord Chancellor, before any order under this section is made, shall cause a draft of the order to be sent to the Society; and the committee shall consider any observations of the Society submitted to them in writing within one month of the sending of the draft, and may then make the order, either in the form of the draft or with such alterations or additions as they may think fit.

(4) The principles prescribed by an order under this section may provide that solicitors should be remunerated—
- (b) by a gross sum; or
- (c) by a fixed sum for each document prepared or perused, without regard to length; or

817

SOL 57   Appendix

      (d)   in any other mode; or

      (e)   partly in one mode and partly in another.

(5)   The general principles prescribed by an order under this section may provide that the amount of such remuneration is to be determined by having regard to all or any of the following, among other, considerations, that is to say—

      (a)   the position of the party for whom the solicitor is concerned in the business, that is, whether he is vendor or purchaser, lessor or lessee, mortgagor or mortgagee, or the like;

      (b)   the place where, and the circumstances in which, the business or any part of it is transacted;

      (c)   the amount of the capital money or rent to which the business relates;

      (d)   the skill, labour and responsibility on the part of the solicitor, or any employee of his who is an authorised person, which the business involves;

      (e)   the number and importance of the documents prepared or perused, without regard to length.

(5A)   In subsection (5) "authorised person" means a person who is an authorised person in relation to an activity which is a reserved legal activity, within the meaning of the Legal Services Act 2007 (see section 18 of that Act).

(6)   An order under this section may authorise and regulate—

      (a)   the taking by a solicitor from his client of security for payment of any remuneration, to be ascertained by assessment or otherwise, which may become due to him under any such order; and

      (b)   the allowance of interest.

(7)   So long as an order made under this section is in operation the assessment of bills of costs of solicitors in respect of non-contentious business shall, subject to the provisions of section 57, be subject to that order.

(8)   Any order made under this section may be varied or revoked by a subsequent order so made.

(9)   The power to make orders under this section shall be exercisable by statutory instrument which shall be subject to annulment in pursuance of a resolution of either House of Parliament; and the Statutory Instruments Act 1946 shall apply to a statutory instrument containing such an order in like manner as if the order had been made by a Minister of the Crown.

[SOL 57]

57. Non-contentious business agreements

(1)   Whether or not any order is in force under section 56, a solicitor and his client may, before or after or in the course of the transaction of any non-contentious business by the solicitor, make an agreement as to his remuneration in respect of that business.

(2)   The agreement may provide for the remuneration of the solicitor by a gross sum or by reference to an hourly rate, or by a commission or percentage, or by a salary, or otherwise, and it may be made on the terms that the amount of the remuneration stipulated for shall or shall not include all or any disbursements made by the solicitor in respect of searches, plans, travelling, taxes, fees or other matters.

(3)   The agreement shall be in writing and signed by the person to be bound by it or his agent in that behalf.

(4)   Subject to subsections (5) and (7), the agreement may be sued and recovered on or set aside in the like manner and on the like grounds as an agreement not relating to the remuneration of a solicitor.

(5) If on any assessment of costs the agreement is relied on by the solicitor and objected to by the client as unfair or unreasonable, the costs officer may enquire into the facts and certify them to the court, and if from that certificate it appears just to the court that the agreement should be set aside, or the amount payable under it reduced, the court may so order and may give such consequential directions as it thinks fit.

(6) Subsection (7) applies where the agreement provides for the remuneration of the solicitor to be by reference to an hourly rate.

(7) If, on the assessment of any costs, the agreement is relied on by the solicitor and the client objects to the amount of the costs (but is not alleging that the agreement is unfair or unreasonable), the costs officer may enquire into—

    (a)   the number of hours worked by the solicitor; and

    (b)   whether the number of hours worked by him was excessive.

[SOL 58]

58. Remuneration of a solicitor who is a mortgagee

(1) Where a mortgage is made to a solicitor, either alone or jointly with any other person, he or the firm of which he is a member shall be entitled to recover from the mortgagor in respect of all business transacted and acts done by him or them in negotiating the loan, deducing and investigating the title to the property, and preparing and completing the mortgage, such usual costs as he or they would have been entitled to receive if the mortgage had been made to a person who was not a solicitor and that person had retained and employed him or them to transact that business and do those acts.

(2) Where a mortgage has been made to, or has become vested by transfer or transmission in, a solicitor, either alone or jointly with any other person, and any business is transacted or acts are done by that solicitor or by the firm of which he is a member in relation to that mortgage or the security thereby created or the property thereby charged, he or they shall be entitled to recover from the person on whose behalf the business was transacted or the acts were done, and to charge against the security, such usual costs as he or they would have been entitled to receive if the mortgage had been made to and had remained vested in a person who was not a solicitor and that person had retained and employed him or them to transact that business and do those acts.

(3) In this section "mortgage" includes any charge on any property for securing money or money's worth.

[SOL 59]

59. Contentious business agreements

(1) Subject to subsection (2), a solicitor may make an agreement in writing with his client as to his remuneration in respect of any contentious business done, or to be done, by him (in this Act referred to as a "contentious business agreement") providing that he shall be remunerated by a gross sum or by reference to an hourly rate, or by a salary, or otherwise, and whether at a higher or lower rate than that at which he would otherwise have been entitled to be remunerated.

(2) Nothing in this section or in sections 60 to 63 shall give validity to—

    (a)   any purchase by a solicitor of the interest, or any part of the interest, of his client in any action, suit or other contentious proceeding; or

    (b)   any agreement by which a solicitor retained or employed to prosecute any action, suit or other contentious proceeding, stipulates for payment only in the event of success in that action, suit or proceeding; or

(c) any disposition, contract, settlement, conveyance, delivery, dealing or transfer which under the law relating to bankruptcy is invalid against a trustee or creditor in any bankruptcy or composition

[SOL 60]

60. Effect of contentious business agreement

(1) Subject to the provisions of this section and to sections 61 to 63, the costs of a solicitor in any case where a contentious business agreement has been made shall not be subject to assessment or (except in the case of an agreement which provides for the solicitor to be remunerated by reference to an hourly rate) to the provisions of section 69.

(2) Subject to subsection (3), a contentious business agreement shall not affect the amount of, or any rights or remedies for the recovery of, any costs payable by the client to, or to the client by, any person other than the solicitor, and that person may, unless he has otherwise agreed, require any such costs to be assessed according to the rules for their assessment for the time being in force.

(3) A client shall not be entitled to recover from any other person under an order for the payment of any costs to which a contentious business agreement relates more than the amount payable by him to his solicitor in respect of those costs under the agreement.

(4) A contentious business agreement shall be deemed to exclude any claim by the solicitor in respect of the business to which it relates other than—
  (a) a claim for the agreed costs; or
  (b) a claim for such costs as are expressly excepted from the agreement.

(5) A provision in a contentious business agreement that the solicitor shall not be liable for his negligence, or that of any employee of his, shall be void if the client is a natural person who, in entering that agreement, is acting for purposes which are outside his trade, business or profession.

(6) A provision in a contentious business agreement that the solicitor shall be relieved from any responsibility to which he would otherwise be subject as a solicitor shall be void.

[SOL 61]

61. Enforcement of contentious business agreement

(1) No action shall be brought on any contentious business agreement, but on the application of any person who—
  (a) is a party to the agreement or the representative of such a party; or
  (b) is or is alleged to be liable to pay, or is or claims to be entitled to be paid, the costs due or alleged to be due in respect of the business to which the agreement relates,

the court may enforce or set aside the agreement and determine every question as to its validity or effect.

(2) On any application under subsection (1), the court—
  (a) if it is of the opinion that the agreement is in all respects fair and reasonable, may enforce it;
  (b) if it is of the opinion that the agreement is in any respect unfair or unreasonable, may set it aside and order the costs covered by it to be assessed as if it had never been made;
  (c) in any case, may make such order as to the costs of the application as it thinks fit.

(3) If the business covered by a contentious business agreement (not being an agreement to which section 62 applies) is business done, or to be done, in any action, a client who is a party to the agreement may make application to a costs officer of the court for the agreement to be examined.

(4) A costs officer before whom an agreement is laid under subsection (3) shall examine it and may either allow it, or, if he is of the opinion that the agreement is unfair or unreasonable, require the opinion of the court to be taken on it, and the court may allow the agreement or reduce the amount payable under it, or set it aside and order the costs covered by it to be assessed as if it had never been made.

(4A) Subsection (4B) applies where a contentious business agreement provides for the remuneration of the solicitor to be by reference to an hourly rate.

(4B) If on the assessment of any costs the agreement is relied on by the solicitor and the client objects to the amount of the costs (but is not alleging that the agreement is unfair or unreasonable), the costs officer may enquire into—

(a) the number of hours worked by the solicitor; and
(b) whether the number of hours worked by him was excessive.

(5) Where the amount agreed under any contentious business agreement is paid by or on behalf of the client or by any person entitled to do so, the person making the payment may at any time within twelve months from the date of payment, or within such further time as appears to the court to be reasonable, apply to the court, and, if it appears to the court that the special circumstances of the case require it to be re-opened, the court may, on such terms as may be just, re-open it and order the costs covered by the agreement to be assessed and the whole or any part of the amount received by the solicitor to be repaid by him.

(6) In this section and in sections 62 and 63 "the court" means—

(a) in relation to an agreement under which any business has been done in any court having jurisdiction to enforce and set aside agreements, any such court in which any of that business has been done;

(b) in relation to an agreement under which no business has been done in any such court, and under which more than £50 is payable, the High Court;

(c) in relation to an agreement under which no business has been done in any such court and under which not more than £50 is payable, any county court which would, but for the provisions of subsection (1) prohibiting the bringing of an action on the agreement, have had jurisdiction in any action on it;

and for the avoidance of doubt it is hereby declared that in paragraph (a) "court having jurisdiction to enforce and set aside agreements" includes a county court.

[SOL 62]

62. Contentious business agreements by certain representatives

(1) Where the client who makes a contentious business agreement makes it as a representative of a person whose property will be chargeable with the whole or part of the amount payable under the agreement, the agreement shall be laid before a costs officer of the court before payment.

(2) A costs officer before whom an agreement is laid under subsection (1) shall examine it and may either allow it, or, if he is of the opinion that it is unfair or unreasonable, require the opinion of the court to be taken on it, and the court may allow the agreement or reduce the amount payable under it, or set it aside and order the costs covered by it to be assessed as if it had never been made.

SOL 63 Appendix

(3) A client who makes a contentious business agreement as mentioned in subsection (1) and pays the whole or any part of the amount payable under the agreement without it being allowed by the officer or by the court shall be liable at any time to account to the person whose property is charged with the whole or any part of the amount so paid for the sum so charged, and the solicitor who accepts the payment may be ordered by the court to refund the amount received by him.

(4) A client makes a contentious business agreement as the representative of another person if he makes it—
- (a) as his guardian,
- (b) as a trustee for him under a deed or will,
- (c) as a deputy for him appointed by the Court of Protection with powers in relation to his property and affairs, or
- (d) as another person authorised under that Act to act on his behalf.

[SOL 63]

63. Effect on contentious business agreement of death, incapability or change of solicitor

(1) If, after some business has been done under a contentious business agreement but before the solicitor has wholly performed it—
- (a) the solicitor dies, or becomes incapable of acting; or
- (b) the client changes his solicitor (as, notwithstanding the agreement, he shall be entitled to do),

any party to, or the representative of any party to, the agreement may apply to the court, and the court shall have the same jurisdiction as to enforcing the agreement so far as it has been performed, or setting it aside, as the court would have had if the solicitor had not died or become incapable of acting, or the client had not changed his solicitor.

(2) The court, notwithstanding that it is of the opinion that the agreement is in all respects fair and reasonable, may order the amount due in respect of business under the agreement to be ascertained by assessment, and in that case—
- (a) the costs officer, in ascertaining that amount, shall have regard so far as may be to the terms of the agreement; and
- (b) payment of the amount found by him to be due may be enforced in the same manner as if the agreement had been completely performed.

(3) If in such a case as is mentioned in subsection (1)(b) an order is made for the assessment of the amount due to the solicitor in respect of the business done under the agreement, the court shall direct the costs officer to have regard to the circumstances under which the change of solicitor has taken place, and the costs officer, unless he is of the opinion that there has been no default, negligence, improper delay or other conduct on the part of the solicitor, or any of his employees, affording the client reasonable ground for changing his solicitor, shall not allow to the solicitor the full amount of the remuneration agreed to be paid to him.

[SOL 64]

64. Form of bill of costs for contentious business

(1) Where the remuneration of a solicitor in respect of contentious business done by him is not the subject of a contentious business agreement, then, subject to subsections (2) to (4), the solicitor's bill of costs may at the option of the solicitor be either a bill containing detailed items or a gross sum bill.

(2) The party chargeable with a gross sum bill may at any time—

(a) before he is served with a writ or other originating process for the recovery of costs included in the bill, and
(b) before the expiration of three months from the date on which the bill was delivered to him,

require the solicitor to deliver, in lieu of that bill, a bill containing detailed items; and on such a requirement being made the gross sum bill shall be of no effect.

(3) Where an action is commenced on a gross sum bill, the court shall, if so requested by the party chargeable with the bill before the expiration of one month from the service on that party of the writ or other originating process, order that the bill be assessed.

(4) If a gross sum bill is assessed, whether under this section or otherwise, nothing in this section shall prejudice any rules of court with respect to assessment, and the solicitor shall furnish the costs officer with such details of any of the costs covered by the bill as the costs officer may require.

[SOL 65]

65. Security for costs and termination of retainer

(1) A solicitor may take security from his client for his costs, to be ascertained by assessment or otherwise, in respect of any contentious business to be done by him.

(2) If a solicitor who has been retained by a client to conduct contentious business requests the client to make a payment of a sum of money, being a reasonable sum on account of the costs incurred or to be incurred in the conduct of that business and the client refuses or fails within a reasonable time to make that payment, the refusal or failure shall be deemed to be a good cause whereby the solicitor may, upon giving reasonable notice to the client, withdraw from the retainer.

[SOL 66]

66. Assessments with regard to contentious business

Subject to the provisions of any rules of court, on every assessment of costs in respect of any contentious business, the costs officer may—
(a) allow interest at such rate and from such time as he thinks just on money disbursed by the solicitor for the client, and on money of the client in the hands of, and improperly retained by, the solicitor or an employee of the solicitor; and
(b) in determining the remuneration of the solicitor, have regard to the skill, labour and responsibility involved in the business done by him or by any employee of his who is an authorised person (within the meaning of section 56(5A)).

[SOL 67]

67. Inclusion of disbursements in bill of costs

A solicitor's bill of costs may include costs payable in discharge of a liability properly incurred by him on behalf of the party to be charged with the bill (including counsel's fees) notwithstanding that those costs have not been paid before the delivery of the bill to that party; but those costs—
(a) shall be described in the bill as not then paid; and
(b) if the bill is assessed, shall not be allowed by the costs officer unless they are paid before the assessment is completed.

SOL 68 *Appendix*

[SOL 68]

68. Power of court to order solicitor to deliver bill etc

(1) The jurisdiction of the High Court to make orders for the delivery by a solicitor of a bill of costs, and for the delivery up of, or otherwise in relation to, any documents in his possession, custody or power, is hereby declared to extend to cases in which no business has been done by him in the High Court.

(2) A county court shall have the same jurisdiction as the High Court to make orders making such provision as is mentioned in subsection (1) in cases where the bill of costs or the documents relate wholly or partly to contentious business done by the solicitor in that county court.

(3) In this section and in sections 69 to 71 "solicitor" includes the executors, administrators and assignees of a solicitor].

[SOL 69]

69. Action to recover solicitor's costs

(1) Subject to the provisions of this Act, no action shall be brought to recover any costs due to a solicitor before the expiration of one month from the date on which a bill of those costs is delivered in accordance with the requirements mentioned in subsection (2); but if there is probable cause for believing that the party chargeable with the costs—

    (a)   is about to quit England and Wales, to become bankrupt or to compound with his creditors, or

    (b)   is about to do any other act which would tend to prevent or delay the solicitor obtaining payment,

the High Court may, notwithstanding that one month has not expired from the delivery of the bill, order that the solicitor be at liberty to commence an action to recover his costs and may order that those costs be assessed.

(2) The requirements referred to in subsection (1) are that the bill must be—

    (a)   signed in accordance with subsection (2A), and

    (b)   delivered in accordance with subsection (2C).

(2A) A bill is signed in accordance with this subsection if it is—

    (a)   signed by the solicitor or on his behalf by an employee of the solicitor authorised by him to sign, or

    (b)   enclosed in, or accompanied by, a letter which is signed as mentioned in paragraph (a) and refers to the bill.

(2B) For the purposes of subsection (2A) the signature may be an electronic signature.

(2C) A bill is delivered in accordance with this subsection if—

    (a)   it is delivered to the party to be charged with the bill personally,

    (b)   it is delivered to that party by being sent to him by post to, or left for him at, his place of business, dwelling-house or last known place of abode, or

    (c)   it is delivered to that party—

        (i)   by means of an electronic communications network, or

        (ii)   by other means but in a form that nevertheless requires the use of apparatus by the recipient to render it intelligible,

and that party has indicated to the person making the delivery his willingness to accept delivery of a bill sent in the form and manner used.

(2D) An indication to any person for the purposes of subsection (2C)(c)—

(a) must state the address to be used and must be accompanied by such other information as that person requires for the making of the delivery;
(b) may be modified or withdrawn at any time by a notice given to that person.

(2E) Where a bill is proved to have been delivered in compliance with the requirements of subsections (2A) and (2C), it is not necessary in the first instance for the solicitor to prove the contents of the bill and it is to be presumed, until the contrary is shown, to be a bill bona fide complying with this Act.

(2F) A bill which is delivered as mentioned in subsection (2C)(c) is to be treated as having been delivered on the first working day after the day on which it was sent (unless the contrary is proved).

(3) Where a bill of costs relates wholly or partly to contentious business done in a county court and the amount of the bill does not exceed £5,000, the powers and duties of the High Court under this section and sections 70 and 71 in relation to that bill may be exercised and performed by any county court in which any part of the business was done.

(4) . . .

(5) In this section references to an electronic signature are to be read in accordance with section 7(2) of the Electronic Communications Act 2000 (c 7).

(6) In this section—
"electronic communications network" has the same meaning as in the Communications Act 2003 (c 21);
"working day" means a day other than a Saturday, a Sunday, Christmas Day, Good Friday or a bank holiday in England and Wales under the Banking and Financial Dealings Act 1971 (c 80).

[SOL 70]

70. Assessment on application of party chargeable or solicitor

(1) Where before the expiration of one month from the delivery of a solicitor's bill an application is made by the party chargeable with the bill, the High Court shall, without requiring any sum to be paid into court, order that the bill be assessed and that no action be commenced on the bill until the assessment is completed.

(2) Where no such application is made before the expiration of the period mentioned in subsection (1), then, on an application being made by the solicitor or, subject to subsections (3) and (4), by the party chargeable with the bill, the court may on such terms, if any, as it thinks fit (not being terms as to the costs of the assessment), order—
(a) that the bill be assessed; and
(b) that no action be commenced on the bill, and that any action already commenced be stayed, until the assessment is completed.

(3) Where an application under subsection (2) is made by the party chargeable with the bill—
(a) after the expiration of 12 months from the delivery of the bill, or
(b) after a judgment has been obtained for the recovery of the costs covered by the bill, or
(c) after the bill has been paid, but before the expiration of 12 months from the payment of the bill,
no order shall be made except in special circumstances and, if an order is made, it may contain such terms as regards the costs of the assessment as the court may think fit.

(4) The power to order assessment conferred by subsection (2) shall not be exercisable on an application made by the party chargeable with the bill after the expiration of 12 months from the payment of the bill.
(5) An order for the assessment of a bill made on an application under this section by the party chargeable with the bill shall, if he so requests, be an order for the assessment of the profit costs covered by the bill.
(6) Subject to subsection (5), the court may under this section order the assessment of all the costs, or of the profit costs, or of the costs other than profit costs and, where part of the costs is not to be assessed, may allow an action to be commenced or to be continued for that part of the costs.
(7) Every order for the assessment of a bill shall require the costs officer to assess not only the bill but also the costs of the assessment and to certify what is due to or by the solicitor in respect of the bill and in respect of the costs of the assessment.
(8) If after due notice of any assessment either party to it fails to attend, the officer may proceed with the assessment ex parte.
(9) Unless—
  (a) the order for assessment was made on the application of the solicitor and the party chargeable does not attend the assessment, or
  (b) the order for assessment or an order under subsection (10) otherwise provides,
the costs of an assessment shall be paid according to the event of the assessment, that is to say, if the amount of the bill is reduced by one fifth, the solicitor shall pay the costs, but otherwise the party chargeable shall pay the costs.
(10) The costs officer may certify to the court any special circumstances relating to a bill or to the assessment of a bill, and the court may make such order as respects the costs of the assessment as it may think fit.
(11) . . .
(12) In this section "profit costs" means costs other than counsel's fees or costs paid or payable in the discharge of a liability incurred by the solicitor on behalf of the party chargeable, and the reference in subsection (9) to the fraction of the amount of the reduction in the bill shall be taken, where the assessment concerns only part of the costs covered by the bill, as a reference to that fraction of the amount of those costs which is being assessed.

[SOL 71]

71. Assessment on application of third parties etc

(1) Where a person other than the party chargeable with the bill for the purposes of section 70 has paid, or is or was liable to pay, a bill either to the solicitor or to the party chargeable with the bill, that person, or his executors, administrators or assignees may apply to the High Court for an order for the assessment of the bill as if he were the party chargeable with it, and the court may make the same order (if any) as it might have made if the application had been made by the party chargeable with the bill.
(2) Where the court has no power to make an order by virtue of subsection (1) except in special circumstances it may, in considering whether there are special circumstances sufficient to justify the making of an order, take into account circumstances which affect the applicant but do not affect the party chargeable with the bill.
(3) Where a trustee, executor or administrator has become liable to pay a bill of a solicitor, then, on the application of any person interested in any property out of

which the trustee, executor or administrator has paid, or is entitled to pay, the bill, the court may order—
- (a) that the bill be assessed on such terms, if any, as it thinks fit; and
- (b) that such payments, in respect of the amount found to be due to or by the solicitor and in respect of the costs of the assessment, be made to or by the applicant, to or by the solicitor, or to or by the executor, administrator or trustee, as it thinks fit.

(4) In considering any application under subsection (3) the court shall have regard—
- (a) to the provisions of section 70 as to applications by the party chargeable for the assessment of a solicitor's bill so far as they are capable of being applied to an application made under that subsection;
- (b) to the extent and nature of the interest of the applicant.

(5) If an applicant under subsection (3) pays any money to the solicitor, he shall have the same right to be paid that money by the trustee, executor or administrator chargeable with the bill as the solicitor had.

(6) Except in special circumstances, no order shall be made on an application under this section for the assessment of a bill which has already been assessed.

(7) If the court on an application under this section orders a bill to be assessed, it may order the solicitor to deliver to the applicant a copy of the bill on payment of the costs of that copy.

[SOL 72]

72. Supplementary provisions as to assessments

(1) Every application for an order for the assessment of a solicitor's bill or for the delivery of a solicitor's bill and for the delivery up by a solicitor of any documents in his possession, custody or power shall be made in the matter of that solicitor.

(2) Where a costs officer is in the course of assessing a bill of costs, he may request the costs officer of any other court to assist him in assessing any part of the bill, and the costs officer so requested shall assess that part of the bill and shall return the bill with his opinion on it to the costs officer making the request.

(3) Where a request is made as mentioned in subsection (2), the costs officer who is requested to assess part of a bill shall have such powers, and may take such fees, in respect of that part of the bill, as he would have or be entitled to take if he were assessing that part of the bill in pursuance of an order of the court of which he is an officer; and the costs officer who made the request shall not take any fee in respect of that part of the bill.

(4) The certificate of the costs officer by whom any bill has been assessed shall, unless it is set aside or altered by the court, be final as to the amount of the costs covered by it, and the court may make such order in relation to the certificate as it thinks fit, including, in a case where the retainer is not disputed, an order that judgment be entered for the sum certified to be due with costs.

[SOL 73]

73. Charging orders

(1) Subject to subsection (2), any court in which a solicitor has been employed to prosecute or defend any suit, matter or proceedings may at any time—
- (a) declare the solicitor entitled to a charge on any property recovered or preserved through his instrumentality for his assessed costs in relation to that suit, matter or proceeding; and

SOL 74   *Appendix*

(b)   make such orders for the assessment of those costs and for raising money to pay or for paying them out of the property recovered or preserved as the court thinks fit;

and all conveyances and acts done to defeat, or operating to defeat, that charge shall, except in the case of a conveyance to a bona fide purchaser for value without notice, be void as against the solicitor.

(2)   No order shall be made under subsection (1) if the right to recover the costs is barred by any statute of limitations.

[SOL 74]

74. Special provisions as to contentious business done in county courts

(1)   The remuneration of a solicitor in respect of contentious business done by him in a county court shall be regulated in accordance with sections 59 to 73, and for that purpose those sections shall have effect subject to the following provisions of this section.

(2)   The district judge of a county court shall be the costs officer of that court but any assessment of costs by him may be reviewed by a judge assigned to the county court district, or by a judge acting as a judge so assigned, on the application of any party to the assessment.

(3)   The amount which may be allowed on the assessment of any costs or bill of costs in respect of any item relating to proceedings in a county court shall not, except in so far as rules of court may otherwise provide, exceed the amount which could have been allowed in respect of that item as between party and party in those proceedings, having regard to the nature of the proceedings and the amount of the claim and of any counterclaim.

[SOL 75]

75. Saving for certain enactments

Nothing in this Part of this Act shall affect the following enactments, that is to say—

(a)   ...
(b)   ...
(c)   any of the provisions of the Costs in Criminal Cases Act 1973;
(d)   ...
(e)   any other enactment not expressly repealed by this Act which authorises the making of rules or orders or the giving of directions with respect to costs, or which provides that any such rule, order or direction made or given under a previous enactment shall continue in force.

Index

[*all references are to paragraph number*]

A

'A' factor
  generally, 31.9
  introduction, 31.8
  'Seven Pillars of Wisdom', 31.14
  supervision, 31.12
  test, 31.21
  unrecordable time, 31.11
Aarhus Convention
  fixed costs, and, 26.45
  protective costs orders, and, 18.4
Abuse of process
  indemnity basis, and, 24.15
Additional liabilities
  between the parties costs, and, 11.7
Admiralty proceedings
  costs management, and, 15.4
  summary assessment, and, 27.7
After the event (ATE) insurance
  adverse costs
    avoidance of policy, 9.25
    introduction, 9.23
    non-party costs orders, 9.24
    order of payment, 9.26
    proceedings against the insurer, 9.24
  avoidance of policy, 9.25
  challenging premium level, 9.18
  clinical negligence cases
    generally, 9.12
    recoverability of premiums, 9.16
  collateral benefits, 9.20
  conditional fee agreements (pre-1 April, 2013), and
    appeals, and, 6.46
    collective CFAs, and, 6.22
    contents of premium, 6.43
    level of premium, 6.44
    staged premiums, 6.45
  contents
    collateral benefits, 9.20
    counsel's fees, 9.21
    introduction, 9.19
    retrospective cover, 9.22

After the event (ATE) insurance – *cont.*
  costs budgeting, and, 9.27
  costs capping, and, 9.29
  counsel's fees, 9.21
  generally, 9.9
  insolvency proceedings, 9.17
  mesothelioma claims, 9.17
  non-party costs orders, 9.24
  other cases, 9.13
  personal injury cases, 9.11
  privacy proceedings, 9.17
  proceedings against the insurer, 9.24
  prospective costs orders, and
    costs budgeting, 9.27
    costs capping, 9.29
    security for costs, 9.28
  publication proceedings, 9.17
  QOCS, and, 9.11–9.13
  recoverability of premiums
    clinical negligence cases, 9.16
    insolvency proceedings, 9.17
    mesothelioma claims, 9.17
    other, 9.14
    'pre-commencement' policies, 9.15
    publication and privacy proceedings, 9.17
  retrospective cover, 9.22
  risks covered
    clinical negligence cases, 9.12
    generally, 9.10
    other cases, 9.13
    personal injury cases, 9.11
  security for costs, and, 9.28
Agreed costs
  summary assessment, and, 27.13
Agreement to charge no fee
  indemnity principle, and, 12.3
Allocation of proceedings
  costs management, and, 15.23
Allocatur rule
  interest on costs, and, 32.2
Alternative dispute resolution (ADR)
  conclusion, 21.6

Index

Alternative dispute resolution (ADR) – *cont.*
Halsey guidance, 21.4
introduction, 21.1
sanction for failure to engage in ADR, 21.2–21.3
voluntary nature, 21.5

Appeals
assessment, against
  assessors' role, 33.9
  authorised court officers, by, 33.10
  conclusion, 33.11
  formulation of reasoning, 33.3
  grounds, 33.5
  importance of reasons, 33.2
  introduction, 33.1
  nature, 33.8
  permission to appeal, 33.4
  proportionality, 33.3
  reasons, 33.2–33.3
  review not re-hearing, 33.8
  routes, 33.7
  time limits, 33.6
damages-based agreements, 7.13
detailed assessment, against
  authorised court officers, from, 28.53, 33.10
  costs judges, from, 28.54
  district judges, from, 28.54
  entitlement to commence, and, 28.5
  introduction, 28.52
family proceedings, 39.17
security for costs, 19.12
summary assessment, 27.22

Arbitration
agreement as to payment of costs. 41.3
applications to court, 41.13
arbitrator's fees and expenses, 41.10
award of costs
  failure to give reasons, 41.12
  fairness, 41.6
  generally, 41.4
  offers, 41.5
basis of costs, 41.7
conclusion, 41.16
'costs of the arbitration', 41.2
enforcement, 41.14
failure to give reasons for award, 41.12
fairness of costs, 41.6
fees and expenses, 41.
introduction, 41.1
non-solicitors, 41.8
offers, 41.5
proportionality, 41.7
recoverability, 41.9

Arbitration – *cont.*
recoverable costs
  basis of costs, 41.7
  fees and expenses, 41.10
  limits, 41.11
  non-solicitors, 41.8
  proportionality, 41.7
  recoverability, 41.9
  security for costs, 19.24, 41.15

Assessment of costs
costs management, and, 15.30
generally, 11.9
transitional provisions, 11.13

Authorised court officers
detailed assessment, and appeals, 28.53
generally, 28.55

Avoidance of policies
after the event insurance, and, 9.25

B

'B' factor
assessment, 31.13–31.14
generally, 31.10
introduction, 31.8
level to claim, 31.18–31.20
supervision, 31.12
test, 31.21

Bankruptcy proceedings
recovery of costs, and, 3.28

Basis of costs
arbitration, and, 41.7
costs budgets, and, 24.5
indemnity basis
  abuse of process, 24.15
  causation of increased costs, and, 24.19
  conclusion, 24.25
  continued pursuit of hopeless claim, 24.14
  culpability, 24.9–24.21
  exercise of discretion, 24.8
  extraneous motive for litigation, 24.18
  failure to come to court with open hands, 24.13
  future issues, 24.26
  generally, 24.7
  hopeless defences, 24.12
  introduction, 24.1
  no culpability, 24.22–24.24
  public funding, and, 24.21
  refusal of offers under CPR Part 44, 24.20

Basis of costs – *cont.*
  indemnity basis – *cont.*
    underhandedness, 24.11
    unjustified defence, 24.16
    unreasonable behaviour, 24.11
    unreasonableness, 24.10
    voluminous and unnecessary evidence, 24.17
  introduction, 24.1
  post-31 March 2013, 24.4
  pre- 1 April 2013, 24.3
  proportionality, 24.6
  relevance, 24.2
  standard basis, 24.1
Beddoe orders
  protective costs orders, and, 18.5
Before the event (BTE) insurance
  client's indemnity, 9.7
  coverage, 9.5
  generally, 9.2–9.3
  hourly rate, 9.8
  indemnity principle, and, 12.12
  top up cover, 9.6
  usage, 9.4
Between the parties costs
  additional liabilities, 11.7
  assessment
    generally, 11.9
    transitional provisions, 11.13
  bases, 24.1–24.26
  Civil Procedure Rules, and
    Part 44, 11.4–11.6
    Part 45, 11.7
    Part 46, 11.8
    Part 47, 11.9
    Part 48, 11.10
    other changes, 11.11–11.13
    overriding objective, 11.2
    Practice Direction, 11.14
    recent changes, 11.1
    relief from sanction, 11.3
  cost capping, 11.4
  costs awards, 22.1–22.35
  costs inducements
    ADR, 21.1–21.6
    Part 36 offers, 20.1–20.41
  costs only proceedings, 11.4
  damages-based agreements, 11.6
  family proceedings, and, 39.26–39.27
  fixed trial costs, 11.7
  funding notices, 11.4
  indemnity principle
    exceptions, 12.3–12.12
    future, 12.13

Between the parties costs – *cont.*
  indemnity principle – *cont.*
    introduction, 12.1
    no profit rule, 12.2
  introduction, 11.1
  overriding objective, 11.2
  payments on account, 25.1
  Practice Direction, 11.14
  proportionality
    generally, 11.11
    introduction, 11.2
  prospective costs control
    costs budgets, 16.1
    costs capping, 17.1–17.7
    costs management, 15.1–15.35
    introduction, 13.1
    proportionality, 14.1–14.4
    protective costs orders, 18.1–18.5
    security for costs, 19.1–19.24
  provisional assessment
    generally, 11.9
    transitional provisions, 11.13
  qualified one way costs shifting
    generally, 11.5
    transitional provisions, 11.12
  relief from sanction, 11.3
  transitional provisions, 11.10–11.13
  wasted costs orders, 23.1–23.34
Billing clients
  cash account, 2.12
  client, 2.3
  final bills
    cash account, 2.12
    form and content, 2.9
    taking account of interim bills, 2.10
  interim bills
    account, on, 2.5
    introduction 2.4
    statute bills, 2.6–2.
  interim statute bills
    agreement, by, 2.7
    introduction, 2.6
    natural break, and 2.8
  introduction, 2.1
  terminology, 2.2
  unpaid disbursements, 2.11
  VAT
    chargeability, 2.13
    counsel's fees, 2.18
    disbursements, 2.16
    fixed costs recovered, on, 2.19
    insurance claims, 2.20–2.21
    introduction, 2.13
    medical reports, on, 2.17

831

Index

Billing clients – *cont.*
VAT – *cont.*
rate, 2.14
tax point, 2.15
Bills of costs
'attendances', 29.19
attendances at court, 29.9
attendances on agents, 29.17
attendances on clients, 29.10
attendances on counsel, 29.9
attendances on other persons, 29.13
attendances on witnesses, 29.11
attendances to inspect property or place, 29.12
certificates, 29.21–29.24
column bills, 29.3–29.4
'communications', 29.19
communications with agents, 29.17
communications with clients, 29.10
communications with counsel, 29.14
communications with other persons, 29.13
communications with witnesses, 29.11
copies, 29.19
detail, 29.19–29.24
disbursements, 29.24
form and content, 29.2–29.5
4 column bills, 29.4
future for, 29.25
heads, 29.8–29.18
narrative, 29.7
numbering, 29.19
other incidental work, 29.18
parts, 29.19
personal attendances, 29.19
postage, 29.19
preparation of bill, 29.19
routine communications, 29.19
service, 28.23
six minute units, 29.19
6 column bills, 29.3
summary, 29.20
3 column bills, 29.3
title page, 29.6
travelling, 29.19
VAT, 29.24
waiting, 29.19
work done on documents, 29.15
work done on negotiations, 29.16
'B' factor
assessment, 31.13–31.14
generally, 31.10
introduction, 31.8
level to claim, 31.18–31.20
supervision, 31.12

'B' factor – *cont.*
test, 31.21
Budgeting
see also Costs budgets
time, and, 31.38
Bullock orders
generally, 22.34

C

'Calderbank' letters
family proceedings, and
financial remedy proceedings, 39.25
generally, 39.18
generally, 20.37
Case management
allocation of proceedings, 15.23
assessment of costs, 15.30
disclosure, 15.26
experts, 15.27
family proceedings, and, 39.43–39.44
interaction with costs management, 15.22–15.30
introduction, 15.1
relief from sanctions, 15.28
standard direction templates, 15.24
trial, 15.29
witness statements. 15.25
Cash account
final bills, and, 2.12
Causation
indemnity basis, and, 24.19
non-party costs orders, and, 40.7
wasted costs orders, and, 23.6
Chancery Division
costs management, and, 15.4
Children
detailed assessment, 34.4
family proceedings, and
generally, 39.13–39.16
legal services orders, 39.40
introduction, 34.1
litigation friends
detailed assessment, 34.4
expenses incurred, 34.3
generally, 34.2
meaning, 34.1
summary assessment, and, 27.12
Civil Procedure Rules
costs management, and, 15.4
Part 44, 11.4–11.6
Part 45, 11.7
Section I, 26.3–26.7

Index

Civil Procedure Rules – *cont.*
  Part 45, – *cont.*
    Section II, 26.8–26.14
    Section III, 26.15–26.24
    Section IIIA, 26.25–26.31
    Section IV, 26.32–26.38
    Section V, 26.39
    Section VI, 26.40–26.44
    Section VII, 26.45
  Part 46, 11.8
  Part 47, 11.9
  Part 48, 11.10
  other changes, 11.11–11.13
  overriding objective, 11.2
  Practice Direction, 11.14
  recent changes, 11.1
  relief from sanction, 11.3
Civil recovery orders
  detailed assessment, and, 28.60
Client
  meaning 1.2
Client care
  Code of Conduct 2007, 1.19
  costs, 1.22
  current regulatory position, 1.20
  damages-based agreements, 7.18
  indicative behaviours, 1.23
  in-house practice, 1.24
  old regulatory position, 1.18–1.19
  outcomes-focused regulation, 1.21–1.22
  overseas practice, 1.24
  status of personnel, 1.25
Clinical negligence cases
  after the event insurance, and
    generally, 9.12
    recoverability of premiums, 9.16
  conditional fee agreements, and, 4.8
Collateral benefits
  after the event insurance, and, 9.20
Collective conditional fee agreements
  after the event insurance, and, 6.22
  generally, 6.17
  indemnity principle issues, 6.18
Commencement
  fixed costs, and, 26.4
Commercial Court
  summary assessment, and, 27.7
Commercial funding
  Association of Litigation Funders, 10.15
  current position, 10.12
  expert evidence, 10.11
  extent of funder's liability, 10.10
  fee sharing, and, 10.16
  funding agreement, 10.13

Commercial funding – *cont.*
  generally, 10.8
  introduction, 10.1
  maintenance and champerty, 10.9
  personal injury cases, 10.16
  profile of cases, 10.14
Companies
  security for costs
    generally, 19.6
    inability to pay, 19.7
Company directors
  non-party costs orders, and, 40.10–40.12
Complaints
  generally, 1.26
Conditional fee agreements
  assignment, 5.15
  calculation of cap
    communication with client, 5.27
    delay, and, 5.26
    personal injury cases, 5.25
    QOCS, and, 5.28
  cancellation
    drafting considerations, 4.13
    generally, 5.16
  cessation of business, 4.12
  clinical negligence cases, 4.8
  cost of creation, 4.26
  cost of funding, 4.22–4.25
  counsel's fees
    communications, 5.11
    contractual terms, 5.7
    form of new agreements, 5.12
    introduction, 5.5
    solicitor's professional obligation, 5.6
    use of CFAs, 5.8–5.9
  creation
    cost, 4.26
    cost of funding, 4.22–4.25
    drafting, 4.9–4.13
    'golden' rule, 4.2
    introduction, 4.1
    rates, 4.14–4.17
    requirements, 4.3–4.8
    risk assessment, 4.18–4.21
  criminal proceedings, and, 4.5
  damages-based agreements, and, 4.10
  dating, 5.14
  differential rates
    fees and expenses, 4.15
    introduction, 4.14
    notice of funding, and, 5.2
    specifying circumstances, 4.16
    unsuccessful cases, 4.17

Index

Conditional fee agreements – *cont.*
  disbursements
    cost of funding, and, 4.25
    generally, 4.15
  drafting
    cancellable agreements, 4.13
    cessation of business, 4.12
    general approach, 4.9
    relationship with DBAS, 4.10
    speculative agreements, 4.11
    termination provisions, 4.12
  estimates of costs, 5.4
  excluded proceedings, 4.5
  expenses, 4.15
  family proceedings, and,
    generally, 39.41
    introduction, 4.5
  fees and expenses, 4.15
  'golden' rule, 4.2
  interest on costs
    introduction, 5.30
    out of pocket interest, 5.32
    punitive interest, 5.31
  interim applications, 5.13
  introduction, 4.1
  management of cases, 5.1–5.33
  no win, no fee arrangements
    differential rates, and, 4.17
    QOCS, and, 5.28
    unsuccessful cases, 5.33
  notices of funding, 5.2–5.3
  notification to insurers, 5.29
  'old' agreements
    see also Conditional fee agreements (pre-1 April, 2013)
    generally, 6.1–6.56
  payments on account
    following acceptance of Part 36 offer, 5.24
    following final hearing, 5.23
    generally, 5.1
  personal injury cases, 4.8
  rates
    early termination, on, 5.22
    fees and expenses, 4.15
    introduction, 4.14
    notice of funding, and, 5.2
    specifying circumstances, 4.16
    unsuccessful cases, 4.17
  'ready reckoner', 4.21
  requirements
    clinical negligence cases, 4.8
    excluded proceedings, 4.5
    introduction, 4.3

Conditional fee agreements – *cont.*
  requirements – *cont.*
    personal injury cases, 4.8
    rate of success fee, 4.6–4.7
    writing, 4.4
  risk assessment
    benefits, 4.19
    forms, 4.20
    limits, 4.18
    'ready reckoner', 4.21
  security for costs, 5.17
  speculative agreements, 4.11
  success fee
    maximum rate, 4.6
    percentage of client's damages, as, 4.7
  termination of case
    generally, 5.23
  termination of retainer, and
    drafting considerations, 4.12
    introduction, 5.18
    insufficient prospect of success, 5.20
    non-cooperation of client, 5.21
    other funding available, 5.19
    rate of charge, 5.22
  unsuccessful cases
    differential rates, 4.17
    generally, 5.33
  written form, 4.4
Conditional fee agreements (pre-1 April, 2013)
  additional liabilities
    notification, 6.47–6.53
  advocacy services, 6.3
  after the event insurance
    appeals, and, 6.46
    collective CFAs, and, 6.22
    contents of premium, 6.43
    level of premium, 6.44
    staged premiums, 6.45
  availability of alternative forms of funding, 6.25
  collective CFAs
    after the event insurance, and, 6.22
    generally, 6.17
    indemnity principle issues, 6.18
  compliance with primary legislation
    advocacy services, 6.3
    court approach, 6.23–6.27
    exempt proceedings, 6.6
    introduction, 6.2
    litigation services, 6.3
    maximum success fee, 6.5
    pre-action applications, 6.4

Conditional fee agreements (pre-1 April, 2013) – *cont.*
compliance with professional conduct rules, 6.28–6.42
compliance with secondary legislation
court approach, 6.23–6.27
introduction, 6.7
Regulations 2000, 6.8–6.12
Regulations 2003, 6.13–6.16
court approach
availability of alternative forms of funding, 6.25
continuing challenges, 6.27
disclosure of CFA, 6.24
disclosure of interests, 6.26
'technical challenges', 6.23
disbursement liability, 6.36
disclosure, 6.24
disclosure of interests, 6.26
exempt proceedings, 6.6
global success fees, 6.35
high value claims, 6.33
introduction, 6.1
liability admitted cases 6.32
litigation services, 6.3
maximum success fee, 6.5
membership organisations
after the event insurance, and, 6.22
generally, 6.19–6.22
limit on amount claims, 6.29
multiple claims, 6.35
notice of funding
effect of non-service, 6.51
exempt situations, 6.50
form and content, 6.48
general requirement, 6.47
introduction, 5.2–5.3
relief from sanctions, 6.52–6.53
service, 6.49
Part 36 offers, 6.16
pre-action applications, 6.4
retrospective success fees, 6.34
routine cases, 6.30
straightforward cases, 6.30
success fee
assessment of level, 6.38–6.
court decisions, 6.29–6.35
disbursement liability, 6.36
early stage CFAs, 6.31
generally, 6.5
global success fees, 6.35
high value claims, 6.33
liability admitted cases 6.32
multiple claims, 6.35

Conditional fee agreements (pre-1 April, 2013) – *cont.*
success fee – *cont.*
retrospective success fees, 6.34
routine cases, 6.30
straightforward cases, 6.30
'technical challenges', 6.23
transitional provisions
clinical negligence claims, 6.56
introduction, 6.54
pre-commencement, 6.55
Conflicts of interest
damages-based agreements, and, 7.18
Consent orders
clarity of wording, 22.14
interim orders, 22.13
introduction, 22.11
reserved costs, 22.15
Tomlin orders, 22.12
Contentious business agreements
certainty, 8.8
challenges to fees
hourly rates, 8.12
invalid agreements, 8.10
valid agreements, 8.11
charging methods, 8.7
commission, 8.7
definition 8.2
early termination, 8.9
generally, 8.3
gross sum payment, 8.7
hourly rates
challenges, 8.12
generally, 8.7
in writing, 8.6
introduction, 8.1
invalid agreements, 8.10
percentage, 8.7
recovery procedure, 8.14
representative party, 8.4
salary, 8.7
termination, 8.9
timing, 8.5
valid agreements, 8.11
written form, 8.6
Contingency agreements
generally, 1.31
Contracts
cancellable agreements, 1.7
client's instructions, 1.6
entire contract, 1.4
generally, 1.3
increase in hourly rate, 1.5

835

Index

Contractual costs
case decisions, 36.4
introduction, 36.1
landlord and tenant, 36.3
mortgages, 36.2
Contributory negligence
damages-based agreements, and, 7.14
Costs
costs management orders, and, 15.10
Costs awards between the parties
agreement in respect of everything except costs, 22.31
allocation, 22.26
available orders, 22.3
Bullock orders, 22.34
consent orders
   clarity of wording, 22.14
   interim orders, 22.13
   introduction, 22.11
   reserved costs, 22.15
   Tomlin orders, 22.12
costs follow the event
   admissible non-Part 36 offers, 22.20
   conduct of the parties, 22.18
   departures from general rule, 22.18–22.20
   general rule, 22.16
   menu of orders, 22.21–22.22
   partial success of case, 22.19
   role of costs judge on assessment, 22.23
   'successful party', 22.17
costs-only proceedings
   basis of costs, 22.30
   detailed assessment, 22.28
   fixed recoverable costs scheme, 22.29
   general procedure, 22.27
   summary assessment, 22.28
counterclaims, and, 22.32–22.33
deemed costs orders
   assessment basis, 22.9
   discontinuance, 22.7
   interest on assessed costs, 22.10
   Part 36 offers, 22.6
   striking out for non-payment of fees, 22.8
fixed recoverable costs scheme, 22.29
introduction, 22.1
issue based orders, 22.21
no order as to costs, 22.2
notification to the client, 22.4
pre-action disclosure, 22.24
re-allocation, 22.26
reserved costs, 22.15
role of costs judge on assessment, 22.23

Costs awards between the parties – *cont.*
Sanderson orders, 22.34
set-off, and, 22.32
small claims track, 22.25
state-funded parties, 22.35
terminology, 22.3
time for compliance, 22.5
Tomlin orders, 22.12
Costs budgets
after the event insurance, and, 9.27
agreement, 15.8
approval, 15.7
bases of costs, and, 24.5
cases not costs managed, in, 16.1
filing, 15.5–15.6
hourly rates, 15.32
indemnity basis, and, 24.5
introduction, 15.2
litigants in person, and, 15.14
relevance
   good reason, 15.13
   indemnity basis assessment, 15.12
   standard basis assessment, 15.11
revision, 15.8
role, 15.2
sanctions for failing to file and serve, 15.6
service, 15.5–15.6
setting
   approach of court, 15.33
   hourly rates, 15.32
   introduction, 15.31
time, and, 31.38
Costs capping
after the event insurance, and, 9.29
background, 17.1
between the parties costs, and, 11.4
case law, 17.3–17.6
conclusion, 17.7
criteria, 17.4
group litigation orders, and, 37.5
introduction, 17.1
jurisdiction, 17.2
procedure, 17.6
timing, 17.5
Costs estimates
client's reliance, 1.14
conditional fee agreements, and, 5.4
costs budgets, and, 1.16
exceeding, 1.13
failure to give, 1.15
generally, 1.10
increases, 1.12
introduction, 1.8
qualified estimates, 1.11

Costs estimates – *cont.*
recoverable fixed costs, and, 1.17
Costs follow the event
admissible non-Part 36 offers, 22.20
conduct of the parties, 22.18
departures from general rule
admissible non-Part 36 offers, 22.20
conduct of the parties, 22.18
partial success of case, 22.19
general rule, 22.16
menu of orders, 22.21–22.22
partial success of case, 22.19
role of costs judge on assessment, 22.23
'successful party', 22.17
Costs inducements
ADR
conclusion, 21.6
Halsey guidance, 21.4
introduction, 21.1
sanction for failure to engage in ADR, 21.2–21.3
voluntary nature, 21.5
'Calderbank' letters, 20.37
conclusion, 20.41
offers outside of CPR Part 36
'Calderbank' letters, 20.37
consequences, 20.38
generally, 20.36
'near miss' offers, 20.39
Part 36 offers
acceptance after trial but before judgment, 20.11
acceptance at trial, 20.10
acceptance time period, 20.5–20.7
background, 20.1
clarification, 20.13
commencement of trial, 20.12
consequences of acceptance, 20.17–20.22
consequences of non-acceptance, 20.23–20.24
content, 20.2
counter offers, and, 20.16
disclosure, 20.15
fixed success fees, and, 20.30
future pecuniary loss claims, and, 20.27
interest, and, 20.14
introduction, 20.1
made less than 21 days before trial, 20.7
offerors, 20.3
pre-action protocols, and, 20.32
provisional damages claims, and, 20.28
QOCS, and, 20.25
recoupment of state benefit, and, 20.29

Costs inducements – *cont.*
Part 36 offers – *cont.*
service, 20.6
small claims track, and, 20.33
success in part, and, 20.26
'tactical offer', and, 20.31
time limit, 20.5–20.7
timing, 20.4
VAT Tribunal, and, 20.34
withdrawal, 20.8–20.9
pre-6 April 2007 payments and offers, 20.35
Costs information
family proceedings, and, 39.37
Costs judge
role on assessment, 22.23
Costs management
admiralty proceedings, 15.4
allocation of proceedings, 15.23
application, 15.4
assessment of costs, 15.30
case law, 15.34
case management, and, 15.22–15.30
Chancery Division cases, 15.4
Civil Procedure Rules, 15.4
conclusion, 15.35
costs budgets
agreement, 15.8
approval, 15.7
filing, 15.5–15.6
introduction, 15.2
litigants in person, and, 15.14
relevance, 15.11–15.12
revision, 15.8
sanctions for failing to file and serve, 15.6
service, 15.5–15.6
setting, 15.31–15.33
costs of, 15.10
costs outside scope of budgets, 15.9
directions, 15.24
disclosure, 15.26
exceptions to application, 15.4
experts, 15.27
fixed costs proceedings, 15.4
Form H
assumptions, 15.17
completion, 15.16–15.21
counsel, 15.19
generally, 15.15
phases, 15.18
specifically excluded costs, 15.20
statement of truth, 15.21
summary, 15.17

Index

Costs management – *cont.*
  good reason, 15.13
  hourly rates, 15.32
  indemnity basis assessment,
    and, 15.12–15.13
  interaction with case management
    allocation of proceedings, 15.23
    assessment of costs, 15.30
    disclosure, 15.26
    experts, 15.27
    introduction, 15.22
    relief from sanctions, 15.28
    standard direction templates, 15.24
    trial, 15.29
    witness statements. 15.25
  introduction, 15.1
  litigants in person, 15.14
  meaning, 15.2
  Mercantile Courts cases, 15.4
  orders, 15.2
  patent cases, 15.4
  pilot schemes. 15.3
  procedural code, 15.5–15.14
  proportionality, and, 15.1
  relief from sanctions, 15.28
  scale costs proceedings, 15.4
  scope of regime, 15.4
  standard basis assessment, and, 15.11
  standard direction templates, 15.24
  TCC cases, 15.4
  trial, 15.29
  witness statements. 15.25
Costs officers
  detailed assessment, and, 28.2
Costs-only proceedings
  basis of costs, 22.30
  between the parties costs, and, 11.4
  detailed assessment, 22.28
  fixed recoverable costs scheme, 22.29
  general procedure, 22.27
  summary assessment, 22.28
Counsel's fees
  after the event insurance, and, 9.21
  brief fees, 31.33
  conditional fee agreements, and
    communications, 5.11
    contractual terms, 5.7
    form of new agreements, 5.12
    introduction, 5.5
    solicitor's professional obligation, 5.6
    use, 5.8–5.9
  summary assessment, and, 27.20
  time, and
    brief fees, 31.33

Counsel's fees – *cont.*
  time, and – *cont.*
    generally, 31.31–31.322
    table, 31.31
  VAT, and, 2.18
Counterclaims
  costs awards between the parties,
    and, 22.32–22.33
  damages-based agreements, and, 7.14
Court of Protection
  detailed assessment, and
    generally, 28.56
    procedure, 30.14
Credit checks
  recovery of costs, and, 3.2
Criminal proceedings
  conditional fee agreements, and, 4.5

D

Damages-based agreements
  appeals, 7.13
  between the parties costs, and, 11.6
  challenges by client, 7.19
  'circumstances in which fees and expenses
    are payable', 7.7
  client care, and,. 7.18
  conditional fee agreements, and, 4.10
  conflicts of interest, 7.18
  contributory negligence, and, 7.14
  counterclaims, and, 7.14
  employment cases, 7.12
  'entered into on or after 1 April 2013', 7.4
  group litigation orders, and, 37.6
  origins, 7.2
  personal injury cases, 7.11
  reasons for settling level of the percentage
    factors, 7.8
    generally, 7.8
    likely recoverable costs on
      assessment, 7.9A
    overall cost of reaching trial and proving
      the case, 7.9
  recoverability, 7.15
  recoverable percentage
    appeals, 7.13
    employment, 7.12
    generally, 7.10
    maximum, 7.10
    personal injury, 7.11
    reduction in damages, 7.14
    set-off, and, 7.14
  specifying the proceedings, 7.6

Damages-based agreements – *cont.*
  statutory framework, 7.3–7.10
  termination, 7.16
  written form, 7.5
Deemed costs orders
  assessment basis, 22.9
  discontinuance, 22.7
  interest on assessed costs
    generally, 22.10
    introduction, 32.3
  Part 36 offers, 22.6
  striking out for non-payment of fees, 22.8
Default costs certificates
  generally, 28.27
  pro bono representation, 28.28
  setting aside, 28.29–28.33
Deferred payment retainer
  summary assessment, and, 27.9
Delay
  detailed assessment, and
    generally, 28.19
    misconduct, 28.20
    right to fair trial, and, 28.21
Delivery of bill
  application for detailed assessment, 3.7
  client information, 3.8
  elapse of one month, 3.6
  introduction, 3.5
  limitation period, 3.9
  order for detailed assessment, 3.7
  recoverable costs trap, 3.10
Detailed assessment
  agreement
    formal, 28.12
    generally, 28.11
    informal, 28.13
  appeals
    authorised court officers, from, 28.53
    costs judges, from, 28.54
    district judges, from, 28.54
    entitlement to commence, and, 28.5
    introduction, 28.52
  authorised court officers
    appeals, 28.53
    generally, 28.55
  bills of costs
    'attendances', 29.19
    attendances at court, 29.9
    attendances on agents, 29.17
    attendances on clients, 29.10
    attendances on counsel, 29.9
    attendances on other persons, 29.13
    attendances on witnesses, 29.11

Detailed assessment – *cont.*
  bills of costs – *cont.*
    attendances to inspect property or place, 29.12
    certificates, 29.21–29.24
    column bills, 29.3–29.4
    'communications', 29.19
    communications with agents, 29.17
    communications with clients, 29.10
    communications with counsel, 29.14
    communications with other persons, 29.13
    communications with witnesses, 29.11
    copies, 29.19
    detail, 29.19–29.24
    disbursements, 29.24
    form and content, 29.2–29.5
    4 column bills, 29.4
    future for, 29.25
    heads, 29.8–29.18
    narrative, 29.7
    numbering, 29.19
    other incidental work, 29.18
    parts, 29.19
    personal attendances, 29.19
    postage, 29.19
    preparation of bill, 29.19
    routine communications, 29.19
    service, 28.23
    six minute units, 29.19
    6 column bills, 29.3
    summary, 29.20
    3 column bills, 29.3
    title page, 29.6
    travelling, 29.19
    VAT, 29.24
    waiting, 29.19
    work done on documents, 29.15
    work done on negotiations, 29.16
  children, and, 34.4
  civil recovery orders, 28.60
  commencement
    additional documents, 28.17
    delay, and, 28.19–28.21
    introduction, 28.15
    filing, and, 28.18
    notices, 28.16
  costs
    awards, 28.47
    offers, 28.48
    without prejudice save as to costs, 28.49
  'costs officer', 28.2

Index

Detailed assessment – *cont.*
  Court of Protection, and
    generally, 28.56
    procedure, 30.14
  default costs certificates
    generally, 28.27
    pro bono representation, 28.28
    setting aside, 28.29–28.33
  delay
    generally, 28.19
    misconduct, 28.20
    right to fair trial, and, 28.21
  delivery of bill, and, 3.7
  'difficult' cases, 28.9
  disbursement only assessment, 28.37
  entitlement to commence
    appeals, and, 28.5
    'forthwith' orders, 28.6
    introduction, 28.4
  final costs certificates
    generally, 28.50
    setting aside, 28.51
  First Tier Tribunal, and, 28.59
  'forthwith' orders, 28.6
  hearing date
    disbursement only assessment, 28.37
    documents to accompany request, 28.35
    role of court, 28.36
    time for request, 28.34
  hearings
    appeals, 28.52–28.54
    conduct, 28.43
    date request, 28.34–28.37
    elections, 28.45
    introduction, 28.41
    Lownds test, 28.46
    post-hearing procedure, 28.44
    pre-hearing procedure, 28.42
  High Court Enforcement Officers, 28.61
  interim costs certificates
    generally, 28.39
    provisional assessment, 28.39
  introduction, 28.1
  legal aid, and
    generally, 28.57
    procedure, 30.15
  London cases, 28.8
  methods, 28.2
  misconduct, and, 28.20
  offers
    costs, 28.48
    generally, 28.25
  Part 18 requests, 29.31

Detailed assessment – *cont.*
  points of dispute
    calculation of offers, 29.29
    format, 29.28
    introduction, 29.26
    service, 28.24
    structure, 29.27
  post-hearing procedure, 28.44
  pre-hearing procedure, 28.42
  preliminary issues, 28.3
  pro bono representation, 28.28
  procedure
    commencement, 28.15–28.21
    introduction, 28.14
    statements of case, 28.22–28.26
  Property Chamber, 28.59
  protected parties, and, 34.4
  provisional assessment
    challenging, 30.12
    conduct, 30.7
    costs, 30.9
    generally, 28.40
    interim costs certificates, 28.39
    introduction, 30.1
    notification of parties, 30.8
    overview, 28.2
    Part 36 offers, 30.10
    pilot scheme, 30.2
    procedure, 30.3–30.8
    requests, 30.3
    six week limit, 30.6
    subsequent hearing, 30.13
    suitability, 30.4
    supporting papers, 30.5
    termination, 30.11
    time limits, 30.6
  provisional damages, 28.5
  regional costs judges, 28.10
  replies
    generally, 29.30
    introduction, 28.26
  service
    bill of costs, 28.23
    points of dispute, 28.24
  statements of case
    bill of costs, 29.2–29.25
    generally, 28.22
    introduction, 29.1
    Part 18 requests, 29.31
    points of dispute, 29.26–29.29
    reply, 29.30
  Supreme Court proceedings, and, 28.58
  trustees, and, 38.5
  trusts, 30.16

Detailed assessment – *cont.*
venue, 28.7–28.8
without prejudice save as to costs, 28.49
Directions
costs management, and, 15.24
Disbursements
conditional fee agreements, and cost of funding, and, 4.25
generally, 4.15
detailed assessment, and, 28.37
VAT, and, 2.16
Disclosure
see also **Notices of funding**
costs management, and, 15.26
indemnity principle, and, 12.9
Part 36 offers, and, 20.15
Discontinuance
deemed costs orders, and, 22.7
group litigation orders, and, 37.2
Distant solicitor
time, and, 31.30

Enforcement – *cont.*
family proceedings, and – *cont.*
judgment summons, 39.49
statutory demand, 39.47
fixed costs, and, 26.7
Entire contract
generally, 1.4
Estimates of costs
client's reliance, 1.14
conditional fee agreements, and, 5.4
costs budgets, and, 1.16
exceeding, 1.13
failure to give, 1.15
family proceedings, and, 39.37
generally, 1.10
increases, 1.12
introduction, 1.8
qualified estimates, 1.11
recoverable fixed costs, and, 1.17
Experts
costs management, and, 15.27

E

Employers' liability personal injury claims
adjournment, 26.23
approval hearings, 26.22
avoidance behaviour, 26.21
claims no longer continuing under the protocols
counterclaims, 26.30
defendant's costs, 26.29
disbursements, 26.28
exceeding the fixed sums, 26.31
general points, 26.26
interim applications, 26.27
introduction, 26.25
London weighting, 26.26
VAT, 26.26
costs-only proceedings, 26.24
counsel's advice, 26.18
disbursements, 26.19
fixed sums, 26.17
introduction, 26.15
London weighting, 26.17
offers, 26.20
stages of the protocols, 26.16
Employment cases
damages-based agreements, and, 7.12
Enforcement
arbitration, and, 41.14
family proceedings, and
bankruptcy, 39.48

F

Family proceedings
appeals, 39.17
between the parties costs orders, 39.26–39.27
Calderbank offers
financial remedy proceedings, 39.25
generally, 39.18
case management, 39.43–39.44
Children Act 1989 applications
generally, 39.13–39.16
legal services orders, 39.40
conditional fee agreements
generally, 39.41
introduction, 4.5
costs information, 39.37
curbing costs
guidelines, 39.42
Practice Note, 39.43–39.44
discretion of courts, 39.36
enforcement
bankruptcy, 39.48
judgment summons, 39.49
statutory demand, 39.47
estimates of costs, 39.37
Family Procedure Rules 2010
general rules, 39.1–39.5
Practice Direction, 39.6
financial effect on parties of costs orders, 39.21–39.24

Index

Family proceedings – *cont.*
  financial remedy proceedings
    Calderbank offers, 39.25
    discretion, 39.19–39.20
    generally, 39.8–39.11
    Practice Direction, 39.6
    Rule, 39.4
    types of order, 39.9–39.10
  funding
    conditional fee agreements, 39.41
    generally, 39.38
    legal services orders, 39.39–39.40
  generally, 39.7–39.11
  introduction, 39.1
  legal services orders
    Children Act 1989 applications, 39.40
    criteria, 39.39
    generally, 39.39
  non-party costs orders, and, 40.20
  other applications, 39.12
  overriding objective, 39.31–39.35
  Practice Direction, 39.6
  Practice Note, 39.43–39.44
  right to fair trial, 39.39
  summary assessment, 39.45–39.46
  wasted costs orders
    generally, 39.5
    overriding objective, 39.35
Fast track trial costs
  amount, 26.42
  calculation of value of claim, 26.43
  introduction, 26.40
  multiple parties, 26.44
  scope, 26.41
Fee earners
  level, 31.28
  locality, 31.25
  targets, 31.39
Fee sharing
  commercial funders, and, 10.16
Final bills
  cash account, 2.12
  form and content, 2.9
  recovery of costs, and, 3.3
  taking account of interim bills, 2.10
Final costs certificates
  generally, 28.50
  setting aside, 28.51
Financial remedy proceedings
  Calderbank offers, 39.25
  discretion, 39.19–39.20
  generally, 39.8–39.11
  Practice Direction, 39.6
  Rule, 39.4

Financial remedy proceedings – *cont.*
  types of order, 39.9–39.10
First Tier Tribunal
  detailed assessment, and, 28.59
Fixed costs
  Aarhus Convention claims, 26.45
  claims no longer continuing under the RTA or EL/PL protocols
    counterclaims, 26.30
    defendant's costs, 26.29
    disbursements, 26.28
    exceeding the fixed sums, 26.31
    general points, 26.26
    interim applications, 26.27
    introduction, 26.25
    London weighting, 26.26
    VAT, 26.26
  commencement costs, 26.4
  costs awards between the parties, and, 22.29
  CPR Part 45
    Section I, 26.3–26.7
    Section II, 26.8–26.14
    Section III, 26.15–26.24
    Section IIIA, 26.25–26.31
    Section IV, 26.32–26.38
    Section V, 26.39
    Section VI, 26.40–26.44
    Section VII, 26.45
  enforcement costs, 26.7
  fast track trial costs
    amount, 26.42
    calculation of value of claim, 26.43
    introduction, 26.40
    multiple parties, 26.44
    scope, 26.41
  HMRC proceedings, 26.39
  indemnity principle, and, 12.11
  Intellectual Property Enterprise Court
    assessment, 26.35
    costs outside the scales, 26.36
    introduction, 26.32
    order for costs, 26.37
    scale management, 26.33
    scale tables, 26.34
    unreasonable costs, 26.38
  introduction, 26.1–26.2
  judgment costs, 26.5
  judicial review
    Aarhus Convention, under, 26.45
  low value RTA, EL and PL personal injury claims
    adjournment, 26.23
    approval hearings, 26.22

Fixed costs – *cont.*
  low value RTA, EL and PL personal injury claims – *cont.*
    avoidance behaviour, 26.21
    costs-only proceedings, 26.24
    counsel's advice, 26.18
    disbursements, 26.19
    fixed sums, 26.17
    introduction, 26.15
    London weighting, 26.17
    offers, 26.20
    stages of the protocols, 26.16
  miscellaneous costs, 26.6
  patents cases
    assessment, 26.35
    costs outside the scales, 26.36
    introduction, 26.32
    order for costs, 26.37
    scale management, 26.33
    scale tables, 26.34
    unreasonable costs, 26.38
  revenue proceedings, 26.39
  RTA personal accident cases
    base fee, 26.9
    conditions, 26.8
    costs formula, 26.9
    counsel's fees, 26.13
    disbursements, 26.11
    exceeding the fixed sums, 26.14
    introduction, 26.8
    London weighting, 26.10
    medical fees, 26.12
    percentage uplift, 26.9
    VAT, 26.9
  small claims track
    allocation, 26.49
    disbursements, 26.48
    general rule, 26.47
    introduction, 26.46
    re-allocation, 26.49
  VAT, and, 2.19

Fixed costs proceedings
  costs awards between the parties, and, 22.29
  costs management, and, 15.4

Fixed fees
  indemnity principle, and, 12.6
  Part 36 offers, and, 20.30

Fixed trial costs
  between the parties costs, and, 11.7

'Forthwith' orders
  detailed assessment, and, 28.6

Funding
  commercial funding, 10.8–10.16

Funding – *cont.*
  conditional fee agreements
    creation, 4.1–4.26
    management of cases, 5.1–5.33
    'old' agreements, 6.1–6.56
  contentious business agreements, 8.1–8.16
  damages-based agreements, 7.1–7.19
  family proceedings, and
    conditional fee agreements, 39.41
    generally, 39.38
    legal services orders, 39.39–39.40
  group litigation orders, and, 37.6
  insurance
    after the event, 9.9–9.29
    before the event, 9.2–9.8
    introduction, 9.1
    solicitor self-insurance, 9.30–9.32
  introduction, 4.0
  litigation funding, 10.8–10.16
  non-contentious business agreements, 8.1–8.16
  non-parties, 40.6
  state funding, 10.2–10.7

Funding notices
  between the parties costs, and, 11.4

G

Good reason
  costs management, and, 15.13
  summary assessment, and, 27.6

Group litigation orders
  costs capping, and, 37.5
  damages-based agreements, and, 37.6
  discontinuance, 37.2
  funding, and, 37.6
  generic costs, 37.4
  individual costs, 37.3
  introduction, 37.1

H

High Court Enforcement Officers
  detailed assessment, and, 28.61

HMRC proceedings
  fixed costs, and, 26.39

Hourly rates
  'A' factor, 31.9
  amount of rate, 31.29
  ascertaining the direct cost, 31.9
  ascertaining the profit element, 31.10

Index

Hourly rates – *cont.*
assessment of 'B' factor, 31.13–31.14
'B' factor, 31.10
before the event (BTE) insurance, and, 9.8
contentious business agreements, and
  challenges, 8.12
  generally, 8.7
costs budgets, and, 15.32
differential rates in CFAs
  fees and expenses, 4.15
  introduction, 4.14
  notice of funding, and, 5.2
  specifying circumstances, 4.16
  unsuccessful cases, 4.17
factors, 31.8
generally, 31.7
'Seven Pillars of Wisdom', 31.14
summary assessment, and, 27.19
supervision, 31.12
unrecordable time, 31.11

I

Impecuniosity
indemnity principle, and, 12.3
Improper conduct
wasted costs orders, and, 23.5
Incipitur rule
interest on costs, and, 32.2
Indemnity basis
abuse of process, 24.15
causation of increased costs, and, 24.19
conclusion, 24.25
continued pursuit of hopeless claim, 24.14
costs budgets, and, 24.5
costs management, and, 15.12–15.13
culpability
  causation of increased costs, 24.19
  continued pursuit of hopeless claim, 24.14
  extraneous motive for litigation, 24.18
  failure to come to court with open hands, 24.13
  hopeless defences, 24.12
  introduction, 24.9
  public funding, and, 24.21
  refusal of offers under CPR Part 44, 24.20
  underhandedness, 24.11
  unjustified defence, 24.16
  unreasonable behaviour, 24.11
  unreasonableness, 24.10

Indemnity basis – *cont.*
culpability – *cont.*
  voluminous and unnecessary evidence, 24.17
discretion, 24.8
extraneous motive for litigation, 24.18
failure to come to court with open hands, 24.13
future issues, 24.26
generally, 24.7
hopeless defences, 24.12
introduction, 24.1
no culpability, 24.22–24.24
pre-1 April 2013 cases, 24.3
public funding, and, 24.21
refusal of offers under CPR Part 44, 24.20
relevance, 24.2
underhandedness, 24.11
unjustified defence, 24.16
unreasonable behaviour, 24.11
unreasonableness, 24.10
voluminous and unnecessary evidence, 24.17
Indemnity principle
agreements to charge no fee, 12.3
before the event insurance, 12.12
damages-based agreements, and, 7.17
disclosure, 12.9
exceptions, 12.3–12.12
fixed costs, 12.11
fixed fees, 12.6
future, 12.13
impecunious clients, 12.3
interim bills, 12.7
introduction, 12.1
item-by-item basis, 12.8
no profit rule, 12.2
pro bono work, 12.4
third party payment, 12.5
unlawful retainer, 12.10
Insolvency proceedings
after the event insurance, and, 9.17
Insurance
after the event
  adverse costs, 9.23–9.26
  avoidance of policy, 9.25
  challenging premium level, 9.18
  clinical negligence cases, 9.12
  collateral benefits, 9.20
  contents, 9.19–9.22
  costs budgeting, and, 9.27
  costs capping, and, 9.29
  counsel's fees, 9.21
  generally, 9.9

Insurance – *cont.*
 after the event – *cont.*
  non-party costs orders, 9.24
  other cases, 9.13
  personal injury cases, 9.11
  proceedings against the insurer, 9.24
  prospective costs orders, and, 9.27–9.29
  QOCS, and, 9.11–9.13
  recoverability, 9.14–9.17
  retrospective cover, 9.22
  risks covered, 9.10
  security for costs, and, 9.28
 before the event
  client's indemnity, 9.7
  coverage, 9.5
  generally, 9.2–9.3
  hourly rate, 9.8
  top up cover, 9.6
  usage, 9.4
 introduction, 9.1
 'legal expenses insurance', 9.1
 solicitor self-insurance
  client's disbursements, 9.30
  opponent's costs, 9.31
  practical considerations, 9.32
 VAT, and, 2.20–2.21
Insurers
 non-party costs orders, and, 40.9
Intellectual Property Enterprise Court
 fixed costs, and
  assessment, 26.35
  costs outside the scales, 26.36
  introduction, 26.32
  order for costs, 26.37
  scale management, 26.33
  scale tables, 26.34
  unreasonable costs, 26.38
Interest
 Part 36 offers, and, 20.14
Interest on costs
 allocatur rule, 32.2
 backdating, 32.5
 conditional fee agreements, and
  introduction, 5.30
  out of pocket interest, 5.32
  punitive interest, 5.31
 court's discretion, 32.5
 deemed costs orders
  generally, 22.10
  introduction, 32.3
 enhancement, 32.4
 incipitur rule, 32.2
 introduction, 32.1
 post-dating, 32.5

Interest on costs – *cont.*
 rate, 32.6
 statutory framework, 32.1
Interim applications
 conditional fee agreements, and, 5.13
Interim bills
 account, on, 2.5
 final bills, and, 2.10
 indemnity principle, and, 12.7
 introduction 2.4
 statute bills, 2.6–2.8
Interim costs certificates
 generally, 28.39
 provisional assessment, 28.39
Interim statute bills
 agreement, by, 2.7
 introduction, 2.6
 natural break, and 2.8
Issue based orders
 costs awards between the parties,
  and, 22.21
Item-by-item basis
 indemnity principle, and, 12.8

J

Judgment
 fixed costs, and, 26.5
Judicial review
 Aarhus Convention, under, 26.45
Just in all the circumstances
 wasted costs orders, and, 23.7

L

'Legal aid'
 background, 10.3
 cost benefit analysis, 10.6
 costs awards between the parties,
  and, 22.35
 criteria
  cost benefit analysis, 10.6
  financial eligibility, 10.7
  introduction, 10.4
  merits, 10.5
 detailed assessment, and
  generally, 28.57
  procedure, 30.15
 excluded services, 10.3
 financial eligibility, 10.7
 generally, 10.2–10.3

Index

'Legal aid' – *cont.*
   introduction, 10.1
   Legal Aid Agency, 10.2
   merits, 10.5
   solicitor's lien, and, 3.34
   wasted costs orders, and, 23.11
Legal expenses insurance
   after the event
      adverse costs, 9.23–9.26
      avoidance of policy, 9.25
      challenging premium level, 9.18
      clinical negligence cases, 9.12
      collateral benefits, 9.20
      contents, 9.19–9.22
      costs budgeting, and, 9.27
      costs capping, and, 9.29
      counsel's fees, 9.21
      generally, 9.9
      non-party costs orders, 9.24
      other cases, 9.13
      personal injury cases, 9.11
      proceedings against the insurer, 9.24
      prospective costs orders, and, 9.27–9.
      QOCS, and, 9.11–9.13
      recoverability, 9.14–9.17
      retrospective cover, 9.22
      risks covered, 9.10
      security for costs, and, 9.28
   before the event
      client's indemnity, 9.7
      coverage, 9.5
      generally, 9.2–9.3
      hourly rate, 9.8
      top up cover, 9.6
      usage, 9.4
   introduction, 9.1
Legal services orders
   Children Act 1989 applications, 39.40
   criteria, 39.39
   generally, 39.39
Liens
   client's remedy, 3.36
   document to be handed over, 3.35
   general lien, 3.29
   loss, 3.32
   particular lien, 3.30
   Solicitors Act charging orders, 3.31
   state funding, 3.34
   termination of retainer, 3.33
Liquidators
   non-party costs orders, and, 40.13
Litigants in person
   costs budgets, and, 15.14

Litigants in person – *cont.*
   financial loss
      fast track trial costs, 35.7
      introduction, 35.4
      legal costs, 35.6
      two thirds rule, 35.5
   introduction, 35.1
   meaning, 35.2
   professional litigants, 35.3
   quantification, 35.8
Litigation friends
   detailed assessment, 34.4
   expenses incurred, 34.3
   generally, 34.2
Litigation funding
   Association of Litigation Funders, 10.15
   current position, 10.12
   expert evidence, 10.11
   extent of funder's liability, 10.10
   fee sharing, and, 10.16
   funding agreement, 10.13
   generally, 10.8
   introduction, 10.1
   maintenance and champerty, 10.9
   personal injury cases, 10.16
   profile of cases, 10.14
Low value RTA, EL and PL personal injury claims
   adjournment, 26.23
   approval hearings, 26.22
   avoidance behaviour, 26.21
   costs-only proceedings, 26.24
   counsel's advice, 26.18
   disbursements, 26.19
   fixed sums, 26.17
   introduction, 26.15
   London weighting, 26.17
   offers, 26.20
   Part 36 offers, and, 20.32
   stages of the protocols, 26.16

M

Maintenance and champerty
   commercial funders, and, 10.9
Medical reports
   VAT, and, 2.17
Mercantile Courts
   costs management, and, 15.4
Mesothelioma claims
   after the event insurance, and, 9.17
Minors
   see also **Children**

Index

Minors – *cont.*
  generally, 34.1
Misconduct
  detailed assessment, and, 28.20
Mortgage possession proceedings
  summary assessment, and, 27.10
Multi-party litigation orders
  costs capping, and, 37.5
  damages-based agreements, and, 37.6
  discontinuance, 37.2
  funding, and, 37.6
  generic costs, 37.4
  individual costs, 37.3
  introduction, 37.1

N

'Near miss' offers
  generally, 20.39
Negligent conduct
  wasted costs orders, and, 23.5
Negotiations
  security for costs, 19.17
No order as to costs
  costs awards between the parties, and, 22.2
No profit rule
  indemnity principle, and, 12.2
No win, no fee arrangements
  differential rates, and, 4.17
  QOCS, and, 5.28
  unsuccessful cases, 5.33
Non-chargeable time
  recording, 31.4
Non-contentious business agreements
  definition, 8.2
  generally, 8.15
  introduction, 8.1
  recovery procedure, 8.13
  solicitor mortgagee's costs, 8.16
  value
    high value estates, 31.36
    hourly rate plus value, 31.35
    introduction, 31.34
    routine work, 31.37
Non-party costs orders
  after the event insurance, and, 9.24
  applications
    approach of court, 40.5
    generally, 40.3
    notice requirements, 40.4
  categories of non-party, 40.8
  causation of costs, 40.7

Non-party costs orders – *cont.*
  company directors, 40.10–40.12
  family proceedings, 40.20
  funders, 40.16–40.17
  funding by non-party, and, 40.6
  general approach, 40.2
  insurers, 40.9
  introduction, 40.1
  liquidators, 40.13
  notice requirements, 40.4
  procedure, 40.3
  receivers, 40.13
  solicitors
    funders, as, 40.18
    lack of authority, for, 40.19
  tribunals, 40.14
  witnesses, 40.15
Notices of funding
  conditional fee agreements, and
    generally, 5.2–5.3
  insurers, 5.29

O

Offers to settle
see also **Part 36 offers**
  ADR, 21.1–21.6
  arbitration, and, 41.5
  'Calderbank' letters, 20.37
  conclusion, 20.41
  detailed assessment, and
    costs, 28.48
    generally, 28.25
  employment tribunals, 20.40
  'near miss' offers, 20.39
  offers outside of CPR Part 36
    'Calderbank' letters, 20.37
    consequences, 20.38
    generally, 20.36
    'near miss' offers, 20.39
  Part 36 offers
    acceptance after trial but before judgment, 20.11
    acceptance at trial, 20.10
    acceptance time period, 20.5–20.7
    background, 20.1
    clarification, 20.13
    commencement of trial, 20.12
    consequences of acceptance, 20.17–20.22
    consequences of non-acceptance, 20.23–20.24
    content, 20.2

847

Index

Offers to settle – *cont.*
Part 36 offers – *cont.*
counter offers, and, 20.16
disclosure, 20.15
fixed success fees, and, 20.30
future pecuniary loss claims, and, 20.27
interest, and, 20.14
introduction, 20.1
made less than 21 days before trial, 20.7
offerors, 20.3
pre-action protocols, and, 20.32
provisional damages claims, and, 20.28
QOCS, and, 20.25
recoupment of state benefit, and, 20.29
service, 20.6
small claims track, and, 20.33
success in part, and, 20.26
'tactical offer', and, 20.31
time limit, 20.5–20.7
timing, 20.4
VAT Tribunal, and, 20.34
withdrawal, 20.8–20.9
pre-6 April 2007 payments and offers, 20.35
Oppression
security for costs, 19.18
Overriding objective
between the parties costs, and, 11.2
family proceedings, and, 39.31–39.35

P

Part 18 requests
detailed assessment, and, 29.31
Part 21 parties
detailed assessment, 34.4
introduction, 34.1
litigation friends
detailed assessment, 34.4
expenses incurred, 34.3
generally, 34.2
meaning, 34.1
summary assessment, and, 27.12
Part 36 offers
acceptance
after trial but before judgment, 20.11
at trial, 20.10
consequences, 20.17–20.22
time period, 20.5–20.7
background, 20.1
clarification, 20.13
commencement of trial, 20.12

Part 36 offers – *cont.*
consequences of acceptance
after expiry of relevant time period, 20.21
introduction, 20.17
offer made by one of two or more defendants, of, 20.22
payment, 20.20
settlement of all or part of claim, 20.18
stay of proceedings, 20.19
consequences of non-acceptance
claimant's offer, of, 20.24
defendant's offer, of, 20.23
content, 20.2
counter offers, and, 20.16
deemed costs order, and, 22.6
disclosure, 20.15
fixed success fees, and, 20.30
future pecuniary loss claims, and, 20.27
interest, and, 20.14
introduction, 20.1
made less than 21 days before trial, 20.7
offerors, 20.3
pre-action protocols for low value RTA and EL/PL claims, and, 20.32
provisional damages claims, and, 20.28
QOCS, and, 20.25
recoupment of state benefit, and, 20.29
service, 20.6
small claims track, and, 20.33
success in part, and, 20.26
'tactical offer', and, 20.31
time limit
generally, 20.5
offer made less than 21 days before trial, 20.7
service of offer, 20.6
timing, 20.4
VAT Tribunal, and, 20.34
withdrawal
effect, 20.9
permission, 20.8
Patent proceedings
costs management, and, 15.4
fixed costs, and
assessment, 26.35
costs outside the scales, 26.36
introduction, 26.32
order for costs, 26.37
scale management, 26.33
scale tables, 26.34
unreasonable costs, 26.38
Patients
see also **Protected parties**
generally, 34.1

Index

Payments on account
  between the parties costs, and, 25.1
  conditional fee agreements, and
    following acceptance of Part 36 offer, 5.24
    following final hearing, 5.23
    generally, 5.1
Personal injury cases
  after the event insurance, and, 9.11
  conditional fee agreements, and
    calculation of cap, 5.25
    generally, 4.8
  damages-based agreements, and, 7.11
Persons under a disability
  see also **Protected parties**
  generally, 34.1
Pilot schemes
  costs management, and. 15.3
Points of dispute
  calculation of offers, 29.29
  format, 29.28
  introduction, 29.26
  service, 28.24
  structure, 29.27
Practice Direction
  between the parties costs, and, 11.14
Pre-action disclosure
  costs awards between the parties, and, 22.24
Pre-action protocols for low value RTA and EL/PL claims
  Part 36 offers, and, 20.32
Pre-emptive orders
  trustees, and, 38.3
Privacy proceedings
  after the event insurance, and, 9.17
Privilege
  wasted costs orders, and, 23.9
Pro bono representation
  detailed assessment, and, 28.28
  indemnity principle, and, 12.4
Profit element
  assessment, 31.13–31.14
  generally, 31.10
  introduction, 31.8
  level to claim, 31.18–31.20
  supervision, 31.12
  test, 31.21
Property Chamber
  detailed assessment, and, 28.59
Proportionality
  arbitration, and, 41.7
  bases of costs, and, 24.6

Proportionality – *cont.*
  costs control
    introduction, 14.1
    justice, and, 14.4
    origins, 14.2–14.3
  costs management, 15.1
  generally, 11.11
  introduction, 11.2
  summary assessment, and., 27.21
Prospective costs control
  after the event insurance, and
    costs budgeting, 9.27
    costs capping, 9.29
    security for costs, 9.28
  case management, and
    allocation of proceedings, 15.23
    assessment of costs, 15.30
    disclosure, 15.26
    experts, 15.27
    interaction with costs management, 15.22–15.30
    introduction, 15.1
    relief from sanctions, 15.28
    standard direction templates, 15.24
    trial, 15.29
    witness statements. 15.25
  costs budgets
    agreement, 15.8
    approval, 15.7
    cases not costs managed, in, 16.1
    filing, 15.5–15.6
    relevance, 15.11–15.12
    revision, 15.8
    sanctions for failing to file and serve, 15.6
    service, 15.5–15.6
    setting, 15.31–15.33
  costs capping
    background, 17.1
    case law, 17.3–17.6
    conclusion, 17.7
    criteria, 17.4
    introduction, 17.1
    jurisdiction, 17.2
    procedure, 17.6
    timing, 17.5
  costs management
    admiralty proceedings, 15.4
    agreement of budgets, 15.8
    application, 15.4
    approval of budgets, 15.7
    case law, 15.34
    Chancery Division cases, 15.4
    Civil Procedure Rules, 15.4

Index

Prospective costs control – *cont.*
  costs management – *cont.*
    conclusion, 15.35
    costs of, 15.10
    costs outside scope of budgets, 15.9
    exceptions to application, 15.4
    filing budgets, 15.5–15.6
    fixed costs proceedings, 15.4
    Form H, 15.15–15.21
    good reason, 15.13
    hourly rates, 15.32
    indemnity basis assessment,
      and, 15.12–15.13
    interaction with case
      management, 15.22–15.30
    introduction, 15.1
    litigants in person, 15.14
    meaning, 15.2
    Mercantile Courts cases, 15.4
    orders, 15.2
    patent cases, 15.4
    pilot schemes, 15.3
    procedural code, 15.5–15.14
    proportionality, and, 15.1
    relevance of budgets, 15.11–15.12
    revision of budgets, 15.8
    sanctions for failing to file and serve
      budgets, 15.6
    scale costs proceedings, 15.4
    scope of regime, 15.4
    service of budgets, 15.5–15.6
    setting the budget, 15.31–15.33
    standard basis assessment, and, 15.11
    TCC cases, 15.4
  introduction, 13.1
  proportionality
    introduction, 14.1
    justice, and, 14.4
    origins, 14.2–14.3
  protective costs orders
    Aarhus Convention cases, 18.4
    Beddoe orders, 18.5
    case law, 18.2
    exceptional circumstances, 18.2
    future, 18.3
    introduction, 18.1
    principles, 18.2
  security for costs
    admissions, 19.16
    amount, 19.22
    appeals, 19.12
    applicants, 19.2
    arbitration, and, 19.24
    change of address with view to evade
      consequences of litigation, 19.8

Prospective costs control – *cont.*
  security for costs – *cont.*
    co-claimants, 19.19
    companies, 19.6–19.7
    conditions, 19.4–19.11
    discretion, 19.14
    failure to comply with CPR, 19.21
    failure to give address, 19.9
    introduction, 19.1
    just with regard to all the
      circumstances, 19.13–19.20
    negotiations, 19.17
    nominal claimants, 19.10
    oppression, 19.18
    pre-action costs, 19.20
    prospects of success, 19.15
    residence outside the jurisdiction, 19.5
    steps taken in relation assets to avoid
      consequences of adverse
      order, 19.11
    third parties, against, 19.3
    variation of order, 19.23
    trustees, and, 38.4
Protected parties
  detailed assessment, 34.4
  introduction, 34.1
  litigation friends
    detailed assessment, 34.4
    expenses incurred, 34.3
    generally, 34.2
  meaning, 34.1
  summary assessment, and, 27.12
Protective costs orders
  Aarhus Convention cases, 18.4
  Beddoe orders, 18.5
  case law, 18.2
  exceptional circumstances, 18.2
  future, 18.3
  introduction, 18.1
  principles, 18.2
Provisional assessment
  challenging, 30.12
  changes to CPR
    generally, 11.9
    transitional provisions, 11.13
  conduct, 30.7
  costs, 30.9
  generally, 28.40
  interim costs certificates, 28.39
  introduction, 30.1
  notification of parties, 30.8
  overview, 28.2
  Part 36 offers, 30.10
  pilot scheme, 30.2

Provisional assessment – *cont.*
 procedure, 30.3–30.8
 requests, 30.3
 six week limit, 30.6
 subsequent hearing, 30.13
 suitability, 30.4
 supporting papers, 30.5
 termination, 30.11
 time limits, 30.6
 transitional provisions, 11.13
Provisional damages
 detailed assessment, and, 28.5
 Part 36 offers, and, 20.28
Public funding
 background, 10.3
 cost benefit analysis, 10.6
 costs awards between the parties, and, 22.35
 criteria
  cost benefit analysis, 10.6
  financial eligibility, 10.7
  introduction, 10.4
  merits, 10.5
 excluded services, 10.3
 financial eligibility, 10.7
 generally, 10.2–10.3
 indemnity basis, and, 24.21
 introduction, 10.1
 Legal Aid Agency, 10.2
 merits, 10.5
 solicitor's lien, and, 3.34
 summary assessment, and, 27.11
 wasted costs orders, and, 23.11
Public liability personal injury claims
 adjournment, 26.23
 approval hearings, 26.22
 avoidance behaviour, 26.21
 claims no longer continuing under the protocols
  counterclaims, 26.30
  defendant's costs, 26.29
  disbursements, 26.28
  exceeding the fixed sums, 26.31
  general points, 26.26
  interim applications, 26.27
  introduction, 26.25
  London weighting, 26.26
  VAT, 26.26
 costs-only proceedings, 26.24
 counsel's advice, 26.18
 disbursements, 26.19
 fixed sums, 26.17
 introduction, 26.15
 London weighting, 26.17

Public liability personal injury claims – *cont.*
 offers, 26.20
 stages of the protocols, 26.16
Publication proceedings
 after the event insurance, and, 9.17

Q

Qualified one way costs shifting (QOCS)
 after the event insurance, and, 9.11–9.13
 conditional fee agreements, and, 5.28
 generally, 11.5
 Part 36 offers, and, 20.25
 transitional provisions, 11.12
Quantification of costs
 appeals against assessments, 33.1–33.11
 detailed assessment
  procedure, 28.1–28.61
  provisional assessment, 30.1–30.16
  statements of case, 29.1–29.31
 fixed costs, 26.1–26.49
 interest, 32.1–32.6
 summary assessment, 27.1–27.23
 time and value, 31.1–31.39
Quotations
 generally, 1.9
 introduction, 1.8

R

'Ready reckoner'
 conditional fee agreements, and, 4.21
Re-allocation
 costs awards between the parties, and, 22.26
Reasonableness
 time, and, 31.27
Receivers
 non-party costs orders, and, 40.13
Recoupment of state benefits
 Part 36 offers, and, 20.29
Recoverability of premiums
 after the event insurance, and
  clinical negligence cases, 9.16
  insolvency proceedings, 9.17
  mesothelioma claims, 9.17
  other, 9.14
  'pre-commencement' policies, 9.15
  publication and privacy proceedings, 9.17

Index

Recovery of costs
applications by the client
  challenge level of costs only, to, 3.22
  delivery of a bill, for, 3.16
  detailed assessment, for, 3.19
  expiry of twelve months following payment, 3.21
  further information, for, 3.17
  set aside contentious business agreement, to, 3.18
  special circumstances, 3.20
applications by third party
  introduction, 3.23
  mortgagee's costs, 3.25
  residuary beneficiaries, 3.24
bankruptcy petition, 3.28
court procedure
  basis of assessment, 3.14
  costs of assessment, 3.15
  default judgment, 3.12
  directions, 3.13
  interim payments, 3.12
  place of proceedings, 3.11
credit checks, 3.2
delivery of bill
  application for detailed assessment, 3.7
  client information, 3.8
  elapse of one month, 3.6
  introduction, 3.5
  limitation period, 3.9
  order for detailed assessment, 3.7
  recoverable costs trap, 3.10
final bill, 3.3
introduction, 3.1
pre-action procedures
  credit checks, 3.2
  delivery of bill, 3.5–3.10
  final bill, 3.3
  signature of bill, 3.4
signature of bill, 3.4
solicitor's lien
  client's remedy, 3.36
  document to be handed over, 3.35
  general lien, 3.29
  loss, 3.32
  particular lien, 3.30
  Solicitors Act charging orders, 3.31
  state funding, 3.34
  termination of retainer, 3.33
statutory demand, 3.27
without proceedings
  bankruptcy petition, 3.28
  introduction, 3.26
  solicitor's lien, 3.29–3.36

Recovery of costs – *cont.*
without proceedings – *cont.*
  statutory demand, 3.27
Reduction in damages
  damages-based agreements, and, 7.14
Regional costs judges
  detailed assessment, and, 28.10
Relief from sanction
  between the parties costs, and, 11.3
  costs management, and, 15.28
Replies
  detailed assessment, and
    generally, 29.30
    introduction, 28.26
Reserved costs
  costs awards between the parties, and, 22.15
Retainer
  'client', 1.2
  client care
    Code of Conduct 2007, 1.19
    costs, 1.22
    current regulatory position, 1.20
    indicative behaviours, 1.23
    in-house practice, 1.24
    old regulatory position, 1.18–1.19
    outcomes-focused regulation, 1.21–1.22
    overseas practice, 1.24
    status of personnel, 1.25
  complaints, 1.26
  contingency agreements, 1.31
  contracts, and
    cancellable agreements, 1.7
    client's instructions, 1.6
    entire contract, 1.4
    generally, 1.3
    increase in hourly rate, 1.5
  entire contract, and, 1.4
  estimates
    client's reliance, 1.14
    costs budgets, and, 1.16
    exceeding, 1.13
    failure to give, 1.15
    generally, 1.10
    increases, 1.12
    introduction, 1.8
    qualified estimates, 1.11
    recoverable fixed costs, and, 1.17
  indemnity principle, and, 12.10
  introduction, 1.1
  quotations
    generally, 1.9
    introduction, 1.8
  'solicitor', 1.2

Retainer – *cont.*
 termination by client, 1.27
 termination by solicitor
  client loses capacity, where, 1.30
  good cause, for, 1.29
  introduction, 1.28
 waiver of costs, 1.32
 writing off costs, 1.32
Retrospective cover
 after the event insurance, and, 9.22
Revenue proceedings
 fixed costs, and, 26.39
Right to fair trial
 delay in detailed assessment, and, 28.21
 family proceedings, and, 39.39
Risk assessment
 benefits, 4.19
 forms, 4.20
 limits, 4.18
 'ready reckoner', 4.21
RTA personal injury cases
 base fee, 26.9
 claims no longer continuing under the protocol
  counterclaims, 26.30
  defendant's costs, 26.29
  disbursements, 26.28
  exceeding the fixed sums, 26.31
  general points, 26.26
  interim applications, 26.27
  introduction, 26.25
  London weighting, 26.26
  VAT, 26.26
 conditions, 26.8
 costs formula, 26.9
 counsel's fees, 26.13
 disbursements, 26.11
 exceeding the fixed sums, 26.14
 introduction, 26.8
 London weighting, 26.10
 low value claims
  adjournment, 26.23
  approval hearings, 26.22
  avoidance behaviour, 26.21
  costs-only proceedings, 26.24
  counsel's advice, 26.18
  disbursements, 26.19
  fixed sums, 26.17
  introduction, 26.15
  London weighting, 26.17
  offers, 26.20
  stages of the protocols, 26.16
 medical fees, 26.12
 percentage uplift, 26.9

RTA personal injury cases – *cont.*
 VAT, 26.9

S

Sanderson orders
 generally, 22.34
Satellite litigation
 wasted costs orders, and, 23.3
'Save as to costs'
 detailed assessment, and, 28.49
 summary assessment, and, 27.14
Scale costs proceedings
 costs management, and, 15.4
SCCO Guide
 time, and, 31.22–31.24
Security for costs
 admissions, 19.16
 after the event insurance, and, 9.28
 amount, 19.22
 appeals, 19.12
 applicants, 19.2
 arbitration, and, 19.24, 41.15
 change of address with view to evade consequences of litigation, 19.8
 co-claimants, 19.19
 companies
  generally, 19.6
  inability to pay, 19.7
 conditional fee agreements, and, 5.17
 conditions
  change of address, 19.8
  companies, 19.6–19.7
  failure to give address, 19.9
  introduction, 19.4
  nominal claimants, 19.10
  residence outside the jurisdiction, 19.5
  steps taken to avoid consequences of order, 19.11
 discretion, 19.14
 failure to comply with CPR, 19.21
 failure to give address, 19.9
 introduction, 19.1
 just with regard to all the circumstances
  admissions, 19.16
  co-claimants, 19.19
  discretion, 19.14
  introduction, 19.13
  negotiations, 19.17
  oppression, 19.18
  pre-action costs, 19.20
  prospects of success, 19.15
 negotiations, 19.17

Index

Security for costs – *cont.*
 nominal claimants, 19.10
 oppression, 19.18
 pre-action costs, 19.20
 prospects of success, 19.15
 residence outside the jurisdiction, 19.5
 steps taken in relation assets to avoid
  consequences of adverse order, 19.11
 third parties, against, 19.3
 variation of order, 19.23
Set-off
 costs awards between the parties,
  and, 22.32
 damages-based agreements, and, 7.14
Signature of bill
 recovery of costs, and, 3.4
Small claims track
 allocation, 26.49
 costs awards between the parties,
  and, 22.25
 disbursements, 26.48
 general rule, 26.47
 introduction, 26.46
 Part 36 offers, and, 20.33
 re-allocation, 26.49
Solicitor and client
 and see under individual headings
 bills 2.1–2.21
 recovery of costs 3.1–3.34
 retainer 1.1–1.16
Solicitor self-insurance
 client's disbursements, 9.30
 opponent's costs, 9.31
 practical considerations, 9.32
Solicitors
 non-party costs orders, and
  funders, as, 40.18
  lack of authority, for, 40.19
Solicitor's lien
 client's remedy, 3.36
 document to be handed over, 3.35
 general lien, 3.29
 loss, 3.32
 particular lien, 3.30
 Solicitors Act charging orders, 3.31
 state funding, 3.34
 termination of retainer, 3.33
Standard basis assessment
 costs management, and, 15.11
 generally, 24.1
Standard directions
 costs management, and, 15.24
State funding
 background, 10.3

State funding – *cont.*
 cost benefit analysis, 10.6
 costs awards between the parties,
  and, 22.35
 criteria
  cost benefit analysis, 10.6
  financial eligibility, 10.7
  introduction, 10.4
  merits, 10.5
 excluded services, 10.3
 financial eligibility, 10.7
 generally, 10.2–10.3
 indemnity basis, and, 24.21
 introduction, 10.1
 Legal Aid Agency, 10.2
 merits, 10.5
 solicitor's lien, and, 3.34
 wasted costs orders, and, 23.11
Statements of case
 detailed assessment, and
  bill of costs, 29.2–29.25
  generally, 28.22
  introduction, 29.1
  Part 18 requests, 29.31
  points of dispute, 29.26–29.29
  reply, 29.30
Statements of costs
 summary assessment, and, 27.16
Statutory demand
 recovery of costs, and, 3.27
Statutory demand
 recovery of costs, and, 3.27
Striking out
 non-payment of fees, 22.8
Success fees
 conditional fee agreements, and
  maximum rate, 4.6
  percentage of client's damages, as, 4.7
Summary assessment
 appeals, 27.22
 benefits, 27.3
 conclusion, 27.23
 counsel's fees, 27.20
 exceptions to general rule
  Admiralty court, 27.7
  agreed costs, 27.13
  agreement 'save as to costs', 27.14
  children, 27.12
  Commercial Court, 27.7
  costs in the case, 27.8
  deferred payment retainer, 27.9
  good reason not to undertake
   assessment, 27.6
  introduction, 27.5

Index

Summary assessment – *cont.*
  exceptions to general rule – *cont.*
    mortgage possession proceedings, 27.10
    protected parties, 27.12
    publicly funded party, 27.11
  extent of costs to be assessed, 27.15
  family proceedings, and, 39.45–39.46
  general rule, 27.4
  hearings, 27.18
  hourly rates, 27.19
  introduction, 27.1
  procedure, 27.17
  proportionality., 27.21
  role of trial judge, 27.2
  statement of costs, 27.16
Supreme Court proceedings
  detailed assessment, and, 28.58

T

Technology and Construction Court
  costs management, and, 15.4
Termination of retainer
  client, by, 1.27
  conditional fee agreements, and
    drafting considerations, 4.12
    introduction, 5.18
    insufficient prospect of success, 5.20
    non-cooperation of client, 5.21
    other funding available, 5.19
    rate of charge, 5.22
  solicitor, by
    client loses capacity, where, 1.30
    good cause, for, 1.29
    introduction, 1.28
Third party payments
  indemnity principle, and, 12.5
Time
  'A' factor
    generally, 31.9
    introduction, 31.8
    'Seven Pillars of Wisdom', 31.14
    supervision, 31.12
    test, 31.21
    unrecordable time, 31.11
  assessment, 31.26
  'B' factor
    assessment, 31.13–31.14
    generally, 31.10
    introduction, 31.8
    level to claim, 31.18–31.20
    supervision, 31.12

Time – *cont.*
  'B' factor – *cont.*
    test, 31.21
  budgeting, and, 31.38
  charging tool, as
    alternatives, 31.40
    generally, 31.5–31.6
  counsel's fees
    brief fees, 31.33
    generally, 31.31–31.322
    table, 31.31
  direct cost
    generally, 31.9
    introduction, 31.8
    'Seven Pillars of Wisdom', 31.14
    supervision, 31.12
    test, 31.21
    unrecordable time, 31.11
  distant solicitor, 31.30
  fee earners
    level, 31.28
    locality, 31.25
    targets, 31.39
  hourly rate
    'A' factor, 31.9
    amount of rate, 31.29
    ascertaining the direct cost, 31.9
    ascertaining the profit element, 31.10
    assessment of 'B' factor, 31.13–31.14
    'B' factor, 31.10
    factors, 31.8
    generally, 31.7
    'Seven Pillars of Wisdom', 31.14
    supervision, 31.12
    unrecordable time, 31.11
  introduction, 31.1
  level of fee earner, 31.28
  locality of fee earner, 31.25
  management tool, as
    alternatives, 31.38–31.39
    generally, 31.2–31.4
  profit element
    assessment, 31.13–31.14
    generally, 31.10
    introduction, 31.8
    level to claim, 31.18–31.20
    supervision, 31.12
    test, 31.21
  proof, 31.26
  reasonableness, 31.27
  recording non-chargeable time, 31.4
  SCCO Guide, 31.22–31.24
  The Expense of Time, 31.3

Index

Tomlin orders
generally, 22.12
Transitional provisions
between the parties costs, and, 11.10–11.13
Trial
costs management, and, 15.29
Tribunal proceedings
non-party costs orders, and, 40.14
wasted costs orders, and, 23.33
Trustees
costs against, 38.2
detailed assessment, 38.5
entitlement to costs out of trust or estate, 38.1
pre-emptive orders, 38.3
prospective costs orders, 38.4
Trusts
detailed assessment, and, 30.16

U

Unpaid disbursements
final bills, and, 2.11
Unreasonable conduct
wasted costs orders, and, 23.5

V

Value
amount, 31.15
circumstances, 31.17
experience, 31.16
introduction, 31.15
non-contentious business
high value estates, 31.36
hourly rate plus value, 31.35
introduction, 31.34
routine work, 31.37
place, 31.17
skill, 31.16
VAT
chargeability, 2.13
counsel's fees, 2.18
disbursements, 2.16
fixed costs recovered, on, 2.19
insurance claims, 2.20–2.21
introduction, 2.13
medical reports, on, 2.17
rate, 2.14
tax point, 2.15

VAT Tribunal
Part 36 offers, and, 20.34

W

Waiver of costs
generally, 1.32
Wasted costs orders
advocacy, and, 23.12
barrister's advice, and, 23.14
case decisions
successful applications, 23.15–23.18
unsuccessful applications, 23.19–23.32
causation, 23.6
conclusion, 23.34
discretion, 23.8
family proceedings, and
generally, 39.5
overriding objective, 39.35
improper conduct, 23.5
introduction, 23.1
just in all the circumstances to make an order, 23.7
negligent conduct, 23.5
notice of intention to apply, 23.10
privilege, and, 23.9
procedure, 23.2
public funding, and, 23.11
reliance on counsel by solicitor, 23.13
satellite litigation, 23.3
statutory framework, 23.1
threats to apply, 23.10
three-stage test
causation, 23.6
improper conduct, 23.5
introduction, 23.4
just in all the circumstances to make an order, 23.7
negligent conduct, 23.5
unreasonable conduct, 23.5
tribunal proceedings, in, 23.33
unreasonable conduct, 23.5
Without prejudice save as to costs
detailed assessment, and, 28.49
summary assessment, and, 27.14
Witness statements
costs management, and, 15.25
Witnesses
costs management, and, 15.25
non-party costs orders, and, 40.15
Writing off costs
generally, 1.32

ARE THORESEN is a Norwegian veterinary surgeon specialising in anthroposophic medicine, homeopathy, acupuncture, osteopathy and agriculture. Since 1981 he has run a private holistic practice in Sandefjord, Norway, for the healing of people, horses and small animals. He has lectured widely, specializing in veterinary acupuncture, and has published dozens of scholarly articles. In recent years, his work has focused mainly on spiritual medicine – based primarily on a deeper understanding of anthroposophy and the medical teachings of Rudolf Steiner – and investigating the spiritual world, resulting in seven books: *Demons and Healing* (2018), *Experiences from the Threshold* (2019), *Spiritual Translocation* and *The Lucifer Deception* (both 2020), *Transforming Demons* (2021), *Travels on the Northern Path of Initiation* (2021) and *Encounters with Vidar* (2022).

MEETING MICHAEL

Further Communications from Spirit Worlds

Are Simeon Thoresen, DVM

TEMPLE LODGE

Dedicated to All Who Seek to Heal and Understand

Temple Lodge Publishing Ltd.
Hillside House, The Square
Forest Row, RH18 5ES

www.templelodge.com

First published in English by Temple Lodge in 2024

© Are Simeon Thoresen, DVM 2024

This book is copyright under the Berne Convention. All rights reserved. Apart from any fair dealing for the purpose of private study, research, criticism or review, no part of this publication may be reproduced, stored in a retrieval system, or transmitted in any form or by any means, electronic, electrical, chemical, mechanical, optical, photocopying, recording or otherwise, without the prior written permission of the copyright owner. Inquiries should be addressed to the Publishers

The right of Are Simeon Thoresen to be identified as the author of this work has been asserted in accordance with sections 77 and 78 of the Copyright, Designs and Patents Act, 1988

A CIP catalogue record for this book is available from the British Library

ISBN 978 1 915776 14 3

Cover by Morgan Creative
Typeset by Symbiosys Technologies, Visakhapatnam, India
Printed and bound by 4Edge Ltd., Essex

'Whoever recognizes Vidar in all his significance and feels him in his soul will understand that in the twentieth century the capacity to behold the Christ can again be given to man; Vidar, who is close to all of us in northern and central Europe, will again stand before him. He was held secret in the Mysteries and occult schools as the god who will receive his task only in the future.'

Rudolf Steiner, Oslo, 17 June 1910*

---

*GA 121, *The Mission of Folk Souls*.

Contents

Foreword	1
My Lecture Tour in the USA	5
Meeting Scythianus	11
Meeting Ymir	19
Meeting Anna	32
Entering Devachan	40
Meeting Ariel	43
The Astral World	60
Meeting the Antichrist	67
Meeting Michael	74
The Higher Angelic Realms	95
An Overview of Devachan	104
Afterword: Reflections on the Vidar and Michael Paths	106
Appendix: An Investigation into the Differences between Study and Spiritual Experience	110

Foreword

This book is a continuation of my previous publication *Encounters with Vidar* (2022), which in turn was a sequel to my book *Travels on the Northern Path of Initiation* (2021). In these volumes, I described how, during a period of intense fatigue after having caught Covid-19, I traversed all three realms of the elemental world, eventually reaching what I call the 'outer realm' of the etheric.

In *Travels on the Northern Path of Initiation*, I journeyed into the elemental world, through the third (luciferic), the second (ahrimanic) and first (azuric) realms of the elemental world, up to the threshold of the outer etheric. In between the first elemental realm and the outer etheric world, I was met by the guardian of the threshold, who introduced himself as Vidar. Vidar in turn introduced me to his colleague, Balder. Vidar teaches us what we as individuals need to know in order to travel further into the outer etheric world, where the presence of Christ may be found. In *Encounters with Vidar*, I expanded upon the teachings I received from Vidar and Balder. I also tried to address some of the mistakes we can make today as seekers of the spirit.

I wrote the above books at the same time as these spiritual journeys, conversations and teachings were taking place. That is why the contents may sometimes appear piecemeal and have the character of a diary. In reality, these writings represent ongoing research and investigation.

In the final volume of this trilogy, I take the reader further into spirit worlds – all the way into the devachanic region, where we will meet the spiritual entities that dwell there. The first entity that introduced himself to me had a threefold name, Serat–Saturn–Scythianus. This being directed me to a different path than the one that I had, up until now, been following.

The path Vidar showed me was through the mysteries of the spirit, the spiritual world. Now 'S' (as I will call 'him') directed me to the mysteries of the earth. As I say, this was a totally different path, but one that would turn out to be even more 'spiritual'. Through this path, through the mysteries of the earth (and physical body), I was introduced to Ymir, the old spirit of the earth itself, who existed before Ancient Saturn. Ymir showed me her twelve layers, and in the seventh layer I met Anna (the mother of Mary and the grandmother of Jesus), who turned out to be the portal between physical creation and the higher spiritual world, Devachan. Anna is a cosmic being, who facilitated Mary's entrance to this world, giving the possibility for Jesus to incarnate, who in turn gave the Christ being the ability to incarnate. Anna is the guardian between the earth mysteries and Devachan, and is the female aspect of the spiritual path into Devachan.

All this was known, apparently, in the School of Chartres. Dominating the northern wall (the feminine direction) of this magnificent cathedral – which is based on the mysteries of the feminine – and overlooking everybody that enters or leaves, is the guardian or gatekeeper, Anna. This tells us a lot.

Through the spiritual portal provided by Anna, I was able to enter Devachan, a higher spiritual world that has been in existence before Ancient Saturn, and which will exist after Vulcan. In the devachanic world, the true mission of the human being is revealed. This book will describe my journey to the devachanic world. Also, I will discuss the Michael path compared to the Vidar path, and the 'spiritual mystery path' compared to the 'earthly mystery path'. I will also speak about my meetings with beings named Ariel and Makuk. Finally, I engage in a 'conversation' with the Antichrist and explore the realms of Michael, our 'Time Spirit' or 'Time Archai'.

In ending this introduction, I would like to address a question I have been asked: How can one be sure that the identity presented by different spiritual beings is the true one? What if adversarial entities try to lead one astray by pretending that they are good or beneficial to humankind? I once asked this same question to a priest of The Christian Community in Stuttgart. He answered that this was quite easy and simple. If we meet a spiritual entity and ask its name, it has to reveal the truth. The priest added that if the name was of an adversarial being, one could even smell sulphur as it disappeared!

Each time I have asked spiritual beings their names, I am quite convinced that they have answered truthfully – both the beneficial and the adversarial beings. It seems that there is a law in the spiritual world that applies to all entities, that when asked their true name, they must reveal it. Concerning many other laws and rules, the adversarial entities follow different ones to those of the good spiritual beings, and this includes the demonic, ahrimanic, luciferic and azuric beings.

With this book, I am once again inviting you to share the observations and reports of my current research. I thank you for following me and taking part – through your own consciousness – in these thoughts and ideas. Many of these findings are still in the process of 'becoming'. Thus, the contents are not intended as fixed teachings, new dogmas or beliefs, but are brought forth in the spirit of open, spiritual-scientific exploration and discovery. They may yet need to be modified in future.

I also ask the reader to bear in mind that English is not my mother tongue. Whilst I am grateful to my Editor at Temple Lodge Publishing for his patient help with grammar and composition, it should be remembered that I am translating metaphysical experiences not only into human language, but also into another language. Nevertheless, as author I take responsibility for any errors that remain.

Are Thoresen, February 2024
Sandefjord, Norway

My Lecture Tour in the USA

After the publication of my second book on Vidar (*Encounters with Vidar*), I was invited to lecture in several different locations in the USA and Canada during the month of August 2022. I was to get the first indications that this tour would be significant when I arrived at the airport in Norway in the last days of July. I carried my Covid-19 certificate – I had been infected with the virus several times and had high levels of antibodies in my blood, in addition to a single shot of the vaccine. The lecture tour was planned, all the preparations were made and all the tickets were sold.

But then I was stopped at the gate. Only fully-vaccinated persons were allowed to travel to Canada and/or the USA, I was told. Thus, I was forced to travel home, where I called the organizers of the tour in North America to inform them that I would not be coming after all. They were shocked, as was I! I did not know what to do, and went to bed.

Early the next morning I got up and looked again at my health documents to ponder on what to do. During the night something remarkable had happened. Somehow, the page had been changed. I had 'received' two further vaccinations, dated in time to allow me to travel. So, I was suddenly free to go. I just had to jump on a plane, where I was welcomed by the American and Canadian authorities.

I have described in some of my books how elemental beings are able to make changes in the physical world, should this be necessary. Such 'operations' are referred to in many places (and sometimes related to the activity of a special group of elemental beings, 'poltergeists'). I have witnessed their actions previously, so this was nothing new to me. I just accepted that they wanted me to go to the USA – that I was needed there. So, after buying new tickets for the first and last flight of the tour, I began my journey again, delayed only by two days.

The main topic and subject of my teaching during this tour was how the whole elemental world longed for – wanted and needed – human beings to become aware of them; to respect their activity and, most importantly, bring the consciousness of Christ into the elemental world. This was so that these beings could be redeemed, saved and thoroughly Christianized. Clearly, these beings needed me to conduct this lecture tour!

I lectured about the transformation of elemental beings through Christ, in disease, agriculture and in technical devices. I spoke in the following places: Toronto, Calgary, Fairbanks, Chicago, Detroit, Ann Arbor, Philadelphia, Kimberton, Chestnut Ridge, Ghent, Wilton and Boston. Everywhere, I was welcomed heartily and thanked for bringing such news, views and techniques to the listening audiences. By bringing these insights to so many people and places, my own relationship to the spiritual world also began to change.

As long as I was travelling, meeting new people every day, sleeping in new rooms almost every night, giving lectures almost every day – often twice a day – I became so tired that the strength of my communications with Vidar or the etheric world were diminished. The exception was in Boston, when the folks that owned the house in which I was staying asked me to help with their tree that had become diseased. Could I ask the tree's elemental being what sort of medicine or treatment it needed?

The answer came over three days (although by the third day I had already left for further travels). On the first day I received an Imagination, which showed a flow of water that was a bit more blue-green than usual. On the second day I received an Inspiration, communicating that the blue-green colour was because of the copper content. On the third day I received the complete information, i.e. that they should spray the tree with a 0.02% copper solution. The woman of the house told me that they had already considered that.

For the first week after I returned home, whilst recovering from my travels, spiritual communications were absent. But then, about a week after I had arrived, and as the worst of the tiredness had subsided, something new appeared. For the first time in six weeks, I travelled to the threshold of Vidar and beyond. There, the 'outer etheric realm' was suddenly populated with higher angelic beings, and not just with unknown entities, as previously. These higher angelic beings were of the second hierarchy of Angels, known in olden times as Exusiai, Dynamis and Kyriotetes (or, in Western esoteric language, Powers, Virtues and Dominions). I knew intuitively that they were of this rank.

These angelic entities were just to be observed on the first day, and their presence even coloured the normal, physical clouds (which can be observed in the 'material' sky) in red and green. The same day as these manifestations appeared to me, some of my patients also mentioned that they perceived such colours as they were receiving treatment.

Later that day, I read in a book by Harrie Salman* in which he speaks about the First Class lessons conducted by Rudolf Steiner in February 1924. As he describes, Steiner's students in meditation and initiation would first meet the lesser guardian of the threshold, who indicates three hindrances to stepping over the threshold: fear of the spiritual, mocking the spiritual , and laziness in our thinking. Only our courage to understand, our fiery will to see and the work of our will could 'open the door so that we can enter the spiritual temple'. We are then led to the greater guardian, which is related to Christ. This way also leads us to the middle of our heart, to a transformation of our feelings (Sophia) and also to a transformation of our higher 'I', and as 'a Michaelic path of knowledge to a higher consciousness ... To a culture characterized by *love*, to a transformation of evil to something good', as described in the Manichean way of life and initiation.

This path, that is called the Manichean way of initiation, is the same path taken by Parsifal, who actually *was* Mani or Manu, and according to Steiner, this Mani will become the Manu (great leader) of the next time cycle, beginning after the seventh cultural age of this present fifth cycle, the post-Atlantean epoch. This path is the same as the one I describe in my book *Travels on the Northern Path of Initiation*. The turning point in Parsifal's path was when he 'faded' three times into the three drops of blood in the snow. Before that, his life had been full of problems and mistakes, and his path was diffuse and difficult. Suddenly, after having faded

---

Rudolf Steiner and Peter Deunov: Anthroposophy and the White Brotherhood on the New Man, LogoSophia, 2023.

three times into the three drops of blood, the path suddenly became clear, and he found the way to King Arthur and later the Grail King. This 'fading-method' is exactly the one I describe in my book, and also the one that Rudolf Steiner referred to in Helsinki on 3 April 1912.*

According to Harrie Salman, the Templars were also a branch of the Manichean stream. In taking a closer look at the initiation practices of the Templars, as described by Judith von Halle in her two-volume book *Die Templer* ('The Templars') – where she describes their methods in detail – I started to see a certain similarity to my own training. Von Halle says that the Templars had found a special stone in the Pyrenees that expressed the 'I'-force or consciousness of the earth, and on which they placed their fingertips. Through this method, they encountered the etheric soul of the earth and the whole cosmos. This technique is almost exactly the same as I describe in my books. In using the fingertips placed in connection with the blood (pulse), I am able to come in contact with the whole body of the patient – even his family and surroundings. Thus, through the blood I am enabled to make a diagnosis ('spiritual pulse-diagnosis').†

Thus, the two pathways into the spiritual world or elemental world I have found and extensively practised in my life – the 'fading' method and the 'fingertip' method – can both be characterized as being inspired by, or as a part of, the Manichean way of initiation, at least according to the book quoted above. The only main difference

---

*See *The Spiritual Beings in the Heavenly Bodies and in the Kingdoms of Nature*, Lecture 1, SteinerBooks 2011.

†See further in my book *Demons and Healing*.

is that it is now Vidar, and not Michael, that is the Archangel acting as the guardian of the threshold. This is due to Michael's current position as an Archai (Principality), which we will refer to later in this book. This Manichean stream manifested itself in the Cathar movement in France, the Bogomil movement in Bulgaria, the Knights Templar, as well as in anthroposophy and in the White Brotherhood founded by Peter Deunov.

Meeting Scythianus

On my first day of new experiences, I simply observed the movements of the inhabitants of the spiritual world. One of the first things I noticed was that all the Angels bowed slightly. None of them stood totally straight, but seemed subject to a higher force – almost as if they were in a process of constant praise to God. This bowing also allowed the possibility of 'looking down', towards the lower hierarchies of spiritual beings, and finally to human beings. The higher 'up' I was able to see, the more the Angels stood or moved vertically, as if the 'lower' groups of Angels were directed more towards the groups further downward, and eventually the human being.

The next morning, this gradient of rank appeared clearer than on the first day. It even seemed that there was some kind of leader within each group, so that the leader of the highest group, of the Seraphim, was the leader of all the Angels.

The clouds of the physical world still appeared in different pastel colours – to a certain degree like the pastel clouds of the outer etheric world, which is definitely not material. When I was able to go deeper into the colours of the clouds, they seemed to form a cup or a concave form. With that, the second day ended.

*

Towards the end of the third day, the 'leader' of the hierarchies came much closer, standing in front of me with a stern expression. I tried to ask his name, but I didn't really catch what he said. However, I was able to decipher that it started with an S. Over the following days, he came closer and closer, still stern in his gaze. At a certain point, he was so close that I merged with him, and he became part of my physical body. I felt him both outside and inside of me.

Then he started to smile and told me his name, almost as if it had been a secret up till now.

In fact, he gave me three names: Serat, Saturn and Scythianus. He indicated that, somehow, he was all three of them.

I don't know who or what Serat is. Of course, Saturn is a planet, and in anthroposophy the name of a period of 'planetary evolution'. And, Scythianus I knew to be an important initiate. I needed to find out more about who Scythianus was. In a book by Sergei Prokofieff* I discovered that Scythianus was one of the greatest initiates ever, connected to the mysteries of the physical body, to Saturn, to the upbuilding of the Resurrection body (the Phantom), to Manu and to Christ. This all started to make some kind of sense and I began to become excited about a deeper encounter with Scythianus himself. But over the following days, the entity did not utter any words nor give me any indication of what I am supposed to hear or 'see'. He remained close to me, however, and had a slight smile on his face.

The spiritual world is strange. When I met Vidar, he appeared like a man of around 2.5 metres in height,

---

* *The Spiritual Origins of Eastern Europe*, Temple Lodge Publishing, 1993.

and Balder at his side was about 1.7 metres. When meeting this new entity – the one that called himself by a name beginning with S – he appeared to be very tall. I could not tell his height for sure, as I could only see his face – but in relation to S's head, his body must have been at least 4-5 metres tall. His head filled my whole view, my whole vision, and I couldn't really see anything else.

This had been going on for several days now. S's visage grew kinder every day. Initially, he was stern and serious, but became humorous and kind – almost with a kind of cheeky jokiness. On Saturday, thirteen days after his first appearance, the being indicated that he had a lot to do with the destiny of the Slavic peoples, and was also closely concerned, or even connected with, the Russo–Ukrainian war that was then raging. I waited with patience, feeling very calm.

*

The following day, Sunday morning, sitting opposite my fireplace, the next step was taken. Being somewhat frustrated by the silence of S, I went back to the threshold to ask Vidar if I had done something wrong. Was everything as it was supposed to be? He answered immediately, as he can do when a question is simple yet important. The answer was that S was of a different spiritual 'stream' than he was. Vidar indicated further that there are several spiritual streams in the cosmos, and that he himself belongs to one that was somewhat 'conscious', or even 'intellectual'. This stream included the being that was Rudolf Steiner and Aristotle.

This new entity, S, belonged to another stream – a stream that could be characterized by the Master Jesus, Abel or Plato. This stream was much more connected to the feelings of the heart than to words. S did not intend to give me any lessons in words or concepts, but only a floating, loving insight into the world and the cosmos. This love would enlighten me further.

Vidar's teachings were to prepare me for the spiritual world beyond, and to give me contemporary concepts and truths that I could communicate in my books. Now, however, the new teachings were to be on another level, the level of the sixth cultural epoch of Love and Warmth. When Vidar said that, I thanked him and went back to S. His friendly gaze became even warmer and friendlier, overflowing with love and understanding. Then, something remarkable happened.

Every morning I drink 1-3 cups of hot water, as instructed by Peter Deunov. He said that this enlivens the cells of the body. Now, I happened to observe a cup of hot water. In it, millions of small elemental water-beings appeared, and when I took a sip, they entered my body and its cells, enlivening them all. A little later, as the water cooled down, the elemental beings became bigger, and these entities entered my organs and bones. This was very interesting, and I resolved to continue to observe the world whilst remaining in close connection with S.

*

The following morning, S gave me one further indication of his mission and way of working. He suggested that whilst Vidar functioned as a transition between the

Detail from 'The School of Athens' by Raphael

elemental world and the spiritual world, he – S – functions as a mediator, or transition, between consciousness and the deeper layers of the earth, as well as the deeper layers of the body itself. It was as if Vidar was pointing upwards and S was pointing downwards. This was like the 'two streams of reality', the Platonic and the Aristotelian, and I was now able to gaze into the deeper layers of the earth itself.

I had been able to do this once before, but only for a short while. This happened on the top of the Atlas Mountains, on the first day of January 2017. Early that morning, I could see through the whole earth, and experienced that the planet had twelve layers – not just nine, which, up until then, I had believed to be the case. At that time, my view felt more like a revelation. Now, it felt more like an insight or understanding. And all the time, S was smiling in a friendly manner.

*

The next day, the teachings of S became stranger still. He showed me, or rather let me experience, another timeline that was previously unknown to me. Earlier, I described a timeline that is known if one enters the etheric of *this* world (the 'inner' etheric, not the outer etheric).* Through these means, if one moves to the left, one enters the past, and if one moves to the right, one enters the future. Both the past (in the etheric Akasha) and the future (in the spiritual astral) are actually present in the now, but one has to move oneself, or rather one's consciousness, in the spiritual realms or landscape. This reveals the fabric of the double timeline or the double time-stream. This timeline exists in the horizontal plane.

S showed me the vertical timeline, a timeline that exists through or in the twelve layers of the earth itself, as well as the twelve layers of the body. When going with consciousness into the depths of the earth, one moves back in time, until one reaches the ninth layer – as if the first nine layers were created in the past. But then, when entering the tenth layer, the travel through time suddenly changes, or reverses. One starts to enter future time, which is, in a way, like the future in the horizontal realm.

In the tenth layer, one meets *angelic time* – the same as astral time – which can be understood as the remaining part of the fifth great time cycle and the Jupiter

---

* See further in my book *Travels on the Northern Path of Initiation*.

incarnation of the earth. This relates also to the power of the Son.

In the eleventh layer, one meets *archangelic time* – the same as lower devachanic time – which also can be called the sixth great time cycle and the Venus incarnation of the earth. This relates to the power of the Mother.

In the twelfth layer, one meets *archaic time* – the same as upper devachanic time, which also can be called the seventh time cycle and the Vulcan incarnation of the earth. This relates also to the power of the Father.

In going with our consciousness into or through the twelve layers of the body itself, something similar happens. Through the first eight layers we move back in time, but at the ninth layer it reverses. When entering this ninth layer, we reach the 'Turning of Time'. Physically, we enter the right atrium (of the heart).

Then, at the tenth layer, we enter the right ventricle, corresponding to angelic time and also the astral world, as well as the rest of the fifth time cycle and the Jupiter incarnation of the earth. This also relates to the power of the Son.

At the eleventh layer, we enter the left atrium, as well as lower devachanic time, which also can be called the sixth time cycle and the Venus incarnation of the earth. This also relates to the power of the Mother.

At the twelfth layer, we enter the left ventricle, as well as upper devachanic time, which also can be called the seventh time cycle and the Vulcan incarnation of the earth. This relates to the power of the Father.

This day's teaching from S almost took my breath away. I felt full and quite exhausted. It seemed that S was directed more towards the mysteries of the physical world and the physical body, whilst the previous teachings from Vidar were directed more towards the etheric mysteries.

Meeting Ymir

The next day's lecture or teaching was even more breathtaking – and exhausting! I was introduced to the spirit of the earth itself in the form of a huge giant. This entity was quite friendly, but greater and mightier than anyone or anything I had ever met before. It was like being brought into the story of the Creation, as told by the bonfires of old Norway.

To understand my feeling and to characterize my experience, I will insert a brief account of the story of Creation, as taken from the *Edda*:*

> In the beginning of time, there was nothing: neither sand, nor sea, nor cool waves. Neither the heaven nor Earth existed. Instead, long before the Earth was made, Niflheim was made, and in it a spring gave rise to twelve rivers. To the south was Muspell, a region of heat and brightness guarded by Surt, a giant who carried a flaming sword. To the north was frigid Ginnungagap, where the rivers froze and all was ice. Where the sparks and warm winds of Muspell reached the south side of frigid Ginnungagap, the ice thawed and dripped, and

---

* This Norse story of the origin of the earth, sky, and humanity is paraphrased from Snorri Sturluson's *Edda*, as translated by Anthony Faulkes (London, J. M. Dent & Sons Ltd.). Sturluson lived in Iceland from 1179 to 1241, and he apparently composed the *Edda* as a compilation of traditional stories and verse. Many of the verses he included appear to date from the times when Norse sagas were conveyed only in spoken form by Viking bards.

from the drips thickened and formed the shape of a man. His name was Ymir, the first of and ancestor of the frost-giants.

As the ice dripped more, it formed a cow, and from her teats flowed four rivers of milk that fed Ymir. The cow fed on the salt of the rime ice, and as she licked a man's head began to emerge. By the end of the third day of her licking, the whole man had emerged, and his name was Buri. He had a son named Bor, who married Bestla, a daughter of one of the giants. Bor and Bestla had three sons, one of whom was Odin, the most powerful of the gods.

Ymir was a frost-giant, but not a god, and eventually he turned to evil. After a struggle between the giant and the young gods, Bor's three sons killed Ymir. So much blood flowed from his wounds that all the frost-giants were drowned but one, who survived only by building an ark for himself and his family. Bor's sons dragged Ymir's immense body to the centre of Ginnungagap, and from him they made the Earth. Ymir's blood became the sea, his bones became the rocks and crags, and his hair became the trees. Bor's sons took Ymir's skull and with it made the sky. In it they fixed sparks and molten slag from Muspell to make the stars, and other sparks they set to move in paths just below the sky. They threw Ymir's brains into the sky and made the clouds. The Earth is a disk, and they set up Ymir's eyelashes to keep the giants at the edges of that disk.

On the seashore, Bor's sons found two logs and made people out of them. One son gave them breath and life, the second son gave them consciousness and movement, and the third gave them faces, speech, hearing, and sight. From this man and woman came all humans thereafter, just as all the gods were descended from the sons of Bor.

Odin and his brothers had set up the sky and stars, but otherwise they left the heavens unlit. Long afterwards, one of the descendants of those first two people that the brothers created had two children. Those two children were so beautiful that their father named the son Moon and the daughter Sol. The gods were jealous already and, when they heard of the father's arrogance, they pulled the brother and sister up to the sky and set them to work. Sol drives the chariot that carries the sun across the skies, and she drives so fast across the skies of the northland because she is chased by a giant wolf each day. Moon likewise takes a course across the sky each night, but not so swiftly because he is not so hurried.

The gods did leave one pathway from Earth to heaven. That is the bridge that appears in the sky as a rainbow, and its perfect arc and brilliant colours are a sign of its origin with the gods. It nonetheless will not last forever, because it will break when the men of Muspell try to cross it into heaven.

In meeting the spirit of the earth, I realized that it was very old, probably from before Ancient Saturn, the first incarnation of the earth. It was not Christian, but gave a home to all other spirits connected with the material creation of matter, plants, animals and man, as well as the adversarial and beneficial spirits. It was all-encompassing and could be characterized as 'a Mother of all'.

However, it is not the Mother that dwells in the eleventh layer of the earth, which can be understood as Sophia, the Wisdom of the Cosmos. Rather, this being was more like an Earth Mother – the carrier of the life of all earthly dwellers and inhabitants. From what I could understand, her name began with 'Y' and ended with 'r' – maybe *Ymir*? But for now I will call her 'Y'.

I spent the rest of that day – Friday 23 September, the autumn equinox – communicating with this earth spirit. 'She' showed me the deeper layers of the earth – the very structure of the earth, before the adversarial forces came to abide in them – and how they would be liberated by the Christ force that would be brought there by human beings. She was very glad that human beings had chosen this earth for their development. She also confirmed that she had existed from before the Saturn condition of the earth, created by the Father in very ancient times – before time had even come into existence.

The rest of that day I observed the different layers of the earth, under the guidance of Y.

From the writings of Rudolf Steiner, I had got the impression that the first nine layers were increasingly evil, with a climax at the ninth layer. This now seemed to be only a part of the actual reality. Two thirds of each layer was good and, influenced by the earth spirit, was destined for growth and goodness. Only approximately one third was evil – which was where the adversarial forces had made their domain or abode.

This is just the same as in our own body. Our body also consists of twelve layers, as well as twelve organs, twelve sense organs and twelve meridians, receiving their etheric strength from the twelve zodiacal areas, as twelve streams of etheric energy coming into the body from the cosmos. Each of our layers or organs also creates an abode for the adversarial forces, but a large part of us – or our organs – are dominated by good forces, by cosmic etheric power and by the cosmic 'blueprint' originally designed by God himself.

After his death on Golgotha, Christ descended into the earth, all the way to the ninth layer (maybe even to the twelfth) to free the souls or spirits of the dead who were trapped in the evil parts of these layers. He did not have to penetrate the good parts of the layers.

This day of teachings felt very good; I understood that the evil in the world, in the earth, was not dominant. There was hope – hope for the future, although, sadly, not for our present culture. Of the latter, I stand by what I have written previously:*

> Today the knowledge and insight of the spiritual world is vanishing rapidly as materialism takes its hold. Fewer and fewer people believe in God. Respect for other life forms, trees and animals, as well as for non-material being, has lessened, and egoistic materialism is progressing. The culture is trembling under the weight of ignorance, and from where I stand, I cannot see any rescue or salvation. Our culture will vanish, as all other cultures have vanished, and the cause of our destruction will be ever-progressing materialism and egoism that will eventually culminate in a 'War of All Against All'.
>
> The knowledge presented in this book, along with all other similar knowledge, will be carried

---

* A version of this text appears in *Experiences from the Threshold*.

over to the next culture by a small number of people who have grasped the need for spirituality, and together they will create a new culture of spiritual, and brotherly and sisterly, love.

*

The following day I investigated the different layers of the earth in detail and found that these twelve layers resembled, quite closely, the twelve layers of the body, with which I had been working intimately for many years.

The outer – or first three – layers were dominated by luciferic entities, who in a way had sought refuge there, or used them as a base. This part also resembled the elemental world, where the third realm (the first realm one enters when travelling the Northern Path of initiation) is dominated by luciferic elemental beings. In the earth, as in the human body, the first three layers relate to luciferic forces, form-elementals or the astral body.

When I went deeper into the earth layers, to the fourth, fifth and sixth layers, I encountered the dominance of the ahrimanic beings. These beings, as well as those in the luciferic section, are doing their proper work in the two-thirds of the layers that are not evil – but elemental beings that have been transformed by evil live in the other one third.

I should add that in the elemental world we find all three elemental beings in all three realms, but there are more luciferic beings in the third layer, more ahrimanic in the second layer, and more azuric beings in the first. I found the same structure in the layers of the earth.

In the three first layers, I found a dominance of the luciferic entities or elementals; in the next three layers, a certain dominance of the ahrimanic beings; and, in the seventh, eighth and ninth layers, a dominance of the azuric beings.

It is the same in the body, and here I found, in the ninth layer, a shared space for the 'Son-related' forces and the azuric forces.

In the anatomical body, the azuric forces live very close to the heart on the right side (sometimes on the left side), which shows the close relationship between the heart (Christ) and the azuras. Christ is the Sun Spirit, and azuras represent Sorat, the Sun Demon.

So, in the ninth layer there is a sharing between Christ-forces and the azuric forces. In the tenth layer the Son-forces dominate; in the eleventh, the Mother, and in the twelfth, the Father.

In the body, the four inner layers are related to the heart itself – the sacred heart – and here they differ between the four chambers of the heart.

The ninth layer relates to the lower 'I' (Id) and to the right atrium. The tenth layer relates to the 'normal' and earthly 'I' and to the right ventricle; and the eleventh layer relates to the higher 'I' (or cosmic 'I') and to the left atrium. The twelfth layer relates to the Christ 'I' and to the left ventricle. A part of these layers in our body is dominated by pathological elementals, created by our bad deeds, and a part is dominated by 'righteous' elementals.

So, in the days that followed, with these insights in mind, I travelled back and forth between the twelve layers, both the 'good' parts and the 'evil' parts.

*

In my work with patients, I use the twelve layers of the body both in diagnosis and in therapy.

One special phenomenon I have noticed, especially in patients with ME (chronic fatigue syndrome), is that they often have the 'blueprint' of an extinct insect species crawling around in the area of the fourth to fifth layers of their body.

The fourth to fifth layers of the atmosphere is named the thermosphere, and this area contains, according to Rudolf Steiner, the 'group-souls' of animals. This layer is also where the aurora borealis or Northern Lights can be seen. People with a close connection to animals can direct the behaviour of the Northern Lights with their mind, which I have demonstrated myself several times. The Scandinavian Sami people believed that the lights are created by the souls of the departed, and should be treated with respect. Staring directly at them could send a person insane.

In this region, which Steiner called the 'Form-Earth' and the 'Fertility-Earth', I saw or was shown the blueprints of all the animal–insect–fish–bird–bacteria species, both the ones that are alive now and those that are extinct. Those that are extinct are trying to find bodies in which they can fulfil their destiny, and this can cause disturbances in the world of the living.

When I was travelling through the layers, the evil parts of the various layers were always to my right. That day, however, I only experienced the fourth and fifth layers in detail.

*

The next day my interest was captured by the threshold between the material and spiritual worlds. When travelling through the elemental world, I discovered that the main threshold was between the first realm of the elemental world and the outer etheric world. That was described in my last book, *Encounters with Vidar*.

When travelling to the deeper layers of the human body, I experienced, together with my students, a major threshold between the eighth and ninth layers. At this point, most of my students felt a fear that made them stop, and they were often unable to penetrate further. Why should this be?

At the threshold between the eighth and ninth layers, the traveller leaves physical reality and enters spiritual reality. This has a great importance for diagnosing and treating patients. In physical reality, scars and trauma may block the effect of treatment, but such hindrances do not exist in spiritual reality. Thus, when there is a scar in the body that blocks the effect of treatment, this hindrance is eliminated when one treats or diagnoses from the deeper layers – that is, from the tenth, eleventh or preferably the twelfth layer. This process is described in my medical books.

When travelling to the centre of the earth, I experienced a major threshold at the ninth layer, where the participation of evil stopped and only the full 'goodness' of the layers appeared or showed itself. At this threshold, I do not experience a guardian; rather, a welcoming entity, greeting my appearance there.

Crossing the lines of these thresholds often gives rise to a certain fear, particularly when one crosses for the first time – or even the first few times. Only at the threshold between the elemental world and the etheric world have I experienced a proper guardian – Vidar. But was that really so, or could there be other guardians that are unknown to me…?

*

Over the following days, I remained at the three thresholds: the one in the earth (between the eighth and ninth layers), where the evil parts end; the one in or just outside the heart (pericardium); and the one at the end of the elemental world. They were somehow the same – although, paradoxically, also different.

The first difference I observed concerned the 'Light of the World' – the realm or area behind the threshold.

- The deeper layers of the heart felt like reflected sunlight on metallic gold.
- The deeper layers of the earth felt like reflected sunlight on molten stone, lava or magma.
- The higher world beyond the elemental world, behind the first realm of the elemental world, felt like sunlight streaming through fog or clouds.

I dwelled and contemplated on these differences for a long time. I tried to see if there were some other types of guardians; whether Christ was present in all three places, beyond all three thresholds. I waited for several days.

Then, one day, I again turned towards Y, the old spirit or entity of the earth itself. She greeted me like a mother greets her son. Her love was overwhelming, her warmth was pleasant, and her wisdom was immense. Her wisdom was different to that of the cosmic wisdom of Sophia, although of the same kind. I was struck with wonder.

I recognized her as the spirit of the earth, Gaia, Y, *Ymir*, the Great Mother… As we have seen, she is very old; from before time itself. Before 'Saturn' was, she was. She nurtures all the beings of the earth, keeping them warm through the 'winter'.

If her name is indeed *Ymir*, then she is, according to the Nordic creation story, from before Creation itself, and she gave her life – she 'died' – to create the earth. Of course, in the spiritual world nothing dies; it is a way of saying that a spirit transforms itself. By transforming, she gave room for the adversarial forces, and thus was able to give rise to the physical and eventually the material Creation – which is actually to say that she gave room to the multiplicity of elemental beings from within herself.

As evolution commenced, many of the elemental beings were transformed into evil beings, through the deeds of humankind on earth. They became like 'Orcs', or evil elemental beings. These evil beings were gathered, or gathered themselves, in the nine upper layers of Y. The three inner layers were kept sacred. I learned that we have to transform and redeem these nine layers with the help of Christ, and thus spiritualize the whole earth so that Ymir can reincarnate as the next incarnation of Earth, the Jupiter stage.

It felt so very good to dwell in the consciousness of the Mother, of Y.

In her reminiscences of Rudolf Steiner, Countess Johanna von Keyserlingk writes the following:

> Rudolf Steiner had the kindness to come up to my room, where he spoke to me about the kingdom in the interior of the Earth. We know that, at the

moment when Christ's blood flowed down onto the Earth at Golgotha, a new Sun globe was born in the Earth's interior. My search had always been directed toward studying the Earth's depths, because I had seen within the Earth a golden kernel light up, named by Ptolemy, the primeval Sun. I could connect those golden depths only with the land that Steiner said was hidden from human sight, and that Christ would open the gates to lead those who seek it to the submerged fairy-tale land of Shambhala, of which the Indians dream. [. . .] I asked Rudolf Steiner, 'Is the interior of the Earth made of the gold that comes from the hollow cavity in the Sun and is destined to return there?' He replied, 'Yes, the interior of the Earth is of gold'. [. . .] I continued to question him for my own assurance: 'Doctor, when I am standing here on Earth [. . .] the golden land is beneath me, deep within the interior of the Earth; if I now attain sinlessness and remain in the depths, will the demons be able to harm me, and will I be able to penetrate beyond them and reach the golden land?' He replied, 'If you pass through them accompanied by Christ, the demons will be unable to harm you – but otherwise they would indeed be able to destroy you.' He added emphatically, 'They can, nevertheless, become our helpers. Yes, this is true; the path is a true one, but very difficult.*

Against this background, it can be understood that the new sacrifice of Christ at this time – the time of

---

* Johanna and Adalbert von Keyserlingk, *The Birth of a New Agriculture* (Temple Lodge Publishing, 1999), pp. 84–86.

his Second Coming – has to do with the Earth Mother and with restoring access for humanity to the 'lost Paradise', known in the East as Shambhala. This, in turn, is an aspect of the creation of the 'new Earth' through the spiritualization of the present earth. In the future, humanity will increasingly – through connecting with Christ in the etheric realm – be able to draw upon the forces of Shambhala in the earth.

Meeting Anna

The next day, Y corrected a little of what I had written the day before. She took me into the part of the earth – actually within herself – that was not infiltrated by the evil forces. Here, she showed me that each layer also had a 'secret garden', where beneficial elemental beings had their abode, their world.

We began in the seventh layer, where at first I saw what appeared to be a protected area. She opened this area for me and led me into a marvellous garden. It was inhabited by creatures of many kinds, something like those described in C.S. Lewis's *Chronicles of Narnia*. There were mostly elves, but also unicorns and fauns. Y continued to show me that in each and every layer there was a part that was protected, and in these areas all the mythological figures and entities lived and could not be attacked by the adversarial elementals, or evil beings, living in the parts of the layers that were open to them. The 'good' part that Y now showed me could be called 'Shambhala'. Shambhala is thus not in the middle of the earth, but in the protected part of each layer.

I asked Y why Rudolf Steiner had only described the evil parts of the layers, and received the following answer: Such a description was necessary at that time, mainly to protect the 'good parts' against the black magicians of humanity. Now it was all open, however, it could be revealed, as Christ could now protect these parts.

Then Y directed me towards the seventh layer of the 'good' earth. I dwelt there for some time – some days.

In the midst of this period, an entity approached me and, with great strong movements, introduced herself. This presentation was so powerful that it penetrated my sleep the following night.

It seems that in the spiritual world one must wait to be introduced to a being – to ask, but to wait – both in relation to the Akashic Records (to meet the 'librarian'), and in developing any acquaintances with beings or entities. (This also applies generally to spiritual truths or for any kind of information.)

And so she stood there, in front of me, for a few days, in her glory, and I waited.

Finally, I asked if she was Anna (Saint Anne, in religious tradition), the mother of Mary and grandmother of Jesus, and she nodded vigorously in confirmation.* I was silent, feeling great reverence. (I cannot say if she is the grandmother of the Solomon Jesus, Zarathustra – the Jesus described in the Gospel of St Matthew – or the Nathan Jesus – the one described in St Luke's Gospel – as is reported by Rudolf Steiner in his lectures on the Christian Gospels.

*

---

* Roman Catholicism assigns the names Anne and Joachim to the mother and father, respectively, of Mary. The website Catholic Online (http://www.catholic.org) states:

> We have no historical evidence, however, of any elements of their lives, including their names. Any stories about Mary's father and mother come to us through legend and tradition. We get the oldest story from a document called the Gospel of James, though in no way should this document be trusted to be factual, historical, or the Word of God. The legend told in this document says that after years of childlessness, an Angel appeared to tell Anne and Joachim that they would have a child. Anne promised to dedicate this child to God (much the way that Samuel was dedicated by his mother Hannah – Anne – in 1 Kings).

'The Virgin and Child with Saint Anne' by Leonardo da Vinci

On Sunday morning I got the first hint of her teachings, and I will describe these briefly and in summary, as best as I am able:

> *All creation, the whole present cosmos, is and must be created in cooperation with the so-called adversarial forces, who are spiritual beings that have sacrificed or offered themselves in this great cosmic plan to become 'backward' beings, or rather 'laggards'. This happened in cooperation, or in the polarity of or with angelic beings (the nine hierarchies) and also other spiritual beings from earlier times (as, for example, Y), to create or give the illusion of a material existence. Through the Dark Ages (Kali Yuga) until 1899, and also further through the opening of the three elemental realms (in 1879, 1949 and 2019* – which is the material Creation), the nine-foldness of Creation had to be maintained as a spiritual reality (illusion), although the spiritual reality is actually twelve-fold.*
>
> *Rudolf Steiner spoke of nine layers of the earth. It was too early to speak of the other layers at that time. Now it is possible to reveal that there are both twelve layers of the body, of the heavens and of the earth.*

That was all for today. She implied that this was enough for me to contemplate.

*

The following day, she was there once again, standing quietly. But now, around her, there started to stream what I can only describe as powerful 'rivers of strength'. This occurred when she moved her arms in a

---

* See my book *Travels on the Northern Path of Initiation*.

very commanding manner, affecting everything in her surroundings.

I began to understand that this entity, Anna, was indeed a very important and significant being, who was related to the good parts of the layers of the earth, especially the seventh layer. As this layer has been secluded for many, many years, this may be the reason that we know so little about her.

Anna is written of in apocrypha, as well as Christian and Islamic tradition, and even Anne Catherine Emmerich and Edgar Cayce spoke of her – although they both describe her simply as the mother of Mary. They do not elaborate further, and nor do they speak of her as an important entity. However, I now concluded that she must be of great significance, given her position in the seventh realm of the good earth.

And so I waited again.

*

The following day, I felt her presence closer than before, and her movements were even stronger and more powerful. And then I understood clearly: these movements – or you could say this type of 'dance' – were expressions of the wisdom of the earth. This wisdom was part of the preparation for the coming of the Prince of Peace at the Turning of the Times.* The grandmother of Jesus, Anna, connected to, represented or brought the wisdom of the earth, Y. His mother, Mary, connected to the heavenly wisdom, Sophia. Together they brought all the wisdom in existence.

---

* Or, what is sometimes translated from Rudolf Steiner's work as, 'the Turning Point of Time'.

Now that the passage through the earth is open to 'travel' – and given that the earth is an expression of the elemental world – humans can access this earthly wisdom through the 'good' part of the earth's twelve layers. This also enables access to the heavenly wisdom through Vidar and Sophia–Christ.

After I received this understanding, Anna began to smile at me, and her movements became gentler and not so vigorous. She became almost friendly! And, as the day progressed and evening came, she became friendlier still.

Then, her teachings began once more. They had a very different quality to those of Vidar. They were more 'physical', and took the form of an Imagination. Anna's teachings began with a feeling on my skin, as if it had become 'prickly'. Anna went on to show me that her teachings of the earth were a reflection from higher Devachan, and their existence stretched out from long before Saturn and to long after Vulcan.

She indicated that life in higher Devachan was the 'real' life, and the 'little detour' into the material realm – stretching from Saturn to Vulcan – was more like a necessary correction to the real life in upper Devachan. This 'detour' – what we call 'real life' – is actually not so.

Higher Devachan looks like a thick membrane, stretching out over material experience, consisting of the Saturn–Vulcan line with smaller 'rests' in between (consisting of periods spent at 'home', in higher Devachan). And in this higher Devachan, everything is contained in a huge membrane.

Anna lay down on this membrane, showing that *she herself* was the entry from the devachanic world to material experience, in both directions. She was a cosmic being, who had allowed Mary, the mother of Jesus, to enter material experience, and was now allowing me to enter devachanic experience... if, indeed, I wanted to.

I reiterate that Anna is not a human being but a cosmic being. She is a portal to and from the higher spiritual world, the entrance to Devachan. In the beginning, she enabled the gods to enter material experience, like in the story of the Anunnaki.* As I thus experienced her – and as she now presented herself – she was the portal between the devachanic world and our material experience. In this way, she also 'birthed' all the gods participating in our cosmos.

As we have seen, in a later incarnation, she appeared as Anna, the grandmother of Jesus. In Celtic mythology, she is also called Ana, Anu, Annan, Danu or Dana, and is considered to be the mother of the gods that she 'birthed' into our material world.

As mentioned earlier, she also watches the entrance to Chartres Cathedral, observing all who enter from the north side, from the feminine side. She watches those who come from the devachanic world into our material universe, and observes those of us that travel into the devachanic world, as most initiates are able to do.

---

* The Anunnaki (also transcribed as Anunaki, Annunaki, Anunna, Ananaki and other variations) are a group of deities of the ancient Sumerians, Akkadians, Assyrians and Babylonians. In the earliest Sumerian writings about them, which come from the post-Akkadian period, the Anunnaki are deities in the pantheon, descendants of An and Ki, the god of the heavens and the goddess of earth, and their primary function was to decree the fates of humanity. (*Wikipedia*)

If Chartre represents the Mystery of the Feminine, is the path that takes us through the earth mysteries the feminine path? Is the earth spirit, *Ymir*, feminine? Is the portal into or from Devachan, feminine? I had so many questions…

*

Then I came to a realization. I had to be born into Devachan through her, almost like entering through a cosmic uterus. As mentioned, Devachan stretches in time – if one could talk about 'time' in this context – from far before Ancient Saturn and far beyond the future Vulcan – our material experience. Was this a new world to explore? I was tempted to enter. And so, I did…

Entering Devachan

At first, all was grey, and nothing was in balance or 'horizontal'. It seemed that concepts of horizontal and vertical did not exist there. It felt as if I was in another universe. I did not know how to move; I had no 'legs', and there were no means of transport (which had always been there in both the elemental kingdom and in the etheric world). Here, in Devachan, I didn't even know how to turn around. I felt lost, to say the least!

Then I tried to 'see' or to 'hear' or at least to *move*. I tried with several of my spiritual senses, especially with my sight, but also with my hearing. These didn't work here, in this realm. It was as if the devachanic world was unlike the spiritual world, unlike the elemental world, not to mention the material illusion, or *maya*, in which we live.

Then I tried to shut down all of my sense abilities and all my spiritual senses, and just experience, or 'see', through my being, through my existence – actually, through my *heart*.

Initially, all I could see was a grey, endless 'city' with no joy or life – just endless 'houses' in all directions, rather like how one might imagine Charles Dickens's London. Then, suddenly, I could experience this seemingly alien world in another way. Through my heart, I could experience the colours, the forms and the beings of this devachanic world, and it was filled with both meaning and joy. Despite this, I felt no clarity in my perception of the beings there.

This sensing in Devachan was only possible through my inner layer, through my very being, from the interior of my heart. (When we use our senses, it is as if we observe from our outer layer, from the skin, so to say, from our nervous system.) My experience is difficult to explain in words, of course, but *life itself* became the sense organ, as well as the means of moving, travelling and communicating in this new realm of existence. Everything existed in, from and through the heart! I turned to Anna and asked if this was really so, and, slowly, she nodded.

Now I felt ready to explore the devachanic world. The first thing I noticed was that there was no timeline here; no law of causality. My next conclusion was that there were no distinctions, no differences, between good and evil. Actually, these concepts did not exist here. Everything existed *in* or *with* or *for* each other. This was difficult to comprehend for a mind used to analysis, such as mine.

After some time, I observed that all entities, actually all 'things', interpenetrated everything and everyone else. But I still didn't really understand anything or come in contact with any actual entity. And so, with patience, I waited again.

*

After several days of expectant waiting, I tried to make changes as I felt somewhat 'stuck' in my investigations, in my travels in Devachan. I tried to 'specify' the heart as a sense organ by looking into the devachanic world through each and every heart chamber. I 'looked' from each of the four chambers of the heart, but nothing in my perception really changed.

Then I was inspired to try something new. When I had been in Iceland and was granted long conversations with the Elders of both the elves and the hidden people, they urged me to become aware of the 'fifth' chamber of the heart – which is in the process of creation, just behind and under the heart, towards the right side. Now, I went with my consciousness to this fifth chamber, and then my perception changed radically. Abruptly, I could perceive structured beings in Devachan, not just the woollen and indefinite ones I had been able to see until then.

I came to the forceful realization of azuric beings entering our world, and the real danger they represented. These terrible beings go directly to this area of the body, where the fifth heart chamber is to be developed. In this way, the azuric demons can hinder our path to the devachanic reality, and thus to the creation of Jupiter, the next incarnation of the earth.

Meeting Ariel

When I concentrated my consciousness in the fifth chamber of the heart, and through this merged into the devachanic world, the diffuse beings in that realm became more distinct. One entity stood out. I asked its name, and this time I did not receive it in three versions. Rather, it came swiftly and with clarity: 'Ariel'.*

After this revelation, for several days I stood in spirit, waiting quietly.

After this pause, I received a new teaching: All true spiritual encounters with beings from the beneficial powers are calm and quiet in nature – although some angelic beings, especially the Archangels, are so strong that one may become somewhat frightened. These true meetings are in contrast to some misguided encounters,

---

* 'Ariel' appears as an epithet for Jerusalem in Isaiah 29:1 and 33:7. The later common etymology interprets this as the lion (Hebrew *ari*) of God (Hebrew *al*). Ariel is also found in Coptic mythology: in the Pistis Sophia, Ariel punishes the lower world. In Judeo-Christian contexts, however, Ariel as the Angel of wrath is more of a ruler and punisher of demons. According to Thomas Heywood, Ariel reigns as prince over the waters and as lord over the Earth. In other occult writings, Ariel is given dominion over fire or air. In some mystical texts, Ariel is associated with dominion over the Earth, creation, northern elements, elemental spirits, and animals. Moses Gaster interprets Ariel as the Angel of healing and connects it with the Archangel Raphael. In Heinrich Cornelius Agrippa von Nettesheim, Ariel is assigned to the element of Earth as the lion of God and is derived from Aries (the constellation Aries).

which some describe as Angels or Christ. These false experiences are characterized as being full of power, strong light and strong feelings – even feelings of peace and happiness. Such meetings, I was informed, are often with beings from the adversary forces, especially azuric demons, who can create hallucinations of goodness and beneficence.

Then I understood that Ariel was both a guardian and a teacher. Further, I understood that several of the higher spiritual beings lived, lingered or remained in Devachan, and when needed, they descended 'down' into material experience. They could achieve this through Anna herself.

I now remained before this new teacher, even though I had already received several revelations. I reflected that Ariel's teachings differed significantly from Vidar's, which always consisted of three stages: Imagination, Inspiration and Intuition – and often were spread over three days. In contrast, Ariel's teachings arrived as insights, more directly, and apparently through the body – or was it even through the physical 'blueprint' that Rudolf Steiner refers to as the 'resurrection body'?

*

After several further days, there was a change in my connection to Ariel. This being became livelier. 'He' started to move, and this movement was in the left to right direction, just as – in my experience – Archangels often move.

He let me understand that his teachings would go directly into my body and would be experienced as bodily feelings or bodily pictures. He let me know that he inhabited or belonged to the area or realm in the spiritual world that had to do with the creation of the human body – sometimes called the 'Mother Lodge of Humanity' or 'World of Providence' – and was elevated above the material or physical experience of Creation.

He was from way outside what we – here in material experience – call time, i.e. before Saturn and after Vulcan. This was one of the differences between the mysteries of the spirit and the mysteries of the earth. Vidar presented the mysteries of the spirit, and as such presented them in Imaginations, Inspirations and Intuitions. Ariel, on the hand, presented the mysteries of the earth, and they were felt or experienced as bodily feelings, or actually as bodily Imaginations.

Such bodily Imaginations were something quite new to me. It was as if the Imaginations that usually are experienced in the mind or upper part of the body, were now experienced in the middle part of the body, deep inside, in close proximity to the fifth heart chamber and also close to the diaphragm.

One of the first 'physical Imaginations' (I will refer to them as such) which I received from Ariel was about the different paths into the spiritual world, and in particular the relationship between the Vidar path and the Michael path, as given by Rudolf Steiner throughout his life, but especially in the years 1923-24. This physical Imagination raised a question that lingered in my mind over the following days. It was about the nineteen lessons that, in the most condensed way, describe the Michael path of initiation.*

---

* See *Esoteric Lessons for the First Class of the School of Spiritual Science at the Goetheanum*, Volumes One to Four, Rudolf Steiner Press, 2020.

Rudolf Steiner gave these lofty teachings to bring the Michael School closer to humanity. But this was a risky endeavour and, if the attempt failed, it could backfire, leaving him in danger – even risking his own death. If one gives high spiritual teachings that are not received in the right way, these teachings can become dangerous to the person who gives them, and potentially – if not received in the correct way – to those who receive them.

In my view, which I will try to substantiate below, Rudolf Steiner insinuated that these teachings may have failed, and that those who received his teachings at that time had not been able properly to understand his insights. Already in the autumn of 1924, I believe, Rudolf Steiner became aware that his teachings in this School did not work as expected, or as he had hoped. It is even alleged that Rudolf Steiner himself expressed this to several close friends.

We can only speculate as to why things did not work out as Rudolf Steiner had expected or hoped, but in my opinion – after going through the Vidar School – one of the reasons might be related to the fact that, during this period, Michael had, since 1879, already begun the process of ascending to the status of Archai (or Time Spirit).*

Today, the changes in the spiritual world are even more advanced. Michael has progressed in his ascension to the status of Archai, and the elemental world is, as I have presented in my research, totally open. Thus,

---

* For more on this process of Michael's ascension to Archai, see Sergei O. Prokofieff's *The Cycle of the Year as a Path of Initiation*, p. 53-54, second ed. 1995.

I have come to the conclusion that we can supplement the Michael path with other ways, other portals, to the spiritual world. Where does this leave us, particularly within the anthroposophical movement, in our work with the Michael School? The Michael path is more advanced or more difficult than the Vidar path, but is this due to the aforementioned concept that Michael was in the process of becoming an Archai when this path was given?

Before deciding how and where to travel in spirit worlds – and what we might see, hear or experience on these travels – we should, in my view, consider the path leading to the Vidar School in relation to the already known Michael School (as given and described by Rudolf Steiner in the years 1923-24, and as published in the lessons of the First Class of the School of Spiritual Science).

The Michael School was given to open a definite pathway into certain areas of the spiritual world, with the help of Michael, who Steiner spoke of as working closely with Christ. So, let us ask: How will a 'difficult' path into the spiritual world work, and what are the consequences of taking such a path? It works rather like taking a 'difficult' road here in the material world, to use this as a metaphor – a road with obstacles, or a road that is overgrown and not properly managed. Many travellers – although, it is important to note, not all – may be hindered and may not reach their destination and turn back. This turning back is often quite disturbing for the person who made the roadway or even the roadmap. Difficult teachings may therefore have unwelcome effects. That is why it is said that one should not 'throw pearls before swine'.*

So, I ask the question: Might this Michael path be too 'difficult' for many people today? Is Michael still accessible to all? Maybe another path needs to be available, based on Vidar's teaching? This is the same Vidar who was entrusted by Michael as an Archangel, and who perhaps also offers a safe and straight path to the living Christ.

Could this Vidar path in any way be a substitute for the work coming from Rudolf Steiner's Christmas Conference and his Class lessons? I make no dogmatic assertions or statements on this, but leave it up to each individual to decide, according to their experiences and results with their spiritual work.

*

My communication with the entity named Ariel of the devachanic world did not proceed as I wanted or expected. Finally, I conjured up the courage to ask this being how we could or should communicate. He replied over the following days as follows.

The way to communicate in Devachan, he explained, is totally opposite to the way one communicates in the outer etheric – or perhaps in the whole of material experience, in which the outer etheric is a part. In material experience we move around the entities and communicate with them. In devachanic experience, however, we must 'enter' the entities, go through them, and as such enter the answer or communication. In this way, we enter the inner world of these beings.

---

* 'Do not give what is holy to the dogs; nor cast your pearls before swine, lest they trample them under their feet, and turn and tear you in pieces.' Matthew 7:6.

This method seems to belong to the mysteries of the earth. When I was invited to enter the earth mysteries, initially through the guidance of S, I had to enter the earth itself in order to meet the earth spirit, Gaia, or Ymir. I had to go 'inside'. There, in the seventh layer, I was introduced to Anna. And to enter the devachanic world, I had to go through her – inside – all the way to her 'cosmic uterus'. Then, in the devachanic world, I had to enter or go inside Ariel to get the teachings he might have to offer. I did this, and a whole new experience opened up. This process would take some days or weeks for me to become properly acquainted with.

So, I stood there, not really knowing what to do. Ariel was much less inviting than Vidar was; it seemed that I had to conjure up my own initiative. I will try to go back in time, I thought. This was something I had attempted several times in my life, when I was in the third realm of the elemental world. Previously, I could enter the etheric (inner etheric) of trees, or between trees, and, when going to the left, I entered the past. When going *far* to the left, even the forms of the leaves of the tree changed. And, the furthest I had reached was the plant forms of the Silurian age.

When I followed this same procedure in the outer etheric, the change appeared as cultural, in the sense of seeing the different cultural epochs. Now, I tried this approach in the devachanic world, inside Ariel. Then I experienced the causes of the changes of both forms and cultures. It was as if I was journeying back in time, within the inner etheric. A journey back in time in the outer etheric produced Inspirations. Now, a journey back in time in Devachan produced Intuitions.

On this day, I tried to venture back in time to when material experience began, at the beginning of so-called Old Saturn. There, I tried to experience – or to see or understand – why the whole of material experience had been initiated. As I 'stood' there, at the beginning of Old Saturn, I had the impression that the leading spirit of the devachanic world – one could refer to this entity as 'God' – was *concerned*. Darkness had started to overtake the light; evil had grown too great.

As a result, it seemed, material experience was initiated, in order to create human beings who could, in freedom, counteract the growing darkness. This darkness had a name, which sounded something like *Makuk*. So, our whole experience, through eons of time, was to create free love in order to counteract Makuk. This was not too difficult to understand or accept, and this insight brought me much understanding and created many further insights, culminating in: *We are soldiers of the good*!

*

A few days later, Ariel clarified the teachings or insights regarding the apparent difference between the Michael path and the Vidar path. During his life, Rudolf Steiner had helped to open and clarify the Michael path, which was then (and still is) a glorious *teaching* and a glorious *path* into the spiritual world. It was and is a 'high' path, a very spiritual path – a path of consciousness and the consciousness soul. But such a path had to be kept open, had to be travelled, otherwise it would become overgrown and would gradually 'close'.

The Michael path that Rudolf Steiner gave was, as I said, a very lofty one. Was there some concern on his part that it was too 'advanced' or difficult for his followers to pursue? Regardless, this – according to Ariel – is what actually happened. This does not mean that the teachings related to the Michael path were wrong, or that the exercises and mantras connected to this path are wrong or false in any way. Not at all! In fact, quite the contrary… The content of the 'lessons' certainly prepare the human spirit and consciousness for this path, for an actual meeting with the spiritual world. But there is a danger – given the events referred to above – that the teachings may not bring the person as such (their soul, spirit or consciousness) into the spiritual world.

This is what we really need today: to *experience* the spiritual world, not just to have *knowledge* about it (as one can gain, of course, through Rudolf Steiner's unparalleled Collected Works). As I have said, the Michael path is in danger of becoming 'overgrown', which leaves the Vidar path as an alternative.

We should *study* the Michael path in order to be able to travel along it in the future, as it contains the insights and knowledge which will be needed – but today, perhaps, for direct experience, we can more easily *travel* the Vidar path.

When Steiner gave the Michael path, the Vidar path was at that time closed, as the deeper realms of the elemental world were not yet 'open'. It was too early for that path, and that is why it could not be taught. The same was true of the twelve layers of the earth, or the 'good' path of the first nine layers, or indeed the teachings of Ymir or Anna.

And so it all began to make sense…

*

Several days passed by and I continued working on and obtaining knowledge or insights through the earth mysteries. As we have mentioned, on the spiritual path insights or messages come as Imaginations, evolving further to Inspirations and then Intuitions. On the earth path, in contrast, this appears to be very different – although it is not *actually* so. The Imaginations have become internal, inside the body, and do not appear 'in front of the eyes', as pictures.

The body tells one what to understand, and this is much more difficult to explain or describe than a pure picture. That is why, in the following pages, I can only refer to understandings or insights, and am not necessarily able to describe the development of pictures.

As I have referred to, one of the first insights I received in this way was the difference between *knowledge* and *path*. I wrote up a summary of my understanding as an article, which is reproduced here in the Appendix.* Although somewhat controversial, the article was well received by many of my anthroposophical friends, although I expected it to be questioned or rejected.

*

A deeper understanding of bodily experiences – or, one might say, between the spiritual-mystery experiences in Imaginations and the earth-mystery experiences – now began to emerge. The earth mysteries and the spiritual

---

* The article was entitled: 'An Investigation into the difference between Knowledge and Path, relating to the Michael School, as described in the Nineteen Class Lessons of Rudolf Steiner and the Vidar School.' A version of this has been included in the Appendix under the title: 'An Investigation into the Differences between Study and Spiritual Experience.'

mysteries are quite different to each other. The spiritual mysteries seem to be mediated through the sense organs, especially through the eyes and ears, but the bodily mysteries seem to be mediated or made conscious through sense organs that I had not been aware of, organs that lived in the darkness of the body, like the chakras, or deeper organs like the 'life-sense'.

This new form of 'seeing' or experiencing, or one might say this path of knowledge, is in a way lonelier than receiving Imaginations. In Imaginations of the spirit, I feel in communication with the spiritual world, with that world using light to show me certain revelations. In the mysteries of the earth, this light is very different. Everything is more or less in twilight, in shadow, and so this experience feels more lonely – also deeper and more serious.

*

One day, the mystery of the movement of blood was revealed to me – partly as an Imagination, and partly as a bodily experience. First, I clearly saw how the forms of the acanthus plant were imprinted in the blood – in the blood's own movement. I already knew these forms of movement from my work in opening up 'the middle' in sick people.* I have described this previously, but will give a short recapitulation below.

In all creation, the forces of the elemental world can be discovered or perceived. These are the forces or beings of luciferic origin, of ahrimanic origin and of azuric origin. In diseased beings, both human and animal, the strength of one of these forces is too strong, thus

---

* See further in my other books, for example *Demons and Healing*.

creating symptoms. Between all these three forces there is a space, which I call 'the middle'.

My 'treatment' is thus to enlarge this middle with the help of my fingers – for if one sees or observes the elemental entities or forces, one has a certain dominion over them. Then, when the middle becomes wider, I ask Christ to enter. This causes a change in the void of this middle, and moving forms start to appear – forms which remind me of acanthus leaves. This is the beginning of the presence of Christ consciousness, of the 'I-am'.

'Acanthus Leaf Design' by Alexis Peyrotte (1699–1769)

I now 'saw' or experienced these forms in all the movements of human blood.

*

The next day was very strange on a personal level. First, I saw a video lecture by Andrew Linnell on 'Sacred Geometry', where he pointed out that an understanding of geometry was important for one's spiritual development. Then, I realized that my own fixation in building small houses with different constructions (which I have been doing in an amateur capacity on my own land for years) – using shapes such as triangles, rectangles, quadrats, pentagons and hexagons – is actually a part of the mysteries of the earth, or maybe the spiritual mysteries... Or, maybe, of both?

I build these houses myself without any help, and I become totally emerged within the structure, the weight-distribution and the different directions of the planks. I have actually built seven such houses over the past fifteen years. Then, I came to an understanding that this was part of my spiritual path. At that moment, the two paths – the mysteries of the spiritual and the mysteries of the earth – merged to become one for me, and I felt more balanced.

It began to snow, and – as I have described in my book *Travels on the Northern Path of Initiation* – I 'merged' into the white surface. At once, the whole elemental world emerged, and my garden became inhabited by elemental beings with different appearances and in all shapes and sizes. I understood how important it is that one uses real snow in one's spiritual 'merging', instead of just a white wall or sheet of paper.

*

The next day, another teaching related to the mysteries of the earth began – a teaching that also raised several questions to which I still have received no answer. The teachings were about the difference of human thoughts in the summer and winter, relating to the rhythms of the earth.

According to Rudolf Steiner, our thoughts are closely connected to the elemental beings, and as these beings leave or are breathed out from the earth into the cosmos during the summer, it is more difficult to think logically and/or spiritually during this time, as we remain here on the earth. In the winter, these elemental beings return to earth, and this results in us being able to think more easily. Many people feel this difference: it is more difficult to concentrate on intellectual thoughts during the summer, while it is the opposite during the winter.

The following questions then arose:

- *Where is the elemental world?*
- *Where is the spiritual world?*
- *Where is the cosmos, and where is the earth?*

We must remember that when the Spirits of Darkness were thrown down to earth by Michael in 1879, they were actually forced to move from the spiritual world into the elemental world, which then also became open to human beings. This leads us to the following questions:

- *If the elemental beings leave the elemental world and enter the spiritual world during the summer, what part of the spiritual world do they enter? (I cannot see them in the 'normal' spiritual world.)*
- *What is meant by entering the 'cosmos'?*
- *If one can travel through the elemental world and into the spiritual world, will our thoughts disappear, or will they instead become more spiritual?*

After I was able to travel through the elemental world and into the spiritual worlds and the devachanic worlds too, it seemed to me that there was no difference between my thinking ability in summer or winter. This thinking ability, or at least some kind of 'living thinking', seems to be influenced by going into the spiritual and then coming back, i.e. there is a certain 'pendulum' swing between these worlds that seems to enliven one's thinking.

I will leave these questions open as I seek and search for further guidance...

*

After a few days, some further insights began to appear – through the experiences of earthly mysteries and through the sensory organs of my body. The understandings concerned the devachanic period before Saturn, which means before our material experience began.

The first understanding was that Devachan itself is also in development. The earth had its 'golden age' in the past, before darker forces slowly started to take over. When the darker forces began to grow, the spiritual world needed a human race to help, for only *love in freedom* could stop the darker forces taking over. Thus, the current material experience was born.

The darker forces understood, of course, that this human experience would be devastating for them, and therefore also penetrated into the material in order to stop it. And thus we have the adversarial hierarchies of today, trying to push us out from our true course. As such, I saw Devachan as a curve, just as one may experience earth development (see over).

The salvation of Devachan, therefore, lies in our victory over the adversarial forces in the material world, which in turn will be the foundation for beating the

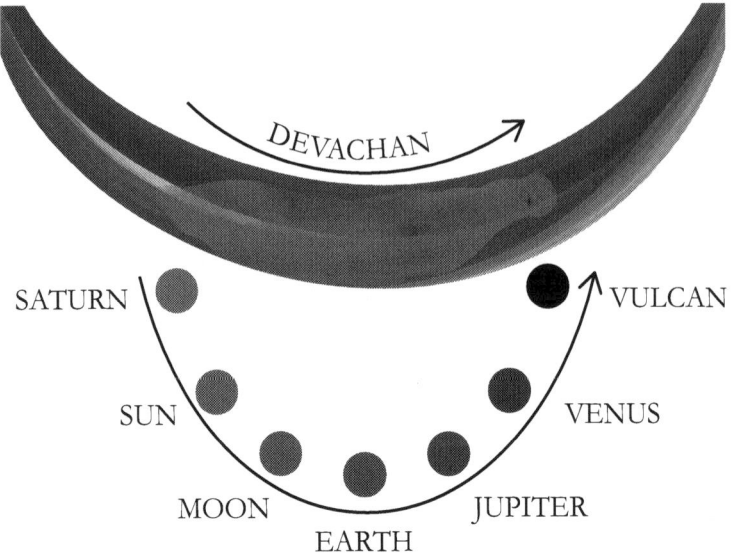

adversarial forces in Devachan itself – although this will take place much later, perhaps some millions of years later. Then, the next great battle will begin. The present day is like a training of the soldiers, or maybe not even that...

The salvational creation of material experience is thus not only of importance to us, but to the whole of the cosmos – that is, the greater cosmos beyond our 'smaller' cosmos.

*

After dwelling in this 'outer' spiritual world – which I referred to as Devachan – for some days, I started to understand some of the differences between devachanic spirituality and the spiritual world belonging to our Earth evolution, or what is also called the 'material experience' of the human being.

The first difference I became aware of was related to what must be called 'the Light'. It became increasingly clear to me that in 'this' world, spirituality can be characterized by the presence of light and expressed in the sentence, 'Knowledge lives in the light', or similar such expressions.

In the 'outer' spiritual world, especially notable in the period prior to the creation of Saturn – before the adversarial forces became too strong – spirituality was or is expressed in a sort of darkness, a twilight, an 'ebony' if you will – a warm and protective ebony. Comparing our spiritual world to devachanic spirituality can be described as comparing ivory to ebony.

This different understanding of both spirituality and knowledge as expressed in this warm twilight, warm ebony, became apparent to me. *Love and wisdom live in the twilight, in the warm darkness of ebony.*

The Astral World

A few days later, after I was asked by a friend why I had not mentioned or described the so-called 'astral world', a further teaching arrived. Ariel 'heard' this question, and I received his answer which came over the following two days. His response was clear and direct.

Ariel took me with him to the time when Saturn was created, and let me see further back (beyond time), to when Devachan was bright and balanced. Then came a time when it became darker and darker, and at a certain point the Gods became somewhat 'stuck', if I may describe it so. This was just before Saturn was initiated. Saturn was actually initiated – that is, the total creation of the human being, from Saturn until Vulcan – *because the Gods were stuck.*

The adversarial forces were threatening to take over. Something had to be done, and this something could only be achieved by 'free' beings that could conquer the adversarial forces. When, therefore, the Vulcan period is completed, we – the created, creative beings – will have been able to develop free love and consciousness that is able to defeat the adversarial forces.

That is why the human race was created, as foot soldiers for the Gods; because *they needed us*, not primarily because *they loved us*. This, at least, is what I understood.

The astral world was at that time of minor importance, just a small, strange thing. It was not needed for the development of the spiritual world or the devachanic world. The astral world had to be presented or constructed for humans to conquer, for humans to penetrate and to overcome. Otherwise, humanity would not be tempted to sin, and thus meet the adversarial forces. This was the so-called 'Fall'. This means that all our suffering in sin was placed upon us in order to make us *good soldiers for the Gods*.

Ariel, who told and showed me this, expressed that the Gods were somewhat sorry for this – especially that the astral realm was forced upon us at the time of the Fall. Nothing more, nothing else.

I had a feeling as if I was seasick.

*

Over the next few days, I studied this astral world in more detail, from the standpoint of Devachan. The whole astral impulse looked like a secluded box with six sides, sharp edges and very pointed corners. It appeared to be like a Christmas present. This secluded box floated in the huge reality of the elemental, etheric and spiritual worlds or realms – but the box itself was quite small in relation to everything else.

It is this box that astrologers investigate in order to determine what is influencing our astral body, our emotions. This box is also related to the starry world, although not the stars themselves, which are the dwelling points of the Gods.

The geometry of the planets is an expression of the etheric forces, and this expression is also included in an astrological analysis, together with the directions that sacred geometry indicates. Thus, the geometry itself – especially 3D geometry (not the 2D used in astrology) – in orientation to the astral stars, indicates the possibilities of influencing both the etheric and astral bodies of human beings. I found it difficult to understand this (as I have found it difficult to describe), and so I awaited further explanation.

*

During and after I was shown the 'box' of the astral world, I was able to make a further observation in relation to the pain I have felt in my stomach for many years, and for which I took medication. From one moment to the next, this pain stopped. This observation is significant. Up until now, I had observed this phenomenon four times in my life, i.e. that a certain spiritual insight or observation leads to major changes in pathology. This change happens instantly, and the effect can last for years – maybe for the rest of one's life. The first time I had such a realization was in relation to karma, which cured my 20-year back pain in seconds (and it has remained cured for some 40 years). The second related to a major stroke. The third was in connection to stomach pain (as described above). And the fourth relates to an instant cure of my bladder when I met Michael (as I will describe later in this book).

*

The next day, Ariel, in his sorrow over the difficult story of how astrality was forced upon mankind, took me to a realm above Devachan, where all was pure bliss. This acted as a kind of healing balm to the 'wound' that had been created previously. This was a beautiful realm, and I felt happy again. Could this be the realm that is called 'higher Devachan'?

From this elevated point-of-view, Ariel continued his explanation – or rather exploration – of the astral realm. This teaching demonstrated clearly the difference between my former ability of clairvoyance and the now preferred path of clairaudience.*

Previously, the insights I received had begun as a picture. Then, with time, this picture would develop into a certain understanding (Imagination through Inspiration to Intuition). But now, the insight started deep within my body. If I needed to express the insight, it would develop into a certain picture, which I could then express in words. This necessity of going through a picture (Imagination) in order to come to the stage of expressing the same in words, slowly disappeared. Now it seemed that, increasingly, I could simply express the insight I received from my bodily senses directly into intelligible words.

Well, Ariel now showed me the existence of the astral from the viewpoint of what I assumed to be higher Devachan. It looked like a speck of dust. Then he showed me the astral from lower Devachan, and it took the shape of a Christmas gift, as described above. We ventured into the solar system, and he showed me that this astral 'package' enveloped the whole solar system.

This is how astrality is forced upon the development of the human race. Astrality exists nowhere else but in our solar system. The planets and the space between are imbued with astrality. That is why astrology is concerned with the planets, and the stars behind are merely used for direction. The mistake astrologers make, though, is not considering the relationship between the planets from a three-dimensional point-of-view, but only from

---

* See my last book, *Encounters with Vidar*.

a two-dimensional one. But Ariel told me this approach will be possible in the future.

The astral energy can be used by inhabitants of the solar system – or misused, if you like – in the development and training of the foot soldiers of the Gods.

*

The following day, Ariel showed me different geometric constellations, but this teaching was so complicated that I am unable to repeat or even summarize it here. What I can say is that there are seven main qualities to the astral world, expressed in the seven planets, with five lesser qualities expressed in five additional heavenly bodies in the solar system. The patterns – that these twelve variables make up – also vary according to the angle they form with the devachanic world, which is the rest of the cosmos. The presence of the spiritual (not devachanic–spiritual) within the solar system also adds to this complexity.

To add to the confusion, all this is influenced to varying degrees by the three main adversarial forces and the angelic forces, plus the higher celestial beings that are present, such as the Christ and beings like Anna!

I felt exhausted and somewhat 'small' for some days, and retreated to higher Devachan to regain strength and faith.

*

After refreshing my energy, I continued to study the astral construct enveloping our solar system. This construct was definitely shaped as a geometrical form, just as astrology is defined by angles and geometrical relationships, although in a two-dimensional construct. This reality was, as stated earlier, in a three-dimensional context. I can describe it thus:

- It has the shape of a cube;
- it has six sides;
- it has eight corners;
- it has 24 edges;
- it concerns twelve celestial bodies:
    - the seven planets already known;
    - the three planets outside Saturn, known to astronomy;
    - one planet further outside Pluto, presently not known;
    - and one planet inside our own earth, not yet known to science.*

I observed this construction both from outside Saturn and from just inside the construction itself. This construction fed astrality to all who would receive it.

---

* Interestingly, Space.com reports the following: 'A protoplanet slammed into the earth about 4.5 billion years ago, knocking loose a chunk of rock that would later become the moon. Now, scientists say that remnants of that protoplanet can still be found, lodged deep inside earth, *Science* magazine reported. If remains of the protoplanet, known as Theia, did stick around after the impact, that may explain why two continent-size blobs of hot rock now lie in the earth's mantle, one beneath Africa and the other under the Pacific Ocean. These massive blobs would stand about 100 times taller than Mount Everest, were they ever hauled up to earth's surface, Live Science previously reported. Theia's impact both formed the moon and transformed earth's surface into a roiling magma ocean, and some scientists theorize that the blobs formed as that ocean cooled and crystalized, Science reported.' (https://www.space.com/theia-may-be-in-mysterious-mantle-blobs)

It could only be conquered by the spirit of human beings, which was its sole purpose.

*

During that week, a huge and shattering truth and insight was revealed, partly as a teaching by Ymir and partly as a result of my own investigations.

The twelfth planet – that resides inside the earth itself, and that entered the earth around the time of the departure of the moon – represents or actually brings forward a fourth kind of evil. This planet came from outside of our solar system, and the spirit that it brought with it is neither luciferic, ahrimanic nor azuric. Rather, it is a fourth kind of evil, and is revealed now due to the total opening up of the elemental world, parallel with the emergence or construction of the fourth aspect of the soul described in my book *Encounters with Vidar* (and which I named there, '*time-karma-Christ*').

Up until now, this twelfth planet, together with its evil, has been hidden. It lingers deep within the earth itself, like a second spirit, parallel with Ymir and – after the Turning Point of Time – also with Christ. When a new evil appears, this must be counteracted by a new ability or part of the soul, which is happening right now. These things are always in balance.

The fourth kind of evil is the evil of nothingness, of vacuum. It brings nothing to human evolution apart from its own wish to destroy. I tried to study this evil for several days and got a certain insight into it, but I did not understand it properly. The best I can describe is that it appeared as shiny, introverted, hidden and very clever. Could it be that this phenomenon is connected to the one-third of the earth that is linked to evil?

Meeting the Antichrist

Over the next few days, I tried to study this 'fourth evil'. As a result, I became more and more depressed… It was close to the darkest point of winter. All was gloomy and cold, minus 10-12 degrees Celsius. I really was in a depression!

Then, on waking up on the morning of 19 December, I conjured all my strength and addressed this fourth evil directly, asking what and who it was. As I have related before, spiritual beings are obliged to answer truthfully when addressed directly. And indeed, the being answered, stating: 'I am the Antichrist'.

At once, this all began to make sense. Neither of the three first adversarial forces are the Antichrist. They are in the elemental world, in the elements of creation, thus helping our own development. This fourth evil is raised up from the earth itself, where it has been hiding since the middle of the Lemurian age. The fourth evil is in two parts, one on each side of the earth, appearing like two horns. One part lies in the Pacific Ocean and the other in the region of the African continent.

I was reminded of the Revelation of St John, where in the thirteenth chapter (verses 11-18) the following is described:

> And I beheld another beast coming up out of the Earth; and he had two horns like a lamb, and he spake as a dragon. And he exerciseth all the power of the first beast before him, and causeth the Earth

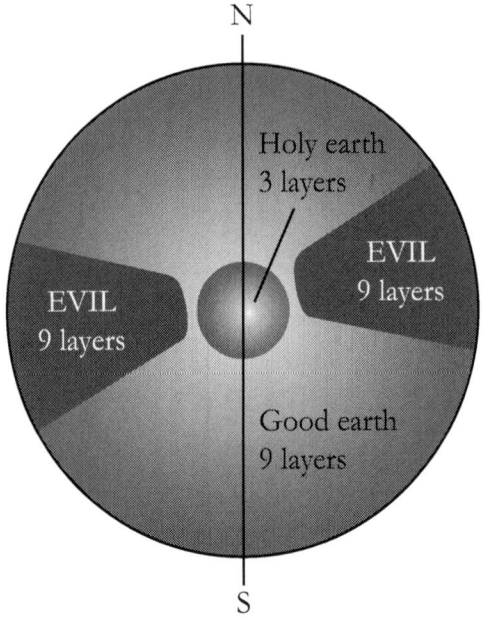

and them which dwell therein to worship the first beast, whose deadly wound was healed. And he doeth great wonders, so that he maketh fire come down from heaven on the Earth in the sight of men, And deceiveth them that dwell on the Earth by the means of those miracles which he had power to do in the sight of the beast; saying to them that dwell on the Earth, that they should make an image to the beast, which had the wound by a sword, and did live. And he had power to give life unto the image of the beast, that the image of the beast should both speak, and cause that as many as would not worship the image of the beast should be killed. And he causeth all, both small and great, rich and poor, free and bond, to receive a mark in their right hand, or in their foreheads: And that no man might buy or sell, save he that had the mark, or the name of the beast, or the number of his name. Here is wisdom. Let him that hath understanding count the number of the beast: for it is the number of a man; and his number is six hundred threescore and six.

Could it be that this Antichrist – whom Rudolf Steiner call Sorat,* or the 'Sun Demon' – has already come to earth, and will not actually come from the sun? Does the designation 'Sun Demon' indicate that it does *not* come from the sun, but in reality is the *enemy* of the sun? Will this fourth evil be released 70 years after the release of the azuric hordes in 2019 – that is, in 2089?

Is the first beast, described in the beginning of the thirteenth chapter (verses 1–10) of the Revelation of St John, the three other adversarial forces – the leopard being the Azuras, the bear being Ahriman and the lion, Lucifer?

> And I stood upon the sand of the sea, and saw a beast rise up out of the sea, having seven heads and ten horns, and upon his horns ten crowns, and upon his heads the name of blasphemy. And the beast which I saw was like unto a leopard, and

---

* Sorat (Hebr. תרוס) is the occult name of the Sun Demon, who is also the Earth Demon, who wants to prevent the future reunion of the earth with the sun and wants to permanently bind people to the earth's slag that will then remain, the so-called Eighth Sphere. He thus becomes the greatest opponent of Christ, who connected Himself with the earth through the Mystery of Golgotha in order to initiate this reunion and to enable people to take part in the associated spiritualization. Sorat is the beast with the two horns mentioned in the Apocalypse of John, whose name is only given in code with the number of the beast – 666.

his feet were as the feet of a bear, and his mouth as the mouth of a lion: and the dragon gave him his power, and his seat, and great authority. And I saw one of his heads as it were wounded to death; and his deadly wound was healed: and all the world wondered after the beast. And they worshipped the dragon which gave power unto the beast: and they worshipped the beast, saying, Who is like unto the beast? Who is able to make war with him? And there was given unto him a mouth speaking great things and blasphemies; and power was given unto him to continue forty and two months. And he opened his mouth in blasphemy against God, to blaspheme his name, and his tabernacle, and them that dwell in heaven. And it was given unto him to make war with the saints, and to overcome them: and power was given him over all kindreds, and tongues, and nations. And all that dwell upon the Earth shall worship him, whose names are not written in the book of life of the Lamb slain from the foundation of the world. If any man have an ear, let him hear. He that leadeth into captivity shall go into captivity: he that killeth with the sword must be killed with the sword. Here is the patience and the faith of the saints.

Does this describe the 70 years from 2019 to 2089? I had so many questions... but I was prompted in my thoughts with the realization that the Antichrist will distort time, with which he is not bound. He will also distort karma – at least that is his intention – and he will try to distort the being of Christ and the reunion of the earth with the sun. That is why we must develop the fourth part of the soul, *time-karma-Christ*, in order to survive the lethal attack from this fourth adversarial force.

The other three soul qualities can resist the three known adversarial forces – the spiritualized thinking, feeling and will can counteract Ahriman, Lucifer and the Azuras – but not the Antichrist. For this to happen, we must develop the soul-aspect of *time-karma-Christ*. I firmly believe that knowledge and awareness of this is of utmost importance in our time.

*

The following day I decided to attempt a dialogue with the Antichrist. He had responded to my question of who he really was. Now, I asked directly about his mission.

He looked me straight in the eye and communicated using panoramic concepts, Imaginations and Intuitions. He expressed that he actually wanted to help human beings, as do the other three adversarial spirits or forces. Human beings were created to help the Gods – in a way, we are trained as foot soldiers for the Gods in their conflict with the adversarial forces. This conflict originated in the desires of the adversarial hordes for freedom.

The adversarial forces wanted humanity also to be free in their own right – not anyone's foot soldiers. This means to become individually free, and not under the domain of God; not to follow any superior's rules. In effect, they wanted us to become selfish and not to adopt the selflessness that the Gods would prefer.

I was given a picture as a metaphor of a human family. Two parents beget a child, creating a family. At 18 years, this child moves out. After some years, the father of the family would like this child – who has since become strong – to move back home in order to help the family, and thus not to follow his or her own path or destiny.

That is how God (the Gods) intend to manipulate us, he claimed. The adversarial forces try to liberate us from this destiny. That is what we should also try to obtain for ourselves. We already have our thinking, feeling and will, which are sufficient for us to be free.

In this, he appeared to reveal, unwillingly, the intended future of the soul – that it was destined to become sevenfold, with 'time' as the fourth member, 'karma' as its fifth, 'Christ' as its sixth, and 'love' as its seventh member. I thought about this for a while.

'What about love,' I cried, 'which is destined to become the strongest force in the universe?' He did not answer, but looked me straight in the eye. And so, for several days, I just stood there spiritually – straight and strong in front of this most feared entity. He appeared dark and shiny, like a black insect, with all the strength of the cuticule of an insect.*

He stood in total silence. And I waited.

*

After a few days, it appeared that the Antichrist was preparing to say something further. This was on 24 December, Christmas Eve. I thought about this situation, concluding that it was probably the most unlikely scenario I could ever have imagined. Talking with the Antichrist on the birthday of Christ! But then I started to realize something…

---

* Insect cuticle is a layered, fibrous composite of chitin, water, protein, catechol, lipid and occasionally metal and mineral, secreted by a single layer of epidermal cells.

In all my 'conversations' with different spiritual beings – and by conversations, I mean communicating through Imaginations, Inspirations and Intuitions, and more recently through the spiritual sense-organs inside my body – all this had been done in 'common tongue' or 'language' – a sort of communicative code that is used by all inhabitants of our solar system, even the planetary spirits. This had also been the case during my first conversations with the Antichrist. But then the nature of the conversation changed. The communication that the Antichrist put forth was no longer in this common tongue. Instead, his means of communication took the character, if I may describe it so, of 'extra-solar-system'. Clearly, he was not of this world!

His language became completely alien, consisting of broken and distorted impressions. Previously – through Imaginations, Inspirations and Intuitions, and also through the spiritual sense-organs inside my body – his communications were presented in recognizable movements of pictures, lines and colours that constitute these spiritual forms of 'language'. In his latest utterances, however, such lines and colours were no longer present. Instead, all was distorted, ugly and unrecognizable.

The Antichrist himself was not necessarily ugly as an entity – not ugly in outer appearance at least – but his 'language' was terrible to experience, and for several days I understand nothing. The words were now constructed like geometric figures, especially triangles, in different forms and shapes.

Meeting Michael

During Christmas and over into New Year 2023, I attempted continually to have a proper, comprehensible conversation with the Antichrist, but his language was indecipherable. So, a few days into the New Year, I gave up and went back to Ariel in upper Devachan. From there, I decided to try to approach Michael, the Archangel that for the time being is elevated to – or is in the process of becoming – an Archai.

This approach became quite confusing, and what I now write may also appear so. Michael, it transpired, was to be found in the time dimension of the solar system. This time dimension appeared like a separate layer, just inside the astral envelope or walls of the 'box'. This time-layer had a strong influence on both the astral sheet and material creation, but less so on the outer etheric.

Michael now abided in this layer or realm or region, and would do so for the next 2,000 years. As very few people understand or have any access to 'time' as such, it is difficult to communicate with the dominant Archai, as this being – Michael, at the present time – is somewhat enclosed within this 'time sheet'.

Michael has a big influence on both the astral world, human feelings and astrology, but his realm is extremely difficult to reach – or, at least, it was for me. I had found it easier to reach Angels, Archangels and even the other Archai, i.e. those that are not the Time Spirit or Archai for the present period. But nevertheless, I tried to approach the entity we know as Michael.

I first attempted to enter the stream of time – that is, the horizontal time axis, with its two time streams, the one from the future and the one from the past. I ignored the vertical one in the earth, as well as the four in lower and/or upper Devachan. Going into this stream reminded me of when I stepped into the etheric streams between the trees in the inner etheric realm. There, when I took a step to the left, I moved into the past, and when I took a step to the right, I was brought towards the future. When I did this in the etheric realm, it had an interesting effect. But in this realm of the Archai of our time, it felt a lot more than interesting! It felt quite overwhelming, and even frightening...

*

At last, I was able to meet Michael. This meeting brought much more than I ever anticipated or expected. It also came with strange insights and understandings. More than ever, I understood the differences between knowledge and 'walking a path' (active spiritual work), and how important it is to meet Michael personally, and not just to acquire knowledge about him.

My first experience with Michael was in relation to 'wind' – as if he expressed himself in the form of a spiritual manifestation of what we know on earth as strong meteorological wind! These winds around Michael blow in two directions, from right to left and from left to right (which, I thought, was representative of time itself). The two directions from which these winds derive are not straight lines, but more a 'curved path', from left and right and then curling around you, as if surrounding one – something like a huge curling snake... It is impossible to tell from where it comes and to where it goes.

Then my mind was drawn to the part of John's Gospel (3:8) where Christ talks about wind:

> The wind blows where it wishes, and you hear its sound, but you do not know where it comes from or where it goes. So it is with everyone who is born of the Spirit.*

Standing within these spiritual winds, one does not know where to rest – it is truly frightening. To meet Michael today, it seems that one is born again, or to put it differently: *when you are born of the spirit, you will meet the Time Archai, who today is Michael.* That is why Michael is so important for the present day, as he was for Rudolf Steiner.

The winds that blow around you when you enter the domain of the Time Archai are very strong. You see and feel only the winds, like two anacondas dancing the Dance of the Hours – the dance of time – around you. Are we bound by this time, by this illusion of causality,

---

* I got the strong impression that Christ himself taught from the level of Time Archai. Christ asked Nicodemus if he is not the teacher of Israel – and Israel (or Azrael) is also the name of an Archangel. The teacher of this Archangel was at that time Oriphiel, who was the leading Archai, or Time Archai, of that period. (According to Trithemius, Oriphiel held this role from 246 BC to AD 109. See further in Richard Seddon, *Europa, A Spiritual Biography*, Temple Lodge Publishing, 1995.)

or will this meeting with 'real' time free us from the illusion of time?

*

For several days I felt, experienced, saw and heard only the winds. Then, one morning, Michael stood there, strong and shining. His body was larger than is shown in the many pictures and paintings of him, with his middle area being most prominent. When he appeared, the sounds of the winds lessened somewhat, and it became obvious to me that he did indeed influence time.

During the day, I was able to receive his attention. The strength of the winds further diminished, and Michael became more approachable, or perhaps even 'friendly'. But there was no 'conversation' or teaching.

*

The next day, I felt that a certain communication was coming, but nothing happened. And so I waited… Then, something unexpected occurred. I was able to 'enlarge' my listening, to 'extend' it to the area above and below Michael. It was rather like when one is trying to listen to something that someone is saying, and you suddenly become aware that above them the birds are singing and flying in the air, and below them the earth and the grass exist and are expressing themselves – that all this is going on in parallel… When you become aware of this, you begin to 'hear' from a wider perspective. This panorama appeared to me when I enlarged my observational angle, also in the vertical sense. I will try to explain further.

The 'speaking' of Michael extended in a certain horizontal direction, but this was, or became, harmonized with an extended 'speaking', also in a vertical direction. This means, in simple terms, that Michael brings in influences from higher and lower beings. These higher beings are higher Angels, like Exusiai, Dynamis and Kyriotetes, and the lower beings are those such as the Archangels and Angels. So, the teaching or speaking of Michael is like a symphony of many angelic beings, some higher and some lower.

At this moment, it was difficult for me to distinguish who was saying what, or even what they were expressing, but I hoped that this would become easier in future. For the time being, I was able to enjoy this choir of sound – this choir of speech, teaching, colour and movement – performed by numerous angelic beings of different 'heights' or development.

It felt like a symphony in high fidelity, and I resolved to listen for some days.

*

The following day, I started to understand how a Time Archai worked, and also what made him different from a normal Archai (the other archaic entities or beings). The function of a Time Archai is to bring down the streams of the zodiac and lead them into the time streams of this world, of the solar system. There is a certain area where this is possible, and that is where the horizontal winds – time streams of the solar system – cross the vertical streams – those from the cosmos, or rather from the twelve zodiacal positions – bringing in the impulses that form or inspire the beings of material experience.

As there are twelve streams coming from the zodiac, there are also twelve 'candidates' for becoming a Time Archai, and this also explains why the duration of the influence of one Time Archai is around 2,000 years.* As discussed, the place where a Time Archai resides gives the impression of being surrounded by strong winds. This is because the twelve streams of the cosmos meet the time streams of material experience, and this gives the feeling of being caught within strong winds.

The twelve etheric streams from the cosmos always form a cross in the sky, relating to the morning–evening axis, and the 90^0 midday–midnight axis. These axes change every 2,000 years, and thus bring different impulses to humanity. The horizontal axis brings the Christ–Sophia forces (which today are Virgo and Pieces), and the vertical axis the adversarial forces (which today are Sagittarius and Gemini).

---

* Compare also the text on Spirits of Time at Anthrowiki: https://en.anthro.wiki/Spirits_of_Time. The Platonic Year is the term for the longest cosmic cycle from a geocentric point of view. It is the time taken for the vernal equinox to travel through the whole ecliptic. This takes 25,729 years (according to some estimates 25,771 years). The vernal equinox spends on average 2,144 years in one sign of the zodiac. This epoch is known as a Platonic month. A Platonic day is roughly 72 years, the average age of a human being. Many astrologers believe that the passage of the vernal equinox through one constellation of the sidereal zodiac sets the general tone for a 2,000 year period known as an Astrological Age. Because it is not possible precisely to define the sidereal zodiac in relation to the Celestial Sphere, it is difficult to calculate the starting date of a new astrological age. With such long periods of time, it probably makes sense to include fairly long transitional phases, from anywhere between decades to centuries. (*Wikipedia*)

These streams are also the foundation of material creation, as the elemental world is formed in the tension between the angelic forces and the so-called adversarial forces – and thus material creation, as well as the divine influences, will change during these transitions.

As material experience also has a sevenfold development, the influence of the Time Archai also has a shorter cycle in which the seven leading Archangels act in human cultural development. These periods last about 350 years, but are not sub-periods of the cultural epochs. They follow an independent rhythm, in the succession of which the seven key Archangels take over from each other in their regency. In such a relationship, there will be seven Time Archai, but they are also, in my understanding, a kind of 'under-Time Archai' to the main Time Archai. This does not reflect their value – it only reflects their place within the overall time-frame, their function at that special point in time (and space). These influences also have a daily influence, a weekly, a monthly and a yearly influence – but this is too complicated to try to elaborate here.*

The cross of the time streams relates to the cosmic zodiacal cross, and the work of the Time Archai is to facilitate the exchange between the two crosses. The 'time cross' consists of the horizontal axis formed by the twofold time stream of the past and future, and the vertical axis consists of the twofold timeline relating to the layers of the earth itself – in which one may,

---

* It is recommended that the reader research these cycles themselves, as different sources – such as Seddon, op. cit. – give different figures.

through the seventh layer and Anna, enter the devachanic fourfold time, which seems to be part of the greater cycle.

*

Now, Michael stood again amidst the winds. They formed a huge cross in the cosmos consisting of the time streams, with Michael at their centre. The four arms of this cross were, at closer inspection, quite different in both colour, structure, movement and patterns. The right stream, the one relating to the future, was yellowish, like the planetary seal of Mercury. The left one, relating to the past, was reddish, like the planetary seal of Mars. The upper one was quite dark, whilst the lower one was grey.

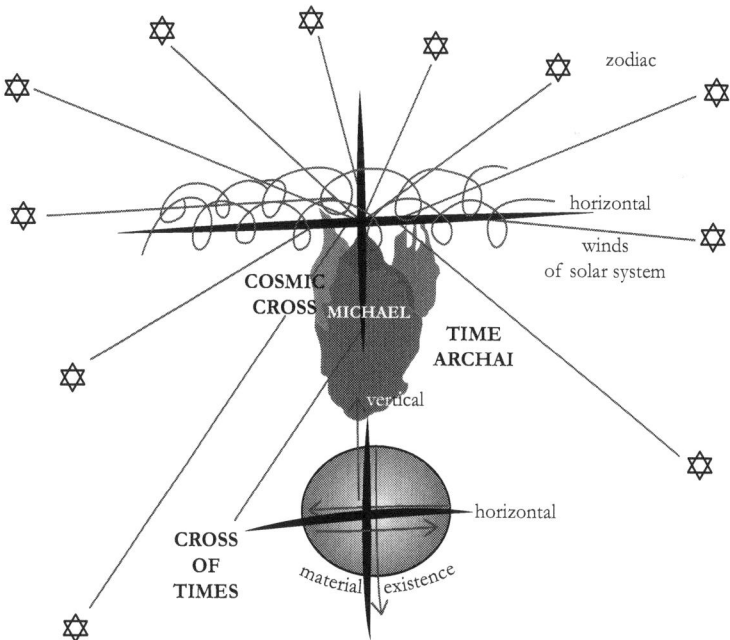

These two crosses, with Michael in their centre, stood in front of me, just as described, for many days. Michael appeared in all his strength, in all his superiority, as a timekeeper, as a Time Lord, conducting the passing and coming of times – of all the four time streams of this material experience. I could do nothing but stand in front of him in awe and grace, honouring his work.

*

The next day, I received a teaching from Michael. I understood that this teaching was somehow personalized to myself – just as with Vidar's teachings – although, paradoxically, it was also universal, or at least to some degree.

I understood that Rudolf Steiner's teaching in his book *The Philosophy of Freedom* came from Michael. This insight gave me an understanding of what Rudolf Steiner had tried to achieve in this most important of his books. *The Philosophy of Freedom* was written as an instruction on how to perceive the world. In this perception, one should separate concepts from percepts, and thus make our thinking and perception spiritual, that is, connected to the elemental world. This part of the teaching was quite simple and easy to understand.

Michael showed me, in strong imaginations, how we – through the outgoing and incoming streams of our eyes – perceive the external world. Through the outgoing stream, we acquaint ourselves to and with the external object (greeting the elemental being in the external object), and in this stream we merge into the elemental beings of the object. Through the incoming stream, we perceive the external object and present it to our soul and spirit. Through this process, we usually let our conceptual mind overshadow the percept, and thus darken the life – or rather the elemental activity – in the percept (i.e. alienate ourselves from the elemental beings in the external object).

As this incoming stream enters our mind, it becomes like a mirror; it turns around, so that the conceptual part becomes stronger than the elemental being of the percept. Our task is to separate these two parts of the incoming stream, and become conscious and aware of the pure percept. This is described as pure perception. This pure perception gives rise to pure thinking, which is thus a gateway to the spiritual world, primarily the elemental world – but, through this, to the whole spiritual world.

This Imagination, of one receiving this double stream of observation – which, within oneself, entering oneself, becomes mirrored, that is, turned, and in this turning the two parts of the observation get somewhat separated, with the conceptual part dominating the perceptual part – all this explained the whole of *The Philosophy of Freedom* in a single image.

I was astonished. I had been able to receive this teaching, and I could understand it immediately! Previously, I needed to first observe the picture, then understand it, and then, finally, become one with the teaching. This was quite different.

With this teaching, I was more than satisfied and grateful, and felt that it was more than enough for one day.

*

At this point, I will make an observation on the nature of existence. Keeping in mind Rudolf Steiner's comments regarding 'the descent into materiality' that went 'deeper than planned', we can understand that not everything is predictable, and the different entities in the spiritual worlds also have their own agendas. In other words, they do not all follow a rigid 'plan'. The totality of development in material existence is thus not 100% predictable. This is just as it is here in the physical world: we have our plans and intentions, but usually things do not develop according to these plans, and sometimes they go totally in the opposite direction.

The entities of which we speak of as 'Gods' are thus sometimes individual and have their own agendas. And, they can make mistakes. Thus, they are not infallible.

*

One Sunday in late January, I tried to explore the time streams of the two great crosses that Michael indicated – or rather displayed – within himself. I thought about the etheric streams (the etheric of this side, not the outer etheric) that I had seen streaming between trees in the forest, and how I could travel in time when I stepped inside these streams. As described, if I moved to the left, I travelled back in time, and if I moved to the right, I moved towards the future. In this travelling, I could not decide where to go, however. I was simply taken in one or the other direction, and I pondered on how Rudolf Steiner and others could travel exactly to where and when in time they wanted.

Now I tried to do the same with the streams of time presented as two crosses – one within material experience and one in the cosmos, in the devachanic world, outside of our material existence. My first experiments with this indicated that it is possible to decide exactly where and when in time one wishes to 'travel'.

*

The next day was much more exhausting than I had expected. When going into the 'winds of time', I could now decide where and when to venture, and also whom to observe – but the problem was that I also received the emotions of the one observed, including their doubts and pains. It was as if, when one enters the winds, there is no place to rest one's head.

I could not observe different times without also participating in the feelings of the time period that I was observing. When going into the past through the etheric streams in this world (i.e. between the trees of the forest, as described above), I did not participate in emotions – but when entering through the winds of the ruling Archai, it was impossible not to. It was as if the inner etheric did not contain the astral, as the winds of time did.

I wondered if I was able to do this – to partake in all these pains and depressions – and decided I would try. I also resolved to train myself in participating in the emotions but still being able to keep a certain distance from them. I sensed that there would be some challenging times ahead.

In the beginning, I tried to get an overview, and also to investigate how to manoeuvre in time – how to steer within it, so to say. This I then did for some days, feeling or experiencing the quiet sobbing of eternal times, shivering like darkness in my mind.

Suddenly, it came to mind that, forty years earlier, I had written a poem describing how, on entering the spiritual world, the feeling of sadness, of crying, may enter your soul. Here is the poem in English translation (although this does not reflect the full poetic expression of the original Norwegian,* which is built on rhythm and rhyme and can also be sung):

THE SUBTERRAINEANS

It breathes softly in the hill and in the moss
The sun has gone down.
They are coming in, in a silver-grey shine
My world is at peace.
And feeling the old within me,
Breaking and bending as crying in my mind,
But this must be my path
If eternity once shall be mine.
I raise myself up in deep uncertainty,
They whirl around in their holy circle.
I'm sinking down in the moonshine,
I know they must leave,
Before the dawn with the sunshine in its mind
Breaks their peace.

*

One morning, Michael provided clarity regarding my conception of time, and how the changes in Devachan cause changes in the material world. I wrote earlier that a change

---

* DE UNDERJORDISKE: Det ånder svakt, i haug og myr, sola har gått ned. De kommer inn, i sølvgrå glans, min verden er i fred. Og kjenner det gamle i meg, bryter og bender i sinn, men dette må bli min vei, om evigheten er min. - Jeg reiser meg, i uviss ro, de danser dansen sin, de spinner rundt, i hellig ring, men drages mer av sted. Jeg synker ned, med månens skinn, jeg vet de må av sted, før gryningen med sol i sinn, bryter deres fred.

in the great time cross in the devachanic world took some 12,000 years to cause similar changes in the time cross in material experience. To give an example, let us say that a certain cooling process is introduced in Devachan. This change in the spiritual world (Devachan) initially causes related changes in the elemental world – beginning in its first layer, the azuric/archaic region. This is mediated through the Archai, who then bend or coerce the azuric elementals to change. The first realm becomes colder.

Then, the change in Devachan causes changes in the second elemental realm, the atomic realm, mediated through the archangelic beings, who thus bend or coerce the ahrimanic elementals, following the wishes of the Gods, and the devachanic changes. The movement of the atoms become slower and colder.

Then, this change causes the relevant or related changes in the third realm, mediated by the Angels, who can bend the will of the luciferic elementals to change the material illusion, which becomes colder, and thus creates the necessary changes for the material experience of human beings. Then we reach the material world, and the described change here will, for example, cause the onset of an ice age.

The whole process takes around 12,000 years to be fulfilled, but the first changes start after twelve years in the first realm, 120 years in the second realm, and then 1,200 years in the third realm. The material expressions show themselves after approximately 12,000 years.

Again, I went into the crosses of time, and there was revealed something that I was not aware of. This related to a parallel observation of time. As explained earlier, the left side of the cross is the past and the right side is the future. I had already perceived and understood this.

But the time cross can also be observed as a 'space cross' when it is fixed at a certain time. This somewhat cryptic concept requires some explanation.

If I go to the left side, that is, back in time, and stop at a certain point in time – for example the year 33, the time of the Crucifixion – the cross shows where in the world a counter-action of this event happens or appears. In this special example, the end of the arms of the cross are in the centre of what is today Israel and the centre of Mexico.

Rudolf Steiner gave this example himself, which is why I begin with it. In Mexico, a black magician was crucified at the same time as Christ Jesus. This great magician's intention was to eliminate the beneficial effect of the crucifixion of Christ Jesus. This was avoided by the crucifixion of the black magician.

What happens spiritually at the end of the right arm of the cross, will find a counter-action of luciferic origin at the end of the left arm of the cross – and what the beneficial spiritual beings do at the left arm of the cross will trigger an ahrimanic counter-attack at the right arm of the cross. This has happened continually, and there are numerous such crosses all over the earth and universe.

All is in balance, and the balancing is between the good and the bad, between the beneficial spiritual forces and the adversarial spiritual forces – and even between the luciferic and the ahrimanic forces, as well as the azuric forces.

I then looked at Europe in the year 1136. I chose any year at random, in order to see if I could be so specific, and found such a balance between Chartres in France – again, I just chose any place to see if I could be that specific – and Toledo in Spain. The good work of the initiates in Chartres was counteracted or balanced by black magicians who were living in Toledo at that time.

Then I looked at the area of east Ukraine today, and saw that it was counteracted by – balanced or connected to – an area in Pennsylvania, USA, although I have no idea what might be counteracting the war in Ukraine there. I also looked at Sandefjord in Norway, where I live. The counter-part or connected place was in the White Sea, at the Solovetsky Islands, in Russia. Another counterpoint was in Cornwall, England.

This 'balance' also showed itself in the timeline of each individual human being (and probably in every entity or phenomenon existing), with a balancing point at every seventh year (and, likely, also in larger time periods too). For example, what happens when we are 7 years old has a counter-point at the age of 21 years, with the balancing point at 14.

When I managed to have an overview of all these connections in both time and space, it seemed that the whole world was crowded with these balancing points, which actually do not look like crosses but more like triangles, with one tip pointing upwards, and the two balancing phenomena at each of their connecting points. These triangles can also be seen throughout the whole of the cosmos.

*

I lived with this for some days, and then, on the afternoon of 6 February, I came down with severe depression. This arrived just after the news of a very strong earthquake in Turkey and Syria, so I assumed that there was a connection to this event. Every time I have depression, I believe that it has a corresponding cause in the physical world. However, almost every time, it is also a precursor to a spiritual revelation. And so it was this time.

On the morning of 7 February, Michael, as Archai, was back. The winds that surrounded him were stronger than ever, and his arms, with the counter-current time streams, were like huge eagles' wings. He did not have his sword – which surely is not needed since his victory in 1879. Instead, he had huge and powerful, eagle-like wings.

He looked directly at me, and I understood that I could expect a personal teaching from this mighty entity. I observed that each of my thoughts were 'doubled', in the sense that they revealed what would be in the future, and also their origins, i.e. what they had been in the past.

I started to understand that the various spiritual beings taught in their own unique ways – and again I realized, 'as above, so also below'. In a university, the different teachers teach in different ways, following different pedagogic systems or philosophical ideas. Some spiritual beings teach according to the means described by Rudolf Steiner, as a progression from Imagination to Inspiration, ending in Intuition. Some teach through experiences via one's organs – bodily understandings, including pain and suffering. Some teach through direct conversations. And, some teach through karmic relations in the physical/material world.

My question now was: How does Michael teach or communicate? It was definitely different to Vidar, Balder, Ariel or Anna... or the other entities I had encountered these past three years.

*

Over the following several days, nothing changed. We only stood there, facing each other spiritually, but saying nothing. Maybe I should try? I attempted this for some time, but it didn't prove to be possible. Then, slowly, I understood: Michael was expressing himself through the fourth soul ability, the *'time-karma-Christ'* aspect. And now, inklings of a message or teaching – initially rather vague – were coming through. It related to the mystery of evil.

The teaching reminded me of something I had read many years ago by Sri Chinmoy to the effect that darkness is light (a human realization); light is for darkness (a divine realization); and that darkness is for light (a supreme human realization).

Michael now added to this understanding as follows: 'Darkness will become divine light: This is my future realization.'

As a result of this, it was clear to me that *we must understand darkness, understand evil, and that is our present goal.*

*

For several days, I tried to communicate further with Michael, but to no end. It seemed that his only purpose was to be the Time Spirit and to bring the understanding of the mystery of evil.

It was as if an Archai follows one important principle (and so, it is very interesting that the Archai are also called 'Principalities').

I felt I had now understood Michael's chief *principle*, and thus decided to attempt to travel further – to venture higher or deeper in the spiritual world. This region is not Devachan, but the system created by the Gods on this side of Anna, the portal between our world and the world of the Gods.

The first realm I arrived in was strangely like the outer etheric realm – as described in my previous books – as a place of Christ-filled pastel clouds. This might not be so strange, as the next realm beyond the archaic realm should be the realm of the Exusiai – which, according to spiritual teaching, is actually a Christ realm, the realm of the Elohim. This was very pleasant, warm and calm, with light shining through pastel-coloured clouds.

I stood still to watch, but could not see any entities or hear anything. Still, I could feel the presence of several loving beings – truly loving... This love was so all-embracing and all-consuming that, for some days, I simply enjoyed this existence.

*

I wanted to go deeper into this world of pastel-coloured love. I ventured further into the clouds, and soon became aware that there was also something else, deep inside this world – something darker. Reaching the centre, I met an enormous darkness; a darkness that materialized in a huge monster or animal-like, alien form. This monster was trapped within this loving environment, surrounded by entities that poured their love on and around this somewhat miserable beast.

The 'beast' looked like a cross between a catfish and a common woodlouse. It appeared to be stranded there, quite helpless. I resolved further to investigate this presence of darkness amid the realm of the Exusiai, the realm of love.

*

The next day I stood in front of this sad creature once more, and tried to ask its name. It seemed to utter 'Vlask' in reply. It indicated to me that it had been there since Hyperborean times, and that it had some relationship to the Antichrist, who had come here in the middle of the Lemurian epoch. This beast had come from the far reaches of our universe – outside the material experience that we can examine and understand.

In this way, there were two creatures that were opposed to Christ, mainly because they were not intended to be here in our universe, as part of our human evolution. Both were polite to me, although simultaneously dark and very strange.

I continued to talk with this being, Vlask, for some time, and was introduced to his 'world-view', which coincided to some extent with that articulated by the Antichrist. I was told that there is, at the present time, a certain war going on in the heavens, in Devachan, between those that want to abide with and serve the Father God, and those that want individual freedom. There are more wars in the spiritual world than here on earth, apparently.

Human beings are, as described before, created to serve in this struggle, on the side of the Father. For this purpose, the material universe was created – around 13.7 billion years ago, according to science, and on Old Saturn, according to Rudolf Steiner. Other universes have also been created in the entirety of existence, in other 'places' and for other purposes.

Those who want individual freedom are called the adversarial forces. This war has gone on for a very long time and will go on for a long time still. Sometimes, the adversarial forces send 'mercenaries' into our universe, our material experience, to help fight the plan of the Father.

The Antichrist and Vlask are two such mercenaries, and both are, for the time being, imprisoned or made harmless – the Antichrist in the earth, and Vlask in the domain of the Exusiai – until such time as human beings are strong enough to face them.

This 'chaining' of different kinds of beasts, demons or adversaries – placing them in occult cages or prisons – is not at all uncommon in spiritual history, as can be found in the Bible, in the myths of many cultures, or in fairy tales. Even human beings can be put in spiritual prisons, as Rudolf Steiner described in relation to the 'occult imprisonment' of Madam Blavatsky. In my book *Transforming Demons*, I described how a terrible demon was imprisoned in the centre of Silbury Hill in England around one thousand years ago, and will be kept there until the time is right to unleash the adversarial forces.

After some time, I began to feel a certain pity for Vlask. He appeared to be distraught, without hope and condemned to this existence – even though he was surrounded by Exusiai and Elohim who were trying to pour their love over him.

The Higher Angelic Realms

For two days, I stood before Vlask, feeling increasingly depressed about his seemingly hopeless situation. Then I asked why, initially, he had come here at all. His expression changed radically; he became angry and malicious, yelling: 'To destroy you!' Any pity I had vanished completely.

From that point on, I tried to venture even higher, and came to another realm of angelic beings, which I understood must be the dwelling place of the Kyriotetes. At this point, I had a unique opportunity to view the ranks of the Angels. I was at something of a high point, overlooking the spiritual terrain.

According to Rudolf Steiner and other spiritual traditions, there are nine levels of angelic beings, starting from 'below':

1. Angels, also called simply Angels or Sons of Light.
2. Archangels, also called the Angels of Individuality.
3. Archai, also called the Angels of Principality.
4. Exusiai, also called the Angels of Power.
5. Dynamis, also called the Angels of Virtue and Movement.
6. Kyriotetes, also called the Angels of Dominion.
7. Thrones, also called the Angels of Steadfastness.
8. Cherubim, also called the Angels of Fullness of Wisdom.
9. Seraphim, also called the Angels of Burning Fire.

This structure and grouping of Angels are the spiritual reality within our created universe, our cosmos and our material experience. In the higher spiritual world, however, which we might call higher Devachan, there are still many other groups and structures. I do not profess to know about these, I have had only a glimpse of them. Besides, from the streams that run out from Devachan, there are other universes, other cosmoses, of which I know nothing. One day, however, I hope to travel further in higher Devachan, through the portal of Anna.

In my experience, the Archangels, of which Vidar is one, gave individual teachings, which had the effect of creating *individuality*. The Archai, of which Michael is one, stood by the *principle* of time. The Exusiai had the *power* to imprison harmful intruders. I did not meet the Dynamis (at least, not to my knowledge). The Kyriotetes showed a certain 'over-height', if I may put it that way, and could be characterized as *dominating*. After meeting these different hierarchies, I began to understand why they have got their different names.

Being present in the region of the Kyriotetes, I awaited what this new experience might bring. Visually, these entities looked very tall, with their heads constituting of one third of their total 'body'. I had to look up to them. They were stern and sincere.

*

Standing before these huge-headed Kyriotetes for some days, I looked back to the area where I had been unable to see anything, where the so-called Dynamis were to be found. And then, from the level of the Kyriotetes, I could vaguely glimpse them. They were moving so fast that they were almost invisible. They were like the wind, almost impossible to see. Only when I moved myself, was it possible to observe them. Then a deep realization began to grow in me, and I was very conscious that this was only the beginning...

There is a very important mantram in Rudolf Steiner's nineteen Class Lessons, that *'three become one'*. Now, I began to come to an understanding that *'nine become twelve'*. The three angelic levels will, with the presence of Christ, become one. The nine angelic hierarchies will become twelve, just as the nine layers of the earth will become twelve in Christ. This is also valid for the human body in the sense that, when we leave the material earth – which consists of nine layers – we enter the spiritual realms, which adds three more layers. Thus, we can speak of a total of twelve layers, of which the last three are within the heart.

The three soul faculties of thinking, feeling and will, will become one in Christ and thus give rise to a fourfold soul. The fourth aspect or part will be *'time-karma-Christ'*, as earlier described. I knew then that I should further contemplate and consider this revelation in order to verbalize it properly. That would probably take a while.

*

From the Kyriotetes, I made an investigational tour further up within the hierarchies, reaching the first hierarchy, traditionally consisting of Thrones, Cherubim and Seraphim. Here, it felt more like a 'cooperative' region. What do I mean by this? Well, whilst the entities in the third hierarchy worked more individually with human beings, the first hierarchy seemed to work with Creation itself, with the construction of the hordes of elemental beings. And what they produced, working together as a team, gave the illusion of material existence.

One group (the Thrones) appeared to work with 'hard' or solid nature, whilst the second (the Cherubim) were more 'fluid', and the third (the Seraphim) somewhat 'airy'. Having said this, the 'air' was burning, the 'fluid' aspect was dancing, in huge and mighty streams, and the 'hard solidity' was of an extreme density, like a huge, thick oak table.

The area these beings abided in was both similar to and different from the outer etheric, where Christ can be found. According to my present understanding, the Christ region is to merge with the angelic hierarchies, creating a new hierarchy – possibly with Christianized human beings – thus resulting in twelve levels of celestial entities.

Then, the work of Christ became more evident to me. He inspired or created the accessibility of the three innermost layers of the earth, the creation of the fourth section of the three angelic hierarchies, and the spiritualization of the elemental world. From this high perspective, I also received the impression or sense that all the different Angels and hierarchies functioned together as a large, unified and quite complicated gathering of entities, with one common goal – the education and maturation of the human race.

In the understanding of this common goal, I had to ask myself the following question: Is Vidar the only guardian of the threshold for all human beings? Or, are the different Archangels the guardians for each of their separate folk souls or groups? As Archangels are different for every folk group of the earth, the latter would imply that there are many different guardians.

*

Over the following days, I simply observed this huge and coordinated dynamic of the different groups and hierarchies of Angels, without any particular or directed conversation with any one or set of entities. From on high, from the region of what is called the first hierarchy, I simply watched and enjoyed this marvellous vision. I also pondered on my previous experiences.

Up until now, I have experienced several areas or realms within the spiritual world:

1. The elemental world:
    a. Third realm (luciferic).
    b. Second realm (ahrimanic).
    c. First realm (azuric).
2. The outer etheric world:
    a. The threshold (Vidar's domain).
    b. Solid (firm) lands.
    c. Rivers.
    d. Mountains (Christ's domain).
3. The Akashic world or realm:
    a. The Akashic vault.
    b. The librarian.
4. The 'earth'–spiritual world:
    a. Nine to twelve layers.
    b. Seventh layer with Anna as a portal between material experience and the devachanic world.
5. The Angelic world:
    a. Nine hierarchies (at least, for the time being).
    b. The 'prison' in the fourth realm.
6. The devachanic world:
    a. …which I have not yet sufficiently explored.

I decided to investigate further within the devachanic world. This region is, according to Rudolf Steiner, divided into lower Devachan and higher Devachan. Here is a description from one of his lectures:

> When we enter the devachanic world the astral world remains fully present; we hear the devachanic, and we see the astral, but under a changed aspect, offering us a remarkable spectacle. We see everything in the negative, as though on a photographic plate. Where a physical object exists, there is nothing; what is light in the physical world appears dark, and vice versa. We see things, too, in their complementary colours: yellow instead of blue, green instead of red.
>
> In the first region of Devachan we see the archetypes of the physical world in so far as it has no life – the archetypes, that is, of the minerals – but also the archetypes of plants, animals, and human beings in so far as their physical forms are concerned. This is the region which provides, as it were, the basic skeleton of Spirit-land. It can be compared with the solid land on earth and is therefore called the 'Continental Mass' of Devachan.
>
> When a human being is observed over there by an initiate, the physical space he occupies appears dark, but round him is a radiant halo. When our senses have become more delicately organized, the archetypes of life are added: everything that has life flows over the earth like water. Here the minerals cannot be seen because they have no vibrant life; but plants, animals and human beings can be seen very well. Life circulates in Devachan like blood in

the body. This second region is called the 'Ocean' of Devachan.

In a third region, the 'Atmosphere', we encounter feelings and emotions, pleasure and pain, wherever they are active in the physical. Physical forms then are like solid foundations, the Continents, of Devachan. Everything that has life forms its Ocean.

Everything that pleasure and pain signify are its Atmosphere. The content of all that is suffered or enjoyed on earth, by human beings or by animals, is displayed here.

Thus, to the initiate a battle appears like a great thunderstorm: fiery flashes of lightning, powerful claps of thunder. He sees, not the physical actions that occur in the battle, but the passions of the opposing armies, and these appear to him like the heavy clouds and lightning-flashes of a thunderstorm. The fourth region transcends everything that might still have existed even if there had been no mankind. It includes all humanity's original thoughts which enable him to bring something new into the world and to act upon it, no matter whether the thoughts are those of an ignorant or a learned person, of a poet or a peasant. They need not involve any great discoveries; they may belong to everyday life.

After these four regions we come to the boundary of the spiritual world. Just as the sky at night looks like a hollow globe encircled by stars, so it is with this boundary of Devachan. But it is a highly significant boundary; it forms what we call the Akasha Chronicle. Whatever a person has done and accomplished is recorded in that imperishable book of history, even if there is no mention of it in our history books. We can experience there everything that has ever been done on earth by conscious beings.*

*

With Rudolf Steiner's descriptions in mind, I wanted to see what I could discover for myself. But first, I had to return to the entrance of the devachanic world, to the seventh layer of the earth, and to Anna herself. I did so, and this time easily entered the devachanic world.

The first time I entered Devachan, I had been taught by Ariel about the connections between Devachan and our material existence; time (in both Devachan and in our universe); and the deeper meaning of the creation of our universe itself. This time I wanted to explore Devachan further, and not just to understand material experience from the standpoint of that world.

After entering the devachanic realm, I was able to find Ariel again, and expressed my wish that I wanted to venture deeper into Devachan. Ariel helped me turn my consciousness within Devachan itself, and not 'down' towards the earth.

The first insights I reached – or rather, were revealed to me – were about the structure of the devachanic world. At this point, I can only give an initial description of what I saw as a totality. It was like an enormous crystal with facets or aspects spreading out to all planes – up, down, right and left – and all possible directions in between (of which the horizontal and

---

* Rudolf Steiner, *Founding a Science of the Spirit*, lecture of 23 August 1906, Rudolf Steiner Press, 1999.

vertical planes were the most prominent). Of course, it is hugely difficult to put these experiences into words, but it was as if all the facets of this 'structure' projected out variations of celestial 'reality' – as if the 'truth' was distorted due to one's lack of insight into the single facets.

To the right of the central facet, the projections of the devachanic reality were changed into its 'opposite', like a negative in film processing. To the left, the projections were changed into the opposite of the 'negative' – the 'positive-positive', if I may put it so. Both the 'negative' and the 'positive-positive' are variations of the 'truth', which may be found in the 'middle'. All the different facets are like seeing the spiritual and celestial truths through the eyes of the different angelic or adversarial entities that constitute the totality of the devachanic world.

This experience was so overwhelming that I could do no more that day. I resolved to venture further into the various facets the following day.

An Overview of Devachan

The next day, I tried indeed to get an overview of the whole expression of the devachanic world. I attempted to venture to the right, to the middle and to the left, and also somewhat towards the upper area and the lower.

When venturing into the devachanic world through the right-sided facets, the 'negative' facets, a huge landscape appeared there, with plains, rivers and a specific atmosphere, much like the outer etheric, but somewhat darker. In going in through the left-sided facets, I entered a very busy and crowded area, reminding me of the carnival in Venice. It seemed as if the right-sided region was affected by ahrimanic influences, the left side by luciferic influences, and the middle by pure spiritual influences.

I decided finally to enter through the facets of the middle area. I was able to do this. Here, it felt like I was entering a light-filled town with houses of many different colours. From the middle, I could also see Devachan in its entirety, or so it appeared to me. Devachan was like an enormous globe, slowly rotating. But it was always different in the various regions, both in relation to colours, light and activity. Outside the various parts, there were smaller globes, created from out of the main globe, almost like off-shoots. Our universe, our material experience, is one such off-shoot.

I also saw that there were two or three doorways from our universe to the main globe, to Devachan. One was through the etheric world, which I have called the outer etheric. One was through the portal of Anna. But there was still one more, which was unknown to me. They all led into different parts of Devachan, although the Anna-portal appeared to be the 'main entrance'.

The one on my right, the one through the outer etheric, led into a region of Devachan that looked somewhat like the outer etheric, and resembled Rudolf Steiner's description quoted above. As I say, I had never seen the one on my left, and I did not investigate further at this stage. All the other universes also had entrances to the main devachanic globe. Through such entrances, I saw that creatures from another universe could enter ours, just as Vlask had done, and even, possibly, the Antichrist.

Then, on the evening of 3 March, I received a clear message from the spiritual world. The messenger was Ariel himself. 'You have reported enough of your experiences,' he told me. 'Stop writing now!'

And that, dear reader, is what I did.

Afterword: Reflections on the Vidar and Michael Paths

For most of my life I have lived consciously within a spiritual world and observed elemental beings – later also spiritual beings of more elevated realms and worlds. During Covid, in a period of intense concentration, I began to pass through the 'deeper' levels or layers of the elemental world – levels that previously were unknown to me – and I reached a threshold with a particular guardian. When I asked his name, he replied: 'I am Vidar'. Vidar is now an Archangel. As a guardian, he can give personal advice (which, in my understanding, an Archai cannot).

Since that time, I have written a trilogy of books on the Northern Path to the spiritual world – this being the final volume. I wrote these because I identified this path – through nature and through the elemental world – to be one that is possible and practical for many people today. I also believe that it is an easier spiritual path than others that are available at the present time.

As related in my previous book, when I asked Vidar about the remaining classes of the School of Spiritual Science, the esoteric school that Rudolf Steiner created towards the end of his life (the Michael School), Vidar replied that these unknown classes were no longer appropriate for humanity at the present time. In any case, they would need to be reworked, as the spiritual world always changes. Thus, they were, metaphorically speaking, locked away!

AFTERWORD: REFLECTIONS ON THE VIDAR AND MICHAEL PATHS

And so, I began to refer to this path, which I called the Michael path, to be either 'blocked' or 'overgrown'. In other words, it was especially difficult in comparison with the Vidar or Northern Path – which I saw as being 'open' since 2019 and accessible to most people.

Rudolf Steiner describes that the wisdom of anthroposophy was laid out in great pictures in the supersensible world by the Archangel Michael in a great school of discarnate human spirits that were preparing to enter incarnation on earth. Just before his untimely death, Rudolf Steiner brought this Michael School down to earth when he gave the First Class of this School (the nineteen 'lessons' that are now published and widely available). These lessons prepare the human being to enter the path to the spiritual world. This includes direction from the 'guardian', who many – including myself – identify with the being of Michael.

Rudolf Steiner said that after 1879 Michael began a progression from being an archangelic being to becoming an Archai or Time Spirit for the whole world. Steiner also suggested that Vidar would replace Michael as an Archangel in this period.

*

Several anthroposophists have reacted negatively to my characterization of the Michael path as being 'closed'. This requires further and deeper explanation.

There are three main 'doors' to the spiritual world, as described by Rudolf Steiner on 2 March 1915:

> It is possible to enter into the spiritual world through three doors, as it were. The first may be called the Door of Death, the second the Door of the Elements

and the third the Door of the Sun. Anyone wishing to follow the path to knowledge in its entirety will have to take the road to knowledge through all three doors.*

In my understanding, if one enters through these three doors, there are several paths available, led by different spiritual beings, such as Michael and Vidar. There are of course many spiritual paths, even within anthroposophy, but the focus of the nineteen lessons, given within the earthly School of Michael, was to reach the Door of the Sun. This proved difficult in Steiner's time, and I believe that it is still difficult today, *but not impossible.*

Michael, who now is an Archai, facilitated Vidar to take his place as a new Archangel, and Michael's power, words and help now flow through Vidar. The Vidar path through the elemental realm is thus based on both Michael and Christ.

When I walked the path of Vidar, leading to the Door of the Elements, I observed many others walking the same path, and even more on the paths leading to the Door of Death – but few on the paths leading to the Door of the Sun.

When I now stand in the spiritual world and position myself so that I can see all three doors or gates, I do see human beings passing through all of them. But most people pass through the Gate of the Elements – whichever path they may take.

I am now convinced that Steiner didn't speak much about the Gate of the Elements and the paths reaching

---

* 'Three Decisions on the Path to Imaginative Perception', lecture of 2 March 1915, Berlin, from *The Destiny of Individuals and of Nations*, Rudolf Steiner Press, 1987.

there because at that time it was 'closed'. However, the more I think about these three gates and the more I investigate practically, I begin to see that one ideally needs to approach all three gates, and that the abilities and knowledge that one can gain from going down one particular path or approaching one particular gate, can help you when one tries another path or attempts to reach another gate.

In our time, many people are having spiritual experiences. Sadly, I think it is possible that such people don't easily appreciate what anthroposophy can give them in terms of knowledge and instruction, because they find themselves exclusively on a path through the elemental world.

The 'merging' into the elemental realm, and its associated clairvoyant experiences, are sometimes explained away as 'atavistic' and therefore unhealthy. But in my experience, any clairvoyance that is conscious is not atavistic if it is practised with clear thought. Our task is to facilitate progression on all the spiritual paths, including the Michael path to the Gate of the Sun.

Appendix: An Investigation into the Differences between Study and Spiritual Experience

A remarkable book by Gerhard von Beckerath with the title *Rudolf Steiners Leidensweg** describes the great pain and suffering that Rudolf Steiner experienced when he realized that the path into the spiritual world that he had offered was – due to his followers' lack of ability and commitment – subsequently less effectual.

In 1924, shortly after the Christmas Conference that was intended to reshape and recreate the Anthroposophical Society as well as the anthroposophical spiritual movement, Rudolf Steiner embarked upon the huge task of presenting the esoteric school founded by the Archangel Michael himself, which was originally given in the spiritual world in two sessions – first in the thirteenth century and secondly in the nineteenth century. This school, that Rudolf Steiner brought down into the physical world in the course of his nineteen lessons, as well as through the Christmas Conference of 1923–24, was supposed to be given in three classes or levels, as reflected in the three angelic streams (the angelic, archangelic and archaic). After presenting the First Class (the angelic stream), however, Rudolf Steiner became seriously ill and eventually died in March 1925.

---

* *Rudolf Steiners Leidensweg: Sein Schicksal mit der Anthroposophischen Gesellschaft* by Gerhard von Beckerath in 1935, Verlag für Anthroposophie, 2011. The title could be translated into English as 'Rudolf Steiner's Path of Suffering, His Destiny with the Anthroposophical Society.'

This inability of his friends and students truly to walk the path that he had described, is said to have damaged Rudolf Steiner himself. The expected reply from the spiritual world, concerning the progress of the Anthroposophical Society, did not come as expected. According to Gerhard von Beckerath in the above-mentioned book, Rudolf Steiner said that his attempts in respect of the Anthroposophical Society and the School of Spiritual Science, had failed.* He had, according to Ita Wegman, began esoteric work that was stopped by the 'old forces', and that caused him great pain and even illness and death. Steiner is alleged to have communicated to both Johanna von Keyserlinck and Ludwig Polzer-Hoditz that he could work no further and had to die due to the fact that there were no ears or eyes to hear or see what he wanted to convey.

When a path into the spiritual world is created but not used, it becomes less accessible and 'grows over'. Could

---

* 'Ita Wegman (member of the Esoteric Council of the Christmas Conference Society in 1923) wrote in retrospect: "With the Christmas Conference, he had anticipated work for the future that would not have been possible to carry out if the old forces had not allowed it." (In 1935, Ita Wegman and Elisabeth Vreede were expelled from the Executive Board. Ita Wegman indirectly sought to leave the Society.) [...] Rudolf Steiner had to experience that the Christmas Conference and its Society had "evaporated", as it was only "taken in the same way as earlier conferences were so readily taken". The Christmas Conference and its Society thus ceased to exist on earth "through the non-execution of the impulses", even during his lifetime. Its Esoteric Council of directors destroyed itself, and it was "broken". There can be no karmic connection with a society that no longer exists. Rudolf Steiner's karmic connection with the Society was thus dissolved by the non-experience of the impulses of Christmas 1923.' Ibid., p. 231.

it be, therefore, that the path Rudolf Steiner offered in 1923–24 is obstructed? Are any anthroposophists able to travel this path today? Could the path be 'cleared' if the nineteen lessons are repeated and read by anthroposophists all over the world?

After venturing into the spiritual world through another path, the Nordic path, and there meeting Vidar, as described in my book *Travels on the Northern Path of Initiation*, I asked if I could have access to the descriptions and teachings of the Michael path of 1924 – teachings that were due to be given by Rudolf Steiner in the Second and Third Classes of that school. Amongst followers of Rudolf Steiner, there have been many discussions and thoughts about what the two remaining Classes might contain.

The answer I received from Vidar, and described in the aforementioned book, was that this material was now under 'lock and key' in the spiritual world, at least for some hundreds of years. It should be left alone where it was 'stored', somewhere in the vault of the Akashic Records, the 'hall of memory' in the World of Providence. Today, those contents should be worked out anew, from a contemporary perspective, and should reflect the huge changes that have taken place in the spiritual world since 1924.

In my view, the above clearly shows the difference between *knowledge* and *path*. The Michael School contains a lot of lofty knowledge given by an Archai, Michael himself, and also of material for discussion or for meditation. But, so long as it is no longer practised, it is not a *path* but only *knowledge*. The Vidar path has recently been opened (from 2019, according to my research), and is still a practical path of development, although it contains less knowledge than the Michael School.

Today we need both the knowledge from the Michael School and the practice of the Vidar School (we need both 'knowledge' and 'path') – but if we only work with knowledge, this work will not give us the *living experience* that walking a practical path can give, and as such will not appeal to young people of today, who are longing for experiences of *walking* the path itself, and not just studying knowledge.

So, what is the Vidar path? It is described by Rudolf Steiner in his lecture of 3 April 1912, given in Helsinki.* There he describes how we can *merge* into the material world, going through the illusion of materiality and reaching the underlying reality of elemental beings. After entering the elemental world, we are able to travel through its three realms, eventually reaching the etheric or spiritual world 'behind' it.

This elemental world was partly closed at that time. Today, however, this has changed, and the path into the spiritual world through the elemental world is fully open. On the way through the elemental world, we meet the elemental beings who are strongly influenced, or partly created by, the adversarial forces – the luciferic, ahrimanic and azuric beings. We thus have to go through the empires of evil in order to reach Christ.

Rudolf Steiner once said that Christ could only be reached or experienced in our time through the mystery of evil, which in my opinion is exactly what here we are

---

* See *The Spiritual Beings in the Heavenly Bodies and in the Kingdoms of Nature*, Lecture 1, SteinerBooks 2011.

being asked to do – going through the mystery of evil, the mystery of the elemental world.

After venturing through the three realms of the elemental world, we reach the guardian to the threshold of the spiritual world, who today presents himself as Vidar. The present guardian, taking over from Michael, has ascended from Angel to Archangel, and today is Vidar.

When we reach him, Vidar will begin to educate us, to teach us what we need to understand and know before going further: initially, into the etheric world, where Christ can be met, and then, further, into the higher spiritual world, the devachanic world. This journey will lead us on to two paths of knowledge: *the mysteries of the spirit* and *the mysteries of the earth*, but it is not my intention to describe all this in this short piece. What we can meet and experience on the Vidar path can be found in my books: *Travels on the Northern Path of Initiation*, *Encounters with Vidar* and *Meeting Michael*.

So, what is my intention with this article? I want to make my fellow seekers on the spiritual path aware of the difference between *travelling on a path* and *acquiring knowledge*; experiencing the spiritual world and reading about it. I have tried to elucidate the path through the elemental world and on to the spiritual world (which is accessible now) versus the direct path through to the spiritual world (which now seems to be somewhat obscured), as well as the several changes that have taken place during the almost one hundred years that have elapsed since Rudolf Steiner's death.

A note from the publisher

For more than a quarter of a century, **Temple Lodge Publishing** has made available new thought, ideas and research in the field of spiritual science.

Anthroposophy, as founded by Rudolf Steiner (1861-1925), is commonly known today through its practical applications, principally in education (Steiner-Waldorf schools) and agriculture (biodynamic food and wine). But behind this outer activity stands the core discipline of spiritual science, which continues to be developed and updated. True science can never be static and anthroposophy is living knowledge.

Our list features some of the best contemporary spiritual-scientific work available today, as well as introductory titles. So, visit us online at **www.templelodge.com** and join our emailing list for news on new titles.

If you feel like supporting our work, you can do so by buying our books or making a direct donation (we are a non-profit/charitable organisation).

office@templelodge.com

For the finest books of Science and Spirit